Pathology of Wildlife and Zoo Animals

Pathology of Wildlife and Zoo Animals

Second Edition

Edited by

Karen A. Terio
Zoological Pathology Program
College of Veterinary Medicine
University of Illinois
Brookfield, IL, United Stated

Denise McAloose
Pathology Department
Zoological Health Program
Wildlife Conservation Society
Bronx, NY, United States

Judy St. Leger
Department of Population
Medicine and Diagnostic Sciences
College of Veterinary Medicine
Cornell University
Ithaca, NY, United States

ELSEVIER

ACADEMIC PRESS
An imprint of Elsevier

Academic Press is an imprint of Elsevier
125 London Wall, London EC2Y 5AS, United Kingdom
525 B Street, Suite 1650, San Diego, CA 92101, United States
50 Hampshire Street, 5th Floor, Cambridge, MA 02139, United States

ISBN 978-0-443-18437-6

For information on all Academic Press publications
visit our website at https://www.elsevier.com/books-and-journals

Cover Images: The elements of wildlife pathology include macroscopic and microscopic manifestations of health and disease that are influenced by intra- and interspecies, environmental, anthropogenic and many other factors. The images on the cover reflect a few important select disease processes in terrestrial and aquatic species. Original artwork by Dr. Maris Daleo. Tissue images include *Monodon baculovirus* in a shrimp (Courtesy of Roxanna Smolowitz, Roger Williams University), SARS-CoV-2 in situ hybridization in the lung of a snow leopard (Courtesy Martha Delaney, University of Illinois Zoological Pathology Program), HPAI meningoencephalitis in a Dalmatian pelican (Courtesy of Denise McAloose, Wildlife Conservation Society), *Emydomyces testavorans* infection in a GMS stained section of shell of a Northwestern pond turtle (Courtesy of Karen Terio, University of Illinois Zoological Pathology Program) and HPAI pancreatitis in a Dalmatian pelican (Courtesy of Denise McAloose, the Wildlife Conservation Society).

Publisher: Mara Conner
Acquisitions Editor: Simonetta Harrison
Editorial Project Manager: Lindsay Lawrence
Production Project Manager: Vijayaraj Purushothaman
Cover Designer: Matthew Limbert

Typeset by STRAIVE, India

Dedication

Dedicated to all those we have lost during, since, and to the SARS-CoV-2 (COVID-19) pandemic.

Contents

Contributors

Tawfik Aboellail Colorado State University, Fort Collins, CO, United States

Dalen Agnew Michigan State University, Lansing, MI, United States

Anibal G. Armién California Animal Health and Food Safety Laboratory System, Davis Branch, CA, University of California, Davis, CA, United States

Abigail Armwood North Carolina State University, Raleigh, NC, United States

Neel Aziz Smithsonian National Zoological and Conservation Biological Institute, Washington, DC, United States

Lineke Begeman Erasmus University Medical Centre, Department of Viroscience, Rotterdam, The Netherlands

Josephine Braun San Diego Zoo Wildlife Alliance, San Diego, CA, United States

Kathy A. Burek-Huntington Alaska Veterinary Pathology Services, Eagle River, AK, United States

Rachel E. Burns San Diego Zoo Wildlife Alliance, San Diego, CA, United States

Caitlin E. Burrell University of Georgia, Athens, GA, United States

Alvin C. Camus University of Georgia, Athens, GA, United States

José Luiz Catão-Dias Department of Pathology, School of Veterinary Medicine and Animal Sciences, University of São Paulo, São Paulo, Brazil

Tylis Y. Chang Zucker School of Medicine, New Hyde Park, NY, United States

Molly E. Church University of Pennsylvania, School of Veterinary Medicine, Philadelphia, PA, United States

Natália Coelho Couto de Azevedo Fernandes Centro de Patologia, Instituto Adolfo Lutz, São Paulo, Brazil

Kathleen M. Colegrove Zoological Pathology Program, College of Veterinary Medicine, University of Illinois, Brookfield, IL, United States

Kenneth J. Conley Disney's® Animals, Science and Environment, Lake Buena Vista, FL, United States

Rocio Crespo North Carolina State University, Raleigh, NC, United States

Martine de Wit Florida Fish and Wildlife Conservation Commission, Saint Petersburg, FL, United States

María del Carmen Carmona Muciño Africam Safari Zoo, Puebla, Mexico

Martha A. Delaney Zoological Pathology Program, College of Veterinary Medicine, University of Illinois, Brookfield, IL, United States

Daniela Denk SeaWorld Abu Dhabi, Yas Island, Abu Dhabi, United Arab Emirates

Shannon L. Donahoe Sydney School of Veterinary Science, Faculty of Science, The University of Sydney, Camperdown, NSW, Australia

Gerry Dorrestein Pathology Laboratory NOIVBD, Vessem, The Netherlands

Mary Duncan Saint Louis Zoo, Saint Louis, MO, United States

Ana María Henao Duque Zoopath Veterinary Pathology Laboratory-Cali Zoo, Cali, Colombia

Lisa L. Farina College of Veterinary Medicine, University of Florida, Gainesville, FL, United States

Heather Fenton Australian Registry of Wildlife Health, Taronga Conservation Society, Mosman, NSW, Australia

Antonio Fernandez Institute of Animal Health and Food Safety, University of Las Palmas de Gran Canaria, Canary Islands, Spain

Monique S. França Pathology at Johnson and Johnson Innovative Medicine, San Diego, CA, United States

Salvatore Frasca University of Connecticut, Storrs, CT, United States

Patricia M. Gaffney San Diego Zoo Wildlife Alliance, San Diego, CA, United States

Michael M. Garner Northwest ZooPath, Monroe, WA, United States

David J. Gasper University of Wisconsin-Madison, Madison, WI, United States

Omar Gonzales-Viera California Animal Health and Food Safety Lab, Davis Branch, Davis, CA, United States

Nicole L. Gottdenker Department of Pathology, College of Veterinary Medicine, University of Georgia, Athens, GA, United States

Jane Hall Australian Registry of Wildlife Health, Taronga Conservation Society Australia, Mosman, NSW, Australia

Sushan Han Denver Zoo Conservation Alliance, Denver, CO, United States

Damien Higgins Sydney School of Veterinary Science, Faculty of Science, The University of Sydney, Camperdown, NSW, Australia

Ursula Höfle SaBio (Health and Biotechnology) Working Group, National Game and Wildlife Research Institute, Cuidad Real, Spain

Charlotte Hollinger Inotiv, Kalamazoo, MI; Wildlife Conservation Society, Bronx, NY, United States

Emma H. Hooijberg University of Pretoria, Onderstepoort, South Africa

Elizabeth W. Howerth University of Georgia, Athens, GA, United States

Carles Juan-Sallés Noah's Path, Elche, Alicante, Spain

Rebecca A. Kagan United States Fish and Wildlife Service, Ashland, OR, United States

M. Kevin Keel University of California, Davis, CA, United States

Jörg Kinne Central Veterinary Research Laboratory, Dubai, United Arab Emirates

Thijs Kuiken Erasmus University Medical Centre, Department of Viroscience, Rotterdam, The Netherlands

Elise LaDouceur Joint Pathology Center, Veterinary Pathology Service, Silver Spring, MD, United States

Jennifer A. Landolfi Zoological Pathology Program, College of Veterinary Medicine, University of Illinois, Brookfield, IL, United States

Julia S. Lankton United States Geological Survey-National Wildlife Health Center, Madison, WI, United States

Jeann Leal Universidade Federal da Paraíba UFPB, Areia, Paraíba, Brazil

Eric D. Lombardini Pre-Clinical Global, Boston Scientific Corporation, Marlborough, MA, United States

Linda J. Lowenstine University of California, Davis, CA, United States

María Ángeles Jiménez Martínez Veterinary Faculty, Complutense University of Madrid, Madrid, Spain

Kerstin Mätz-Rensing German Primate Center, Leibniz Institute for Primate Research, Göttingen, Germany

Denise McAloose Pathology Department, Zoological Health Program, Wildlife Conservation Society, Bronx, NY, United States

Rita McManamon University of Georgia, Athens, GA, United States

Alexandria Mena SeaWorld Parks and Entertainment, San Diego, CA, United States

Melissa A. Miller Marine Wildlife Veterinary Care and Research Center, California Dept. of Fish and Wildlife, Santa Cruz, CA, United States

Emily Mitchell Department of Paraclinical Sciences, University of Pretoria, Pretoria, South Africa

Maria Morell Institute of Terrestrial and Aquatic Wildlife Research, University of Veterinary Medicine Hannover, Büsum, Hannover, Germany

Pedro Enrique Navas-Suárez University of São Paulo, São Paulo, Brazil

Aleksija Nemanis Swedish Veterinary Agency, Uppsala, Sweden

Nicole M. Nemeth University of Georgia, Athens, GA, United States

Alisa L. Newton ZooQuatic Laboratory, LLC, Baltimore, MD; OCEARCH, Park City, UT, United States

Sally Nofs Menzies Institute for Medical Research, University of Tasmania, Hobart, TAS, Australia

Mark S. Okihiro California Department of Fish and Wildlife, Wildlife and Fisheries Division, Fisheries Branch, Vista, CA, United States

Francesco C. Origgi Institute of Microbiology, Department of Environment Construction and Design, University of Applied Sciences and Arts of Southern Switzerland, Mendrisio, Switzerland; Department of Veterinary Sciences, University of Messina, Messina, Italy

Robert J. Ossiboff Aquatic, Amphibian, and Reptile Pathology Service, College of Veterinary Medicine, University of Florida, Gainesville, FL, United States

Allan P. Pessier Washington State University, Pullman, WA, United States

Stephen Raverty Ministry of Agriculture and Food, Animal Health Center, Abbotsford, BC, Canada

Drury R. Reavill Zoo/Exotic Pathology Service, Citrus Heights, CA, United States

Emily M. Reinhardt University of Connecticut, Storrs, CT, United States

Bruce Rideout San Diego Zoo Wildlife Alliance, San Diego, CA, United States

Alison Righton San Diego Zoo Wildlife Alliance, San Diego, CA, United States

David Risco Neobéitar, Cáceres, Spain

Carlos E. Rodríguez Disney's® Animals, Science and Environment, Lake Buena Vista, FL, United States

Wendi Roe Massey University, Palmerston North, New Zealand

Jamie L. Rothenburger Faculty of Veterinary Medicine, University of Calgary, Calgary, AB, Canada

David Rotstein Marine Mammal Pathology Services, Olney, MD, United States

Marie-Pierre Ryser-Degiorgis Centre for Fish and Wildlife Health, University of Bern, Bern, Switzerland

Catherine M. Shilton Northern Territory Department of Industry, Tourism and Trade, NT, Australia

H.L. Shivaprasad California Animal Health and Food Safety Laboratory System, Tulare Branch, University of California, Tulare, CA, United States

Ursula Siebert Institute of Terrestrial and Aquatic Wildlife Research, University of Veterinary Medicine Hannover, Büsum, Hannover, Germany

Dale A. Smith University of Guelph, Guelph, ON, Canada

Roxanna Smolowitz Roger Williams University, Bristol, RI, United States

Jessica M. Snyder Department of Comparative Medicine, School of Medicine, University of Washington, Seattle, WA, United States

David Spratt CSIRO, Australian National Wildlife Collection, Canberra, ACT, Australia

Judy St. Leger Department of Population Medicine and Diagnostic Sciences, College of Veterinary Medicine, Cornell University, Ithaca, NY, United States

Nicole I. Stacy University of Florida, Gainesville, FL, United States

Ilse H. Stalis Independent Scholar, Long Island, NY, United States

Nancy L. Stedman Busch Gardens Tampa, Tampa, FL, United States

Jennifer Steinberg Mid-Atlantic Veterinary Pathology, LLC, Baltimore, MD, United States

Johan Steyl Department of Paraclinical Sciences, University of Pretoria, Pretoria, South Africa

Mark F. Stidworthy International Zoo Veterinary Group, Keighley, West Yorkshire, United Kingdom

Simone Stoute California Animal Health and Food Safety Laboratory System, Trulock Branch, University of California, Turlock, CA, United States

María José Navarrete Talloni Ross University School of Veterinary Medicine, Basseterre, Saint Kitts

Karen A. Terio Zoological Pathology Program, College of Veterinary Medicine, University of Illinois, Brookfield, IL, United States

John Trupkiewicz Northwest ZooPath, Monroe, WA; School of Medicine, Johns Hopkins University, Baltimore, MD, United States

Edwin J.B. Veldhuis Kroeze Erasmus University Medical Centre, Department of Viroscience, Rotterdam, The Netherlands

Tabitha C. Viner United States Fish and Wildlife Service, Ashland, OR, United States

Arno Wünschmann University of Minnesota, St. Paul, MN, United States

Foreword

When I opened the impressive first edition of *Pathology of Wildlife and Zoo Animals*, I was struck by the breadth of topics covered and the array of insights shared. It invited me to enjoy and explore an impressive wealth of information. Now, just a few years later, the tireless editors have delivered their second edition, which builds upon the earlier foundation and expands its scope.

This updated version is, first and foremost, a testament to the editors and the many authors who have once again made time and effort to update and supplement their respective chapters, incorporating advances and discoveries that have been made since the first edition's release. The book covers a vast range of species, including mammals, birds, reptiles, amphibians, fish, and invertebrates, each with their unique set of conditions and diseases. It provides extensive and detailed insight into their respective pathologies, both in human care and in the wild, making it an invaluable resource for those involved in wildlife pathology, health, and management.

Each chapter provides an overview of the most relevant details of a compilation of non-infectious, infectious, genetic, neoplastic, and, often most fascinating, miscellaneous diseases. This includes symptoms, etiologies, and diagnostic highlights, all accompanied by high-quality photographs and eSlides that aid in understanding and diagnosis. An invaluable highlight of this collection appears in the initial chapters, one of which delivers detailed and practical background information on necropsy examinations of wildlife. Experts provide guidelines for sample collection, storage, histology, documentation, and reporting. This is crucial baseline information in disease diagnosis and research and an integral tool for wildlife pathology capacity building and bridging.

The first edition arrived at our desks in 2018. The world has fundamentally changed since then. COVID-19, a wildlife-origin pandemic responsible for over 700 million cases and over 7 million human deaths, immeasurable suffering, and social turmoil paired with direct and indirect global economic devastation on the order of tens of trillions of US dollars, affected every one of us and our respective communities. Simultaneously, we saw several Ebola virus disease outbreaks while mpox (monkeypox) marched across the globe. Highly pathogenic avian influenza (HPAI) poses a massive threat to global avian biodiversity as the virus potentially acquires novel traits with each new mammal outbreak. African swine fever continues to spread, impacting communities and wildlife across Asia.

The present interpandemic phase is defined by multiple tightly interconnected crises—climate emergency, biodiversity loss, emerging infectious diseases, and global inequities. There is no question that humanity's destructive relationship with nature and the insatiable demand for ever-more resources has dramatically increased the risk of future zoonotic, specifically wildlife-origin, pandemics, consequential panzootics, and a myriad of non-infectious threats for all species.

We know that stopping disease spread after a successful spillover is extremely difficult and often impossible on our tightly interconnected and entangled home planet. This makes prevention based on a solid knowledge base all the more important. Critically, we must remain grounded in science; be agile in the face of ambiguities, uncertainties, and false equivalences; and be able to navigate conflicting priorities and agendas, mobilizing our substantial knowledge to prevent disease from pummeling our planet.

While our ability to appropriately prevent, prepare for, and respond to future disease threats builds on the work and wisdom of early wildlife pathology trailblazers, solidly researched and consistently updated volumes like this one provide a critical knowledge base for generations to come. Evidence-based decision-making can be jeopardized by highly polarized, politicized, and entrenched discourses, and we all must remain focused on the root-cause responses that will mitigate the emergence of novel pathogens, the reemergence of old disease foes, and the ever-increasing threats from pollution.

I have yet to read the entire book, but I refer to it frequently for information I need to better tackle daily work demands, and invariably, while turning a page, I find myself diving down a new and fascinating avenue of wildlife pathology. Enjoy, explore, and, most importantly, stand up for evidence- and science-based decision-making. Every day.

Chris Walzer
Wildlife Conservation Society—Global

Preface

Since the publication of the first edition of this book, emerging and known diseases have caused significant mortality events and pandemics in both animals and humans. The SARS-CoV-2/COVID-19 pandemic has impacted the global community and is a stark reminder of the interconnectedness of animal and human health. Intra- and interspecies transmission of highly pathogenic H5N1 avian influenza virus has caused mass mortality across several continents, affecting not only birds but mammals as well. Anthropogenic factors, including those related to changes on our ever-warming planet, continue to have significant impacts across countless vertebrate and invertebrate species and increase wildlife-human interactions. As the global landscape continues to change, understanding the many diverse and dynamic factors that drive health and disease in wildlife and zoo animals has never been more important. This applies to individuals and populations alike because of the role these animals play as sentinels of disturbed environments and their underlying factors.

The *Pathology of Wildlife and Zoo Animals*, second edition, continues the exploration of diseases in nondomestic species laid out in the first edition. The updated content in the main text and online Digital Content includes new images, eSlides, and extended reference lists. An introductory chapter that discusses the role wildlife pathologists play in disease investigations provides historical context and underscores the importance of wildlife and zoo pathology as a bridge between animal and human health and a number of other important sectors, including policy. The textbook offers a comprehensive exploration of the diseases that affect these species with the goal of equipping students, researchers, and practitioners with the knowledge and tools necessary to identify and contextualize diseases and disease processes.

As with the first edition, the creation of this book absolutely took a village. As we take a moment to thank those who made this textbook possible, we first want to recognize all of the authors who contributed to this project and acknowledge the many challenges and sacrifices they made in order for this project to be realized. To them, we offer our sincere and deep-felt thanks. We also want to recognize the contributions of Dr. Drury Reavill, Dr. Marie-Pierre Ryser-Degiorgis and Dr. Patricia Gaffney. All were remarkable people whose legacy extends well beyond the chapters in this book. We also want to remember those we've lost during and to the COVID-19 pandemic. Our lives have been permanently impacted by both your presence and, now, your absence.

To our colleagues at Academic Press/Elsevier; we extend our thanks to our Project Manager, Lindsay Lawrence, and our Acquisitions Editor, Simonetta Harrison, for carrying this project from conception through production. We would also like to thank our Production Managers, Kumar Anbazhagan and Vijayaraj Purushothaman, for helping to get us over the finish line.

We would also like to thank Maris Daleo for providing the beautiful original artwork on the cover. We also thank Drs. Ilse Stalis and Lisa Harrenstien for lending their critical eyes to helping us edit the index.

Lastly, and always most importantly, to our families. Your love and continued support of this project and our crazy careers—including late nights, lost weekends, and time away, among many other distractions that pull us away from you—is something that all of us are continually amazed by and eternally grateful for. Without you, none of what we do would be as enjoyable, possible, or fulfilling. So to Dan and Abigail Winter, Tylis Chang, and Marc Kratzschmar, you are our hearts and souls. Thank you, thank you, thank you.

Chapter 1

Role of the Veterinary Pathologist in Emerging Infectious Diseases

Thijs Kuiken, Lineke Begeman, and Edwin J.B. Veldhuis Kroeze
Erasmus University Medical Centre, Department of Viroscience, Rotterdam, The Netherlands

INTRODUCTION

The discovery, diagnosis, and study of emerging infectious diseases benefit from the knowledge and skills of veterinary pathologists, especially those with expertise in wildlife pathology. This is because veterinary pathologists have expertise in multiple aspects of infectious disease emergence, including interspecies interaction, route of transmission, and comparative disease presentation. An emerging infectious disease in a novel host species, including humankind (*Homo sapiens*), often is caused by an etiological agent originating from another species, the reservoir. The spillover of this etiological agent requires interaction between reservoir host and novel host, directly or via an intermediate or stepping-stone host. The interspecies interactions that allow infectious diseases to spill over between species are well known to and understood by zoo and wildlife pathologists, be it in the zoo setting (e.g., Aujeszky's disease from pigs to brown bears (*Ursus arctos*),[48] in free-living wildlife (e.g., avian cholera from snow geese [*Chen caerulescens*] to other waterfowl species[35]), or at the animal-human interface (e.g., tularemia from free-living European hares [*Lepus europaeus*] to humans[37]). The next step after interaction between reservoir host and novel host is the actual transmission of the etiological agent. Again, veterinary pathologists are familiar with the different possible routes of transmission, for example, via air into the respiratory tract, via food into the gastrointestinal tract, or via a bite wound into the nervous system. Once in the novel host, the etiological agent can cause disease. The pathological presentation of this disease may differ in several ways from that in the reservoir host, including character, distribution, duration, and severity. It is particularly at this stage that veterinary pathologists, and especially wildlife pathologists, whether they had an official training or not, can prove their worth. Because the pathology of only a small proportion of diseases of wild animals is well described or even known, wildlife pathologists are trained to carefully and systematically describe the characteristics of a lesion, to select subsequent diagnostic analyses as needed, and by comparison of their observations with what they know about the pathology of other diseases in other species, to arrive at a possible, probable, or definitive disease diagnosis. These skills and knowledge, which are unique to veterinary pathologists, make them a valuable, if not crucial, member of a team investigating emerging infectious diseases.

Veterinary pathologists have a role to play in each of the different phases of an emerging infectious disease event, from its initial emergence to its eradication or management. Some of these roles are as follows. Veterinary pathologists can be part of the team that discovers the etiological agent that causes an emerging infectious disease. Veterinary pathologists can improve the diagnosis of the emerging infectious disease by careful characterization of its pathology in the novel host. Veterinary pathologists can help to test candidate vaccines and therapeutic agents against the emerging infectious disease by helping to develop and use a suitable laboratory animal model that accurately mimics the disease, or a significant part of it, in the novel host. At all phases of an emerging infectious disease event, veterinary pathologists can generate knowledge about the emerging infectious disease by investigating its pathogenesis in laboratory animal models and by comparing this with the pathogenesis in both the novel host and the reservoir host. This knowledge can be generated for its own sake and for the purposes of managing and preventing the emerging infectious disease. These different roles of veterinary pathology in the research, management, and prevention of emerging infectious diseases will be discussed in more detail below.

For some of the roles indicated above, veterinary pathologists need to switch from a diagnosis-centered mindset to a more research-centered mindset. Training programs for veterinary pathology typically focus on diagnosis. Trainees develop encyclopedic knowledge on pathological features of diseases and learn to apply this knowledge to cases of morbidity or mortality in order to rapidly and accurately diagnose known diseases. However, this training sometimes can inhibit scientific curiosity. For many activities

Pathology of Wildlife and Zoo Animals. https://doi.org/10.1016/B978-0-443-18437-6.00040-7

related to emerging infectious diseases, a questioning and critical approach is required, with more focus on aspects that are puzzling and not understood than those that fit with existing knowledge. Also, innovation and creativity often are required to use knowledge of veterinary pathology to develop methods, if possible quantitative, to test hypotheses about characteristics. Therefore in addition to maintaining their phenomenal diagnostic skills, veterinary pathologists should nurture the questioning and creative parts of their minds to improve their research skills.

ROLES OF VETERINARY PATHOLOGISTS IN RESEARCH, MANAGEMENT, AND PREVENTION OF EMERGING INFECTIOUS DISEASES

The examples provided in the text and table are selected to illustrate different ways in which veterinary pathologists can contribute to solving issues related to emerging infectious diseases. They do not represent all such potential examples. Many of these investigations were championed by veterinary pathologists, collaborating in a team of scientists with various backgrounds.

Discovering the Etiology

An important and exciting role for veterinary pathologists in an emerging infectious disease event is discovering its etiology. In some cases it may even turn out to be a non-infectious etiology. Without pathology, the body of a human or other animal that has died of an unknown disease often is a black box.

Veterinary pathologists can help guide further analyses by characterizing the main lesions, even if this does not consist of a precise identification of the etiological agent involved. In general, veterinary pathologists are able to broadly classify lesions as more likely to be caused by viral, bacterial, protozoal, fungal, or metazoal infections or by a non-infectious agent. For example, pulmonary lesions characterized by infiltration of neutrophils and lytic necrosis put bacterial infection higher on the differential diagnosis, while infiltration with lymphocytes and epithelial desquamation point more toward a viral infection. In this way veterinary pathologists can provide valuable help in restricting the choice of tissue samples to be tested and the types of etiological agents to be tested for. In some cases pathological examination reveals characteristic lesions that allow more or less precise identification of the etiological agent involved. For example, the detection of both nuclear and cytoplasmic eosinophilic inclusions in a lesion is highly characteristic of morbillivirus infection. There are several examples of emerging infectious diseases in which the contribution of veterinary pathologists was critical in discovering the etiological agent (Table 1.1).

One example is the diagnosis of a chronic disease of unknown cause that had been observed since 1967 in captive mule deer (*Odocoileus hemionus hemionus*) in Colorado, USA. In 1980 two scientists published a report on this disease, which they identified as chronic wasting disease.[47] Clinically, deer affected by chronic wasting disease showed listlessness, progressive weight loss, and depression, leading to emaciation and death. Clinical evaluation of samples of blood, urine, and central spinal fluid showed no deviations from normal clinical chemical or hematological values except low specific gravity of urine suggestive of diabetes insipidus. Serologically, deer were negative for toxoplasmosis, leptospirosis, brucellosis, infectious bovine rhinotracheitis, bovine virus diarrhea, and bluetongue. Toxicological analysis of the liver and kidney showed normal levels of selenium, lead, copper, mercury, and molybdenum. No consistent lesions were present at gross autopsy except emaciation and the marked excess fluid, mixed with sand or gravel, in the rumen.

The clue to the cause of chronic wasting disease came from histopathology. Histopathological examination of an extensive selection of postmortem tissues showed consistent lesions restricted to the brain and spinal cord. These consisted of spongiform changes in the neuropil, intracytoplasmic vacuoles in neurons, neuronal degeneration, astrocyte hyperplasia, and fibrillar hyperplasia, characteristic of the pathological diagnosis spongiform encephalopathy (Fig. 1.1). Based on these histopathological results, the researchers concluded that spongiform encephalopathy was specific for chronic wasting disease and resembled spongiform encephalopathies in humans and other animals. Eleven years later, together with other scientists, they identified the cause of chronic wasting disease as a scrapie-like prion, by immunolocalization of scrapie amyloid protein (PrP27–30) in plaques observed in brain tissues of Rocky Mountain elk (*Cervus elaphus nelsoni*) and hybrids of mule deer and white-tailed deer (*Odocoileus virginianus*) naturally affected with chronic wasting disease.[9] Chronic wasting disease is increasing in prevalence and range, thereby threatening the survival of cervid populations in North America, and now is also known to occur in Europe.[43]

A second example is the discovery of the cause of mortality of vultures in the Indian subcontinent. Starting in 1990, numbers of oriental white-backed vultures (*Gyps bengalensis*) declined by more than 95%. Across the subcontinent of India, steep declines were observed in other vulture species, as well. Affected vultures were lethargic and had drooping necks before they died. Deaths of vultures appeared to spread across breeding and roosting colonies over a period of several years.[4] Potential explanations for this mortality event were (1) a rapid reduction in food availability, (2) an epidemic of infectious disease, and (3) poisoning.[4] The first of these possible causes was discounted by the studies of

TABLE 1.1 Summary of relative recent major emerging diseases or outbreaks in which veterinary pathologists played key roles in diagnosing the cause, improving the diagnosis, testing candidate vaccines and therapeutic agents, or investigating the pathogenesis

Year and Location of Disease Emergence	Emerging Disease	Main Role of Veterinary Pathologist in Characterization of Emerging Disease (Reference)
1967 USA	Chronic wasting disease Mule deer	Identified a novel spongiform encephalopathy as cause of CWD, infectious pathogen presumed[9,47]
1986 United Kingdom	Bovine spongiform encephalopathy	Identified BSE fibrils in brains similar to scrapie-associated fibrils by EM and protein sequencing. Koch's postulates fulfilled[12]
1988 North/Baltic Seas	Morbilliviral pneumonia Harbor seals	Serology and EM suggested morbillivirus infection as cause, confirmed by typical microscopic morphology, IF and virus isolation[14,26,27]
1989 USA	Abortions Cattle	Identified protozoan *Neospora caninum* as cause of abortion by IHC and excluded other pathogens[41]
1990 Spain	Morbilliviral pneumonia Striped dolphins	Identified morbillivirus as cause of disease, by typical microscopic morphology and IHC, EM, isolated and cultured virus[5,6]
1990s India	Diclofenac-induced mortality Vultures	Identified diclofenac by chromatography and mass spectroscopy in carrion as cause of kidney necrosis and death. Koch's postulates fulfilled[21,24]
1990s Tasmania	Tasmanian Devil's Facial Tumor Disease	Identified transmissible neuroendocrine tumor by IHC and karyotyping and excluded virus by EM. Koch's postulates fulfilled[17,18,31]
1993–4 Tanzania Russia	Canine distemper Felids	Identified and characterized Canine Distemper Virus by IHC and serology as cause of mortalities in free-ranging lions[34] and in a tiger from Russian far east[32]
1994 Australia	Respiratory disease Horses and Humans	Characterized Hendra virus from horses and man by IHC, EM, culture, PCR and showed viral endotheliotropism, excluded other pathogens. Koch's postulates fulfilled[22,23]
1997 Eurasia	Influenza H5N1 Humans	Tested vaccine and antiviral therapy to prevent CNS invasion[38]
1998 USA	Domoic Acid toxicity Sea lions	Established domoic acid to cause disease and specific CNS lesions, coincide with algal (*Pseudo-nitzschia* sp.) blooms, excluded other pathogens[10,11,36]
1998 Australia	Chytridiomycosis Amphibia	First descriptions of chytridiomycosis causing die-offs of free-ranging amphibians[3] contemporaneously, pathologists and scientists in the USA were describing *Batrachochytrium dendrobatidis*[28]
1999 USA	West Nile virus Wild and domestic animals and humans	Identified mosquito-borne WNV as cause of encephalitis by IHC, ISH, EM, isolated and genetically sequenced virus[16,39]
2002 East Asia	Severe acute respiratory syndrome Humans	Identified SARS-CoV1 and characterized the pathogenesis in NHP model by IHC, EM. Koch's postulates fulfilled in NHP model [7,15]
2005 Thailand	Influenza H5N1 virus Tigers	Identified HPAIV H5N1 by IHC as lethal etiology and tiger-to-tiger transmission after feeding raw chicken carcasses[40]
2007 North America	White-nose syndrome Bats	Identified invasive fungus during hibernation as etiology. Koch's postulates fulfilled[20]
2010 Europe	Chytridiomycosis Amphibia	Identified *Batrachochytrium salamandrivorans* as cause, characterized pathology, and improved diagnosis in salamanders and newts to distinguish between *Bsal* and *Bd*[8,19,42]
2020 The Netherlands	Influenza H5N1 Wild carnivores	Improved diagnosis of highly pathogenic avian influenza (HPAI) in free-living carnivores (red fox, polecat, badger, otter) by examination of CNS[46]
Not appropriate	11 zoonotic virus infections humans	Compared pathogenesis in reservoir hosts and human host[2]

EM, electron microscopy; *HPAIV*, highly pathogenic avian influenza virus; *IF*, immunofluorescence; *IHC*, immunohistochemistry; *ISH*, in situ hybridization; *NHP*, nonhuman primate.

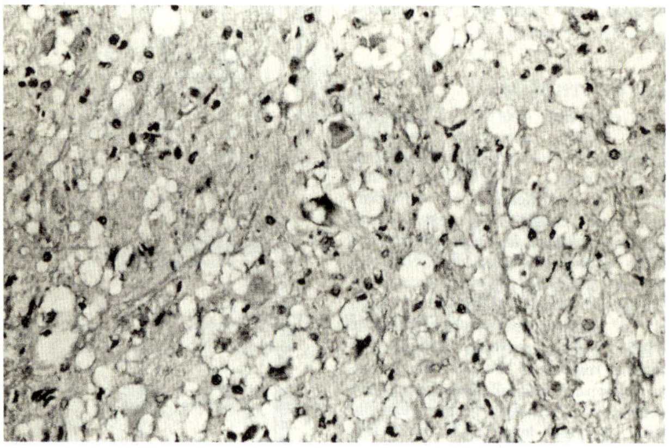

FIGURE 1.1 Chronic wasting disease. Spongiform encephalopathy in mule deer with chronic wasting disease. *(Image reprinted by permission from Williams ES, Young S. Chronic wasting disease of captive mule deer: a spongiform encephalopathy. J Wildl Dis. 1980;16:89–98.)*

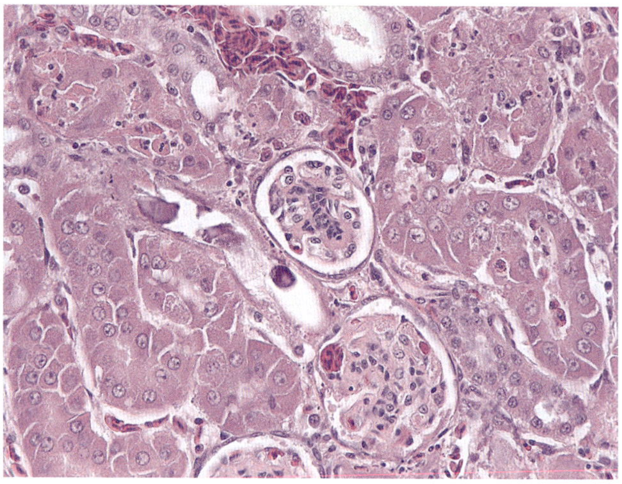

FIGURE 1.2 Diclofenac poisoning in vultures. Acute necrosis of the proximal convoluted tubules (top left of photograph) of the renal cortex in an oriental white-backed vulture due to poisoning with the NSAID diclofenac. *(Photo Courtesy of C. Meteyer, National Wildlife Health Center.)*

Prakash and colleagues,[30] who showed that animal carcasses remained abundant throughout India and that, of several hundred carcasses recorded during nationwide surveys in 2000, less than 5% had attendant vultures.

Pathological analysis of vultures helped to investigate the two remaining causes. Between 2000 and 2002, autopsies were performed on 259 vultures.[21,24] Regarding the possibility of infectious disease, evidence of inflammation indicating an infectious disease was not consistently present in the tissues of the vultures examined. Tissues tested negative for avian influenza virus, infectious bronchitis virus, and West Nile virus by PCR. No viruses were isolated in cell cultures inoculated with samples of kidney, spleen, lung, or intestine.

Regarding the possibility of poisoning, 219 of 259 vultures had grossly apparent urate deposits, characteristic of visceral gout, on the surface of internal organs, most commonly kidney, liver, spleen, lung, and heart. The differential diagnosis of visceral gout included dehydration. However, histologically, there was severe, acute necrosis of the proximal convoluted tubules in the renal cortex (Fig. 1.2), indicating acute renal disease. Toxicological analysis ruled out acute renal intoxication from cadmium, lead, and mercury as well as general intoxication by carbamates, organophosphate or organochlorine pesticides, and polychlorinated biphenyls. The question remained: what caused the acute renal disease?

After discussing this question, the scientists proposed that the renal pathology in the vultures pointed to an anthropogenic toxicant that targeted the kidneys and perhaps a nonsteroidal antiinflammatory drug (NSAID).[33] Because the primary food source of the vultures was dead livestock, they designed a questionnaire that was used to gather information on a large number of pharmaceutical products sold for use in livestock in India and found one candidate that fit the epidemiological pattern: the NSAID diclofenac. Indeed, diclofenac was detected in the kidney tissues of affected vultures by liquid chromatography and mass spectroscopy. Administration of diclofenac, both directly and by feeding meat of treated livestock, resulted in the deaths of all treated vultures.[24] Therefore pinpointing acute renal tubular necrosis as the main pathological diagnosis, followed up by good detective work and solid research, was instrumental in identifying the etiological agent of this mass mortality of vultures.

A third example is the discovery of the cause of a mass die-off of striped dolphins (*Stenella coeruleoalba*) in the Mediterranean Sea. Between July and November 1990, more than 400 striped dolphins were stranded dead on the Mediterranean coast of Spain. The differential diagnosis of this mass stranding included an infectious disease epidemic, by-catch, intoxication, acoustic trauma, and extreme weather events. By pathological examination of freshly dead carcasses, it was possible to pinpoint the cause. Scientists examined 58 dolphin carcasses.[5,6] Grossly, they observed pneumonia and encephalitis. By histology, they diagnosed diffuse interstitial pneumonia (Fig. 1.3), which was consistent with a viral cause. More detailed examination of these lesions revealed both syncytia as well as eosinophilic nuclear and cytoplasmic inclusions, which were characteristic of morbillivirus infection. Additional evidence was morbillivirus antigen expression by immunohistochemistry, detection of paramyxovirus-like particles in cytoplasm of affected cells by electron microscopy, and isolation of a morbillivirus by virus culture.

Did these dolphins die of the same phocine distemper virus that had decimated the North Sea population of harbor seals (*Phoca vitulina*) 2 years earlier? No, the cause of this striped dolphin die-off was subsequently identified as a

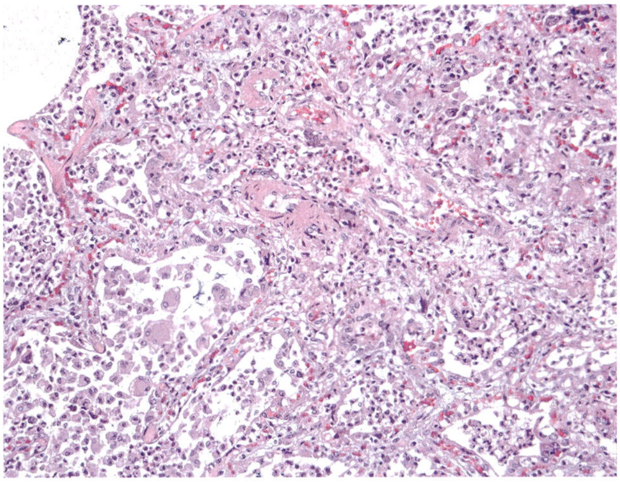

FIGURE 1.3 Cetacean morbillivirus in striped dolphins. Diffuse histiocytic bronchointerstitial pneumonia with viral inclusions (not evident in this magnification) in lung of striped dolphins characteristic of infection with Cetacean morbillivirus. *(Photo Courtesy of M. Domingo, Universitat Autònoma de Barcelona.)*

separate morbillivirus, dolphin morbillivirus.[44] This virus was later classified within the species *Cetacean morbillivirus*.[13] Thus, by a combination of expert pathological and virological analyses, the researchers were able to diagnose dolphin morbillivirus as the cause of this mass mortality in striped dolphins.

A final example is the discovery of the cause of mortality of California sea lions (*Zalophus californianus*) in 1998–2000 along the central Californian coast, USA. More than 400 sea lions were found dead, and many others were stranded alive with neurological signs. At the onset of the first mass stranding, a long list of differential diagnoses was considered, including bacterial diseases such as leptospirosis, viral diseases such as phocine distemper, and intoxications from heavy metals, industrial chemicals, or algal blooms.[11]

Scientists performed clinical and pathological examinations on 70 sea lions and on one northern fur seal (*Callorhinus ursinus*) from the beginning of this mortality event.[10] Stranded live animals were in good physical condition, and all showed similar neurological signs. These included seizures, ataxia, head weaving, depression, and abnormal scratching behavior. Despite supportive treatments, many sea lions died.[10] Bacteriological, virological, and histopathological analyses did not provide evidence for a viral or bacterial cause. There was no evidence of chemical pollutants based on normal brain cholinesterase levels nor of lead intoxication based on blood lead levels below the detection limit.[11,36]

Several strands of evidence pointed toward algal toxin poisoning, but this diagnosis had never been made before in marine mammals, and the epidemiological, clinical, and toxicological analyses were suggestive but not conclusive. The mass mortality event coincided temporally and spatially with a bloom of the diatom (*Pseudonitzschia australis*; marine algae/phytoplankton) associated with the production of a neurotoxin, domoic acid (DA). DA was found in high levels in anchovies and sardines, upon which sea lions were known to feed. The abnormal scratching behavior of the affected sea lions was reminiscent of the scratching behavior documented in mice and pelicans poisoned with DA. However, DA only was found in serum, urine, or feces of a minority (between 2 and 8 animals) of the 48 live-stranded sea lions that died and were autopsied.[36]

Again, histopathological characterization was critical for pinpointing the cause of mortality. All 48 autopsied sea lions had zonal vacuolation of the hippocampal neuropile (see 23.1 in Pinnipediae), which was most severe in the anterior ventral hippocampus.[36] These lesions are similar to those identified in mice, macaques, and humans exposed to DA. The consistency and characteristic features of these histopathological results, together with the oceanographical, epidemiological, clinical, and toxicological data, strengthened the diagnosis of DA intoxication as the cause of this mass mortality event in California sea lions.[11,36]

Improving the Diagnosis

Autopsies often are starting points for establishing the cause of death of an animal. Veterinary pathologists can help to improve our ability to diagnose an emerging infectious disease by carefully and comprehensively characterizing the pathological changes that it causes in the novel host.

This can be illustrated by the studies of *Batrachochytrium salamandrivorans* (*Bsal*) in amphibians. Scientists first identified this novel chytrid fungus as a cause of mass mortality of fire salamanders in the Netherlands and Belgium and subsequently in other salamander and newt populations in Europe.[19] In contrast to the better-known *Batrachochytrium dendrobatidis* (*Bd*), which causes die-offs in all groups of amphibians (including salamanders and newts) and occurs worldwide,[3,28] *Bsal* specifically targets salamanders and newts in Europe and has only been reported in scattered locations. Therefore to be able to determine its distribution and possible spread and hopefully slow it down or even prevent it, it is key to be able to distinguish the pathology of *Bsal*-induced disease from that of *Bd*-induced disease and to have sensitive methods of diagnosis of chytridiomycosis in salamanders and newts that are able to distinguish between *Bsal* and *Bd*.[8] This is what the researchers set out to do.

Regarding pathology *Bsal* consistently induced multifocal superficial erosions and deep ulcerations in the skin all over the body.[19] At the periphery of the erosions were keratinocytes with eosinophilic necrosis and marginated nuclei. Each of these keratinocytes contained one centrally located thallus, the majority being segmented (colonial thalli). The intraepidermal organisms reacted positively by immunohistochemistry using a primary antibody to *Bsal*. In contrast to this marked skin ulceration caused by *Bsal* infection, *Bd* infection typically induces epidermal hyperplasia and hyperkeratosis.

Regarding differentiating between *Bsal* and *Bd*, the researchers compared various methods of diagnosis. In samples from fire salamanders (*Salamandra salamandra*) infected either naturally or experimentally with *Bsal* they compared histopathological examination, immunohistochemical detection of *Bsal* thalli, and a duplex real-time PCR for *Bsal* and *Bd*. They found that the *Bsal* and *Bd* species–specific duplex real-time PCR allowed detection of both pathogens on noninvasively collected amphibian samples and could easily be implemented by diagnostic laboratories. Besides allowing rapid and reliable detection of *Bsal* and *Bd*, this PCR had the advantage over histopathology and immunohistochemistry of being able to simultaneously identify and better quantify *Bsal* and *Bd* in amphibian samples. The primary antibodies to *Bsal* and *Bd* did not differentiate between the two. For samples stored for months to years in neutral-buffered formalin, however, histopathology in combination with immunohistochemistry was preferred as this storage method did not affect tissue morphology nor the ability of the immunohistochemical staining technique to detect *Bsal*, whereas PCR-based methods did not work because tissue lysis and/or DNA damage. Overall, the researchers advocated the use of molecular diagnostic tools such as the duplex real-time PCR in conjunction with techniques that are able to show colonization and/or pathology such as histopathology and immunohistochemistry, whenever possible.[42]

Through these studies, scientists provided the diagnostic molecular and pathological tools to accurately diagnose *Bsal*-induced chytridiomycosis in salamanders and to prevent introduction and spread of *Bsal*. These are crucial elements in action plans for *Bsal* in order to try to protect native salamander populations and salamander species.

Another illustration is the more recent study of highly pathogenic avian influenza (HPAI) in free-living carnivores. Since 2021, a variant of HPAI virus, the H5N1 2.3.4.4b clade of the goose/Guangdong lineage of H5, appears to persist in free-living wild birds.[29] As a consequence, this H5N1 virus occasionally spills over to free-living mammals, predominantly those that predate on wild birds or scavenge their carcasses.[1] Influenza in mammals has historically been considered primarily a respiratory disease. Therefore diagnosis focuses on the respiratory tract: swabs from nose and throat for virological diagnosis in live animals and respiratory tract samples including trachea and lung for virological and pathological diagnosis in animals found dead. However, many free-living mammals suspected of H5N1 virus infection presented with neurological signs, suggesting central nervous system (CNS) disease. Is sampling that is restricted to the respiratory tract adequate to diagnose this new HPAI variant?

To help answer this question, scientists examined a series of carnivores infected with HPAI virus H5N1 virus of the 2.3.4.4b clade.[46] These consisted of 11 animals: 6 red foxes (*Vulpes vulpes*), 3 polecats (*Mustela putorius*), 1 badger (*Meles meles*), and 1 otter (*Lutra lutra*). These cases had been identified through targeted surveillance of euthanized wild carnivores with antemortem neurological signs or through general wildlife disease surveillance on wild carnivores found dead and with evidence of viral infection during routine autopsy. Tissue samples from the respiratory tract, gastrointestinal tract, CNS, and other major organs were subjected to histological and immunohistochemical analyses for the presence of lesions and the expression of influenza virus antigen, respectively.

All 11 carnivores showed a mild to severe, nonsuppurative meningitis, encephalitis, or meningoencephalitis, which was associated with H5N1 virus infection based on positive RT-PCR for influenza virus in all cases and colocalized with influenza virus antigen expression in neurons of many cases. Surprisingly, none of the cases had respiratory tract disease associated with H5N1 virus infection, although a few had influenza virus antigen expression in a few epithelial cells or macrophages in the lungs, or nasal mucosa. Similarly, none of the cases had disease associated with H5N1 virus infection in other extra-respiratory organs, with the exception of a single otter, which had influenza virus antigen expression in a few mononuclear and epithelial cells of the pancreas. Interestingly, some of the animals in this study were negative by influenza virus RT-PCR of anal and throat swabs, and HPAI diagnosis would have been missed without additional brain samples.

In this study systematic histopathological examination was critical in establishing that pathological changes from infection with this novel H5N1 virus in carnivores may be restricted to the CNS. The results of combined pathological and virological analyses also show that reliable diagnosis of this virus infection in carnivores submitted for autopsy requires sampling of the brain.

Testing Candidate Vaccines and Therapeutic Agents

If an emerging infectious disease cannot be prevented, vaccines and therapeutic agents can be part of the tools to lower the burden of disease. The evaluation of the safety and efficacy of such vaccines and therapeutic agents often requires a suitable animal model that accurately mimics the disease, or at least that part of the disease that one wishes to temper. Veterinary pathologists can help to develop and use such animal models.

For example, veterinary pathologists have contributed their skills to develop a ferret model for central nervous system (CNS) disease in humans from highly pathogenic avian influenza (HPAI) H5N1 virus infection, and its use to test the efficacy of vaccination or oseltamivir, an inhibitor of the influenza virus's neuraminidase.[38] HPAI H5N1 is a poultry disease that occasionally spills over to people and typically causes severe respiratory tract disease. However, the most common extra-respiratory complication of HPAI H5N1, both in humans and in other mammals, is severe or even fatal encephalitis, indicating the neurotropic potential of this virus across a wide range of mammalian species. The pathogenesis of influenza virus-induced encephalitis was shown to involve virus entry by the olfactory route: from the olfactory mucosa in the nasal cavity via the olfactory nerve and into the olfactory bulb and the rest of the brain.[45] However, while vaccination and prophylactic oseltamivir had been designed to reduce morbidity and mortality from respiratory tract disease in humans, the efficacy of these prophylactic strategies on virus replication in the olfactory mucosa and subsequent CNS invasion were unknown.

Scientists therefore investigated the efficacy of vaccination and prophylactic oseltamivir in a ferret animal model. Eighteen ferrets were divided among three groups of six ferrets each: a control group, a vaccination group, and an oseltamivir-treated group. Ferrets in the two treatment groups were inoculated intranasally with HPAI H5N1 virus while lying on their backs to enhance the exposure of the olfactory mucosa and so reproduce virus-induced encephalitis. At specified time points after inoculation, ferrets were euthanized and autopsied, and tissues were collected for virological and pathological analyses.

While all control ferrets developed moderate to severe virus-induced encephalitis, the brains of all vaccinated ferrets were virus free and only mildly inflamed. In contrast, virus was detected in the brain of the majority of the oseltamivir-treated ferrets with encephalitis of comparable severity to that in control ferrets. Therefore the results of this study showed that vaccination prevented virus from spreading to the CNS, while HPAI H5N1 virus in oseltamivir-treated ferrets was able to spread to the CNS via the olfactory nerve, resulting in a meningoencephalitis comparable in severity to that in control ferrets.

This study illustrates the value of veterinary pathology to virological studies of HPAI H5N1. Veterinary pathological analysis not only generated a research question (that is, the distinction between CNS disease and respiratory disease from HPAI H5N1 virus infection) but also made it possible to determine how two different prophylactic strategies inhibited the invasion of HPAI H5N1 virus into the brain in different degrees.

Investigating the Pathogenesis

Understanding the pathogenesis of a disease, that is, its origination and development, is at the core of the scientific discipline of pathology. Veterinary pathologists can generate knowledge on pathogenesis at all stages of an emerging infectious disease event, from its initial emergence to its eradication or management. Pathogenesis can be studied during the life (in vivo) or after the death (*postmortem*) of the novel host, the reservoir host, or a lab oratory animal, as well as in systems involving tissues (ex vivo, organoid) and cells (in vitro). The goals of generating knowledge of pathogenesis can be for the sake of knowledge itself, as well as for improving the management and prevention of emerging infectious diseases.

An example where several of these goals come together is provided by a study comparing the pathogenesis of 11 zoonotic viruses (rabies, macacine alpha herpes-, West Nile, Puumala orthohanta-, monkey pox, Lassa, H5N1 highly pathogenic avian influenza, Marburg, Nipah, Middle East respiratory syndrome, and simian/human immunodeficiency viruses) in the reservoir hosts (for which they coined the term "orthopathogenesis") with that in the novel, human host ("neopathogenesis").[2] The main goal of this study was to determine whether insights in development of severe human disease can be derived from a comparison between orthopathogenesis and neopathogenesis of some relatively well-studied zoonotic virus infections.

Through extensive literature review and taking the light microscopical viewpoint of a veterinary pathologist, the authors compared the following aspects of the pathogenesis of the 11 virus infections: (1) attachment to receptors on exposed host cells, (2) replication in host cells at the entry site, (3) release from host cells with subsequent dissemination within the host to other replication sites, (4) replication in host cells at sites that are important for virus amplification, and (5) excretion from the host to enable transmission to a new host.

From this comparison between orthopathogenesis and neopathogenesis of 11 arbitrarily selected zoonotic virus infections, they derived three main insights. First, most aspects of the pathogenesis in humans were remarkably similar to those in reservoir species. This is an important finding because it can help us to predict aspects of infections of potential zoonotic pathogens. Second, those aspects where ortho- and neopathogenesis differed helped to explain the severe outcome of infection in people. Third, even for the relatively well-known viral diseases selected, knowledge of the pathogenesis in reservoir hosts was sorely lacking. If these knowledge gaps are resolved, we will be better able to explain the severe disease from these virus infections in people.

Aside from the main goal stated in this study, there are additional possible goals for studying orthopathogenesis.

Better understanding of orthopathogenesis can lead to the recognition of specific virus characteristics and their selective advantages in the reservoir host in which it evolved. Also, orthopathogenesis may inform us about how the virus will behave in its novel host: because a newly emerged virus is at an early stage of adaptation in its novel host, knowledge of the pathogenesis in its reservoir host, where it is at an advanced stage of coevolution, will inform us about the possible future route of evolution that the virus will take in its novel host. In summary, this example illustrates how a veterinary pathology viewpoint can shed light both on the evolution of a virus infection in its reservoir host and on its manifestations after spilling over into a new host.

CONCLUSIONS AND FUTURE PERSPECTIVES

The role of a veterinary pathologist in dealing with emerging infectious diseases has varied from discovering the etiology and improving diagnoses to the detection of cures and might be expanded beyond this in the future. *That* veterinary pathologists can play an important part in protecting society from emerging infectious diseases is clear. These are turbulent times, and how the future will develop is becoming less certain. In the best-case scenario Artificial Intelligence will do the job faster and better, and veterinary pathologists can make themselves useful in other ways. In the worst-case scenario, due to worldwide trends, emerging infectious diseases like COVID-19 are expected to occur more frequently, so there is need for further involvement by veterinary pathologists.

While veterinary pathologists are used to providing a supporting role in collaborative research, several of these examples show that they also can and do (and maybe should) take the lead in research on emerging infectious diseases, including the role of principal investigator. We suggest that the following tools and assets ought to be featured more in veterinary pathology training: pattern recognition and a comparative approach are critical for discovering the etiology for classes of etiologies and to guide ancillary tests to identify the cause of disease. It is possible that in the future more patterns will be recognized, aided by development of new diagnostics for immune and cellular responses. In addition, the knowledge of what are 'background lesions' for common species, including free-ranging animals – or in other words – the further understanding of what is within normal limits (see Unique Features sections of the taxa-focused chapters) is invaluable to interpret both histopathology and ancillary test results. Veterinarians are in a unique position to educate others about the importance of species differences in health and disease. For improving the diagnosis, it is essential that pathologists collaborate with experts in the field and learn about other techniques that might be combined to make diagnosis easier and quicker in the future. There is a need to think beyond the diagnosis toward the implications that the histopathological findings might have for the management of the disease. Pathologists can, and do, more than diagnose disease. By looking at the tissues and combining findings with available knowledge, pathologists can provide much more information. Pathologists should integrate the histopathology findings with clinical history, other tests performed to investigate how the animal may have become infected, possible routes of transmission, and how emergence of the disease fits with ongoing environmental changes, anthropogenic activities, or both. For testing candidate vaccines and therapeutic agents, pathologists should continue to do careful and detailed observations that they have been so well trained and be active participants in the scientific process. We encourage veterinary pathologists to come up with hypotheses, maybe even beyond the research question of the experiment (e.g., treatment A is associated with less infection, but more eosinophils than treatment B – suggestive of an allergic immune response). Pathologists should be involved in method development to quantify observations, making it possible to test hypotheses and to make observational data more robust, for example, looking at genetic and protein levels in tissues. Trends and dose responses might not always be statistically significant. Here, it is important to realize that the low numbers of animals in experimental groups and the unpredictability of certain subtle findings increase the chances of such statistically nonsignificant, but potentially biologically relevant, trends. These might still be relevant to pursue and mention, especially if the trends fit with mechanistic explanations. Novel imaging and machine learning techniques might aid in this hypothesis testing, and it seems logical for veterinary pathologists to learn to work with these programs. For investigating the pathogenesis, pathologists should go beyond the knowledge of anatomy and histology of the hosts, to integrate the natural history of the host. The knowledge of the animal's behavior and interactions with its environment should be integrated with the knowledge of the pathogenesis of the emerging infectious disease to gain novel insights. Overall, when dealing with emerging infectious diseases, we encourage veterinary pathologists to occasionally take the time to step back for a while and think outside the box.

Beyond emerging infectious diseases, veterinary pathologists also have a role to play in other major threats to human society, especially the climate crisis and biodiversity loss, which partly share the same underlying causes as emerging infectious diseases. Veterinary pathologists, in general, and zoo and wildlife pathologists, in particular, have the appropriate background, interest, and expertise to help to deal with these major challenges. Because their work entails comparing diseases among species, including humans, and operating at the interface between nature conservation, animal husbandry, and public health, they are exquisitely prepared to balance the health and other needs of humans with those of domestic animals, wildlife, and ecosystems. In essence, this is a One Health approach, as recently defined by the interdisciplinary One Health High-Level Expert Panel.[25] In this way veterinary pathologists will not only be able to help prevent future emergence of infectious diseases but also contribute to attaining a sustainable society.

References

Full references for this chapter can be found online at https://doi.org/10.1016/B978-0-443-18437-6.00040-7.

Chapter 2

Wildlife Necropsy

Denise McAloose[a], Kathleen M. Colegrove[b], and Alisa L. Newton[c,d]

[a]Pathology Department, Zoological Health Program, Wildlife Conservation Society, Bronx, NY, United States; [b]Zoological Pathology Program, College of Veterinary Medicine, University of Illinois, Brookfield, IL, United States; [c]ZooQuatic Laboratory, LLC, Baltimore, MD, United States; [d]OCEARCH, Park City, UT, United States

INTRODUCTION

From the Greek for *pathos* (suffering) and *logia* (study of), pathology is the study of diseases, their cause or etiology, pathogenesis, and the structural and functional changes (clinical signs or symptoms) they produce. Clinical (or laboratory medicine; see "Introduction to Comparative Clinical Pathology" chapter) and anatomic pathology are the two main branches of this specialized medical field. In veterinary medicine, a postmortem/necropsy examination on a dead animal provides a significant opportunity to understand normal anatomy, biology, and the causes and various manifestations of disease. The aim of this chapter is to provide basic information for prosectors and pathologists performing necropsy examinations and disease investigations on nontraditional species both in individual animals and in outbreak situations.

Thorough postmortem examinations capture all of the information that is relevant to the death of an animal or group of animals. When considered narrowly, results provide information about a single animal. More broadly and depending on the context, the compilation of historical and animal-related data forms the basis of understanding disease and the impacts of pathogens at the individual, population, species, and ecosystem levels. These data may be locally, regionally, or internationally relevant for wildlife, zoo, livestock and companion animals, and for human public health. Necropsies are performed for many reasons. These include (but are not limited to) health and disease surveillance programs; characterization of normal and abnormal gross and morphologic anatomic features; establishing baseline health parameters and normal reference intervals; identification of the cause/s of morbidity and mortality in individual animals, groups, or populations; contribution of data and biomaterials to short and long-term research; determination of the effectiveness of medical or husbandry interventions or mitigation activities; collection of forensic information necessary in legal proceedings and prosecution (see "Wildlife Forensic Pathology" chapter), and in teaching and training. In conservation efforts, necropsy data is also important in rescue, recovery, reintroduction, and translocation programs. For example, to provide information about disease presence/absence in assurance colonies or relocation animals as well as in endemic populations and sympatric species to prevent unintended disease transmission. Results can also be used in the establishment of protected areas or to influence policy decisions (e.g., habitat use and resource extraction).

Protocols and procedures for laboratory or field-based necropsies for a number of terrestrial or aquatic, domestic, and nondomestic species are available from a variety of sources. These include governmental, nongovernmental, and university or zoo-based biology, veterinary medicine, and conservation organizations. Historically and to date, much of this information and these resources have been available in textbooks or printed manuals. Related information can now also be accessed online, with some of it in the form of instructional videos. A few examples of online protocols are those that are available through The World, European, or American Association of Zoos and Aquariums (WAZA, EAZA, and AZA, respectively), Taxon Advisory Group (TAG) or Species Survival Plan® (SSP) Programs, and the International Union for the Conservation of Nature (IUCN), though there are many others.[2,13,16,18,21,24,25,26,35,41] Some provide general instructions while others are taxon, genus, or species based. Some of this information, including a basic necropsy form, and vertebrate (carnivore) and invertebrate (several groups) tissue checklists, can also be found in the Digital Content for this chapter.

NECROPSY BASICS

Safety

Personal safety, the safety of the team, biosecurity, and communication are of paramount importance during any necropsy examination regardless of scope or scale. The complexity and coordination of these activities will differ

Pathology of Wildlife and Zoo Animals. https://doi.org/10.1016/B978-0-443-18437-6.00031-6

between an individual animal death and a disease outbreak or mass mortality event. But the basic principals are the same.

It is important, especially in field situations, to evaluate local environmental conditions to identify problems that pose safety risks, such as cold or rain, rising tides or suboptimal animal position, or presence of local scavengers or predators. In some situations, creative solutions or enlisting help from others (e.g., heavy machinery to move or position an extremely large carcass) will correct the situation and allow an examination to be safely performed. In others, selective or abbreviated necropsy procedures or abandoning the necropsy altogether may be the best and safest solution. Considering the safety of the media or the public and their pets is also important. Addressing this may require restricting access to the necropsy site with pylons, tape, or rope and providing restricted access signage or designated personnel for crowd control; local law or other wildlife enforcement agency presence may be necessary to protect the team and/or the public under certain circumstances.

Whether in a laboratory or the field, the best outcomes involve following standard protocols and having backup plans. This is equally important when responding to a mass mortality event or performing a necropsy on an individual animal in a laboratory. You can't plan for all possibilities. However, it's important to work out the details of your situation-based needs and communicate the plan and contingency plans to those involved in the necropsy as well as to those providing logistical, technical, medical, and financial support.

Roles and Responsibilities

In advance of a necropsy, it is important to assign roles and responsibilities. Standard tasks include identification of a lead prosector/pathologist and individuals or teams to collect morphometric data; perform prosection; label and manage samples (diagnostic, research, archival); perform photodocumentation; manage data/data sheets (record, upload, and organize) and ensure fulfillment of all protocols; ensure participant safety; ship samples; perform cleanup, disinfection, and carcass disposal; and manage communication within and between necropsy investigative team members and stakeholders, including the media. The least complicated situations are those involving a single animal, prosector and support staff in a controlled setting, like a necropsy suite or a laboratory. More complicated situations, for example, large carcass necropsies (e.g., whales, elephants), confiscation events, or outbreak situations or mass mortality events, will involve a greater number of people from highly trained experts to unskilled volunteers, so a more formal and structured set of guidelines, protocols, reporting structure, and chain of command is needed to ensure everyone's safety and successful outcomes (see "Mass Mortality Events and Outbreaks" section). Organizations may find it helpful to have a dedicated necropsy team in place or on-call to manage and coordinate necropsies of very large animals.

Carcass Management

Consideration and plans for carcass disposal should be coordinated well in advance of the necropsy or as standard operating procedures for institutions. In some situations, carcasses may be left in place for natural processes to occur (e.g., scavenging, decomposition) or arrangements may need to be made for burial or relocation. Preplanning aids in ensuring adequate availability of removal equipment and disposal containers, especially for very large animals, and, in the case of euthanasia, will allow decisions to be made that mitigate the risk of unintended secondary exposure/intoxication.

Regulations and Permissions

Just as there are laws and regulations that govern the jurisdiction and handling of wildlife and zoo species, permissions to perform postmortem examinations and collect and distribute tissue samples are also required. The two are not always the same, even in zoo settings. It is therefore important to understand under whose jurisdiction (e.g., native peoples, local to central governmental authorities, wildlife agencies) a species is managed and what permissions are needed to perform necropsy examinations. In addition, the conditions of a permit may require the holder to provide reports or data to the permitting organization. It is therefore important not only to obtain permission to perform necropsies but also to understand and comply with the restrictions and requirements associated with that permission.

If at any point before, during or after a necropsy examination or mortality investigation, there is suspicion or confirmation of a locally, regionally, or internationally reportable/notifiable disease [e.g., World Organization for Animal Health (WOAH)] or if criminal activities are suspected, appropriate authorities must be immediately contacted and activities coordinated. In legal or forensics cases, collection of data and samples using appropriate chain of custody and documentation (see "Wildlife Forensic Pathology" chapter) is also necessary. In addition, structured reporting systems exist in every country. It is important to be familiar with those systems that apply in the jurisdiction in which you work. Direct contact with regulatory agencies or online information will provide the most up-to-date information about those diseases that are listed and reportable (e.g., https://www.woah.org/en/what-we-do/animal-health-and-welfare/animal-diseases/-2023).

NECROPSY EXAMINATION

Preparation

Before a postmortem examination begins, it's worth remembering that *you* are the most valuable asset in the process. All conclusions drawn in a mortality investigation rest on the information and samples you and your team collect. Best intentions and memory are no substitution for the collection of written and photographic documentation. Also, since it is uncommon to retain an entire carcass for sampling after death, the postmortem examination is your first, and perhaps only, opportunity to collect circumstantial evidence and diagnostic samples, and to permanently record the information from which relevant death factors and baseline information are established. Each person involved in the necropsy must take personal responsibility for their actions. Essential in all of this is focus and the engagement of your brain and critical senses: sight, smell, touch, and hearing. Outcomes are as good as the data you collect and the focus you bring to the tasks. What you do during and after the examination will make the impossible, possible.

The tools needed to perform a necropsy examination will vary based on the number, size, and shape of the carcass/es (e.g., dissecting scope for very small animals, hoists, cranes, and power tools for large animals). Box 2.1 contains a general list of necropsy supplies, tools, and equipment that can be modified to appropriately address the species and circumstances at hand.

The Essentials

All mortality investigations, regardless of whether they involve a single animal or a large mortality event, share a set of consistent elements: an iterative diagnostic plan (Fig. 2.1) that applies evidence against a set of case definitions (Box 2.2) and differential diagnoses to establish a cause of death. Central in this process is the necropsy examination. Providing specific necropsy protocols for all taxa and for every situation is beyond the scope of this chapter. However, all necropsy procedures follow a similar, basic format.

Box 2.1 Necropsy Examination Supplies, Equipment, Tools, and Personal Protective Equipment (PPE)

Documentation

- Forms: necropsy protocol and form, morphometrics data sheet, tissue sample checklist, human interaction form, notebook/paper [regular and waterproof (e.g., Rite in the Rain®)].
- Labeling: tape, laundry tags with metal clips, pencils, waterproof marking pens and pencils (to label samples that will be immersed in liquid fixatives), Tyvek®.[a]
- Photodocumentation: digital or film camera, batteries, memory cards, labels (include in every image).

Safety and basic PPE

- Clothing: gloves (surgical, cut-proof), mask/respirator [surgical, N95, powered air purifying respirator (PAPR), masks with integrated face shield], eye protection (goggles, face shield), surgical scrubs, lab coat, coveralls, rubberized overalls, aprons, boots, gloves, caps (head cover), weather or task appropriate clothing (cold, rain, etc.).
- Disinfection: sponges, dish soap, scrub brushes, bactericidal and virucidal disinfectant, bleach, alcohol (70% ethyl alcohol).
- Other: biological safety cabinet, first aid kit, communication link (e.g., satellite phone), SDS (safety data sheets), MSDS (material safety data sheets).

Tools

- Cutting: scalpel blades, scalpel handles, knives (6″ or 8″ blade, flensing, etc.), saws (e.g., hack saw, reciprocating saw), band saw, axe/hatchet, knife sharpeners,[b] scissors (small and large), rongeurs, bone shears, loppers (hedge clippers), chisel/wedge (e.g., T-shaped), mallet/hammer, cutting boards.
- Tissue handling: forceps, meat hook.
- Containers: rigid leakproof wide-mouth spill-proof screw top containers (various sizes), zip-top plastic bags and Whirl-paks® (various sizes), serum tubes for fluid, blood and urine collection, aluminum foil, Teflon® bags, cryovials, sterile vials/containers/bags, sterile needles and syringes (various sizes), trochar (various sizes).
- Morphometrics (metric): ruler, calipers, tape measure, scales.
- Sterilization: sterile instruments, matches or propane torch and searing blade/spatula, isopropyl alcohol for flaming instruments.
- Culture: sterile swabs, urine cups, bags, culture transport media/tubes (for bacteria, virus).
- Tissue fixation: 10% neutral buffered formalin, 4% glutaraldehyde or other EM fixative in small screw-top vials, isopropyl alcohol (for ecto and endoparasites, cytologic preparations, etc.).
- For genetic/molecular diagnostics (aliquot into small screw-top vials): 20% DMSO/saturated saline solution (genetics), RNA-later® or TRIzol® (molecular diagnostics).
- Lighting: headlamp, flashlight, portable lighting, batteries, light bulbs, generator, extra bulbs and fuel.
- Cold chain: ice chest, ice packs, refrigerator, freezer (−20°C, −80°C, dry ice, liquid nitrogen), absorbent packing materials.

Continued

Box 2.1 Necropsy Examination Supplies, Equipment, Tools, and Personal Protective Equipment (PPE)—cont'd

- Laboratory equipment: microscope [for field settings: field adapted (mirror as a light source), car battery adapted power source, generator], centrifuge.
- Other: ropes/straps/chains, string/suture, parafilm, glass slides and slide boxes, water supply/source (for cleaning/clean-up), plastic tarps, plastic tape/ropes to cordon off necropsy site, garbage bags, biohazard bags, disinfectant, bleach, sponges/scrub brushes, paper towels, portable generator (for electric powered saws, refrigerators, etc.).

Additional equipment for small carcasses (<5 g)

- Magnification: dissecting microscope, magnification headband or surgical loupe.
- Cutting: microdissection forceps, scissors.

Additional equipment for megavertebrate (bison, gaur, whale, elephant, etc.)

- Cutting: flensing knives, reciprocating saw, large hammer/mallet and chisel, appropriate sharpeners, shovel.
- Tissue handling: large meat hooks, gaff hook.
- Morphometrics (metric): 20 m long (min).
- Mobility (for carcass burial or moving carcass, appendages, etc.): hoist/crane with large capacity mounted scale.
- Containers: large sealable containers (e.g., vials to garbage cans).
- Other: thick rope/chain (min of 20 m long), block and tackle.

This is a general guide that may be tailored to meet the specific situational needs of a particular necropsy examination/s.

[a]When labeled with permanent marker or pencil, it may be placed in liquid (e.g., formalin) for identification purposes.

[b]For standard knives, recommend hand-held sharpeners not sharpening stones since the use of stones requires significant expertise and inexperienced operators may dull rather than sharpen cutting instruments (e.g., knives, scissors).

Adapted from Munson L. Necropsy procedures for wild animals. In: White L, Edwards A, eds. *Conservation Research in the African Rain Forests—A Technical Handbook.* New York, NY: The Wildlife Conservation Society; 2000:196–217 and Woodford M, Keet D, Bengis R. Procedures for wildlife veterinarians and field biologists. In: Woodford M, ed. *Published Jointly by the Office International Des Epizooties, Care for the Wild and the Veterinary Specialist Group/Species Survival Commission of the World Conservation Union (IUCN), Paris, France;* 2000.
http://www.seadocsociety.org/wp-content/uploads/Orca-necropsy-protocol-FINAL-May-15-2014.pdf.

Collection and Documentation of Relevant Historical Information

In addition to information that will be collected from the carcass, it is also important to collect and record relevant clinical and death-related information. This includes basic (e.g., species, breed, gender, age/age group, birth and death dates) and medical information about the subject animal, contact animals, and other sympatric contact species (wildlife/vermin, humans, pets, livestock, etc.). A complete data set also includes health and disease-related epidemiological (number of deaths, spatial/temporal events, etc.) and environmental information. Carefully examine the area and collect information about the location in which an animal was found dead (e.g., GPS location, weather, environmental conditions, changes in husbandry practices, evidence of a struggle) to record those things that may be relevant to the circumstances of an animal's death. For example, weather patterns or environmental sampling of soil/sediment or water in aquatic settings may be important in die-offs of multiple or large numbers of animals and/or when there are deaths across species or taxa. These and other types of information may be of importance as forensic evidence in a legal case (see "Wildlife Forensic Pathology" chapter).

Personal Protective Equipment

Personal protective equipment (PPE) includes gloves, surgical masks or other respirators, aprons, boots, and dedicated clothing (e.g., coveralls and surgical scrubs) (Box 2.1). Some combination of PPE should be worn during all necropsy procedures, cleanup, and carcass disposal. The type of PPE needed is informed by a risk assessment that takes into account the common diseases in a particular and sympatric species or in a collection or region. In some situations, prophylactic vaccinations are also needed to safely perform necropsy procedures. In addition to basic PPE, eye protection should be worn during venomous animal necropsies or to protect against splash risk. Additional attention should be paid to proper PPE any time equipment that can aerosolize tissues or pathogens (e.g., high-pressure hose, drills, and saws) is used. Properly fitted N95 respirators, half- or full-face respirators with appropriate cartridges, or powered air purifying respirator (PAPR) should be worn when performing necropsy examinations on animals that pose a risk as carriers of zoonotic diseases and those that can be transmitted via aerosol routes (e.g., SARS-CoV-2 susceptible species, tuberculosis in great apes and elephants). Rabies vaccination or proof of protective titer should be standard if working with carnivores and other mammals. A necropsy should not be performed in cases where the risk to human health exceeds what one can be protected against with the available PPE or vaccination status. Veterinarians and human health professionals are typically responsible for making decisions about PPE and necessary vaccinations based on assessments that rely on an understanding of specific and relative risk, knowledge of infectious diseases, institutional standard operating protocols, and best practice protocols (e.g., United States Centers for Disease Control (USCDC), US Department of Agriculture, WOAH).[22,28]

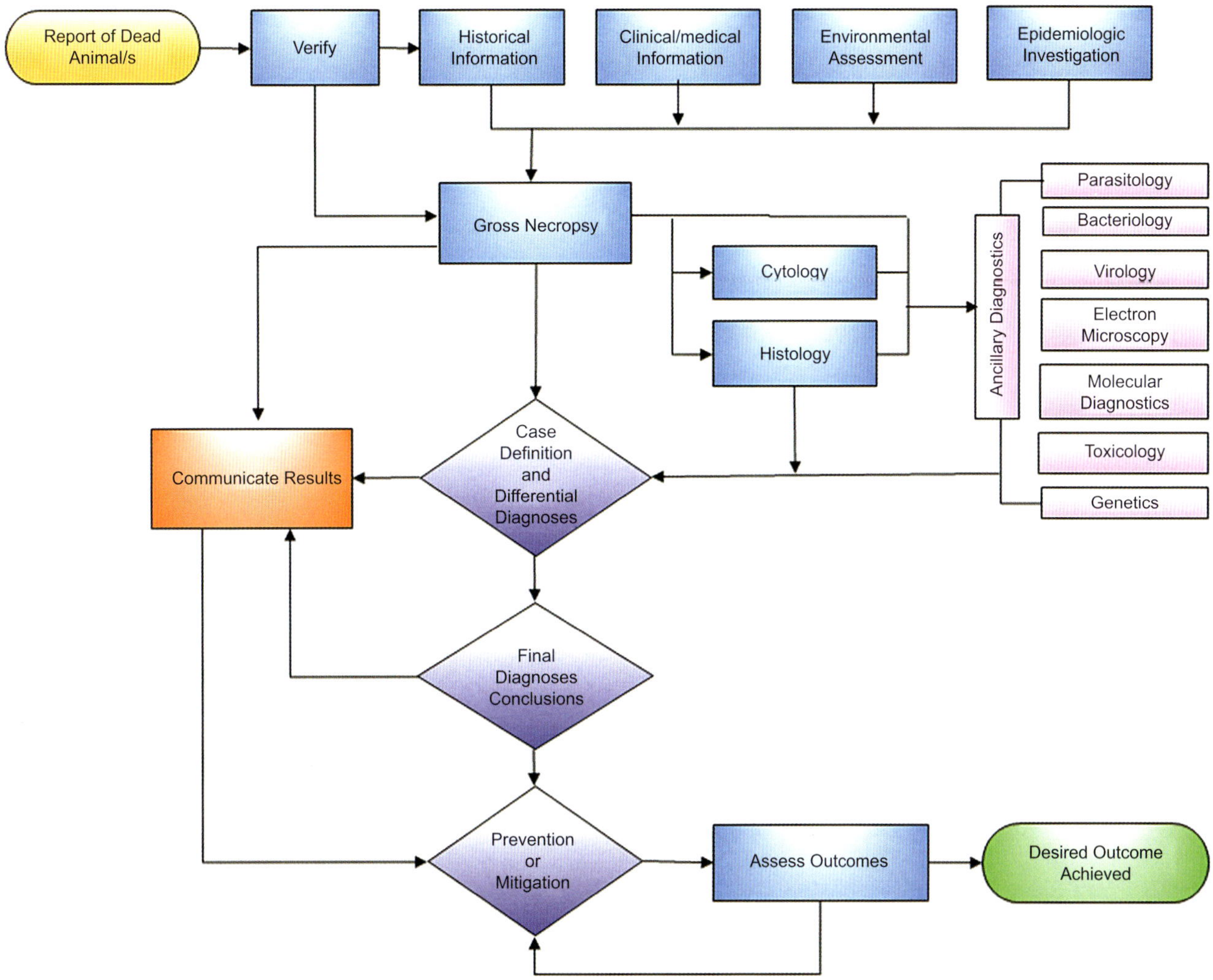

FIGURE 2.1 Mortality investigation flow chart. All mortality investigations include a consistent set of standardized activities. These include collection of information related to the circumstances of an animal's death, gross necropsy and sample collection for histology and ancillary diagnostic testing, and establishing a set of differential diagnoses and case definitions against which test results are applied in order to arrive at final diagnoses and conclusions. Integral in this process is collection of information and documentation of all findings, and clear and consistent communication with all relevant stakeholders.

Box 2.2 Case Definition

A **case definition** is a mechanism for categorizing animals as having or not having a disease. It is based on a clearly stated set of criteria that includes history (e.g., epidemiological, temporospatial, medical, husbandry), signalment (e.g., species, age, sex), clinical signs, and necropsy, histology, or other diagnostic test results. Fulfillment of the criteria determines whether an animal is considered a "case." In some situations, multiple case definitions may be used.

A case definition should contain criteria for "inclusion" as well as criteria for "exclusion." For example, if the case definition requires that affected animals are neonates, an animal in another age class (e.g., juvenile, subadult, adult) would be excluded and not classified as a case. Sometimes the absence of critical information is sufficient for exclusion of a case from a series. For example, if sex is important, then absence of documentation of the sex of the carcass can result in exclusion of an individual as a case even if all other information is of high value.

Cases can be classified using the following criteria:

Confirmed: Fulfills all criteria for inclusion, including positive laboratory results.

Probable: Fulfills all or most of the criteria for inclusion but lacks laboratory confirmation.

Possible: Has some but not all criteria for inclusion and lacks laboratory confirmation.

Negative: Does not fulfill the criteria for inclusion and lacks laboratory confirmation.

External Examination

Under ideal conditions, a necropsy is performed and samples are collected from freshly dead animals. If necropsy is delayed, carcasses should be refrigerated or kept cool on ice for several hours to a few days. Freezing should be an option of last resort as the freeze/thaw cycle will alter the color of many tissues, kill many pathogens (e.g., bacteria and fungi), and ice crystal formation can significantly limit the value of histology. However, under some circumstances, for example, when a necropsy cannot be safely, thoroughly, or adequately performed in a timely manner, a necropsy can be performed on previously frozen carcasses, though the results may not be optimal.

A good rule of thumb before performing a necropsy procedure is to familiarize yourself with the normal anatomy, basic biology, and natural history of the species with which you will be working (e.g., normal diet, fecal consistency, color patterns, sexual dimorphism, common and zoonotic diseases) as this will help in your understanding of what you see, from anatomy to gastrointestinal content to comparative histologic features. This is particularly important when performing a necropsy on a species that you are not familiar with and will inform activities related to personal protective equipment, risk, and findings. Reviewing taxa-specific chapters in this text and the Digital Content of this chapter, other reference textbooks, online resources, and contacting colleagues with species-specific expertise to discuss normal and abnormal findings are all important resources that can be consulted prior to performing a postmortem examination. It is also worth remembering that while every taxa and species has unique features, familiarity with one often serves as a helpful reference for another. For example, familiarity with the anatomy of well described and studied domestic or farmed species is directly applicable to the anatomy of most nondomestic mammals, fish, reptiles, and birds. Invertebrates provide a greater challenge. This is an incredibly diverse group that includes more than 99% of all animal species on the planet. Over the past decade, there has been increased interest in this group, and expertise and resources have increased. Recommendations below and in the invertebrate chapter in this book (see "Invertebrates" chapter) provide basic anatomy and an approach to invertebrate necropsy.[10,13,30]

Written descriptions of visual observations (animal, physical environment, etc.) remain the cornerstone of a necropsy examination. Two important general rules of thumb during a necropsy are to describe what you see and limit descriptions to factual observations. For this, it is worth remembering that you describe all sorts of things every day and the language you use to describe what you see during a necropsy examination is no different than the language you use during everyday conversations. Remember, a description should include those details that allow someone who did not witness the necropsy to create an accurate mental image of what you saw. Also, remember that a description is not an interpretation or a diagnosis. Those are best left to the pathologist to formulate based on your description and other documentation (e.g., photographs). Basic necropsy descriptions include information about location, number, color, size, shape, distribution (e.g., focal, regionally extensive, diffuse), consistency and texture (soft, firm, hard), and odor. Weights of organs are important to establish normal reference intervals and for organs thought to be too large or too small. Size measurements should be taken in three dimensions (length × width × height/thickness). For example, "dozens of pinpoint to 5 mm in diameter, soft, light tan nodules that contain thick, white material on cut section are present throughout the liver." Differentials based on this description include abscesses, hyperplasia, or cancer. The description is the same regardless of the pathologist's final diagnosis, which is based on a combination of the gross description as well as histology and ancillary diagnostics.

For photodocumentation, always include a size reference (a ruler is best) and animal-related information (at a minimum, identification number and date). Rulers and identification tags should be placed next to and should not obscure the lesion. Take pictures from both a wide and close perspective to capture relevant size as well as fine details of the lesion. All information from the necropsy and collected samples should be entered into logs and secure databases developed for this purpose. This is important for a whole host of reasons, including rapid data retrieval and collation during mortality events, and data and sample sharing for immediate and future research projects. Information should be as detailed as possible to ensure that important information is not forgotten or lost.

All necropsy examinations should be systematic and follow standardized procedures and protocols. This ensures that the carcass is thoroughly examined, standardized information for generating baseline information and documenting abnormalities is captured, and a complete set (or sets) of data and samples are collected for baseline, diagnostic, archival, and research purposes. There are several common approaches to a necropsy examination. One is not necessarily better than another and your preferred method may vary depending on the species and specific circumstances. Regardless of your method, of paramount importance is that you work safely and have a consistent and systematic strategy for organ/tissue and organ system examination. This is especially important when working in large teams or in complicated situations (e.g., mass mortality event). It is also important to work "cleanly" to maximize visibility and minimize contamination of the worksite and exposure risk. Where applicable, standardized necropsy procedures and/or sampling/research-associated requests (e.g., institutional biomaterials or research requests) and those that are required by government agencies should also be followed. When practical, fulfilling biomaterial requests allows valuable, additional information to be generated.

The external examination is an assessment of the tissues outside of the main body cavities. The exam begins with an overall assessment of the animal's body/nutritional condition, direct measurement or estimation of carcass weight, collection of morphometric data, photodocumentation, and assessment of the state of decomposition. Every carcass, regardless of its state of decomposition, can provide valuable scientific data. In many field situations, access to fresh carcasses is the exception rather than the rule. Environmental conditions (e.g., high humidity, high temperatures, or dense foliage in natural or recreated settings) and interval from death to discovery all affect the degree of decomposition. Tissue degradation varies by tissue type and affects the reliability of diagnostic test results. It is therefore important to document the degree of carcass decomposition, otherwise known as carcass condition, to contextualize observations and diagnostic test results. This is a subjective assessment and a relatively common scheme that categorizes postmortem carcass condition from excellent (freshly dead) to mildly autolyzed, moderately autolyzed, severely autolyzed, or skeletonized/desiccated/mummified. In the marine mammal community, several protocols employ a similar carcass condition coding system (Box 2.3, Figs. 2.2A–D).[7] Assessment of carcass condition is particularly useful in low-resource or time-restricted settings to identify those carcasses that will be the most diagnostically valuable and to direct sampling and subsequent testing.

In vertebrates, a systematic external (external to the body cavities) examination with the animal in lateral or dorsal recumbency includes evaluation of normal and abnormal findings in the structures of the head (e.g., eyes, ears, nares, bill, beak, teeth, cere, gills, tonsil); the outer covering of the body (e.g., skin, hair, feathers, scales, claws, hooves); the subcutis, glands (e.g., periorbital, salivary, scent, salt, perianal, uropygial); external reproductive system tissues; the musculoskeletal system (including assessing muscle mass and one to multiple joints and bone marrow); and the peripheral nervous system. It also includes objective (measured) and/or subjective assessment of fat stores (e.g., subcutaneous, blubber, cavitary, coelomic). In some cases, skeletal preps (e.g., warm water maceration) and radiographs or other imaging modalities [e.g., radiography, computed tomography (CT), magnetic resonance imaging (MRI)] may expose or aid in documenting lesions. Written and photographic documentation of all notable normal and abnormal findings should be recorded and representative sample/s from all tissues and all lesions should be collected and placed into 10% neutral buffered formalin (NBF) for histology or collected as appropriate for ancillary diagnostic testing (Table 2.1). Box 2.4 and the Digital Contents include examples of general forms and tissue checklists for vertebrate and invertebrate necropsies.

Internal Examination

In vertebrates, the internal examination often begins by making a ventral midline incision from the symphysis of the mandible to the inguinal region and reflection of the skin and limbs. This is followed by opening (1) the abdomen or coelom by removing the abdominal/coelomic wall distal to the xiphoid and ribs, (2) the thorax by cutting the diaphragm (when present), checking for negative

Box 2.3 Carcass Condition Codes

Code 1: Freshly dead/euthanized animal.

Code 2: Mild decomposition. The body is intact. The skin is firm with little drying or wrinkling of the skin, eyes, and mucous membranes. The eyes are clear. The abdomen/gastrointestinal tract is not distended with gas. The tongue, penis, and anus do not protrude.

Code 3: Moderate decomposition. Intact carcass but there is distention (bloating) of the abdomen due to gas accumulation. The tongue, penis, and rectum may be protruding, especially in large animals. The skin is dry and the hair may epilate easily with manual handling. The organs are discolored but are largely identifiable.

Code 4: Advanced/severe decomposition. Intact, collapsed carcass. The hair and skin slough easily with manual handling. Some internal organs are intact and identifiable but others (e.g., gastrointestinal tract) are extremely friable or liquefied.

Code 5: Mummified or skeletal remains. Often all that remains are portions of dried skin, hair, and/or skeletal muscle over bones.

This scheme primarily describes assessment of vertebrates. A similar scheme can be applied with some modifications to invertebrates. Carcasses in condition code 1 or 2 are the most valuable for sample collection for histology and ancillary diagnostic testing. Many tests can also be run on samples collected from code 3 and 4 carcasses but caution should be used when interpreting results. A limited number of diagnostic tests, for example, skeletal morphometrics and some molecular diagnostic tests, can be run on code 5 carcasses; many caveats apply.

Adapted from Geraci J, Lounsbury V. In: Geraci JR, Lounsbury VJ, eds. *Marine Mammals Ashore: A Field Guide for Strandings*. 2nd ed. Baltimore, MD: National Aquarium of Baltimore; 2005.

FIGURE 2.2 Carcass condition in large whales based on methods by Geraci and Lounsbury.[7] These or similar schemes are available or can be created and applied to other species. Code 1. Animals are freshly dead (not shown). (A) Code 2. Minke whale. The carcass is intact, not bloated, and the skin is firm with little wrinkling or drying. Note the heavy equipment being used to move and bury the carcass (truck on right; backhoe on left). (B) Code 3. Melon-headed whale. The carcass is intact but bloated, the penis is extruded, the skin easily sloughs when handled, and has already sloughed off of large parts of the carcass. (C) Code 4. Southern right whale. The carcass is intact but collapsed. (D) Code 5. Southern right whale. The carcass is desiccated and what remains covers remnants of the remaining skeletal structure. *(Parts C, D: Photos Courtesy of the Southern Right Whale Health Monitoring Program.)*

pressure in the thorax, and removing the ribs unilaterally or bilaterally (depending on position), and (3) the pericardial sac to expose the heart. Some modifications will be necessary based on species-specific anatomy. It is standard practice to open these cavities and assess the general body condition of the carcass (e.g., cavitary fat stores, fat body size, size of the liver) and examine all organs and organ systems in situ prior to tissue removal or sampling. As for external examinations, knowledge of domestic animal anatomy is directly applicable and serves as a guide for the identification of tissues and organs in many nondomestic species. It is standard practice when opening these cavities to immediately collect sterile samples (e.g., heart blood, abscesses, cavitary fluid, fungal plaques) for ancillary diagnostics, archiving, and to perform in situ photodocumentation before handling the organs. If care is taken to open the cavities with minimal handling, the risk of contaminating diagnostic samples is minimized and collection of tissues for archival purposes is more easily managed.

TABLE 2.1 Diagnostic Tests and Sample Collection, Handling, and Storage[a,b]

Diagnostic Tests	Sample Collection	Sample Handling	Sample Storage
Histology[c]: the microscopic examination of tissue in order to study the manifestation of disease. It is a valuable adjunct to gross observation as many diseases may look similar macroscopically	• Collect samples from all organs and any tissue that looks abnormal (relative to adjacent tissue) • Collect a representative sample from each part of the organ/lesion • Limit sample thickness to 0.5 cm ($1\ cm^3$) for good tissue fixation • For abnormal tissue, collect junction between normal and abnormal • Label important items or samples that came from a specific location • All samples from a necropsy may be placed in a single container	• Place tissue in 10% neutral buffered formalin (NBF) in leak-proof container • 1 part tissue:10 part formalin ratio for good fixation	• Short term: room temperature • Long term: room temperature • DO NOT: FREEZE
Electron microscopy[d]: higher magnification to resolve pathogens and tissue structure than can be achieved with routine light microscopy. Options include transmission, scanning, or negative stain electron microscopy	• Tissue should be no greater than $1\ mm^3$ • Common fixatives include glutaraldehyde or osmium tetroxide	• Place tissue into fixative in leak-proof container • 1 part tissue:10 part fixative ratio for good fixation	• Short term: refrigerate • Long term: refrigerate • DO NOT: FREEZE
Cytology: the microscopic examination of fluids and cells. It may be performed immediately and aids in making a diagnosis. Cytologic review may occur locally or at an off-site laboratory	• Fluids: direct smear of fluid in thin film onto slide OR spin sample to make cell pellet (cell poor sample) that is then smeared onto a slide • Tissues: touch prep: gently blot on a paper towel to remove blood, then gently touch tissue to slide surface • Tissues: crush prep: collect small tissue sample onto slide; cover with coverslip/cover glass; flatten tissue	• Fluids, touch prep: air dry, fix (e.g., methanol), stain (e.g., Diff-Quik®, Gram's, Wright's Giemsa, acid-fast); affix coverslip • Crush prep: immediately directly view unstained tissue/cells; affix coverslip	• Short term: clean, sturdy, crush-proof, bug/vermin-proof slide box or container • Long term: clean, sturdy, crush-proof, bug/vermin-proof slide box or drawer • DO NOT: EXPOSE AIR DRIED SLIDES TO FORMALIN
Microbiology: the study of microorganisms including bacteria and fungi. Typical tests include bacterial or fungal culture. The most commonly sampled sites are obvious lesions (e.g., abscess) or the intestinal tract	• Clean and decontaminate sampling site if contaminated (e.g., alcohol, heated blade) • Use sterile instruments/techniques; minimizes risk of contamination • Instruments: swab, needle plus syringe, scalpel blade, skin punch, trochar, etc.	• Place the sample/swab in appropriate medium for the desired test (e.g., culture media for aerobic or anaerobic bacteria or fungus)	• Short term (<1 week): Room temp OR refrigeration • Long term (>1 week): −80°C, liquid nitrogen (but may kill target pathogens) • DO NOT: FREEZE BLOOD OR SAMPLES FOR VIBRIO OR LEPTOSPIRE CULTURE
Virology: The study of viruses. Typical tests include viral culture or PCR. Common samples include lesions or organs such as lung, liver, spleen, lymph nodes, intestinal tract/feces, and brain	• Clean and decontaminate sampling site if contaminated (e.g., alcohol, heated blade) • Use sterile instruments/techniques; minimizes risk of contamination • Instruments: swab, needle plus syringe, scalpel blade, skin punch, trochar	• For virus isolation, place swab/tissue in viral transport medium (VTM); for molecular diagnostics, place fresh sample in sterile collection vile (no media) • No transport media, no refrigeration available: place 1 part tissue: 5–10 parts 50% buffered glycerin	• Short term (<48 h): refrigerate • Long term: fresh tissue, VTM: −80°C or liquid nitrogen • DO NOT: EXPOSE SAMPLES TO DRY ICE REFRIGERANT

Continued

TABLE 2.1 Diagnostic Tests and Sample Collection, Handling, and Storage[a,b]—cont'd

Diagnostic Tests	Sample Collection	Sample Handling	Sample Storage
Parasitology: the study of parasites and parasitism. Depending on the species, a wide variety of external (ectoparasites) and internal parasites (endoparasites) may be present. Collection allows for parasite identification and documentation. Some parasites also harbor infectious agents, such as bacteria and viruses, and may be used for PCR analysis or virus isolation	• Ectoparasites: collect grossly visible parasites. Swab or scrape ears or skin for small parasites (may not be grossly visible) and smear material as a thin film with or without mineral oil onto a slide • Endoparasites: collect grossly visible parasites. Collect fresh feces in a clean container or bag	• Place grossly visible parasites in 70% alcohol (ethanol) or formalin for identification • Examine scrapes under oil immersion microscopy immediately • Examine feces using standard procedures (e.g., gross appearance, direct smear, fecal floatation [for metazoan parasites]; glycerin sedimentation [for amoeba])	• Short term: fresh feces: refrigerate • Short and long term: visible parasites: 70% ethyl alcohol or formalin; room temperature
Toxicology: toxins are important to consider, particularly in the acute death of an animal or group of animals with few clinical signs. Toxins may be natural/biological (e.g., poisonous plants, algae), agricultural/chemical (e. g., herbicides, pesticides, rodenticides), or industrial (e.g., lead, arsenic, waste from mining). Because many toxins are unstable and degrade with time, rapid testing, appropriate handling and short and long-term storage are essential	• A large volume of fresh tissue is often needed (50–500 g of sample). • Tissues useful in a toxin screen: o Liver o Kidney o Brain o Stomach or intestinal content o Fat o Eye (aqueous and vitreous humor) o Blood o Urine o Hair, flight feather, baleen	• Place tissues in clean, leak-proof containers with no chemical preservatives • Note: aluminum and plastic may leach and interfere with tests, so collection of two sets of tissues in different container types is recommended when possible • Filter paper (e.g., Whatman® 903) can be used for heavy metal analysis	• Short term: refrigeration or freezing for transport is essential • Long term: −20°C, −80°C, liquid nitrogen
Molecular diagnostics: typically performed for genetic or pathogen testing or discovery. Target tissues usually include lesions and/or liver, kidney, lung, spleen, brain, pancreas, gonads, lymph nodes, conjunctiva, mucocutaneous junctions of the oral cavity or genital tract, fluids/secretions, hair, or skin	• Small sample volume needed (1 g or less of tissue, 0.1 mL of blood or a clean dry swab) • Sample site should be clean and sterile • Wear gloves; use sterile instruments and techniques to collect samples and minimize risk of contamination • Instruments typically used for sampling: swab, needle and syringe, scalpel blade, skin punch, trochar	• Place sample in sterile container • Options: fresh frozen; viral transport media (VTM) RNAlater®, RNAlater-ICE®, ethanol, DNA/RNA Shield®, TRIzol® (check preference of your laboratory, research project, etc.) • Blood or fluids: may be dried on a piece of filter paper (e.g., Whatman® FTA for DNA/RNA-based infectious disease; Whatman® 903 for proteins)	• Short term (see manufacturer's recommendations): Fresh: −80°C; RNA later®: (<1 day) room temp (37°C), (<1 week) refrigeration (4°C), (>1 week) −80°C; filter paper (<1 m), room temperature • Long term (see manufacturer's recommendations where appropriate): fresh frozen, VTM, RNA later®, DNA/RNA Shield®, filter paper: −80°C

[a]This list includes several common test categories and general recommendations for sample collection, handling, and storage. Specific best practice protocols should be consulted whenever possible in the development of your in-house standard operating procedures for sample management.
[b]Recommended for labeling: pencil for glass slides; for bags, tags, and labels - permanent marker or other appropriate inks that are resistant to alcohol, formalin, and other solvents and that will not peel off with refrigeration or ultralow temperatures (e.g., 20°C, −80°C, liquid nitrogen) that are routinely used in necropsy or cytologic procedures or tissue storage/handling and transport. It is also useful, in addition to labeling the container/bag, to place a labeled tag (resistant ink) in the container (e.g., cut pieces of Tyvek®).
[c]Tissues collected for histology are typically placed in 10% neutral buffered formalin (NBF). For NBF and tissues placed in other fixatives, it's critical to maintain a minimum ratio of 1 part tissue to 10 parts fixative to ensure proper fixation. Salinity and pH of fixative are important considerations for marine invertebrates. For invertebrates, 10% formalin or sea-water formalin are adequate fixatives in most cases. Specialized fixatives are preferred for some species, such as zinc-formalin (Z-fix) for corals and Davidson's for soft-bodied invertebrates (cnidaria). Decalcification can cause significant artifactual changes in the heavily mineralized tissues of some species, particularly corals and echinoderms. If the structure of the skeleton of these animals is critical for evaluation, the pathologist should consider submitting tissues for processing in a mineralized state (e.g., through plastic embedded and sectioning at a bone pathology laboratory) or a slow method of decalcification should be chosen (e.g., using the acetic acid in Davidson's solution or pH balanced EDTA). Enrobing the delicate tissues of coral species in alginate prior to decalcification has proven an excellent method for protecting the delicate surface polyps while retaining skeletal structural features. Formic acid and hydrochloric acid may be used for decalcification but tissue disruption through gas bubble formation should be recognized and not interpreted as a lesion. For species with a chitin-based ectoskeleton, Pyreni's solution may be used to soften the chitin and provide softening of tissue prior to sectioning.
[d]Recipes for marine invertebrate-specific electron microscopy fixative are recommended.

Box 2.4 Examples of Gross Anatomy and Sample Collection Checklists[a,b]

Vertebrate: Carnivore

	Skin		Gall bladder
	Umbilicus (neonates)		Pancreas (with and without intestine)
	Skeletal muscle		Spleen
	Peripheral nerve (e.g., sciatic)		Adrenal glands (both)
	Bone/bone marrow (e.g., femur, rib)		Kidney (include cortex, medulla, pelvis)
	Salivary gland		Ureters (both)
	Lymph nodes		Urinary bladder
	Tongue, tonsils, salivary gland/s		Urethra
	Thyroid glands (both)		Accessory glands
	Parathyroid glands		Testes/epididymis (bisect longitudinally)
	Trachea		Penis
	Esophagus		Ovaries
	Heart (free walls, septum, valves)		Uterus
	Lung (each lobe and bronchus)		Cervix
	Thymus		Vagina
	Diaphragm		Tonsil
	Stomach (cardia, fundus, pylorus)		Brain (whole or cut along midline)
	Small intestine (with and without pancreas)		Spinal cord
	Large intestine, cecum		Pituitary gland
	Liver (with and without gall bladder)		Eye

Invertebrate: Echinoderm

Body wall		**Structures on aboral surface**	
	Central disc (aboral and oral surfaces)		Mouth
	Rays (aboral and oral surfaces)		Peristome
	Interradius (space between the rays)		Ambulacral groove
	Endoskeleton (dermal ossicles and spines)		Ambulacral spines
	Coelomic epithelium		Tube feet
Structures on oral surface			Terminal tentacles
	Madreporite		Pigmented spot (eye)
	Anus	**Internal structures**	
	Spines		Esophagus
	Pedicellaria		Stomach
	Papulae		Pyloric cecae
			Gonad (ovary/testis)

[a]See the Digital Content for additional general necropsy forms and information on invertebrate necropsy and gross anatomy, histology and tissue checklists for arthropods, cnidarians, echinoderms, bivalve and gastropod mollusks, and cephalopod mollusks.

[b]Note: These are general lists that should be modified to reflect the anatomy of the species being necropsied and any additional specific needs (e.g., research requests, specific pathogen screening). For example, collection of tissues from bovids and camelids should include sections from each gastric compartment; fish sampling would include gill and anterior kidney. Collect one to multiple samples of normal and abnormal tissue and junctions between normal and abnormal. For bilateral organs (e.g., kidneys, adrenal glands), samples should be collected from each. Multiple parallel cuts should be made in solid organs to assess the parenchyma prior to sampling. Tubular organs should be opened along their long axis and examined prior to sampling. Tissue samples for histology should not be greater than 0.5 cm thick and should be placed in 10% neutral buffered formalin (NBF) for fixation (1 part tissue:10 parts NBF). In some animals (small animals less than ~5 g; those that have fragile tissues like arthropods, etc.), perfusion of the body cavity/ies or fixation whole with subsequent postfixation dissection may be preferable to gross dissection of the fresh animal.

Three common approaches for performing detailed tissue dissection and sample collection after in situ examination are (1) removal of tissues in the thoracic cavity/cranial coelom and the abdominal cavity/caudal coelom en bloc, (2) removal of tissues by organ or organ system, or (3) a combination of the two. Preference as to which approach is taken, whether some modification of these is chosen, and the order of the examination is up to the prosector. Important considerations include the species, situational needs, known or suspected diseases, and general rates of autolysis (e.g., removal, dissection, and sampling of gastrointestinal tract tissues might occur before thoracic cavity tissues to minimize autolysis in diagnostically important tissues). In a very fresh carcass, you may want to consider examining the gastrointestinal tract later in the process to reduce contamination of the body cavities with content. Regardless of the approach, a thorough, systematic examination is essential. As for the external examination, written and photographic documentation of all notable normal and abnormal findings should be recorded and representative sample/s from all tissues and all lesions should be collected and placed into 10% NBF for histology (Box 2.e1) or collected as appropriate for ancillary diagnostic testing and archiving. See Box 2.4, Table 2.1, and the Digital Content for sample collection, storage, and handling recommendations and for general forms and tissue checklists.

CARNIVORES

Depending on carcass size, carnivores are placed in dorsal or lateral recumbency. After completion of the external examination and sample collection, in situ examination, and collection of sterile samples and samples for ancillary diagnostics or archiving, the internal organs are removed and more thoroughly examined (see Box 2.4 and Digital Contents).

In the thoracic cavity, examine and remove the "pluck" en bloc. The "pluck" includes the tongue, oropharynx, and esophagus; thyroid and parathyroid glands; larynx, trachea, lungs, thymus (if present), pulmonary lymph nodes, and the heart and pericardial sac. Prosectors may elect to remove and collect samples from the thyroid and parathyroid glands prior to removing the pluck as they are relatively small and easily overlooked or "lost" during a necropsy. Examine and collect samples of tongue, thyroid glands, parathyroid glands, thymus (if present), and lymph nodes. Open the esophagus along its length noting mucosal changes (e.g., discoloration, erosion, ulceration, proliferation) and intraluminal content. Open the trachea along its length and the bronchi and major vessels into the lungs noting mucosal changes and any abnormal intraluminal content. Palpate the lungs, noting their consistency, any nodules and color and consistency of fluid that drains from cut surfaces; note whether lung samples sink (abnormal) or float in formalin (normal). Open the chambers of the heart and vessels (e.g., aorta, vena cava, pulmonary artery) noting any inconsistencies in the heart valves, clotted blood (postmortem clots are smooth and shiny; thrombi are dull and may be adherent), and/or defects in the walls. Collect samples of all of the above tissues. Heart samples should include papillary muscles, valves, and vessels.

The organs in the abdomen include the gastrointestinal tract (stomach, intestines, and liver), hematopoietic tissues, endocrine tissues, and the urogenital tract, which may be removed en bloc then further dissected by the organ system after removal. An alternate approach is to examine, remove, dissect, and collect tissue samples by organ system: endocrine (paired adrenal glands); hematopoietic [spleen, lymph nodes (e.g., ileocecocolic, mesenteric)]; digestive system—liver and gall bladder with or independent of the stomach, pancreas, cecum, and small and large intestines; and urogenital system (kidneys, ureters, urinary bladder, urethra, ovaries, uterus, cervix, vagina, intra-abdominal testes, accessory sex glands). Prosectors may elect to remove and collect samples from the adrenal glands prior to removal of other tissues as they are relatively small and can be difficult to identify in animals that are overconditioned or if the abdomen becomes contaminated by blood or gastrointestinal contents during tissue/organ removal. Prosectors may also choose to linearize the intestinal tract to allow better visualization of the abdominal organs in situ prior to removal of other organs or organ systems.

Examination of the abdominal viscera occurs by longitudinally bisecting (e.g., adrenal glands, lymph nodes, kidneys, gonads) or making a series of parallel cuts (e.g., pancreas, liver spleen) in solid tissues and opening tubular or saccular structures (e.g., stomach, intestine gall bladder, urinary bladder, uterus) along their length. Paired endocrine and other bilateral organs (e.g., ovaries, kidneys) should always be examined independently and in relation to the contralateral organ; samples should be collected from each of these paired organs. Note any inconsistencies or abnormalities in the tissues and the content and consistency of intraluminal materials or fluids.

There are two basic approaches to the examination and removal of the brain and pituitary gland. After disarticulating the head at the atlantooccipital joint and removing the skin and soft tissues of the head, the first and simplest is to longitudinally bisect the head along midline and remove the bisected halves of the brain. Many pathologists prefer, when practical, removal of the intact brain. This is done by removing a roughly rhomboid-shaped skull cap through a set of bilateral, parasagittal cuts that extend rostrally from the lateral aspect of the occipital condyles and are connected by a coronal cut in the rostral parietal bone. Removal of connective tissues overlying the sella turcica and adjacent bone is often necessary for exposure and collection of the pituitary gland.

RUMINANTS AND OTHER HERBIVORES

Ruminant necropsies are often performed in left lateral recumbency such that the approach to the abdomen places the rumen deep to other parenchymal organs such as the liver and spleen. There are pros and cons to left or right lateral recumbency for necropsy of these species and often the choice depends on personal prosector preference, the position of the carcass if found dead, and options for repositioning the carcass. The procedure is otherwise like that for carnivores.

Special mention of anthrax, a serious zoonotic bacterial disease that is caused by the bacterium *Bacillus anthracis* and can be fatal in humans, is of particular concern during ruminant necropsies though it should be remembered that in addition to wild and domestic ruminants, other species (e.g., horses, zebra, wild pigs, carnivores) can also be infected.[18,41] The most common gross finding is bloody discharge from body orifices (similar findings can be seen in several other disease processes, for example hemorrhagic diseases in primates or swelling of the neck and head in nondomestic carnivores). Another finding that should raise the suspicion of anthrax is a swollen/enlarged spleen that exudes tarry, black blood on the cut section. If anthrax is suspected, it is recommended that the necropsy not proceed and the body should not be moved (to minimize bacterial sporulation) until appropriate authorities are notified for their input and recommendations about how best to proceed and testing to confirm or exclude its presence is completed. If possible, guarding the carcass to prevent scavenging while diagnostic testing is being performed is recommended. Rapid screening of multiple cytologic preparations of blood (ear nick from the downside ear or coronary band are recommended) or aspirates from the thoracic cavity or tissue (submaxillary lymph node, spleen) should be air dried, fixed in methanol, and stained with polychrome methylene blue (McFadyean reaction) or Giemsa stain. The bacteria are large bacilli with square ends and a clear or pale halo; they must be differentiated from *Clostridium* spp., which have rounded edges and no halo (Figs. 2.3A, B; see "Introduction to Comparative Clinical Pathology" and "Bovidae, Antilocapridae, Giraffidae, Tragulidae, and Hippopotamidae" chapters). Bacterial culture can be performed for confirmation, but cytology will typically provide a quicker preliminary result. If positive, it is generally recommended that the necropsy not proceed, and local, state, regional or national governmental agencies responsible for animal and public health and safety should be immediately contacted. Recommendations for disposal of infected carcasses should follow recommendations from regulatory agencies and can include burial, burning or a combination of the two.[6,39]

MARINE MAMMALS

Marine mammal anatomy and necropsy are similar to that of terrestrial mammals, especially carnivores, with some notable exceptions.[2,24–26] This includes a blowhole in cetaceans that is the proximal end of the respiratory tract and opens rostral to the brain on the dorsal surface of the head, and hemidiaphragms in manatees and dugongs that along with the gastrointestinal tract, must be removed prior to in situ examination of the lungs (see "Cetacea," "Pinnipediae," and "Sirenia" chapters). For whales, flensing knives are useful and often necessary to remove the skin and thick blubber layer.

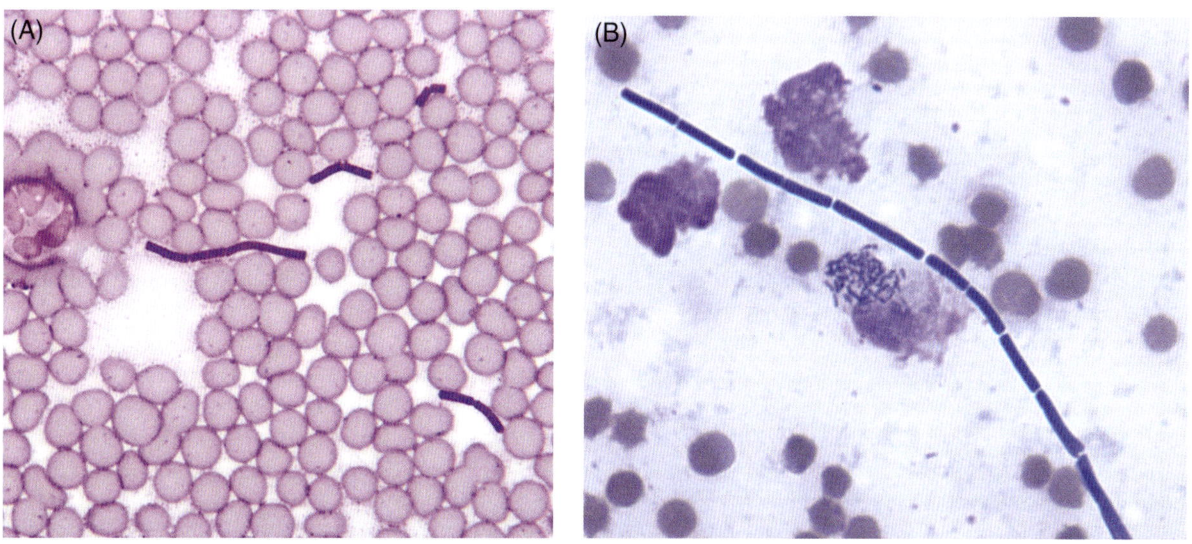

FIGURE 2.3 *Bacillus anthracis* and *Clostridium* sp. in blood films. (A) Anthrax bacteria can be seen as individual organisms or in chains. They are large (up to 10 μm long), blunt ended (appear rectangular), and are surrounded by a thin, clear halo in cytologic preparations. (B) They must be differentiated from *Clostridium* spp. bacteria, which will proliferate after death and are also large but have rounded edges and do not have a halo. Modified Wright's Giemsa. *(Part A: Photo Courtesy of University of Illinois Zoological Pathology Program; Part B: Photo Courtesy of J. Steyl, University of Pretoria, South Africa.)*

BIRDS

In birds, wetting the carcass or ventrum with water or disinfectant is a quick solution for managing the flight of feathers that are released when opening a carcass. Upon opening the carcass, assess the air sacs for transparency/opacity, fluids, or plaques. There are three common approaches for the removal and examination of the organs in birds: removal of all tissues from the tongue to the cloaca en bloc; removal of tissues en bloc except for the lungs, kidneys, adrenal glands, and gonads (located adjacent to the vertebral column on the ventral surface of the kidneys at their proximal end); or removal of the organs ventral to dorsal with the animal in dorsal recumbency. Note that the thyroid glands are often located near the thoracic inlet and base of the heart adjacent to the common carotid arteries. For the gastrointestinal tract, pancreas is often collected and attached to one or both adjacent lengths of the duodenum.

TURTLES

The most challenging aspect of a turtle necropsy is entering the coelom. This is most easily achieved with the animal in dorsal recumbency to facilitate the removal of the plastron by cutting the marginal bridge that connects the carapace (upper/dorsal shell) to the plastron (bottom/ventral shell) (see "Chelonia" chapter). Scissors, rongeurs, loppers, saws, a machete, or similar tools are needed to cut the soft to boney (species dependent) marginal bridge. The scapula and pelvis must then be reflected in order to allow maximal exposure of the coelomic viscera.[4,33,35] Necropsy examination is otherwise like that in carnivores.

SNAKES

Before beginning a snake necropsy, it is important to know if the snake is venomous or nonvenomous since venom remains viable after death. For carcass submission, it is recommended that the head of venomous snakes be taped or otherwise secured such that the mouth is not free to open and the snake should be placed in a completely secured, see-through container for transport to minimize the risk of accidental envenomation. Information about the species and antivenin should be attached or submitted with the animal/container. In addition to basic PPE, protocols should include mandatory eye protection and guidelines for contacting emergency/first responders, an envenomation center/emergency hospital/zoological institution/others that maintain antivenin stocks, and an emergency hospital with expertise in managing envenomation patients. For venomous snakes, the recommended first step in the necropsy procedure after confirming that the snake is dead by observation through the see-through container is to remove the head and place it into formalin. Formalin deactivates many of the proteins and peptides in snake venom; however, it is not known if formalin deactivates all venom proteins, and tissue fixation may be incomplete in deep tissues, especially in large heads.[15,35] Extreme caution and necropsy by experienced pathologists/prosectors should be the rule rather than the exception when handling the heads of venomous snakes to prevent injury. Most snake necropsies are performed with the animal in dorsal recumbency. A ventral incision along the length of the body is most easily accomplished at the junction between the dorsal and ventral scales but can also be made along ventral midline. Snakes have a single thyroid gland that is located immediately orad to the heart. There is variability in the symmetry and functionality of lungs among snake species. For example, boas and pythons have two lungs (left, right) while colubrids, viperids, and elapids have a rudimentary left lung; the left is absent or vestigial in some species (see "Serpentes" chapter).[36] The hemipenes in snakes are located in the subcutis, just caudal to the cloaca. The remainder of the necropsy is like that in carnivores.

LIZARDS

The approach to a lizard necropsy is like that of carnivores with the exception that the kidneys are located deep within the pelvis. The skin in Crocodilia, especially on the head and back, can be very thick and difficult to cut or dissect free from underlying bone (see "Crocodilia" and "Lacertilia" chapters). Reciprocating or other saws are useful for cutting these tissues and are often necessary for brain removal in larger animals. For venomous lizards, knowledge of the types and location of venom production should inform appropriate precautions and protocol development.

FISH

An important component of a fish necropsy is cytologic examination of the skin and external mucus layer, fin tissue, and gills (see "Osteichthyes" and "Chondrichthyes" chapters); squash preparations and cytologic examination of lesions in fresh tissues are also valuable. Necropsy is typically performed in lateral recumbency and is like that in carnivores. Sharks are covered by placoid scales (dermal denticles). These structures create an armor-like surface that is difficult to cut. Insertion of a blade under the skin and cutting from deep to superficial rather than superficial to deep is recommended for reflection of the skin. For electric eels, the use of rubberized gloves is important as these fish retain their charge, which dissipates over time, after death. In some, especially small fish, sampling of the anterior (head) and posterior kidneys is often easiest by collecting a transverse section across the body that includes kidney and adjacent skeletal muscle and vertebrae. Necropsy of chondrichthes and osteichthes is similar with the exception of the cartilaginous skeleton in the former, which makes cutting the skeleton easier. The spines of some fish and rays are venomous. Knowledge of the types and location of venom production in venomous fish species (e.g., spines of stonefish) should inform appropriate precautions and protocol development.

INVERTEBRATES

As previously stated, invertebrates present a challenge at necropsy due to the myriad species and body types represented by this vast group of animals. Critical to gross and histologic evaluation, as with all species, is first reviewing basic textbooks and journal publications on anatomy, biology, and published diseases,[5,8,10,11,13,14,19,31,34] and through this, generating a checklist of organs and organ systems for consistent gross and histologic evaluation. As a guide, see Box 2.4 and the Digital Content for general comments related to the unique anatomy of invertebrates, suggestions for external and internal examinations, and tissue checklists for arthropods, cnidarians, echinoderms, bivalve, gastropod, and cephalopod mollusks. In some species, gross evaluation will significantly disrupt tissue architecture. This is particularly true of arthropods and bivalve mollusks. In these groups, limited external evaluation of the ectoskeleton (carapace) and soft tissues, and collection and examination of hemolymph and cytologic (wet mount) evaluation of the gills and external abnormalities are recommended prior to fixing the animal entire for histologic evaluation. Special considerations need to be discussed with a histology laboratory to determine what methods are available to soften exoskeletons (e.g., decalcification, chitinase). For many marine invertebrate species with calcium-based skeletons, typical decalcification solutions (hydrochloric acid, formic acid) will cause a rapid reaction and gas production which can disrupt tissues and be disconcerting to inexperienced histology technicians. Less aggressive decalcification protocols are recommended.

NEONATES

There are a number of notable, normal features to be aware of when performing necropsies on neonates. In mammals and birds, these include skeletal muscles that will generally be lighter in color than in adults; a soft, light tan/orange liver (due to hepatocellular glycogen stores); low amounts of subcuticular or cavitary fat; and a very soft brain even in a very fresh carcass. In mammals, the ductus arteriosus and foramen ovale may be probe patent, the umbilicus may be raw/unhealed and the urachus may be patent, the gonads and adrenal glands may appear relatively large, and if the carcass was refrigerated after death, the lenses may become opaque white (rather than translucent). In birds, an egg tooth and (occasionally prominent) pipping muscle are seen. Sections of the pipping muscles should be sampled as they are susceptible to degenerative changes that can impact hatching. Birds and reptiles will have a yolk sac. In term hatchlings, the yolk sac should be internalized and may occupy a majority of the caudal coelom. In birds, it is generally absorbed within several days and retention could reflect underlying disease. In preterm fetuses, the yolk sac may be completely external to partially external. The bursa of Fabricius and thymus may be poorly cellular at hatch in altricial (relatively featherless, require paternal care) or precocial (covered with down feathers, require limited paternal care) birds (species dependent); both will develop after hatch then slowly regress over time. As in their domestic counterparts, important abnormal findings to look for and document in nondomestic species include the presence of congenital abnormalities or defects (e.g., cleft palate, extra or missing limbs or limb deformities, cardiac septal defects), and excessive numbers of squames, meconium, yolk, or other foreign material in the lungs.

EGGS

Approach to the necropsy of an egg begins with the external examination and collection of morphometric information of the egg as a whole (e.g., weight, straight length and width, longitudinal and transverse circumference). In species with an air cell, a hole is made into the blunt end of the egg (away from the developing embryo/fetus) to assess the content and examine the air cell (it should be empty; swab samples for ancillary diagnostics of accumulated fluid or collection of abnormal materials or membranes can be collected at this point), to measure shell thickness, and to assess and the condition of the egg membranes (color, intact, pipped, etc.). The hole is then enlarged and the membranes are cut to assess the egg for fertilization (e.g., presence of blood spot and embryonic membranes) and the developmental stage and position of a developing fetus/embryo. References that describe developmental stages and appropriate and lethal embryonic positions for hatching are available for some domestic birds and may serve as a valuable resource when assessing eggs from domestic and potentially some nondomestic species.[29] Embryonic development and staging tables are also available for many nonavian species.[9,12,20,40]

VERY LARGE ANIMALS

For large animals like elephants or whales, removal of the "pluck" or other organ systems en bloc may not be practical due to their large size. Organs will therefore need to be removed independent of each other. Heavy equipment, for example, a backhoe, may be useful in maneuvering the carcass, limbs, or head and for removing and/or weighing of organs, limbs, or the head, and during burial (Fig. 2.2B). Standard safety practices should be undertaken when this type of equipment or machinery is utilized.

SMALL ANIMALS

For small animals (<5 g), a dissecting microscope or head loupe is an incredibly helpful tool. External examination with the animal in dorsal or lateral recumbency is the same as that for other species as is the internal examination. After examination and sampling of tissues for ancillary diagnostics and nonformalin tissue archiving, tissues can be removed separately for fixation. In cases where the intestines are very small and it is impractical to open them, injection of the unopened gastrointestinal tract with fixative (e.g., neutral buffered formalin) may improve preservation of the mucosal lining. In some cases (e.g., animals <1–3 g), the best option after opening the body cavity/ies may be to fix the entire animal in formalin due to its small size. Alternatively, a small incision in the body cavities to allow fixative to contact the internal organs or through injection of the body cavities with fixative can be performed.

MASS MORTALITY EVENTS AND OUTBREAKS

Necropsy evaluation is a critical component in any outbreak or mass mortality event. These events typically involve an elevated number of deaths over baseline (Fig. 2.1, Box 2.5). Good communication and coordination of all activities are essential in order to manage the increased complexity due to the expanded scope and scale that are inherent in an outbreak/mortality event. In many countries, local and national plans or other strategies may have been developed and are activated in outbreak situations. Familiarity with those that exist in your region is important as valuable resources that aid or are necessary in your response and investigation may be available through these programs. For example, in the United States, the Incident Command System (ICS) is a management strategy that was created to address the demands and complexities inherent in emergency response or natural disaster events. The ICS is used by federal agencies and its adapted use can provide organizational structure to outbreak and mass mortality event response and investigations, and for postmortem examinations of large animals or complex cases such as beach stranded large whales.

In addition to activities that are part of routine necropsies, critical components in mortality events and outbreak investigations include: (1) identifying an investigative team and resources, (2) verifying that an outbreak is occurring, (3) establishing a case definition(s) and categorizing cases using results of necropsy examination, histology, ancillary diagnostic test results, etc., to establish the scope of the outbreak, (4) establishing baseline information about the event and disease (e.g., temporospatial, species, gender, and age class), (5) examining the descriptive epidemiologic features of the cases to generate hypotheses, (6) testing hypotheses and performing additional analysis as needed, (7) implementing control measures (when possible), and (8) communicating findings and maintaining surveillance.[27,37] For any event, it is also important to quickly determine if a disease (e.g., infectious, toxic) with significant zoonotic or animal health impacts (e.g., anthrax, Ebola, highly pathogenic avian influenza) or illegal activities are present and to engage and work together with the appropriate agencies and authorities in the investigation (Box 2.5). It is also important that the investigative team have a broad range of expertise (epidemiology, population biology, pathology, clinical medicine, genetics, bacteriology, virology, toxicology, molecular diagnostics, species and environmental expertise, etc.) to help guide the investigation and interpret findings, other information/data and evidence.

Pathologists often work closely with epidemiologists during outbreak or mass mortality event investigations to

Box 2.5 Unusual Mortality Event/s

When compared to historical disease epidemiology, indicators of possible unusual morbidity or mortality events or possible criminal activities include:

- Increased morbidity and/or mortality over baseline.
- Increased magnitude or severity of disease.
- A temporal or spatial change in disease occurrence.
- Difference in species, age, or sex distribution of affected animals.
- Detection of disease in nonendemic or new areas.
- Suspicious activity at site of morbidity or mortality.
- Signs of toxicity unrelated to natural conditions.
- Reappearance of a previously eradicated or controlled disease.
- Presumptive or confirmed laboratory diagnosis of notifiable high consequence or foreign animal disease.
- Simultaneous or progressive outbreaks in multiple locations with no epidemiological link.

construct case definitions (Box 2.2), characterize the epidemiologic features of the outbreak (time frame, geographic borders, species affected, etc.), develop hypotheses on the cause of the event, analyze data, recommend ancillary diagnostic testing, and ultimately, to make recommendations for control measures and/or long-term surveillance. Epidemiologic characteristics of the outbreak along with historic data are important in establishing that mortality and disease prevalence are higher during the outbreak than is noted normally in the population. Sometimes, when working with free-ranging populations, the background prevalence of disease has not been well established. In these instances, using background data on closely related populations or populations with similar geographic ranges may be helpful. These historical data and necropsy findings guide decisions related to first-tier ancillary diagnostic testing (those deemed most relevant and immediately important). For example, in small cetacean unusual mortality events along the eastern and southern coasts of the United States, testing for cetacean morbillivirus and marine biotoxins are often high on the list of tests to be considered due to historical occurrence. First-tier diagnostic tests and histopathology findings guide additional ancillary diagnostic testing.

Approaches to necropsy evaluation and sampling during an outbreak or large-scale mortality event are like those utilized during routine, thorough systematic diagnostic necropsies. Establishing a sampling and necropsy strategy as soon as possible at the outset of the outbreak or mortality event is crucial for obtaining the most complete sample set and the best data to aid the investigation. Sampling and necropsy protocols should incorporate appropriate tissue and sample collection for evaluation of not only the first-tier disease concerns but also a broad range of alternative disease processes. Collecting multiple sets of archived tissues that will be available for a wide range of ancillary diagnostic tests is recommended. Often with large-scale mortality events, it is not practical or logistically possible to examine or sample every carcass. As such, focusing thorough diagnostic efforts on the freshest carcasses, especially those that die earliest in an event, may yield the most reliable data. In addition, plans should be established at the outset of the investigation for carcass management, database and sample management, storage and sharing of data, and communication with all relevant stakeholders, including governmental or other relevant agencies.

CLEANUP AND CARCASS DISPOSAL

As occurs during a necropsy procedure, PPE (gloves, masks, eye protection, etc.) should be worn during cleanup and carcass disposal. Blood and residual tissues should be removed from instruments and work surfaces with soap and water after which they should be disinfected. The best choices for disinfection have broad antimicrobial properties (e.g., VirkonS®, 0.5% sodium hypochlorite, 10% bleach, 70% alcohol, borax). Depending on the disinfectant (e.g., bleach), instruments may need to be rinsed after disinfection to prevent corrosion. It should also be remembered that all disinfectants are not created equally, so one should be knowledgeable about the product being used, understand its efficacy against different pathogens, use those that will be most broadly efficacious, and adhere to manufacturers' recommended contact times for adequate disinfection.[1,38] This is particularly important for pathogens that pose significant human and animal health risks. Of note in this regard are prions, which are resistant to a number of common disinfectants and other forms of destruction. Current, best practice protocols for disinfection and carcass management should be obtained and implemented in consultation with governmental agencies for suspected or confirmed cases.

Whether in a laboratory or a field setting, carcass and necropsy waste and sharps and infectious materials disposal should be performed in a manner that minimizes environmental contamination and exposure of the necropsy team or domestic or wild animals to disease agents. When considering options and performing disposal, it is also important to be aware of and adhere to relevant organizational, local, national, and international regulations and protocols for carcass and medical waste disposal, which may involve consultation with local environmental and health agencies. Options may be limited due to a number of factors including personal safety, location, size of a carcass/es, environmental conditions, location relative to water sources/catchments/wells, ability to safely move/relocate a carcass for disposal generally and within a reasonable timeframe (this may be logistically and politically complex and challenging), financial limitations, animal welfare, and willingness, ability, and permission of contracting facilities or landfills to accept carcasses and necropsy waste.

Options for carcass and tissue disposal include incineration/burning, burial, rendering, landfill, fermentation or biocremation (alkaline hydrolysis), which may be possible on-site or by external contractors, as well as mounding, composting, and natural decay at the site of death and necropsy. A number of detailed descriptions of several of these options are available and include those of national and international governmental and nongovernmental regulatory agencies.[6,23,28,39] On-site incineration/burning or burial may be the most practical option in some cases, for example, in mass mortality events. In known or suspected anthrax cases, incineration/burning or burial at a depth of 2m (to prevent scavenging) and coverage of the carcass with an unbroken layer of lime ($Ca(OH)_2$) have been described. In hot conditions, if neither is possible due to environmental conditions or the size or number of carcasses, guards can be posted to protect the site and carcasses from scavengers/predators for 4 days in tropical conditions, by which time it is reported that the anthrax bacilli will be destroyed by pH changes and putrefactive bacteria in the carcass, which limit its ability to sporulate.[41] If prion disease is suspected, a 2001 report suggested rendering, incineration, and alkaline hydrolysis (all under certain conditions) as the most reliable technologies at that time for reducing the infectivity of the organism.[28,39] From an environmental perspective, efforts should be made to minimize the volume of nonanimal waste (e.g., aprons, masks, gloves, plastic bags) that is generated and burned, buried, or transported. It is worth remembering that pathogens may persist, proliferate, and pose extended human and animal exposure risks in unburned, buried, and closed plastic bags.

SAMPLES

Collection

Even when a cause of death seems obvious, it's important that interpretations and conclusions drawn during a mortality investigation are informed by all relevant factual information. This process is sequential and iterative with gross necropsy and cytology being typically followed by histology. All directly affect decisions about ancillary diagnostic tests that are subsequently performed and are needed to confidently arrive at final conclusions (Fig. 2.1). In addition to a thorough, systematic necropsy examination, collection of a suite of tissues for histology and ancillary diagnostic testing and tissue archiving (for immediate or future testing) should be standard in all necropsy procedures (see Table 2.1, Box 2.4, and Digital Content for general forms and tissue checklists for vertebrates and invertebrates).

It may not always be possible to collect all of the recommended samples from each animal, but the more consistently these goals can be achieved and reports generated, the greater the chance of accurately identifying diseases and disease trends. It is also worth noting that even in this era of molecular diagnostic and other advanced test modalities (e.g., high throughput sequencing, metagenomics), the importance of collecting a complete set of tissues for histologic examination as a driver of ancillary diagnostics and for contextualizing diagnostic tests cannot be overstated.

Checklists for collection of "standard" sets of tissues are a component of all necropsy protocols (Box 2.4 and Digital Content for general forms and tissue checklists for vertebrates and invertebrates). These lists may be a simple tissue list or be more complicated and include instructions about where, why, and how many samples are to be collected from each organ/tissue/site for histology, ancillary diagnostics, or archiving. As a general rule, both grossly normal and abnormal (including junctions between normal and abnormal) tissue samples should be collected. Sample collection should always be performed with the expectation that they will be submitted for subsequent histology, ancillary diagnostic testing, and archiving and should therefore be optimally handled and stored to allow immediate or future testing (Table 2.1).

Pathologists are often asked about the minimum set of tissues that should be collected or that are necessary to make a diagnosis. The difficulty in answering that question lies in the fact that limiting access to samples limits the ability to identify and contextualize lesions, identify relevant concurrent diseases, and arrive at the best and most informed conclusions. The goal therefore should, whenever possible, be to collect a complete set of normal and abnormal tissue samples that can be processed and tested immediately or archived until this becomes possible. In situations where collection of a complete set of tissues is not possible, a "limited" set of high-value tissues should be collected. This will vary by species and circumstance but should include those tissues that are most likely to be most diagnostically relevant (common sites of infection, toxicity, etc.). Generally, these are major organs and include heart, lung, liver, kidney, spleen, brain/spinal cord, stomach, and small and large intestines. Targeted tissue collection can also focus on obvious gross lesions or sample collection for specific suspect diseases (e.g., brain, lung, tracheobronchial lymph nodes for suspected canine distemper testing; liver, kidney, fat, stomach content, spleen, hair, brain for toxicology screening; urine, and feces for domoic acid testing). Minimally invasive, closed-body, targeted biopsy sampling may be an option to consider in animals suspected of being infected with high-risk zoonotic pathogens (e.g., rabies, yellow fever and other arboviruses).[3,17,32] In some situations, environmental samples should also be collected. Collecting a limited set of tissues may provide a diagnosis. However, it may reflect an incomplete picture of the cause, contributing, or predisposing factor/s in death. In addition, and especially because tools like PCR are so sensitive, some organisms might be detected but their role in death or an outbreak may not be appropriately contextualized in the absence of other diagnostics or complete tissue examination. In worst-case scenarios, this can lead to misinterpretations and false conclusions that drive inappropriate husbandry, medical management, or other animal or outbreak-related activities and mitigation strategies.

Handling, Management, and Storage

Sample handling and storage are critically important to optimize the value of diagnostic samples and contributions to animal health, management, surveillance, and conservation outcomes (Table 2.1). All samples should be labeled (collection date, species, tissue/sample, identification number at a minimum), logged into a database (paper, electronic, etc.), and properly stored to ensure the successful management of what are incredibly valuable resources. Recommended for labeling is pencil for glass slides, and permanent marker or other appropriate inks that are resistant to alcohol, formalin, other solvents, refrigeration, and ultralow temperatures (e.g., −20°C, −80°C, liquid nitrogen) for bags, tags, and labels that are routinely used in necropsy and cytologic or histologic processing, and for tissue storage, handling, and transport. It is also useful, in addition to labeling the container/bag, to place an appropriately labeled tag in the container (e.g., cut pieces of Tyvek®); specific tissues of interest can also be labeled with laundry tags or pieces of suture and so noted in necropsy reports for later reference.

Tissues collected for histology are typically fixed in 10% neutral buffered formalin (NBF) (Table 2.1, Box 2.e1). For both NBF and alternate fixatives, it is critical to maintain a minimum ratio of 1 part tissue to 10 parts fixative to ensure proper fixation. It is also important to know the penetration limits of tissues being fixed. For example, tissues fixed in formalin for histology should not be more than 0.5 cm thick and tissues in glutaraldehyde for electron microscopy should not be more than 1 mm^3. Because fixed tissues become less pliable, wide-mouthed containers should always be used for ease and safety of later removal. For medium- to large-sized mammals, fixing the brain in a large container separate from other tissues may be necessary for adequate fixation. Institutions should work closely with the diagnostic laboratory that performs their histologic evaluations to ensure tissues are being collected, stored, and shipped in appropriate containers or leak-proof bags. It is also important to have adequate numbers of appropriately sized sample containers on hand and available during necropsy examinations.

Special handling is needed for invertebrate tissues. Fixation, decalcification, and dechitonization are critical for preparing invertebrate tissues for histologic evaluation. Salinity and pH of fixatives are important considerations for marine invertebrates. In most cases, 10% formalin or sea-water formalin are adequate fixatives. Specialized fixatives are preferred for some species, such as zinc-formalin (Z-fix) for corals and Davidson's for soft-bodied invertebrates (cnidaria). Recipes for marine invertebrate-specific electron microscopy fixative are available and recommended. In addition, decalcification can cause significant artifactual changes in the heavily mineralized tissues of some species, particularly corals and echinoderms. If the structure of the skeleton of these animals is critical for evaluation, the pathologist should consider submitting tissues for processing in a mineralized state (e.g., plastic embedding and sectioning at a bone pathology laboratory) or a slow method of decalcification should be chosen (e.g., acetic acid in Davidson's solution or pH balanced EDTA). Enrobing the delicate tissues of coral species in alginate prior to decalcification has proven an excellent method for protecting the delicate surface polyps while retaining skeletal structural features. Formic acid and hydrochloric acid can be used for decalcification in invertebrates but tissue disruption through gas bubble formation can occur and should not be interpreted as a lesion. For species with a chitin-based ectoskeleton, Pyreni's solution can be used to soften the chitin and provide soft, supple tissue for sectioning.

A set of general guidelines for handling and storage of samples for several of the most common diagnostic tests (microbiologic, virologic, parasitologic, toxicologic, molecular diagnostic, electron microscopic) is presented in Table 2.1. With the exception of formalin-fixed tissues, most diagnostic samples need to be refrigerated or frozen to prevent degradation and to maintain the viability of pathogens. This often includes establishing and maintaining a cold chain, which can be particularly challenging in field settings.[23] Equipment used for samples should not be used for holding food, beverages, or other medical materials like pharmaceuticals. Basic equipment typically includes refrigerators (4°C ice chests/boxes, ice/gel packs, etc.), freezers (e.g., 0°C, −20°C, −80°C; should not have automatic defrosting ("frost-free") options), and containers appropriate for standard and ultralow storage (e.g., cryovials). Other options include dry ice and liquid nitrogen (LN_2) [e.g., dewars (vacuum flasks), dry shippers (vapor shippers)]. Use and handling of samples stored at ultralow temperatures should be carried out with caution as exposure can cause severe burns. Dry ice and LN_2 vapor can cause asphyxiation and should be used only in ventilated spaces or containers to prevent explosion. Rubberized "o-rings" used in some cryovials/containers should only be used if rated for use for ultralow archiving to prevent contraction with freezing and explosive expansion when being brought to room temperature. For each of these ultralow applications, handling should be limited to trained personnel. For the electrical equipment that is being used to power refrigerators, freezers, or other storage methods, a consistent source of power is an absolute necessity. In all settings, backup sources of electricity, often in the form of a generator, are essential. In less developed settings, multiple generators may serve as both primary and backup power sources. For samples stored in dry ice or LN_2, monitoring and "topping-off" the ice or liquid should be part of routine management protocols. Scheduled monitoring and documentation of temperatures should also be included in standard operating procedures to ensure consistent and continual storage temperatures. Whenever possible, samples should be immediately stored, and frozen samples should be handled in ways that minimize freeze–thaw cycles.

Two significant sample management challenges are storage space and sample curation. With proper storage, tissues can be a valuable diagnostic resource for many years. Understanding and managing space, power needs, ventilation, duration of research projects, chain of custody, and other aspects of short- and long-term storage of samples all need to be considered. Plans and protocols for the curation of sample storage and movement (e.g., for diagnostic testing, archiving, to fulfill research requests) must also be developed and implemented. This includes managing the information related to samples and movement of the samples themselves. Sample information has historically been maintained in paper logs. However, many people, even in remote locations, now use electronic databases or software that needs to be regularly updated and should be regularly backed up to prevent catastrophic data loss. In some situations, for example, large, active sample repositories or zoo collections with large sample inventories, dedicated staff may be needed to curate and manage sample acquisition and disposition, databases, and sample retrieval and shipment. Archived inventories should be reviewed regularly, for example, every 6 months, to update additions and remove tissues that become redundant.

Transport and Permits

With the exception of some governmental and university-based laboratories, many organizations do not have the in-house resources or expertise to perform the complete suite of tests that are run as components of a complete necropsy/postmortem examination. Shipment of samples to external laboratories is therefore often necessary. In general, labeled samples should be placed in an appropriate leak-proof primary container (e.g., screw-top vials with parafilmed seals; zip-top bags, heat-sealed bags) and a secondary leak-proof container (e.g., zip-top bags) with appropriate absorbent; external packaging should also be leak proof and be sturdy enough to withstand the rigors of shipment (cardboard box with styrofoam liner, etc.). All sample shipments must comply with local, state/provincial, federal/national, and international laws and regulations to ensure human and animal safety and sample viability. In the United States, shipment of biological samples requires completion of specialized training offered through the US International Air Transport Association (IATA) and the US Department of Transportation (DOT); similar requirements apply in other countries. Special packaging, handling, and labeling may be required, for example, for shipments containing dry ice or LN_2.

For shipment to local laboratories, contacting the laboratory to discuss specific package labeling, submission forms, test requests, and minimum sample volumes is of value to ensure the best results. For international shipments, export and import permits to ship samples, and compliance with treaties and regulations concerned with the international movement of samples from endangered or managed species [e.g., CITES, US NOAA (marine mammals), US Centers for Disease Control (primates, rodents, civets, bats), US Department of Agriculture (USDA)] may also be required. The regulations and paperwork to import and export samples differ in every country, and processing can be complicated and time consuming. However, it is essential to understand the process and pay attention to the many associated details to be in compliance with permitting and shipping requirements. Failure to do so knowingly or unknowingly can result in significant fines, import/export permit refusal, delays resulting in sample degradation, confiscation, and return or destruction of samples by border patrol agents. It is a good idea, whenever possible, to maintain a duplicate set of samples at your facility as an assurance against problems that might occur during shipping. It is also important to recognize that diagnostic laboratories typically discard tissues, glass slides, or paraffin blocks after a certain retention time and many claim ownership of all submitted specimens. Written agreements, established prior to submission, may be required to ensure return, retention, or ownership of your samples.

DATA MANAGEMENT AND SHARING

All historical data, pathology reports, ancillary diagnostic test results, photographs, and information related to sample curation that are generated during a necropsy must be organized and managed. In many settings, this is currently achieved using a combination of paper and electronic documents, databases, and more recently, phone apps. Technological advances will likely push all activities related to documentation, even that done in the field, onto electronic platforms in the not-too-distant future, and the use and development of robust systems for secure data storage and backups is and will continue to be a necessity. These databases contain important baseline and disease-related information and lists of available samples, all of which are hugely valuable in retrospective and prospective projects focused on wildlife ecology, health monitoring and disease surveillance, and conservation projects for animals in zoos and other managed and free-ranging settings. Shared access to these valuable resources should be the rule rather than the exception, for example, through collaborative research projects, biomaterials requests, and contributions to other programs such as the Association of Zoos and Aquariums and European Association of Zoos and Aquaria Species Survival Plan (SSP), Taxon Advisory Group (TAG) and European Endangered Species (EEP) Programs. Whether retrospective or prospective, inclusion and participation of wildlife veterinary pathologists to prevent misinterpretation (a common problem in data review by nonpathologists) and guide early engagement in the development of protocols that involve necropsy and sample collection from free-ranging wildlife or zoo animals is essential.

COMMUNICATION

Communication, both during a necropsy or active mortality investigation and after, is critical and results should be shared regularly with stakeholders (veterinary staff, keepers, curatorial staff, collaborators, indigenous people, wildlife biologists, public, governmental agencies, etc.) as they become available and interpretations and conclusions are drawn. Immediate communication of test results is also specifically important for medical treatment and/or mitigation of disease in contact wild or domestic animals, staff, or others (e.g., public health networks) and for directing additional ancillary diagnostic testing. As mentioned earlier, suspicion or confirmation of listed, reportable or emerging or pandemic diseases must be immediately shared with appropriate local, national, and international authorities who may need to participate in the mortality investigation and disease management. The format for the communication will vary depending on need and may include sharing of complete or summary pathology reports; peer-reviewed publications; presentations at professional conferences or community meetings; development or iterative revision of protocols, practical technical manuals, and other teaching materials; or other forms of communication with the media including interviews and articles in the lay press. For high-profile cases or outbreaks with high public interest, it may be of value to identify a communications lead or team to field questions and comments that are informed by the pathologist and necropsy and investigative teams. Creating a list of key talking points to be used by the communications team is beneficial to guide messaging, especially when conveying complex scientific information to the public. Care should also be taken to only communicate confirmed data or to qualify all results as preliminary until they are confirmed.

CONCLUSIONS

A necropsy examination is a foundational skill for pathologists. Findings drive ancillary diagnostics and are the first step in identifying, discovering, characterizing, and contextualizing infectious and non-infectious processes as the causes of morbidity and mortality. A standardized approach and plan for information and sample collection are applicable across species and essential to ensure you have what is needed to draw appropriate conclusions. Collaboration and communication across disciplines allows us to tackle even the most complicated situations.

ADDITIONAL RESOURCES

Eros C, Marsh H, Bonde R, O'Shea T, Beck C, Recchia C, Dobbs K, Turner M, Lemm S, Pears R, Bowater R (2007) Procedures for the Salvage and Necropsy of the Dugong, 2nd ed. The Great Barrier Reef Marine Park Authority, Townsville, Queensland, Australia.

Fanelli A, Awada L, Caceres-Soto P, Diaz F, Grillo T, Gizo I, Hamilton K, Rolez CL, Melens P, Morales R, Mur L, Muset S, Nake L, Thompson L, Wannous C, Tizzani P. Sensitivity of an international notification system for wildlife diseases: A case study using the OIE-WAHIS data on tularemia. Zoonoses Public Health. 2022 Jun;69(4):286-294. https://doi.org/10.1111/zph.12916.

Farris S, Squires M, Ridgley F, Lavergne E, Serota M, Mazzotti F (2013) Necropsies of Reptiles: Recommendations and Techniques for Examining Invasive Species. Davie, FL.

Geering W, Penrith M, Nyakahuma D (2001) Disposal Procedures. In: *Manual on Procedures for Disease Eradiction by Stamping Out*. Food and Agriculture Organization of the United Nations. Food and Agriculture Organization of the United Nations. Emergency Prevention System for Transboundary Animal and Plant Pests and Diseases, Rome, Italy, p 130.

Gestal C, Pascual S, Guerra A, Fiorito G, Vieites JM (eds) (2019) Handbook of Pathogens and Diseases in Cephalopods. Springer Open, Cham, Switzerland.

Gulland EM, Diefauf LA, Whitman KL (eds) (2018) CRC Handbook of Marine Mammal Medicine. CRC Press, Boca Raton, FL.

Kuisma E, Olson SH, Cameron KN, Reed PE, Karesh WB, Ondzie AI, Akongo M-J, Kaba SD, Fischer RJ, Seifert SN, Muñoz-Fontela C, Becker-Ziaja B, Escudero-Pérez B, Goma-Nkoua C, Munster VJ, Mombouli J-V (2019) Long-term wildlife mortality surveillance in northern Congo: A model for the detection of Ebola virus disease epizootics. Phil Trans R Soc B 374: 20180339. http://doi.org/10.1098/rstb.2018.0339.

LaDouceur EEB (2021) Invertebrate Histology. LaDoucer EEB (ed) Wiley-Blackwell, Hoboken, NJ.

Munson L (2000) Necropsy Procedures for Wild Animals. In: *Conservation Research in the African Rain Forests—A Technical Handbook*. White L, Edwards A (eds) The Wildlife Conservation Society, New York, NY, p 196–217.

Pugliares K, Bogomolni A, Touhey K, Herzig S, Harry C, Moore M (2007) Marine mammal necropsy: An introductory guide for stranding responders and field biologists.

Raverty S, Duignan P, Jepson P, Morell M (2018) Marine Mammal Gross Necropsy. In: *CRC Handbook of Marine Mammal Medicine*, 3rd ed. Dierauf L, Gulland F, Whitman K (eds) CRC Press, Boca Raton, FL, p 249–265.

Rideout BA (2012) Investigating embryo deaths and hatching failure. Vet Clin North Am Exot Anim Pract 15:155–162.

Terrell S, Stacy B (2007) Reptile Necropsy Techniques. In: *Infectious Diseases and Pathology of Reptiles*. Jacobson E (ed) CRC Press Taylor & Francis Group, Boca Raton, FL, p 219–256.

Villarroel, A., 2015. Outbreak investigations. In *Practical Clinical Epidemiology for the Veterinarian*. John Wiley & Sons, Ltd, Chichester, UK, p 109–119.

Villarroel A (2015) Outbreak Investigations. In: *Practical Clinical Epidemiology for the Veterinarian*, First. John Wiley & Sons, Inc, Ames, p 109–119.

References

Full references for this chapter can be found online at https://doi.org/10.1016/B978-0-443-18437-6.00031-6.

Chapter 3

Wildlife Forensic Pathology

Tabitha C. Viner[a], Rebecca A. Kagan[a], and María José Navarrete Talloni[b]

[a]United States Fish and Wildlife Service, Ashland, OR, United States; [b]Ross University School of Veterinary Medicine, Basseterre, Saint Kitts

INTRODUCTION

Veterinary pathologists working on disease diagnosis and herd health are increasingly asked to work on forensic cases. Common sense is the most valuable asset for a diagnostic pathologist presented with these cases. The cause and effect events that occur when an animal dies under questionable legal circumstances generally occur on a macroscale with very little hidden in molecular, physiologic pathways or microscopic, infective particles. The well-trained pathologist already has the toolbox to reconstruct the events in a legal investigation into an animal's death.

Great emphasis is placed upon data and documentation in a forensic case. As in diagnostic pathology, history is an essential component of the investigation. In a forensic case this must be followed by the detailed documentation of every step in the postmortem examination process, including documenting who was involved and what was discovered. While recording this much detail may seem unnecessary or onerous, simple omissions or mistakes may dictate the difference between effective testimony and inadmissible information. Specific and accurate record keeping is essential to a successful outcome.

This chapter aims to provide the pathologist with a broad overview of wildlife forensics and the ways individuals and corporations may negatively impact free-ranging animals. Readers should come away with a better understanding of the gross and microscopic characteristics of the various types of trauma and the appropriate steps to take to assist law enforcement in the pursuit of poisoning, gunshot, or industry-related cases. This should help give the wildlife pathologist confidence to recognize and process forensic evidence and to demystify the field of veterinary forensic pathology. For the intrepid learner, further, expansive resources in human and animal forensics and taphonomy are provided.

TRAUMA

The hallmarks of **blunt force trauma** include lacerations, abrasions, and contusions.[22,65] **Lacerations** of soft tissue occur when the elastic capacity of that tissue is overwhelmed by a force applied to the surface. The tissue loses its cohesive properties and the individual cells or fibers dissociate. In the skin a laceration will have a u-shaped profile, with deeper portions of the wound appearing rounded (Fig. 3.1A). Because of the heterogeneous nature of skin and the differing elastic qualities of dermal structures (blood vessels, collagen, etc.), tissue bridging is often seen at the deep portions of skin lacerations (Fig. 3.1B). Lacerations may also occur in other organs; for example, the liver of passerines is often lacerated on impact with windows.

Abrasions of the skin result from a force applied roughly parallel to the surface of the skin. The superficial layers of the epidermis are scraped away exposing deeper layers of the epidermis or dermis. Abrasions are often hyperemic and may be patterned, depending on the shape of the force applied to the skin surface.

Contusions are simply bruises or the extravasation of red blood cells within a tissue. Contusions may be present in any organ affected by blunt trauma and result from the rupture of capillaries and larger blood vessels. Petechiae, ecchymoses, and hematomas are all manifestations of contusions.

The impact of a moving vehicle on a living body can produce all three hallmarks of blunt force trauma—lacerations, abrasions, and contusions. It is important to note, however, that a diagnosis of blunt force trauma does not require the presence of all three lesions. Blunt force trauma may also result in bone fractures. In general, high-velocity forces produce fractures that are highly comminuted in nature. The impact of a wind turbine blade or a speeding bullet often produces long bone fractures with greater than 10 bony fragments. In contrast, fractures due to collision with fences or slow-moving vehicles are simple or only mildly comminuted. In these cases of accidental trauma fractures are usually unilateral.[32]

Incisions due to sharp force trauma are the result of the application of a sharp object to the body. Incisions due to sharp force trauma differ from blunt force lacerations in that the wound profile of a sharp force incision is v-shaped with a defined angle at the base of the lesion (Fig. 3.2A). All components of the organ (collagen, blood

Pathology of Wildlife and Zoo Animals. https://doi.org/10.1016/B978-0-443-18437-6.00004-3

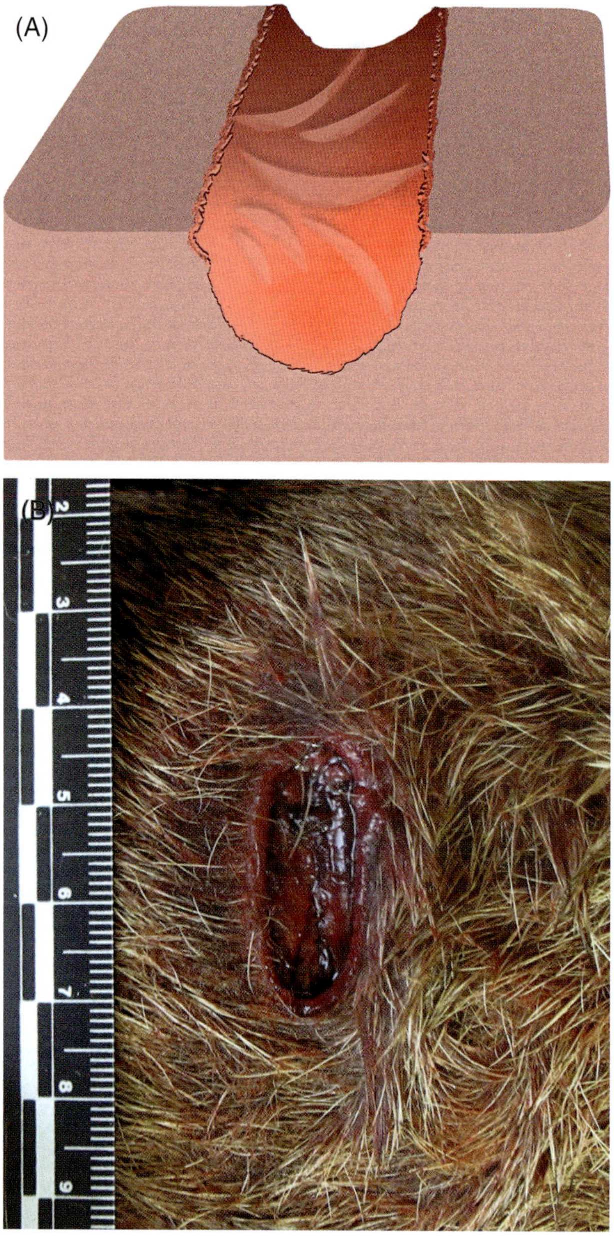

FIGURE 3.1 Laceration in the skin. Blunt forces may overwhelm the elastic capacity of the skin, resulting in separation of the epidermis and underlying soft tissues. (A) This separation exhibits a broad, u-shaped profile, often with bridging by tissues with greater elastic capacity. (B) Impact by a vehicle caused the skin of this mountain lion to tear. The skin edges are rounded and the base of the wound is broad and irregular.

vessels, etc.) are cut at the same plane, regardless of their physical properties (Fig. 3.2B). Scalpels, knives, scissors, and ice picks are examples of tools that may cause sharp force injury in a body. Nails and exposed fence wire may cause an initial sharp force wound followed by tearing or degloving of the skin if the tissue remains attached to the item while the animal is still in motion. Sharp-edged arrowheads can also produce this type of trauma by slicing the skin and organs, as well as producing a complete perforation that runs through an animal.[75]

Injuries associated with **inter/intraspecific aggression** result from the weapons that animals use against each other including claws/talons/hooves and teeth. The lesions produced from the use of these weapons may have characteristics of blunt and/or sharp force trauma. The teeth of mammalian predators are technically blunt objects but are often applied with enough force to cut through the skin. Even flat bones, such as the skull, can be punctured by the canine teeth of a predator (Figs. 3.3A, B). Grabbing and holding the prey/adversary results in deep muscle lacerations underneath somewhat inconspicuous skin punctures. Tears in the musculature can be dramatic, appearing radiographically as large defects in the soft tissue (Fig. 3.4). Grossly, muscle is lacerated and markedly hemorrhagic. Particularly aggressive attacks may result in penetration of the abdominal body wall or intercostal muscles, resulting in pneumothorax.

In addition, the pattern of injury is generally dictated by the behavior and social nature of the species in question. For example, fighting bald eagles will use their talons to grab at the chest or head of an opponent. Birds that succumb during an attack will have deep puncture wounds into the muscle, often perforating the sternum or skull. These penetrating wounds should be differentiated from gunshot wounds through careful exploration of the wound tracts. Talon wounds are often multiple and will have a finite depth and will not be associated with gross or radiographically evident metal particles.

Most aquatic vertebrates (sea otters, pinnipeds, sirenians, cetaceans, turtles, and crocodilians), as well as fishes that feed close to the surface or in shallow waters, are susceptible to **vessel-strike injuries (VSI)**.[10,26,42] Trauma incurred from VSI may include the blunt impact of hulls or keels, or sharp cuts made by propeller blades and skegs. Wounds caused by propeller blades are chop wounds, evenly spaced and generally parallel, and commonly s-shaped or curved, featuring skin incisions or comminuted fractures of the affected bones (see Chapter 24: Sirenia). In the case of turtles the injuries show crushing of the carapace or plastron, skin lacerations, and soft tissues or organ exposure. Propeller incisions may extend into the body cavities causing damage to the gastrointestinal (GI) tract, kidneys, and lungs. Sharp trauma to the lungs may result in pneumothorax or subsequent pyothorax. In the case of manatees, because of the physical division between the two sides of the thoracic cavity, pathologic processes that follow sharp incision into the chest cavity are often unilateral. While dolphins commonly manage to evade ship and boat interactions, the dorsum of whales may be partially or fully transected by a large ship propeller.

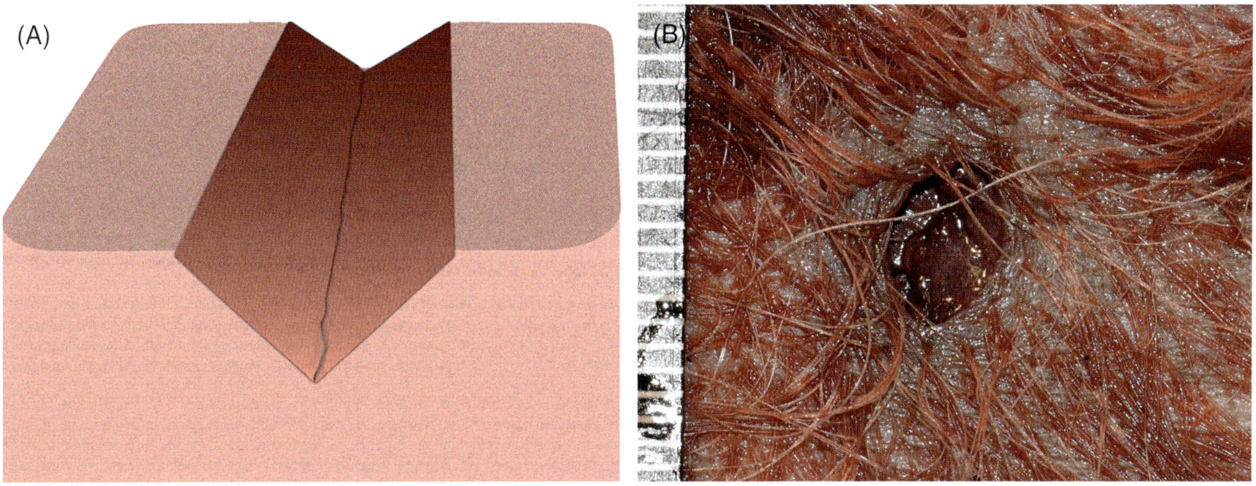

FIGURE 3.2 Sharp incision of the skin. An incision made by a blade or other sharp instrument results in a v-shaped groove and clean wound edges. (A) In contrast to lacerations, the wound edges of the skin of this wolf were cut by a broadhead arrow. (B) The muscle below also exhibits sharp angularity.

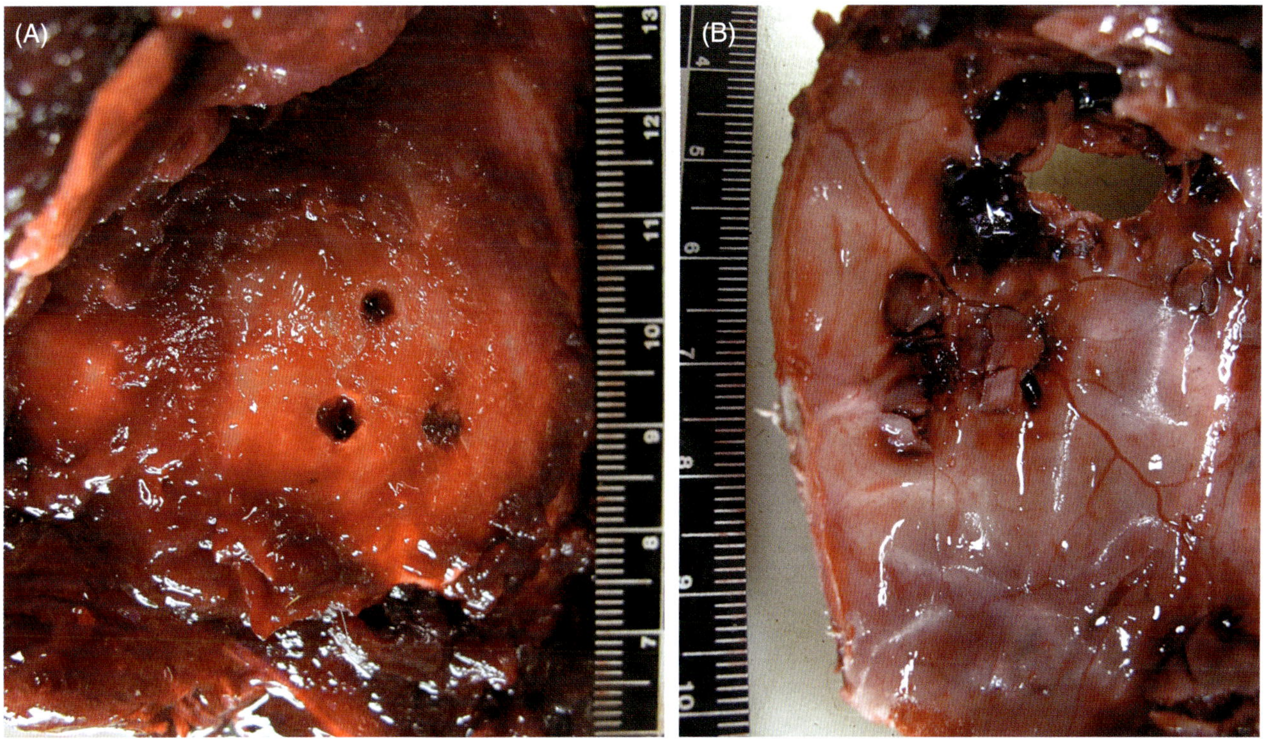

FIGURE 3.3 Skull punctures in a Mexican gray wolf. The skin is reflected to demonstrate tooth punctures due to interspecific conflict. The canine teeth of the combatant punctured the skull of this wolf multiple times. Wounds such as these must be differentiated from gunshot wounds. In cases of tooth or talon punctures, beveling of the bones will be consistently directed inwardly with no exit patterns. (A) Concave surface of the skull showing perforations with clean margins. (B) Inner, convex surface of the skull with beveling and inward protrusion of skull fragments.

The skin of heavily pigmented marine mammals will camouflage hematomas resulting from the impact of ship hulls. Methodical flensing of the skin and outer soft tissues will reveal hemorrhage in the deeper tissues of traumatized whales and manatees.

Communication towers and power lines have become ubiquitous in the 21st century; their impact on free-ranging birds and bat populations is significant. Birds and bats may collide with these structures due to low visibility or attraction to lighted portions of the structure. It is estimated that 6.8 million birds die in the United States and Canada each year due to collision with communication towers.[44] Impacts with power lines result in the deaths of between 8 and 57 million birds annually in the United States.[46]

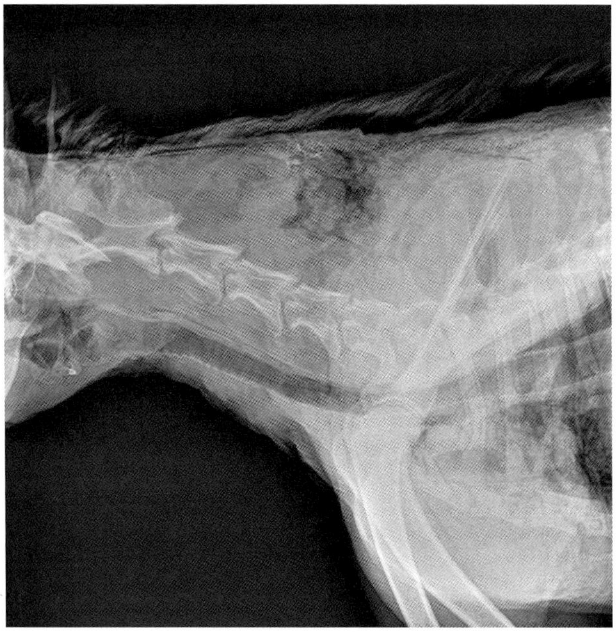

FIGURE 3.4 Cervical muscle tears in a gray wolf. This lateral radiograph of a gray wolf that died of interspecific conflict shows air cavitation in the muscle on the dorsal aspect of the neck due to tearing by the canine teeth of a combatant. When animals grab and hold with teeth or talons, the more elastic skin may puncture and stretch while tearing of the underlying, less elastic muscle may occur.

Entanglement is another important anthropogenic cause of mortality and trauma in aquatic animals. Entrapment of mammals within fishing nets, pipes, or floating debris may result in death due to drowning. The postmortem findings may differ depending on the species (see Chapter 22: Cetacea, Chapter 23: Pinnipediae, Chapter 24: Sirenia, and Chapter 34: Chelonia). If the animal is not found dead with the entrapping material, the skin may also exhibit evidence of rope abrasions or circumferential compression in the blubber layer.

Electrocution occurs often due to contact with electrical wires but may also occur when a body is struck by lightning. The extent of damage in electrocution injury depends on the voltage encountered, the duration of exposure, and the conductive capacity of the body part contacting it. Classic lesions of power line electrocution include singed feathers/hair and burns on the apteric/glabrous skin, and in raptors often at the distal ends of wings and legs[35] (Figs. 3.5A, C–D). Affected hair and feather barbs appear curled and melted and are often discolored brown to black. These changes may be visible to the naked eye or highlighted with the use of an alternate light source[86] (Fig. 3.5B). Electrothermally damaged keratin glows orange-red when light of a wavelength of 530–560 nm is applied and the tissue is observed through a red filter. Signs of electrocution may be subtle, involving one or few feathers on the wings or head or small areas of the feet.

In severe cases of electrocution the skin loses elasticity and is easily torn during necropsy. There is often separation of the skin from the underlying tissues. The musculature appears dry and is discolored light brown to tan, resembling cooked meat. Histologic changes in the skin include smudging of dermal collagen, loss of differential staining, and elongation of epithelial nuclei[19] (Fig. 3.5E).

Avian mortality at **solar facilities** was first described in 1986 by McCrary.[47] Morbidity and mortality at solar facilities takes three forms: blunt force impact trauma, stranding, and feather burns. Photovoltaic panels and heliostats (mirrors) present a reflective surface that can mimic sky or water, similar to the deceptive reflections from windows on a building. Bony trauma that may result from impact with these surfaces includes broken wings or skulls. Deceleration injury, including lacerations of the liver and heart-base vessels resulting in hemocoelom, are other common manifestations of solar panel impact. Obligate waterfowl, such as grebes, may attempt to land on the water-like surface only to become suddenly grounded. Without the body of water these birds require to take flight, death due to exposure, dehydration, and starvation may ensue.

Solar power tower facilities use hundreds or thousands of mirrors that reflect the sun's rays onto a central tower, heating fluid into steam that in turn runs turbines. Concentration of the radiant flux from the mirrors can create a visible glow around the tower. Contact with a surface is required for radiant flux to be translated into heat. Insects that are attracted to the glow—and the birds that follow them—provide that contact, resulting in thermal burns to the feathers. Feather singeing in these cases is often diffuse and symmetrical, in contrast to the feather burns of power line electrocution. The solar aspect of the burns is demonstrated in the differential effect of the radiant flux on darker and lighter areas of the feathers. The darker portions of a feather absorb more solar heat and are thus more affected by singeing and warping than white portions, which are more reflective of the sun rays. Birds that do not die as a direct result of the burns suffer from loss of feather function and subsequent inability to obtain food, flee from predators, and to thermoregulate. Birds with extensive feather burns but little to no soft tissue damage may be found emaciated.[36]

Estimates of avian **wind turbine mortality** in the United States range between 20,000 and 573,000 birds per year. Based on the data from several studies, Loss et al.[45] calculated that 3–7 birds die at each monopole wind turbine in the contiguous 48 states each year. The most commonly affected bird is the golden eagle but bald eagles, hawks, and passerines are also affected. Additionally, tree roosting bats (see also Chapter 25: Chiroptera) may mistake the monopoles for roosting habitat, approaching the turbines while the blades are still rotating.[15]

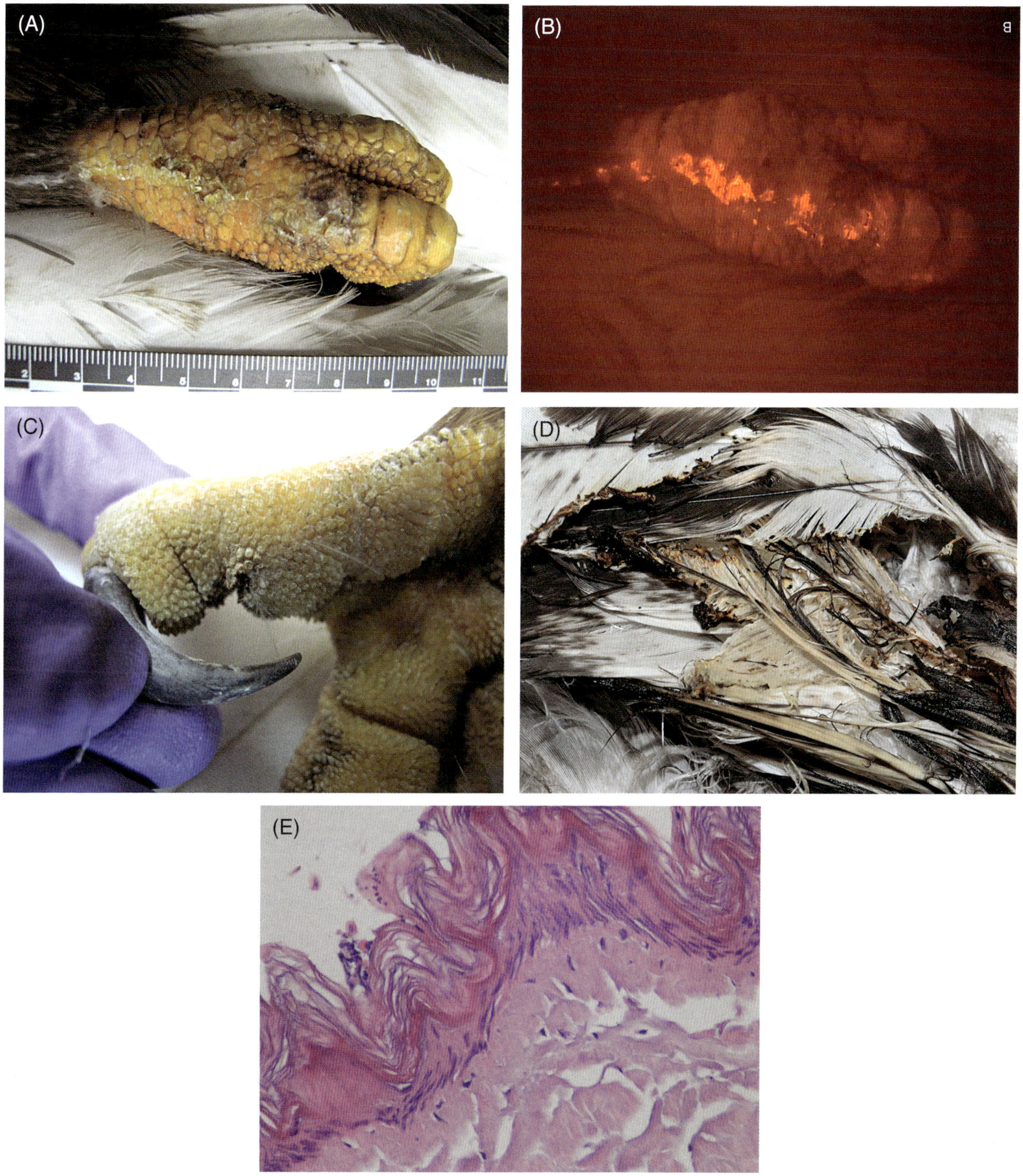

FIGURE 3.5 Electrothermal burns in bald (A, B) and golden (C–E) eagles. An electrical current from a power line passed through and over the left foot of this bald eagle. (A) Keratin over the dorsal aspect of the foot is dry and peeling away from the underlying skin and dermis. The more distal portions of the wound are charred and blackened. (B) Electrothermal damage to the foot is highlighted with the application of an alternate light source at 560 nm and a red viewing filter. Affected areas photoluminescence bright orange. (C) Only a small patch of skin on the plantar surface of the digit at an interphalangeal joint was affected by the electrothermal energy. The skin appears pale at the periphery and brown or black toward the center of the lesion. (D) The feather barbs of this subadult bald eagle are curled, discolored, and blackened due to contact with a powerline. (E) Histologically, electrocution causes smudging of the dermal collagen and elongation of the basal epidermal nuclei resulting in a "picket fence" appearance. *(Parts A and B: Reprinted with permission from Viner TC, Kagan RA, Johnson JL. Using an alternate light source to detect electrically singed feathers and hair in a forensic setting. Forensic Sci Int. 2014;234:e25–e29.)*

FIGURE 3.6 Wind turbine strike in a golden eagle. The high energy impact of a wind turbine blade often results in limb amputation, decapitation, or bi- or trisection of the body of larger birds. In this case, the tail was amputated, and the trunk was bisected at the thoracosynsacral junction. The dashed lines indicate the areas of complete transection.

The speed at the tips of the blades may reach 180 mph. Lesions seen in animals struck by wind turbine blades are usually those of severe blunt force trauma. Leg and/or wing amputation, evisceration, and decapitation are common presentations in large birds that are struck by turbine blades (Fig. 3.6). If the body remains intact, bone fractures are usually multiple and comminuted. Closer to the nacelle (hub) of the turbine, the blade speed is slower and the physical damage is less dramatic. In these cases the history or location of the body when found dead can be married with the postmortem findings to build the case for wind turbine mortality.

Most bats that are found dead in proximity to wind turbines have traumatic lesions. Most common are fractures of the wings and skull (Fig. 3.7A). Soft tissue injuries often involve muscular bruising and diaphragmatic ruptures with herniation of abdominal organs into the thorax or subcutaneous space (Fig. 3.7B). Barotrauma, identified by concurrent rupture of the pulmonary capillaries, has been described as an additional concern but likely plays a minor role in bat-wind turbine mortalities.[69]

Traps and snares have a long history of use in wildlife as pest control, food, fur, and other animal products. Even when legally used, it is often difficult to exclude non-target animals, resulting in "by-catch" of animals that may be legally protected species. In assessing the legality of a trapping incident, it is important not just to consider the trapper's intent but also the season, the type of trap, any modifications, whether it was properly placed and baited, and whether it was adequately monitored. Depending on your jurisdiction, a crime may have been committed if the trapper failed to meet requirements based on the above criteria.

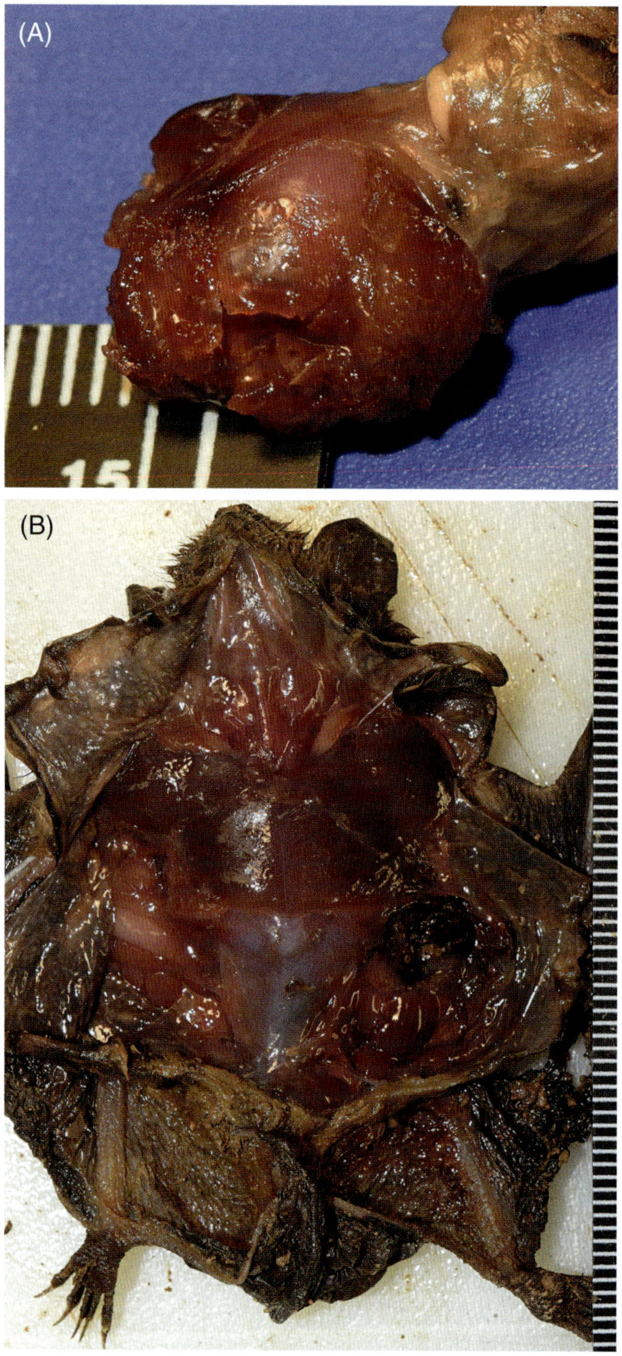

FIGURE 3.7 Wind turbine strike in a gray bat. (A) Injuries commonly seen in bats that are struck by wind turbine blades include fractures of the wings, skull, and spine. (B) Tears in the diaphragm and body wall result in herniation of abdominal organs into the thoracic cavity or subcutaneous space.

A snare trap is a loop of wire or cable attached to a stationary object. The wire constricts when a part of the animal (usually a limb or head) passes through it and pulls the loop closed. A stop or breakaway may be added to the trap to limit the amount of closure. This is done to exclude certain

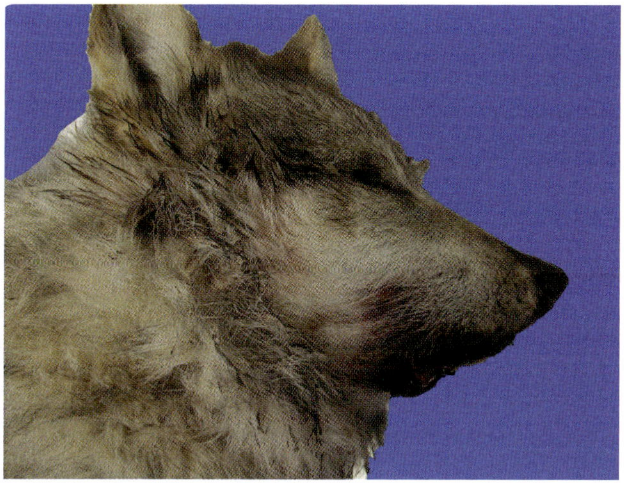

FIGURE 3.8 Snare trap trauma of the head and neck of a Mexican gray wolf. This wolf was found with a snare trap around the neck. The snare had been removed prior to submission. There was constriction of the soft tissues at the caudal end of the cervical spine with severe facial and cervical congestion and swelling. The cause of death was strangulation from the snare.

types of animals (generally larger animals such as deer) and may be a legal requirement in some areas. Snare traps cause lacerations, swelling, and eventual local and distal tissue necrosis[75] (Fig. 3.8). If the snare is tight enough to obstruct cervical arterial blood flow, there may be rapid death with no visible lesions.

Leghold traps may be smooth edged or toothed. They can cause fractures of the phalanges or metacarpals of mammals and the leg bones of birds. If toothed, there may be patterned lacerations[14,24] (Figs. 3.9A, B).

Animals found in traps may have also been shot, bludgeoned, or otherwise injured if the trap was of a restraining type such as a leghold trap. Animals that have escaped or attempted to escape may have limb injuries or show signs of struggle such as broken teeth or nails.[24] Other types of restraint traps include cage traps, pit traps, nets, and sticky traps. Direct injury and escape-induced injuries may or may not be present (Fig. 3.10).

Conibear (aka body grip) traps are kill traps, meaning the trap is intended to cause rapid death from cervical compression. When an animal passes through a set trap, they hit a trigger, which causes the trap to forcefully and rapidly close, ideally around the chest. Unmodified body grip traps are designed with a small gap to prevent injury to the trapper. Because of this gap, these traps suffocate an animal rather than fracture the cervical spine and most often leave no signs of injury. Modifications to body grip traps in the form of clamping bars are more likely to cause hemorrhage, or cervical or cranial fracture. These modifications are not available commercially in the United States and may be legally prohibited.[34,60,61]

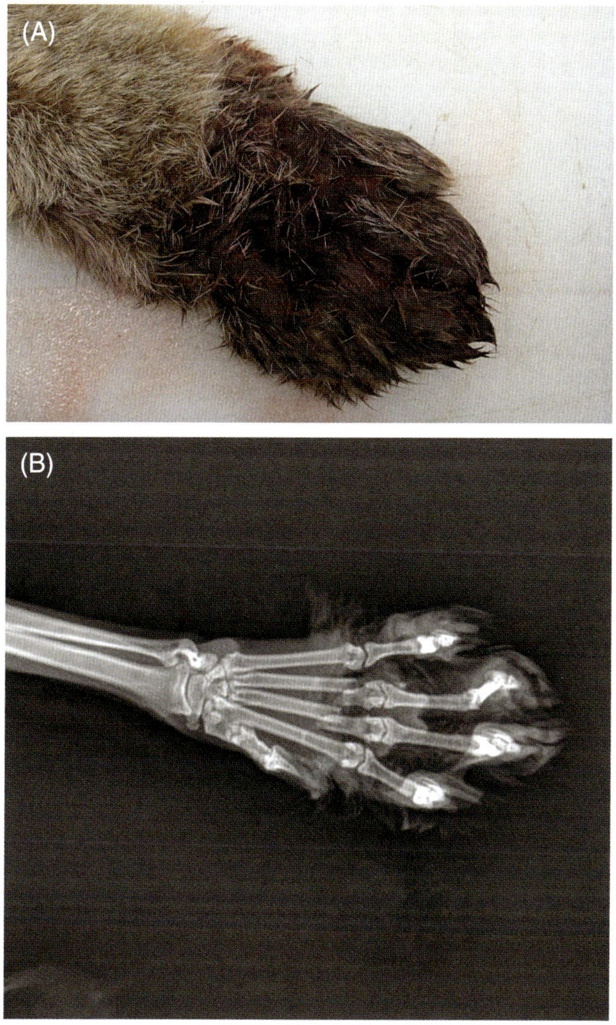

FIGURE 3.9 Leghold trap trauma of the left forefoot of a Canada lynx. (A) This lynx was found dead in a leghold trap. Trauma in the left forefoot related to the leghold trap included fractures, hemorrhage, soft tissue inflammation, and necrosis. (B) Radiographs were important in documenting fractures of the left third and fourth metacarpal bones.

In evaluating suspected entrapment cases, stomach contents may also be saved for identification of bait. If the trap is submitted with the animal, consider the possibility of fingerprint analysis and handle and preserve the trap accordingly.

INDUSTRIAL HAZARDS

Industrial **mining** processes generate wastewater and solids that are often disposed of or stored on-site. These open storage ponds can be up to several acres in area and, depending on the industry in question, may contain toxic chemicals that are dangerous to wildlife. To comply with federal laws prohibiting the killing or harming of wildlife (e.g., the Migratory Bird Treaty Act in the United States), mining companies often use physical or psychological deterrents to keep animals away from these pits. These

FIGURE 3.10 Cage trap trauma on the rhinotheca of a red-tailed hawk. Trauma related to the trap includes keratin abrasion and staining with green paint from a homemade cage trap. These wounds were consistent with self-injury from escape attempts. Other findings related to escape attempts included abrasions and green staining of the skin along the dorsal edge of the orbit and leading edges of the carpi, and broken tail feathers.

deterrents, however, may fail or break down allowing access to these chemicals.

Toxic materials may involve the object of the mining efforts or chemicals used in the extraction of materials from the substrate. **Cyanide** is still in use in US mining operations to leach gold from ores.[13] The effluent from these leach heaps is collected in ponds for eventual reuse. Birds and mammals are exposed through ingestion or inhalation. Other mining materials, such as lead, may be absorbed into the surrounding soil, water, or vegetation, which then acts as toxin sources.[12]

Gross and microscopic postmortem changes will vary depending on the compound involved. In peracute death no pathologic changes may be visible and effective pursuit of the cause of death may be based on ancillary diagnostics. Because of its somewhat volatile nature and short half-life, postmortem testing for cyanide may yield ambiguous results.[58] Conversely, heavy metal concentrations in the liver or kidneys of free-ranging animals may be reflective of feeding activities in the days or weeks prior to death and not indicative of events immediately preceding death or a die-off. Because of these variables, investigations into wildlife mortality at mine sites should involve collection of water and/or soil samples in the area of the die-off. These samples can provide supportive evidence in cases of wildlife mortality events at mining sites when testing of the animals themselves is difficult or inconclusive.

Industrial waste may present physical rather than metabolic hazards. In some locations oil fields produce water that is highly enriched in **dissolved salts**, most often sodium chloride.[8] This super-saline water is disposed of in evaporation ponds. Salts in this water rapidly crystallize on the plumage of waterfowl that land on unprotected ponds (Fig. 3.11). The weight of the salt crystals and disruption of the feathers prevent flight and disrupt the thermoregulatory function of the plumage. Birds become "trapped" in the pond and die due to thermoregulatory imbalance, starvation, and/or dehydration.

The oil industry presents two hazards to wildlife: **oil pits** or **ponds** containing waste or unrefined oil, and **oil spills** from drilling or conveyance operations. Similar to mining industries, containment ponds are used at oil production facilities and are attractive to waterfowl as well as ground-foraging bird species.[79] Mammals, reptiles, and amphibians are less frequently found in oil and waste ponds. The National Oceanic and Atmospheric Association (NOAA) in the United States have tracked oil spill incidents since 1985, recording an average of 110 spills per year (range 15–197) in the United States (https://incidentnews.noaa.gov/).

Acute effects of oil exposure in birds are typically physical in nature. Unlike water, hydrocarbon products are not repelled by a bird's natural body oils and tend to mat the feathers and spread deeply into the plumage. Disruption of the external thermoregulatory apparatus leaves

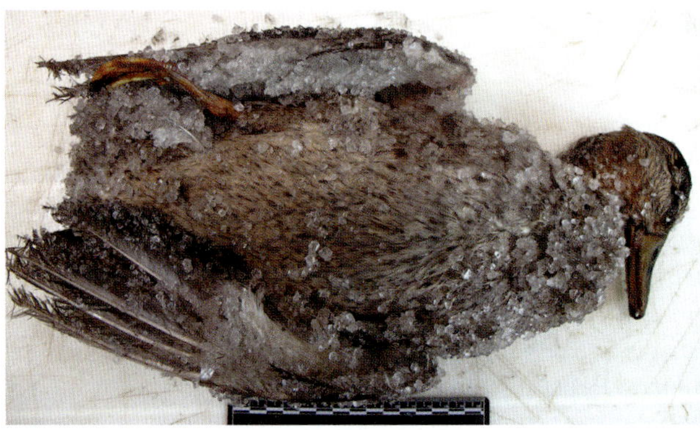

FIGURE 3.11 Salt encrustation of a mallard. This mallard landed in a super-saline pond. The feather barbs provided a nidus for crystallization of the minerals in the pond. Disruption of the plumage and the weight of the crystals prevented the bird from flying away. Subsequent dehydration, starvation, and possible thermal imbalances can lead to death in these cases.

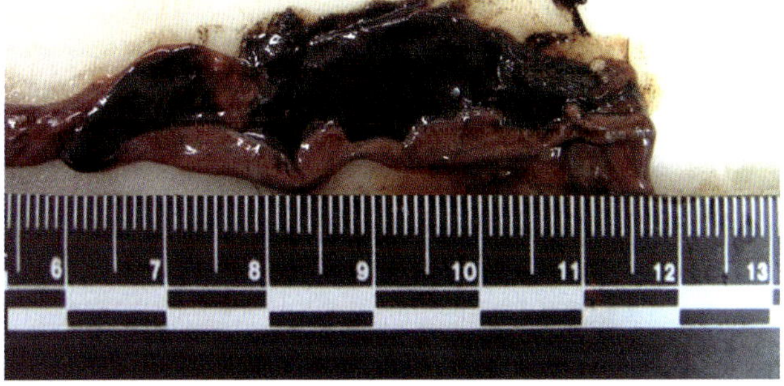

FIGURE 3.12 Crude oil ingestion by a northern gannet. Thick oil is present in the intestine of this bird, supporting the fact that the bird was alive and actively ingesting the oil prior to death. Mixing the contents with water highlights the oily morphology of the substance and can help differentiate this from digested blood.

birds susceptible to environmental temperatures, leading to either hypothermia or hyperthermia. Due to the absorptive nature of dark crude, oil spills in open bodies of water are often much warmer than ambient temperature. In these cases typical changes of hyperthermia or thermal damage to the skin may be seen. Similar to electrothermal injury, the skin may be friable and inelastic, and the underlying muscles soft and tan.

The presence of oil in the respiratory or GI tract (Fig. 3.12) helps to substantiate antemortem exposure to the oil but it is not necessary for a diagnosis. Oil presence shows that the animal was actively breathing and/or eating/drinking while in the oil and did not die of other causes prior to contact with the substance. The long-term effects of exposure and ingestion of hydrocarbon products by birds primarily include decreased reproduction and egg hatchability.[90] Marine vertebrates (cetaceans in particular) are vulnerable to the effect of oil spills, and their exposure is usually through direct contact and/or ingestion (direct ingestion or prey) (see Chapter 22: Cetacea). Exposure of marine mammals to crude oil spills may result in adrenal cortical atrophy, lymphoid depletion, and bacterial pneumonia.[84] These histologic findings occurred with more frequency in dolphins exposed to oil from the Macondo Prospect well in the Gulf of Mexico than in stranded dolphins unassociated with an oil spill. Careful postmortem examination and history can help to differentiate oil-exposed marine life from animals involved in other unusual mortality events caused by harmful algal blooms or disease, such as Morbillivirus or brucellosis.[43]

Each natural oil source has a unique molecular composition. Thus oil from the plumage of a recovered bird can be compared against oil taken from the source of a spill. This source matching or "oil fingerprinting" can be used to definitively link wildlife mortalities to a specific oil spill event.

BALLISTICS

The vast array of ammunition and firearms available today is mind-boggling; however, in wildlife cases there are particular types of **firearms** that are more commonly used. Brief descriptions as they pertain to the necropsy examination as well as general wound characteristics and collection techniques that can be applied broadly to projectile-related deaths are presented below.

Rifle and shotgun projectiles may be simple, such as an air rifle pellet, or may be part of a larger unit, such as a rifle or shotgun cartridge (Figs. 3.13A–C). Cartridges are composed of a casing with a bullet or pellets at one end and a primer at the other end. When the firing pin strikes the primer, the bullet or pellets and wad explode out of the casing and through the gun barrel.

Caliber is the measurement of the diameter of the bullet or barrel (Fig. 3.14A). Rifling refers to the ridges (lands and grooves) that spiral along the inner surface of the rifled gun barrel (Fig. 3.14B). The patterning of striations, scratches, and other unique imperfections left on the bullet as it passes through the barrel is what allows spent bullets to be matched to a suspect firearm or air rifle.[9,21,50]

Though other firearms, such as handguns and machine guns also have rifling, the term "rifle" typically refers to long-barreled firearms commonly used for hunting. Ammunition is either centerfire or rimfire. The distinction is made based on where the primer is located, either at the center or around the rim of the cartridge base. Centerfire bullets with partial metal-jacketing are designed to expand and fragment on contact with a body resulting in maximum damage and the radiographically visible "lead snowstorm" of small metal fragments (Fig. 3.15).[21,78] Some of these bullets are designed with plastic tips to provide a weak point to better ensure bullet fragmentation while preserving aerodynamics. Shotguns are smooth-barreled firearms. Projectile types vary, from slugs to pellets of various sizes. Within a shotgun shell, pellets can vary in size and composition.

An **air rifle**, as the name suggests, is a rifle that uses compressed air or a carbon dioxide cartridge as propellant. Air rifles have rifled barrels, whereas other air-powered guns, such as BB guns, have smooth barrels. Though velocities do not approach those of centerfire or rimfire bullets, pellets can still easily cause injury and death[4,51] (Fig. 3.16).

Harpoons, darts, crossbows, or spear guns are considered weapons that fire sharp objects or **arrows**. Arrow wounds may need to be differentiated from gunshot wounds, particularly if there is a question as to whether an animal was killed with the appropriate weapon for the

FIGURE 3.13 Ballistic projectiles. (A) From left to right: Cutaways of 12 gauge shotgun shells, 9 mm Lugar cartridge and test-fired bullets, 0.22 long rifle or handgun cartridge and test-fired bullet, 0.223 Remington cartridge and test-fired bullet, 3 examples of Nosler plastic-tipped bullets, and 4 detached tips of various colors. (B) Radiograph of previous image demonstrating differences in the appearance of components in radiographic evaluations. (C) From left to right: plastic-tipped air rifle pellet 0.177 gauge, pointed tip air rifle pellet 0.177 gauge, wadcutter air rifle pellet 0.177 gauge, sabot air rifle pellet 0.177 gauge (below, pellet with plastic collar removed).

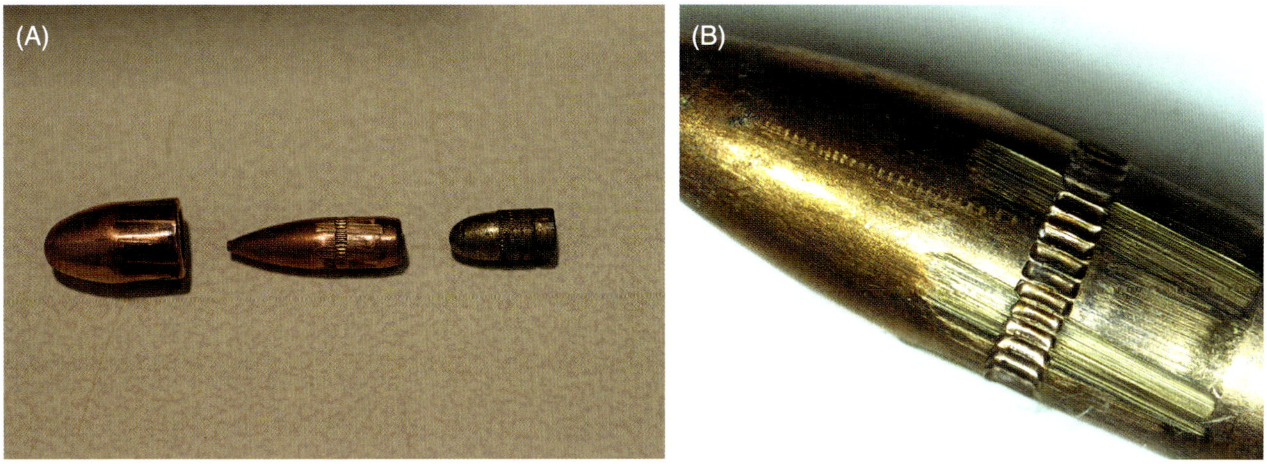

FIGURE 3.14 Bullets. (A) Variously sized rifle bullets demonstrating rifling along the length of the projectile. L to R: 9mm Luger, 0.223 Remington, 0.22 long rifle or handgun. (B) 0.223 Remington close-up to demonstrate the pattern of rifling from the gun barrel.

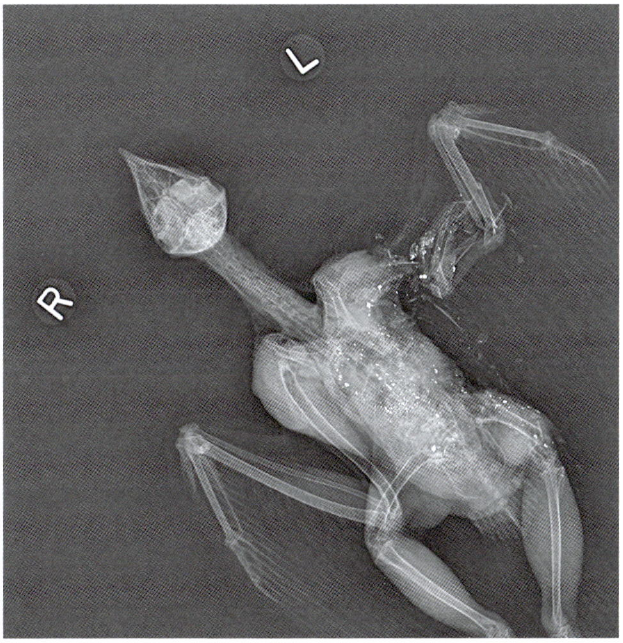

FIGURE 3.15 Gunshot injury in a rough-legged hawk. This radiograph demonstrates multiple bone fractures due to a gunshot injury to the left wing, leg, and body cavity. Fragmentation of the bullet as it passed through the body caused a radiographic "lead snowstorm" appearance of ballistic fragments. These changes are consistent with gunshot injury from a high-velocity firearm.

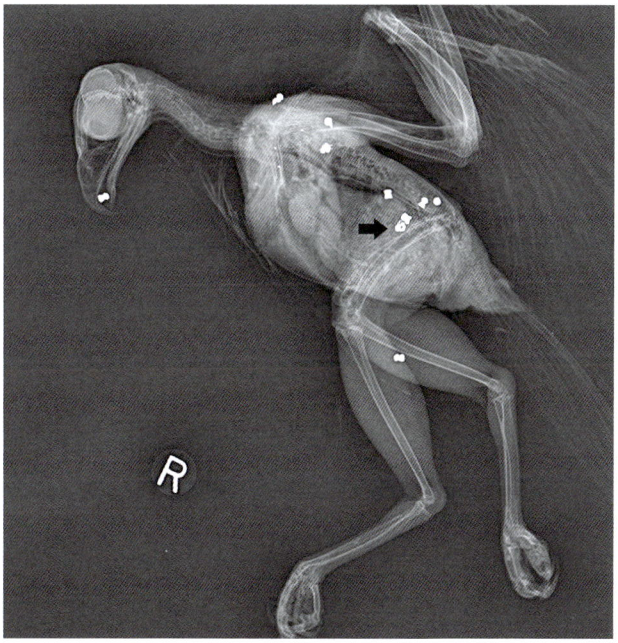

FIGURE 3.16 Air gun pellets in a red-tailed hawk (cadaver study). Eleven different air rifle pellets of a variety of compositions and brands (including lead, tin, and copper) were fired. This image shows various degrees of pellet deformity and fragmentation that may occur. Only a few pinpoint fragments are left in a shot to the shoulder. An end-on pellet at the proximal extent of the femur *(black arrow)* has a radiolucent center consistent with a plastic-tipped pellet.

hunting season. These types of weapons provide a speed of action that is similar to the objects used in stabbings and their mechanism of injury has relevant differences.[5] There are two main types of arrows: field-tip and broadhead. Field-tips are conical, creating entrance wounds that are very similar to those of a bullet.[30] Broadhead arrows have blades, which cause radiating incisions at the entrance site. When skin edges are pulled together, a y- or x-shaped incision may become apparent (Figs. 3.17A, B).

Broadhead arrows are designed to cut and incise tissue directly along the wound path, in contrast to cavitated wound tracts resulting from high-velocity firearms.[78] A fishing speargun is very similar in shape and function to a crossbow or an arch, and therefore the lesions are very similar to the ones observed in broadhead arrows[5] (Figs. 3.18A–D). When dealing with arrows found in the body, be mindful of the potential for fingerprint analysis and treat the evidence accordingly.

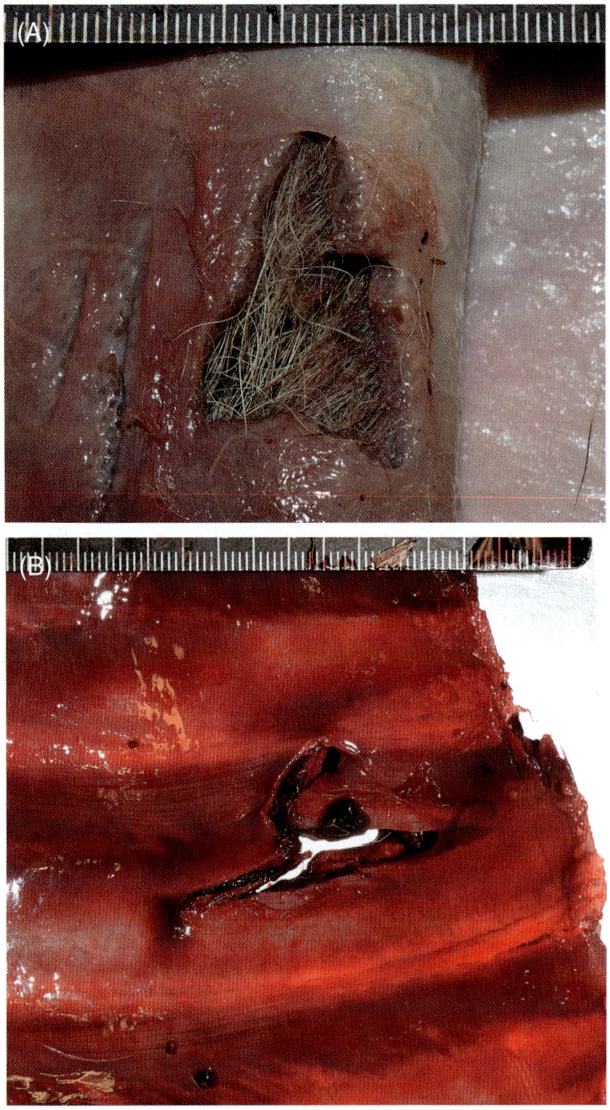

FIGURE 3.17 Arrow wounds in a gray wolf. (A) This tri-radiate incision in the skin was made by a 3-bladed broadhead arrow. Arrowheads cause incisions in soft tissue and bone that have sharp edges and are meant to cause clean kills through exsanguination. (B) The arrow also cut through the ribs and intercostal muscles and caused damage to the heart and lungs.

There are several main **mechanisms of bullet injury**. Lacerations, crushing, and fractures are caused by direct contact with the bullet and bullet fragments. Temporary or cavitated wound tracts from the bullet wake result in stretching and crushing of tissue. Projectiles from higher velocity firearms cause greater cavitation. Secondary types of injury are bullet embolization (particularly with small projectiles, such as shot or air rifle pellets) and secondary infection. Lead toxicity secondary to retention of lead bullets or fragments within the soft tissues does not seem to occur.[33,41]

Do not underestimate an animal's mobility immediately after being hit. Though some wounds are immediately incapacitating (e.g., one that severs the spine), adrenaline and forward momentum may allow an animal to travel some distance before dropping.[21,39]

Entrance wounds are classically "tidier" than **exit wounds**, often forming discrete holes that can sometimes be difficult to find. There may be a thin rim of abrasion along the margin of the hole where the bullet is pushed inward. A maximum bullet diameter can be determined from holes in bone; however, due to the flexible nature of skin, the entrance wound measurement should never be used to determine caliber.[28] If bullets ricochet or hit an intermediate target, such as the folded wing of a perched bird, destabilization of the bullet can cause the entry wound to be irregular. Though uncommon in wildlife, contact wounds—created when the muzzle of the firearm is placed directly against the body as it is fired—will also create an atypical entrance due to expulsion of gasses during the discharge.[18]

Because bullets tumble and break apart in the body, exit wounds are generally (but not always) larger than entrance wounds and can be any shape. Surface tears around the margin of the wound are caused by expulsion of the bullet or fragment and suggest an exit wound.

Beveling of bone can help to determine the direction of bullet travel. Force applied to bone on one side will cause beveling (or angling) of the fracture fragment on the other side of the bone. Visibly, there is a smaller hole on the outside (entrance side) of the bone and beveling of the bone cortex on the exit surface. This is particularly obvious in skull wounds but can also be appreciated in other flat bones and sometimes long bones (Figs. 3.19A, B). As in blunt force trauma, Puppe's rule applies: fractures propagated from the exit wound will not be able to cross those of the entrance wound.[62]

Determination of range of fire or the distance the victim is from a shooter is often a component of human forensic case investigation but is less useful in wildlife cases. Most free-ranging animals will naturally try to stay some distance away from humans. Beyond 0.5–1 m (2–3 ft; as will be the case in most shootings of free-ranging animals), signs, such as singeing, soot, and stippling will not occur.[21] If these signs are present, they indicate a close-range shot and may support a claim of self-defense (the animal was charging) or indicate that the animal was shot while captive (or trapped).

In cases involving shotguns evidence of impact from the wadding and/or shell indicates close range. As pellets leave the muzzle of the gun, they spread out; therefore an animal that is hit at closer range will likely have far more pellets in the body than one that is hit from further away. It is not recommended that distance be more than generally determined by pellet spread as there are many variables involved.

Examination and collection techniques for ballistic investigations involve multiple steps. Radiographs or computed tomography (CT) should always be taken in suspected gunshot cases for diagnosis, retrieval of fragments,

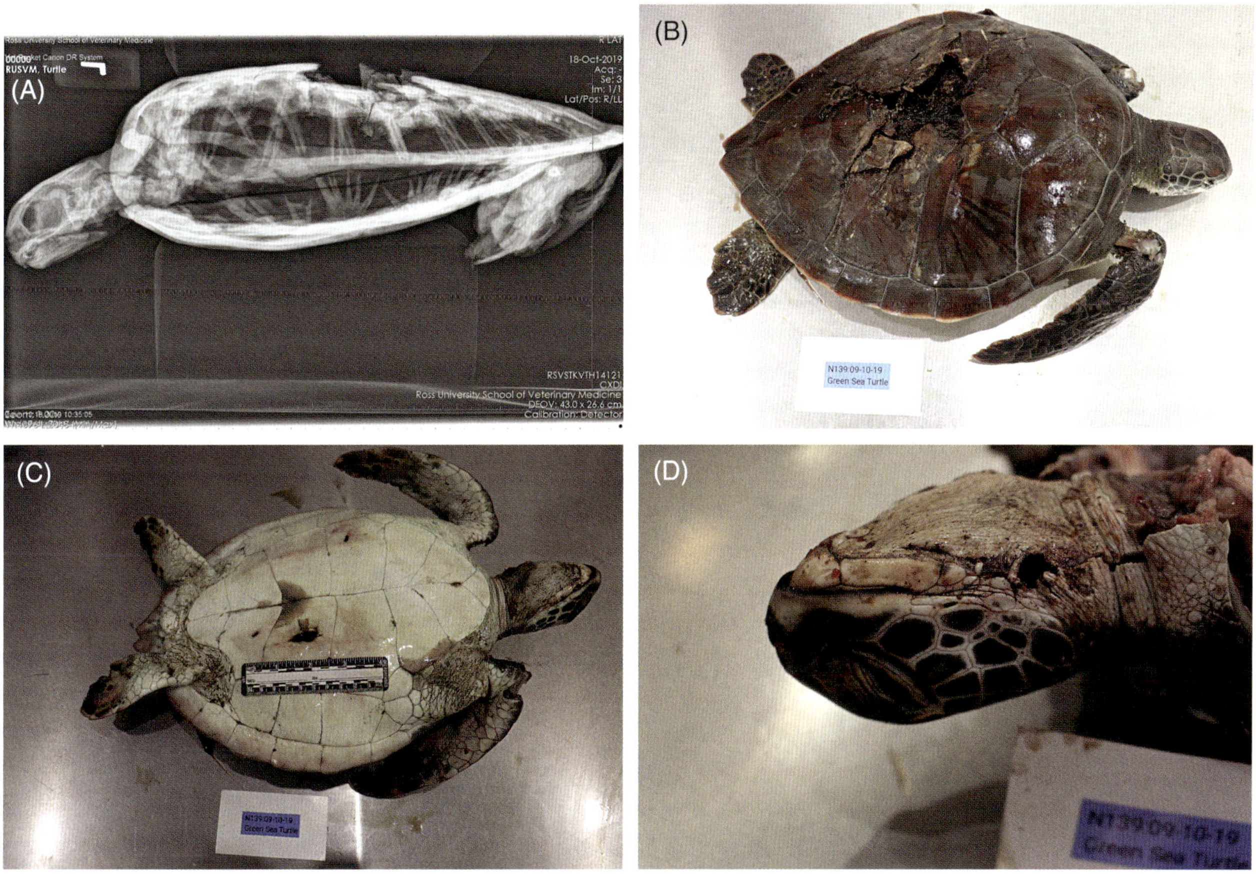

FIGURE 3.18 Speargun injury in a green turtle. (A) Right lateral, whole body view radiograph image, illustrating the damage to the carapace in this animal (*arrow*). (B) The suspected entry wound from the spear pole shot is present in the carapace. Fragmented scutes in the carapace were the results of a through-and-through injury (*arrow*). (C) An irregularly shaped polygonal wound (*asterisk*) in the plastron was the suspected exit wound. (D) Three other linear lesions, each less than 1 cm and suspected to also have been due to the spear pole, were present in the neck. Also present was a circular penetrating lesion with regular margins.

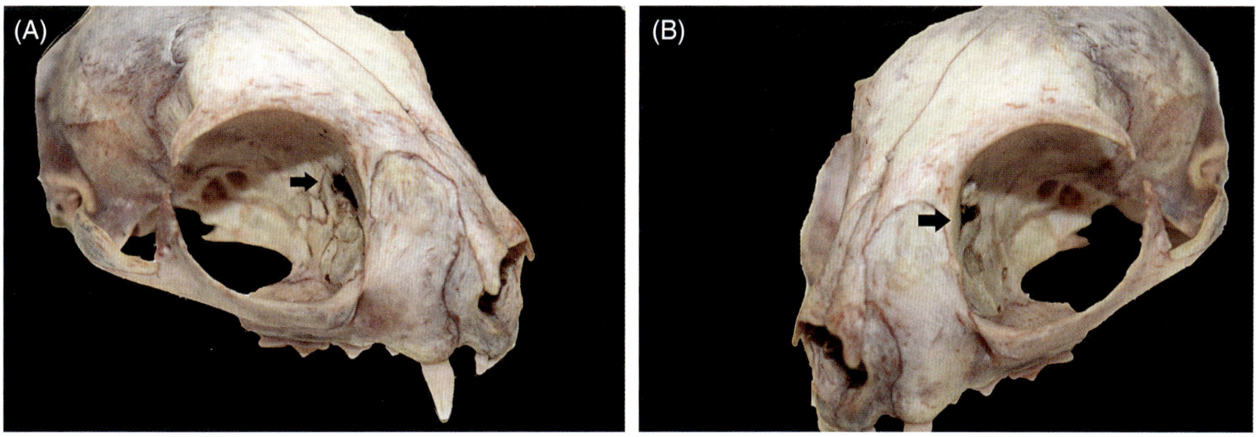

FIGURE 3.19 Air gun injury in a domestic cat. This is a through-and-through shot from an air rifle. (A) The entry wound is an irregularly round hole in the orbital bones with an inwardly beveled, right orbital fracture (*arrow*). (B) An exit wound on the left side consists of an outward protrusion of bone fragments (*arrow*).

and determination of bullet direction (Fig. 3.20). Estimation of bullet caliber from radiographs is not accurate as measurements may differ with distance from and angle to the beam.[37]

While photographing the body, probes can be used to demonstrate the wound path. It is recommended that the wound path be examined and documented before putting in probes, as an aggressive prober can inadvertently alter

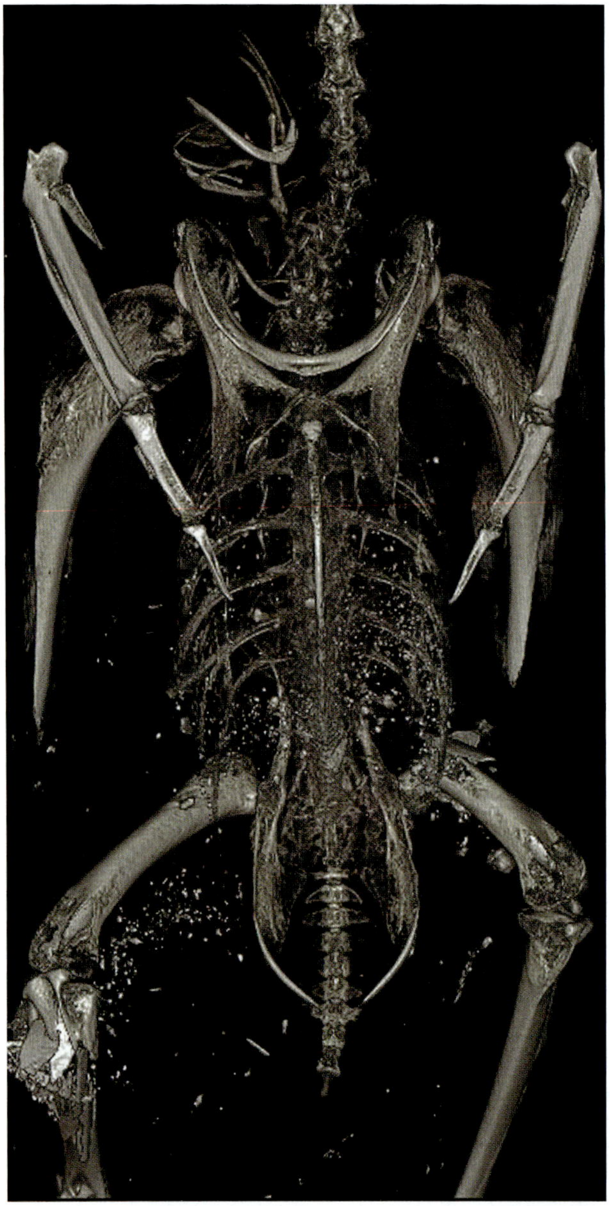

FIGURE 3.20 Computed tomography of gunshot trajectory in a bald eagle. Computed tomography can supplement or replace traditional radiographs and can be useful in defining the path of a projectile through the body. Numerous, small bullet fragments are arrayed from the fractured proximal extent of the right tibiotarsus to the fractured proximal extent of the left femur.

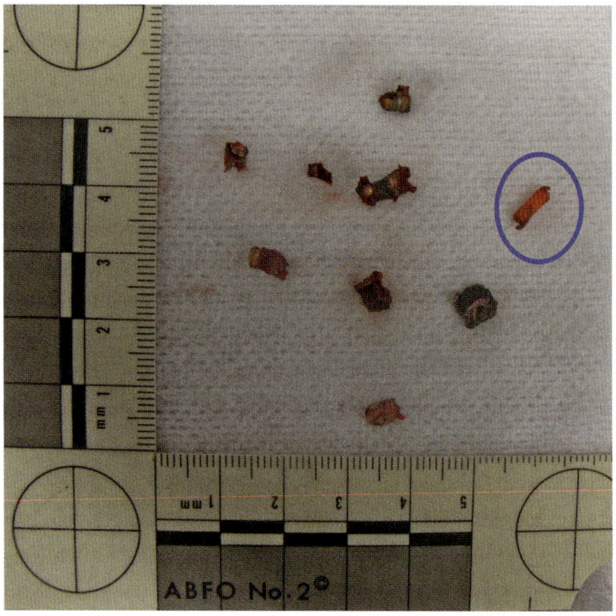

FIGURE 3.21 Bullet fragments from a gray wolf. These bullet fragments were retrieved from the wound tract. Circled is the stem from a plastic tip. Other fragments consist of pieces of bullet jacket and lead core material.

the wound path or create a new one. If the body is to be photographed with a probe in place, the area should also be photographed before the probe is inserted. Close-ups should be preceded by shots taken at an intermediate distance so the location can be appreciated in relation to body landmarks.[18]

Bullets and bullet fragments should be collected gently (preferably with plastic forceps) so as not to damage any rifling that may be present. If found, be sure to also retrieve and analyze plastic tip fragments (Fig. 3.21). Shotgun injuries may result in hundreds of pellets in the body. If there are large numbers of pellets in the body, be sure to retrieve a representative sample, including multiple sizes if present. Fragments should be photographed with a scale so that size may be inferred. They may be lightly rinsed with water and rubbing alcohol.[28] Do not store them wet in non-breathable packaging (e.g., plastic) or markings may degrade. Optimally, fragments are stored in small paper envelopes but vials may be preferable, if fragments are very small.

As lead projectiles pass through skin, feathers, or bone, trace amounts of lead may be imparted onto the tissue. If skin or other tissue is to be tested for trace metal, tissue should be stored frozen. A square section of skin around and including the bullet hole can be laid or pinned on paperboard or another firm material for preservation. Tissues should otherwise be minimally handled without washing or immersion in formalin.[28] Instrumentation, such as x-ray fluorescence, which is used for elemental analysis, may be limited in the size of the specimen that can be accommodated. Determine the best way to preserve and submit such items to the laboratory ahead of time.

Gunshot residues consist of both organic and inorganic particles originating from the primer, propellant, gun barrel, and bullet.[16] Gunshot residue analysis is commonly discussed in human cases as a means of identifying the shooter and/or distance from the victim. In wildlife cases it is unusual for the shooter to be that close to the victim, though it may occur if an animal is charging or is trapped. Residue should not be used to determine either the range of fire or exit versus entrance.[20,57] If residue collection is desired, tape lifts are an easy and readily available method.[16] The tape lift technique involves lightly pressing the sticky side of a piece of tape on the area to be sampled. The tape is then lifted off and the collected material protected by either folding the tape on itself or placing it onto a base such as packing tape or an index card. Hair from around the wound tract may also be collected.

DROWNING

Investigation of suspect **drowning** cases can be a challenging endeavor. Many factors, including water composition and extent of decomposition, must be considered and signs may be nonspecific, subtle, or absent.[48] Analysis requires detailed gross and microscopic examinations, a thorough medical history, crime scene data, and eyewitness accounts. With all this information, diagnosis of drowning is still largely a diagnosis of exclusion. First, it must be established that the animal died in the water and was not dead prior to entering the water. Other primary causes of death must also be eliminated. Further, taphonomic changes due to maceration, decomposition, and scavenging must be differentiated from antemortem lesions that may indicate struggling or entrapment.

Fluid in the lungs and stable froth in the trachea are commonly seen in animals that drown. Microscopic changes may include hemorrhage or edema within the alveoli, aspirated foreign material, and distension of alveoli. However, these changes are not pathognomonic for drowning as they may occur with shock or disease. During the process of drowning, water may be swallowed as well as inhaled. Excessive amounts of swallowed water may result in tears of the stomach wall or vomition with possible aspiration. The presence of foreign material in the bronchi or alveoli/parabronchi can support a diagnosis of drowning, as this material would not reach deeper lung tissue by passive movement. Aspiration is required to propel foreign substances into the lungs. The air sacs of a bird are connected to their pneumatic bones via small ostia. An actively breathing bird may inhale water that makes its way into the medullary space of pneumatic bones. Passive transfer of water into the bones is unlikely to occur in birds that die of other causes in the water, unless decomposition is such that the pneumatic fossae are exposed. In the case of sirenians suspected of drowning, water, foreign materials, or foam in the respiratory tract is absent. Dugongs, for example, may experience laryngospasms that cause asphyxiation or the paralysis of the larynx due to carbon dioxide in the circulating blood (see Chapter 24: Sirenia).

POISONING

Poisoning in this context (as opposed to environmental toxicants) refers to primary or secondary intoxication of wildlife from deliberate or accidental misuse of pesticides, pharmaceuticals, or other dangerous chemicals. Whether a pesticide has been used legally or illegally is contingent upon jurisdiction, as laws outlining pesticide application vary among countries, states, and provinces. Wildlife poisoning may be prosecuted based on the species involved or, in the case of unintentional pesticide poisonings, whether the directions for use were properly followed.

The poisons described below generally have a global distribution though laws and regional accessibility may differ. For example, sodium **monofluoroacetate (also known as 1080)** has been largely phased out in the United States but remains available in livestock collars; however, it is still a regulated pesticide used in oral baits in New Zealand and Australia.[53] **Carbamates** are still available worldwide though they are banned in the UK and heavily restricted in the United States. They remain one of the most common types of illegal poisons used in African countries due to their relatively low expense and availability for agricultural use. As an agricultural pesticide, carbofuran is still legal throughout Africa though national laws vary on use in wildlife.[56] **Strychnine** is banned in European nations but can still be purchased for use as a below-ground bait in the United States[70,82] (Fig. 3.22).

Some poisons are a common cause of problems in some locations but are virtually unheard of in others. The nonsteroidal anti-inflammatory drug **diclofenac** was a devastating cause of mortality in vultures in South Asia through accidental intoxication from scavenging medicated

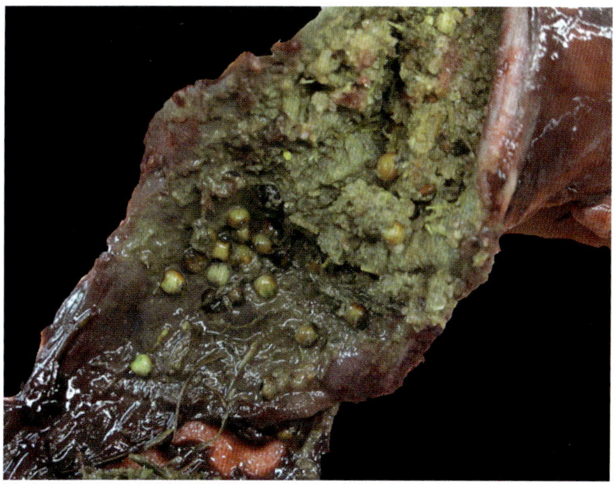

FIGURE 3.22 Content in the ventriculus from a poisoned Canada goose. This goose had ingested strychnine-laced seed gopher bait. The bait was applied aboveground—in violation of pesticide application regulations—where it could be accessed by the goose. This particular bait is green tinged and a small amount of the artificial coloring can still be seen.

livestock.[14] The **avicides** Starlicide and 4-aminopyridine (Avitrol) were developed specifically to target invasive European starlings and they may be uncommon outside of North America. In contrast, the narcotic pesticide **alpha-chloralose** is used in the United Kingdom and New Zealand as an avicide, but as of the writing of this second edition is still in the EPA approval process in the United States.[83] A complete review of all poisons implicated in illegal wildlife deaths is deserving of its own book, so the reader is encouraged to research the laws and trends specific to their area. Described below are those considered to be of the most globally applicable importance as well as some with which the authors, working in the United States, have had the most experience. An overview of these poisons, including necropsy findings and recommended test samples, can be found in Table 3.1.

One of the main challenges with poisoning cases is a lack of specific gross or histologic findings. Death of one or multiple animals with no obvious signs of trauma or disease raises suspicion of poisoning, especially if there is an evidence of recent feeding. The most effective toxicology testing requires some idea of which toxin to test for. Screening for the most commonly encountered poisons may be helpful in cases with no history and no suspects. Saving portions of the stomach contents for genetic identification of species may help to identify the bait.

It is important to establish contact ahead of time with a diagnostic or criminal laboratory that can process your samples. For legal casework, it is ideal if the laboratory is accredited, meaning that it has been assessed by a third party (accrediting body) for quality control, quality assurance, and personnel requirements. Worldwide there are numerous laboratory accreditation services. A good place to start searching for accredited labs is through the International Laboratory Accreditation Cooperation (ILAC) which oversees accreditation bodies for reference and diagnostic laboratories and operates under ISO (International Organization for Standardization) standards (ilac.org). Keep in mind that ILAC does not include all accrediting bodies. Make sure that all outgoing evidence is accompanied by a chain of custody form and that the receiving laboratory understands how to handle a chain of custody and evidence seals. In legal cases opposing counsel must always have the opportunity for their experts to examine the evidence. To that end, a representative portion of the tested sample should be saved if possible.

Organophosphates and **carbamates** have similar mechanisms of action. Organophosphates differ from carbamates in that they have prolonged effects. Carbamates are shorter-acting, increasing the chances that an animal with a low dose exposure will survive.[27,52]

Because of their potency and persistence, baited carcasses can kill many animals including both those that feed on the bait (direct intoxication) and predators or scavengers that feed on the primary victims (indirect intoxication). Crime scene findings may include multiple animals found dead in one area, dead insects on the body, the presence of a scavenged carcass, and body position consistent with terminal convulsions. Depending on the compound, the poison may be visible in the form of dark granules or white or purple powder; however, it also may be applied in a liquid form.

Ideally, collection and testing of both stomach contents and brain is used to make a diagnosis. Detection of the toxin in the stomach contents in conjunction with low acetylcholinesterase levels in the brain is confirmatory. Practically speaking, it is enough to detect poisons in the stomach contents or skin since the evaluation of acetylcholinesterase in the brain presents challenges in species for which reference ranges have not been determined. Decomposition can falsely raise levels while prolonged freezing can falsely reduce levels.[72,85] If acetylcholinesterase levels are to be evaluated, a hemisection of brain should be collected, frozen, and tested as quickly as possible.[52]

These poisons seem to be remarkably hardy in the environment and even when the animal is heavily scavenged or devoid of stomach contents, traces may still be detected on the oral cavity, esophageal, or gastric mucosa. Foot skin from birds can be positive, indicating percutaneous exposure. Birds seem to be especially sensitive to these poisons and may absorb a fatal dose from just perching on a baited carcass.

As poisoning can result from inhalation or percutaneous absorption, it is important to take precautions when working with or transporting cases. Always use personal protective equipment (PPE) and work in a well-ventilated environment.

Avitrol is the trade name for **4-aminopyridine**, a restricted-use pesticide designed to frighten away flocks of nuisance birds by causing distressed behavior in the affected members. It also kills them. The poison is often formulated in cracked corn bait. Effects are mitigated by dose, time of day, and temperature. Colder weather and early morning feeding are associated with greater mortality.[49,73] 4-aminopyridine toxicity should be suspected if multiple granivorous birds are found dead. Corn may be found in the stomach contents or around the site. A clinical sign may be distress behavior, which can result in blunt force or other trauma noted clinically and during the necropsy. Because the mechanism of action is blockage of potassium channels, there are no characteristic gross or microscopic lesions.

Starlicide is the trade name for **DRC-1339**.[81] This is a slow-acting toxin used to kill blackbirds, starlings, pigeons, corvids, and gulls. Starlicide is billed as causing a "painless death," presumably because the birds are too sick to show outward signs of distress. Secondary toxicity has not been demonstrated.[17,23]

TABLE 3.1 Toxic Agents That Impact Wildlife

Toxin	Selected Chemical and Trade Names (Note This Is Not a Complete List)	Mechanism of Action	Clinical Signs	Necropsy Findings	Test Samples
Organophosphates and carbamates	Terbufos Diazinon Chlorpyrifos Malathion Fenthion Famphur—Warbex Methomyl Aldicarb Carbofuran—Furadan	Inhibition of the neurotransmitter, acetylcholinesterase	Convulsions, salivation, lacrimation, urination, defecation, dyspnea, emesis	May be evidence of recent feeding. Otherwise, no specific findings	Stomach and/or crop contents, vomit, skin from the feet, oral mucous membranes (toxin testing). Brain (acetylcholinesterase testing)
Strychnine		Inhibition of the neurotransmitter, glycine	Weakness, muscle spasms, convulsions	May be baited grain in stomach or ventriculus, otherwise no specific findings	Stomach contents, vomit, urine. Liver, kidney, bile (less sensitive)
4-Aminopyridine	Avitrol	Enhanced acetylcholine release via blockage of voltage-gated potassium channels	Depressed and/or distressed behavior	Secondary trauma, may be ingested bait (usually corn)	Ventriculus and crop contents
DRC-1339	Starlicide	Damages heart and kidneys—mechanism unknown	Depressed behavior	Possibly gout but more often no specific findings	Feces, liver, muscle, lung, kidney, ventriculus contents
Zinc phosphide		Corrosive to the respiratory and gastrointestinal tract	Lethargy, weakness, disorientation, diarrhea/hematochezia, emesis/hematemesis	Systemic congestion, hyperemia and/or hemorrhage of the respiratory and/or gastrointestinal tract, may be a fishy or garlic odor	Stomach or ventriculus contents. Consider submitting unopened organ to reduce phosphine gas dissipation
Pentobarbital	Euthasol	Neurodepressant; activates GABA and inhibits glutamate	Depression, weakness, ataxia	May be evidence of recent meal, otherwise no specific findings	Liver, stomach and/or crop contents, urine, feces
Anticoagulant rodenticide	Brodifacoum Bromadiolone Chlorphacinone Difenacoum Diphacinone Warfarin	Inhibits coagulation by blocking recycling of vitamin K_1	Uncontrolled bleeding results in a variety of possible clinical signs	Hemorrhage; gross and/or microscopic	Liver

Continued

TABLE 3.1 Toxic Agents That Impact Wildlife—cont'd

Toxin	Selected Chemical and Trade Names (Note This Is Not a Complete List)	Mechanism of Action	Clinical Signs	Necropsy Findings	Test Samples
Bromethalin		Neurotoxicity via blockage of ATP production	Muscle tremors progressing to seizures, hyperthermia, hyperexcitability	Central nervous system (CNS) white matter vacuolization	Fat, liver, kidney and brain. Stomach contents if acute onset
Cholecalciferol		Causes hypercalcemia	Depression, weakness, anorexia, polyuria, polydipsia, dehydration, hematemesis	Gastric hemorrhage, soft tissue mineralization	Kidney, bile
Sodium monofluoroacetate	Compound 1080	Inhibits the Krebs-citric acid cycle	Hyperexcitability, convulsions, vomiting, salivation, urination, and defecation	No specific findings	Stomach contents and/or vomit. Urine, kidneys (less sensitive)
Cyanide	M-44	Tissue hypoxia through inhibition of oxidative phosphorylation	Tachypnea, hypotension, convulsions, coma	Possibly bright red mucous membranes, almond odor, gastric hemorrhage	Whole blood. Heart, liver are less sensitive
Metaldehyde		Decreases inhibitory neurotransmitter, GABA	Tremors, seizure, hyperthermia	Formaldehyde odor, no other specific findings	Stomach contents

Starlicide can be a frustrating poison to diagnose as it may be eliminated as rapidly as 2h after ingestion whereas death occurs anywhere from 3 to approximately 50h after ingestion.[17]

Sampling of feces, liver, muscle, lung, and kidney may yield the best results.[23] More often, diagnosis is made based on scene findings (large numbers of dead birds, often of the above-mentioned sensitive species, with or without baited grain, or seeds present) and, possibly, evidence of renal dysfunction. Gross and histologic lesions may or may not be present. When present, necropsy findings include visceral gout. Histologically, there may be acute renal hemorrhage and/or tubular necrosis. Distension and rupture of the gall bladder has also been reported.[17]

Anticoagulant rodenticides pose well-known dangers to target and nontarget species.Animals may be directly poisoned by eating the bait or secondarily by eating poisoned rodents. Because vitamin K_1 and clotting factors must be exhausted before coagulopathy develops, there is a lag time between toxin ingestion and clinical effects. In dogs this period can be as long as 5 days.[55] Animals, particularly raptors, may have low levels of toxin but otherwise be clinically normal; therefore the diagnosis must be made taking into consideration both the presence of the toxin and clinical or necropsy findings.[1,77] Suggestive postmortem findings include fat and muscle depletion; visceral pallor; and black, tarry GI contents. Disproportionate subcutaneous or body cavity hemorrhage may occur from otherwise minor trauma and may result in hemorrhagic shock in subclinically affected animals. Anticoagulant intoxication may be diagnosed antemortem via evaluation of clotting times and analysis of blood, vomitus, or bait. Liver is the best sample for postmortem diagnosis of AR toxicosis.[55]

Zinc phosphide is used primarily for rodent control. It is sold as a powder or formulated into bait pellets or grain. In an acidic environment, such as the stomach, zinc phosphide is hydrolyzed to phosphine gas, which is directly toxic. Because moisture also triggers effects, some animals, such as waterfowl, are more vulnerable. Studies suggest a low risk of secondary toxicity but it has been reported in dogs.[40,76]

Because zinc phosphide dissociates into a gas, the poison is lost with exposure to air. In suspect cases stomach contents should be sealed into an air-tight container and frozen as soon as possible. Detection of elevated levels of zinc in the stomach contents provides indirect evidence of zinc phosphide toxicity and may be the only remaining indicator, if the gas has dissipated.[59] Be aware of personal safety when working with zinc phosphide cases and only open the carcass in a well-ventilated environment.

Cyanide is still used as coyote bait in M-44 "coyote getters." The M-44 is a spring-loaded device that ejects cyanide powder. When exposed to moisture in the mouth and nose, the powder forms hydrogen cyanide gas. Ingredients in the M-44 capsules include an inert marker that is either visible to the naked eye or fluoresces under ultraviolet light.[66] Free-ranging animals may also be exposed to cyanide through mining operations (see "Industrial Hazards" section). Diagnosis of cyanide toxicity relies largely on the field investigation and (in M-44 cases) the presence of the marker. Ideally, cyanide will also be detected in the blood. In the absence of available whole blood, blood-filled organs (e.g., the heart and liver) can be tested but may be falsely negative.[89]

Additional important poisons, such as **bromethalin, strychnine, cholecalciferol, pentobarbital,** and **sodium monofluoroacetate**, are listed in Table 3.1.

FORENSIC TAPHONOMY AND TIME OF DEATH DETERMINATION

Taphonomy is the study of how bodies decay. Forensic taphonomy is the application of taphonomy to a crime scene with the aim of estimating the time since death and differentiating injury from postmortem alterations. In forensic cases, and in wildlife in particular, bodies may not be discovered until long after death; however, skeletonized remains can still provide valuable information.

Death has been described as a process rather than an event, one that can last decades or longer as remains decay back to their elemental state.[80] In the first hours after death bodies progress through the stages of **algor, livor,** and **rigor mortis**. Algor mortis is cooling of the body. Livor mortis is dependent blood pooling. In furred or feathered creatures livor mortis may be difficult to appreciate externally but blood pooling—evidenced by darker or redder coloration—can be observed in the subcutis and organs on the dependent or "down-side". Within the first 8–12h, livor mortis can shift if the body is moved. It eventually becomes "fixed" as blood leaks from the vasculature into the surrounding tissues.[22] Rigor mortis is the stiffening of muscles with exhaustion of ATP. Rigor mortis typically starts a few hours after death, reaching a maximum point at approximately 12h postmortem. Rigor then disappears as the body continues to decay and is gone 24–36h postmortem. The onset and period of rigor may be hastened by elevated body or environmental temperature at the time of death. Conversely, reduced muscle mass or low temperature may delay the onset of rigor mortis.

There are also ways in which bodies may be "naturally preserved" and these should be kept in mind when evaluating postmortem change and time of death. Decomposition rates vary widely when bodies are desiccated, immersed in water, buried, and/or frozen. The decomposition degree may be estimated in a subjective way by odor, color, corneal changes, skin sloughing, blood pooling in different areas, body temperature, and degree of bloating.[67,71] Bodies that are large, cylindrical, and well insulated, such as whales, may putrefy and autolyze faster at low hydrostatic pressures. Further variations in the decomposition rate may be observed due to aquatic salinity and water temperature.[31] Often, increased internal pressure from bloating will result in extrusion of the soft tissues of the mouth or reproductive tract in these species. The bloated carcass in water may experience inappropriate transport or dissection (mechanical stress), and therefore the internal organs and body fluids may be suddenly ejected ("carcass explosion"). In these cases only fluids are ejected; there is no evidence for the inclusion of skeletal elements in such "explosions". Thus the scattering of bones over a specific area is not observed.[64] Also, putrefaction gasses play a major role with respect to the buoyancy and drift behavior of vertebrate carcasses in shallow marine (as well as lacustrine) environments. The carcasses of many marine mammals, such as those of whales, cannot be observed because most of them will never surface or strand, having released their bloat at sea and sunk.[11] Gathering and recording the data of postmortem hydrostatic pressure, temperature, and scavenging rate conditions in aquatic animals is fundamental to define the cause of death in these animals.[2]

Weathering of bones causes cracking, flaking, and splintering, typically along the fibers and along the long axis in long bones. Bone weathering, taking place over years to decades, was described in Behrensmeyer's classic 1978 article; starting with cracking and progressing through five stages to eventual splintering and dissociation[6,7] (Figs. 3.23A, B). Caution should be used in assigning specific death times based on weathering but this can give the investigator a rough idea of chronicity. Small and nonmammalian species may differ somewhat and buried bones also will not weather predictably.

Determining the time since death in both humans and nonhumans is an area of ongoing research. In animals time since death determination may be more difficult for a number of reasons. Controlled cadaver studies have not been done for most species, so data, such as body cooling rate, is not available. As an added complication, wildlife bodies are often not discovered until after the immediate postmortem period. Nor are they always treated in a way in which data, such as body temperature, can be accurately recorded (e.g., they may be picked up and immediately put in a freezer).

Scene clues are often heavily relied upon in human and domestic animal forensic cases. When was the mail last collected? How old is the food in the refrigerator? When did someone last see this person alive? Unfortunately, in cases involving free-ranging animals, eyewitness accounts are rarely available and are not always trustworthy. Radio collar tracking data, if available, should be acquired and applied to the case investigation.

Forensic entomology is another method of estimating the time of death. Time of death (or more accurately, time of colonization) is determined in this way by collecting maggots, determining the species, reviewing the literature to find the appropriate growth temperature studies, and back-calculating using weather data from the site where the body was found. Because results are most accurately determined from the oldest developing stage (which may be pupae), good collection technique is critical (Fig. 3.24).

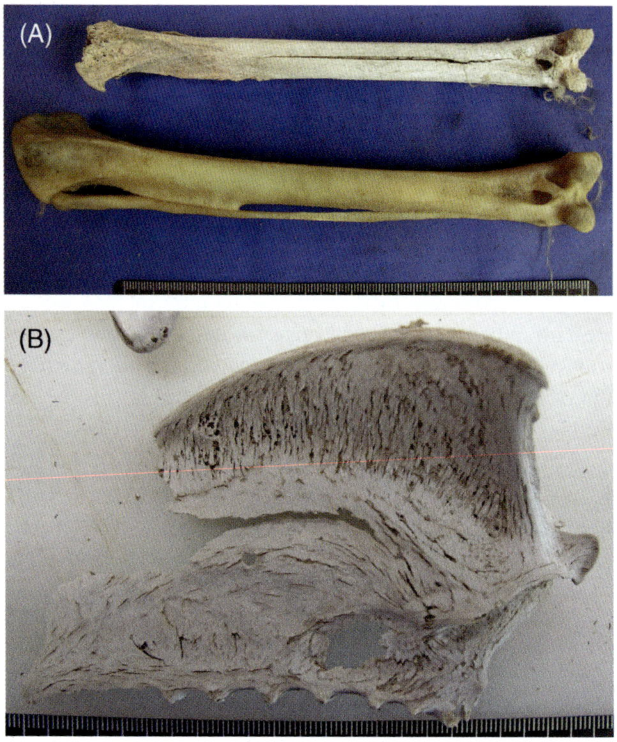

FIGURE 3.23 Weathering of bones from golden eagles. Exposed to the sun and elements, bones begin to crack and fragment as they pass through Behrensmeyer's (1978) five stages of weathering. (A) Pictured are tibiotarsi from two different golden eagles that have undergone natural decomposition and removal of soft tissues. The smaller bone at the top of the image has additionally been weathered and exposed to the sun, resulting in longitudinal cracks and pitting of the bone surface at the epiphyses (stage 1). (B) Over time, the bone will begin to flake apart (stage 5).

Evaluating Skeletonized Remains

The pathologist may be dismayed when presented with **skeletal remains**; however, there is valuable information hidden in even the most poorly preserved bodies. Clues as to the cause of death may be found in fracture patterns. There may be bullet fragments or evidence of osteomyelitis or other bone alterations (Figs. 3.25A, B).

When remains are skeletonized, mummified, or otherwise markedly decomposed, radiographs can be used to help document which bones are present, relative sizes, the presence of fractures, fracture types, and the presence (or absence) of trace metal. Advanced imaging techniques,

FIGURE 3.24 A maggot mass. Necrophagous flies are extremely efficient, and they use collective group feeding, forming a maggot mass that is commonly found on carrion.

such as CT scanning and scanning electron microscopy (SEM) may also be of value.

SEM can reveal details of bone injury, such as knife or other tool marks but may require a skilled analyst for interpretation. Careful evaluation of fracture patterns can help to determine the cause of death. Additional clues, such as fracture location and scene information, can help to narrow down the cause. Vehicle collision, for example, commonly results in fractures of the pelvis of quadrupeds.[32] Both radiography and table-top reconstruction of the skeletal elements are useful during the examination. Removal of remaining soft tissues can help with analysis (see below).

Species identification is often necessary to determine the legal course of action as well as aiding in cause of death determination. A specialist may be needed to make determination from some bones, though larger bones and skulls may not present much of a challenge. Bone, bone marrow, or soft tissue remnants can be used for molecular analysis.

Postmortem Scavenging

Postmortem alteration of the body by scavengers is common in wildlife cases. Carrion consumption by scavengers is a relevant part of terrestrial and aquatic food webs. However, there are differences in these two different ecosystems, usually between the structure and functioning of the scavenging communities. In general, **scavenger trauma** is not accompanied by hemorrhage and findings are characteristic of the particular scavenger(s) involved. Identification of the culprit will help differentiate scavenging and depredation. Scene clues, such as tracks and the presence of scat, should be included in the evaluation. DNA swabs of bite marks can also be used to determine species.

Bite mark measurements can be taken and compared to reference standards.[54] Patterning of punctures arranged in groups of three or four is suggestive of raptor talon marks in cases of predation or fight wounds. Avian scavengers (typically vultures, bald eagles, and/or corvids) can make short work of a carcass. Scratches on bone, frayed tendons and ligaments, skin pulled down around the limbs resulting in a "case-skinned" (a carcass processing method by which the skin is removed by pulling it down and peeling it off like a sock) look and tissue loss around orifices and pre-existing wounds are all suggestive of avian scavengers.[3,63] Parallel grooves on the ends of bones are consistent with rodent scavenging whereas punctures and pit-like damage to bones, particularly at the ends, are suggestive of canid feeding.[29,38]

Determining whether a fracture occurred in the perimortem or postmortem period can be challenging when analyzing skeletal remains. As a general rule, discoloration around the fractured area that encompasses both the fracture and cortical surface is considered likely to be perimortem (Fig. 3.26). Nonsplintering fracture patterns, including spiral, green-stick, and butterfly fractures, are suggestive of relatively moist (therefore perimortem) bone. Caution should be used in drawing firm conclusions from these findings as bone moisture loss is variable, depending on the type of bone and the environment.[87,88]

GROSS AND HISTOLOGIC NON-LESIONS

Submission of carcasses for veterinary forensic pathology examination is not always prompt. Animals may be buried and exhumed after a period of time, or the carcass may be frozen to reduce the effects of decomposition. Freezing in a commercial freezer causes ice crystals to form within cells and organs, resulting in tinctorial and architectural changes that must be "read through," if possible to formulate an opinion on a case. Damage to organs can vary with species and not all changes described here may be present.

The destruction of cell membranes and leakage of cytoplasmic contents caused by **freezing** is grossly most visible in and around the blood vessels. Hemoglobin, normally intracellular, leaks out of red blood cells and vessel walls causing smudge-like staining of the surrounding soft tissues. This ill-defined discoloration can mimic bruising and must be differentiated from real hemorrhage.[68] Subtle differences in color, location, or edge definition can help to discriminate between the two entities and are outlined in Table 3.2.

The greatest challenge in the evaluation of frozen–thawed carcasses is histologic interpretation of tissues. Cellular fluid shifts as a result of freezing may rupture the plasma membrane of individual cells leaving only the nuclei intact. This effect is especially noticeable in blood vessels where the cells are transformed into heterogeneous, proteinaceous material with free nuclei. Small-caliber vessels (e.g., meningeal vasculature) leak the free

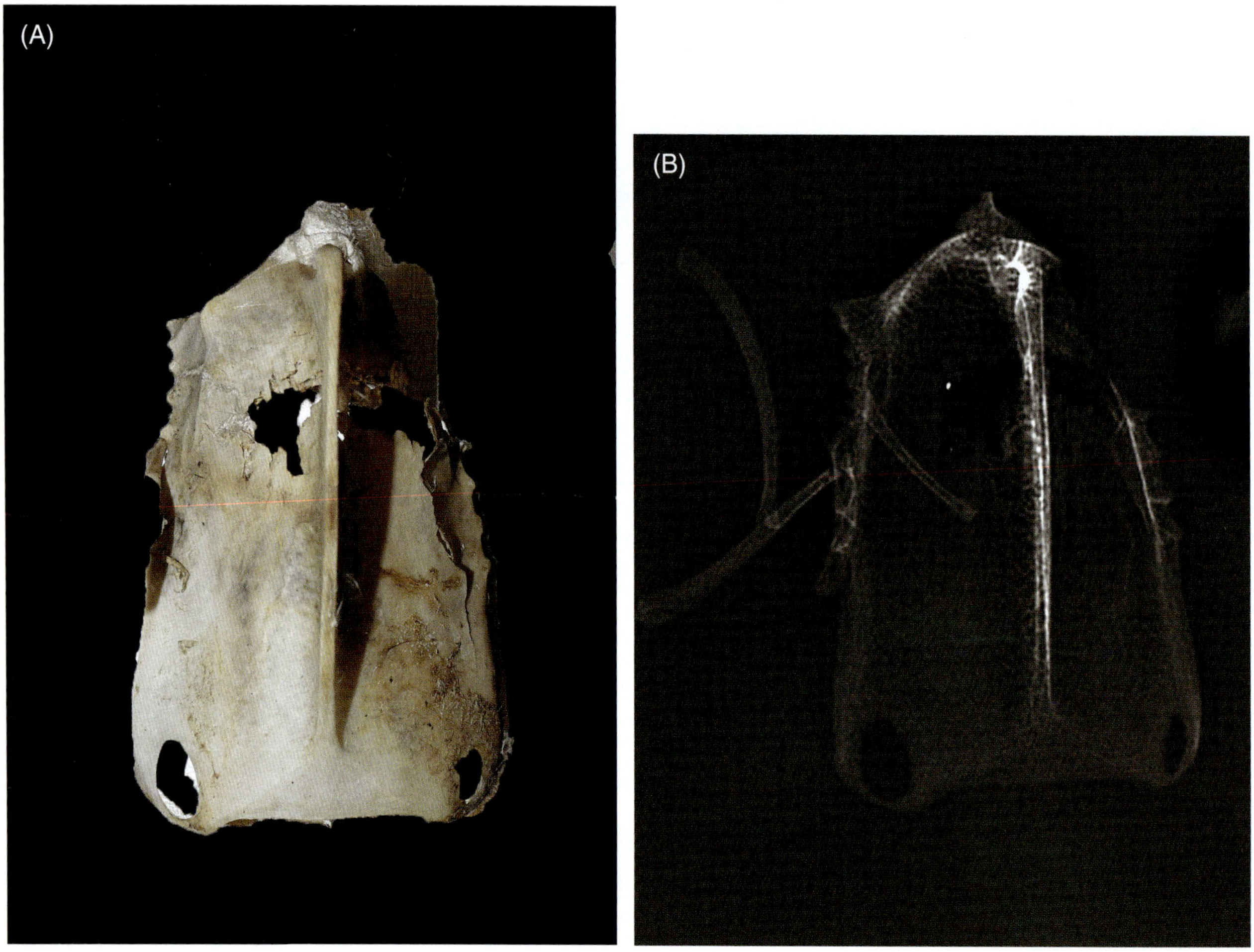

FIGURE 3.25 Decomposition in the sternum of a red-tailed hawk. This bird was one of numerous raptors found in various stages of decomposition on private property. This particular hawk was in an advanced state of decomposition with only a partial skeleton and a small amount of feathered skin remaining. (A) There are fractures on either side of the sternum that, on the left, form a transversely elongated hole. Fractured edges on both sides protrude to the left. (B) In this radiograph, the fracture shown in (A) is visible as a roughly transverse area of radiolucency across the sternum. At the right edge of the fracture, there is a pinpoint radiodensity. The gross features of the fracture in conjunction with the presence of metal is consistent with a gunshot wound.

hemoglobin resulting in diffuse, red discoloration of the organ surface (Fig. 3.27A). The cerebellum characteristically takes on an intense, red hue. The intensity of the change is not indicative of antemortem congestion or hemorrhage and is merely an effect of freezing. Larger crystals may alter the microscopic architecture of organs. The parenchyma of most organs is grossly of normal shape and color, and the animal may otherwise appear well preserved. Microscopically, freezing causes ice crystals to form throughout the parenchyma, often resulting in linear and haphazardly arranged areas of cell clearing, especially in the liver (Fig. 3.27B).

The postmortem decomposition of lipids and formation of **adipocere** is well described in the human literature.[22] Adipocere develops in wet and low-oxygen environments from the breakdown of adipose tissue into glycerin and fatty acids. A calcium mineral component often makes adipocere gritty. The most common manifestation of adipocere in animals is on the liver capsule and diaphragm of mammals. The color and texture are similar to that of visceral gout but the change is not widely distributed and only present on markedly decomposed carcasses.

SKELETAL PREPARATION TECHNIQUES

There are certain instances when purposely rendering a body into a skeleton is helpful in the analysis. Situations may include the examination of mummified remains in which bones are obscured by desiccated soft tissue, the need to differentiate chronic from acute bone injury, evaluation of areas that are difficult to visualize when fleshed (e.g., the cervical spine), and reconstruction of skeletal fractures. There are multiple ways in which a body can be skeletonized, including mere exposure to the environment;

FIGURE 3.26 Evidence of vital reaction in weathered bones. This synsacrum and pelvis from a red-tailed hawk has undergone natural decomposition, which removed all the soft tissues. The cranial extent of the synsacrum is raggedly fractured. Dark discoloration of the cranial half of the bone suggests that this fracture occurred while the animal was alive. Analysis of fracture fragments revealed trace amounts of lead, indicating that the bird had been shot through the back.

however, in forensic cases security of the body and related evidence should be a priority.

A hot water bath can be used to quickly strip bones of soft tissue. This is most easily done using a hot plate or a crock pot. Temperatures should be monitored and water kept to a simmer as boiling the bones can create artifact. A simple procedure using baking soda and detergent followed by household ammonia has been described.[25] When removing the bones from the hot water, transition gently as sudden changes in temperature can cause cracking.

Insects do a fabulous job of cleaning skeletons. In the wild, maggots are the chief remover of soft tissues, aided by cadaver beetles. Dermestid beetle colonies are used in museums for the preparation of reference specimens. If space, time, and resources are available for starting and maintaining a beetle colony, this method is highly recommended. Biological supply companies sell starter kits and there are numerous references available online regarding beetle care. Skeletonization in this way is gentle, fast, easily monitored, and readily secured.

Water bath maceration is a slow but easy and gentle method whereby bodies are immersed in water until the soft tissues fall off. Odor control and time are considerations with this method. The addition of enzymes, such as detergents, can speed the process.[74]

If desired, bones can be gently whitened by immersing them in hydrogen peroxide. Exposure to sunlight will expedite the process. Dilute bleach can also be used but more cautiously as it may result in flaking of the bone surface. Whitening of bone is generally unnecessary for casework.

DOCUMENTATION AND REPORTING

Postmortem examinations are, by necessity, destructive processes. The original piece of evidence in the veterinary forensic pathology case—the animal carcass—is left irreversibly damaged. In a legal sense, this makes the verification of findings and the presentation of evidence for court proceedings physically impossible. A good set of **photographs** to compliment the final report, however, allows the virtual verification of findings and immortalizes the normal and abnormal changes that were seen during the necropsy. The photographs become the official piece of evidence that can be presented to a judge or jury. Effective photography requires attention to detail and several techniques in order to fully capture those findings.

TABLE 3.2 Differences Between Freeze–Thaw Artifact and True Hemorrhage in Soft Tissues

	RBC Freeze Artifact	Real Hemorrhage
Color	Red-brown	Bright red
Definition	Hazy, smudged edges	Edges more sharply defined
Location	Around blood vessels	Anywhere

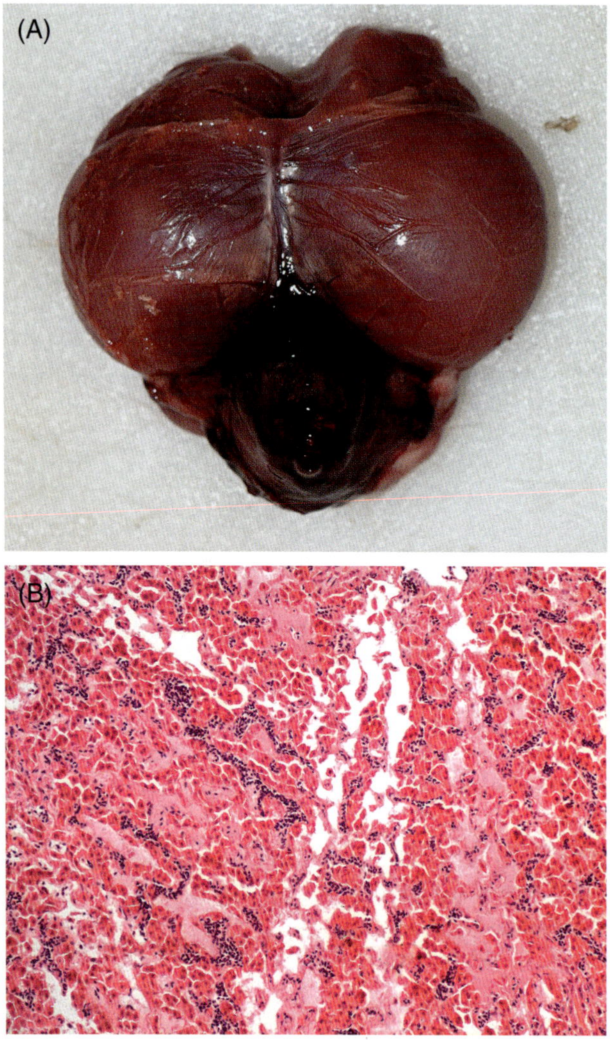

FIGURE 3.27 Freeze artifacts in tissues from a golden eagle that was frozen prior to necropsy. (A) Ice crystals cause red blood cells to rupture. Free hemoglobin then leaks into the surrounding tissues. The external surface of the cerebellum is often diffusely discolored red in the frozen and thawed carcass. (B) Ice crystals also formed in the liver during the freezing process. While the liver appeared grossly normal, the microscopic sections show the marked disruption of architecture where the crystals formed. Hepatic cords have been displaced by the ice crystals and hepatocytes are occasionally individualized or lysed. Additionally, red blood cell plasma membranes have ruptured in the vessels and sinusoids, leaving only nuclear remnants and proteinaceous material.

Photodocumentation should begin by photographing the carcass in its packaging. Photographic documentation of the packaging shows the integrity of the evidence as it was received and can support the chain of custody in the case. Images of the external surface of the animal, including body position, should follow to show how the animal appeared upon receipt.

Photographs of normal as well as abnormal findings constitute a comprehensive set of images. Photographing the organs in situ and ex situ can help to lead an external investigator through what was seen during the necropsy. Photographs taken in error (accidental shots or images that are less-than-perfect) should be maintained and not deleted. Gaps in the image set's metadata sequential numbering may pique suspicion and questioning by a defense team in the legal case, which can be avoided by keeping all images taken, regardless of quality.

Long range as well as close-up images of lesions can help an external observer see and locate an area of interest on an animal. Fig. 3.28A shows a close-up image of a hematoma in the subcutis of a bald eagle. Few clues are present in the image to help localize the lesion to a particular body site. However, when paired with a long-range shot (Fig. 3.28B), the viewer can more easily see the location and position of the bruise.

Rarely are lesions of interest displayed two-dimensionally on the body. The natural topography of the body presents both focus and lighting challenges to the photographer. The use of a camera-mounted or ancillary flash or light source can help reduce the obscuring effect of shadows. **Illumination** from a point directly above the subject, as opposed to side lighting, provides diffuse lighting that appears more natural to the eye. An extra light source can be used to help light narrow spaces, such as the lumen of a trachea or deep body cavities. Care should be taken to reduce reflections or glare from wet tissues or surfaces.

Composition of images should be simple with minimal distractions. Pictures should be taken at right angles to the plane of interest to reduce the impacts of distortion in an image. A right-angled forensic ruler may be displayed in the picture (Fig. 3.29). The targets on the ruler indicate the amount of distortion, and the gray and white coloration allow for white balance correction.

The subject of a picture should fill the frame of the image but not overlap the edges. The areas of highest interest in the image should be in the sharpest focus and, ideally, centered within the frame. Blood, necropsy instruments, and extra body parts can be a distasteful sight and can distract from what the pathologist is trying to show in a photograph. Additionally, images showing excess blood or dirt may be prevented from being admitted as evidence because they may be seen as inflammatory or offensive to some jurors. To increase the effectiveness of your images, the area in question should be as clean as possible. Images may be taken prior to cleaning for documentary purposes, but an image of the uncompromised lesion should be taken for potential use in court. Eliminating extra material from an area of interest can also highlight a lesion, making it easier to see and explain to a judge or jury. Prior to cleaning an area, however, ensure that all necessary samples, such as bite mark swabs and touch impressions, have been taken.

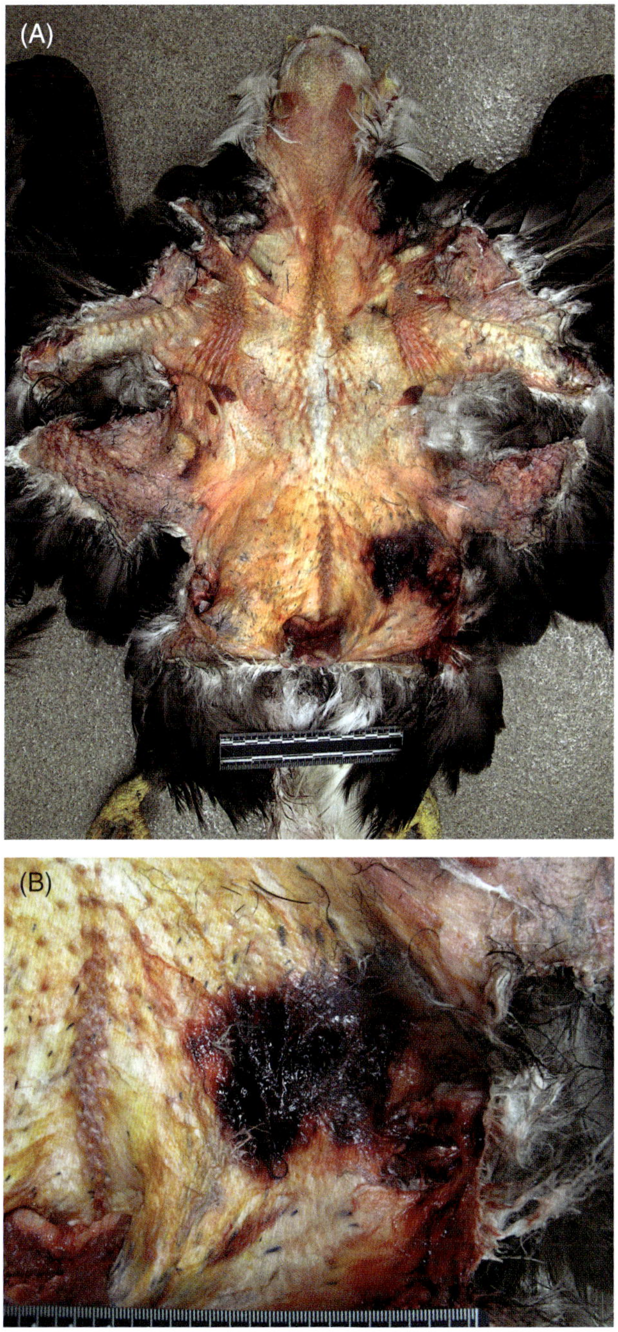

FIGURE 3.28 The reflected skin from a bald eagle demonstrate the value of scope and composition in forensic photographs. (A) A photograph taken from a distance allows the viewer to visually locate a lesion in the anatomy of the animal. In this case, the subcutis near the left hip is darkly discolored. (B) Closer photographs reveal greater detail about the lesion in question. Here, dark discoloration is shown to be a subcutaneous hematoma.

REPORT CONTENT

The **final veterinary forensic pathology report** is a lasting and often public document containing all of the findings in a forensic postmortem examination. It encapsulates everything that was seen and done and by whom. This report

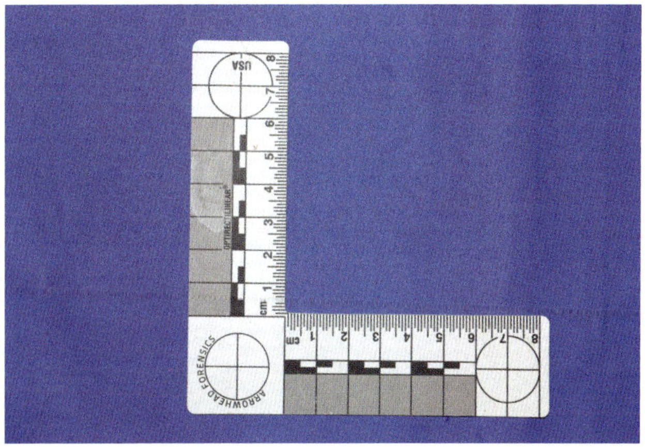

FIGURE 3.29 Forensic ruler or American Board of Forensic Odontology bite mark (ABFO) scale. This ruler features a metric scale, round targets, and a middle gray reference band. The round targets help the viewer evaluate the angle at which the surface was photographed. Ideally, the targets should be round without any oval distortion. The gray reference helps to create consistency and authenticity in color photography.

may be seen by a judge, jury, and clients; it may be subject to freedom of information requests and, thus, must be understandable to laypeople. Technical terms should be used in the narratives but plain language interpretation should be provided in the case summary so that an average person can understand the conclusions and opinions drawn by the pathologist.

Information that should be included in a final forensic pathology report is listed in Table 3.3. The report should be signed by the pathologist who oversaw or performed the necropsy. Review of the report by another veterinary pathologist provides further legal support of the findings. The reviewer checks for technical and nontechnical errors and ensures that the conclusions drawn by the veterinary pathologist who performed the necropsy are sound and logical. Documentation of the review stays in the case file or digital folder.

PRESERVATION OF EVIDENCE

Physical evidence, such as carcasses and samples, must be retained by either the pathology laboratory or the submitting agency until the case is adjudicated or otherwise closed.

The destruction of evidence prior to court proceedings can be viewed as a suspicious act. Additionally, the defense may wish to retest certain samples to ensure the validity of results. If no sample is available for testing, results of the original analysis may become inadmissible in court. Photographs, radiographs, notes taken during necropsy, and other case-related documentation should be retained indefinitely in either a clearly labeled case file folder or digitally in a secure drive that is regularly backed up.

TABLE 3.3 Elements in a Forensic Veterinary Pathology Report

Report title including discipline (e.g., veterinary pathology report)
The name of the person or agency that submitted the animal(s)
The case number and any other identifiers associated with the case
Date of submission or receipt by the veterinary pathologist
Date of the necropsy
History and examination(s) requested
Statement of the examinations conducted (e.g., radiographs, photography, and dissection)
Signalment, weight, body condition, and postmortem preservation of the animal
Narrative of the necropsy findings
Summary of the case with a clearly stated opinion as to the cause of death of the animal(s)
Disposition of the carcass and any samples taken—whether retained or returned to the submitting party
List of the samples taken for additional diagnostics
Results of tests performed on samples, the date of the report, and the name of the scientist who produced the test report

SUMMARY

National, state, territorial, and tribal laws exist to minimize the negative impacts of humans upon animals. Inevitable violations of these laws must be pursued with a multidisciplinary approach that involves law enforcement, legal professionals, and individuals knowledgeable of animal physiology and pathology. The ability to differentiate disease from misdeeds and nature from neglect helps the wildlife veterinarian inform and build cases against law violators. Thoroughly documenting and providing the forensic pathology findings to law enforcement can result in the successful prosecution of a legal case and mitigate future violations.

ADDITIONAL RESOURCES

American Academy of Forensic Sciences Standards Board. *Veterinary Forensic Postmortem Examination Standards* (170). https://www.aafs.org/academy-standards-board. 2022

Cooper JE, Cooper ME. *Wildlife Forensic Investigation: Principles and Practice*. CRC Press, Taylor & Francis Group; 2013.

Di Maio VJM. *Gunshot Wounds: Practical Aspects of Firearms, Ballistics, and Forensic Techniques*. 2nd ed. CRC Press; 1999.

Dolinak D, Matshes EW, Lew EO. *Forensic Pathology: Principles and Practice*. Elsevier/Academic Press; 2005.

Haglund WD, Sorg MH, eds. *Forensic Taphonomy: The Postmortem Fate of Human Remains*. CRC Press; 1997.

Huffman JE, Wallace JR, eds. *Wildlife Forensics: Methods and Applications*. Wiley-Blackwell;2011.

Merck M, ed. Veterinary Forensics: Animal Cruelty Investigations. 1st ed. Blackwell Pub; 2007.

Sincerbox SN, DiGangi EA, eds. *Forensic Taphonomy and Ecology of North American Scavengers*. Academic Press; 2018.

Sriraman PK. *Wildlife necropsy and forensics*. CRC Press; 2021.

Veterinary Pathology, Special Focus: Veterinary Forensic Pathology. McDonough SP, McEwen BJ, guest eds. Volume 53 Issue 5, September 2016.

References

Full references for this chapter can be found online at https://doi.org/10.1016/B978-0-443-18437-6.00004-3.

Legal note: The findings and conclusions in this article are those of the authors and do not necessarily represent the views of the U.S. Fish and Wildlife Service.

Chapter 4

Introduction to Comparative Clinical Pathology

Nicole I. Stacy[a], Charlotte Hollinger[b,c], and Emma H. Hooijberg[d]
[a]University of Florida, Gainesville, FL, United States; [b]Inotiv, Kalamazoo, MI, United States; [c]Wildlife Conservation Society, Bronx, NY, United States; [d]University of Pretoria, Onderstepoort, South Africa

INTRODUCTION

Clinical pathology focuses on the diagnosis and understanding of disease through laboratory analysis of blood and other body fluids, as well as nonarchitectural tissue samples. Assessing and supporting the quality of laboratory results and their interpretation is also included. Laboratory testing is used antemortem to evaluate health status, to identify disease, and to monitor progress after treatment, and is an important component of the diagnostic workup. Clinicopathological data from healthy animals is used to establish reference intervals and provides an insight into species physiology. Laboratory testing on samples obtained postmortem can similarly provide useful diagnostic information to elucidate pathology with judicious consideration of death-related limitations and artifacts. While some features of clinical pathology are variable across species, many basic principles are universal, and an understanding of both fundamental tenets and unique features is important to best arrive at evidence-based conclusions.

This chapter provides a broad overview of fundamental aspects of clinicopathological data acquisition and interpretation across commonly available diagnostic laboratory testing modalities, highlights unique features of clinical pathology in different taxa and/or species, and discusses the role and implementation of clinical pathology in postmortem assessment. Clinical pathology of nondomestic species could fill a book in and of itself. For this reason, this chapter has extensive materials in the form of digital content. Chapter subsections are divided by major clinical pathology assessment types (i.e., hematology, hemostasis, chemistry, urinalysis, cytology) and for all major analytes with an assessment type, content is organized into general physiology and interpretation, followed by information for mammals and nonmammalian vertebrates (as available), and overarching quality assurance considerations (preanalytical and analytical) as applicable. The chapter concludes with a discussion of laboratory testing during postmortem examinations, zoonotic risk management, and caveats for field, forensic, and unusual mortality events.

BASIC PRINCIPLES OF CLINICAL PATHOLOGY ACROSS SPECIES

This section addresses details on the establishment of reference intervals as well as aspects on methodology and quality assurance and control (see digital content).

COMPARATIVE HEMATOLOGY

Physiology and Interpretation of Changes in Disease

Hematological analysis (i.e., assessment of the cellular components of blood) is an essential diagnostic tool as screening test, part of wellness examinations and patient monitoring, or for diagnosis in clinically ill patients. Critical foundational information is found in the digital content.

Mammals

The principles and methodologies of hematological analysis and interpretation are more similar than different within the diverse group of mammalian species.[34,104] Thus, it comes as no surprise that many animal species are used as experimental models for human disease and toxicological studies, and that methodologies generally apply across mammals, including nondomestic species.

Mammalian erythrocytes are anucleated, biconcave, and round (discocytes) in most species (Fig. 4.e1; Table 4.e1).[4–6] Similar to human erythrocytes, Canidae have central pallor spanning about one-third of the center of the cell, the extent of which is not observed in other taxa. Uniquely among mammals, camelidae in health have

Pathology of Wildlife and Zoo Animals. https://doi.org/10.1016/B978-0-443-18437-6.00033-X

oval-shaped erythrocytes (elliptocytes, ovalocytes) that occasionally exhibit Cabot rings (ring- or figure eight-shaped thin structures that are considered to be remnants of mitotic spindles) that are of unknown clinical significance and occasionally also have scattered densely eosinophilic hemoglobin crystals on blood films (Fig. 4.1).[236] Additional species with a low number of hemoglobin crystals recognized on blood films in health include cats and dogs,[104] and have been observed by the authors in clinically normal aye-aye, pangolin, capybara, and agouti. Some cervidae species form drepanocytes (fusiform-shaped RBC, a.k.a. sickle cells) in vitro due to changes in pH and oxygen tension (Fig. 4.2). Neonates may exhibit significant poikilocytosis in health, including acanthocytes, echinocytes, and/or schistocytes. In some young wild and domestic ruminant species, including roan antelope, this is suspected to be associated with a unique hemoglobin molecule at that life stage[30,181,222,236] (Fig. 4.3). Erythrocytes of suidae and rats are prone to echinocyte formation (echinocytes; crenated erythrocytes) in vitro likely due to lower intracellular ATP[222] (Fig. 4.4). Low percentages of Heinz bodies (focal pink precipitates of hemoglobin along the cell membrane resulting from oxidative injury) are normal in healthy felidae, sirenia, rhinoceros, marmosets, and tamarin (Fig. 4.5); mechanisms for the occurrence of Heinz bodies in these species include a higher susceptibility to oxidative damage, lower efficiency of the spleen to remove affected erythrocytes, and/or higher concentrations of oxidative content in their diet.[68,104,106,251] Therefore, including Heinz body counts (visualized by new methylene blue [NMB] stain; described below) in these species as part of routine CBCs can be helpful for establishing baselines for individuals and for monitoring during disease. If there is evidence of Heinz body formation in other mammalian species, these should be quantified. Mild rouleaux formation (erythrocytes arranged as a stack of coins) may be

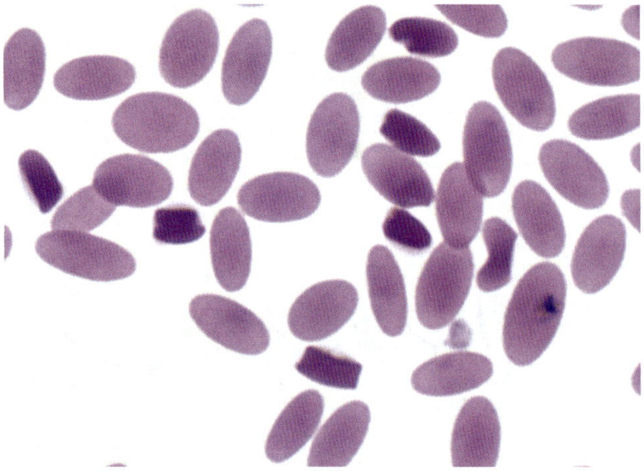

FIGURE 4.1 Elliptocytes (*oval-shaped* erythrocytes) and hemoglobin crystals in a healthy llama. Wright-Giemsa (WG).

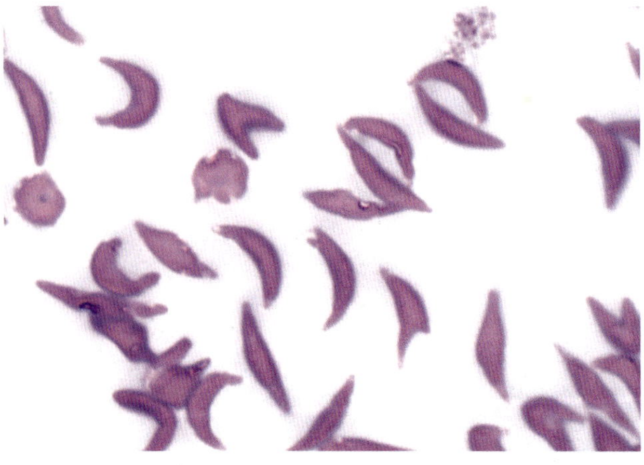

FIGURE 4.2 Blood film from a healthy white-tailed deer with *spindle-shaped* sickle cells (drepanocytes). A low number of sickle cells exhibit variably sized, irregular, nonstaining refractile drying artifact (no clinical significance). WG.

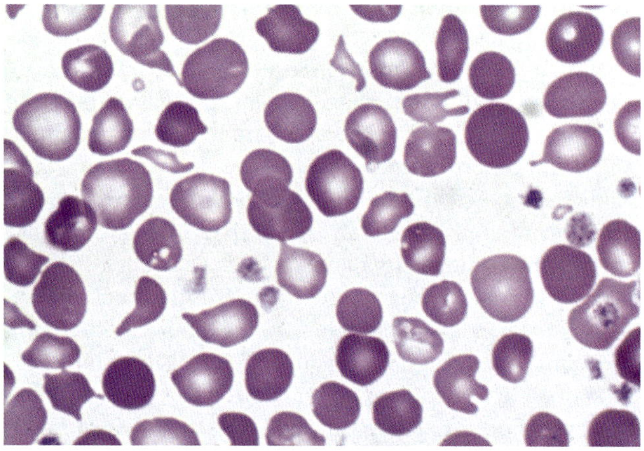

FIGURE 4.3 Blood film from a healthy 3-month-old calf: poikilocytosis (variable erythrocyte shapes). WG.

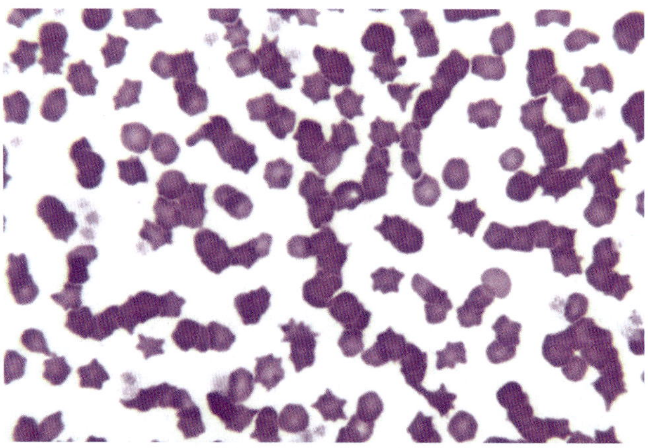

FIGURE 4.4 Blood film from a healthy domestic pig: rouleaux (erythrocytes grouped together like a stack of coins) and echinocytes (crenated erythrocytes). WG.

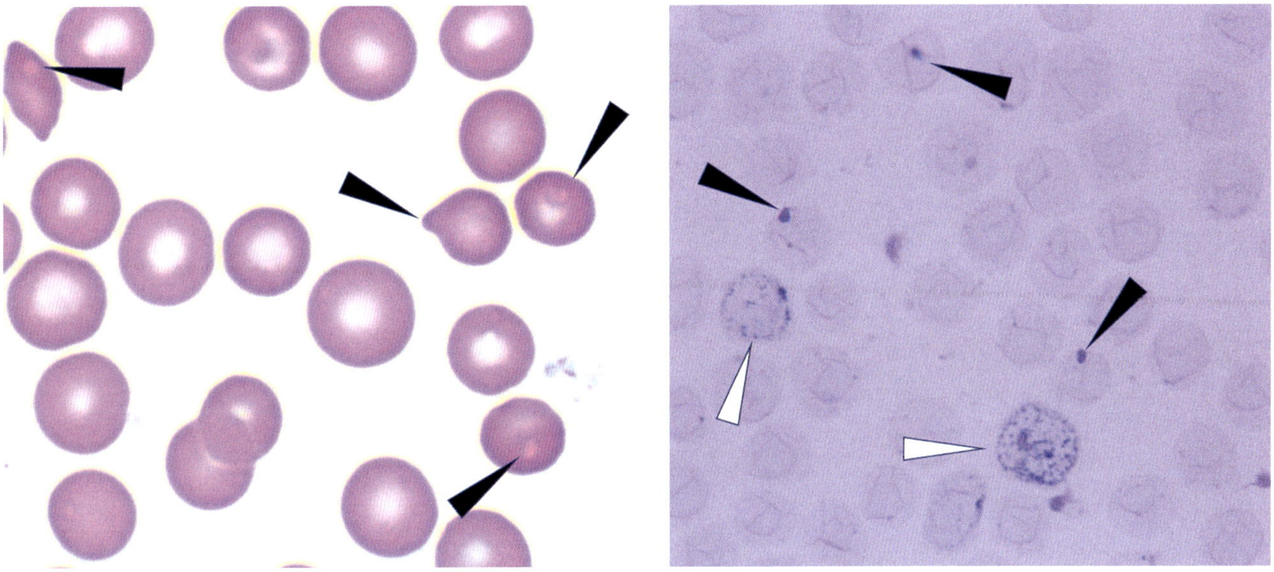

FIGURE 4.5 Blood film from a Florida manatee containing erythrocytes with Heinz bodies (denatured precipitated hemoglobin). Healthy manatees can have up to 20% Heinz bodies (*black arrowheads*). WG (*left*). New methylene blue stain (*right*) showing also reticulocytes (*white arrowheads*).

observed in equidae and other ungulates (e.g., elephants), felidae, and suidae.[30] Increased rouleaux formation is typically seen with inflammation. In domestic mammals, it has been associated with lymphoproliferative neoplasia.[104] Nucleated red blood cells (nRBC) may be normal in health in some taxa (e.g., New World monkey species, koala, dolphin) and should always be quantified (see below).[30,222]

Most mammalian species have low numbers of circulating **reticulocytes/polychromatophils**, which are observed as mild anisocytosis and polychromasia in health. The terms reticulocyte and polychromatophil are often used interchangeably; however, on routinely stained blood films, not all reticulocytes have enough RNA content to be observably polychromatophilic (termed polychromatophils). Reticulocytes are identified using NMB stain (Fig. 4.5); the preparation of smears for reticulocyte evaluation is explained below. Given the short life span of reticulocytes in most mammalian species, all NMB-positive staining RBCs are commonly termed "reticulocytes" without differentiating into various forms. However, the reticulocytes of cats and their relatives are commonly subclassified into punctate and aggregate cells, as the relative proportion of these forms provides information about the duration and progression of the regenerative response in felids.[104,222]

The degree of the **erythroid regenerative response** (Figs. 4.6, 4.7) is driven by the underlying cause(s) and

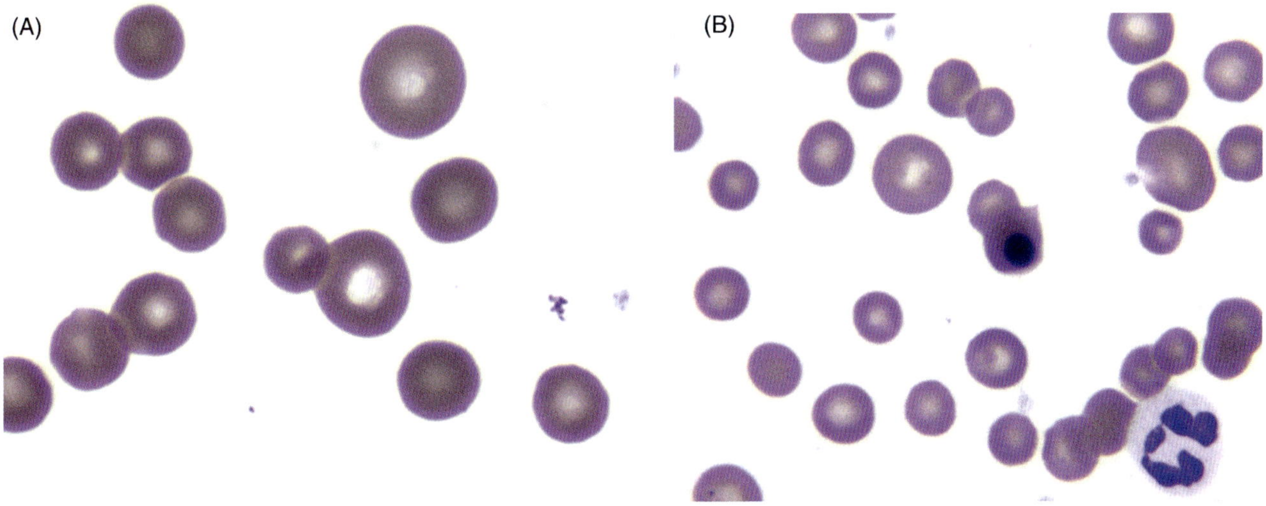

FIGURE 4.6 Blood film from a black rhinoceros with regenerative anemia (PCV 14%) showing morphological features of erythroid regeneration (anisocytosis with macrocytes) in species that do not normally release polychromatophils (similar to horses). (A) Erythrocytes exhibiting marked anisocytosis (variation in size) but no polychromasia. (B) Erythrocytes exhibiting marked anisocytosis but no polychromasia and a single nucleated red blood cell. WG.

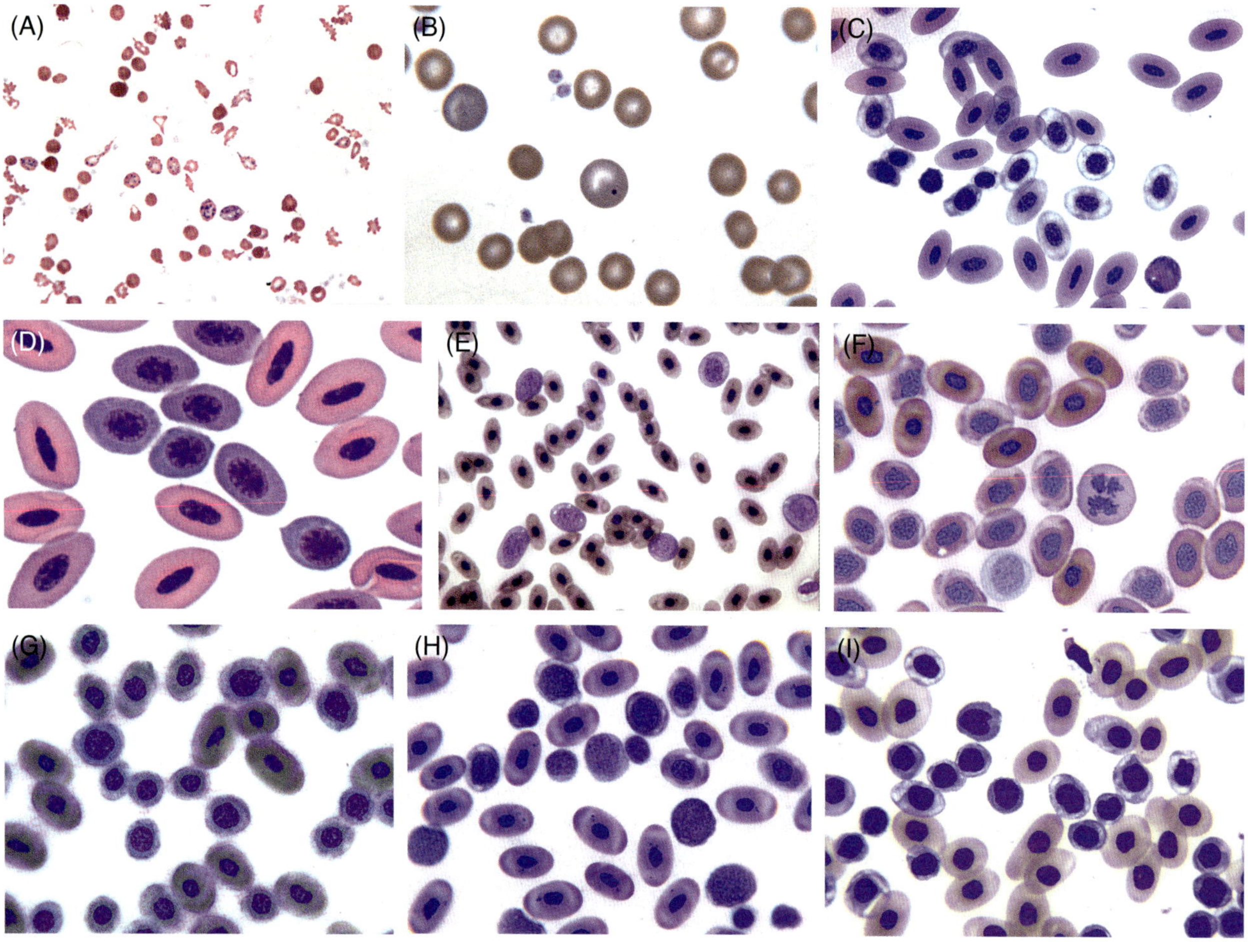

FIGURE 4.7 Morphological features of erythroid regenerative responses in various species. (A) Sable antelope with PCV 9% and erythrocytes exhibiting anisocytosis, polychromasia, and basophilic stippling; in addition, hypochromasia and poikilocytosis suggest iron deficiency anemia. (B) Domestic dog with anisocytosis, polychromasia, and a Howell-Jolly body. (C) Frogmouth with anisocytosis and polychromasia (note: anisocytosis reflects immature erythrocytes being smaller in size compared to polychromatophils in nonmammals, the opposite of anisocytosis due to macrocytosis commonly seen with regenerative responses in mammals). (D) Spoonbill with anisocytosis and polychromasia. (E) Freshwater turtle with regenerative response including anisocytosis, polychromasia, and circulating early erythroid precursors (basophilic, more rounded cells) (F) Green turtle with anisocytosis, polychromasia, and a mitotic figure. (G) Green turtle with regenerative response (anisocytosis, polychromasia, basophilic stippling). (H) Loggerhead sea turtle with anisocytosis, polychromasia, late- and early-stage precursors. (I) Cownose ray with anisocytosis, polychromasia, mitotic figure, early-stage precursors). Red blood cell sizes are not directly comparable among images. WG.

duration of anemia (e.g., hemolysis, hemorrhage, bone marrow depression), intrinsic factors (e.g., age, reproductive, nutritional status, lifespan of erythrocytes), and extrinsic factors (e.g., medications, husbandry, diet, season). Mammals with shorter erythrocyte life span often have more circulating reticulocytes (Table 4.e1), and may exhibit occasional Howell–Jolly bodies and metarubricytes, for example, monkeys and rodents.[30] However, some members of the clade of ungulates do not exhibit overt circulating reticulocytes during phases of regenerative responses to anemia. For example, the horse lacks a prominent reticulocyte release into circulation, even in the face of prominent bone marrow hyperplasia.[104] While the RBC changes may be too subtle to be recognized by blood film evaluation in horses, they can be detected by hematology analyzers.[14] The authors have also observed no overt polychromasia in response to anemia in other ungulate species, including rhinoceros, elephant, tapir, and manatee (Fig. 4.6). Black rhinoceros suffering from hemolytic anemia therefore do not show polychromasia, although the presence of nRBCs has been reported and may indicate an erythroid regenerative response.[57,158] Thus, the observation of anisocytosis and rare nRBC in anemic horses and some other ungulate species can be the only morphological evidence of erythroid regeneration. These findings are often accompanied by a progressively increasing PCV over time. Bone marrow evaluation and new methylene blue staining of bone marrow cytology specimens could support the hematological assessment of anemic domestic horses and their relatives but is rarely performed, given the potential procedural risk. Red blood cell indices and blood film evaluation provide the basis for the evaluation of erythroid regenerative response in other mammalian species. In the erythroid response of ruminants, basophilic stippling and increased nucleated red blood cells (rubricytosis) may be particularly prominent[30] (Fig. 4.7A). Bone marrow evaluation may be indicated in the assessment of anemic mammals with unexplained or persistent cytopenias, or presence of abnormal circulating cells, among other possible indications.[104]

Mammalian leukocytes (Fig. 4.e2) are the basis of the general understanding of leukocyte morphology and kinetics, as described above. Female mammals may exhibit a distinct **sex chromatin lobe** (Barr body; inactivated X chromosome) attached to a terminal lobe of neutrophil nuclei.[104] For some species, such as rabbits, elephants, hyraxes, and manatees, the literature is inconsistent as to whether these species have heterophils versus neutrophils. In most such species, no differences in cellular composition or function are documented to substantiate the terminology argument based on morphological features, and the use of heterophil/neutrophil terminology is a matter of convention or preference rather than substance. One notable unique leukocyte cell type in mammals is the **Foa-Kurloff cell** or **Kurloff cells** in guinea pig, capybara, and paca. This is a mononuclear cell with a focal membrane-bound inclusion composed of mucopolysaccharides that may represent up to 4% of the WBC differential, with higher numbers being possible during pregnancy (Fig. 4.8).[70]

Leukogram responses in inflammatory disease have been well studied in domestic mammal species[30,104]: dogs have a large bone marrow storage pool and thus respond with prominent leukocytosis in disorders or diseases of increased peripheral tissue demand; in contrast, cattle have a very small storage pool and often initially present with neutropenia after an inflammatory stimulus. Leukocyte responses similar to cattle have been observed in dolphins, elephants, rhinoceros, and manatees. These species often develop neutropenia rather than prominent leukocytosis with the onset of inflammatory disease. A leukocytosis of >20,000/μL in these species is an exception and is often observed with chronic-active systemic inflammation.

Mammalian platelets are morphologically very similar across species (Fig. 4.e3). Felidae may exhibit a few larger platelets (macroplatelets) in health, and their platelets tend to aggregate more rapidly in vivo compared to other species (Fig. 4.e3A). Platelet activation often occurs in vitro in mammals, with morphological features including degranulation, filopod formation, and centralized granules.[104] Since platelet clumps can falsely lower automated platelet clumps, blood film review is indispensable for recording their presence and may indicate the need for a manual hemocytometer count.

Nonmammalian Vertebrates

In contrast to mammals, **nonmammalian erythrocytes** are nucleated, generally larger, and elliptical in shape, although they may be rounded in some teleost species (Fig. 4.e4). The centrally placed nucleus is round to oval with clumped chromatin. Variably sized distinct basophilic inclusions and/or a few small clear vacuoles in the erythrocyte cytoplasm may be observed in erythrocytes of most nonmammalian species. These are considered of no clinical significance, and presumptively represent degenerate organelles presumably of mitochondrial or other cytoplasmic organelle membranal origin or artifact (Figs. 4.9, 4.e4A–E, G–H).[5,6,34] Hemoglobin crystals of unknown

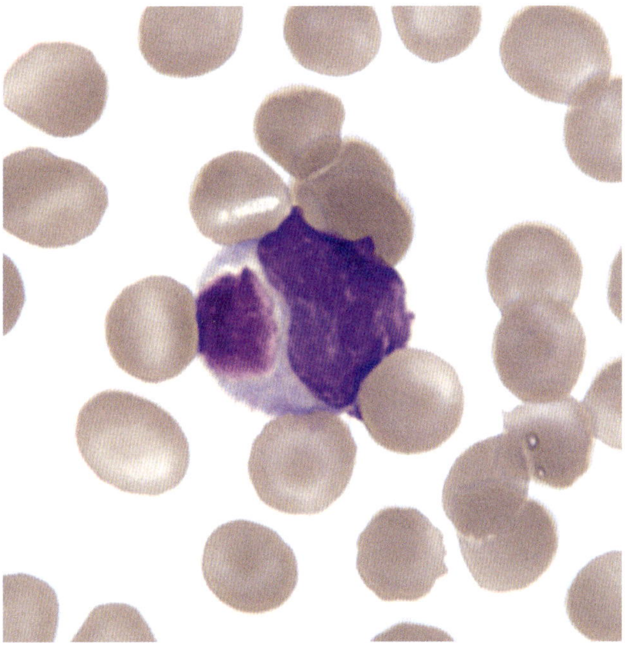

FIGURE 4.8 Guinea pig Kurloff cell. Presumptive lymphocyte with granular magenta-colored inclusion. WG.

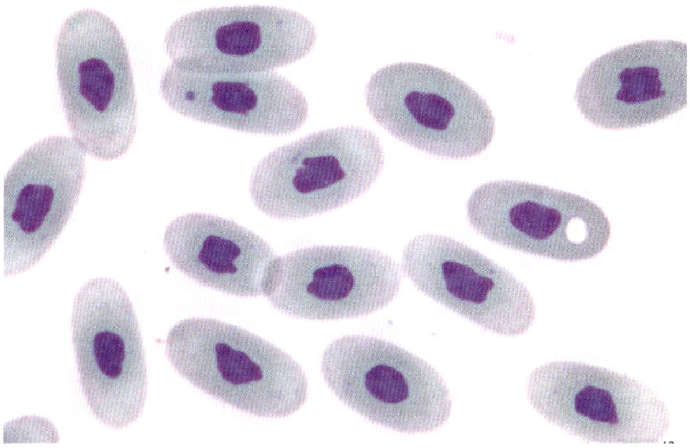

FIGURE 4.9 Normal erythrocytes from a box turtle. Small distinct basophilic inclusions and/or vacuoles are most consistent with degenerate organelles or artifact.

clinical relevance have been documented in various lizard species (e.g., green and rhinoceros iguana).

Nonmammalian vertebrates frequently have a low number of circulating reticulocytes/polychromatophils in health. In the assessment of the **erythroid regenerative response in nonmammalians**, a spun packed cell volume (PCV), plasma colour, total protein, and blood film evaluation provide the most diagnostically useful information. In addition to changes in PCV, blood film evaluation in nonmammals is the only means to determine whether an anemia is regenerative or nonregenerative. In birds, enumeration of only ring-form reticulocytes—those with a continuous or discontinuous circumferential ring of reticulum (precipitated RNA) around at least half of the nucleus is suggested to have strong correlation with polychromatophils and improved intraobserver reproducibility; this strong correlation deems reticulocyte counts unnecessary as reticulocytes can be assessed by routinely stained blood films as polychromatophils.[128] Red blood cell morphological changes can be quite prominent in a regenerative response in nonmammals and may include the following: increased anisocytosis and polychromasia, basophilic stippling (especially in chelonians), presence of early erythroid precursors of various stages, micronuclei, and mitotic figures (Figs. 4.7C–I). The presence of early erythroid stages, such as the rubriblast, should be interpreted cautiously in nonmammals, since misdiagnosis of leukemia is possible, especially in an early regenerative response with release of immature precursors but lack of synchronous maturation all the way to mature erythrocytes (Fig. 4.7). The interpretation of such morphological observations needs to be made in the context of any other relevant clinical and laboratory findings.

Nonmammalian leukocytes are largely morphologically and functionally similar and classified according to mammalian schemes. Reptiles have a unique leukocyte type, the **azurophil** for which cellular lineage is still controversial (Fig. 4.10). While the azurophil is cytochemically similar to the monocyte in chelonians and crocodilians, this cell type is peroxidase positive and more comparable to mammalian neutrophils in snakes and some lizard species; in snakes and lizards, azurophils should be reported as a separate leukocyte category in the WBC differential.[4,44,97,160]

The nonmammalian equivalent of the mammalian platelet is the nucleated **thrombocyte** (Fig. 4.e3). Thrombocytes are oval or elliptical with a small amount of cytoplasm that may contain a few fine pink granules on Romanowsky-stained blood films. They may even exhibit pseudopods after activation but can be more rounded, especially in reptiles. Thrombocytes of birds and reptiles have phagocytic capability, although the clinical significance of this phenomenon is unknown.[60] The authors have observed thrombocytosis in nonmammals with inflammatory disease and marked thrombocytopenia in several reptile cases with consumptive coagulopathy.

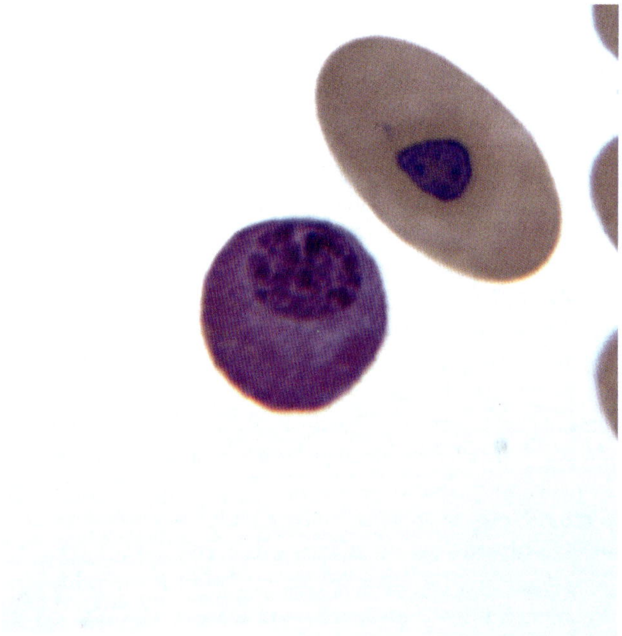

FIGURE 4.10 Normal reptile azurophil from a rainbow boa.

Preanalytical and Analytical Consideration

Blood film evaluation is an important quality control step in every CBC and provides information about cell morphology that no other testing can provide, such as RBC poikilocytosis (abnormal RBC shapes) or inclusions, WBC morphological changes (e.g., neutrophil or heterophil left-shift, toxic change, inclusions), abnormal cells in circulation (i.e., leukemia), and infectious agents (free or within any cell lineage including erythrocytes, leukocytes, or platelets/thrombocytes).

In the presence of inflammatory stimulation, morphological changes of neutrophil/heterophil toxic change and granulocyte lineage left shift occur across all vertebrate species and are recognized by blood film evaluation (Fig. 4.11).[218] **Toxic change** is observed as a spectrum of morphological changes in neutrophils or heterophils that result from accelerated production and reduced maturation time; these changes include presence of Doehle bodies (focal basophilic aggregates of retained rough endoplasmic reticulum within the cytoplasm), increased diffuse cytoplasmic basophilia, cytoplasmic foaminess, cytoplasmic degranulation/vacuolation, or abnormal granulation. There is a lack of standardization of toxic change grading schemes in the literature as well as a lack of, or inconsistent correlation, of such schemes with clinical outcomes. Rather than subjective numerical scales, communicating the degree of toxic change as "absent," "present," or "severe" along with a description of the associated morphological changes may prove more informative given the documented clinical relevance in domestic

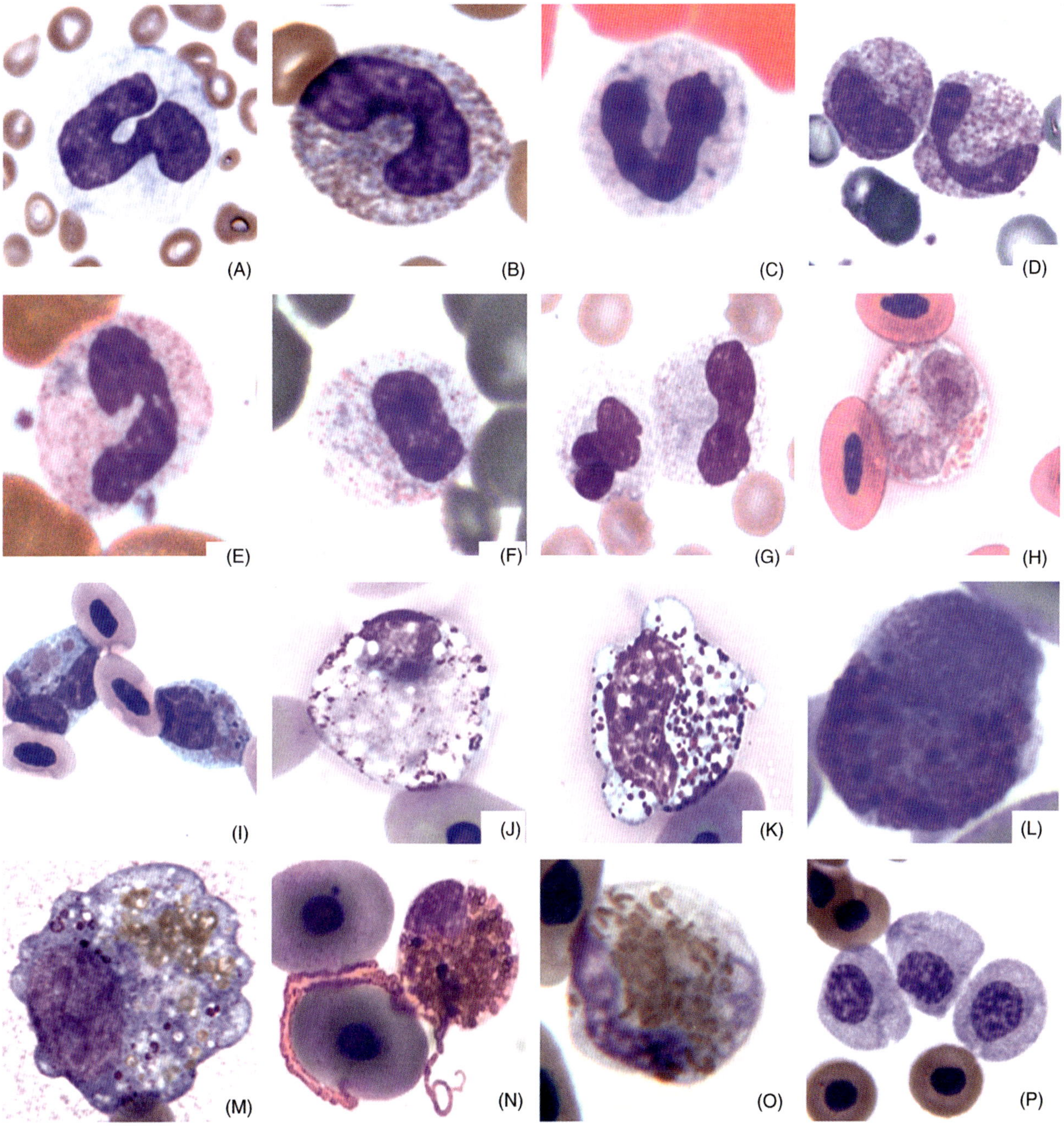

FIGURE 4.11 Image composite of the morphology of neutrophil/heterophil toxic change (Doehle bodies, cytoplasmic basophilia, foaminess/vacuolation, degranulation, and/or abnormal granulation) and left shift (increased circulating immature stages) in various taxa. The severity of toxic change and left shift is determined across a population. (A) Domestic cat band neutrophil. (B) Pygmy macaque band neutrophil. (C) Guadalupe fur seal band neutrophil. (D) Florida manatee metamyelocyte (*left*) and segmented heterophil (*right*). (E) Lowland gorilla band neutrophil. (F) African elephant metamyelocyte. (G) Black bear segmented neutrophil (*left*) and band neutrophil (*right*). (H) Red-tailed hawk immature heterophil. (I) Cockatoo mature heterophil (*left*) and immature heterophil (*right*). (J) American alligator mature heterophil. (*K*) *Sulcata* tortoise immature heterophil. (L) Green turtle immature heterophil. (M) Green iguana immature heterophil. (N) Monitor lizard with immature heterophil exhibiting whip-like projections (considered associated with inflammation or less likely artifact). (O) Chameleon immature heterophil. (P) Blue catfish immature neutrophils. Leukocyte sizes are not directly comparable among images. WG.

mammals.[218,219] With severe toxic change, differentiation of some cell lineages (e.g., granulocytes and monocytes) may be difficult and affect the accuracy of the WBC differential. Toxic change is nonspecific and may be associated with various inflammatory cytokine stimuli, including non-infectious, infectious, neoplastic, or metabolic derangements, or effects from various drugs; resolution of toxic change is considered a good prognostic indicator.[104] Toxic morphological changes are often accompanied by concurrent release of earlier precursors, termed "left shift," due to the convention of arranging earlier stages to the left of later stages on hematopoietic diagrams. **Left shift** is a drift toward immaturity in a cell line and is defined as the release of immature myeloid precursors to a greater concentration than observed in health in a given species (i.e., a low number of band neutrophils can be present in healthy carnivores, but when increased, there is a left shift). Left shifts are typically organized, with more "late stage" and progressively fewer "early stage" precursors (i.e., band cells > metamyelocytes > myelocytes >promyelocytes), though severe responses can shift further back (left) in the maturation pyramid. Like toxic change, left shifts are most frequently associated with inflammatory stimuli and both the extent of the left shift and the absolute cell concentrations provide information about how the body is responding to cellular demand. Both left shift and toxic change provide information about the severity of inflammatory stimulation and state of bone marrow storage and proliferation cell pools and often occur concurrently. When independently present, it is more common for left shifts to occur without toxic change, whereas toxic change without some degree of left shift is uncommon in both mammalian and nonmammalian vertebrates.

Infectious agents of erythrocytes include hemoparasites, bacteria, fungi/yeast, and viruses, many of which are species-specific; it is important to recognize the often pathognomonic morphology in order to determine the clinical significance since several organisms may not be pathogenic, especially in noncaptive animals (Fig. 4.12). Furthermore, morphological identification provides guidance for additional diagnostic testing (e.g., blood culture, PCR, electron microscopy). It is important to differentiate red blood cell inclusions (especially in nonmammalian vertebrates, see below) and drying artifact in any species from infectious agents. Artifacts in blood films include drying or smearing artifacts, staining precipitate, cell lysis as seen with humidity/condensation, or exposure to formalin fumes.

Infectious agents of leukocytes include bacteria, fungi, viruses, and hemoparasites (Fig. 4.13). Incidental findings

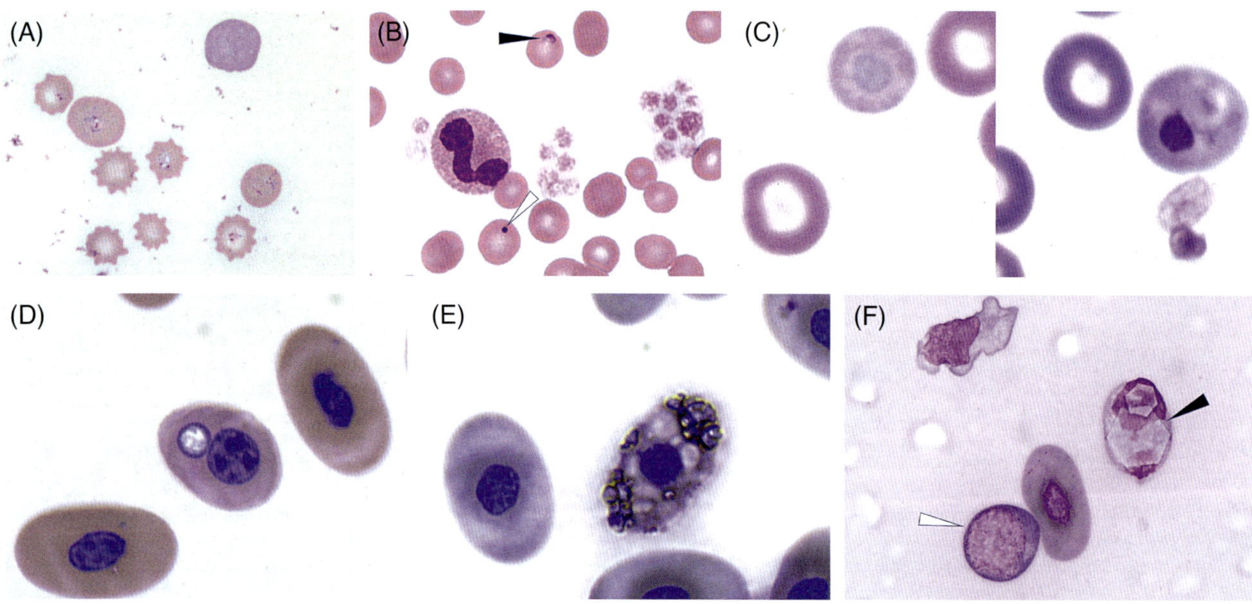

FIGURE 4.12 Examples of infectious agents and inclusions in erythrocytes of nondomestic species. (A) Virginia opossum: multiple small, pale basophilic intracytoplasmic pyriform organisms consistent with *Babesia* sp., anisocytosis and polychromasia (regenerative anemia), and echinocytes (crenated erythrocytes). (B) Intracytoplasmic *Cytauxzoon* sp. (*black arrowhead*) in a North American bobcat, erythrocyte with Howell-Jolly body (*white arrowhead*) below eosinophil and platelet clumps. (C) Canine distemper viral inclusions in erythrocytes of a domestic dog, seen as pale purple indistinct inclusions with WG (*left image*), but distinct pink staining with Diff-Quik™) (*right image*). (D) Large distinct purple granular *Anaplasma* sp. inclusion in an erythrocyte of a gopher tortoise. (E) Variably sized, irregular, refractile drying artifact (no clinical significance) in an erythrocyte of a diamondback terrapin; drying artifact should not be confused with a hemoparasite. (F) Fer de lance snake with hexagonal snake erythrocyte viral (SEV)-inclusions (*black arrowhead*) in an erythrocyte precursor, with PCV 14% and evidence of erythroid regeneration (early-stage erythrocyte precursor; *white arrowhead*). Red blood cell sizes are not directly comparable among images. WG. *(Part C: Photo Courtesy of J. Harvey, University of Florida.)*

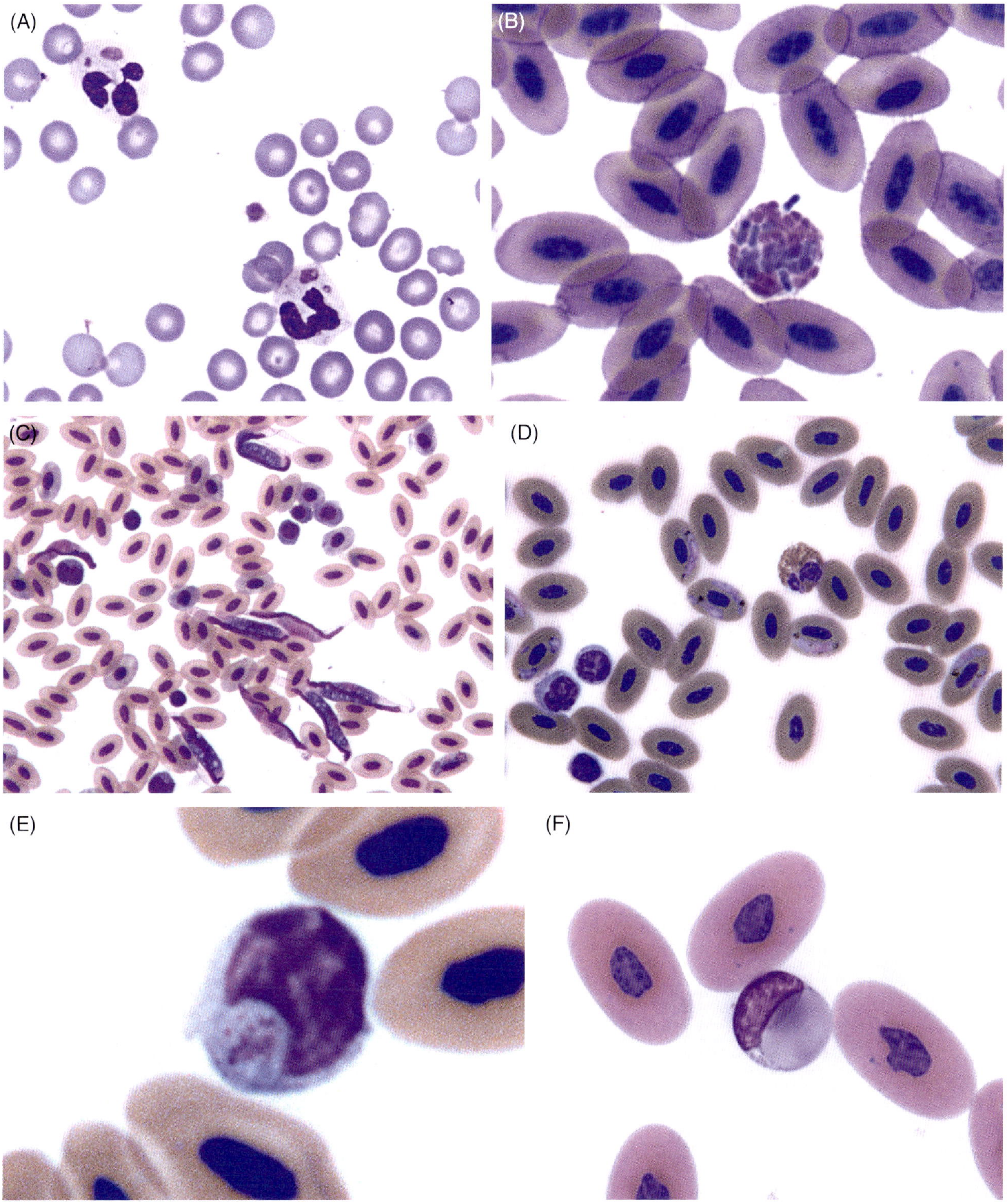

FIGURE 4.13 Examples of infectious agents and inclusions in leukocytes of various species. (A) domestic dog: two neutrophils with canine distemper viral inclusions. Diff-Quik™. (B) Blood film from an injured red-tailed hawk: heterophil with phagocytized bacilli. (C) Blood film from a mallard duck with *Leukocytozoon* sp. gametocytes and a single intraerythrocytic *Hemoproteus* sp. gametocyte (*lower right corner*). It is still unconfirmed whether *Leukocytozoon* sp. gametocytes develop in leukocytes or erythrocytes. (D) Blood film from a Greater Flamingo with intraerythrocytic stages of *Plasmodium* spp. (E) Blood film from an Oropendola with intracytoplasmic inclusion consistent with *Isospora* spp. in lymphocyte. (F) Blood film from an emerald tree boa: arenavirus-associated inclusion body disease inclusion in lymphocyte. Blood cell sizes are not directly comparable among images. All images: WG unless otherwise indicated.

in leukocytes that need to be differentiated from infectious agents include morphological features of immaturity and toxic change, drying artifact, staining variability in granules (e.g., birds), and phagocytized diatoms in fish.[62]

Specifics of blood film preparation are provided in the digital content.

Mammals

Hematology analyzers have historically been designed for humans and only more recently for cross-species application. It is of general importance to properly calibrate **hematology analyzers** to the unique features of the species of interest and choose the appropriate species setting based on mean cell volume (MCV) (Table 4.e1); however, some species pose additional challenges. For example, the rigid RBC membrane of camelidae precludes accurate RBC enumeration and measurements by hematology analyzers,[56] and size differences in RBCs and platelets may result in inaccurate platelet counts, as seen frequently in cats given the similar size of RBC and platelets.[104,241] The cytometric methods used in many point-of-care automated analyzers cannot differentiate some leukocyte types accurately (band neutrophils, eosinophils, basophils, and sometimes monocytes).[30,140] Since the hematocrit (Hct) data provided by the analyzer is a calculated value from mean cell volume and RBC count, it is affected by errors in these measurements and can be inaccurate in comparison to a properly performed manually spun packed cell volume (PCV). A spun PCV is easily obtained and should be part of every complete blood count (CBC). In mammals, circulating nucleated RBCs (nRBC), as can be observed with regenerative anemia or various causes of inappropriate rubricytosis, cannot be differentiated from WBCs by automated analyzers and thus nRBC are included in the total nucleated cell count. In the presence of nRBC, the WBC count generated by the analyzer needs to be corrected using established formulas, and the absolute number of nRBC per microliter calculated (e.g., #/100 WBC with calculation of the absolute number in #/μL using standardized formulas).[104]

For the laboratory assessment of erythroid regenerative responses in mammals, reticulocytes are technically identified in supravital new NMB-stained preparations by observation of increased precipitated, punctate or reticulated, basophilic-staining RNA[222] (discussed above). **Reticulocyte staining** is performed using equal volumes of blood and NMB stain, which are incubated together for 10–20 min, and then used to prepare blood films. Reticulocytes are counted as number per 1000 erythrocytes frequently by use of a Miller's disc counting reticle.[104] The smear count is converted to a percentage and the obtained raw count is then corrected for the degree of anemia (i.e., corrected reticulocyte count), using the PCV or HCT of the patient divided by the mean normal PCV or HCT for the species, multiplied by the raw reticulocyte count.

Heinz bodies are quantified by NMB stain by counting the proportion of Heinz bodies during a 100 red blood cell count and reporting the percentage.

Nonmammalian Vertebrates

In nonmammals, because all blood cells (WBC, RBC, and thrombocytes) are nucleated, the use of automated hematology analyzers is not possible and *only* **manual methods** are suitable for hematological analysis in nonmammalian species. In order to avoid the high degree of imprecision associated with manual methods, consistency and standardization of blood sample collection, handling, processing, and laboratory analysis are key elements to reducing variability and in obtaining the most accurate and reliable hematological results. Manual WBC quantification methods include the manual WBC count (e.g., Natt-Herricks, Phloxine B), and a WBC estimate from the blood film. It is important to note that both a manual WBC count in conjunction with a WBC estimate from the blood film are beneficial in the identification of potential laboratory error. A major challenge in manual WBC quantification of nonmammalian species is the morphological differentiation of lymphocytes from thrombocytes by hemocytometer and blood film review, since they are similar in size and shape. This differentiation is especially challenging in reptiles and is comparatively easier in birds, fish, and amphibians, given the more distinct features of both cell types in these taxa.

Rounded **thrombocytes** have been observed with delayed blood film preparation, and the preparation of fresh blood films is recommended to avoid this and other delays in processing artifacts. Since thrombocytes tend to clump in heparinized blood samples, resulting in variably sized thrombocyte clumps in hemocytometer and blood films, their quantification is inaccurate and should be considered as a subjective semiquantitative estimate (i.e., adequate, increased, decreased).

Comparative Hemostasis

Physiology and Interpretation of Changes in Disease

The major role of hemostasis across vertebrate taxa is to maintain physiological flow through blood vessels. This is achieved by balanced interactions of promotors and inhibitors, including cells, proteins, and other intra- and peri-vascular components (e.g., calcium, collagen). Abnormalities of hemostasis result in hemorrhage or thrombosis. The major mechanisms involved in vertebrate hemostasis are vasoconstriction, platelet/thrombocyte plug formation (primary hemostasis), coagulation to form a stable fibrin clot (secondary hemostasis), and fibrinolysis to permit reendothelialization and reestablishment of blood flow. The traditional "cascade" model of coagulation portrays a series of sequential reactions converting inactive zymogens to active enzymes and ultimately producing cross-linked fibrin to form a stable clot. This model is useful to describe in vitro coagulation as it occurs in routine plasma-based coagulation tests (i.e., prothrombin time [PT] and partial thromboplastin time [PTT]) and has been a mainstay of our understanding of coagulation for more than 50 years. In recent years, however, increasing recognition of the complexity of hemostasis has produced more holistic models of hemostasis that better incorporate the interactions and roles of coagulation factors, cells and subcellular components (e.g., cell-based model of blood coagulation), and other in vivo circulatory and structural dynamics (e.g., intravital microscopy).[155]

Foundational review of hemostasis and hemostatic testing are beyond the scope of this chapter and several excellent overviews have been published elsewhere.[214,222] Table 4.1 provides a limited summary of common in vitro tests of hemostasis with abbreviations, primary influencing factors, and brief comments. In general, in vitro assays of hemostasis have largely been developed in mammals (humans) and applied to other species. These include tests of primary hemostasis (e.g., buccal mucosal bleeding time [BMBT], platelet aggregometry), secondary hemostasis (e.g., PT, PTT), and global hemostasis (e.g., thromboelastography, TEG).

Mammals

As described in mammals, **primary hemostasis** yields a platelet plug formed when platelets adhere to sites of vascular injury through exposed tissue collagen and von Willebrand factor (vWF), aggregate to one another through fibrinogen and vWF bridges, and secrete regulatory factors (e.g., ADP, Factor V, Fibrinogen, Thromboxane A_2).[222] The platelet fibrinogen receptor, glycoprotein IIb-IIIa, plays a critical role in mammalian platelet aggregation and genetic deficiencies of the receptor cause Glanzmann's thrombasthenia with associated bleeding tendency. Primary hemostasis resulting in the formation of the platelet plug facilitates secondary hemostasis during which the coagulation system sets up a stable fibrin clot. Disorders of primary hemostasis resulting in hemorrhage can occur with decreased platelet concentration (thrombocytopenia) or abnormal function (thrombopathia); an increased tendency to thrombosis is possible with increased platelet concentration (thrombocytosis) but is clinically rare.[171]

Secondary hemostasis of mammals involves protein and nonprotein coagulation factors and cellular components. Abnormalities in coagulation can result from inherited defects (e.g., coagulation factor deficiencies) or acquired disease conditions. Acquired lesions are often multifactorial: examples include liver disease resulting in diminished coagulation factor production, and consumptive processes (i.e., sepsis, localized or disseminated intravascular coagulation) causing increased coagulation factor utilization or inactivation. The pattern of coagulation assay prolongation in these conditions depends on the dynamics of disease and assay properties. In mammals, the tissue factor pathway is thought to initiate thrombin generation, whereas the intrinsic pathway via activated factors IX, VIII, and X leads to coagulation propagation.[214] While some naturally occurring factor deficiencies cause risk for clinically significant in vivo bleeding and associated in vitro test prolongations (e.g., factor IX, PTT prolongation), others cause in vitro test prolongation without associated clinical bleeding tendency (e.g., factor XII deficiency [Hageman trait], PTT prolongation in domestic cats, marine mammals, some dogs). It is also possible for therapeutic interventions to result in evidence of coagulopathy, such as with large volume plasma-poor transfusions or medications that interfere with coagulation factors (e.g., heparin) or with platelet function (e.g., nonsteroidal inflammatory drugs).

Vitamin K is an essential nutrient for mammals, and deficiency or antagonism can manifest as hemorrhage. Vitamin K antagonism can result from intended therapy (e.g., warfarin) or unintended intoxication (e.g., anticoagulant rodenticides). The mechanism of action for these compounds is to inhibit the enzymatic recycling of oxidized vitamin K back to its reduced form. The reduced form of vitamin K is necessary for carboxylation reactions involved in activating factors II, VII, IX, and X, and the absence of reduced vitamin K prevents normal assembly of coagulation complexes. In mammals, factor VII has the shortest half-life, and vitamin K antagonism or deficiency typically manifests first by prolongation of PT, followed by PTT prolongation and risk for clinical bleeding. There is recognized species variation in sensitivity to anticoagulant rodenticides as well as broad differences in toxicity risk between compounds.[188] The underpinnings for this variation require further investigation, and may involve genetic, metabolic, and other factors. Susceptibility to hemorrhage can additionally be affected by concurrent disease and other stressors, as well as exposure type (e.g., single-dose acute, repeated subacute). Differences in sensitivity to clinical intoxication appear particularly high across birds, with raptors at relatively higher risk.[87,169]

In order to maintain healthy circulatory flow, procoagulant systems are balanced by inhibitors and fibrinolytic activity. **Inhibitors of coagulation** in mammals include tissue factor pathway inhibitor, antithrombin, protein C, and other proteins. Fibrinolysis works to degrade fibrin though activation of plasmin, which is promoted by tissue plasminogen activator and protein C, and inhibited by plasminogen activator inhibitor. Other substrates degraded by activated plasmin include fibrinogen and activated factors V and VIII, thus contributing to anticoagulant effects. Deficiency of fibrinolysis promotes thrombosis.

TABLE 4.1 Limited Summary of Common in Vitro Tests of Hemostasis with Major Tested Hemostatic Components and Other Considerations

Test Name, Abbreviation, and Sample Type	Major Hemostatic Component(s)	Comments
Buccal mucosal bleeding time (BMBT) (buccal mucosa)	Primary hemostasis • Platelets • vWF (Other potential influences: Hct, vascular disease, fibrinogen)	• Point-of-care screening employs a standardized small cut in the buccal mucosa to assess primary hemostasis[a] • Relatively insensitive, requires standardized methods • Infrequently applied in nondomestic animals; comparative reference times in healthy individuals of many species are not established • Not indicated in thrombocytopenic animals as it will be prolonged
Activated coagulation time (ACT) (whole blood)	Secondary hemostasis: intrinsic and common coagulation pathways • Intrinsic: Factors XII, XI, IX, VIII • Common: Factors X, V, II (prothrombin), I (fibrinogen) (Other potential influences: interferents, medications)	• Point-of-care screening that activates contact factors using particulates (e.g., diatomaceous earth, kaolin) • Insensitive (less sensitive than PTT) and requires standardized method of implementation • Infrequently applied in nonmammalian species, especially in those with limited intrinsic (contact) pathway activity (see text), though the use of tubes containing diatomaceous earth is reported in birds
Activated partial thromboplastin time (PTT) (citrated plasma)	Secondary hemostasis: intrinsic and common coagulation pathways - Intrinsic: Prekallekrein, factors XII, XI, IX, VIII - Common: Factors X, V, II, I (Other potential influences: interferents, medications)	• Laboratory method stimulating the intrinsic pathway by addition of excess phospholipids, a surface activator (e.g., kaolin, silicates), and Ca^{2+}, and measuring time to clot detection (methods of detection vary) • Requires species-specific consideration of reagents, method validation, and reference intervals (see text) • Generally prolonged compared to mammals in taxa lacking intrinsic factors (e.g., birds, see text)
Prothrombin time (PT) (citrated plasma)	Secondary hemostasis: extrinsic and common coagulation pathways • Extrinsic: Factor VII • Common: Factors X, V, II, I (Other potential influences: interferents, medications)	• Laboratory method stimulating the extrinsic pathway by addition of reagent (Thromboplastin reagent) containing tissue factor, phospholipid, and Ca^{2+}, and measuring time to clot detection (methods of detection vary) • Varying clotting times with different thromboplastin reagents • Requires species-specific consideration of reagents, method validation, and reference intervals (see text)
Fibrinogen concentration (modified Clauss assay) (citrated plasma)	Conversion of fibrinogen (factor I) to fibrin (Other potential influences: interferents, medications)	• Laboratory method assessing fibrinogen to fibrin conversion by addition of thrombin and measuring time to clot detection (thrombin time); Clauss method uses high thrombin concentration and compares to a reference standard curve to determine fibrinogen concentration[b] • Requires species-specific consideration of reagents, method validation, and reference intervals (see text)
Thromboelastography (TEG) (whole blood)	Global hemostasis (primary, secondary, and fibrinolysis)	• Viscoelastic assay of clot formation kinetics from platelet aggregation to fibrinolysis • Limited data in nonmammalian species • Requires consideration of preanalytical factors and taxon/ species-specific reagents (see text)

[a]*Bleeding times assessed by cuticle and skin cuts are generally less standardized and prone to involve secondary as well as primary hemostasis, complicating interpretation.*
[b]*Fibrinogen concentration can also be semiquantitatively measured by the heat-precipitation method using plasma, but the analytical sensitivity and precision of this method are lower than the modified Clauss method; fibrinogen measured by heat precipitation is applied predominantly for the detection of inflammation (positive acute phase protein response) (see Protein section in text).*
Note: Factor III is tissue factor, factor IV is Ca^{2+}, and factor VI is unassigned. Factor XIII, which is activated by thrombin and acts to cross-link and stabilize fibrin, is not assessed by these assays. Additional taxa-specific information and assay considerations are discussed in the text.
Sources: Stockham S, Scott M. Fundamentals of Veterinary Clinical Pathology. 2nd ed. Ames, IA: Blackwell Publishing Ltd; 2008; Russell KE, Heatley JJ. Avian Hemostasis. In: Brooks MB, Harr KE, Seelig DM, Wardrop KJ, Weiss DJ, eds. Schalm's Veterinary Hematology. Hoboken, NJ: Wiley-Blackwell; 2022:1073–1078.

Increased fibrinolysis occurs in a variety of disease conditions, such as sepsis, intravascular coagulation, and envenomation.[222]

Nonmammalian Vertebrates

Hemostasis in nonmammalian vertebrates has been variably assessed across taxa. Some components, including vitamin K-dependent thrombin-catalyzed conversion of fibrinogen to fibrin, are identified across all vertebrates.[55,65,185] The evolution of a coagulation cascade, including factor X activation of prothrombin to thrombin and fibrinogen cleavage to fibrin, appears to have occurred prior to jawless fish (e.g., lamprey), the earliest extant vertebrates.[65,185] Tissue factor activation is the predominant coagulation pathway of most vertebrates, including birds, reptiles, amphibians, and fish.[31,65] Factors XI and XII are naturally markedly reduced or lacking in the plasma of nonmammals, correlating to in vitro prolongations of traditional PTT assays without in vivo bleeding tendency.[65,172,185,195,209] Recent whole genome sequencing identified parts of the kallikrein-kinin (intrinsic) system (prekallikrein and high-molecular-weight kininogen) in lobed-finned fish, ancestors to all land vertebrates, and origination of factor XI as a duplication of the prekallikrein gene in early mammals.[185]

Birds share much of the same hemostatic and hemodynamic physiology of mammals. In primary hemostasis, avian thrombocytes like mammalian platelets contain dense granules, microtubules, and mitochondria, and respond to primary platelet agonists, including collagen and thrombin. Differences from mammalian platelets include lower ADP content, fewer glycoprotein IIb-IIIa receptors, and generally weaker, less efficient, and sluggish thrombocyte plugs forming thin (single-layered sheets) over damaged areas rather than the shear-resistant aggregates of mammals.[199,203] Birds also employ an in vivo coagulation cascade (secondary hemostasis) similar to mammals, with many of the same identified factors, including tissue factor, factors VII, IX, and X, prothrombin, fibrinogen, and inhibitors (protein C, antithrombin, etc.).[31,55] Most factors are produced by the liver. Birds lack prekallikrein, factors XI and XII and therefore lack contact activation of the intrinsic pathway. Instead, activation of the limited avian intrinsic pathway (factor IX) is achieved by cross-activation from the extrinsic pathway (factor VIIa/TF complex). In comparison to mammals, avian bleeding times are generally prolonged and traditional PTT tests requiring contact activation often extend beyond the upper limit of detection, thereby negating clinical utility for the detection of avian coagulopathy.

Birds are prone to vitamin K deficiency which is promoted by factors including a generally short digestive tract, feeding of low-fat diets, and high supplementation of competitive inhibitors of vitamin K, including vitamin E and some therapeutics (e.g., sulfonamide antibiotics and coccidiostats). A vitamin-K-responsive coagulopathy associated with elevated vitamin E has been described in pink-backed pelicans.[174] As in mammals, vitamin K deficiency and antagonism (e.g., due to rodenticide intoxication) are associated with prolonged PT.[31,245] Inadvertent secondary rodenticide exposures can occur through scavenging, ingestion of intoxicated prey, or accidental consumption of bait. Proximity to human populations and rodent-rich diet are risk factors for such intoxications, but they do not explain all occurrences. Further studies assessing additional exposure pathways and detailed toxicokinetics are needed, especially as anticoagulant rodenticides pose a conservation concern to avian populations.[87,169]

Hemostasis in reptiles and amphibians is less well-defined and subject to species-specific testing considerations. In general, the clotting times of reptiles and amphibians are prolonged compared to mammals. Limited data in these species support the presence of thrombocyte adhesion and spreading (primary hemostasis), and a coagulation pathway comparable to the mammalian extrinsic pathway, in which tissue factor mediates thrombin generation and fibrinogen to fibrin conversion.[55,141,216] Factors VIII, IX, and/or XII have been detected in low concentrations in some reptiles; these may provide positive feedback of the extrinsic pathway, but the intrinsic pathway is weak and PTT samples are commonly prolonged or fail to clot.[172,216] As assessed in sea turtle species, factors II and X correspond to mammalian equivalents antigenically and functionally.[216] Nontraumatic hemorrhage due to suspected coagulopathy is rarely reported in reptiles and amphibians, but based on conserved mechanisms, diagnostic considerations are akin to other taxa.

Piscine thrombocytes have been shown to share similar agonist responsiveness and receptor function to mammalian platelets.[248] The majority of intrinsic, extrinsic, and common pathway factors have been demonstrated to have structural and functional homology in teleosts, including fibrinogen, prothrombin, factors V, VII, VIII, IX, X, XII, and inhibitory factors, such as antithrombin III and protein C.[231] The whole blood clotting times of fish vary and may relate to factor activities; the whole blood of some teleosts clots faster than mammals, while other fish, such as lampreys and dogfish, are slow or may clot only weakly, even given prolonged time.[64,231] Functional tissue factor activity has been demonstrated in jawless, cartilaginous, and bony fish, and tissue factor-driven coagulation (extrinsic pathway) is considered the primary activating system of fish.[55,65] Due to the similarities in hemostatic physiology, pathological considerations are considered similar in fish as in mammals. Vitamin K requirements appear to be variable among fish species, but hemorrhage has been reported with deprivation and exposure to vitamin K antagonists (e.g., warfarin).[231]

The hemostatic systems of invertebrates function to help control fluid loss through injuries in the presence of an open circulatory system but are also tightly linked to functions of the innate immune system and wound healing.[36,142] The exact constituents of the hemostatic systems are different from vertebrates and vary between invertebrate groups, but the predominant outcomes include plasma protein coagulation/gelatinization and blood cell aggregation. Horseshoe crabs have a long history of scientific study and the *Limulus* amebocyte lysate assay is employed commercially for endotoxin detection. The hemostatic series of horseshoe crabs involves two proteinase cascades in which factors released from hemocytes in response to a stimulus (e.g., lipopolysaccharide) result in the conversion of a soluble protein, coagulogen, to a gel of linked coagulin polymers. Hemostasis in crustaceans including crayfish and shrimp employs clotting lipoproteins that form long cross-linked polymers with similar effect. Proteins involved in gelatinization of plasma in arthropods vary, including transport proteins and more specialized factors (e.g., fondue protein in *Drosophila*). Similar to the cross-linking activity of activated factor XIII in mammals, the stability of the polymerized clot in multiple invertebrates (e.g., horseshoe crabs, shrimp, *Drosophila*) is bolstered by the activity of transglutaminases released from hemocytes. Transglutaminase activity is the only factor recognized to be homologous across invertebrate and vertebrate hemostasis,[142] and assays employed for vertebrates (e.g., PT, PTT) are not applicable to invertebrates.

COMPARATIVE CLINICAL CHEMISTRY

Blood chemistry panels are part of routine diagnostic assessment in health and disease. Ideally, such testing is performed with species-validated instruments and methodology, species-specific standards, and minimization of sources of error (preanalytical, analytical, and postanalytical). Interpretation of clinical chemistry results is best performed in conjunction with species-specific and laboratory/method-congruent reference intervals. When these are not available, as is common in nondomestic species, analysis and interpretation require consideration of the analytic accuracy of results, utility of available references, and possible confounding variables, including preanalytical factors, analytical methodology, handling-related stress, medications or anesthetics, diet, circadian and seasonal patterns, age and reproductive status, and various environmental factors. Amphibians, reptiles, and fish are particularly influenced by environmental conditions, including temperature, humidity, water quality, population density, etc. Storage and freezing, often needed for samples collected in field settings, can further complicate the acquisition of reliable results as analytes vary in stability to storage and freeze/thaw cycles. To obtain the best diagnostic results, it is useful to become familiar with species-specific features and suitable methods, consider all potential influencing factors, incorporate these into research protocol development and clinical assessment, and institute consistent protocols for sampling, analysis, and interpretation.

The normal color of serum or plasma in vertebrates ranges from clear to yellow, with the latter due to the presence of carotenoids in herbivores. Plasma in some amphibians, like tree frogs and the giant salamander, is blue.[75] Antarctic icefishes lack erythrocytes and hemoglobin, and their plasma and whole blood have the same appearance: colorless to pale blue.[170]

Analytes vary in sensitivity to delayed analysis and variable storage conditions, which is necessary to consider in the interpretation of chemistry data. A comprehensive list of the effects of tube type, storage time and temperature, interferents and biological factors on a wide range of analytes in various nondomestic species, based on published data, is available.[26] In vitro hemolysis and delayed separation of serum or plasma from cellular blood components can result in spurious changes of analyte concentrations, including increases in potassium and phosphorus and decreases in glucose. These effects are also species dependent, as differences exist in cellular composition and metabolism. Species with higher documented intraerythrocytic potassium concentration include primates, pigs, horses, sea turtles, fish, rabbits, and rodents; species with lower intraerythrocytic potassium include most dogs, cats, seals, ferrets, and most cattle.[103] However, within taxa, there may be variation between species in the type and extent of artifactual changes in potassium concentration (e.g., different bird species).[98] Furthermore, within some species, individuals vary in having either low or high erythrocyte potassium concentrations based on polymorphisms affecting the activity of the sodium-potassium ATPase pumps, as has been documented in different breeds of sheep and some Japanese-origin dogs.[103] Early erythroid precursors are relatively potassium rich and increased numbers of these cells as well as reticulocytes produced in the presence of accelerated erythropoiesis also have higher intraerythrocyte potassium concentration. A similar pattern of increasing potassium concentration and decreasing glucose concentration due to delay in the separation of heparinized plasma from cells occurs in macaws and Burmese pythons.[100] Thus, delayed sample processing should be avoided and, when it occurs, it needs to be considered in result interpretation.

Blood chemistry analysis is performed commonly on serum from mammals and heparinized plasma from birds, reptiles, amphibians, and fish. The predominant use of plasma in nonmammalian vertebrates relates to animal size and difficulty of sampling, as well as greater sample volume yield when plasma is used, and avoidance of prolonged and unpredictable clotting in reptiles, amphibians, and fish that can affect chemistry results. Relevant differences in plasma versus serum chemistry results include differences in protein and potassium concentrations, which are discussed in more detail below. For plasma samples, blood collection into lithium, ammonium, or sodium heparin is suggested. Collection into sodium- or potassium-EDTA or potassium-heparin salts may result in confounding electrolyte data and EDTA may falsely lower results for calcium and enzymes that use divalent cations as a cofactor, like ALP and CK.[37,159] The order of blood draw and blood tube filling during blood sampling into multiple blood collection tubes must be considered to avoid introducing artifacts from tube to tube by accidental transfer of anticoagulant substance; EDTA-tubes are filled last for the reasons mentioned.[46]

The following sections review commonly assessed blood biochemistry analytes across taxa. Table 4.2 contains recommended panels for various taxa. Boxes 4.e1–4.e9 (adapted and modified from Ref. 222) briefly summarize common considerations for various chemistry abnormalities.

Total Serum/Plasma Protein and Albumin

Physiology and Interpretation of Changes in Disease

Proteins constitute an abundant component of plasma and have physiological roles that include facilitating metabolism, shuttling substances, maintaining oncotic pressure, and participating in immune responses. Blood proteins contain the vast majority of nitrogenous material in the blood. The main site of protein production is in hepatocytes, with the exception of immunoglobulins that are produced by cells of the humoral immune system (lymphocytes). Plasma total protein and serum total protein are inherently different due to the absence of clotting factors, including fibrinogen, in serum. Globulins constitute all nonalbumin proteins ([total protein] − [albumin] = [globulins]). The concentration of total proteins varies between species, with nonmammalian taxa tending to have lower concentrations compared to mammals. Albumin has not been identified or is present in very low concentrations in several species of elasmobranchs, with the albumin-like fraction presumably representing lipoproteins.[134,157,213] Common considerations for alterations in blood protein concentrations are summarized in Box 4.e1.

Preanalytical and Analytical Considerations

Contamination of blood samples by other body fluids within sample collection sites, such as by extravascular fluid, lymph, or spinal fluid in reptiles, amphibians, or fish—may artifactually decrease blood total protein and affect electrolyte concentrations.[88,144]

The recommended methods of measurement vary between species with some assays failing to maintain accuracy across taxa. Measurement of plasma **total protein** (TP) concentration by **refractometry** is based upon the proportional increase of light bending (refraction) to the concentration of solids in solution. The TP scale of refractometers is calibrated to assume that changes in measurement are due to alterations in TP concentration and that the concentrations of other solids are stable (therefore TP measurement rather than total solids or TS). Despite most refractometers being calibrated to human plasma proteins rather than animals, differences in the application of this scale between mammalian species are generally clinically insignificant.[222,226] Besides changes in protein concentration, refractometry is also affected by the presence of other soluble solids, and significant variations in the concentrations of other solids, including glucose, urea, electrolytes, and lipoproteins, can alter the refractometric reading. Furthermore, solutes of anticoagulants can also increase the refractive index if present in sufficient concentration in comparison to blood (i.e., marked underfilling of anticoagulated blood collection tubes). Pigmentation of plasma/serum from hemolysis and hyperbilirubinemia may make visualization of refractometric reading more difficult but is not expected to cause artifactually increased TS.[93,222] In birds, reptiles, and fish (especially elasmobranchs), refractometric measurement of total protein is inaccurate and commonly overestimates total protein. If performed, the measurement should be considered an estimate.[145,148]

Total protein measured by chemistry analyzers commonly employs the **Biuret reaction**, in which a colour change is measured spectrophotometrically after copper binds to peptide bonds.[222] This reaction is proportional to protein concentration, although some proteins may contribute to a larger degree. As with most assays based on colour change, hemolysis may cause mild positive interference. **Albumin** concentration at most veterinary labs is measured by the **bromocresol green (BCG) dye-binding reaction** in which the spectrophotometric colour change due to BCG-binding is proportional to albumin concentration. There is variability of the BCG-binding affinity among species, and interferences can include binding to globulins (including fibrinogen), hemolysis, triglycerides, and even the presence of plasma anticoagulants, depending on the assay.[222,223] BCG dye-binding reportedly yields inaccurate results in birds, reptiles, rats, white rhinoceros, dolphins, and manatees, and has limited assessment across amphibian and fish species.[15,23,51,116,166] For mammals, use of the BCG method together with method-specific-reference intervals is common. For nonmammalian vertebrates, however, protein electrophoresis with measurement of total protein by Biuret reaction and subsequent calculation of the albumin fraction is recommended, and use of the BCG method is not advised.[47,98,146,150,166] In the presence of hemolysis sufficient to cause red plasma discoloration, the results of biuret total protein and BCG albumin reactions are reported as invalid in amphibians.[75]

TABLE 4.2 Suggested Chemistry Analyte Panels for Various Taxa. Crosses in black text represent core analytes, crosses in orange text represent additional analytes, selected based on clinical findings and results of the core panel.

Analyte	Avian	Reptile	Elasmo-branch	Teleost	Marine mammal	Ungulate	Rodent/ Rabbit	Other mammal
Total Protein	X	X	X	X	X	X	X	X
Albumin					X	X	X	X
Globulin					X	X	X	X
A/G Ratio					X	X	X	X
Acute phase protein[a]				X	X	X	X	X
Protein electrophoresis	X	X	X	X			X	X
Blood urea nitrogen	X	X	X	X	X	X	X	X
Creatinine	X				X	X	X	X
Uric acid	X	X	X	X				
ALT					X		X	X
AST	X	X	X	X	X	X	X	X
GLDH		X				X		
CK	X	X	X	X	X	X	X	X
ALP		X			X	X	X	X
GGT	X	X	X	X	X	X	X	X
LDH	X							
Bile acids	X					X	X	X
Total Bilirubin						X	X	X
Amylase	X	X						
Lipase	X	X						X
Glucose	X	X	X	X	X	X	X	X
Cholesterol	X		X	X	X		X	X
Triglycerides	X	X	X	X	X	X	X	X
Sodium	X	X	X	X	X	X	X	X
Potassium	X	X	X	X	X	X	X	X
Chloride		X	X	X	X	X	X	X
Calcium	X	X	X	X	X	X	X	X
Phosphorus	X	X	X	X	X	X	X	X
Magnesium		X			X	X		X
Co2	X	X	X	X	X	X	X	X

[a]Select only if a specific acute phase protein assay has been validated for the species and is available.

Furthermore, the bromocresol purple (BCP) dye-binding reaction typically employed by human clinical pathology laboratories is not reliable across species, even within mammals.[222] Researchers and clinicians should be aware of the testing methods used by their laboratory and the relevance or limitations to their species of interest.

Protein Electrophoresis

Physiology and Interpretation of Changes in Disease

Pre/analytical aspects of protein electrophoresis are presented in the online content. Typical electrophoretic regions across taxa are summarized in Box 4.e2.

Mammals

The protein electrophoretic patterns of domestic mammals are described in detail elsewhere[67,162,222] and are the basis for common electrophoretic pattern interpretation across mammalian taxa, though differences are increasingly recognized.

Nonmammalian Vertebrates

Comprehensive information regarding protein electrophoresis in nonmammalian species is available in a recent review paper.[47] Applications of protein electrophoresis in these taxa include the accurate quantitation of albumin and the detection of an acute phase or chronic inflammatory response. Additionally, low albumin concentrations and albumin-to-globulin ratio (A/G) have been shown to be prognostic for decreased survival in loggerhead turtles and birds with aspergillosis.[47] Lastly, protein electrophoresis can be used to identify monoclonal-appearing bands; however, these are not always diagnostic for the presence of paraproteins as narrow globulin peaks can occur in inflammatory diseases, in hemolytic samples, and in egg-laying birds.

In comparison to mammals, birds tend to have higher concentrations of prealbumin and lower concentrations of gamma globulins.[47,49] The composition of the prealbumin peak has not been determined, but this peak has been documented to decrease in falcons with aspergillosis and increase in stressed geese.[47] In several groups of birds (psittaciformes, falconiformes, columbiformes), a prominent alpha globulin peak has been documented on agarose gel electrophoresis and found to be composed of apolipoprotein A-I, a negative acute phase protein involved in cholesterol homeostasis.[197,198] This component must be considered when interpreting avian plasma electrophoresis, as unlike many of the positive acute phase proteins in the alpha region, the peak of Apo A-I is not indicative of inflammation, and in fact, would be expected to decrease with inflammation. Increases in alpha, beta, or gamma globulins occur in a range of infectious diseases in avian species. Additionally, a prominent monoclonal-appearing beta globulin peak has been reported in some birds during egg laying and may relate to transferrin concentration.[50] Examples of common avian electrophoretic tracings are displayed in Fig. 4.14. Familiarity with species-specific and reproductive stage-related electrophoretic patterns as well as development of appropriate reference intervals are recommended to best assess and interpret protein electrophoresis results.

Protein electrophoresis is also used to assess protein fractions in reptiles, amphibians, and fish.[47] Causes for dysproteinemias are generally expected to be consistent with other taxa. However, electrophoretic composition and response in these animals require further study and marked variations in fraction patterns are recognized between species and across taxa.[47,85] Notable physiological highlights in these animal groups include reported physiological increases in protein concentrations during metamorphosis of amphibian tadpoles to adult frogs, and during overwintering periods[75]; seasonal, sex, and reproductive-stage-associated differences in protein concentration and plasma protein electrophoresis of some reptiles, including estrogen-induced hyperproteinemia[47,168]; seasonal fluctuation of electrophoretic patterns in teleosts, and significant contributions of vitellogenin to TP in gravid females[90]; and significant contribution of lipoproteins to electrophoretic fractions of elasmobranchs, in which albumin may be negligible.[47,48] In terms of changes in disease, a decrease in albumin and an increase in beta globulins may occur in diseased or debilitated reptiles.[47] Across all taxa, the correlation of electrophoretic results to extrinsic and intrinsic factors and clinical findings is important during interpretation, and repeated assessment of electrophoresis over time may yield additional insights into health and disease.

Acute Phase Proteins

Physiology and Interpretation of Changes in Disease

Acute phase proteins (APP) are proteins whose production is commonly linked positively or negatively to inflammatory stimuli through cytokine effects. The concentrations of positive APP increase and the concentrations of negative APP decrease with inflammation. The composition of major and minor APP as well as the rate of change is variable between species, but typically APP changes are noted within one to several days, in comparison to immunoglobulin or complement changes that may take weeks. APPs are sensitive indicators for the diagnosis of systemic inflammation but are not specific as to the inciting cause. Therefore, they need to be interpreted in the context of clinical findings and other diagnostic tests.

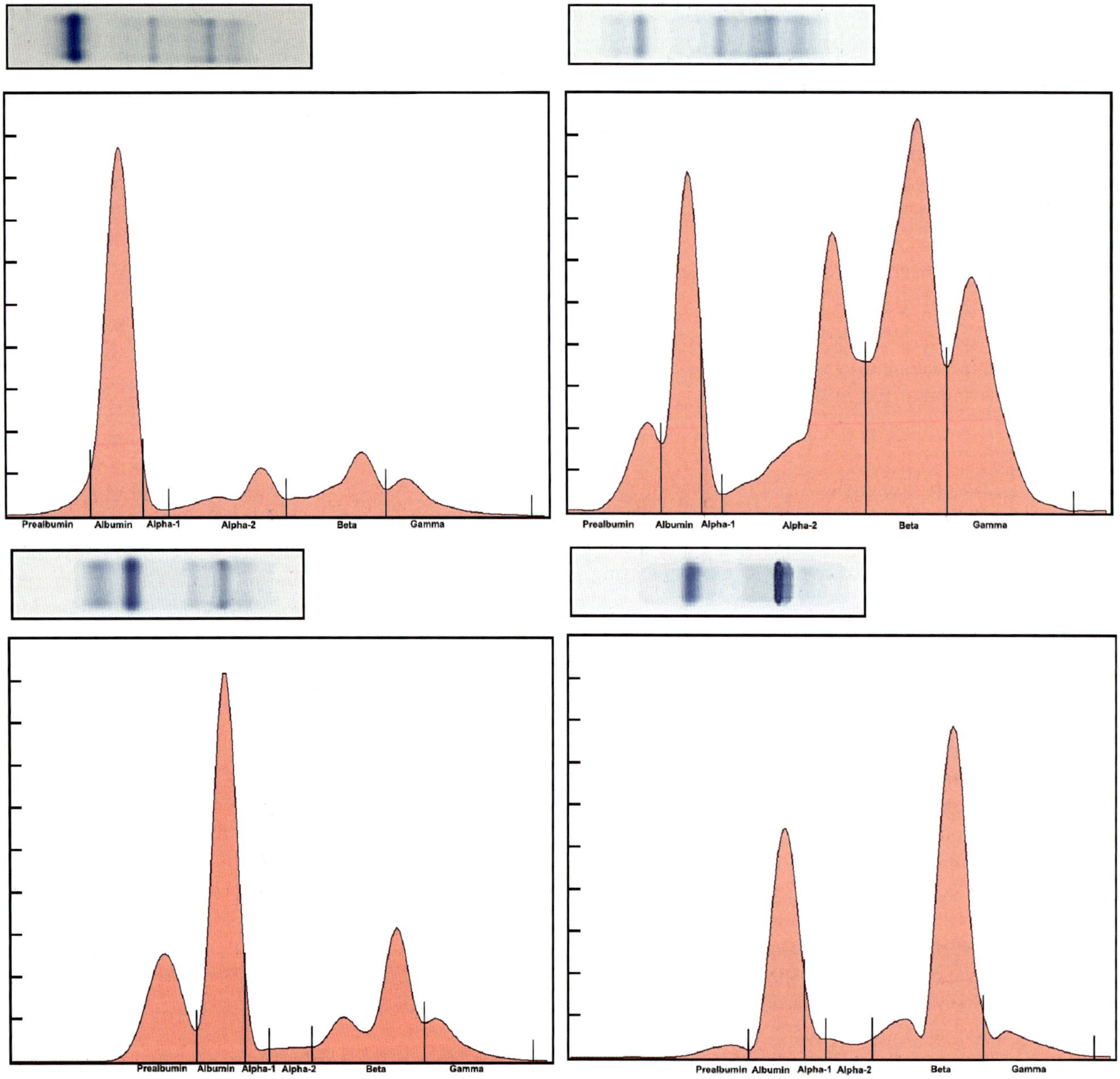

FIGURE 4.14 Examples of plasma protein electrophoretic gels and associated electrophoretic tracings. (A) Healthy African penguin with a prominent albumin band on the electrophoretic gel and associated peak on the electrophoretic tracing. (B) African penguin with chronic-active inflammatory disease due to aspergillosis showing prominent alpha 2, beta, and gamma peaks; the albumin percentage (area under the curve) is reduced; a prealbumin peak is also present. (C) Healthy *Pionus* parrot. (D) Egg-laying *Pionus* (D) demonstrating a prominent beta globulin peak. Differences in the size of the prealbumin peak may be individual variation. *(Photos Courtesy of C. Cray, University of Miami).*

Mammals

Serum amyloid A (SAA) is an important positive APP in nondomestic equids, some ruminants, rhinoceros, elephants, manatees, and elephant seals[113]. Haptoglobin is a minor or moderate APP in several nondomestic ungulates and shows promise as a marker of subclinical inflammation associated with infectious diseases like bovine tuberculosis.[113] C-reactive protein (CRP) is a major acute phase protein in dogs, pigs, and primates but does not appear to be a useful marker of inflammation in many other mammalian species. Fibrinogen is a ubiquitous but minor positive APP and increases up to twofold in inflammatory disease. Albumin is the most well-known negative APP, although it is not usually used as a specific indicator of inflammation as there are often many additional causes for hypoalbuminemia in a sick patient. Examples of other APPs that have been identified in domestic animals but have not been well investigated across mammals include ferritin, ceruloplasmin, alpha-1 acid glycoprotein, transferrin, and apolipoprotein A-I. The concentration of some APPs can also change due to stimuli that are not directly inflammatory, including pregnancy, lactation, or drug effects (e.g., glucocorticoids, phenobarbital) in mammals.[222]

Nonmammalian Vertebrates

APPs identified in chickens include SAA, alpha-1 acid glycoprotein (AGP), ceruloplasmin, fibrinogen, and (ovo) transferrin, which is a positive APP in birds.[176] An increase in fibrinogen has been documented in a range of avian species with bacterial infections and this acute-phase protein may be useful as a routine screening test for the detection of inflammation in birds.[107] Serum amyloid A and fibrinogen have been identified as positive APPs in reptiles, although fibrinogen does not appear to be a useful indicator of inflammation in red-eared sliders.[161,193] Hemoglobin binding protein has shown some promise as diagnostic test in chelonians.[59,74] Acute phase proteins identified in fish include fibrinogen, haptoglobin, α-2 macroglobulin, ceruloplasmin, hepcidin, warm acclimation protein 65 kDa, ntelectin, SAA, CRP, and serum transferrin, which is a positive APP as in birds.[193] Haptoglobin and CRP have shown promise as markers of inflammation in some shark species and SAA could be a potentially valuable biomarker of bacterial infection in fish.[35,122] Fibrinogen increases during hibernation in some amphibians; other APPs have not been studied in amphibians to date.[75]

Preanalytical and Analytical Considerations

Commercially available APP assays for mammalian SAA are antibody based, and therefore species specific. It is imperative that an SAA assay be validated for the species of interest. SAA assays have successfully been used in several ungulate species, elephants, dolphins, and northern elephant seals.[114] Haptoglobin measurement is usually performed using a colorimetric assay, which is not species-specific. Hemolysis will cause falsely low results because haptoglobin present in the specimen forms complexes with free hemoglobin and is not measured. Human-specific CRP assays are suitable for use in primates and have been validated in northern elephant seals.[211] Nonimmunological methods (modified Clauss methods or the heat precipitation method) are preferred for the measurement of fibrinogen. Patient-side APP tests are popular in companion animal medicine, but so far, none of these devices have been validated for nondomestic species. Assays for the measurement of APPs in nonmammalian vertebrates are not commercially available.

Renal

Physiology and Interpretation of Changes in Disease

Kidneys function in the excretion of wastes, retention of physiological solutes, and water balance, although in some nonmammalian species, other tissues and organs are also integral to maintaining homeostasis (e.g., gills in fish). Renal chemistry analytes are influenced by prerenal, renal, and postrenal factors. The analytes traditionally captured in routine veterinary chemistry panels (e.g., urea, creatinine, uric acid) are insensitive biomarkers of renal disease and require a significant loss of renal reserve (~75%) prior to recognition of biochemical abnormalities. In toxicological pathology studies, research, and safety assessment applications, other biomarkers of nephrotoxicity, such as glutathione-S-transferase alpha, kidney injury molecule 1, clusterin, and cystatin C may provide more diagnostic sensitivity or mechanistic information. However, these are incompletely studied across taxa and are not routinely used in the clinical setting.[180]

Azotemia is defined by the presence of increased nonprotein nitrogenous compounds in the blood. The predominant end product of protein metabolism varies among taxa from ammonia to urea to uric acid. **Symmetric dimethylarginine (SDMA),** a methylated form of the amino acid arginine, is produced by nucleated cells at a constant rate and excreted via the kidneys. Blood concentrations are therefore dependent on glomerular filtration rate and may be altered in renal disorders. Across taxa, **other blood chemistry abnormalities** that may be detected in renal disorders include changes in serum albumin, potassium (K) and phosphorus (P) concentration, acid-base status, magnesium, and anemia. Potassium concentration often increases in birds and mammals with reduced renal excretion, but hypokalemia in chronic renal failure is also common secondary to multifactorial effects (e.g., diminished intake, renal wasting, and cellular shifts). Similarly, with reduced GFR, inorganic P concentration often increases. However, ruminants may not develop hyperphosphatemia due to the relatively small renal role in P excretion compared to alimentary losses (saliva, rumen).[222] Hypocalcemia with renal failure is also common, although hypercalcemia can occur in species that commonly secrete renal calcium (e.g., horses), and calcium status depends on other endocrine factors (e.g., parathyroid and vitamin D levels) as well. Renal disease-associated anemia is multifactorial and involves factors, such as decreased renal erythropoietin production, shortened erythrocyte lifespan, and poor nutritional status.

General causes of altered concentration of nitrogenous wastes and SDMA in blood are summarized in (Box 4.e3).

Mammals

Assessment of renal health is typically based on the concentrations of blood urea nitrogen, creatinine, symmetric dimethylarginine (SDMA) (some species), and electrolytes in conjunction with urinalysis including assessment of urine concentrating ability by measurement of **urine specific gravity (USG).** The degree of normal urine concentrating ability varies between mammalian species, with higher USG expected in kidneys adapted to water conservation. For example, in the face of simple dehydration common USG values in mammals are expected to be >1.040 in cats, >1.030 in dogs, and > 1.025 in cattle and horses.[222] Besides renal impairment, extrarenal factors that can affect USG include decreased hepatic urea production (reduced medullary concentrating gradient), antidiuretic hormone influence (e.g., nephrogenic or central diabetes insipidus), solute excesses (e.g., glucosuria), endogenous and exogenous corticosteroid excess, and medications (e.g., diuretics).

Creatine phosphate is a high-energy molecule used in muscles and **creatinine** is the nonprotein nitrogenous product of creatine degradation. Creatinine enters plasma after degradation of creatine or creatine phosphate in muscle fibers. Baseline creatinine concentration can be noted to vary with degree of muscling (muscle mass). In mammals, creatinine is often considered one of the best measures of glomerular filtration rate (GFR) because it is freely filtered in the kidney with very little secretion and no resorption. In contrast, **urea** (UN) is affected by resorption and nonrenal influences, such as from the alimentary tract, particularly in equids and ruminants in which gastrointestinal excretion and microbial urea metabolism can cause significant effects.

SDMA concentrations are not influenced by muscle mass, and in dogs and cats, SDMA may detect a decrease in GFR earlier than creatinine.[201] SDMA has been investigated as a marker of renal function in cheetahs and tigers; in the former species SDMA has not been shown to correlate with GFR but correlates variably with creatinine,[200,244] while in tigers, positive correlations between SDMA and creatinine, and SDMA (and creatinine) and the degree of histological renal damage, has been demonstrated.[9,165] SDMA was more sensitive in detecting renal disease, compared to creatinine, in some but not all tigers and cheetahs with chronic kidney disease.[135,165] The utility of SDMA for detecting and monitoring kidney disease in nondomestic mammals still requires extensive research.

Nonmammalian Vertebrates

In birds, creatine is excreted by the kidney before being converted to creatinine; therefore, measurement of **creatinine** has low utility in birds. Although creatine concentration might provide useful information in birds, veterinary laboratories do not routinely test for creatine. Birds are able to concentrate urine to only a limited degree[110] and the predominant end product of nitrogen metabolism in birds is **uric acid** (UA). Concentrations of uric acid tend to be higher in adults than in juveniles. Uricotelism allows water conservation, and uric acid is less toxic than urea or ammonia, allowing the accumulation of UA in eggs to be tolerated during embryogenesis.[148] UA is synthesized mainly in the liver and is predominantly excreted (~90%) by active renal tubular secretion. This secretion is largely independent of hydration status; not depending on urine flow or water resorption, and exceeding GFR. Uric acid is not sensitive for prerenal azotemia, and tends to increase only late in renal disease. Increased UA production occurs with catabolism of proteins from meat ingestion or breakdown of protein tissues, and fasting prior to blood collection is helpful in meat-eating birds.

Hyperuricemia can result in the precipitation of urates as crystals (monosodium urate monohydrate) in body cavities, within organs, and in synovial joints (see Chapter 31: Galliformes and Columbiformes). This precipitation indicates imbalanced UA production and elimination. Contributing factors include high protein intake (dietary; in excess of capacity for tubular secretion), or more commonly, renal functional insufficiency causing diminished tubular secretion. Interestingly, the degree of hyperuricemia occurring in raptorial and piscivorous birds postprandially may exceed levels expected to result in urate precipitation, and the mechanisms preventing deposition during health are not known.[148]

Birds have very low concentrations of plasma urea such that **UN** is not a reliable indicator of renal function. As in mammals, increased protein catabolism (i.e., dietary) results in more hepatic urea formation. Although not sensitive for renal disease in birds, UN may have value in the detection of prerenal azotemia as, similar to mammals; it undergoes glomerular filtration and tubular reabsorption that is related to urine flow. With dehydration, more urea is reabsorbed in the avian kidney and this may be diagnostically informative.[148] When severe dehydration leads to or develops secondary to renal function abnormalities, hyperuricemia is often concurrent.

Nitrogenous wastes in reptiles are variable. As a group, reptiles lack elongated loops of Henle and do not concentrate urine above the osmolarity of blood.[110] Some species have evolved to excrete uric acid, which, as in birds, aids water conservation as well as in ovo embryonic survival. Marine and freshwater turtles produce and excrete ammonia and urea, crocodilians ammonia and uric acid, amphibious turtles mainly urea, and tortoises, snakes, and other terrestrial reptiles mainly uric acid.[125] UN and creatinine concentrations are generally not sensitive indicators of renal disease in reptiles, especially nonureotelic species in which concentrations of urea and creatinine may be very low. Unlike other uricotelic terrestrial reptiles with low blood UN concentration, higher UN concentrations have been documented in chelonians but the possible causes for this are unconfirmed. As in birds, terrestrial uricotelic reptiles may develop hyperuricemia with or without gout due to progressive renal insufficiency, though other causes (e.g., protein intake, dehydration) may also play a role. SDMA has been measured in clinically normal Hermann's tortoises, but no further information on the utility of SDMA as a renal biomarker in reptiles is available.[139]

The predominant form of nitrogenous waste in amphibians depends on species and life stage. In general, aquatic stages or species commonly excrete ammonia through the kidneys (ammonotelism), skin, and gills, while terrestrial stages or species excrete urea (ureotelism). However, amphibians may be uricotelic, such as some species of adult tree frogs that utilize uric acid to enable more water conservation. Some species of frogs switch from ammonia to urea excretion in response to environmental water abundance or deficiency, respectively. Thus, it is necessary to understand and test the appropriate waste product form. Furthermore, amphibians are highly tolerant of elevations in blood urea as it is used to enable cutaneous water absorption.[75] Because amphibian kidneys do not concentrate urine beyond the osmolality of plasma, USG is a marker of plasma osmolality rather than an indicator of renal concentrating ability in these species.[75] Generalized edema (anasarca) is a common sequela to renal disease in amphibians and may reflect glomerular protein loss or tubular electrolyte wasting (e.g., hyponatremia).

Kidneys as well as gills, function in the excretion of nitrogenous wastes in fish but are also key to ion excretion, acid-base status, and osmoregulation. Most teleosts are ammoniotelic and have only small amounts of blood UN.[90] However, urea is the main nitrogenous end product in marine elasmobranchs, few teleosts, and coelacanths.[33] Gills are heavily involved in ammonia and urea hemostasis and elevated concentrations may reflect branchial rather than kidney disease. Furthermore, environmental and water quality factors affect homeostasis, and some species of fish (e.g., lungfish) change from ammonotelism in water to ureotelism when out of water. Fish vary in osmoregulation strategies. Freshwater teleosts that are hyperosmolar to their environment have high GFR and urine production, while physiological mechanisms in saltwater teleosts act to conserve water and excrete NaCl. Elasmobranchs utilize urea and trimethylamine produced from the liver to maintain high body osmolarity, allowing for high water influx across the gills and high GFR. It has been postulated that SDMA may be a useful marker of renal function in those fish species that have glomeruli. SDMA has been documented in the plasma of brook trout, using the LC-MS method but further information as to the usefulness of SDMA in detecting renal disease in fish is not yet available.[189]

Preanalytical and Analytical Considerations

Blood **UN** is generally measured via a urease reaction that splits urea into NH_3 or NH_4^+, although the exact methods vary. This test measures the nitrogenous portion of the urea, not actually the urea compound (CH_4N_2O). **Creatinine** is also measured by various assays depending on system and manufacturer (i.e., enzymatic hydrolysis, complex formation [Jaffe's reaction]). As stated previously, differences in methodology are one of the many reasons the results and reference values generated by different laboratories and different analyzers should be compared with caution, and species-, laboratory-, and method-specific reference intervals are optimal. Units of measure are used variably across the world. Furthermore, different assays are sensitive to different interferences, including from hemolysis, chromogens, and medications. For these reasons, it is best that the users research and communicate with the testing laboratory to understand the assay details and potential complicating factors. The gold standard measurement method for **SDMA** is LC-MS, and a commercial immunoassay is available which has thus far only been validated for use in cheetahs and has shown good agreement with the LC-MS method in tigers.[52,244]

Enzymology

Physiology and Interpretation of Changes in Disease

Commonly assayed enzymes are listed in Table 4.3, along with abbreviations used in the following text.[7] Baseline enzyme activities reflect physiological rates of enzyme plasma entry (rate or production and release) and clearance (inactivation or elimination). Enzyme activities increase when rates of entry increase or removal decreases. Enzymes vary in tissue and cellular locations, and an understanding of such distributions is needed to interpret findings. Enzyme activities in sequestered and external-facing tissues, such as the brain and kidney, respectively, do not generally contribute to circulating enzyme activities. For example, although ALT, AST, and ALP are documented in the kidney of multiple species, in most situations these enzymes are shed in urine rather than into blood. Enzyme activities are variably affected by organ mass, cellular damage, rates of synthesis, intestinal absorption, cellular proliferation, and clearance. Circulating half-lives also vary between different enzymes and species. Table 4.3 summarizes major diagnostically significant enzyme activities with tissue sources, rough estimates of half-life, and relevant notes for birds and mammals. Less is known about specific enzymology in reptiles, amphibians, and fish, although some consistency in tissue distributions and subsequent interpretations of changes in enzyme activity are generally accepted (e.g., elevations in CK activity with muscle damage).

Cytosolic and mitochondrial enzymes, such as **ALT**, **AST**, **ID**, **LDH**, **GLDH**, and **CK**, are often considered "leakage" enzymes since their activity often increases in circulation due to active reversible or irreversible cell damage. The largest increases in these enzyme activities are frequently seen with necrotizing and inflammatory

TABLE 4.3 Summary of Commonly Assessed Blood Enzymes Activities with Diagnostically Relevant Tissue Sources, Cellular Location, and Half-Lives

Name	Abbreviation	Major Tissue Sources	Cellular Location	Half-Life	Comments
Alanine transaminase	ALT	Hepatocytes Skeletal myocytes (erythrocytes in birds)	Cytoplasm	2–3 days canine 3–4 h feline ~15 h avian	Hepatocytes of cattle, horses, pigs, and guinea pigs, and some birds have little ALT, limiting usefulness as a marker of hepatocellular damage. Additionally, in birds, the T1/2 of ALT is short, and elevations can occur with hemolysis. ALT has low activity and tissue specificity in reptiles.
Aspartate transaminase	AST, GOT	Hepatocytes Skeletal myocytes Cardiac myocytes Erythrocytes	Cytoplasm Mitochondria	<1 day canine ~1 h feline ~7 h avian ~7 days equine ~1 day cattle	Generally sensitive for hepatocyte or skeletal muscle damage in most species. Interpret at minimum with CK to improve specificity for hepatic disease.
Alkaline phosphatase	ALP	Hepatocytes Biliary epithelial cells Osteoblasts	Cell membrane	~3 days canine L-ALP <8 h feline L-ALP	Birds have low liver ALP; osteoblastic ALP activity is predominant. ALP has low tissue specificity in reptiles.
Creatine kinase	CK	Skeletal myocytes Cardiac myocytes Brain	Cytoplasm	<2 h canine ~3 h avian	
Gamma-glutamyltransferase	GGT	Biliary epithelial cells Hepatocytes	Cell membrane	~3 days equine	Birds and reptiles have low hepatic GGT. Guinea pigs have higher GGT than other rodents.
Glutamate dehydrogenase	GLDH, GDH, GMD	Hepatocytes	Mitochondria	~8 h dog <24 h equine ~14 h bovine <1 h avian	GLDH is a useful and specific indicator of hepatocellular damage in many species (including horses, cattle, dogs, birds, some reptiles, and amphibians). However, the assay is less frequently available. In birds, the T1/2 is short.
Iditol (sorbitol) dehydrogenase	ID, SDH	Hepatocytes	Cytoplasm	~4–5 h canine ~3–4 h feline ~12–24 h equine	Tissue specific, but relatively short in vitro stability (hours at room temperature).
Lactate dehydrogenase	LDH, LD	Hepatocytes Skeletal myocytes Cardiac myocytes Erythrocytes	Cytoplasm	<6 h canine <1 h avian	Limited diagnostic utility due to lack of tissue specificity and short T1/2.

Sources: Allison RW. Laboratory evaluation of the liver. In: Thrall MA, Weiser G, Allison RA, Campbell T, eds. Veterinary Hematology and Clinical Chemistry. 2nd ed. Ames, IA, USA: Wiley-Blackwell Publishing Ltd; 2012:401–424; Campbell T. Clinical chemistry of common nondomestic mammals, birds, reptiles, fish, and amphibians. In: Thrall MA, Weiser G, Allison RW, Campbell TW, eds. Veterinary Hematology, Clinical Chemistry, and Cytology. 3rd ed. Hoboken, NJ, USA: John Wiley & Sons; 2022:569–632; Lumeij JT, Westerhof I. Blood chemistry for the diagnosis of hepatobiliary disease in birds: a review. Vet Q. 1987;9:255–261; Lumeij T. Avian clinical biochemistry. In: Kaneko J, Harvey J, Bruss M, eds. Clinical Biochemistry of Domestic Animals. New York, NY: Academic Press; 2008:839–872; Stockham S, Scott M. Fundamentals of Veterinary Clinical Pathology. 2nd ed. Ames, IA: Blackwell Publishing Ltd; 2008.

conditions that cause cell lysis. When cell injury is not active, such as in cholestatic disease or neoplasia, increases in these leakage enzymes may be minor.

Hepatic functional insufficiency is often not associated with detectable alterations in enzyme activities because the tissue cell injury is past, and the elevations in enzyme activities with portosystemic shunting are also generally mild. Regardless of tissue source, low enzyme activities are often not diagnostically significant. Such values may occur with enzyme degradation, effects from interferents, or in vitro inhibition. However, cases in which decreased activities may be of value include decreased organ tissue mass and specific toxicities, such as decreased cholinesterase activity in organophosphate toxicity.

The most diagnostically useful enzymatic markers for hepatocellular injury vary across species and depend on laboratory test availability and sample handling parameters.

Box 4.e4 reviews selected pathophysiological considerations for increased blood enzyme activities in all taxa.

Mammals

ALT is a fairly specific indicator of liver disease in canids, felids, lagomorphs, rodents (apart from guinea pigs), and primates, but is not useful in equids, ruminants, and porcines due to low activity in hepatocytes.

AST is generally considered to be a sensitive marker of hepatocellular disease in many mammals but has poor specificity for the liver as it is present in many other tissues, most notably muscle.

For assessment of hepatocellular injury in horses, **ID** has greater diagnostic sensitivity than ALT and greater diagnostic specificity than AST and LDH, but the stability is relatively short with loss of enzyme activity detected by 24 h in both room temperature and refrigerated samples (recovery 59% and 87%, respectively).[120] ID is also the most sensitive and specific marker of hepatocellular damage in rats.[71]

GLDH is a mitochondrial enzyme present in high concentrations in the centrilobular zones of the liver. GLDH is liver specific, and elevations are considered to indicate irreversible hepatocyte injury. The sensitivity of GLDH for detecting hepatocellular injury is moderate to good and varies depending on the underlying process but it is higher than that reported for ID. GLDH is a stable enzyme, and due to its stability and high specificity and sensitivity, it is used in many countries as the enzyme of choice for evaluating hepatocellular injury in ruminants and equids.[111]

ALP is present in the cell membranes of many tissues, but only a few have enough activity to play a role in circulating ALP activity. Most mammals have two isoenzymes of ALP: an intestinal ALP and a tissue nonspecific ALP. The tissue nonspecific ALP undergoes posttranslational modification to liver, bone, and kidney isoforms. The L-ALP isoform of hepatocytes and biliary epithelium and the B-ALP of osteoblasts contribute to circulating ALP, while the intestinal ALP and kidney ALP generally contribute little due to external release, apart from intestinal ALP in rats.[71] Rabbits have three ALP isoenzymes, producing two different isoenzymes of ALP in the liver and kidney in addition to intestinal ALP. Humans and great apes have placental ALP that is absent in other mammals.[86,222] ALP isoforms can be differentiated and quantified by ancillary techniques, including selective inhibition, thermostability, and immunological assays.

The most common cause of increased ALP activity in mammals is increased L-ALP from hepatocytes and biliary epithelium in association with cholestasis. Bile acids increase L-ALP production and membrane release.[222] Some chemical substances, including corticosteroids and phenobarbital, also induce the production of L-ALP. Additionally, domestic and wild canids have a unique ALP isoform, C-ALP, which is closely related to I-ALP, and is produced by hepatocytes when stimulated by corticosteroids. Increases in B-ALP contribute to hyperphosphatemia in animals with elevated osteoblastic activity, such as with skeletal growth in young animals, bone neoplasia, osteolysis, or fracture repair. ALP also increases in cervids during osteoblastic phases of antler growth and calcification. In domestic cats, B-ALP is increased with hyperthyroidism. The diagnostic sensitivity of ALP for cholestatic disease varies across species, being relatively high in dogs, moderate in cats and cattle, and poor in horses.

GGT also increases with cholestasis, biliary hyperplasia, and chemical induction. Compared to ALP, GGT is considered to have greater diagnostic sensitivity for cholestasis and biliary disease in animals like horses and cattle.[222]

In muscle cells, creatine kinase (**CK**) catalyzes the transfer of phosphate (PO_4^{3-}) between creatine and ADP/ATP. The reaction is reversible: when energy is low, it allows storage as creatine phosphate; when ATP is needed, it catalyzes the transfer of phosphate to ADP to form ATP. "Creatinine" kinase is a common misnomer for the enzyme. Isoenzymes of CK occur in brain, cardiac and skeletal muscles, and mitochondria of many tissues. CK activity in serum has been reported to be higher than activity in plasma (2.5 times higher in healthy dogs); this difference may relate to CK release from platelets and may cause mild elevations in CK.[3] Elevation in CK activity is specific to muscular damage and will not increase with liver injury alone. With muscle damage, compared to AST, CK activity usually peaks higher but returns to baseline quickly (within hours to days), whereas AST activity peaks lower but remains elevated longer (days). Increased CK can also occur with in vitro hemolysis, not because of CK release but because of other components that interfere with enzymatic assays (e.g., glucose-6-phosphate, adenylate kinase, ATP).[222] Myocardial-specific markers, including the CK-MB isoenzyme or cardiac troponins, are increasingly being studied. Limited data has recently become available for validation and clinical applications of cardiac biomarkers in marine mammals.[32,127,164] Muscle leakage enzymes (AST, CK) are often elevated in wild animals not adapted to handling. Marked tissue cell lysis, as can occur with rhabdomyolysis, may be accompanied by hyperkalemia, hyperphosphatemia, and hyponatremia, due to shifting of intra- and extracellular ions, as well as hypocalcemia due to the influx of calcium into damaged tissue.

Amylase and **lipase** are cytoplasmic enzymes included on some chemistry panels. Amylase activity can include pancreatic, intestinal, and possibly hepatic components.

Salivary amylase has been documented in pigs, guinea pigs, rats and other rodents, and a variety of primates (humans, various macaques, collared mangabey, various baboon, gorilla, and orangutan species). Notably, chimpanzees and bonobos have low salivary amylase activity and squirrel monkeys may be the only primates with negligible salivary amylase.[22] Lipases are produced by gastric, pancreatic, hepatic, and other cells, including adipocytes and myocytes; diagnostic assays have varied in analytical specificity for pancreatic lipase over other tissue sources. While pancreatic acinar damage is a cause for hyperamylasemia and hyperlipasemia, the diagnostic sensitivity for pancreatitis in various mammals is low and pancreatitis may be present without detectable increases in amylase or lipase activity. Furthermore, amylase and lipase are both cleared by the kidneys and decreased renal clearance may cause or contribute to elevated serum activity.[222] Immunoassays specific for pancreatic lipase have been developed for dogs and cats (canine PLI, feline PLI). The extent to which these offer value, if any, to nondomestic species is not yet determined and is likely to be limited to closely related species.

Nonmammalian Vertebrates

In birds, **GLDH** is considered a strong and specific diagnostic marker of hepatocellular injury, but the assay is less commonly available across diagnostic laboratories and short half-life in birds may result in missing elevations depending on the timing of injury and assessment.[148] GLDH is found in mitochondria, and thus GLDH activity will only increase with severe liver damage.[80] Increased **AST** activity has been reported as a sensitive, useful, and routinely assessed marker of liver disease in a variety of bird species, but the diagnostic specificity of this enzyme for liver disease is low as it is also present in skeletal and cardiac muscles and erythrocytes.[91,108,147] AST activity in conjunction with muscle-specific **CK** activity is commonly assessed to distinguish liver and muscle damage and elevated AST without CK suggests hepatocellular disease. However, because AST has a longer half-life than CK, it is also possible for CK to return to baseline more quickly with muscle damage.[98] **ALT** has moderate sensitivity and poor specificity for hepatic disease in birds, and is not commonly assayed.[80]

The L-ALP activity of birds is very low, and **ALP** activity in avian blood is predominantly a reflection of osteoblastic activity associated with skeletal growth, skeletal pathology, and egg shell deposition,[98,148] thus the diagnostic sensitivity of ALP for cholestatic disease is low in birds. **GGT** is considered to have good specificity and better sensitivity than ALP for cholestasis and biliary disease in birds.[148]

Amylase and **lipase** have not been well-studied in birds and there is no information regarding their usefulness in the diagnosis of pancreatitis.

Much less is known about enzymology of lower vertebrates and the sensitivity of such species to intrinsic factors and assay conditions must be considered during interpretation. However, elevations in **ALT, AST, GLDH, and LDH** activities may occur with acute hepatocellular injury, and must be differentiated from increases due to other tissue sources.[33,90] **GGT** activity is low to undetectable in reptiles and does not appear to be a useful biomarker.[108]

Amylase and lipase have been studied only to a small extent in reptiles, fish, and amphibians and clinical applications may be limited.[220]

Preanalytical and Analytical Considerations

Enzyme assays are mostly spectrophotometric and can be influenced by factors affecting light transmission (e.g., hemolysis, bilirubinemia, bilirubinemia, or lipemia). Furthermore, cellular disruption including hemolysis can release cellular enzymes and other constituents to the sample (i.e., AST or ATP). The stability of enzymes is variable between analytes, species, and under different storage and physiological conditions. For projects in which sample collection and storage are expected, a review of published data for analytes of particular interest is suggested prior to collection (e.g., Ref. 26).

Routine blood chemistry enzyme analyses measure enzyme activities, not concentrations. Differences in methodology based on reaction type, substrates, and reaction conditions can result in different measured values. Less commonly, immunological assays may be available to measure the concentration of a protein (e.g., canine PLI and TLI), but these are less readily available and more species-specific.

The effects of routine measurement of enzyme activities at 37°C must be considered in poikilothermic animals, as some enzymes in cold-adapted species may be inactivated at such temperatures. Instrumentation temperature is a source of variation to control if comparisons are made between the same or different individuals over time. Furthermore, high concentrations of ammonia may interfere with assay methods and result in falsely high values of ALT and AST, as demonstrated in rainbow trout.[54]

Liver Function

Physiology and Interpretation of Changes in Disease

Common analytes assessed as markers of hepatic function include bilirubin (or biliverdin), bile acids, and ammonium concentrations.

Bile acids are produced from cholesterol by the liver, excreted in the bile, and are involved in the emulsification and digestion of lipids in the intestine. In species both with and without a gall bladder, bile acids increase postprandially and most excreted bile acids are reabsorbed by enterohepatic recirculation. Common bile acids are variable by taxa and include cholic acid, chenodeoxycholic acid, and deoxycholic acid.[222]

Ammonium (NH_4^+) concentration is another measurement of liver function that is routinely assessed in plasma. Although this test may also be referred to as "ammonia" concentration, at physiological pH the majority exists in the ionized, NH_4^+ form.[222] Under physiological conditions, ammonium is produced mainly in the intestine, enters portal blood, is taken up by hepatocytes, and used for the urea cycle and protein catabolism.

Mammals

In healthy mammals, heme from the turnover of erythrocytes and other heme-containing proteins results in the formation of unconjugated **bilirubin**, which complexes with albumin, is taken up by hepatocytes, undergoes conjugation reactions, and is excreted into the bile. The transport of conjugated bilirubin to bile canaliculi is the rate-limiting step of bilirubin excretion. Some conjugated bilirubin also enters the blood from hepatocytes, is freely filtered by the glomerulus and is excreted in urine.

Total bilirubin in plasma consists of unconjugated (Bu) bilirubin, conjugated (Bc) bilirubin, and delta-bilirubin. Delta bilirubin is formed by the covalent binding of some conjugated bilirubin to albumin. In health, the majority of total bilirubin is unconjugated bilirubin complexed with albumin. Bu and Bc are rapidly removed from plasma; Bu mainly by hepatic, and Bc by both hepatic and renal clearance. Delta-bilirubin clearance is linked to albumin, and therefore has a longer half-life.[222] Delta bilirubin is usually negligible but increases with increased Bc concentration, as in obstructive cholestasis.[133]

Hyperbilirubinemia occurs with prehepatic, hepatic, or posthepatic factors. The paramount prehepatic condition is hemolysis, which causes Bu formation that exceeds hepatic uptake or excretion. Fasting may also result in hyperbilirubinemia due to liberated fatty acids interfering with the uptake of Bu by hepatocytes, which is especially common in horses. Hepatic hyperbilirubinemia results from decreased hepatic functional mass which causes reduced uptake or conjugation of Bu and/or decreased excretion of Bc. Posthepatic hyperbilirubinemia results from decreased excretion of Bc, which may be due to intra- or posthepatic obstructive cholestasis or functional cholestasis associated with sepsis. Bilirubin concentration should be assessed in conjunction with erythrocyte mass and other measures of hepatobiliary health. Persistent hyperbilirubinemia due to hereditary deficiencies of bilirubin transport (e.g., Dubin-Johnson syndrome, Gilbert syndrome) have been reported in humans, Southdown and Corriedale sheep, Eisai hyperbilirubinemic rats (EHBR), and golden lion tamarins.[207] Though these conditions are benign, the bilirubin concentrations may be striking and must be differentiated from clinically relevant liver disease.

Bile acid concentration increases are due to decreased hepatic function or portosystemic shunting, causing reduced clearance from portal blood, or cholestasis, resulting in decreased biliary excretion. Although bile acid concentration may increase mildly (generally <2 times) postprandially and with gall bladder contraction, significantly increased concentration is a sensitive and specific indicator of hepatobiliary disease.

Hyperammonemia may result from decreased ammonium clearance from portal blood due to liver insufficiency, portosystemic shunting or urea cycle dysfunction, increased ammonium production (e.g., urea toxicosis in ruminants, intestinal disease in horses), or increased ammonium intake via administration (e.g., NH_4Cl, anhydrous NH_3).

With marked hepatic functional insufficiency (liver failure), in addition to hyperammonemia and increased bile acids hypocholemia, analytes produced by the liver may also become deficient. These include albumin and most other nonimmunoglobulin proteins, glucose, cholesterol, urea, and coagulation factors.

Nonmammalian Vertebrates

The major product of heme degradation in birds, reptiles, and amphibians is **biliverdin**. In mammals, biliverdin is further converted by biliverdin reductase to bilirubin which may provide evolutionary cytoprotective advantages at relatively low concentrations.[208] Unlike mammals in which yellow-brown plasma and urine discoloration suggests bilirubinemia and bilirubinuria (icterus), yellow or orange discoloration of avian plasma is more often normal and due to dietary carotenoids (e.g., flamingo). In birds, hyperbilirubinemia causing icterus is rare, and when present, may reflect the enzymatic conversion by microorganisms or nonspecific reductases.[148] More commonly, the presence of liver disease in birds can be detected by green serum/plasma (biliverdinemia) and excretion of green urates (biliverdinuria). The plasma color of reptiles varies from clear to yellowish (carotenoids in herbivores) to greenish-yellow (riboflavin and carotenoids in pythons) or, in disease (hemolysis, liver disease, starvation), medium to dark green (biliverdinemia).[33]

The normal color of amphibian plasma ranges from clear to yellow and then yellow to blue.[75] Bile pigments in fish include bilirubin and biliverdin depending on species, and serum may be light yellow or light green with increases of one or the other, respectively.[33]

Bile acids are sensitive and specific indicators of liver dysfunction in birds. Testing should be performed after a fasting period of 12–24 h in raptors, but fasting is not necessary in psittacines.[99] Healthy birds are reported to have higher plasma **ammonia** in comparison to mammals. The diagnostic utility of ammonium assessment in birds with hepatic disease is uncertain and warrants further study.

Current limited data suggest no clear diagnostic decision levels using fasting or tolerance test values.[98,148]

There is little information concerning bile acid and ammonium changes in hepatic disease in other nonmammalian vertebrates.

Preanalytical and Analytical Considerations

Hemolysis and lipemia may cause positive or negative interferences in **bilirubin** measurement, depending on the assay. Furthermore, bilirubin is sensitive to ultraviolet (UV) light and the concentration in a sample may decrease up to 50% in 1 h with direct sunlight.[222]

Bilirubin assays vary in their measurement of total or fractions of bilirubin. Some blood chemistry tests measure total bilirubin and "direct" bilirubin (Bd), which is a combination of conjugated and delta bilirubin [Bd = Bc + Bdelta]; indirect bilirubin (Bi) concentration is then calculated by subtracting direct from total bilirubin and this is equal to Bu [Bi = total bilirubin − Bd = Bu]. Fewer, dry-chemical, blood chemistry systems measure total bilirubin, Bu, and Bc; and some research systems (e.g., high-performance liquid chromatography) can measure multiple specific fractions including delta bilirubin. When values for Bu (indirect) and Bc (or direct) bilirubin are provided in addition to total bilirubin concentration, the rule of thumb patterns of hyperbilirubinemia are hemolysis Bu > Bc; decreased hepatic function Bu > Bc or Bc > Bu depending on the remaining function, underlying cause, and presence of concurrent cholestasis; and cholestasis Bc > Bu but these are generalizations and combinations of influencing factors may be present.

Ammonium is unstable and suboptimal handling of samples will give falsely high or low results. Blood must be drawn into tubes with EDTA or heparin anticoagulant; the tubes must be spun and plasma harvested as soon as possible, and analyzed immediately. If plasma cannot be assayed immediately, it can be frozen at -20°C for 1–2 days.[222]

Bile acids are measured by spectrophotometric reactions or immunoassays that can be influenced by sample hemolysis or lipemia. The use of immunoassays (often optimized for humans) may be further affected by limitations of antibody cross-reactivity between taxa.

There is no assay commercially available for the measurement of **biliverdin.**

Monovalent Electrolytes

Physiology and Interpretation of Changes in Disease

Interpretation of the monovalent electrolytes sodium (Na), chloride (Cl), and potassium (K) (Boxes 4.e5 and 4.e6) is generally consistent across taxa, although physiological ranges vary. Electrolyte fluxes in fish are discussed further later.

Sodium is a major, osmotically active electrolyte of plasma and extracellular fluid in most taxa and must be interpreted in light of hydration status. Sodium concentration is generally regulated by response to blood volume through the renin-angiotensin-aldosterone system, and by response to blood osmolality via thirst centers and antidiuretic hormone response. **Chloride** is the major extracellular anion of most taxa and is influenced by Na and HCO_3^- concentrations (acid-base status). The acid-base integration of Cl relates to influences from the maintenance of electroneutrality and linked exchange of HCO_3^- and Cl across some cell membranes, such as gastric parietal cells and some components of the nephron.[222]

Potassium is the major intracellular cation in most cells due to the activity of Na-K-ATPase pumps. K concentration is affected by acid-base status and is best interpreted in light of acid-base balance. Depending on the type of process and concurrent conditions, the effects of acid-base status on potassium concentration include K shifts out of or into cells opposite of H^+ excess or deficiency, and K loss with renal excretion of anions such as lactate and ketone bodies. Acid-base alterations due to respiratory mechanisms are similar to the effects of metabolic disturbances, but the net effects on K concentration are smaller and often not clinically recognized.[12,222] Insulin also promotes the movement of K into cells. The overall effects on K concentration are commonly multifactorial and normokalemia in the face of acidemia or alkalemia may mask the presence of overall increases or decreases in total body potassium.

Mammals

The general principles discussed above apply to mammals.

Nonmammalian Vertebrates

Salt glands, present in some marine and desert birds and reptiles as well as in elasmobranchs (rectal gland), play osmoregulatory roles. In marine species, these glands enable the excretion of a majority of Na and Cl with diminished reliance on renal clearance. Plasma osmolality and hormones play a role in the rate of secretion.[25] Salt glands in marine birds are activated upon exposure to osmotic stress and atrophy when exposed to prolonged freshwater.[76,110]

In addition to renal control, osmoregulation in amphibians is also affected by electrolyte and water resorption in the urinary bladder, across the skin, and in the lymph hearts of frogs.[75] These mechanisms enable water conservation in addition to renal control, as kidneys in amphibians are not able to concentrate urine above plasma osmolality. Electrolytes in amphibians are also more sensitive to environment conditions such as acidic substrate that can alter cutaneous ion transporters. Amphibians are generally able to tolerate wide ranges of electrolytes and osmolality shifts. Over-wintering amphibians and reptiles undergo a variety of alterations in proteins, carbohydrates, acid-base, and electrolytes to control ice crystal formation and lower tissue freezing point to constrain ice formation to extracellular compartments.[75,124,242]

Electrolyte regulation in fish is influenced significantly by osmoregulatory and environmental factors. Furthermore, the gills play a significant role in electrolyte balance. Due to osmotic loss of water to the environment, marine teleosts ingest saltwater and excrete excess monovalent ions primarily across the gills. Marine elasmobranchs employ urea, trimethylamine oxide, and NaCl to regulate internal fluid balance, and utilize the rectal gland as a major site of salt secretion. Freshwater teleosts and elasmobranchs are hypertonic to the environment and excrete large volumes of dilute urine to balance osmotic water uptake; electrolyte uptake and reabsorption through the gills and kidney also aid in the maintenance of ionic and acid-base homeostasis. Like other species, potassium is largely absorbed via the intestine of fish. Alterations in potassium and sodium concentration may arise from intestinal, renal, or cutaneous effects, and acid-base status.[33,90,95] Stress alone may result in electrolyte and acid-base shifts, including acidemia, hyponatremia, hypochloremia, and hyperkalemia.[90]

Preanalytical and Analytical Considerations

Na, Cl, and K are generally measured via indirect or direct potentiometry. Using indirect potentiometry, the sample is diluted using a predetermined dilution factor; if the sample does not meet normal criteria for this dilution (e.g., marked hyper- or hypoproteinemia, lipemia, or the sample is prediluted due to low sample volume), electrolytes may be spuriously increased or decreased (e.g., electrolyte exclusion effect).

As discussed earlier, **pseudohyperkalemia** often results from artifactual increases in potassium concentration that may occur due to in vitro hemolysis in species with relatively high erythrocyte K concentration in comparison to extracellular fluid. Other causes of pseudohyperkalemia include contamination by K-EDTA or K-heparin anticoagulant, delayed separation from cells, or marked thrombocytosis. Serum K concentration is generally mildly elevated compared to plasma due to platelet/thrombocyte K release during clotting.

Bivalent Electrolytes

Physiology and Interpretation of Changes in Disease

Box 4.e7 provides an overview of common considerations for altered calcium concentration. **Total calcium** (tCa) consists of free and bound subsets, commonly in a ratio of 1:1 to 1:2. The majority of bound calcium is complexed to anionic proteins, mainly albumin, but also globulins to a small extent. A small percentage of calcium in health is bound to nonprotein anions, including citrates, lactate, and phosphates. Nonbound, **free calcium (fCa)** is the physiologically active component and the concentration of fCa is tightly regulated within a narrow range through physiological mechanisms, including the production of parathyroid hormone (parathyroid glands in mammals, birds, reptiles, and amphibians); calcitonin (parathyroid gland C cells in mammals, reptiles, amphibians, and cells of the ultimobranchial gland in birds); stanniocalcin in fish (corpuscles of Stannius); and vitamin D3 (cholecalciferol). The active form of vitamin D in most mammals is 1,25-dihydrocholecalciferol, for which the final activation reaction is hydroxylation catalyzed by alpha-1 hydroxylase in the kidney. The biological activity of 1,25-dihydroxycholecalciferol has been shown to be much greater than that of 25-hydroxycholecaliferol or cholecalciferol.[222] In general, PTH production is stimulated by low fCa and acts to counteract hypocalcemia by increasing Ca and P mobilization from bones, increasing Ca absorption in the intestine, and both decreasing Ca and increasing P excretion by the kidneys. Vitamin D also enhances Ca and P absorption in the intestine and liberation from bone, and increases Ca resorption in the kidney. Calcitonin production is stimulated by elevated fCa and acts to decrease mobilization from bone and to decrease resorption of Ca and P at the kidneys. The term "ionized calcium" (iCa) is commonly used when referring to free calcium, but since all calcium in the body is ionized Ca^{2+} (both bound and free forms), the term "free calcium" (fCa) is more accurate when referring to the unbound, physiologically regulated, portion.

The concentration of fCa is tightly controlled. It is not uncommon for abnormalities in total calcium concentration to occur without corresponding shifts in fCa. The most common example of this is a shift in blood protein concentration that affects bound but not free Ca. Similar to in vitro effects, fCa is also sensitive to acid-base shifts in vivo, which alter Ca binding to anions; acidemia tends to increased fCa and alkalemia results in the opposite.

The **phosphorus** concentration (P) (total inorganic phosphorus) of blood is composed of a combination of forms: PO_4^{-3}, HPO_4^{2-}, $H_2PO_4^-$, and H_3PO_4. Like calcium, blood phosphorus exists in free and bound forms (approximately 1:1), and the bound portion is complexed to nonprotein cations or, to a lesser extent, protein cations.[222] Like calcium, P concentration is largely maintained via absorption in the intestine (promoted by vitamin D), resorption in the bone (promoted by PTH), and excretion by the kidneys (promoted by PTH). Other contributing factors include insulin, which promotes P entry into cells, and growth hormone, which promotes renal resorption of P. Common findings in young animals are increased P and ALP (bone isoform) relative to adults. Box 4.e8 provides an overview of common considerations for altered P concentration.

All taxa are susceptible to metabolic bone disease resulting from **secondary hyperparathyroidism**. It is especially common in reptiles and amphibians with improper husbandry parameters, and in new world monkeys, particularly marmosets, who have uniquely high requirement for dietary vitamin D3.[61,143] Nutritional secondary hyperparathyroidism results from improperly balanced dietary calcium and phosphorus, deficient dietary vitamin D, or insufficient UV light exposure. Young, growing, animals are commonly affected. In large amphibians, excess dietary vitamin A from the ingestion of rodents is also reported as a potential cause of absolute or functional hypovitaminosis D.[75] Renal secondary hyperparathyroidism occurs as a consequence of chronic renal failure in a variety of species, generally in older animals, in which renal disease causes Ca:P imbalance stemming from P retention and hypocalcemia secondary to diminished active vitamin D. This pattern of disease is uncommon in species that typically excrete large amounts of Ca via the kidneys, including horses and rabbits, as decreased GFR in these animals on standard diets more often results in hypercalcemia not hypocalcemia.[18,222] In both nutritional and renal secondary hyperparathyroidism, the body responds to decreased fCa and often increased P by increasing PTH production, stimulating bone calcium resorption, intestinal absorption, renal Ca retention, and renal P excretion, in an attempt to correct the imbalance. Depending on animal age, duration, and severity, these metabolic derangements may lead to a spectrum of osteopenic bony abnormalities, including rickets, osteomalacia, fractures, and fibrous osteodystrophy.

Magnesium (Mg), like calcium, consists of ionized magnesium (Mg^{2+}) in free and bound portions (approximately 1:1) with the bound portion complexed primarily to anionic proteins and to a lesser degree to nonprotein anions. Free Mg (fMg) is the physiologically regulated portion. Magnesium is absorbed in the intestinal tract (including the rumen in ruminants and distal small intestine and colon in monogastrics) and absorption is promoted by vitamin D. Mg is present mainly in bones and soft tissues. The effects of protein concentration on total magnesium concentration are similar to those of total calcium and its protein-bound fraction. Major regulators of fMg are changes in gastrointestinal absorption and renal excretion.

Mammals

Equine kidneys lack 1-alpha-hydroxylase, have low levels of hydroxylated cholecalciferol in plasma, and do not rely on vitamin D to the same extent in the regulation of calcium.[27] Whether this is also the case in nondomestic Equidae is not known.

Nonmammalian Vertebrates

Birds uniquely manage calcium requirements to produce eggs surrounded by hard, calcified shells. The majority of this calcium derives from intestinal absorption as well as from resorption of a medullary bone reservoir that is deposited in response to gonadal hormones.[53] During egg laying, birds may develop physiological estrogen-induced total hypercalcemia due to increased calcium binding to elevated yolk proteins (globulins).[98] Dietary excess of calcium and vitamin D are also reported as a common cause of hypercalcemia in birds.[148] Reproductive disease in birds can cause either hypercalcemia or hypocalcemia (e.g., egg binding or chronic egg laying).[98] A hypocalcemic syndrome of unknown origin that is associated with decreased fCa and neurological signs is described in African Gray parrots.[221] The calcium metabolism of reptiles is similar to birds. Hypercalcemia due to protein-bound calcium also occurs in female reptiles during folliculogenesis.[33] Detailed assessment of calcium and phosphorus homeostasis and abnormalities are limited in amphibians but include seasonal and life-stage alterations.[75] As with other electrolytes, blood calcium concentrations in fish are influenced by environmental factors and shifts across the gills. Fish lack parathyroid glands but Stanniocalcin produced by the corpuscles of Stannius acts against hypercalcemia.

Preanalytical and Analytical Considerations

Calcium assays are expected to be inaccurate in the presence of anticoagulants that bind Ca (EDTA, citrate). Heparinized plasma may be used with some total calcium assays, but excess heparinization can cause false decreases, especially affecting fCa measurement. Total calcium concentration is influenced by plasma protein concentrations due to the large protein-bound fraction. Therefore, total calcium should always be interpreted in relation to protein concentration, particularly albumin concentration. Adjustment formulas for tCa based on TP or albumin concentrations are reliable only for the circumstances in which they were based (preanalytic and analytical factors)[98,222] and these formulas do not take into account other bound calcium; the calculated adjusted values are "ballpark" estimates, and thoughtful clinical interpretation of protein and calcium is more helpful to assess whether protein variation alone or in combination with other factors reasonably explains hypo/hypercalcemia.

Free calcium (fCa), measured by ion-selective electrodes, provides a direct measurement of the concentration of physiologically active calcium. It is possible to measure fCa on serum or heparinized whole blood or plasma with consideration of sample collection and handling conditions. Heparin is the only anticoagulant suitable for use when performing fCa measurements, but excess heparin risks falsely decreased fCa concentration. The use of commercially available calcium-titrated heparin tubes is therefore recommended. If such tubes are not available, the preparation of collection devices with heparin concentration of <15 U/mL of blood is suggested.[222] Additional preanalytical factors that need to be considered when performing fCa testing are the need for anaerobic handling and separation of serum or plasma from cells as soon as possible (within 1 h after collection) as fCa is sensitive to pH changes: in aerobic conditions, loss of CO_2 to air raises the pH and artifactually decreases fCa; and with delay in separation of plasma or serum from cells, cellular metabolism lowers the pH and artifactually increases fCa. Additionally, there is evidence that silicone serum separator tubes result in mildly artifactually elevated fCa due to the release of calcium and acid from the tube gel.[136]

Assays for **phosphorus** may be affected by in vitro hemolysis or delayed separation due to the release of intracellular P, and also by hyperbilirubinemia which can falsely increase or decrease values depending on methodology.

Iron

Physiology and Interpretation of Changes in Disease

Iron is an essential component of hemoglobin and several enzymes and is also a cofactor for enzymes involved in the citric acid cycle. Free iron causes oxidant damage, and iron in blood (serum or plasma) is bound to the carrier protein apotransferrin, to form transferrin. As transferrin, iron is transported throughout the body but most circulating iron is utilized in erythropoiesis.[102] Most of the iron in blood originates from the breakdown of hemoglobin by macrophages; less than 3% originates from the diet. Iron is stored in tissues (mostly spleen, bone marrow, liver, and enterocytes) as ferritin or hemosiderin; the former is also present in low concentrations in blood. The protein hepcidin plays a key role in iron metabolism, by causing degradation of ferroportin, a transport protein responsible for iron export from cells. This results in decreased iron transport out of enterocytes, hepatocytes, and macrophages, thereby reducing iron availability to other cells like erythroid progenitor cells.[89] Inflammatory cytokines result in an increase in hepcidin and ferritin (positive APPs) and a decrease in transferrin (negative APP) with the result that iron is sequestered and less available for erythropoiesis, resulting in a nonregenerative anemia. Anemia caused by restricted iron availability (either iron deficiency or restriction) takes some weeks to develop and is dependent on the lifespan of the currently circulating, non–iron-deficient erythrocytes.[102] In inflammation (iron-restriction), hepcidin and ferritin will be increased, and serum or plasma iron and transferrin decreased. There may be a microcytic hypochromic nonregenerative anemia. With true iron deficiency, there is also anemia, hypoferremia, and low transferrin, as well as decreases in ferritin and hepcidin.[89]

Iron overload disorders are characterized by hemochromatosis and hemosiderosis which is the accumulation of iron in tissues, particularly the liver, which leads to oxidative cellular damage. Iron overload may be primary (because of a genetic disorder), or secondary, due to excessive iron availability. The latter mechanism is more common in veterinary species; causes include excessive dietary iron bioavailability, repeated blood transfusions, and chronic hemolytic anemia.[102]

Mammals

A decrease in serum iron secondary to inflammation has been reported in several nondomestic mammal species, including white rhinoceros, and manatees, where transferrin saturation was also low.[105,117] Many species of nondomestic mammals have been reported to suffer from iron overload disorder, but this disease has been most well-documented in black and Sumatran rhinoceros, tapirs, lemurs, dolphins, and pinnipeds.[39] Most of these species probably have natural diets with low iron bioavailability and have subsequently not evolved mechanisms that could prevent them from accumulating excess tissue iron because of captive diets with high iron bioavailability.

Nonmammalian Vertebrates

Iron overload disorder occurs in birds and has been reported in captive frugivorous and insectivorous species (mynahs, toucans, birds of paradise) that naturally have a diet low in iron.[45,109,212] Serum iron, total iron binding capacity and transferrin saturation have been used to monitor response to dietary manipulation in Raggia birds of paradise.[109]

Detailed assessments of iron homeostasis and abnormalities in reptiles, amphibians, and fish are limited.

Preanalytical and Analytical Considerations

Iron overload disorder is usually confirmed by histopathological examination of tissues, but in some cases, blood markers of iron metabolism have been measured.[39] Serum or plasma-free iron measurement is widely available and can be determined across taxa but is not specific as increases or decreases occur in many conditions. Sample hemolysis can artefactually increase serum or plasma iron concentration. Ideally, serum or plasma iron results should be compared to reference intervals derived from healthy free-living individuals of a species, but these are often not available.

Transferrin and transferrin saturation are indirectly measured by determining total iron binding capacity, which is a measure of the total amount of iron that can be carried in blood. Total iron binding capacity methods are not species-specific and are carried out in many commercial laboratories. Measurement of serum ferritin is done by ELISA, using species-specific assays. These are not commercially available, and ferritin has only been measured in very few nondomestic mammals (lemurs, black and white rhinoceroses, northern fur seals, and tapirs).[39]

Glucose

Physiology and Interpretation of Changes in Disease

Glucose is regulated by multiple factors including intake, insulin, and "diabetogenic" (insulin-antagonistic) hormones including glucagon, catecholamines, glucocorticoids, and growth hormone. Insulin promotes cellular glucose utilization and storage. Diabetogenic hormones variably stimulate gluconeogenesis (e.g., steroids, glucagon) or glycogenolysis (e.g., catecholamines, glucagon), interfere with cellular glucose uptake (e.g., growth hormone, steroids), or decrease insulin release and cellular glucose utilization (e.g., catecholamines). Physiology of glucose homeostasis is similar across taxa, although the concentrations of circulating components vary.

Common considerations for altered blood glucose concentration are listed in Box 4.e9. Many cases of hyper- or hypoglycemia are multifactorial and have a combination of underlying mechanisms. For example, in sepsis, hypoglycemia may develop through increased host or organism cellular glucose utilization, hypoperfusion-associated shifts to anaerobic glycolysis, diminished hepatic gluconeogenesis, and other factors. While insulin acts to decreased glucose concentration, hormones that shift toward hyperglycemia include growth hormone, glucagon, glucocorticoids, catecholamines, stomatostatin, and progesterone (diabetogenic hormones). In a series of gastric neuroendocrine carcinomas positive for stomatostatin by immunohistochemistry, severe hyperglycemia was reported in 2 out of 5 cases.[194] Animals not adapted to handling, including wildlife, commonly show significant hematological effects of stress and excitement, which can include hyperglycemia.

Some species may naturally enter states of starvation during periods such as winter, migration, and egg incubation. During starvation, triglycerides are a major source of energy, and the sequence of progression is typically utilization of ingested food with glycogen storage followed by metabolism of triglycerides and lastly by catabolism of proteins for gluconeogenesis. During extreme starvation, protein metabolism may result in increased circulating nitrogenous compounds (e.g., uric acid in birds).

Prolonged hyperglycemia or hypoglycemia may be further documented by assessment of **fructosamine** and **glycated hemoglobin**, which are complexes of glucose to proteins and hemoglobin, respectively. The half-life of these complexes is weeks for proteins and up to months for hemoglobin, but this duration varies between species and is altered depending on pathological states affecting protein or erythrocyte concentrations and half-lives.

Mammals

The general principles discussed above apply to mammals.

Nonmammalian Vertebrates

Amphibians and reptiles tend to have low blood glucose concentration in comparison to birds and mammals. In birds, plasma insulin concentrations are generally lower than mammals, while glucagon concentrations are higher and play a larger role in avian physiology. Birds vary in their dependence on glucagon (granivorous birds) or insulin (carnivorous birds) and diabetes mellitus in birds may be due to glucagon excess, insulin deficiency/resistance, or a combination of factors.[148] The control of glucose in reptiles appears primarily regulated by insulin, to which environmental and seasonal influences play a larger role than in other taxa.[220]

Some over-wintering reptiles and amphibians develop physiological hyperglycemia by the accumulation of low-molecular-weight carbohydrates including glucose. This can be documented in blood or as glucosuria during hibernation or soon after thawing.[75,242] There is little information concerning glucose metabolism in fish.

Preanalytical and Analytical Considerations

In general, serum or plasma used for **glucose** analysis should be separated from cells soon after collection to avoid artifactual decreases occurring from cellular glycolysis. The rate of cellular glucose utilization varies between species and cellular conditions, including temperature and leukocyte concentrations. Although avian erythrocytes utilize very little glucose in lieu of other substrates, the effects of delayed separation can still be identified.[100] The erythrocytes of pigs have notably low glucose utilization due to the absence of functional glucose transporters on mature adult erythrocytes (present only on reticulocytes and neonatal erythrocytes). Instead, erythrocytes from adult pigs employ inosine as a major energy substrate.[103] Since sample types used for glucose measurement often vary (whole blood, plasma, or serum) and some analyzers calculate plasma glucose concentration even if whole blood is used as the sample, it is important to investigate and understand the methodology used and to report results as whole blood, plasma, or serum glucose concentration as appropriate. Hemolysis, icterus, and lipemia can interfere with some assays for glycated hemoglobin and fructosamine and yield erroneous results.

Lipids

Physiology and Interpretation of Changes in Disease

Although body lipids include a wide variety of forms (sterols, fatty acids, phospholipids, etc.), the most frequently assessed components are **cholesterol** and **triglyceride** concentrations in blood. Both cholesterol and triglycerides are commonly transported in blood within lipoprotein complexes, and their concentrations are generally influenced by increased or decreased production, increased or decreased utilization (cellular uptake, lipolysis), and alterations in clearance.

Mammals

Hyperlipidemia (including hypercholesterolemia and hypertriglyceridemia) occurs postprandially with increased production (physiological or high-fat diets), as well as with decreased utilization (e.g., hypothyroidism, lipase deficiencies) and decreased clearance (e.g., obstructive cholestasis). Some conditions, such as hypercholesterolemia (and sometimes hypertriglyceridemia) in nephrotic syndrome or hypertriglyceridemia (and sometimes hypercholesterolemia) in diabetes mellitus, likely reflect multiple concurrent mechanisms. Hypocholesterolemia is less common and not of great clinical significance but may occur with enteropathies, malnutrition, or decreased functional hepatic mass (e.g., liver failure, portosystemic shunting).

Nonmammalian Vertebrates

Excessive accumulation of fat in the liver (hepatic steatosis) may occur in birds fed high carbohydrate diets, especially if there are concurrent protein or vitamin deficiencies. In captivity, some reptile and amphibian species are prone to hypercholesterolemia and hypertriglyceridemia in conjunction with obesity and corneal lipidosis.[75] Fish are generally hyperlipidemic and hypercholesterolemic in health compared to mammals, although concentrations are influenced by physiological and environmental factors.[90]

Acid-Base Status

Physiology and Interpretation of Changes in Disease

Acid-base status is maintained mainly by respiratory and metabolic (especially renal) mechanisms in most terrestrial vertebrates, and additionally by the gills and environmental factors in fish. Acid-base status has key effects on blood buffering, electrolyte concentrations, and the physiological function of many body systems.

The respiratory effect on acid-base status is fairly immediate, whereas the renal response may take several days. Respiratory effects on acid-base status reflect ventilation, with hypocapnia resulting from increased alveolar ventilation (hyperventilation) causing respiratory alkalosis, and hypercapnia from decreased alveolar ventilation (hypoventilation) causing respiratory acidosis. A variety of underlying mechanical and functional, intra- and extrapulmonary defects may influence ventilation, including nervous system disease, drugs/medications, myopathies, airway lesions, body cavity disease, and lesions of the gas-exchange surfaces of the lung.

Metabolic effects on acid-base status result from increasing hydrogen ion (H^+) or decreasing bicarbonate ion (HCO_3^-) content in the blood. Increased H^+ may result from excess generation of H^+ (lactic acidosis and ketoacidosis) and/or decreased H^+ excretion (renal failure). HCO_3^- may be lost through the alimentary (e.g., diarrhea) and renal systems. Although acid-base measurement is less commonly utilized in nondomestic species and a complete discussion of pathophysiological mechanisms is beyond the scope of this chapter, an understanding of acid-base shifts and effects on other analytes (i.e., electrolytes, as previously discussed) is recommended to arrive at well-rounded conclusions.

Mammals

The general principles discussed above apply to mammals.

Nonmammalian Vertebrates

Although reference intervals are available for birds, reptiles, and fish, understanding of pathophysiological mechanism is limited to date.

Preanalytical and analytical considerations

Blood gas instruments measure hydrogen ion and the partial pressures of oxygen and carbon dioxide and calculate other values, including bicarbonate ion concentration, base excess, and percent oxygen saturation. These instruments are often designed for human use and include assumptions about body temperature (typically run at 37°C) and calculation variables that do not hold true across species. Correction of blood gas values for body temperature is controversial. Especially in poikilotherms, correction can aid in understanding in vivo acid-base status and is considered best practice in aquatic reptiles.[123] Serial monitoring of values in nonpoikilotherm species reported at 37°C may aid in understanding changes in blood gas values over time without confounding influences of temperature correction.[222] Other analytic factors that significantly influence blood gas variables include delay in analysis, air exposure, and excess heparinization. Immediate analysis with prevention of air exposure is recommended. The net result of exposure to air or excess heparinization yields artifactually increased pH, decreased partial pressure of carbon dioxide, and decreased bicarbonate ion.[119,222] To accurately assess the oxygenation of blood, arterial blood must be sampled.

URINALYSIS

General and pre-/analytical considerations are presented in the online content.

Urine sediment evaluation includes a qualitative and semiquantitative review of epithelial cells, crystals, leukocytes, erythrocytes, casts, and microorganisms following routine techniques of evaluation and general principles of interpretation.[177,222] Sediment evaluation can provide valuable evidence for the diagnosis of blood loss, inflammation, infections (e.g., bacteria, fungi, nematodes), and metabolic disease. Inflammation is the most frequently diagnosed condition of the urinary tract across taxa of nondomestic species. In addition to routine wet mount preparations, dry mount cytology of the urine sediment can be used to provide further diagnostic benefit of cellular elements, such as for the detection of bacteriuria and for detailed evaluation of leukocytes and epithelial cells.[229,246] Atypical cytological features in urine epithelial cells need to be cautiously interpreted due to the potential for nonneoplastic reactivity as well as morphological changes related to deterioration and/or urine scalding. For example, cellular atypia in the presence of increased leukocytes may reflect a reactive tissue response to an inflammatory process, but in other cases may reflect inflammation in conjunction with neoplasia. Review by experienced cytologists is recommended and correlation to clinical information and other ancillary diagnostics (e.g., imaging, histology) may be necessary for a definitive diagnosis.

Mammals

Although urine is typically clear to minimally cloudy in health across taxa, cloudy urine may be normal in some species, for example, in Asian elephants as well as horses and other equids, rabbits, and guinea pigs due to normally occurring calcium carbonate crystals [184,249]. Rabbits and possibly other lagomorphs may have brown or red urine due to the presence of urochromes, which are pigments of dietary or endogenous origin.[151]

Notably, for urine chemistry analysis, a trace amount of urine protein may be observed in most mammalian species, especially in voided or catheterized samples or in concentrated urine with specific gravity >1.035. However, a greater amount of protein is reported in the urine of small exotic mammals (e.g., chinchilla, rat, ferret).[66,72] This may be associated with increased protein from sampling (i.e., voided or catheterized samples) or urine alkalinity in many small mammals. In chinchillas, false positive high urine protein dipstick results were associated with high urine specific gravity and quantitative protein analysis by chemistry analyzer is recommended.[66] Urine generally tests negative for bilirubin other than male dogs (and likely related species) and ferrets with urine specific gravity >1.025; positive results indicate liver disease, posthepatic cholestasis, or hemolytic disease[121]). Mild bilirubinuria may be seen in horses with anorexia. Positive dipstick results can be confirmed with the Ictotest method.[222] Urine from healthy animals should be negative for glucose, ketones, and hemoprotein; considerations for positive urine dipstick and concurrent hyperglycemia include stress or diabetes mellitus. Positive urine ketones indicate starvation, fasting, or diabetes mellitus. The reagent strip for hemoprotein detects hemoglobin, intact red blood cells, and myoglobin; thus, positive results need to be interpreted in context of urine collection technique, urine specific gravity, pH, serum or plasma muscle enzyme activities, sediment findings, and the patient's hematocrit and plasma colour (e.g., few red blood cells are normal in urine depending on collection technique but frequent red blood cells suggest hematuria; concurrent red/pink plasma in a sample with positive hemoprotein indicates hemoglobinuria). The ammonium sulfate precipitation test can help in differentiating myoglobin and hemoglobin.[222]

It is important to be aware of normal urine sediment findings in certain species, especially crystals or incidental findings that may be observed in healthy animals. For example, elephants and rhinoceros, similar to horses, rabbits, and guinea pigs, typically have variable numbers of calcium carbonate crystals (Fig. 4.15A); elephants may also have magnesium ammonium phosphate, calcium oxalate dihydrate, amorphous phosphate, and infrequently mucus.[94,177,249] Calcium oxalate dihydrate crystals can be seen in healthy animals or with prolonged urine storage (Fig. 4.15B). Spermatozoa are a normal finding in male animals (Fig. 4.15B). Low numbers of magnesium ammonium phosphate crystals are often observed in carnivores in health, but increased numbers may be associated with urease-producing bacteria. Since manatee urine is collected voided, it often includes diatoms and rhabditoid nematode larvae as incidental findings from skin contamination (Figs. 4.15C–D). Amorphous crystals tend to be phosphates in alkaline or urates in acidic urine and can be seen in healthy mammals (Figs. 4.15A, C).

Nonmammalian Vertebrates

The urate portion of bird and reptile urine is typically white or cream-colored, accompanied by variable amounts of mucus and/or fluid. Green discoloration of urates indicates excretion of biliverdin and is abnormal as it suggests cholestasis.[98]

Since the renal concentration ability is limited in birds or lacking in reptiles and amphibians, specific gravity cannot be used diagnostically in these species.[75,98,110]

In birds, urine pH varies from acidic to alkaline, and varies with reproductive activity, phases of hypoxia during diving in birds, and other factors.[225] Extrinsic factors influencing urine pH have been reported in tortoises (e.g., season, hibernation). Acidic urine in herbivorous chelonians may be associated with metabolic acidosis from chronic starvation.[84]

In nonmammalian species with a cloaca, the urinary, digestive, and reproductive tracts are in close approximation and cytological findings need to be interpreted while

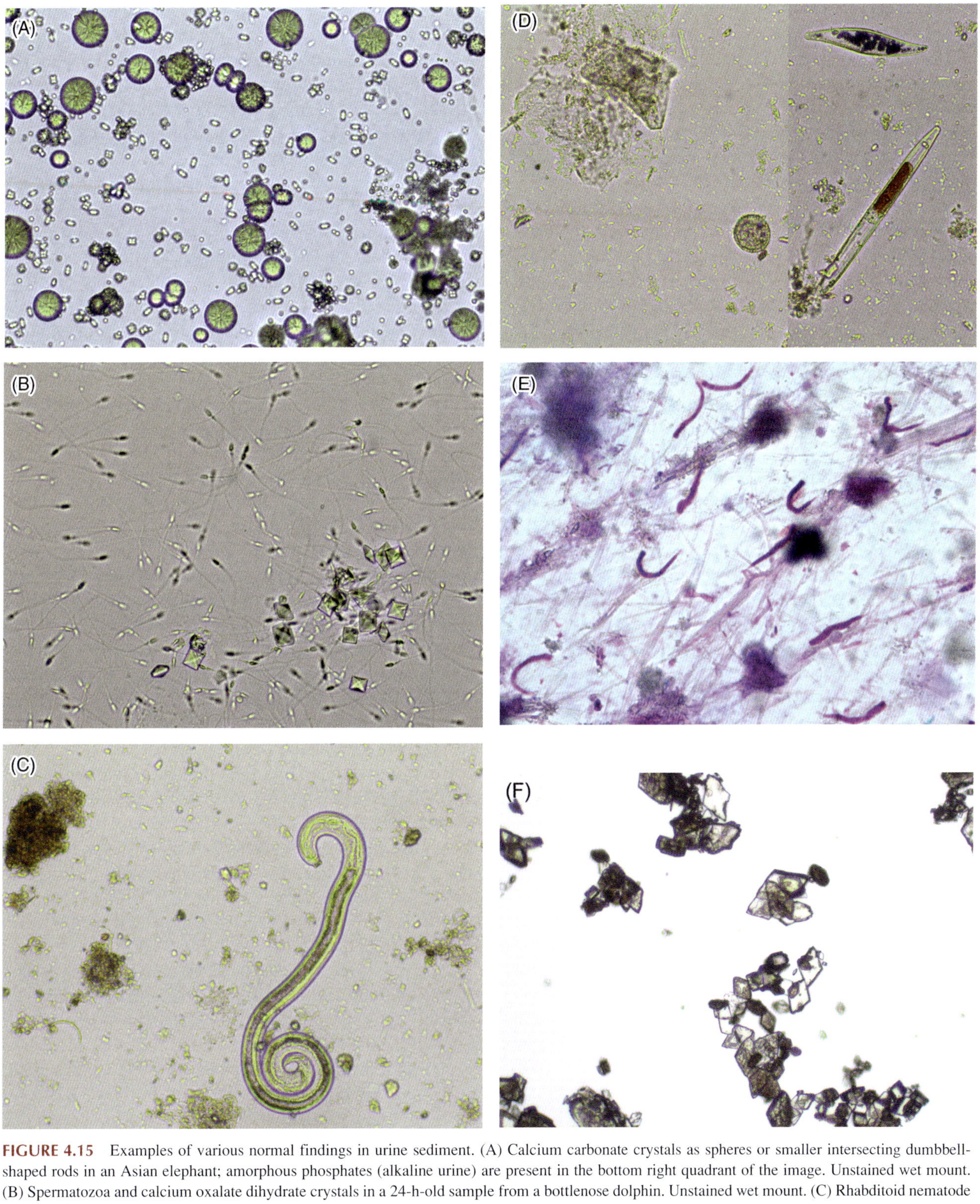

FIGURE 4.15 Examples of various normal findings in urine sediment. (A) Calcium carbonate crystals as spheres or smaller intersecting dumbbell-shaped rods in an Asian elephant; amorphous phosphates (alkaline urine) are present in the bottom right quadrant of the image. Unstained wet mount. (B) Spermatozoa and calcium oxalate dihydrate crystals in a 24-h-old sample from a bottlenose dolphin. Unstained wet mount. (C) Rhabditoid nematode larva and amorphous phosphates (alkaline urine) in manatee urine. Unstained wet mount. (D) Image composite showing three pleomorphic diatoms (*round, diamond-shaped, elongated*) in manatee urine taken at same magnification. Squamous epithelial cells in upper left corner. Unstained wet mount. (E) Needle-like urate crystals and spermatozoa in desert tortoise urine. Modified Giemsa-stained direct smear. (F) Uric acid crystals in a green turtle with acidic urine recently fed a high protein fish diet. Unstained wet mount.

keeping this (potentially confounding) anatomy in mind. In these species, mixed colonic flora, digestive tract parasites, and diatoms in aquatic species are observed in voided samples more commonly than in mammals. Urates are commonly present in birds and reptiles and may be observed as diamond-shaped or needle-like crystals (Figs. 4.15E–F). Amorphous crystals can also be present in healthy nonmammals; however, whether these are phosphates versus urates or other chemical composition has not been determined. The importance in recognition of amorphous crystals lies in differentiation from bacteria. Urine cytology may also show evidence of active reproductivity (i.e., presence of spermatozoa) (Fig. 4.15E).

COMPARATIVE CYTOLOGY

Physiology and Interpretation of Changes in Disease

The principles of cytology sampling and evaluation are consistent across taxa. However, unique organism physiology, anatomy, and disease processes need to be considered. A comprehensive cytological evaluation by a trained specialist (veterinary clinical pathologist) typically includes evaluation and categorization of lesions into pathological processes (i.e., inflammation, neoplasia); identification of potential causative factors (e.g., infectious agents, foreign material); interpretation of cumulative findings (i.e., synthesis into a cytological diagnosis); and provision of useful information to contextualize findings amid the current state of scientific knowledge and recommendations for additional diagnostic tests. Cytology requires minimal equipment, minimal time and effort for sample collection and processing, and provides near real-time diagnostic information. It can be used as a stand-alone test or in combination with histology and other ancillary diagnostics to formulate a definitive diagnosis. As cytological preparations are relatively easy to make and are inexpensive, they are readily accessible to most investigators and clinicians. Based on expertise and situational factors (i.e., hospital or field, patient diagnosis or research), the goals of cytology for nonspecialist clinicians or investigators can range from confirmation of basic sample quality prior to diagnostic laboratory submission (i.e., cellular or acellular, cells intact or lysed), to categorization of main pathological process (i.e., inflammatory or noninflammatory, screening for infectious agents), guiding empiric treatment decisions, or even complete definitive diagnosis and interpretation. Cytology adds value at each of these levels when judiciously performed with recognition of limitations and the "art" that comes with expertise and experience; when possible, collaboration with a skilled clinical pathologist typically yields the most information and diagnostic insight.

Assessment of sample quality is a fundamental first step to achieve useful cytological information. This includes assessment of sample yield on the cytological preparation (i.e., sample is present on the slide), evaluation of whether the sample is likely to be representative of the intended lesion (i.e., not simply contamination), and screening for sufficient intact cell preservation and adequate staining to achieve diagnostic insight. Samples that are entirely acellular consist of material other than the targeted sample (e.g., only surface squamous epithelial cells, glove talc, ultrasound lube, or blood rather than lesion cells), and/or are composed of only lysed cells are unlikely to be diagnostically representative and may warrant re-sampling. Understained samples can be restained as needed. The question of representative sampling is especially important in antemortem fine-needle aspirates in which the sampling may be a relatively "blind stick" (in comparison to postmortem sampling, during which the tissue of origin can be more definitively isolated). The whole slide should be scanned to get an impression of overall cell distribution. Systematic approaches to cytological evaluation are described in various resources.[186,237] Expected cytologic findings in health for selected tissues are summarized in Table 4.4.

The main purpose of cytological evaluation is to categorize a lesion into one or more major pathological processes (Tables 4.e2 and 4.e3).[21,58,179,190,234] Common lesions include inflammation (± identification of infectious agents) or tissue proliferation (hyperplasia, neoplasia). Mixed samples, consisting of both inflammatory and tissue cells, often require nuanced assessment, as the cytological features of inflammation with reactive cellular changes (atypia) and cellular proliferation (e.g., neoplasia) with associated inflammation can overlap. Other core categories of pathological processes identified on cytology include squamous cyst formation (e.g., follicular cysts or follicular neoplasms), necrosis, and fluid accumulations (Tables 4.e2 and 4.e3).[178,186]

Mammals

Cytology is applied widely in mammals for both inflammatory and neoplastic disease processes, and much of our general understanding of and recommendations for cytology (as above) are derived from mammals. Notably, acute inflammation in mammals is often neutrophilic, myeloperoxidase-dependent, and pus forming. Pigs and cattle have comparatively low myeloperoxidase concentrations and thus have less capability to liquefy inflammatory exudates.[1,160]

Nonmammalian Vertebrates

Cytology is increasingly being applied in nonmammalian vertebrates as a relatively quick and low-cost diagnostic method. In these taxa, inflammatory lesions are more

TABLE 4.4 **A Condensed Guide to Expected Findings in Health for Selected Tissues Commonly Sampled for Cytology**

Organ/Origin	Cytologic Findings	Comments
Skin/Shell/ Adnexa	Well-differentiated squamous epithelial cells; possible variable numbers of mixed bacteria (superficial flora/ contaminants). Also possible in skin: basilar epithelium, sebaceous epithelial cells, follicular components, and melanocytes.	Diatoms are a common incidental finding in aquatic species.
Respiratory tract	Upper (nasal cavity and nasopharynx): Well-differentiated squamous epithelial cells and/or ciliated columnar epithelial cells, possible rare inflammatory cells consistent with resident leukocytes. Lower (trachea, lung): Ciliated columnar respiratory epithelial cells, goblet cells, small amount of mucus, rare resident macrophages.	Fungal structures suggestive of environmental contaminants in herbivores, for example, *Alternaria* sp. Commensals include bacteria with *Simonsiella*-like morphology across taxa Diatoms are a common incidental finding in aquatic species. Ciliates, flagellates, and few yeasts, in the absence of inflammation, are considered normal findings in upper respiratory samples from cetaceans.
Upper gastrointestinal tract	Oral cavity/esophagus/nonglandular stomach: well-differentiated squamous epithelium. Glandular stomach: gastric columnar mucosal epithelium (may exfoliate as honeycomb-appearing sheets) ± Mixed bacilli (normal flora), ± small amount of mucus, ± food particle fragments.	Oropharyngeal commensals include bacteria with *Simonsiella*-like morphology across taxa.[42] Esophageal cytology across taxa is not completely elucidated; based on histologic composition, ciliated columnar epithelium may be seen (e.g., reptiles, amphibians).
Fecal	Frequent mixed bacilli, possible rare cocci and spore-forming bacteria; few yeast or protozoa (e.g., *Balantidium, Blastocystis* sp.); digesta (cellulose or amorphous material); well-differentiated squamous epithelial cells; columnar colonic epithelial cells.	Diatoms are a common incidental finding in aquatic species. Melanin pigment granules and pale-yellow long needle-like crystals of probable dietary origin can be seen in marine mammals. Urinary crystals may be seen (e.g., urates common in birds and reptiles). Normal flora yeast in low numbers may include *Cyniclomyces guttulatus* (rodents, lagomorphs, canids).
Lymphoid organs	Predominance of small well-differentiated lymphocytes (often >80%–90%) with low numbers of intermediate and large lymphocytes; few plasma cells, macrophages, or other leukocytes may be present (typically <3%–5%).	Spleen may also include extramedullary hematopoiesis and reticuloendothelial stroma; rare mast cells, and hemosiderin-laden macrophages. Thymus may also include mast cells and few thymic epithelial cells.
Liver	Hepatocytes (low number may be binucleated in health); occasionally few biliary epithelial cells.	Considerations for hepatocellular-associated pigment include hemosiderin, lipofuscin, and bile.

Sources: Campbell TW, Grant KR. Exotic Animal Hematology and Cytology. 5th ed. Hoboken, NJ, USA: John Wiley & Sons, Inc.; 2022; Raskin RE, Meyer DJ, Boes KM. Canine and Feline Cytopathology: A Color Atlas and Interpretation Guide. 4th edition. St. Louis, MI: Saunders Elsevier; 2022.

frequently diagnosed compared to neoplasia. Features of inflammatory responses, including composition, duration, and degree of inflammation, vary across species and are driven by factors including inflammatory cell types, cellular functions, and body thermoregulation strategies. The type of cellular inflammatory response and concentration of myeloperoxidase and hydrolytic enzymes affect the occurrence of tissue liquefaction and pus formation versus granuloma formation.[1] Granulocytes with low or that lack myeloperoxidase cannot produce liquefaction, and inflammation results in caseous granuloma formation.[217]

Similar to mammals, neutrophils in amphibians and fish contain variable amounts of myeloperoxidase and their antimicrobial function is mainly employed through oxygen-dependent mechanisms.[28,160,187,230] In particular, the inflammatory responses of fish mimic mammals in that neutrophils are the first WBC type to appear (within 10h in mammals and up to 24h in fish), followed by macrophages (within 12–48h in mammals and within 24h in fish).[1,227] The difference in timing of leukocyte tissue migration in mammals and fish may be associated with differences in body temperature. Heterophils of some species (e.g., rabbits, guinea pigs, some birds, and reptiles) lack or have low concentrations of myeloperoxidase and appear to depend mainly upon oxygen-independent mechanisms such as lysozyme for microbial killing.[160,182,187] Exceptions include chicken and turkey heterophils, which have bactericidal activities comparable to mammalian neutrophils.[160]

The detailed impact of species-specific differences in leukocyte granule contents and host defense mechanisms are not fully understood to date. Common types of inflammation and related etiologies across taxa are briefly presented in Table 4.e2.

Cytology of neoplasms in nondomestic species follows diagnostic and classification algorithms as in mammals, as discussed above and summarized in Table 4.e2.

APPLICATION OF CLINICAL PATHOLOGY TESTING DURING POSTMORTEM EXAMINATIONS

Clinicopathological sampling in postmortem examinations offers benefits similar to those discussed above, including ease of collection and potential for rapid diagnostic information. In addition to typical antemortem considerations of species differences, sample handling, methodology, and other sources of error, postmortem clinical pathology testing (e.g., hematology, biochemistry, cytology, and urinalysis) must also take into account peri- and postmortem factors, including antemortem therapeutic interventions (e.g., fluids, resuscitation attempts), time since death, and tissue preservation conditions. Conclusions must be reached prudently amid the complexity of physiological, circumstantial, and environmental factors. Any available data from the animal in health and/or before death should be considered in the overall evaluation of all clinical and pathological findings. As with antemortem samples, those collected during postmortem examinations should be handled and processed appropriately and in a timely manner to prevent additional artifactual changes. Examples of potential informative postmortem findings include detection of hemoparasites or other infectious agents, azotemia, or malignancy.

Postmortem Hematology

Blood collection is difficult from the peripheral veins of carcasses due to clotting, blood settling, and loss of blood pressure. In fresh carcasses, heart blood can be collected via a needle inserted into the heart or by lacerating a major artery (e.g., axillary, femoral) and collecting the pooled blood within the associated tissue pocket. Caution is warranted when assessing routine CBC measurands (e.g., PCV, Hgb, WBC, etc.) as these are unlikely to reflect antemortem results due to rapid postmortem effects, including cell settling and lysis. However, in conjunction with other antemortem and necropsy findings, a blood film (whole blood and/or buffy coat) may reveal qualitative morphological changes soon after death (<6–12h,[126]) or the presence of parasitemia/other infectious agents, which can be detectable for up to several days by light microscopy. Postmortem whole blood samples can also be used for molecular diagnostics, even in the presence of hemolysis. In postmortem samples of all three cell lines, erythrocytes appear to be the most resistant to autolysis. However, despite the presence of autolytic changes in leukocytes (e.g., vacuolation, cell lysis), platelets, or thrombocytes, morphological identification of infectious agents in a postmortem blood film may still provide useful diagnostic information.

Postmortem Biochemistry

Postmortem samples offer a proxy for antemortem results to provide evidence in the cumulative diagnosis of metabolic diseases and intoxications. However, lacking species-specific and postmortem-related validated methods, detailed knowledge of postmortem changes and factors affecting the rate and degree of such changes, and applicable control data, interpretations/conclusions must be reached with caution.

Death causes a multitude of physiological consequences in biological organisms, including rapid and marked metabolic and structural alterations. Some effects, such as cessation of tissue oxygenation, changes in enzymatic reactions, cellular degeneration, and loss of blood buffering systems, result in changes that are detectable by biochemical testing.[63] A notable macroscopically apparent change in delayed postmortem blood samples is the development of hemolysis, characterized by red-discolored plasma and serum. Depending on the analyte, methodology, and degree of hemolysis, released erythrocyte components (e.g., potassium) and/or colour interference can result in minor to diagnostically significant artifactual changes in analyte values. Other identified postmortem biochemical changes in humans and animals (e.g., rats, pigs, deer, dogs) include diminished blood and/or fluid (CSF or intraocular) pH, glucose, sodium, and chloride concentrations, and increases in blood and/or fluid lactate, formate, hypoxanthine, uric acid, phosphorus, ammonia, magnesium, potassium, and reduced-form nicotinamide adenine dinucleotide (NADH) concentrations.[29,63,183,204–206]

In both human and animal species samples investigated for utility in postmortem chemistry or toxicology have included blood, urine, cerebrospinal fluid, aqueous humor, vitreous humor, gastric contents, and a wide variety of solid body tissues. Ocular fluids have increased in popularity for postmortem assessment as they represent filtrates of the blood, are located in the relatively protected ocular environment, and are fairly easy to collect. Vitreous humor has proven more stable than aqueous humor and less susceptible to autolytic change, but the viscous consistency of vitreal fluid can require additional preanalytical steps such as centrifugation or heating that may also slightly alter analytic results.[19,24,29,233] Contamination of vitreal fluid with retinal cells has also been recognized in humans as a potential source of falsely increased potassium concentration.[19] In comparison to ocular fluids, uncontaminated CSF samples are generally more difficult to obtain. As nonblood samples for biochemical analyses are less common in routine laboratory analysis, standardized and validated methods are generally not available. Furthermore, each sample type (matrix) has the potential for specific postmortem, preanalytical, and analytical effects, further complicating interpretation.

Studies of postmortem biochemical testing have long focused on the determination of postmortem interval (PMI). Although limited support has been garnered for such assessment, particularly for the use of potassium concentration assayed in vitreous humor,[228] the compiled human and animal findings indicate variation and inconsistency in rate and degree of changes among species, sample types, and analytes. Additionally, differences in investigative methodology, incomplete assessment of circumstantial and environmental factors (e.g., temperature, humidity, body composition, underlying disease, bacterial composition, extent of putrefaction), and issues of practicality limit confidence in routine biochemical markers as reliable indicators of PMI. Less common analytes, such as hypoxanthine, ammonia, and NADH, have more recently shown a positive linear association with PMI unaffected by putrefaction, and such tests may hold promise for PMI determination in the future.[29,63] Overall, however, conclusive interpretative guidelines for specific analytes, species, and postmortem situations have not been reliably established, and there is a trend to move away from one-dimensional attempts to calculate the time of death from postmortem chemistry.[19,29]

Postmortem biochemistry on blood or body fluids has also been evaluated as a surrogate for the diagnosis of mortality-associated metabolic disease. Such reports have included identification of ketone bodies in blood, ocular fluids, CSF, or urine of humans with alcohol- or diabetes-related ketoacidosis; evaluation of cardiac troponin I in blood of humans with acute myocardial injury or electrocution; evidence of hypomagnesemic tetany by vitreous and/or aqueous humor testing as a cause of sudden death in cattle and sheep; and assessment of azotemia using aqueous humor in humans, California sea lions, cattle, dogs, and cats.[19,96,153,202] Notably, the tested analytes may be sensitive to factors such as environmental conditions, degree of analyte elevation, and postmortem delay in sample acquisition: in horses, the variability of vitreous humor UN and creatinine concentrations increased with progressive elevations in temperature and time; and in cats and cattle, postmortem serum and aqueous humor concentrations of UN varied as a function of UN concentration itself.[96,154] Additionally, methods are often unvalidated in a specific sample type or species. In both animals and man, results of postmortem chemistry are limited and sometimes conflicting, and further work is needed to evaluate preanalytic and analytic factors and to more completely assess the use of postmortem samples for the diagnostic interpretation of metabolic derangements. As in antemortem cases, proper interpretation of findings must take into account all available evidence.

Postmortem Cytology

The collection of postmortem cytological samples is guided by species and gross postmortem findings. Taxa-specific sampling guides are available[34,83,210,253] As with antemortem cytology, postmortem sampling can help guide ancillary diagnostics or provide an early identification of disease processes prior to processing for histopathology. Examples of postmortem cytology and correlative histology are provided in Figs. 4.16–4.25.

As with antemortem testing, the purpose of postmortem cytological sampling is to obtain a well-spread, representative sample of the organ or lesion. The best sampling technique depends on factors, including the lesion type, organ, and species. While fine needle biopsy or aspirate techniques are commonly used in live animals, tissue preparations, such as imprints, squash preparations, and smears, are more commonly employed in postmortem samples. **Tissue imprints** are made by obtaining a small fragment of the desired tissue (<0.5 cm diameter), gently blotting that sample on absorbent material to remove excess tissue fluid and blood, and gently touching the sample in serial successive locations along a glass slide to form imprints that can be air-dried, methanol-fixed, and stained. Blotting is an important step to diminish the presence of excess blood that might interfere with assessment. A gentle touch is needed to reduce artifactual cell lysis.

For softer, semisolid, or liquid samples, **squash preparations** or squash-smear preparations may also be useful and are created by placing a small sample (i.e., drop) on one slide, gently placing a second slide or coverslip on top of the first, and either leaving it in place (squash) or gently pulling the slides apart horizontally (smear). The slides may be placed in either longitudinally parallel or perpendicular positions relative to one another for the smear. The "squashing" of the sample between the two slides should be relatively gentle (only as much pressure as needed to sufficiently spread the sample) as excess pressure, particularly with the smear technique, may contribute to cell lysis. The goal is to obtain a thin layer (i.e., monolayer) of cellular material. Swabs, brushes, or edged instruments (e.g., blades, spatulas) can also be used to transfer material from organs or lesions (e.g., bone marrow or plaque lesions) to a slide: the swab or brush is gently rolled or "painted" along the slide; the edged instrument is gently touched to the slide surface. If necessary, a secondary technique of squash or smear can be employed to better spread the transferred material.

Although postmortem bone marrow cytology may be worth a sampling attempt in fresh carcasses, cell

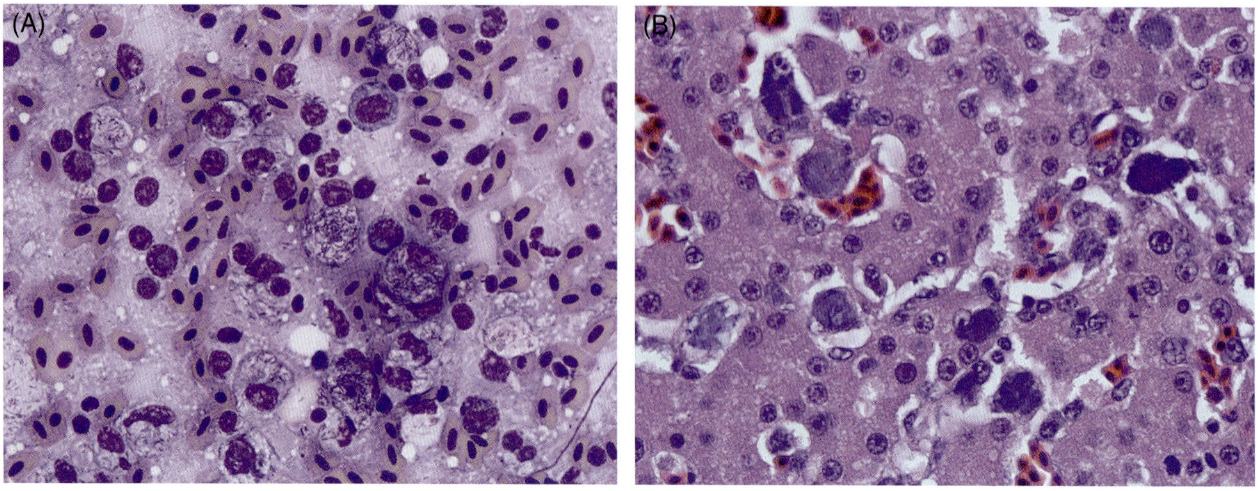

FIGURE 4.16 Cytological (A) and histological (B) correlate: *Erysipelothrix rhusiopathiae* infection in the liver of a pied avocet with mild liver enlargement. (A) Histiocytes with abundant intracellular slender rod-shaped bacteria in liver tissue imprint. Modified WG. (B) Bacteria-laden histiocytes within hepatic sinusoids. Bacterial morphology is histologically indistinct. Hematoxylin & Eosin (H&E). Culture identified *Erysipelothrix rhusiopathiae.*

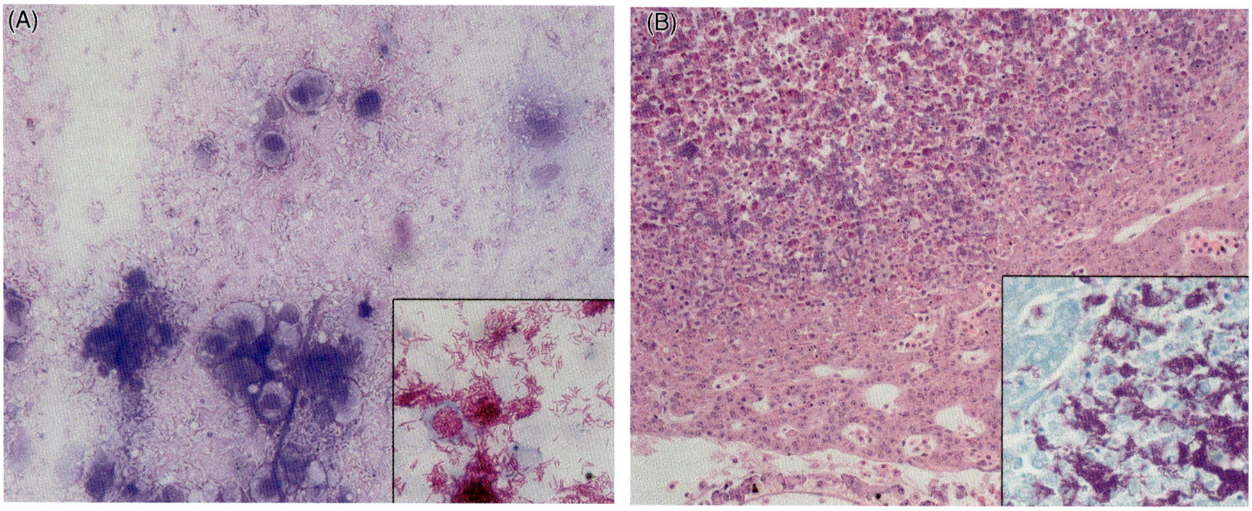

FIGURE 4.17 Cytological (A) and histological (B) correlate: mycobacteriosis in a northern seahorse. (A) Abundant extracellular and lesser histiocyte-phagocytized nonstaining rod-shaped bacteria typical of mycobacteriosis in a coelomic nodule. Modified WG. Positive acid-fast staining (inset). Ziehl-Neelsen acid fast. (B) Necrotizing hepatitis with intralesional bacteria (H&E) that are acid-fast stain (inset). Ziehl-Neelsen acid fast. *(Photo Courtesy of A. Newton, Wildlife Conservation Society.)*

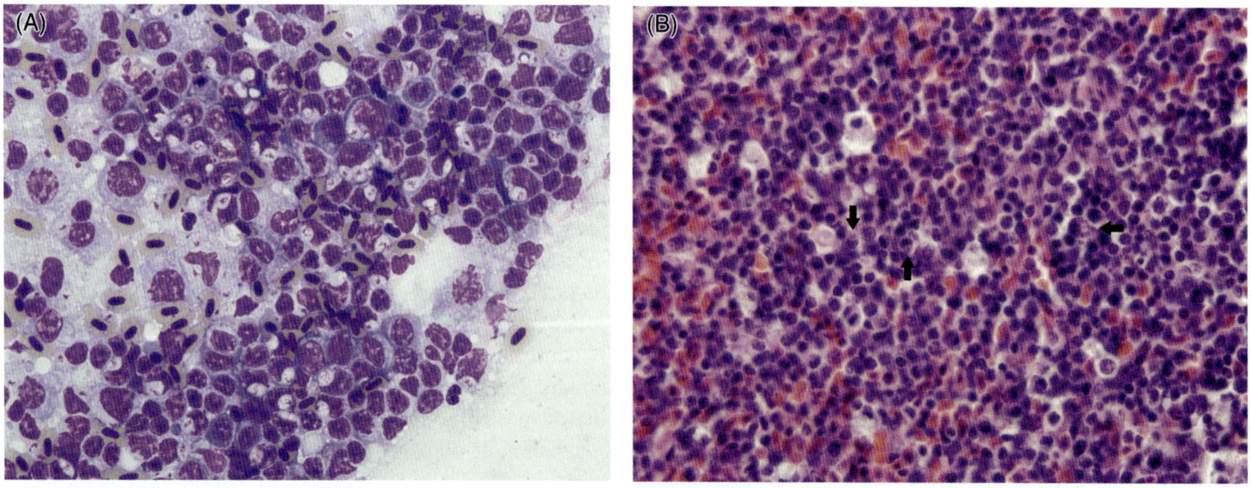

FIGURE 4.18 Cytological (A) and histological (B) correlate: systemic isosporosis (atoxoplasmosis) in a house sparrow with moderate splenic enlargement. (A) Individual and multiple protozoal trophozoites within moderate numbers of mononuclear cells (predominantly lymphocytes) in a splenic tissue imprint. Leukocytes show mild postmortem deterioration. Modified WG. (B) Protozoal organisms within scattered mononuclear leukocytes. H&E. Molecular testing confirmed detection of *Isospora* sp.

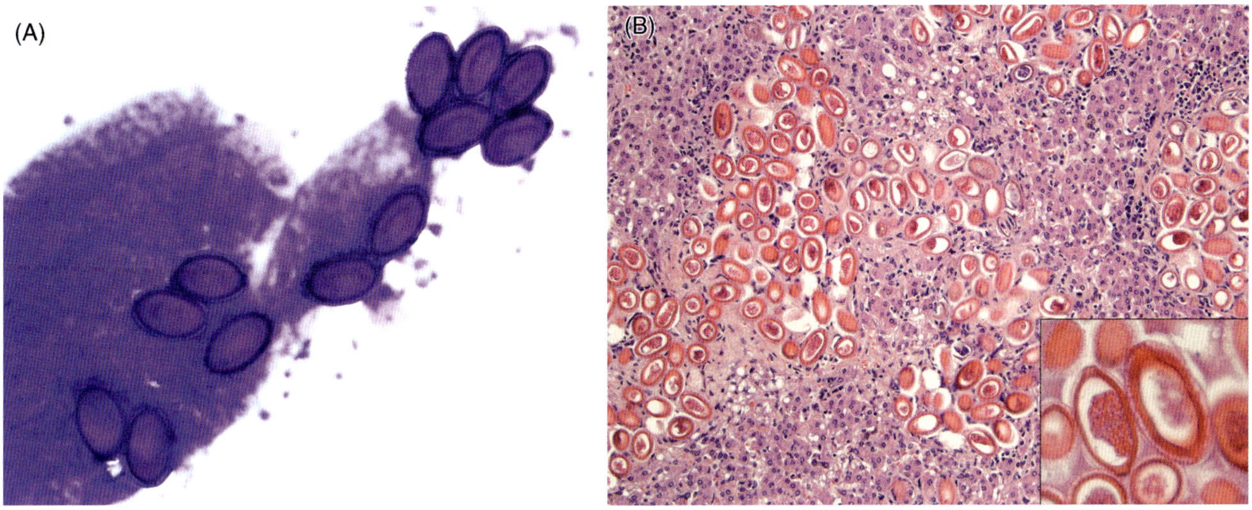

FIGURE 4.19 Cytological (A) and histological (B) correlate: capillariasis in a black-tailed prairie dog with serpiginous pale foci in the liver. (A) Numerous oval, slightly three-dimensional, nematode ova with a thick refractile wall and bipolar opercula plugs, typical of *Calodium hepaticum* (previously *Capillaria hepatica*) in liver tissue imprint. Hepatocytes are poorly discerned in the background. Modified WG. (B) Hepatitis with numerous intralesional nematode eggs with features comparable to the cytology. Depending on orientation and plane of section, bipolar opercula plugs are visible (inset). H&E.

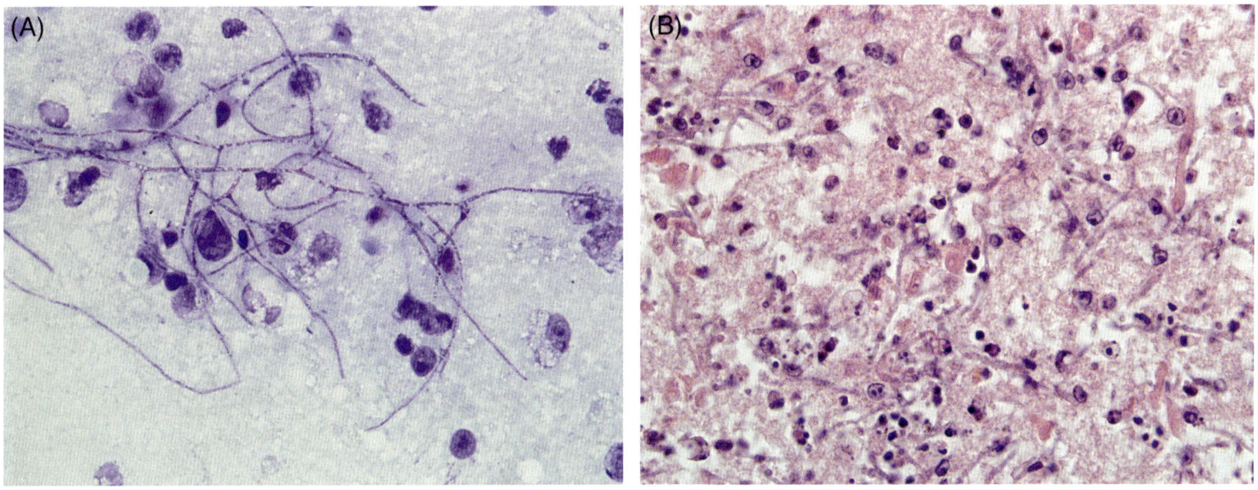

FIGURE 4.20 Cytological (A) and histological (B) correlate: fungal encephalitis (systemic aspergillosis suspected) in a thick-billed parrot. (A) Branching, nonpigmented, parallel-walled, septate hyphae, with scattered macrophages containing phagocytized basophilic material (suggestive of myelin), and few lymphocytes in tissue imprint of a focal pale and firm region in the brain. Modified WG. (B) Fungal encephalitis with pale-staining basophilic hyphae and inflammatory encephalomalacia with gitter cells. H&E.

deterioration limits utility in delayed samples. The postmortem bone marrow preparations are typically made by using rolled or **paint-brush** techniques in which a swab or paint-brush is gently touched/rolled along the marrow and then along a slide; wetting the utensil with uncontaminated plasma or body fluid from the carcass seems to help with intact cell yield. These touch preparations generally lack bone marrow particle morphology as seen in aspirates (i.e., stromal and hematopoietic aggregates), hindering assessment of marrow cellularity. Taxa-specific collection sites must also be considered. Although antemortem sampling considerations for weighing procedural risks and diagnostic benefits do not need to be considered during necropsy, postmortem sampling may benefit from being collected from sites that are accessible antemortem for comparative purposes. While the proximal appendicular or axial skeleton is a common collection site in many mammals, elephants, cetaceans, and sirenians have limited appendicular bone marrow stores and require axial skeletal sampling which is made more difficult by large body architecture and substantial vertebral bone. In birds, the tibiotarsus and keel are generally used as many other bones are

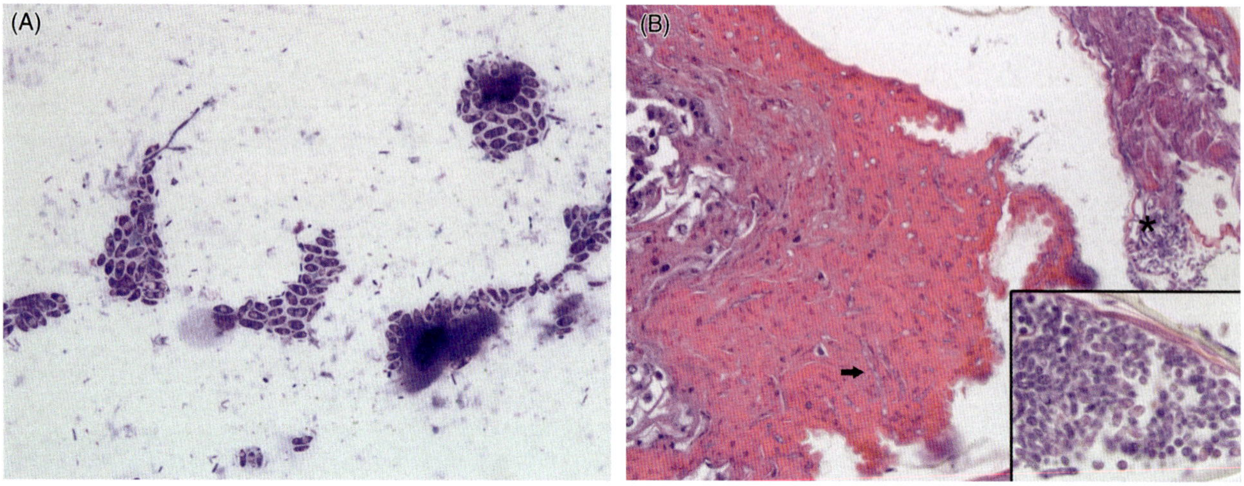

FIGURE 4.21 Cytological (A) and histological (B) correlate: fungal keratitis in a tentacled snake. (A) Numerous ovoid fungal conidia and rare narrow, parallel-walled hyphae amid moderate numbers of mixed extracellular bacteria and a small amount of clumped keratinaceous debris in a superficial squash preparation of a raised skin crust. Modified WG. (B) Crusting epidermitis with numerous embedded fungal hyphae (*arrow*), and at the right of the image, superficial fungal conidia (*asterisk*, inset) and bacteria. H&E. Molecular analysis of the lesion identified *Paranannizziopsis australasiensis*.

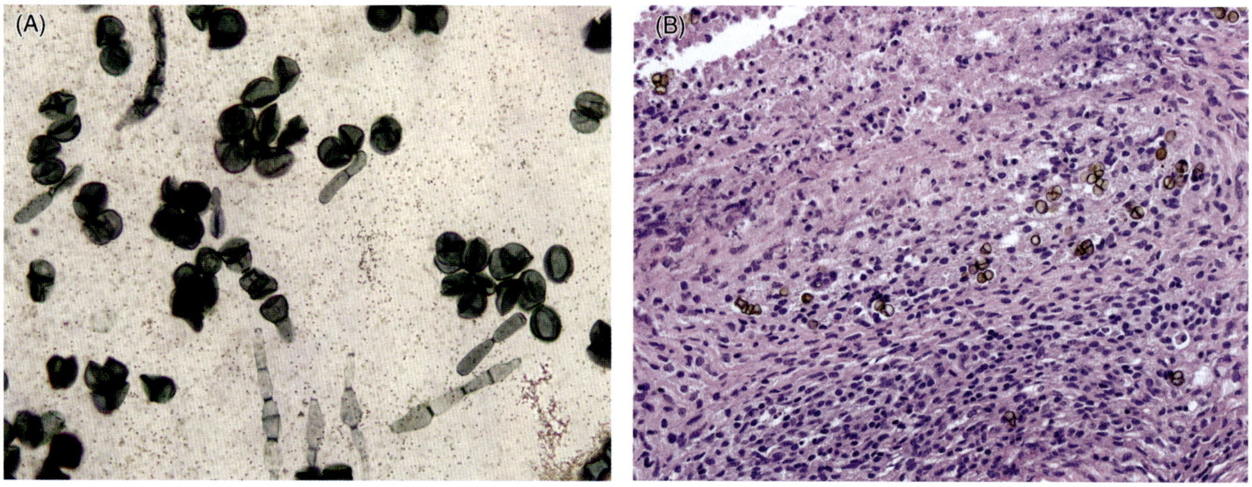

FIGURE 4.22 Cytological (A) and histological (B) correlate: chromomycosis in a giant marine toad. (A) Numerous pigmented fungal structures (sclerotic bodies and hyphae) amid abundant extracellular scattered fine brown melanin granules in a tissue imprint of brown to black dermal nodules. Modified WG. (B) Pigmented fungal structures (sclerotic bodies) within a chronic inflammatory nodule. H&E.

pneumatized and do not contain marrow elements. In reptiles, collection sites include, for instance, the proximal femur or tibia of legged lizards and pelvis or the carapace/plastron bridge in chelonians, but fibrous reptilian marrow tends to exfoliate poorly even with open postmortem access and scraping. Interpretation of hematopoiesis in nonmammals also requires increased consideration of extramedullary hematopoietic sites that play a larger (sometimes primary) hematopoietic role in these taxa (e.g., liver and spleen in birds; liver, spleen, thymus in reptiles; spleen, liver, and kidney in amphibians, which can exceed the bone marrow contribution; anterior kidney and spleen in teleost fish; and epigonal, Leydig organ, and spleen in elasmobranchs who lack a bone marrow contribution),[34,217,254] (see Chapter 38: Amphibia, Chapter 39: Osteichthyes, and Chapter 40: Chondrichthyes). As with other samples, factors important to consider for the interpretation of postmortem bone marrow cytology include time since death, type and duration of illness, therapeutic measures, etc.

Similar to antemortem samples, postmortem effusion body fluids are commonly assessed for color, transparency, and total protein (refractometer), and then used to make cytological preparations, with retention of residual material for potential additional analysis (e.g., chemistry, toxicology). Fluid sample processing for cytology can include a variety of techniques, such as direct smears, squash preparations, flow-back preparations, or concentrated sediment smear preparations, depending on the composition of the

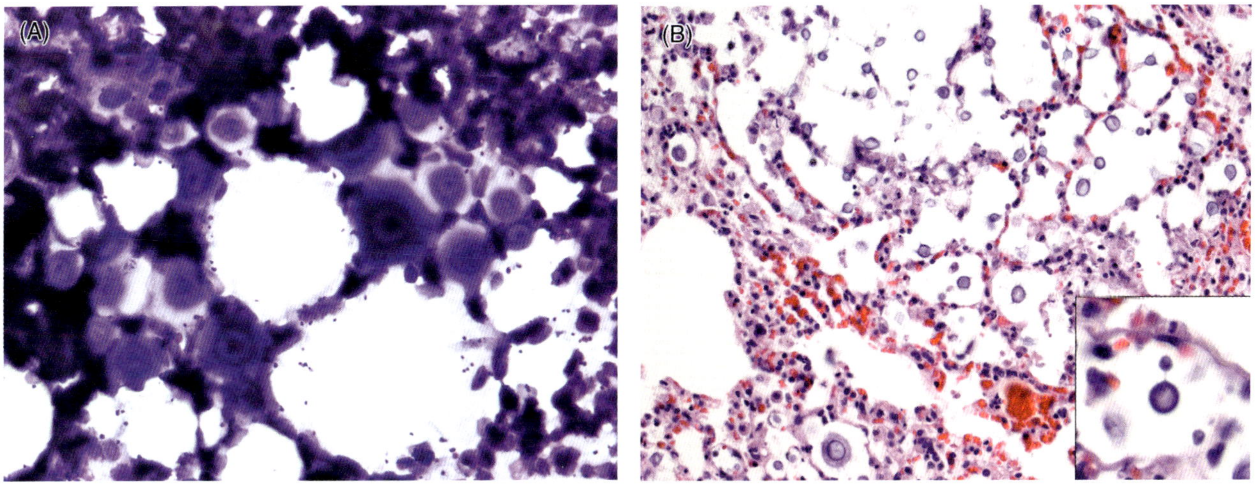

FIGURE 4.23 Cytological (A) and histological (B) correlate: cryptococcosis in a bushy-tailed jird. (A) Multiple *Cryptococcus* sp. yeast amid blood in a tissue imprint of mottled dark red lungs. Yeasts are variably sized with a nonstaining capsule. Narrow-based budding is not obvious in this field. Modified WG. (B) Cryptococcosis yeast filling and distorting pulmonary alveoli with little associated inflammation. Rare yeast demonstrates narrow-based budding (inset). H&E.

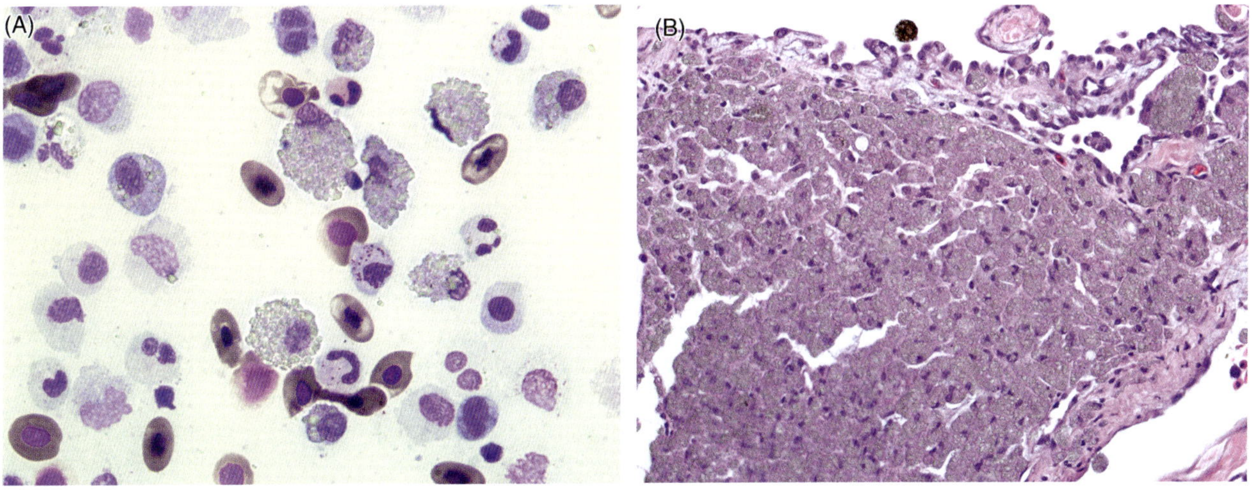

FIGURE 4.24 Cytological (A) and histological (B) correlate: barium coelomitis in a frilled lizard. (A) Direct smear of coelomic fluid shows increased mixed inflammatory cells (inflammatory exudate) with phagocytized refractile barium material in macrophages. Modified WG. (B) Numerous macrophages in the mesentery containing phagocytized refractile barium material and mild surface mesothelial hypertrophy. H&E.

fluid. If the fluid is cloudy and suggestive of moderate to high cellularity, direct smears, or squash preparations may yield the best cell spread. **Direct smears** are made in a method similar to a blood film, with a stationary slide and a spreader slide. The flow-back preparation is made similarly, but rather than pushing the spreader slide off of the stationary slide at the end of the push, it is abruptly lifted prior to reaching the end of the stationary slide, and the stationary slide is angled backward to allow the remaining fluid to "flow back" into the body of the smear, creating a concentrated end region (rather than a feathered edge). A similar technique is the line smear concentration technique with which the smearing of the fluid is performed as for a direct smear, but then the spreader is slide stopped and lifted (while the slide is kept level), resulting in a concentrated line at the end of the smear (rather than a flow back). If the fluid is clear and appears of low cellularity, the concentration of the sample may yield a cytologically more informative sample with more cellular material and thus additional cellular detail and the possibility to identify infectious agents, especially when present in low numbers. Direct smears should be made prior to concentration since they provide information about true cellularity of the sample. Cytocentrifuges (cytospin) are very helpful for obtaining **concentrated preparations** of fluid samples. However, since cytocentrifuges may not be readily

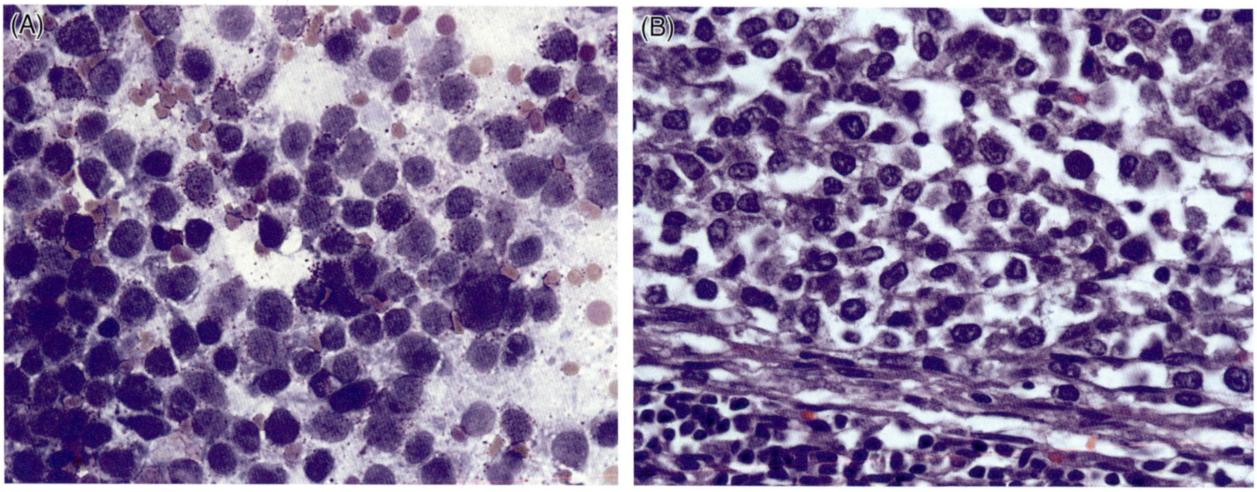

FIGURE 4.25 Cytological (A) and histological (B) correlate: mast cell tumor in a four-toed hedgehog. (A) Many discrete neoplastic mast cells with mild anisokaryosis and low to moderate numbers of distinct cytoplasmic granules in a tissue imprint of a subcutaneous mass. Cells show mild postmortem deterioration, including cytoplasmic dissolution. Modified WG. (B) Round cell neoplasm (mast cell tumor, top of image) effacing a lymph node. The neoplastic round cells have mildly granular cytoplasm, but distinct cytoplasmic granules are not apparent histologically as they are on cytology. The residual nodal architecture with small lymphocytes is apparent histologically (bottom of image). H&E.

available, a "Do-It-Yourself" cytospin composed of two paperclips, filter paper, and a cut off tuberculin syringe presents an excellent alternative.[186] With sufficient volume, another option is to centrifuge the sample in a standard centrifuge, remove the majority of the supernatant, and resuspend the remaining pellet by flicking the tube to create concentrated sediment smears or squash preparations. Additional descriptions of methods for cytological sample preparation are described in veterinary cytology texts.[156,186]

While cytological samples are commonly prepared to assess specific pathological findings (e.g., foci, masses, effusions), in some species the routine collection of cytological samples as part of the complete necropsy examination, even in the absence of apparent lesions is helpful for disease surveillance screening and to direct additional diagnostic testing, especially for agents more easily recognized on cytology (even postmortem) than on routine histology. In birds, routine tissue imprints from liver, spleen, and lung provide a valuable screening method for hemoparasites (e.g., *Plasmodium, Hemoproteus, Leukocytozoon,* systemic *Isospora*) and species-targeted concerns such as systemic coccidiosis in cranes. In fish, routine collection of wet-mount gill, fin, and skin/mucus preparations are a core component of both antemortem and postmortem sampling to investigate abnormalities including parasitic, fungal, and bacterial infections and air embolism (see Chapter 39: Osteichthyes).[34,112,175] In brief, small wedges are cut from the primary lamellae of the gills and the edge of the fin, and the superficial mucus and epidermal layers are obtained by gently scraping along the body with an edged instrument (blunted blade, scissors, or glass slide); these samples are placed on a glass slide with a drop of water (tap water, tank water, or saline as available and appropriate for fresh and marine species) and coverslipped. The wet mount preparations are immediately microscopically examined with a lowered condenser to promote the refraction of light around the sample components. Unstained samples of fish organ tissues, including liver, anterior and posterior kidney, and gastrointestinal tract contents, can also be prepared with the "squash" technique and examined for evidence of disease, including granulomas or parasites.

For any lesion or tissue, preparation of multiple cytology slides (i.e., at least 3) can be advantageous, with one or more representative slides stained for cytological evaluation and the remainder of slides available for additional diagnostics. Examples of ancillary diagnostics are listed in Table 4.e4.[40,149,192] For cytological assessment, preparations are typically air dried and then stained with Romanowsky type-based stains, such as Diff-Quik® or other Wright-Giemsa. Heat fixation is unnecessary and potentially damaging for most cytology samples. Fecal smears or gastrointestinal contents are best stained in separate stain containers to avoid contamination of nonfecal specimens. Careful dipping or a gentle stream of distilled water is sufficient to remove excess stain. Changes in the pH of the rinsing water, such as can occur with tap water, may affect staining characteristics of cytology samples. This and other factors, such as stain concentration and duration of staining, should be considered when staining results are suboptimal. Understained samples can be restained as needed. Stained slides may be cover-slipped and archived. For long-term storage or archiving of unstained preparations, the slides should be methanol fixed (10 min) and can be stained at a later time (months to years); storage of unstained slides without methanol fixation results in staining artifacts within weeks. It is essential to keep cytological preparations (blood, tissue, or fluid preparations) away from formalin fumes prior to methanol fixation, since these fumes partially fix and thereby alter the staining quality and cellular appearance. Frequently observed postmortem artifacts in cytology samples include autolysis and bacterial contamination and/or proliferation. Common postmortem bacteria include anaerobes, such as *Clostridium* sp., that quickly disseminate throughout tissues. In the absence of correlative findings (e.g., inflammation and/or intracellular bacteria to support infection), postmortem artifacts must be judiciously assessed to avoid over/misinterpreted.

ZOONOTIC RISK MANAGEMENT

As with all clinical and postmortem procedures, collection and handling of clinicopathological samples poses some degree of hazard for the associated personnel. Basic rules of protection apply to minimize human exposure and the risk of infection with potential zoonotic pathogens (see Chapter 1: Role of the Veterinary Pathologist in Emerging Infectious Diseases). A general understanding of the conditions on the list of potential rule outs when working with a certain species is of utmost importance. Zoonotic risk is higher in cases of infectious disease. However, as terminal non-infectious events (e.g., trauma) can occur secondarily to reduce physical ability from infectious concerns, an element of caution is always recommended.

Examples of infectious diseases with zoonotic potential from various nondomestic species include infections from bacteria (e.g., tuberculosis, anthrax, brucellosis, leptospirosis), fungi (e.g., aspergillosis, blastomycosis), parasites (i.e., *Cryptosporidium* sp., *Giardia* sp.), and viruses (i.e., rabies, West Nile virus). One example of notable importance because of zoonotic risk and the potential role of cytology in decision-making during postmortem assessment is anthrax septicemia in high-risk locations and species. Using appropriate precautions for handling samples from a suspected case, vegetative anthrax bacilli (*Bacillus anthracis*) can be identified in smears of blood and exudate from animals dying of septicemia as large bacilli sometimes in short chains with blunted square ends and a variably apparent capsule (Fig. 4.26). Multiple cytology stains are effective for visualization (e.g., modified Wright's stain, Gram stain, methylene blue) (see Chapter 1: Role of the Veterinary Pathologist in Emerging Infectious Diseases). Other common large postmortem bacilli, *Clostridium* sp., have rounded ends. The vegetative anthrax bacilli are destroyed by postmortem putrefaction and are expected to become less apparent with longer postmortem interval. Anthrax sporulation occurs upon exposure to oxygen and results in long-lived and highly resistant spores. As in all situations of biosecurity concern, a safety-first policy is key, and postmortem examination of

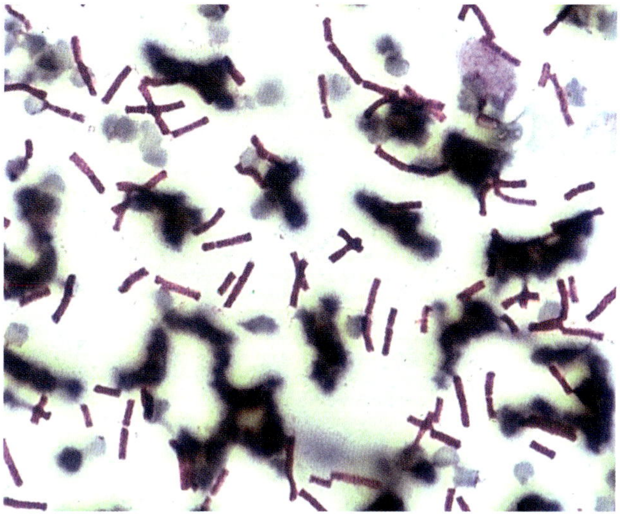

FIGURE 4.26 Postmortem collected blood film from a kudu showing abundant monomorphic large plump bacilli consistent with Anthrax bacilli (*Bacillus anthracis*) in a blood smear from a kudu. Giemsa stain. *(Photo Courtesy of R. Bengis, Department of Agriculture, Forestry and Fisheries, South Africa.)*

anthrax cases is not recommended due to the risk of zoonotic infection and air exposure enhancing sporulation and environmental contamination.[20,167] Typical gross pathological observations associated with anthrax septicemia include incomplete or lack of rigor mortis, presence of blood-stained fluid from body openings, petechiae and ecchymoses, dark nonclotting blood, and edema; suggested samples, if collected, include whole blood for blood smears and blood culture in fresh carcasses, and nasal swabs or hoof lamellae in older carcasses.[20] Anthrax is considered a notifiable disease by the World Organization for Animal Health (WOAH, founded as OIE, https://www.woah.org/en/what-we-do/animal-health-and-welfare/animal-diseases/) and in the United States by the Department of Agriculture (https://www.aphis.usda.gov/aphis/ourfocus/animalhealth/nvap/NVAP-Reference-Guide/Animal-Health-Emergency-Management/Notifiable-Diseases-and-Conditions).

NOTES ON CLINICAL PATHOLOGY SAMPLING IN THE FIELD, FOR FORENSICS, AND IN UNUSUAL MORTALITY EVENTS

Please see online content.

Conclusion

Clinical pathology encompasses a broad range of testing that combines an understanding of the principles of laboratory testing, quality assurance, physiology, and pathology. While many pathophysiological mechanisms and resultant findings are translatable across taxa, it is critical to consider species-specific nuances during the entire testing cycle, from sampling to interpretation. With the ongoing evolution of laboratory testing, the role of clinical pathology in health screening and disease diagnosis in zoo and wildlife species will continue to expand.

ADDITIONAL RESOURCES

Allender, M.C., Fry, M.M., 2008. Amphibian hematology. Vet. Clin. North Am. Exot. Anim. Pract. 11, 463–480.

Arnold, J.E., Camus, M.S., Freeman, K.P., Giori, L., Hooijberg, E.H., Jeffery, U., Korchia, J., Meindel, M.J., Moore, A.R., Sisson, S.C., Vap, L.M. and Cook, J.R., 2019. ASVCP Guidelines: Principles of Quality Assurance and Standards for Veterinary Clinical Pathology (version 3.0). Vet. Clin. Pathol. 48, 542–618.

ASVCP Quality Assurance and Laboratory Standards Committee, 2010. Guidelines for the determination of reference intervals in veterinary species and other related topics. Am. Soc. Vet. Clin. Pathol. Available from: https://www.asvcp.org/page/QALS_Guidelines.

Brooks, M.B., Harr, K.E., Seelig, D.M., Wardrop, K.J., Weiss, D.J., 2022. Schalm's Veterinary Hematology, seventh ed. John Wiley & Sons, Hoboken, NJ, USA.

Campbell, T.W., Grant, K.R., 2022. Exotic Animal Hematology and Cytology, fifth ed., John Wiley & Sons, Inc., Hoboken, NJ, USA.

Friedrichs, K.R., Harr, K.E., Freeman, K.P., Szladovits, B., Walton, R.M., Barnhart, K.F., Blanco-Chavez, J., 2012. ASVCP reference interval guidelines: determination of de novo reference intervals in veterinary species and other related topics. Vet. Clin. Pathol. 41, 441–453.

Fudge AM, 2000. Laboratory Medicine, Avian and Exotic Pets. WB Saunders Co., Philadelphia, PA, USA. pp. 47–55.

Jacobson, E., Garner, M.M., 2021. 2nd ed., Infectious Disease and Pathology of Reptiles. CRC Press, Boca Raton, FL, USA.

Kaneko JJ, Harvey JW, Bruss ML, 2008. Clinical Biochemistry of Domestic Animals. sixth ed. Elsevier, Burlington, MA. pp. 351–378.

Mader, D., 2006. Reptile Medicine and Surgery. Elsevier Science, St. Louis, MO, USA.

Reavill, D., Roberts, H., 2007. Diagnostic Cytology of Fish. Vet. Clin. North Am. Exot. Anim. Pract. 10, 207–234.

Sharkey, L.C., Radin, M.J., Seelig, D.M., 2021. Veterinary Cytology. Wiley-Blackwell, Hoboken, NJ.

Stockham, S., Scott, M., 2008. Fundamentals of Veterinary Clinical Pathology, second ed. Blackwell Publishing Ltd., Ames, IA.

Thrall, M.A., Weiser, G., Allison, R.W., Campbell, T.W., 2022. Veterinary Hematology, Clinical Chemistry, and Cytology. third ed., John Wiley & Sons, Hoboken, NJ, USA.

Valenciano A.C., Cowell, R.L, 2020. Cowell and Tyler's Diagnostic Cytology and Hematology of the Dog and Cat. fifth ed. Mosby Elsevier, St. Louis, MO.

References

Full references for this chapter can be found online at https://doi.org/10.1016/B978-0-443-18437-6.00033-X.

Chapter 5

Bovidae, Antilocapridae, Giraffidae, Tragulidae, and Hippopotamidae

Emily Mitchell[a], Johan Steyl[a], and David J. Gasper[b]

[a]*Department of Paraclinical Sciences, University of Pretoria, Pretoria, South Africa;* [b]*University of Wisconsin-Madison, Madison, WI, United States*

INTRODUCTION

The families Bovidae, Antilocapridae, Giraffidae, Tragulidae, and Hippopotamidae represent a large and diverse group of ruminant and pseudoruminant artiodactyls. Full common and scientific names as well as conservation status (where known) in these taxa are presented in the digital content Table 5.e1. Clinical pathology and unique anatomical features, as well as dental formulae, details about urolithiasis, selected toxicosis, neoplastic conditions, bacterial infections, fungal diseases, and ecto and endoparasites (see Tables 5.e2–5.e9) and references are also available in the digital content. Bovidae is the largest family in this group, with 8 subfamilies, 50 genera, and approximately 140 species. The family Tragulidae includes 3 genera and at least 10 species of chevrotains or mouse deer and includes the smallest living ungulates. The family Giraffidae includes 2 genera: the giraffe and okapi. The family Hippopotamidae includes 2 genera: the common and the pygmy hippo. The family Antilocapridae contains only pronghorn antelope. Free-ranging pronghorn are restricted to Central and Western North America. Records of disease in this group are derived primarily from disease surveillance in free-ranging or farmed populations, particularly where these animals inhabit the wildlife-domestic animal interface. Many of these species are susceptible to infectious diseases subject to international trade restrictions (e.g., foot and mouth disease) or domestic animal disease control legislation (e.g., brucellosis and tuberculosis). Disease susceptibility in those species that are extinct in the wild, or critically endangered, is largely obtained from zoos or other captive breeding institutions. Disease records for the Tragulidae and Hippopotamidae are relatively scarce.

NON-INFECTIOUS DISEASES

Nutritional

Anthropogenic activities increasingly impact the nutritional ecology of free-ranging herbivorous wildlife and result in the escalation of many nutritional imbalances. Protein, energy, and macromineral deficiencies commonly cause ill thrift and mortality in captive animals, especially browsers.[40] Consumption of protein-binding **tannins** by browsers (especially obligate browsers like kudu and giraffe) can result in a "starvation syndrome" as a result of reduced digestive enzyme function and protein bioavailability.[25,69,382] No lesions specific for tannin-induced maldigestion have been reported. Depleted adipose reserves, manifesting as **serous atrophy** of cardiac coronary, renal pelvic and bone marrow adipose tissue (Fig. 5.1), and variable **skeletal muscle atrophy** are common.[80,328,423] These nonspecific findings necessitate the investigation of both direct nutritional factors (quantitative and qualitative) and other debilitating diseases (e.g., see dental attrition, endo-/ecto parasitism, and peracute mortality syndrome). Certain types of tannins have positive antioxidant, -parasitic, and -bloat effects and should not be completely excluded from the diet of browsing ruminants.[69]

Micronutrient (trace minerals, vitamins and amino acids) deficiencies occur in captive browsers offered a limited variety of feed and free-ranging animals with restricted habitats or in regions with known mineral deficiencies.[96] **Copper deficiency** is well described in domestic ruminants and has been documented in most taxa in this chapter.[96] Primary deficiency is associated with a lack of dietary copper; however, secondary deficiencies are associated with feed components such as molybdenum that impair copper uptake. In nondomestic species lesions associated with copper deficiency include anemia; generalized poor body condition, diarrhea, dermatopathy, and neonatal degenerative neuropathy with generalized weakness and posterior paresis (enzootic ataxia/swayback); developmental defects including cerebral, cardiovascular, and skeletal malformation; and reduced fertility and abortion. In contrast to hydrocephalus, the brain cavities are formed by loss of tissue (hydrancephaly) rather than expansion of the ventricles, and the calvarium is usually morphologically unaffected (Fig. 5.2). An "**unsteady gait disease**" characterized by pica, dyskinesia, and unsteady gait due to secondary copper

Pathology of Wildlife and Zoo Animals. https://doi.org/10.1016/B978-0-443-18437-6.00005-5

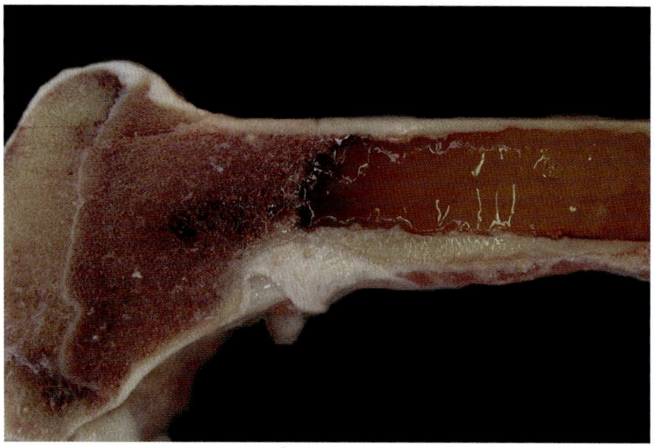

FIGURE 5.1 Serous atrophy of bone marrow adipose tissue in an adult nyala bull. The bone marrow cavity in a sagittal sectioned femur is filled with clear, pale yellow jelly-like adipose tissue.

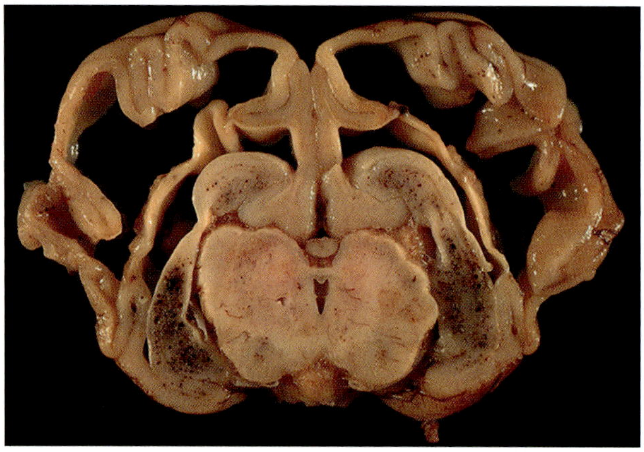

FIGURE 5.2 Hydranencephaly due to copper deficiency in a stillborn impala. Coronal section of a formalin fixed brain. Bilaterally, there is nearly complete replacement of the cerebral hemispheres by large, cavitary spaces enclosed by thin layers of remnant cortical tissue. *(Photo Courtesy of San Diego Zoo Wildlife Alliance.)*

deficiency has been reported in free-ranging Tibetan gazelles.[350] Dermatopathy, secondary to impaired tyrosinase activity, is characterized by progressive bleaching of the hair coat and straightening or coarsening of hair fibers and is more noticeable in dark-coated animals. Although the pathogenesis is unclear, in utero and neonatal copper deficiency causes a neuropathy characterized by neuronal degeneration, cerebral white matter malacia, and Wallerian degeneration in the spinal cord. Severity is dependent on the age at onset. As in domestic ruminants, spinal cord Wallerian degeneration is characterized by bilaterally symmetrical white matter spongiosis affecting the dorsolateral and ventromedial funiculi (eSlide 5.e1).[299] Mineral analysis of liver should be evaluated to support or confirm suspected diagnoses. Normal ranges of hepatic copper (wet mass, WM) vary among species. Reference ranges are reported for bison (25–100 ppm WM), blackbuck (50–100 ppm WM), blesbok (7.8–21.9 ppm WM (wet mass)), eland (5–65 ppm WM), giraffe (1.8–45 ppm WM), hippopotamus (143 ppm WM), nilgai (70–230 ppm WM), and oryx (4–8 ppm WM).[299,314] Hepatic mineral screening for copper-antagonizing agents such as molybdenum could add insight into the underlying mechanism of copper deficiency. Copper deficiency may also increase susceptibility to gastrointestinal parasitism, including abomasal nematodes such as *Haemonchus* spp.[116]

Selenium, with or without concurrent **Vitamin E** deficiency, is reported to be the most common cause for "**nutritional myopathy**" in most mammals and is characterized by oxidative injury in many tissues, but most prominently in striated muscle. It is specifically mentioned as a cause of death in juvenile Kirk's dik-dik.[423] The primary lesion of nutritional myopathy is multifocal, polyphasic myofiber degeneration with variable fiber mineralization. In contrast, exertional myopathy (see "Metabolic" section) often exhibits a monophasic pattern and retains damaged myofiber basal laminae and satellite cells that facilitate myofiber regeneration and repair.[80] Myoglobinuria is not a classic feature at necropsy. Selenium deficiency may potentiate exertional myopathy and complicate the interpretation of the underlying pathomechanisms of lesions at necropsy. Regions with selenium-deficient soils are distributed on all continents. Secondary deficiency due to ingestion of selenium-antagonizing elements, especially sulfur, is also common, and lesions cannot be distinguished from those of primary deficiency.[362] A definitive diagnosis of selenium deficiency depends on mineral analysis. Reference ranges of hepatic selenium have been reported for bison (0.25–0.5 ppm WM), blackbuck (0.27–0.46 ppm WM), eland (0.2–0.35 ppm WM), giraffe (0.26–0.63 ppm WM), hippopotamus (1.02 ppm WM), nilgai (0.30–0.88 ppm WM), and oryx (0.2–0.71 ppm WM). Analytic screening of liver tissue for copper, selenium, molybdenum, manganese, and zinc provides very useful data when investigating deficiencies or toxicities of these elements.[314]

Osteoporosis of undetermined etiology has been described in the skulls of free-ranging mountain sheep in the southwestern United States.[42] Possible etiologies include dietary deficiency of copper or magnesium and increased demand for calcium due to pregnancy or lactation. "**Stiffness of the Extremities Disease**" is a recently described condition associated with low dietary phosphorus in the eastern Asian yak; a similar disease has been reported in water buffalo.[351] The mandible, scapulae, sacrum, ilium, and ribs are most severely affected and exhibit varying degrees of osteoporosis and brittle fragility with thinning of the cortical bone and expansion of the marrow cavity into the epiphyses. Additional gross lesions include nonmineralizing expansion of the epiphyses of long bones, with concurrent diaphyseal bowing, which are similar to those observed in **hypophosphatemic rickets**.

Dietary **calcium and phosphorus imbalances** also occur in captive and free-ranging animals in areas overlapping with selenium-deficiency regions. These imbalances often result in metabolic bone diseases such as **rickets** (growing animals) and **osteomalacia** (adults). Osteomalacia in Tibetan gazelle is characterized macroscopically by bone softening and pathological fractures. In growing animals disturbed endochondral ossification results in widening of growth plates and poorly mineralized cartilage in the epiphyses.[174] Calcium- and/or phosphorus-deficient free-ranging animals may exhibit pica, and the subsequent osteophagia may predispose them to botulism. Geophagia may result in high or even toxic intake of minerals like iron and manganese.

Metabolic

Ruminal acidosis is widely reported in captive nondomestic ruminants. The clinical presentation is similar to that described in domestic ruminants and commonly occurs secondary to dietary grain excess. Lesions include rumen papillary blunting with epithelial hyperplasia and parakeratosis; acute or chronic rumenitis, with or without intra-epithelial pustules. Secondary bacterial and/or fungal rumenitis is often due to *Trueperella pyogenes*, *Fusobacterium necrophorum*, or *Mucor* spp. **Hepatic abscessation** and **laminitis** are common in captive animals, particularly springbok, deer, and pronghorn.[404] Serious serum mineral imbalances (high phosphate, low magnesium and calcium) have been recorded in captive wild ruminants on high-concentrate diets. Seizures, tetany, and laminitis were observed in kudu, eland, nyala, bongo, and giraffe.[252] In pronghorn acidosis resulted in hemorrhagic foci in the thalamus and brainstem and cortical laminar necrosis resembling thiamine-responsive polioencephalomalacia.[412]

Ruminal **tympany/bloat** is common although published reports in nondomestic ruminants are rare. The condition has been described in a captive eland and as a secondary complication to many conditions such as intoxications.[25,111] Clinical bloat, especially during chemical immobilization, frequently occurs and can be attributed to nonsternal recumbency and/or use of alpha-2-agonists in the chemical capture cocktail.[18,156] Bloat-induced respiratory compromise can be rapidly fatal in animals immobilized using respiratory depressants such as opioid derivatives.

Urolithiasis, with or without obstruction, occurs frequently in these taxa, and diet has been identified as a significant contributing factor (see Table 5.e3). Similar to domestic ruminant species, the sigmoid urethral flexure and urethral process in males are common sites of obstruction, but uroliths can occur anywhere in the urinary tract. Calcium carbonate uroliths are most commonly reported in giraffe and wildebeest, and elevated dietary phosphorus and high ratios of feed concentrate to forage have been suggested as possible contributing factors.[253] Gross and microscopic lesions are referable to chronic irritation or obstruction. Important sequelae include hydronephrosis, azotemia, and urinary bladder rupture.

Hypothyroidism has been described in individual captive animals including a reticulated giraffe and African forest buffalo.[11,227] Generalized mild alopecia and epidermal hyperkeratosis, nonregenerative anemia, decreased serum thyroid hormone levels, and conformational deformity of the hooves were noted. The thyroid gland was hyperplastic with lipofuscin-laden follicular epithelial cells and the skin showed hyperkeratotic acanthosis with proliferating and ectatic sweat glands. Other endocrine organs including the pituitary gland were normal.

In "**Peracute mortality syndrome**" captive giraffes are suddenly found dead, with poor body condition and serous atrophy of fat in the cardiac coronary groove and bone marrow but no other histological lesions.[119] The condition is associated with malnutrition, principally inadequate dietary energy. Sudden environmental stressors like cold weather, especially at night, can precipitate peracute terminal collapse and death, most likely as a result of hypothermia and/or hypoglycemia.[312] This syndrome, as well as significant differences in dental wear exhibited by captive and free-ranging giraffes, illustrates the incomplete understanding of the dietary requirements of *Giraffidae*, and, most likely, other wild browsers.[71]

Cardiovascular disease is rarely described in free-ranging animals but has been recognized as a poorly characterized disease in captivity. A syndrome of recurrent pregnancy-associated **congestive heart failure** has been described in okapi.[395] The syndrome is of uncertain etiology and presents, in mid-gestation, with left-sided heart failure. Globoid cardiac enlargement with severe left atrial dilatation, left atrial epicardial and mural myocardial hemorrhage, pericardial hemorrhage, and pleural effusion occurs. Microscopic changes include severe left atrial fibrosis and acute hemorrhage, bilateral ventricular fibrosis, arteriosclerosis of the great vessels, and chronic passive hepatic and pulmonary congestion.

A review of mortality in captive pygmy hippopotamuses, ranging from 3 to 30 years of age, revealed that **cardiovascular disease** accounted for 17% of all deaths. Infection with encephalomyocarditis virus was the primary cause (see "Infectious Diseases" section); however, dilated cardiomyopathy, **myxomatous valvular degeneration** (**endocardiosis**), and myocardial degeneration and fibrosis were also documented.[115]

Exertional rhabdomyolysis ("**capture myopathy**") may be the most significant and well-described metabolic disease affecting animals in this chapter and has been reviewed in detail in nondomestic ruminants.[295] It has also been referred to as "transit or translocation myopathy," "capture stress," "stress myopathy," "diffuse muscular degeneration," "idiopathic muscle necrosis" and even as "white muscle disease" or "muscular dystrophy."[51,156,295] Exertional rhabdomyolysis (ER) was originally described as a fatal sequela to capture in free-ranging ruminants and has subsequently been described in capture and noncapture related morbidity and mortality in a broad range of free-ranging and captive nondomestic bovids, including, giraffe, hippos, Kirk's dik-dik, okapi, and pronghorn.[115,295,392,414,423] The clinical presentations and lesions of ER have been classified into four manifestations: hyperacute (capture shock syndrome), acute (ataxic myoglobinuric syndrome), subacute (ruptured muscle syndrome), and chronic (delayed peracute syndrome).[156,363] Clinical presentations and lesions may be influenced by numerous factors including environmental temperature and humidity, cardiovascular fitness, nutritional status, comorbidities, and degree and duration of the exertion.[51,248] Considering the antioxidant protective function of essential trace elements like copper and selenium against muscle fiber injury, deficiencies in these elements have been postulated to increase susceptibility to exertional rhabdomyolysis.[51]

The hyperacute syndrome of ER is uncommon and primarily characterized by severe metabolic alterations presenting as weakness or muscle stiffness, ataxia, hyperthermia, metabolic acidosis-induced hyperkalemia, and shock with sudden death occurring less than 6 h following an exertional event. Gross musculoskeletal lesions may be absent or nonspecific, consistent with terminal shock. Microscopic lesions include acute coagulative necrosis of individual or small groups of skeletal (Fig. 5.3A) and cardiac myofibers (Fig. 5.3B). Granular renal tubular casts resulting from myoglobinemia are often present, although not always macroscopically evident in the urine. These changes may be accompanied by microscopic foci of acute necrosis in most tissues susceptible to acute hypoxia and acidosis.

The acute syndrome (6–24 h) is most common, characterized by acute skeletal and cardiac myodegeneration and necrosis resulting in weakness and ataxia. In contrast to selenium deficiency (see "Nutritional" section), the basement membranes of myofibers are significantly injured, releasing large amounts of myoglobin, with resultant myoglobinemia and myoglobinuria. Myoglobinuric nephrosis and renal failure may occur hours to days after the initial insult. Gross lesions may be evident in the myocardium, but gross lesions in the axial and appendicular skeletal muscles, and kidneys are always present. The main muscle groups affected are the quadriceps and gastrocnemius muscles, although muscles of the neck, extensors of the front limbs, and diaphragm may show lesions in severe cases. Acute myocardial and skeletal muscle lesions are soft, red to black, congested, or hemorrhagic with regions of pallor or linear, pale streaking (Fig. 5.4A). With time, affected muscles may become firm and variably pale, or tan in color. White streaks representative of dystrophic mineralization can also be present. Kidneys are swollen and dark brown; a darker brown band may be present within the cortex. Urine is dark brown to burgundy in color. Myoglobinuria is distinguished from hemoglobinuria by a positive ammonium sulfate precipitation test on the urine.

Microscopic lesions in the cardiac and skeletal muscle are often monophasic and characterized by hemorrhage, hypereosinophilic myofiber swelling, loss of cross-striations, contraction band or retraction cap formation, sarcoplasmic vacuolation or fragmentation, and nuclear pyknosis (Figs. 5.4B–E). Type II and modified type I skeletal myofibers are primarily affected; however, all types may be affected in severe cases. Neutrophilic inflammation may be present in acutely affected regions. With time, there is increased infiltration by mononuclear inflammatory cells (mainly macrophages), myofiber satellite cell proliferation, and fibrosis. Myofiber dystrophic mineralization may be present. Renal lesions typically show widespread

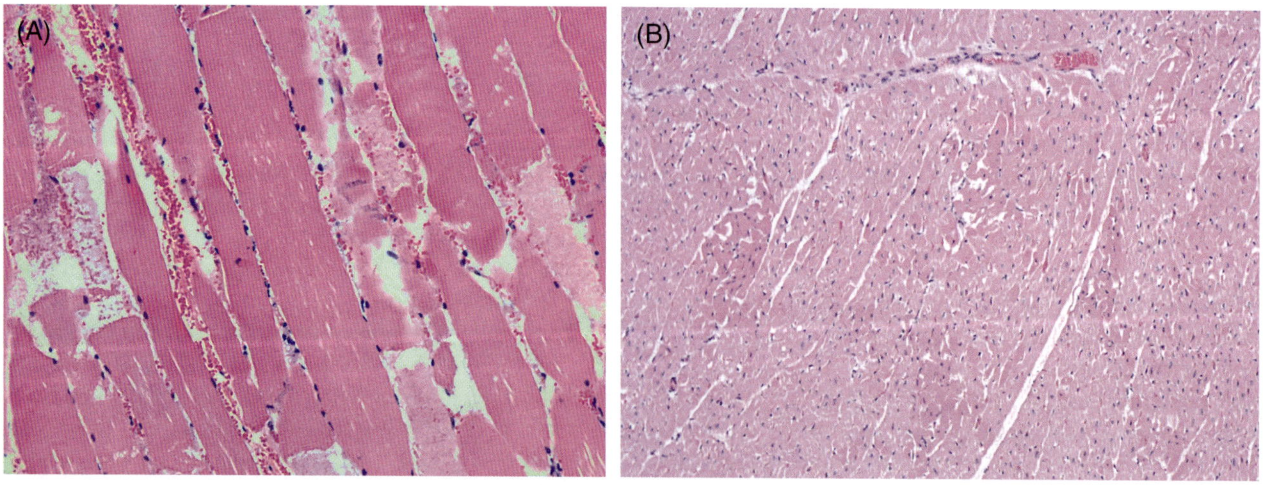

FIGURE 5.3 Hyperacute exertional rhabdomyolysis in a waterbuck bull. (A) The quadriceps muscle shows monophasic striated myofiber swelling, loss of cross striation, fragmentation, sarcoplasmic leakage, and perimesial hemorrhage. (B) Regional cardiac myofibrillar hypereosinophilic degeneration.

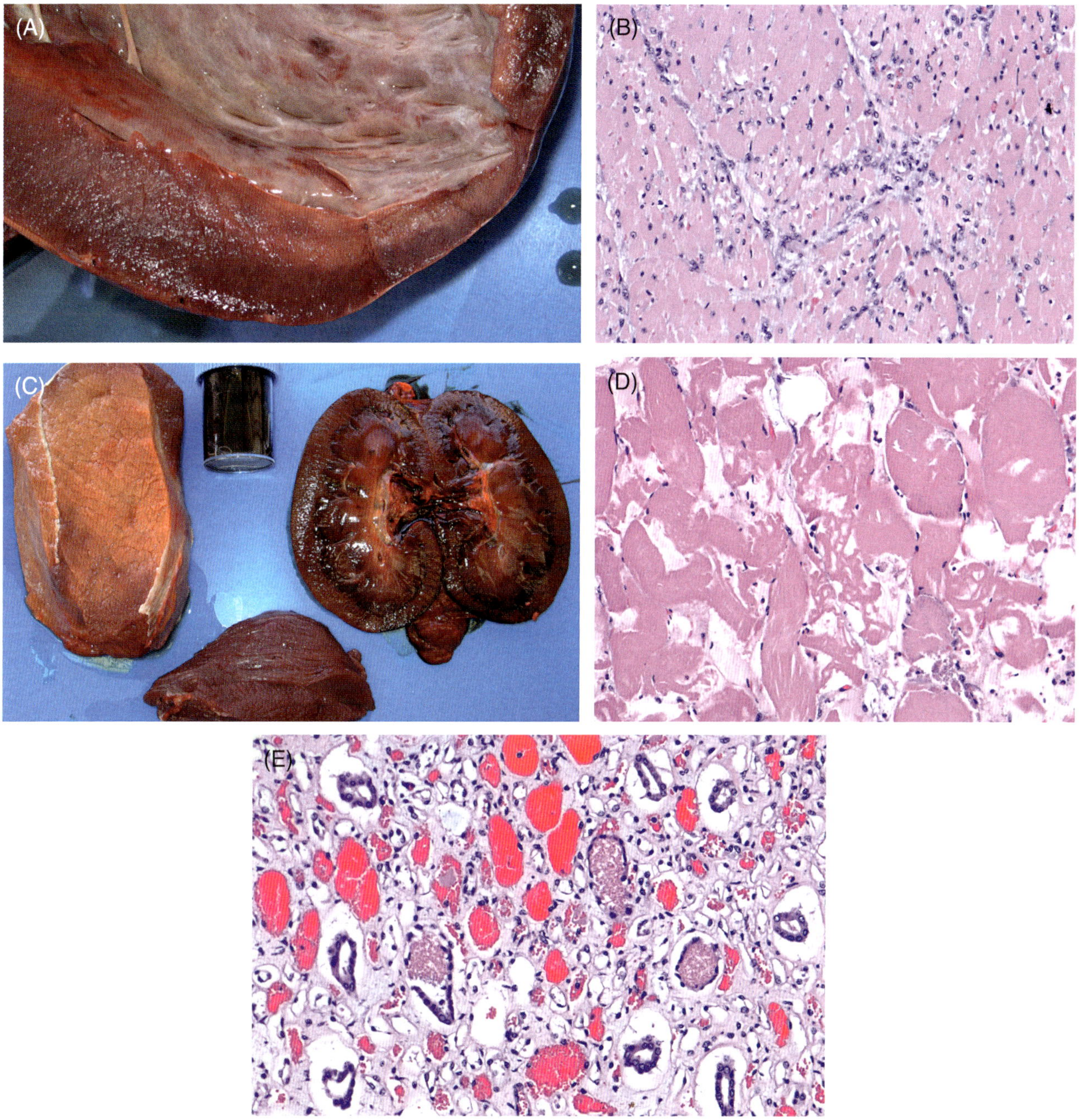

FIGURE 5.4 Acute exertional rhabdomyolysis in a sable antelope bull. (A) Sagittally dissected myocardium showing subendocardial to mid myocardial pallor. (B) Vasocentric radiating myofiber necrosis, myofibrillar depletion, and early repair. (C) Extensive pallor indicative of myonecrosis in the semimembranosus muscle. Compare with normal muscle at the bottom of the image. Transected kidney shows marked medullary congestion and/or myoglobinuria, as indicated by the dark red color of the urine (separate container). (D) Striated muscle showing fiber loss, swelling, and fragmentation and loss of striations. (E) Multiple renal medullary collecting ducts containing granular pigment (myoglobin) casts.

cortical tubular ectasia, degeneration, and necrosis. Tubules frequently contain hypereosinophilic hyaline and granular casts. Often, only medullary granular casts in collecting ducts are evident without significant or sublethal cortical tubular injury. Immunohistochemical labeling for myoglobin is necessary to differentiate granular pigment from hemoglobin.

The subacute syndrome may present 24–48 h following the exertion event. Gross lesions are similar to those in the acute syndrome with the addition of rupture and extensive hemorrhage along regions of necrosis. The appendicular weight-bearing muscles and those subject to greater loads are typically affected including unilateral or bilateral rupture of the gastrocnemius muscle, resulting in hyperflexion of the

FIGURE 5.5 Subacute exertional rhabdomyolysis in a subadult blue wildebeest. Bilateral hock hyperflexion resulting from gastrocnemius, semitendinosus, and semimembranosus muscle rupture.

hock joint with the hind quarters often dropped to a sitting position (Fig. 5.5). Microscopic lesions are more extensive with entire muscle bellies and muscle groups affected. Inflammatory and regenerative myofiber changes may be evident and characterized by sarcolemmal proliferation with increased sarcoplasmic basophilia, nuclear centralization, multinucleation, and nuclear rowing. Lesions in the diaphragm help differentiate true ER from recumbency-induced pressure necrosis, in susceptible muscle groups, in animals that showed protracted recumbency prior to death.

The chronic syndrome of ER is uncommon and often referred to as the "indefinite phase" of the exertional rhabdomyolysis complex. Animals that survive the acute syndrome with or without significant clinical abnormalities may die suddenly following a subsequent stressor, such as recapture. In affected animals there are few gross lesions and the microscopic lesions in the heart and skeletal muscle are less severe than those described in the acute syndrome. Variable interstitial myocardial and skeletal muscle fibrosis may be observed (Fig. 5.6), suggestive of significant previous sublethal injury, potentially reducing myocardial resilience to subsequent stressors and resulting in acute heart failure. Other myocardial insults like cardiotoxicosis, ischemia, and postinflammatory events should be ruled out.

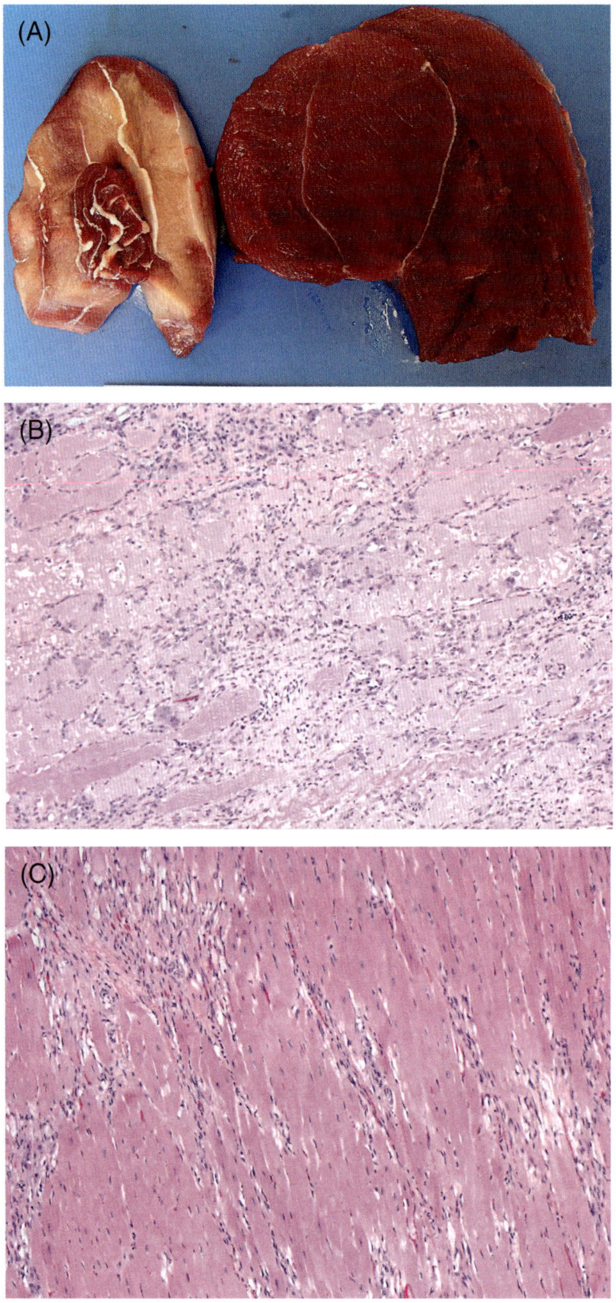

FIGURE 5.6 Subacute to chronic exertional rhabdomyolysis in a waterbuck cow. (A) Transected gastrocnemius muscle (left) showing widespread pallor (compared to normal quadriceps femoris muscle on the right) indicative of muscle necrosis with myoglobin depletion and active repair. (B) Advanced monophasic skeletal striated muscle repair characterized by sarcolemmal nuclear proliferation, interstitial fibroplasia, and mild mononuclear leukocytosis. (C) Multifocal, vasocentric myocardial replacement fibrosis.

Toxic

Toxicoses in captive and free-ranging wildlife may result from exposure to a number of natural and anthropogenic agents including toxic plants and animals, agricultural and industrial toxins, and intentional poisonings secondary to illegal hunting (see Table 5.e4). Although toxic plants occupy an important ecological niche in most range lands, plant toxicoses are infrequent. Evolutionary pressures have likely resulted in avoidance behaviors and physiological adaptations that enable some ruminant species to ingest plants that would otherwise be toxic.[253] However, phytotoxicosis has been recorded sporadically in wild ruminants following relocation to unfamiliar habitats. Intentional fatal pesticide toxicoses due to illegal hunting of African antelope have recently been reviewed and represent an ongoing threat to free-ranging animals.[276]

A wide variety of nondomestic ruminants may be affected by toxin producing bacteria (see "Bacteria" section).

Congenital/Genetic

The prevalence of **polycystic kidney disease (PKD)** in captive pygmy hippopotamuses is reported to exceed 35% in adult and geriatric animals.[115] Some cases are accompanied by elevated BUN and creatinine. Though most affected animals have been female, both sexes are affected.[265] The relatedness of some affected individuals suggests a familiar pattern of inheritance, and the condition shares similarities with autosomal dominant polycystic kidney disease in cats and humans. However, specific genetic mutations have not yet been identified. Grossly, few to myriad cysts, up to several centimeters in diameter, are distributed throughout the renal cortex and medulla, often effacing normal parenchyma.[317] Histologically, ectatic tubules and cysts are lined by cuboidal to flattened epithelium, with occasional epithelial hyperplasia or dysplasia. Bowman's capsules are dilated and membranous glomerulopathy has been described. Cysts are surrounded by interstitial fibrosis and may contain cell debris or pale, eosinophilic fluid (Fig. 5.7). In some cases cysts are also described in the liver (biliary cysts), duodenum (cystic Brunner's glands), thyroid glands, and/or pancreas, but these appear to be clinically insignificant.

A lethal, **neonatal polycystic kidney disease** has also been described in captive springbok, affecting up to 18% of live births in one captive population.[177,317] In these cases there was no sex predilection and inheritance was most consistent with an autosomal recessive pattern. Grossly, affected animals exhibited bilaterally symmetric nephromegaly. Cysts were approximately 1–3 mm in diameter and were most frequent in the outer medulla but also present in the inner medulla and cortex. In one report springbok with polycystic kidneys were also diagnosed with **lysosomal storage disease**.[317]

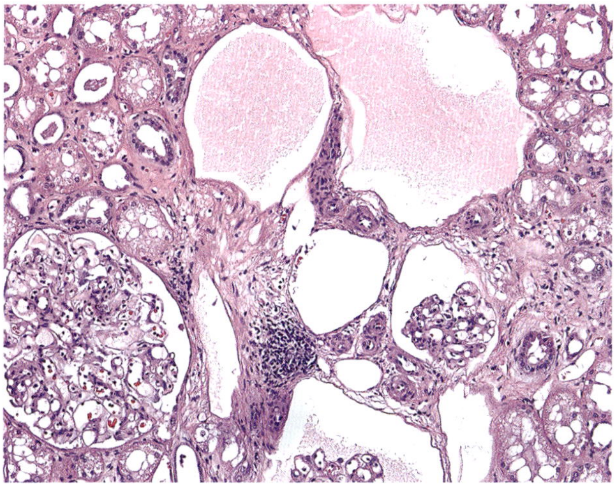

FIGURE 5.7 Polycystic kidney from a pygmy hippopotamus. Dilated renal tubules and Bowman's capsules are scattered throughout the renal cortex. Cysts contain pale, eosinophilic fluid. *(Photo Courtesy of San Diego Zoo Wildlife Alliance.)*

Fetal developmental anomalies and malformations, such as ***amorphous globosus***, **arthrogryposis**, **cyclopia**, **laryngeal atresia**, ***schistosomus reflexus***, and segmental **colonic aplasia**, occur in various captive nondomestic hoofstock species (Fig. 5.8). The underlying cause of most malformations is not known. Malformations are most likely spontaneous, rather than hereditary.[324,349,423] Bilateral **cryptorchidism** and **hypospadias** is recorded in an adult free-ranging giraffe.[247] Hyperplastic **goiter** with signet ring cells is recorded in an emaciated giraffe calf. Thyroid follicles were lined by single to multiple layers of hyperplastic epithelial cells, sometimes forming papillary projections. Most of the follicles were collapsed with no colloid. Signet ring cells with clear small to large cytoplasmic vacuoles and peripheralized nuclei were multifocally present.[394]

Age-Related/Degenerative

Degenerative joint disease is common in captive, aged bovids, giraffes, okapi, and hippopotamuses, and may be a significant factor in decisions to euthanize geriatric animals. Any joint may be involved, including the atlanto-occipital joint.

Due to challenges in diagnosis, **laminitis** may be an underdiagnosed cause of lameness in captive nondomestic

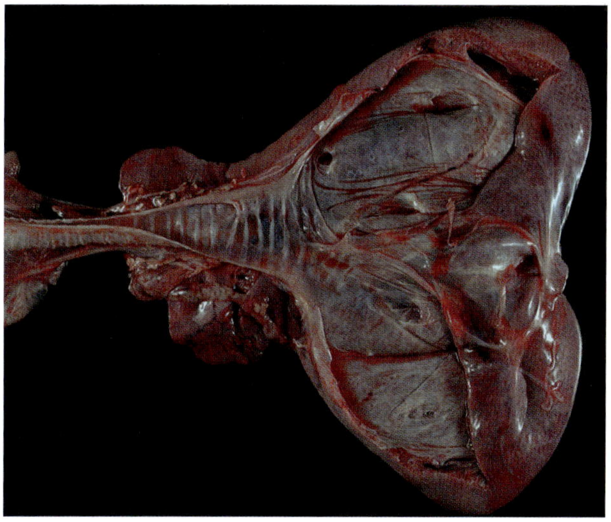

FIGURE 5.8 Complete laryngeal atresia in a stillborn Indian blackbuck. A congenital failure of formation of the laryngeal opening leads to chronic accumulation of pulmonary secretions and progressive ectasia of pulmonary spaces during development. *(Photo Courtesy of San Diego Zoo Wildlife Alliance.)*

hoofstock. Laminitis with pedal bone rotation and protrusion of the third phalanx through the sole has been described in multiple species, including takin, eland, greater Malayan chevrotain, and giraffe.[404] Advanced imaging of the limbs may aid in antemortem and postmortem evaluation of bony and soft tissue changes in the hooves; however, postmortem examination of parasagittal sections through the foot is often necessary to identify subclinical or mild cases of laminitis.

Dental attrition and wear (Fig. 5.9) result in inadequate mastication and poor body condition mainly in captive, geriatric bovids; malocclusion is a common cause of disease in captive and free-ranging ruminants and, in particular, predisposes them to mixed bacterial infections and chronic bacterial alveolar osteomyelitis.[199] Free-ranging North American Dall's and bighorn sheep have a high prevalence of dental disease, with a high rate of concomitant lumpy jaw-type lesions.[167] Significant differences have been noted in the extent and severity of dental attrition in free-ranging and captive giraffes, with dietary differences resulting in the functional classification of free-ranging giraffes as browsers and captive giraffes as grazers.[71] Cheek teeth of captive giraffe tend to have a rounded cusp, with only the first molar commonly having a blunted cusp. This contrasts with a predominance of sharp cusps on the cheek teeth of free-ranging giraffe. These differences suggest that traditional captive diets may have been overly abrasive. The resulting dental alterations have been further implicated as a contributing factor in giraffe acute mortality syndrome (see "Metabolic" section). There are several reports of canine tooth overgrowth and **malocclusion** in both pygmy and common hippopotamuses.[115]

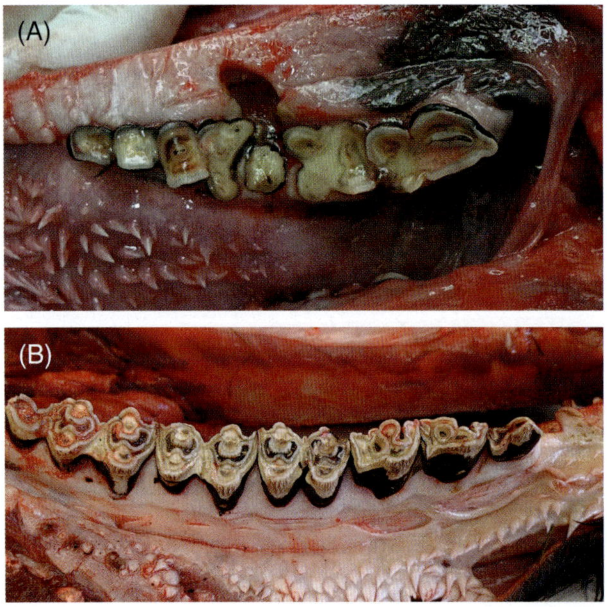

FIGURE 5.9 Severe molar and premolar dental attrition in a senile sable antelope. (A) The maxillary dental arcade shows significant abrasion of the hypsodont tooth's enamel ridges exposing mainly much softer cementum and dentine, significantly decreasing the dental abrasive ability. Simultaneously, a decrease in crown height to the level of the gingiva predisposed to traumatic injury and focal ulceration. (B) Normal African buffalo mandibular dental arcade for comparison. Note the presence of prominent enamel ridges for effective shredding of vegetation in the center of the typical hypsodont molar tooth.

Inflammatory Non-Infectious

Chronic tubulointerstitial nephropathy has been described in captive okapi housed in European and North American zoos.[34,151] Histologic lesions include tubular degeneration and atrophy, with interstitial fibrosis, mononuclear cell interstitial nephritis, thickening of tubular basement membranes, and hyaline and granular cast formation. Glomeruli are often affected to a lesser degree, with thickening of Bowman's capsule basal laminae, periglomerular fibrosis, and focal glomerular obsolescence. The character of the lesions suggests primary tubular insult, and although unproven, toxicosis from ingestion of plants such as willow (*Salix* spp.) has been suggested as a possible etiology.

Multicentric vesicular and ulcerative dermatopathy has been described in captive common hippos.[159,364] Gross lesions are erosive and ulcerative and may be widely distributed, especially affecting the axillary, inguinal, perineal, and limb regions. Seasonality has been described.[78] Microscopic changes include dermal edema, epidermal necrosis with individual keratinocyte necrosis, and neutrophilic dermal vasculitis with thrombosis. The etiopathogenesis has not been determined. A primary pathogen has not been identified; mixed bacteria have been cultured from the lesions. Lesions respond to various combinations of increased water salinity, broad-spectrum antibiotics, and immunosuppression with pentoxifylline.[364]

Abdominal and perirenal fat saponification and necrosis occur in nondomestic bovid species in zoos. Grossly, small to large, individual to coalescing masses of necrotic and mineralized fat may be present in the mesentery, retroperitoneum, or omentum. Masses may be yellow, hemorrhagic, or white and are gritty on cut section. Histologically, lesions include necrosis and mineralization of adipocytes, sometimes accompanied by histiocytic and mixed steatitis. Rarely, osseous metaplasia and even extra-skeletal osteosarcoma can occur. Secondary ureteral or gastrointestinal obstruction can occur, and the presence of masses in the pelvic canal can lead to dystocia.[324] Lesions can be spontaneous, especially in obese animals, but association with endophyte-infected fescue has been reported in cervids.

Miscellaneous/Multifactorial

Trauma is a common cause of death in captive nondomestic hoofstock. Trauma from conspecifics or other species sharing an exhibit can include blunt force fractures, bites, and goring wounds. Environmental trauma from running into obstacles is a potential cause of death for African antelope including Kirk's dik-dik[423] as well as giraffe. Gore wounds can elicit significant internal hemorrhage and can be associated with pneumothorax, myocardial laceration, or gastrointestinal perforations and secondary infections. Frequently, there is surprisingly minimal external evidence of a perforating wound despite the presence of severe, fatal internal trauma. Free-ranging giraffes are susceptible to electrocution when caught in range fires close to electricity lines. Neonatal trauma, typically inflicted by a parent, is a significant cause of death in captive pygmy and common hippopotamuses.[115]

Amyloidosis refers to group of diseases characterized by the accumulation of **amyloid**, an amorphous, hyaline, extracellular proteinaceous material, in various tissues. The condition is diagnosed relatively frequently in captive nondomestic hoofstock. Published reports include both wild and captive caprids, captive gazelles, pronghorn, and captive bongo,[13,122,149,194,220,236,325,401,420] but presumably any member of *Bovidae* is susceptible. In cases where amyloid has been characterized in nondomestic bovids, only reactive systemic or secondary (AA) amyloidosis has been recognized.

Clinical signs vary depending on the amyloid distribution, but animals are often emaciated, and there may be clinical evidence of renal or hepatic failure. Tissues most frequently involved include liver, kidney, spleen, gastrointestinal tract (lamina propria and submucosa), adrenal glands (cortical interstitium), thyroid gland, exocrine pancreas, lymph nodes, salivary glands, and blood vessels in multiple sites.[122,325] Grossly, the liver may be firm and more prone to fracture and hemorrhage. Kidneys may be pale and waxy (Fig. 5.10). If glomeruli are involved, the cortex may have a granular texture. Amyloid in the gastrointestinal tract may cause generalized or segmental, submucosal thickening similar to Johne's disease. Amyloid is hyaline, pale eosinophilic, and often induces atrophy and loss of adjacent tissues. In the liver amyloid tends to be deposited initially in the space of Dissé and then extends along sinusoids, inducing hepatic cord atrophy. Renal amyloidosis most often involves both medullary and glomerular

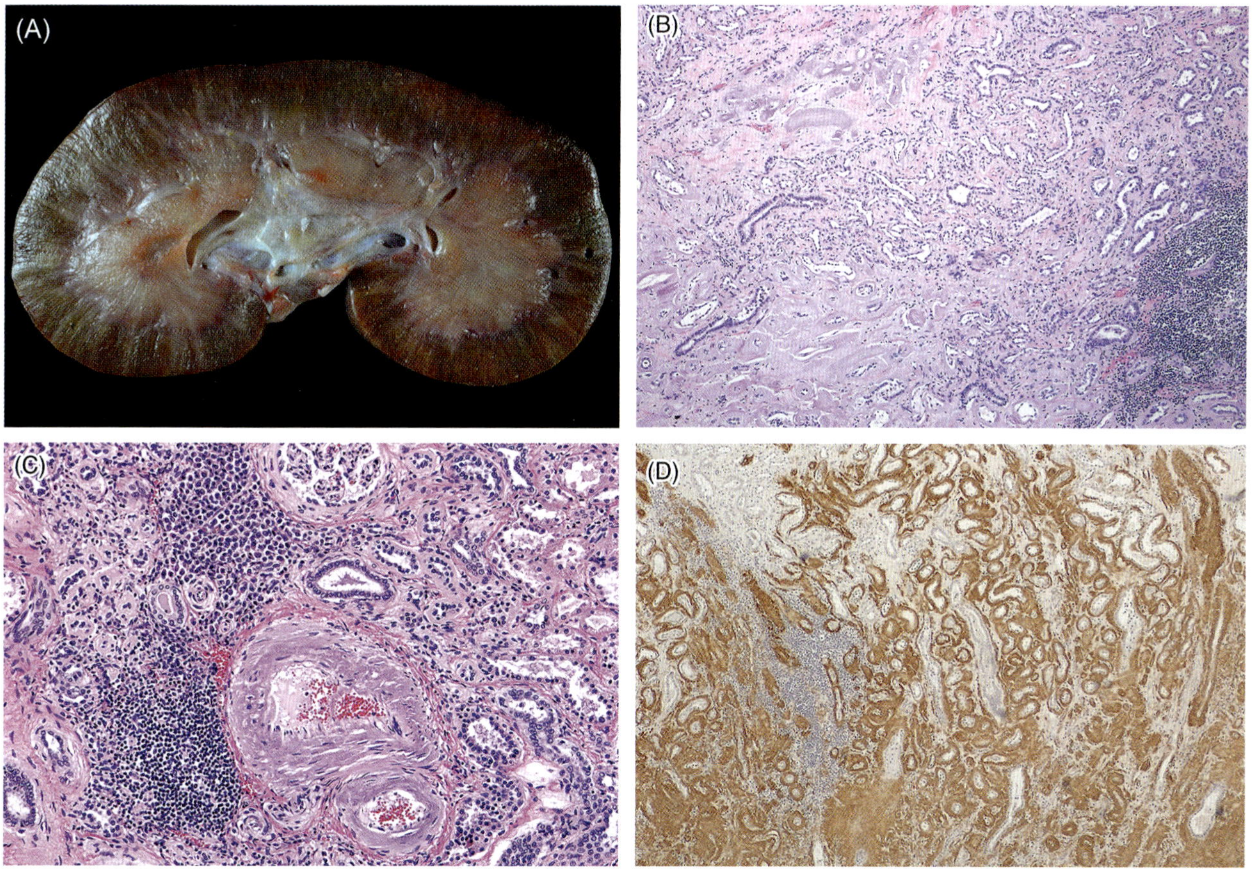

FIGURE 5.10 Renal medullary and interstitial amyloidosis in an adult bongo antelope. (A) The cortical surface is uneven and there are streaks of pallor extending from medulla to cortex. (B) Amorphous, hyaline extracellular material expands both the interstitium, widely separating remaining collecting ducts and tubules. There are patches of lymphoplasmacytic interstitial nephritis. (C) Arterial walls, in the kidney and other tissues, were disrupted by accumulation of amorphous, hyaline material. (D) AA amyloid in the renal interstitium and in blood vessels throughout the kidney. AA amyloid immunohistochemistry. *(Parts A–D: Photos Courtesy of P. Gaffney, University of California, Davis.)*

deposition, sometimes also involving cortical interstitium and blood vessels. In a series of captive Dorcas gazelles renal amyloidosis was exclusively medullary and glomeruli were spared.[325] When severe, renal amyloidosis may be accompanied by papillary or renal crest necrosis, presumably secondary to ischemia.[325] Renal tubular dilation and protein casts may also be present. Amyloid in blood vessels causes mural thickening and sometimes reduced luminal diameter.

Rectal prolapse and **stenosis** have been reported in captive okapi in the Netherlands and North America.[34] This is often accompanied by dilation of the orad rectum and colon and accumulation of hard dry fecal material (obstipation). Excessive grooming by dams with insertion of the elongated tongue into the rectum is a proposed etiology. This behavior can result in rectal erosion, ulceration, and eventual strictures.

Chronic pancreatic disease of unknown etiology is described in captive lesser kudu.[132] Lesions were limited to pancreatic acinar cell depletion, ductular hyperplasia, and fibrosis. Temporary distal limb swelling due to periosteal bone proliferation (**hypertrophic osteopathy**) is described in captive adult female takin. The condition resolved postpartum.[234]

A variety of **non-infectious factors** have been **associated with abortion, stillbirth, or neonatal death**. These include nutritional imbalances, immobilization, or anesthesia of pregnant dams; maternal neglect; failure of passive antibody transfer; advanced maternal age; fetal malformations; algal toxins and toxic plants; trauma; twinning in some species; or hereditary/genetic causes.[324,423] **Dystocia** is relatively common in captive bovids and may be related to maternal nutritional status, fetal malposition, or other causes. Fetal lesions after fatal dystocia often include nonspecific indicators of stress, such as external meconium staining and meconium and squamous cell aspiration in the lungs.

Neoplastic

Neoplasia is uncommonly reported in this group (see Table 5.e5) and has only recently been recognized as a threat to conservation in a few species.[241]

Novel **delta papillomaviruses** are reported in association with skin lesions in giraffe and okapi.[316,384] Typical viral cutaneous papillomas and fibropapillomas have been reported in pronghorn antelope.[368,369] Many benign papillomavirus-induced lesions regress spontaneously, but malignant transformation can occur in response to incompletely understood, host and environmental conditions.[369] *Rupicapra r. rupicapra* papillomavirus 1 (RrupPV1) was isolated from a large, nasolabial papilloma in a free-ranging Alpine chamois; the virus was most closely related to a papillomavirus associated with squamous cell carcinoma in domestic sheep (OaPV3).[246]

Tumors are classified, based on the proportion of epithelial and mesenchymal components, as papillomas, fibropapillomas, or fibromas. Grossly, typical papillomas are exophytic, hyperkeratotic, verrucous nodules that may be pigmented or nonpigmented. Histologically, hyperplasia of keratinocytes in the strata spinosum and granulosum; cell swelling and cytoplasmic pallor (koilocytes); orthokeratotic and parakeratotic hyperkeratosis; prominent, eosinophilic, cytoplasmic granules; and occasional amphophilic, intranuclear viral inclusion bodies are described. Fibropapillomas contain a much greater mesenchymal component and are covered by variably hyperplastic and hyperkeratotic epithelium that typically extends thin rete pegs into the underlying tissue. The mesenchymal component is composed of well-differentiated or anaplastic fibroblastic cells arranged in bundles, whorls, or storiform patterns. The tumor must be differentiated from other tumors of the dermis and from granulation tissue.

Equine sarcoid-like fibropapillomas associated with BPV-1 and BPV-2 have been reported in an African buffalo, giraffe, and sable antelope in South Africa.[383,407] Giraffe had coalescent, verrucous to nodular masses variably involving large proportions of the trunk, neck, head, and/or limbs (Fig. 5.11). The sable antelope had smaller, focal nodular lesions on the distal hind limb, shoulder, and lip; in one case a limb mass recurred after excision. Histologically, epidermal changes typical of papillomaviral infection included marked acanthosis with angular rete pegs, orthokeratotic hyperkeratosis, and koilocytes. Epidermal ulceration was variable. The dermis was replaced by pleomorphic fibroblastic cells arranged in whorls and bundles that displaced adnexa (Fig. 5.12). Viral inclusions are not typically histologically evident. Various bovine papillomaviruses have been detected in skin and mucous membranes of clinically healthy chamois and mouflon,[340] which demonstrates that infection is not always associated with disease.

FIGURE 5.11 Disseminated fibropapillomatosis associated with bovine papillomavirus infection in a free-ranging giraffe. Grossly, there are multiple, coalescent, verrucous to nodular cutaneous masses involving much of the neck, which resemble equine sarcoids. *(Photo Courtesy of R. Bengis, Department of Agriculture, Forestry and Fisheries, South Africa.)*

macrophages, contribute to the vasculitis.[339] The inflammation may be triggered by infected endothelial cells. In bison infected with OvHV-2 disease progresses very rapidly and is characterized by prominent gastrointestinal mucosal necrosis and minimal lymph node hyperplasia. OvHV-2 was detected in the skin of free-ranging bighorn sheep skin with alopecia; hyperpigmentation and crusting of skin on the face, pinnae, distal limbs, perineum and tail; and granulomatous mural cutaneous folliculitis.[357]

Cross-species transmission of **equid herpesvirus-1** in captive Thompson's gazelles is recorded.[191,334] The gazelles became lethargic, recumbent, with head shaking and seizures. Gross lesions were limited to meningeal congestion. Nonsuppurative meningoencephalitis was associated with neuronal necrosis and intranuclear eosinophilic and amphophilic inclusion bodies. **Equid herpesvirus-9** was originally isolated from a Thomson's gazelle with encephalitis.[191] It has been associated with disease in a giraffe with encephalitis.[187]

Contagious ecthyma, contagious pustular dermatitis, or **orf** is a highly contagious zoonotic disease caused by infection with **orf virus (ORFV)**, genus *Parapoxvirus*. Orf occurs worldwide in domestic sheep and goats, and many nondomestic members of the subfamily *Caprinae* are susceptible including free-ranging blackbuck, bighorn sheep, chamois, Dall's sheep, mountain goats, captive and wild muskoxen, Sichuan takin, Himalayan tahr, and serow.[147,210,336,348,370,376] Outbreaks of contagious ecthyma occur in both captive and free-ranging populations, and morbidity rates correlate with population densities. Isolated cases of parapoxvirus-associated disease have also been reported in noncaprid species, such as blackbuck, mountain gazelle, and steenbok, and it is likely that the host range is broad.[348] Experimental infection causes mild disease in pronghorn, though natural infection has not been reported.[211]

Infection most likely occurs through defects in skin or mucous membranes. Virus in sloughed skin crusts in the environment may be infectious for prolonged periods and may be a source of indirect infection. ORFV is epitheliotropic and causes proliferative, necrotizing, and vesicular lesions on the skin and mucous membranes (Fig. 5.15). The muzzle, lips, and gingiva are most often involved, but lesions also occur on the tongue, periocular region, distal limbs, interdigital cleft, heel bulb, mammary gland, vulva, prepuce, and skin of the dorsum. Severe infections can inhibit eating and serve as a portal for secondary infections. Lesions progress from papules to flat vesicles to pustules that rupture and become covered by very thick crusts. Young animals are usually most severely affected, and infection may be fatal if lesions interfere with feeding or lead to secondary infections including cellulitis, stomatitis, aspiration pneumonia, or mastitis. Histologically, orf is characterized by marked epidermal proliferation, swelling of stratum spinosum cells, intra-epidermal vesicles and pustules, and dense superficial crusting. Intra-cytoplasmic, eosinophilic viral inclusion bodies may be present. Dermal lesions include edema and neutrophilic and mixed superficial inflammation. Molecular testing may be necessary to distinguish ORFV lesions from those caused by other parapoxviruses, such as bovine popular stomatitis virus and pseudocowpox virus, and *Capripoxvirus*. **Pseudocowpox virus** infection in a bison was associated with raised hyperkeratotic proliferative nodules on the ventral skin due to marked epidermal hyperplasia, parakeratotic hyperkeratosis, keratinocyte ballooning degeneration, and eosinophilic intracytoplasmic viral inclusions in keratinocytes.[353]

Viruses in the genus *Capripoxvirus* cause sheep and goat pox in small ruminants, and lumpy skin disease (LSD) in cattle. These viruses are serologically cross-reactive but have differing, minimally overlapping geographic

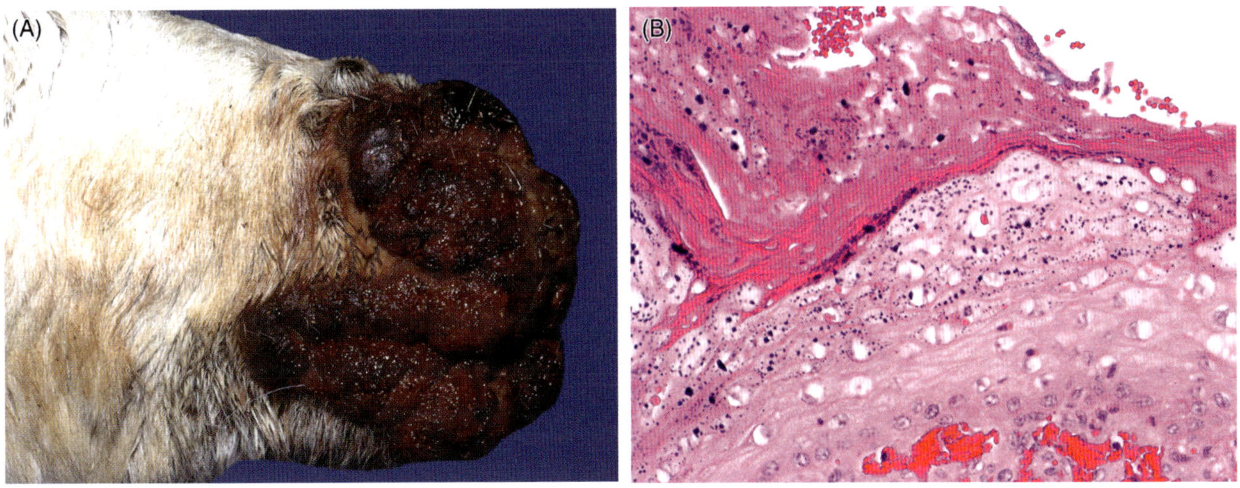

FIGURE 5.15 Proliferative and necrotizing dermatitis on the muzzle of a Dall's sheep infected with orf parapoxvirus. (A) Epithelial hyperplasia with degeneration and necrosis. (B) Vacuolar degeneration of subcorneal epithelial cells with eosinophilic intracytoplasmic inclusion bodies characteristic of parapoxvirus infection. *(Parts A, B: Photos Courtesy of University of Illinois, Zoological Pathology Program.)*

distribution. There are no reports of sheep or goat pox in nondomestic ruminants. LSD in cattle is a WOAH-listed, reportable disease caused by **lumpy skin disease virus, LSDV**. The virus is found in much of Africa and has recently been identified in the Republic of Georgia, Greece, India, the Middle East, Russia, and Turkey.[283,292] Unconfirmed LSD was reported in Arabian oryx in Saudi Arabia. A broad serosurvey of African wildlife revealed a low seroprevalence in giraffe, impala, kudu, reedbuck, springbok, and waterbuck.[158] Experimental infection of giraffe and impala calves resulted in fatal, systemic disease with proliferative and necrotizing dermatitis and panniculitis similar to classical LSD in cattle; African buffalo calves and adult wildebeest showed no clinical signs.[421] Gross lesions in the giraffe and impala included widespread dermal nodules up to 10 cm in diameter that turned black and ruptured with the release of serosanguinous fluid. The disease progressed over several weeks to include diffuse lymphadenopathy, multifocal mucosal and lingual ulceration, and necrosis leading to anorexia, severe weight loss, and death. Similar dermal nodules have been seen in LSD natural infections in springbok (Fig. 5.16). Microscopic lesions were predominantly characterized by epidermal proliferation, vacuolar degeneration and necrosis with chronic neutrophilic panniculitis and myositis, and varying degrees of acute and chronic dermal vasculitis and microthrombosis (eSlide 5.e3). Epithelial cells in and around the lesions contained intracytoplasmic inclusion bodies, and intranuclear inclusion bodies were present within affected skeletal myofibers or stromal cells. Macrophages within the lesions contained both intracytoplasmic and intranuclear inclusion bodies. Additional lesions included necrotizing esophagitis and lymphadenitis, mild multifocal acute hepatocellular necrosis, and suppurative bronchopneumonia. Definitive diagnosis is based on PCR, immunohistochemistry, and/or electron microscopy (Fig. 5.17).

A small number of mainly delta papillomaviruses, including **bovine papillomavirus 1 (BPV 1) and BPV 2**, have been shown to cause proliferative epithelial lesions and benign neoplasms in the skin and mucous membranes of the gastrointestinal and urinary tracts of nondomestic bovids (see “Neoplastic” section).[91]

FIGURE 5.16 Cutaneous infarcts on the head of a white color mutant springbok with Capripoxvirus infection (lumpy skin disease). Multifocal to coalescing round raised hairless scabs of necrotic skin.

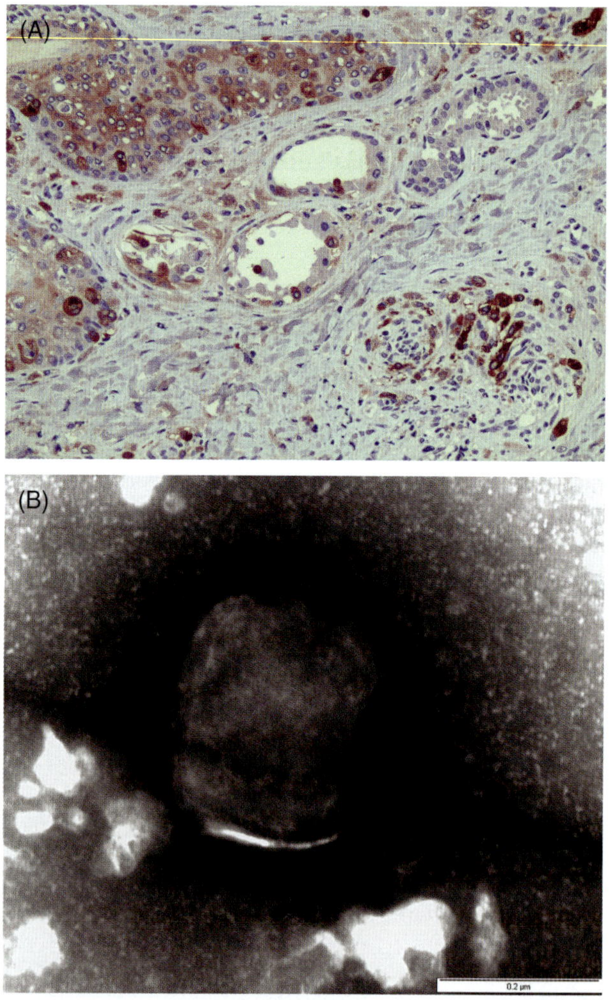

FIGURE 5.17 Lumpy skin disease in a springbok. (A) Positive immunohistochemical labeling of cytoplasmic viral antigen in epidermal keratinocytes and perivascular leukocytes. (B) Transmission electron microscopy image showing a pox viral particle in necrotic skin debris from a cutaneous scab using negative staining (see eSlide 5.e3). *(Photo Courtesy of the Electron Microscopy Unit, Faculty of Veterinary Science, University of Pretoria.)*

RNA Viruses

Bluetongue virus (BTV) causes **bluetongue (BT)**, a reemerging vector-borne hemorrhagic fever of ruminants. Biting midges (*Culicoides* sp.) act as biological vectors in tropical and temperate areas worldwide; and Cervinae may act as reservoirs (see Chapter 6: Cervidae). Vertical and oral transmission are less important in the disease epidemiology. There are at least 26 recognized BTV serotypes with considerable genetic variation among naturally occurring strains within each serotype.[225] The virus is endemic throughout tropical and subtropical regions. In the last decade there have also been numerous virus incursions into Northern Europe.

The severity of disease varies significantly with virus serotype and strain, host species, and breed.[224] For example, the strain of BTV-8 that emerged in northern Europe in 2006 is particularly pathogenic in multiple species and is associated with increased rates of trans-placental infection compared to most wild-type field strains. The disease is most severe in domestic sheep, with similar but more subtle lesions in cattle and nondomestic ruminants. During the BTV-8 Europe outbreak, significant disease was recorded in American and European bison, Alpine and Siberian ibex, muskoxen, mouflon, and yaks.[337] In 2018 Iberian ibex and domestic ruminant mortalities were caused by reemergence of BTV-4 strains.[140] Large outbreaks of BT have also been reported in pronghorn in North America.[375]

Bluetongue disease is characterized by widespread vascular injury, with the most significant lesions in the upper gastrointestinal tract, lungs, and skin. Clinically, animals are febrile, with variable combinations of erosions and ulcers in oral mucosa, muzzle, and teats; conjunctivitis; serous, mucous, or hemorrhagic nasal discharge; subcutaneous edema of the head, neck, and limbs; and hyperemia of the coronary band. Some animals may exhibit only lameness, and in some cases sudden death may occur with no other signs.[338] Cyanosis, edema, and ulceration of the lips, dental pad, and lateral tongue (from which the disease gets its name) are uncommon but have been observed in captive yaks.[239] Pulmonary edema can be severe. Grossly, there may be pleural and pericardial effusion and hemorrhage in multiple sites including upper gastrointestinal serosa and mucosa, omasal/abomasal orifice, lymph nodes, subcutis, and the subintima of the pulmonary artery. Acute hemorrhagic enteritis is reported in captive yaks dying from BT. Histologic lesions may include acute to chronic myocardial and skeletal muscle necrosis and hemorrhage. Small-vessel changes may be mild, consisting only of endothelial cell hypertrophy, and there may also be disseminated intravascular coagulation.[224] Prior to the emergence of the BTV-8 strain affecting northern Europe in 2006, transplacental transmission of BTV was uncommon in natural infections. Infected pregnant animals may abort, and fetuses infected early in gestation may be born with congenital defects including hydranencephaly.

Epizootic hemorrhagic disease virus (EHDV), like BTV, is a *Culicoides*-vectored orbivirus capable of infecting and causing hemorrhagic fever in domestic and wild ruminants. EHDV is an especially significant disease in cervids (see Chapter 6: Cervidae) but disease is also reported in cattle, and sporadic outbreaks have been reported in other species, such as bighorn sheep and yak.[272]

Foot and mouth disease (FMD) is a highly contagious disease of artiodactyls characterized by high morbidity but low mortality. FMD is a WOAH-listed, reportable disease that is endemic in large regions of Africa, Asia, and South America. Its importance lies in the economic implications of production losses in infected livestock, and significant trade restrictions imposed when **foot and mouth disease virus (FMDV)** is present. Conservation activities, such as translocation and reintroduction programs, can be severely limited by FMDV detection in species of concern.

There are seven FMDV serotypes (O, A, C, SAT 1, SAT 2, SAT 3, and Asia 1), and infection with one serotype does not confer immunity to another. Each serotype contains several antigenic subtypes or strains that can vary considerably in virulence. Numerous case reports describing both natural and experimental infection in nondomestic artiodactyls have been reviewed.[19,20,306,341,396] Outside of sub-Saharan Africa, infection of captive and free-ranging wildlife with FMDV tends to coincide with FMD outbreaks in domestic animals, with virus typically spreading from livestock to wildlife.[374] In sub-Saharan Africa, the African buffalo is the only nondomestic species conclusively shown to transmit virus and exhibit a prolonged carrier state, playing a role in the maintenance and spread of disease. Outbreaks of FMD in the Kruger National Park, South Africa, have occurred, often in impala, and viral sequencing has demonstrated that the source is often African buffalo. Infection in African buffalo is usually subclinical, though typical oral vesicles and ulcers have been reported. There have been no documented FMDV infections in the common hippopotamus and a serosurvey in an endemic FMDV region (Kruger National Park, South Africa) detected no evidence of exposure in this species. Reserve perimeter fencing and livestock vaccination are used to prevent spread between wildlife (primarily buffalo) and livestock.

The detailed pathogenesis of FMD in domestic cattle has been extensively reviewed.[19,215] Briefly, FMDV is transmitted via direct contact with body secretions or vesicular fluid of infected animals, or indirectly via aerosol or contaminated fomites. Long-distance airborne spread can occur. Primary replication occurs in the follicle-associated epithelium of the pharyngeal mucosa-associated lymphoid tissue, in pulmonary alveolar epithelium, as well as at the portal of entry and in regional lymph nodes (if infection occurs via skin abrasion). Viremia leads to widespread

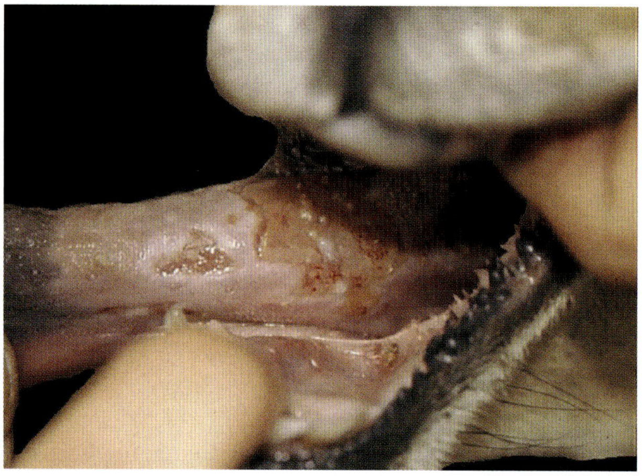

FIGURE 5.18 Foot and mouth disease in an impala. The extruded tongue shows multifocal irregularly shaped postvesicular ulceration. *(Photo Courtesy of R. Bengis, Department of Agriculture, Forestry and Fisheries, South Africa.)*

epithelial dissemination, and lesions develop in areas of mechanical or physiologic stress. The incubation period can range from 1 day to 2 weeks. Early lesions consist of mucosal or epidermal swelling, blanching, or reddening. Vesicles occur on the dental pad and tongue (Fig. 5.18); coronary band and interdigital region; mammary gland; and ruminal pillars. Vesicles begin in the stratum spinosum as epithelial degeneration and necrosis, ballooning degeneration, and intercellular edema that progress to form fluid-filled spaces whose surfaces often slough to form erosions. Vesicles can be very large, and vesicle fluid harbors large quantities of virus. In young animals the virus can be myotropic, leading to myocardial degeneration, necrosis, myocarditis, and sudden death. Free-ranging mountain gazelles in Israel showed mortality rates of 10%–50%.[36]

Clinical signs and gross lesions in nondomestic species are similar to those in domestic livestock. Significant FMD-associated disease has been documented in blackbuck, impala, mouflon, mountain gazelles, saiga, and urial.[175] Pronghorn have been shown to be susceptible experimentally.[323] Affected animals are often lame, may display arched backs, shake their feet, and lag behind the herd. Piloerection has been reported in impala, presumably reflecting fever, and hoof wall defects may be present for several months in recovered animals. Gross findings include vesicles that progress to erosions and ulcers on the mouth and feet, especially affecting the dental pad and tongue; coronitis; and pale streaking of the myocardium. During outbreaks in mountain gazelles, sloughing of horns and degeneration and necrosis in both cardiac and skeletal muscle were also seen. Pancreatic necrosis and pancreatitis are also described in natural outbreaks and experimental infections in mountain gazelles, and in experimentally infected pronghorn antelope.

Differentials for the gross lesions of FMD in wild artiodactyl species include the relevant vesicular viral diseases vesicular stomatitis, swine vesicular disease, and vesicular exanthema in suids. Other etiologies of mucosal and cutaneous ulceration must also be considered, including bluetongue virus, epizootic hemorrhagic disease virus, malignant catarrhal fever virus, mucosal disease/BVDV, photosensitization, and exposure to irritant or caustic chemicals. As FMD is a WOAH-listed disease, diagnosis of suspect cases should be referred to the relevant regulatory authorities. Assays including virus isolation, detection of viral antigen by ELISA, and identification of viral nucleic acid by reverse-transcriptase PCR can be performed at accredited laboratories.

Death due to **encephalomyocarditis virus** (ECMV) infection has been documented in multiple species in zoological settings, including pygmy hippopotamuses, Thomson's gazelle, and an oryx-addax cross.[318,399] Outbreaks of ECMV infection tend to occur across multiple mammalian taxa and are associated with sudden death. Lesions of acute to subacute, necrotizing pleocellular encephalitis and myocarditis (Fig. 5.19) are accompanied by pericardial effusion, pulmonary edema, or other signs of acute heart failure. Myocardium is the most useful tissue for laboratory confirmation via virus isolation or PCR. Rodent control is critical for management in zoo settings.

Border disease virus (BDV-4) has been associated with mass mortality events in free-ranging chamois since 2001.[16,126] Phylogenetic analysis indicates that infection of Pyrenean chamois, with an Ovid BVD-4a clade, occurred in the early 1990s.[222] In contrast to BDV infection in domestic sheep, natural, postnatal BDV infection in chamois is usually associated with a high mortality rate, though some populations appear to be resistant to disease.[233] Persistent infection with BDV has not been definitively documented in chamois, though vertical transmission has been confirmed experimentally.[59] Abortions in experimentally infected pregnant Pyrenean chamois have been implicated as one factor in population decreases in this species.[235] Clinically, affected animals exhibit depression and decreased flight distance, or are found dead. Gross lesions include emaciation, patchy to widespread alopecia, and epidermal hyperpigmentation.[232] Secondary bacterial pneumonia and subcutaneous abscesses may also be present. Cerebral spongiosis, perivascular edema, neuronal degeneration and necrosis, glial nodules, and mononuclear perivascular cuffing occur. Cutaneous follicular atrophy, excessive tricholemmal keratin deposition (flame follicles), telogenization of follicles, acanthosis, melanosis, and orthokeratotic hyperkeratosis may be seen. Other reported lesions include lymph node cortical necrosis and histiocytosis. Pestiviral antigen may be present in splenic and lymph node macrophages; epithelial cells of skin, kidney, or rumen; and in cells in the brain and bone marrow. Diagnosis of BDV in chamois can be confirmed via RT-PCR, virus isolation, antigen-capture ELISA, or virus neutralization tests.

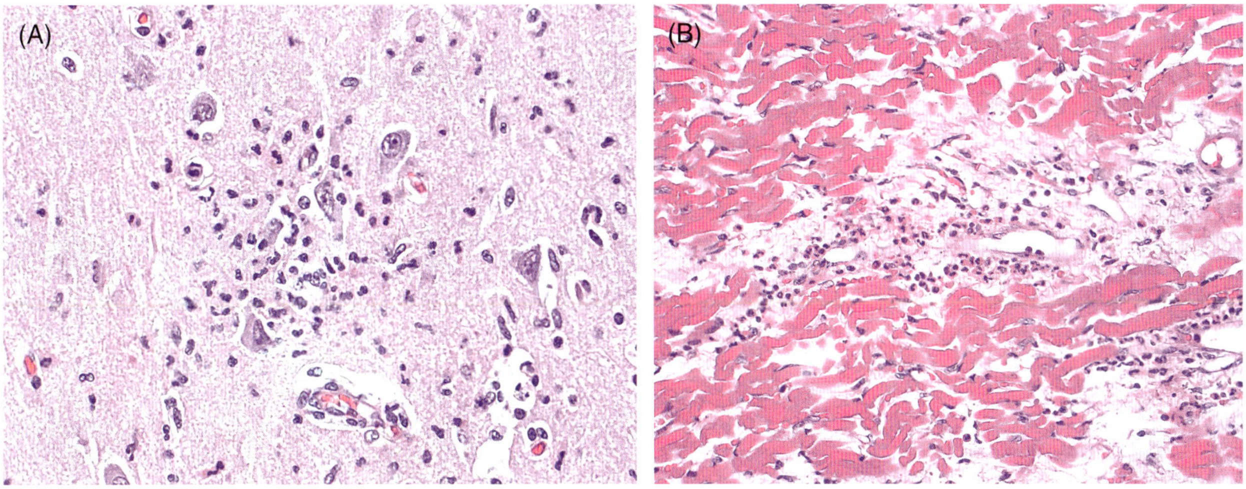

FIGURE 5.19 Encephalomyocarditis virus in a pygmy hippopotamus. (A) Small numbers of lymphocytes, plasma cells, and neutrophils infiltrate the neuropil. (B) Moderate numbers of neutrophils with fewer plasma cells and lymphocytes infiltrate the myocardium, with associated edema and mild hemorrhage. *(Photo Courtesy of M. Garner, Northwest ZooPath.)*

Most nondomestic ruminants are thought to be susceptible to **Bovine Viral Diarrhea Virus (BVDV)** infection. Persistent infection has been demonstrated in unthrifty American bison, while viruses isolated from eland were considered spillover infections from cattle.[157,391] On the other hand, captive persistently infected symptom-free mousedeer demonstrated vertical transmission of the virus as well as transmission to cattle.[379] Immunosuppression due to BVDV infection has been implicated in the pathogenesis of periodic epizootics of bacterial pneumonia and/or in bighorn sheep and mountain goats.[416] BVDV1a was isolated from the spleen of an affected mountain goat kid with bacterial pneumonia and necrotizing mesenteric lymphadenitis. Positive immunohistochemical labeling was seen in various cells in multiple organs in two captive mountain goat kids, one of which died suddenly due to enteritis of unknown cause; the other died due to bacterial pneumonia.[266]

Peste des petits ruminants (PPR) is a WOAH-listed, reportable disease that primarily affects domestic goats and sheep with spillover into nondomestic ruminants. PPRV is closely related to **rinderpest virus**, and the two morbilliviruses share common antigenic determinants. Significant clinical disease and mortality have been reported in nondomestic Caprinae (Afghan markhor, argali, bharal, Barbary sheep, Capra, and Nubian ibex) and several gazelle species (including Dorcas, goitred, and Thomson's gazelle), as well as impala, bushbuck, gemsbok, and springbok.[1,35,128,195,216] An outbreak of PRR in 2016–17 resulted in catastrophic losses of various free-ranging species, including the critically endangered Mongolian saiga; the magnitude of the outbreak may have been influenced by the immunologically naive status of the population.[216] Genetic analysis suggests that the virus spilled over from livestock in China and was circulating for at least 6 months before the outbreak occurred.[31] Syncytia associated with gingival erosions along the tooth margins and multiple foci of hepatocellular and biliary epithelial necrosis (with or without eosinophilic intracytoplasmic and intranuclear viral inclusions) are described.[313] Serologic evidence of exposure has been documented in several other species, including free-ranging Grant's and Thompson's gazelle, impala, and wildebeest in Tanzania.[182,226]

PPR is an acute disease, with an incubation period of 3–10 days.[281] Animals are febrile with serous to mucopurulent oculonasal discharge and often have pneumonia and hemorrhagic diarrhea. Oropharyngeal erosion, ulceration, and pseudomembrane formation may be seen (Fig. 5.20). Gastrointestinal lesions can include pseudomembranous esophagitis and hemorrhagic enterocolitis. Severe lymphoid necrosis is often evident in the

FIGURE 5.20 Peste des Petits Ruminants (PPR) in an Arabian mountain gazelle. Severe widespread palatal epithelial erosion and ulceration. *(Photo Courtesy of J.W. Eksteen, Kroonstad Animal Hospital, South Africa.)*

gut-associated lymphoid tissue (GALT), with associated intestinal mucosal ulceration in regions overlying Peyer's patches. In peracute cases in gazelles lesions may be limited to abomasal ulcers, perhaps with fibrinous peritonitis. In contrast to rinderpest (see below), animals with PPR may also have fibrinonecrotizing tracheitis, fibrinous pleuritis, and cranioventral lung consolidation. Histologically, additional lesions can include lymph node and splenic lymphocytolysis and lymphoid depletion, as well as random acute hepatic necrosis. Bronchointerstitial pneumonia may be accompanied by necrosis of large and small airway epithelium, type II pneumocyte hyperplasia, and alveolar epithelial syncytial cells. Syncytial cells may be also present in oral mucosa, liver, and lymphoid tissue. Eosinophilic, intranuclear, and intracytoplasmic viral inclusion bodies may be found in syncytia, in epithelial cells in affected tissues, and in monocyte–macrophage lineage cells in lymphoid tissues.

Laboratory confirmation of PPRV infection is by competitive ELISA, immunocapture ELISA, PCR, or virus isolation. Virus neutralization tests are considered to be the gold standard but are time consuming. PPRV will cross-react with rinderpest virus in agar gel immunodiffusion and complement fixation tests. Depending on lesion distribution, differentials include caprine pleuropneumonia, pasteurellosis, bluetongue, bovine viral diarrhea, and foot and mouth disease. Immunosuppression due to lymphoid depletion can lead to secondary infections that can complicate the diagnosis.

Rinderpest, a morbillivirus closely related to PPRV, was declared to be globally eradicated in 2011.[181] Rinderpest was a severe and highly fatal disease of cattle, with a potentially wide host range in other artiodactyls.[345] Confirmed outbreaks of rinderpest were documented in several wild bovid species including the highly susceptible African buffalo, common eland, and lesser and greater kudu.[24,201,203] Clinically, animals had diarrhea, oculonasal discharge, and blindness. Gross lesions are similar to PPR with less respiratory system involvement: mucosal erosion and ulceration, most frequently in the oropharynx, abomasum, and colon. Linear submucosal congestion and hemorrhage along the longitudinal folds of the colonic and cecal mucosa (tiger striping) are frequently cited, though this is a nonspecific finding. Lymph nodes are edematous and hemorrhagic, and Peyer's patches are necrotic with ulceration of the overlying intestinal mucosa. Necrosis, congestion, and inflammation may also be present in the urogenital tract. Ocular lesions in some wild bovids and giraffes tend to be more pronounced than in domestic cattle and include corneal opacity due to keratoconjunctivitis and corneal ulceration, uveitis, and cataract formation.[331] Dermatitis and plaque-like hyperkeratosis are described in African buffalo, and swollen joints and tenosynovitis in lesser kudu, though these lesions may reflect secondary infection. Histologic lesions include mucosal epithelial cell necrosis, erosion, and ulceration; lymphocytic to mixed-cellular enterocolitis; and hepatic, splenic, renal tubular, and salivary and bile duct epithelial necrosis. Syncytial cells with or without nuclear and cytoplasmic inclusion bodies may be present in affected mucosal epithelium and lymphoid tissue.

Rabies virus infection is a disease of carnivores (see Chapter 9: Canidae, Ursidae, and Ailuridae); arteriodactylids are largely infected as spillover dead-end hosts. However, regular catastrophic epizootics of the disease occur in free-ranging greater kudu in Namibia, with losses of up to 70% of animals. Kudu are primarily infected with jackal viral strains.[231] Horizontal transmission, through communal browsing of thorny bushes that cause oral lesions, is thought to contribute to the epidemiology of the disease in this social antelope.[23] Diagnosis is by detection of the antigen using immunofluorescent or immunohistochemical tests.

Bacteria

Mycobacteria are nonmotile, nonspore-forming coccobacilli. **Bovine tuberculosis (BTB)**, caused by *Mycobacterium bovis*, is a WOAH-listed, reportable, zoonotic disease with potentially significant impacts on trade, wildlife conservation, and human public health. It is found worldwide in domestic cattle, but can also affect wild and captive nondomestic ruminants, and other wildlife species.[93] Clinical cases of BTB have become less common in cattle since the development of eradication programs in many countries, and regulatory attention has focused increasingly on wildlife. The presence of *M. bovis* in a population can limit wildlife translocations or reintroduction programs, having conservation effects that extend beyond direct impacts of disease on mortality or reproduction rates.[411] Lesions indistinguishable from BTB can be caused by other species in the *M. tuberculosis* complex, such as *M. caprae*, in European ungulates and *M. tuberculosis* in several species of nondomestic bovids as well as free-ranging giraffe.[144,163] Antemortem diagnosis in nondomestic ruminants, especially for screening of clinically healthy animals, can be challenging because the classic intradermal tuberculin test lacks sensitivity and specificity. Several direct and indirect diagnostic tests have been recently developed.[213]

Globally, the management of *M. bovis* at the wildlife–domestic animal interface is challenging, complex, and sometimes controversial. While many species are susceptible to infection, only a proportion can act as maintenance hosts capable of transmitting disease. Reported BTB reservoir species include the African buffalo (South Africa), lechwe (Zambia), white-tailed deer (United States), and wood bison (Canada).[250,271] Greater kudu may act as maintenance hosts at high population densities but are usually considered spillover hosts.

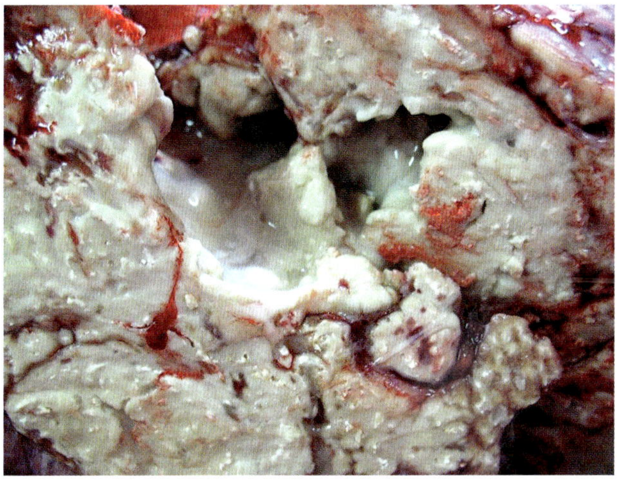

FIGURE 5.21 Bovine tuberculosis necrogranuloma in a nyala, caused by infection with *Mycobacterium bovis*. A region of central liquefactive necrosis is surrounded by areas of caseous necrosis. These may be encircled by a fibrous capsule (see eSlide 5.e4). *(Photo Courtesy of R. Bengis, Department of Agriculture, Forestry and Fisheries, South Africa.)*

Chronicity and severity of disease and characteristics of lesions depend on infectious dose, type of immune response by the host, and the tissue inoculated. Lesions may be rare and require a painstaking search in subclinical cases where BTB is suspected. African buffalo, which act as reservoir hosts, may maintain subclinical infections for several years prior to manifesting clinical signs that include weight loss, cough, and dull coat.[321] Lesions, as in cattle, are primarily respiratory and include classic, encapsulated granulomas in the lungs and lymph nodes of the thorax and head, particularly mediastinal and retropharyngeal (Fig. 5.21; eSlide 5.e4). Lesions in the mesenteric lymph nodes in lechwe occur in some cases. Both captive and free-ranging greater kudu exhibit "almost pathognomonic" swelling in the parotid and submandibular areas, often with draining fistulae, thought to be due to innoculation of oral lesions in these browsers.[190] Hematogenous dissemination can occur.

Histologically, granulomas have a central region of necrotic cellular debris or caseous material, surrounded by a layer of activated, epithelioid macrophages and multinucleated giant cells. The outer rim often contains lymphocytes and a fibrous connective tissue capsule, depending on species and chronicity. The organisms stain positively with Gram, but acid-fast stains such as Ziehl-Neelsen and Fite Faraco may be more sensitive. Lesions may be paucibacillary, so definitive diagnosis may require culture of lesions in appropriate media. PCR on fresh or paraffin-embedded tissue is also available but sensitivity will depend on the number of bacilli in the sample.

Infection with ***Mycobacterium avium* subsp. paratuberculosis (MAP)** causes **paratuberculosis** or **Johne's disease**. Johne's disease is characterized by chronic, granulomatous enteritis and progressive wasting that principally affects ruminants but can affect a wide range of captive and free-ranging, nondomestic ruminant species.[61,160,213,329,358,409] Johne's disease and the presence of MAP infection in a herd can impede animal transfers between institutions. This can have important implications for captive breeding efforts in support of endangered or threatened species.

The epidemiology of MAP in captive ruminants is assumed to be similar to that observed in domestic cattle and sheep. Fecal–oral transmission is the primary route of infection, and young calves or lambs are much more susceptible than adults. In utero infection and neonatal exposure via milk or colostrum are also possible.[403] Clinical signs include chronic wasting and diarrhea, but the latter may be intermittent. Submandibular or generalized edema and effusion and serous fat atrophy may be present in hypoproteinemic animals.[406] Gross lesions in advanced cases include intestinal mucosal thickening, particularly distal small intestine, mesenteric lymphadenomegaly, and mesenteric lymphangitis. In nondomestic ruminants lesions may be mild relative to classic Johne's disease in cattle, and in early cases there may be few gross changes.

Microscopic lesions of Johne's disease include granulomatous enterocolitis, mesenteric lymphadenitis, and lymphangitis, with intracellular, acid-fast bacilli (Fig. 5.22; eSlides 5.e5 and 5.e6). Johne's disease in domestic sheep may be tuberculoid (paucibacillary), or lepromatous (pleuribacillary), but lesions in nondomestic ruminants may not always fall clearly into one category. The tuberculoid form is characterized by infiltrates of epithelioid macrophages, multinucleated giant cells, and lymphocytes, necrosis with or without mineralization, and often few positive bacteria on acid-fast stains.[405] Similar lesions in bighorn sheep may be pleuribacillary. The lepromatous form is characterized by more diffuse lamina proprial infiltration of epithelioid macrophages with numerous intracellular acid-fast bacilli. This form has been described in captive scimitar-horned oryx and takin.[160,264] Both forms can be accompanied by villus blunting and fusion, wide separation of intestinal glands, and transmural inflammatory cell infiltrates.

In early or mild cases, especially if paucibacillary, aggregates of epithelioid macrophages may be uncommon and small, and acid-fast staining and examination of multiple sections are required for diagnosis. Fecal culture using specialized mycobactin-supplemented media is generally considered to be the gold standard for diagnosis of MAP infection in live animals. Fecal culture is reported to have specificity approaching 100%; however, false-positive results can occur due to "pass-through" effect if an individual animal ingests large numbers of organisms from a contaminated environment.[371] Though more sensitive than other antemortem tests (including IS900 PCR), fecal culture may not identify all infected animals, especially in early stages where shedding may be minimal or intermittent.[269]

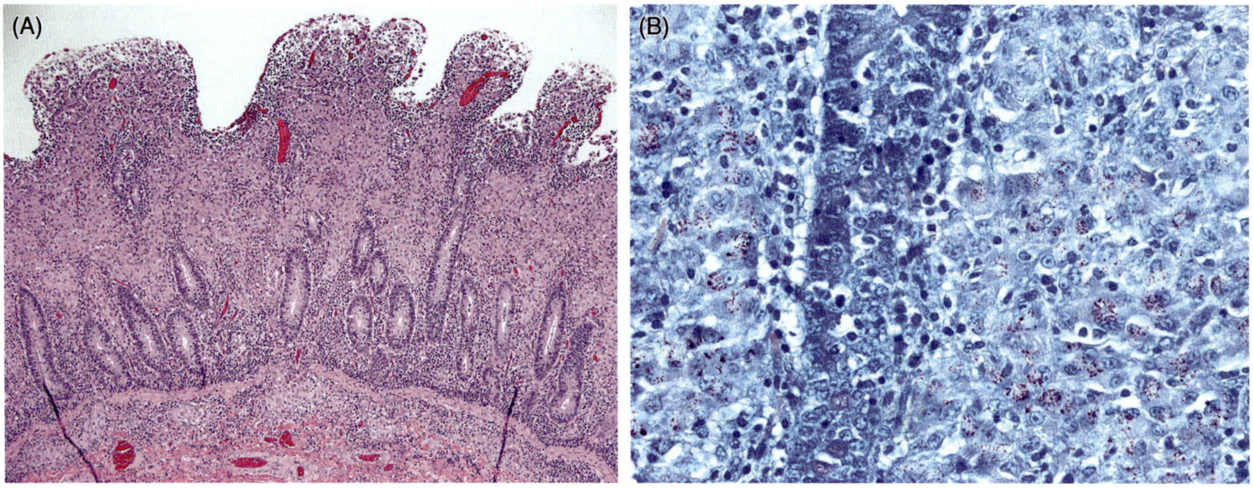

FIGURE 5.22 Granulomatous enteritis in a urial with Johne's disease caused by infection with *Mycobacterium avium* subsp. *paratuberculosis*. (A) The mucosal lamina propria is severely expanded by sheets of epithelioid macrophages which separate and replace the crypts, and there is blunting and fusion of villi. Inflammation extends to the muscularis mucosa. (B) Macrophages and multinucleated cells contain clusters of red-staining, acid-fast bacilli consistent with mycobacteria. Ziehl-Neelsen. (see eSlides 5.e5 and 5.e6). *(Parts A, B: Photos Courtesy of San Diego Zoo Wildlife Alliance.)*

Isolation of organisms or PCR from intestinal or lymph node tissue samples can provide a definitive diagnosis and may be more sensitive than histologic examination in early cases. As with fecal culture, isolation from tissue samples requires technical expertise, specialized media, and prolonged incubation time. More than one test is recommended to confirm the diagnosis.[329]

Anthrax is a WOAH-listed, reportable, zoonotic bacterial disease with worldwide distribution. It is primarily a disease of herbivores, though any mammalian species is potentially susceptible. Ecological and epidemiological factors that contribute to outbreaks of disease in wildlife are reviewed elsewhere.[173] The causative agent, ***Bacillus anthracis***, is a gram-positive, spore-forming, bacterial rod that gains entry to a host via skin or mucous membrane defects, ingestion, or inhalation. Infection by spores is followed by rapid replication in regional lymph nodes, bacteremia, and toxemia. Massive elaboration of exotoxins leads to widespread, acute edema, hemorrhage, and terminal shock. Death may occur within hours in highly susceptible species. Carcasses often become bloated very quickly due to accelerated putrefaction, and they may exhibit petechial and ecchymotic hemorrhages and leak hemorrhagic fluid from body orifices (Fig. 5.23A).

If anthrax is suspected based on the appearance of the carcass, a blood smear from distal tail or ear (areas likely to be least autolyzed) should be examined. Bacilli form pairs or chains in vivo and have a characteristic clear capsule with blunt to squared ends with a Giemsa stain (Fig. 5.23B, see also Chapter 2: Wildlife Necropsy).[33] Necropsy should not be performed without appropriate personal protective equipment due to biosafety concerns. Gross lesions are those of septicemia and include generalized edema, splenomegaly with a soft, gelatinous texture, and multicavitary effusions. Confirmation of the diagnosis typically relies on bacterial culture from fluid draining from body orifices or an unstained blood smear. Serology and molecular techniques are available but usually limited to research settings. Clinical samples and laboratory cultures of anthrax require biosafety level 3 facilities.

Anthrax has been reported in numerous wild artiodactyl species, including wood bison in northern Alberta and Northwest Territories, Canada.[267] Greater kudu, nyala, waterbuck, and roan antelope are reportedly more susceptible than other species, such as the African buffalo, and consistently exhibit higher levels of bacteremia in outbreaks in Southern Africa.[73,81,223] Mortality due to anthrax is also high in hippopotamuses when compared to other species, and sporadic mass outbreaks occur in hippos in Uganda and Zambia.[155,377] Facultative carnivorous behavior has been proposed as a contributing factor to increased mortality and large-scale outbreaks in hippos.[108] Poverty and hunger in communities in or around wildlife preserves contribute to the significant zoonotic risk associated with these outbreaks.[215]

Hemorrhagic septicemia (HS) is an acute, epizootic, highly fatal disease of ruminants caused by infection with certain serotypes of ***Pasteurella multocida***.[279] Hemorrhagic septicemia is a WOAH-listed and reportable disease that occurs in Asia, Africa, the Middle East, southern Europe, and North America.[352] The disease is most common in water buffaloes and domestic cattle but has been reported in African buffalo, bighorn sheep, American bison, chamois, muskoxen, pronghorn, and various cervid species (see Chapter 6: Cervidae). *P. multocida* serotype B was implicated in a catastrophic mortality event in free-ranging saiga antelope in central Kazakhstan in 2015 in which more than 200,000 adults died over a period of 3 weeks.[204] Warm

FIGURE 5.23 Anthrax in an African buffalo. (A) Blood-stained fluid drains from mouth and nares with marked facial swelling. (B) Large bacterial rods with square ends and an eosinophilic capsule. Giemsa. *(Photo Courtesy of R. Bengis, Department of Agriculture, Forestry and Fisheries, South Africa.)*

humid conditions, congregation of migratory animals, and unknown other factors may have triggered the outbreak.[204,327] *P. multocida* is a common component of the normal nasopharyngeal flora of many ruminants and usually acts as an opportunistic pathogen (e.g., as a component of bighorn sheep pneumonia). However, in cases of hemorrhagic septicemia, serotypes B2, E2, or, occasionally, other serotypes of *P. multocida* can act as primary pathogens and cause rapidly fatal septicemia with case fatality rates approaching 100%. *P. multocida* has many virulence factors that likely play a role in the rapid progression of disease, though endotoxin may be the most significant.

Hemorrhagic septicemia is typically a peracute to acute disease, and animals may be found dead or die less than 24h after initial clinical signs. Clinically, there is fever, excessive salivation, dyspnea, recumbency, and sometimes diarrhea that may be hemorrhagic. Grossly, petechial and ecchymotic hemorrhages may be present in the subcutis, serosa, endocardium, lymph nodes, and muscle (Figs. 5.24A, B). In cases of longer duration there may be prominent subcutaneous edema in the submandibular, cervical, and brisket regions. The lungs may be congested or consolidated, with fibrinous pleuritis or pneumonia and variably hemorrhagic effusions in body cavities. Histologically, lesions are those of septicemia and include acute hemorrhagic gastroenteritis, fibrinohemorrhagic interstitial pneumonia, pulmonary edema, and randomly distributed foci of acute inflammation in multiple sites. Diagnosis may be suspected based on epidemiologic and postmortem findings and on the presence in blood smears or impression smears of subcutaneous pharyngeal edema fluid of short bacilli (Fig. 5.24C). Bacteria are gram negative and show bipolar staining with methylene blue. Confirmation of hemorrhagic septicemia requires culture of blood, heart swab, or bone marrow, followed by identification of capsular serotype (A, B, D, E, or F) and somatic type (1–16). Molecular techniques are also available to identify specific serotypes.

Pneumonic pasteurellosis can be caused by infection by gram-negative bacteria in the family *Pasteurellaceae*. *P. multocida*, *Mannheimia haemolytica*, and *Bibersteinia trelahosi* have been reported in nondomestic ruminants.[251] Infection is usually opportunistic and disease tends to occur in association with primary viral or mycoplasmal infections, trauma, stress, or parasitic infections (see bighorn sheep pneumonia).[251,327] Lesions include hemorrhagic to suppurative bronchopneumonia and fibrinous pleuritis (Fig. 5.25).

Contagious caprine pleuropneumonia (CCPP) and **contagious bovine pleuropneumonia (CBPP)** are WOAH-listed, reportable diseases caused by *Mycoplasma* sp. bacteria belonging to the "mycoides cluster."[278] CCPP, caused by ***Mycoplasma capricolum*** subsp. ***capripneumoniae (Mccp)***, is a severe disease that exclusively affects the respiratory system. CCPP is primarily a disease of domestic goats, but mortality has also been reported in captive, nondomestic caprids (Nubian ibex, Laristan mouflon), gazelles (sand gazelles, gerenuk), an Arabian oryx

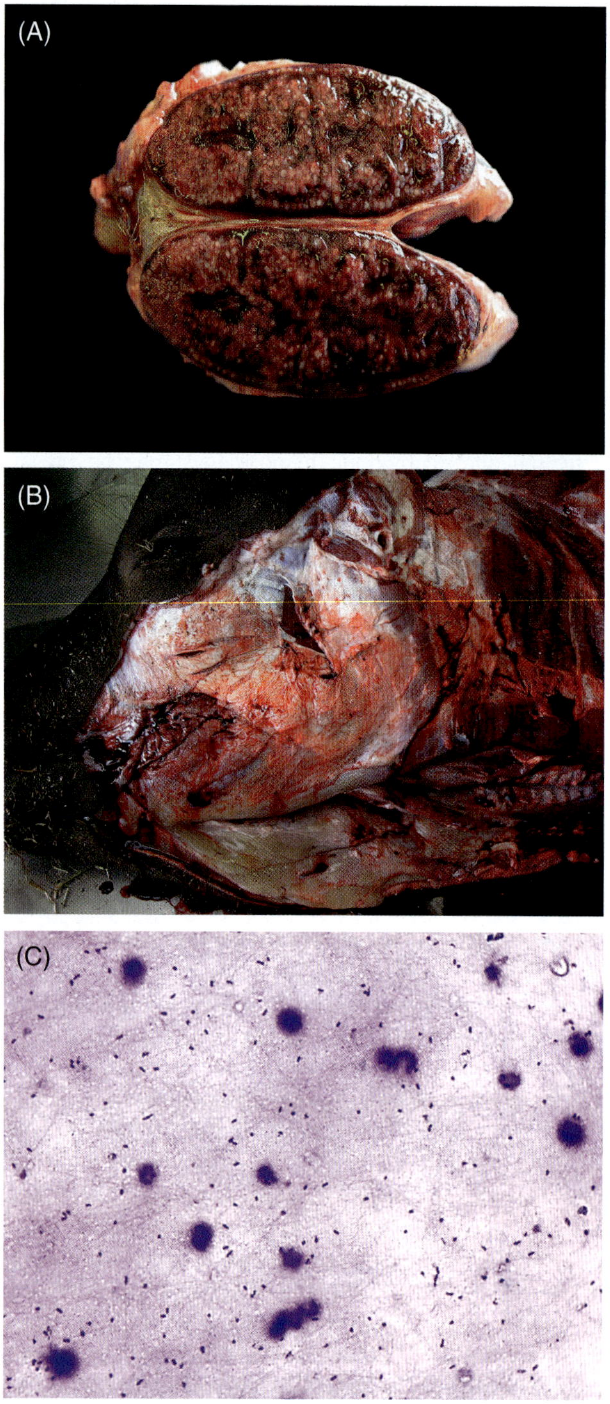

FIGURE 5.24 Hemorrhagic septicemia in an African buffalo. (A) Marked lymphoid hyperplasia and hemorrhage in a retropharyngeal lymph node. (B) Submandibular subcutaneous protein-rich edema. (C) Impression smear from B exhibiting innumerable short (1 to 1.3 μm) bacterial rods often showing bipolar staining. Many extravasated degenerating erythrocytes and amorphous eosinophilic material (protein-rich edema) are seen in the background. Diff-Quik®.

in the Arabian Peninsula, free-ranging Tibetan antelope in China, and markhors in Tajikistan.[15,66,268,290,422] CBPP, caused by ***Mycoplasma mycoides* subsp. *mycoides***, is

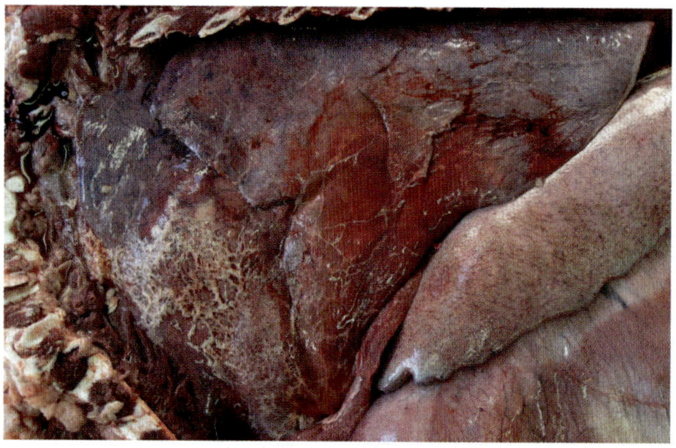

FIGURE 5.25 Pneumonic pasteurellosis in an African buffalo. Fibrinous pleuropneumonia with consolidation of the cranioventral lung lobes and a fibrin mat on the pleura.

primarily a disease of bovids of genera *Bos* and *Bubalas* and has not been widely reported in nondomestic wildlife species. Both diseases are characterized by pleuropneumonia with a thick layer of pleural fibrin and abundant, straw-colored pleural effusion. Lungs are often dark red to gray with lobular areas of consolidation and there may be a granular texture on cut section. Fibrinous pericarditis is often present. Pulmonary sequestra may form in chronic cases and have been reported in gazelles with CCPP. Histologically, serofibrinous to chronic fibrinonecrotic pleuropneumonia and interstitial pneumonia with prominent edema and neutrophilic to lymphocytic interstitial and alveolar infiltrates is seen. There may be vasculitis with thrombosis and coagulative necrosis. Differential diagnoses include *Mannheimia haemolytica* infection, PPR, or other mycoplasmas. Other organisms from the "mycoides" cluster can also cause similar disease (Fig. 5.26). Definitive diagnosis can be achieved using PCR. Culture requires enriched media and prolonged incubation time. Serologic testing is also available, though some cross-reactivity can occur with other *Mycoplasma* species.

Mycoplasma bovis is associated with lung lesions in free-ranging pronghorn. Lesions include fibrinous pleuropneumonia, fibrinonecrotic and suppurative bronchointerstitial pneumonia, and variably mineralized caseonecrotic foci. Concurrent *Trueperella pyogenes* and *Mannheimia* and *Histophilus* spp. co-infections and lungworm infections are inconsistently found.[228] In farmed bison lameness with swollen joints, dyspnea, coughing, and poor body condition are reported.[50]

Population-limiting, **epizootic bronchopneumonia of bighorn sheep (bighorn sheep pneumonia)** has been documented in North America since early in the 20th century. Primary etiology or etiologies have long been debated. At varying times, lungworm infection (*Protostrongylus* sp.), *Pasteurellaceae* (*Bibersteinia trelahosi*, *Mannheimia haemolytica*, *Pasteurella multocida*), respiratory viruses (bovine parainfluenza-3, bovine respiratory

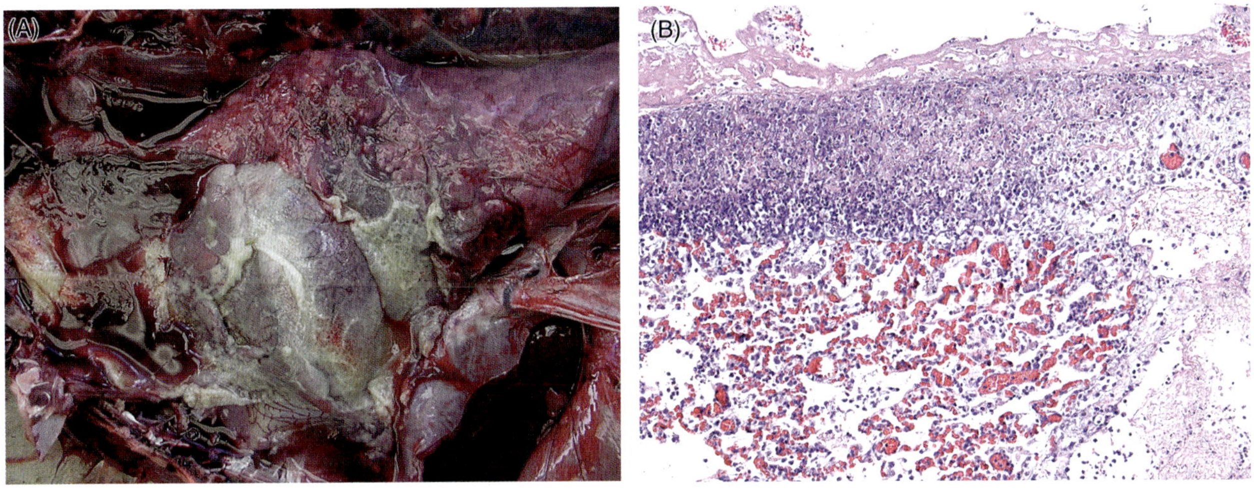

FIGURE 5.26 Fibrinous pleuropneumonia and epicarditis in a Vaal rhebok infected with a mycoides-cluster *Mycoplasma* sp. (A) The ventral lung lobe is consolidated and the pleural and pericardial surfaces covered by a pale yellow fibrinous exudate. (B) The pleural surface and subpleural alveoli are obscured by a band of fibrin that entraps large numbers of neutrophils and cellular debris. The inflammation extends along interlobular septa to the right of the image. *(Parts A, B: Photos Courtesy of San Diego Zoo Wildlife Alliance.)*

syncytial virus), and *Mycoplasma ovipneumoniae* have been proposed as underlying causes and/or contributing factors.[39,63,251,332] *Pasteurellaceae* including leukotoxin-positive (lkt +) *M. haemolytica* and *B. trelahosi* are often isolated from pneumonic bighorn sheep, and lesions compatible with lkt-induced leukocytolysis may be present.[136,415,418] Several recent studies have provided support for a primary role of *M. ovipneumoniae*.[39] Epizootic bighorn sheep pneumonia has been linked to contact with domestic sheep[63] and occasionally other domestic ruminants including cattle.[415] In a naive population animals of all ages may be affected and mortality rates can be greater than 90%. Subsequently, disease may be maintained in a population and pneumonia-related mortality mainly occurs in lambs.

The disease is characterized by acute to chronic, fibrinous, suppurative, or necrotizing sinusitis and bronchopneumonia sometimes accompanied by pleuritis (Fig. 5.27). Ventral regions of lung are dark red, firm, and sink in formalin. Chronic lesions include abscesses, bronchiectasis, and pleural adhesions. Histologically, lesions include acute fibrinous or necrotizing bronchopneumonia and pleuritis; acute hemorrhage; bronchiolar epithelial hyperplasia; lymphocyte cuffing around airways and extending into adjacent tissue; neutrophilic exudate in lumina of bronchioles, bronchi, and some alveoli; and alveolar histiocytosis. Bacterial colonies may be present. In some cases there is necrosis of neutrophils and macrophages with formation of "oat cells" (necrotic neutrophils with distinctive streaming hyperchromatic nuclei). Concomitant lesions can include tracheitis, rhinitis, sinusitis, and suppurative otitis media with osteomyelitis.

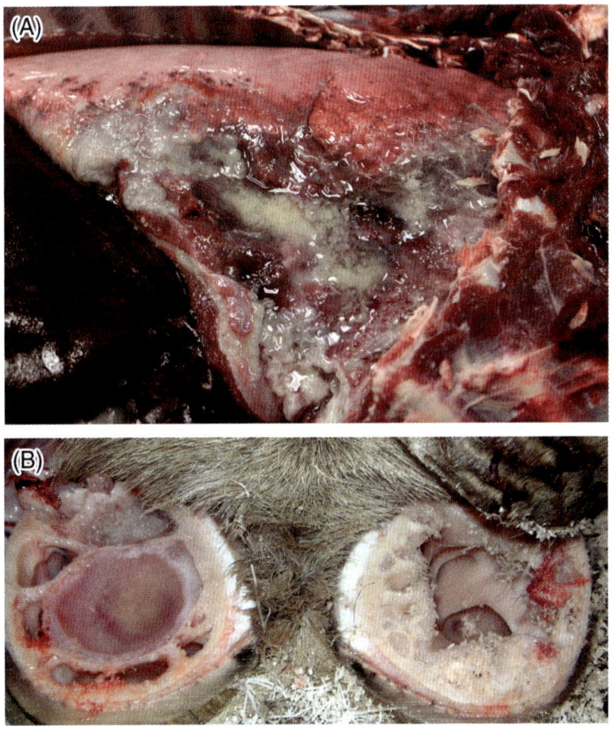

FIGURE 5.27 Epizootic bronchopneumonia of bighorn sheep. (A) Fibrinosuppurative exudate covers the consolidated, middle, and cranioventral lung. (B) Purulent material within the cornual diverticulum of the frontal sinus (visible when the horns are removed). *(Parts A, B: Photos Courtesy of Washington Animal Disease Diagnostic Laboratory.)*

Outbreaks of **infectious keratoconjunctivitis (IKC)** have been described in Alpine and Pyrenean chamois, ibex, other European caprinae, and wild bighorn sheep in North America.[17,137,240] In European species ***Mycoplasma conjunctivae*** has been identified as a primary etiologic agent, though other bacteria may be present.[137] *M. conjunctivae* is transmitted readily through aerosol, direct contact via ocular secretions and exudate, or indirectly by insects. Histologically, there is mixed mononuclear conjunctival infiltrate. Over days to weeks, lesions progress to keratitis with mononuclear to neutrophilic, limbic and corneal infiltrates, edema, and neovascularization accompanied by conjunctival lymphofollicular hyperplasia. Corneal edema, erosion, ulceration, and stromal necrosis can progress to perforation in advanced cases, with anterior synechiae and staphyloma formation. Corneal fibrosis may be present in animals that survive. Death during outbreaks of IKC is usually secondary to blindness with lesions of trauma, predation, or emaciation.[17,240] Laboratory confirmation of *M. conjunctivae* keratoconjunctivitis is through mycoplasma-specific culture techniques on conjunctival swabs, or by PCR.

In an outbreak of IKC in bighorn sheep in North America, inclusions consistent with **Chlamydia** sp. were identified on histologic examination and *Chlamydiaceae* were confirmed by immunofluorescent antibody testing on conjunctival swabs.[245] Lesions were similar to those described in mycoplasma-associated IKC. Sporadically, other organisms, such as ***Moraxella ovis***, ***Staphylococcus aureus***, ***Corynebacterium* sp.**, and ***Rickettsia conjunctivae*** are detected in cases of IKC.

Brucellosis is a zoonotic disease caused by gram-negative, intracellular bacteria in the genus **Brucella**. The primary *Brucella* species that cause disease in the nondomestic *Bovidae*, *Giraffidae*, *Antilocapridae*, and *Hippopotamidae* are *B. abortus* and *B. suis*, with fewer reports of *B. melitensis*.[139] Experimental infection with *B. ovis* has been investigated in captive mouflon and bighorn sheep but natural infection has not been recognized in wild populations. All four *Brucella* species are recognized as notifiable pathogens listed by the WOAH. *B. abortus* is a significant endemic pathogen in free-ranging American bison in the western United States and Canada, and in free-ranging African buffalo in sub-Saharan Africa. *B. abortus* has been isolated from a broad range of nondomestic bovid species. *B. suis* biovar 4 has been identified in musk oxen; *B. mellitensis* has been identified in free-ranging chamois, ibex, yaks, sable antelope, captive Arabian and scimitar horned oryx, and Neumann's and sand gazelles.[219,360] *B. mellitensis* has not been identified in free-ranging wildlife in the United States.

Reproductive lesions may include abortions, placental retention (as observed in bison), placentitis, and significant hemorrhage that may also involve the fetus. Intercotyledonary exudate or peri-cotyledonary inflammation may be observed, and fetal membranes may be edematous. Metritis and orchitis are commonly observed and can be accompanied by localized or widely disseminated acute and chronic abscesses. Abscesses may be liquefactive or caseating, and mineralization of affected tissues often progresses with chronicity. Polyarthritis, tenosynovitis, and bursitis have also been reported (Fig. 5.28).[219]

Microscopically, large numbers of gram-negative, intracellular bacteria occur within chorionic trophoblasts with or without lymphohistiocytic, necrotizing, or suppurative placentitis. In aborted fetuses mild to severe fibrinosuppurative bronchopneumonia and multisystemic hemorrhages may be prominent. Microscopic findings in adult animals reflect subacute to chronic, suppurative inflammation and abscessation observed grossly in the reproductive tract or other organs. Large numbers of gram-negative bacteria may be present within regions of necrosis and within macrophages in affected areas.[322] Granulation tissue and fibrosis may be associated with chronic and mineralized lesions. Definitive diagnosis requires bacterial isolation or PCR in tissues or clinical samples.[139] Regulatory agencies may also require specific laboratory testing including biotyping.

Heartwater (cowdriosis) is a WOAH-listed, reportable disease caused by the tick-borne, obligate intracellular, gram-negative rickettsial agent ***Ehrlichia ruminantium***. Heartwater is endemic throughout sub-Saharan Africa, offshore African islands, and in some areas of the Caribbean.[12] It is transmitted by ticks of the genus *Amblyomma*. Though the disease is less significant in wild ruminants, both natural and experimental infections have been demonstrated in several African ungulate species.[302]

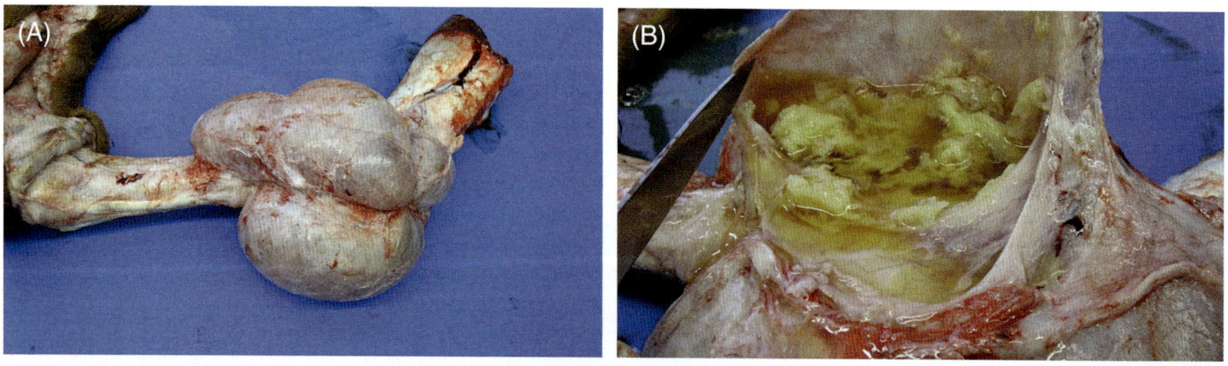

FIGURE 5.28 Carpal hygroma in a sable antelope due to *Brucella mellitensis*. (A) A large multilocular hygroma obscuring the carpus. (B) Serofibrinous exudate filling the hygroma. The joint cavity was not affected in this case.

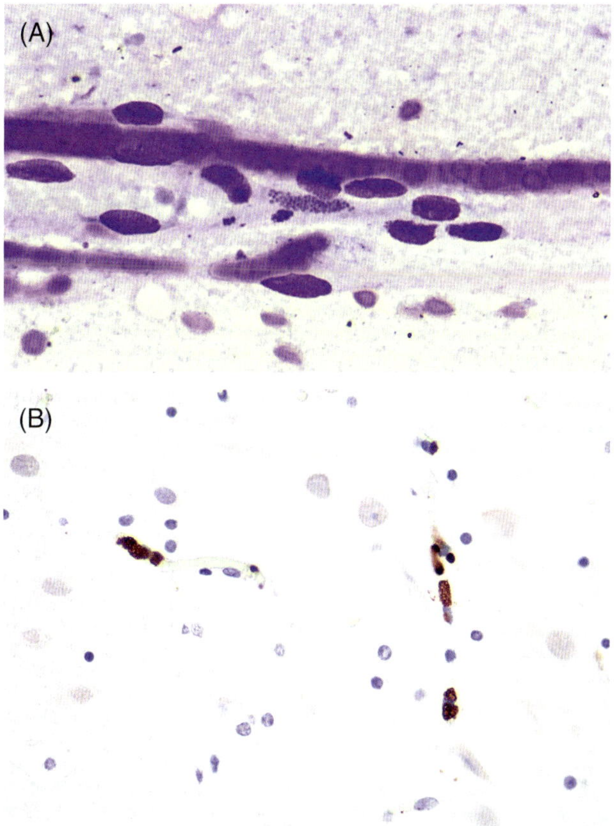

FIGURE 5.29 Heartwater in a springbok. (A) A colony of tiny basophilic *Ehrlichia ruminantium* organisms in the endothelial cell cytoplasm in a brain smear. Diff-Quik®. (B) Positive immunohistochemical labeling of *E. ruminatium* colonies in capillary endothelial cells.

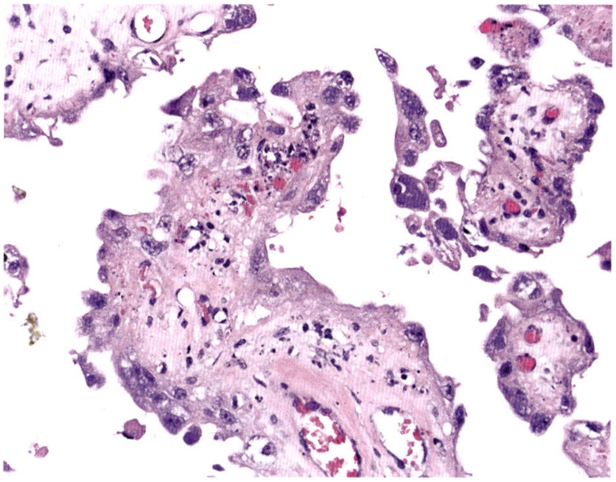

FIGURE 5.30 Placental villi from an Eastern white-bearded gnu with coxiellosis. Multifocal villus necrosis, sloughing of the trophoblast layer, and dense granular colonies of intracytoplasmic bacteria in individual trophoblasts are present (see eSlide 5.e7). *(Photo Courtesy of San Diego Zoo Wildlife Alliance.)*

Heartwater-associated mortality has been reported in naturally infected eland, springbok, steenbok, lechwe, and sitatunga. Individuals of several other species have been experimentally infected and may act as reservoirs without exhibiting significant disease.[302,303] Diagnosis is by demonstration of the organism in endothelial cell cytoplasm in a brain smear or by immunohistochemistry (Fig. 5.29).

Coxiella burnetii is a gram-negative, obligate intracellular bacterium that can cause reproductive losses in nondomestic ruminants. **Coxiellosis** or **Q fever** is a WOAH-listed, emerging, reportable zoonotic disease with a very low infectious dose in humans and suspected cases should be managed using appropriate biosecurity. *Coxiella*-associated placentitis, abortion, and stillbirth have been reported in several species of captive *Bovidae*, including bongo, dama, Dorcas and Cuvier's gazelle, giraffe, greater kudu, sable and roan antelope, water buffalo, and waterbuck.[74,131,141,221,408] It is likely that most ruminants are susceptible to infection and disease.

Late-gestation fetal losses or weak neonates are seen in nondomestic ruminants, as in cattle, sheep, and goats. Fetal lesions are nonspecific and secondary to fetal distress: alveolar meconium aspiration and increased alveolar squamous cells. Grossly, the placenta may be within normal limits, or there may be pale discoloration or mineralization of cotyledons. Histologically, multifocal to widespread necrosis of cotyledonary villi and intercotyledonary trophoblasts is associated with the accumulation of necrotic cell debris, fibrin, and neutrophils between villi (Fig. 5.30; eSlide 5.e7).[141] Lymphoplasmacytic and neutrophilic infiltrates may be present in the interstitium. Trophoblasts contain finely granular, pale basophilic to gray aggregates of organisms that can distend the cytoplasm. Organisms can be stained using modified Ziehl-Neelsen, Stamp, Gimenez, or Macchiavello stains. Important differential diagnoses include *Brucella* and *Chlamydia* spp.

Chlamydiaceae are gram-negative, obligate, intracellular zoonotic bacteria that cause reproductive loss, infectious keratoconjunctivitis, and several other clinical syndromes in artiodactyls. **Chlamydia (Chlamydophila) abortus** is a common cause of placentitis and abortion in domestic sheep and goats. Though there are few confirmed case reports of *Chlamydia*-associated abortion in nondomestic ruminants, *C. abortus* DNA or anti-*C. abortus* antibodies have been detected in several free-ranging ungulate species.[169,170,218,311,335] Given the potentially wide host range, *C. abortus* should be considered as a differential for fetal loss in nondomestic ruminants and appropriate biosecurity should be enacted in suspect cases.[324] In domestic ruminants vasculitis is described as a prominent feature of chlamydial placentitis that distinguishes it from coxiellosis, but this may not be relevant in nondomestic ruminants. Chlamydial organisms have been identified in lesions of IKC in bighorn sheep in North America, and in other species in Europe (see above).

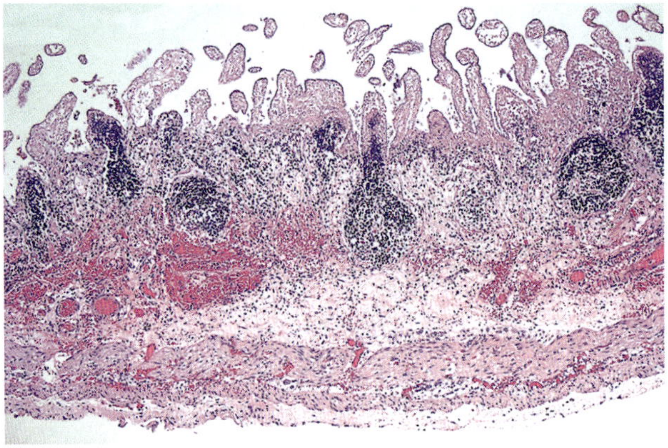

FIGURE 5.31 Necrohemorrhagic enteritis in a neonatal urial infected with *Clostridium perfringens* type C. There is transmural necrosis and submucosal hemorrhage and edema. Ghosts of necrotic villi are lined by bacilli. *(Photo Courtesy of San Diego Zoo Wildlife Alliance.)*

A wide variety of nondomestic ruminants may be affected by enteric infection with one or more toxin-producing ***C. perfringens*** types; however, significant disease is most frequently associated with alpha and beta toxins produced by ***C. perfringens* type** C, which causes hemorrhagic enteritis. Lesions in neonatal nondomestic bovids are comparable to those described in domestic ruminants and include mural acute necrotizing enteritis with ulceration and peritonitis. Diagnosis may be made based on the clinical history and clinical signs, and gross and histological findings with isolation of *C. perfringens* type C from jejunal lesions (Fig. 5.31). Toxin identification through ELISA, or identification of the toxin-producing genes, is required for definitive diagnosis, as *C. perfringens* is a normal inhabitant of gastrointestinal flora. *C. perfringens* **type A** has been implicated as the cause of fatal **hemorrhagic enteritis** in a neonatal hippopotamus. *C. perfringens* type A has also been associated with fatal cases of necrotizing and hemorrhagic enteritis in a variety of neonatal bovids.

Malignant edema** or **gas gangrene is caused by *C. septicum*. Other clostridial species including *C. novyi* may produce similar disease. Most species of nondomestic *Bovidae*, *Antilocapridae*, *Tragulidae*, and *Hippopotamidae* are susceptible. Disease usually results from contamination of skin wounds. Clinical signs include sudden death, sometimes preceded by febrile disease. Gross and microscopic findings include traumatic injury, subcutaneous and muscle edema, and necrosis with or without crepitation or gas bubbles with large numbers of large bacterial rods. Diagnosis is based on clinical history, signs, and identification of bacteria within tissues. Confirmation is achieved through isolation of relevant *Clostridium* species from lesions and fluorescent antibody assay for toxins on tissue impressions.

Blackleg or **clostridial myositis** is caused by *Clostridium chauvoei*. Blackleg is poorly characterized in animals other than domestic cattle. Most species of nondomestic *Bovidae*, *Antilocapridae*, and *Tragulidae* may be susceptible; susceptibility in *Hippopotamidae* is not known. In wildlife contamination of deep wounds has been suggested as a potential route of infection.[253] Clinical signs may include peracute death, or sudden onset of muscle stiffness, anorexia, and depression rapidly leading to death. Gross findings include significant regions of muscle necrosis and hemorrhage with or without crepitation. Ancillary diagnostics include culture, Gram stain, immunohistochemistry, fluorescent antibody testing for toxins, and PCR.

Other significant bacterial diseases that are less frequently reported in nondomestic *Bovidae*, *Antilocapridae*, *Giraffidae*, *Hippopotamidae*, and *Tragulidae* are summarized in Table 5.e6 in the digital content.

Fungi

Fungal infections are sporadically reported in nondomestic *Bovidae*, *Antilocapridae*, *Giraffidae*, *Hippopotamidae*, and *Tragulidae* (see Table 5.e7). Disease most often occurs in individual animals, and outbreaks affecting herds are uncommon.

Metazoa

Gastrointestinal nematodes may be clinically significant in captive settings, especially on permanent pasture or when naive animals are exposed to novel parasites through translocation or reintroduction programs. The presence of parasites does not necessarily imply disease, and other factors should be evaluated when assessing the significance for an individual animal or a population. A detailed inventory of gastrointestinal nematodes reported in free-ranging ruminants of North America is available.[164]

The **abomasal trichostrongyles** are among the most important and common gastrointestinal parasites in the *Giraffidae* and nondomestic *Bovidae*. **Haemonchosis** is caused by infection with ***Haemonchus*** sp. nematodes, including ***H. contortus*** and ***H. placei***. Larvae and adults are hematophagous, and heavy worm burdens are associated with significant anemia and hypoproteinemia. The life cycle is direct, and environmental parasite loads can build up rapidly on feces-contaminated pastures in warm, moist conditions, or where there is regular watering. Copper deficiency may predispose ruminants to gastrointestinal parasitism. A wide range of species is susceptible, including giraffe, nondomestic caprids (Caspian urial, mouflon, Sudan Barbary sheep), and various antelope.[133]

Grossly, mucous membranes, muscles, and viscera are pale and there may be subcutaneous or submandibular edema, hydrothorax, and ascites due to hypoproteinemia. Diarrhea is usually absent unless there is coinfection with other gastrointestinal helminths.[373] Abomasal contents may be hemorrhagic. Adult *Haemonchus* sp. females are 20–30mm long and can be identified in abomasal contents by the "barber pole" appearance imparted by the white uterus spiraling around the blood-filled gut. The trichostrongyle *Ashworthius sedemi*, a parasite of Asiatic cervids, can cause similar disease in free-ranging European bison.[206]

Ostertagia ostertagi in cattle, ***Teladorsagia circumcincta*** in sheep, related nematodes of genera **Teladorsagia**, **Marshallagia**, **Spiculopteragia**, and **Camelostrongylus** in other nondomestic ruminants can be the primary cause of an **Ostertagiosis**-like syndrome.[114,119,142,143,152,202,208,209,287,296,359,381] Clinically, affected animals may be emaciated and hypoproteinemic, and abomasal pH may be elevated due to a loss of parietal cells in the metaplastic abomasal mucosa. Grossly, the abomasal mucosa may be irregularly thickened by small nodules, each of which may contain small, protruding nematodes detectable if examined under magnification. Lesions of chronic abomasitis, glandular hyperplasia, and mucous metaplasia are due to larval development within abomasal glands. Quantification of worm burden can be performed on abomasal contents. Adult and larval worms are small (<1.5 cm long) and brown and may require magnification for detection. On histology, adults have longitudinal ridges visible in cross-section and are smaller than the morphologically similar *Haemonchus* spp.

Trichuris sp. nematodes, commonly called **whipworms**, infect the cecum and, less frequently, colon of ruminants and other species. Heavy infections in captive, nondomestic ruminants can cause mucohemorrhagic typhlocolitis and associated chronic diarrhea, hypoproteinemia, and poor body condition. *Trichuris* spp. are hematophagous, but anemia is a less important component of trichuriasis than inflammation and protein loss. Grossly, the intestine may contain hemorrhagic or blood-flecked contents, and there may be mucosal erosion or thickening. Worms are visible on the mucosal surface. Histologically, sections of anterior ends of adult nematodes may be found embedded in the mucosa, and there may be erosion, glandular hyperplasia, mixed inflammatory cell infiltrates, and superficial hemorrhage.

Eleaophorosis is caused by infection with ***Elaeophora schneideri*** (superfamily *Filarioidea*, family *Onchocercidae*). Infection is usually incidental in definitive mule- and black-tailed deer hosts, but in aberrant hosts, such as Barbary and bighorn sheep, *E. schneideri* can cause severe, ulcerative to crusting facial skin lesions, pneumonia, rhinitis, and meningomyelitis associated with the presence of degenerating microfilaria.[14,49,297] Skin lesions can be both necrotizing and proliferative, may be complicated by secondary infection, and should be differentiated from poxvirus infection (contagious ecthyma). Histologically, affected skin, lung and brain show mixed, eosinophilic to granulomatous inflammation, and intact or degenerating microfilaria may be present within foci of necrosis. Thrombosis and direct vascular damage may lead to ischemic necrosis. Adult nematodes up to 10 cm long are most often present in the carotid arteries but may be found in any large artery. Significant proliferative arterial intimal lesions are described in elk (see Chapter 6: Cervidae); arterial lesions in sheep appear to be less pronounced.

Elaeophorosis caused by ***E. sagitta*** (***Cordophilus sagittus***) has been described in African bovids including greater kudu, blackbuck, bushbuck, nyala, eland, and African buffalo.[307] Adult *E. sagitta* worms inhabit the heart chambers, coronary arteries, and large and small branches of pulmonary arteries including those in caudal regions. Gross lesions include coronary artery aneurysms that contain several adult worms (Fig. 5.32). Histologic findings include myocardial fibrosis, arterial mural fibrosis and smooth muscle hypertrophy, lymphoplasmacytic and eosinophilic arteritis, and periarteritis; villous intimal proliferation is seen in pulmonary arteries. Occasional mineralized parasites are present in nodular foci of fibrosis, and small numbers of microfilaria may be present in the myocardium or other tissues.

Filarial cutaneous infection in free-ranging giraffe in East Africa, "**giraffe skin disease**", results in raised, firm edematous nodules that progress to hairless crusted fissured hyperkeratotic crusted plaques (10–15 cm dia.).[112,153,214,262,402] Prevalence in Ruaha National Park, in Tanzania, may exceed 90% and is significantly higher in adults. Lesions are most frequent on the palmar aspects of the carpi and are usually bilateral. Less commonly, the skin of the brisket region, hindlimbs, or vulva may be affected; occasional lesions are present in distal forelimbs and hindlimbs. Histologically, in Nubian giraffe, orthokeratotic hyperkeratosis, epidermal acanthosis with long irregular rete pegs, epidermal erosions with serocellular

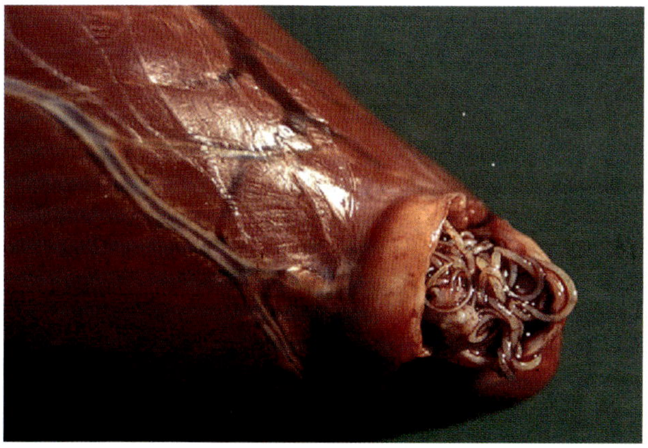

FIGURE 5.32 Elaeophorosis in a greater kudu. Coronary artery aneurysm filled with *E. sagitta* nematodes. *(Photo Courtesy of Pathology Section, Faculty of Veterinary Science, University of Pretoria.)*

crusts, and corneal pustules are present.[153] Eosinophilic granulomatous and lymphoplasmacytic dermatitis are associated with L3 nematode larvae in the epidermal and follicular epithelium and microfilariae in dermal vacuoles. Chronic lesions consist of dense granulation tissue, depigmentation, and fibrosis. The nematodes are genetically similar to *Stephanofilaria* spp. and *Brugia malayi*. Research regarding the insect vector and the significance of fungal hyphae and spores, described in some cases,[197] is needed. There may be an association with certain soil features.[214]

Disease associated with the meningeal worm ***Parelephostrongylus tenuis*** (superfamily *Metastrongyloidea*, family *Protostrongylidae*) is reported in several wild and captive bovids and in pronghorn. The definitive host of *P. tenuis* is the white-tailed deer, in which infection is incidental or associated with minimal lesions. Susceptible aberrant hosts include American bison, bighorn sheep, blackbuck, bongo antelope, eland, pronghorn, sable antelope, scimitar horned oryx, and sitatunga.[95] Lesions are associated with nematode migration in the spinal cord and brain and include foci of malacia, hemorrhage, Wallerian degeneration, and meningoencephalitis often with eosinophilic infiltrate.[315,355,398] Adult nematodes may be present in white or gray matter or on meninges. Eggs and developing larvae may be associated with granulomatous interstitial pneumonia.

Parelaphostrongylus odocoilei are small, thread-like worms that can be found in the skeletal muscle of various cervid and bovid species. Like *P. tenuis*, they are transmitted through ingestion of gastropod intermediate hosts.[180,212] In Dall's sheep encephalomalacia and myelomalacia have been associated with migrating worms, eggs, and larvae. Granulomatous interstitial pneumonia and myositis, sometimes associated with hemorrhage, are also reported.

Verminous pneumonia caused by other nematode species including ***Dictyocaulus* spp.**, ***Pneumostrongylus* spp.**, ***Muellerius* spp.**, and ***Protostrongylus* spp.** may manifest as interstitial pneumonia or pulmonary granulomas.[67,130] These conditions could be mistaken for tuberculosis in regions where these parasites are endemic. Secondary bacterial infections can occur, and lungworms have been implicated as a component of multipathogen pneumonias, such as bighorn sheep pneumonia (see above).[38]

Clinically significant endoparasites or those that pose a zoonotic risk are listed in Table 5.e8 of the digital content.

Protozoa

***Babesia* sp.** parasites (order Piroplasmida) of mammals are obligate intracellular, tick-borne, apicomplexan protozoa that almost exclusively infect erythrocytes. Zoonotic babesiosis is described as an emerging disease. Clinically significant or fatal babesiosis has been reported sporadically in a variety of nondomestic bovid species, including sable antelope in Africa (*Babesia* sp. (sable)) and European chamois (*Babesia divergens/capreoli*).[166,286,300] Lesions consistent with babesiosis include hemoglobinuria, enlarged spleens due to red pulp hyperplasia, and erythrocytic piroplasmosis (Fig. 5.33). Similar findings were reported in small numbers of giraffe in South Africa infected with a novel *Babesia* species, though a link between infection and death was not definitively established.[285] *B. ocodoilei*, which usually causes disease in cervids, can cause fatal babesiosis in captive muskoxen and has been identified in clinically healthy bighorn sheep.[344]

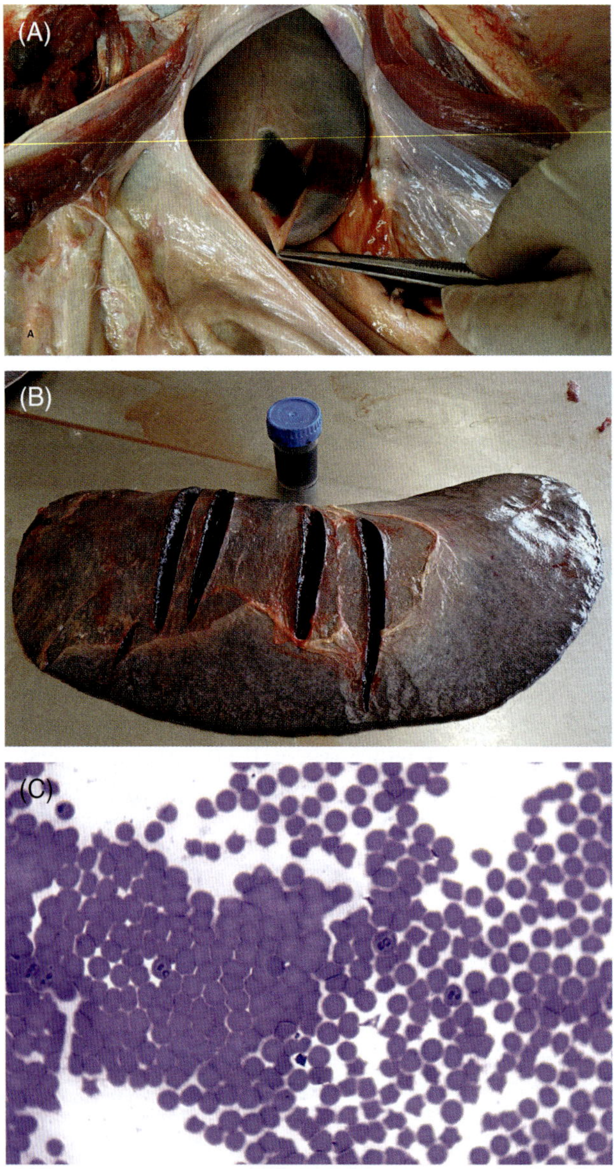

FIGURE 5.33 Babesiosis infection in a sable antelope. (A) Hemoglobinuria. (B) Splenomegaly resulting from intravascular hemolysis-induced red pulp hyperplasia; urine in the container depicts hemoglobinuria. (C) Blood smear with paired *Babesia* sp. (sable) piroplasms in erythrocytes. Diff-Quik®.

Exposure of ruminant calves (<9-month-old) to infected vectors in areas where babesiosis (and also other hemoprotozoa) infections are endemic will result in infection and immunity from clinical disease. Lack of exposure during this period will increase the risk of clinical disease; hence a recent history of re-location is an important epidemiological component to consider in disease investigation.[300] In affected sable antelope had been recently imported to South Africa from Germany; it is possible their naïve status increased their susceptibility to infection and clinical disease.

Histologically, marked congestion of capillaries, with erythrophagocytosis and hemosiderin accumulation in macrophages occurs in multiple tissues. There may be renal hemoglobin casts, tubular degeneration, and tubular epithelial hemosiderosis. Hepatic hypoxic lesions are characterized by centrilobular sublethal hepatocellular degeneration or necrosis. Canalicular cholestasis may be seen. Using a Giemsa or HE stain, organisms are visible as pale basophilic structures in erythrocytes and may be more prominent in renal glomerular and pulmonary alveolar capillaries. Babesia morphology depends on the species. On blood smears or impression cytology of organs, sporozoites and merozoites may be located centrally or near the erythrocyte margin, and organisms may be single, or paired and oriented at obtuse angles. Single forms may be pyriform, round, or elongated. Intra-erythrocytic organisms also include *Theileria* spp., *Cytauxzoon* spp., and *Anaplasma* spp. Confirmation of babesiosis is based on evidence of extravascular hemolysis in association with significant numbers of infected erythrocytes. Identification of the *Babesia* species involved requires PCR.

Babesiosis may be distinguished from anaplasmosis by the lack of hemolysis in the latter.

Piroplasmids of the genus **Theileria** are tick-vectored, apicomplexan hemoparasites that infect both leukocytes and erythrocytes in mammalian hosts. *Theileria* are transmitted by ixodid ticks.[41] The most economically significant and well-studied members of the genus are ***T. parva***, subspecies of which cause **East Coast fever** (cattle to cattle transmitted) and **corridor disease** (African buffalo to cattle transmitted) in sub-Saharan Africa. *T. annulata* causes **tropical theileriosis** in North Africa and Asia. Vectors for *T. parva* are *Rhipicephalus appendiculatus*, *R. zambesiensis*, and *R. duttoni*; *Hyalomma* species transmit *T. annulata*. Bovine theileriosis is a WOAH-listed and reportable disease. The parasite life cycle, pathogenesis, and diagnosis in domestic animals are complex.[41,230,277,366] The pathogenesis and pathology in artiodactyls is reviewed by Clift et al.[76] Historic cases of **cytauxzoonosis** in several African ungulate species are now generally attributed to Theilerial infection.[270]

Clinical infection with *Theileria* has been reported in giraffe and in numerous bovid species, including African buffalo, American bison, blesbok, blue wildebeest, bushbuck, common eland, Grant's gazelle, gray duiker, klipspringer, kudu, nyala, reedbuck, roan, sable tsessebe, and waterbuck.[53,138,145,171,200,230,270,285,286,291,393] Infection of free-ranging wildlife is mostly subclinical (endemic) but disease is most common in roan and sable antelope. Disease due to *Theileria* sp. (sable) infection has been reported in association with stressful events such as translocation of naïve animals of all ages and exposure of young calves to high infected tick burdens.[365] Clinical findings include lethargy, fever, anemia, and icterus. Dyspnea, diarrhea, or neurologic signs may be terminally present. Gross findings include prominent, generalized lymphadenomegaly, splenomegaly, pulmonary edema, and petechial and ecchymotic hemorrhage on serosal surfaces and in the renal and adrenal cortices. Both intra- and/or extravascular hemolysis may occur, with or without hemoglobinuria (Fig. 5.34).

Histologically, there is widespread vasocentric atypical mononuclear leukocytic infiltration and proliferation, with many leukocytes containing cytoplasmic theilerial schizonts; leukocytolysis and necrosis; and thrombosis and hemorrhage. Kidney, gastrointestinal tract, liver, lung, heart, adrenal gland, and brain are particularly affected (eSlide 5.e8). *Theileria* sp. schizonts (in Koch's bodies) in blast-transformed, atypical mononuclear leukocytes (monocytes and/or lymphocytes) are visible in aspirates of lymphoid tissue or in blood smears. Koch's bodies containing multiple schizonts may be extracellular if cells rupture during smear preparation. Depending on the host, stage of infection, and the species of *Theileria*, schizonts may be rare in circulating lymphocytes but large numbers of piroplasms may be visible in erythrocytes. Blood smears from most healthy ungulates in endemic regions will show scant but notable theilerial piroplasms in erythrocytes. Hence, a diagnosis of clinical theileriosis depends on the presence of schizonts, which indicate active infection and clinical disease. PCR is required for *Theileria* species identification. Immunohistochemical labeling has been developed for detecting theilerial schizonts in infected leukocytes.[77]

Trypanosomes are insect-vectored protozoal hemoparasites that parasitize many species, including humans. Significant disease has not been reported in free-ranging African ruminants in endemic areas, although they may be parasitized (**trypanosomiasis**) and are reservoirs for the agents of tsetse fly-transmitted bovine **trypanosomosis** or **nagana** (*Trypanosoma congolense*, *T. vivax*, *T. brucei brucei*), a WOAH-listed, reportable disease.[21,280,385] Giraffe and hippopotamuses are also important hosts for tsetse flies (*Glossina* spp.). There are many species of pathogenic and nonpathogenic trypanosomes. Species identification using molecular techniques is more reliable than parasite morphology in blood smears.

Toxoplasma gondii is a coccidian protozoan parasite with worldwide distribution. Serosurveys have identified

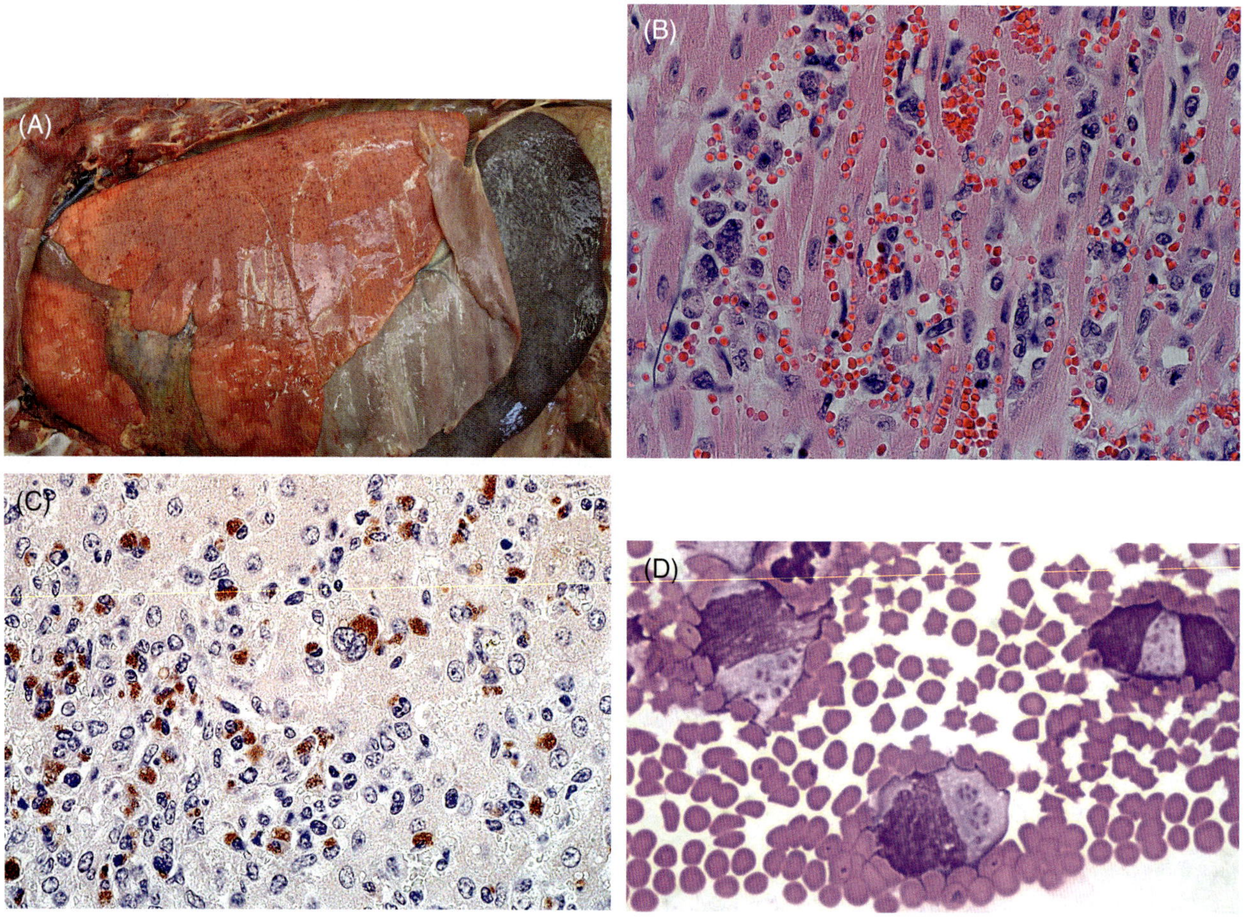

FIGURE 5.34 Theileriosis (*Theileria* sp. (sable)) in a roan antelope calf. (A) Pleural petechiation with severe splenomegaly resulting from hemolysis. (B) Interstitial myocardial infiltration of theilerial schizont infected blast-transformed mononuclear leukocytes. (C) Positive immunohistochemical labeling for *Theileria* sp. schizonts in mononuclear leukocytes. (D) Blood smear with cytoplasmic theilerial schizonts (Koch's bodies) in blast transformed mononuclear leukocytes; single theilerial piroplasms are occasionally present in erythrocytes. Diff-Quik®. (see eSlide 5.e8). *(Photo Courtesy of Pathology Section, Faculty of Veterinary Science, University of Pretoria.)*

exposure to *Toxoplasma* in many species of *Bovidae* including free-ranging giraffe and common hippopotamuses.[22,326] *Toxoplasma*-associated disease can include placentitis, abortion, early neonatal death, pneumonia, or disseminated, fatal disease.[102] Bovid species susceptible to **toxoplasmosis** include bison, dik-dik, gazelles (dama, Cuvier's, slender-horned), gerenuk, and saiga.[56,105,185,258,346,367] Pronghorn are highly susceptible to experimental infection.[104] Abortion and neonatal death have been reported in muskoxen and nilgai.[86,346]

In acute, disseminated infections, *Toxoplasma* sporozoites from ingested oocysts, or bradyzoites from ingested tissue cysts, invade intestinal epithelial cells and multiply asexually as tachyzoites that spread to mesenteric lymph nodes and systemically via blood and lymphatics. Gross lesions include mesenteric lymphadenopathy and lymph node necrosis; hemorrhagic or catarrhal enteritis; hemorrhagic to fibrinous peritoneal effusion; serosal and endocardial petechiae and ecchymoses; pulmonary edema and hemorrhage; and pale foci of necrosis in tissues, notably liver, spleen, or kidney. Histologically, necrosis and variably intense mononuclear to mixed inflammatory cell infiltrates are present in multiple tissues, sometimes centered on blood and lymphatic vessels (necrotizing vasculitis). Necrosis is usually due to intracellular tachyzoites, but tissue cysts may also be present. In transplacental infections dams may remain asymptomatic and lesions may be limited to the placenta, where there is necrosis associated with the presence of tachyzoites. In domestic sheep and goats, gross findings include dark red cotyledons with multiple, small, white foci of necrosis. Histologically, cysts or tachyzoites may be present in the placenta, and in some cases there are lesions in fetal heart, lungs, brain, or other tissues.[324] Tachyzoites are approximately 6 µm long, often curved, may be uninucleate or binucleate, and may be present individually, paired, or in clusters in tissue sections or impression smears. Tissue cysts vary in size and are most often found in brain, muscle, and eye but can occur anywhere. Note that tissue cysts may be found in clinically healthy animals, not associated with inflammation; in these cases cysts could serve as a source of infection to humans who may eat game. Cysts are round to elongate and may be up to 60 μm long. Diagnosis can be confirmed by immunohistochemistry or PCR.

Neospora caninum is an apicomplexan protozoan parasite closely related, and morphologically similar, to *Toxoplasma gondii*.[99] Infection has been associated with stillbirth in lesser kudu. Lesions include nonsuppurative encephalitis, with gliosis, and myocarditis.[304] Tissue cysts may be present and can be identified using immunohistochemistry. Placental necrosis with protozoal zoites may also be present.[324]

***Sarcocystis* sp.** protozoa usually have a 2-host life cycle with carnivore definitive hosts. Many species of *Sarcocystis* have been identified, and most are relatively species-specific to their mammalian or avian intermediate hosts.[102] For many *Sarcocystis* sp. of wildlife, the life cycle and definitive hosts are not known. *Sarcocystis* infection in wild *Bovidae*, *Giraffidae*, and *Hippopotamidae* is usually not clinically significant.[67] Large tissue cysts may be observed histologically, or sometimes grossly, in skeletal and cardiac muscle of African buffalo (Fig. 5.35).[101] In domestic ruminants systemic infection may be associated with acute hemorrhage, vasculitis, and necrosis in multiple tissues.

***Cryptosporidium* sp.** oocysts have been identified in fecal smears stained with modified Ziehl Neelsen in hand-reared captive artiodactyls exhibiting clinical diarrhea (Fig. 5.36).[161] Cryptosporidial infection of the small intestine, and less commonly in the large intestine, associated with minimal inflammation has been described for captive addax, blackbuck, fringe-eared oryx, sable antelope, and scimitar-horned oryx.[386]

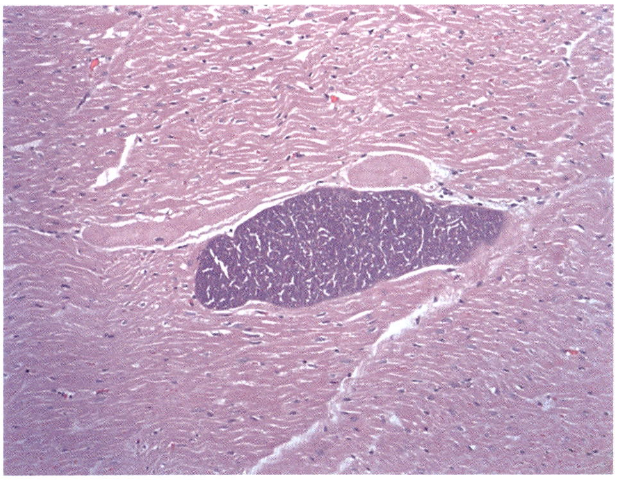

FIGURE 5.35 Sarcocystosis in a springbok. A large thin-walled sarcocyst filled with basophilic crescentic metrocytes distorts a myocardial myocyte.

Hepatic coccidiosis has been reported in chamois and red lechwe, characterized by mild portal inflammation and fibrosis.[54,400] ***Intestinal coccidiosis*** has been described for African and water buffalo, kudu, and nilgai in intensive conditions. Histomorphologically various developmental stages of *Eimeria* species result in subtotal infection of jejunal and ileal mucosae, and organisms may obliterate villous structures (Fig. 5.37).[103,106,347,387]

Amoebiasis (especially *Entamoeba* sp.) and **giardiasis** have been reported in captive artiodactylids with diarrhea and positive fecal analysis. However, mortality and necropsy findings attributed to these are yet to be described. Clinical **besnoitiosis** has not been reported in free-ranging Bovidae; however, *Besnoitia* sp. cysts were described in the intimal layers of the venous vasculature, particularly the jugular veins, of clinically healthy blue wildebeest and impala in the Kruger National Park, South Africa.[242]

Ectoparasites

Dermatitis and pruritis associated with **sarcoptic mange** (etiology: ***Sarcoptes scabiei***) has been reported in over 100 mammalian species. Outbreaks have been reported in free-ranging animals (Fig. 5.38). For example, sarcoptic mange was diagnosed in several giraffe in Kenya in 2010.[5] **Psoroptic mange** associated with ***Psoroptes* sp.** infection is moderately common in some populations of free-ranging bighorn sheep.[64] Lesions can include alopecia, erythema, papule formation, lichenification, crusting, and acanthosis on legs, back, shoulders, neck, and muzzle. Diagnosis is through identification of mites and eggs on skin scraping.

Wild ruminants and hippopotamuses may have a wide range of ectoparasites including ticks, mites, lice, rectal leeches, and flies. The clinical significance of many of these infestations is variable and often unclear. A brief summary of clinically significant ectoparasites or those that pose a zoonotic risk is presented in the digital content (Table 5.e9).

Other Infectious Organisms

Transmissible spongiform encephalopathies (TSEs) of ungulates include chronic wasting disease (see Chapter 6: Cervidae), **scrapie**, and **bovine spongiform encephalopathy (BSE)** (WOAH-listed, reportable diseases). Bison, bighorn sheep, mountain goats, and pronghorn can be affected by chronic wasting disease.[87] **Scrapie**, which has been recognized in domestic sheep and goats for hundreds of years, also occurs naturally in mouflon and possibly other species of wild *Caprinae*. TSEs reported in several species of captive, nondomestic ruminants housed in zoological gardens in the United Kingdom were coincident with the epidemic of BSE in the 1980s.[88,179,196] Cases of TSE in nyala, greater kudu, gemsbok, Arabian oryx, scimitar-horned oryx,

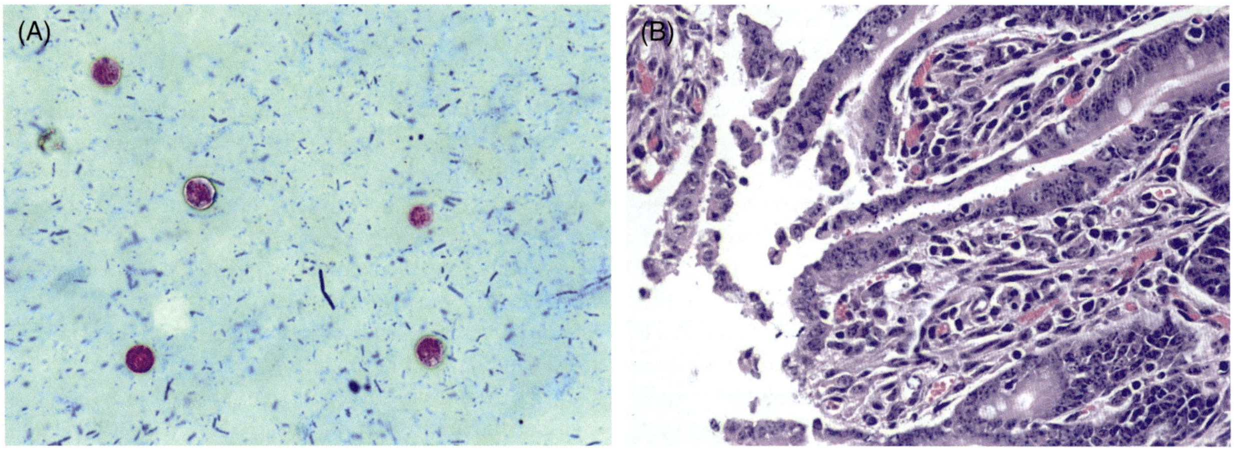

FIGURE 5.36 Cryptosporidiosis in a 3-week-old impala lamb. (A) Fecal smear with acid-fast positive *Cryptosporidium parvum* oocysts measuring 5.4 to 5.8 µm in diameter. A florid mixed bacterial population is also present. Modified Ziehl Neelsen. (B) Numerous small spherical basophilic cryptosporidial trophozoites adhered to the jejunal enterocyte brush border.

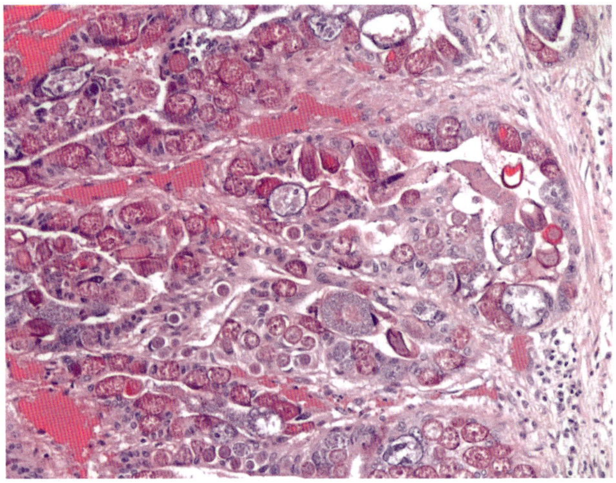

FIGURE 5.37 Intestinal coccidiosis in a buffalo cow. Subtotal effacement of jejunal enterocytes by innumerable cytoplasmic coccidian organisms in various stages of development.

FIGURE 5.38 Sarcoptic mange in a blue wildebeest. The skin is thickened, fissured, and hyperplastic and hyperkeratotic with scales.

eland, and American bison are now recognized as BSE acquired through ingestion of affected ruminant-derived meat and bone meal in feed.[354] During this BSE epidemic, greater kudu appeared to be highly susceptible. Mouse bio-assay studies demonstrate a very wide tissue distribution of the BSE agent in kudu, and horizontal transmission of the BSE agent during the epidemic has been proposed.

Diagnosis is through demonstration of typical **neuronal vacuolation**, which is most often present in the medulla oblongata at the level of the obex in both BSE and scrapie. Specialized instruments for sampling the obex via the foramen magnum have been developed for both cattle and sheep and may be appropriate for testing nondomestic ruminants. However, samples from multiple regions of brain may be required to identify "atypical" scrapie, which affects the cerebellum and only minimally involves the medulla.[282] Vacuolation of neuron cell bodies may be accompanied by neuropil vacuolation and astrogliosis. Immunohistochemistry can be used to confirm the presence of PrP^{sc} and to detect PrP^{sc} prior to induction of vacuolar change in the brain. Other methods such as western immunoblot and ELISA can be used as rapid tests, but these should be followed by confirmatory tests. There are no reliable tests for detecting BSE or scrapie in a live animal.

ADDITIONAL RESOURCES

Allsopp, B.A. Heartwater—*Ehrlichia ruminantium* infection. OIE Rev. Sci. Tech. 2015; **34**, 557–568.

Bengis RG, Frean J. Anthrax as an example of the One Health concept. OIE Rev. Sci. Tech. 2014;**33** (2):593–604.

Breed D, Meyer LC, Steyl JC, Goddard A, Burroughs R, Kohn TA. Conserving wildlife in a changing world: Understanding capture myopathy—A malignant outcome of stress during capture and translocation. Conservation Physiology. 2019;**7**:coz027.

Clauss M. Tannins in the nutrition of wild animals: a review. Zoo animal nutrition. 2003;**2**:53–89.

Clift SJ, Collins NE, Oosthuizen MC, Steyl JCA, Lawrence JA, Mitchell EP. The Pathology of Pathogenic Theileriosis in African Wild Artiodactyls. Vet Pathol. 2020;**57** (1).

Godfroid, J., Garin-Bastuji, B., Saegerman, C., Blasco, J.M. Brucellosis in terrestrial wildlife. Rev. Sci. Tech. 2013; **32**, 27–42.

Lecu, A., Ball, R. Mycobacterial infections in zoo animals: relevance, diagnosis, and management. Int. Zoo. Yb. 2011; **45**, 183–202.

OIE, 2022. Manual of Diagnostic Tests and Vaccines for Terrestrial Animals. Available from: https://www.woah.org/en/what-we-do/standards/codes-and-manuals/terrestrial-manual-online-access/.

Pinto, A.A., 2004. Foot-and-mouth disease in tropical wildlife. Ann. N.Y. Acad. Sci. 1026, 65–72.

Roller M, Hansen, S, Knauf-Witzens, T. et al., *Mycobacterium avium* subspecies *paratuberculosis* infection in zoo animals: a review of susceptibility and disease process. Frontiers in Veterinary Science, 2020;**7**:572724.

Shivachandra, S.B., Viswas, K.N., Kumar, A.A., 2011. A review of hemorrhagic septicemia in cattle and buffalo. Anim. Health Res. Rev. 12, 67–82.

References

Full references for this chapter can be found online at https://doi.org/10.1016/B978-0-443-18437-6.00005-5.

Chapter 6

Cervidae

Elizabeth W. Howerth[a], Nicole M. Nemeth[a], and Marie-Pierre Ryser-Degiorgis[b,*]

[a]*University of Georgia, Athens, GA, United States;* [b]*Centre for Fish and Wildlife Health, University of Bern, Bern, Switzerland*

INTRODUCTION

This chapter concerns the family Cervidae, which has four recognized subfamilies: Cervinae (*Dama*, *Axis*, *Cervus*, and *Elaphurus*), Hydropotinae (*Hydropotes*), Muntiacinae (*Elaphodus*, *Muntiacus*, and *Megamuntiacus*), and Odocoileinae (*Odocoileus*, *Blastocerus*, *Ozotoceros*, *Hippocamelus*, *Mazama*, *Pudu*, *Alces*, *Rangifer*, and *Capreolus*). Older classifications also included Moschinae (musk deer; *Moschus* sp.), which is now considered to have a familial ranking, Moschidae. Cervid species are naturally free ranging throughout North America, South America, and Eurasia, on most associated continental islands, and North Africa, and free-ranging populations have been established elsewhere, for example, in New Zealand. Some cervids are kept in captivity or farmed, for example, red deer and white-tailed deer. Reindeer have been semidomesticated.[239] Some cervids are considered extinct in the wild (e.g., Pere David's deer), endangered (e.g., Calamian deer), or vulnerable (e.g., marsh deer) due to various factors, including limited host range due to geographical isolation, hunting pressure, habitat destruction, human settlement, and agricultural expansion. Other free-ranging populations, for example, European roe deer and wapiti, are widespread (http://www.iucnredlist.org/).

NON-INFECTIOUS DISEASES

Nutritional

Nutritional myopathy (vitamin E/selenium deficiency) occurs in both captive and free-ranging cervids; cardiac and skeletal muscle lesions are similar to those in other species.[322] Gross findings include white or chalky streaks in cardiac and/or skeletal muscle. Microscopically, polyphasic myodegeneration, necrosis, and mineralization are seen. Selenium levels can be assessed by measuring concentrations in whole blood, serum, and tissue (liver). Vitamin E levels can be assessed by measuring the concentration of circulating plasma alpha-tocopherol, with levels $<3.5\,\mu mol/L$ being associated with deficiency,[70] or through measurement of concentrations in liver.

Osteopathy co-occurring with selenium deficiency has been described in Patagonian huemul.[99] Affected deer have lesions in mandibular, maxillary, and appendicular bones that include eroded alveoli and perforated maxillary and mandibular bone, exostoses on several bones, and deformed hooves or phalanges.[101] Selenium deficiency, which impairs bone growth and causes periodontitis in ruminants, is thought to be the cause.

Älvsborg (wasting) syndrome in free-ranging Swedish moose likely has a multifactorial etiology with malnutrition/starvation as the major cause. Potential contributing factors are overpopulation, phytotoxic substances, soil pH imbalances, and viral infections.[33,110,112,266] Associated dietary mineral imbalances, assessed via tissue levels in affected moose, may include low copper and associated increased molybdenum, and abnormal levels of other metals including iron, zinc, and cadmium. Grossly, moose are in poor body condition or are emaciated with loss of musculature and serous atrophy of fat. Fragile skin, achromotrichia, corneal opacity, cataract, abomasal wall edema and hyperemia, thin intestinal walls, splenic atrophy, an edematous "flabby" heart, and osteoporosis may also be seen.[110,266] Histologically, edema, hemorrhage, erosions, and ulcers may be present in the mucosa of the oral cavity, esophagus, rumen, or abomasum. Dilation of lymphatic vessels in the abomasum and Brunner's glands in the duodenum is often marked. Additional findings may include a small liver and spleen, splenic hemosiderosis, dilated heart walls, nephrosis, and cerebellar abiotrophy. Definitive diagnosis of this condition is difficult but can be supported by gross findings, low hepatic copper levels, and lack of evidence of other underlying diseases.

Urolithiasis has been reported in a number of cervids and is mostly a disease of captive animals. Factors such as diet, early castration, and dehydration likely contribute to its occurrence similarly to domestic ruminants. Multiple stone types (e.g., calcium oxalate, calcium carbonate, magnesium carbonate) have been described[142,200,351] (Fig. 6.1).

*Deceased.

Pathology of Wildlife and Zoo Animals. https://doi.org/10.1016/B978-0-443-18437-6.00051-1

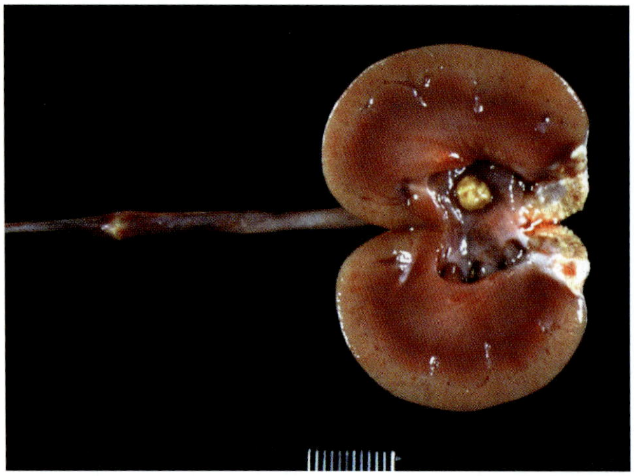

FIGURE 6.1 Urolithiasis in the urinary system of a white-tailed deer. Calculi are present in the cranial pole, renal pelvis, and ureter, both of which are dilated. *(Photo Courtesy of J. Wright, North Carolina State University and Noah's Arkive.)*

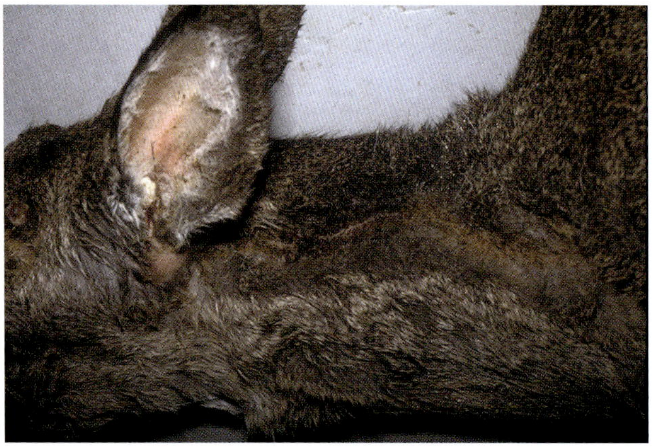

FIGURE 6.2 Lightning strike in the skin of a white-tailed deer. There is a serpentine trail of singed hair along the neck beginning just caudal to the pinna.

Lightning strike can cause mortality in individual animals, but mass mortality events from a single strike also have been reported in herds of cervids standing on water-soaked ground. Gross lesions may not be present, particularly in mass mortality events. However, in individuals, animals may have a track of singed hair from hoof to head (Fig. 6.2). Diagnosis can be supported by finding a Lichtenberg figure etched into the ground or on the skin of victims.[291] Lichtenberg figures reflect electric discharge that can create large, fern-like or branching patterns in the skin (burn injury) or grass surrounding the point of lightning strike.

Metabolic

Ruminal acidosis is a fermentation disorder characterized by abnormally low pH. It reflects a microbial imbalance with ruminal absorption of volatile fatty acids. The disease course may vary from peracute with sudden death to subacute with mild clinical signs. It can be the driver of a wide range of secondary health problems affecting other organs.[144] Ruminal acidosis and eventual **bloat** from grain overload or ingestion of other feeds rich in carbohydrates occur both in captive and free-ranging cervids. Abrupt diet changes and/or inappropriate feed, such as large amounts of bread, often initiate the disease in farmed deer. Adult, and especially dominant, animals seem to be prone to developing ruminal acidosis, likely because they are at particular risk for eating too large a portion of pellets or supplementary food.[142,296] In the wild, supplemental feeding with inappropriate types or amounts of food in winter is suspected to be the main cause.[253] Gross findings in acute cases include rumen hypotony and hydrorumen. Signs of dehydration and diarrhea and secondary bacterial infections of the internal organs, laminitis, and cerebrocortical necrosis may also be observed. Microscopically, the typical picture is a parakeratosis-rumenitis-liver abscess complex. In farmed deer, diagnosis of ruminal acidosis is based on a ruminal pH <6 and/or histological lesions of the ruminal mucosa, such as parakeratosis, hyperkeratosis, and small pustules.

All cervid species are susceptible to **exertional rhabdomyolysis** (capture myopathy; see Bovidae, Antilocapridae, Giraffidae, Tragulidae, and Hippopotamidae). The pathogenesis is poorly understood but disease is associated with vigorous muscle activity, such as stressful events (e.g., pursuit, capture, restraint, transport).[41] The severe muscle damage, rhabdomyolysis, and myoglobinuria that ensues may be associated with increased catecholamine release and can be fatal due to cardiogenic shock, renal failure, metabolic disorders, or chronic cardiac damage.[240,305] Gross lesions are often bilateral but can be asymmetrical and may include pale foci (acute) or delineated streaks (chronic) in the heart and skeletal muscles (limbs, shoulder, pectoral, intercostal); hemorrhage, edema, and rupture of muscles may also occur (Fig. 6.3). Microscopic lesions may include hemorrhage, monophasic degenerative and necrotic myofibers interspersed with normal myofibers, and small numbers of macrophages and neutrophils.[23] Necropsy immediately following death followed by histopathology is a key to diagnosis, and knowledge of history, clinical signs, and clinical pathology can help distinguish exertional rhabdomyolysis from toxic or nutritional etiologies.[64]

Toxic

Multiple cervid species are susceptible to **fluorosis** due to ingestion of excessive amounts of fluoride.[100,238,294] Exposure can be via water, forage contaminated by industrial effluents, vegetation contaminated with high fluoride-content soil by rain splash or wind, or volcanic ashes

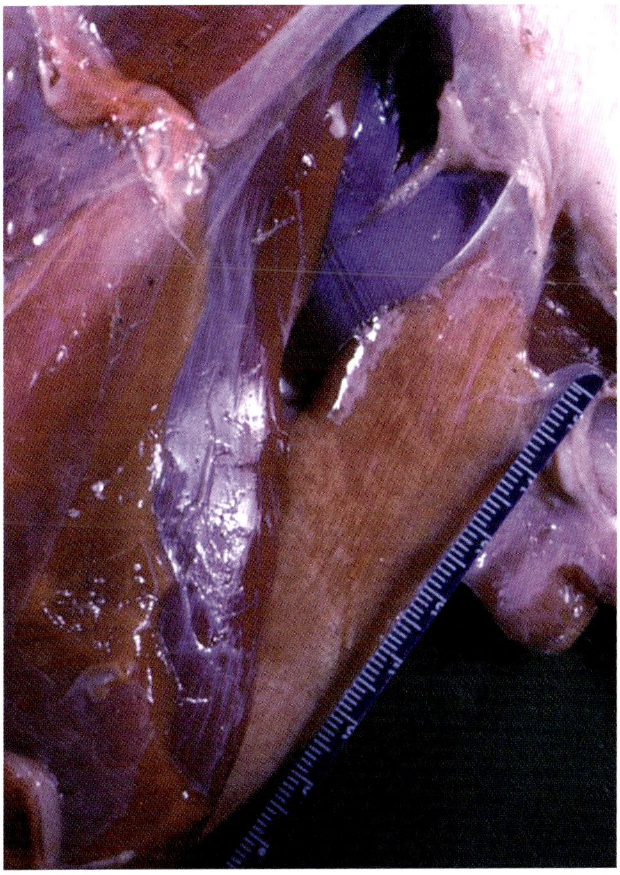

FIGURE 6.3 Exertional rhabdomyolysis in a white-tailed deer. White streaking of thigh muscle is due to myonecrosis and edema within fascial planes. *(Photo Courtesy of W. Chapman, College of Veterinary Medicine, University of Georgia and Noah's Arkive.)*

(tephra). Changes in teeth typical of fluorosis are enamel hypoplasia, breakage, pitting, blackening, and excessive wear. Diagnosis is made by characteristic gross lesions in teeth and bones, and high levels of fluoride in bone.

In general, cervids are not poisoned by most plants that are toxic to domestic ruminants. However, toxicity has been attributed to a variety of other plants in cervids.

Lichen intoxication due to the ingestion of tumbleweed shield lichen (***Xanthoparmelia chlorochroa***) causes red urine, ataxia, and muscle weakness that rapidly progresses to recumbency and eventually death due to dehydration or predation in free-ranging North American elk (hereafter referred to as "wapiti").[52] Gross findings are muscle pallor and streaking, particularly of the muscles of ambulation of the pelvic limbs. Microscopically, degenerative myopathy with varying degrees of myofiber degeneration, rupture and necrosis, mononuclear cell infiltration, and sometimes mineralization and regeneration are seen. Diagnosis is by typical lesions and detection of the lichen in the ruminal contents. The specific toxin in the lichen is currently unknown.

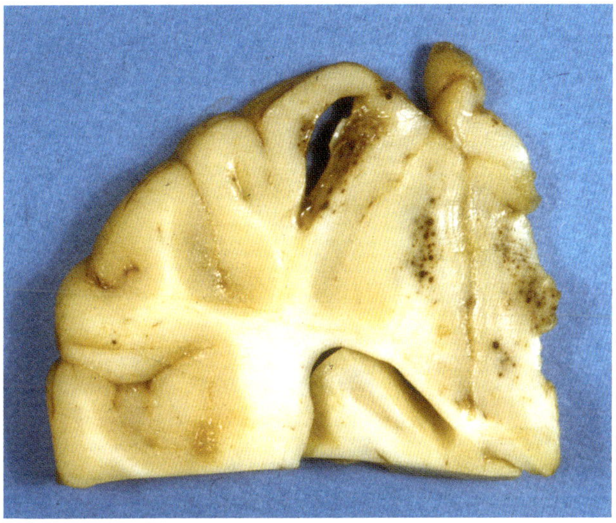

FIGURE 6.4 Leukoencephalomalacia due to fumonisin toxicosis in a white-tailed deer. Characteristic lesions are malacia, cavitation, and hemorrhage in the white matter, as well as edema resulting in swelling of the gyri.

Bracken fern (*Pteridium aquilinum*) **toxicosis** was suspected in several captive fallow deer with lesions similar to those in cattle with enzootic hematuria. Specifically, lesions included hematuria and urinary bladder lesions, such as hemangiosarcoma, hemangioma, transitional cell carcinoma, and chronic cystitis.[285] Intestinal adenocarcinomas in distal ileum, cecum, and proximal colon were linked to bracken fern ingestion in a herd of farmed sika deer[176] (see also "Neoplastic" section and Fig. 6.10).

A variety of **other plant toxicoses** have been described. *Wedelia glauca*, a native South American plant, was responsible for centrilobular hepatocellular necrosis and death in captive axis deer-fed contaminated alfalfa.[121] **Tall fescue toxicosis** resulting in large masses of necrotic abdominal fat has been reported in cervids, including Eld's deer.[345] **Mycotoxicosis** due to **fumonisin toxicosis** characterized by leukoencephalomalacia has been reported in a white-tailed deer[159] (Fig. 6.4).

Congenital/Genetic

A wide range of congenital anomalies affecting different organs have been reported in cervids. Anophthalmia, microphthalmia, aphakia, and other ocular abnormalities occur in free-ranging populations of white-tailed deer, European red deer, and Canadian moose[2,46,117,119,193,234] (Fig. 6.5).

Hypotrichosis is reported in white-tailed deer and characterized macroscopically by sparse hair development on the ventral thorax, rear legs, muzzle, around the eyes, and inside the ears. Remaining skin surfaces may be alopecic, thin, and hyperpigmented.[357] **Ventricular septal defects**

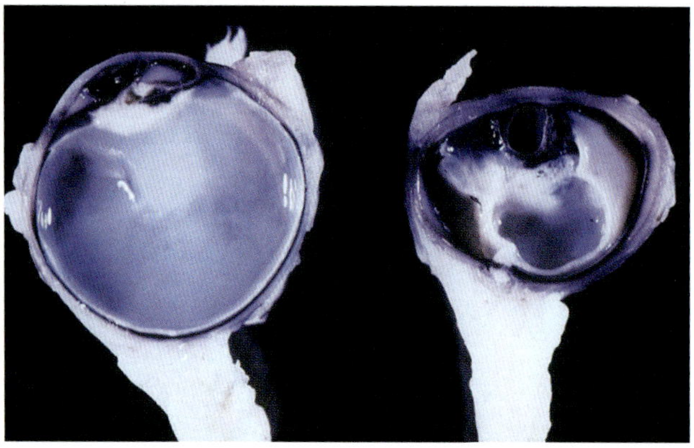

FIGURE 6.5 Congenital microphthalmia in a white-tailed deer. The eye on the left is normal while the right eye is small. Although not obvious here, the small eye also has anterior segment dysgenesis, a condition which includes a number of anomalies of the anterior segment due to disordered development and/or improper separation of the lens from the overlying cornea. *(Photo Courtesy of K. Carmichael, College of Veterinary Medicine, University of Georgia and Noah's Arkive.)*

occasionally occur in fallow deer and roe deer.[2,253,296] **Diverticulosis** was described in rectum and colon of captive fallow deer[44] and esophagus of roe deer.[253] **Diaphragmatic hernia**, **palatoschisis**, **biliary atresia with marked cholestasis**, and **diffuse dissecting fibrosis** have been observed in farmed fallow deer.[296] **Polycystic kidneys** regularly occur in roe deer and are also reported in white-tailed deer with no known genetic or other cause[248,253] (Fig. 6.6).

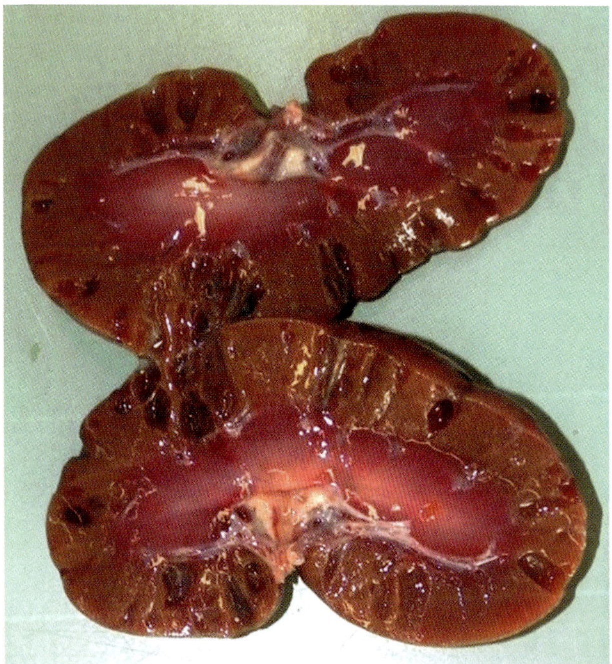

FIGURE 6.6 Polycystic kidneys in a roe deer. Multiple, variably sized tubular cysts are present throughout the renal cortex. *(Photo Courtesy of the Centre for Fish and Wildlife Health, University of Bern.)*

Various skeletal and dental anomalies are reported in farmed fallow deer, white-tailed deer, red deer, roe deer, and caribou.[1,2,166,178,227,228,247,253,276,296,297,310] **Intersex**, linked to SOX9 duplication, occasionally occurs in European roe deer; phenotype ranges from antlered females to true hermaphrodites.[191,246] Testicular abnormalities (e.g., **cryptorchidism, hypogonadism**), with or without antler abnormalities (from total absence to poor calcification or velvet retention) are reported in numerous cervid species, such as roe deer, wapiti, white-tailed deer, red deer, and reindeer.[203] Partial or complete **albinism** occasionally occurs in roe deer.[309]

Age-Related/Degenerative

Degenerative joint disease (DJD) is seen more in aged and captive cervids; and it can occur secondary to trauma or infectious disease. In free-ranging moose, DJD may be associated with poor nutrition early in life and increases susceptibility to predation.[252] Gross lesions may include excessive articular wear with sclerosis following cartilage loss, pitted articular surfaces, and periarticular bony exostoses and osteophytes. In moose, the coxofemoral joint is most commonly affected, as are intervertebral and lower back-lumbosacral; less severe lesions occur in foot (all limbs) and other distal rear limb joints.[198] In white-tailed deer, the distal femur and proximal tibia are more commonly affected and have focally rough articular surfaces, matted cartilage, and articular clefts.[343]

Abnormal tooth wear and sharp forage wedged into the gum can facilitate bacterial invasion of the mandibular bone, culminating in lytic lesions, and chronic mandibular osteomyelitis in all cervid species (see information on *Actinomyces bovis* infection under section, "Bacteria") (Fig. 6.7). Abnormal tooth wear may be related to coarse and abrasive food as well as dental fluorosis.[190]

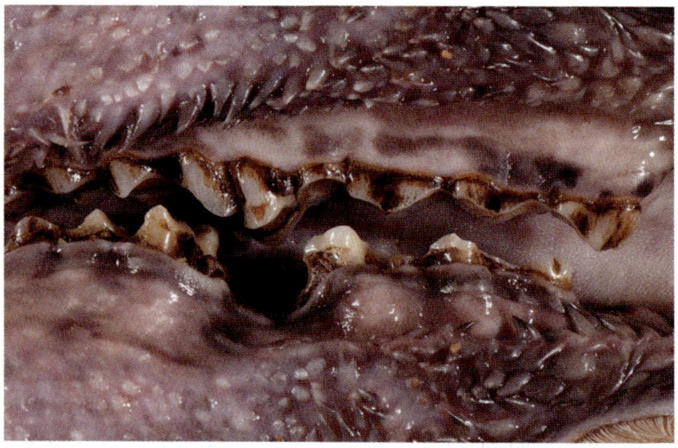

FIGURE 6.7 Dental disease in a white-tailed deer. There is severe tooth wear, an absent premolar, and irregular occlusal surface with multiple points and hooks. *(Photo Courtesy of the Southeastern Cooperative Wildlife Disease Study.)*

Aberrant antlerogenesis ("double" or "triple heads") is the development of ectopic accessory (i.e., supernumerary) pedicles or accessory antlers in the frontal, nasal or parietal bones. This condition occurs in all cervid species except rangiferines (i.e., *Rangifer* spp.) and is more commonly reported in roe deer, white-tailed deer, and moose than red or fallow deer. Antler deformities or malformations may also result from cryptorchidism (i.e., lack of circulating androgens) or continued exposure of dams to estrogenic compounds, repeated skull or pedicle trauma, malnutrition (e.g., copper deficiency, low dietary protein, poor foraging conditions), infectious agents (e.g., epizootic hemorrhagic disease virus), and genetic factors (see "Congenital/Genetic" section).[35,36,109,126] Lack of circulating androgens in castrated male deer associated with sarcoma-like proliferation of soft bone tissue is referred to as "perukes" or "wig antlers." These may be especially distinctive in roe deer, in which castrated bucks retain the velvet permanently. This condition is consistent with the broader designation of **antleroma**, a neoplasm that may be associated with abnormal levels of circulating testosterone (e.g., hypogonadism, cryptorchidism, castration, toxic insult) and originates from one or more antler tissues (e.g., bone, cartilage, fibrous connective tissue, velveted skin). Antleromas have been documented in numerous free-ranging and captive cervid species, including white-tailed deer (Fig. 6.8; eSlide 6.e1), fallow deer, roe deer, mule deer, moose, and others. Presentation is variable and consists of distortion of one or more antler tines, which are covered by velvet. Larger tumors may protrude internally and compress the brain. Histopathology also varies. In castrated fallow deer, skin outgrowths on antlers consist of collagen fibers with interspersed fibroblasts and no bone, or cartilage. Antleromas reported in roe deer and moose contain intradermal bone, and in white-tailed deer, the neoplasm consists of bony masses with scattered fibrous connective tissue and chondrocytes with areas of necrosis and hemorrhage.[179,233]

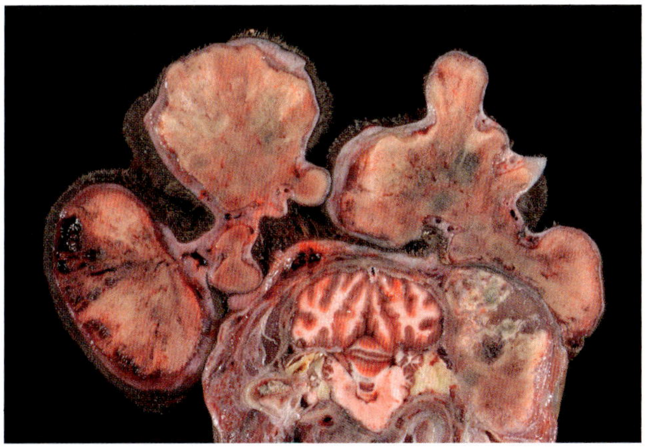

FIGURE 6.8 Antleroma in a white-tailed deer. Irregular bony masses along the antlers consist of mature and immature bone, and mature to mineralized cartilage with scattered fibrous connective tissue. Masses contain multifocal necrosis and hemorrhage (see eSlide 6.e1). *(Photo Courtesy of the Southeastern Cooperative Wildlife Disease Study.)*

Antler malformations that vary from twisted, knobby, stunted antlers to rotated and ventrally deviated beams with thin cortices and a hollow central core have been associated with low serum or hepatic copper levels in white-tailed deer and tule elk.[126,180] A condition known as "**cactus buck**" refers to males whose antlers have short, aberrant points for several or more years. The condition, as well as, other antler malformations are associated with testicular atrophy and hypoplasia, which purportedly could be due to dietary deficiencies and toxins (i.e., estrogenic substances). Testicular hemorrhage, seminiferous tubular necrosis, epididymitis, orchitis, and hypospermia, and less commonly, necrotizing vasculitis, edema and more chronic lesions of fibrosis, mineralization, and seminiferous tubular loss were observed in mule deer that also tested PCR and serologically positive for epizootic hemorrhagic disease and bluetongue viruses.[109]

Perennial velvet antlers occur due to inadequate circulating androgens or traumatic head injury and have permanent soft antler tissue growth over the entire antlers. In castrated fallow deer bucks, histology includes an osseous portion of lamellar or woven bone and lack of cartilage or intradermal bone.[179]

Inflammatory Non-Infectious

Hoof lesions have been described in captive and free-ranging cervids. Non-infectious causes may be multifactorial and include nutritional imbalance, endophyte toxicity, and mineral deficiency.[137]

Neoplastic

Neoplasia is neither species specific nor specific to captive or free-ranging cervids. However, aged captive animals may contribute to an increased number of cases detected in this group, and environmental exposures to known carcinogens within a given region or habitat may be a predisposing factor in the tumor development in some species or settings (e.g., intestinal adenocarcinoma in sika deer, see below). Roe deer seem to be particularly prone to neoplasia even in free-ranging conditions.[253]

Cutaneous fibroma is most commonly reported in white-tailed deer, roe deer, and European moose but has been documented in a variety of cervids. Caused by **fibroma viruses (FV)** (family *Papillomaviridae*, genus *Deltapapillomavirus*), the tumor may regress with time.[195] One or more, up to 25 cm diameter, smooth to verrucose, pedunculated to pendulous, dermal masses are generally most numerous on the head and neck but may include the oral mucosa and legs[86] (Fig. 6.9).

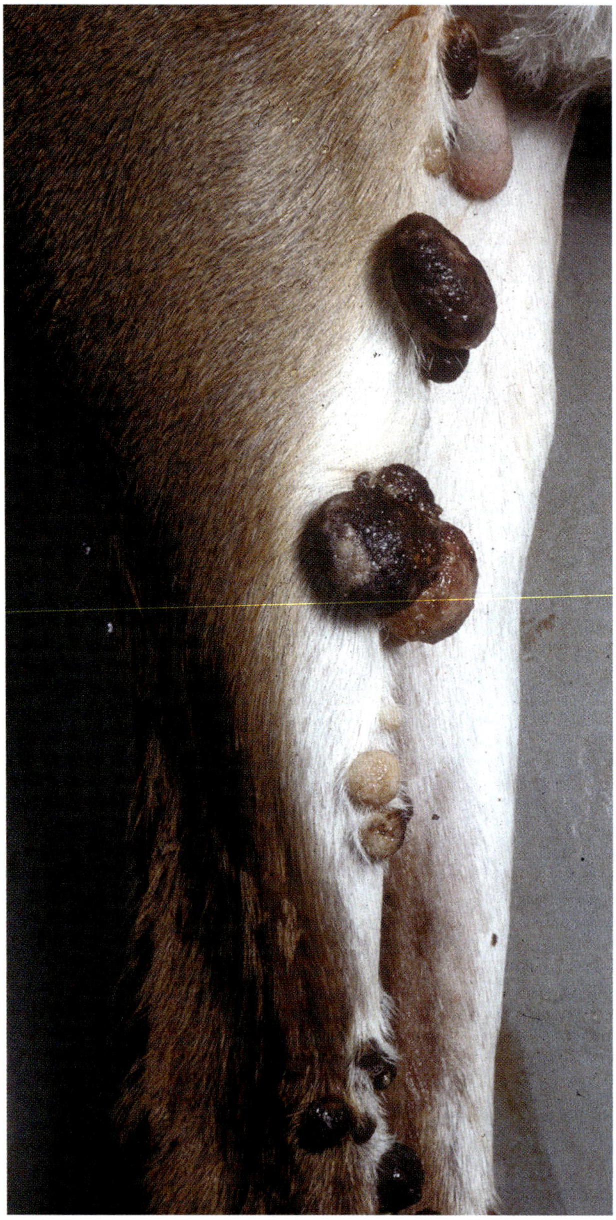

FIGURE 6.9 Cutaneous fibromas on the leg of a white-tailed deer. Multiple, variably sized, white to brown/black, pedunculated dermal masses are present in the skin. Tumors are associated with deer fibroma virus (DFV) and can regress with time. *(Photo Courtesy of the Canadian Wildlife Health Cooperative.)*

Microscopically, the neoplasm is nonencapsulated and consists of neoplastic fibroblasts supported by a collagenous matrix that is overlain by epidermal hyperplasia with rete pegs and hyperparakeratosis. Evidence of viral cytopathic effect includes cytoplasmic keratohyaline granules in the stratum granulosum, koilocytes, and inconsistent, pale, eosinophilic to amphophilic, intranuclear inclusions.[293,311,328] Tumor regression may involve perivascular lymphocyte infiltration and hyalinization of the connective tissue matrix. Confirmation of papillomavirus infection may be made with indirect immunofluorescence, immunohistochemistry (IHC), and PCR of fresh-frozen tumor tissue, including affected epithelium (i.e., containing viral DNA).[86,313] Monoclonal antibody panels can aid in partial typing of viruses by immunohistochemistry.[312]

Lymphoma has been reported in a variety of cervid species, including white-tailed deer, roe deer, red deer, and moose, and is likely the most common tumor in cervids.[253] CD3+ T cell alimentary/abdominal lymphoma is most commonly reported; some cases have central nervous system involvement.[198] Immunohistochemistry can be used to identify neoplastic cells as being of lymphocytic origin.[185,223] Other central nervous system tumors reported in white-tailed deer include **oligodendroglioma**[128] and **astrocytoma**.[154]

Myxoma has been reported in moose and wapiti. It presents as a nonencapsulated, invasive, amorphous, or cauliflower-like mass. Microscopically, the tumor is composed of fusiform to stellate cells within a mucinous to myxomatous stroma with moderate numbers of mitotic figures and scattered collagen bundles.[198,298]

A novel low-grade **intestinal adenocarcinoma** has been described in farmed sika deer. Single or multiple neoplasms in the intestinal mucosa and submucosa consist of poorly circumscribed, plaque-like, pale-gray thickenings in the ileocecocolic junction, proximal colon, cecum, and ileum. Neoplastic cells were well-differentiated, locally infiltrative, and most commonly of tubular morphology, followed by mucinous and mixed. Sika deer with intestinal adenocarcinoma grazed on pasture with **bracken fern**, which was associated with **urinary bladder hemangiosarcoma**, **hemangioma**, **transitional cell carcinoma**, and **chronic cystitis in the urinary bladder** of six captive fallow deer[176,285] (Fig. 6.10; see also "Toxic" section).

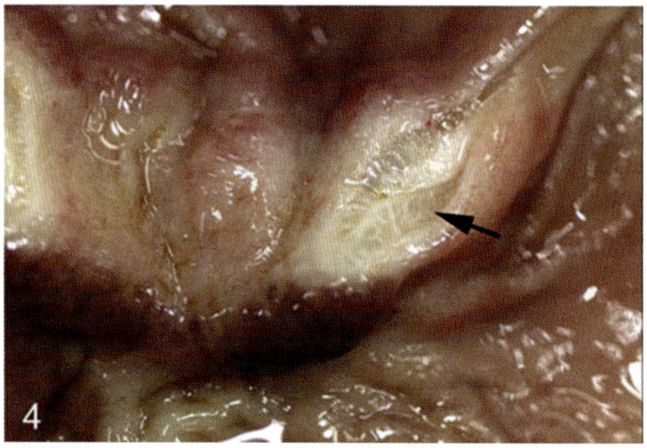

FIGURE 6.10 Intestinal adenocarcinoma, mucinous type, in a farmed sika deer. The intestinal wall at the ileocecal junction is markedly thickened by an infiltration of cystic structures *(arrow)*. Tumors occurred in deer grazing on bracken fern that also had multiple lesions in the urinary bladder. *(Reprinted with permission from Kelly PA, Toolan D, Jahns H. Intestinal adenocarcinoma in a herd of farmed sika deer (Cervus nippon): a novel syndrome. Vet Pathol. 2015;52:193–200.)*

Rarely reported neoplasms include **ovarian teratomas** in a Père David's deer[354] and free-living Iberian red deer,[149] **cementoblastoma** in a red deer,[177] **hepatocellular carcinoma** in a white-tailed deer,[254] **embryonal nephroma** in a free-ranging wapiti,[299] **oral squamous cell carcinoma** in a free-ranging red deer,[326] **benign and malignant melanomas** and **pigmented schwannomas** in captive fallow deer,[161,279] and **malignant melanoma** in red deer.[286] **Antleromas** are documented in numerous cervid species (see above aberrant antlerogenesis). A range of neoplasms originating from various tissues have been reported in roe deer.[253]

INFECTIOUS DISEASES

DNA Viruses

Susceptibility to **malignant catarrhal fever (MCF) virus** (family *Herpesviridae*, subfamily *Gammaherpesvirinae*, genus *Macavirus*) varies among cervids, with Père David's deer and white-tailed deer being highly susceptible; wapiti, moose, reindeer, and roe and red deer are also susceptible, and fallow deer relatively resistant. Other species reported with MCF virus infections include mule deer, axis deer, sika deer, Indian hog deer, barasingha or swamp deer, rusa deer, Chinese water deer, red brocket deer, Eld's deer, Sunda sambar, Malayan sambar, and Reeve's muntjac.[145] Cervids infected with ***Ovine gammaherpesvirus* 2 (OvHV-2)**, the most common etiology, often experience acute death (1–3 days postinfection). Other herpesviruses within this subfamily have also been reported to cause disease including **alcelaphine herpesvirus 1** and **2** (**AlHV-1** and **AlHV-2**, respectively), **caprine HV-2**, and **MCF virus-white-tailed deer**.[53,186,244] MCF-white-tailed deer has a suggested designation of caprine HV-3 due to recent identification of a goat reservoir; it was first isolated from diseased white-tailed deer in a zoo and caused disease in red brocket deer and reindeer.[209–212]

Transmission of OvHV-2 is most commonly via contact with aerosolized nasal secretions. The pathogenesis of MCF has not been fully elucidated and varies for the different MCF-causing viruses, but increased cytotoxicity or failed regulation of T lymphocytes (e.g., CD8+ and regulatory) and autoimmunity appear to play a role. Open reading frame (ORF) 25 expression in tissues of some species increases with disease progression and severity.

Few grossly evident lesions may be seen in cervids that undergo sudden death. These animals may have hemorrhagic gastroenterocolitis and oral (including pharyngeal and laryngeal), nasal, and tracheobronchial mucosal lesions, which may include combinations of hyperemia, inflammation, hemorrhage, edema, necrosis, ulceration, and pseudomembranes. Alimentary lesions become more severe with disease progression and may include ulcers on the gingiva, dental pad, palate, buccal papillae, tongue, esophagus, abomasum, rumen, and intestine. Hemorrhage, inflammation, and erosions may be seen in the esophagus, abomasum, intestine, and rectum. Arterial walls may be thickened and tortuous, as has been observed along the ruminal serosa, hepatic portal regions, and renal interlobar and lobular and arcuate arteries of naturally infected axis deer.

The classic **head-and-eye form of MCF** in cattle is less pronounced in cervids, who may also have an intestinal form. In the latter, affected animals may be dehydrated and emaciated with serous to mucopurulent nasal discharge, crusting of the skin over the muzzle and other areas, such as around the mouth, eyes, perineum, and feet, and even truncal, generalized exanthema, and cutaneous hyperemia in the abdominal, thoracic, inguinal, and perineal regions. Feces may be blood stained and scant. Generalized lymphadenopathy is common and pulmonary edema and emphysema may occur. Chronically infected animals may have bronchopneumonia. Ocular lesions may include conjunctivitis, ophthalmia, unilateral or bilateral corneal opacity/edema, and fibrin or purulent material in the anterior chamber. Urinary tract lesions (e.g., mucosal hyperemia, edema, inflammation, and occasional hemorrhage, erosion, and ulceration) are more common in bison and cattle than in cervids. Other possible gross findings include hemorrhage in the uterus and gall bladder mucosa, hepatomegaly with a reticular pattern, and pale, raised foci in kidneys and heart (Fig. 6.11A). Chronic disease in sika deer with caprine HV-2 infection is characterized by vascular lesions accompanied by granulomatous mural folliculitis in the skin over the dorsum.[54]

The histopathologic hallmark of MCF is lymphocytic vasculitis and perivascular inflammation of medium-sized arteries and veins, which is often less widespread in cervids and other wildlife species than cattle (Fig. 6.11B). Vascular lesions may also include medial fibrinoid necrosis and thrombosis. Expansion, effacement, and fragmentation of artery walls by irregular mesenchymal cells and lymphocytes have been seen in axis deer, in which larger arteries have intimal spindle cell proliferation and collagen deposition. Lymph nodes often exhibit lymphoid hyperplasia in paracortical and interfollicular areas, and lymphoid proliferation with mononuclear cell infiltration is also seen in the spleen and the vasculature in the kidney, liver, adrenal glands, and the central nervous system. Inclusion bodies are absent[145,211,244,331] (eSlide 6.e2).

A presumptive diagnosis of MCF can be made based on the collective history of contact with potential reservoir host species as well as clinical, gross, and histopathology findings. Definitive diagnosis depends on detection of viral DNA in tissue. Multiplex PCR can confirm and distinguish the different MCF-causing viruses, and qPCR is recommended for the diagnosis of MCF caused by OvHV-2. Virus

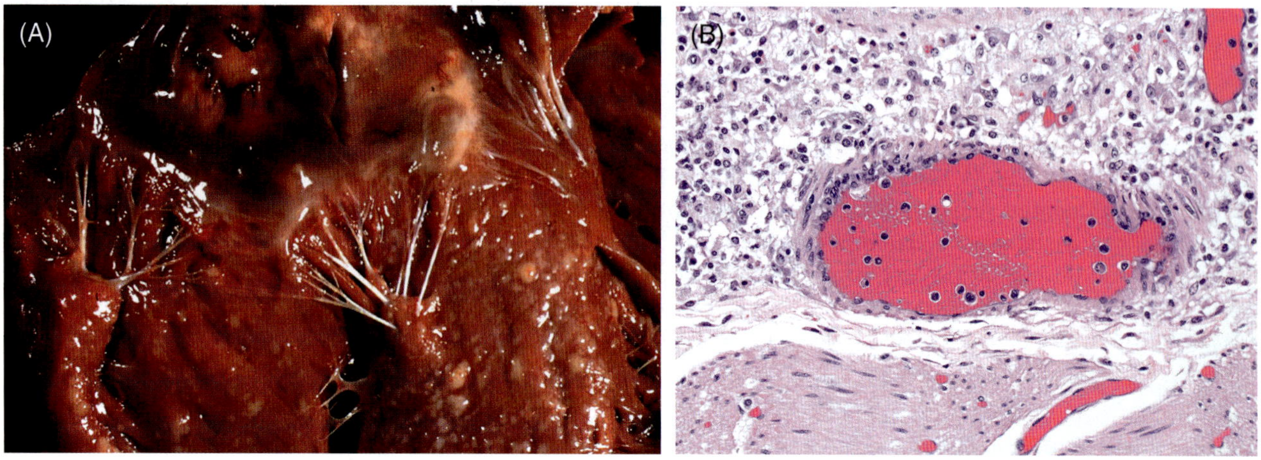

FIGURE 6.11 Malignant catarrhal fever (MCF) in deer. (A) Multifocal myocarditis characterized by multiple white, raised foci composed of infiltrating lymphocytes. (B) Small intestine from a white-tailed deer with malignant catarrhal fever. Vascular or perivascular inflammation are characteristic findings in MCF. A linear aggregate of lymphocytes is present within the tunica media of a vein. Lymphocytes and plasma cells are also scattered within the surrounding submucosa (see eSlide 6.e2). *(Part A: Photo Courtesy of METRO, Ohio State University and Noah's Arkive.)*

isolation can also be performed. Common preferred samples for virus isolation and PCR are peripheral blood leukocytes (i.e., noncoagulated blood) and lymphoreticular tissues.[244]

Adenovirus hemorrhagic disease of deer (family *Adenoviridae*, genus *Atadenovirus*—closely antigenically related to ovine adenovirus-7, goat adenovirus-1, and bovine adenovirus-7) was first described in 1993 in California. Mule deer are the most susceptible species, with Columbian black-tailed deer apparently affected more frequently than Rocky mountain mule deer. Additional cervids reported with fatal infections include white-tailed deer and moose.

Adenoviruses are primarily transmitted by inhalation and the fecal-oral route. The virus that causes adenovirus hemorrhagic disease in deer is endotheliotropic, and although vasculitis is primarily seen in the lungs and alimentary tract, systemic infection may occur. The most consistent gross finding is pulmonary edema; intraluminal intestinal hemorrhage is seen but is less commonly observed. Similar to the orbiviral infections caused by bluetongue viruses (BTV) and epizootic hemorrhagic disease viruses (EHDV), damage to endothelial cells can lead to disseminated intravascular coagulopathy. Upper alimentary tract disease can be diffuse or localized, with ulcerative necrotizing stomatitis involving the lips, pharynx, tongue, and less often, the forestomachs. Endothelial damage leads to thrombosis and ischemic necrosis, inflammation, and secondary bacterial invasion (e.g., abscessation) of the oral cavity and subsequent anorexia and wasting; osteomyelitis may also occur.

Histopathology includes vascular lesions (e.g., vasculitis, endothelial hypertrophy or necrosis, fibrinoid necrosis) in the lungs and alimentary tract. Eosinophilic to amphophilic, often indistinct, endothelial intranuclear inclusion bodies are seen but can be rare and transient. In fatally infected animals, vascular lesions can also occur in the brain, pulmonary artery, spleen, kidney, uterus, urinary bladder, and lymph nodes. Other lesions attributed to vascular damage include pulmonary edema, upper alimentary tract ulceration, and intestinal mucosal hemorrhage. Lymphoid depletion and necrosis may occur in multiple lymph nodes (e.g., retropharyngeal, mandibular, parotid, and cranial cervical). Histopathology in localized infections is nonspecific and includes focal necroulcerative oral lesions, most notably in the pharynx.[348–350] Bronchiolitis associated with adenovirus infections has also been documented in red and fallow deer. The latter was associated with **bovine adenovirus-6**.[292,301]

Gross and histopathology findings in adenoviral hemorrhagic disease of deer can be similar to bluetongue and epizootic hemorrhagic disease; endothelial intranuclear inclusion bodies associated with vascular lesions are consistent with adenovirus infection. Definitive diagnosis is made with virus isolation, immunohistochemical staining of affected tissues for antibody against bovine adenovirus-5 or deer adenovirus, or PCR.[199] Transmission electron-microscopic examination of affected tissue can aid in diagnosis.[347]

Poxviruses of cervids include **deer poxvirus** (subfamily *Chordopoxvirinae*, family *Poxviridae*, genus *Cervidpoxvirus*) in mule deer, white-tailed deer, and black-tailed deer, **moosepox virus** in moose, an unclassified **cervidpoxvirus** in reindeer, and **parapoxvirus** (genus *Parapoxvirus*) of red deer and semidomesticated reindeer.[9,19,153] In addition, a novel poxvirus related to **capripoxvirus** has been reported in captive pudu.[171] Reports of all three viruses in cervids are more common among young captive or semidomesticated animals, including herded and zoo reindeer, a captive herd of white-tailed deer, black-tailed deer in rehabilitation, and moose calves at a zoological garden. In experimentally infected mule deer, white-tailed deer, caribou, and moose, **contagious ecthyma** (i.e., orf; caused by the parapoxvirus, orf virus), derived from infected noncervid and cervid hosts (e.g., bighorn sheep, Dall sheep, reindeer), led to mild proliferative oral lesions that subsequently regressed at the mucocutaneous junction of the lips.[196,355]

Transmission of cervidpox and parapox viruses is by animal-to-animal contact and contaminated fomites by which the virus enters via breaches in the keratin layer of the skin. Contagious ecthyma outbreaks in reindeer may be associated with aspects of domestication, such as supplemental feeding and corralling. Poxvirus lesions are often associated with morbidity rather than mortality and regress without scarring. However, parapoxvirus-infected farmed red deer and semidomesticated reindeer have high mortality rates, and cervid poxvirus-infected black-tailed and white-tailed deer fawns have been euthanized or died due to severe disease. Additional opportunistic infections (e.g., bacterial) are suspected to be involved in the death of red deer. Zoonotic transmission and the development of cutaneous lesions have occurred in humans associated with both orf virus infections in deer[192] and a novel deer-associated parapoxvirus in white-tailed deer.[275]

Gross lesions of parapoxvirus infection in red deer include pustular dermatitis on the muzzle, face, ears, and sometimes on the neck and limbs. Histopathology may include hyperkeratotic and hyperplastic dermatitis with intraepithelial pustules, ulcers, and intracytoplasmic eosinophilic inclusion bodies in keratinocytes.[153,273,324]

Gross lesions of cervidpoxvirus infection in deer species may include keratoconjunctivitis and proliferative, ulcerative, and crusty skin lesions on the face, neck, pinna, legs, feet, ventral thorax, abdomen, and vulva of reindeer and mule deer. Proliferative skin lesions are prominent on the abdomen of captive white-tailed deer fawns; and erythema, alopecia, papules, crusting, and ulceration have been reported on the neck (**raw neck syndrome**) of black-tailed deer fawns in a rehabilitation setting (Fig. 6.12). Epithelial ulceration on the hard palate has been observed in zoo reindeer and mule deer; on the lips, dorsal tongue, and buccal cavity in white-tailed deer; and on the tongue in mule deer. White-tailed deer fawns can also have transmural ruminal ulcers and mule deer fawns may have ulcerative dermatitis and hoof wall separation at the hind coronary bands.[16] Captive moose calves had well-demarcated, crust-covered, raised, 0.5–1.5 cm diameter plaques or nodules, sometimes with purulent discharge, on the skin of the face (primarily adjacent to medial and later canthus) (Fig. 6.13A) and fetlocks.[9]

Histopathology of cervidpoxvirus infection includes epidermal hyperplasia with hydropic change and prominent rete ridges, erosions and ulcers, occasional pustules in the epidermis and superficial dermis, mixed inflammatory infiltrates, and intracytoplasmic, eosinophilic inclusion bodies in keratinocytes. Similar inclusion bodies have been reported in the ruminal mucosal epithelium of white-tailed deer fawns. Eyes of infected mule deer fawns have keratoconjunctivitis with neovascularization, epithelial ulceration, and neutrophils and fibrin in the anterior chamber.[19,27,338] Lesions in moose infected with moosepox virus include subacute necrosis of hair shafts and adnexal glands without distinct inclusions. Suppurative, eosinophilic dermatitis and hemorrhage, along with mixed, superficial to dermal bacteria, have been described. More chronically infected moose calves had thick crusts with proteinaceous exudate, stratum spinosum hyperplasia, orthokeratotic hyperkeratosis, and papillary dermal fibrosis with lack of adnexa and lymphoplasmacytic and granulomatous dermatitis (Figs. 6.13B, C).[9]

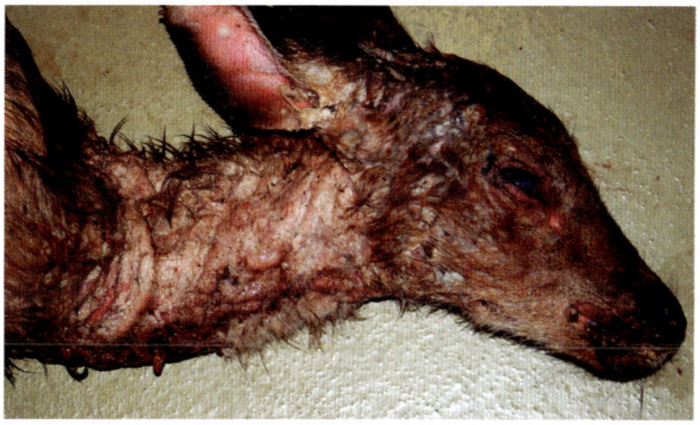

FIGURE 6.12 Cervidpoxvirus infection in the skin of a black-tailed deer fawn. Multifocal erythema, alopecia, papules, and ulceration are present on the neck (raw neck syndrome) and head. Secondary pyoderma contributes to the disheveled hair coat. *(Photo Courtesy of R. Bildfell, Veterinary Diagnostic Laboratory, College of Veterinary Medicine, Oregon State University.)*

The diagnosis of pox and parapoxviruses is made by identification of characteristic gross lesions and histopathology (e.g., intracytoplasmic, eosinophilic inclusion bodies). Viral infection can also be confirmed by PCR and virus isolation or negative-contrast electron microscopy. Past infections can be detected via serum neutralization testing, ELISA, complement fixation, and immunodiffusion.[273]

Red deer cervid alphaherpesvirus 1 (CerHV-1), **(rangiferine) cervid alphaherpesvirus 2 (CerVH-2)**, and **elk herpesvirus 1** (family *Herpesviridae*, subfamily *Alphaherpesvirus*, genus *Varicellovirus*) are closely related to Bovine alphaherpesvirus 1 and have been detected either serologically or by virus isolation from a variety of cervids including white-tailed deer, caribou/reindeer, red deer, wapiti, and mule deer.

CerHV-1 was first isolated from farmed red deer fawns during an outbreak, but alone it likely does not cause

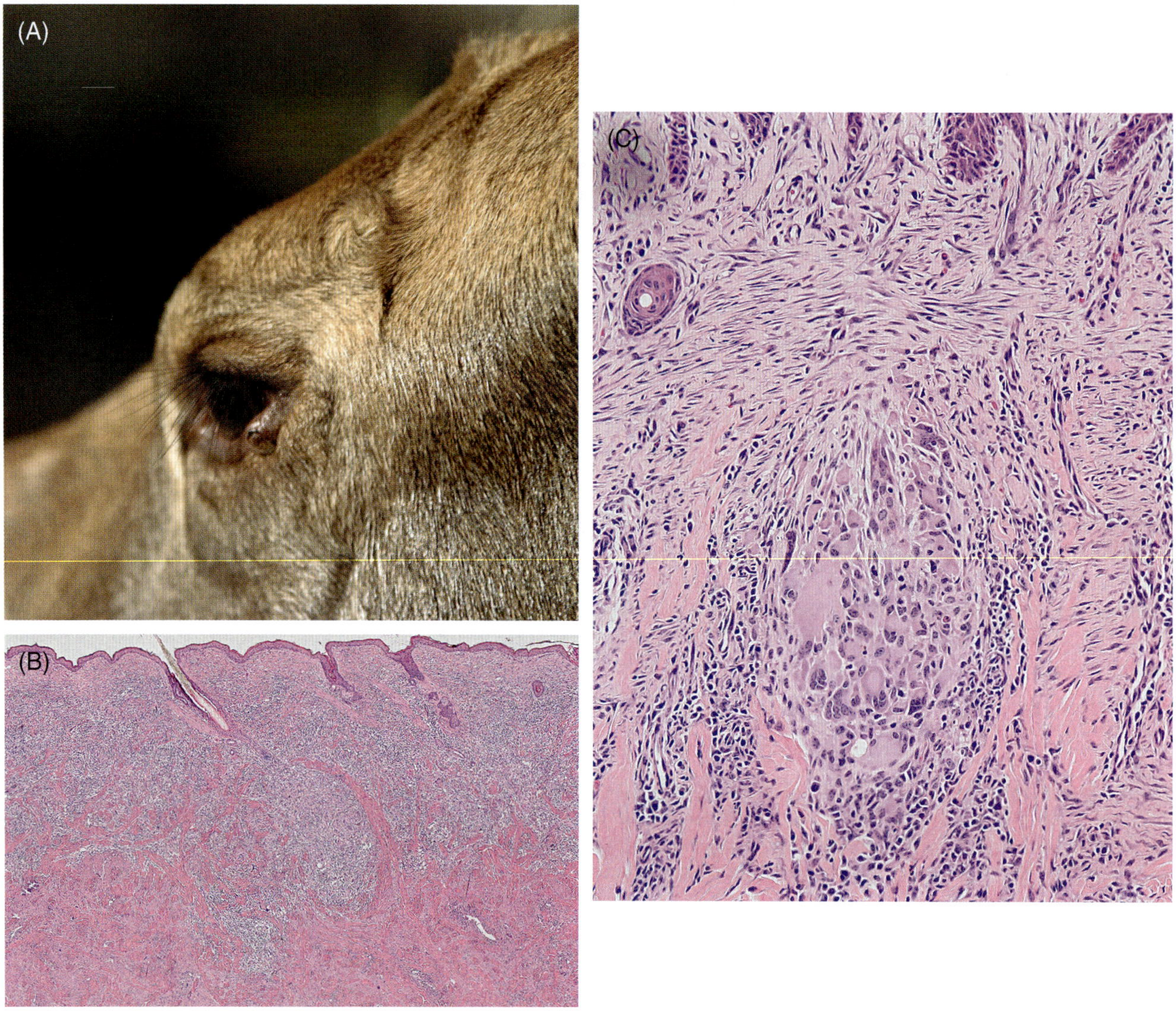

FIGURE 6.13 Moosepox dermatitis in a moose calf. (A) The lateral canthus of the eye has a small crusty subacute lesion. (B) A chronic skin lesion from the fetlock has marked coalescing granulomatous inflammation that expands the dermis and effaces the adnexa. (C) A layer of granulation tissue underlies a hyperplastic epidermis with rete pegs evident. An adnexa is replaced with granulomatous inflammation surrounded by lymphocytes and plasma cells. *(Parts A–C: Photos Courtesy of A. Armién, California Animal Health and Food Safety Laboratory System, University of California-Davis.)*

significant disease and is more commonly reported in young animals. Lesions in red deer fawns include conjunctivitis, upper eyelid edema, hypopyon, corneal opacity, and ocular and nasal discharge.[162,267]

CerHV-2 or rangiferine herpesvirus was first isolated from vaginal secretions of a healthy reindeer in the 1980s.[84] CerHV-2 has been associated with outbreaks of keratoconjunctivitis in semidomesticated reindeer, characterized by corneal opacity, eyelid and periorbital edema, and ocular exudate (sometimes with hemorrhage) that leads to crusted, discolored periocular fur. In severe cases, disease progresses to panophthalmitis with corneal perforation and secondary bacterial infection. Morbidity may be high in a herd but many individuals recover. In a past outbreak in reindeer, a small proportion of affected animals were euthanized due to severe clinical signs.[323] CerHV-2 is thought to play a role in multifactorial diseases including neonatal death and abortion, and the "reindeer respiratory disease complex."[56] Histopathology may include vitreal hemorrhage and fibrin deposition, fibrinous uveitis, conjunctivitis with edema, neutrophilic infiltration of the third eyelid lacrimal glands, corneal epithelial necrosis, edema, hemorrhage, and anterior synechiae.

Elk HV-1 was isolated from the semen of a healthy wapiti and it is not known to cause disease in this species.[66]

Virus isolation and PCR can be used to detect alphaherpesviruses, and are useful in confirming CerHV infection from swab samples from suspicious lesions in the eyes, nose, vagina, or semen. Latent infections may require additional amplification steps by PCR.[56,323]

RNA Viruses

Cervid orbiviruses include **bluetongue viruses (BTV)** and **epizootic hemorrhagic diseases viruses (EHDV)** [family *Reoviridae*, genus *Orbivirus*; World Organization for Animal Health (WOAH)-listed disease-multiple species concern] and affect numerous cervids. Reported naturally acquired, fatal infections of BTV and/or EHDV in cervids are limited to free-ranging native and captive native and nonnative cervids in North America. White-tailed deer are most susceptible, but disease has occurred in other cervids species including mule deer, black-tailed deer, wapiti, Rocky Mountain elk, axis deer, fallow deer, sika deer, yaks, Pere David's deer, and moose[272] and more recently reindeer.[320] Orbiviral disease is generally considered mild in wapiti, and absent in moose and hog deer (as reviewed in Refs. 148,158). As of 2021, 27 serotypes of BTV and seven serotypes of EHDV are recognized, although the number of BTV serotypes may increase as putative new serotypes and strains have been described.[272,277,289] Infection occurs via the bite of *Culicoides* spp. midges and is followed by virus replication in macrophages of draining lymph nodes, virus entry into circulation (viremia is usually 7–10 days duration in experimentally infected deer[352]), and dissemination to secondary replication sites (e.g., spleen) where it replicates in endothelial cells, mononuclear phagocytic cells, and lymphocytes.[221] Endotheliotropism leads to vascular damage, activation of the coagulation and fibrinolytic systems, thrombosis, hemorrhage, and disseminated intravascular coagulation. Viral evasion of neutralizing antibodies due to virus association with erythrocytes can lead to prolonged viremia[158] (Fig. 6.14A).

Gross lesions associated with BTV and EHDV-induced disease in naturally and experimentally infected white-tailed deer manifest early as pulmonary and subcutaneous edema generally in the head, neck, and along fascial planes. Similar lesions are seen but less commonly in the omentum, mesentery, joints, and cranium, as well as pleural, peritoneal, and pericardial effusions. With disease progression, multisystemic hemorrhage is seen and often involves the alimentary tract [e.g., oral mucosa, abomasum—especially the pyloric region, rumen—involving papillae and pillars (Fig. 6.14B), and intestine] and heart, but may also occur in the testes and less commonly, the diaphragm, kidney, urinary bladder, fascia, subcutis, lymph nodes, lung, pleura, peritoneum, thymus, salivary glands, and adrenal glands. Hemorrhage in the tunica media of the pulmonary artery is a near-pathognomonic lesion (Fig. 6.14C). As disease progresses, alimentary tract erosions, ulcers, and necrosis may occur in the gingiva, buccal papillae, tongue (Fig. 6.14D), frenulum, palate, forestomachs, and abomasum, and focal necrosis in the heart, lung, tongue, abomasum, skeletal muscles, kidney, and salivary glands develops. Hemorrhagic gastrointestinal contents, congested lymph nodes, splenomegaly, and corneal opacity may also occur.[221] Hoof and feet lesions may include coronitis, coronary band hemorrhage, and laminitis. Chronically affected animals may have muscle atrophy, rumen scarring, and hoof sloughing, cracking, and overgrowth.[157,158] During EHDV outbreaks, mule deer naturally infected with EHDV-2 develop inflammation, hemorrhage, edema, and necrosis of the testes and epididymis. Animals with chronic testicular lesions may develop associated proliferative antler growths and retained velvet.[109]

Histopathology of BTV and EHDV in white-tailed deer includes vascular congestion, hemorrhage, thrombosis, and necrosis, which may be prominent in the alimentary tract and heart, but can also occur in large blood vessels, lung, kidneys, adrenal gland, spleen, thymus, lymph nodes, and skin adjacent to the coronary band (Fig. 6.14E). Necrotic and hemorrhagic foci may contain mineralization. Smaller blood vessels may have swollen endothelial cells, mild lymphocytic infiltration, and perivascular neutrophils. Alimentary tract epithelium may exhibit spongiosis, leukocyte infiltration, and necrosis that may progress to sloughed and ulcerated epithelium; this can be prominent in the buccal and ruminal papillae. Lymphoid depletion, erythrophagocytosis and hemosiderosis may be seen in spleen and lymph nodes, and megakaryocyte hyperplasia may be seen in the bone marrow[157,158] (eSlide 6.e3).

A high suspicion of orbiviral hemorrhagic disease in cervids is based on gross lesions, especially hemorrhage at the base of the pulmonary artery, although lesions can be subtle in peracute and acute stages. Infection is confirmed through PCR or virus isolation of whole blood (in an anticoagulant, such as EDTA or heparin), spleen, lung, lymph node, or bone marrow (refrigerated, not frozen) prior to testing. PCR requires less stringent storage conditions and can be used to determine serotype and serogroup of the infecting virus, but cannot rule out prolonged or persistent virus infections.[173]

White-tailed deer with **rabies** may present with a distinctive gross pattern of skin lesions (e.g., alopecia, ulceration, swelling, hyperemia) suggestive of excessive head rubbing (Figs. 6.15A, B).[336]

White-tailed deer experimentally infected with **severe acute respiratory syndrome coronavirus 2 (SARS coronavirus-2)** developed subclinical infections, shed virus in nasal secretions and feces, transmitted the virus to contacts, and developed neutralizing antibodies.[249] Coronavirus antibodies have also been detected in free-ranging deer indicating natural infection. Free-ranging WTD may also serve as a wildlife reservoir in some deer populations based on PCR findings.[40,42] To date, disease or pathology due to natural SARS-CoV-2 has not been reported in deer.

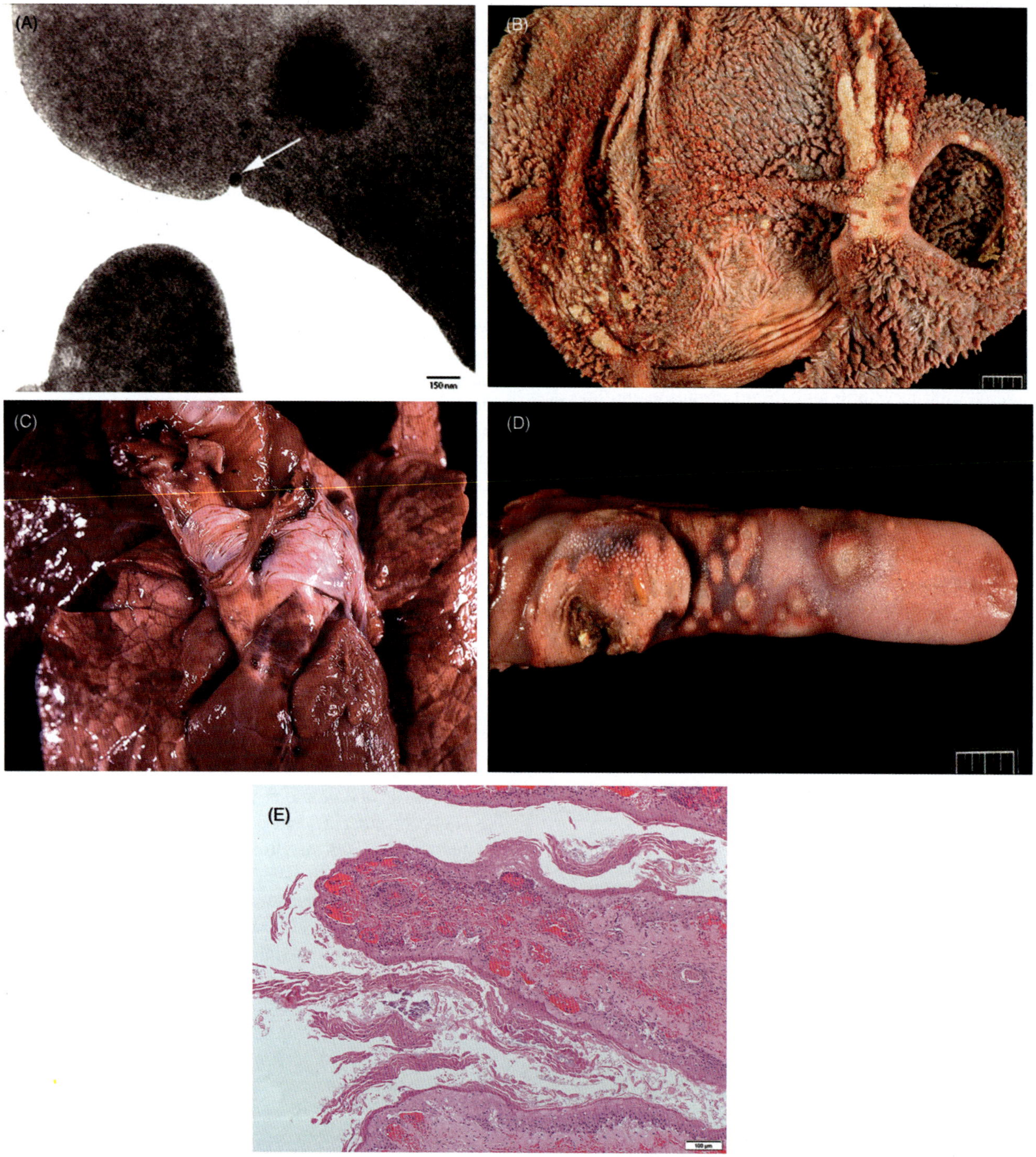

FIGURE 6.14 Orbivirus epizootic hemorrhagic disease virus (EHDV) infection in white-tailed deer. (A) These viruses replicate primarily in microvascular endothelial cells and have a long period of viremia as viral particles circulate hidden from the immune response in pits on the erythrocyte surface. A viral particle is present on the surface of this erythrocyte *(arrow)*. TEM. (B) The rumen has characteristic hemorrhage, erosion, ulceration, and necrosis that appears as white areas that is especially prominent on the pillars throughout the rumen. Similar changes occur with bluetongue virus infection. (C) Hemorrhage on the adventitial surface of the pulmonary artery is a characteristic lesion for these viral infections. (D) Ulcerative, necrotic glossitis is a common lesion in subacute disease. (E) Ruminal papillae have microvascular damage and hemorrhage. Minimal inflammatory cell infiltration is common and there are no characteristic inclusion bodies. Microvascular damage results in necrosis and ulceration of mucosal surfaces (see eSlide 6.e3). *(Parts B–D: Photos Courtesy of the Southeastern Cooperative Wildlife Disease Study; Part E: Photo Courtesy of J. Rothenburger, Canadian Wildlife Health Cooperative.)*

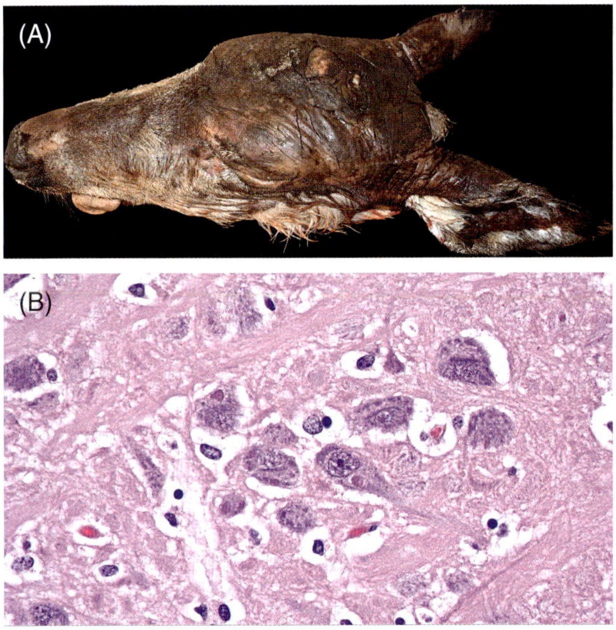

FIGURE 6.15 Rabies in a white-tailed deer. (A) Severe alopecia, ulceration, erythema, and inflammation over the frontal and lateral aspects of the skull, in the periorbital region, and on the muzzle. (B) Microscopically, numerous, eosinophilic, cytoplasmic inclusions are evident in neurons in the cerebrum at the level of the hippocampus. *(Parts A, B: Photos Courtesy of the Southeastern Cooperative Wildlife Disease Study.)*

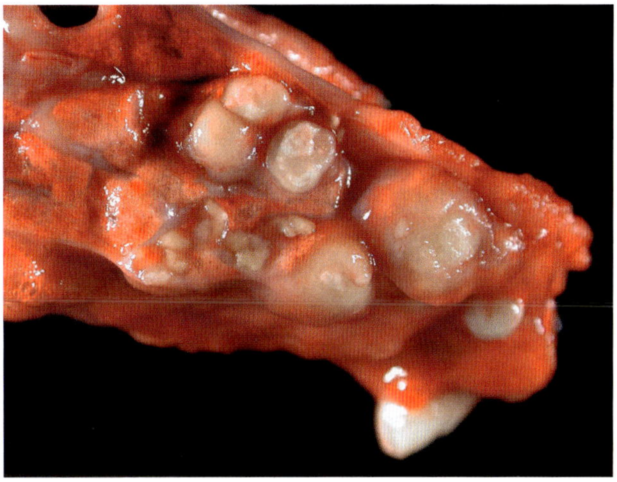

FIGURE 6.16 Tuberculosis in the lung of a white-tailed deer. The majority of the pulmonary parenchyma is replaced with firm, pale nodules due to *Mycobacterium bovis* infection. *(Photo Courtesy of M. Palmer, Agricultural Research Services, United States Department of Agriculture.)*

Additional information on viral diseases of cervids including **pseudorabies, foot and mouth disease, bovine viral diarrhea, rabies, eastern equine encephalitis** (eSlide 6.e4), **Schmallenberg, West Nile, and Rift Valley fever viruses** (eSlide 6.e5) are described in the digital content for this chapter (Table 6.e1).

Bacteria

Cervids are susceptible to **bovine tuberculosis** (bTB; WOAH-listed reportable disease) caused either by ***Mycobacterium bovis*** or ***M. caprae***. Transmission is commonly through an oral or respiratory route. White-tailed deer, wapiti, fallow deer, and red deer are often maintenance hosts and the roe deer is typically a spillover host. Mycobacteria can be shed by oral, nasal, or fecal routes and in urine. Cervids may shed *Mycobacteria* spp. for years in the absence of clinical signs.[98] Red deer have been identified as supershedders[282]; susceptible and resistant red deer breed-lines exist.[220]

Cervids tend to develop severe tuberculous lesions in the lungs, medial retropharyngeal, tracheobronchial, mediastinal and mesenteric lymph nodes, ileocecal junction, and tonsils (Fig. 6.16). Dissemination to multiple body cavities and tissues is uncommon and generally affects fawns less than 6 months old. In *M. caprae*-infected red deer, medial retropharyngeal lymph nodes are most commonly affected, followed by mesenteric lymph nodes, either exclusively or in combination with other sites. Thoracic lesions may extend to both the visceral and parietal pleural surfaces. Thoracic lesions due to *M. bovis* are more frequent in fallow deer, while *M. caprae*-associated intestinal lesions are common in free-ranging red deer. Lesions range from miliary, white foci to 10 cm diameter coalescing tubercles with central caseous necrosis. Abscesses are typically observed in red deer infected with *M. caprae*. Red deer are frequently observed with enlarged superficial lymph nodes that may develop draining fistulas.[96,98,118] Microscopic changes consist of encapsulated granulomas with central necrosis, sometimes with calcification, a peripheral zone of epithelioid macrophages, Langhans or irregularly shaped giant cells, lymphocytes, plasma cells, and neutrophils. Acid-fast bacilli vary in number, and similar to *M. tuberculosis* complex infections in other species, may be rare or not observed. Following the presumptive diagnosis at necropsy, demonstration of *M. bovis* or *M. caprae* by PCR or culture followed by molecular typing is required for definitive diagnosis; bacterial isolation remains the "gold standard" for confirmation of tuberculosis.

Mycobacterium kansasii, a nontuberculosis zoonotic bacterium, was isolated from a white-tailed deer with lesions similar to *M. bovis*.[102] The deer had fibrinous pleuropneumonia with mineralized nodules seen histologically as multifocal to coalescing granulomas with few acid-fast bacilli. Kidney also had foci of histiocytes with acid-fast bacteria. Diagnosis was by PCR.

Paratuberculosis (Johne's disease, WOAH-listed reportable disease), caused by ***M. avium*** subsp. **paratuberculosis** (MAP), is a chronic disease resulting in weight loss and diarrhea. Clinical disease has been reported in a wide range of captive and free-ranging cervid species including reindeer, axis deer, wapiti, tule wapiti, roe deer, white-tailed deer, and red deer.[103,214,253]

On postmortem examination, animals often are emaciated with diarrhea-stained perineum and hindlimbs. Internal gross lesions are not always evident and early cases may be missed. When present, lesions are mainly in the distal jejunum and ileum, ileocecal junction, and associated mesenteric lymph nodes. They may resemble those induced by *M. bovis* or *M. avium* subsp. *avium* but the classic picture consists of diffuse thickening of the intestinal mucosa, sometimes with visible transverse folds. In free-ranging deer, gross lesions may be limited to enlarged mesenteric lymph nodes with mild intestinal thickening. Lymph nodes are pale and edematous with indistinct corticomedullary junctions, occasionally with caseous exudate. Mesenteric lymphatic vessels may be prominent and resemble thickened cords. Systemic infection with disseminated lesions in lung and liver may occur.[214]

Microscopically, Johne's disease is characterized by diffuse granulomatous inflammation in the small intestinal mucosa and associated lymphatics and lymph nodes. Occasionally, organized granulomas composed of macrophages, Langhans multinucleated giant cells and lymphocytes are noted in the liver, lung, and associated lymph nodes. Granulomas may contain necrotic foci in severe cases; mineralization is rarely seen in cervids.[214] Variable numbers of acid-fast bacilli are observed in macrophages and Langhans giant cells due to different immune responses that result in two distinct disease forms: multibacillary (i.e., lepromatous) and paucibacillary. The multibacillary form is characterized by numerous intracellular bacteria and epithelioid macrophages, a weak cell-mediated immune response, and the presence of serum antibodies. In contrast, in the paucibacillary form, intracellular bacteria are not consistently seen with routine or special stains in the intestinal mucosa but can be found in lymph nodes or Peyer's patches, and the host develops a strong cell-mediated immune response with increased lymphocytes but a poor antibody response.[45,214] As in bTB infection, pathological changes are suggestive but not diagnostic for paratuberculosis and confirmation by culture and/or PCR is required.

Necrobacillosis results from infection with ***Fusobacterium necrophorum***, a commensal bacterium of the mammalian alimentary tract. It acts as an opportunistic pathogen requiring a physical defect in the protective epithelium of the skin or alimentary tract to invade tissues and cause disease. There are two subspecies or biotypes of *F. necrophorum*: ***F. necrophorum necrophorum* (biotype A)**, which is most pathogenic and typically associated with necrobacillosis, and ***F. necrophorum funduliforme* (biotype B)**, which predominates in the digestive tract. Necrobacillosis is a serious health problem in semidomesticated tundra reindeer,[138] farmed fallow deer,[296] and farmed white-tailed deer.[131] Major epizootics have been reported in free-ranging white-tailed deer, mule deer, and wapiti[344]; disease has also been documented in red deer, sambar,[206] and roe deer.[253]

Necrobacillosis may manifest as a digital form (sometimes referred to as "foot rot," not to be confused with *Dichelobacter nodosus* infection) or oral form (i.e., lumpy jaw), but other organs may be primarily or secondarily affected. In general, the stress of crowding and inadequate nutrition, including vitamin A, E, and selenium deficiencies, are risk factors.[140] Crowding in muddy and feces-contaminated (i.e., humid, warm) areas predisposes to both digital and oral necrobacillosis and skin abrasions serve as a portal of pathogen entry. Digital trauma due to stony or sharp substrate or coarse pasture and trauma to the oral mucosa and gingiva from highly abrasive food material, as well as dental eruption, are believed to create portals of entry leading to the oral form.[138,139,206]

Once *F. necrophorum* penetrates the subepithelium, multiplication and toxin production occurs, with toxins (e.g., leukotoxin and endotoxin) causing damage to local host tissues. Resulting bacteremia may lead to dissemination to internal organs. Extension to adjacent bones and joints or soft tissues may occur. Coinfection with other bacteria (e.g., *Trueperella pyogenes*, *Escherichia coli*, and *Staphylococcus aureus*) enhances the growth and pathogenic potential of *F. necrophorum*.

Oral lesions may involve zones of necrosis on the buccal surfaces, tongue, pharynx, gingiva, or periodontal tissues. In farmed fallow deer, the base of the tongue is typically severely affected and associated with obvious swelling of the mandibular region (Figs. 6.17A, B). Oral necrobacillosis frequently leads to secondary foci of infection in the lower alimentary tract, especially the rumen and in the liver. Secondary lesions in lung (through aspiration) and brain (through tissue necrosis and extension) may also occur. Digital lesions include swelling of the foot and cutaneous sinus tracks with sparse discharge of pus. Tendons, tendon sheaths, joints, and periosteum of the digital bones are often affected. As in the oral form, secondary foci may develop. In reindeer, skin, genital, udder, and eye infections have been recorded.[138] Lesions may be focal or multifocal and consist of a zone of necrotic tissue sharply demarcated from adjacent living tissue, which may be bordered by a zone of edema and severe inflammation with fibrosis and granulation tissue. Cellulitis, vasculitis, and thrombosis are common. Inflammation may lead to liquefaction, sloughing of necrotic tissue from cutaneous and mucosal surfaces, and abscess formation. Infected bone undergoes both lysis and reactive proliferation, resulting in deformity and enlargement. Joint infection may result in cartilage loss and periarticular bone formation. Filamentous, gram-negative bacteria are present within or at the periphery of

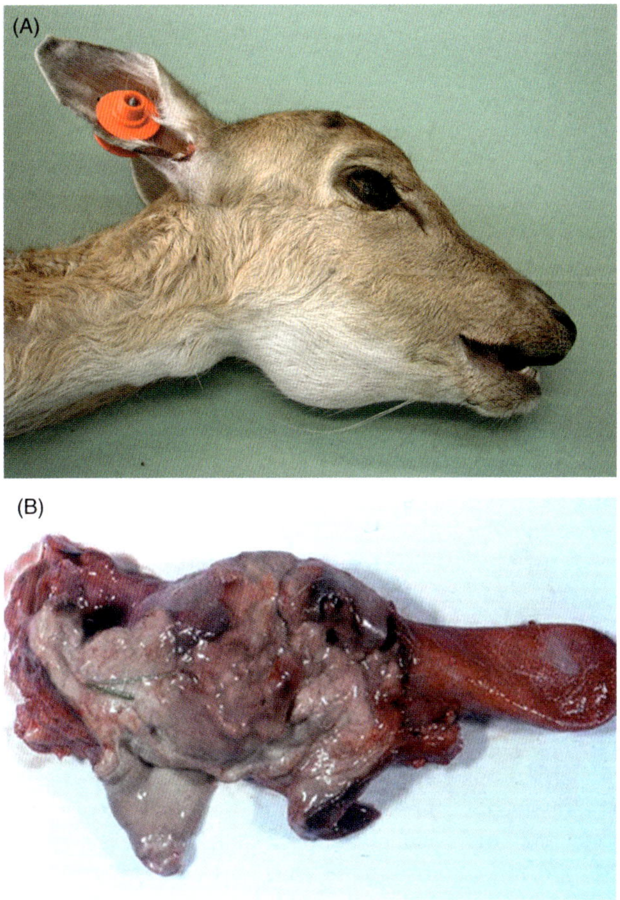

FIGURE 6.17 Necrobacillosis due to *Fusobacterium necrophorum* in the jaw of a fallow deer. (A) Typical intermandibular swelling due to lesions in the tongue. (B) The base of the tongue is obscured and replaced by a large swelling and extensive tissue necrosis that is associated with infection. *(Part A: Photo Courtesy of the Centre for Fish and Wildlife Health, University of Bern; Part B: Photo Courtesy of the Centre for Fish and Wildlife Health, University of Bern.)*

necrotic foci. Anaerobic culture confirms infection. Fluorescence in situ hybridization is a useful alternative to bacterial culture.

Intracranial abscessation-suppurative meningoencephalitis complex, also known as cranial abscess disease, may include grossly evident purulent material within the cranial vault (e.g., well-defined abscesses, diffuse suppurative meningoencephalitis, or a combination) or may have only microscopic suppurative lesions in the brain or cerebral meninges.[61] This complex is primarily an important cause of morbidity and mortality in free-ranging white-tailed deer across portions of the United States and Canada.[20,172] It is also commonly seen in roe deer in Europe.[253] Adult males are most commonly affected and disease is more prevalent in areas with a high buck-to-doe ratio.[49] Disease occurs during breeding season (fall to winter for white-tailed deer and spring to summer for roe deer) when behaviors, such as antler sparring and rubbing, as well as antler casting occur and predispose to trauma and/or facilitate bacterial transmission. Brain infection results from the opportunistic invasion of cuts, abrasions, or injured pedicles by pyogenic bacteria (usually *T. pyogenes*), followed by further penetration into the skull along bone sutures. *T. pyogenes* is a common resident bacterium of white-tailed deer, especially in males. The reported interregional differences in bacterial and disease prevalence may be due to varying virulence factors of *T. pyogenes* and to environmental factors, such as the aridity of the climate.[21]

Grossly, one or both cerebral hemispheres may be affected. Abscesses typically range from 1 to 3 cm diameter and contain purulent material surrounded by an intense inflammatory reaction (Fig. 6.18; eSlide 6.e6). Examination of skulls may reveal necrosis, erosion, and pitting of the cranial bones. These lesions are localized on the external surfaces of the parietal and frontal bones, usually on or near the antler pedicles. Erosions often extend into the cranium through sutures between the frontal bones or between the frontal and parietal bones.[61]

Bacterial pneumonia has been reported in numerous species of farmed (e.g., axis deer) and wild cervids (e.g., white-tailed and black-tailed deer, mule deer) and can represent opportunistic infections following periods of stress or predisposing disease (e.g., *Dictyocaulus viviparous*).[229] Bacteria cultured from affected lungs most commonly include *Truperella pyogenes*, *Pasteurella multocida*, and *Bibersteinia trehalosi*, but may also include *Mannheimia hemolytica* and *Mycoplasma* spp. Gross lesions vary from more discrete focal to multifocal abscesses (Fig. 6.19) to multifocal to coalescing inflammation with variable fibrinous adhesions and associated effusion and hemorrhage (Fig. 6.20). Microscopically, foci of suppurative to pyogranulomatous inflammation may be surrounded by granulation tissue and scattered lymphocytes and plasma cells if more chronic (eSlide 6.e7).

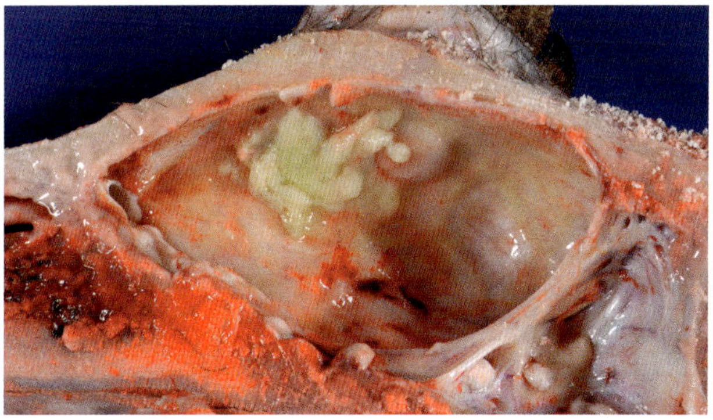

FIGURE 6.18 Intracranial abscess of a white-tailed deer. Infection was due to *Trueperella pyogenes*. An accumulation of purulent material is located dorsally within the vault. *(Photo Courtesy of the Southeastern Cooperative Wildlife Disease Study.)*

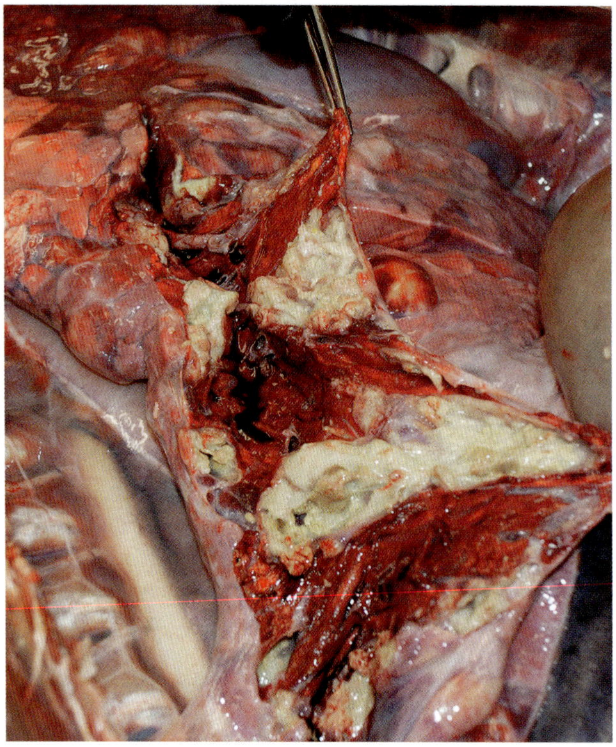

FIGURE 6.19 Bacterial (*Trueperella pyogenes*) pneumonia in a white-tailed deer doe. Multiple, large abscesses replace portions of the pulmonary parenchyma. This deer also had abscesses in the kidneys and lymph nodes. *(Photo Courtesy of the Southeastern Cooperative Wildlife Disease Study.)*

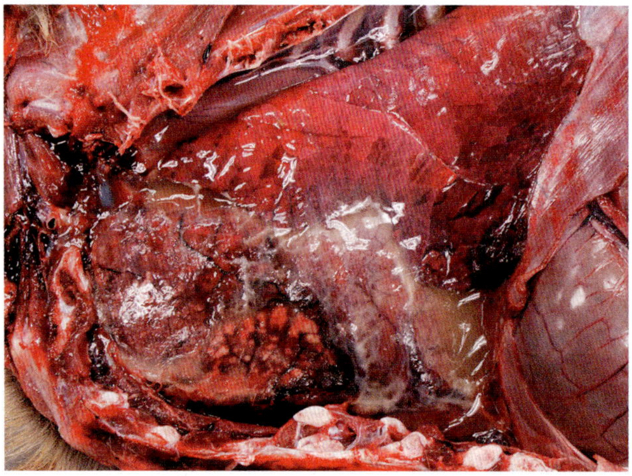

FIGURE 6.20 Bacterial (*Trueperella pyogenes* and *Bibersteinia trehalosi*) pneumonia in a white-tailed deer. A large portion of the cranial lung lobe is consolidated and contains multifocal hemorrhage and fibrin with pleural adhesions. Thoracic, fibrinous effusion also was present. *(Photo Courtesy of the Southeastern Cooperative Wildlife Disease Study.)*

Mandibular osteomyelitis (also referred to as **actinomycosis** or **lumpy jaw**) has been reported in various species including *Odocoileus* spp., moose, caribou,[342] red deer, fallow deer,[13] and roe deer.[190] Mandibular osteomyelitis can be induced by various agents, such as *Actinomyces*

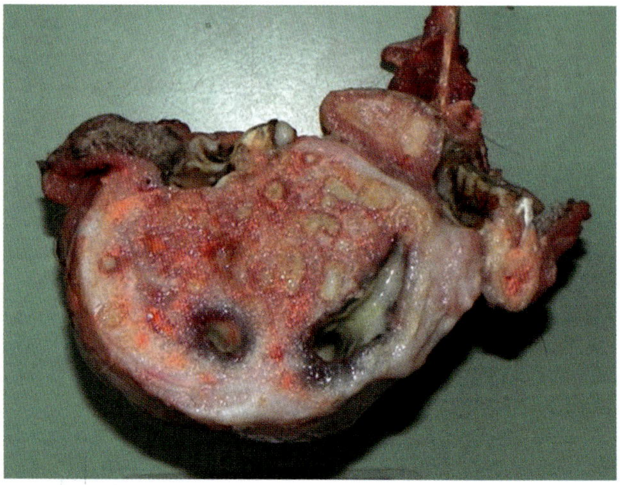

FIGURE 6.21 Actinomycosis-like mandibular osteomyelitis in a roe deer. Cross section of mandible with bony destruction and expansion by extensive fibrosis and multiple tracks of purulent exudate. *(Photo Courtesy of the Centre for Fish and Wildlife Health, University of Bern.)*

bovis[302] and *T. pyogenes*.[253] Infection is facilitated by abrasions or breaks in the skin, trauma to the oral mucosa (e.g., by rough feed, impacted vegetation, or abnormal tooth wear or misalignment), and through the dental alveoli. Mandibular lesions can result in difficulty in mastication and progressive loss of body condition. Stress, poor diet, suboptimal environmental conditions, karst habitat (limestone base), dental fluorosis, replacement of deciduous dentition, and high population density may contribute to disease. Species and sex differences in disease frequency have been attributed to differences in foraging habits and range use strategies. Gross lesions include a deformed and swollen mandible, sometimes with purulent exudate from fistulas and tooth loss (Fig. 6.21; eSlide 6.e8). Pus-filled tracks may contain pale, miliary, slightly gritty, up to 3 mm diameter foci called "sulfur granules" (i.e., grossly visible *A. bovis* colonies). In red and fallow deer, changes have been observed mainly in the region of molar teeth.

Microscopically, localized pyogranulomatous inflammation is surrounded by granulation tissue, mononuclear cell infiltration, and fibrosis. In chronic mandibular osteomyelitis, suppurative necrosis with progressive destruction of trabeculae, and proliferation of periosteal new bone is also seen. Diagnosis of actinomycosis is through positive bacterial culture for *A. bovis*, histology, and/or cytologic examination of sulfur granules. Granules can be examined cytologically in a drop of 10% KOH on a microscope slide and gently crushing them by applying pressure on the coverslip (unstained), or by a Gram-stained smear. Colonies of *A. bovis* consist of filamentous and branching gram-positive bacteria.

Brucellosis caused by *Brucella abortus* (WOAH-listed reportable disease) occurs in multiple species and is an important disease in free-ranging wapiti in the Greater

Yellowstone Area of the United States. Transmission occurs via contact with fetal fluids, vaginal exudates, and aborted fetuses. However, because wapiti usually consume the placenta before it touches the ground, placentas from infected wapiti are unlikely to be the source of environmental contamination. In wapiti, abortion, stillbirth, and infertility in females and orchitis and epididymitis in males are the most common clinical signs, and swollen joints and lameness may be seen. The potential impact of infection with *B. abortus* in moose is controversial, and white-tailed and mule deer appear relatively resistant to infection.[219,319]

Gross lesions in wapiti are neither dramatic nor diagnostic. Early stages consist of enlarged, edematous lymph nodes. The most significant lesions include necrotizing placentitis with purulent exudate in females and testicular swelling and edema in males.[219] *Brucella*-infected hygromas (carpal bursitis), fluid-filled swellings of the carpus surrounded by a thick, moist, rough, red-brown capsule and sometimes containing opaque, straw-colored fluid and fibrinous debris, are frequent. Aborted wapiti fetuses may have serosanguinous fluid in the subcutis and within body cavities, interlobular pulmonary edema, and meconium in the abomasum. In moose, pleuritis, pericarditis, peritonitis, lymphadenitis, enlarged kidneys and testis, and joint lesions are reported.[319]

Microscopically, acute infections in wapiti are characterized by lymphadenomegaly due to an increase in the size and number of lymphoid follicles. The enlarged nodes can contain many neutrophils and some eosinophils. Fetal lungs have focal to diffuse infiltrates of lymphoreticular cells around bronchioles and associated vessels, and suppurative to fibrinopurulent bronchopneumonia. In the placenta, the placentome and intercotyledonary epithelium may contain necrosis and mixed inflammation with numerous bacteria in necrotic chorionic and trophoblastic cells. Giant cells and epithelioid macrophages are present in the uterus, and distended uterine glands contain necrotic debris and neutrophils. In male wapiti, lesions are mild and consist of focal to multifocal testicular necrosis and focal to diffuse mononuclear inflammation in the prostate, seminal vesicles, and ampullae. Epididymitis and orchitis may also be observed. Hygromas may have roughened surfaces covered with fibrin tags, edema, and infiltration by neutrophils, plasma cells, macrophages, and occasional giant cells in the underlying tissues. Synovitis and arthritis also occur.[219,319]

Rangiferine brucellosis refers to brucellosis in reindeer and caribou caused by ***B. suis*** **biovar 4** (WOAH-listed reportable disease in multiple species) throughout the Arctic region, Siberia, Canada, and Alaska. Orchitis, epididymitis, bursitis, synovitis, large hygromas colloquially called "**caribou knees**," and metritis-abortion syndromes (retained placentas and/or excessive bleeding) have been observed. *Brucella suis* biovar 4 may also infect moose.[125,319]

Brucella ovis causes epididymitis in red deer and is associated with reduced semen quality and lowered fertility.[219] The epididymis may be grossly enlarged. Stags may present with nonsuppurative epididymitis, vesiculitis, prostatitis, and ampullitis, with neutrophil exudation in glandular ducts. Spermatic granulomas and epithelial hyperplasia with intraepithelial cyst formation also occur.[269]

Bacterial culture, isolation, and PCR provide the definitive diagnosis of brucellosis.

Clostridiosis involves a variety of disease syndromes caused by numerous *Clostridium* spp.: ***C. perfringens*** **type D (enterotoxemia)**, ***C. chauvoei*** **(blackleg)**, ***C. septicum***, ***C. sordelli*** and ***C. novyi*** **(malignant edema)**, ***C. tetani*** **(tetanus)**, ***C. hemolyticum*** **(bacillary hemoglobinuria)**, and ***C. piliforme*** **(Tyzzer's disease)**.[34,219] Clostridiosis is sporadic in farmed deer. Enterotoxemia affects mostly young red and fallow deer but enteric infections associated with diarrhea have also been reported in roe deer.[253] *C. perfringens* has been isolated from farmed fallow deer and red deer with rumen acidosis, enteritis, and/or colitis.[141,296] Blackleg (myositis and myonecrosis) or malignant edema (gas gangrene) has been described in adult red deer and wapiti, reindeer, and white-tailed deer in association with traumatic lesions. Tetanus has been reported in red deer and fallow deer, especially in bucks after castration with rubber rings, and has been suspected in other deer species. Infection with *C. hemolyticum* in free-ranging wapiti is often associated with giant liver fluke infection. Tyzzer's disease has been reported in white-tailed deer fawns.

Animals with acute to peracute enterotoxemia are often in good nutritional condition, with straw-colored pericardial effusion, cardiac petechiae, abomasal, intestinal, and hepatic congestion, soft and gelatinous kidneys, and watery feces in the small intestine. Blackleg is characterized grossly by muscle swelling and crepitus, usually in association with deep puncture wounds and hemorrhage. Microscopically, skeletal muscle hyalinization with mild necrosis and degeneration, neutrophilic infiltration, and thrombosis are seen.[327] Malignant edema consists of skin gangrene with associated edema of subcutaneous and connective tissues. In wapiti, emphysema of extremities, intramuscular hemorrhage, edema of internal organs, and hemorrhagic exudates from body orifices have been observed. *C. piliforme* isolated from a neonatal white-tailed deer with a history of diarrhea had no significant gross lesions. Microscopically, infection was associated with multifocal, random, coagulative hepatocellular necrosis with mixed inflammation in necrotic foci. Multifocal necrosis of myocardiocytes and Peyer's patches were also present, and intracytoplasmatic filamentous bacteria morphologically consistent with. *C. piliforme* were present with Warthin-Starry staining were seen in areas of necrosis.

For clostridial enterotoxemia, bacteria isolation and toxin identification by ELISA or PCR for toxin genes from intestinal content can provide a definitive diagnosis; fluorescent antibody testing is most often used to confirm infections that produce blackleg and malignant edema; Tyzzer's disease is best confirmed by PCR. Bacterial culture of lesions may also aid in diagnosis.

Dermatophilosis is a skin disease characterized by crust formation, which may be confused with mange. Among cervids, dermatophilosis has been documented in roe deer,[65] mule deer,[204] and white-tailed deer.[236] It is caused by the aerobic gram-positive actinomycete ***Dermatophilus congolensis***. This bacterium exists in two morphologic forms (hyphae and zoospores) and is characterized by having a fungus-like life cycle and morphology. Dormant, immature zoospores resistant to desiccation and high temperature are transmitted (from diseased animals or healthy carriers) via direct contact, ectoparasites, or thorn bushes. When mature, spores become motile, infective, and may initiate infection in damaged skin. Humid environments (e.g., damp hair coat), minor skin trauma, and ectoparasite infestation may facilitate infection.[303] Following dermal invasion, bacterial growth induces epidermal hyperplasia and local inflammation, resulting in exudation of serum and neutrophils from underlying dermal vessels. Progressively thicker masses of exudate and shed epidermis form and entrap the hair coat.

Disease severity and lesion distribution vary. Mild, multifocal to coalescing erythema with crust formation to severe, extensive, exudative lesions may be seen in affected skin. Crusts are most common on the back, tail, and carpal and tarsal regions and less common on the udder, scrotum, neck, and head. Severely affected animals become emaciated. Crusts are often easily detached from the skin, leading to moist hairless foci, which may be hyperemic and covered with exudate. In white-tailed deer, alopecia, erythema, and crusting are usually present on the face, ears, and distal limbs (Fig. 6.22A). Survivors (usually captive) may have scarring of the face and legs and scarring and cropping of the ears. Characteristic microscopic lesions consist of thick laminated crusts of alternating layers of keratin and degenerated neutrophils. This is associated with intralesional *D. congolensis* bacteria along with variable epidermal hyperplasia and neutrophilic and mononuclear cell infiltration in the underlying dermis (Fig. 6.22B). *D. congolensis* infection is diagnosed by its pathognomonic morphology and characterized by 2–6 parallel rows of cocci that look like railroad tracks and stain positive with Grocott, Giemsa, and Gram stains. Identification of the organism can also be made in direct smears or culture of detached crusts.

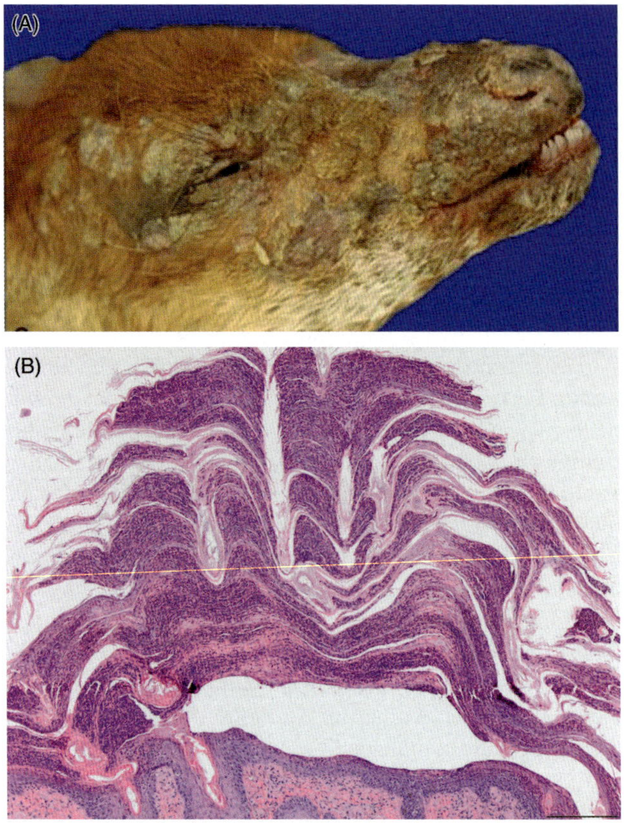

FIGURE 6.22 Dermatophilosis due to *Dermatophilus congolensis* infection in the skin of a white-tailed deer fawn. (A) Characteristic facial dermatitis includes crusting and alopecia on the muzzle, periocular region, and pinnae. (B) Characteristic histopathology includes a hyperplastic epidermis covered by a thick laminated crust that consists of alternating layers of dense keratin and exudate composed of necrotic cell debris and degenerated streaming neutrophils. Characteristic branching chains of coccoid gram-positive bacteria are most obvious in the keratin layers. *(Part A: Photo Courtesy of the Southeastern Cooperative Wildlife Disease Study.)*

Treponeme-associated elk hoof disease was first described in lame, free-ranging Roosevelt elk in Washington State (United States) that were copper and selenium deficient.[47,137] Since first described, it has spread all over the Northwestern and Western United States, including Rocky Mountain elk. The disease is characterized by elongated, deformed hooves with capsular breakage and sole ulcers. Hoof elongation can be exacerbated by disuse from pain and lack of normal mobility and wear. In some cases, fracture of the long pastern bone (proximal phalanx) may occur. Gross lesions consist of asymmetrically elongated (up to 20 cm long), curved hooves with capsular breakage and sloughing and sole ulcers; proliferative granulation tissue may be present (Fig. 6.23). Microscopic lesions may include coronary band, heel bulb, and interdigital ulcers, suppurative inflammation, epithelial hyperplasia, separation and undermining of the hoof capsule and heel-sole junction, and deeply invading intralesional spirochetes. A grading scheme of gross lesions has been described based in part on severity and distribution of ulceration and hoof overgrowth and deformation.[136] *Treponema* sp. has been consistently observed within lesions of elk of various ages, and can be detected by culture or PCR. Silver stain highlights organisms microscopically (Fig. 6.24) and an immunohistochemistry assay for use in its identification has been developed but is not widely commercially available.[136]

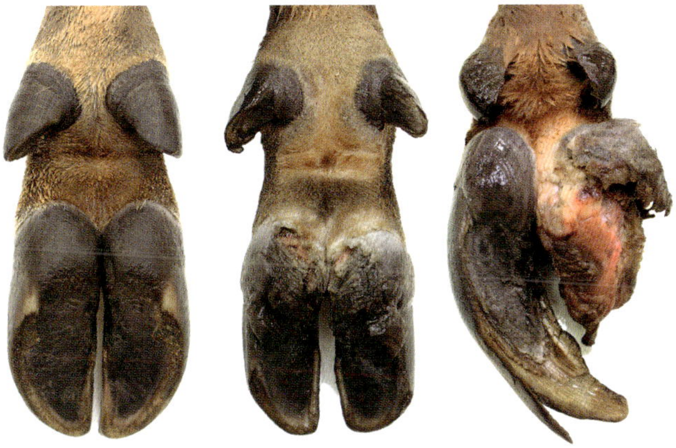

FIGURE 6.23 Treponeme-associated hoof disease of elk. Hooves from three in different elk: normal on left with increasing severity grade from middle to right. Early disease manifests as sole ulcers, that can progress to deeper and more widespread ulceration, inflammation of underlying tissue, and sloughed or overgrowth hoof capsules (middle photo), and eventually severe overgrowth, deformed (curved) regrowth including granulation tissue, sloughed (lost) hoof tissue, and ulcerative dermatitis (right photo). Laminar necrosis and thrombosis may also be present. *(Photo Courtesy of K. Taylor, Washington Animal Disease Diagnostic Laboratory and Department of Veterinary Microbiology and Pathology, College of Veterinary Medicine, Washington State University.)*

FIGURE 6.24 Treponeme-associated hoof disease of elk. Large, thick filamentous opportunistic organisms are present with the densely knotted and fine *Treponeme* spirochetes throughout the background and extending deeper into the lamina. Steiner's silver stain. *(Photo Courtesy of S. Han, Colorado State University.)*

Hypertrophic osteopathy of the metacarpal and metatarsal bones with exuberant deposition of periosteal new bone is sporadic in roe deer and, as occurs in other species, is presumed to develop secondary to intrathoracic masses. Bone lesions in cervids have typically been associated with lung lesions, such as fungal pyogranulomas or large bacterial abscesses (*Streptococcus*, *Staphylococcus*, *Fusobacterium* spp.[253]; Fig. 6.25).

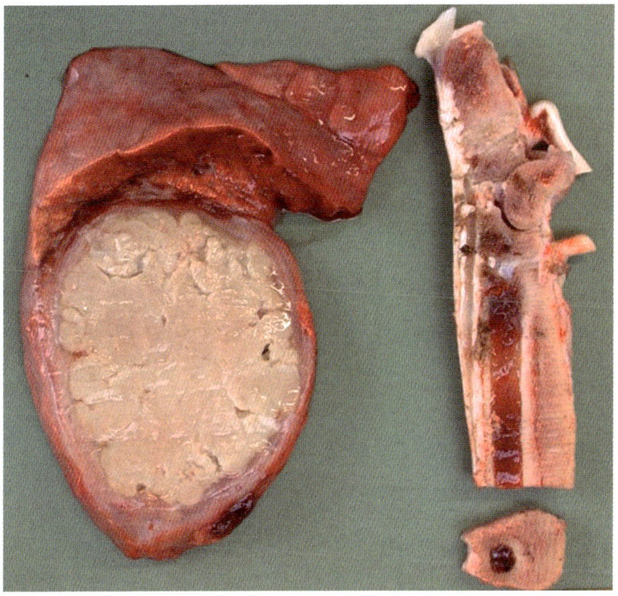

FIGURE 6.25 Hypertrophic osteopathy in a roe deer with a pulmonary abscess. Severe periosteal bony proliferation caused marked thickening along the cortical surfaces of the bone. A large abscess is in the lung on the left. *(Reprinted with permission from Pewsner M, Origgi FC, Frey J, Ryser-Degiorgis M-P. Assessing fifty years of general health surveillance of roe deer in Switzerland: a retrospective analysis of necropsy reports. PLoS One. 2017;12(1).)*

FIGURE 6.26 Chronic bacterial cellulitis, so-called Bullwinkle deer disease, in a white-tailed deer. The muzzle is greatly swollen due to chronic bacterial cellulitis associated with infection with *Mannheimia granulomatis*. *(Photo Courtesy of Southeastern Cooperative Wildlife Disease Study.)*

Multiple cases of "**Bullwinkle deer**" disease have been described in white-tailed deer with swollen and deformed muzzles, which gives them the appearance of the cartoon moose "Bullwinkle" who had a large muzzle, throughout the United States (Fig. 6.26). The condition appears to be a chronic bacterial infection of the skin and subcutis of the muzzle. Histologically, there is nodular and sclerosing pyogranulomatous and eosinophilic dermatitis and cellulitis with Splendore-Hoeppli reaction and intralesional gram-negative coccobacilli. The causative organism/s has been difficult to culture, but ***Mannheimia granulomatis*** has been isolated from several cases.[174] *M. granulomatis* has also been reported in roe deer with keratoconjunctivitis, stomatitis and glossitis, pleuropneumonia, and cervical cellulitis/abscess.[31]

Additional bacterial diseases of cervids including **yersiniosis (pseudotuberculosis), salmonellosis, pasteurellosis** (including hemorrhagic septicemia), **anthrax**, **listeriosis**, **leptospirosis**, **colibacillosis** (including colisepticemia), **heartwater**, **erysipelas**, **caseous lymphadenitis**, and infections without associated disease signs in cervids (***Anaplasma***, ***Rickettsia***, ***Ehrlichia***), are described in the digital content for this chapter (Tables 6.e2 and 6.e3).

Fungi

In general, fungal diseases are not common, but there are case reports of various fungal diseases occurring in individual captive and free-ranging animals (Table 6.1). In roe deer, fungal infections are sporadically but regularly

TABLE 6.1 Other Fungal Diseases Reported in Cervids

Disease	Etiology	Gross Lesions	Histologic Lesions	Cervidae Species Affected	Key to Diagnosis
Mucormycosis	Multiple spp. Often not identified in reports[59,284]	Pulmonary (fallow), gastric (sika); often secondary to other disease, such as ruminal acidosis	Gastric: neutrophilic infiltration and hyphal infiltration of the mucosa, hyphal invasion of vessels with associated necrosis	Sika deer (captive); fallow deer (free-ranging)	Fungal hyphae 5–20 μm, irregular contours, haphazard branching that is random, infrequent septations; angioinvasive
Aspergillosis with hypertrophic osteopathy	*Aspergillus fumigatus*[91]	Encapsulated nodules in lungs and other organs; thickening of cortical bone in a variety of bones	Granulomas with fungal hyphae and periosteal hyperostosis	Elk (farmed)	Fungal hyphae with uniform width, 3–6 μm in diameter, septate, dichotomously branched; conidial head may form in cavitary lesions
Conidiobolosis	*Conidiobolus incongruus*[122]	Disseminated white nodules throughout thoracic and abdominal viscera	Central necrosis, neutrophils, eosinophils plasma cells, and macrophages	Red deer (farmed)	Fungal hyphae that are poorly stained, 6–25 μm in diameter with nonparallel walls, infrequent septa, and random branches and eosinophilic and granulomatous inflammation
Cryptococcosis	*Cryptococcus gatti*[245]	Pale tan masses in tracheobronchial lymph nodes and lung filled with yellow gelatinous material, yellow gelatinous material obliterating ethmoturbinates, and small pitted lesions in brain	Cystic spaces and adjacent tissue with yeast with mucinous capsules and a granulomatous inflammatory response	White-tailed deer (free-ranging)	Yeast 15–33 μm in diameter, mucinous capsule, and narrow-based budding. Some yeast may be dematiaceous and stain for melanin
	Cryptococcus neoformans[8,134]	Meningoencephalitis, pneumonia, cellulitis	Cystic spaces and adjacent tissue with yeast with mucinous capsules and a granulomatous inflammatory response	Wapiti (free ranging); Barasignga	Yeast 15–33 μm in diameter, mucinous capsule, and narrow-based budding. Some yeast may be dematiaceous and stain for melanin

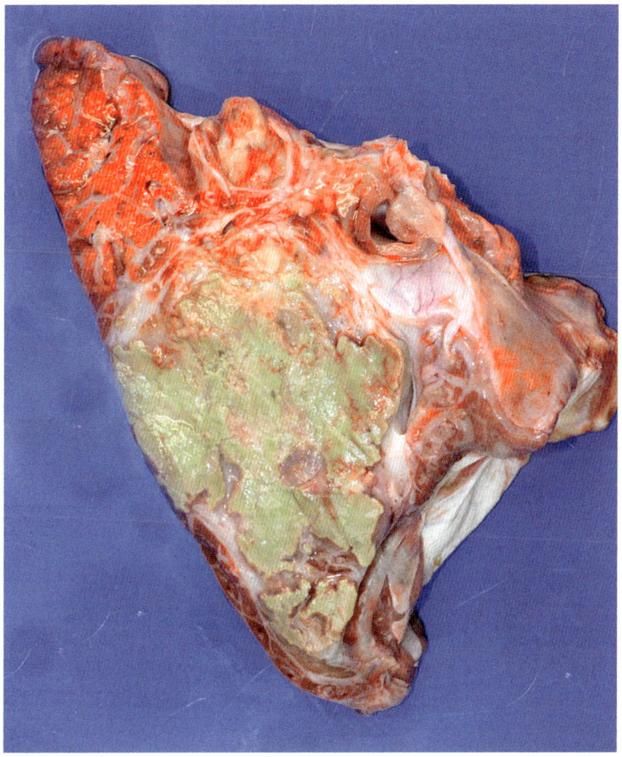

FIGURE 6.27 Green lung syndrome due to an unidentified fungus in a white-tailed deer. A large portion of the lung lobe is replaced by firm, green material surrounded by a fibrotic capsule (see eSlide 6.e9). *(Photo Courtesy of the Southeastern Cooperative Wildlife Disease Study.)*

observed. They are most commonly seen in the lungs (pneumonia) and due to ***Aspergillus*** **sp**., or in the head in the form of actinomycosis-like proliferations of soft tissue and bone and due to infections with *Aspergillus* sp. or occasionally ***Mucor*** **sp**.[253] **Green lung syndrome** is a recently described condition in which prominent, green-hued, compact lesions replace portions of the lungs; similar, less commonly described locations are in the forestomach wall of white-tailed deer (Fig. 6.27; eSlide 6.e9). Lesions are often limited to one lobe but large portions may be affected. Microscopically, areas of severe, chronic eosinophilic and pyogranulomatous inflammation and necrosis contain myriad intralesional fungal-like elements that variably stain with Grocotts methenamine silver. Thus far, no predominant organism has been identified[335] (eSlides 6.e9 and 6.e10).

Metazoa

The **meningeal worm** of white-tailed deer is ***Parelaphostrongylus tenuis***. White-tailed deer are the natural host for the parasite and infection in them is generally meningeal and subclinical, but massively infected deer can have neurologic signs.[6] In aberrant hosts, which include almost all other cervids, such as black-tailed deer, mule deer, fallow deer, moose (referred to as "moose sickness"), caribou,

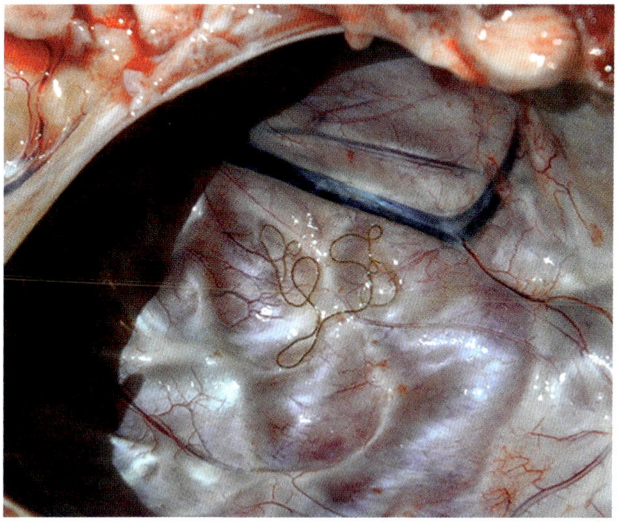

FIGURE 6.28 *Parelaphostrongylus tenuis* (meningeal worm) in a white-tailed deer. In the natural host, these long, slender, red meningeal worms can be found in the subdural space, along the dura, and free within venous sinuses. The dura mater has been removed and the parasites in this image are present on the calvarial bone. *(Photo Courtesy of the Southeastern Cooperative Wildlife Disease Study.)*

reindeer, wapiti, and red deer, *P. tenuis* infects nervous tissue (instead of the meninges) and may cause clinical neurologic disease that culminates in paralysis and death. Infection in these cervids occurs where their range overlaps that of white-tailed deer. In white-tailed deer, worms may be found anywhere in the cranial subdural space but favor the tentorium cerebelli, falx cerebri, and diaphragm sellae (Fig. 6.28). They may also be partially embedded in the dura mater near the intercavernous sinus or, rarely, in the leptomeninges or neuroparenchyma. Worms may be free within the dorsal venous sinus and other sinuses. Worms on the dura mater may be associated with yellow hemorrhagic exudate or there may be small hematomas in the dura. Eggs enter the venous sinuses and pass to the lung, where they develop into L1 larvae, so tiny discolored spots may be distributed throughout the lung, especially under the pleura. Microscopically, lesions in the spinal cord associated with developing worms are mild. The main lesions are in the cranium, where collections of eggs or larvae in the dura or walls of the sinuses are surrounded by fibrosis and inflammation consisting of eosinophils, lymphocytes, histiocytes, and giant cells. Adult worms embedded in the dura are surrounded by necrotic debris and similar inflammatory cells. Microscopically, numerous, scattered nodules of granulomatous inflammation surrounding one to several eggs are present in the alveolar capillaries; larvae may be in alveoli.

Some cervids (e.g., moose and caribou) are especially susceptible to infection and ingestion of even a few worms may result in fatal neurologic disease. In cervids other than white-tailed deer, larvae migrate from the gastrointestinal tract into the dorsal horn of the spinal cord, and continue to migrate in the spinal cord and to the brain resulting in tracks with intralesional, fifth-stage larvae in the brain or spinal cord. They are surrounded by necrotic cell debris and mixed inflammation of eosinophils and histiocytes or are present in the meninges surrounded by granulomatous inflammation[183] (eSlide 6.e11). In moose that survive the initial phase of the infection, adult worms in the subdural space may subsequently invade the neuroparenchyma and cause parenchymal damage and death of the host. Granulomatous nodules have also been detected in the leptomeninges of some species.[6]

Parelaphostrongylus andersoni, the muscle worm of white-tailed deer, resides adjacent to or partially within vessels in the musculature of the hind body of white-tailed deer.[6] Eggs are deposited in the vasculature and emboli can be seen in alveolar capillaries where first-stage larvae develop. Inflammation is similar to that with *P. tenuis*. Moderate infections are well tolerated. The longissimus dorsi muscle is the best location to find adults. Natural infections of ***P. odocoilei*** only occur in black-tailed and mule deer. Adults are found in the connective tissues of muscles and eggs are present in the lungs, similar to *P. andersoni*.[262] Infection has been reported, though rarely, as the primary cause of death in black-tailed deer, but is considered a potentially important infection in mule deer populations.

There are three species of ***Elaphostrongylus*** that normally infect cervid hosts. These parasites migrate into the brain and spinal cord, with clinical signs of ataxia, paralysis, or blindness. Pneumonia may also occur due to large numbers of eggs and migrating larvae in the lungs. Disease occurrence depends on the species of *Elaphostrongylus* and host. ***E. cervi*** is common in red deer. ***E. rangiferi*** is found in reindeer and may cause disease in its normal reindeer host or aberrant hosts, such as moose.[197]

The arterial worm, ***Elaeophora schneideri***, is a filarid nematode found most commonly in the common carotid or internal maxillary arteries.[146] Mule deer and black-tailed deer are considered natural hosts and infection does not cause clinical disease. However, the parasite is pathogenic in other cervids causing a condition called "clear-eyed blindness" in elk and moose; muzzle and ear necrosis, antler deformities, and cerebral necrosis in elk; and intimal proliferation in the carotid artery of moose. White-tailed deer may develop feed impactions. Unlike these species where clinical disease is associated with adult worms in vessels, sika deer may have facial dermatitis due to the microfilaria. Diagnosis is by finding the adult worms in vessels or sheathed microfilaria in the dermis of the skin of the head.

Most helminths in the gastrointestinal tract of cervids are **trichostrongyles**, such as ***Haemonchus***, ***Ostertagia***, ***Trichostrongylus***, ***Apteragia***, ***Nematodirus***, and ***Cooperia*** **spp.**[257] Disease due to these parasites, as well as gross and histopathologic findings, is similar to that in other ruminants and occurs in both free-ranging and captive animals.

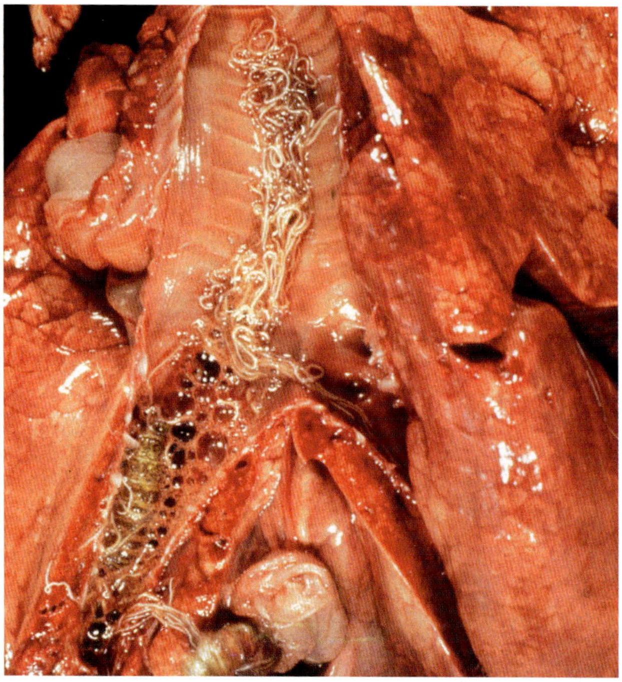

FIGURE 6.29 Lungworm infection due to *Dictyocaulus* sp. in a wapiti. Numerous slender, white nematode parasites are present within the tracheal and bronchial lumens. *(Photo Courtesy of the Canadian Wildlife Health Cooperative.)*

Several **lungworms** infect both captive and free-ranging cervids and can be associated with disease when present in large numbers.[6] ***Dictyocaulus viviparous*** probably occurs in most cervids. Adults are found in bronchioles or bronchi (Fig. 6.29); bronchitis and peribronchitis with mucus, neutrophils, mononuclear cells, and eosinophils along with eggs and larvae in the smaller air passages may be seen microscopically. ***D. filaria*** is occasionally diagnosed in cervids. Other lungworms that are generally less pathogenic in cervids include ***Protostrongylus* spp.**, ***Orthostrongylus macrotis*** (mule deer, moose), and ***Varestongylus* spp**.

White-tailed deer are considered the natural host of ***Fascioloides magna*** (the **large American liver fluke**). The fluke is most common in the southeastern United States but also occurs in western regions of North America, where mule deer, black-tailed deer, and wapiti are the definitive host. Disease is rare in white-tailed deer, but fatalities are reported in wapiti.[260] Moose can be infected but do not develop disease and are considered dead-end hosts. The parasite, which is reddish-brown and 4–10 cm in length, resides in encapsulated cysts in the liver parenchyma. Flukes that do not penetrate the liver may be found dead in granulomas in the lung and lymph nodes. Generally,

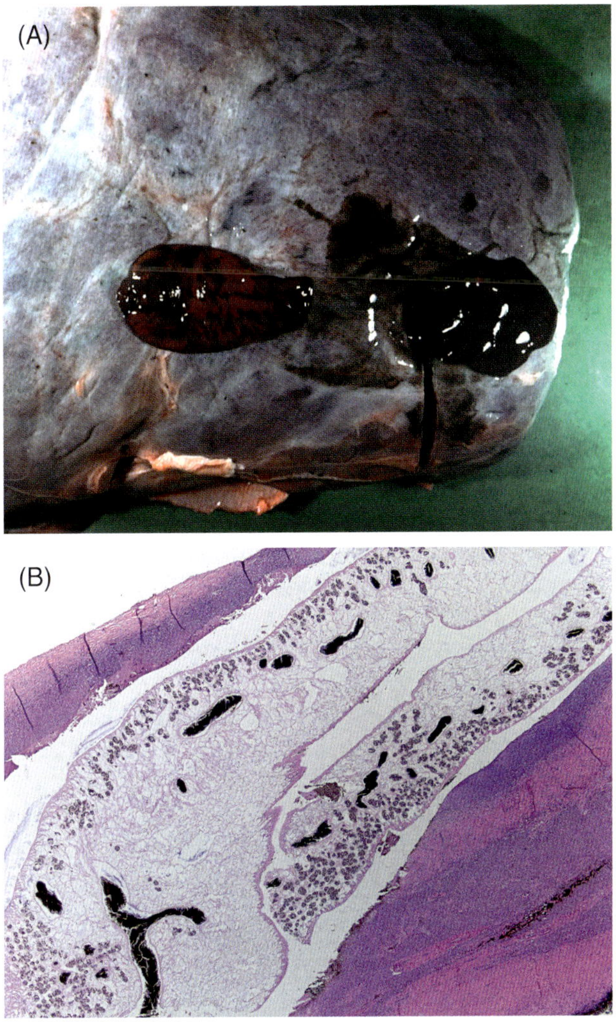

FIGURE 6.30 Large liver fluke infection due to *Fascioloides magna* infection in a white-tailed deer. (A) The flukes are present within cystic spaces in the liver that are surrounded by fibrous connective tissue and give the parenchyma a multifocal, nodular appearance. Thick black fluid fills the cysts. (B) Cysts typically contain two flukes. Characteristic black pigment is present within the flukes in this cyst. The surrounding fibrous tissue capsule is infiltrated by inflammatory cells and also contains black pigment. (see eSlide 6.e12). *(Part A: Photo Courtesy of A. Prestwood, College of Veterinary Medicine, University of Georgia and Noah's Arkive.)*

two flukes are found in each liver cyst and the cysts communicate with the bile ducts, so eggs may be found in the feces. A distinctive black pigment (a hematin compound from partial digestion of red blood cells and fluke excreta) is associated with infection and may be in or on the mesentery, lymph nodes, hepatic tissue and other organs (Figs. 6.30A, B; eSlide 6.e12). Fatalities in wapiti are associated with extensive liver damage due to migration tracks of numerous tiny flukes (5–10 mm) or from exsanguination into the abdominal cavity due to rupture of the hepatic capsule or hepatic vein.

The **small liver fluke**, ***Fasciola hepatica***, inhabits the bile ducts of domestic ruminants and has been infrequently reported in several cervids including white-tailed deer, wapiti, red deer, roe deer, and fallow deer.[253,330] Clinical disease is not apparent in natural infections. Histologically, bile duct hyperplasia is expected in infected animals.[104]

Protozoa

Theileria* spp.** (phylum Apicomplexa, class Aconoidasida, order Piroplasmida, family Theileridae; WOAH-listed reportable disease in cattle) are tick-borne hemoprotozoa. ***T. cervi infection has been observed in numerous cervid species originating from multiple continents, with clinical manifestations varying by species.[57,156,188] In addition to *T. cervi*, ***T. tarandirangiferis*** has been associated with high mortality in reindeer in Russia.[115] Additional *Theileria* spp. reported in cervids (e.g., roe deer, sika deer, and red deer) include ***T. uilenbergi***, ***T. luwenshuni***, ***T. capreoli***, and ***T. ovis***, and ***Theileria* OT3**.[213]

The lone star tick (*Amblyomma americanum*) is the definitive host and biologic vector of *T. cervi*, and introduces sporozoites into the vertebrate host while feeding. Sporozoites invade lymphocytes, where schizogony and both cell and parasite division occur.[181] The parasites then disseminate to lymphatics and lymphoid tissues and schizonts undergo merogony, which can occur in the liver, spleen, and lymph nodes. Merozoites are released and infect erythrocytes, where piroplasms develop. Hemoglobin and coproporphyrin leak from parasitized erythrocytes resulting in anemia and icterus.[156,346]

T. cervi is often asymptomatic in naturally infected cervids unless the animal has a heavy parasite load, malnourishment, concurrent disease or immunosuppression, or is in areas of high deer population densities. In these cases, disease may be more prevalent in fawns.[115,188] Rarely, gross lesions include poor nutritional condition with serous atrophy of fat, anemia (pale mucous membranes and skeletal muscles), jaundice, and involuted or enlarged spleens. Some mule deer fawns develop an acute fatal hemolytic crisis and have multiple gray foci on the hepatic capsular surface and parenchyma or an enlarged brown spleen.[57,181,274,346] Ovoid, ~125 × 150 µm protozoa may be seen histologically within the cytoplasm of mononuclear cells in the liver, lymph nodes, and spleen. Mononuclear cells containing meronts may be hypertrophied with enlarged and elongated nuclei and the cytoplasm may contain 1–2 µm, pinpoint to aggregated, basophilic merozoites. When infected host cells rupture, zoites are released and incite pyogranulomatous inflammation, hemorrhage, necrosis, and subsequent fibrosis. Additional findings may include lymph node and splenic lymphoid depletion, and splenic, hepatic, pulmonary, and renal hemosiderosis.[274,346]

Diagnosis is based on microscopic visualization of intraerythrocytic piroplasms via Giemsa-stained blood smears or schizonts in lymph node impression smears or aspirates.[188] Intraerythrocytic piroplasms are pleomorphic. Shapes may resemble a comma, signet-ring, paperclip (bipolar), pear, rod, or are ovoid, round, or anaplasmoid. All have a small, dark red to purple-staining, eccentric nucleus and basophilic cytoplasm.[115,156] However, parasitemia is generally low (1%–5% of erythrocytes), potentially leading to false negative test results.[181] Developing meronts and merozoites may appear in tissues (e.g., liver, lymph node, spleen) as granular to 1–2 μm basophilic structures within the cytoplasm of hypertrophied monocytes. Host cells may rupture, releasing zoites, and inciting pyogranulomatous inflammation.[346] Nested PCR can be used to distinguish *Theileria* spp. from other hemoparasites, and molecular typing can be used for further characterization.[57,135] Serology can be performed by indirect fluorescent antibody test (IFAT), which can distinguish between piroplasms of different species.[188,263]

Babesiosis is caused by ***Babesia* spp.** (phylum Apicomplexa, class Aconoidasida, order Piroplasmida, family Babesiidae; WOAH-listed reportable disease in bovine). Similar to *Theileria* spp., they are ixodid tick-borne hemoprotozoa. Fatal ***B. odocoilei*** infection has been reported in captive reindeer and wapiti.[150,337] Many cervid species have asymptomatic *Babesia* spp. infections, due to a number of different *Babesia* spp.[37,57,259] Reproduction of piroplasms in erythrocytes is by binary fission and leads to cell destruction and hemoglobin leakage, resulting in anemia, hemoglobinuria, and icterus. The resulting loss of oxygen-carrying capacity in the blood can be fatal.[155]

Naturally acquired *Babesia* spp. infection in cervids rarely leads to clinical disease. Gross findings may also be absent or may include emaciation, hematuria, anemia, icterus, splenomegaly, hepatomegaly, and gall bladder distention by dark, viscous bile, and abomasal and intestinal mucosal edema with subserosal hemorrhage.[155] Intraerythrocytic paired pyriform bodies may be evident histologically, but lesions are minimal and may include splenic erythrophagocytosis and hepatic bile accumulation.[155,181] A fatally infected reindeer had pigmentary nephrosis and tubular necrosis associated with intravascular hemolysis and hemorrhages in cardiac and skeletal muscles.[17]

Diagnosis of *Babesia* spp. is by microscopic visualization via Giemsa-stained blood smears or histopathologic identification of the intraerythrocytic paired pyriform bodies, which form an acute angle at the connection of their pointed ends. *B. odocoilei* and others are often located adjacent to the margin of the erythrocyte[181] (Fig. 6.31). In Giemsa-stained preparations, the parasite cytoplasm is blue with a red nucleus toward the pointed end. However, blood smears may yield a false negative due to low parasitemia in chronic infections, and PCR can confirm the diagnosis with species determination by sequencing of a hypervariable 18S rRNA region.[337] Diagnosis is supported by the presence of known tick vectors on the host or with known distribution in the region.[155]

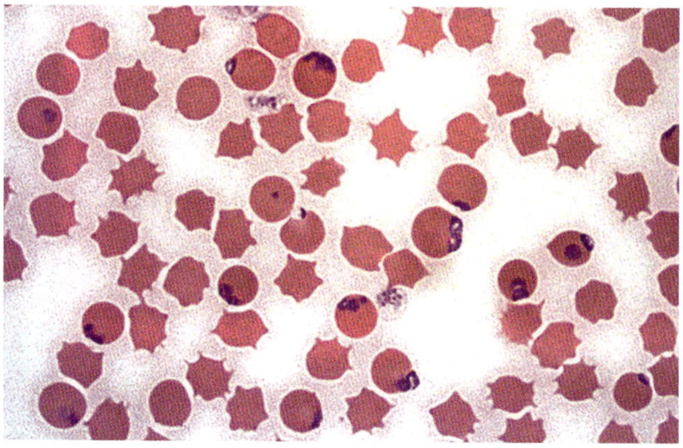

FIGURE 6.31 *Babesia odocoilei* piroplasms in the erythrocytes of a reindeer. *(Photo Courtesy of the Toronto Zoo.)*

Coccidiosis in cervids is most commonly due to ***Eimeria* spp.** (Phylum Apicomplexa, class Conoidasida, order Eucoccidiorida, family Eimeriidae), of which at least 25 species have been reported in cervids.[168,181,261] *Eimeria* spp. are obligate intracellular parasites that reproduce both asexually (merogony) and sexually (gamogony) in the epithelial or endothelial cells of the host intestine; gamogony results in a resistant oocyst that is shed in feces. After ingesting sporulated oocysts, exsporulation occurs in the gut and resulting sporozoites migrate and penetrate the epithelial cytoplasm or, rarely, the nucleus. Nuclear divisions (i.e., schizogony) produce merozoites that cause cell rupture and then infect adjacent cells.[181] Epithelial damage can cause clinical disease through disruption of digestion and absorption, and in more severe cases by altering villous and crypt architecture thereby permitting secondary infections. Infection is often subclinical but its effects vary due to sporulated oocyst traits (e.g., number, strain, species) and host factors (e.g., age, sex, species, infection site, nutritional and immune status, feeding habits, coinfections). Captive settings that may involve high densities of cervids and stressful scenarios (e.g., nonnative cohorts, human presence) may lead to increased oocyst shedding and facilitate infections and outbreaks.[82,168] Coccidia can also occur outside of the intestinal tract; various stages of *Eimeria* sp. have been found in the epididymis of a wapiti.[160]

Coccidiosis in cervids is usually not pathogenic under natural conditions. However, emaciation, dehydration, and watery to bloody diarrhea may occur, and younger animals are likely more susceptible. The asexual or sexual stage of the parasite can be seen microscopically in epithelial cells with standard hematoxylin and eosin staining.[82]

Diagnosis is made by microscopic visualization of oocysts in the feces by the direct centrifugal flotation technique using sucrose or zinc sulfate ($ZnSO_4$) solutions or aqueous (w/v) potassium dichromate. Oocyst morphology can help differentiate species, however, salt preparations may distort oocysts.[261] Molecular methods, such as PCR and random amplified polymorphic DNA (RAPD), can also provide species identification. Histology and electron microscopy may reveal distinguishing morphology of apicomplexans, class Conoidasida, which includes a characteristic apical complex with at least one polar ring, rhoptries, micronemes, and a complete conoid. The sporulated oocysts of *Eimeria* spp. have four sporocysts, each of which has two sporozoites.[82,168,181]

Sarcocystis spp. are tissue-inhabiting protozoa (class Apicomplexa, order Eucoccidiida, family Sarcocystidae). They are often nonpathogenic (i.e., encyst in muscle) in cervids, which serve as intermediate hosts, but massive infections can result in myositis. Many cervid species can be infected and those known in wild ruminants, as of 2015, are provided in Ref. 72 with additional cervid-specific information in Ref. 25. Cervids are infected after ingestion of sporocysts shed in carnivore feces, and asexual development occurs as part of the two-stage parasitic life cycle. Sporozoites excyst in the small intestine, schizogony (meronts) occurs in endothelial cells, and schizonts rupture, releasing merozoites that can enter muscle cells and initiate the formation of bradyzoite-filled sarcocysts within the host cell cytoplasm.[74,181]

Gross lesions are usually absent except with heavy infections, which may produce pale streaks in muscle.[181] Reports in elk have multiorgan serosal petechiae, and pericardial and pleural effusion. Mule deer experimentally infected with ***S. hemionilatrantis*** had perivascular necrosis, mononuclear cell and neutrophilic inflammation, and edema.[168] Histologically, sarcocysts are found in myofibers in cardiac, skeletal (especially tongue and diaphragm), and smooth muscle (e.g., esophagus), and are round, ovoid, elongated, or ellipsoid. Their thick walls are smooth, striated, or hirsute or branching. They range in size from 10 μm to 5 mm. Cyst contents appear as myriad vermiform or crescent-shaped bradyzoites.[168,181] Diagnosis is supported by the presence of immature and mature sarcocysts at the same developmental stage. Electron microscopy may reveal characteristic features such as conoidal rings, polar rings, a conoid pellicle, subpellicular microtubules, micropores, and micronemes. IHC (using anti-*S. neurona* antibodies) may aid in diagnosis. *Sarcocystis* spp. can be distinguished from *Toxoplasma* spp. by the presence of a parasitophorous vacuole that separates the latter from the host cell cytoplasm.[74]

Besnoitiosis is caused by tissue-inhabiting protozoa, ***Besnoitia* spp.** (class Apicomplexa, order Eucoccidiida, family Sarcocystidae). Disease occurs in reindeer, barren ground caribou, and mule deer. Infections have also been reported in roe deer.[12,113,165] Transmission to intermediate hosts (e.g., cervids) may be direct or via mechanical vectors such as biting flies.[124] Merozoites may develop in endothelial cells and subsequent merogony occurs in a variety of tissues, with a predilection for connective tissue and fibroblasts, in which intracellular cysts may form.[207]

Besnoitia spp. infections are often inapparent, but in some species, such as reindeer and caribou, chronic infections may result in characteristic, ≤1 mm diameter, firm, spherical, glistening, white cysts in the skin, subcutis, and underlying tissues. Grossly evident skin lesions on the limbs may be prominent over the metatarsus and metacarpus, and extend deep to the panniculus, epimysium, periosteum, and associated fascia.[12] Cysts may also occur in the skin of the face (periocular and rostral), as well as scrotal regions. Cysts may cause thickening and ulceration in the nasal and pharyngeal mucosa, and ocular scleral and palpebral conjunctiva. Cysts in connective tissue of muscle, spleen, kidney, visceral serosa, bone, and testicle can appear as linear chains within blood vessel walls. Severe infections may cause thickened, hyperpigmented skin leading to ulceration and fissure formation, and secondary infections with associated alopecia, crusting, exudation, hyperkeratosis, inflammation, and superficial hemorrhage. Male sterility is a potential sequela.[78,124,207]

Histologically, large, approximately 250–350 μm diameter cysts may be seen in connective tissue of numerous organs including the tunica media of blood vessels, which may cause occlusion of vascular lumens. Cysts have a thick, hyaline, extracellular capsule that surrounds the parasitized host cell and contains multiple, peripheralized, flattened nuclei and a central mass of myriad bradyzoites within a parasitophorous vacuole. An irregular, clear space (processing artifact) often separates the host cell from the surrounding capsule.[124] Inflammation is usually absent but cyst rupture or degeneration, or associated skin damage may invoke macrophage, multinucleated giant cell, and lymphocyte inflammation.[207]

In cervids (i.e., intermediate hosts), gross lesions and histopathology, including bradyzoite-filled cysts exclusively in connective tissues, are strongly supportive of *Besnoitia* spp. infection. Real-time PCR assays have been developed to detect and quantify ***B. besnoiti*** parasite load in cattle and have potential utility in caribou, although specific molecular markers for some large animal *Besnoitia* spp. are currently lacking.[78,243,287] Electron microscopy reveals characteristic structures that may be shared among *Besnoitia* spp. and related apicomplexan protozoans (e.g., *T. gondii*), such as a conoid, rhoptries, micronemes, dense granules, mitochondrion, the nucleus, lipid bodies, and amylopectin. Further, tachyzoites and bradyzoites of ***B. tarandi*** are ultrastructurally similar to other *Besnoitia* spp. (e.g., lack of enigmatic bodies).

Immunohistochemistry for *B. tarandi* has been performed[76] but is not widely available in diagnostic laboratories.

Trypanosoma* spp.** (class Zoomastigophora, order Kinetoplastida, family Trypanosomatidae; WOAH-listed reportable disease in multiple species) are flagellated hemoparasites that have been reported in many captive and free-ranging cervid species. Species documented in numerous cervid species include ***T. cervi and ***T. evansi***; ***T. theileri*** has been diagnosed in roe deer and red elk; and ***T. mazamarum*** has been reported in brocket deer.[68,168,182,237] Tabanid biting flies transmit *Trypanosoma* spp. to cervids and other mammals. Upon infection, trypomastigotes circulate in peripheral blood and enter macrophages and other cells of the reticuloendothelial system in the heart and skeletal muscle, where amastigotes multiply by binary fission and then spread throughout the body.[187] Deer keds (*Lipoptena mazamae*) can also transmit *T. cervi*.[97] Trypanosomes have evolved immune escape mechanisms and may cause immunosuppression. Otherwise, host effects are often regarded as nonpathogenic. However, disease varies by host species, as acute, fatal infections are reported in *Axis* sp. deer and sambar deer, and chronic effects in axis deer and Sunda sambar include anemia, weight loss, and abortion (reviewed in Ref. 68). Hog deer with fatal *T. evansi*-associated neurologic disease have nonsuppurative meningoencephalitis including perivascular cuffs of lymphocytes and plasma cells. *T. cervi* caused lymphoplasmacytic choroid plexitis in white-tailed deer[208] and late-stage abortion in an infected hog deer.[325] As hemoflagellates, trypanosomes are elongated and laterally flattened. *T. theileri* is approximately 60–70 µm long (range 25–120 µm) with a long, pointed posterior end, medium-sized kinetoplast, prominent undulating membrane, and free flagellum. *T. cervi* are of similar length (40–83 µm) and morphology but have a shorter free flagellum (10 µm long) and a more posterior large granular nucleus, located anterior to the relatively large kinetoplast.[97,181] *T. evansi* is more slender with a small subterminal kinetoplast, central nucleus, and thin posterior extremity.[68]

Trypanosomiasis can be diagnosed by microscopic visualization of parasites in blood or impression smears of spleen and liver. With heavy infections, trypanosomes in peripheral blood hematocrit tubes may be visible in the plasma above the buffy coat layer.[181] More sensitive methods include PCR or blood culture, and IHC of the brain of animals with neurologic disease.[168,237,325]

Neospora* spp.** are obligate intracellular apicomplexan protozoal parasites (class Conoidasida, subclass Coccidiasina, order Eucoccidiorida, family Sarcocystidae) that form tissue cysts. Detection in cervids has been mostly via serological surveys in healthy animals with rare disease detection, primarily in young animals or fetuses. Organisms may be detected histologically via hematoxylin and eosin staining, with tachyzoites and cysts in brain, spinal cord, lung, kidney, liver, and other tissues, but histology is less sensitive than other methods of detection and should be performed in conjunction with immunohistochemistry (polyclonal antibodies are commercially available and more sensitive) or PCR.[300] *N. caninum* can be difficult to differentiate from *T. gondii* histologically. Transmission electron microscopy can distinguish ***N. caninum from *Toxoplasma gondii* by specific morphology of tachyzoites and cysts but is less commonly used. PCR is an excellent confirmatory diagnostic tool for *N. caninum* (specific markers include repetitive *Nc5* gene and the internal transcribed spacer 1) and is often performed on brain or other useful tissues: skeletal muscle, heart, liver, kidney, spinal cord, and lung.[71] The testing of more than one brain section and multiple tissue types by PCR increases the detection likelihood.[74,85] Additional information on *Neospora* and *Toxoplasma* spp. in cervids is listed in Table 6.e4 in the digital content.

Ectoparasites

Larval mites of the family **Trombiculidae** (genera ***Eutrombicula***, ***Neotrombicula***) and others are found on the skin of a variety of cervid species (Fig. 6.32) and are usually nonpathogenic but may cause pruritis. Larvae attach to the host, puncture its skin with its chelicerae, and inject saliva (via the buccal aperture) into the epidermis that digests host tissue and forms a stylostome (i.e., feeding tube) in the dermis.[216] Routine surveys of free-ranging, white-tailed deer reveal infestations of ***N. richmondi*** that are associated with facial dermatitis, alopecia, excoriation, and serous exudate with orange, granular material most prominent around the mouth and eyes. The mites and associated inflammation may lead to occlusion of the lacrimal ducts. Histopathologic lesions include larvae on the surface of the hyperkeratotic, acanthotic epidermis and within crusts,

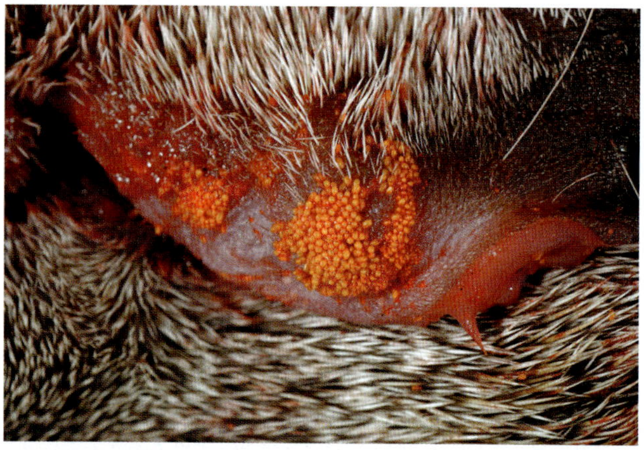

FIGURE 6.32 Dermal trombidiosis of the head and ears of a white-tailed deer. Clusters of these orange/red mites are present in the skin adjacent to the mouth. Deer can develop an allergic reaction to the larval mite (i.e., chigger; *Eutrombicula* or *Neotrombicula* sp.) infestation. *(Photo Courtesy of the Southeastern Cooperative Wildlife Disease Study.)*

with severe plasma cell infiltration of the underlying dermis, folliculitis, and furunculosis. Additional reports include trombiculid mites clustered in the axilla of roe deer, and on the skin of sika deer with no gross pathology.[315,332] Larvae appear as round-bodied, orange-pale mites with six legs and a stylostome by light microscopy on wet mounts or histology.[216] For taxonomic identification, specialized entomology or diagnostic laboratories are required.

***Demodex* spp.** (i.e., hair follicle mites) are obligate parasites (class Arachnida, subclass Acari, order Trombidiformes, family Demodicidae) and reside in the skin of numerous cervid species on several continents. A thorough review of *Demodex* spp. on ungulates is provided by Izdebska and Fryderyk.[163] *Demodex* spp. mites inhabit the hair follicles and do not cause pathology under normal circumstances. However, in rare cases, they may cause local or generalized skin disease. Predisposing factors for clinical demodicosis are likely multifactorial and may include host immunologic or genetic factors, immunosuppression due to substandard environmental conditions, and concurrent infections (e.g., lungworm) or endocrinopathies. In a retrospective review, white-tailed deer over 2 years of age were more likely to have demodicosis than younger deer, and most of the detections were in the fall and corresponded to the rut.[236] Gross findings in severe cases may include emaciation and patchy to generalized alopecia (especially involving the head, neck, limbs, and trunk), hyperemia, hyperpigmentation, lichenification, and less commonly, cutaneous nodules (Fig. 6.33; eSlide 6.e13). Histopathology includes dilated hair follicles and sebaceous glands that are filled with mites, variably thick superficial serocellular crusts, acanthosis, hyperkeratosis, mixed cell (lymphocytes, plasma cells, macrophages, multinucleated giant cells, neutrophils) dermatitis, suppurative folliculitis and furunculosis, and perifollicular, lymphoplasmacytic inflammation; inflammation may also be absent.[63,67,120]

Mites can be viewed microscopically on wet mounts of skin scrapings or by histopathology. They are easily identified due to their characteristic cigar-shape.[63,120] Dissection and assessment of mite morphology and anatomy via phase-contrast microscopy can aid in speciation.[163] Molecular methods (i.e., PCR) have been used to differentiate and characterize *Demodex* species in other animals based on available genetic sequences.[24]

Dermacentor albopictus (i.e., "winter ticks"), a one-host parasite, infests numerous North American cervids, with moose being the most susceptible to heavy and potentially fatal infestations. The ticks may over-winter and readily feed on numerous free-ranging cervid species. They have also been documented on captive reindeer.[333] Larvae infest cervids in the late summer/early fall, when the ticks molt into nymphs and adults that overwinter. Female ticks engorge late in the winter, mate, and leave the host to lay eggs. Tick numbers can reach tens of thousands on moose (up to 178,000) who have much higher mean densities than other cervid species, and cause winter coat damage due to excessive grooming or rubbing with a characteristic pattern on the neck and shoulders in the early stages and later, the withers and tail[123,198,281] (Fig. 6.34). Chronic effects may include poor nutritional condition, alopecia and hair breakage, anemia (pale liver, heart, and kidneys; decreased hepatic iron concentration), and hydropericardium. Death may occur due to severe anemia and exhaustion. Calves are especially vulnerable to emaciation and anemia, and may show neurologic signs.[151,353] Wapiti, mule deer, and

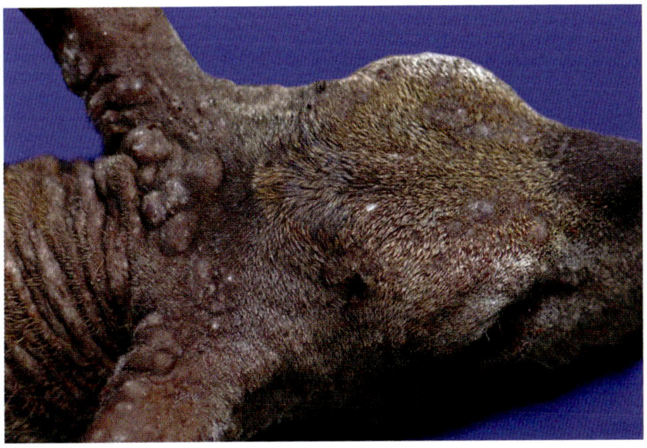

FIGURE 6.33 Demodicosis dermatitis in a white-tailed deer. Patchy alopecia and severe distension of the hair follicles by *Demodex* sp. mites cause multifocal nodules in the skin. (see eSlide 6.e13). *(Photo Courtesy of the Southeastern Cooperative Wildlife Disease Study.)*

FIGURE 6.34 Winter tick disease in due to *Dermacentor albipictus* infection in the skin of a moose. Infection is associated with coat damage and alopecia along the neck and shoulders. *(Photo Courtesy of Ontario Ministry of Natural Resources and Forestry.)*

white-tailed deer may have mild alopecia, which has been reported at the base of the neck, and rare mortality is reported in elk.[4] In one extreme case, a captive reindeer had over 400,000 ticks with no gross lesions. Diagnosis is by the finding of high densities of *D. albopictus* on moose during late winter at above 60 degrees north latitude in North America. Blood spots may be evident in the snow where engorged ticks ruptured on recumbent animals.[151]

***Cephenemyia* spp.** (order Diptera, family Oestridae, subfamily Oestrinae) are referred to as nasal, pharyngeal, head, or throat bots because the first instar larvae develop in the nasal cavity following deposition of larval packets in the nostrils of the host by the female fly. The second and third instar larvae are found in retropharyngeal pouches. Infestations occur in a number of North American and European cervid species (Fig. 6.35). These larvae are obligate parasites and feed upon secretions and cell debris. Infestations of *Cephenemyia* larvae vary (up to 100 larvae/host) but generally do not cause significant disease.[51] However, in severe cases, roe deer with a history of lethargy and coughing may suffocate if they have larvae in the nostrils and/or main stem bronchi.[253] First instar larvae may cause local eosinophilic and mastocytic inflammation at the penetration site, and mechanically enlarged retropharyngeal recesses and cause epithelial erosion, fibrosis, and inflammation.[48] Keratitis caused by the first instar larvae of *C. trompe* has been reported in reindeer.[265] Aberrant *Cephenemyia* spp. larval migration may lead to penetration of the cranium or other sites, as reported in a mule and roe deer with cerebral abscesses, and a wapiti with a damaged eustachian tube.[107,167,253] In addition, adult *Cephenemyia* spp. activity may elicit fly deterrence behavior leading to decreased feeding and poor nutritional condition in reindeer and roe deer; associated reduced milk yield also occurs in the former[50] (reviewed in Ref. 280). Diagnosis is made by gross observation of bot larvae in the nasal cavities or pharyngeal pouches.[51] First instars are small (1–2 mm long) and challenging to identify, but third instars (20–35 mm long) can be speciated. Ideally, specimens are allowed to pupate and mature into adults for definitive identification.[50]

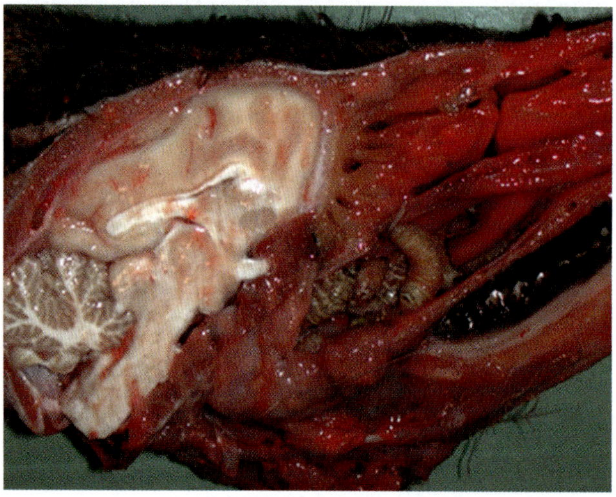

FIGURE 6.35 Botflies in the nasopharynx of a roe deer. The nasal cavity contains multiple white fly larvae. Severe infestations can result in suffocation. *(Photo Courtesy of the Centre for Fish and Wildlife Health, University of Bern.)*

Deer keds are hematophagous biting flies (***Lipoptena*** and ***Neolipoptena* spp.**; order Diptera, suborder Brachycera, infraorder Muscomorpha, family Hippoboscidae, subfamily Hippoboscinae). They shed their wings once they find a host, on which they remain nearly year-round. Pupae drop from the coat after copulation, which occurs on the host. Deer keds can spread from doe to fawn during suckling.[62] White-tailed deer are infested by numerous species (***L. cervi*, *L. mazamae*, *L. depressa***, and ***N. ferrisi***), which is in part determined by geographic distribution. In addition, *L. cervi* is widely distributed on moose in Norway and roe deer in continental Europe; *L. mazamae* is reported on exotic deer in North America (axis, fallow, sika); *L. depressa* and *N. ferrisi* are common on mule deer; and *L. depressa* are common on black-tailed deer.[3,62] Infestations can reach thousands of keds, but associated pathology is rarely reported. However, eczema and alopecia have been attributed to deer keds in red deer, and semidomesticated reindeer and fallow deer with massive infestations have been reported with growth retardation. In an epizootic of alopecia with severe infestations of deer keds in free-ranging moose, affected animals were in poor nutritional condition with generalized lymphadenopathy and severe, bilateral alopecia with crusting dermatitis on the cheeks, ventral neck, lateral thorax, abdomen, and less severe alopecia on the dorsal neck, back, and limbs. Histopathology in severe cases revealed orthokeratotic hyperkeratosis, epidermal hyperplasia with rete ridges, and perivascular dermatitis (dominated early by eosinophils, followed by lymphocyte, macrophage, and plasma cell infiltration). The skin of some moose also had epidermal ulcerations and micropustules, flame follicles, serocellular crusts, and dermal fibrosis. These lesions may culminate in starvation and hypothermia as the proximate cause of death, and blood loss as a contributing factor.[222] Deer keds can be diagnosed by the presence of flat, dark brown, small, wingless flies in the hair coat.[3]

Additional metazoan, protozoal, and ectoparasitic diseases of cervids are described in the digital content (Table 6.e4).

Other Infectious Organisms

Chronic wasting disease (CWD) is a transmissible spongiform encephalopathy reported in wapiti, white-tailed deer, mule deer, moose, and reindeer. It has been reported in both captive and free-ranging populations.[14,22] Survival

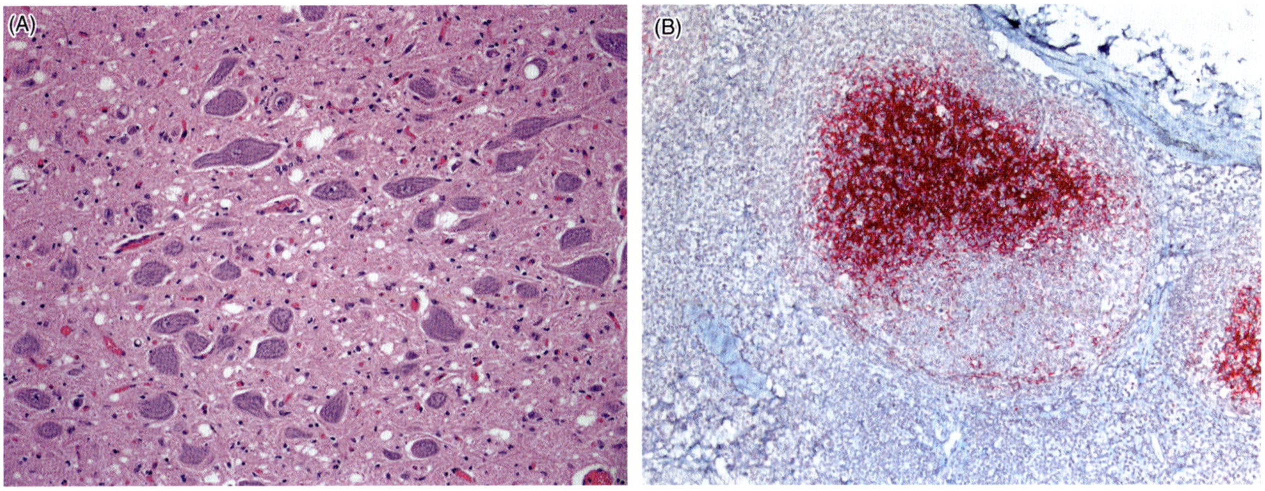

FIGURE 6.36 Chronic wasting disease due to prion proteins in the brain of an elk. (A) Characteristic lesions are neuronal vacuolation and astrocytosis. Histologic lesions in the brain start in the dorsal vagal nuclei which are in the obex region of the brain stem. (B) Prion protein is first seen in lymphoid tissue and then in nuclei in the brain stem in the area of the obex. IHC:alkaline phosphatase technique with red chromogen (see eSlide 6.e14). *(Part A: Photo Courtesy of T. Spraker, Colorado State University; Part B: Photo Courtesy of the Southeastern Cooperative Wildlife Disease Study.)*

of deer and wapiti herds in North America is currently threatened by the increasing geographic distribution and prevalence of CWD among cervid populations. The pathogenesis of CWD involves conversion of endogenous cellular prion protein (PrP^{C}) in tissues, such as lymph nodes and brain, to an abnormal, protease-resistant form (PrP^{CWD}) after exposure to an infectious dose of the abnormal form.[339] Under natural conditions, cervids are infected via the oral and nasal mucosa by contact with infected animals or by indirect environmental exposure associated with food consumption or scraping during rut.[230] The disease is readily transmitted horizontally from infected to susceptible cervids and residual infectivity in contaminated environments appears to help sustain epidemics. Clinical signs include progressive weight loss, behavioral changes, increased drinking and urination, and excessive salivation (http://www.nwhc.usgs.gov/publications/fact_sheets/pdfs/cwd/CWDFactsheet2007.pdf). However, the disease is present and can be diagnosed long before clinical signs occur.

Oropharyngeal lymphoid tissue is thought to be the earliest site of PrP^{CWD} accumulation[152] and abnormal prion protein PrP^{CWD} accumulates early in gut-associated lymphoid tissue (GALT). Early diagnosis can be made by detecting PrP^{CDW} in lymphoid tissues, such as GALT, lymph nodes, such as the medial retropharyngeal lymph node, and tonsil by IHC or an enzyme-linked immunosorbent assay (ELISA).[231] Gross and microscopic changes are not seen in these tissues. Later in the progression of the disease, the abnormal protein can be detected in the brain. Accumulation of the protein starts in the dorsal vagal nucleus, which is in the region of the obex and is the brain region of choice for IHC or ELISA testing.[306] PrP^{CWD} has also been detected in feces from infected deer and Rocky Mountain elk using amplification techniques (e.g., protein misfolding cyclic amplification)[258,317] (Tennant et al., 2022). There are no gross lesions in the brain and prion protein accumulates prior to the detection of microscopic lesions. When microscopic lesions develop, they include vacuoles in the nerve cell body and axons, neuronal degeneration and loss, astrocytosis/gliosis, and production of amyloid plaques; inflammation is not a feature of infection[340] (Fig. 6.36A; see eSlide 6.e14). The medial retropharyngeal lymph node and obex region of the brain are considered the most sensitive and best samples for postmortem testing[133] (Fig. 6.36B). Human-assisted movement of live cervids should be discouraged, but herd-level antemortem testing, including tonsillar and/or rectal biopsies with prion detection by IHC or a real-time quaking-induced conversion assay (RT-QuIC), has been used to lower the risk of disease spread when translocating or exporting animals.[143,231] Abnormal prions may also be detected in saliva, urine, blood, feces, cerebral spinal fluid, nasal brushings, and skin biopsies by RT-QuIC, which is being evaluated for diagnostic and surveillance utility but is not currently a recommended diagnostic test[93,132] (eSlide 6.e14).

ADDITIONAL RESOURCES

Cohen, B.S., Belser, E.H., Killmaster, C.H., Bowers, J.W., Irwin, B.J., Yabsley, M.J., Miller, K.V., 2015. Epizootiology of cranial abscess disease in white-tailed deer (*Odocoileus virginianus*) of Georgia, USA. J. Wildl. Dis. 51, 609–618.

das Neves, C.G., Roth, S., Rimstad, E., Thiry, E., Tryland, M., 2010. Cervid herpesvirus 2 infection in reindeer: a review. Vet. Microbiol. 143, 70–80.

Donahoe, S.L., Lindsay, S.A., Krockenberger, M., Phalen, D., Slapeta, J., 2015. A review of neosporosis and pathologic findings of *Neospora caninum* infection in wildlife. Int. J. Parasitol. 4, 216–238.

Pewsner, M., Origgi, F.C., Frey, J., Ryser-Degiorgis, M.-P., 2017. Assessing fifty years of general health surveillance of roe deer in Switzerland: a retrospective analysis of necropsy reports. Plos One 12 (1), e0170338.

Rivera, N.A., Varga, C., Ruder, M.G., Dorak, S.J., Roca, A.L, Novakofski, J.E., Mateus-Pinilla, N.E., 2021. Bluetongue and Epizootic Hemorrhagic Disease in the United States of America at the Wildlife-Livestock Interface. Pathogens. 10, 915.

Speck, S., 2012b. Dermatophilus infections. In: Gavier-Widén, D., Duff, J. P., Meredith, A. (Eds.), Infectious Diseases of Wild Mammals and Birds in Europe. Wiley Blackwell, Oxford, pp. 439–440.

Sundberg, J.P., Williams, E.S., Hill, D., Lancaster, W.D., Nielsen, S.W., 1985. Detection of a papillomavirus in cutaneous fibroma of white-tailed and mule deer. Am. J. Vet. Res. 46, 1145–1149.

Williams, E.S., Young, S., 1993. Neuropathology of chronic wasting disease of mule deer (*Odocoileus hemionus*) and elk (*Cervus elaphus nelsoni*). Vet. Pathol. 30, 36–45.

References

Full references for this chapter can be found online at https://doi.org/10.1016/B978-0-443-18437-6.00051-1.

Chapter 7

Camelidae

Omar Gonzales-Viera[a], Tawfik Aboellail[b], and Dalen Agnew[c]

[a]*California Animal Health and Food Safety Lab, Davis Branch, Davis, CA, United States;* [b]*Colorado State University, Fort Collins, CO, United States;* [c]*Michigan State University, Lansing, MI, United States*

INTRODUCTION

The Camelidae family (generally falling in the suborder Tylopoda and the order Cetartiodactyla) consists of the Old and New World camelids. The three Old World (OW) camelids are the dromedary, the domestic Bactrian camel, and the wild Bactrian camel. Dromedary camels inhabit arid and semiarid areas of the Arabian Peninsula and northern Africa. Wild Bactrian camels inhabit high plains in China and Mongolia and are genetically distinct from the domesticated Bactrian camels. The lineage of New World (NW) camelids has been more controversial, though most now agree it consists of the genus *Lama* and a separate genus *Vicugna*. The llama belongs to *Lama glama* and the guanaco belongs to *L. guanicoe*. The alpaca and vicuña belong to *Vicugna pacos* and *V. vicugna*, respectively. Llamas and alpacas do not live in the wild anymore since they have been domesticated 6,000 to 7,000 years ago.[162] Both species inhabit the dry, arid, and cold Andean highlands at elevations ranging from 2,300 to 4,000 m above the sea level (a.s.l.). On the other hand, vicuñas and guanacos are found in the wild, inhabiting semiarid grasslands and plains at elevations between 0 and 5,000 m a.s.l. Interestingly, guanacos can be seen in coastal deserts with high temperatures. Interbreeding between NW species and between OW species produces fertile crosses, but to date, fertile young have not been produced by crosses between NW and OW camelids.[47] Only the free-ranging population of wild Bactrian camel is listed by the International Union for the Conservation of Nature (IUCN) as critically endangered. Bactrian camels held in zoos are generally of domestic origin. Vicuña were listed as vulnerable, but they have now been classified as least concerned thanks to the conservation measures. The anatomical and physiological features of NW and OW camelids allowed their adaptation to harsh environments, which was used by native populations to use them as pack animals as well as milk and meat source. This chapter focuses on diseases of camelids, most of which have been described in domesticated and zoo animals in their native environments or elsewhere in the world. Where a disease has also been described in free-range animals, it will be specifically mentioned.

NON-INFECTIOUS

Nutritional

Nutrition-related diseases are relatively common in NW camelids maintained in captivity (zoos and hobby pets). While arguably not true "ruminants," both OW and NW camelids are able to utilize coarse forage through anaerobic fermentation in C-1 and C-2. New World camelids have evolved in the presence of highly variable food resources, with a "feast and famine" routine—periods of high dietary input followed by very low input. Because of this physiologic adaptation, **over feeding** in captive and domesticated NW camelids is common and can lead to significant consequences such as obesity and hyperthermia. Maintaining and monitoring proper body condition in camelids can be challenging. Due to their heavy fleece, NW camelids may be emaciated and managers will not recognize this unless they are palpated. In OW camels, there is not a similar "feast and famine" syndrome, but these animals are well adapted to storing nutrients and maintaining conditions with poor-quality feeds. The hump stores fat and will increase and decrease in size and firmness depending on the nutritional condition.[46]

Obesity, the accumulation of excessive amounts of fat in the subcutis, abdomen, and retroperitoneal cavity, is a common problem in captive OW and NW camelids. Without carefully managing animal weight and incorporating lower-quality feeds and "lean" periods of nutrition, animals are prone to developing obesity and secondary complications, including exacerbating osteoarthritis, infertility, and susceptibility to hyperthermia.

At necropsy, **starvation** is usually recognized by the absence of abdominal, retroperitoneal, pericardial, subcutaneous, or marrow fat. If mobilization of fat is rapid, it may present as serous atrophy in any of these fat depots. The bone marrow is typically the last to be depleted (Fig. 7.1).

Pathology of Wildlife and Zoo Animals. https://doi.org/10.1016/B978-0-443-18437-6.00009-2

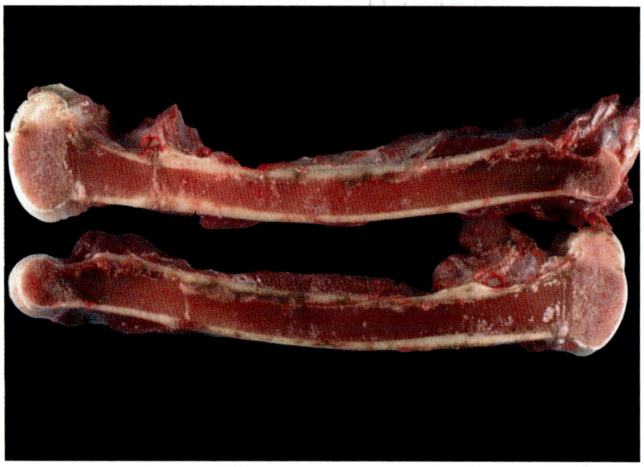

FIGURE 7.1 Serous atrophy of fat in a llama. Bone marrow fat is considered the last adipose tissue depot to be depleted during starvation. Serous atrophy of fat is yellow to red, translucent, and gelatinous. This contrasts to normal firm white to tan marrow fat (not shown).

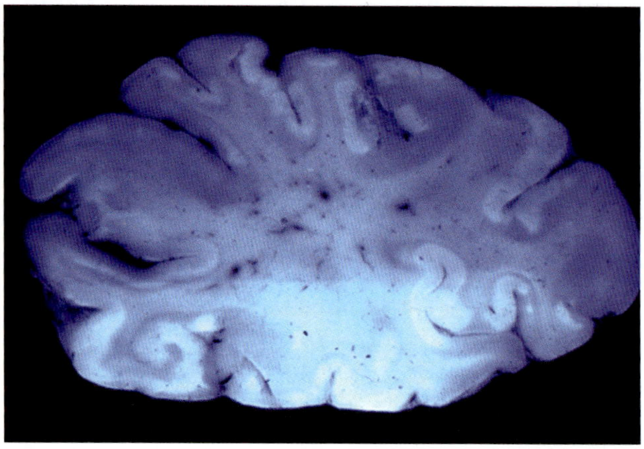

FIGURE 7.2 Polioencephalomalacia in a llama. Necrotic tissue autoflouresces with ultraviolet light. In this image, which was taken with ultraviolet light illumination, areas of necrosis can be seen as bright white fluorescent areas in the cortical gray matter.

Testing of bone marrow for percentage of fat has been validated in other species but not in camelids, though a very low percentage of marrow fat (<10%) is a likely indicator of poor nutritional condition and starvation in camelids.[91]

Polioencephalomalacia (PEM; syn. cerebral cortical necrosis) in ruminants is frequently associated with **thiamine (vitamin B1) deficiency** resulting from dietary changes that lead to acidosis, such as the feeding of carbohydrate-rich grain diets that encourage the proliferation of thiaminase-producing bacteria. Thiaminase-containing plants, some anthelmintics (e.g., amprolium), and excessive sulfur intake may also cause PEM. Although this condition is rarely diagnosed in camelids, in both NW and OW, the disease is like that described in ruminants.[75] Grossly, diffuse swelling and multifocal necrosis of the cerebral cortex are seen in the superficial gray matter with occasional extension into the underlying white matter. Yellow areas of discolored necrotic tissue will autofluoresce under ultraviolet light (Fig. 7.2). Histologically, bilateral deep laminar necrosis of the neurons, edema of the superficial laminae, and gliosis are characteristic. While necrosis is most typically seen in the cerebrum, it can also be found in the cerebellum. Gliosis will vary with the age of the lesion, being more prominent as the lesion matures. Very chronic lesions may contain cystic cavities.[88,165] Bilateral laminar necrosis on histology is confirmatory of PEM. Knowing the levels of carbohydrate in the diet or the administration of anthelmintic, and measuring the levels of sulfur in water source would be possible approaches to establish a diagnosis.

NW camelids are particularly prone to **hyperglycemia**. Poor glucose tolerance, partial insulin resistance, and low concentrations of circulating insulin are characteristic in these species and are likely a consequence of evolutionary factors that allow them to survive on a uniquely nutrient-poor diet.[26] Besides primary metabolic diseases, the effects of these energy disturbances likely exacerbate or predispose to other diseases, including sepsis. Environmental, metabolic, or social stress and the release of corticosteroids are likely to lead to persistent hyperglycemia. Dehydration is the most common and clinically significant sequela to persistent hyperglycemia, glucosuria, and glucose diuresis. This, in turn, can be associated with hypernatremia, metabolic acidosis, and azotemia. While these signs may mimic diabetes mellitus, true diabetes mellitus is rare in NW camelids.[26]

Sick NW camelids, particularly crias, may develop **hypoglycemia**, though this is less commonly noted than in other species. Several diseases, including hepatic disease, sepsis, and *Mycoplasma haemolamae* predispose to its development.[26] It is also associated with hypothermia and failure of passive transfer in compromised crias that have not nursed.[164]

Hepatic lipidosis and **hyperlipemia** are also relatively common conditions seen in NW camelids. Grossly, the liver is enlarged with rounded edges, and it is pale tan to yellow with a greasy texture (Fig. 7.3). In rare cases, hepatic changes may be so severe as to cause the liver to float in formalin or water. Histologically, changes are diffuse with large, clear, distinct cytoplasmic vacuoles and peripheralized nuclei in hepatocytes. Pregnant and lactating camelids are prone to lipidosis in a manner similar to ketotic ruminants with inadequate carbohydrates in the diet.[26] In other NW camelids, stressful events and/or hepatotoxins are more likely responsible. Secondary hepatic encephalopathy associated with lipidosis has been described in llamas. Alzheimer type II astrocytosis in the cerebrum and to a lesser extent in the hippocampus, thalamus, and cerebellum are present in affected animals.[113] Decreased astrocytic immunohistochemical labeling intensity for glial fibrillary

FIGURE 7.3 Hepatic lipidosis in a llama. Characteristic findings include yellow to light tan, locally extensive to diffuse discoloration of the liver, a greasy friable texture, and rounded lobe margins. It is common in NW camelids, particularly in pregnant and lactating camelids when dietary carbohydrates are inadequate. *(Photo Courtesy of A. Tibary, Washington State University.)*

acid protein is seen in the brains of neurological llamas with hepatic dysfunction. In addition, distinct from ruminants, fat can accumulate in renal tubular epithelium **(renal lipidosis)** under similar circumstances as hepatic lipidosis and lead to kidney dysfunction in camelids.

Vitamin E and **selenium** are important antioxidants that typically act synergistically via glutathione peroxidase across the taxonomic spectrum. As in domestic ruminants, deficiencies—**hypovitaminosis E** or **selenium deficiency**—in camelids can be due to selenium-deficient soil (some geographic areas have low selenium) or vitamin E deficient (mature or inadequately stored) forage. Vitamin E is a fat-soluble nutrient that cannot be stored in the liver, so daily ingestion is required. With very severe deficiency, cardiac and skeletal muscle degeneration and necrosis develop; mineralization may be present in areas of necrosis. Heavily used muscles such as the intercostals, diaphragm, and tongue are commonly affected. Cardiac necrosis can lead to sudden death; important differentials for sudden cardiac death due to cardiomyocyte necrosis are oleander or ionophore toxicity, both of which are relatively common in captive camelids. It is likely that other plants throughout the natural ranges of camelids and ornamental plants used in zoos have similar cardiotoxic effects, such as Lily of the Valley, avocado leaves, Christmas rose, and sea onion.[95] In addition to feed and tissue level testing (liver, serum, or plasma) for selenium and vitamin E, toxicologic testing by liquid chromatography with tandem mass spectrophotometry (LC–MS–MS) may be required to rule out toxins such as oleandrin and ionophores in gastrointestinal (GI) content or urine and feed, respectively, when myocardial necrosis is seen. Old World camelids appear to be more sensitive to selenium toxicosis and toxicity can develop with lower doses than are required to cause disease in their livestock counterparts. In general, polyphasic myonecrosis/degeneration is more suggestive of nutritional myopathy than toxic myopathy, which is often monophasic. Captive/zoo camels should be limited to 4–6 mg/day for adult camels. Selenium toxicosis, if fed >8 mg/day for an adult camel, can result in hyperkeratosis of the food pads (Figs. 7.4A, B)

Fat necrosis in camelids is associated with vitamin E and selenium deficiency, rapid weight loss, and some toxins (such as endophyte-infected fescue), particularly in overweight animals with excessive intra-abdominal fat. In dromedary camels, **hump fat necrosis** is also anecdotally associated with vitamin E and selenium deficiency and causes histologic lesions like those seen in other species, including saponification and mineralization.

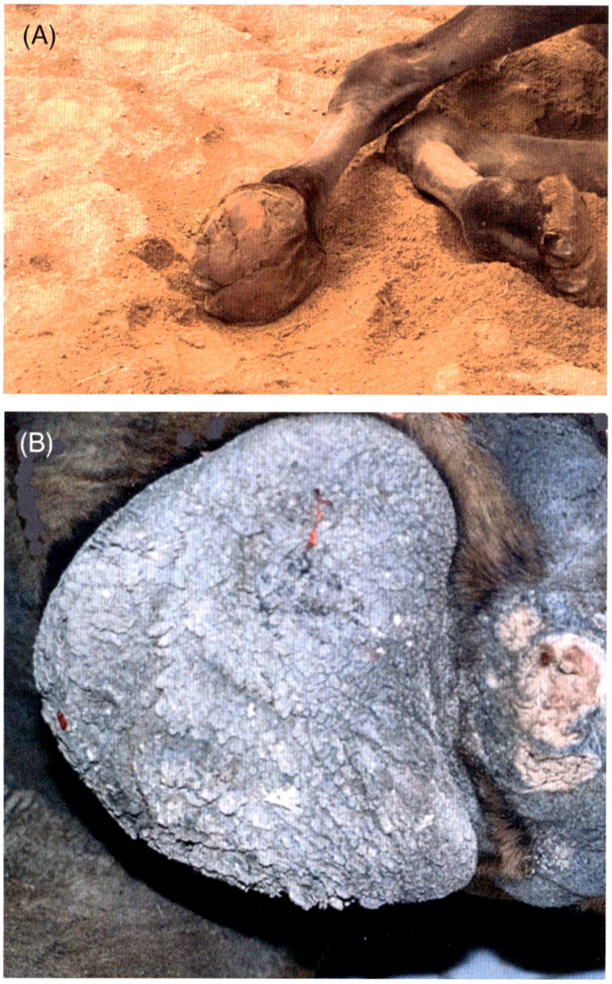

FIGURE 7.4 Hyperselenosis in the footpads (skin) of a Bactrian camel. (A) Characteristic findings include marked thickening and hyperkeratosis of the foot pads, which are also multifocally cracked in this animal. (B) The pedal surface of the foot pad is thick and rough. *(Parts A, B: Photo Courtesy of M. Dioli.)*

Goiter has been described in OW camelids in Africa and in the United States. **Iodine deficiency** has been described in fetal, neonatal, and adult camelids.[6,34] In adult animals, goiter has been characterized by enlarged thyroids with variable follicular dilation. Follicles are filled with copious colloid and lined by plump cuboidal and sometimes columnar epithelium. Clinically, disease has been associated with low serum thyroid hormones (T3, T4) as well as normocytic normochromic anemia and infertility. **Congenital goiter** is typically reported in late gestation with significant hypotrichosis (Fig. 7.5).

As occurs in domestic livestock, **zinc-responsive dermatosis** has also been reported in NW camelids.[50,127] Black NW adult camelids seem predisposed, and lesions are typically noted on the face, ventral abdomen, lateral thorax, and inguinal region with variable pruritus. Lesions consist of alopecia, dermal thickening, scaling, and hyperpigmentation (Fig. 7.6A). Microscopically, variably thick orthokeratotic hyperkeratosis and perivascular lymphoplasmacytic, eosinophilic, and histiocytic dermatitis are observed (Fig. 7.6B; eSlide 7.e1). A notable difference in the condition between camelids and domestic animals is that skin changes in camelids are orthokeratotic hyperkeratosis while that in domestic ruminants, dogs, and pigs is parakeratotic.[133] Low plasma zinc levels (<0.5 mg/dL) may be associated with these lesions.

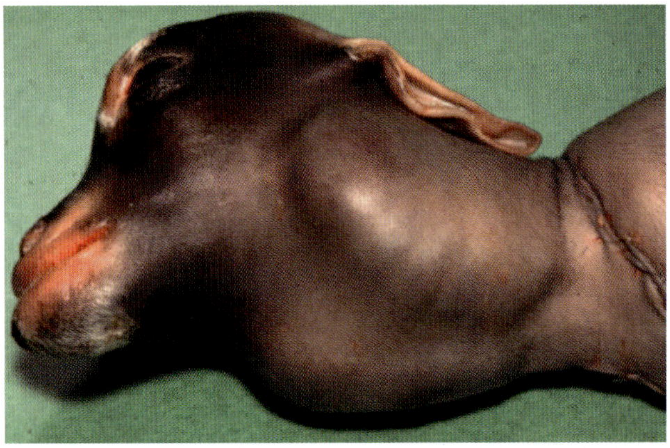

FIGURE 7.5 A stillborn dromedary camel with goiter. Severe bilateral enlargement of the thyroid glands is seen in congenital goiter. There is also significant hypotrichosis and thickening of the skin due to myxedema. Hypotrichosis and myxedema are often associated with goiter. *(Photo Courtesy of M. Dioli.)*

Toxic

Oleander (*Nerium oleander*) is one of the most toxic shrubs that affects NW camelids and other mammals. The toxic metabolite is oleandrin, which is a cardiotoxic glycoside. The lethal dose of either green or dried leaves when the animal ingests them is 225 mg/kg. For a 150 kg llama or alpaca, it would mean less than 35 g of leaves. The low number of leaves necessary to kill an animal makes this an important poisonous plant of camelids.[44] At necropsy, epicardial and endocardial hemorrhages with myocardial necrosis (Figs. 7.7A, B; eSlide 7.e2) are usually reported as well as hemorrhages in the GI tract. On histology, monophasic myocardial necrosis and degeneration are typically seen adjacent to areas of hemorrhage. Liver, GI content, urine, and serum are good samples for the diagnosis of oleander toxicity by liquid chromatography with tandem mass spectrometry.

Consumption of **black oil sunflower seeds (*Helianthus annuus*)** is considered to be fatal in alpacas and llamas from two different premises from Missouri and Washington.[101] Accidental ingestion of large volumes of sunflower seeds caused high mortality with renal tubular necrosis, azotemia, and gastroesophageal ulceration (Figs. 7.8A–C), although the mechanism of GI mucosal and renal tubular damage is unknown.

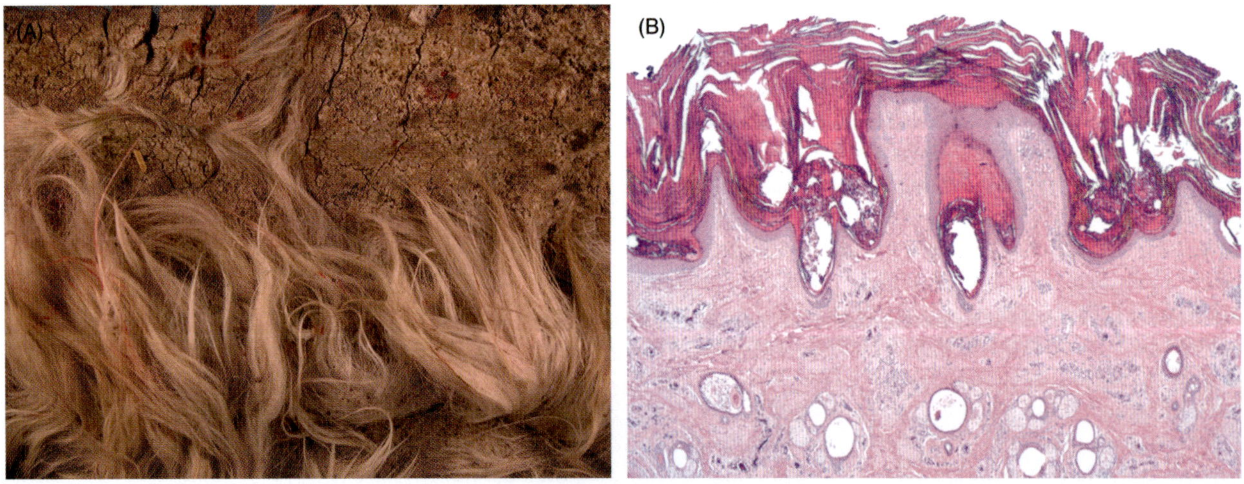

FIGURE 7.6 Zinc-responsive dermatosis in a llama. (A) This condition is seen in NW camelids on the face, ventral abdomen, lateral thorax, and inguinal region. It is characterized by alopecia, dermal thickening, scaling, crusting, and variable hyperpigmentation. (B) Markedly thickened orthokeratotic hyperkeratosis, abundant keratin accumulation, and hair follicle distention are predominant features. Diffuse perivascular lymphoplasmacytic, eosinophilic, and histiocytic dermatitis are also often present (see eSlide 7.e1).

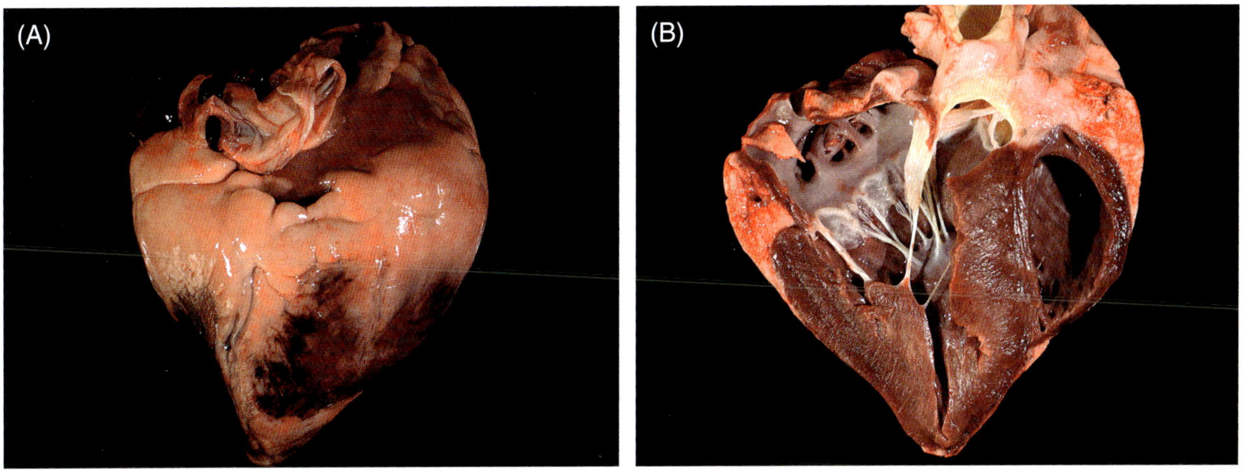

FIGURE 7.7 Oleander toxicosis in an alpaca. (A) Multifocal to coalescing epicardial hemorrhage is seen in the myocardium of the ventricular walls. (B) Associated pallor of the papillary muscle and interventricular septa corresponded histologically (not shown) to necrosis (see eSlide 7.e2).

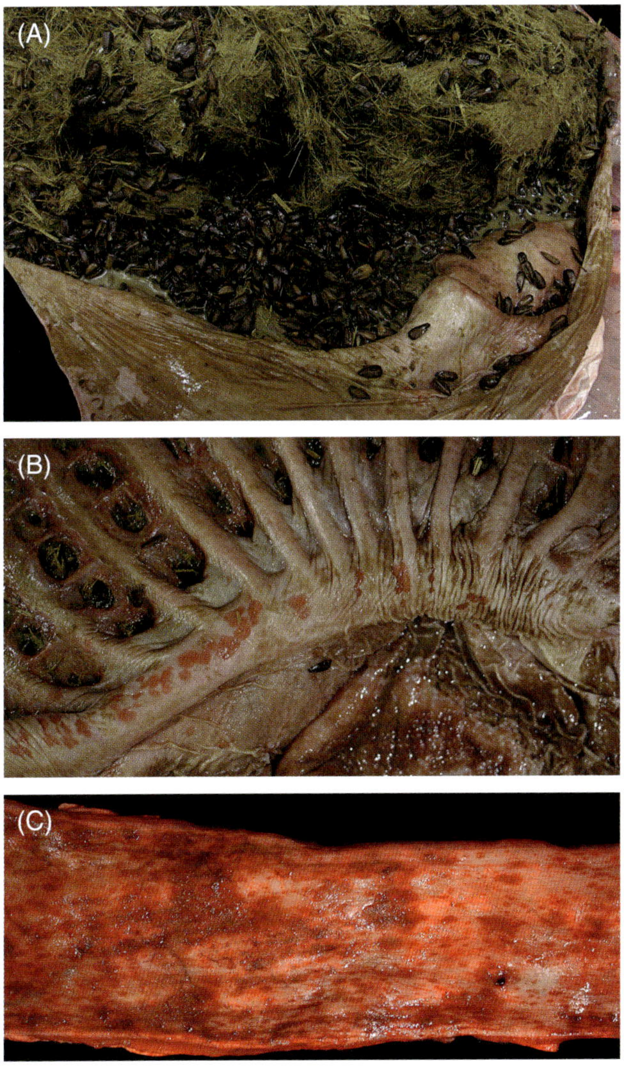

FIGURE 7.8 Black oil sunflower seed toxicosis in an alpaca. (A) Numerous unmasticated sunflower seeds are present in the distended C-1 from an alpaca. (B) Multifocal to coalescing ulcerations in the mucosa of the transversal pillar of C-1. (C) Toxicity can also be associated with severe acute ulceration of the esophageal mucosa. *(Part A: Photo Courtesy of A. Mete, California Animal Health and Food Safety Laboratory System, University of California, Davis.)*

Congenital/Genetic

In North America, NW camelids have been selectively bred for many generations, often from a small founder population, leading to the presence of heritable defects. Teratogens and viral infections as well as nutritional deficiencies are likely additional contributing factors. Fowler has generated an extensive list of defects in NW camelids that have been reported in the literature or have been personally observed or communicated.[45] Of these, the most common are **angular limb deformities, polydactyly, hemivertebra**, and **cranial dysgenesis**. **Congenital deafness** associated with pure white haircoat, and blue iris pigmentation is seen in alpacas. Diagnosis of this condition is usually confirmed by brainstem auditory-evoked potential assessment.[59]

Since crias are obligate nasal breathers, **choanal atresia**, characterized by the presence of a membranous or bony partition between the nasal cavity and the pharynx, is often a lethal defect[45,60] (Figs. 7.9A–D). In some cases, atresia may be incomplete and lead to partial respiratory obstruction. Choanal atresia is often coupled with other facial and skull deformities and appears highly heritable (Figs. 7.9A, B). **Cleft palate** in NW camelids does not appear to be associated with choanal atresia but is typically also a lethal defect (Fig. 7.10). **Pituitary (Rathke's pouch) cyst, wobbler syndrome, wry face, arrhinia, synophthalmia (cyclopia),** and **brachygnathia** have also been rarely identified in camelids (Johnson, Holt and Aboellail; unpublished data) (Figs. 7.11A–I).

Reproductive tract defects are also common in NW camelids. In a Peruvian study, 10% of the animals studied had one or more abnormalities.[45,141] Defects include uterine fistulae or agenesis/hypogenesis of the uterus or vagina (with secondary mucometra and dilation), intersexes, cryptorchidism, penile hypoplasia, or malformation, retained penile frenulum, and testicular cysts.

In contrast, fewer congenital defects have been reported in OW camels, possibly due to less intensive breeding programs or under-reporting of congenital malformations. Reported conditions include **ophthalmic defects,**

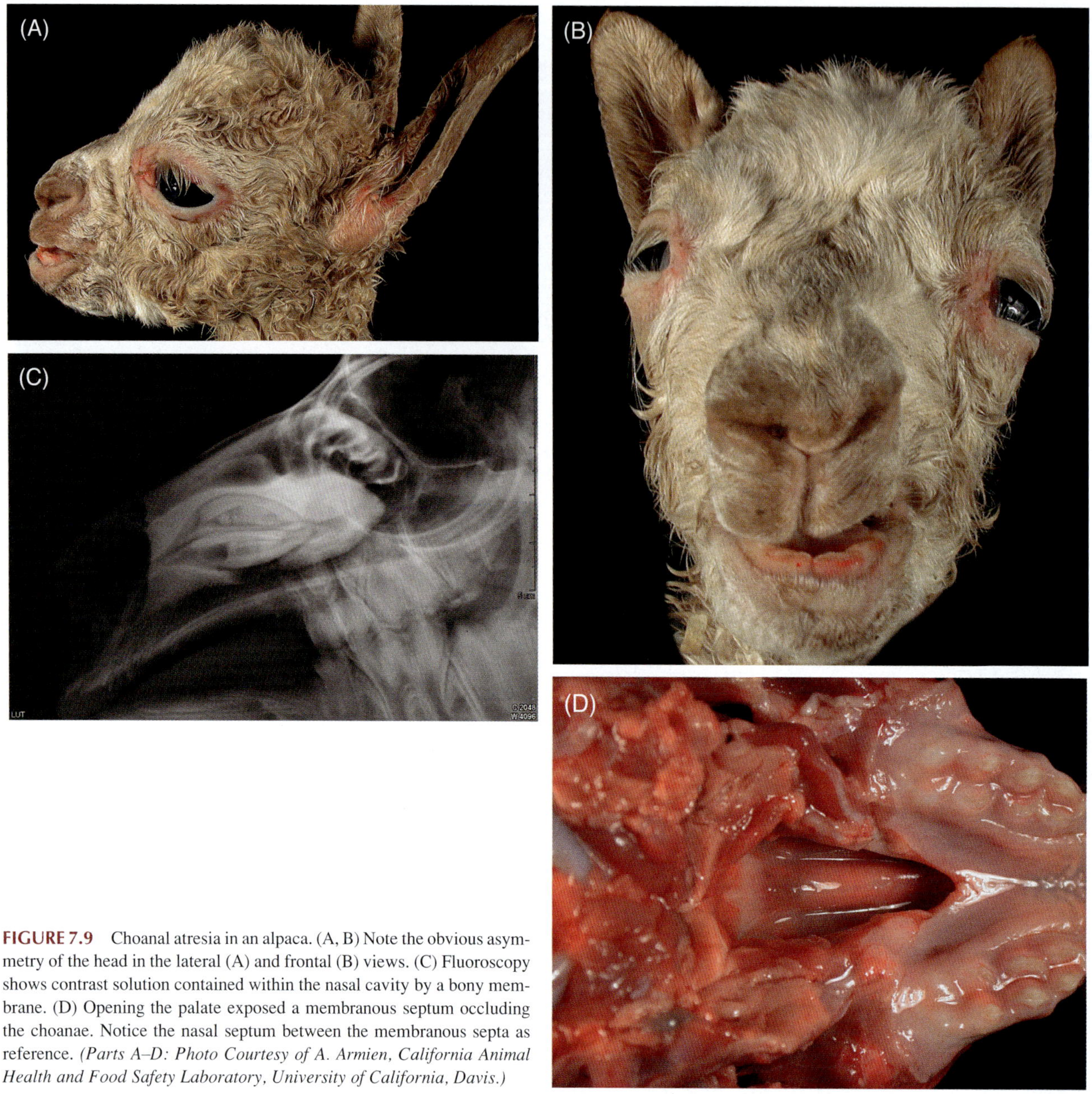

FIGURE 7.9 Choanal atresia in an alpaca. (A, B) Note the obvious asymmetry of the head in the lateral (A) and frontal (B) views. (C) Fluoroscopy shows contrast solution contained within the nasal cavity by a bony membrane. (D) Opening the palate exposed a membranous septum occluding the choanae. Notice the nasal septum between the membranous septa as reference. *(Parts A–D: Photo Courtesy of A. Armien, California Animal Health and Food Safety Laboratory, University of California, Davis.)*

ventricular septal defects, polydactyly, cleft palate, atresia ani, hernia, and **growth retardation**.[20,106,128] In a survey of abattoir samples from 850 female dromedary camels in Nigeria, only 0.12% had cervical hypoplasia or any other congenital reproductive abnormality.[119]

Age-Related/Degenerative

Osteoarthritis and **degenerative joint disease** are common in NW and OW camelids, though no diseases unique to these species have been reported. In older animals, overextension of the metacarpophalangeal or metatarsophalangeal joint (also called **dropped fetlocks**) has been well-described. Copper deficiency in conjunction with zinc excess rather than degeneration or trauma is a potential etiology.[118] It is likely that in other animals, angular limb deformities and poor conformation, possibly coupled with obesity and lack of exercise, lead to degenerative joint diseases such as carpal, tarsal, or hip joint disease (Fig. 7.12) commonly reported later in life.

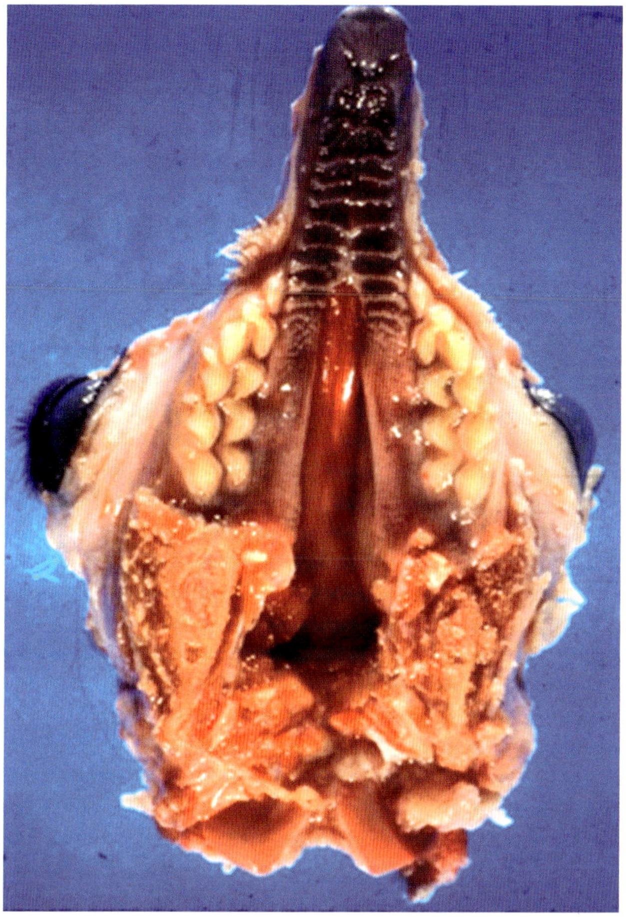

FIGURE 7.10 Cleft palate (palatoschisis) in a newborn alpaca. Note the failure of fusion of the palatine processes resulting in communication between the oral and nasal cavities. This can lead to aspiration during nursing and aspiration pneumonia.

Ulceration and secondary infection of the sternal callosity (Fig. 7.13) occurs in all species of camelids in captivity and is probably related to lameness and prolonged recumbency in older animals. These lesions are characterized by chronic fibrosis, neovascularization, and granulation tissue, as well as pockets of suppurative inflammation. In OW camels, a bony protuberance underlies this callosity, and it can become chronically inflamed, infected, and undergo bony remodeling and irregular proliferation secondary to the extension of the overlying inflammatory process.

Abnormal tooth wear is another commonly observed age-related change, particularly in NW camelids. Congenitally deformed dental structures can lead to elongated teeth, uneven wear, broken teeth, and laceration of the oral mucosa; this can occur even at a young age. Heredity is likely involved in deformities of the oral cavity as it is in other mammalian species. Clinical monitoring and management are critical because emaciation and chronic bacteremia can result from impaired mastication.

Inflammatory Non-Infectious

Autoimmune thyroiditis has been reported in dromedary camels in association with hypothyroidism and thyroid hormone autoantibodies.[13,14] It has also been associated with *Trypanosoma evansi* infection, a possible predisposing factor, and lowered fertility in males may also be associated. Further investigation into this condition is needed to better elucidate the mechanisms involved and the associated pathology.

Ovarian hydrobursitis is a unique condition described in dromedary camels. It is reported in up to 15% of females at slaughter, and in up to 33% of infertile camels.[11,145] The disease may be unilateral or bilateral and is most often seen in the left ovary, though studies vary in the specific percentages of unilateral versus bilateral lesions. Grossly, the bursa is usually mildly to markedly distended by hemorrhagic fluid (Fig. 7.14). Histologically, there is degeneration and hyperplasia of the bursal epithelial lining; infiltration of the lining by variably intense aggregates of lymphocytes, plasma cells, and macrophages; cysts within the epithelial lining; multiple variably expansive regions of hemorrhage; and intralesional hemosiderin-laden macrophages. The high estrogen content of the bursal fluid in this condition suggests it is a mixture of hemorrhage and follicular fluid (the estrogen content of the distended bursa tends to be much higher than serum). While a specific cause has not been proven, over 86% of camels with hydrobursitis have serological evidence of *Chlamydia abortus* infection, suggesting this organism may be involved in the pathogenesis.[10] Oxytetracycline, an antichlamydial drug, returns some of the fertility to affected animals, further supporting the theory that this is, at least in part, an infectious disease syndrome.

Miscellaneous/Multifactorial

While OW camels are very resistant to **hyperthermia**, NW camelids, perhaps because of their evolutionary development at high altitudes and cool temperatures, are more susceptible. A history of high environmental temperatures, high humidity, physical activity, social stress, dehydration, and/or fever in a heavily fleeced down, moribund, or dead NW camelid is highly suggestive of hyperthermia or heat stress. Tissues will rapidly autolyze under these conditions, so prompt postmortem inspection is critical to make a diagnosis. Lesions suggestive of hyperthermia include petechial hemorrhages, hyperemia, thrombosis, pulmonary edema and congestion, neuronal necrosis, hepatic necrosis, and renal tubular necrosis.[51] **Hypothermia** can also be a concern in NW and OW camelids kept in northern climates where frostbite can occur; it primarily affects the ears. Some NW camelids have congenitally shortened ears ("gopher ear") that can be distinguished from healed frostbite by the presence of

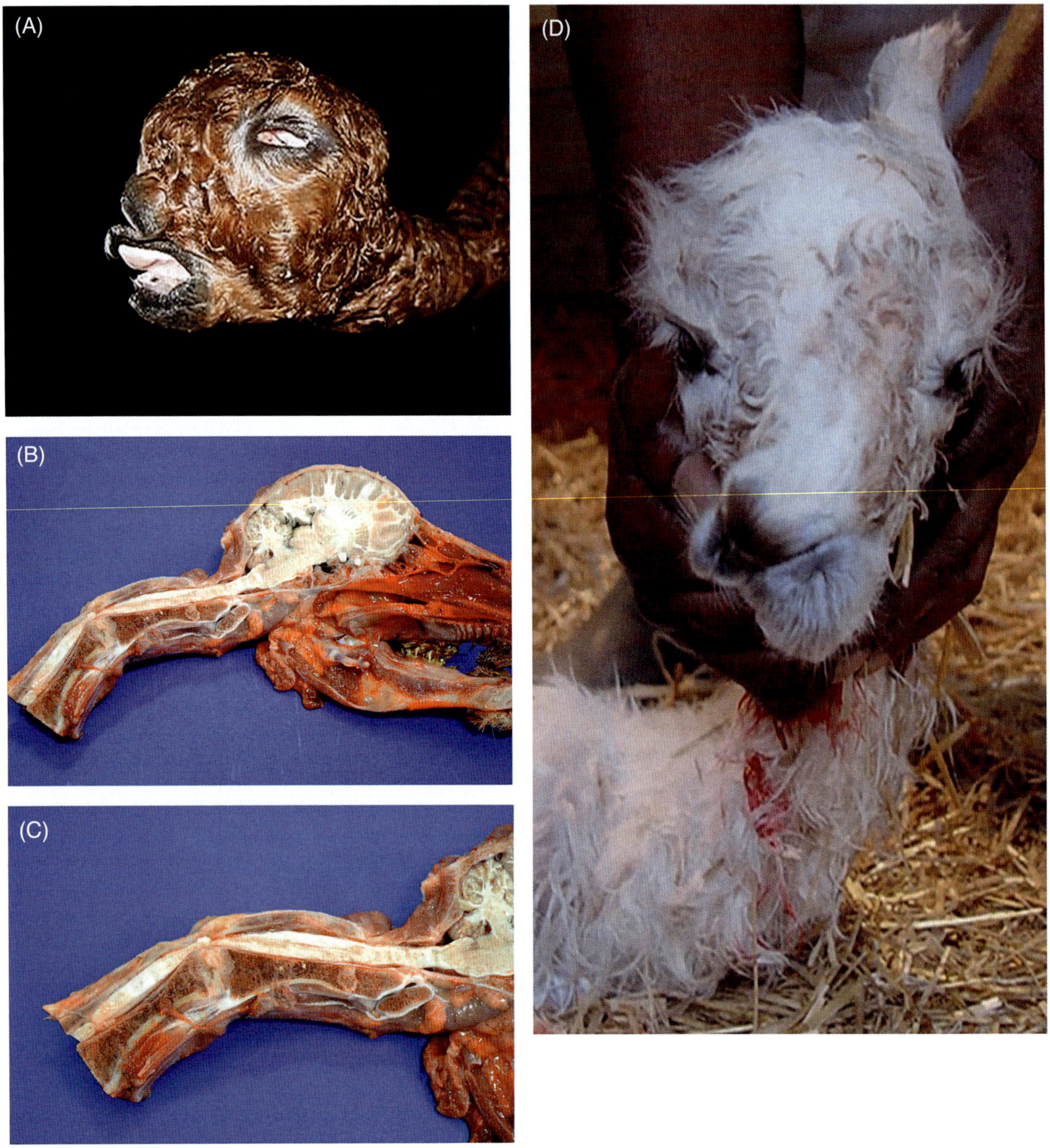

FIGURE 7.11 Congenital facial deformities in alpaca (A–C), llama (D, H), and camel (I). (A) Facial deformity, anotia, and arrhinia. (B, C) Severe stenotic myelopathy at the level of C1–2 resulted in progressive neurological symptoms. (D, E) Rostral deviation of the maxillae. (F) Brachygnathia superior and absence of the nares (arrhinia). (G) Synophathalmia (cyclopia) with brachygnathia and protrusion of the tongue. (H) Severe prognathism. (I) Vertebral scoliosis and lordosis (wry neck). *(Part A: Photo Courtesy of L. W. Johnson, Colorado State University.)*

(Continued)

fibrosis and alopecia at the squared margin of the latter, distinct from the haired and tapered tip of the former.[51]

Camelids can develop many of the same non-infectious skin diseases that are seen in other livestock. Two relatively unique conditions occur in NW camelids: **idiopathic superficial necrolytic dermatitis** and **zinc-responsive dermatosis** (see above). Idiopathic superficial necrolytic dermatosis is characterized by nonpruritic, waxing and waning erythematous, and crusting vesicles that are typically present in the axilla, inguinal area, ventral abdomen,

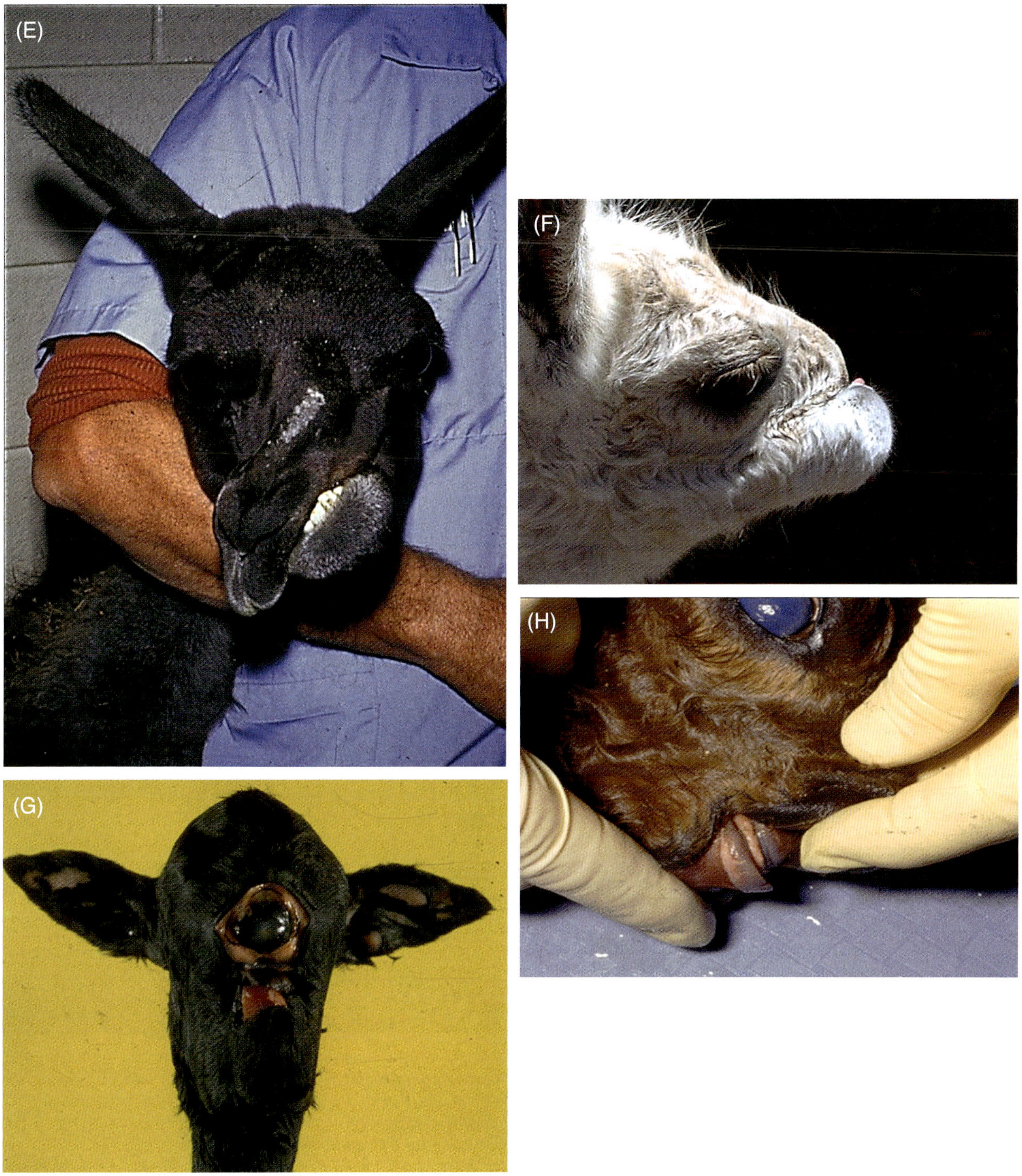

FIGURE 7.11—Cont'd

(Continued)

distal extremities, perineum, and on the face. Intracorneal pustules and suppurative epidermitis with ballooning degeneration are associated with orthokeratotic and parakeratotic hyperkeratosis. These lesions sometimes form the characteristic laminar distribution of "red, white, and blue" lesions: superficial parakeratosis ("red") overlying keratinocyte vacuolation and spongiosis ("white"), and basilar epidermal cell hyperplasia ("blue") like those seen in other species. An etiology has not been determined but it may be multifactorial.[50,127]

FIGURE 7.11—Cont'd

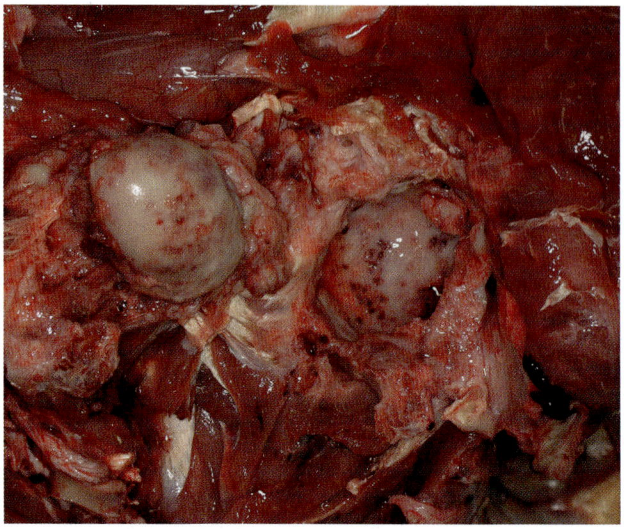

FIGURE 7.12 Degenerative joint disease in a llama. Multifocal to coalescing ulceration is present in the articular cartilage and acetabulum of the hip joint. Additional changes include the formation of bone spurs and synovial proliferation. Degenerative joint disease is a common skeletal disease in geriatric camelids. *(Photo Courtesy of L. Woods, California Animal Health and Food Safety Laboratory, University of California, Davis.)*

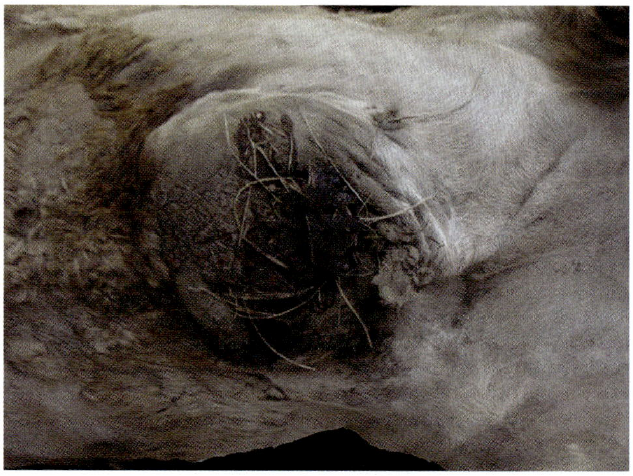

FIGURE 7.13 Ulceration of the sternal callosity in a dromedary camel. The sternal callosity (pedestal) contains a large central cavitation with hemorrhage that is covered by crusted, serocellular debris. In older NW and OW camelids, this condition is due to prolonged recumbency or poor substrates. In OW camels, a bony protuberance underlies this callosity and extension of inflammation can lead to osteomyelitis.

Gastric ulceration (C-1 through C-3) is a common lesion that can be a primary disease process or develop in association with other disease states.[51] In C-1 and C-2, ulcerations are typically noted along the margins of saccules, primarily in the glandular regions (Fig. 7.15A). In C-3, ulcers are typically linear along the longitudinal pleats in the proximal four-fifth but are focal and deeper in the last one-fifth (truly glandular area of C-3). Expansive ulcers may occur at the margin of compartments C-2 and C-3. These are most often seen in neonates with hypothermia, sepsis, or during periods of social or metabolic stress. Ulcers in the distal part of C-3 are the most likely to perforate, leading to peritonitis and rapid death (Figs. 7.15B, C). Underlying causes of ulcer development are poorly understood and likely multifactorial. Social and environmental stresses and associated catecholamine secretion have been suggested. Hypercortisolism is associated with acid hypersecretion in the gastric lumen of other species and is presumed to also be the case in camelids. Duodenal reflux and gastric vasoconstriction may also be factors in stressed animals. There is no acid secretion in C-1, so ulceration in this area cannot be explained by this mechanism. Other lesions that might be associated with

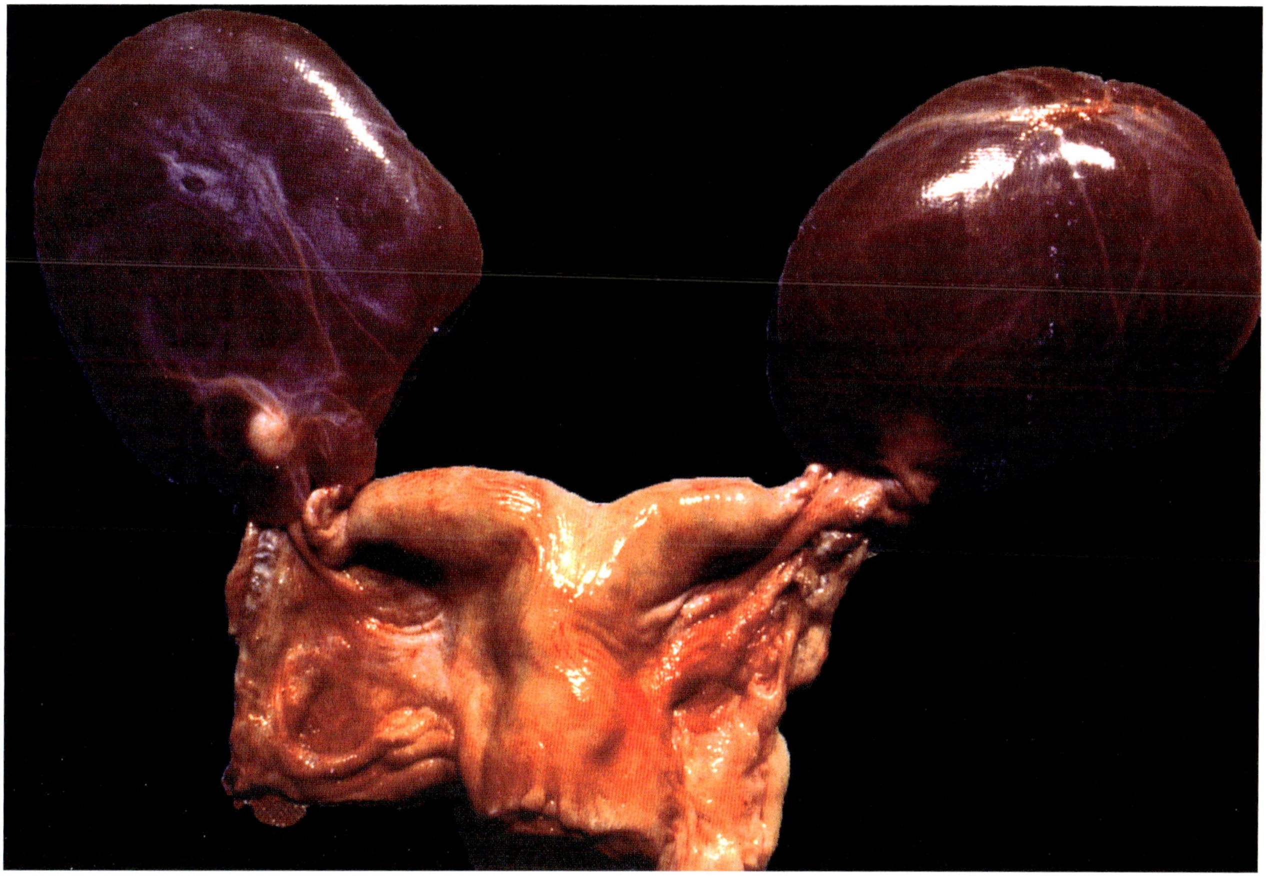

FIGURE 7.14 Ovarian hydrobursitis in a dromedary camel. This is a unique condition seen exclusively in dromedary camels. The ovarian bursa is mildly to markedly distended and filled with a mixture of follicular fluid and hemorrhage. The condition may be unilateral or bilateral. *(Photo Courtesy of A. Tibary, Washington State University.)*

stress-induced gastric ulceration are lymphoid atrophy and adrenal cortical hyperplasia; the latter used to build a case for stress-induced disease when ulceration is noted at postmortem examination.

Megaesophagus, an excessive dilation of the esophagus, is occasionally observed in llamas and alpacas with different degrees of severity (Fig. 7.16). Salivation is the most common clinical sign, but regurgitation due to dysphagia might be noted. In crias, megaesophagus has been associated with the presence of vascular ring anomalies.[102] A study that described megaesophagus in 15 llamas identified organophosphate toxicity to be the cause in one and vagal neuropathy or degenerative myopathy of unknown cause in 3; etiologies in the remaining animals were undetermined.[156]

Phrenic nerve neuropathy and diaphragmatic paralysis have been diagnosed in alpacas with variable degrees of dyspnea and respiratory distress.[23,151] Affected animals were not genetically related and less than 1 year of age. Microscopically, denervation atrophy of the diaphragmatic myofibers is seen along with axonal degeneration of both phrenic nerves characterized by dilation of the myelin sheaths with necrotic debris and macrophages (digestion chambers) and occasional swollen axons (Fig. 7.17A; eSlide 7.e3). Besides the diaphragm, collection of the whole extension of the phrenic nerves as well as other nerves (e.g., sciatic, vagus, etc.) from the same animal for comparison (Fig. 7.17B) are needed for accurate diagnosis. The cause and possible pathogenesis of this neuropathy is unknown.

Several other, somewhat common necropsy findings have been reported in camelids. One is **endogenous lipid pneumonia**. One report noted that it was confined to the dorsal surface of the diaphragmatic lung lobes (Fig. 7.18A). Typical histologic findings were similar to what occurs in other species and consisted of clusters of foamy macrophages with cholesterol clefts in the subpleural alveoli (Fig. 7.18B).[66] Another noted finding in a second report was subacute to chronic **thrombosis of the splenic vein** (Fig. 7.19). The thrombi were not associated with clinical signs and no evidence of an underlying infectious cause was seen.[68]

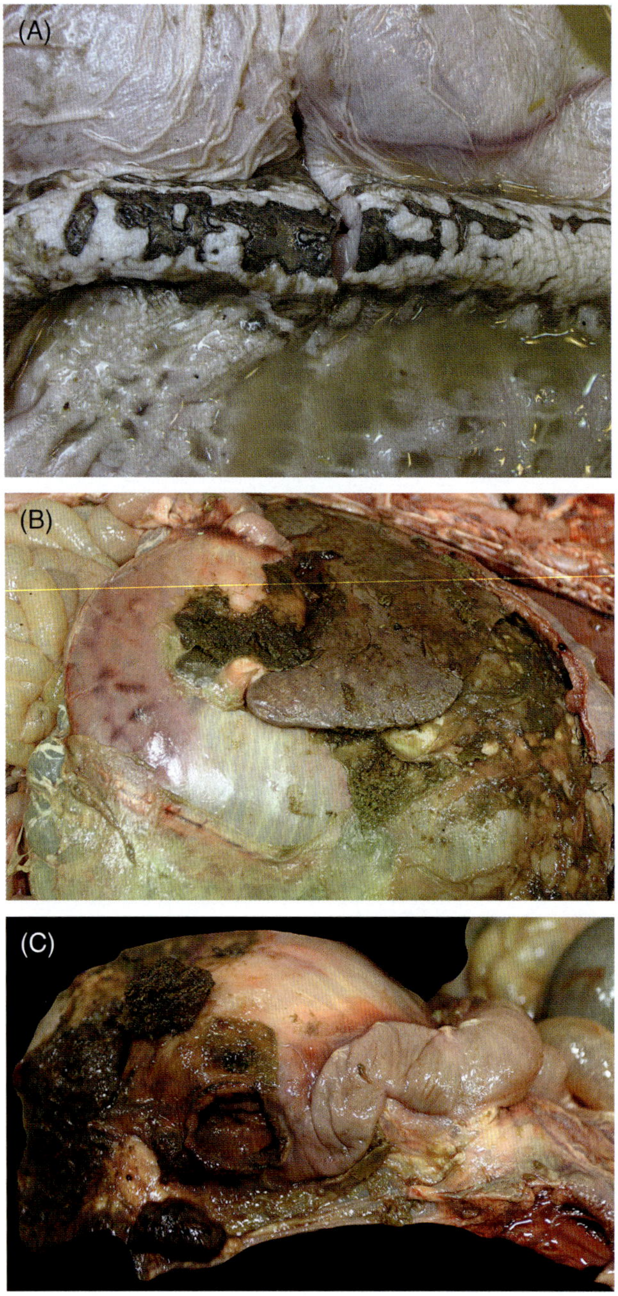

FIGURE 7.15 Gastric ulceration in an alpaca and llama. (A) Multifocal to coalescing, variably and irregularly shaped, sharply demarcated ulcers with cavitation and gray/black discoloration are present along pillars and at the margin between saccules in the glandular regions. Compartments C-1 and C-2 are most commonly affected. (B) Gastric ulceration presents grossly as regionally extensive red to purple discoloration in the wall of C-3. In this case, it was associated with intraperitoneal gastric content. (C) The latter was due to transmural ulceration, seen here as focal, crateriform cavitations in the gastric wall, with perforation of the ulcers in the distal (caudad) portion of C-3.

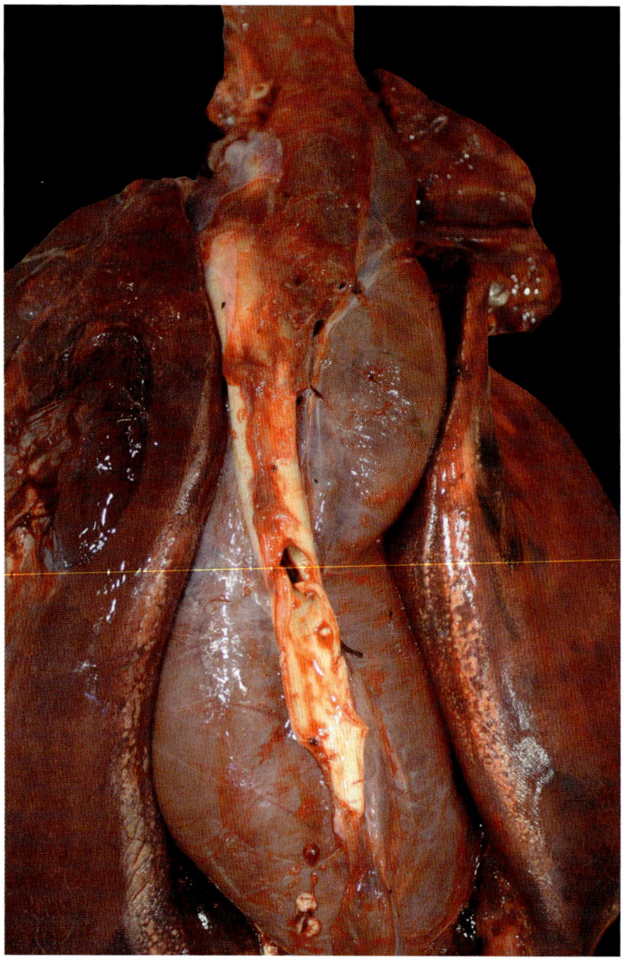

FIGURE 7.16 Megaesophagus in an alpaca. The esophagus is markedly dilated and appears as a saccular structure between the lungs and below the aorta. The causes of the megaesophagus are numerous in camelids; most are idiopathic.

Neoplastic

Neoplasia is not common in camelids. Relatively few reports exist in the literature, though two reviews of cases in diagnostic laboratories put the prevalence at 8.5% for all camelid submissions and 11% and 4.9% in llamas and alpacas, respectively.[4,51,152] It is unlikely that this is due to decreased reporting in these species, as there is considerable veterinary and abattoir attention to camelids.

Of the reported cancers, the most common is **lymphosarcoma (syn. lymphoma)**.[3,27,79,97,122,135,136,149] Immunohistochemistry using antibodies against CD3 (for T-cells), PAX-5, BLA36, CD79α (for B-cells), CD18, and synaptophysin are useful in differentiating lymphoid neoplasms in alpacas and llamas.[3] Lymphomas that do not label with either T- or B-cell markers may be non-B, non-T-cell lymphomas, or primitive malignant round cell tumors (PMRCT).[3,97] Interestingly, three cases (one llama and two alpaca) were reported from Canada in animals 6–7 months of age; other reports also describe young camelids (especially alpacas) with disseminated lymphosarcoma.[3,79] Except for bovine leukemia virus-induced lymphosarcoma, confirmed by in situ hybridization and serology in a 13-month-old alpaca,[92] no underlying etiologies or

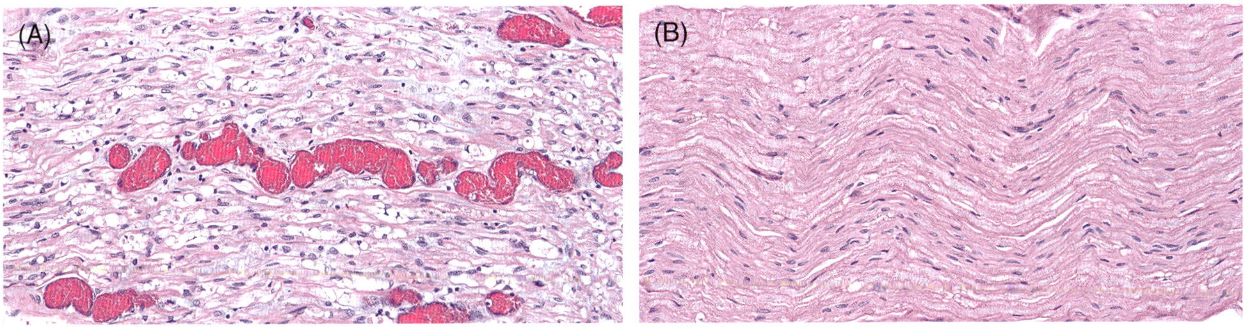

FIGURE 7.17 Phrenic nerve neuropathy and diaphragmatic paralysis in alpacas. (A) The nerve contains numerous digestion chambers and macrophages that contain phagocytized myelin debris. Blood vessels are congested and lined by plump (reactive) endothelium. (B) Normal sciatic nerve from the same alpaca (A) collected for comparison (see eSlide 7.e3).

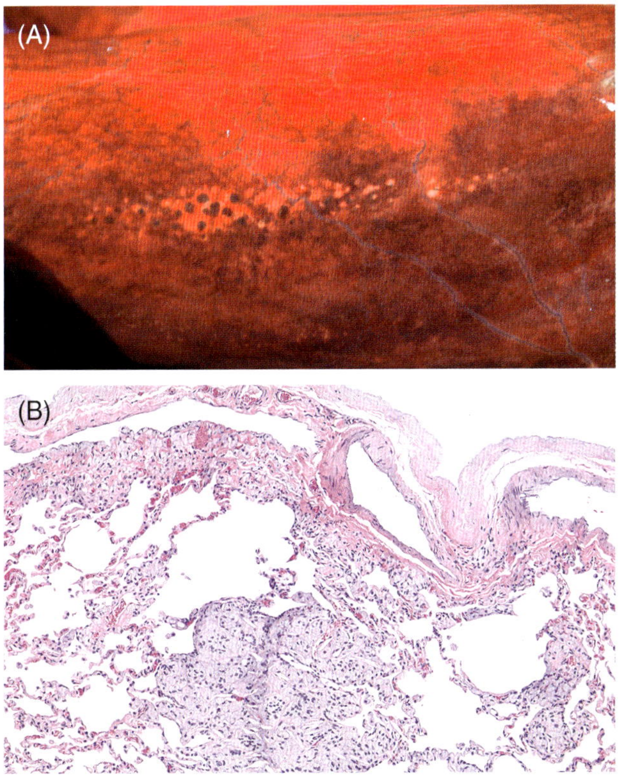

FIGURE 7.18 Endogenous lipid pneumonia. (A) Multiple, tan-white foci of discoloration are present in the subpleural tissue. (B) This corresponds histologically to aggregates of foamy macrophages with intralesional cholesterol clefts that form nodules in the subpleural alveoli.

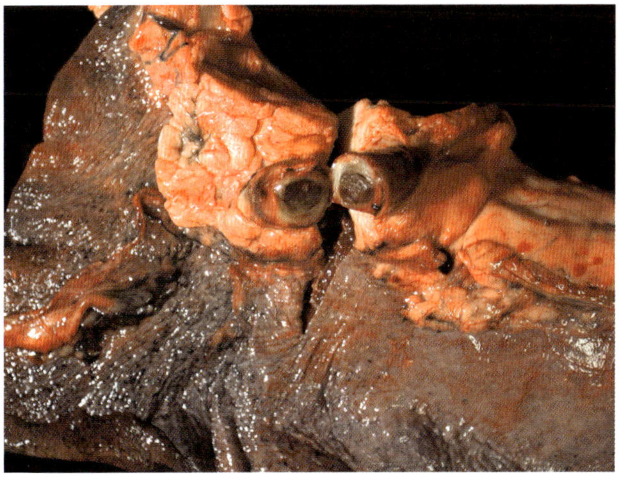

FIGURE 7.19 Thrombus in the splenic vein in a llama. Transverse section of the splenic vein showing the complete occlusion of the lumen by a subacute or chronic thrombus.

predisposing factors have been confirmed in lymphosarcoma (or any other cancer) of camelids. No trend toward any particular type of lymphosarcoma has been confirmed in camelids.

Mucocutaneous fibropapillomas are common in NW camelids. One study demonstrated papillomavirus in multiple mucocutaneous fibropapillomas.[132] Efforts have been made to look for viruses in other tumors via electron microscopy without success.

Other cancers that have been reported in camelids are listed in Table 7.1.

TABLE 7.1 Reported Neoplasms in Camelidae

Neoplasm	Species of Camelid
Epithelial tumors	
Ameloblastoma[80]	Llama
Basal cell carcinoma[8]	Dromedary camel
Cholangiocarcinoma[142]	Llama, alpaca
Cutaneous squamous cell carcinoma[121]	Llama, alpaca
Gastrointestinal adenocarcinoma[90]	Llama, alpaca, Bactrian camel
Gastrointestinal squamous cell carcinoma[30]	Llama, alpaca
Mammary carcinoma[19,22]	Llama, alpaca, dromedary camel

Continued

TABLE 7.1 Reported Neoplasms in Camelidae—cont'd

Neoplasm	Species of Camelid
Nephroblastoma[35]	Guanaco
Ocular medulloepithelioma (nonteratoid)[131]	Llama
Pulmonary carcinoma[22,115]	Llama, dromedary camel
Renal cell carcinoma[155]	Dromedary camel
Retinoblastoma[56]	Llama
Trichoepithelioma (multiple)[140]	Alpaca
Uterine adenocarcinoma[89]	Llama
Mesenchymal tumors	
Ameloblastic odontoma[138]	Llama
Astrocytoma[58]	Llama
Chondrosarcoma[82]	Dromedary camel
Embryonal rhabdomyosarcoma[64]	Alpaca
Fibromas/fibropapillomas[132,152]	Llama, alpaca
Fibrosarcoma[71]	Llama, alpaca
Hemangiosarcoma[67]	Llama
Lipoma[152]	Llama
Meningioma[104]	Bactrian camel
Ossifying fibroma[99]	Llama
Osteosarcoma[148]	Dromedary camel
Peripheral nerve sheath tumor[87]	Dromedary camel
Rhabdomyosarcoma[64,169]	Dromedary camel
Uterine leiomyosarcoma[69]	Alpaca
Round cell and other tumors	
Granulosa cell tumor[16]	Llama
Histiocytic sarcoma[104]	Bactrian camel
Interstitial cell tumor (ovary)[152]	Alpaca
Malignant melanoma[105]	Alpaca
Mast cell tumor[93]	Llama
Melanocytoma[105]	Llama
Mesothelioma[163]	Dromedary camel
Neuroendocrine tumor[130]	Alpaca
Ocular melanoma[73]	Alpaca
Pituitary adenoma[29,62]	Llama, alpaca
Seminoma[9]	Dromedary camel
Teratoma[74,103]	Alpaca, dromedary camel

INFECTIOUS DISEASES

Viral, bacterial, fungal, algal, and parasitic diseases have been described in NW and OW camelids. Prion diseases have not been diagnosed in camelids, and camelids are considered low risk for bovine spongiform encephalopathy (BSE), though no studies have looked at transmission and only one has examined prion protein gene polymorphisms in camels.[166]

DNA Viruses

Equine *Herpesvirus* (EHV-1), a DNA virus in the *Herpesvirus* family, has been frequently reported in NW and OW camels, and has been experimentally reproduced in llamas from an alpaca isolate.[76] Infection in camelids likely develops as a "spill-over" from infected cohoused equids. In camelids, clinical signs are typical of equine neurological herpes viral infection. Edema, neuronal necrosis, and perivascular and meningeal lymphocytic inflammation are usually present.[76] Eosinophilic intranuclear inclusion bodies are typically present in endothelium and neurons. Clinical neurologic signs include blindness, which appears particularly common. Histologic lesions include necrotizing retinitis with retinal detachment, hemorrhage, choroiditis, vitritis, and optic neuritis characterized by edema and gitter cell infiltration.[117] Abortion has not been reported in camelids due to EHV-1. Vaccination has been proposed but there are no published studies on the effectiveness in the prevention of infection or disease.

Camelpox is caused by **Camelpox virus**, a DNA virus in the *Poxviridae* family and *Orthopoxvirus* genus. It is reportable to the World Organization for Animal Health (WOAH) and is the most common viral disease in dromedary and Bactrian camels. As in some poxviruses from domestic species, camelpox is zoonotic and capable of infecting immunocompetent humans, though it is more likely to cause severe disease in immunocompromised individuals. With electron microscopy this virus appears as brick-shaped, enveloped particles with rounded edges measuring 360 µm × 270 µm × 250 µm. Parapoxviruses can be differentiated by their more ovoid shape than camelpox. Camelpox has been experimentally reproduced in NW camelids, though it is not present in the Americas or Australia. Transmission is usually due to direct exposure through skin abrasions or inhalation of infected secretions from lesions, milk, saliva, blood, ocular or nasal secretions, or urine. Arthropods, including flies and ticks, are suspected vectors. Upon infection, the virus replicates locally, producing a primary erythematous lesion. The infection then spreads to local lymph nodes and a leukocyte-associated viremia develops.[18] During this time virus can be identified in skin, turbinates, lungs, and lymphoid organs. A few days after viremia begins, widespread secondary skin lesions (typical pox) appear and persist for 2–3 days until viremia subsides. Like other poxviral diseases, camelpox virus encodes multiple genes that modulate the host immune response by interfering with interferon (IFN), proinflammatory cytokines [Interleukin (IL)-1b, IL-18 and tumor necrosis factors (TNFs)], chemokines, and complement.[39]

Severity of disease is dependent on viral strain, the immune status of the infected host, and the age of the host (younger animals are much more susceptible). Typical pox lesions begin as erythematous macules that develop into papules and vesicles and progress to pustules that rupture, leading to crust formation (Figs. 7.20A, B). Lesions typically affect the head, nostrils, ear margins, and eyelids (Fig. 7.20C). Disseminated pox lesions are rarely described (suspected to be related to stress by weaning or poor husbandry) but have been reported in other areas besides of the skin, like trachea, lung, and eyes.[2] Mucous membranes, including the lips, and oral and nasal cavities, can be affected and infection may also spread to other parts of the body, including the genitalia and mammary gland. In the most severe cases, proliferative necrotizing lesions in the oral cavity or respiratory or GI tracts may lead to death. Healing can take up to 4–6 weeks and can leave scars or cause multifocal alopecia.[18] Clinical signs in these animals include hypersalivation, anorexia, mucopurulent nasal and ocular discharge, and diarrhea. Abortion can also occur in pregnant females. Death is often due to secondary bacterial infection and subsequent septicemia.

Histopathology is typical for poxviral lesions: cytoplasmic swelling and ballooning degeneration of keratinocytes in the stratum spinosum of the epidermis with subsequent rupture and vesicle formation, lymphoplasmacytic perivascular cuffing and infiltration of neutrophils and eosinophils in the dermis, and prominent epithelial hyperplasia at the border of the epidermal lesions. It is notable that, unlike many other poxviral lesions, inclusions are not a diagnostic feature of poxvirus infection in camels. Lung lesions are characterized by bronchial epithelial cell proliferation with vacuolation, proliferative alveolitis and bronchiolitis, necrosis, and in older lesions, fibrosis. Immunohistochemistry, in situ hybridization, and PCR can all be used to confirm the diagnosis, and transmission electron microscopy (TEM) on scabs or tissue can be used to differentiate the brick-shape of the orthopoxvirus from clinically relevant differentials, importantly contagious ecthyma ("Orf;" a parapoxvirus), papillomatosis (papillomaviruses) and insect bites. After infection, immunity to camelpox is life-long.

Parapoxvirus infection (also called **contagious ecthyma**, **contagious pustular dermatitis**, or **"Orf"**), caused by a double-stranded DNA virus in the family *Parapoxviridae*, is well-recognized in both OW and NW camelids. The epidemiology, pathogenesis, and lesions are very similar to those seen in small domestic ruminants.

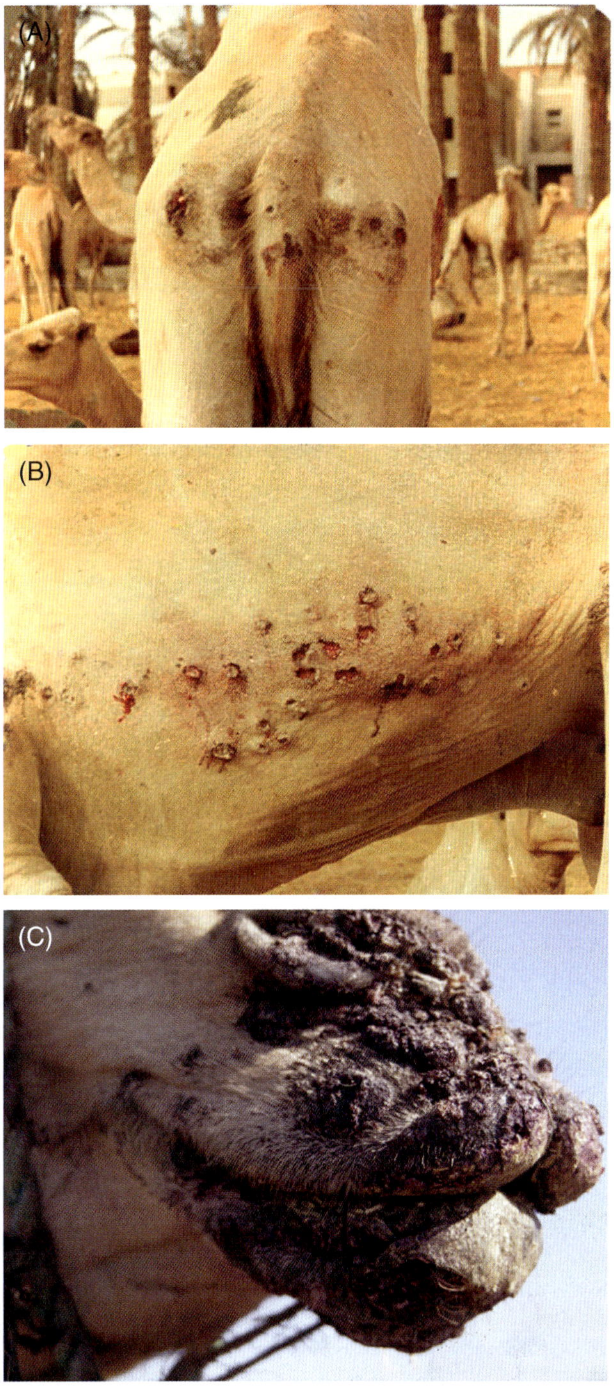

FIGURE 7.20 Cutaneous camelpox virus infection in dromedary (A) and Somali (B, C) camels. (A, B) Lesions in the skin begin with erythematous macules that develop into papules and vesicles. These progress to pustules that rupture and become covered by crusts. Lesions often affect the head, nostrils, ear margins, and eyelids. (C) Camelpox affects the mucous membranes, including the lips and oral and nasal cavities, making it difficult for animals to eat or nurse. *(Part C: Photo Courtesy of A. Tibary, Washington State University.)*

Typical lesions are characterized by marked epidermal proliferation with superficial necrosis that is most common on the oral commissures and lips of crias and calves, as well as on the teats and perineum of dams. Microscopically, there is swelling and vacuolation of keratinocytes, epidermal proliferation, suppurative inflammation, and a marked accumulation of crusts. Eosinophilic intracytoplasmic inclusions appear by 2–3 days postinfection and persist for 3–4 days. Lesions in camelids tend to persist longer than those seen in goats and sheep. Transmission is likely associated with shared housing with small ruminants. This is a zoonotic disease, and there are reports of humans also infecting llamas (anthroponotic).[49]

Fibropapillomatosis, caused by a DNA virus in the family *Papillomaviridae*, is relatively common in OW camels. In dromedary camels, it is associated with two identified papillomaviruses, *Camelus dromedarius* papillomavirus types 1 (CdPV1) and 2 (CdPV2).[150] Lesions present as large, round, cauliflower-like, pedunculated masses that, in their mature stage, are unlikely to be confused with other lesions but, in early stages, may be confused with parapox or camelpox lesions. Microscopically, fibropapillomas are exophytic, papillary projections with hyperkeratosis, acanthosis, ballooning degeneration, downgrowth rete ridges, and dermal fibrosis. Fibropapillomas in NW camelids (both alpaca and llamas) are associated with a unique papillomavirus that shares only 73% homology to bovine papillomavirus.[132] These lesions resemble equine sarcoid, with a fibroblastic proliferation in the dermis, hyperplasia of the overlying epithelium, and prominent tendrils of rete ridges. Inclusions are seen in the fibropapillomas of neither OW nor NW camelids.

RNA Viruses

Bovine viral diarrhea virus (BVDV) is an RNA virus in the *Flaviviridae* family and *Pestivirus* genus. Two BVDV genotypes, BVDV-1 and BVDV-2, each with multiple subtypes, are recognized. Many ruminant species, including OW and NW camelids, can be infected. Clinical signs include immunosuppressive and respiratory and reproductive tract diseases. In utero, infection can lead to the birth of persistently infected (PI) offspring. In NW camelids, in utero infection with subsequent abortion or birth of PI offspring are the most described outcomes. Infection in alpaca has been associated with abortion and BVDV-positive fetal tissues, or chronically ill, stunted, and immunosuppressed crias that may be seronegative, but positive for BVDV by PCR or virus isolation.[24,98,108] Other outcomes include stillbirth, neonatal death, and congenital deformities; chronic respiratory or GI disease have also been reported. Disease has not been described in OW camelids, but both serological response and molecular presence of BVDV-1 in both Bactrian and dromedary camels confirm exposure.[2,57,78] Both BVDV type 1 and 2 were isolated from a dromedary camel herd that experienced reproductive failure and perinatal congenital anomalies. Experimental infection of dromedary camels resulted in C-3 ulceration and diarrhea and one of the infected camels was euthanized due to poor body condition (Figs. 7.21A–C).[2,168]

A new substrain of BVDV (BVDV-1q) was first identified in Bactrian camels and later identified in dairy cattle and swine in China.[57]

Pathogenesis has been investigated in experimental infections in both OW and NW camelids. An experimental BVDV-1 infection study in alpacas less than 2 years of age showed lymphocytopenia and a decrease in peripheral lymphocytes.[137] GI-associated lymphoid tissue (GALT) B-cells were depleted, while parafollicular and submucosal T-cells remained. Viral antigen was identified by immunohistochemistry in GALT of ileum, proximal colon, and C-3; it was also variably present in other lymphoid tissues. Histologic depletion of lymphoid tissues correlated with the presence of antigen. Viremia was present in all infected alpacas by day 5 and as early as day 3 postinfection. Nasal shedding was present in all infected alpacas. None of these experimentally infected alpacas were clinically ill.

Antigen distribution has been reported in PI alpacas and is widespread in neurons, endothelial cells, and vascular tunica media myocytes, less in epithelium of multiple organs, and still less in lymphoid tissues.[70] Macrophages were also strongly labeled in the intestinal submucosa and medullary sinuses of lymph nodes. Antigen distribution was similar in reproductive organs to that seen in cattle, suggesting possible venereal transmission and impact on fertility. Unlike most bovids, antigen was only rarely found in the bone marrow.[2]

Peste des petits ruminants virus (PPRV) is the *Morbillivirus* of small ruminants (similar in pathogenesis and lesions to the now eradicated rinderpest in bovines) and has been reported in dromedary camels in the Middle East and East Africa. Common clinical signs include fever, dehydration, oral ulcers, diarrhea, lymphadenopathy, dermatitis, corneal ulceration, conjunctivitis, and respiratory distress. Pulmonary congestion, edema, and pneumonia; hepatic necrosis; renal tubular necrosis; and multifocal necrotizing dermatitis and mucosal ulcerations are common necropsy findings. One of the most remarkable clinical signs observed in camels is keratitis and conjunctivitis, which has also been seen in wild small ruminants.[86,170]

Foot and mouth disease (FMD) (also called **hoof and mouth disease**) is caused by **food and mouth disease virus,** a member of the genus *Aphthovirus* in the family *Picornaviridae*. There are 7 types and over 60 subtypes. This is a WOAH reportable disease that has major impacts on livestock trade worldwide. Susceptibility varies markedly between species of camelids, making the role of camelids in the epidemiology and regulation of this disease controversial. Dromedary camels are considered essentially resistant to FMD infection, with no confirmed reports of natural infection and no seroconversion, even

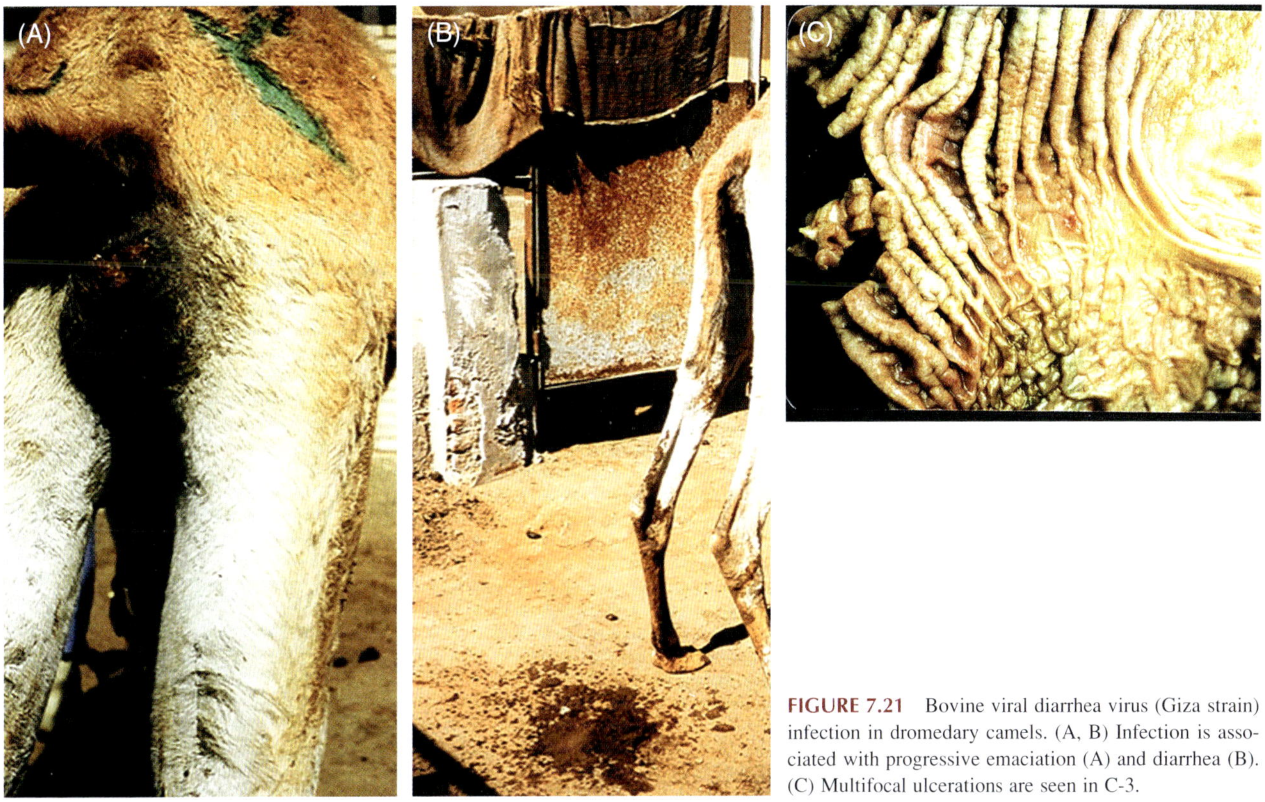

FIGURE 7.21 Bovine viral diarrhea virus (Giza strain) infection in dromedary camels. (A, B) Infection is associated with progressive emaciation (A) and diarrhea (B). (C) Multifocal ulcerations are seen in C-3.

in the face of ongoing outbreaks in closely associated cattle.[160] Bactrian camels, in contrast, are susceptible and may play a small but significant role in the epidemiology of the disease in Central Asia. NW camelids can contract the disease both naturally and experimentally, but they are significantly less susceptible than cattle and sheep and stop shedding 14 days postinfection; they do not remain carriers after infection.[158] Lesions are similar to those described in ruminants: vesicle formation followed by erosions and ulcers on the tongue, dental pads, oral cavity, palate, lips, nostril, coronary band, interdigital space with undermining of the footpads, teats, and the pillars of C-1 and C-2.

Vesicular and erosive lesions similar to those of FMD are present in **vesicular stomatitis (VS),** a disease caused by an RNA virus in the *Rhabdoviridae* family (*Vesiculovirus* genus) that has a broader host range than FMD. This is an internationally recognized, reportable disease, but it is not a WOAH-listed reportable disease. New World camelids likely play a very small to no role in the epidemiology of the disease. Only one natural case of VS has been reported in the literature in NW camelids and none in OW camelids. Experimental infections require inoculation directly into the dorsum of the tongue and produce localized vesicles, fever, and recumbency.[49]

Borna disease virus (*Bornaviridae* family, *Bornavirus* genus) affects multiple taxa. Infection and disease are often associated with horses and donkeys in central Europe, but it has also been identified in captive NW camelids.[15] The virus is likely tick transmitted, but oral transmission has also been demonstrated in horses. Infection in llamas and alpacas can cause anorexia and weight loss, loss of libido, and stretching convulsions.[81] No gross lesions are apparent, but histologically, there is nonsuppurative encephalomyelitis, neuronal degeneration, gliosis, and perivascular cuffing in the retina, brain, cervical spinal cord, and trigeminal ganglion. The neurons in the hippocampus might have pathognomonic eosinophilic intranuclear inclusion bodies called Joest-Degen inclusions. The frequent infection of NW camelids by Borna disease virus, but not other animals on European farms, suggests a special susceptibility in NW camelids, allowing them to be sentinels of the occurrence of Borna disease virus in endemic regions.[96]

Eastern equine encephalomyelitis virus (EEEV) and **West Nile virus (WNV)** infection, both caused by arboviruses (*Alphavirus* and *Flavivirus,* respectively), have been reported in NW camelids. Infection with either causes very similar clinical signs and disease. Fever, seizures, ataxia, lethargy, recumbency, and a high mortality rate are characteristic; differentials include listeriosis, rabies, polioencephalomalacia, and meningeal nematodiasis. Gross lesions are inapparent. Histologic lesions in both include polioencephalitis with typical lymphocytic perivascular cuffs, neutrophilic infiltration (mostly in EEEV), gliosis, satellitosis, multifocal neuronal and glial necrosis, and edema (Fig. 7.22). For EEE, immunohistochemistry and in situ hybridization will demonstrate intracytoplasmic neuronal and glial antigen and RNA.[110] Immunohistochemistry highlights abundant WNV in the neurofibers as well as in neuronal and glial cytoplasm.[38]

Bluetongue virus (*Reoviridae* family, *Orbivirus* genus) infection is a vector-borne disease that is predominantly transmitted by biting midges (*Culicoides* genus). It was historically thought that camelids were resistant to bluetongue virus; however, sporadic cases and outbreaks in camelids from Europe and the United States demonstrated that they are susceptible. Serotype susceptibility to bluetongue virus in camelids is important in creating immunoprophylaxis plans. Reported clinical signs are sudden onset of weakness, recumbency, and respiratory distress. Grossly, pulmonary congestion and edema as well as hydrothorax and hydropericardium with subendocardial and myocardial hemorrhages are observed (Fig. 7.23A). Multifocal hemorrhage expands the myocardium predominantly adjacent to the papillary muscles of the left ventricle.[12,111] Typical bluetongue virus lesions seen in sheep, such as ulceration of the oral and esophageal mucosa, subcutaneous edema of the head, cervical muscles, surrounding the ligamentum nuchae and subintimal hemorrhages of the pulmonary artery, are not seen in NW camelids.[61,111] Histologically, there is alveolar edema and congestion with rare and mild lymphocytic vasculitis (Fig. 7.23B). Clinical signs, postmortem findings, molecular tests using spleen, lung, or other bloody tissues, and ELISA are important for the diagnosis of bluetongue virus since acute infections will demonstrate positive molecular but not positive ELISA results. Also, this virus is capable of attaching to erythrocytes for several months, so a positive molecular diagnosis has to be correlated with typical clinical and postmortem findings to interpret it as a cause of death.

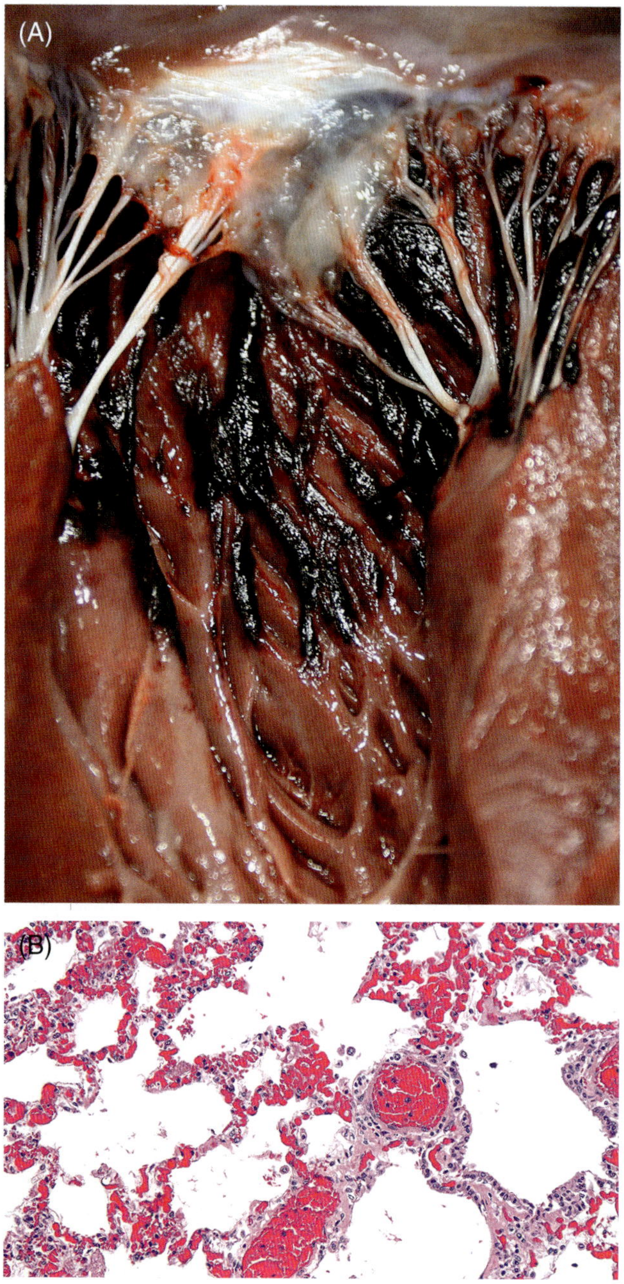

FIGURE 7.23 Bluetongue virus infection in the heart (A) and lung (B) of a llama. (A) Hemorrhage in the tongue or other tissues is characteristic. In this case, there are severe subendocardial hemorrhages in the left ventricle. (B) Pulmonary lesions consistent with vascular/endothelial cell damage include congestion, intralveolar fibrin, and perivascular lymphocytes surrounding a small caliber blood vessel.

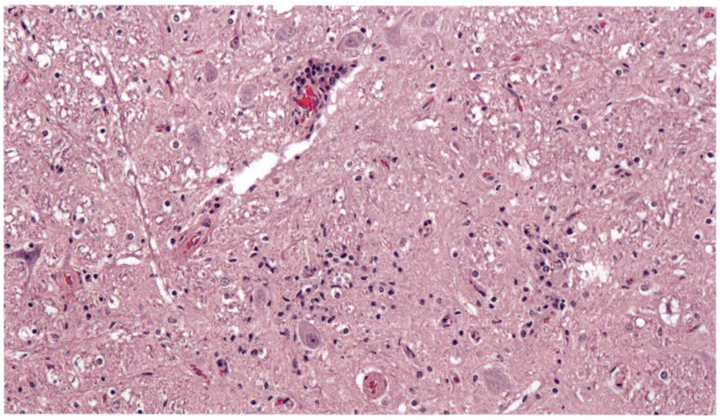

FIGURE 7.22 West Nile virus (WNV) infection in the brain of a llama. Characteristic lesions associated with viral infection include lymphoplasmacytic perivascular cuffs and multifocal areas of gliosis in the neuropil. Inclusions are generally not seen with WNV infection.

In neonatal and young NW camelids, like young domestic ruminants, **rotavirus** (family *Reoviridae*) **and coronavirus** (family *Coronavirdae*) infections are likely very common. One North American serosurvey suggested up to 98% prevalence of rotavirus on some farms and another report suggested

42% of diarrhea in young NW camelids was due to coronavirus infection.[85] Rotavirus and coronavirus have also been reported in alpaca crias with diarrhea in the Peruvian highlands.[94] Like cattle, predisposition to disease is likely linked to failure of passive transfer of maternal antibodies. Clinical signs and gross and histologic lesions are also similar to those domestic cattle. Diarrhea in young OW camelids associated with both rotavirus and coronavirus have been reported as well; however, gross and or histologic lesions have not been described. These are likely similar to those seen in cattle.

Middle eastern respiratory syndrome (MERS; also called **camel flu)** is caused by **MERS coronavirus (MERS-CoV)** and is a WOAH reportable disease. Clinical signs in humans vary from mild to severe and life-threatening. Dromedary camels likely play a role in transmission to humans, though how this occurs is currently unclear.[167] Serosurveys suggest the virus has been circulating in camels since the 1980s, well before the disease was recognized in humans in 2012. The virus appears to infect and spread among young camels more readily than older camels, but there are no documented lesions in camels. The virus is significant for the substantial risk it poses to humans exposed to camels shedding the virus. Experimental infections of alpacas with MERS-CoV showed successful infection, replication, and transmission of the virus.[7] Mild to moderate upper respiratory tract infection with epithelial hyperplasia and lymphohistiocytic and neutrophilic infiltration is seen in the lamina propria or within the upper respiratory epithelium. The lower respiratory tract is minimally affected by this virus in dromedary camels, llamas, and alpacas. In contrast to the minimal or no clinical signs in camelids, MERS-CoV causes focal hemorrhagic and necrotizing pneumonia with extrapulmonary manifestations due to the excessive production of cytokines and chemokines in humans.[143]

A **coronavirus** was detected by TEM and PCR as the cause of an outbreak of respiratory disease in alpacas in California in 2007.[32] Disease was primarily seen in pregnant animals but was also identified in other demographics. Gross lesions included pulmonary congestion, edema and pleural effusion. Histologically, diffuse, acute bronchointerstitial pneumonia centered on terminal airways was characterized by fibrin deposition and hyaline membranes accompanied by variable epithelial necrosis and regeneration and mixed inflammation (Fig. 7.24). Subsequent outbreaks have not been reported, but additional sequence analysis indicates the virus is closely related to the common human coronavirus (HCoV) 229E.[33]

Bacteria

Johne's disease is caused by ***Mycobacterium avium*** subsp. ***paratuberculosis*** (see Bovidae, Antilocapridae, Giraffidae, Tragulidae, and Hippopotamidae; Cervidae). Infection has been reported in alpacas, llamas,

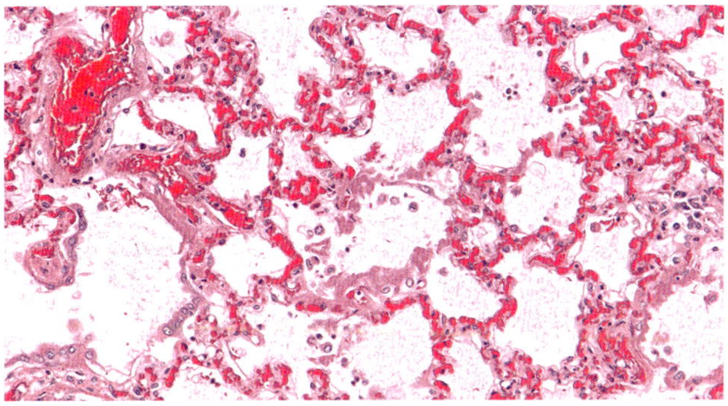

FIGURE 7.24 Coronavirus infection in the lungs of an alpaca. The alveolar septa are congested and alveoli are multifocally lined by hyaline membranes. Other changes include alveolar histiocytosis and type II pneumocyte hyperplasia.

dromedary, and Bactrian camels. Clinically, animals present with severe weight loss (often difficult to detect visually due to heavy fleece) and diarrhea. Gross and histologic lesions are identical to those in cattle, with mild to marked expansion of the intestinal lamina propria and mesenteric and ileocecocolic lymph nodes by macrophages containing few to numerous acid-fast bacteria.[144] In a group of alpacas infected in Australia, extensive granulomatous lesions extended to the lungs, liver, and peritoneal lymphatics.[120]

Tuberculosis (*Mycobacterium tuberculosis, M. bovis,* and other members of the ***M. tuberculosis*** **Complex**), is a WOAH-listed reportable disease that is widely reported in OW camelids. Infection is less often documented in NW camelids, though cases have been reported in llamas co-housed with *M. bovis*-infected cervids.[49] Lesions in camelids are most often associated with the respiratory tract and affect the lungs and associated lymph nodes, though disseminated disease has been reported. Inflammation is granulomatous with characteristic tubercles composed of a central core of caseous necrosis and mineralization surrounded by macrophages, lymphocytes, plasma cells, neutrophils, and a dense fibrous capsule. Unlike infection in cattle, giant cells are rarely present in OW camelids but may be seen in NW camelids infected with *M. microti* infection. Lesions contain very rare, acid-fast positive bacilli; some lesions in NW camelids have had more numerous intralesional organisms.[107,159] Antemortem diagnostics, as in many nondomestic species, is problematic and at this time, there is no definitive and reliable test.

Brucellosis, caused by ***Brucella*** **spp.**, a gram-negative, facultatively intracellular coccobacillus, is common in OW camels and is a WOAH-listed reportable disease (see Bovidae, Antilocapridae, Giraffidae, Tragulidae, and Hippopotamidae). Zoonotic transmission typically occurs via the consumption of unpasteurized milk. In dromedary camels, brucellosis is primarily caused by ***B. melitensis*** and rarely by ***B. abortus***.[157] Brucellosis is less common in NW camelids, though outbreaks in alpacas and zoonotic transmission have occurred. Abortion is the most common and significant clinical presentation, though this is less common in camels than in small ruminants and cattle. Placental retention is also uncommon in camels, likely due to the different structures of their placenta. Necrosis, suppurative inflammation, and fibrosis of the chorion in more chronic cases are typical. Organisms are often abundant and can be found in trophoblasts lining the chorion.

Reproductive tract infections are a very common source of infertility and poor production in both NW and OW camelids. In NW camelids, ***Escherichia coli*** and ***Streptococcus equi*** subsp. ***zooepidemicus*** are common causes of endometritis. Infection is associated with endometrial ulceration, neutrophilic infiltrates, intraglandular aggregates of debris and pus, and in chronic cases, fibrosis, glandular atrophy, and infertility. An endometrial biopsy scoring system similar to that used in horses has been developed and can provide valuable treatment and prognostic information.[114] While abortion can be due to a variety of genetic, management, and nutritional factors, infectious diseases of all kinds have been estimated to cause from 10% to more than 70% of abortions in some areas.[146] Of these infections, **leptospirosis**, **toxoplasmosis**, **neosporosis** (see "Protozoa" section) and **chlamydiosis** have been frequently diagnosed in llamas and alpacas while in dromedary camels, **brucellosis** and **trypanosomiasis** (see "Protozoa" section) are more common in the Middle East and Africa. ***Leptospira* serovar *icterohaemorrhagiae*** and ***Leptospira* serovar *pomona*** are the most prevalent in alpacas and vicuñas from the Peruvian Andean region.[126] Mastitis is rare in South American camelids but can be very common in dromedary and Bactrian camels. It is most often subclinical and chronic, with nonsuppurative inflammation and fibrosis being the most common histologic lesions.

Alpaca fever (*la fiebre de las alpaca*), caused by ***S. equi*** subsp. ***zooepidemicus***, is an important disease of alpacas that is associated with low morbidity and very high mortality (up to 100%).[84] It is seen primarily in Peru and South America but has also been reported in North America. Acute forms are usually seen in young animals that develop high fever, depression, and recumbency. Polyserositis is a significant common finding and systemic effects include fibrinous peritonitis and pleuritis (Figs. 7.25A, B), as well as fibrinosuppurative meningitis. Chronic abscesses are more common in adults in which there may be an environmental source of infection. Transmission is likely by direct contact or oral routes, and more severe forms are generally associated with environmental, nutritional, or social stress.[49,84] Similar disease has recently been described in OW camelids in North America and in abattoirs in Africa, suggesting that this may be a more widespread disease among camelids than previously thought.[5,139]

Colibacillosis is a significant cause of morbidity and mortality in NW camelid neonates (and presumably OW camelids, as well, though this is less well documented).[164] ***E. coli*** can lead to neonatal diarrhea and sepsis, is often associated with a failure of passive transfer of maternal antibodies, and the disease is most commonly reported in animals less than 7 days of age. When combined with viral or other factors, older animals may be affected. Clinical signs and lesions are typical of those in domestic ruminants with the same disease, though enteropathogenic and enterotoxigenic strains have not been identified in camelids. Late in the disease, hemorrhagic enteritis may be present. Typical enteric lesions include piecemeal necrosis, increased neutrophil, lymphocyte, and plasma cell infiltration of the lamina propria, crypt abscesses, and variably adherent gram-negative coccobacilli adhered to the surface of enterocytes.

Neonatal mortality in alpacas from the Peruvian highlands is led by the "neonatal diarrheic complex", from which ***Clostridium perfringens* type A** is commonly present, but other pathogens are frequently detected such as *Eimeria*

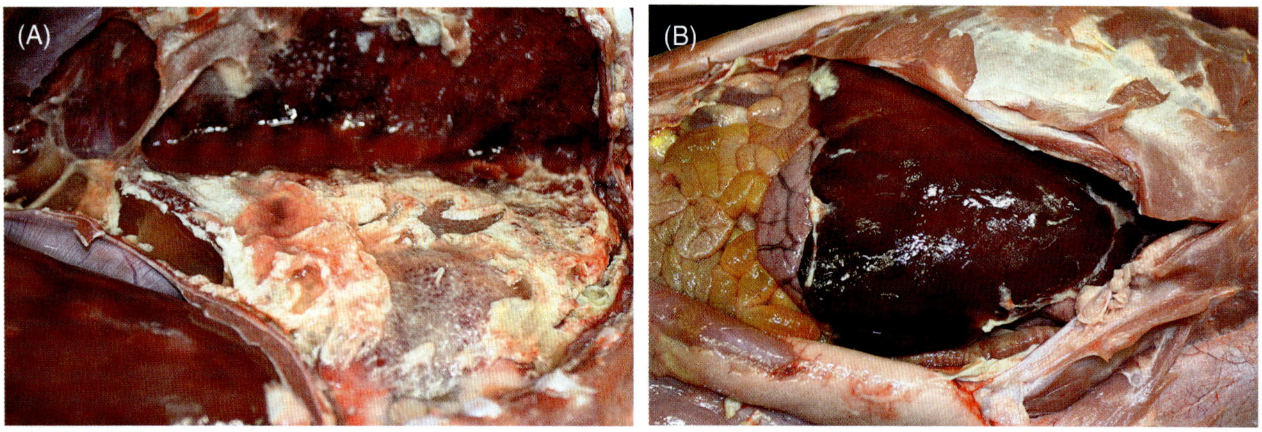

FIGURE 7.25 Alpaca fever (la fiebre de las alpaca) due to *Streptococcus equi* subsp. *zooepidermicus* infection in an alpaca. (A) Infection is associated with fibrinosuppurative pleuritis and hyperemia of the parietal pleura. (B) Fibrinous serositis and peritonitis with fibrin tags can be seen in the abdominal cavity. *(Photo Courtesy of A. Mete, California Animal Health and Food Safety Laboratory, University of California, Davis.)*

macusaniensis that is considered a predisposing factor, *Giardia duodenalis* and coronavirus.[125] *C. perfringens* type A has been implicated in enterotoxemia of young and adult OW camelids.[161] Racing camels appear more commonly affected than less intensively trained animals. Death often occurs at 2–3 weeks of age as colostral antibodies wane and before naturally acquired antibodies increase in circulation.[49] Sudden death, bloating and constipation, or neurologic signs are common; diarrhea is not a feature without concurrent infections with other bacteria. Gross lesions include edema, congestion, and hemorrhage of the lungs, lymph nodes, and thymus, distension of the intestine by gas or fluid, and congestion of the saccules of C-1 and C-2. Congestion of kidney, intestines, and other organs is also common, and myocardial necrosis and degeneration have been reported.[40] Microscopically, there is diffuse collapse of the intestinal architecture with severe hemorrhage and necrosis, and the superficial mucosa is lined by gram-positive bacilli (Fig. 7.26). The second most common cause of neonatal mortality in alpacas in the Peruvian highlands are acute pneumonias caused by viral and bacterial interactions due to *Pasteurella multocida*, *Mannheimia haemolytica*, respiratory syncytial virus, and parainfluenza-3, which is similar to the causes of pneumonia in other young ruminants.[123]

***C. perfringens* type C infection** is well described in young NW camelids. It causes similar signs to those of *C. perfringens* type A enterotoxemia, but diarrhea is more common. In both types A and C, enterotoxemia, pulmonary and tissue edema, cerebral edema, and neuronal necrosis are present. In all cases of clostridial enteritis, isolation of the organism from the ileum is not sufficient for confirmation of disease since this bacterium can be part of the normal microbiota. Culture should be coupled with appropriate lesions, histologic evidence of intralesional gram-positive bacilli with morphologically consistent with *Clostridium* spp., and demonstration of the toxin using ELISA or identification of the toxin-producing gene by PCR.

***C. perfringens* type D** enterotoxemia has also been suspected in camelids with lesions similar to those in cattle and sheep. Serosal hemorrhage, prominent fibrin clots in the pericardial sac, and pulmonary and cerebral edema are strongly suggestive.

M. haemolamae is a highly endemic, hemotropic, epierythrocytic mycoplasma of llamas and alpacas that binds to the surface or red blood cells and has been detected in camelids in Europe and North and South America.[65] It is similar to the hemotropic mycoplasma organisms reported in cattle, pigs, and cats.[147] These are small, 0.5 µm diameter bacteria that lack a cell wall and are present at the periphery of red blood cells on blood films (Fig. 7.27). They are usually round but can be ring shaped or even linear. Electron microscopy suggests the organism does not actually invade the RBC but rather indents the membrane.[116] Infection is likely transmitted by insects, though specific vectors have not been identified. Vertical transmission has been established, with crias showing infection at 1 day of age before colostrum ingestion. Weight loss, lethargy, and anemia, which can be life-threatening, are commonly reported, though the pathogenesis of infection is not well documented. It is likely that this organism predisposes to other diseases. Blood films can be diagnostic, but infection levels may be very low or cyclic. PCR tests are available and are particularly valuable in animals with low-level infections.

Fungi

While numerous fungal diseases, including dermatophytosis and aspergillosis, have been reported in captive and free-range NW and OW camelids, **coccidioidomycosis**

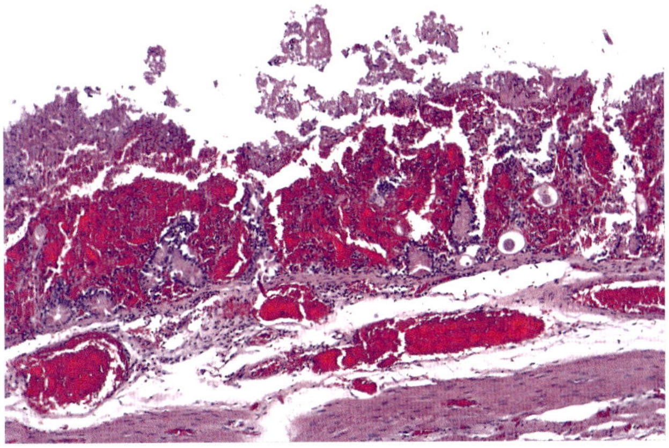

FIGURE 7.26 Clostridiosis in the intestines of an alpaca. Severe hemorrhage and superficial necrosis are typical of the bacterial infection. Coinfection with *Eimeria macusaniensis* (multiple life stages in the crypts) was also seen in this case. *(Photo Courtesy of R. Perales, Universidad Nacional Mayor de San Marcos.)*

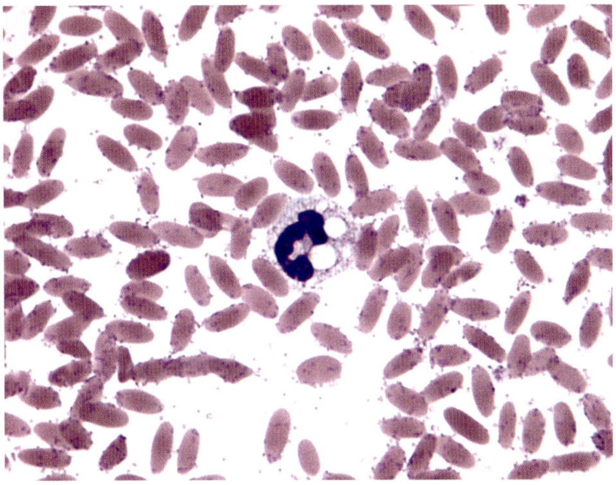

FIGURE 7.27 *Mycoplasma haemolamae* in an alpaca. Numerous, up to 1 µm in diameter, coccoid and ring-shaped bacteria (basophilic in this blood film) are present on the surfaces of and between the characteristic, ovoid camelid erythrocytes. Modified Wright stain. *(Photo Courtesy of M. Scott, Michigan State University.)*

due to infection with ***Coccidioides immitis*** or ***C. posadasii*** is likely the most significant fungal disease in NW camelids. NW camelids appear to be very susceptible to this infection. It is acquired via inhalation of infectious arthrospores. The fungus is endemic in desert areas of southern California, the San Joaquin Valley, Utah, Nevada, Arizona, New Mexico, and Texas in the United States; as well as desert areas of northern Mexico, Northwest Venezuela, Argentina, Bolivia, and Paraguay, Honduras, Guatemala, and Nicaragua.[49] Coccidioidomycosis usually presents as a respiratory disease. However, disseminated disease is common and can cause lesions affecting the neurological system, heart, liver, kidney, skin, bone, eyes, or lymphoid tissues; clinical signs reflect the location of the lesions.[31,54] Granulomas or granulomatous inflammation surrounding intralesional fungal arthroconidia, spherules, and endospores are characteristic and inflammation can coalesce into large masses that efface organs (Figs. 7.28A, B; eSlide 7.e4). Inflammation surrounding freshly ruptured spherules may be suppurative. Multinucleated giant cells may also be present. In its parasitic form, infective arthroconidia transform into immature spherules (10–20 μm in diameter) that mature into large, up to 200 μm diameter mature spherules (or sporangia) filled with 2–5 μm endospores.

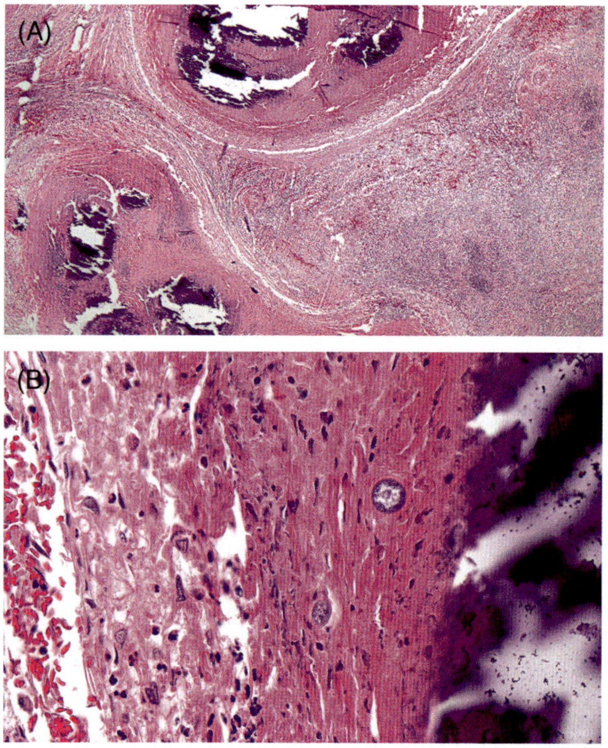

FIGURE 7.28 Coccidioidomycosis in the lung of an alpaca. (A) Numerous mineralized small- and medium-sized nodular granulomas obliterate the pulmonary parenchyma. (B) High magnification of a mineralized granuloma with two fungal spherules in an area of necrosis (see eSlide 7.e4).

Sporangia subsequently rupture, releasing endospores into adjacent tissue that form new spherules, or if they are released into the environment, the endospores form mycelia. Aerosolized endospores can be a significant zoonotic threat, especially if appropriate precautions are not followed.[25]

Cryptococcosis was reported in a llama in the United States that developed ulcerative cheilitis and subsequent dissemination to the lungs and internal lymph nodes[1] (Figs. 7.29A–C).

Metazoa

Parasitic diseases of OW and NW camelids have a significant impact on meat, milk, and fleece production in South America and Africa and on the management of breeding herds and individuals in North America and Europe. Parasites of camelids have been reviewed by Fowler.[52] A few of the most important will be discussed here.

Lamanema chavezi, an intestinal trichostrongylid nematode, causes severe clinical signs, including high mortality in NW camelids.[52] Its range was previously restricted to South America, but recent reports of infections in North America and New Zealand indicate the range has expanded with exportation of camelids from endemic areas.[83,100] The life cycle includes ingestion of larvae, which then penetrate the intestinal wall and migrate to the liver and lung. Maturation occurs after migration up the trachea, re-ingestion, and migration back to the small intestine. Migration through the intestinal wall leads to severe catarrhal and hemorrhagic enteritis. Hepatic lesions include coagulative necrosis, hemorrhage, and multiple small abscesses, which if the animal survives will become fibrotic and mineralized. The mountain viscacha, a South American rodent, is considered the definitive host. How this infection might be maintained outside of South America is unknown.[52]

Other trichostrongyles that infect camelids are similar to those of small ruminants and cattle. These include ***Ostertagia*** spp. and ***Haemonchus contortus.*** Both cause significant disease. Ostertagiasis causes similar proliferative gastritis as in domestic ruminants. *H. contortus* causes blood and protein loss through traumatic damage to the C-3 mucosa, which leads to ill-thrift and weight loss in addition to severe anemia and hypoproteinemia. Haemonchosis causes multifocal to diffuse micro-ulcerations in the intestinal mucosa, as well as anemia-related hepatic centrilobular degeneration and necrosis and tissue hypoxia. Another important trichostrongyle from camelids is ***Camelostrongylus mentulatus*** that causes proliferative lesions in the pyloric region of C-3.

Whipworms, especially ***Trichuris* spp.**, are also a significant source of unthriftiness in NW and OW camelids. Ova are remarkably resistant to environmental degradation, and thus *Trichuris* spp. are a particular problem in intensively managed camelids such as in small zoo exhibits. They have a direct life cycle. Eggs passed in the feces

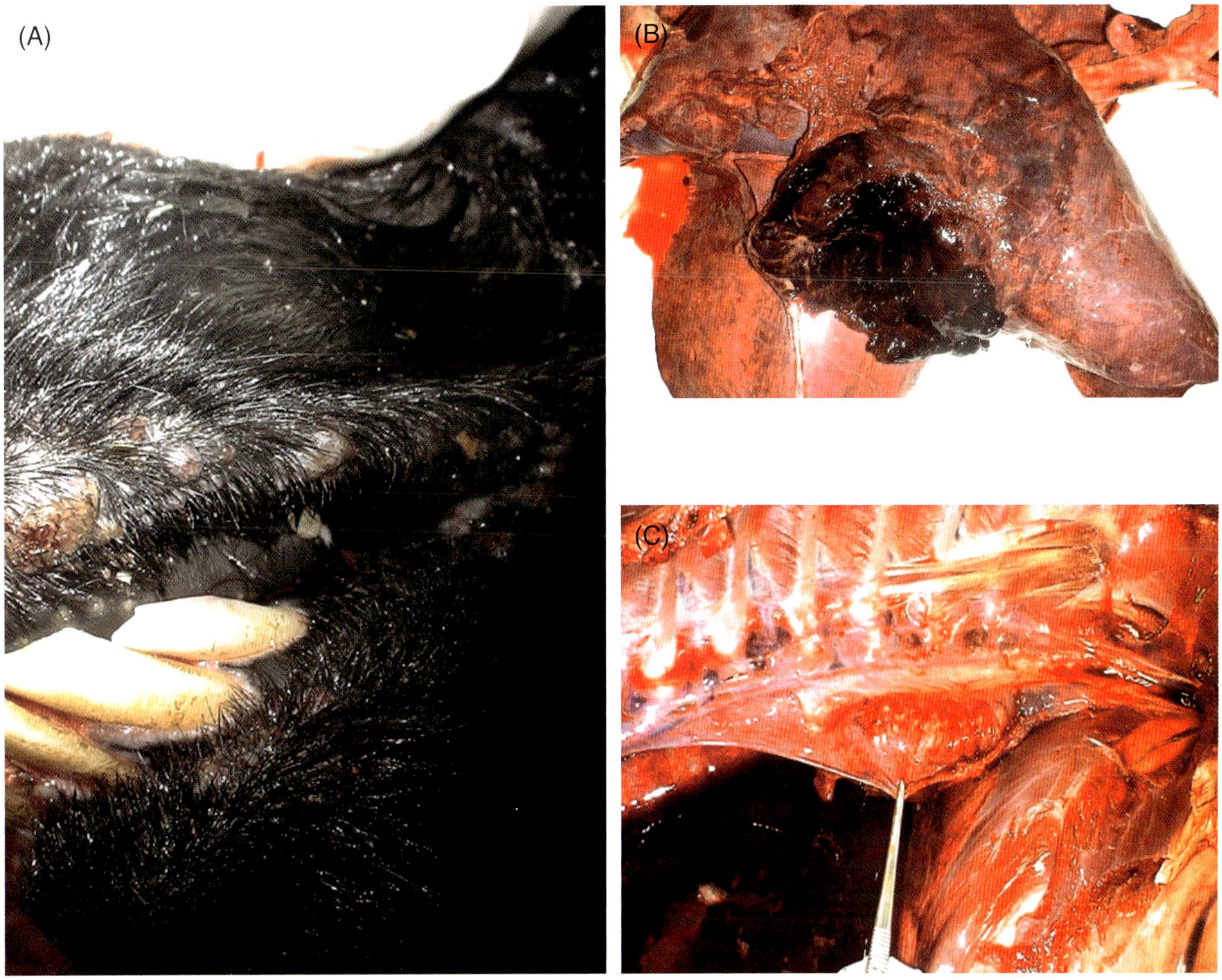

FIGURE 7.29 Disseminated cryptococcosis in the skin (A), lungs (B), and lymph nodes (C) of a llama. (A) Infection in the skin can present as exophytic, blanched nodules or as ulcera. (B) Dissemination to the lungs can be associated with hemorrhage or pneumonia. (C) Infection can be associated with markedly enlarged lymph nodes. Affected tissue can be soft to firm depending on the severity and stage of infection.

become infective on pasture/substrate within 3 weeks. Upon ingestion, the eggs hatch and larvae invade the small intestinal mucosa, mature, and then migrate to the cecum and colon to complete their life cycle. Whipworms have a characteristic tapered appearance; the slender tapered end burrows into the intestinal mucosa. Catarrhal and hemorrhagic enterocolitis are typically seen grossly and histologically.

Parelaphostrongylus tenuis, the **meningeal worm** (see Cervidae), and aberrant larval migrans are significant differential diagnoses for neurologic disease in camelids in North America. The white-tailed deer is the definitive host for the meningeal worm and the life cycle is complex.[52] In the host, adult worms live in the veins of the dura mater where the eggs and larvae enter the blood vessels, flow to the lungs, and infiltrate the alveoli. They are then expectorated, swallowed, and are defecated into the environment. Terrestrial snails and slugs ingest the larvae, which are subsequently ingested again by the host. Larvae released from the snail in the host's stomach penetrate the peritoneum and migrate to the spinal cord via the spinal nerves. Larvae mature in the gray matter dorsal horns and adults then migrate to the brain via the subdural space and finally through the dura mater to the venous sinuses. In deer, the definitive host, there are few lesions and little clinical disease. However, in aberrant hosts, such as NW camelids, damage to the neurologic system is often severe and clinical signs reflect sites of damage. The presence of peripheral eosinophilia and increased eosinophils in the cerebrospinal fluid is variable.

There are typically none to very rare visible gross lesions in *P. tenuis* infections, and nematode larvae are only very rarely identified in the meninges of the brain. Microscopic lesions include lymphoplasmacytic perivascular cuffs, axonal degeneration, hemosiderin-laden macrophages, eosinophilic infiltrates, meningitis, mineralization, hemorrhage, multinucleated giant cells, glial scars (linear tracts of glial cells), linear tracts of necrosis and mixed inflammation consistent with parasitic migration, and typically upon exhaustive histologic sectioning and examination, cross sections of nematodes.[36] Histologically, nematodes are usually degenerate, 100–200 mm in diameter, with a prominent thin, smooth cuticle, coelomyarian musculature, accessory hypodermal chords, and a GI tract composed of multinucleate cells. A reproductive tract is often present and contains sperm or thin-shelled eggs.

Liver flukes, *Fasciola hepatica* and *Dicrocoelium dendriticum*, are commonly found in NW camelids and cause subclinical and clinical disease. Acute massive infections can lead to bile stasis, liver degeneration and necrosis, hepatic insufficiency, and death. More commonly, chronic infections are characterized by bile duct obstruction and bile stasis with cholangitis and biliary epithelial hyperplasia.[72] Anemia and hypoproteinemia are common, and an unthrifty condition and particularly poor fleece quality develops. Grossly, hepatic fibrosis and ascites are prominent with adult forms of flukes (Fig. 7.30). Regenerative hepatic nodules, mineralized foci, and abscesses may also be present. Histologically, in chronically infected animals, the liver contains multifocal to regionally extensive and occasionally coalescing areas of abundant fibrosis with biliary hyperplasia that surrounds regenerative nodules. Neutrophils, lymphocytes and plasma cells, and some macrophages are free in the parenchyma, periportal, and in the lumina of bile ducts. Flukes and their ova are often found in affected areas in the bile ducts and within the parenchyma. Characteristic features of trematodes are their tegument, lack of body cavity, and the presence of both gonads (hermaphrodites). Flukes found in camels can be differentiated by their size and the presence of spines. *Fasciola* spp. are the largest flukes in camelids (cm vs mm) and have spines on their tegument. *F. hepatica* ova are large (120–140 mm) and lack a miracidium. *D. dendriticum* ova are small (40 mm × 25 mm) oval, have a golden wall, are operculated, and contain a miracidium. Hepatic trematodiasis in NW camelids can be associated with severe intimal and adventitial thickening and mild to moderate medial thickening in the pulmonary arteries that resemble pulmonary arterial hypertension in humans with hepatic parasites.[72]

FIGURE 7.30 Fasciolosis in the liver of a llama. These trematode parasites are characteristically leaf-shaped, flat, and tan and become apparent after serially sectioning the liver. *(Photo Courtesy of California Animal Health and Food Safety Laboratory, University of California, Davis.)*

Cutaneous pelodera (syn. **rhabditic dermatitis**) was diagnosed in a 4-year-old llama. Infection caused severe dermatitis with intensive pruritus, localized alopecia, and coalescing cutaneous nodules that had a central pore. Histological lesions showed severe periadnexal eosinophilic dermatitis and the presence of characteristic nematodes with a rhabditiform esophagus (Aboellail, unpublished data).

Protozoa

In camelids, ***Eimeria macusaniensis***, an apicomplexan coccidian parasite that primarily affects young NW camelids, is considered the most significant protozoal infection in this age group. At least five other species, ***E. alpacae, E. lamae, E. ivitaensis, E. punoensis,* and *E. peruviana***, also infect NW camelids, but none have been associated with the severe clinical signs seen in *E. macusaniensis* infection.[28,52] *E. macusaniensis* is by far the largest camelid coccidian. Its oocytes are 81–107 µm × 61–80 µm and tissue meronts are greater than 100 µm in diameter. Other *Eimeria* are typically a quarter of this size.[28,52] In OW camelids, four species of *Eimeria* have been characterized: ***E. bactriani, E. cameli, E. dromedarii***, and ***E. pellerdyi***. Similar to *E. macusaniensis, E. cameli* is the largest by far, with oocytes that are three to four-fold larger than other species. All have been associated with severe disease, particularly in young camels.[77]

Clinical signs of ***E. macusaniensis*** infection include lethargy, weight loss, anorexia, diarrhea, and in some cases sudden death. The number of organisms in a given infection is important and heavily contaminated pastures can be a source of high infective doses and a significant factor in clinical outcomes. Coinfections are also likely important, since the destruction of enteric epithelium, particularly in the small intestinal crypts, can predispose to other infections, particularly with clostridial organisms.[124] Histologically, the density of meronts and gamonts within the mucosa of the jejunum and ileum is often remarkable and they may efface the parenchyma leaving little remaining unaffected tissue (Figs. 7.31A, B; eSlide 7.e5). Necrosis and loss of villi are the primary lesions, though in more chronic cases, severe fibrosis in the lamina propria may occur. The colon and duodenum are rarely affected. Coccidia are typically most dense in the deep crypts, but in heavy infections, parasites occupy the entire villus. Despite massive infections, fecal examinations for *E. macusaniensis* are often negative, at least initially, in many animals.

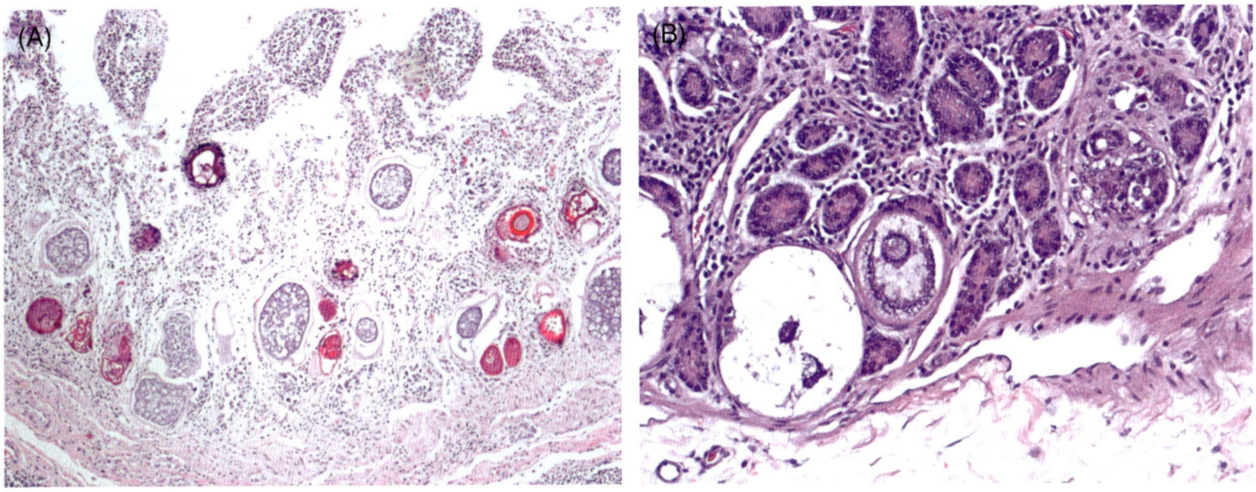

FIGURE 7.31 *Eimeria macusaniensis* in the ileum of a llama. (A) Myriad, round, coccidian parasites of multiple life stages are present in the intestinal mucosa of this llama. These include meronts, microgamonts, macrogamonts, and oocysts. In severe infections, the organisms can completely efface the intestinal architecture. (B) An area of fibrosis in the lamina propria (right side of the image) is associated with multiple degenerating coccidial schizonts and cysts in the lamina propria. Others have caused compression or ruptured into intestinal crypts (see eSlide 7.e5).

Cryptosporidia are important apicomplexan protozoal pathogens in both young NW and OW camelids. They are an important cause of neonatal diarrhea that must be differentiated from other causes of neonatal diarrhea, including bacterial endotoxemia or enterotoxemia, coccidiosis, coronavirus, rotavirus, or BVDV.[52,164] Fecal smears and histologic sections can be examined with carbol-fuchsin modified Ziehl Neelsen stain or periodic acid-Schiff reaction, for which the parasite will be positive. Villus atrophy is the primary histologic feature, with small 4–5 μm diameter basophilic round organisms that are intracellular but extracytoplasmic, and are typically most apparent superficially in cells at the tips of small intestinal villus epithelium.

Trypanosomiasis (also known as **nagana** or **surra**) is a severe disease of OW camels that affects productivity in regions where these animals are kept. ***Trypanosoma evansii*** causes surra in camels and is found in Southeast, East, and Central Asia, as well as North Africa. *T. evansii* has been reported in South America but has not been identified in NW camelids. The tsetse fly (*Glossina* spp.) is the intermediate host and after feeding on the blood of an infected animal, the organisms multiply in the fly's hindgut and are deposited as metamorphosed trypanosome in the salivary secretions when the fly takes its next blood meal. Two syndromes have been reported in camels: acute and chronic form. In acute surra, there is fever, anemia, edema, and weakness that progresses to paralysis. Pulmonary edema may lead to secondary pneumonia. Abortion and caseous mastitis have also been reported. Death usually occurs within weeks and there are many parasites in blood smears (Fig. 7.32). In the more chronic form, there is intermittent fever, anemia and edema, and progressive loss of condition. Between febrile episodes, finding organisms in blood smears may be challenging, so sample collection should coincide with periods of fever.

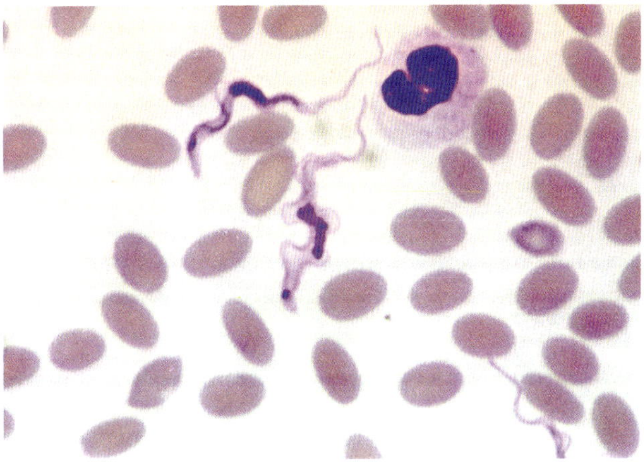

FIGURE 7.32 *Trypanosoma evansii* (surra) in the blood of a camel. The organism is characterized by its spindloid body, undulating membrane along one side, anterior flagellum, and kinetoplast. To detect organisms in blood smears, samples should be collected during periods of fever. Organisms may be very rare during the intervening afebrile periods. Wright-Giemsa stain. *(Photo Courtesy of A. Tibary, Washington State University.)*

Trypanosoma brucei and ***T. congolensis*** are the causes of human sleeping sickness and can also infect camels. However, they are not a primary cause of surra in camels, though these organisms may prevent camels from being kept in the so-called Tsetse Belt of Africa. *T. brucei, T. congolense,* and *T. vivax* have been well described in cattle; however, lesions in camels have not been characterized in the literature. It would be expected that lesions might be similar, and consist of interstitial pneumonia, myocardial degeneration and myocarditis, bone marrow atrophy, lymphadenitis with follicular hyperplasia followed by atrophy and sclerosis, thymic atrophy, and renal membranoproliferative glomerulonephritis, and interstitial fibrosis.[154] In cattle, organisms are most often seen in the liver and the cerebral cortex, though they may be found in any organ.[154] PCR assays have been developed and can be used in camels.

Toxoplasma gondii and ***Neospora caninum*** have been reported to cause abortion in NW camelids. ***T. gondii*** might cause hepatomegaly and hydrocephalus on gross examination in aborted alpaca crias while mild nonsuppurative meningoencephalitis and hydromelia with intralesional protozoal cysts are present histologically.[37] ***N. caninum*** induces nonsuppurative meningoencephalitis, perivascular cuffing, and glial nodules in aborted llamas and alpacas in Peru.[134]

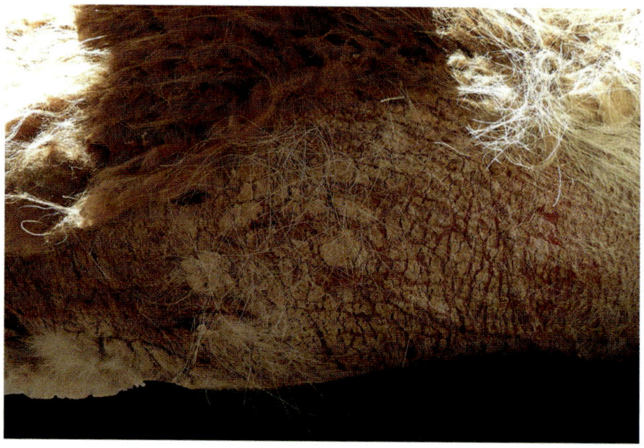

FIGURE 7.33 Sarcoptic mange due to *Sarcoptes scabiei* infestation in the ventral skin of a vicuña. Infection is associated with severe alopecia with thick, gray to white, crusty, erythematous skin. *(Photo Courtesy of L. Gomez-Puerta, Universidad Nacional Mayor de San Marcos.)*

Ectoparasites

New World camelids are aberrant hosts of the deer and sheep **nasal bot, *Oestrus ovis*;** the **deer bot** (*Cephenemyia* sp.) is more common.[52,55] Mortality has not been reported, but a significant mucosal reaction in the nasopharynx consisting of granulomatous inflammation surrounding embedded larvae is common. Sneezing, coughing, and nasal discharge are associated with dyspnea since camelids are obligate nasal breathers. Old World camels are primary host of the camel bot, ***Cephalopina titillator***, which can lead to significant morbidity and rare mortality.[112] Nasopharyngeal mucous membranes are congested and edematous and covered by adherent layers of hemorrhage, mucus, and fibrin. Histologically, the mucosa is hyperplastic and infiltrated by lymphocytes, plasma cells, macrophages, and eosinophils; fibrosis develops in more chronic cases. Goblet cell hyperplasia and dilation of submucosal secretory glands may also be present. In rare cases, larvae may migrate through the ethmoturbinates, resulting in parasitic meningitis and death.

Mange mites infect NW and OW camelids. **Sarcoptic mange (*Sarcoptes scabiei*, scabies)** is the most common and has the greatest impact on fiber quality and animal health. Sarcoptic mange is considered host specific and is probably not zoonotic. It has the potential to decimate wild populations of NW camelids, as has occurred in a National Park in Argentina, where vicuñas and guanacos died due to infection after it was introduced by domestic llamas.[42] Cutaneous lesions are often localized to the skin of the ventral abdomen, inguinal region, and limbs, which are alopecic, erythematous, and have crusts and fissured skin (Fig. 7.33).[63]

Camelids can also be infected with **Psoroptic (*Psoropotes* spp.)** and **chorioptic (*Chorioptes bovis*) mange.**[52] Psorioptic and chorioptic mange mites are not considered host specific, can be transmitted across species, and may be zoonotic. Chorioptic mange is characterized by crusty papules and pustules primarily on the limbs and ventrum, with hyperkeratosis and frequent secondary bacterial infection. Thickening of the skin is a differentiating feature of sarcoptic mange. Psoroptic mange produces moist papules progressing to crusts, on the shoulders, back, sides, and tailhead, as well as in the ears with subsequent fiber loss. Dermatitis induces pruritus, but there is little thickening of the skin. Chorioptic mange is less common, but lesions resemble psoroptic mange. All forms of mange are reportable in the United States; they may also be reportable in other jurisdictions.

Algae

Algal enteritis caused by **blue-green algae, *Chlorella* spp.**, was the reported cause of granulomatous ileitis in a dromedary camel. The organisms were intralesional, 5–11 μm encapsulated, spheroidal to ovoid, and characterized by a single nucleus and slightly basophilic cytoplasm.[109]

ADDITIONAL RESOURCES

Aboellail TA, Waugh M, Harvey A, Fisher J, Vilander AC. Neoplasia and Proliferative Lesions of New World Camelids: A Systematic Literature Review and Retrospective Study of Cases Submitted to Colorado State University From 1995 to 2020. Front Vet Sci. 2021 Oct 22;8:743498.

Cebra, C., Anderson, D. E., Tibary, A., Van Saun, R.J., Johnson, L. W., 2014. Llama and alpaca care: Medicine, Surgery, Reproduction, Nutrition, and Herd Health. Sounders, St. Louis, MO.

Fowler, M., 2010. Medicine and surgery of camelids. 3 ed. Wiley-Blackwell, Ames, IA.

References

Full references for this chapter can be found online at https://doi.org/10.1016/B978-0-443-18437-6.00009-2.

Chapter 8

Suidae and Tayassuidae

María Ángeles Jiménez Martínez[a], María del Carmen Carmona Muciño[b], David Risco[c], Nicole L. Gottdenker[d], and David J. Gasper[e]

[a]*Veterinary Faculty, Complutense University of Madrid, Madrid, Spain;* [b]*Africam Safari Zoo, Puebla, Mexico;* [c]*Neobéitar, Cáceres, Spain;* [d]*Department of Pathology, College of Veterinary Medicine, University of Georgia, Athens, GA, United States;* [e]*University of Wisconsin-Madison, Madison, WI, United States*

INTRODUCTION

The suids, *Suidae* and *Tayassuidae*, primarily species of the tropics and subtropics, have a worldwide distribution on all continents except Antarctica. The true wild boars were indigenous to Europe and Asia and are the ancestors to the domestic pig with which it shares the same scientific name, *Sus scrofa*. Wild boars can also survive in temperate climates and have been introduced to the Americas and many islands. In many areas, they have interbred with domestic pigs and formed considerable populations of feral suids.

Wild suid populations are relatively hardy and most disease research has been focused on their potential as a reservoir for diseases of concern for commercial pig production. High densities and gregarious behavior of wild boar increase the chance of disease transmission. Many wild boar populations have overlapping ranges with commercial swine operations, which increases the risk for transmission among these groups. Within the *Suidae*, the Togian Island babirusa, pygmy hog, Visayan warty pig, and the Javan warty pig are listed as endangered or critically endangered. Within the *Tayassuidae*, only the Chacoan peccary is endangered, and the white-lipped peccary is vulnerable in many areas throughout its range. For all suids, habitat loss, hunting, and hybridization are important threats to conservation.

NON-INFECTIOUS DISEASES

Nutritional

Obesity is not uncommon in captive suids. It predisposes to arthritis and may negatively impact reproduction.[226]

Metabolic

Hyperthermia is not uncommon in both the Suidae and Tayassuidae, and can be fatal. Beyond elevated environmental temperatures in areas with inadequate ventilation, shipping, anesthesia, and stressful situations can precipitate hyperthermia. Capture and restraint are common causes of mortality, possibly exacerbated by hyperthermia, as is capture myopathy.[11,12,226]

Piglets are sensitive to **hypothermia**. Unlike other mammals, young suids lack brown adipose, which hinders their ability to thermoregulate. Domestic pigs, European wild boars, Bornean bearded pigs, warthogs, and red river hogs all have deletions within the *uncoupling protein 1* (UCP1) gene.[17] UCP1 is expressed exclusively in brown fat and located along the inner membrane of mitochondria where it catalyzes protein leakage that results in the release of heat. The European wild boar shivers when cold and builds farrowing nests, which may be behavioral adaptations to more temperate climates.

Gastric ulceration in the stomach is not uncommon. Ulcer may occur in both glandular and nonglandular portions but are more frequent in the latter, near the pars esophagea. In this area, lesions vary from initial hyperkeratosis, evolving into fissures and ulcers. Death may occur due to acute gastric hemorrhage. Other clinical signs are poor growth, especially with chronic ulcers. The cause is attributed to alterations in feed and occasionally have been associated with stress. Ulceration of the glandular stomach is contrarily associated with infectious disease.[160]

Congenital/Genetic

Rare congenital anomalies have been described. These include duplication of the pelvis and rear extremities suggestive of ischiopagus, a type of conjoined twin, in Chacoan peccaries,[16] and cyclopia and limb deformities in collared peccaries.[87]

Age-Related/Degenerative

In both captivity and in the wild, nondomestic suids are generally hardy animals and develop common age-related degenerative conditions. **Arthritis** is most common in joints

Pathology of Wildlife and Zoo Animals. https://doi.org/10.1016/B978-0-443-18437-6.00014-6

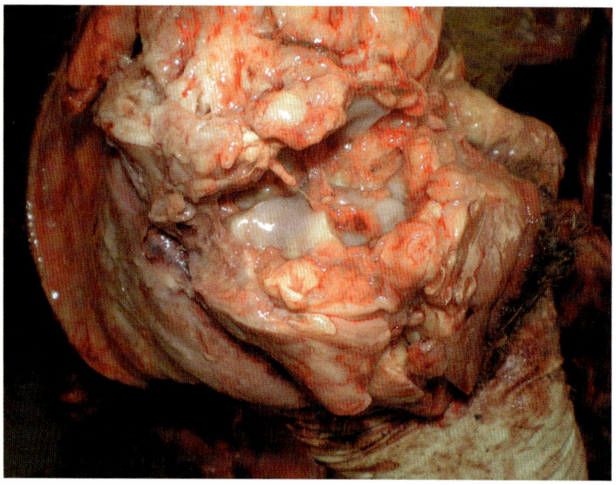

FIGURE 8.1 Arthritis in the femur of a red river hog. Arthritis with marked proliferation of the joint capsule. *(Photo Courtesy of Madrid Zoo, Madrid, Spain.)*

of the distal appendicular skeleton (Fig. 8.1), although intervertebral disc disease also occurs. In captivity, hoof problems such as overgrowth of horn of the claw, cracking, or separation, with secondary complications such as septic laminitis or abscesses, may occur. **Periodontal disease** can also occur and primarily affects the cheek teeth. Studies of wild red river hog and warthog skulls noted an increase in the prevalence of periodontal disease with age in red river hogs but not warthogs, presumably due to their more abrasive diet.[254] **Polycystic kidney disease** has also been reported in peccaries and is presumed to be acquired rather than genetic.[12] It can be found in other suid species (Fig. 8.2). Similar to domestic species, **cystic endometrial hyperplasia** (Fig. 8.3; eSlide 8.e1) can develop in nondomestic suids.[230] Cyst epithelium lacks vacuoles typical of progesterone stimulation, thus the pathogenesis is presumed similar to that of domestic pigs where cystic endometrial hyperplasia is associated with prolonged estrogen stimulation (endogenous or exogenous).

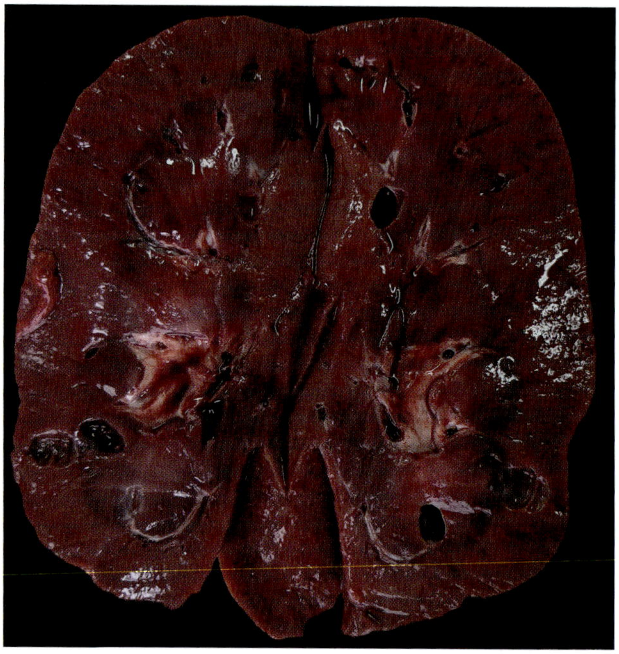

FIGURE 8.2 Polycystic kidney in a wild boar. The renal parenchyma contains multiple cysts predominantly in the cortex. *(Photo Courtesy of Madrid Zoo, Madrid, Spain.)*

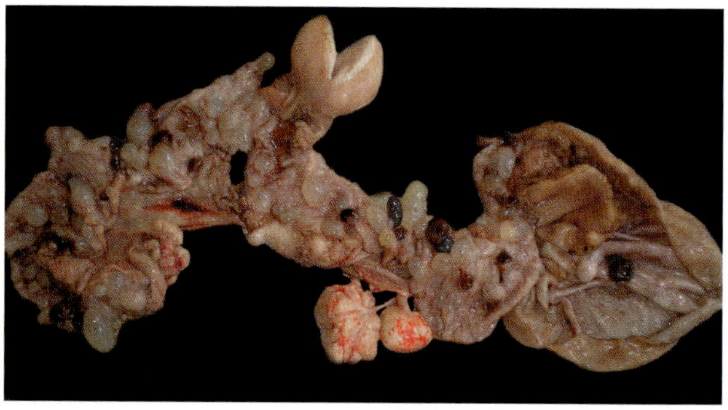

FIGURE 8.3 Cystic endometrial hyperplasia in a wild boar. Bilaterally the uterine horns are enlarged and dilated. Both horns are opened and contain numerous variably sized cystic structures that expand the endometrium. In suids, this condition is associated with prolonged estrogen stimulation. The horn on the left additionally contains a leiomyoma within the wall (see eSlide 8.e1). *(Photo Courtesy of Madrid Zoo, Madrid, Spain.)*

Miscellaneous/Multifactorial

Both **intestinal** and **gastric volvulus** occur in a variety of suids (Fig. 8.4).[12] **Trichobezoars** are somewhat common in captive-collared peccaries.[12] Most cases are incidental findings at necropsy but they may be associated with hyporrhexia and weight loss. Bezoars occur in adults without apparent sex predilection. They tend to be single, round to stomach-shaped (Fig. 8.5) and can fill most of the gastric lumen. Bezoars are composed almost entirely of compacted hair but may contain areas of mineralization. Mild to severe erosions of the gastric mucosa can occur. Low fiber diets may predispose to the formation of bezoars, and excessive grooming due to stress is contributory.

Neoplastic

Neoplasms are infrequently described in suids and include lymphosarcoma, intestinal carcinoma, squamous cell carcinoma (Fig. 8.6), uterine adenocarcinoma (see eSlide 8.e2), uterine leiomyosarcoma, pheochromocytoma, plasmacytomas, hemangiomas, and mammary carcinomas.[12,42,158,226]

INFECTIOUS DISEASES

Much of the research on infectious disease in wild suids is focused on diseases of concern for domestic pigs or potential

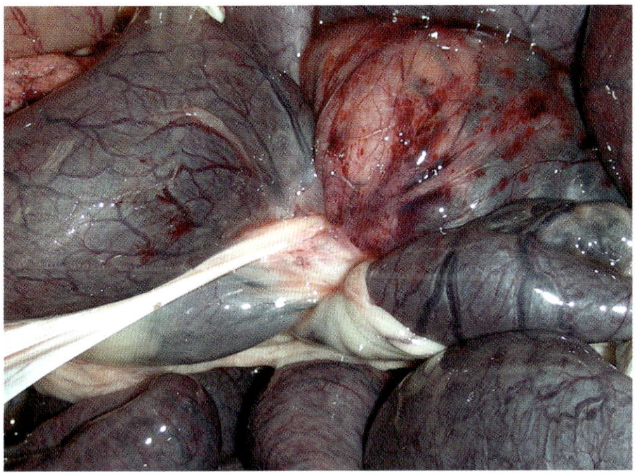

FIGURE 8.4 Intestinal volvulus in a red river hog. Torsion at the root of the mesentery occurs in multiple species of suids. In this case, 360-degree torsion around the cranial mesenteric artery (*center of the image*) is associated with distension and congestion of the small intestine and multifocal serosal hemorrhage.

FIGURE 8.5 Trichobezoars from the stomachs of collared peccaries. Note that some bezoars take on the shape of the stomach lumen.

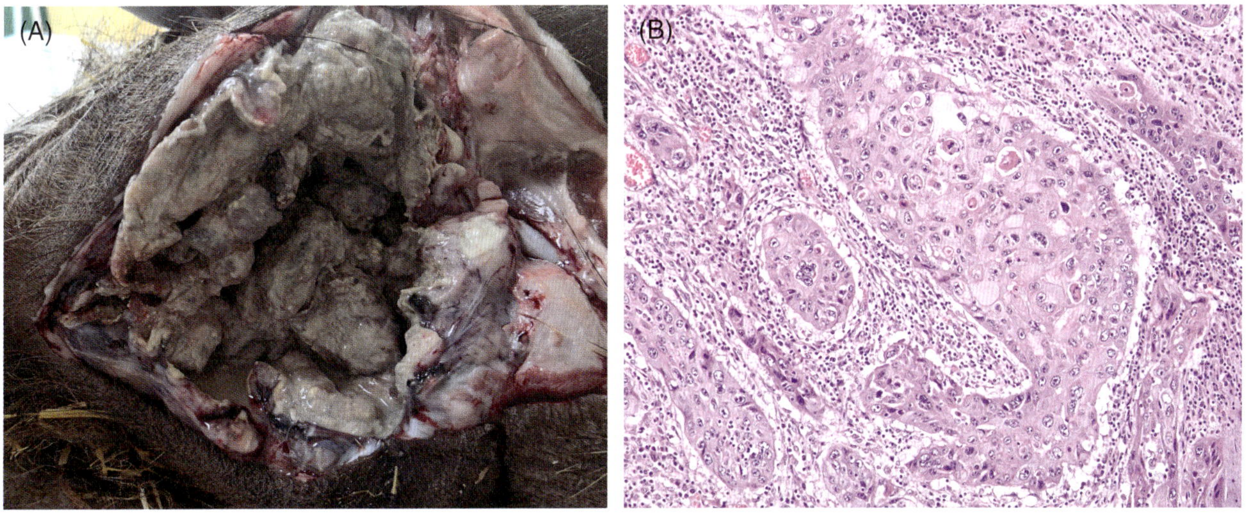

FIGURE 8.6 Submandibular fistula with underlying infiltrative squamous cell carcinoma in a European wild boar. (A) Submandibular fistula with putrid and necrotizing myositis and cellulitis with underlying, infiltrative squamous cell carcinoma. (B) Detail of large, polygonal neoplastic cells with defined borders, abundant cytoplasm, marked anisokaryosis, and apoptotic cells. Inflammatory cells consist mainly of lymphocytes and neutrophils. *(Part A: Photo Courtesy of Madrid Zoo, Madrid, Spain.)*

zoonoses. Wild suids are important reservoirs for maintenance and transmission of several viruses and bacteria of concern to livestock and human health.[142,195] Supplemental Table 8.e1 contains a list of multihost pathogens that have been isolated in wild suids, which may serve as reservoirs or develop mild or subclinical disease.

DNA Viruses

African swine fever (ASF) is a devastating, WOAH listed, notifiable systemic disease in domestic pigs, feral pigs, and wild boars. Morbidity and mortality may reach 100% in naïve populations of domestic pigs. The disease is caused by a DNA arbovirus (family *Asfaviridae*) with multiple genotypes from different geographic regions. **ASF virus (ASFV)** originated in sub-Saharan Africa where it remains endemic through an ancient sylvatic cycle among warthogs and soft ticks of the genus Ornithodoros.[82] Disease and mortality rarely occur in natural hosts, however, the introduction of the virus in naïve domestic pig or wild boar populations causes a variably severe viral hemorrhagic fever with high mortality.[172] In 2007 the virus was introduced in Georgia through imported contaminated pig products from Eastern Africa or Madagascar (2007), and has hence spread to the Trans-Caucasian region, reached the Russian Federation, spread into the European Union in 2014, and later reached China (2019) spreading throughout the Asian Pacific region, followed by Dominican Republic and Haiti in 2021.[20,82,203,252] In the past 13 years ASF has therefore gained an unprecedented spread, causing a panzootic of great economic impact.

In endemic regions with natural hosts, such as warthogs and bushpigs, ASFV is transmitted by Ornithodorus ticks.[57,82,179] Where no ticks are involved, ASFV is transmitted by direct contact with infected animals, contaminated fomites, feces, and ingestion of contaminated feed.[82,179]

Warthogs, bushpigs, and *Ornithodorus* species of ticks are natural hosts of ASFV. A sylvatic cycle involving wild suids and soft ticks of the *Ornithodorus moubata* complex has been established in East and Southern Africa, and possibly the Indian Ocean region.[57] Warthogs are susceptible to infection but do not develop clinical signs of disease.[111] The majority of warthogs are infected during the first few weeks of life and develop a generalized viremia, sufficient to infect ticks that feed on them. The virus then remains in high numbers for a limited period in systemic lymph nodes. Adult warthogs remain infected but remaining viral load is low. Ticks in warthog burrows feed on infected animals and become infected.[97,98] Transmission to domestic pigs in endemic areas is by tick bites and direct contact.[82,203]

Bushpigs are also suspected to be natural reservoirs of ASFV, but the role they play in the ecology of ASF is considered of lesser importance than the warthog.[178,179] Natural infection is known to occur without clinical signs, and blood levels in infected bushpigs are high enough to infect both ticks and domestic pigs. However, unlike warthogs, bushpigs do not live in burrows that facilitate tick exposure. In Madagascar, where ASF was introduced in the late 1990s and evolved into an enzootic disease, bushpigs were sporadically infected but, to date, do not appear to play a role in maintenance and transmission of ASF.[178,179,192] Although the frequency and prevalence of infection are lower than in warthogs, bushpigs should be considered possible reservoirs and a source of transmission of ASFV in endemic countries in east and southern Africa.[98] Peccaries were not susceptible to ASFV in experimental studies.[51]

The role of other wild pigs in the epidemiology of ASF is often obscure and variable, particularly because of differences between the four wild suid genera (*Phacochoerus*, *Potamochoerus*, *Hylochoerus*, and *Sus*), taxonomy, distribution, and ecology.[97] For instance, though ASF infection in warthogs and bushpigs in endemic regions is asymptomatic, experimental studies have proven that the Eurasian wild boar is highly susceptible to acute disease similar to domestic pigs. Additional experimental studies demonstrated transmission occurring directly between wild boars and domestic pigs.[68] Furthermore, infection of free-ranging wild boars and deaths caused by ASF has been observed in Russia.[74,82] High mortalities of the vulnerable bearded pig (*Sus barbatus*) have been reported in Malaysia since 2021, raising the concern that outbreaks could push the 11 native wild pig species in Southeast Asia to the edge of extinction.[62,129] The disease has been found to establish self-sustaining cycles within wild boar populations, when earlier it was thought to be self-limiting in wild boar unless sustained by spillover from domestic pigs.[82,117,172]

Up to 24 genotypes of ASFV have been identified in Africa,[29] where genetic diversity is promoted through the sylvatic cycle, and it is not sustained outside of these regions.[246] Virulence or pathogenicity is not related to genotype, and international genotypes are solely restricted to types I and II. Virus strains in Europe and Asia are of genotype II and show high virulence for both domestic pigs and European wild boars under experimental conditions.[19]

Depending on the virulence of the strain, the clinical course of the disease can be peracute, acute, or subclinical/chronic in both domestic pigs and wild boars.[19,21,68] Highly virulent strains cause peracute disease with up to 100% mortality within 7–10 days. In experimental studies, Eurasian wild boars with peracute disease show clinical signs of lethargy, depression, anorexia, diarrhea, respiratory distress, and high fever; and less commonly, epistaxis and neurological signs.[21,200] Often in the peracute course of the disease, the only gross findings are enlarged hemorrhagic lymph nodes. The full range of gross lesions includes enlarged, hemorrhagic, and edematous lymph nodes, hemorrhagic gastritis, pulmonary edema, renal cortical petechiae, and ecchymoses (Figs. 8.7 and 8.8). Hepatogastric and renal lymph nodes are usually most severely affected. Rarely gallbladder wall edema and renal infarcts occur. Splenomegaly has often been mentioned as a characteristic feature of ASF, but in experimental conditions, it is rarely observed. Skin lesions have not, to date, been reported in natural or experimental disease in boars; however, because their skin is dark colored, they are difficult to assess.

Moderately virulent strains lead to acute disease with up to 30%–70% mortality. Clinical signs include high fever, anorexia, fatigue, nonspecific respiratory and gastrointestinal signs, and abortions. Domestic pigs with ASF have concurrent thrombocytopenia and leukopenia.[205] Thrombocytopenia generally occurs at the end of the acute phase.[244] Gross findings are similar to peracute disease. Histologically, in addition to systemic hemorrhages and edema, experimental studies in domestic pigs revealed a lymphohistiocytic interstitial pneumonia. However, in wild boars the inflammation pattern was of a bronchopneumonia, showing hypertrophy of the bronchial epithelium and alveoli, bronchioles, and bronchi filled with lymphocytes and cell debris.[19,205] Additional histological findings shared in acute and chronic disease caused by certain strains are lymphoid follicle hypertrophy with numerous apoptotic lymphocytes and tingible body macrophages (Figs. 8.7 and 8.8; eSlides 8.e3 and 8.e4).

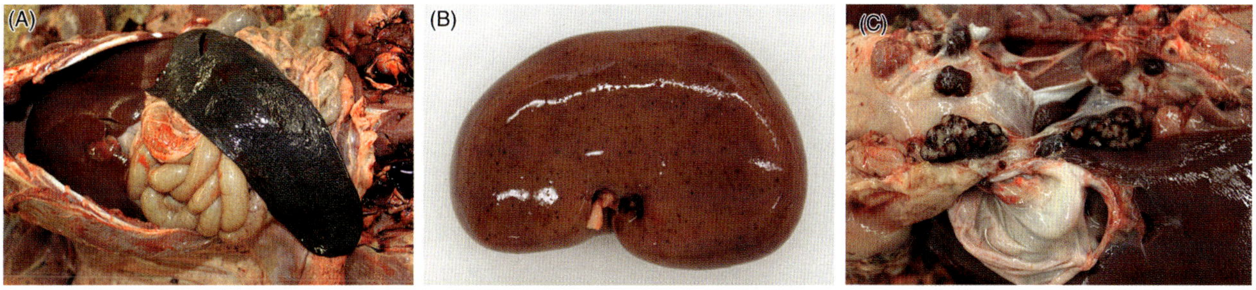

FIGURE 8.7 Experimental African swine fever virus (ASFV) infection in a European wild boar. (A) Splenomegaly, while not pathognomonic, is more common in ASFV than with classical swine fever infections. (B) Renal petechiae can occur in infections with both African swine fever and classical swine fever virus infections and cannot therefore be used as a differentiating feature (see also Fig. 8.11C). (C) Hemorrhagic lymphadenitis. *(Parts A–C: Photos Courtesy of the Friedrich Loeffler Institute.)*

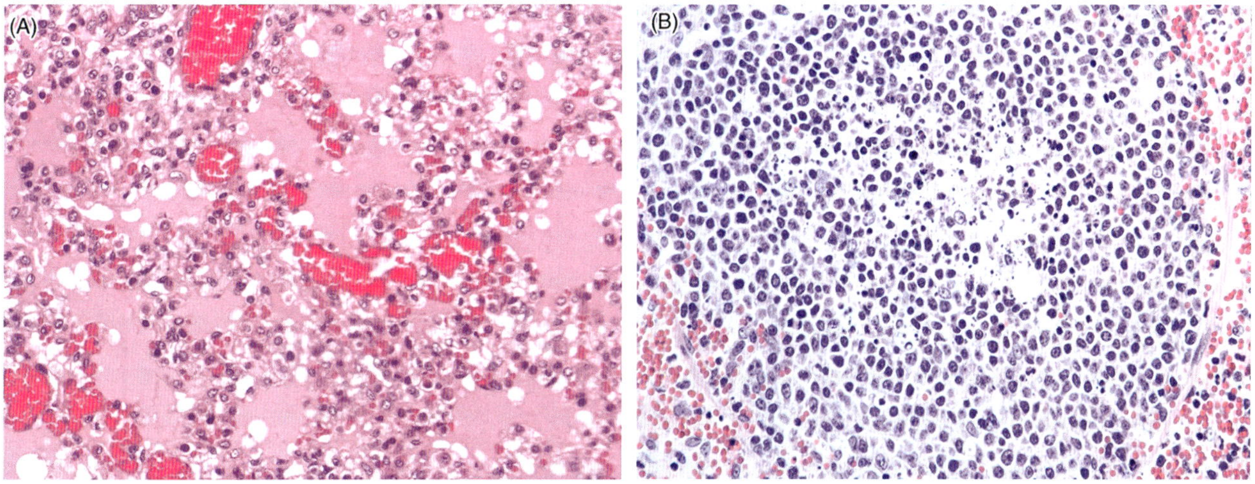

FIGURE 8.8 African swine fever virus infection in a domestic pig. (A) Diffuse pulmonary edema with flooding of alveolar air spaces by edema and foamy macrophages. There is hemorrhage within alveolar septae. (B) Necrosis of lymphocytes within a lymph node with central cellular debris and apoptotic bodies. Surrounding the follicle are extravasated erythrocytes (see eSlides 8.e3 and 8.e4). *(Parts A, B: Photos Courtesy of the Department of Compared Anatomy and Anatomic Pathology. Veterinary Faculty of the University of Cordoba, Spain.)*

Low virulent strains show subclinical and chronic courses with nonspecific clinical signs and low mortality.[19] An autoimmune component may accompany chronic disease, and lesions resulting from immune-complex deposition, particularly in the kidneys (membranoproliferative glomerulonephritis) and lungs, may occur.[173]

PCR for the p72 gene and immunofluorescence antigen detection in tissue smears offer rapid diagnostic options. Sections of tonsil, lymph node, kidney, and/or spleen, as well as blood samples, are the most useful for diagnosis.

Like in domestic pigs, the ASF virus replicates in wild boars, mainly in monocytes and macrophages. In the domestic pig, the monocyte/macrophage lineage is a key viral target, and subsequent production of cytokines, such as TNF-α, IL-1α, IL-1β, and IL-6 are critical to the development of disease. However, in experimental studies, wild boar did not respond to infection with the production of these cytokines.[94] These proinflammatory cytokines, rather than direct viral effects, are thought to be the cause of clinical signs and endothelial cell activation that triggers the clotting cascade as well as lymphocyte apoptosis.[76,172] Further studies are needed to understand the disease mechanism in wild boar. In contrast to domestic pigs, CD8+ T cells have been found to play a crucial role in survival and are activated in wild boar postinfection.[94]

Suid herpesvirus 1 (SuHV1) (syn. Pseudorabies virus [PrV]) is an *Alphaherpesvirus* and the causative agent of **Aujeszky's disease** or **pseudorabies**. This disease is WOAH (formerly OIE) listed and notifiable. It has been eradicated from domestic pigs in North America, New Zealand, and some countries in Europe. However, globally, it remains an important disease of domestic pigs and is a disease for which wild boars are considered reservoir hosts. Seroprevalence in wild boars varies geographically and in some regions viral infection may be endemic.[113,147,149,196,239,263] Rare seropositivity has also been noted in peccaries in the southwestern United States. SuHV-1 has been identified by serology and PCR in postmortem samples from feral swine, white-lipped and collared peccaries in El Pantanal, Brazil where the origin of the virus is unknown. In these species, it is generally present in its latent form (serologically nonreactive).[45,53,162] Multiple genotypes are present in wild boar populations, some of which suggest transmission from domestic or feral pigs while others are distinct from strains found in domestic pigs inhabiting the same geographic region. Oral, nasal, and venereal shedding are suggested based on virus isolation studies with higher viral DNA loads in genital swabs.[149,193] Piglets and fetuses of seropositive sows may have both anti-SuHV antibodies and virus suggesting vertical transmission.[162,240] Taken together, these findings support the importance of venereal transmission in adult boars.

The pathogenesis of SuHV infection in wild boars is incomplete as outcome of experimental infection has varied and disease has generally been milder than what is noted in domestic pigs. To what extent this is due to SuHV strain factors, host adaptation, or earlier maturation of CD8 T cells in wild boars is uncertain.[149,163] While young domestic piglets (<7 days) are highly susceptible, high mortality has not been described in similar age wild suids; disease in young wild suids may be underrepresented due to the difficulty in recovering carcasses from very young wild animals. In domestic pigs, primary replication occurs in nasal and oropharyngeal mucosa with viral latency within ganglia. Clinical disease and lesions, when described, appear to be more common and more severe in juvenile boars (4–18 months) than adults.[77,208] Similar to domestic pigs, neurologic signs and lesions are more common in younger boars. Gross lesions in natural outbreaks include lymphadenomegaly and meningeal congestion. Histologically, nonsuppurative encephalitis with neuronal necrosis, lymphoplasmacytic perivascular cuffs, and edema have been described (Fig. 8.9). Intranuclear inclusions are variably noted but the virus can be demonstrated by immunohistochemistry (IHC) and immunofluorescence in sections of tonsil and brain. In adult domestic pigs, respiratory signs are a hallmark of SuHV infection but they have not been described in natural infections in boars. Mild sneezing, nasal discharge, and conjunctivitis have been described after experimental infection of wild boars with strains of SuHV isolated from wild boars.[132] However, wild boars treated with immunosuppressant drugs develop severe pleuropneumonia as do wild boars inoculated experimentally with some domestic pig SuHV strains. Further research is needed to better understand the pathogenesis of these varying strains in the differing hosts.

Porcine circoviruses (PCV) occur as two genotypes, PCV1 which is considered nonpathogenic in domestic pigs and PCV2, the cause of **postweaning multisystemic wasting syndrome (PMWS)**. PCV1 has been detected by PCR and serology in wild boar populations but not associated with disease, similar to domestic pigs.[49,231] PCV2 seroprevalence in European wild boar populations is generally high, suggesting the virus is endemic.[196,32] Some populations have evidence of exposure without evidence of disease,[182] but other populations have had histologic lesions and increased piglet mortality. In some cases, disease has followed the confinement of wild animals, suggesting potential cofactors in pathogenesis.[61] Affected animals are generally young (4–10 months).

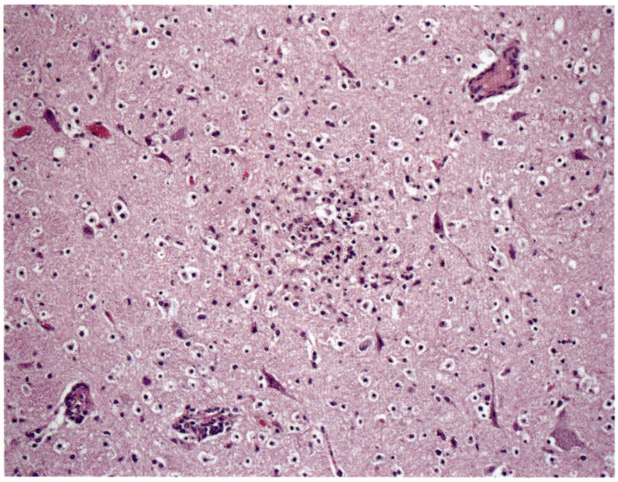

FIGURE 8.9 Suid herpesvirus infection encephalitis in a pig. Necrotizing nonsuppurative encephalitis is associated with neuronal necrosis and perivascular cuffing. Eosinophilic intranuclear inclusions may be present but are generally more easily seen with immunohistochemical labeling.

Lesions in wild boars are similar to those described in domestic pigs and include emaciation with adipose and skeletal muscle atrophy, lymphoid depletion with lymph node and Peyer's patch sinus histiocytosis (Fig. 8.10A), lymphoplasmacytic interstitial pneumonia, and mild erosive enterocolitis associated with mononuclear inflammatory cell infiltrates.[60,89,125,209,218,242]

Characteristic basophilic, botryoid intracytoplasmic, and less common intranuclear inclusions are present within macrophages (Fig. 8.10B). Diagnosis can be confirmed by immunohistochemical labeling of PCV antigen in macrophages, lymphatic endothelial cells, and in gut-associated lymphoid tissue. PCR is also available, although as the virus appears to be endemic in wild suids in some areas, identification of viral DNA may not imply causality. Furthermore, in some studies, PCV2 DNA could not be detected even in seropositive boars with histologic lesions of PMWS.[91] Other lesions, such as hyperplasia of bronchial-associated lymphoid tissue, may be due to secondary infections with *Mycoplasma* sp. or *Pneumocystis* sp.[30] Populations of boar with higher PCV2 prevalence also have a higher prevalence of *Mycobacterium bovis* infection, and more severe and more generalized disease.[185,189] However, findings in other studies have been less clear, suggesting that the interaction between PCV2 infection and potential coinfections may be more complex

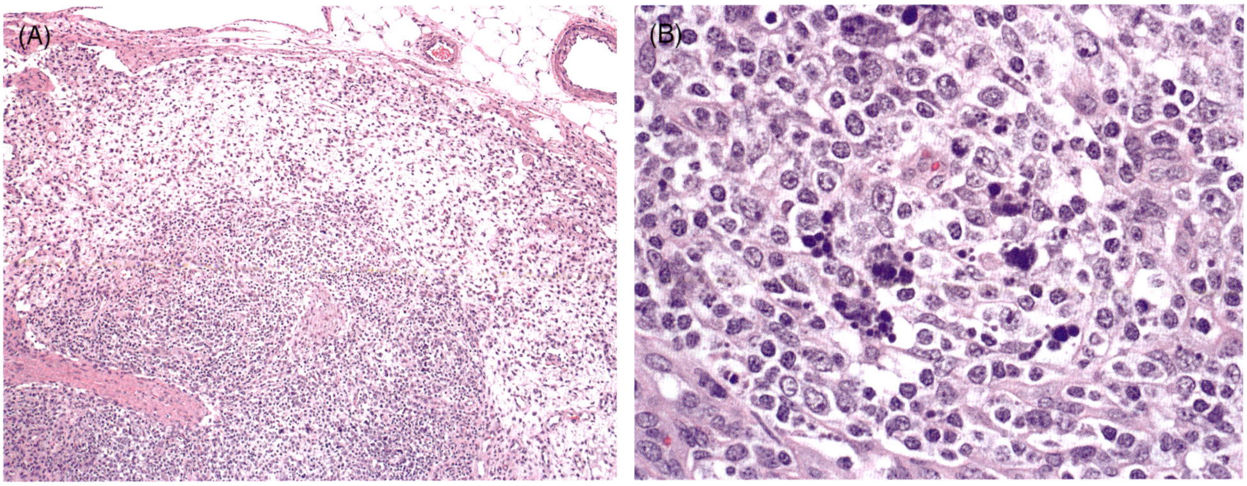

FIGURE 8.10 Porcine circovirus type 2 infection in a lymph node in a pig. (A) Sinuses contain increased numbers of histiocytes and there is diffuse lymphoid depletion. (B) Intrahistiocytic, basophilic, intracytoplasmic inclusions have a characteristic botryoid appearance.

than simply PCV2-induced immune suppression.[52] Despite this, recent experiences have shown that vaccination against PCV2 targeting young wild boar reduces TB severity in this species.[186] In addition to boars, PCV2 has been identified by PCR in tissues from white-lipped and collared peccaries in Brazil but no lesions have been described.[53] While in one study, collared peccaries lacked antibodies to PCV2.[71] absence of detectable titers can be noted in domestic pigs as well and in peccaries this could be due to viral or host factors or potential geographic differences since PCV2 antibodies were detected by ELISA in a collared peccary and feral pigs in Colombia.[151]

A novel papillomavirus, *Sus scrofa* **papillomavirus 2 (SsPV2)** has been associated with cutaneous fibropapillomas in a wild boar from Germany.[125] The distribution of this virus in wild suids is unknown.

RNA Viruses

Classical swine fever (CSF), also called **hog cholera**, is caused by an enveloped RNA virus of the *Pestivirus* genus (family *Flaviviridae*) and is WOAH notifiable. The natural hosts for **CSF virus (CSFV)** are members of *Suidae*. The disease causes major economic losses in domestic pig populations. The Eurasian wild boar is also susceptible and plays an important role in the transmission and epidemiology of the disease in Europe.[777,194] The wild boar is not a classic reservoir but is capable of maintaining infection over long periods of time.[9] Wild African *Suidae*, such as the common warthog and the bushpig, have been experimentally infected.[61,70]

Transmission routes are similar in the domestic pig and wild boar. The virus is spread by direct contact between infected animals and indirectly by contact with contaminated food, feces, and carcasses. Naïve populations are infected accidentally by indirect or direct contact with infected domestic pigs or wild boars.[146] CSF outbreaks with highly virulent strains of CSFV in small populations of wild boars can be self-limiting.[146,777] However, there is concern in Europe about the wild boar becoming a dangerous reservoir in large populations where CSFV tends to become endemic and persist for years.[146,167,777] Animal density also plays an important role in maintenance of the virus, since a higher turnover rate increases the number of younger susceptible pigs.[146] A study in France supported the idea that denser populations with more young favored the persistence of CSF. Additionally, populations of wild boar occupying larger forests had a higher incidence of CSF than populations occupying smaller spaces.[195]

The pathogenesis is not fully understood but involves cytopathic effects of the virus on lymphoreticular cells and macrophages, the vascular endothelium, and epithelial cells. Virus enters through mucous membranes, replicates in the tonsillar epithelium, and spreads to cervical lymph nodes. Depending on strain virulence (classified as high, moderate, or low virulent) and the interaction with the host, CSF may occur as an acute, subacute, chronic, or inapparent disease. Acute infections of CSF cause hemorrhagic fever, but subacute, chronic, and subclinical forms of the disease also occur depending on virulence of the strain and the immune response of the pig.[146,167] Clinically, animals with acute disease show anorexia, lethargy, fever, and leukopenia.[167,190] Hemorrhages in the periphery of lymph nodes and renal petechiae are common findings. Splenic infarction is considered almost pathognomonic for CSF. The spleen is not enlarged, so infarcts stand out within the parenchyma.[146] Experimental studies in common warthogs and bushpigs demonstrated intra-species transmission.[61,70] Bushpigs develop similar clinical signs and lesions as domestic pigs. Consistent lesions include necrotizing and ulcerative

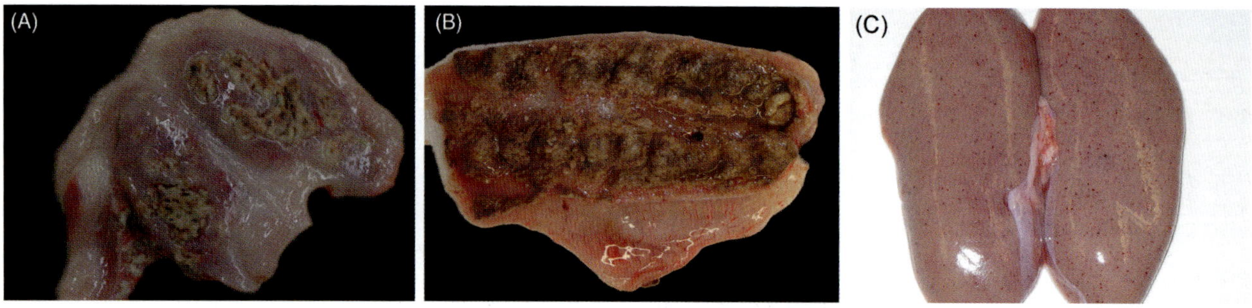

FIGURE 8.11 Experimental classical swine fever virus infection in a European wild boar. (A) Necrotizing tonsillitis. (B) Necrotizing enterocolitis with button ulcers due to infarction of capillaries. (C) Renal petechiae are similar in gross appearance to those in ASFV infection (see also Fig. 8.7B). *(Parts A-C: Photos Courtesy of the Friedrich Loeffler Institute.)*

enteritis, suppurative rhinitis, and pneumonia (Fig. 8.11). Renal petechiae can also be present, similar to ASFV infection (Fig. 8.7B). Histologically, affected animals also have lymphoid necrosis and depletion. Perivascular cuffs in the brain and multiple, random, organs can be seen in surviving animals. In contrast, infected warthogs may be clinically normal. Histologically, warthogs may have rare or subtle lesions consisting of lymphoplasmacytic infiltrates in various organs and rare perivascular cuffing.[72]

In the chronic form of the disease in domestic pigs, colonic and cecal mucosal defects ("button ulcers") are characteristic findings. These lesions are characterized by sharply delineated, circular areas of hemorrhage with necrosis of submucosal and mucosal lymphoid tissue. These centers are filled with cellular debris and fibrin; when removed, they leave a sharp, sunken, discoid ulcer.[146]

Clinical and gross findings for CSF are very similar to ASF making them clinicopathologically indistinguishable. Histologically, minor differences can be extrapolated from the literature, particularly with subacute and chronic forms of the disease, but are hardly sufficient for differentiating between the two diseases. In CSF, vascular lesions particularly in lymph nodes, consist of fibrinoid necrosis and vasculitis, and are often accompanied by a more overt inflammatory response than in ASF. Perivascular cuffing is also described in chronic persistent forms of CSF. Additionally, the suppurative rhinitis and pneumonia seen in some species of wild suids infected with CSF, differs from the lymphohistiocytic interstitial pneumonia in domestic pigs or the bronchopneumonia in wild boars with the acute to chronic disease of ASF. Button ulcers in the large intestine are uncommon in chronic forms of ASF. Also, endemic wild suid species in Africa do not develop overt disease when infected with ASF, contrary to CSF, where both warthogs and bushpigs develop similar clinical and pathological signs as domestic pigs.

Given the impossibility of differentiating CSF from ASF due to the lack of specificity of the clinical signs and lesions, the WOAH requires laboratory testing of antibodies or the virus itself to confirm CSF. According to the WOAH Manual of Diagnostic Tests,[253] the virus must be cultured and isolated in pig kidney cell lines or other CSFV permissive cell lines. Serological tests that detect CSFV antibodies are useful in herds suspected of being infected 21 days prior.

Control measures for domestic pigs in the European Union (EU) are based on culling infected or suspicious animals and temporary animal movement restrictions; emergency vaccinations may be employed when uncontrollable spread of CSF is suspected/occurs.[146] In wildlife, detection of infection is more difficult and control measures are focused on reduction of population densities, intensive diagnostics, and good hygiene practice during hunting season. In some parts of Germany, these measures have been supplemented by oral vaccination combined with hunting efforts that target young wild boars.[167]

In addition to carnivores, **canine distemper virus (CDV)** (genus *Morbillivus*) can cause encephalitis in collared peccaries.[6] Clinically affected animals appear blind, reluctant to move and develop myoclonus. Lesions are restricted to the brain with variable neuronal necrosis, astrocytosis, and perivascular cuffs. Intracytoplasmic eosinophilic inclusions are present within neurons. CDV antigen can be demonstrated within neurons by IHC. Serologic surveys suggest peccaries are enzootically infected[157] although much is unknown about disease epidemiology.

Rinderpest virus has caused fatal disease in wild suids. Natural infection has been reported in warthogs, bush pigs (red river hogs), and giant forest hogs in Africa,[4] and been experimentally reproduced in peccaries.[66] Lesions include extensive ulceration with overlying diphtheritic membranes in the stomach and over Peyer's patches in the small intestine and cecum, although they can occur in other areas within the gastrointestinal tract. Congestion is common in the gastrointestinal tract, kidneys, and urinary bladder. Regions of cyanosis or mucosal ulceration may be seen in the skin, tongue, trachea, and larynx. The myocardium may be pale. Rinderpest was eradicated in the wild in 2011 but is still a WOAH-listed notifiable disease.

Foot and mouth disease (FMD), caused by an *Aphthovirus*, **foot and mouth disease virus**, is a severe, highly contagious disease of a wide variety of cloven-hoofed species, including domestic suids. The disease has been eradicated in some countries and the geographic distribution of disease has decreased; however, it is still present in South America, Africa, Asia, and parts of the Middle East and Eastern Europe. There are seven different serotypes but how this relates to pathogenicity in wild suids is unknown. While the virus can survive within the oropharynx in persistently infected hoof stock, viral persistence has not been shown in suids. Some serologic surveys in wild boar have identified seroconversion in some, but not all, outbreaks related to livestock.[2,59] However, to date, how and if FMD unfolds naturally in wild boars or free-ranging pigs is unclear. Classic FMD lesions are cutaneous and mucosal vesicles with ulceration. Associated clinical signs depend on lesion site. In Suidae, the most common clinical sign is lameness due to vesicles and ulceration along the coronary band between the claws.[130] Secondary bacterial infection is common. Vesicles also occur on the snout and to a lesser extent in the mouth. Disease in wild boars appears to be less severe than in domestic pigs based on experimental studies.[30]

Ebola viruses (EBOV), family *Filoviridae*, can cause fatal hemorrhagic diseases in humans and nonhuman primates (also see New World and Old World Monkeys and Apes). In 2008 investigation into an unusually severe outbreak of respiratory disease and abortions in domestic pigs suggestive of a pathogenic strain of porcine reproductive and respiratory syndrome virus identified coinfection with **Reston EBOV (REBOV)**.[10] In this initial study, REBOV antigen localized to the capsular surface of lymph nodes while PRRSV was present within germinal centers. Both viruses co-localized to the lung and were associated with interstitial pneumonia and necrotic debris within alveoli. Subsequent studies have found PRRSV and REBOV coinfections in other populations of domestic pigs.[164] In experimental studies in pigs, Reston EBOV replicates within internal organs and can be shed from the nasopharynx without evident clinical signs[136]; experimental infection of domestic pigs with Zaire EBOV (ZEBOV) causes severe neutrophilic and histiocytic pneumonia and respiratory distress.[112,155]

Fruit bats are thought to be the natural reservoir for EBOV. However, given the known susceptibility of domestic pigs to EBOV and the sympatric distribution of wild suids in regions where previous EBOV outbreaks have occurred, some have suggested suids as either an incidental host or potentially an important part of EBOV disease ecology. Serosurveys in domestic swine farms infected with REBOV demonstrated wide variation in seroprevalence from 3% to 70% for REBOV.[206] REBOV can be transmitted from suids to humans and experimental studies suggest that suids can also transmit ZEBOV to nonhuman primates.[249,250] Virus neutralization assays to test for ZEBOV have recently been optimized for use in swine.[170]

Swine influenza is caused by a number of **influenza A** viruses, further characterized as H1N1, H1N2, and H3N2, which include avian and human influenza viruses. Coinfection increases the opportunity for viral reassortment and the development of increased pathogenicity. Additionally, avian and human subtypes were identified in wild suids, including pandemic H1N1 2009, classical H1N1, H1N2, and H3N2.[1,18,37,65,101,168,213,233,32,242] In one study, H1N2 virus was associated with bronchopneumonia in boars; however, it is unknown if the virus was contributory, as IHC was negative for influenza A and positive for *Mycoplasma hyopneumoniae*.[18]

Bacteria

The *Mycobacterium tuberculosis* complex (MTC) includes mycobacteria capable of causing **tuberculosis (TB)** in countless wildlife species, including suids. The wild boar is generally considered as a spillover or dead-end host for tuberculosis and similar observations have been noted in other wild suids.[46,108] Generalized lesions and mortality due to tuberculosis are low and infection usually occurs in wild boar populations in contact with livestock and wild ungulates.[108,151] Recent epidemiological studies show this is true for wild boar in central and northern Europe. However, the scenario differs in the Spanish Mediterranean ecosystems where evidence suggests wild boars are natural reservoirs able to maintain TB infection and transmit disease to other species.[8,138,151]

Mycobacterium bovis (bTB) and other genotypes of the MTC are prevalent in wild boar in the Spanish Mediterranean region.[8,151] Infection is high in dense populations and juvenile piglets play an important role in disease epidemiology. Studies show that up to one-third of juvenile wild boars may become infected during the first months of life and are more likely to develop larger, generalized lesions.[8,138] Additionally, bTB was found to be the cause of death in approximately 30% of the wild boars. The proportion of wild boars with generalized bTB lesions decreased with age as did deaths associated with bTB.[8] These results contrast with results obtained in other parts of Europe, where the death rate due to bTB is 3%.[108]

In wild boar, bTB lesions can either be localized in one anatomical region or found affecting several organ systems. In one study, head lymph nodes (LNs), particularly the mandibular, were the sites most frequently and consistently affected, often being the only site of disease. Other affected organs include lung, thoracic LNs, liver, spleen, and mammary gland.[138] Both respiratory and food-borne infections seem plausible in this species, given the distribution of lesions. In most cases, mycobacteria enter the oral mucosa and disseminate to tonsils and mandibular LNs. Wild boars with extensive lesions and systemic infection are potentially capable of excreting mycobacteria through several

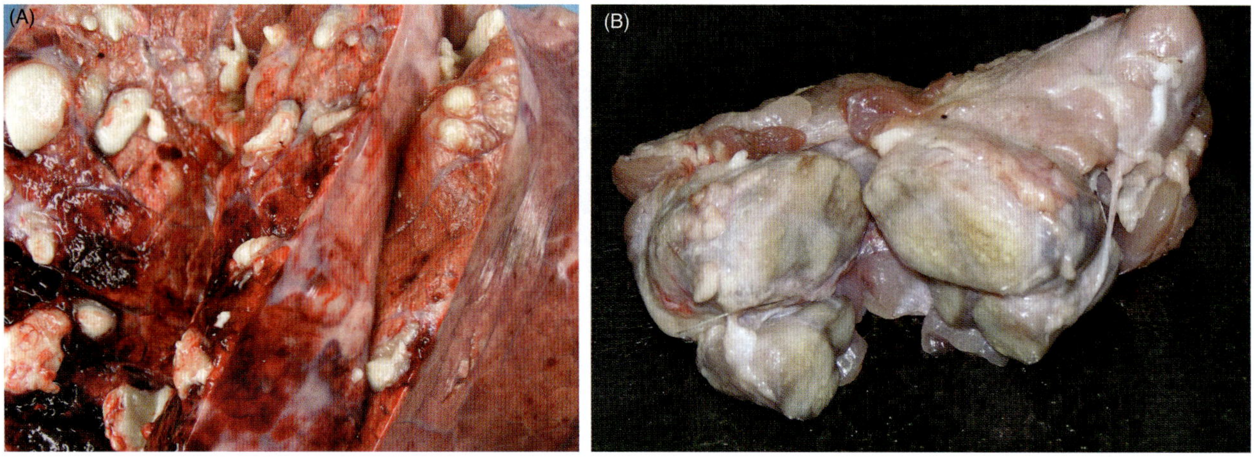

FIGURE 8.12 Mycobacteriosis due to *Mycobacterium bovis* in a European wild boar. (A) Multiple pulmonary granulomas are characteristic lesions in tuberculous mycobacterial infections. (B) Granulomatous lymphadenitis. *(Parts A, B: Photos Courtesy of José Ángel Barasona García-Arévalo, SUAT-VISAVET, Department of Animal Health of the Complutense University of Madrid.)*

routes, mainly respiratory and gastrointestinal.[8,138] *M. bovis* excretion in wild boar is also positively correlated with the existence of coinfections with PCV2 and *Metastrongylus* spp.[188]

Gross lesions in boar with bTB resemble those of tuberculosis in other species, and consist of focal to multifocal to miliary, variably sized granulomas (Fig. 8.12). Microscopically, granulomas are characterized by nodular aggregates of epithelioid macrophages, multinucleated giant cells, surrounded by lymphocytes, plasma cells, and macrophages (Fig. 8.13). Larger or more chronic granulomas are additionally surrounded by fibrosis and contain central areas of necrosis with multifocal mineralization. Necrosis and mineralization are more extensive in generalized cases of TB, when more than one organ is affected. Acid-fast bacilli are consistently found in larger numbers in pulmonary granulomas, in contrast to lymph nodes where intralesional mycobacteria are scant to absent.

Diagnosis of bTB is made by detection and identification of mycobacteria of the MTC in affected organs. Bacterial culture, molecular diagnostics, and enzyme-linked immunosorbent assays (ELISAs) are invaluable tools for diagnosis and detection of exposure.[23,25,75] Histochemical detection of acid-fast bacteria or immunohistochemical labeling of intracellular or intralesional mycobacteria within or near granulomas or necrosis, are suggestive of the disease; however, false negative results may occur when bacteria are rare. Additionally, classic and real-time polymerase chain reaction (PCR), with genome reamplification are sensitive, specific, and fast diagnostic tools for TB and are of use in the wild boar.[75]

Brucellosis in wild swine is typically caused by ***Brucella suis*** although ***B. abortus*** and ***B. melitensis*** are also reported to cause mild disease. *B. suis* biovar 1 has been detected in populations of collared peccaries in Venezuela and wild boars in Croatia. In these populations, few animals develop disease but the populations are considered an important reservoir for infection. In contrast, biovar 2 is associated with disease outbreaks in European wild boars.[73,127] Disease in wild swine is similar to that in domestic swine with oral or venereal transmission followed by lymphadenitis and bacteremia. Granulomatous lesions with central coagulative necrosis surrounded by suppurative exudate are typical and can be found in male (Figs. 8.14A, B) and female genitalia, bone, joints, mammary gland, lymph nodes (Fig. 8.14C), spleen, liver, kidney, bladder, and brain. Uterine lesions consist of focal to coalescing miliary abscesses and granulomas in endometrium and placenta. Histologically, uterine glands are dilated and filled with neutrophils and mucus, and there

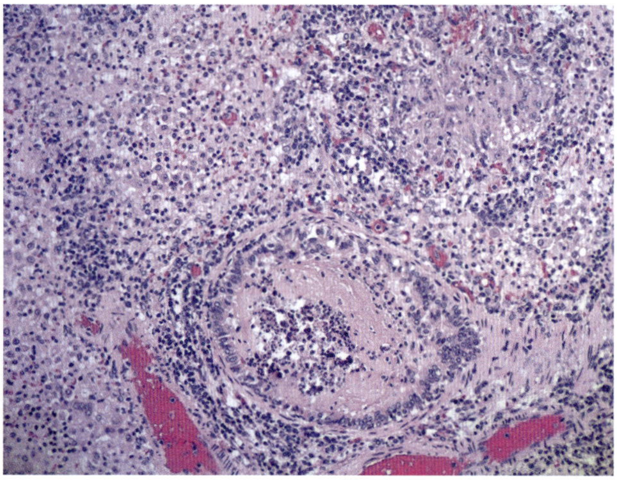

FIGURE 8.13 Mycobacteriosis due to *Mycobacterium bovis* in a free-ranging warthog. Granulomatous pneumonia with both more diffuse granulomatous inflammation as well as organizing granulomas. *(Photo Courtesy of E. Mitchell, University of Pretoria.)*

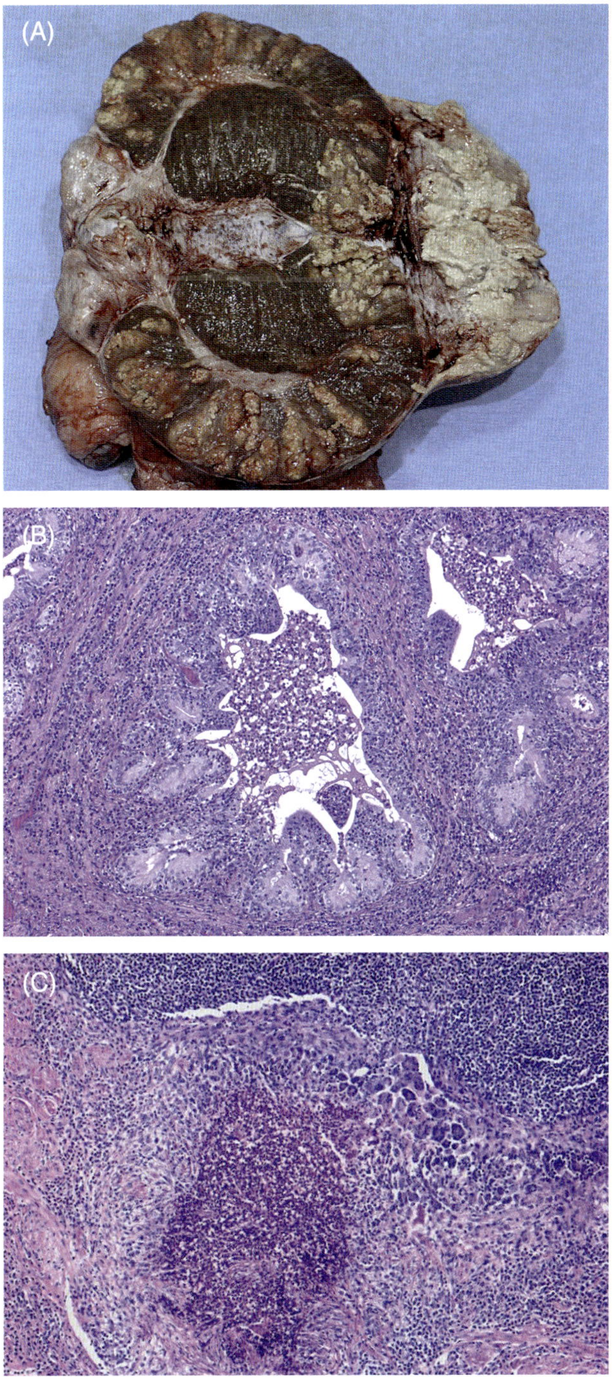

FIGURE 8.14 *Brucella suis* in feral *Sus scrofa*. (A) The epididymis contains numerous coalescing granulomas and granulomatous inflammation. (B) Granulomatous infiltrates within the seminal vesicle mixed within proteinaceous fluid and cell debris. (C) Lymphadenitis with a large granuloma rimmed by multinucleated giant cells. *(Part A: Photo Courtesy of T. LaBranche, University of Georgia.)*

is a loss of endometrium. The fetal placenta is congested and there is multifocal to coalescing hemorrhage, edema, and a mucopurulent exudate with numerous bacteria. Aborted fetuses are usually autolyzed with serosanguinous fluid in body cavities, subcutaneous tissues, and abnormal stomach content. Vertebral lumbar osteomyelitis and fibrinopurulent synovitis are common lesions in juveniles and in chronic disease. Serological cross-reaction has been reported with *Yersinia* O:9, *Escherichia coli* O:17, and some serobiovars of *Salmonella*, which complicates diagnosis based solely on serology.[73,131,207]

Erysipelothrix rhusiopathiae is a gram-positive bacillus with a wide geographic distribution and host range. It is the cause of **porcine erysipelas** and is capable of causing outbreaks in other domestic and wild animals. Pigs are a reservoir for infection, particularly for other pigs, since *E. rhusiopathiae* can persist for long periods in lesions and persists in the tonsils, bone marrow, and intestine of healthy swine.[190] Aggregation and high wild boar density can influence spread.[24,187,195] Seroprevalence has been detected in wild boars in Japan and coastal Mediterranean countries, such as Greece and Spain; it is also likely to be present in other populations of wild boars.[41,134,187,212,241]

Porcine erysipelas occurs in swine of all ages but young animals (up to 12 months) and pregnant sows are most susceptible. Disease can be acute, causing septicemia and death. Mild and chronic forms are characterized by well-demarcated skin necrosis, endocarditis, and polyarthritis.[190] Gross lesions in acute or chronic disease include multifocal to diffuse petechiae, erythema, and/or cyanosis (Fig. 8.e5). Histologically, vasculitis with microthrombi, intralesional bacteria and associated leukocytic infiltrates, mural fibrinoid necrosis, and hemorrhage occur throughout the body but are most remarkable in the kidneys, urinary bladder, and skin (Fig. 8.15). Multifocal mononuclear infiltrates may be seen in the kidneys,[187] and lymphoplasmacytic infiltrates, microthrombi, as well as alveolar edema and congestion, may be seen in alveolar septae. Corneal edema and panuveitis with fibrinoid vasculitis have also been described. Lesions in wild boars reported by Risco et al.[187] were similar to those of porcine erysipelas and erysipelas in other species. Septicemia in wild boars due to erysipelas has also been reported in Japan.[257]

Brachyspira is gram-negative, anaerobic spirochetes, adapted to occupy specialized niches in the large intestines of animals, including swine. Two important entities in domestic swine are caused by ***Brachyspira hyodysenteriae*** and ***B. pilosicoli***, known as **swine dysentery (bloody scours)** and **porcine intestinal spirochetosis**, respectively.[84] *B. hyodisenteriae* is a highly hemolytic bacteria, capable of causing episodes of mucohemorrhagic diarrhea that can result in high morbidity and mortality rates, which cause important economic losses for domestic swine farms. Information about *Brachyspira* infection in wild boars or in other wild suid genera is limited. Several studies in free-ranging wild boar in Sweden and Spain failed to demonstrate *Brachyspira* in fecal samples, which seemed to indicate that these spirochetes are not a natural part of the

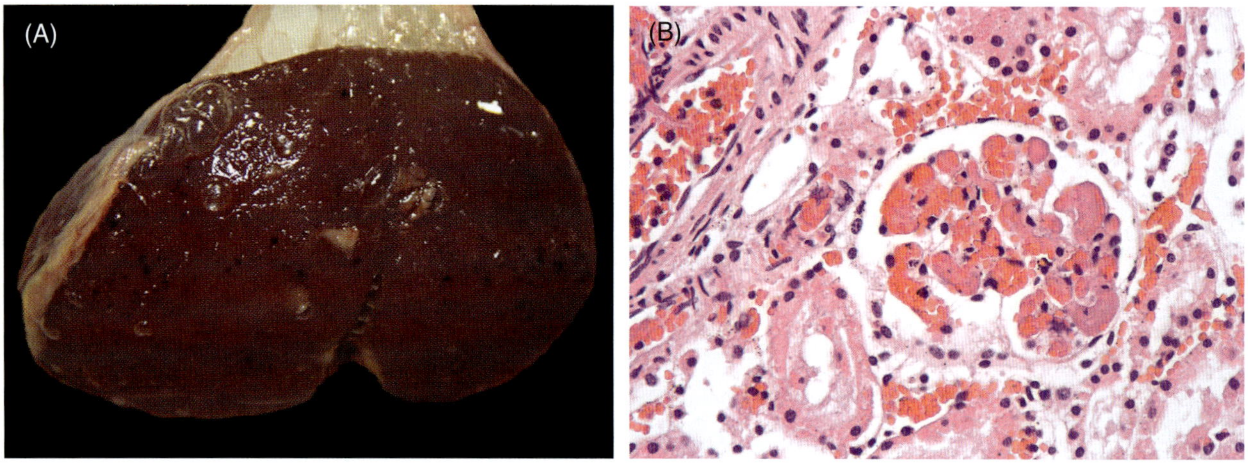

FIGURE 8.15 *Erysipelothrix rhusiopathiae* nephritis in a wild boar. (A) Renal petechiae cannot be differentiated from petechiae due to other causes in suids (see also Figs. 8.7B and 8.11C). (B) Hemorrhage is present in the renal interstitium surrounding the glomerulus. Note the ischemic epithelium of the proximal tubules and fibrin thrombi within the glomerular capillaries. *(Parts A, B: Photos Courtesy of T Risco Perez, INGULADOS S.L.)*

intestinal flora.[95,161,236] However, in one study in Spain that included 40 wild boars from several intensive-growth farms, *B. intermedia*, *B. pulli*, and *B. hyodysenteriae* were isolated and associated with clinical disease in three animals.[236] In addition to documenting infection, this study also supported the hypothesis that weakly hemolytic species, such as *B. intermedia* and *B. pulli*, have pathogenic potential in nondomestic suids. Details of the lesions in wild boar were not specified. Several species of *Brachyspira*, including a highly hemolytic strain of *B. hyodysenteriae*, have also been identified in feral pigs in Western Australia (information on associated disease or lesions has not been described).[169]

Mycoplasma hyopneumoniae is the cause of mycoplasmal pneumonia, also known as **enzootic pneumonia**, of swine. *M. hyopneumoniae* is a small bacterium that lacks a cell wall and adheres to the ciliated epithelium of large airways and to a lesser extent, the bronchioles. Most mycoplasmas contain superantigens in their cell membranes that induce polyclonal lymphoid proliferation and characteristic lymphoid aggregates adjacent to airways.[119] Seroprevalence of *M. hyopneumoniae* has been reported in wild boars in several parts of Europe, including Switzerland, Greece, and Spain.[36,114,115,215,239] Most of these studies do not have associated serology with disease. However, in one comprehensive study on the prevalence of *M. hyopneumoniae* in wild boars in Switzerland, mild to moderate lymphoplasmacytic and histiocytic bronchointerstitial pneumonia, primarily affected the cranioventral region of the lung (Fig. 8.16A).[13] Lesions were consistent with the acute and chronic stages of the disease described in domestic pigs. These include acute disease with multilobular, patchy red to dark red areas affecting less than 50% of the lungs versus more chronic lesions in which a higher percentage of the pulmonary parenchyma is affected by dark red to gray, atelectatic areas. Histologically, infected boars had hyperplastic bronchial associated lymphoid tissue (BALT) with reactive lymphoid follicles, interstitial infiltrates of plasma cells; lymphocytes and histiocytes, alveolar and bronchial neutrophilic infiltrates, and alveolar edema were also, but less commonly, seen (Fig. 8.16B). The amount of inflammation, number of neutrophils, and hyperplastic BALT increased with the chronicity. Juvenile and young adult boars are more susceptible to infection, which corresponds to serological reports of wild boars in Italy, and also to the age range of susceptibility in the domestic pig. While wild boars are frequently viewed as the reservoir for disease in domestic pigs, in this study, it was instead more likely for wild boars to become infected by contact with domestic pigs. Based on this, the role of wild boars as reservoirs for *M. hyopneumoniae* transmission to domestic pigs is considered minor.

Several species of ***Chlamydophila*** have been reported both serologically and histologically in wild boars[93,32]; however, its significance as a pathogen is uncertain. In one study, *C. psittaci* was the most prevalent species; *C. abortus* and *C. suis* were less commonly identified. Chlamydial inclusions in the lung were present in smooth muscle cells around bronchioles, the endothelium and interstitium of alveolar walls as well as the myometrium of the uterus.[93] In Spain, a fatal case of bilateral keratoconjunctivitis caused by *C. suis* and affecting a 2-month-old piglet has been also reported.[187] However, the significance of chlamydial infection in wild boar is currently uncertain.

Clostridia are large, spore forming, strictly anaerobic to oxygen-tolerant anaerobic bacteria. *Clostridium perfringens* types A and C, and *C. difficile* are the principal enteric clostridial pathogens of domestic swine. All subtypes of *C. perfringens* can be commensals in the intestines of many animal species; therefore, diagnosis must be made

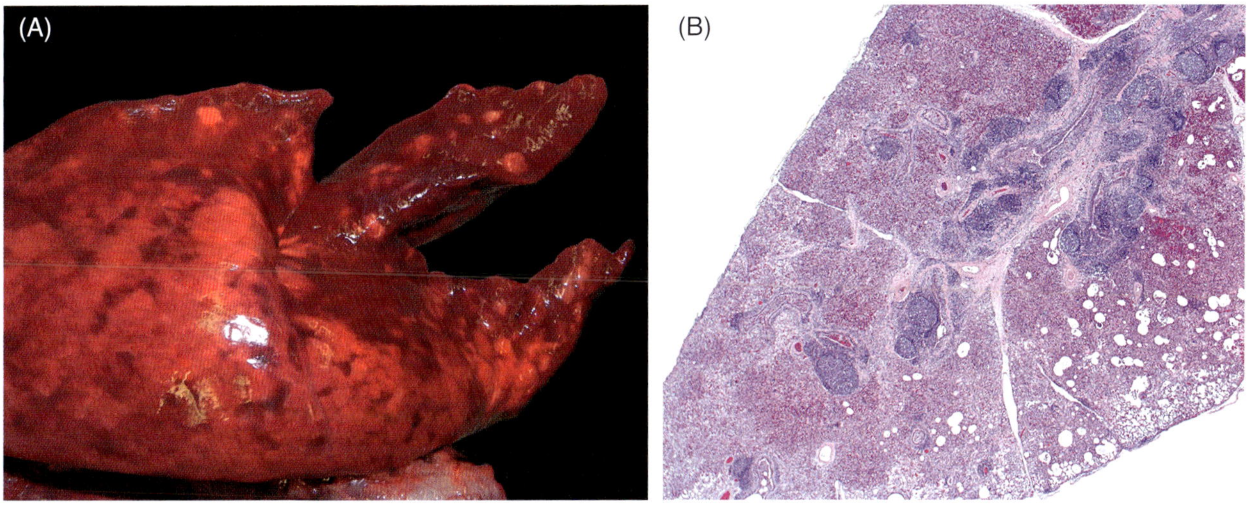

FIGURE 8.16 *Mycoplasma hyopneumoniae* infection (enzootic pneumonia) in the lungs of a wild boar. (A) Infection causes bronchopneumonia with consolidation that often has a cranioventral pattern. Affected parenchyma will be firm, often discolored dark red, and demarcation between affected and less or unaffected adjacent tissue may be apparent. (B) Infection is airway oriented and regionally, the airways and the alveolar interstitium contain numerous inflammatory cells. Bronchial-associated lymphoid tissue is hyperplastic. With chronicity, the severity of infection increases. *(Parts A, B: Photos Courtesy of the Centre for Fish and Wildlife Health, University of Bern.)*

by detecting the toxin in association with consistent lesions and in the absence of other potential pathogens. Standard tests for toxin detection include real-time PCR for toxin-encoding genes (more commonly alpha-toxin gene and enterotoxin gene). Detection and toxin typing in formalin-fixed tissue samples have also been developed.[255]

C. perfringens type A is part of the normal intestinal flora in pigs but can cause enteric disease in neonatal and occasionally weaned pigs. The pathogenesis of *C. perfringens* type A in pigs is poorly understood but it is likely to be multifactorial. *C. perfringens* is a known cause of enteritis in a variety of wild animal species and has been reported in wild Suidae, such as wild boar and pygmy hogs.[52,72,123,214] In an outbreak occurring in wild boar in China, *C. perfringens* type A and C were isolated by serological and molecular assays, and found to be the cause of death. Gross lesions consisted of extensive hemorrhage in lungs, intestine, and inguinal lymph nodes.[123] Hemorrhagic enteritis associated with *C. perfringens* type A has also been reported in a research and breeding center for pygmy hog in Assam, India. Lesions were similar to those reported in the domestic pig.[214] Hemorrhagic and necrotizing enteritis with diffuse small and large intestinal edema occurred in another adult, female pygmy hog at the Assam State Zoo, in Assam, India. Molecular assays confirmed *C. perfringens* type A infection.[52] *C. perfringens* type C occurs worldwide and causes necrotizing and hemorrhagic enteritis that is often fatal in domestic swine. Hallmark histologic lesions are deep mucosal necrosis and hemorrhage with emphysema in the small intestine that sometimes also affects the colon. It is a primary pathogen and usually causes more severe disease in young animals.

C. difficile is a cause of enteritis in neonatal pigs that usually develops between 1 and 7 days of age. Hallmark lesions include edema of the mesocolon, suppurative inflammation of the colonic lamina propria, and mesenteric lymph node edema.[72] *C. difficile* has been isolated from feral pig feces in areas with intensive commercial swine production in Canada, and feral pigs could be reservoirs.[228]

Wild suids are important reservoirs for ***Salmonella*** species and infection is reported worldwide, particularly in wild boars and pygmy hogs. Wild boars may be asymptomatic carriers and maintenance reservoirs in areas with sufficient boar density.[176,32] Transmission occurs directly or indirectly by contact with contaminated water, food, feces, or other infected individuals. Several acute, fatal outbreaks have occurred in both adult and young (<1 year of age) pygmy hogs.[176] Only adults showed clinical signs, which included generalized loss of body condition, anorexia, high fever, loose feces that contained mucus and blood, weakness, and tremors. All animals had necrohemorrhagic enterocolitis with multifocal, discrete areas of necrosis in the cecum and colon. Additionally, the liver, epicardium, and pericardium contained areas of hemorrhage, and the spleen was enlarged. Cerebral congestion was also a consistent feature. **Enterotoxigenic *Salmonella enteritidis*** (***Salmonella enterica enteritidis***) that carried fimbrial genes for colonization of enterocytes was isolated in other pygmy hog outbreaks, ***Salmonella typhimurium*** has been isolated.[176]

Haemophilus parasuis, the cause of **Glasser's disease** or porcine polyserositis and arthritis syndrome, causes fibrinosuppurative inflammation of one to several serosal surfaces (Fig. 8.e6) and joints; it may also cause pneumonia or meningoencephalitis.[47] Antibodies to *H. parasuis* have been reported in wild boars in Slovenia, Spain, and Germany,[159,182,239] but information on disease development in this species or other wild suids is limited. Fatal infections have been reported in free-ranging wild boars in central Spain. Clinical signs included respiratory distress and weakness; however, there was only one documented fatality, suggesting a low mortality rate.[50] The one young boar that died had fibrinous bursitis, interstitial bronchopneumonia, interstitial nephritis with hemorrhage, and mild meningitis. *H. parasuis* was cultured and confirmed with molecular methods. The polyserositis and serofibrinous exudates usually associated with infection in domestic pigs were lacking in the boar.[47,50]

Pasteurella multocida is a gram-negative bacillus with five known serovars, A, B, C, D, and E. Coinfection with toxin-producing strains of *P. multocida* and other bacterial pathogens are the cause of **progressive atrophic rhinitis (PAR)** in domestic pigs and sporadic outbreaks of septicemia can occur. The disease usually affects 6–12 week-old pigs and causes sneezing, nasal discharge, unilateral epistaxis, and nasal deformity.[34] Serovars A and D are known causes of pneumonia in weaner and finisher pigs but are rarely primary pathogens. Predisposing factors, such as environmental conditions or coinfection with *Mycoplasma hyopneumoniae*, porcine herpesvirus 1, or other respiratory pathogens are considered necessary for disease development in domestic pigs.[34,180] Serologic evidence of *P. multocida* infection without disease has been reported in peccaries.[53] However, both PAR and systemic pasteurellosis have been reported in wild boars.[102,185] In cases of systemic pasteurellosis, boars were found dead with epistaxis but had no clinical signs prior to death. Gross lesions consisted of subcutaneous edema in the ventral region of the neck, generalized congestion, and fibrin within the peritoneal cavity and pericardial cavity. Histologically, multiple organs were congested and multifocal hemorrhage and thrombosis were seen (Fig. 8.17).

Streptococcus suis is an important bacterial pathogen in swine that has worldwide distribution. Infection causes a variety of diseases, including sepsis, meningitis, polyarthritis, bronchopneumonia, and endocarditis. Over 30 serotypes are recognized, but serotypes 1 and 2 are generally of greater significance in domestic pigs. Serotype 1 infects suckling pigs and is not pathogenic for other species, while serotype 2 causes disease in weaner and feeder pigs and humans. Serotype 7 can also be zoonotic.[47] Feral swine and wild boars are known carriers of several *S. suis* serotypes, some of which are potentially zoonotic.[9,14,186,201,211] Several cases of *S. suis* infection in hunters in contact with wild boars have occurred.[14] However, not much is known about the clinical and pathological significance of infection in wild boars. A single case of fatal *S. suis* serotype 2 infection was reported in a wild boar in Spain. The animal was young (approximately 2 weeks old) and had suppurative bronchopneumonia and meningitis with hemorrhage (Fig. 8.18), which is similar to lesions that are described in domestic pigs.[186] Other organs, such as the kidneys, had congested vessels, microthrombi, and hemorrhages, consistent with septicemia.

Fungi

Coccidioides immitis is a dimorphic fungus that is endemic to the southwestern United States, Northern Mexico, and

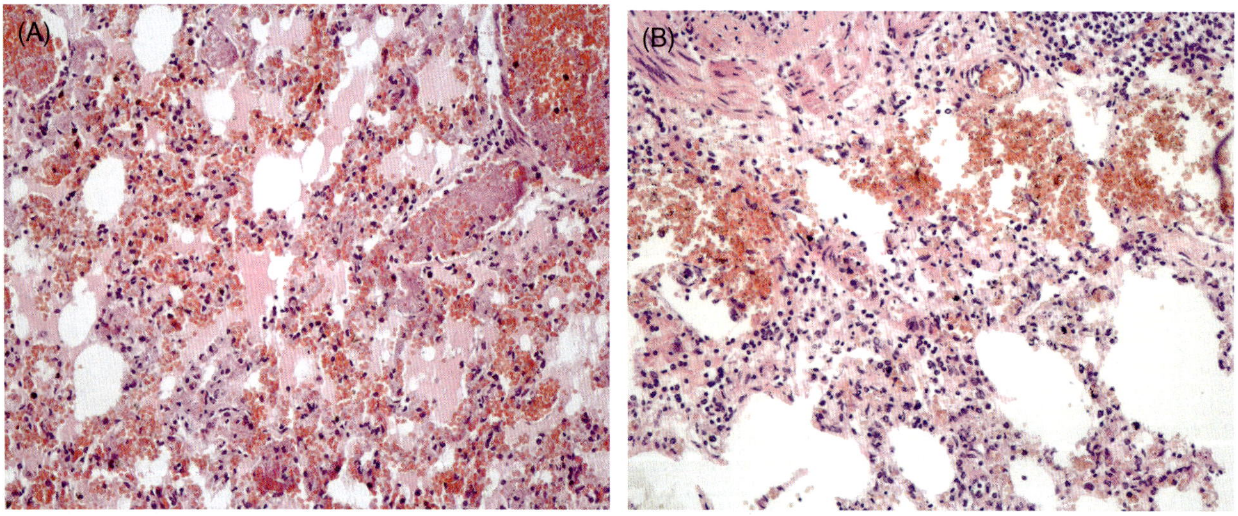

FIGURE 8.17 Systemic pasteurellosis in a wild boar. (A) Several capillaries are occluded by fibrin thrombi. Alveolar septae are hemorrhagic and capillaries contain fibrin, neutrophils, and numerous erythrocytes. Alveolar spaces or filled with edema fluid. (B) The alveolar interstitium contains sheets of extravasated erythrocytes and infiltrates of neutrophils. *(Photo Courtesy of D. Risco Perez, INGULADOS S.L.)*

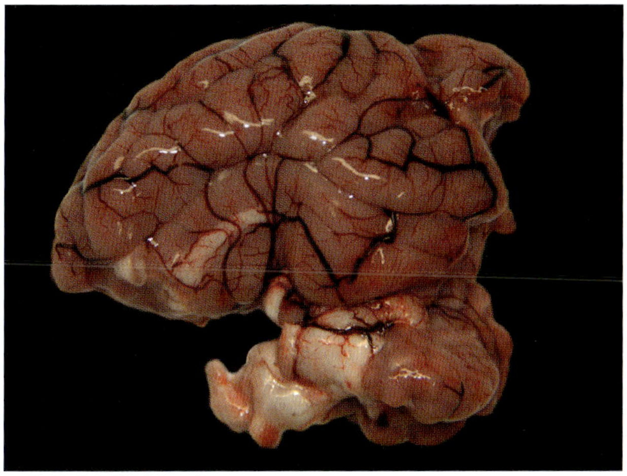

FIGURE 8.18 *Streptococcus suis* in a wild boar. Acute meningitis and hemorrhage with meningeal vascular congestion. *(Photo Courtesy of D. Risco Perez, INGULADOS S.L.)*

some countries in Central and South America. **Coccidioidomycosis** has been reported in collared peccaries but despite the species, rooting behavior and the range overlap with the agent, few exposed individuals develop clinical signs and lesions. Nodular to miliary pyogranulomatous or granulomatous pneumonia is most common, but lesions may be multisystemic and involve regional lymph nodes. Microscopically, necrotic foci contain fungal spherules in different stages of maturation and are surrounded by pyogranulomatous or granulomatous inflammation that matures into organized granulomas or pyogranulomas. Intralesional basophilic, 10–80 µm diameter, round to oval and double contour walled spherules with multiple, round to oval, 2–5 µm, basophilic, PAS and GMS positive endospores support the diagnosis. Infection has also been associated with a neurological-like disorder in Texan Peccaries (incoordination, muscle tremors, and convulsive seizures), however neither gross nor histologic lesions were reported.[66,126] High antibody titers may be suggestive of active disease. Mature sporangia may be up to 200 µm diameter, a feature that is helpful in differentiating them from blastomycosis and histoplasmosis.[34,104]

Metazoa

A wide variety of parasites have been described in free-ranging suids. Many of these are important pathogens of domestic swine but appear to be asymptomatic in wild suids. Table 8.e2 in the digital content contains additional parasites known to infect wild suids.

Gastric metazoa include ***Hyostrongylus rubidus***, a trichostrongylid nematode that infects wild boars and warthogs, and ***Ascarops strongylina***, a spirurid nematode that infects wild boar and pygmy hogs.[15,63,103,133] The life cycle for both is indirect and they use an arthropod intermediate host (dung beetle for *Ascarops*). *Hyostrongylus* burrows into the stomach wall causing ulceration and gastric nodules with increased mucus production while *Ascarops* causes catarrhal gastritis. Hemorrhage due to infection with either parasite may cause anemia, anorexia, and melena. Diagnosis is by identification of eggs in feces or examination of mucosal scrapings at necropsy for immature larvae.

The most important small intestinal parasite of suids is ***Ascaris suum***, which has been identified in wild boars and collared peccaries.[63,133] Outbreaks have been reported in captive peccaries with clinical signs of respiratory distress and coughing. This nematode parasite has a direct life cycle. Pigs are infected by ingesting eggs with second-stage larvae or infected paratenic hosts (e.g., earthworms). The larvae migrate to the liver via the portal circulation and molt into third-stage larvae that then travel via the venous system to the lungs, where they are coughed up, swallowed, and mature into adults. Visceral migration of larvae can be associated with mucosal necrosis, hemorrhage, healing fibrosis, and lymphoplasmacytic inflammation (white areas called "milk spots") (Fig. 8.e7; eSlide 8.e5); heavy larval infection in the lungs is associated with pulmonary edema and acute necrosuppurative pneumonia. Diagnosis is by identification of eggs or worms in feces or by the presence of larvae within the lungs.

Globocephalus sp. are globally distributed hookworms. *G. urosubulinatus* is common in domestic pigs in the western hemisphere and Europe; it has also been reported in free-ranging feral populations of wild boars and peccaries. *G. versteri* more commonly occurs in African bushpigs, and *G. samoensis* and *G. longemucronatus* are more common in wild boars in southeast Asia and Japan.[63,210,238] The life cycle is direct. Following larval ingestion or percutaneous infection, migration can be extensive and cause myocardial and or pulmonary inflammation before returning to the gastrointestinal tract. The adults are hematophagous and utilize specialized mouthparts to adhere to and cause ulceration of the small intestinal mucosa resulting in enteritis and blood loss. Diagnosis is by the identification of eggs in the feces by flotation.

Strongyloides ransomi infects the intestine of wild boars and is pathogenic in suckling piglets. Vertical transmission occurs during suckling after which the parasites tunnel into the base of small intestinal villi, resulting in blood loss, anemia, and diarrhea in piglets. Infection in adults is asymptomatic. Diagnosis is most commonly made by identification of eggs in feces collected from the rectum, mucosal scrapings at necropsy, or recovery of immature worms using the Baermann technique.

Macrocanthorhyncus hirudinaceus, the thorny-headed worm, infects the ileum and caudal small intestine of wild boar.[63,133,226] The life cycle of this nematode parasite is indirect with arthropod intermediate hosts. While eggs are passed in the feces, diagnosis via floatation is difficult because the eggs do not float, but they can be detected by direct fecal smear or sedimentation. Infection may cause multifocal chronic granulomatous enteritis with the formation of intramural granulomas that can be identified along the serosal surface. Nodules rarely perforate.

Onchocerciasis dermatitis and lymphadenitis have been described in free-ranging Suidae in many countries; however, descriptions are largely limited to parasite morphology without description of gross or microscopic lesions.[234,235] Transmission is typically mechanical via arthropod vectors, including Simulium flies. In *O. dewittei japonica* infection of Japanese wild boars, adult filarial worms are encysted in fibrous subcutaneous nodules in the distal limbs and the fat bodies of the foot pads. Microfilariae are more commonly found in the skin of the ears, back, tail, and less frequently on limbs. In contrast, *O. tacaokai* microfilariae and filarial adults infest the skin of ears, neck, and back; these adult parasites are not associated with cutaneous nodules. Coinfection with *O. dewittei japonica* and *O. tacaokai* has been reported in Japanese wild boars. In *O. ramachandrini* infection of African warthogs, microfilariae are present in the skin of the neck, back, and feet, and filarial adults are found coiled in the subcutis of the distal limbs. Diagnosis is by identification of larvae or adults in skin sections. Larvae will emerge from sections of skin placed into saline.

Trichinellosis is an WOAH listed, reportable disease with worldwide distribution. While several species of *Trichinella* may infect wild suids, *T. spiralis* and *T. pseudospiralis* are most commonly reported.[156,260] No gross lesions and only mild histologic inflammation have been noted in infected wild boars, and infection is of importance because of the zoonotic potential (Fig. 8.19). Real-time PCR of muscle tissue is a sensitive diagnostic tool.[80]

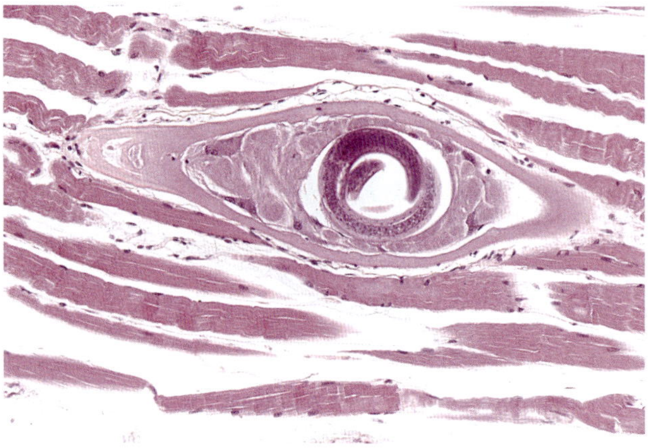

FIGURE 8.19 *Trichinella* sp. encysted in the skeletal muscle of a European wild boar. Note the hypertrophied nurse cell and the characteristic esophageal stichosome within the nematode. *(Photo Courtesy of D. Risco Perez, INGULADOS S.L.)*

Protozoa

Trypanosomiasis is recognized as a clinically significant vector-borne, protozoal disease of humans and livestock, with distinct agents and disease entities recognized in Africa and Central and South America. Subclinical infection may be regionally common in free-ranging *Suidae*, and warthogs, bushpigs, red river hogs, giant forest hogs, peccaries, and wild boars are important in the epidemiology due to the preference of insect vectors to feed on them. These suid species are reservoirs for trypanosomes pathogenic to humans and domestic animals, including ***Trypanosoma congolense***, ***T. simiae***, ***T. suis***, ***T. evansi***, ***T. brucei***, and ***T. cruzi***. Warthogs are probably the main wild reservoirs of *T. simiae*, which causes acute fulminating infection in domestic pigs in Africa; peccaries are important in the maintenance of *T. cruzi* and *T. evansi* in Brazil; and red river hogs are important in the epidemiology of human sleeping sickness in Africa caused by *T. brucei*.[40,88,118,174]

Suids generally do not develop severe disease as a result of trypanosome infection but they can maintain parasitemia for a long period of time.[40] In white-lipped peccaries, *T. evansi* causes severe anemia, weight loss, and immunosuppression that predisposes to opportunistic infections. In juvenile peccaries, disseminated intravascular coagulation develops secondary to the deposition of immune complexes in the heart, liver, brain, and kidneys.[88] Encephalitis due to *T. suis* infection has been reported in some species.[118] Disease severity in suids is variable, with *T. brucei* and *T. congolense* typically only causing mild disease, and *T. suis* and *T. simiae* often causing more clinically significant disease in suids.[174] Presence of intrasarcoplasmic protozoal pseudocysts in the myocardium with or without mild myositis and inflammation can occur with *T. cruzi*,[43] but it can be difficult to see amastigotes in tissue sections. Intralesional trypanosomes, hemorrhages, and extravascular hemolysis may be seen in some organs. The splenic parenchyma may contain large numbers of histiocytes with marked periarteriolar sheath lymphoid hyperplasia, lymphoid depletion, and/or hemosiderosis due to extravascular hemolysis. Cerebral endothelial hypertrophy and intralesional organisms may also be present.

Free-ranging wild boars, warthogs, and bushpigs are a reservoir for ***Babesia trautmanni*** and ***B. perroncitoi***; however, clinical diseases are best described in wild boars. *Babesia* spp. are typically transmitted between *Suidae* and *Tayssuidae* by ticks in the genera *Rhipicephalus*, *Boophilus*, and *Dermacentor*. *B. trautmanni* in domestic pigs and wild boars results in a chronic hemolytic disease with clinical signs including emaciation, pyrexia, anorexia, icterus, and hemoglobinuria. Gross lesions may be limited to poor body condition or emaciation, splenomegaly, and hepatomegaly, with or without polycavitary effusion, and multisystemic hemorrhages. Microscopic lesions reflect hemolytic disease with hemoglobin nephrosis, and degeneration and necrosis with a pattern suggestive of hypoxia in multiple tissues. Diagnosis is based on identification of the parasite in erythrocytes. *B. gibsoni*-like organisms have been detected by PCR in wild boar in Asia but have not yet been linked to disease.[139]

Cryptosporidium suis and ***C. scrofarum*** are ubiquitous in domestic pigs and wild boars, and all *Suidae* and *Tayassuidae* appear to be susceptible to infection.[66,106,116,154,191,197] *Cryptosporidium* spp. have a direct life cycle, and while most naturally occurring infections are subclinical, infection has infrequently been associated with mild enteritis and diarrhea in suids. The protozoa are most commonly identified in young animals; adults appear to have a moderate degree of resistance to infection. Zoonotic transmission is the most significant feature of infection in these species, and transmission to human and nonhuman primates by wild boar, warthogs, and bushpigs is a primary concern.[165]

Additional pathogenic and commensal protozoa are presented in Supplemental Tables 8.e2 and 8.e3.

Ectoparasites

Sarcoptes scabiei* var *suis, the cause of **sarcoptic mange** in domestic pigs, have also been reported in free-ranging wild boars, warthogs, and peccaries.[66,83] Lesions are similar to those described in other species and include alopecia and crusting dermatitis with lichenification that primarily affects the ventrum and extremities (Figs. 8.20A, B). Histologically, parakeratotic hyperkeratosis with intralesional mites and serocellular crusts are common findings (Fig. 8.20C). Dermal inflammation is primarily eosinophilic. Diagnosis is by skin scrapes.

Free-ranging *Suidae* and *Tayassuidae* are hosts to numerous tick species, and infestations are not uncommon in captive animals. In free-ranging animals, tick burdens may be extremely high, and most *Suidae* and *Tayassuidae* have developed behavioral adaptations, such as wallowing that rubs ticks off of the skin and commensal relationships with other animals that scavenge ticks. Tick infestation is most commonly associated with chronic multifocal dermatitis; it is not generally associated with systemic disease. Ticks are more important due to their roles as vectors for infectious disease transmission, such as in the pathogenesis of ASF.

Additional information on ticks and other ectoparasites is provided in Supplemental Table 8.e2.

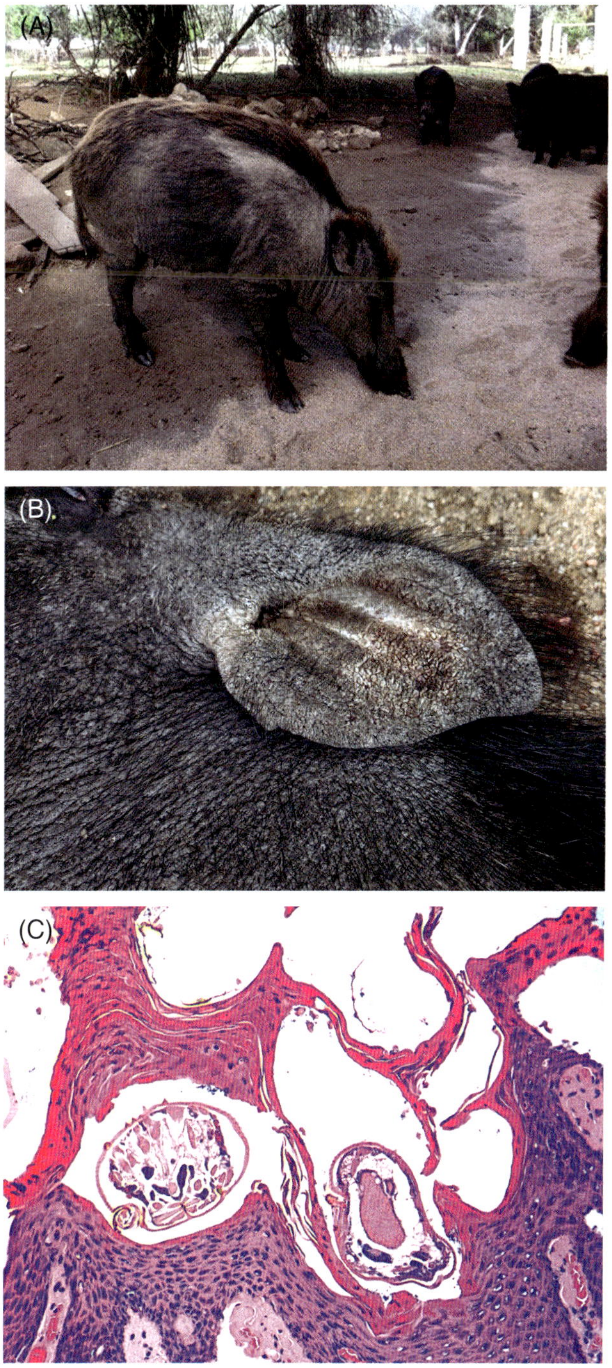

FIGURE 8.20 Scabies due to *Sarcoptes scabeii* infection in a wild boar (A and B) and babirusa (C) Gross lesions in wild boar with alopecia, hyperkeratosis, and hyperplasia of the skin (A) and pinnae (B). Histologically, there is marked epidermal hyperplasia and parakeratotic hyperkeratosis associated with intracorneal mites in a babirusa. *(Part A: Photo Courtesy of D. Risco Perez, INGULADOS S.L; Part B: Photo Courtesy of the Wildlife Ecology & Health Group—Universitat Autònoma de Barcelona; Part C: Photo Courtesy of E. Mitchell, University of Pretoria.)*

ADDITIONAL RESOURCES

Gottdenker, N., Bodmer, R., 2004. Hutchins, M. (Ed.), Grizimek's Animal Life Encyclopedia, 16, second ed., Gale, Farmington Hills, Michigan, pp. 291–301.

Martin-Hernando, M.P., Höfle, U., Vicente, J., Ruiz-Fons, F., Vidal, D., Barral, M., Garrido, J.M., de la Fuente, J., Gortazar, C., 2007. Lesions associated with *Mycobacterium tuberculosis* complex infection in the European wild boar. Tuberculosis 87, 360–367.

Pikalo, J., Zani, L., Hühr, J., Beer, M., Blome, S. 2019. Pathogenesis of African swine fever in domestic pigs and European wild boar—Lessons learned from recent animal trials. Virus Research 271: 197614.

Ruiz-Fons, F., 2017. A review of the current status of relevant zoonotic pathogens in wild swine (*Sus scrofa*) populations: changes in modulating the risk of transmission to humans. Transbound. Emerg. Dis. 64, 68–88.

Sutherland-Smith, M., 2015. Suidae and tayassuidae (wild pigs, peccaries). Miller, R.E., Fowler, M.E. (Eds.), Fowler's Zoo and Wild Animal Medicine, 8, Elsevier Saunders, St. Louis, pp. 568–584.

References

Full references for this chapter can be found online at https://doi.org/10.1016/B978-0-443-18437-6.00014-6.

Chapter 9

Canidae, Ursidae, and Ailuridae

M. Kevin Keel
University of California, Davis, CA, United States

INTRODUCTION

Members of the family *Canidae* include approximately 35 species of carnivores that are found on every continent except Antarctica. They include wolves, coyotes, jackals, dingoes, foxes, and other caniform or dog-like animals (Table 9.e1).[1] They are generally cursorial species and range in size from the fennec fox to the gray wolf. The canids generally exhibit long-term monogamy, a high degree of parental investment, long periods of neonatal dependency, and relatively large litters. Some are cooperative hunters, such as gray wolf and African wild dogs, whereas other species, often small canids, such as red foxes, tend to be solitary foragers. Most species demonstrate some degree of omnivory. Conservation concerns include depredation by humans, habitat loss, depletion of prey species, and the emergence of infectious diseases.

There are eight extant species of bears (Table 9.e2) in the family *Ursidae*. Most bear species are omnivorous, but the polar bear is mostly carnivorous and the giant panda is almost exclusively a bamboo eater. Bears have long played a role in human culture and mythology. They are also prominent symbols in wildlife conservation and are indicators of very real problems associated with planetary climate-related changes. Other conservation concerns include habitat loss and the illegal wildlife trade, particularly the trade in bear bile for traditional medicinal purposes. Many species are also hunted for meat, which may pose a risk for zoonotic disease transmission (e.g., trichinellosis).

The *Ailuridae* family includes just a single species, the red panda, which was previously placed phylogenetically within either *Procyonidae* or *Ursidae* families. Recent molecular evidence supports its placement in its own family within the caniform carnivores or as a member of the mustelid clade.[2,93] Recent genetic evidence also suggests there are two species of red panda, the Chinese red panda (*A. styani*) and the Himalayan red panda (*A. fulgens*).[202] The red panda is currently listed as endangered, and its population is declining.[175] Loss of habitat, unregulated hunting (for medicine, pelts, and meat), increasing interaction with domestic dogs, and interspecies transmission of canine distemper virus (CDV), to which the species is uniquely susceptible, are significant conservation concerns.

NON-INFECTIOUS

Nutritional

Natural food preferences vary greatly among wild canids. Some species are predominately carnivorous, whereas others are omnivores or even insectivores. Selecting appropriate diets for captive animals with unusual natural food preferences can be challenging. Maned wolves in captivity were historically fed raw meat, but their natural diet is diverse and includes an abundance of soft fruit. The administration of a predominately meat-based diet is thought to be the source of many problems in maned wolves, which include chronically soft stools, reduced intestinal transit time, and poor nutritional condition. The diet of captive maned wolves is further complicated by a high prevalence of cystine uroliths (described in greater detail later) and attempts to manage them by diet. Dietary changes to reduce cystinuria have resulted in taurine deficiency, and this is hypothesized to be related to poor reproductive success.[69]

Rickets in wild Arctic foxes is associated with classic lesions, which include enlargement of costochondral junctions (the "rachitic rosary") and delayed endochondral ossification that affects different bones to varying degrees. Associated gross lesions include widening of physes, angular limb deformities, and metaphyseal fractures.[307] Osteodystrophy variably occurs in the trabecular and cortical bone of the metaphyses and ossification centers. Growth plates may be thickened, and the thinner trabeculae in metaphyses may be partially replaced by fibrous connective tissue. Rickets has also been reported in captive red foxes, a captive coyote pup, and a captive crab-eating fox.[15,387] Rickets and fractures have been reported in polar bears, as well.[228,252] Vitamin D levels were found to be low in some cases, and this may have been a predisposing factor. Taurine deficiency may also contribute to rickets because it is an essential amino acid that is necessary for the

Pathology of Wildlife and Zoo Animals. https://doi.org/10.1016/B978-0-443-18437-6.00038-9

conjugation of ursodeoxycholic acid. Ursodeoxycholic acid, in turn, assists in the absorption of fat-soluble vitamins. Studies comparing taurine levels among captive and free-ranging polar bears found significantly higher taurine in free-ranging polar bear cubs and in milk from polar bears, in comparison to commercial formulas.[67] Deficiency in hand-reared individuals was therefore suspected as a possible contribution to this disease. Myocardial disease was not reported in the affected individuals.

Nutritional secondary hyperparathyroidism has been described in wild, juvenile red foxes and is suspected, but not proven, in African wild dogs.[383,387] The etiology was suspected to be nutritional in all cases, but more detail was not available. Affected red fox kits are typically emaciated with enlargement of the distal epiphyses of the radii, ulnae, and femurs, and the costochondral junctions of the ribs. Bone cortices are thin and pliable. Histopathology confirms **fibrous osteodystrophy** characterized by complete replacement of some regions of the cortices with loosely arranged fibroblast and collagen bundles admixed with thin spicules of woven bone that are lined by variable numbers of osteoblasts (new bone formation), osteoclasts (bone resorption), and occasional Howship's lacunae (bone resorption). Fibrous osteodystrophy can also occur as a comorbidity with rickets in wild canids.[85]

Thiamine (vitamin B1) deficiency has been reported in farm-raised foxes worldwide, and it has also been described in wild red foxes.[308,309] The disease is called **Chastek paralysis**, after the Chastek farm in Minnesota, USA, where the disease was first noted. It occurs when foxes are fed certain types of raw, whole fish containing thiaminase, which causes the degradation of thiamine in food.[180] All carnivores require a dietary source of thiamine, and deficiency is associated with bilaterally symmetric hemorrhage and necrosis of the brain and brainstem. Poliomalacia (gray-matter necrosis) may occur in the middle laminae of the occipital and temporal cortex, but lesions are most striking in the periventricular gray matter.[55] Bilateral lesions are most consistent in the caudal colliculi and vestibular nuclei, but these may also be accompanied by lesions in other periventricular nuclei. Malacia may also be seen in the lingula and nodulus of the cerebellum. Myocardial degeneration, hepatic lipidosis, and nephrosis are also reported. Myocardial necrosis is typically more severe in the right ventricle and may be manifested grossly as pale, flabby myocardial muscle.

Other nutritional diseases in these groups of animals include **hyperostotic bone disease**, which has been reported in a single captive group of red pandas.[260] Excessive dietary vitamin A was suspected as the underlying cause but was not confirmed. In addition, **zinc-responsive dermatosis** with secondary pyoderma has been reported in a captive red wolf. Acanthosis and orthokeratotic and parakeratotic hyperkeratosis affected the skin of the footpads and distal limbs of the wolf. Superficial ulcerative dermatitis and furunculosis also occurred and were associated with secondary bacterial infection.[226]

Metabolic

Cystinuria is a relatively common disease in maned wolves and appears to be present in both wild and captive animals. Cystinuria in humans and domestic dogs is an autosomal recessive condition characterized by defective transport of cystine, lysine, ornithine, and arginine across the renal tubular epithelium. Affected maned wolves have dibasic aminoaciduria with excretion of cysteine and lysine (variable excretion), a condition that is more similar to that of domestic dogs than humans.[46] A primarily meat-based diet for captive maned wolves has been hypothesized to predispose them to urolith formation.[68] However, whether genetic, anatomic, or other factors are involved, remains to be fully determined. Cystine uroliths in maned wolves may be present in the renal pelvis, ureters, urinary bladder, or urethra (Fig. 9.1). Chronic cystitis, nephritis, obstruction of the urinary tract, and death are reported sequelae.

Exertional myopathy is less commonly encountered in wild canids than in other species, but it has been reported in association with leg-hold traps.[402] The lesions are similar among various taxa and include monophasic necrosis of striated muscle that may be pale and less compliant than unaffected tissue. Grossly this may appear as regions of muscular pallor with variable swelling, that are slightly firm, and dry.

Toxic

Malicious poisoning of wild canids often involves acute-acting compounds, particularly carbamates and

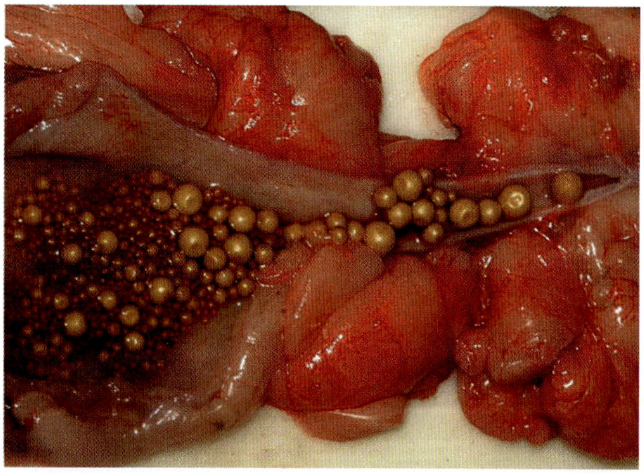

FIGURE 9.1 Cystine urolithiasis in the urinary tract of a maned wolf. Numerous, variably sized, round, yellow calculi fill, and obstruct the urethral lumen. *(Photo Courtesy of the Smithsonian Institution's National Zoo and Conservation Biology Institute.)*

organophosphates. Gross lesions may not be evident with **organophosphate** or **carbamate poisoning**; however, bait material (e.g., chicken, hamburger, and eggs) in the stomach, dead animals in good nutritional condition, and mortality of multiple individuals near an obvious bait site provide evidence of acute intoxication. Secondary intoxication of other species (e.g., vultures) or mortality among nontarget species may also be seen. Unintended poisoning and unlikely scenarios occasionally occur. For example, a Mexican gray wolf died of organophosphate toxicity after ingesting a cow's ear tag that was impregnated with diazinon as a parasite deterrent.[220]

Organochlorine pesticides (**OCP**) and **polychlorinated biphenyls** (**PCB**) are persistent compounds with relatively low toxicity, but with a tendency for bioaccumulation that is particularly evident in aquatic ecosystems. Residues are very common in wildlife. It is widely held that OCPs and PCBs can have subclinical effects on immune, endocrine, and reproductive function, but it is often difficult to associate lesions with the presence of such contaminants.[367] Because these compounds accumulate in fat, rapid mobilization of lipids in adipose tissue can result in sudden, high-level exposure to pesticides.[190] Heavy use of dieldrin and heptachlor epoxide, both organochlorine insecticides, is thought to have been the cause of significant fox mortality in British red foxes in the 1950s. The insecticides were used as seed protectants, and secondary intoxication through ingestion of granivorous prey was considered a possible route of exposure. The necropsy findings and tissue concentrations of the compounds were consistent with those seen after experimental administration of lethal doses. Lesions included emaciation, centrilobular hepatic lipidosis with variable centrilobular necrosis, and lipid accumulation in the tubules of the kidney.[38] Captive Arctic foxes chronically exposed to OCPs and PCBs at levels consistent with natural exposure tend to develop thyroid lesions including follicular cysts, C-cell hyperplasia, and remnant embryonic ducts.[368]

As apex predators that consume large quantities of blubber from other high trophic species, polar bears are at high risk for bioaccumulation of environmental contaminants. Multiple studies have confirmed elevated levels of varying **PCBs** as well as elevated **mercury** (see Cetacea) in polar bears. High levels of PCBs and other **organohalogen contaminants** (**e.g., organochlorine pesticides**) have been negatively correlated with reproductive and thyroid hormone levels, as well as suppression of immune function.[111,157,184,234] Elevated organohalogen and mercury levels have been correlated with microscopic lesions, including Ito-cell hypertrophy, hepatic steatosis, leucocytic infiltrates in the liver and kidney, renal tubular hyaline casts, and thickening of glomerular capillary walls.[364–366] However, the correlation of individual contaminants to specific lesions is complicated by the presence of multiple contaminants within individual bears, temporal trends over study periods, and age-related changes. Pathologists should be aware of the potential confounding impact of these toxic compounds on polar bear health and histopathology.

Poisoning by other compounds may result in lesions similar to those seen in domestic dogs. For example, **ethylene glycol** ingestion leads to renal nephrosis with large numbers of birefringent (by polarized light) calcium oxalate crystals in renal tubules. Crystal deposition can also occur in cerebral vessels or in the Virchow-Robbins space. **Anticoagulant rodenticide intoxication** is associated with severe hemorrhage, often in the subcutis, or in body cavities.[33,155,400] Even mild trauma, such as bumps or small sharps injuries, can result in suffusive hemorrhage or hematomas in animals with anticoagulant rodenticide intoxication. Secondary toxicity may occur through the ingestion of affected prey species. Because of this, animals are likely to prey on or scavenge dead rodents or other target animals are more likely to be affected. **Strychnine** can cause death quickly, without significant lesions, but the convulsions associated with this poison can result in self-inflicted trauma and myoglobinuria. Nonspecific lesions, such as serosal and mucosal hemorrhage, and passive congestion of various parenchymal organs, have been reported.[274] **Cholecalciferol poisoning** has been described in captive foxes and is associated with deposits of calcium that are evident on microscopic evaluation of the kidneys and, in some cases, skeletal muscle, gastric mucosa, bronchi, and larger blood vessels.[191] Gross lesions are not evident. **Melarsomine toxicity** was implicated in the death of a red panda that died within 24 h after treatment for *Dirofilaria* sp. infection using melarsomine dihydrochloride. At necropsy, marked pulmonary edema was presumed to be secondary to direct pulmonary cellular cytotoxicity.[303] For additional descriptions of the wide range of poisons known to affect canids and other wildlife, the reader is directed to several excellent reviews.[32,47,328]

Congenital/Genetic

It is likely that congenital or genetic defects, especially those that have a lethal phenotype, occur more frequently in the wild or captive settings than are represented in the small number of published case series or case reports. Several of the more notable conditions are described below.

Congenital hypotrichosis occurs in several species of foxes. Gray foxes with this lesion have been described from a small region in the southeastern United States. Only the guard hairs tend to be missing in affected animals, and the foxes usually appear to be clinically normal in other respects. The appearance is that of a lightly haired or nearly hairless fox. The disease seems to have a low prevalence and the effect on the fitness of the population is largely unknown. An absence of guard hairs has also been reported in Arctic foxes and the red fox population of south Finland, where they have been given the name "Samson foxes."[192,395] The condition has been associated with decreased fitness (e.g., developmental delay in cubs, higher mortality). **Gingival fibromatosis** (syn. **Hereditary hyperplastic gingivitis**), an autosomal recessive trait, is periodically reported in farm-reared silver foxes.[78] The condition has a much higher prevalence in males and is associated with superior fur quality. Although silver foxes are derived from a North American subspecies of red fox, the disease was first identified in Sweden around 1940 and until recently was only recognized in Europe. The proliferation may progress rapidly and result in the displacement of teeth and ultimately encapsulation of teeth by the proliferative tissue. Proliferative gingiva is often soft, bleeds easily, and may be associated with entrapped food. Thickening of the gingiva is predominately due to the proliferation of submucosal collagen, but the tissue can become inflamed. Lesions matching those of gingival fibromatosis in silver foxes have also been observed in a single wild red fox.[351]

Mandibular hypoplasia or **micrognathism** with a high prevalence of malocclusion was reported in an isolated population of red foxes in the Netherlands.[45] This was presumed to be due to a recessive trait associated with a genetic bottleneck in the population. **Supernumerary teeth**, both bilateral and unilateral, are occasionally observed in nondomestic canids.[187,312] **Failure of tooth development** may also occur and may be difficult to differentiate from tooth loss. Tooth absence may be due to impaction or failure to erupt. For example, maxillary tooth impaction was diagnosed in a young maned wolf after it was presented for what turned out to be infraorbital abscesses and ulcers.[319] Other developmental anomalies of teeth in these taxa include **extra roots, malposition, malformation**, and **hypoplasia**. **Craniomandibular osteopathy** has been described in a young, free-living golden jackal.[271] This is a rare condition that is heritable in dogs; the pathogenesis is unknown. New bone formation was evident in the endosteum and periosteum of the mandibular bone and the distal aspects of the hind limbs.

Congenital hypothyroidism was associated with varying degrees of dwarfism in red pandas.[135] A presumptive diagnosis of nongoitrous, congenital hypothyroidism was made in 75% of cubs from a common sire and two closely related dams. Thyroid glands were hypoplastic with little to no colloid. **Congenital hyperplastic goiter** and **cretinism** were documented in a 16-month-old black bear that was captured at 8 months of age.[124] This individual had significant growth retardation, clumsiness, and facial dysmorphism.

A condition known as **cardiorespiratory syndrome** has been described in farmed red foxes since the 1980s.

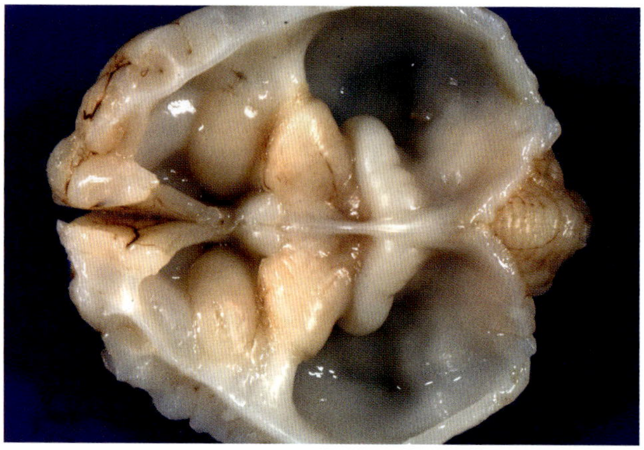

FIGURE 9.2 Hydrocephalus in a red fox kit. The lateral ventricles are greatly expanded and the cortices are extremely thin remnant. This was a congenital lesion identified in a very young individual.

Affected animals have been described in Poland and Canada and are principally affected by right-sided heart failure at a young age. Postmortem examinations of affected 2- to 6-month-old foxes in Poland consistently revealed tricuspid valve dysplasia with right-sided atrioventricular dilatation.[305]

Congenital internal **hydrocephalus** has been seen in red fox kits from one region of the eastern United States (Keel, unpublished data). Kits were up to 5 weeks of age and had severe distension of the lateral ventricles (Fig. 9.2), expansion of the fontanelles, and, in one instance, percutaneous transudation of fluid. Minimal lymphocytic meningitis has been variably identified. Immunohistochemistry for canine parvovirus and canine distemper virus have been negative, but the possible involvement of congenital or neonatal viral infection has not been completely ruled out. A 2-month-old red fox from Italy diagnosed with internal hydrocephalus had associated neutrophilic periventricular and choroidal inflammation thought to be due to a bacterial infection.[267]

Lafora-like disease has been reported in a mature fennec fox with a 6-month history of seizures.[200] Lafora disease in humans is an autosomal recessive condition characterized by the intracytoplasmic accumulation of insoluble complex glycoprotein polymers (Lafora bodies) in neurons and various cells in other tissue types. No gross lesions were present in the fox. Histologically, amphophilic to basophilic, periodic acid-Schiff (PAS) positive, diastase-resistant inclusions consistent with Lafora bodies were present in the perikaryon of affected neurons, macrophages, myocardiocytes, and occasionally hepatocytes. Ultrastructurally, inclusions were variably electron dense, fibrillary to granular, and lacked a delimiting membrane.

GM-1-gangliosidosis has been reported in free-ranging black bears in the northeastern United States.[300] Affected bears are lethargic and in poor body condition with

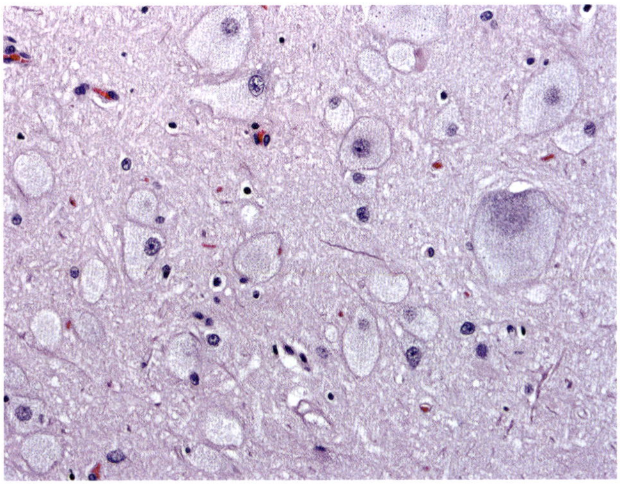

FIGURE 9.3 G_{M-1}-gangliosidosis in the brain of a black bear. Storage material causes neuronal hypertrophy with foamy cytoplasm. *(Photo Courtesy of S. Jennings, Cummings School of Veterinary Medicine, Tufts University.)*

intention tremors of the head and body, as well as generalized ataxia and forelimb hypermetria. Circulating mononuclear cells contain rare cytoplasmic vacuoles. Histologically, neurons in the brain and peripheral ganglia, including the retina, are enlarged and have foamy cytoplasm (Fig. 9.3). Some cells in the brain and retina also contain small eosinophilic granules. Renal proximal tubular epithelium, hepatocytes, and chondrocytes show similar changes. By electron microscopy, neurons contain enlarged secondary lysosomes filled with lamellated membranes. Decreased acid β-galactosidase activity was consistent with GM1-gangliosidosis. The β-galactosidase-1 (GLB1) gene from an affected bear contained a homozygous recessive mutation within a highly conserved region of the gene. These genetic findings and regional geographic distribution of cases suggest an increased frequency of a founder mutation in this black bear population.

Heritable **blindness** in a captive population of gray wolves has been associated with various ocular abnormalities including cataracts, retinal dysplasia, and abnormal blood vessels in the retina and vitreous.[243]

Age-Related/Degenerative

Some incidental lesions encountered in domestic dogs can be seen in other nondomestic canid species. For example, pulmonary **osseous metaplasia** is commonly encountered in many taxa and is typically more common in aged animals. As with many aged animals of other taxa, aged canids and ursids may also develop **dental wear and attrition** as they age. **Periodontal disease** is not well documented in nondomestic canids but likely has a similar pathogenesis to that described for domestic dogs. The influence of diet on the occurrence of periodontal disease has been speculated. However, in at least one species, the African wild dog, a natural diet did not appear to be protective.[373] Dental disease including fractures, excessive wear, and loss of teeth that increase in severity with age has been reported in free-ranging giant pandas, and fractured canines are common in captive ursids.[218] In giant pandas, wear is symmetrical and most severe in the incisors consistent with mechanical grinding; tooth loss is most common in the premolars. In some pandas, lack of appropriate mastication has led to large segments of intact bamboo within the intestines leading to ulceration and perforation. Interestingly, periodontal disease is not noted commonly in this species, possibly due to flavonoid compounds in bamboo that act as an antiseptic and to the fibrous nature of bamboo acting as a mechanical "brush" preventing the formation of plaque.

Temporomandibular joint osteoarthritis (TMJ-OA) and dental lesions have been common in skeletal studies of polar bears, black bears, and brown bears from museums.[77,404,405] Adults were more commonly affected with TMJ-OA. Common dental lesions include dental fractures, periapical disease, periodontitis, and dental fractures. Nearly 1/3 of the brown bears had some degree of malocclusion. A significant number of black bears had one or more small vestigial teeth with crowns that did not protrude above the adjacent alveolar bone.

Age-related and degenerative musculoskeletal problems are reported for many species and may be more prevalent among older animals in captivity. Canids, ursids, and red pandas tend to be long-lived in captivity and develop many age-related degenerative conditions. **Degenerative joint disease** is common in aged, captive ursids. Lesions are generally most severe in the appendicular skeleton. Individuals in the wild are less likely to reach the advanced age associated with degenerative joint disease, but it does occur.[407] Similarly, **spondylosis deformans** is age-related and more likely to occur in captive animals (Fig. 9.4). Larger species are more likely to be affected, but smaller species (e.g., red fox) can also be affected.[189] **Hypertrophic osteodystrophy**, a metaphyseal lesion typically associated with necrosis, neutrophilic infiltrates, and modeling of bone in fast-growing large-breed dogs, has also been described in captive red wolves.[173] As in dogs, radial, ulnar, and tibial bones are most often affected but lesions have also been described in the metaphyses of multiple long bones.

End-stage renal disease is a common finding and causes for euthanasia in captive bears, but it has been best described in polar bears.[242] Clinical signs are like those of domestic species and include inappetence, lethargy, polyuria, and polydipsia. Biochemical abnormalities include azotemia, anemia, hyperphosphatemia, and isosthenuria. Histologically, lesions include membranous and/or proliferative glomerulopathy and although suspected, the role of immune complex deposition in the pathogenesis has not been fully investigated. Tubular protein, interstitial

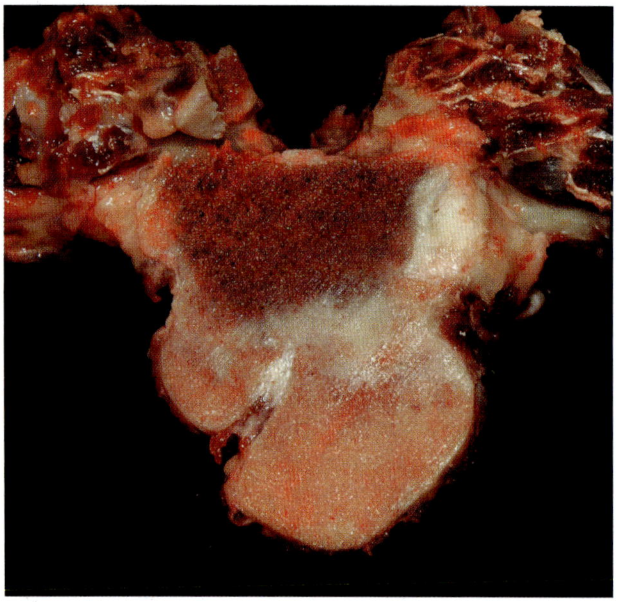

FIGURE 9.4 Spondylosis in the vertebral column (transverse section) of an aged grizzly bear. Characteristic features are irregularly shaped, smooth, dense, proliferative bone along the ventral aspects of the vertebra. Note the irregular ventral border and band of thick, white tissue along the ventral aspect of the marrow space suggestive of concurrent, chronic osteomyelitis. *(Photo Courtesy of C. Rodriguez, Wildlife Conservation Society.)*

fibrosis, and inflammation are common concurrent findings. In free-ranging polar bears, similar lesions, including diffuse glomerular capillary wall thickening, mesangial deposits, hyalinization of tubular basement membranes, and interstitial fibrosis, have been reported and were correlated with age.[365] After controlling for age, glomerular mesangial deposits, glomerular capillary wall thickening, medullary hyaline casts, and interstitial fibrosis were also correlated with organohalogen contaminants (see "Toxic" section), and the responsibility of long-term toxin exposure in the development of renal disease in free-ranging polar bears may be possible.

Cardiac degenerative disease is rarely reported in free-ranging canids. The diseases described include **valvular dysplasia** in a red wolf and **dilated cardiomyopathy** in African wild dogs.[1,246] Valvular dysplasia has also been frequently noted in African wild dogs. In captive red pandas, cardiovascular and renal diseases are reportedly common, age-related findings in adult and geriatric animals.[109] Myocardial fibrosis and hypertrophic cardiomyopathy are the most common diagnoses; renal disease was not characterized beyond nephritis.

Miscellaneous/Multifactorial

Interspecific and intraspecific conflicts are a common cause of **trauma** among canids.[407] Smaller canids may be prey to larger canids or other predators. Conversely, species that prey on much larger game may suffer injuries from their prey. For example, fractures attributed to prey species are common in wolves, but fractures are uncommon in coyotes due to differing prey and hunting strategies. Both gray wolves and coyotes have been observed with sticks lodged transversely between the carnassial teeth (Fig. 9.5).[187,407] In most cases, the sticks are associated with mechanical ulcers in the associated palatine mucosa.

FIGURE 9.5 Oral trauma and papillomatosis in a coyote. A foreign body (stick) is lodged between the carnassial teeth of a coyote. This individual also had multiple oral papillomas *(arrowheads)*.

Among wild canids, nutrition is often prey dependent and **starvation** can be a significant mortality factor with a higher rate of mortality in pups that are dependent on dams for nutrition and may have inadequate nutritional reserves.[285] **Nutritional insufficiency** may predispose such animals to other, more proximal, mortality factors such as predation or parasitic or infectious diseases. Some species, such as the Arctic fox, are subject to loss of fat reserves with reduced prey availability in winter, but captive individuals also voluntarily reduce their food intake during the winter.[153] Mobilization of fat in wild animals poses some risk of the release of stable, lipid-soluble contaminants, such as organochlorine pesticides, from adipose tissue (also see "Toxic" section).

Loss of sea ice due to climate-related changes will greatly impact hunting opportunities for polar bears. This raises concerns about **starvation** in free-ranging polar bears, who build energy stores needed to sustain them during ice-free periods on land. Earlier ice melts create a timing mismatch between the presence of ice and the presence of ringed seal pups, the preferred polar bear prey. It is uncertain if polar bears can alter their foraging behavior to take advantage of land-based food sources, such as snow geese and caribou, but if they can, land-based activities may increase conflicts with humans or domestic animals or the transmission of diseases from terrestrial carnivores to the bears.[19,177]

Gastric dilatation and volvulus (GDV) have been reported a number of times in captive red wolves, in several maned wolves, and in one red panda. As in domestic dogs, conformation may be a predisposing factor and death is a common outcome.[1,354] Additionally, wolves may develop GDV immediately after a struggle associated with capture and, dietary factors have been suggested but not conclusively proven as predisposing to GDV. The presentation of GDV in red pandas may be similar to that in domestic canids.[347] Clinical signs include nonproductive retching, blanched mucous membranes, and abdominal distension with tympany. Gas distension with a displacement of the pylorus cranial to the body of the stomach, rather than ventral and caudal to it, is a characteristic finding. Various degrees of volvulus, typically around the root of the mesentery and occasionally also involving the spleen, are seen and are associated with congestion, hemorrhage and, depending on duration, mucosal or transmural necrosis.

Seasonal alopecia occurs in both captive and free-ranging ursids and red pandas. Lesions in spectacled bears include atrophic hair follicles in the telogen phase and perifollicular fibrosis.[211] While no cause has been identified, perivascular infiltrates of mast cells and eosinophils suggest an allergic component. Clinically, animals have progressive alopecia, pruritis with seborrhea, and lichenification. Secondary yeast infections are not uncommon. In red pandas some, but not all, cases are associated with dermatophyte infection (see "Fungi" section). **Denning-associated alopecia** has also been described in free-ranging black bears.[87] In these cases, lesions are primarily around the eyes, muzzle, neck, and thorax and are characterized by dry scaly skin with hair loss. Lichenification is also rarely seen. Mange was proposed as a differential in some cases; however, no cause was identified on scrapings or histopathology.

Bilateral, asymmetric, patchy alopecia in polar bears, characterized as loss of guard hairs and thinning of the undercoat of the head and neck regions (Fig. 9.6), has been described in free-ranging animals in Alaska (United States).[19] Histologically, affected bears have greater numbers of telogen follicles, follicular, and epidermal hyperkeratosis, and thickened, irregular, trichilemmal keratin. Some bears also have crusting lesions or broken hair shafts suggestive of pruritis. Males and subadults are more likely to be affected, and, in contrast to black bears, lesions are not associated with denning. A cause is currently unknown. Affected bears had lower body condition scores than unaffected bears, although whether this was causal or an effect of hair loss is uncertain. Alopecia has also been described in captive polar bears, although seasonal allergies, self-trauma, and ectoparasites are often implicated.[382]

Captive polar bears housed in warmer climates can develop **green fur** due to algal overgrowth within the hollow hair shaft of outer guard hairs.[250] The hairs of the

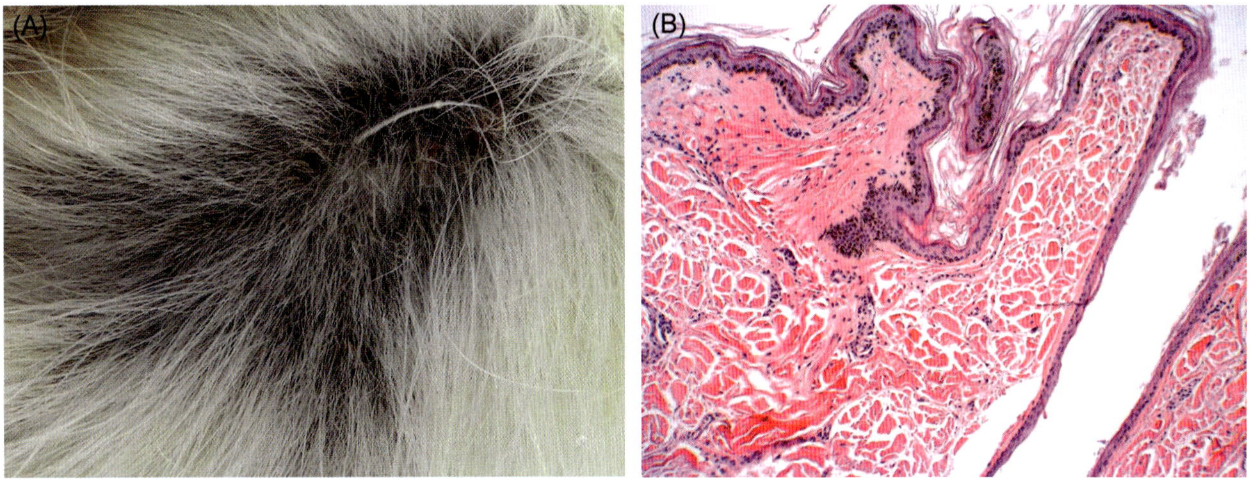

FIGURE 9.6 Alopecia in a wild polar bear. (A) Alopecia is characterized grossly by loss of guard hairs and thinning of the undercoat. (B) Histologically, many hair follicles are in the telogen phase with follicular and epidermal hyperkeratosis. *(Photo Courtesy of K. Burek Huntington, Alaska Veterinary Pathology Services.)*

undercoat are thinner and not hollow and thus are not similarly affected. Isolated algae are compatible with freshwater cyanophytes and are likely contaminants from the captive environment.

Embedded **porcupine quills** are not uncommon in wild canids that prey on porcupines. Associated complications (e.g., ocular trauma, pain, acute and chronic inflammation with draining tracts, and opportunistic infection) can be associated with significant morbidity or even mortality in affected predators.

Spongy degeneration of white matter in the central nervous system of young captive silver foxes has been associated with large cytoplasmic vacuoles in oligodendrocytes, vacuolation due to separation of myelin lamellae at intraperiod lines, demyelination, and myelin edema.[185] The disease tends to spare axons and neurons, but lesions are associated with astrocytic hypertrophy and with astrogliosis and axonal remyelination in the late stages of the disease. Both the brain and spinal cord are affected, and lesions are bilaterally symmetrical. Clinical disease in some foxes was stabilized or even improved. A cause is currently undetermined.

Cystic endometrial hyperplasia (**CEH**) and **pyometra** are commonly identified in aged red wolves and African wild dogs.[1,354] Although spontaneous disease occurs, it is thought that the use of synthetic progestin contraceptives, such as melengestrol acetate, may increase the risk for severe CEH and mineralization.[296] Treatment with a deslorelin implant alone increases the risk for pyometra. However, this effect is reduced by treatment with oral megestrol acetate immediately before and after placement of the deslorelin implants. The lesions are typical of those in domestic dogs and are characterized by proliferative endometrial glands that are distended by variably mucinous secretions (Fig. 9.7).

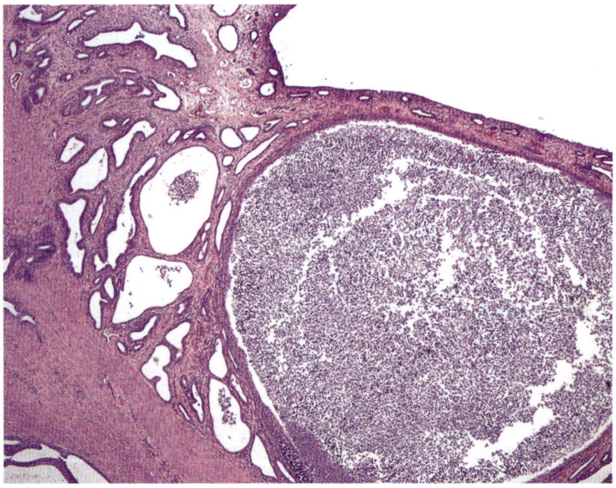

FIGURE 9.7 Cystic endometrial hyperplasia and pyometra in the uterus of an African wild dog. Numerous endometrial glands are enlarged, tortuous, and branching. Some contain luminal neutrophils. This dog had purulent material within the lumen of the uterus as well. *(Photo Courtesy of K. Terio, University of Illinois Zoological Pathology Program.)*

Systemic amyloidosis associated with elevated serum amyloid A is well documented in free-ranging island foxes.[158] It has been reported to affect as many as one-third of the population and poses a risk to the conservation and recovery of this endangered species. Older foxes in one population had an increased prevalence, and amyloidosis has been positively correlated with nephritis and captivity. This may suggest a multifactorial pathogenesis that could include a genetic predisposition and exogenous risk factors. Amyloid deposition in affected island foxes is present in nearly every tissue except the brain and is particularly common in the kidneys, oral tissues, and spleen. In the kidney, amyloid diffusely expands the renal interstitium (98% of cases) and less commonly the glomeruli (56% of cases). Submucosal deposition in oral tissues can result in a bosselated appearance, partly due to hyperplastic mucosal epithelium. Splenic amyloid deposition can produce large nodules that replace a major portion of the parenchyma.

Descriptions of amyloidosis in nondomestic canid species, other than island foxes, are relatively rare but undoubtedly occur. Red foxes in Sweden have been reported with amyloidosis due to serum amyloid A, and two of 25 randomly selected cases were positive.[332] Both individuals had severe inflammatory lesions that were considered to be the instigating factor in the development of amyloidosis. A report of thyroid C-cell carcinoma with amyloidosis has been described in a single red fox.[196] As occurs in humans, amyloidogenic fibrils in the fox were presumed to be derived from calcitonin.

A single free-ranging grizzly bear has been described with **tracheobronchopathia osteochondroplastica.**[339] This lesion is characterized by nodular aggregates of mature cartilage that protrude into the lumen. It is a very rare lesion in animals and is primarily described in human literature. It was thought to be incidental in the bear (death was due to a vehicle collision).

Neoplastic

With a few notable exceptions, tumors in nondomestic canids and ursids are described in a small number of case reports and reviews. They likely underrepresent the true prevalence and variety of tumors as most species have not been extensively studied in the wild, and it is assumed that most types of tumors seen in domestic dogs have the potential to occur in wild species. Several neoplastic diseases with relatively high prevalence are described below.

Ceruminous gland tumors in the Santa Catalina Island subspecies of the island fox occur at a prevalence that is one of the highest known for any tumor in a wild population. In one survey, 52.2% of the foxes had ceruminous gland tumors and of those, 34.8% were malignant.[393] The

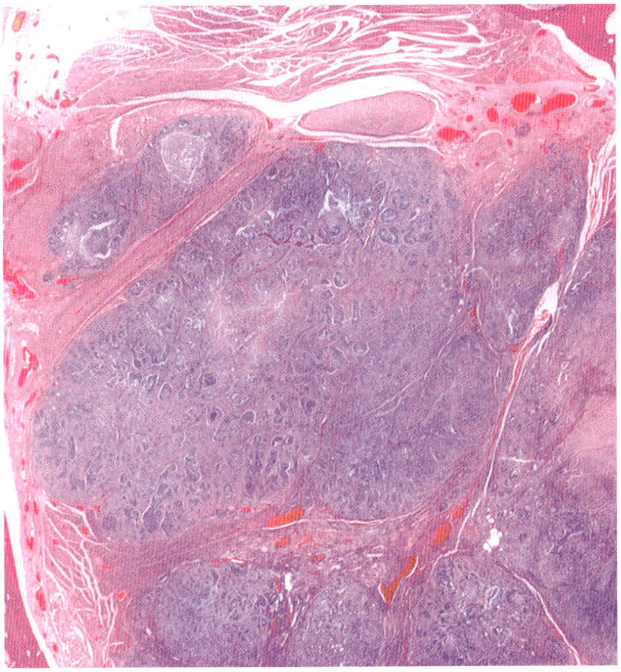

FIGURE 9.8 Invasive ceruminous gland adenocarcinoma in the ear of an island fox. These tumors are seen in the Santa Catalina Island subspecies and are associated with infestation by the ear mite, *Otodectes cynotis*. *(Photo Courtesy of L. Munson, William R Pritchard Veterinary Medical Teaching Hospital.)*

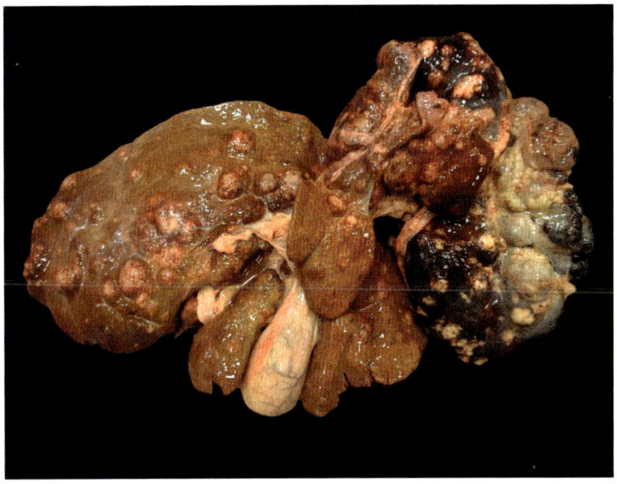

FIGURE 9.9 Cholangiocellular (biliary) carcinoma in the liver of a polar bear. Tumors are typically multicentric, variably sized, nodular, tan masses that replace, compress, and may locally invade the adjacent parenchyma. Red discoloration in this tumor was associated with multifocal and regionally extensive hemorrhage (contrast this with the hepatocellular carcinoma in Fig. 9.10). *(Photo Courtesy of C. Rodriguez, Wildlife Conservation Society.)*

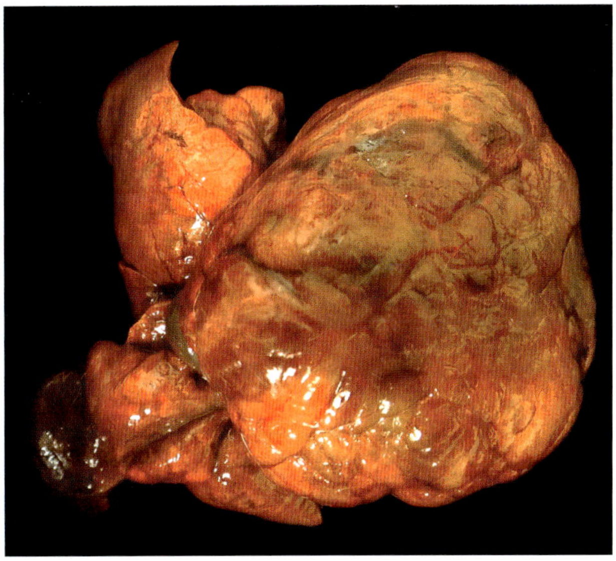

FIGURE 9.10 Hepatocellular carcinoma in the liver of a fennec fox. Tumors most commonly consist of a single variably sized mass that expands and replaces the parenchyma (contrast this with the cholangiocellular carcinoma in Fig. 9.9). *(Photo Courtesy of K. Conley, Wildlife Conservation Society.)*

adenocarcinomas consist of epithelial cells arranged in tubules, acini, or sheets that invade the deep dermis and sometimes induce desmoplasia (Fig. 9.8). They have been reported to invade the tympanic bullae and other bones of the skull or to metastasize to regional lymph nodes or the lungs. These types of tumors have been associated with chronic inflammation due to infestation by the ear mite *Otodectes cynotis*. The predisposition appears to only be seen in Santa Catalina Island foxes as foxes on other islands in the Channel Islands archipelago are unaffected even though ear mites are also present in those populations. The reason for the greater susceptibility of the Santa Catalina Island foxes is unknown, but hypotheses include an inappropriate immune response among this subspecies.

Hepatic tumors are relatively common in ursids and fennec foxes. These include **biliary adenoma** in a grizzly bear; **biliary adenocarcinoma** in sloth bears, grizzly bears, and polar bears (Fig. 9.9); **hepatoma** in a Kodiak bear and a polar bear; and **hepatocellular carcinoma** in polar bears, Hokkaido brown bears, and fennec foxes (Fig. 9.10).[12,53,146,178,279,291] Grossly and histologically, these neoplasms resemble hepatic neoplasms in other species. Biliary neoplasms can be extrahepatic and may be associated with the main bile duct, or they may be intrahepatic and multifocal throughout the parenchyma; carcinomatosis has been reported.[149] Histologically, they are composed of irregularly shaped glandular structures frequently surrounded by abundant fibrous connective tissue. Primary hepatocellular neoplasms are intrahepatic and composed of neoplastic hepatocytes of varying degrees of differentiation. They are reported to metastasize to local lymph nodes and adrenal glands.[279] The cause for this apparent predisposition for hepatobiliary cancer is unclear, although chronic inflammation, chronic viral infection, or dietary carcinogens have been raised as possible etiologies.[53,149,255,321] Bears used for **bile farming** commonly have cholecystitis and cholelithiasis, because hygienic practices are often not followed during the collection of bile. They also have a high incidence of hepatic and gall bladder neoplasia.[255]

Malignant lymphoma was a common tumor in one survey of red wolves.[1] Lymphomas are a diverse group of lymphoid neoplasms that arise in various locations outside of the bone marrow. Phenotypic classification is often lacking in reports concerning nondomestic canids, but T-cell lymphoma was confirmed in a raccoon dog.[318] This was a case of multicentric lymphoma with a leukemic phase, which resulted in diffuse hepatomegaly, splenomegaly, and lymph node metastases. Two red wolves diagnosed with lymphoma had multicentric tumors. A captive brown bear was diagnosed with large B-cell lymphoma of both kidneys and metastases to the central nervous system.[27] The histopathology of lymphomas is generally similar among species and has been reviewed extensively in other texts.

Mexican gray wolves appear to be predisposed to developing **nasal carcinoma** (Fig. 9.11).[342] These tumors were described as sinonasal, squamous-cell carcinomas. Other than a single report of adenosquamous carcinoma in a red fox, nasal tumors seem to be underrepresented in other species.[154]

Ovarian tumors occur with relatively high frequency in maned wolves.[298] The most common tumor is **dysgerminoma**, a neoplasm considered very rare in other species. These germ cell tumors are typically solid, but they can appear multinodular and contain cystic cavities. They are usually gray and may contain variable amounts of hemorrhage. Dysgerminomas are densely cellular, and they are composed of sheets of primitive round cells with distinct cell borders, moderate amounts of eosinophilic granular cytoplasm, and vesiculate nuclei. The cells are supported by scant connective tissue, or they may be subdivided into lobules by thin connective tissue septa. Neoplastic cells have moderate anisokaryosis and anisocytosis; mitotic figures may be common. All are potentially malignant and can metastasize to regional lymph nodes or within the abdominal cavity. **Granulosa cell tumors** have also been diagnosed in captive maned wolves and in captive African wild dogs.[214] They are the most common sex-cord stromal tumors among domestic dogs and likely occur in a variety of other canid species. Neoplastic cells often palisade or form follicular structures, some of which can be cystic. In other cases, the cells may be more closely packed into sheets, cords, trabeculae, or nests. Some tumors have cells arranged in tubules that resemble Sertoli cell tumors, as has been described in a maned wolf. Other ovarian tumors reported in nondomestic canids include **papillary cystadenoma** in an African wild dog and **papillary carcinoma** in a Syrian golden jackal.[130]

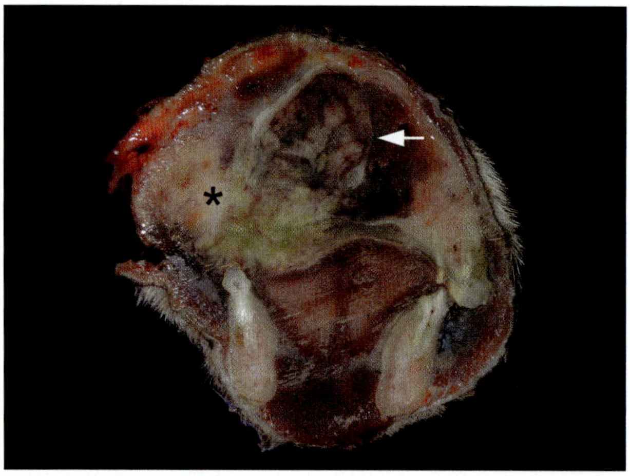

FIGURE 9.11 Nasal carcinoma in a Mexican wolf. Unilateral carcinoma *(asterisk)* within the nasal cavity has deviated from the nasal septum *(arrow)* and extended through the maxillary bone. *(Photo Courtesy of K. Terio, University of Illinois Zoological Pathology Program.)*

Despite their frequency in domestic dogs, mammary tumors are rarely reported in nondomestic canids. **Malignant mammary tumors** have been described as singleton cases in a Mexican gray wolf, a maned wolf, and a red fox.[138,160,215] **Tumors of sweat glands** (apocrine gland tumors) are also occasionally identified in canids. These include adenoma in a gray fox and carcinoma in a silver fox.[66] Cutaneous, apocrine gland tumors have been described in captive African wild dogs as well.[292] Females were overrepresented, and tumors tended to be malignant with a location on the dorsal midline.

Hemangiosarcoma is not commonly reported in nondomestic canids, but it has been described in several African wild dogs.[246] Although, the heart is the most common site, tumors in the liver and retroperitoneal space have also been seen.

Osteosarcoma has been described in red wolves and maned wolves.[1,286] In one case, extraosseous osteosarcoma in a maned wolf developed at a vaccination site, although no certain association could be made.[329] **Hypertrophic osteopathy** associated with metastatic **osteosarcoma** in a red wolf was attributed to a heart-worm infestation.[1] **Multilobular tumor of bone** has been described in the calvarium of an African wild dog.[280] A black-backed jackal was diagnosed with an osteoma, but further information regarding location or histologic description was unavailable.[66] **Synovial cell sarcoma** with invasion and lysis of the adjacent bones has been described in the stifle of a maned wolf.[89]

Oral squamous cell carcinoma in Malayan sun bears has been associated with a novel gammaherpesvirus, Ursid herpesvirus 1.[245,301] These neoplasms behave as oral, squamous cell carcinomas do in other species, with local invasion and recurrence; metastasis has not been reported. Regardless of species, these tumors are characterized by invasive nests of neoplastic, squamous epithelium surrounded by the scirrhous response.

Thyroid gland tumors have been sporadically reported in a number of bear species.[146,376] **Both follicular adenoma** and **cystadenoma** have been described. In one American black bear, a thyroid gland cystadenoma was concurrent with colloid goiter, and both were considered to contribute to hypothyroidism in that individual.[376] Metastatic thyroid gland carcinoma with epithelial-to-mesenchymal transition has been described in a brown bear.[336]

INFECTIOUS DISEASES

DNA Viruses

Canine adenovirus 1 (family *Adenoviridae*; syn. **Canine infectious hepatitis virus**) causes disease in a wide variety of canids and bears. Lesions in coyotes, wolves, and various fox species are like those described in dogs, although gross lesions in foxes may not be as apparent.[71,168,397,410] If present, gross lesions include petechial and ecchymotic hemorrhages on serosal surfaces, lymph node edema and hemorrhage, gall bladder edema, hepatomegaly, liver mottling, and increased liver fragility. Hepatosplenomegaly and adrenomegaly also occur.[396] Vasculitis and disseminated hemorrhage can be seen in the brain. Many lesions are secondary to vascular damage. However, renal and hepatic lesions occur due to direct infection of epithelial cells. Renal lesions can include glomerulonephritis.[396] Intranuclear viral inclusion bodies, which are acidophilic or basophilic, are present in endothelial cells, hepatocytes, biliary epithelium, and adrenal epithelium. Hepatic necrosis in dogs is typically periacinar (centrilobular), but this pattern may not be recognized in foxes.[168] In addition to canids, canine adenovirus 1 can infect sloth bears, black bears, and brown bears.[161,241,361] Histologically, lesions resemble those in dogs, including periacinar and hepatocellular necrosis with intranuclear viral inclusion bodies. A serological survey of free-ranging bears revealed that up to 29% of grizzly bears in Alaska were seropositive for antibodies to the virus.[413]

Canine parvoviral (**CPV**) disease in some wild canids is identical to that seen in domestic dogs. It is particularly well documented in congeneric species such as coyotes and wolves. Although most species of canids seem to be susceptible to infection, clinical disease and mortality are variable. For example, red foxes may develop mild leukopenia and shed canine parvovirus in feces but seem to be resistant to clinical disease.[24,51] Disease attributed to CPV has been reported in various captive canids, including coyotes, crab-eating foxes, dingos, maned wolves, and raccoon dogs.[24] Clinical signs include anorexia, vomiting, bloody diarrhea, and depression.[216] Lesions are similar to those of CPV observed in domestic dogs with necrosis of small intestine crypts and mucosal collapse, lymphoid depletion (e.g., spleen and lymph nodes), and pulmonary hemorrhage. **Blue fox parvovirus** and **raccoon dog parvovirus** are similar to CPV and are named after the host from which they were isolated.[24] Lesions are minimal in association with these viruses.

Not all parvoviruses cause an acute systemic disease. An example is **red panda amdoparvovirus**. This recently described virus is highly prevalent (50%) in zoo-housed red pandas and the animals are presumed to be persistently infected for life.[6] To date, the disease has only been associated with viral infection in a single red panda that had pyogranulomatous peritonitis, pancreatitis, and myocarditis.[7]

Pseudorabies, also known as **Aujeszky disease**, mad itch, or infectious bulbar paralysis, is caused by **suid herpesvirus 1** (family *Herpesviridae*). Feral swine and European wild boar are the natural reservoir hosts (see Suidae and Tayassuidae), but many species, including domestic and nondomestic canids, can be incidentally infected. Central nervous system signs in wild canids are related to nonsuppurative encephalitis with intranuclear, eosinophilic, inclusion bodies.[230] Among canids, the most significant differential diagnoses are rabies and canine distemper virus infections (see below). Suid herpesvirus 1 can also infect a variety of bear species.[23,350,412] While many wild bear populations show serologic evidence of exposure, clinical disease has only been reported in captive animals, possibly because the disease is highly fatal and thus rarely encountered in wild bears. Clinical signs include tremors, excessive salivation, pruritis, self-mutilation, and anorexia. Reported gross and histologic lesions have been nonspecific or related to self-mutilation.

Equine herpesviruses can infect polar bears and fatal neurologic disease has been reported in captive bears. Infection with equine herpesviruses 1 and 9 has been confirmed. The virus was thought to have been transmitted via contact with a fomite rather than direct contact with an infected equid. Lesions consist of nonsuppurative encephalitis with neuronal necrosis and gliosis.[181,349] Intranuclear basophilic inclusions are present in glial cells and neurons.

Nonsuppurative encephalitis of unknown etiology has been reported in bears from the state of Nevada, United States. Several viruses, including a novel gammaherpesvirus, have been identified in the affected bears, but causation still remains uncertain.[5,37]

Papillomaviruses are nonenveloped, double-stranded DNA viruses are associated with cutaneous and mucosal papillomas or plaques in a variety of mammals, birds, and reptiles. Although numerous papillomaviruses have been identified in domestic dogs, the diversity of those reported in wild carnivores is limited. **Oral papillomatosis** of coyotes is the most commonly recognized papillomavirus-associated disease among wild canids (Fig. 9.5). Extensive lesions can interfere with swallowing or cause secondary infections that may contribute to morbidity or even death. However, most cases are self-limiting and clinically insignificant. ***Lambdapapillomavirus* 2** has been identified in oral papillomas from domestic dogs, coyotes, and gray wolves, and possible transmission between these species has been suggested.[235] The exophytic lesions occur almost exclusively on the oral mucosa, but infrequently, smaller papillomas may be observed on the nasal planum or ocular palpebrae. The tumors are usually nonpigmented and papilliform to cauliflower shaped. The basal-cell layer is hyperplastic, and the stratum spinosum and granular-cell layer may be expanded with many cells that have swollen, eosinophilic to lightly basophilic cytoplasm (ballooning degeneration), and variably enlarged or shrunken nuclei. Enlarged nuclei have central pallor and marginated chromatin. If present, the viral inclusions are intranuclear and basophilic.[341]

RNA Viruses

Canine distemper virus (CDV) (*Morbillivirus* genus; *Paramyxoviridae* family) is one of the most globally important pathogens of domestic canids and nondomestic terrestrial carnivores (*Canidae*, *Felidae*, *Hyaenidae*, *Mustelidae*, *Procyonidae*, *Ursidae*, and *Viverridae* families). It can also have significant population-level effects in some species of marine mammals (family *Phocidae*).[108,257] The virus is epitheliotropic and neurotropic and can be found in cells in the respiratory, gastrointestinal, urinary, and nervous systems and in lymphoid, neuroectodermal, and mesenchymal cells including vascular fibroblasts.[62] Clinical signs are related to the stage and location of infection with respiratory, neurologic, and cutaneous signs typically being most clinically apparent. However, clinical signs are highly variable and may be influenced by virus strain, environmental conditions, host age, and immune status. Species susceptibility varies widely. For example, mortality in gray foxes appears to approach 100%, and although clinical distemper in red foxes appears to be relatively high in some parts of the world, it appears to be relatively rare in southeastern North America.[99,249,254]

Like other morbilliviruses, CDV is highly contagious, and it is spread principally by aerosol droplets, infected body fluids (e.g., nasal and ocular secretions and urine), or feces. Virus entry into host cells is mediated by specific viral receptors.[257] Signaling lymphocyte activation molecule family member 1 (SLAM/F1 or CD150) is expressed by subsets of thymocytes, dendritic cells (DCs), macrophages hematopoietic stem cells, and T- and B-lymphocytes. Nectin cell adhesion molecule 4 (nectin-4, syn poliovirus receptor-related 4) is expressed on the basolateral aspect of epithelial cells at the adherens junction complexes. Nectin-4 is also expressed by endothelial cells and keratinocytes. Both receptors are important in the pathogenesis of disease. Viral entry and dissemination are critically dependent upon CD150, but nectin-4-mediated transmission appears to be essential for virus transmission.

Acute lesions in nondomestic canids are like those in domestic dogs. Bronchointerstitial pneumonia and rhinitis with purulent nasal or oculonasal discharge are common. Bronchioles often contain suppurative exudate with patchy necrosis and attenuation of bronchiolar epithelium. Alveolar septa are expanded by mononuclear cell infiltrates, and alveoli may be filled with mixed inflammatory cell infiltrates and fibrin. Intraalveolar histiocytes and multinucleated histiocytes can be common. Viral epithelial cell syncytia may also occur but may be less evident than histiocytic syncytia. Eosinophilic, cytoplasmic, and viral inclusions are most commonly identified in syncytia, macrophages, and bronchiolar epithelial cells. Secondary bacterial infections are common and can complicate the identification of syncytia and viral inclusions.

Systemic spread and the development of neurologic disease may follow the acute development of pneumonia in domestic dogs or may be present as part of the primary clinical disease in susceptible animals.[378] This distinction has not been well characterized in other species. In the nervous system, infection often develops first in perivascular macrophages, astrocytes, and choroid plexus epithelium. Lesions in white matter and gray matter may occur concurrently or independently of each other. White-matter demyelination is the hallmark nervous system lesion in CDV infections, but it is not pathognomonic

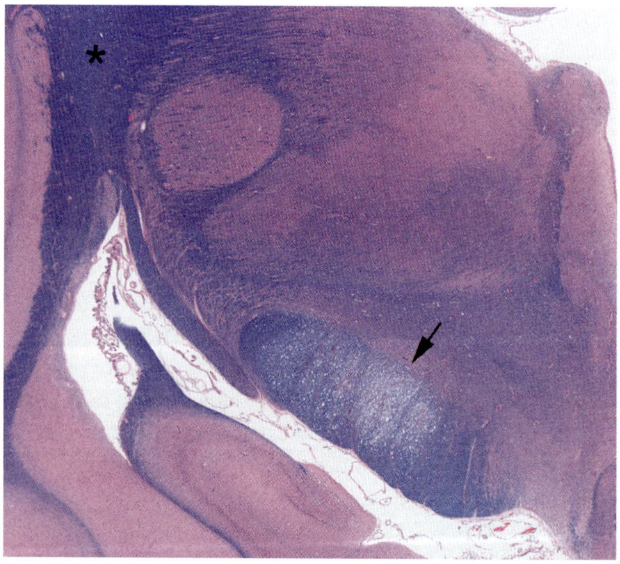

FIGURE 9.12 Canine distemper virus infection in the brain of a coyote. Canine distemper virus infections are frequently associated with demyelination. The demyelinated white matter at one focus in the ventral aspect of the internal capsule is rarefied in this section *(black arrow)* and contrasts sharply with other white matter tracts visible in this image *(asterisk)*. Luxol fast-blue stain (also see eSlide 9.e1).

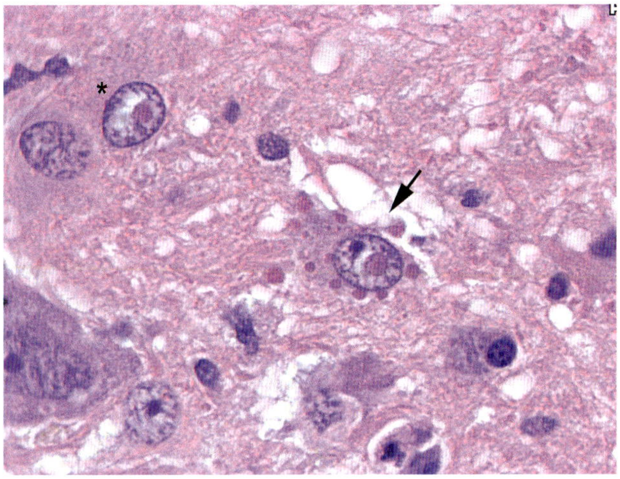

FIGURE 9.13 Canine distemper virus infection in the brain of a coyote. One neuron contains a single eosinophilic intranuclear inclusion *(asterisk)*. Another neuron *(arrow)* contains multiple intracytoplasmic inclusions and an intranuclear inclusion (also see eSlide 9.e1).

(Fig. 9.12). The demyelination may be multifocal or patchy, and it tends to be most severe in the cerebellum, periventricular white matter adjacent to the fourth ventricle, rostral medullary velum, optic tracts, and spinal cord. The neuropil may be vacuolated with loss of myelin and preservation of axons in early lesions. Gray matter lesions can be seen in the cerebral cortex, cerebellum, brainstem, or spinal cord. Nuclear or cytoplasmic, eosinophilic, viral, inclusion bodies will be present in astrocytes, other glial cells, or neurons (Fig. 9.13; eSlide 9.e1). Some authors have suggested that intranuclear inclusions are more common in neurons, but others describe abundant, intracytoplasmic inclusions with few intranuclear inclusions. Neuronal necrosis and nonsuppurative encephalitis are associated with infection, but early lesions may have minimal inflammation.

Marked hyperkeratosis is the classic, cutaneous CDV lesion, and the characteristic footpad thickening is the source of the colloquial name, "**hard pad disease**." Other sites in which hyperkeratosis may be seen are the nasal planum or hairy skin. Syncytial cells and both intracytoplasmic and intranuclear inclusion bodies may be present in the epidermis. Small numbers of lymphocytes may infiltrate the epithelium. Secondary infection with *Dermatophilus congolensis* in hyperkeratotic footpad lesions can be encountered in animals living on soil. Pustular dermatitis may occur secondary to the bacterial infection of affected-haired skin.

Other clinical signs of CDV can include diarrhea, especially in the early stages of acute infection in some species, since viral infection and replication occur in epithelial cells along the gastrointestinal tract. Inclusion bodies may be common but are highly variable. Secondary enteric infections with agents such as *Cryptosporidium* spp. or attaching and effacing *Escherichia coli* may exacerbate gastrointestinal signs. Lymphoid necrosis and depletion are also typical in animals with CDV infection; premature thymic atrophy is common in neonates. Subsequent immunosuppression is a significant sequela and is associated with opportunistic bacterial, protozoal, and concurrent viral infections. Certain conditions, such as metaphyseal osteosclerosis, dental anomalies (e.g., enamel hypoplasia), and facial "ticks" occur in domestic dogs but have not, to date, been reported in wild canids.[26,36]

CDV may be highly suspected based on clinical signs and lesions. These include identification of cytoplasmic inclusions in stained conjunctival epithelial cells or lymphocytes (blood film) or identification of characteristic histologic lesions, which variably include cytoplasmic and nuclear inclusions in syncytia and in cells of epithelial (e.g., lung and urinary bladder), mesenchymal, lymphoid, or nervous system tissues. Immunohistochemical labeling (IHC), in situ hybridization (ISH), PCR, and DNA sequencing can be used to confirm the diagnosis and differentiate it from other causes of respiratory and nervous system diseases including rabies, which can occur as a coinfection and may not be easily differentiated from CDV with routine hematoxylin and eosin (HE) stains.

As with other members of the order Carnivora, free-ranging populations of ursids have serologic evidence of exposure to CDV.[108] Gross and histologic lesions in the few reported clinical cases are similar to those for canids.[88,326] In polar bears, an inverse relationship between CDV titer and total leukocytes has been noted, suggesting the bears may be immunologically compromised even after clearing CDV infection.[232] Those strains isolated from ursids have had genetic similarities to strains circulating regionally in other carnivores, both wild and domestic.[88,183] In addition to exposure to CDV, free-ranging polar bears in multiple studies had serologic evidence of exposure to phocine distemper virus (PDV), phocine morbillivirus (PMV), or dolphin morbillivirus infections (DMV).[65,164,232] All of the individual polar bears that had titers to PDV, PMV, or DMV also had titers to CDV. In all cases, titers to CDV were higher, and it is possible that these results represent assay cross-reaction rather than true exposure to multiple morbilliviruses from infected prey species.

In addition to natural disease, **vaccine-induced CDV** infections have been described in red pandas vaccinated with modified-live CDV vaccines.[52,134,209] Disease and lesions in vaccine-induced cases are similar to those in natural infections and have characteristic viral inclusions. In cases where virus isolation has been attempted, viral isolates are consistent with the attenuated strain rather than virulent wild-type virus.

Highly pathogenic avian influenza (HPAI), better known for the devastating impacts it has on birds, has recently been confirmed as the cause of disease in various

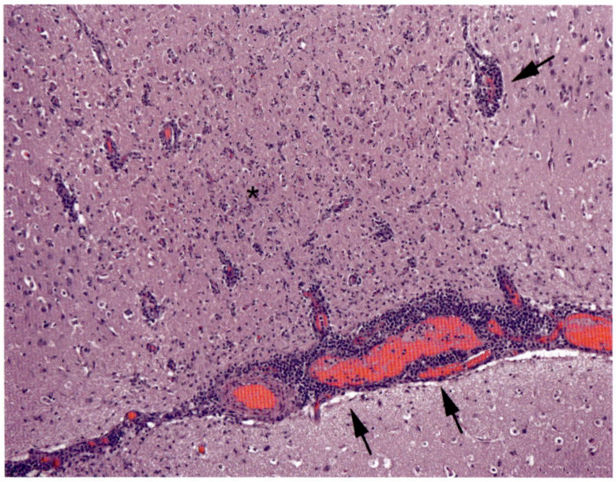

FIGURE 9.14 Highly pathogenic avian influenza infection in the brain of a red fox. Many lymphocytes and plasma cells expand the Virchow-Robbins space around blood vessels *(arrows)* and occasionally extend into the neural parenchyma. Neuronal necrosis, gliosis, and karyorrhectic debris are also evident *(asterisk)*. *(Photo Courtesy of B. Elsmo, Wisconsin Veterinary Diagnostic Laboratory.)*

wild carnivores, including black bears, brown bears, coyotes, and red foxes.[16,380] Neurologic signs and histologic evidence of disease have been observed in the carnivores in North America infected with H5N1 clade 2.3.4.4b viruses (Fig. 9.14), and neurotropism of the virus has been described in red foxes in Europe.[42] The expansion of the epidemic in North America beginning in the Fall of 2022 was associated with many more carnivore cases than had been described in previous HPAI outbreaks, suggesting the host specificity of the virus continues to change.

Rabies is caused by a number of lyssaviruses (family *Rhabdoviridae*), but **rabies virus** is the type species. It is a globally significant disease, and it is often locally or regionally reportable. Species susceptibility varies for different strains of these viruses, but it is generally accepted that all warm-blooded mammals, including humans, are susceptible.[340] Infection causes progressive encephalomyelitis that is lethal once clinical signs become apparent. Transmission typically occurs by inoculation of virus-infected saliva from a rabid animal into tissue or less commonly, through exposure of open wounds or intact mucous membranes to infected saliva, salivary glands, or brain tissue. Dogs and wild canids can both be infected and serve as important reservoirs for the virus. Infection can have significant local and regional population-level effects on these and other species. The variant of rabies that is adapted to dogs readily infects other wild canid species, and spillover of variants from other wildlife reservoirs (e.g., mongoose variant and raccoon variant) also occurs. A cycle of so-called street rabies in feral or poorly managed dog populations provides a significant source of infection for wildlife, particularly where the ranges of domestic and wild canids overlap.

The classic signs of rabies range from depression and listlessness, the "dumb" form, to extreme aggression, the "furious" form. However, a variety of neurologic signs can occur. In animals exhibiting aggression, the disease can progress to a paralytic stage caused by progressive damage to motor neurons. The latter stage is variably characterized by incoordination (typically associated with hind limb paralysis), difficulty swallowing (most commonly recognized as pseudoptyalism), and paralysis of facial muscles. Infection does not directly result in gross lesions, but bite wounds, tooth fractures, and gastric foreign bodies may be seen in association with aggression and pica. Aggression promotes transmission of the virus through inoculation of bite wounds with infected saliva in susceptible animals. At bite wounds, the virus invades neuromuscular junctions and ascends peripheral nerves to the spinal cord and subsequently to the central nervous system (CNS). From the CNS, it is distributed centrifugally to the salivary glands and other tissues, including but not limited to the adrenal glands, nasal mucosal epithelium, and corneas.[22] The incubation period is unpredictable and can range from weeks to months. The clinical course of the disease is usually acute but can be as long as 10 days.

Microscopic lesions include nonsuppurative encephalomyelitis, ganglioneuritis, adrenalitis, and sialadenitis. Perivascular, lymphoid cuffing and multifocal gliosis may be present but are variable, and inflammatory responses can be minimal.[55] Lesions are generally more severe from the pons to the hypothalamus and in the cervical spinal cord. Rabies and related lyssaviruses produce intracytoplasmic, eosinophilic, viral inclusions (Negri bodies) in neurons. They are reported to be most common in the hippocampus of carnivores but may also be abundant in other regions including, but probably not limited to, cortical neurons, basal nuclei, thalamus, and caudal colliculi.[18] The size and number of inclusions are highly variable; the latter tends to be inversely correlated with inflammation. Neuronal degeneration can be severe in carnivores, and degenerate neurons typically lack inclusion bodies.

Diagnosis is typically based on a combination of clinical signs, serology, histologic lesions, and identification of viral antigen in tissues by immunologic assays. A fluorescent antibody test is commonly used to identify the virus in unfixed samples of the fresh brain that includes the medulla oblongata and cerebellum. Immunohistochemical assays can also demonstrate viral antigen (Fig. 9.15; eSlide 9.e2). PCR and DNA sequencing can provide confirmation and specific information about viral strains, and, in addition to the brain, can also be run on nonneural tissues including salivary samples or skin.

The negative impacts of rabies-associated mortality on wild populations are largely speculative, and many populations seem to be stable, even with endemic rabies virus. However, there are some reports of dramatic mortality

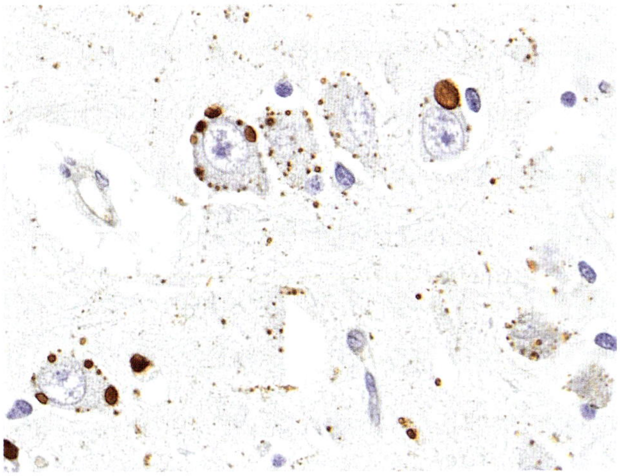

FIGURE 9.15 Rabies virus infection in the brain of a gray fox. Viral inclusions (Negri bodies) are present (brown) in the cytoplasm of the hippocampal neurons. The inclusions in this case were labeled using an immunohistochemical assay based on an antibody raised against the rabies virus. IHC (also see eSlide 9.e2).

rates; up to 90% of bat-eared fox deaths in one study were attributed to rabies.[262] Additionally, species or populations that are already endangered may be at greater risk of rabies-associated mortality, and spillover from other domestic dog reservoirs can be important. For example, the extirpation of a significant African wild dog population in the Serengeti ecosystem was attributed to rabies, and repeated rabies mortality events are a significant conservation threat in the Ethiopian wolf.[224,327] The risk of rabies in small or isolated populations of endangered animals may warrant parenteral vaccination using inactivated virus or subunits. However, caution is advised using attenuated vaccine strains of rabies virus, because they have caused clinical rabies in wild canids.[201]

Several reports describe the infection of captive Asian black bears and sun bears with the picornavirus, **foot and mouth disease virus**.[306] The lesions were similar to those in other species and included vesicle formation on foot pads, in the oral cavity, and on the nasal planum with subsequent sloughing of skin (Fig. 9.16).

Bacteria

In addition to bacterial infections described below, positive serology for a number of additional bacterial pathogens has been reported in a wide range of wild canid species (Table 9.e3).

Bacteria in the genus ***Mycobacterium*** have a broad host range and are widely distributed. ***Mycobacterium bovis*** is particularly important, and it is the causative agent of bovine tuberculosis, a chronic, progressive disease that is a listed reportable disease by the World Organization for Animal Health (WOAH) (see Bovidae,

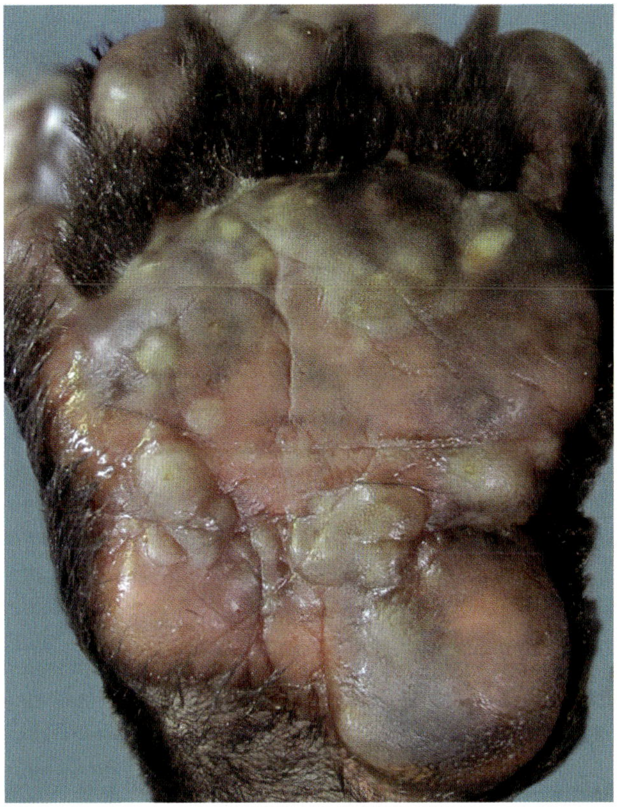

FIGURE 9.16 Foot and mouth disease virus in the foot of an Asiatic black bear. Infection in bears is associated with the development of large vesicles on the footpads. Vesicles can also develop on the nasal planum and in the oral cavity in affected bears. *(Photo Courtesy of K. Officer and Animals Asia.)*

Antilocapridae, Giraffidae, Tragulidae, and Hippopotamidae). Infection is characterized by granulomas and granulomatous inflammation in the lymph nodes, lungs, or other tissues. As the name implies, this is typically a disease in ruminants, both domestic and free ranging. However, infection has been described in various carnivore species including red foxes, gray foxes, coyotes, wolves, wild dogs, and captive fennec foxes.[31,56,195,276,287,390,406] Infection with *M. bovis* has also been reported in free-ranging black bears in regions where deer are infected, as well as in sloth bears held in substandard conditions.[48,389]

Transmission to wild carnivores is thought to occur through ingestion of infected prey or carcasses. In many reports, gross lesions are not seen, and detection of infection relies on histology, acid-fast stains, and other ancillary diagnostic tests. When present, lymphadenomegaly with granulomas and granulomatous lymphadenitis of the mesenteric lymph nodes is often the most common finding. Other lesions can include more generalized infections and lymphadenomegaly, as well as granuloma formation and granulomatous inflammation in the lungs, bronchial or other thoracic cavity lymph nodes, or other tissues. Granulomatous infiltrates typically contain low but variable numbers of acid-fast-positive bacilli, variable amounts of intralesional mineralization, and necrosis that is often centralized in the granulomas. Disseminated infection with no gross or histologic lesions has also been reported in red foxes.[277] The role of carnivores in bovine tuberculosis transmission is unclear. In one study, oral and nasal secretions and feces from *M. bovis*-positive coyotes suggested minimal shedding.

Infections with several **other mycobacterial species** have been reported in wild carnivores. These include *M. avium* subsp. ***paratuberculosis*** (paratuberculosis; Johne disease) infection in red foxes and a coyote; *Mycobacterium avium* subsp. ***avium*** in a jackal ***M. avium, M. intracellulare***, and ***M. kansasii*** in coyotes; ***M. caprae* subtype Lechtal** in a red fox; and ***M. tuberculosis*** in captive sloth bears.[11,270,278,315,374,390] Red foxes often lacked gross lesions of *M. avium* subsp. *paratuberculosis* infection and infection was confirmed with acid-fast stains, bacterial culture, or PCR. If lesions were present, they generally occurred in mesenteric lymph nodes and mucosa-associated lymphoid tissue and consisted of histiocytic and granulomatous inflammation and granuloma formation with few to numerous, intrahistiocytic, acid-fast bacilli. Lesions associated with *M. tuberculosis* infection in sloth bears were confined to the lungs and regional lymph nodes and were multifocal granulomas that often had central caseous necrosis. Histologic descriptions of mycobacteriosis were not provided for the jackal or coyotes.

The gold standard for mycobacterial diagnosis is bacterial culture. However, the likelihood of detection and confirmation increases with the use of a combination of diagnostic techniques. A thorough systematic necropsy with cytologic and histologic examination, acid-fast staining, and bacterial culture are routine initial diagnostics. The use of confirmatory ancillary assays, such as the multiantigen print immunoassay (MAPIA), rapid lateral-flow antibody tests (RLFAT), PCR, DNA sequencing, spoligotyping, or the analysis of variable-number tandem repeats of mycobacterial interspersed repetitive units (MIRU-VNTR) can greatly facilitate diagnostic or surveillance efforts.

Anthrax is a WOAH-listed, reportable bacterial disease of global concern. The causative agent is ***Bacillus anthracis*** **(BA)**, a large, 1- to 9-μm-long, 3- to 5-μm-wide, spore-forming, gram-positive, rod-shaped bacterium that has truncated ends (giving it a rectangular appearance) and a characteristic clear capsule. It can be seen in blood smears as individual bacilli or in long chains (see Wildlife Necropsy). The disease is most common in ruminants and humans, though most mammals are thought to be susceptible (see also Chapter 5: Bovidae, Antilocapridae, Giraffidae, Tragulidae, and Hippopotamidae, Chapter 6: Cervidae, and Chapter 10: Felidae). Carnivores can become infected through ingestion or inhalation of spores in infected carcasses. Proteins produced by the bacteria (protective antigen, edema factor, and lethal factor) combine to produce edema toxin and lethal toxin, which cause tissue swelling (edema) and death. Exposure to BA appears to be quite high in some populations of carnivores. For example, in one serologic study of black-backed jackals in Etosha National Park in Namibia, the prevalence of antibodies to BA was 95%, and antibodies were present in all animals over 6 months of age. This suggested exposure at a young age and survival despite infection.[29]

Though carnivores are commonly thought to be relatively resistant, death due to anthrax has been reported and mortality rates can be significant, especially in small populations and endangered species. Outbreaks have been reported in African wild dogs in both Zambia and Tanzania.[90,381] Clinical signs in pups have included swelling of the soft tissues of the jaws, lips, gums, and the ventral aspect of the neck. Adults had periocular swelling. Some members of both age groups developed ocular discharge, and blood oozed from the swollen lips and gums of some pups. Several affected animals developed wart-like lesions on the face that were suspected, but were not proven to be, anthrax related. Gross lesions included extensive liquid and clotted blood in the subcutis of the face, submandibular and thoracic regions, and similar accumulations of blood in the thoracic and abdominal cavities. Pulmonary congestion as well as marked lymphadenomegaly of superficial nodes was also present. In this outbreak, animals across age groups were affected but mortality occurred only in the pups.

Salmon poisoning due to ***Neorickettsia helminthoeca*** infection has been reported in captive sun bears and in natural and experimental infection in coyotes.[145,159] *N. helminthoeca* is a bacterium that infects the trematode ***Nanophyetus salmincola***. Carnivores are commonly infected with *Nanophyetus* by ingestion of one of the intermediate hosts of the trematode, typically a fish. After ingestion, metacercariae excyst from the tissues of the prey item, attach to the intestinal mucosa, and introduce the bacteria into the host. Clinical signs include vomiting, diarrhea, lethargy, and lymphadenomegaly that typically develop within 5–7 days of infection as a result of the bacterial infection. Thrombocytopenia and hypoalbuminemia may also be seen. Diagnosis of *Nanophyetus* ova in feces allows for a presumptive diagnosis, however, *Neorickettsia* infection may precede fecal shedding of the fluke ova and produce false-negative results. Lesions in canids include granulomatous inflammation in the gastrointestinal tract, spleen, and lymph nodes. Infected bears may develop eosinophilic gastritis and enteritis with erosion and attenuation of villi. Hemorrhage and granulomatous inflammation, although common in other species, have not been described in bears. Although free-ranging black bears are native to regions (e.g., the Pacific Northwest of the United States), where *Nanophyetus* is present and disease is endemic, significant disease has not been described in them, leading to speculation that black bears have developed resistance.

A second *Neorickettsia* sp., known as the ***Stellanchasmus falcatus*** **(SF)** agent, has also been associated with salmon poisoning disease in dogs.[314] Experimental inoculation of grizzly bears with *N. helminthoeca* and the SF agent resulted in clinical disease, but only the SF agent was recovered from sick bears.[333]

A few additional bacterial infections are worth mentioning. These include sporadic reports of **borreliosis** caused by spirochetes in the genus *Borrelia*. Disseminated borreliosis has been described in a wild red fox.[208] Gross lesions included areas of cutaneous erythema. Microscopic lesions include lymphoplasmacytic, neutrophilic, and eosinophilic dermatitis. Other histologic lesions include lymphoplasmacytic and neutrophilic inflammation in the liver, kidney, and conjunctiva. The bacterium has been isolated from multiple tissues. Diagnostic options include serology, histology (with silver stains), darkfield microscopy, electron microscopy, bacterial culture, and PCR. ***Clostridium piliforme***, the causative agent of **Tyzzer's disease**, was implicated in the death of a red panda with necrotizing colitis and multifocal hepatic necrosis with microabscesses.[247] Both lesions included characteristic intracytoplasmic, interlacing stacks of bacteria.

Fungi

It is likely that wild carnivores are susceptible to infection and disease by the same wide range of fungal pathogens as their domestic counterparts. It is also likely that animals with immature immune systems or those that are immunodeficient or immunosuppressed are at a higher risk for opportunistic infection with resident or infectious fungal species.

Dermatophytoses are fungal infections of the epidermis or hair follicles caused by keratinophilic dermatophytes. Species of *Microsporum* and *Trichophyton* are common causes of infection and disease in domestic dogs. Clinical signs include focal, multifocal, or coalescing, characteristically round foci of alopecia and scaly skin, with or without crusts; more generalized hair loss can be seen in severe cases. Histologically, acanthosis with orthokeratotic hyperkeratosis is associated with 2- to 4-μm-diameter, basophilic arthrospores and 2- to 7-μm-wide, thin, parallel-walled hyphae in hair shafts or hair follicles. Invasion of the hair shaft is called an endothrix infection, whereas exothrix (or ectothrix) refers to infections confined to the hair surface. The disease is often more severe in young or immunosuppressed animals. ***Trichophyton mentagrophytes*** were associated with generalized partial alopecia and dermatitis in red foxes.[236] In another survey, a single fox (unknown species) was diagnosed with ***M. canis*** infection.[358] An atypical *T. mentagrophytes* infection has also been diagnosed in two captive African wild dogs. Affected dogs had severe, generalized alopecia. Histologic lesions included acanthosis, hyperkeratosis, hyperpigmentation, and sebaceous gland atrophy.[386] Dermatophytosis associated with ***M. gypseum*** is an important cause of skin disease in red pandas and causes lesions similar to those in other species, including alopecia, crusting, and hyperkeratotic dermatitis.[226] Lesions typically develop in warm seasons on the tail, extremities, muzzle, and ears. Some of the more severely affected individuals develop furunculosis and cellulitis. Rare cases have also been reported in captive bears.[310] Black bears in California (United States) are occasionally diagnosed with dermatophytosis and most of the cases have yielded cultures of ***T. equinum***.[80] Direct microscopy of skin scrapes and hairs, histology, or culture (Sabouraud's dextrose agar) confirms the diagnosis of dermatophytosis.

Microsporidia are unicellular, obligately intracellular, fungal parasites. Several species are important in animals, including those in the genus ***Encephalitozoon***. Infection is rarely described in nondomestic canids, but it has been reported in a litter of captive African wild dogs with concurrent CDV infection and in a litter of Arctic foxes.[384,388] In both litters, histologic lesions were seen in the brain and kidney (similar to infection with *E. cuniculi* in rabbits). The fox kits also had dissemination of *Encephalitozoon* to the spleen, liver, mesenteric lymph nodes, lungs, and urinary bladder. The organisms were ovoid, 0.8×2.3 to $1.6 \times 3.0\,\mu m$, and those in the brain and kidneys were associated with microgranulomas, lymphocytic, and histiocytic inflammation. Organisms in the fox kidneys were occasionally associated with necrosis. The organisms are best seen with Gram or Periodic acid-Schiff (PAS) stains, which also highlight their somewhat characteristic polar cap.

Coccidioidomycosis, also known as "valley fever," is a fungal disease caused by infection with *Coccidioides immitis* and *C. posadasii*. These fungi are endemic to several western and southwestern states in the United States and parts of Mexico and Central and South America. Infection with aerosolized spores (arthroconidia) typically causes respiratory disease; however, disseminated infections can occur. In tissues, arthroconidia develop into spherules that, when mature, contain numerous endospores. Fungal size varies based on developmental stage: arthroconidia are ~4 μm in diameter, immature spherules are 5–30 μm in diameter and mature sporangia are 30–200 μm in diameter and contain 2- to 5-μm-diameter uninucleate, endospores (sporangiospores). Histologically, the sporangia have a characteristic, 4- to 5-μm-thick, double-contoured, refractile wall (Fig. 9.17). Infection is often associated with pyogranulomatous inflammation with epithelioid macrophages, multinucleated giant cells, lymphocytes, plasma cells, and neutrophils. Serologic evidence of infection has been reported in an engendered San Joaquin kit fox.[283]

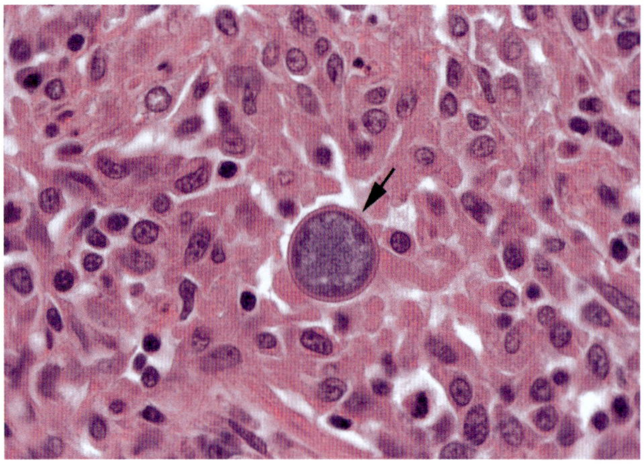

FIGURE 9.17 *Coccidioides immitis* infection in the lung of a coyote. Spherules are the asexual stage of the fungus and are characterized by a thick, refractile wall. This image shows a young spherule *(arrow)* without endospores. More mature endospores contain 2- to 5-μm-diameter spherules that reproduce by asexual division. Numerous epithelioid macrophages surround the spherule in this image.

Systemic infections with ***Blastomyces dermatitidis*** and ***Histoplasma capsulatum*** are rarely observed in both captive and free-ranging bears, but they are seen somewhat frequently in coyotes in certain parts of the United States (Terio, unpublished data). Lesions resemble those in other carnivores.[128,194]

Parasites

Numerous case reports and wide geographic studies describe parasite diversity in wild canids.[10,73,140,152,171,363,401] The vast majority are also seen in sympatric domestic dogs and many parasites have a host range that includes noncanid carnivores, as well. Several notable parasitic diseases in wild canids are described below.

Metazoa

Oslerus osleri (formerly ***Filaroides osleri***) (superfamily Metastrongyloidea; family *Filaroididae*) is a nematode that infests the trachea, tracheal bifurcation (the most commonly reported site) and mainstem bronchi. Infection occurs in domestic and nondomestic canids in North and South America, Europe, Africa, Asia, and Australia, with reports in coyotes and wolves in Canada and the United States, the hoary fox and maned wolf in South America, and dingoes and foxes in Australia.[60,104,113,125,126,210,355,392] Unlike other metastrongyles, *O. osleri* has a direct life cycle. Larvae hatch from eggs laid in the respiratory tract, are coughed up, swallowed, and then shed in feces. Infection and transmission occur via ingestion of infective larva in saliva (e.g., through maternal grooming), in vomitus (e.g., regurgitant feeding; the primary transmission mechanism

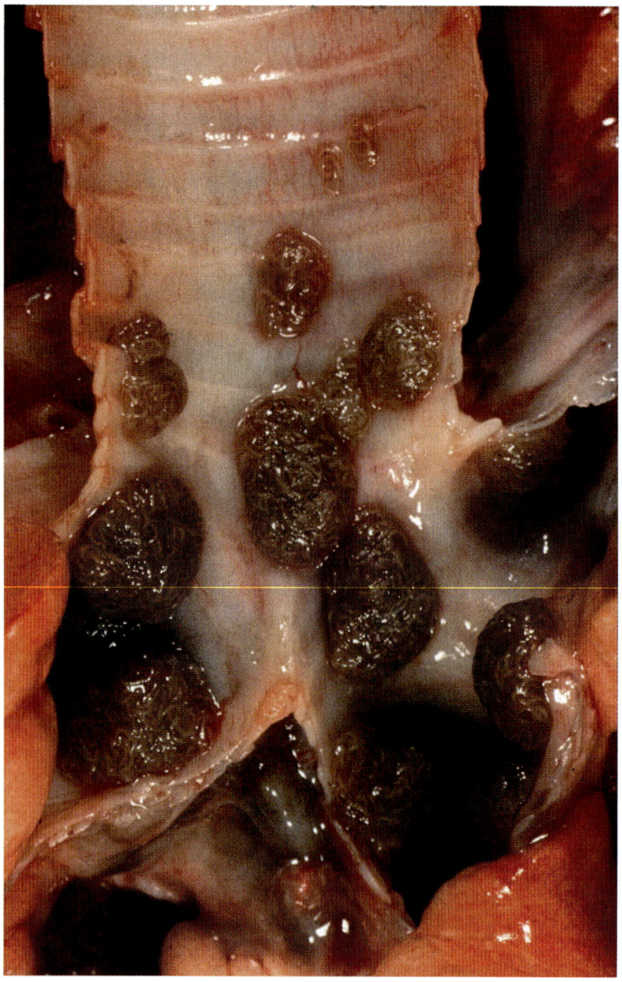

FIGURE 9.18 *Oslerus osleri* infestation of the trachea and bronchi of a coyote. Adult nematodes expand the mucosa and appear as nodules composed of tangles of slender worms. *(Photo Courtesy of R. Fairley, Western College of Veterinary Medicine.)*

in wild canids), or from the environment. Infection may be asymptomatic or associated with a chronic, nonproductive, dry cough. In the case of severe infection or large nodules, it may cause dyspnea or death. Lesions include 1- to 10-mm-diameter, firm, gray to pink, foci of mucosal thickening, and sessile nodule formation in the trachea, bronchi, or lungs (Fig. 9.18). Nodular expansion of the lamina propria is characterized by intralesional, coiled nematodes, and larva that are surrounded by mixed histiocytic, lymphoplasmacytic, eosinophilic, and granulomatous inflammation with variably mature fibrous connective tissue. The adult parasites have a smooth cuticle, a pseudocoelom, coelomyarian musculature, a multinucleate intestine, and larvated eggs in adult females. Fecal flotation is a useful diagnostic tool, but it is not sensitive enough to detect low numbers of worms. Zinc sulfate-solution flotation is preferred over the Baermann technique to identify eggs in fecal samples. A diagnosis can also be made by identification of characteristic gross and microscopic lesions in biopsy or necropsy samples, as well as PCR with DNA sequencing.

The definitive hosts for ***Crenosoma vulpis*** (superfamily Metastrongyloidea; family *Crenosomatidae*), also called the fox lungworm, include foxes, wolves, raccoons, and dogs in northeastern North America and various countries in Europe.[35,82] Infection has also been reported in badgers, wolverines, and black bears. In one study, an undesignated ***Crenosoma*** sp. was described in 43% of wild red pandas in Nepal.[83,265,359] Adults reside in the trachea, bronchi and bronchioles, and larvae are released into the airways. They are then coughed up, swallowed, and shed in the feces. Mammals acquire the parasites when they consume intermediate hosts (i.e., snails or slugs) that contain infective larvae. Clinical signs include rhinotracheitis, bronchitis, coughing, and nasal discharge. Gross lesions consist of single or multiple areas of parenchymal consolidation with intralesional adult nematodes. The females are up to 15 mm long. Microscopic changes in small bronchi and bronchioles can include: severe epithelial and glandular hyperplasia; marked, bronchiolar, mucous metaplasia; smooth muscle hypertrophy; submucosal eosinophilic and mononuclear cell infiltrates; and pyogranulomatous to eosinophilic bronchopneumonia with involvement of alveoli. Adult parasites have crenated folds on their anterior ends. Larvae have a blunt, rounded anterior end, and a finely tapered posterior end with a slight bend; they lack a dorsal spine (see *Angiostrongylus vasorum* below). Similar respiratory tract lesions are also seen in infections with ***Eucoleus aerophilus*** (previously ***Capillaria aerophila***; family *Capillariidae*). This is a larger (adult females 20–40 mm long), oviparous (embryonated, operculated eggs), aphasmid lungworm of wild carnivores including foxes. It is present in many parts of the world. Some studies suggest, it is more commonly found embedded in the mucosa and higher in the respiratory tree (i.e., trachea and bronchi) than *C. vulpis*, which classically occurs in small bronchi and bronchioles the caudal lung lobes. *E. aerophilus* can also be found more widely, and parasites are often free in bronchial and bronchiolar lumina.[304] Larvae of *C. vulpus* and *E. aerophilus* are identified using the Baermann technique, zinc-sulfate fecal flotation, or from tracheal washes. Gross and histologic features of the adults and larvae (especially tail morphology) are diagnostic and allow differentiation between these two nematodes and other pulmonary parasites.

Members of the genus **Angiostrongylus** (superfamily Metastrongyloidea) are intravascular nematode parasites with a wide host range that includes canids, and occasionally humans, in Europe, Asia, Africa, Australia, and North and South America.[372] Some species are geographically limited, but others are regionally or globally widespread. *Angiostrongylus* species have an indirect life cycle. Infection of definitive hosts occurs when they ingest infective larvae from the environment or from infected gastropods (slugs or aquatic or terrestrial snails). The gastropods are themselves infected when first-stage larvae (passed in the feces of the definitive hosts) penetrate their muscular foot. Also known as the "French lungworm," ***A. vasorum*** affects a wide range of wild canid species including red foxes in Europe and Canada, wolves in Europe, coyotes in Canada, golden jackals in Serbia, pampas foxes in Bolivia, and hoary and crab-eating foxes in Brazil; it has also been reported in captive red pandas.[20,44,106,116,141,166,167,251,356,372] In one study in Ireland, the prevalence of *A. vasorum* in red foxes was 39.9% and it is considered an emerging health problem in many regions within Europe.[282] *A. vasorum* in Europe and ***A. raillieti*** in South America were recently consolidated into a single species based on taxonomic revision. However, molecular studies suggest that *A. vasorum* in South America may represent a different species.[217] Adult worms reside in the right atrium and ventricle of the heart and pulmonary arteries or branches; larvae and eggs are present in the lungs.[372] Several other *Angiostrongylus* species in wild canids include ***A. gubernaculatus*** (syn. ***Angiocaulus gubernaculatus***) in an island fox (Channel Islands, United States), *A. raillieti* in a crab-eating fox (Rio de Janeiro, Brazil), and another incompletely characterized *Angiostrongylus* sp. in a crab-eating fox (Brazil).[86,137,217,394]

Animals infected with *Angiostrongylus* species may be asymptomatic, or they may have mild respiratory signs (e.g., coughing). Grossly, the lungs contain multifocal, variably sized, firm nodules that may cause obliteration of the pulmonary parenchyma. Adult nematodes may be encountered in pulmonary arteries or the right ventricle. In the lungs, they induce eosinophilic and lymphoplasmacytic arteritis with intimal fibromuscular thickening and medial smooth muscle hypertrophy and hyperplasia. Intraluminal nematodes may be associated with thrombi that range from acute fibrin aggregates to more mature recanalized lesions. Adult parasites have a smooth eosinophilic cuticle, thin coelomyarian, polymyarian musculature, and an intestinal tract lined by a few, multinucleated cells. Microscopically the pulmonary lesions vary from smaller aggregates of eosinophils and mononuclear cells to more extensive granulomatous foci containing eosinophils and neutrophils that are oriented around nematode eggs and larvae. Eggs are ovoid, 50–60 μm in diameter, and contain larvae or morulae. Lesions may also include sloughed epithelial cells and mucus in the bronchioles and bronchi may contain sloughed epithelial cells and mucus. Aberrant larval migration can lead to granuloma formation in various other tissues (e.g., brain, kidney, lymph nodes). Diagnosis is made by identification of adult parasites with characteristic gross and histologic features in the right ventricle or pulmonary arteries, identification of larvae with a characteristic "kinked" tail and small, dorsal spine in fecal samples using the Baermann technique, fecal flotation, or cytology of bronchiolar lavage samples or tissue touch preparations (impression smears).

Dirofilaria immitis is a spirurid, filarial, nematode parasite (family *Onchocercidae*) that causes **heartworm disease**. Domestic and nondomestic canids are definitive hosts and other species including bears, red pandas, sea lions, ferrets, and occasionally humans can be infected.[13] Transmission is via mosquito vectors. The parasite occurs in temperate, subtropical, and tropical regions in North and South America, Europe, the Middle East, Asia, and Australia, where the primary mosquito vectors (*Culex, Aedes, Anopheles*, and *Culiseta* species) are present. The most significant lesions typically occur in the pulmonary arteries and lungs. Low-level infections may be asymptomatic, or they may be associated with cough or exercise intolerance. More severe infections can cause pulmonary hypertension and symptoms associated with right-sided congestive heart failure. Adult worms up to 30 cm in length are typically present in the pulmonary arteries and in the right chambers of the heart (Fig. 9.19A). Histologic changes in large to small pulmonary arteries include villous endarteritis characterized by frond-like, intimal proliferations with or without disruption of the internal elastic lamina; hypertrophied endothelium; intimal infiltrates of mononuclear cells and eosinophils; medial smooth muscle hypertrophy and hyperplasia with disruption of the elastic fibers; mural collagen deposition; and luminal narrowing (Fig. 9.19B). Intraluminal adult parasites have a thin, eosinophilic cuticle, prominent, lateral cords, an internal, lateral a thin, eosinophilic cuticle; prominent, lateral cords; an internal, lateral cuticular ridge; tall, coelomyarian muscle; and an intestinal tract lined by a few, multinucleated, epithelial cells. Lesions include vascular thrombi; worm-fragment emboli; periarteritis with pulmonary, interstitial fibrosis and mononuclear cell and eosinophil infiltration; vascular, mural, medial hypertrophy, and hyperplasia (suggestive of hypertension); and marked hypertrophy of smooth muscle layers in terminal bronchioles and alveolar ducts.[225] Circulating microfilaria may be present with or without associated inflammation (e.g., pneumonitis, glomerulonephritis). Caval syndrome can occur due to heavy intracardiac infections that result in obstruction of the right atrioventricular valve.[290] Infection in bears is typically less severe than in canids. However, vascular endothelial proliferation can occur in red pandas, and death has been reported in that species.[109,303] Diagnosis can be confirmed by the identification of circulating microfilaria in a direct blood smear, by the modified Knott test, or by membrane filtration. The microfilariae lack a sheath, have a pointed cephalic end, and a straight tail with a pointed end. Antigen testing (e.g., ELISA and lateral flow immunoassay), PCR and DNA sequencing, radiography, and echocardiography are also useful antemortem assays.[281] Adult *D. immitis* in histology sections can be differentiated from *A. vasorum* by virtue of the intrauterine microfilaria (as opposed to eggs for *A. vasorum*), a more well-developed coelomyarian muscle fibers and a smaller intestine.

Domestic dogs and other carnivores (e.g., fox and wolves) are also the definitive host for ***Dirofilaria repens***, which causes subcutaneous and occasionally intraabdominal,

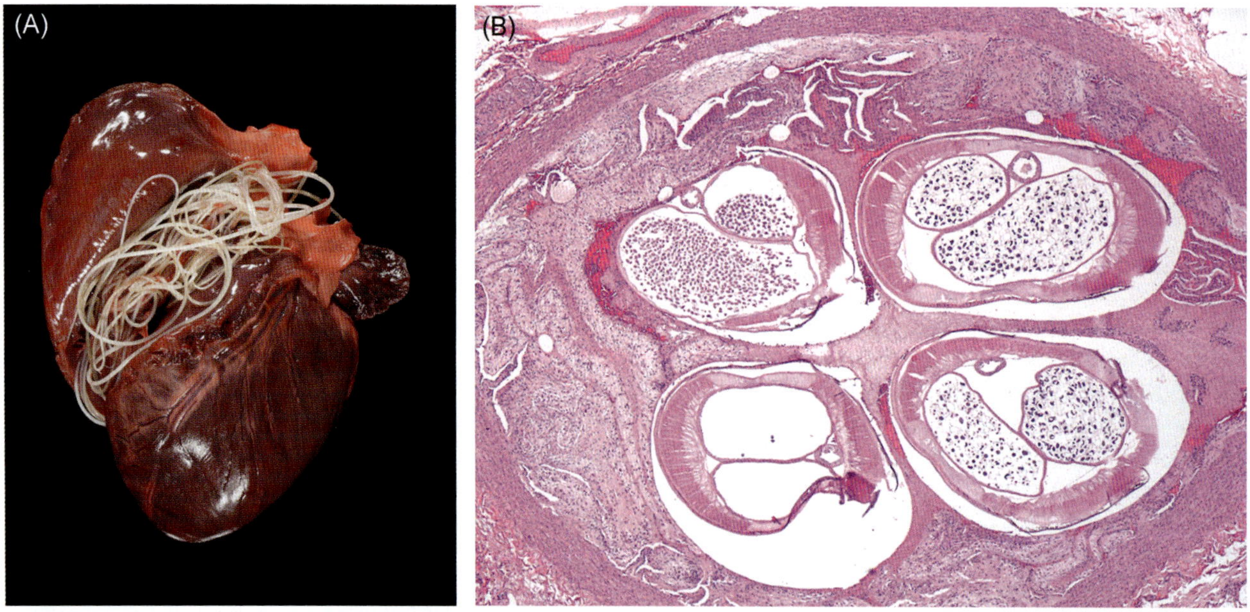

FIGURE 9.19 Heartworm infection in a coyote due to *Dirofilaria immitis*. (A) Adult parasites are present in, and occlude, the right ventricle. Worms were also present in the lumen of the pulmonary artery and its branches. (B) Infection is associated with severe endarteritis. Intimal inflammation is evident and the lumen is partially obstructed by proliferative, villous fibrous connective tissue. The uteri in the worm sections (right side of the image) are distended by abundant, thin-shelled, larvated eggs (i.e., sheathed microfilaria).

lymph node, or intestinal infections.[75,151] In humans, it has also been associated with subcutaneous nodules and ocular infections. Distribution is limited to the Old World (southern and eastern Europe, Asia Minor, central Asia, and Sri Lanka). Morphologic features of the parasite include longitudinal cuticular ridges and coelomyarian muscle.

Spirocerca lupi (order Spirurida; family *Spirocercidae*) can cause fibrotic aortic and esophageal nodules and aortic aneurysms in wild carnivores (e.g., foxes, wolves, coyotes, black-backed jackals, and bush dog). It has a world wide distribution but is generally found in tropical and subtropical areas.[9,40,50,147,264,272,295,331] ***Spirocerca vulpis*** is a parasite of red foxes in Europe that has recently been distinguished from *S. lupi*.[338] The lesions of *S. vulpis* differ from those of *S. lupi* in that the nodules tend to be in the stomach, and the species seems to be host adapted to foxes.[272] The life cycle of *S. lupi*, and presumably that of *S. vulpis*, includes coprophagous beetles as intermediate hosts and a variety of paratenic hosts such as poultry, lizards, and rodents. Outcomes vary from asymptomatic infections to sudden death. Infected animals may develop a cough and can be weak or lethargic. Other clinical signs include retching, regurgitation, or difficulty swallowing. Developing larvae are present in the aortic wall, typically in the caudal thoracic region, and they may cause mural thickening and aneurysmal dilation of the aorta (Fig. 9.20A). Microscopically the lesions include neutrophilic, eosinophilic, or lymphoplasmacytic inflammation, smooth muscle and elastic fiber degeneration with fibroplasia, hemorrhage, mineralization, and heterotopic bone formation (Fig. 9.20B). Sudden death in infected animals most commonly occurs secondary to aneurysm rupture and fatal hemothorax. Adults typically reside in mural esophageal nodules, or occasionally in gastric nodules, where they create necrotic tracts with inflammation (e.g., eosinophils, lymphocytes, plasma cells, neutrophils, and macrophages), fibroplasia, and fibrosis. Esophageal lesions are most often fibrotic rather than granulomatous. Adults are also occasionally reported at other sites (e.g., lung, omentum). Aortic lesions due to larvae are commonly encountered in the absence of esophageal or gastric nodules.[50] Chronic inflammation due to *S. lupi* can lead to mural hyperplasia or dysplasia, and some have been associated with neoplastic transformation (e.g., fibrosarcoma, osteosarcoma).[337] The adults, like most spirurids, are red and they are large. Females are 6–7 cm long and males are 3–4 cm long. Histologically, the parasites have a smooth, thick cuticle, coelomyarian-polymyarian muscle, prominent lateral cords, and an intestine lined by uninucleate columnar epithelium. The pseudocoelom is filled with eosinophilic fluid. Infection can be detected by fecal floatation, but sensitivity is reportedly increased by using a modified sugar solution, the Markovics and Medinsi technique.[331] Other diagnostic options include esophageal

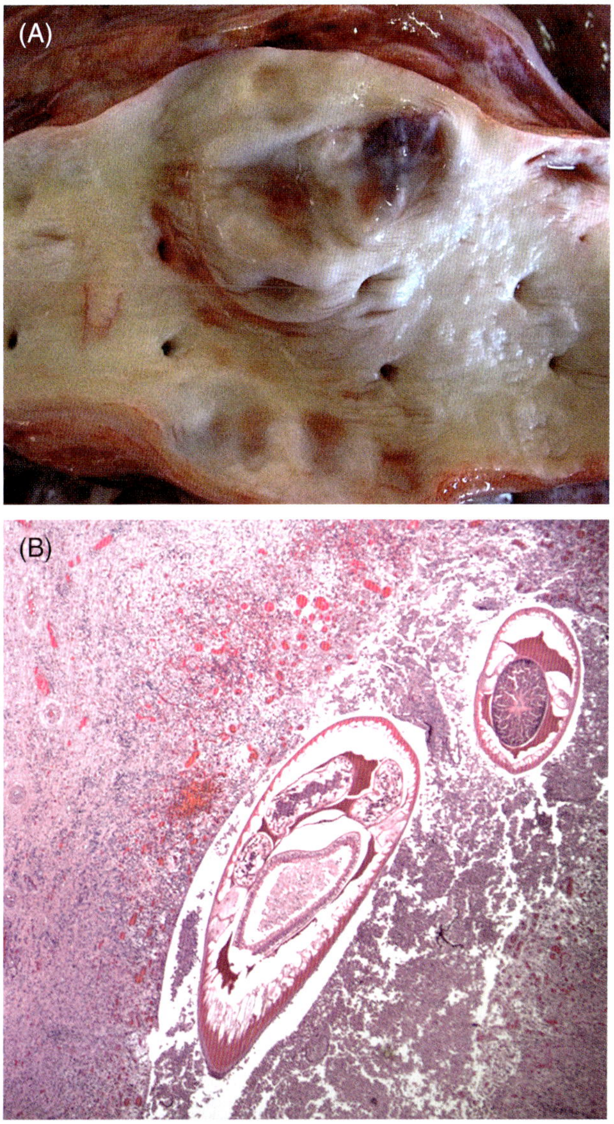

FIGURE 9.20 *Spirocerca lupi* infection in an African wild dog. (A) Developing larvae in the vascular wall (often the descending thoracic aorta) incite local inflammation and fibrosis and are associated with nodular, mural thickening, and aneurysmal dilatation. (B) Adults in the esophagus are associated with chronic esophagitis with mural fibrosis and granulation tissue formation and mural and luminal mixed inflammation (macrophages, plasma cells, lymphocytes, and eosinophils). Characteristic features of the parasite are a smooth, thick cuticle, coelomyarian, polymyarian muscle, prominent lateral cords, and an intestine lined by uninucleate columnar epithelium. *(Photos Courtesy of E. Mitchell, National Research Foundation of South Africa.)*

endoscopy, radiography, and computed tomography, identification of characteristic aortic and esophageal nodules with intralesional worms, and PCR.

The giant kidney worm, ***Dioctophyma renale*** (order Ascaridida; family *Dioctophymatidae*), occurs globally in mammals, primarily in mink and dogs. Infection in wild canids has been reported in wolves, coyotes, jackals, and foxes, and it has been identified with a relatively high prevalence in free-ranging maned wolves in Brazil.[61,186]

Transmission occurs through ingestion of encysted larvae in infected raw fish, frogs, or annelid worms. Adult *D. renale* are notably dark red and are the largest nematode known to infect carnivores. Females are up to 103 cm long and 12 mm in diameter; the smaller males are up to 40 cm long. Complications of infection include hematuria, ureteral obstruction, and hydronephrosis. Renal failure can develop in animals with bilateral disease. However, unilateral infections are more common and the right kidney is typically affected. Gross findings include asymmetric kidneys, and the affected kidneys may be atrophied, enlarged, or hydronephrotic. In severe cases, there may be near complete destruction and loss of the renal parenchyma with only a thin rim of residual tissue and capsule due to progressive pressure atrophy of the renal parenchyma as the parasite grows. Compensatory enlargement may develop in the unaffected kidney. Histologic changes include parenchymal collapse, fibrosis, and multifocal lymphoplasmacytic to mixed lymphoplasmacytic, eosinophilic, and neutrophilic inflammation. Transitional cell hyperplasia and epithelial degeneration may occur in the renal pelvis. Aberrant migration can lead to an intraperitoneal location that can cause peritonitis and adhesions.[54] Morphologic features of the parasites include a thick cuticle, polymyarian-coelomyarian muscle, ventral nerve cords, uninucleate columnar cells in the intestine, and intralesional or intraluminal eggs that are ~75 × 40 μm, round to oval, have a thick capsule, and are larvated or embryonated. Diagnosis is typically made by identifying these large, characteristically red worms and associated renal parenchymal loss on radiography, ultrasound, exploratory laparotomy, or gross necropsy. Thick-shelled, pitted, barrel-shaped, embryonated eggs with slightly flattened ends are seen microscopically in urine sediment.

The urinary bladder can also be affected by nematode infestation. ***Pearsonema* spp.** (syn ***Capillaria* spp.**, order Enoplida; family Capillariidae) have been identified in red foxes, wolves, and various mustelids in central Italy.[131] Infestation was associated with eosinophilic cystitis in approximately half of the affected individuals. Other researchers further identified the parasites affecting red foxes in Bosnia and Herzegovina as ***Pearsonema plica***.[197]

Trichinella spp. occur on all continents but Antarctica, infect a wide range of mammalian, avian, and reptilian hosts, and cause the disease **trichinellosis** (syn. **trichinosis**), a WOAH-listed, reportable disease. There are two clades. The encapsulated clade infects mammals and includes ***Trichinella spiralis*** (affecting carnivores and omnivores worldwide), ***T. nativa*** (affecting holoarctic terrestrial and marine carnivores), ***T. britovi*** (affecting temperate palearctic carnivores) ***T. murelli*** (affecting temperate Nearctic carnivores), ***T. nelsoni*** (affecting African predators and scavengers), and ***T. patagoniensis***. The nonencapsulated clade primarily infects mammals and birds and includes ***T. pseudospiralis*** (cosmopolitan distribution infecting mammals and reptiles), ***T. papuae*** (Papua New Guinea and Thailand), and ***T. zimbabwensis*** (eastern Africa).[17,322] *Trichinella* species are ovoviviparous, nematodes. Zoonotic transmission occurs through the consumption of larvae in undercooked meat including pork, meat from various wild animals (e.g., bears), particularly carnivores and scavengers, and marine mammals. Wildlife are an important reservoir and source of zoonotic infection. The most notable features of infection with species in the encapsulated clade are the formation of "nurse cells," markedly modified skeletal muscle cells containing multiple nuclei within which is a coiled, intrasarcoplasmic larvae with a thick (up to 10 μm), eosinophilic capsule (Fig. 9.21).[411] Nurse cell formation in the absence of capsule formation and more diffuse myofibrillar degeneration is seen in nonencapsulated *Trichinella* species.[28,41] Larvae are 25–30 μm in diameter; have a thin, eosinophilic cuticle; coelomyarian, polymyarian musculature; bilateral hypodermal bands; and a stichosome. Myofiber necrosis and variably intense infiltrates of leukocytes and eosinophils may surround affected cells or larval migration tracks. Infection occurs with relatively high frequency in foxes, wolves, and bears. In addition to *T. spiralis*, the classically

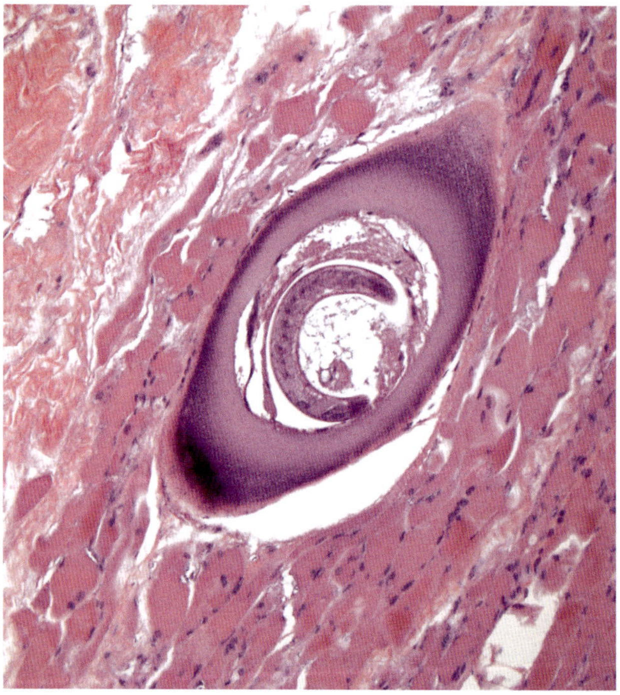

FIGURE 9.21 Trichinellosis in the skeletal muscle of a bear. A larval *Trichinella* sp. parasite is encysted in a nurse cell, which typically forms when the larvae invade a striated myofiber. The larva is completely encapsulated within the sarcoplasm of the nurse cell. The dedifferentiated nurse cell eventually forms a collagen capsule, visible here as a thin, margin of eosinophilic tissue enveloping the nurse cell.

described species, *T. britovi* and *T. pseudospiralis* have been described in wild canids, and *T. nativa* has been described in Arctic bears. Elevated muscle enzymes or blood eosinophilia support the diagnosis. Larvae can be identified in muscle by histology or trichinoscopy, the practice of squashing muscle between two slides and identifying the capsules and larvae by microscopy. PCR with DNA sequencing is also useful in confirming infection and species. Detection of nonencapsulated species is enhanced through the use of enzymatic tissue digestion techniques (e.g., various pepsin solutions), which are reportedly more sensitive for these *Trichinella* sp. than traditional trichinoscopy. In the recent past, tissue digestion has contributed to more widespread detection of these parasites.

***Ancylostoma* spp.** (order Rhabditida; family Ancylostomatidae), also known as hookworms, is a common concern for canid species and has been described as a concern for the management of endangered red wolves.[49] This is a particular concern for pup mortality and is manifested clinically by severe anemia.[95] These small intestinal nematodes are generally evident on fecal examination.

Eyeworms of the genus ***Thelazia*** have been described in a variety of wild canids and bears.[311] Lesions frequently include mild conjunctivitis with hyperemia and mucopurulent discharge; corneal ulceration and corneal edema can also occur. *Thelazia* spp. are found worldwide and many affect multiple host species. The intermediate hosts are lacrymophagous drosophilid or muscid flies. Flies ingest L1 larvae while feeding on ocular secretions. They mature to the L3 stage on the fly before being transmitted to another host. Aberrant migration within the eye occurs occasionally. Some species demonstrate weak zoonotic potential.

Protozoa

In addition to several of the more commonly described protozoal diseases described below, see Table 9.e4 for an extended list of protozoal infections in nondomestic canids.

Hepatozoon is a genus of apicomplexan parasites containing over 300 species. Hematophagous arthropods are the definitive host and the site of parasite sexual development (gametogony). Ticks, mites, lice, and tsetse flies are all vectors for various *Hepatozoon* spp. Asexual development occurs in a vertebrate host and infection begins with ingestion of an infected arthropod or vertical transmission to a fetus.[163] Sporozoites enter various host tissues where they undergo merogony. Merozoites are released, enter leucocytes, and begin a stage of gametogony. The lifecycle is completed when another arthropod ingests leucocytes containing gamonts. Blood infection and local or disseminated tissue infection have been reported in a number of wild canids in Asia, Africa, Europe, and South and North America. Affected species include black-backed jackals, African wild dogs, golden jackals, crab-eating foxes, red foxes, Pampas gray foxes, and coyotes.[4,129,136,156,172,237,401] Clinical signs are often absent, especially in mild infections, and will vary depending on affected tissues (which may depend on infective species) and age. Some suggest higher morbidity and mortality in young animals. Infection can cause anemia or leukocytosis. Lesions include pulmonary infection with interstitial or bronchopneumonia, or myocardial infection with necrotizing myocarditis. The disease can be associated with labored breathing. Of the various *Hepatozoon* species, meronts of ***H. canis*** have a characteristic "wheel spoke" appearance with a central eosinophilic nucleus and peripheral, basophilic merozoites; they are more commonly found in peripheral blood, spleen, and bone marrow. Meronts of ***H. americanum*** have a central core surrounded by concentric layers of mucinous material giving them a laminated "onionskin" appearance; they may more commonly be found in the skeletal and cardiac muscles. Inflammation, when present, is typically mild with infiltrates of lymphocytes and plasma cells. Mixed lymphoplasmacytic, neutrophilic, histiocytic, or pyogranulomatous inflammation can also be seen. An additional interesting lesion attributed to *H. americanum* infection in domestic dogs and experimentally infected coyotes, is mild to severe, disseminated, periosteal proliferation of woven bone.[238] Diagnostic options include blood smears, serology, PCR, and DNA sequencing. Intracellular protozoal cysts or zoites may be identified by cytology, biopsy, or histology (Fig. 9.22; eSlide 9.e3) with routine staining, or with immunohistochemical methods.

Piroplasmosis is caused by apicomplexan hemoprotozoa in four related genera, ***Babesia***, ***Theileria***,

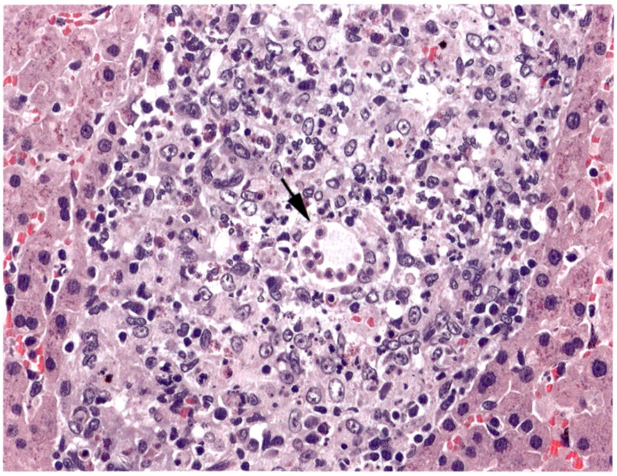

FIGURE 9.22 Hepatic hepatozoonosis due to infection with *Hepatozoon canis*. A developing schizont *(arrow)* is associated with mixed inflammation dominated by epithelioid macrophages (also see eSlide 9.e3).

Cytauxzoon, and ***Rangelia***. Like *Hepatozoon* spp., the lifecycle of these parasites includes an invertebrate definitive host, such as ticks, and a vertebrate intermediate host. Infection has been detected serologically, on blood smears, through histology and electron microscopy, or through molecular methods in several canid and ursid species in the wild or in zoos in North and South America, Europe, Asia, and Africa.[10,81,171] In general, infection is often considered an incidental finding. However, it has occasionally been associated with morbidity and mortality. Immunosuppression or other stressors are thought to be predisposing factors in the development of clinical disease. Clinical signs of **babesiosis** (e.g., *B. annae*, ***B. canis***) include anemia and icterus with variable thrombocytopenia or evidence of hemolysis. Gross lesions can include hepatomegaly, splenomegaly, and petechial or radiating tracts of hemorrhage in the kidneys. Microscopic lesions may include erythrocyte sludging in capillaries, disseminated intravascular coagulation, hemoglobinuric nephrosis, microvascular necrosis, and ischemia of various tissues, but these are inconsistent findings.[76,81,300,320] The pathogenesis is still poorly understood, but host immune factors are suspected to play a role in the diversity of lesions. Identification of intraerythrocytic parasites with characteristic morphology is diagnostic. Tetrad or cruciform merozoites are unique to *Babesia* spp., and the trophozoites have a ring formation. ***Babesia microti*** are 1–3 μm, whereas ***B. canis*** and ***B. vogeli*** are much larger (2.5–4.5 μm).[363] PCR and electron microscopy are additional diagnostic options, but speciation may not be possible.

Rangeliosis, caused by ***Rangelia vitalii***, has many synonyms, such as bleeding ears (nambi-uvu), bleeding plague (peste de sangue), and yellow fever of dogs (febre amarela dos caes). It is a disease of South American domestic dogs, but has also been reported in wild crab-eating and pampas foxes.[105,148,150,362] The protozoa are 2–4 μm in diameter and are round, oval, or tear-dropped shaped. They infect erythrocytes, leukocytes, and endothelial cells and can also be seen free in plasma. The parasites cause hemolysis, petechial or ecchymotic hemorrhage, hematemesis, gastrointestinal luminal hemorrhage, bloody diarrhea, anemia, icterus, fever, and persistent bleeding from the nose or venipuncture sites. Severe bilateral hemorrhage along the lateral margins of the pinna is a characteristic and highly suggestive sign of infection. Hematological findings are consistent with hemolysis and include regenerative or nonregenerative anemia, spherocytosis, and Howell-Jolly body formation. Thrombocytopenia and leukopenia, leukocytosis, or a leukemoid reaction are also common. Single or paired merozoites are present in erythrocytes. Gross lesions, in addition to hemorrhage, include hepatomegaly, splenomegaly, and lymphadenomegaly. Numerous zoites are found in leukocytes and endothelial cells in multiple sites, including the liver, spleen, heart, lymph nodes, and bone marrow.[148] Diagnosis is based on a combination of findings from hematology, cytology, gross pathology, and histopathology, with confirmation by PCR.

The apicomplexan ***Toxoplasma gondii*** is a ubiquitous protozoal parasite that affects a wide range of warm-blooded vertebrate hosts including birds, humans, and marine mammals. Felids are the definitive hosts (see Felidae). It is transmitted to intermediate hosts by ingestion of infective oocysts in feces, consumption of protozoal cysts containing bradyzoites in tissues, or transplacental transmission of tachyzoites. Infection in canids and ursids is generally asymptomatic, and many species may be serologically positive without developing clinical signs. However, fatal infections have been reported in a number of species in the wild and in captivity. This includes infections in red, gray, Arctic, sand, Fennec and Blanford's foxes, giant pandas, and brown bears.[117,119,120,240,261,313,369,382] Mild focal infection with intralesional tachyzoites or protozoal cysts can be associated with mild lymphoplasmacytic inflammation or gliosis (nervous tissue). Severe, disseminated infection, more often containing intralesional or intracellular tachyzoites than protozoal cysts, is typically associated with necrosis. Common sites of infection include the liver, spleen, heart, intestine, and lymph nodes, though any site (e.g., brain, lung, pancreas, skeletal muscle, and eye) can be affected. Diagnosis is based on histological identification of protozoal cysts or tachyzoites, coupled with IHC or PCR.

Enteric infections with ***Cryptosporidium*** sp. have been described in wild canids. They are usually associated with immunosuppressive conditions such as infections with canine distemper virus.[99] The protozoal trophozoites are small and are often evident in apical vacuoles on the surface of epithelial cells in the gastrointestinal tract (Fig. 9.23).

Protozoa in the genus ***Leishmania*** (order Trypanosomatida) are an important cause of cutaneous and visceral disease in a number of species including canids and humans. These are obligately intracellular protozoa that infect monocytes and macrophages and are transmitted to mammalian reservoir hosts (typically canids) by sandflies (*Phlebotomus* species in the Old World; *Lutzomyia* species in the New World).[371] The protozoa have a two-host lifecycle in which motile, flagellated promastigotes in the invertebrate are injected into the vertebrate host, then phagocytosed by mononuclear cells. Within phagocytic cells, promastigotes lose their flagellum to become nonmotile amastigotes that undergo asexual replication. Cell rupture releases amastigotes that infect other cells. Infection has been detected in a number of wild and captive canid species and has generally been asymptomatic.[94,294] However, **cutaneous** and **visceral leishmaniasis** have been reported in captive bush dogs, crab-eating foxes, golden jackals, hoary foxes, and maned wolves.[63,97,193,259] Gross lesions include emaciation, anemia, multifocal hemorrhage, cavitary effusion, hepatomegaly, splenomegaly,

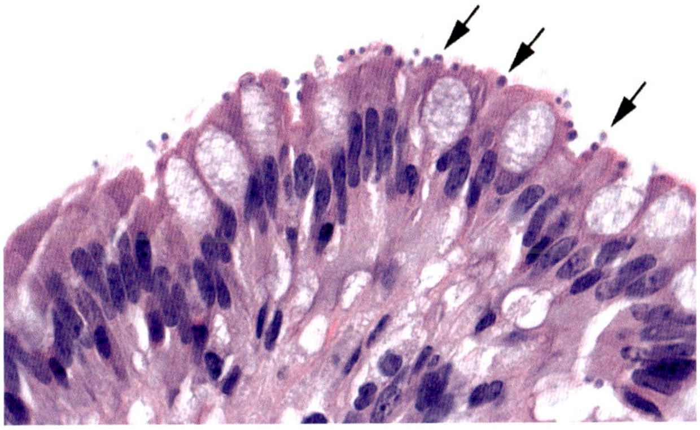

FIGURE 9.23 *Cryptosporidium* sp. infection secondary to canine distemper virus infection in a gray fox. Cryptosporidiosis is a common secondary infection in animals suffering from distemper. Numerous, small, round, basophilic protozoa are present along the apical surface of the enterocytes *(arrows)*. Although the parasites may seem to be situated on the luminal surface of enterocytes, they are actually confined within the cell membrane.

lymphadenomegaly, and thickening of the intestinal wall. Ulcers or crusts on the skin or mucosal surfaces may also occur. Histologically, histiocytic infiltrates with intracytoplasmic amastigotes may be local (e.g., skin, adrenal gland) or disseminated (e.g., liver, spleen, lymph nodes, kidney, lung, and gastrointestinal tract). Membranoproliferative glomerulonephritis, vasculitis, lymphoid depletion, and disseminated intravascular coagulation have also been described. In addition to histology, cytology, IHC, EM, and PCR can be used to support or confirm the diagnosis.

Fatal **sarcocystosis** has been reported in American black bears, grizzly bears, and polar bears, and cases have been associated with multifocal necrotizing hepatitis, variable nonsuppurative encephalitis, and intramuscular sarcocysts.[100,248,414] In British Columbia, Canada, ***Sarcocystis canis*** was identified in 69% of black bears and grizzly bears with hepatic and neuronal sarcocystosis, and *S. canis* was the etiologic agent associated with the death of the bears in which it was identified.[248] Bears can also serve as intermediate hosts of *Sarcocystis* spp., and ***S. arctosi*** and ***S. ursusi*** were identified in the skeletal muscles of brown bears and black bears, respectively.[92,118,121] Protozoa in many infections have not been speciated, and further work is needed to determine the true pathogenic potential of the different species. In fatal cases with liver disease, animals may be icteric with edema of the gall bladder and hemorrhages in multiple sites. Histologically, hepatic necrosis and nonsuppurative encephalitis with intralesional protozoa, morphologically consistent with *Sarcocystis* spp., is the most consistent finding.

Systemic sarcocystosis has also been blamed for the death of two red panda cubs in Michigan, United States.[415] Lesions included lymphoplasmacytic and histiocytic pneumonia, encephalomyelitis, and myocardial infection. Sequencing indicated 99% homology to ***S. neurona*** and ***S. dasypi***.

Cutaneous ***Neospora caninum*** tachyzoites and concurrent ***Toxoplasma gondii*** infection in a red fox were also associated with severe hyperplastic dermatitis with parakeratosis, keratinocyte necrosis, and extensive dermal necrosis and mineralization.[123]

Ectoparasites

Sarcoptic mange, caused by the burrowing mite ***Sarcoptes scabiei***, is a widespread, contagious, pruritic disease of numerous species, including various canids and ursids. These include multiple species of foxes and wolves, coyotes, dingos, jackal, raccoon dogs, and black bears.[43,143,348] Two, often overlapping, forms of skin disease, hyperkeratotic and alopecic, are recognized. The first is thought to generally occur in naïve or immunocompromised individuals whose immune status prevents them from mounting an inflammatory response appropriate to clear infection. These animals develop mild to marked, focal to diffuse epidermal hyperplasia, spongiosis, and hyperkeratosis with variably thick, fissured crusts (some up to 1.5 cm thick) with numerous intralesional mites (Fig. 9.24). Dermal inflammation, if present, is typically more lymphoplasmacytic or lymphohistiocytic than eosinophilic. The second form, the so-called hypersensitivity form, is associated with mild, local to generalized alopecia and subsequent thickening and lichenification of the skin. It is presumed to develop in immunocompetent animals who mount strong type I (immediate) and type IV (delayed) type hypersensitivity reactions to the mites.[316] In addition to alopecia, infection in these animals is associated with mild to intense dermal eosinophilic and mast cell inflammation with fewer lymphocytes and plasma cells and few intralesional mites.

Mange can affect isolated individuals, but it has also been associated with outbreaks in several species. It has occasionally been considered the main cause of the local population declines or local extinction and is therefore considered a conservation threat in some small populations or endangered species.[96,370] Variability in lesion type and severity across species has been noted and could have a genetic component, though this remains to be more completely investigated. Behavioral changes, presumably associated with discomfort (e.g., pruritis) or debilitation, particularly in the hyperkeratotic form of the disease, have been associated with reduced hunting and reproductive success. Increasingly severe infection is associated with progressive debilitation, emaciation, and mortality. Diagnosis is based on the identification of characteristic parakeratotic hyperkeratosis with variable crusts, seborrhea,

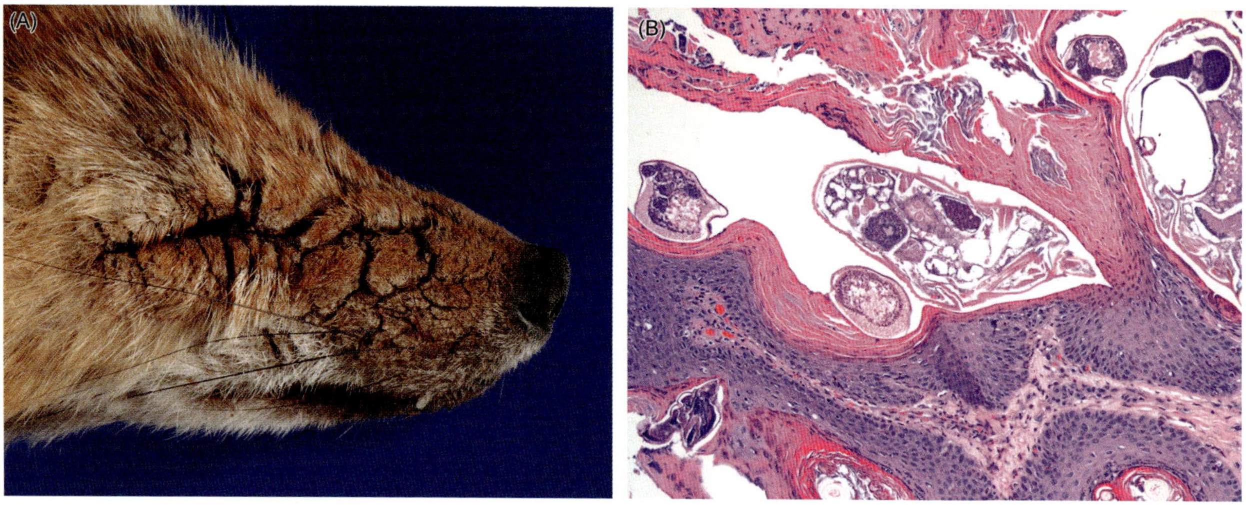

FIGURE 9.24 Sarcoptic mange due to *Sarcoptes scabei* infection in a red fox. (A) Severe hyperkeratosis is associated with partial alopecia and the formation of deep fissures. (B) Numerous mites migrate through the keratin layers of the thickened epidermis. Note the absence of significant inflammation.

alopecia; hyperpigmentation, and lichenification. Mites or oval, 150 × 175–200 × 250 μm eggs are typically found in the hyperkeratotic crusts or the superficial epidermis. Serology and IHC are available and may be valuable in animals with atypical presentations or low-level infection.

Otodectic mange, a form of otitis externa caused by the so-called ear canker mite ***Otodectes cynotis***, occurs worldwide in numerous species. In farmed Arctic foxes and silver foxes, it can have significant economic impacts due to pruritis and scratching that damages the pelt.[182] Affected wild species include red foxes, gray foxes, Arctic foxes, crab-eating foxes, and the endangered Santa Catalina Island fox.[98,254,256,297,393] In the latter, a study between 2001 and 2008 identified infestations in almost the entire population. Ceruminous gland hyperplasia and subsequent tumor development (adenomas and carcinomas) occurred in about half of the foxes over 4 years of age (see Neoplastic). Gross findings across species include excessive cerumen accumulation in the external ear canal, and open, crusty, skin wounds (typically on the back or base of the ear) with associated scabs and crusts that develop secondary to self-trauma. Otitis externa, which varies in severity based on the level and chronicity of infection, consists histologically of variably severe hyperkeratosis and epidermal hyperplasia with nonburrowing, intracorneal mites or eggs that are associated with variable mixtures of lymphocytes, plasma cells, histiocytes, neutrophils, eosinophils, and mast cells. Variably severe ceruminous gland hyperplasia, with or without dysplasia, is also seen. As already mentioned, sequelae include proliferative, exophytic, benign, or malignant (may be locally invasive or metastatic) ceruminous gland tumors. Opportunistic infections may develop in traumatized skin. Diagnosis is typically by histology or microscopic identification of the mites in skin scrapes or cerumen samples dissolved in 4% NaOH. PCR is also available.

Pentastomids, also commonly called "tongue worms" due to the resemblance of some species to a mammalian tongue, are flattened, elongate, arthropods that infest the upper respiratory tracts of a number of reptiles, birds, and mammals, including humans. **Nasal pentastomiasis** due to ***Linguatula serrata*** occurs in domestic and nondomestic canids, the definitive hosts. Infection is typically asymptomatic and is often identified incidentally at necropsy; however, sneezing, coughing, nasal discharge, and dyspnea are described in domestic dogs. In wild canids, the infection has been reported in Andean foxes (culpeo) and Patagonian foxes (chilla) in Chile, in wild dogs (dingos) and red foxes in Australia, and in red foxes in Herzegovina.[152,198,357] The presence and number of parasites may increase with age, and differences in infestation rate and intensity between species have been suggested to have a dietary component, due to the likelihood of ingesting infected intermediate hosts.[357] It is unclear whether detection might be underrecognized since nasal cavities are not always regularly examined.

Pentastomid nymphs are characteristically C-shaped, and adults have cuticular annulations that give the parasites a pseudo-segmented appearance. Histologic features of the adults include a thick, eosinophilic, chitinous cuticle with multiple, pore-like, refractile, eosinophilic, sclerotized openings. The latter is a unique feature of pentastomes that can be highlighted with Movat's pentachrome stain. The parasites also have two pairs of cephalic hooks next to the mouth, annular bands, striated muscle, and aggregates of large, eosinophilic cells that form acidophilic glands adjacent to the digestive tract. Diagnosis of pentastomiasis is typically based on the identification of the characteristic parasites on gross necropsy examination. Histology and PCR and DNA sequencing support the gross diagnosis and in the case of the latter, can be used to speciate the parasite.

ADDITIONAL RESOURCES

Anderson RC. Filarioid Nematodes. In: Samuel WM, Pybus MJ, Kocan AA, eds. Parasitic Diseases of Wild Mammals. John Wiley & Sons, Ltd. 2001:342–356.

Appleton JA, Blum LK, Gebreselassie NG. Nematoda: Trichinella. In: Immunity to Parasitic Infection. John Wiley & Sons, Ltd. 2012:275–285.

Derocher AE. Polar Bears: A Complete Guide to Their Biology and Behavior. Johns Hopkins University Press 2012.

Ebani VV, Nardoni S, Mancianti F. Arthropod-Borne Pathogens in Wild Canids. Vet Sci. 2023:10 (2):165.

Fairbrother A, Hoff G, Locke LN, eds. Non-Infectious Diseases of Wildlife. Iowa State University Press 1999.

Gardiner CH, Poynton SL. An atlas of metazoan parasites in animal tissues. Armed Forces Institute of Pathology 2006.

Loots AK, Mitchell E, Dalton DL, Kotzé A, Venter EH. Advances in canine distemper virus pathogenesis research: a wildlife perspective. J Gen Virol. 2017;98 (3):311–321.

Macdonald DW, Sillero-Zubiri C. The biology and conservation of wild canids. Oxford University Press. 2004.

Padilla LR, Hilton CD. Canidae. In: Fowler's Zoo and Wild Animal Medicine, Volume 8. 2015:457.

Samuel WM, Pybus MJ, Kocan AA, eds. Parasitic Diseases of Wild Mammals. John Wiley & Sons, Ltd. 2001.

Williams ES, Barker IK, eds. Infectious Diseases of Wild Mammals. Iowa State University Press 2001.

References

Full references for this chapter can be found online at https://doi.org/10.1016/B978-0-443-18437-6.00038-9.

Chapter 10

Felidae

Karen A. Terio[a], Denise McAloose[b], and Emily Mitchell[c]

[a]*Zoological Pathology Program, College of Veterinary Medicine, University of Illinois, Brookfield, IL, United States;* [b]*Pathology Department, Zoological Health Program, Wildlife Conservation Society, Bronx, NY, United States;* [c]*Department of Paraclinical Sciences, University of Pretoria, Pretoria, South Africa*

INTRODUCTION

Nondomestic felids are susceptible to many of the same diseases as their domestic counterparts. Infectious diseases uncommon in domestic cats, such as canine distemper virus, however, can cause episodic mortalities in wild and captive felids. The solitary nature of some species may help limit transmission of infectious diseases; however, expansion of human populations with their domestic animals provides increased opportunities for contact, making naïve wild felids especially vulnerable to cross-species infectious disease transmission. The illegal wildlife trade, human–wildlife conflict (e.g., retribution killing), as well as urbanization and fragmentation of habitats all contribute to mortality and decline of wild populations.[8,122,250,345] Many felid populations may also be vulnerable to the impacts of climate change whether due to direct loss of habitat, changes in habitat suitability, changes in prey base, or climate-induced alterations in infectious disease dynamics.[46,47,234,244,396] Over 80% of all extant felids have decreasing wild populations (Table 10.e1); however, intensive, multifaceted conservation efforts that include habitat restoration and breeding programs have not only halted declines but resulted in increasing populations of the previously critically endangered Iberian lynx (www.IUCNRedList.org). In addition to understanding disease risks to wild populations, understanding unique disease concerns of ex situ populations will be critical for ensuring healthy, viable populations for future reintroduction and rewilding efforts to be successful.

NON-INFECTIOUS DISEASES

Nutritional

Cranial malformations and secondary neurologic disorders have been diagnosed in young (typically <2 years) African lions worldwide. All affected lions have been captive at the time of diagnosis though some were wild born.[15] Similar lesions have been reported in cheetahs with thickening of the parietal bones.[77,312] Common presenting signs include incoordination, ataxia, head tilt, star-gazing, and or opisthotonus. Characteristic gross lesions include diffuse thickening of the bones of the skull, osseous tentorium cerebelli, dorsal arch of the atlas with narrowing of the foramen magnum, secondary cerebellar herniation, and dilation of the ventricles and central canal of the spinal cord (Fig. 10.1). Affected lions may have compression of the anterior spinal cord. Histologically, the bones of the skull are thickened due to large, irregular islands of fibrocartilage and woven bone that contain retained cartilage cores. Osteoclasts are only rarely noted. In the brain and spinal cord, there is histologic evidence of hemorrhage, rarefaction, and thinning of the cerebellar folia with loss of all cell layers, as well as gliosis. Wallerian degeneration is most prominent in the cervical spinal cord and in cheetahs, it may be accompanied by syringomyelia. The leptomeninges are often thickened by fibrous connective tissue.

Low serum and/or hepatic vitamin A levels have been reported in many lions with cranial malformations. Similar lesions have also been documented in other species with vitamin A deficiency, although lesions have not been reproduced in domestic cats[109] and the pathogenesis is not well understood. In **hypovitaminosis A**, bone production occurs at sites of normal resorption, and the paucity of osteoclasts within lesions suggests a potential link, though the role of vitamin A in bone development and maintenance is not completely understood. Treatment in many clinical cases has been unrewarding, although improvement has been noted in mild cases treated with excessive vitamin A supplementation and in other cases following surgical reduction of bone.[212,323] Whether the poor response to replacement therapy is due to the chronicity of lesions or other undetermined causes is uncertain. Affected lions are often fed diets similar to those of other captive felids that do not develop these lesions. Many reports include multiple cubs from the same litter suggesting genetics may contribute. A separate myelopathy associated with hypovitaminosis A with diffuse spinal cord white matter

Pathology of Wildlife and Zoo Animals. https://doi.org/10.1016/B978-0-443-18437-6.00006-7

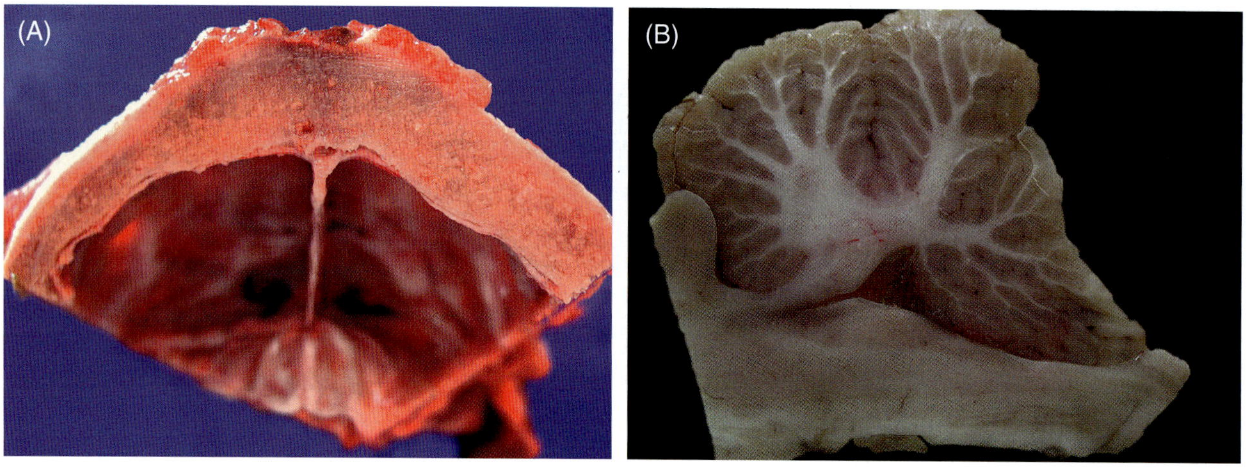

FIGURE 10.1 Cranial malformation in an African lion presumed secondary to hypovitaminosis A. (A) Diffuse thickening of the parietal bones of the skull. (B) Herniation of the cerebellar vermis in a lion with osseous cranial malformations. There is caudal dorsoventral flattening of the cerebellar vermis resulting in a "wedge-shaped" appearance.

degeneration in the absence of bone lesions has been diagnosed in a captive lion.[200] In this case, dilated myelin sheaths, reactive astrocytosis, and Wallerian degeneration were present throughout the spinal cord with bilaterally symmetrical demyelination in the fasciculus cuneatus of the cervical spinal cord and peripheral white matter of all segments. Within gray matter, there were rare neurons with chromatolysis and neurofilament accumulation.

Toxic

Toxic levels of **mercury**, specifically monomethyl and inorganic mercury, have been found in tissues from free-ranging Florida panthers and in hair from museum specimens collected as early as the 1890s.[256,296] Mercury levels are high within the United States Everglades ecosystem, and exposure of panthers is likely through predation of piscivorous prey, particularly raccoons.[13,84] It is hypothesized that these contaminants contribute to some of the reproductive, endocrine, and immune system problems noted in this species.[94,296] High levels of mercury have also been found in hair samples from jaguars in areas near gold mining operations in Brazil and have been proposed as a potential conservation threat.[209] Interestingly, higher concentrations of monomethyl mercury have been detected in fur from coastal Amur tigers than in those from inland regions.[282] The source was presumed to be coastal marine atmospheric fog.

Captive carnivores fed animals that have been shot are susceptible to **lead toxicity.**[259] Affected animals show various neurological symptoms; pathological findings, apart from lead shot fragments in the gastrointestinal tract, are unremarkable.

Hypervitaminosis D has been described in Iberian lynx under managed care receiving a diet with excessive levels of vitamin D.[190] Affected lynx were hypercalcemic, hyperphosphatemia with increased plasma calcidiol and decreased parathyroid hormone. Toxicity resulted in renal failure in most lynx as well as severe extraskeletal mineralization of the kidneys and to a lesser extent cardiovascular tissues.

Urbanization of landscapes and exposure to **anticoagulant rodenticides** have been associated with immune suppression and susceptibility to opportunistic infections such as cases of notoedric mange in bobcats (see "Ectoparasites" section).[319,320] While anticoagulants can cause direct toxic effects on a variety of nontarget wildlife species, cryptic effects are less well recognized but may be more pervasive. In bobcats, reduction in neutrophils but elevated lymphocyte counts have been identified as well as cytokine alterations that suggested both immune suppression and stimulation.[320] Gene expression profiles have identified differential expression of genes known to be impacted by warfarin in apparently healthy bobcats with evidence of rodenticide exposure.[103] Results of these studies suggest that although they may not be direct causes of mortality, exposure to rodenticides should be evaluated within urban landscapes as factors in species decline and susceptibility to disease.

Congenital/Genetic

As a result of habitat loss, geographic isolation from other populations of pumas, and small population numbers, in the United States, the Florida panther has become increasingly inbred.[294] This may be a primary contributing factor in several congenital defects noted in free-ranging panthers including **atrial septal defects** and **cryptorchidism. Ostium secundum atrial septal defects** have been diagnosed in up to 18% of the panthers and are thought to have been the cause of death in 9% of studied animals.[39,68]

Bilateral cryptorchidism, which can be a heritable trait, has also been described in Florida panthers. Depending on the study, the prevalence varied from 49% to 54%[39,199]; however, since Texas cougars were introduced into the population in 1995 to increase genetic diversity, none of the progeny resulting from this introduction have been cryptorchid.[199] Other wild, free-ranging nondomestic felid species in which cryptorchidism has also been reported include lion, cougar, Eurasian lynx, cheetah, fishing cat, ocelot, black-footed cat, Amur leopard, and jaguar.[149] In male felids, phimosis, the inability to extrude the penis from the prepuce, has been described in a group of related, juvenile Amur leopards. This condition develops secondary to narrowing of the preputial orifice, can be congenital or acquired (e.g., trauma), and the clinical signs (e.g., pollakiuria, stranguria, dribbling urine, vocalization) vary depending on the severity of the obstruction.[155]

Ocular colobomas are common in snow leopards in zoos in Europe and the United States.[10,106,116,309] The condition has not been reported in free-ranging snow leopards. Ocular colobomas are defects in one or more structures of the eye. Those that occur in the eyelids, cornea, iris, ciliary body, choroid, retina, and/or optic disc result from failure of the embryonic fissure to close correctly during embryogenesis. Clinical signs and sequelae depend on the type and extent of the coloboma/s. With eyelid colobomas, this can include pain, epiphora, blepharospasm, corneal abrasion or ulceration due to trichiasis, or exposure keratitis with corneal clouding, pigmentation, and neovascularization. The most commonly recognized manifestation in snow leopards is unilateral or bilateral eyelid colobomas with **agenesis.** This is seen as a small to extensive region in which the lid is not present. All cases to date have occurred in the upper lid/s (Fig. 10.2) in areas where there is a direct transition from the periocular skin to the bulbar conjunctiva. Other described lesions include **microphthalmia** or **anophthalmia**, and choroidal or optic nerve colobomas. **Persistent pupillary membranes**, slender, pigmented, fibrovascular strands of remnant fetal membranes that extend from the iris to the anterior lens capsule, posterior surface of the cornea, or other parts of the iris, are relatively common concurrent lesions. Histologic lesions can include choroid hypoplasia, retinal cysts, retinal dysplasia (retinal folds), and cupping or cystic ectasia of the optic nerve. Multiple hypotheses to explain the colobomas in snow leopards have been proposed but, to date, none have been proven.[10,116] A heritable component is suspected, given the high percentage of affected leopards that have an affected sibling, parent, or grandparent.[106] Ocular colobomas have also been described in captive cheetah and cougar.[25,71,165]

Single, multiple, and multilocular **biliary cysts** are common postmortem findings in African and Asian lions as well as African leopard, clouded leopard, snow leopard, jaguar, tiger, mountain lion, and cheetah (Fig. 10.3).[18,108,192,278] Cysts vary in size (some are quite large) and contain clear to dark green bile or bloody fluid. Histologically, cysts are lined by flattened to low cuboidal epithelium supported by a few layers of fibrous connective tissue. Although cysts have also been noted in the kidney in a few cases, mutations in *PKD1* (*polycystic kidney disease* 1) have not been identified and the vast majority of lions with hepatic cysts do not have concurrent renal cysts.[18,108]

FIGURE 10.2 Ocular colobomas in a snow leopard. There is an abrupt transition from haired skin to the bulbar conjunctiva due to agenesis of the eyelids. Lesions develop in the upper lids and often affect the central (seen here) and/or lateral aspects of the lids. *(Photo Courtesy of the Wildlife Conservation Society.)*

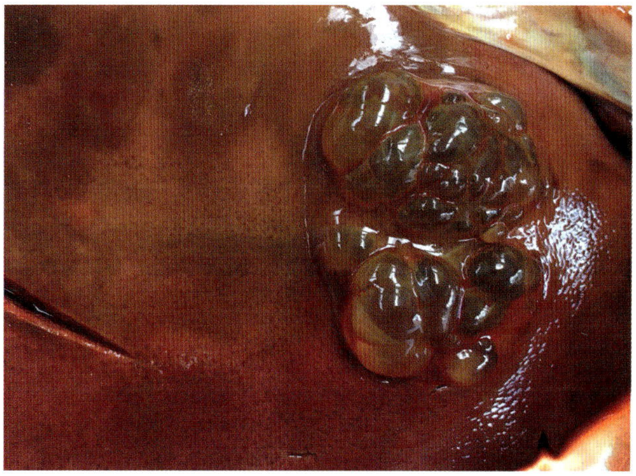

FIGURE 10.3 Multiple biliary cysts in a lion. Within the hepatic parenchyma and raising the capsular surface are multiple thin-walled cystic structures with green fluid (bile). This is a common and incidental finding.

Pallas' cats in North American zoos have been diagnosed with **polycystic kidneys**. Affected cats are generally adults at the time of diagnosis. Cysts are confined to the kidneys, and appear as multiple, large (up to 3 cm diameter), fluid-filled tubular cysts lined by low cuboidal to attenuated

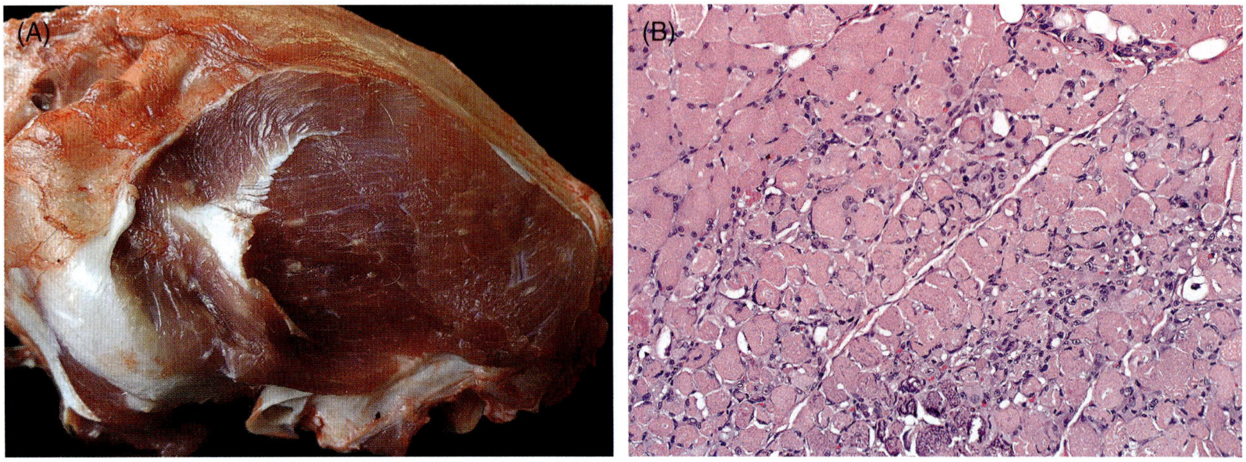

FIGURE 10.4 Myopathy in a free-ranging African lion. (A) Atrophy of the masseter muscles. (B) Polyphasic myopathy with some swollen and or fragmented myofibers, multifocal myofiber mineralization, and evidence of regeneration with central myofiber nuclei and prominent satellite cells. *(Part A: Photo Courtesy of R. Bengis, Department of Agriculture, Forestry and Fisheries, South Africa.)*

epithelium in the renal cortex. Because several affected cats have been siblings or closely related, a genetic cause is suspected.

A syndrome of severe **polyphasic skeletal muscle necrosis** in free-ranging lion cubs in South Africa has been reported.[73] Affected cubs are unable to walk, with greatly reduced muscle mass, and are abandoned by the pride. Skeletal muscle fibers are variably swollen and eosinophilic with loss of cross striations, fragmented and/or mineralized with scattered regenerative fibers (Fig. 10.4). Because only cubs were affected, and since no evidence of infectious or parasitic disease has been found, a genetic cause is suspected. However, the condition is not similar to known myopathies in domestic cats.

Congenital vestibular disease, described in Sumatran tigers in zoos in Australasia, has variable presentations, including head tilt, circling, ataxia, strabismus, and/or nystagmus. Clinical signs are present at or noted shortly after birth, and in all reported cases resolved by 2 years of age.[141,387] A genetic cause is suspected since all affected animals have been closely related and share a common ancestor. Descriptions of histologic lesions remain to be reported.

Cystinuria and **cystine urolithiasis** have been reported in captive servals.[43] There was no association with diet, sex, or reproductive status. Evaluation of pedigree suggested a possible autosomal recessive pattern of inheritance, although parentage was incomplete. Evaluation of genes (e.g., *SLC3A1* and *SLC7A9*) associated with disease in other species may be warranted.

Age-Related/Degenerative

Spondylosis (spondylosis deformans; ankylosing spondylosis) is a degenerative disease of the vertebral column that has been described in a variety of captive, nondomestic felids.[173] Commonly considered a disease of large cats, spondylosis can also occur in medium and small cats. Rare cases have been identified in free-ranging felids. The spectrum of lesions varies from small bone spurs on the lateral or ventral edges of the articular surfaces of vertebral bodies to large, bony bridges or callus-like proliferations of bone between adjacent vertebral bodies and sclerosis of vertebral end plates (Fig. 10.5). Primary factors important in its development include **intervertebral disc degeneration**, disc herniation with intervertebral space collapse, and degeneration of the articular facets of the vertebral bodies. Affected animals may be asymptomatic or show decreased

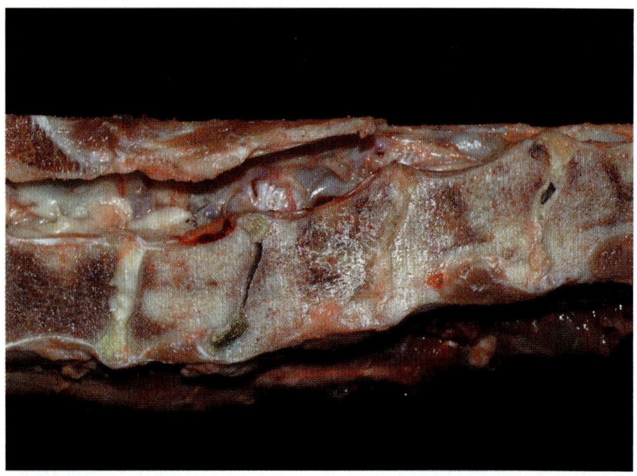

FIGURE 10.5 Intervertebral disc degeneration and spondylosis in an African lion. Collapse of multiple disc spaces is associated with dorsal extrusion of degenerate disc material into the vertebral canal (spinal cord has been removed). Bridging spondylosis, seen as smooth, nodular proliferative bone between adjacent vertebrae, is present along the ventral aspect of multiple vertebrae and there is increased bone (end plate sclerosis) in multiple affected vertebral bodies.

activity, weakness, ataxia, proprioceptive deficits, or paresis. While lesions can occur anywhere along the vertebral column, those in the lumbar spine and lumbosacral junction are most common. Axonal damage and Wallerian degeneration in the spinal cord are associated with disc herniation and compression but can be due to osteophyte formation, and compressive lesions may occur in spinal nerves due to bone expansion.

Similar to domestic cats, chronic renal disease, often of undetermined cause, is a common cause of morbidity and mortality in aged felids. While there are some species-specific diseases (e.g., glomerulonephritis in Iberian lynx, glomerulosclerosis in cheetahs, amyloidosis in black-footed cats) and infectious causes (e.g., *Leptospirosis*) that are discussed in other sections, all nondomestic felids can develop **tubulointerstitial nephritis** with interstitial inflammation and fibrosis with tubular degeneration.[5,218,254,255,393] Primary glomerular disease appears to be less common, but glomerulosclerosis is often present in areas of interstitial fibrosis. In addition to traditional biomarkers, serum or plasma measurement of symmetric dimethylarginine (SDMA) may have utility for antemortem diagnosis of renal function in nondomestic felids.[5,232,384]

Inflammatory Non-Infectious

Progressive membranous glomerulonephritis has been described in wild and captive Iberian lynx.[145] Lesion progression occurs with age irrespective of gender. Histologically, diffuse membranous glomerulonephritis is associated with rare sclerotic glomeruli. Glomerular basement membranes are thickened by increased amounts of laminin, fibronectin, and type IV collagen, even in the early stages. Ultrastructurally, in addition to thickened basement membranes, foot processes are effaced in some cases and electron-dense deposits compatible with immune complexes are present in the intramembranous spaces. Positive immunogold staining for IgM and IgG but not IgA is present in these complexes. In addition to glomerular lesions, interstitial lymphoplasmacytic infiltrates are present and increase with severity of the glomerular lesions. Interstitial fibrosis is also seen in more severe cases. Proteinuria and low urine specific gravity, consistent with glomerular dysfunction, as well as azotemia and hyperphosphatemia, are seen in some cases. Underlying viral infections are not associated with the glomerular lesions and thus a genetic predisposition has been hypothesized.

Palatine clefts adjacent to the upper fourth premolar, and occasionally third premolar, occur in both free-ranging and captive cheetahs, as well as many other felids, to accommodate the large maxillary premolars.[203,334,400] Although previously described as palatine erosions, more recent research that included histologic evaluation of lesions suggests that the term **palatitis** is more appropriate. The inflamed ulcerated palatine epithelium, often impacted with food material or grass, may be edematous, and have microabscesses or granulomatous gingivitis. The palatine bone may perforate, forming an oronasal fistula (Fig. 10.6) which is usually lined by intact noninflamed epithelium.[334]

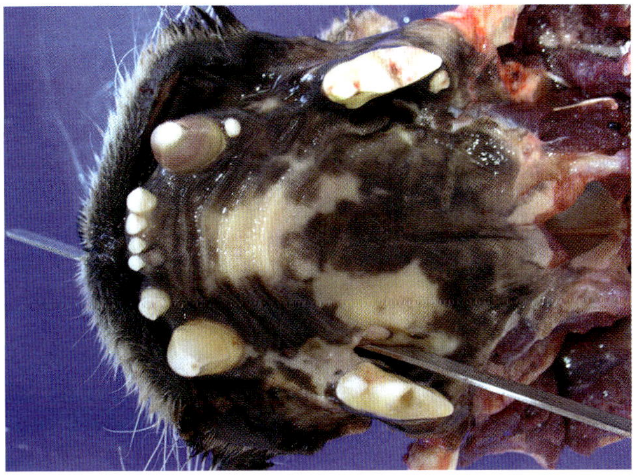

FIGURE 10.6 Palatitis with oronasal fistula in a cheetah. Medial to the fourth premolar are bilateral erosions and cavities in the palate. These erosions can become impacted with food and or foreign material and infected with subsequent inflammation and perforation of the palatine bone. A probe is present demonstrating a fistula extending from the erosion on the right side into the nasal cavity.

Similar to domestic cats, various other skull, dental, and oral lesions are documented in captive and free-ranging felids.[1,59,60,90,157,174,279,301,314,317,333] The most common findings include **dental trauma**, particularly **fractures** of the canine teeth, **periodontitis, gingivitis, resorptive lesions, attrition, caries, calculus,** and the development of benign or malignant neoplasms (see "Neoplastic" section). In wild lynx, pathologic lesions noted in some cats were considered of sufficient severity to impact hunting success.[59,60] In one study, the accumulation of calculus and periodontal disease were more common in captive than in wild lions and tigers. Soft, ground meat diets rather than bone-in meats or whole carcasses were suggested as a possible cause for this outcome.

Chronic lower airway disease, that resembles feline **asthma**, is common in captive African black-footed cats and lions. Macroscopically, collapsed lung lobes and airways filled with pale green mucoid material are seen in black-footed cats (Fig. 10.7) while occasionally large subpleural bullae are common in lions. Histological findings include variable mixtures of bronchiolar mucous gland hyperplasia, bronchitis, hyperplasia of terminal bronchiolar smooth muscle, and alveolar emphysema.

Miscellaneous/Multifactorial

Veno-occlusive disease (VOD), also diagnosed as sinusoidal obstructive syndrome or sinusoidal fibrosis, is

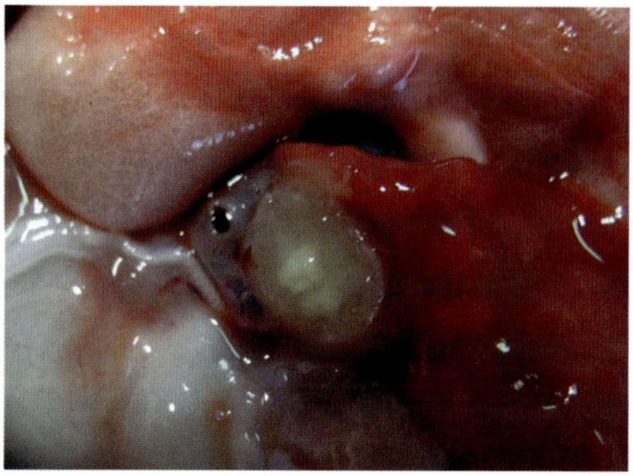

FIGURE 10.7 Lower airway disease in a black-footed cat. Bronchi are filled with thick pale green to white mucoid material with regional atelectasis of the pulmonary parenchyma. More normal pulmonary parenchyma is at the top and left of the photo.

described in cheetahs and snow leopards.[18,114,247] Disease is characterized by progressive accumulation of collagen in perisinusoidal spaces (space of Disse) and subintimal spaces of central and sublobular veins. The result is progressive attenuation or obliteration of vascular lumina with collapse of regional hepatic cords, atrophy, and loss of hepatocytes surrounding affected vessels and subsequent fibrosis (Fig. 10.8; eSlide 10.e1). Subendothelial damage, thought to be the primary lesion, is evident ultrastructurally. Chronic cases can progress to cirrhosis. In severe acute cases, there is centrilobular to massive hepatic necrosis with hemorrhage. Clinically, affected cats can be asymptomatic or have signs of hepatic failure, including hypoproteinemia, serous or chylous abdominal effusion, elevated liver enzymes (aspartate aminotransferase, alanine aminotransferase), and icterus.[358] In rare cases, cheetahs have exhibited neurologic signs presumably due to hepatic encephalopathy. A dietary cause for VOD, including phytoestrogens (a cause of the disease in humans) or elevated dietary vitamin A, has been proposed but not proven.[114,322] The diagnosis of VOD in a few wild cheetahs seems to refute the hypothesis that captive diets are causative.[245] Some cheetahs, not all of which have VOD, have multiple portosystemic shunts.

Among felids, **amyloidosis** has been described in cheetahs, black-footed cats, and caracals although the pathogenesis differs.[115,271,357] Captive black-footed cats have a very high prevalence of **systemic AA amyloidosis**; a single, free-ranging juvenile black-footed cat with small amounts of amyloid has been described but the prevalence in and impact on the wild population is unknown.[357] Amyloid deposition occurs most commonly in the renal medulla, glomeruli, spleen, and gastrointestinal tract (Fig. 10.9A; eSlide10.e2). Affected black-footed cats commonly develop amyloid-associated renal failure at a young age (around 4 years). Amyloidosis is not associated with chronic inflammation and heritability estimations suggest it may have a genetic component. Similarly, amyloidosis in caracals affects primarily the glomeruli with splenic, intestinal lamina propria, glossal submucosa, and exocrine pancreas less commonly affected.[115] In caracals, amyloid was confirmed type AA by immunohistochemistry. No cause could be determined although genetics were suspected.

In cheetahs, amyloid fibrils are commonly deposited in the renal medulla and can be associated with papillary necrosis[271]; less common is deposition in the hepatic perisinusoidal space (space of Disse; Fig. 10.9B) and intestinal lamina propria. These amyloid deposits contain full serum amyloid A (SAA) protein as well as fragmented AA protein.[17,147] Unlike in black-footed cats, concurrent

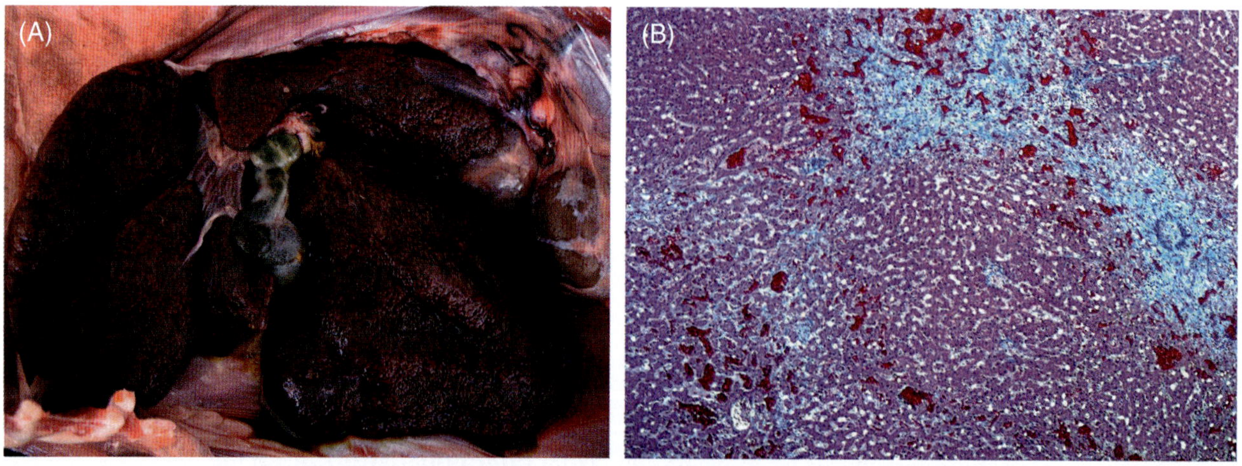

FIGURE 10.8 Veno-occlusive disease (perisinusoidal fibrosis) in a cheetah. (A) Chronic veno-occlusive disease with fibrosis, parenchymal loss, and nodular hyperplasia consistent with cirrhosis. (B) Complete occlusion of a central vein with broad bands of fibrosis extending from central regions into the adjacent parenchyma (Masson's trichrome) (see eSlide 10.e1).

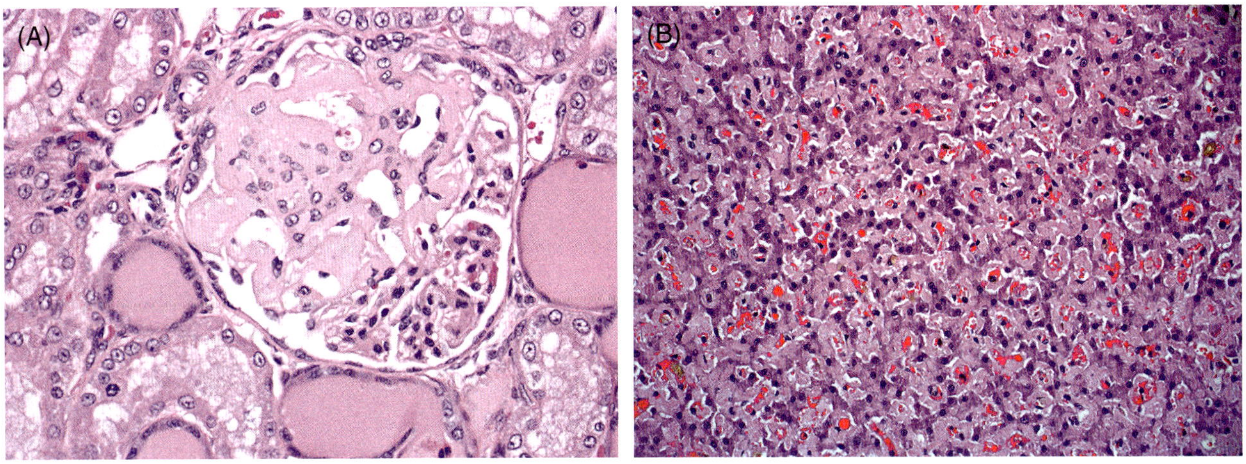

FIGURE 10.9 Amyloidosis in nondomestic felids. (A) Segmental amyloid deposition within the glomerulus of a black-footed cat obliterates capillary and urinary spaces with loss of podocytes. Adjacent tubules contain abundant brightly eosinophilic homogenous proteinaceous material consistent with protein leakage from the damaged glomerulus (see eSlide10.e2). (B) Hepatic amyloid deposition within the space of Disse of a cheetah with thin, fragmented hepatic cords.

inflammation, such as gastritis or nephritis, is common in cheetahs; however, not all cheetahs with inflammation develop amyloidosis. The amino acid sequence of the cheetah AA protein is similar to the sequence in domestic cat breeds with familial amyloidosis, suggesting a structural predisposition and genetic component to disease pathogenesis. Additionally, single nucleotide polymorphisms have been identified in genes regulating transcription of SAA1, suggesting a possible genetic role in increased production of SAA.[398] Different alleles of this gene have been associated with differing SAA serum concentrations in cheetahs, but no significant allelic difference has been identified in the presence or absence of amyloidosis in tissues at necropsy.[102] Furthermore, if amyloidosis was solely due to a genetic predisposition, we would expect to find amyloidosis in free-ranging cheetahs. To date, this has not been documented,[245](Terio and Mitchell, unpublished) suggesting that disease development is multifactorial. In some species, AA amyloidosis is thought to be transmissible and AA fibrils have been identified in cheetah feces.[399] Epidemiological studies of amyloidosis in captive cheetahs that control for concurrent inflammation have not found evidence of amyloid transmission in captive North American facilities.[213] In addition to systemic amyloidosis, **beta-amyloid deposits** have been described in the brains of some cheetahs. A subset of these had neurofibrillary tangles suggestive of Alzheimer's disease.[321] While cognitive declines were suggested in one case, this is difficult to assess in wild animals, and the advanced age of affected cats suggests that it is unlikely to be of significance to in situ conservation.

Glomerulosclerosis is a primary renal disease of captive cheetahs that is only rarely documented in free-ranging cheetahs.[236,241,245] Early lesions typically are arranged in fine radial rays extending outward from the corticomedullary junction. Glomerular basement membranes are segmental to globally thickened with individual globally sclerotic (obsolescent) glomeruli (Fig. 10.10A).[23] Some Bowman's capsule and tubular basement membranes are similarly thickened. Glomerular inflammation and synechia are lacking. Associated lesions include tubular atrophy, cortical atrophy, and cortical and medullary interstitial fibrosis with lymphoplasmacytic inflammation. In a study of lesions from the North American captive population, ultrastructurally, glomerular membranes contain homogeneous and fibrillar material that lacks electron-dense deposits (Fig. 10.10B) and severe lesions have fusion of podocyte foot processes. Thickened membranes are histochemically positive for collagen, glycoproteins, and reticulin and immunopositive for the advanced glycosylation end products, pyrraline and pentosidine.[23] In a study from the European captive population,[372] lesions were variably immunopositive for collagen type IV, fibronectin, IgG, and anticomplement C3. In this study, some glomeruli had evidence of membranous or mesangioproliferative glomerulonephritis, however, immune deposits could not be confirmed by transmission electron microscopy.

Several mechanisms have been suggested in the development of glomerulosclerosis, but none have been confirmed. The presence of advanced glycosylation end products within glomerular basement membranes of cheetahs parallels lesions in diabetic nephropathy of humans; however, affected cheetahs are generally not diabetic. Stress-induced hypercorticoidism with subsequent hyperglycemia and or glomerular hypertension with capillary basement membrane damage has also been proposed.[23,354] As the lesions are similar to old rat nephropathy, which is linked in some rat strains to high protein diets, the large proportion of skeletal meat in the diet of captive cheetahs (in contrast to wild cheetahs that eat more of the whole carcass) may contribute to disease

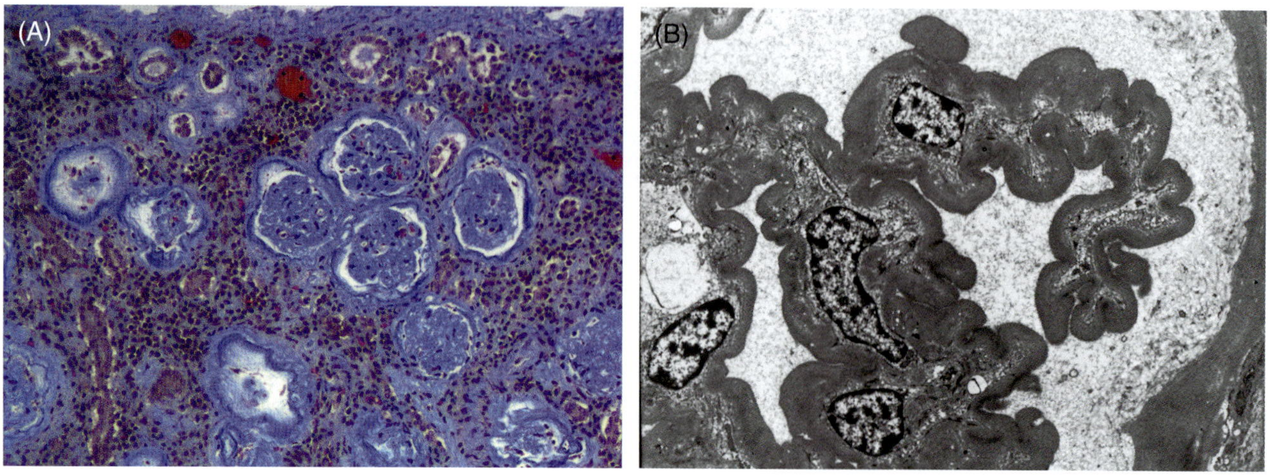

FIGURE 10.10 Glomerulosclerosis in a cheetah. (A) Variably thickened glomerular capsules and loops with obsolescent glomeruli are accompanied by tubular atrophy and thickened tubular basement membranes as well as multifocal lymphoplasmacytic interstitial nephritis. Masson's trichrome. (B) Basement membranes are diffusely thickened by homogeneous material. There is loss of podocytes and endothelial cells and the membranes have collapsed around the remaining mesangial cells. Transmission electron micrograph. *(Reprinted with permission from* Bolton LA, Munson L. Glomerulosclerosis in captive cheetahs (*Acinonyx jubatus*). *Vet Pathol.* 1999;36:14–22.*)*

predisposition in captivity. A genetic predisposition is possible but unlikely since wild cheetahs, which share the species' renowned homogeneity, generally lack these lesions.[245,261] The role of antigen–antibody deposition requires further investigation as these findings have been inconsistent across populations and studies.[372] The lesions also resemble chronic renal disease that can be experimentally induced by ischemia in domestic cats.[227,331]

Since the mid-1970s, sporadic cases of **oxalate nephrosis** have been reported in captive cougars, jaguars, leopards, and most commonly, cheetahs.[228] Affected cats are found dead or present in acute renal failure with severe azotemia, and poly- to anuria. Crystals may be noted in the urine. Histologically, characteristic birefringent crystals forming rosettes are present in renal tubules and associated with epithelial necrosis, tubular dilation, and cellular casts (Fig. 10.11). In more chronic cases, there may be tubular regeneration. Crystals are positive for calcium, and in some cases, basement membranes are mineralized. Ethylene glycol toxicosis has been suspected in some cases but never proven and cases have occurred in areas of the world that do not use antifreeze. Additionally, approximately 10% of the North American and South African captive cheetah populations have histologically evident oxalate crystals at the time of death, making widespread toxicity unlikely. In cheetahs, there is no association between risk factors such as concurrent renal disease (amyloidosis, glomerulosclerosis, interstitial nephritis) or gastritis and the severity of oxalate nephrosis. Some of the most severe cases occur in young cheetahs suggesting an inherent predisposition, although the pathogenesis remains uncertain.

In 1996 **leukoencephalopathy** was diagnosed in cheetahs at multiple captive breeding facilities from all geographic regions of North America.[30] This disease had not

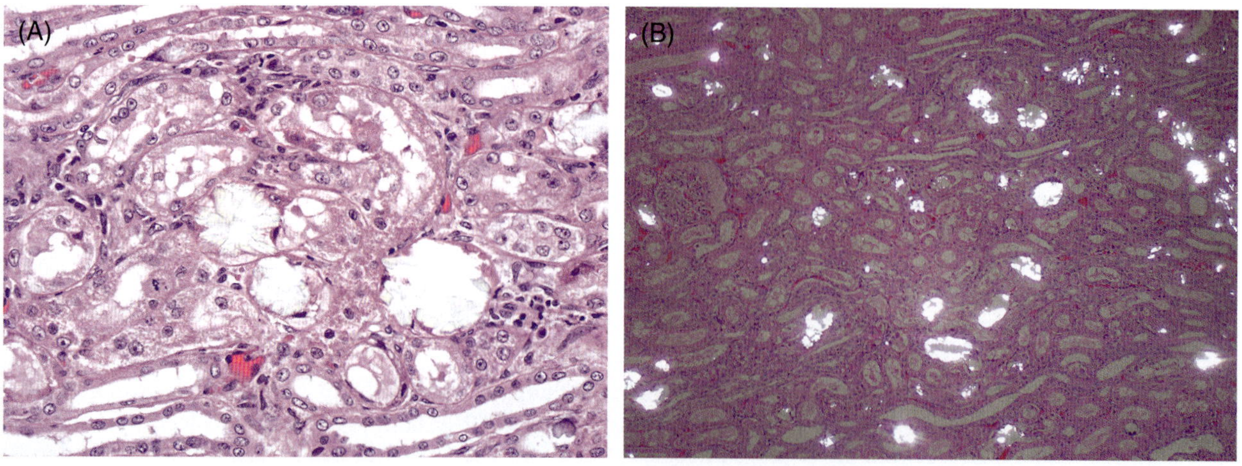

FIGURE 10.11 Oxalate nephrosis in a cheetah. (A) Crystals form rosettes within tubules and are associated with tubular epithelial necrosis. (B) Numerous crystals are evident throughout the cortex upon polarization of routine hematoxylin–eosin stained slides.

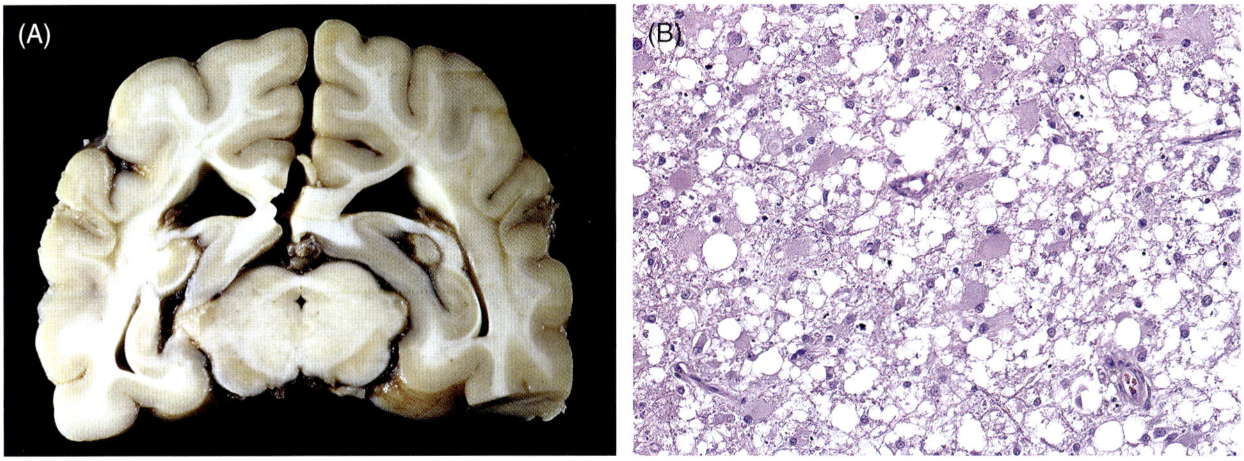

FIGURE 10.12 Leukoencephalopathy in a cheetah. (A) Bilateral cavitation and graying of the white matter in the centrum semiovale, corona radiata, and internal capsule with enlargement of the lateral ventricles. (B) Marked neuropil vacuolation with myelin and axonal loss and numerous astrocytes. Astrocytes have abundant eosinophilic granular cytoplasm.

been diagnosed in captive cheetahs previously but was similar to lesions in a tiger and two captive Florida panthers. Affected cheetahs were older (median and mean age of 12 years) and presented with ataxia, suspected blindness, altered behavior, disorientation, proprioceptive deficits and, in some cases, seizures. At necropsy, lesions included bilateral cavitation within the white matter of the centrum semiovale and corona radiata and to a lesser extent, the internal capsule with hydrocephalus (Fig. 10.12A). Histologically, the disease was characterized by numerous reactive astrocytes with varying degrees of myelin vacuolation and degeneration, as well as cavitation (Fig. 10.12B). Some astrocytes were large, with abundant granular eosinophilic cytoplasm that had reduced GFAP staining with vacuolation, karyomegaly, and multiple nuclei while other astrocytes were more typical gemistocytic astrocytes with abundant GFAP positive cytoplasm. Gitter cells contained abundant myelin debris in severe lesions while in less severe lesions, axons remained intact suggesting that myelin loss preceded axonal degeneration. The cause of leukoencephalopathy has not been determined. Given the lesions, lack of viral evidence, and epidemiology of the outbreak, exposure of cheetahs to some common neurotoxin through feed, pharmaceuticals, or vaccines was considered the most likely cause. No cases have been identified in North American captive cheetahs since 2005; however, cases have been identified in wild-caught captive cheetahs in Namibia (Mitchell, unpublished). More recent reports describe similar disease in two Sumatran tigers and a snow leopard.[125,153,377] Histologically, these cases had severe, extensive leukomalacia with gliosis, large, reactive, gemistocytic astrocytes, and intralesional foamy macrophages (gitter cells); internal hydrocephalus ex vacuo was also described in the tiger in Germany.

A **myelopathy** has been described in captive-born cheetahs housed in Europe and the United Arab Emirates.[324,380,381] Age of onset varies from 2.5 months to 12 years with a rapid and fatal course. Onset is acute with ataxia, paresis, and variable progression to dragging of the hind limbs. Histologically, there is bilaterally symmetric degeneration of the white matter tracts of the spinal cord. Myelin loss exceeds that of axonal loss suggesting a primary demyelinating disorder. Gliosis and degeneration of the Purkinje and granular cells of the cerebellum also occur. The pathogenesis is unknown, although copper deficiency has been suspected.

In 2018 panthers and bobcats in Florida, the United States were seen with clinical signs of **feline leukomyelopathy** consisting of rear-limb weakness, incoordination, difficulty walking, and emaciation.[66] Many of the affected cats have been cubs with multiple litter mates affected. White matter degeneration of the ventral and lateral funiculi of the spinal cord with dilation of myelin sheaths and axonal loss are the most consistent histologic findings (Fig. 10.13).

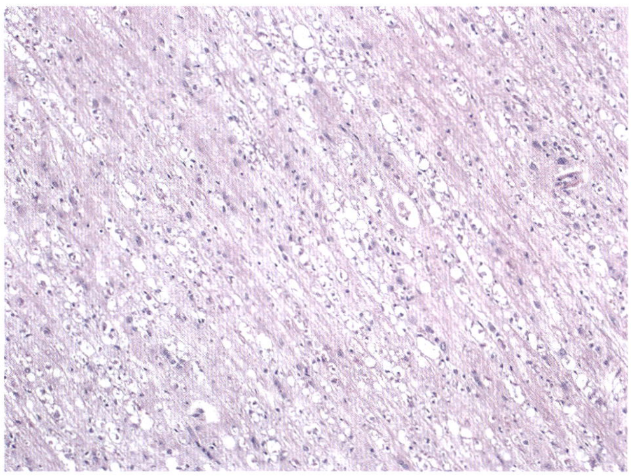

FIGURE 10.13 Myelopathy in a Florida panther. Dilation of white matter sheaths, loss of axons, and digestion chambers within white matter tracts of the spinal cord lateral funiculi. *(Photo Courtesy of N. Nemeth, University of Georgia.)*

Peripheral nerves are not affected. To date, testing for infectious, nutritional, and toxic causes has not yielded a definitive etiology.

Endometrial hyperplasia (EH) occurs spontaneously in a number of species of zoo felids and is significantly more common in **melengestrol acetate (MGA)** contracepted animals[238,242] It is characterized grossly by mild to marked expansion of the endometrium with or without macroscopic cyst formation (Fig. 10.14A). Histologically, there is proliferation of endometrial glands with adenomatous projections (adenomatous hyperplasia), dilated cystic glands (cystic hyperplasia) (Fig. 10.14B), or expansion of the endometrium by irregular folds and/or polypoid projections. Hyperplastic glands are lined by crowded, densely packed, well-differentiated cuboidal to columnar epithelial cells that are asynchronous with the estrous cycle stage. In cystic glands, epithelial cells may be attenuated. In more severe cases, there may also be endometrial fibrosis and endometrial or stromal mineralization.[325] Presence of glands in the myometrium (adenomyosis) can also occur and should not be mistaken for a neoplastic process in the absence of other criteria of malignancy. Hyperplasia develops earlier and is significantly more common in MGA-treated animals; risk is increased with increased treatment duration (>72 months). In addition to MGA treatment, age is an additional risk factor while parity is associated with decreased risk. The pathogenesis of MGA-associated EH is unclear; however, exposure of estrogen primed endometrium to prolonged progestins, the latter inducing secretory differentiation of epithelial cells, is suspected.

Though relatively uncommon, **gastric dilatation with or without torsion** has been reported in several species (snow leopard; Amur, Malayan, and Sumatran tigers; lion; jaguar) of nondomestic felids in North American zoos.[4] Some have been found dead with no premonitory signs. In others, clinical signs included lethargy and/or abdominal distension with or without vomiting. As in other species, prognosis is poor even with rapid diagnosis and intervention. Histologic changes in the torsed tissue can include transmural congestion, hemorrhage, edema, necrosis, and vascular thrombosis or evidence of disseminated intravascular coagulopathy. Radiographs are an important tool and can be diagnostic if the characteristic, gastric gas distension is seen. A cause for its development is often not identified.

Neoplastic

Nondomestic felids develop many of the same types of benign and malignant tumors described in domestic cats. The vast majority of these have been described in captive animals and many in single case reports. Several excellent retrospective mortality studies of nondomestic felids include tumor lists and there are multiple well-described case series of more common neoplasms.[58,69,72,77,80,127,138,170,208,211,231,254,255,261,269,289,306,364,395]

Lymphoma, while it can occur in any nondomestic felid, is most common in older (>14 years) African lions. In one study,[128] the majority of tumors were T cell (CD3+ CD79a−) with primary splenic involvement. Liver and lymph nodes were also commonly affected. Tumors in these lions were not associated with feline leukemia virus (FeLV) or feline immunodeficiency virus (FIV) infection, although there is a case report of an FIV-associated lymphoma in a lion.[283] In contrast, a single Namibian cheetah has been diagnosed with FeLV-associated multicentric T cell lymphoma.[202]

Multiple myeloma has been diagnosed in an African lion, jaguar, snow leopard, cheetah (Terio, unpublished), and several tigers.[69,158,207,284,363] Lesions in these species

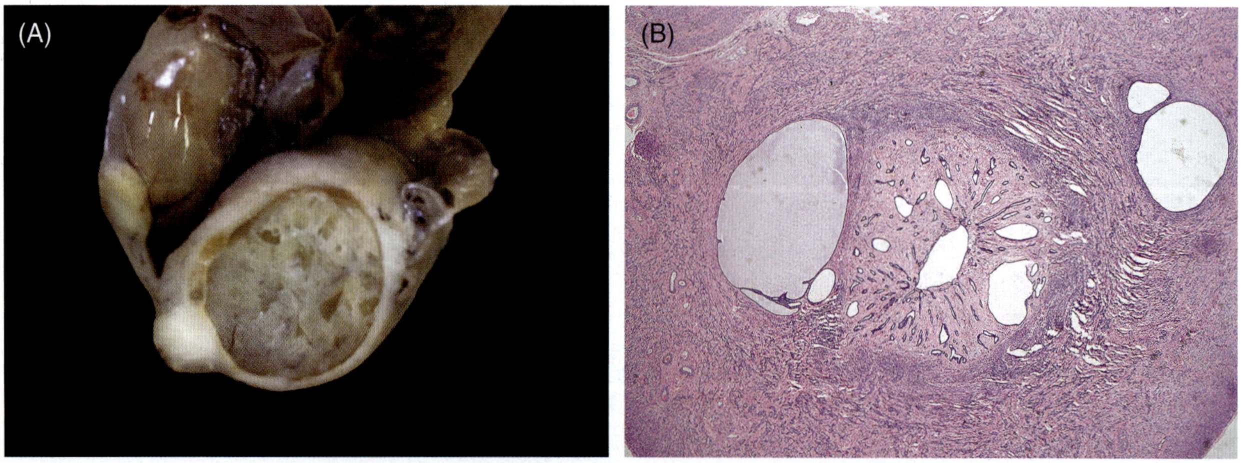

FIGURE 10.14 Cystic endometrial hyperplasia in felids. (A) The uterine lumen is filled with endometrium that is greatly expanded by small to large cystic spaces. (B) Endometrial glands are mildly to markedly dilated and lined by variably attenuated epithelium. Some glands contain pale basophilic mucinous material. A few glands are present within the myometrium (adenomyosis).

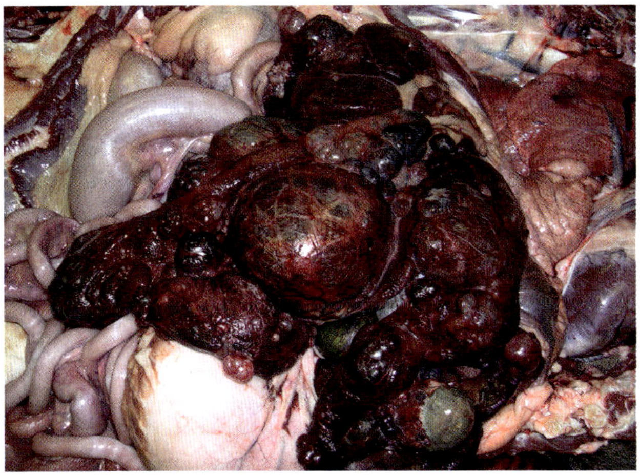

FIGURE 10.15 Biliary carcinoma in an African lion. Numerous variably sized, solid to cystic structures, some of which are filled with dark green fluid (bile), efface the normal hepatic parenchyma.

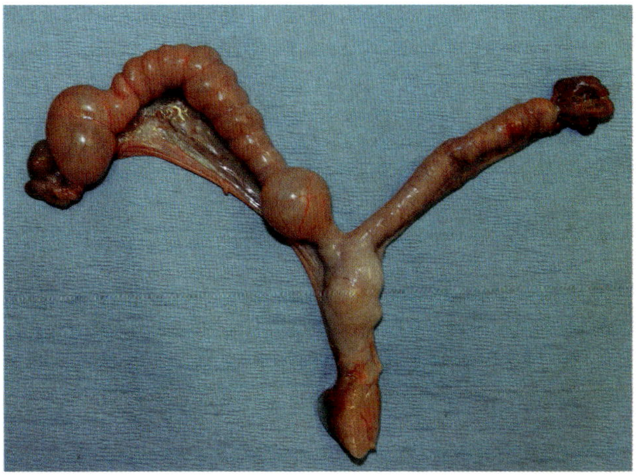

FIGURE 10.16 Leiomyoma and hydrometra in a serval. This multinodular mass expands the wall of the uterine body. Obstruction to the outflow of uterine secretions has resulted in fluid retention and dilation of one uterine horn.

resemble the disease in domestic animals including intraosseous neoplasms with osteolysis and bone destruction, monoclonal gammopathy, excretion of light chains in the urine, and sheets of neoplastic plasma cells in the bone marrow or soft tissues that have immunoreactivity to multiple myeloma oncogene 1 (MUM1), CD79a, and other B cell markers. Spread to extramedullary sites appears to be common in nondomestic felids.

Biliary adenomas and **biliary adenocarcinomas** (Fig. 10.15) have been reported in lions and leopards.[18,182,306] In most cases, these tumors are cystic rather than solid. Whether these are related to the high prevalence of biliary cysts (Fig. 10.3) in these species is not known.

Leiomyomas and **leiomyosarcomas** are benign and malignant smooth muscle tumors, respectively. They can originate in smooth muscle anywhere in the body. In zoo felids, they appear to be relatively common in the female genital tract.[51,207] Leiomyomas have been diagnosed more commonly in large cats than small cats, which may reflect true incidence or collection bias. Tumors have been reported more commonly in the uterus than in the ovary or broad ligament; cervical leiomyomas have also been reported. Tumors may be single or multiple, and appear grossly as expansile or pedunculated, fleshy to firm, light pink/tan to tan/white masses (Fig. 10.16). In the uterus, they appear to originate more commonly from the intramyometrial smooth muscle than from subserosal or subendometrial smooth muscle. Tumors are often well-demarcated, but they can blend with adjacent normal tissue or have edges that vary from being well to poorly defined but noninvasive. Leiomyomas are composed of dense, interwoven bundles and fascicles of well-differentiated, elongate cells with moderate amounts of fibrillar, eosinophilic cytoplasm, and oval to elongate, normochromatic, vesicular nuclei with a single nucleolus. Mitoses are typically absent or rare (<1 per 10 high-power fields). Tumoral necrosis may be present but inflammation is uncommon and metastasis has not been described.

Leiomyosarcomas are less common in zoo felids than leiomyomas. They are poorly demarcated, invasive, firm, multilobulated masses that often replace normal architecture. Cells have mild to moderate cellular pleomorphism with anisocytosis, anisokaryosis, occasional multinucleation, and one to a few nucleoli. Mitosis are much higher than in leiomyomas and are often common (several to over 20 per 10 high-power fields). Tumor necrosis and inflammation are more common in leiomyosarcomas than in leiomyomas, and foci of tumoral mineralization, ossification, and local or distant metastasis may be present. Immunohistochemical labeling is useful in differentiating leiomyomas and leiomyosarcomas from other mesenchymal tumors, and will be positive for vimentin, smooth muscle actin, and/or desmin though labeling with the smooth muscle-specific markers may be less common in leiomyosarcomas than leiomyomas.

Possible complications of either tumor include obstruction along the reproductive tract and in leiomyosarcomas, adhesions between the tumor and adjacent structures due to invasive growth.[51,326] In a review of the reproductive histories of 52 zoo felids with uterine leiomyomas, there was no association between progestin contraception and tumor presence; in five zoo felids with leiomyosarcoma, four had been treated with the progestin contraceptive melengesterol acetate (MGA); however, an association between the two was not significant and further investigation was deemed necessary.[51]

Mammary gland (MG) carcinoma has been diagnosed in many species of large and small zoo felids.[126,170,207,211]

More common reports in large cats may reflect collection bias; however, the relatively high incidence of these cancers in jaguars may represent a true genetic predisposition. Tumors typically develop in older animals. In general, and as occurs in domestic cats, MG cancers in zoo felids are typically aggressive, have multiple patterns, and many have local, distant, or occasionally widespread metastasis at the time of diagnosis.[211]

MG carcinomas typically present as variably sized, single or multiple, demarcated to invasive tumors in one or more of the mammary glands. Tumors are typically composed of firm, dense, light tan to light pink tissue. Ulceration of overlying skin may be present (Fig. 10.17A). Histologically, the majority contain multiple morphologic patterns.[126,211] Most cancers have some areas of tubulopapillary growth with formation of tubular, solid, cribriform, comedone, and less commonly mucinous patterns (Fig. 10.17B). Regardless of morphologic pattern, tumors are typically multinodular to multilobular with scant to small amounts of fibrous or fibrovascular stroma although moderate intertubular stroma may be seen in tubular patterns. Focal to multifocal inflammation, typically lymphoplasmacytic, neutrophilic, or mixed, may be present but is not a consistent finding; necrosis may also be present but is not as common as intratubular necrotic cellular debris.

Tumors vary from well to poorly differentiated based on anisocytosis, anisokaryosis, and chromatin pattern, with poorly differentiated tumors having more marked degrees of anisocytosis and anisokaryosis, a vesicular rather than uniform chromatin pattern, and multiple rather than single nucleoli. In general, most MG carcinomas in zoo felids are moderately to poorly differentiated and of high grade based on the percentage of tubule formation, mitotic count, and nuclear morphology (using criteria of Elston and Ellis[89]). Metastasis is common and most often to local and distant lymph nodes, lung, and liver. As occurs in domestic cats, MG carcinomas in zoo felids are more commonly positive for progesterone receptors than estrogen receptors, though testing has only been performed on a small number of animals.[211]

While MG carcinomas occur spontaneously, a significant increase in incidence is seen in zoo felids on continuous, long-term (>3 years) MGA contraception.[126] Use of this potent synthetic progestin in implant form (subcutaneous or intramuscular) began in the 1970s and continued until concerns about its association with mammary and uterine cancers and cystic endometrial hyperplasia arose in the 1990s.[126,238,242] Though not completely understood, concurrent exposure to endogenous estrogen and this potent progestin is suspected as a predisposing factor in MGA-associated MG carcinogenesis. Felids are seasonally polyestrous and typically ovulate only if bred (induced ovulation), though spontaneous ovulation has been documented in some species (see "Unique Features in Digital Content" section). When not bred, they are exposed to recurrent peaks of estrogen as waves of follicles develop and then undergo atresia.[240] Because MGA does not suppress ovulation,[160] treated cats are likely exposed to concurrent intermittent high levels of endogenous estrogen (during folliculogenesis) and continuous exposure to this potent, synthetic progestin. In women, dogs, and cats, cumulative endogenous and exogenous steroid hormone exposure are risk factors in MG carcinogenesis.[240] This combination is also suspected as a predisposing factor in carcinogenesis in zoo felids as estrogen and progesterone promote MG interlobular duct and lobuloalveolar growth, respectively.[211] Interestingly, there are no significant differences in spontaneous and MGA-associated MG cancers,

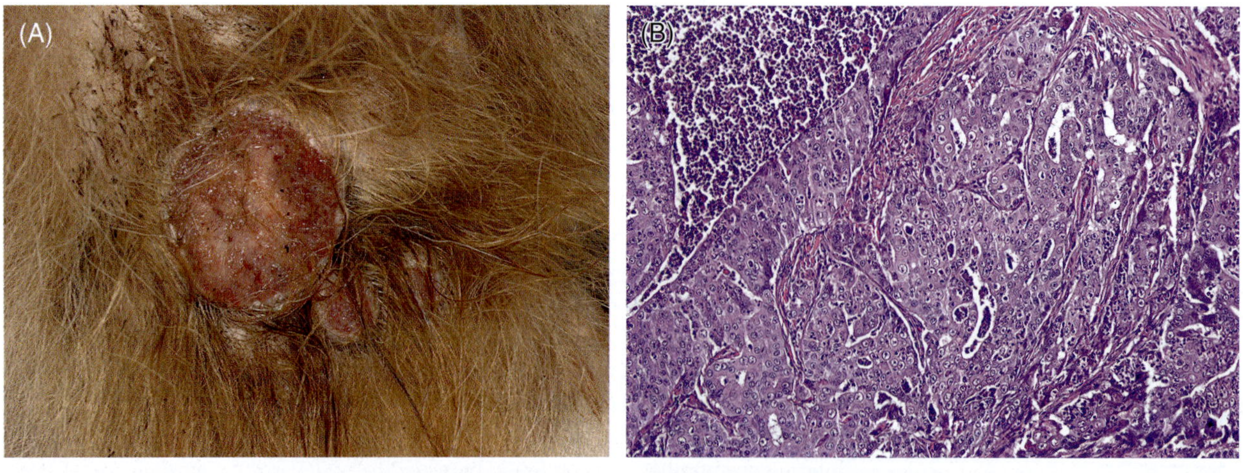

FIGURE 10.17 Mammary carcinoma in a melengestrol acetate (MGA) contracepted tiger. (A) Multiple raised pink to tan mammary masses with ulceration of the overlying skin. (B) Multiple histologic patterns are common in mammary gland carcinomas. In this case, there are areas of comedone, solid, and tubular growth separated by small amounts of fibrous stroma. Intratubular inflammation and necrotic debris are also present. *(Parts A, B: Photos Courtesy of the Wildlife Conservation Society.)*

including the age of development, morphology, grade, or occurrence of metastasis. In either group, MG cancer may be associated with hypercalcemia or anemia.

Spontaneous or MGA-associated **endometrial neoplasia** are also described in a number of species of zoo felids and appears to be significantly more common in MGA-treated animals.[238,242] **Endometrial carcinoma** in zoo felids is similar to that in other species. On gross examination, tumors present as firm, unencapsulated, light tan to white, demarcated to poorly demarcated thickened areas within the uterine wall. Histologically, there are proliferative neoplastic endometrial cells that form papillary intraluminal growths.[242] Invasion of clusters of neoplastic endometrial cells into the adjacent stroma is seen.

Captive fishing cats have a high incidence of **transitional cell carcinomas** (TCC).[177,207,344] Mean age in affected cats is 10.8 years and no sex predilection has been noted. Both captive-bred and wild-caught captive-held cats have developed these tumors. However, it is unknown if this disease occurs in wild fishing cats. Tumors commonly develop in the trigone region of the bladder. Histologically, lesions are primarily mural, although exophytic papillary proliferation does occur in about half of the cases (eSlide 10.e3).[177] Carcinoma in situ occurs in, presumably, early lesions. Invasive portions of the neoplasms express cyclo-oxygenase 2 (COX-2) and therefore treatment with COX-2 inhibitors may slow disease progression.[177] A dietary component to the pathogenesis has been suspected because in captivity cats have commonly been fed commercial carnivore diets consisting primarily of beef with little fish. Nutritional studies found significantly increased levels of saturated fatty acids in cats with TCC compared to cats without tumors.[204] As some of these fatty acids are correlated with TCC in humans, recommendations have been made to decrease beef and increase fish content of captive diets. In addition to fishing cats, transitional cell carcinomas have been described in snow leopards.[393]

Among nondomestic felids, only jaguars commonly develop **ovarian papillary cystadenocarcinomas.**[24,237] This tumor is a significant cause of mortality in captive adult and geriatric jaguars affecting breeding age as well as geriatric cats but its prevalence in wild jaguars is unknown.[137] Diagnosis is common at necropsy. Affected jaguars will have multiple, variably sized cystic masses, many of which are bilateral with spread along the uterine horns and abdominal carcinomatosis (Fig. 10.18A). Histologically, tumors are composed of papillary projections lined by neoplastic polygonal cells overlying a thin connective tissue stroma (Fig. 10.18B). Cysts formed by neoplastic cells contain proteinic fluid and or hemorrhage. An important differential is peritoneal mesothelioma. In jaguars, the occurrence of this tumor has not been correlated with progestin (MGA) contraception.[160] Interestingly, jaguars also have a high prevalence of mammary cancer,[38,237] suggesting that abnormalities in genes, such as *BRCA* may be important in the pathogenesis.

Pheochromocytomas, an uncommon neoplasm in domestic felids, are common in clouded leopards.[62,362] Many neoplasms are solitary expansile masses; however, a few have invaded the vena cava causing fatal hemoabdomen. By electron microscopy, neoplastic cells contain both epinephrine and norepinephrine granules. Retinal detachment, in some cases, suggests that tumors may be functional.

Splenic myelolipomas are a common benign neoplasm of cheetahs occurring in both captive and wild cats,[236] and have been described in African lion, jaguar, and snow leopard.[158,284,363] Masses are commonly multiple throughout the parenchyma (Fig. 10.19A) and can be greater than 2 cm in diameter. Masses are composed of

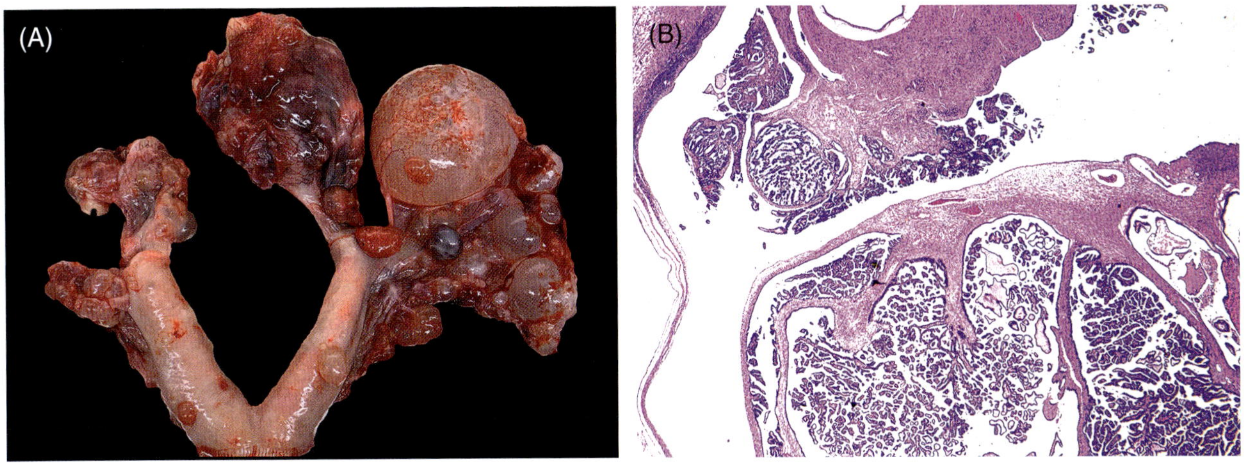

FIGURE 10.18 Ovarian cystadenocarcinoma in a jaguar. (A) The ovaries are effaced by a numerous, variably sized, cystic, tan, red, or purple nodular masses. Local metastasis is seen along the uterine serosa. (B) The cystadenocarcinoma is composed of multiple cystic spaces with irregular papillary projections lined by fronds of neoplastic epithelial cells. *(Parts A, B: Photos Courtesy of S. Corner, Michigan State University.)*

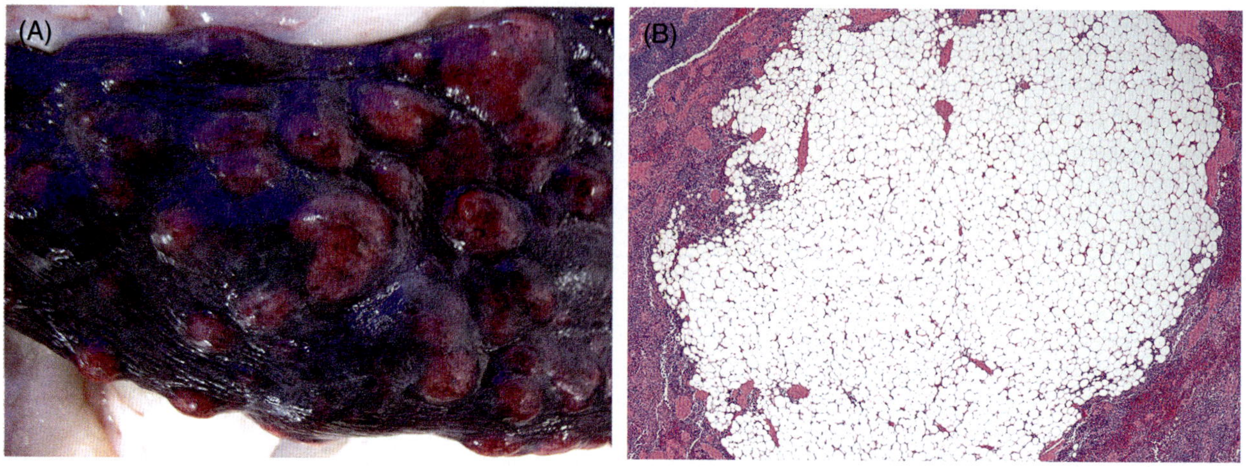

FIGURE 10.19 Splenic myelolipomas in a cheetah. (A) Myelolipomas in cheetahs can be single but are often multiple and appear as variably sized, unencapsulated, well-demarcated, expansile masses that are light to dark red/purple. (B) Myelolipomas consist of variable combinations of mature adipose tissue and hematopoietic tissue from all three cell lines.

well-differentiated adipose with variable amounts of admixed hematopoietic tissue (Fig. 10.19B). These masses may be mistaken for malignant tumors on ultrasound but are incidental and benign.

Meningiomas, a somewhat common finding in domestic cats, have also been reported in one large retrospective study of brain lesions in nondomestic felids. In that study, it was seen in leopards, lions, tigers, and cougars. No additional details about the specific tumor types were provided.[377]

Sublingual **papillomas** associated with papillomavirus infection are relatively common in all felids (see "DNA Viruses" section). Snow leopards can also develop oral squamous cell carcinomas, cutaneous plaques, papillomas, and cutaneous squamous cell carcinoma, which have also been associated with papillomaviruses.[207,393]

In a study of oral cavity lesions, **fibromatous epulis** of periodontal ligament origin, was the most common odontogenic tumor reported in lions, jaguars, a bobcat, cheetah, and serval. These tumors arise from the tooth-associated gingiva and consist of ligament-like stroma (dense collagen surrounding stellate mesenchymal cells) within which are variable amounts of cemento-osseous matrix, and odontogenic epithelium arranged as nests or branching cords lined by columnar cells.[317]

Mesothelioma, a relatively uncommon tumor in animals, has been reported in lion, tiger, jaguar, snow leopard, clouded leopard, and cheetah.[207,362] These tumors can form plaques, nodules or papillary fronds on the parietal and serosal surfaces in the body cavities. There are three histologic types: epithelial, sarcomatous, and biphasic. Its variable morphology can pose a diagnostic challenge. Immunohistochemical markers helpful in the diagnosis include panytokeratin, vimentin, Wilms tumor 1 (WT1), and calretinin.

Malignant **melanomas** and **melanocytomas** have been described in lions, tigers, and snow leopards.[288] While melanocytomas were present in the skin, malignant melanomas were noted in the oral cavity, skin, nasal planum, lip, and iris. For poorly pigmented neoplasms, Melan-A, PNL-2, and TRP-2 immunohistochemistry were utilized for confirmation and each was utilized to confirm histogenesis on an individual case. In these cases, tumors were often aggressive and metastasis was noted at necropsy in many cases. Of the tiger cases, the majority were white tigers suggesting a predilection.

Cheetahs appear to have an increased incidence of **mast cell tumors** and **mastocytosis.**[42] Masses are generally benign and some have been described to regress which has led to questions about potential insect bite hypersensitivity; however, others are distinct masses. Histologically, tumors are low grade and well-differentiated although granules may stain poorly with routine stains and toluidine blue. Giemsa stains may highlight granules more consistently in this species. It is important to rule out underlying herpesviral skin lesions which can also have large numbers of eosinophils and mast cells but which should also have a large plasmacytic component and intraepithelial inclusions (see "DNA Viruses" section). Evaluation of mutations in the KIT proto-oncogene in cheetahs and domestic cats failed to identify specific association between mutations and histologic tumor grade.[348]

INFECTIOUS DISEASES

Numerous surveys have documented exposure of free-ranging felids to a variety of pathogens without evident disease suggesting that they are generally well-adapted to their presence. However, many felid habitats have become highly fragmented, increasing the risk of disease due to perturbation of normal host–parasite relationships and exposure to novel pathogens.[243] While the solitary and reclusive nature of felids decreases intraspecific contact rates, fragmentation of habitat results in increased contact with other species, particularly domestic dogs and cats, exposing felids to pathogens for which there may not be individual or population-level immunity. For example, FeLV does not appear to be endemic in most free-ranging felid populations but has resulted in mortalities after introduction threatening the Iberian lynx and Florida panther.[35,216,252] As regional and pathogen-specific variables are also contributory, continued surveys for pathogen exposure combined with pathologic investigations are needed for accurate risk analysis.[48,118,280,303,368] In captivity, despite vaccination and generally excellent medical care, increased animal density facilitates transmission.

DNA Viruses

All felids are susceptible to **feline herpesviruses. Feline alphaherpesvirus-1 (FeHV-1),** the cause of **feline viral rhinotracheitis,** is associated with upper respiratory signs, including sneezing, nasal discharge, as well as ocular signs such as conjunctivitis, ocular discharge, keratitis, and corneal ulcers. Transmission is direct and through saliva and ocular and nasal secretions. Infected animals develop lifelong latency in sensory neurons and recrudescence can occur, often preceded by any of a number of stressors.[196] Many wild felid populations have serologic evidence of exposure to feline herpesviruses, although disease has not often been described.[243] Vaccination with modified live vaccines may, in some circumstances, induce disease in nondomestic felids[292]; transmission from domestic cats is also suspected.[201,390]

Severe upper respiratory tract disease with corneal ulcers and/or keratitis have been noted in both Pallas' cats and cheetahs infected with FeHV-1. Infected captive cheetahs have also been diagnosed with unusual skin lesions due to FeHV-1.[152,246,310] This has not been noted in wild cheetahs despite evidence of seropositivity,[239,245,360] but has been seen in wild-caught, captive-held cats.[101] Lesions present as ulcers and often plaque-like proliferative lesions on the face and/or forelimbs, or sites in contact with lacrimal (tear) and salivary gland secretions (Fig. 10.20A). Biopsy is essential to rule out neoplasia and differentiate from cowpox lesions. Histologically, there are dense dermal infiltrates of eosinophils, plasma cells, and mast cells. The intact epithelium adjacent to the ulcer may be hyperplastic (Fig. 10.20B; eSlide 10.e4). Eosinophilic nuclear inclusions can be found in the intact epithelium of the skin, adnexa, and rarely in salivary gland ducts.

Immune responses to FeHV-1 in cheetahs have been studied in vitro using peripheral blood mononuclear (lymphocyte and monocyte) cell (PBMC) cultures.[223,224] In these studies, cheetah PBMC as a group had significantly lower proliferative responses than domestic cats to FeHV-1, although individual responses varied widely. Interestingly, the cheetah PBMC responses improved when they were first cultured with interleukin-2 (IL-2), a TH-1 cytokine critical to an effective cell-mediated immune response. Other studies have found low levels of IL-2 in captive cheetahs as opposed to wild counterparts suggesting that ineffective immune responses may play a role in disease susceptibility.[352]

Gammaherpesviruses (GHV) have also been described in several species of nondomestic cats.[191,197,368] This includes *Lynx rufus* GHV1 and *Lynx rufus* GHV2 in bobcats, *Puma concolor* GHV1 and *Lynx rufus* GHV1,

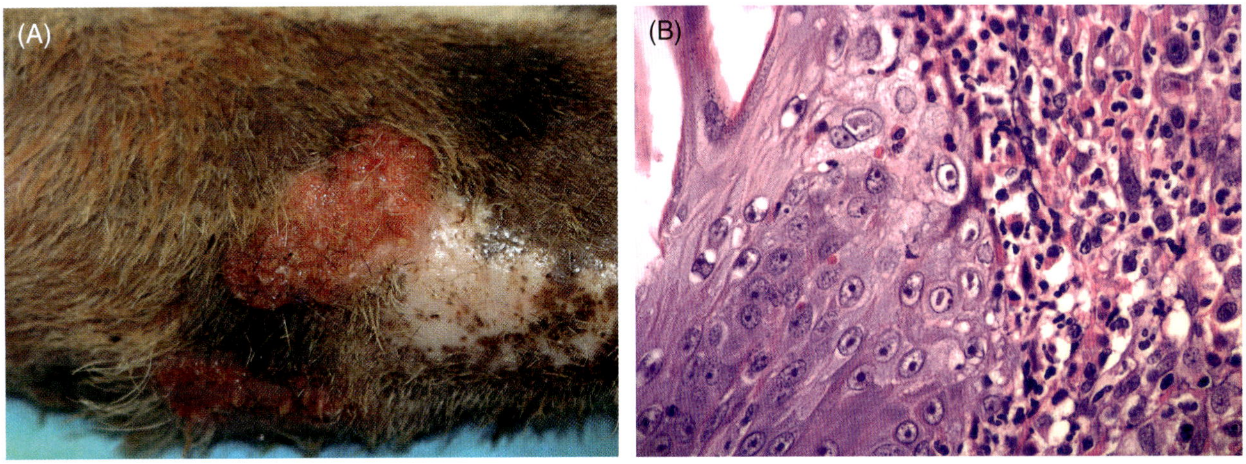

FIGURE 10.20 Feline herpesviral dermatitis in a cheetah. (A) Lesions present as ulcerated, raised plaque-like areas. The forelimb is a common site. Similar lesions can be noted on the face. (B) The epidermis is thickened and epithelial cells contain characteristic eosinophilic to amphophilic intranuclear inclusions. Dermal inflammation is typically mixed and, as is this case, can include numerous eosinophils and plasma cells. Mast cells as well as neutrophils in regions with ulceration may also be present (see eSlide10.e4).

Felis catus GHV 1 in leopard cats, *Panthera leo* GHV1 in a lion, and *Leopardus pardalis* GHV1 in ocelots. Latent infection is often in lymphocytes, and as with alphaherpesviruses, reactivation and recrudescence can occur, and infection with multiple different gammaherpesviruses in individual animals has been reported. Associated disease remains to be described.

Necrotizing and ulcerative facial skin lesions in cheetahs can also be caused by **cowpoxvirus** infection.[332] This infection is typically seasonal and cheetahs that also develop respiratory distress may die. Histologically, ulcerative necrotizing cheilitis and nasal dermatitis are associated with hydropic degeneration of keratinocytes containing large angular eosinophilic intracytoplasmic poxviral inclusions (Fig. 10.21). Exposure to cowpoxvirus infection has also been documented in jaguarundi and Eurasian lynx.[176,369]

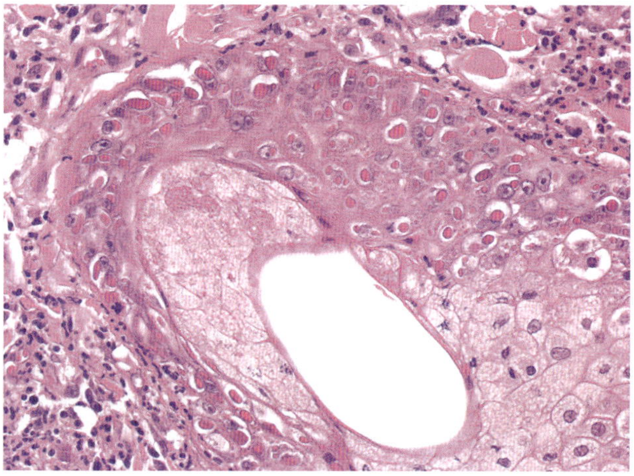

FIGURE 10.21 Cowpoxviral dermatitis in a cheetah. There is swelling and hydropic degeneration of epithelial cells, many of which contain large eosinophilic intracytoplasmic viral inclusions and neutrophilic inflammation within the surrounding dermis. *(Photo Courtesy of A. Pocknell, Finn Pathologists.)*

Parvoviruses, single-stranded DNA viruses, are multihost pathogens that naturally infect a wide range of carnivores including Felidae. The host range of the feline subgroup is poorly defined. Felids are susceptible to **carnivore protoparvovirus 1** which includes ***feline panleukopenia virus* (FPV), *canine parvovirus* 2 (CPV 2)**, and ***raccoon parvovirus*** variants, all of which have been associated with disease in felids. Little is known about the epidemiology of these viruses in nondomestic species, including their role in free-ranging hosts as natural reservoirs and the importance of bidirectional transmission between and among domestic and wild carnivores. Carnivore breeding or housing facilities may be particularly prone to recurrent disease outbreaks even if vaccination in late pregnancy using an inactivated vaccine is practiced.[178] The virus is hardy, particularly in cool, moist, shaded environments, so even solitary, widely dispersed, free-ranging carnivores may be exposed at marking sites, latrines, or play trees shared with other carnivores.

Clinical signs and pathological lesions are rarely documented in wild felids.[83,178,304,336] Differing outcomes depend on the timing of infection and may include fever, reduced litter size, neonatal deaths, severe lymphoid depletion and leukopenia, cerebellar hypoplasia, and intestinal disease in older cats characterized by intestinal hemorrhage, villous atrophy and collapse, necrosis and regeneration of crypt epithelium, and opportunistic bacterial catarrhal to necrohemorrhagic enteritis (Fig. 10.22). Eosinophilic to amphophilic, nuclear inclusions in intestine and viral shedding are more common early in the course of disease. In addition to histology, fecal electron microscopy (most useful in early infections), immunohistochemistry,

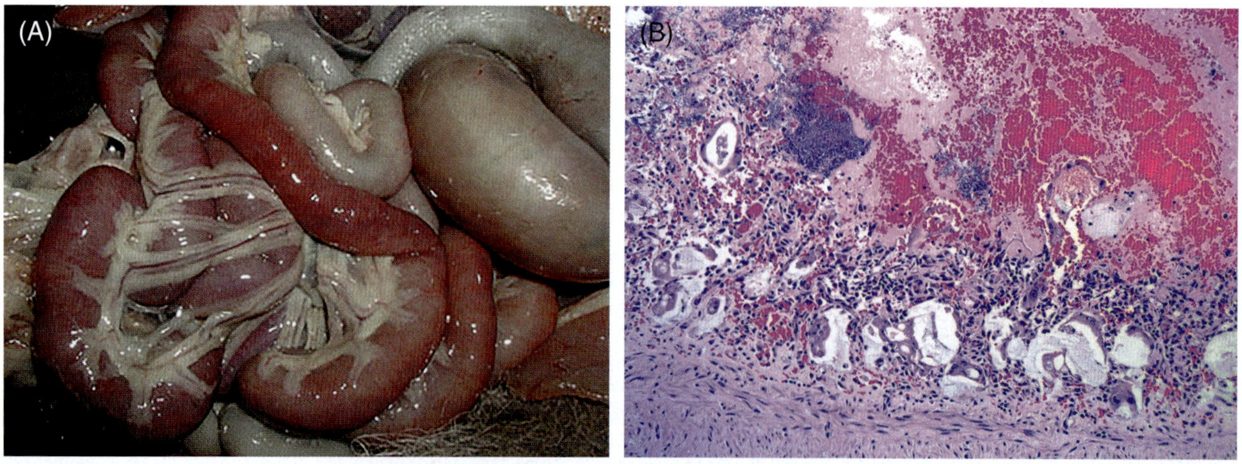

FIGURE 10.22 Carnivore protoparvovirus 1 (feline panleukopenia virus) enteritis in a cheetah. (A) The small intestines are segmentally congested and flaccid. (B) Characteristic lesions include severe, acute crypt epithelial cell necrosis with villous blunting and villous collapse. Intraluminal and intralesional hemorrhage and opportunistic bacterial colonization are common.

virus isolation, or PCR are diagnostic. Disease caused by CPV in felids may be milder than that caused by FPV.

***Papillomaviruses* (PV)** are nonenveloped, circular, DNA viruses in the family Papillomaviridae.[235,289] Most PVs are species specific and many have specific tissue tropism. In nondomestic felids, papillomaviruses have been detected and are thought to play a role in the development of oral and cutaneous papillomas with transformation in some cases causing squamous cell carcinomas and sarcoids.[235] Basilar epithelial cells are the primary site of infection, and viral replication and assembly occur in maturing epithelial cells.

Oral papillomas associated with **PuPV-1** (also known as UuPV-1) are a relatively common tumor of snow leopards in many zoo collections[151,343]; to our knowledge, they have not been detected in free-ranging snow leopards. They typically arise as single to multiple, variably sized, nonpigmented plaques and or papillary exophytic masses commonly on the ventral surface of the tongue (Fig. 10.23A). Characteristic histologic lesions include mild to marked mucosal hyperplasia and hyperkeratosis with expansion of the suprabasilar layers of the epithelium and evidence of orderly maturation. A sharp line of demarcation is typically present between lesional and nonlesional tissue. In papillary lesions, underlying fibrous stroma may contain mild inflammation. Viral cytopathologic changes occur in the granular cell layer and include formation of large epithelial cells containing abundant gray-blue, granular cytoplasm, prominent, large basophilic, cytoplasmic keratohyalin granules, and koilocytes (large cells with perinuclear clearing). Inclusions are typically rare and infection can be confirmed by immunohistochemistry (IHC) (Fig. 10.23B), in situ hybridization (ISH), electron microscopy, or by PCR. PV-associated oral papillomas have also been reported in several captive and wild Asian lions and captive bobcats, Florida panthers, cheetahs, clouded leopard, and an Asian desert cat.[335,342,343]

In contrast to other nondomestic felids and similar to oral papillomas, oral **squamous cell carcinomas** (SCC) are more common in snow leopards than in other species.[207] Oral SCCs in snow leopards may be due to PV-induced hyperplasia and transformation but SCC are less commonly PV IHC positive than are benign papillomas. However, there can be an inverse relationship between the degree of cellular dysplasia and positive immunolabeling and the absence of immunolabeling should not be taken to infer the lack of PV involvement.[235] Tumors are commonly sublingual (similar to papilloma location) but can occur in other locations in the oral cavity. A relationship between those that appear to arise in tonsillar crypts or salivary gland and PV infection is unclear. The morphology is typical of SCC in other species.

Cutaneous plaques, **papillomas**, or squamous cell carcinomas (SCC) have also been occasionally reported in captive snow leopards; some have been associated with **PuPV-2** infection.[151,343] They can occur in any location but are more often described on the face and forelimbs and less commonly on the feet. They present as flat to slightly raised, sharply demarcated, focal areas of transepidermal and (occasionally) follicular hyperplasia with marked thickening of the stratum granulosum. Individual or small clusters of koilocytes in the upper stratum spinosum, and scattered cells with large, cytoplasmic, basophilic, keratohyalin granules in the stratum granulosum (consistent with viral cytopathology) are also seen. Some have been positive for PV by IHC. Research is needed to better understand the pathogenesis and predilection for transformation in snow leopards.

Sarcoids have been described in a group of captive African lions and a captive mountain lion but can occur

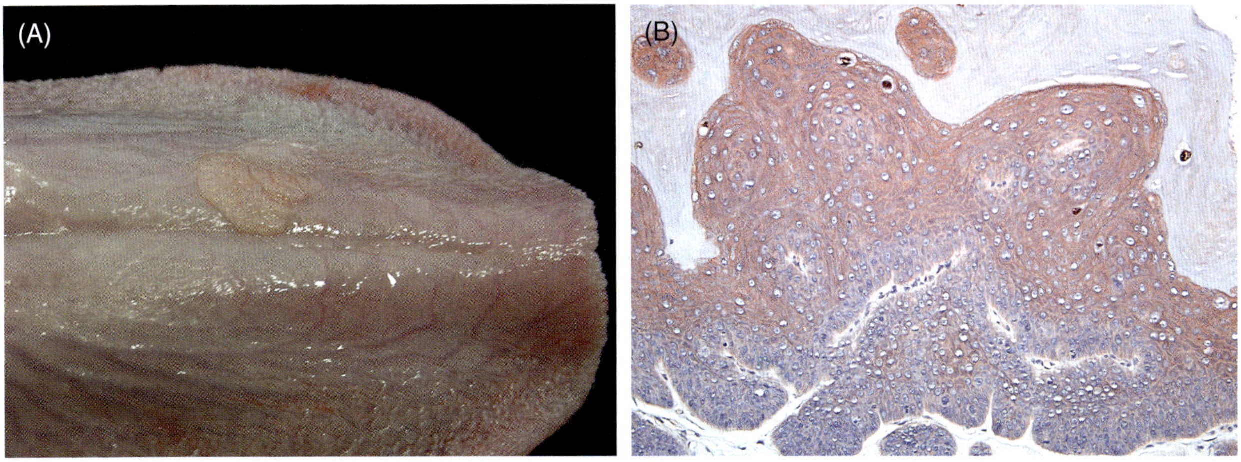

FIGURE 10.23 Oral papilloma on the tongue of a snow leopard. (A) Two nonpigmented, raised, flat plaques are present just lateral to midline on the ventral surface of the tongue. (B) Mucosal epithelial hyperplasia and hyperkeratosis with intense cytoplasmic staining within individual koilocytes. Papillomavirus immunohistochemistry. *(Part A: Photo Courtesy of the University of Illinois Zoological Pathology Program.)*

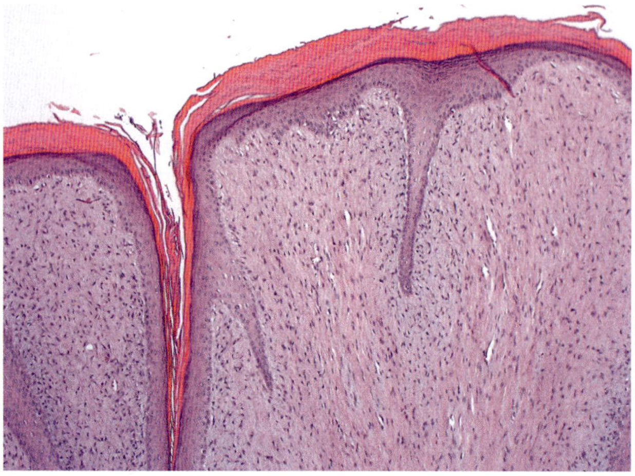

FIGURE 10.24 Sarcoid in a tiger. Moderate epidermal hyperplasia with narrow pointed, deep rete pegs and the dermis greatly expanded by streams and bundles of spindle-shaped to stellate cells. In felids, many sarcoids are PCR positive for feline sarcoid-associated papillomavirus.

in other species.[264,316] In the African lions, tumors develop as single or multiple perioral masses while in the mountain lion the lesions are between the lip and nasal planum. Tumors in both have gross and histologic features typical of sarcoids including densely packed spindle to stellate cells overlain by hyperkeratotic epidermis with long rete pegs; a close association between proliferative fibroblastic cells and the overlying epidermis is a characteristic feature (Fig. 10.24). Tissue from the affected lions and mountain lion has been PCR positive for PV with 96%–97% sequence similarity to feline sarcoid-associated papillomavirus (FeSarPV). Surgical resection was not associated with recurrence in the African lions but regrowth occurred a year after surgical resection in the mountain lion.

RNA Viruses

Canine distemper virus **(CDV)** has become a significant disease concern for captive and wild felids.[6,21,74,110,186,188,217,251,265,287,297,318,341,351,376] Interestingly, retrospective studies of archived tissues have identified CDV infection in captive lions and tigers prior to the first described outbreak, suggesting that CDV infection in felids is not a new phenomenon.[249] CDV lesions in felids are similar to those in other carnivores (see Canidae, Ursidae, and Ailuridae) including bronchointerstitial pneumonia (Fig. 10.25A) and nonsuppurative encephalitis (Fig. 10.25B) with slight variations in manifestation that may be due to viral strain, tropism, and pathogenicity as well as host immunity.[189] Coinfections may also be important in the outcome of viral infection, although they are not required for fatal infection.[244]

In wild felids, CDV infection has commonly been attributed to unvaccinated domestic dogs.[297] In the Serengeti ecosystem, serological data from outbreaks that occurred in the 1990s suggested spillover from dogs as the source of wildlife outbreaks. However, after the implementation of a vaccination program, exposure in lions became asynchronous from those in dogs, which highlighted the importance of other sympatric carnivores, such as hyena, as reservoirs or a source of infection.[376,385] Additionally, a more recent study suggests that a more complicated suite of noncanid-adapted CDV strains, rather than spillover of a single canid strain, may have been important drivers of these outbreaks.[258] Once infected, whether there is cat-to-cat spread is uncertain, though models predict that lions may be able to sustain an epidemic due to their social nature.[64]

Whether felids are naturally susceptible to CDV, similar to other carnivores, or accidental spillover hosts susceptible

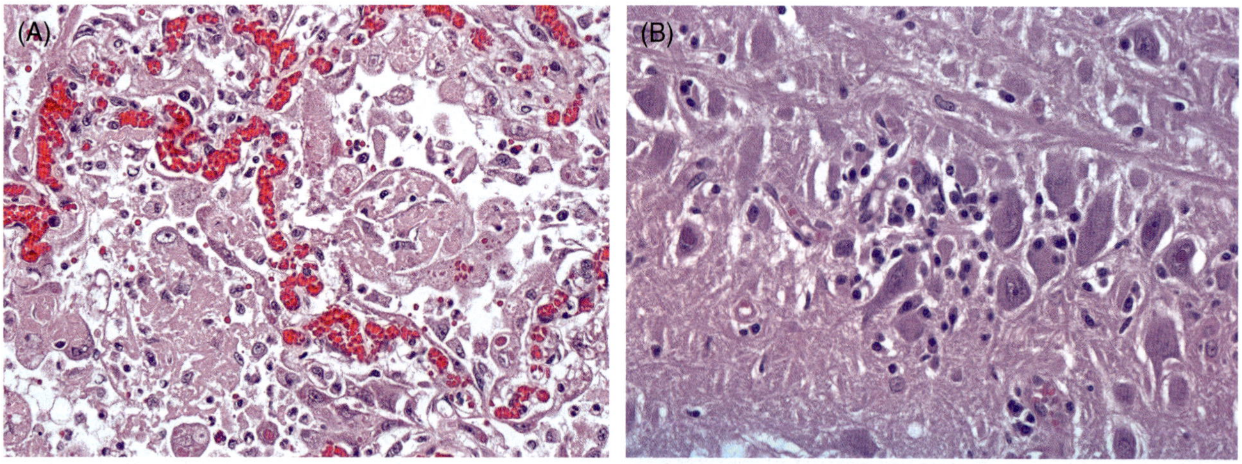

FIGURE 10.25 Canine distemper virus bronchointerstitial pneumonia and encephalitis in two tigers. (A) Alveoli are lined by plump Type II pneumocytes and syncytial cells with characteristic intracytoplasmic and intranuclear inclusion bodies. Alveolar lumens contain numerous foamy macrophages and sloughed epithelial cells. (B) Multiple neurons and astrocytes contain eosinophilic, intranuclear inclusion bodies with mild nonsuppurative, lymphoplasmacytic inflammation.

to only some viral strains is an area of active research. All wild felid populations tested to date have some degree of seropositivity, suggesting that exposure and infection are widespread.[243] Analysis of multiple sites on the CDV-H gene which encodes the hemagglutinin protein important for binding to host receptors have found no consistent amino acid sequence to identify those strains that infect felids.[353] Additional research is needed to better understand CDV infection dynamics in felids as well as the ecology of disease spread in multihost ecosystems.

Feline enteric coronavirus **(FCoV)** infection can cause mild enteritis as well as fatal **feline infectious peritonitis (FIP)** in nondomestic felids.[339] Infections in the latter include cases in a serval, bobcat, puma, cheetahs, European wildcats, tiger, and mountain lion. Lesions and clinical signs of FIP in nondomestic felids are similar to those described in domestic cats and include high protein effusions within the abdominal and or thoracic cavities and/or pyogranulomatous inflammation centered on vessels (Fig. 10.26); bilateral panuveitis has also been described.[248] In 1983 an epidemic of FIP in a captive cheetah population caused >60% mortality.[93,133] It was hypothesized that lack of heterogeneity at MHC loci encoding peptides that mediate immune responses to pathogens was an important factor in the high mortality rate.[261] However, subsequent studies and comparison with data from wild cheetah populations that also lack MHC heterogeneity suggest that extrinsic environmental factors and possibly stress may also be significant in disease pathogenesis.[245,354] IHC is considered one of the best diagnostic options in establishing a diagnosis of FIP since PCR and serologic tests do not discriminate between forms of the virus and because antibody titers can be unpredictable.[339]

Nondomestic felids in managed care were among the first animals diagnosed with **severe acute respiratory syndrome coronavirus 2 (SARS-CoV-2)** infection.[3,14,70,98,159,171,210,229,359,382] In most cases, infection appears to be from humans, as variants are most closely related to local human strains. However, cat-to-cat transmission has been confirmed experimentally in domestic cats and suspected in some cohoused nondomestic felids.[14,105,117,171,226,327] Lion-to-human transmission has also been suspected.[327] Why nondomestic, but not domestic felids which have higher contact rates with people, have been disproportionately affected is an area of ongoing research. Shedding in nondomestic felids, assessed through RT-PCR of viral RNA, in both nasal swabs and in feces can be prolonged (>4 weeks). To date, surveys of free-ranging puma and bobcat have failed to find evidence of infection.[78,300]

As in humans, clinical disease severity has differed among individual felids infected with the same viral variant and disease severity has differed among viral variants. Multiple snow leopards and a lion infected with the Delta variant succumbed to infection after a rapid and severe disease course.[81] Clinical cases have been reported in tigers, lions, pumas, snow leopards, lynx, and fishing cats.[394] Clinical signs vary from mild upper respiratory disease to severe pneumonia, respiratory distress, and death. Some cats that survived initial viral insult succumbed to secondary infections, predominately fungal. Gross lesions include congestion, edema, and ulceration of the mucosal epithelium of the nasal cavities, trachea, and mainstem bronchi. In acute cases, pulmonary edema was marked and interlobular septae were edematous and grossly evident (Fig. 10.27A). Histologically and cytologically, direct viral damage to the respiratory tract including nasal, tracheal, bronchial, and alveolar epithelium have been observed and associated with viral proteins and RNA by IHC and ISH, respectively (Fig. 10.27B).[81,210] Airway lumens contain abundant cellular debris, neutrophils, foamy macrophages, hemorrhage, fibrin, and varying numbers of mixed bacteria. Within alveoli there is abundant fibrin exudation and interstitial edema is prominent (Fig. 10.27C; eSlide 10.e5).[81] Areas of necrosis and abscessation due to secondary fungal infections were present in more chronic cases. Animals that survived initial infection but succumbed, after recovery, to another disease process, have had regionally extensive areas of interstitial fibrosis and bronchiolitis obliterans (M. Delaney, personal communication).

Feline calicivirus is a common, usually self-limiting, infection in free-ranging and captive felids. Characteristic lesions are vesicles and ulcers primarily on the tongue, oral, and nasal mucosa; they are less common on the foot pads.[154] Infection is highly contagious and can thus spread quickly. Clinical infection has occurred after vaccination with modified-live vaccines in nondomestic felids.[292] Nondomestic felids are also susceptible to systemic calicivirus infections, which can lead to sloughing of glossal epithelium and death.[129] Dual-strain killed vaccines may be protective.[127]

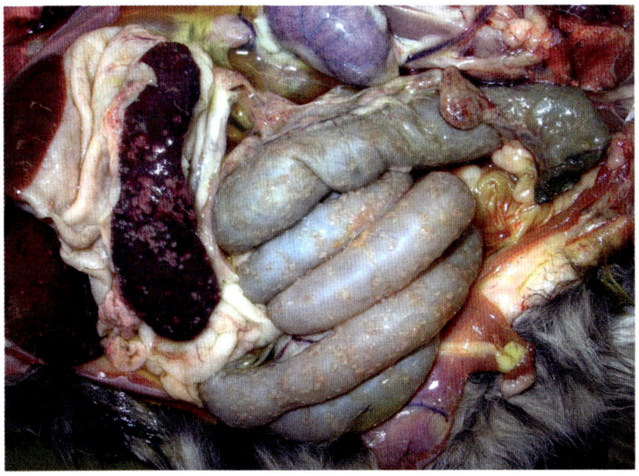

FIGURE 10.26 Feline infectious peritonitis in an African black-footed cat. Large numbers of small pale tan granulomas cover the serosal surfaces of the spleen and intestine and a mild fibrinous ascites is present.

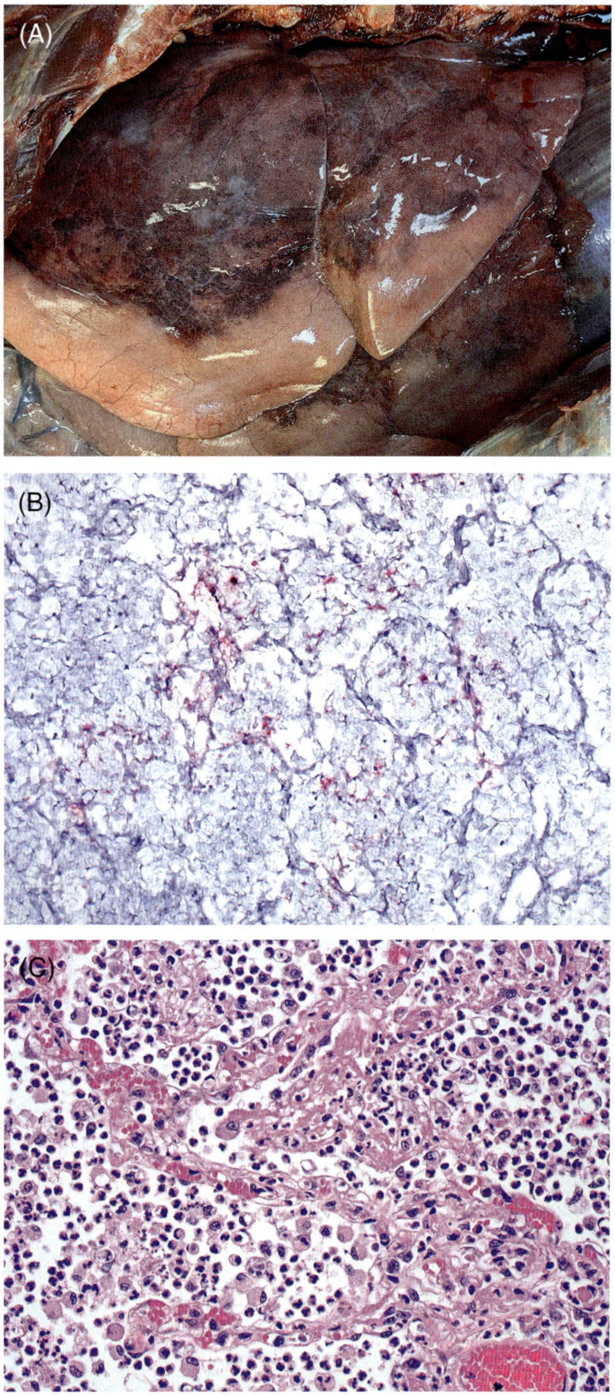

FIGURE 10.27 SARS-CoV-2 infection in a lion (A, C) and snow leopard (B). (A) Regional consolidation and edema of the pulmonary parenchyma. Interlobular septae are prominent and widened due to edema. (B) In situ hybridization using a probe against the SARS-CoV-2 neuraminidase gene with signal along alveolar septa and within macrophages and cell debris in the alveoli. (C) Alveolar septae are widened by edema and fibrin and are partially denuded of pneumocytes. Rare remnant pneumocytes are plump (reactive). Alveoli contain clumped fibrin, macrophages, cell debris, and neutrophils (see eSlide10.e5). *(Part A: Photo Courtesy of J. Odani, University of Hawai'i.)*

Avian influenza caused by an **influenza type A virus** of the Orthomyxoviridae family is a known pathogen of domestic and nondomestic felids. Multiple experimental studies in domestic cats have shown that cats exposed to various influenza A viruses can become infected and develop disease, and that viral spread between cats can occur.[135,175,291] Virus is shed through aerosolized respiratory secretions and feces. In 2003–2004 captive clouded leopards, leopard cats, and tigers died of **highly pathogenic avian influenza (HPAI) H5N1** after being fed infected chicken and quail.[161] In one zoo, there was evidence of tiger-to-tiger viral spread.[361] Additional cases of H5N1 in captive tigers have occurred in subsequent outbreaks[131,139]; H5N1 infection and death has also been reported in a lion.[53] In 2022 H5N1 influenza A virus clade 2.3.4.4b was implicated in the deaths of free-ranging nondomestic felids including bobcat and puma.[88,394] Source of infection was presumed to be wild birds. Clinical signs in affected felids vary with viral strain but include high fever, respiratory distress, and neurological signs including loss of fear of humans. At necropsy there is pulmonary consolidation and hemorrhage with tracheal and bronchial exudates. Histologically, cats can have pulmonary hemorrhage, edema, and necrosis of alveolar walls as well as meningoencephalitis, myocarditis, and necrotizing hepatitis (Fig. 10.28). In addition to H5N1 viruses, nonfatal H1N1 viruses have been recovered from cheetahs with clinical respiratory signs during multiple unrelated outbreaks.[65,373] In one of these outbreaks, evidence suggests that cheetahs were infected by caregivers.[373]

Feline immunodeficiency virus **(FIV)** is prevalent worldwide and many species of felids, both free-ranging and captive, have serological evidence of exposure.[243] Of the nondomestic felids, FIV infection has been most extensively studied in African lions. In contrast to domestic cats, there is controversy as to whether disease is associated with FIV infections in nondomestic felids. Only a single case of FIV-associated lymphoma has been described.[283] Alterations in lymphocyte subsets with depletion of CD4+ T cells occur in both captive and free-ranging lions and pumas infected with FIV.[40,293,295] In wild lions, different FIV(Ple) subtypes are associated with different mortality rates.[367] There is conflicting evidence of differential susceptibility to other infectious agents in FIV infected felids.[29,193,274,330,347] In addition to lions, mountain lions (pumas), bobcats, Pallas' cats, leopards, and cheetahs can be infected with their own lentiviruses or those of other felids.[12,31,36,44,45,180,366]

Feline leukemia virus **(FeLV)** is not endemic in most wild felid populations[243]; the exception is free-ranging wildcats where infection in enzootic and seroprevalences can reach 81%.[82,132,184,221] This may be due to increased contact between wildcats and feral domestic cats. Similarly, FeLV exposure has been linked with fragmented landscapes

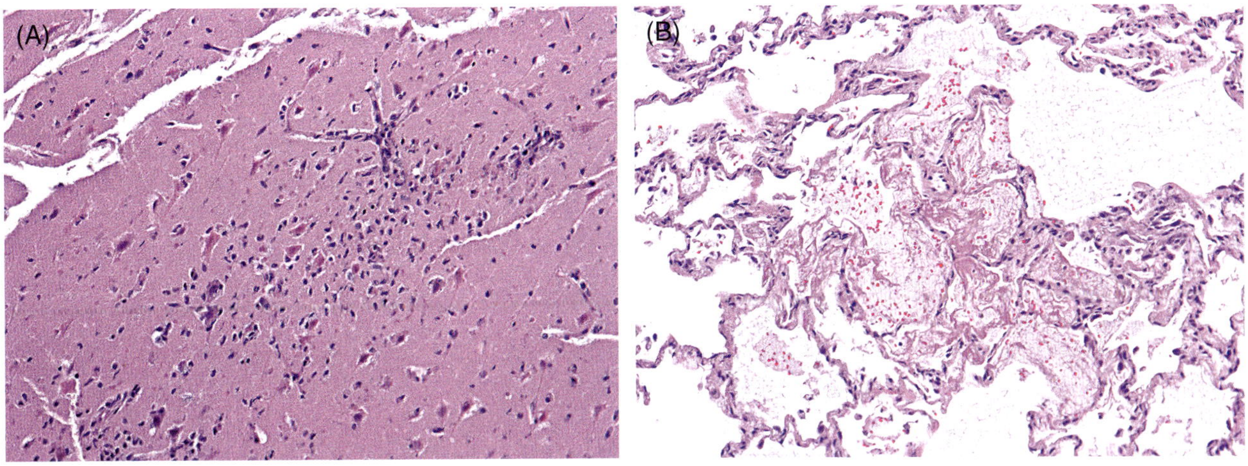

FIGURE 10.28 H5N1 influenza A in a free-ranging bobcat. (A) Meningoencephalitis associated with influenza infection with neuronal necrosis, gliosis, and perivascular inflammation. Fracturing of parenchyma in the section is due to freeze–thaw artifact. (B) Acute necrotizing interstitial pneumonia with fibrin, edema, and hemorrhage within alveolar spaces. *(Parts A, B: Photos Courtesy of E. Elsmo, Wisconsin Veterinary Diagnostic Laboratory.)*

in populations of guiña.[303] While many cases have been linked to spillover from domestic cats, there is evidence for cat-to-cat transmission among puma.[277] FeLV has been associated with clinical disease and mortality in free-ranging Iberian lynx, Florida panthers, and a mountain lion.[67,107,111,144,215,216] and has been associated with lymphoma in a single captive cheetah.[202] Viremia has been linked to anemia, lymphopenia, and lymphadenopathy, and death in some cats with associated sepsis. In the Iberian lynx and Florida panther, viral sequence analysis indicates similarity to strains previously described in domestic cats, suggesting possible cross-species transmission.[35,215,277,303]

***Bluetongue* (BT) virus serotype 8** has been associated with disease in captive Eurasian lynx-fed stillborn or aborted fetuses from farms with confirmed BT infection.[143] Anemia, subcutaneous hematomas, petechial hemorrhages, pulmonary congestion, edema, and death were noted macroscopically, and histologic lesions included vasculitis, edema, and reactive endothelial cells. Free-ranging felids have evidence of infection with bluetongue virus but disease has not been reported.[2] Transmission is presumed from infected prey; transmission by biting midges, as occurs in ruminants, has not been ruled out.

Bacteria

Both captive and free-ranging felids are susceptible to **tuberculosis** due to mycobacterial infections, mainly by ***Mycobacterium bovis* (bTB)**. Disease has been described in free-ranging African lions, leopards, cheetahs, Iberian lynx, and bobcats.[7,28,37,56,162,168,205,219,222,276,311] Wild African lions, leopards and cheetahs are affected with the same spoligotype and variable number tandem repeat (VNTR) genotypes isolated from sympatric buffalo, while free-ranging lynx in Spain are infected with spoligotypes identical to local cattle, wild boar, and fallow deer. These findings suggest that cats are infected from prey. Free-ranging cheetahs, leopards, and lynx are considered spillover hosts. Lions may be either maintenance or spillover hosts.[206] The long-term effects of bTB on free-ranging lion populations are uncertain.[56,99,378] Seroconversion of necropsy staff and keepers has been reported, emphasizing the importance of appropriate personal protective equipment.[134,231,383]

Clinical signs, antemortem tests, and clinical pathology in felids have been reviewed by several groups.[75,194,222,378] The pathogenesis of mycobacterial infections is determined by multiple host factors, including host immunity, coinfections, genetics, population density, contact rates, longevity, and behavior (including gregariousness, denning, scavenging, and grooming behavior); pathogen virulence and genetics; and ecological factors including feral cat, host, and prey geographic distribution.[55] The evidence for the role of FIV in the pathogenesis of bTB in free-ranging lions or lynx is contradictory.[195,274,330,347] Increased susceptibility to bTB or recrudescence of latent infections is related to in-breeding in free-ranging lions,[365] and recent episodes of stress or concurrent disease in captive felids.[205,231]

Lesions are similar among different felid species and disease is often advanced. External findings include emaciation, decubital ulcers, cutaneous alopecia, corneal opacity, poorly healed bite wounds, and elbow and hock hygromas (Fig. 10.29). Internally, there can be localized or systemic granulomatous inflammation within the lung (Fig. 10.30A) and lymph nodes or other sites such as the spleen, kidney, eye, or adrenal glands. Pulmonary cavitations (<4 cm diameter) caused by bronchiectasis with luminal mucus are most common in the diaphragmatic lung lobes. These lesions can rupture and result in pneumothorax

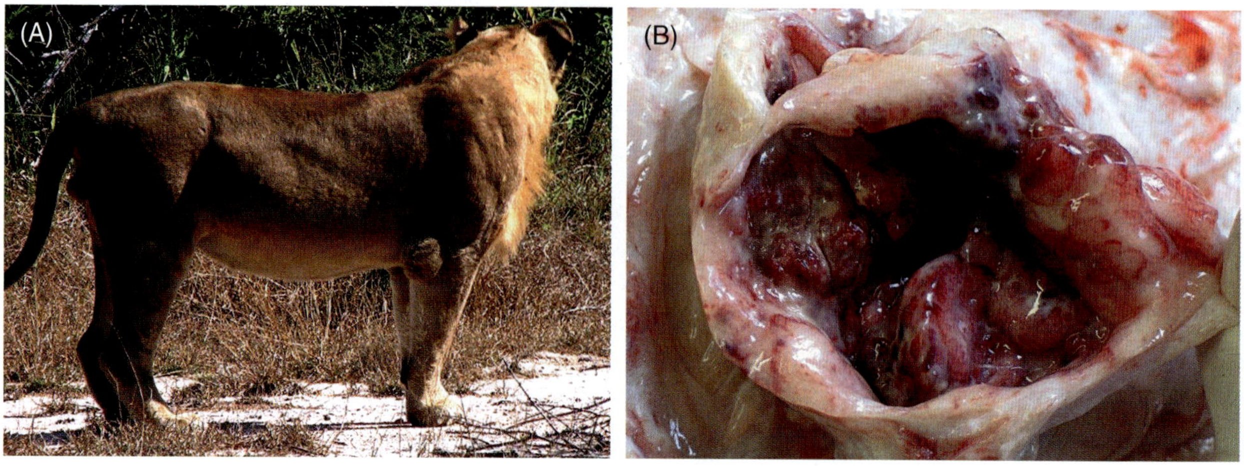

FIGURE 10.29 Mycobacterial hygroma in a lion. (A) Infection with *Mycobacterium bovis* can lead to distended, sac-like structures (hygromas), over the elbow and hock joints. (B) The opened joint capsule is markedly thickened and there is a chronic, hemorrhagic synovitis within the elbow joint. *(Parts A, B: Photos Courtesy of R. Bengis, Department of Agriculture, Forestry and Fisheries, South Africa.)*

as well as hydrothorax and pleural adhesions. Arthritis and proliferative osteitis are other common findings. Histologically, granulomas, multinucleate giant cells, necrosis, and mineralization are rare. Instead there are expansile, nonencapsulated aggregates of macrophages, epithelioid cells, lymphocytes, plasma cells, and neutrophils along bronchioles and in alveolar walls (Fig. 10.30B). While acid-fast bacteria may be numerous in airways, lung, and joint exudates, bacteria are rare in pulmonary macrophages (Fig. 10.30C). Lymphoid atrophy is common. Granulomatous lymphadenitis, panophthalmitis, choroiditis, uveitis, conjunctivitis (with or without retinal detachment), pleuritis, mural enteritis, and renal medullary amyloidosis may be present, and osteomyelitis, myositis, and periostitis may be accompanied by secondary hypertrophic osteoarthropathy. Granulomatous arthritis may be associated with joint capsule mineralization and ossified fragments in the joint space.[162,168]

Cases of bTB have also been described in captive cheetahs, leopards, lions, lynx, pumas, snow leopards, and tigers.[92,148,172] Clinical diagnostic and pathological features are similar to those described in free-ranging felids. Disease due to *M. avium* has been described in a captive Bengal tiger[54] and *M. caprae* has been isolated from a captive adult Amur tiger.[179] Nontuberculous mycobacteria (NTM) have also been isolated from the oropharynx of clinically healthy captive pumas in Argentina.[364]

Typical lesions and culture results should rule out other bacterial infections unless prolonged antibacterial therapy has been administered prior to death. Mycobacterial isolation from affected tissues, the gold standard diagnostic test, is slow and insensitive. Acid-fast stains, immunohistochemistry, and molecular tests may provide valuable interim results while waiting for culture. Speciation, typing (spoligotyping and VNTR), PCR, and DNA sequencing on fresh or paraffin-embedded tissues may be valuable for determining the source of infection, ruling out NTM infection, and assessing the risk to personnel.[63,262] A commercial cat interferon-gamma release assay using mycobacterial antigen-stimulated blood has shown promise for antemortem diagnosis of *M. bovis*-infected lions and cheetahs.[119–121]

Carnivores worldwide are susceptible to **anthrax**, caused by the gram-positive endospore-forming multihost bacterium ***Bacillus anthracis***. Infection is typically through ingestion of infected prey. Confirmed and suspected cases have been described in captive and free-ranging Amur leopard, bobcat, caracal, cheetah, clouded leopard, golden cat, jungle cat, leopard, leopard cat, lion, lynx, ocelot, panther, puma, serval, and tiger.[285,370,395] Serological titers in free-ranging lions correlate with geographic areas of anthrax incidence and periods of higher rainfall. Regular exposure may be protective since high mortality occurs in lions after periods of low anthrax incidence whereas nonscavenging felids, such as cheetahs, are highly susceptible to fatal disease.[113,140,181,233,285,371]

Infection occurs through skin defects, ingestion, or inhalation of spores that gain entry through oropharyngeal or intestinal tract epithelia, or pulmonary alveolar membranes. Spores in macrophages or mast cells are transported to the major draining lymph nodes of the neck or intestinal tract where they germinate, replicate, and produce toxins that then spread systemically causing septicemia. The toxin complex consists of two separate protein components (edema and lethal toxins) and their shared cell receptor binding protein (protective antigen). Severe edema at the site of entry and draining lymph nodes are caused by the edema toxin and are modulated by macrophage production of cytokines, including interleukin-1 (IL-1), tumor necrosis factor-alpha (TNF-α), IL-6, and IL-8; while lethal toxin

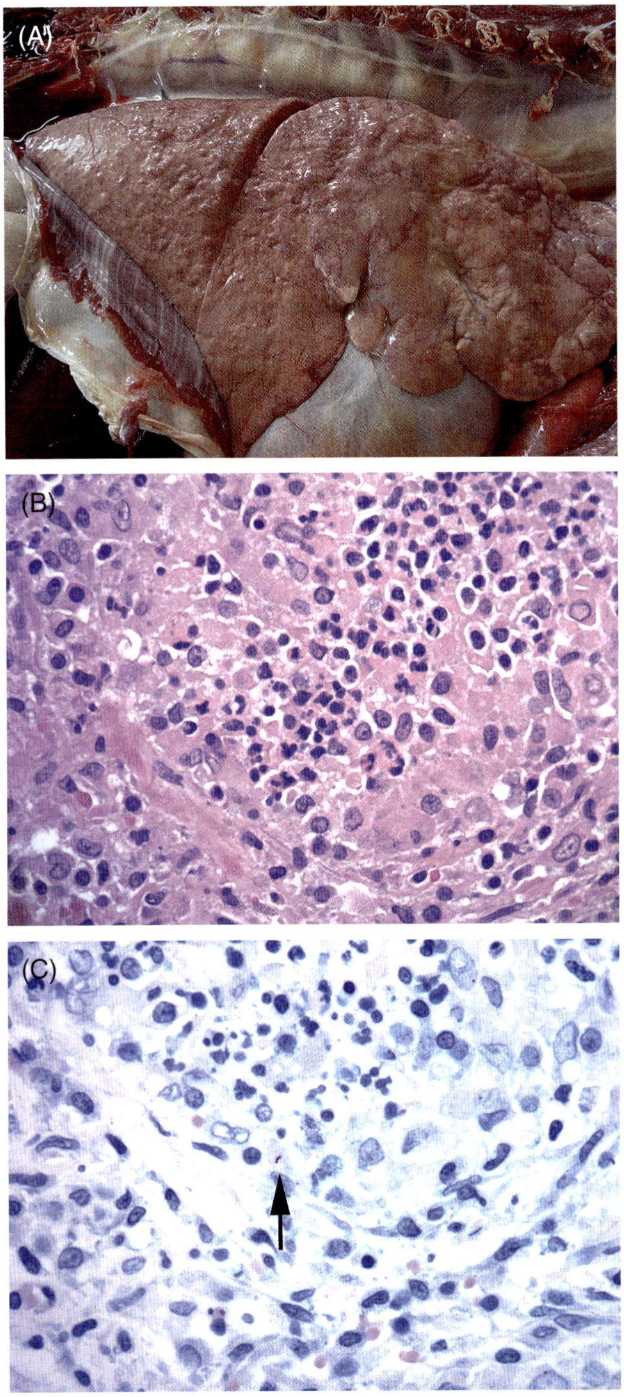

FIGURE 10.30 *Mycobacterium bovis* pneumonia in an African lion. (A) The lungs contain multiple collapsed areas and bronchiectasis as well as more diffuse areas of granulomatous inflammation. (B) Characteristic granulomatous inflammation with aggregates of epithelioid macrophages, fewer lymphocytes, and plasma cells. (C) Rare, intracellular, acid-fast bacilli (*arrow*) are present within lesions, typical of *Mycobacterium tuberculosis* complex infections. Ziehl-Neelsen. *(Part A: Photo Courtesy of Lin-Marie de Klerk-Lorist, South African Department of Agriculture, Forestry, and Fisheries.)*

inactivates mitogen-activated protein kinases and inhibits IL-1β expression with resultant cell necrosis, reduced phagocytosis and altered blood clotting. Death is due to edema, shock, acute renal failure, and central nervous system-mediated anoxia.[123,136]

Disease in felids is usually mild, transient, and localized with head and neck subcutaneous edema (Fig. 10.31), lethargy, and anorexia; acute death may occur. Marked lip, gingival, submandibular and lingual edema, serosanguinous ascites, marked edema of the gastric and/or intestinal wall, pancreatic congestion and edema, multifocal acute agonal subepicardial and endocardial hemorrhages, and rapid putrefaction of the spleen, liver, kidney, and parotid salivary glands are described in fatal cases. Histologically, abundant fibrin-rich edema fluid containing large numbers of neutrophils and variable numbers of spore-forming bacilli expand the affected subcutaneous adipose tissue and gastric wall. Fibrinoid vasculitis may be present in the skin and renal blood vessels; acute necrotizing and hemorrhagic lymphadenitis and necrotizing enteritis are also seen. Asphyxiation and aspiration of gastric contents may occur.

Differentiation from cytotoxic snake bites is necessary. To reduce environmental contamination, carcasses of suspected cases should not be opened to limit exposure, including mechanical transfer of spores by arthropods. Carcasses should be disposed of by incineration, rendering, or burial. Cytological preparations of tissue smears, fluids aspirated from edematous lesions, or tissue lesions stained with Giemsa stains show variable numbers of basophilic blunt-ended, bacterial rods with a well-developed capsule that may form short chains (see Wildlife Necropsy). Bacterial morphology may be altered after death. Organisms in affected tissues may not stain with Gram stains

FIGURE 10.31 Anthrax cellulitis and edema in an African lion. There is marked swelling of the head and neck due to edema and cellulitis. *(Photo Courtesy of R. Bengis, Department of Agriculture, Forestry and Fisheries, South Africa.)*

if antibiotics have been administered before death, but silver stains usually remain positive. Bacteria may be cultured from unstained blood, tissue, or exudate smears. Molecular tests for anthrax are largely used for human epidemiological studies and research. As anthrax is a WOAH-listed, reportable zoonotic disease, appropriate protective measures should be taken.

Helicobacter sp. are spiral bacteria that commonly colonize the stomachs of free-ranging and captive felids.[166,167] *Helicobacter* sp. are generally an incidental finding with only rare individual cases of gastritis reported in most felid species.[115,156,315] The exception is cheetahs in which, historically, the majority (>95%) of captive cheetahs have associated lymphoplasmacytic gastritis.[85,86,236] Affected cheetahs lose weight, may exhibit chronic vomiting and regurgitation, and may pass undigested food in the feces. Grossly, stomachs are thickened with a reddened cobblestone appearance; some cases have loss of rugae and atrophy (Fig. 10.32A). Histologically, there are large numbers of lymphocytes and plasma cells within the lamina propria as well as intraepithelial (CD3+ CD8α+) T cells (Fig. 10.32B; eSlide 10.e6).[356] Small to large numbers of globular leukocytes may also be present within gland epithelium. Severe cases are characterized by large numbers (>70%) of lamina proprial activated B (CD79a+ CD21–) and plasma cells.[356] Some lymphoid aggregates along the base of the lamina propria have follicular differentiation. Parietal cell necrosis, intestinal metaplasia, and or mucosal atrophy are present in some cases. Cheetahs with gastritis can develop secondary amyloidosis.[271] Wild cheetahs are also infected with *Helicobacter* sp. and the types of *Helicobacter* are similar to those in captive cheetahs suggesting that bacteria alone do not cause disease and the importance of host factors in disease development.[355] With improvements in management, prevalence in North American captive cheetah populations has been declining (Terio, unpublished).

Salmonella sp. can cause enteritis in felids but culture of *Salmonella* sp. from feces should be interpreted with caution as the bacteria can be shed from asymptomatic cats.[57,374] *Salmonella*, primarily *S. typhimurium*, was isolated from >90% of samples from a variety of felids, including snow leopards, leopards, cougars, caracals, and servals. The source of *Salmonella* has been presumed to be contaminated feed as *Salmonella* have been isolated from dietary meat.

Leptospirosis can occur in free-ranging felids and those in managed care.[220,340] Leptospira infection has been best studied in free-ranging mountain lions and bobcats. In these cases, the most common serovar is *Leptospira interrogans* serovar Pomona. Infected cats can develop lymphocytic interstitial nephritis similar to that described in other species.

Plague, due to infection with ***Yersinia pestis***, occurs in felids primarily due to consumption of infected rodents. Disease is uncommon in captive settings as food sources are managed, but it can occur, especially with "opportunistic" predation on wild rodents. Lesions include necrotizing and suppurative pneumonia, splenitis, hepatitis, and lymphadenitis with intralesional gram-negative coccoid bacilli. Disease has also occurred in free-ranging Canada lynx where lesions were predominately in the lungs.[389]

Domestic and nondomestic felids are also susceptible to infection with ***Yersinia pseudotuberculosis,*** which has been described in several species, including a cougar, Amur tiger, lions, and cheetah.[18,268,281,349,392] Rodents and birds are considered reservoirs for the bacterium, and transmission to felids is generally through ingestion of contaminated prey. Acute disease is associated with enterocolitis, sepsis, and death with no premonitory signs or within a few days of the development of clinical signs. Lethargy, anorexia, diarrhea, respiratory distress, and ataxia have been described in the subacute to chronic form of the

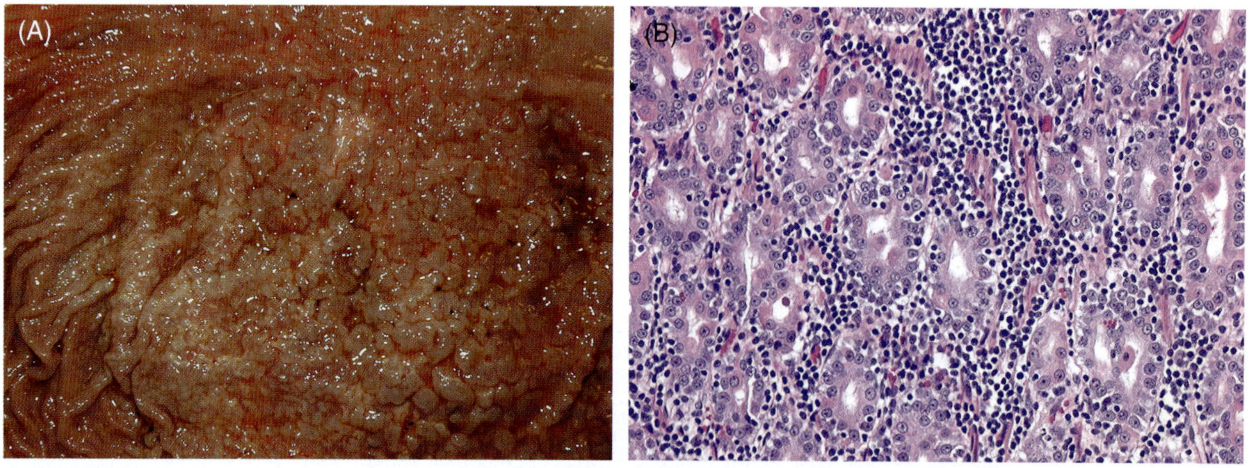

FIGURE 10.32 *Helicobacter*-associated gastritis in a captive cheetah. (A) The gastric mucosa is markedly thickened and reddened with a cobblestone appearance. (B) Large numbers of plasma cells and lymphocytes are present within the lamina propria and within the gland epithelium. Individual epithelial cells are necrotic and sloughed into gland lumens (see eSlide10.e6).

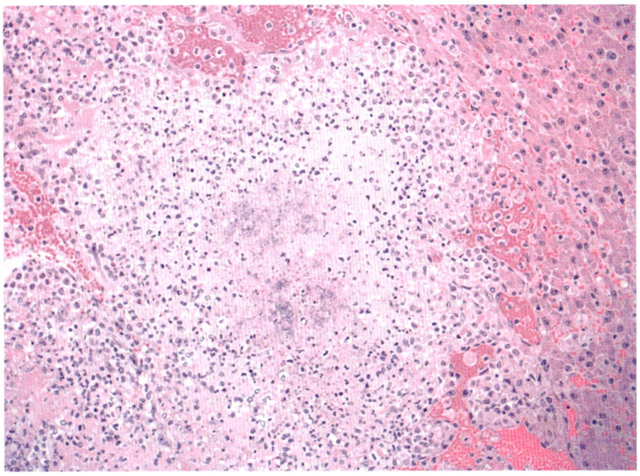

FIGURE 10.33 *Yersinia pseudotuberculosis* hepatitis in a cheetah. Necrosis and pyogranulomatous inflammation within the liver with intralesional bacteria. Similar lesions are also common in the spleen (see eSlide 10.e7).

disease. Infection in nondomestic felids has generally been in the liver and spleen; infection in the lung, kidney, and lymph nodes has also been reported. Lesions are typically variably sized, demarcated, tan nodules in the parenchyma of one or more tissues. This corresponds histologically to pyogranulomatous inflammation with necrosis and intralesional gram-negative coccoid bacilli (Fig. 10.33; eSlide 10.e7). Associated lesions can include local or systemic thrombi, and cavitary effusions, serosal/pleural inflammation, and sepsis. Differentials include tuberculosis and other bacterial infections such as *Y. pestis* and *Francisella tularensis*, all of which are zoonotic, or neoplasia. Gram and Giemsa stains highlight the bacteria; bacterial culture or molecular testing is confirmatory.

Hemoplasmas (hemotropic *Mycoplasma*) have been reported in multiple wild felids. Their contribution to disease outcome is unclear since, in some cases, association between infection and disease has not been found while in other cases they have been identified in mixed infections.[107,244]

Fungi

Cryptococcosis, caused by the dimorphic fungus ***Cryptococcus gattii,*** has been described in captive and free-ranging cheetahs.[142,225] Inhalation of infectious basidiospores may be common in cheetahs because of their habit of using play trees and aerosolizing the organism through slapping the ground with their forelimbs when threatened. The outcome of infection depends on host immune status, fungal virulence factors including capsule thickness, and the capacity to combat host oxidative processes.[183] While the predisposition of the king cheetah coat variant to clinical disease suggests an inherited susceptibility in this subtype,[19,22] neither the limited genetic variation of the cheetah nor underlying viral infections are predisposing.[224,225] The role of chronic stress in captive cheetahs in the susceptibility to this disease is uncertain.[354]

Ulcerative skin nodules and discrete to coalescing white, light gray to tan, soft to gelatinous space occupying foci in the nose (Fig. 10.34A) and underlying bone or oral mucosa, lungs, pleura, meninges, brain, spinal cord, kidney, optic nerve, and retrobulbar adipose tissue as well as draining lymph nodes may be present. Histologically, uninucleate thin-walled spherical to oval yeast (2–20 μm diameter) with single, narrow-based budding and a wide clear capsule are pathognomonic. Inflammation is variable and may include macrophages (with phagocytosed organisms), multinucleate giant cells, lymphocytes, and plasma cells. Lesions are often present on the ventral brain surface and are associated with the meninges, ventricles, and spinal cord central canal (Fig. 10.34B). Cytology of lesions or centrifuged cerebrospinal fluid using

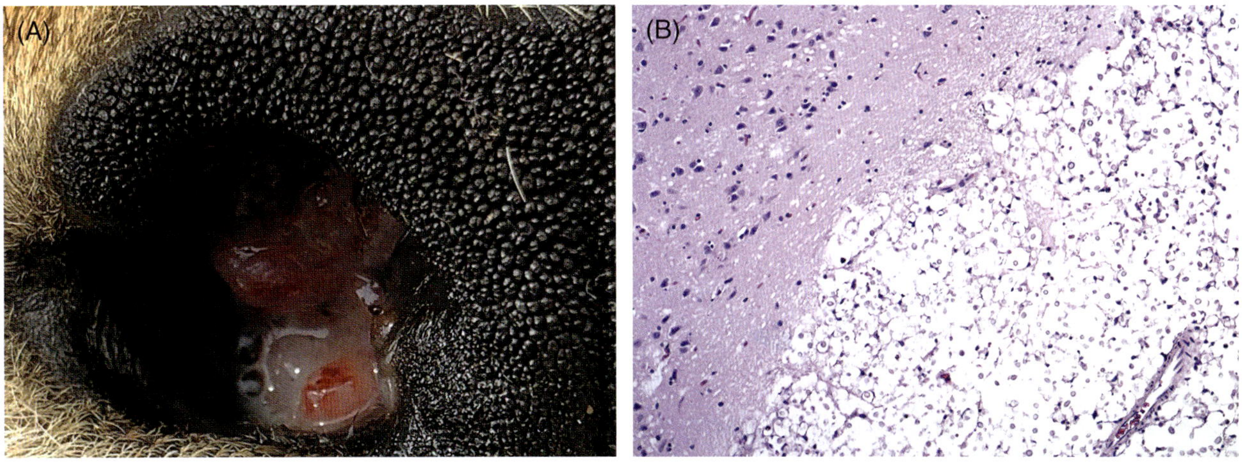

FIGURE 10.34 Cryptococcosis in a cheetah. (A) Small white to red glistening cryptococcal granulomas within the nostril with a scant mucoid nasal exudate. (B) Sheets of round to oval fungi with thick clear capsules efface the cerebral neuropil. Minimal inflammation is present.

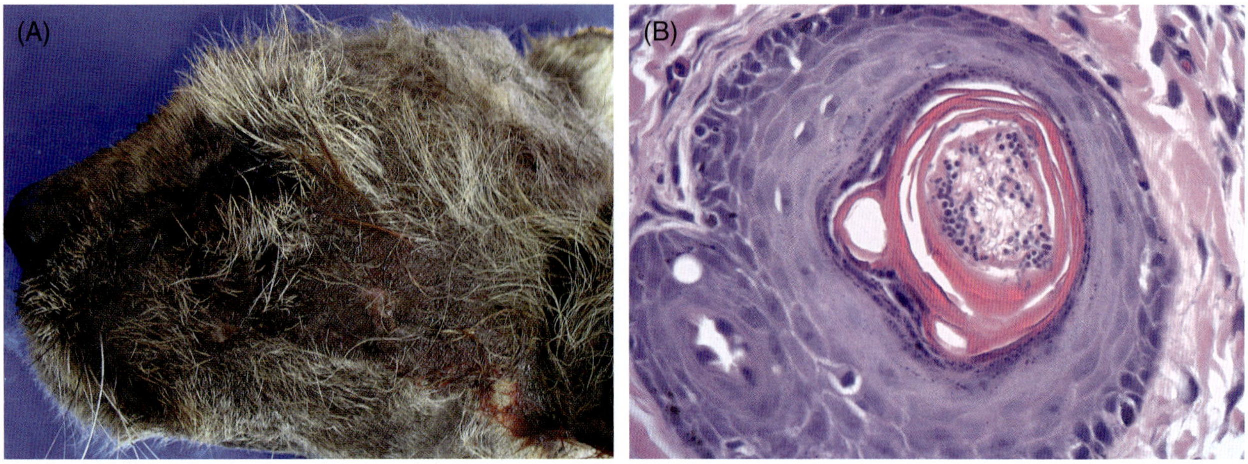

FIGURE 10.35 Dermatophytosis in a cheetah. (A) The facial skin is thickened due to hyperkeratosis and marked alopecia is present. (B) Arthrospores invade the hair shaft and follicular keratin.

Romanowsky-type stains is diagnostic, as is histology, since *Cryptococcus* is distinctive and is the only fungus with a mucinous capsule that stains with Alcian blue and Mayer's mucicarmine. Genotyping by various molecular methods has replaced serotyping.[183]

Dermatophytosis due to ***Trichophyton mentagrophytes or Microsporum*** spp. can occur in any felid, captive or wild, and causes alopecia, orthokeratotic, hyperkeratosis, and pyoderma (Fig. 10.35A).[299] The 2–3 µm, basophilic arthrospores will be present in affected hairs and hair follicles (Fig. 10.35B). Infection may be associated with epidermal hyperplasia and mixed dermal inflammation. Cytology of plucked hairs, histology, special stains (GMS, PAS), or culture using specialized media are used diagnostically. Topical treatment may be effective.[346] In cheetahs, topical antifungal treatments are recommended as griseofulvin has been associated with lethargy, diarrhea, and bone marrow suppression.[379]

A few case reports describe infection with the dimorphic fungus ***Histoplasma capsulatum*** in leopards, tigers, and snow leopards.[49,91] Clinical signs in affected animals have been nonspecific. Gross findings include lymphadenomegaly and multifocal to coalescing nodules in the lung, liver, or spleen; many of the reported cases have been systemic. Associated inflammation is generally histiocytic to granulomatous or pyogranulomatous. Numerous macrophages contain few to many, 2–4 µm diameter yeast with a 1–2 µm basophilic nucleus, that show narrow-based budding and are highlighted with GMS or PAS staining; IHC and PCR can be diagnostic.

Metazoa

Felids can be infected with a wide variety of parasites, some of which are associated with disease and others that are asymptomatic in immune-competent hosts. Additional selected metazoan and protozoan are described below and presented in the Digital Content (Table 10.e2).

Ollulanus tricuspis is a gastric trichostrongylid that occurs worldwide with cases recorded in nondomestic felids including captive cheetahs, lions, and tigers[26,52,61,130] and free-ranging wild cats, pumas, and tigers.[27,198] Infected free-ranging pumas have been seropositive for FeLV, which may have predisposed to infection.[144,290] Clinical signs include vomiting, inappetence, intermittent bloody or mucoid diarrhea, and ill-thrift. Grossly, cats have an uneven gastric mucosal surface, small raised pale pink plaques, hemorrhage, and/or areas of ulceration. Histologically, abundant globule leukocytes and multifocal lymphoplasmacytic gastritis with fibrosis and glandular atrophy, as well as moderately hyperplastic basilar mucosal-associated and submucosal lymphoid tissue are reported. Mucus cell hyperplasia is noted in some cases. Nematode sections may be rare but occur in the surface mucus and gastric pits (Fig. 10.36). Since the life cycle is completed in the stomach, infections may be overlooked due to the small size of the nematode (c. 1 mm long), virtual absence of adults, larvae or eggs in feces, and often small numbers of parasites in the mucosa. In cheetahs, histological differentiation from *Helicobacter*-associated gastritis is necessary.

Protozoa

Toxoplasma gondii infects captive and wild felids but is only rarely associated with disease. In contrast to most felids, captive Pallas' cats are uniquely susceptible to infection with *Toxoplasma*, which results in high neonatal mortality.[163] Common lesions in neonatal Pallas' cats include necrotizing encephalitis, pneumonia, and/or hepatitis (Fig. 10.37; eSlide 10.e8). Necrotizing to granulomatous inflammation may also occur in other tissues such as the spleen, kidney, and adipose. In aged Pallas' cats,

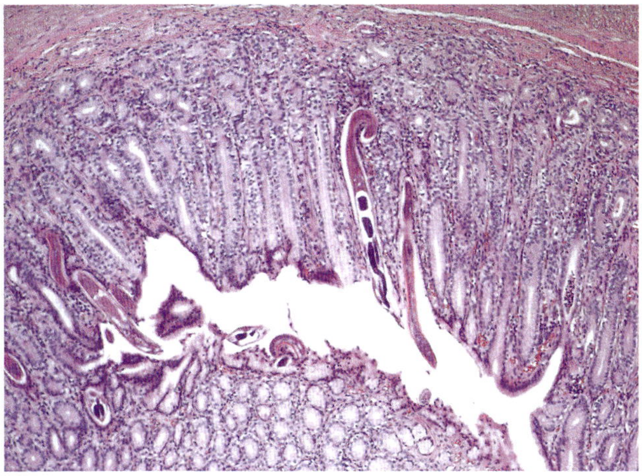

FIGURE 10.36 Gastritis associated with *Ollulanus tricuspis* in a tiger. Lymphoplasmacytic gastritis is associated with embedded trichostrongyle nematodes. *(Reprinted with permission from* Breuer W, Hasslinger MA, Hermanns W. Chronic gastritis caused by *Ollulanus tricuspis* (Leuckart, 1865) in a tiger. *Berl Munch Tierarztl Wochenschr.* 1993;106(2):47–49.*)*

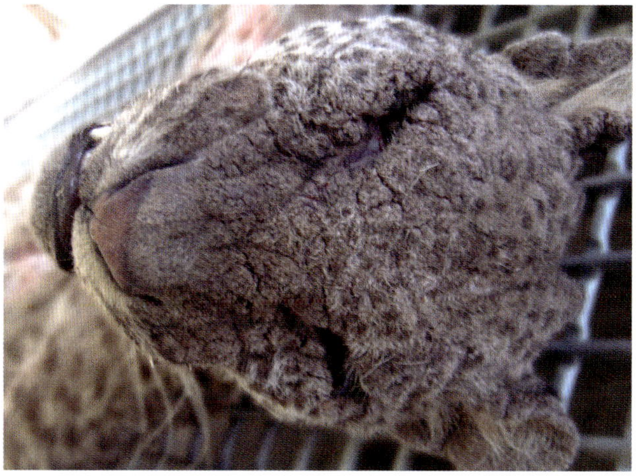

FIGURE 10.38 Sarcopic mange in a free-ranging leopard. The facial skin is thickened due to epidermal hyperplasia and hyperkeratosis. *(Photo Courtesy of R. Bengis, Department of Agriculture, Forestry and Fisheries, South Africa.)*

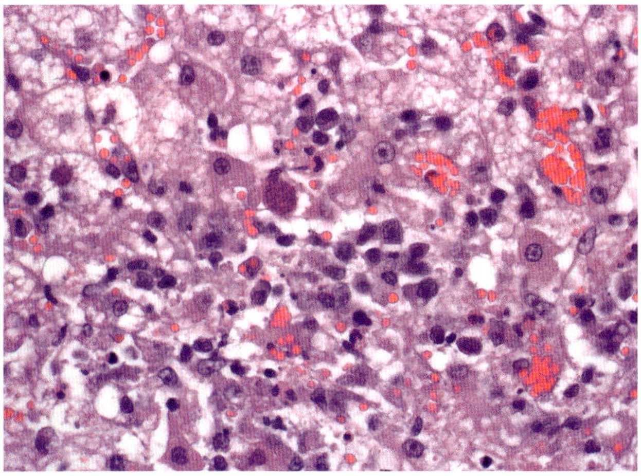

FIGURE 10.37 *Toxoplasma gondii* hepatitis in a Pallas' cat. Multiple areas of hepatic necrosis and neutrophilic inflammation are associated with protozoal zoites (see eSlide 10.e8).

disease manifests primarily as encephalomyelitis. In these cats, disease may be related to declining immune function and recrudescence of latent infection. A survey of wild Pallas' cats in Mongolia found that <13% are seropositive and organisms were not seen in the feces or in tissues.[34] The increased severity of the outcomes in Pallas' cats suggests they may have evolved without exposure to this parasite. Others have suggested that immune deficiencies may contribute to the susceptibility.[164] Rare cases have also been noted in juvenile cheetahs with pyrexia, tachypnea, and disseminated infection.[187]

Ectoparasites

Some parasites, such as *Sarcoptes scabeii* and *Notoedres cati*, have negatively impacted wild populations of felids.[124,263,302,313,319,320,337] In some cases, exposure to anticoagulant rodenticide is a predisposing factor for disease. The lesions and pathogenesis of mange in felids are similar to those of other species. Affected cats have hair loss, hyperkeratosis, and epidermal hyperplasia with variable excoriation and crusting (Fig. 10.38). Mites can be identified on skin scrapes but differentiation between *Notoedres* and *Sarcoptes* spp. can be difficult, although *Notoedres* are generally smaller with a dorsal anus.

Other Infectious Organisms

Several captive cheetahs in Europe developed **spongiform encephalopathy** presumably from eating contaminated meat during bovine spongiform encephalopathy (BSE) outbreaks.[11,87,169,185,273] Lesions include vacuolation in the neuropil and neurons of the midbrain, thalamus, and hypothalamus. The obex, brainstem, and cerebellum were affected to a lesser degree. Lesions were associated with deposition of prion protein (PrP^{Sc}) within glial cells and neurons. Prion gene (*PNRP*) sequences are similar among cheetahs and other felids that lack evidence of disease,[338] such as mountain lions who prey upon deer and elk affected by chronic wasting disease (CWD). To date there is no evidence of transmission of CWD to mountain lions and they appear to digest prions that they consume.[16,391] The role of dosage as well as differences between BSE and CWD are unknown in disease pathogenesis.

ADDITIONAL RESOURCES

Meli ML, Cattori V, Martinez F, et al. Feline Leukemia Virus and other pathogens as important threats to the survival of the critically endangered Iberian lynx (*Lynx pardinus*). *Plos One*. 2009;4: e4744–e4744.

Munson L, Terio KA, Kock R, et al. Climate Extremes Promote Fatal Co-Infections during Canine Distemper Epidemics in African Lions. *PLoS ONE*. 2008;3: e2545–e2545.

Munson L, Terio K, Lane E, Robert N, Courchamp F. Wild Felid Diseases: Conservation Implications and Management Strategies. In: MacDonald D, Loveridge A, eds. *Biology and Conservation of Wild Felids*. Oxford: Oxford University Press; 2010:237–259.

Munson L, Terio KA, Worley M, Jago M, Bagot-Smith A, Marker L. Extrinsic factors significantly affect patterns of disease in free-ranging and captive cheetah (*Acinonyx jubatus*) populations. *Journal of Wildlife Diseases*. 2005;41: 542–548.

Roelke-Parker ME, Munson L, Packer C, et al. A canine distemper virus epidemic in Serengeti lions (*Panthera leo*). *Nature*. 1996;379: 441–445.

Ryser-Degiorgis M-P, Ryser A, Bacciarini LN, et al. Notoedric and Sarcoptic Mange in Free-ranging Lynx from Switzerland. *Journal of Wildlife Diseases*. 2002;38: 228–232.

References

Full references for this chapter can be found online at https://doi.org/10.1016/B978-0-443-18437-6.00006-7.

Chapter 11

Mustelidae and Mephitidae

Alison Righton[a], Kathy A. Burek-Huntington[b], Melissa A. Miller[c], and Stephen Raverty[d]

[a]San Diego Zoo Wildlife Alliance, San Diego, CA, United States; [b]Alaska Veterinary Pathology Services, Eagle River, AK, United States; [c]Marine Wildlife Veterinary Care and Research Center, California Dept. of Fish and Wildlife, Santa Cruz, CA, United States; [d]Ministry of Agriculture and Food, Animal Health Center, Abbotsford, BC, Canada

INTRODUCTION

The family Mustelidae, comprising the stoats, polecats, mink, fishers, wolverines, weasels, martens, badgers, tayra, grison, ferrets, and otters, is the largest family within the order Carnivora. Skunks were traditionally considered as mustelids, however, recent molecular evidence places them in the family Mephitidae.[310,382] Both Mustelidae and Mephitidae are within the Musteloidea superfamily with Procyonidae and Ailruidae. Globally, mustelids and mephitids are free ranging in terrestrial, fresh water, and marine environments, are kept as pets, farmed, used as laboratory animals, and managed in zoological settings. In the Mustelidae family, European mink (*Mustela lutreola*) are listed as critically endangered and multiple species including the giant otter (*Pteronura brasiliensis*), black-footed ferret (*Mustela nigripes*), sea otter (*Enhydra lutris*), marine otter (*Lontra felina*), and southern river otter (*Lontra provocax*) are listed as endangered according to the International Union for Conservation of Nature.[195] Aside from extensive literature published on free-ranging sea otters, there is less information known about free-ranging morbidity and mortality in mustelids and mephitids that is not directly correlated with human interactions. Most of the information on disease processes and syndromes affecting members of this group comes from individuals housed under human care. This chapter contains an introduction to the natural history, normal anatomy, physiology, and common diseases and syndromes affecting members of Mustelidae and Mephitidae.

NON-INFECTIOUS DISEASES

Nutritional

An association between neurologic disease and **thiamine deficiency** (Chastek paralysis) in mink was first noted in the 1940s, particularly in mink fed diets containing high levels of thiaminase-containing fish.[313] Additional reports include a single captive river otter[385] and an incident in New Zealand, where farmed ferrets were fed a diet of raw eggs.[203] However, it is likely that this condition could affect any member of the taxa. The pathogenesis of thiamine deficiency is complex and is likely related to the role of this nutrient as a cofactor for pyruvate decarboxylase and transketolase synthesis, with lesion distribution closely paralleling transketolase activity in the brain. Decreased activity of ATP-dependent sodium and water transport results in intraneuronal swelling that progresses to neuronal necrosis with vascular degeneration, edema, and hemorrhage. Clinical disease is characterized by anorexia and malaise followed by incoordination, paralysis, convulsions, and death within 2–3 days.[14] Gross lesions of thiamine deficiency in mustelids resemble those in other carnivores and consist of bilaterally symmetrical hemorrhage and necrosis of the periventricular gray matter, caudal colliculi, and vestibular nuclei. Histologically, vacuolar degeneration and necrosis of neurons with accompanying gliosis are often most apparent in nuclei within the lateral geniculate bodies, caudal colliculi, and red nuclei. In one report, lesions were present in adult male mink, but not adult females or kits. A flaccid, dilated right cardiac ventricle and fatty degeneration of the liver were also identified in one case.[313] In North American river otters, cases of thiamine deficiency with typical clinical signs can manifest as diverse gross and histologic pathologic changes, including no histologic findings,[14] dilated cardiomyopathy,[17] or bilaterally symmetrical axonal degeneration with spheroids in the nucleus cuneatus and gracilis of the medulla oblongata, pons, and obex.[385]

Vitamin E deficiency was also described in farmed mink in the 1940s with intermittent reports in other mustelids,[154] including young ferrets fed diets high in improperly stored fish, horsemeat, or polyunsaturated fatty acids.[203] This disease primarily affects healthy young, farmed mink in the summer and fall, and may present with no premonitory signs.[154] As an important antioxidant, vitamin E protects cellular membranes against oxidative

Pathology of Wildlife and Zoo Animals. https://doi.org/10.1016/B978-0-443-18437-6.00032-8

processes initiated by free radicals. Affected mustelids suffer from steatitis and skeletal and cardiac myodegeneration. Mink can also develop microcytic, normochromic anemia.[428] Gross lesions justify the early name associated with this condition, "**yellow fat disease**." At necropsy, the abdominal skin appears thickened and doughy, and may exude a thin fluid.[124] Oxidized abdominal and subcutaneous fat is firm and yellow-brown (Fig. 11.1A), and a fishy odor may permeate from the cadaver in severe cases. Histologic lesions include widespread adipocyte necrosis with multifocal saponification, dystrophic mineralization, and mixed suppurative and histiocytic inflammation (Fig. 11.1B). In severe cases, entire fat lobules may be replaced by inflammatory cells with cytoplasmic acid-fast positive ceroid pigment. Skeletal and cardiac muscle may have varying degrees of myofiber degeneration, swelling, cytoplasmic vacuolation, fragmentation, contraction band necrosis, and fibrosis, with proliferation of satellite nuclei. Some mustelid species, including badgers, otters, and black-footed ferrets, have been reported to be susceptible to exertional myopathy or rhabdomyolysis associated with vitamin E deficiency.[225]

Vitamin A deficiency is relatively common in captive sea otters, likely due to dietary deficiency or individual otter noncompliance when administered oral vitamin A supplements. Clinical signs vary from none to lethargy, anorexia, and failure to thrive. Grossly, multifocal and coalescing, 1–2 mm irregular bony nodular proliferations (hyperostoses) are visible on the internal surface of the calvarium. In contrast, a linear multinodular band of mineralized dura and arachnoid mater of the falx cerebri is a common incidental finding in older wild sea otters; in the absence of the above-mentioned calvarial hyperostoses, this condition should be distinguished from vitamin A deficiency. Histologically, multifocal to diffuse nodular hyperplasia of the lamellar calvarial bone creates a generalized wavy profile. This lamellar bone contains variably sized lacunae, dilated vascular channels, and diminished osteoclast activity. Additional histologic changes due to hypovitaminosis A in otters include squamous metaplasia of nasal seromucous glands with secondary bacterial rhinitis.[422] **Pyridoxine (vitamin B6) deficiency** has been documented in mink.[203] Males demonstrate testicular atrophy, aspermia, and degeneration. Absorption sterility occurs in females.

Imbalances in levels of **calcium, phosphorus, and vitamin D**, especially when exposure to ultraviolet light is minimized, can result in a variety of pathologies in mustelids. In ferrets, **fibrous osteodystrophy** (nutritional hyperparathyroidism) is associated with feeding an all-meat diet without calcium supplementation.[203] Affected ferrets are reluctant to move, cannot support their own weight, and exhibit forelimb abduction. Long bones are soft and pliable, and fractures may be present. The parathyroid glands are hyperplastic. **Salt toxicity** has been observed in ferrets fed a diet composed entirely of salted fish. Affected ferrets are significantly depressed, with periodic choreiform dyskinesia 24–96 h postingestion of excessively salty diets; death occurs shortly thereafter. Pathology is restricted to the brain and is characterized by diffuse neuropil edema with coning of the cerebellar vermis and nonsuppurative eosinophilic meningitis.[203]

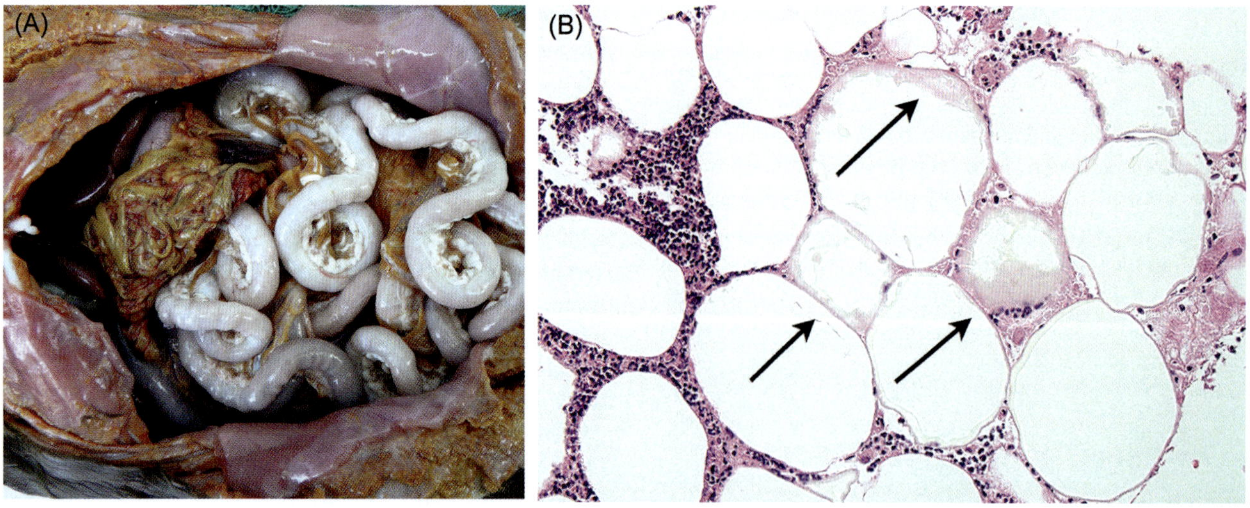

FIGURE 11.1 Hypovitaminosis E in a smooth-coated otter and an American mink. (A) Gross presentation of subcutaneous steatitis and abdominal fat necrosis in a smooth-coated otter with vitamin E deficiency. Subcutaneous and abdominal adipose tissue stores are diffusely orange-brown with multiple nodular, more hyperemic areas of mesenteric necrosis. (B) Steatitis within the mesentery of a mink suffering from hypovitaminosis E. The interstices between adipocytes contain lymphocytes, plasma cells, and macrophages. There is saponification of several adipocytes (*arrows*). *(Part A: Photo Courtesy of T. Kasantikul, Zoological Park Organization and Faculty of Veterinary Science, Mahidol University.)*

Metabolic

Nephrolithiasis and **urolithiasis** are well-documented entities in several species of mustelids, including Asian small-clawed otters (ASCOs), cape clawless otters, Eurasian otters, mink, ferrets, North American river otters, sea otters, and wolverines. Information regarding urolithiasis and nephrolithiasis in otters and wolverines is summarized in Table 11.e1.

The pathophysiology is multifactorial, with complex interplay of metabolic, nutritional, dietary, gastrointestinal and urinary tract microbiome, age-related, and genetic factors. In farmed mink, struvite urolithiasis exhibits clear seasonal patterns, with pregnant females affected in the spring, and male kits affected in the fall. In gravid females, stones may hinder parturition, resulting in dystocia and death, while in males; urethral obstruction may result from urolith-mediated obstruction at the distal os penis.[154] Concurrent *Staphylococcus intermedius* cystitis is common in affected mink.

ASCOs have a high incidence of calcium oxalate uroliths with fewer uric acid and ammonium urate stones. In a survey of ASCOs under human care in North America, 66% exhibited nephrolithiasis, and 23% had cystolithiasis.[55] Both kidneys were affected in 84% of cases.[55,56] Females and males are equally affected and risk factors include older age and increased dietary calcium. Hypercalcemia with hyperoxaluria during food consumption and fasting contributes to a 1:1 calcium-to-oxalate ratio, which may facilitate urolith formation. In contrast, increased dietary crude protein and phosphorus are associated with a reduced risk of stone formation.[338,339,492] Dietary risk factors have also been identified in Eurasian otters, which present predominantly with ammonium urate calculi. Urolith formation has been correlated with elevated purine intake, protein quality and digestibility, and renal uric acid excretion.[375] In contrast, no evidence of dietary involvement was found for nephrolithiasis in another captive cohort of Eurasian otters with a 67% incidence of stone formation and a genetic predisposition was proposed.[36] In free-ranging North American river otters, uric acid[157] and calcium phosphate nephroliths have been documented with no associated renal or lower urinary tract lesions.[303]

Freshwater otters with nephrolithiasis often have bilateral calculi observed in minor calyces throughout the kidneys (Fig. 11.2A), focal or extensive scarring of reniculi, and variable hydronephrosis and hydroureter (Fig. 11.2B). Histopathology often reveals crystals within medullary tubules, renal pelvises, or ureters with associated medullary and cortical atrophy, fibrosis, and scant mononuclear inflammation (eSlide 11.e1). Secondary suppurative pyelonephritis or tubulointerstitial nephritis is occasionally observed. Hyperplasia of the urethral mucosa and hypertrophy of the tunica muscularis may also occur secondary to urolith obstruction. Uroliths within the urinary bladder are similar in appearance to nephroliths (Fig. 11.2C) but are not as common as nephroliths in freshwater otters.

In a study of free-ranging wolverines, ammonium acid urate with magnesium ammonium phosphate and calcium phosphate uroliths were associated with unilateral nephrolithiasis in 87.5% of surveyed animals.[311] In northwestern Canada, the prevalence of renal calculi in wolverines was 8.9%, but the pathogenesis of urolith formation and potential impacts on health are unknown.

In a review of 408 uroliths from pet ferrets, 202 (67%) of uroliths were struvite, 61 (15%) were cystine, and 43 (11%) were calcium oxalate. Uroliths were most common in males (73%), prevalence increased with age, and most (77%) were retrieved from the bladder. When diets high in cereal proteins were fed to pregnant or lactating jills, 5%–10% had uroliths, and many required emergency surgery to remove a single large or multiple small struvite uroliths.[203] Unlike mink, dogs, and cats, ferret struvite uroliths are not associated with bacterial infections.[309]

Gastric erosions or bleeding mucosal ulcers with "coffee-ground" melena are common in mustelids and are similar grossly and microscopically to published descriptions of gastritis in humans.[268] Secreted hydrogen ions and enzymes are necessary to break down nutritional components into absorbable amino acids, carbohydrates, and fats. Altered pH can disrupt digestion, resulting in dysregulation of gastric pH. In humans, this dysregulation is often associated with physiologic stress, leading to gastric inflammation (stress-induced gastritis), mediated by elevated cortisol and histamine.[268]

In affected mustelids, erosions are grossly visible in the pyloric mucosa (Fig. 11.3A) and gastric body (Fig. 11.3B), and they often have a surface coat of dark brown to black, partially digested blood. Gastric erosions or ulcers are extremely common in stranded sea otters. In a recent study of 500 sea otters assessed for this condition, 59% (295/500) had gastric erosions or ulcers, and 47% (237/500) had lesions severe enough to be a contributing cause of death.[277] Erosions and ulcers were twice as common in otters in poor or emaciated nutritional condition at necropsy (56%; 144/258), when compared to otters in fair, good, or excellent nutritional condition (28%; 27/238).[277] Gastric erosions and ulcers may be a nonspecific expression of severe stress, inanition, emaciation, poor health, and/or NSAID administration, and/or *Helicobacter* sp. mucosal bacterial infections in sea otters and other mustelids.[277] Although a causal relationship has been proposed for the development of gastric inflammation and erosions or ulcers in older ferrets and sea otters with confirmed *Helicobacter mustelae* or *H. enhydrae* infection, respectively,[24,128,402] a direct causal link has not been established.

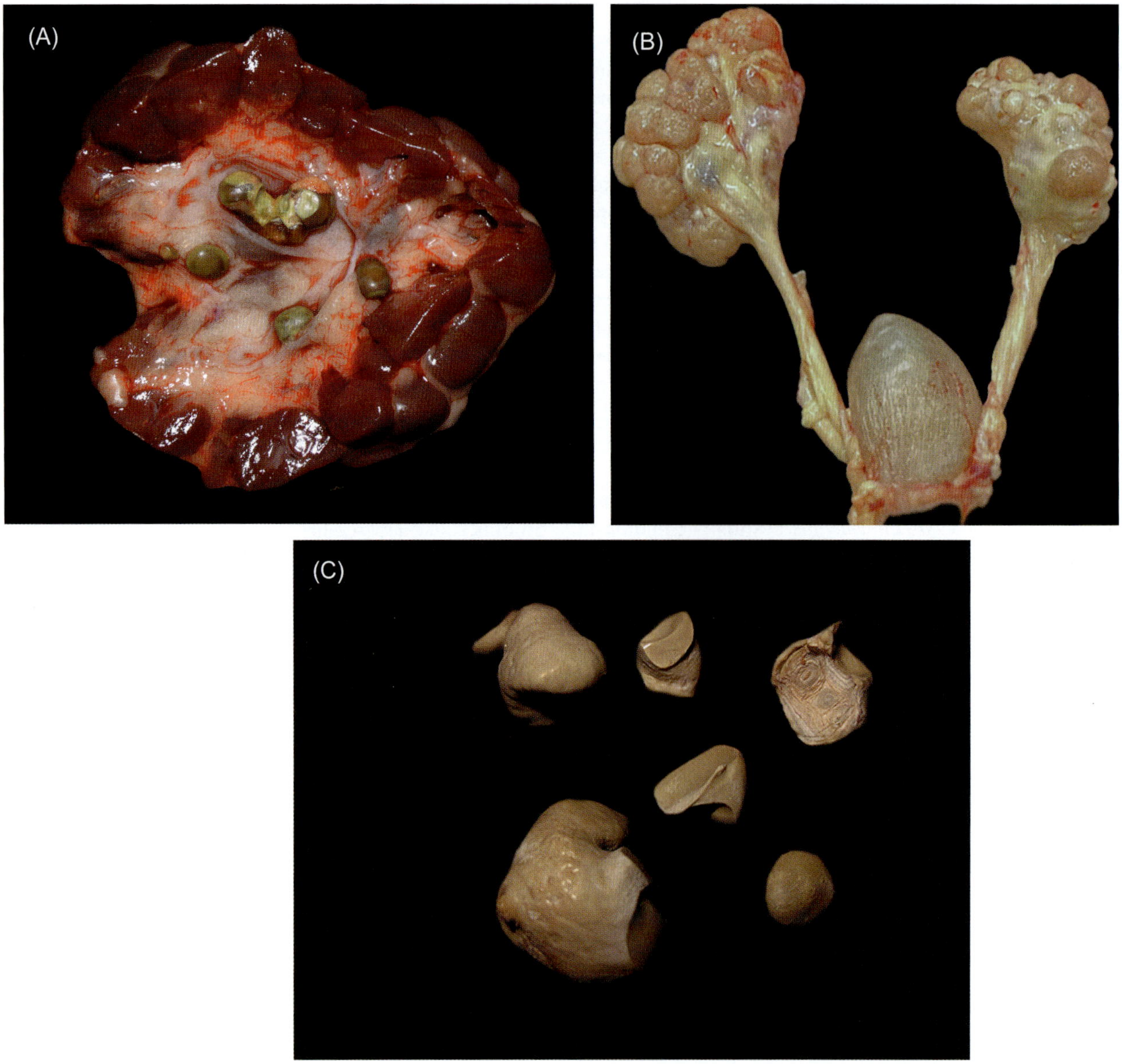

FIGURE 11.2 Nephrolithiasis and urolithiasis in otters. (A) Nephroliths in the kidney of an Asian small-clawed otter. Multiple green to brown, smooth, and occasionally branching nephroliths are present within the major calyces with secondary hydronephrosis and fibroplasia. (B) Kidneys, ureters, and urinary bladder in an Asian small-clawed otter with bilateral nephrolithiasis. There is atrophy, fibrosis, and loss of multiple individual or contiguous reniculi. The capsular surfaces of less affected reniculi are variably cobblestoned and the ureter on the right is mildly dilated. No uroliths were present within the urinary bladder of this individual (see eSlide 11.e1). (C) Uroliths from the urinary bladder of a Cape clawless otter. *(Part A: Photo Courtesy of K. Terio, University of Illinois Zoological Pathology Program. Part B: Photo Courtesy of Omaha's Henry Doorly Zoo and Aquarium. Part C: Photo Courtesy of B. Rideout, San Diego Zoo Wildlife Alliance.)*

Histologically, these gastric lesions are typically erosions, not true ulcers, characterized by wedge-shaped areas of mucosal coagulation necrosis and hemorrhage, often with one or more tiny fibrin thrombi within subtending blood vessels. In the gastric body, these lesions often concentrate along the outer edges of rugal folds. While rarely serving as a primary cause of death, gastric erosions or ulcers in sea otters often contribute to morbidity and mortality through progressive, severe blood loss, and occasionally, acute submucosal vascular rupture. In ferrets, these lesions are often concentrated in the pylorus and may result in full-thickness ulceration, perforation, and severe or fatal hemorrhage.

Adrenal-associated endocrinopathy (AAE) is a common metabolic condition in middle-aged and older ferrets. In intact ferrets, seasonal hypothalamic secretion

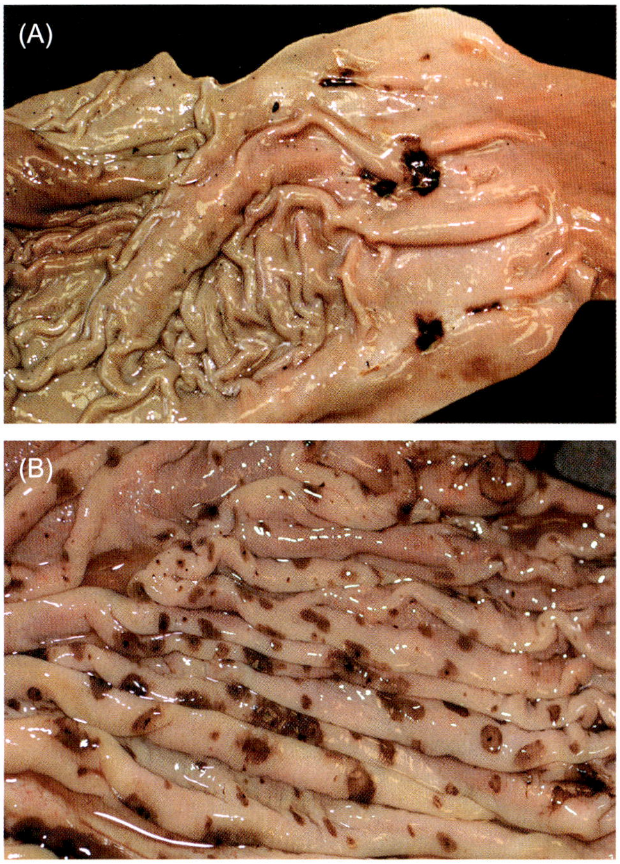

FIGURE 11.3 Gastric ulcers in sea otters. (A) Multiple, occasionally coalescing, dark red to brown depressed ulcerations are present throughout the pyloric mucosa of this northern sea otter. (B) Throughout the entire mucosal surface of the gastric body, but especially along the outer edges of the gastric rugae, there are multiple variably sized red to dark red, occasionally depressed gastric erosions and ulcerations in this southern sea otter. *(Part A: Photo Courtesy of Alaska Veterinary Pathology Services. Part B: Photo Courtesy of the California Department of Fish and Wildlife.)*

of gonadotropin- releasing hormone triggers secretion of luteinizing hormone and follicle-stimulating hormone to prepare the ovaries and testes for reproduction. Early castration or ovariohysterectomy can cause chronic luteinizing hormone secretion due to loss of negative gonadal feedback, resulting in hyperplastic and potentially neoplastic proliferation of primitive adrenocortical cells, resulting in AAE.[31,487]

AAE is characterized by adrenal hypersecretion of sex steroids, not cortisol. One or more sex steroids, including androstenedione, dehydroepiandrosterone sulfate (DHEA), estradiol, or 17-a-hydroxyprogesterone (17-OHP) will be elevated, whereas cortisol levels are usually normal.[487] Serum tests for these compounds, especially in neutered animals, can confirm the diagnosis. Because excessive cortisol is only rarely encountered in ferrets with this condition, the term "Cushing's disease" is inappropriate. Male and female ferrets are equally affected.[487] Proliferative adrenocortical lesions may be observed in animals less than 1 year of age and can range from multiple hyperplastic nodules to invasive carcinoma. Precise classification of the resulting tumors is not always possible nor necessary; except for a myxoid variant of carcinoma, tumor metastasis is uncommon and occurs late in disease progression.[337]

Clinical signs of AAE in ferrets include various patterns of truncal alopecia, vulvar swelling in spayed females, life-threatening bone marrow suppression in both sexes, stranguria or pollakiuria in castrated males due to prostatomegaly, and increased sexual behavior in neutered animals. Bilaterally symmetric **truncal alopecia** often begins caudally and progresses cranially due to inhibitory effects of excess estrogen secretion on hair follicles (Fig. 11.4). It may begin as seasonal alopecia that eventually does not resolve after 2 to 3 seasonal cycles.[487] Pruritus, characterized by erythema or papules on the dorsal neck or interscapular area, and muscle atrophy may be present along with alopecia. Mucopurulent vaginal discharge may be associated with vulvar hypertrophy secondary to hyperestrogenism, and some females develop secondary vaginitis.[487] Hyperestrogenism-associated bone marrow suppression may present as a nonregenerative anemia with a hematocrit $<25\%$, nucleated red blood cells, neutropenia, and thrombocytopenia. Spontaneous hemorrhage can occur when platelet counts fall below 20,000 cells/μL. Neutered males with adrenal hypersecretion of sex steroids due to AAE can develop sterile **prostatic cysts** and prostatomegaly, leading to partial or complete urinary obstruction that requires emergency intervention.[487] Concurrent disease may include prostatitis, urinary tract infection, and urolithiasis.[357] Although mechanisms of pathogenesis are unclear, high levels of circulating sex steroid hormones appear to stimulate proliferation of

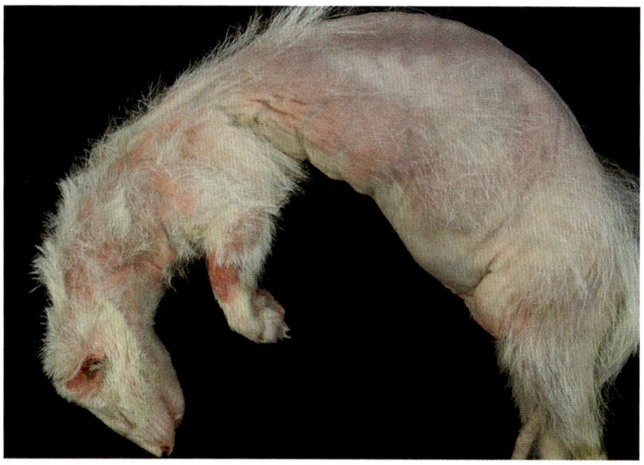

FIGURE 11.4 Truncal alopecia in a domestic ferret with adrenal-associated endocrinopathy. Patchy to diffuse alopecia and multifocal erythema are present on the lateral trunk, dorsum, ventrum, and tail. *(Photo Courtesy of T. Donovan, Schwarzman Animal Medical Center.)*

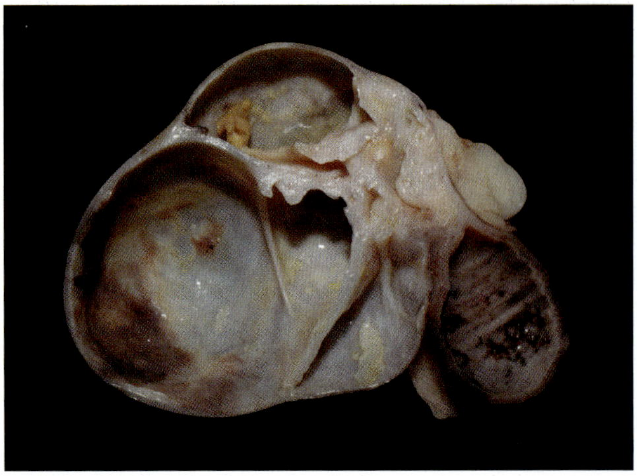

FIGURE 11.5 Prostatic cysts in a domestic ferret. Large multiloculated cysts expand the prostate gland and compress the adjacent trigone of the urinary bladder (bottom right). Cyst walls are thin, multifocally mottled brown to tan (congestion, inflammation, squamous metaplasia), and have scant tan to yellow adherent keratinaceous and purulent luminal material. Small amounts of hemorrhage and congestion are also present in the urinary bladder mucosa. *(Photo Courtesy of C. Bertram, Freie Universität Berlin.)*

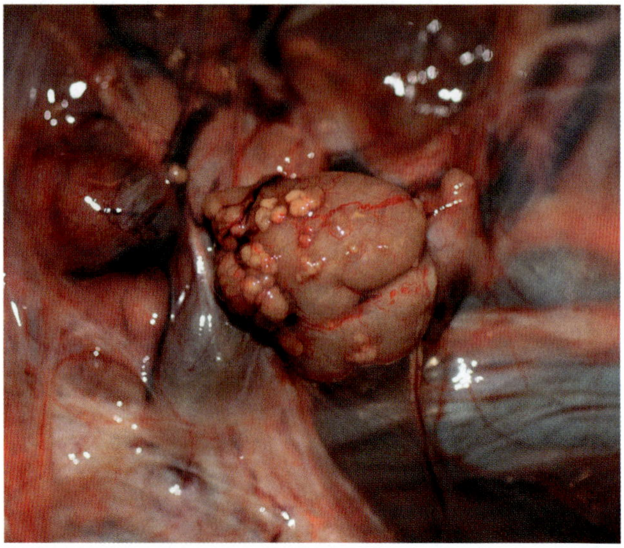

FIGURE 11.6 Capsular and ectopic adrenocortical nodular hyperplasia in a southern sea otter. Multiple, variably sized, pale yellow to tan masses are present on the cortical surface of the adrenal glands as well as the adjacent retroperitoneal capsule. This is a common incidental finding in adult and older adult sea otters and domestic ferrets. *(Photo Courtesy of the California Department of Fish and Wildlife.)*

prostatic tissue, resulting in squamous metaplasia, keratinization of prostatic glandular epithelium, and accumulation of purulent exudate, forming large cysts (Fig. 11.5).[95] AAE-associated prostatic disease may be either cystic or hyperplastic, and clinical signs usually resolve 24–48 h after surgical removal of the affected adrenal gland.[487]

Hyperestrogenism is common in intact female ferrets that are not bred or artificially stimulated to ovulate. Estrus lasting more than 1 month is associated with hyperestrogenism which can lead to bone marrow depression. Other potential causes of estrogen toxicosis in ferrets include cystic ovaries, an estrogen-secreting ovarian remnant, and ovarian neoplasia.[487] In intact females, recent estrus and clinical signs suggestive of anemia are sufficient for presumptive diagnosis. In spayed ferrets, abdominal ultrasound may help to discriminate between an ovarian remnant and AAE. Protracted hyperestrogenism may also cause melena, petechial or ecchymotic hemorrhages, metritis, pyometra, or vaginitis.[487]

Adrenocortical nodular hyperplasia is common in sea otters and becomes increasingly visible with age. Unlike ferrets, no significant clinical impacts have been reported. Although nodules of hyperplastic adrenocortical tissue often develop on the adrenal surface and in adjacent tissues (Fig. 11.6), the adrenal glands are not grossly enlarged. Histologically, these nodular aggregates are composed of well-differentiated cortical epithelial cells, often with mild to moderate cytoplasmic vacuolization.

Diabetes mellitus has been reported in a domestic ferret and a captive black-footed ferret.[29,65] Clinical signs include lethargy, dehydration, alopecia, muscle twitching, hindlimb paresis, weight loss, polyuria, polydipsia, polyphagia, and or diarrhea. Clinical chemistry and urinalysis reveal hyperglycemia, glycosuria, and possibly ketonuria. Gross lesions are often absent; histologic changes may be absent or can include diffuse vacuolation of pancreatic islet of Langerhans cells with PAS-positive material that can be digested with amylase pretreatment (putative glycogen). Cytoplasmic immunoreactivity for insulin can confirm these vacuolated cells are beta cells. Renal tubular epithelium may also be vacuolated, and can contain similar PAS-positive material that digests with amylase.[29] Histopathology may be unremarkable in cases of diabetes due to lack of insulin release or diminished effectiveness of secreted insulin.[65]

Exertional rhabdomyolysis has been diagnosed in North American river otters during translocation efforts.[174] In cases of exertional rhabdomyolysis, chemistry reveals elevated aspartate aminotransferase (AST) and creatinine kinase. Symptoms can develop 1–2 days postcapture or other exertional activity and range from subclinical to depression, lethargy, hypothermia, and dehydration. At necropsy, linear bands of pale pink to tan skeletal muscle are visible in the subscapularis, rectus abdominis, quadriceps femoris, laryngeal muscles, or multiple muscle groups concurrently. Histopathology often reveals acute to subacute skeletal myofiber necrosis and myophagia. Associated lesions may also include gastrointestinal hemorrhage, bacterial pneumonia, and clostridial enteritis.

Toxic

Table 11.e2 summarizes documented causes of toxicity in mustelids and skunks. **Oil spills** are especially devastating to mustelids living in aquatic environments. The Exxon Valdez oil spill in Prince William Sound, Alaska in 1989 was of particular significance to sea otters. Up to 4,000 sea otters perished following that event, and in many areas, population recovery remains incomplete. Due to the wide variation in chemical composition of refined and unrefined petroleum products, clinical disease and lesion patterns can vary greatly depending on the spill type, the extent of post-spill weathering, and the dose and route of exposure. Also, due to the dark pelage of many mustelids and skunks and the volatile nature of many petroleum products, visual confirmation of oil exposure can be challenging in some cases (Fig. 11.7).

A single episode of petroleum exposure can cause both acute and chronic disease. Acute clinical signs include pelage matting and wetting, hypothermia, emaciation, anorexia, depression, seizures or tremors, melena, and dyspnea. Bloodwork from live, acutely exposed animals may reveal anemia, hypoglycemia, increased blood urea nitrogen (BUN) and total protein, elevated liver enzymes, hypoxemia, and acidosis. Mild pelage oiling with weathered tar patties is relatively common for southern sea otters due to the presence of numerous natural oil seeps offshore; when the fouling is focal and mild, healthy animals can often remove these tar patties via grooming and survive. However, in some cases, pelage fouling can be fatal, or can contribute to morbidity.

Because tar patties are often present on beaches, it is important to discern between antemortem and postmortem surface contamination. Antemortem tar patties are usually embedded deep into the pelage and are often fragmented into small clumps due to grooming activity (Fig. 11.8A). Postmortem contamination of carcasses by tar patties is usually characterized by superficial attachment of large, round-edged, tar patties to the outer hair tips, with lack of patty distortion and fragmentation due to grooming.

FIGURE 11.8 Tar patties on sea otter pelage. Photo (A) shows antemortem tar patties that are deeply embedded and broken into small fragments throughout the affected area of pelage due to grooming activity. Photo (B) represents a postmortem tar patty where the tar is more superficial, lacks distortion or separation into smaller fragments, and has large amounts of seagrass adherent to and enmeshed into the patty. *(Parts A, B: Photos Courtesy of the California Department of Fish and Wildlife.)*

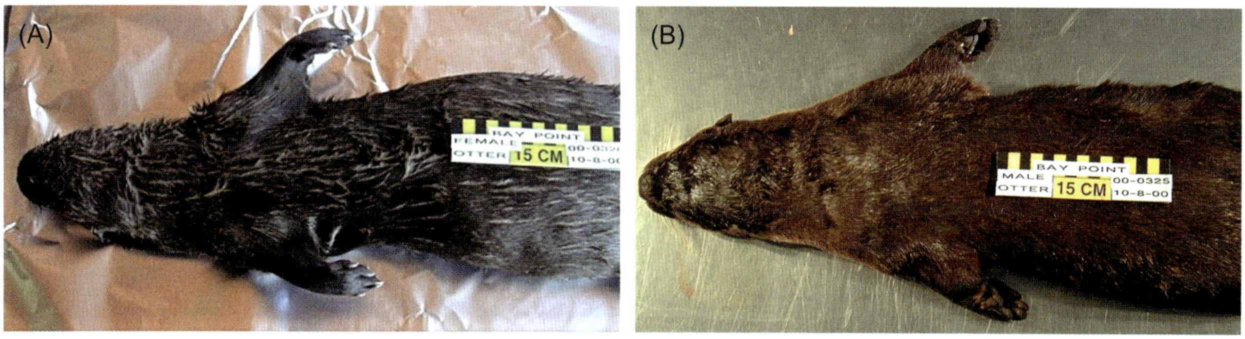

FIGURE 11.7 Petroleum on the pelage of North American river otters. These river otters were found dead after a petroleum pipeline break resulted in severe wetland contamination by a refined, highly volatile petroleum product. Otter (A) was photographed soon after carcass recovery and the pelage is diffusely wet-looking due to extensive surface contamination with the petroleum product. Otter (B) was photographed after much of the highly volatile product had vaporized from the pelage surface, although the animal still had a strong petroleum smell. Depending on the interval, the pelage may be relatively dry and grossly unremarkable. *(Parts A, B: Photos Courtesy of the California Department of Fish and Wildlife.)*

In addition, large amounts of seagrass, kelp, or other material are often enmeshed into adherent postmortem tar patties; this material is usually quickly removed by live animals (Fig. 11.8B).

The most common and immediate pathologic effects of petroleum exposure in mustelids and skunks are pelage matting and wetting, which impairs air trapping and thermoregulation, resulting in hypothermia and death. In addition to surface contamination, associated gross lesions can include skin, ocular and oral chemical burns, physical obstruction of the eyes, nares, and mouth, chemical esophagitis and gastroenteritis, secondary stress-related gastric erosions or ulcers, melena, erosive tracheitis and bronchitis, pneumonia, pulmonary emphysema, cardiotoxicity, neurotoxicity, and necrotizing hepatitis.[246,358,434]

Potential chronic effects following oil exposure include bone marrow suppression, disrupted blood cell function, cardiotoxicity, neurotoxicity, oxidative damage, immune, reproductive, and endocrine impairment, impaired stress responses, and tumor induction.[246,358,434] In a recent review, a progression of impacts of oil exposure was noted across animal taxa, from molecular and cellular effects that manifest as organ dysfunction, to systemic effects that compromise fitness, growth, reproductive potential, and survival.[434] Adverse health effects can form a suite of signs and host responses that can result in multiorgan system failure.[434] Histologic lesions are relatively nonspecific, encompass the above-listed conditions, and often include centrilobular hepatic necrosis, as well as lipidosis of periportal hepatocytes and renal tubules. Definitive diagnosis is achieved through forensic petroleum fingerprinting of source material and contaminated skin or hair, digesta, tissue, or body fluids, and ruling out other potential causes of morbidity or mortality.

In general, refined petroleum products that have been enriched for shorter, more volatile petroleum compounds are more acutely toxic, while crude and/or weathered products are more likely to cause death via physical fouling and hypothermia. Petroleum products that are especially high in volatile compounds, benzenes, or additives are often more acutely lethal due to high toxicity and concurrent dermal, ocular, respiratory (Fig. 11.9), and gastrointestinal exposure. Risk to personnel and equipment exposure to hazardous compounds must be carefully considered when collecting, transporting, and necropsying oiled animals. The toxicology of oil in vertebrates was recently reviewed following the response to the Deepwater Horizon oil spill.[434] Readers are encouraged to read this overview of gross and microscopic pathology related to oil exposure.

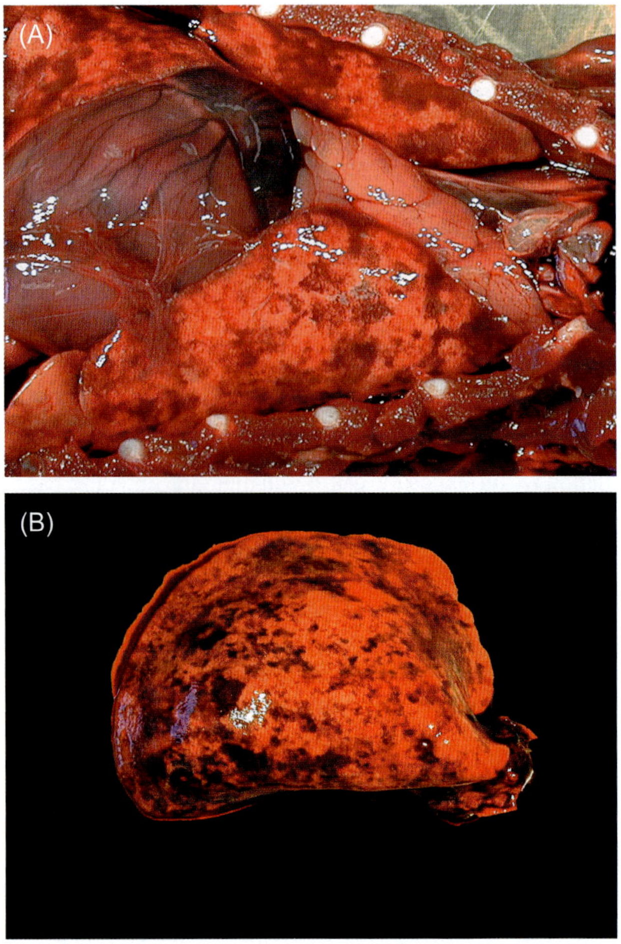

FIGURE 11.9 Pulmonary lesions associated with inhalation of volatile petroleum products in North American river otters. These are the same otters with external petroleum contamination seen in Fig. 11.7. Both otters have severe diffuse pneumonitis characterized by multiple foci of pulmonary atelectasis (A) and hemorrhage (B). *(Parts A, B: Photos Courtesy of the California Department of Fish and Wildlife.)*

Anticoagulant rodenticide poisoning has emerged as a significant concern for conservation and management of nontarget wildlife, including mustelids and mephitids. Species known to be affected include fishers, marine and European otters, American mink, beech martens, cape clawless otters, honey badgers, American badgers, European polecats, least weasels, stoats, and striped skunks.[5,87,126,134,240,250,343,352,364,394,426] Mustelid species can be directly targeted with anticoagulant bait deployment; however, they are often poisoned secondarily by eating targeted prey species such as rodents and rabbits.[5] Anticoagulant rodenticides interfere with synthesis of clotting factors, resulting in hemorrhage and death. Vitamin K is essential for synthesis of prothrombin and other clotting factors. Anticoagulant rodenticides antagonize vitamin K_1-epoxide reductase in the liver, causing gradual depletion of vitamin K, and consequently, vitamin K-dependent clotting factors. Two basic forms of anticoagulant rodenticide exist: first-generation anticoagulants (warfarin, chlorophacinone, and diphacinone) and second-generation anticoagulants (brodifacoum, bromadiolone, flocoumafen, difenacoum, and difethialone) which are more acutely toxic. Second-generation anticoagulants are widely used for rodent control in the United States and other countries. Their superior potency is related to a greater affinity for vitamin K-epoxide reductase. Classical gross and microscopic lesions consist of hemorrhages and abnormal blood clotting. Transplacental and lactational transfer of anticoagulants have been described in fishers[364] and multiple biologic samples from mustelids worldwide have tested positive for anticoagulants without documented disease.[87,240,250,343,394]

Microcystin, a potent cyanotoxin, has been identified as a cause of death in southern sea otters in central California (United States).[274] These toxins are produced by freshwater and estuarine cyanobacteria (*Microcystis aeruginosa*), which were formerly classified as "blue-green algae". ***Microcystis aeruginosa*** super-blooms with toxin production are relatively common in warm, thermally stratified, nutrient-enriched freshwater lakes, ponds, streams, and rivers globally, especially during warmer months. Several studies suggest that the frequency and severity of these bloom events may be increasing worldwide in relation to climate change and anthropogenic manipulation of watersheds.[256,465] Risk factor analysis in one study indicated that coastal nutrient loading (nitrate, phosphate silicate, ammonium, and urea) was a significant predictor of microcystin concentrations within affected watersheds.[146] Once produced, microcystins are relatively stable and can persist in water, sediment, and biota for days to months. In addition, these toxins can flow downstream and biomagnify within food chains, posing broader health risks.[147,274]

The cyanobacterium *M. aeruginosa* often floats in the water as small clusters (Fig. 11.10A) that have been described grossly as resembling chunks of floating broccoli. Extensive blooms can turn affected water bodies into a bright, fluorescent green color that resembles paint (Fig. 11.10B). Although microcystins can be toxic via aerosol and transdermal exposure, the most common route of exposure is ingestion of contaminated water or food. Microcystins are toxic to all living things, including plants. These toxins can cycle enterohepatically between the liver, bile, and gastrointestinal tract, prolonging toxicity. The liver is especially vulnerable to microcystin toxicity because hepatocytes possess a carrier-mediated transport system that facilitates microcystin uptake. Microcystins severely impair cytoskeletal function, resulting in loss of hepatocyte adhesion, necrosis, and apoptosis.

Clinical and clinicopathological signs of microcystin intoxication can include diarrhea, vomiting, weakness, recumbency, icterus, elevated liver enzymes, coagulopathy, and death. Grossly, livers of affected sea otters are swollen, friable, and hemorrhagic (Fig. 11.11A), and affected animals have mild to severe icterus.[274] Microscopic lesions include hepatocellular vacuolation, cellular disassociation, necrosis, apoptosis, and hemorrhage

FIGURE 11.10 Microscopic and gross photos of the toxic cyanobacterium *Microcystis aeruginosa*. Photo (A) represents a microscopic view of colonies of *M. aeruginosa*. Photo (B) is a jar of bright green, opaque, cyanobacterial surface scum collected at Pinto Lake, California during an *M. aeruginosa* super-bloom in 2007, when dead, microcystin-positive sea otters were first detected in adjacent nearshore marine waters. This sample contained one of the highest concentrations of the potent hepatotoxins called microcystins ever detected in the United States: 2.9 million ppb. Microcystin was also detected in the Pajaro River located downstream from Pinto Lake to within 1 km of the ocean, and in freshwater discharges into Monterey Bay. *(Part A: Photo Courtesy of B. Rosen, Florida Gulf Coast University. Part B: Image reprinted with permission from Miller MA, Kudela RM, Mekebri A, et al. Evidence for a novel marine harmful algal bloom: cyanotoxin (microcystin) transfer from land to sea otters. PloS One. 2010;5(9).)*

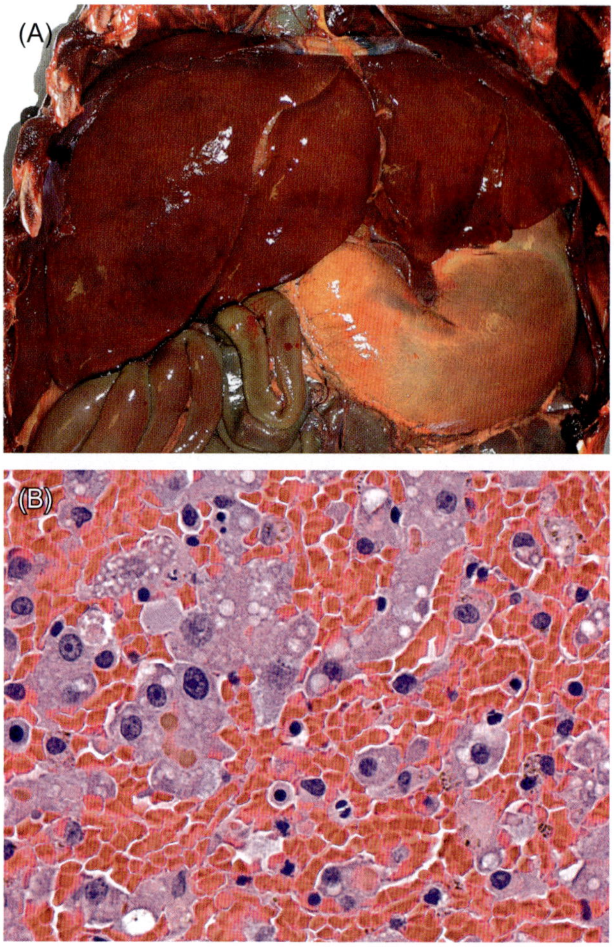

FIGURE 11.11 Microcystin intoxication in a southern sea otter. There is moderate diffuse hepatomegaly and icterus (A). The hepatic parenchyma had a mottled appearance grossly and was soft and friable. Microscopic examination of the liver (B) from the same otter reveals severe hepatocellular cytoplasmic vacuolation, necrosis, apoptosis, and parenchymal hemorrhage. The orderly arrangement of hepatic sinusoids was disrupted, especially in the centrilobular and mid-zonal regions, and clumps of bile are visible within remaining canaliculi (within the clump of hepatocytes at center left, and near the far lower left edge of the photo). *(Parts A, B: Photos Courtesy of the California Department of Fish and Wildlife.)*

(Fig. 11.11B).[274] Although the liver is a key target, the heart, kidneys, and other tissues are also affected, and can exhibit similar pathology.

At present, diagnosis and characterization of microcystin-associated pathology in animals are hindered by several obstacles, including diverse microcystin congeners with variable chemical properties and toxicity, challenges with precise toxin detection postingestion, an unknown toxic dose for sea otters, and limited access to validated commercial diagnostic assays. Because they bind covalently (strongly) to host tissues, these toxins can be difficult to detect and quantify in body fluids and tissues following ingestion. Although testing of water and bloom material for microcystins is relatively straightforward, only specialized biochemical laboratories with proven expertise in microcystin detection should be used for assessment of tissues and body fluids from suspect intoxication cases. The most sensitive and specific tests available, such as liquid chromatography/mass spectrophotometry (LCMS) should be used for tissue analyses, and the testing protocol should include analysis of matching unexposed and microcystin-spiked tissues from the same animal species. A newer biochemical assay that has shown promise for microcystin detection from tissues of exposed animals is the MMPB (2-methyl-3-methoxy-4-phenylbutyric acid) technique, which was used to assess total (bound and free) concentrations of the cyanobacterial hepatotoxins microcystins and nodularins in livers of bottlenose dolphins in Florida (the assay cannot discern between these toxin groups).[43] Although commercial assays such as ELISAs may be helpful for detection of microcystin in water and bloom material, they appear to be unreliable for microcystin detection in animal tissues.[43] Another promising but minimally available diagnostic assay is immunohistochemistry (IHC). Although some research laboratories have developed and validated IHC assays to assess microcystin binding in animal tissue, these assays are not validated for wildlife species and are not generally commercially available. Given these constraints, diagnosis is often based on a holistic approach to investigation of potential cyanobacterial intoxication events that combines biochemical testing of water, bloom material, and scum from the suspect water body, clinical examination and clinical pathology for any affected live animals, phycological analysis of bloom material to identify the cyanobacterial species, and gross necropsy, histopathology and strategic testing of suspect animal tissues or body fluids. Suspect microcystin cases should also be screened for leptospirosis, as lesions can overlap.

Domoic acid (DA) intoxication is a cause of neurotoxicity, toxic cardiomyopathy, and heart failure in sea otters.[48,229,239,276,277,287,288] Produced by marine diatoms (e.g., *Pseudo-nitzschia australis*), DA is a potent neurotoxin that acts on the central nervous system (CNS), cardiovascular system, and other tissues that express kainic acid glutamate receptors (see also Chapter 23: Pinnipediae). In a recent study, DA toxicity was determined to be the primary cause of death in 9% of southern sea otters and was a primary or contributing cause of death in 20%.[277] Domoic acid was detected in biological samples from 26% (19/72) of Alaskan sea otters necropsied from 2002 through 2012,[48] and was detected concurrently with saxitoxins in another study,[239] although toxin detection was rarely associated with characteristic pathology. In multiple studies, DA exposure was shown to be a significant risk factor for cardiovascular pathology in sea otters,[229,277] and crab and clam-based diets posed a higher risk for cardiomyopathy development.[287] Pyura (sea squirts or tunicates) are efficient accumulators of DA along the Chilean coast and could serve as a source of exposure for native marine otters.[251] Due to the widespread distribution of this toxin globally, concentration within food chains, and difficulty in toxin identification, many cases of DA toxicosis likely go undiagnosed.

Southern sea otters with acute, fatal DA toxicosis typically present with neurological signs, severe diffuse vascular congestion, and multifocal microscopic hemorrhages (microhemorrhages) in the brain, spinal cord, cardiovascular system, and eyes. The congestion and microhemorrhages are associated with detection of high concentrations of DA in postmortem urine or gastrointestinal content, and often precede the development of histologically apparent CNS lesions.[276] In the CNS, pathology is most common in the hippocampus, olfactory, entorhinal, and parahippocampal cortex, periventricular neuropil, and ventricles. In subacute cases, hippocampal pyramidal neurons may be segmentally swollen and brightly eosinophilic, consistent with DA toxicosis in other animals and humans; however, many sea otters with fatal DA toxicosis instead have segmental bands of shrunken, hyperchromic necrotic pyramidal neurons in the same location, possibly because this "classical" lesion of DA toxicosis is relatively transient in sea otters. Circumventricular organs are also important DA targets, and microscopic examination of the pituitary gland, area postrema, other circumventricular organs, and both eyes facilitates confirmation of acute DA toxicosis.[276]

DA-associated histologic lesions are also common in the heart and may be more severe than the CNS lesions in animals with previous sublethal DA toxicosis. Progressive cardiomyocyte loss and arteriosclerosis occur in the left ventricular free wall, papillary muscles, cardiac apex, atria, and coronary arterioles.[276] Diagnosis of DA toxicosis is based on a preponderance of evidence approach that considers all findings from the clinical history, necropsy, histopathology, and biochemical testing. Because toxin clearance can be very rapid (e.g., hours), a low or negative biochemical test does not necessarily exclude DA as a cause of death. Cases of chronic DA toxicosis often present with cardiovascular pathology that is more severe than the CNS pathology; however, the lesions at both anatomic sites are relatively quiescent, reflecting previous damage.

Saxitoxins are a diverse group of neurotoxic biotoxins composed of saxitoxins, neosaxitoxin, gonyautoxins, and decarbamoylsaxitoxin. They are produced by marine dinoflagellates (e.g., *Alexandrium* spp.) and some freshwater cyanobacteria (e.g., *Cylindrospermopsis* sp.). As with other biotoxins, saxitoxins can biomagnify within food chains. One of the most potent natural toxins, saxitoxins act on voltage-gated sodium channels of neurons, causing paralysis and asphyxiation. Saxitoxin has been detected in biologic samples from northern and southern sea otters; one northern sea otter was thought to have died from saxitoxin poisoning, and two other boat-strike fatalities were potentially associated with saxitoxin toxicity, including possible chronic low-level exposure.[48] Due to their mechanism of action, there are no definitive gross or microscopic lesions. Diagnosis is achieved through assessment of the case history, toxicological tests, and exclusion of other cases of death.

Botulism is a paralytic disease caused by a potent neurotoxin that blocks presynaptic release of acetylcholine in peripheral nerve endings at neuromuscular junctions. Botulinum neurotoxin is produced during exponential growth of the anaerobic bacterium ***Clostridium botulinum***. The usual source of the toxin is decaying carcasses or vegetable. Based on their serological properties, botulinum neurotoxins are classified into types A through G; types C and D cause disease in animals.[245] The specificity of motor neuron surface receptors varies for different botulinum toxins, and animal species differ in susceptibility. Fur production animals such as mink are especially prone to large-scale botulism outbreaks that can affect thousands of individuals.[245] Botulism in mink is usually caused by type C strains of *C. botulinum* that produce toxins in chopped raw meat or fish, but type A and E strains are occasionally involved. Extensive mortality of unvaccinated mink can occur within 18–96 h of consuming toxin-laden food. Clinical signs of botulism are characterized by lassitude, acute muscular incoordination, and stiffness, progressing to limb paralysis, dyspnea, and death, with mortality as high as 100%.[154,160,245] Death is usually due to respiratory or cardiac paralysis. Milder cases may exhibit anorexia, and animals may have a worse outcome following repeated toxin exposure.

Ferret deaths from *C. botulinum* type C and type D exposure have been reported with increased susceptibility in commercially raised animals; type D *C. botulinum* is considered the most pathogenic. Clinical signs of botulism toxicity in ferrets include muscle incoordination and stiffness that progresses to respiratory paralysis and death.[172,376] No published reports of botulism have been described in skunks. Traditional diagnostic tests include the mouse inoculation test or response to botulinum antitoxin. PCR assays are also available to identify *C. botulinum* type C bacteria and the type C botulinum toxin.

Although numerous studies document detection of **anthropogenic chemicals**, such as **pesticides**, in mustelid and skunk tissues,[208,295] linking exposure with morbidity and mortality is often challenging. In studies of southern sea otters, animals that died from infectious diseases, neoplasia, and emaciation contained higher concentrations of

DDTs, butylins, perfluorooctanoic acid, and perfluorooctane sulfonic acid than those that died from trauma or other unknown causes.[208,209,295] Virtually all mustelids and skunks are presumed to be exposed to anthropogenic chemicals via their environment and diet, including organochlorine pesticides, herbicides, PCBs, and fire retardants. These compounds are extremely diverse in terms of their structure, persistence, and routes and mechanisms of toxicity. All organ systems can be affected, and potential negative impacts have been reported for immune function, reproduction, and endocrine homeostasis. However, correlations between these contaminants and specific diseases have not been determined. Due to the global diversity, pervasiveness and environmental persistence of many anthropogenic chemicals, as well as their ability to biomagnify within food webs and pass to neonates via transplacental and lactational transfer, more research is greatly needed. Given the indirect and often relatively obscure mechanisms of toxicity for most of these compounds, potential health impairment is usually reported as nonspecific impacts on animal weight, reproduction,[25,26] survival, immune competence, or other factors.[241]

Descriptions of specific gross and microscopic pathology associated with anthropogenic chemical exposure in mustelids and skunks are relatively limited. Eurasian otters exhibiting retinal dysplasia have significantly higher concentrations of **dieldrin** (an organochlorine pesticide), and lower concentrations of vitamin A than otters with normal retinas. Microscopic lesions include retinal folds and rosettes, retinal detachment, and retinal pigmented epithelial hypertrophy.[477] Confirmation of exposure is achieved via biochemical testing of tissues and body fluids.

Congenital/Genetic

A variety of congenital diseases occur sporadically in mustelid species. Many documented congenital diseases or abnormalities are listed in Table 11.e3. Color dilution is associated with multiple congenital or genetic diseases in mink and ferrets. The high prevalence of nephrolithiasis (see "Metabolic" setion) in some mustelid species suggests the possibility of shared heritable metabolic abnormalities; however, this has not been proven.

Age-Related/Degenerative

Age-related diseases are common in mustelids. These include **periodontal and dental disease**, **degenerative joint disease**, and **intervertebral disc disease**. The prevalence of periodontal disease in sea otters and beech martens is estimated to be 74.4% and 47.7%, respectively.[412,485] In sea otters, marked tooth wear, fractures, tooth root abscesses, and tooth loss are common in older

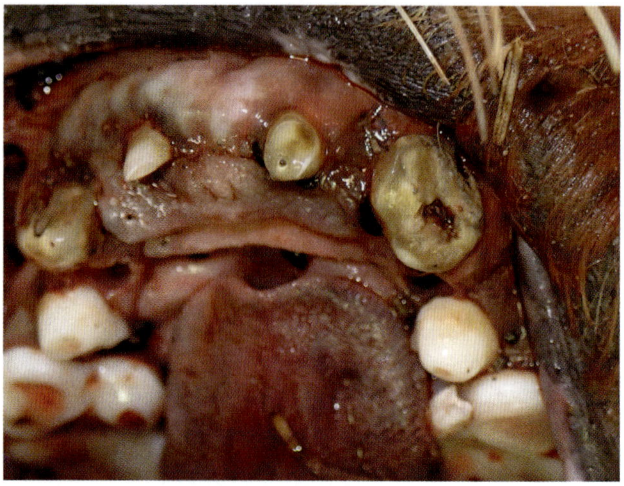

FIGURE 11.12 Dental disease and attrition in a northern sea otter. Some incisors are missing, and remaining incisors and canine teeth are significantly worn or fractured just above the gingival margin with pulp cavity exposure in a northern sea otter. Adjacent gingiva is hyperemic and hyperplastic. *(Photo Courtesy of Alaska Veterinary Pathology Services.)*

animals. Severe periodontal disease in older wild otters (Fig. 11.12) may result in emaciation, secondary bacterial infection, and can be a primary or contributing cause of death.[277,485] Review of >1,000 archival southern sea otter skulls revealed that temporomandibular joint osteoarthritis was common, affecting 4.1% of archival specimens; this may reflect in part their hard diet of bivalves, snails, and crabs.[13] In captive black-footed ferrets, dental disease increases with age and is thought to be related to an unnaturally soft diet.[11] Aged American badgers and wolverines also often have marked dental attrition, thought to be a byproduct of masticating dirt during consumption of ground-dwelling rodents.

Thoracolumbar **intervertebral disc herniation** is an important differential for paraparesis in striped skunks and ASCOs.[228,367] Though perhaps not an aging change, another differential for hind limb paralysis in a mustelid would be **fibrocartilaginous emboli**, which have been documented in an 8-year-old captive tayra, associated with unilateral gray and white matter malacia in the lumbar spinal cord. Affected spinal veins may contain multiple, pale basophilic, amorphous emboli with lacuna-like structures consistent with cartilage.[361]

Age-related ocular disease has been described in ferrets and mink. **Cataracts** are common in ferrets. Progressive cataract formation has been reported in genetically unrelated populations of laboratory ferrets, and in one population, 47% of animals were affected. In some populations, persistent fetal vasculature has been implicated as a cause of cataracts in young ferrets, and was shown to be inherited as an autosomal dominant defect.[456] In most cases, cataracts develop in middle age and progress over time.[247] **Central retinal degeneration**, characterized by degeneration of rods, cones, and outer retinal layers, is also relatively common in older ferrets and mink.[162] In ferrets, retinal degeneration may accompany cataract formation. Older ferrets may also spontaneously develop glaucoma with buphthalmia. Chronic corneal edema is seen in older mink (8–11 years), resulting from **spontaneous degeneration of the corneal endothelium**. Royal pastel color variant females are predisposed, and the condition is persistent, bilateral, and rarely associated with ulceration.[165,456]

Various presumed congenital **cystic urogenital lesions** have been described in older ferrets and sea otters.[95,242] The incidence of renal cysts (Fig. 11.13) ranges from 32% to 63% of ferrets; cysts can be single or multiple and are generally unilateral. Rarely, this condition presents as polycystic kidney disease[95] and may be associated with other concurrent developmental anomalies, end-stage kidney disease, or mesangial glomerulopathy.[198] Urethral gland distention and cyst formation occur in older female mink, leading to pressure atrophy and urethral wall degeneration.[166] Predisposing causes are unknown. Chronic interstitial nephritis is commonly found in geriatric ferrets including interstitial nephritis, pyelonephritis, glomerulonephritis, and immune complex-mediated glomerulonephropathy.[95] As with dogs, cats, and humans, older age is a common risk factor for mustelid nephrolithiasis (see "Metabolic" section). It is also common for adult and aged sea otters to develop one or multiple thyroid cysts, hyperplastic thyroid nodules, or benign neoplastic thyroid masses (Fig. 11.14).

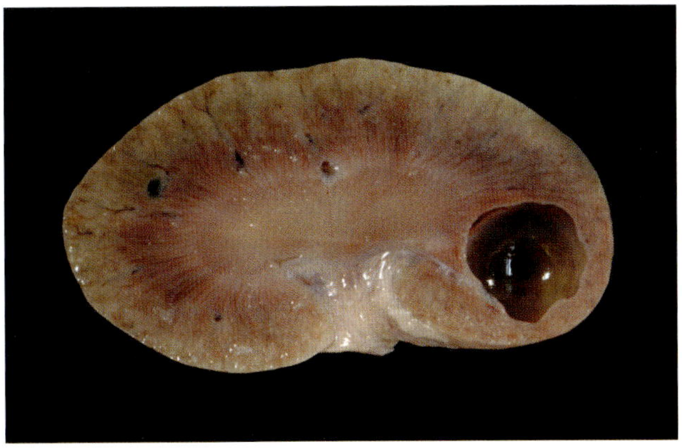

FIGURE 11.13 Renal cyst in a domestic ferret. A single, discrete, cyst is present within the renal cortex with gross compression of the adjacent cortex and medulla. The pale tan to brown renal cortex and pale streaking of the medulla are suggestive of concurrent chronic kidney disease. *(Photo Courtesy of C. Bertram, Freie Universität Berlin.)*

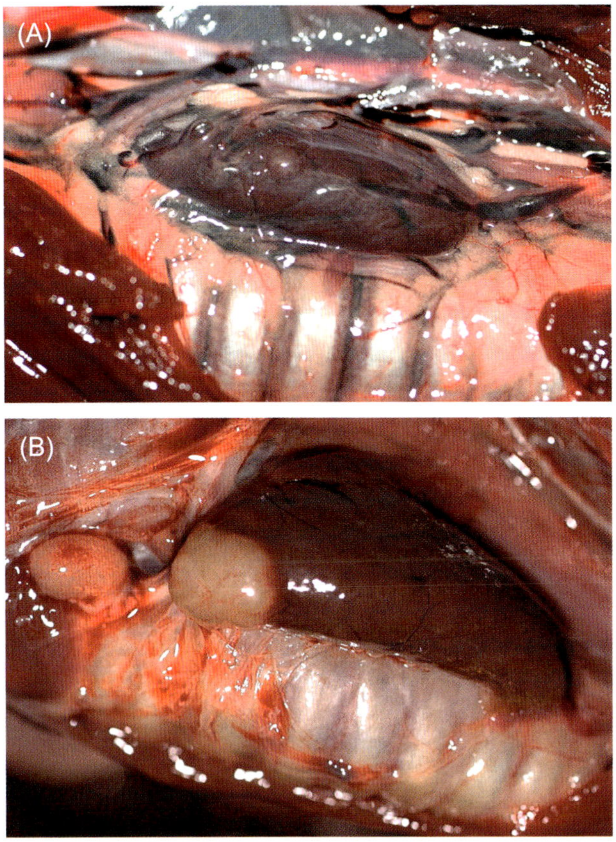

FIGURE 11.14 Thyroid gland lesions in older sea otters. (A) A small cyst is present at the capsular surface, and (B) a larger, discrete, tan cystic mass containing nodules of hyperplastic or benign neoplastic thyroid epithelium. A normal pink elliptical parathyroid gland is also visible at the top left in both photos. *(Parts A, B: Photos Courtesy of the California Department of Fish and Wildlife and The Inner Otter—Sea Otter Care and Science Resource Center.)*

Inflammatory Non-Infectious

Several inflammatory diseases with a possible underlying **immune-associated** etiopathogenesis have been reported over the past 25 years in ferrets. **Inflammatory bowel disease (IBD)** is common in middle-aged and older ferrets. IBD is characterized histologically by intestinal mucosal lymphocyte infiltration, epithelial damage, and marked villous blunting and fusion. The etiology of this condition in ferrets remains unknown, however, a complex interaction between dietary proteins that are not part of their natural diet, gastrointestinal microbiota, and the mucosal immune system is suspected.[50] **Eosinophilic gastroenteritis** is associated with systemic hypereosinophilia and variably severe, predominantly eosinophilic inflammatory infiltrates in multiple tissues. Documented lesions include mural eosinophilic infiltrates within the intestine, occasionally resulting in effacement of the muscular tunic, eosinophilic gastritis, eosinophilic interstitial pneumonia, eosinophilic vasculitis, and eosinophilic granulomas with Splendore-Hoeppli material in the mesenteric lymph nodes, liver, mesentery, and choroid plexus. The etiology is not known, but an allergic response is suspected.[129]

An **idiopathic systemic myofasciitis and/or myositis** affecting 5 to 24-month-old ferrets has been documented. Characteristic findings include severe pyrexia and neutrophilia, as well as infiltration of skeletal muscles (almost always including the esophagus) with severe suppurative to pyogranulomatous infiltrates in the absence of demonstrable infectious agents.[139,140] Serum creatine kinase levels are rarely elevated in affected individuals and muscle biopsy is required for antemortem diagnosis. Idiopathic myofasciitis was initially thought to be associated with a specific ferret distemper vaccine; Fervac-D. While the incidence of this condition declined following removal of this vaccine from the market, additional cases have been identified and the inciting cause remains unknown.[192] **Masticatory myositis** of presumed immune-mediated etiology has been documented in mink. The condition is characterized by myodegeneration and myonecrosis of the temporal and masseter muscles, with a prominent eosinophilic infiltrate and diffuse subcutaneous edema over the dorsal head.[296] Mink masticatory muscles possess unique type 2M fibers, suggesting a similar pathogenesis to canine masticatory myositis. Rare cases of **erythema multiforme** and other autoimmune skin diseases have been identified in domestic ferrets with possible association to underlying adrenal disease.[123]

Endogenous lipid pneumonia is a common, incidental gross and histologic finding in older mustelids (otters, ferrets) that is not associated with any underlying infectious or cardiovascular disease process. It is characterized by variably sized, often subpleural, intra-alveolar aggregates of foamy macrophages and giant cells with variable numbers of intervening intralesional cleft-like colorless spaces or cholesterol crystals. Grossly, these aggregates appear as pale pink to white, slightly raised, and occasionally coalescing subpleural foci affecting one or more lung lobes (Fig. 11.e31). Severe, fatal primary endogenous lipid pneumonia has been reported in a young domestic ferret; large portions of alveolar spaces and entire sections of lung were replaced by cholesterol crystals, histiocytic inflammation, and septal fibroplasia with associated pleural effusion.[331]

Miscellaneous/Multifactorial

Cardiomyopathy comprises a heterogeneous group of disorders which typically manifest as conduction defects and/or mechanical dysfunction of the myocardium and may present with ventricular hypertrophy or dilatation. The underlying pathogenesis may originate in the heart or may represent sequelae from other systemic processes. Cardiomyopathies are common in specific bloodlines of ferrets, older striped skunks,[28] sea otters, ASCOs, and North American river otters under human care.[367] In the domestic ferret, it is the most common cause of cardiac disease, is considered congenital or heritable, and may present from 6 months of age to geriatric individuals. While both dilatative and hypertrophic forms have been identified; the dilatative form is more common in ferrets (Fig. 11.15A) with unilateral or biventricular cardiac involvement and occasional eccentric ventricular hypertrophy. Pulmonary edema with thoracic and abdominal effusion and chronic passive hepatic congestion may occur. The pathogenesis of ferret cardiomyopathy is unresolved.

Histologic lesions associated with dilated cardiomyopathy in mustelids may be subtle or obvious and can include variation in myofiber size, myocyte degeneration, necrosis or compensatory hypertrophy, and variable interstitial edema or fibrosis (Figs. 11.15B–D). Myofiber disarray is not commonly encountered. In contrast, hypertrophic cardiomyopathy typically presents with enlarged myocardial fibers with prominent vesiculated nuclei and myocyte disarray. Interstitial fibrosis is prominent with occasional myocardial necrosis, intramural coronary vascular medial hypertrophy, and lymphohistiocytic infiltrates. Mild lesions are often first identified in perivascular areas throughout the ventricles, and to a lesser extent, the interventricular septum. Foci of myocardial fibrosis are more common in older animals of all mustelid species and affected animals often show little or no outward evidence of cardiac decompensation. The degree of fibrosis observed on histopathology often does not closely correlate with the degree of clinical impairment. Concurrent evidence of cardiac insufficiency or congestive heart failure, such as chronic hepatic congestion, pulmonary siderosis, pulmonary edema, and ascites, is helpful in corroborating the histologic diagnosis of cardiomyopathy.

Cardiomyopathy was identified as a common primary or contributing cause of death in southern sea otters in two mortality reviews.[229,277] In contrast, this condition is uncommon in northern sea otters. Grossly, affected animals exhibit variably dilated, orange/white-mottled myocardium, often with evidence of left and/or right heart failure (Fig. 11.16). As with ferrets, the extent of myocardial dilation, inflammation, and fibrosis may not always correspond with the severity of clinical disease. On histopathology, sea otter cardiomyopathy may lack inflammation or display mild to severe lymphocytic to pleocellular myocarditis with progressive myofiber loss and replacement

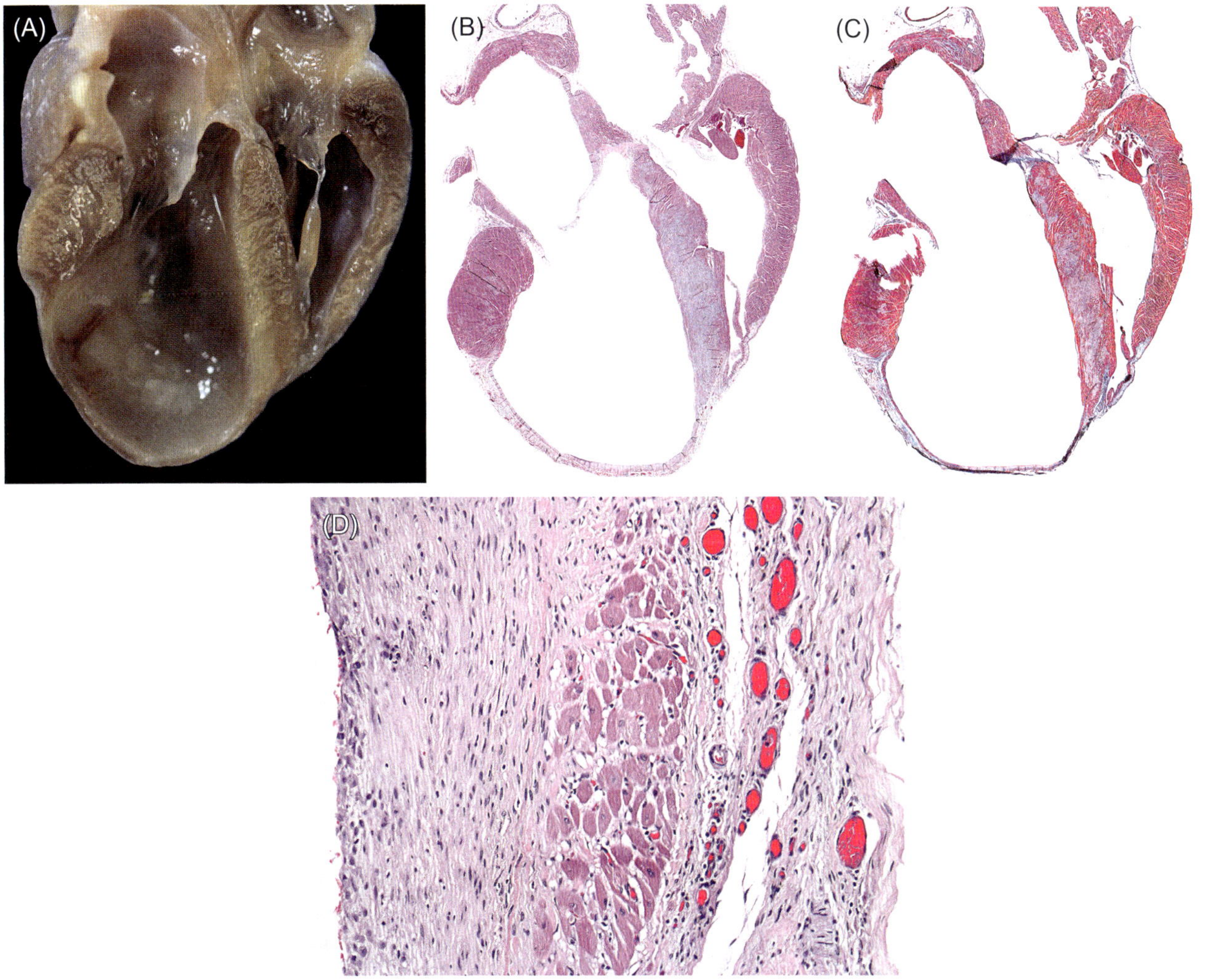

FIGURE 11.15 Dilated cardiomyopathy in the heart of a ferret. (A–C) Severe left ventricular dilation with cardiomyocyte loss, mural attenuation, and fibrous connective tissue replacement. Masson's Trichome (C). (D) The ventricular wall is markedly thinned with extensive cardiomyocyte loss and fibrous tissue replacement. Many remaining cardiac myocytes are atrophied and compressed. *(Parts A–D: Photos Courtesy of T. Donovan, Schwarzman Animal Medical Center.)*

fibrosis; the amount of associated inflammation is likely dependent on the primary inciting cause of cardiomyopathy (protozoal, toxic, or ischemic). *Sarcocystis neurona* or *Toxoplasma gondii* infection, and sublethal domoic acid (DA) toxicosis are significant risk factors for development of cardiomyopathy in prime-age and aged sea otters.[229,277,288] Additional information regarding predisposing toxic and protozoal causes of sea otter cardiomyopathy is mentioned in their respective sections in this chapter. ASCOs most commonly have gross and histologic lesions consistent with biventricular congestive heart failure, whereas North American river otters generally have lesions consistent with right-sided heart failure.[367]

In free-ranging mustelids, **trauma** is a significant cause of mortality, with vehicular strikes, entanglement, entrapment, ballistic trauma, interspecific predation, and conspecific aggression and infanticide among the most common causes of traumatic death. Rural development and habitat disruption, fragmentation, and destruction have had profound effects on forest specialists such as fishers, pine martens, American martens, and other endangered or threatened species in more geographically confined regions. Vehicle strikes tend to occur along waterways, dykes, or on high-traffic roadways and bridges.[344] In many instances, animals present with catastrophic blunt force injuries, comminuted fractures of the calvarium, vertebral column, ribs, and long bones, particularly toward the cranial

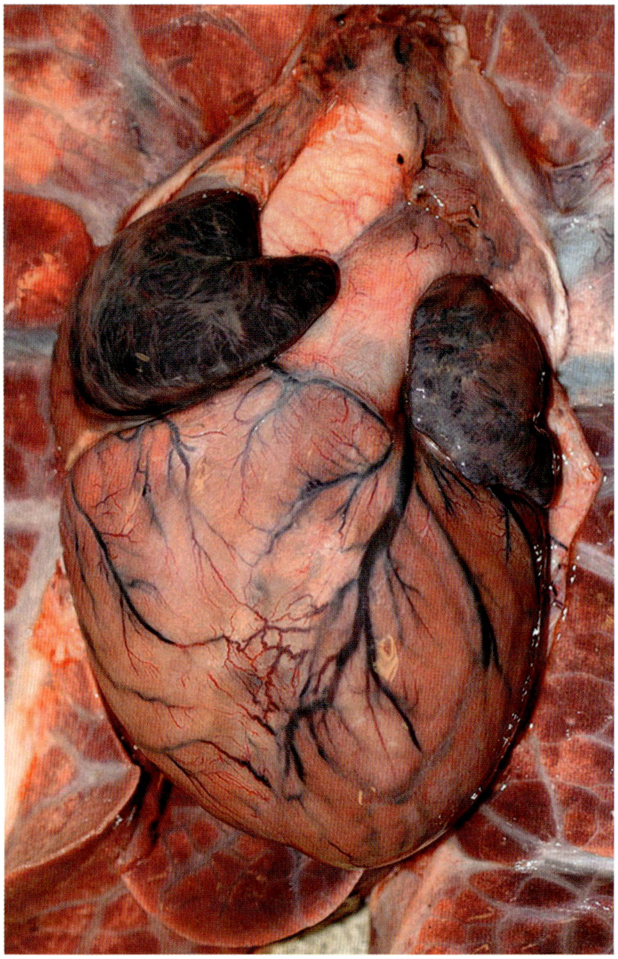

FIGURE 11.16 Cardiomyopathy in a southern sea otter. Biventricular pallor and orange discoloration, dilation of the right ventricle, and congestion of coronary vessels are present. Pulmonary septal fibrosis is also visible and is secondary to left heart failure. *(Photo Courtesy of the California Department of Fish and Wildlife.)*

portion of the torso. Vehicular trauma has been identified as the most common cause of death for stoats and polecats in the United Kingdom.[410]

Anthropogenic injuries inflicted on sea otters include incised or chopped cutaneous wounds associated with propeller strike, blunt force trauma due to vessel strike (Fig. 11.17A), gunshot, and entrapment with drowning in crab and fish gear or other anchored structures.[176,277] Similar reports of human interactions have occurred with marine otters with evidence of demersal gillnet entanglement.[252,266,343] Anthropogenic trauma has also been documented in Eurasian, neotropical, and giant otters, martens, badgers, and adult wolverines.[353,370] In some geographic areas, habitat destruction has contributed to mustelid depredation on farms, leading to poaching, trapping, and local genetic inbreeding effects.[1,122,333]

In addition to human-associated trauma, predation, cannibalism, infanticide, abandonment, and intraspecific

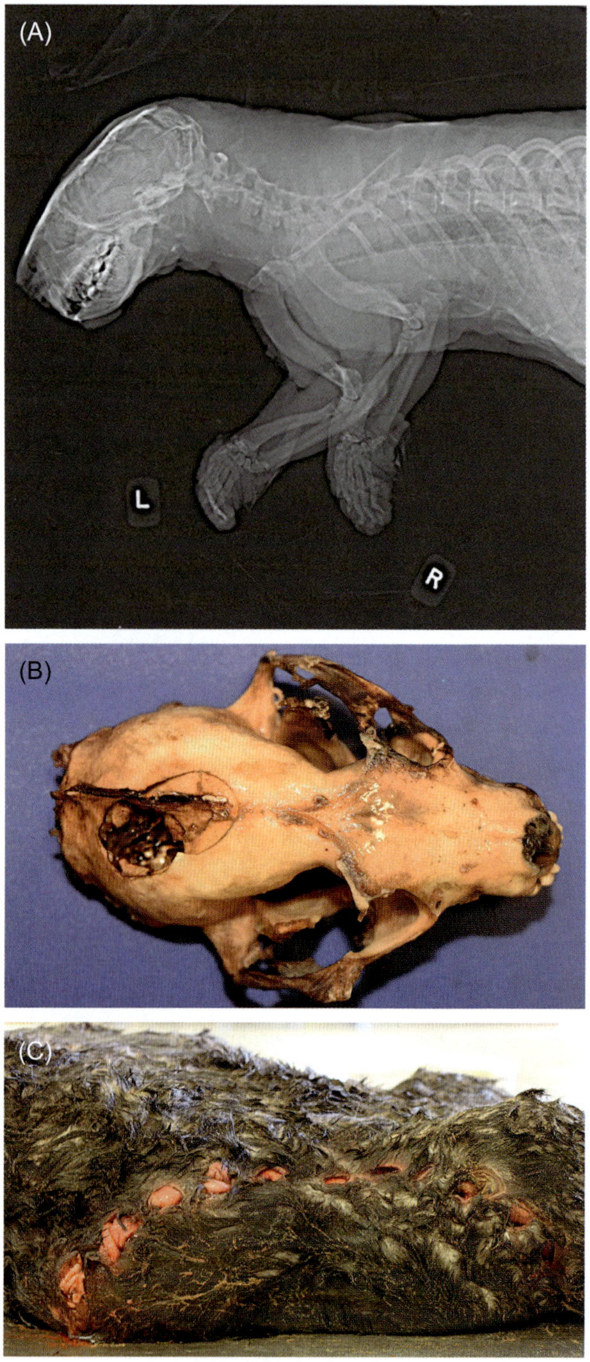

FIGURE 11.17 Traumatic injuries in southern sea otters and a pine marten. (A) The skull of this subadult southern sea otter is fractured in multiple places due to blunt force trauma from a boat strike. (B) The dorsal aspect of this pine marten skull contains a large hole that is inwardly beveled, indicative of crushing forces, as from a predator. Other puncture wounds and muscle tears were present (not pictured) in this animal to support the diagnosis of predator trauma rather than a gunshot wound. (C) A curvilinear arrangement of regularly sized full-thickness lacerations is present over the body wall consistent with a shark bite injury in this southern sea otter. *(Part A: Photo Courtesy of R. Kagan, National Fish and Wildlife Forensic Laboratory. Part B: Photo Courtesy of T. Viner, National Fish and Wildlife Forensic Laboratory. Part C: Photo Courtesy of the California Department of Fish and Wildlife.)*

aggression are also significant contributors to morbidity and mortality in specific mustelid species, such as giant and marine otters, Pacific martens, and juvenile wolverines.[107,266,291,334,417] Gross findings include partially consumed or buried carcasses and depredated remains. The gross appearance (Fig. 11.17B) and anatomic location of bite wounds (e.g., punctures, depressed skull fractures, or consumption of most or the entire skull) may facilitate differentiation of predation events by badgers, cougars, feral and domestic cats, raptors, and coyotes. Molecular screening of feces has confirmed predation of least weasels by domestic cats; this innovative approach may be valuable to identify common predators.[342,431,472] Domestic cat predation exceeded deaths from vehicular trauma in weasels almost 3:1 in the UK.[410] Common predators of black-footed ferrets include coyotes, owls, and badgers.

Predation of sea otters by bald eagles, sharks, and killer whales has been observed in different regions along the Pacific coast of North America.[10,366,467] These cases are confirmed following recovery of floating sea otter remains from predation events or from eagle nests. Shark attacks are not invariably lethal, but bite wounds may be difficult to interpret. Presence of sequential or multifocal stab-like wounds (which may or may not have a curvilinear alignment) with intralesional parallel bone scratches due to contact with serrated edges of shark teeth or embedded serrated tooth fragments confirms diagnosis of white shark (*Carcharodon carcharias*) bite (Fig. 11.17C).[6,277,444] To date, predation by other shark species has not been reported in sea otters.

Traumatic wounds associated with conspecific attacks are also commonly reported in sea otters.[277] In male southern sea otters, healed baculum fractures (Fig. 11.18), fractures or luxations of the digits of the hind feet, and trauma to the testicles with associated orchitis and epididymitis are observed as sequela to a lifetime of territorial fighting.[277,326] Percutaneous punctures attributed to bite wounds to the nasal planum (Fig. 11.19), face, and neck, as well as colonic and vaginal perforation, can be inflicted by males upon females during the breeding season. Mating-associated nose wounds are often more severe in southern sea otters than in other otter populations.[171,277]

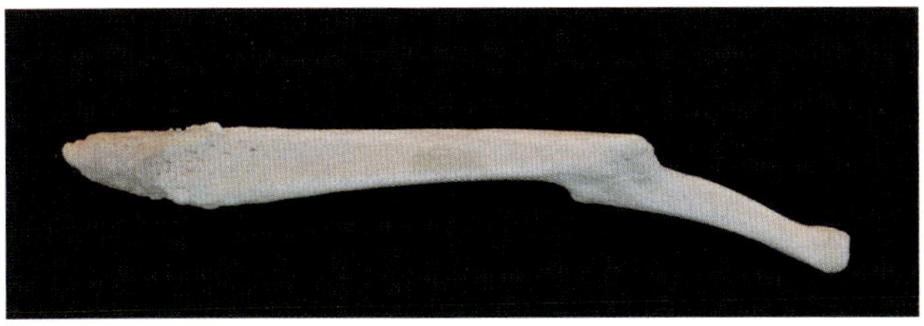

FIGURE 11.18 Healed subluxated baculum fracture in an adult male southern sea otter. These lesions are commonly associated with intraspecific fighting. *(Photo Courtesy of A. Reed, California Department of Fish and Wildlife.)*

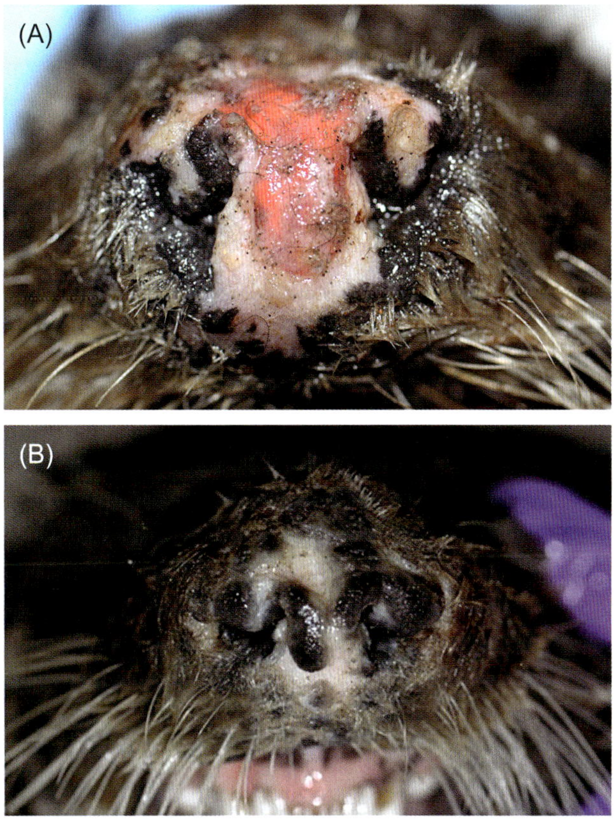

FIGURE 11.19 Copulation-associated nasal wounds in female southern sea otters. Subacute (A) wounds are partially scarred and ulcerated with loss of nasal pigmentation and early deformity of the nasal planum. Chronic (B) wounds have significant scarring, shrinking, and depigmentation of the nasal planum and surrounding skin. *(Parts A, B: Photos Courtesy of the California Department of Fish and Wildlife.)*

Systemic or regional **amyloidosis** has been identified in mustelids and skunks, but documentation is most complete in mink in which serum amyloid A (SAA) protein has been sequenced. In one review, tissues from 24% of 68 mink >3 years old were amyloid-positive, compared with 3% prevalence for 68 mink <3 years. SAA is most common in the cervical salivary glands, but is also detected in the liver, spleen, kidney, heart, adrenal, tongue, stomach, intestine, pituitary, and pancreas.[391] A study of black-footed ferrets from eight US zoological institutions revealed SAA in the renal glomeruli and other tissues in association with proteinuria and emaciation.[139,140] The condition in black-footed ferrets has been associated with a lack of genetic heterogeneity for captive populations and supervening disease.[139,140]

Individual case reports of amyloid light-chain (AL) amyloidosis of undetermined pathogenesis have been reported in beech martens.[37,383] On gross examination, firm white nodules are present throughout the heart and kidney along with splenomegaly, hepatomegaly, and pallor or miliary white foci throughout affected tissues (Fig. 11.20A). Amyloid deposits are observed in renal

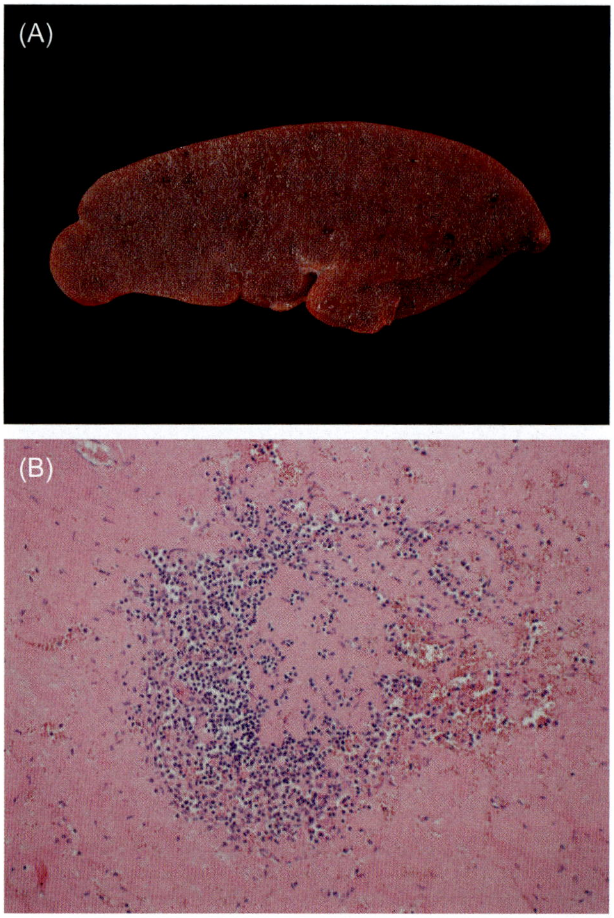

FIGURE 11.20 Splenic amyloidosis in a beech marten. (A) In the cut section, the splenic parenchymal margins are rounded and the parenchyma is diffusely mottled red to pink with a waxy appearance. (B) Large aggregates of amyloid are present throughout the splenic red and white pulp, partially obscuring a lymphoid follicle. *(Parts A, B: Photos Courtesy of C. Bertram, Freie Universität Berlin.)*

glomeruli, coronary and renal vessels, splenic parenchyma (Fig. 11.20B), and hepatic sinusoids, where they may be associated with degeneration of intervening hepatic cords. Congo red stain with 5% potassium permanganate confirms AL amyloid. Similar gross and microscopic findings of amyloidosis have also been reported in a skunk, badger, and Eurasian otter.[113,199,383] In martens, amyloidosis has been associated with endo and ectoparasitism, and as a sequela to septicemia.[2] A striped skunk presented with generalized alopecia of the pelvic and thoracic limbs, cutaneous papules, and hepatorenomegaly with no predisposing or preexisting pathology identified in examined tissues.[113] In addition to vascular, renal, hepatic, splenic, and lymph node amyloidosis, histopathology revealed cutaneous periadnexal amyloid deposition with epidermal atrophy, and AA amyloidosis was confirmed by Congo red staining and immunohistochemistry for mc4 antibody. A wolverine exhibited a unique amyloid-beta (Aβ)-associated cerebral angiopathy with neuritic plaques, neurofibrillary tangles, and granulovacuolar degeneration, similar to lesions associated with Alzheimer's disease in humans.[369]

Hemorrhagic enteropathy is frequently identified in river and sea otters and captive ASCOs.[367] Grossly, the gastric and or intestinal lumen is filled with clotted blood. Histologically, this condition is characterized by prominent blood pooling in the intestinal mucosa and submucosa, with minimal damage to the mucosa; associated bacterial infection has not been observed.[236] Considering the absence of opportunistic pathogens in such cases, underlying stress, shock, and/or cardiac insufficiency should be considered, depending upon case findings. Additionally, while not associated with hemorrhagic enteropathy, due to their long small intestine and small localized mesenteric root, sea otters with gastrointestinal disease are predisposed to development of intestinal **volvulus**.

Pododermatitis/fur animal epidemic necrotic pyoderma (FENP) is a common and important problem in farmed mink. Genetic factors, reproductive status, immune status, various bacteria, and environmental parameters such as cold weather and management are possible contributors. This disease was first recognized in the US and Canada where disease onset coincided with use of seal byproducts in mink feed. Acute disease presents as multifocal footpad and nail bed ulcers, and chronic infections are characterized by footpad hyperkeratosis and lesions on other parts of the body, particularly the legs and facial skin. Systemic pathology such as pleuritis and bronchopneumonia are also observed. Histologic changes include orthokeratotic hyperkeratosis, folliculitis, and dermatitis. Cases with partial or complete epidermal ulceration may contain neutrophilic infiltrates, and gram-positive cocci and microabscesses may be observed in intact skin. Subcutaneous suppurative vasculitis and thrombosis may also be present. Opportunistic *Staphylococcus* spp. (including *S. canis*, *S. intermedius*, and *Staphylococcus delphini*), *Streptococcus* sp. and *Arcanobacterium phocae* are common isolates.[70,201] *Arcanobacterium phocae* alone or in conjunction with *Streptococcus* spp. can reproduce FENP in inoculated mink (Fig. 11.21),[307,308] and *A. phocae* has also been reported as an opportunistic pathogen secondary to bite wounds, resulting in abscesses and systemic infections in sea and Eurasian otters.[202,306]

Starvation resulting from suboptimal nutrition can manifest as generalized emaciation, and reflect changes in prey availability, habitat disruption, competition with introduced or sympatric species, high population density, and other exogenous or endogenous stressors. These phenomena may contribute to insufficient caloric intake or suboptimal nutritional quality, may impair homeostasis, and can affect individuals or entire populations. Failure to acclimate has also been reported during efforts to

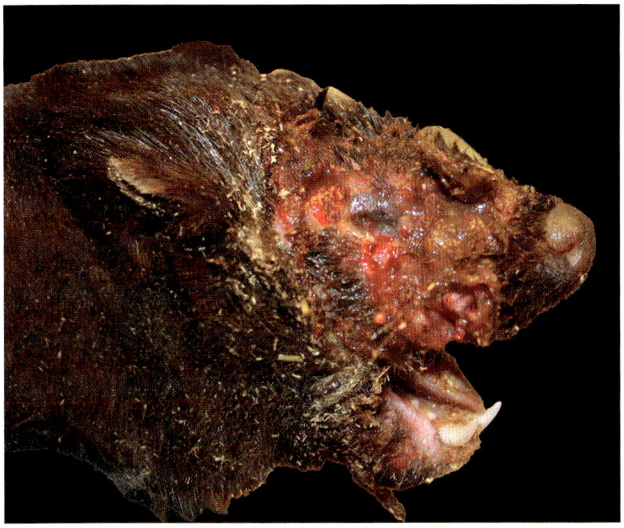

FIGURE 11.21 *Arcanobacterium phocae* infection in an American mink. Severe necroulcerative dermatitis and cellulitis affect almost the entirety of the facial skin. These types of lesions occur in fur animal epidemic necrotic pyoderma (FENP). *(Photo Courtesy of T. Kare Jensen, National Veterinary Institute of Denmark, Technical University of Denmark.)*

reintroduce black-footed ferrets into compatible habitat from captive propagation programs. In both free-living mustelid and mephitid populations and managed care facilities, perinatal or neonatal starvation is common and may be secondary to in utero malnutrition, where the dam has sustained an insufficient nutritional plane for fetal development, especially during the last trimester. Other potential causes include postpartum maternal separation, loss, mismothering, or neglect. Congenital malformations, dystocia, infections, and nonviable perinates may also contribute to poor body condition.

Affected neonates and pups may exhibit hypothermia, hypoglycemia, anemia, paresis, seizures, or obtundation. Gross lesions may include generalized emaciation, dehydration, matted fur, an empty gastrointestinal tract, and diffuse hepatic and pancreatic atrophy.[277,473] Microscopically, cytoplasmic condensation of adipocytes, myocyte atrophy, hepatic lobular collapse and hemosiderosis, and diffuse pancreatic atrophy with zymogen depletion from exocrine pancreatic acinar cells may support the gross findings. Insufficient colostrum consumption may predispose neonates to emaciation due to an enhanced risk of secondary or opportunistic infections.

End lactation syndrome (ELS) in southern sea otters is characterized by moderate to severe emaciation in females dying during late pup care or postweaning that is not attributable to a concurrent, independent disease process.[75] Peak ELS prevalence occurs in late spring, possibly reflecting the population trend toward fall/winter pupping in southern sea otters. Increasing age and number of pregnancies are associated with a higher risk of fatal ELS. In one retrospective study, ELS accounted for 58% (106/183) of reproductive-age female sea otter deaths as a primary or contributing cause of death.[277] Although the proportion of ELS females is highest in areas with dense sea otter populations, cases have been recovered throughout the range, suggesting that death from ELS is associated with, but not solely caused by, population-level resource limitation.[75]

Neoplastic

Neoplasms are common in mustelids and mephitids, and many resemble the breadth, variety, appearance, and biological behavior of tumors observed in domestic mammals. Table 11.e4 summarizes reported neoplasms in several mustelid species and skunks, and Figs. 11.e2–11.e6 show the gross and histologic appearance of several common mustelid neoplasms.

Lymphoma is one of the most common neoplasms in adult and geriatric mustelids. Gastrointestinal lymphoma, predominately of T-cell origin, is common in freshwater otters and can behave similarly to epitheliotropic lymphoma described in domestic cats, with progressive transition between lesions compatible with inflammatory bowel disease and lymphoma. Neoplastic small lymphocytes are present within the lamina propria and occasionally infiltrate the epithelium, crypts, or submucosa[359]; the mesenteric lymph nodes, spleen, and liver are documented sites of metastasis. Large cell intestinal lymphoma and lymphoma of granular lymphocytes resulting in transmural effacement of the gastrointestinal wall (Fig. 11.e32), in some cases with perforation and abdominal metastasis, are also reported.[23]

Lymphoma is the most common hemolymphatic neoplasm in domestic ferrets and often presents as widespread multicentric lymphoma involving multiple lymph nodes and/or visceral organs (Fig. 11.22). No age or sex

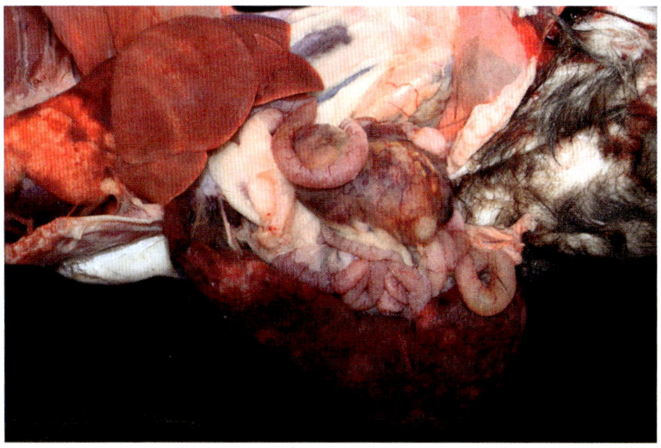

FIGURE 11.22 Lymphoma in a ferret. The enlarged, rounded, and red to tan mottled spleen, enlarged liver, and enlarged mesenteric lymph nodes demonstrate the "classical" presentation of this neoplasm in ferrets. *(Photo Courtesy of F. Del Piero, Louisiana State University.)*

predilections have been identified and clinical signs generally reflect the range of organs affected. Ferrets may have no clinical signs or may be lethargic, anorexic, and weak with diarrhea or respiratory signs. Nonregenerative anemia is the most common hematological abnormality; hypercalcemia is rare. Primary cutaneous and gastric lymphomas have also been described in ferrets.[476]

Other types of lymphoma described in mustelids and mephitids include splenic marginal zone B-cell lymphoma,[423] peripheral lymphoma,[7] thymic/mediastinal lymphoma,[216,248] primary cerebral T-cell lymphoma,[435] pancreatic lymphoma (Fig. 11.e33), and Hodgkin's lymphoma.[416] Recently, **mustelid gammaherpesvirus-2 (MuGHV-2)** was associated with B-cell lymphoma in an older European mink. Neoplastic syncytia had visible intranuclear inclusion bodies confirmed with electron microscopy to be MuGHV-2. Neuritis and a range of neoplasms have been identified in older MuGHV-2 and MuGHV-3 positive mink with unknown viral association or induction.[301]

Adrenal cortical adenomas and carcinomas are the second most common tumors in older castrated or spayed domestic ferrets. Dysregulation of the negative feedback mechanism for GnRH and resultant LH release leads to a persistent elevation of LH contributing to hyperplastic proliferation and neoplastic transformation of adrenal cortical cells. More information regarding the pathogenesis of adrenal neoplasms in domestic ferrets is present within the Metabolic Diseases section of this chapter. Adrenocortical adenomas are discrete, nodular, often >1 cm diameter proliferations of well-differentiated, often vacuolated, adrenocortical cells (Fig. 11.e34). Affected adrenal glands are grossly enlarged and cellular pleomorphism is mild with <1 mitotic figure per 10 high powered fields (HPF). Occasionally, affected and unaffected ferret adrenal glands may contain large dilated cystic spaces filled with mucinous material (Fig. 11.23) that are distinct from the neoplasm. These are suspected to be of biliary origin, given the close proximity of the right adrenal gland and liver.[454] Adrenocortical carcinomas are also large (Fig. 11.24) with similar histologic features as adenomas, except that mitotic figures are greater than 1 per 10 HPF and other features of malignancy such as metastasis, vascular invasion, and necrotic foci are present. Spindle cell and myxoid variant adrenocortical carcinomas have also been described.[16] Pheochromocytomas and abdominal neuroendocrine carcinomas of unknown origin, both frequently associated with wide-spread metastases, are much less common in domestic ferrets, but have been described in North American river otters and ASCO under human care.[367,387]

Insulinoma is the most common neoplasm in domestic ferrets and has been observed in at least one North American river otter under human care.[367] Clinical signs include weight loss and hypoglycemia. Grossly, the pancreas contains one to multiple up to 1 cm diameter red-brown, firm masses (Fig. 11.25A). Histologically, insulinomas are expansile or rarely infiltrative, and are composed of packets, nests, and cords, of well-differentiated cuboidal to columnar cells with eosinophilic granular cytoplasm, rare mitotic activity, and scant supportive fibrovascular stroma (Fig. 11.26B). Immunohistochemistry for synaptophysin, chromogranin A, and insulin facilitate diagnosis.[16]

Chordomas (neoplasms originating from remnants of the embryonic notochord) are common slow-growing

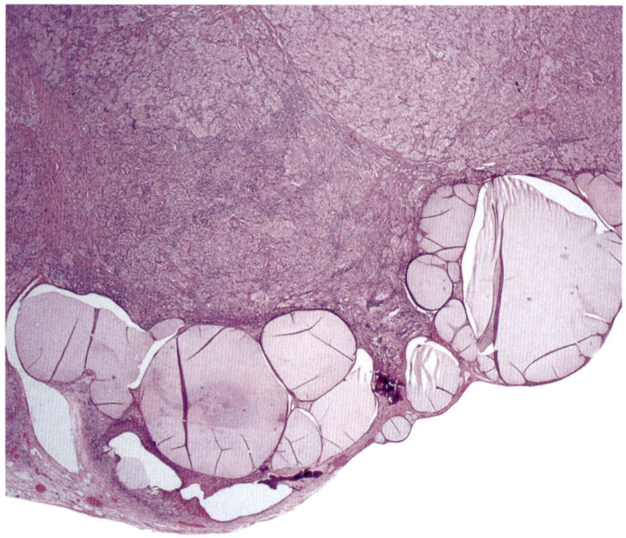

FIGURE 11.23 Adrenal gland with cortical cysts and a cortical adenoma in a domestic ferret. Low magnification view of an adrenal cortical adenoma displaying the multinodular cellular arrangement of neoplastic cells distorting the normal architecture of the adrenal gland. Multiple, peripheral cystic dilations filled with mucinous material are present in the outer cortex of the affected gland but are not a part of the neoplasm (possible biliary origin). (see Fig. 11.e34). *(Photo Courtesy of T. Donovan, Schwarzman Animal Medical Center.)*

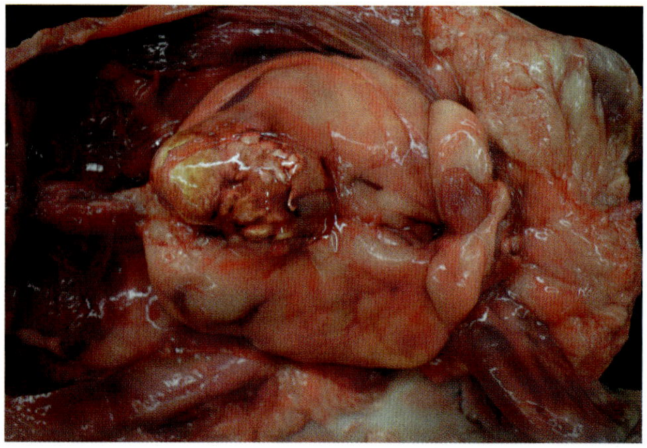

FIGURE 11.24 Adrenal cortical carcinoma in a domestic ferret. The left adrenal gland is effaced by a large multinodular, tan to pale yellow mass. *(Photo Courtesy of C. Bertram, Freie Universität Berlin.)*

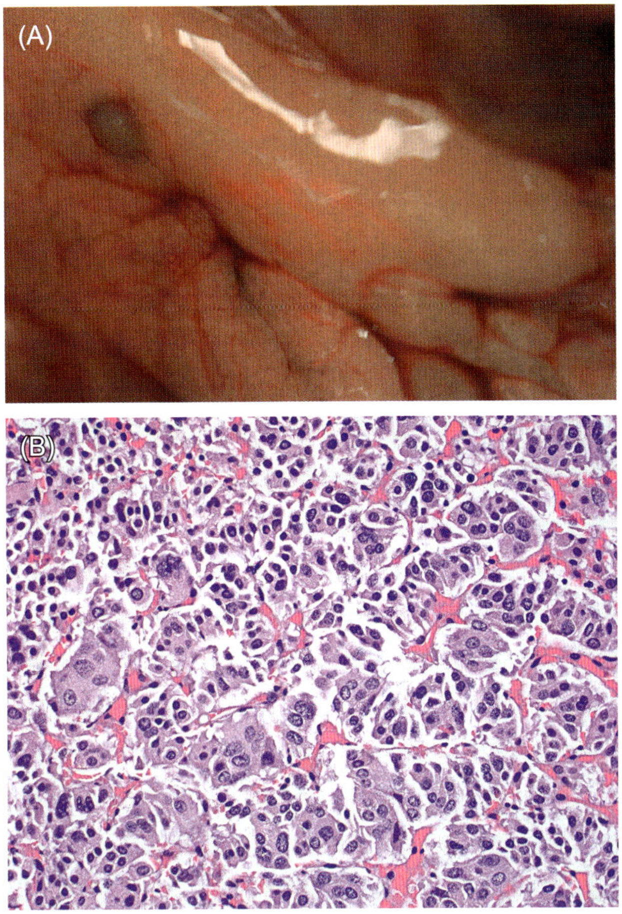

FIGURE 11.25 Insulinoma (pancreatic islet beta cell carcinoma) in the pancreas of a domestic ferret. (A) A small, tan, smooth, nodular mass is present in the left limb of the pancreas. (B) High magnification shows packets of neoplastic cells have moderate amounts of granular, amphophilic cytoplasm and round to oval, pleomorphic nuclei. *(Parts A, B: Photos Courtesy of T. Donovan, Schwarzman Animal Medical Center.)*

neoplasms of domestic ferrets that typically arise from the tail tip near the last vertebral body but occasionally occur elsewhere in the vertebral column. Grossly, chordomas present as slow-growing, discrete, firm to semifirm, multinodular, often partially alopecic masses that are firmly attached to the vertebral column. Female ferrets are twice as likely to develop chordomas compared to males.[389] Histologically, these neoplasms often exhibit three distinct zones that are circumferentially arranged (Fig. 11.26A); lobules of vacuolated polygonal cells (physaliferous cells) surround a deeper cartilaginous matrix, with a central core of trabecular bone (Fig. 11.26B). Physaliferous cells are large, closely packed, polygonal cells with distinct borders and single or multiple colorless cytoplasmic vacuoles (Fig. 11.26C). Along the outer margins of lobules, physaliferous cells are smaller with a microvacuolated eosinophilic

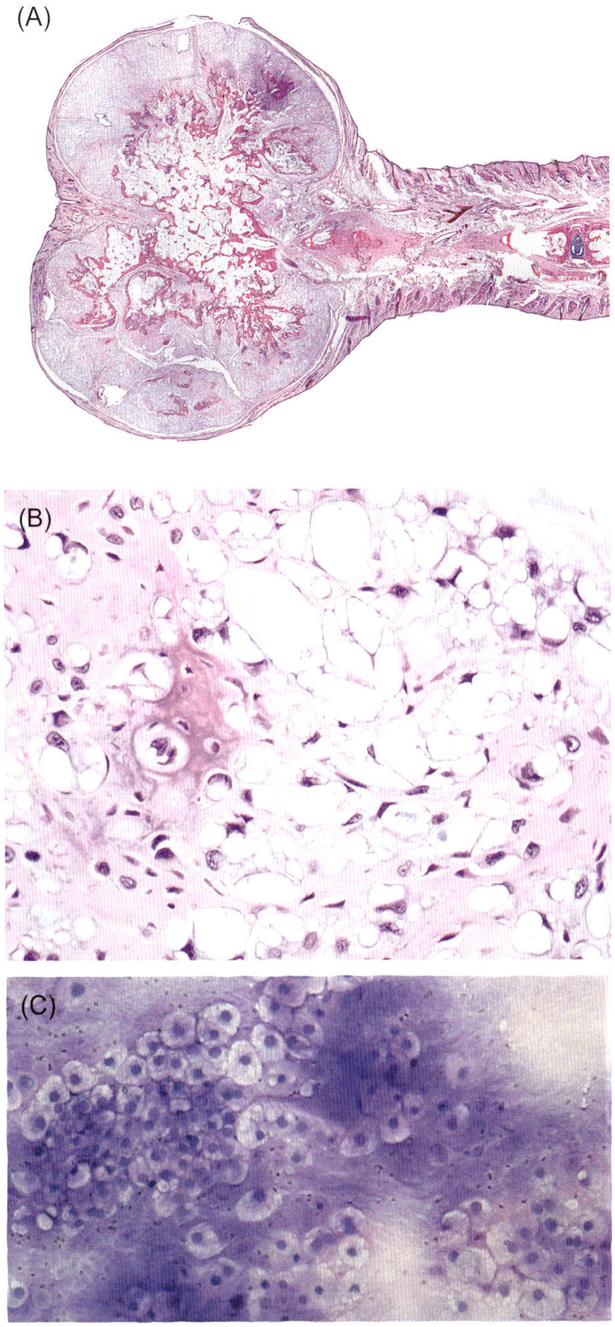

FIGURE 11.26 Chordoma in the tail tip of a ferret. (A) Expanding the tail tip is a large mass comprised of lobules of large, vacuolated (physaliferous) cells surrounding cartilage with a central core of bone. (B) Higher magnification of the mass with vacuolated (physaliferous) cells and central bone formation. (C) Fine needle aspirate of chordoma with individual or clusters of large, vacuolated, polygonal cells with condensed chromatin and no visible mitotic figures among a myxomatous matrix. Diff-Quik. *(Parts A, B: Photos Courtesy of T. Donovan, Schwarzman Animal Medical Center. Part C: Photo Courtesy of Omaha's Henry Doorly Zoo and Aquarium.)*

cytoplasm. Nuclei are round to oval with finely stippled chromatin and occasional prominent nucleoli. Mitotic figures are rare. These cells are supported by a thin fibrous stroma with abundant mucinous extracellular matrix that blends with the deeper cartilaginous and osseous layers.[106] Unlike chordomas in other species, ferret chordomas are rarely metastatic and complete excision of tail tip chordomas is usually curative.[389]

Cutaneous **mast cell tumors** are relatively common in middle age to older ferrets and often resemble the discrete, well-differentiated, superficial dermal cutaneous mast cell tumors of domestic cats. There is no anatomic or sex predilection. Masses are composed of sheets and rows of heavily granulated, minimally pleomorphic mast cells with few intervening eosinophils (eSlide 11.e2). Mitoses are rare. Masses may be multifocal but are usually restricted to one area (i.e., ears) and metastasis is not described. Toluidine blue stain reveals numerous intracytoplasmic metachromatic granules.[324]

Malignant melanomas have been documented in the Eurasian otter and ASCO. These neoplasms often arise in the skin of extremities including the tail, toe, and nose as raised, fast-growing, darkly pigmented masses. Histologically, masses are composed of pleomorphic polygonal to spindloid cells with moderate to large amounts of cytoplasmic melanin pigment, pronounced anisokaryosis, anisocytosis, and a high mitotic rate. Foci of necrosis and hemorrhage are common. All documented cases of melanomas in Eurasian otters and ASCO recurred after initial excision and exhibited widespread metastasis at necropsy. Common sites of metastasis include the lymph nodes, liver, spleen, and lung.[133,367]

Leiomyoma of the reproductive tract of a souther sea otter (Fig. 11.e35) and trichoblastoma in a domestic ferret (Fig. 11.e36) have also been seen.

INFECTIOUS DISEASES

DNA Viruses

Multiple **herpes viruses** infect mustelids. **Suid alphaherpesvirus-1** is the causative agent of **pseudorabies** (Aujeszky's Disease) in farmed mink that were fed infected pigs, resulting in severe, rapidly fatal seizures.[298] Inhalation exposure may also occur, as cases were documented 2–3 days after porcine fecal slurry was sprayed on pastures surrounding a mink farm.[219] In experimentally infected mink, the prepatent period ranged from 72 to 96 h and was followed by a short illness characterized by hypersalivation, vomiting or retching, dyspnea, depression, coma, and death in all infected mink, often within 24 h of exhibiting initial clinical signs.[354] The prevalence of Aujeszky's disease in wild mustelids is low. In naturally occurring pseudorabies outbreaks in the Netherlands, mortality varied from less than 1%–15%, with all age groups being equally susceptible.[219] Gross lesions include widespread petechiae and ecchymoses, empty gastrointestinal tracts, moderate splenomegaly, and pulmonary edema. Histologically, pseudorabies infection in mink differs from the nonsuppurative encephalitis that is associated with this disease in other species,[219] and is typically characterized by fibrinoid degeneration of arteries, veins, and smaller vessels in the central nervous system (CNS), myocardium, and oropharynx. Affected vessels have endothelial karyorrhexis, luminal thrombi and neutrophils, and perivascular fibrin and hemorrhage. There may be perivascular tissue necrosis with mild status spongiosis in the CNS, or glomerular hemorrhage with secondary hemoglobinuric nephrosis. Infected mink also exhibit jejunal crypt necrosis or lymphocytolysis in the gastrointestinal-associated lymphoid tissue, lymph nodes, and spleen. In contrast, ferrets usually display more classical lesion patterns, including nonsuppurative encephalitis with focal neuronal necrosis, perivascular cuffing, microgliosis, and eosinophilic intranuclear inclusions in neurons and glial cells.[219] Ferrets also develop facial pruritus, which is typical of the disease in other species.[221]

Mustelid gammaherpesvirus-1 (MusGHV-1), also referred to as badger herpesvirus-1 or mustelid herpesvirus-1, is frequently detected via PCR from genital swabs collected from wild badgers in the United Kingdom. Both horizontal and vertical transmission are suspected.[213,449] The proportion of PCR-positive animals ranges from 55% to 100% with reactivation of the virus at the beginning of the winter mating season in males. Females with active infections have a significantly lower pregnancy rate and decreased reproductive success.[449,450] Nonspecific gross lesions in infected females include small tan nodules or vesicles on the vaginal mucosa or around the cervix, mucosal hyperemia or congestion, and punctate dark brown mucosal foci. Gross lesions in infected males include small tan nodules or vesicles on the outer urethral mucosa. Notably, these gross lesions are not always associated with active infection. Histologic changes include vaginal, cervical, or urethral epithelial and lymphoid hyperplasia with variable submucosal mixed inflammation. In one male badger, basophilic to amphophilic intranuclear inclusions were identified within the superficial penile base epithelium in the absence of gross lesions.[449] **Fisher herpesvirus-1**, a virus closely related to MuGHV-1, was detected in dermal ulcers on the muzzle and plantar pads of a male fisher. Histologically, the ulcer borders contained clusters of epithelial cells with basophilic to amphophilic intra nuclear inclusions; a pan-herpesvirus nested PCR polymerase assay and DNA sequencing confirmed infection.[136]

Mustelid herpesvirus-2, a gammaherpesvirus, was first identified in oral ulcers from northern sea otters stranding during the 1989 Exxon Valdez oil spill. Clinical disease was presumed incidental and likely stress related. This virus, which appears to be closely related to MusGHV-1, was detected in affected mucosal tissues from 29 otters after the Exxon Valdez oil spill, and in nasal swabs from 83 live-captured northern sea otters with and without visible ulcers from the Kodiak Archipelago. In addition to oral and sublingual ulcers with adjacent proliferative or hyperplastic epithelium, lymphoplasmacytic conjunctivitis and corneal ulcers with intralesional epithelial intranuclear inclusions were also identified.[451] **Herpes simplex**, an alphaherpesvirus, was identified as the cause of nonsuppurative and necrotizing encephalitis with neuronal and glial inclusions in a striped skunk. Experimental intravenous inoculation of skunks results in cellular necrosis that is most severe in the liver and adrenal glands.[73]

Aleutian mink disease virus (AMDV), a **carnivore amdoparvovirus**, is probably the most well-known virus of mink as it causes **Aleutian mink disease** (AMD) or **mink plasmacytosis**, a chronic wasting syndrome characterized by high mortality and reduced reproduction in farmed mink. AMDV infection has been reported in most mustelid species including American mink, domestic ferrets, European mink, European badgers, European polecats, Eurasian otters, pine martens, least weasels, stoats, and wolverines.[496] Closely related carnivore amdoparvoviruses cause similar disease in other species including striped skunks, raccoons, dogs, cats, genets, and red pandas.[3,4,61,496] Some studies suggest that AMDV is the most common cause of death in juvenile mink and the second most common cause of death in farmed mink throughout their production cycle.[483] AMDV infection may also be contributing to the population decline of wild European mink.[490]

The first cases of AMDV in mink that occurred in North America were primarily in the gunmetal gray shades of mink with similar color to Aleutian foxes. Although all color phases of mink may be infected, color-dilute varieties genetically related to Aleutian strains appear to be particularly susceptible. Given the enhanced susceptibility of color-dilute strains, lesions associated with AMDV infection were initially presumed to be genetic, and mink were shipped worldwide before a viral etiology was confirmed. The clinical course of AMDV in Aleutian strains of mink may be as rapid as 4 months, with environmental stress often accelerating disease progression. For ferrets and other color strains of mink, a 2-year subclinical phase is common, and healthy carriers may have a significant role in spread of AMDV to other captive care facilities and the environment by acting as reservoirs.[59]

AMDV can be spread horizontally through feces, urine, and saliva, or vertically from dams to offspring. Oral infection appears to be most common, although airborne, iatrogenic (e.g., through blood sampling) transmission, and vector-borne infections have also been described.[496] Virions are environmentally persistent and can survive high temperatures and chemical treatments making eradication difficult.[61] Chronic viral infection over months to years results in plasmacytosis and hypergammaglobulinemia, increased antibody titers, and high levels of circulating antigen–antibody complexes. These immune complexes deposit in systemic basement membranes, resulting in type-III hypersensitivity reactions.[234] Clinical signs in adult mink include weight loss, anorexia, polydipsia, lethargy, diarrhea, melena, spontaneous abortion, and acute death. Grossly, multifocal hemorrhage may occur due to vascular damage, thrombocytopenia, and coagulation disorders related to severe monoclonal gammopathy. Histologically, adult mink often have immune-complex related glomerulonephritis, arteritis with fibrinoid necrosis, and lymphoplasmacytic hepatitis and nephritis; affected animals are especially susceptible to death due to glomerulonephritis. Acute, severe hematuria may distend the bladder, resulting in a "cherry-like" appearance. In mink kits, infection is acute and manifests clinically and pathologically as respiratory distress due to fulminant interstitial pneumonia with type II pneumocyte hyperplasia resulting in death 3 weeks postinfection.[61,151] Clinical signs in ferrets differ from other mustelids, as infected individuals often develop posterior ataxia, progressive paresis, cachexia, malaise, and melena.[471]

Skunk amdoparvovirus (proposed as carnivore amdoparvovirus-5) infection results in a similar disease course to AMDV in mustelids, and adults are more likely to be infected than juveniles. Skunk amdoparvovirus infection appears to be an emerging disease that is becoming endemic in California, USA with high prevalence across North America.[3,151,234] Gross lesions include ascites or hydrothorax, splenomegaly, diffuse renal pallor (Fig. 11.27), and multifocal hemorrhage. Histologically, there is multisystemic lymphoplasmacytic inflammation (interstitial nephritis, myocarditis, hepatitis, meningoencephalitis, pneumonia, and splenitis), encephalomalacia with cerebral microangiopathy, and arteritis with or without fibrinoid necrosis in multiple organs (eSlide 11.e3). Glomerulonephritis is less common in striped skunks than in mustelid AMDV infections.[151,234]

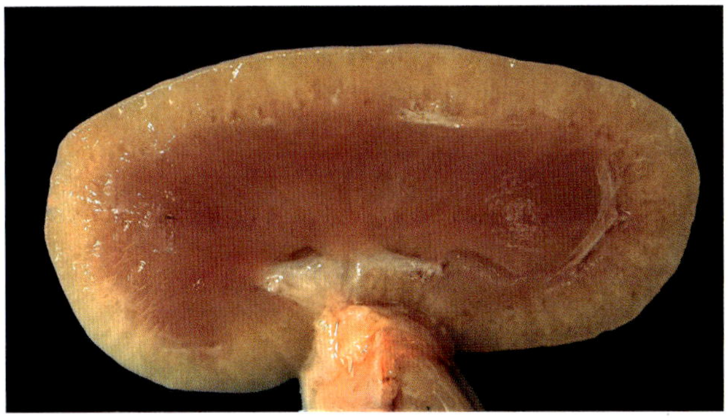

FIGURE 11.27 Amdoparvovirus infection in a striped skunk. The affected kidney has generalized pallor with poor corticomedullary distinction, tan cortical and medullary streaking, and prominent vasculature (see eSlide 11.e3). *(Photo Courtesy of L. Bourque, Canadian Wildlife Health Cooperative.)*

Mink viral enteritis is caused by mink enteritis virus (MEV), a parvovirus closely related to feline panleukopenia virus (FPV) and canine parvovirus 2 (CPV-2). All three viruses are members of the carnivore protoparvovirus-1 species, and MEV and FPV are so closely related they may be variants of the same virus. One report identified MEV as the third most common cause of death in ranched mink, and the second most common cause of death in mink 4–10 months of age.[483] Clinical signs develop 4–14 days postexposure and manifest as diarrhea, anorexia, squinting, vomiting, and depression. Histologically, intestinal crypts are dilated, and villi are fused or blunted. Crypt epithelial cells are variably necrotic or have large nuclei with prominent nucleoli and many crypts are filled with neutrophils and exfoliated necrotic epithelium. Some enterocytes have nuclei with marginated chromatin and smudged basophilic intranuclear inclusions. The lamina propria is infiltrated by eosinophils and lymphocytes, and necrosis of leukocyte progenitors and lymphocytes in gut-associated lymphoid tissue and lymphoreticular organs are common. Partial to complete sloughing of the intestinal mucosa may be observed 7–8 days postinfection.[18,483] As with feline panleukopenia, younger animals are more severely affected. Asymptomatic carriers can excrete virus for over a year postinfection,[38] and cross-species infection may occur, with mink susceptible to feline panleukopenia virus, and raccoons susceptible to mink enteritis virus.[18,424] Vaccination is protective and is typically administered to kits at approximately 6–8 weeks of age. **Feline panleukopenia virus** has been detected via PCR in American river otters, Eurasian otters, skunks, domestic ferrets, beech martens, and free-ranging honey badgers[54,60,424,461] although viral-associated clinical disease was not detected in river otters, martens, or skunks. Ferrets demonstrated disease after in utero infection and free-ranging honey badgers had diarrhea. North American river otters may act as maintenance hosts for FPV.[60,461]

Canine parvovirus-2 (CPV-2) infection was identified in zoo collections of neotropical[109] and Asian small-clawed otters, sometimes in conjunction with *Clostridium difficile*.[466] Additional reports of canine parvovirus infection have been documented in Eurasian badgers[20] and stone martens.[424] Case presentation and pathology (Fig. 11.28) resemble descriptions of mink viral enteritis. Infection results in inappetence, lethargy, vomiting, severe diarrhea, and hematochezia.[150] CPV-2 was detected via serology in free-ranging American badgers and wolverines[116] and by PCR in Eurasian otters and beech martens.[54,461] A novel endemic protoparvovirus has been identified in southern sea otters and two American pine martens, but no disease associations were established.[60,411]

Canine adenovirus-1 infection has been described in striped skunks and a captive Eurasian river otter, resulting in fatal hepatitis.[210,323] The skunks were about 2 weeks old

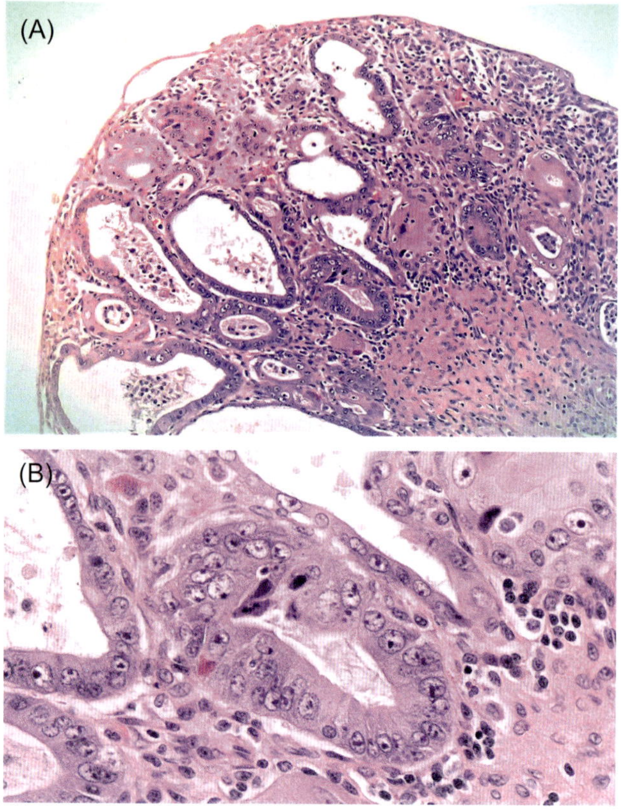

FIGURE 11.28 Canine parvovirus infection in a Eurasian badger. (A) Intestinal villi are markedly blunted and fused, and crypts are dilated and lined by a single layer of variably attenuated, hypertrophic, or necrotic epithelium with central karyorrhectic debris and neutrophils. (B) Occasionally, nuclei of crypt epithelial cells have smudged basophilic intranuclear inclusion bodies. *(Parts A, B: Photos Courtesy of A. Barlow, Wildlife Network for Disease Surveillance.)*

and were from a single litter. Clinical signs included nasal discharge, depression, and acute death. The Eurasian river otter was 10 years old with signs of chronic anorexia, weight loss, and diffusely swollen liver lobes with many scattered 1 mm dark brown foci. Histologically, multiple random areas of hepatocellular necrosis contained deeply acidophilic to basophilic intranuclear inclusions within hepatocytes, biliary and renal tubular epithelium, and adrenal cortical epithelial cells. Interstitial pneumonia associated with intranuclear inclusions in respiratory epithelium and secondary bacterial bronchopneumonia also occurred. A closely related virus, **skunk adenovirus-1**, was identified in skunks with similar clinical and pathologic findings. It is not clear if the initial report[210] represented spillover from canine adenovirus-1 or was actually the unique skunk adenovirus-1.[39,227] Skunk adenovirus-1 has also been associated with fatal interstitial pneumonia in domestic ferrets.[316] A novel **mastadenovirus**, closely related to otarine adenovirus-1, was found in a 4-month-old North American river otter. This animal presented with severe palmar, plantar, and oral ulcers, interstitial pneumonia, and pericardial and pleural effusion. Oral and skin lesions were associated histologically with necrotizing vasculitis and endothelial intranuclear viral inclusions. Diffuse alveolar damage with no visible inclusion bodies was noted in lung tissue.[80,152]

***Enhydra lutris* papillomavirus-1 (ElPV-1)** has been associated with multifocal raised, variably pigmented, flat-topped, 2–5 mm diameter, **papillomas** on the gingiva and lips of southern sea otters[300] (Fig. 11.29A). Microscopically, lesions are composed of moderate epithelial hyperplasia accompanied by marked parakeratotic hyperkeratosis, hypergranulosis with irregularly shaped large keratohyalin granules, moderate ballooning degeneration (koilocytes), and mild spongiosis. Within the cells of the stratum spinosum and granulosum are small, round to oval, lightly basophilic intranuclear viral inclusions as well as large, finely granular, pale eosinophilic to amphophilic intracytoplasmic inclusions consistent with keratin aggregates (Figs. 11.29B–C). Both intranuclear inclusions and intracytoplasmic keratin granules label via immunohistochemistry using antibodies raised to bovine papillomavirus. Cutaneous viral papillomas have been observed in a Eurasian badger[118] (Fig. 11.30) and a pine marten (eSlide11.e4). A single case of fibroepithelial papilloma or acrochordon in a stone marten (*Martes foina)* was associated with a *Dyonupapillomavirus* classified as MfoiPV1.[232]

Poxviral dermatitis occurred in two sea otter pups, one northern and one southern, with slightly raised, pink, ulcerated, nodular masses. In the northern sea otter, the mass was located on the caudal abdomen and in the southern pup, two similar masses were present on the upper lip. Histologically, these lesions were composed of areas of marked epidermal papillomatous hyperplasia with prominent rete pegs and nests extending into the deep dermis. Epithelial cells displayed ballooning degeneration and necrosis along with scattered intracytoplasmic eosinophilic inclusions typical of Bollinger bodies. These lesions were self-limiting and were not associated with systemic illness.[453]

RNA Viruses

Coronaviruses are best described in the mink and ferret, where they can be clinically silent, or can cause epizootics characterized by rapid spread of respiratory disease (betacoronaviruses) or diarrhea (alphacoronaviruses) throughout naïve animal colonies. In late 2019, a novel betacoronavirus, **severe acute respiratory syndrome coronavirus 2 (SARS-CoV-2, the virus that causes COVID-19)**, was identified in Wuhan, China as a cause of severe acute respiratory disease in humans that resulted in a global pandemic. Presumably due to reverse zoonoses, infection with this

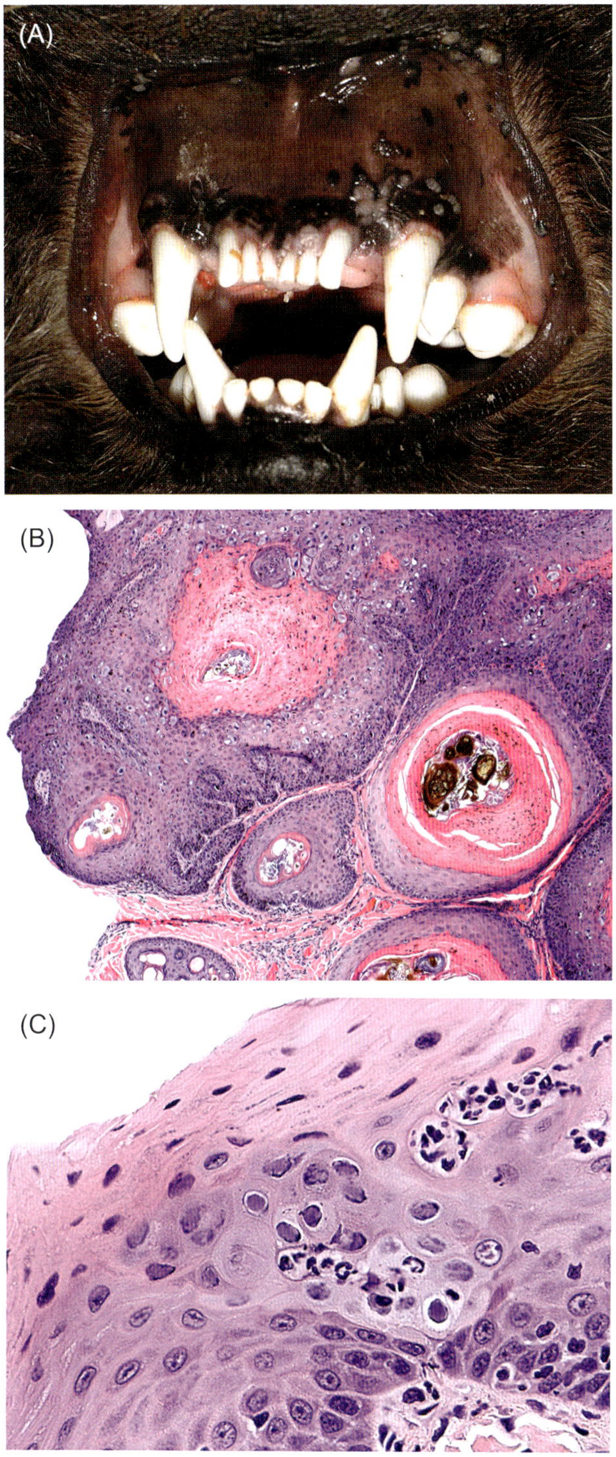

FIGURE 11.29 *Enhydra lutris* papillomavirus-1 infection in a sea otter. (A) Clusters of white raised plaques are present throughout gingival and buccal mucosa. (B) Marked oral mucosal hyperplasia with clusters of koilocytes, intracellular edema, and hyperkeratosis. (C) Characteristic smudged intranuclear viral inclusions with a colorless halo are present within koilocytes along with a few intervening pustules. *(Parts A, B: Photos Courtesy of the California Department of Fish and Wildlife.)*

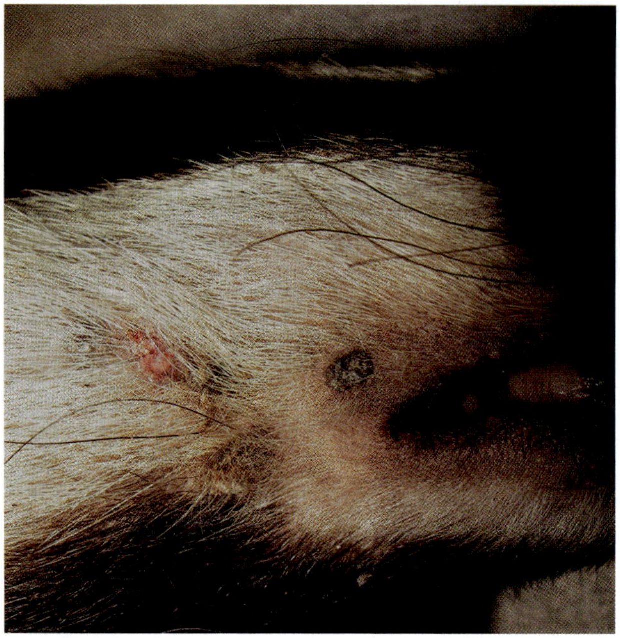

FIGURE 11.30 Cutaneous viral papilloma in a Eurasian badger. Multiple, irregular, tan to brown, and occasionally ulcerated, papillary growths are present in the perioral and right cheek skin. A novel papillomavirus was identified within the lesions. *(Photo Courtesy of D. Couper, RSPCA West Hatch, Somerset, England.)*

novel coronavirus was subsequently identified as a cause of morbidity and mortality in mustelids worldwide. Viral transmission occurs through inhalation of respiratory droplets from closely housed animals or infected caretakers.[404] American mink are particularly susceptible to severe morbidity and mortality,[368] with up to 55% adults dying in one farm outbreak.[110] Housing design flaws, overcrowding, and poor sanitation may contribute to rapid disease spread and development of viral variants at some mink farms.[96]

Clinical disease and mortality from natural and experimental infections of SARS-CoV-2 have been described in farmed and free-ranging American mink, domestic ferrets, European mink, and ASCOs.[96,368,404,497] Adult mink >1 year of age appear to be more susceptible to clinical disease and most animals are in good nutritional condition.[284] Clinical signs include inappetence, lethargy, fever, sneezing, ocular discharge, and coughing.[110,284,368] In severe cases, animals exhibit disorientation and increased abdominal respiratory effort.[69] Subclinical infections are common, comprising 70% of mink in one report.[347]

At necropsy, predominant gross lesions include nasal discharge and diffusely dark red, wet, consolidated lungs (Fig. 11.31A).[110,368] Histologic changes in individuals with severe acute disease include degeneration and expansion of alveolar septa by fibrin and mononuclear cells. Septa are sometimes lined by hyaline membranes (diffuse alveolar damage) (Fig. 11.31B) and type II pneumocyte hyperplasia with or without early fibroblast proliferation and organization.[368] Alveolar spaces are often filled with mononuclear inflammatory cells, desquamated necrotic respiratory epithelial cells, edema, fibrin, many foamy alveolar macrophages, and variable numbers of neutrophils.[110,368]

In acute infections, bronchioles are most commonly affected, with epithelial necrosis and syncytia (Fig. 11.31C). However, bronchial, tracheal, and nasal respiratory epithelium can also frequently exhibit ciliary loss, cytoplasmic swelling, and cell degeneration associated with variable numbers of neutrophils. Rhinitis is one of the most common histologic findings in mink with congestion, edema, lymphoplasmacytic to neutrophilic infiltrates, suppurative mucoid exudate, and in some cases, admixed foreign debris and bacteria (Fig. 11.31D).[368] Subacute to chronic bronchiolitis with lymphoplasmacytic inflammation, squamous metaplasia of the bronchiolar epithelium, bronchiolar smooth muscle hypertrophy, extension of bronchiolar epithelium into alveolar spaces (bronchiolization), and fibroblast proliferation (Fig. 11.31E) are common in animals with fatal infections.[368] In mink, pulmonary arterioles have varying degrees of mural necrosis with transmural edema, lymphocytes, plasma cells, and thrombosis (Fig. 11.31F; eSlide 11.e5).[69,110,284,368]

Severe lymphoplasmacytic perivasculitis and vasculitis have been described in experimental infections in domestic ferrets,[404] but pulmonary vascular fibrinoid necrosis is not as common in mustelid infections as in human cases of SARS-CoV-2. Similar respiratory disease and pathology have been documented in domestic ferrets experimentally infected with SARS-CoV-1.[427] Although striped skunks are experimentally susceptible to SARS-CoV-2 infection and seroconvert, no clinical disease and minimal nonspecific histologic lesions such as increased nasal and pulmonary perivascular lymphoplasmacytic infiltrates have been described.[131]

Confirmation of SARS-CoV-2 infections in mustelids can be obtained via PCR, immunohistochemistry, and in situ hybridization. As the virus tends to replicate preferentially in upper respiratory or nasal epithelium, PCR of nasopharyngeal swabs can be optimal to confirm infection[284,368]; however, younger mink may only be positive on colorectal swab PCR.[368] Similar to humans, SARS-CoV-2 antigen is present in epithelial cells of the nasal conchae, trachea, bronchi, bronchioles, alveoli, and alveolar macrophages via immunohistochemistry (Fig. 11.31G).[368] However, in early infections, only the upper respiratory tract may demonstrate positive immunolabeling.[284] Positive immunohistochemistry labeling has also been described in the ferret intestine.[404] In situ

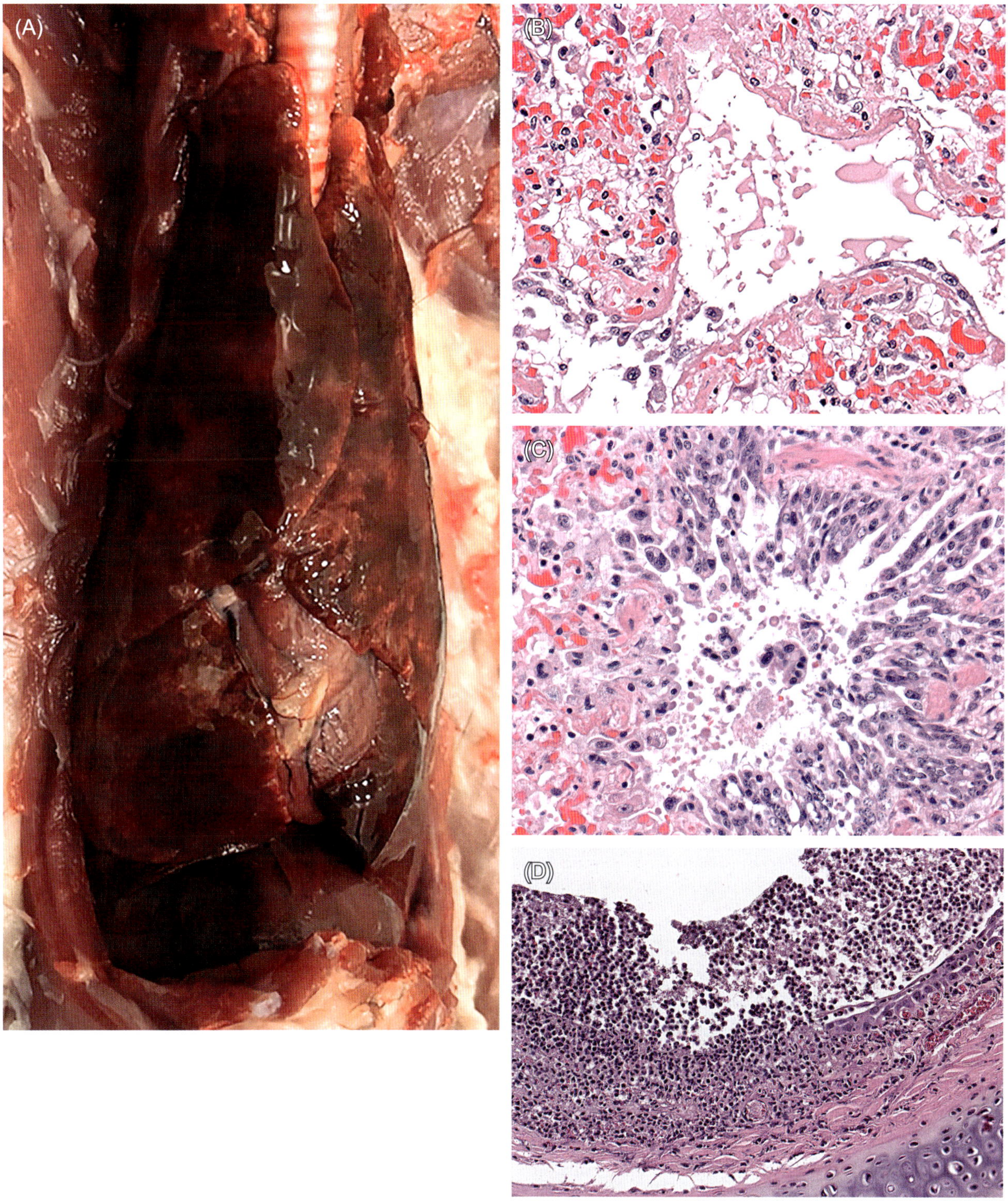

FIGURE 11.31 SARS-CoV-2 infection in American mink. (A) The pulmonary parenchyma is diffusely dark red. (B) Diffuse alveolar damage is evident with hyaline membranes lining alveoli along with segmental type II pneumocytes and vascular congestion. (C) Bronchiolitis with degeneration, necrosis, and sloughing of the hyperplastic bronchiolar epithelium and syncytia formation. (D) Nasal turbinate with ulceration, mucosal attenuation, loss of cilia, suppurative inflammation, and fibrinosuppurative exudate.

(Continued)

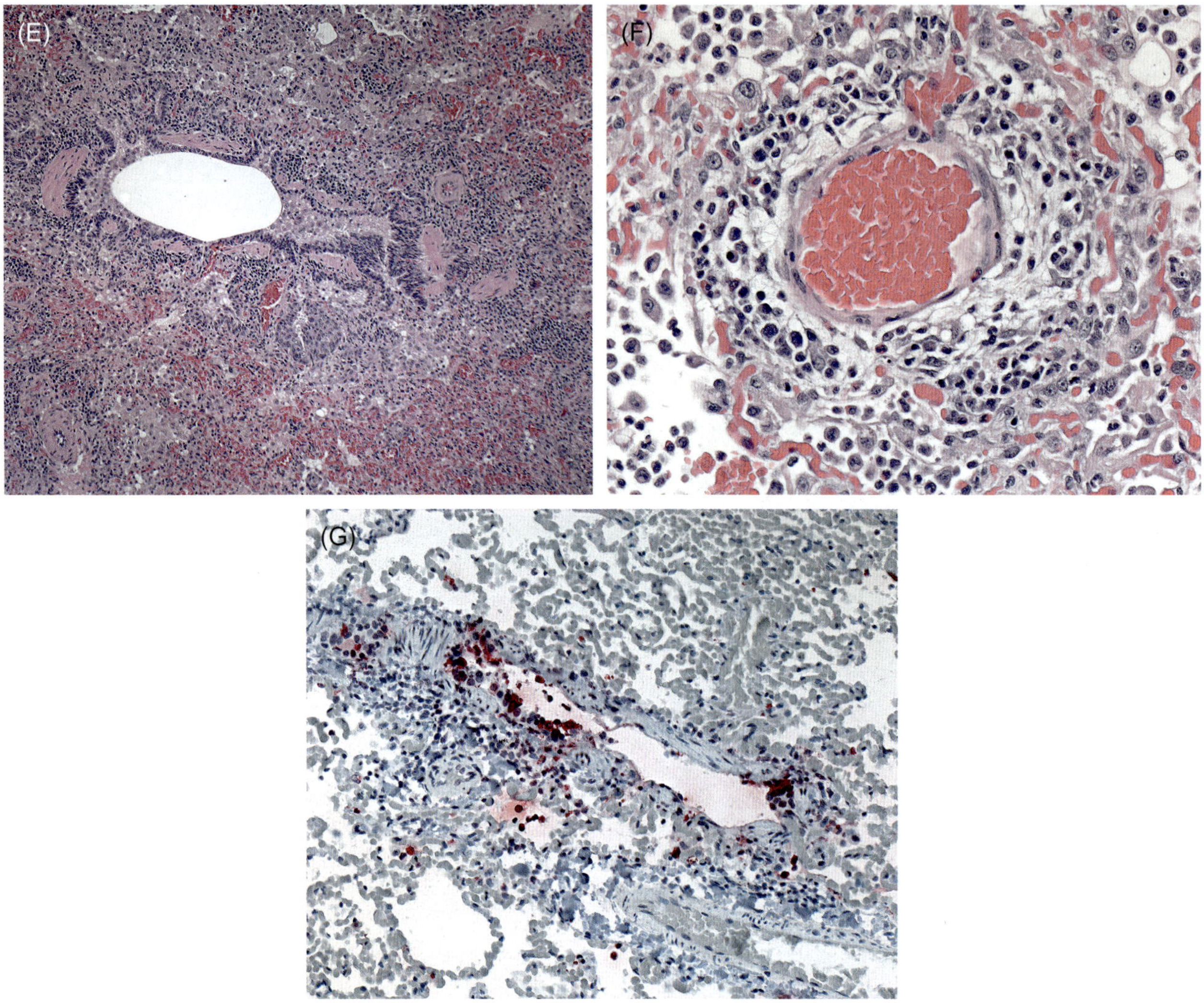

FIGURE 11.31, CONT'D (E) Chronic bronchiolitis with peribronchiolar mononuclear inflammation, smooth muscle hypertrophy, hyperplastic bronchiolar epithelium extending into alveoli (bronchiolization), and squamous metaplasia of the bronchiolar epithelium. (F) Pulmonary vessel with perivascular edema and lymphoplasmacytic inflammation. (G) Lung immunohistochemistry for SARS-CoV-2 with immunolabeling of bronchiolar epithelium, scattered alveolar macrophages, and a few type II pneumocytes (see eSlide 11.e5). *(Part A: Photo Courtesy of Center for Disease Control and Prevention (CDC) Field Team. Parts B–G: Photos Courtesy of the Infectious Diseases Pathology Branch, CDC.)*

hybridization using FFPE tissue revealed abundant positive staining in nasal exudate and epithelial cells within the upper respiratory tract with less involvement of the lower respiratory tract.[110,368]

Diseases associated with alphacoronaviruses include **epizootic catarrhal gastroenteritis** in mink and **epizootic catarrhal enteritis (ECE)** in ferrets. In mink, epizootic catarrhal gastroenteritis is most common in dark-colored juveniles, and outbreaks are more prevalent during fall. While younger ferrets are often asymptomatic, older animals commonly exhibit severe disease.[161,427] Anorexia, vomiting, and mucoid to bright green diarrhea that can result in reduced body condition and fur quality may affect up to 100% of naive individuals, but mortality tends to be low and lesions are restricted to the gastrointestinal tract. Grossly, affected sections of jejunal mucosa are hyperemic and thin.[497] Histologically, viral cytopathic effects are most severe for enterocytes at the tips of intestinal villi, resulting in marked villous blunting and fusion, followed by lymphocytic infiltration into the lamina propria.[161] In ferrets, these lesions may be apparent up to 8 months postinfection.

Ferret systemic coronavirus (FRSCV), similar to feline infectious peritonitis, has been correlated with ferret enteric coronavirus (ECE) infection. While unproven, the pathogenesis of feline enteric coronavirus infection is thought to be characterized by viral mutation within infected individuals that facilitates persistence within macrophages, and a similar etiopathogenesis is likely for ferrets. Most ferret cases resemble the "dry form" of feline infectious peritonitis (FIP), with fewer cases resembling the "wet" form.[138] Clinical and gross findings include marked polyclonal hyperglobulinemia, hepatosplenomegaly, and renomegaly, often with multifocal to coalescing white to tan nodules and plaques (granulomas) in multiple visceral organs, especially the mesentery, spleen, liver, kidney, lung, and lymph nodes.[161] Granulomas may have central areas with necrosis or suppurative inflammation and, similar to FIP, the lesions may localize around small blood vessels in the tunica adventitia and media.[427,497] Because feline coronavirus anti-gp70 antibody cross-reacts with both the ferret enteric and systemic coronaviruses, immunohistochemistry can demonstrate antigen in macrophages of the lung, liver, kidney, spleen, lymph nodes, intestine, heart, brain, bone marrow, pancreas, and other tissues, facilitating diagnosis.[427,436]

Morbillivirus (family *Paramyxoviridae*) strains have been implicated in mustelid infections worldwide. **Canine distemper virus (CDV)**, phocine distemper virus (PDV), and cetacean morbillivirus (CMV) have all been associated with disease in mustelids, although CDV is the most prominent and clinically significant, and is the only morbillivirus described in skunks.[153,231,290,321,327,438] CDV infection has been associated with morbidity and mortality in American badgers,[116] ferret badgers,[74] river, Eurasian, giant, and sea otters,[207,290,321,384,442] martens,[317] fishers,[212] black-footed ferrets,[480] skunks,[97,231] and weasels.[439] Veterinarians must remain vigilant for distemper outbreaks following modified live virus vaccine administration in wildlife, as some modified live vaccines developed for use in dogs and cats have caused fatal disease in black-footed ferrets and European mink.[63,430]

Clinical signs of CDV in mustelids are similar to those in canids and include anorexia, oculonasal mucopurulent discharge (Fig. 11.32A), multifocal dermal hyperkeratosis (commonly called 'hard pad disease') (Figs. 11.32B, C), emaciation, severe pruritus, intermittent diarrhea, pneumonia, and rapid death.[290,480] Clinical signs in mustelids

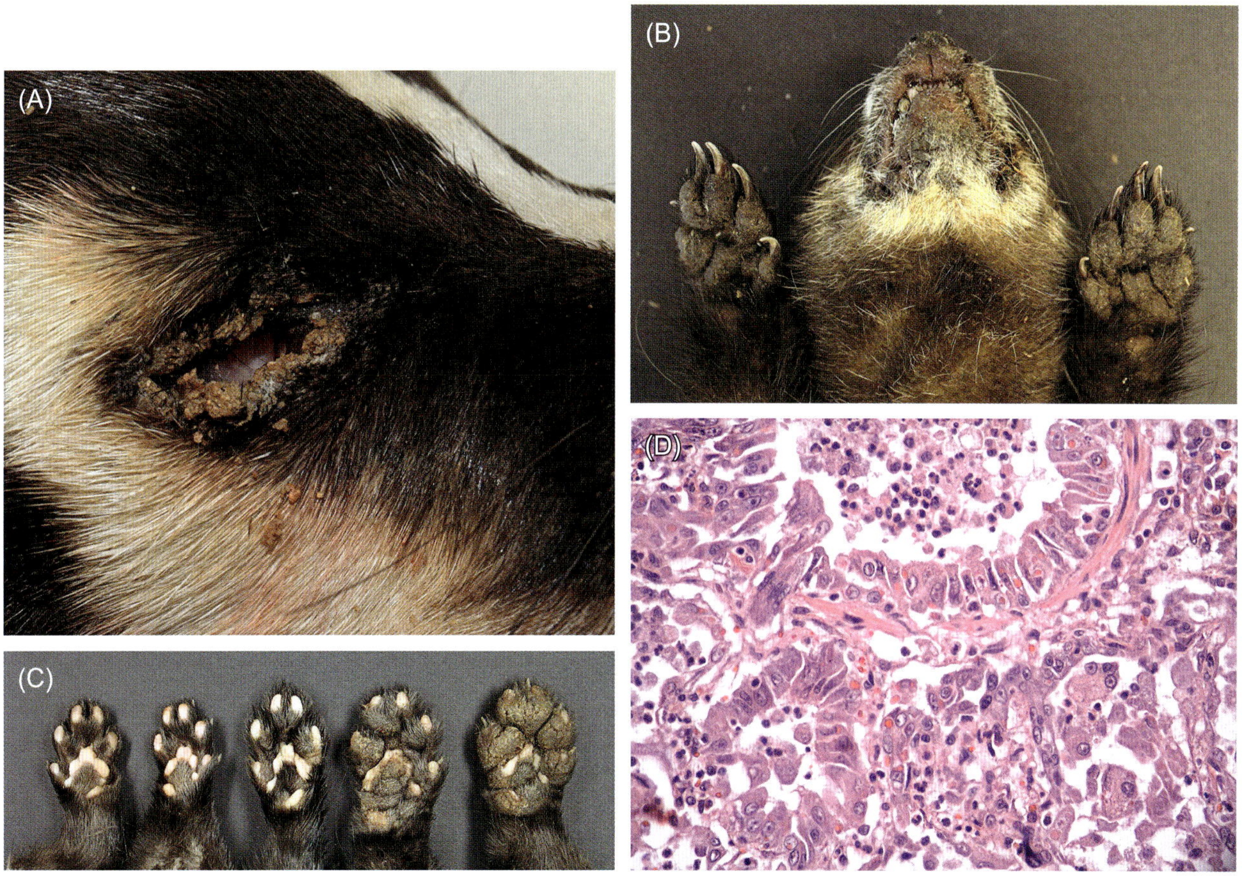

FIGURE 11.32 Canine distemper virus infection in various mustelids. (A) American badger with abundant mucopurulent ocular discharge resulting in wetting, clumping, and tan to brown discoloration of the periocular fur. (B) European polecat with marked, diffuse hyperkeratosis of the footpads, nasal planum, and perioral skin. (C) Footpads from farmed American mink with similar hyperkeratosis of varying severity from mild (left) to severe (right). (D) American badger with marked hypertrophy and hyperplasia of bronchiolar epithelial cells along with occasional syncytia formation, type II pneumocyte hyperplasia in adjacent alveolar spaces, and numerous intracytoplasmic and intranuclear eosinophilic viral inclusions. (see eSlide 11.e6). *(Parts A, D: Photos Courtesy of T. Viner, National Fish and Wildlife Forensic Laboratory. Parts B, C: Photos Courtesy of T. Kare Jensen, National Veterinary Institute of Denmark, Technical University of Denmark.)*

often progress faster than in dogs, with ferret-adapted strains causing death in as few as 12 days. Microscopic lesions include bronchointerstitial pneumonia, alveolar syncytia (Fig. 11.32D), mild to moderate meningoencephalitis, and variable numbers of intracytoplasmic and intranuclear eosinophilic inclusions in epithelium, syncytial cells, neurons, and glial cells (eSlide 11.e6). In one retrospective study, lesions in northern sea otters were restricted to the central nervous system.[442] Due to the rapid disease course, sick animals may be seronegative.[290,480] Conventional CDV immunohistochemistry and PCR are available in most diagnostic laboratories for viral detection. Concurrent bacterial, protozoal, fungal, and viral infections are common due to CDV-associated immunosuppression.[231,442] CDV seropositivity without clinical disease has been reported in wolverines, polecats, and sea otters[90,340,442]; CDV RNA (PCR-positive) northern sea otters were identified without histologic evidence of morbilliviral disease.[442]

Phocine distemper virus (PDV) has been detected by PCR in northern sea otters with rare inclusion bodies identified by electron microscopy in the liver and pancreas. No significant lesions were observed aside from putative viral-associated vasculitis and encephalitis.[48] Sea otter PDV seroprevalence can be as high as 40% in some areas of coastal Alaska.[153] Cetacean morbillivirus (CeMV) infection has been documented in Eurasian otters on the Italian coast, presumably due to scavenging on stranded dolphins or infected invertebrate reservoirs during a cetacean morbillivirus epizootic. There was evidence of bronchopneumonia or bronchointerstitial pneumonia, and positive immunohistochemical labeling of CeMV localized to vascular walls and alveolar epithelial cells. Sections of lung and kidney from affected otters were PCR-positive for CeMV. The proximate cause of death for all CeMV-infected otters was vehicular trauma.[321]

Although reports are sparse for some species, all mustelids are presumed to be susceptible to **rabies** virus infection. In many parts of the world, mustelids and skunks serve as important reservoirs for host-adapted rabies viral strains, especially striped skunks in the United States and ferret badgers in Southeast Asia.[72] Based on anecdotal reports, some eastern US state monitoring programs have detected rabies-positive river otters. However, many cases are not documented in the peer-reviewed literature, which leads to the misperception that rabies is uncommon in this species. Affected animals are often described as being unusually aggressive or tame, and will often bite or approach humans prior to capture, euthanasia, and rabies testing. Rabies was diagnosed in a river otter intended for a reintroduction program, illustrating the importance of pathogen screening in wildlife prior to transportation and release efforts.[393] Clinical signs in mustelids and skunks with rabies virus infection are similar to other rabid mammals and include emaciation, hypersalivation, abnormal mentation, and decreased nociception. Histologic lesions are most prominent in the brainstem and include nonsuppurative meningoencephalomyelitis, variable white matter spongiform degeneration, and neurons containing intracytoplasmic eosinophilic inclusion bodies (Negri bodies). Sialadenitis and adrenal necrosis may also be observed in ferret badgers.[77] Although vaccination has dramatically reduced the case prevalence in zoos and other captive care facilities, any unusual behavior or sudden death, especially involving unvaccinated animals, necessitates consideration of rabies, in part due to significant human health risks and the potential for exposure of other animals.

Influenza viruses are zoonotic pathogens with a broad host range that includes migratory and production birds, canids, horses, marine mammals, and mustelids. Ferrets are susceptible to both type A and B influenza strains and are commonly used as influenza research models.[498] Free-ranging striped skunks and farmed mink have been implicated as potential conduits for influenza viral amplification and spread between humans and other animals. Based on the necropsy findings and experimental infection, both groups may shed influenza virus for extended periods, facilitating transmission.[41,115,374] Clinical signs of influenza A and B infection in mustelids and mephitids include malaise, serous nasal discharge, sneezing, sudden death, and lower respiratory tract disease due to viral infection or secondary bacterial infections. Influenza A encephalitis has also been reported in a stone marten with a diffuse nonsuppurative panencephalitis, perivascular cuffing, multifocal gliosis, neuronal necrosis, and focal necrosis of pancreatic islet cells.[223] Sea otters can be seropositive for influenza A infection with no associated clinical disease or pathology.[243]

In mustelids and mephitids infected with **H1N1 and H7N9 influenza viruses**, clinical signs include fever, oculonasal exudate, coughing, sneezing, and anorexia. At necropsy, affected individuals have suppurative rhinitis and pulmonary consolidation. Histologic features include necrotizing bronchointerstitial pneumonia with intraluminal bronchiolar accumulation of cellular debris, edema, and fibrin (eSlide 11.e7).[41,373] A reverse zoonosis outbreak of H1N1 on an American mink farm in 2019 only affected kits <14 days old; lesions included moderate to severe acute neutrophilic, histiocytic, and lymphocytic interstitial pneumonia.[78] In contrast, infection with the **H5N1** variant of highly pathogenic avian influenza was associated with antemortem neurologic signs (e.g., seizures, ataxia, and lack of fear response) in a fisher and stone marten, but only dyspnea in striped skunks. Histologic findings in the H5N1-positive striped skunks were predominately interstitial pneumonia, lymphoid depletion, hepatic necrosis, and rarely necrotizing meningoencephalitis, whereas necrotizing nonsuppurative meningoencephalitis was more common in other infected mammalian species.[114]

Numerous **enteric viruses** have been associated with diarrhea in mustelids. **Rotaviruses** have been identified in young ferrets and mink, with group A and C strains infecting ferret kits. Gross lesions include diarrhea, dehydration, and thin-walled, gas and fluid-filled intestines. Histologic changes are typical of rotavirus infections in other species and include superficial atrophic enteritis with degeneration and necrosis of villous tip enterocytes and sloughing of necrotic epithelium.[486] Preweaning diarrhea syndrome of mink kits in Europe is purported to be caused by mink **calicivirus or astrovirus**. Gross findings include evidence of dehydration and increased greasy apocrine gland secretions on the cervical skin, paws, and tail (e.g., "greasy kit syndrome"). Histologic findings include enterocyte vacuolization with atrophy, fusion of jejunal villi, and atrophy of colonic glands. Superficial colonization and occasional proliferation of gram-positive bacterial cocci within enterocytes of the duodenum and colon, and variable mononuclear or neutrophilic infiltrates within the lamina propria are common.[167]

Astrovirus infection has been identified as the cause of a neurological condition in Scandinavian mink known as **shaking mink syndrome**. Brain homogenates from infected mink produce similar signs when inoculated into healthy animals and metagenomic sequencing identified an astrovirus.[35] Infected animals exhibit tremors, ataxia, and a staggering gait with no apparent gross lesions. Histologic changes are most prominent in the caudal brainstem and cerebellum and include diffuse and multifocal cerebellar gliosis, particularly in the molecular layer along the junction with the Purkinje cell layer, with segmental degeneration or loss of Purkinje cells. Multifocal gliosis, neuronal degeneration and necrosis, satellitosis, neuronophagia, and mononuclear and occasional neutrophilic perivascular cuffing are observed in the gray matter of affected areas. In more severe cases, moderate mononuclear infiltrates are observed in the meninges and choroid plexus along with occasional white matter vacuolization.[143] Astroviruses have also been associated with neurological disease in cattle, pigs, and humans.

Bacteria

Mycobacteria are common pathogens in zoo settings and infection has also been documented in free-ranging mustelids. Feral domestic ferrets in New Zealand are reservoirs of ***Mycobacterium bovis***; while they can amplify and transmit the disease to conspecifics and cattle, the frequency of cattle transmission is unknown.[53] Wild badgers are recognized as major reservoirs of *M. bovis* in the United Kingdom, Ireland, France, and Spain.[34,261] Despite formidable challenges, vaccination is considered the best long-term preventative measure to date; studies with captive badgers demonstrated that vaccination with *M. bovis* bacillus Calmette Guerin (BCG) and heat-inactivated *M. bovis* vaccine (HIMB) induced protective immunity when delivered by various routes.[82] In badgers, lesions associated with *M. bovis* infection include caseating granulomas with dystrophic mineralization, ranging from 1 to 10 mm diameter in the lungs (Fig. 11.33), submandibular, retropharyngeal, bronchial, or mediastinal lymph nodes; peri-hepatic and mesenteric lymph nodes are less commonly affected.[34] Infections may be mild or limited to microgranulomas which are easily missed during gross necropsy. In contrast to badgers, where pulmonary granulomas dominate, hepatic granulomas (and microgranulomas) are common in ferrets, and lymph node granulomas are common in both species.[346] *M. bovis* infection has been

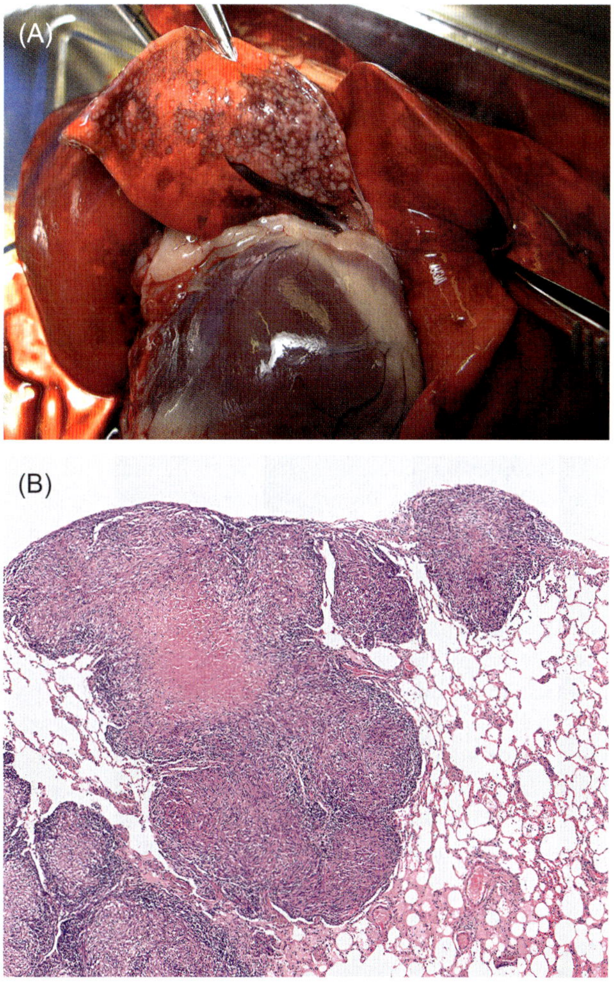

FIGURE 11.33 Experimental infection with *Mycobacterium bovis* in the lungs of Eurasian badgers. (A) Multiple discrete and coalescing tan to gray granulomatous foci are present throughout the lung lobes. (B) Multiple coalescing and occasionally caseating granulomas are present throughout the lung and pleura. *(Parts A, B: Photos Courtesy of Animal and Plant Health Agency, UK (Crown Copyright 2024).)*

reported in multiple species of zoo and farmed mustelids, including an ASCO with disseminated disease.[237] Transmission is likely due to chronic fecal shedding by other exhibit animals or when fur farms feed potentially contaminated feedstuffs. On histology, variable numbers of acid-fast bacilli are present in necrotic areas within granulomas. Bacterial isolation requires selective media and long-term culture at 37°C. Spoligotyping and mycobacterial interspersed repetitive units-variable number of tandem repeats (MIRUVNTR) typing can be used to confirm a diagnosis and determine the *M. bovis* genotype.[237] PCR using the heat shock protein gene, *hsp65*, is a rapid, sensitive, and specific method to identify slow-growing mycobacteria.[94]

Infection by numerous subspecies of **atypical mycobacteria** has been detected in mustelid and mephitid species, often without evidence of clinical disease. ***Mycobacterium avium*** ssp. ***paratuberculosis*** (MAP) infections, the cause of Johne's disease, have been identified in various free-ranging mustelids with sparse bacteria-associated pathology, suggesting that consumption of infected prey or carrion may create passive or asymptomatic carriers. MAP infection has been detected in Eurasian badgers, Eurasian otters, and beech martens in Portugal[260]; stoats, weasels, and European badgers in Scotland; and skunks in Wisconsin, USA.[81,371] Eurasian otters infected with MAP have nonspecific chronic retropharyngeal and mesenteric lymphadenitis.[371] *Mycobacterium celatum* and *Mycobacterium genavense* infections were associated with focal or disseminated granulomas in pet ferrets.[94,346] Pulmonary microgranulomas associated with *Mycobacterium kumamotonensis* and *M. avium* complex infections have been described in stoats and weasels in the United Kingdom.[410]

Mycoplasma infections described in mustelids and skunks include outbreaks of pneumonia in domestic ferrets due to an organism most closely related to *Mycoplasma molare* and polyarthritis in free-ranging striped skunks attributed to an organism similar to *Mycoplasma fermentans/caviae*.[137,220] One ferret outbreak occurred in a commercial pet vendor facility where 95% of infected animals were clinically ill, with acute mortalities in 6–8-week-old ferrets, and chronic respiratory disease in older individuals; stress may have precipitated or exacerbated the severity of infection. Animals developed a dry cough, hemoptysis, dyspnea, sneezing, and conjunctivitis. Gross lesions included multiple, widespread, tan to gray, somewhat firm nodules centered on airways. Histologic changes were characterized by moderate bronchointerstitial pneumonia, catarrhal bronchitis, severe bronchial-associated lymphoid tissue (BALT) hyperplasia, perivascular lymphoid cuffing, and diffuse pulmonary congestion.[220] *Mycoplasma* sp.-infected free-ranging striped skunks in Massachusetts were emaciated with alopecia, pustular dermatitis of the neck and shoulders, and purulent osteoarthritis affecting two or more joints. Many (68%) animals also had positive serum titers to canine distemper virus and clinical signs consistent with CDV infection.[137]

Multiple bacteria have been associated with outbreaks of diarrhea in mustelid species. ***Clostridium perfringens* type A** gastroenteritis has been documented in young black-footed ferrets and is attributed to dietary change and overeating.[263,390] Affected animals display marked abdominal distention, dyspnea, and cyanosis. Myriad gram-positive bacilli line the necrotic gastrointestinal mucosa and extend into the gastric glands and intestinal crypts on histopathology. Lymphoid necrosis may also be present in gut-associated lymphoid tissue and lymphoid organs. *C. perfringens* type A can be isolated from gastric and small intestinal contents, and the toxin can be identified via PCR or a mouse protection assay. **Enterotoxigenic *Escherichia coli*** infection has been associated with an outbreak of fatal diarrhea in captive black-footed ferrets. Clinical signs include acute death, dehydration, anorexia, and mucoid diarrhea in adults and kits. *Escherichia coli* can be isolated from rectal swabs, stomach, small intestine, liver, kidney, and spleen consistent with septicemia. Gross lesions are subtle and may include gastric and intestinal mucosal erythema. Histologically, lesions vary from none to mild congestion of gastric and intestinal mucosa and large numbers of gram-negative bacteria on the surface of intestinal villi. Contaminated food is the suspected source of infection.[40]

Infection by ***Campylobacter jejuni***, a gram-negative, spiral-to-curved, microaerophilic bacterium, is associated with diarrhea in mink and ferrets. However, this bacterium can also be cultured from the feces of asymptomatic ferrets.[263] Infection can cause self-limiting diarrhea that may be watery, with occasional blood and mucus. Infection in pregnant females may result in fetal resorption, abortion, or delivery of premature kits. Infection is most severe in kits, especially when associated with concurrent cryptosporidiosis.[263] Experimental challenge results in mild to moderate enterocolitis with neutrophilic infiltration of the lamina propria. Laboratory confirmation of infection can be completed via fecal culture and PCR.

Streptococcal infections have been reported in several mustelid species, often as part of mixed bacterial flora infecting wounds and bite injuries. Lancefield group C *Streptococcus zooepidemicus* infection has been associated with valvular endocarditis in ferrets, whereas *S. bovis* has been detected in mink with verrucous endocarditis.[328] Because distinct bacterial species can have overlapping biochemical properties and subgroups (e.g., *S. bovis* and *S. lutetiensis* sp. nov. described below), molecular characterization of isolates is advised. Numerous cases of bacterial discospondylitis have been reported from mink farms and *Streptococcus* sp. is most commonly isolated from these lesions. Affected animals exhibit posterior paresis associated with suppurative intervertebral discospondylitis, myelitis, periosteal proliferation, and vertebral osteomyelitis.

Streptococcus lutetiensis sp. **nov**, a member of the ***S. bovis/equinus*** group (was formerly identified as *S. infantarius* subsp. *coli*), is an important pathogen of northern sea otters and was the cause of death for 44% of 780 northern sea otters necropsied between 2002 and 2012.[48,185,349] A similar bacterial strain infects southern sea otters at much lower incidence.[84] Primary infection with *S. lutetiensis* sp. nov or other *Streptococcus* spp., known as "strep syndrome", occurs in prime-age animals with no preexisting health conditions.[48,84,85] Acute infections of animals in good nutritional condition can result in fatal meningoencephalitis (eSlide 11.e8), while chronic infections often present with septic valvular endocarditis of the left atrioventricular and/or aortic valves (Fig. 11.34A). A dilated, thin-walled left ventricle, massive pulmonary edema, thoracic effusion, and multifocal thromboemboli with coagulation necrosis and inflammation in the myocardium, kidneys (Fig. 11.34B; eSlide 11.e9), spleen, or descending aorta (secondary to saddle thrombus) may also be observed. Oral exposure is the likely route of infection as 31% of animals with gastrointestinal contents cultured were positive for *S. bovis/equinus* species, including primarily *S. lutetiensis*, rare *S. bovis* nonspecified, *S. gallolyticus* ssp. *gallolyticus*, and *S. gallolyticus* ssp. *pasteurianus*.[48] Coinfection with multiple species of ***Bartonella*** sp. has been identified in both northern and southern sea otters, but infection prevalence is not correlated with the observed valvular lesions.[66] Another beta-hemolytic streptococcus, *Streptococcus equisimilis*, recently recategorized as ***Streptococcus dysgalactiae*** ssp. ***equisimilis***, has been associated with mandibular and subcutaneous abscesses, necrotizing suppurative pneumonia, meningoencephalitis, myocarditis, and acute pyothorax in wild striped skunks. Infection is attributed to intraspecific trauma incurred during the mating season.[193]

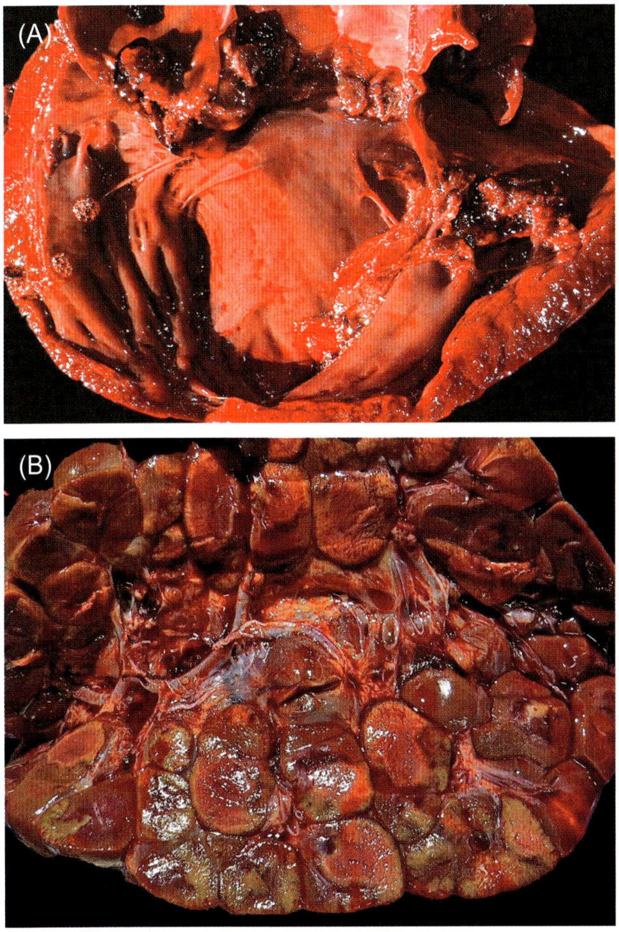

FIGURE 11.34 *Streptococcus lutetiensis* sp. nov infection in northern sea otters. (A) Large, irregularly nodular, vegetative aggregates of tan to red friable material are adhered to the aortic and left atrioventricular valves. (B) Discrete pink to tan to brown, wedge-shaped foci representing acute to subacute renal infarctions are present within multiple reniculi (see eSlides 11.e8 and 11.e9). *(Parts A, B: Photos Courtesy of Alaska Veterinary Pathology Services.)*

Streptococcus phocae and related beta-hemolytic streptococci have been identified as one of the most common opportunistic pathogens of southern sea otters. *S. phocae* prevalence was 30% in fresh dead otters examined from 2004 to 2010,[22] and, alone or in conjunction with *S. lutetiensis*, was the cause of "strep syndrome" in as many as 8%–10% of northern sea otters. Lesions associated with *S. phocae* infection include infected skin wounds, pneumonia, lymphadenopathy, sepsis, abscess formation, and meningoencephalitis (Fig. 11.35A). Histologically, lesions include suppurative or necrosuppurative inflammation in one or more tissues associated with gram-positive cocci in chains or clusters.[22,437] Otters with *S. phocae* infection often have preexisting conditions that presumably contribute to debilitation or immunosuppression such as emaciation, trauma, severe endoparasitism, or concurrent bacterial or viral infections.[437] *S. phocae* may also be associated with mixed bacterial infections in farmed mink with pleuropneumonia (Fig. 11.35B) and pododermatitis.[304]

Mustelids have recently been identified as natural hosts of ***Staphylococcus delphini***, one of three members of the *S. intermedius* group. Mink, badgers, and ferrets are consistently culture-positive for *S. delphini* group A.[158] *Staphylococcus delphini* infection in ferret and mink kits causes **hypersecretory diarrhea** known as "sticky kit disease".[142] Grossly, multiple loops of bowel are distended by mucoid material. In mink, concurrent hypersecretion of cervical apocrine glands results in sticky pelts. In some mink farms, infections can result in 30%–40% mortality of kits. Histologically, there is no evidence of inflammation or substantial alteration of mucosal morphology, although gram-positive cocci colonize segments of the small intestine. PCR or enzyme-linked immunosorbent assay (ELISA) can be used to detect *Staphylococcus*

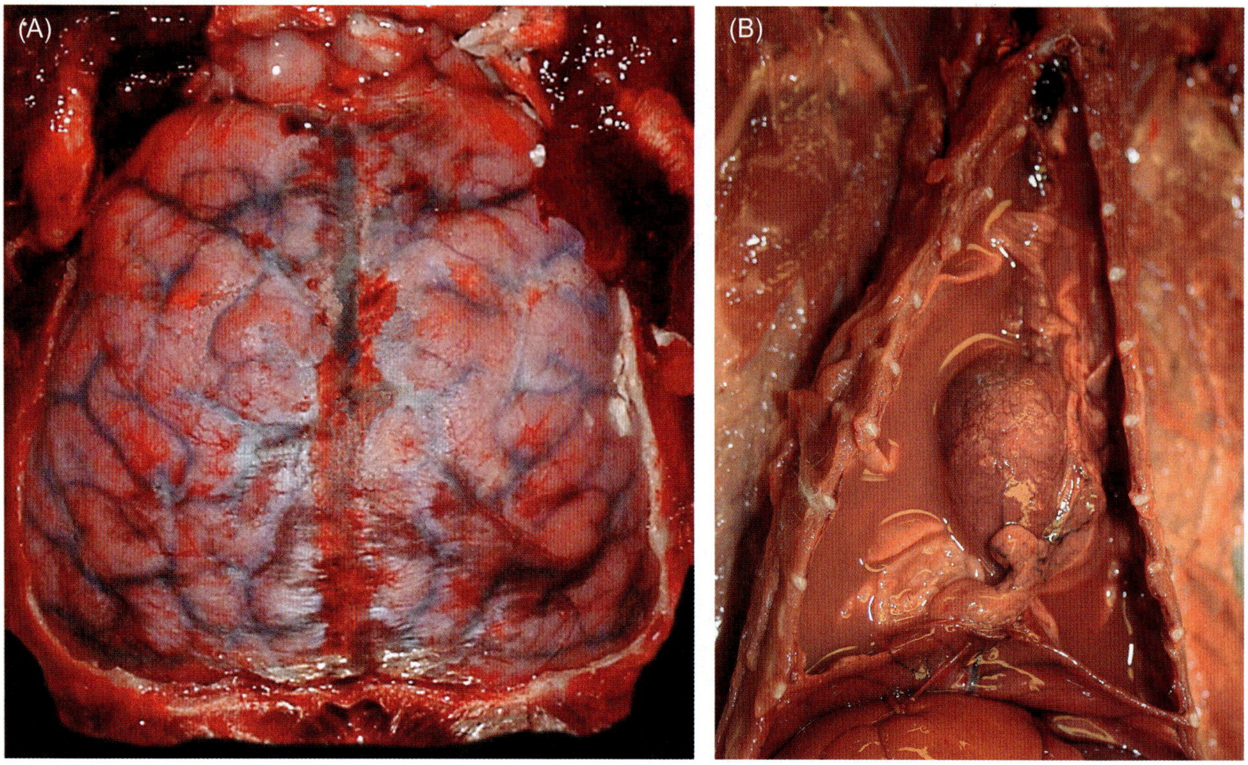

FIGURE 11.35 *Streptococcus phocae* infection in a northern sea otter and American mink. (A) The meninges are partially opaque, and the subdural space contains abundant pale red to tan, cloudy fluid. Vasculature is diffusely congested with multifocal petechial and ecchymotic hemorrhages. This individual had a mixed infection of *S. phocae* and *S. lutetiensis*. (B) A marked, opaque, pink to red pleural exudate consistent with pleural empyema fills the thoracic cavity. Additional lesions may be present including pododermatitis or necroulcerative facial dermatitis and cellulitis, all part of the fur animal epidemic necrotic pyoderma syndrome. *(Part A: Photo Courtesy of C. Goertz, Alaska Sea Life Center. Part B: Photo Courtesy of T. Kare Jensen, National Veterinary Institute of Denmark, Technical University of Denmark.)*

enterotoxins A and E.[142] *Staphylococcus delphini* is also a frequent opportunistic pathogen in cases of pododermatitis in farmed mink.[70] In mink, bacteria that are phenotypically consistent with *S. intermedius* have been isolated from cervical adenitis in kits and pyometra, vaginitis, cystitis, pyelonephritis, pleuritis, peritonitis, and mastitis in adults.[190,414] Earlier reports of *S. intermedius* infections in mink were likely *S. delphini*, as almost all pathogenic *Staphylococcus* sp. infections in mink have been confirmed as *S. delphini*.[158,463]

Helicobacter mustelae is a ubiquitous bacterial pathogen of ferrets that has also been detected in stoats.[125] In ferrets, *H. mustelae* is associated with atrophic and lymphocytic gastritis; affected ferrets are used as models for human *H. pylori* infection.[128] Almost all North American ferrets are infected with *H. mustelae* by weaning age. Infection is lifelong, the severity of *H. mustelae* gastritis increases with age, and infections may become clinically apparent when animals are stressed due to concurrent disease.[186] Infection in mustelids results in clinical hypochlorhydria, vomiting, and diarrhea; most commonly in animals that are >3.5 years old. While not generally fatal, chronic *H. mustelae* infection contributes to debilitation in older ferrets. Microscopic lesions are most common in the pylorus where the pH is optimal for bacterial colonization and consist of lymphoplasmacytic gastritis with marked attenuation of gastric glands. *H. mustelae* infection has not been causally linked to gastric ulcers, a common condition in ferrets. Silver stains demonstrate the presence of extracellular spiral bacteria within the mucus layer. An enteropathogenic *Helicobacter* species, *H. cholecystus*, has been identified in domestic ferrets in association with hepatobiliary disease.[312]

Helicobacter enhydrae sp. **nov** has been isolated from southern sea otters.[24,402] Gross pathology varies from no lesions to a potential but unproven link with gastric erosions and ulcers. Histological changes range from mild cystic degeneration of gastric glands to severe mucosal erosions and ulcers. Slightly curved to spiral-shaped bacilli can be visualized in the superficial mucosa and the epithelial surface of gastric glands of infected animals using silver stains. The mucosa and submucosa can also contain small numbers of neutrophils, macrophages, and lymphocytes, with occasional lymphoid aggregates and well-defined lymphoid follicles. Because gastric erosions and ulcers are extremely common in sick sea otters, including those with and without *H. enhydrae* sp. nov infection, studies focused on confirming or refuting causal associations are needed. While no direct correlation has been reported to date, diagnosis of gastric ulcers at necropsy was significantly associated with *Helicobacter* qPCR detection from gastric mucosa.[24]

Hypermucoviscous (HMV) and non-hypermucoviscous ***Klebsiella pneumoniae*** strains, and other *Klebsiella* spp. are occasionally associated with infected wounds, lymphadenomegaly, abscesses, pneumonia, meningoencephalitis (eSlide 11.e10), endocarditis, and septicemia in southern sea otters and other mustelids.[71] Histologically, suppurative or necrosuppurative inflammation in one or more tissues is associated with numerous short bacterial rods. Infection by HMV *Klebsiella* strains is characterized by intralesional bacterial rods that are surrounded by a thick, colorless, outer capsule (Fig. 11.36).

Pseudomonas aeruginosa can cause epizootics of necrotizing hemorrhagic pleuropneumonia and septicemia in farmed mink, referred to as **fatal hemorrhagic pneumonia**. First described in 1953, this highly contagious bacterium spreads rapidly across affected farms and occurs most commonly during the fall.[154] Peracute death is common, and losses may approach 75%. Kits are routinely vaccinated at 6–8 weeks, although the duration of protective immunity is short at approximately 90 days.[483] Gross necropsy reveals cranioventral bronchopneumonia with copious bloody fluid on cut surface. Microscopic lesions include prominent transmural vasculitis, alveolar hemorrhage, and abundant bacteria within alveoli and capillaries. Similar lesions occur in the tracheobronchial and hilar lymph nodes. Hemolytic *E. coli* is a primary differential for this type of bronchopneumonia in mink, although

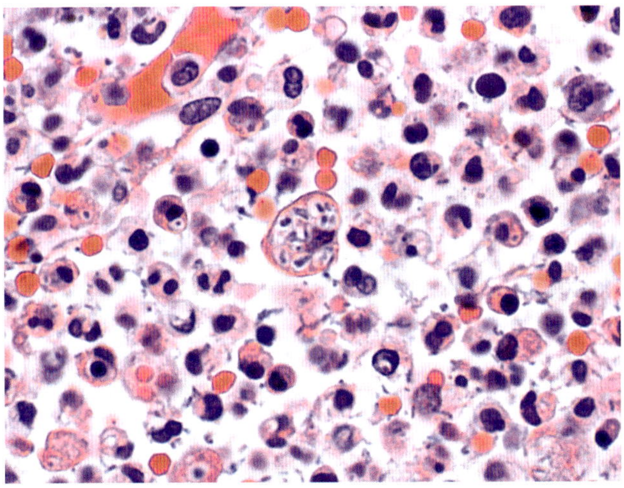

FIGURE 11.36 Hypermucoviscous (HMV) *Klebsiella pneumoniae* pneumonia in a southern sea otter. Marked suppurative and mildly histiocytic inflammation along with few erythrocytes fills alveolar spaces. Associated with the inflammatory infiltrate are large numbers of intracellular or extracellular bacilli surrounded by a colorless capsule. *(Photo Courtesy of the California Department of Fish and Wildlife.)*

E. coli-associated pneumonia often presents with a more diffuse pattern, with fewer bacteria and more alveolar edema than hemorrhage.[377]

Acinetobacter baumannii infection has been associated with an outbreak of fatal, acute pneumonia in American mink and septicemia in a captive European mink. Clinical signs include tachypnea and tachycardia, and gross lesions include severe bilateral acute hemorrhagic pneumonia, pleural effusion, multisystemic congestion, and moderate splenomegaly. Histologically, severe suppurative and fibrinohemorrhagic bronchopneumonia is associated with numerous extracellular coccobacilli that often occlude alveolar spaces, and are also visible within the cytoplasm of alveolar macrophages.[57] Affected individuals may also have necrotizing arteritis.[284] In mustelids, other bacteria associated with primary pneumonia or secondary to fungal or viral infections include ***Pasteurella multocida, Bordetella bronchiseptica*** (eSlide 11.e11), *E. coli*, and *Actinomyces* spp. Gross and microscopic lesions resemble other causes of bacterial bronchopneumonia with varying associated lymphadenitis, pleuritis, and septicemia.[39]

Serologic evidence of ***Leptospira*** spp. infection has been documented at relatively high frequency in multiple mustelid species and skunks, including European and American mink, European polecats, pine martens, stone martens, and striped skunks.[88,179,283,355] High seropositivity in mustelids and skunks likely reflects high prevalence of infection in rodent prey.[283] Although striped skunks are commonly *Leptospira* antigen positive in the United States and Canada, infections are often asymptomatic.[401] In one study *L. pomona* and *L. canicola* infections in skunks were associated with tubulointerstitial nephritis.[88] In contrast, 68% (15/22) of American badgers were positive via immunohistochemistry, and bacterial detection was associated with mild nephritis.[116] *Leptospira interrogans* seroprevalence in sea otters is comparatively low, ranging from 0% to 2% in live, wild, northern, and southern sea otters, respectively, and 10% in sick, stranded otters.[169] Sea otter deaths due to leptospirosis are rare, but occasional clustering of cases occurs. Gross lesions may be absent or include cyanosis, renal swelling, congestion, pale streaks on the cut surface of renal lobules, renal petechiae, hematuria, dehydration, lymphadenopathy, pulmonary congestion, and adrenal hemorrhage or congestion.[224] Microscopically, there is moderate to severe lymphoplasmacytic tubulointerstitial nephritis. Silver stains demonstrate intratubular spirochetes, whereas immunohistochemistry localizes leptospiral antigens in renal tubules and the interstitium.

Lawsonia intracellularis infection is associated with **proliferative bowel disease** in ferrets, especially young males. In contrast with hamsters, pigs, and horses, where the ileum is preferentially affected, *L. intracellularis* infection in ferrets is primarily associated with colonic pathology, although it can also affect the small intestine. Clinical signs include lethargy, anorexia, weight loss, dehydration, and diarrhea. Diarrhea is often blood-tinged, may be green and mucoid, and rectal prolapses may also occur. Grossly, the colonic mucosa is thickened and corrugated due to marked hyperplasia of colonic enterocytes. Histologically, marked colonic mucosal hyperplasia is accompanied by reduced goblet cells and a marked lymphocytic infiltrate in the lamina propria. Silver stains demonstrate numerous bacteria within the apical cytoplasm of proliferating enterocytes.[263]

Sylvatic plague, caused by *Yersinia pestis,* is an important cause of mortality for endangered black-footed ferrets, threatening species recovery.[490] The primary prey of black-footed ferrets are prairie dogs; prairie dog colonies and associated fleas serve as *Y. pestis* reservoirs. As part of in situ conservation programs, prairie dog colonies located where black-footed ferrets are concentrated or released are dusted with pesticides to control fleas and reduce disease risk. Plague-infected ferrets have histologic lesions similar to infected wild felids including prominent vasculitis, necrohemorrhagic inflammation, and perivascular accumulations of bacilli in submandibular and mesenteric lymph nodes.[478] Vasculitis also occurs in the skin and lungs, causing subcutaneous hemorrhage and pulmonary edema, respectively. Myriad peri- and intravascular bacilli may be observed in other organs. The Siberian polecat is a suitable model for plague in black-footed ferrets.[481] In New Zealand, a *Y. pestis*-like bacterium has also been detected by PCR in feral stoats with a high prevalence of idiopathic pneumonia.[265] Although American badgers are considered susceptible and some populations are seropositive, clinical signs and pathology are not reported.[351] The related bacterium *Yersinia pseudotuberculosis* has been isolated from Eurasian otters, martens, polecats, and mink with no associated lesions.[305]

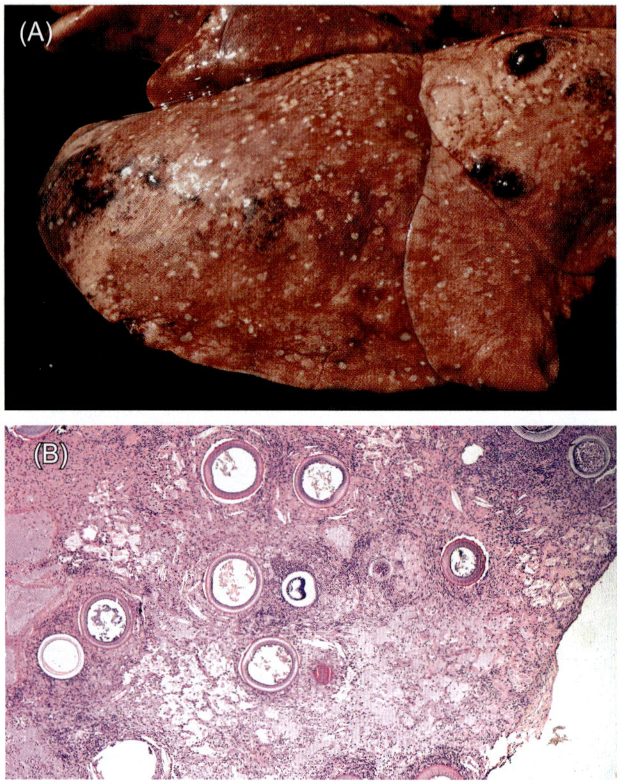

FIGURE 11.37 Adiaspiromycosis in Eurasian otters. (A) Grossly affected lungs contain multiple, often coalescing, tan to white pinpoint nodules throughout the lung lobes. (B) Histologically, pulmonary architecture is obscured by multiple large yeasts and associated chronic granulomatous and lymphoplasmacytic inflammation. Yeasts have a brightly eosinophilic outer cuticle, wider pale basophilic inner layer, and central clearing or flocculent basophilic granular material. Adjacent alveolar spaces contain edema, cholesterol clefts, and scattered neutrophils. *(Part A: Photo Courtesy of V. Simpson, Wildlife Veterinary Investigation Centre. Part B: Photo Courtesy of A. Barlow, Wildlife Network for Disease Surveillance.)*

Fungi

Fungal infections in mustelids predominantly involve **dimorphic fungi** and dermatophytes. **Adiaspiromycosis,** caused by infection with ***Emmonsia* spp.,** has been reported in free-ranging mustelids in the United Kingdom and Europe.[254,378,410] ***Coccidioides*** spp. infections have been described in stranded sea otters along the California coast[189] and a river otter in Arizona, USA.[218] **Histoplasmosis** (*Histoplasma capsulatum*)[47] and paracoccidioidomycosis (*Paracoccidioides brasiliensis*)[365] have also been described in northern sea otters and a greater grison, respectively. Gross lesions in dimorphic fungal infections in mustelids range from multiple, gray to tan, 0.5–1 mm, sharply demarcated to often coalescing nodules to complete effacement of affected tissue. Histologically, dimorphic fungal infections cause granulomatous to pyogranulomatous inflammation, with fungal yeast stages surrounded by necrosis, peripheral lymphoplasmacytic infiltrates, and fibrosis. *Emmonsia* spp. (Fig. 11.37) are commonly associated with inflammation of the lungs and intrathoracic lymph nodes,[378] while coccidioidomycosis (Fig. 11.38) and histoplasmosis (Fig. 11.39) infections are often more widely disseminated, involving the lungs, lymph nodes, liver, spleen, heart, brain, skin, and eyes.[47,189,489] Dimorphic fungi are generally discernable histologically and identifiable to genus with routine hematoxylin and eosin stains (eSlide 11.e12); diagnosis is often confirmed with culture or PAS (eSlide 11.e13) and GMS stains. Due to the zoonotic potential of these dimorphic fungi, culture outside of highly experienced fungal specialty laboratories is not recommended. Freezing of infected carcasses or tissues does not inactivate these fungi. Concurrent bacterial infections are not uncommon in cases of fungal pneumonia in mustelids.

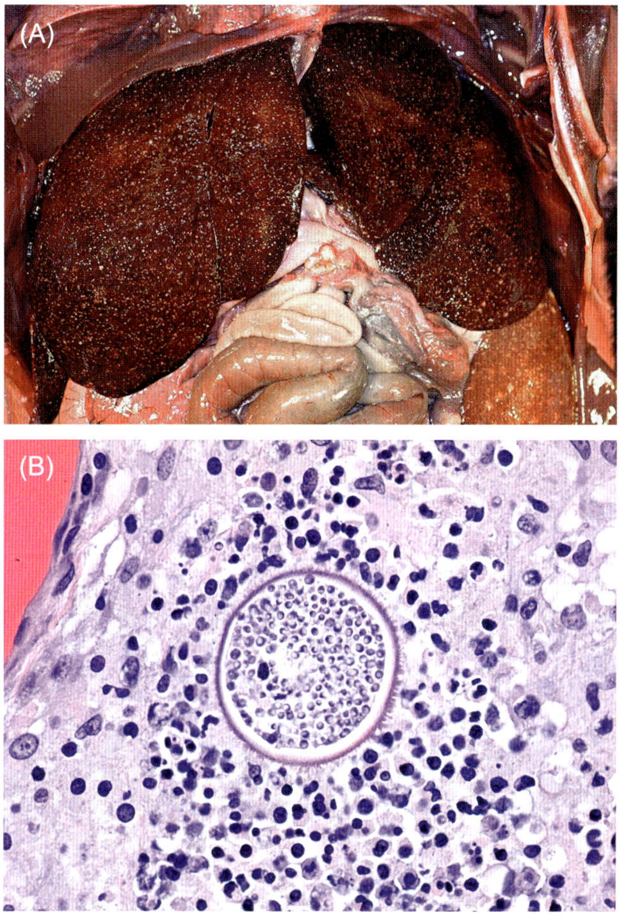

FIGURE 11.38 Coccidioidomycosis in a southern sea otter. (A) Multiple, occasionally coalescing, variably sized gray to tan granulomas are present throughout the liver and spleen due to infection with *Coccidioides immitis*. (B) High magnification view of a spherule containing numerous endospores of *C. immitis* or, less likely *C. posadasii* surrounded by pyogranulomatous inflammation and minute surface projections of Splendore-Hoeppli material in the mediastinum. *(Parts A–C: Photos Courtesy of the California Department of Fish and Wildlife.)*

Dermatophytosis (ringworm infection) has been extensively described in farmed American mink, where it can cause significant pelt damage and economic loss to producers. Clinical disease generally affects kits, with fewer cases reported in mature adults.[320] Gross lesions include patchy alopecia and scaling, most commonly along the dorsum.[32] Histopathology is characterized by chronic hyperplastic dermatitis with suppurative to pyogranulomatous folliculitis, furunculosis, intracorneal pustules, and secondary bacterial pyoderma or cellulitis. Fungal hyphae and arthrospores are visualized histologically with routine hematoxylin and eosin or PAS, often associated with affected follicles and hair shafts. Organisms identified in outbreaks of dermatophytosis in mink include *Trichophyton equinum* and *Trichophyton* sp.[32,320] *Microsporum canis* has also been associated with ringworm infection in Eastern spotted skunks.[159] One case of encephalitozoonosis (*Encephalitozoon cuniculi*) has been described in a black-footed ferret kit resulting in pyogranulomatous tubulointerstitial nephritis, pneumonia, and encephalitis.[440] Table 11.e5 summarizes reported fungal infections in mustelids and skunks.

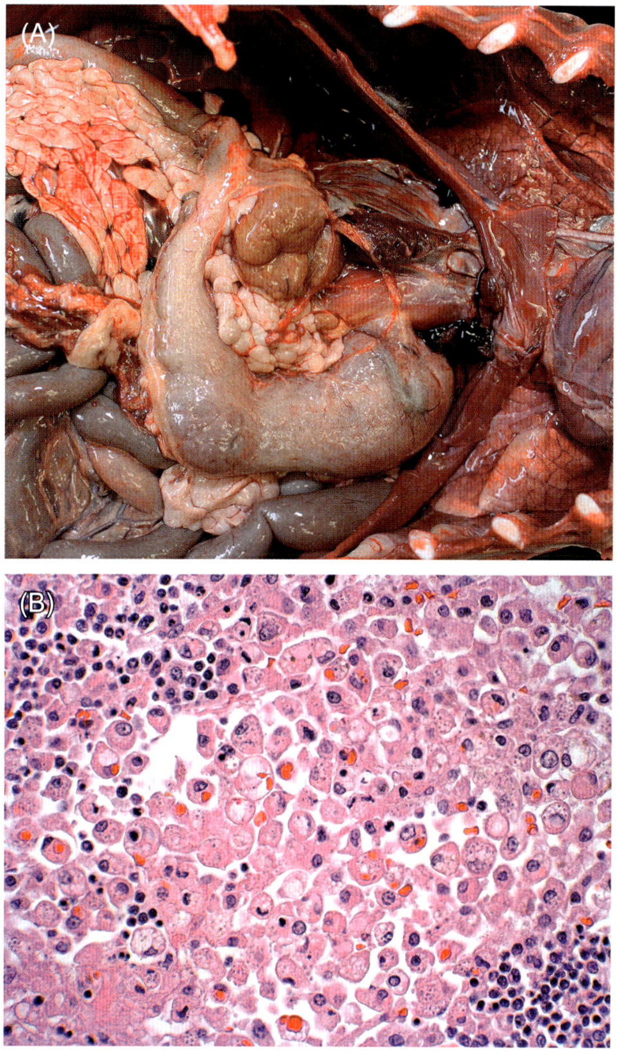

FIGURE 11.39 Histoplasmosis in a northern sea otter. (A) Massive gastric lymphadenomegaly due to lymphadenitis is present near the lesser curvature of the stomach (center of image). (B) High magnification (40×) view of the affected lymph node reveals a marked granulomatous lymphadenitis with most of the lymph node replaced by macrophages. Many macrophages contain multiple, intracytoplasmic round, deeply eosinophilic yeasts surrounded by a clear capsule. (see eSlides 11.e12 and 11.e13). *(Parts A, B: Photos Courtesy of Alaska Veterinary Pathology Services.)*

Metazoa

Various parasites have been identified in Mustelidae and Mephitidae. Some are unique to the family, but several also occur in other wildlife species. Table 11.e6, available in the digital content, summarizes published reports of metazoan

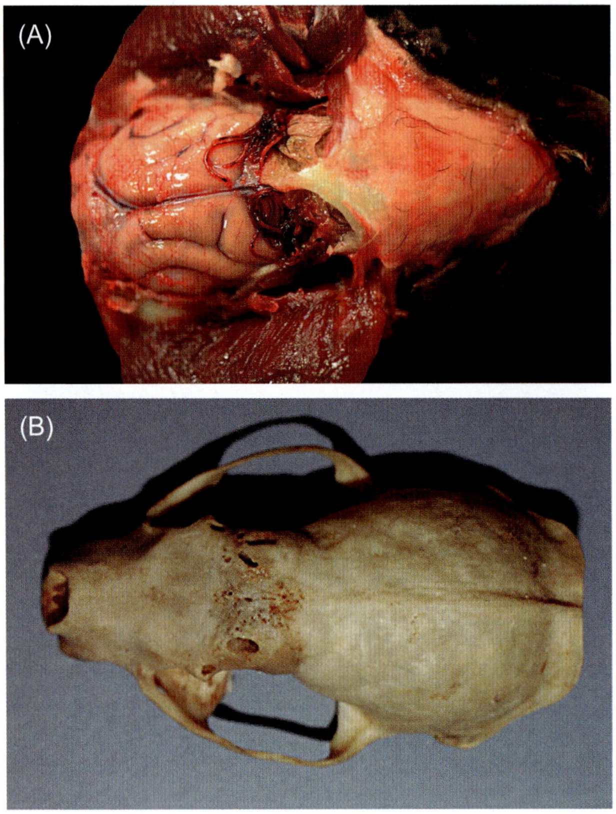

FIGURE 11.40 *Skrjabingylus* sp. infection in a striped skunk and American Mink. (A) Many thin, long, red nematodes are present bilaterally within the frontal sinuses of a striped skunk. (B) Bilateral perforations and bony remodeling of the frontal bones in an American Mink with *Skrjabingylus* sp. infection. *(Part A: Photo Courtesy of E. Edmondson, Colorado State University. Part B: Photo Courtesy of R. Tumlison, Henderson State University, Arkansas.)*

and protozoan parasites in mustelids and skunks. A common metazoan parasite in the nasal and frontal sinuses of mustelids is ***Skrjabingylus*** sp. In addition to aggregates of thin, bright red, nematodes within affected sinuses (Fig. 11.40A), moderate to severe infections can cause substantial remodeling, erosion, fenestration, and ruptured fontanelles in the frontal bone and sinuses (Figs. 11.40B and 11.e37A–E). Despite the severity of these lesions, *Skrjabingylus* infection is not considered a significant mortality factor in mustelids and skunks. The life cycle of this parasite includes a mollusk intermediate host, and shrews or small rodents as paratenic hosts. *Skrjabingylus lutrae* is found in river otters, *S. nasicola* infects mink, stoats, and European pine martens,[180,406] and *S. chitwoodorum* infects skunks.

Dioctophyme renale is another metazoan parasite with highly variable infection of mustelids globally. The parasite has a predilection for the right kidney, with nematodes more rarely detected free within the abdominal cavity or loosely attached to the serosa of abdominal viscera.[111] Depending on the stage of infection, the infected kidney may be enlarged with cystic change and numerous intralesional parasites, or, may progress to multifocal renicular atrophy, polycystic disease, and fibroplasia in more chronic cases. Bilateral involvement may be fatal; however, compensatory hypertrophy of the left kidney is more common. Histopathology may include renal capsular fibrosis with mesothelial hypertrophy, chronic inflammation with giant cell formation, parasitic ova, necrosis, and dystrophic calcification. The parasite has a complex life cycle with an annelid intermediate host and fish, crayfish, and frog paratenic hosts; infective ova are shed in the urine and mustelids acquire infection through consumption of infected prey. Infections are commonly reported in mink, river otters, martens, weasels, and wolverines. Another parasite with a tropism for the lower urinary tract is ***Pearsonema* spp.**, which has an earthworm intermediate host. In beech martens, infections are characterized grossly by urothelial to transmural hemorrhage, congestion, and hematuria; histopathology is characterized by hemorrhagic and eosinophilic cystitis.[112]

The metazoan ***Capillaria plica*** infects beech martens with a predilection for the urinary bladder. This parasite also features an annelid intermediate host and canids as definitive hosts. Infections may be asymptomatic, or animals may present with pollakiuria, dysuria, and hematuria.[336] Other ***Capillaria*** spp. infections reported in mustelids, include *C. aerophila* in the trachea and bronchi of martens, *C. procyonis* in the esophagus of striped skunks,[52] and *C. hepatica* in the liver of long-tailed weasels and sea otters.[273,419] As with some apicomplexan parasites, sea otter *C. hepatica* infections represent another example of land-to-sea pathogen transfer, with infections often detected during or soon after the rainy season.

Gastrointestinal nematodiasis is common in mustelids. Regional and host subspecies variations in ***Pseudoterranova decipiens***, *P. azarasi*, and *Anisakis* spp. infections between northern and southern sea otters have been attributed to diet preference, with northern sea otters being more piscivorous, and thus more likely to be infected, compared to their southern counterparts. Increases in population density along with human-caused increased access to fish waste presumably also contribute to the prevalence of clinical cases in northern sea otters (personal communication). Infections are typically detected at necropsy, with adults free within the lumen or insinuated within the mucosa of the alimentary tract.[257,277] ***Baylisascaris*** spp. is another nematode reported in skunks and mustelids, particularly of the clade Guloninae.[381] Initially, *Baylisascaris* spp. speciation was based on parasite morphology, then by definitive host, and most recently through molecular characterization. Lesions associated with infection vary according to the parasite and host species (host adaptation), the stage of infection, and exposure route. Infections present as visceral larval migrans, neural larval migrans, or ocular larval migrans, with encapsulated white-brown elongated sinuous ascarids with tapering tails in the lung, brain, eye, abdominal cavity or free within the gastrointestinal tract. In North America, fishers, martens, and wolverines are typically infected with *B. devosi* and *B. columnaris* is more commonly detected in striped skunks.[214] In these species, there is gastrointestinal tract tropism and few gross findings. Encephalitis with serpiginous parasitic tracts and associated granulomatous inflammation, putatively due to *Baylisascaris* spp. or *Paragonimus* spp., have been rarely diagnosed in sea otters.[277,473]

Trichinella spiralis has also been reported in numerous mustelids with parasites observed in the diaphragm, tongue, and skeletal muscles.[183] Histologically, lesions typically contain encapsulated parasites in nurse cells with variable eosinophilic and granulomatous inflammation, often associated with degeneration and necrosis of the infected muscle fiber. As with *Baylisascaris* spp., molecular analysis of *Trichinella* spp. has expanded the number of species. Mixed cryptic infections are reported in Canadian high arctic wolverines.[398,400]

Cardiopulmonary helminth infections with a predilection for the right ventricle, pulmonary artery, and lungs have been reported in Eurasian and North American otters due to ***Dirofilaria immitis***,[330,418] *D. spectans* in neotropical otters,[420] ***Crenosoma* sp.** (Fig. 11.41) in Eurasian badgers and American mink, ***Angiostrongylus*** sp. in long-tailed weasels, giant otters, European badgers (Fig. 11.42), tayra, and stoats,[86,145,329,406] and *Perostrongylus* (*Aleurostrongylus*) *falciformis* (Fig. 11.43) in Norwegian badgers.[91] Patent *D. immitis* infections have been described in Eurasian otters.[330] With more severe infections, animals may present with valvular prolapse, cardiac insufficiency, consolidation of the diaphragmatic lung lobes, and reactive hilar lymph nodes. Microscopically, L1 parasites are present in terminal airways often associated with granulomatous inflammation, while adults are most common within pulmonary arteries and right ventricle.

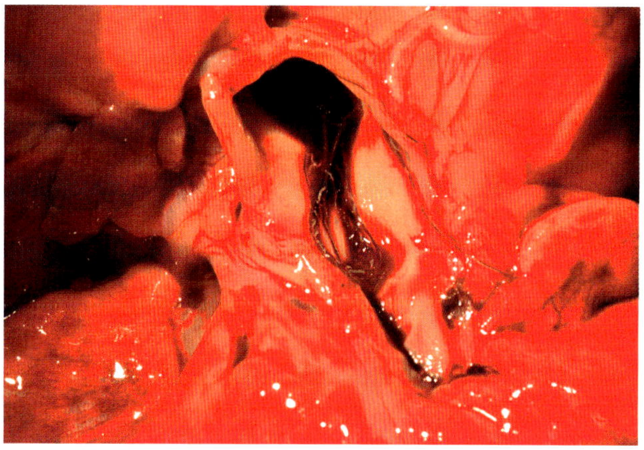

FIGURE 11.42 *Angiostrongylus daskalovi* in a European badger. Multiple, slender, red nematodes are present within the pulmonary artery. *(Photo Courtesy of A. Mihalca, University of Agricultural Sciences and Veterinary Medicine of Cluj-Napoca, Romania.)*

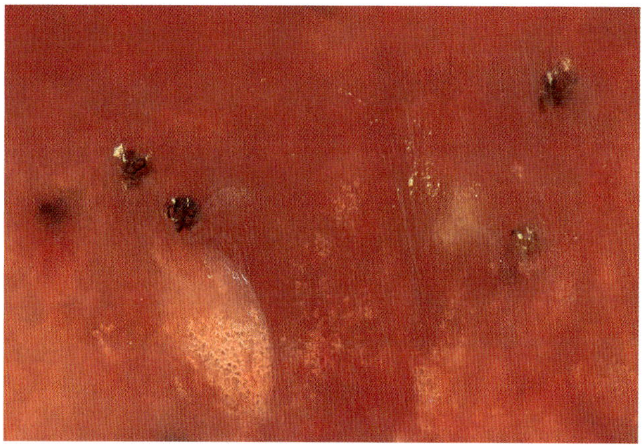

FIGURE 11.43 *Perostrongylus* (*Aleurostrongylus*) *falciformis* in the lungs of a Eurasian badger. Dark brown, thin, coiled adult nematodes are visible on the pleural surface. *(Photo Courtesy of A. Mihalca, University of Agricultural Sciences and Veterinary Medicine of Cluj-Napoca, Romania.)*

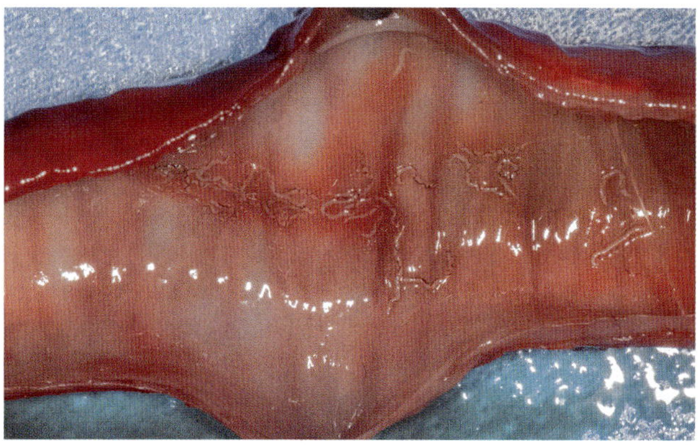

FIGURE 11.41 *Crenosoma petrowi* in the trachea of a European Badger. Adult nematodes are present as thin, white to brown, variably coiled nematodes overlying the tracheal mucosa. *(Photo Courtesy of A. Mihalca, University of Agricultural Sciences and Veterinary Medicine of Cluj-Napoca, Romania.)*

Subcutaneous nematodiasis in mustelids and mephitids is associated with infection by ***Filaria* spp.**, *Dracunculus* spp., and *Dirofilaria* spp., *Filaria martens* is associated with disease in mustelids in the United States, Europe, and Japan, *F. taxideae* is described in American badgers and skunks,[130] and *Dracunculus* sp. (possibly *D. insignis*) is described in fishers in the United States.[79,318] Lesions associated with *F. martens* infection are most common in the subcutis along the flank, perineum, medial, and caudal aspect of the thigh and dorsum, and rarely over the stifles, elbows, or hips. Characteristic microscopic lesions include superficial perivascular dermatitis and crust formation with vesicles that contain embryonated ova, epidermal erosion, ulceration, and granulomatous lymphangitis. Free or encapsulated *Dracunculus* spp. or *Dirofilaria* spp. parasites may be detected in the subcutis and fascia (Fig. 11.44). *Dracunculus insignis* has been reported in fishers, mink, and

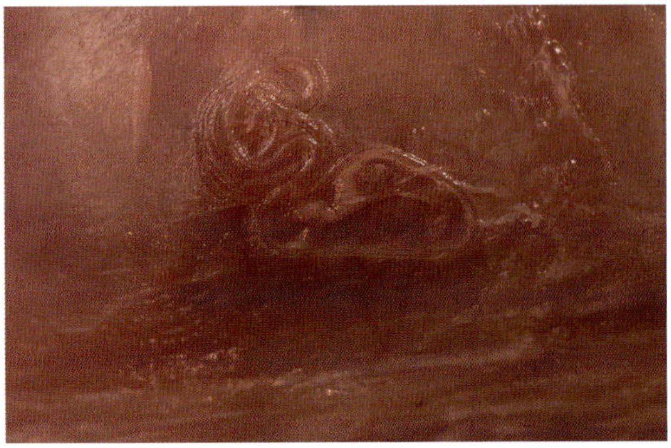

FIGURE 11.44 *Dirofilaria repens* in the subcutis of a Eurasian badger. A coiled adult nematode is present within the subcutaneous tissues with no significant associated gross inflammation. *(Photo Courtesy of A. Mihalca, University of Agricultural Sciences and Veterinary Medicine of Cluj-Napoca, Romania.)*

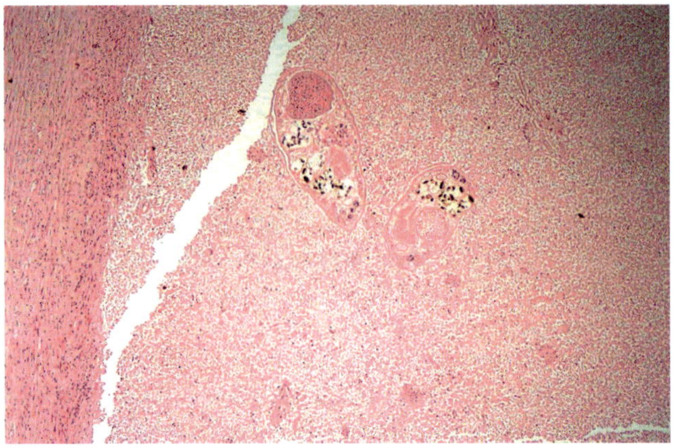

FIGURE 11.46 *Pseudamphistomum truncatum* in the gallbladder of a Eurasian otter. Luminal flukes with characteristic male and female gonads and pigmented ova are present within the gallbladder lumen. Although gallbladder mucosal epithelium has sloughed due to autolysis, fibroplasia of the gallbladder wall is evident along the left side of the photo. *(Photo Courtesy of A. Barlow, Wildlife Network for Disease Surveillance.)*

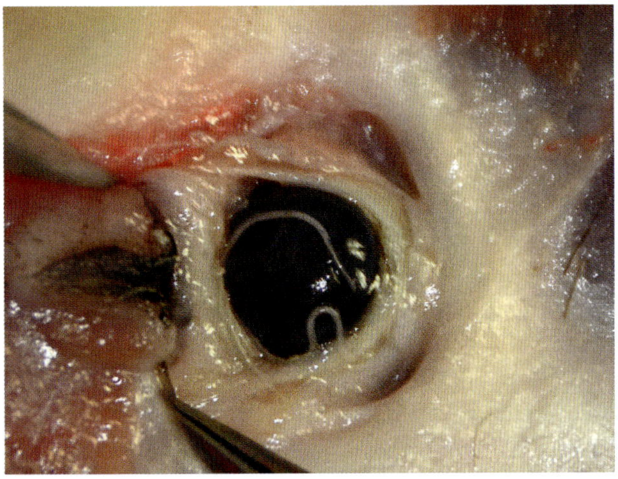

FIGURE 11.45 *Thelazia callipaeda* in a beech marten. A thin, partially coiled, white to tan nematode is present within the ocular conjunctival tissues and overlying the corneal surface. *(Photo Courtesy of A. Mihalca, University of Agricultural Sciences and Veterinary Medicine of Cluj-Napoca, Romania.)*

domestic ferrets; *Dracunculus medinensis* in domestic ferrets; *Dirofilaria lutrae* in North American river otters; and *Dirofilaria repens* in Eurasian badgers. ***Thelazia callipaeda*** infests the eye surface and associated ocular tissues and causes keratoconjunctivitis in beech martens (Fig. 11.45).[392]

Several trematode infections have been reported in mustelids, with distinct anatomic distributions. Environmental factors such as rainfall, terrain, and distribution of mollusk intermediate hosts contribute to the geographic distribution of these parasites. Gallbladder and extrahepatic bile duct infections with ***Pseudamphistomum truncatum*** are reported in American mink and Eurasian otters. Gross lesions include a distended or collapsed gallbladder with thickened walls, serous or mucoid bile, fibrotic bile ducts, and regional lymphadenopathy. Histologically, mucosal hyperplasia, laminar fibrosis of the smooth muscle layers, and nonsuppurative and eosinophilic inflammation may be apparent (Fig. 11.46). More severe cases can exhibit eosinophilic cholangiohepatitis, fibrosing cholangitis with ductular and ductal cholestasis, and hepatocellular individualization and degeneration.[178,386,408] *Orthosplanchnus fraterculus* was reported in the gallbladder of northern, but not southern sea otters and is associated with chronic cholecystitis, cholestasis, and stenosis of the common bile duct. Trematodes are typically identified in the gallbladder and bile ducts. Similarly, the trematode *Opisthorchis felineus* has a predilection for the gallbladder, intrahepatic bile ducts, and pancreatic ducts of Siberian weasels; freshwater snails and fish serve as intermediate hosts.[218]

Subcutaneous and skeletal muscle infections by the trematode *Alaria alata* are reported in European mink.[433] The definitive hosts are red foxes (*Vulpes vulpes*), while mustelids are considered paratenic hosts. Grossly, organisms present as pinpoint white to tan foci scattered throughout the skeletal muscles (Fig. 11.47A). Organisms may also be identified on wet mount cytology of affected muscle (Fig. 11.47B). Frontal nasal sinus and nasal cavity infections with the trematode ***Troglotrema acutum*** are reported in Eurasian and European badgers, European polecats and pine martens, American mink, beech martens, and Eurasian otters.[180] As with *Skrjabingylus* spp. infections in mustelids and mephitids, parasite-associated hyperostosis, osteolysis, and bone fenestration are common, and tend to be more severe and widespread with *T. acutum* infections. Due to a more robust skull, lesions tend to be less extensive and severe in badgers compared to

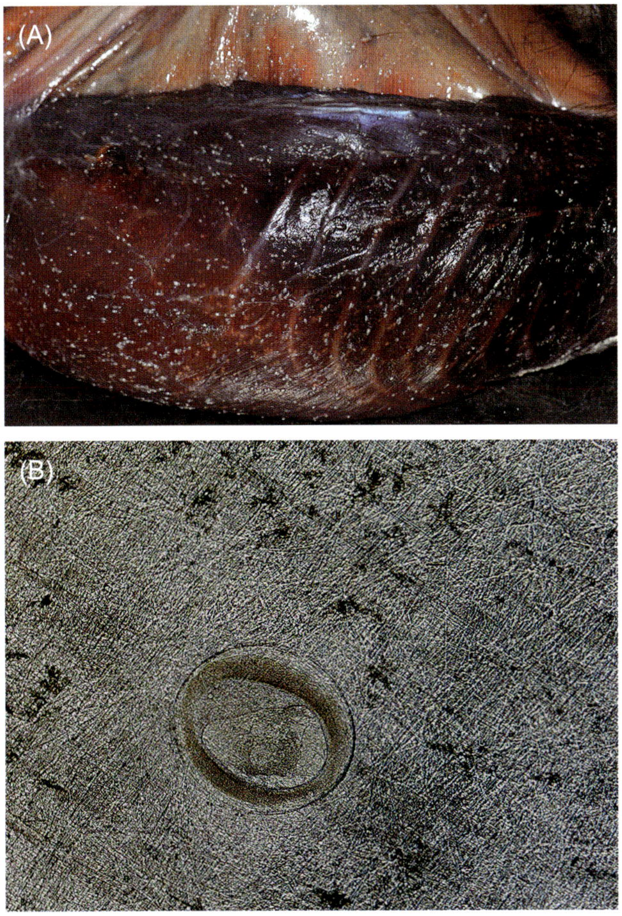

FIGURE 11.47 *Alaria alata* in the muscle of a European mink. (A) Multiple pinpoint white foci are present throughout the body wall skeletal muscle representing encysted parasites. (B) Mesocercariae of the parasite are present within the muscular tissue on wet mount cytology. *(Parts A, B: Photos Courtesy of A. Mihalca, University of Agricultural Sciences and Veterinary Medicine of Cluj-Napoca, Romania.)*

polecats.[259,363] The life cycle of *T. acutum* is complex, with a prosobranch snail first intermediate host, and an amphibian second intermediate host. In sea otters, gastrointestinal *Microphallus pirum*, *M. nicolli*, and *Plenosoma minum* can be detected grossly, primarily in postweaned animals, as white or black punctate foci throughout the intestinal mucosa.[257,356] On histology, these parasites are often seen within variably dilated intestinal crypts.

Incidental infections or individual case reports of **cestodiasis** in mustelids have been documented including *Spirometra erinacei* in skeletal muscle of a Siberian weasel,[238] *Taenia twitchelli* and *Taenia* spp. in intestine of hunter-harvested wolverines in Canada,[468] and *T. martis* in intestine of beech martens.[93] In contrast, infections with **acanthocephalans**, which are highly modified parasitic rotifers, often cause more significant gastrointestinal and systemic disease. These parasites are widespread among pinnipeds, but only recently described in mustelids.[48,156] *Corynosoma strumosum* infection has been detected in marine foraging mink and river otters. Infection has been associated with enteric perforations and secondary septic peritonitis in mink, however, no significant lesions have been described in infections in river otters.[395,396] In sea otters, *C. enhydri* is a common intestinal acanthocephalan with no or minimal associated pathology.[277,395] In contrast, dense ***Profillicollis altmani*** and *P. kenti* infections can cause nutrient depletion, and *P. altmani* is a common cause of intestinal perforation and fatal peritonitis in southern sea otters.[156,264,277]

Pentastomes are enigmatic parasitic arthropods with rodents and other terrestrial mammals as intermediate hosts, and snakes as definitive hosts. Infections are acquired through ingestion of water or vegetation contaminated with snake feces, saliva, or infected snakes. Infections are generally asymptomatic, however, large parasite loads may result in intestinal obstruction, secondary bacterial invasion, or abscessation. Mustelid infections are rare, but with the advent of more specific mitochondrial and ribosomal molecular studies, subpatent infections may be more readily detected. Infection with *Sebekia mississippiensis* nymphs has been reported in a river otter.[218,319]

Protozoa

All mustelids and skunks can serve as intermediate hosts for the protozoan parasite ***Toxoplasma gondii*** by consuming infected intermediate hosts, paratenic hosts (e.g., snails or mussels), sporulated oocysts, or via transplacental infection in utero.[98] Ingested *T. gondii* can then leave the intestine and spread systemically. As host immunity develops, rapidly multiplying tachyzoites convert to bradyzoite-filled tissue cysts, resulting in chronic infection. Encysted bradyzoites may persist for life, and reactivation of infection (recrudescence) can occur when host immunity wanes.[98] Concurrent infection (e.g., viral infection or sepsis) may trigger recrudescence, or worsen the impact of acute infection.[97,155]

Systemic *T. gondii* infection can cause significant morbidity and mortality (toxoplasmosis) in mustelids and skunks, especially free-ranging southern sea otters, captive black-footed ferrets, and farmed mink.[51,222,269,443] Clinical disease is most often observed in captive, fetal, young, or immune-suppressed animals, and individuals with concurrent viral, bacterial, or parasitic infections. Clinical signs range from asymptomatic to fatal, and may include anorexia, depression, fever, seizures, tremors, paresis, paralysis, corneal edema, blindness, aggression, myalgia, ataxia, reduced bladder tone, obtundation, and coma.[98,279,443] Gross lesions include lymphadenomegaly, splenomegaly, interstitial pneumonia, abortion, stillbirth, and neonatal mortality.[98,271] Congenital brain malformation has been reported in association with severe *T. gondii* infection in a neonatal sea otter.[271] Protozoal-associated

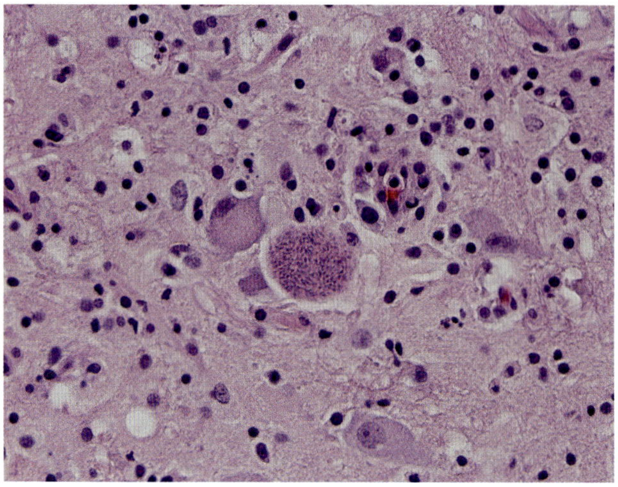

FIGURE 11.48 *Toxoplasma gondii* encephalitis in southern sea otters. On high magnification, cysts are composed of multiple, basophilic to eosinophilic, crescent-shaped, bradyzoites with moderate numbers of lymphocytes and plasma cells forming nodular inflammatory infiltrates in the adjacent neuropil. *(Photo Courtesy of the California Department of Fish and Wildlife.)*

reproductive failure has been reported in sea otters and mink,[271,397] and pregnancy may trigger parasite recrudescence with maternal *T. gondii* serum antibody spikes.[397] Acute fatal toxoplasmosis affecting apparently healthy adult and subadult southern sea otters was recently reported, and all cases were affected with the same rare strain of *T. gondii* (COUG strain). All affected animals exhibited extremely high parasite loads and striking diffuse, severe, grossly apparent protozoal steatitis.[278]

Histologically, mustelids and skunks with toxoplasmosis commonly have nonsuppurative meningoencephalomyelitis with intralesional tachyzoites and thin-walled, angular-edged tissue cysts (Fig. 11.48). The inflammation is dominated by small lymphocytes, with fewer histiocytes and plasma cells, often forming large inflammatory nodules dispersed randomly throughout the brain and spinal cord, with variable necrosis, gliosis, cavitation, and dystrophic mineralization. Tissue cysts are often concentrated along the edges of inflammatory nodules, and prominent perivascular cuffs surround blood vessels near areas of parasite proliferation. The cysts (approximately 50–300 μm diameter) are characterized by a thin, almost invisible outer cyst wall surrounding thousands of short, stout (6 × 2 μm), tachy- or bradyzoites. Individual tachy- and bradyzoites are often slightly curved, with a pointed anterior end, rounded posterior end, and eosinophilic cytoplasm. Lesions outside of the CNS include myositis, myocarditis, steatitis, enteritis, interstitial pneumonia, necrotizing hepatitis, adrenalitis, pancreatitis, thymitis, placentitis, lymphadenitis, and splenitis. Histologically, parasites are most common in the brain, heart, skeletal muscle, adrenal gland, lung, lymph nodes, and spleen.

Although the genus *Sarcocystis* contains several species and strains reported from mustelids and skunks, the most clinically important is ***Sarcocystis neurona***. The life cycle resembles that of *T. gondii*, except that new world opossums (*Didelphis virginiana* and *D. albiventris*) are the definitive hosts. Sporocysts are immediately infective when defecated, and repeated fecal shedding can occur. Ingested sporocysts undergo asexual division to produce merozoites, which penetrate host cells, and proliferate to form schizonts and tissue cysts. Although opossums are distributed throughout much of North and South America, reports of *S. neurona*-associated mustelid mortality are concentrated along the Pacific coast of North America, where nonnative Virginia opossums (*Didelphis virginiana*) have been introduced by humans.[148,279,443] Seroreactivity is sometimes reported in mustelids from areas where opossums are absent or less common, suggesting other modes of exposure such as consumption of avian intermediate hosts, or infection by closely related parasites. Terrestrial runoff appears to transport *T. gondii* oocysts and *S. neurona* sporocysts to estuarine and coastal marine waters, where invertebrate prey can ingest and accumulate the parasite.[12,275]

Fatal *S. neurona* infections are common in southern sea otters and have also been reported in American mink, fishers, and striped skunks. In a recent study,[277] *T. gondii* and/or *S. neurona* infection was a primary or contributing cause of death for 20% of >540 examined southern sea otters. *Toxoplasma gondii* and/or *S. neurona* were detected in 78% of otters in the same sample, suggesting that subclinical infection is common. *Sarcocystis neurona* is more common as a primary cause of southern sea otter death than *T. gondii*.[277]

Sarcocystosis is difficult to distinguish from toxoplasmosis clinically. Gross lesions in mustelids and skunks are nonspecific and can include conjunctival chemosis, subcutaneous petechia, congestion, lymphadenomegaly, splenomegaly, mottled myocardium, pericardial effusion, pulmonary edema, muscle atrophy, bladder distension, interstitial pneumonia, brain swelling and herniation, abortion, stillbirth, and neonatal mortality.[45,99–101,104,148,212,222,231,270,272,277,279,397,443] As with *T. gondii*, immature and subadult animals, and those concurrently infected with other parasites and viruses may be more susceptible to severe clinical disease.[148,212,272]

On histology, sarcocystosis is often distinguishable from toxoplasmosis based on the cellular composition of the inflammatory infiltrate, even when parasites are sparse. *Sarcocystis neurona*-associated inflammation is more pleocellular and diffuse, containing histiocytes, lymphocytes, plasma cells, neutrophils, and occasionally, eosinophils. In contrast with the large, widely spaced inflammatory nodules characteristic of *T. gondii* infection in brain tissue, *S. neurona* brain infection is characterized by miliary

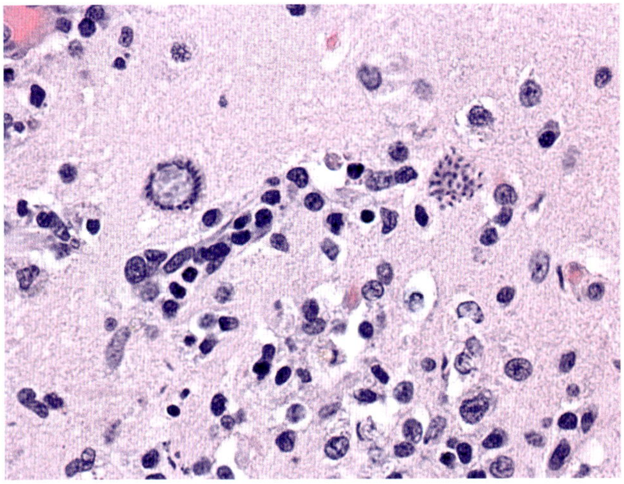

FIGURE 11.49 *Sarcocystis neurona* meningoencephalitis in a southern sea otter. On high magnification of the cerebellar molecular layer, an excellent location to search for parasites, an early schizont (upper right), a rosette-form schizont (upper left), and free merozoites (lower left) are present with intervening pleocellular inflammation. *(Photo Courtesy of the California Department of Fish and Wildlife.)*

clusters of inflammatory cells that could be confused with viral meningoencephalomyelitis. Although parasites are often easiest to visualize in the cerebellar and hippocampal molecular layer and brainstem, inflammation and parasites are usually distributed throughout the brain and spinal cord, along with the heart, skeletal muscle, lung, lymph nodes, and other tissues. Parasite-associated vasculitis and placentitis have also been reported.[45,148,397]

Although infected sea otters often have severe meningoencephalomyelitis, parasites are often hard to visualize on hematoxylin and eosin tissue sections and immunohistochemistry can help confirm the diagnosis. Detection of merozoites and schizonts, especially rosette-form schizonts in brain tissue on histopathology provides strong preliminary evidence for *S. neurona* (or *Sarcocystis* sp.) infection (Fig. 11.49). Additionally, microscopic examination of skeletal muscle and heart often reveals developing sarcocysts. *Toxoplasma gondii* tachyzoites and bradyzoites are slightly stouter and more eosinophilic on hematoxylin and eosin tissue sections than *S. neurona* merozoites and bradyzoites, which tend to have more basophilic cytoplasm. *Sarcocystis neurona* tissue cysts are most common in cardiac and skeletal muscle (Fig. 11.50A). In contrast with *T. gondii*, *S. neurona* tissue cysts are rare in neuropil; when present they are typically smaller, round, more basophilic, and have a thicker cyst wall with small surface projections (Fig. 11.50B).[270]

Other *Sarcocystis* spp. are distinguished from *S. neurona* via differences in parasite morphology, tissue distribution, pathology, immunohistochemistry, and PCR. Definitive speciation is not possible without molecular or ultrastructural characterization. Reports of mustelid and skunk infection by other *Sarcocystis* spp. include wolverines, beech martens, Eurasian otters, and striped skunks infected with *Sarcocystis* sp., *S˙ kitikmeotensis*, *S. kalvikus*, *S. lutrae* or *S. rileyi*.[68,101,103,350,370,474] In most cases, sarcocysts are noted incidentally in skeletal or cardiac muscle. *Sarcocystis* spp. and *S. neurona* infections can occur concurrently.[273,277] Skunks serve as the definitive host of *S. rileyi* with high sporocyst production.

Enteric coccidia are diverse, often species host-specific, and mainly or entirely infect the gastrointestinal tract. Key genera infecting skunks and mustelids include ***Eimeria* spp.**, ***Isospora* spp.**, and ***Cryptosporidium* spp.** Infection can be asymptomatic, or cause fever, vomiting,

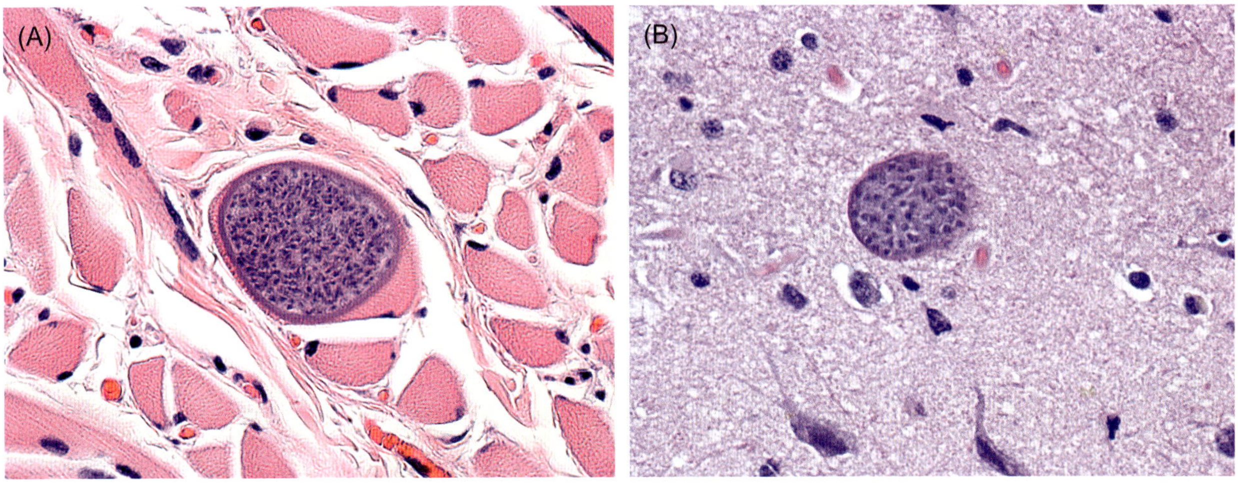

FIGURE 11.50 Tissue cysts of *Sarcocystis neurona* in a southern sea otter. Mature tissue cysts in skeletal muscle (A) have multiple basophilic lunate bradyzoites surrounded by a thick cyst wall with prominent outer surface projections. In comparison, cysts of *S. neurona* in the brain (B) are smaller, round, and more basophilic than *T. gondii* cysts with a thicker cyst wall and in some cases, small surface projections or pegs. *(Parts A, B: Photos Courtesy of the California Department of Fish and Wildlife.)*

diarrhea, enteritis, intestinal hemorrhage, and death. Younger, stressed, sick, and immunosuppressed animals in captive environments are more likely to display clinical signs. Outbreaks often occur in wild or captive settings when feces containing environmentally resistant oocysts accumulate. *Eimeria* spp. and *Isospora* spp. include several important opportunistic pathogens of mustelids and skunks, including *E. furonis, E. ictidea, E. hiepei, E. mustelae, E. melis*, and *E. vison* with most infections being host specific. Affected animals may be asymptomatic or display a range of symptoms including bloody diarrhea, anorexia, emaciation, rectal prolapse, dehydration, and death.[222] Histologically, gamonts of enteric coccidia are visualized within the apical cytoplasm of jejunal and ileal enterocytes, accompanied by variable lymphoplasmacytic inflammation within the lamina propria. Occasionally, epithelial sloughing, villous blunting, lamina propria hemorrhage, and granulocytic inflammation can occur.[325,413,479] Potential concurrent conditions include enteric *L. intracellularis* infection and anorexia-associated hepatic lipidosis.[325]

Biliary coccidiosis secondary to *Eimeria* sp. or *Eimeria ictidea* infection has been diagnosed in neotropical otters[362] and domestic and black-footed ferrets.[386] Infection can be subclinical or associated with jaundice, abdominal distension, pain, emaciation, and anorexia. Biliary coccidiosis is characterized grossly by hepatomegaly, dilated firm bile ducts, and a thickened gallbladder wall. Histopathology reveals cystic proliferation of mucosal glands, fibroplasia, granulomatous inflammation, hepatic biliary tubular dilation, and epithelial hyperplasia.[325] On histopathology, asexual and sexual parasite reproduction can be visualized in the gallbladder or bile duct epithelium, and oocysts may be present in the lumen. *Isospora* sp. infection was also associated with diffuse thickening of the common bile duct in mink that grossly resembled neoplasia.[386]

Fecal shedding of ***Cryptosporidium*** spp. has been detected in asymptomatic river otters in Washington,[144] and sick captive black-footed ferrets with concurrent mixed *Eimeria* sp., *Cryptosporidium* sp., and bacterial infections,[42] where the pathogenicity of *Cryptosporidium* sp. relative to the other concurrent infections was not clear. Some *Cystoisospora* spp. species can form extraintestinal tissue cysts, each containing a single organism. Enteric *Cystoisospora* spp. infection has been described in ASCOs, European otters, and steppe polecats.[445,448,491] Affected ASCOs have green diarrhea, anorexia, and vomiting, some animals die, and numerous sporulated oocysts (putative *C. rivolta*) can be observed on fecal floatation. At necropsy, there is intestinal mucosal necrosis and aspiration pneumonia.[445]

Infections with flagellated protozoa are occasionally reported in mustelids and skunks. Many species are commensals or opportunistic pathogens, and asymptomatic infections are likely under-recognized. Asymptomatic fecal shedding of ***Giardia*** spp. has been detected in river otters in Puget Sound, Washington[144] and *Giardia duodenalis* assemblage E infection has been associated with diarrhea in European badger cubs.[19] Coinfections of *Giardia* sp., *Salmonella* sp., and *Eimeria* sp. may also occur. Suspected or confirmed subclinical trypanosome infections have been reported in river otters and tayra.[120,218,380] Experimental ***Trypanosoma cruzi*** infection of striped skunks results in focal to diffuse chronic granulomatous myocarditis, with amastigotes visible within myocardial fibers. Microgranulomas may be present in the esophageal, gastric, and proximal intestinal muscularis. In these experimental infections, lesions resolved 46 days postexposure in some animals, and skunks were hypothesized to serve as reservoir hosts.[92]

Leishmania infantum infection or seroreactivity has been reported in American and European mink, domestic ferrets, Eurasian otters, and European pine martens.[58,62,149,222,460] Reports of clinically significant *L. infantum* infections are restricted to mustelids housed in Spain, including a domestic ferret, farmed American mink, and Eurasian otters. Clinical signs include nonpruritic, erythematous, and edematous papular skin lesions, lymphadenomegaly, splenomegaly, and hyperglobulinemia. Hematological changes include increased ALT, ALP, GGT, hyperproteinemia, hyperglobulinemia, proteinuria, increased haptoglobin and ferritin. Histologically, amastigotes can be observed within macrophages from splenic aspirates or affected tissues.[460] Secondary *P. aeruginosa*-associated hemorrhagic pneumonia has been described in mink kits with primary leishmaniasis.[222]

Diagnosis of protozoal infections in live mustelids and skunks is often based on demonstration of elevated or rising antibody titers, or clinical response to antiprotozoal therapy. For enteric coccidia, diagnosis is facilitated by fecal floatation, microscopic examination, immunofluorescence, or PCR. For flagellates, some stages can be visualized on fresh wet mounts, Wright-Giemsa stains, or immunohistochemistry. Due to chronic persistence of tissue cysts, *T. gondii* and *S. neurona* seropositivity is suggestive of infection, not simply prior exposure. Where possible, serological findings should be corroborated with results from histopathology, immunohistochemistry, parasite isolation, and PCR. Muscle biopsy, MRI or CT, and PCR from serum or cerebrospinal fluid may be attempted for especially valuable or rare individuals.

Ectoparasites

Ectoparasites have been detected on a variety of mustelid species worldwide. In subtropical and tropical areas, abiotic factors, such as altitude, latitude, and exposure to susceptible host species, contribute to parasite recruitment and infestation.[249] Clinical signs may be inapparent or lesions may be attributed to direct injury by the parasites, host hypersensitivity, secondary bacterial involvement, and other factors. Many ectoparasites are also biological or mechanical vectors of pathogenic bacteria, rickettsia, viruses, protozoa, and helminths. These arthropods comprise a broad range of organisms, including lice (*Phthiraptera* and *Mallophaga*; biting lice, and *Anoplura*; sucking lice), biting flies (*Insecta*), bot and warble flies (*Diptera*), ticks (*Arachnida*), and mites (*Acaridae*). Members of these groups may feature a complex life cycle with extended periods of time off a host as facultative feeders, or may be obligate and host-specific parasites.

The composition of tick fauna may vary by geographic region and typically includes members of genera *Amblyomma*, *Ixodes*, *Rhipicephalus*, and *Dermacentor*.[249] Host response to tick bites may include hypersensitivity reactions, and with more severe infestations, anemia, and death. Some ticks elaborate neurotoxins in salivary secretions, secondary bacterial infections are common, and ticks can be important biological and mechanical disease vectors. *Ixodes* spp. infestations have been detected on badgers from Belgium, the Netherlands, and Switzerland, in association with detection of *Borrelia burgdorferi*, *Anaplasma phagocytophilum*, and rare *B. miyamotoi* infections. In Belgium and the Netherlands, a survey of European pine martens and European polecats detected varying levels of infection by the tick-transmitted pathogens *A. phagocytophilum*, *Neoehrlichia mikurensis*, and *B. burgdorferi*.[187] Massive infestations of *Ixodes rugicollis* and *Trichodectes retusus* were observed in stone martens in Austria with few clinical signs or gross lesions.[462]

Most fleas reported in mustelids have been identified through live capture surveys or necropsy reviews. These parasites have a broad host range and tend to infest animals that regularly return to burrows or nests. Associated clinical signs include pruritus, erythema, and anemia. In mustelids, individual case series have recorded *Paraceras* spp. infesting badgers and martens, *Oropsylla hirsute* in black-footed ferrets, *Neopsylla bidentiadformis* in Steppe polecats, and *Ctenocephalides felis* and *Oropsylla actomys* in striped skunks.[2,108,267,282] Flea infestation was implicated in plague transmission in black-footed ferrets and Steppe polecats; based on extrapolation from other wildlife, livestock, and companion animals, these parasites are also known to transmit tularemia, the tapeworm *Dipylidium caninum*, swine poxvirus, and rabbit myxomatosis. Animals may remain asymptomatic with large ectoparasite infestations or may develop flea bite dermatitis or flea allergy dermatitis.

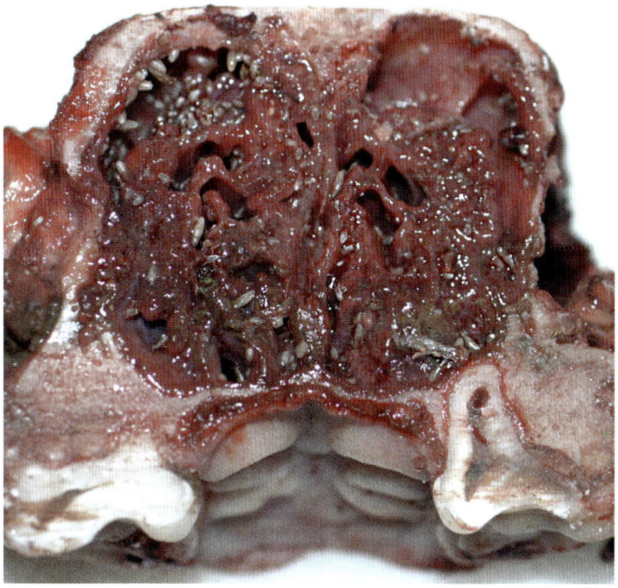

FIGURE 11.51 Nasal acariasis in a southern sea otter. Multiple tan to brown nasal mites are present within the extensive nasal turbinate network. Mite infestation is accompanied by diffuse mucosal hyperemia and brownish catarrhal exudate. *Halarachne halichoeri* mite infestation can contribute to secondary nasal or lower respiratory bacterial infections. *(Photo Courtesy of the California Department of Fish and Wildlife.)*

Mite infestations have been reported in multiple mustelids with a broad geographic and host range. Sarcoptic mange due to ***Sarcoptes scabiei*** has been documented in badgers and martens; clinical signs and gross pathology include diffuse dermal crusting and alopecia of the ears, torso, tail and occasionally flanks.[2,253] Histopathologic features include epidermal hyperplasia, hyperkeratosis, and fibrinosuppurative exudate with numerous embedded mites.[2] In southern sea otters, nasopulmonary acariasis due to infection with ***Halarachne halichoeri*** mites may be subclinical, or may result in mucopurulent to hemorrhagic nasal exudate, rhinitis, tracheobronchitis, turbinate erosion and lysis, and secondary bacterial infections (Fig. 11.51). Mite infestation is more common in older otters, and individuals are more than three times more likely to be infected if they have a prior history of captive care or had close proximity to harbor seals within 10 days of death.[403] In contrast, beech martens infested with *Lynxacarus mustelae*, stone marten with *Trichodectes retusus*, and wolverine with *Otodectes cynotis* infections had few clinical signs aside from an occasional dull hair coat.[462,484]

Biting and sucking lice, which tend to be host specific and may have regional specialization in individual hosts, are transferred through intimate contact between animals and persist indefinitely in the host pelage. Louse infections have been identified in striped skunks in North America.

Other Infectious Organisms

Transmissible mink encephalopathy (TME) is a spongiform disease diagnosed in farmed mink in numerous countries. First identified in Brown County, Wisconsin in 1947,[173] the most recent US outbreak occurred in 1985. While the origin of TME is unconfirmed, a potential source was consumption of prions in meat of infected downer or dead cattle.[244] In mink, the disease spreads through biting or cannibalism, with an incubation period of 6–12 months, and a clinical course of 2–8 weeks. Affected animals exhibit ataxia, aggression, nest soiling, and self-mutilation. Histologically, there is marked spongiform degeneration and cerebral astrocytosis, especially in the frontal cortex and less pronounced in the caudal brain.[194] Ferrets and mink have been used as a model for prion disease research with experimental challenge via intracranial inoculation. Although mink- and ferrets-fed tissues from downer and dead cows have resulted in TSE outbreaks, natural prion infections have not been reported.[332,405] Variations in host-encoded PrP^c and TSE strains may account for variable susceptibility.[425]

Algal epibionts, organisms that live on the surface of other organisms, have been documented on aquatic, marine, and terrestrial vertebrates where they may be opportunistic, commensal, or symbiotic. Although anecdotal, observations of red algal growth on the pelage of wild southern sea otters in California date back to the 1980s and there are rare reports from live northern sea otters in British Columbia, Canada. Algal growth is most common on hairs around the tail and perineum ($n=44$, 88% of 50 cases with descriptions), hind limbs (42%), front limbs (22%), lateral body wall near the water line (20%), ventral chest and abdomen (16%), dorsum (10%) and head or neck (10%).[30] Microscopic examination and molecular characterization confirmed the causative alga as ***Acrochaetium secundatum***[30] (Fig. 11.52A). Visual monitoring of live, free-ranging sea otters with visible algal growth on the pelage (Fig. 11.e38A) and assessment of nutritional condition suggest that the epizoic red algal growth does not significantly affect animal survival. Algal growth is most common on guard hairs and is largely restricted to the portion of hairs protruding above the insulation layer (Fig. 11.e38B), minimizing compromise of thermal integrity of the pelage.[30] However, it might still be possible for animals with severe algal burdens to suffer some compromise, including increased energy expenditure due to excessive grooming. Algal growth on the pelage appears to be most prominent during periods with greater solar radiation. Continued assessment of the condition is advised.

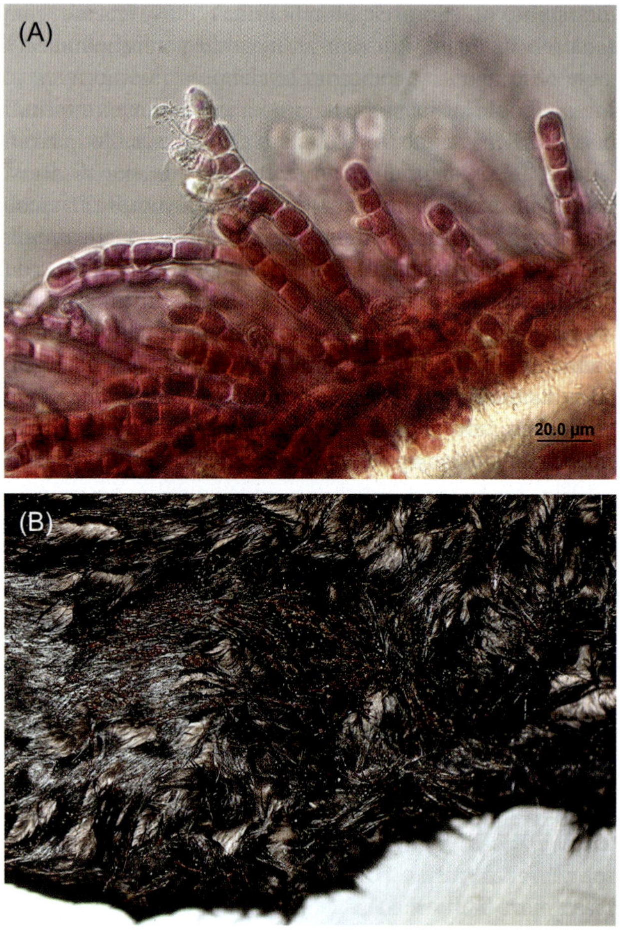

FIGURE 11.52 *Acrochaetium secundatum* growth on the pelage of a wild southern sea otter. (A) Microscopic view of *A. secundatum* growing on a hair from a sea otter pelt. Bright red, linear, septate algal structures grow off the outer surface of the hair shaft. (B) Subtle dark red to red brown growth of algae present on the pelage can easily be missed at necropsy. *(Part A: Photo Courtesy of B. Rosen, Florida Gulf Coast University. Part B: Photo Courtesy of the California Department of Fish and Wildlife.)*

Because of the dark brown color of sea otter fur, epibiotic red algal growth can easily be missed, especially on dead animals with surface contamination by sand or other material (Fig. 11.52B). Prior freezing of the carcass or contact of the pelage with fresh water can cause the algae to rupture, making detection even more difficult. An ultraviolet (Wood's) lamp can be used to optimize algal detection on the pelage of live or dead sea otters and help distinguish between epizoic red algal growth and petroleum contamination; on live animals and fresh carcasses, the epizoic alga fluorescence bright orange when illuminated with a Wood's lamp (Fig. 11.53A), while many (but not all) petroleum products fluoresce pale gray blue (Fig. 11.53B).

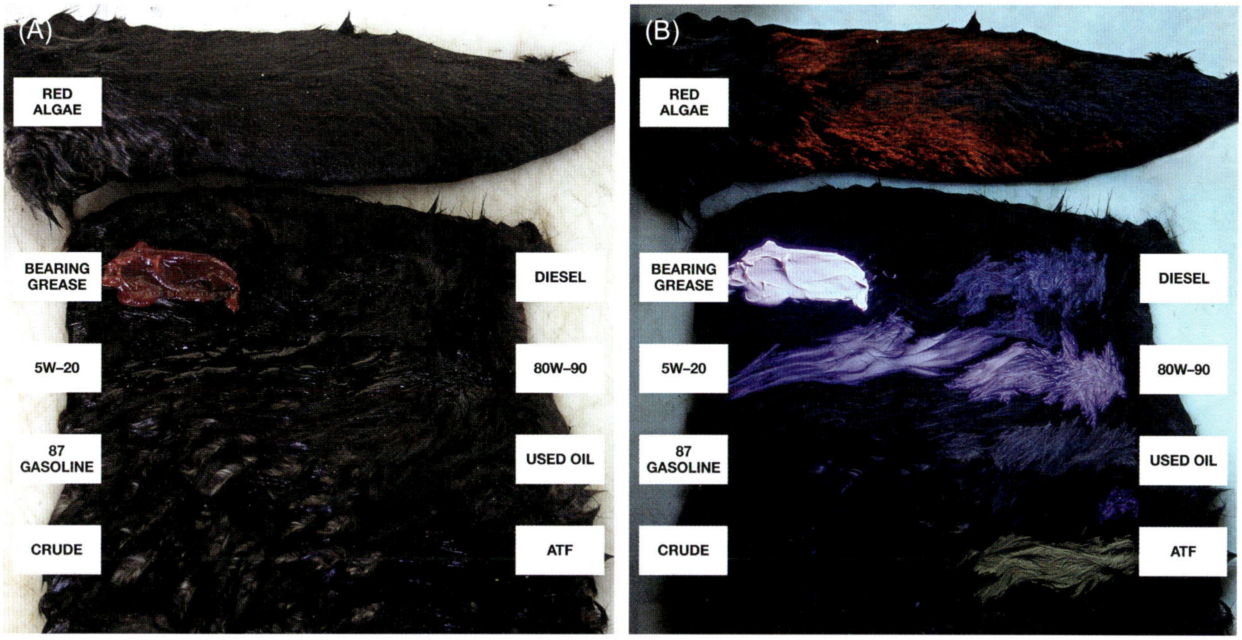

FIGURE 11.53 Ultraviolet light differentiation of epizootic red algae growth and petroleum product contamination on the pelage of southern sea otters. A commercial UV light (i.e., Wood's lamp) can be used to optimize detection of some cryptic contaminants on the pelage of live animals and minimally decomposed carcasses. (A) Contamination of a sea otter pelage by epizoic red alga and various petroleum compounds, as viewed under natural light. (B) The same pelage samples viewed under ultraviolet illumination using a Wood's lamp. The epizoic alga fluoresces bright orange, while many (but not all) petroleum products fluoresce pale gray blue, light tan, or white. Key to the products used in these images: Left: Bearing grease, 5W-20 motor oil, 87 octane gasoline, and unrefined crude oil. Right side: Diesel fuel, 80W-90 motor oil, used automotive oil, and automatic transmission fluid (ATF). *(Parts A, B: Photos Courtesy of A. Reed and S. Kam, California Department of Fish and Wildlife.)*

ADDITIONAL RESOURCES

Alex CE, Watson KD, Schlesinger M, et al. Amdoparvovirus-associated disease in striped skunks (*Mephitis mephitis*). *Vet Pathol*. 2023: 03009858231173468.

Burek-Huntington KA, Gill VA, Berrian AM, et al. Causes of mortality of northern sea otters (*Enhydra lutris kenyoni*) in Alaska from 2002 to 2012. *Front Mar Sci*. 2021; 8:105.

Miller MA, Moriarty ME, Henkel L, et al. Predators, disease, and environmental change in the nearshore ecosystem: mortality in southern sea otters (*Enhydra lutris nereis*) from 1998 to 2012. *Front Mar Sci*. 2020; 7:582.

Quesenberry KE, Orcutt CJ, Mans C, eds. *Ferrets, rabbits, and rodents: clinical medicine and surgery*. 4th ed. Elsevier; 2021.

Stout AE, Guo Q, Millet JK, de Matos R, Whittaker GR. Coronaviruses associated with the superfamily Musteloidea. *MBio*. 2021;12: e02873–20.

References

Full references for this chapter can be found online at https://doi.org/10.1016/B978-0-443-18437-6.00032-8.

Chapter 12

Procyonidae, Viverridae, Hyaenidae, Herpestidae, Eupleridae, and Prionodontidae

Molly E. Church[a], Sushan Han[b], Caitlin E. Burrell[c], and Karen A. Terio[d]

[a]*University of Pennsylvania, School of Veterinary Medicine, Philadelphia, PA, United States;* [b]*Denver Zoo Conservation Alliance, Denver, CO, United States;* [c]*University of Georgia, Athens, GA, United States;* [d]*Zoological Pathology Program, College of Veterinary Medicine, University of Illinois, Brookfield, IL, United States*

INTRODUCTION

This chapter covers the diseases and pathology of multiple taxonomic groups within the order Carnivora with particular focus on Procyonidae and several of the Feliformia carnivores. There is considerable overlap in disease susceptibility between these groups and with Felidae and Canidae (see Canidae, Ursidae, and Ailuridae and Felidae).

The procyonids include the raccoons, kinkajous, coatis, ring-tail cat, and olingos. These groups are characterized by masked faces and banded tails and are native to the western hemisphere. With the exception of the near threatened olinguito, procyonids exist in stable numbers.

Feliformia carnivores include the Hyenidae (spotted, striped, and brown hyenas as well as aardwolf), Viverridae (civets, binturongs, genets, and linsang), Prionodontidae (a monogeneric family of linsangs), Eupleridae (includes fossa, civets, mongoose, and falanouc), and Herpestidae (includes mongoose and meerkats).[202] A notable feature of the feliform carnivores is a double-chambered auditory bulla composed of two bones joined by a septum.[67] Many of these species are listed as vulnerable or endangered due to habitat loss and hunting (see Table 12.e1). Hyaenidae are largely restricted to Africa, the Arabian Peninsula, and Asia as far east as Bhutan. Although they have a similar quadrupedal, digitigrade form to members of the suborder Caniformia in the family Canidae, modern taxonomic revisions place Hyaenidae in the suborder Feliformia. Hyenas have a powerful build with somewhat long forelimbs. Hyena fur is coarse and lacks an undercoat, unlike most of the canids.

The majority of knowledge about disease pathogenesis for these species has been biased toward studies of raccoons in part due to overlapping habitat and disease susceptibility with domestic dogs and cats and because raccoons have adapted to life within urban and suburban environments, which increases potential interactions with pets and people.[72,110] For the feliform groups in this chapter, susceptibility to and potential reservoir status of several species for diseases, such as avian influenza virus and coronaviruses (e.g., SARS), have recently received renewed scientific interest.[194] Several species are commonly displayed in zoological facilities.

NON-INFECTIOUS DISEASES

Nutritional

The species in these groups are omnivorous and eat a wide variety of items from whole vertebrate prey to invertebrates and eggs. Because the nutritional composition of many of the dietary items of wild populations is unknown, providing a variety of food items in managed care settings is critical for ensuring a balanced diet. **Taurine** should be a component of diets, as inadequate intake may be a contributing factor in dilated cardiomyopathies noted in meerkats and possibly mongooses.[122] **Obesity** is not an uncommon finding in urban and suburban raccoons that have ample access to human or domestic animal food sources.

As can occur in domestic carnivores, **pancreatitis** has been reported in captive civets and meerkats.[116,140] Lesions are similar to those in domestic species and include interlobular and peripancreatic neutrophilic inflammation and pancreatic and peripancreatic fat necrosis. Obesity, high-fat diets, hyperlipidemia, and dietary indiscretion are considered risk factors for the development of pancreatitis in these species.

Severe **hydroxyapatite deposition** was reported on the footpads, periarticular regions, and in viscera in a collection of binturongs. Deposition resulted in progressive immobility

Pathology of Wildlife and Zoo Animals. https://doi.org/10.1016/B978-0-443-18437-6.00028-6

and emaciation.[181] The affected binturongs demonstrated hyperphosphatemia and elevated serum vitamin D, though the underlying cause was not determined. Primary renal disease was not ruled out, but affected binturongs had normal serum urea and creatinine levels.

Metabolic

Metabolic disease has been reported infrequently in this group. Pancreatic **islet amyloidosis** is a common postmortem finding in both free-ranging and captive raccoons (Fig. 12.1). An association between this finding and diabetes, as occurs in some other species, has not been established in raccoons.[11,89] Amyloid deposition is characteristic, with abundant, amorphous, homogeneous eosinophilic material in the islets of Langerhans that exhibits birefringence in Congo-red stained sections examined under polarized light.

A **seasonal dermatopathy** was reported in female fossa. It is characterized by symmetrical trunk alopecia with variable pruritis. Lesions are most common in spring and summer and correlate with thickening and enlargement of the uterus. Because of this relationship, the dermatopathy is proposed to be due to hormonal changes associated with reproductive activity. Lesions resolve and reoccur in individuals.[159]

Toxic

Nontarget exposure to **rodenticides** is the most commonly encountered type of exposure to toxins in this group of carnivores, though in some cases, intentional poisoning may occur if they are perceived as pests. Both anticoagulant toxins and those that uncouple oxidative phosphorylation are often manufactured into green-blue pellets or cakes that may be identified in the gastric or intestinal contents of poisoned individuals. Anticoagulant rodenticides include first-generation (e.g., warfarin, diphacinone, chlorophacinone, and coumachlor) and second-generation (e.g., brodifacoum, bromadiolone, and difethialone) compounds. These rodenticides inhibit vitamin K reactivation, thus preventing the production of vitamin K-dependent clotting factors. In one long-term, unpublished survey of free-ranging raccoons in Northern California, up to 80% of raccoons tested from a rehabilitation center over 10 years were positive for rodenticides (M. Piazza and K. Kacmarcik, personal communication). Other studies in raccoons[47,179] and other free-ranging wildlife species[73,160] have also documented frequent rodenticide exposure, indicating an important risk. Gross and histologic lesions include hemorrhage in multiple anatomic sites, including the subcutis, brain, pericardial, thoracic, and abdominal cavities, and within the gastrointestinal tract. Definitive diagnosis is made by detecting anticoagulant rodenticides by high-performance liquid chromatography–mass spectrometry (HPLC-MS/MS) in liver, serum samples, whole blood, or gastrointestinal contents.

Nonanticoagulant rodenticides, like those that uncouple oxidative phosphorylation (e.g., bromethalin), frequently do not cause gross lesions. Death occurs secondary to an influx of fluid into the central nervous system (CNS) with an associated increase in intracranial pressure and subsequent compression of cardiovascular and respiratory centers in the brainstem. Histologically, there is marked vacuolar degeneration of the white matter that corresponds to splitting of myelin fibers by edema (Fig. 12.2).[15] There is variable but often marked astroglial and oligodendroglial gliosis. Definitive diagnosis is made by detection of desmethylbromethalin (a metabolite of

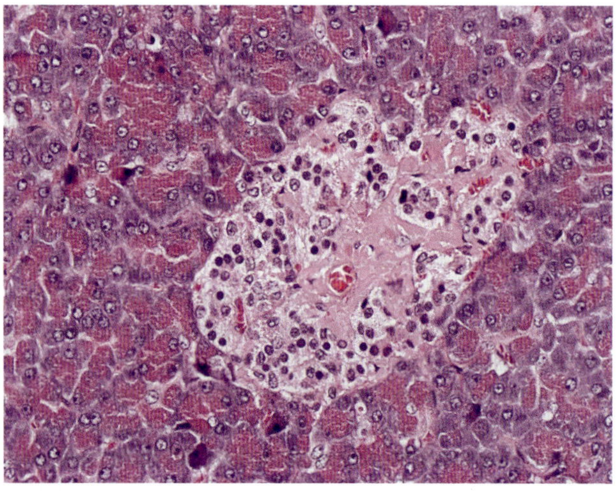

FIGURE 12.1 Pancreatic islet amyloidosis in a raccoon. In some raccoons, pancreatic islets contain small to moderate amounts of acellular, amorphous, eosinophilic fibrillar material consistent with amyloid. There is no evidence that this accumulation is associated with abnormal hormone production or other physiologic or metabolic diseases.

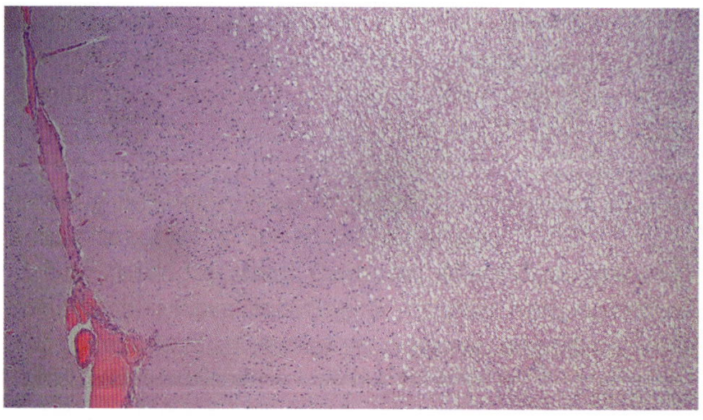

FIGURE 12.2 Bromethalin toxicosis in a raccoon. There is vacuolar degeneration corresponding to myelin fiber splitting and edema with clear demarcation between the affected cerebral white matter and relatively unaffected neuropil. Gross lesions are frequently not apparent in cases of bromethalin toxicosis.

bromethalin) by HPLC-MS/MS in brain, fat, liver, or gastrointestinal contents.

Elevated tissue mercury levels have been reported in Egyptian and Javan mongooses but there was no clinical or histologic evidence of toxicity.[97,132] It has been suggested that mongooses have elevated thioredoxin reductase activity, an intrinsic detoxification pathway.

Ingestion of pennies resulting in **zinc toxicosis** has been described in a captive striped hyena.[2] The animal had severe, nonregenerative anemia, jaundice, and gall bladder edema. Multifocal thrombosis was evident microscopically and associated with multifocal necrosis. The liver demonstrated centrilobular hemosiderosis, cholestasis, Kupffer cell hyperplasia, and single-cell necrosis. Multifocal tubular necrosis in the kidney was associated with intraluminal casts and lymphoplasmacytic, tubulointerstitial nephritis.

Mongooses, whose diet can include snakes, have developed complex mechanisms to avoid the toxic effects of **snake venom**. Examples include toxin-neutralizing proteins in the serum that inhibit venom metalloproteinases and myotoxins, as well as genetic adaptation of the nicotinic acetylcholine receptor that prohibits binding of α-neurotoxins.[190]

Successful treatment of **Heinz body hemolytic anemia** associated with **leek toxicosis** has been reported in a South American coati. Heinz bodies, ghost cells, eccentrocytes, and neutrophilic Döhle bodies were identified on blood smear and were similar to reports of leek toxicity in other species.[103]

Lily toxicosis was described in a case report of a juvenile meerkat. Associated lesions included acute renal tubular necrosis and renal failure. This is similar to lily toxicosis in domestic cats, but affected meerkats also had massive hepatic necrosis and pulmonary edema, which may be unique features in this species.[148]

Congenital/Genetic

Albinism occurs in raccoons and is linked to two independently inherited alleles.[121] Crosses of albino raccoons and normal-colored raccoons produce offspring with a reddish-brown hair coat, referred to by some as "cinnamon," "leucistic," or "red" raccoons. **Congenital follicular dysplasia** has been reported in related litters of coatis.[143] Coatis were born alopecic with only very short dark hairs and with cutaneous lichenification, crusting, and scaling. Histologically, the most significant lesion was premature cornification within the cells of the hair shaft.

Most other congenital/genetic diseases in this group have been reported only in individual case reports. A case of **lysosomal storage disease**, **sphingomyelin lipidosis** (**Niemann-Pick disease**), was described in a raccoon with vacuolated neurons and glial cells in the CNS and macrophages in numerous different organs (liver, tongue, trachea, tonsils, spleen, lymph nodes, small and large intestines, and lungs). Cytoplasmic vacuoles did not stain with Sudan black, oil red O, or PAS, and ultrastructurally they were composed of concentrically laminated, myelin-like figures with a regular periodicity.[188] An uncharacterized **glycogen storage disease** has been reported in a captive coati with progressive neurologic disease. Histologically, neurons, hepatocytes, and macrophages had distended cytoplasm containing finely granular material that ultrastructurally is composed of membranous structures within lysosomes.[34]

Other congenital lesions in raccoons are reported rarely and include single case reports of forelimb and tail absence,[94] tail duplication,[189] diaphragmatic hernia,[169] renal hypoplasia,[129] dental abnormalities,[112] and duplicated pelvises and hind limbs (M. Church, unpublished). In meerkat, congenital atlanto-occipital dysplasia with cranial displacement of the dens into the vertebral canal of C1 (Evans unpublished, 2018) as well as megaesophagus due to a persistent right aortic arch (Han, unpublished) have been diagnosed.

Age-Related/Degenerative

Aged animals have potential problems with **dental wear, dental attrition**, and **periodontal disease** as occurs in other taxa. **Dietary indiscretion** can often result in damaged teeth and gastrointestinal foreign bodies in striped hyenas (R. Wack, personal communication). Severe lymphoplasmacytic and suppurative periodontitis is described in a dwarf mongoose which was associated with fibrous hyperplasia and generalized gingival enlargement and loss of alveolar bone, responsive to extensive dental extraction.[18]

Similar to domestic carnivores, there are many documented cases of both **hypertrophic** and **dilated cardiomyopathy** in aged dwarf and banded mongoose and meerkats (Fig. 12.3).[66,95] The pathogenesis is not known. Dilated cardiomyopathy with myocardial degeneration and fibrosis appears to be common in captive mongooses. Detailed population studies with a focus on characterization of pathology and heritability are currently underway. In meerkats histological features of cardiomyopathy include perivascular and replacement fibrosis that is most prominent in subendocardial tissue and within papillary muscles. Myocardial fibrosis with and without dilation has been identified as a common cause of sudden death in binturongs. It is most often diagnosed through necropsy and histologic examinations and can include clinical congestive heart failure with bicavitary effusion.[127]

Chronic renal disease, characterized by lymphoplasmacytic interstitial nephritis and fibrosis, are common

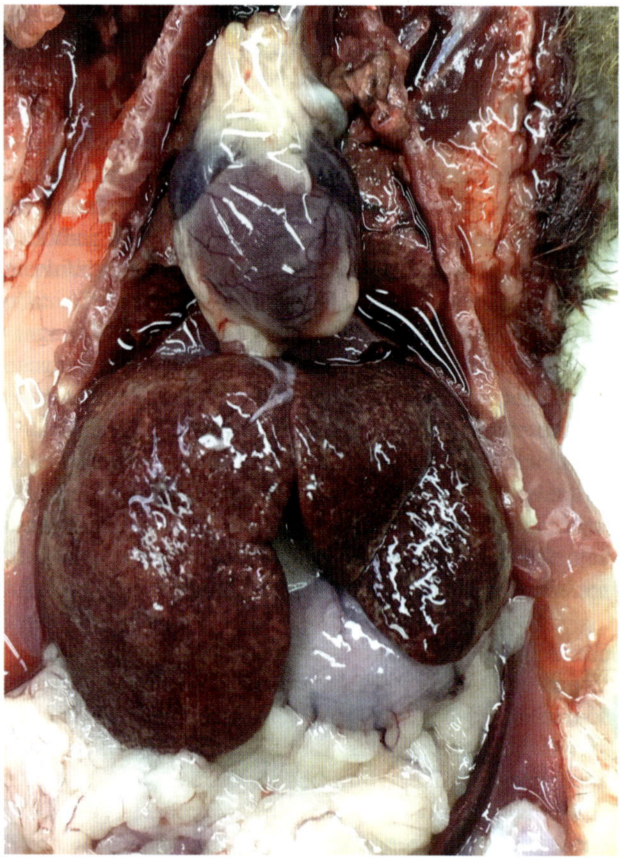

FIGURE 12.3 Cardiomyopathy and secondary hepatopathy in a dwarf mongoose. There is hypertrophic cardiomegaly with secondary hepatomegaly and parenchymal mottling due to noninflammatory, cardiogenic, bridging hepatic fibrosis.

findings in aged dwarf and banded mongoose, meerkats, and binturongs. It occurs as a primary condition or in conjunction with chronic myocardial degeneration. Renal insufficiency in binturongs can be challenging to diagnose clinically as serum biochemistry and electrolyte levels do not necessarily correlate with advanced lesions diagnosed on postmortem examination, suggesting a lack of currently appropriate reference values for this species.[127]

Intervertebral disc disease has been reported in coatis.[130,139] Similar to domestic animals, clinical signs included hindlimb paresis and ataxia and magnetic resonance imaging (MRI) demonstrated thoracolumbar extradural spinal cord compression. Related histologic lesions included dilated myelin sheaths, spheroids, myelinomacrophages, astrogliosis, endothelial cell swelling, and perivascular lymphohistiocytic inflammation in affected spinal cord segments.

In aged coatis, PAS-positive, diastase-resistant, **intraneuronal cytoplasmic globules** are noted within neurons of the peripheral autonomic ganglia.[41] There is no evidence of concurrent neurodegeneration and globules appear to be an incidental aging change.

Miscellaneous/Multifactorial

Vascular mineralization in the leptomeninges (overlying cerebrum, cerebellum, and spinal cord), choroid plexuses, and capillaries is common in the brains of adult free-ranging, especially aged, raccoons similar to domestic geriatric carnivores and herbivores. It appears multifocally as round deposits of deeply basophilic, refractile material arranged in concentric rings in the walls of small capillaries and randomly throughout the neuropil (Fig. 12.4). While these mineralized foci are sometimes associated with mild inflammation and fibrosis, the lack of associated clinical signs and lack of disruption of surrounding tissue make these foci of mineralization an incidental finding.[84]

Another commonly encountered lesion in the CNS of raccoons is **neuroaxonal degeneration**, which is seen consistently in the gracilis and cuneate nuclei of the medulla and is characterized by abundant vacuolated neurons and axonal spheroids. This lesion has no associated inflammation or clinical signs and has been reported in young and adult free-ranging and captive raccoons.[85,89,90]

Perivascular eosinophilic cuffs, primarily in the gray matter of the cerebrum, are not uncommon in free-ranging adult raccoons. Their significance is unknown, but these changes in the brain are found in raccoons without neurologic signs.

Cystic endometrial hyperplasia, characterized by glandular dilation with intraluminal proteinaceous material and mild endometrial inflammation, is commonly seen in captive adult female raccoons.[85,89] **Polycystic kidneys**, in which cysts of varying sizes lined by a single layer of tubular epithelium that expanded and destroyed the renal cortex and medulla and were associated with lymphocytic

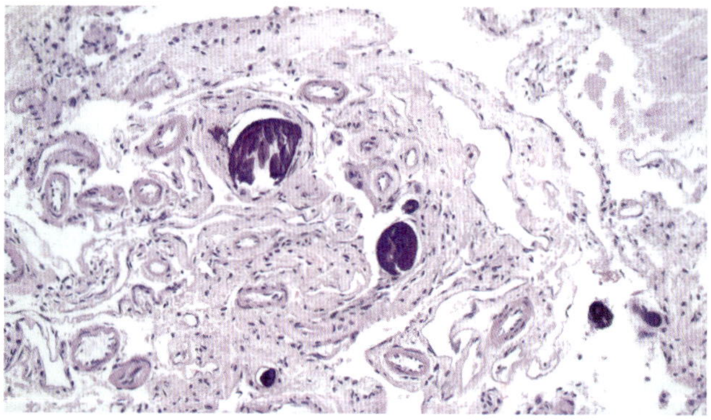

FIGURE 12.4 Meningeal vascular mineralization in a free-ranging raccoon. Meningeal capillaries are variably dilated and lumina contain mineralized foci (psammoma bodies), some of which are associated with segmental endothelial loss. These are typically considered an incidental finding in raccoons.

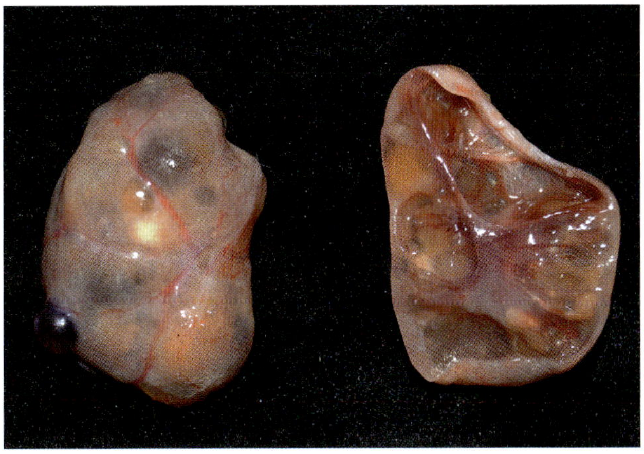

FIGURE 12.5 Polycystic kidney disease in a meerkat with renomegaly and severe cortical and medullary loss from multiple, fluid-filled cysts. Multiple, variably sized cystic spaces expand and replace the normal renal parenchyma.

inflammation, have been reported in a captive-aged raccoon.[88] Single or multiple and extensive **renal cysts** are occasionally diagnosed in dwarf mongooses and meerkats (Fig. 12.5).

Neoplastic

As might be expected, neoplasms in procyonids and feliform carnivores have similar gross and histologic features, disease progression and prognosis to those commonly diagnosed in domestic cats and dogs. Reported neoplasms in these groups include lymphoma (Fig. 12.6); other round cell and mesenchymal sarcomas,[3,21,119,124]; and carcinomas of the urinary bladder, gastrointestinal tract (including anal sac glands), mammary gland, hepatobiliary, pulmonary, and renal systems.[3,19,21,31,35,46,49,63,79,87,89,93,98,109,120,125,127,137,175,176,183,191,192] A single case of uterine adenocarcinoma in a captive coati associated with implantation of the contraceptive **melengestrol acetate** has also been reported.[32]

Thyroid neoplasms have been reported in raccoons in the United States and Germany.[126] In Germany, the incidence of thyroid disease was 64.3% (18/28) and 77.5% (31/40) in two necropsy surveillance studies of captive raccoons. Age likely plays a role in the oncogenesis, as none of the 37 free-ranging juvenile raccoons in one of the studies had thyroid disease. Carcinomas were most frequent (15/31, 48.4%), some of which metastasized (similar to dogs); functional follicular hyperplastic lesions (7/31, 22.6%) or adenomas (4/31, 12.9%) (similar to cats) were less common.

Neuroglial brain tumors associated with high levels of a novel **raccoon polyomavirus (RacPyV)** have recently been described in raccoons in the western United States.[36,37,53] These tumors are consistently found in the olfactory tract and bulb of raccoons, and in some cases are so extensive that they extend rostrally into the nasal cavity (Fig. 12.7). Histologically these tumors have a varied appearance, ranging from bundled spindloid cells to sheets of polygonal cells interrupted by serpentine areas of necrosis with a pseudopallisading pattern[75] (Fig. 12.8; eSlide 12.e1). Confirmation of causality in oncogenesis remains to be confirmed. However, the viral oncogenic gene, Large T antigen, is highly transcribed in tumor tissue relative to host,[24] and the RacPyV genome is maintained in cultured tumor cells and the transformed phenotype is lost if the viral genome is inhibited from binding to the host genome in vitro.[37] This suggests RacPyV has an important

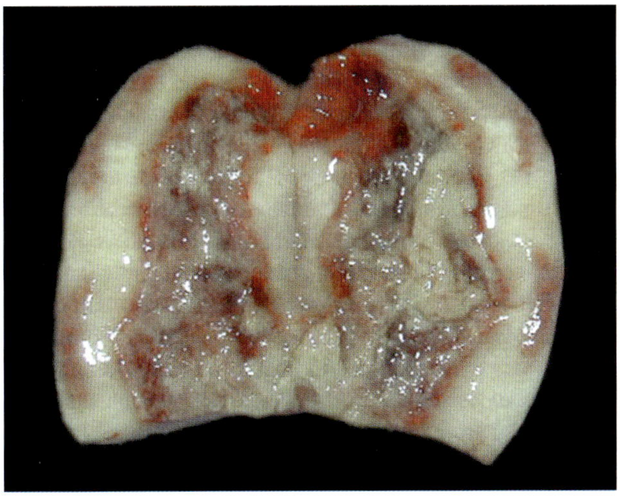

FIGURE 12.6 Uterine lymphosarcoma in a free-ranging raccoon. Longitudinal section of the uterine body with exposure of the lumen is shown. The uterine wall and mucosa are expanded by infiltrative soft, pale tan tissue that thickens and obscures the normal layers of the uterine wall. This is a common tumor of free-ranging and captive raccoons.

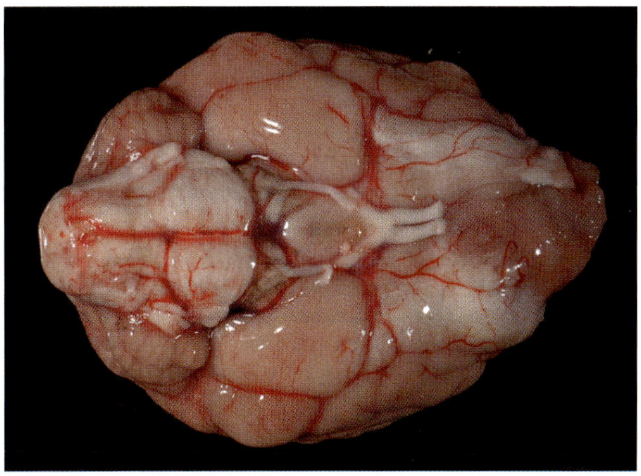

FIGURE 12.7 Neuroglial brain tumor in raccoon. An unencapsulated, poorly demarcated, soft, pale tan/gray mass distorts the normal architecture of the left olfactory bulb. These tumors are consistently found in this location. In severe cases, they can extend into the nasal cavity.

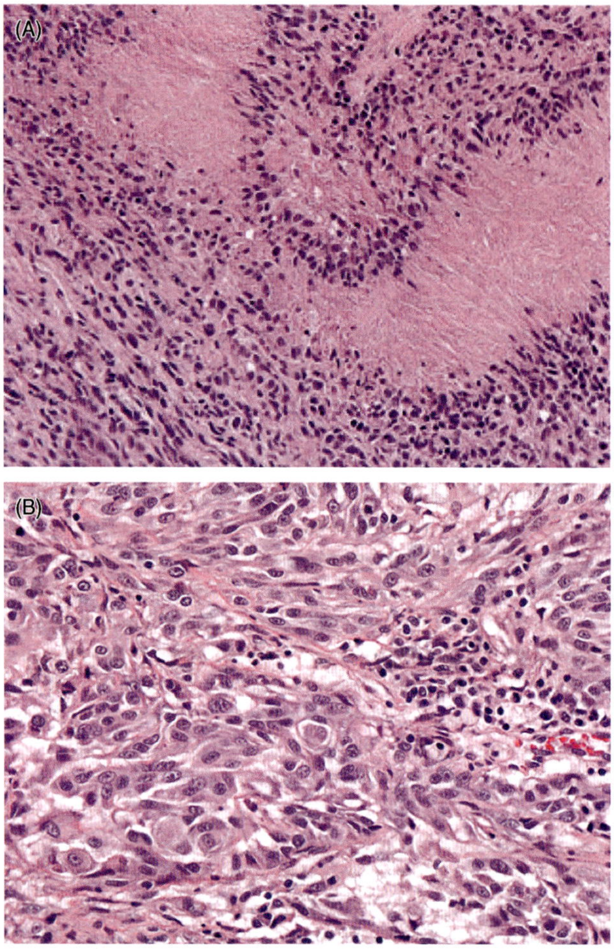

FIGURE 12.8 Neuroglial brain tumor in a raccoon. (A) A characteristic feature is a prominent pseudopalisading pattern of neoplastic cells. (B) Cells vary in morphology from sheets of polygonal cells to streams of spindloid cells (see eSlide 12.e1).

role in the promotion of neoplasia in neural stem cells. Serological surveys have confirmed viral exposure in free-ranging raccoons with and without tumors across the United States and Canada, suggesting additional factors may play a role in the oncogenesis of these tumors.[36]

Renal carcinoma in binturongs is well described and is associated with metastatic disease.[31,109,127,183] In one case, both the dam and sire of an affected animal died from complications of metastatic renal carcinoma.[183] The tumors in binturongs resemble those in other species. Grossly, they can be focal or extensive and replace normal parenchyma (Fig. 12.9). Histologically, they are composed of nests of well-differentiated to pleomorphic, cuboidal to polygonal (or spindle shaped, in the case of the sarcomatoid variant) cells sometimes arranged in irregular tubules. Pulmonary metastasis was present in all of the reported cases; metastases also occur in the mediastinum, pericardium, parietal

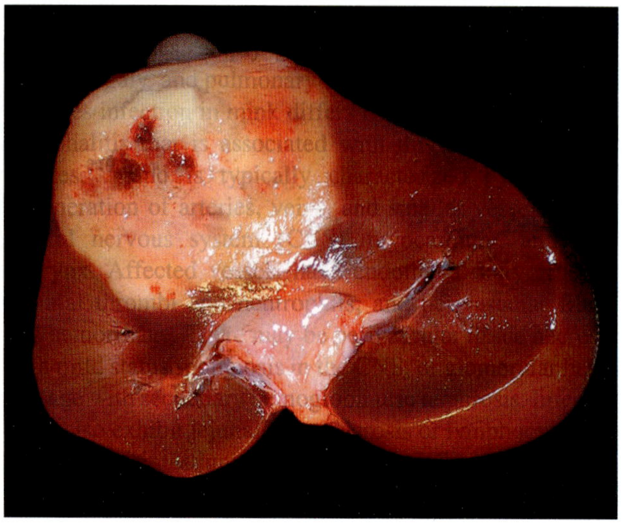

FIGURE 12.9 Renal carcinoma in a binturong. A focal, well-demarcated, tan mass within the cortex replaces the normal parenchyma. Metastasis is common, even in cases with a focal primary tumor. *(Photo Courtesy of C. Rodriguez, Wildlife Conservation Society.)*

pleura, muscles of the thoracic wall, adrenal glands, liver, spleen, and pancreas.

Oral and **genital papillomas** have been identified in a population of wild spotted hyenas at the Masai Mara Game Reserve in Kenya.[142] Spontaneous regression occurred in some animals. The microscopic and gross appearance was similar to oral and cutaneous papillomas in other species. Characteristic features included thickened, hyperplastic epithelium with prominent stratum spinosum, thickened keratin layers, and the formation of wide, discrete rete pegs. Scattered koilocytes were present in the stratum granulosum and ballooning degeneration was seen in some cells. Virus isolated from hyenas was most closely related to the lambdapapillomaviruses of cats.

INFECTIOUS DISEASES

Procyonids and feliform carnivores are ubiquitous in their environments and share habitat and environmental resources with other nondomestic and domestic carnivores and humans. Because several of the species in this chapter act as reservoirs for a number of important multispecies or zoonotic pathogens, for example, raccoons (e.g., canine distemper virus and rabies) and civets (e.g. SARS coronavirus), surveys for pathogens that may be harbored or vectored by these species have been active areas of investigation. Unfortunately, less research has focused on the potential effects of these pathogens on their hosts. The descriptions below are primarily those for which disease is observed in wildlife species.

DNA Viruses

Infection with several viruses in the family *Parvoviridae*, including feline parvovirus, canine parvovirus, and mink enteritis virus, occur in raccoons.[9,13,106] The majority of **parvovirus** disease outbreaks are due to infections with **carnivore protoparvovirus-1** (CPPV-1). Outbreaks in raccoons are primarily due to racoon parvovirus, a unique virus characterized to be phylogenetically positioned between canine parvovirus 2 (the initial virus that emerged in the 1970s) and 2a (one of the three subtypes, 2a, 2b, and 2c, that have emerged more recently). However, raccoons can be infected with all subtypes of canine parvovirus.[9] Disease manifestation with any of these parvoviruses consists of lymphoid depletion, lymphoid necrosis, and necrohemorrhagic enteritis with sloughing of intestinal crypt epithelium and mucosal collapse ("radiomimetic" lesions) (Fig. 12.10; eSlide 12.e2).[106] Parvoviral disease is anecdotally more severe in young raccoon kits than in adults, similar to disease in domestic animals. Several species of civets and genets can also be infected with canine and feline parvoviruses with similar gastrointestinal lesions and lymphoid depletion described.[28,54,194] In a survey of vehicle-killed and hunted free-ranging mongoose in Portugal, almost 60% of tested individuals harbored feline parvoviral DNA.[58] However, given extremely low recovery of viral DNA and the lack of pathologic lesions, it is unclear if this species is truly susceptible to parvoviral-induced disease or is merely susceptible to infection and has the potential to act as a virus reservoir and/or vector. In nonraccoon species, the carnivore protoparvovirus-1 **feline panleukopenia virus** was identified as the cause of acute diarrhea in a banded linsang; infection was associated with fecal shedding of the virus[102] and cause of an outbreak of hemorrhagic gastroenteritis in farmed, small Indian civets in Thailand.[28]

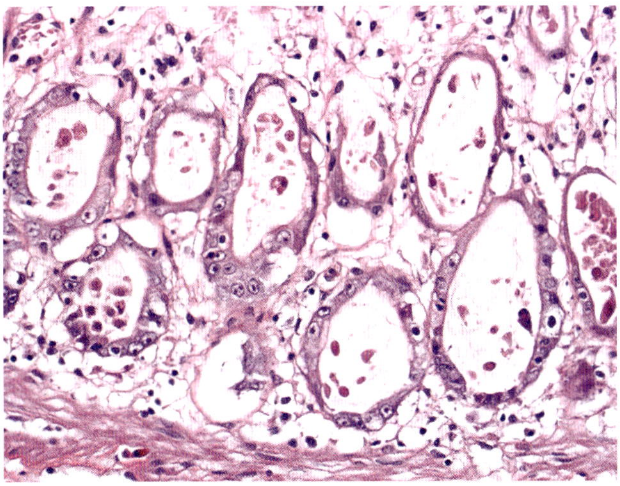

FIGURE 12.10 Parvoviral enteritis in the small intestine of a raccoon. Lesions resemble parvovirus infection in other species and include crypt dilation, crypt epithelium attenuation and necrosis, and villous collapse. Sloughed necrotic cells can be seen within the lumen of distended crypts and associated neutrophils and edema are present in the adjacent lamina propria (see eSlide 12.e2).

Aleutian disease is caused by **Aleutian disease virus** (syn. **Carnivore amdoparvovirus 1**). It is a highly contagious parvoviral disease that naturally occurs in mustelids, especially mink and ferrets (Mustelidae and Mephitidae), and other free-ranging carnivora including skunks, foxes, and raccoons. Antibodies to the virus have been detected in raccoons in North America, and experimental seroconversion has been demonstrated but disease is not often observed, suggesting raccoons as a potential reservoir for the virus.[145] Free-ranging common genets in France have also been shown to harbor antibodies to this parvovirus, but active disease has not been demonstrated.[70]

Captive mongooses in Germany are reportedly extremely susceptible to **cowpox virus**, with infection rates in some outbreaks as high as 100%.[115,170] Rodents are thought to be the virus vectors. Gross lesions include multifocal cutaneous and mucocutaneous vesicles, pustules, and ulcerations that correspond histologically to epithelial cell necrosis with eosinophilic, intraepithelial cytoplasmatic viral inclusion bodies and associated inflammation. Pinpoint to coalescing foci of necrosis are also seen in the lymph nodes, liver, spleen, and gastrointestinal tract. These gross and histologic lesions are typical of cowpox lesions reported in other species that are susceptible to infection.[170]

Pseudorabies due to **Aujeszky's disease virus** (**ADV**; syn. **Pseudorabies virus**, **suid herpesvirus 1**) has been reported in raccoons and hyenas.[178] It is an important differential diagnosis in suspect rabies cases, as animals infected with either virus can present with a variety of neurological signs. Gross lesions in raccoons are usually not evident or are associated with self-trauma from severe pruritus that is frequently seen clinically. Histologic findings include nonsuppurative encephalitis with intranuclear, eosinophilic viral inclusion bodies in neurons and myocardial, lymphoid, and pancreatic necrosis. While 30 years ago it was thought that raccoons were historically thought to be a reservoir for ADV, however, it is now suspected that feral swine and European wild boar are the main reservoirs.[20,48] Raccoons also appear to be susceptible to other unique herpesviruses,[91] possibly their own, although these are less well characterized and concurrent immunosuppression secondary to other disease processes, for example, infection with canine distemper virus (CDV), may be important in disease manifestation.

Oral and **genital papillomas** similar to lambdapapillomavirus in felids have been identified in a population of wild spotted hyenas at the Masai Mara Game Reserve in Kenya.[142] Gross lesions were typical of oral-cutaneous papillomas in other species and appeared to spontaneously regress. Histological findings included thickened, hyperplastic epithelium with peaks of hyperkeratosis; wide, discrete rete pegs; and scattered koilocytes with ballooning degeneration.

RNA Viruses

Rabies virus is a single-stranded RNA virus in the *Lyssavirus* genus and the *Rhabdoviridae* family. Raccoons are consistently the most reported rabid wildlife species in the eastern United States, and rabies infection is considered enzootic in populations in some eastern states. In addition to causing disease in raccoons, the raccoon strain of rabies is also the most common strain diagnosed in rabid domestic companion and production animals in the United States (in cases where variant is reported).[123] In 2021 most cases of rabies reported in the United States (91.5%) were in wildlife species, with a little over 60% of those positive cases occurring in bats and raccoons.[123] Raccoons infected with rabies typically display a wide range of neurologic clinical signs including, but not limited to, diurnal activity, ataxia, head pressing, paralysis, and aggressiveness.[167] There are no gross lesions directly associated with rabies virus infection, though affected animals may have superficial cutaneous lesions (i.e., contusions), tooth fractures, or gastric foreign bodies as a result of ataxia, incoordination, aggression, and pica. Histopathologic findings include polioencephalomyelitis with perivascular cuffing by lymphocytes and plasma cells throughout cerebrum, brainstem, and spinal cord.[84,93,167] In some cases intracytoplasmic eosinophilic viral inclusion bodies (Negri bodies) can be seen (Fig. 12.11). Inclusions are reported to be most common in the hippocampus of carnivores but they may also be abundant in other regions including but probably not limited to cortical neurons, basal nuclei, thalamus, and the caudal colliculi.[50] The gold standard for diagnosis of rabies virus infection (in any species) is direct staining with fluorescent antibody.[48,84,167] Domestic dogs and other nondomestic canid and carnivore species can serve as reservoirs for rabies virus in different geographic locations; mongooses are recognized reservoirs in the Caribbean.[17,168,173,194,203]

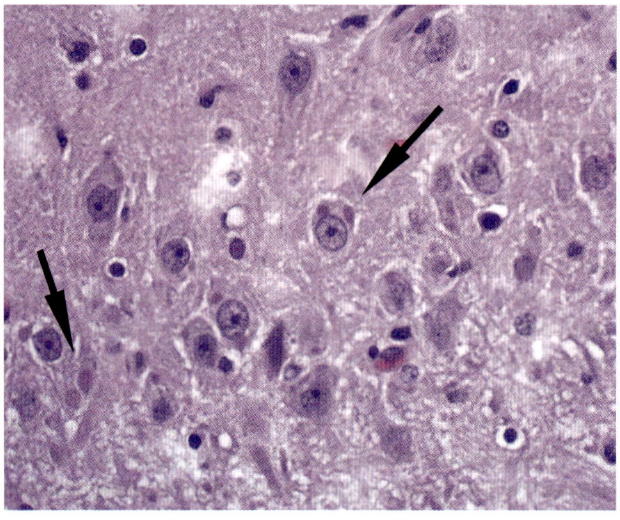

FIGURE 12.11 Rabies viral encephalitis in a raccoon. Neurons contain characteristic intracytoplasmic viral inclusion bodies (Negri bodies; *arrows*). Infection in this case was associated with minimal inflammation. *(Photo Courtesy of D. McAloose, Wildlife Conservation Society.)*

Mongooses, genets, and civets are important vectors of rabies in Africa, where a mongoose rabies virus biotype, distinct from the canid biotype, has been identified in Zimbabwe and South Africa.[141,168] The yellow and slender mongoose of South Africa and Zimbabwe, respectively, are the principal vectors of this biotype, where the species has been shown to be capable of sustained transmission of the virus. The mongoose rabies biotype is responsible for up to 44% of confirmed rabies cases in South Africa.[141]

Canine distemper virus (CDV) is a single-stranded RNA virus in the *Morbillivirus* genus, *Paramyxoviridae* family. It is an extremely important multihost pathogen that affects all carnivores.[158,196] Case reports and periodic outbreaks of canine distemper have been reported in most of the groups included in this chapter. Some species, like raccoons, are reservoirs and possible sources of newly evolved strains.[165] Disease manifestation is similar to that seen in domestic dogs, and young animals are most susceptible.[196] Evidence of infection occurs in many different organ systems, notably the respiratory, gastrointestinal, lymphoid, and nervous systems, and clinical signs can therefore vary and include oculonasal discharge, coughing, diarrhea, anorexia, and concurrent infections. Blepharoconjunctivitis, rhinitis, and footpad hyperkeratosis can also be seen.[16,196] Neurologic signs are frequently encountered, and rabies is an important differential diagnosis. The most common histopathologic findings are bronchointerstitial pneumonia (Fig. 12.12A; eSlide 12.e3) and generalized lymphoid depletion (Fig. 12.12B). Nonsuppurative encephalitis with neuronal necrosis and demyelination can be seen anywhere in the brain but is often most prominent in the cerebellar white matter. Nonsuppurative polioencephalitis has also been described in farmed civets in Thailand, where a unique civet-adapted Asia-4 lineage canine distemper virus was implicated.[154,182] This specific strain causes pathologic lesions similar to those seen in other species including severe catarrhal enteritis, pulmonary edema, and necrotizing hepatitis with footpad and nasal planum hyperkeratosis.[154] Eosinophilic intracytoplasmic and intranuclear inclusion bodies can be seen in many different cells including epithelial cells and viral syncytial cells in many different organ systems (e.g., enterocytes, pneumocytes, transitional epithelium of the urinary bladder, neurons, and astrocytes). Lesions in captive binturongs in the United States and Korea included interstitial pneumonia, lymphoid depletion, and footpad hyperkeratosis but interestingly,

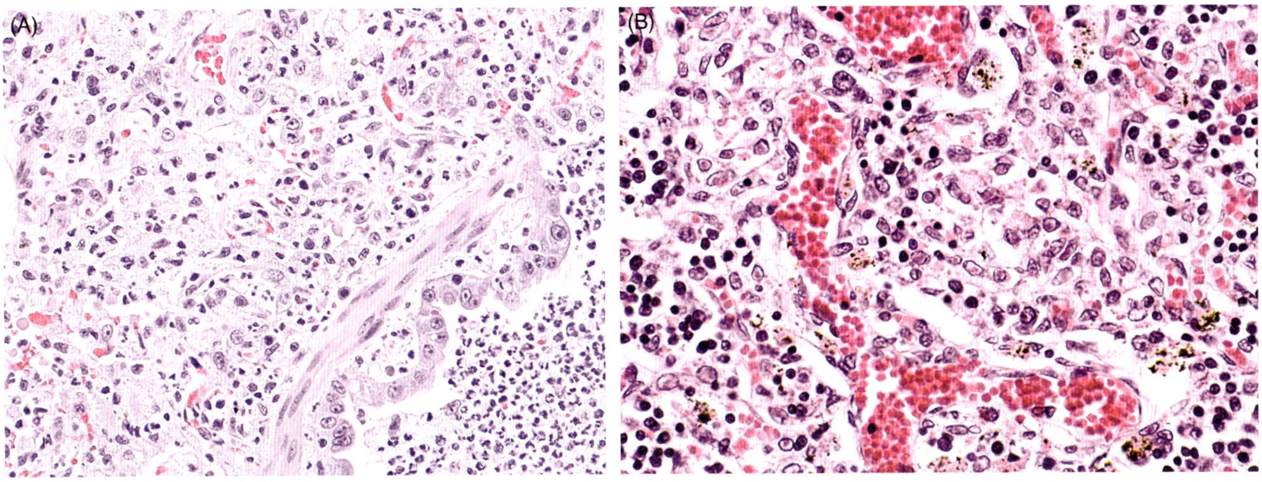

FIGURE 12.12 Canine distemper virus (CDV) infection in the lung of a raccoon (A) and thymus of a hyena (B). (A) Bronchiolar epithelium contains numerous intracytoplasmic and intranuclear eosinophilic inclusions and rare syncytia. Alveolar type II pneumocyte hyperplasia with intranuclear and intracytoplasmic inclusions, and intraalveolar and intrabronchiolar neutrophils and foamy macrophages are also seen (B) There is diffuse lymphoid depletion and numerous histiocytes contain prominent intranuclear eosinophilic inclusions (see Fig. 12.15; eSlide 12.e3).

they did not develop CNS disease.[30,100] Opportunistic infection with commensals or less common pathogens including *Toxoplasma* sp.,[60] *Sarcocystis* sp.,[114] *Neospora* sp.,[117] *Clostridium piliforme*,[198] and rabies[93] can occur as comorbidities related to lymphoid depletion and immunosuppression.

As previously mentioned, viverrid species have been reported to be susceptible to a large number of viruses, raising the concern that they can serve as vectors for zoonotic disease. For example, masked palm civets were implicated as the source of **severe acute respiratory syndrome coronavirus-1 (SARS-CoV-1)** in humans.[29,44,74,118] Civets are sensitive to illness from this virus, and given the short duration of associated pathological lesions, they likely represent an amplification host rather than a reservoir host. In an experimental study in masked palm civets, viral RNA was detected by in situ hybridization (ISH) in the lung, small intestine, and cerebrum and was associated with interstitial pneumonia, lymphoid depletion, and neuronal degeneration.[199]

SARS-CoV-2 has been detected in a number of nondomestic animals including groups of binturongs, coati, and spotted hyenas in zoos in North America.[8] Clinical signs varied among individuals and included hyporexia, lethargy, mild fever, coughing, serous nasal discharge, and diarrhea that in all cases resolved with supportive care. Infection was traced back in most cases to infected animal handlers but appeared to spread within groups of animals. In North America, SARS-CoV-2 does not appear to amplify in raccoons.[23,71]

Several viverrid species are also susceptible to **influenza virus** infection, which is transmissible between animal species as well as zoonotic. Severe interstitial pneumonia, in the absence of other infectious agents, was reported in a captive binturong infected with H1N1 influenza A during a global pandemic in 2009.[171] An H5N1 outbreak was observed in captive Owston's civets and palm civets,[194] which caused respiratory and neurologic disease, and death in affected individuals.[166]

Bacteria

It is assumed that all of the species in this chapter can be infected with mycobacteria, and some species, particularly free-ranging mongoose and meerkats, appear to be particularly susceptible. Studies of **mycobacteriosis** in meerkats identified ***Mycobacterium tuberculosis*** or ***M. bovis***, members of the ***M. tuberculosis* complex**, as causative organisms. However, more recent molecular analyses have identified a novel tuberculous mycobacterial species, ***M. suricatta***, in meerkats and the closely related ***M. mungi*** in mongooses.[4,6,151] It is unclear if these novel and less common mycobacterial species represent the main mycobacterial pathogen in these species or if they are present in addition to *M. tuberculosis* and *M. bovis*. Larger epidemiological studies are required to answer that question. ***M. microti*** has also been reported in meerkat.[149] Epidemiological studies in wild meerkats suggest that mycobacterial infections spread locally within clusters of closely interacting individuals within a mob, particularly among grooming individuals.[56,152] The pattern of lesions suggests infection is via inhalation. Social behaviors are also thought to be important in the transmission of *M. mungi* in banded mongoose. High levels of *M. mungi* organisms are found in anal gland and nasal planum of mongoose, suggesting olfactory-related behaviors as being important in transmission.[7] Infections have also been reported in civet and genet, and spoligotyping identified *M. bovis* with similarity to buffalo strains suggestive of spillover infections.[107]

Lesions in meerkats are typical of tuberculous mycobacterial infections and include characteristic multifocal to coalescing granulomas in the lung, liver, lymph nodes, spleen, kidneys, and skin.[57] Histologically, granulomas contain numerous epithelioid macrophages with few lymphocytes. Multinucleated giant cells and neutrophils, as well as fibrosis, are rare findings. Acid-fast bacilli are common within lesions, but macrophages may contain only a single bacillus. In mongoose, *M. mungi* lesions are common in the skin of the nasal planum and nasal cavity (Fig. 12.13). In mongoose, lesions in the lungs were adjacent to vessels and thus fit a pattern of hematogenous spread rather than exposure through inhalation.[5,7] Similar to meerkats, fibrosis is minimal and central caseous necrosis within granulomas is variable. In contrast, reported *M. bovis* lesions in mongoose consist of granulomas with central necrosis and mineralization in the lungs and lymph nodes; intralesional bacilli are rare.[25]

Both *M. bovis* and the nontuberculous ***M. avium*** have been detected in the tissues of free-ranging raccoons.[134,197] In experimental inoculation studies, raccoons developed granulomatous lymphadenitis and characteristic granulomas.[150] Though not a significant disease in free-ranging raccoons, given their wide distribution and contact opportunities with domestic animals, raccoons could be a reservoir species for *M. bovis*.[197]

Additional case reports of mycobacteriosis include ***M. intracellulare*** infection in a captive binturong[1] and ***M. goodii*** in a spotted hyena.[187] In the latter, infection caused bronchial lymphadenomegaly and severe, multifocal and regionally extensive, pneumonia with pulmonary consolidation. Histologically, multifocal to coalescing, pyogranulomatous inflammation with multinucleated giant cells and intralesional acid-fast positive bacilli were present.

FIGURE 12.13 *Mycobacterium mungi* rhinitis in a mongoose. The nose is a common site for infection with this bacterium in mongoose. Ulcerative lesions may be present in the nasal cavity, skin of the nares, or on the hard palate. *(Photo Courtesy of C. Sanderson and K. Alexander, Department of Fish and Wildlife Conservation, Virginia Tech.)*

Raccoons are both susceptible to and may serve as a reservoir for a number of leptospire species serotypes. Several serologic surveys in the United States demonstrate that raccoons commonly have circulating antibodies to a number of ***Leptospira interrogans*** serovars including grippotyphosa, autumnalis, hardjo, and icterohemorrhagiae.[105,157,163] Some studies have confirmed nephritis in association with the bacteria by using PCR and immunohistochemistry (IHC), indicating that raccoons are susceptible to disease and do not simply serve as asymptomatic reservoirs.[61,86,113] Leptospiral infections have also been reported in mongoose and genets, and mongooses are a reservoir host for *Leptospira* sp. in the US Virgin Islands.[43] *L. interrogans* serovar icterohemorrhagiae is the most commonly identified strain.[133,135] Histologic lesions are similar to those in other species and consist of lymphoplasmacytic interstitial nephritis with fibrosis and tubular ectasia (Fig. 12.14A; eSlide 12.e4). Intralesional organisms can be identified in silver-stained sections and with IHC (Fig. 12.14B).

Severe swelling of the face, head, and neck; marked mandibular lymphadenomegaly and abscessation with rupture and drainage, respiratory distress, ocular discharge, ataxia, lethargy; and occasionally death were the hallmark findings of an outbreak of ***Streptococcus equi*** subsp. ***ruminatorum*** (Lancefield group C) in spotted hyenas in the Ngorongoro Crater in Tanzania.[96,177] Histologically, suppurative inflammation and edema with intralesional colonies of gram-positive cocci were seen in the soft tissues and nodes in the head/neck. Internal draining or aspiration with necrotizing and suppurative pneumonia and similar intralesional bacteria has also been described. The bacterium was cultured from affected and unaffected hyenas, suggesting a possible carrier state. It has also been cultured from Burchell's zebra, which are a prey species and the suspected source. Clinical signs in the hyenas were similar to a disease colloquially called "strangles" in domestic horses caused by the related bacterium *Streptococcus equi*.

Yersinia pseudotuberculosis is a gram-negative bacillus, and infection has been described in captive meerkats with necrotizing enteritis, hepatitis, and splenitis.[138] Inflammation was predominately neutrophilic and surrounded two morphologic forms of the bacteria: bacilli and larger globular bodies that were positive for *Y. pseudotuberculosis* O4 antigen by IHC.

An unclassified mycoplasma species causes epiphysitis and periostitis in captive and rehabilitating raccoons.[193] Clinical findings and gross lesions include joint swelling and abscessation of the extremities. **Mycoplasmosis** can be successfully treated with long-term antibiotic administration, even though with resolution, infection has the potential to cause permanent skeletal deformities if it occurs

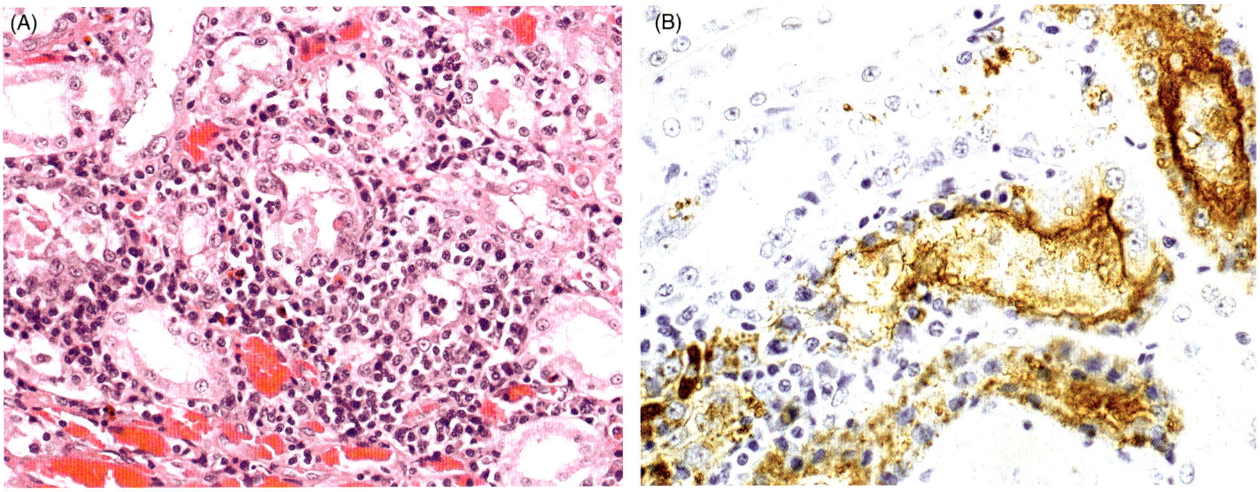

FIGURE 12.14 Leptospiral nephritis in a raccoon. (A) Tubulointerstitial lymphoplasmacytic nephritis centered on renal tubules is associated with tubular epithelial necrosis. (B) Antibodies against leptospiral antigens demonstrate strong positive chromogenic (brown) labeling in tubular epithelial cells and tubular lumens. IHC. (see eSlide 12.e4).

in a young, growing animal. One study detected *Mycoplasma* sp. in conjunctival swabs of healthy, free-ranging raccoons, so it is unknown if disease is the result of opportunistic infection of a commensal organism.[155]

Anthrax, due to ***Bacillus anthracis***, is a significant infectious, zoonotic disease. It is typically associated with ungulates, though most mammals, including carnivores that prey upon infected animals or carcasses, are thought to be susceptible. It is a large, spore-forming, gram-positive, rod-shaped bacterium with characteristic blunt ends and a clear capsule (refer to Chapter 2: Wildlife Necropsy; Chapter 4: Introduction to Comparative Clinical Pathology; and Chapter 5: Bovidae, Antilocapridae, Giraffidae, Tragulidae, and Hippopotamidae). Infection has been reported in captive genet.[101] Lesions included systemic hemorrhages as well as hemorrhage around the larynx and trachea, similar to what has been described in felids. Histologically, hemorrhagic necrosis of the intestinal villi was associated with bacterial invasion. Lymph node and splenic sinusoids contained numerous characteristic bacilli. Ingestion of contaminated meat was the presumed route of infection in the genets.

An outbreak of melioidosis due to systemic infection with ***Burkholderia pseudomallei*** was identified in a collection of slender-tailed meerkats that presented with deep ulcerative and suppurative dermatitis, particularly of the distal limbs and tail. Infection was associated with septicemia and suppurative thrombosis of the lungs, portal vein, and liver. Less commonly seen was nephritis and enteritis, suggesting this species is highly susceptible to infection.[156]

Fungi

Fungal disease is rarely reported in these species. Case reports include **histoplasmosis** (***Histoplasma capsulatum***) in two raccoon kits with disseminated, systemic infection and granulomatous pneumonia, lymphadenitis, splenitis, enteritis, nephritis, encephalitis, and thymitis with intrahistiocytic yeast[38]; ***Trichophyton mentagrophytes*** alopecic dermatitis in a captive adult female spotted hyena and her two cubs[82]; and fatal meningoencephalitis due to ***Encephalitozoon cuniculi*** in captive meerkats.[158]

Metazoa

There are a number of metazoan parasites described in the procyonids and feliform carnivores without associated disease. Wicker et al.[194] provide an excellent review of Viverridae pathogens and the reader is referred to that article for more information regarding parasites infecting that group.

Baylisascaris procyonis is a nematode parasite that is commonly found in the small intestines of free-ranging raccoons. Raccoons are the natural host and infection in aberrant hosts, including other mammalian and avian species as well as humans, results in larval migrans. *B. procyonis* infection has been reported in pet kinkajous in the United States and in free-ranging kinkajous in Colombia.[108,147] A related roundworm, ***B. potosis***, was recently described in captive kinkajous. Its zoonotic potential, if any, is unknown.[184] ***Physaloptera rara*** and ***Gnathostoma procyonis*** are also common in raccoons. Infection with these gastric nematodes is also typically asymptomatic, but ulcers may be seen in heavily parasitized individuals.[48]

The subcutaneous nematode of raccoons, ***Dracunculus insignis***, can cause edema and inflammation in the subcutis and skeletal muscle fascia, and extrusion of the female worm through the skin may result in an ulcerative wound.[48] These worms must be distinguished from larvae of the subcutaneous tapeworm ***Spirometra mansonoides***, which are smaller and flatter than *D. insignis*. *Spirometra* sp. infection was reported in a group of captive meerkats with subcutaneous, intramuscular, peritoneal, and pleural cysts, containing cestode larvae (sparganosis), with reactive pyogranulomatous inflammation. Lesions were fatal in one case with intrathoracic rupture of cyst, showing meerkats to be possible paratenic hosts for this parasite.[128]

Species of *Angiostrongylus* are metastrongyle nematode parasites that are found in the cardiovascular and respiratory system of definitive hosts and utilize (typically) a gastropod as an intermediate. ***Angiostrongylus dujardini*** infection has been described in a group of captive meerkats.[64,76] Affected animals had congested lungs and yellow foci in the lungs and liver. Histologically, inflammation in the lungs surrounded parasite eggs and larvae and adults were present in pulmonary arteries. Inflammation in the liver was variable. As cases occurred concurrently with other animals housed at the facility, consumption of infected gastropod intermediate hosts was presumed.

Ancyclostoma sp. and ***Arthrocephalus*** sp. have been reported in mongooses, civets, hyenas, and raccoons.[174] Raccoons have also been reported to be infected with ***Necator***, ***Arthrostoma***, and ***Uncinaria*** spp. While local tissue damage, reduced growth rates, and anemia have been reported in other species, reports of associated disease in the groups covered in this chapter are limited.

Eleven species of **acanthocephalans** have been reported in free-ranging North American raccoons (*Arythmorhynchus frassoni*, *Centrorhynchus conspectus*, *Centrorhynchus spinonsus*, *Centrorhynchus* sp., *Echinorhynchus gadi*, *Moniliformis* sp., *Neoechinorhynchus cylindratus*, *Oligacanthorhynchus tortuosa*, *Plagiorhynchus cylindraceus*, *Profilicollis botulus*, *Southwellina hispida*, and *Macracanthorhynchus ingens*). *M. ingens* is the only species for which the raccoon is the definitive host.[162] This particular acanthocephalan can also infect kinkajous. Despite this thorny-headed worm's large size, it does not commonly cause disease in raccoons. *O. tortuosa* have been reported to cause severe ulceration at the site of attachment in isolated cases.[164]

Protozoa

Meerkats appear to be highly susceptible to ***Toxoplasma gondii*** infection and multiple outbreaks with high mortality have been described in captive groups.[14,26,42,55,104] Clinical signs include respiratory distress and neurologic signs. The lungs are congested and fail to collapse, corresponding histologically to interstitial pneumonia. Pale foci that correspond to areas of necrosis occur in multiple sites, commonly the lung, spleen, and lymph nodes. Histologically, necrosis with neutrophilic to pyogranulomatous inflammation surrounds free and intracytoplasmic, individual and clusters of zoites in tissues with gross evidence of infection as well as those that appear grossly normal. Associated hepatitis and nonsuppurative encephalitis are not uncommon in animals with toxoplasmosis. Infection can be confirmed by immunohistochemistry or molecular methods. Infection is often presumed to occur through ingestion of feed contaminated with cat feces that contain infected oocysts (cats are the definitive host). Typing of isolates from one outbreak in slender-tailed meerkats identified clonal type III lineage, although the types present within potential source populations were not described.[14]

In raccoons infected with CDV, ***Toxoplasma gondii***, ***Neospora caninum***, and ***Sarcocystis neurona*** (Fig. 12.15) are not uncommon secondary opportunistic infections. Seroprevalence of exposure to *Toxoplasma gondii* is relatively high in free-ranging raccoons and lower ringtails in North America, with seropositivity rates of about 80% and about 20%, respectively, depending on the geographical location.[84,180] *T. gondii* has also been transmitted readily in an experimental setting to raccoons.[84] Lesions associated with naturally occurring *T. gondii* infection in raccoons are most often the result of infection with a coinfection with the immunosuppressive pathogen, canine distemper virus, and include presence of tissue cysts in the liver and tachyzoites in the brain.[84] Pyogranulomatous pneumonia can also be seen in raccoons with CDV and *T.* coinfections and is histologically similar to changes described in other species.

Leishmania infantum infection has been reported in free-ranging genets and Egyptian mongoose but disease and associated clinical signs have not been described.[52,78]

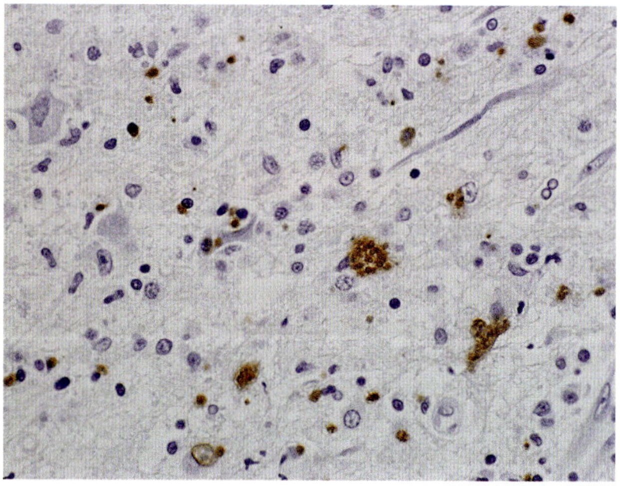

FIGURE 12.15 *Sarcocystis neurona* encephalitis in a raccoon concurrently infected with canine distemper virus (CDV). Protozoal infections are common secondary infections in raccoons with CDV infection. In the center of the image strong positive antibody labeling for *S. neurona* highlights a mature schizont with merozoites that are peripherally arranged in a characteristic pattern. IHC.

Serum biochemistry abnormalities, such as elevated globulins, have been reported in some individual animals but are inconsistent findings.[131]

Enteric coccidia including two species of ***Eimeria*** (***E. nuttali*** and ***E. procyonis***) and one species of ***Isospora*** have been described in the raccoon.[59] Though the significance of these pathogens on a population level is unclear, marked enteritis associated with parasitism contributed to mortality in raccoons in published cases.

Free-ranging raccoons have a high seroprevalence in antibodies to ***Trypanosoma cruzi***, though associated disease is extremely rare.[200]

Babesia sp. have been reported in coatis, raccoons, genets, spotted hyenas, civets, and meerkats.[10,194] Some individuals have evidence of splenomegaly and anemia associated with infection. Coinfections with other hemoparasites such as ***Hepatozoon*** sp. in hyena and ***Cytauxzoon*** sp. in meerkat have also been identified by PCR, but the clinical impact of these infections was not described.

Additional protozoa of hyenidae are presented in Table 12.e2.

Ectoparasites

All species in this chapter are susceptible to **sarcoptic mange (scabies)**, a zoonotic disease caused by the mite ***Sarcoptes scabei***. Lesions have been described in native raccoons in the United States[68] and also in introduced invasive populations in Germany.[161] Affected animals are severely pruritic and have regional alopecia and crusting dermatitis. Diagnosis is usually made by identifying mites on skin scrapes, though they can also be identified histologically in sections of skin. In addition to *Sarcoptes* sp., ***Notoedres*** sp. infections have been identified in free-ranging and captive civets.[144,146] Affected civets were pruritic and had severe alopecia and crusting lesions. Histologically affected skin was acanthotic and hyperkeratotic with intralesional mites. Hair follicles were parakeratotic and many were devoid of hair. Dermal inflammation was mild. Diagnosis can be confirmed by histopathology and microscopic evaluation of skin scrapings.

Visceral pentastomiasis due to ***Armillifer armillatus*** nymphal migrans has been diagnosed in a zoo-housed, brown[81] and a striped hyena[51]; *A. armillatus* and ***A. moniliformis*** have also been reported in aardwolf[33] and *A. moniliformis* has been reported in wild and captive civets and binturong.[194] Snakes are the definitive and canids are the intermediate hosts. Encysted nymphs in the striped hyena were present throughout the omentum, mesentery, diaphragm, intestines, liver, spleen, kidneys, and urinary bladder, where they were surrounded by a thin fibrous capsule and mild, chronic or neutrophilic (intestinal) inflammation. Diagnosis of pentastomiasis is typically based on the identification of the characteristic parasites on gross necropsy examination. Histology and PCR with DNA sequencing support the gross diagnosis and in the case of the latter, it can be used to speciate the parasite.

Fleas are a common, generally species-specific parasite. Free-ranging raccoons in the United States are frequently infested with *Orchopeas howardi*, though they can also be infested by fleas from the *Ctenocephalides* genus.[99,136] Raccoons are also frequently infested with the **louse *Trichodectes octomaculatus***. Infestation by these ectoparasites rarely causes disease. Species of **Ixodes** tick are important vectors for *Babesia* sp. infections in genets and raccoons.[10]

Other Infectious Organisms

Raccoons have been experimentally infected with three types of transmissible spongiform encephalopathies including transmissible mink encephalopathy, sheep scrapie, and chronic wasting disease.[89] Results suggest susceptibility to transmissible mink encephalopathy and sheep scrapie but resistance to chronic wasting disease. No natural cases have been reported.

ADDITIONAL RESOURCES

Hamir, A.N., 2011a. Pathology of neurologic disorders of raccoons (*Procyon lotor*). J. Vet. Diagn. Invest. 23, 873–884.

Hamir, A.N., 2011b. Spontaneous lesions in aged captive raccoons (*Procyon lotor*). J. Am. Assoc. Lab. Anim. Sci. 50, 322–325.

Hamir, A.N., Hanlon, C.A., Niezgoda, M., Rupprecht, C.E., 2001.The prevalence of interstitial nephritis and leptospirosis in 283 raccoons (*Procyon lotor*) from 5 different sites in the United States. Can. Vet. J. 42, 869–871.

Ruprecht, C.E., Stohr, K., Meredith, C., 2001. Rabies. In: Williams, E.S., Barker, I.K. (Eds.), Infectious Diseases of Wild Mammals. Iowa State Press, Ames, pp. 3–36.

Wicker, L.V., Canfield, P.J., Higgins, D.P., 2017. Potential pathogens reported in species of the family *Viverridae* and their implications for human and animal health. Zoonoses Public Health 64, 75–93.

Williams, E.S., 2001. Canine distemper. In: Williams, E.S., Barker, I.K. (Eds.), Infectious Diseases of Wild Mammals. Iowa State Press, Ames, pp. 50–59.

References

Full references for this chapter can be found online at https://doi.org/10.1016/B978-0-443-18437-6.00028-6.

Chapter 13

Prosimians

Denise McAloose[a] and Ilse H. Stalis[b]

[a]Pathology Department, Zoological Health Program, Wildlife Conservation Society, Bronx, NY, United States; [b]Independent Scholar, Long Island, NY, United States

INTRODUCTION

Prosimians are medium- to small-bodied primates in the suborders Strepsirhini and Haplorhini. The taxonomy of prosimians has been debated for decades and some consider the term "prosimian" taxonomically obsolete. However, until resolved and because the term prosimian is still commonly used to describe animals in these two suborders, its use will be maintained in this chapter. Within the Strepsirhini are two infraorders, the Lemuriformes and Lorisiformes, and within the Haplorhini are the Tarsiiformes of Southeast Asia. The Lemuriformes group includes five families of lemurs endemic to Madagascar: Cheirogaleidae (five genera; includes fork-marked, dwarf, and mouse lemurs), Lemuridae (five genera; includes bamboo, ring-tailed, ruffed, and true lemurs), Lepilemuridae (one genus; includes sportive lemurs), Indriidae (three genera; includes indris, sifakas, and woolly lemurs), and Daubentoniidae (one genus and species, the aye-aye). The Lorisiformes group includes the *Lorisinae* (slow and slender lorises) from India, Southeast Asia, western Indonesia and Africa, and the galagos (bushbabies) and pottos of Africa.[286] See Table 13.e1 for an expanded list of prosimians and their conservation status.

Most prosimians occur in areas of high biodiversity and species endemism. Habitat loss due to anthropogenic factors, which include development and natural resource extraction (e.g., land clearing for cultivation, logging and/or forest fragmentation, mining), is a major conservation threat for many species, and hunting for bushmeat, harvesting for traditional medicine, and collection of animals for the local and international pet trade also contribute to declines.[225,252] In a 2020 assessment by the International Union for the Conservation of Nature (IUCN), lemurs were recognized as the most endangered primate group in the world with 103 of 107 (98%) known species listed as threatened with extinction; 33 and 45 are listed as critically endangered or endangered, respectively. Of the four species not listed as threatened, two (the gray-brown mouse lemur and gray mouse lemur) were listed as least concern and in the remaining two (the Groves' dwarf lemur and the Boraha mouse lemur), their conservation status remained to be established (listed as data deficient).[142]

Prosimians in free-range and captive settings are affected by a number of non-infectious and infectious disease processes. Those that are more commonly encountered are described in this chapter. Less common diseases, single case descriptions, or reports in which a pathogen has been detected but for which no associated disease has been described or for which limited information is available are listed in the Digital Content: Table 13.e2, Neoplasia in Prosimians; Table 13.e3, Viruses and Viral Antibodies in Prosimians; and Table 13.e4, Reported Metazoa, Protozoa, and Ectoparasites in Prosimians. Also present in the Digital Content are descriptions of unique anatomic features in prosimians.

NON-INFECTIOUS DISEASES

Nutritional

Iron overload, also known as **hemosiderosis**, is the intracellular accumulation of iron in the absence of other tissue damage. It is a common finding in a number of lemur species in captive settings.[22,73,118,119,261,286,287,288,292] Diet has been suggested as one of several possible etiologic factors due to an absence of iron accumulation in newborns and animals less than 6 months of age. It can be mild to abundant and the severity appears to be somewhat species dependent. Iron accumulation in prosimians usually occurs first in the small intestine (particularly duodenum) and is followed by storage in the liver, spleen, and bone marrow. In severe cases, iron accumulation may result in brown discoloration of the tissue (Fig. 13.1A). Hepatic deposition occurs first in the cytoplasm of Kupffer cells with subsequent accumulation in hepatocytes. Iron appears histologically as granular to globular material that stains brown with routine hematoxylin and eosin stains and blue with Prussian blue stains (Fig. 13.1B). Intestinal accumulation often starts in macrophages at the tips of villi (Fig. 13.1C; eSlides 13.e1 and 13.e2) but can eventually fill macrophages throughout

Pathology of Wildlife and Zoo Animals. https://doi.org/10.1016/B978-0-443-18437-6.00021-3

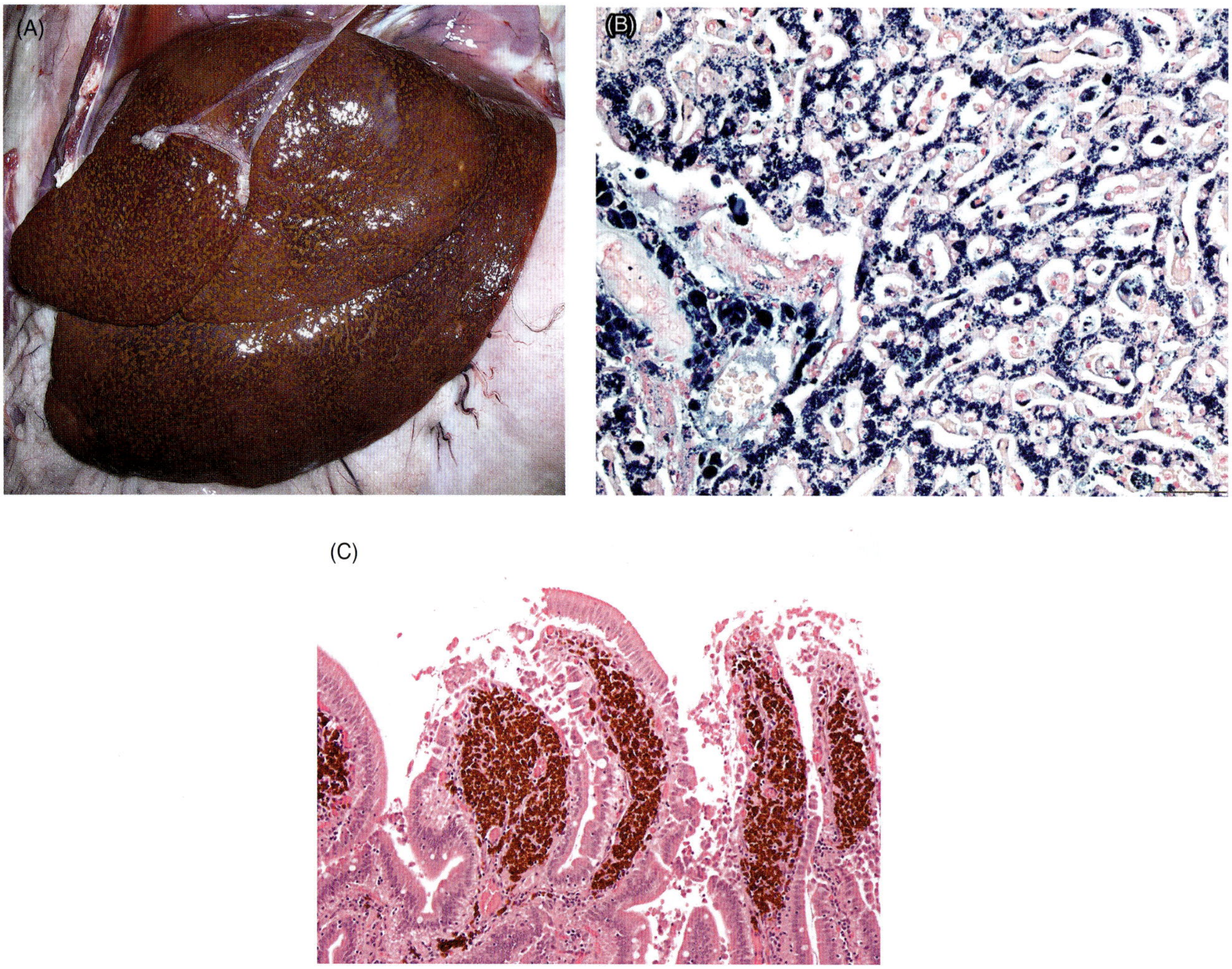

FIGURE 13.1 Iron overload (hemosiderosis) in the liver of a black-and-white ruffed lemur (A) and lesser mouse lemur (B) and the small intestine of a black-and-white ruffed lemur (C). (A) The parenchyma of the liver is discolored and mottled brown and tan due to severe iron accumulation. (B) Hemosiderin appears histologically as granular to clumped intracytoplasmic material. It stains brown with routine hematoxylin and eosin and blue with Prussian blue staining. Hemosiderosis is not associated with hepatocellular necrosis, degeneration, fibrosis, biliary hyperplasia, or nodular regeneration, which are typical changes associated with hemochromatosis. Prussian blue. (C) Iron can accumulate in multiple sites including the small intestine. In this case, the lamina propria within the intestinal villi is moderately expanded by numerous hemosiderin-laden macrophages (see eSlides 13.e1 and 13.e2). *(Part A: Photo Courtesy of K. Terio, University of Illinois Zoological Pathology Program.)*

the lamina propria. In severe cases, excessive iron accumulation in the liver can result in **hemochromatosis** characterized by fibrosis, hepatocyte necrosis, hepatocyte fatty degeneration, bile duct hyperplasia, and/or nodular regeneration. Hemochromatosis is much less common than hemosiderosis.

Abundant hepatic iron accumulation is a known risk factor for benign or malignant neoplasms in humans. A similar association in prosimians has been suggested in some publications but not confirmed in other retrospective reviews. Recent studies of multiple lemur species have not found a relationship between clinical disease and excessive hepatic hemosiderosis or high liver iron concentrations (up to 12,800 ppm dry weight).[22,118,177,261,287]

Dental disease is common in captive Strepsirrhines, especially pygmy and slow lorises and Senegal bushbabies.[211,256,270] Gross lesions include mild to severe deposition of dental calculus, severe gingivitis or periodontitis characterized as a variably wide band of bright pink to red discoloration and swelling along the gingival margin, tooth wear, fracture or loss of single or multiple teeth, and abscessed sulci or tooth roots (Fig. 13.2). Inflammation and tissue response depend on chronicity and other factors such as opportunistic bacterial infection. Earlier, more acute inflammatory lesions may appear primarily as neutrophilic inflammation with or without necrosis. More chronic lesions can include abscesses with an active, neutrophilic and necrotic core surrounded by a fibrous capsule or regionally extensive mixed neutrophilic, lymphoplasmacytic and histiocytic inflammation. Fibroplasia or fibrosis can be seen in chronic lesions, and focal or extensive osteomyelitis can be present in more severe cases. Facial

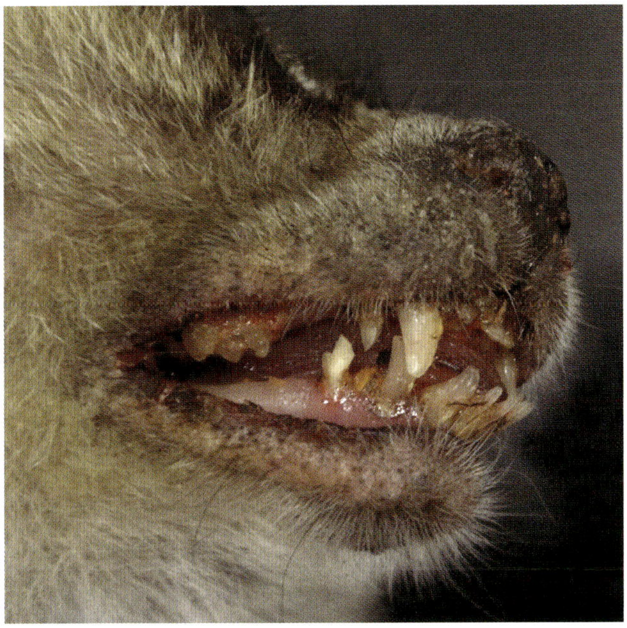

FIGURE 13.2 Chronic dental disease in a lesser slow loris. Gross lesions include mild deposition of dental calculus, chronic gingivitis with pink to red discoloration and hemorrhage along the gingival margin, and multifocal tooth wear, fracture, and loss.

swelling can be a feature of any type of dental disease but is more likely to be present in cases with more significant/severe disease.[99,192] Extension to ocular structures or the brain with resultant suppurative encephalitis or meningoencephalitis can occur. One or a number of concurrent factors, including sugar-rich diets, mechanical trauma due to food items, genetic factors, or viral or bacterial infections, are thought to contribute to the development of dental disease in prosimians. Slow loris diets with a high proportion of fruit and little to no plant exudates such as gums, which are a major component of the natural diet, may promote the development of dental disease.[52] Notable bacteria that have been isolated from dental lesions in prosimians include *Sebaldella termitidis* and *Trueperella pyogenes*.

Choleliths, the development of which is also often associated with dietary factors, have been reported with low frequency in several species of prosimians. These include detection of choleliths of undetermined composition in the greater galago, black lemur, and a slender loris, and cholesterol liths in a group of slender lorises.[47,122,212,270] In the lorises, affected animals were typically 11–13 years of age. Clinical signs were absent or included inappetence, emaciation, abdominal pain, lethargy, or vomiting. One to numerous stones, up to 1 cm in diameter, were present in the gallbladder; none were described as being in the common bile duct. Associated tissue changes ranged from none to thickening of the gallbladder wall with marked

FIGURE 13.3 Choleliths in the gallbladder of a slender loris. Multiple, variably sized, smooth concretions (liths) are present in the gallbladder. Sequelae to the liths and inspissated bile include secondary chronic cholecystitis (seen as thickening and white discoloration of the gallbladder wall) and choledochitis.

cystic dilation of the gland, mild, multifocal, lymphoplasmacytic inflammation, and occasional intrahepatic, periductular fibrosis (Fig. 13.3). A high-cholesterol diet or lack of dietary components found in naturally foraged food in the wild (e.g., gum) may be predisposing factors in the development of choleliths in prosimians in managed care. Ultrasound may aid in antemortem cholelith detection.

Metabolic

Diabetes mellitus has been reported in several species of middle-aged to aged, captive prosimians.[149,286] Associated gross and histologic lesions have not been reported, and obesity is not necessarily a predisposing factor. Because the diagnosis may be complicated by a hyperglycemic stress response in prosimians, antemortem diagnostics should include baseline and long-term monitoring of serum fructosamine or glycated hemoglobin (HbA1c).

Toxic

Seasonal alopecia, also known as "**bald lemur syndrome**", has been described in wild ring-tailed lemurs in the Berenty Reserve in Madagascar.[24,74,75,144,145] The condition is due to consumption of mimosine (β-[*N*-3-hydroxy-4(1H)-pyridone] α-amino propionic acid; l-mimosine), a toxic, water soluble, plant amino acid (chemically similar to tyrosine) of the *Leucaena leucocephala* tree.[7,77,164] Toxin is present in higher concentration in the seeds and in mature leaves and stems. Native to Mexico, Central America, and northern South America, the plant was introduced into some regions in Madagascar in the 1990s to provide a fast-growing, high-protein food source for domestic ruminants. The plant is toxic when eaten in high concentration (<5%–10% dry weight in non-ruminant animals) and causes alopecia, poor body condition or poor weight gain, and infertility. Effects resolve with withdrawal of the toxin. Alopecia in ring-tailed lemurs is associated with increased *Leucaena* consumption (over 50% of diet in some groups) during the dry season (May to September) when other food resources become scarce. The distribution of hair loss and severity varies from mild, with partial alopecia of the trunk and tail, to severe with complete loss of fur on the trunk, legs, and tail and near complete hair loss on the face and feet (Fig. 13.4). Affected animals may lose body condition, and the most severely affected animals may be lethargic. Mortality has been reported in adults. Infant mortality may also occur due to an inability of neonates to cling to alopecic dams.[74,137]

FIGURE 13.4 Seasonal alopecia ("bald lemur syndrome") in a wild ring-tailed lemur. Severe alopecia occurs due to consumption of the seeds, leaves, and other parts of the *Leucaena leucocephala* tree (seen in the background behind the lemur). The plant contains mimosine, a toxic plant amino acid that is thought to interfere with metabolic pathways associated with hair growth. *(Reprinted with permission from Crawford G, Puschner B, Affolter V, Stalis I, Davidson A, Baker T, Tahara J, Jolly A, Ostapak S. Systemic effects of Leucaena leucocephala ingestion on ringtailed lemurs (Lemur catta) at Berenty reserve, Madagascar. Am J Primatol. 2015;77 (6):633–641.)*

Affected animals have mild to moderate orthokeratotic hyperkeratosis and follicular arrest with the majority of hairs in the telogen (resting) phase of the hair growth cycle.[75] The latter is characterized by decreased numbers of hair bulbs and narrow outer root sheaths. Broken hair shafts are seen in some telogen follicles that contain hair. A decrease in the anagen-to-telogen ratio is seen. Changes in the skin are not associated with inflammation, systemic effects, or organ toxicity, though some affected animals may have elevated packed cell volume, albumin, α-tocopherol, and free-thyroxine. The mechanism of mimosine toxicity is not completely understood but may include inhibition of DNA, RNA, and protein synthesis and several metabolic pathways that result in cell cycle arrest and inhibition of cell division. In the ring-tailed lemur the toxin is thought to induce rapid progression of hair growth from the anagen to the telogen phase of the growth cycle with subsequent inhibition of a new cycle of hair growth.

Plants in the *Solanaceae* (nightshade) family contain **solanine**, a toxic glycoalkaloid that inhibits cholinesterase enzymes including acetylcholinesterase. Mechanisms of action include termination of synaptic transmission at neuromuscular junctions and synapses.[230] Its primary sites of action are in the gastrointestinal and nervous systems. Signs of intoxication include vomiting, diarrhea, cardiac dysfunction, respiratory depression, muscle tremors, paralysis, and death. Consumption of hairy nightshade (*Solanum sarrachoides*) in prosimians at a zoo in the western United States caused anorexia, depression, lethargy, ataxia, diarrhea, and/or death. Histologic lesions include severe hemorrhagic enteritis and typhlitis.[87]

Avocado toxicity has been suspected as an infrequent cause of death in captive aye-ayes.[248] The toxin, persin, is present in the leaves and bark of the avocado tree. It is also present in the pits, skin, and possibly the fruit. Associated lesions include severe pericardial effusion, myocardial pallor, and acute myocardial contraction band necrosis. Massive catecholamine release and/or other stressors are thought to be important in the pathogenesis. Variable toxin levels in the plant/fruit elements and individual susceptibility, even in animals that have previously been fed avocados, may determine clinical outcomes. Diagnosis may be suspected by finding the plant/fruit or detection of the toxin in stomach contents.

Congenital/Genetic

A number of congenital and possibly heritable conditions have been described in captive black-and-white ruffed lemurs. Pairings between related lemurs have produced offspring with developmental abnormalities including scoliosis (thoracic, lumbosacral, and/or lumbar vertebral column), kinked tails, bilateral exencephaly of the frontal bones of the skull, cleft palate, pectus excavatum, hypoplasia or aplasia of a portion of the lung, *cor biloculare* (two-chambered heart), hydroureter, and hydronephrosis.[20,22] Karyotypic abnormalities were not identified in affected animals and the mode of inheritance was not established through pedigree analysis.

Age Related/Degenerative

Cataracts are reported in individuals and are a relatively common finding in certain groups of captive prosimians.[19,61,109] As in other species, causes vary and can include metabolic, traumatic, heritable, or degenerative processes. They can also develop secondary to ocular infectious, inflammatory, or neoplastic diseases. Blindness can be the end result in animals with severe, complete cataracts. Grossly, they appear as small or focal to diffuse opacities in one or both lenses (Fig. 13.5A). Cataractous lesions in prosimians are similar to those in other species and include subcapsular epithelial cell hyperplasia with or without fibrous metaplasia; disorganization, separation, and swelling of lens fibers with or without the formation of bladder cells (swollen lens fibers with retained nuclei); lens fiber degeneration with the formation of globular protein aggregates (Morgagnian globules); and fragmentation of lens fibers with the formation of clefts or vacuoles (Fig. 13.5B). Corneal opacities have also been described in gray mouse lemurs from two colonies that had a high incidence of cataracts.[5] Histologic examination of the eyes of one animal revealed xanthomatous corneal inflammation.

Amyloid-β (Aβ) peptide, a major component of senile plaques and one of the hallmarks of Alzheimer's disease in humans, can be seen in the brains of aged mouse lemurs.[241] Cytoplasmic accumulations of Aβ in mouse lemurs have been described in neurons in the parietal cortex, hippocampus, and caudate nucleus. It appears as diffuse or densely packed, spherical vesicles with routine hematoxylin and eosin staining; the material may be more easily seen using immunohistochemical labeling (4G8 monoclonal antibody directed against the 17–24 residues of Aβ). Some animals may also have extracellular Aβ deposition. Mouse lemurs with tissue Aβ accumulation may also have elevated plasma Aβ. Accumulation of amyloid in the brain of mouse lemurs has been associated with decreased cognitive function.[247]

As in many mammal species, zoo and free-range prosimians can develop **degenerative arthritis** or **inflammatory diseases of the joints.** Degenerative lesions appear to be much less common in prosimians than in other Old World primates, and osteoarthritis appears to be more common in captive than free-range prosimians.[238,239] Osteoarthritis in prosimians is like that in other species and includes erosion of the articular cartilage, osteophyte formation, and sclerosis in subchondral trabecular bone. It is only occasionally associated with an infectious process. Degenerative lesions described in galagos and members of the Lemur, Lepilemur, Cheirogaleus, Indri, and Nycticebus groups include isolated erosion in the articular cartilage of the metacarpal and metatarsal joints or bones of the knee or elbow joints; described lesions in galagos are typically limited to a single joint.

Miscellaneous/Multifactorial

Chronic renal disease is a common finding in prosimians.[17,32,45–47,61,206] It most often occurs in middle-aged to older animals. Lesions vary in their severity and distribution and are characterized by interstitial fibrosis, cystic tubules, and nonsuppurative, interstitial, or tubulointerstitial inflammation (Figs. 13.6A, B; eSlide 13.e3). Focal and diffuse proliferative (increased number of glomerular endothelial and mesangial cells with or without increased

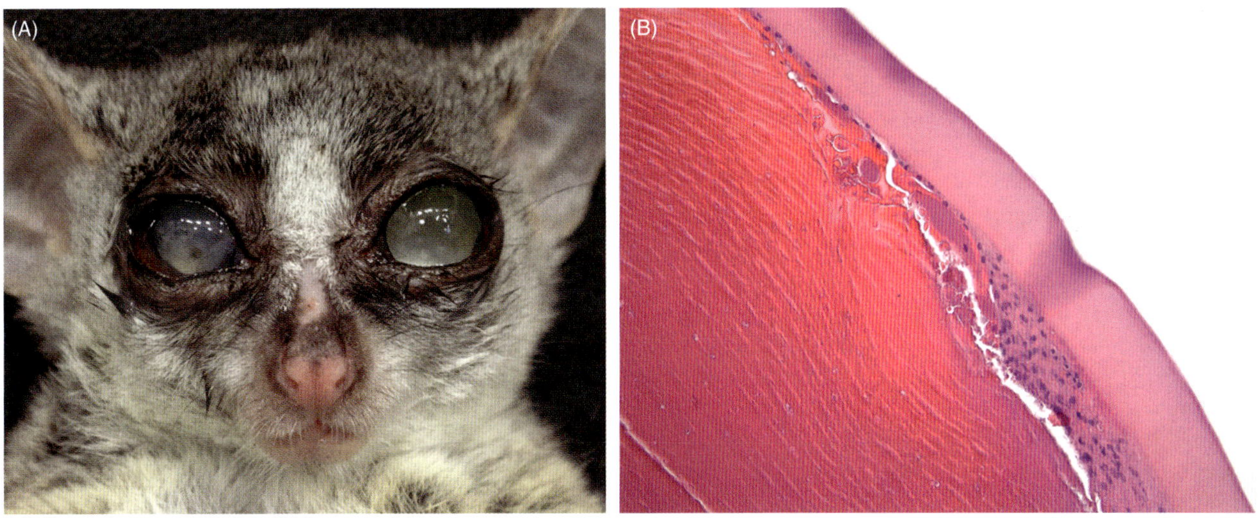

FIGURE 13.5 Cataracts in an aged bushbaby. (A) The lenses are bilaterally diffusely opaque. (B) Histologic lesions include subcapsular epithelial cell hyperplasia and fibrous metaplasia, and lens fiber degeneration with Morgagnian globule formation.

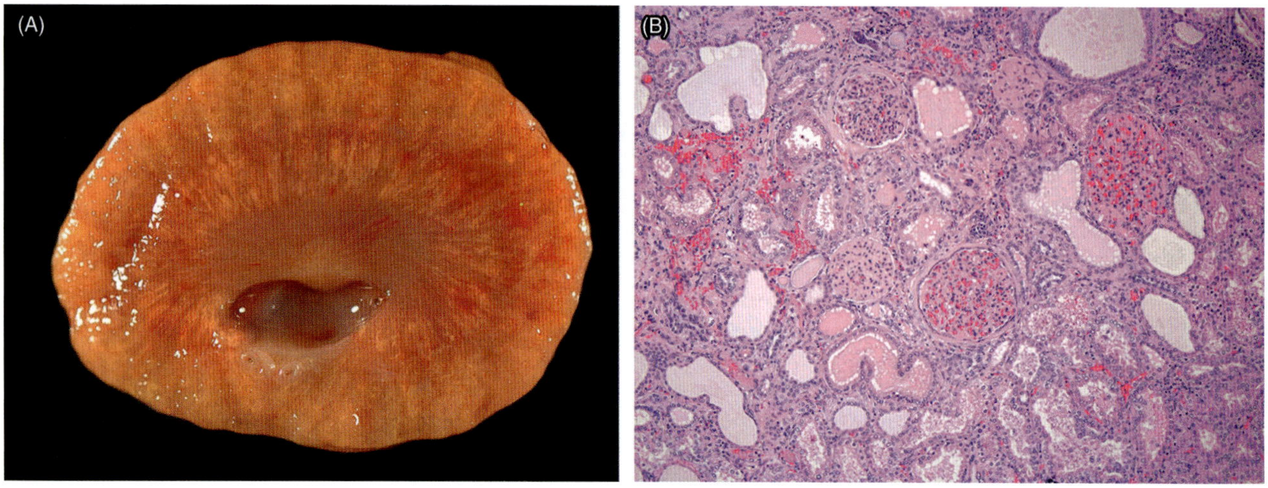

FIGURE 13.6 Chronic renal disease in a pygmy slow loris. (A) The kidney has an undulating capsular surface and linear or coalescing areas of light tan discoloration radiate peripherally from the renal pelvis (which is mildly dilated) and corticomedullary junction. (B) Histologic lesions include severe, (predominantly) lymphocytic and histiocytic tubulointerstitial inflammation, membranoproliferative glomerulonephritis with synechia, and fibrosis of Bowman's capsule and the interstitium. Many tubules are dilated and some contain proteinaceous fluid (see eSlide 13.e3).

mesangial matrix), membranous (thickened glomerular basement membranes, with or without mesangial proliferation and without cellular proliferation), or membranoproliferative (all of the above) glomerulonephritis may also be seen. Changes are typically associated with glomerular endothelial cell hypertrophy, variable presence of adhesions between the parietal and visceral layers of Bowman's capsule (synechia), and occasional enlarged or shrunken, fibrotic (sclerotic) glomeruli. Fibrous osteodystrophy and pathologic fractures have been reported as sequelae to chronic renal disease in gray mouse lemurs, and associated uroliths (calcium oxalate monohydrate and dihydrate; mixed calcium carbonate and calcium phosphate) have been reported in Alaotran gentle lemurs.[17,61] Additional sequela in this latter group of animals included renal and extrarenal metastatic soft tissue mineralization and renal interstitial amyloidosis. An underlying cause is often not identified, especially in chronic, "end stage" renal disease. Perret attributed the development of lesions in mouse lemurs to the stress of captivity, but analysis was only based on histologic evaluation so this factor has not been fully assessed.[206]

Polycystic nephropathy has been described in adult slender lorises.[210] Kidneys in affected animals have undulating capsular surfaces that correspond to underlying, variably sized (up to 0.4 cm diameter) cortical cysts. Cysts are lined by cuboidal to attenuated tubular epithelium and are associated with variable pericystic fibrosis and scattered tissue mineralization. Cystic changes are also associated with moderate lymphocytic and histiocytic interstitial nephritis, and mild to severe proliferative or membranous glomerular changes that, when severe, are associated with fibrosis or sclerosis. An underlying etiopathogenesis (e.g., genetics, stress, infection) has not been established.

A syndrome of **necrotizing enterocolitis** and **typhlitis** has been described in ring-tailed, red-ruffed, and mongoose lemurs.[197] Affected lemurs presented with abdominal pain, lethargy and anorexia, and some developed diarrhea. Findings included mucosal to transmural necrosis of the small intestine, cecum, and/or large intestine. Attempts to identify a specific infectious cause were unsuccessful. Localized or multifocal ischemia was proposed as a possible factor.

Gastric pneumatosis (emphysematous gastritis; pneumatosis cystoides intestinalis) has been described in black-and-white ruffed lemurs, red-ruffed lemurs, and a Bengal slow loris.[27,53,196] It is characterized by focal or regionally extensive expansion of the stomach wall by large, multifocal to coalescing, cyst-like spaces (Fig. 13.7A, eSlide 13.e4). Expansion of the wall by gas may be apparent radiographically as a variably thick, radiopaque expansion of the wall along the outer edge of the greater curvature of the stomach (Fig. 13.7B). Clinical signs are nonspecific and may be absent or include lethargy, weakness, inappetence, vomiting, and weight loss or sudden death with no premonitory signs. Histologically, numerous, variably sized, coalescing, cyst-like spaces expand the submucosa. Mild multifocal inflammation with various cell types (primarily eosinophilic to histiocytic, lymphoplasmacytic, or mixed with occasional multinucleated giant cells) in the presence or absence of fibrosis may be seen in the interstitium between the cystic spaces (Fig. 13.7C). The cause of this condition has not been identified. Intralesional infectious agents, including gas-forming bacteria, fungi, or parasites, have not been described. Some have suggested that diets high in carbohydrates and sugars may be a contributing factor in lesion development, but this remains to be proven.

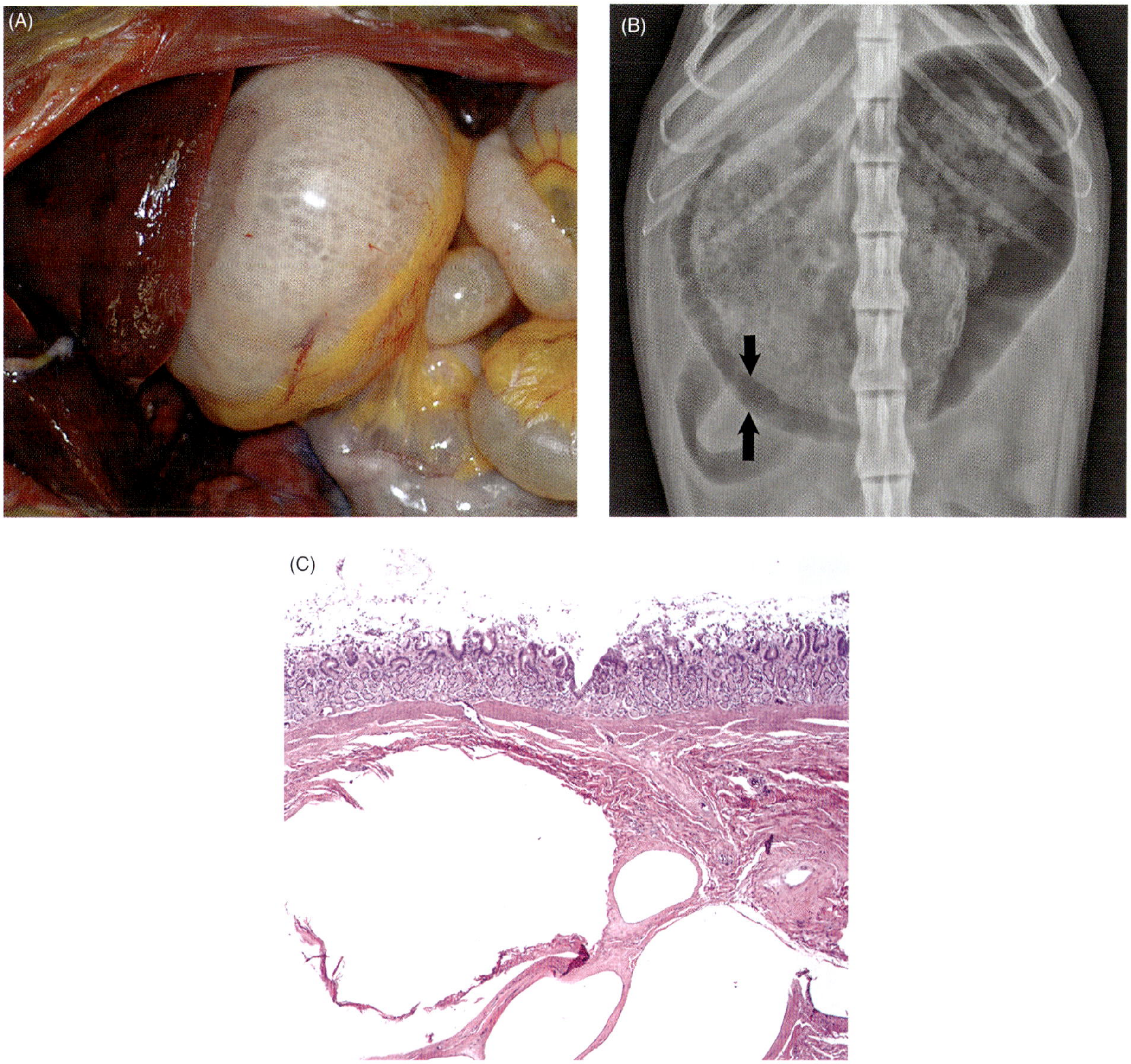

FIGURE 13.7 Gastric pneumatosis in a slow loris. (A) The stomach is distended and the wall has a reticulated pattern. (B) Expansion of the wall by gas may be apparent radiographically as a variably thick, radiopaque expansion of the wall along the outer edge of the greater curvature of the stomach (area between arrows). (C) Expansion of the wall corresponds histologically to numerous, variably sized, and often large cystic spaces that markedly expand the submucosa. These are not true cysts as they do not have an epithelial lining. They have not been associated with intraluminal bacteria (see eSlide 13.e4). *(Part B: Radiograph Courtesy of Wildlife Conservation Society.)*

Multifocal pyogranulomatous osteomyelitis has been described in captive red-ruffed lemurs and is associated with lameness and skeletal muscle atrophy. The disease is characterized radiographically as multifocal lytic lesions with cortical thickening and sclerosis in the diaphysis of the long bones of the fore and hind legs.[16] Histologic lesions in affected areas vary with chronicity. Earlier lesions contain woven trabecular bone, neutrophilic inflammation, and small foci of necrotic trabecular bone. More chronic lesions contain numerous foci of necrotic bone, decreased production of woven trabecular bone, and chronic inflammation consisting of macrophages, lymphocytes, plasma cells, and few neutrophils, occasional granuloma formation, and fibrosis. Clinical signs may resolve, and associated radiographic lesions that transition from lytic to sclerotic may be present with long-term antiinflammatory treatment. Lesions are not associated with an infectious process, and the histologic lesions and response and resolution rather

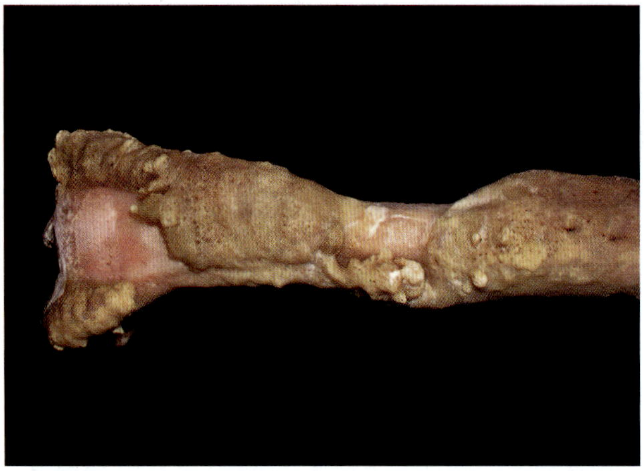

FIGURE 13.8 Hypertrophic osteoarthropathy in a black lemur. The periosteum is multifocally and markedly expanded by thick, irregular, deposits of new bone formation (see eSlide 13.e5). *(Photo Courtesy of K. Terio, University of Illinois Zoological Pathology Program.)*

than progression with antiinflammatory treatment have similarities to reports of chronic, recurrent, multifocal osteomyelitis in humans. A cause for the condition is not known.

Periarticular hyperostosis (PH) has been described in black lemurs.[150] Lesions are characterized by proliferative, periosteal, new bone formation that is primarily associated with the metaphyses. Periarticular osteophyte formation, degeneration of articular cartilage, and periarticular fibrosis surrounding joint capsules can be present in advanced cases. The condition in black lemurs is associated with chronic renal disease and is histologically similar to hypertrophic osteoarthropathy (Fig. 13.8; eSlide 13.e5). The latter typically affects the diaphysis while PH generally affects the metaphysis.

Endometritis appears to be a relatively common finding in female prosimians in zoos. In one review, it was diagnosed in 10.1% of female reproductive tracts submitted to the Reproductive Health Surveillance Program (RHSP).[188] **Adenomyosis,** the presence of endometrial glands in the muscular layers of the uterus and/or perimetrium, and high numbers of atretic follicles were also identified as relatively common findings in lemurs by the RHSP. Adenomyosis was associated with advanced endometrial hyperplasia.

Neoplastic

A wide variety of **neoplasms** have been described in prosimians. In an excellent review, Remick et al. provide a comprehensive list of spontaneous neoplasms in captive born or wild-caught, captive prosimians.[228] Their list includes tumors in five families of prosimians: Lemuridae, Cheirogaleidae, Galagidae, Lorisidae, and Indriidae. In this summary, our experience, and that of others, the most common tumors in prosimians occur in the liver.[18,232,300] Tumors of the reproductive tract, hematopoietic system, skin, endocrine organs, and urinary system are less common. Several of the more common tumors are listed below. See Table 13.e2 for a more complete list of neoplasms that have been reported in prosimians.

Among prosimians, **hepatocellular tumors** have been described most commonly in lemurs. In general, hepatocellular tumors are more common than biliary tumors, and carcinomas are more common than adenomas.[18,22,228,232,300] Benign and malignant tumors are like those in other species with several notable differences. As in other species, **hepatocellular carcinomas** can be variably sized, single to multiple, invasive masses (Fig. 13.9; eSlide 13.e6). In contrast to other species, they have marked hepatocellular pleomorphism and anisocytosis with low numbers of, often atypical, mitotic figures. In one study of 14 cases, the mitotic count was higher in animals with metastasis than those without metastasis.[232] The most common histologic features of hepatocellular carcinomas in this group of cases were a trabecular pattern and multinucleated cells. Metastasis was relatively common and typically involved the lungs. Other reported metastatic sites include the stomach, heart, kidney, and mediastinum. Hepatocellular carcinomas can be immunohistochemically labeled with hepatocyte paraffin 1, which can be used to differentiate them from cholangiocellular tumors. Hepatocellular adenomas differ from carcinomas in that they are composed of a proliferation of well-differentiated hepatocytes. They lack normal architectural structures, such as portal tracts, and may contain areas of cystic degeneration filled with blood or proteinaceous fluid. Necrosis may be seen in benign or malignant tumors. Inflammation is not a common feature in the tumors or in adjacent, nonneoplastic tissue.

A cause for the development of hepatic tumors in prosimians has not been identified. Hepadnaviruses, known to cause hepatocellular carcinomas in humans and a number of other mammal and bird species, have not been linked to hepatocellular tumor development in prosimians.[22,195,300] Additionally, there are no histologic lesions in prosimians to suggest hepatitis B or C viruses or aflatoxin B1 as predisposing factors. Because of the high incidence of hemosiderosis in captive prosimians, some have suggested it as an underlying causal factor. Others have found no association between the two and despite changes to feeding diets lower in iron, hepatic neoplasms remain a common finding in lemurs (C. Williams, personal communication).

Lymphoma and **leukemia** are also relatively common tumors in prosimians.[228] Tumors can develop in many sites including the liver, spleen, kidneys, lymph nodes, bone marrow, lung, small intestine, pancreas, testicles, and skin (Figs. 13.10A, B). Cell morphology is variable. Immunophenotyping and tumor predilection or predisposing disease (e.g., possible herpesvirus-associated tumorogenesis in a captive slow loris) has been rarely reported in prosimians.[216,266] Further studies are needed to establish any causal links.

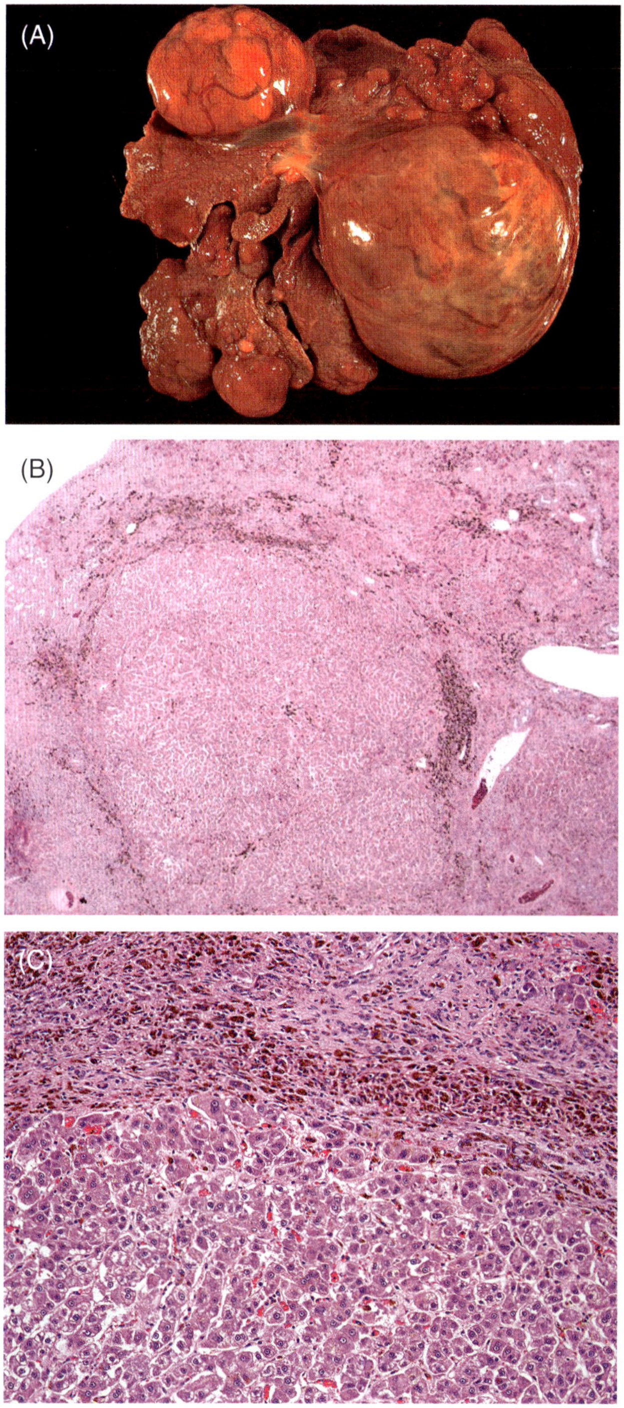

FIGURE 13.9 Hepatocellular carcinoma in the liver of a black-and-white ruffed lemur. (A) Multiple, variably sized, unencapsulated nodular masses expand and replace normal parenchyma (Note: this lemur had concurrent cirrhosis, characterized grossly by multiple foci of nodular regeneration and intervening firm, fibrotic contracted parenchyma). (B) The tumor is composed of demarcated, nodular aggregates of neoplastic hepatocytes arranged in cords and disorganized clusters. (C) The masses compress adjacent nonneoplastic parenchyma (upper area of image), which in this case shows evidence of hepatic damage including hepatocyte degeneration and drop-out, fibrosis, and aggregates of hemosiderin-laden macrophages (see eSlide 13.e6).

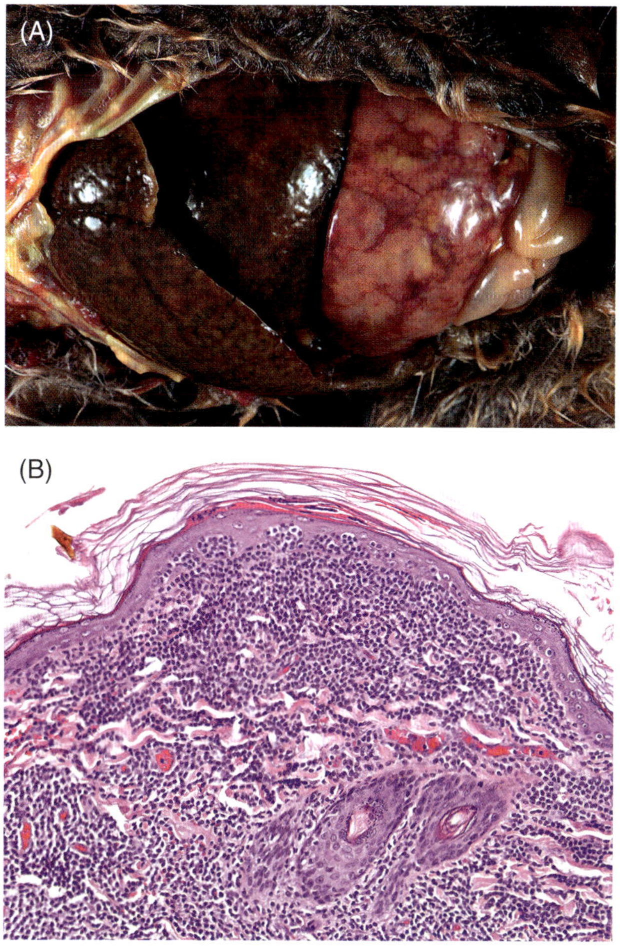

FIGURE 13.10 Lymphoma in a pygmy slow loris (A) and a fat-tailed lemur (B). (A) The liver and spleen are markedly enlarged and have mottled coloration due to neoplastic infiltrates. (B) This tumor is a cutaneous epitheliotropic lymphoma consisting of sheets of mildly pleomorphic, neoplastic lymphocytes that expand the subcutis and dermis, surround adnexa, and infiltrate and form small aggregates in the epidermis (consistent with Pautrier's microabscesses).

INFECTIOUS DISEASES

In addition to virus-associated diseases described below, see Table 13.e3 in the Digital Content for a list of viruses or antibodies to viruses that have been detected in prosimians but for which associated disease has not been reported.

DNA Viruses

Herpesviruses are among the most ubiquitous viruses on the planet. Infections in natural hosts are typically asymptomatic except during periods of immunosuppression. Interspecies infection can be associated with significant disease and even death. Endogenous herpesviruses have been described in the tarsier (TsyrHVL; *Tarsius syrichta*

Herpesviridae-like virus) and aye-aye (DmadHVL; *Daubentonia madagascariensis Herpesviridae*-like virus), and a herpesvirus has been detected in circulating lymphocytes in the slow loris.[15,293] Novel herpesviruses have also been found in a pygmy slow loris and gray mouse lemurs with lymphoproliferative disease, although a relationship to disease has not conclusively been established.[165] As occurs in gorillas, interspecies transmission of human herpes simplex virus type 1 (HSV-1) can cause significant disease in prosimians.[116]

Herpesvirus encephalitis and **meningoencephalitis** have been described in captive and privately owned black-and-white ruffed and ring-tailed lemurs and pottos.[141,154,162] Clinical signs included lethargy, weight loss, decreased coordination, lameness or paralysis, seizures, ptyalism, and death.[154] Gross lesions can be absent or consist of multifocal hemorrhage or malacia (Fig. 13.11A; eSlide 13.e7). Histologic lesions vary in severity and are typical of many viral encephalitides. They include multifocal, nonsuppurative inflammation with mononuclear cells (lymphocytes, plasma cells, and macrophages) and perivascular cuffs and gliosis (Fig. 13.11B; eSlide13.e7). Other lesions include hypereosinophilic, necrotic neurons, malacia, demyelination, and intranuclear, eosinophilic inclusions (Fig. 13.11B; eSlide 13.e7). Human alphaherpesvirus 1 (HHV-1; syn. herpes simplex virus 1 [HSV-1]) infection with encephalitis, intralesional Type A inclusions typical of alphaherpesvirus infection (Fig. 13.11C), and oropharyngeal ulceration was diagnosed with PCR and DNA sequencing in two privately owned ring-tailed lemurs and one privately owned black-and-white ruffed lemur (H. Hauck, personal communication). Virus isolation and molecular characterization via PCR testing and/or DNA sequencing are confirmatory.

Acute viral hepatitis associated with **hepadnavirus-like** particles has been reported in two captive black-and-white ruffed lemurs.[294] Hepadnaviruses are small viruses in the *Hepadnaviridae* family that have partially double-stranded and partially single-stranded circular DNA. Infection in nonhuman primates may be asymptomatic or as in other species, it can result in hepatitis, hepatic necrosis, hepatic degeneration, and benign hepatomas or hepatocellular carcinoma. In the two lemurs, clinical findings included lethargy, weakness or sudden collapse, anemia, icterus, hypoproteinemia, and/or bilirubinemia. Grossly, the livers were enlarged, mottled light tan, and soft. There are no pathognomonic histologic lesions but disorganization of hepatic cords with or without hepatocellular enlargement and vacuolization, multifocal anisokaryosis or binucleation, scattered hepatocellular degeneration or necrosis, canilicular bile stasis, and intrasinusoidal neutrophilic aggregates are suggestive of infection. Viral inclusions are typically not seen in hepadnavirus infections, and they were not seen in the affected livers from the lemurs.

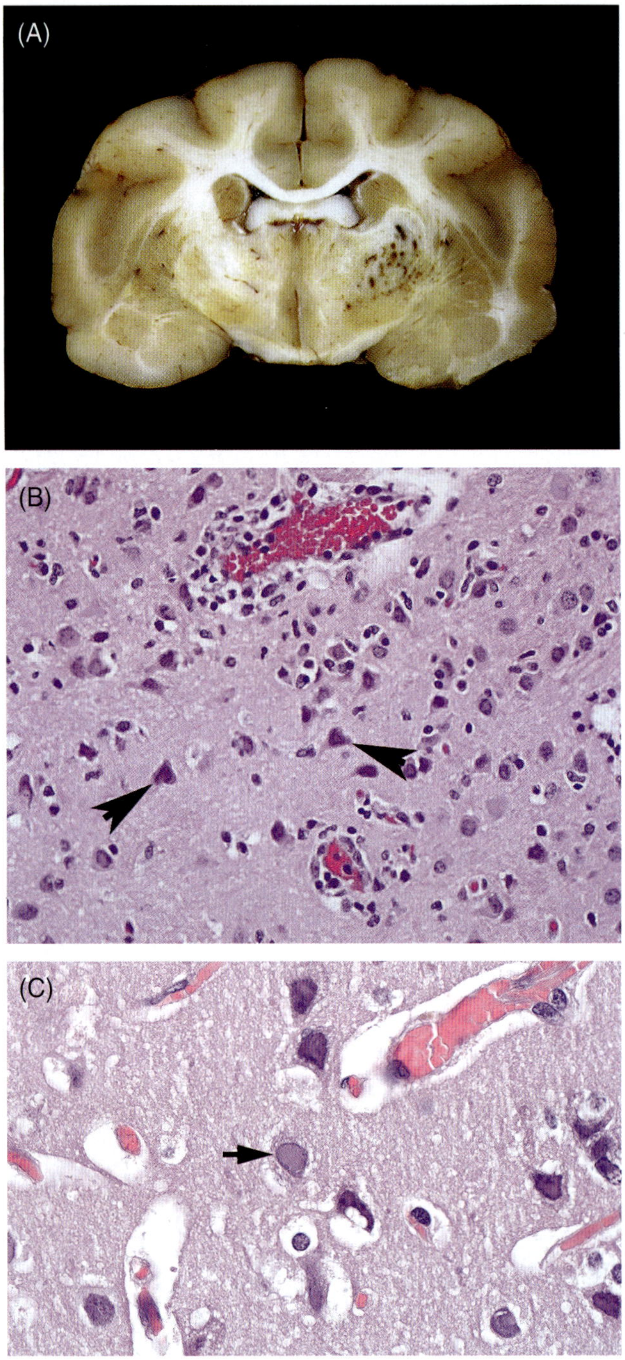

FIGURE 13.11 Encephalitis due to herpes virus infection in black-and-white ruffed lemurs. (A) Multiple small areas of red/brown discoloration consistent with hemorrhage and necrosis are present in the gray matter. (B) Nonsuppurative, lymphoplasmacytic inflammation with perivascular lymphoid cuffs, gliosis, and eosinophilic intranuclear inclusions in astrocytes, glial cells, or neurons (*arrows*) are characteristic of herpesviral infection. (C) Type A inclusions (*arrow*), intranuclear and eosinophilic with margination of the chromatin, are characteristic of herpesviral infection. Human alphaherpesvirus-1 infection was confirmed in this case (see eSlide 13.e7). *(Parts A, B: Photos Courtesy of William R Pritchard Veterinary Medical Teaching Hospital; Part C: Photo Courtesy of H. Hauck, College of Veterinary Medicine, University of Florida.)*

Parvoviruses are small, nonenveloped DNA viruses. The family is divided into two subfamilies: the *Parvovirinae* viruses that infect vertebrates and the *Densovirinae* viruses that infect invertebrates. The near-complete genome of a parvovirus has been described in a slow loris. It has been tentatively named slow loris parvovirus 1 (Sl.L-PV-1).[59] In addition to being the first parvovirus detected in this species, the infected animal had persistent viremia and developed histiocytic sarcoma. Primary gross lesions were marked splenomegaly, mottled coloration throughout the liver with multifocal areas of pallor, and pancreatic and mediastinal lymphadenomegaly. Round to polygonal neoplastic cells with moderate anisocytosis and anisokaryosis infiltrated and effaced normal splenic and lymph node architecture and formed portal and centrilobular aggregates in the liver. Mitotic figures were common and multinucleated cells and erythrophagocytosis were also seen. Numerous tissues including the liver, spleen, kidney, lung, and intestine were PCR positive for the virus, as was serum and whole blood collected up to 8 years prior to death. A direct oncogenic relationship between viral infection and the development of the tumor was suspected but not confirmed.

RNA Viruses

Encephalomyocarditis viruses (EMCV) have worldwide distribution and cause significant neurologic and cardiac disease in a number of wild and captive nonhuman primate (NHP) species including prosimians.[60] These small, nonenveloped, single-stranded, RNA viruses are members of the *Picornaviridae* family. Rodents, and in particular rats and mice, are considered natural hosts and reservoirs, and infection in other species can be asymptomatic or cause encephalitis, paralysis, myocarditis, diabetes, orchitis, or reproductive disorders. Viral shedding occurs via feces and urine, and consumption of contaminated food, water, or infected carcasses are sources of infection for other species.

Outbreaks of EMCV in captive prosimians have been reported in Italy (black, ring-tailed, red-ruffed, white fronted lemurs), Australia (ring-tailed lemur), and the United States (black-and-white ruffed lemurs).[57,226,283] Clinical signs vary from sudden, unexpected death in asymptomatic animals to a short period of weakness and lethargy prior to death. The heart is typically the most significantly affected organ. Gross lesions include cardiomegaly, multifocal or diffuse myocardial pallor, and epicardial petechial or ecchymotic hemorrhage; pulmonary congestion, hemorrhage, or edema; serous, serofibrinous, or serosanguineous hydropericardium or thoracic effusion; and ascites. Myocardial necrosis, cardiomyocyte hydropic degeneration, and multifocal lymphoplasmacytic epi- and/or myocarditis with or without scattered neutrophilic inflammation or necrotizing vasculitis are seen histologically (Figs. 13.12A, B). Neurologic lesions, such as neuronal necrosis, satellitosis, and perivascular meningeal hemorrhage, are less common. Ancillary diagnostics useful in confirming infection include immunohistochemical labeling for EMCV, electron microscopy, and virus isolation. Heart is the preferred diagnostic sample, though virus can also be isolated from liver, spleen, kidney, intestine, skeletal muscle, lung, blood, pancreas, or feces.

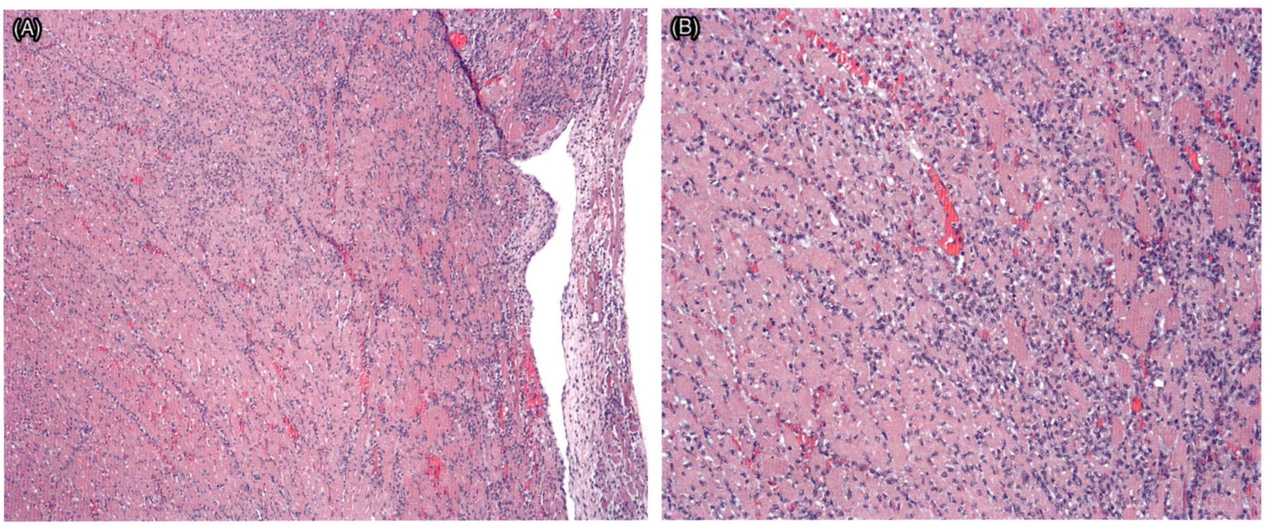

FIGURE 13.12 Encephalomyocarditis virus in a ring-tailed lemur. (A) The epicardium and myocardium are infiltrated and expanded by moderate to severe inflammation. (B) Inflammation is primarily lymphoplasmacytic and associated with acute myocardiocyte degeneration and necrosis characterized by cellular swelling, loss of muscle cross-striations, and hypereosinophilia. *(Parts A, B: Photos Courtesy of M. Garner, Northwest ZooPath.)*

Bacteria

Prosimians are susceptible to a variety of bacterial infections but do not seem to have an increased susceptibility to any particular bacterial agents.

Numerous bacteria can cause enteric disease in prosimians (Fig. 13.13). Several that have been described in captive prosimians include *Campylobacter jejuni*, *Yersinia pseudotuberculosis, Yersinia enterocolitica*, enteropathogenic *Escherichia coli*, *Salmonella typhimurium, Shigella* spp., *Klebsiella pneumoniae*, *Listeria monocytogenes*, and *Clostridium difficile*.[38,62,64,130,285] Several of these bacteria have also been detected in feces from free-ranging lemurs.[43] In addition to enteric disease, fibrinous and necrotizing bronchopneumonia due to *K. pneumoniae* or mixed bacterial infections have been reported in a collared lemur and slow lorises, and it was identified as the cause of neutrophilic and lymphocytic meningitis with intralesional gram-negative bacilli in a group of black-and-white ruffed lemurs.[245] In the latter, gross lesions included opaque white discoloration of the meninges and multifocal petechiae in the parenchyma of the brain. While also generally associated with enteric disease, *Y. pseudotuberculosis* has been reported as a cause of osteomyelitis in a ring-tailed lemur.[280]

Trueperella pyogenes infection by clonally related organisms was reported in three gray slender lorises housed at a single institution.[192] One had a facial abscess. A second died and had catarrhal purulent rhinitis, sinusitis, pneumonia, and pericarditis. The third also developed fatal septicemia secondary to a facial abscess. Additional bacteria were isolated in the second and third cases, but *T. pyogenes* was considered the primary pathogen. *T. pyogenes* has also been isolated from a galago with abscesses secondary to injuries from a fight.[242]

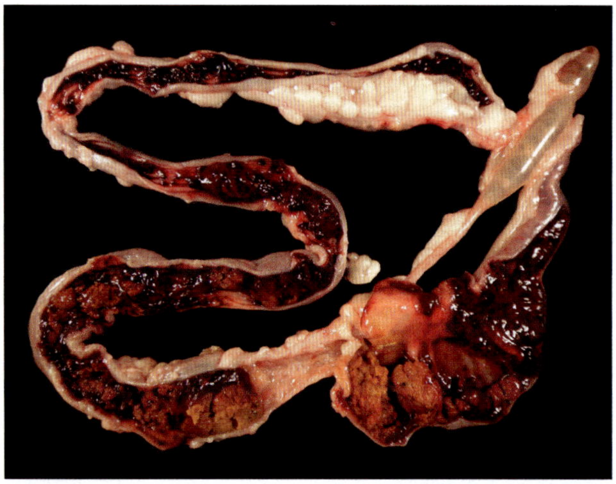

FIGURE 13.13 Necrotizing bacterial enterotyphlocolitis in a ring-tailed lemur. *Enterococcus faecium* was isolated from the intestinal tract and was associated with marked, necrotizing inflammation with abundant intraluminal necrotic debris and hemorrhage. *(Photo Courtesy of A. Newton, Wildlife Conservation Society.)*

Systemic **listeriosis** due to infection with ***L. monocytogenes*** has been reported in gray mouse lemurs.[124] Histologic lesions in one were similar to infections in other species and included necrotizing and pyogranulomatous inflammation in liver, spleen, bone marrow, adrenal gland, and brain. Bacteria were highlighted with immunohistochemical labeling. ***Citrobacter* sp., *S. typhimurium***, and ***Streptococcus* sp.** have been isolated in cases of bacterial septicemia in several species of prosimians.[22,62,122,286]

Ehrlichiosis is a tick-borne, bacterial infection caused by *Ehrlichia* species in the Rickettsiales order, Anaplasmataceae family. These small, gram-negative, obligate intracellular bacteria cause disease in both animals and humans. In the United States, infection with ***Ehrlichia chaffeensis*** occurs predominantly in the south-central and mid-Atlantic regions. Transmission is typically through the bite of infected lone star tick (*Amblyomma americanum*) nymphs and adults; the white-tailed deer is the main reservoir host for the bacterium. In humans and domestic dogs, bacteria primarily infect monocytes, though lymphocytes, myelocytes, and neutrophils can also be infected. In 2001, an outbreak of naturally acquired ehrlichiosis occurred in the south-eastern United States in a lemur colony in which animals had summer access to a fenced, mixed pine and deciduous hardwood forest.[289] Affected lemurs developed anorexia, lethargy, fever, and enlarged lymph nodes. Thrombocytopenia and leukopenia (lymphopenia and neutropenia) were common. Hyperbilirubinemia, azotemia, and proteinuria were also seen. Confirmation of infection included identification of morulae of *Ehrlichia* in lymphocytes and monocytes from lymph node aspirates, morula-like inclusions in cultured leukocytes, and positive PCR amplification.

Like ehrlichiosis, **anaplasmosis** is a vector-borne, bacterial infection. The causative bacteria (Order: Rickettsiales; Family: Ehrlichiaceae) infect host erythrocytes (e.g., ***Anaplasma marginale***) or neutrophils (e.g., ***Anaplasma phagocytophilum***) and are generally transmitted by ticks, especially those of in the genus *Ixodes*. A single case of granulocytic anaplasmosis has been described in a captive ring-tailed lemur.[1] The lemur presented with anorexia, weakness, epistaxis, and an uncoordinated gait. The animal was thrombocytopenic with no other hematological or biochemical test abnormalities. Round, basophilic morulae were seen in the cytoplasm of infected neutrophils in peripheral blood smears (Giemsa staining), and *A. phagocytophilum* infection was confirmed in blood with PCR testing and DNA sequencing. Two adult female *Ixodes ricinus* collected from this lemur were infected with *A. phagocytophilum*. Antibodies to *A. phagocytophilum* in the absence of clinically apparent disease have also been reported in a managed group of free-range blue-eyed black, black-and-white ruffed, and ring-tailed lemur.[296] Exposure in each species reflects the presence of competent vectors in the environment.

Several other tick-vectored bacterial infections have been described in wild, free-ranging lemurs in Madagascar. Affected species include critically endangered indris and diademed sifakas. Those identified during health surveys have included ***Babesia* sp., *Borrelia* sp.,** and ***Neoehrlichia* sp.**[170,218]

As occurs in many species, prosimians are susceptible to mycobacterial infections. ***Mycobacterium tuberculosis*** has been described in ring-tailed, ruffed, black, black-and-white ruffed, and mongoose lemurs.[160,167,168] *M. tuberculosis* infection results in the formation of granulomatous inflammation or granulomas characterized by epithelioid macrophages, lymphocytes, plasma cells, and variable numbers of multinucleated giant cells with caseous necrosis, and (typically few) intralesional acid-fast positive bacilli. In general, the lungs are the most common site of infection in primates. Infection in prosimians has been described in the lungs, intestines, thoracic cavity, liver, spleen, lymph nodes and bone. ***M. xenopi***, described in a captive black-and-white ruffed lemur, produced similar histologic lesions that were extrapulmonary and present throughout the mesentery and in a mesenteric lymph node.[275]

Francisella tularensis, which causes **tularemia**, is a small, gram-negative coccobacillus that can infect prosimians. Infection can be transmitted by ticks, deer flies, or exposure to infected rodents and rabbits. Systemic disease has been described in captive ring-tailed and black-and-white ruffed lemurs.[54] Consistent gross lesions include marked splenomegaly, microabscess formation in the liver and spleen, and bronchopneumonia. As occurs in many other species, tularemia-associated splenic and liver microabscesses have a multifocal and random distribution and are composed of roughly nodular aggregates of neutrophilic inflammation with central necrosis. Bacterial culture or PCR is confirmatory, and vector control and antibiotic treatment are important to prevent or mitigate disease transmission and progression, respectively.

Fungi

Coccidioidomycosis (Valley fever, San Joaquin Valley fever) is caused by ***Coccidioides immitis*** or ***C. posadasii***, which are dimorphic, saprophytic fungi found as mycelia in warm, arid regions of the southwestern United States, Mexico, and parts of Central and South America. Infection is typically via inhalation of aerosolized spores, called arthroconidia (or arthrospores). In tissue, arthroconidia develop into large, 30–200 µm diameter spherules with a 2–4 µm thick, refractile, double-layered wall. The spherules undergo endosporulation to produce 2–5 µm diameter endospores that upon release infect adjacent tissue. Primary infections typically occur in the lungs or (less commonly) through breaks in the skin, and disseminated disease may subsequently develop. Associated inflammation is often granulomatous or pyogranulomatous with granuloma and/or abscess formation. Diagnostic options include cytologic preparations, histology, immunohistochemical labeling, PCR, and in situ hybridization. Infection described in a captive ring-tailed lemur was typical of pulmonary and disseminated coccidioidomycosis.[48] Gross and histologic lesions included pleural and pulmonary abscesses, abscess formation in the liver and spleen, granulomatous inflammation in the retina, and characteristic intralesional coccidioidal spherules and endospores.

Blastomyces dermtitidis is a dimorphic fungus found in the eastern and mid-western United States. Blastomycosis has been reported in a red-ruffed lemur.[237] Characteristic 7 µm to15 µm diameter yeasts with broad-based budding were seen in cytologic preparations of a fine needle aspirate of a lung mass, and in histologic sections of lung, tracheobronchial lymph nodes, spleen, kidney, liver, brain, and eye. Associated inflammation consisted of macrophages and neutrophils, and multifocal foci of necrosis and hemorrhage.

Reports of other fungal infections are rare. These include case reports of disseminated **cryptococcosis** in a mongoose lemur and oral **candidiasis** due to ***Candida albicans*** in lemurs treated with long-term antibiotics.[286] Another report described crusting dermatitis and alopecia in the skin around the commissures of the mouth, eyes, and along the neck, crusting of the skin of a forefoot, and thickened perineal skin in a Coquerel's sifaka with **dermatophytosis** due to ***Trichophyton mentagrophytes*** infection.[209] Histologic lesions included lymphoplasmacytic perivascular dermatitis with folliculitis, furunculosis, and intralesional fungal hyphae. Confirmation of the diagnosis was made through a combination of skin scrapes, histology, and fungal culture using dermatophyte test medium. Treatment was initiated due to concerns about possible intraspecies, interspecies, or zoonotic transmission.

Parasites

A variety of metazoan, protozoan, acanthocephalan, and ectoparasites are reported in prosimians. Treatment in the absence of clinical disease may be performed in captive settings as part of standard practices rather than in response to specific clinical signs or the development of lesions. Several parasites in which significant disease has been described in groups of animals or that are notable for management purposes are described below. Other parasites that have been less frequently described or that have been detected in the absence of reported disease are listed in Table 13.e4.

Metazoa

Numerous nematode parasites can infect wild and captive prosimians. These include oxyurids, strongyles, spirurids, and trichurids.[3] Notable diseases are associated with several spirurid nematodes.

Gongylonema* sp.** are thread-like nematodes of birds and mammals that are transmitted by insects (typically beetles and cockroaches). Infection is seen occasionally in captive prosimians. As in other species, infections are typically seen in the tongue or esophagus (Fig. 13.14; eSlide 13.e8). Adult worms are present in the superficial, stratified squamous epithelial mucosa where they form characteristic serpiginous tracts. They generally are not considered pathogenic, but associated mild inflammation can occasionally be seen.[286] One publication reported an association between esophageal ***G. pulchrum infection and squamous cell carcinoma in a vari.[30]

Gastric ***Physaloptera* spp.** infections are occasionally reported in captive prosimians.[286] Infection is generally asymptomatic even when parasites are present in large numbers (Fig. 13.15). However, they can occasionally cause mild or significant disease. For example, there is a single case report of *Physaloptera*-associated gastritis in a ruffed lemur, and a black lemur died due to gastric perforation and suppurative peritonitis resulting from ***P. sibirica*** infection.[22,23]

Spirocerca lupi aortic aneurysm formation, rupture, and death have been reported as a cause of mortality in captive Malagasy lemurs and an aye-aye.[29,71] The spirurid nematode ***Pterygodermatites nycticebi*** was first described by Monnig in a slow loris in 1920.[138] It is generally not considered a pathogen in prosimians but may have contributed

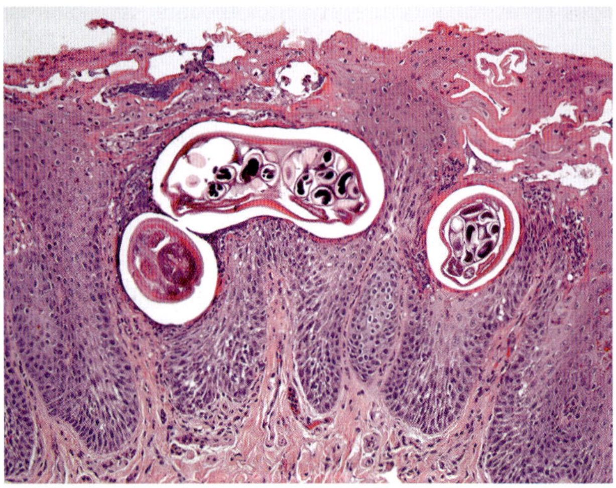

FIGURE 13.14 Gongylonema in the tongue of a slow loris. Adult parasites are embedded in the mucosa. Characteristic features of this spirurid nematode include coelomyarian musculature, a large intestinal tract with few uninucleate, cuboidal cells, and thick-shelled, embryonated eggs. Mild mixed inflammation is present in the adjacent tissue (see eSlide 13.e8).

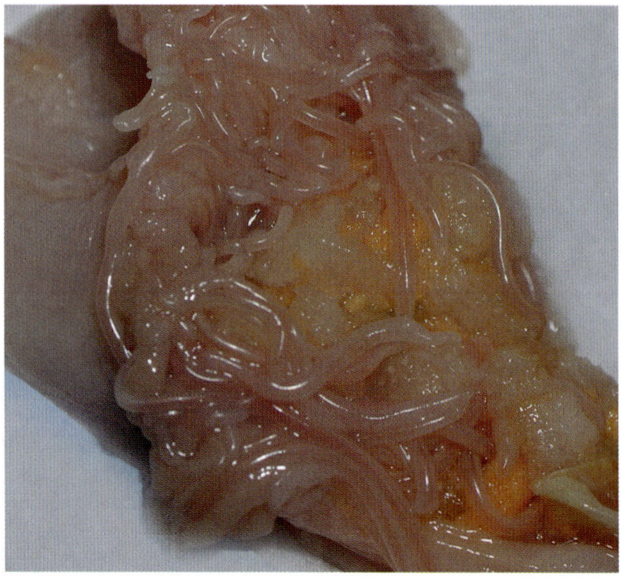

FIGURE 13.15 Physaloptera in the stomach of an eastern lesser bamboo lemur. Numerous elongated, light pink nematode parasites are present in the stomach. Even when severe, infection is typically asymptomatic. *(Photo Courtesy of D. Haring, Duke Lemur Center.)*

to death in an aye-aye and galagos at zoos in Germany and Japan, respectively.[246,254] Control of this parasite should include pest management, as cockroaches carry infective *P. nycticebi* larvae.[187,299]

As occurs in many species, lemurs are susceptible to encephalitis caused by larval migration of the ascarid nematode ***Baylisascaris procyonis***. The definitive host is the North American raccoon, which has been introduced into Europe, Russia, and Japan. Infection in raccoons is asymptomatic and aberrant hosts are infected through ingestion of eggs shed in the environment. Larvae can cause severe and sometimes fatal nervous system disease in nondefinitive hosts including mammals and birds. Published reports in prosimians include cases in black-and-white ruffed and white-headed lemurs.[55,143] The ruffed lemurs had areas of malacia with Gitter cells in multiple sites in the brain. Characteristic 40–60 µm diameter parasites were seen in only one animal. This animal also had a granuloma with similar parasites in the mesentery. The white-headed lemurs had granulomatous meningoencephalitis with tracts of white matter necrosis and intralesional nematode larvae. Granulomatous and eosinophilic inflammation and intralesional nematode larvae were also found in lungs, kidney, spleen, heart, and jejunum. Characteristic features of the larval parasites include a smooth, eosinophilic cuticle, lateral alae, prominent lateral cords, and coelomyerian musculature (Fig. 13.16). Larvae were confirmed to be *B. procyonis* based on histologic appearance and PCR.

Nervous system nematodiasis has also been described in a red-ruffed lemur infected with ***Angiostrongylus***

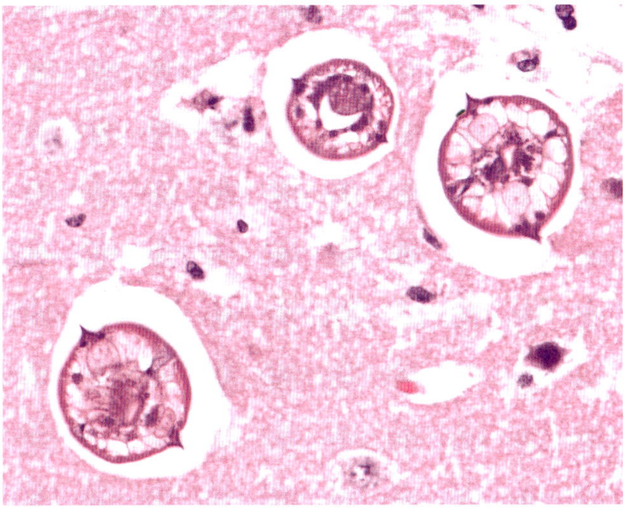

FIGURE 13.16 Visceral larval migrans due to *Baylisascaris procyonis* in the brain of a white-headed lemur. Multiple sections of larval nematodes are present in the brain. Characteristic features include a smooth, eosinophilic cuticle, lateral alae, prominent lateral cords, and coelomyerian musculature. *(Photo Courtesy of M. Jimenez Martinez, Zoo and Wildlife Pathology Service of the Complutense University Veterinary Hospital, Madrid.)*

cantonensis.[231] The rat is the definitive host. Clinical signs included hind limb paresis and head tilt. Adult nematodes were present in the subarachnoid space and neuropil of the cerebellum and brainstem and in the subdural space in the thoracic spinal cord. Characteristic features of the parasites included a smooth, eosinophilic cuticle, prominent lateral cords, coelomyarian musculature, a pseudocoelom, and intestinal tract lined by multinucleated cells. Associated inflammation included eosinophils, neutrophils, macrophages, multinucleated giant cells, and lymphocytes; meningeal inflammation was predominantly eosinophilic. *A. cantonensis* was suspected based on the appearance of the organism and was confirmed with PCR. *A. cantonensis* is endemic in Southeast Asia and the Pacific islands, but the lemurs in these reports and other animals were housed in the United States.

Cysticercosis is caused by infection with metacestodes (larval tapeworms) of ***Taenia*** species. The normal life cycle of this parasite typically involves a carnivore definitive and rodent intermediate host. In nondefinitive hosts, infection may be local or multifocal and extensive and can cause spectacular lesions with tissue expansion, abdominal distension, or organ compression due to the formation of large, cystic structures in the subcutis or body cavities. The cysticercus is a fluid-filled, bladder-like structure containing a single attached inverted scolex. Multiple, double rows of hooks (***Taenia saginata*** is the exception; scolex has four large suckers but lacks a rostellum and rostellar hooks) are present on the rostellum, calcareous corpuscles are present within the parenchyma (often most numerous in the scolex and neck), and the bladder may be internally subdivided by fine connective tissue septa.[113] Associated chronic inflammation and peripheral fibrosis may be associated with cyst formation.

Cysticercosis has been described in case reports in ring-tailed, black, and white-headed lemurs at zoos in Europe and the United States, and we have seen an additional case in a red-ruffed lemur (Figs. 13.17A, B; eSlide 13.e9). Infections are associated with cysts containing myriad cysticerci and associated inflammation in the subcutis, pleural cavity, lungs, or abdomen.[4,81,178,207,243,255] PCR targeting the mitochondrial cytochrome oxidase 1 gene (COX1) has confirmed

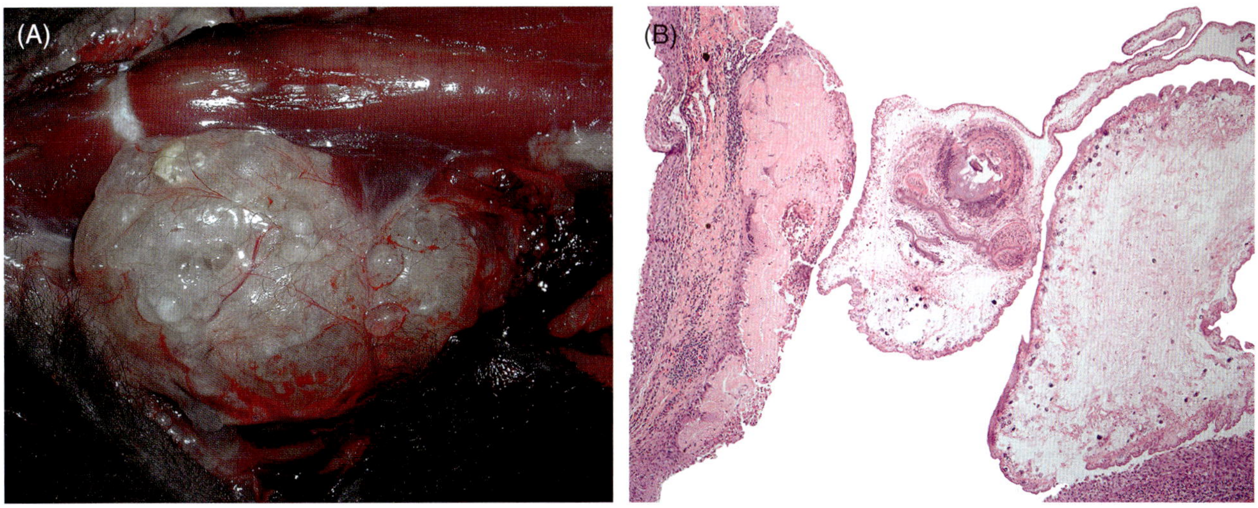

FIGURE 13.17 Cysticercosis due to metacestodes of *Taenia crassiceps* in the subcutis of a red-ruffed lemur. (A) This large, multinodular, multicystic mass was present lateral to the trachea in the subcutis of the neck. (B) The larval cestode has a characteristic single scolex and calcaneous corpuscles (right side of the image). The wall of the parasitic cystic space (left side of the image) is composed predominantly of macrophages (see eSlide 13.e9).

Cysticercus longicollis (larval form of ***Taenia crassiceps***) and ***Taenia martis*** infections in some of the lemurs. In the black lemur, infection caused a large, fluctuant swelling in the subcutis over the back.[97] This corresponded to a large, multiloculated, demarcated, encapsulated cystic structure containing myriad luminal cysts, each of which contained a single, characteristic, developing scolex. Similar lesions were seen in the abdominal and thoracic cavities, and infection replaced a portion of a lung lobe and caused displacement of abdominal organs. *C. longicollis* was diagnosed based on the morphologic features of the parasites. A source of infection was not identified in any case and each of the lemurs died due to complications of its infection.

Hydatid cysts are also produced by larval cestodes but are the result of infection with the genus ***Echinococcus***. They are large, unilocular or multilocular, fluid-filled cystic structures lined by an outer membrane and inner germinal epithelium from which brood capsules containing one to multiple, small inverted scolices (called protoscolices due to their small size) form and remain attached to or bud from and float freely within the cyst.[113] Hydatid cysts, in which a diagnosis has been based on typical metacestode morphology and examination of scolices and hooks, have been described in a captive greater galago and ring-tailed lemurs.[82,161,201,272,273] Disease in the galago and some lemurs caused marked abdominal distension. Multiple hydatid cysts were seen in the galago and cysts were present in multiple sites including the liver, lungs, thoracic and abdominal cavities, and/or lymph nodes in the lemurs. Cysts ranged in size and were often large, up to 9 cm diameter. One, a large, uniloculated hydatid cyst in the caudal thorax of an infected lemur, caused lateral displacement of the heart and atelectasis and atrophy of the right lung lobe

An unusual case of disseminated ***Hymenolepis*** cestodiasis with atypical larval parasite morphology was described in a ring-tailed lemur.[76] The parasites replaced a majority of the hepatic parenchyma, which was firm and contained multifocal to coalescing, demarcated, unencapsulated mottled red to tan-brown-white masses. This corresponded histologically to areas of pyogranulomatous hepatitis and necrosis with numerous, intralesional, round to ellipsoid to irregularly serpiginous structures that were 200 µm in width and 400 µm in length, had a thin eosinophilic tegument, and contained lightly eosinophilic fibrillar stroma supporting numerous, round, basophilic cells. Similar parasites were also seen in the lung and a mesenteric vein. The parasites lacked characteristic cestode features (e.g., calcareous corpuscles, proglottid segmentation, scolices). Given the unusual morphology of the parasites, PCR and DNA sequencing were performed and confirmed infection with ***Hymenolepis nana***. This is a zoonotic cestode parasite that can infect a wide range of species and has occasionally been reported in nonhuman primates including ring-tailed lemurs.[174]

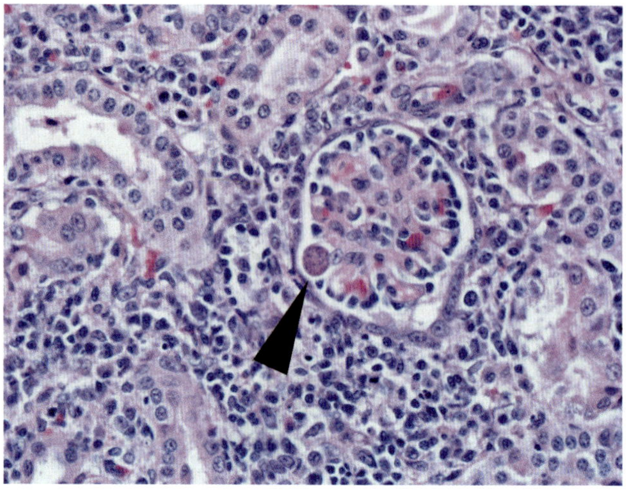

FIGURE 13.18 Toxoplasmosis in the kidney of a red-ruffed lemur. A focal protozoal cyst (*arrowhead*) containing numerous bradyzoites is present in a glomerulus. The surrounding renal interstitium contains moderate, nonsuppurative, lymphoplasmacytic inflammation.

Protozoa

As occurs in many mammals including other primates, prosimians are susceptible to infection with the protozoal parasite ***Toxoplasma gondii***. Infections may be localized or systemic with mononuclear cell infiltrates, necrosis, and intra- and extracellular, 1–3 µm tachyzoites, or parenchymal cysts (Fig. 13.18). Outcomes of infection in pregnant dams can include placentitis, abortion, stillbirth, disseminated fetal infection, or neonatal death; however, offspring of persistently infected dams may be born uninfected since transplacental infection does not always occur.[22,40,129,262] The tissue distribution and severity of infection and inflammation often vary from one organ to another. Lesions in the placenta may be mild, consist of edema and histiocytic infiltrates, and protozoal organisms may be infrequent or absent. Immunohistochemistry, electron microscopy, and PCR are useful ancillary diagnostic tests to confirm the identity of the organism and to help in the identification of tachyzoites.

Amoebic encephalitis causes significant, life-threatening disease and has been diagnosed in a number of species including humans and primates.[229] In prosimians, necrotizing meningoencephalitis due to ***Acanthameoba* T4 genotype** caused the death of a captive black-and-white ruffed lemur after a sudden onset and short course of lethargy, recumbency, and seizures.[111] Hematologic evaluation revealed neutrophilic leukocytosis and lymphopenia. Gross and histologic lesions, which were limited to the frontal and parietal lobes of the cerebrum, consisted of severe, multifocal hemorrhage, necrosis, and (predominantly) neutrophilic inflammation and fibrin that expanded Virchow-Robin's space and the leptomeninges. Within affected areas were numerous, intralesional, double-layered amoebic cysts and trophozoites. The latter were 15–30 μm in diameter, filled with abundant eosinophilic, vacuolated cytoplasm, and contained a central round nucleus with a prominent central nucleolus. Amoeba were highlighted with periodic-acid Schiff (PAS) staining and positive immunohistochemical labeling for amoebic lectins. PCR for 18S ribosomal RNA and DNA sequencing confirmed the identity of the organism. The brain was the only affected organ. Infection was thought to have occurred through the nasal mucosa and water was the likely source of infection.

Cryptosporidiosis has been reported in several species of free-range lemurs and several species of captive lemurs and sifakas.[39,65,221,277] In the latter, clinical signs varied from none to mild intermittent anorexia, lethargy, and abdominal bloat with or without soft or diarrheic feces in individuals or groups of animals. Deceased Coquerel's sifakas diagnosed with enteritis were reported to have lymphoplasmacytic infiltrates in the mucosa of the intestine or mucosal edema.[62]

Several ***Plasmodium*** species, including ***P. bucki, P. coulangesi, P. foleyi, P. girardi, P. lemuris, P. malagasi***, and ***P. percygarnhami***, have been reported in wild lemurs but to date, clinical disease has not been described.[96,169,173] Asymptomatic ***Trypanosoma cruzi*** infection has been identified in ring-tailed lemurs in a free-range, captive collection.[125] Blue-eyed black, black-and-white ruffed, and ring-tailed lemurs from the same collection were seropositive for the organism.

***Giardia* spp.** and ***Entamoeba* spp.**, important enteric zoonotic pathogens that can cause diarrheal disease (with or without hemorrhage) and death in nonhuman primates, have been detected in the feces of several species of wild and zoo-housed prosimians.[26,70,220,277,286] Detection of *E. histolytica* in diarrheal feces from wild lemurs in Madagascar has been described, but whether infection was the cause of disease was not established.[220]

Ectoparasites

Parasitic mites in the ***Demodex*** genus are the cause of **demodicosis**, a noncontagious skin disease resulting from the over proliferation of these (typically) nonpathogenic, host-specific parasites.[123] The parasites can vary in size but are typically 100–300 μm in length and approximately 30–50 μm in diameter with a tapered posterior end. They have six to eight short, jointed appendages that are positioned toward the anterior end of the body and have an eosinophilic, chitinous exoskeleton. Associated disease most commonly occurs in young, old, or immunocompromised hosts. Clinical presentation may be local or generalized. Gross lesions may include alopecia, crusty or erythematous skin, and the formation of plaques, nodules, or pustules with or without comedones. Demodicosis has been described in a group of captive Senegal bushbabies.[166] Disease consisted of multifocal, up to approximately 0.8 cm diameter, raised nodules and papules on the haired (dorsal) and unhaired (ventral) surfaces of the ear pinnae (Fig. 13.19A) or (rarely) the legs. Nodules contained thick, waxy, white material consistent with sebum. Numerous mites consistent with *Demodex* sp. were present in cytologic preparations and in histologic sections (Fig. 13.19B). Lesions were limited to the dermis and consisted of single or large groups of intrafollicular and dermal *Demodex* sp. mites surrounded by amorphous eosinophilic material, variable

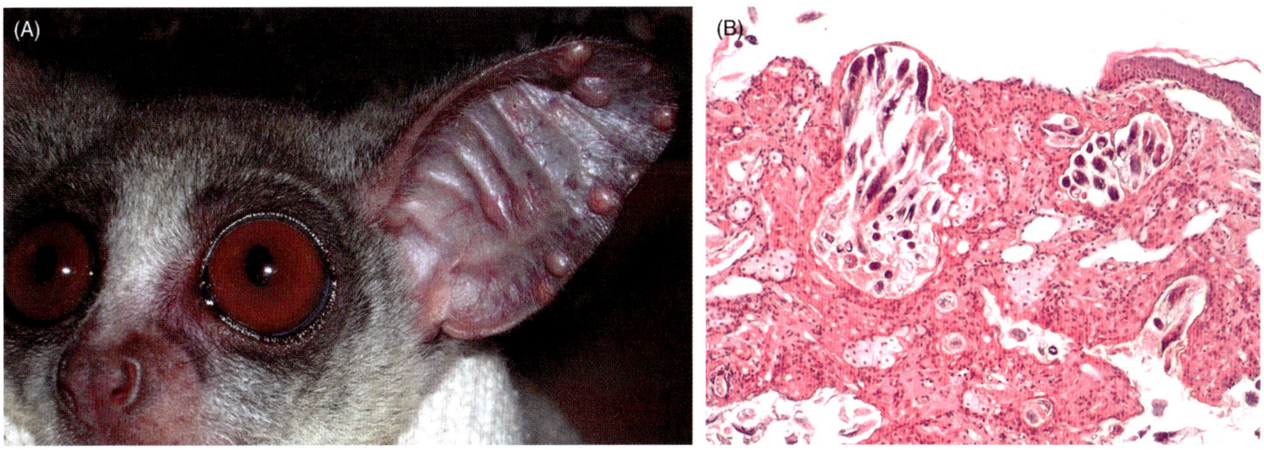

FIGURE 13.19 Demodicosis in the skin of a Senegal bushbaby. (A) Numerous, variably sized, nonpruritic, papular and nodular lesions are present on the ear pinna. (B) Papules and nodules are the result of marked dilation of hair follicles by groups of elongate mites that are morphologically consistent with *Demodex*. Characteristic features include a chitinous exoskeleton and short, jointed appendages. Hair follicles are surrounded by mixed neutrophilic and lymphocytic inflammation; similar inflammation surrounds extrafollicular mites in the dermis (presumed secondary to hair follicle rupture and parasite release into the dermis). *(Images reprinted with permission from Kuznetsova E, Vysokikh A, Bourdeau P. First description of demodicosis in 12 galagos (*Galago senegalensis*). Vet Dermatol. 2012;23(1).)*

numbers of sebocytes, and variable numbers of lymphocytes and neutrophils. Characteristic features of the mites included a thin, eosinophilic, chitinous exoskeleton, short, jointed appendages and striated muscle.

ADDITIONAL RESOURCES

Remick AK, Van Wettere AJ, Williams CV. Neoplasia in prosimians: case series from a captive prosimian population and literature review. Vet Pathol. 2009. 46(4):746–72.

Robveille C, Cullen JM. Hepatocellular carcinomas in captive prosimians. Vet Pathol. 2022. 59(6):1012–1021.

Williams CV. Prosimians. In: Fowler's Zoo and Wild Animal Medicine, volume 8. Miller ER, Fowler ME (eds). 2015. WB Saunders; 2015.

References

Full references for this chapter can be found online at https://doi.org/10.1016/B978-0-443-18437-6.00021-3.

Chapter 14

New World and Old World Monkeys

Kerstin Mätz-Rensing[a], Natália Coelho Couto de Azevedo Fernandes[b], and Linda J. Lowenstine[c]

[a]German Primate Center, Leibniz Institute for Primate Research, Göttingen, Germany; [b]Centro de Patologia, Instituto Adolfo Lutz, São Paulo, Brazil; [c]University of California, Davis, CA, United States

INTRODUCTION

New World monkeys (NWMs) are naturally distributed from Mexico through the Central and South American tropics and subtropics (Table 14.e1).[185] Most are listed as threatened, endangered, or critically endangered by the International Union for the Conservation of Nature (IUCN). They are a complicated taxonomic group with disputed systematics that are constantly being refined, or sometimes confused, by molecular studies.[233] NWM were historically divided into two main families, the small, squirrel-like arboreal Callitrichidae (marmosets and tamarins) and the larger-bodied Cebidae. Newer evidence has split Cebidae into Cebidae and Atelidae. The best-studied species, in terms of medicine and pathology, include the common marmoset, the squirrel monkey, and the owl monkey, which have all been used extensively as laboratory animals.

Old World monkeys (OWM) are found in tropical, subtropical, and temperate regions of Asia and Africa (Table 14.e2). Many are threatened, endangered, or critically endangered, though some species are still abundant and have become "peridomestic," dwelling in close proximity to humans (e.g., rhesus macaques, Hanuman langurs). Normal habitats are forests, grasslands, and wooded savannas. Some species are highly arboreal. Others spend much of their time on the ground and retreat to the trees for safety, while a few are nearly entirely ground living. OWM are divided into two subfamilies, the Cercopithecinae or cheek-pouched monkeys, and the Colobinae, or leaf-eating monkeys. Members of both subfamilies are found in Africa and Asia. As is the case for NWM, systematics and molecular techniques are being used to refine relationships, particularly within the macaques, baboons, guenons, and colobine monkeys. The best-studied species, in terms of medicine and pathology, include rhesus macaques and cynomolgus monkey groups.

Information that follows focuses on notable non-infectious and infectious diseases in NWM and OWM. Notable unique anatomic features are described in the Digital Content (Table 14.e3).

NON-INFECTIOUS DISEASES

Nutritional

Multiple **metabolic bone diseases (MBD)** have been described in NWM and OWM. Common forms of MBD in NWM and OWM include **rickets and osteomalacia**, both of which are disorders characterized by a failure of osteoid mineralization. Rickets occurs in immature animals before epiphyseal closure; osteomalacia is a similar disorder in mature animals. Rickets in young growing animals affects growing cartilage and bone; osteomalacia in mature animals affects bone and not cartilage. Although primarily seen in NWM, rickets and osteomalacia are also seen in young OWM and apes housed indoors in the absence of full-spectrum sunlight.[189] In general, vitamin D requirements vary with species of primate and must be in the correct form (D3). NWM require higher vitamin D in diets compared to OWM. Clinical presentations manifest as distorted limbs, kyphosis (Fig. 14.1), fractures of the long bones, and thickening of the maxillary and mandibular bones. Bones in both conditions are soft and often easily manually cut. Enlargement/widening and deformation of epiphyseal growth plates due to impaired chondrocyte development and failure to mineralize growth plate cartilage is a common gross finding in young animals with rickets; this will not be seen in older animals with osteomalacia due to MBD. Osteopenia with folding fractures of ribs and long bones is seen in older animals. Extensive bone resorption with increased osteoclastic osteolysis, impaired mineralization of newly formed osteoid, thinning of bone cortices and trabeculae, and medullary fibrosis are seen histologically in both rickets and osteomalacia.

Fibrous osteodystrophy of the skull and mandible as a result of both spontaneous and experimental vitamin D3 deficiency (hypovitaminosis D) has been described in capuchins, squirrel, and woolly monkeys. Grossly, the calvarium, maxilla, and mandibles are swollen, and the teeth are loosely attached or easily shed. Histologically, there is osteoclastic resorption of bone and expansion and replacement of bone

Pathology of Wildlife and Zoo Animals. https://doi.org/10.1016/B978-0-443-18437-6.00018-3

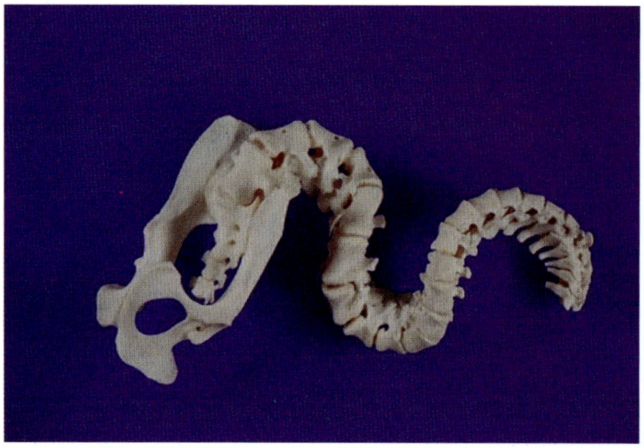

FIGURE 14.1 Metabolic bone disease in a douc langur. The vertebral column is disfigured due to excessive convex kyphotic and inward lordotic curvature.

by myxomatous fibrous connective tissue within which are embedded thin, poorly mineralized bony trabeculae.

Vitamin C is an essential dietary nutrient because NWM and OWM lack the enzyme L-gulono-lactone oxidase. Due to its significant role in collagen synthesis and stability, **hypovitaminosis** C, produces **scurvy**. Gross lesions include gingival bleeding and gingivitis, loosening of the teeth, cutaneous petechiation, and periarticular and subperiosteal hemorrhage with epiphyseal and costochondral fractures.[154] Cephalohematomas, due to subperiosteal hemorrhage in the outer calvarium, are pathognomonic in squirrel monkeys.[209] Anemia is a consistent finding likely related to the role of ascorbic acid in iron and folic acid absorption. Though much more common historically in pet monkeys, outbreaks of scurvy can occur due to errors in the formulation of commercial diets. Histologically, hypovitaminosis C osteopathy is characterized by osteopenia, microfractures, subperiosteal hemorrhage, and short stout metaphyseal trabeculae. In young animals, the growth plates are broad and hypocellular with a paucity of primary spongiosa, which though mineralized, is not properly ossified.[68] The pathogenesis involves defective collagen synthesis leading to abnormal osteoid formation, weakening of periodontal ligaments, and increased vascular fragility. As a consequence, inadequate strength of subephyseal bone leads to microfractures to macrofractures (e.g., slipped epiphysis; Salter-Harris fractures). Vascular fragility leads to petechiation or occasionally more massive hemorrhage.

Vitamin E-responsive **hemolytic anemia** and **necrotizing myopathy** have been described in owl monkeys, marmosets, and gelada.[125,132] The requirement for vitamin E appears to be high in owl and capuchin monkeys and callitrichids. Macrocytic hemolytic anemia may be the first clinical sign. Pathologically, scattered polyphasic necrosis of skeletal muscle and the myocardium with varying degrees of fibrosis and regeneration are found.[156] The lesions are accompanied by increased medullary and extramedullary hematopoiesis and erythroid hyperplasia. Binucleated rubricytes are also seen.[25]

Iron overload (hemosiderosis) is one of the most common postmortem findings in captive callitrichids. It is also seen, but less frequently, in free-living marmosets. The disease is defined by an excessive accumulation of ferrous (iron) pigment in Kupffer cells and/or hepatocytes in the liver. It is likely caused by excessive dietary intake in conjunction with enhanced enterocyte uptake of iron. Ferritin and serum iron concentrations and the transferrin saturation percent are convenient, noninvasive methods of assessing the severity of hemosiderosis in callitrichids.[248] Serum liver enzyme values are usually normal. The most frequent macroscopic change is brown discoloration of the liver. Histologically, granular brown-yellow, Prussian blue positive material is present in hepatocytes, Kupffer cells, and bile duct epithelium (Fig. 14.2). Iron overload may be associated with increased susceptibility to infections. In common marmosets, there is a correlation between dietary iron, the severity of hemosiderosis, and early death.[183] Hepatocellular finely granular to globular, green-brown pigment accumulation is also present in familial Dubin–Johnson-like disease in golden lion tamarin (see below).[236] Pigment can be differentiated from hemosiderosis with special stains, which will be iron positive in hemosiderosis but negative in Dubin–Johnson-like disease.

Cholelithiasis has been reported in several NHP species, including black lion and golden lion tamarins, silvery and Wied's marmosets, squirrel monkeys, owl and African green monkeys, and baboons.[134,151,247] Animals can be asymptomatic or develop associated icterus and/or diarrhea, elevated bilirubin may be detected, and liths may be seen radiographically or with ultrasonography

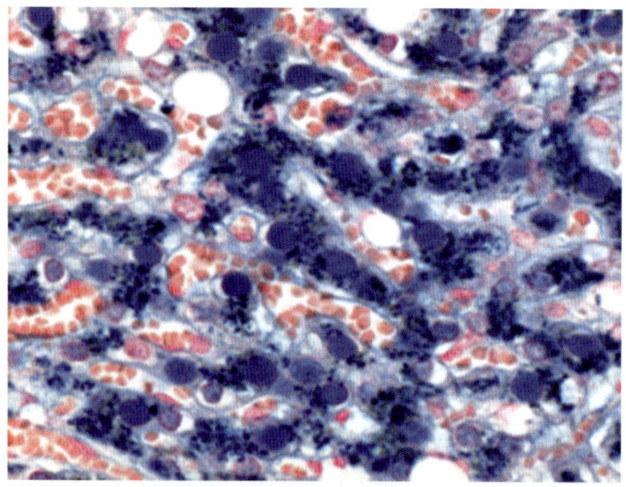

FIGURE 14.2 Severe hemosiderosis in a common marmoset. Hepatocytes and Kupffer cells throughout the liver contain a massive amount of intracytoplasmic and intranuclear iron. Prussian blue.

FIGURE 14.3 Cholelithiasis in a common marmoset. Mild thickening and distension of the gall bladder epithelium and multiple small liths and concretions in the gall bladder.

(depends on lith composition). Gross findings include mild to severe thickening and distension of the gall bladder wall and single to multiple liths in the gall bladder or common bile duct (Fig. 14.3). Histology can include varying degrees of gallbladder epithelium hyperplasia, hepatic pericholangiolar fibrosis, and acute to chronic hyperplastic cholecystitis or choledochitis. Compositional analysis in some cases has identified cystine or cholesterol and protein as components of the liths.

Metabolic

Type 1 and type 2 diabetes mellitus (DM) occur in OWM and NWM species; type 2 DM is most common.[36] Clinical features of diabetes are polyphagia, polydipsia, and polyuria. In most species, type 2 DM is associated with obesity. The clinical features in nonhuman primates (NHP) are similar to those in humans. Before developing overt diabetes, NHPs have a period of obesity-associated insulin resistance with compensatory insulin secretion. Pathological changes in NHP pancreatic islets are also similar to those seen in human diabetics. Initially there is hyperplasia of the islets with abundant insulin production typically followed by beta cell apoptosis, relative alpha cell hyperplasia, and replacement of islets with islet amyloid polypeptide (amylin peptide).[96] Microscopically, islet amyloid appears as homogenous, extracellular eosinophilic material that is congophilic and exhibits apple-green birefringence under polarized light. Islet amyloidosis is most commonly seen in macaques and baboons.[113] Type 1 DM has been reported in cynomolgus monkeys but at a lower frequency than type 2 DM.[269]

Nutritional fatal fasting syndrome, also known as **fatal fatty liver syndrome**, occurs primarily in obese female macaques.[30] It has also been seen in African cercopithecines housed in zoos. It is an acute metabolic disorder associated with anorexia, depression, and substantial (up to 30%) weight loss. The clinicopathologic hallmark of the disease is the development of azotemia. At death, animals still have abundant fat. The most consistently found lesions at necropsy are enlarged yellow or tan livers due to severe hepatocellular lipidosis along with necrosis of abdominal fat.[47] Renal tubular lipidosis, most conspicuous in proximal nephrons, is seen along with necrosis of renal tubular epithelium and pancreatic zymogen granule depletion or acinar cell dissociation.

Amyloidosis is a disease caused by the accumulation of abnormal, insoluble amyloid proteins in tissues. The most common form, **reactive secondary amyloidosis**, is especially prevalent in aged macaques, baboons, and common marmosets.[20,113,163] The disease is characterized by the extracellular deposition of amyloid-A (AA), an amorphous hyaline substance, associated typically with a chronic inflammatory process that induces elevated levels of the amyloidogenic precursor molecule. Amyloid has a uniform ultrastructural appearance with a characteristic protein folding structure and nonbranching fibrils that are deposited in several tissues. In macaques, the disease is strongly associated with chronic enterocolitis and diarrhea. Correlation with indwelling catheters and medical devices, tuberculosis, or retroperitoneal fibromatosis/simian retroviral infection (see below) is also observed.[245] Amyloid deposition frequently occurs in the liver, spleen, adrenal glands, and intestine. Hepatic involvement may result in marked hepatomegaly. Microscopically, deposition of homogenous, fibrillar, acellular eosinophilic material is found in affected organs. In the liver, the process starts in the space of Disse (perisinusoidal space). In the small intestine, which is typically more severely infected than the colon, early deposition is seen in the lamina propria. In the spleen, amyloid is often deposited at the periphery of the periarteriolar lymphoid sheaths. Other organs are also involved in severe cases. The material is birefringent with Congo-red staining and immunoreactive with commercially available anti-AA antibodies.

Congenital/Genetic

A condition similar to **Dubin–Johnson syndrome** in humans was described in two groups of closely related, adult female golden lion tamarins.[236] Affected animals were asymptomatic but had persistent conjugated hyperbilirubinemia, and defective secretion of nonbile acid anions from hepatocytes to bile. The livers of affected animals were dark brown to black with a prominent reticulated pattern. Abundant, finely granular to globular, green-brown, hepatocellular cytoplasmic pigment was seen in the livers of affected animals. The distribution was generally centrilobular and mid-zonal but was diffuse in severe cases. The pigment was positive with Fontana and lipofuscin-ferric ferricyanide reduction stains and negative with Perl's Prussian blue (iron), Hall's (bilirubin), and lipofuscin stains. It appeared as pleomorphic, variably electron-dense, membrane-bound structures with electron microscopy. In humans, this autosomal recessive disorder is due to a defect in the transfer of nonbile acid organic anions from hepatocytes into bile. This results in a retention of conjugated hyperbilirubinemia and the characteristic black appearance of the liver. The mode of inheritance in the tamarins was not established, but a heritable component was suspected through pedigree analysis of two familial groups over three successive generations.

Cystic liver disease is sporadically seen in cotton-top tamarins. The disease is similar to polycystic liver disease in humans (PCLD) and seems to have a genetic etiology. There are no clinical symptoms associated with the disease. Macroscopically, multiple fluid filled cysts of variable size and number are found in the liver (Fig. 14.4A). Histologically, the cysts are lined by cuboidal to flat epithelium, are partly filled with serous fluid, and some are surrounded by pericystic fibrosis (Fig. 14.4B). The cysts originate from irregular, ectatic, cystic, and intrahepatic bile ducts and are called **von Meyenburg complexes, biliary hamartoma**, or **ductal plate malformations**. Von Meyenburg complexes tend to be multiple, frequently paraportal or subcapsular, small nodules, and benign tumor-like malformation in the liver.

Inflammatory Non-Infectious

Wasting marmoset syndrome (WMS) describes several disease entities that are common in captive marmosets and tamarins. To date, no specific cause has been identified.[208] Described conditions include some or all of the following symptoms: severe weight loss, generalized weakness, muscle atrophy, intermittent to chronic diarrhea, and anemia.[282] Alopecia and paralysis of the hind legs, also called cage paralysis, often accompanies the disease. The syndrome is also often associated with **chronic lymphocytic enteritis (CLE)**, **inflammatory bowel disease (IBD)**, or chronic renal disease. CLE is a distinct form of IBD. The etiology of WMS is unknown but dietary factors may play a role. For example, protein deficiency has been implicated because improved body condition has been noted when dietary protein content has been increased in some species,[76] and intestinal mucosal atrophy with decreased villous height, malabsorption, and weight loss has been reported in low protein diets in others, for example, owl monkeys (data from New England Primate Center). The US National Research Council recommends dietary protein content on a dry matter (DM) basis of 15%–22% for adult monkeys, whether NWM or OWM. Dietary allergens, such as gluten-sensitive enteropathy as they occur in man, have been suggested as having a possible role. In humans, gluten-containing diets lead to an increase of antibodies to gliadin (AGA), tissue transglutaminase (tTG), deamidated gliadin (ADGA), and glycoprotein 2 (AGP2A). Animals with increased AGA and tTG titers have more frequent diarrhea on gluten-containing diet.[139,235]

One of the most common and important clinical signs of WMS is chronic diarrhea. This is related to severe maldigestion and malabsorption and leads to chronic protein or vitamin deficiencies, all of which are thought to be important contributing factors in the development of WMS. The most prominent lesions are present in the small intestine and consist of lymphocytic enteritis with villous atrophy and hyperplasia of the glandular epithelium (Fig. 14.5A; eSlide 14.e1). Large numbers of predominantly CD3 lymphocytes infiltrate the lamina propria and enteric epithelium.[162] Other changes can include chronic colitis, hepatocellular hemosiderosis, and severe chronic tubulointerstitial nephritis and mesangioproliferative or membranoproliferative glomerulopathy (Figs. 14.5B, C), all of which can be involved

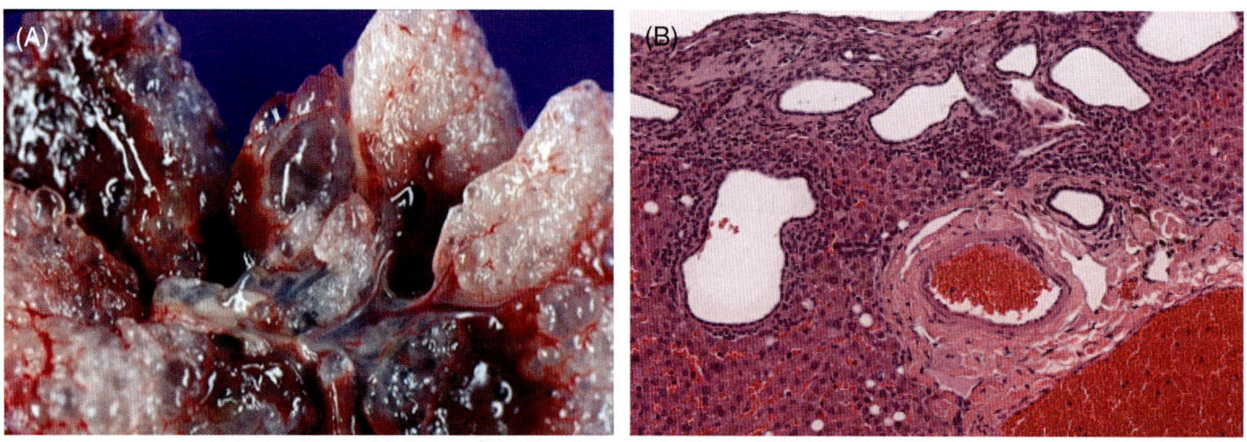

FIGURE 14.4 Cystic liver in a cotton-top tamarin. (A) The liver parenchyma is diffusely expanded by variably sized cysts and mild pericystic fibrosis. (B) The cysts are von Meyenburg complexes consisting of disorganized, ectatic bile ducts, and caniculi with cylindric epithelium.

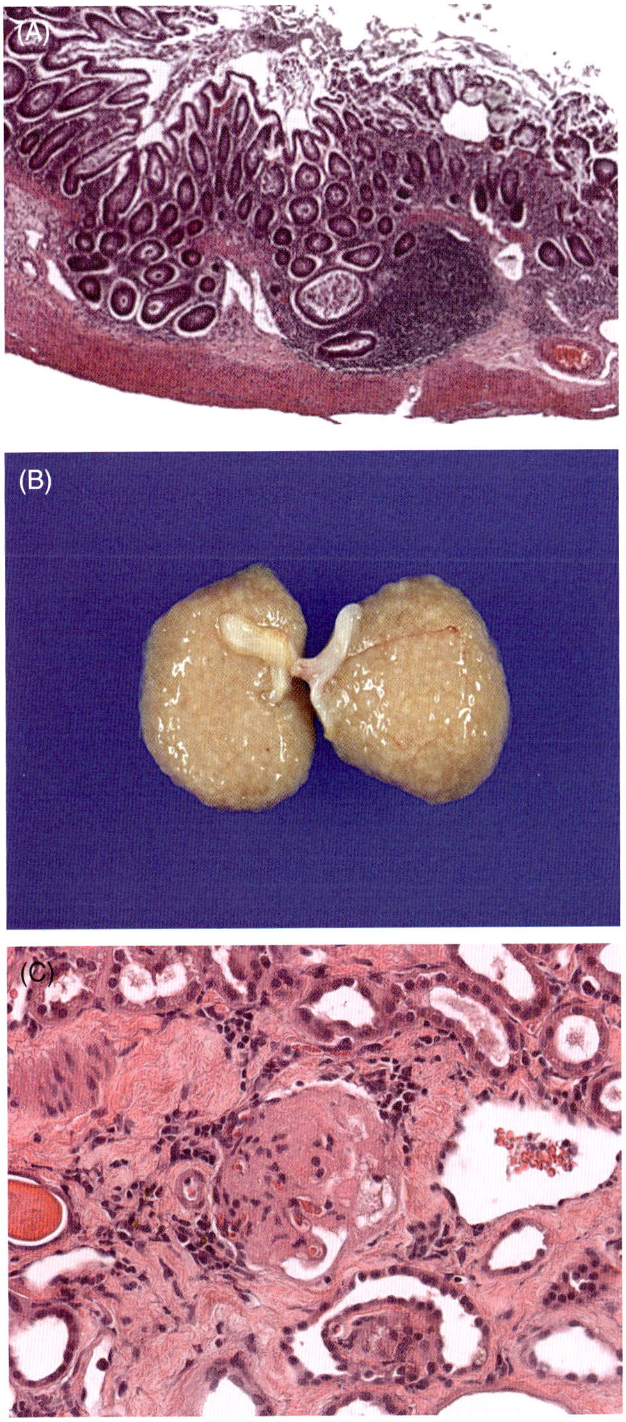

FIGURE 14.5 Wasting marmoset syndrome in a common marmoset. Common lesions occur in the intestine and kidneys. (A) Lymphocytic enteritis with dilatation of crypts and crypt abscesses. The mucosa is infiltrated with lymphocytes; the altered crypts are filled with cell detritus, fibrin, and inflammatory cells. (B) The kidneys are bilaterally enlarged, discolored pale tan, and have an irregular granulated surface consistent with chronic progressive nephropathy. (C) Altered kidneys with severe fibrosis and sclerosis of glomeruli and the interstitium (see eSlide 14.e1).

in disease progression.[24] The symptoms of WMS may differ slightly from colony to colony depending on environmental factors, gastrointestinal infections, and the genetic background of the animals. Whenever possible, causative factors should be investigated to clearly define the underlying cause/s of disease. In general, the term WMS should be avoided because it lacks specificity.

Chronic colitis occurs in captive macaques and baboons. The pathogenesis is complex and involves repeated enteric infections, gut microbiome dysbiosis, dysregulation of mucosal defenses, environmental stress, and possible hypersensitivity to dietary antigens. Clinical signs are recurrent to persistent liquid diarrhea leading to severe dehydration and weight loss. The cecum and ascending colon are often most severely affected. Microscopically, lymphoplasmacytic colitis occurs with numerous crypt abscesses, and herniation of glands into the submucosa, especially in areas of gut-associated lymphoid tissue may also be seen. Erosions and ulcers may be seen but to a lesser extent. The ulcerative form closely mimics chronic colitis in humans.[221]

A wasting-type syndrome in colobus, similar to **sprue** in humans, has been described in colobus. The condition is the result of an inability of the intestines to absorb nutrients. This leads to diarrhea, hypoproteinemia, and the subsequent development of effusions and edema including colonic mural edema. Histologically there is blunting and fusion of intestinal villi. The disease can be progressive and fatal. Reformulation of commercial biscuits to include more fiber and protein and less cereal, especially wheat, as well as feeding browse, seems to have decreased the occurrence of this condition in animals under managed care.

Endometriosis is characterized by the presence of ectopic, most frequently intraabdominal and occasionally lymph node endometrial tissue that undergoes cyclic changes under the influence of estrogen and progesterone. Implantation of normal endometrial tissue in ectopic locations can occur through retrograde menstruation, lymphatic transport, or stem cell proliferation.[114,121] The disease is the most common reproductive disorder in menstruating OWM, affecting 30% or more of sexually mature female rhesus and cynomolgus macaques in some colonies.[2,284] The incidence increases with age and there may be a genetic predisposition to disease development. The disease is only sporadically seen in other OWM and NWM monkeys.[55] Endometriosis appears as variably sized, single or multiple, soft, red-brown masses adhered to serosal or peritoneal surfaces. Multiple cysts filled with brownish fluid composed of blood components frequently develop at sites of implantation (Figs. 14.6A, B). These so-called chocolate cysts are accompanied by extensive fibrosis, hemosiderosis, and adhesion to adjacent organs. The adhesions may cause compression and subsequent obstruction of the colon or ureters. Foci of endometriosis can occur throughout the

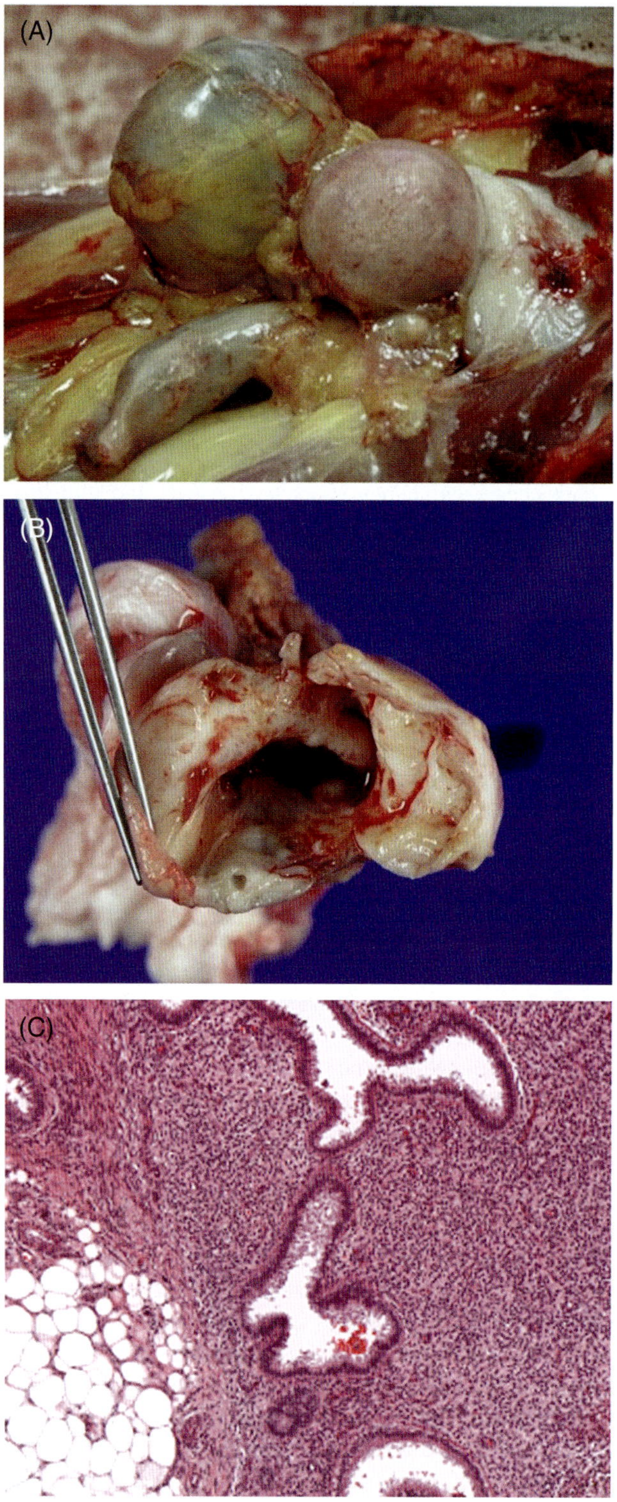

FIGURE 14.6 Endometriosis in a rhesus macaque. (A) Characteristic gross lesions are cysts with hemorrhagic content (chocolate cyst) in the pelvic region adjacent to the uterus (cyst to the left of the uterus). (B) The opened cyst is thick-walled, filled with blood, and lined by endometrial tissue on the inner surface. (C) Endometrial tissue growth on serosal and peritoneal surfaces forms characteristic endometrial glands surrounded by endometrial stroma (see eSlide 14.e2).

abdomen and affect multiple organs but are most frequent in the pelvic cavity. Extraperitoneal endometriosis has been described, but only rarely so, in the lung. Histologically, endometriosis consists of variably shaped and sized glands lined by endometrial epithelium that are surrounded by spindle-cell stroma; glands and stroma occur in varying proportions (Fig. 14.6C; eSlide 14.e2). Clinical signs depend on the location and extent of the lesions. Infertility is a frequent consequence when endometrial tissue obstructs the oviduct or uterus, or leads to extensive adhesions and obstruction of the pelvic organs.

Neoplastic

Colonic adenocarcinoma is common in cotton-top tamarins.[40] Clinical signs include diarrhea, weight loss, intestinal obstruction, and palpable intraabdominal masses. In some cases, neoplastic transformation appears to develop as a sequela to chronic or repeated episodes of colitis. Histologically, the tumors are typically anaplastic, mucinous adenocarcinomas that infiltrate the colonic mucosa transmurally (Fig. 14.7). Neoplastic glands are composed of typical signet ring cells that contain cytoplasmic mucin globules and a centrically placed nucleus. The neoplastic cells are cytokeratin positive and produce mucin, which is deposited within dilated ducts and glands and can be highlighted with PAS-reaction and Alcian blue staining. Lymphatic invasion and dissemination to regional lymph nodes are common.[155] The disease is only observed in captivity, suggesting a link to environmental or other health or husbandry-related factors in managed care settings.[275] In contrast to tamarins, intestinal adenocarcinomas in marmosets tend to arise at the duodenal/jejunal junction producing a "napkin ring" like constriction due to scirrhous response. Intestinal adenocarcinomas are the most common

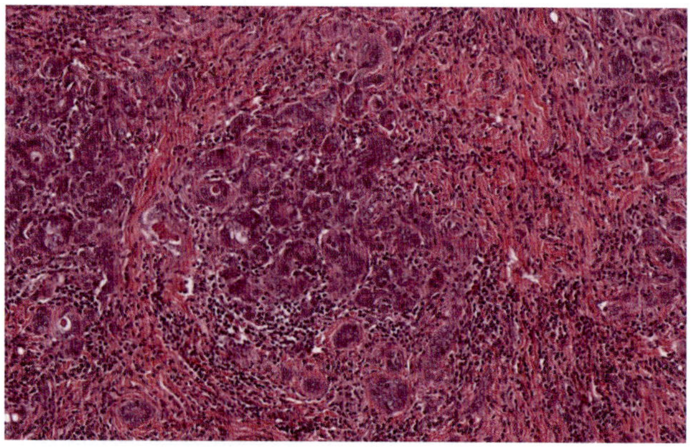

FIGURE 14.7 Colonic carcinoma in a rhesus monkey. Tumors are composed of transmural, epithelial cell proliferations within the lamina muscularis. Neoplastic epithelial cells are cuboidal with large, round vesicular nuclei, and moderate eosinophilic cytoplasm. Mitoses can be common and tumor nodules can be surrounded by mixed inflammatory infiltrate.

cause of neoplasia-related morbidity and mortality in aged common marmosets.[182] Intestinal carcinomas are the most frequent neoplasia in older rhesus macaques in whom they tend to occur in the cecum, ileocecal junction and transverse colon.[242] Chronic enteritis and colitis may play a role in pathogenesis and *Helicobacter* spp., which adhere to enterocyte apical cytoplasm and upregulate proinflammatory cytokines, are being investigated as a possible cofactor.[149]

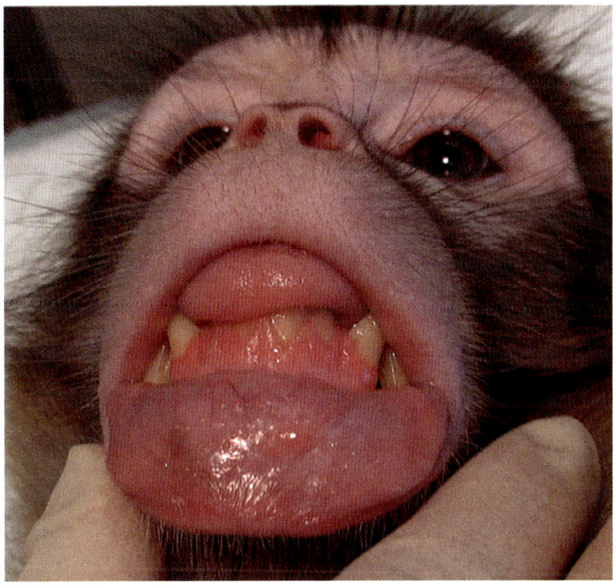

FIGURE 14.8 Herpes B infection in a rhesus macaque. Characteristic vesicles on the oral mucosa and lips.

INFECTIOUS DISEASES

NWM and OWM species are highly susceptible to a variety of viral infections. Some exist as latent or subclinical infections in their natural hosts but can cause severe disease or death when interspecies transmission occurs. The alphaherpesviruses, Macacine herpesvirus 1 (B-virus) and Herpes simplex virus (Human herpesvirus 1) are zoonotic and important in both NHP and human health as infection can lead to severe fatal disease in the accidental hosts. Descriptions of important herpes infections (Table 14.e4) and several other notable viral infections are below; several others [e.g., rabies virus, Ebola (ebolaviruses), Dengue virus, West Nile virus, virus-associated genital ulcers, venereal papillomas, macaque parvovirus, canine distemper virus] are listed in Table 14.e5.

DNA Viruses

Alphaherpesviruses

Several important species-specific, human-derived, and zoonotic herpesviruses infect nonhuman primates. Arguably one of the most important in OWM is **B virus (BV) (syn. Macacine herpesvirus 1, Herpesvirus simiae, herpesvirus B, monkey B virus, herpes B, Cercopithecine herpesvirus 1)**. BV is a zoonotic alphaherpesvirus that is enzootic in all species of Asian macaques. According to current knowledge, long-tailed macaques from the Mauritius Islands appear to be free of the agent. There is a high rate of seropositivity (70%–100% by adulthood) in conventionally reared, captive macaques. BV causes lifelong persistent infection. Virus is latent in the trigeminal and lumbosacral ganglia.[195] Intermittent reactivation and virus shedding may occur during periods of stress. Virus is shed in oral and genital secretions. Transmission occurs through bites, scratches, and copulation (venereal transmission). The clinical signs are analogous to herpes simplex virus (HSV) infection in humans. Disease in macaques is usually mild or asymptomatic. Lesions include the formation of vesicles and ulcers on the oral mucosa and lips (Fig. 14.8). They can also occasionally develop on the conjunctiva. Disseminated infections are rare but have been reported in young or debilitated animals. The mucosal alterations are characterized by ballooning degeneration of epithelial cells with progression to vesicle formation. The presence of multinucleated syncytial giant cells with typical eosinophilic intranuclear viral inclusion bodies on the edge of areas of necrosis is considered pathognomonic. BV can disseminate to the liver, lung, central nervous system, and other organs leading to severe necrotizing inflammation. Interspecies transmission can cause fatal disseminated disease. Owl monkeys, marmosets, African green monkeys, Barbary macaques, bonnet monkeys, gibbons, and DeBrazzás monkeys are reported to be susceptible.[157,261] An asymptomatic course of interspecies transmission has been described in capuchin monkeys housed together with macaques.[52] B virus is a zoonotic disease and more than 40 fatal human cases have been described since the virus was first described in 1932.[70] Humans can be infected by bites or scratches, mucous membrane contact with splashed body fluids, or during laboratory work while handling monkey cell culture or blood components. Infected humans can develop lethal encephalomyelitis. Facilities that house macaque species should have a BV prevention and control plan.[212] For BV detection, antigen (e.g., PCR to detect viral DNA) or antibody (e.g., ELISA/Western blot, multiplex Colony Surveillance Assays such as CSA, Intuitive Biosciences, Madison, WI) based techniques are used.[130] Virus isolation is also possible, but requires access to BSL3 or BSL4 laboratories. The results of antibody-based diagnostic tests need to be interpreted with care. The generation of antibodies in infected animals can be delayed and antibody levels can fall below the limit of detection, either of which can produce false negative results. Cross-reactivity to antibodies raised against HSV is also possible and can produce false positive results in antibody-based assays. Important differential diagnoses are other herpes infections, including HSV (see next).

Human **herpes simplex virus (HSV 1 and 2)** can infect and cause disease in both OWM and NWM. Infection in OWM most often causes mild, localized mucocutaneous lesions. This differs significantly from infection in NWM, including marmosets, howler monkeys, titi monkeys, and tamarins, who are highly susceptible to and often develop severe, often fatal, disease.[22,126,146] Transmission from humans to monkeys can subsequently result in monkey-to-monkey transmission leading to epizootics with high morbidity and mortality. Lesions may be local or generalized with severe mucosal involvement. In NWM, infection often presents initially as oral, lingual (Fig. 14.9A), and genital vesicles and ulcers and is followed by the development of necrotizing meningoencephalitis and multiorgan necrosis if disseminated disease occurs.[102,177] Histopathologic changes consist of mild to severe vesicle formation and ulceration of oral, lingual, and genital mucous membrane squamous epithelia. The cells undergo ballooning degeneration and acantholysis. Necrotic plaques may extend from the submucosa to the mucosal surface. Multinucleated syncytia and intranuclear, eosinophilic inclusion bodies, which are best found at the ulcer margins, are histopathologic hallmarks of the infection (Fig. 14.9B; eSlide 14.e3). Nonsuppurative meningoencephalitis occurs after the acute phase of infection. Strict separation of NWM from infected humans with acute herpetic lesions and use of personal protective equipment is required to prevent interspecies transmission.

The **simian varicella virus (SVV; syn. Cercopithecine herpesvirus 9)** causes natural erythematous disease in OWM. SVV belongs to a group of closely related viruses including the **Liverpool vervet virus, Medical Lake macaque virus**, and **Delta herpesvirus**; all are able to induce a similar exanthemous herpetic rash. Infections occur naturally in a number of OWM species. Especially macaques are responsible for sporadic epizootics. Infection with SVV is lifelong and periodic reactivation may occur. The virus is genetically and antigenically similar to human herpesvirus 3, commonly known as varicella-zoster virus, which causes chicken pox. Transmission occurs by close contact and via the respiratory route. The primary lesion is a diffuse inguinal rash that spreads centripetally, progressing to vesiculoulcerative dermatitis of the trunk, face, and extremities. Vesicles may extend to mucocutaneous junctions and the oral mucosa but usually spare the soles and palms.[115] The lesions are pruritic and there is often evidence of self-excoriation. Cutaneous lesions progress from papule to vesicle to crust. Lesions of all stages may be present at the same time. Microscopically, cutaneous lesions are vesicles containing variably sized acantholytic epithelial cells with intranuclear eosinophilic viral inclusions, consistent with herpes virus. Syncytia are seldom seen. The infection can disseminate, especially under immunosuppressive conditions, and cause widespread necrosis in various organs.[96] The disease must be differentiated from B virus infection (see above), which normally does not induce extensive cutaneous involvement, and morbillivirus infection (measles; see below), in which the rash is not vesicular. Transmission of SVV to humans has not been documented.

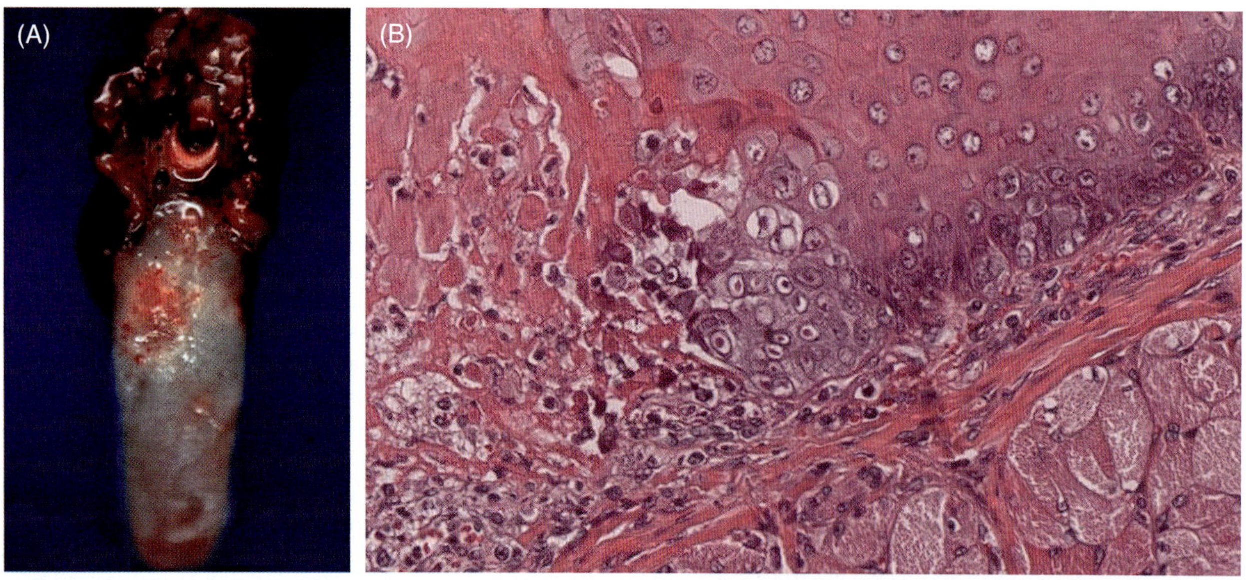

FIGURE 14.9 *Herpes simplex* infection in the tongue of a common marmoset. (A) Multifocal to coalescing, variably sized erosions, and ulcers are present on the dorsal surface of the tongue. (B) Characteristic lesions include necrotizing glossitis with ballooning degeneration of epithelial cells and intranuclear eosinophilic inclusion bodies at margins of ulcers (see eSlide 14.e3).

Betaherpesviruses

Simian cytomegaloviruses (CMV) are described in a variety of NWM and OWM species. Infection in the natural adult host is usually latent and asymptomatic. Clinical signs are most often seen in stressed or immunodeficient individuals. CMV is a common opportunistic infection in simian immunodeficiency virus (SIV) or simian retrovirus (SRV) infected macaques (see below). After reactivation/recrudescence, generalized infection may occur and result in necrotizing meningitis, neuritis, enterocolitis, and interstitial pneumonia. Less frequently, infections are associated with necrotizing sialoadenitis, arteritis, hepatitis, or orchitis. In the intestinal tract, gross lesions consist of raised hyperemic papules (Fig. 14.10A). This corresponds histologically to neutrophilic inflammation with intralesional vascular and stromal cytomegalic cells. The histopathologic hallmarks are karyomegaly and large basophilic intranuclear inclusions (sometimes called "owl's eye" cells) (Fig. 14.10B). Also occasionally seen are poorly defined, eosinophilic granular intracytoplasmatic inclusion bodies within mesenchymal cells or epithelia.[11] There is no known zoonotic potential.

Gammaherpesviruses

The **rhesus lymphocryptovirus (RhLCV)** or **macacine herpesvirus 4** (also known as **cercopithecine herpesvirus 15**) is a gammaherpesvirus in the *Lymphocryptovirus* genus. More than 50 distinct simian lymphocryptoviruses have been isolated from OWM and NWM species.[67] All are related genetically to the type strain human herpesvirus 4, the Epstein Barr virus (EBV). Like EBV, distinct lineages of RhLCV have been identified. They differ in the sequence of their EBNA genes. Epizootiology of lymphocryptovirus infection in OWM closely parallels human infection, which occurs in young animals and is usually asymptomatic. By the age of 2 years, all animals harbor the virus leading to a 100% prevalence of the virus in macaque colonies.[81] In immunodeficient animals, especially SIV-infected rhesus macaques, RhLCV infection may progress to malignant lymphoma, which is similar to the development of non-Hodgkin lymphoma in HIV-infected AIDS patients. Most common are B cell lymphomas that usually develop in extranodal locations such as the gastrointestinal tract, the central nervous system, the nasal cavity, or in the periorbital tissue.[172] These lymphomas are classified as centroblastic, immunoblastic, large cell, or Burkitt-like lymphomas.[127] A proliferative epidermal lesion termed **oral hairy leukoplakia** commonly observed in the oral mucosa and esophagus can be associated with RhLCV infection. Oral hairy leukoplakia is characterized by pale, swollen acanthocytes that produce a white raised patch or plaque/s on mucosal surfaces in the oral cavity.[13] Acidophilic Cowdry type A intranuclear herpesviral inclusions are present in cells in the middle and superficial mucosal epithelial layers. Another important member of this group is the **marmoset lymphocryptovirus** or **Callitrichine herpesvirus 3**, which can induce B-cell lymphoma of the intestinal tract and associated lymph nodes.[22,46] Outbreaks and individual cases have occurred in zoo and laboratory settings. There is no known zoonotic potential.

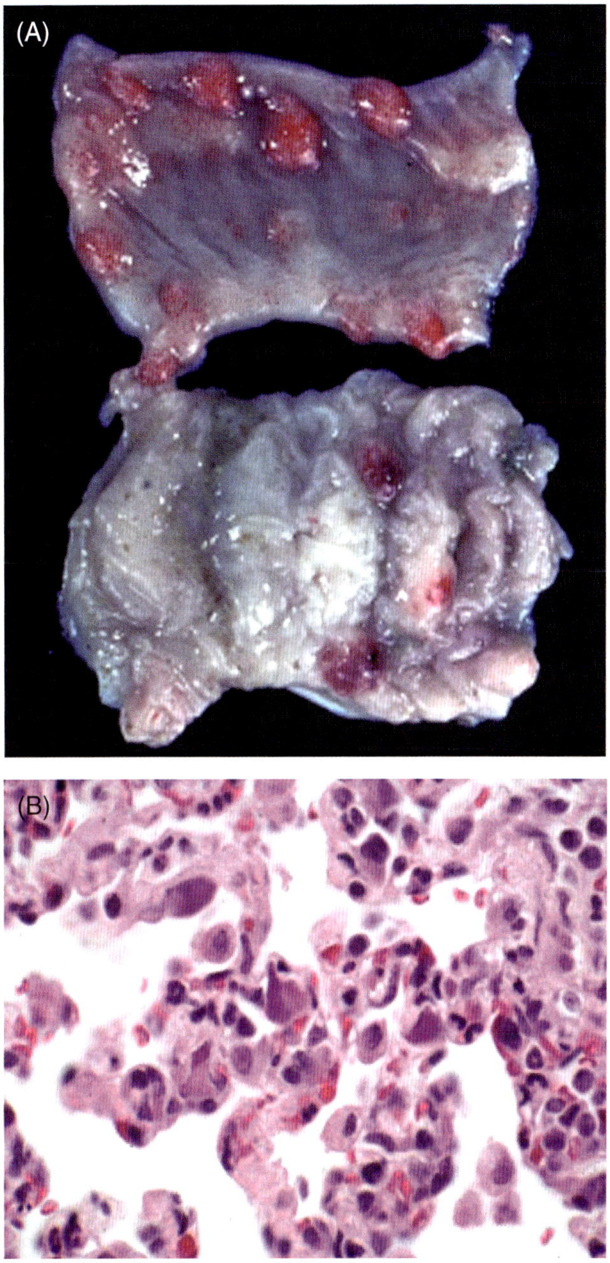

FIGURE 14.10 Cytomegalovirus infection in the intestine and lung of a rhesus macaque. (A) Enterocolitis of the small intestine (above) and colon (below) with prominent raised reddened papules in the mucosa. (B) Lung lesion with large, basophilic intranuclear "owl's eye" inclusions in infected pneumocytes accompanied by mild interstitial mononuclear cell inflammation.

Herpesvirus saimiri [HVS; syn. Saimiriine herpesvirus 2, Herpes saimiri (HVS)] is a T-lymphotropic, gamma-2 herpesvirus that is closely related to human herpesvirus 8 (HHV-8), which is also known as Kaposi's sarcoma-associated herpesvirus (KSHV). HVS is an enzootic infection in squirrel monkeys in whom it causes no recognized disease.[75] Serological surveys have shown that more than 95% of adult squirrel monkeys are infected. This differs from acute lymphoproliferative disease, which has been described in other susceptible species such as tamarins, owl monkeys, and marmosets.[117] The virus in these species persists in lymphoid cells and leads to abnormal lymphoid proliferation with hyperplastic or neoplastic infiltrates in multiple organs including the gastrointestinal tract, spleen, liver, and kidney; associated gross lesions include splenomegaly and/or peripheral and visceral lymphadenomegaly. Neoplastic cells are primarily CD3- and CD8-positive T lymphocytes that are often present along blood vessels. Leukemia may be seen in blood smears. HVS has been naturally transmitted to callitrichids in zoo settings. Therefore, squirrel monkeys should not be held in mixed exhibits with marmosets and tamarins. There is no known zoonotic potential.

Rhesus rhadinovirus (RRV) or **Macacine herpesvirus 5**, and other macaque rhadinoviruses including **retroperitoneal fibromatosis-associated herpesvirus (RFHV)**, are closely related to HHV-8/KSHV (see above). RRV infection in rhesus macaques is usually asymptomatic and occurs early in life. Seroprevalence in surveys of infected colonies is often greater than 90%. RFHV infections are frequently associated with natural or experimental simian retrovirus infection immunosuppression (see below) and may lead to the development of mesenchymal neoplasia involving the gastrointestinal tract and mesentery [retroperitoneal fibromatosis (RF)] or the skin [subcutaneous fibromatosis (SF)]. Tumors share histopathological similarities with Kaposi's sarcoma[8] and present as multiple, firm pale nodular masses in the abdomen or abdominal viscera, or the subcutis or myofascial planes, respectively. In the gastrointestinal tract, early lesions of RF are often present at the ileocecal junction and tumor subsequently infiltrates the mesentery, mesenteric lymph nodes, and gastrointestinal tract causing expansion and replacement of normal tissue. Neoplastic cells are spindle-shaped and arranged in variably thick intersecting fascicles with prominent and varying degrees of collagen production. Tumor cells are vimentin, desmin, and smooth muscle actin positive.[263]

Kaposi's sarcoma herpesvirus [KSHV; syn. Human herpesvirus (HHV-8)] is a human herpesvirus that causes Kaposi's sarcoma. The tumor is an endothelial neoplasm of the dermis, oral cavity, and intestinal organs. A KSHV-like tumor was described in a mantled guereza that developed circumscribed, flattened to fissured, nodular masses on the upper and lower mucosal surfaces of the gingiva and lips.[95] Histologically, the masses consisted of spindle cells in a collagen-rich, fibrovascular stroma. Similar fibrovascular tissue was identified in mandibular and axillary lymph nodes, and endothelial-lined cyst-like spaces were seen adjacent to one of the mandibular nodes. Most of the spindle cells were IHC-positive for CD31 (endothelial cells) and approximately 50% were positive with antibodies to KSHV latent nuclear antigen. A novel rhadinovirus, **Colobine gammaherpesvirus 1 (CbGHV1)**, that had high sequence homology to KSHV was identified in lesional tissue and suspected to have been involved in tumorigenesis.

Other DNA Viruses

Poxviruses are large, brick to ovoid-shaped enveloped DNA viruses. Four members of the pox virus family belonging to the *Orthopoxvirus* and *Yatapoxvirus* genera can cause disease in nonhuman primates. Outbreaks with **mpox** (previously **monkeypox**) virus, **cowpox virus**, **Yaba monkey tumor virus**, and **Yaba-like disease virus** are infrequently reported in NWM and OWM species. In addition, nonhuman primates are susceptible to **variola virus** and **vaccinia virus** and represent important animal models for human infection.[120] All pox virus infections have a zoonotic potential.

Mpox virus is a World Organization for Animal Health (WOAH) listed, notifiable, zoonotic disease caused by an orthopoxvirus that is closely related to human smallpox virus. Human infections with mpox virus occur most often in Africa, but infections have been diagnosed sporadically in other countries. The virus has received attention due to a global outbreak that was identified in early 2022 and by September 2023 had caused almost 100,000 laboratory-confirmed cases and more than 150 deaths; the highest number of cases have been in the North and South America and Europe (https://www.who.int/docs/default-source/coronaviruse/situation-reports/20230919_mpox__external-sitrep-28.pdf?sfvrsn=2654b9d2_5&download=true; accessed 21 September 2023). Human-to-human infection is the most common, though zoonotic transmission continues to be of concern. Mpox is the most important poxviral disease in OWM and NWM monkeys.[5] OWM species seem to have a higher risk of infection than NWM. Nevertheless, outbreaks can occur in NWM and infection can be zoonotic.[202] Transmission occurs by aerosols or direct contact and leads to generalized disease characterized by circular maculopapular to nodular lesions of the skin that undergo vesiculation and pustule formation. Central collapse and umbilication of vesicles and pustules leads to classic "pock" lesion. Pustules crust over and drop off, leaving small scars. A rash develops and spreads centripetally starting at the face and extremities (Figs. 14.11A, B). Histologically, skin lesions are initially composed of proliferative acanthosis with ballooning degeneration of keratinocytes and spongiosis that progress to

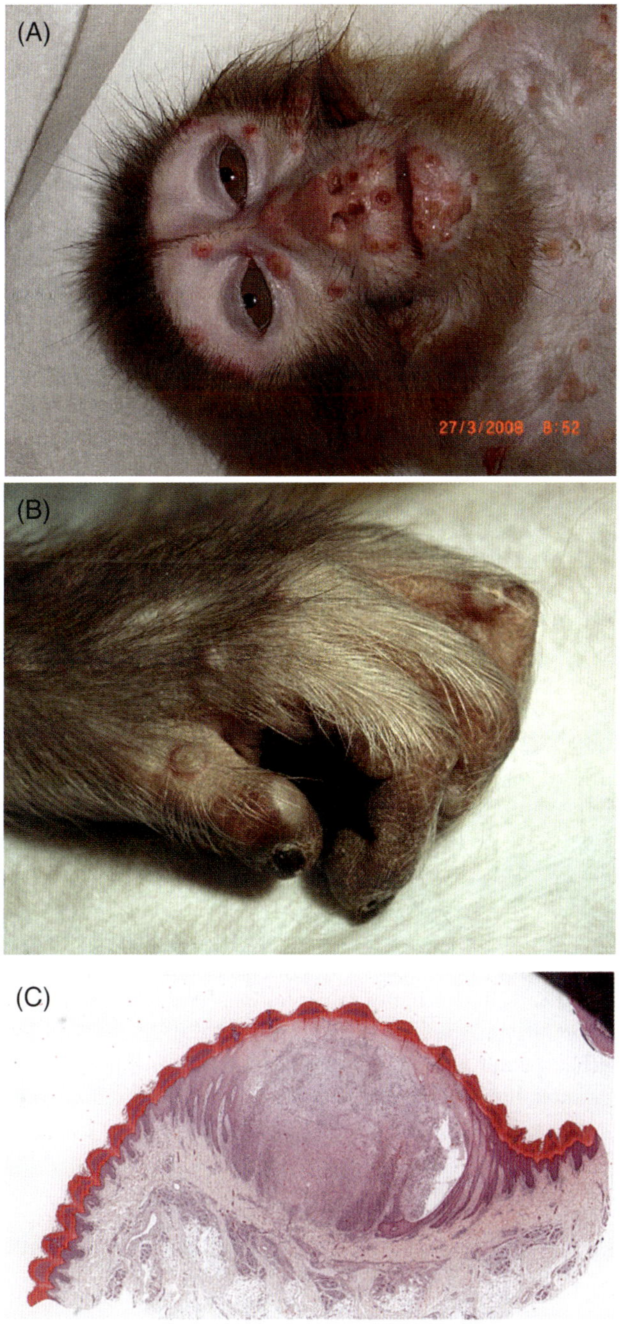

FIGURE 14.11 Mpox virus infection in the skin of a rhesus macaque. (A) Skin rash with classic umbilicated pox lesions on the face and (B) multiple, exophytic, papular nodules with a light tan, occasionally umbilicated center, and a reddened base on the hand and fingers are characteristic of pox lesions. (C) The epidermis is hyperplastic and markedly, focally expanded by vesicular to pustular inflammation.

vesicle formation within a few days (Fig. 14.11C). Intracytoplasmic eosinophilic inclusions are sparse. Respiratory tract involvement is characterized by hemorrhagic necrosis. After several weeks, most affected animals recover spontaneously. Severe cases with respiratory tract involvement may be fatal.

Cowpox virus, a zoonotic orthopoxvirus that is enzootic in rodents across Europe and can be found as far east as Central Asia and western Russia, induces similar, but generally less severe, lesions to those of mpox infection (see above). Rodents are the reservoir host and do not develop disease. Zoonotic transmission is through handling infected rodents (especially voles) or domestic cats that have been infected by rodents. NWM, especially callitrichids, seem to be highly susceptible to infection. Infection leads to vesicular, hemorrhagic dermal lesions that typically develop on the face, scrotum/labia (Fig. 14.12A), and palmar or plantar surfaces. Histologically, skin lesions show epidermal acanthosis, vesiculation, hemorrhage, and necrosis with typical pox-like intracytoplasmic inclusions (Fig. 14.12B; eSlide 14.e4). The disease course is often fatal due to systemic spread and severe inflammation followed by uncontrolled excessive release of inflammatory mediators like cytokines (cytokine storm) which leads to death.[173,174]

Poxviruses belonging to the genus ***Yatapoxvirus***, which includes **Yaba monkey tumor virus** and **Yaba-like disease virus** (previously **Tanapox virus**), are causative agents of cutaneous lesions in humans and monkeys. Transmission to humans has occurred in monkey handlers through accidental needle sticks. A unique feature among poxviral infections is that **Yaba monkey tumor virus** affects histiocytes rather than epithelial cells and induces histiocyte proliferation and the production of subcutaneous tumor-like masses. Natural infections with the Yaba monkey tumor virus can occur in rhesus monkeys, baboons, and vervet monkeys[28,63]; transmission to humans has (rarely) been reported. The disease is characterized by single or low numbers of rapidly growing subcutaneous nodules typically on the head and limbs including the digits.[31] Histological hallmarks are large pleomorphic histiocytes that form nonencapsulated, infiltrative masses. Large eosinophilic, intrahistiocytic, and intracytoplasmic inclusions are a key finding. Regression, which usually occurs over the course of a few months, is associated with erosion, ulceration, and formation of multinucleated giant cells. The infection is not observed in NWM. The **Yaba-like disease virus** is a contagious skin disease in macaques and humans. Infection produces multiple crusted macules on the face and arms that heal in 3 to 4 weeks and resemble mpox virus infection that is arrested at the papular stage, though umbilication can occur. Histologically, lesions are characterized by epidermal proliferation and ballooning degeneration. Swollen epithelial cells contain eosinophilic intracytoplasmic inclusion bodies.[99]

Hepatitis B virus (HBV) belongs to the genus *Orthohepadnavirus* and is a major cause of human hepatitis. Transmission occurs via infected blood, saliva, and semen. Mother-to-infant transmission may play a critical role in human infections. HBV can cause persistent infection leading to chronic hepatitis and hepatocellular carcinomas.

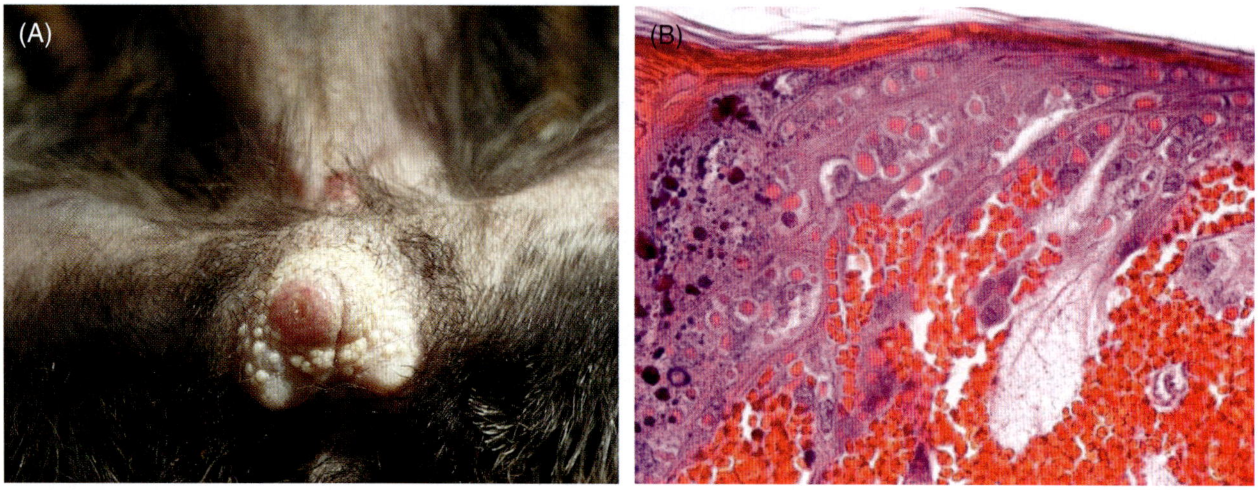

FIGURE 14.12 Cowpox infection in skin and external genitalia of a female marmoset. (A) The skin of the labia has a round, focal area of ulceration, and red discoloration with central umbilication typical of pox virus infection. (B) Hallmarks of infection are acanthosis, acantholysis, and intraepidermal vesicles with associated hemorrhage. Eosinophilic intracytoplasmic, globular, inclusion bodies (Guarnieri bodies) are a typical finding (see eSlide 14.e4).

Human HBV infection can be transmitted to macaques and chimpanzees. Cynomolgus monkeys are highly susceptible.[138] Each group of nonhuman primates appears to have a distinct strain of hepatitis B virus.[215] Natural infections with species-specific strains have been identified in orangutans, gibbons, gorillas, chimpanzees, macaques, baboons, and woolly monkeys. Pathologic findings in nonhuman primates are comparable to those that develop in humans and consist of nonsuppurative, periportal inflammation with lymphocytic infiltration. Progression to cirrhosis and hepatocellular carcinoma, as occurs in humans, has not been documented in naturally infected nonhuman primates. The zoonotic potential of monkey-specific strains for humans is unknown. The use of a serologic panel to detect HBV surface antigen, HBV surface antibody, HBV core antibody, and IgM against core antibody is recommended in the detection of and differentiation between acute and chronic infections or immunity.

RNA Viruses

Yellow fever (YF) is a flaviviral infection caused by the **yellow fever virus (YFV)**. The Flaviviridae are small, positive-stranded, nonsegmented, enveloped RNA viruses of arthropods. In addition to YFV, the family includes several other important mosquito-vectored viruses (e.g., Dengue fever, Japanese encephalitis, West Nile virus, and Zika viruses). YFV can cause fatal hepatitis in humans and is one of the most important viral infections of free-living NWM. Originating in West Africa, the virus is now endemic in Sub-Saharan Africa and South America. Mosquitoes of the genera *Aedes* and canopy mosquitos in the *Sabethes* and *Haemagogus* genera are the primary vectors. The virus has an incubation period of 3–6 days with initial replication in lymph nodes and hepatocytes; peak viremia in rhesus monkeys is 3–7 days and higher viral loads are related to a worse prognosis and death.[71]

YF has two established cycles: urban and sylvatic (or jungle). The urban cycle exists in African and Asian monkeys, African monkeys are relatively resistant to clinical disease. However, Asian monkeys and NWM are highly susceptible. Epidemic cycles of the sylvatic form have been described in the Americas. In the Americas, the cycle is maintained in rainforest biomes that are endemic in the Amazon rainforest, with sporadic circulation in almost all of South America. Outbreaks of the disease in wild monkeys are related to human infections. Because of this, Brazil has a national surveillance program to detect outbreaks in primates. Detection in the primates informs public health and the implementation of preventive vaccination in humans. As an example, in response to a large outbreak of YF in wild monkeys in new areas of Brazil in 2016–17, vaccination coverage and protection were provided to more than 20 million people.[53]

Devastating outbreaks, which can cause population-level effects, have been reported in wild NWM.[188] The disease is especially severe in howler monkeys, owl monkeys, titis, and tamarins, while marmosets, spider, woolly, squirrel, and capuchin monkeys suffer lower mortality.[56,224] The most characteristic gross lesion is yellow discoloration of the liver with or without pinpointing to larger areas of red discoloration and reduction in the size of the liver (due to necrosis; see later). Icterus, pulmonary hemorrhage, and peritoneal effusion may also be seen. Classical YF lesions are characterized histology by marked diffuse, lytic hepatocellular necrosis with acidophilic, apoptotic (Councilman-Rocha Lima) bodies. Other changes can include microvesicular hepatocellular vacuolation (lipid type) and less commonly or consistently, multifocal hepatic hemorrhage; mild multifocal neutrophilic and/or mononuclear cell infiltrates, splenic lymphoid depletion, and renal tubular necrosis.[73,144] The pattern of gross and histologic lesions follows viral infection in which initial replication in macrophages is followed by viremia with virus dissemination to many organs and tissues; in the liver, Kupffer cell infection and necrosis precede hepatocellular necrosis.

Antemortem diagnosis of YF is based on detection of the virus with quantitative reverse transcription PCR and/or isolation of the virus in cell culture. Characteristic histologic lesions coupled with specific immunohistochemical assays are also diagnostic (Figs. 14.13A, B). An exception is in marmosets and tamarins who may not develop characteristic lesions and in whom PCR is therefore the most reliable diagnostic tool.

Kyasanur forest disease (KFD), also called **monkey fever** and caused by **Kyasanur Forest disease virus (KFDV)**, is a tickborne, flaviviral hemorrhagic disease that is an important cause of epizootic mortality in black-faced (Hanuman) langurs and bonnet macaques in southern India.[110] Deaths in monkeys often presage human outbreaks, which have become an almost annual occurrence; an effective vaccine is available for use in humans. In experimental infections, mortality can be as high as 100% in langurs and 82% in bonnet monkeys, while rhesus macaques become viremic but do not die.[240] Clinical signs include fever, anorexia, and diarrhea. Monkeys examined during outbreaks of KFD have multiorgan hemorrhage. Hemorrhage is not a pronounced feature in experimental infections in langurs or bonnet monkeys. Instead, there is hepatic fatty change and random hepatocellular necrosis. Kupffer cells and macrophages in spleen and lymph nodes contain erythrocytic and leukocytoclastic debris. Encephalitis, mild myocardial necrosis, and myocarditis are seen in some experimentally infected monkeys. Nephrosis is seen in naturally infected monkeys.

Another flavivirus that has recently received attention due to severe developmental disease that it can cause in human fetuses is **Zika virus (ZK)**. The virus has been detected in free-ranging adult *Callithrix* spp. living near urban areas in Brazil where the virus is endemic. No histological lesions or viral antigen was seen in tissues and titers in infected animals were low.[260] In contrast to natural infection, experimental infections in pregnant marmosets inoculated with a Brazilian strain of the virus resulted in abortion and fetal malformations including microcephaly, which has been seen in human infections. Interestingly, dams were asymptomatic after inoculation and during pregnancy.[237]

Callitrichid hepatitis is an acute, fatal epizootic disease of NWM including tamarins and marmosets. It is caused by infection with **lymphocytic choriomeningitis virus (LCMV)**, an arenavirus of mice. Transmission occurs through contact with or ingestion of infected murid rodents. The house mouse is the major reservoir. Clinical signs include dyspnea, weakness, and jaundice. Coagulopathy may be apparent. At necropsy, hepatosplenomegaly (Fig. 14.14A), pleural and pericardial effusions, jaundice, and subcutaneous and intramuscular hemorrhage are characteristic. Histopathologically, multifocal hepatic necrosis with prominent hepatocellular swelling, acidophilic degeneration, and acidophilic body formation is a consistent finding (Fig. 14.14B). Nonsuppurative, lymphoplasmacytic meningitis with perivascular lymphoid cuffs may accompany the liver lesions. Venocentric lymphocytic infiltrates may be

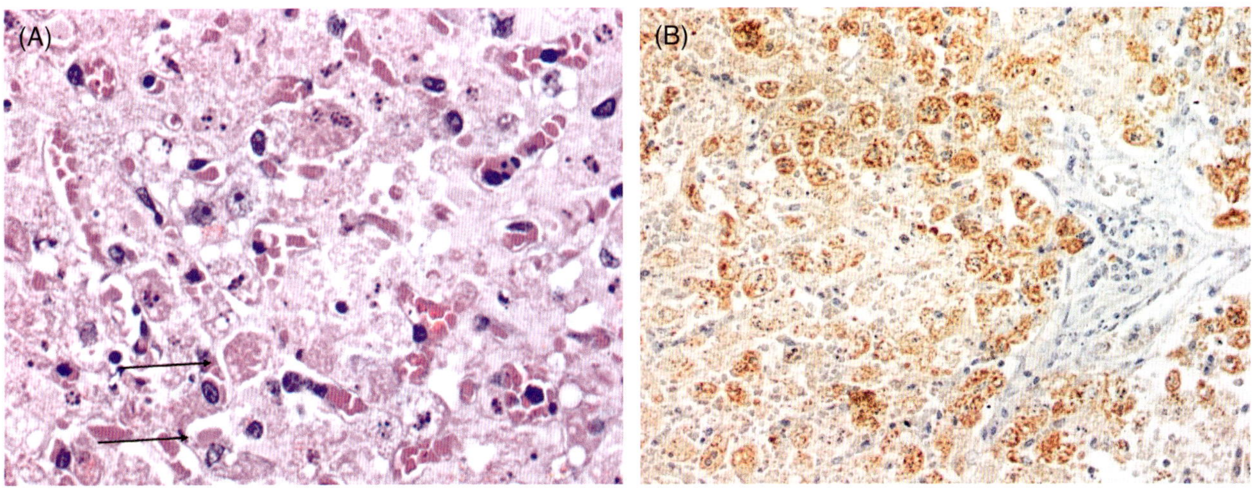

FIGURE 14.13 Yellow fever in *Sapajus* sp. (A) Liver with necrosis and apoptosis of hepatocytes and presence of round eosinophilic bodies *(arrows)*, microvesicular steatosis, note the absence of inflammation. (B) Yellow fever virus antigen in the cytoplasm of hepatocytes detectable by immunohistochemistry (IHC).

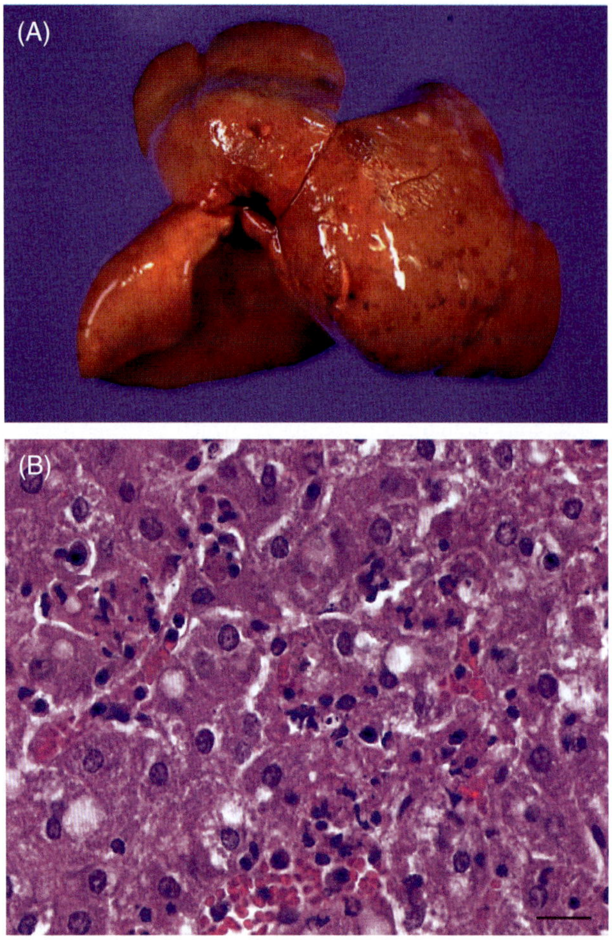

FIGURE 14.14 Callitrichid hepatitis in an emperor tamarin. (A) There is severe hepatomegaly with multifocal hemorrhage. (B) Characteristic histologic findings are diffuse hepatocellular degeneration with lipidosis, Ito cell hyperplasia, isolated, rounded, shrunken, hypereosinophilic, apoptotic hepatocytes (Councilman body formation), and intracytoplasmic acidophilic inclusions.

seen in other organs. Extrahepatic lesions are usually milder than those that develop in the liver.[186] LCMV is a zoonotic agent. Seroconversion of veterinarians in contact with infected marmosets has been described.

Hepatitis A viruses (HAV) are members of the genus *Hepatovirus* in the family *Picornaviridae*. Serologic surveys have documented infection in marmosets, owl monkeys, *Cebus* and howler monkeys, macaques, African green monkeys, and chimpanzees.[226] Though cross-species infections may occur naturally and experimentally, nonhuman primates are hosts to their own strains of hepatoviruses. Transmission occurs by the fecal-oral route. Seroconversion and elevation of liver enzymes are usually the only clinical signs of infection. Infection leads to lifelong immunity and does not lead to chronic hepatitis. Histopathologic findings are periportal and parenchymal, mononuclear inflammatory cell infiltrates, slight hepatocellular degeneration, and Kupffer cell hyperplasia. Nonhuman primate HAVs are potentially zoonotic, and monkeys can become infected with human strains.

Measles virus, a morbillivirus in the *Paramyxoviridae* family, has been described in many OWM and NWM species. Humans are the natural hosts of the virus, and monkeys are highly susceptible to infection and disease development.[8,158] Infected OWM develop maculopapular exanthema and dry or scaly desquamative dermatitis. The rash is most pronounced on the ventral body surface and generally spares the plantar and palmar surfaces of the feet and hands. Typical **Koplik spots**, small white foci rimmed by a red border in the oral mucosa, are infrequently seen. Severe cases may progress to respiratory illness typically manifested in OWM as giant cell interstitial pneumonia. Abortions and stillbirths occur. In NWM the disease may be more severe leading to epizootics with high morbidity and mortality. The characteristic exanthema is lacking and necrotizing gastroenteritis is the predominant lesion. Histologically, multinucleated giant cells in various tissues, including lymph nodes (Warthin-Finkeldey cells), and eosinophilic intracytoplasmic or intranuclear inclusions in epithelial and syncytial cells are hallmarks of (but are not pathognomonic for) measles virus infection (Fig. 14.15). Note that these are rare in the skin lesions. Measles is strongly immunosuppressive, and opportunistic infections (e.g., cytomegalovirus infection, bacterial respiratory infections) are common. Note that the immunosuppressive effects of measles virus infection may result in false negative intradermal tuberculin skin testing after infection.

Simian hemorrhagic fever viruses (SHFV) are a group of highly divergent viruses in the family *Arteriviridae*, genus *Simartevirus*.[140] The natural hosts of these viruses are thought to be African cercopithecines

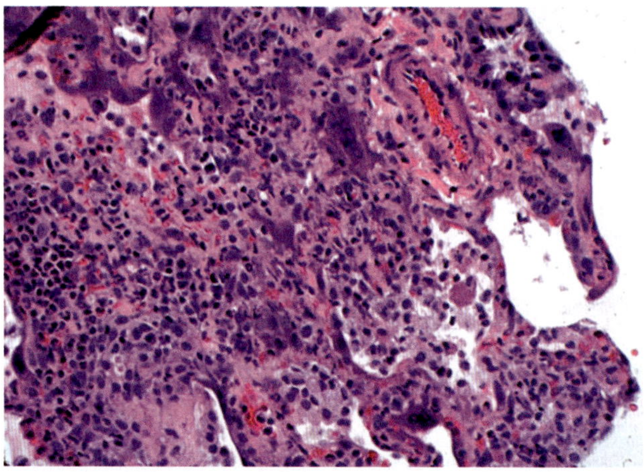

FIGURE 14.15 Measles infection in the lung of a rhesus macaque. Marked interstitial pneumonia is characterized by expansion and infiltration of the interstitium by mononuclear cells and syncytial cell formation.

(e.g., baboons, patas monkeys, African green monkeys) and colobines.[29] Natural transmission in African monkeys occurs during social interactions. Cross-species transmission has led to severe outbreaks of fatal hemorrhagic disease in captive macaque colonies. Strict separation of African and Asian OWM monkeys is essential. Transmission in colony situations is likely iatrogenic through tattooing or reuse of tuberculin needles and from animal to animal by direct contact. In the natural host, disease is usually subclinical or transient and leads to persistent infection. Macaques develop febrile illness with lethargy, anorexia, facial edema, cutaneous petechia, subcutaneous and retrobulbar hemorrhage, melena, and epistaxis. Clinical pathology is indicative of a consumptive coagulopathy with decreased platelets, abnormal clotting times, elevated liver enzymes, and hypoalbuminemia. Hematuria and proteinuria are present. Necropsy findings include splenomegaly, duodenal mucosal hemorrhage beginning abruptly at the pylorus, colonic hemorrhage, and cutaneous/subcutaneous hemorrhages. Histology reveals hemorrhage and evidence of disseminated intravascular coagulopathy (DIC) in multiple organs with fibrin thrombi throughout the spleen. A lack of hepatocellular necrosis distinguishes SHFV from Ebola virus disease (EVD).

Retroviruses

Simian immunodeficiency viruses (SIV) are a group of closely related, zoonotic, lentiviruses that occur naturally in some African primates including guenons, baboons, mangabeys, vervets, chimpanzees, and gorillas. Natural hosts are persistently infected, develop antibody and cellular immune responses, and generally remain asymptomatic for life. In rare cases, infection may progress to **simian acquired immunodeficiency syndrome (SAIDS**; see below) depending on host or environmental factors.[198] In contrast, Asian macaques are highly susceptible to SIV infection and develop fatal immunodeficiency that shares similarities with human immunodeficiency virus (HIV) infected human AIDS (acquired immunodeficiency syndrome) patient disease. SIV is not a naturally occurring infection of wild macaques in their countries of origin. The first epizootics in macaques were likely related to contact with cohoused, naturally infected sooty mangabeys. Once within a group, SIV can be transmitted horizontally in macaques, spreading rapidly within a colony and causing epidemics of immunodeficiency.[161] Thus, mixed species exhibits containing both Asian and African primates should be discouraged. SIV infects CD4+ lymphocytes and macrophages and causes profound and progressive CD4 T-cell decline in macaques. Immunosuppression is followed by lymph node and splenic lymphoid depletion, lymphopenia, opportunistic infections, and the development of (typically B-cell) lymphoma, similar to what occurs in human AIDS. Because of the great similarities to human disease, SIV-infected macaques are widely used as animal models for AIDS pathogenesis and vaccine studies. Experimental transmission is by oral, genital, rectal, or parenteral inoculation. A combination of species specific, viral, and environmental factors influence disease progression.[212]

The initial clinical signs of **SIV-induced AIDS-like disease** in macaques are a macular rash and lymphadenopathy in the first weeks after infection. Opportunistic infections and virus-induced lesions develop in a variety of organ systems.[12] Gross lesions associated with primary infection include marked lymph node and splenic lymphoid follicular and interfollicular hyperplasia that progresses to lymphoid depletion. Furthermore, atrial and pulmonary thrombosis, proliferative arteriopathy, and SIV-induced enteropathy are typical SIV-associated lesions (Figs. 14.16A, B). SIV-induced enteropathy is characterized by CD4 T-cell loss, villous blunting, and atrophy accompanied by crypt hyperplasia. SIV-induced arteriopathy is characterized by extensive medial and intimal smooth muscle proliferation of medium-sized arteries, which when severe, may contribute to atrial and pulmonary thrombosis. A characteristic histologic finding of primary SIV retroviral infection is giant cell disease with large multinucleated syncytia in multiple organs including lymph nodes, lungs, intestines, and the central nervous system.[150] Opportunistic infections, most commonly with cytomegalovirus, SV40, *Mycobacterium avium* subspecies *intracellulare*, *Pneumocystis carinii*, *Trichomonas* spp., *Candida* spp., and *Plasmodium* spp., typically develop when the absolute CD4 T-cell count drops below 300–400 cells/mm^3. SIV is not oncogenic, but lymphoid neoplasms are common and associated with rhesus lymphocryptoviruses. Due to the significance of SIV-related diseases in primates and their zoonotic potential, SIV prevention and control plans should be implemented in facilities housing potentially infected animals.

Simian retroviruses (SRV) (formerly known as **Type D retroviruses**) belong to the subfamily *Orthoretrovirinae*. They are endemic in several species of Asian macaques and langurs and like SIV, they can cause an immunodeficiency syndrome called **simian AIDS (SAIDS)**. Natural infections occur in wild, laboratory, and zoo-housed primates. SRV-SAIDS differs in several aspects from SIV-induced immunodeficiency. Several serotypes are recognized but SRV-1 and SRV-2 are the serotypes of greatest clinical significance. SRV-1 is most common in rhesus macaques while SRV-2 is most common in pig-tailed and cynomolgus macaques. Animals may carry both or possibly multiple serotypes.[148] Infection is transmitted horizontally by fighting and sexual contact, vertically from infected dam to neonate, and once introduced, it spreads rapidly within a group. SRV has broad cell tropism and infects B,

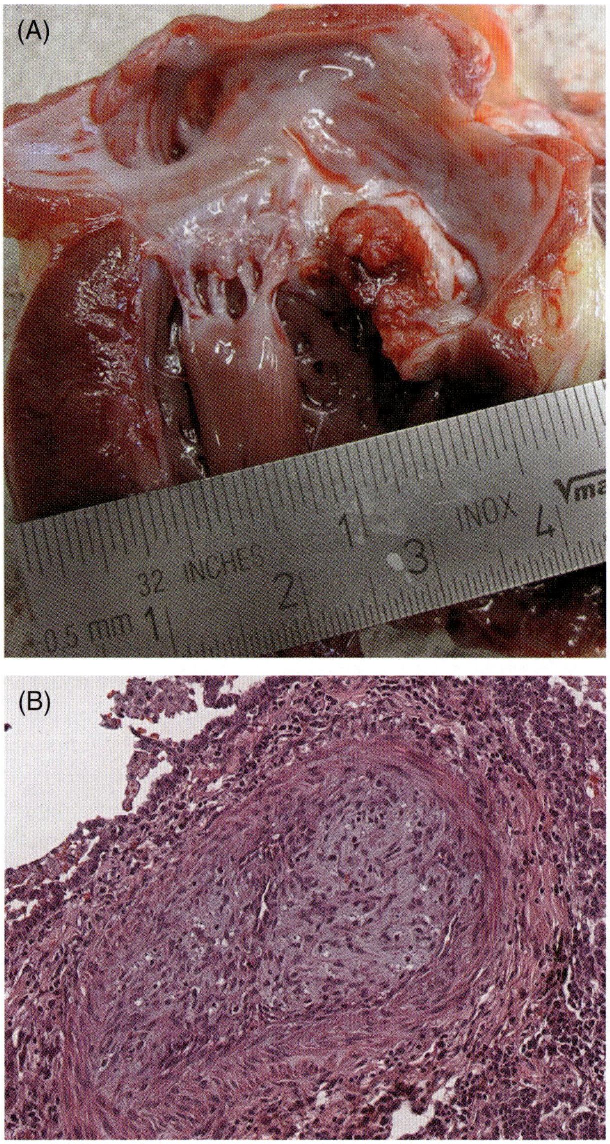

FIGURE 14.16 SIV-induced lesions in the cardiovascular system of a rhesus macaque. (A) Severe endocarditis with thrombosis. (B) Arteriopathy with severe hyperplasia of the media and intima of small pulmonary vessels.

CD4 and CD8 T cells, macrophages, hematopoietic progenitor cells, and epithelial cells.[142]

Animals may harbor SRV without developing clinical disease. The infection will progress to SAIDS in only a subset of animals. The clinical course ranges from subclinical carrier status to rapidly developing fatal immunosuppressive disease. Clinical findings include persistent diarrhea, weight loss, and hematologic abnormalities, such as anemia and pancytopenia. Gross findings include marked lymphoid hyperplasia progressing to lymphoid depletion and immunosuppression with opportunistic infections.[98,159] The spectrum of opportunistic infections differs slightly from that seen in SIV-induced immunodeficiency. In SRV-infected animals, lymphopenia as well as neutrophilia develop and because of the latter, animals are more susceptible to infections with pyogenic bacteria and **noma**, a rapidly progressive, polymicrobial infection of the oral cavity that results in necrotizing and ulcerative gingivitis and stomatitis with osteonecrosis. *Mycobacterium avium*, *Pneumocystis carinii*, and SV40 infections are common in SIV immunodeficiency and are less frequently seen in SRV-SAIDS. Other common opportunistic pathogens, like disseminated cytomegalovirus infection, candidiasis, or cryptosporidiosis are seen in both SRV or SIV infection.[159] SRV-SAIDS is often accompanied by **retroperitoneal (RF)** or **subcutaneous fibromatosis (SF)**, an aggressive fibroproliferative disorder reminiscent of Kaposi's sarcoma in patients with HIV-induced AIDS. The zoonotic potential of SRV is unclear. While seroconversion has been reported in a few primate handlers, there have been no reports of clinical effects.

Simian T-lymphotropic viruses, or **simian T-cell leukemia viruses (STLVs)**, are commonly found in numerous OWM species. Three different groups exist that are closely related to their human counterpart, human T-lymphotropic virus (HTLV-1, -2, -3). In most cases, infection is asymptomatic and not associated with clinical disease or pathology.[160] Disease is reported in African green monkeys and baboons. In these species, infection is associated with lymphoproliferative disorders such as malignant lymphoma and leukemia.[112] There is evidence that HTLVs are derived from STLVs and that multiple cross-species transmissions have occurred.

Bacteria

Bacteria and fungi are ubiquitous in the environment. Some are commensals, and some are pathogens that can induce serious clinical disease in NHP. They may infect large numbers of animals leading to epizootics or cause disease in individual animals. Diagnosis requires culture and identification by biochemical, serological, or molecular biological tests. Mycobacterial infections historically have been and continue to produce significant disease in OWM facilities. Mycotic infections occur less frequently than bacterial in NHPs in managed care. Notable bacterial and fungal infections are described below and information related to several additional mycotic infections is listed in Table 14.e6.[243]

Tuberculosis is caused by infection with bacteria in the ***Mycobacterium tuberculosis complex*** **(MTBC)**. These include ***Mycobacterium tuberculosis***, ***M. bovis*** (cause of bovine tuberculosis), ***M. africanum*** (parts of Africa), ***M. canetti***, and ***M. microti***. *M. tuberculosis* is the most common cause of tuberculosis. Infection with each of these bacteria can be zoonotic or anthropozoonotic, and all are easily transmitted from humans to NHPs who are known to be highly susceptible to mycobacterial infection. Mycobacteria are intracellular, gram-variable bacteria that are best visualized with acid-fast stains like Ziehl Neelsen or Fite Faraco stains. Transmission occurs by aerosolization or ingestion of the organism. *M. tuberculosis* is the primary cause of disease in captive primates, especially in laboratory settings.[87,88,176,197] However, the disease is reported in zoo animals as well.[216] A few reports also describe tuberculosis in wild, free-living baboons in peri-urban environments in African countries.[200] There are great differences in the susceptibility among nonhuman primate species. The most severe outbreaks have been described in OWM, whereas NWM seem to be more resistant.[23] Once established in a colony, the disease spreads rapidly to co-housed animals leading to epizootics with high morbidity and mortality. Tuberculosis in OWM commonly results in a rapidly progressive, debilitating disease. Latency is not considered a common feature but cynomolgus monkeys may develop a more clinically silent disease course.[153] The most common clinical signs are coughing and weight loss. Clinical disease is directly related to the severity and distribution of gross lesions. The hallmark of tuberculosis is the typical tubercle. Grossly, these are variably sized, firm, yellow-white, or grayish nodules. If these nodules grow, they tend to coalesce forming lesions with a central cavity containing necrotic debris. Respiratory disease is a common site of infection. Tubercles may extend into the bronchus system, the trachea, or thoracic cavity. Tracheobronchial lymph nodes can be greatly enlarged with apparent caseation on cut section (Fig. 14.17A). Advanced stages of the disease are characterized by dissemination to extrapulmonary sites (Fig. 14.17B) including the spleen, kidney, liver, lymph nodes, or any other site of the body. Dissemination to other organs varies among monkeys and does not necessarily correlate with the severity of lung involvement. Histologically, tubercles are typical granulomas containing central acellular necrotic debris or proteinaceous material, which is seldom mineralized (Fig. 14.17C; eSlide 14.e5). The central cores are surrounded by a cuff of epithelioid macrophages, lymphocytes, plasma cells, and multinucleated giant cells, usually of the Langhans type. In contrast to other animal species, a fibrous capsule is usually not found in nonhuman primates. The number of acid-fast bacilli in granulomas can vary considerably, but often there are only a few intralesional organisms. Confirmatory testing is usually through a combination of different test methods based on serology, radiographs, histology, microbiology, and molecular methods. In captive settings, periodic intradermal tuberculin skin testing and serology are often used to monitor animals for infection. It is advisable to use combinations of different test methods to confirm the diagnosis of tuberculosis since a positive result has implications for both the animals (e.g., quarantine, long-term treatment, euthanasia) and the humans who care for them.[32,88]

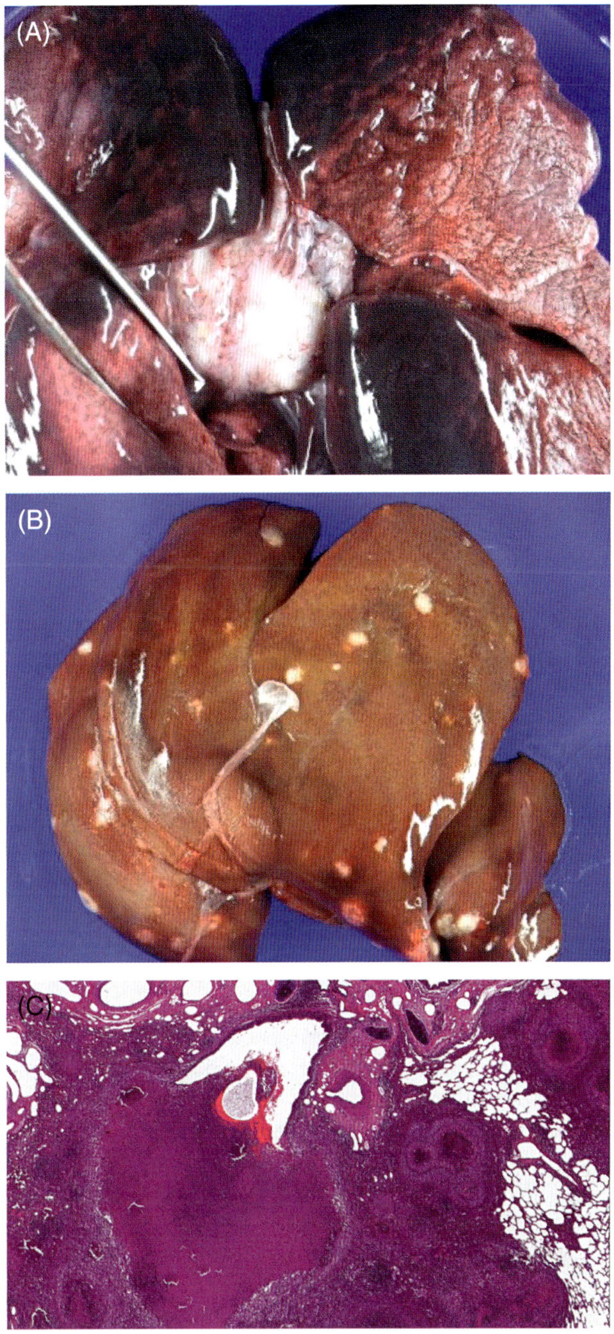

FIGURE 14.17 Tuberculosis due to *Mycobacterium tuberculosis* in a rhesus macaque. Infection and characteristic granuloma formation can be seen in many tissues, although the lung is the most common site of infection. (A) The bronchial lymph nodes may be severely hyperplastic and enlarged. (B) Multiple, well-demarcated, crème colored granulomas are present in the liver parenchyma. (C) Classic tubercle lesions (here in the lung) are characterized by central areas of necrosis surrounded by a cuff of granulomatous inflammation that in this case is confluent with an airway (see eSlide 14.e5).

Mycobacterium kansasii and other environmental, atypical mycobacteria are not normally associated with clinical disease but may cause false positive intradermal tuberculin skin tests in NWM and OWM. Infections occur by inhalation. An established infection may lead to granulomatous inflammation of the pulmonary lymph nodes that are comparable to tuberculosis. The disease is most commonly reported and problematic in squirrel monkeys[26] but is also described in rhesus macaques.[119]

The ***Mycobacterium avium* complex (MAC)** group of bacteria includes ***M. avium*** and ***M. intracellulare***. These are common environmental bacteria that can induce severe systemic infection in immunodeficient animals or localized infections in normal NHP.[105,170] The induced lesions are referred as mycobacteriosis and must be differentiated, through bacterial culture or molecular methods, from tuberculosis (MTBC infections). Infected animals may have a false positive reaction in intradermal tuberculin skin testing.

Paratuberculosis (syn. Johne's disease) is caused by ***M. avium* subspecies *paratuberculosis***. The disease is most commonly described in ruminants (see Chapter 5 Bovidae, Antilocapridae, Giraffidae, Tragulidae, Hippopotamidae and Chapter 6: Cervidae). Infected animals develop progressive diarrhea and weight loss. The infection can be acquired from contaminated water, soil, or paratenic insects.[168] Both NWM and OWM are susceptible to infection but the disease is less common in NWM. The disseminated form of MAC is most often seen in animals with simian AIDS. Disease expression and severity are directly correlated to special SIV strains. It typically causes intestinal lesions that are characterized by diffusely rugose or multinodular, firm, thickened mucosa near the ileocecocolic junction or involving the entire colon and small intestine (Fig. 14.18A). Draining lymphatics are often thickened and adhesions are formed between enlarged mesenteric lymph nodes and the intestinal serosa. In the localized form, enlarged mesenteric lymph nodes may be the only finding.[103] The respiratory tract is rarely the primary site of infection.[206] Microscopically, the intestinal mucosa is infiltrated by sheets of large epithelioid histiocytes (Fig. 14.18B). Giant cells, granulomas, or typical tubercles are infrequent. Mesenteric lymph nodes are enlarged with effacement by epithelioid macrophages. Large numbers of intrahistiocytic, acid-fast positive bacilli are seen with acid-fast stains (Fig. 14.18C). In localized infections independent of simian AIDs, typical caseating granulomas in infected lymph nodes cannot be morphologically discriminated from tuberculosis. Definitive diagnosis requires isolation of the organism by culture, or PCR and subsequent DNA sequencing.

Mycobacterium leprae is the causative agent of **leprosy**, a chronic granulomatous disease affecting the skin, dermis, and peripheral nerves. Transmission occurs via the respiratory route or through infected skin lesions. Natural infections are reported in sooty mangabeys and cynomolgus monkeys.[181,267] Nodular thickening of the skin and peripheral nerves may lead to paralysis or deformities of the face, hands, and feet. Lesions occur predominantly in cooler areas of the body, such as the ears, nose, scrotum, and tail. Histologically, inflammation is predominantly histiocytic and contains intralesional acid-fast bacilli. Confirmation is by bacterial culture and molecular methods.

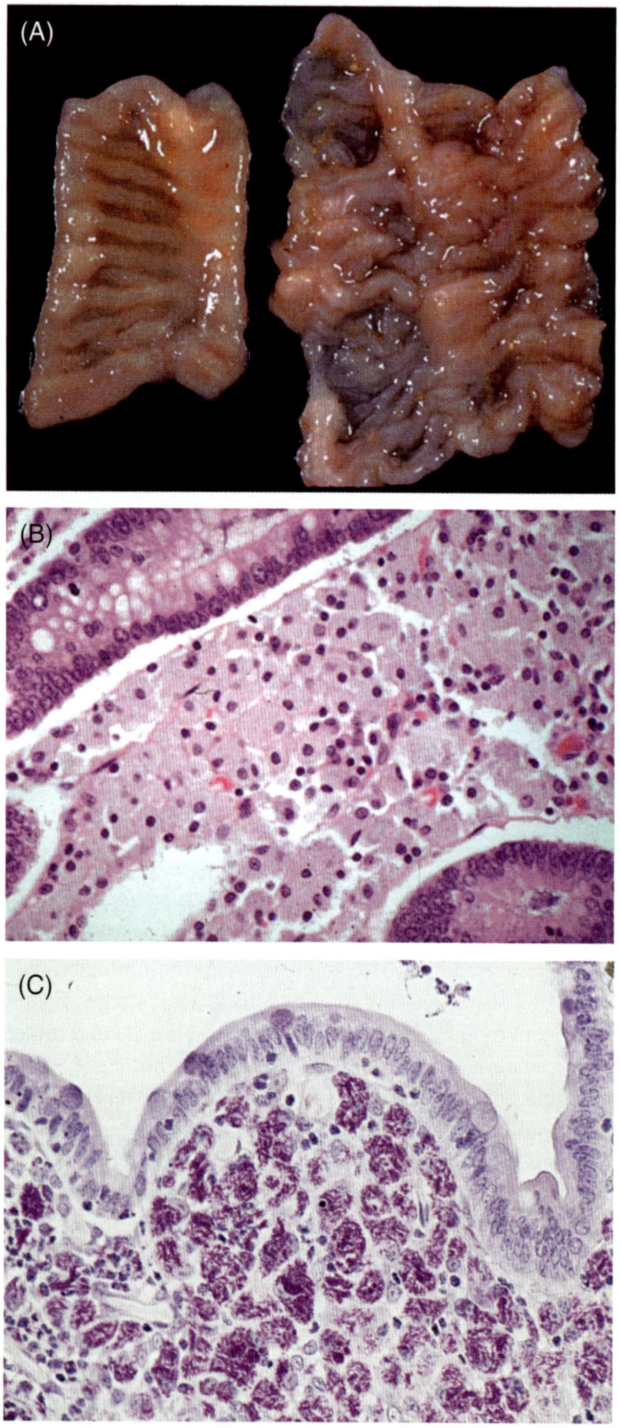

FIGURE 14.18 Intestinal *Mycobacterium avium* infection in a rhesus macaque. (A) The ileum and colon are diffusely thickened and the mucosa has a wrinkled appearance. (B) Sheets of large, epithelioid macrophages expand the mucosal lamina propria. (C) Myriad, intracellular, acid-fast-positive bacteria are present in epithelioid macrophages. Ziehl Neelsen.

Shigellosis is caused by ***Shigella*** bacteria. The bacteria are divided into four serogroups: ***S. flexneri***, ***S. dysenteriae***, ***S. boydii***, and ***S. sonnei***. All are gram-negative enteroinvasive bacteria that induce dysentery, most commonly in OWM. Lesions of shigellosis are limited to the large intestine (cecum and colon); the small intestine is spared. Lesions may be focal or diffuse and are characterized by edema, hemorrhage, erosion, ulceration, and pseudomembrane formation (Fig. 14.19A). Histologically, it is characterized by necrotizing colitis, often with crypt abscess formation, gland herniation into gut-associated lymphoid tissue (GALT), and superficial mucosal neutrophilic exudation[129] (Fig. 14.19B). Ulceration can be pronounced in chronic cases. The diagnosis is confirmed by bacterial culture. Shigellosis is a zoonotic disease that is easily transmitted from monkeys to humans.[145] *S. flexneri* also causes a separate syndrome of ulcerative, gum-line gingivitis in macaques. Gingival forms may or may not be associated with dysentery. Gingivitis can persist for a prolonged period of time and infected animals therefore can serve as carriers.[6] Differential diagnosis includes acute necrotizing, ulcerative gingivitis (ANUG) which affects the interdental papilla and is associated with immunodeficiency. Reactive arthritis may follow shigellosis.[266]

Salmonellosis, usually caused by the gram-negative bacteria ***Salmonella enteritidis*** or ***S. typhimurium***, is typically rare in established primate colonies. Infection and disease can be sporadic or epizootic, and most cases occur after importation or during quarantine. Asymptomatic carriers are common. The most common clinical sign is diarrhea, which varies from watery to bloody or mucoid depending on bacterial virulence factors and host response. Affected animals rapidly become moribund and die.[258] The most severe microscopic alterations are found in the mucosa of the ileum and colon.[220] The dominant finding is necrotizing and suppurative enterocolitis. Bacteria invade epithelial cells and the lamina propria leading to systemic dissemination and septicemia resulting in pyogranulomas or histiocytic nodules in the liver. Salmonellosis is a zoonotic disease that is easily transmitted from monkeys to humans.

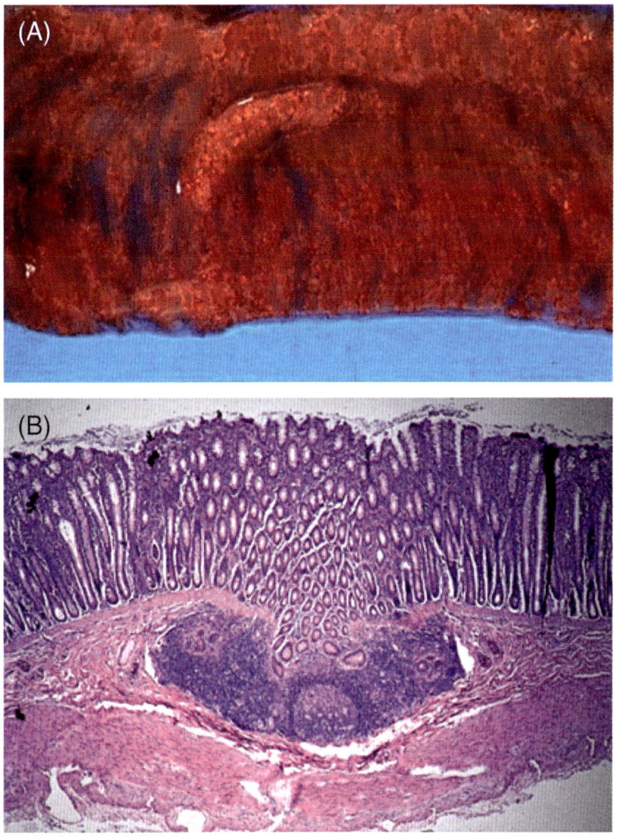

FIGURE 14.19 Shigellosis in the large intestine of a rhesus macaque. (A) Diffuse erosion and hemorrhages in the colonic mucosa. (B) Subacute colitis with diffuse inflammation in the lamina propria and crypt herniation into gut-associated lymphoid nodules.

Escherichia coli, a gram-negative bacterium, can also cause diarrheal disease in OWM or NWM. Disease can be due to **enteropathogenic *E. coli* (EPEC), enterotoxigenic *E. coli* (ETEC), enterohemorrhagic *E. coli* (EHEC), enteroinvasive *E. coli*** (EIEC), and/or **diffuse adherent *E. coli* (DAEC)**. EPEC infection is common among marmoset and tamarin species, especially cottontop tamarins. These species tend to develop idiopathic colitis and concurrent EPEC infection is assumed to be a co-factor in disease progression.[169] Clinical disease is frequently characterized by acute hemorrhagic diarrhea (Fig. 14.20), but more frequently chronic, intermittent, or progressive diarrhea may be seen. Primary histologic findings include neutrophilic colitis with crypt epithelial cell hyperplasia and attenuation of surface epithelium. Key diagnostic features are adherent bacteria on the apical surface of the mucosal epithelial cells leading to effacement of normal epithelial cell architecture.[262] The surface

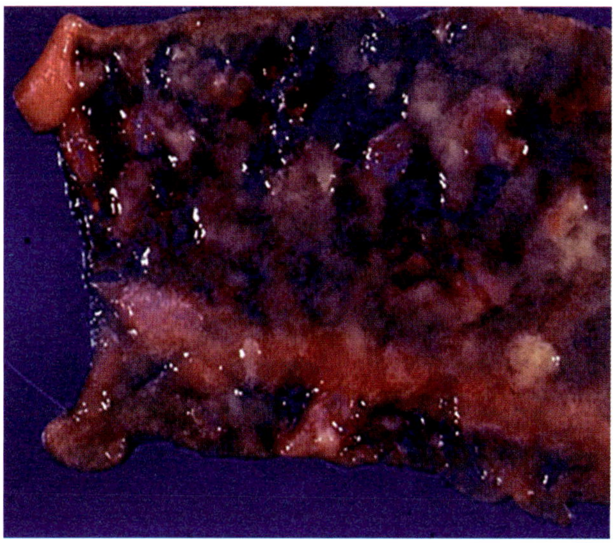

FIGURE 14.20 Enterotoxigenic *Escherichia coli* in a Titi monkey. Infection is associated with severe hemorrhagic enteritis. Small erosions on the mucosa are covered with clotted blood.

epithelium may become hyperplastic with increased mitotic activity, loss of goblet cells, and loss of microvilli; grossly, the mucosal surface may have a cobblestone appearance. In severe cases, the ileum and distal jejunum may be colonized as well.

Yersinia enterocolitica and ***Y. pseudotuberculosis*** are the two most common and important causes of bacterial **yersiniosis** in OWM and NWM. The clinical symptoms and lesions induced by both species are principally comparable, but disease induced by *Y. enterocolitica* is usually more severe.[23] Reservoir hosts are wild birds and rodents, and transmission occurs primarily by the fecal-oral route through ingestion of contaminated feed or water. Infections with both agents are reported in a variety of OWM and NWM species, but are most frequent in marmosets, tamarins, and squirrel monkeys.[80,118,192] OWM outbreaks have also occurred in breeding colonies of macaques[166,218] and African green monkeys.[250] The disease can become endemic in zoo settings.[19] Acute morbidity and mortality is typical but chronic debilitating disease occurs as well. Affected monkeys are often found dead without premonitory signs of disease. Clinical signs, when present, can include diarrhea, depression, and dehydration. Yersiniosis can also be associated with abortion and stillbirth.[218] The disease has a seasonal pattern with most infections occurring during the wet and/or cold seasons. Characteristic lesions are most often present in the gastrointestinal tract, liver, and spleen (Fig. 14.21A). Ulcerative enterocolitis, often centered on lymphoid patches, is the primary lesion and the compromised mucosal barrier is the portal through which bacteria enter the systemic circulation. The resulting bacteremia causes hepatomegaly, splenomegaly, and/or mesenteric lymphadenomegaly. The characteristic gross lesions in the liver and spleen are multifocal, small, demarcated, unencapsulated, yellowish-white foci. This corresponds histologically to multifocal to coalescing areas of necrosuppurative hepatitis and splenitis, or in more chronic cases, granulomatous inflammation, with intralesional lobulated colonies of gram-negative bacteria.[191] The histologic appearance, which is similar to but distinct from the tubercle-like nodules that are characteristic of tuberculosis, underlies the common name "pseudotuberculosis" in *Y. pseudotuberculosis* infections. The presence of large, intralesional, lobulated colonies of gram-negative bacteria in areas of inflammation is highly characteristic of yersiniosis (Fig. 14.21B; eSlide 14.e6). Bacterial culture confirms the diagnosis. Yersiniosis is a zoonotic disease.

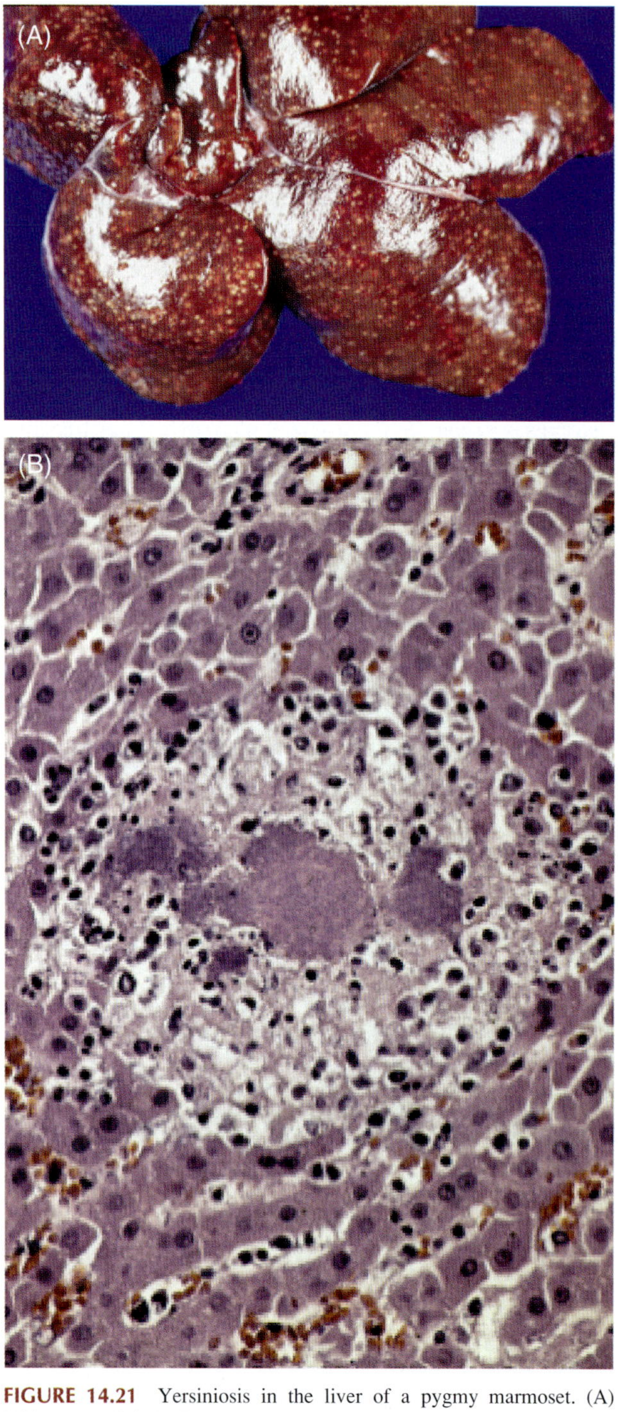

FIGURE 14.21 Yersiniosis in the liver of a pygmy marmoset. (A) Miliary, small foci of yellow-white discoloration are present throughout the enlarged liver. (B) Large, botryoidal bacterial colonies with associated hepatic necrosis, and moderate, mixed inflammation are characteristic (see eSlide 14.e6).

Tularemia is a highly contagious, zoonotic disease that is transmissible by inoculation, ingestion, or inhalation of the aerobic, gram-negative coccobacillus ***Francisella tularensis***. It is a reportable disease in some regions (e.g., the United States). Several subtypes (biovars) of the bacterium, each with differing levels of virulence, exist. These include: *F. tularensis tularensis* (Type A) which is the most virulent and is found only in North America; *F. t. holoarctica* (Type B) found in North America, Europe and Asia; *F. t. mediaasiatica*; and *F. t. novicida*. Cottontail rabbits and ticks are the primary reservoirs for

Type A.[278] Transmission is by the bite of an infected tick or by handling or ingestion of infected animals. Type B is primarily spread by rodents and squirrels, aquatic rodents (e.g., muskrats, beavers), or contaminated water and food and causes a milder form of the disease. In general, tularemia can show a seasonal pattern. Nonhuman primates housed outdoors may be at a higher risk of exposure than those housed indoors due to a higher risk of exposure to infected reservoir animals, vectors, or contaminated water or environmental substrates. Regardless of the route of infection, tularemia is a serious and often fatal disease. The course of infection depends on the virulence of the bacterial strain, the portal of entry, the extent of systemic involvement, and the immune status of the host. There are several reports of naturally acquired tularemia in nonhuman primates.[34,69] Most of the reported cases are observed among NWM and lemurs in parts of the United States where the disease is endemic. In Europe, only a few cases have been reported with the largest outbreak being in outdoor-housed cynomolgus monkeys and marmosets in Germany.[175] Additional cases are reported from Switzerland and Hungary.

The main forms of tularemia in both humans and nonhuman primates are ulceroglandular (skin, mucous membranes, draining lymph nodes), pneumonic (lung), typhoidal (gastrointestinal tract), and septicemic. Clinical signs of the disease are nonspecific and resemble respiratory tract infections or gastroenteritis caused by other common bacterial or viral pathogens. Large lingual ulcers or abscesses in the tongue, and severe submandibular lymphadenopathy are prominent findings in some cases. Macroscopically, small granulomas that cannot be grossly differentiated from other forms of granulomatous inflammation (e.g., yersiniosis) are present multifocally in multiple organs (Fig. 14.22A). Histologically, pyogranulomatous lesions in liver (Fig. 14.22B), spleen, respiratory tract, and lymph nodes are characteristic but not pathognomonic. The demonstration of the bacteria in histological sections can be difficult, even with special stains. Diagnostic confirmation can be challenging as managing suspect samples requires special handling, Biosafety Level 3 laboratory facilities, and special media and conditions for optimal growth. Confirmatory tests include bacterial culture, PCR, immunohistochemistry or direct fluorescent antibody testing, and MALDI-ToF mass spectrometry.

Bordetellosis is caused by the pleomorphic, motile gram-negative, coccobacillus ***Bordetella bronchiseptica***. The organism is considered to be part of the normal commensal bacteria of oropharyngeal mucus membranes, and is present in many domestic and wild animal species. It has a special affinity for the respiratory tract, becoming attached to the ciliated epithelium of the airways. Transmission occurs by direct contact, through the air (coughing or sneezing; respiratory route), or via contaminated fomites. Clinical disease produced by *B. bronchiseptica* alone or in combination with other microorganisms, is associated with stress induced by shipping, poor body condition, overcrowding, or immunosuppression. The disease may affect OWM and NWM, though it is more severe and more frequent among NWM. In marmosets, characteristic signs include mucopurulent nasal discharge and pneumonia with dyspnea, coughing, and pyrexia. A marked difference in age susceptibility to severe illness and pneumonia exists.[10] Young marmosets are highly susceptible and may die shortly after infection. The bacteria induce purulent bronchopneumonia. Histologically, bronchioles are filled with

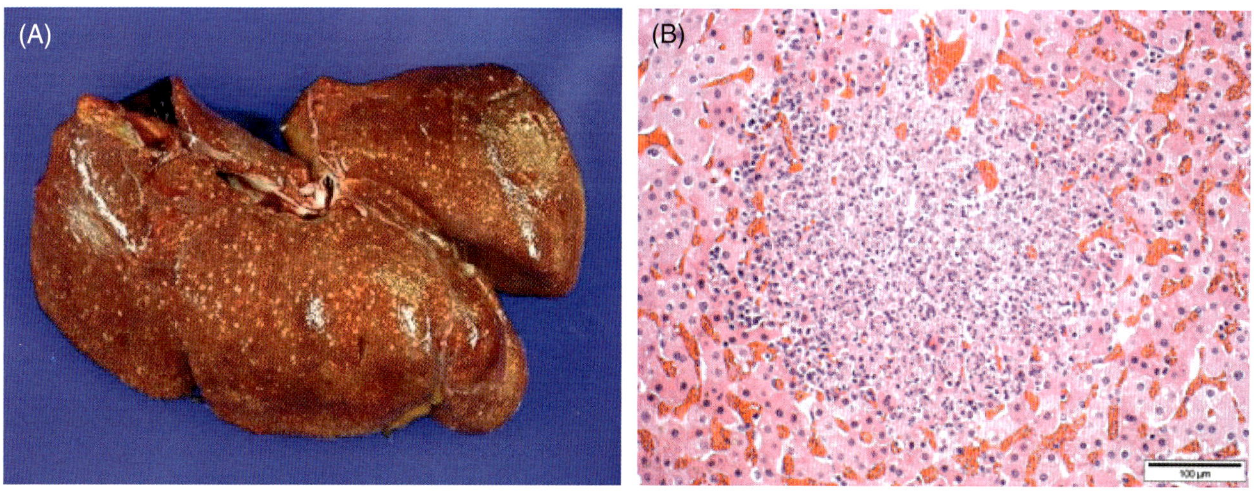

FIGURE 14.22 Tularemia in the liver of a long-tailed macaque. (A) The liver contains miliary, small, white foci. (B) These correspond to multifocal areas of acute hepatic necrosis and associated lymphohistiocytic inflammation. As in this image, bacteria may be difficult to see with routine hematoxylin and eosin or special stains.

necrotic cellular debris, polymorphonuclear leucocytes, and macrophages. Bronchiolar epithelium and vascular components can become necrotic.

Klebsiella pneumoniae is a nonmotile, encapsulated gram-negative bacillus that is part of the normal fecal and oral flora. However, some forms can cause significant morbidity and mortality in a variety of OWM and NWM species. Invasive, abscess-forming strains of *K. pneumoniae* associated with the so-called **hypermucoviscosity (HMV) phenotype** are becoming more important. Transmission occurs by oral or respiratory routes. Infection can also be acquired from environmental sources. Disease is frequently associated with concurrent *Pasteurella multocida* infection and affects animals debilitated by stress or concurrent disease. Infants and juveniles are highly susceptible. Clinically, *K. pneumonia* may induce pneumonia, necrotizing and suppurative typhlitis and or typhlocolitis, and septicemia.[204] Multiple abdominal abscesses are seen after infection with HMV strains in African green monkeys. Infection is associated with peritonitis, mesenteric adhesions, and multisystemic abscessation.[265] The bacterium is also involved in the pathogenesis of airsacculitis.[92] The pattern and distribution of lesions depend on the primary organ system involved. At necropsy, there is often diffuse enteritis and/or pneumonia. Microscopically, lesions consist of suppurative inflammation and necrosis with small, intralesional bacilli that are often surrounded by a clear halo. In more long-standing disease, fibrinopurulent and histiocytic exudates are often seen and may be associated with pneumonia, pleuritis, or peritonitis.

The "**bloody nose syndrome**" of rhesus and cynomolgus macaques, previously thought to be due to infection with *Branhamella (Moraxella) catarrhalis*, has more recently been shown to be induced by the gram-negative, oxidase-positive, aerobic diplococcus ***Moraxella macacae***.[268,273] The syndrome is most frequently observed in winter, may be associated with low humidity, can spread easily between animals in close quarters, and is often self-limiting. Clinical signs are characterized by epistaxis and periorbital edema accompanied by upper respiratory tract signs. Lesions consist of mucohemorrhagic rhinitis with gram-negative diplococci in the exudate. Diagnosis is confirmed through culture and biochemical methods. Differentiation between *M. macae* and *B. catarrhalis* can be made based on their beta-lactamase reactivity.

Campylobacteriosis is caused by ***Campylobacter fetus***, ***C. jejuni***, and ***C. coli***. These are curved, spiral "gull-winged"-shaped, "S"-shaped, or rod-shaped, gram-negative bacilli. Transmission occurs by the fecal-oral route. Clinical asymptomatic carriers are common, and the organism typically resides in the biliary tract. *Campylobacter* is widely distributed among OWM, particularly among macaques.[128] The bacteria have a variety of virulence factors including adhesins, hemagglutinins, flagella, endotoxins, and cholera-like enterotoxins. The disease varies from mild secretory watery diarrhea to severe dysentery with or without fecal blood and mucus according to strain and toxins produced. Campylobacteriosis can also be associated with abortion in primates[15] and the bacteria are part of the multifactorial chronic colitis syndrome of macaques. Small intestinal and colonic mucosal hyperemia, roughening, and edema (Fig. 14.23A) correspond histologically to neutrophilic colitis with crypt abscesses (Fig. 14.23B). *C. jejuni* also affects the small intestine and can cause villous blunting. Curved rods with typical morphology can be seen in the altered mucosa and in crypt abscesses using silver stains.

Helicobacter and related species are curved to spiral-shaped, gram-negative, silver-positive bacilli that infect primates and may induce gastrointestinal disease. Some represent normal commensal organisms that typically only cause disease in immunocompromised animals.[74] The most important member of this group in OWM and NWM is ***Helicobacter pylori***. In macaques, transmission occurs by the oral route early in life from mother to offspring, and 90% of animals are infected in the first year of their

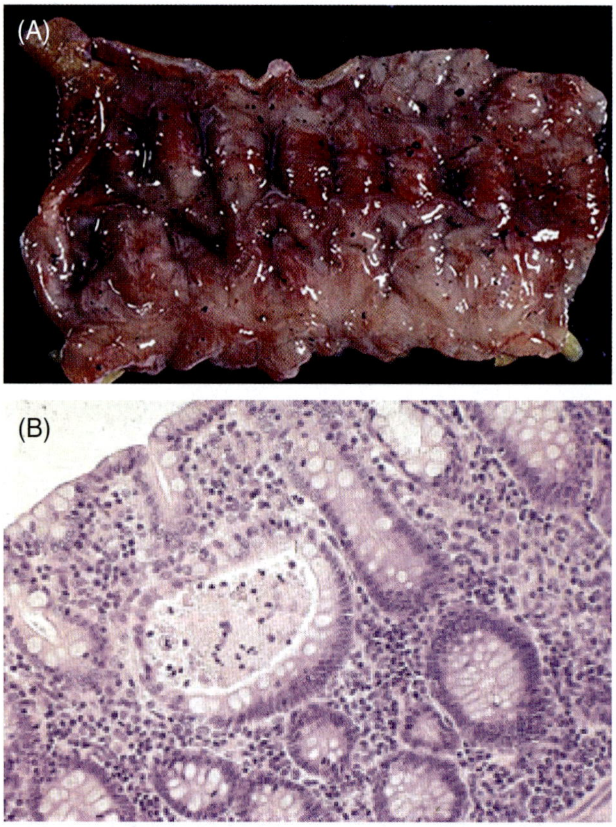

FIGURE 14.23 Campylobacteriosis in the large intestine of a rhesus macaque. (A) The inflamed mucosa is edematous, reddened, and swollen. (B) Severe colitis is associated with mixed inflammation and crypt abscess formation in the colonic mucosa.

life.[65,249] Clinical symptoms are usually not observed but occasional vomiting may occur. At necropsy, focal reddening or erosions of the gastric mucosa are occasionally seen. The antrum is most commonly infected. Histologically, chronic active gastritis is present.[65] Mononuclear cell infiltration in the gastric mucosa, superficial erosions, and epithelial hyperplasia may be evident. The bacteria are best visualized with silver stains (e.g., Warthin-Starry stain), which highlight the slightly curved, rod-shaped, loosely coiled organisms associated with the gastric epithelium in the antral mucosa. The bacteria must be differentiated from ***Gastrospirillum hominis*-like organisms (GHLO)** also called ***H. heilmannii***. GHLO are nearly ubiquitous in rhesus monkeys. They are mainly found in the fundus of the stomach. The bacteria are larger than *H. pylori*. They are spiral shaped, tightly coiled with 6–8 coils per cell, and located in surface mucus, the lumen of the gastric pits, and in parietal cells. Normally, there is no disease associated with GHLO infection. Colonic helicobacters have an affinity for the brush border of enterocytes and may play a role in the chronic colitis syndromes of macaques as they have been shown to upregulate proinflammatory cytokines.[149]

Lawsonia intracellularis is a curved, gram-negative, cell-associated bacterium that induces proliferative enteritis in several mammalian species. Principally a disease of swine, it has been diagnosed in macaques with chronic diarrhea and weight loss.[143] Infected animals develop proliferative ileitis characterized by segmental thickening of the distal ileum with markedly thickened mucosa.[133] This corresponds histologically to mucosal expansion with lymphohistiocytic infiltrates, distortion of the crypts, blunting and fusion of villi, and microabscesses in the intestinal glands. Mucosal ulceration and necrosis may occur. The intracellular curved rods within infected epithelial cells are best identified with silver stains or with specific immunohistochemical methods.

Treponematosis (syn. treponemiasis) refers to a group of nonvenereal diseases caused by spiral-shaped bacteria in the genus Treponema. Infection *and* disease occur in free-living baboons in Africa that are infected with ***Treponema pallidum*** (Figs. 14.24A, B). Original reports are from West Africa and involve animals with small ulcerated lesions on the face and muzzle. More recently, destructive lesions involving the perineum, ischial callosities, and external genitalia have been identified in olive baboons in Tanzania.[101,135,246] Histologically the lesions consist of acanthosis with deep dermal rete and varying degrees of ulceration. Dermal inflammation ranges from mixed mononuclear cell infiltration with large numbers of T-lymphocytes to granulomatous lesions. Ulcers are often filled with neutrophils and serocellular crusts. Intralesional treponemes are best highlighted with silver stains. Based on molecular characterization, the baboon organism is related to ***T. pallidum pertenue***, the agent of **yaws** (causes disfiguring lesions in skin and bones), rather than ***T. p. pallidum***, the agent of human syphilis.

Leptospirosis, caused by bacteria in the genus Leptospira, is an uncommon, naturally acquired infection in monkeys. Most reports describe the disease in NWM species.[9,201] Transmission is most often via contact or ingestion of infected rodents. Clinical disease is associated with jaundice, anemia, hemorrhage, abortion, and hematologic evidence of nephritis. Single infections or outbreaks, for example, in capuchin monkeys rescued from homes and housed in a wildlife rehabilitation center in Colombia, have been reported.[257] Interstitial nephritis in combination with necrotizing tubulonephritis or alone is a frequent histologic finding. Marked hepatocyte dissociation and pulmonary hemorrhage may also be seen.

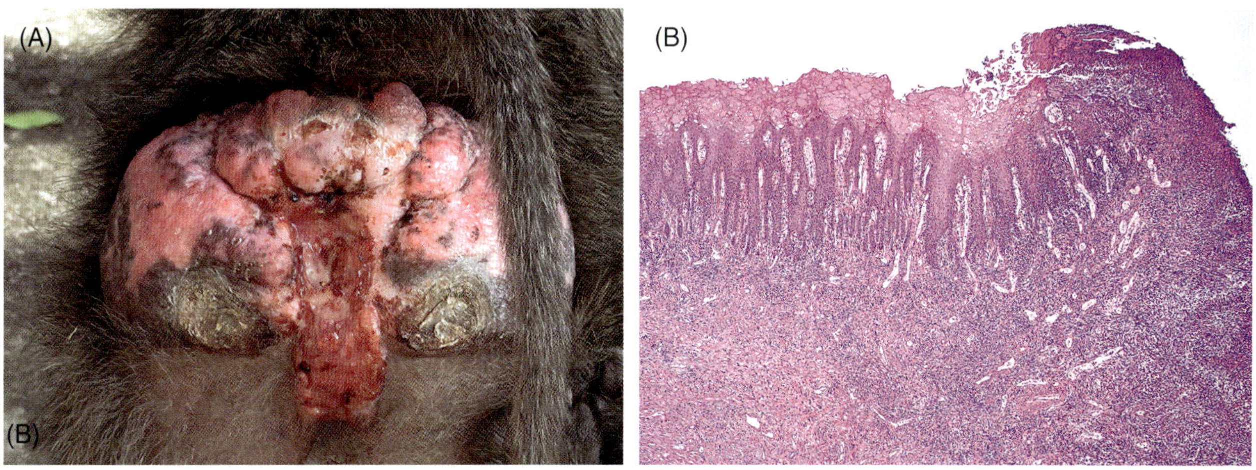

FIGURE 14.24 Treponematosis in a female baboon. (A) Ischial callosities and external genitalia with severe ulcerative lesions. (B) This is associated with varying degrees of dermal ulceration and hyperkeratosis. *(Parts A, B: Photos Courtesy of Gombe Ecohealth Program.)*

Listeria monocytogenes causes **Listeriosis**, an important food-borne disease that can affect a large range of species. The agent is a gram-positive, facultative anaerobic, rod-shaped bacterium that invades cells and replicates in the cytoplasm. Clinical disease has been infrequently reported in NHP. Most cases in NHP show symptoms of reproductive failure including abortion, stillbirth, and neonatal death due to septicemia or cerebral complications.[41] Affected dams usually show little or no clinical evidence of disease. The disease can become endemic in group-housed, colony-reared animals. Infection in a free-living *colobus guereza* in Kenya caused widely distributed liver necrosis and intestinal hemorrhage.[137] However, lesions are most often only seen in the placenta and the aborted fetus and consist of diffuse fibrinopurulent placentitis and multiple discrete foci of necrosis in the liver and lung of the fetus or infected adults (Figs. 14.25A, B). Gram-positive rods are easily identified in necrotic tissue.

Tetanus is a disease caused by infection with ***Clostridium tetani***, a spore-forming, gram-positive, obligate anaerobe found in the soil. Contamination of wounds (e.g., bites, frostbite) is the portal of infection.[93,210] The type and location of the wounds can influence the risk of infection.[251] Disease is recognized in OWM and NWM and can be problematic in free-ranging and outdoor-housed animals. Clinically affected animals exhibit several characteristic early signs of tetanus toxin toxicity including torpor, reluctance to interact with others, bipedal locomotion, hopping and falling, piloerection, and progressive stiffness. In advanced cases, the classical triad of trismus (lockjaw), opisthotonus, and status epilepticus occurs. The disease is usually fatal in 1 to 10 days due to respiratory paralysis and exhaustion. The clinical symptoms are induced by the neurotoxin tetanospasmin, which is produced by *C. tetani* during vegetative growth of bacteria in the body. Tetanus toxin blocks the release of inhibitory neurotransmitters [glycine and gamma-aminobutyric acid (GABA)] at neuromuscular junctions, resulting in skeletal muscle spasms, continuous stimulation of skeletal muscle, and the development of rigor. In most cases, the clinical signs are diagnostic. Morphologic changes are limited to secondary processes, such as aspiration pneumonia. An immunization program for tetanus prophylaxis is advisable for institutions housing nonhuman primates outdoors.[131]

Clostridium difficile is commonly found in the gastrointestinal tract of various NHP species. Healthy animals carry the gram-positive bacteria as a part of their normal intestinal flora. Disruption of the normal flora may induce overgrowth of the bacteria. The bacterium is one of the most important causes of antibiotic-associated diarrhea in OWM and NWM. Prolonged antibiotic treatment with beta-lactam antibiotics or periods of stress or chronic diseases, such as diabetes in NWM, predisposes to opportunistic infections that may precede the onset of diarrhea. Cytotoxin produced by the bacteria causes severe pseudomembranous colitis. Infected animals develop profuse watery diarrhea that may contain blood and fibrin.[217] Microulceration of the colonic mucosa with exudation of fibrin and neutrophils

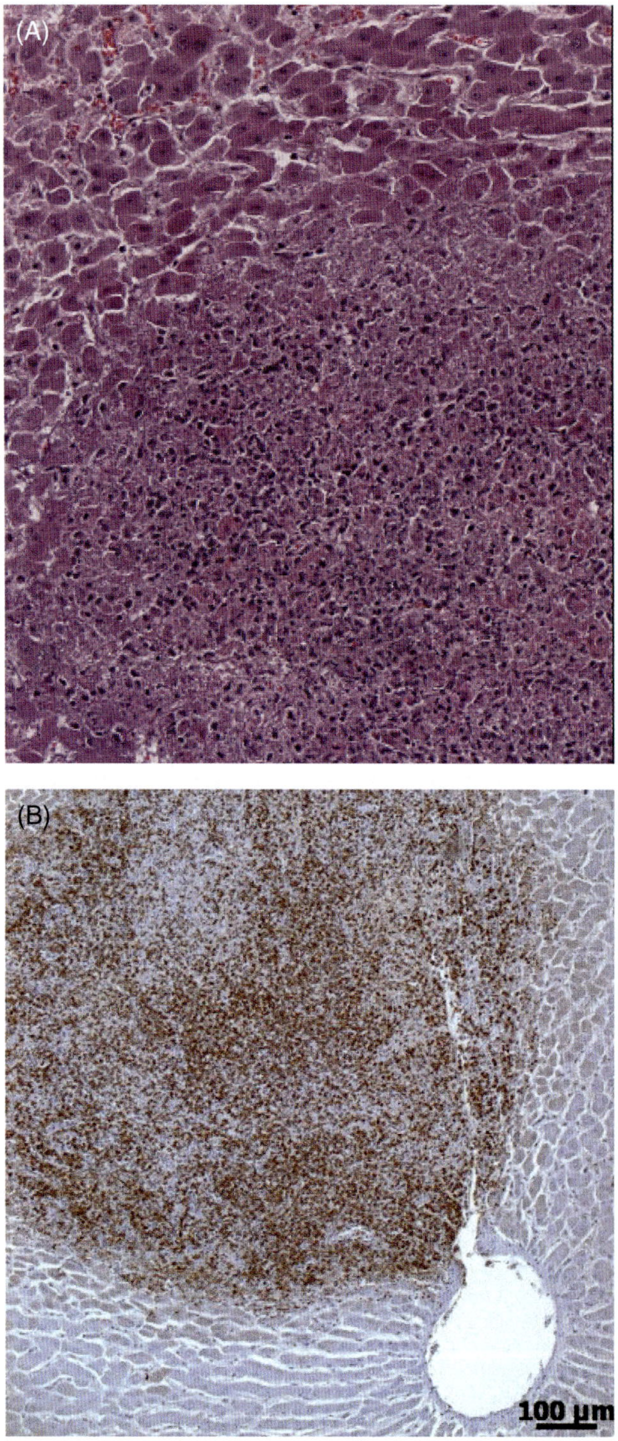

FIGURE 14.25 Listeriosis in the liver due to *Listeria monocytogenes*. (A) A large area of necrosis is present in the parenchyma. (B) Detection of *Listeria monocytogenes* antigen within the demarcated necrotic foci by immunohistochemistry.

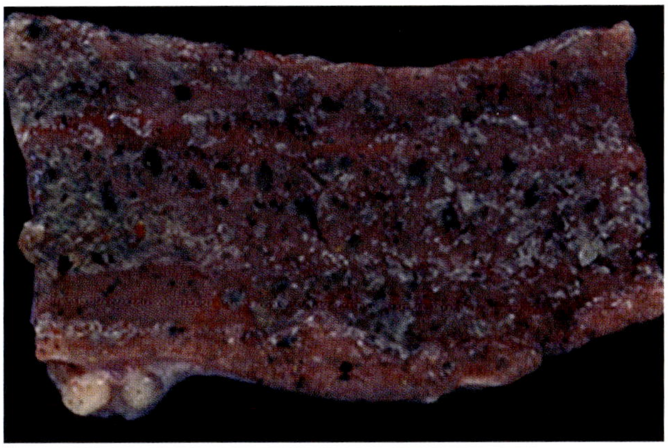

FIGURE 14.26 *Clostridium difficile* induced pseudomembranous colitis and antibiotic-associated diarrhea in a rhesus macaque. Multifocal microulceration, colon mucosal hyperemia, fibrin exudation, and pseudomembrane formation covers the mucosal surface.

that spread to cover the mucosa with flat pseudomembranes is characteristic and diagnostic (Fig. 14.26). The adjacent mucosa is heavily inflamed and hyperemic.

Clostridium perfringens is associated with **acute gastric dilatation** or **"bloat syndrome"** of nonhuman primates. Bloat syndrome can occur after overeating and excessive drinking, following anesthesia, or from gastric dysbiosis induced by antimicrobial treatment.[194] Clinically diseased monkeys present with abdominal distension induced by gastric dilatation (Fig. 14.27). The stomach and small intestine are filled with large amounts of gas and fluid.[18] The disease is observed among cercopithecine and colobine OWM and NWM. Acute gastric dilatation is a medical emergency. If untreated, animals die due to acute shock, most often related to hemodynamic collapse secondary to vascular compression or compromised flow due to pressure on thoracic organs and the vascular system.

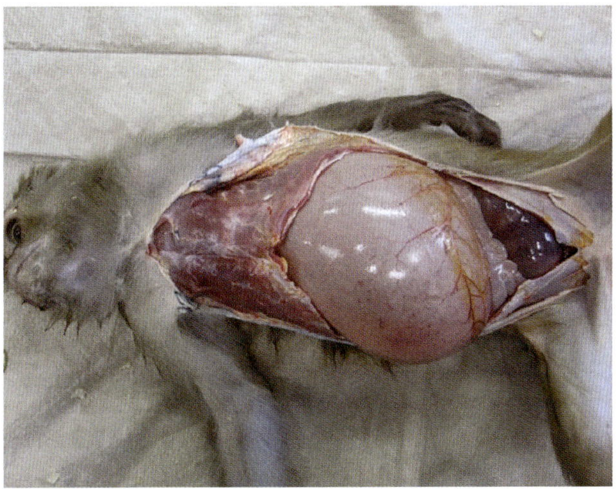

FIGURE 14.27 Bloat syndrome in a rhesus macaque. Severe abdominal distension induced by gastric dilatation.

Most common microscopic lesions are hyperemia, hemorrhage, and necrosis of the gastrointestinal mucosa. *C. perfringens* can be isolated from gastric contents.

Streptococcus pneumoniae, a gram-positive, encapsulated coccoid bacterium, is the most common cause of meningitis in NHP. The bacterium is present on the nasopharyngeal mucosa and can be carried by clinically asymptomatic animals and humans who represent a source of infection for NHP.[94] Transmission occurs through aerosols or by contaminated fomites. Stress-related factors such as capture, transportation, or quarantine predispose to infection. Neurologic signs include ataxia, nuchal rigidity, and coma.[91] *S. pneumoniae* can also cause respiratory disease that results in focal outbreaks or develops secondary to viral infections.[285] Infected animals develop signs of bronchopneumonia, which may be followed by the development of arthritis or meningitis. At necropsy, lung lesions range from acute inflammation with hyperemia and congestion or exudative and fibrinopurulent bronchopneumonia (Fig. 14.28A). Pleuritis is very common. In cases with meningitis, the disease course is rapidly progressive. A dull white to yellow exudate on the brain surface is suggestive of inflammation (Fig. 14.28B) which is microscopically characterized by fibrinopurulent meningoencephalitis. Sequela includes the formation of fibrin thrombi and infarcts. The causative diplococci are easily seen on Gram-stained smears of exudate.

Streptococcus equi* subspecies *zooepidemicus belongs to the beta-hemolytic Lancefield group C streptococcus. It is found in the nasopharynx and respiratory tracts of healthy horses and cattle and can infect many different species. It primarily causes disease in foals and young horses. Transmission of the bacteria is through respiratory droplets or direct contact. Outbreaks among NHPs are sporadically reported and have been described in both NWM and OWM. Transmission in zoo-housed callitrichids may be via ingestion of infected horse meat being fed to other species in mixed exhibits[231] or from keepers in contact with horses inadvertently carrying the organism into primate exhibits.[179] Free-living or peridomestic monkeys in Indonesia have been infected through exposure to infected pigs.[222] Infected animals die within a few days of infection. Reported disease includes bronchopneumonia, pleuritis, epicarditis, endocarditis, and meningitis with intralesional gram-positive cocci. Confirmation is through bacterial culture.

Staphylococcus aureus is a common commensal bacterium of the skin and nasopharyngeal mucosa of NHPs. Infection has been reported in several species. The agent gains entry through bite wounds or surgical sites. Beta-lactam resistance is common and this and **methicillin-resistant staphylococcal infections (MRSA)** are important contemplations when considering treatment options.[14] The agent induces pustular dermatitis in young animals, and vaginal infections in adults.[207] Wound infections result in

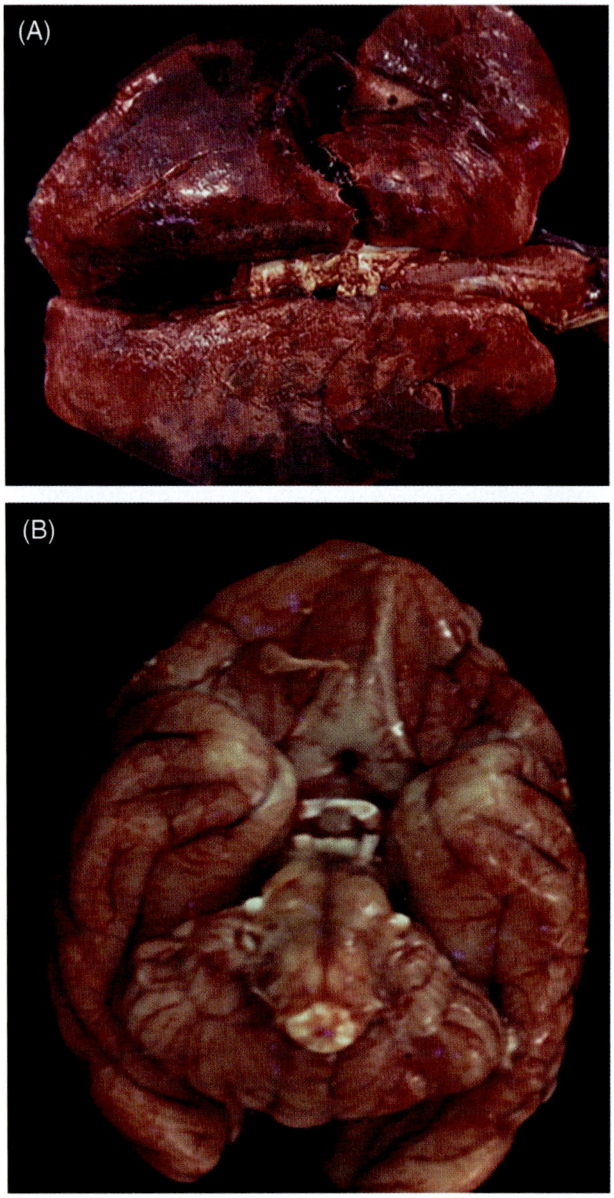

FIGURE 14.28 *Streptococcus pneumonia* infection in a rhesus macaque. Common manifestations occur in the lung and central nervous system. (A) The lung is diffusely congested and the parenchyma is expanded and firm due to fibrinopurulent bronchopneumonia. It does not collapse upon removal. (B) The leptomeninges are hyperemic, dull, thickened, and multifocally opaque white suggestive of suppurative meningitis.

cellulitis, lymphangitis, and abscessation. Sequela to bacteremia includes visceral abscesses, endocarditis, pericarditis, septic shock, or arthritis. Vegetative valvular endocarditis may produce septic emboli and infarcts in various organs. Staphylococcal abscesses and cellulitis typically contain thick creamy pus. Histologic lesions consist of fibrinopurulent exudate with masses of intralesional, gram-positive cocci. Chronic abscesses may develop a rim of macrophages.

Fungi

Mycotic infections occur less frequently in nonhuman primates than bacterial diseases (Table 14.e6). The most commonly diagnosed disease is **candidiasis** (Figs. 14.29A, B), which may involve mucosal surfaces, mucocutaneous junctions, and/or the skin.[283]

Some fungal infections, for example, **enterocytozoonosis** or **pneumocystosis** induced by ***Pneumocystis jirovecii*** or ***P. carinii***, are typically opportunistic infections in immunosuppressed macaques (Figs. 14.30A–C).[136,167,244,277] Some agents, such as ***Cryptococcus neoformans***, ***Coccidioides immitis***, ***Histoplasma***

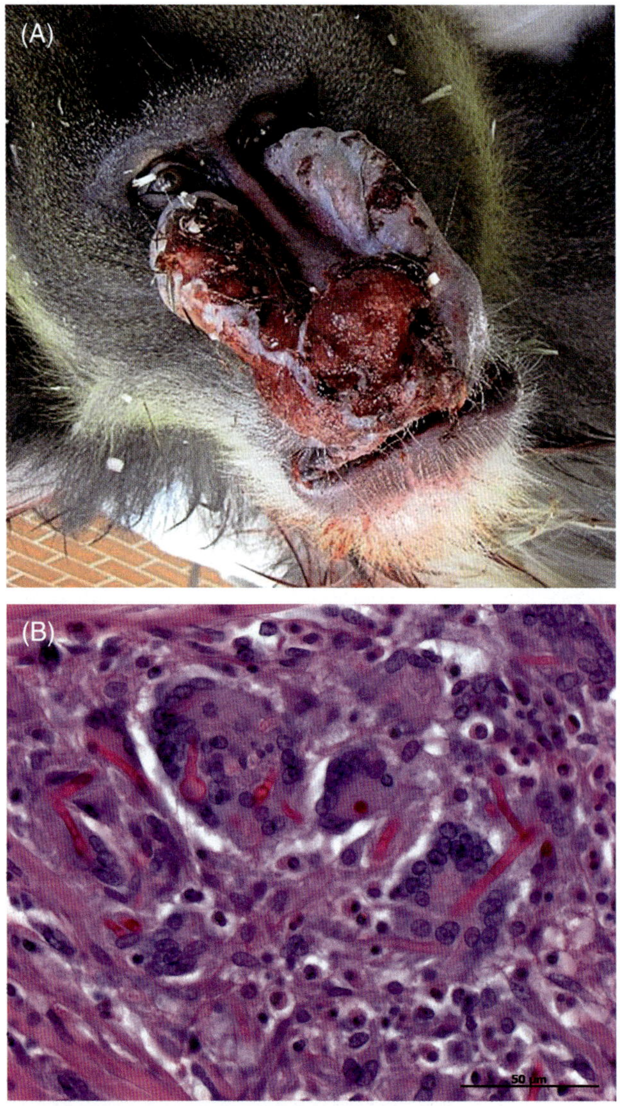

FIGURE 14.29 Candidiasis infection in a mandrill. (A) Severe ulcerative skin lesions on the bridge of the nose, the nasal planum, and the mucocutaneous junctions. (B) Granulomatous inflammation in the subcutis with fungal hyphae and blastospores in the cytoplasm of multinucleated giant cells. PAS reaction.

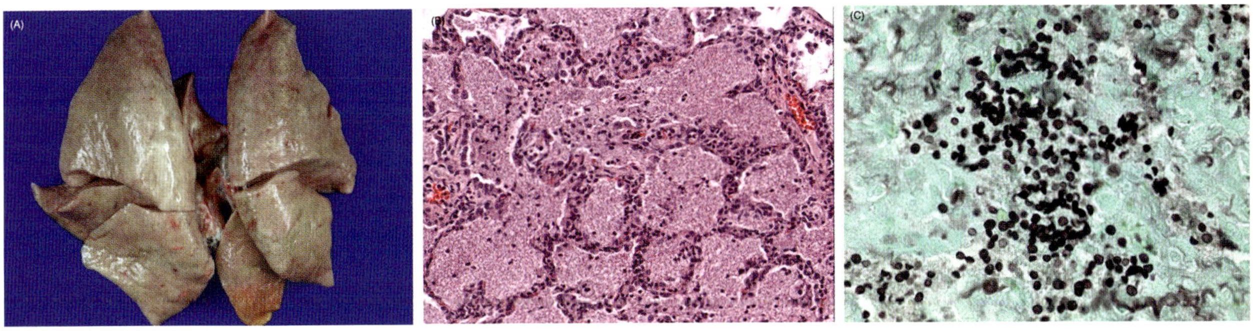

FIGURE 14.30 *Pneumocystis carinii* infection in the lung of a rhesus macaque. (A) The lungs fail to collapse, are discolored pale tan, and have a firm, meaty consistency on palpation. (B) Interstitial pneumonia is characterized by a moderate histiocytic cell infiltrate. Myriad free and intrahistiocytic, eosinophilic *P. carinii* cysts and trophozoites give the cells and infiltrate a foamy appearance. (C) Silver staining highlights intralesional 4–6 μm cysts. Grocott silver.

capsulatum var. ***capsulatum***, and ***Paracoccidioides brasiliensis***, tend to induce systemic infections,[17,38,50,104,123,272] while others like ***Sporothrix schenckii*** and ***H. capsulatum*** var. ***duboisii*** are usually confined to the skin.[33,51] Several cases of cryptococcosis have been described in NHP including squirrel monkeys,[219] common marmoset,[123] and several old world monkey species.[104,184,196] In disseminated cryptococcosis, nearly any tissue can be affected.[104] The most common sites of infection are the lungs and the central nervous system *C. neoformans* is neurotropic and frequently spreads to the central nervous system from the respiratory tract either hematogenously or via direct extension through the cribriform plate. Related inflammation is suppurative to necrotizing and the yeasts are 4–15 μm diameter and surrounded by a thick capsule, colorless with HE staining and PAS reaction (Figs. 14.31A, B) that is mucicarmine-positive.

Aspergillosis infection is also most common in the lungs but disseminated infection may also occur (Fig. 14.32).[276] *Encephalitozoon cuniculi* is limited to NWM.[97,122] An overview of mycotic infections in NWM and OWM is presented in Table 14.e6.

Metazoa

Parasitic infections of NWM and OWM are frequently observed, especially in wild-living populations. An exhaustive description of the wide spectrum of parasites to which NWM and OWM are susceptible is beyond the scope of this text. Notable parasites in these species are described below and listed in Table 14.e7.[253]

Hydatidosis, also known as **echinococcosis**, is induced by the larval stages of cestodes belonging to the genus

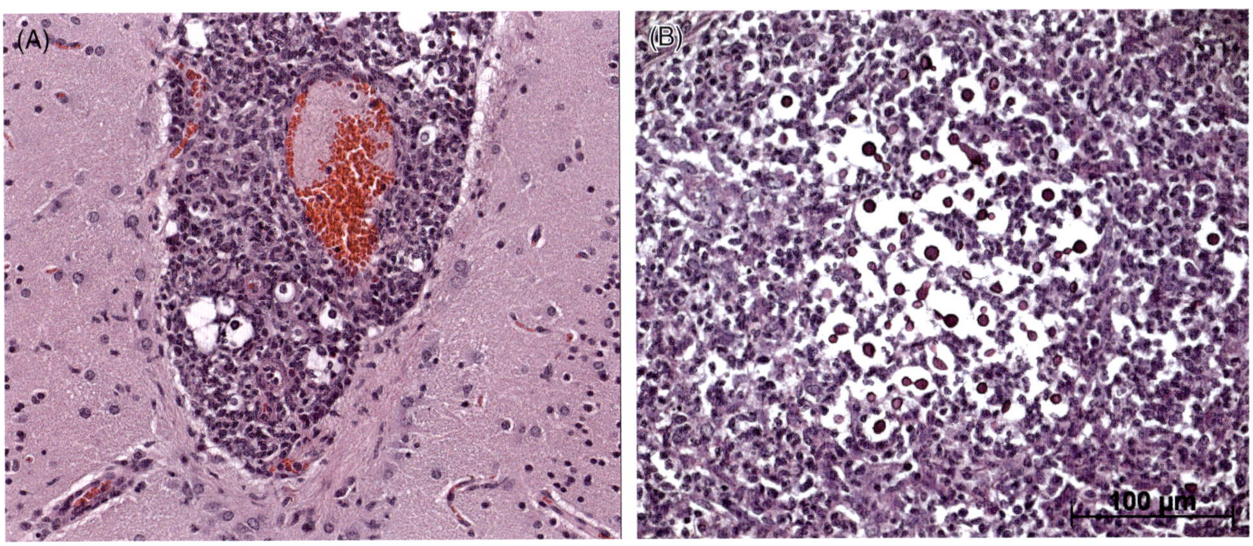

FIGURE 14.31 Cryptococcosis in the central nervous system of a white-lipped tamarin. (A) Suppurative meningitis and encephalitis with typical yeasts of 4–15 μm diameter. (B) The yeasts are surrounded by a thick capsule that is colorless with hematoxylin and eosin staining and the PAS.

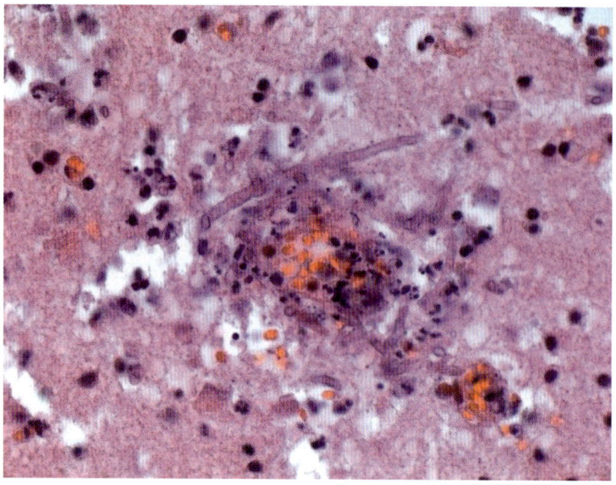

FIGURE 14.32 *Aspergillus fumigatus* infection in the brain of a putty-nosed monkey. Angiocentric lytic necrosis is associated with perivascular fungal hyphae that are parallel-walled, septate, and have dichotomous branching.

Echinococcus. The definitive hosts are canids. Hydatid cysts caused by infection with ***E. granulosus*** are described in a number of NWM and OWM species.[90,171] Hydatid cysts are large and unilocular. The characteristic hyalinized cyst wall is finely laminated and amphophilic. The inner layer of the cyst is lined by germinal epithelium, which produces numerous brood capsules containing protoscolices and calcareous corpuscles. Protoscolices have armed rostella with acid-fast hooklets and four suckers. Cysts may be surrounded by fibrosis. Hydatid cysts may be located in the abdominal or thoracic cavity, liver, lung, and retrobulbar space. Cyst rupture releases protoscolices that seed body cavities and tissues that can lead to the development of massive numbers of daughter cysts. Associated inflammation is generally absent or mild. The closely related ***E. ortleppi*** is described in langurs and gibbons in Asia.[205]

Echinococcus multilocularis hydatid cysts are reported with increasing frequency in OWM. This fox tapeworm is widely distributed in the northern hemisphere. The sylvatic life cycle involves red foxes as a final definitive host and small rodents as intermediate hosts. Many mammal species including humans and nonhuman primates can be infected accidentally. Reports from Switzerland,[7] Germany,[259] and Japan[227] indicate that the disease is a threat to nonhuman primates living in zoological gardens or other outdoor facilities. The fox tapeworm larvae differ from other hydatid cysts by forming multiple small budding cysts resembling alveoli (Fig. 14.33A). The larvae propagate by budding from the germinal membrane to form multilocular cysts with poorly defined borders that invade neighboring host tissue like a tumor. Cysts are composed of a typical laminated membrane with a germinal layer producing numerous protoscolices (Fig. 14.33B; eSlide 14.e7). Sterile cysts with no protoscolices can occur but are rare. Granulomatous inflammation is generally present around the cysts. Cyst size depends on its age, location within the host, and host reaction. Large cysts often develop necrotic centers. Ruptured cysts may release protoscolices that implant on other tissues producing daughter cysts. Cyst rupture can lead to an anaphylactic reaction. Pathologic changes produced by ***E. vogeli***, a tapeworm in South and Central America, produce polycystic hydatid cysts that are intermediate between those of the unilocular hydatid and alveolar hydatid cysts.

Cysticercosis is due to infection with larval cestodes in the family Taeniidae. Cysticerci are round, thin-walled fluid-filled cysts containing a single invaginated larval

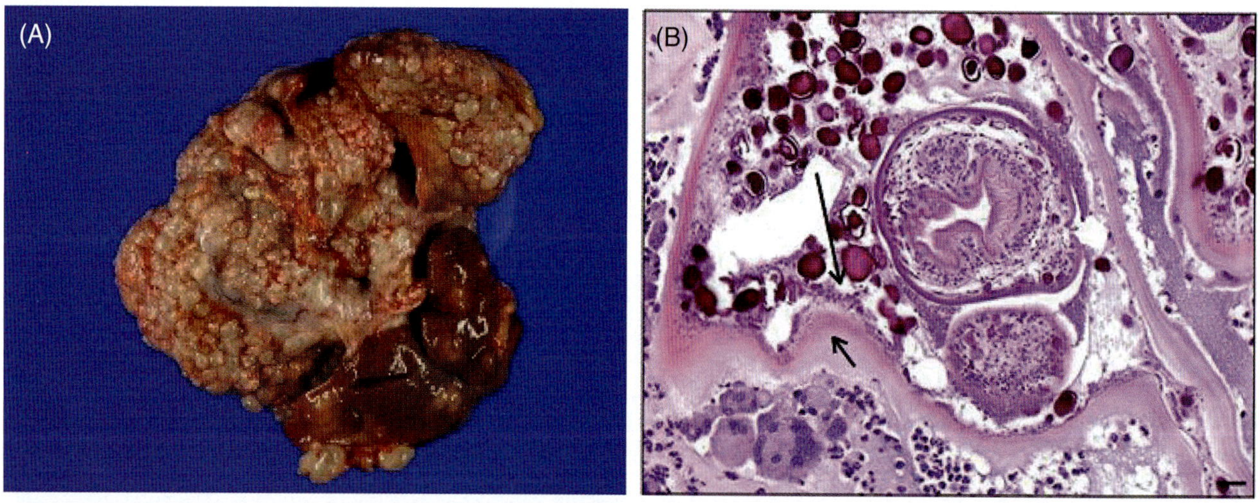

FIGURE 14.33 *Echinococcus multilocularis* infection in the liver of a long-tailed macaque. (A) The liver is markedly expanded and the parenchyma is replaced by multiple cystic to firm, nodular masses. (B) The cysts of the developing parasites have characteristic architecture consisting of an outer laminar layer, inner germinal layer *(arrows)*, and multiple inner protoscolices and mineralized bodies within a brood capsule (see eSlide 14.e7).

scolex with four suckers. Cysts are generally found in the peritoneal or pleural cavity[109,264] or in the muscle.[21] Infections are also described, but less commonly, in the central nervous system or subcutaneous locations.[279] Clinical symptoms are related to mass effect, as the cysts can grow quite large. Some species, such as ***Taenia crassiceps***, multiply asexually in tissue producing multiple cysts (Figs. 14.34A, B) containing characteristic cysticerci.[21,165] **Cysticercosis** must be differentiated from **coenurosis** which develops due to infection with larval forms of ***Multiceps multiceps*** or ***M. serialis***. Canids are their definitive hosts. Coenurus infection can be differentiated from cysticercosis as the former has a thicker-walled cyst with a germinal layer that produces multiple scolices.

Platynosomum sp. is a trematode parasite with worldwide distribution. It has a heteroxenous life cycle with more than one intermediate host, and mammals, including New World Primates, as definitive hosts.[203] Callitrichidae are susceptible to infection with adult forms found in the bile ducts. Histologic lesions range from mild to severe and are associated with neutrophilic, eosinophilic, and histiocytic inflammation that is centered on bile ducts and reactive fibrosis (Fig. 14.35). In severe cases, bile ducts are hyperplastic to dysplastic and bridging fibrosis and cirrhosis may be found. An association between infection and hepatocellular carcinoma was suggested in a marmoset.[61]

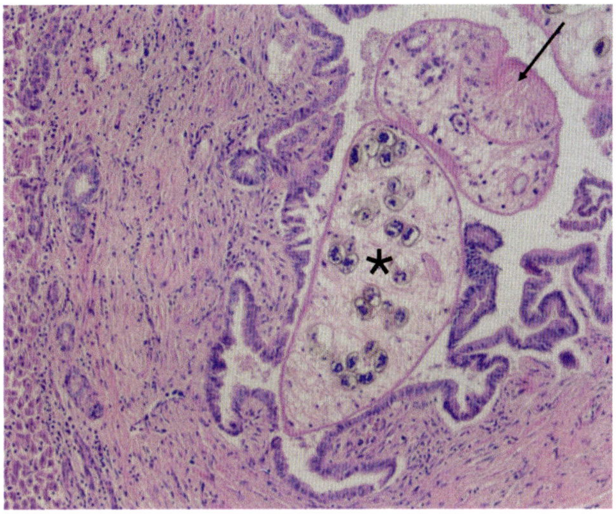

FIGURE 14.35 Intrahepatic biliary trematodiasis in a marmoset. An adult trematode *(asterisk)* is present in the lumen of a dilated bile duct. The *arrow* indicates the parasite's sucker. There is moderate to severe periductular fibrosis, with ductular reaction and mild inflammation.

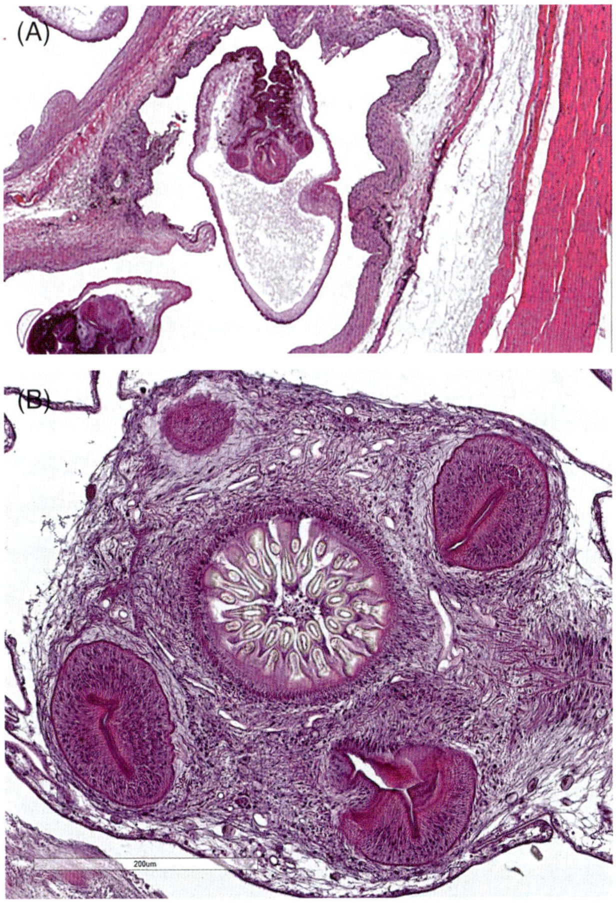

FIGURE 14.34 Cysticercosis induced by cysts of *Taenia crassiceps* in a *Nilgiri langur*. (A) Multichambered cysts containing protoscolices are found in the muscle. (B) The scolices have four suckers and a typical rostellum with 16 hooks.

Parasites in the phylum Acanthocephala (at least 1000 species), can be found in a wide range of hosts including fishes, reptiles, birds, and mammals. Beetles and cockroaches are intermediate hosts for vertebrates who are the definitive hosts. Parasites in the *Prosthenorchis* genus, for example, ***Prostenorchis elegans***, have been reported in NWP. The parasites have a thick hypodermis and lack a body cavity, mouth, and digestive tract. Infections have been reported in lion tamarins and marmosets in both wild and captive settings. The parasite invades the muscular layer of the intestine causing transmural enteritis with nodular leiomyositis (Figs. 14.36A, B). Perforation and peritonitis have resulted in death in severe cases. Chronic infections may predispose to malnutrition, infections, and sepsis.[39,58,254]

Protozoa

NWM are highly susceptible to **toxoplasmosis** caused by the apicomplexan protozoan *Toxoplasma gondii*.[62,72,124] The organisms are found worldwide and felids are the only known definitive hosts. Most reported epizootics in NWM have occurred among captive squirrel monkeys.[37] The most common transmission route is by ingestion of feed that is

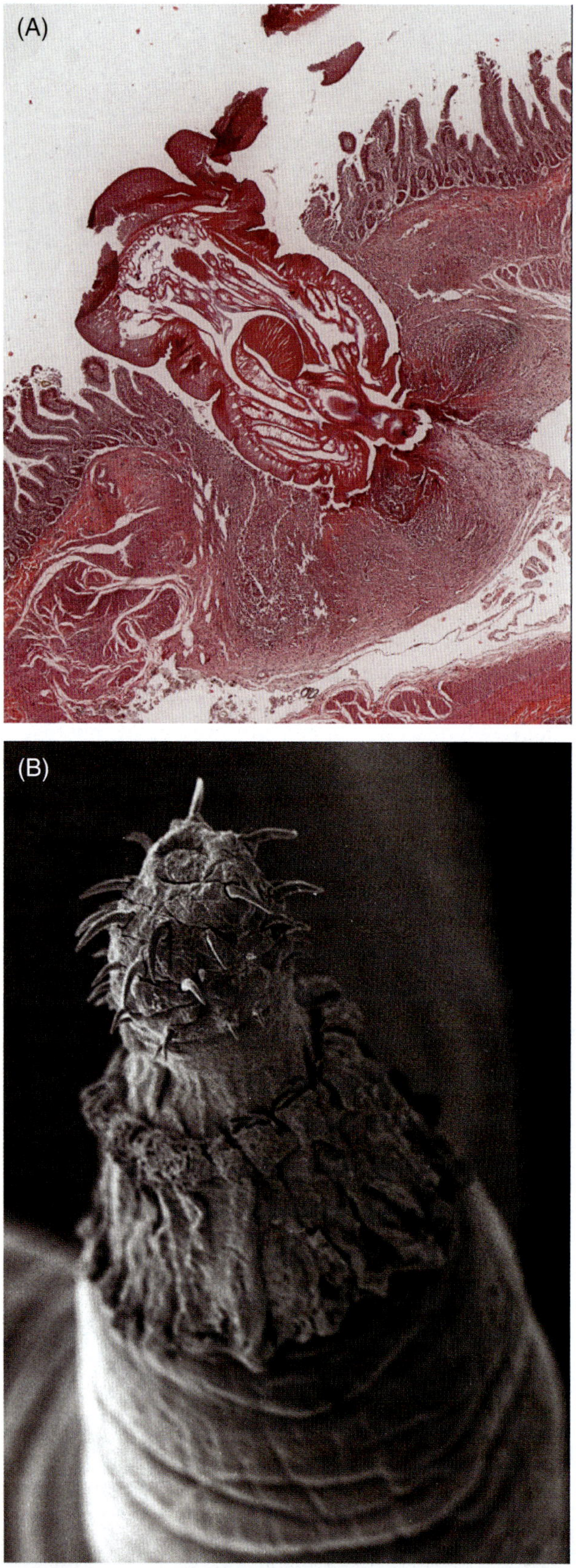

FIGURE 14.36 Acanthocephaliasis induced by *Prostenorchis elegans* in a coppery titi monkey. (A) The head of the thorny-headed parasite is deeply embedded in the intestinal wall inducing severe ulcerative and granulomatous enteritis. (B) Electron microscopic picture demonstrates thorns on the head surface of the parasite.

contaminated with infective oocysts. Other sources are infected rodents caught and eaten by NHP or infected meat, insects, and earthworms that serve as paratenic hosts. Horizontal transmission among infected animals is possible.[84] Primates in the wild are rarely seropositive.[86] Clinical signs are nonspecific and include malaise and dyspnea. The most common postmortem findings are pulmonary congestion and severe pulmonary edema (Fig. 14.37A), splenomegaly, and mesenteric lymphadenitis. The most prominent histopathological findings are multifocal necrotizing hepatitis, lymphadenitis, interstitial pneumonia, and necrotizing splenitis. Mixed inflammation is present in altered organs. Tachyzoites are common in necrotic lesions while tissue cysts containing bradyzoites are encountered in more chronic lesions (Fig. 14.37B). In OWM with inapparent infections, tissue cysts are most common in the brain.

Ectoparasites

Pulmonary acariasis is induced by **lung mites**, which are obligate endoparasites of the respiratory tract. They feed on host pulmonary epithelial cells, erythrocytes, and lymph. Infections can be caused by several species of the genus ***Pneumonyssus***, with ***Pneumonyssus simicola*** being most commonly identified.[116] Lung mite infections have been reported in more than 20 OWM species. The prevalence of lung mites in wild-caught monkeys is approximately 100%, whereas NHPs born and raised in captivity have a low incidence of infection.[83] Suggested transmission routes include inhalation of infected aerosols, ingestion during grooming, or through fecal-oral routes. Lung mites in general do not cause any clinical symptoms, but they can predispose to pulmonary bacterial infection due to impaired mucociliary clearance. In addition to pneumonia, serious complications from lung mite infection include pneumothorax or pulmonary arteritis that can lead to fatal outcomes. There is an apparent predilection for infection in the cranial lung lobes, which might be related to aerogenous transmission or higher oxygen content in these areas.[3] Reliable, standardized in vivo tests are not currently available for antemortem diagnosis. Therefore, and with respect to subclinical course of infection in most cases, prophylactic or metaphylactic anthelminthic treatment (e.g., ivermectin) is administered in captive settings. Most inflammatory changes seem to be reversible after mite elimination; however, chronic mite-induced pulmonary changes, such as bronchiectasis, pigment accumulation, or smooth muscle hypertrophy persist, making affected animals more susceptible to other lung diseases.

Macroscopically, lung mite infections appear on the lung surface and throughout the parenchyma as multifocal gray-yellow, thin to irregularly thick walled, 1–7 mm in diameter cysts containing air, serous fluid or, in case of inflammatory reaction, mucopurulent exudate or necrotic caseous debris. Besides pulmonary lesions, tracheobronchial lymph nodes can contain dark brown to black

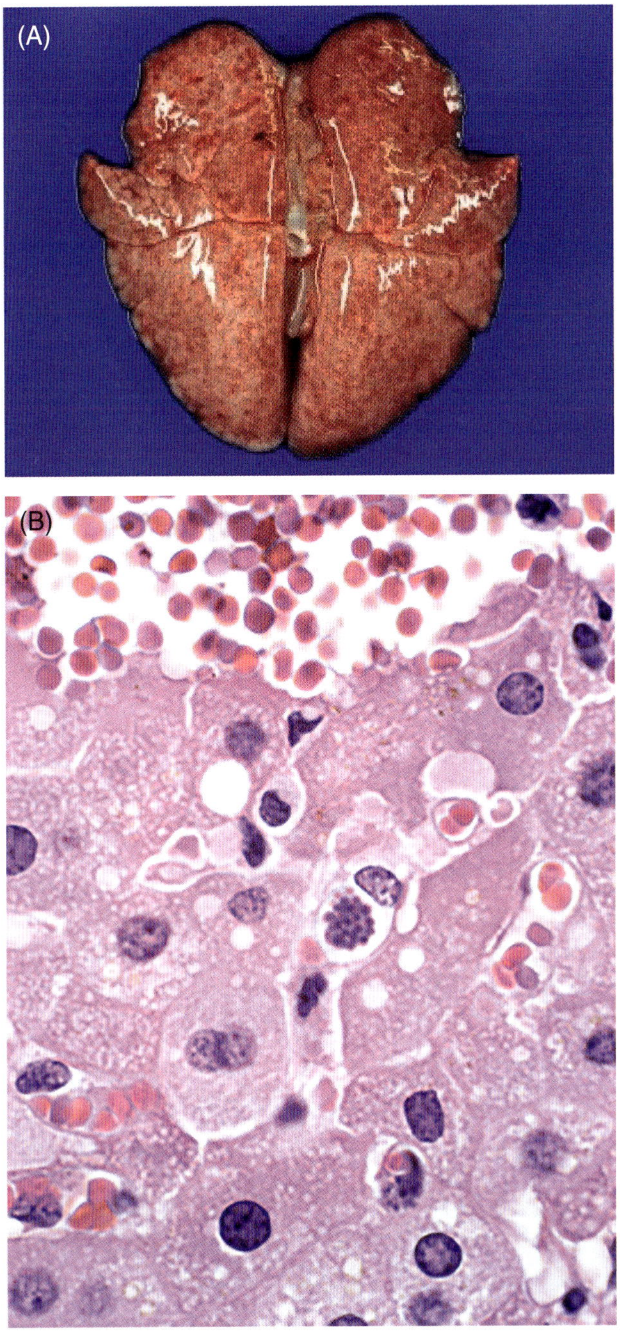

FIGURE 14.37 Toxoplasmosis in a squirrel monkey. (A) Infection-associated pneumonia can result in firm lungs that are edematous with multifocal areas of hyperemia. (B) In the liver, protozoal infection can be associated with hepatocellular necrosis. In this case, a necrotic hepatocyte is expanded by an intracytoplasmic toxoplasma cyst (center of the field).

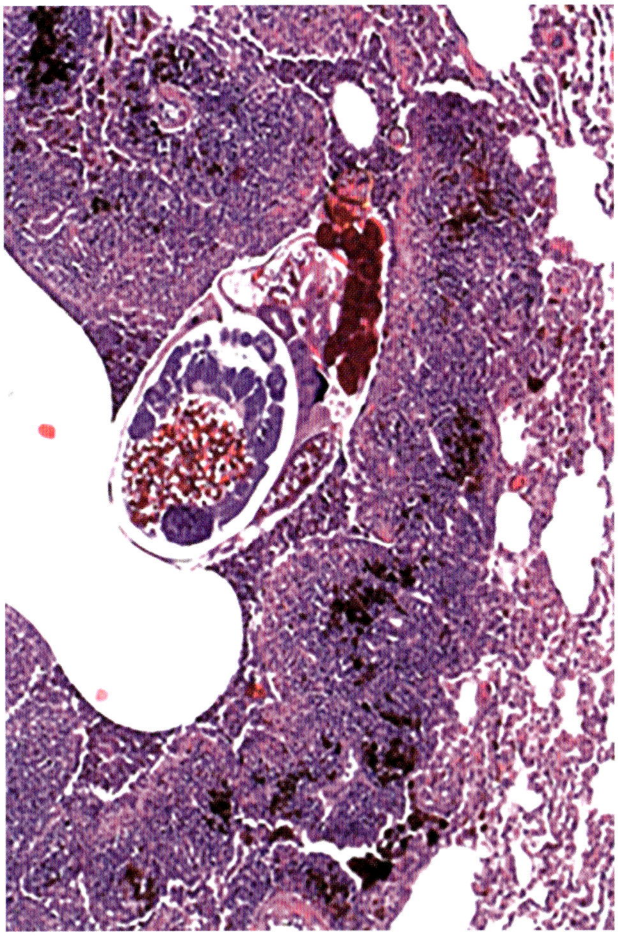

FIGURE 14.38 Lung mite infection in a rhesus macaque. A large, intrabronchiolar mite is associated with granulomatous bronchopneumonia that markedly expands the bronchiolar wall. Intense mixed inflammation consists of epithelioid macrophages, multinucleated giant cells, lymphocytes, and plasma cells admixed with granular brown to black mite pigment. Characteristic features of lung mites in this image include a chitinous exoskeleton, jointed appendages, striated muscle, a body cavity, gut segments, and a uterus.

discoloration due to mite pigment accumulation, and fibrinous pleural adhesions can develop.

Histologically, cystic lesions are characterized by variable numbers of mite sections, that are most often located in bronchial or bronchiolar walls (Fig. 14.38) and are surrounded by mucus, cellular debris, and granulomatous inflammation consisting of epithelioid macrophages, multinucleated giant cells (foreign body type), eosinophils, neutrophils, fibroblasts, and collagen deposition. Bronchiolar ectasia, alveolar edema, BALT (bronchiolar associated lymphoid tissue) hyperplasia, and pulmonary arteritis with medial hyperplasia can also be seen. A striking histologic feature of lung mite infection is the typical intralesional golden brown, birefringent, granular pigment that accumulates in areas of peribronchiolar and peribronchial inflammation, macrophages or vessel walls; the pigment can also often be readily detected macroscopically. This pigment represents mite excrement, which contains at least three different components. The majority consists of large granules of hemosiderin or lipofuscin and a lesser amount of needle-like crystals and small granular or amorphous products of mucinous substance. Large parts of the pigmented material are identified as silicates. The pigmented material can be found even when living mites are no longer present in the lung.

OTHER INFECTIOUS ORGANISMS

Visceral **pentastomiasis** is an unusual parasitic zoonotic disease caused by larval stages of several species of pentastome parasites that form a unique phylum with characteristics of both arthropods and annelids.[44] Pentastomes are regarded as parasitic crustaceans within the family of arthropods, whose final hosts are carnivore reptiles, birds, and mammals. Intermediate hosts are herbivore vertebrates (e.g., fish, rodents, ungulates). Pentastomiasis usually does not cause any clinical symptoms or pathological changes and may be an incidental finding during surgery or necropsy. The c-shaped parasites can be found in various visceral organs including the mesentery, serosal surfaces, liver, and lung. Histologically, parasitic cysts are lined by a thin, 15–20 μm hyaline membrane, likely generated by the parasite, and surrounded by variably thick fibrous tissue.[178] Characteristic features of the parasite are a pseudosegmented body, two pairs of chitinous hooks flanking the mouth, a chitinous cuticle, fascicles of striated muscle in the body wall, and large acidophilic glands (Fig. 14.39).

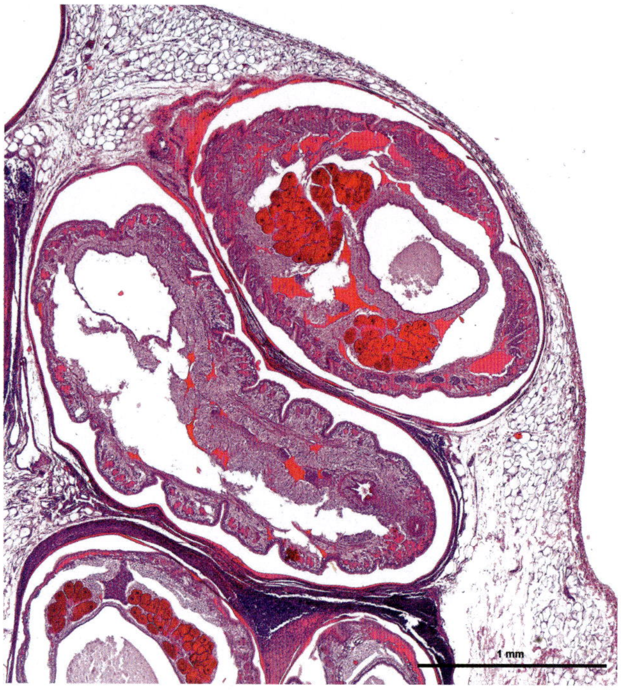

FIGURE. 14.39 Pentastomiasis in a cynomolgus monkey. Histology of typical c-shaped parasites within the mesentery. The parasitic cysts are lined by a thin, 15–20 μm hyaline membrane. Characteristic features of the parasite are a pseudosegmented body, a chitinous cuticle, fascicles of striated muscle in the body wall, and large acidophilic glands.

ADDITIONAL RESOURCES

Mittermeier, R.A., Rylands, A.B., Wilson, D.E., 2013. Handbook of the Mammals of the World: Primates. Lynx Editions, Barcelona.

Pereira, A.H., Vasconcelos, A.L., Silva, V.L., Nogueira, B.S., Silva, A.C., Pacheco, R.C., Souza, M.A., Colodel. E.M., Ubiali, D.G., Biondo, A.W., Nakazato, L., Dutra, V. 2022. Natural SARS-CoV-2 infection in a free-ranging black-tailed marmoset (*Mico melanurus*) from an urban area in mid-west Brazil. J Comp Pathol. 194, 22–27.

Simmons, J., Gibson, S., 2012. Bacterial and mycotic diseases of nonhuman primates. In: Abee, C.R., Mansfield, K., Tardif, S., Morris, T. (Ed.), Nonhuman Primates in Biomedical Research. Elsevier, 105–172.

Strait, K., Else, J.G., Eberhard, M.L., 2012. Parasitic diseases of nonhuman primates, In: Abee, C.R., Mansfield, K., Tardif, S., Morris, T. (Eds.), Nonhuman Primates in Biomedical research. Diseases. Elsevier, 197–297.

References

Full references for this chapter can be found online at https://doi.org/10.1016/B978-0-443-18437-6.00018-3.

Chapter 15

Apes

Linda J. Lowenstine[a], Rita McManamon[b], Karen A. Terio[c], and Sushan Han[d]
[a]University of California, Davis, CA, United States; [b]University of Georgia, Athens, GA, United States; [c]Zoological Pathology Program, College of Veterinary Medicine, University of Illinois, Brookfield, IL, United States; [d]Denver Zoo Conservation Alliance, Denver, CO, United States

INTRODUCTION

Apes are Old World primates found in Southeast Asia and Africa. The group includes the gibbons or small apes (family Hylobatidae) and the great apes (family Hominidae): bonobos ("pygmy" chimpanzees), (common) chimpanzees, gorillas, and orangutans. Many ape species are found in zoos, including several of the 17 species of gibbons, common chimpanzees, bonobos, western lowland gorillas, and Bornean, Sumatran, and, possibly, Tampanuli orangutans. In terms of diseases and pathology, the common chimpanzee is the best studied of the apes, primarily due to large laboratory colonies, zoo collections, and longstanding field studies.[204,363,380,406,576,608,694] Most literature on diseases of other apes concerns zoo-housed animals; however, there is an expanding literature on health and diseases of free-living western lowland and eastern (mountain and Grauer's) gorillas and chimpanzees.[39,42,105,106,238,260,287,288,323,393,394,424,460,498,651,652,727] Only a few health studies of free-living and range country sanctuary-housed bonobos, orangutans, and gibbons have been published.[186,339,441,739,740,745,771]

All ape species are endangered or critically endangered[294] (Table 15.e1). Threats to wild populations include: hunting for bush-meat trade; exploitation for pet trade; accidental entrapment in snares set for other animals; habitat loss due to deforestation for firewood, commercial logging, mining, and palm oil plantations; wildfires created by human activities, climate change and nonanthropogenic factors including predation by large carnivores, especially leopards, and even, in rare cases, other apes, and lightning strikes.[122,377,656] African ape populations are also threatened by diseases, such as Ebola virus, anthrax-like infection, and respiratory diseases.[387,392,658,734]

Extensive review of ape special anatomy is beyond the scope of this chapter. The reader is referred to physical anthropology/primatology texts and refereed literature.[10,209,215,231,409,674,683] Notable gross and histologic features that impact interpretations by pathologists are noted in the following sections and in the Digital Content that accompanies this chapter.

NON-INFECTIOUS DISEASES

Nutritional

A syndrome of **regurgitation and reingestion** is described in zoo-housed gorillas, bonobos, chimpanzees, and orangutans.[449] It can be mitigated by increasing the fiber, the amount of browse in the diet, or opportunities to feed.[197,393] Damage to dental enamel is the only described pathological correlate.

In apes, **protein deficiency** is occasionally seen in captive or rehabilitation settings but is much less commonly reported than in New World (NW) monkeys.[478] The protein requirement of apes is similar to that of Old World (OW) monkeys: ~14% on a dry matter basis in young chimpanzees, and less in adults.[518] Protein deficiency in orphaned, young lowland gorillas is associated with chronic alopecia along the back and legs, hair color change from black to gray or brown, poor growth, weight loss, normocytic normochromic anemia, and hypoalbuminemia. Similar signs can occur secondary to severe parasitism even in adult animals.[542]

Both **vitamin D deficiency** (**rickets**, **osteopenia**, **metabolic bone disease**) and **vitamin C deficiency** (**scurvy**, **scorbutic rickets**) can occur in apes in managed care. Like all other nonhuman primates, apes lack functional L-gulono-γ-lactone oxidase rendering them unable to synthesize vitamin C.[155] Vitamin D deficiency can occur in gorillas and chimpanzees housed in indoor exhibits without sufficient natural sunlight. Maternal deficiency can lead to metabolic bone disease in nursing infants.[36,310,454] Lesions in apes with deficiencies of vitamins C and D are similar to those described in NW and OW monkeys and other vertebrates (see Chapter 14: New World and Old World Monkeys).

Obesity is a common problem in many zoo gorillas and orangutans and in chimpanzees in both zoos and colonies.[101,561] **Metabolic syndrome** in female chimpanzees is linked to obesity and likely also exists in other apes.[667] Clinical parameters include increased abdominal fat (measured by increased waist/umbilical

Pathology of Wildlife and Zoo Animals. https://doi.org/10.1016/B978-0-443-18437-6.00046-8

girth circumference), increased fasting blood glucose and triglycerides, decreased high-density lipoproteins, and elevated blood pressure.[174,723] Morbid obesity is associated with diabetes mellitus, heart disease, and stroke in adult orangutans.[43,247,410]

Iron overload occurs in zoo-housed apes. As is the case in some species of monkeys and birds (refer to Chapter 33: Passeriformes, Caprimulgiformes, Coraciiformes, Piciformes, Bucerotiformes, and Apodiformes), it is most common in folivorous and frugivorous species (gorillas, orangutans, and gibbons). It is rare in free-living mountain gorillas except in neonates and in starvation or chronic illness causing hepatic atrophy.[413] Grossly, there may be orange/brown discoloration of the duodenal mucosa and bronze discoloration of the liver. Histologically, hemosiderin is present in tingible body macrophages in the duodenal villous lamina propria, hepatocytes (periportal to panlobular), and Kupffer cells. Note that pigment commonly seen in centrilobular hepatocytes of mountain gorillas is lipofuscin. Susceptible species have likely evolved enhanced mucosal absorption (no mucosal blockade) to compensate for natural diets high in tannins and other phytochemicals that bind iron in the intestinal lumen. Molecular physiology of ape iron balance has not been investigated. Chronic disease (e.g., mycobacteriosis) or captive diets high in ascorbic acid and low in fiber, which enhance iron absorption, can lead to iron sequestration and cause, or exacerbate, iron overload.[479]

Metabolic

The literature on endocrine disorders of apes is sparser than that on NW and OW monkeys.[29] **Type II diabetes mellitus (T2DM) and insulin resistance** are associated with obesity in female chimpanzees and orangutans but are occasionally reported in nonobese chimpanzees.[8,43,574] Type II diabetes, defined by elevated fasting glucose and hemoglobin A1C, occurs in about 1% of all laboratory-housed chimpanzees and 3.7% of chimpanzees over 30 years of age.[439] Values suggestive of prediabetes are fasting glucose (FG) 106 to 119 mg/dL and glycated hemoglobin (HbA1c) 5.1%–5.2%. Those suggestive of diabetes are FG greater than or equal to 120 mg/dL and HbA1c, greater than 5.3% Elevated FG levels are also noted in about 6.7% of zoo-housed orangutans with levels suggestive of prediabetes and diabetes when evaluated in concert with fasting insulin levels.[232] Overweight adults, both male and female, are usually affected. However, weight or age alone is not statistically correlated with fasting insulin and glucose levels in orangutans. Renal dysfunction with decreased glomerular filtration rate, a common complication of diabetes in other species, is also reported in diabetic chimpanzees.[575] Histological islet lesions, such as altered ratios of hormone-producing cells (decreased insulin-producing beta cells and increased glucagon-producing alpha cells) and accumulation of amylin-type amyloid in the islets of Langerhans, as are reported in diabetic humans, OW monkeys, felids and other species, are not common in apes.[29,659] Islet lesions reported in apes include: islet cell hyperplasia with decreased insulin in a chimpanzee with clinically impaired glucose tolerance and hyperglucagonemia; islet amyloidosis noted in a single chimpanzee in a mortality review from a research colony; and swelling and vacuolation of islet cells in a chimpanzee with diabetic hyperosmolar state (glucose >500 and HbA1c 12%).[72,363,574]

Though less common than diabetes, **hypothyroidism** and **hyperthyroidism** have been reported in gorillas, orangutans, and a chimpanzee.[187,372,679] Congenital thyroid dysfunction due to thyroid dysgenesis, determined by ultrasonography and radionucleotide uptake scintigraphy, is reported in two infant orangutans (a 9-month Bornean and a 6-week Sumatran).[187] Clinical signs include delayed growth, poor development, and thermoregulation. Thyroid-stimulating hormone (TSH) levels are markedly elevated and serum thyroxine and free T4 levels markedly decreased when compared to age-matched controls. Supplementation with levothyroxine results in clinical improvement. In an adult female western lowland gorilla, reported clinical signs of hypothyroidism include weight gain, altered appetite, lethargy, and behavioral abnormalities including anxiety, elevated TSH, and decreased T4 and free T4. Autoimmune thyroiditis is discounted as a cause of ape hypothyroidism based on a preliminary study.[4] The histology of ape thyroids in health and disease is not well described, and thyroids should be examined and weighed routinely at necropsy.

Congenital/Genetic

All great apes have a diploid number of 48 chromosomes (humans have 46). The number of chromosomes in gibbons varies from 2N = 38 to 52. **Chromosomal defects** in apes that are usually lethal in infancy include **trisomy 22** (analogous to human trisomy 21, Down syndrome) in chimpanzees and a Sumatran orangutan, and **trisomy 17** (analogous to trisomy 18 in humans) in a bonobo.[9,41,382,434] In addition to fetal death, phenotypic characteristics suggesting karyotypic anomalies include low birthweight, delayed developmental milestones, weakness, poor nursing, dull mentation, prominent epicanthic folds, syndactyly, protrusion of the tongue, abnormal pinnae, flat facial plane, spinal anomalies, and heart defects. Congenital hypothyroidism can have similar signs.[187] Karyotypic anomalies identified in adults include mosaic **triple X** in a female gorilla (similar to XXX Turner syndrome) with obesity and low fertility, a **3q deletion** (analogous to human deletion 4q) in a male gorilla with abnormal face and short limbs, and trisomy 22 in an adult female chimpanzee with stunted growth and atrial septal defect.[58,275,382]

Syndactyly (interdigital webbing) is the most common congenital defect in eastern gorillas, occurring in >50% of

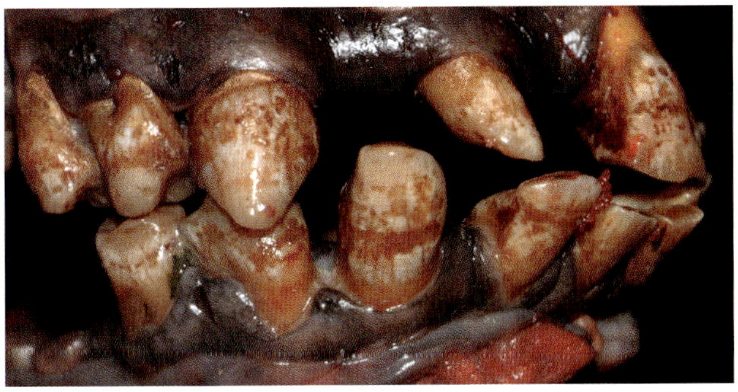

FIGURE 15.1 Linear enamel hypoplasia and pitting of enamel in a geriatric orangutan. Hypoplasia is characterized by horizontal bands of decreased thickness or absent enamel. Tooth loss, gingival regression, and periodontal disease are also present.

mountain gorillas.[471] Webbing between the second and third toes is normal in the siamang gibbon, hence the species name *syndactylus*. Bilateral absence of the terminal phalanx of the hallux in a lowland gorilla is also reported.[274]

Linear enamel hypoplasia (LEH) and **localized enamel hypoplasia of the primary canines (LHPC)** are the most common **dental anomalies** in apes in captivity and in the wild.[252] Defects occur in either deciduous or permanent teeth and may involve one or multiple teeth (Fig. 15.1). The pathogenesis involves transient disruption of enamelogenesis due to internal stressors, such as disease, or environmental stressors, such as climatic events that disrupt food availability.[238] Synthesis of several studies confirms that, in general, orangutans are most affected followed by chimpanzees and gorillas, though in sympatric western lowland gorillas and chimpanzees, the gorillas are more often affected. The influence of fluoridation of city water on enamel dysgenesis in zoo-housed apes has been suspected but not proven. A maxillary lateral incisor enamel defect (MLID), which differs from linear enamel hypoplasia in that it is nonlinear, is described in orangutans, chimpanzees, and gorillas, occurring in 30% of 136 skulls examined.[251]

Several congenital **cardiac defects** are described in apes including atrial and ventricular septal defects in orangutans, univentricular heart in an orangutan, ventricular septal defects in a premature chimpanzee and a gorilla, and coarctation of the aorta in a gorilla.[46,107,230,443,705] None are species specific or of population concern.

Congenital facial dysplasias including abnormal flattening of the facial profile and nose, abnormal shape, size and positioning of the nostrils, and cleft nostrils are described in 16 of 66 wild chimpanzees living near an agricultural area.[356,358] Baboons in the area are also affected. A variety of pesticides and other chemicals are used in the area. It is hypothesized, but not proven, that chemically induced, maternal endocrine disruption during pregnancy leads to defects in offspring.

Palate defects in two mountain gorillas, a lowland gorilla, and a gibbon seem to be isolated cases without known genetic or environmental predisposition.[520,630,701]

A Chiari defect of the skull (herniation of portions of the cerebellum through the foramen magnum) and syringomyelia are described in a male, zoo-born, chimpanzee with behavior abnormalities.[653] An 11-year-old male sanctuary-housed former research chimpanzee with progressive behavioral and motor abnormalities and liver failure was determined by MRI to have extensive symmetrical leukodystrophy involving frontal lobe white mater and corpus callosum with genetic abnormalities compatible with human juvenile onset X-linked adrenoleukodystrophy.[120]

Age-Related/Degenerative

The most significant degenerative diseases of captive ape populations are **cardiovascular disease, renal disease,** and **arthritis**. **Neurodegenerative diseases** also occur in apes and are important for comparison with human dementias. Additional information on the pathology of aging in apes can be found in a review.[411]

Cardiovascular disease (CVD) is a significant factor in the deaths of zoo and laboratory-housed apes, being reported in approximately 45% of bonobos, 41% of gorillas, 38% of zoo chimpanzees, >50% of colony chimpanzees, and 20% of orangutans.[43,239,247,375,380,410,438,442,617,676,677] Gibbons are also affected but the prevalence is unknown.[53,336] Cardiovascular disease, often incidental to cause of death, is also recognized in free-living chimpanzees and eastern gorillas.[323,519,694]

Normal heart weights (mass) (HW) and measurements are not established for the apes. Data from a limited number of zoo-housed gorillas indicate that adult male heart mass ranges from 449 to 1,600 g.[40] Adult male gorilla hearts without overt heart disease range from 449 to 720 g; hearts with mild fibrosis or hypertrophy 720 to 820 g; and hearts from gorillas with confirmed heart disease range from 725 to 1,600 g. The heaviest hearts are dilated as well as hypertrophied. Heart weights of laboratory chimpanzees, from colonies with a high prevalence of sudden cardiac death, range from 195 to 450 g (HW/BW [body weight] 0.39% to 0.94%) for females, and 275 to 660 g (HW/BW 0.46%–1.06%) for males.[375]

The only biomarker that seems to correlate with clinical heart disease in gorillas is brain-natriuretic protein (BNP).[483] BNP as well as troponin 1, some markers of fibrosis/collagen synthesis and degradation (procollagen carboxyl-terminal telopeptide [ICTP] and procollagen III amino-terminal propeptide [PIIINP]) may have utility in diagnosing cardiovascular disease in chimpanzees.[176] In gorillas, low TNF-alpha has been suggested as a possible predictive marker for cardiovascular morbidity.[164] In gorillas, adiposity in males and altered gut microbiome are thought to be contributory, or indicators for, the development of cardiovascular disease.[138,361]

The most frequent cardiac lesion in all four great ape taxa is **myocardial fibrosis**.[373,375,410,611,676,677] Also seen are **aortic dissection**, **arteriosclerosis**, **atherosclerosis**, and **valvular disease**.[335,614] Less common is a condition called **arrhythmogenic right ventricular cardiomyopathy**, which is reported in occasional chimpanzees and a bonobo.[88,704] Because of the importance of CVD in the apes, special cardiac necropsy protocols have been developed in both the United States and Europe.[229,776] These protocols standardize data collection and should be followed to enable development of consensus terminology in line with best practices for human and veterinary cardiology.[624]

Sudden spontaneous and perianesthetic deaths, presumably from fatal arrhythmias, are often associated with **myocardial fibrosis**. This is typically seen in adult male gorillas and chimpanzees in their mid-20s and older, though it can occur in younger males and females, as well as in orangutans. Grossly, affected hearts may have pale streaks or patches of tan or gray myocardial discoloration that corresponds histologically to fibrosis or steatosis. Often also noted is increased cardiac mass, frequently due to left ventricular hypertrophy (Fig. 15.2), and, less often, globoid enlargement with biventricular dilation. Female gorillas, orangutans, and some chimpanzees may have extensive fibrosis without left ventricular hypertrophy. Histologically, there can be multiple patterns of fibrosis including dissecting fibrosis (Fig. 15.3; eSlide 15.e1), often radiating from, or surrounding, intrinsic coronary arteries, replacement fibrosis where portions of the myocardium lack myocytes and they are "replaced" by dense collagen, and interstitial fibrosis. The differing types of fibrosis may be present in any of the four chambers including the trabeculae carneae and papillary muscles and in some cases multiple patterns are present in the same heart. Although replacement fibrosis can occur due to infarction or large vessel blockage, in apes the extent of the lesion is more suggestive of smaller vessel involvement. Fibrosis is seen in both

FIGURE 15.2 Cardiomegaly and left ventricular hypertrophy in a common chimpanzee. Transverse short axis section taken 3 cm from the apex of the heart. There is hypertrophy of the left ventricular wall with narrowing of the lumen.

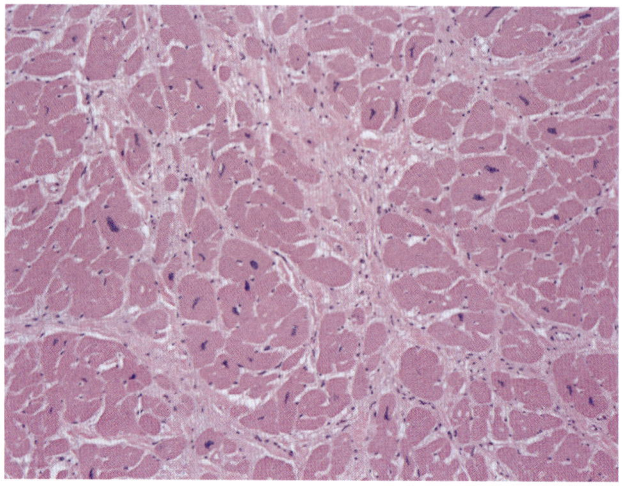

FIGURE 15.3 Myocardial fibrosis in a geriatric common chimpanzee. The interstitium is mildly to moderately expanded by fibrosis. Fibrosis is associated with atrophy of some and hypertrophy (indicated by karyomegaly) of other entrapped myocardiocytes (see eSlide 15.e1).

hypertrophied and dilated hearts and is usually associated with little or no inflammation. When present, inflammation is centered on individual myofibers and may be lymphocytic, histiocytic, or neutrophilic (in order of frequency). Myofiber hypertrophy with karyomegaly (representing polyploidy) and atrophy of entrapped myofibers can occur in the same heart. Although polyploidy normally increases with age, the polyploid nuclei in affected hearts are often distorted as well as enlarged.

Fibrosis is a common endpoint in many disease processes, and it is plausible that the pathogenesis of myocardial fibrosis in apes is heterogeneous. Various names are applied to the syndrome (fibrosing cardiomyopathy, idiopathic myocardial fibrosis, interstitial myocardial fibrosis, hypertrophic cardiomyopathy, and dilated cardiomyopathy). The latter two designations may be different stages of the same process, as dilated ape hearts are usually heavy, suggesting prior hypertrophy.[625] Left ventricular hypertrophy and arteriosclerosis of intramural coronary arteries (Fig. 15.4A; eSlide 15.e2) suggest that **hypertension** may be involved in the pathogenesis of ape myocardial fibrosis.[265,410,448,681] Documenting antemortem indirect blood pressure measurement in apes is challenging. Anesthetics can alter blood pressure, and measurements without anesthesia require special training of keepers and apes and validation of equipment and techniques.[127,497] Reference ranges for chimpanzees have been established under Telezol sedation.[175]

Factors linked to hypertension in humans may also affect adult apes. These include age-associated increased arterial stiffness, chronic inflammatory conditions, abnormal sympathetic nervous system activity, psychosocial stress, metabolic syndrome, age-associated increased salt sensitivity, and chronic renal disease. Some of these factors have been

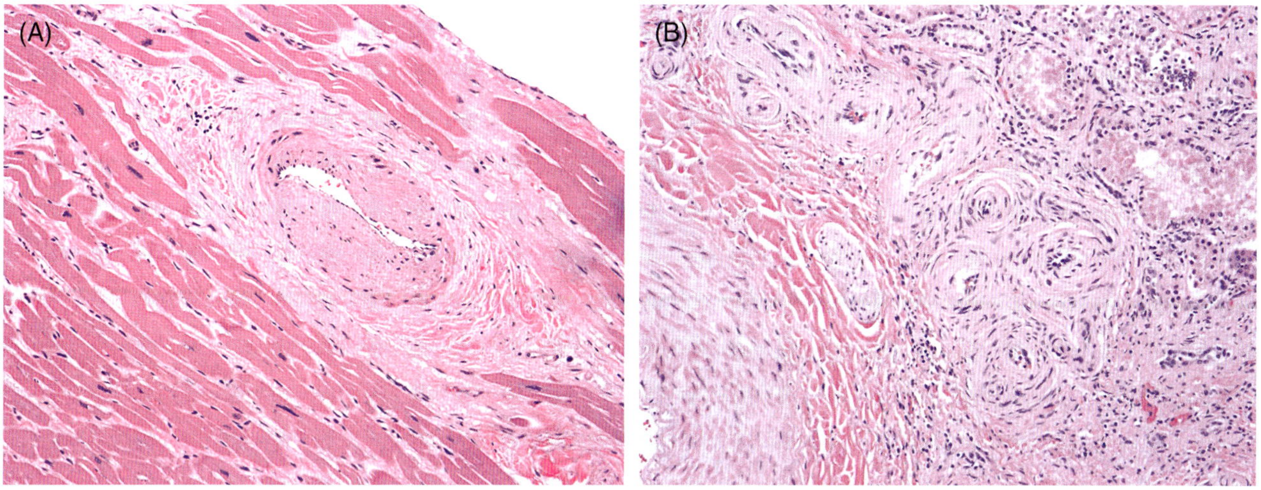

FIGURE 15.4 Arteriosclerosis in the common chimpanzee. (A) In an intrinsic coronary arteriole, changes are characterized by medial smooth muscle hypertrophy and hyalinization. (B) Concentric, hypertrophy of the tunica media with luminal narrowing to occlusion is seen in several renal arterioles (see eSlide 15.e2).

investigated in chimpanzees, in which age >30 years and obesity in females, but not males, are associated with hypertension.[174] Experimentally, high salt in chimpanzees' diet increases blood pressure in some but not all individuals.[139] Chronic renal disease is common is older chimpanzees and orangutans (see below).

Hyaline arteriosclerosis and arteriolosclerosis cause functional narrowing of the lumen of vessels due to thickening of the vessel wall by medial smooth muscle hypertrophy and/or hyalinization, often accompanied by thickening or splitting of the internal elastic membrane of intramural (intrinsic) coronary arteries and arterioles of the heart. These vascular changes result in localized ischemia. This is the dominant vascular change in gorilla and bonobo hearts. It is present, but inconsistent, in chimpanzee and orangutan hearts. The pathogenesis of hyaline arteriolosclerosis involves leakage of serum proteins, including IgM and complement factors, through the endothelium into the media.[203]

Systematic examination of vessels in other organs (e.g., eyes, kidney, and brain) to corroborate small vessel hypertensive changes (Fig. 15.4B) are important goals in ape necropsies.[506] Although inflammatory myopathies due to viruses (e.g., encephalomyocarditis virus, coxsackieviruses), protozoa (e.g., trypanosomes), and undetermined causes occur in apes, the preponderance of ape heart disease is not inflammatory at the time of diagnosis.[131,718]

Arrhythmogenic right ventricular cardiomyopathy is a recently recognized and uncommon form of presumed hereditary cardiomyopathy in chimpanzees and bonobos.[88,429,704] This condition, originally described in humans, is also well described in dogs and cats.[127] Marked fatty or fibro-fatty replacement of the right ventricular free wall myocardium is the defining lesion (Fig. 15.5), but left ventricular involvement also occurs.

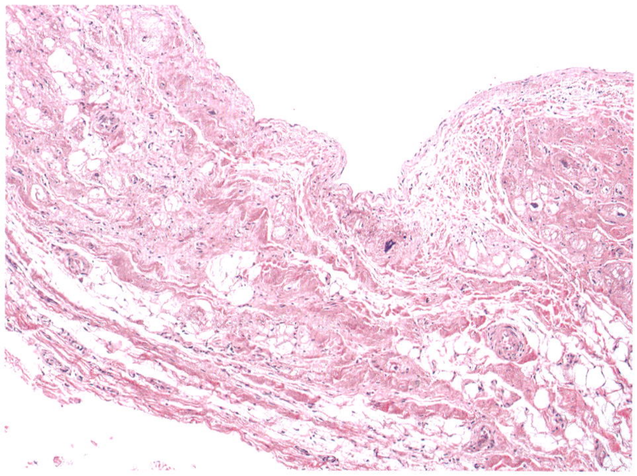

FIGURE 15.5 Arrhythmogenic right ventricular cardiomyopathy in a chimpanzee. The right ventricular myocardium is replaced by fibrosis and mature adipocytes. There is karyomegaly of remaining myocytes.

The fibrofatty replacement begins in the subepicardium and extends to the endocardium as the disease progresses leading to dilation and thinning of the right ventricular free wall. Mononuclear inflammation is mild but may be involved in the pathogenesis. The pathogenesis in humans involves genetic mutations interfering with desosomal structure or function leading to sometimes fatal arrhythmias.[773] The prevalence of underlying genetic factors in this condition in great apes is, as yet, unknown.

Thoracic aortic dissection is the second most common cardiovascular lesion in gorillas, and is an important problem in bonobos but rare in chimpanzees and orangutans.[247,335,410] The reasons for this discrepancy, such as possible structural, physiological, or inflammatory differences between ape species, are being examined by the Great Ape Heart Project using criteria established for evaluation

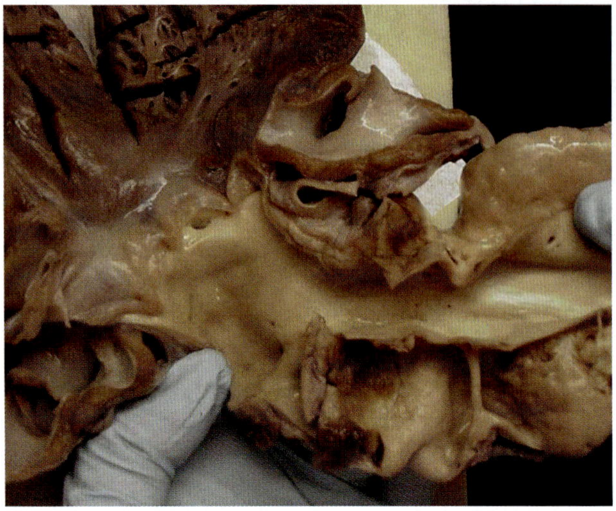

FIGURE 15.6 Aortic dissection in a bonobo. Dissection in the wall of the ascending aorta causes the appearance of a "double-barreled" vessel. Note the smooth intima in the true lumen and the roughened lining of the false lumen. *(Photo Courtesy of San Diego Zoo Wildlife Alliance.)*

of the aorta in humans.[243,673] Dissections generally begin in the ascending aorta (Debakey type II, Stanford type A) just distal to the aortic valves, or within an aortic sinus. Aortic dissection occurs with or without aneurysm. Dissections may result in a "double-barreled aorta" with blood in both the true lumen and within the mural defect (Fig. 15.6). Death is due to pericardial hemorrhage and tamponade, acute adventitial hemorrhage that disrupts conduction pathways, or hemothorax. Histologically, in a few cases, there is "cystic mucinous medial degeneration" characterized by disorganization or loss of the parallel orientation of elastin fibers, loss of smooth muscle cells, and accumulation of amorphous, Alcian blue positive, ground substance (mucopolysaccharides and glycosaminoglycans) between remaining lamellae (Fig. 15.7A). Inflammation is sometimes seen in the aortic adventitia in areas of dissection, and the vasa vasorum sometimes exhibit arteriolosclerosis (Fig. 15.7B), changes that may be involved in the pathogenesis.[2,221] However, mild adventitial inflammation can be a nonspecific aging change in humans. In humans, the key risk factors for dissection are older age, hypertension, atherosclerosis, infective and noninfective aortitis, pregnancy, and genetic or acquired connective tissue disorders (e.g., Marfan syndrome, matrix metalloprotease polymorphisms, scurvy, mineral deficiencies). These result in an imbalance between dynamic wear and repair mechanisms that weakens the aortic wall. Although risk factors in apes have not been conclusively determined, hypertension may play a role in bonobos. Aortitis may predispose to dissection in people, however, idiopathic lymphocytic and granulomatous aortitis have been seen in gorillas without dissection.[455]

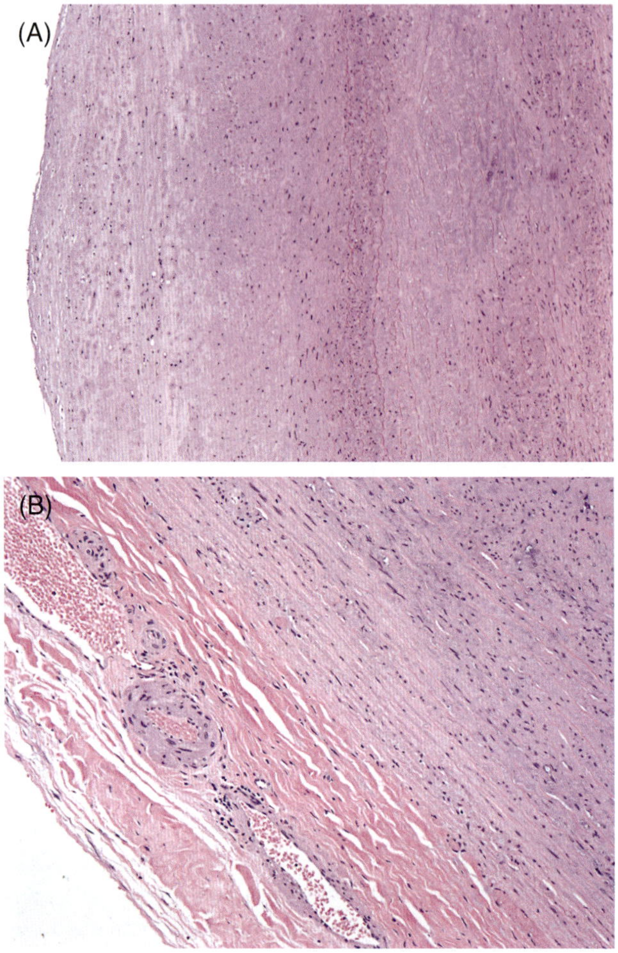

FIGURE 15.7 Degenerative aortic lesions in apes. (A) Mucinous medial degeneration and atherosclerosis in a lowland gorilla. The intima of the aortic arch is markedly expanded by an atherosclerotic intimal plaque to the left of the elastic lamina. Mucinous median degeneration and elastic fiber fragmentation are present on the right. (B) Arteriolosclerosis in vaso vasorum of the aorta (small vessels on left side of image) in an orangutan. There is also moderate mucinous degeneration of the aorta (right side of image).

Coronary atherosclerosis was once common in captive apes, but it is now only seen occasionally in very old apes that lived under outmoded husbandry conditions.[438] Atherosclerosis is more common in large elastic vessels, especially the abdominal aorta, iliac arteries, and aortic arch, than in coronary arteries. It can be detected by clinical ultrasound but is more frequently discovered after death.[24] Grossly, atherosclerosis ranges from mild fatty streaks (also occasionally seen in wild mountain gorillas) to complex atheromas with raised ulcerated plaques and "hardening" of the vessel (Fig. 15.8A). Histologically, lipid-laden foam cells that accumulate in the intima are accompanied by varying degrees of fibrosis, necrosis, and mineralization (Fig. 15.8B). Medial arteriosclerosis with mineralization (analogous to human Monckeburg's medial calcific arteriosclerosis) is also seen. The pathogenesis of atherosclerosis

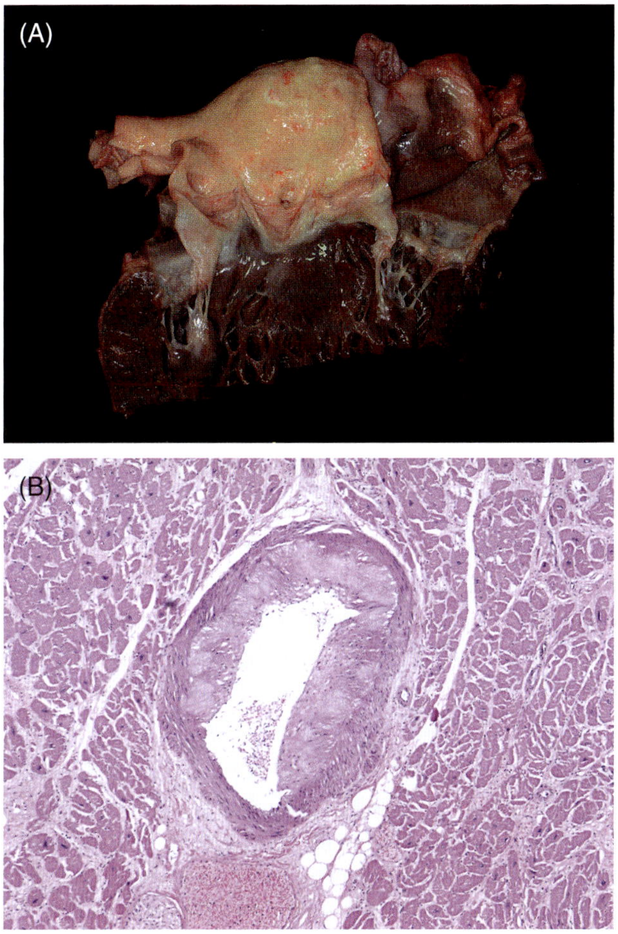

FIGURE 15.8 Atherosclerosis in a lowland gorilla. (A) Multifocal, irregular thickening, and pale yellow discoloration in the aorta is most apparent above the valve leaflet furthest to the right. (B) Characteristic histologic features in an intramural coronary artery include expansion of the intima by eosinophilic material, macrophages, and cholesterol clefts. *(Part A: Photo Courtesy of San Diego Zoo Wildlife Alliance.)*

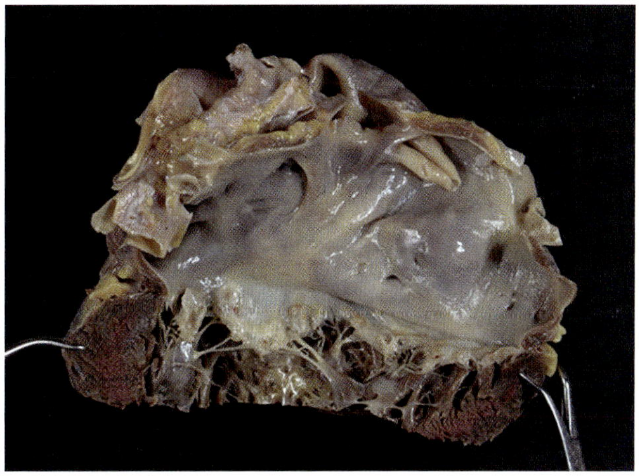

FIGURE 15.9 Valvular myxomatous degeneration in a geriatric chimpanzee. Heart (formalin fixed) with nodular myxomatous degeneration of the mitral valve in a geriatric male chimpanzee with congestive heart failure. The slight orange discoloration of the valve margins is due to small foci of hemorrhage and hemosiderosis withing the myxomatous nodules which is common in chimpanzees. *(Photo Courtesy of Anatomic Pathology, William R. Prichard Veterinary Medical Teaching Hospital, University of California, Davis.)*

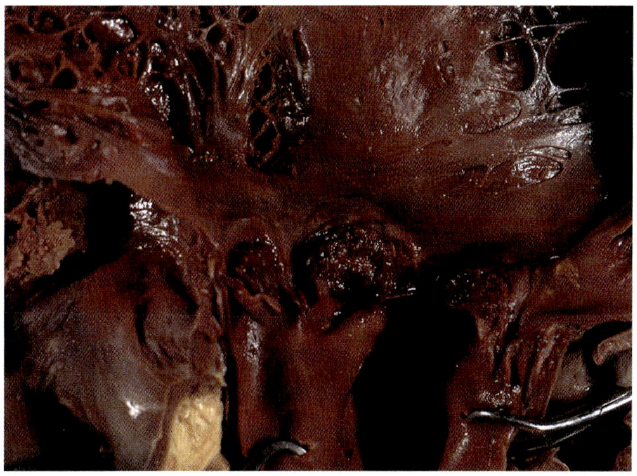

FIGURE 15.10 Vegetative valvular endocarditis in a mountain gorilla. Heart (formalin fixed) with vegetative valvular endocarditis in an older free-living silverback (male) mountain gorilla with severe cardiomegaly. Colonies of gram-negative rods were present in the vegetations. The dark discoloration of the heart is due to hemoglobin imbibition and advanced autolysis, often a problem in ape field necropsies. *(Photo Courtesy and Copyright of Gorilla Doctors.)*

has been well studied using NW and OW monkeys as models. It is a complex process, involving alterations in the endothelium and arterial intercellular matrix, as well as chronic inflammation and dyslipidemia.[626,738] Published values for cholesterol in laboratory chimpanzees and zoo and wild gorillas and orangutans are generally high compared to norms in humans.[24,607,667] In chimpanzees, high-density lipoproteins (HDLs) decrease, and low-density lipoproteins (LDL) increase with age.

Degenerative and **infective valvular diseases** can occur in captive and wild apes. Echocardiograms in affected zoo-housed gorillas occasionally detect mild mitral regurgitation and mitral valve myxomatous degeneration (endocardiosis) may be seen at necropsy in geriatric free-living mountain gorillas, zoo-housed chimpanzees, and orangutans (Fig. 15.9).[411,482] Orange to *red* discoloration of the valves can occur with small foci of hemorrhage and hemosiderosis and are common in chimpanzees. In cases of vegetative endocarditis, which most often involves the aortic and mitral valves, both gram-positive cocci and gram-negative bacilli have been identified in free-living mountain and captive lowland gorillas (Fig. 15.10).

As occurs in many species, **renal disease** is common in apes as they age. The nature and causes underlying ape renal disease are as yet undetermined. Based on a review of the North American SSP pathology databases, the most common diagnosis in all apes is "chronic interstitial nephritis." Glomerular lesions are frequently reported as well, and proteinuria, suggesting glomerulopathy, occurs in laboratory chimpanzees beginning at about 25 years of age.[724] Serum urea and creatine increase in males after 25 years and females in their late 30s, but are not associated with clinical renal failure. Urinary protein creatinine ratios also increase with age and suggest impaired renal function.[374] Chronic renal disease with evidence of protein in tubules has also been described in wild chimpanzees.[411] Advanced chronic interstitial nephritis with glomerulopathy is the most common lesion described in adult and old adult orangutans from North American zoos and was identified as a primary cause of death in 15.7% of orangutans necropsied.[247] An interesting aspect of ape renal disease is the statistical correlation with cardiovascular disease described in chimpanzees and known as "cardiorenal syndrome."[97] This syndrome is characterized by glomerular sclerosis and tubulointerstitial fibrosis in association with myocardial fibrosis (Fig. 15.11). Statistically, heart disease and renal disease are also linked in orangutans.[247,410] Chronic renal disease can compound heart disease due to increased resistance in the renal vascular bed. Conversely heart disease can lead to renal disease through poor perfusion or from primary hypertension leading to glomerular and renal small vessel sclerosis.

Senile cortical neuronal loss, **senile amyloid plaques**, and **cerebral amyloid vasculopathy** occur in apes, but *tau* protein deposition (a hallmark of human Alzheimer disease) is rare, though reported, in chimpanzees.[165,166,208,299,411] Frontal lobes and hippocampus are most affected. Cerebral and meningeal arteriolosclerosis are common, especially in chimpanzees with suspected hypertension.[515] Strokes ("cerebrovascular accidents") have been seen in all species of apes.[53,303,742]

All the ape species suffer from various forms of **arthritis** as they age, and chronic arthritis with uncontrollable pain is often a cause for euthanasia in geriatric apes in managed care.[411] In the wild, posttraumatic, monoarticular degenerative joint disease is common.[312,313] However, both managed and free-living apes, especially chimpanzees, and gorillas, appear to suffer from both osteoarthritis and non-infectious, inflammatory, seronegative (nonrheumatoid) polyarticular arthritis with spinal involvement, similar to spondyloarthropathy of humans.[313,581] Additionally, mycoplasma-associated, rheumatoid-like arthritis and a single case of juvenile reactive arthritis secondary to shigellosis are described in zoo-house gorillas.[480,501] Factors contributing to development of osteoarthritis in humans that may also contribute in apes include obesity, sedentary lifestyle, genetics, and injury.

Lesions of **osteoarthritis** in apes, similar to those in other species, include cartilage loss on weight-bearing surfaces with sclerosis and eburnation of the exposed subchondral bone in one or more joints, including articular facets of the spine, and osteophyte formation (Fig. 15.12). Uniarticular changes or rupture of tendons or ligaments indicate posttraumatic arthritis. Joint capsule thickening and osteophyte production in response to joint instability can be marked. Cartilage defects in inflammatory arthritis, such as spondyloarthropathy, are more often at joint margins and there may be mineralization and osteophyte production associated with the joint capsule insertion and ligaments and tendons. In the spine, bridging spondylosis involving the margins of the annulus fibrosis can be quite severe, especially in the lumbar area. Independently, degenerative disk disease of the spine can also lead to spondylosis, as can osteoarthritis, which often affects facets as well as vertebral

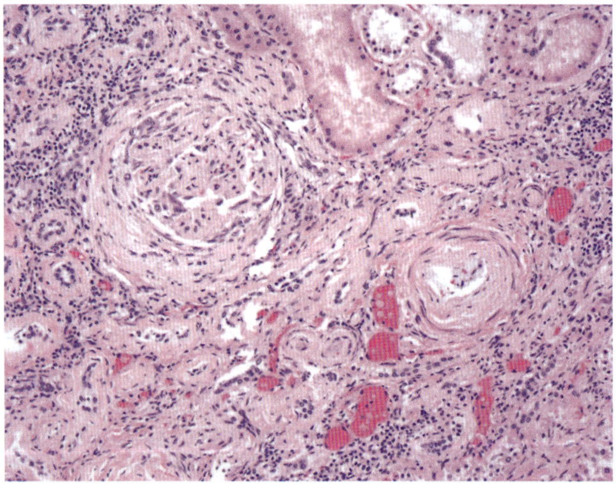

FIGURE 15.11 Cardiorenal syndrome in common chimpanzee. Chronic renal lesions suggestive of hypertension secondary to cardiac disease include glomerular sclerosis and arteriolar concentric medial hypertrophy with luminal narrowing. Also present is chronic interstitial nephritis.

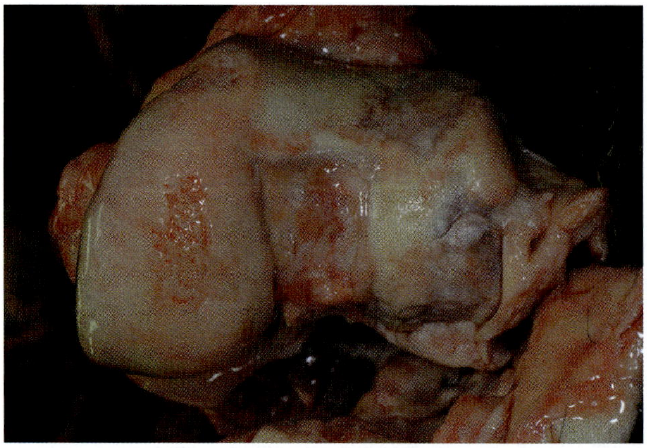

FIGURE 15.12 Osteoarthritis of the femorotibial joint in a western lowland gorilla. Degenerative changes include severe multifocal loss of cartilage, loss and remodeling of the condyle, and osteophyte formation. *(Photo Courtesy of San Diego Zoo Wildlife Alliance.)*

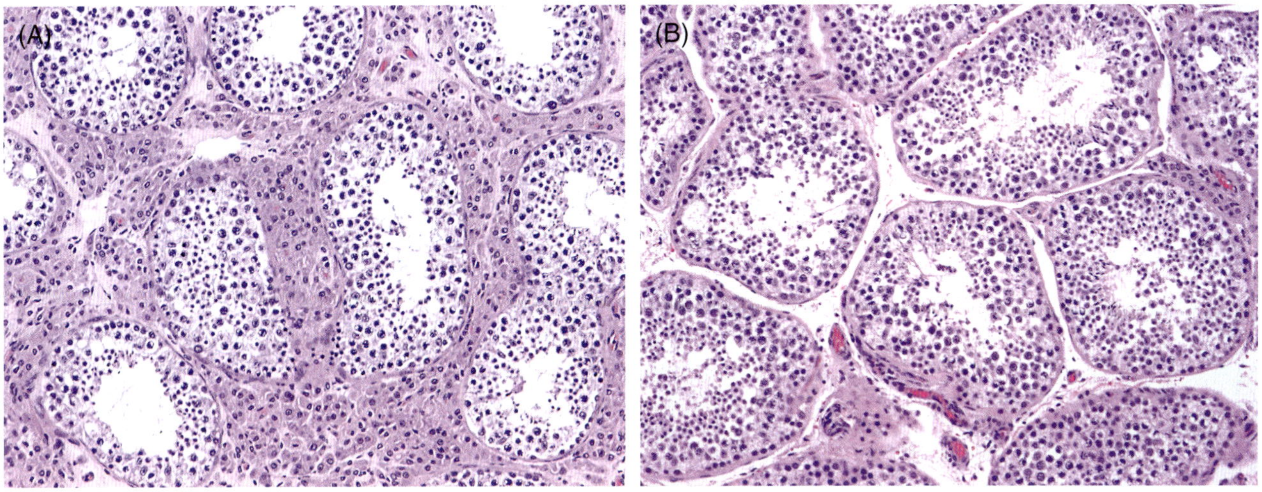

FIGURE 15.13 Testes in a western lowland gorilla (A) and chimpanzee (B). (A) Testicular hypoplasia with abundant interstitial cells and hypospermiogenesis is common in gorillas in the wild and captivity and is thought to be within normal limits. (B) Chimpanzees have larger seminiferous tubules and little interstitial tissue in comparison to the gorilla.

bodies. Another form of spinal arthritis called **diffuse idiopathic skeletal hyperostosis (DISH)**, a condition of elderly humans, is also thought to occur in apes.[404] DISH differs from spondyloarthropathy in that it primarily effects the attachment sites of ligaments and tendons (entheses) especially those of the spine.[417] Distinguishing between different forms of osteoarthritis relies on detailed examination of the skeleton before and after defleshing, and is facilitated by radiographic or MRI imaging prior to dissection.

Miscellaneous/Multifactorial

Trauma is an important cause of morbidity and mortality in free-living gorillas and chimpanzees but also occurs in group-housed, managed settings. Infants and juveniles are most at risk in the wild and in zoos.[261,380,676] Accidents occur, but conspecific trauma is especially common. Conspecific causes include infanticide, in which a male kills an infant that is not his offspring, which is common in gorillas and chimpanzees and infrequent in orangutans and bonobos. Infanticide is diagnosed by history and evidence of bite wounds, which are often extensive.[343] Intergroup aggression can result in traumatic deaths in adult animals due to immediate injuries or secondary infections. Snare entrapments are also a common cause of morbidity and mortality in all African apes and can result in loss or distortion of digits and entire distal limbs. In free-living bonobos, frequent limb abnormalities are ascribed to trauma, not congenital lesions.[326] Long bone, vertebral and pelvic fractures suggestive of trauma from falls also occur in African apes.[312,694]

Reproductive tract lesions in male chimpanzees include seminiferous tubule degeneration and lymphocytic infiltrates in the prostate gland. **Benign prostatic hyperplasia (BPH)** has been described in chimpanzees at "late middle age."[666] BPH is not reported in the other apes, perhaps because accessory sex glands are not always examined clinically or after death. **Testicular hypoplasia** with hypospermatogenesis and increased interstitial cells is a common finding in western lowland and mountain gorillas[179,201] (Fig. 15.13) and may be physiological, not pathological. Seminiferous tubule atrophy can occur in any ape with chronic illnesses.

Ovarian atrophy, uterine leiomyomas (see Neoplastic), **adenomyosis**, and **endometrial atrophy** are the most common lesions in the reproductive system of female chimpanzees.[91] Although adenomyosis is diagnosed in all apes, including wild mountain gorillas, true endometriosis is very rare but reported in chimpanzee, gorilla, and orangutan.[35,150,360,480,515] Cases of pelvic, retroperitoneal, and abdominal inflammation and adhesions due to carcinomatosis are seen in gorillas and must be distinguished from other causes of chronic peritonitis in apes, such as appendicitis and diverticulitis (see also discussion in "Bacteria" section and Fig. 15.32). Estrogen and progesterone receptor immunohistochemistry may be useful in identifying a reproductive tract origin of carcinomatosis.

Complications of pregnancy are not uncommon in the apes. Negative outcomes include abortion, stillbirth, prematurity, small-for-gestational-age neonates, neonatal death, and maternal death (Table 15.e2 for details of ape reproduction).

All OW primates including apes have hemochorial placentas with endovascular trophoblast extension into the maternal decidua and remodeling of the spiral arteries. The gorillas, chimpanzees, and humans, but not gibbons, differ in that, additionally, there is deeper decidual invasion and interstitial trophoblast extension that extends into the myometrium.[82–85,119,544,545] Complications associated with this type of placentation include: maternal preeclampsia; placental abruption (premature separation with retroplacental hemorrhage); placenta previa (placental disk adjacent to or covering the internal cervical os); placenta accreta (abnormally deep invasion of the placenta into the uterine wall causing placental retention at parturition); placental retention due to other causes; and fetal growth restriction (small for gestational age fetus).

Preeclampsia has been suspected in gorillas and chimpanzees,[23,82,736] though the pathogenesis is incompletely understood, defective remodeling of the uterine spiral arteries and inadequate decidual invasion may play a role. Abnormal placental circulation leads to maternal hypertension mediated through glomerular endothelial swelling and altered renal perfusion. Proteinuria and edema are clinical features. Pregnancy-associated seizures suggestive of **eclampsia** are reported in a gorilla. Expected placental lesions are those suggestive of hemodynamic alterations and include increased fibrin deposition, maternal floor infarction, calcification, villous basement membrane thickening, hypermature villi, and increased syncytial knots.[143,206] **Placenta accreta** (PA) is described in cases of periparturient deaths in an orangutan and a chimpanzee with placental retention, both with massive hemorrhage.[307] PA is more common in older multiparous women and both reported that apes fit that phenotype. Findings include difficulty in removing the placenta from the uterus and histological evidence of deep myometrial invasion. **Retained placenta** without verified accreta is also reported in a primiparous bonobo, which survived a postparturient infection after manual extraction of the placenta.[242] **Placenta abruption** leads to marked occult retroplacental hemorrhage or overt vaginal bleeding if the cervical os is open. Fetal and maternal death can occur if delivery is not procured. Placental infarctions are often present. **Placenta previa** is reported in an orangutan and is suspected in other apes.[341] **Ascending uterine/placental infection** in gorillas may occur because female gorillas allow copulation during the last part of gestation. This can result in weak or stillborn infants with "infected amnion syndrome" or "in utero pneumonia" characterized by intraalveolar squames with bacteria and associated neutrophils (Fig. 15.14). Chorioamnionitis leading to fetal pneumonia has also been observed in an orangutan.[39] Placentitis can also occur via hematogenous spread from the mother and resultant infection of the fetus.[419] Numerous viruses, bacteria, and protozoans are known to cause transplacental infections in humans and animals, and *Salmonella heidelberg* is reported as a cause of septic abortion in a gibbon.[700]

Other pregnancy complications in apes are umbilical cord accidents such as knots, excessive twisting, and strangulation by the umbilical cord.[39]

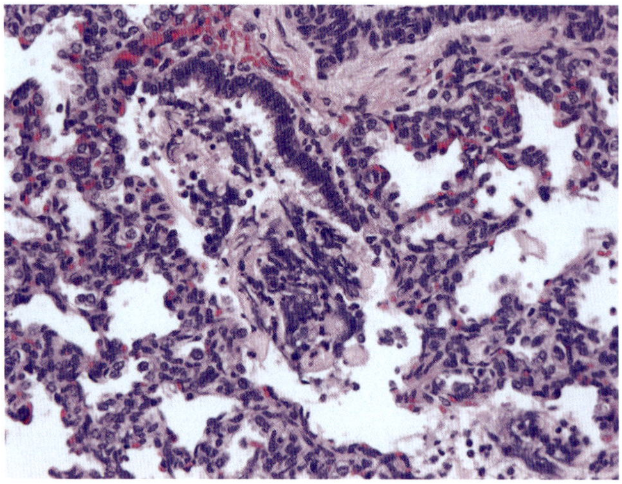

FIGURE 15.14 In utero pneumonia (infected amnion syndrome) in a premature stillborn western lowland gorilla. The condition is characterized by intraalveolar squames and neutrophils in bronchioles and alveolar spaces.

Stillbirths, defined as delivery of a fetus that died in utero, occur in 10% of orangutan births, 12% of chimpanzee births, and 13% of gorilla births in AZA-accredited zoos, most commonly from pregnancies of older females.[591] Other authors define stillbirths as being delivered dead after the age of viability, which in chimpanzees is given as greater than 208 days (mean gestation period 228±9.9 days), while abortions are defined as being expelled dead before viability.[752] In the latter study, 18% of pregnancies resulted in abortions, stillbirths, or death in the first days of life. Prematurity contributes to postnatal mortality; only one of 17 infants born at <208 days gestation lived to adulthood. An additional 8.6% of pregnancies resulted in full-term infants that died in the perinatal period. Causes of death include sepsis, meconium aspiration, viral pneumonia, and meningitis, but most are of undetermined cause. Placental examination is often not performed. In humans, diagnosis of the cause of stillbirth is greatly increased when the fetus receives a full autopsy and further increased when placenta is also examined.[227]

Twinning, which is fairly common in chimpanzees and infrequent in other apes, often results in death of one or both of the infants due to prematurity, low birth weight, or inadequate maternal care.[173] A better understanding of complications of pregnancy in apes will only happen if clinicians and pathologists successfully encourage fetal and placental examination. Protocols are available for examination modeled on published protocols for examination of human stillbirths and placentas.[1,327,328]

Infant mortality limits population growth in both captive settings and in the wild. Trauma, including infanticide, is an important cause in wild gorillas; infectious diseases and maternal issues, such as illness or hypogalactia, also contribute.[261] **Prematurity** is an important cause of

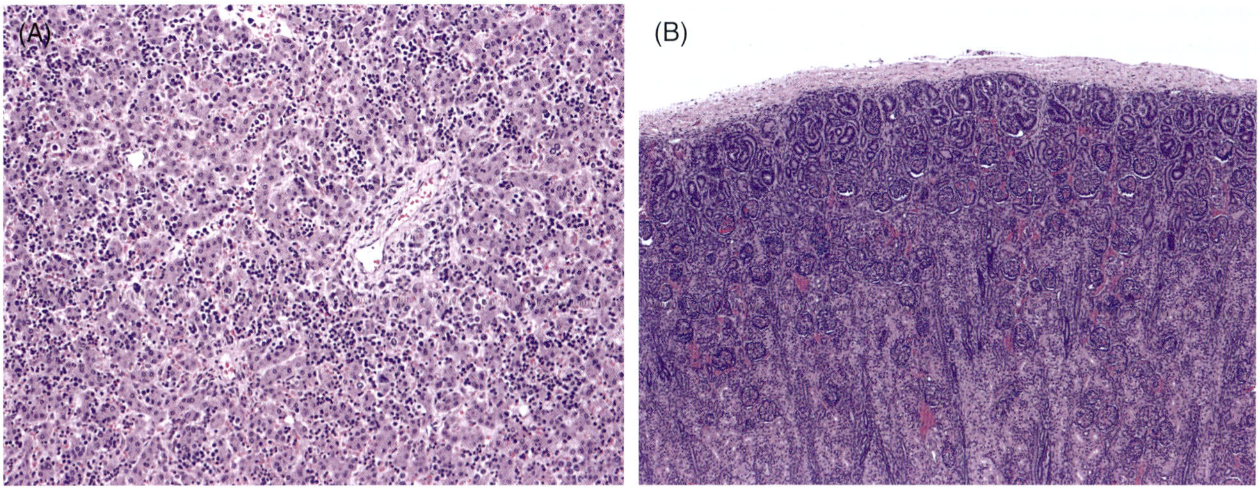

FIGURE 15.15 Histologic features of prematurity in a western lowland gorilla. (A) Findings include abundant hepatic extramedullary hematopoiesis in the liver. (B) Renal glomerulopoiesis characterized by fetal glomeruli (small, hypercellular glomerular tufts and prominent peripheral nuclei) and immature tubules (may be dilated, branching, (nephropoiesis), which are coiled and lined by small, cuboidal epithelial cells with closely spaced nuclei).

perinatal deaths in all the apes.[752] Gross evidence of prematurity includes low birth weight and open fontanelles (Fig. 15.e1). Histological evidence of prematurity, based on data from human neonates, includes: abundant liver extramedullary hematopoiesis (EMH) (Fig. 15.15A) and hepatocellular glycogen; glomerulopoiesis (nephropoiesis is present up to 2 years) (Fig. 15.15B); low degree of pulmonary alveolization; presence of intraalveolar fluid; and the adrenal fetal zone. Evidence of in utero distress or hypoxia in a stillbirth includes meconium staining of the pelage, keratinization and mineralization of thymic epithelium, abundant intraalveolar squames and/or meconium, neuronal ischemic necrosis, and in utero involution of the fetal adrenal cortex.

Alopecia in apes and other nonhuman primates is often a concern in both managed and free-living animals and has multiple etiologies including seasonality, pregnancy, nutrition (e.g., malnutrition), behavior (e.g., hair plucking), or infections (e.g., internal and external parasites), while endocrine alopecias (e.g., due to adrenal or thyroid dysfunction) is not well described in apes.[514] Hair plucking during both self- and allo-grooming is common in bonobos in managed care.[60] Clinical and pathological evaluation, using techniques standard for domestic species, is advocated for macaques and should also be used for hair loss in apes.[415,514]

A progressive neurological disorder consistent with a **leukoencephalopathy** has been sporadically identified in gibbons within North American zoological collections. Animals have generally been young and multiple species of gibbon are affected. Lesions are predominately within the white matter and include asymmetrical foci of hemorrhage, cavitation, and collapse of the parenchyma due to loss of white matter (Fig. 15.16A). Histologically, cerebral white matter from the centrum semiovale to corona radiata may have vacuolation and or cavitation of the parenchyma with numerous gemistocytic astrocytes (Fig. 15.16A), gitter cells and, in some areas, small foci of hemorrhage. Vessels may be surrounded by lymphocytes and plasma cells. While lesions have similarities to the leukoencephalopathy associated with JC polyomavirus in humans or Simian virus 40 lesions in macaques,[730] no viral cause has been identified to date and the pathogenesis remains unknown.

Neoplastic

Reports of **neoplasms** occurring in apes have been recently reviewed.[67,411] In general, neoplasms, especially carcinomas, do not seem to be as common in apes as in humans and some other primates.[720] A notable exception is the prevalence of female reproductive malignancies in lowland gorillas, which are rarely seen in other apes.[411] Another exception is gastric adenocarcinoma in free-living mountain gorillas which is associated with chronic parasitic gastritis.[475] Additional reported neoplastic and hyperplastic conditions include Brunner's gland hyperplasia in chimpanzees, a Brunner's gland adenocarcinoma in a gorilla, a mammary adenocarcinoma in a male orangutan, and a malignant melanoma in a mountain gorilla.[81,324,602,732] Leukemias and lymphomas (due to gibbon ape leukemia virus), once common in gibbon colonies, now appear to be rare.[66] Hemolymphatic system malignancies seem to be more common in gorillas and orangutans than chimpanzees and may have an association with viruses such as lymphocryptoviruses (EBV-like) and retroviruses (HTLV-1).[20,26]

Testicular tumors are rare and include a mixed Sertoli-Leydig cell tumor in chimpanzees and interstitial cell adenomas in gorillas.[411] Leiomyomas occur in as many as 30%

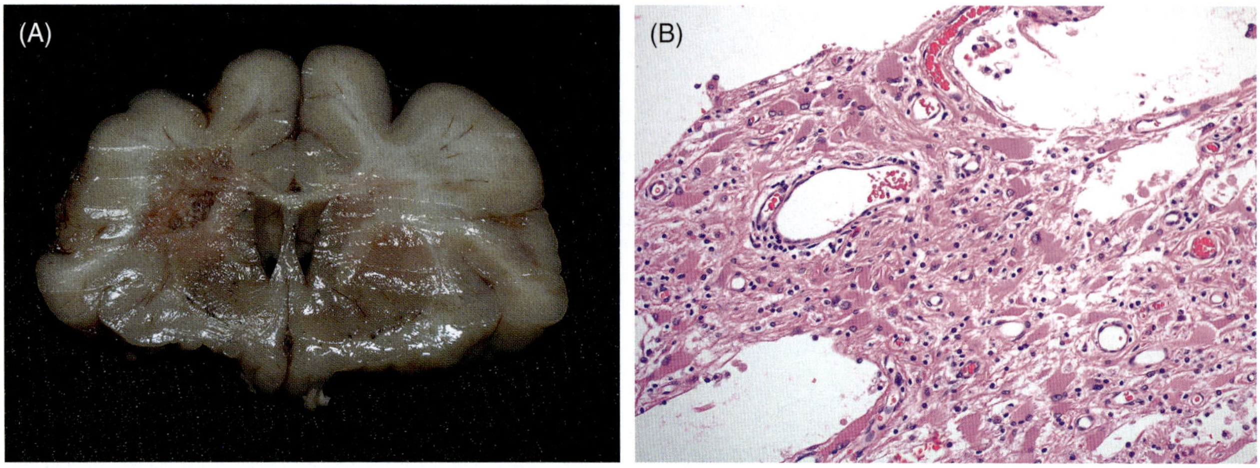

FIGURE 15.16 Leukoencephalopathy in a Javan gibbon. (A) Cavitation of white matter within the corona radiata and centrum semiovale of the left hemisphere. (B) Cavitated spaces contain few gitter cells and are rimmed by remnant white matter. There are numerous gemistocytic astrocytes and smaller numbers of lymphocytes within affected areas. *(Part A: Photo Courtesy of University of Illinois Zoological Pathology Program.)*

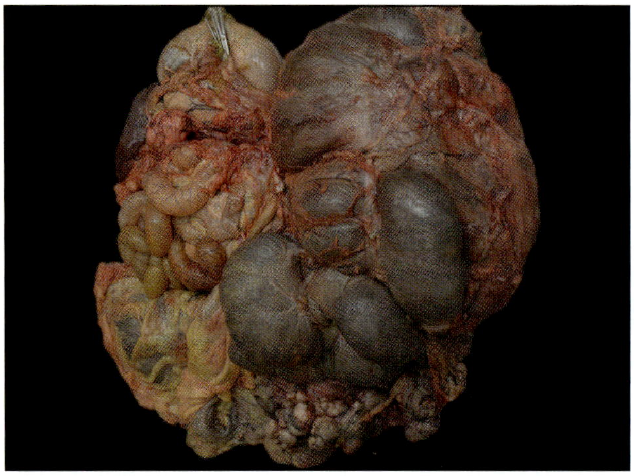

FIGURE 15.17 Uterine adenocarcinoma and carcinomatosis. Abdominal viscera (ex situ) with adhesions and carcinomatosis due to uterine adenocarcinoma from a geriatric female western lowland gorilla in managed care. *(Photo Courtesy of Anatomic Pathology, William R. Prichard Veterinary Medical Teaching Hospital, University of California, Davis.)*

of laboratory chimpanzees, but are less frequently reported in gorillas, orangutans, bonobos, and gibbons.[67,363] **Malignancies of the female reproductive tract** are unusually frequent in zoo-housed gorillas.[411] They include cervical, uterine (Fig. 15.17), and ovarian adenocarcinomas and a uterine leiomyosarcoma.

INFECTIOUS DISEASES

Apes are host to several unique **viruses** and are also susceptible to viruses of humans and OW monkeys. Human contact is a risk factor for both apes in captivity and groups of apes in the wild that are habituated for research and tourism or whose habitat is encroached upon by other human activities. Isolation and serology were the initial means by which the virome of apes was studied.[318–322,649–651] Serologic surveys show that both captive and free-living gorillas, chimpanzees, and orangutans carry antibodies against many viruses including herpesviruses, adenoviruses, paramyxoviruses, pneumoviruses, orthomyxoviruses, and enteroviruses.[70,339,639,749,750] Molecular techniques now allow for virus and antibody detection in noninvasively collected samples, such as feces, urine, and saliva.[349,458,637] This has revolutionized our ability to screen wild populations for "novel" viruses and to explore viral biome diversity and cross-species transfer of pathogens. In addition to notable viral infections in apes discussed below, Table 15.e3 has an extended list of viruses that have been detected in apes.

DNA Viruses

Poxviruses

Monkeypox virus infection (MPXV, genus *Orthopoxvirus* to be renamed "Mpox virus") is a World Organisation of Animal Health (WOAH) listed reportable, zoonotic disease that is endemic to rainforests of central and western Africa where African rope squirrels and other rodents are likely reservoir hosts.[180,709] Natural and experimental infection is reported in free-living and rehabilitant chimpanzees in Africa and in captive chimpanzees, orangutans, gorillas, and a gibbon in a European zoo.[535,537,541] Though monkeypox has not yet been described in bonobos, their range includes endemic areas.[293] The virus is transmitted by direct contact, ingestion, inhalation, or, possibly via arthropod vectors such as flies. Three forms of monkeypox virus infection occur in nonhuman and human primates: (1) benign upper respiratory signs followed by cutaneous eruptions; (2) mucous membrane lesions, facial and laryngeal edema, and death due to asphyxia; and

(3) bronchopneumonia. Incubation in apes is less than 10 days. Initial signs are variable but can include acute death or respiratory signs such as purulent nasal discharge and cough without cutaneous eruptions; cutaneous eruptions without systemic illness; vesicles on the soles of the feet, back of hands, face and oral mucous membranes; and, in chimpanzees in Africa, nodular lesions on the face and abdomen. The disease can be fatal in zoo-housed orangutans, chimps, and gibbons with as few as 4 days from onset to death.

Eruptions begin as papules and progress to vesicles and pustules that ulcerate and become covered with scabs. These typical "pocks" have raised margins and depressed centers. Secondary bacterial infections occur; opportunistic *Staphylococcus aureus* infection is common in vesicles and systemically. Lesions typically start on the face, lips, and oral cavity and spread over the whole body including the soles of feet. Lymphoid tissues become infected and monocyte-associated viremia leads to systemic disease. Histologically, based on infections in OW monkeys and zoo-housed orangutans, chimpanzees, and gibbons, epithelial proliferation and intra- and intercellular edema with small, inconspicuous, eosinophilic, intracytoplasmic inclusion bodies involve the epidermis and hair follicles. In humans, but not in apes, intranuclear pseudoinclusions and multinucleated keratinocytes are reported. Internal lesions can include fibrinonecrotic bronchopneumonia and proliferative lesions in the oral cavity and gastrointestinal tract, the latter associated with diarrhea. As of the recent international spread of human infections, WHO has issued guidelines for monkeypox testing.[762] In the United States, the Centers for Disease Control and Prevention is the WHO Collaborating Centre for Smallpox and Other Poxvirus Infections.

Molluscum contagiosum (MC) is a human disease caused by a poxvirus of the genus *Molluscipoxvirus*. Molluscipoxvirus-like infection is reported in laboratory chimpanzees.[152] The prevalence is unknown and it is uncertain if MC in chimpanzees is due to a human or chimpanzee-specific virus. Transmission in humans is by direct contact with skin lesions and fomites or through sexual contact. The route of transmission in chimpanzees is unknown, but reported cases are all juveniles. Lesions consist of small papules with depressed centers on the face, eyelids, and inguinal area that regress spontaneously in several weeks. Sebum-like material, expressed from the lesions and examined cytologically with Wright's stain, contains squamous cells with large, basophilic, cytoplasmic inclusion bodies. Histological lesions consist of marked focal acanthosis and hypergranulosis with both elevation and infolding of the epidermis creating a "cup-shaped" lesion. Huge amphophilic to eosinophilic, finely granular cytoplasmic inclusions (molluscum bodies) expand the cytoplasm of cells and are pathognomonic.[295] A single molluscum-like lesion was seen as a serendipitous histological finding on the breast of a free-living female mountain gorilla (Fig. 15.18). The prevalence of molluscum virus in ape populations is yet to be investigated.

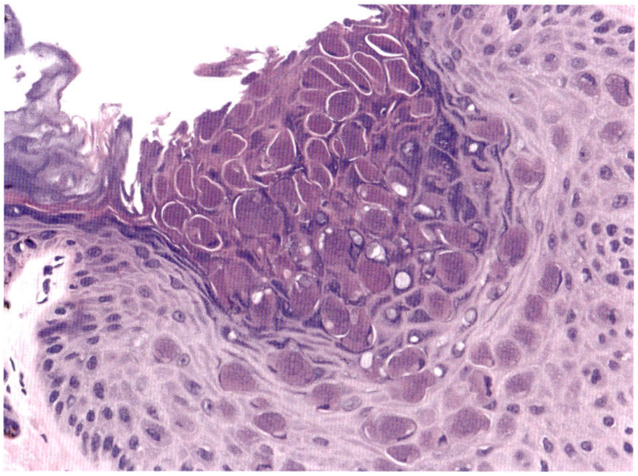

FIGURE 15.18 Molluscum contagiosum (poxvirus) infection in a mountain gorilla. Cutaneous lesion of suspected molluscum contagiosum from the breast area of an adult female free-living mountain gorilla. This small cup-shaped (inverted papilloma-like) lesion contains squamous cells filled with large eosinophilic granular inclusions. This was a serendipitous finding. *(Photo Courtesy and Copyright of J. Bosco Noheri for Gorilla Doctors.)*

Alphaherpesviruses

Human herpesvirus-1 and **2 (HHV-1** and **2; Herpes simplex types 1** and **2)** are alphaherpesviruses that may cause disease in apes. Serological reactivity to HHV-1 and HHV-2 is common in captive western lowland gorillas, chimpanzees, orangutans, and gibbons.[162,163,376,595] Molecular evidence of similar viruses found in wild chimpanzees, bonobos, and western gorillas shows that are all of HHV-2-like lineage suggesting that HHV-2 was originally a zoonotic virus.[618,639,745]

Nonfatal, human-acquired **HHV-1 infection** in juvenile eastern gorillas consists of malaise, anorexia, and gingival and labial vesicles with ulceration[216] (Fig. 15.19A). Biopsy reveals epithelial necrosis and vesicle formation with syncytial giant cells containing clear to smudgy intranuclear inclusions (Fig. 15.19B). The oral lesions are identical to those of B-virus infection in macaques. Fatal disease in infant zoo-housed apes is characterized by labored breathing and/or vomiting and diarrhea with rapid progression to death.[268,338]

Disseminated cutaneous eruptions and vesicles involve the face, chest, and limbs, and there may be ulceration throughout the gastrointestinal tract mucosa. Gross lesions also include interstitial pneumonia with or without pericardial effusion, and hepatosplenomegaly. Histologically, multifocal necrosis in lungs, liver spleen, lymph nodes, heart, and brain are associated with Cowdry type A inclusion bodies. Syncytia with inclusions are most

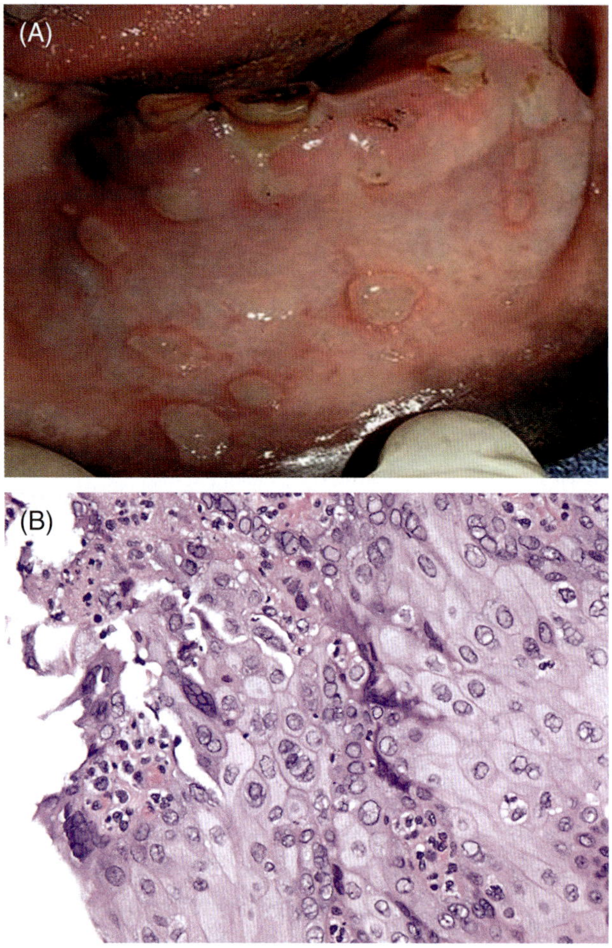

FIGURE 15.19 Human herpesvirus-1 in an eastern lowland gorilla. (A) Multiple, papular, vesicular, and ulcerative lesions are present in the buccal mucosa. (B) Syncytia with intranuclear viral inclusion bodies characteristic of HHV-1 infection are present in the hyperplastic epithelium at the margin of an oral vesicle. *(Parts A, B: Photos Courtesy and Copyright of Gorilla Doctors.)*

common in the skin, upper gastrointestinal (GI) tract, and lung. Cases may be isolated or may occur in the face of illness in adult members of the group.

Fatal infections in gibbons occur as both individual cases and outbreaks.[376,464,645] Oral ulcers and vesicles may occur in otherwise healthy animals. Other animals die of encephalitis with or without prodromal oral lesions. Additional clinical signs include lethargy, loss of normal aggression, cranial nerve deficits, paresis, and paralysis ending in coma or seizures necessitating euthanasia. Cerebral edema and congestion may be appreciated grossly. Encephalitis is characterized histologically by foci of necrosis, nonsuppurative inflammation, cuffing, and multifocal gliosis. Cowdry type A intranuclear inclusions are observed in glial cells and neurons (Fig. 15.20) in some, but not all, cases.

As much as 42% of colony-housed chimpanzees have antibodies cross-reactive with HHV-1 or HHV-2. Naturally occurring **HHV-2-like virus-infection** can cause genital ulcers in colony-housed bonobos and chimpanzees.[436] More recent evidence suggests that these infections may be due to an HHV-2-like virus of chimpanzee not human origin.[639,745]

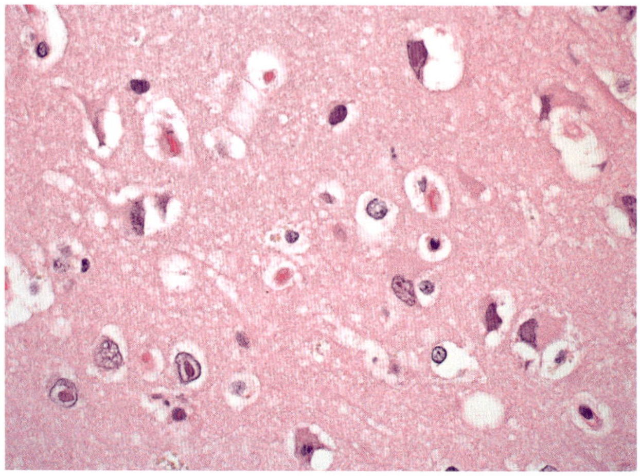

FIGURE 15.20 Human herpesvirus-1 encephalitis in a gibbon. Encephalitis associated with HHV-1 infection with neuronal necrosis and characteristic intranuclear inclusions within glial cells and neurons.

Additionally, the only verified ape-specific simplex-like alphaherpesvirus isolated from a chimpanzee, **chimpanzee alphaherpesvirus (ChHV)**, is related to HHV-2 not HHV-1.[621] Oroesophageal ulceration characterized by a red rim with a "whitish" center resolved in 2–3 weeks.[416] Histopathology is not described; however, the cytopathic effect (CPE) in Vero cells includes necrosis without syncytia.

Varicellovirus (syn: varicella-zoster; HHV-3; VZV; chicken "pox"; shingles), a human alphaherpesvirus, can infect apes.[228] Serosurveys demonstrate exposure in chimpanzees, gorillas, orangutans, and gibbons.[318] Infection may be lifelong with latency in sensory ganglia and reactivation due to intercurrent diseases or stress (called "shingles" in people). Natural primary infections, similar to "chicken pox" of children, occur in infant or juvenile captive chimpanzees and gorillas, and experimental work confirms susceptibility of chimpanzees.[103,270,486,747] Cutaneous eruptions occur all over the body, especially the trunk, and are characterized by small vesicles that ulcerate and become covered by scabs (Fig. 15.21). Lesions heal without scarring unless pruritis-associated excoriation occurs. Histology in apes is similar to that in humans and OW monkeys (see New World and Old World Monkeys), namely epidermal vesicles containing variably sized acantholytic cells with intranuclear inclusions. Systemic signs of disease, such as pneumonia, are not reported in apes. Recrudescent infection in an adult female gorilla consisted of typical cutaneous lesions in a single dermatome and detectable virus in ganglia of that dermatome, similar to "shingles" in humans.[430]

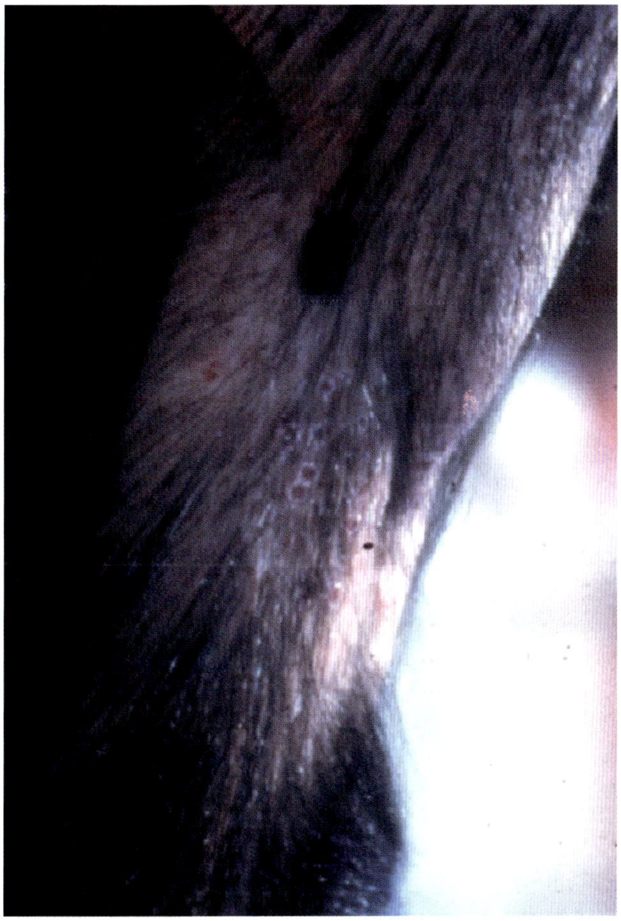

FIGURE 15.21 "Chicken pox" (varicella zoster) infection in the skin of a chimpanzee. Infection is characterized by multiple, small, round, shallow ulcers resulting from rupture of vesicles. *(Photo Courtesy of L. Gage.)*

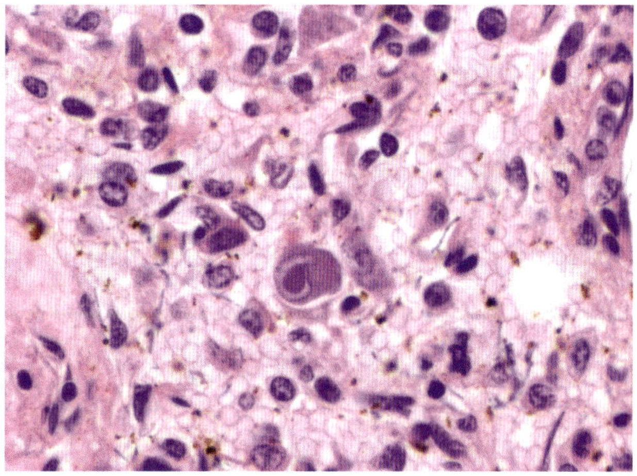

FIGURE 15.22 Cytomegaloviral pneumonia in an infant gorilla. Presumed cytomegalovirus recrudescence in an infant Eastern lowland gorilla confiscated from poachers and in sanctuary care. This megalocyte with "owl's eye" intranuclear and ill-defined eosinophilic cytoplasmic inclusion typical of cytomegalovirus infection. *(Photo Courtesy and Copyright of E. Kambale for Gorilla Doctors.)*

Betaherpesviruses

Cytomegalovirus (*CMV*) (similar to humans HHV-5) and *Roseolovirus* (similar to HHV-6 and HHV-7) are members of the *Betaherpesvirinae* subfamily. Cytomegalovirus infections are usually clinically silent, but can cause serious disease in fetuses, infants, and immune-compromised hosts. Infection is lifelong. Recrudescence occurs due to stressors, such as other viral infections and immunosuppressive drugs. Antibodies reactive with chimpanzee CMV are present with high prevalence in captive western lowland gorillas, chimpanzees, orangutans, and gibbons, and in wild eastern gorillas.[318,595,749,750,764] In addition, CMVs have been detected by molecular methods in wild western chimpanzees, wild mountain gorillas, wild and zoo-housed western lowland gorillas, and zoo-housed orangutans with evidence of cross-species transmission between gorillas and chimpanzees.[386,618,639] Clinical disease and deaths are infrequent. Clinical signs of CMV infection are nonspecific including anorexia, abdominal pain, or respiratory signs. Histologically, infection is characterized by the presence of "cytomegalic cells" with karyomegaly, amphophilic, Cowdry type A intranuclear inclusions, and ill-defined granular acidophilic intracytoplasmic inclusions, similar to lesions in other primates (Fig. 15.22). Cytomegalic cells are seen in alveolar septa with interstitial pneumonia and throughout the GI tract (including salivary glands), liver, lymphoid organs, brain, eye, and adrenal glands. Lesions are frequently necrotizing and CMV is one of the few viruses associated with neutrophilic inflammation. However, in acute disseminated infection in chimpanzees, cytomegalic cells may be present without inflammation. Reported associated conditions in chimpanzees include experimental multiple sclerosis, adenovirus hepatitis, and ulcerative colitis.[129,706,764] Although molecular detection is common, lesions of CMV are rare in wild eastern gorillas, but have been seen in one wild and one confiscated infant in association with localized and systemic bacterial infections with lymphoid depletion (Fig. 15.22).

Gammaherpesviruses

Gamma-1 lymphocryptoviruses (EBV-like viruses) (subfamily *Gammaherpesvirinae*) are found only in primates and are cell associated, lymphotropic, and/or epitheliotropic; they are also potentially oncogenic. The type virus is human Epstein-Barr virus (EBV, HHV-4). Serosurveys show that all apes, including wild mountain gorillas, wild chimpanzees, and wild and semicaptive orangutans, have a very high prevalence of antibodies cross-reactive with EBV.[319,320,339,396,749,750] EBV-like viruses are also detectable by PCR amplification from peripheral blood mononuclear cells, feces, or saliva in all species of apes, including gibbons.[168,169,212,213,367,638] Cross or shared infection with the same lymphocryptovirus can occur in orangutan and chimpanzee, but there is no known zoonotic potential.

The pathogenicity of chimpanzee, bonobo, gibbon, and orangutan EBV-like lymphocryptoviruses is not well established. Histological lesions suspected to be associated with mountain gorilla lymphocryptovirus include oroesophageal leukoplakia, generally in adult gorillas. In infants, follicular hyperplasia in tonsils and lymph nodes, and pulmonary lymphoid hyperplasia (lymphofollicular bronchitis and lymphocytic interstitial pneumonia) are seen[638] (Fig. 15.23A). Leukoplakia is characterized histologically by acanthosis, intracytoplasmic edema (ballooning) in cells of the upper stratum spinosum, and intranuclear inclusions (Fig. 15.23B). These lesions are similar to those reported

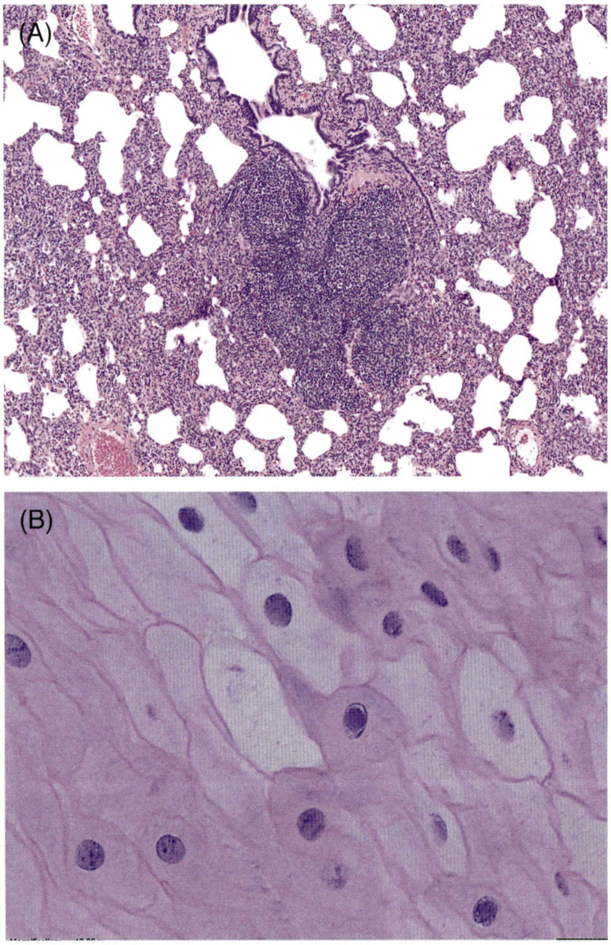

FIGURE 15.23 Gorilla lymphocryptovirus infection in a mountain gorilla. (A) Pulmonary lymphoid proliferation (follicular bronchiolitis and lymphocytic interstitial pneumonia) is thought to be due to a primary lymphocryptovirus infection. (B) Epithelial hyperplasia with cytoplasmic edema with intranuclear inclusion bodies. The lesion is similar to leukoplakia described in humans and macaques with lymphocryptovirus infection. Electron microscopy confirmed the presence of herpesviral particles and PCR confirmed this infant was infected with GorLCV. *(Photo Courtesy and Copyright of J. Bosco Noheri for Gorilla Doctors.)*

in immunosuppressed OW monkeys with lymphocryptovirus infection.[86] To date there is no proven association between LCV and lymphomas in apes.

Papillomaviruses

Papillomaviruses (family *Papillomaviridae*) are strongly tropic for epithelial surfaces and become latent in basal epithelial cells causing lifelong infections that can be detected by molecular methods.[279] Papillomaviruses can be asymptomatic and several different ones may be carried by a single host species including chimpanzees and gorillas.[11] When pathogenic, they cause epithelial hyperplasia with characteristic intracytoplasmic edema (ballooning degeneration, koilicytes). Complete virions are formed in the outer stratum spinosum and stratum corneum and result in smudgy amphophilic intranuclear inclusion bodies. **Oral epithelial hyperplasia**, also known as **oral papillomatosis**, is seen in both chimpanzees and bonobos.[281,680,716] Clinical disease, generally self-limiting, and lesions seen on routine health exams may be associated with recrudescence during stressful events. Horizontal transmission occurs, possibly through wounding during social interactions. Grossly, lesions on the gums and oral mucosa of the tongue, lips, and cheeks are single or multiple, pale pink, plaques or nodules with cobble-stone to villiform surfaces (Fig. 15.24A). Histologically, they are characterized by acanthosis with irregular branching rete pegs (Fig. 15.24B; eSlide 15.e3). There is little cytoplasmic pallor or intracellular edema and inclusion bodies are often inapparent. Immunohistochemistry using antibodies directed toward a papillomavirus group antigen highlights the virions (Fig. 15.24C) and PCR can be used to detect latent infections.

RNA Viruses

Respiratory Disease Viruses

Respiratory infections are a common cause of morbidity and mortality in apes in the wild and in managed care and are second only to trauma as a cause of death in some wild populations.[70,144,145,178,236,237,331,349,406,519,531,634,658] Infections range from self-limiting upper respiratory "colds" to severe lower airway disease and pneumonia often complicated by superimposed bacterial infections. Causative viruses include those of human origin, as evidenced by natural outbreaks and experimental infections. These include: **rhinoviruses** (*Picornaviridae*, genus *Enterovirus*); **human metapneumovirus** (*Pneumoviridae*, genus *Metapneumovirus*); **respiratory syncytial virus** (*Pneumoviridae*, genus *Orthopneumovirus*); **measles virus** (*Paramyxoviridae*, genus *Morbillivirus*); **parainfluenza viruses** (*Paramyxoviridae* genera *Respirovirus* and *Rubulavirus*); **influenza viruses** (*Orthomyxoviridae*, genera *Influenzavirus A*, *Influenzavirus B*, *Influenzavirus C*); and **coronaviruses** (*Coronaviridae*, genus *Betacornoavirus*). Respiratory DNA viruses (e.g., adenoviruses, bocaviruses) are also

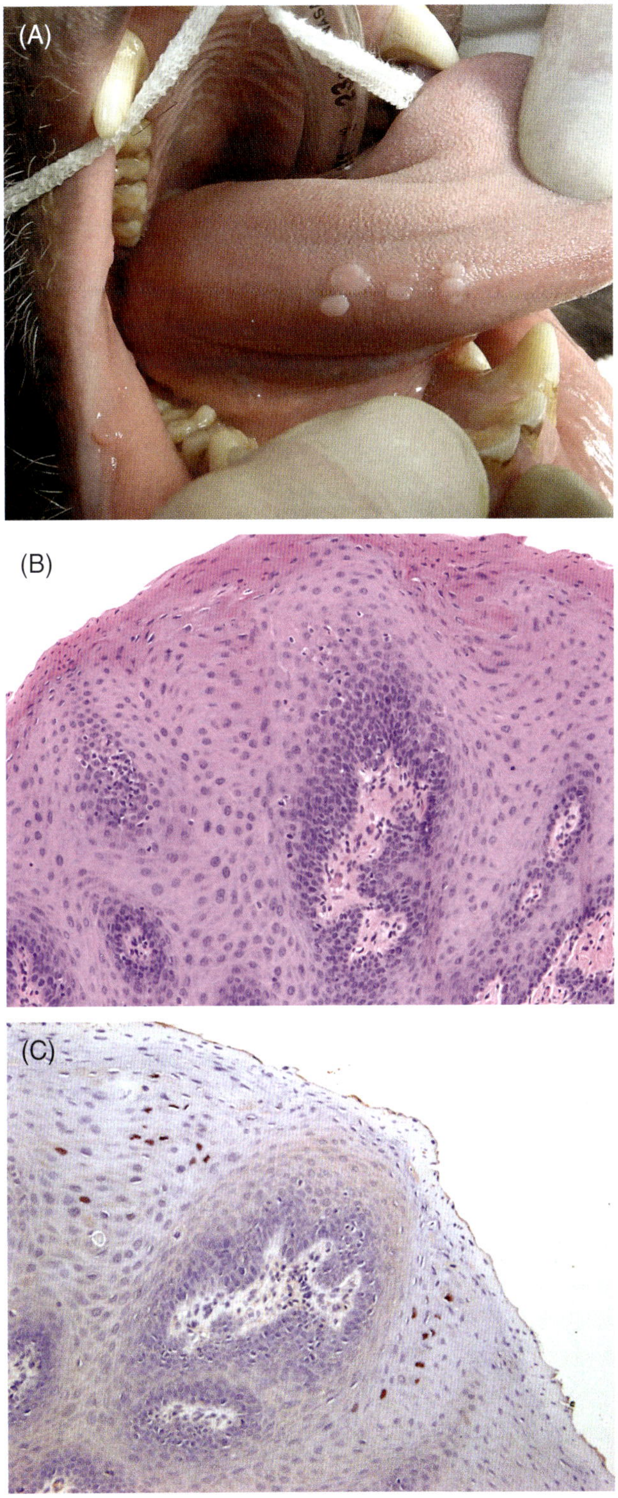

FIGURE 15.24 Chimpanzee papillomavirus infection in the oral cavity of a chimpanzee. (A) Multifocal, demarcated, slightly raised foci of epithelial hyperplasia (papillomatosis) are present on the tongue and buccal mucosa. (B) Histologic lesions are characterized by acanthosis with mild hyperplasia of the basal epithelial cell layer. Several smudgy or clear nuclei are seen, but karyomegaly or unequivocal intranuclear inclusions are not (see eSlide 15.e3). (C) Papillomavirus antigen is demonstrated by immunohistochemistry using antibodies directed toward papillomaviral group-specific antigens. *(Part A: Photo Courtesy of R. Wack, University of California, Davis, Zoological Medicine Service.)*

suspected to have reverse zoonotic potential. Definitive diagnosis of respiratory viral infections relies on histology and virus detection by immunohistochemistry and/or molecular techniques. There are commercially available antibodies for immunohistochemistry for RSV, human metapneumovirus, human parainfluenza viruses 1–3, measles, mumps, and others, as well as published sequences for primers for molecular amplification and in situ hybridization. In addition to postmortem samples, chewed plant material (wadges) and feces can be used for antemortem detection.

First described in humans in 2001, **human metapneumovirus (HuMPV)** (genus *Metapneumovirus*) is a leading cause of upper respiratory disease, bronchitis, and pneumonia in children and adults worldwide. It is a reverse zoonosis. Naturally occurring, sometimes fatal infections are recognized in chimpanzees in the wild and in zoos as well as in free-living, human-habituated mountain gorillas.[178,331,531,634,711] Serologic testing suggests widespread exposure to HuMPV in captive apes. Keeper to chimpanzee to keeper transmission is documented. This virus predisposes to *Streptococcus pneumoniae* and other bacterial pneumonias, but viral infection alone can be fatal. Clinical signs include cough and nasal discharge. Grossly, lungs may be firm, mottled, and fail to collapse when the chest is opened (Fig. 15.25A). Histologic lesions include bronchointerstitial pneumonia with loss of cilia and epithelial necrosis, alveolar epithelial necrosis and fibrin exudation, type II pneumocyte hyperplasia, and airway and alveolar histiocytes (Figs. 15.25B, C). "Smudge cells" may be seen, however there are no true syncytia and no inclusion bodies. Although death due to HuMPV occurs in naturally infected apes, experimentally naïve chimpanzees experience only self-limiting nasal discharge.

Respiratory Syncytial Virus (RSV; genus *Orthopneumovirus*, family *Pneumoviridae*, "chimpanzee coryza agent"), first identified in laboratory-housed chimpanzees, is a reverse zoonosis with humans as natural hosts.[59] The infection can be fatal in infants and elderly humans. RSV causes upper respiratory disease in humans, providing ample opportunity for aerosol transmission to apes in captivity and the wild. Serology reveals antibodies in mountain gorillas and the highest seroprevalence of any of the common human respiratory viruses in zoo-housed great apes.[70,749] RSV respiratory disease outbreaks are documented in zoo-housed chimpanzees, gorillas, a gibbon, orangutans, and in free-living, human-habituated mountain gorillas.[433,652,685,710] In an outbreak in free-living, western lowland gorillas, the strain of RSV was similar to that causing human community-acquired respiratory disease.[237]

Chimpanzees present with sneezing, rhinorrhea, catarrh, and cough (tussis) that often progress to bronchitis and pneumonia. Severity is highly variable and fatal infections are reported, usually in association with concurrent streptococcal pneumonia. In chimpanzee outbreaks, the most severely affected animals exhibit lethargy, anorexia, and

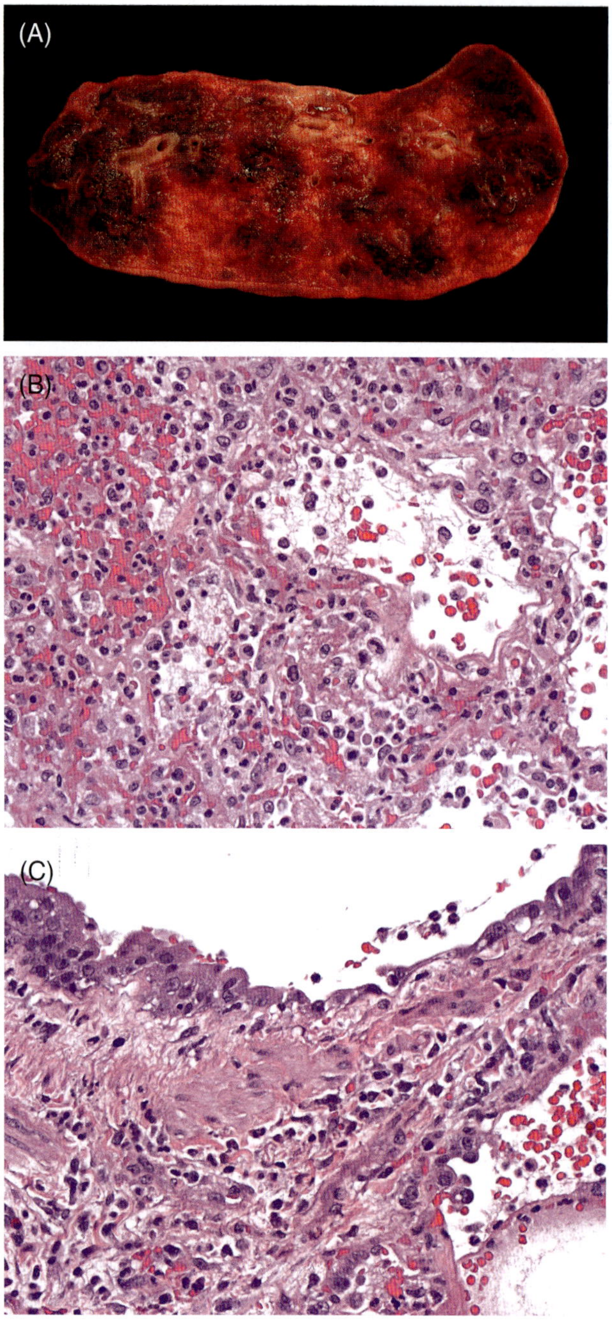

FIGURE 15.25 Human metapneumovirus (HuMPV) infection in the lung of a common chimpanzee. (A) Infection is characterized by nonspecific patchy discoloration and consolidation. (B) Bronchointerstitial pneumonia including acute alveolar septal damage and fibrin exudation is typical of many respiratory viral pneumonias, including HuMPV. Neutrophils may be responding to secondary bacterial infection. (C) Bronchiolar damage includes characteristic epithelial deciliation *(on left)* and attenuation *(on right)*. *(Parts A–C: Photos Courtesy of University of Illinois Zoological Pathology Program.)*

weakness. Reddening of the nasopharyngeal mucosa, enlarged tonsils and regional cervical and tracheobronchial lymph nodes, pulmonary edema, and patchy to panlobar pulmonary consolidation with fibrinous pleural adhesions are features in fatal cases. Histologically, there is a typical viral bronchointerstitial pattern with small airway epithelial necrosis or proliferation and occasional multinucleated syncytia with intracytoplasmic acidophilic inclusion bodies. Suppurative bronchopneumonia is present if there is secondary bacterial infection, in which case viral "foot prints" may be largely effaced.[236] Crusting of the nares and respiratory distress are reported in a gibbon. Histologically, skin lesions are multifocal superficial and follicular acanthosis and hyperkeratosis with rare epithelial syncytia and cytoplasmic eosinophilic inclusions. Diagnosis is based on histology and detection of the virus by PCR or IHC. Measles giant cell bronchointerstitial pneumonia is a main differential diagnosis, but in measles both intranuclear and intracytoplasmic inclusions are present and the GI tract is often affected.

Measles (rubeola) virus (family Paramyxoviridae, genus *Morbillivirus*) is a highly contagious, potentially fatal, infection of humans that occurs as periodic epidemics in children and unvaccinated adults. It has a global distribution but is largely eradicated in developed countries due to vaccination. All monkeys and apes are susceptible, with high morbidity and mortality ranging from 0% to 100% depending on the species (see New World and Old World Monkeys). Infection is via the respiratory route. The virus is trophic for dendritic cells, monocytes, lymphocytes, epithelia, and nervous system. Initial infection involves replication in both T and B lymphocytes with secondary infection of dendritic cells followed by epithelia and nervous system.[132] Affected epithelia include the upper and lower respiratory tract, skin, and GI tract. Due to lympholysis, measles virus causes profound immune suppression causing secondary viral and bacterial infections. Death is usually due to diarrhea or pneumonia and, occasionally, CNS complications. In humans, four forms of measles encephalitis are recognized: acute primary measles encephalitis, acute postmeasles encephalitis, measles inclusion body encephalitis, and subacute sclerosing panencephalitis.[189] The latter two can occur months to years, respectively, after initial infection.

Although apes are presumed to be susceptible to measles infection and vaccination is recommended, there are few published reports of the clinical or pathological signs in these species. A wide-reaching respiratory disease outbreak in wild mountain gorillas in Rwanda in 1988 is presumed to have been due to measles and secondary mycoplasma pneumonia, based on rising serologic titers and histological findings in young animals.[262] Although a characteristic truncal rash, facial edema, and chemosis are common in OW monkeys, none was seen in gorillas. Lung lesions are of typical bronchointerstitial pneumonia, centered on small bronchi and bronchioles with necrosis of epithelium and filling of lumens by mucus and cell debris. The lesion extends into the interstitium as acute alveolar damage with exudation of protein-rich edema

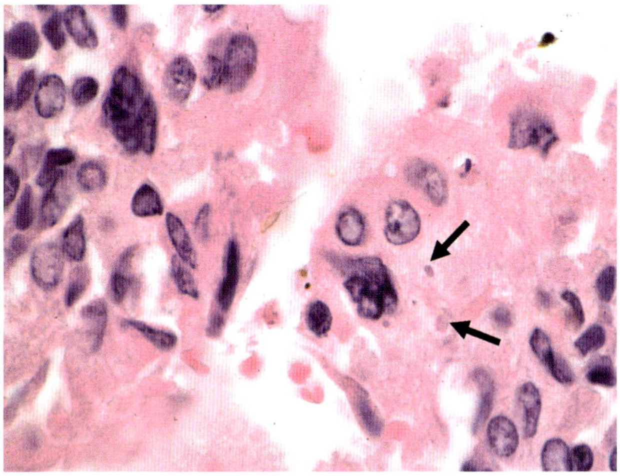

FIGURE 15.26 Viral airsacculitis in a mountain gorilla. Syncytia containing both intranuclear and intracytoplasmic inclusion bodies *(arrows)*. Electron microscopy confirmed paramyxoviral nucleocapsids suggestive of measles virus infection.

fluid, increased numbers of alveolar macrophages, and sloughed cellular debris. Syncytia are present in the parenchyma or airways (giant cell interstitial pneumonia) and air sac epithelium (Fig. 15.26). Both intranuclear and intracytoplasmic inclusion bodies are seen in syncytia and other cells. Lymphoid depletion with germinal center syncytia (Warthin-Finkeldy cells) may be seen.

Parainfluenza viruses (PIV) types 1, 2, 3, and 4 (human: HPIV-1, HPIV-2, HPIV-3, respectively; Syn: Paramyxovirus PMV; family *Paramyxoviridae*; genera *Respirovirus* [HPIV-1 and HPIV-3] and *Rubulavirus* [HPIV-2 and HPIV-4]) are the most prevalent and significant of the human parainfluenza viruses.[170] HPIV-1 and HPIV-2 are often associated with laryngotracheitis, while HPIV-3 is often associated with bronchitis and bronchiolitis. HPMV/HPIV-1 infects chimpanzees and replicates in lower airway epithelium.[633] Disease varies from asymptomatic with virus replication and shedding to fatal pneumonia in infant chimpanzees. In a small outbreak of HPMV/HPIV-2 in three zoo-housed gorillas, reported clinical signs are serous progressing to purulent rhinorrhea, cough, and lethargy. Hemorrhagic laryngotracheitis and pulmonary consolidation with abscessed lymph nodes are the main lesions present in the one gorilla that died.[112] Histologically, PMV-1 may form large multinucleated syncytia without inclusion bodies in respiratory epithelium. In the gorilla, interstitial pneumonia with type II pneumocyte hypertrophy is largely obliterated by bacterial suppurative pneumonia. Gibbons are susceptible to spontaneous infection with PMV-3 (synonymous with PIV-3) occurring as respiratory disease outbreaks.[428] Mortality is not reported. As is the case for many of the respiratory viruses, PIV infections are often complicated by bacterial pneumonia with either gram-positive or gram-negative organisms.[308]

Mumps virus, a *Rubulavirus* related to HPIV-2, causes "mumps"; humans are the natural host. The typical childhood disease consists of fever, aches and pains, sore throat, and parotid gland inflammation that causes swelling of the face and angle of the jaw.[583] Respiratory signs are also present because the respiratory tract is the portal of entry. Infection of the CNS occurs commonly but is largely asymptomatic, though clinical aseptic meningitis can occur. Orchitis is common in puberal and adult males. Antibodies against mumps virus are detected in gorillas, chimpanzees, and orangutans.[339,763] Natural disease in chimpanzees mimics the childhood syndrome with sialadenitis (parotiditis), and pharyngeal and palatine erosions. Histologically, there is interstitial and periglandular edema with perivascular and interstitial mononuclear cell inflammation and necrosis of acinar and ductal epithelium. Postinfection scarring of the parotid gland may occur. Mumps virus causes cytoplasmic inclusions and syncytia formation in tissue culture, but syncytia are not reported in clinical cases.[400]

Influenza viruses (family *Orthomyxoviridae*; genus *Influenzavirus*) including **Influenza A** and **B**, have a broad host range with avian, swine, equine, canine, marine mammals, and human (primate) strains; genetic recombination can result in shifts of host range. Antibodies to influenza A and B are common in serosurveys of apes in captivity and in the wild.[70,318,749,771] Chimpanzees can be infected experimentally. Susceptibility to different strains depends on sialic acid cell surface receptors in the upper and lower respiratory tract and the surface proteins of the virus.[202] There is debate about the infectivity of human influenza viruses in nonhuman primates in general.[128] However, illness is often reported in ape collections at times when "flu" is circulating in the human population and positive serology has been detected in chimpanzees, orangutans, and gibbons. Flu-like disease can be caused by a variety of respiratory viruses and serology or molecular detection should be used to confirm the causative etiology.[248] Morbidity in colony-housed gibbons infected with influenza A2 Hong Kong 68 includes both respiratory and gastrointestinal signs (e.g., lethargy, anorexia, serous to mucous nasal discharge, cough, and diarrhea or constipation).[306] At necropsy gibbons have pneumonia, most severe in caudal lobes, with consolidation and hemorrhage. Fibrinous bronchointerstitial pneumonia is seen histologically with epithelial necrosis and sloughing in terminal bronchioles and filling of alveoli by fibrin, edema, erythrocytes, and macrophages. Bronchiolitis obliterans may be present focally. Secondary bacterial infections result in superimposed, suppurative bronchopneumonia. The absence of syncytia or inclusion bodies and presence of severe alveolar hemorrhage may help to differentiate "flu" from other fatal viral respiratory infections.

Coronaviruses

Coronaviruses (family Coronaviridae) detected in great apes include HCoV-OC43 in chimpanzees and SARS-CoV-2 in gorillas.

HCoV-OC43 (genus *Alphacoronavirus*) an alphacoronavirus is a human virus that is one of the causes of the "common cold." It was determined to be the cause of a mild respiratory disease outbreak in a well-studied population of chimpanzees in the Tai Forest.[538] Fecal PCR ruled out other possible respiratory viruses. Clinical signs were sneezing and coughing.

Severe Acute Respiratory Disease Coronavirus-2 (SARS-CoV-2 [COVID-19]) (genus *Betacoronavirus*) is a zoonotic virus and the cause of a major human pandemic. Infection has been documented in zoo-housed gorillas in the United States and Europe, but not in human-habituated, free-living gorillas.[214,342,493] Based on ACE2 receptor sequence identity, all apes as well African and Asian cercopithecines, are predicted to be susceptible to SARS-CoV-2 infection.[444] No mortalities have been reported, and only a few geriatric animals have been seriously ill, one of which had apparent carditis as well as pneumonia.[342] To date, no gorillas have died from their infection. The source of infection has been caregivers. Signs are those of a nonspecific respiratory illness: decreased activity to severe lethargy, coughing, rhinorrhea, and decreased appetite. Fecal PCR testing is used for diagnosis when pharyngeal or nasal swabs are not possible. Rapid lateral-flow tests can be used for sedated apes in field situations. It is unknown whether lesions in apes would be similar to those in experimentally infected macaques, which mimic, to varying extent, those of humans.[573] Gross and histologic features of human SARS-CoV-2 are those of severe bronchointerstitial viral pneumonia with acute alveolar damage and evidence of enhanced coagulation such as capillary and small vessel fibrin thrombi.[25] Differentiation from other viral pneumonias requires demonstration of the virus in antemortem and/or postmortem samples.[354] SARS-CoV-2 can be detected in type I and II pneumocytes by IHC. There are several commercially available antibodies for IHC and in situ hybridization probes.

Picornaviruses (Family Picornaviridae)

Encephalomyocarditis virus (EMCV; genus *Cardiovirus*) was first isolated from a gibbon in captivity in Florida that died of congestive heart failure.[80] The virus occurs in many regions of the world, but with discontinuous distribution. In the United States, natural infections in zoo animals are much more common in the southeast and south. Reservoir hosts are rodents, but the virus has broad host range including nonhuman primates, swine, carnivores, elephants, and other exotic hoofstock. The transmission route is orally from food or water contaminated by infected rodent feces and from ingestion of raw meat from an infected animal. The virus has broad tissue tropism but clinical signs are attributable to viral damage to beta cells of the pancreatic islets (causing type I diabetes, primarily in rodent models), myocardial fibers causing myocarditis, and infection of CNS tissues causing meningoencephalitis and/or myelitis. Placental infection and fetal death may occur in nonhuman primates and swine. Although EMCV causes cell death, pathogenicity is also linked to the host inflammatory response.

Spontaneous disease can occur in outbreaks in zoos and primate colonies.[76,289,770] Affected species include lemurs, squirrel monkeys, baboons, siamangs, orangutans, chimpanzees, and bonobos.[309,562,743] Clinical signs associated with cardiac lesions include sudden death or congestive heart failure and pulmonary edema. Infected bonobos may present with coughing, mimicking respiratory virus infection. Placentitis and abortions are reported in baboons but not in apes. Gross lesions include pericardial effusion and severe pulmonary edema. Cardiomegaly and pale foci in the myocardium may be seen. Histologically, inflammation in the heart is lymphoplasmacytic and histiocytic, associated with myofiber degeneration and necrosis and variable amounts of hemorrhage. In affected bonobos, neutrophils may also be present (Fig. 15.27). The main differential diagnoses are coxsackievirus carditis, Chagas disease (trypanosomiasis), and toxoplasmosis. Encephalitis, best described in squirrel monkeys, is characterized by cerebrocortical neutrophilic and lymphocytic inflammation with perivascular cuffing and loss of neurons in cerebral gray matter. Definitive diagnosis relies on virus detection by culture or PCR.

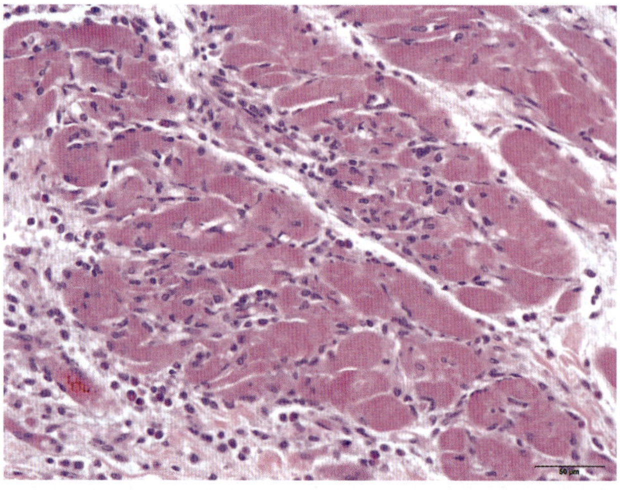

FIGURE 15.27 Encephalomyocarditis virus myocarditis infection in a bonobo. Characteristic myonecrosis, myophagia, interstitial edema, and mixed inflammation are seen in the heart.

Enteroviruses (Family Picornaviridae)

Humans are host to over 70 different **enteroviruses** (genus *Enterovirus*) and many others are found in cattle, swine, and nonhuman primates. These small, single-stranded RNA viruses often have affinity for gastrointestinal and respiratory tracts but also cause cutaneous, myocardial, or nervous system lesions. Well-known enteroviruses that have been reported in apes include **rhinoviruses**, **poliovirus**, and **coxsackieviruses**.[158,257,258,457,459,588] The prevalence and effect of these viruses in apes have been incompletely explored. Studies in free-living apes have shown that enteroviruses can cocirculate in apes and humans.

Rhinoviruses (genus *Enterovirus*) cause a large proportion of the "common cold" in humans. A respiratory disease outbreak of **Rhinovirus C**, which is the most pathogenic of the human cold viruses, is described in a chimpanzee sanctuary (detected in blood samples) and in a group of free-living chimpanzees in Uganda.[158,615] One wild infant chimpanzee died. Fecal PCR was used to follow the outbreak in the Uganda chimps and rhinovirus was also detected in swabs collected from the dead infant. Lesions are not described. The cadherin-related family member 3 CDHR3-Y529 allele, which increases susceptibility to severe rhinovirus infections and asthma in human infants, is present in all chimpanzee populations studied to date.

Coxsackieviruses are divided into two groups, **Coxsackie A** and **B**, each with several serotypes. Positive serology is seen in several species of nonhuman primates including orangutans.[339] Deaths reported in apes include neonatal and adult chimpanzees infected with Coxsackie B5 and orangutans infected with Coxsackie B4.[505] Diarrhea and respiratory signs occur. Gross lesions include pneumonia, visceral congestion, and cardiac enlargement with pale streaking. Histologically, lymphocytic myocarditis consists primarily of CD3-positive T-cells with occasional granulocytes in the interstitium centered on necrotic myofibers. Meningitis and hepatic necrosis may also be seen. The lesions resemble those of EMCV infection and virus detection is needed to differentiate the two. Coxsackievirus infection may predispose to streptococcal pneumonia. Transmission is most often fecal–oral. Infection of tonsils and regional lymph nodes results in secondary hematogenous spread to lungs, meninges, liver, and heart. Chronic intestinal carriage of coxsackieviruses has been noted in experimentally infected, clinically normal chimpanzees, raising concerns about zoonotic transmission.[445] Coxsackie-like enteroviruses can be detected by culture and PCR in feces of wild and human-habituated gorillas in western Africa without identified clinical signs.[6]

Poliomyelitis virus (genus *Enterovirus*) can affect all "simian" primates, but not prosimians. Humans are the primary host. The natural infection in apes is primarily a nonparalytic enteric disease. CNS involvement is reported in less than 1% of chimpanzee cases as well as in orangutans and gorillas.[151] After experimental infection, laboratory chimpanzees may be asymptomatic, though brain and cord lesions may be present in some asymptomatic individuals.[49] Natural infection in free-living chimpanzees can lead to motor neuron signs including paresis and paralysis with secondary skeletal asymmetries.[465,754] Poliovirus transmission is fecal–oral and, occasionally, respiratory. Virus multiplies in the oropharynx and intestinal mucosa, spreads to lymphoid tissues [especially tonsils, cervical lymph nodes, gut-associated lymphoid tissue (GALT), and mesenteric lymph nodes], and then becomes viremic leading to CNS infection. Fecal shedding may persist for months. Neurological lesions include nonsuppurative, polioencephalomyelitis with neuronal swelling, chromatolysis, neuron cell death, and neuronophagia. Inflammation is transiently neutrophilic becoming lymphocytic. Myelitis is centered on anterior (ventral) horns of the spinal cord and medulla. Occasional glial nodules are present in the motor cortex. Direct intranasal inoculation leads to CNS lesions in the olfactory lobes, presumably by direct neuronal transport.[717] Oral polio vaccination of chimpanzees in sanctuary settings confirms antibody protection lasting many years.[474]

Additional polio-like enteroviruses have been detected by fecal serology in wild western lowland gorillas, and enteroviruses are implicated in acute flaccid paralysis resembling polio (**enterovirus, C99**) in a chimpanzee, and in facial paralysis in a western gorilla (**Enterovirus C** [strain Ibou002]).[6,257,258,457,620]

Hepatitis Viruses

Hepatitis A (HAV) (family Picornaviridae, genus *Hepatovirus*), a picornavirus transmitted via fecal–oral route, is found worldwide. Humans are the probable natural host, but hepatitis A can be spread from human to primate, primate to primate, and primate to human, and there are nonhuman primate host-specific strains of HAV.[572] Serological evidence of natural infection can be found in zoo-housed gibbons, chimpanzees, orangutans, and gorillas and chimpanzees can be experimentally infected.[551,598] Spontaneous infection is usually subclinical, detected only with clinical chemistry, but fatal infections can occur. Acute lesions in chimpanzees consist of multifocal, predominately periportal, individual hepatocyte necrosis, and acidophilic bodies (apoptotic hepatocytes) with moderate lymphocytic inflammation accompanied by mononuclear periportal hepatitis that breaches the limiting plate and bridges between portal areas. Lesions typically resolve in less than a month. Liver enzymes, especially serum glutamic-pyruvic transaminase (SGPT), also called alanine aminotransferase (ALT), increase dramatically but transiently. HAV does not establish persistent infections.

Hepatitis B (**HBV**; family Hepadnaviridae; genus *Orthohepadnavirus*) occurs worldwide in humans and produces chronic active infection leading to cirrhosis and hepatocellular carcinoma. Human hepadnaviruses are used experimentally to infect chimpanzees as a model for HBV vaccine development. Nonhuman primate HBV strains also exist in gibbons, orangutans, gorillas, and chimpanzees in captivity and in the wild.[52,511,596,599] The prevalence of HBV exposure and chronic infection in chimpanzees in the US zoos is about 2%, and must be taken into account in management decisions.[440] Ex-captive orangutans in Indonesia have a 59% prevalence of antibodies to HBV and of these almost 11% are chronic active carriers.[739] Seroprevalance may be high (up to 88.6%) in captive breeding programs of gibbons and some of these gibbons may be positive for HBV DNA and surface antigen. HBV DNA can also be detected in saliva. Infected gibbons have higher levels of ALT than gibbons without active infection.[513] Ape-to-ape transmission can occur within and between species of apes. The zoonotic potential of ape HBVs is unclear; however, there is documented transmission between a zoo-housed gorilla and a keeper.[399] In apes, acute HuHBV infections are usually subclinical. Histologically, there is diffuse, panlobular "sinusoidal lining cell activation" and multifocal random individual hepatocyte necrosis (sometimes becoming confluent) with a few attendant lymphocytes. Acidophilic bodies (apoptotic hepatocytes) and multinucleated hepatocytes may be present, and elevation of liver enzymes, especially serum glutamic-oxaloacetic transaminase (SGOT) (also called aspartate amino transferase or AST), is seen. Histological lesions in chronic HBV carrier chimpanzees consist of marked lymphocytic periportal hepatitis without disruption of the limiting plate, and scattered "ground glass" hepatocytes in which hepatitis B surface antigen (HBsAg) can be demonstrated by immunohistochemistry.[698] Liver cancer and cirrhosis, common sequelae in humans, are unusual in experimentally infected chimpanzees.[552]

Filoviruses (Hemorrhagic Fever Viruses)

African Ebola viruses (genus *Ebolavirus*; family Filoviridae) are WOAH-listed reportable, zoonotic viruses that are highly pathogenic for humans, apes, and other primates. Of the five recognized subtypes of African Ebola, three have been associated with outbreaks: Ebola Zaire, Ebola Sudan, and Ebola Bundibugyo. Bats are suspected, but not proven, reservoir hosts. Devastating outbreaks of African Ebola with population effects occur in chimpanzees, bonobos, and western lowland gorillas in equatorial Africa.[392,734] Transmission is through direct contact with blood or other secretions from infected individuals, or ingestion of infected tissues. Experimental disease (rhesus and cynomolgus macaques) consists of a 2- to 6-day incubation period, early fever and petechial rash (in rhesus), depression, dyspnea, diarrhea, blood from the rectum and vagina, lymphopenia, and thrombocytopenia. Gross lesions include hemorrhage in the skin, lungs, liver, and spleen, splenomegaly, and interstitial pneumonia. Histologically, hemorrhage, hepatic lipidosis, hepatic necrosis with basophilic cytoplasmic inclusions, and lymphoid necrosis are reported in humans and experimentally infected primates. Macrophages are infected and inflammation is often lymphocytic and histiocytic. In the only description of naturally occurring Ebola in a chimpanzee, gross lesions suggestive of "hemorrhagic septicemia" were seen.[765] Histological lesions include multifocal hepatitis, multinucleated hepatocytes throughout the parenchyma, and Kupffer cell hyperplasia. Eosinophilic intracytoplasmic inclusion bodies are seen in Kupffer cells but not hepatocytes. Lymphoid depletion with lymphocytolysis and hemorrhage is seen in the spleen along with "fibrinoid necrosis" in red pulp, especially in the marginal zones of the periarteriolar lymphoid sheaths. Splenic macrophages contain cytoplasmic inclusions, similar to those in the Kupffer cells. The presence of Ebola virus is confirmed with immunohistochemistry, for which reagents are available commercially.[401] Serological studies suggest some apes survive exposure to Ebola viruses.[392] Both IgG and IgM antibodies can also be detected in chimpanzee and gorilla fecal samples allowing for noninvasive epidemiological studies.[458] This suggests that Ebola exposure has not been uniformly fatal in these populations. Naïve populations may be candidates for an Ebola vaccination strategy being developed specifically for great apes.[735,757]

Retroviruses

Many **exogenous retroviruses** (family Retroviridae) have been identified in apes. These include: the gibbon ape leukemia virus (GALV, genus *Gammaretrovirus*); primate T lymphotropic viruses (PTLVs; genus *Deltaretrovirus*), simian immunodeficiency viruses (SIVs, genus *Lentivirus*); and foamy viruses (genus *Spumavirus*).[540] Most of these are homologous to human retroviruses and cross-species transmission has occurred historically and in recent times. Some appear to be apathogenic (e.g., foamy viruses). All the retroviruses have the propensity to cause chronic infections with integration of DNA provirus into the host genome. Some are oncogenic.

Gibbon ape leukemia virus (GALV) has historically been associated with outbreaks or hemolymphatic malignancies in gibbons in managed care.[332] Virus is shed in the urine and feces, although the mode of transmission is uncertain. Antibodies against GALV can be identified in gibbons without evidence of disease, but both natural and experimental infection can result in disease.[632] Because GALV lacks viral oncogenes, latency may be prolonged and pathology variable depending on when and where proviral integration occurs. Accordingly, age at onset of illness varies, and disease course can be prolonged (up to 3 years). Conditions associated with GALV infection in gibbons include malignant lymphoma, lymphoblastic leukemia, myelogenous (granulocytic) leukemia, and "osteopetrosis" of long bones. In cases of granulocytic leukemia, there is marked circulating granulocytosis with infiltration of bone marrow, liver, lymph nodes, and spleen. Infiltrated tissues often have a greenish hue (chlorosis). Disease due to GALV is no longer prevalent in the US SSP populations of gibbons, and, although antibodies are still present in some apes, it is thought that this disease may no longer be of concern.[66]

Simian T-lymphotropic viruses (STLV, syn. Primate T-lymphotropic viruses PTLV) (Retroviridae, Orthovirinae, *Deltaretrovirus*) have been identified in gorillas, chimpanzees, bonobos, and gibbons.[19,311,430,712] Human PTLVs are thought to have been originally derived from apes and zoonotic transmission has occurred from bites from gorillas. Several disease syndromes are recognized in humans. Pathogenicity in gorillas has been suggested by reports of immunosuppression in naturally infected gorillas and reports of T-cell lymphoma or leukemia in zoo-housed gorillas.[20,47,383,554]

Simian immunodeficiency viruses (SIVcpz, SIVgor) (genus *Lentivirus*) occur in a variety of nonhuman primates, including chimpanzees and western lowland gorillas, but not (to date) in free-ranging eastern lowland and mountain gorillas, bonobos, gibbons, or orangutans.[19,123,181,182,334,398,500,585,600,601,739] Two strains of SIVcpz exist, SIVcpz*Ptt* infecting the central chimpanzee subspecies and SIVcpz*Pts* infecting the eastern chimpanzee. Wild chimpanzees of the other two subspecies (*Pan troglodytes verus* and *P. troglodytes ellioti*) are not naturally infected.[623] SIVgor appears to have arisen from cross-species transmission of SIVcpz*Ptt*.[687] Cross-species transmission of SIVcpz*Ptt* into humans is likely the origin of HIV-1 strains in phylogenetic Group M, the strain associated with the AIDS pandemic, as well as strains in Group N.[205,334] In contrast, SIVgor appears to be the origin of HIV-1 Group O and P strains.[123] Vertical transmission has been demonstrated for SIVcpz and SIVgor and horizontal transmission proven for SIVcpz and suspected for SIVgor.[181,333] Population-wide studies indicated negative effects on population growth, survival, and reproduction in free-ranging eastern chimpanzees with SIVcpz.[584]

Pathologic changes are reported in SIVcpz*Pts* and SIVcpz*Ptt*-infected eastern chimpanzees.[182,333,694] Significantly lower numbers of CD4+ lymphocytes are present in lymphoid tissues in comparison with controls. In some chimpanzees, CD4+ T-cell loss is selective, while other lymphocyte populations remained intact. Marked depletion of cortical and paracortical T and B lymphocytes in lymph nodes (Fig. 15.28) and marked lymphoid depletion in the spleen, consistent with end-stage infection or AIDS-like disease, may be seen. At death, chimpanzees may be emaciated with profound sarcopenia and hepatocellular atrophy. In all infected chimpanzees, hyaline deposits within splenic periarteriolar lymphoid sheaths, a nonspecific marker for chronic immune activation, is noted with severity paralleling the loss of lymphocytes. To date, SIVcpz*Pts* infections have not been associated with malignancies. Opportunistic infections likely occur but are difficult to prove due to rapid autolysis in apes found dead in the wild. Destabilization of the gut microbiome at the end-stage disease (months prior to death) but not early in infection has been documented from antemortem fecal samples suggesting the possibility of enteric dysbiosis with immunodeficiency.[32] Decreased peripheral blood CD4+ T-cell count has been described in a sanctuary-housed SIVcpz*Ptt*-infected chimpanzee with frequent infections with various pathogens.[182]

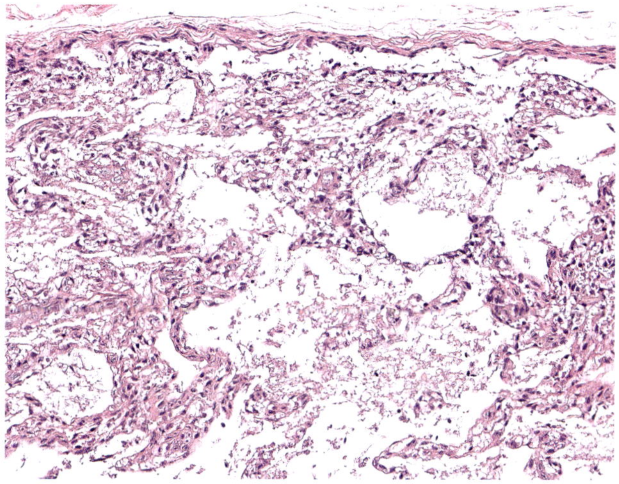

FIGURE 15.28 Natural SIVcpz infection in a chimpanzee. Infection is associated with marked lymphoid depletion that results in a "cobweb" appearance in the lymph nodes. *(Photo Courtesy of Gombe Ecohealth Program.)*

Bacteria

See Table 15.e4 for a list of additional bacteria and fungi reported from apes.

Streptococcus pneumoniae (syn. ***Diplococcus***, ***Pneumococcus***), a gram-positive, diplococcal bacterium, is a common cause of pneumonia, meningitis, and pericarditis in apes, as is the case in monkeys. Vaccination is recommended. Predisposing factors include young age, immune suppression, and underlying viral infections, such as parainfluenza-3, human metapneumovirus, and human respiratory syncytial virus.[308] Reports include pneumonia and meningitis in a juvenile wild chimpanzee, pneumonia and meningitis in chimpanzees and gorillas, respiratory disease including sinusitis and airsacculitis in gibbons, bonobos, lowland gorillas, and orangutans, and tonsillitis in a captive gorilla.[96,236,654,710,713] Numerous strains and serotypes of *S. pneumoniae* have been described in humans and animals. In many cases, strain types of *S. pneumoniae* recovered from captive apes are similar to those identified in their human caretakers, raising the concern for anthropozoonotic and zoonotic transmission.[685] Although nasopharyngeal carriage of *S. pneumoniae* occurs commonly in humans, it is not thought to be a natural pharyngeal inhabitant in other primates. However, oropharyngeal carriage has been demonstrated by PCR in sanctuary-housed chimpanzees, which suggests that ape-specific strains may exist. The extent to which strain type is associated with pathogenicity remains uncertain.[472]

Pulmonary lesions include suppurative bronchopneumonia and/or pleuropneumonia with fibrinous adhesions in the thoracic cavity (Fig. 15.29A). There may be abscesses within the lung and thoracic cavity and inflammation may efface large areas of pulmonary parenchyma (lobar pneumonia). Histologically, lesions contain numerous degenerate neutrophils, fibrin, fewer macrophages, and intra- and extracellular gram-positive cocci, which can be identified antemortem by bronchoalveolar lavage (Fig. 15.29B). In cases with pleuritis or pericarditis, similar inflammation coats the diaphragm, thoracic wall and pericardium. While some reports mention an interstitial component, this may be due to viral coinfection rather than the bacteria.

Other *Streptococcus* species, such as *Strep pyogenes* and *Strep angiosus* may be associated with suppurative bronchopneumonia and chronic airsacculitis in orangutans.[292,426]A single case of necrotizing cervical fasciitis in a zoo housed gorilla was associated with an alphahemolytic *Streptococcus* and a *Prevotella* sp.[5] **Necrotizing fasciitis** can also be caused by the gram-positive *Clostridium septicum* and other bacteria.[195]

Klebsiella pneumoniae, a gram-negative bacillus, is, as in OW and NW primates, a common component of ape respiratory disease, including airsacculitis and pneumonia. Although it can be a primary pathogen, especially if it carries the hypermucoviscosity phenotype, it is often secondary to viral infections or is a coinfection with *P. pneumoniae*. The capsule of *Klebsiella* can inhibit neutrophil chemotaxis and phagocytosis and, although pyogenic, flooding of respiratory bronchioles and alveoli with activated macrophages is common.[528] Due to the capsule, the presence of a clear zone around intra- and extracellular bacilli on impression smears and in sections can be a clue to the presence of *Klebsiella* as is Gomori's methenamine silver (GMS) staining.

The gram-negative coccobacillus ***Pasteurella multocida*** can be cultured from chimpanzees with pneumonia and airsacculitis.[348]

Bacterial meningitis is a sequel to streptococcal and other pneumonias, especially in infant apes. Lesions are similar to those in OW monkeys and are characterized by meningeal vessel congestion, reddened or dull white to yellow leptomeninges with exudates present in sulci and pooling at the base of the brain around the pituitary. Histologically, there are dense infiltrates of degenerate neutrophils within the leptomeninges, ventricles and extending into the parenchyma. gram-positive cocci are seen in most cases. Macrophages, some with intracytoplasmic cocci, and lymphocytes may be present depending on the chronicity of infection. Necrotizing vasculitis and fibrin thrombi are common. Most reported cases of meningitis also have pneumonia, suggesting the respiratory tract as the primary site of infection. Note that in acutely affected infants, leptomeningeal edema may be the only lesion, though bacteria can be identified with H&E, Giemsa, or Gram stains.

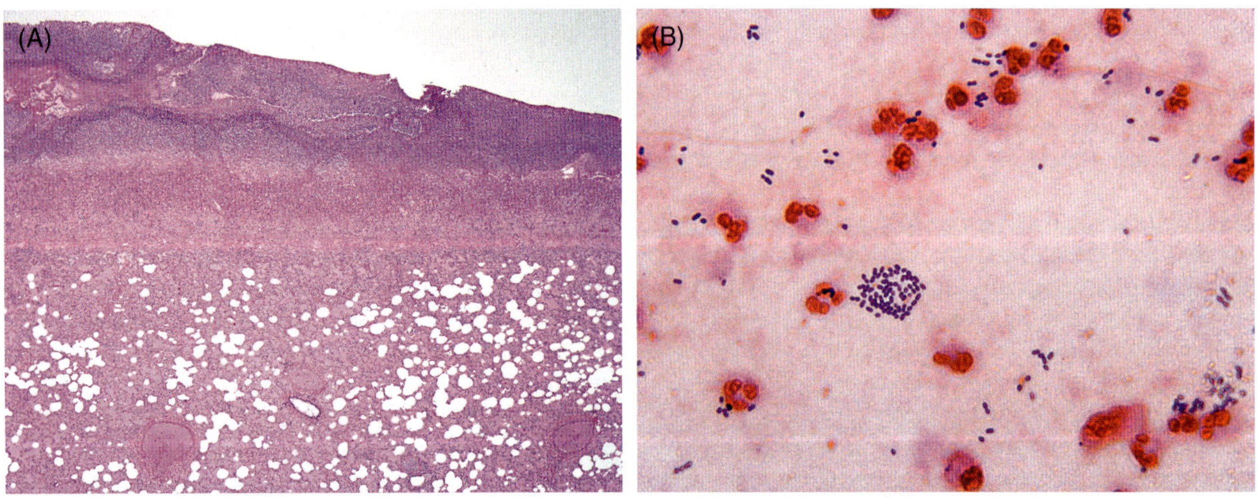

FIGURE 15.29 Suppurative pleuropneumonia due to *Streptococcus pneumoniae* in a chimpanzee (A) and gorilla (B). (A) The pleural is markedly thickened by fibrin and neutrophils *(top of image)* and there is consolidation of the underlying lung parenchyma. (B) Typical gram-positive diplococci can be seen in bronchoalveolar lavage cytology. *(Part A: Photo Courtesy of Gombe Ecohealth Program.)*

Initial diagnosis can be based on cytology (bacterial morphology in exudates) and gross or histologic findings, but confirmation requires culture or molecular detection. Multilocus strain typing of *S. pneumoniae* can provide additional information on molecular epidemiology. As many cases have concurrent viral infections, evaluation for common viral respiratory viruses is warranted.

Meningitis in apes can also be caused by the gram-positive coccus ***Staphylococcus aureus***. Most cases appear to arise from opportunistic infection of wounds with septicemic spread to the brain or extension from otitis. Lesions in the brain are similar to those described for *S. pneumonia*, but encephalitis with regional necrosis and gliosis appears to be more common with *S. aureus*. Gram-negative bacterial causes of meningitis in apes include *Klebsiella pneumoniae* and *Proteus mirabilis*.[298]

Staphylococcus aureus strains can be shared between humans and chimpanzees in sanctuary settings and **methicillin-resistant *Staph aureus* (MRSA)** can be carried by chimpanzees in managed colonies.[250] Suppurative arthritis due to *S. aureus* has been reported in a single neonatal orangutan, and necrotizing cellulitis leading to sepsis and death in a sanctuary-housed western lowland gorilla.[282,492]

Apes have ethmoid, sphenoid, frontal, and maxillary sinuses, of which the maxillary sinuses are generally the largest.[350–352,553,557] The maxillary sinus surrounds the roots of the "cheek teeth" in gibbons, chimpanzees, and gorillas (but not orangutans) and dental disease may extend into the sinuses in these species. The use of MRI and CT has enhanced our understanding of the three-dimensional anatomy of the sinuses and their role in respiratory disease. **Sinusitis** is reported in all apes but is best studied in orangutans in which species it may be the precursor to infection of air sacs and lungs.[675] Mastoid bone air spaces may also be inflamed/infected (mastoiditis).

Airsacculitis has been described in all of the ape species. It is most common and severe in orangutans, possibly due to the extensive and pendulous nature of the ventral "throat sac" and the extensive associated fat pad causing traction on the air sac. Airsacculitis is often not an isolated entity in orangutans, but occurs as part of the orangutan respiratory disease syndrome (ORDS) which includes sinusitis, mastoiditis, airsacculitis, bronchiectasis, and pneumonia.[13,197,661,675] Computerized tomography (CT) examination (both ante- and postmortem) is recommended for better determination of the extent of involvement of pneumatized intraosseous spaces.[668] Disease occurs in all age groups except infants. Prevalence seems to be increased in males, hand-reared apes, and Bornean (vs. Sumatran) orangutans. Affected orangutans are more often related to others with respiratory disease than to healthy individuals, suggesting a possible genetic component. The constellation of lesions in orangutans is suggestive of cystic fibrosis, a hereditary disease of humans.[675] A potentially lethal variant of the cystic fibrosis gene has been detected in orangutans, further pointing to a genetic predisposition.[690]

In chimpanzees with airsacculitis, some studies have found equal sex distribution while others (possibly confounded by a larger male population) noted a male bias.[362] In all species, the most common presenting signs include cough, halitosis, and purulent nasal discharge. A variety of gram-negative and gram-positive bacteria are identified, often as mixed infections with *Pseudomonas aeruginosa* the most reported isolate. An initiating role of viral respiratory infections has been suggested.

Gross and histologic lesions are similar across species, but the extent of pulmonary and sinus involvement is greatest in orangutans. Exudate can be present bilaterally or unilaterally and distension of the air sacs may be seen externally. Exudates may be seen in the ostia, which extend from the lateral saccules of the larynx. Exudate varies from pale or dark tan to yellow and is often viscous (Fig. 15.30A). Air sac walls are thickened and opaque in severe chronic cases, and septation and compartmentalization may develop due to fibrosis. Histologically, the air sac epithelium is thickened with loss of cilia and squamous metaplasia (Fig. 15.30B). The subepithelial stroma is thickened by edema or fibrosis with variable inflammation including formation of lymphoid follicles. Aspiration of purulent material and secondary bronchopneumonia occur most frequently in orangutans.

Periodontal disease with gingivitis, gingival recession, bone loss, and osteomyelitis is common in apes in managed care and in the wild and increases with age in both settings.[411,605] Gross lesions are obvious and include dental plaque, swelling, erythema, rounding and retraction of gingival margins, and ulceration over apical gingiva in younger animals (Fig. 15.31). The same lesions are superimposed on thickened, corrugated fibrotic gingiva in older animals.[12] Skeletal remains of apes give testimony to periodontal bone loss.[407] Histologically, inflammation is most severe in the sulcal epithelium and is largely neutrophilic, and the degree of gingival epithelial hyperplasia increases with duration.[529] Although there is a positive correlation of periodontal disease and cardiac disease in humans, this association was not found in a small study of 12 sanctuary-housed chimpanzees.[559]

Peritonitis, intrabdominal and retroperitoneal abscesses are reported to be frequent in gorillas and there are case reports in orangutans, a subadult female chimpanzee, and a juvenile male chimpanzee.[92,484,488,547] In one study, 8.5% of captive western lowland gorillas had suspected intraabdominal abscesses of which 90% were mostly in overweight, reproductively inactive adult females.[488] Affected animals are often lethargic and anorexic with recto-genital discharge and constipation. Abdominal distension is noted in some cases. Abscesses

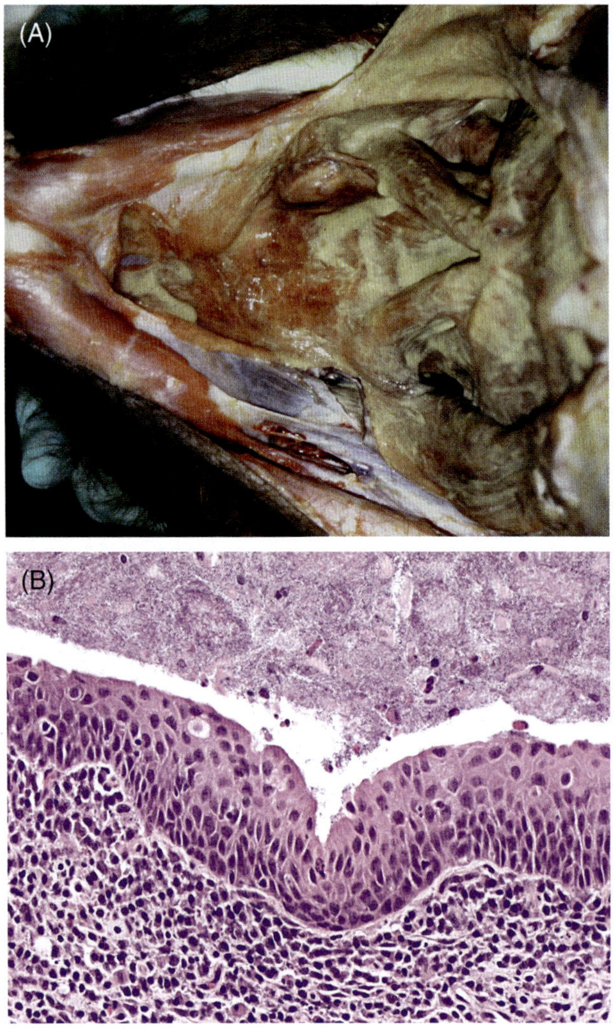

FIGURE 15.30 Severe chronic airsacculitis in an orangutan. (A) There is thickening and opacification of the air sacs and accumulation of a tan exudate. (B) The air sac epithelium is markedly thickened and has undergone squamous metaplasia. Subjacent, dense lymphoplasmacytic inflammation is present and a luminal exudate consists primarily of bacteria and necrotic cellular debris with occasional intact neutrophils.

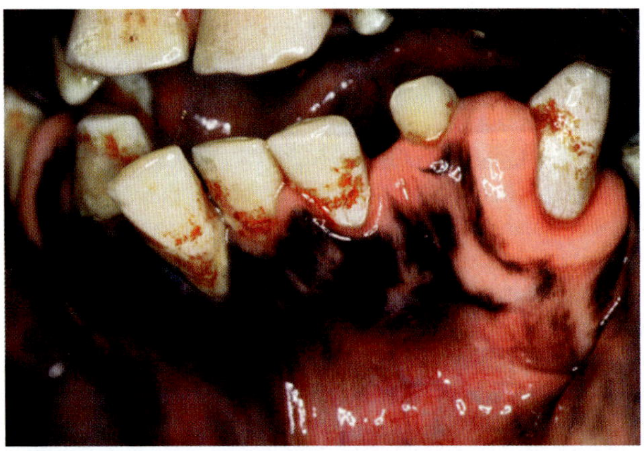

FIGURE 15.31 Periodontal disease and stomatitis in an orangutan. Both severe gingival regression and proliferation of the gingival mucosa are seen. *(Photo Courtesy of L. Braswell and Zoo Atlanta.)*

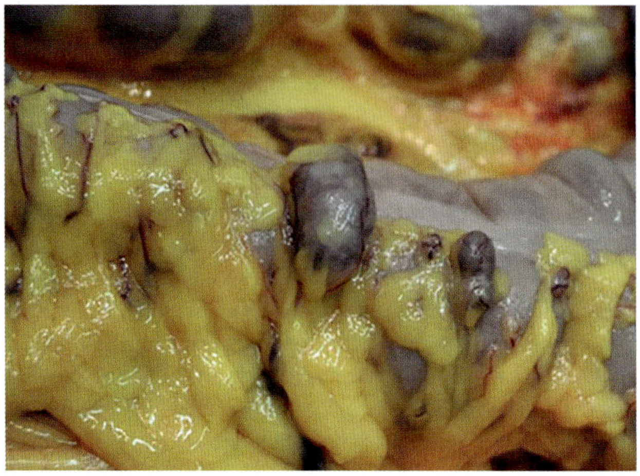

FIGURE 15.32 Diverticulosis and mid diverticulitis in an obese geriatric female orangutan in managed care. Note the distended saccules and congestion of serosal vessels. This was an incidental finding. *(Photo Courtesy of William R. Prichard Veterinary Medical Teaching Hospital, University of California, Davis.)*

typically occur in the caudal abdomen and intrapelvic region and can involve the perirenal retroperitoneum and abaxial muscles. Lumbar, perirectal or perivaginal fistulas may develop. Extensive fibrous adhesions are common regionally among the reproductive, distal gastrointestinal, and urinary tracts, often making it difficult to discern their origin. Regional peritonitis due to gastrointestinal perforation, transmural appendicitis, or diverticulitis (Fig. 15.32) is often suspected.[121,475] Typhlitis with rupture is reported in a zoo-housed gorilla and colonic rupture is occasionally seen in mountain gorillas.[697] Since female gorillas are disproportionately affected by retroperitoneal abscesses, a reproductive tract origin is also plausible. Various bacteria isolated include *Escherichia coli*, *Morganella morganii*, *Fusobacterium necrophorum*, *Klebsiella pneumoniae*, hemolytic *Streptococcus* and *Staphylococcus* spp., *Bacteroides* sp., *Enterococcus* sp., *Hemophilus* sp., *Proteus* sp., and *Actinomyces* sp. Note that **intraabdominal adhesions** are common in all wild and captive adult apes, even in the absence of overt peritonitis (Fig. 15.33).

Bacterial enterocolitis is common in apes, often presenting as diarrhea with lethargy and anorexia sometimes culminating in death.[247,442,676] While the majority of cases of ape enteritis are attributed to bacteria, such as ***Salmonella*** spp., ***Yersinia* spp.**, ***Campylobacter*** spp., and ***Shigella* spp.**, parasitic and viral causes must also be considered (though the latter are less well documented). *Campylobacter*, *Salmonella*, and *Shigella* spp. have all been identified in free-ranging, human-habituated mountain gorillas, primarily in sub- and young adults, frequently in the absence of diarrhea.[509] *Shigella* spp., primarily *flexneri*, are common isolates from clinical cases of diarrhea in zoo-housed apes.[31,40,478,501] Transmission is fecal–oral. In gibbons, persistent infection with periodic shedding of *S. flexneri* complicates eradication. Lesions of shigellosis in apes have not

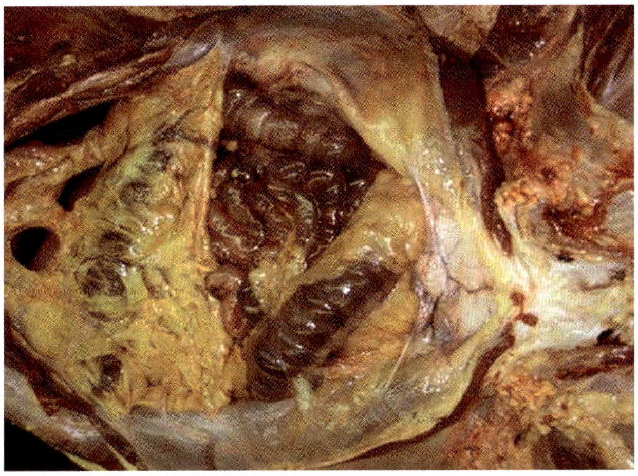

FIGURE 15.33 Abdominal viscera in situ from an adult female orangutan in managed care. Multiple adhesions between visceral, omentum, colon, and body wall. Adhesions such as these are of undetermined etiology and are common in adult apes. *(Photo Courtesy of San Diego Zoo Wildlife Alliance.)*

been well described but are expected to be similar to those noted in OW monkeys (i.e., colonic dysentery, hemorrhagic colitis). Diagnosis is based on culture and or PCR. *Salmonella* spp. can also cause acute hemorrhagic enterocolitis with septicemia especially in young, aged, or immunocompromised individuals.[42,530] *Salmonella enteritidis* can occur in outbreak form and can cause fatal necrotizing and hemorrhagic typhlocolitis in siamangs. Both *Campylobacter coli* and *C. jejuni* can be isolated from apes with diarrhea.[539] Natural gastroenteric infection with *Helicobacter* spp. is reported but pathogenicity is not documented.[191]

A ***Sarcina* sp.** (Candidatus *Sarcina troglodytae*) (family Clostridiaceae) has been identified by metagenomics and culture to be associated with mortality in chimpanzees in a Sierra Leon sanctuary.[527] Clinical signs included anorexia, neuromuscular weakness, ataxia, seizures, vomiting, and abdominal distension, lending the name "epizootic neurological and enteric syndrome." Clinical course is variable but the syndrome is 100% fatal. Gross lesions include gastric dilation (bloat), pulmonary edema, systemic congestion suggestive of "shock," and thymic and pancreatic hemorrhage. Microvascular neutrophil margination is seen histologically. Brains have not been examined. *Sarcinia* spp. are gram-positive cocci that occur in tetrad packets. Though sometimes part of GI flora, *S. ventriculi* has been associated with gastroparesis and gastritis in humans and bloat in infant ruminants.[3,167]

Mycobacterial infections can occur in any ape, as well as in NW and OW monkeys (Chapter 14: New World and Old World Monkeys). All are susceptible to ***Mycobacterium tuberculosis* (Mtb)** and other members of the Mtb complex, for example, ***M. bovis*** and ***M. africanum***.[108,669] Humans are the main/natural hosts of Mtb. Transmission is by aerosolization or through fomites contaminated with human sputum and exudates. Oral infection can occur, especially in infant apes fed unpasteurized milk from cows or goats. The source of infection is not always evident, but transmission from caretakers, visitors, conspecifics, or other animals in a collection is described. There is a perception that apes are less susceptible than OW monkeys, but this may be a false impression as fewer apes are in studied wild populations or in captive situations than are monkeys like African green monkeys or rhesus macaques.[597] In early days of ape zoo-keeping, this was one of the most common causes of mortality.[240] When glass-fronted enclosures separate apes from human visitors, the incidence of tuberculosis (TB) in apes drops precipitously. At the Philadelphia Zoo, 43% of 7 ape deaths from 1901 to 1920 were due to TB, while from 1921 to 1975, only 11% of 98 apes necropsied had TB (most occurring early in the time period).[647] Given the global rise of tuberculosis, zoo collections are still vulnerable.[728] Chimpanzees held in research facilities have been infected, but apes (especially orangutans) kept as pets, confiscated from illegal ownership, and housed in sanctuaries or rehabilitation centers with close human contact, are at especially high risk.[71]

Mtb is still considered one of the greatest reverse zoonotic health concerns for human-habituated wild ape populations.[761] Fortunately, no cases have been identified during 30 years of necropsies of habituated wild mountain gorillas or by necropsy or PCR-based screening over the past 10 years in the well-studied chimpanzees of Gombe.[406,694,760] However, a wild chimpanzee from Côte d'Ivoire, killed by a leopard, was determined to be infected with *Mycobacterium africanum*.[108]

Apes with tuberculosis may have anorexia, weight loss, and respiratory signs, such as dyspnea and cough. The lungs are the primary site of infection with drainage to local lymph nodes and systemic spread to the liver, spleen, kidney, and other organs in advanced cases. At necropsy, lesions are similar among the *M. tuberculosis* complex bacteria (e.g., *M. bovis*) and to those noted in OW monkeys. When the route of bacterial transmission is respiratory; lesions consist of multifocal to coalescing granulomatous pneumonia with multinucleated cells and varying degrees of central caseation and mineralization. Air sac involvement has been described in orangutans with concurrent lung lesions. Oral route of infection can result in widespread granulomatous inflammation in abdominal viscera. Both intrahistiocytic and extracellular acid-fast positive bacilli can be identified in lesions, but infections are commonly paucimicrobial.[627] In the chimpanzee infected with *M. africanum*, caseous tuberculoid granulomas with Langerhans-type giant cells were found in liver, spleen, and mesenteric lymph nodes but not lungs. Acid-fast bacteria were abundant (pleurimicrobial) and readily visualized using Ziehl-Neelsen stain.

Postmortem diagnosis is based on the characteristic lesions, presence of acid-fast positive bacilli and, ideally, microbiological or genetic characterization.[759] Tuberculin skin tests are not always reliable for antemortem diagnosis as these tests can be falsely negative due to stage of disease (both early or advanced tuberculosis) and overall immune function.[136] False positive results occur due to cross-reaction with environmental nontuberculous *Mycobacterium* spp. (especially in orangutans) or vaccine adjuvant (in vaccinated animals). Serologic tests are unreliable for identifying latent disease. A gamma-interferon stimulation assay has been developed for use in nonhuman primates, but multiple tests may be necessary for definitive antemortem diagnosis. Fecal PCR is also an option.[391,759]

Nontuberculous mycobacteria, including the ***M. avium*** complex organisms, can also infect apes. As in monkeys, underlying immune suppression should be suspected. These organisms are often environmental and can be found in municipal tap water. Histiocytic enteritis, resembling Johne disease of cattle, is one manifestation as reported in a siamang gibbon as well as in immune-compromised macaques. These infections are usually pleurimicrobial and organisms may be better stained with Fite-Faraco than with Ziehl-Neelsen acid-fast stain.[479]

Naturally occurring **leprosy** caused by ***Mycobacterium leprae*** can occur in chimpanzees, and gibbons are susceptible experimentally.[276,288,390,733,741] The organism cannot be cultured on artificial media, but can be detected in lesions and environmental samples by PCR. Transmission route in apes is unknown, but in humans, close contact, infected aerosols, environmental reservoirs, and traumatic inoculation are established routes.[61] To date, all naturally occurring cases have been in wild or wild-caught chimpanzees and no cases have been reported in in-contact apes born in captivity. In the past it has been difficult to study *M. leprae* infections due to the prolonged latency period and lack of accurate diagnostics for subclinical infections. Lesions can appear in captive chimpanzees decades after infection with organisms originating in country of capture.[682] PCR detection is beginning to overcome these difficulties and fecal PCR has helped understand the extent of leprosy in wild populations. Lesions in apes are similar to those in humans and occur primarily on the face, ears, and fingers but also on body and genitalia. Early lesions are areas of cutaneous erosion that progress to coalescing dermal nodules. Histologically, the lesions are of the lepromatous, not tuberculoid, type, composed of activated macrophages/histiocytes with fewer lymphocytes, plasma cells, and neutrophils and no organized granulomas. While inflammation can be diffuse, nodules characteristically surround deep dermal nerves and there is an overlying subepidermal "grenz" zone lacking inflammatory cells.[288,390,427] Numerous acid-fast bacilli are present within histiocytes (multibacillary/pleurimicrobial) with lower numbers in nerves. In a few cases, systemic spread of intrahistiocytic acid-fast bacilli has been noted. Some lesions are complicated by self-trauma. Diagnosis is based on the characteristic histological lesions, PCR testing of skin, internal organs or feces, and serology. Leprosy bacilli stain more intensely with Fite-Faraco than Ziehl-Neelsen stains.

A condition similar to tertiary **yaws** in humans is described in a population of wild western lowland gorillas in the Republic of Congo, in chimpanzees in Guinea, west Africa, and in multiple species of cercopithecine monkeys in east Africa, all areas where human yaws is endemic.[395,408,469,774] The disease is caused by the spirochete ***Treponema pallidum pertenue*** and related subspecies. In humans, tertiary yaws causes "gangosa" (syn. Rhinopharyngitis mutilans) with severe destruction of the palate and nasopharynx, and "gondou," which is characterized by bony proliferation of the maxilla. Sixteen percent of gorillas in affected areas have severely disfiguring destructive facial lesions similar to gangosa. Lesions also develop on the chest and shoulders with increasing age. Male gorillas are more affected than females, and affected males are less likely to be an alpha silverback. *T. p. pertenue* is transmitted by direct contact with infected lesions and by biting flies. Periostitic lesions suggestive of yaws osteitis have been seen in museum skeletons from chimpanzees and gorillas collected in west Africa, and gondou-like proliferative lesions on a gorilla skull are reported.[111,408] Camera "trapping" is used to identify lesions in free-living chimpanzees. The lesions are similar to the facial lesions described in gorillas, with extensive destruction of the lips. Molecular characterization of the organism classified it as ***T. pallidum endemicum***.

Histological lesions described in a wild chimpanzee are similar to those described in humans and baboons, namely, acanthosis and spongiosis with deep rete pegs, intraepidermal pustules, and spirochetes (demonstrated with silver stains or IHC using anti-*Treponema* antibodies) leading to ulceration.[452] There is dense plasma cell infiltration in the dermis. Bone lesions are of proliferative periostitis. Diagnosis is based on gross lesions, histology, and molecular detection.[481]

Tularemia caused by ***Francisella tularensis***, a gram-negative coccobacillus endemic in parts of the Northern Hemisphere, is reported in gibbons, orangutans, and a gorilla, as well as in monkeys in managed care.[74,337] Both Type A (lagomorph-associated) and Type B (rodent-associated) strains are identified, which emphasizes the importance of rodent/lagomorph control in primate exhibits. Clinical signs are variable and similar to disease in other species. Pyogranulomatous inflammation in affected tissues (e.g., oral cavity, skin, lymph nodes, and lungs) and terminal bacteremia without inflammation are described. Diagnosis of tularemia requires identification of the bacteria. They are often poorly demonstrated with tissue Gram stains. PCR and immunohistochemistry are available ancillary diagnostic tests. Isolation can be complicated due to slow growth and culture overgrowth by concurrent bacteria or contaminants. *Francisella tularensis* is zoonotic. It is a Tier 1 Select Agent in the United States and is reportable to the Centers for Disease Control and Prevention.

Anthrax-like disease is an important emerging cause of fatalities in chimpanzees and gorillas in Africa and is WOAH-listed reportable.[345,387] Classical anthrax, caused by *Bacillus anthracis*, affects a wide variety of animals and is frequent in African wildlife. The disease resembling anthrax in free-living African apes is caused by ***Bacillus cereus*** **var.** ***anthracis***, a species of *B. cereus* that has acquired anthrax plasmids.[344] Affected apes present with sudden death or weakness and vomiting. Hemorrhage in multiple sites throughout the body, notably in the lung and intestines, is present at necropsy. Additionally, the lungs contain areas of edema and emphysema. Histologically, plump, gram-positive bacilli are present intra- and extravascularly in all examined tissues, consistent with acute bacterial sepsis (Fig. 15.34). Extreme biosafety precautions should be taken during any necropsy investigation of possible anthrax in wild apes as Ebola occurs in similar geographic regions and is also associated with hemorrhage.

Fungi

Dermatophytosis (ringworm) is occasionally reported in apes, including gibbons and siamang, gorillas, chimpanzees, and orangutans in zoos and chimpanzees in sanctuaries and in the wild.[18,156,446,507,526,642,689] Lesions are variable

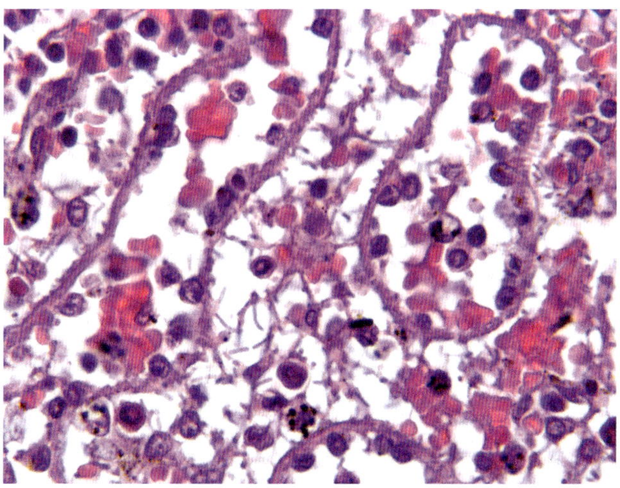

FIGURE 15.34 *Anthrax*-like infection in a chimpanzee. Myriad *Bacillus cereus* biovar *anthracis* bacilli are present in the splenic sinusoids. *(Photo Courtesy of the Taï Chimpanzee Project.)*

including papular, crusting lesions on the face around the nostrils, focal or multifocal hair loss, and scaling of the skin over the entire body. Lesions may be pruritic. Histologic lesions are similar to those in domestic animals with varying degrees of acanthosis and hyperkeratosis of the epidermis and hair follicles, and fungal elements in surface keratin, hair follicles, and hair shafts. PAS and GMS stains help highlight the fungal elements in tissue sections. Differential diagnoses include dermatophilosis and ectoparasitism. Reported organisms include *Microsporum caninum*, *Micosporum gypseum*, *Microsporum audouinii*, and *Trichophyton rubrum*. Definitive diagnosis relies on morphology by KOH-cleared direct microscopic examination, culture on Sabauroud's agar (which is actually less sensitive), PCR or MALDI-TOF (Matrix-assisted laser desorption/ionization-time of flight mass spectrometry), which is a reliable alternative to ribosomal RNA gene sequencing for identification of fungi and bacteria of clinical significance.[266]

Reports of **Candidiasis** due to *Candida* spp. Are uncommon in the literature on zoo and free-living apes but not uncommon in actuality. Oroesophageal and tracheal candidiasis is noted in free-living and zoo-housed gorillas, chimpanzees, and orangutans. As in other species, generalized disability, immune dysfunction, and antibiotic treatment are likely predisposing factors. Gross and histological mucous membrane lesions include plaques of hyperkeratosis with superficial ovoid yeasts and pseudohyphae oriented perpendicular to the epidermal layers causing retention of the stratum corneum. A rare manifestation is focal nodular proliferation of the gastric mucosa as reported in a perinatal gibbon.[589]

Coccidioidomycosis, due to ***Coccidioides immitis*** or ***C. posadasii***, is a significant threat to captive chimpanzees and gorillas in endemic areas of the southwestern United States.[40,278,347,603] These two fungal species are separated by geography, cultural characteristics, and molecular techniques. *C. immitis* is endemic to Mexico and the San Joaquin Valley of California, while *C. posadasii* is endemic elsewhere in California, Arizona, Texas, Mexico, and South America. Both have a broad host range including many nonhuman primate species. Infection is via inhalation of the small (<10 µm), airborne, arthroconidia from saprophytic hyphae in the soil. Once inhaled into the lungs, the conidia become larger spores or spherules that divide by endosporulation. Rupture of the spherules leads to endospore dissemination in the lungs and throughout the body by macrophage transport. Clinical signs include weight loss and cough. Lesions are typical of "valley fever" in other species and include focal, multifocal, or diffuse pyogranulomatous pneumonia with characteristic spherules readily recognized by size (10–200 µm), thick, double-contoured, refractile, hyaline walls and endospores (2–5 µm). Dissemination to brain, eye, bone, and other locations frequently occurs.

Hypertrophic osteoarthropathy can occur in infected chimpanzees. Antifungal therapy is sometimes successful if the diagnosis is made early enough.

Cryptococcus neoformans and ***C. gattii*** cause **cryptococcosis**, a serious systemic infection that has been described in humans and a variety of domestic and wild animals including occasional reports in apes.[394,451] *C. neoformans*, found worldwide, is thought to be mainly opportunistic, while *C. gattii*, found initially in tropical and subtropical areas and the Pacific Northwest of North America, also infects immune-competent hosts. *C. gattii* can be found on many species of trees and its geographic range is thought to be expanding.[660] Clinical disease and pathologic lesions are similar in both infections. Signs in a zoo-housed gorilla from Europe include marked weight loss, progressive lethargy, and coughing followed by neurological signs of nystagmus, ataxia, and tremor leading to euthanasia. Gross and histological lesions include cloudy meninges and cystic degeneration of periventricular white matter. Large numbers of cryptococcal spores surrounded by a thick, clear capsule are accompanied by predominantly nonsuppurative inflammation. The fungus is positive with silver stains; mucicarmine highlights the polysaccharide capsule.

See Digital Content Table 15.e4 for additional fungi.

Metazoa

There are numerous coprological studies of all the apes, both in the wild and in managed care.[54,100,148,155,188,211,212,249,304,335,346,347,466,468,477,496,556,558,565,610]

Routine parasitological examinations of free-living apes as well as use of molecular detection are expanding the understanding of the diversity of ape parasites. Below are several with known pathogenicity. Table 15.e5 in the Digital Content includes a more comprehensive list of parasites identified in apes.

Pinworms indigenous to their natural hosts are found in wild gorillas, chimpanzees, and orangutans.[517] These parasites have a direct life cycle and live in the large intestine, including the vermiform appendix (which is present in all apes). Pinworm infections are usually subclinical; however, the human pinworm ***Enterobius vermicularis*** can be fatal in captive chimpanzees.[768] Gross lesions include colonic mucosal roughening and nodularity and draining tracks in the perianal skin. Histologically, *E. vermicularis* may be seen in the lumen of the ileum, cecum, and colon, within nodules in the mucosa, in regional lymphatics, mesenteric lymph nodes, the liver, and lungs.

Strongyloidiasis is an important, sometimes fatal disease of young apes in captivity. *Strongyloides* eggs or larvae are frequently identified in coprological studies of apes in captivity and in the wild.[40,140,346,365,366,410,435,480,708] Two species can parasitize hominids: ***Strongyloides fuelleborni*** (OW monkeys and apes natural hosts) and ***S. stercoralis*** (human natural host). Infection occurs when filariform larvae in the environment penetrate skin or mucous membranes, enter circulation, and travel through the heart to the lungs where they undergo further development in alveoli. Developing larvae are then coughed up, swallowed, and penetrate the wall of small intestine where they develop into parthenogenic adult females that live in the crypts and produce eggs that are shed in the feces (*S. fuelleborni*) or hatch in the lumen to release larvae that are shed in the feces (*S. stercoralis*). Although infections are often asymptomatic, *S. stercoralis* can be associated with mortality due to hyperinfection in which larvae penetrate the colon wall instead of being shed in the feces (autoinfection), resulting in heavy parasite loads. Ivermectin has reduced the incidence, but hyperinfection with *S. stercorialis* continues to be an occasional cause of death in zoo-housed infant and juvenile orangutans. Clinical signs include cough, dyspnea, diarrhea or constipation, ileus, weight loss, and death. Gross lesions are often limited to pulmonary hemorrhage (Fig. 15.35A), but there may also be hemorrhagic or ulcerative enteritis with mesenteric lymphadenopathy and pericarditis. Histologically, larvae can be seen in the lung where they are associated with acute interstitial pneumonia and multifocal mild to severe hemorrhage (Fig. 15.35B). They also may be found in the lumen, crypts, and intestinal lymphatics as well as in mesenteric lymph nodes. Adult worms are small and found in the intestinal glands/crypts (Fig. 15.36; eSlide 15.e4). Complications include toxemia/septicemia due to bacteria accompanying larval migration.

Chronic proliferative gastritis associated with **trichostrongyles** is seen in mountain gorillas from the Virunga massif and can lead to chronic debilitation and death.[475,542] Grossly, there is diffuse or nodular thickening of the rugal folds, which is more severe in the pyloric region and is accompanied by variable ulceration (Fig. 15.37). Histologically, hypertrophied and hyperplastic gastric glands often exhibit severe dysplasia. A gastroduodenal carcinoid and gastric adenocarcinomas have been identified. In the Bwindi Impenetrable Forest, a syndrome of weight loss, poor or discolored hair coat, and general debilitation has been linked to severe **intestinal strongyloidiasis**.[542]

Oesophagostomiasis, due to nodular worms (***Oesophagostomum* spp.**), is a common, sometimes clinically important, parasitic disease of apes, especially chimpanzees in range countries.[695] ***O. stephanostomum*** is the species most often identified in chimpanzees and gorillas. ***O. bifurcum*** can also infect African apes, creating concern for zoonotic transmission.[357] Orangutans and gibbons are infected by ***O. aculeatum***.[769] The worms live and cause lesions in the colon and adjacent mesentery. The life cycle is direct and there is no systemic migration. Ingested larvae pass to the colon where they penetrate the mucosa forming a

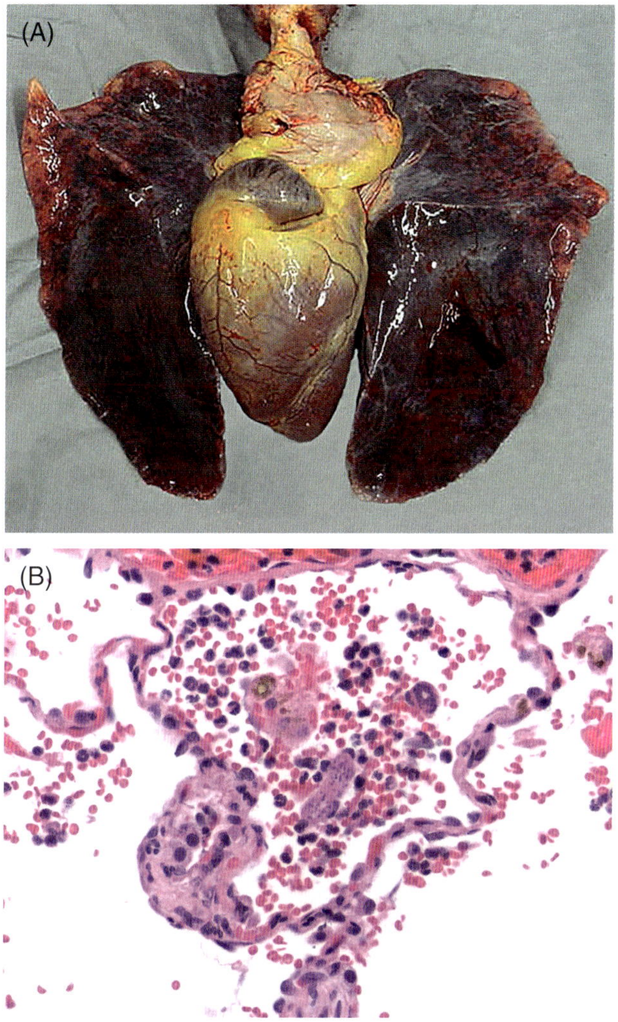

FIGURE 15.35 Pulmonary strongyloidiasis in an orangutan. (A) Infection and hemorrhagic pneumonia are associated with diffuse red discoloration and failure of the lungs to collapse. (B) Intraluminal larval nematodes are associated with hemorrhage and neutrophilic inflammation. *(Parts A, B: Photos Courtesy of T. Schoeb, University of Alabama, Birmingham.)*

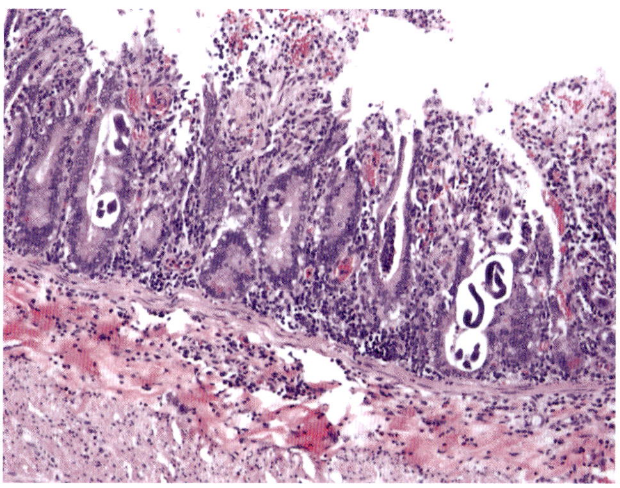

FIGURE 15.36 *Strongyloides stercoralis* colitis in a chimpanzee. Parasites within colonic crypts are associated with inflammation and necrosis (see eSlide 15.e4).

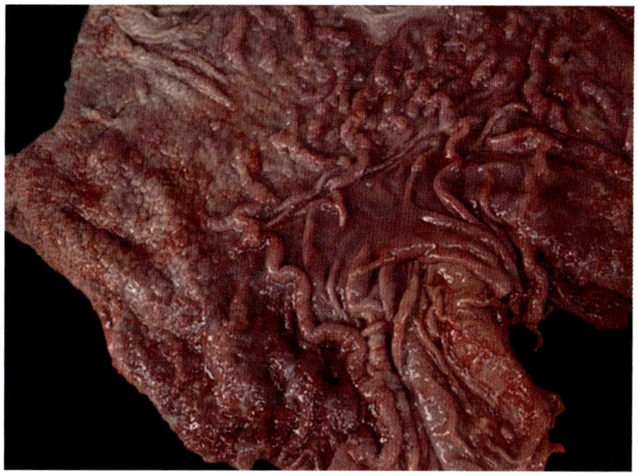

FIGURE 15.37 Stomach with severe, chronic proliferative gastritis in a free-living adult mountain gorilla. Note the rugal thickening in the body of the stomach and cobblestone appearance of the mucosa. This lesion has been associated with chronic trichostrongyle infection. *(Photo Courtesy and Copyright of J. Bosco Noheri for Gorilla Doctors.)*

submucosal or mural nodule. In primary infections, mural lesions are small and heal after the fourth stage larvae exit into the lumen. However, when a host is reinfected, larvae may become trapped in the nodule by immune-mediated inflammation. Clinical signs range from none to severe diarrhea, weight loss, anemia, hypoproteinemia, and mortality. Grossly, white, yellow, red, or black 2–5 mm nodules stud the colonic serosa and mesentery (Fig. 15.38A). Viable larvae may be seen in some nodules. Inflammation varies from necrotizing and hemorrhagic to granulomatous and can become suppurative due to secondary bacterial invasion (Fig. 15.38B; eSlide 15.e5). The overlying mucosa may be normal, thickened, covered by a diphtheritic membrane, or contain small ulcers. Rupture of the nodules or transmural extension of inflammation is associated with peritonitis and mesenteric/omental adhesions. If hypoproteinemia is present, hydropericardium and hydrothorax can occur. Antemortem differentiation of the ova and larvae of hookworms and nodular worms in fecal samples is not always possible and PCR techniques have been developed for more reliable diagnosis.[722]

Capillariasis, due to ***Calodium hepatica*** (syn. ***Capillaria hepaticum***), is found worldwide. Many species, including nonhuman primates, can be infected; rodents are the natural host. Infection occurs via ingestion of eggs from environments contaminated with decomposed liver tissue or the feces of a predator or scavenger that has eaten an infected animal. Infection is identified at necropsy, and in mountain gorillas it is usually incidental but occasionally severe in infants.[225] Eggs with thick walls and bipolar caps (opercula) and adults with aphasmid morphology (hologonic, stichosome, bacillary bands, and characteristic eggs) are found in tracks in the liver parenchyma, often near

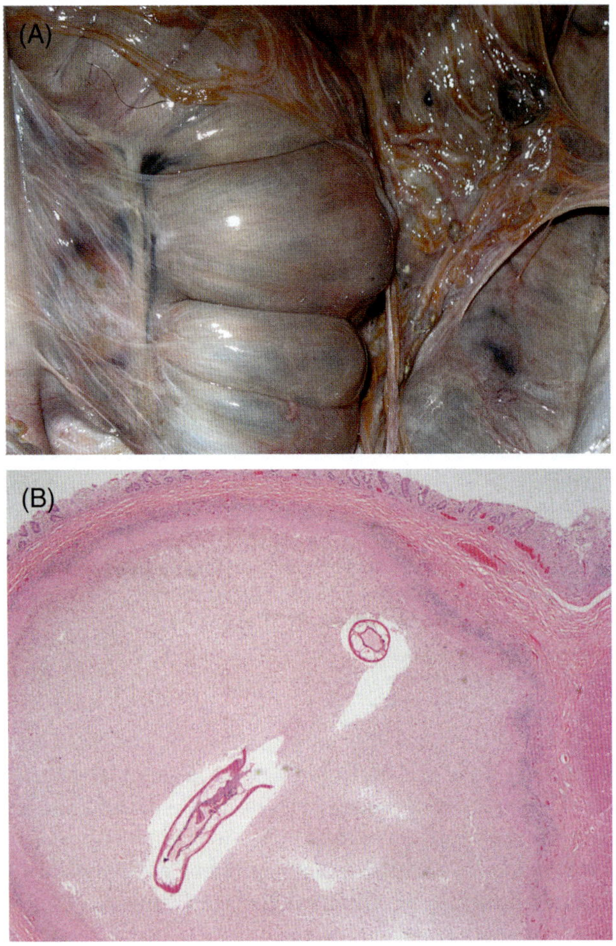

FIGURE 15.38 Oesophagostomiasis in free-living chimpanzee (A) and mountain gorilla (B). (A) Multiple variably sized black nodules within the colonic wall and within the mesentery. Nodules contain inflammation and larvae. (B) A large, demarcated, mural nodule composed of chronic granulomatous and suppurative inflammation contains sections of intralesional spirurid nematodes (see eSlide 15.e5).

portal areas. Reaction to the parasites varies from none to marked granulomatous hepatitis (Fig. 15.39). ***Capillaria brochieri***, an enteric parasite, is reported to cause fatal diarrhea in bonobos.[314]

Mammomonogamus laryngeus is a syngamid nematode that is related to the gapeworms of birds. It has a broad host range and is endemic to Southeast Asia, the Caribbean, and Brazil. Infections accompanied by respiratory signs and death are documented in a semiwild population of Sumatran orangutans in Indonesia.[193] The life cycle is unknown but is likely direct with fecal–oral transmission. Eggs and occasional whole nematodes are coughed up, swallowed, and shed in feces. Grossly, large numbers of parasites occur in upper and lower airways of young orangutans dying with dyspnea and lethargy. Microscopic lesions are not described in orangutans, but in affected bovids, mucosal

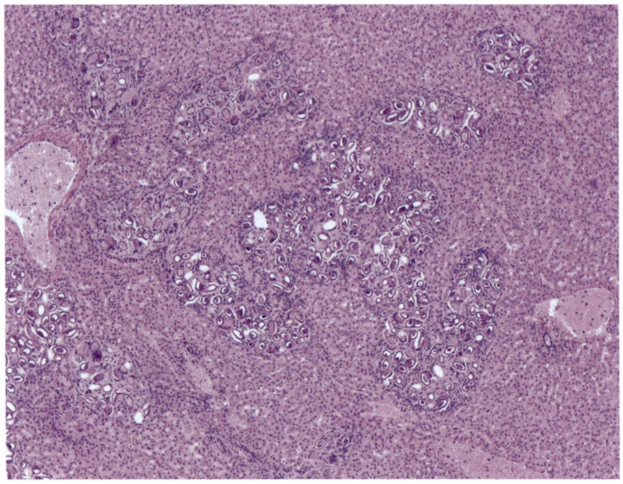

FIGURE 15.39 Hepatic capillariasis in a mountain gorilla. Severe granulomatous inflammation surrounds nematode eggs laid in serpiginous tracts created by female worms.

edema and polypoid proliferations are found at nematode attachment sites. Eggs of *Mammomonogamus* sp. are also identified in feces of western lowland gorillas in the Central African Republic, but lesions are not yet described.[89]

Larval migrans, due to ***Baylisascaris* sp.** and ***Angiostrongylus* sp.**, can be a problem in zoo-housed gibbons and orangutans.[177,249,447] Definitive hosts of *Baylisascaris* spp. include raccoons (*B. procyonis*) and skunks (*B. columnaris*). The infection is found throughout the United States. Parasite migration tracts in the brain are grossly malacic with or without hemorrhage. Necrosis, spongiosis, and nonsuppurative inflammation are present histologically. The larval nematodes, with characteristic prominent lateral alae, can be surprisingly hard to find in brain sections. Systemic vascular involvement is not a feature.

Parastrongylus (Angiostrongylus) cantonensis and ***P. costaricensis***, parasites of rats with a snail intermediate host that have been introduced and become endemic in the southeastern United States, can cause larval migrans in zoo-housed gibbons, orangutan, owl monkeys, and callitrichids and in native wildlife including raccoons and opossums. *P. cantonensis* generally causes pulmonary angiostrongyliasis or eosinophilic meningoencephalitis and/or myelitis. *P. costaricensis* causes abdominal angiostrongyliasis. Inflammation is usually eosinophilic and necrotizing or granulomatous and the nematodes are usually visible in vessels of lung or lumen of heart (*P. cantonensis*) or in mesenteric vessels (*P. costaricensis*).

Apes are host to a variety of **cestodes (tapeworms)**, some of which, especially larval forms, are highly pathogenic. Of greatest concern is **hydatid disease** due to ***Echinococcus granulosa*** (dog definitive host) or ***E. multilocularis*** (fox definitive host). Individuals with the former parasite may be infected in countries of origin, but gorillas and chimpanzees with the latter are infected after importation as *E. multilocularis* only occurs naturally in the Northern Hemisphere (including portions of Asia). Mortality from hydatid disease is reported in gorillas in zoos in the United States, Europe, and Japan.[40,564,744] Gross lesions consist of abdominal enlargement due to uni- or multiloculated cysts that contain many daughter brood capsules and daughter cysts. Cysts can be diagnosed antemortem by ultrasound, and aspiration for cytology and PCR. Differential diagnoses include hepatic abscesses due to bacteria or amoeba. Rupture of the cyst(s) may lead to seeding of the peritoneum by the asexually dividing larvae, causing florid, chronic, effusive peritonitis. The liver may be replaced by necrosis and cystic, caseous, or fibrotic nodules. Renal involvement may occur. Histologically, the wall of hydatid cysts due to *E. granulosa*, has a characteristic hyaline, laminated appearance while this layer in *E. multilocularis* cysts is narrower and lacks distinct striation.[566] Inflammation ranges from suppurative to granulomatous. Hydatid disease differs from **cysticercosis** in that the cysticercus is a single thin-walled cyst with single invaginated larval scolex in CNS or viscera.[590] An unusual, fatal case of larval cestodiasis due to ***Versteria*** sp. (mustelid carnivore definitive host) occurred in a juvenile orangutan with a short clinical course of lethargy, anorexia, and cough and gross lesions of hepatosplenomegaly, pulmonary hemorrhage, and pericardial and thoracic effusions.[220] On histology, the liver contained uni- and multilocular, unencapsulated, thin-walled cysts containing marginated, multinucleated parasitic protoplasm. Similar cysts were present in the lungs and spleen, often in blood vessels.

Intestinal tapeworms of the genera ***Bertiella*** and ***Anoplocephla*** are commonly seen in apes.[147,517] Wild mountain gorillas, especially older adults, are often heavily parasitized by ***A. gorilla***.[635] These animals are often in poor condition, but whether the tapeworms are opportunistic or pathogenic is unclear.

Protozoa

The most important protozoan disease of captive apes, especially gorillas, is **balantidiasis** due to ***Balantidium coli***. Amoebic infections are also common and vary from incidental enteric colonization (e.g., ***Entamoeba dispar***) to disseminated fatal disease (e.g., ***Balamuthia mandrillaris***). ***Plasmodium*** spp. infections are common in free-living western gorillas, chimpanzees, orangutans, and gibbons, but clinical malaria is infrequently documented. Other protozoan infections, such as **cryptosporidiosis**, **cyclosporidiosis**, and **giardiasis**, have been detected by microscopic and molecular fecal examinations, but rarely associated with clinical disease.[489,593,753]

Balantidium coli is a ciliated protozoan with worldwide distribution and broad host range including apes, other nonhuman primates, and humans.[496] Molecular studies have resulted in the suggested reassignment of the primate parasites to the genus ***Neobalantidium***.[548] *B. coli*-like ciliates are identified in free-ranging, sanctuary, zoo, and colony-housed apes, though infection is much more common in captive than in free-ranging populations.[339,550] *Balantidium* is generally considered commensal but can become invasive when animals are immune compromised, stressed, or suffering from other causes of diarrhea (e.g., salmonellosis). Morbidity and mortality are described in western lowland gorillas, chimpanzees, and orangutans in zoos and in semicaptive (rehabilitant) orangutans.[379] Captive diets high in starch may allow for overgrowth of the organisms.[610] The most common gross lesion is ulcerative typhlocolitis with hyperemia and hemorrhage in the colon and cecum and hemorrhagic diarrhea (dysentery).[379] Histologically, the protozoa are present in the lumen, on the mucosal surface, within ulcer beds, and throughout the wall (Figs. 15.40A, B). The trophozoite stage has two nuclei and

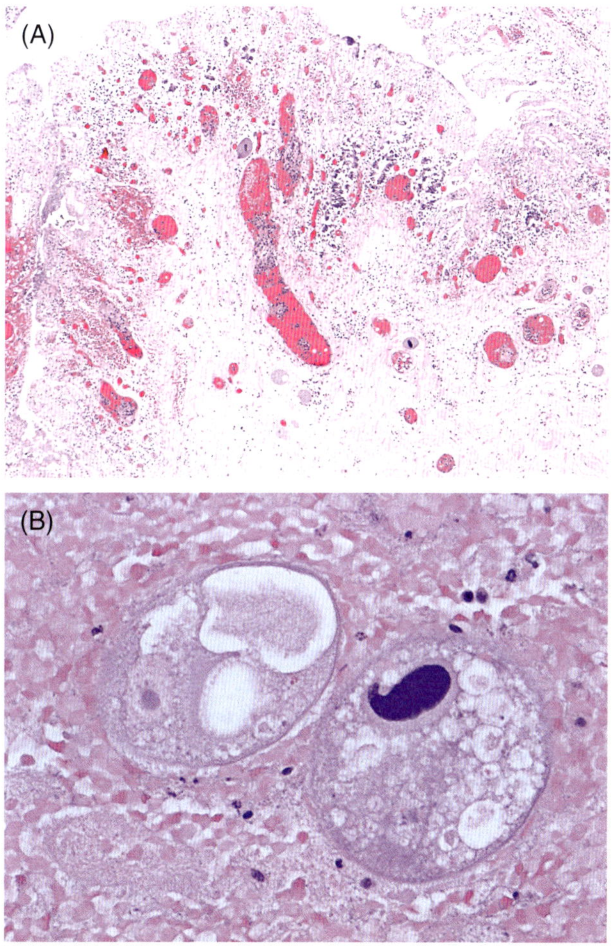

FIGURE 15.40 Invasive balantidiasis with hemorrhagic typhlocolitis in a western lowland gorilla. (A) Round to ovoid amoeboid, amphophilic protozoa with a prominent curved macronucleus are present in the mucosa and submucosa of a western lowland gorilla. Infection and inflammation are associated with multifocal necrosis and hemorrhage. (B) Higher magnification of two encysting *Balantidium* sp. trophozoites in a chimpanzee.

the cyst stage has a large, curved, or horseshoe-shaped macronucleus. These and the distribution of circumferential cilia (holotrich) are useful morphologic features that aid in differentiating them from commensal ciliated protozoans, such as ***Troglodytella gorillae*** and ***T. gabonensis*** in wild gorillas and ***T. abraesarti*** in chimpanzees, bonobos, gorillas, and siamangs in captivity.[549]

Cerebral and systemic **amoebiasis** due to ***Balamuthia mandrillaris***, a free-living, soil-dwelling, leptomyxid amoeba with global distribution, is a serious emerging infection of gorillas, and, less frequently, other apes (orangutans, gibbons), monkeys, and humans.[77,188,219,263,432,568] It accounts for almost 3% of gorilla deaths in the North American SSP population and is described in zoo-housed apes in Europe and Australia. The presumed route of infection is inhalation of airborne cysts from contaminated soils or, rarely, through skin wounds. Brain lesions may be visible grossly as foci of malacia, congestion, or hemorrhage in both gray and white matter without specific geographic distribution (Fig. 15.41A). Histologically, the infection may be necrotizing and/or granulomatous and is often vasocentric with associated fibrinoid vasculitis (Figs. 15.41B, C; eSlide 15.e6). Organisms often cluster in Virchow-Robin spaces, but immunological staining reveals wider distribution. In some cases, granulomatous inflammation and fibrosis, which cause mass lesions suggestive of neoplasia, are present in liver, kidney, pancreas, and lymph nodes. Differential diagnosis includes cerebral amoebiasis due to *E. histolytica* or other, free-living amoeba (e.g., *Naegleria* or *Acanthamoeba* spp.). Amoebae are identified on the basis of nuclear morphology and shape and size of the karyosome. *E. histolytica* has a distinctive large karyosome. *Balamuthia* and *Acanthamoeba* are sometimes difficult to differentiate, though *Balamuthia* sometimes has a double endosome (Fig. 15.41C). Immunohistochemistry and PCR differentiation are available through the United States Centers for Disease Control (CDC) and some diagnostic laboratories. Serologic testing may be useful for screening in sanctuary settings.

Intestinal amoebae are often detected on routine fecal examinations of both managed and free-living apes or PCR amplification from feces.[305,364,450] Amoeba identified include ***Entamoeba dispar*** and ***E. histolytica***, ***Entamoeba*** spp., ***Dientamoeba*** *fragilis* and other *Dientamoeba* spp., and *Iodoamoeba* sp. Although ***E. histolytica*** is considered to be the most pathogenic of the enteric amoeba, it is less often detected in apes than in OW monkeys. Lesions of *E. histolytica* include ulcerative colitis with "flask-shaped ulcers" expanding into the submucosa and amoebic hepatic and cerebral abscesses. ***Dientamoeba fragilis*** is recognized as an emerging human pathogen causing chronic irritable bowel-like symptoms and may also be pathogenic in gorillas.[378] Signs include persistent diarrhea, without blood,

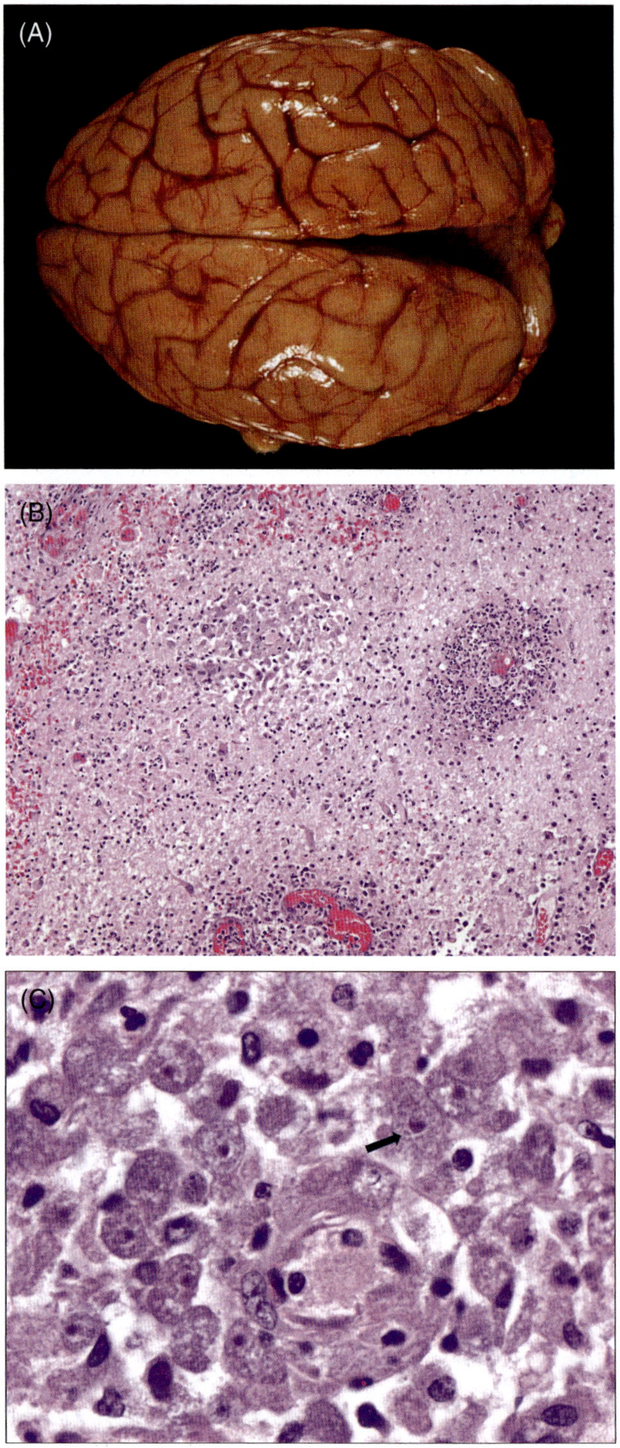

FIGURE 15.41 Amoebic encephalitis due to *Balamuthia mandrillaris* in an orangutan. (A) A focal area in which the parenchyma appears roughened and the vasculature is indistinct is present in the left occipital lobe near the longitudinal fissure. (B) Infection causes regionally extensive malacia and necrotizing vasculitis. (C) Numerous amoeboid protozoa form a thick perivascular cuff (see eSlide 15.e6). A characteristic double endosome is present in one amoeba *(arrow)*.

and weight loss that resolves with appropriate treatment, such as metronidazole.

American trypanosomiasis or **Chaga disease**, caused by ***Trypanosoma cruzi***, occurs in captive nonhuman primates, including gibbons and chimpanzees, housed outdoors in endemic areas of the southern United States and Central and South America.[51,631] Transmission is percutaneous via feces from the triatomid ("kissing bug") vector contaminating the bite wound, or through ingestion of infected bugs or their feces. Transplacental transmission and congenital infection are described in marmosets and baboons but not in apes. Heart failure is the most common presentation in all primates. Megaesophagus, common in humans, is not described in nonhuman primates. Gross lesions include cardiomyopathy, myocardial necrosis or fibrosis, pulmonary edema, and effusions. Histologically, there is primarily mononuclear myocarditis and myocardial necrosis. The amastigote forms of *T. cruzi* are seen in pseudocysts in heart, skeletal muscle, and reticuloendothelial cells. They are distinguished from other protozoans that might infect the myocardium (e.g., *Toxoplasma*) by the presence of both nucleus and bar-shaped kinetoplast. Trypomastigotes or trypanosomes are rarely seen in blood smears from infected chimpanzees.

African trypanosomes, ***T. brucei brucei***, ***T. brucei rodiensis***, and ***T. brucei gambiensis*** are the cause of animal and human sleeping sickness. It is known that chimpanzees (as well as vervet monkeys and other nonhuman primates) can be experimentally infected with African trypanosomes resulting in CNS infection.[28] Clinical signs include piloerection, ataxia, and hind limb paresis. Recent molecular epidemiological studies using serum, postmortem tissues, or feces from wild, sanctuary-housed, and European zoo chimpanzees have revealed that natural infections occur in free-living chimpanzees.[304] Lesions of human cerebral trypanosomiasis include meningeal vessel congestion, choroiditis, meningitis, lymphocytic and plasmacytic cuffing, gliosis, and vasogenic edema. Trypanosomes can be detected in the CSF.[98] Trypanosomiasis should be included in the differential diagnoses of meningoencephalitis in wild western chimpanzees.

Leishmania* sp.** DNA has been detected by fecal PCR in wild gorillas in Cameroon though there is debate about this finding.[245] However, a confirmed case of **visceral leishmaniasis** cause by the amastigote parasite ***Leishmania infantum is described in a single infant western lowland gorilla from a zoo in Brazil.[702] Clinical signs include weight loss, alopecia, and lymphadenomegaly of axillary and inguinal lymph nodes. Amastigotes in lymph node aspirates, positive serology, and positive PCR on whole blood confirm the diagnosis.

Malaria due to ***Plasmodium* spp.** is one of the globally most important diseases of humans, causing both lifelong morbidity and mortality in tropical and subtropical areas of the world. Nonhuman primates are naturally infected with malarial parasites often identical or closely related to human pathogens, which may have arisen as zoonoses.[546] Cross-species, mosquito-mediated transmission is well documented.[423,503] Identified species include: *P. rodhaini* in chimps and western gorillas (similar to *P. malariae*, zoonotic); *P. reichenowi* in chimps and western gorillas (ancestor of *P. falciparum*; not zoonotic); *P. schwetzi* in chimps and western gorillas (similar to *P. vivax*; zoonotic); *P. gaboni* and *P. lomamiensis* sp. nov. in bonobos; *P. pitheci*, *P. sylvaticum*, *P. inui*, *P. cynomolgi*, and human *P. vivax* in orangutans; and *P. hylobati*, *P eylesi*, *P. youngei*, and *P. jeofferyi* in gibbons.[56,183,402,565]

Plasmodia require an arthropod vector, usually an anopheline mosquito. Infection in apes is generally subclinical, but in gibbons, it may be associated with mild cyclic depression due to anemia. In confiscated orangutans "malaria-like symptoms" have been abrogated by antimalarial treatment. Examination of thick blood smears by experienced lab personnel for intraerythrocytic gametes and/or schizonts is necessary for diagnosis and examination of multiple repeat samples may be needed. Speciation can sometimes be made on morphology alone. PCR is available for differentiation of related species. Fecal PCR screening is used for noninvasive detection in populations of free-living apes. Gross pathology, based on experimental studies, includes hepatosplenomegaly (especially if there is extraerythrocytic schizogony). Histologically, hepatic and splenic extramedullary hematopoiesis, Kupffer cell hyperplasia, and expansion of the mononuclear phagocytic system cells in the spleen are associated with granular, black pigment (hemozoin) in phagocytic cells. Hemozoin is refractile, birefringent, and does not stain with iron stains (e.g., Perl's). It must be differentiated from acid hematin, a fixation artifact. Cerebral malaria, a severe consequence of human malaria due to sludging of parasitized red cells in brain capillaries, is reported in monkeys but not apes. Glomerulopathy, characterized by membranous thickening leading to nephrotic syndrome, is seen in affected humans but not reported in experimentally infected chimpanzees.[105]

Ectoparasites

Mange, due to ***Sarcoptes scabei* (scabies)**, is reported in free-living gorillas and chimpanzees and can cause death in infants.[226,316,406] Skin lesions include alopecia associated with florid hyperkeratosis and occasionally nodule formation. Histologically, mites are visible in deep epidermal tunnels. Morbidity and mortality are associated with alopecia, anorexia, and weight loss. The source of the mites may be humans or other animals. ***Pangorillalges gorillae*** is a common, gorilla-specific mite that can also cause mange in mountain gorillas.[521] It is a much less serious infection than scabies, although alopecia and pruritis may

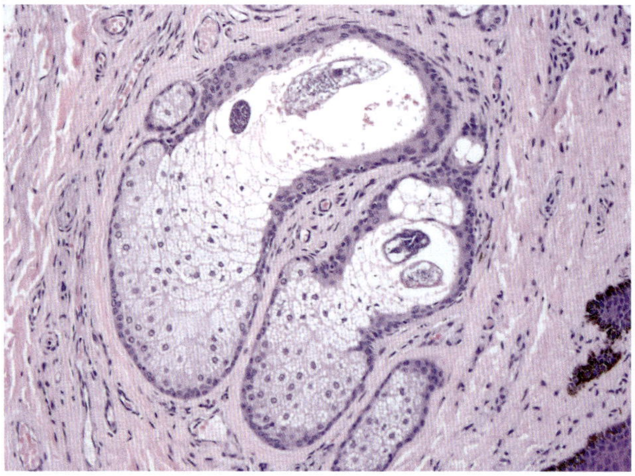

FIGURE 15.42 Prepuce from a young silverback western lowland gorilla in managed care. Demodex mites are present within distended sebaceous gland ducts. This is an incidental finding.

be present. ***Psorergates*** mange in siamang causes pruritis, scaley skin, and multifocal alopecia.[17] Histologically, mites are present just beneath or within the stratum corneum. Whole mites are needed for speciation or mite-specific DNA can be amplified from host feces.

Demodex-like mites are seen incidentally in sebaceous glands and ducts of mountain and western lowland gorillas and are likely present in all apes (Fig. 15.42). Though small sebaceous glands are associated with hair follicles over the entire body, very large sebaceous glands are present only on the brow ridge, cheeks, lips, and genital area. Ducts of these complex sebaceous glands provide habitat for *Demodex* mites. Eccrine sweat glands, found over the entire body in gorillas and chimps (except lips, nose, male prepuce, and female labia), and apocrine sweat glands (associated with some hair follicles and scent glands, such as the sternal scent glands of gibbons and orangutans and the axillary organ of African apes) are not affected.[671]

Respiratory mites are found in the nasal cavity, sinuses, larynx, trachea, or lungs. ***Pneumonyssus oudemansi*** is described in chimps and gorillas and ***Rhinophagia pongicola*** in the frontal sinus of orangutans.[184,185] In zoo-housed gorillas and bonobos, respiratory mites are occasionally seen during tracheal intubation or on nasal secretion cytology.

Pediculosis (lice infestation) in free-living chimpanzees and gorillas can be especially severe in young or debilitated animals. ***Pthirus gorilla***, a blood-sucking louse related to the human pubic louse, is found all over the body in gorillas.[563] Nits (eggs) and adults are readily visible grossly. Chimpanzees can be heavily infested with the blood-sucking louse, ***Pediculus schaeffi***, which is related to the human head louse.[269]

ADDITIONAL RESOURCES

Abee CR, Mansfield K, Tardif S, Morris T, eds. *Nonhuman Primates in Biomedical Research*. 2 ed. San Diego: Academic Press; 2012.

Cooper JE, Hull G. Gorilla pathology and health. Academic Press. 2017.

Lowenstine LJ. A Primer of Primate Pathology: Lesions and Nonlesions. *Toxicologic Pathology*. 2003;31: 92–102.

Lowenstine LJ, McManamon R, Terio KA. Comparative Pathology of Aging Great Apes. *Veterinary Pathology*. 2016;53: 250–276.

Weston Murphy H. "Great Apes" In Fowler's Zoo and Wild Animal Medicine, Vol. 8. Elsevier. 2015. pgs. 336–354.

References

Full references for this chapter can be found online at https://doi.org/10.1016/B978-0-443-18437-6.00046-8.

Chapter 16

Proboscidae

Jennifer A. Landolfi[a] and Daniela Denk[b]

[a]Zoological Pathology Program, College of Veterinary Medicine, University of Illinois, Brookfield, IL, United States; [b]SeaWorld Abu Dhabi, Yas Island, Abu Dhabi, United Arab Emirates

INTRODUCTION

The taxa Proboscidae includes three living species of elephants classified in two genera, *Loxodonta* and *Elephas*. Elephant species are generally referred to by their current geographic distribution. African elephants include *Loxodonta africana*, the African savanna elephant, and *Loxodonta cyclotis*, the African forest elephant. The Asian elephant species is *Elephas maximus*; subspecies include *E. m. indicus*, *E. m. maximus*, *E. m. sumatranus*, and *E. m. borneensis*, the Bornean pygmy elephant. African savanna elephants are listed as endangered and African forest elephants are listed as critically endangered by the International Union for the Conservation of Nature (IUCN), and both are in CITES Appendix I except for populations in Botswana, Namibia, Zimbabwe, and South Africa, which are in CITES Appendix II. Asian elephants are endangered CITES Appendix I species[139] (www.cites.org; www.iucn.org). Pathologic lesions in these two closely related genera are similar for many diseases; however, important differences in prevalence and susceptibility, and to some extent lesions, do exist. No notable differences in pathology between African savanna and African forest elephants have been documented, thus discussion of pathology in African elephants will be applicable to both *Loxodonta* species. In the absence of data to indicate otherwise, all Asian elephant subspecies are presumed to have similar susceptibility and manifestations of disease.

Please refer to the chapter's digital content for a discussion of some unique normal anatomic features of Proboscidae.

NON-INFECTIOUS DISEASES

Age-Related/Degenerative

Foot lesions are the most common pathologic abnormality of captive elephants. Prevalence is higher in Asian elephants, but lesions occur in African elephants as well. Though not as well documented in free-ranging elephants, foot disease does occur, and **pododermatitis** has been recognized in several wild African elephants in Kruger National Park in South Africa.[60] Optimal elephant foot health is best achieved in a natural environment with ample space, diverse topography, and native substrates including wetlands, pasture, and wooded areas. Provision of such habitat is often challenging in a managed-care environment, and several factors may contribute to the development of foot disease in captive elephants. These include neglect of regular nail and sole trimming and care, prolonged exposure to a wet or soiled environment, artificial substrates, sedentary lifestyle, conformation, and trauma. Physical abnormalities in foot structures (e.g., nails, sole, and cuticles) due to overgrowth, trauma, or other mechanical disturbances may predispose to foot lesion development. Secondary opportunistic bacterial and/or fungal infection of damaged foot structures may develop. Bacterial pathogens include enteric genera, such as *Aeromonas*, *Citrobacter*, *Morganella*, *Salmonella*, *Proteus*, *Klebsiella*, and *Escherichia* sp. Fungi may include *Fusarium*, *Penicillium*, and *Cladosporium* species.[34,75] Mixed infections occur frequently.

Grossly, lesions often involve overgrowth of one or more of the keratinized structures of the foot (sole, nails, and cuticles). **Overgrown soles** have an irregular contour with plantar grooves, ridges, and pockets (Fig. 16.1). Pododermatitis may develop and manifest as soft, discolored, and macerated sole keratin with edema and inflammation of underlying tissues (Fig. 16.2). **Toenail overgrowth** may result in deformed nails and/or lateral growth that can penetrate adjacent skin or the sole. Associated reactive fibroplasia or granulation tissue formation at the affected site presents as white, friable material.[34] Overgrown cuticles harden and may crack forming hangnails (Fig. 16.3). Cuticle overgrowth may be associated with thickening and hardening of the internail skin resulting in "interdigital callus" formation. Cuticle lesions can also lead to compression and obstruction of sweat gland ducts that are present in high concentration at this location and result in formation of fluid-filled pockets ("blisters" or "blebs") beneath affected cuticles.[130]

Pathology of Wildlife and Zoo Animals. https://doi.org/10.1016/B978-0-443-18437-6.00011-0

FIGURE 16.1 Sole overgrowth in the foot of an Asian elephant. The overgrown sole is characterized by an irregular plantar contour with grooves, ridges, and pockets. *(Photo Courtesy of R. Isaza, University of Florida College of Veterinary Medicine.)*

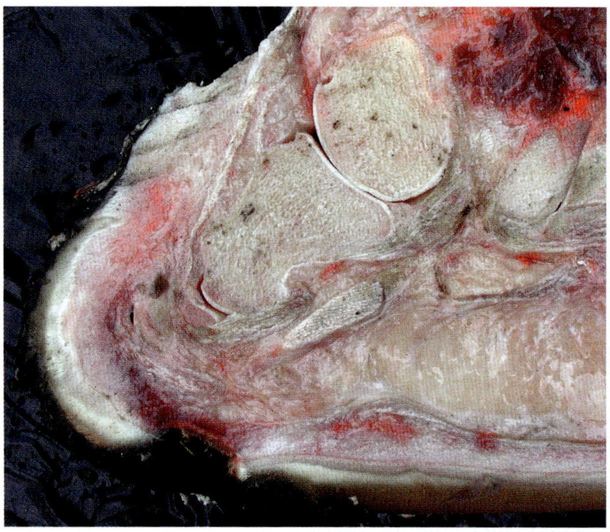

FIGURE 16.2 Pododermatitis in an Asian elephant (sagittal section). The rostral plantar sole has a focal, depressed, cavitated defect. Underlying pedal soft tissues are thickened, wet, and multifocally reddened due to edema, inflammation, and hemorrhage. *(Photo Courtesy of R. Isaza, University of Florida College of Veterinary Medicine.)*

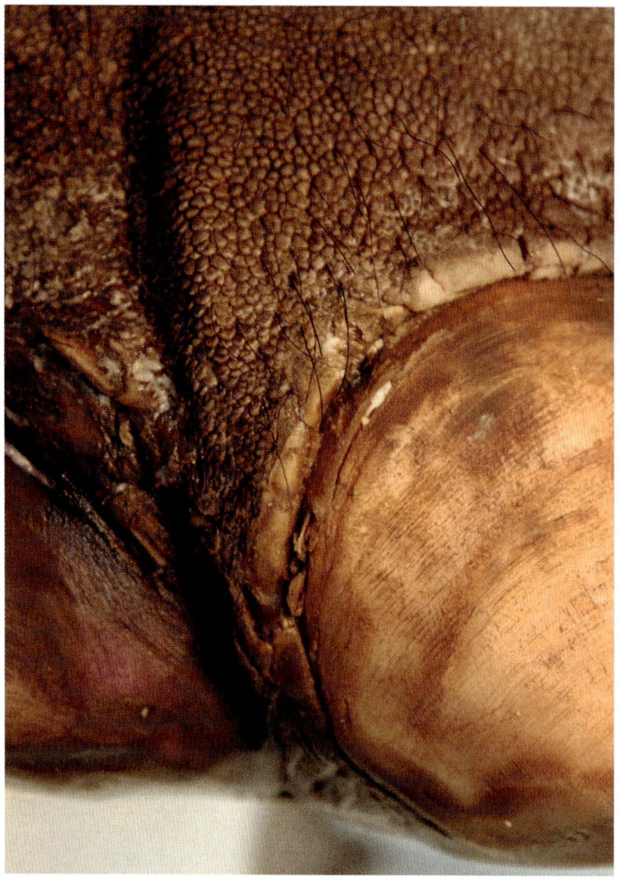

FIGURE 16.3 Cuticle overgrowth in the foot of an Asian elephant. Cuticular skin is irregularly thickened and cracked due to overgrowth. *(Photo Courtesy of R. Isaza, University of Florida College of Veterinary Medicine.)*

Toenail cracks are another common foot abnormality of captive elephants. Overgrowth and trauma are key contributors. Cracks may be horizontal or vertical; vertical cracks occur more often. A crack may extend from the cuticle distally or originate from the nail margin (Fig. 16.4). Cracks can be superficial or deep to the level of the corium.[34,130]

Overgrowth or damage of the keratinized foot structures described earlier commonly results in secondary infections that complicate lesions. Infection may manifest as generalized pododermatitis with variable amounts of associated exudate ranging from serous to thick, caseous, and fetid. Alternatively, infection may result in the formation of a localized **abscess** that presents as a relatively discrete pocket of fluid or exudate bordered by granulation tissue and/or fibrosis. Exudate is typically thick, turbid, and tan-yellow to brown. Surrounding soft tissue is thickened, tough, and white indicating fibroplasia (granulation tissue and fibrosis). Overgrown or damaged foot structures overlying an abscess frequently have a defect with irregular margins through which exudate may drain (Fig. 16.5).

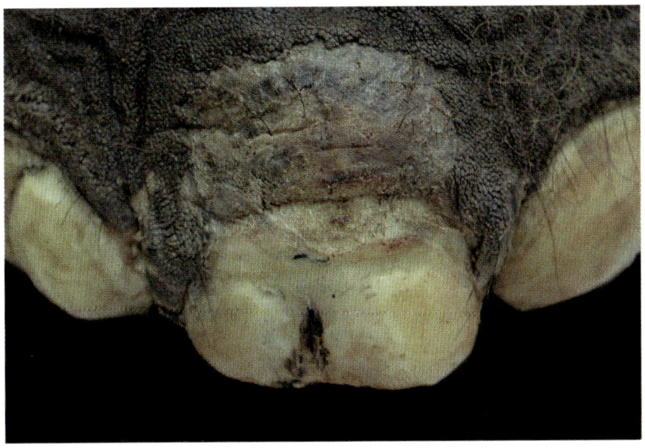

FIGURE 16.4 Nail crack in the foot of an African elephant. The hindlimb third digit nail has a longitudinal crack that extends proximally from its plantar aspect. Regionally, cuticular skin is eroded with a pale tan, discolored, depressed, dull, and granular surface. *(Photo Courtesy of San Diego Zoo Wildlife Alliance.)*

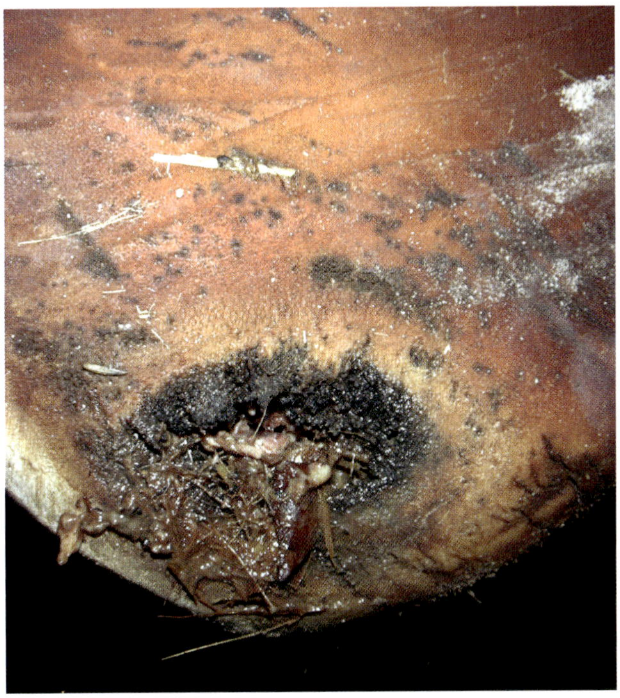

FIGURE 16.5 Sole abscess in the foot of an Asian elephant. The rostral plantar aspect of the sole is ulcerated with irregular, ragged, black-brown discolored margins, and exposure of underlying soft tissue. Small amounts of opaque, tan-brown, mucoid exudate drain from and cover the defect.

In an affected foot, lesions are typically focal, but multiple feet may be involved. Histologic features of foot lesions will vary depending on chronicity. Changes include hyperplasia and hyperkeratosis of affected epithelium with regions of ulceration and associated inflammation. Some degree of fibroplasia ranging from granulation tissue to mature fibrosis may also be present.

Osteoarthritis and **degenerative joint disease** are common causes of morbidity in captive Asian and African elephants. Lesions are typically seen in mature, older animals. The pathogenesis is multifactorial, and factors include one or more of the following: age-related deterioration, mechanical trauma, repetitive loading stress due to hard environmental substrates, lack of appropriate exercise, excessive body weight, and poor conformation. Fore limbs are generally more commonly and severely affected than hind limbs due to greater weight bearing.[34,79] Gross lesions may include cartilage erosion/loss, thickening/fibrosis and/or mineralization of joint capsule and periarticular soft tissues, periarticular bony proliferation (osteophytosis), and/or subchondral bone sclerosis (eburnation) (Figs. 16.6A–C). Histologically, variably severe chronic inflammation and fibrosis of articular/periarticular soft tissues occur. Characteristic gross lesions combined with compatible history and radiographic abnormalities are diagnostic.

Malocclusion and associated dental disease are common in aged captive elephants (Fig. 16.7). Failure of worn cheek teeth to fracture and shed properly can prevent forward progression of other cheek teeth leading to crowding, malalignment, and malocclusion. Impaired forward progression of cheek teeth may delay or inhibit delamination of cheek teeth crowns leading to malformed or fused teeth and chronic dental disease. Entrapped foreign material may subsequently result in periodontitis. Grossly, maloccluded arcades show irregular wear, abnormal curvature to the plane of migration, and fusion of affected teeth.[24,151] Chronic periodontal disease can lead to osteomyelitis and extensive remodeling of the associated maxillary or mandibular bone. A consequence may be weight loss and decrease in body condition.

Arteriosclerosis and **atherosclerosis** were first documented in free-ranging African elephants in the 1960s and 1970s during examination of the cardiovascular systems of randomly culled animals in Kenya, Uganda, and Zambia.[23,84,88] The precise cause is unknown; however, postulated contributing factors include mechanical stress in affected vessels, medial anoxia, high serum calcium, and advancing age. Correlations with serum lipid levels have not been found. Aorta, coronary arteries, and aortic branch arteries are most commonly affected. Grossly, vessels have a thickened, rough, discolored endothelial surface with discrete, white, firm, raised plaques. Histologically, two morphologic variations occur. The first is medial sclerosis, which involves calcified fibrotic plaques in the inner tunica media with smooth muscle atrophy[86] (Fig. 16.8). The second, intimal atherosclerosis, is characterized by intimal plaques composed of lipid accumulation, lymphocytic inflammation, mineralization, and necrosis.[85] Arterio- and atherosclerosis are not associated with physical disruption of affected vessel walls. These specific vascular lesions

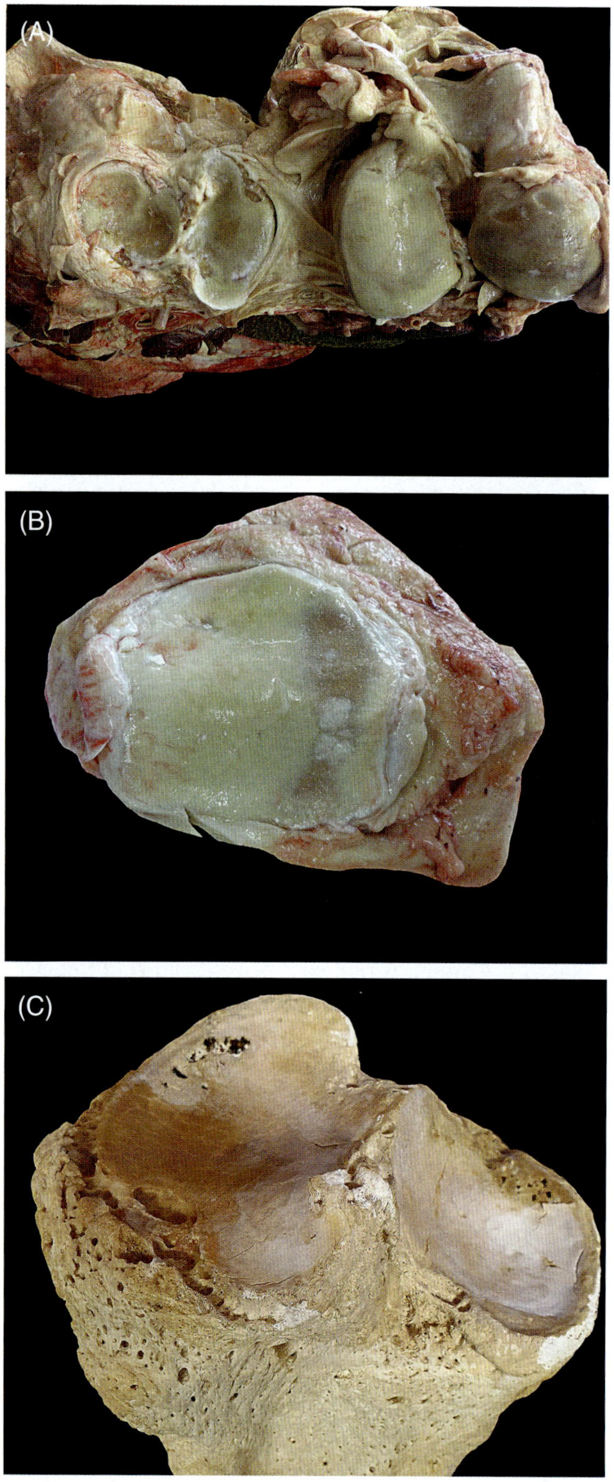

FIGURE 16.6 Chronic osteoarthritis (degenerative joint disease) in joints of an Asian elephant. (A) Femoral (right) and tibial (left) condyles have loss and erosion of articular cartilage, mild marginal osteophytosis, and thickening of the joint capsule and periarticular soft tissues. (B) The patella has marginal osteophytosis and extensive cartilage erosion and loss with exposure of subchondral bone. (C) A skeletal preparation of the proximal tibial has extensive bony remodeling and sclerosis of medial (left) and lateral (right) condyles.

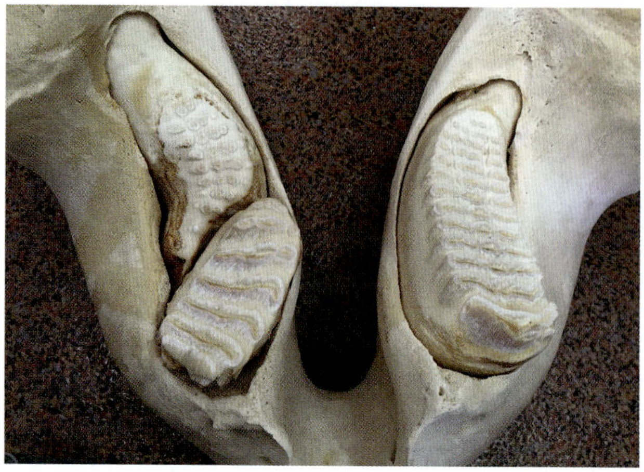

FIGURE 16.7 Molar malformation and impaction in the mandible of an Asian elephant (skeletal preparation). Mandibular molars have unilateral impaction with malformation and partial fusion due to impaired forward progression of cheek teeth, with resultant malocclusion. *(Photo Courtesy of J. Dagostino, Oklahoma City Zoo.)*

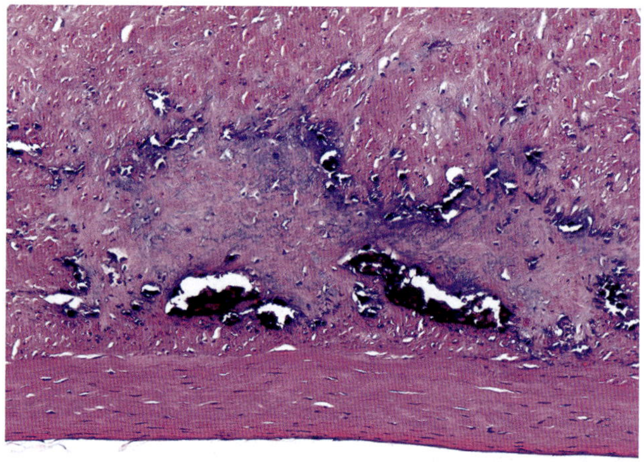

FIGURE 16.8 Arteriosclerosis in the aorta of an African elephant. The aortic media is disrupted and expanded with deposits of deeply basophilic, fragmented mineral. *(Photo Courtesy of A. Falkenau, Ludwig Maximilian University, Munich, Germany.)*

have not been detected in more recent mortalities; however, aortic aneurysms have been described in elephants in the absence of underlying arterio- or atherosclerosis; aneurysmal rupture may cause sudden death.

Degenerative cardiac disease is recognized in aged captive African, and to a lesser extent, Asian elephants. In some animals found dead or moribund in the absence of any premonitory signs, degenerative cardiac lesions are the only significant finding; death may be attributable to induced arrhythmias. Gross changes include heart enlargement and/or chamber dilatation. Histologically, multifocal myocardial degeneration and fibrosis dominate. Degeneration is characterized by swollen myofibers with pale, vacuolated sarcoplasm, and variable loss of cross striations. Cardiomyocyte dyskaryosis with anisokaryosis and karyomegaly is also evident histologically. Fibrous tissue is hypocellular, collagen rich, and dense and ranges from perivascular to more extensive myocardial replacement. The inciting cause for these changes is unknown, and identification is complicated by the chronicity of the lesions. Hypotheses include hypoxic damage and previous myocarditis, though no evidence of any specific underlying infectious cause has been confirmed.

Pulmonary fibrosis is a common finding in aged captive Asian and African elephants. Caudal or caudodorsal portions of the lungs are most frequently affected with a patchy, nodular or more extensive, subpleural and regional distribution. Grossly, affected lung is consolidated, firm, and pale tan to white (Fig. 16.9A). Fibrosis may be associated with regions of emphysema or formation of pseudocysts in the pulmonary parenchyma. The interstitium has abundant hypocellular, collagen-rich, dense fibrous tissue with distortion or replacement of normal architecture (Fig. 16.9B). Fibrous tissue may multifocally contain low to moderate numbers of lymphocytes and plasma cells. Entrapped bronchial glands can be hyperplastic and ectatic with luminal mucoid material. Mineralization of the affected interstitium and/or associated vessels is common. No features suggestive of hypersensitivity/allergic pneumonitis (i.e., alveolar eosinophilic hyaline membranes) have been detected. The cause is unknown. The lesions may represent an age-related change or chronic response to previous insult or injury and share some morphologic features with equine multinodular pulmonary fibrosis, a condition associated with the presence of equine gammaherpesviruses (i.e., EHV-5). In one elephant case, elephantid herpesvirus DNA was detected in the affected lung via PCR, but a causative relationship has yet to be confirmed.

Chronic interstitial nephritis is seen in aged captive elephants and presumed to be a degenerative, age-related disease. Previous insults or infections are potential inciting factors; however, chronicity of lesions generally precludes identification of primary cause(s). Affected kidneys are red/brown to pale grossly and have rough, granular, or slightly pitted subcapsular cortical contours. Cortical parenchyma may have pale cortical to medullary streaking on cut section and/or few, variably sized, clear fluid-filled cystic structures. Histologically, the renal interstitium multifocally has variable amounts of hypocellular, collagen-rich, dense fibrous connective tissue with separation and replacement of renal parenchyma. Affected areas may have low numbers of lymphocytes and plasma cells. Tubules are decreased in number and atrophic; some may be cystic. Glomerular changes are generally limited to areas of interstitial fibrosis and range from mild segmental membranous change to

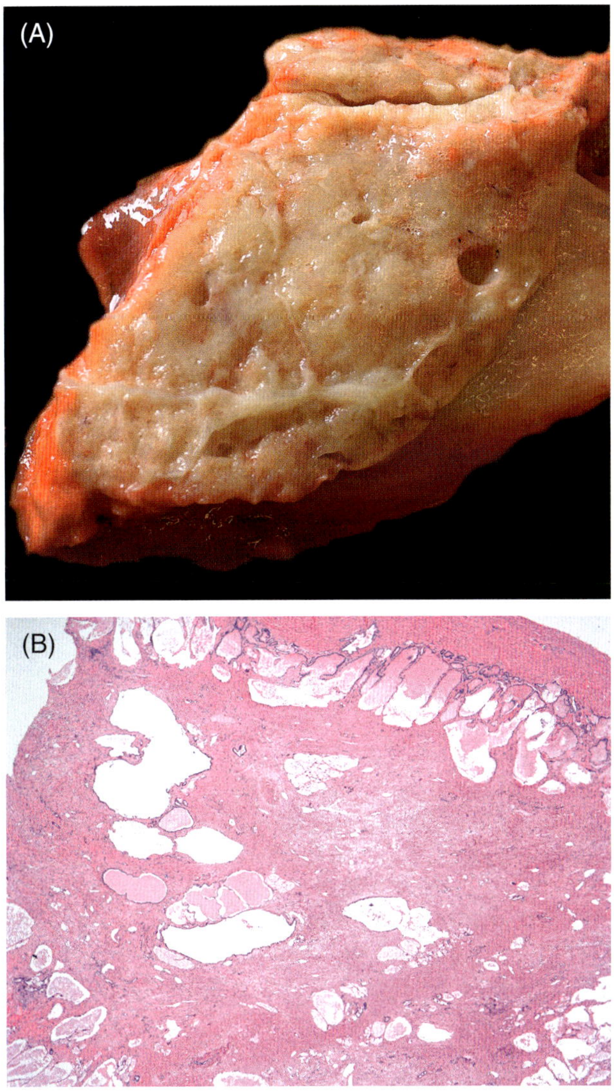

FIGURE 16.9 Pulmonary fibrosis in an Asian elephant. (A) The pulmonary parenchyma is consolidated, firm, and pale tan to white. (B) The interstitium has abundant hypocellular, collagen-rich, dense fibrous tissue with distortion and replacement of normal architecture. *(Part A: Photo Courtesy of S. Citino, White Oak Conservation; Part B: Photo Courtesy of M. Garner, Northwest ZooPath.)*

sclerosis. Interstitial fibrosis may also be evident in the medulla, and variable numbers of medullary and fewer cortical tubules may contain mineral (Fig. 16.10).

Endometrial hyperplasia is common in captive, nulliparous, aged females. Similar to carnivores, pathogenesis of endometrial hyperplasia in elephants is associated with a normal hormonal environment including both prolonged estrogen and progesterone phases. Estrogen primes the endometrium for maximum effect of prolonged progesterone exposure during the luteal phase. Size and number of glands increase under estrogenic influence, and secretory function increases with progesterone exposure. Irreversible

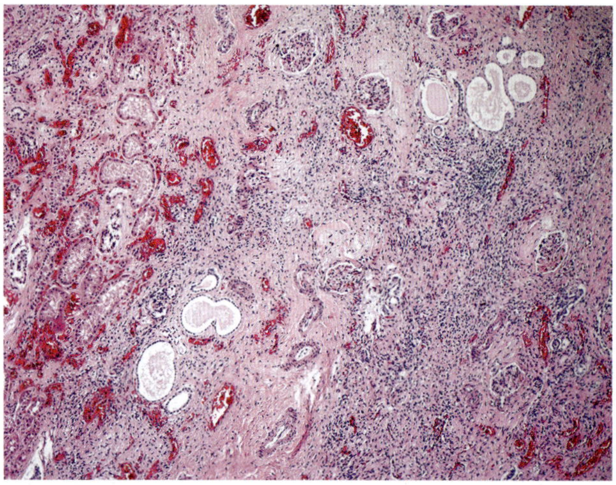

FIGURE 16.10 Chronic interstitial nephritis in an Asian elephant. The renal interstitium has extensive fibrosis and scattered lymphoplasmacytic inflammation with wide separation and replacement of tubules. Some remaining tubules are ectatic and filled with luminal proteinic material. Glomeruli are variably sclerotic or obsolescent.

damage to the endometrium occurs in the absence of pregnancy-associated endometrial remodeling. Lesions may impact fertility but are otherwise typically clinically silent. Grossly, the endometrium is thickened with redundant folds and multiple irregular polyploid projections and/or thin-walled, fluid-filled cystic structures (Fig. 16.11). Histologically, endometrial glands are irregular, dilated, and sometimes tortuous. Gland epithelium ranges from attenuated to hyperplastic. Mild endometrial fibrosis and mild lymphoplasmacytic inflammation may be present.[2,102]

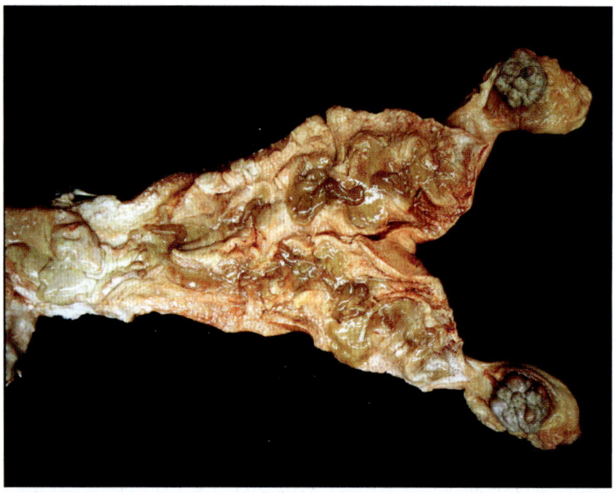

FIGURE 16.11 Cystic endometrial hyperplasia in an Asian elephant. The endometrium is thickened and rugose with several pale tan, gelatinous, polyploid projections. *(Photo Courtesy of University of California, Davis William R. Pritchard Veterinary Medical Teaching Hospital.)*

Miscellaneous/Multifactorial

A condition termed **flaccid trunk paralysis** or **floppy trunk syndrome** was recognized in free-ranging African savanna elephants in Zimbabwe in 1989. Older bulls were affected, and the condition developed slowly over a period of months. The cause was not confirmed but investigations suggested toxicity associated with the plant *Heliotropium ovalifolium* or lead as a possibly contributory toxin. No gross lesions have been documented. Histologically, peripheral nerve degeneration is seen in the trunk with degenerative changes noted first in the distal portions. Diagnosis is based on peripheral nerve degenerative changes in the trunk and a compatible clinical history of trunk dysfunction/paralysis.[64]

Gastric ulcers are common gross necropsy findings in elephants (Fig. 16.12). Ulcers can range in size and appearance from pinpoint shallow, dark red to black depressions in the gastric mucosa to large, coalescing, 2–3 cm diameter, almost full-thickness defects. Ulcers may be acute with sharply demarcated edges or chronic with thick rounded margins. Evidence of mucosal hemorrhage is often seen. Possible underlying causes of gastric ulcers in many species include NSAID administration alone or in combination with antiinflammatory steroid administration, "stress" or endogenous glucocorticoid release, infectious agents, and chronic renal disease, but definitive causation is often elusive.

Intestinal mechanical or physical disturbances occur in both captive African and Asian elephants, with a higher incidence in African elephants. Mechanical or physical disturbances include **impaction**, **obstruction**, and **torsion/volvulus**. The large intestine, particularly the cecum, is most commonly affected. Impacted material is often a

FIGURE 16.12 Gastric ulcerations in an African elephant. Multiple, well-demarcated, dark-red to black depressions consistent with mucosal ulceration are evident.

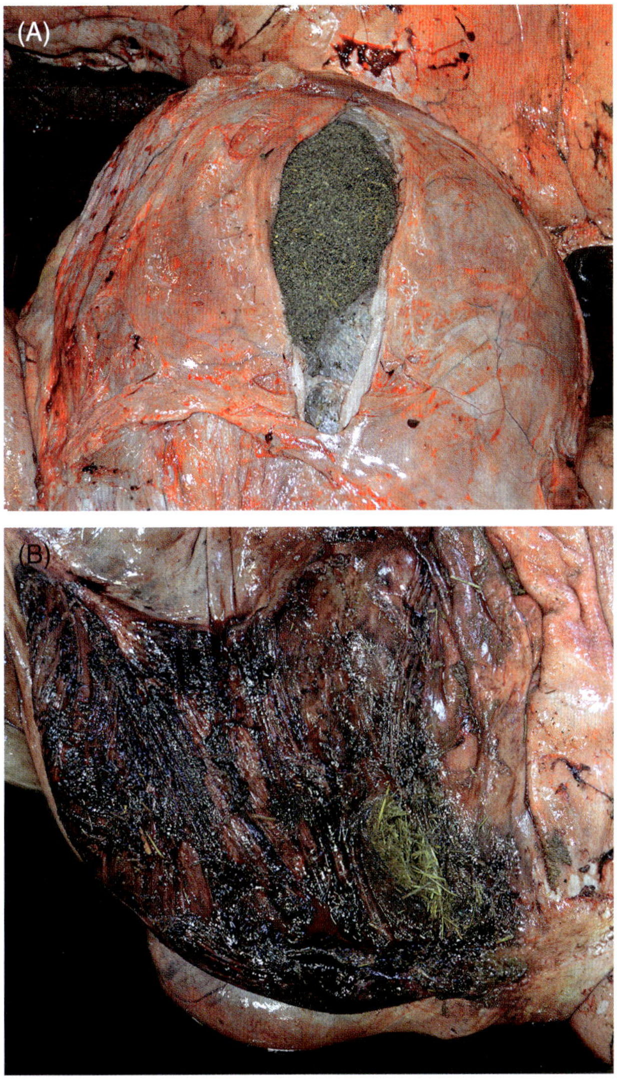

FIGURE 16.13 Cecal sand impaction and rupture in an African elephant. (A) The cecum is filled with coarse, granular environmental substrate (sand). (B) The apex of the cecum has an irregular full-thickness rent (rupture) through which digesta protrudes. The surrounding serosal surfaces are covered with admixed hemorrhage, fibrin, and free fragments of digesta.

combination of environmental substrate (i.e., sand) and feedstuff (Fig. 16.13A). Environmental substrate may be inadvertently ingested along with food items, and over time, this material collects and settles within the alimentary tract, especially the cecum. Feed placement (e.g., on enclosure substrate vs in suspended feeders) can be a factor in pathogenesis. Cecal impaction may be uncomplicated or result in torsion of the impacted segment. A possible sequel to impaction is intestinal wall rupture and septic peritonitis (Fig. 16.13B).

Nonsteroidal antiinflammatory drugs (NSAIDs) are commonly used to treat osteoarthritis and foot lesions in elephants, and nephrotoxicity is the suspected cause of **renal medullary** and/or **papillary necrosis** in elephants. Other possible causes include renal hypoxia, dehydration, pyelonephritis, and chronic urinary tract obstruction [urinary calculi (calcium carbonate) infrequently occur in these species]. Gross lesions vary from thin yellow/white streaks in the medulla to radiating areas of necrosis or even infarction of medullary tissue. Histologically, gross lesions correspond to areas of necrosis, inflammation, and often mineralization of medullary tubules and associated connective tissue.

Neoplastic

Published reports of neoplasia in elephants are few and consist predominantly of single animal case reports (Table 16.1). Elephants have been cited as an example of Peto's paradox, the observation that the incidence of cancer at the species level does not correlate with the number of cells in an organism. Studies describe a lower-than-expected rate of neoplasia in elephants given their large size and extended lifespan.[1,45,49,133,137,154,156] Several recent reports have proposed cancer resistance in elephants mediated by multiple copies (retropseudogenes) of the tumor suppressor gene TP53 in the African elephant genome with a resulting increase in the cellular apoptotic response following DNA damage.[1,41,154,156,158,159] However, confirmatory evidence of a causal link between extra TP53 copies and cancer suppression in elephants has not yet been documented.[1,14] It is likely that elephant neoplasia is underreported in the published literature. In fact, information gleaned from a retrospective analysis of archival postmortem data has shown that the prevalence of neoplasia in captive elephants is similar to humans and other captive zoo mammals.[8,68,115] Despite the limited literature, the practical experience of veterinarians, pathologists, and caretakers indicates neoplasia is an important factor in health and welfare.

Myometrial leiomyomas, benign smooth muscle tumors, are by far the most common tumors identified in elephants.[51,68,98] They are typically found in nulliparous, aged Asian elephants and present as variably sized (usually 1–10 cm diameter), discrete, unencapsulated, firm, tan-white, and nodular masses in the uterine wall (Fig. 16.14A). Multiple masses are often present, and numerous coalescing leiomyomas may markedly distort uterine architecture (Fig. 16.14B).[68,98,117,132] Histologically, leiomyomas are composed of intersecting bands of well-differentiated neoplastic smooth muscle. Neoplastic cells are immunoreactive for vimentin, desmin, and smooth muscle actin; immunolabeling for estrogen receptor has also been demonstrated.[68,117] No vascular invasion or metastasis to indicate malignancy has been documented.

TABLE 16.1 Published Reports of Neoplasia in Elephants

Neoplasm	Site Affected	Species	References
Ovarian cystadenoma	Ovary	African	Hoby et al.[52]
Eccrine carcinoma	Foot skin	Asian	Grabarević et al.[43]
Fibroma	Trunk skin	African	Stevens et al.[152]
Leiomyoma/ fibroleiomyoma	Uterus	Asian	Landolfi et al.[68] Sapundzhiev et al.[132]
Adenoma	Uterus	Asian	Landolfi et al.[68]
Adenocarcinoma	Uterus	Asian	Landolfi et al.[68] Laricchiuta et al.[72]
Anaplastic carcinoma	Uterus	Asian	Landolfi et al.[68]
Peripheral neuroectodermal tumor (PNET)	Uterus	Asian	Landolfi et al.[68]
Angiosarcoma	Uterus	Asian	Landolfi et al.[68]
Epithelial dysplasia	Conjunctiva	Asian	Fraunfelder et al.[37]
Phaeochromocytoma	Adrenal gland	African	Bonar et al.[9]
Fibrosarcoma	Foot skin	All	Brown et al.[13] Liu et al.[76]
Odontoma	Tooth	African	Raubenheimer et al.[121]
Dermoid cyst	Neck skin	African	Ritchie et al.[125]
Papilloma, fibropapilloma	Skin, vulva, uterus	All	Basson et al.[7] Jacobson et al.[56] Munson et al.[102] Robinson and Meier[127] Sundberg et al.[155]
Carcinoid	Liver[a]	Not specified	Werle et al.[161]

[a]*Presumed metastatic; primary site of neoplasia not determined.*

These tumors are typically incidental findings at necropsy. Significance resides in the potential of myometrial leiomyomas to impact fecundity and thus influence conservation efforts in managed care.[68]

Uterine adenocarcinomas are the second most common neoplasm in elephants and have been described in aged, nulliparous Asian elephants. Pathogenesis is uncertain, but the contribution of hormone influence, body condition, and physical activity levels have been considered. Distant and widespread metastasis has been documented in most cases. Adenocarcinoma is frequently detected concurrently with myometrial leiomyomas, and in such cases may be grossly indistinguishable from the smooth muscle neoplasms. Uterine adenocarcinomas are poorly demarcated, multinodular, firm masses that range from 12 to 30cm in diameter with distortion of the uterine wall and elevation of the endometrial surface (Fig. 16.15A). Masses may have foci of necrosis evident as cystic or cavitated areas containing yellow-tan, mucoid to friable material. Histologically, neoplastic cells are arranged in solid nests, acini or irregular, elongate, tortuous glandular structures, sometimes with luminal papillary projections (Fig. 16.15B). Supporting stroma is typically abundant and densely collagenous. Neoplastic cells are positive for pan-cytokeratin and negative for vimentin. In some cases, neoplastic cells have positive intranuclear immunoreactivity for estrogen receptor[68,72] (Fig. 16.15C). Uterine malignancies other than adenocarcinoma have also been described in captive Asian elephants and include **anaplastic sarcoma**, **peripheral neuroectodermal tumor (PNET)**, **angiosarcoma**, and **anaplastic carcinoma**.[68] Female reproductive tract

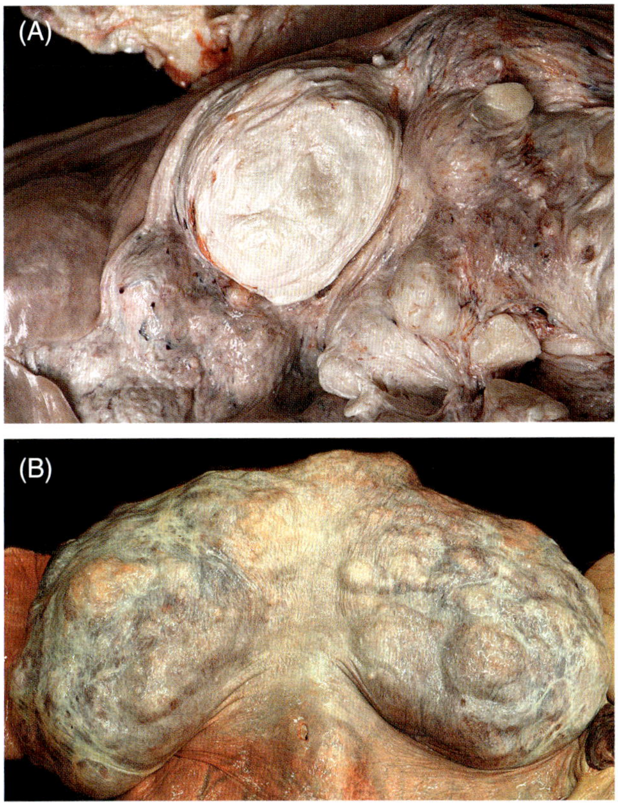

FIGURE 16.14 Uterine leiomyoma in an Asian elephant. (A) The myometrium is expanded by a nodular, white, firm, and fibrous mass. (B) Bilaterally, the uterine horns are enlarged and distorted by numerous, coalescing myometrial leiomyomas. *(Part A: Photo Courtesy of F. Uzal, University of California, Davis William R. Pritchard Veterinary Medical Teaching Hospital. Part B: Photo Courtesy of L. Lowenstine, University of California, Davis William R. Pritchard Veterinary Medical Teaching Hospital.)*

serosal fibropapillomas are common incidental lesions in aged elephants (Fig. 16.16). Polyploid or papillary fibrous projections may be present on serosal surfaces of the ovary, oviduct, uterus, mesovarium, and/or mesometrium.

INFECTIOUS DISEASES

DNA Viruses

Elephant endotheliotropic herpesvirus (**EEHV**) causes **EEHV hemorrhagic disease** (**EEHV-HD**), an acute, multisystemic, often fatal hemorrhagic syndrome that affects both managed and free-ranging elephants worldwide. EEHV-HD is the leading cause of death in zoo-managed Asian elephant calves in North America and Europe as well as a significant cause of calf mortality in Asian elephant range countries; consequently, infection has profound implications for population growth and sustainability.[6,30,53,73,77,108,109,118,123,131,147,153,164,166] Elephant endotheliotropic herpesviruses are double-stranded DNA herpesviruses (*Herpesviridae* family; *Proboscivirus* genus).

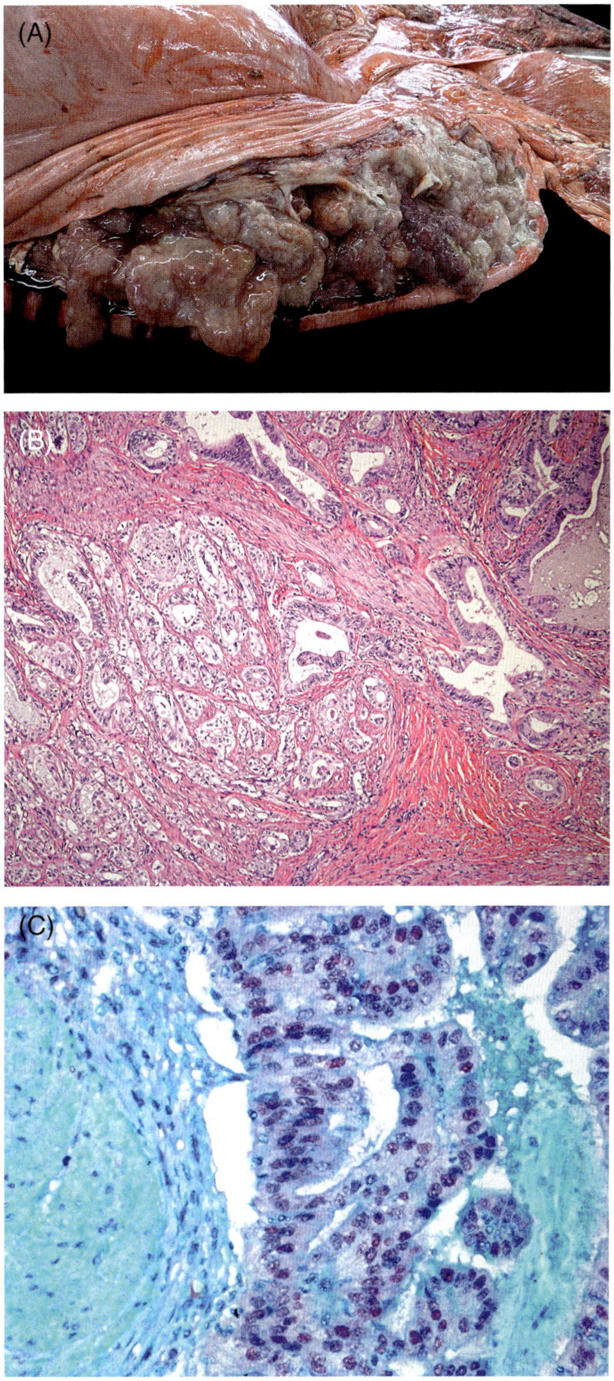

FIGURE 16.15 Uterine adenocarcinoma in an Asian elephant. (A) The endometrium is irregularly thickened by a firm, pale tan to pink, multilobular, and cystic mass. (B) The mass is composed of neoplastic epithelial cells arranged in haphazard glandular structures supported by dense fibrous stroma. (C) Neoplastic epithelial cells have nuclear immunolabeling for estrogen receptor [(ER) immunohistochemistry and hematoxylin counterstain]. *(Part C: Photo Courtesy of San Diego Zoo Wildlife Alliance.)*

Research indicates these viruses should be classified into a unique subfamily (*Deltaherpesvirinae*) representing an intermediate branch between *Betaherpesvirinae* and *Gammaherpesvirinae*. EEHVs have a unique arrangement of

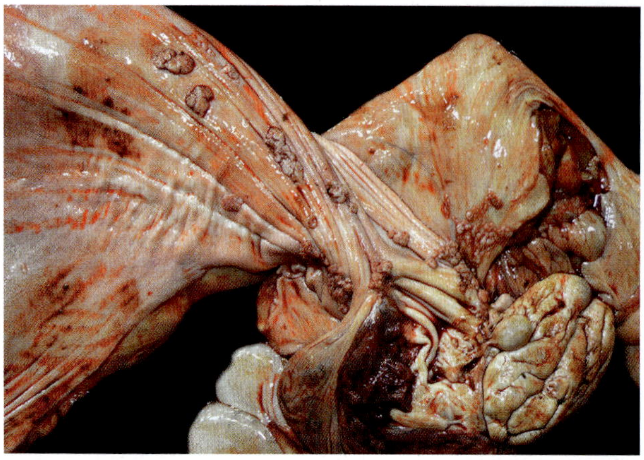

FIGURE 16.16 Reproductive tract serosal fibropapilloma in an African elephant. The uterine serosa has a few short, polyploid, tough, white projections consistent with fibropapillomas.

the core domain gene blocks and retention of TK, RRBm, and OBP proteins, which are absent in other betaherpesviruses.[77,124] Several types or species of EEHVs exist (EEHV1a, EEHV1b, EEHV2, EEHV3, EEHV4, EEHV5, EEHV6, EEHV7), and multiple strains of each EEHV type can cause disease. Molecular studies examining virus genotypes in elephant tissues have shown that *Proboscivirus* infection is nearly ubiquitous in adult, clinically normal elephants. Adult Asian elephants are naturally infected with EEHV1a, EEHV1b, EEHV4, and EEHV5; African elephants harbor EEHV2, EEHV3, EEHV6, and EEHV7. Asian elephants may occasionally shed EEHVs in trunk wash secretions in the absence of disease. In African elephants, EEHVs have been detected in skin, lung, and/or saliva of clinically normal animals.[150,168] Most cases of lethal EEHV-HD in Asian elephants are due to EEHV1a.[122,123] Other virus types associated with lethal disease include EEHV1b, EEHV3, EEHV4, and EEHV5. In African elephants, lethal disease has occurred with EEHV2, EEHV3, and EEHV6.[5,22,30,32,39,109,114,122,123,148,162] Asian elephant calves, typically between 2 and 8 years of age, are most susceptible to EEHV-HD. Maternal antibody appears to provide protection against disease given the absence of EEHV-HD in animals less than 1 year of age. Research has also demonstrated that EEHV1 antibodies wane in Asian elephant calves during the first 2 years of life.[38] Importantly, serologic evidence indicates that EEHV-HD results from primary infection rather than reactivation of latent virus.[38,119] Though more needs to be learned about the mechanisms governing susceptibility to EEHV-HD, current knowledge indicates that significant EEHV-related disease develops when primary infection of a calf is not controlled by normal innate cellular and humoral immune responses.[38,77]

In cases of EEHV-HD, viral antigens have been detected in epithelia of the alimentary tract and salivary glands indicating the possibility of initial epithelial infection at these sites.[48,62] Dissemination occurs with the circulation of virus-infected blood monocytes.[47,62,149] Transmission of infection to vascular endothelial cells is facilitated by monocyte-leukocyte adhesion molecules (i.e., PCAM-1). Viral infection and replication result in endothelial cell damage. Direct virus-mediated injury is considered the most likely mechanism; induction of apoptosis has also been considered. Consequent vascular compromise with increased permeability leads to widespread hemorrhage and edema. Endothelial damage is associated with increased expression of PCAM-1 and von Willebrand factor (vWF) with activation of the clotting cascade and upregulation of inflammatory cytokines including IL-1b, IL-2, IL4, and TNF-α. Coagulopathy ensues, likely contributes to further endothelial damage, and leads to formation of microthrombi and thromboemboli in vessels consistent with disseminated intravascular coagulation.[47,114] Most EEHV types primarily affect capillary endothelium (EEHV1a/b, EEHV2, EEHV4, and EEHV5); however, endothelial infection and damage in arterioles and venules are typical in cases of EEHV3.[39,114] Cardiac vasculature is severely affected in most cases. Death is the result of hypovolemic shock and/or cardiac and other organ failure. Hypovolemic shock is produced by both blood loss via hemorrhage and systemic cytokine release (cytokine storm) that facilitates circulatory collapse. Acute myocardial hypoxic damage may also impact contractility and conduction and contribute to cardiac dysfunction.[47,77,114]

Gross lesions in EEHV-HD are manifestations of vascular damage secondary to viral endothelial cell infection and, to a lesser degree, cytokine-mediated activation of systemic inflammation. Lesions are similar in Asian and African elephants; virus type, as opposed to affected host species, has the greatest impact on disease manifestations.[30,77,122] Though character, extent, and severity of gross findings may vary depending on the EEHV type involved, the time course of disease and therapeutic measures, the principal lesions of EEHV-HD include lingual cyanosis (Fig. 16.17); widespread edema affecting the trunk, head, neck, limbs and dependent abdomen, intestinal wall, mesentery and mesenteric root, omentum, and lungs; pericardial effusion; cardiac petechial to extensive ecchymotic hemorrhages (Fig. 16.18A); and visceral petechial hemorrhage.[77,122]

In general, histologic features of EEHV-HD may include, in order of decreasing frequency, perivascular to interstitial microhemorrhages with interstitial edema and inflammation consisting of low numbers of heterophils, macrophages, and/or lymphocytes (Fig. 16.18B); endothelial cell hypertrophy, sloughing or necrosis (Fig 16.18C); endothelial cell nuclear inclusions ranging from eosinophilic to amphophilic or basophilic and typically smudged with variable margination of nuclear chromatin (Figs. 16.18D, E; eSlide 16.e1); vascular mural edema

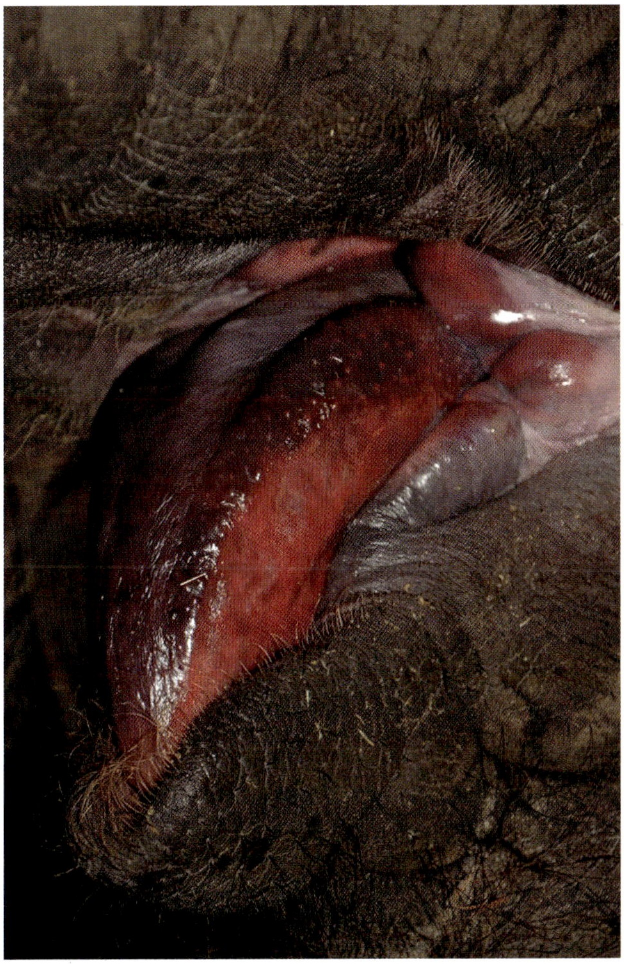

FIGURE 16.17 Herpesviral glossal hemorrhage and edema due to elephant endotheliotropic herpesvirus infection in an Asian elephant. The tongue is swollen, and the mucosa is discolored red to purple (cyanosis) indicative of interstitial congestion, hemorrhage, and edema due to virus-mediated endothelial damage. *(Photo Courtesy of J. Cracknell, International Zoo Veterinary Group.)*

and fibrinoid degeneration; and microthrombosis. Histologic lesion distribution generally parallels location of gross lesions. However, some variation in distribution is noted for the different recognized types of EEHV. Vascular lesions in EEHV1 infection are predominantly detected in the heart, tongue, and liver and may be restricted to capillaries. Other histologic changes include mild hepatocellular degeneration and oral/laryngeal mucosal ulceration or necrosis.[47,114,122] With EEHV3 infection, the distribution of affected tissues is broader and includes kidney, uvea, splenic capsule, and alimentary tract as well as heart and liver. Affected vessels include capillaries, venules, small veins, arterioles, and small arteries.[30,39] In EEHV4 cases, hemorrhage, edema, and mild inflammation affect the alimentary, respiratory, and cardiovascular systems.[47,77,148] Broad distribution and vascular disease are also seen with EEHV5 infection in which endothelial cell nuclear inclusions are seen in capillaries of heart, lung, liver, adrenal gland, and oral mucosa. Hepatic inflammation and necrosis are also features of EEHV5 infection.[5,22,162] Ultrastructurally, endothelial cell inclusions contain 70–90 nm, hexagonal herpesviral particles. Enveloped particles may be visible budding from the nuclear membrane.[27] Viral protein and nucleic acid have also been localized to endothelial cell nuclei in cases of EEHV-HD using immunohistochemistry and in situ hybridization assays.[18,48,62,63]

A combination of earlier listed findings in susceptible animals should strongly suggest EEHV-HD. Differential diagnoses for gross and histologic lesions in the absence of endothelial cell intranuclear inclusions or detectable EEHV protein or nucleic acid include bacterial sepsis, other systemic infections, or coagulopathies (e.g., toxins, autoimmune mediated, and parasitic). Infection with encephalomyocarditis virus (EMCV; see below) is a particularly important rule out due to the prominence of cardiovascular lesions in both infections. Notable differences between the two infections are more prominent myocardial degeneration and necrosis, less severe hemorrhage, and absence of viral inclusions in ECMV.

In clinically healthy, adult African elephants, EEHV2, EEHV3, EEHV6, and/or EEHV7 are associated with nonlethal, generally incidental cutaneous papillomas and nodular foci of lymphoid hyperplasia in lung and distal urogenital mucosa.[56,82,87,103,122,123,168]

Herpesvirus-associated lymphoid nodules in asymptomatic, clinically healthy, adult African elephants occur in the lung and/or distal urogenital tract mucosa. One or more of the EEHVs carried by African elephants (EEHV2, EEHV3, EEHV6, and EEHV7) may be present in the lesions. In the lung, nodules appear as discrete, up to 1–2 cm diameter, pale pink-tan, soft to firm foci (Fig. 16.19A). Urogenital mucosa lesions present as single to multiple, 2–10 mm diameter, raised, red nodules. Histologically, nodules are composed of multiple large lymphoid follicles with prominent central germinal centers (Fig. 16.19B). In pulmonary nodules, the lymphoid follicles surround syncytial epithelial cells containing eosinophilic nuclear inclusion bodies (Fig. 16.19C). No viral inclusions are present in urogenital mucosa nodules, however, herpesviral DNA is detectable via PCR.[87,103,122,123,168]

Herpesvirus-associated cutaneous papillomas, also noted in otherwise healthy, captive, and free-ranging African elephants, present as variably sized, often large, nodular, firm masses distributed predominantly on the trunk (Fig. 16.20A). Histologically, lesions are endophytic, inverted papillomas with branching and anastomosing trabeculae of hypertrophic and hyperplastic epithelial cells that extend into the dermis (Fig. 16.20B). Affected epithelial cells contain eosinophilic to amphophilic, herpesviral nuclear inclusions (Fig. 16.20C), and virions morphologically compatible with herpesvirus are evident via electron

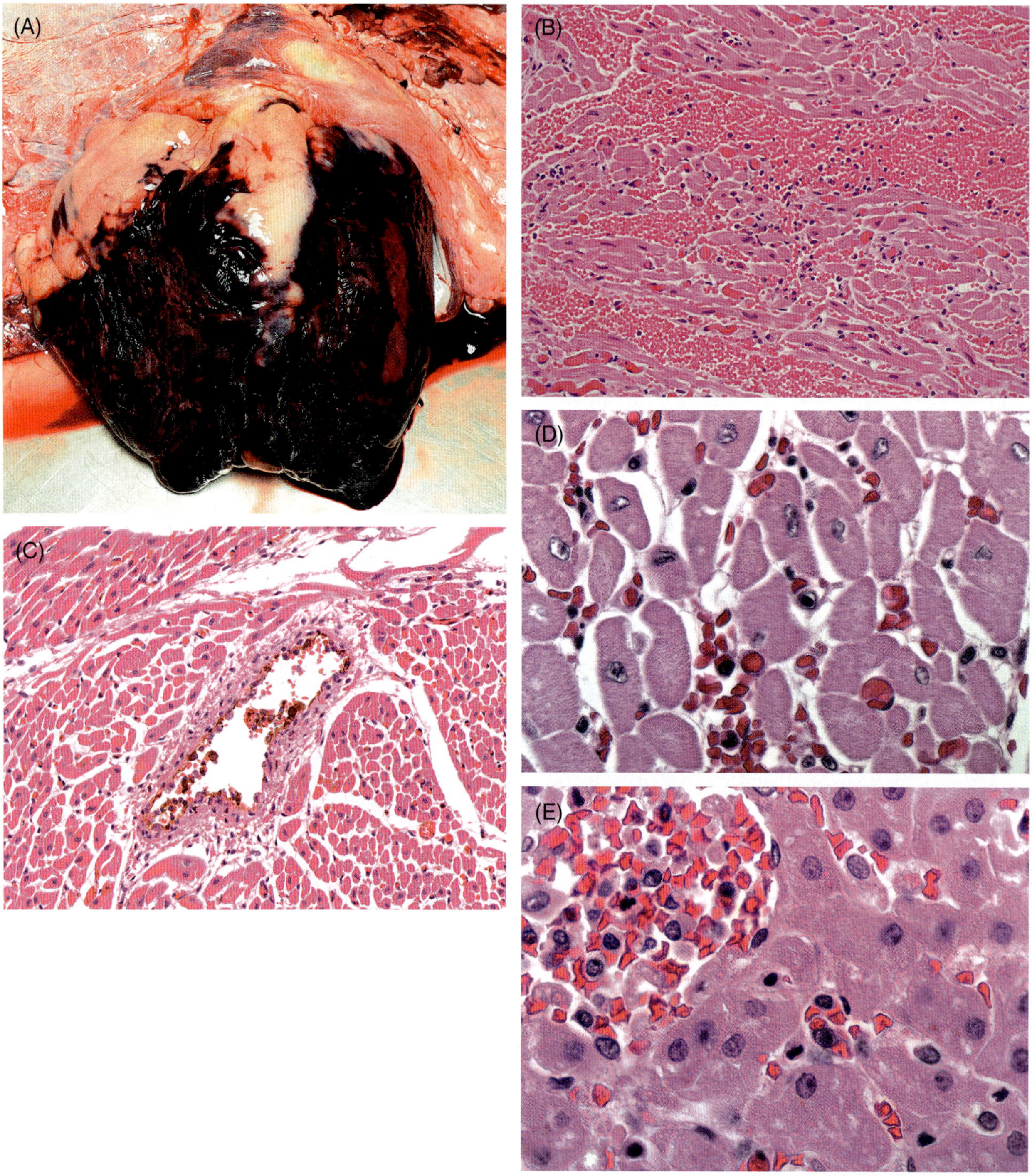

FIGURE 16.18 Elephant endotheliotropic herpesvirus hemorrhagic disease in an Asian elephant. (A) The epicardium is diffusely discolored dark red as a result of severe hemorrhage due to viral infection. (B) The myocardial interstitium is expanded by hemorrhage (see eSlide 16.e1). (C) Endothelial cells highlighted by brown immunolabeling for von Willebrand factor are enlarged and swollen with rounded nuclei containing viral inclusion bodies. Von Willebrand immunohistochemistry with HE counterstain. (D) Enlarged vascular endothelial cells in the myocardium have marginated nuclear chromatin and a discrete intranuclear eosinophilic inclusion (see eSlide 16.e1). (E) Few hepatic sinusoidal endothelial cells have similar enlarged nuclei with inclusions. *(Part A: Photo Courtesy of J. Cracknell, International Zoo Veterinary Group.)*

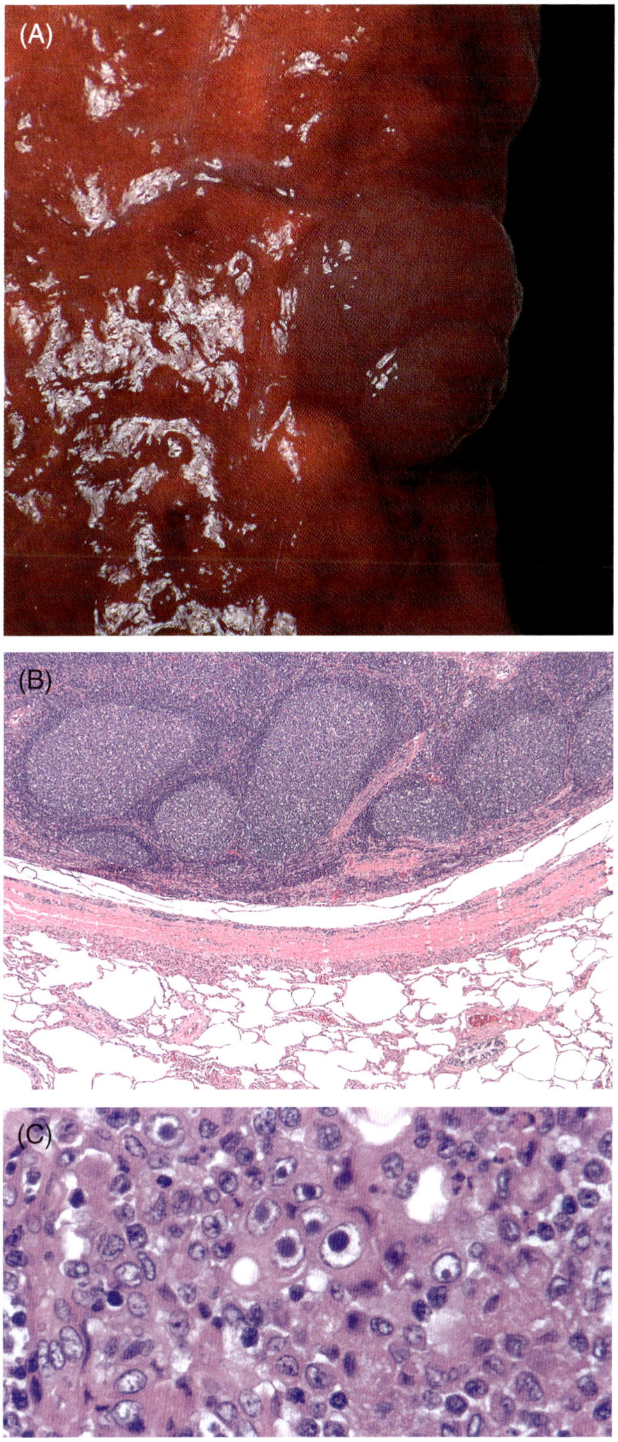

FIGURE 16.19 Elephant endotheliotropic herpesvirus-associated pulmonary lymphoid nodule in an African elephant. (A) The lung contains a discrete, soft, pink nodule with elevation of the pleura. (B) The parenchyma is focally replaced and displaced by a discrete nodular aggregate of lymphoid tissue with numerous follicles containing germinal centers. (C) Epithelial cells within pulmonary lymphoid nodules have eosinophilic, intranuclear herpesviral inclusions. *(Parts A–C: Photos Courtesy of Omaha's Henry Doorly Zoo and Aquarium.)*

microscopy.[56,82,122] With chronicity, many lesions become ulcerated with attendant dermal fibrosis and inflammation. Though EEHV DNA is present in the lesions, contribution of infection to pathogenesis is uncertain as elephant polyomavirus (AelPyV-1) and elephant gamma herpesvirus DNA are also detectable via PCR.[82,113,122] Lesions may be the result of coinfection with multiple viruses.

Elephant pox is a serious, contagious disease caused by an orthopoxvirus (*Poxviridae* family, *Orthopoxvirus* genus). Infections have primarily been reported in elephants in European zoos.[65,116] Infected rodents, which may represent the definitive host for the virus, serve as a reservoir for elephant infection. Transmission occurs via direct contact with infected rodents or contaminated fomites. Infection can be zoonotic and close contact with an infected elephant is an important risk factor. Manifestations of disease vary widely and range from mild local or generalized cutaneous infection to systemic, fatal illness. Virus infects epithelial cells of the skin and mucous membranes.[126] In mild cases, gross lesions are often distributed on the skin of the trunk, temples, perineum, and vulva. In more severe cases, lesions also develop in the mucous membranes of the conjunctiva, tongue, oral cavity, pharynx, esophagus, trachea, intestinal mucosa, and the pleural and peritoneal linings of the thorax and abdomen. Affected skin is covered by a few to multiple, 1–2 cm diameter, coalescing, gray-yellow papules with central reddening that progress to vesicles, pustules, and ulcerations. Ulcers are common on affected mucous membranes (Fig. 16.21). Toenail and slipper corneum may be affected. Lesions include tissue necrosis and sloughing (Fig. 16.22).[65] Histologically, lesions are typical of orthopoxvirus infection and characterized by epithelial hyperplasia with intracellular swelling (ballooning degeneration) and intracytoplasmic, eosinophilic inclusions. Epithelial necrosis and consequent ulceration occur with progression of disease. Diagnosis is supported by characteristic gross and histologic lesions with demonstration of epithelial intracytoplasmic inclusion bodies and/or virions typical of poxvirus infection via electron microscopy. Phylogenetic analysis can aid in establishing transmission chains.[66]

RNA Viruses

Encephalomyocarditis virus (EMCV) is a single-stranded RNA virus in the *Picornaviridae* family. Infection occurs in both captive and free-ranging Asian and African elephants.[46,67,128,136,157,160] Infection in elephants is most often associated with sudden death in the absence of promontory signs. EMCV is reported and causes encephalitis and myocarditis in a wide range of species. Rodents are believed to be the primary host and serve as a reservoir

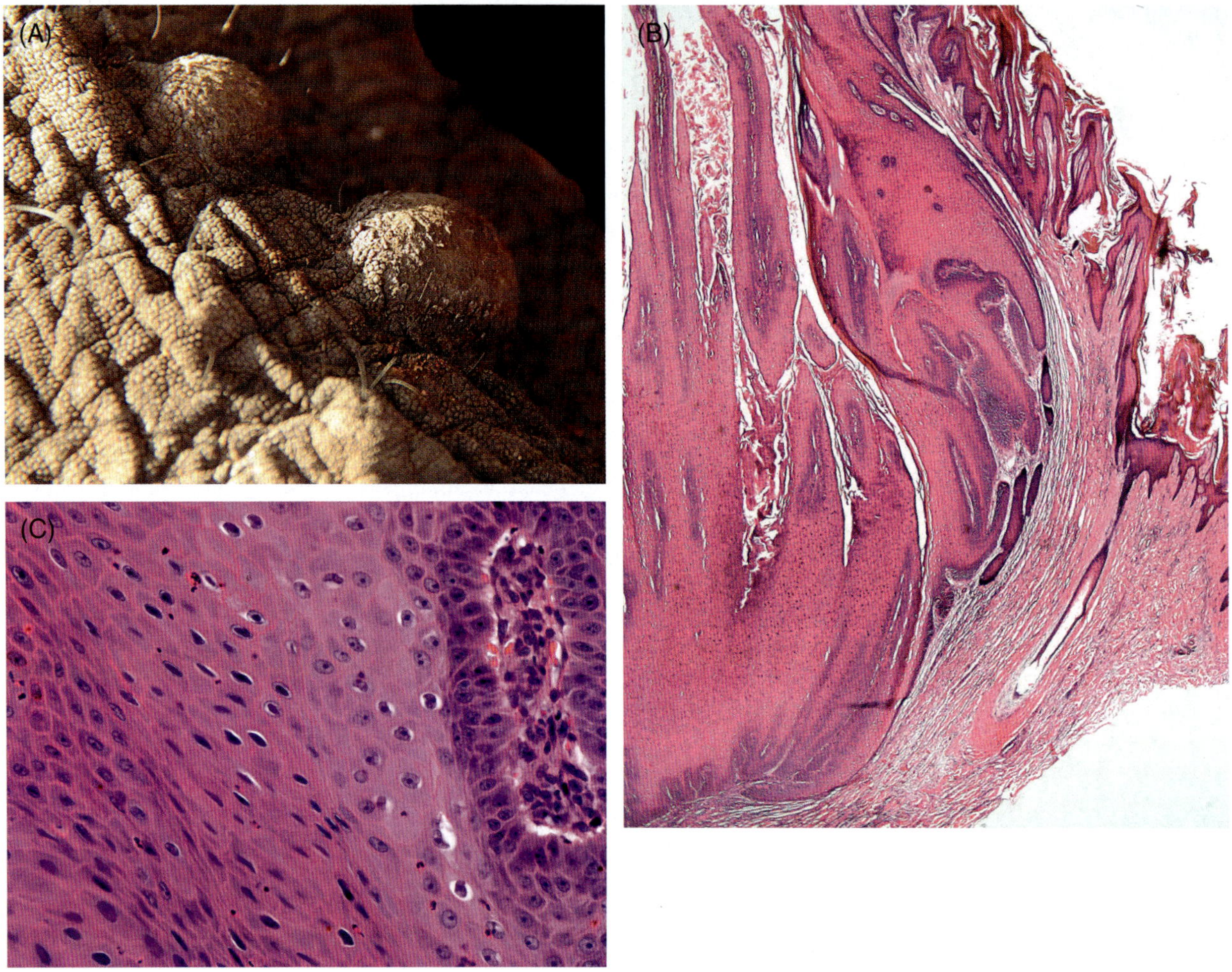

FIGURE 16.20 Elephant endotheliotropic herpesvirus-associated cutaneous papilloma in an African elephant. (A) Skin of the trunk is expanded by a discrete, firm, raised nodular mass. (B) Histologically, the mass is an endophytic, inverted papilloma with branching and anastomosing trabeculae of hypertrophic and hyperplastic epithelial cells that extend into the dermis. (C) Epithelial cells contain intranuclear herpesviral inclusions with chromatin margination. *(Parts A–C: Photos Courtesy of V. Pearson, Fox Chase Cancer Center.)*

of infection; infected rodent carcasses and feces facilitate transmission and spread.[128,157] The virus is believed to be principally cardiotropic, and infection results in acute myocarditis with cardiac dysfunction leading to death. Gross lesions in cases of clinical EMCV infection may include pericardial and/or peritoneal transudate effusion, epicardial and myocardial hemorrhage, multifocal myocardial pallor, and multicentric mucosal and serosal petechial hemorrhage (Figs. 16.23A, B). Multifocal myocardial degeneration and necrosis with hemorrhage and edema are noted histologically (Fig. 16.23C). Mild, lymphoplasmacytic to heterophilic myocarditis may also be present[67] (Fig. 16.23D). Given clinical signs and gross and histologic lesions are similar to those of EEHV-HD herpesvirus infection is a primary differential. Diagnosis requires identification of EMCV via PCR or virus isolation in tissues from affected elephants in concert with necrotizing myocarditis and related lesions.

Foot and Mouth Disease (**FMD**; also **Hoof and Mouth Disease**), though most commonly recognized as an infection of domestic, cloven-hoofed animals, can infect and cause disease in many nondomestic species including elephants. Due to its importance in agricultural industries around the world, FMD is a reportable disease locally, regionally, nationally, and internationally. Disease is caused by a single-stranded RNA virus (*Aphthovirus* genus) that is highly contagious, and transmission occurs via aerosol, direct contact, vectors, or exposure to contaminated feed or water.[50] Infection has been primarily documented in captive elephants though rare cases in free-ranging elephants exposed to infected ruminants have been reported.[16,54,120] Gross lesions in adult elephants are typically confined to the oronasal mucous membranes and feet. Multiple, thin-walled, clear fluid-filled vesicles may be present on the tongue, palate, cheeks, and mucosa of the trunk. Rupture of vesicles leads to mucosal ulceration. Skin

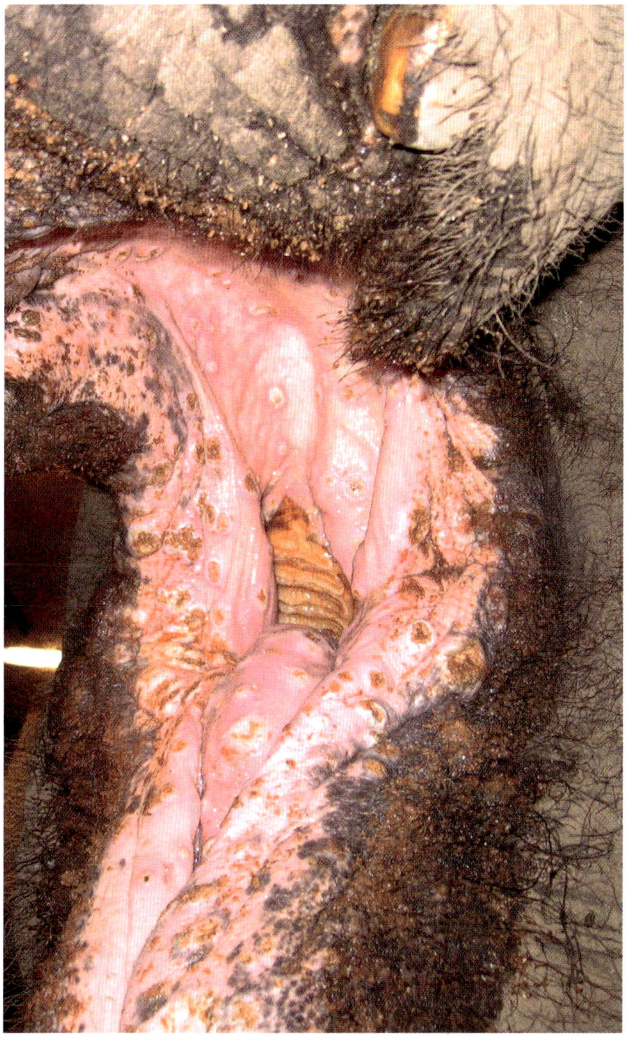

FIGURE 16.21 Poxviral stomatitis in an Asian elephant. The oral mucosa is covered by many discrete, raised, round, white to yellow foci typical of elephant poxviral infection. *(Photo Courtesy of H. Luten, Royal Burgers' Zoo, Arnhem.)*

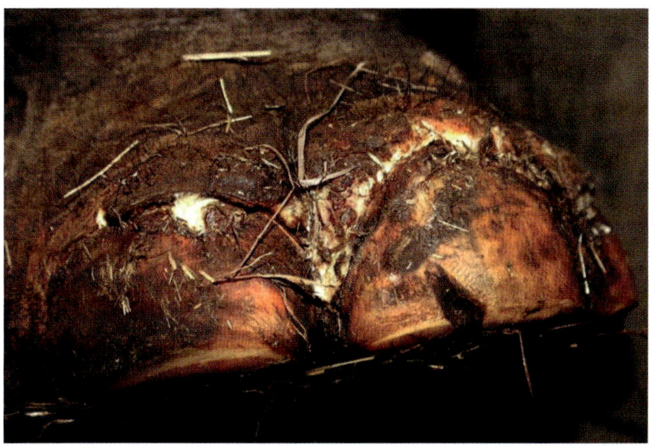

FIGURE 16.22 Poxviral pododermatitis an Asian elephant. The cuticular skin is cracked with irregular depressed defects (ulceration). Ulcerated areas are covered by dry granular, red-brown to yellow crusts. *(Photo Courtesy of H. Luten, Royal Burgers' Zoo, Arnhem.)*

around the toenails and slipper margins is swollen and reddened, and the slipper may slough with disease progression.[16,35,54] Histologically, intact vesicles consist of intraepithelial clefts filled with proteinic material and/or heterophils. Ruptured vesicles correspond to areas of ulceration with associated inflammation. Young elephants may have systemic infection with lymphocytic myocarditis and erosive enteritis.[138] FMD is the only vesicular disease to which elephants are susceptible. Because the vesicular stage of the disease may go easily undetected, FMD should be considered as a possible differential for any ulcerative or inflammatory lesions involving the oral cavity, trunk, and feet. FMD is a World Organization for Animal Health (WOAH) listed reportable disease, and suspected cases should be promptly reported to appropriate state and/or federal regulatory veterinary agencies.

Bacteria

Tuberculosis is one of the most important infectious diseases of captive and managed elephants worldwide.[4,31,81,92,95,110,141] Recent evidence has also confirmed infection in free-ranging animals in India.[17,95,96,107,165] Tuberculosis is a locally, nationally, and WOAH reportable disease, and elephant infections have significant public health implications. The majority of elephant infections are due to ***Mycobacterium tuberculosis*** (**Mtb**). Rare cases of disease with other *Mtb*-complex organisms, such as ***M. bovis*** and other nontuberculous mycobacteria, occur.[10,93,96,112] Humans are the natural and reservoir hosts for *M. tuberculosis*. Both zoonotic and anthropozoonotic infections have been reported in humans and elephants.[42,89,95,105,107,111,165,167] Tuberculosis occurs in both Asian and African elephants; however, disease is much more prevalent in Asian elephants.[35,99,100,110] Transmission is primarily via exposure to aerosolized infectious respiratory particles. In elephants, infection most often results in pulmonary disease with lungs and thoracic lymph nodes being the predominant tissues affected. Disseminated infection involving abdominal lymph nodes, the reproductive tract, and bone also occurs.[94]

Reasons for elephant susceptibility to this infection are unknown. Research suggests that altered host immunity likely plays a role in disease development and progression.[69,70] Tuberculosis in elephants is typically a chronic, subclinical disease. Many cases are diagnosed at the time of necropsy. Mycobacteria are acid-fast positive, obligate intracellular pathogens that infect and persist in host macrophages. Gram staining is variable though often positive. To survive in macrophages, the bacteria utilize intracellular survival strategies including inhibition of phagolysosomal fusion, evasion of macrophage-generated reactive oxygen and nitrogen free radicals, and induction of host-derived antiinflammatory mediators (e.g.,

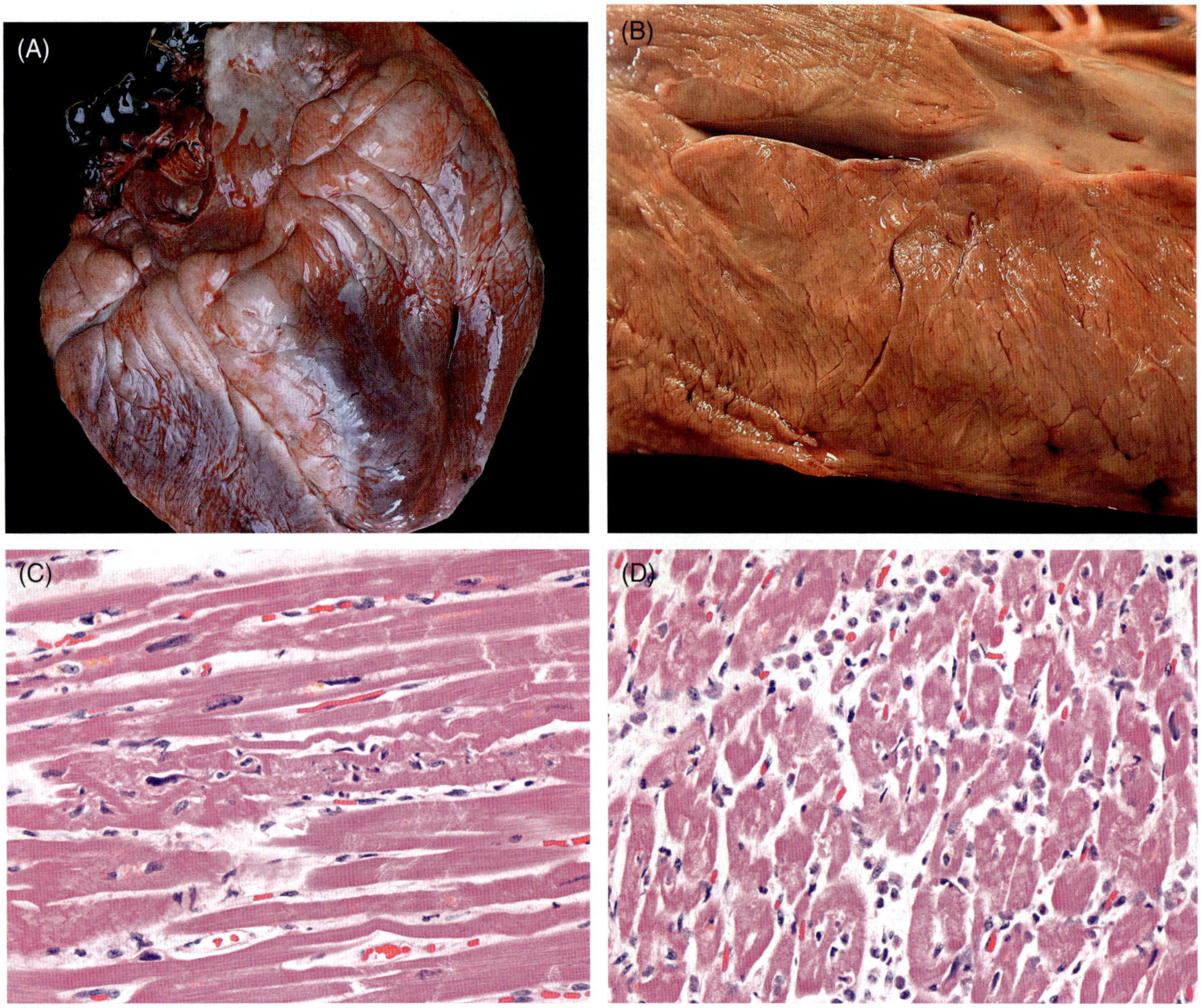

FIGURE 16.23 Encephalomyocarditis virus infection in the heart of an African elephant. (A) The myocardium is flaccid and multifocally mottled dark red to tan due to hemorrhage and myocardial degeneration secondary to viral infection. (B) Cut sections of myocardium have irregular pale tan streaking. (C) Cardiac myofibers have smudged, clumped, and fragmented sarcoplasm with loss or disruption of cross striations indicative of necrosis. Necrotic myofibers are infiltrated with several macrophages and other necrotic leukocytes. (D) Myocardial interstitium is expanded with small amounts of colorless space (edema) and has low numbers of heterophils, macrophages, and lymphocytes. *(Part A: Photo Courtesy of B. Lamglait, Réserve Africaine de Sigean; Parts B–D: Photos Courtesy of Disney's Animal Kingdom.)*

interleukin-10). These mechanisms allow mycobacteria to avoid immune surveillance and establish infection in susceptible individuals. Infected elephants have been shown to have alterations in hepatic iron-regulating hormones (i.e., hepcidin, ferroportin). Iron sequestration that occurs in macrophages as a component of critical innate immune mechanisms is inhibited via depression of hepcidin and elevation of hepatic ferroportin in infected elephants. These hormone alterations appear to counteract iron restriction in infected elephants to provide iron for erythropoiesis while decreasing iron availability for mycobacterial replication.[129] Both innate immune mechanisms and the character of the subsequent adaptive immune response influence disease progression and outcome. Thus, mycobacterial infection in elephants may result in active and possibly disseminated disease or more commonly localized, clinically silent latent disease.[71]

Gross lesions in the lung and thoracic lymph nodes manifest as few, isolated, or many variably sized, possibly coalescing, firm, yellow-white nodular granulomas. Tracheal granulomas can be present. Granulomatous inflammation associated with active disease is often extensive and characterized by coalescing, caseous to cavitated nodules resulting in pulmonary consolidation (Fig. 16.24A). In addition to the pulmonary parenchyma, inflammation may involve adjacent airways and produce luminal mucoid

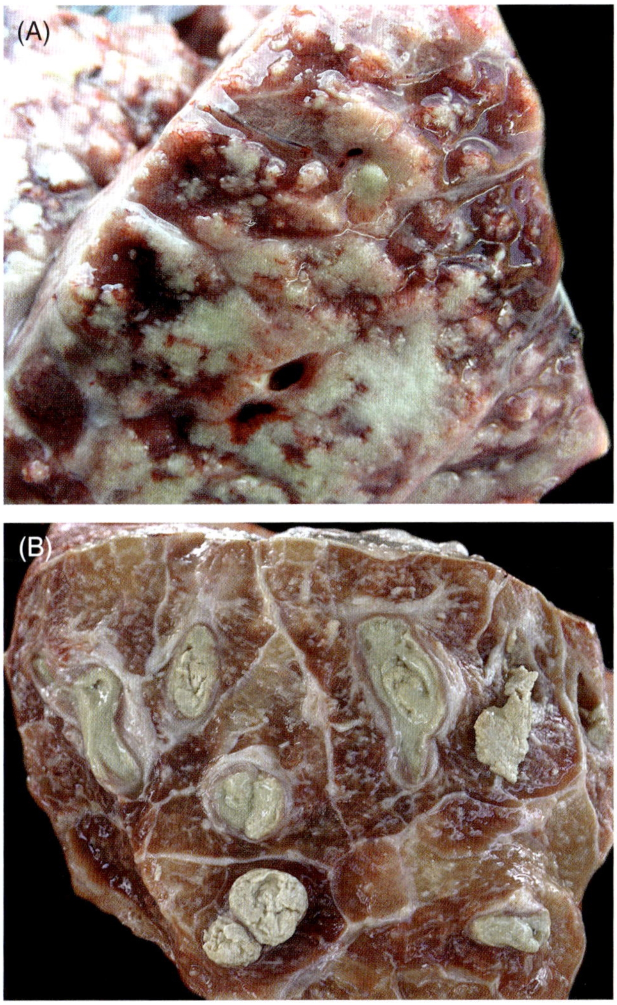

FIGURE 16.24 Pulmonary tuberculosis in an Asian elephant. (A) The pulmonary parenchyma is obscured by numerous, coalescing, roughly nodular, firm, yellow-white, caseous foci of granulomatous inflammation with distribution and character compatible with "active disease." (B) The lung contains several discrete granulomas with a core of caseous to mucoid exudate that are surrounded by a tough white fibrous wall consistent with "latent" *Mycobacterium tuberculosis* infection. *(Part A: Photo Courtesy of S. Mikota, Elephant Care International; Part B: Photo Courtesy of S. Citino, White Oak Conservation.)*

to mucopurulent exudate.[35,99,100] Similar granulomatous lesions may be present in other affected nonthoracic tissues.[25] Extrapulmonary lesions may occur alone or with concurrent thoracic disease. Few isolated, small, chronic granulomas may be indicative of latent disease (Fig. 16.24B). These granulomas are often firm due to fibrosis and central mineralization.

Histologically, the degree of organization associated with the granulomatous inflammation varies widely from case to case. Lesions compatible with both latent and active disease, as recognized in human and nonhuman primate *M. tuberculosis* infections, can be seen in elephant lungs. Latent granulomas are discrete collections of central macrophages, giant cells, and scant cellular and abundant mineralized debris surrounded by concentric layers of mature fibrous connective tissue (Figs. 16.25A, B; eSlide 16.e2). Walls of these granulomas typically contain moderate to high numbers of CD3 positive lymphocytes. Active granulomatous inflammation is poorly organized and characterized by large central areas of necrosis and some mineralization surrounded by low to moderate numbers of epithelioid macrophages and rare multinucleated giant cells (Figs. 16.26A, B; eSlide 16.e3). Lymphocyte numbers within active lesions vary, but most lymphocytes are CD20 positive B cells. In both latent and active cases, granulomas contain few to very rare acid-fast positive bacilli.[71]

Probable infection can be inferred given compatible gross and histologic lesions with intralesional acid-fast bacilli. For definitive diagnosis and to rule out other nontuberculous mycobacteria, demonstration of *M. tuberculosis* organisms or DNA within granulomatous lesions via culture and/or PCR of affected fresh tissue is required. Given the potential for zoonosis and serious human disease, elephant necropsy procedures should include proper precautions.[89,105,107,167] Appropriate respiratory and other personal protective equipment should be utilized by all involved/exposed staff, at least until the thorax is determined to be free of disease by experienced veterinary professionals supervising the procedure.[20]

Salmonellosis in elephants presents primarily as enteritis though septicemic infection is possible. Reported cases have been seen in animals managed in human care, and most infections occur in association with episodes of immunosuppression secondary to environmental and biological stressors, such as overcrowding, transport, or parturition.[35,57,80,83] Salmonella are motile, gram-negative bacilli found in the gastrointestinal tracts of a wide variety of both poikilothermic and homeothermic species, where they may be commensal and part of the normal flora. Development of disease in a new susceptible host depends on exposure and infectious dose, host resistance to colonization of the gastrointestinal tract, and *Salmonella* serovar.[101] Cases of elephant salmonellosis have included infection with ***S. enterica serovars Typhimurium***, ***Dublin***, ***Enteriditis***, ***Montevideo***, and ***Muenchen***.[35,80] Pathogenic infection results in damage to ileal and proximal colonic enterocytes and/or binding to intestinal microfold cells overlying gut-associated lymphoid tissue (GALT) that facilitates infection of tissue macrophages. Infection limited to the intestine by the host immune response results in enteritis. *Salmonella* serovars capable of surviving and proliferating in macrophages can disseminate and cause systemic disease.[101]

Grossly, distal small intestinal and/or colonic mucosa is reddened or discolored and overlain by variable amounts of adherent, friable, tan-yellow to green-black, fetid material

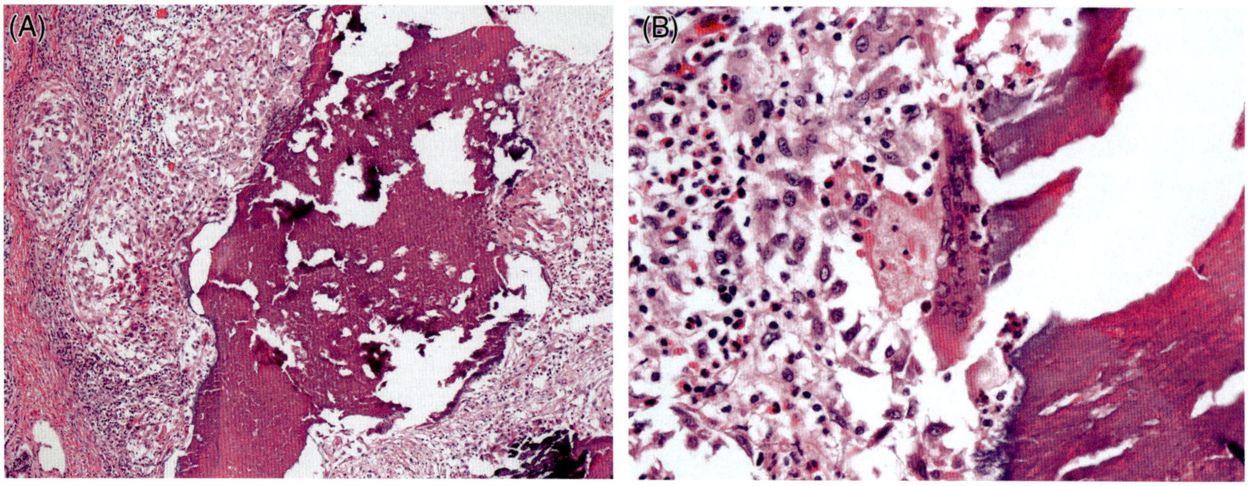

FIGURE 16.25 "Latent" pulmonary tuberculosis in an Asian elephant. (A) Inflammation is characterized by discrete granulomas centered on cores of cellular and mineralized debris surrounded by several concentric layers of mature fibrous connective tissue with many lymphocytes. (B) Granulomatous inflammation contains many large epithelioid macrophages, fewer multinucleated giant cells, and variable numbers of other inflammatory cells (see eSlide 16.e2).

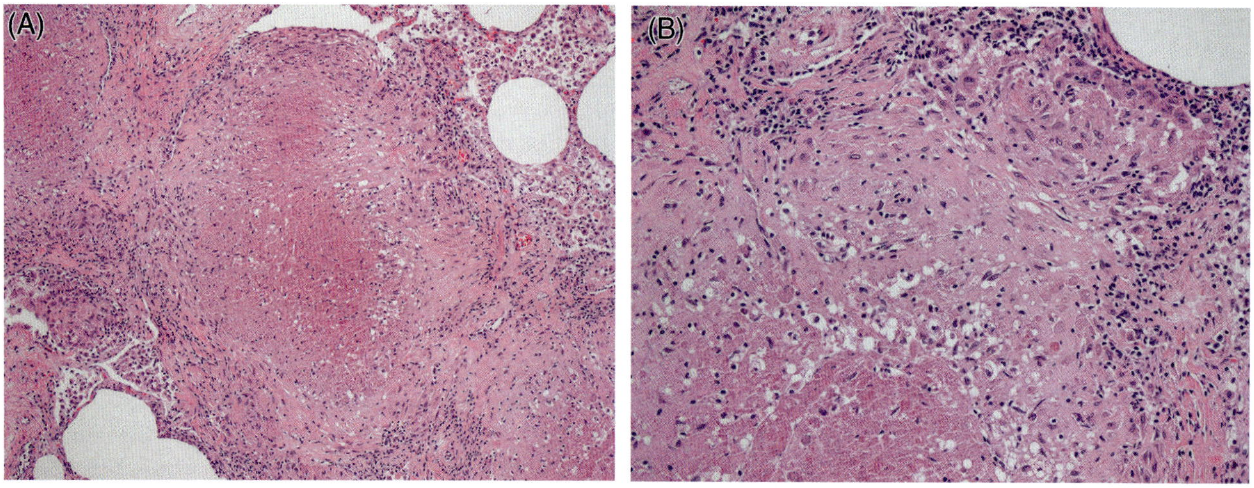

FIGURE 16.26 "Active" pulmonary tuberculosis in an Asian elephant. (A) In characteristic lesions, granulomatous inflammation is poorly organized and characterized by large central areas of necrosis surrounded by relatively small amounts of granulomatous inflammation and minimal fibrosis. (B) Inflammation is composed predominantly of macrophages, some of which are epithelioid, rare multinucleated giant cells, and variable numbers of lymphocytes (see eSlide 16.e3).

(Fig. 16.27). In cases of systemic infection, affected viscera, such as the liver and spleen may contain multiple, pinpoint to larger, pale tan-yellow foci of necrosis and inflammation.[83,163] Histologically, there is multifocal to coalescing erosion and ulceration of the mucosa overlying GALT. Mucosal defects are overlain by adherent fibrin, cell debris, and degenerate heterophils admixed with bacteria.[80] GALT contains pyogranulomatous inflammation with intrahistiocytic small, gram-negative bacilli. Visceral lesions in cases of sepsis consist of randomly distributed foci of necrosuppurative inflammation. Diagnosis is confirmed based on the combination of gross and histologic lesions and *Salmonella* positive bacterial culture of feces and/or other infected tissue.

Elephants are susceptible to **anthrax**, a WOAH listed reportable, zoonotic, and highly fatal disease. Infection is caused by the aerobic, gram-positive, endospore-forming bacterium ***Bacillus anthracis***.[40] Most cases have occurred in free-ranging elephants though rare zoo infections have been documented.[21,35,106] Epizootics are seasonal and most often associated with hot, dry weather following heavy rains or floods. Routes of infection may include respiratory aerosols, ingestion, or wound contamination. Macrophages phagocytose vegetative or spore forms of the organism. Bacterial replication results in macrophage lysis and dissemination. *B. anthracis* produces a potent exotoxin that causes increased capillary permeability and coagulopathy. Disease is typically peracute or acute with no promontory

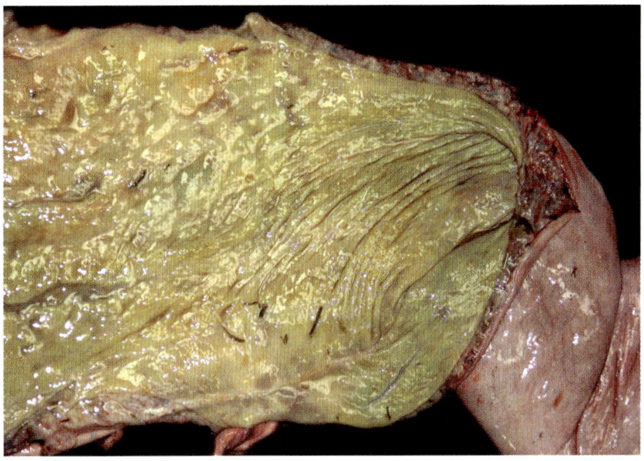

FIGURE 16.27 Salmonellosis in the small intestine of an Asian elephant. Characteristic lesions include necrotizing inflammation in which the intestinal mucosa is overlain by variable amounts of adherent, friable, tan-yellow material (fibrinonecrtoic pseudomembrane). *(Photo Courtesy of F. Uzal, University of California, Davis William R. Pritchard Veterinary Medical Teaching Hospital.)*

signs.[40] As in other mammals, lesions in elephants reflect toxin-mediated vasculopathy. Hemorrhagic fluid exuding from body orifices is a characteristic feature of infection. Subcutaneous tissues, especially of the head and neck, have marked edema (see Felidae). Blood is dark, thick, and clots poorly. Widespread hemorrhage is evident in the heart and on serosal and mucosal surfaces of the gastrointestinal tract. The spleen is enlarged, dark, and meaty. Anthrax exotoxin inhibits rigor mortis thus the carcass may be flaccid or have delayed rigor.[21,40,106] A history of peracute death and compatible appearance of the carcass should suggest possible anthrax infection. Such carcasses should not be opened to prevent contamination of the environment and exposure of humans and other animals. Blood smears from infected animals contain high numbers of sporulated, gram-positive bacilli (see Wildlife Necropsy). Histological lesions can include splenic reticuloendothelial necrosis, endothelial cell necrosis, and hemorrhagic lymphadenitis. A diagnosis of anthrax is presumed given compatible history, carcass presentation, and detection of sporulated bacilli in blood smears.

Clostridial species rarely cause disease in elephants. Infection with ***Clostridium septicum*** may result in fatal toxemia similar to that seen in domestic species (e.g., cattle, horses, small ruminants). Bacterial contamination of cutaneous wounds results in infection and subsequent toxemia. In addition to necrotizing myositis at the site of primary infection, lesions include widespread visceral edema and hemorrhage. Given the character of the visceral lesions and because some clostridial species are spore formers, anthrax would be an important differential and must be ruled out. Both ***C. perfringens*** and ***C. difficile*** infection may result in enteritis that ranges from hemorrhagic to fibrinonecrotic, and associated enterotoxemia can be fatal.[24,134] Fatal coinfection with *C. perfringens* and EEHV4 in an adult Asian elephant has been reported and was characterized by multifocal necroulcerative enterocolitis and multicentric hemorrhage. The initial/primary infection could not be confirmed; however, clostridial enteritis may have led to recrudescence of latent EEHV4 infection with EEHV-HD.[19] ***C. tetani*** is infrequently reported in elephants. Infection is associated with sudden death in the absence of lesions, like cases of tetanus in other species.[24,134]

Parasites

Information related to several parasites reported in elephants is presented in Table 16.e1.

ADDITIONAL RESOURCES

Fowler, M.E., 2006. Foot disorders. In: Fowler, M.E., Mikota, S.K. (Eds.), Biology, Medicine, and Surgery of Elephants. first ed. Blackwell Publishing, Ames, pp. 271–290.

Guntawang, T., Sittisak, T., Kochagul, V., Srivorakul, S., Photichai, K., Boosri, K., Janyamethakul, T., Boonprasert, K., Langkaphin, W., Thitaram, C., Pringproa, K. 2021. Pathogenesis of hemorrhagic disease caused by elephant endotheliotropic herpesvirus (EEHV) in Asian elephants (*Elephas maximus*). Sci. Rep. 11, 12998.

References

Full references for this chapter can be found online at https://doi.org/10.1016/B978-0-443-18437-6.00011-0.

Chapter 17

Perissodactyla

Mary Duncan[a], Patricia M. Gaffney[b,*], and Pedro Enrique Navas-Suárez[c]
[a]*Saint Louis Zoo, Saint Louis, MO, United States;* [b]*San Diego Zoo Wildlife Alliance, San Diego, CA, United States;* [c]*University of São Paulo, São Paulo, Brazil*

INTRODUCTION

Perissodactyla are hoofed ungulates with an odd number of digits. Body weight is carried on the central digit of the limb. The group includes the *Equidae*, *Tapiridae*, and *Rhinoceridae* families. The taxonomy and conservation status of these groups are listed in the Digital Content (Table 17.e1). Likewise, details of unique anatomic features, including a list of comparative dentition (Table 17.e2) and related images (Figs. 17.e1–17.e5), and lists of reported neoplasms (Table 17.e3), serologically detected pathogens (Table 17.e4), endoparasites (Table 17.e5), and ectoparasites (Table 17.e6) that expand descriptions of diseases that follow below, can be found in the Digital Content.

NON-INFECTIOUS DISEASES

Nutritional

Iron overload disease occurs in captive black and Sumatran rhinoceros, but not free-ranging or recently collected rhinoceros.[74,108,136] The mechanisms underlying iron overload in susceptible perissodactyls are currently not well understood, nor is the precise significance of the condition to overall health. Iron overload continues to be an area of active research and the pathogenesis is likely multifactorial. **Hemosiderosis** is the presence of iron in a tissue, while **hemochromatosis** indicates high levels have altered histologic structure.

After absorption by duodenal enterocytes, iron is exported from the cells into the circulation (transferrin) by ferroportin. Ferroportin production is controlled by the hormone hepcidin, which is secreted by hepatocytes. Increased serum iron and the presence of interleukin-6 (proinflammatory cytokine) cause hepcidin release. Circulating hepcidin binds ferroportin and prevents iron uptake. When hepcidin levels are high, iron remains in enterocytes and is shed in feces. Hepcidin production is increased under certain conditions, such as inflammation from infection or autoimmune disease.[42,141] Hepcidin production is down regulated in iron deficiency and with increased erythropoiesis to allow for increased iron absorption. Hepcidin deficiency leads to increased iron absorption and subsequent deposition in hepatocytes, cardiomyocytes, and endocrine glands.

Hemochromatosis, or primary iron overload, in humans is generally a heritable disorder. Mutation of the hepcidin or hemojuvelin genes results in earlier iron deposition in the myocardium or endocrine organs, while mutation of the transferrin receptor-2, ferroportin, or hemochromatosis genes results in milder later onset accumulation of iron, particularly in the liver. The hemochromatosis gene (*HFE*) encodes a protein that complexes with the transferrin receptor; abnormal gene expression leads to increased iron absorption by enterocytes. The HFE protein with transferrin receptor-2 protein can also affect hepcidin expression. Hemochromatosis gene nucleotide mutations have been identified across rhinoceros species, but a protein polymorphism has been found only in the black and not the white, greater one-horned or Sumatran rhinoceros. The S88T protein polymorphism is in the portion of the HFE protein that interacts with the transferrin receptor and may have aided iron uptake in the black rhinoceros with a low iron diet.[9,133]

Black and Sumatran rhinoceroses are browsers while white and greater one-horned rhinoceroses are grazers. The diet of wild black rhinoceroses is rich in compounds that chelate iron (tannins, phytates, polyphenolic compounds), forming insoluble complexes that pass through the gastrointestinal tract unabsorbed.[136] In captivity, lower levels of dietary iron-binding compounds may increase the amount of dietary iron that is available for absorption. Ascorbate (vitamin C) increases iron absorption, so ascorbate rich produce can increase iron uptake. The bioavailability of iron cannot be predicted simply from the measurement of total dietary iron since absorption is affected by other dietary elements (e.g., tannins, vitamin C).[170] The black rhinoceros appears genetically adapted for iron absorption from a naturally low iron diet, and overload

*Deceased.

Pathology of Wildlife and Zoo Animals. https://doi.org/10.1016/B978-0-443-18437-6.00019-5

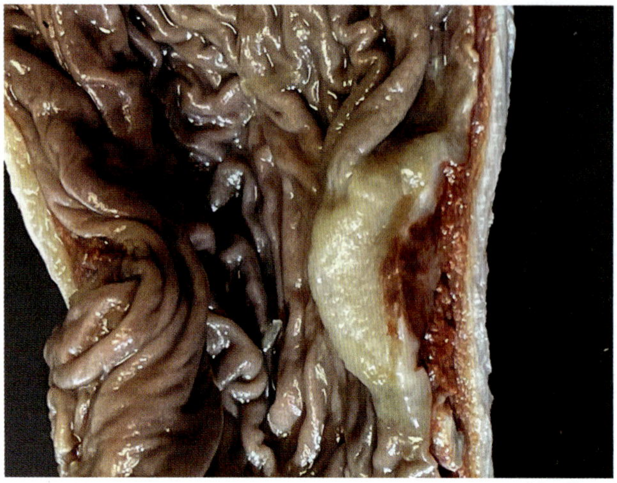

FIGURE 17.1 Hemosiderosis (iron overload) in the duodenum of a black rhinoceros. Iron accumulation appears grossly as red/brown deposits in the submucosa, particularly in the region of the common bile duct opening. This corresponds histologically to hemosiderin-laden macrophages (see Fig. 17.2; eSlides 17.e1 and 17.e2).

develops with loss of homeostasis in captivity when the diet has high iron bioavailability.[9,42]

In captivity, iron accumulation is typically seen in hepatocytes and histiocytes of the spleen, liver, small intestine (Fig. 17.1; eSlides 17.e1 and 17.e2), and lungs of black rhinoceros, the histiocytes of the spleen, and other organs in greater one-horned rhinos, and the histiocytes of the intestines, liver, and lung in white rhinos.[133] The presence of hemosiderin in histiocytes is consistent with secondary iron overload.[44] The captive white and greater one-horned rhinoceros do not demonstrate ferritin and tissue iron elevations suggestive of iron storage.[105] In hepatic hemochromatosis (iron accumulation associated with cellular degeneration, necrosis, and fibrosis) both parenchymal and reticuloendothelial cell lines are affected and cellular and tissue structure is altered. Initially, hemosiderin is deposited in periportal hepatocytes but over time increased iron staining occurs throughout the hepatic lobule and is prominent in both hepatocytes and Kupffer cells (Figs. 17.2A–D; eSlides 17.e1 and 17.e2); in hemochromatosis, periportal fibrosis develops (Figs. 17.2C, D). Iron accumulation in the bone marrow may cause myelodysplasia; marrow is hypocellular and is often fibrotic with sheets of hemosiderin-laden macrophages.[170]

The ability to assess and monitor the development of iron overload in living animals has been an ongoing quest. Serum haptoglobin levels are similar in both white and black rhinoceros suggesting iron accumulation in the black rhinoceros is unlikely secondary to hemolysis. Hepatic nonheme iron in black rhinoceroses in captivity has been shown to be significantly higher than that in white rhinoceroses, and the level in black rhinoceroses increased with age.[160,170] Measurement of serum ferritin using an enzyme-linked immunoabsorbent assay (ELISA) with antibodies to equine ferritin, which cross-react with rhinoceros ferritin, is thought to be a good estimate of total body iron stores in rhinoceros.[170] A Sumatran rhinoceros species-specific ferritin assay was developed due to inconsistent results with the equine assay. Ferritin concentration was found to vary between individuals and did not correlate well with iron overload or illness.[161] The Sumatran rhinoceros ferritin assay was subsequently validated for use in black rhinoceros. In one study, as in Sumatran rhinoceros, serum ferritin concentrations varied between individual black rhinos, and there were no correlations between serum ferritin and sex, season, age, or length of time in captivity. Ferritin is an acute-phase protein that is released in response to inflammatory and neoplastic processes; it is not specific to iron overload disease. The usefulness of serum ferritin to monitor iron overload in browsing rhinoceros appears limited.[191] Transferrin iron binding capacity has been used as a marker of body iron status, but accurate measurement has been difficult in rhinoceros.[151] Nontransferrin-bound labile plasma iron has also been assessed as a potential biomarker of iron overload, but while all Sumatran rhinos positive for labile plasma iron showed clinical signs of overload, positive black, white, and greater one-horned rhinoceros appeared healthy.

Iron overload is associated with a variety of diseases. Free iron catalyzes the production of hydroxyl free radicals, and iron overload disease may therefore predispose black rhinoceros to hemolytic anemia, mucocutaneous ulcerative diseases, and stress intolerance.[136,170] Iron accumulation disrupts physiological functions and increases susceptibility to many infectious diseases. At normal levels, iron is captured by the host and bound to protein, which deprives pathogens of this essential cation. However, organisms such as mycobacteria thrive in iron-rich environments and leukocyte phagocytic and bactericidal properties are compromised with iron excess. The influence of iron overload in infectious disease is further complicated by the form of overload (primary or secondary) that is present, with histiocytes as the main iron storage cell in secondary overload.[69,160]

Further research is needed into iron overload in tapirs as the topic has been studied less extensively. Captive tapirs develop higher serum ferritin and increased transferrin saturation (lowland > Baird's > Malayan) relative to free-ranging Baird's tapirs.[12,137,146] Hemosiderosis has not been described in free-ranging lowland tapirs.[124]

Microcytic, hypochromic iron deficiency anemia has been reported in captive Malayan tapir neonates from two institutions.[52] Both had unremarkable neonatal examinations and blood parameters but developed anemia by 1 to 2 months of age. Notable hematologic findings were anisocytosis, polychromasia, and hypochromia with acanthocytes, schistocytes, and poikilocytes. Stores of iron are

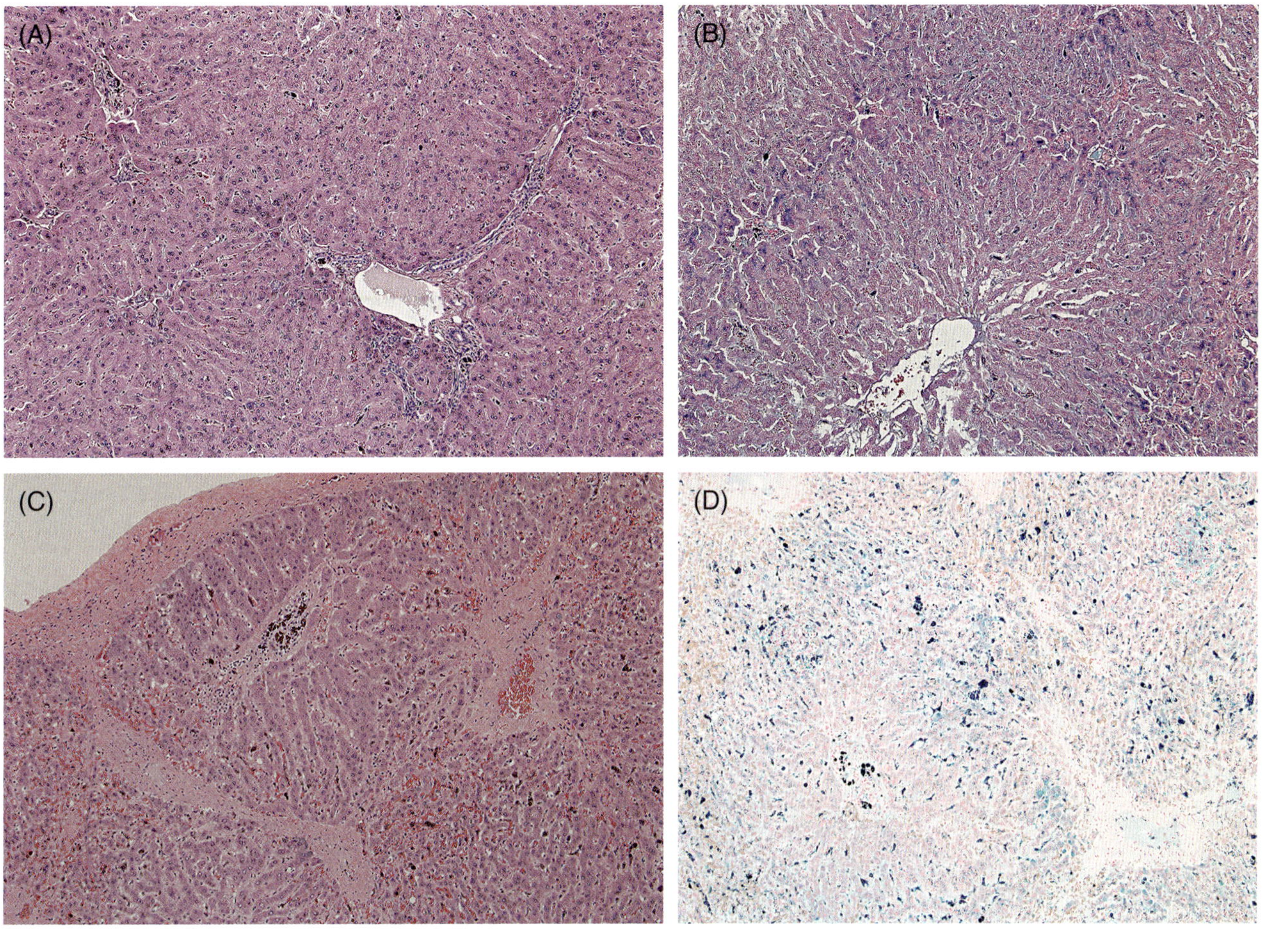

FIGURE 17.2 Mild hemosiderosis (iron overload) (A, B) and mild hemochromatosis (C, D) in the livers of black rhinoceros. (A) Mild iron accumulation appears as granular brown intracytoplasmic pigment in intrasinusoidal Kupffer cells. (B) Intracellular iron is highlighted with special stains. Perl's iron/Prussian blue. (C) Moderate iron accumulation is associated with increased numbers of periportal, pigment-laden Kupffer cells and periportal and perivascular fibrosis. (D) With special staining, the accumulation of iron can be seen to be more widespread in mild hemochromatosis than mild hemosiderosis. Perl's iron/Prussian blue (see Fig. 17.1; eSlides 17.e1 and 17.e2).

low at birth, milk has low iron content, growth is occurring rapidly, and limited access to iron sources, such as soil, may all be factors in the development of iron deficiency in these young animals.

Copper deficiency has been diagnosed in tapirs.[60,61] Although the diet may appear to have adequate levels, interaction between copper and other trace elements, such as iron, zinc, sulfur, or molybdenum, may be responsible for low bioavailability. Stillbirths and light haircoat, seen in other species with copper deficiency, have been seen in tapirs with low serum copper levels.

Hypovitaminosis D is of interest across multiple species. This fat-soluble vitamin is involved in many metabolic pathways and disease processes. Circulating 25-hydroxyvitamin D is the best indicator of vitamin D in the body. Vitamin D_3 is at least 3 times more effective than vitamin D_2 in maintaining 25-hydroxyvitamin D values. Ergocalciferol (vitamin D_2) comes from the diet, plant material. Cholecalciferol (vitamin D_3) is endogenously produced in the skin on exposure to ultraviolet B radiation. Levels vary with distance from the equator and levels of sunlight and are reduced by barriers such as glass. Relative to free-ranging black rhinoceros (Zimbabwe), two individuals in managed care in North America had low circulating 25-hydroxyvitamin D. Exposure to sunlight was shown to be highly significant, despite dietary supplementation. Of note is that in other hindgut fermenters, bone growth and calcium homeostasis appear to occur independently of vitamin D.[132]

Vitamin E (alpha-tocopherol) is important in maintaining membrane integrity relative to oxidant stress. **Hypovitaminosis E** as a cause of hemolysis has been reported in primates, rats, and horses. It also causes low bile salt secretion with reduced solubilization of dietary fat. It has been implicated in **hemolytic anemia** and **encephalomalacia** in black rhinoceros and **myopathy** in a Brazilian tapir.[193] Free-ranging black rhinoceros are reported to have significantly higher plasma vitamin E than those in managed care. The lower levels in the latter are suggested to be due to lack of access to fresh browse/grazing.[23] One study showed poor efficacy of absorption of vitamin E with most dietary supplement formulations, which may also contribute to lower vitamin E in managed care settings.[142] Grazers and browsers may have different dietary requirements for vitamin E. Higher vitamin E has been measured in the leaves relative to the twigs of the same browse plants in Zimbabwe.[22] Vitamin E increases with chlorophyll degradation during plant maturation and would be expected to be higher in photosynthesizing leaves.

Degenerative myelopathy in captive Przewalski's horses is associated with hypovitaminosis E due to low dietary alpha-tocopherol in feed.[86] Clinical presentation varies from mild ataxia to wide-based gait and stance, and uncoordinated movements of the hind limbs. Gross or histologic lesions in cardiac or skeletal muscle are not noted. The most significant lesions are degeneration, demyelination, and astrogliosis of the ventral and lateral funiculi in the caudal cervical and cranial thoracic spinal cord. Axons in the gray matter throughout the cord are swollen and degenerate. Plasma alpha-tocopherol concentration in affected horses is less than 0.03–0.08 mg/dL (normal >0.5 mg/dL; < 0.3 mg/dL is considered deficient). Similar demyelinating lesions are seen in humans with vitamin E malabsorption due to chronic liver disease and laboratory animals fed vitamin E-deficient diets.

Metabolic

Exertional rhabdomyolysis (capture myopathy) is a potential problem in all perissodactyls but the condition is of greatest concern during and in association with capture and translocation of rhinoceros.[97,149] Lesions are consistent with those seen in other species (see Bovidae, Antilocapridae, Giraffidae, Tragulidae, Hippopotamidae and Cervidae).

Toxic

A number of chemical and biological toxins have been reported in nondomestic Perissodactyls. Most are single case reports or series (Table 17.1). In one notable toxicological study of free-living lowland tapirs in Brazil, carbamates (aldicarb, carbaryl), organophosphates (diazinon, malathion, mevinphos), pyrethroids (deltamethrin, permethrin), and toxic metals (cadmium, lead) were detected in a variety of tissues, but no significant effects were linked to these findings.[99]

Congenital/Genetic

Adenomatous hyperplasia of the placenta (Fig. 17.3) has been seen in greater one-horned rhinoceros. Nodules are composed of variably sized glandular or cystic structures lined by plump cuboidal to low columnar epithelium or epithelial fronds with various amounts of fibrous stroma. The change has not been associated with placentitis or abortion.

Familial degenerative myelopathy was reported in 8 of 17 Plains zebra foals, the progeny of one stallion and two mares.[116] Foals developed progressive ataxia between 4 and 7 months of age. This was not associated with radiographic or postmortem findings to indicate narrowing of the vertebral canal which can result in spinal cord damage. Histologically, degeneration of ascending and descending tracts with bilaterally symmetrical demyelination was found in the spinal cord at the lateral and ventral funiculi. The entire cords were examined in two zebra and lesions were seen in the cervical, thoracic, lumbar, and sacral regions. Lesions also extended cranially to the medulla oblongata. Small blood vessels in the lesions showed fibrosis in the adventitia. No histologic lesions were found in the brains.

Skeletal, cardiac, and neurologic congenital defects have been reported in nondomestic perissodactyls. A **schistosomus reflexus-like malformation** was reported in a white rhino calf that was aborted at approximately 13 months of gestation.[83] The fetus exhibited spinal inversion and a ventral midline abdominal defect with externalization of the abdominal viscera. Computed tomography revealed severe scoliosis, spiral spinal rotation, multiple vertebral anomalies, hypoplasia of multiple bones of the skull, and mild prognathism. In cattle, autosomal recessive inheritance is suspected to cause the defects based on case clustering around certain bulls. The white rhino dam was captive born and had successfully delivered two normal calves sired by the same bull prior to the abortion of the affected fetus.

Congestive heart failure as a result of **persistent truncus arteriosus**, and **ventricular and atrial septal defects** has been reported in a neonatal black rhinoceros.[156] Hypoplastic pulmonary arteries and secondary right and left ventricular dilation were also noted. Persistent truncus arteriosus is characterized by a single arterial trunk arising from the ventricular outflow tract to supply the pulmonic, coronary, and systemic circulations. A septum dividing the truncus arteriosus into the aorta and pulmonary artery failed to form. The cause of persistent truncus arteriosus may be multifactorial. It is more common in certain breeds and certain human families in which it may have monogenetic inheritance. Another black rhinoceros calf that died less than 12 h after birth had congenital defects including **cardiac septal defects**, **cleft palate**, and **patent foramen ovale**.[85]

Atresia ani is a congenital embryological anomaly in which the hindgut fails to fully communicate with the perineum. The anus may be either stenotic or imperforate. Atresia ani may appear alone or in combination with rectovaginal or rectovestibular fistula. Atresia ani has caused

TABLE 17.1 Toxins Reported in Nondomestic Perissodactyls

Family/Species	Toxic Agent	Comments
Equidae		
Plains zebra[126]	Organophosphate	Postdeworming ($n = 1$)
Grevy's zebra[188]	Red maple *Acer rubrum*, includes hybrid trees. Seasonal—wilted/dried leaves, gallic acid, precise toxin unknown	Hemolytic anemia, methemoglobinemia, (oxidative damage erythrocytes) $n = 2$ females. Heinz body formation, leukocytosis, icterus, hemoglobinuric nephrosis, paracentral hepatic necrosis
Przewalski's horse[114]	Sycamore seed *Acer pseudoplanatus* Sycamore/sycamore maple. Toxin is hypoglycin A	Equine atypical myopathy $n = 1$ foal. Sudden onset neck stretching, progressive weakness. Increased creatine kinase, increased aspartate aminotransferase, mild increase in inorganic phosphorus, neutrophilia. Brown urine—increased protein and hemo/myoglobin Multiple acyl co-A dehydrogenase deficiencies were indicated. Fatty acid oxidation in muscle mitochondria Rhabdomyolysis striated and cardiac muscles
Rhinoceridae		
Black rhinoceros[97]	Organophosphate	Degenerative myopathy, $n = 2$
Black rhinoceros[39]	Vitamin D_3	Diet compounding error, diet 13–38 times expected vitamin levels; 3/10 individuals developed hypercalcemia and partial anorexia, and died over a 5-month period; of these 2/3 had mildly elevated serum 25-hydroxyvitamin D_3, 1/3 marked increase liver 25-hydroxyvitamin D_3
Rhinoceros, including white rhinoceros[105,108]	Blue-green algae *Microcystis aeruginosa*	Periodic algal blooms associated with high organic matter and low water levels. Random massive acute hepatic necrosis
Black rhinoceros	Creosote from coal tar, cresols, associated wood preservatives	20 wild-caught in creosote treated pen for >1 month. 4/20 became lethargic, anorexic, anemic, icteric. Increased aspartate transaminase, bilirubinemia, dark brown urine. Two of the four died with icterus, ulcers oral cavity and stomach, large cholestatic (green) liver, and erythroid hyperplasia of marrow at necropsy. Acute periacinar liver necrosis. Post translocation five more died with similar liver lesions[78] Two captive rhinoceros developed protracted anemia and died, ulcerative skin lesions, possible icterus, suspected creosote toxicity. Degenerative hepatopathy with green/brown pigmentation. White rhinoceros similarly housed unaffected. Fencing old telephone poles[165]
Black rhinoceros[140]	Anthracycline (doxorubicin) chemotherapy	Juvenile treated for acute lymphoblastic leukemia. Developed congestive heart failure with ventricular dilatation, congestion of lung, liver, and spleen. Doxorubicin-induced cardiomyopathy identified in humans

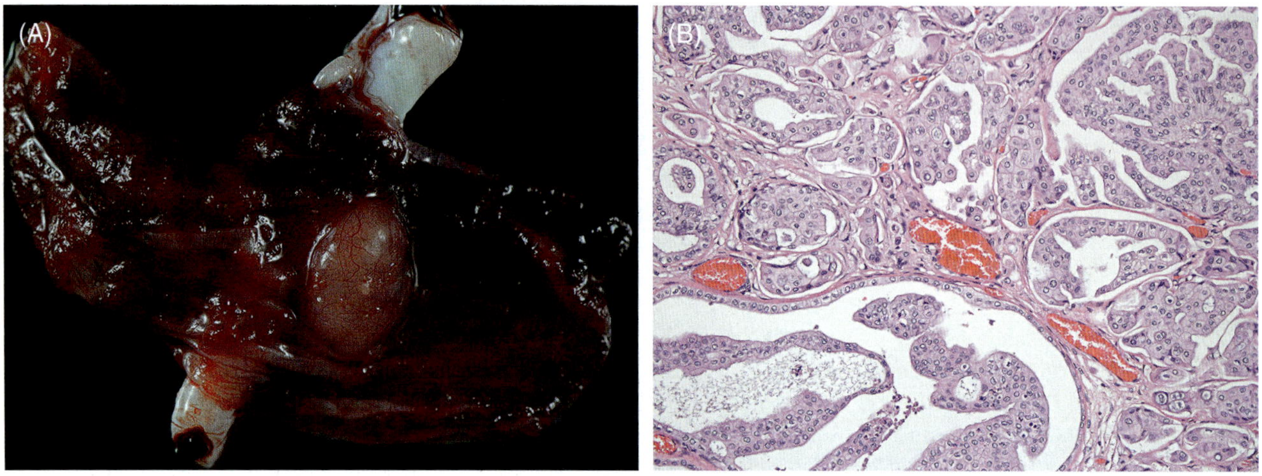

FIGURE 17.3 Adenomatous hyperplasia in the placenta of a greater one-horned rhinoceros. (A) These can be grossly inapparent or appear as nodular foci. Tissue in this case is light pink/tan and the surface is covered by multiple prominent vessels. (B) Nodules are composed of variably sized glandular or cystic structures lined by cuboidal to low columnar epithelium or epithelial fronds with various amounts of fibrous stroma.

death in neonatal tapirs; abdominal distention has been seen, but associated fistulae have not been reported.[60] Other congenital abnormalities in perissodactyls include **albinism** in a free-living lowland tapir in Brazil,[178] **leucism** in a lowland and a mountain tapir,[129,176] and coccygeal scoliosis, unilateral microphthalmia, and optic nerve hypoplasia in a Grevy's zebra foal (Figs. 17.4A–C).

Age-Related/Degenerative

Histologic features of **chronic renal disease** in perissodactyla are similar to those in other species and include mononuclear interstitial nephritis, fibrosis, tubular dilation, and tubular degeneration and loss (Fig. 17.5A) and pyelonephritis (Fig. 17.5B). Secondary changes, such as uremic mineralization in the lungs (Fig. 17.5C) or stomach (Fig. 17.5D) and gastric ulcers (Fig. 17.5E), are seen. As with sequelae to renal disease in many species, blood and urine analysis can be used for clinical monitoring. Management with a high-energy, low-protein diet may help.[105]

Degenerative joint disease is frequently seen in older animals under managed care. In a review of the osteopathology of rhinoceros, skeletons and cadavers from 27 rhinoceroses of various species were examined.[158] Twenty-two individuals had bone lesions in at least one limb. Six categories of lesions were identified with the following frequencies: enthesopathies 20/27, osteoarthritis 15/27, bone remodeling 12/27, osteitis/osteomyelitis 3/27, fracture 3/8, and subluxation 3/8. Entheses are the sites of ligament and tendon attachment to the bone surface. Response to injury at these sites includes new bone deposition. Enthesopathies were present at the dorsoproximal phalanges in particular. Osteoarthritis was most common at the distal interphalangeal joints.

Dental disease is common in aged perissodactyls. Tooth loss may be associated with tartar accumulation and gingivitis due to insufficient browse in captive diets in rhinoceros.[108] Dental disease, **mandibular and maxillary osteomyelitis**, and **facial abscesses** are common in aged tapirs. The lesions are thought to be initiated by rough feed that causes mechanical trauma and enables pathogen entry. Surveys of dentition in museum specimens have indicated markedly different patterns of wear and lesions between Malayan and lowland tapir species. Malayan tapirs are selective browsers and feed on soft vegetation. In skulls and mandibles from Malayan tapirs, dental lesions in one study were identified in 52% of the captive and 6% of the free-ranging tapirs.[20] Lesions were most common at the second, third, and fourth premolars and the first molar teeth and consisted of resorption with loss of radiodensity extending into the tooth at the cementoenamel junction. The subgingival location suggested a metabolic rather than an external etiology. Mandibular teeth were more commonly affected than maxillary teeth. Age and captivity were positively correlated with lesion development. In a study of the skulls of lowland tapirs, 24% had lesions with similar prevalence in free-ranging and captive animals.[177] The most common problems were dental fracture with associated periapical reaction in 15% of skulls, periapical reaction without dental pathology in 13% of skulls, and dental fracture with no periapical reaction in 4% of skulls. One skull from a free-ranging lowland tapir had caries-like lesions. The lesions in the lowland tapirs were most consistent with dental trauma. Increasing age was positively correlated with lesion development.

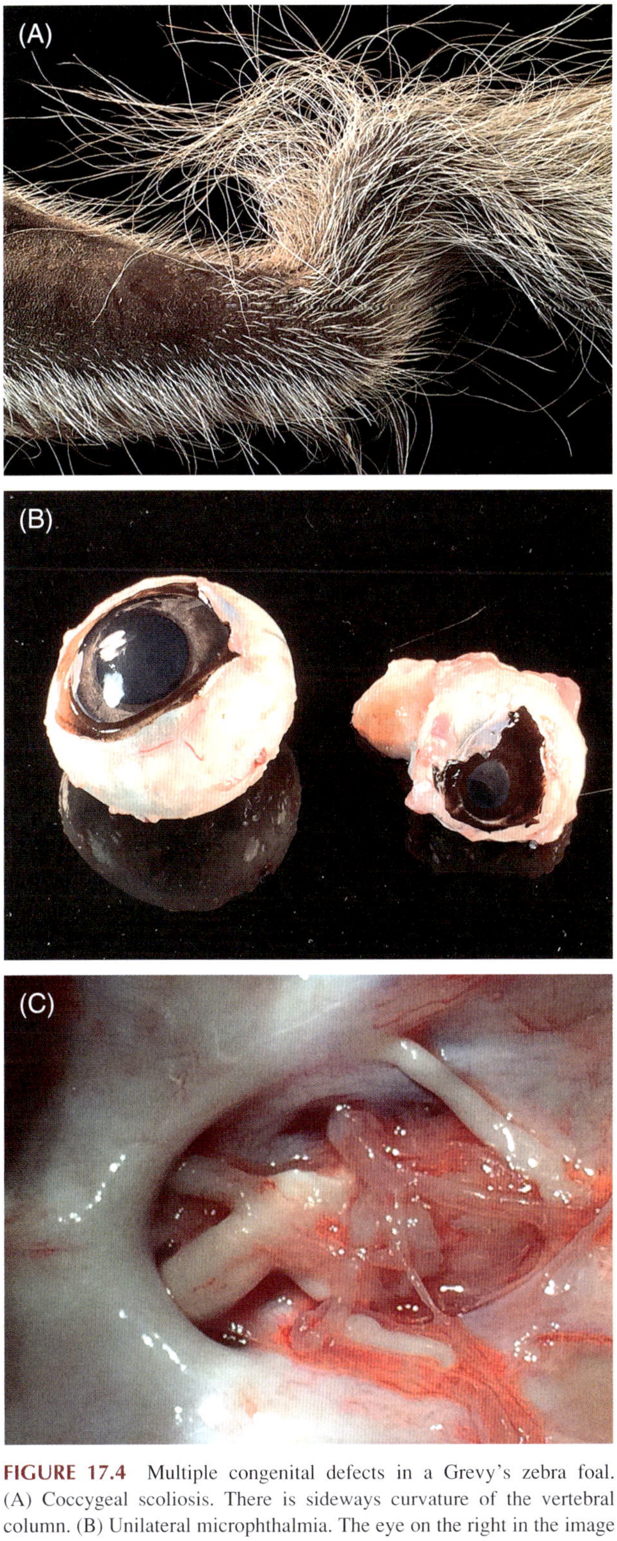

FIGURE 17.4 Multiple congenital defects in a Grevy's zebra foal. (A) Coccygeal scoliosis. There is sideways curvature of the vertebral column. (B) Unilateral microphthalmia. The eye on the right in the image is significantly smaller than the normal contralateral eye. (C) Unilateral optic nerve hypoplasia. In this image of the optic chiasm, the abnormal left optic nerve (upper left) is notably smaller than the normal right optic nerve (lower left).

Inflammatory, Non-Infectious

The density of hair follicles is low in rhino skin and there is an increased prominence of epidermal rete ridges. The dermis is composed of thick, coarse dense collagen (reticular dermis). When hemorrhage occurs in the superficial dermis, erythrocyte exocytosis through the epidermis may result in subcorneal hemorrhage or hemorrhagic crusts.[119]

A variety of skin diseases affect captive black rhinoceros. Some have frequent episodes of ulcerative skin/mucocutaneous disease alone or in association with other disease problems.

These include **superficial necrolytic dermatopathy** (also known as **superficial necrolytic dermatitis, necrolytic migratory erythema, vesicular and ulcerative dermatopathy, metabolic epidermal necrosis, mucosal and cutaneous ulcerative syndrome, hepatocutaneous syndrome**, and **ulcerative skin disease**). With this condition, individuals develop lesions of variable severity.[118] Animals of all ages can be affected. Epidermal plaques progress to vesicles or pustules that erode or ulcerate. Lesions are generally bilateral, possibly symmetrical, with a tendency to form at pressure points. Common sites are the lateral body surfaces, coronary bands, tail tip, and ear margins. Oral and nasal lesions are seen and may be more persistent. Affected rhinoceroses appear depressed, anorexic, and lose weight. Hypoalbuminemia, hypocholesterolemia, and decreased hematocrit may be found reflecting loss of albumin and blood through the skin. Similar gross and histologic lesions in domestic species are associated with hyperglucagonemia or hypoaminoacidemia. Low circulating amino acids disrupt epidermal homeostasis causing ulceration or increased susceptibility to injury. Glucagonoma, diabetes mellitus, hepatic disease, hyperadrenocorticism, and hypothyroidism have been linked to the skin lesions in domestic species. Hypoaminoacidemia was not confirmed in a study of black rhinoceros in which monthly plasma levels were assessed.[26]

Histologically, lesions have a characteristic layered "red-white-blue" appearance: superficial parakeratosis with eosinophilic keratinocytes retaining nuclei (red) overlying hydropic degeneration in the stratum spongiosum, spongiosis, and intraepidermal vesicles resulting in pallor (white), and acanthosis of the deep stratum spongiosum (blue). Intracytoplasmic inclusions seen in some cases are keratin intermediate filament aggregates as determined by electron microscopy. Dermal inflammation develops following ulceration.[119]

Epidermal exfoliation may be a variant of superficial necrolytic dermatopathy and is associated with shedding of superficial hyperkeratotic epidermis from the thorax

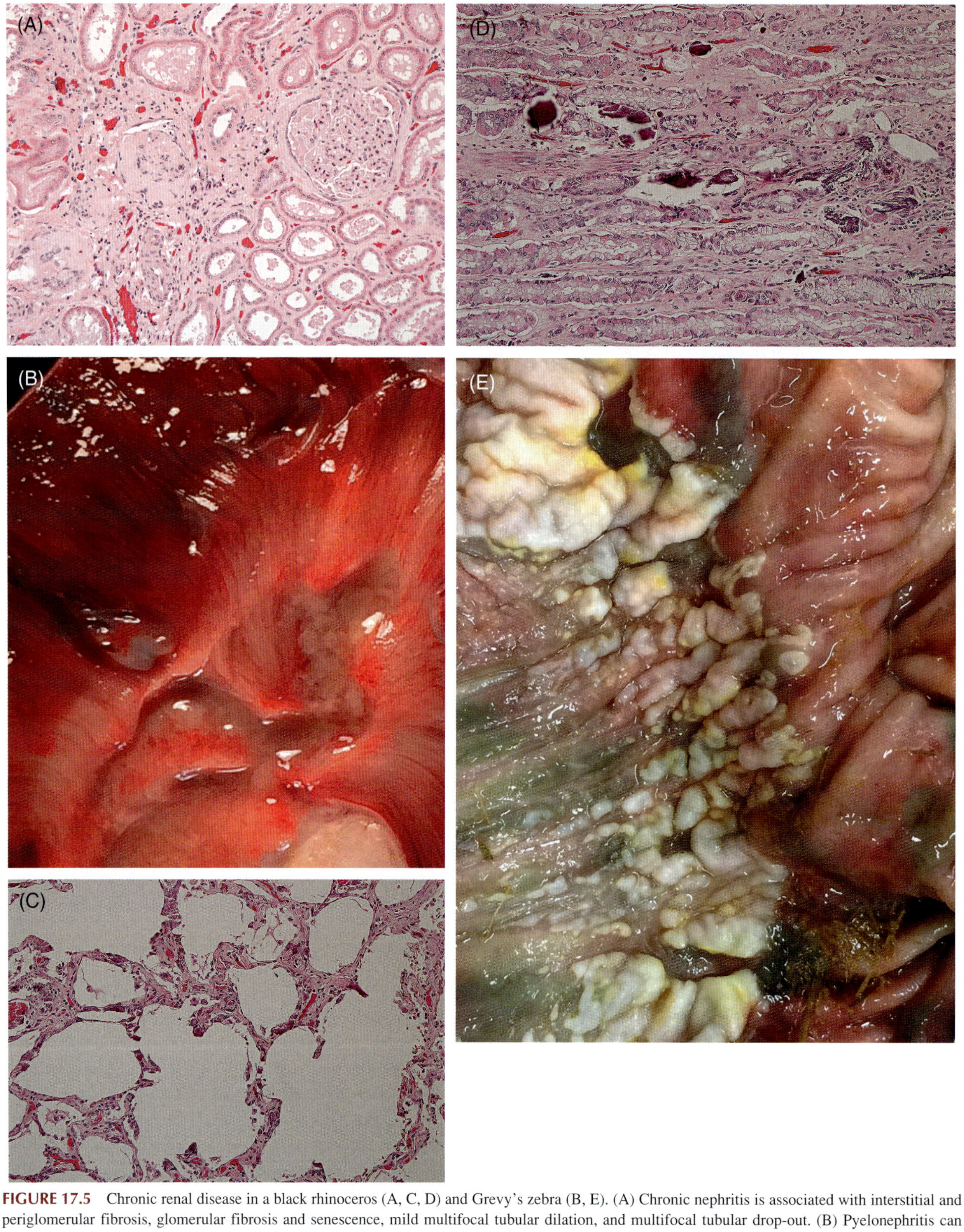

FIGURE 17.5 Chronic renal disease in a black rhinoceros (A, C, D) and Grevy's zebra (B, E). (A) Chronic nephritis is associated with interstitial and periglomerular fibrosis, glomerular fibrosis and senescence, mild multifocal tubular dilation, and multifocal tubular drop-out. (B) Pyelonephritis can appear grossly as fine tan streaks that extend from the renal pelvis to the cortex. Complications of chronic renal disease can include (C) pulmonary and/or (D) gastric mineralization (basophilic deposits) secondary to uremia and (E) gastric ulceration at the margo plicatus.

and flanks that leaves a shiny gray surface. Intraepidermal pustules are seen occasionally with no further epidermal degeneration (pustules are not a component of superficial necrolytic dermatitis). Rarely, coagulative necrosis of the epidermis may occur in association with neutrophil accumulations and colonies of cocci. Dermal inflammation is seen in these lesions.

Nodular collagen degeneration with dystrophic mineralization alone or in association with chronic ulcers has been reported in the rhinoceros. Macrophages and multinucleate giant cells may be associated with these lesions. The etiology is currently unknown. Similar changes are seen in horses with arthropod-induced injury and dogs with hyperglucocorticoidism.[119]

Eosinophilic granulomas have been described in eight captive black rhinoceros. Affected sites are the oral and nasal cavities and the skin (frequently behind the prehensile upper lip). Gross lesions appear as proliferative, exophytic masses that are prone to trauma. Histologically, the epidermis is proliferative, acanthotic, and often ulcerated. The dermis contains prominent infiltrates of eosinophils, collagen degeneration with flame figures, and marked neovascularization.[119,145] Lesions can resolve spontaneously, and corticosteroid and cryotherapy have been used to resolve lesions effectively. The condition does not appear to be seasonal or occur in related individuals. It can recur and may be seen concurrently with superficial necrolytic dermatopathy. The etiology is currently unknown, but hypersensitivity as a primary cause is suspected. Recent cases in rhinoceros have responded to antihistamines.[10]

Seasonal allergic dermatitis has been reported in related greater one-horned rhinoceros. The lesions were eosinophilic and consistent with a hypersensitivity reaction. Intradermal skin testing and allergen-specific IgE serum testing (the latter not validated for rhinoceros) identified specific allergens for hyposensitization treatments, which resulted in a positive response.[14]

Vesicular skin disease is described in tapirs. Lesions develop most commonly in the dorsal cervical and lumbosacral areas and resolve spontaneously.[37,61] Erythematous papules form, coalesce, rupture, and slough leaving ulcers. Vesicles are subepidermal and may contain neutrophils and eosinophils. The overlying epidermis is spongiotic with superficial to full-thickness necrosis. Follicles are unaffected. A narrow zone of degenerate collagen is present in the superficial dermis below the vesicles; perivascular edema and hemorrhage are present in the dermis. The histologic features are consistent with a number of conditions **including bullous pemphigoid, dermatitis herpetiformis**, junctional form of **erythema multiforme**, and the acute hemorrhagic form of **staphylococcal hypersensitivity**. However, lesion distribution and disease epidemiology in tapirs is inconsistent with those found in other species. Sloughing and repair occur approximately a week after lesions first develop. Resolution may be associated with fluctuating hormone levels in females.[194] Viral particles have not been identified by electron microscopy in vesicular fluid and virus was not isolated from equine fibroblast culture. In one survey, tapir (Malayan, lowland, Baird's) with similar clinical histories had biopsies consistent with vesicular skin disease. In Baird's tapir, the lesions were more lateral and pruritic. Many animals also exhibited neurologic signs, such as hindlimb ataxia, lameness, weakness, and syncopal episodes during periods of skin disease. Many affected tapirs also have chronic intermittent respiratory infections.

A survey of free-ranging Baird's tapir in Costa Rica identified **depigmentation of undetermined etiology** (also known as **leukoderma** or **vitiligo**) along the dorsum, perineum, inguinal, and perioral regions.[54] Histologically, the epidermis was hyperplastic with little pigmentation. Mild mononuclear cell deposits were present in the superficial dermis and there were perivascular eosinophils. Hairs were generally in catagen or telogen stage. Histologic lesions were consistent with hypersensitivity dermatitis; a specific etiologic agent was not identified but a response to biting insects was suspected. In a somewhat similar case, a captive black rhinoceros developed nonpigmented foci initially at the nares and lips and subsequently multifocally on the limbs, head, and trunk. Multifocal hypopigmentation of the basal layer of the epidermis with pigmentary incontinence and perivascular lymphohistiocytic dermatitis were seen. Phototherapy was used to induce repigmentation.[173]

In the 1980s and 1990s, 44 episodes of **acute intravascular hemolytic anemia** occurred in 36 captive black rhinoceros in North America, Europe, and Japan. Among them, 23 individuals died (~64%).[108] Early cases were characterized by hemoglobinuria, lethargy, and severe anemia. Rare animals survived episodes with supportive care.[27,109] Investigated causes of hemolytic anemia included autoimmune hemolytic anemia, equine infectious anemia, equine viral arteritis, copper toxicosis, vitamin E deficiency, and clostridial infection.[108] Mortality due to hemolytic anemia occurs in free-ranging rhinoceros in association with babesiosis and trypanosomiasis; however, none of these infections were confirmed as contributing to hemolytic anemia in affected captive black rhinoceros. Leptospirosis was diagnosed in many of the affected captive black rhinos and may have played a role in the pathogenesis of hemolytic anemia. Iron deposition in parenchymatous organs is a general finding, originally thought to be evidence for hemolytic anemia (see iron overload disease discussed earlier).[136,139] The relationship between hemolytic anemia and iron accumulation in tissues is still poorly understood.[170] Histologic findings in affected rhinos included renal tubular casts with red blood cells and protein; hepatic centrilobular fatty change and necrosis were also seen.

Hemolytic syndromes in humans are commonly caused by **enzyme defects** that impair metabolic pathways required to neutralize oxidants in red blood cells. These include hereditary deficiency of glucose 6 phosphate dehydrogenase as the most common defect and bacterial or viral infections that lead to secondary oxidant stress.[138] Several factors, including 6 phosphate dehydrogenase, erythrocyte catalase activity, erythrocyte adenosine kinase activity, and catalase and glutathione *S*-transferase, have been suggested in the pathogenesis of hemolytic syndrome cases in black rhinoceros. However, to date, no definitive pathogenesis has been elucidated to explain cases of hemolytic anemia in rhinos. Unlike humans, erythrocyte glucose 6 phosphate dehydrogenase activity in affected black rhinos is normal.[139] Erythrocyte catalase activity levels are similar in black rhinos with or without hemolytic anemia and are similar among black, white, greater one-horned, and Sumatran rhinos.[141] Low erythrocyte adenosine kinase activity (important for maintaining ATP levels),[139] hypophosphatemia (inhibits glycolysis, reducing the amount of available ATP),[107] and low erythrocyte ATP levels (relative to humans) have all been suggested as possibly predisposing red cells to oxidative stress and hemolysis.[139] Low ATP in rhino red blood cells may be rate limiting in the antioxidant activity of the hexose monophosphate shunt, which is important in the neutralization of oxidative byproducts and may contribute to mucocutaneous ulcerative disease and acute episodic hemolysis.[138] Recent cases have been rare; though further investigation of oxidative stress in black rhinoceros may inform procedures for conservation transportation events.[149]

Suspected **neonatal isoerythrolysis (syn. Hemolytic icterus)** has been reported in Baird's tapirs.[184] Neonatal isoerythrolysis is a form of immune-mediated hemolytic anemia in which maternal antibodies to red cell components inherited from the sire bind to and destroy the neonate's red blood cells. The dam develops an immune response to fetal red blood cells during her first pregnancy with an individual sire, so isoerythrolysis only occurs in subsequent offspring. Clinical signs include lethargy, inappetence, icterus, and red urine. Hemolysis in tapir foals fed equine colostrum has been reported and may be due partly to natural antibodies to erythrocytes of a different species. It is unknown whether equine colostrum is an appropriate replacement for tapir colostrum.

Idiopathic hemorrhagic vasculopathy syndrome has been reported in captive black rhinoceros in the United States. No sex bias or tendency to develop in wild-caught versus captive-born animals was noted. Signs included locally extensive swelling of the neck and limbs, lethargy, laminitis, and sloughing of the nails. Nonhemolytic anemia is seen; hematocrit and serum phosphorus fall rapidly in affected animals. Histologically, neutrophilic vasculitis in the superficial limb dermis, proliferative vasculitis in the oral mucosa, and extravasation from small vessels suggestive of immune-mediated disease are seen. Extensive testing to identify an infectious agent has been unproductive, and the cause remains unknown.[122,141] Equine purpura hemorrhagica, a sporadic noncontagious allergic reaction to streptococcal or viral antigens, appears to be the most similar condition.

Hoof lesions, such as **laminitis** and foreign bodies at the white line resulting in abscess formation, have been reported in captive equids including Somali wild ass.[135] In captive rhino high-protein diets have been suggested as a cause of laminitis.[108] Additional predisposing conditions may include excessive feeding of concentrates, enteritis, chronic renal failure, and acute intravascular hemolytic anemia. Laminitis occurs when the blood supply to the corium is interrupted. In severe cases, the union between the horny and sensitive laminae breaks down, progressing to separation of the nail at the coronary band. Signs of laminitis are lameness and pain or discharge at the coronary band. The affected nail may remain while the new nail grows in and displaces it. The nail may remain attached at the sole with total separation at the coronary band.[182]

Greater one-horned rhinoceroses are prone to developing **chronic foot disease**. The species lives in river plains and they are "hoof-walkers;" the outer rim of the hoof wall is denser and carries the animal's weight. In captivity on hard substrate, foot problems tend to develop especially if animals become overweight. Chronic foot disease was found to affect nearly 25% of the captive population in a 1996 international survey. Male rhinos were twice as likely as females to have chronic foot disease. The condition usually occurs in one or both hind feet, but all four feet are vulnerable. The hoof wall breaks down/cracks with increased stress on the more delicate central pad, and hematomas of the pad may be seen. Fissures develop in the pad behind the middle toe and excessive tissue growth occurs between the toes. Cracks in the pad initially appear as ragged edges of overgrown sole. Trauma promotes separation and scarring results. Secondary bacterial infection of the lesions may occur. A majority of cases also have overgrowth of the toenails, although it is not certain whether this is a cause or a consequence of pad separation.[182]

Though not specifically reported in the literature, many non-infectious chronic inflammatory diseases are seen in aged individuals; over time lesions degenerate and mineralize. Chronic degenerative vascular diseases, including **arteriosclerosis** (Fig. 17.6) and **atherosclerosis**, are seen in many aged captive perissodactyls. These changes are dissimilar to vitamin D_3 toxicosis (described earlier) and unrelated to vascular rupture. **Coronary artery**

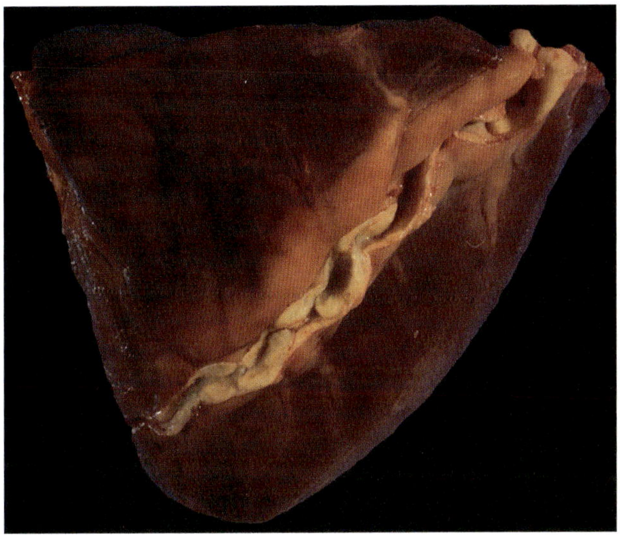

FIGURE 17.6 Arteriosclerosis in a Malayan tapir. There is moderate transmural thickening of the cardiac arteries in the interventricular groove of the heart. *(Photo Courtesy of Toronto Zoo.)*

aneurysms, with or without rupture and hemopericardium, have been described in black rhinoceroses.[73,108] Histological lesions include ulceration of the intima with superficial/luminal granulation tissue. Cardiac **Purkinje fiber degeneration, necrosis, and mineralization** have been described in free-ranging black rhinoceroses.[76] The cause of the lesions is undetermined. Sites of **chronic skin trauma** can develop fibrosis and mineralization; some have been associated with the development of calcinosis circumscripta (Fig. 17.7).

Gastrointestinal tract disease is a common source of problems in the domestic horse and similar conditions, including **colonic impaction, esophageal rupture,**

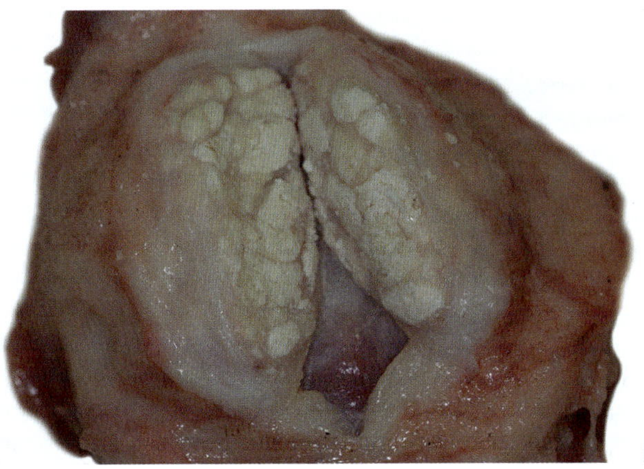

FIGURE 17.7 Calcinosis circumscripta due to chronic trauma in the skin of a white rhinoceros. Characteristic features are the nodular deposition of mineral in soft tissues such as the dermis and subcutis.

intestinal perforation, serositis, ulcerative colitis, and **colonic torsion**, have been seen in a number of nondomestic perissodactyls (Figs. 17.8A–F). Dilation of the caudal esophagus in one report was associated with regurgitation in a black rhinoceros whose symptoms resolved on a soft diet.[154] In another study, a third of adult tapir deaths involved the gastrointestinal tract.[60] Findings included **intestinal volvulus, gastric and colonic impactions, colonic incarceration, oropharyngeal abscess/oral necrobacillosis, oropharyngeal trauma, acute pancreatitis**, and **eosinophilic enterocolitis**. Sand and shavings have caused colic and intestinal impaction in *Equidae* and *Tapiridae*, especially newborns.[194] Some gastrointestinal problems occurred due to overconsumption, especially if food lacks fiber as the stomach is relatively small.[194] **Rectal prolapse** has been seen in perissodactyls, most frequently in tapirs in association with poor quality dietary roughage and lack of access to water and consequent dehydration. Prolapse has also been reported in rhinoceros.[1,7] Prolonged tenesmus, constipation, diarrhea, intestinal protozoal parasites, dystocia, urinary obstruction, proctitis, rectal tear, and space-occupying lesions have been associated with rectal prolapses. Rectal prolapse was associated with marked edema and submucosal fibrosis in a greater one-horned rhino.[7]

Pseudodiverticulosis with jejunal rupture and peritonitis has been reported in a captive-born Przewalski's horse.[112] **Enteroliths** have also been described in nondomestic perissodactyls. They develop when mineral deposits in layers around a nidus. Diet acidification may decrease formation. Tapir fecal enteroliths are composed of hydrated phosphates, vivianite (Fe), and newberyite (Mg) rather than struvite (struvite is more common in horses).[121]

Corneal cloudiness due to **keratitis** in captive tapirs may be associated with excessive light exposure, trauma, or herpesvirus.[58,60,155,194] The nocturnal nature of Malayan tapirs and the dense vegetation of the habitat of these and Sumatran rhino may predispose these species to keratitis in the captive setting when managed with less shade and encouraged to shift their wake/sleep cycle.

Miscellaneous/Multifactorial

Fracture of cervical vertebra and femur has occurred during courtships of Somali wild ass.[135] Traumatic lesions during breeding also occur in many species of rhinoceros.[111] **Bite wounds** are common in *Equidae*. **Cervical fractures** occur with some frequency in captive zebras after collision with enclosure fencing.[172] Animals are often found dead at the fence with no history of any illness. Diagnosis is straightforward radiographically and on gross examination based on the cervical hemorrhage and disrupted vertebrae (Fig. 17.9). **Rhinoceros horn avulsion**

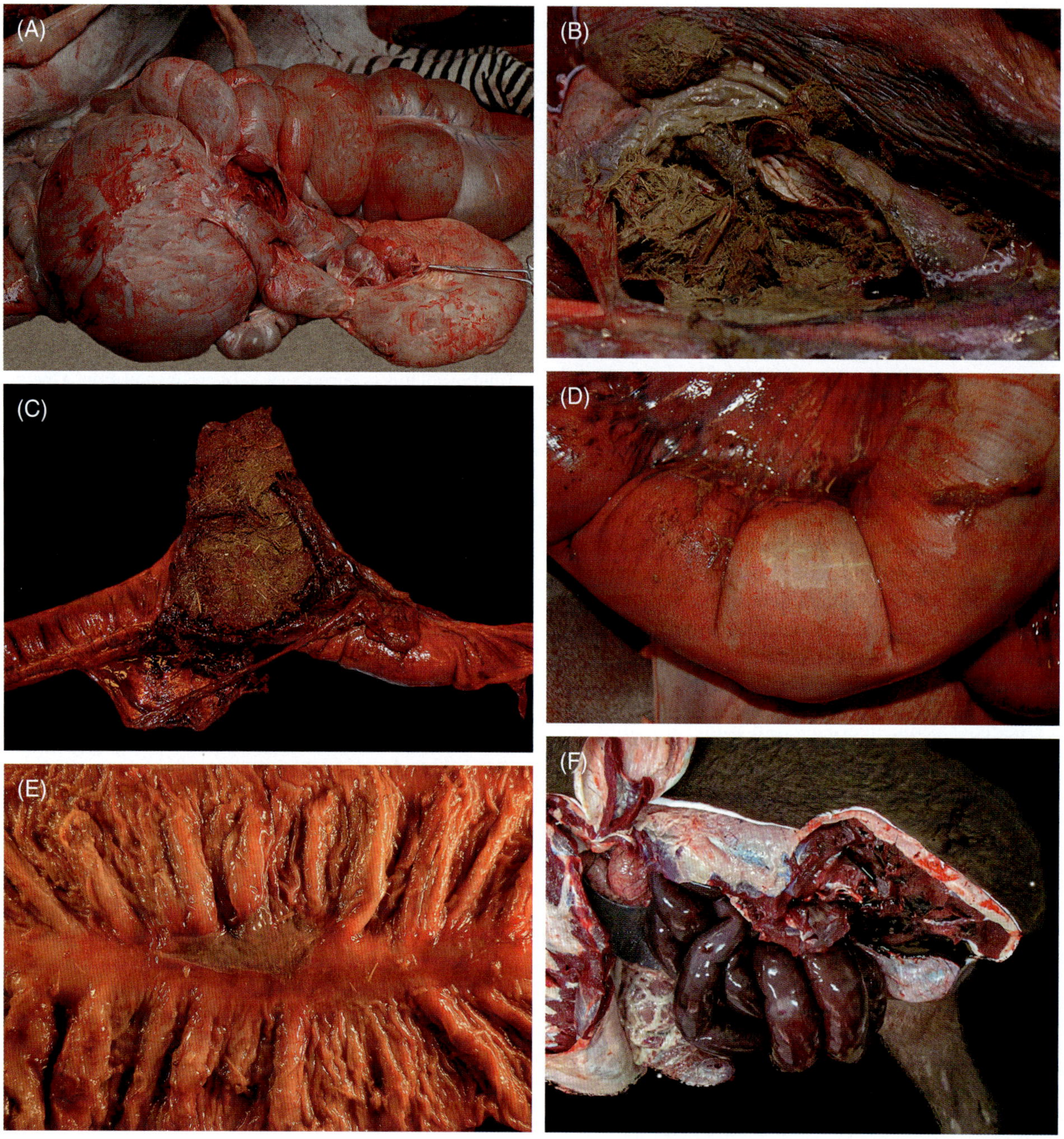

FIGURE 17.8 Gastrointestinal tract disease in perissodactyls. (A) Colonic impaction at the diaphragmatic flexure in a Grevy's zebra. (B) Esophageal rupture with intrathoracic gastrointestinal content at the mediastinum (left lung is reflected dorsally) in a Grevy's zebra. (C) Intestinal perforation in a Grevy's zebra. (D) Acute fibrinonecrotic serositis in a Grevy's zebra. (E) Focal ulcerative colitis in a Grevy's zebra. (F) Colonic torsion in a lowland tapir. *(Part F: Photo Courtesy of Zoológico de Sorocaba, Brazil.)*

can occur if horizontal poles are used for fencing. **Poacher-related trauma** from dehorning and snares are common in wild rhinoceros. **Interspecies aggression** may occur between male tapirs housed together. Death of free-ranging tapir occurs frequently through collision with **motor vehicles**.[65,98]

Obesity is a common concern in all captive equids that receive a rich diet and get relatively little exercise.[126]

Severe decubitus ulcers with necrosis have been described in white rhinoceros. One neonate with mild angular limb deformity developed ulcers secondary to reduced mobility. Particular susceptibility to pressure necrosis in the skin of neonatal rhinos has been suggested.[41] In another report, a rhino developed ulcerations due to trauma associated with cement flooring during winter housing.

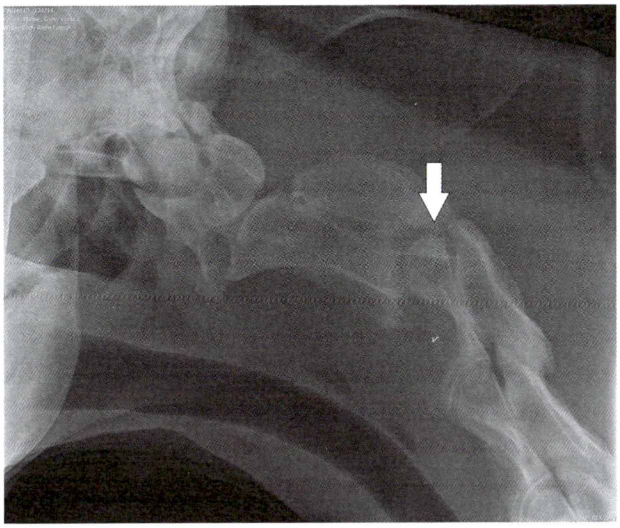

FIGURE 17.9 Cervical vertebral column trauma in a Grevy's zebra. Radiograph showing C2 vertebral displacement with facet fracture.

Ischemic distal tail necrosis after presumed trauma and entrapment of the extremity necessitated amputation in a black rhinoceros (Figs. 17.10A, B).

Leukoencephalomalacia has been reported in young, female black rhinoceroses in three different zoos; two of these animals had clinical hyperglycemia.[68,110] The calves developed sudden onset lethargy, later becoming recumbent and comatose. One calf developed epistaxis and hyperresponsiveness. Cloudy meninges, yellow-green foci in the frontal lobes, and subdural hemorrhages were seen grossly. White matter cavitation was seen in coronal sections. Histologically, edema, hemorrhage, neutrophil infiltration, and focal necrotizing vasculitis were present. The cerebrum was most severely affected. The pattern of encephalomalacia was similar to that seen in horses that ingest moldy feed contaminated with the fungus *Fusarium moniliforme*, which produces several toxins including fumonisins. No mold toxin was identified in feed in the affected rhinos. The fourth case developed acute neurologic signs: circling, head tilt, head pressing, continuous vocalization, hypermetria, ptyalism, and apparent blindness.[68] Computed tomography scans showed areas of reduced density in the cerebrum and thalamus, ventricular dilation, and dense areas in the caudate nuclei compatible with hemorrhage. Grossly, cavitating necrosis of the white matter was seen in the cerebral hemispheres. Histologically, demyelination was the main lesion with minimal reaction. In one case, the dam had been severely traumatized in the last trimester of pregnancy with possible compromise of blood circulation to the fetal brain. Increased age of the dam, trauma, and/or excess maternal iron stores were suggested as possible contributing factors.[105]

Neoplastic

The number and frequency of neoplasms occurring in perissodactyls may not be fully represented in the published literature. **Leiomyoma** of the female reproductive tract and cutaneous **sarcoids** are described below. Other reported neoplasms in nondomestic perissodactyls are listed in Table 17.e3.

Leiomyoma, a benign smooth muscle tumor, is found with relative frequency in the reproductive tract of female aged zoo species, especially greater one-horned rhinoceros. In one clinical study, 100% of greater one-horned rhinoceros over 12 years of age had reproductive tract masses on ultrasound examination.[53] The number and size of masses increase with age. The masses are particularly common at the cervix and in the vagina. They can be single or multiple and are generally well demarcated, white, and

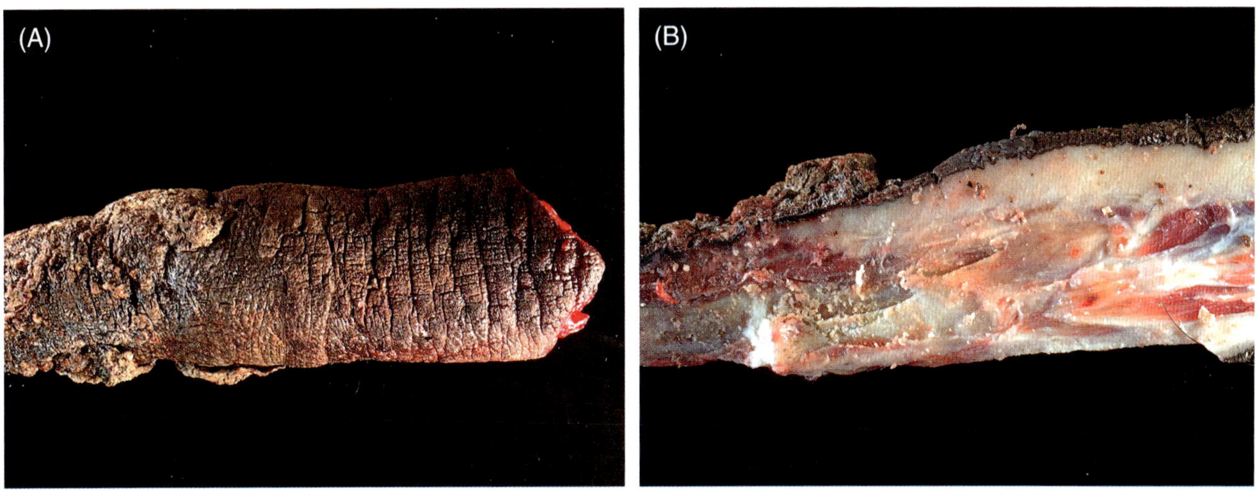

FIGURE 17.10 Ischemic necrosis (undetermined cause) of the distal tail of a black rhinoceros. (A) Depressed distal surface with crusted dried serocellular deposits to the left of the image. (B) Central segmental pallor of the striated muscle is consistent with necrosis.

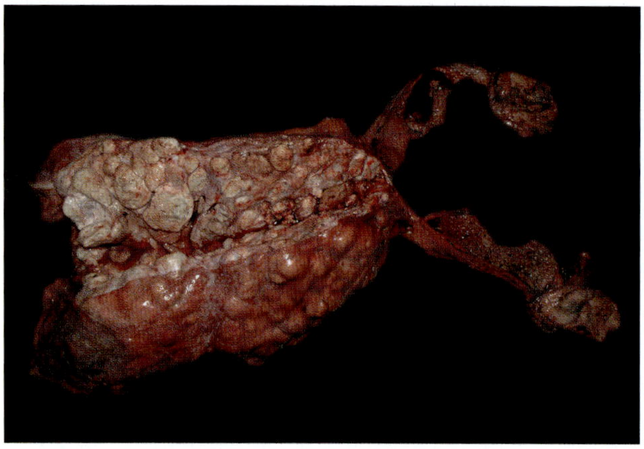

FIGURE 17.11 Uterine leiomyomas in a greater one-horned rhinoceros. The uterine body, which is opened longitudinally along its long axis, is markedly expanded by numerous, variably sized, firm, demarcated, coalescing, light tan masses that obscure the uterine lumen (see eSlide 17.e3). *(Photo Courtesy of the Wildlife Conservation Society.)*

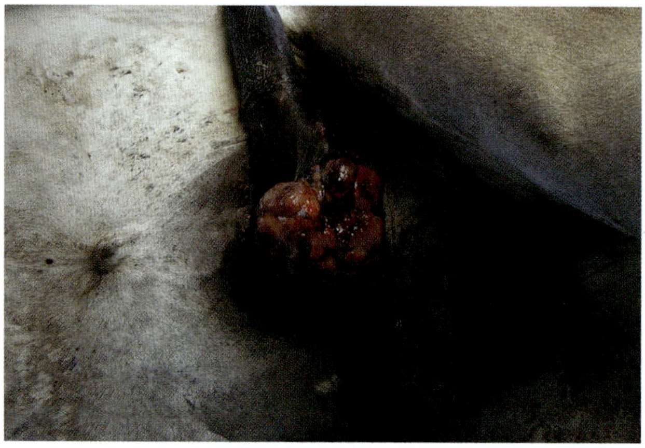

FIGURE 17.12 Preputial sarcoid in a Somali wild ass. These are exophytic, flat to fleshy tumors that may, as in this case, have an ulcerated surface (see eSlides 17.e4–e6). *(Photo Courtesy of S. Citino, White Oak Conservation.)*

firm. The wall of the uterus can be markedly distorted (Fig. 17.11; eSlide 17.e3) and cause compression or dislocation of adjacent abdominal viscera. The average gestational length of the white rhinoceros is 16 months. In the wild, a single offspring is produced every 2–3 years, so the uterus of a wild female is generally under the influence of progesterone. Increased progesterone levels are protective against the development of leiomyomas, while acyclicity has been suggested to predispose captive rhinos to their development.[53,190]

Sarcoids are unencapsulated, poorly demarcated fibroblastic skin tumors that invade locally and may recur after surgery, but that do not metastasize. Spontaneous regression has been reported in young individuals with mild lesions. Sarcoids are composed of spindle cells within the dermis that are arranged in streams and whorls and form well to poorly demarcated masses. The overlying epidermis is frequently hyperplastic with long rete pegs. Equine sarcoids associated with **bovine papillomavirus (BPV) types 1, 2, and 13** infection have been identified (see "DNA viruses" section). Sarcoids have been reported at the eyelid, nose, and inguinal region of a captive Plains zebra with BPV type 1. Virus was demonstrated in the tumor with in situ hybridization of the inguinal mass.[88] Sarcoids at the ear, prepuce, and mammary gland in Somali wild asses have also been described and are also thought to be virus-associated (Fig. 17.12; eSlides 17.e4–e6).[135] A survey of sarcoid cases in African equids in European zoo collections suggests prevalence varies with equid species. A genetic predisposition has been suggested in domestic horses and captive Somali wild ass; in the latter this is because the population is descended from 17 wild-caught individuals.[189]

Sarcoids have also been reported in wild Cape mountain zebra in two reserves in South Africa. The prevalence was high (up to 53%) and lesions were most frequent on the ventrum and legs.[93] Free-ranging zebra populations with greater genetic diversity seem less likely to be affected by sarcoids.[164] Two cases of sarcoid at the pinna of captive Baird's tapir with possible association with BPV type 1 have been reported.[70] Papillomas have been identified on rhinos. An association with viral infection has not, to date, been identified.

INFECTIOUS DISEASES

The literature on infectious diseases of nondomestic perissodactyls consists primarily of single case reports or case series of disease and serology of captive animals. In addition to descriptions of notable diseases below, see Table 17.2 (Viral infections), Table 17.3 (Mycotic infections), and Tables 17.e4–17.e6 (Serologic evidence of infection, Endoparasites, and Ectoparasites, respectively) for lists of additional reported infections. The World Organization for Animal Health (WOAH) listed, reportable diseases of specific global concern for equids include African horse sickness, Western and Venezuelan equine encephalitis viruses, equine influenza, equine herpesvirus-1, equine infectious anemia, equine viral arteritis, vesicular stomatitis virus (VSV), glanders (*Burkholderia mallei*), contagious equine metritis (*Taylorella equigenitalis*), equine piroplasmosis, and dourine (*Trypanosoma equiperdum*). A recent outbreak of VSV in a group of captive white rhinos demonstrated their susceptibility to this virus. Nondomestic equids are also assumed to be susceptible to Borna virus.[186]

TABLE 17.2 Viral Infections Reported in Nondomestic Perissodactyls

Family/Species	Viral Pathogen	Comments
Equidae		
Grevy's zebra	Equine herpesvirus 1 (EHV-1)[48,192] Unusual alphaherpesvirus—lacks strong host specificity	Captive, abortion, perinatal foal mortality, virus neutralization, DNA restriction endonuclease analysis Clinical disease rare, zebra considered natural host, recommend separation from other susceptible species[84]
	Equine herpesvirus 9 (EHV-9)[48,84] (closely related to EHV-1) Unusual alphaherpesvirus—lacks strong host specificity	Captive, 3/7 severe acute disease, 2 deaths, fibrinonecrotic pneumonia, hyaline membranes, nested PCR herpesvirus in lung Clinical disease rare, zebra natural host, Grevy's possible reservoir, recommend separation from other susceptible species
Onager, zebra[115]	Equine herpesvirus-1	Captive, abortion onager (multifocal necrosis spleen and liver), suspected myelitis zebra
Nondomestic equids[148]	African horse sickness	Captive, epizootic in Europe with mortalities following introduction wild-caught zebra. Plains zebra considered the natural host for the virus, rarely develops signs. *Culicoides* spp. are insect vectors
Tapiridae		
Lowland and Malayan tapirs[155]	Foot and mouth disease virus	Captive, Europe interdigital lesions
Malayan and mountain tapirs[155]	Herpesvirus	Captive, fever, keratitis, erosion to ulceration at mucocutaneous junction 3/6 mortalities
Malayan tapir[45]	Equine herpesvirus 1 (EHV-1), presumptive	Captive, pregnant female died, hemorrhage and necrosis multiple organs
Malayan tapir[181]	Encephalomyocarditis virus	Captive, unexpected death 5 month female
Rhinoceridae		
Black rhinoceros[45]	Herpesvirus	Captive (same collection as tapir with EHV-1), one death with gastrointestinal tract ulcers, 3 cases cutaneous ulcers. Herpesvirus on electron microscopy
Black and white rhinoceros[155]	Poxvirus	Captive, vesicles and pustules
White and greater one-horned rhinoceros	Vesicular stomatitis virus	Captive, United States Swellings to ulcers at nostrils, lips, and oral cavity. Cracks, ulcers, sloughing tissue at feet

DNA Viruses

Sarcoids (see "Neoplastic" section) in horses have been associated with **bovine papillomavirus types 1, 2, and 13** infection. Viral DNA has been detected in the nuclei of the neoplastic mesenchymal cells of zebra sarcoids.[88] Late viral antigens are not found in sarcoids. Neutralizing antibodies to virus are not found in horses, suggesting infection is nonproductive in this (probable) nonpermissive host. Mature viral particles are found in the upper keratinizing layers of these epithelial tumors in bovids; the immune system response results in limited progression of the lesion. The mechanism of transmission of bovine papillomavirus to zebras has not been identified. Sarcoids do not produce virions, but viral DNA and RNA may be detected, and BPV type 1 and 2 major transforming protein E5 is expressed. Of 12 skin lesions from Somali wild ass at 10 European collections, BPV type 1 was detected in eight and BPV type 2 was detected in one by PCR.[189]

RNA Viruses

Perissodactyls are susceptible to **rabies virus** infection (may be locally or regionally reportable) and, similar to other susceptible mammal species, they can die after developing clinical neurologic disease. Infection has been reported in a zebra foal housed at a safari lodge in Kenya

TABLE 17.3 Mycotic Infections Reported in Nondomestic Perissodactyls

Species/Family	Fungal Pathogen	Comments
Equidae[186]	*Trichophyton*	
Grevy's zebra[24,186]	Phaeohyphomycosis	Captive, mortality
Przewalski's horse[174]	*Coccidioides immitis*	Captive, granulomatous lesions, respiratory system predisposed, males overrepresented, intraspecies aggression suggested
Tapiridae[155]	*Microsporum gypseum, M. canis, Trichophyton tonsurans*	Dermatophytosis. Alopecia in a mountain tapir
Lowland tapir[91,155]	*Coccidioides immitis*	Wild caught from endemic area. Case report, captive. Pulmonary granulomata in association with amebic meningoencephalitis
Rhinoceridae		
White rhinoceros[155]	*Trichophyton mentagrophytes*	Group housed after recent capture
White rhinoceros[5]	*Malassezia pachydermatis, Candida parapsilosis*	Captive, pruritus
Black rhinoceros[155]	*Aspergillus fumigatus, Absidia corymbifera*	Pulmonary aspergillosis
Black rhinoceros[187]	Aspergillosis most frequent	Survey captive United States, $n=8$, pneumonia at necropsy, often concurrent disease, 2/8 mycobacteriosis
Black rhinoceros[185]	*Coccidioides immitis*	Captive, United States, chronic progressive lameness

that was suspected to have been bitten by a feral dog. Approximately 1 month later, the foal died after experiencing neurological disease for 2 days. Rabies was confirmed by direct fluorescent-antibody testing.[82]

Fatal encephalomyocarditis virus (EMCV) infection has been identified in a tapir. White foci and streaks in the myocardium correspond histologically to areas of cardiomyocyte degeneration and mineralization, and interstitial lymphoplasmacytic and occasionally neutrophilic inflammation (Figs. 17.13A, B). Mild to moderate congestion of the lungs and hepatic centrilobular vacuolar degeneration may also be seen. Infection is confirmed by PCR.[181] Several fatal cases of EMCV have also been described in rhinoceroses.[108] While lesions have not been specifically described, viral presence has been confirmed in the heart, spleen, and other viscera. Virus is excreted in the feces and urine of rodents, the natural hosts for the virus. Management of the virus in captive settings should focus on pest control.

Vesicular stomatitis virus (VSV) has been diagnosed in southern white and greater one-horned rhinoceroses as the cause of significant clinical disease (personal communication, P.M. Gaffney). PCR testing confirmed VSV-New Jersey strain infection. Gross findings were localized to mucous membranes and mucocutaneous junctions of the head, or the nails and coronary bands of limbs. On the nostrils, lips, and in the oral cavity, lesions varied from swellings to erosions to ulcers. At the distal limbs, lesions progressed from cracks to ulcers to separation and sloughing of foot pads (Fig. 17.14A). Histologic appearance of sloughed or debrided tissue revealed inflamed to necrotic epidermis, intercellular edema, and cellular dissociation (spongiosis) of the stratum spinosum, progressing to individualized, necrotic keratinocytes within enlarging vesicles. Formation of pustules following necrosis, and secondary bacterial infection, may occur (Fig. 17.14B). Some individuals seroconverted, and VSV nucleic acid was detectable in lesions for at least 6 weeks after the development of gross lesions.

Bacteria

Sporadic ***Mycobacterium bovis*** and ***M. tuberculosis*** (WOAH-listed reportable diseases) infections have been reported in captive rhinoceros.[103,108,155] Most cases have been subclinical and only diagnosed during postmortem examination. Advanced cases may present with chronic progressive respiratory disease. Isolated cases of *M. bovis*

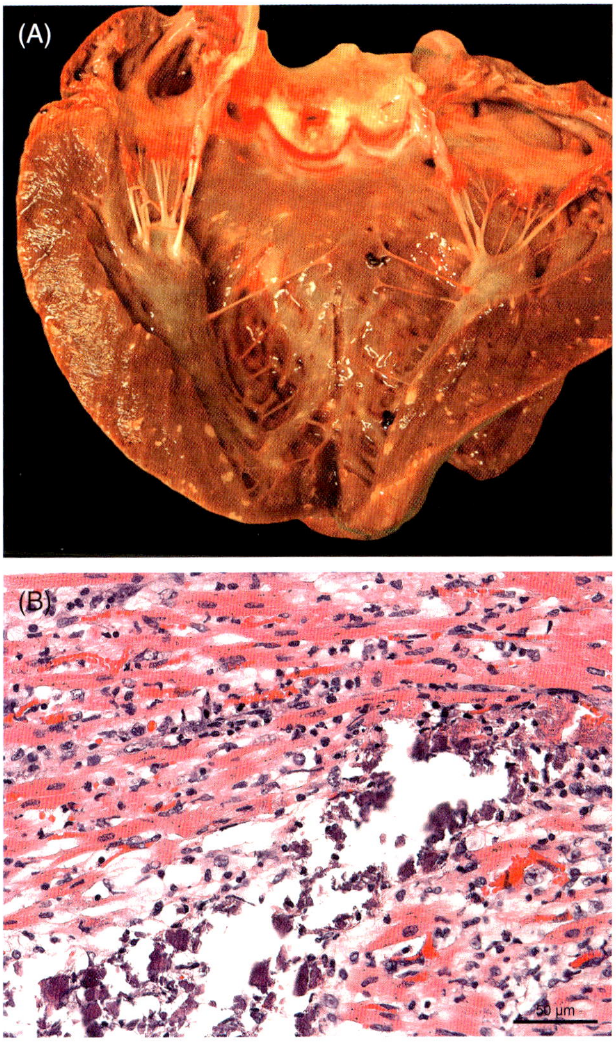

FIGURE 17.13 Encephalomyocarditis virus infection in the heart of a tapir. (A) Multifocal white discoloration is present throughout the myocardium. (B) Discoloration corresponds to foci of myocardial degeneration, mineralization, and mild inflammation. *(Reprinted with permission from* Vercammen F, Bosseler L, Tignon M, Cay AB. Encephalomyocarditis virus in a captive Malayan tapir *(Tapirus indicus). Open Vet J.* 2017;7(2):100–103.*)*

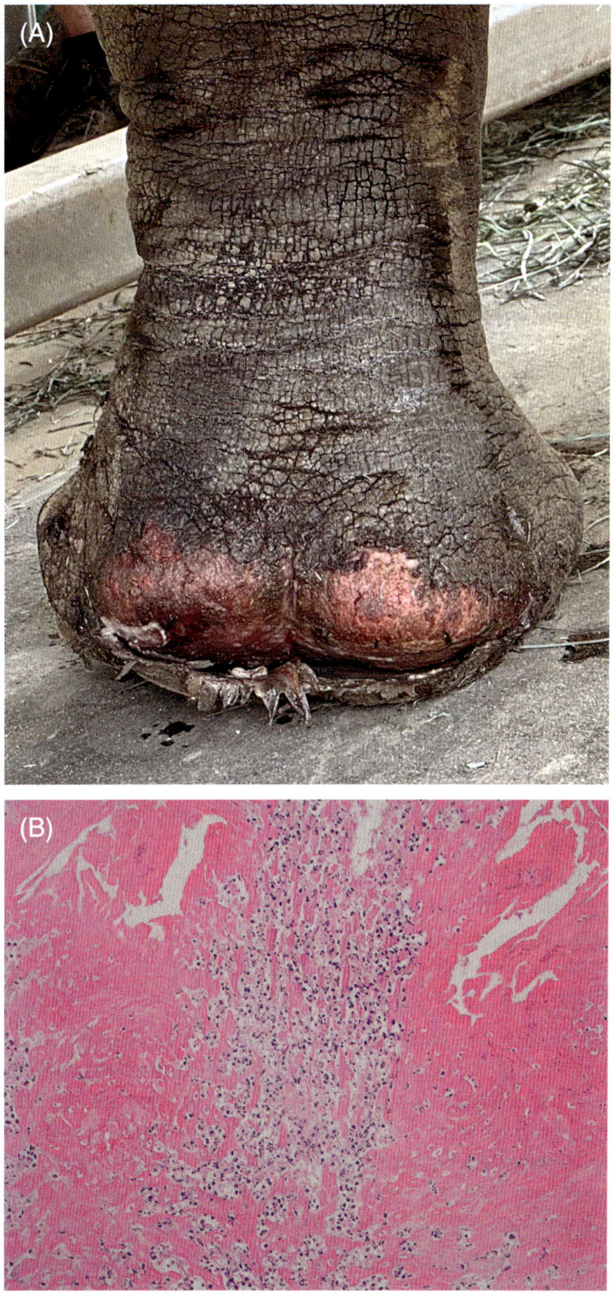

FIGURE 17.14 Vesicular stomatitis virus in the skin of a southern white rhinoceros. (A) Locally extensive ulceration is present along the junction between the skin of the foot and the foot pad. (B) This corresponds to severe epidermal necrosis and neutrophilic inflammation with abundant pyknotic and karyorrhectic debris.

or *M. tuberculosis* have also been reported in free-ranging rhinos.[34,106] The presence of granulomas in the lungs and lymph nodes should raise suspicion for mycobacteriosis (Fig. 17.15; eSlides 17.e7 and 17.e8). Diagnosis is confirmed by bacterial culture and PCR analysis. ***M. orygis*** was recently described in the ***Mycobacterium tuberculosis* complex**. Infections have occurred in East Africa, the Arabian Peninsula, and South Asia. It was reported in a free-ranging Indian rhinoceros in Nepal with pulmonary granulomata and confirmed by molecular methods.[175] A greater one-horned male rhino that was captive-born in India and imported to the United States as a juvenile had a period of lethargy and inappetence as a young adult. A year later the problem recurred and progressed to pelvic limb paresis, ataxia, and gluteal and epaxial muscle atrophy. Fibrohistiocytic to pyogranulomatous inflammation centered at the sublumbar lymph nodes and extending into a sciatic nerve and the cauda equine was identified on gross examination. Lymph node culture and polymerase chain reaction isolated *M. orygis*.[90]

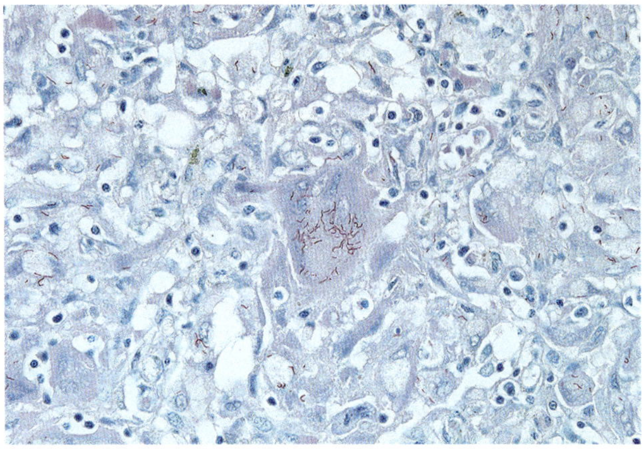

FIGURE 17.15 Pulmonary mycobacteriosis in a black rhinoceros. Histiocytic pneumonia is associated with numerous intracytoplasmic, acid fast positive bacilli in multinucleate giant cells and histiocytes. Ziehl-Neelsen (see eSlides 17.e7 and 17.e8).

Mycobacterium pinnipedii has been isolated from captive pinnipeds worldwide and in the Southern Hemisphere from wild individuals. A captive Malayan tapir, previously housed adjacent to sea lions, developed respiratory signs and weight loss. Sputum was positive for acid-fast bacilli. Necrogranulomas were present in the lungs and mesenteric lymph nodes at necropsy. Spoligotyping was positive for *M. pinnipedii*. Three additional tapirs were seropositive and subsequently euthanized. Granulomatous lymphadenitis was found, and bacterial culture and spoligotyping confirmed *M. pinnipedii* in these three animals.[63] Three lowland tapirs at a Brazilian zoo died with pulmonary granulomas.[120] Polymerase chain reaction of bronchoalveolar lavage samples from two animals was positive for a *M. tuberculosis* complex strain commonly associated with regional human cases of tuberculosis.

Disease caused by the ***Mycobacterium avium–intracellulare* complex** bacteria is becoming more prevalent, including lesions of lymphadenitis, pulmonary granulomata, and disseminated infections. Infections in Perissodactyla include those in a breeding pair of captive-born lowland tapirs from different origins. Both were initially negative within intradermal skin testing. However, both developed mild respiratory signs within a few months of their pairing, and nontuberculous mycobacteria were present among bacteria recovered from sputum. Clinical signs resolved completely in the male with antibiotic treatment, while the female continued to have mild signs. A female offspring born a year later developed respiratory signs at 3 weeks; similar respiratory signs developed in a male offspring born a year later. The female calf recovered with antibiotic treatment but signs progressed in the male calf, which was euthanized. Necropsy revealed chronic bronchopneumonia. Bacterial culture was positive for ***M. avium* subspecies *hominissuis***.[95] The remaining tapirs showed weight loss and respiratory disease, and *M. avium* ssp. *hominissuis* was cultured from bronchoalveolar lavage and identified by PCR testing. Following a long treatment course the group tested negative. The source of mycobacteriosis was thought to be environmental.[94]

Paratuberculosis or **Johne's disease** caused by ***M. avium* subspecies *paratuberculosis*** is a WOAH-listed reportable disease that is typically seen in ruminants. Clinical signs include progressive weight loss, debilitation, and occasionally death. Infection causes histiocytic to granulomatous ileitis-enteritis. Johne's disease is uncommon in perissodactyls. However, the bacterium has been isolated from the feces of a wild-caught black rhino with diarrhea and weight loss.[13] Due to the conservation value of the individual, antimycobacterials were administered until fecal shedding was no longer detected.

Anthrax, a WOAH-listed reportable, zoonotic disease due to infection with ***Bacillus anthracis***, is a significant problem in the free-living *Equidae* and *Rhinoceridae* in Africa and is a significant threat to wild populations of Grevy's zebra. In periods of drought, increased trauma to the oral cavity from dry or woody browse may increase the risk of spore acquisition. Infected animals may be asymptomatic and die unexpectedly. Bloody discharges from the mouth, nostrils, ears, or anus; rapid bloating of the carcass; incomplete rigor mortis; and dark unclotted blood should raise one's index of suspicion for infection. Differential diagnoses include several other important infectious diseases including African horse sickness, botulism, Rift Valley fever, and snake bite. Vaccination has been conducted in some parks in Kenya and Namibia. Efficacy has not yet been determined.[59,105,186] Transmission of anthrax in herbivores is assumed to occur by ingestion of spores. A recent study found the potential for infection by inhalation during dust bathing was unlikely.[3] Caution should be used in the handling and necropsy of suspect cases due to the risk of zoonotic transmission. Cytology is a recommended early diagnostic test that should be performed prior to other sample collection or necropsy as it may confirm infection with the encapsulated, square-end bacilli in suspect cases and inform risk assessment (see Wildlife Necropsy).

Leptospirosis occurs worldwide, especially in moist environments. Infection is associated with rodent transmission and water contamination. Leptospirosis was associated with early cases of hemolytic anemia in captive black rhinoceros. In one survey, leptospirosis was considered the cause of hemolytic anemia in 9 of 31 cases.[62] At necropsy, red fluid was found in the pericardial sac, peritoneum, and urinary bladder, and marked edema was noted throughout the omentum, lungs, and perirenal tissues. Leptospirosis was implicated by microscopic agglutination test. Blood collected from rats trapped on zoo grounds where two rhinos died from leptospirosis had significant titers for ***Leptospira icterohaemorrhagiae***. Rats were considered to be the vectors in these cases.[27] In another study, antibodies to 8 serovars of ***L. interrogans*** were reported in free-ranging and captive black rhinoceros. The free-ranging black rhinoceros had low antibody levels and serovars varied with region and ecosystem. In other infections, *L. interrogans* was the cause of death in black rhinoceros that died of hemolytic anemia and in another rhinoceros that died as a complication of ulcerative skin disease.[62] Peracute depression, anorexia, hind leg trembling, dysuria, glucosuria, gastrointestinal discomfort, and decreased fecal output in the absence of hemolytic anemia have also been associated with *L. interrogans* infection in captive black rhinos[125] and with abortion in an Indian rhinoceros.[111]

Salmonellosis is a common cause of septicemia and enteritis in rhinoceros and tapir[155] and of abortion and genital tract infection of equids.[186] Asymptomatic shedding may occur and antibiotic resistance may quickly develop with transfer of plasmids between *Salmonella* species. Salmonellosis has been reported in wild black rhinoceros with diarrhea during translocations. ***Salmonella enterica* subspecies *arizonae*** was cultured from three black rhinoceros from one American zoo that died with gastroenteritis and sepsis.[67] Other reported signs of salmonellosis include lethargy, partial anorexia, epistaxis, ulcerative skin lesions, and/or diarrhea in a captive female and two male rhinos. In one report, Tokay geckos released into a building housing rhinos were the suspected source of salmonellosis (but instead *S. eastbourne* was isolated from the intestine of one of two geckos euthanized). Gross lesions of salmonellosis in these rhinos included fibrinous pleuritis, a large lung abscess in one, and resorptive bone lesions in the distal digits. In one survey of salmonellosis conducted across rhinoceros species from 1990 to 1997,[66] 11% of fecal samples were culture positive for *Salmonella* spp. Some animals suspected of having clinical salmonellosis had leukopenia, hemoconcentration, and hyperproteinemia. In four, sepsis and death were attributed to salmonellosis. Monthly surveillance for salmonellosis in a captive, asymptomatic herd of black rhinoceros for nearly 3 years by fecal culture and PCR identified periodic shedding in four of six rhinoceros.[104] This is important from a management perspective as periodic, intermittent testing is often needed to confirm the presence or clearing of infection and shedding. Clinical cases of salmonellosis in black rhino calves have also been reported. Clinical signs can include self-limiting diarrhea or sepsis-related death (death may be related to low transfer of maternal immunity and primiparity).[89]

Other notable bacterial diseases in perissodactyls include **tetanus** due to ***Clostridium tetani*** of a hoof abscess in an onager, and bacterial sepsis secondary to ***C. welchii*** wound infection in a Somali wild ass.[135] Unexpected death, presumably from ***C. perfringens*** enterotoxemia has been reported in white rhinoceros; the main findings were mild abdominal exudate, splenic and multifocal gastrointestinal congestion, and adrenal hemorrhage.[15] ***Streptococcus equi*** causing soft tissue swelling due to suppurative lymphadenitis that is most often associated with infection of the pharynx and respiratory tract, strangles, was the cause of a cluster of deaths in Przewalski's horses reintroduced into Southwestern Mongolia. A harsh winter and wolf predation were suspected as predisposing factors.[159] **Bacterial vegetative valvular endocarditis** was reported in a captive, aged, wild-caught male white rhinoceros that developed immobility and decubital ulcers. There was roughening at the left atrioventricular valve and ventricular endocardium. This corresponded histologically to proliferative inflammation on the valve leaflet (vegetative endocarditis) and associated myocardial degeneration. Fibrosis and degeneration in the liver and chronic nephritis developed as sequelae. ***Streptococcus dysgalactiae*** subsp. ***equisimilis*** and ***S. ovis*** were isolated from multiple sites. The decubitus ulcers were the suspected portal of bacterial infection.[55]

Anaplasma phagocytophilum is a gram-negative rickettsial organism that is transmitted by *Ixodes* spp. ticks. The rickettsial has a tropism for granulocytes and forms morulae in their cytoplasm that appear as granular inclusion bodies. Anaplasmosis has been reported in Przewalski's horses in managed care based on the presence of intracytoplasmic morulae in cytologic or hematologic preparations and PCR assay of whole blood. Signs included dehydration, pyrexia, and lethargy with possible ataxia and contusions.[168]

Coxiella burnetti, the cause of **coxiellosis** (also known as **Q fever**), is most commonly reported in ruminants and most often causes abortion. Infection has been described in captive primiparous white rhinoceroses.[6] The most characteristic histologic lesion is the presence of gram-negative intracellular bacteria in necrotic placental trophoblasts and macrophages. PCR testing of frozen placenta can confirm the presence of *Coxiella burnetti* DNA. Vaginal shedding from infected female rhinos has been shown to last several weeks. Transmission is usually by aerosols. The disease may be subclinical and reproductive failure may suggest an underlying or developing problem. Human cases of coxiellosis are reportable and the zoonotic implications of the disease should be considered. In human infections, acute infection is characterized by phase II antibodies and chronic infection by phase I antibodies on serology. Serological study showed similar phase antibody patterns in rhinoceros dams.

Fungi

Phaeohyphomycosis is an opportunistic, dematiaceous (pigmented) fungal infection caused by a variety of fungal species. These fungal organisms are ubiquitous in soil and woody plants. Systemic phaeohyphomycosis is rare.

Vascular invasion is infrequent and often associated with immunosuppression. Infection usually originates in the respiratory system following inhalation, while cutaneous disease follows wound infection. Infection in a young adult, captive male Grevy's zebra caused sudden onset weight loss, lethargy, and hypothermia.[24] Focal pyogranulomatous pneumonia, myocarditis, and pericardial effusion were associated with intralesional, 3–6 μm diameter moniliform fungal hyphae. Yeast forms were found in the heart.

Coccidioides immitis is a dimorphic fungus that is present in the soil and may become airborne. Transmission is usually by inhalation of arthroconidia. Most infections are asymptomatic and resolve spontaneously. **Coccidioidomycosis** was the leading cause of death in a captive population of Przewalski's horses between 1980 and 2000 that were housed in Southern California, United States.[174] Abscesses, granulomas, pyogranulomas, and/or granulomatous inflammation with intralesional, thick-walled, 20–40 μm diameter, fungal spherules, and internal endospores were often disseminated, but the respiratory system was most often affected (Figs. 17.16A, B). Cases occurred more commonly in males and younger individuals, possibly due to intraspecific aggression and stress in the bachelor herds. Skeletal muscle lesions in the hind quarters were consistent with bite wound contamination or systemic spread to an inflamed vascular area. Multiple genetic lineages of Przewalski's horse were affected, while other equid species at the facility were unaffected. Przewalski's and domestic horses had similar numbers of CD3, CD4, CD5, CD8, and CD21 lymphocyte subsets by flow cytometry. Lymphocyte blastogenesis assay responses to the generic T-cell mitogen concanavalin A were alike but lymphocytes of two Przewalski's horses failed to proliferate in response to coccidioidomycosis antigen, suggesting a poor immune response to this specific pathogen.[174] Coccidioidomycosis was also identified incidentally in pulmonary and lymph node granulomas and in the synovium of a black rhinoceros with chronic progressive lameness. Initial serological testing was negative, though a positive titer was noted at the time of euthanasia.[185]

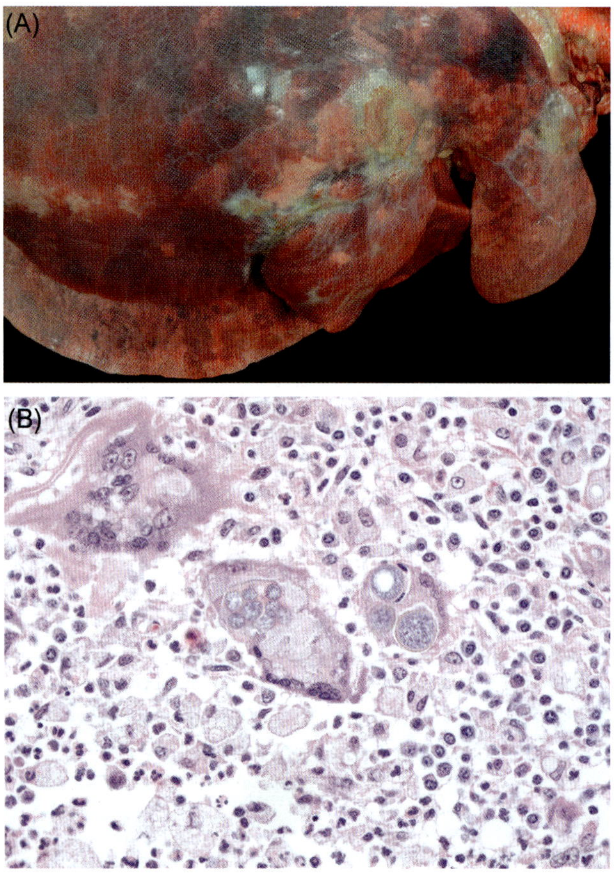

FIGURE 17.16 *Coccidioides immitis* infection in the lung of a Przewalski's horse. (A) Multiple yellow-white pyogranulomas are present throughout the parenchyma. (B) Histiocytes and neutrophils are admixed with lymphocytes and plasma cells. Fungal spherules containing endospores are present within multinucleate giant cells.

Concurrent infection with ***Malassezia pachydermatis*** and ***Candida parapsilosis*** has been described in a white rhinoceros that developed pruritic, erosive lesions up to 20 cm in diameter along the dorsum. Lesions were hyperemic with exudate. Cytologic examination of the exudate was positive for *Malassezia* spp. type yeasts. Biopsy showed exudative epidermitis. Fungal culture confirmed concurrent infection. *Malassezia* spp. have been cultured from the skin of unaffected individuals suggesting it may be part of the normal flora of rhinoceros skin.[5]

Metazoa

Stephanofilaria dinniki has been reported in free-ranging black and white rhinos with crusty, dry raised areas of dermatitis (Fig. 17.17). Histologically, ulcerative dermatitis and lymphohistiocytic nodules with serocellular crusts and intralesional, epidermal adult filarial nematodes are associated with necrosis, dermal granulomas, and mixed inflammation. Acanthosis and rete ridge formation are present at lesion margins. Chronic lesions have thick granulation tissue deposits in the dermis and there is marked inflammation with eosinophils, histiocytes, and multinucleate giant cells. Both adults and microfilariae can be seen in lesions. There is a pattern of seasonal variation in infection, with pruritus and erythema in the summer and resolution in the winter.[75] Crusts fall off to reveal red areas and leave depigmented scars. White rhinos have larger lesions that average 23 cm in diameter relative to 15 cm in black rhinos. Black rhino lesions are typically in the shoulder and ventral thorax, while white rhino lesions are seen at the rump. The life cycle of *S. dinniki* is unknown but a bloodsucking arthropod is required to complete the cycle. Poor body condition and heavy rainfall are thought

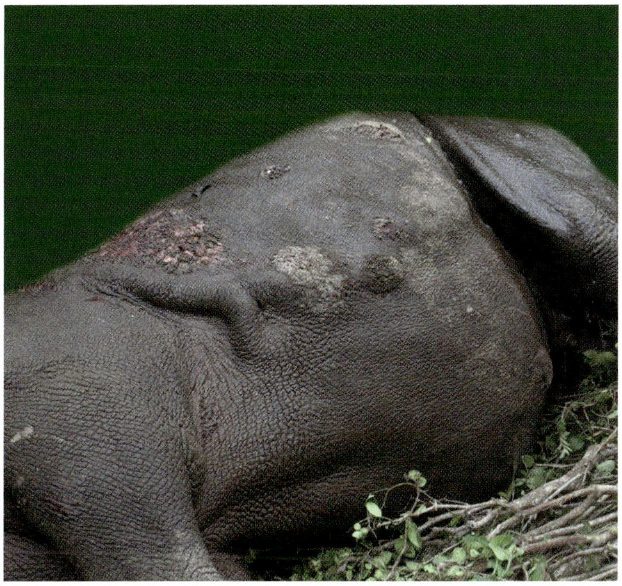

FIGURE 17.17 *Stephanofilaria dinniki* infection in the skin of a black rhinoceros. Irregular raised crusty lesions with multifocal ulceration are present along the lateral body wall. *(Photo Courtesy of S. Chege, Vetinwild.)*

FIGURE 17.18 *Kiluluma longipene* infection in the stomach of a lowland tapir. Numerous slender, white nematodes are admixed within the gastric content. *(Photo Courtesy of the Lowland Tapir Conservation Initiative.)*

to lead to recurrence of the lesions. Red-billed oxpeckers may aid in parasite spread and cause ulceration.[123]

Halicephalobus gingivalis has been reported in a captive female Grevy's zebra that appeared blind and demonstrated abnormal behavior.[57] Bilateral uveitis and associated normocytic, hyperchromic anemia, and elevated total protein and globulin with elevation of beta globulin fraction were noted. Disseminated granulomatous inflammation was found in the kidneys, heart, eyes, uterus, and lymph nodes in association with rhabditiform nematode infection. Cataracts and marked granulomatous inflammation of the ciliary body and choroid with synechiae were also seen. Larvae with adults were associated with the inflammatory infiltrates. Only female adults have been identified in infections suggesting parthenogenic reproduction.

Gastric nematodiasis due to ***Kiluluma longipene*** (Fig. 17.18) and ***Physocephalus*** spp. have been reported in free-ranging tapir. The latter is associated with proliferative mucosal nodules with central cavitations and intralesional slender red worms (Fig. 17.19).

Cestode infections have been identified nondomestic perissodactyla including tapir, Przewalski's horses, and rhinoceros (Figs. 17.20A, B). Reports include descriptions of several species of ***Anoplocephala***. Most reports list these parasites, even when loads are high, as incidental findings.

Schistosomiasis was found in a juvenile captive lowland tapir with a history of occasional bloody diarrhea. Multiple, often periportal, granulomas were present in the liver and eosinophils and multinucleate giant cells were

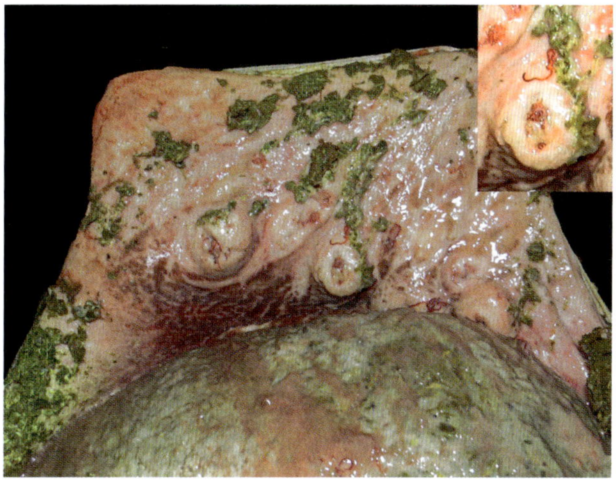

FIGURE 17.19 *Physocephalus* spp. infection in the stomach of a lowland tapir. Multiple raised nodules in the gastric wall contain a central cavitation with slender, intralesional red nematodes. *(Photo Courtesy of the Lowland Tapir Conservation Initiative.)*

associated with pigmented embryonated eggs. Hemorrhagic gastritis and enteritis were seen grossly and many trematode eggs were seen histologically within villi. Trematodes of schistosomiasis are intravascular; eggs are shed in the mesenteric venous plexuses and may backwash into the portal system or penetrate the intestinal villi.[193]

Protozoa

Piroplasmosis is an infectious, tick-borne disease that is caused by ***Babesia caballus*** or ***Theileria equi***. The parasites attack and destroy red blood cells. Infection may occur seasonally and parallels the vector tick lifecycle. Host

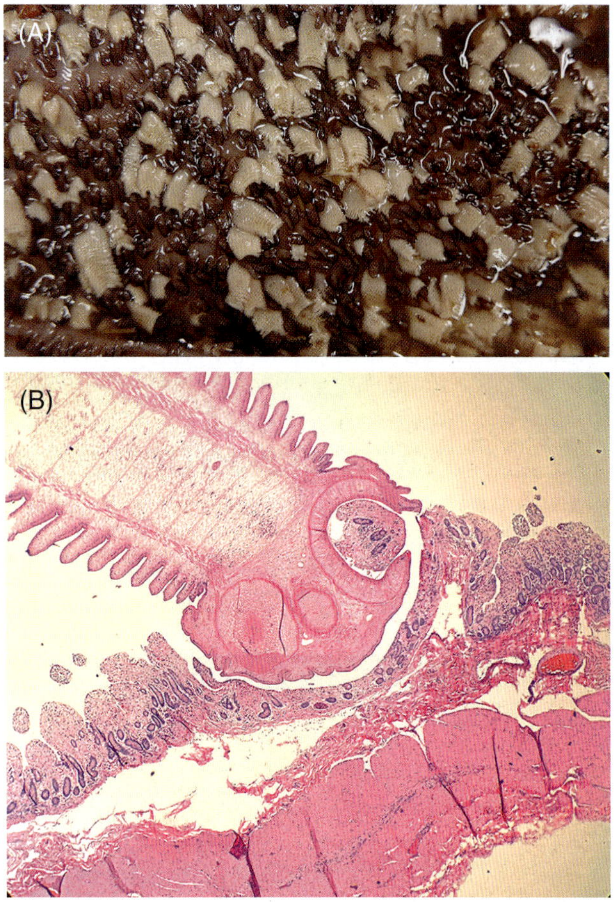

FIGURE 17.20 Enteric cestodiasis in a greater one-horned rhinoceros. (A) Myriad short, white cestodes fill the lumen and are admixed with luminal content. (B) Characteristic features of cestodes include a thick tegument, segmented body, lack of a body cavity and intestine, and multiple suckers on the scolex (head).

translocation may induce disease by spreading pathogens and modulating the effects of stress hormones on the immune system, allowing piroplasm proliferation.

In equids, severe regenerative hemolytic anemia and jaundice are seen in acute cases, while chronic cases may be seen in subclinical carriers. Gross findings include pulmonary edema and splenic congestion. Histologically, renal tubular necrosis and disseminated hemosiderin-laden macrophages are noted. Transplacental infection may cause abortion. Aborted fetuses have high levels of parasitemia. In a mortality survey of Przewalski's horses reintroduced to Southwestern Mongolia, piroplasmosis was implicated as the cause of death of stallions and a stillborn foal.[159] Babesiosis has been reported in a Somali wild ass stallion.[135] In cograzing donkeys and wild Grevy's zebra in Kenya, all zebra tested positive for *T. equi*.[51]

Piroplasmosis in black rhinoceros, caused by ***Babesia bicornis*** and ***Theileria bicornis***, has been shown to be fatal, especially in small and isolated populations in Tanzania and South Africa. In one study, *B. bicornis* was not detected in Kenya, but *T. bicornis* had a 49% prevalence in white rhinoceros and 43% in black rhinoceros. Infections have an increased importance because of conservation efforts related to animal relocations.[134]

Sarcocystis neurona has been isolated from domestic horses with **equine protozoal myeloencephalitis**. The opossum, *Didelphis virginiana*, is the definitive host of *Sarcocystis falcatula* and may be the definitive host of *S. neurona*. Similar lesions have been described in a captive Grant's zebra that developed acute ataxia, weakness, and depression. Cerebrospinal fluid was positive for antibodies to *S. neurona* on Western blot. Perivascular cuffing with mononuclear cells was noted focally at the brainstem. *S. neurona* was identified by immunohistochemical labeling of intralesional merozoites.[96]

Klossiella equi is a coccidian parasite of the renal parenchyma of equids. Its complete lifecycle is unknown. Schizogony (asexual reproduction) is thought to occur in the endothelial cells of Bowman's capsule. Merozoites are released and enter the proximal convoluted tubules to develop into second-generation schizonts. Merozoites are released into the lumen and enter cells of the loop of Henlé, where they develop into microgametocytes and macrogametocytes. When pronuclei of micro and macrogametocytes fuse, a zygote/sporont develops, sporogony occurs, and sporoblasts form that develop into sporocysts. Sporocysts are released into the lumen and pass out in the urine. The cycle is completed when sporocysts are ingested and sporozoites are released into the digestive tract, penetrate the wall, enter the blood stream, and travel to the glomeruli. The coccidian has been described as an incidental finding in sections of kidney from Hartmann's mountain zebra[172] and Przewalski's horses.[159]

Disseminated **neosporosis** was found on histopathologic examination of tissues from an adult, wild-caught white rhinoceros that died unexpectedly. At necropsy, hepatomegaly with multifocal necrosis, erosions to ulcerations of the stomach, and enlargement of the mesenteric lymph nodes were seen. Histologically scattered foci of necrosis with tachyzoites were present in liver, adrenal cortex, kidney, and intestine. Immunohistochemistry and polymerase chain reaction confirmed infection with ***Neospora caninum***.[171] Parasite transmission may be horizontal (ingestion) or vertical (transplacental). A source of infection was not identified but was likely through ingested oocysts. Abortion is the most common sign of infection in the intermediate host and cases of abortion have been reported in a Malayan tapir and a white rhinoceros.[147,163] Definitive hosts (dogs) most commonly have paresis and paralysis of the hindlimbs.

Necrosuppurative meningoencephalitis due to infection with ***Naegleria fowleri*** has been reported in a captive South American tapir. Inflammation was associated with perivascular ameboid trophozoites. Confirmation was by monoclonal antibody testing. In humans, infection is associated with water contact. Olfactory neuroepithelium is the portal of entry for the organisms.[91]

ADDITIONAL RESOURCES

Giundi, M., 2017. Liver disease in iron overload. In: Saxena, R., (Ed.), Practical Hepatic Pathology: A Diagnostic Approach. Elsevier, pp. 151–165.

Miller, M.A., Buss, P.E., 2015. Rhinoceridae (Rhinoceroses). In: Miller, R.E., Fowler, M.E. (Eds.), Fowler's Zoo and Wild Animal Medicine, 8, Elsevier, Saint Louis, Missouri, pp. 538–547.

Munson, L., Miller, R.E., 1999. Skin diseases of black rhinoceroses. In: Fowler, M.E., Miller, R.E. (Eds.), Zoo and Wild Animal Medicine, Current Therapy. fourth ed. W.B. Saunders Company, Philadelphia, Pennsylvania, pp. 551–555.

Zimmerman, D.M., Hernandez, S., 2015. Tapiridae. In: Miller, R.E., Fowler, M.E. (Eds.), Fowler's Zoo and Wild Animal Medicine, 8, Elsevier, Saint Louis, Missouri, pp. 547–559.

References

Full references for this chapter can be found online at https://doi.org/10.1016/B978-0-443-18437-6.00019-5.

Chapter 18

Monotremes and Marsupials

Shannon L. Donahoe[a], Jane Hall[b], David Spratt[c], and Damien Higgins[a]

[a]*Sydney School of Veterinary Science, Faculty of Science, The University of Sydney, Camperdown, NSW, Australia;* [b]*Australian Registry of Wildlife Health, Taronga Conservation Society Australia, Mosman, NSW, Australia;* [c]*CSIRO, Australian National Wildlife Collection, Canberra, ACT, Australia*

INTRODUCTION

The marsupials (metatheria) and monotremes (monotremata) are an extremely diverse group that includes approximately 340 species across 25 families. They occupy a wide range of habitats in the Americas, Australia, and Papua New Guinea. Their behavioral adaptations range from social to solitary, nocturnal to diurnal, and specialist to generalist. The majority are either cryptic or remote. This has shaped our understanding of disease in the group with most of our knowledge being in either relatively common species that are encountered in proximity to humans or in the few species that are the subject of active threatened species recovery programs. It is safe to assume that we have only scratched the surface in identifying the range of infectious agents and non-infectious processes that affect the taxa and understanding their significance in disease.

In Australia, where the majority of this group is endemic, many species are in serious decline (see Digital Content Table 18.e1). As specialized small mammals, they are under significant pressure from predation by introduced carnivores such as cats, foxes, and dogs, habitat fragmentation and degradation, land clearing, and extreme weather events such as drought, fire, and floods. Our understanding of the role of disease in these declines, either as a primary or compounding factor, is limited even in the most studied and iconic species, such as the koala. As in many countries, the field of wildlife disease is growing due to investment in training and capacity, developments in technology, and the awareness raised by the emergence of a few, high-profile diseases, such as devil facial tumor disease (DFTD). However, further research is needed to assess associated risks; particularly as the resilience and isolation of these species are increasingly challenged by changes to their environment. In relation to diseases of importance to livestock or public health, the roles of members of the taxa as potential reservoirs are less predictable than that of their better understood eutherian counterparts.

NON-INFECTIOUS DISEASES

Nutritional

Skeletal deformities consistent with **vitamin D deficiency** have been reported in hand-raised pouch young of a range of species. These commonly manifest as deformities of the rib cage. **Metabolic bone disease**, leading to hip and shoulder dysplasia, has been detected radiographically in young koalas and is associated with being housed in shaded enclosures. Serum calcium, phosphorus, and alkaline phosphatase in these cases are within normal limits.[189]

A vitamin E-responsive **nutritional myopathy**, distinct from exertional rhabdomyolysis (capture myopathy), has been observed in a range of macropod species.[3,16,130,220] Although any muscle can be affected, changes are most common in the pelvic and femoral muscles and to a lesser extent in the muscles of the thorax and spine. Myocardial lesions can also occur and are common in the quokka, in which the disease has been used as an experimental model for myotonia and muscular dystrophy in humans. Gross lesions vary from pale streaks to generalized pallor and atrophy. Tissues may become yellow-brown or friable.[137] Histologically, myocyte lesions are polyphasic (unless an animal is presenting in a very acute stage of disease) and include hyalinization of myofibers with granular or vacuolar degeneration and occasional mineralization. Myofiber swelling with pyknosis of sarcolemmal nuclei is apparent in the early stages in some affected areas, along with proliferation in adjacent viable tissue. In more chronic lesions, inflammation and repair are evident. In a small proportion of cases, there is acute renal tubular necrosis, presumably either as a result of ischemia, myoglobinuria, or both.[137]

Pathology of Wildlife and Zoo Animals. https://doi.org/10.1016/B978-0-443-18437-6.00036-5

Metabolic

Oxalate nephrosis has been reported in several species including the koala, Gilbert's potoroo, swamp wallaby, and scaly-tailed possum.[84,90,222] Its cause is poorly understood. In some cases, it seems likely that deposition of oxalate crystals in renal tubules is secondary to renal compromise rather than the cause. For example, in koalas, mild oxalate nephrosis is commonly considered an incidental finding outside of South Australia. However, in approximately 50% of the koalas in the Mt. Lofty Ranges of South Australia, oxalates cause significant nephrosis and renal failure without other concurrent renal lesions. Based on current evidence the source of oxalate is unlikely to be dietary,[223] and a genetic factor in this highly inbred population appears likely (e.g., primary hyperoxaluria as occurs in humans, dogs, and cats).[238] Affected koalas have tubular loss and dilation, cortical fibrosis, glomerular atrophy, and crystal-associated inflammation.[137,222] Some show a segmental, cortical pattern of fibrosis similar to that described in humans with advanced nephrocalcinosis, which may be the result of tubular obstruction with crystals and subsequent areas of cortical atrophy. Oxalate crystals have a typical morphology and are birefringent under polarized light (Figs. 18.1A, B).[222]

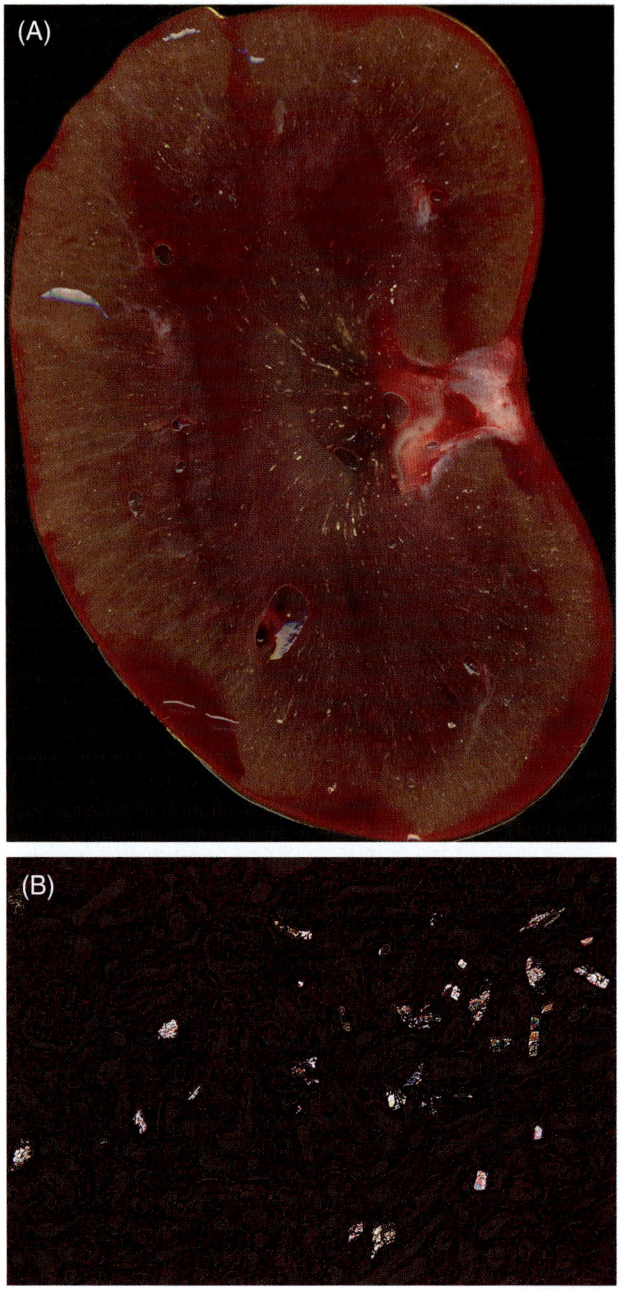

FIGURE 18.1 (A) Oxalate nephrosis in a koala. Oxalate deposits within collecting ducts are visible grossly on the cut surface of a kidney as pale radial corticomedullary streaks. (B) Oxalate nephrosis in a koala. Multicolored birefringent oxalate crystals in tubules and the collecting ducts of the kidney viewed with polarized light. The crystals are arranged in sheaves, rhomboids, and rosettes. *(Parts A, B: Photos Courtesy of N. Speight, Adelaide University.)*

Toxic

Neurotoxicity has been reported in eastern grey kangaroos in Victoria, Australia that grazed on pasture containing ***Phalaris* spp.**[10,239] The toxic compounds in *Phalaris* spp. are indole-derived tryptamine alkaloids, mainly dimethyltryptamine, 5-methoxy-*N,N*-dimethyltryptamine and 5-hydroxy-*N,N*-dimethyltryptamine (also known as **bufotenine**). These are structurally similar to serotonin and competitively inhibit the initial step in the breakdown of serotonin by monoamine oxidase. This causes accumulation of serotonin and other catecholamines, resulting in enhanced response to excitatory inputs. Neurological signs include disorientation, ataxia, wide-based stance, hypermetric gait and generalized muscle tremors, exaggerated tail balancing, and hyper-responsiveness. Grossly, there is prominent greenish-gray discoloration of the gray matter, particularly of the thalamus, brainstem, cerebral cortex, and spinal cord (Fig. 18.2). Microscopically, within neurons, there is accumulation of brown intracytoplasmic pigment that reacts with melanin stains including Fontana-Masson, Schmorl's, and Warthin-Starry. Recent ultrastructure analysis has confirmed the presence of intra-neuronal neuromelanin granules.[239] Similar pigment can also be seen in renal tubular epithelial cells and ganglion cells of the retina and submucosal plexus of the intestines.

Hepatogenous **photosensitization** secondary to **steroidal saponin hepatotoxicity** has been described in eastern grey kangaroos in New South Wales, Australia that grazed on pasture dominated by ***Panicum gilvum***.[233] Affected animals presented with photophobic behavior, blindness, ataxia, lethargy, recumbency, and ear shaking.

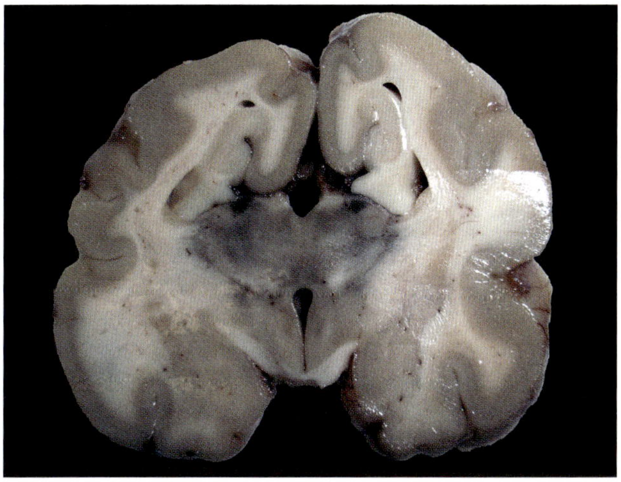

FIGURE 18.2 Phalaris toxicity in the brain of an eastern grey kangaroo. Bilateral prominent greenish-gray discoloration of the thalamus is the result of intraneuronal accumulation of neuromelanin. *(Photo Courtesy of University of Melbourne.)*

Grossly, the kangaroos had subtle to overt skin lesions consisting of excoriations, thickening and scabbing, corneal opacity, and hepatic discoloration. Icterus was seen in more severely affected animals. Histologic lesions varied from mild to severe and included ulcerative and necrotizing dermatitis, cholangiohepatitis with aggregates of acicular colorless clefts in portal areas, corneal edema, and keratitis.

Suspected **pyrrolizidine alkaloid hepatotoxicosis** has been reported in free-ranging southern hairy-nosed wombats in South Australia.[268] Affected animals were predominately juveniles in poor to emaciated body condition. Severe bilaterally symmetrical exudative and necrohemorrhagic dermatitis on the face (particularly on the eyelids and ears), alopecia in sun-exposed sites on the dorsum and lateral body, and icterus were observed (Figs. 18.3A, B). A shrunken nodular and fibrotic liver was identified at necropsy. Histologically, the skin had cutaneous vascular degeneration, thrombosis, and infarction. Changes in the liver were compatible with pyrrolizidine alkaloid hepatopathy in other species and included marked biliary hyperplasia, portal and bridging hepatic fibrosis, nodular regeneration, and hepatocellular megalocytosis and multinucleation (Fig. 18.3C). Toxicological analysis performed on scats and gastrointestinal contents from wombats in the same habitat identified a number of toxic pyrrolizidine alkaloids consistent with ingestion of ***Heliotropeum europaeum***. Classic hepatic lesions of pyrrolizidine alkaloid hepatotoxicosis have also been seen in free-ranging red kangaroos from habitats where pyrrolizidine alkaloid-containing plants occur.[137]

Fluorosis has been observed in a population of eastern grey kangaroos, red-necked wallabies, swamp wallabies, koalas, common brushtail possums, and common ringtail possums, resident near a fluoride-emitting aluminum smelter in southeastern Australia. Overall, eastern grey kangaroo bone fluoride concentrations, determined using a selection of sites across the skeleton, were markedly elevated (up to 20× higher than controls).[112] Lesions were positively associated with bone fluoride concentration and consisted of multiple, large, smooth exostoses over the diaphysis of long bones especially, but not exclusively, the tibia, fibula, and metatarsi (Fig. 18.4). There was also degenerative joint disease including periarticular osteophytosis, articular cartilage erosion/ulceration, synovial hyperplasia, and joint capsule fibrosis, most commonly in the knee, hock, and metatarsophalangeal joints. Articular cartilage changes ranged from small (2–3 mm) pale cysts to more extensive uneven and eventually pale staining of the cartilage with increasing chondrocyte clustering (most severe cases with >5 cells per cluster) and cellular hypertrophy. Bony lesions were typical of exostosis with numerous concentric basophilic reversal lines in cortical bone.[113] Changes in the enamel of affected kangaroos were consistent with fluorosis in other species and included varying degrees of light-brown to almost black staining and surface defects consistent with enamel loss or hypoplasia.[132]

Aluminum toxicity has been implicated in renal failure in free-ranging koalas in the Adelaide Hills, South Australia. The pathogenesis of this syndrome has not been confirmed but affected koalas were exposed to dietary aluminum exceeding that considered nephrotoxic in rats (5 mg/kg bodyweight/day). Affected koalas had polyuria/polydipsia, azotemia, and isosthenuria but routine histological examination of kidneys detected no significant changes. More detailed examination using solochrome azurine and Perl's Prussian blue stains, as well as electron microscopy, confirmed the presence of aluminum in the cytoplasm of proximal convoluted tubules associated with reduced lysosomal and mitochondrial densities. These findings are consistent with aluminum nephrotoxicity in other species.[103]

Congenital/Genetic

Congenital defects in Australian marsupials are rare and are most described in the koala. Documented defects in koalas include **microphthalmia, spinal scoliosis**, and **testicular hypoplasia** or **aplasia**.[40,87,206,238] Testicular lesions have primarily been described in koala populations in the states of South Australia and Victoria that have been subject to marked and sometimes repeated genetic bottlenecks. In one study of a single genetically isolated population on Kangaroo Island, 12.9% of male koalas were reported to have unilateral (7.5% right or 4.5% left) or bilateral (0.9%) testicular aplasia.[206] Microphthalmia and scoliosis occur sporadically across the koala's range.

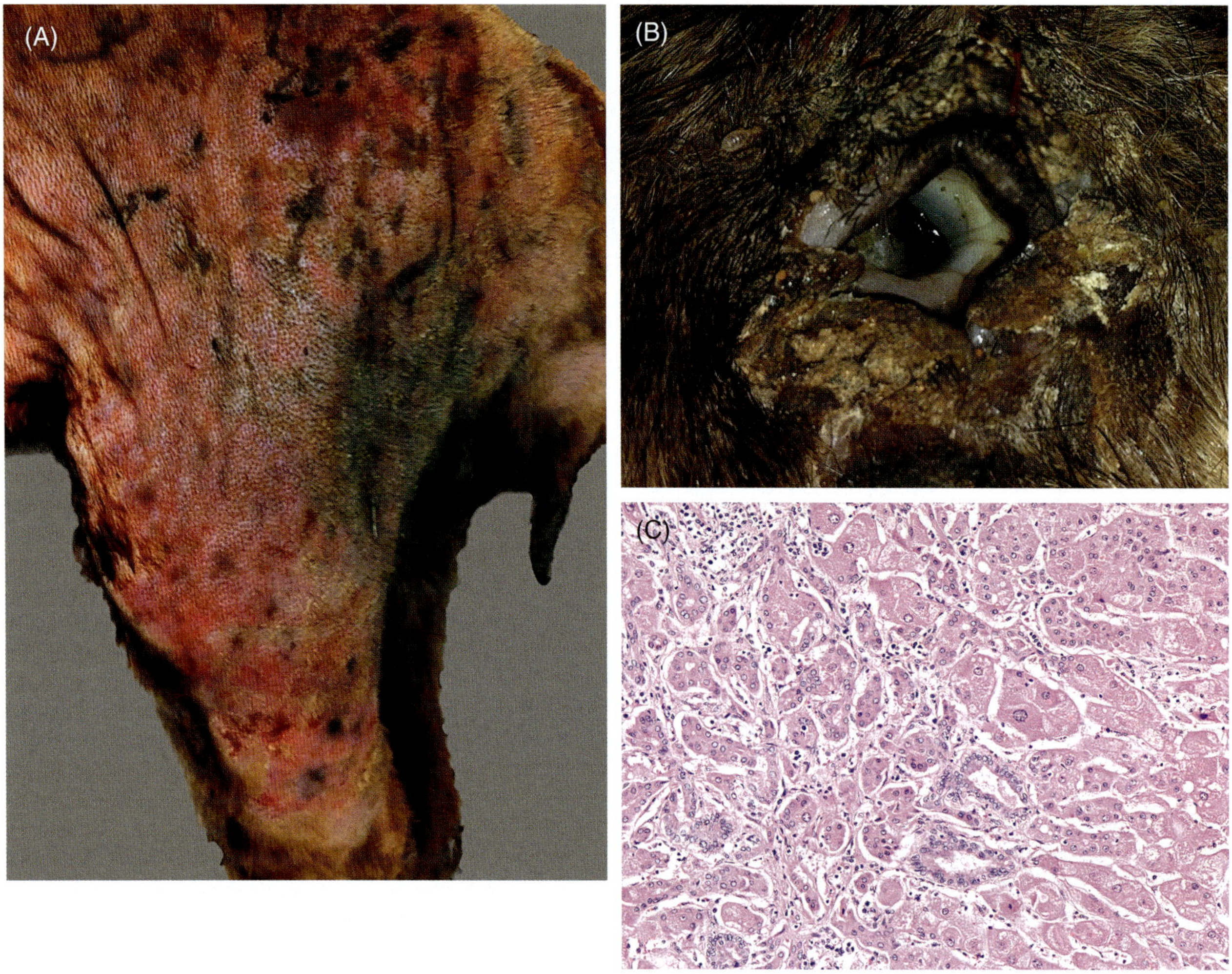

FIGURE 18.3 (A) Photosensitivity due to pyrrolizidine alkaloid toxicity in the skin of a southern hairy-nosed wombat. Necrohemorrhagic and exudative dermatitis and alopecia are the results of cutaneous thrombosis and infarction on the lateral aspect of the hindlimbs. (B) Photosensitivity due to pyrrolizidine alkaloid toxicity in the skin of a southern hairy-nosed wombat. Necrotizing and exudative periocular dermatitis and blepharitis. A louse is present on the fur at the upper left of the image. (C) Pyrrolizidine alkaloid toxicity in the liver of a southern hairy-nosed wombat. Characteristic features include hepatocellular megalocytosis, multinucleation, anisocytosis, anisokaryosis, atypical ductular proliferation, and lymphomonocytic infiltrates. *(Parts A–C: Photos Courtesy of L. Woolford, University of Adelaide.)*

Cryptorchidism and microphthalmia have been reported in male quokkas.[78] Bilateral preaxial **polydactyly** has been described in an eastern grey kangaroo and a western grey kangaroo.[67,257] Congenital **ichthyosis** has been reported in two newly out-of-pouch red-necked wallabies from different pairings of the same parents. Lesions were characterized by severe keratoderma affecting the pads of all feet, diffuse severe scaling of haired skin, failure to thrive, and early mortality.[104]

Age Related/Degenerative

Dilated and hypertrophic **cardiomyopathies** are considered a serious health concern in captive North American opossums older than 2 years of age. It has been suggested that development is related to obesity.[105]

Lesions consistent with **hypertension** have been reported at high prevalence in mature to aged captive western grey kangaroos in a zoo collection in the United States. The etiology is unknown. Lesions are consistent with hypertension in other species and include increased thickness of the renal arteriolar tunica media with smooth muscle hypertrophy and/or hyperplasia, accumulation of extracellular matrix within arterioles, increased vascular tortuosity, and varying degrees of juxtaglomerular hyperplasia (Fig. 18.5). In addition, some animals had severe thickening of arteriolar and arterial walls in tissues other than the kidneys, most often in the central nervous system (cerebrum and spinal cord), myocardium, gastrointestinal-tract walls, mesentery, spleen, thyroid gland, and occasionally the lung. A subset of animals had antemortem multifocal retinal detachment but there were no appreciable vascular changes in the eyes of affected animals.[129]

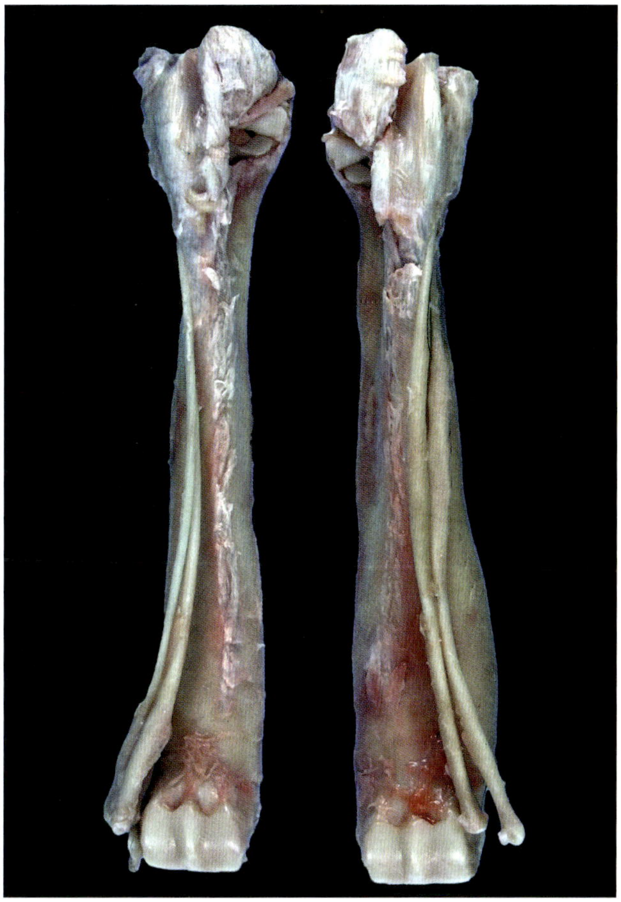

FIGURE 18.4 Fluorosis in an eastern grey kangaroo. Metatarsi IV with skeletal fluorosis showing bilateral irregular hyperostosis, which is particularly pronounced on the medial aspect of the left bone. Metatarsi II and III are also shown and display similar irregularly hyperostotic lesions. *(Photo Courtesy of J. Hufschmid, University of Melbourne.)*

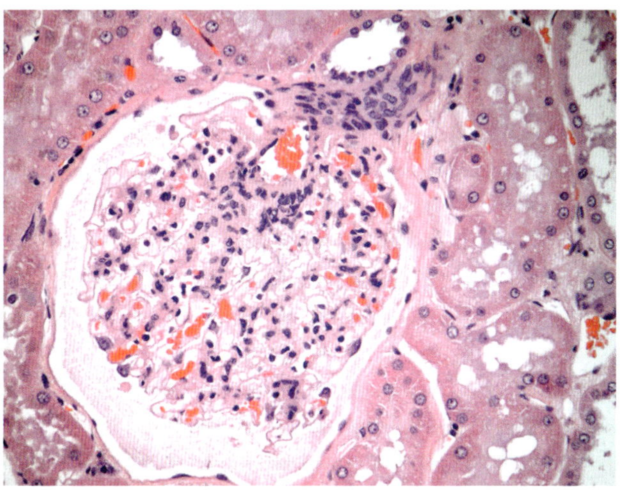

FIGURE 18.5 Arteriolar hypertrophy associated with idiopathic hypertension in an eastern grey kangaroo. Afferent and efferent arteriolar walls are thickened due to smooth muscle hypertrophy/hyperplasia and there is increased vascular tortuosity. *(Photo Courtesy of K. Terio, University of Illinois Zoological Pathology Program.)*

Inflammatory Non-Infectious

Swollen paw syndrome is seen in free-ranging common ringtail possums. The etiology is unknown but it is suspected that the pathogenesis involves avascular necrosis.[126] Postulated causes of compromised microcirculation that may be involved in lesion development include bacterial infection, viruses, thermal injury, electrocution, mycotoxins, plant-derived alkaloids, and photosensitivity. Characteristic lesions consist of edema of the feet, particularly the forefeet, that often progresses to well-demarcated areas of gangrene with loss of the affected feet (Fig. 18.6). Ulceration, exudative dermatitis, crusting, and necrosis of the ears, dorsal nose, and tail tip may also occur. Histologically, coagulative to caseous epidermal necrosis with fibrinoid necrosis and thrombosis of dermal blood vessels is seen. In some cases, there is evidence of repair, with reactive fibroblasts and endothelial cells in adjacent granulation tissue. Lesions are often complicated by secondary infection and inflammation. Animals remain bright, alert, and appetent [Australian Registry of Wildlife Health (ARWH), unpublished data].

A range of dermatological conditions occur in brushtail possums. The most severe and common of these is **exudative dermatitis**, which causes widespread alopecia, exudation, crusting, and ulceration of the skin in common brushtail possums (Figs. 18.7A, B). This occurs most commonly on the face but can also occur on the torso, rump, limbs, or elsewhere on the body. Lesions can appear acute or chronic, with chronic lesions sometimes exhibiting areas of healing and regrowth of fur (K. Rose, personal communication). A range of bacteria may be cultured, and a single or specific etiology has not been established. Anecdotally, the condition is more common in males, particularly those at age of dispersal, and in warm and humid

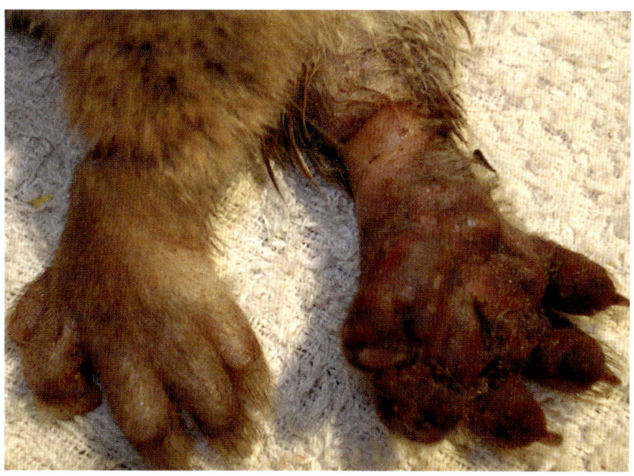

FIGURE 18.6 Swollen paw syndrome in a ringtail possum. Edema and necrosis of the paw distal to a well-demarcated boundary circumscribing the antebrachium. *(Photo Courtesy of D. Spielman, University of Sydney.)*

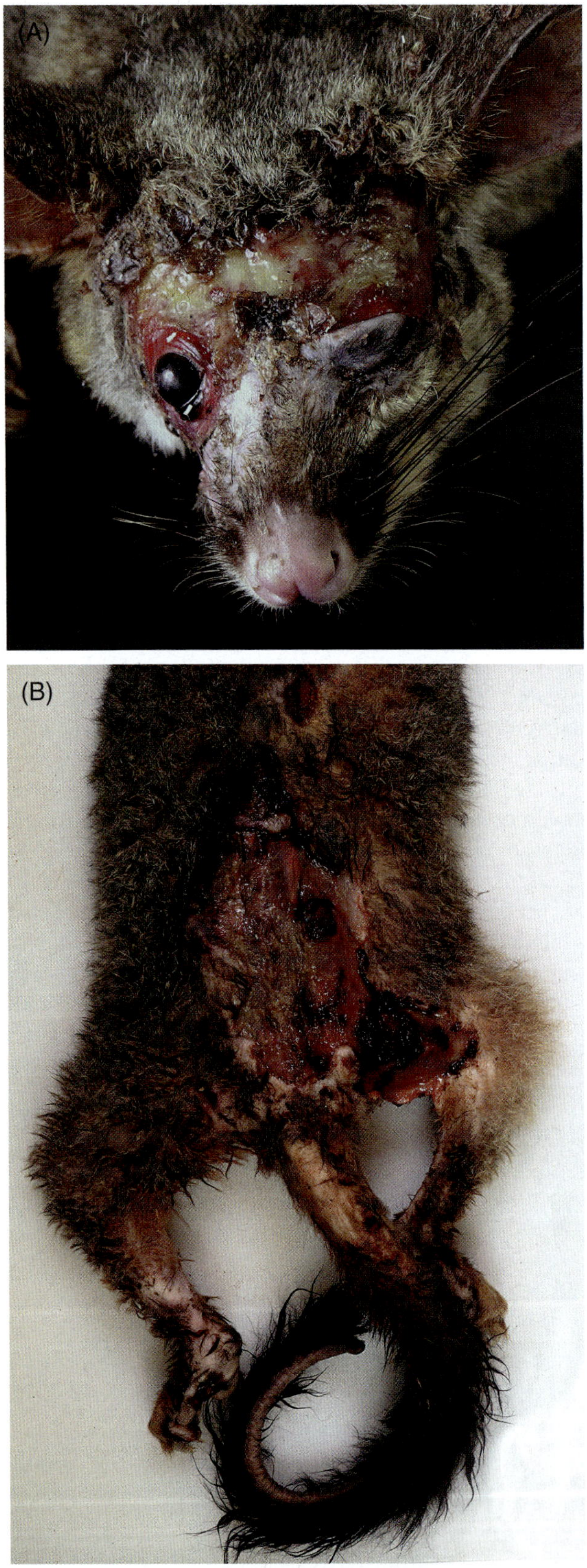

FIGURE 18.7 (A) Facial necrosis due to exudative dermatitis in a common brushtail possum. Exudative ulceration of facial skin. (B) Necrosis of skin overlying rump and limbs due to exudative dermatitis in a common brushtail possum. Exudative ulceration of skin over the rump, flanks, and feet. *(Photo Courtesy of Australian Registry of Wildlife Health.)*

weather suggesting a potential relationship to stress, fight wounds, or maceration of the skin.[126]

A common syndrome, colloquially known as **rump wear,** occurs in mountain and common brushtail possums. The clinical appearance appears to be due to breakage of the fur over the rump and tail base; there is minimal damage to the underlying skin. The etiology of the condition is unknown. There is no recognized sex or seasonal predisposition, though there is an association with increasing age.[114]

Idiopathic progressive renal disease is reported as a relatively common problem in one captive collection of critically endangered mountain pygmy possums, predominantly in animals 6 years of age or older.[205] Histologically, there are glomerular changes ranging from epithelial and mesangial proliferation to atrophy, synechia, and periglomerular fibrosis. Renal tubular degeneration and loss with interstitial fibrosis and lymphoplasmacytic infiltrates are also seen. Affected possums have a high prevalence of associated multisystemic metastatic mineralization, which is most consistently seen in the gastric mucosa, aortic wall, body wall, myocardium, and pulmonary interstitium. A single affected animal also had histological lesions in long bones consistent with renal secondary hyperparathyroidism.

Miscellaneous/Multifactorial

Trauma is an important cause of morbidity and mortality in monotremes and marsupials. The atlanto-occipital joint of macropods is unusually mobile, and it has been suggested this may predispose kangaroos and wallabies to cervical dislocation when startled animals run into stationary objects. Facial trauma, which can occur if they are chased along fences, is also common. Hepatic trauma is a common sequela to fighting between males as a result of kicks to the abdomen. Although koalas commonly fall from heights, particularly when fighting or attempting to mate, they often appear to suffer no apparent ill-effects. However, trauma, including fractures or dislocations of the limbs or rupture of the liver or spleen, can occur. Motor vehicle collisions and predation by dogs, foxes and cats are major causes of mortality in free-ranging marsupials and monotremes in Australia. Lesions are similar to those seen in other mammalian species. **Exertional rhabdomyolysis** (capture myopathy) is particularly common in macropods. Lesions and death due to venomous snake bites have been observed in young koalas, and fatal attacks on koalas by cattle are surprisingly common.[107,122]

Animals surviving **bushfire** may sustain thermal injuries from flames or direct heat of the fire. Burns may be severe, and because thermal damage continues after removal of the heat source, the extent of burns may not be apparent for days. Clinical signs of burn wounds are variable and include singed whiskers, charred fur and flesh, cutaneous erythema and blistering, and widespread necrosis with secondary inflammation and infection. Burns are usually classified by depth of injury as first degree (superficial epidermis), second degree (full-thickness epidermis and partial dermis with sparing of adnexal structures), third degree (full-thickness epidermis and dermis with adnexal destruction), and fourth degree (epidermis, dermis, subcutis, and underlying tissues and bone).[266] In bushfire-injured marsupials including koalas, burns are commonly seen on the limbs,[81] especially the palmar and plantar footpads due to attempts by animals to walk across burning ground to try and escape active flames (Fig. 18.8A). Burns to the head and face and dehydration are also common in marsupials impacted by bushfires. Smoke inhalation is not uncommon in wildlife impacted by bushfires and should be considered in animals with facial burns, singed hair and vibrissae, and respiratory signs.[32] Soot deposition may be encountered on oral and nasal cavity mucous membranes, in the upper and lower airways, and in the upper gastrointestinal tract if swallowed. Pulmonary edema and congestion are common sequelae (Fig. 18.8B). Histologically, dark brown to black particulate matter may be located in the trachea, bronchi and bronchioles. In animals that survive long enough to incite a physiological response, particles may be observed in macrophages adjacent to airways and to a lesser extent within alveolar macrophages.[266] Inhalation of hot gas or air can result in coagulation necrosis of tracheal and bronchial epithelium and laryngotracheal and pulmonary edema. In some instances, shock-associated microvascular thrombosis may be seen.[266] Many animals are susceptible to carbon monoxide poisoning during smoke inhalation which can be lethal. Cherry-red discoloration of blood and tissues (especially musculature) is a feature in animals that succumb to carbon monoxide poisoning. Animals surviving smoke inhalation may be at risk of developing respiratory and cardiovascular disease, immunosuppression, and/or neurological impairment.[201]

Short-beaked echidnas are prone to **orthokeratotic hyperkeratosis**. It is often seen as a feature of chronic proliferative dermatitis which is characterized by alopecia, crusting, and easily fractured brittle quills. Histologic lesions include acanthosis, orthokeratotic epidermal and follicular hyperkeratosis, and pustular to suppurative dermatitis. While no specific etiological agent has been identified, various bacteria, fungi, and mites have been seen in the epidermis and dermis of affected echidnas. These cutaneous changes are probably a nonspecific reaction of echidna skin to chronic insult and/or inflammation.[161]

Endogenous lipid pneumonia is an incidental finding encountered in North American opossums. Multifocal yellow-white subpleural plaques, visible grossly and bilaterally on the dorso-caudal pleura of the lungs, extend into the pulmonary parenchyma as aggregates of foamy macrophages in the alveolar walls and spaces. This change often occurs with minimal disruption of the pulmonary architecture (Fig. 18.9).[39,93] In more severe cases, they form large nodular lesions with cholesterol clefts, localized emphysema, fibrosis, and hyperplasia of type II pneumocytes and perivascular lymphoid tissue. In some cases, there is an association with the presence of the lungworm

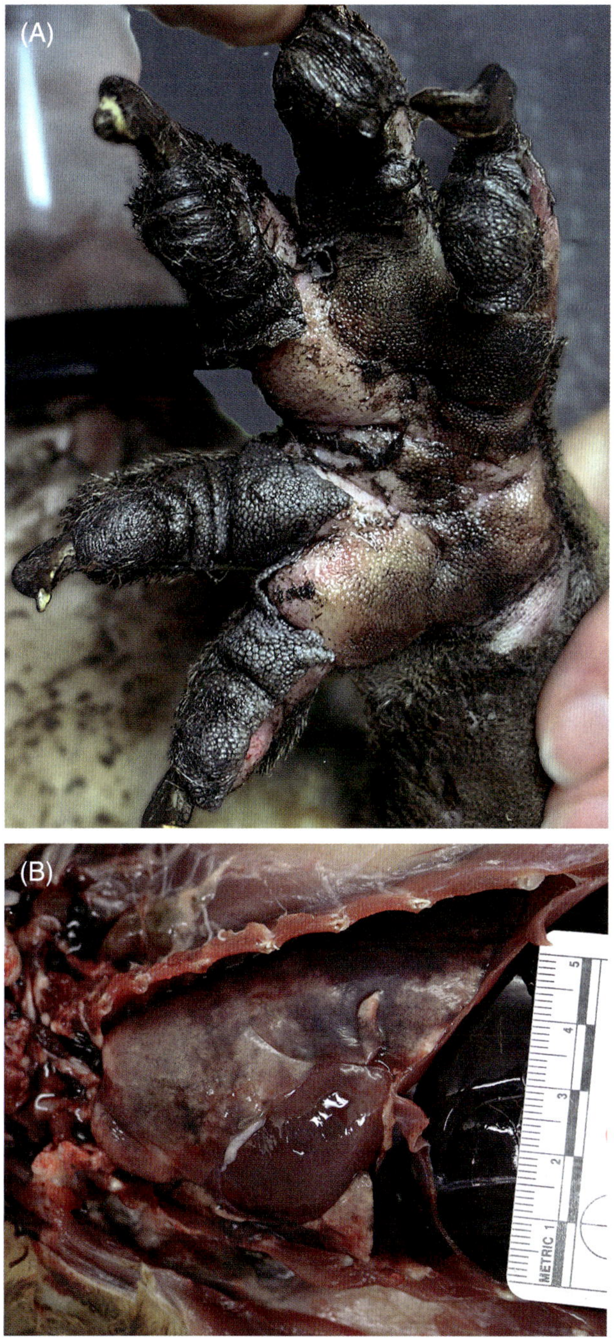

FIGURE 18.8 (A) Bushfire burns on the forepaw of a koala. Partial to full thickness burns of the paws. Note the loss of leather padding, eschar at the ends of the digits, singing of fur, and lifting of skin. This koala walked across burning ground to escape active flames. (B) Smoke inhalation in the lungs of a brush-tailed possum. Pulmonary anthracosis following smoke inhalation during bushfires. The pulmonary parenchyma has a black to gray mottled appearance due to carbon smoke particles accumulating along bronchial mucosa and adjacent tissue. *(Part A: Photo Courtesy of C. Flanagan, Koala Conservation Australia Inc. Part B: Photo Courtesy of Australian Registry of Wildlife Health.)*

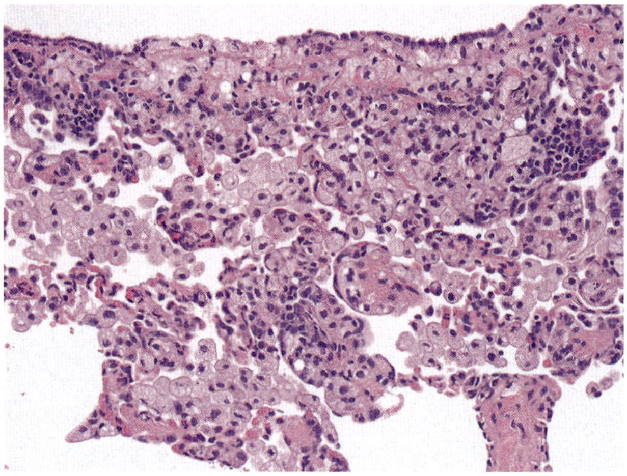

FIGURE 18.9 Endogenous lipid pneumonia in a North American opossum. Large numbers of vacuolated macrophages are present within subpleural alveolar spaces with fewer mixed inflammatory cells. *(Photo Courtesy of K. Terio, University of Illinois Zoological Pathology Program.)*

Didelphostrongylus hayesi. It has been proposed that these parasites induce lipid pneumonia by bronchiolar obstruction.[93]

Neoplastic

A wide range of neoplasms have been reported in marsupials. These have been reviewed by Tong,[242] Ladds,[137] and Canfield et al.[48,49] Of note, dasyurids (see Table 18.e1 for species in this group) are particularly prone to developing neoplasms, especially of epithelial cell origin. This is most common in captive animals, which generally survive well beyond what would be expected in the wild. Several of the most common and important neoplasms are mentioned below.

First detected in 1996 in the far northeast of the state of Tasmania, **DFTD** has since spread across most of the Tasmanian devil's range. The disease is one of the only two known and naturally transmissible tumors of vertebrates. From the conservation perspective, the disease is highly significant and poses local and regional extinction threats. It has affected up to 90% of devils in some populations and at least 50% of devils in populations where it is established. It is almost invariably fatal within 6 months of tumor appearance. Due to its frequency-dependent transmission pattern, the disease maintains prevalence even in populations that have suffered up to 90% declines. As a result, and due to its unusual pathogenesis, it is the subject of extensive research and management programs that have included establishment of a captive insurance population for this (now) threatened species. The condition has not been reported in any other sympatric species.[191]

There are two recognized DFTD cancer cell types: devil facial tumor 1 (DFT1) and devil facial tumor 2 (DFT2). DFT1 tumors are composed of cells possessing the same extensive set of chromosomal rearrangements. Cells lack a Y chromosome suggesting that the tumor arose from a single female Tasmanian devil. Recent minor rearrangements have given rise to several tumor strains and there is preliminary evidence that these strains may differ in their pathogenicity. DFT1 tumors are transmitted as allografts via facial and gingival bite wounds, which are common in this species because of their feeding and other aggressive behaviors. Disease is most common in animals that are 3 years of age. It was originally proposed that transplanted tumor cells evaded rejection due to low major histocompatibility complex (MHC) diversity among the devil population, but this has since been refuted. More recent research has shown that the tumor allograft evades rejection by down-regulation of MHC. This recognition has led to diminishing hopes of identifying genetically resistant devil populations.[191] DFT2, described in five Tasmanian devils from southeast Tasmania in 2015, is completely distinct from DFT1 cytogenetically. Tumor cells contain a Y chromosome, indicating its origin in a male animal.[190]

DFT1 tumors are large, firm, soft tissue masses that commonly contain central areas of ulceration and necrosis (Fig. 18.10A). Lesions are generally well circumscribed, nodular, pale to slightly translucent on cut surface, and they have visible fibrous septae.[145] Tumors appear to originate in the dermis of the facial skin or the submucosal connective tissue in the oral cavity. They are multicentric in about two-thirds of cases. Neoplasms are composed predominantly of well-vascularized, infiltrative, densely cellular, compact proliferations of pleomorphic, anaplastic, round cells that often have indistinct cell borders, high nuclear to cytoplasmic ratio and fibrillar, eosinophilic cytoplasm. The nodular masses are often present within a thin fibrous pseudocapsule. In some cases, cells are arranged in bundles, cords, or streams while in others there are palisades, packets, nests, or sheets (Fig. 18.10B). All neoplastic cells contain a single round central nucleus with a predominantly granular or vesicular nuclear pattern without visible nucleoli. The number of mitoses varies, with about a third of tumors having more than 3 per high-power field. Most cases lack associated inflammation but if present, mild peripheral infiltration with neutrophils, or less commonly lymphocytes, may be noted. Observation of vascular invasion is uncommon but metastases occur in about two-thirds of all cases. Metastasis to regional lymph nodes is reported in slightly more than half of the cases; distant metastases most commonly occur in the lung and spleen.[145] DFT1 neoplasms are negative immunohistologically for cytokeratin, epithelial membrane antigen, von Willebrand factor, smooth muscle actin, desmin, glial fibrillary acid protein, CD16, CD57, CD3, and LSP1, and are positive

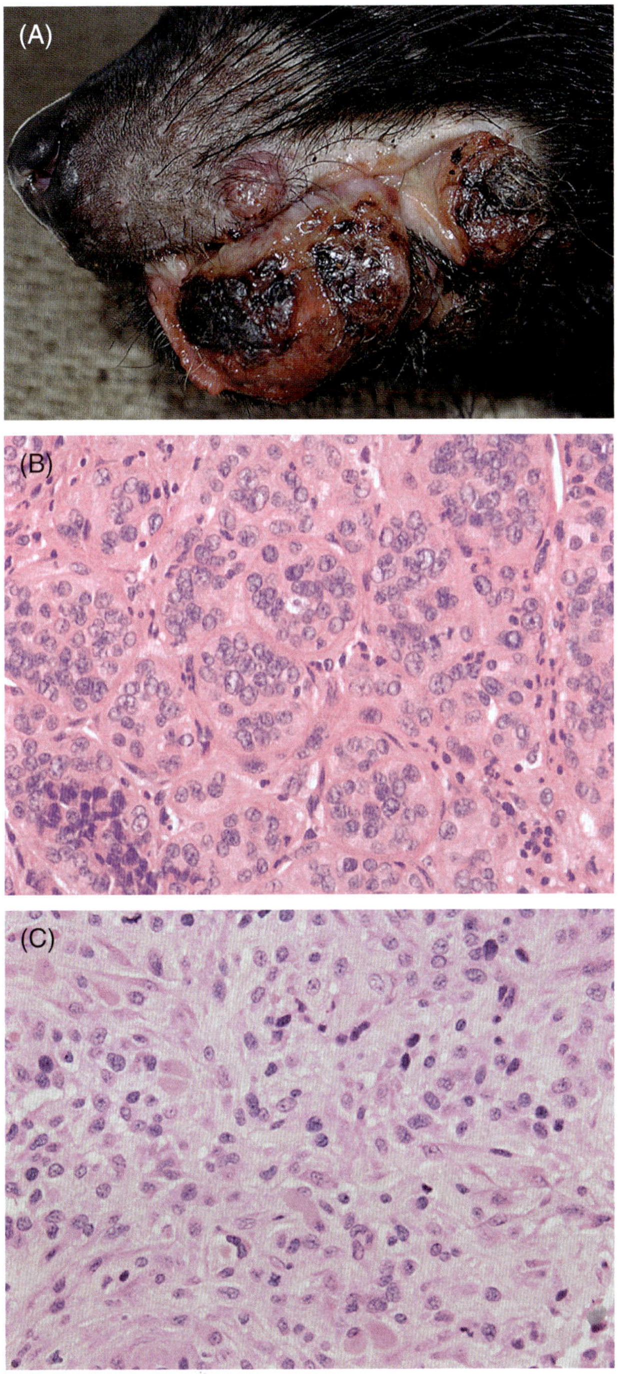

FIGURE 18.10 (A) Devil facial tumor (DFT) in a Tasmanian devil. Variably sized ulcerated masses located around the mouth and chin. (B) Devil facial tumor 1 (DFT1) in a Tasmanian devil. DFT1 comprises pleomorphic, anaplastic, round cells arranged in bundles, cords, or streams; or in palisades, packets, nests, or sheets. (C) Devil facial tumor 2 (DFT2) in a Tasmanian devil. DFT2 comprises pleomorphic, amorphic to stellate or fusiform cells arranged in solid sheets, without the packeting observed in DFT1. (see eSlide 18.e1). *(Part A: Photo Courtesy of Australian Registry of Wildlife Health. Parts B, C: Photos Courtesy of G. Knowles, DPIPWE, Tasmania.)*

for vimentin, S-100, and melan A consistent with a sarcoma. Initial studies further classified the cells as neuroendocrine in origin based on labeling by antibodies against neuron-specific enolase, chromogranin A, and synaptophysin but subsequent immunohistochemical and transcriptomic work has identified the tumors to be of Schwann cell origin.[146] Periaxin, a marker for Schwann cells in peripheral nerves, has been confirmed as the most sensitive and specific marker.[191]

DFT2 tumors are grossly similar to those of DFT1 but histologically they differ in that they are composed of pleomorphic, amorphic to stellate or fusiform cells arranged in solid sheets (Fig. 18.10C; eSlide 18.e1). Also in contrast to DFT1, these tumors lack immunopositivity for periaxin.[190]

With the exception of DFTD, most other reported neoplasms in monotremes and marsupials have been described in koalas, though it is uncertain whether this is due to species predisposition [perhaps associated with retroviral infection (see below)] or due to the disproportionately large number of necropsies that have been conducted in the species. Although no formal prevalence studies have been performed, percentages have been compiled from necropsies at one koala hospital in New South Wales, Australia. Unless noted, percentages listed below are from that population.

Although it can occur in any marsupial species, **lymphoma (lymphosarcoma)** occurs most commonly in koalas (4%–6% of animals necropsied). There is growing evidence of a causative role for **Koala retrovirus** (KoRV) in oncogenesis of the koala.[155,202] With 100% of koalas endogenously infected with KoRV over the northeastern states of New South Wales and Queensland, induction of disease is likely to be complex and this is an active area of research (see Koala retroviruses below).

Common clinical presentations of lymphoma include lymphadenomegaly, weight loss, poor coat condition, or any clinical sign that might be caused by metastasis-induced organ dysfunction. This disease most commonly occurs in middle-aged koalas without sex predilection and progression of disease can be rapid. Connolly et al. reported that 63% of affected animals with lymphoma are leukemic but suggested that this high percentage, in comparison to cats (25%) or dogs (33%), might be associated with later clinical presentation/disease recognition in koalas.[59] Due to this high prevalence of leukemia, hematological examination is a useful diagnostic tool. Abnormal lymphocytes or lymphocytosis are commonly seen and flow cytometric and immunohistochemical methods have been developed to discriminate between T and B lymphocytes using cross-reactive antihuman CD3 and CD79b antibodies. T cell lymphomas are slightly more common (51%). Approximately 25% of tumors failed to react by immunohistochemistry with B or T cell markers. Whether these unidentified phenotypes truly do not express surface markers due to their anaplastic state, originated from CD3-/CD79b (e.g., NK) cells, or failed to label due to poor fixation or autolysis is unknown.[59] Association of clinical presentation or histological phenotype with cell type has not been reported.

Lymphoma in koalas most commonly affects lymphoid tissues (lymph nodes, thymus, and spleen), liver, kidney, and the gastrointestinal and urogenital tracts.[48] Lesions are typical of lymphoid tumors in other species and consist of diffuse sheets of neoplastic cells infiltrating and or replacing normal parenchyma (Fig. 18.11A; eSlide 18.e2).

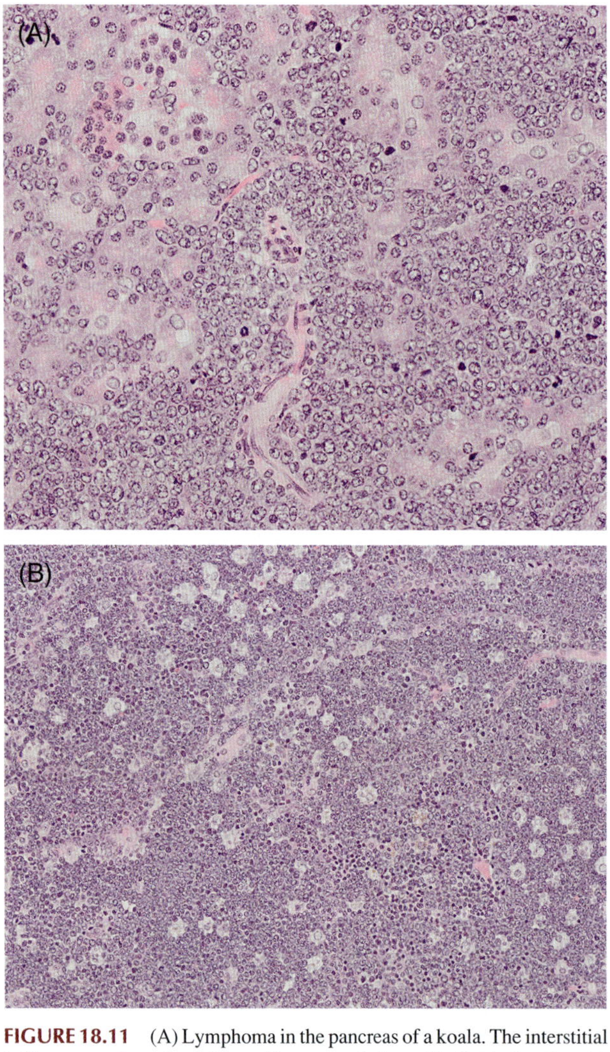

FIGURE 18.11 (A) Lymphoma in the pancreas of a koala. The interstitial infiltrate consists of a fairly uniform population of individual cells with medium-sized, pale, round to ovoid nuclei, and scant cytoplasm. In this case, several mitotic figures are evident. Note the Islet of Langerhans at top left of image. (B) Lymphoma in the lymph node of a koala. Disruption of normal nodal architecture by sheets of neoplastic lymphocytes. The presence of histiocytic cells scattered throughout the neoplasm imparts the typical "starry sky" effect. (see eSlide 18.e2) *(Parts A, B: Photos Courtesy of the University of Sydney.)*

Sheets of lymphocytes commonly exhibit a "starry sky" appearance due to the presence of tingible body macrophages scattered throughout (Fig. 18.11B). Neoplastic lymphocytes can be small, medium, or large, with more than half being medium. Nucleoli are prominent only in larger cell types. In many cases, neoplastic lymphocytes are difficult to differentiate from normal lymphocytes and classification as neoplasia must rest on clinical history and the cell behavior (sheeting, invasion, and exclusion of other leukocyte or lymphocyte types based on morphology and immunohistochemistry).

In koalas, **osteochondromas** present predominantly as craniofacial tumors and generally originate in the bones of the nasal cavity and paranasal sinuses (Fig. 18.12). They have been identified in multiple populations, and in one population they occurred in 0.5% of koalas. The tumors are benign with no reports of metastasis, but they can be quite expansive within the sinuses even if they are not visible externally. Tumors are often lobulated and neoplastic chondrocytes can appear monomorphic, undifferentiated, or sarcomatous. The chondrocytes are generally embedded in cartilaginous ground substance with varying degrees of trabecular ossification but may also be surrounded by vascular and connective tissue septae.[48,137]

Lesions, grossly and histologically consistent with **mesothelioma**, have been reported in approximately 1.5% of koalas and are most common in the abdominal cavity, but some occur in the pericardium, pelvic cavity, and scrotal tunica. Typical clinical signs are nonspecific and include weight loss and death. Grossly, serosal thickening and miliary, white to pale pink or red glistening nodules up to 5 mm diameter are distributed diffusely across the serosal surfaces and extend into the underlying musculature (Fig. 18.13). When in the abdomen, the cavity may contain abundant viscous, tenacious, straw-to-red-colored

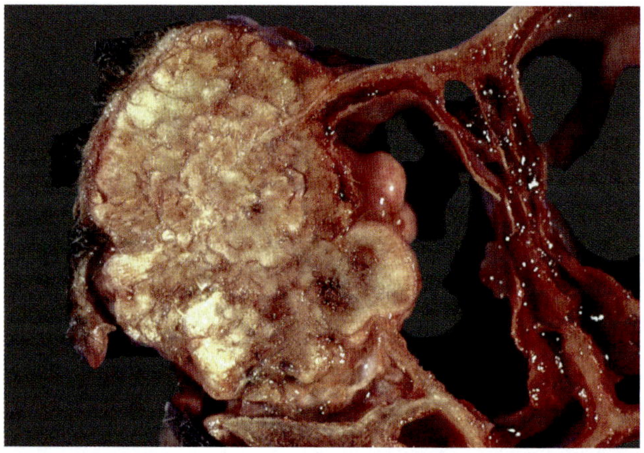

FIGURE 18.12 Osteochondroma in the skull of a koala. Large off-white lobulated tumor involving the zygomatic arch, maxilla, and frontal bone with encroachment upon the left nasal cavity. *(Photo Courtesy of P. Canfield, University of Sydney.)*

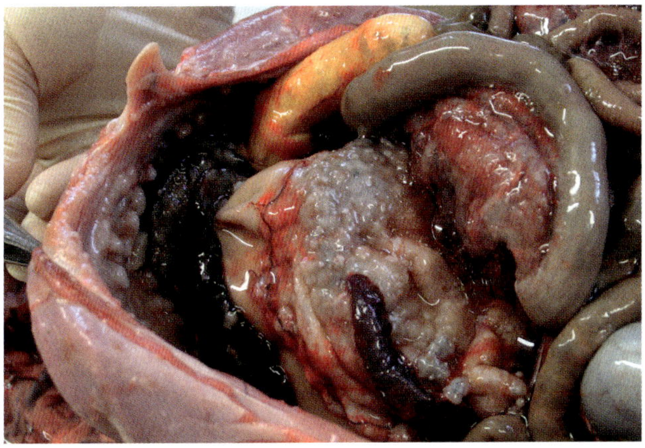

FIGURE 18.13 Mesothelioma in the peritoneal cavity of a koala. Disseminated off-white nodules thickening the mesentery and distributed on the visceral and diaphragmatic peritoneum. The prominent, elongated pale yellow gallbladder at the top of the image is within normal limits for this species. *(Photo Courtesy of C. Marschner, University of Sydney.)*

fluid. Tumors histologically consist of plump, occasionally pleomorphic, fusiform cells interspersed with eosinophilic to lightly basophilic mucinous material (Alcian blue positive, PAS negative) and/or vascular connective tissue. Tumors also often contain less cellular areas composed either of whorls of collagen or a sparse, myxomatous population of cells. Nodules are generally covered by single or multiple layers of fairly uniform round mesothelial cells.[47] Mesotheliomas label for vimentin and cytokeratin immunohistochemically. Recently, a series of tumors in koalas with strong gross and histological similarity to myxomatous mesothelioma were classified as serosal myxosarcoma on the basis of negative immunohistochemical labeling for cytokeratin.[6] It seems likely that a subset of lesions previously classified as myxomatous mesothelioma could meet these criteria.

A range of **soft tissue sarcomas** including myxosarcoma, fibrosarcoma, and osteosarcoma have been described in feathertail gliders and a koala in association with subcutaneous microchip/passive transponder tags.[126,182,242,252] Tumor development is typically detected 2–3 years after placement. Lesions begin as fibrovascular proliferations but progress to locally invasive tumors that extend into the dermis or underlying muscle (Fig. 18.14A). In the koala, the more peripheral and invasive areas are composed of a loose, myxomatous proliferation of stellate cells within a mucinous matrix. Tumors in gliders consist of whorls or streams of pleomorphic fusiform or stellate cells with moderate to abundant cytoplasm, ovoid, vesiculated, moderately anisokaryotic nuclei often with prominent nucleoli, and a moderate number of mitotic figures (Fig. 18.14B). Collagen deposition varies and osteoid is present in some cases (Fig. 18.14C) (ARWH, unpublished).

INFECTIOUS DISEASES

There is a long list of pathogens that marsupial and monotreme species may be exposed to. Due to the geographically isolated nature of Australia, many endemic species have had no, to very limited, exposure to many of these pathogens. The importation of exotic diseases to these and isolated and island populations is a serious risk and reason for the establishment of strict biosecurity and importation regulations. Readers should therefore be informed and open minded when considering pathogens to which these groups may be susceptible and that are beyond those discussed here. It is also important to bear in mind that exotic disease presentation in naïve species may differ from those seen in typical hosts.

DNA Viruses

A range of herpesviruses have been detected in marsupials, but their significance is not yet fully understood. Among the many alpha- and gamma-herpesviruses that have been detected, some have been affiliated with disease. **Macropodid herpesvirus 1 (MaHV-1)**, an alphaherpesvirus, is reported with sudden death and rhinitis, conjunctivitis, pneumonia, cloacal ulceration, and variable splenic, pulmonic, and hepatic necrosis with occasional intranuclear inclusion bodies in captive parma wallabies.[1,89] **MaHV-2**, also an alphaherpesvirus, has caused outbreaks of fatal systemic disease characterized by interstitial pneumonia with alveolar and bronchiolar necrosis, hepatic necrosis, and ulcerative esophagitis, gastritis, cloacitis and dermatitis in gray dorcopsis wallabies.[42,125] **MaHV-3**, a gammaherpesvirus, was described with respiratory disease in a group of free-ranging eastern grey kangaroos;[216,264] and **MaHV-4**, another alphaherpesvirus, was reported in association with respiratory and possibly neurological disease in a free-living eastern grey kangaroo.[250] Multifocal necrotizing hepatitis, adrenalitis, and bronchointerstitial pneumonia were recently described in a koala infected with a novel alphaherpesvirus that had concurrent cryptococcosis. Inclusion bodies were associated with necrosis in respiratory epithelium, pulmonary interstitium, hepatocytes, biliary epithelium, and corticomedullary cells of the adrenal gland.[37] Similar lesions have been reported in greater bilbies infected with an alphaherpesvirus closely related to MaHV-1.[20] Correlations have been found between herpesvirus shedding and absence of pouch young in eastern grey kangaroos, poor body condition in wombats, and detection of *Chlamydia pecorum* and retrovirus in the koala.[232,249] However, it remains unknown whether these relationships are causative, related to an unidentified cofactor or, in the case of koalas and *C. pecorum*, the result of concomitant transmission.

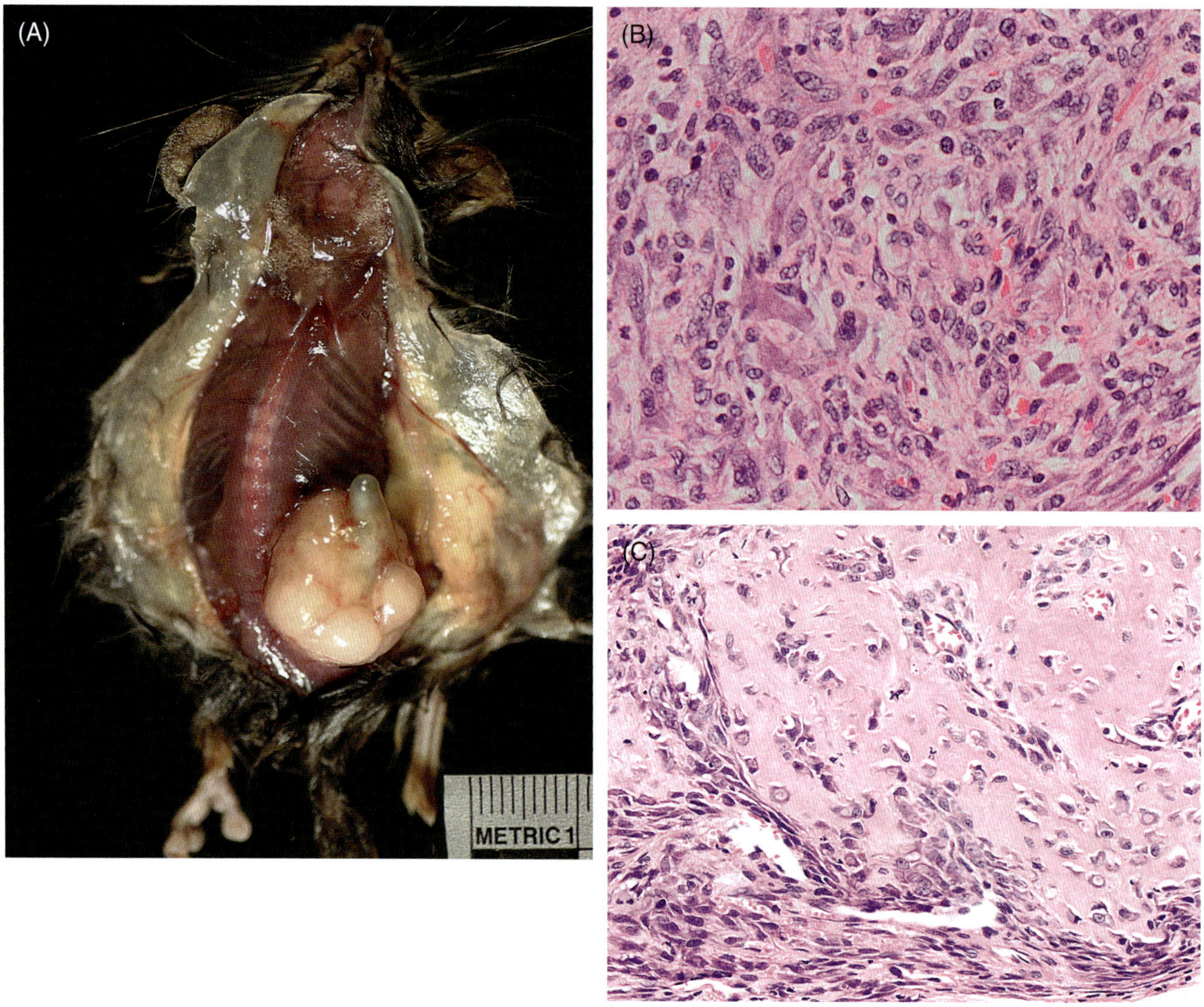

FIGURE 18.14 (A) Microchip-associated sarcoma in a feathertail glider. Off-white lobulated neoplasm located on the left caudolateral dorsal region oriented around a microchip encased with fibrous tissue that projects slightly above the mass. (B) Microchip-associated soft tissue sarcoma in a feathertail glider. Pleomorphic fusiform or stellate cells with moderate to abundant cytoplasm, ovoid, vesiculated, moderately anisokaryotic nuclei often with prominent nucleoli. Cells are surrounded by a modest amount of collagenous stroma. (C) Microchip-associated osteosarcoma in a feathertail glider. A proliferative population of pleomorphic, anisokaryotic mesenchymal cells surrounded by osteoid. *(Parts A, B: Photos Courtesy of Australian Registry of Wildlife Health.)*

Cytomegalic inclusions without associated inflammation or apparent impact on the host have been described in the prostate of the yellow-footed antechinus, agile antechinus, brown antechinus, dusky antechinus, and the brush-tailed phascogale (Fig. 18.15), and the renal tubular epithelium of bettongs, platypus, and pygmy possums.[2,164] These are generally considered to be herpesviral but the morphology of the platypus virus on electron microscopy (EM) was considered more consistent with an adenovirus[262] and no viral particles were demonstrated by EM in pygmy possums. With the exception of the yellow-footed and agile antechinus, in which DNA from a novel gammaherpesvirus was detected in the prostate and other tissues, the inclusions have not yet been characterized genetically.[2,232]

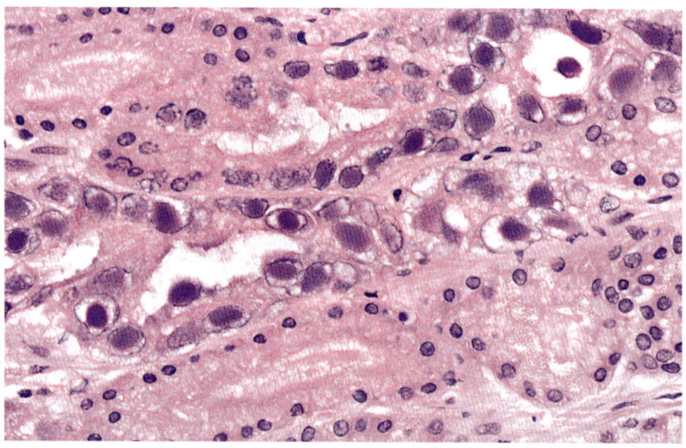

FIGURE 18.15 Idiopathic cytomegalic lesions in the prostate of a brown antechinus. Some tubules are lined by prostatic epithelial cells with nuclear enlargement and large amorphous basophilic intranuclear inclusions. *(Photo Courtesy of Australian Registry of Wildlife Health.)*

Poxviral lesions are most commonly reported in macropods (eastern grey kangaroo, western grey kangaroo, common wallaroo, red kangaroo, tammar wallaby, agile wallaby, swamp wallaby, and quokka).[12,137,156,177,203] Single case reports of poxvirus infection are reported in the common brushtail possum, common ringtail possum, and short-beaked echidna, and an outbreak was recently reported in common ringtail possums.[254] Poxvirus infection typically induces the formation of focal or multifocal, well-demarcated exophytic cutaneous lesions that are generally found in less furred areas of the skin (Fig. 18.16A).[137,203] Grossly, poxviral lesions may resemble those of cutaneous leishmaniasis but the histologic lesions are characteristic. The epidermis is markedly thickened due to acanthosis and hyperkeratosis. Epidermal rete pegs extend into the dermis and the skin is thrown up into a series of papillary projections. Hyperkeratotic debris may accumulate in the folds of the papilliform projections and ulceration with associated suppurative inflammation of the surface epidermis may occur (Fig. 18.16B). Epithelial cells in the stratum spinosum often contain large cytoplasmic vacuoles with large central eosinophilic inclusion bodies ranging from 10 to 20 μm in diameter (Fig. 18.16C; eSlide 18.e3). Lesions may be accompanied by mild mononuclear cell inflammation with varying numbers of neutrophils depending on the degree of ulceration. Electron microscopy can confirm the presence of orthopoxvirus particles, which are oval, approximately 250 nm long and contain a biconcave core.

Bandicoot papillomatosis and carcinomatosis virus 1 (BPCV 1)** and **BPCV 2 are associated with papillomatous and carcinomatous lesions in western barred and southern brown bandicoots, respectively.[18,269] In the case of the western barred bandicoot, disease is an impediment to species recovery efforts. The viruses, which are genetically intermediate between *Papillomaviridae* and *Polyomaviridae*, induce a spectrum of cutaneous and mucocutaneous lesions that resemble the stepwise progression from premalignant stages into invasive squamous cell carcinomas.

In western barred bandicoots, lesions are classified as epithelial hyperplasia, carcinoma in situ, or squamous cell

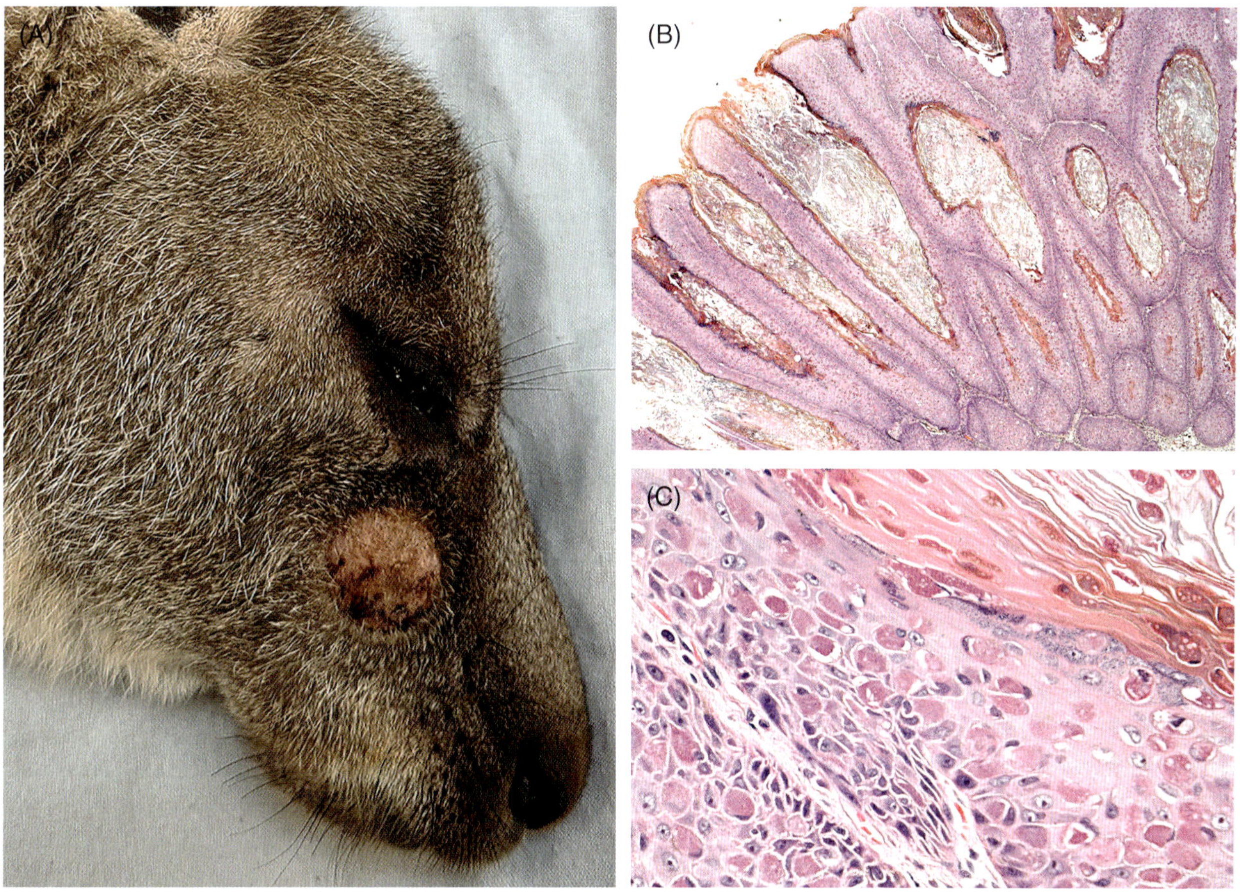

FIGURE 18.16 (A) Poxviral lesion in the skin of an eastern grey kangaroo. Well-demarcated, exophytic, alopecic facial nodule. (B) Poxviral lesions in the skin of an eastern grey kangaroo. Papillary projections, acanthosis, and hyperkeratosis. (C) Poxviral lesions in the skin of an eastern grey kangaroo. Eosinophilic cytoplasmic inclusions originating in the stratum spinosum, associated with marked acanthosis, and hyperkeratosis. (see eSlide 18.e3). *(Parts A–C: Photos Courtesy of Australian Registry of Wildlife Health.)*

carcinoma.[269] Epithelial hyperplasia of the suprabasilar epithelial cells at mucosal sites or keratinocytes in the stratum spinosum and stratum granulosum of the skin is irregular and varies from mild to marked. Hyperplasia of the outer root sheath of hair follicles and the glandular elements of sebaceous and sweat glands, is also seen. Koilocytes and epithelial cells with marginated nuclear chromatin and large amphophilic inclusions are present in some hyperplastic conjunctival lesions. Lesions categorized as carcinoma in situ are confined to the epidermis and have expansion of the stratum spinosum and stratum granulosum with mild to moderate dysplastic changes in keratinocyte morphology and epidermal architecture. Dysplastic changes include disorderly keratinocyte differentiation, suprabasilar mitoses, nuclear hyperchromasia, prominent nucleoli, clear or vacuolated cytoplasm, and mild anisocytosis and anisokaryosis. Lesions categorized as squamous cell carcinoma have typical histological changes including a basement membrane breached by neoplastic cells, moderate to marked cellular pleomorphism, keratin pearl formation, dyskeratosis, and desmoplastic stromal response. Squamous cell carcinomas are locally aggressive and invade the surrounding dermis, subcutis, and even bone. Metastases may be present. Mixed inflammatory cell infiltrates are frequently observed within lesions or in the surrounding subcutis.[269] Virus can be detected by immunohistochemistry using a polyclonal antibody to conserved papillomavirus capsid antigens.

RNA Viruses

Many **arboviruses** that are endemic to Australia and cause disease in humans have marsupials as known or suspected subclinical reservoirs or amplification hosts. Examples include the alphaviruses, **Ross River Virus,** and **Barmah Forest Virus,** as well as the flaviviruses **Murray Valley Encephalitis Virus** and the **Kunjin strain of West Nile Virus**. These viruses are carried by mosquitoes and, in the case of Ross River Virus, also by the biting *Culicoides* spp. midge.[214] The limited data on arboviruses in American marsupials has not yet indicated that they are significant arbovirus reservoirs.

Orbiviral blindness associated with infection by **Wallal** and/or **Warrego viruses** has occurred predominantly in western grey kangaroos and, to a lesser extent, in red and eastern grey kangaroos and the common wallaroo.[82,110] The disease generally occurs in outbreaks and up to 30% of animals in a group or population may be affected. Clinical presentation is consistent with behavioral changes due to blindness and/or encephalitis. In both natural and experimental infection, histological changes include nonsuppurative chorioretinitis and uveitis and, often, encephalitis or meningoencephalitis.[110,194] The latter two can occur without ocular lesions. In acute cases, chorioretinal inflammation and necrosis are severe and may affect most of the retina. With chronic retinitis, segmental retinal atrophy and Wallerian degeneration of the optic nerve and associated tracts in the brain may result. Ultrastructurally, viral particles are icosahedral, 60–70 nm in diameter, and have an electron-dense core and outer shell consistent with orbiviruses.[110,194] Viruses can be isolated in cell culture from a variety of tissues in acute cases. A diagnosis can be achieved based on clinical signs, histological lesions, viral culture, and EM. Immunofluorescence, PCR, and virus-neutralizing serology are research tools that have also been used to diagnose disease outbreaks.

Tammar wallaby sudden death syndrome is reported only in zoo collections and research colonies. The condition has occurred as outbreaks, resulting in the deaths of individuals and up to several hundred animals. Epizootics have been associated with an Orbivirus of the Eubenangee serogroup. When they are observed before death, clinical signs include lethargy, depression, sitting with a drooping head, ataxia, lateral recumbency with muscle fasciculations, and increased respiratory rate. Consistent gross findings are moderate to marked diffuse pulmonary and hepatic congestion, and subcutaneous edema with petechiae along the extremities. Lymphocytolysis in lymphoid germinal centers is also a consistent finding. Affected wallabies may also have moderate to marked hemorrhage in the perirenal retroperitoneal tissues, subcutis, and intermuscular fascia of the hindlimbs, inguinal, ventral thoracic, and dorsal cervical regions. Gastric ulceration and acute periacinar hepatocellular necrosis may also be seen.[198] The gross and histological features of tammar wallaby sudden death syndrome are not specific and are similar to those seen in babesiosis (see below). Following elimination of babesiosis as a differential by examination of renal impression smears and brain squash preparations, the brain, kidney, spleen, heart, and liver should be submitted for viral culture.

Koala retrovirus (KoRV), a gamma-retrovirus, is widespread in Australia and its study is an extremely complex and dynamic field. There is 100% prevalence of KoRV subtype A (KoRV A) in koalas in the northeastern states of Queensland and New South Wales, where it is endogenous (inherited in the germ line) and competent (able to replicate). In southern states, it appears that the virus is not endogenous and is less prevalent than in northern states[86,144]; reported prevalence is 73% in mainland Victoria, 28% on Victorian Islands, and 15% on Kangaroo Island, South Australia.[68,207] There are also defective versions of KoRV A, which lack the *pol* gene and are therefore not replication competent. These versions (known as recKoRV and detected by amplification of *env* or *gag* genes) are endogenous and widespread across both northeastern and southern states. These were not detected by early KoRV studies, which detected KoRV by amplification of *pol* gene. Based on the hypervariable region of the *env* gene, nine KoRV subtypes in three clades (A, B/J, and C–M), have been detected to date, and it appears likely there are more yet to be found.[35,53,128] With the exception of endogenous KoRV A, all are considered to be exogenous or to arise within hosts as a result of recombination events between KoRV A and endogenous retroviral elements. These subtypes only occur in the company of competent KoRV A and their presence in southern states is contentious.[35,128] Transmission of these exogenous subtypes appears to be from mother to young, though the route is still to be confirmed as some subtypes are considered to be not competent. KoRV has been proposed as a potential etiological agent of lymphoma, myeloid leukemia, aplastic anemia, myelodysplasia, and immune dysfunction,[101,270,272] but proving a causative role in these diseases is problematic. Where KoRV A is endogenous and ubiquitous, preliminary evidence exists for an association between greater KoRV proviral or viral loads and lymphoid disorders[237] and chlamydial disease[34,193], and infection with competent KoRV A in free-ranging Victorian populations is associated with ill thrift and disease, and changes to cytokine expression.[142,151] However, lymphoproliferative disorders or chronic inflammation can themselves increase retroviral expression. Association of potentially oncogenic KoRV proviral integrations with neoplasia, and upregulation of oncogenes following in vitro infection are more supportive of KoRV as a causative agent for neoplasia and suggest potential for KoRV integration to dysregulate a range of other genes including those involved in immune regulation.[155,202] A conserved retroviral envelope protein of KoRV which inhibits proliferation and alters cytokine expression by lymphocytes of several species in vitro provides an alternative potential mechanism.[68]

It appears likely that active transcription and integration of KoRV subtypes in general is most associated with disease rather than infection by any particular subtype. Associations between detection of KoRV B by PCR and lymphoma or death of pouch young,[271] changes to cytokine expression in lymphocyte stimulation assays in small cohorts of captive koalas,[151] and with chlamydial disease in free-ranging koalas[192,258] now appear likely to reflect greater KoRV proviral or transcript loads in general rather than presence/absence of KoRV B, as KoRV B commonly exists below detection limits of PCR based on amplicon deep sequencing results. The significance of rec KoRV is unclear, with potential to be protective to infection by KoRV, neutral, or beneficially or detrimentally affect adjacent host genes. Overall, support for involvement of koala retroviruses, particularly exogenous KoRVs, in pathogenesis is developing incrementally. It is likely that advances in genomics, transcriptomics, and immunology will provide additional information about the significance of these viruses in disease development.

A fatal, predominantly neurological disease of introduced common brushtail possums in New Zealand, **Wobbly possum disease (WPV)** is caused by a **nidovirus.**[80] In natural infection, lesions include nonsuppurative encephalitis with perivascular cuffing, diffuse nonsuppurative meningitis, and focal nonsuppurative myocarditis.[181] In experimentally infected animals, the most severe lesions occur in the liver, spleen, and heart. Brain, meningeal, and renal infection can also be seen. Perivascular mononuclear cell infiltrates consist predominantly of lymphocytes and plasma cells with fewer macrophages. Associated lesions include renomegaly, hepatomegaly, splenomegaly, cardiomegaly, and petechial hemorrhage of the gastrointestinal serosa, meninges, and renal capsule.[94]

Nonsuppurative meningoencephalitis occurs in common brushtail possums in Australia and presents with a disease syndrome similar to that of wobbly possum disease in New Zealand. Two different arteriviruses related to the nidovirus responsible for the New Zealand syndrome have recently been identified by meta-transcriptomics in Australian possums though further work is required to fulfill Koch's postulates. Disease presents as nonsuppurative meningoencephalitis.[52] The disease in Australia differs from that in New Zealand by its lack of multisystemic inflammation. In Australian common brushtail possums, lesions consist of multifocal areas of hemorrhage and malacia, nonsuppurative perivascular inflammatory cell cuffing, and fibrinoid vascular necrosis in the meninges and parenchyma of the cerebrum, cerebellum, brainstem, and in some cases, cervical spine (Fig. 18.17A). Retinal atrophy is often evident (Fig. 18.17B) along with mild to marked nonsuppurative choroiditis, retinitis, and optic neuritis with lymphocytic cuffing of associated blood vessels (ARWH, unpublished data).

Bacteria

Necrobacillosis, also called **macropod progressive periodontal disease** or **'lumpy jaw'**, is a common disease of captive macropods that is also occasionally seen in free-ranging macropods and in captive wombats, brushtail possums, sugar, and yellow-bellied gliders, numbat, and kowari.[137,158] The primary bacterium involved in lesion development is ***Fusobacterium necrophorum***, however, a suite of other aerobic and anaerobic bacteria may be involved, including *Nocardia* spp., *Bacillus* spp., *Bacteroides* spp., *Arcanobacterium pyogenes*, *Proteus* sp., *Pseudomonas* sp. and *Clostridium* sp., as well as staphylococci and streptococci. Predisposing factors include overcrowding, nutritional or climatic stress, abrasive or soft forage, facial trauma, and gastrointestinal parasitism.[158] Though the role of abrasive forage suggests that infection occurs via compromised oral mucosal

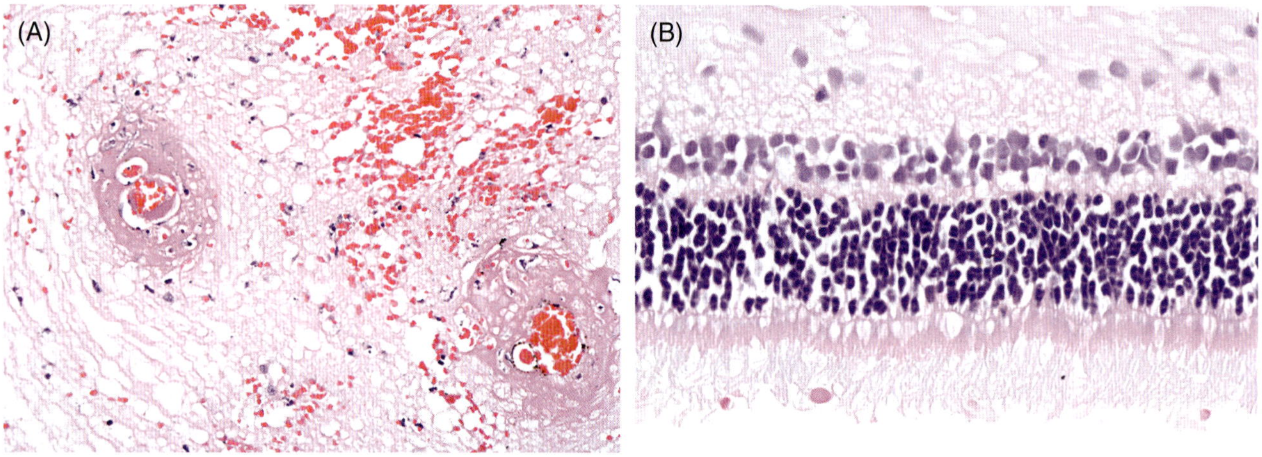

FIGURE 18.17 (A) Arterivirus-associated nonsuppurative encephalitis in a common brushtail possum. Malacia, hemorrhage, and vascular fibrinoid necrosis within the thalamus. (B) Retinal atrophy due to arterivirus-associated nonsuppurative encephalitis and optic neuritis in a common brushtail possum. Atrophy of the ganglion cell layer of the retina. Inflammation is not seen in this image. *(Parts A, B: Photos Courtesy of Australian Registry of Wildlife Health.)*

barriers, the frequency with which visceral foci of infection are found in the lungs, liver, stomach, and intestines in the absence of evident oral lesions suggests a role for hematogenous spread following entry via the gastrointestinal mucosa. Necrobacillosis manifests as edematous or bony swellings in the maxillary and mandibular area, often accompanied by abscessation, gingival and cutaneous ulceration, periodontal disease, and the formation of fistulas and bony sequestra (Fig. 18.18A).[137,158] Ocular and nasal swelling or discharge, salivation, and inappetence are also common. In chronic cases, subperiosteal bony proliferation occurs in adjacent bone (Fig. 18.18B). Infection can spread to involve the ears or central nervous system. Visceral lesions include focal or multifocal nodules of necrosis that can become extensive. Histologically, areas of coagulative to liquefactive necrosis contain masses of tangled filamentous, gram-negative organisms, often accompanied by other mixed bacterial infection/s. These are often peripheral in the lesion, though they can occur throughout. Leukocyte infiltration and surrounding fibroplasia in soft tissues and inflammation, osteolysis, and remodeling with osteoclasts and osteoblasts in bone can also occur. There is often a well demarcated boundary with surrounding normal bone.[137,158]

Tyzzer disease, caused by ***Clostridium piliforme***, has been reported in macropods, koalas, wombats, dasyurids, and brushtail and ringtail possums. Disease occurs mostly in young animals and outbreaks are often associated with stressful events.[137] Grossly, the liver may be enlarged, congested, and mottled with multiple, small (up to 3 mm in diameter), demarcated, gray-white foci scattered throughout the parenchyma. Other reported lesions include irregular pale patches and hemorrhage in the heart, hydropericardium, hydroperitoneum, pulmonary edema, and

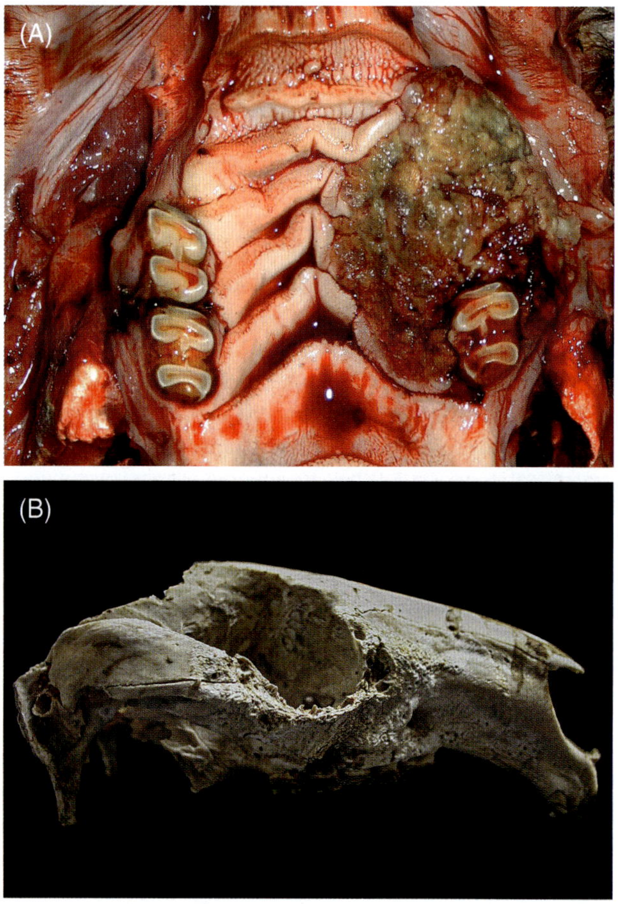

FIGURE 18.18 (A) Macropod progressive periodontal disease ('lumpy jaw') in a red kangaroo. Necrotizing and suppurative inflammation in the left maxillary region associated with molar loss. (B) Macropod progressive periodontal disease ('lumpy jaw') in a swamp wallaby. Osteolysis and periosteal proliferation of the maxillary region above the dental arcade and periorbital bone consistent with osteomyelitis. *(Parts A, B: Photos Courtesy of Australian Registry of Wildlife Health.)*

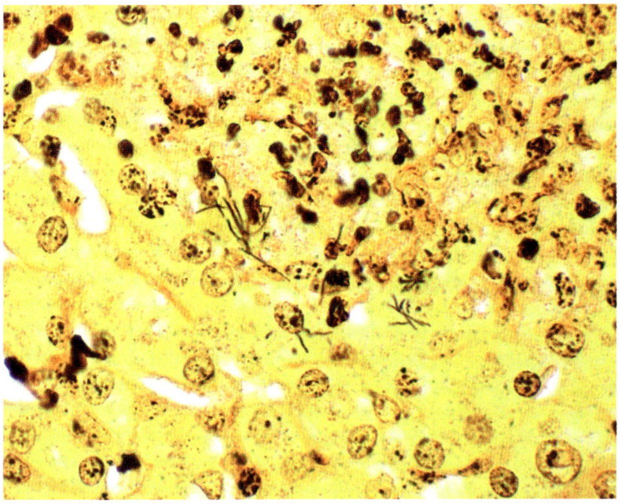

FIGURE 18.19 Tyzzer disease (*Clostridium piliforme*) in the liver of an eastern grey kangaroo. Clusters of argyrophilic filamentous bacteria along the margins of an area of necrosis, Warthin-Starry silver stain. *(Photo Courtesy of the University of Sydney.)*

consolidation of lungs. Gross lesions correspond histologically to multifocal, well-demarcated areas of random hepatic necrosis. Larger lesions and lesions in the myocardium may appear as areas of coagulative necrosis surrounded by a rim of neutrophils. Bacteria are predominantly located at the edges of lesions and appear as slender, faintly basophilic, intracellular bacilli with routine HE staining and as skeins of distinct, often beaded rods with Warthin-Starry and Gomori's methenamine silver staining (Fig. 18.19).[44]

Tetanus due to ***Clostridium tetani*** infection has been reported in a range of macropod species. Clinical signs associated with infection are typical of those in other mammals. **Enterotoxemia** due to ***C. perfringens*** infection in young kangaroos produces translucent, straw-colored fluid in body cavities and petechiation on serous surfaces and of the small intestine.[137]

Marsupials are recognized as being highly susceptible to various species of ***Mycobacteria***. Mycobacterial infections have been reported in members of all Australian marsupial orders in captivity worldwide. Isolated bacteria include: *Mycobacterium bovis* [World Organization for Animal Health (WOAH) reportable], *M. tuberculosis* (WOAH reportable), *M. ulcerans*, *M. scrofulaceum*, *M. paratuberculosis* (WOAH reportable), *M. marinum*, *M. terrae*, *M. asiaticum*, *M. chelonae abscessus*, *M. fortuitum*, *M. chitae*, *M. smegmatis*, *M. avium*, and *M. intracellulare*.[137] In New Zealand, free-ranging common brushtail possums are introduced as the focus of control and research programs because they are a major reservoir for *M. bovis* and culture has also identified *M. avium* subsp. *paratuberculosis* in lymph nodes.[41] The contrasts between the high prevalence of mycobacterial infection in introduced brushtail possums and captive marsupials, as opposed to free-ranging marsupials in Australia, may be related to the level of exposure and host density. Reduction of possum density is an effective tool for management of *M. bovis* in reservoir populations in New Zealand.[166]

The factors affecting the susceptibility of marsupials, as a group, to mycobacterial infection are not completely understood but it appears likely that fundamental biases of the marsupial inflammatory response play a role. This is based on the histopathological differences between these and most other species. Similar to ferrets and badgers, brushtail possums fail to produce the degree of granuloma organization and fibrous encapsulation needed to prevent dissemination of infection by *M. bovis*,[61] and the literature suggests that the same is true of mycobacterial infection, in general, in other marsupial species. Lesions are granulomatous or pyogranulomatous and may have a necrotic center with associated suppurative inflammation. Giant cells are often seen in the granulomatous components but fibrous encapsulation is generally not a prominent feature, if it is present at all (Fig. 18.20A).[41,137] Diagnosis is based on the demonstration of acid-fast organisms with special stains and subsequent identification by PCR and sequence analysis (Fig. 18.20B).

Mycobacterium ulcerans is the causative agent of Bairnsdale, Buruli, or Daintree ulcer in humans and has been reported to occasionally cause focal or multifocal granulomatous ulcerative dermatitis in koalas and brushtail possums. These have also been proposed as reservoirs for the pathogen. The mode of transmission is unknown. Exposure of wounds to contaminated soil, vegetation, or other animals carrying the pathogen has been proposed and mosquitoes are a suspected but unproven vector.[92]

Listeriosis due to ***Listeria monocytogenes*** has been reported in a range of species including antechinus, ringtail and brushtail possum, red-necked wallaby, rufous bettong, and a cuscus.[3,13,45,137] In the antechinus, infection is typically associated with postmating die-offs of males due to physiological exhaustion.[13] Gross lesions include hepatomegaly and multifocal pinpoint to 2 mm diameter pale foci in the hepatic parenchyma. These correspond histologically to areas of necrosis that may contain gram-positive bacilli and infiltrates of neutrophils, histiocytes, and some lymphocytes and plasma cells. Associated lesions include splenic lymphoid depletion and focal necrotic lesions in other tissues such as the myocardium and adrenal glands.[137]

Chlamydiosis is the most common infectious disease of koalas and, along with motor vehicle collisions, displacement, and predation by dogs, has significant population-related impacts, especially in the northern part of its range where the species is listed as Endangered (Environment Protection and Biodiversity Conservation Act, 1999). Disease is typically caused by ***Chlamydia pecorum***. Koalas can also be infected by ***C. pneumoniae*** but it is

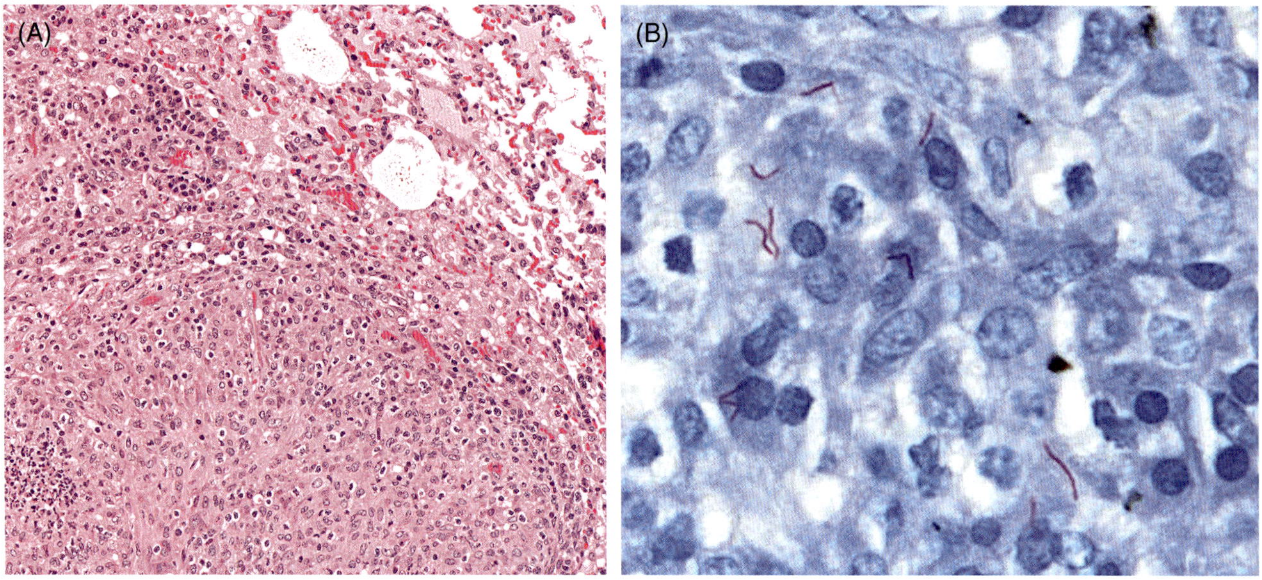

FIGURE 18.20 (A) Granulomatous pneumonia due to *Mycobacterium fortuitum* in a long-nosed potoroo. Minimal fibrous encapsulation, typical of mycobacterial lesions in marsupials. (B) Granulomatous pneumonia due to *Mycobacterium fortuitum* in a long-nosed potoroo. Acid-fast filamentous intracellular rods, Ziehl-Neelsen stain. *(Parts A, B: Photos Courtesy of Australian Registry of Wildlife Health.)*

rarely detected and its role in disease appears to be comparatively minor. *C. pecorum* is present in almost all free-ranging koala populations at prevalence ranging from 10% to 90%. Morbidity varies considerably. In some populations, it approaches 100% while in others, disease in infected animals is uncommon.[183] The drivers of disease expression appear to be multifactorial with anecdotal and association study support for a role for host genetics, chlamydial strain, coinfection, or disturbance and other stressors.[140,193,195,196]

In koalas, *C. pecorum* is transmitted sexually and is also likely to be transmitted during birth, via monocytes in milk, or during pap-feeding (establishment of intestinal flora in young by ingestion of maternal cecal feces). Transfer between anatomical sites can occur. Systemic spread via monocytes and histiocytic cells, as occurs for other chlamydial species, appears likely as inclusions formed by *C. pecorum* are seen with immunohistochemical labeling in submucosal macrophages in diseased tissues and in pulmonary, hepatic, and splenic histiocytes of some koalas.[106]

Most tissue damage in chlamydiosis is caused by the host's response rather than by cell lysis due to chlamydial replication. Infection in the female urogenital tract causes pyometra, metritis, cystitis, and/or ascending pyelonephritis. Commonly, the ovarian bursa fills with fluid to form multiloculated cysts up to approximately 150 mm in diameter (Fig. 18.21A). In these cases, fibrinous and fibrous adhesions between the bursa and ovary may develop. The associated uterus and uterine tube typically become fibrotic and occluded. A recent study has identified modifications of intercellular junctions in the ovarian bursae of female koalas with *Chlamydia*-associated cystic ovarian bursal disease.[176] These include reduced expression of epithelial cell–cell junction markers E-cadherin, β-catenin, and ZO-1, and severe widening of intercellular spaces. It has been postulated that the pathogenesis of abnormal fluid accumulation in the cystic ovarian bursae of infected koalas may involve the alteration of epithelial cell–cell adhesion proteins by chlamydial organisms, leading to epithelial cell separation and reduced absorption or increased leakage of fluid.

Hydroureter and hydronephrosis are common sequelae of fibrotic cystitis and changes associated with renal infection are typical of ascending pyelonephritis in other species, with or without resulting fibrosis.[106] Severe cases of urogenital disease can also produce fibrotic adhesions to any of the pelvic viscera. In male koalas, prostatitis and, rarely, orchitis have also been described.[43,116,127] Abnormalities in seminal fluid and spermatozoal morphology, motility, and viability have been documented in male koalas with severe clinical signs of urogenital chlamydiosis.[115] This may be related to spermatozoal damage and/or loss of function associated with exposure to chlamydial lipopolysaccharide or a consequence of chronic urogenital inflammation.

Ocular lesions range from mild conjunctival hyperemia and edema to epithelial and lymphoid hyperplasia. The nictitating membrane may become markedly thickened and cauliflower-like as a result of edema and lymphoplasmacytic infiltration (Fig. 18.21B). Keratitis and blindness are common sequelae to chronic ocular inflammation. Mucosal lesions vary with severity and chronicity and

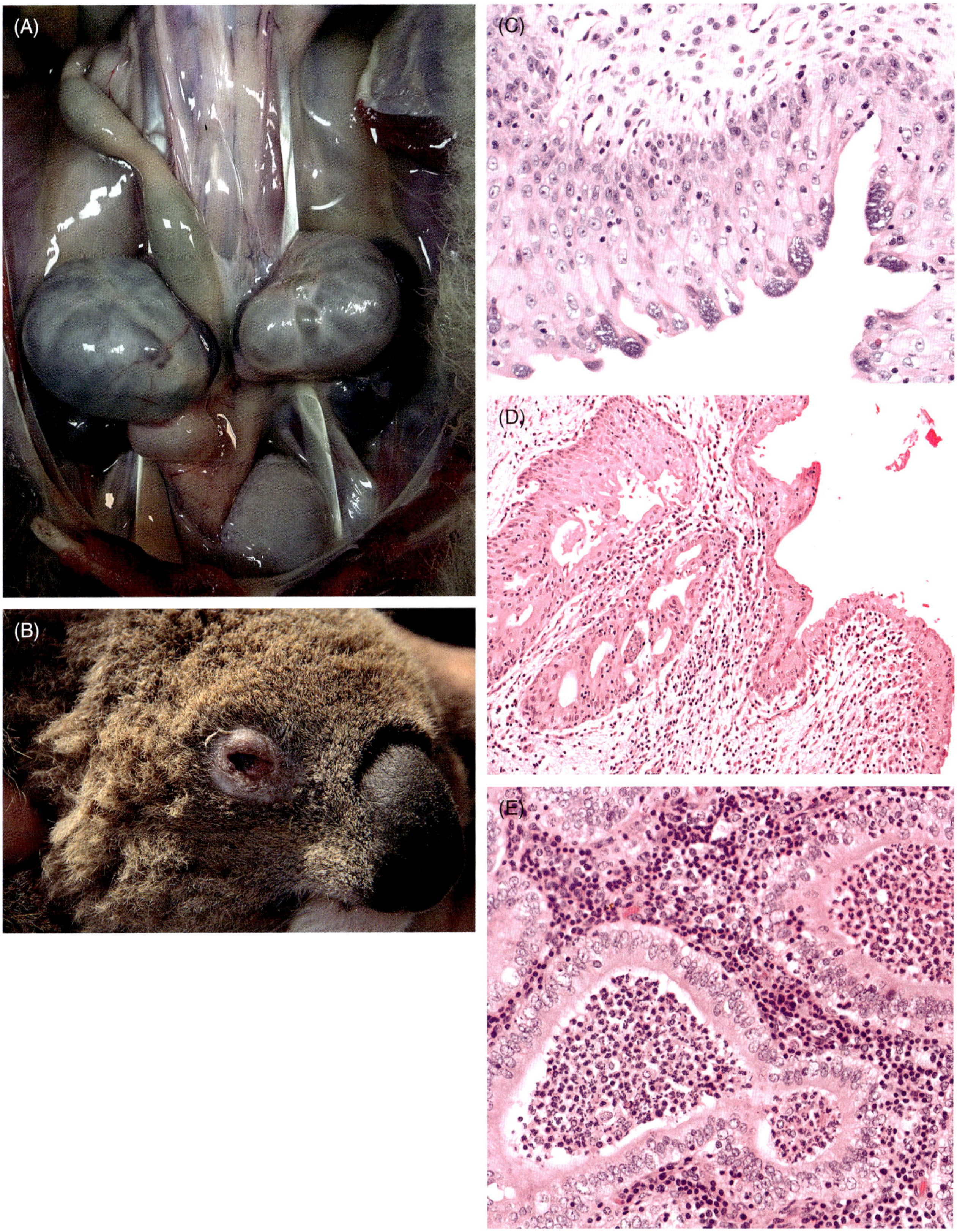

FIGURE 18.21 (A) Parovarian cysts due to chlamydiosis in a koala. Bilateral cystic dilation of the ovarian bursa. (B) Ocular chlamydiosis in a koala. Chronic keratoconjunctivitis with periocular alopecia. (C) Chlamydial cystitis in a koala. Basophilic chlamydial 'inclusions' in superficial epithelium of the urinary bladder. (D) Chronic chlamydial prostatitis in a koala. Moderate epithelial hyperplasia, moderate lymphoplasmacytic infiltration of the submucosa, and mild epithelial lymphocyte transmigration. (E) Active chlamydial metritis in a koala. Marked lymphoplasmacytic infiltration of the lamina propria and suppurative exudation within the lumen of the uterine glands. *(Part A: Photo Courtesy of T. Portas, RSPCA Queensland. Part B: Photo Courtesy of C. Marschner, University of Sydney. Part C: Photo Courtesy of the University of Sydney. Parts D, E: Photos Courtesy of Australian Registry of Wildlife Health.)*

may include villous epithelial hypertrophy and hyperplasia, fibrosis, granulation tissue formation, and squamous metaplasia. Inflammation is most severe in the lamina propria and superficial submucosa and is predominantly lymphocytic with occasional macrophages. Chlamydial elementary bodies appear as cytoplasmic inclusions in epithelial cells that can be seen in some cases. These are highlighted with Gimenez stain (Fig. 18.21C). In active cases with greater numbers of inclusions, transepithelial migration of neutrophils and lymphocytes (often in large numbers) is common. Plasma cells predominate in more chronic, inactive cases (typically immediately beneath the basal epithelium and extending deep into the submucosa) (Figs. 18.21D, E). In chronic cases, extensive fibrosis is common and may occlude the urinary or reproductive tracts or cause marked thickening of the urinary bladder wall. Cases of pneumonia caused by *C. pecorum* are seen sporadically.

Because chlamydiosis is the most common cause of conjunctivitis and urogenital inflammation in koalas, gross and histological signs are strongly suggestive of a diagnosis. Diagnosis can be confirmed by PCR or immunohistochemistry, usually with a genus-specific anti-LPS antibody.[106] Sample selection for IHC should consider that in these cases chlamydial inclusions are predominantly in epithelial cells and are often surprisingly scant in comparison to the degree of damage. Serology is typically an insensitive indicator of chlamydial infection.

Although a range of opportunistic bacteria, for example, *Klebsiella pneumoniae* and *Pseudomonas aeruginosa*, cause respiratory disease in marsupials, ***Bordetella bronchiseptica*** is known for its involvement in outbreaks of often fatal respiratory disease in captive koalas and, to a lesser extent, wallabies, and opossums. It is uncertain exactly what role *B. bronchiseptica* plays in pathogenesis. Due to its role as an opportunist in other mammalian species, it is suspected that stress events, primary infection by a yet-to-be-identified virus, or coinfection with *Mycoplasma* or *Chlamydia* spp. may impair pulmonary defenses and predispose to *B. bronchiseptica* infection. Lesions include severe pulmonary congestion, edema, suppurative bronchopneumonia with hyperplasia of bronchial epithelium, cuboidal metaplasia of alveolar epithelial cells, and fibrinopurulent pleuritis. Bronchopneumonia is accompanied by copious bronchial and tracheal mucopurulent exudate. Bacteria may be seen histologically as clumps of gram-negative bacteria tangled in the epithelial cilia. Disseminated infection can be associated with necrotizing hepatitis, splenitis, and presence of serofibrinous effusion in body cavities.[137]

As for other mammalian species, a range of **opportunistic bacteria** can cause disease in marsupials and monotremes including but not restricted to *Escherichia coli, Aeromonas hydrophila, Pseudomonas, Yersinia, Shigella, Klebsiella,* and *Proteus.* These and other significant bacterial infections are reviewed by Ladds.[137] *Pasteurella multocida* is associated with severe bronchopneumonia in a range of marsupial species, possibly in association with stress episodes, coinfection by *Mycoplasma,* or an unidentified virus. Marsupials appear to be particularly susceptible to fatal septicemia due to *P. multocida* infection resulting from cat bites.[33] Lesions are consistent with septicemia and include peritonitis, bacterial emboli, widespread congestion and hyperemia, and necrotizing lesions with suppurative inflammation in a wide range of tissues. ***Edwardsiella*** sp. is suspected to act as a primary pathogen in captive and free-ranging echidna.[261] Infection causes suppurative or necrotizing bronchopneumonia and serosanguinous pleural exudate. ***Burkholderia pseudomallei*** is reported to cause disease in several marsupial species and the epidemiology and pathology are similar to melioidosis in other mammals.[55] In Australia, cases are restricted to the far north of the country.

Salmonella typhimurium and, to a lesser degree, a variety of other *Salmonella* spp. have been isolated from marsupials or monotremes. They are commonly carried asymptomatically and may be a source of zoonotic disease. This is of particular relevance to the kangaroo and wallaby meat industry and in situations where marsupials are held as pets or when orphaned marsupials, particularly red kangaroos and other macropods, are being hand-raised. When disease does occur in marsupials, it is consistent with salmonellosis in other mammals and primarily causes necrotizing and hemorrhagic enteritis and septicemia.[137]

Infection with ***Leptospira*** spp., predominantly *L. hardjo, L. balcanica,* and *L. pomona* serovars, has been detected in a wide range of marsupial and monotreme species. Marsupial and monotreme species are considered potential reservoirs for some of these bacteria, and associated disease in them is rare.[137] When disease occurs, it causes interstitial nephritis that is generally mild but of variable severity. Leptospires are highlighted in renal tubules with silver stains.

Rickettsia spp. occur in a range of species with disease being mild or asymptomatic. With ticks as the vector, macropods and bandicoots are carriers of ***Coxiella burnetii***, a WOAH-listed pathogen and the cause of Q fever in humans.[62]

Fungi

Cryptococcosis has been reported in both free-ranging and captive koalas. The causative agent is ***Cryptococcus neoformans*** var. ***gattii***. These are saprophytic encapsulated yeasts that are present on a wide range of eucalyptus species and are common in tree hollows of *Eucalyptus camaldulensis* and *E. tereticornis*, two significant food trees for koalas in many regions. The pathogen is environmental rather than contagious and *C. n.* var. *gattii* is also reported to cause disease in Indigenous communities and native and domestic species. The nasal mucosa of koalas is the most common site of colonization and its presence is often not associated with lesions.[134] Disease occurs in a minority of cases, and the factors driving progression to invasion and then clinical disease are poorly understood. The degree of environmental contamination and molecular fungal type may be important. *C. neoformans* var. *gattii* of molecular type VGII appear to be more virulent than VGI organisms.

Cryptococcosis in koalas most often affects the respiratory tract with the lower respiratory tract being more commonly infected (Fig. 18.e17).[135] Dissemination is common, especially to the central nervous system. When upper respiratory tract disease occurs, local extension to surrounding tissues is often observed. Other tissues in which lesions may be seen are (in decreasing order of occurrence) the lymph nodes, gastrointestinal tract, kidneys, spleen, and skin. Histopathological changes are variable and range from large aggregations of cryptococcal organisms with minimal host-response to early neutrophilic response and later pyogranulomatous and granulomatous inflammation (Figs. 18.22A, B; eSlides 18.e4 and 18.e5). Intralesional yeasts are heavily encapsulated and approximately 10–50 µm in diameter. Cryptococcal yeasts commonly colonize the upper respiratory tract of koalas without causing disease and their presence can be an incidental finding. As such, nasal culture is of limited clinical use. Detection of capsular antigen in serum by latex agglutination or lateral flow assays is used as an indicator of invasion and subclinical disease.

Mycotic granulomatous dermatitis, caused by ***Mucor amphibiorum***, is a significant cause of mortality and morbidity in platypus in the state of Tasmania. *M. amphibiorum* causes disseminated disease in anurans. In platypus, ulcerative dermatitis grossly similar to this disease can be caused by other bacteria including *Corynebacterium ulcerans*, *Epicoccum purpurascens*, and *Aeromonas hydrophila*.[149] Interestingly, the organism is common in waterways on mainland Australia, but it does not cause disease in platypus there.

Mucor and other fungi in platypus cause superficial granulomatous dermatitis that progresses to ulcerative nodular or coalescing lesions (Fig. 18.23A). Associated inflammation is composed of T lymphocytes, macrophages, and giant cells (Fig. 18.23B). Epidermal hyperplasia, tissue necrosis with accompanying suppurative inflammation, and fibrovascular proliferation may be present.[58] Lesions can occur in the skin at any site and wounds or tick bites may predispose the skin to infection. Lesions are more common in males, which engage in territorial fighting.[60] Infection can extend into the underlying musculature and disseminate to internal organs, particularly the lungs.

Pulmonary **adiaspiromycosis** occurs in all three species of free-ranging wombats.[64,153,204] The causative agent, a soil saprophytic fungus, was previously described as ***Emmonsia parva***, but recent phylogenetic studies have identified the organism in northern hairy-nosed wombats to be a novel species of the genus ***Emmonsiellopsis***.[64] Recent investigations using molecular diagnostics have confirmed the same species in common and southern hairy-nosed wombats historically diagnosed with

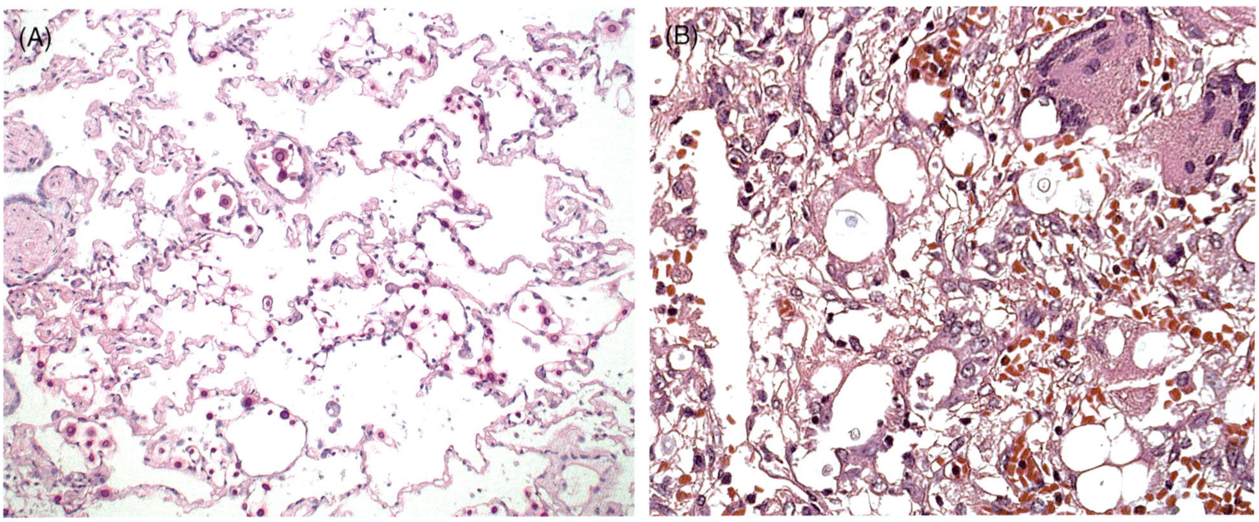

FIGURE 18.22 (A) *Cryptococcus gattii* infection in the lung of a koala. Minimal host response to infection is evident, Periodic acid-Schiff stain. (B) Granulomatous pneumonia due to *Cryptococcus gattii* in a koala. Cryptococcal organisms are clearly evident as mildly refractile ovoid yeasts surrounded by a thick, nonstaining capsule. Organisms are sometimes visible within macrophages or giant cells (as in top right of image). (see eSlide 18.e4 and 18.e5). *(Part A: Photo Courtesy of the University of Sydney. Part B: Photo Courtesy of Australian Registry of Wildlife Health.)*

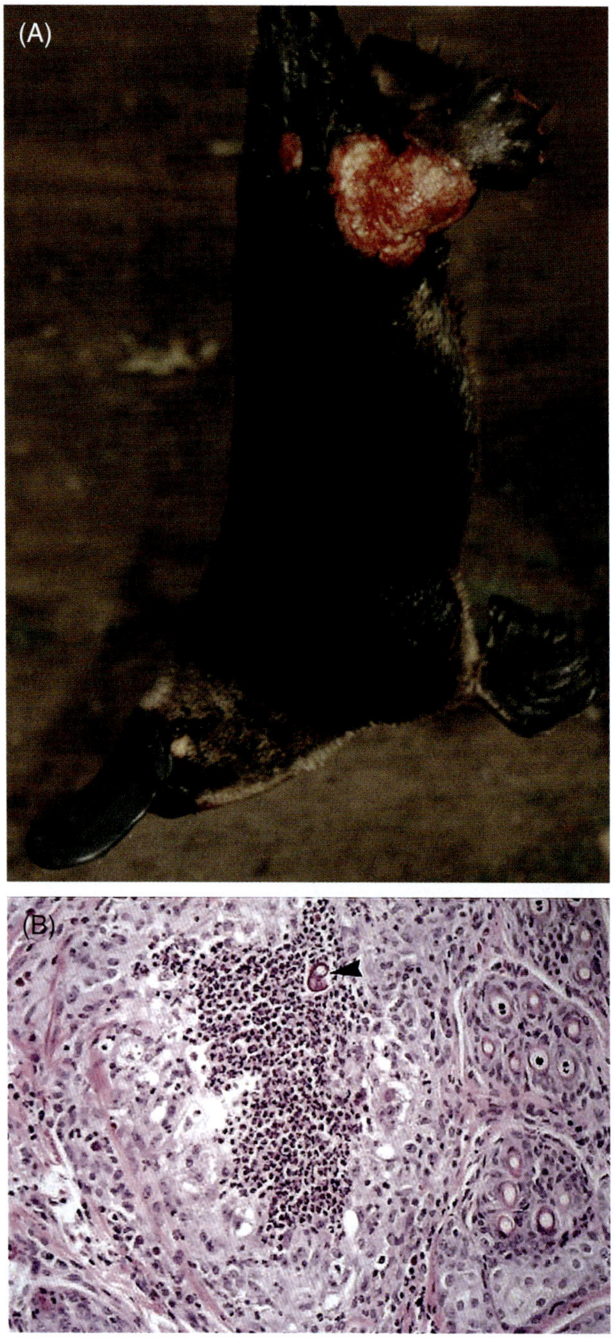

FIGURE 18.23 (A) Dermal ulceration of the hindlimb caused by *Mucor amphibiorum* in a platypus. (B) *Mucor amphibiorum* (*arrowhead*) in a platypus. Dermal granulomatous inflammation. In vivo, the fungus develops as spherical forms containing a number of daughter spherules; no mycelia are seen in tissue sections. By contrast, in vitro, growth consists of aerial aseptate mycelia and sporangia, features typical of the genus *Mucor*. *(Parts A, B: Photos Courtesy of J. Connolly, Charles Sturt University.)*

Emmonsia infection (M. Gimeno and D. Phalen, unpublished data). Burrowing habits of wombats are thought to predispose them to infection via inhalation of the ubiquitous soil-borne conidia (aleuriospores). Infection is usually

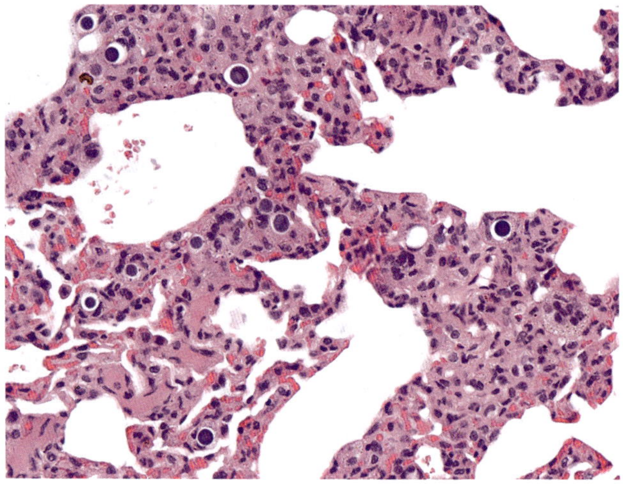

FIGURE 18.24 Pulmonary adiaspiromycosis in a common wombat. Fungal spherules are associated with granulomatous interstitial inflammation in the lungs. Many macrophages and multinucleate giant cells contain coarsely granular to spiculated pigmented material (pneumoconiosis). This is a common incidental finding in wombats due to soil exposure when burrowing. *(Photo Courtesy of D. Phalen, University of Sydney.)*

subclinical but large burdens may be associated with chronic respiratory disease in some animals. At necropsy, ventral lung lobe consolidation and pallor, mucopurulent exudate in larger airways, and variably severe pulmonary edema, pneumonia, fibrosis, and pleuritis may be seen. Gross lesions in subclinical cases may not be evident. Histologic changes include minimal to severe histiocytic pneumonia and fibrosis associated with variably sized, 22–35 μm diameter, intraalveolar and intrahistiocytic adiaspore spherules with a 1.0–1.5 μm wide refractile capsule (Fig. 18.24).

Candidiasis, most commonly caused by ***Candida albicans***, has been seen in many species of Australian mammals including echidnas, macropods, koalas, wombats, and possums. *Candida* yeasts are commensal organisms of the upper gastrointestinal tract, skin, and cloaca and overgrowth of these organisms can result in clinical disease localized to these sites or, in some cases, systemic disease. Clinical candidiasis is most common in hand-reared pouch young, and factors favoring overgrowth include stress, poor hygiene, inappropriate diet, antibiotic therapy, and comorbidities. Pale plaques and ulcerations may be present on the skin around the mouth, lips, tongue, oral cavity, esophagus, and nonglandular stomach. Proliferations of budding yeasts and pseudohyphae are seen adjacent to desquamated epithelial cells and pseudohyphae preferentially invade the superficial layers of squamous epithelium. In more severe infections, invasion of pseudohyphae into deeper tissues is associated with ulceration and neutrophil-rich inflammation.[137] *Candida* stomatitis and gastritis are frequently observed in captive echidnas of any age with concurrent disease or debilitation.

Vesiculopustular and erosive gastritis are seen histologically and resemble lesions of ruminal acidosis in livestock.[243]

Metazoa

In addition to the following descriptions, an overview of monotreme and marsupial parasites associated with pathological lesions is presented in the Digital Content Table 18.e2. A few parasitic infections of note are discussed below.

Infection and disease associated with the rat lungworm, ***Angiostrongylus cantonensis***, is reported in several marsupial species in the wild, zoos, and fauna parks. Infection develops as a consequence of ingestion of terrestrial or aquatic gastropods containing infective third-stage larvae.[227] The reports of debilitating infection with *A. cantonensis* in brushtail possums in Brisbane and Sydney, Australia[148] serve as a reminder that this nematode is well established not only in nonurban areas but also in some parts of these capital cities. Subadult common brushtail possums have been identified as potential biosentinels of this zoonotic agent in the Sydney region.[148,225] The clinical signs of ascending paresis and paralysis and sometimes death in nondefinitive mammalian hosts are induced by migration of larvae from the subarachnoid space into the neural tissues of the spinal cord from where they ascend to the brain. Larvae may be found in the spinal cord (Fig. 18.25A), however often only residual inflammation or Wallerian degeneration are seen (Fig. 18.25B) as the advancing front of late third-, fourth-, or young adult-stage nematodes have reached the brain. Although there may be limited associated inflammation, the nematodes and their tracts can be associated with acute or chronic eosinophilic meningoencephalitis or myelitis, malacia, and hemorrhage. Inflammation may be exacerbated by the death of larvae following treatment with parasiticides.

Most of the large numbers of **cloacinid nematodes** that can be found in the sacculated forestomachs of macropodoids are not pathogenic (Fig. 18.26). However, third-stage larvae of *Parazoniolaimus collaris, Labiosimplex clelandi,* and L. *billardierii* produce granulomas at the junction of the squamous and glandular epithelia.[27,163,213] Third-stage larvae of *Labiosimplex robustus* accumulate in nodules under the epithelial lining of the stomachs of wallaroos, and *Rugopharynx rosemariae* form numerous crateriform lesions on the glandular mucosal surface of the stomach of eastern and western grey kangaroos. The tails of these larvae protrude from each lesion.[5,28] Minute species of *Strongyloides* occur in regions of the cardiac glandular mucosa of macropodids and in heavily infected kangaroos, the gastric mucosa may appear erythematous. Severe infections may be fatal, but these are known only in captive animals.[4,219,265]

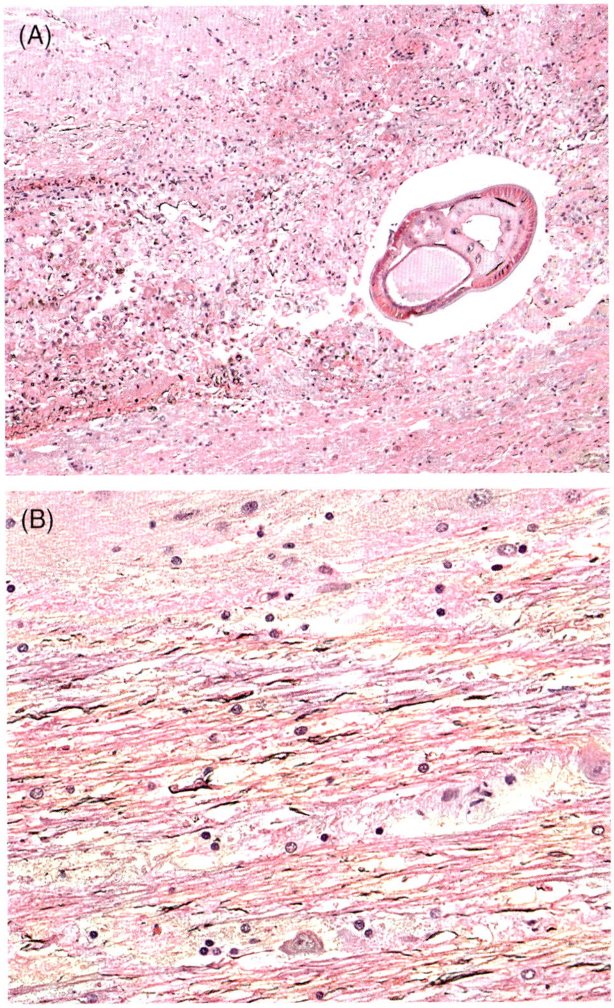

FIGURE 18.25 (A) *Angiostrongylus cantonensis* in the spinal cord of a brushtail possum. Cross section of an adult nematode in the central gray matter of the spinal cord. The nematode is characterized by a thin eosinophilic capsule surrounding coelomyarian-polymyarian musculature and a pseudocoelom containing digestive and reproductive tracts. There is associated malacia and disruption of the normal tissue architecture, as well as replacement by large numbers of macrophages and occasional erythrocytes and polymorphonuclear cells to the left of and around the parasite in this image. (B) *Angiostrongylus cantonensis* in the spinal cord of a brushtail possum. Wallerian degeneration; swollen axons with digestion chambers are scattered throughout the white matter tracts. *(Parts A, B: Photos Courtesy of Australian Registry of Wildlife Health.)*

Multiple species of a large red trichostrongyloid nematode parasite, ***Globocephaloides,*** infect the duodenum and feed on the blood of macropodid hosts. Infection may cause anemia and mortality, especially in juveniles.[3,29] A large hookworm-like nematode, ***Hypodontus macropi***, infects the cecum of wallabies and the cecum and terminal ileum of kangaroos. It feeds on the mucosa and can cause intestinal bleeding. Heavy burdens have the potential to cause anemia and hypoproteinemia.[22] Larvae of the large nematode ***Paramacropostrongylus***

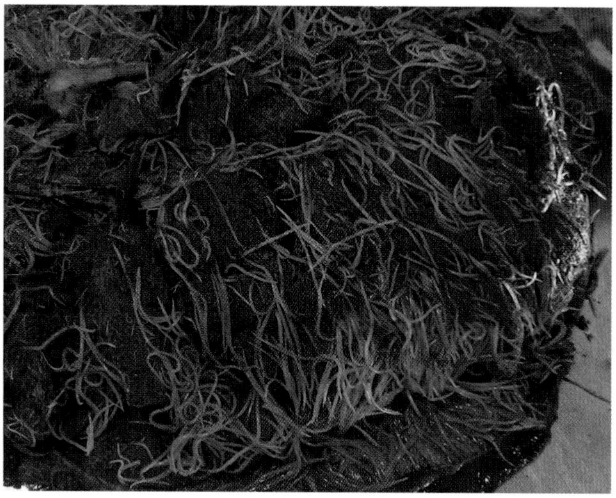

FIGURE 18.26 Cloacinid nematodes in sacculated forestomach of a common wallaroo. *(Photo Courtesy of I. Beveridge, University of Melbourne.)*

toraliformis develop in nodular lesions at the distal tip of the cecum of eastern grey kangaroos.[5]

Many species of kangaroos, wallabies, and rock-wallabies are important intermediate hosts of ***Echinococcus granulosus.*** Hydatid cysts (**hydatid disease**) containing developing larval cestodes have been reported most commonly in eastern Australia, although infection occurs in all States and the Australian Capital Territory.[120,121,136] The preponderance of cysts occurs in the lungs and thoracic cavity rather than the liver, and swamp and black-striped wallabies are more important in the sylvatic cycle than foxes and feral pigs.[120,121] Hydatid cysts have been reported in common wombats only in Victoria.[120] Although historically thought to be distinct strains, subsequent studies have found no detectable genetic differences between *E. granulosus* from wildlife and domestic animals.[111]

Fascioliasis due to infection with ***Fasciola hepatica*** occurs in grazing marsupials in southeastern Australia in areas where their habitat overlaps with that of ruminants. Infection may be severe in the smaller macropods but is reasonably well tolerated in larger kangaroos despite substantial fibrosis around bile ducts.[231] Viable eggs may be passed in the feces. Common wombats develop severe fibrosis of the bile ducts and parenchyma through which flukes have migrated, but viable eggs are not passed in the feces.

Protozoa

Toxoplasma gondii has been reported in most macropodoid species, a wide range of dasyurids and peramelids, brushtail possum, numbat, wombat, koala, platypus, echidna, and opossum.[72,74,76,91,137] Infection has primarily been reported in captive animals but also occurs in free-ranging animals. Vertical transmission of *T. gondii* may occur in chronically infected macropodid and potoroid marsupials.[179] Other sources of infection for omnivorous and insectivorous animals include oocyst ingestion in vegetation or soil and their associated invertebrates, especially earthworms, that have been contaminated with cat feces.[21] The source of infection for carnivores in the wild is ingestion of infected prey. *T. gondii* bradyzoites are common in the muscles of sheep in Australia,[218] and when fed raw sheep meat, small dasyurids in captivity have developed neurological signs and death due to fulminating encephalitis.[7,8] Toxoplasmosis is recognized as a significant cause of mortality in free-ranging eastern barred bandicoots in Tasmania and in a critically small population of this species at Hamilton, Victoria,[143,173] the latter species is thought to be particularly susceptible to infection-associated disease.[174] Recrudescence of latent infection may be a major manifestation of disease in this species.

Clinical signs in marsupials are correlated with species groups.[46,74,76] Animals in all groups exhibit lethargy, inappetence, and nervous signs. Sudden death has been reported in all groups except bandicoots and the bilby.[137] Respiratory distress is documented only in macropodids, wombats, and possums. Diarrhea is reported only in macropodids. Seroprevalence is approximately five times higher at sites with declining eastern quoll populations in Tasmania and there is a negative association between seroprevalence and quoll abundance; however, *T. gondii* infection has not been shown to reduce quoll survival or reproduction.[88]

For diagnosis, predilection sites for *T. gondii* infection include the central nervous system, lung, heart, skeletal muscle, and adrenal gland. Focal disease is common, highlighting the need to assess each of these tissues. In fulminating cases, clusters of *T. gondii* tachyzoites in skeletal and coronary muscle may lead to misinterpretation of disease chronicity as they resemble bradyzoites in tissue cysts. However, on close examination, these "pseudocysts" have no cyst wall. Bradyzoites can remain dormant with little or no host inflammatory response, and individual tissue cysts are commonly found within the muscle and nervous tissue of animals during histologic examination. Unless these cysts are associated with tissue necrosis or inflammation, they most likely represent subclinical infection. In contrast, tachyzoites of *T. gondii* are highly immunogenic, inciting nonsuppurative inflammation and necrosis. Gross postmortem lesions may be inapparent, however, pulmonary congestion, edema, consolidation, hepatosplenomegaly, lymphadenopathy, or multisystemic pale foci may be evident. Histologic examination of tissues from clinically affected animals may reveal any of the following lesions: retinitis, lymphadenitis, hepatitis, myositis, pneumonia, pancreatitis, myelitis, and encephalitis (eSlides 18.e6 and 18.e7).[46,74,76,137] Because it is not possible to unambiguously differentiate between the different tissue cyst-forming coccidia (e.g., *Toxoplasma* sp., *Neospora* spp., *Sarcocystis* spp.) using light microscopic examination of tissue alone, immunohistochemistry for *T. gondii* (see eSlide 18.e8) and/or molecular diagnostics should be conducted in conjunction with histopathology to confirm a diagnosis of toxoplasmosis.

Coccidiosis in marsupials and monotremes is most commonly due to ***Eimeria* spp.**, of which over 40 different species have been reported in macropods alone.[15] Oocysts are commonly identified in feces and coccidian parasite infections are relatively common in healthy captive and wild marsupials and monotremes, but disease is rarely reported. Clinical disease associated with coccidia infection is more commonly encountered in young animals, especially eastern grey kangaroos, western grey kangaroos, red-necked wallabies, common wombats, and echidnas, and it is often associated with comorbidities. Lesions of enteric coccidiosis in monotremes and marsupials are similar to those in eutherian mammals and include necrotizing and hemorrhagic enteritis, mucosal thickening due to the presence of schizonts and micro- and macrogametes in the lamina propria, and mucosal ulceration and villous atrophy associated with rupture and release of parasite life stages (Fig. 18.27A).[137] Gastric hemorrhage and necrosis associated with coccidian organisms may also be seen in western grey and eastern grey kangaroos with severe fulminating enteric coccidiosis.

Coccidiosis in tammar wallabies is characterized by both intestinal and hepatic lesions. Schizogony is initially seen in the small intestine and subsequently occurs in the liver, where merozoites are released into bile ducts and gametogony takes place.[137] Hepatic lesions of coccidiosis in tammar wallabies include coccidian schizonts in hypertrophied cells and associated hemorrhage and necrosis, bile duct dilation and rupture, suppurative and eosinophilic cholangiohepatitis, biliary hyperplasia, periportal fibrosis and accumulations of brown-pigment laden macrophages.[45]

Enteric coccidiosis in the short-beaked echidna may progress to disseminated infection where coccidian development occurs in extraintestinal organs of affected animals. Deep-sequencing molecular technology has implicated ***E. echidnae*** as the cause of **systemic coccidiosis** in the echidna.[212] Gross lesions of systemic coccidiosis depend on the organs involved but generally include congestion, edema, hemorrhage, and necrosis.[137,241] Histologically, immature and mature schizonts and free merozoites are identified most commonly in endothelial cells and the parenchyma of the lungs, liver, heart, kidney, spleen, lymph nodules, and gastrointestinal tract (Fig. 18.27B). The coccidian life stages are associated with variable tissue reaction usually consisting of mononuclear infiltration, necrosis, and less commonly neutrophilic or granulomatous inflammation (eSlides 18.e9 and 18.e10).[75,137] Examination of blood films can be used for antemortem diagnosis of clinical systemic coccidiosis in short-beaked echidnas because animals commonly have parasitemia of more than 30% of monocytes. Very low numbers of intramonocytic *Eimeria* can be detected in up to 50% of healthy echidnas, but large numbers are correlated with clinical disease.[241]

Leishmania sp. have been identified in captive red kangaroos in the Darwin area of the Northern Territory and clinical disease has been described in captive common

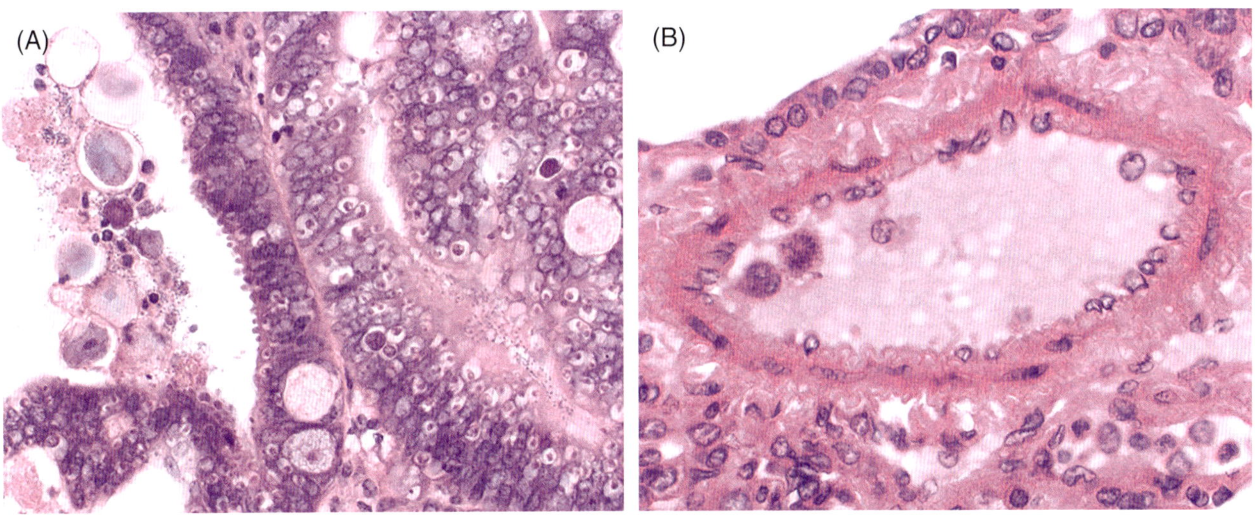

FIGURE 18.27 (A) Enteric coccidiosis and concurrent cryptosporidiosis in a short-beaked echidna. Numerous enterocytes contain various life stages of coccidian parasites. The apical surface of crypt enterocytes is regionally colonized by small round basophilic *Cryptosporidium* parasites. *Eimeria echidnae* and *Cryptosporidium fayeri*, respectively, were identified by molecular diagnostics. This echidna had systemic coccidiosis. (B) Systemic coccidiosis due to *Eimeria echidnae* in the lung of a short-beaked echidna. Disseminating intramonocyte stage of *Eimeria echidnae* in a pulmonary blood vessel. (see eSlides 18.e9 and 18.e10). *(Parts A, B: Photos Courtesy of the University of Sydney.)*

and black wallaroos and agile wallabies in this region.[73,197] Strong evidence suggests that the day-feeding biting midges *Forcipomyia (Lasiohelea)* sp. 1 and possibly *F.* (*L.*) *peregrinator* (Diptera: Ceratopogonidae) are responsible for transmission.[73] The disease manifests as cutaneous lesions on the tail, inner forearms, and hind legs. Affected skin is characteristically thickened or raised; variably encrusted or ulcerative and pale nodules and mildly swollen and variably encrusted lesions can also be present on the distal margins of the pinnae (Figs. 18.28A, B). Gross lesions may resemble poxviral lesions. Histologically, granulomatous inflammation with clusters of macrophages containing single-celled amastigotes with a nucleus and kinetoplast are present (Fig. 18.28C; eSlide 18.e11).[197] Members of the subgenus *Lasiohelea* bite humans and other mammals,[63] warranting further investigation to determine the actual risk of zoonotic transmission of this organism and the risk of propagation of other zoonotic *Leishmania* species that could be introduced by travelers. Opossums are natural hosts for *Leishmania braziliensis* and may develop clinical disease, primarily in the form of ulcers.[91]

Eastern grey kangaroos and agile wallabies may develop fatal **babesiosis** characterized by anemia, lethargy, chronic-ill thrift, and neurological signs. Young, often hand-raised animals are most commonly affected.[65,71] However, since 2014, large outbreaks of mortality have also been documented in free-ranging eastern grey kangaroos of all ages that have undergone significant multifactorial environmental stressors including unnaturally large aggregations due to urbanization, fire, and flood (K. Rose, personal communication). In a review of 38 cases, a piroplasm detected by Donahoe et al. in eastern grey kangaroos from across New South Wales and Queensland was genetically identical (18S rRNA) to ***Babesia macropus*** in eastern grey kangaroos.[71] The parasite had a high degree of similarity to organisms detected in anemic agile wallabies in Northern Queensland and an unnamed *Babesia* sp. in woylies in Western Australia.[65,71,178] The most significant

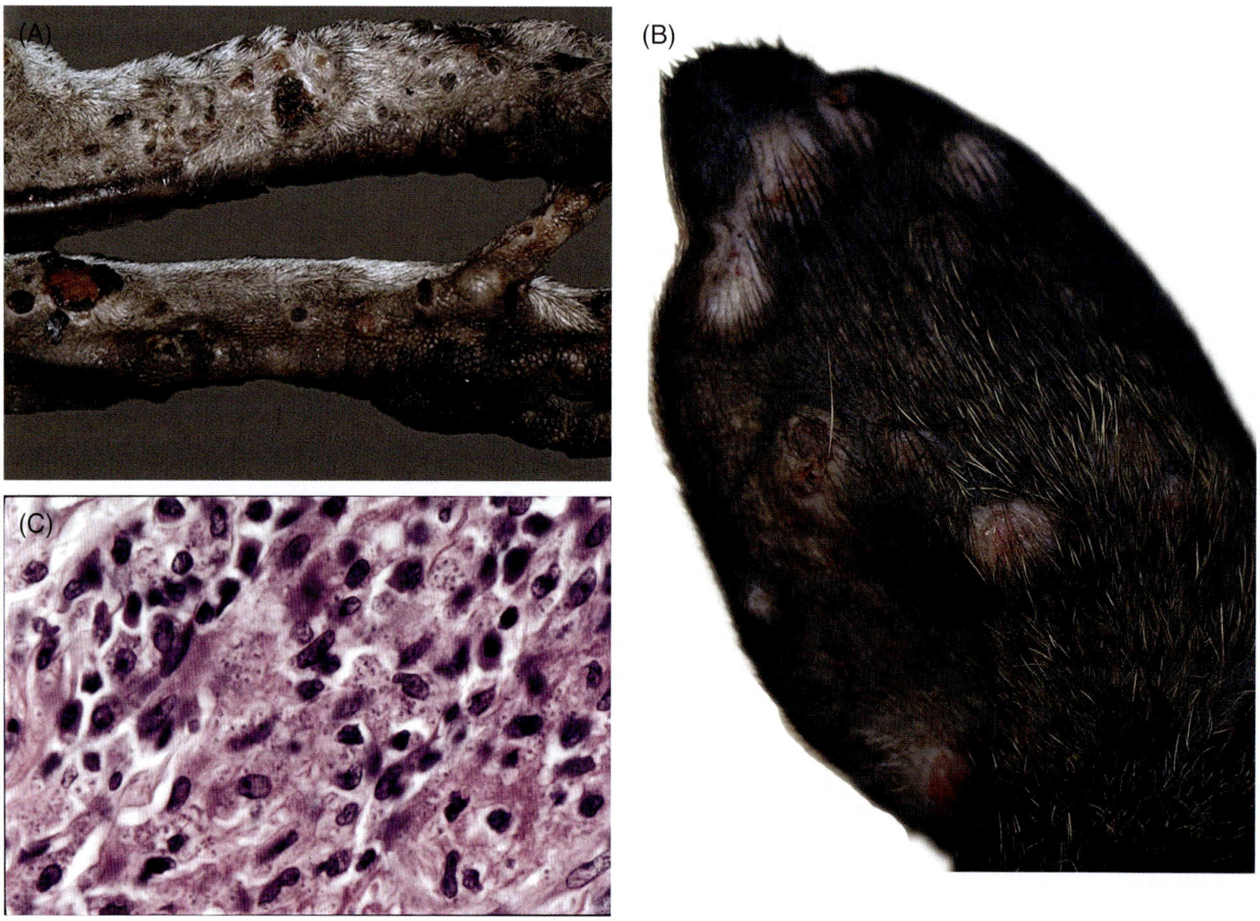

FIGURE 18.28 (A) *Leishmania* sp. in the skin of an agile wallaby. Multiple variably alopecic, ulcerative to crusting, cutaneous nodules over the hind limbs of an affected wallaby. (B) *Leishmania* sp. in the pinna of an agile wallaby. Multiple variably alopecic cutaneous nodules on the pinna, some with central areas of ulceration. (C) *Leishmania* sp. In the skin of a red kangaroo. Clusters of dermal macrophages containing single-celled organisms with a nucleus and often visible kinetoplast. (see eSlide 18.e11). *(Parts A, B: Photos Courtesy of C. Shilton, Northern Territory Berrimah Veterinary Laboratory. Part C: Photo Courtesy of Australian Registry of Wildlife Health.)*

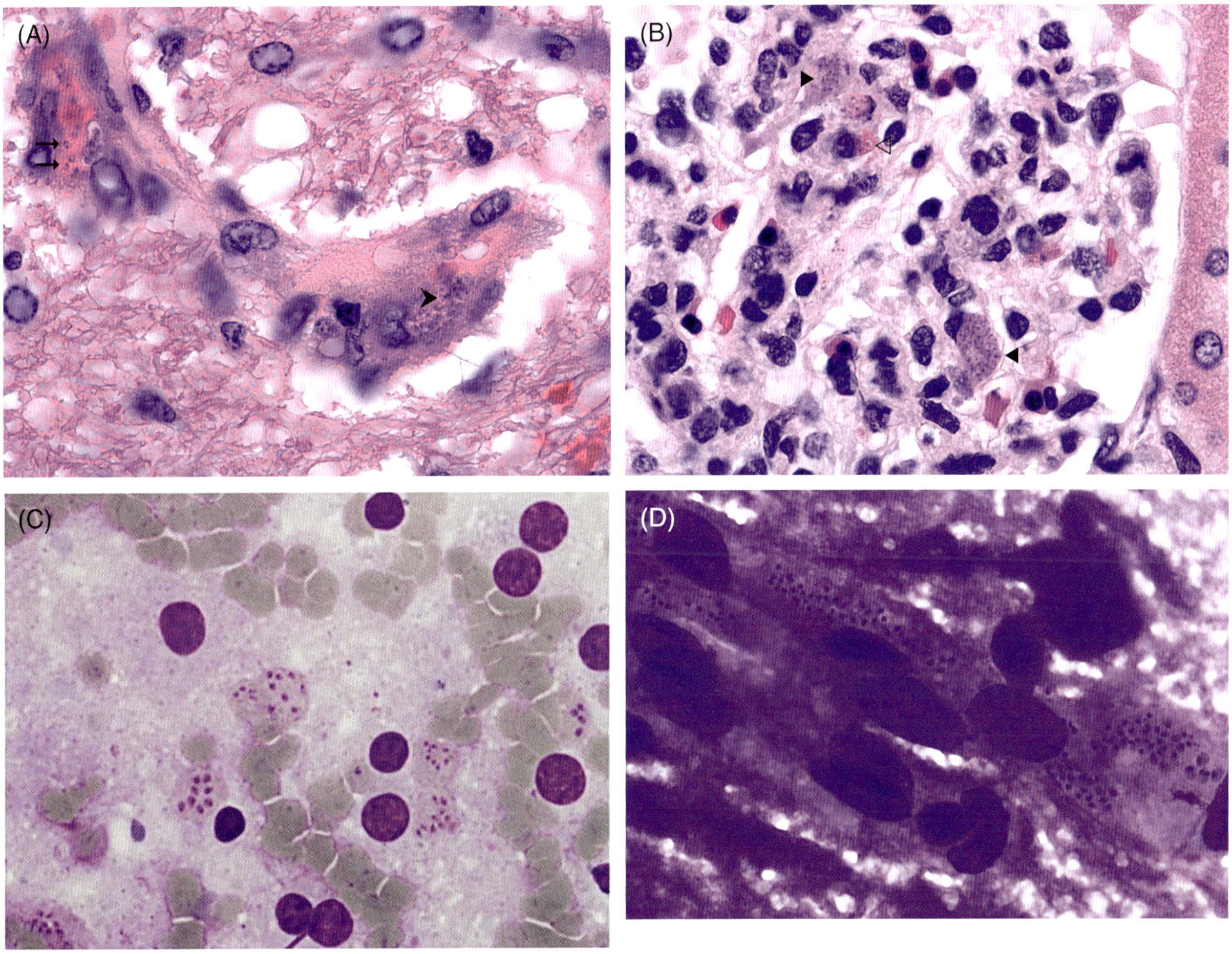

FIGURE 18.29 (A) *Babesia macropi* in the brain of an eastern grey kangaroo. A cerebral blood vessel contains intraerythrocytic (*arrows*) and extraerythrocytic (*arrowhead*) zoites. Focal hemorrhage is present in the bottom left of the image. (B) *Babesia macropi* in the kidney of an eastern grey kangaroo. Clusters of intraerythrocytic (*open arrowhead*) and extraerythrocytic (*closed arrowhead*) zoites are seen in glomerular capillary loops. (C) *Babesia macropi* in an eastern grey kangaroo. Extraerythrocytic clusters of zoites in a renal impression smear, Diff-Quik stain. (D) *Babesia macropi* in the brain of an eastern grey kangaroo. Zoites within capillaries, squash preparation, Diff-Quik stain. *(Parts A–D: Photos Courtesy of Australian Registry of Wildlife Health.)*

findings at necropsy are widespread petechiae and ecchymoses, tissue pallor and edema, and splenomegaly. Parasitized erythrocytes are few or undetectable in peripheral blood but are sequestered in large numbers in small vessels of the kidney and brain where they can be detected histologically (Figs. 18.29A, B; eSlides 18.e14 and 18.e15) or by cytologic evaluation of tissue squash preparations in which distinctive clusters of extraerythrocytic organisms can be seen (Figs. 18.29C, D).

Babesia are seen in southern brown bandicoots and quolls,[56,150,169] echidnas[11] and the Proserpine rock wallaby,[56] but their clinical significance is unknown. *Babesia* spp. infections in Gilbert's potoroo and woylies do not cause apparent disease.[56] Babesiosis is associated with anemia and mortality in agile antechinus[54] and recrudescence of infection has been reported as contributing to postmating mortality in association with anemia, hemoglobinuria, and hemosiderosis in the lung and spleen.[13]

Besnoitia darlingi has historically been considered common and nonpathogenic in opossums. However, it has more recently been recognized as a cause of disease and mortality in a small percentage of North American opossums; predisposing factors have not been elucidated.[83] Granulomatous inflammation and degenerative lesions may be associated with protozoal cysts. Tissue cysts have a mean diameter of 560 μm and are separated from the host tissue by a 5–20 μm thick cyst wall.[85]

There are many native **trypanosome** species in Australia. Most are considered nonpathogenic, although there is growing evidence that some may play a role in disease, especially in compromised hosts. Koalas can be infected with three different trypanosomes: *Trypanosoma irwini,*

T. copemani, and T. gilletti. Preliminary evidence suggests the latter may exacerbate anemia and loss of body condition in association with concurrent disease. The critically endangered woylie or brush-tail bettong can be infected by *T. vegrandis, T. copemani,* and *T.* sp. *H25*. The intracellular amastigote stage of *T. copemani* is associated with inflammation and degeneration of the heart, skeletal muscle, esophagus, and tongue, which are similar to lesions in human patients with Chagas disease. Experimental infection of macropods with *T. evansi*, the agent responsible for Surra, indicates that they are potential reservoirs but *T. evansi* has never been established in Australia.[240]

Ectoparasites

Sarcoptic mange, due to ***Sarcoptes scabei***, occurs in koalas,[38,172,224] free-ranging agile wallabies,[159] and southern hairy-nosed wombats.[175,211] It is the most significant parasitic disease of wombats.[211] Principle lesions of mange in these species include erythema, parakeratosis, alopecia, excoriation, scaling, and fissuring of the parakeratotic crusts and skin, and the latter considered pathognomonic for sarcoptic mange.[209] Inflammation is primarily neutrophilic or lymphocytic with variable numbers of plasma cells consistent with a Th1-dominated Type IV (delayed) hypersensitivity response. In more chronic cases, there is evidence of a Th2-dominated Type I response with infiltration by mast cells and eosinophils.[208]

The **Australian paralysis tick**, ***Ixodes holocyclus*** (Fig. 18.30), parasitizes a broad range of terrestrial and arboreal marsupial hosts along the eastern coast of Australia. It produces holocyclotoxin, a neurotoxin that can induce fatal, ascending flaccid paralysis, ataxia, and respiratory distress in previously unexposed hosts; most commonly young or translocated koalas, possums, and macropods.[50,236] The toxin acts by inhibition of acetylcholine release at neuromuscular junctions.[51] The bite of *I. holocyclus* may also produce a delayed allergic reaction and occasionally anaphylaxis in some humans that consume beef, pork, and lamb. This occurs through transmission of a carbohydrate, galactose-alpha-1,3-galactose, also known as alpha-gal, which is present in both tick bite saliva and red meat. This carbohydrate is not produced by humans and

FIGURE 18.30 A female Australian paralysis tick, *Ixodes holocyclus*. Ventral view, illustrating toothed hypostome and spurs on coxae, scanning electron microscope image. *(Photo Courtesy of Australian Society for Parasitology Inc.)*

other primates and sensitization to alpha-gal is associated with development of specific IgE antibodies to the carbohydrate.[248]

ADDITIONAL RESOURCES

Donahoe SL, Peacock CS, Choo AYL, et al. A retrospective study of *Babesia macropus* associated with morbidity and mortality in eastern grey kangaroos (*Macropus giganteus*) and agile wallabies (*Macropus agilis*). *Int J Parasitol Parasites Wildl*. 2015;4: 268–276.

Fabijan J, Sarker N, Speight N, et al. Pathological findings in Koala Retrovirus-positive koalas (*Phascolarctos cinereus*) from Northern and Southern Australia. *J Comp Pathol*. 2020;176: 50–66.

Ladds PW. *Pathology of Australian Native Wildlife*. Collingwood, VIC: CSIRO Publishing; 2009.

Pye RJ, Woods GM, Kreiss A. Devil facial tumor disease. *Vet Pathol*. 2016;53: 726–736.

Vogelnest, L., Portas, T. *Current Therapy in Medicine of Australian Mammals*. Clayton, Vic: CSIRO Publishing; 2019.

References

Full references for this chapter can be found online at https://doi.org/10.1016/B978-0-443-18437-6.00036-5.

Chapter 19

Lagomorpha

Martha A. Delaney[a], Jessica M. Snyder[b], and Jamie L. Rothenburger[c]

[a]*Zoological Pathology Program, College of Veterinary Medicine, University of Illinois, Brookfield, IL, United States;* [b]*Department of Comparative Medicine, School of Medicine, University of Washington, Seattle, WA, United States;* [c]*Faculty of Veterinary Medicine, University of Calgary, Calgary, AB, Canada*

INTRODUCTION

The order Lagomorpha consists of two families, Leporidae and Ochotonidae. The 11 genera and 54 species in Leporidae include hares (syn. jackrabbits) and rabbits, of which there are over 100 domestic breeds derived from the European wild rabbit. The Ochotonidae are the pikas. This group consists of 1 genus (*Ochotona*), 3 subgenera, and 30 species. Hares and rabbits are indigenous to all continents except Australia and Antarctica. The European wild rabbit or its derivatives have been introduced to most of Oceana. In zoological institutions, domestic rabbits are frequent members of farm and interactive exhibits. As prey species, rabbits and hares are crepuscular, being most active during the twilight hours at sunrise and sunset. Pikas require a more specialized environment than hares and rabbits and are found only at high elevations in cold, semiarid regions of North America, Eastern Europe, and Asia. They are therefore, potentially more susceptible to the impacts of climate change which has led ecologists to use them as indicator species of this phenomenon. Pikas have disappeared from over one-third of their previously known habitat in the United States, and their populations are decreasing globally. Pikas are uncommon in zoos and studies of free-ranging populations are relatively recent, thus information about health and disease in this group is limited.

NON-INFECTIOUS DISEASES

The majority of non-infectious diseases are relevant to domestic breeds and zoo-housed rabbits and hares. Nevertheless, these processes are possible in free-ranging rabbits and hare species as anatomy, physiology, and disease susceptibilities overlap.

Nutritional

Gastrointestinal tract (GI) stasis and ileus are common in domestic rabbits; they are less frequently reported in free-ranging species. The causes of GI stasis are often multifactorial and related to husbandry conditions. Disease can occur following derangements in diet quality (low fiber and high carbohydrate levels) or anorexia due to numerous factors including chronic stress and pain, dysphagia, neoplasia, infection, dental disease (e.g., malocclusion), drug effects, or obstruction. Grossly, there may be gas, fluid, or ingesta/digesta accumulated in segments of the intestine and evidence of obstipation. If ileus persists, associated dysbiosis and overgrowth of toxin-producing bacteria (e.g., *Clostridium*) can result in progressive inappetence, intestinal compromise, enterotoxemia, and potentially fatal hepatic lipidosis.

Gastric trichobezoars (wool block) are common in domestic rabbits (Fig. 19.1) and are associated with excessive grooming, though they may also develop following increased gut transit times or decreased motility. Gross lesions depend on cause, severity, and chronicity of GI stasis. Excessive gas/fluid accumulation with impaction of food, feces, and/or fur may be present. Pressure necrosis, hemorrhage, congestion, edema, and/or evidence of hypovolemic shock and hepatic lipidosis can occur secondarily if obstruction is present. Histologic lesions correlate to gross findings and thus can vary depending on predisposing factors and sequelae to GI disease. Severe cases may be associated with ischemic and pressure necrosis with invasion of the compromised intestinal tunicae by normally commensal intestinal bacteria and associated suppurative inflammation.

Mucoid enteropathy (ME) is a relatively common non-inflammatory, subacute condition of rabbits. It affects weanlings and can cause high morbidity and mortality in this age group. Although the exact pathogenesis is unclear, ME is often associated with GI stasis and low fiber, high carbohydrate diets. Subsequent acidification of cecal contents and dysbiosis causes increases in certain bacterial species. These release excessive toxins (secretagogues) that stimulate intestinal goblet cell hyperplasia and production of abundant mucus. Dysautonomia has been documented in some cases. Colonic and, to a lesser degree, cecal distension

Pathology of Wildlife and Zoo Animals. https://doi.org/10.1016/B978-0-443-18437-6.00029-8

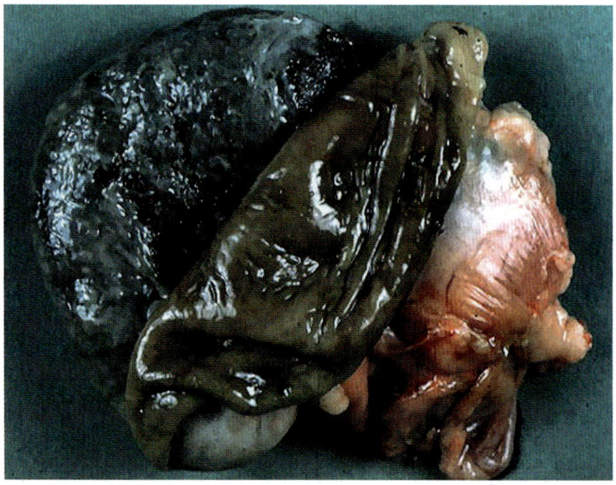

FIGURE 19.1 Gastric trichobezoar in the stomach of a rabbit. The wall of the stomach is reflected to expose a large mass of undigested black hair (the left side of the image) that fills the stomach and is covered with mucus. *(Photo Courtesy of D. Liggitt, University of Washington.)*

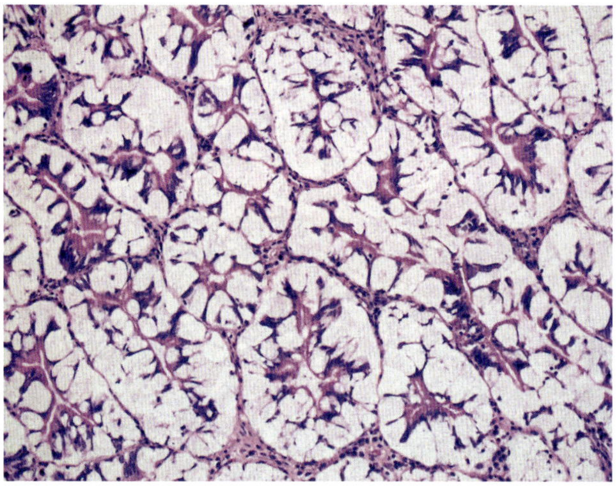

FIGURE 19.2 Mucoid enteropathy in the small intestine of a rabbit. Histologically, there is marked hyperplasia of goblet cells and variable vacuolation and swelling of enterocytes with minimal lamina proprial, heterophilic, and mononuclear inflammation (see eSlide 19.e1). *(Photo Courtesy of K. Terio, University of Illinois Zoological Pathology Program.)*

with abundant luminal colorless, translucent, and gelatinous mucus is the distinguishing gross lesion. Histologically, the colon and ileum have extensive goblet cell hyperplasia and enterocyte vacuolation with mild inflammation consisting of heterophils, lymphocytes, and plasma cells (Fig. 19.2; eSlide 19.e1). In severe cases, erosive and suppurative colitis may also be present. Small intestinal mucosa may have variable neutrophilic inflammation with epithelial loss, villous atrophy, and fusion. Cecal lesions are typically absent. Diagnosis is based on signalment, clinical history, and gross and histologic findings.

Hypervitaminosis D occurs in domestic and zoo-housed rabbits following excessive dietary vitamin D supplementation through inadvertent or improper diet formulation. In acute cases of vitamin D toxicity, lesions include multifocal mineralization of vessels with associated hemorrhage in the GI tract, kidneys, and other parenchymal organs. Grossly, myocardial pallor may occur and depends on the degree of vascular compromise, necrosis, and mineralization. Bone lesions of chronic hypervitaminosis D include osteosclerosis, specifically at the metaphyses, with loss of spongiosa and sparse marrow containing dilated veins and loose connective tissue. Histologically, bones may have severe osteopenia with abundant osteoclasts in acute cases; chronically there is extensive osteoblastic production of a basophilic fibrillary matrix that overlies trabeculae, endosteum, and periosteum. Soft tissue mineralization (vessels and parenchymal organs, such as kidney) may be present. Dietary history, histologic lesions, and derangements in serum calcium and phosphorus are supportive of the diagnosis of vitamin D toxicity.

Hemochromatosis in captive Afghan pikas due to **iron overload** is characterized by tan to brown livers with roughened-to-granular surfaces and in severe cases, multiple nodules separated by thick interconnecting bands of fibrous tissue.[60] Histologically, excessive hemosiderin accumulation (hemosiderosis) in periportal hepatocytes, Kupffer cells, and macrophages is accompanied by hepatocellular degeneration, fibrosis, and bile duct proliferation.

Metabolic

Pregnancy toxemia in lagomorphs is rare and has similar pathogenesis, clinical manifestations, and lesions to those in other species. It is most common in first litter does and can lead to death 1–2 days prior to parturition.[26]

Urolithiasis is common in pet and zoo-housed lagomorphs; its prevalence in wild populations is not known. Secondary lesions include those related to mechanical trauma and blockage. The most commonly reported uroliths are composed of calcium carbonate, which are normally found in the urine in crystalline form. They can be found during gross examination of the urinary bladder or lower urinary tract. Histologically, they are associated with mucosal congestion, hemorrhage, and necrosis at and around the site of urolith deposition/development. Associated cystitis and inflammation of the lower urinary tract as well as bacterial colonization and infection are not uncommon.

Toxic

Antibiotic-related toxicities are important in lagomorphs. Treatment with antibiotics, particularly oral medications that target gram-positive bacteria (e.g., amoxicillin, ampicillin, cephalosporins, clindamycin, erythromycin, lincomycin, and penicillin) disrupt the normal intestinal microflora.

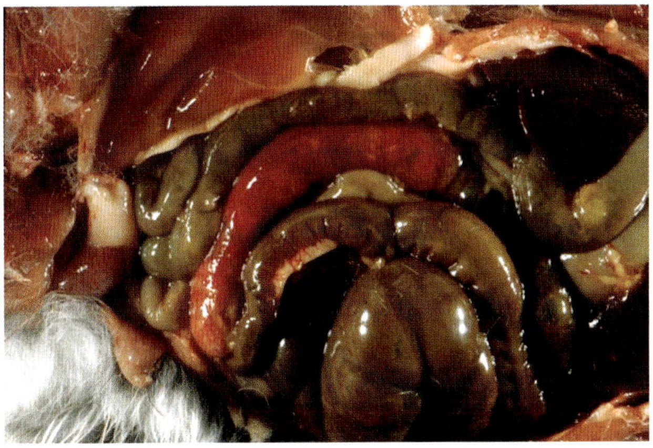

FIGURE 19.3 Antibiotic-associated hemorrhagic typhlitis in the cecum of a rabbit. The cecum is distended and the cecal wall is segmentally and transmurally discolored dark red. The lumen will contain dark red, gelatinous (hemorrhagic) contents. *(Photo Courtesy of K. Terio, University of Illinois Zoological Pathology Program.)*

The resulting dysbiosis permits overgrowth of toxin-producing bacteria, such as *Clostridium* sp., specifically *C. difficile*. Gross lesions may include reddened intestinal serosa and mucosa (Fig. 19.3), variable gas dilatation, mucoid to hemorrhagic luminal contents, and perineal fecal staining. Histologically, mucosal congestion, hemorrhage, and necrosis can be seen throughout the intestinal tract. A history of antibiotic treatment and characteristic lesions support the diagnosis.

***Fipronil* (brand name Frontline®)** is a broad-spectrum insecticide used to treat ectoparasites in dogs, cats, and livestock. Fipronil blocks GABA receptors within the central nervous system leading to increased excitability and is highly toxic to rabbits. Young rabbits are more susceptible to its effects than older rabbits. Clinical, gross, and histologic lesions correlate to clinical manifestations of toxicity including weight loss, seizures, and diarrhea. History of dermal fipronil administration is critical to diagnosis.

Fluorosis in rabbits produces characteristic osseous proliferations along the bones of the extremities and mandible.[11] Severe proliferative gastroduodenopathy accompanies bony lesions. Histologically, severe adenomatous proliferations and branching villous projections are present in the gastric and duodenal mucosa. The reported fluoride levels in bone ash from affected rabbits significantly exceed that in control rabbits (15,000 vs 550 μg/g).

Lead poisoning (syn. Lead toxicosis, plumbism) can occur by the exposure of lead-containing materials in food or water. Clinical manifestations and gross and histologic lesions are similar to those in other species including characteristic renal changes: proteinuria, acute tubular necrosis, and eosinophilic intranuclear inclusion bodies; and cerebral laminar cortical necrosis, hemorrhage, and edema.

Congenital/Genetic

There are numerous reported genetic disorders in domestic rabbits. A complete description is beyond the scope of this chapter.[13] However, those that are more commonly encountered or have pronounced gross or microscopic pathology are described later.

Glaucoma in rabbits is typically a congenital condition that results from goniodysgenesis (abnormal or incomplete development of the filtration angle in the eye). The condition is caused by autosomal dominant inheritance of a defective *bu* gene. Animals carrying this defective gene may have small litter sizes and decreased neonatal survival because there is a close linkage between the *bu* genes with a semilethal locus. Incomplete penetrance has been described in homozygous animals. The disease manifests by 3–5 months of age and results in uni- or bilateral buphthalmia with corneal opacity, mydriasis, and lagophthalmos in severe cases. Histologically, there is absence or underdevelopment of the filtration angle and outflow channels. Sequelae of glaucoma include: cupping of the optic disk; retinal atrophy and degeneration; accentuation of Müller cells and plexiform layers; glial scarring; collapse of uveal trabecular network; atrophy of ciliary processes; attenuation of the sclera and enlarged anterior and posterior chambers; corneal ulceration, edema, metaplasia, striae (breaks in Descemet's membrane); retinal detachment (Fig. 19.4); and lens luxation.

Splay leg is an inherited disease that is variably associated with hip dysplasia but can also result from traumatic injury or housing on slippery floors. Grossly, hind limbs splay laterally and there is variable bowing and valgus of the femur and tibia; the right rear limb is most commonly affected. In cases associated with hip dysplasia, shallow acetabulae, joint capsule thickening, and femoral head subluxation and flattening may be present. Splaying of the forelimbs is infrequent. Diagnosis is confirmed based on

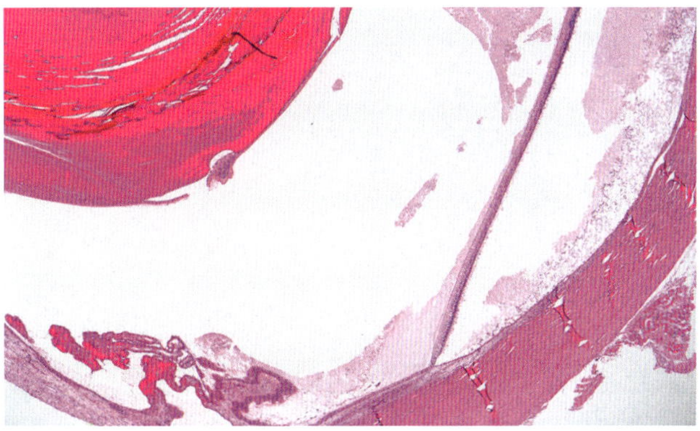

FIGURE 19.4 Retinal detachment in a rabbit with glaucoma. The retina is elevated from the uveal tract, subtended by accumulations of eosinophilic granular to flocculent material. Similar material is found within the posterior chamber and drainage angle.

clinical signs, imaging to rule out traumatic injury, supportive gross lesions and histologic evidence of muscle atrophy secondary to disuse.

Mandibular prognathism (maxillary brachygnathism) is an inherited, autosomal recessive trait most common to dwarf breeds. A genetic mechanism involving the *mp* gene has been proposed as the cause. Abnormal growth of the dorsal and basal skull bones with accompanying displacement of the mandible is evident by 10 weeks of age. With continuous tooth growth, there can be overgrowth of the incisors and cheek teeth with devastating malocclusion. Other forms of malocclusion are common in certain rabbit breeds, though the genetic mechanisms remain unknown. Carrier status can be determined by presence of malocclusion in siblings or offspring.

Age-Related/Degenerative

A common and potentially serious condition in rabbits is **malocclusion** (Fig. 19.5) which develops in part due to tooth anatomy. Leporid teeth are elodont, aradicular, and hypsodont, meaning they are continuously growing, open rooted, and long crowned, respectively. Excessive growth can result from misalignment, insufficient abrasive dietary fiber, or congenital deformities of the skull and jaw (e.g., mandibular prognathism in dwarf breeds). Common sequelae to dental disease in rabbits include buccal/lingual ulcerations and lacerations as well as abscess formation and osteomyelitis. Gross and histologic lesions correlate to clinical disease, which are often best examined with radiographic imaging.

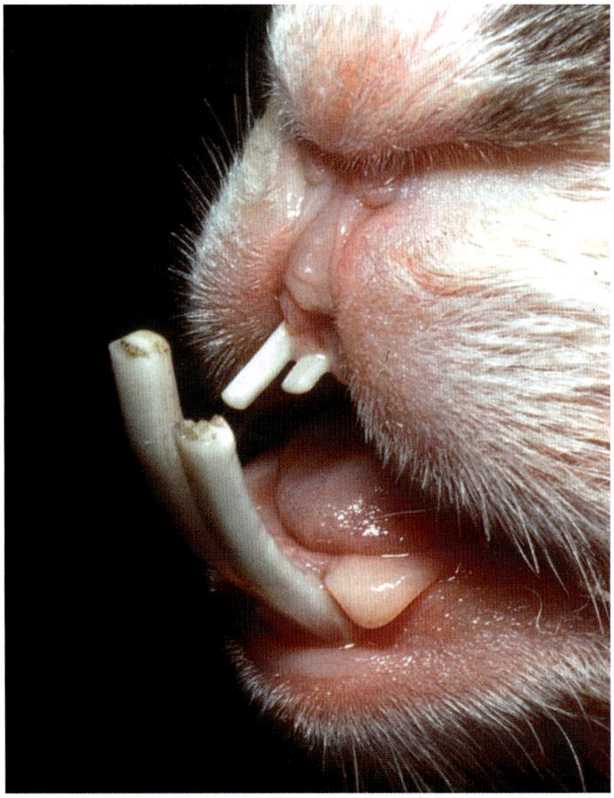

FIGURE 19.5 Malocclusion in an adult rabbit. The maxillary incisors in this rabbit are hypoplastic, being short and thin. Misalignment due to the abnormal teeth prevented normal abrasive wear and allowed the mandibular incisors to grow excessively long. *(Photo Courtesy of R. Kelly, UQ School of Veterinary Science.)*

Odontogenic dysplasia is seen in the incisors of aging rabbits and is often associated with chronic dental disease or trauma.

Dacryocystitis is a common disease in domestic rabbits that involves inflammation and potential infection of the lacrimal sac following obstruction of the nasolacrimal duct, typically in association with dental disease. Grossly, dacryocystitis manifests as copious white to tan, opaque and purulent ocular discharge, and overproduction of tears with periocular staining and irritation. Diagnosis is based on gross findings and concurrent or history of the dental disease.

Hemosiderin-laden cells have been detected in cerebral cortical sections from several older domestic rabbits, in the absence of other substantial CNS lesions. The underlying cause and significance of this finding remains unclear, thus further investigation is warranted, specifically the use of Prussian blue staining to confirm presence of iron (hemosiderin) pigments in brain sections.

Miscellaneous/Multifactorial

Rabbits and hares have a long, fragile skeletal system compared to their heavy musculature so spinal fractures, particularly of the caudal lumbar vertebrae, are common. **Vertebral fractures** are the most common cause of acute onset posterior paralysis in zoo and pet rabbits. Fracture or luxation typically occurs at the seventh lumbar vertebra. Diagnosis can be confirmed radiographically. Gross evidence of fracture with surrounding hemorrhage and tissue maceration is supportive. Histology may be useful in determining the extent of spinal cord and nerve damage. Urine retention due to neurologic dysfunction can cause cystitis, necrosis, and hemorrhage. In free-ranging lagomorphs, trauma is a major cause of death.[29,42]

Heat stroke (hyperthermia) is a serious and potentially fatal condition in Lagomorpha. Pikas are exquisitely susceptible to overheating due to their high basal body temperature. In captivity, strict attention to appropriate environmental conditions is required. Several ongoing studies of free-ranging pikas demonstrate the direct relationship between changes in ambient temperatures (climate-related changes) and the success of pika populations across their natural ranges.

Soiled and wet enclosure floors can cause **hutch burn (syn. Urine scald)**, although it can also occur in animals with incontinence. Perineal and genital regions are irritated, inflamed, and are susceptible to bacterial infection. Grossly, there may be brown or green exudate and hemorrhage in affected areas. Histologically, lesions consist of heterophilic and ulcerative dermatitis with or without bacterial colonization. Lesions can resemble rabbit syphilis (vent disease) caused by *Treponema* sp. and must be differentiated from it using additional testing and (for the latter) demonstration of causative bacteria within the lesions via silver or immunohistochemical stains.

Moist dermatitis (wet dewlap) is a skin condition in female or obese rabbits that affects a heavy fold of skin at the front of the neck (the dewlap). Drinking or excessive salivation (e.g., due to dental malocclusion or disease) moistens the dewlap, leading to irritation and inflammation. There may be alopecia, myiasis, and/or opportunistic infections (e.g., *Pseudomonas aeruginosa*, *Trueperella pyogenes*, *Fusobacterium necrophorum*). Histologic lesions correlate to severity of dermatitis.[101]

Endometrial venous aneurysm occurs in nonpregnant, multiparous does. In other species, arterial rather than venous aneurysms are much more common. Endometrial aneurysms in the rabbit uterus intermittently rupture and bleed into the lumen resulting in urogenital bleeding, anemia, hematuria, and proteinuria. Grossly, multiple dilated, thin-walled, blood-filled endometrial vessels, often accompanied by clotted blood, are seen in the uterine lumen. Microscopically, endometrial stroma is expanded by one to multiple, variably dilated veins containing thrombi and blood (Fig. 19.6; eSlide 19.e2). Ruptured aneurysms can contain necrotic debris and neutrophilic infiltrates. Affected individuals may have concurrent cystic endometrial hyperplasia, which consists of dilated endometrial glands rather than vessels. Venous aneurysms must be differentiated from uterine adenocarcinoma, which is one of the most common neoplasms in domestic rabbits.

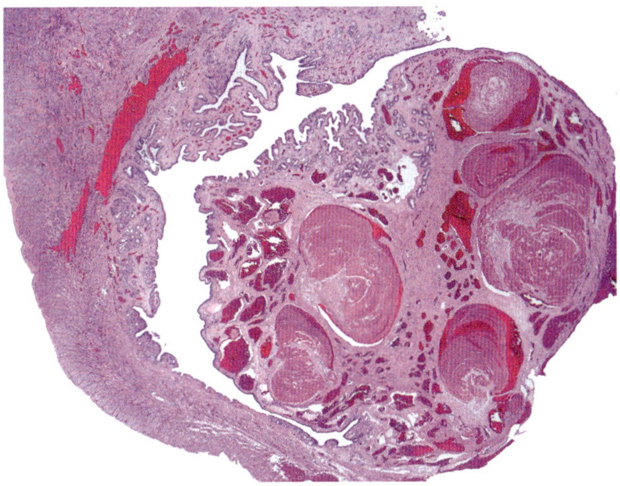

FIGURE 19.6 Endometrial venous aneurysms in a rabbit. The endometrial stroma is expanded by multiple, coalescing markedly dilated veins that contain luminal thrombi and blood (see eSlide 19.e2). *(Photo Courtesy of K. Terio, University of Illinois Zoological Pathology Program.)*

Like other hindgut fermenters, rabbits have a large cecum and colon which can become displaced and entrapped. Hepatorenal ligament entrapment of the intestine has been rarely seen causing obstruction, vascular compromise, and death (personal communication).

Neoplastic

Uterine adenocarcinoma is the most common neoplasia in female domestic rabbits. More than 50% occur in rabbits older than 3 years. Decreased litter size in breeding rabbits may be an indicator of uterine malignancy. Persistent estrogen stimulation in the absence of progesterone opposition may be a factor in the development of these cancers, which are generally slow-growing and often preceded by hyperplastic changes in the endometrium. Grossly, the uterine walls of one or both horns contain multiple nodular to polypoid masses and/or papillary projections (Fig. 19.7A). Histologic features include acinar or tubular structures lined by cuboidal neoplastic epithelial cells supported by myxomatous to moderately collagenous fibrovascular stroma (Fig. 19.7B; eSlide 19.e3). There may also be atrophy of nonneoplastic endometrial glands and abundant luminal mucus. Metastasis to the lung, liver, and peritoneal serosal surfaces (carcinomatosis) is common.

Endometrial and mammary gland hyperplasia are potentially preneoplastic lesions. These lesions can be seen concurrently in intact does older than 3 years. Mammary carcinomas are considered rare in rabbits. A review of mammary tumors in pet rabbits describes several different histomorphologies of epithelial neoplasms; there were no mesenchymal neoplasms in the study population.[83] Bucks may develop testicular tumors, of which **interstitial (Leydig) cell tumors** are the most common. **Seminomas, Sertoli cell tumors,** and **granular cell tumors** have also been reported.[43]

Lymphoma is most common in young rabbits and generally affects visceral organs (liver, spleen, kidney), though it can also arise in the robust GALT (Figs. 19.8A, B). There are also reported cases in free-ranging cottontail rabbits.[54] Leukemia may accompany solid tumors. *Herpesvirus sylvilagus* has been associated with these lesions experimentally; however, the role of this virus in the pathogenesis of spontaneous lymphoma is uncertain.[38] Histologically, neoplasms are composed of lymphoblastic B or T cells. **Thymoma** is rare and does not often result in clinical disease but can induce paraneoplastic syndromes such as exfoliative dermatitis, hypercalcemia, and periodic exophthalmos.[28,97]

There are several reports of renal neoplasms in free-ranging cotton-tail rabbits including **nephroblastoma, adenocarcinoma,** and **urogenital hamartoma.**[18,54]

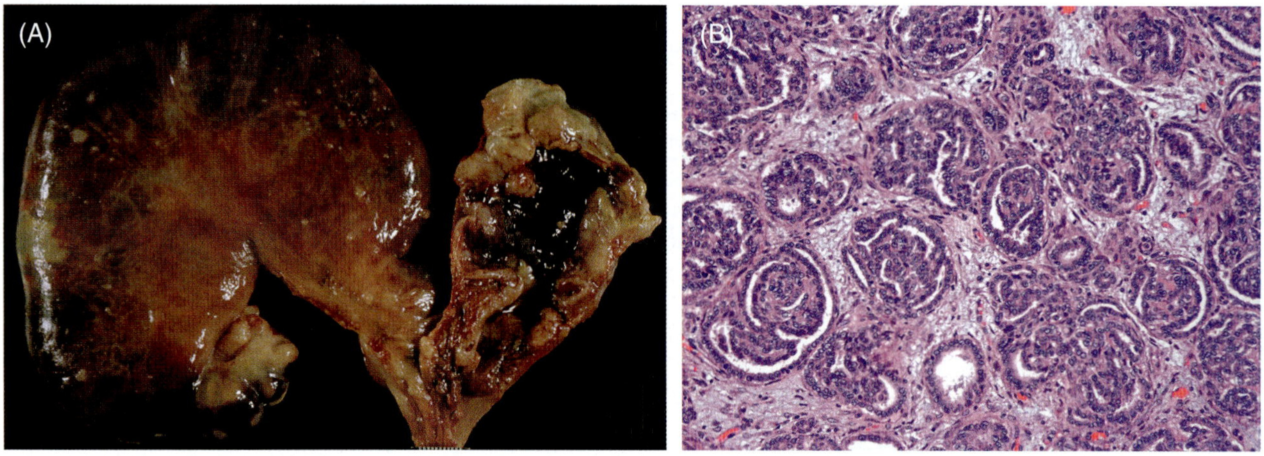

FIGURE 19.7 Uterine adenocarcinoma in a rabbit. (A) The uterine horns are dilated and there is multifocal mural thickening and red/black discoloration. On cut section (right horn), the uterine wall is expanded by multiple nodular to polypoid, pale tan masses, and the lumen contains dark red, gelatinous material. (B) Small neoplastic epithelial cells form small pseudoacinar and tubular structures supported by variably loose to dense fibrous stroma (see eSlide 19.e3). *(Part A: Photo Courtesy of K. Terio, University of Illinois Zoological Pathology Program.)*

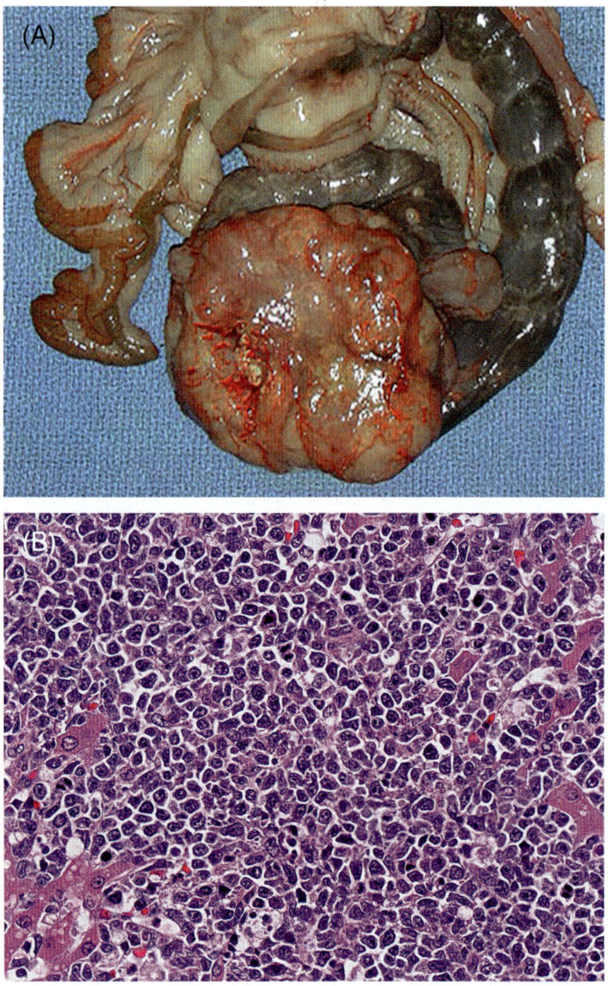

FIGURE 19.8 Cecal (A) and hepatic (B) lymphoma in a rabbit. (A) There is marked regionally extensive, transmural expansion of the cecal wall by a large nodular, exophytic, soft, tan mass typical of lymphoma. (B) Microscopically, masses are composed of dense sheets of large neoplastic lymphocytes among a scant fibrovascular stroma, with wide separation and effacement of the hepatic parenchyma. *(Part A: Photo Courtesy of J. Landolfi, University of Illinois Zoological Pathology Program.)*

Nephroblastomas are also common among free-ranging white-tailed prairie hares (see eSlide 19.e4).[42] **Trichoblastomas** are common, benign cutaneous neoplasms originating from hair follicles. In addition to the virally induced skin neoplasms (Table 19.1), other reported skin tumors in lagomorphs include **squamous cell carcinoma, sebaceous gland carcinoma**, and **histiocytic sarcoma**.[45] There are also rare reports of **ameloblastomas**[59] and **osteosarcoma**[102] in domestic rabbits. Recently, **odontogenic neoplasms** in the cheek tissues were reported in several domestic rabbits (see eSlide 19.e5).[68] Tumors ranged from benign ameloblastomas to ameloblastic carcinoma-like masses but all were limited to the soft tissues of the cheek with no evidence of bone or tooth involvement.

INFECTIOUS DISEASES

DNA Viruses

Myxomatosis, Shope fibroma, and **Shope papilloma** are virally induced, proliferative, cutaneous, and soft tissue lesions in lagomorphs that have similar gross appearance but distinct lesion distribution and histologic characteristics (Table 19.1).

Myxomatosis (bighead disease; mosquito disease) is a WOAH-listed reportable disease that is caused by **myxoma virus** (double-stranded DNA, *Poxviridae* family; *Leporipoxvirus* genus). The virus is transmitted by hemophagus arthropods or through contact with infected fluids. The virus is endemic on several continents and affects both wild and domestic rabbits as well as European and Iberian hares. Strain virulence and host immune status determine clinical signs and lesions: wild rabbits and hares develop skin tumors, while domestic rabbits may also develop systemic signs.[10,33] Gross manifestations involve multiple subcutaneous gelatinous nodules along the face and perineum with variable edema of the eyelids, conjunctiva, and anogenital

TABLE 19.1 Virally Induced Proliferative Lesions in Lagomorphs

Disease	Virus Name	Virus Type	Species Affected	Distribution	External Gross Lesions	Internal Gross Lesions	Histologic Features	Viral Inclusion Location	References
Myxomatosis (bighead; mosquito disease)	Myxoma virus	Poxvirus (*Leporipoxvirus*)	Wild and domestic rabbits; European hares	Face and perineum	SQ edema and gelatinous nodules; ocular and nasal discharge	Splenomegaly; lymph node edema; petechiae and ecchymoses	Hyperplastic epidermis, ballooning degeneration, and mesenchymal (myxoma) cells within myxomatous stroma	Epithelial cells in epidermis; intra-cytoplasmic	Bertagnoli and Marchandeau[10]; DiGiacomo and Mare[23]; Stewart[88]
Shope fibroma	Shope fibroma virus (Rabbit fibromavirus)	Poxvirus (*Leporipoxvirus*)	Wild and domestic rabbits	Face and extremities	Firm dermal masses	Masses in viscera and bone marrow	Proliferation of large fibroblasts and epidermal hyperplasia	Epithelial cells and fibroblasts; intra-cytoplasmic	DiGiacomo and Mare[23]; Lopushinsky and Fay[54]; Pulley and Shively[75]; Prose et al.[74]; von Bomhard et al.[98]
Shope papilloma	Shope papilloma virus (Cottontail rabbit papilloma virus)	Papillomavirus (*Kappa-papillomavirus 2*)	Wild rabbits, hares, and jackrabbits	Head, neck, ears, eyelids	Firm white warts and horn-like growths	None	Epidermal hyperplasia and parakeratotic hyperkeratosis	Epithelial cells of stratum spinosum in epidermis; intracytoplasmic and intranuclear	Lopushinsky and Fay[54]; Shope[86]

SQ, subcutaneous.

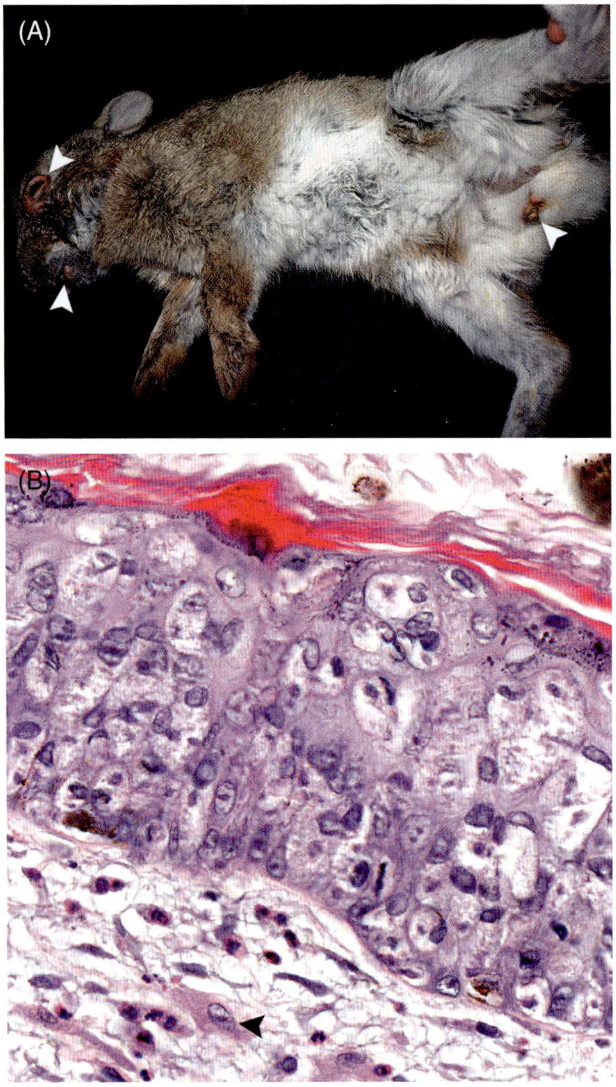

FIGURE 19.9 Myxomatosis in the skin of a wild rabbit. (A) Variably sized, exophytic, gelatinous nodules expand the skin of the lips and perineum. Eyelid edema and hyperemia are characteristic *(white arrowheads)*. (B) Viral infection causes epidermal hyperplasia with ballooning degeneration and intraepithelial and intracytoplasmic eosinophilic inclusions. The dermis contains large undifferentiated stellate, mesenchymal "myxoma cells" *(black arrowhead)*, and heterophilic inflammation. *(Parts A, B: Photos Courtesy of I. Barker, Ontario Veterinary College.)*

regions, as well as conjunctival and nasal discharge (Fig. 19.9A). Internally, systemic lesions include splenomegaly, lymph node edema, and hemorrhage in the heart, kidney, testes, and GI tract.[88] Histologic lesions are characteristic. Subcutaneous nodules are composed of undifferentiated spindloid to stellate mesenchymal cells referred to as myxoma cells within abundant myxomatous stroma. The overlying epidermis is hyperplastic with ballooning degeneration and large, eosinophilic, intracytoplasmic inclusions in the epithelial cells (Fig. 19.9B); viral inclusions are not found in myxoma cells. Lymphoid depletion and necrosis are present in lymph nodes and spleen.[88] There is also necrosis of pneumocytes and hepatocytes. Diagnosis is based on histomorphology and distribution of lesions. Viral inclusions are limited to epithelial cells. Virus isolation, immunohistochemistry (IHC), and PCR are useful for confirmation. PCR of conjunctival swabs is a useful *antemortem* test since affected rabbits may shed virus for up to 30 days. Ultrastructurally, affected tissues contain characteristic, large (300 × 100 nm), brick-shaped poxvirus virions with a smooth surface and dumb-bell-shaped nucleoid.

Rabbit (Shope) fibroma virus infects domestic and wild rabbits. The virus is a large double-stranded DNA virus in the *Leporipoxvirus* family, which also includes squirrel fibroma virus and myxoma virus.[74] Infection produces fibromas (Shope fibromas). Gross lesions consist primarily of firm cutaneous masses (Fig. 19.10A); visceral and bone marrow metastases occur in young rabbits.[98] Histologically, dermal masses are composed of densely arranged, enlarged fibroblasts with abundant pale eosinophilic

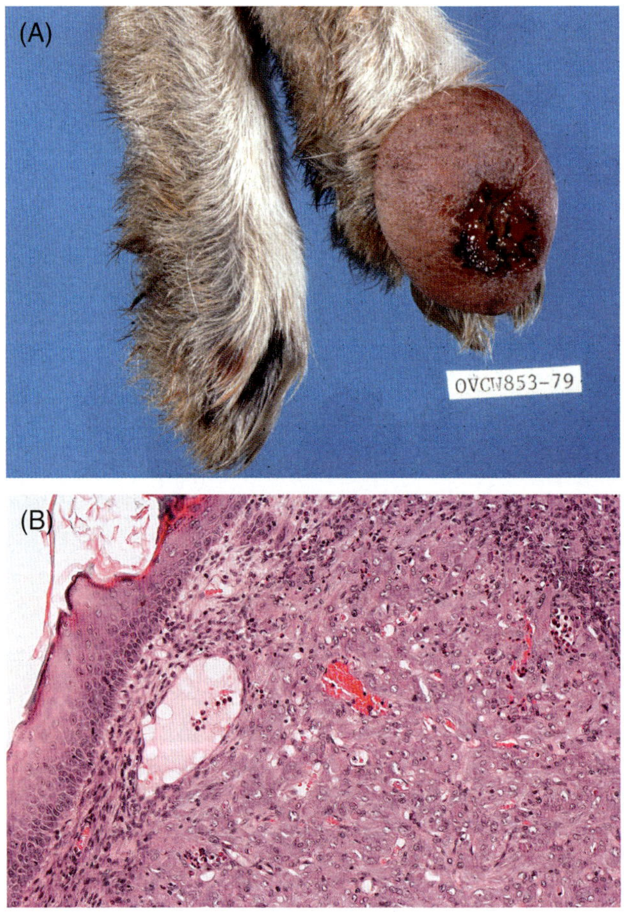

FIGURE 19.10 Rabbit fibroma due to rabbit (Shope) fibroma virus infection in the skin of a rabbit. (A) The soft tissues of the foot are markedly expanded by a large, firm, raised, well-circumscribed, centrally ulcerated cutaneous mass. Affected skin is alopecic and erythematous. (B) Histologically, the hyperplastic epidermis is subtended, and the dermis expanded by a mass composed of enlarged fibroblasts, some of which contain eosinophilic cytoplasmic viral inclusions. *(Part A: Photo Courtesy of I. Barker, Ontario Veterinary College.)*

cytoplasm (Fig. 19.10B). The overlying epithelium is hyperplastic with prominent rete pegs that extend into the fibroblastic mass. Notably, eosinophilic viral inclusions are found within the cytoplasm of both the epithelial cells and fibroblasts.[75] Shope fibromas can be distinguished from myxomatosis because of their presence on the limbs in addition to the face.[23] Transmission between wild and domestic rabbits can occur with exposure to the insect vectors.[75]

Shope papilloma virus, also known as **cottontail rabbit papillomavirus** (CRPV; Kappapapillomavirus 2; double-stranded DNA virus), causes firm, white warts that can progress to "keratinous carcinomas" that resemble horns on the head, neck, pinnae, and eyelids of wild and domestic leporids.[86] Susceptible species include cottontail and brush rabbits, black-tailed jackrabbits, snowshoe hares, and European rabbits. In severe cases, papillomas can impede eating; however, these neoplasms also regress and completely resolve without treatment. The discovery of the causative papillomavirus led to breakthroughs in understanding the pathogenesis of some virus-associated neoplasms. Histologically, masses are similar to other virally induced papillomas and carcinomas.[54] Papillary masses are composed of markedly thickened stratified squamous epithelium with parakeratotic hyperkeratosis supported by central fibrovascular cores. Within the stratum spinosum, epithelial cells are typically swollen and may contain large glassy basophilic intracytoplasmic and, less often, intranuclear inclusions. Immunohistochemical labeling for papillomavirus antigen may be useful to confirm the etiology though molecular testing would be required to determine the specific viral strain.

Rabbit oral papillomavirus (circular double-stranded DNA virus) is known only to affect domestic rabbits, specifically New Zealand white rabbits. As in other species, this virus induces growth of wart-like or pedunculated masses in the oral cavity, often along the ventral aspect of the tongue. Lesions may resolve spontaneously. An important distinction between the oral papillomavirus and Shope papillomavirus is that the latter never occurs in the mouth. There is a report of papillomavirus in a cotton-tail rabbit in which viral inclusions were detected in the Langerhans cells via immunohistochemistry and ultrastructural microscopy.[100]

Rabbit poxvirus (double-stranded DNA virus; Rabbitpox virus), in the *Poxviridae* family and *Orthopoxvirus* genus, causes **rabbitpox**. The virus is transmitted via aerosols and causes papular lesions in the skin and mucocutaneous junctions with variable systemic disease affecting the lungs, spleen, liver, and lymph nodes. Histologically, lesions contain multifocal necrosis with mixed inflammatory infiltrates. Most of the information about this disease comes from biomedical research since rabbitpox is considered a valuable model for smallpox in humans.[3,82]

Like many other species, rabbits can be infected by several **herpesviruses**. There are four described leporid herpesviruses, although natural disease associated with Leporid herpesvirus 1 and 2 have not been reported.[90] Leporid herpesvirus 5 was isolated from lesions in wild Iberian hares (*Lepus granatensis*), however, the hares were coinfected with myxoma virus therefore the role of this herpesvirus in the pathogenesis of the lesions is uncertain.[1]

Leporid herpesvirus 4 is an alphaherpesvirus that affects rabbits of all ages. Clinical presentation resembles myxomatosis with facial swelling and edema, conjunctivitis, ulcerative dermatitis of the periocular and perineal skin, weakness, and abortion. Histologically, there is hemorrhagic dermatitis and panniculitis associated with epidermal degeneration and dermal/subcutaneous vascular necrosis and thrombosis. Characteristic eosinophilic, intranuclear herpesviral inclusions are present in epithelial cells of the epidermis and hair follicles, as well as dermal and subcutaneous mesenchymal cells. Syncytial cells are also present within the epidermis. Hemorrhage and necrosis are found in the heart, lungs, spleen, and lymph nodes.[14]

Leporid herpesvirus 3 (formerly *Herpesvirus sylvilagus*) is a gamma herpesvirus that causes lymphoproliferative disease in juvenile cottontail rabbits.[38,44] Within 10 days of infection, hematologic abnormalities include leukocytosis, monocytosis, and lymphocytosis with atypical lymphocytes. Histologically, there is lymphocytic myocarditis, myositis, and interstitial pneumonia. Splenic lymphoid hyperplasia is followed by lymphoid depletion and hypersplenism (reticuloendothelial cell hyperplasia, hemosiderosis, and extramedullary hematopoiesis) as the disease progresses. Systemic lymphoproliferation presents as enlarged, hyperplastic lymph nodes and perivascular aggregates in various tissues including the lungs, kidneys, GI tract, and salivary glands.

Rabbits are extremely sensitive to **human herpesvirus 1 (HHV-1; herpes simplex virus)**, which causes nonsuppurative meningoencephalitis with viral inclusions in neurons and glial cells. Reports exist of human-to-rabbit transmission with fatal outcomes.[66] HHV-1 infection should therefore be considered as a differential for central nervous system signs in rabbits exposed to humans.

RNA Viruses

Rabbit hemorrhagic disease (RHD), also known as rabbit calicivirus disease (RCD) and viral hemorrhagic disease (VHD), is a fatal disease in rabbits that historically affected adult European rabbits (*Oryctolagus* spp.).[2] It is a WOAH-listed reportable disease caused by **rabbit hemorrhagic disease virus (RHDV)**, which is a member of the *Caliciviridae* family (genus *Lagovirus*). To date, there are two known forms of Type 1 RHDV, **RHDV1** (classical form, also known as ***Lagovirus europaeus* GI.1**) and **RHDVa**. In 2010, the only Type 2 form, **RHDV2** (also known as ***Lagovirus europaeus* GI.2**) was discovered, has a much wider host range infecting many wild rabbit species in addition to domestic rabbits.[80]

RHDV1 was the first serotype to be identified and was isolated from rabbits during a fatal outbreak in China in 1984. RHDV1 has spread worldwide, in some instances intentionally as a rabbit control measure, and is currently endemic in parts of Asia, Europe, Africa, Australia, New Zealand, and Cuba. Several limited outbreaks in the early 2000s in US rabbitries were associated with different virus strains.[62] While classically a disease exclusive to domestic rabbits (*Oryctolagus cuniculus*), RHDV1 has also been rarely reported in dead wild Iberian hares and viral RNA has been detected in rodents, such as wood mice and Algerian mice that share territories with infected rabbits. Maternal antibody provides disease resistance in neonates between the ages of 4 and 8 weeks,[61] and if infected during this time, individuals can develop lifelong immunity. However, 90% of naïve, adult rabbits succumb to the virus with death typically occurring within 1 day following infection. The virus spreads through direct contact, aerosols, vectors, and/or fomites.

RHDV2 emerged as a genetically distinct rabbit **calicivirus** related to Type 1 RHD (RHDV1 and RHDVa) and European brown hare syndrome virus. **RHDV2** is not only genetically distinct but has distinctive antigenic and pathogenic features, notably the extensive host range and age of susceptibility and disease. Unlike RHDV1, RHDV2 commonly infects both domestic and wild rabbits and rabbit kits under 30 days of age.[80] Rabbits infected with RHDV2 may survive longer than those infected with RHDV1; however, in hare species, it causes peracute death. In Europe, there have been epidemics of RHVD2-related disease in wild rabbit populations since immunity to or vaccination against RHDV1 is not protective for RHDV2. These outbreaks primarily affect wild rabbits though spill over into hare species including the brown hare, Italian hare, and Cape hare is documented.[95] The virus has spread globally, likely through human activities.[81] In the spring of 2020, an outbreak of RHDV2 began spreading through native North American lagomorphs including black-tailed jackrabbits, antelope jackrabbits, desert cottontails, mountain cottontails, Eastern cottontail, Riparian brush rabbits, brush rabbits, pygmy rabbits, and also pikas; lesions were similar to affected European species (www.aphis.usda.gov).[52] Since development and implementation, vaccine strategies have been effective in controlling RHDV1[78] and RHDV2.[37]

Clinical presentation, gross and histologic lesions for all RHDV forms are similar and indistinguishable. Together with acute death, clinical presentation is typically hemorrhage from the nares. Gross lesions include serosal, pulmonary, and perineal hemorrhages, hepato- and splenomegaly, and pulmonary congestion, edema, and hemorrhage (Fig. 19.11A). Histologically, there is multisystemic hemorrhage and evidence of disseminated intravascular coagulation, like glomerular thrombosis (Fig. 19.11B), along with acute necrosis in the spleen, small intestinal crypts,

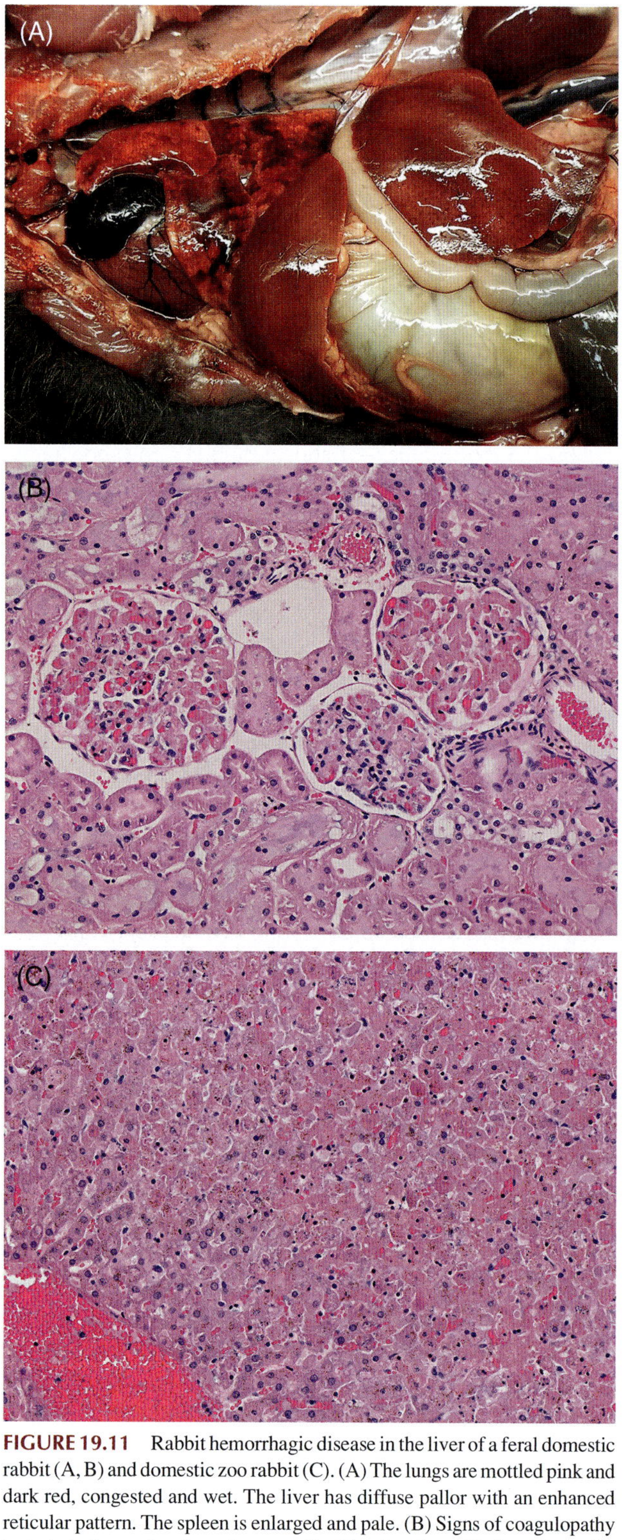

FIGURE 19.11 Rabbit hemorrhagic disease in the liver of a feral domestic rabbit (A, B) and domestic zoo rabbit (C). (A) The lungs are mottled pink and dark red, congested and wet. The liver has diffuse pallor with an enhanced reticular pattern. The spleen is enlarged and pale. (B) Signs of coagulopathy included glomerular thromboses, with dilated tuft capillaries filled with clumped eosinophilic material (fibrin). (C) The liver has massive necrosis with dissociation of hepatic cords. Necrotic hepatocytes are typically rounded and slightly swollen with pyknosis or karyorrhexis (see eSlide 19.e6). *(Parts A–C: Photos Courtesy of H. Snyman, British Columbia Ministry of Agriculture's Veterinary Diagnostic Laboratory.)*

and liver (Fig. 19.11C; eSlide 19.e6). Diagnosis is based on gross and histologic lesions and demonstration of the virus, viral antigen, or antibodies via ultrastructural, immunohistochemical, and serological analyses, respectively; RT-PCR is also available and is rapid and highly sensitive.

A distinct calicivirus, called **Michigan rabbit calicivirus**, was isolated from a rabbitry outbreak in Michigan, United States in 2001.[9] It caused nearly 33% mortality and produced similar clinical signs and lesions as for RHD.

European brown hare syndrome (EBHS) is caused by an RHD-related calicivirus and is characterized by hepatic disease.[34] Gross lesions are limited to hepatomegaly and pallor that correlate histologically with periportal to massive necrosis with minimal periportal inflammation.[31] Virus isolation is difficult, thus histologic lesions coupled with serology and/or PCR are required for diagnosis. There is also evidence that widespread circulation of nonpathogenic lagomorph circoviruses occurs among European lagomorphs.[20]

Lapine rotavirus group A causes variably severe diarrhea in young rabbits. Seroprevalence studies demonstrate exposure in various groups including domestic and wild rabbits and snowshoe hares.[73] The virus can be isolated from feces of healthy individuals. Gross and histologic lesions are comparable to rotavirus infection in piglets and include degeneration of the superficial enterocytes, villous blunting and fusion resulting in atrophy, and mild lymphoplasmacytic inflammation in the lamina propria and submucosa. Disease can be enhanced by coinfection with coliforms, such as *Escherichia coli* and coccidia. Recently, infection of rabbits with **bovine rotavirus** has been reported.[94] "**Rabbit enteritis complex**" is typically associated with a number of organisms, including **rotavirus**, bacteria, and protozoa such as *Clostridium spiroforme*, *C. piliforme*, *E. coli*, *Lawsonia intracellularis*, and *Eimeria* species.[72]

Like other mammals, lagomorphs are susceptible to **rabies**. Paralysis and subsequent death occur within days following the infection. Gross lesions are typically not present and histologic changes in the brain are characteristic and include perivascular lymphoplasmacytic inflammation and eosinophilic, neuronal intracytoplasmic inclusions (Negri bodies).

Influenza virus, specifically low pathogenicity avian influenza H9N2, has been reported in plateau pikas (*Ochotona curzoniae*) in China.[105]

Experiments in rabbits have shown they are susceptible to **SARS-CoV-2** and can shed virus in respiratory secretions, although infections are subclinical.[69]

Bacteria

Within this section are several zoonotic and notable nonzoonotic bacteria that infect lagomorphs, with substantial overlap with those seen in rodents (Rodentia).

Tularemia (also **rabbit fever**) is an important zoonotic disease that occurs worldwide[103]; it is a WOAH-listed reportable disease. Humans acquire infection by exposure to wild rabbits and rodents, which are considered reservoir hosts for the causative pathogen, ***Francisella tularensis***.[93] Muscle from infected hares may harbor the bacteria and undercooked meat is a potential hazard to human consumption.[39] Tularemia is highly contagious and can be acutely fatal to susceptible species, which include hares and New World rabbits. European rabbits appear to be resistant.[16] Tularemia is endemic in Eastern cottontails in the United States and hares in the northern hemisphere.[67] Direct contact, arthropod vectors (ticks), ingestion, and inhalation are all means of transmission. Death occurs within 2 weeks of infection. Gross lesions include spleno-, hepato-, and lymphadenomegaly with pinpoint to small tan/white (necrotic) foci throughout the liver (Fig. 19.12A), spleen, lymph nodes, kidneys, and less commonly in the heart, lungs, and bone marrow. Large caseous granulomas, purulent lymphadenitis, intestinal hemorrhage with Peyer's patch, and GALT necrosis have been reported. Tularemia should also be suspected in cases of found dead, free-ranging lagomorphs with no gross lesions.[39,42] Microscopically, a range of affected organs have multifocal to coalescing lytic necrosis associated with variable types of inflammation ranging from heterophilic to pyogranulomatous to granulomatous depending on chronicity of disease (Fig. 19.12B). There is also vasculitis with and without thrombosis and necrosis of lymphoid tissues in the spleen, thymus (if present), lymph nodes, and intestinal tract. Intralesional aggregates or intrahistiocytic bacteria can be highlighted using immunohistochemical labeling; Gram or other histochemical stains typically do not stain the organisms. Diagnosis is based on gross and histologic lesions and isolation of the organism in biosafety level 3 laboratories or molecular detection of the organism. Lesions can be similar to those of other bacterial diseases, in particular Tyzzer's disease, listeriosis, and yersiniosis. The pattern of lesions may be related to specific strain types; European hares infected by one genotype were more likely to have pneumonia compared to those infected by other genotypes.[50]

Listeriosis is a zoonotic disease of humans and numerous animal species including rabbits and hares. ***Listeria monocytogenes***, the causative agent, is a gram-positive, facultative anaerobic bacillus with many serovars. *L. monocytogenes* is most abundant in temperate climates; however, it is resistant to harsh conditions and is therefore found ubiquitously in the environment. Infections occur following the ingestion of contaminated food or water. There are three main disease syndromes: (1) encephalitis; (2) abortion; and (3) septicemia. Rabbits are most susceptible to abortion and there appears to be tropism for the gravid uterus as adult bucks and nonpregnant does are rarely affected. Systemic disease is possible and manifests

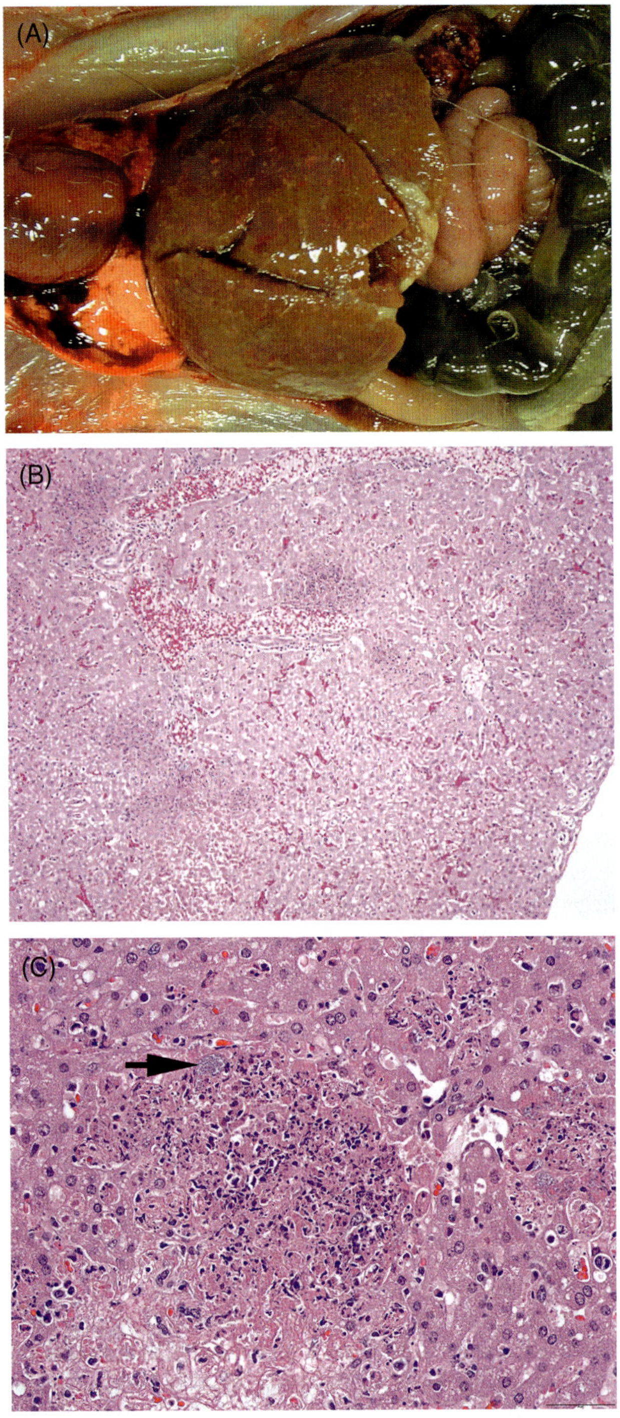

FIGURE 19.12 Tularemia in the liver of wild Eastern cottontail rabbits. (A) The liver is moderately enlarged and pale tan; the capsule is covered by thin yellow fibrin strands that span the intestinal serosa. (B) Multifocal necrosis, both coagulative and liquefactive, is present throughout the hepatic parenchyma. (C) Intralesional and intrahistiocytic bacteria are present *(arrow)* but difficult to see with routine hematoxylin and eosin staining. *(Part A: Photo Courtesy of K. Conley, Wildlife Conservation Society; Part C: Photo Courtesy of D. McAloose, Wildlife Conservation Society.)*

histologically as hepatic necrosis, cerebral and cerebellar microabscesses, and mononuclear leptomeningitis with dense perivascular cuffs. Neuronal necrosis and cranial nerve neuritis may also be seen. Necrotizing and fibrinosuppurative transmural metritis and septicemia attributed to listeriosis have also been reported in female wild hares from Saskatchewan, Canada[79] (Figs. 19.13A, B). Additional findings in hares included fetal maceration, necrotizing myocarditis, hepatic necrosis, and nonsuppurative encephalitis. Bacterial colonies are present in reproductive and systemic lesions as well as fetal tissues from affected does. The diagnosis of listeriosis is based on gross and histologic lesions and demonstration of intralesional, intracellular, gram-positive coccobacilli (Fig. 19.13C). Culture and immunohistochemical stains can also be used to confirm etiology.

Yersiniosis (Pseudotuberculosis) is caused by infection with ***Yersinia pseudotuberculosis*** or ***Y. enterocolitica***. Like listeriosis in leporids, ingestion of contaminated foodstuffs is the main route of infection. Grossly, there is necrosis and granulomatous inflammation in the liver, spleen, and mesenteric lymph nodes that appear as variably sized, yellow to tan caseous nodules. There is also ulcerative and fibrinohemorrhagic typhlitis with abscessation of the GALT and lamina propria. Histologically, large colonies of amphophilic coccobacilli are seen within lesions with routine hematoxylin and eosin (HE) staining. Microabscesses surround bacterial colonies and associated blood vessels may contain fibrin thrombi. Isolation and speciation of the causative bacteria are required to confirm diagnosis.

Pasteurella multocida is a gram-negative bacterium that causes significant disease and mortality in hares and both domestic and wild rabbits. *P. multocida* is the most commonly isolated bacterium from animals with upper respiratory disease, though it is also considered a commensal in healthy individuals. These bacteria can proliferate and spread locally to the lungs, eyes, ears, and brain or systemically through the blood to the heart, reproductive organs, and skin. Thus, there are numerous gross lesions when *P. multocida*-associated disease occurs.

"Snuffles" is the common name for the classic disease caused by *P. multocida* in rabbits. It is characterized by purulent rhinitis and watery or thick, white-to-yellow nasal discharge. The snuffling sounds result from obstruction of the nares and nasal passages by mucus and inflammatory exudates and are the origin of the name. Infection may also extend from the nasal cavity to the ears and occasionally into the brain, resulting in encephalitis and torticollis (wry neck or head tilt). Pulmonary infection results in cranioventral fibrinopurulent pneumonia, pleuritis, and abscessation (Fig. 19.14A). Systemic lesions include endocarditis, endometritis or orchitis, mastitis, and subcutaneous abscesses (Fig. 19.14B) in addition to septicemia. In

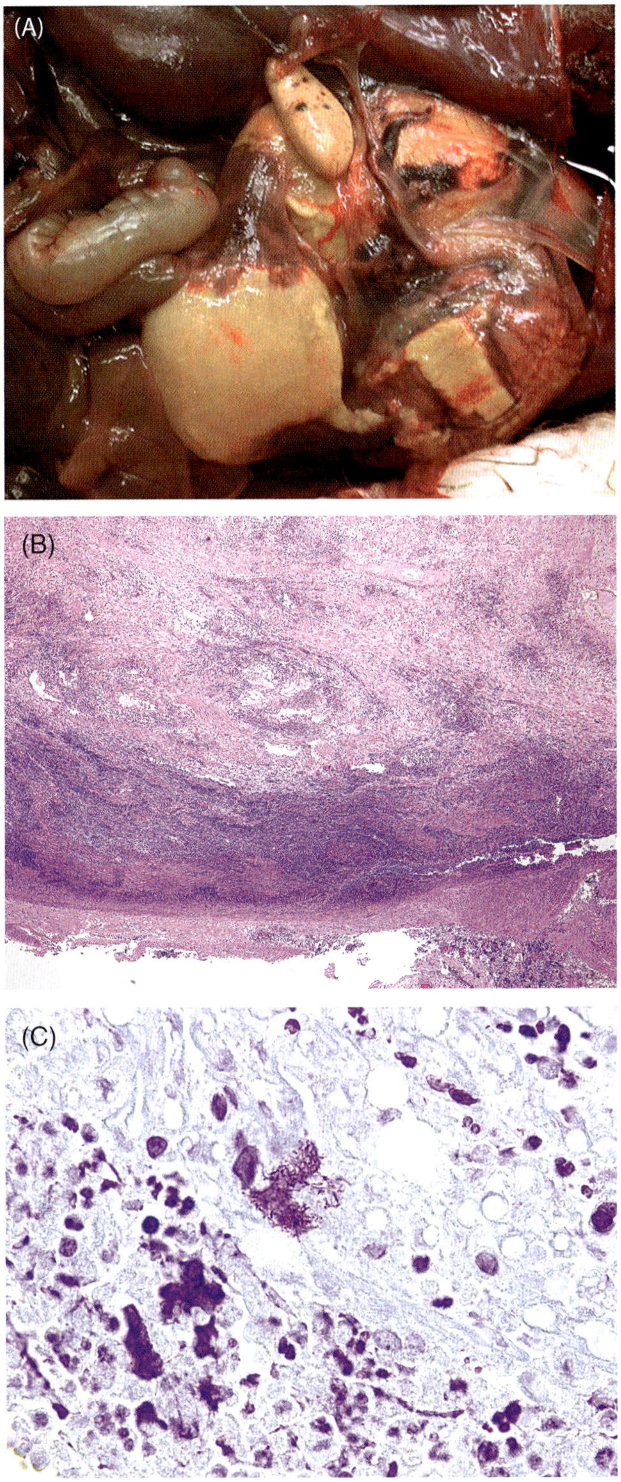

FIGURE 19.13 Listeriosis in the uterus of a wild snowshoe hare. (A) Transmural necrotizing metritis in a gravid doe is characterized by well-demarcated, segmental to circumferential, thickened, pale tan bands. (B) The myometrium is disrupted and replaced by sheets of necrotic heterophils, inflammation, fibrin, and necrotic cellular debris. (C) Aggregates of gram-positive coccobacilli are present among the inflammation and necrotic debris. Gram stain. *(Part A: Photo Courtesy of T. Bollinger, Canadian Wildlife Health Cooperative, Western-Northern Region.)*

chronic cases, atrophic-like rhinitis and/or otitis may occur (Fig. 19.14C). Histology is characteristic of pyogenic bacterial infections and consists of heterophilic inflammation with necrosis and abundant exudate. In more chronic cases, lymphocytes, plasma cells, and multinucleated giant cells with variable fibrosis are typical. Chronic pneumonia will have evidence of type II pneumocyte hyperplasia. Coinfection with *Bordetella bronchiseptica* causes bronchointerstitial pneumonia; mononuclear inflammation is centered on the vessels and airways and is mononuclear. Diagnosis of pasteurellosis is based on characteristic lesions, isolation of the organism with culture, and/or PCR. Other bacteria associated with upper and lower respiratory tract lesions similar to *P. multocida* include *Staphylococcus aureus* and *Klebsiella pneumoniae*.

Proliferative enteropathy caused by ***Lawsonia intracellularis*** in rabbits is similar to the presentation in pigs and ferrets. In rabbits, the disease affects nursing and weanling young rabbits as well as adults. The acute form is most common, though three forms occur: (1) erosive, (2) heterophilic, and (3) proliferative. Lesions are found in the small intestine, cecum, and colon and include eroded, ulcerated, hemorrhagic, and edematous mucosa with fibrinous membranes. In severe cases, the intestinal mucosa is thickened and corrugated with microscopic crypt proliferation. Infection can be detected in affected intestinal mucosa using silver stains and confirmed with immunohistochemistry.

A number of clostridial species are normal commensals in healthy rabbits but can become pathogenic, especially in the face of coinfection with other bacterial or viral pathogens. ***Clostridium spiroforme*** is the most common clostridial infection in rabbits.[19] The bacteria produce Type E iota toxin that causes severe necrotizing enteritis and bloody diarrhea. Grossly, the cecum is dilated and contains abundant luminal liquid green to brown fecal material with multifocal serosal petechiae and ecchymoses. Large numbers of gram-positive bacilli can be detected in necrotic foci with Gram staining. If the animal survives, the intestinal tract may have mucosal atrophy and fibrosis leading to malabsorption and wasting. Other clostridial diseases in rabbits include *C. perfringens*, which causes cecal hemorrhage and edema in young rabbits, and *C. difficile*, which typically leads to ileal (not colonic) lesions following either a dietary change or antibiotic administration.

Tyzzer's disease, caused by ***Clostridium piliforme***, is found in immunocompromised animals of several species but is classically described in foals, rodents, and rabbits. It is rarely reported in snowshoe hares. Lesions are similar to those seen in gerbils and hamsters. Rabbits with Tyzzer's disease commonly develop severe diarrhea with prominent perineal fecal staining. In some cases, there is necrotizing colitis with substantial edema. Linear necrosis of the myocardium is also common in rabbits with Tyzzer's disease.

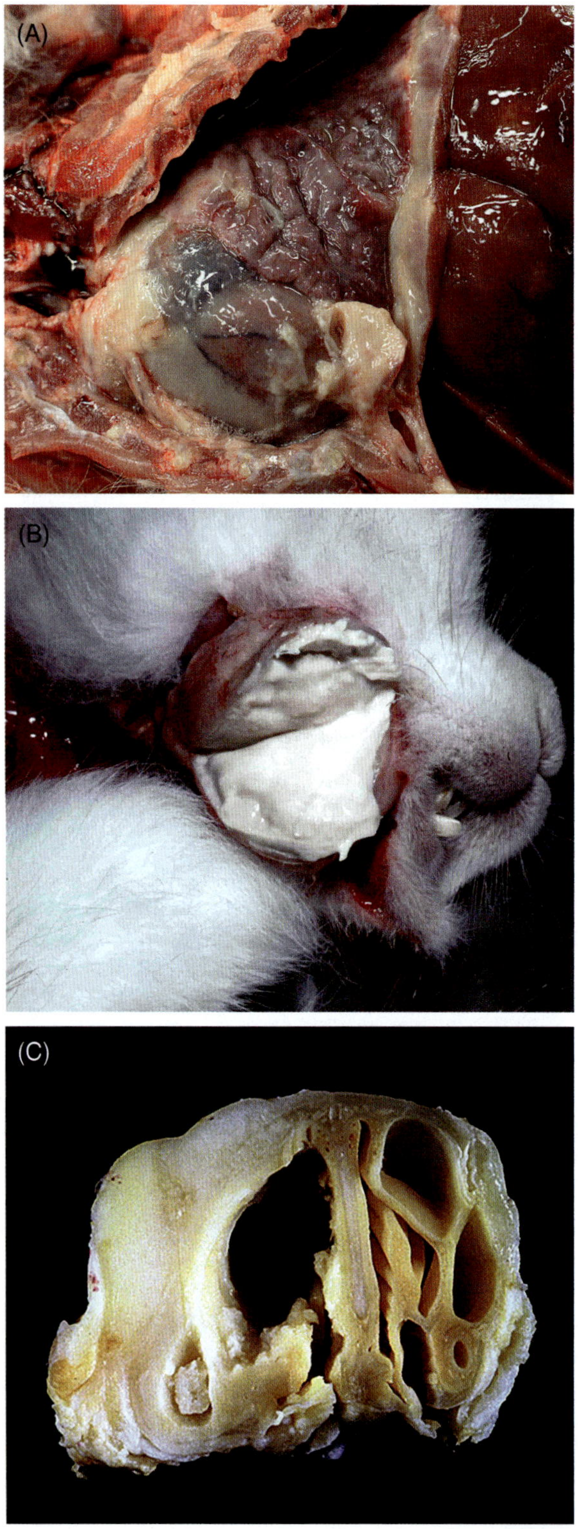

FIGURE 19.14 Pulmonary (A) subcutaneous (B) and nasal and aural (C) pasteurellosis in a rabbit. (A) The thorax contains copious, thick, opaque white fluid and flocculent exudate that obscures the heart and lungs and is adhered to the pleural surfaces. (B) The submandibular region contains a large, well-circumscribed, subcutaneous abscess containing abundant purulent white material. (C) Chronic pasteurellosis can be associated with atrophic rhinitis and otitis. Unilaterally, the nasal turbinates and ear canal are filled with, obscured and obliterated by exudate, chronic inflammation, and fibrosis. *(Part A: Photo Courtesy of J. Davies, University of Calgary; Part B: Photo Courtesy of R. Kelly, UQ School of Veterinary Science; Part C: K. Terio, University of Illinois.)*

The bacteria are long and slender and best highlighted at the edges of lesions with silver stains.

Colicobacillosis refers to disease caused by attaching and effacing (enteropathogenic) ***Escherichia coli*** strains. This is a significant cause of enteritis in rabbitries. Gross lesions include evidence of dehydration (tacky/dry) membranes, perineal staining with watery brown to yellow feces, serosal ecchymoses, cecal and colonic mural edema, and mesenteric lymphadenomegaly. The small intestine typically lacks gross lesions. Histologically, there is ileal mucosal villous blunting and heterophilic inflammation. Superficial mucosal epithelial cells are degenerate and lined by adherent bacilli (Fig. 19.15A). A condition resembling hemolytic uremic syndrome (HUS) may occur in rabbits with *E. coli* enterocolitis.[99] Renal insufficiency and tubular proteinosis are due to glomerular injury manifested as congestion, hemorrhage, and fibrin thrombi (Fig. 19.15B).

Ulcerative pododermatitis, also known as **sore hocks** (a misnomer because this condition affects the soles of the hind feet rather than the hocks), is a bacterial infection that develops secondary to pressure and trauma from wire-floored cages. Heavy breeds of rabbits and obese individuals are more susceptible. Gross and histologic lesions consist of dermal ulcerations and lacerations with heterophilic and mononuclear inflammation and variable bacterial colonization. *Staphylococcus aureus* is commonly isolated from these lesions. Other potentially pathogenic bacteria associated with pododermatitis include *Pseudomonas aeruginosa*, *Trueperella pyogenes*, and *Fusobacterium necrophorum*.

Syphilis, also **called vent disease, treponematosis**, or **venereal spirochetosis**, is a venereal disease of rabbits and hares caused by ***Treponema* sp.**[55] Infected rabbits are used as a model for the human disease; however, the disease is not zoonotic. Syphilis is a self-limiting disease and does not result in mortality. However, it can increase susceptibility to other infections. Asymptomatic carriers, especially bucks, can transmit the disease to multiple does. There have been no demonstrated negative effects on fertility. Gross lesions are limited to the head and perineal region and consist of erythema and edema at the mucocutaneous junctions along the lips, eyelids, nares, and perineum. These may progress to ulcerative dermatitis with serocellular crusts and reactive epidermal hyperplasia. Silver stains highlight spirochete bacilli within the affected areas.

The opportunistic bacterial infection colloquially called **blue breast** is a form of mastitis that affects the mammary tissue of lactating females. These lesions are similar to those described in other species and if untreated can lead to fatal septicemia. Causative bacteria include *Staphylococcus aureus*, *Streptococcus* sp., and *Pasteurella multocida*.

Several other, typically uncommon but important bacterial infections occur in leporids. **Mycobacteriosis** is rare in rabbits, although in biomedical research, rabbits serve as models for latent tuberculosis in humans. ***Mycobacterium***

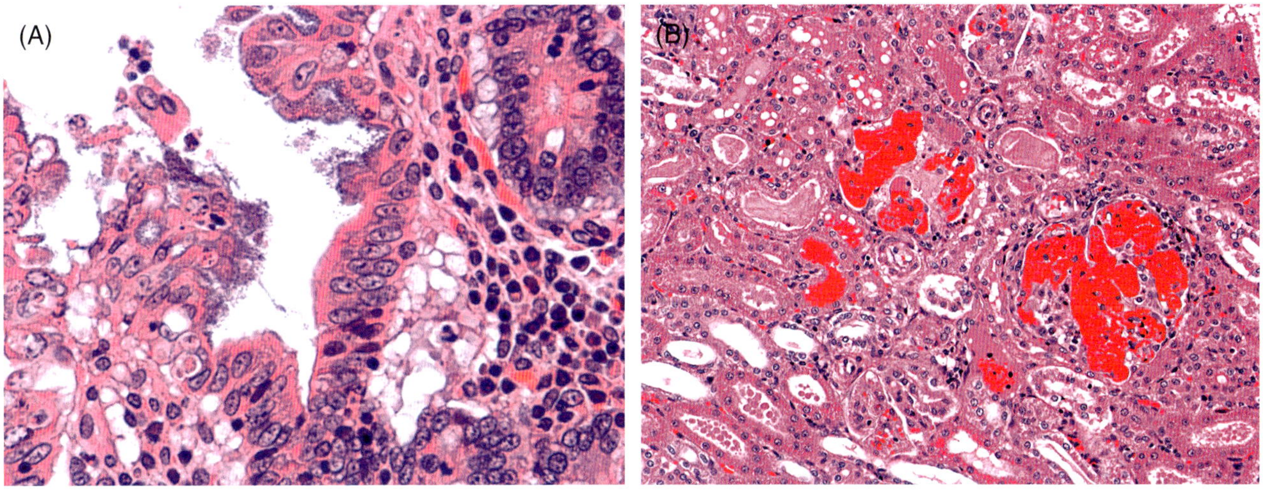

FIGURE 19.15 Colibacillosis in the cecum (A) and associated hemolytic-like uremic syndrome in kidney (B) of a rabbit. (A) The colonic mucosal epithelium is overlain by myriad adherent small bacilli. The epithelium has scattered individual necrotic cells and intraepithelial inflammatory cells. (B) Glomeruli are necrotic and obscured by clumped aggregates of erythrocytes, necrotic cells, and eosinophilic granular to smudged material. Some renal tubules are filled with homogenous to flocculent eosinophilic material (proteinosis).

bovis, ***M. avium***, and ***M. tuberculosis*** have been reported as natural infections in rabbits and hares. ***Dermatophilus congolensis*** is a filamentous bacterium that can cause exudative superficial dermatitis in rabbits similar to that seen in numerous other species. **Lyme disease** caused by ***Borrelia burgdorferi*** has been induced experimentally in rabbits[51]; wild rabbits are considered potential reservoirs for the disease. ***Castellaniella ginsengisoli***, a motile gram-negative bacterium, has been associated with an outbreak of fatal, multisystemic suppurative inflammation, and abscesses in a group of captive Daurian pikas.[92]

Fungi

Ringworm is the most common fungal skin disease in rabbits. It can be caused by ***Trichophyton mentagrophytes***, ***Microsporum canis***, or ***M. gypseum***. These fungal species are zoonotic and can be transmitted to other species. Young and immunocompromised individuals are most susceptible. Well-circumscribed, raised, alopecic red cutaneous lesions are found on the head and ears. Microscopically, the epidermis is hyperplastic and hyperkeratotic with mixed inflammatory cell folliculitis and arthrospores that stain positively with Gomori methenamine silver stain (GMS) and the Periodic acid-Schiff (PAS) reaction. In addition to histopathology, diagnosis can be made by skin scrapings cleared with 10% potassium hydroxide solution (i.e., KOH test) and fungal culture.

Aspergillus fumigatus and ***A. flavus*** can cause disseminated infections in wild rabbits. Lesions are similar to those reported in other species and include necrotic nodules in the spleen, liver, kidney, brain, heart, and intestine. Histologically, nodules are granulomas that consist of central necrosis with intralesional, parallel-walled, dichotomously branching, fungal hyphae surrounded by heterophils, macrophages, multinucleated giant cells, peripheral lymphocytes, and plasma cells. Hyphae are easily visualized with GMS and PAS stains. Conidia may be present if the affected tissues are exposed to air.

Adiaspiromycosis is a rare type of granulomatous pneumonia caused by ***Emmonsia*** sp. Inhalation of fungal spores from soil is thought to be the main route of infection. It has been described in a free-ranging European rabbit, desert cottontail rabbits, American pikas, and laboratory rabbits.[41] Histologically, adiaspores are ~240 μm in diameter with a multilayered wall consisting of a thin outer brightly eosinophilic layer, a thick pale eosinophilic middle layer, and an inner basophilic layer that surrounds a granular basophilic center. Adiaspores are accompanied by mixed inflammatory cells including heterophils, macrophages, multinucleated giant cells, lymphocytes, and plasma cells. GMS and PAS stains are useful to emphasize the adiaspore cell wall. Adiaspores are nonbudding and do not spread systemically.

A common microsporidial disease in rabbits is **encephalitozoonosis** caused by ***Encephalitozoon cuniculi***. These are obligate intracellular parasites that are small (2 μm diameter), oval, and gram-positive. Although this microsporidium is widely reported in domestic rabbits, particularly dwarf breeds, it and other *Encephalitozoon* spp. can also cause disease in hares and other animal species[40,71]; it is also zoonotic. Transmission occurs horizontally through ingestion of spore-contaminated urine and vertically through transplacental infection. Clinically and pathologically, there are three forms of disease: neurologic, ocular, and renal, though disseminated disease can also occur. Gross lesions are not seen in the brain and can be mild in the eyes and kidneys. More apparent lesions can include severe endophthalmitis and cataracts or focal to multifocal, up to 4mm diameter, irregular depressed and pale or reddened regions in the renal cortex. Histologically, the latter correspond to areas of granulomatous interstitial nephritis with associated tubular ectasia and degeneration. Spores are present in vascular endothelial and renal tubular epithelial cells and are best visualized with Gram and Giemsa stains as well as immunohistochemical labeling. Granulomatous encephalitis or meningoencephalomyelitis with intracellular spores may be seen associated with inflammation but characteristic inflammatory lesions are often present without obvious organisms. Diagnosis is most often made after death and is based on histologic lesions and demonstration of organisms.

Protozoa

Many metazoans, protozoans, and ectoparasites have been reported in lagomorphs. Several parasites that cause notable diseases are listed below. Those of lesser importance are listed in the Digital Content (Table 19.e1).

Protozoan parasitic diseases are well documented in rabbits and include both species-specific and zoonotic pathogens. **Coccidiosis** is an important group of protozoal diseases in lagomorphs. Numerous species of *Eimeria* affect either the intestinal tract or liver. Intestinal coccidiosis is important in domestic rabbits and is caused by one of the many species of ***Eimeria*** of which ***E. magna***, ***E. intestinalis***, and ***E. flavescens*** are the most pathogenic. Less pathogenic species include *E. irresidua* and *E. piriformis* (intermediate virulence), and *E. perforans*, *E. neoleporis*, and *E. media* (low virulence). Severity of coccidiosis can be affected by coinfections or other predisposing factors and in turn can predispose individuals to other infectious agents and GI disease.

Hepatic coccidiosis caused by ***Eimeria stiedae*** occurs in both domestic and free-ranging rabbits. It most often affects weanling rabbits since older rabbits can develop immunity. Protozoa target and replicate within biliary epithelial cells. Gross lesions include hepatomegaly with multifocal to coalescing, gray to yellow, raised, circumscribed areas of discoloration (Fig. 19.16A). These correspond histologically to acute to chronic proliferative cholangiohepatitis with bile duct hyperplasia and ectasia; ducts may contain inspissated bile and debris. Hyperplasia can be associated with papillary biliary epithelial projections, and hypertrophic epithelial cells contain intracellular oocysts and gametocytes (Fig. 19.16B). Gallbladder mucosa is similarly hyperplastic and luminal bile may contain oocysts. Diagnosis is based on identification of the organisms in routinely stained sections of affected liver.

Ectoparasites

Psoroptes cuniculi is the most common ectoparasite of rabbits and hares and is most often associated with severe crusting **otitis externa**. *P. cuniculi* is a nonburrowing mite, however, mite saliva and feces cause intense pruritis and inflammatory response resulting in self-trauma, skin damage, and secondary bacterial infections. Gross lesions are usually limited to the ears and start as hyperemia and edema with progression to thickening of the skin and formation of superficial brown crusts that slough into and fill the outer ear canal (Fig. 19.17). Histologically, the

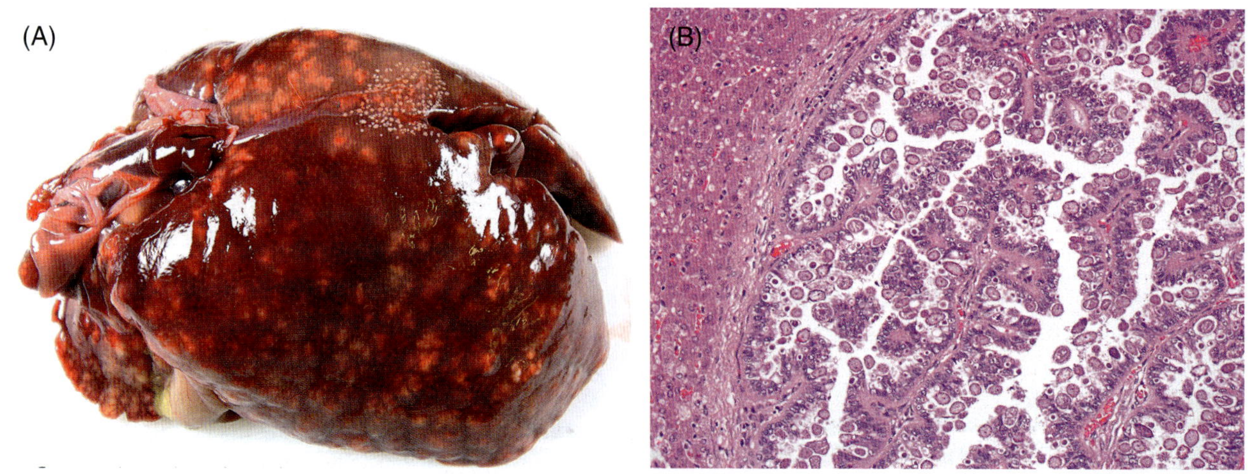

FIGURE 19.16 Hepatic eimeriosis in a rabbit. (A) The hepatic parenchyma contains variably sized, well-demarcated, multifocal random, nonencapsulated areas of tan discoloration. (B) These areas correspond to chronic proliferative cholangiohepatitis with marked bile duct hyperplasia and ectasia. As in this case, hyperplastic biliary epithelium may form papillary projections. Epithelial cells contain intracellular oocysts and gametocytes. *(Part A: Photo Courtesy of C. Legg, University of Calgary; Part B: Photo Courtesy of K. Terio, University of Illinois Zoological Pathology Program.)*

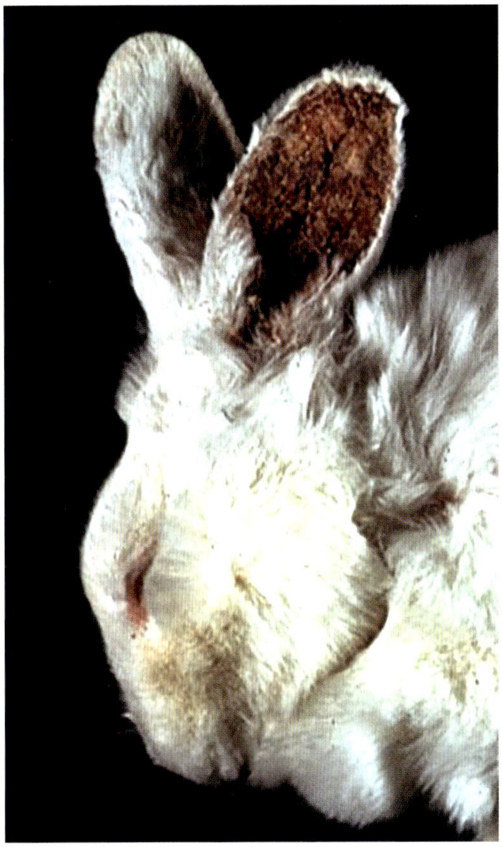

FIGURE 19.17 *Psoroptes cuniculi* in the ear of a rabbit. The ear pinna is markedly thickened and overlain by thick red-brown crusts. The ear canal is filled with exfoliated crusts and exudate. *(Photo Courtesy of D. Liggitt, University of Washington.)*

epidermis is acanthotic with marked parakeratotic hyperkeratosis. Mites and ova may be embedded within adherent and exfoliated keratin. Mites are large (400 × 700 µm) and oval with a chitinous exoskeleton, striated muscle, jointed appendages, an intestinal tract, and gonads. A distinguishing microscopic characteristic of the adult mites is a three-segmented pedicel with terminal caruncle. Eggs are up to 70 µm diameter, oval, and have an anisotropic shell. The disease can spread throughout the body and cause severe cutaneous damage, potentially resulting in death. ***Cheyletiella parasitovorax*** is another mite species affecting rabbits. It is a fur mite that can transmit myxoma virus. *C. parasitovorax* can be distinguished from *Psoroptes* by its smaller size (380 µm long) and large curved palpal hooks around its mouthparts. Fleas and other ectoparasites, while not necessarily implicated in overt disease, are capable of transmitting pathogens such as *Bartonella* sp. between individuals.[15]

ADDITIONAL RESOURCES

Barthold SW, Griffey SM, Percy DH. *Pathology of Laboratory Rodents and Rabbits*. 4th ed. Wiley-Blackwell; 2016.

Manning PJ, Ringler DH, Newcomer CE, eds. *The Biology of the Laboratory Rabbit*. 2nd ed.; 1994.

Miller E, Lamberski N, Calle P, eds. *Fowler's Zoo and Wild Animal Medicine Current Therapy*, vol. 10; 2023.

Quesenberry K, Mans K, Orcutt C. In: James W. Carpenter, ed. *Ferrets, Rabbits, and Rodents Clinical Medicine and Surgery*. 4th ed.; 2020.

Smith MV. *Textbook of Rabbit Medicine*. 3rd ed.; 2022.

Suckow MA, Stevens KA, Wilson RP, eds. *The Laboratory Rabbit, Guinea Pig, Hamster, and Other Rodents*. 1st ed.; 2011.

References

Full references for this chapter can be found online at https://doi.org/10.1016/B978-0-443-18437-6.00029-8.

Chapter 20

Rodentia

Martha A. Delaney[a], Jessica M. Snyder[b], and Jamie L. Rothenburger[c]

[a]Zoological Pathology Program, College of Veterinary Medicine, University of Illinois, Brookfield, IL, United States; [b]Department of Comparative Medicine, School of Medicine, University of Washington, Seattle, WA, United States; [c]Faculty of Veterinary Medicine, University of Calgary, Calgary, AB, Canada

INTRODUCTION

The order Rodentia includes 29 families, 468 genera, and over 2,052 species. Two separate systems exist to classify rodents.[23] The first classification scheme by Brandt[19] is based on morphology of the jaw and skull together with masticatory muscle position and has three suborders: Sciuromorpha (squirrel like), Myomorpha (mouse like) and Hystricomorpha (porcupine like). The second separates rodents into two suborders, Sciurognathi and Hysticognathi, based on the position of incisors and the angle of the jaw.[23] Molecular studies yield seven clades containing three phylogenetic lineages: squirrel-related (Sciuriodea and Gliridae), mouse-related (Anomaluromorpha, Castoridae, Geomyoidae, and Myodonta), and Ctenohystrica. With diverse ecological strategies, rodents are found naturally in a variety of habitats globally and on all continents except Antarctica. The natural history and biology of members of this extensive order exhibit substantial variation. Nevertheless, similarities exist among rodents including dentition, anatomy, and disease susceptibility.

The descriptions of diseases of rodents that follow are not exhaustive but instead highlight the more common findings in species for which there is published literature or knowledge for species managed in zoos and in the wild. A wealth of information also exists regarding the anatomy, physiology, and diseases of rodents used as animal models in biomedical research (i.e., laboratory animals) and is a valuable resource when examining lesser-studied rodent species using a comparative approach.[9,66,139,148] As many rodents are important prey species, diseased or dead rodents are exceedingly difficult to detect in nature using passive surveillance techniques of traditional wildlife disease monitoring. Thus, less is known about the diseases and pathology of free-ranging rodents as compared to larger and, considered by some (but not the authors of this chapter), more charismatic mammals.

NON-INFECTIOUS DISEASES

Nutritional

One of the most well-described nutritional diseases in rodents is **hypovitaminosis C** or **scurvy**. Guinea pigs and a number of other species including capybaras and humans, lack the enzyme L-gulonolactone oxidase which is required for L-ascorbic acid (vitamin C) production; therefore, these species require dietary vitamin C.[35] Vitamin C is necessary for the formation of collagen via hydroxylation of proline/lysine for cross-linking of fibrillar collagen. Hypovitaminosis C results in decreased collagen fibril cross-linking, leading to the increased fragility of blood vessels, cartilage, and osteoid. In affected individuals, gross and histologic lesions are characteristic and include capillary hemorrhage and lack of bone deposition and remodeling as a result of poorly formed and weak collagen. Other gross lesions include evidence of anemia (e.g., pallor, reactive bone marrow), periarticular hemorrhage (Fig. 20.1A), swollen joints, specifically at the costochondral junctions (scorbutic lattice) (Fig. 20.1B), and physeal lesions in young, growing animals (Fig. 20.1C). Characteristic oral lesions include gingival swelling, erythema, hemorrhage, erosion, ulceration, and tooth loss. Histological examination is necessary to confirm and define bone lesions, which include dilation of metaphyseal blood vessels, irregular column formation and hypertrophy of physeal cartilage, calcification of metaphyseal cartilage with reduced or absent osteoid, microfractures with surrounding and subperiosteal hemorrhage, medullary fibrosis, and decreased osteoclastic activity and remodeling.[54] Tooth loss may result from a combination of abnormal odontoblasts, dentin resorption, pulp fibrosis, and a lack of/abnormal alveolar bone remodeling.

Like lagomorphs, some hindgut-fermenting rodents are susceptible to **gut stasis (ileus)** and related diseases if they are not able to obtain enough fiber from the diet

Pathology of Wildlife and Zoo Animals. https://doi.org/10.1016/B978-0-443-18437-6.00025-0

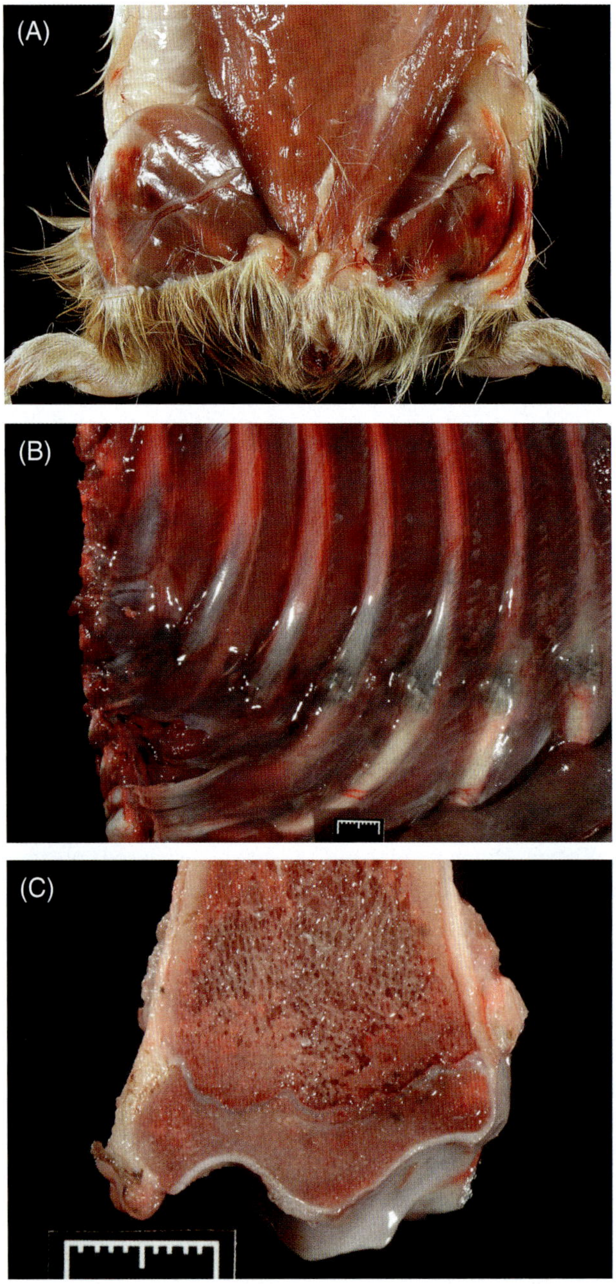

FIGURE 20.1 Hypovitaminosis C (scurvy) in a guinea pig (A) and young capybara (B, C). (A) A common gross manifestation of scurvy in guinea pigs is bilateral, periarticular hemorrhages surrounding the stifle joints. (B) Along the pleural surface of the thoracic wall, there are several enlarged costochondral junctions with surrounding hemorrhage. (C) In cross-section through the proximal tibia, the physis is irregular and the subphyseal trabeculae are absent (necrosis) and or dark red, consistent with infarction. *(Part A: Photo Courtesy of D. Imai-Leonard, University of California, Davis School of Veterinary Medicine; Parts B–C: Photos Courtesy of B. Porter, Texas A&M University.)*

(see Lagomorpha). **Obesity** can be common in rodent species if fed a high-caloric diet and can predispose to diabetes mellitus, cardiac disease, pregnancy toxemia, and hepatic lipidosis in females.

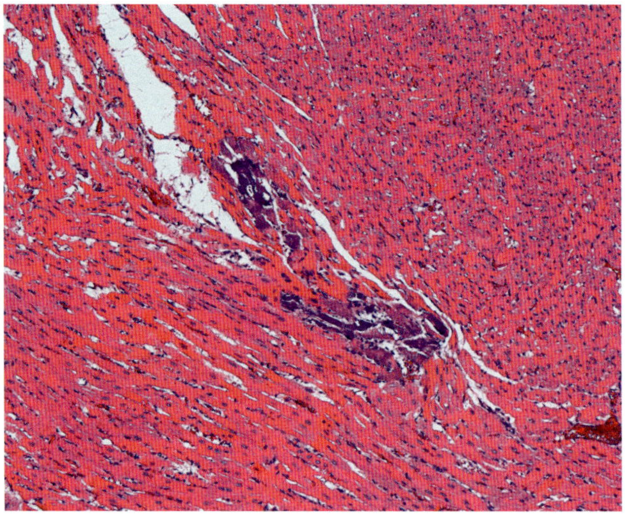

FIGURE 20.2 Dystrophic mineralization in the myocardium of a guinea pig. Several adjacent degenerate cardiac myocytes are obscured and replaced by variable amounts of granular to fragmented, angular deeply basophilic material consistent with mineral.

Metabolic

Dystrophic mineralization is an idiopathic condition that occurs in guinea pigs, hamsters, and free-ranging rats.[128] Mineralization most often affects the **myocardium** (Fig. 20.2) but can also occur in other tissues, such as the diaphragm, tongue, liver, kidney, lungs, cornea, and aorta (see "Congenital/Genetic" section). Though the pathogenesis is not fully elucidated, mitochondria appear to be preferentially mineralized based on ultrastructural studies. Gross lesions may include cardiomegaly with chalky to gritty white material and streaking within the myocardium of the atria and ventricles. Histologically, affected cardiac myocytes are degenerative and/or necrotic with rupture of the sarcolemma. Damaged myocytes will have variable and progressive mineralization until the entire myofiber is densely mineralized. Variably mature fibrosis surrounds individual and clusters of dead and mineralized myocytes. Satellite cells are frequently hypertrophic.

Pregnancy toxemia occurs in overweight, pregnant, and lactating female guinea pigs and hamsters. There are two forms in guinea pigs, a fasting type that results from decreased carbohydrate intake and subsequent mobilization of fat stores, and a circulatory type that occurs when the gravid uterus compresses the aorta and causes reduced blood flow to abdominal organs. Characteristic gross and histologic lesions include hepatic lipidosis and fat deposition in renal tubular epithelial cells. Similarly, hepatic lipidosis and associated metabolic effects occasionally affect overweight nongravid sows and boars.

Urolithiasis is common in guinea pigs, rats, and mice due to high urine pH and calcium concentrations.

Obstructive urolithiasis has also been reported in captive male agoutis resulting in bilateral hydronephrosis, urinary bladder rupture, and peritonitis.[10]

Toxic

Fluorosis is well studied in rats as they are used as models in toxicology studies. Guinea pigs and rabbits are also susceptible, and fluorosis has occurred in free-ranging cotton rats in sites contaminated by petrochemical waste.[115] Pathognomonic lesions occur in teeth and bones. Periosteal hyperostosis occurs first along the medial surface of the metatarsals, then progresses to affect the mandible, metacarpals, and ribs though all bones can be affected in severe, chronic disease. Bones are brittle and chalky-white. Pathological fractures are common, but joints are typically spared. Characteristic, enamel hypoplasia only occurs when fluorosis is present during tooth development. Tooth lesions include dull, dry, and chalky hypomineralized or hypoplastic enamel with pitting, grooves, and discoloration (due to oxidation); excessive attrition affects the incisors first. Irregular wear, malocclusion, and fractures may be present. Histologically, ameloblasts are small while osteoblasts are vacuolated and disorganized. There is abundant globular dentin and hypomineralization of the outer enamel layer. A differential for enamel hypoplasia in young Syrian hamsters is Hamster Parvovirus (HaPV); concurrent testicular hypoplasia and cerebral malacia are present with this viral infection.

Rodenticides containing **cholecalciferol (vitamin D3)** can cause hypervitaminosis D in most species. The deliberate poisoning of free-ranging rats and mice results in inadvertent toxicity of nontarget animals including other rodents, such as squirrels, or carnivorous predators that consume affected carcasses. Multisystemic mineralization occurs prior to death in affected individuals. Unlike other rodents, naked mole-rats, beavers, and woodchucks do not require dietary vitamin D and toxicity can develop if they are fed rodent chow or diets supplemented with vitamin D. Lesions consistent with calcinosis cutis and circumscripta can develop in naked mole-rats fed such diets. Soft tissue calcification can be found throughout the body in severe cases (Figs. 20.3A, B) and, in contrast to dystrophic calcification, is not limited to pressure points. With correction of the diet, resolution of lesions typically follows without treatment in mildly affected mole-rats.[41] In cases of soft tissue or parenchymal mineralization in susceptible rodent species, review of diet and supplementation should be completed to ensure a vitamin D excess or ingestion of vitamin D (calcitriol) analogs (e.g., Solanum, Cestrum) hasn't occurred. Additionally, other causes of metastatic mineralization can be excluded such as severe renal disease and hyperparathyroidism based on clinical, or postmortem examinations.

Anticoagulant rodenticides including warfarin and its derivatives are frequently used to control free-ranging mice and rat populations. These compounds antagonize vitamin K and thus inhibit production of vitamin K-dependent clotting factors (I, II, VII, IX, X). Gross lesions include multisystemic hemorrhage, which may be most common in the subcutis (Fig. 20.4) and periarticular soft tissues. "Bait," some of which are brightly colored (e.g., teal-green), may be present in the stomach and/or intestines. Inadvertent consumption by nontarget wildlife may have fatal consequences.

Hindgut fermenting rodents are susceptible to similar **antibiotic-related toxicities** as rabbits (see Lagomorpha). **Nephrotoxicosis** can be the result of exposure to numerous agents and has been extensively described in laboratory animals. Histologically, acute tubular necrosis is the main finding. For example, **Swainsonine toxicity** occurs following the ingestion of plants from the genus **Swainsona**

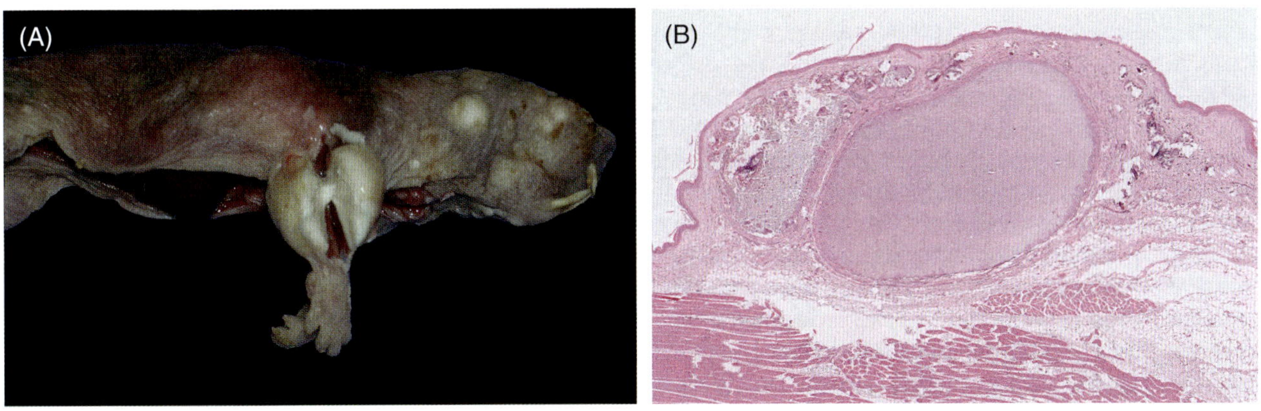

FIGURE 20.3 Calcinosis cutis in a naked mole-rat. (A) The head, neck, and antebrachium have variably sized white dermal and subcutaneous nodules. On cut surface, nodules contain white, opaque, pasty, and gritty material. (B) Nodules are composed of dense accumulations of finely granular basophilic to amphophilic material and lesser large, deeply basophilic fragments (mineral) surrounded by bands of macrophages, multinucleated giant cells, and collagen.

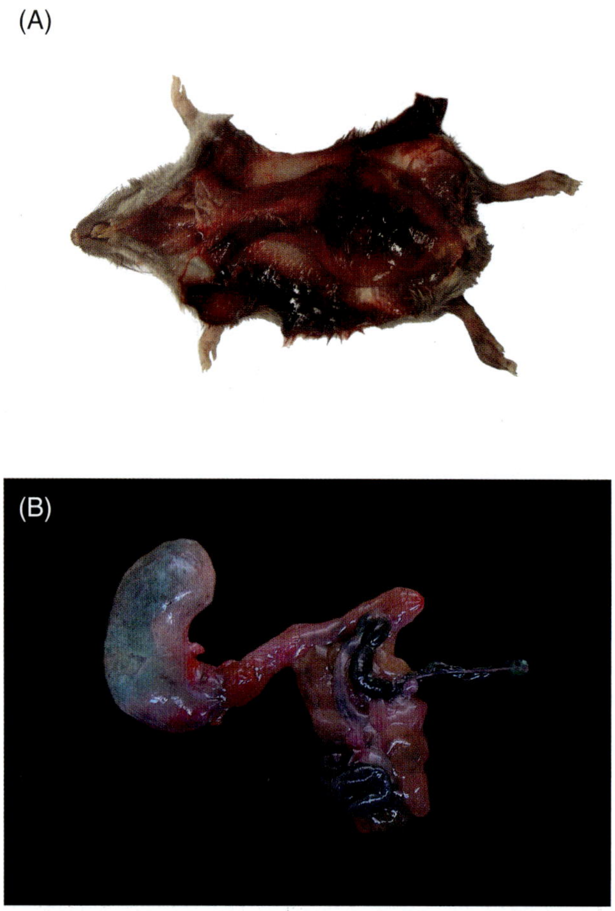

FIGURE 20.4 Hemorrhage due to anticoagulant toxicity in a free-ranging rat. (A) Throughout the body, the subcutis, soft tissues, and muscle have extensive regions of hemorrhage, compatible with anticoagulant toxicity. Internally, there may be cavitary hemorrhage and petechiation and ecchymoses of serosal surfaces. (B) Presence of blue-green digesta and feces in the gastrointestinal tract is confirmatory.

and results in **acquired alpha-mannosidosis** in rats. Histologically, renal tubular epithelial cells are severely and diffusely vacuolated.

Congenital/Genetic

Dystrophic mineralization, also known as **dystrophic cardiac calcinosis,** is considered as a genetic disease in inbred mouse strains (BALB/c, C3H, DBA), particularly in aged individuals.[47] Cardiac mineralization is detected grossly as pinpoint foci to larger irregular crusts composed of white, gritty material within and overlying the epicardium. Histologic lesions vary depending on severity and chronicity with degeneration and necrosis of individual myocytes to extensive regions of myocyte loss and replacement by deeply basophilic granular material. In severe cases, mineralization extends into the myocardium resulting in scar formation and can also be seen in other tissues such as the kidneys and lungs. The prevalence of this disease in free-ranging mice from which these strains originated is unknown.

Diabetes mellitus (DM) in Chinese hamsters is a heritable, autosomal recessive disease.[68] Symptomology and lesions are similar to those found in humans and include hydropic degeneration and degranulation of pancreatic islet β-cells with secondary vascular changes such as arteriosclerosis and glomerulosclerosis. In addition, affected hamsters are predisposed to pancreatic adenocarcinomas. Serum levels of α-2 globulins are a useful indicator of risk for diabetes mellitus development in these populations.

Polycystic kidney disease has been diagnosed in captive populations of Brazilian agoutis.[104] Most affected individuals have no clinical signs and are potentially related genetically. Grossly, affected agoutis have a range of bilateral renal changes including atrophy, roughened granular cortical surfaces, and multiple cysts. Histologically, there is cystic dilatation of renal tubules and Bowman's capsules with mesangial and capsular thickening accompanied by mononuclear interstitial nephritis and fibrosis.

Age-Related/Degenerative

Many degenerative and age-related diseases are described in rodents used in toxicology studies.[9,112] A significant disease confounding these studies is **chronic progressive nephropathy**, which is best described in rats though mice, naked mole-rats, and Australian rodents, particularly the stick-nest rat, are also affected.[40] The exact pathogenesis is unknown; however, genetic factors may play a role. Grossly, kidneys are bilaterally pale and either enlarged or shrunken with numerous pinpoint to large cortical microcysts creating a nodular surface. Histologically, renal tubules are markedly ectatic and tortuous and many contain brightly eosinophilic fluid (proteinosis) (Fig. 20.5; eSlides 20.e1–20.e3). Tubules have a range of degeneration, necrosis, and regeneration with occasional tubular hyperplasia and adenoma formation (predominantly in rats). Glomeruli are variably affected by membranous and/or proliferative changes with thickened basement membranes, sclerosis, and obsolescence. Synechiae, dilated Bowman's capsules, and hypertrophied parietal epithelium are common. The interstitium is expanded by lymphoplasmacytic infiltrates with pigment-laden macrophages, mild hemorrhage, and edema. There is often substantial interstitial fibrosis and in severe cases, infarction. Diagnosis is based on histologic evaluation, though abnormal renal clinical chemistry values may be used to determine renal function if antemortem blood collection is possible.

Cardiomyopathy is common in aged mice, rats, and hamsters; it is particularly common in Syrian hamsters and is reported in woodchucks.[27,121,128] Lesion severity

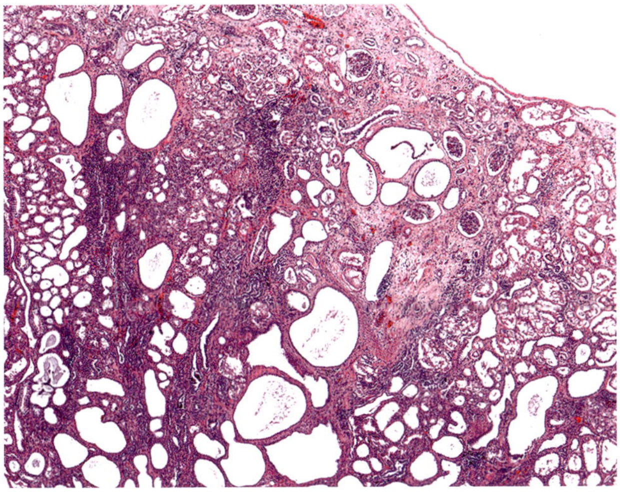

FIGURE 20.5 "Chronic progressive nephropathy" in a naked mole-rat. The renal cortex is undulant with depressed regions. Renal tubules are variably ectatic, atrophied, degenerate, necrotic, or small (atrophy) and separated or replaced by collagenous fibrous tissue. The interstitium contains scattered accumulations of mononuclear cells (primarily lymphocytes and fewer macrophages). Few glomeruli have membranous change, sclerosis, and/or periglomerular fibrosis (see eSlides 20.e1–20.e3).

varies though it appears to increase with age and is more common in males. Though considered strain and thus possibly genetically related in laboratory rats, free-ranging rats develop similar disease.[128] Grossly, there may be cardiomegaly but macroscopic changes are not always apparent. Histologically, myocardial lesions consist of multifocal and perivascular lymphoplasmacytic infiltrates accompanied by fibrosis, myocyte degeneration, and drop out (Figs. 20.6A, B). Trichrome, Movat pentachrome, or other stains that highlight fibrosis are useful in evaluating severity and chronicity. Syrian hamsters develop cardiomyopathy associated with spontaneous atrial thrombosis and subsequent coagulopathy that is often fatal (Fig. 20.6C). Females appear to be affected earlier in life than males.

Like other species, rodents are susceptible to skeletal degeneration and related diseases. Guinea pigs, specifically the Duncan-Hartley breed and Syrian hamsters, frequently develop **degenerative joint disease**. Gross lesions include thickened joint capsules, variable osteophyte formation, roughened cartilage surfaces, eburnation, joint effusion, and synovial proliferation. Depending on which joints are affected and chronicity, crepitus may be present when joints are manipulated. Microscopic examination of joints reveals degeneration, necrosis, and ulceration of the articular cartilage with proliferation and fibrosis of the synovium and joint capsule, periarticular and periosteal bone formation, and remodeling. There may also be variable inflammation and hemorrhage within joint spaces.

Amyloidosis is typically seen in aged animals and is described in detail elsewhere. Amyloid is an insoluble protein that when deposited in soft tissues, can lead to tissue atrophy and dysfunction. Amyloidosis is most common in

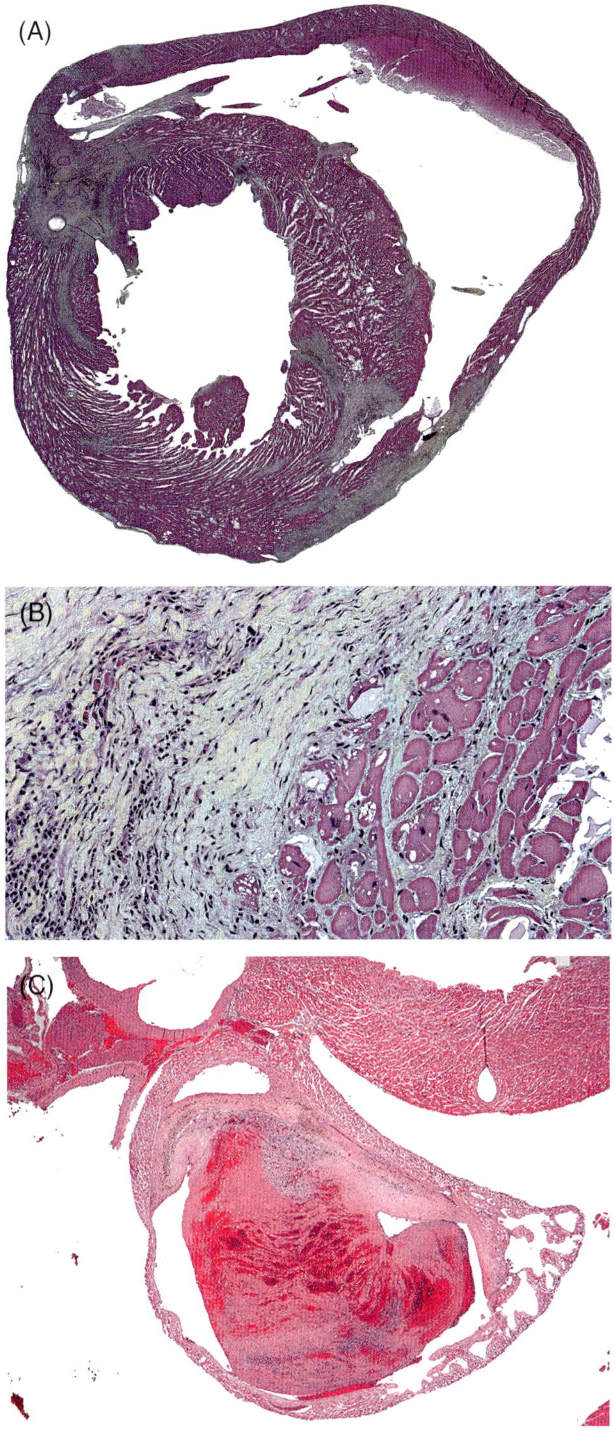

FIGURE 20.6 Cardiomyopathy in a black rat (A, B) and related atrial thrombosis in a Syrian hamster (C). (A) The myocardium has multifocal to regional replacement fibrosis (green-blue staining), Movat pentachrome. (B) Cardiomyocyte degeneration is characterized by cell swelling and loss of cross striations. Cells are separated and surrounded by increased collagenous tissue (blue-green staining) in which are low numbers of lymphocytes and plasma cells. Movat pentachrome. (C). The atrium is dilated and contains a large luminal thrombus that is regionally adherent to the endocardium. The thrombus is composed of laminated layers of fibrin, leukocytes, and erythrocytes with intermixed hemosiderin-laden macrophages. *(Part C: Photo Courtesy of J. Landolfi, University of Illinois Zoological Pathology Program.)*

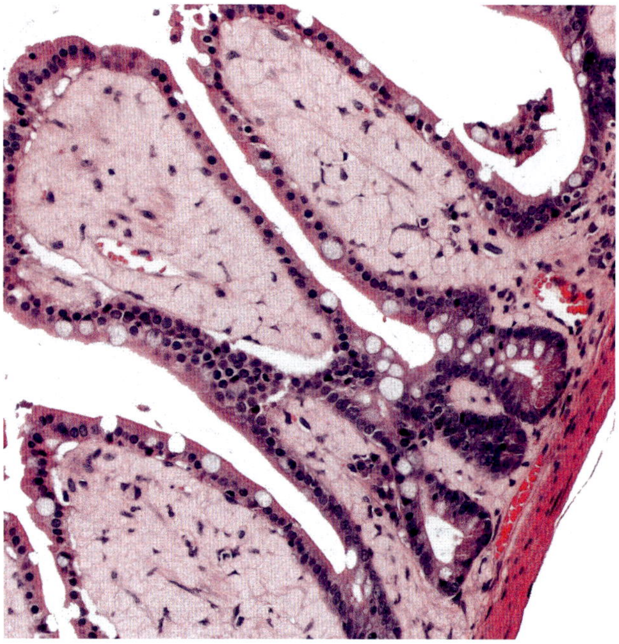

FIGURE 20.7 Intestinal amyloidosis in an aged mouse. Villi are widened and lamina propria expanded by abundant homogenous, pale eosinophilic, extracellular fibrillar to homogenous material consistent with amyloid. There are few lymphocytes embedded within the amyloid. *(Photo Courtesy of C. Pettan-Brewer and P. Treuting, University of Washington.)*

myomorphs (mice, rats, hamsters, gerbils) and is most commonly seen in the kidneys, spleen, adrenals, and liver as well as the gastrointestinal lamina propria (mouse) (Fig. 20.7). There is a higher prevalence in females. Of note is the eosinophilic substance within the nasal planum of aged mice that was once thought to be amyloid. Histochemical and ultrastructural studies have instead shown this material to consist of collagen and complex carbohydrates produced by the nasal gland epithelial cells.[45]

Hemosiderosis, often referred to as **iron overload**, is the accumulation of iron within various tissues, most commonly the liver. It is described in numerous species including primates, birds, rhinoceros, hyraces, and rodents. In one zoological institution, hemosiderosis was present in approximately 65% of adult naked mole rats.[41] Grossly, livers appear bronze. Iron accumulation appears histologically as zonal to diffuse hepatocellular accumulation of brown granular to globular pigment. Pigment is most intense in the periportal hepatocytes. The pigment stains positively with Prussian blue and is not refractile with polarized light (eSlide 20.e4).

Aged mice can develop **intrapulpal denticles**, which are dysplastic tooth-like growths arising in the pulp of incisors. These may result in abnormal wear and/or breakage of the incisors contributing to periodontitis and decreased ability to effectively gnaw.[112]

Neurodegenerative conditions in free-ranging and zoo-housed rodents are understudied. A recent study of zoo-housed naked mole-rats highlights the importance of routinely examining rodent brains histologically.[145] Many degenerative lesions, similar to those described in humans and other animals, were noted in aged naked mole-rats.[145]

Miscellaneous/Multifactorial

Trauma in free-ranging rodents is understudied and presents a challenge to detect since small, injured or dead wildlife are difficult to detect and may not be submitted for diagnostic investigation. Aggressive social interactions between many rodent species may result in bite wounds, which are susceptible to secondary bacterial infections. A study of free-ranging rats identified rare nonlethal trauma in the form of chronic wounds, limb and tail amputations, and healed fractures that may have been caused by snap-type traps.[126] Among rodents submitted to one diagnostic laboratory, trauma was the second most frequent cause of death after infectious/inflammatory causes.[56] Despite their protective quills, porcupines are particularly susceptible to cutaneous trauma and tears due to an exquisitely fragile integument. Secondary infections resulting in pyoderma and persistent ulceration may occur.

Abnormal tooth wear and growth and lead to **malocclusion** (Fig. 20.8), which is common in hystricomorphs, though also seen in brachydonts. It is a significant disease in captivity and may be related to genetics, nutrition and diet, and husbandry practices.[94]

Ovarian cysts are common in rodents. **Cystic rete ovarii** are prevalent in female guinea pigs[152] and can be distinguished from other ovarian cysts as they arise from the rete and compress adjacent ovarian tissue (Fig. 20.9). Other ovarian cystic structures reported in female rodents include bursal cysts, epithelial cysts, follicular cysts, luteal cysts, and paraovarian cysts.[43]

The hemochorial placentation in rodents predisposes them to **trophoblast emboli** (Fig. 20.10). This lesion is best described in chinchillas and is not generally associated with clinical ramifications or gross lesions. Histologically, intravascular trophoblasts are present in the lung, uterus, spleen, liver, and other sites, such as the adrenal glands with no associated inflammation. Trophoblasts are large cells (>100 μm diameter) with abundant amphophilic granular cytoplasm and large nuclei with prominent nucleoli.

Stress dermatosis (or dermatitis) is a condition of several South American rodents including the agouti, acouchis, and capybaras. This syndrome is seen in captive and free-ranging individuals in densely populated groups. It presents as alopecic skin lesions and lacerations along the lumbosacral spine. Stress is thought to increase skin fragility, and intraspecific aggression due to overcrowding may contribute to lesion development. Sun exposure, pruritus, and self-mutilation result in large patches of alopecia with erythema and excoriated skin with loss of the long

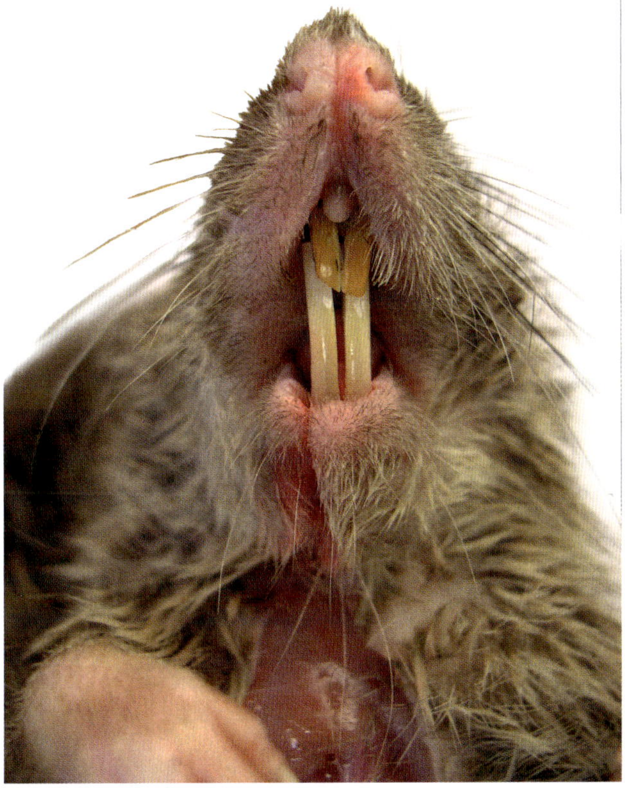

FIGURE 20.8 Malocclusion in a free-ranging rat. The incisors are misaligned with irregular wear of maxillary incisors and overgrowth of mandibular incisors.

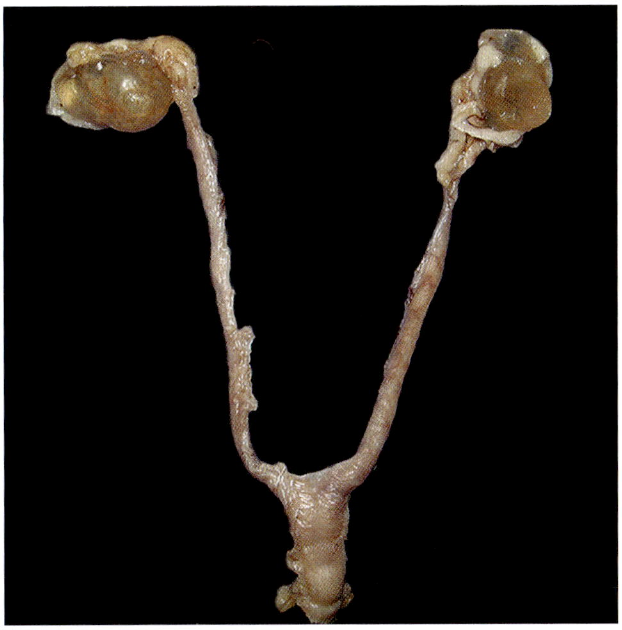

FIGURE 20.9 Cystic rete ovarii in a guinea pig. The ovaries are bilaterally enlarged by multiple, intraovarian rete cysts. *(Photo Courtesy of J. Landolfi, University of Illinois Zoological Pathology Program.)*

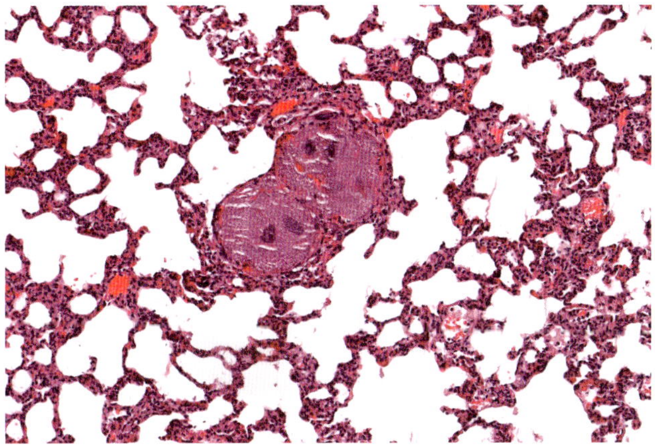

FIGURE 20.10 Intravascular trophoblasts in the lung of a chinchilla. Alveolar capillaries are focally dilated and expanded by intraluminal trophoblasts, which can be up to greater than 100 μm diameter. *(Photo Courtesy of Joint Pathology Conference.)*

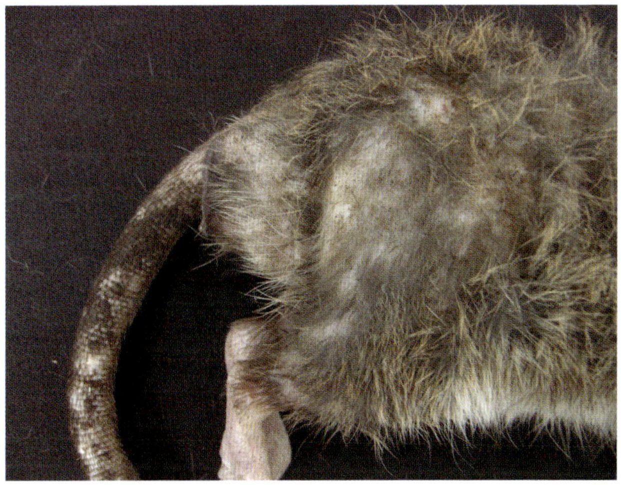

FIGURE 20.11 Stress alopecia (dermatosis) in a free-ranging rat. Large and coalescing regions of pelage are absent or have sparse hairs. The remaining pelage is unkempt.

hairs along the dorsum and tail (Fig. 20.11). Microscopic lesions correlate to severity and chronicity and include variable epidermal degeneration, necrosis, erosion/ulceration, and associated dermatitis with hemorrhage and edema. Diagnosis is based on clinical history, which has demonstrated the absence of a parasitic etiology in skin scrapings and histopathology. A review of dermatitis in several wild rodent species reports that for a large percentage of cases, the underlying cause of alopecia is not determined[119]; thus, additional investigations of the pathogenesis of dermatitis in rodents are warranted.

From 2013 to 2021, a unique pattern of alopecia was documented in related juvenile American red squirrels (*Tamiasciurus hudsonicus*) in New York state, United States during the fall months. Squirrels had widespread, bilaterally symmetric hypotrichosis to complete alopecia involving the entire dorsal and ventral trunk and neck and extending down the legs. Histologic lesions were limited to the skin and included abnormally oriented hair follicles, perforating folliculitis, kinked, bent and malacic hair shafts, and scant crusts. All squirrels grew normal pelage within the next several months. A cause was not determined although genetic anomalies in combination with hormonal or seasonal influences were considered.[65]

Neoplastic

Many neoplasms of inbred mice and rats appear to be strain related. Nevertheless, certain neoplasms occur more frequently among free-ranging and domesticated outbred rodents than in other mammalian orders. Several of them are described below. A list of other common neoplasms in rodent species is listed in Table 20.e1 in the Digital Content.

Naked mole-rats are a popular zoo species and are heralded as animal models for cancer and aging due to their extreme longevity and **cancer resistance**. Genomic sequencing and experimental studies have found potential mechanisms of cancer resistance. High expression of the tumor suppressor gene ARF, a loss of function mutation of the oncogene ERAS, high molecular weight hyaluronan, and related increased cellular contact inhibition have been postulated.[82,101,146] Rare cases of neoplasia have been reported in naked mole-rats and in the related Damaraland mole-rat, which is also long-lived and appears to be relatively resistant to cancer and age-related diseases.[30,39,42,140,143] Other subterranean rodents used as animal models for carcinogenesis research include some members of *Spalacidae* (mole-rats, blind mole-rats, zokors, and bamboo rats). There are currently no reported spontaneous tumors in this group; however, subcutaneous tumors (fibromas and fibrosarcomas) may be induced with the application of carcinogenic compounds.

Odontoma, or more aptly named **odontogenic dysplasia**, is described in several rodent species (rats, mice, guinea pigs, prairie dogs, degus, chinchilla, chipmunk, squirrels, voles[18,76,93]). Most odontomas are not true neoplasms and instead represent dysplastic changes resulting from chronic inflammation or trauma and associated regional tissue response. As there is no demonstrated metastatic potential or criteria for malignancy, these lesions are increasingly identified as hamartomas versus neoplasms (and also referred to as **pseudoodontomas**). Nevertheless, these toothy masses are regionally destructive and almost exclusively affect the incisors (thus the term "**elodontoma**"). Grossly, these odontogenic masses present as nodular swellings around the maxillary or mandibular incisors, or less often premolars or molars (Figs. 20.12A, B). Microscopically, masses are composed of disorganized and dysplastic trabeculae of odontogenic epithelium, enamel matrix, and mineralized enamel, dentin, pulp, and cementum. Teeth outside of the mass may exhibit apical elongation (Fig. 20.12C). In rodents, many odontomas have been described as complex because they contain fully differentiated tooth elements but do not form the tooth-like structures grossly present in compound odontomas seen in other species (e.g., horse). Histologically, rodent odontomas are composed of well-differentiated odontogenic epithelium among an enamel organ with dental pulp mesenchyme and a variably mineralized enamel matrix (Fig. 20.12C). Other odontogenic tumors, such as **odontogenic fibromas** and **ameloblastomas**, have been rarely reported in various rodent species.[48,92,105]

Several endocrine diseases have been documented in various rodent species. They have typically been associated with hormone-producing neoplasms (adrenal, islet, thyroid, parathyroid, and pituitary tumors).

INFECTIOUS DISEASES

Although there are many important infectious and zoonotic diseases of rodents, this section focuses on those that cause significant lesions in rodent species. Additional pathogens, including zoonotic pathogens that occur in rodent species are listed in the Digital Content (Tables 20.e2 and 20.e3). Most zoonotic pathogens carried by rodents do not cause disease or produce only mild lesions in their host species. Several references describe rodent reservoirs of emerging and reemerging zoonotic pathogens.[71,99] Evidence from free-ranging rats suggests that individuals with lesions not attributable to pathogen infection are more likely to carry zoonotic pathogens.[126] Human consumption of wild rodents or rodent-contaminated food is an important factor in the emergence or reemergence of zoonotic pathogens worldwide.[80]

DNA Viruses

Poxviral diseases in rodents are caused by members of *Orthopoxviridae*, which includes **mousepox** and **cowpox**. Both are on the World Organization for Animal Health (WOAH) list of reportable zoonotic diseases. Several species of rodents serve as the reservoir for the cowpox virus, which results in cutaneous lesions in humans. In laboratory mice, mousepox, more commonly known as **ectromelia**, is an important disease, and, like several other viral and bacterial pathogens listed below, is excluded from most specific-pathogen-free (SPF) laboratory facilities.[110] This infection can be acutely fatal with minimal lesions. In the chronic form, gross lesions include crusting lesions of face, legs, and tail, conjunctivitis, an enlarged, swollen, and friable liver and spleen, and in cases of recovery, splenic fibrosis. Necrosis of the kidney, thymus, and lymph nodes

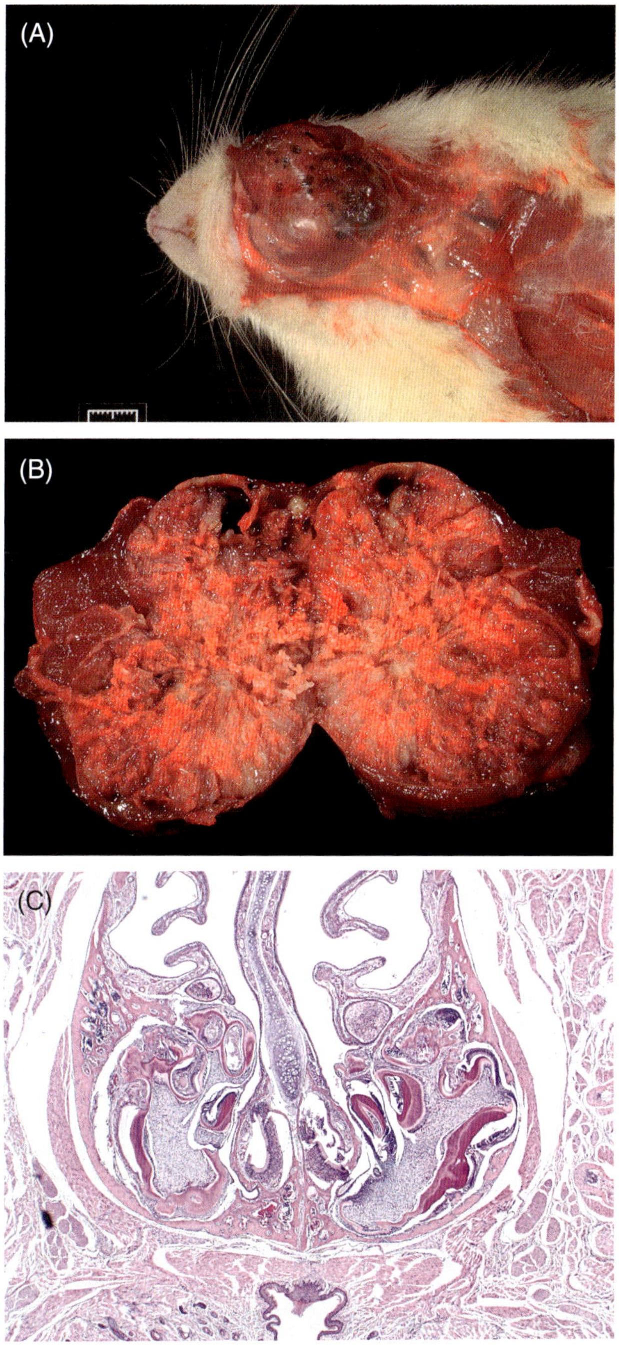

FIGURE 20.12 Odontoma in a rat (A, B) and apical elongation and odontogenic dysplasia in a mouse (C). (A) The ventral and lateral left mandible is expanded and effaced by a nodular, hard, pink, and dark red mass. (B) On cut surface, the mass has radiating bony spicules centered around white, hard, tooth-like structures. (C) The incisors are distorted and misshapen with several coalescing, tooth-like proliferations. *(Parts A–B: Photos Courtesy of B. Porter, Texas A&M University; Part C: Photo Courtesy of D. Imai-Leonard, University of California, Davis School of Veterinary Medicine.)*

may also be grossly evident. The small intestine, urinary bladder, and vagina can have mucosal erosions and hemorrhage. Histologically, skin and mucosal lesions include epithelial ballooning degeneration, hyperplasia, and erosions

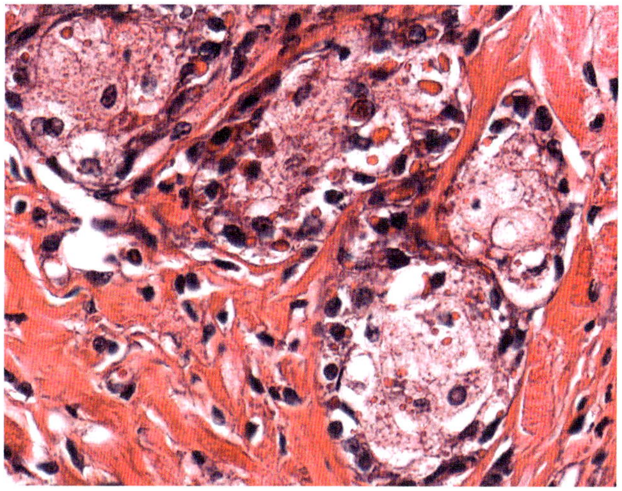

FIGURE 20.13 Ectromelia due to mousepox virus infection in a mouse. Follicular and sebaceous gland epithelial cells are mildly to markedly swollen (degeneration) with one large or multiple small, round to oval, eosinophilic, intracytoplasmic viral inclusion bodies, referred to as "Marchal bodies." *(Photo Courtesy of Joint Pathology Conference.)*

with characteristic intraepithelial, eosinophilic, intracytoplasmic inclusion bodies (Fig. 20.13). Within the parenchymal organs, splenic and hepatocellular necrosis is typically coagulative and multifocal to coalescing. Diagnosis is based on characteristic lesions with viral inclusions. Electron microscopy, PCR, and virus isolation are useful for confirmation of the viral strain. Zoo-housed rodents, such as Patagonian mara (or cavy; *Dolichotis patagonum*), may also develop characteristic lesions and pose a risk to immunocompromised humans.[84]

In 2003, an outbreak of **monkeypox** (*Orthopoxviridae*), a zoonotic virus, occurred in prairie dogs sold as companion animals in the United States.[61] Fatal infections in infected prairie dogs were characterized by necrotizing bronchopneumonia, conjunctivitis, and glossal ulcers. Viruses and viral antigens were detected in multiple tissues and cell types via immunohistochemical staining, PCR, and electron microscopy. Infected humans developed fever and vascular rashes. Forty-seven confirmed and probable cases of monkeypox were reported from six states. The prairie dogs contracted the virus following exposure to infected African rodents. Other rodents that also developed disease and died included: rope squirrels, tree squirrels, Gambian giant rats, brush-tailed porcupine, dormice, and striped mice. Although the animal reservoir(s) have not been confirmed, rodent species such as rope and sun squirrels, giant-pouched rats, and African dormice are thought to maintain the virus in environments of West and Central Africa (cdc.gov). In a subsequent outbreak in 2022, monkeypox infections occurred in over 80 countries with 30,000 infections from human-to-human transmission, mainly through close contact (https://www.cdc.gov/poxvirus/monkeypox/response/2022/).

Fibromas and **fibromatosis** of gray squirrels are potentially fatal conditions caused by **squirrel fibroma virus**. Disease is mainly reported in eastern North America[144] but also occurs in red squirrels and has been reported in a fox squirrel.[157] Transmission is likely through biting arthropods (fleas, ticks, and mosquitoes). Gross and histologic lesions are similar to Shope fibromas in rabbits (both are *Leporipoxviruses*). Multifocal to coalescing tan, firm nodules may cover all parts of the body and begin as alopecic nodules with progression to plaque-like to pedunculated masses (Fig. 20.14A). Systemic disease manifests as renomegaly and multiple pulmonary nodules.[85] Histologically, the skin nodules exhibit marked epidermal hyperplasia with ballooning degeneration of keratinocytes, spongiosis, and intracytoplasmic eosinophilic viral inclusions (Figs. 20.14B, C). Pulmonary nodules correlate histologically to adenomatous hyperplasia with similar viral inclusions. Other microscopic lesions may include atypical mesenchymal proliferation in the liver and seminal vesicles, and renal tubular epithelial hyperplasia with viral inclusions. Diagnosis is based on characteristic lesions and detection of virus by PCR or isolation.

Parapoxvirus is an important cause of disease in red squirrels in the United Kingdom and Ireland, where it is a suspected factor in severe population declines.[8,134,147] The causative agent is referred to as **red squirrel parapoxvirus**. Gray squirrels are likely maintenance hosts since a large percentage of healthy gray squirrels are positive for the virus and their range overlaps with that of red squirrels. Lesions in red squirrels include exudative dermatitis with crusting that is similar to other poxviral diseases. Viral inclusions can be detected histologically within the cytoplasm of infected cells. Electron microscopy and PCR are used to detect the virus.

Papillomaviruses have been well-studied in rodents, mainly for utilization as animal models for human papillomavirus infection.[150] Through advances in cloning methods, novel papillomaviruses have been isolated from proliferative epithelial lesions of free-ranging rodents such as North American porcupine and beaver, in addition to laboratory species like Syrian and Siberian hamsters.[150] As in other species, rodent papillomaviruses cause proliferative benign papillomas of the skin, mucus membranes, or mucocutaneous junctions.

Woodchuck hepatitis virus (**WHV**; *Hepadnaviridae*) is a disease in captive and free-ranging animals.[122] Although not zoonotic, this disease is a valuable animal model for viral hepatitis and carcinogenicity associated with hepatitis B virus infection in humans. It affects both sexes and is typically found in animals greater than 4 years of age. Activated T cells (CD8+) cause cellular and DNA damage leading to persistent hepatitis. Simultaneously, viral DNA integrates into the host genome at loci containing oncogenes, creating the potential for carcinogenic transformation. Hepatocellular carcinomas are reported in 100% of

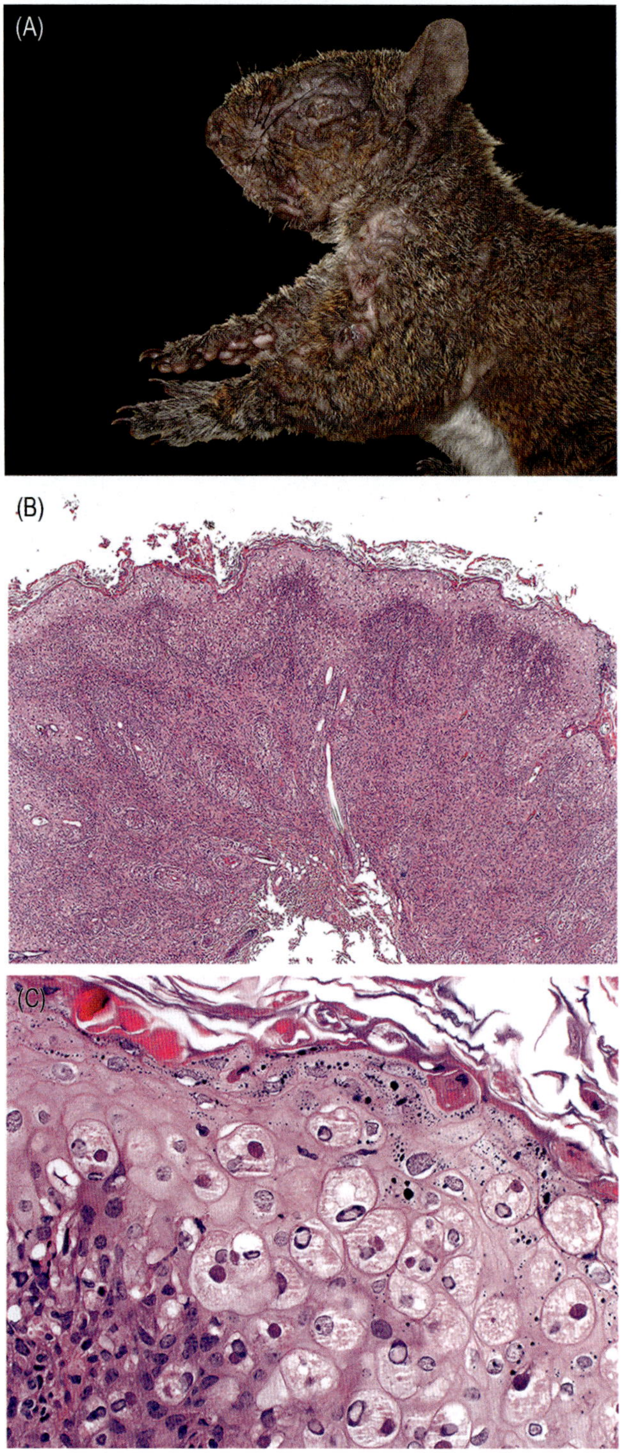

FIGURE 20.14 Fibromatosis in free-ranging squirrels due to squirrel fibroma virus infection. (A) The skin has multiple variably sized dermal plaque-like fibromas (coalescing areas of dermal thickening and alopecia). (B) The epidermis is severely thickened, and the dermis has extensive inflammation. (C) Numerous epithelial cells exhibit ballooning degeneration and contain intracytoplasmic eosinophilic inclusions. The underlying dermis has lymphocytic inflammation. *(Part A: Photo Courtesy of J. Landolfi, University of Illinois Zoological Pathology Program; Parts B, C: Photos Courtesy of D. Campbell, Canadian Wildlife Health Cooperative Ontario-Nunavut.)*

experimentally infected woodchucks. Grossly, most carcinomas are solitary and involve one hepatic lobe. Histologically, neoplastic masses have differing growth patterns: trabecular, adenoid, or solid with and without cystic spaces, hemorrhage, and necrosis. An interesting feature of these tumors is the presence of atypical, bizarre, and giant neoplastic cells. Adenomas may be present and are considered a precursor to carcinomas. Mitotic index and cellular morphology are used to classify tumors as adenomas or carcinomas. The remaining hepatic parenchyma often contains acute and/or chronic hepatitis with neutrophilic to mononuclear inflammation, variable hepatocellular degeneration, necrosis, fibrosis, and biliary degeneration or hyperplasia. Electron microscopic identification of the virus is definitive. Beechey ground squirrels have a similar virus with variable hepatic lesions.[36]

Hamster polyomavirus (HaPyV), a papovavirus, causes cutaneous **trichoepitheliomas** in Syrian hamsters.[53] The virus is transmitted via urine and is highly contagious, with up to 50% morbidity in affected breeding colonies. Early lesions consist of wart-like growths and alopecic nodules of the face and perineum with progression to severe, multifocal to coalescing nodules throughout the body. Histologically, these characteristic skin tumors contain multiple islands of basilar to variably differentiated follicular epithelium and abrupt central keratinization with faded anuclear keratinocytes (ghost cells). Rudimentary hairs may be present and follicular rupture may be associated with inflammation.

Other **polyomaviruses-associated neoplasms** due to squirrel fibrosis virus include other cutaneous tumors, lymphomas, and salivary gland tumors. In virally induced lymphomas, affected tissues are effaced by diffuse sheets of neoplastic lymphocytes among scant stroma. Viral inclusions are not present. In a case of polyomavirus-induced, multicentric lymphoma in an 8-week-old Syrian hamster, the virus was detected in tumor cells, renal tubular epithelial cells, and enterocytes via in situ PCR indicating some degree of tissue tropism.[133] Diagnosis is based on clinical history, gross and histologic lesions, and detection of the virus via PCR and/or electron microscopy.

Lesions and diseases caused by **herpesviruses** are sporadically reported in rodents. Recently, a novel alphaherpesvirus, referred to as **DpAHV**, was discovered as the cause of acute mortality in several zoo-housed Patagonian maras (*Dolichotis patagonum*). Histologic lesions were most severe in the nervous system and included necrotizing encephalitis, myelitis, and ganglioneuritis with characteristic intranuclear inclusions.[21]

RNA Viruses

Mouse hepatitis, caused by **mouse hepatitis virus, MHV** (*Coronaviridae*), is enzootic in some colonies of inbred laboratory mice.[9] It may also occur in nonlaboratory strains of mice in zoo settings. MHV causes polytropic and enterotropic disease syndromes, both of which occur following viral infection, which is highly contagious, by oronasal transmission. Grossly, there is hepatomegaly, splenomegaly, lymphadenomegaly, and lymphadenopathy with jaundice and ascites. There may be gross evidence of intestinal (ileal and cecal) disease with thickened and necrotic mucosa. Multifocal coalescing and random white necrotic foci are seen in the liver, spleen, and lymph nodes. The main histologic lesions include necrotizing hepatitis with intralesional syncytial cells in necrotic foci, and demyelination in CNS in immunodeficient mice. The development of enterotropic disease is age-dependent and may include villous attenuation and syncytial cell formation with eosinophilic intracytoplasmic inclusion bodies in enterocytes, mesenteric lymph nodes, and endothelial cells; granulomatous serositis may also be present. Diagnosis is based on histologic features and immunohistochemical demonstration of viral antigens within lesions. Virus isolation and PCR are used to confirm infection and serology may indicate previous exposure.

During the early months of the **COVID-19** global pandemic, many laboratory rodents were used in studies to elucidate **SARS-CoV-2** pathogenesis and in the development of vaccine and therapeutic strategies.[114] Syrian hamsters are readily infected and can efficiently shed the virus, which made them useful in the investigation of mask mitigation strategies.[26] More recently, studies have focused on animal susceptibility to infection from humans (reverse zoonosis) and potential establishment of animal reservoirs that include peri-domestic rodents such as deer mice and bushy-tailed woodrats.[17,63,74,155]

Sendai virus infection (parainfluenza 1; *Paramyxoviridae)* is common in wild and pet rodents and is actively excluded from laboratory populations of mice, rats, and gerbils.[46] Disease manifestations occur quickly following aerosol inhalation, direct contact, or in utero exposure. The virus targets respiratory epithelium and type II pneumocytes, which may impair ciliary activity and predispose affected individuals to secondary *Mycoplasma pulmonis* and other bacterial infections. Gross lesions include dark purple discoloration, atelectasis, and consolidation of the cranioventral lung lobes and diffuse pulmonary edema. Splenomegaly and lymphadenomegaly are common and if infection occurs in a gravid female, fetal death may occur. Histologically, lesions include necrotizing rhinitis, laryngotracheitis, bronchitis, and bronchointerstitial pneumonia. In immunocompromised animals [e.g., athymic and severe combined immunodeficiency (SCID) mice], intracytoplasmic and intranuclear inclusion bodies may be present. Diagnosis is based on supportive clinical history, histologic findings, and PCR detection of the virus.

Sialodacryoadenitis is a disease specific to rats and is caused by **sialodacryoadenitis virus (SDAV**; *Coronaviridae*). The virus may be enzootic in some research colonies and in pet and free-ranging rats.[46] Virus is shed in nasal secretions and/or saliva. SDAV infects and replicates in the respiratory tract and spreads to salivary and lacrimal glands resulting in secondary lesions of the nasopharynx, respiratory tract, and eyes. Grossly, the parotid and submandibular salivary glands, Harderian glands, exorbital glands, and cervical lymph nodes are enlarged, pale, and edematous. The Harderian glands may exhibit brown pigmentation, which correlates to accumulations of porphyrin within acini. Histologically, there is acute necrotizing adenitis. In chronic cases, there is fibrosis and extensive squamous metaplasia of glandular tissue. There may be rhinitis, tracheitis, bronchitis, and bronchiolitis in severe and chronic cases. Histologic lesions are characteristic and supportive of the diagnosis.

Rabies is rare among rodents, despite their susceptibility to infection. It is likely most rodents do not survive attacks by rabid predators and due to their small size, are unlikely to be found if they do. Groundhogs in North America are the most commonly detected rabid rodent species; infection usually occurs in association with outbreaks of terrestrial rabies in raccoons.[50] Clinical signs and lesions are typical for rabies in other species.

Bacteria

In addition to several zoonotic and notable nonzoonotic bacteria that infect rodents described below, several others that are notable but not associated with disease in their rodent host are listed in the Digital Content (Table 20.e2).

Plague is a WOAH-listed, reportable zoonotic disease that is important to consider when working with rodents. Fleas spread the causative agent, ***Yersinia pestis***, between hosts. Cats and other rodent-eating predators can also contract infection through ingestion and (rarely) inhalation. Infection with this bacterium caused the Black Death pandemic, which caused the deaths of tens, if not hundreds of millions of people during the 14th century in Eurasia and Northern Africa. There are two main host types: enzootic reservoirs (voles and deer mice) and epizootic amplifiers (prairie dogs, rats, and squirrels). With close to 100% mortality, plague outbreaks in prairie dog colonies are a devastating and significant threat not only to this species but to endangered black-footed ferrets, who are susceptible to disease because prairie dogs are their primary food source.[37] There are three disease forms: bubonic (lymphadenomegaly with abscessation), pneumonic (necrotizing pneumonia with fibrinous pleuritis), and septicemic (multifocal necrosis of liver, lung, spleen, kidney, eye, brain).[60] Gross lesions include multifocal pulmonary hemorrhage (Fig. 20.15), hemorrhagic lymph nodes, erythema of axillary and inguinal skin, splenomegaly, and localized hemorrhage and necrosis in the dermis and subcutis (suspected site of flea bite). Histologic lesions of bubonic disease include necrosuppurative lymphadenitis; in the pneumonic form, necrotizing pneumonia; and in the septicemic form, multifocal abscessation. All lesions contain gram- and Giemsa-positive coccobacilli with characteristic bipolar morphology resembling a safety pin. Bacterial culture must be performed at a certified, biosecure laboratory; identification using immunohistochemistry or immunofluorescence is a beneficial diagnostic aid. A 2017 outbreak of urban pneumonic plague in Madagascar highlights the resilience of this ancient pathogen and need for continued surveillance and mitigation in reservoir populations.[151]

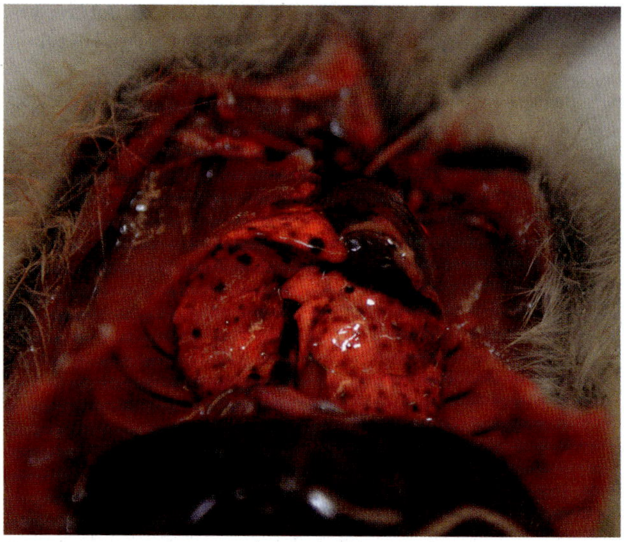

FIGURE 20.15 *Yersinia pestis* infection in the lungs of a prairie dog. Multifocal pulmonary hemorrhage is a common finding in the pneumonic form of plague. *(Photo Courtesy of K. Fox, Colorado Parks and Wildlife.)*

Yersiniosis (pseudotuberculosis) is caused by ***Y. pseudotuberculosis*** and ***Y. entercolitica***. Rodents may serve as carriers of this zoonotic bacterial agent, though some species, including mice, voles, beavers, and muskrats, may succumb to acutely fatal disease or develop more characteristic lesions and symptoms. Gross and histologic lesions include multifocal necrotizing hepatitis and splenitis and/or necrotizing and ulcerative enterocolitis (Fig. 20.16; eSlide 20.e5). Agoutis appear to be uniquely susceptible in zoo-settings (K. Terio, Personal Communication).

Tularemia caused by ***Francisella tularensis*** is an important zoonotic and WOAH-listed reportable disease that is maintained by wild rodents and lagomorphs. Human cases of tularemia have increased in recent years.[163] Disease occurs in many rodents, including mice, beavers, and muskrats, while voles appear to be subclinical carriers and serve as a reservoir.[109,120,160] Lesions in affected species are similar to those described in rabbits

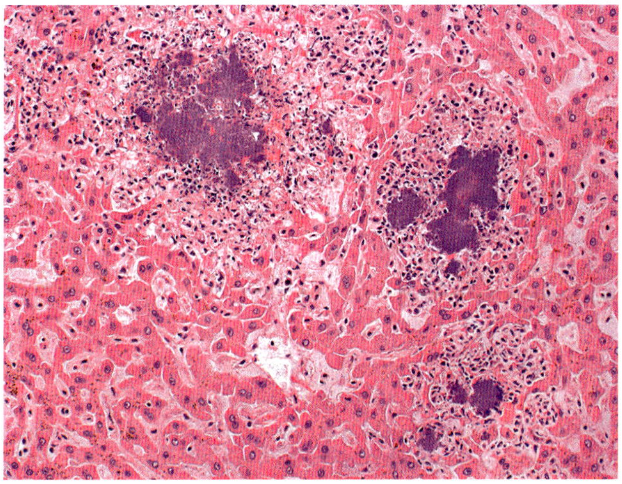

FIGURE 20.16 Yersiniosis in the liver of a free-ranging North American beaver. Bacterial infection causes multifocal, random necrosis throughout the hepatic parenchyma. Large, central colonies of the causative agent, *Yersinia pseudotuberculosis* are surrounded by mild neutrophilic inflammation and necrotic cell debris (eSlide 20.e5). *(Photo Courtesy of D. Campbell, Canadian Wildlife Health Cooperative Ontario-Nunavut.)*

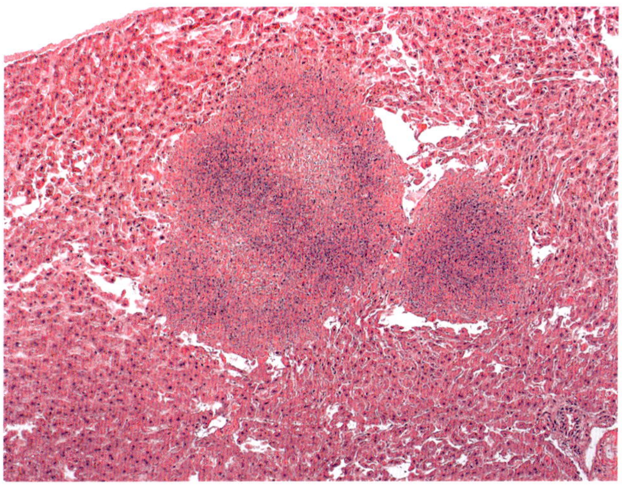

FIGURE 20.17 Tularemia due to *Francisella tularensis* in the liver of a free-ranging North American beaver. Large, random areas of coagulative necrosis are present throughout the hepatic parenchyma. Note the relative lack of inflammation. Unlike yersiniosis, bacterial colonies are not readily detected with routine hematoxylin and eosin staining (eSlide 20.e6). *(Photo Courtesy of D. Campbell, Canadian Wildlife Health Cooperative Ontario-Nunavut.)*

(see Lagomorpha). Miliary to multifocal and coalescing hepatic necrosis with variable inflammation (depending on chronicity) characterize the disease (Fig. 20.17; eSlide 20.e6). Unlike yersiniosis and similar to listeriosis, *Francisella* bacteria are not readily detected with routine hematoxylin and eosin staining but can be identified with special stains.

Listeriosis, caused by ***Listeria monocytogenes***, is a common zoonotic bacterium in rodent species. It can cause variable, yet similar lesions to those that develop in rabbits (see Lagomorpha). The systemic form is most common. During an outbreak in bushy-tailed jirds, death occurred without clinical signs.[142] Likewise, several kangaroo rats in a zoo-housed group died suddenly and had disseminated listeriosis with lesions in the liver, spleen, and brain as well as pneumonia. Diagnosis is made based on characteristic microabscesses in multiple organs including brain, spleen,

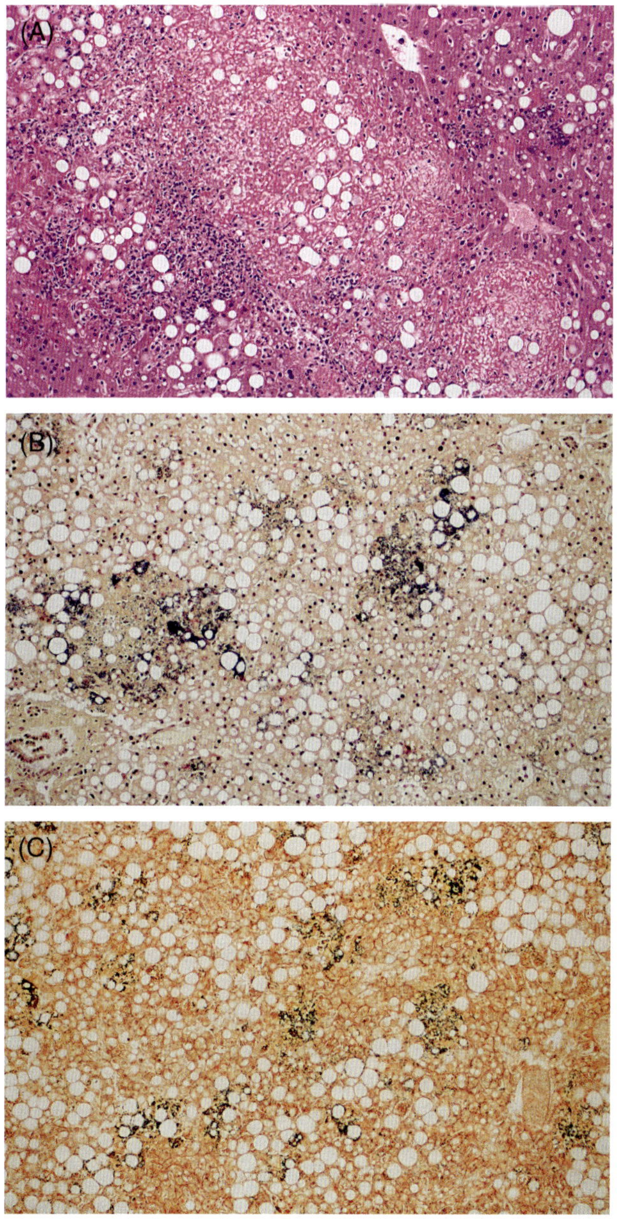

FIGURE 20.18 Listeriosis in a kangaroo rat. (A) The liver has many variably sized to coalescing and extensive regions of acute necrosis with surrounding necrotic and fewer intact neutrophils. Additionally, some hepatocytes have large cytoplasmic lipid vacuoles indicative of lipid mobilization and hepatocellular injury. (B) Gram staining highlights dense aggregates of Gram-positive small bacilli within affected regions. Gram stain. (C) Bacilli are also argyrophilic. Steiner's silver stain.

liver (Fig. 20.18A), and lung with gram-positive and argyrophilic bacilli (Figs. 20.18B, C). Immunohistochemistry, bacterial isolation, and PCR are useful to confirm infection. Though listeriosis rarely occurs naturally, guinea pigs are susceptible and used as animal models for human maternal-fetal listeriosis.

Salmonellosis is reported in guinea pigs, rats, and mice, though any rodent species can be infected and develop disease. ***Salmonella enterica*** **Enteritidis** or **Typhimurium** are the most common zoonotic serovars. Guinea pigs become septicemic and often die acutely. Other rodents may develop gastrointestinal disease that includes diarrhea and abdominal pain prior to systemic spread. In hamsters, salmonellosis must be considered and ruled out in cases of "wet tail," which is typically associated with infection by *Lawsonia intracellularis*. Epidemics of fatal salmonellosis are reported in captive maras (cavies; *Dolichotis patagonum*) with peracute deaths and characteristic gross and histologic lesions.[107] Gross lesions of salmonellosis include cyanosis of mucous membranes and polyserositis with splenomegaly. Hepatic, splenic, and lymph node necrosis are typical, as are multifocal necrosis and button ulcers in the colon with segmental intestinal infarction. Histologic lesions correlate to gross findings and include small foci of hepatocellular necrosis containing cell debris and variable numbers of macrophages, lymphocytes, and fewer neutrophils (paratyphoid nodules), Kupffer cell hyperplasia; granulomatous hepatitis, splenitis, and lymphadenitis; and fibrinonecrotic ileotyphlocolitis. Intestinal infarction can be attributed to vasculitis and thrombosis of the mesenteric or mesocolic vessels. Culture of the causative agent and histologic lesions confirm the diagnosis, which is important as this is a zoonotic pathogen.

Mycobacteriosis is a zoonotic disease that is rarely reported in rodents, though some species may serve as reservoirs. There are reports of mycobacteriosis (***Mycobacterium microti***) in British voles and wood mice,[25] hamsters, and Korean and Richardson's ground squirrels in Spain.

Chlamydophila caviae and less frequently ***C. psittaci*** cause disease in guinea pigs, primarily juveniles. Conjunctivitis, bronchitis, and pneumonia may occur and result in debilitation and death. Histologically, there is heterophilic keratoconjunctivitis and uveitis. Pneumonia has a cranioventral pattern and is heterophilic. Some animals may be asymptomatic. Diagnosis is achieved through cytology of ocular discharge and/or PCR of affected tissues. Immunohistochemistry of tissue sections is also available. This is a zoonotic bacterium and may infect humans.

Bordetella bronchiseptica causes **bordetellosis (epizootic pneumonia)**, a common respiratory disease in guinea pigs that is also reported in free-ranging squirrel populations.[134] Similar disease can occur in other rodents, particularly in the context of immunosuppression or coinfection. Grossly, there is pleuritic and cranioventral pneumonia with discrete, multifocal to coalescing areas of reddish-gray consolidation affecting multiple lobes. Mucopurulent exudate may be present in the nares, nasal passages, trachea, and tympanic bullae. There is often crusting conjunctivitis. Females may have metritis and pyosalpinx. Histologically, lung lesions consist of necrotizing and heterophilic bronchopneumonia with obliteration of the airways. Diagnosis is confirmed with bacterial culture or PCR. Other predisposing pathogens (e.g., adenovirus, *Mycoplasma* spp.) must be ruled out.

Ciliated associated respiratory (CAR) bacillus is an unclassified bacterium that colonizes the ciliated epithelium of the respiratory tract of rats, mice, rabbits, and a variety of other species. This agent acts synergistically with other respiratory pathogens including *Mycoplasma pulmonis*, Sendai virus, coronavirus, *Streptococcus pneumoniae*, *Corynebacterium kutscheri*, *Bordetella bronchiseptica*, *Klebsiella pneumoniae*, or *Pasteurella* spp. Lesions include suppurative to mucopurulent bronchopneumonia, perivascular and bronchiolar lymphoid hyperplasia, rhinitis, and lymphoplasmacytic tracheitis in free-ranging, pet and laboratory rats[6,125] (Fig. 20.19). Coinfection of CAR bacillus and mycoplasmas has been reported in rats and spinifex hopping mice with pneumonia.[96]

Murine respiratory mycoplasmosis is caused by ***Mycoplasma pulmonis***[6] Gross lesions include cranioventral bronchopneumonia with bronchiectasis and mucopurulent exudate in airways with atelectasis of remaining pulmonary parenchyma. Histologic lesions include peribronchial lymphoid cuffing with bronchus-associated lymphoid tissue (BALT) hyperplasia,

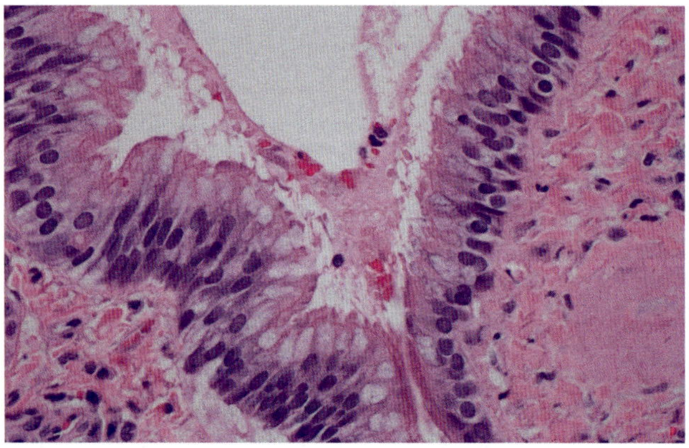

FIGURE 20.19 Cilia-associated respiratory (CAR) bacillus in the bronchus (lung) of a rat. Along the bronchial respiratory epithelium, the cilia are clumped and have adherent, intermixed faintly basophilic bacilli. The bronchial lumen contains homogenous to flocculent, clumped eosinophilic material with enmeshed erythrocytes, few sloughed cells, and aggregates of bacilli.

neutrophilic bronchopneumonia, alveolar type II pneumocyte hyperplasia, and emphysema. Bronchiectasis and peribronchial cuffing are the characteristic findings; however, histology alone cannot differentiate *M. pulmonis* and CAR bacillus respiratory infections.[125] Other lesions associated with *M. pulmonis* include endometritis, salpingitis, and peri-oophoritis. Diagnosis is based on histologic features, culture or PCR detection, and ELISA.

Pasteurellosis, caused by ***Pasteurella multocida***, affects young and immunocompromised rodents, most frequently following direct transmission from the dam. Strain virulence and species determine lesions, which in general, are like those reported in other species, including rabbits (see Lagomorpha). Prairie dogs develop upper respiratory disease and pneumonia, while conjunctivitis is more common in rats and mice. Hamsters have variable susceptibility. In laboratory rodents, *Pasteurella pneumotropica* complex is the most important group of *Pasteurella* species. This complex includes seven new bacterial species, all of which have been reclassified into a new genus *Rodentibacter*.[12] *P. pneumotropica*, now referred to as *Rodentibacter pneumotropicus*, is a common cause of pneumonia in laboratory mice and also affects certain biomedical studies as subclinical infections.[12]

Cervical lymphadenitis, a disease classically described in guinea pigs, is caused by ***Streptococcus equi*** subsp. **zooepidemicus**. Infection is associated with coarse foods that cause traumatic lesions of the oral mucosa; bite wounds may serve as another route of entry. Grossly, the cervical and submandibular lymph nodes are markedly enlarged and contain thick purulent yellow-white to red-gray exudate. Lymph nodes in severely affected and chronic cases may rupture. Histologically, affected lymph nodes are effaced by heterophilic lymphadenitis with central necrosis and peripheral fibrosis. Thoracic lesions include fibrinous pleural adhesions, fibrinosuppurative bronchopneumonia, pleuritis, and pericarditis. Chains of gram-positive cocci are present within lesions. Diagnosis is based on gross and histologic lesions and culture of causative agent.

Tyzzer disease, caused by ***Clostridium piliforme***, is a potentially fatal disease of many rodent species. It has been reported in captive gerbils, hamsters, and spinifex hopping mice.[135] Infection results in a triad of gross lesions including icterus and hepatomegaly with military gray foci, congestion, and edema of the intestine, and lymphadenomegaly with edema and hemorrhage. Sudden death without clinical disease or lesions is possible in peracute cases. Histologically, multifocal random to coalescing hepatic necrosis is surrounded by hemorrhage, heterophils, and macrophages (Fig. 20.20A). At the necrotic margins, hepatocytes may accumulate characteristic criss-crossed bundles of faintly staining intracytoplasmic bacilli that are best visualized with silver stains (Fig. 20.20B).

Proliferative ileitis (wet tail) due to ***Lawsonia intracellularis*** is most commonly reported in hamsters. Similar to this infection in other species, there is segmental thickening of the ileum with serosal nodules. Histologically, there is marked crypt hyperplasia and herniation and villous elongation, hyperplasia, and fusion with variable necrosis and hemorrhage. Silver stains and PCR are useful to confirm infection.

Fungi

A diverse array of ***Pneumocystis*** spp. are carried without clinical signs by a variety of wild rodents; these organisms

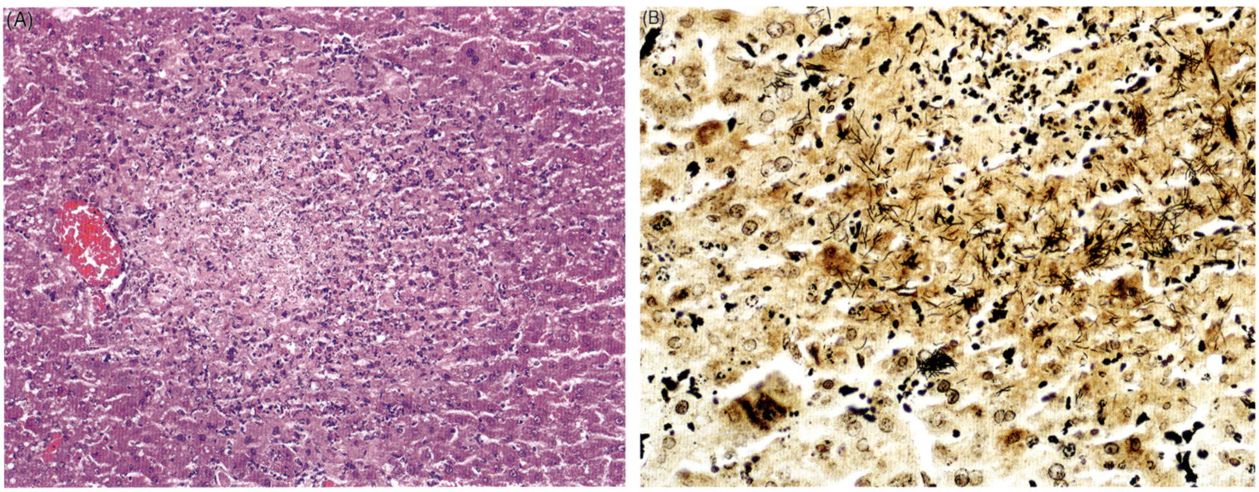

FIGURE 20.20 Tyzzer disease due to *Clostridium piliforme* in the liver of a gerbil. (A) Hepatic parenchyma has multifocal, random necrosis with low numbers of faintly basophilic intracellular bacilli. (B) Individual and clusters of filamentous bacteria stain brown-black. Steiner's silver stain. *(Photo Courtesy of J. Landolfi, University of Illinois Zoological Pathology Program.)*

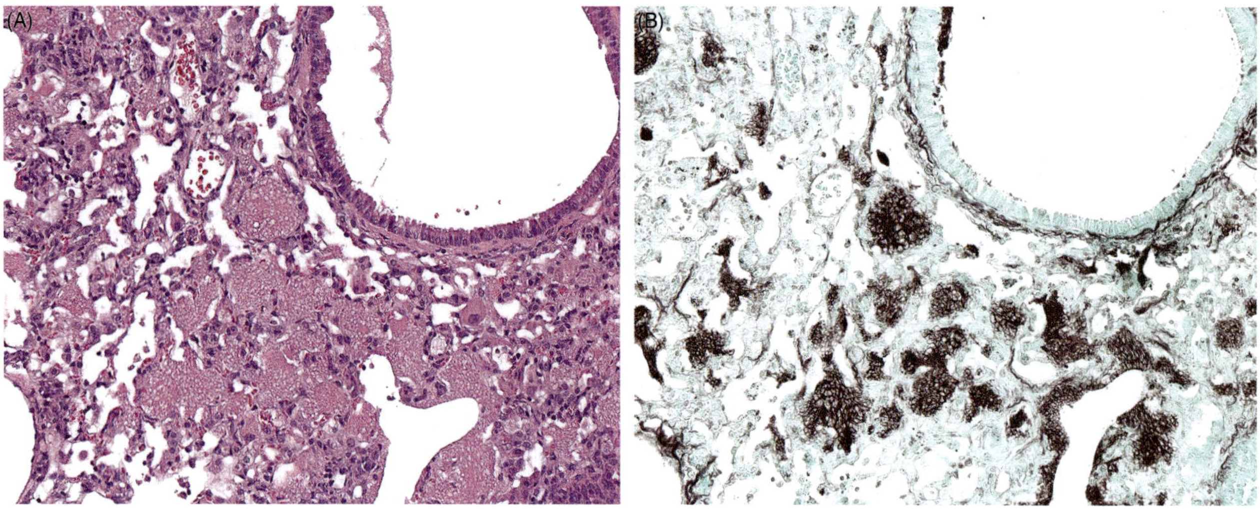

FIGURE 20.21 *Pneumocystis* spp. in the lungs of an immunocompromised mouse. (A) Alveoli are filled with pale eosinophilic material that appears foamy due to infection with this small, yeast-like fungus. Infection is associated with chronic pneumonitis. (B) Silver stains are useful to highlight intralesional and intracellular pneumocystis cysts. GMS (see eSlides 20.e7 and 20.e8). *(Parts A, B: Photos Courtesy of P. Treuting, Seattle Genetics.)*

are thought to have co-evolved with their hosts and are thus highly host-specific. *Pneumocystis* spp. appear as small, 1–8 µm diameter (size is life-stage dependent), eosinophilic round organisms with a small central body. In severe cases, organisms within alveolar spaces and alveolar macrophages can obliterate alveoli; in some cases, they can be associated with interstitial pneumonia (Fig. 20.21A; eSlide 20.e7). Silver stains like Gomori's methenamine silver (GMS) are useful to highlight yeasts (Fig. 20.21B; eSlide 20.e8). This fungus can cause significant opportunistic infections in immunocompromised individuals.

Cryptococcosis (***Cryptococcus neoformans***) can be a significant pathogen in some rodent species. For example, approximately 20% of slender-tailed cloud rats at one zoological institution died with or as a result of cryptococcal pneumonia over a 15-year period.[14] These large, arboreal rodents were likely infected by contaminated substrates and may be a highly susceptible species, similar to some species of nonhuman primates and felids.[141] Lesions included none or mild to severe, chronic granulomatous pneumonia (Figs. 20.22A–D; eSlides 20.e9–20.e11) with or without necrosis and regional lymphadenitis. Disseminated disease can produce subcutaneous nodules (cryptococcoma) and meningoencephalitis. Microscopically, characteristic fungal yeasts with narrow-based budding are present in cytologic preparations and histologic sections. GMS staining can enhance detection of fungal yeasts, which are 5–20 µm diameter and have with a distinct, thick (5–10 µm), nonstaining mucopolysaccharide capsule, the latter of which is best visualized using the PAS reaction and mucicarmine stains (Figs. 20.22B–D; eSlides 20.e10 and 20.e11). Cytology, histology, and special stains can confirm infection; culture and PCR can confirm the fungal species.

Adiaspiromycosis is a systemic fungal disease reported in several captive and free-ranging rodent species. It is caused by the saprophytic fungi *Chrysosporium parvum* (syn. *Emmonsia parva*) and *C. crescens*. Once inhaled, these fungi cause pulmonary granulomas with fungal spherules and conidia. Additional diagnostic tests may be useful, including immunodiffusion and complement fixation. These fungi are zoonotic.

Other systemic mycoses of rodents include infections caused by the endemic (and zoonotic) fungi ***Blastomyces dermatitidis*** and ***Histoplasma capsulatum***. Reported cases include juvenile mice and chinchillas. Immunity to fungal pathogens appears limited in mice.

Dermatophytosis is a cutaneous fungal disease that can be found in any rodent species and is similar to that of other mammalian species. Causative agents include ***Trichophyton mentagrophytes***, ***Microsporum canis***, ***M. gypseum***, and ***Epidermophyton***. Of note, these fungi are zoonotic and can cause skin lesions in other animals. Additionally, ***Sporothrix schenckii***-associated cutaneous disease is reported in several rodent species and can be zoonotic. Dermatitis lesions vary from suppurative and necrotizing to granulomatous and fibrosing depending on chronicity. PAS-positive yeasts are present within macrophages. Characteristic organisms on cytologic preparations of skin scrapes prepared with potassium hydroxide support the diagnosis. During an 8-year period, over 25% of North American porcupines (*Erethizon dorsatum*) examined from states of New England, United States, were severely affected by dermatophytosis, resulting in widespread skin ulcerations, debilitation, and death.[108]

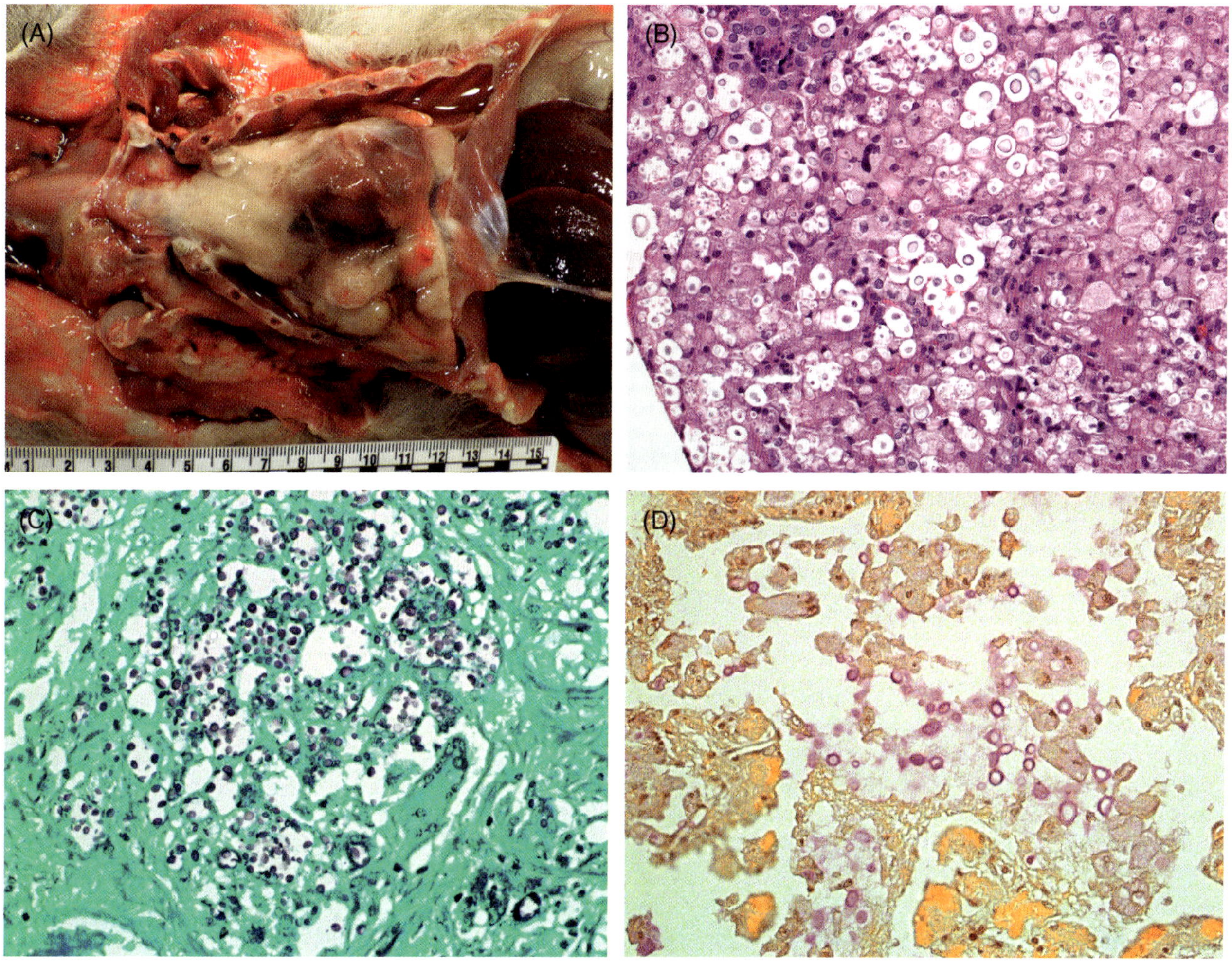

FIGURE 20.22 Cryptococcosis in the lungs of a slender-tailed cloud rat. (A) The lungs and pleura are multifocally obscured by abundant tan material. (B) The pulmonary interstitium is moderately to markedly expanded and distorted by myriad fungal yeasts and associated histiocytic to granulomatous inflammation. Lymphoplasmacytic inflammation is uncommon. Yeast vary in size, have narrow-based budding, and are surrounded by a thick mucopolysaccharide capsule that does not stain with hematoxylin and eosin (B) or silver stains (C; GMS) but can be highlighted with mucicarmine (D) (see eSlides 20.e9–20.e11). *(Part A: Photo Courtesy of K. Conley, Wildlife Conservation Society; Parts B–D: Photos Courtesy of D. McAloose, Wildlife Conservation Society.)*

Metazoa

Many species of free-ranging rodents are susceptible to fatal neural larval migrans associated with the raccoon nematode ***Baylisascaris procyonis*** (Fig. 20.23; eSlide 20.e12). Significant clinical disease is often related to parasite migration in the brain, where they incite encephalitis that is characterized by low numbers of eosinophils, gliosis, spheroids, and rarefaction or necrosis of the neuropil. Encephalitis attributed to larval migrans is a frequent diagnosis among rodents submitted to diagnostic laboratories, although larvae may not be visible in all cases.[56]

Calodium hepaticum (syn. ***Capillaria hepatica***) affects a broad host range worldwide, including rodents and humans. It is a ubiquitous parasite of globally invasive Norway and black rat populations.[124] Gross lesions consist of multifocal to coalescing, tortuous, white tracks in the liver parenchyma (Fig. 20.24A). Histologically, the hepatic parenchyma is replaced by nodular and or serpiginous tracts of necrosis, lymphoplasmacytic and granulomatous inflammation, and fibrosis with intralesional adults and bioperculate eggs (Fig. 20.24B; eSlide 20.e13). Eggs are approximately $40 \times 60\,\mu m$ with a 10-μm thick bioperculated shell and contain either an eosinophilic morula or granular material. Intralesional eggs are generally more common than adult nematodes, the latter having characteristic features including stichosome and hypodermal (bacillary) bands. Eggs are only released from the liver following carcass decomposition or ingestion by a predator/cannibalistic conspecific. Therefore, diagnosis is based on the presence of the nematode and/or eggs in the liver and not fecal parasitology.

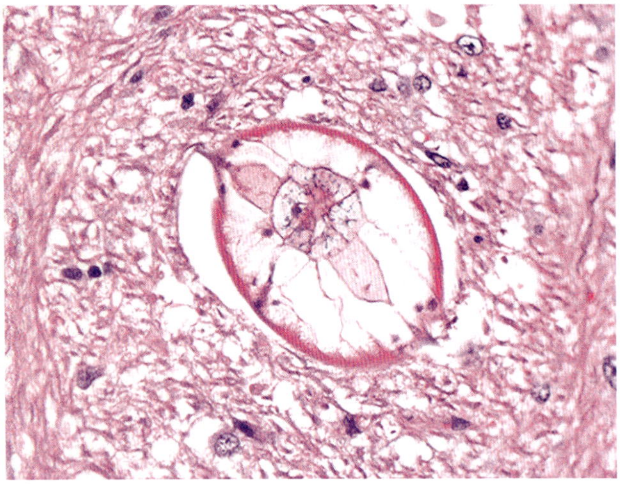

FIGURE 20.23 Visceral larval migrans due to *Baylisascaris procyonis* in the brain of a free-ranging groundhog. The neuropil contains a larval nematode with prominent lateral alae and lateral cords, a thick cuticle, and coelomyarian musculature (see eSlide 20.e12). *(Photo Courtesy of J. Landolfi, University of Illinois Zoological Pathology Program.)*

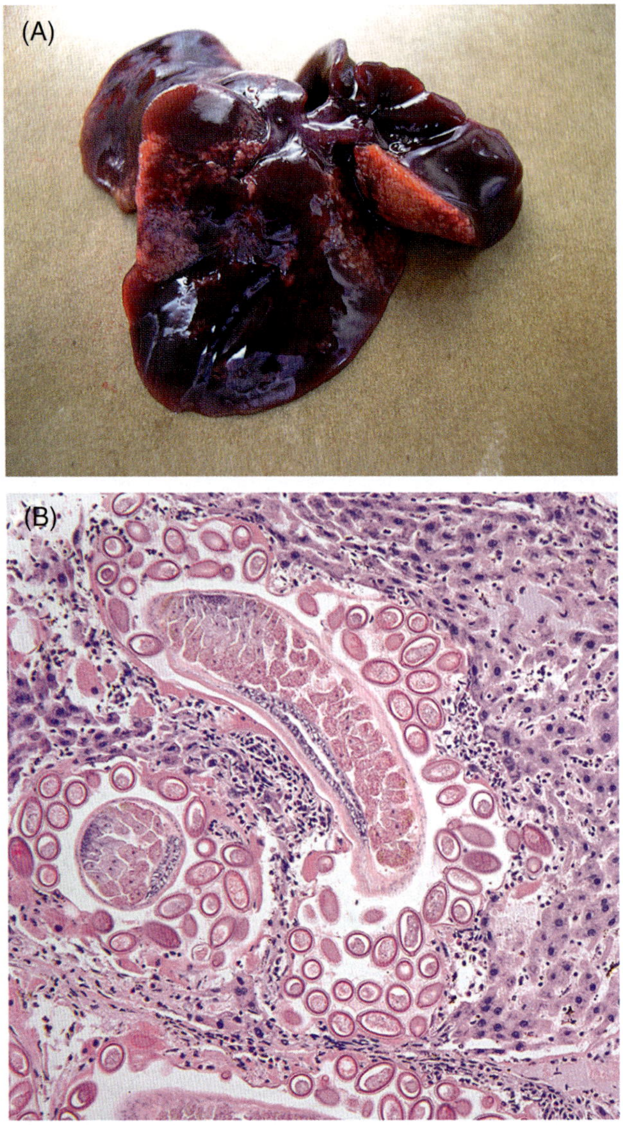

FIGURE 20.24 *Calodium hepaticum* (*Capillaria hepatica*) in the liver of a free-ranging Norway rat. (A) Multiple liver lobes contain extensive tortuous, white nematode tracks. (B) Adult nematodes and eggs compress, distort, and replace hepatic parenchyma. Eggs are characteristically bioperculate. Inflammation is lymphoplasmacytic to granulomatous and may be associated with fibrosis in chronic infections (see eSlide 20.e13).

Free-ranging rodents are the intermediate hosts for ***Echinococcus multilocularis***, a zoonotic tapeworm for which carnivores such as foxes, wolves, coyotes, and domestic dogs are common definitive hosts. Although the parasite is present throughout the northern hemisphere, it is considered an emerging pathogen in North America and has been identified with increasing frequency in several Canadian provinces. Intermediate host species vary by geographical location but are mainly mice and vole species. Rodents acquire infection by consuming eggs, which then develop into the metacestode stage. Gross lesions consist of multifocal to coalescing cystic structures that contain yellow flocculent material. The liver is most often affected but cysts can also spread throughout the abdomen (Fig. 20.25A). Histologically, parasitic cysts are lined by a thick eosinophilic outer membrane (highlighted by PAS reaction) and thin, inner basophilic germinal epithelial layer that contains calcareous corpuscles (Fig. 20.25B). Intact or ruptured brood capsules with protoscoleces may be present. Protoscoleces are approximately 100 μm in diameter with an inner sucker and rostellum that features multiple birefringent hooks. The presence of protoscoleces is indicative of a fertile infection, meaning it is a suitable intermediate host to infect the definitive host. Cysts may be accompanied by variable tissue necrosis, fibrosis, and inflammation.[5,149]

Ectoparasites

Rodents, particularly squirrels, are prone to **notoedric mange**.[4,32,136] Mites burrow into the stratum corneum inducing marked hyperkeratosis, which results in patchy alopecia, and crusting and thickening of the skin (lichenification), which can appear yellow to gray. Gross lesions are predominantly found on the ears, nose, tail, external genitalia, inguinal and perianal regions, and feet though they can cover the entire body in severe cases. Peripheral lymphadenopathy may also be seen. Skin scrapings and fecal floats are useful in examining mite features and the presence of eggs, respectively. Histologically, there is marked epidermal hyperplasia with parakeratotic

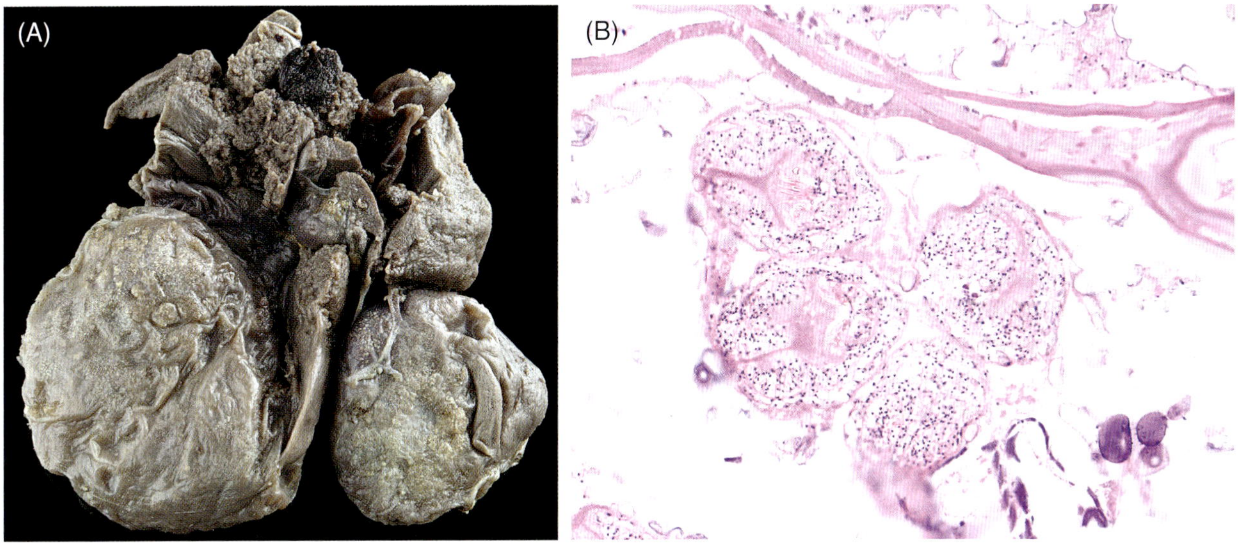

FIGURE 20.25 *Echinococcus multilocularis* in a free-ranging muskrat. (A) The liver (formalin-fixed) has multifocal to coalescing cystic structures that contain yellow flocculent material. (B) Within cysts, protoscoleces are approximately 100 μm diameter with an inner sucker and rostellum that features multiple birefringent hooks. *(Part A: "Pathology of Wildlife and Zoo Animals Second Edition," fig.1a from Trudeau MJ, Goldsmith D, Letain C, Stamler S, Rothenburger JL. Muskrats (Ondatra zibethicus) are competent intermediate hosts of* Echinococcus multilocularis *in North America. J Wildl Dis. 2023;59(4):684–693.)*

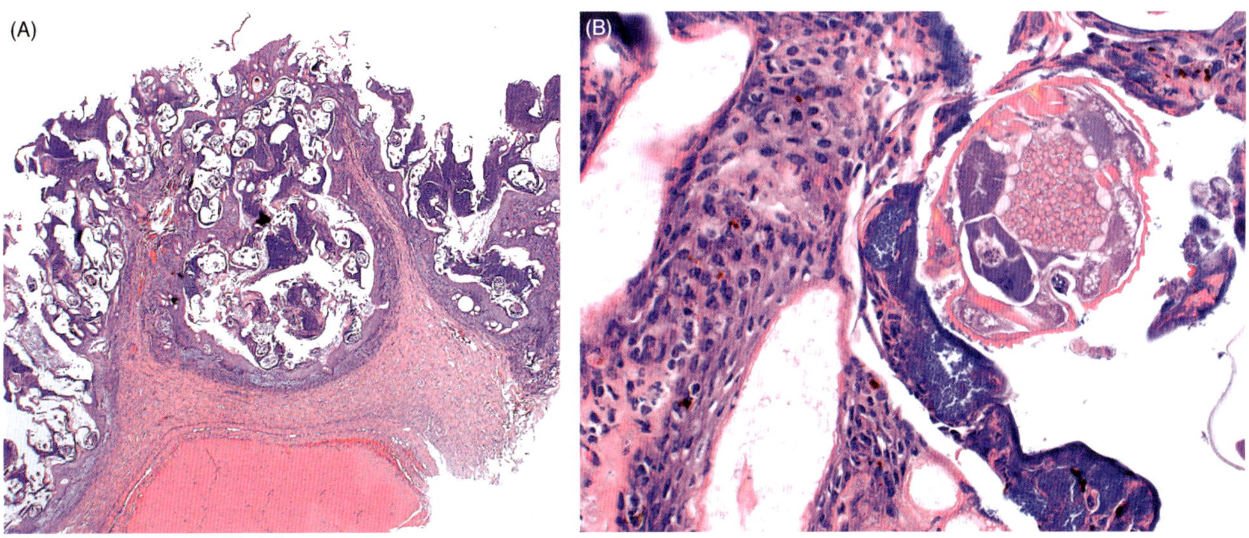

FIGURE 20.26 Notoedric mange in a free-ranging gray squirrel. (A) Severe hyperkeratosis and epidermal hyperplasia contain multiple large mite tunnels. (B) Cross-section of a mite adjacent to colonies of cocci, layers of parakeratotic hyperkeratosis, and neutrophilic cellular debris. *(Parts A, B: Photos Courtesy of D. Campbell, Canadian Wildlife Health Cooperative Ontario-Nunavut.)*

hyperkeratosis forming caps over tunnels containing mites with superficial, perivascular eosinophilic dermatitis (Figs. 20.26A, B). Mites are approximately 250 μm diameter with typical arthropod features: chitinized exoskeleton, striated muscles, and jointed appendages. Specific features, which may not be seen in histologic sections, include a dorsally located anus and appendage extensions with suckers. Secondary bacterial infections are common. Additional notable parasites of rodents are listed in the Digital Content (Table 20.e3).

ADDITIONAL RESOURCES

Barthold. S.W., Griffey, S.M., Percy, D.H. Pathology of Laboratory Rodents and Rabbits. 4th Edition. West Sussex, England: Wiley Blackwell. 2016. pp. 384.

Harkness, J.E., Turner, P.V., VandeWoude, S.V., Colette L. Wheler, C.L. Harkness and Wagner's Biology and Medicine of Rabbits and Rodents. 5th Edition. Ames, Iowa: Wiley-Blackwell. 2010. pp. 480.

Miller, E., Lamberski, N., Calle, P. Fowler' s Zoo and Wild Animal Medicine, Current Therapy. Volume 10. W.B. Saunders. 2023. pp 832.

Quesenberry, K., Mans, C., Orcutt, C., Carpenter, J.W. Ferrets, Rabbits, and Rodents Clinical Medicine and Surgery. 4th Edition. St. Louis, MO: Elsevier. 2020. pp. 656.

Suckow, M.A., Stevens, K.A., Wilson, R.P. The Laboratory Rabbit, Guinea Pig, Hamster, and Other Rodents, 1st Edition. San Diego, CA: Academic Press, an Imprint of Elsevier. 2012. pp. 1288.

Treuting, P.M., Dintzis, S.M., Montine, K.S. Comparative Anatomy and Histology: A Mouse, Rat, and Human Atlas. 2nd Edition. St. Louis: Academic Press, and imprint of Elsevier. 2018. pp. 570.

References

Full references for this chapter can be found online at https://doi.org/10.1016/B978-0-443-18437-6.00025-0.

Chapter 21

Xenarthra, Erinacoemorpha, Some Afrotheria, and Pholidota

Dalen Agnew[a], Sally Nofs[b], Martha A. Delaney[c], Jamie L. Rothenburger[d], and José Luiz Catão-Dias[e]

[a]*Michigan State University, Lansing, MI, United States;* [b]*Menzies Institute for Medical Research, University of Tasmania, Hobart, TAS, Australia;* [c]*Zoological Pathology Program, College of Veterinary Medicine, University of Illinois, Brookfield, IL, United States;* [d]*Faculty of Veterinary Medicine, University of Calgary, Calgary, AB, Canada;* [e]*Department of Pathology, School of Veterinary Medicine and Animal Sciences, University of São Paulo, São Paulo, Brazil*

INTRODUCTION

This chapter covers some of the smaller and more unique orders kept in zoos, as pets, and that are managed as wildlife. Though generally poorly understood and researched relative to other species, these species are increasingly common in zoos due to their unique anatomy and behavior. In the wild, they often play an important role in insect control and soil aeration and disruption. Animals discussed in this chapter include those in the following Orders: Xenarthra (armadillos, sloths, and anteaters), Erinacoemorpha (includes moonrats and hedgehogs; *Atelerix* sp., *Erinaceus* sp., and others), Tubulidentata (aardvarks), Hyracoidea (hyraxes), Afrosoricida (tenrec and golden moles), Macroscelidea (elephant shrews), Scandentia (tree shrews), and Pholidota (pangolins). Among these, pangolins are among the most trafficked wildlife species, which is causing enormous declines in wild populations and captive breeding insurance programs have been started for some species.

NON-INFECTIOUS DISEASES

Nutritional

Cardiomyopathies have been reported in multiple species in this chapter, most notably anteaters, hedgehogs, and rock hyrax. The pathogenesis in these species has not been fully elucidated. In captive anteaters, the dilated form of cardiomyopathy has been reported (Fig. 21.1A), but the true prevalence may not be as common as reports suggest.[23] Cardiomyocytes are thin, individualized, and separated by edema or connective tissue, with some wavy fibers. Fibrosis may be present. Nutritional factors have been suggested,[4,23,82] and some preliminary data suggests the condition is associated with taurine deficiency as lesions in anteaters are similar to the dilated cardiomyopathy that is seen in domestic animals and associated with taurine deficiency.[110] In hedgehogs, cardiomyopathy incidence may be high (up to 38% of mortality) and males appear to be overrepresented.[92,94] Subjectively determined cardiomegaly, globoid distortion of the heart, mild left ventricular dilation, white myocardial streaking, hepatomegaly, pulmonary edema and congestion, and hydrothorax are the most commonly reported gross lesions, though some cases have no grossly evident abnormalities. Histologically, there can be myocardial fibrosis and edema, as well as left ventricular myocardial degeneration, necrosis, atrophy, hypertrophy, and myofiber disarray. Acute and chronic passive congestion of the lungs and liver are also seen. A high incidence of cardiomyopathy, primarily dilative, is also reported in rock hyrax.[59] This is not associated with any specific cause or predilection of sex or age, though little specific research has been done. Thickened, hypertrophic distorted, and vacuolated myofibers with pleomorphic and karyomegalic nuclei are found in the ventricular walls, along with multifocal myocardial fibrosis, edema, and myofiber necrosis (Fig. 21.1B; eSlide 21.e1). Mild inflammation composed of neutrophils and macrophages is sometimes present in damaged areas of the heart. Affected individuals may have pleural and/or pericardial effusion as well as pulmonary and multiorgan congestion suggestive of cardiac insufficiency.

Iron overload is well-described in hyraxes, anteaters, and sloths but not in other species covered in this chapter. These diseases are presumably diet-related, though genetic influences are likely also involved.[6,8,22] Concurrent diseases including parasitism and infections may play a role.[77] In affected hyrax, the liver is diffusely copper-colored and may be enlarged. If there is progression to hemochromatosis, evidence of chronic degeneration with fibrosis, biliary hyperplasia, and nodular hyperplasia (cirrhosis) is seen. Multicentric diseases including gastric hemorrhage and

Pathology of Wildlife and Zoo Animals. https://doi.org/10.1016/B978-0-443-18437-6.00026-2

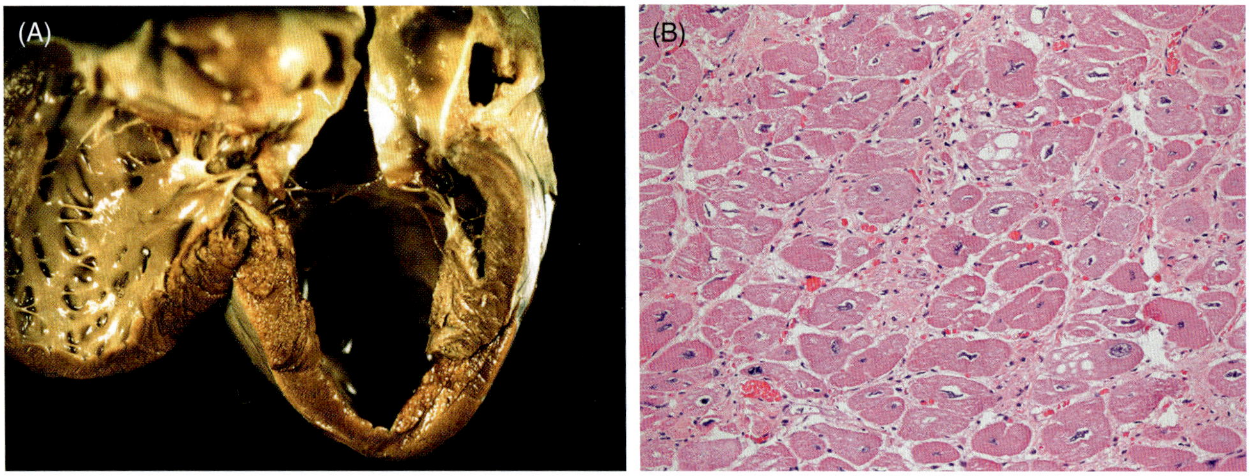

FIGURE 21.1 Dilated cardiomyopathy in a giant anteater (A) and rock hyrax (B). (A) The myocardium of both ventricles is thin and flaccid and the lumen is markedly dilated. (B) Myofibers have variable degeneration, fragmentation, vacuolation, and loss. Fibers are separated by interstitial fibrosis and edema. Nuclei are pleomorphic and karyomegalic (see eSlide 21.e1). *(Photo A: Courtesy of M. Garner, Northwest ZooPath.)*

hemorrhagic enteritis may accompany liver disease. Microscopically, hepatocytes contain increased cytoplasmic iron, with the most abundant accumulations occurring in periportal hepatocytes (Fig. 21.2). In hemochromatosis, iron pigments are associated with variably severe portal fibrosis, hepatocellular vacuolar change, and degeneration. Less frequently, iron accumulates in intestinal mucosal epithelium, macrophages (small intestine lamina propria, lymph nodes, spleen), and adrenal cortical cells.[38] Prussian blue staining of affected tissues is useful in detection and assessing the degree of iron accumulation. Mineral analysis of both fresh and formalin fixed paraffin embedded tissue can be performed to document hepatic iron levels.

Pancreatic islet fibrosis has been described in hyrax from different facilities and lineages, although many of the affected animals have been related.[40,41] Some, but not all, animals present with diabetes mellitus and hyperglycemia. An enlarged, firm pancreas may be identified on gross examination. Fibrosis of the pancreatic islets is the primary histologic feature. Increases in the overall size of islets may include hypercellularity and increased numbers of alpha, beta, and delta cells identified by immunohistochemistry. Fibrosis encircles the islets and extends into the adjacent parenchyma. In severe cases, fibrosis disrupts or effaces the islets and endocrine cells are separated and scattered within the fibrous tissue (Fig. 21.3; eSlide 21.e2). In a large percentage of these cases, islet cells have vacuolated cytoplasm or are necrotic. Pancreatic duct proliferation is also commonly observed near affected islets. Amyloid is

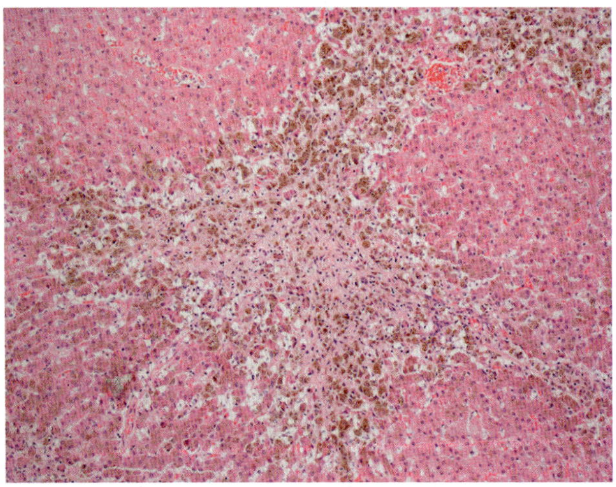

FIGURE 21.2 Hepatic hemochromatosis in a rock hyrax. Iron pigment, which appears light to dark brown with hematoxylin and eosin staining, is present within hepatocytes, most abundantly in periportal hepatocytes. It is associated with portal fibrosis and hepatocellular vacuolar change.

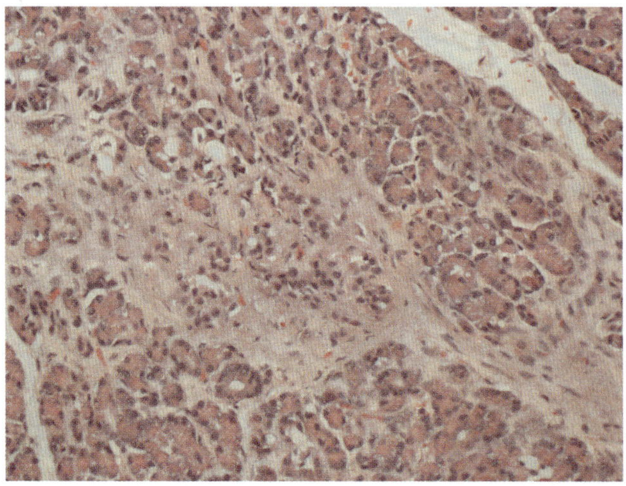

FIGURE 21.3 Pancreatic islet fibrosis in a rock hyrax. Fibrosis is associated with hypercellularity and increased numbers of endocrine cells. In more severe cases of pancreatic islet fibrosis, fibrosis disrupts or effaces the islets, with endocrine cells scattered among the fibrous tissue (see eSlide 21.e2). *(Photo Courtesy of M. Garner, Northwest ZooPath.)*

not a feature of pancreatic islet fibrosis. Histologic examination of the pancreas is important for confirmation, though clinical data including hyperglycemia and elevated creatine phosphokinase, amylase, and lipase are strongly suggestive.[40] Low-fiber diets have been associated with this lesion, but the pathogenesis is uncertain.

Vertebral hyperostosis has been reported in captive collared (lesser) and northern tamandua. The disease is characterized clinically by paresis, bladder distention, reduced mobility, and rigid kyphosis that cannot be reduced during anesthesia.[24] Massive hyperostosis up to 1 cm in depth and limited to the ventral and lateral aspects of the thoracic, lumbar, and cranial coccygeal vertebra is evident radiographically. Coalescing osteophytes lead to fusion of the entire vertebral column (Fig. 21.4). Mineralization of the tracheobronchial tree may also be visible. The disease is progressive. Gross lesions reflect the radiographic findings and histologically the proliferative lesions are composed of irregular cartilage, fibrocartilage, collagen, and poorly organized woven bone with arrest lines and interspersed active bone marrow. Variable mineralization of the myocardium, splenic capsule, tunica media of the aorta and other arteries, and renal medullary tubular basement membranes are also present. It is associated with membranoproliferative glomerulonephritis in some animals, but the relationship is unclear. Plasma calcium, phosphorus, and vitamin D levels appear to be within normal limits (normal reference ranges for this species have not been determined). Liver vitamin A levels are considered higher than most domestic species and reduction in the rate of lesion development appears to be associated with reductions in vitamin A and vitamin D levels in the diet, suggesting that these may be part of the pathogenesis.

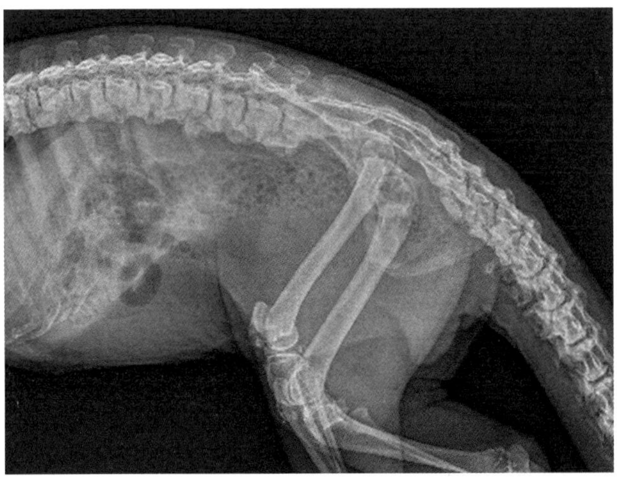

FIGURE 21.4 Hyperostosis in a tamandua. Radiographs show bony proliferation (up to 1 cm in thick), limited to the ventral and lateral aspects of the thoracic, lumbar, and cranial coccygeal vertebra. Osteophyte proliferation can fuse the entire vertebral column. *(Photo Courtesy of M. Tocidlowski, Houston Zoo.)*

Hypovitaminosis K is anecdotally reported in anteaters and armadillos and is associated with spontaneous hemorrhage at mucocutaneous junctions and extremities.[4] Hematuria is also reported. Insectivore species that rely exclusively on termites and ants in the wild may be particularly susceptible to deficiency in captivity, where commercially produced insects and diets may lack natural components rich in vitamin K. Care should be taken to include supplemental vitamin K in all diets. In addition, anteaters and armadillos on antibiotic therapy may be particularly at risk.

In the pet trade and in captive/managed facilities, **obesity** can be a significant concern. The low metabolic rate, state of torpor seen in some species (e.g., hedgehogs), and the lack of good dietary data on many species contributes to this problem. In contrast, the specific diet of pangolins has made sourcing food items in captivity problematic.

Metabolic

Addison's disease (hypoadrenocorticism) has been described in Hoffmann's two-toed sloth[57] and a giant anteater (S. Nofs, unpublished data). Clinical signs include respiratory distress, dehydration, dysphagia, and hypotension. Clinical pathology findings include decreased serum sodium:potassium ratio, low baseline cortisol, and decreased ACTH and reduced aldosterone in stimulation tests (as compared to a cohoused healthy male Hoffmann's two-toed sloth). Histopathology confirmed marked adrenal cortical atrophy. Notably, intranuclear inclusions were described in hepatocytes and mucosal epithelial cells of the oropharynx, esophagus, and stomach; high throughput viral sequencing and immunohistochemistry for selected viral agents did not identify an etiology.

As in other species, energy imbalance resulting in rapid mobilization of fat from adipose stores can result in **pregnancy toxemia** in captive rock hyrax, typically in older and multiparous females. Grossly, atrophy of fat and hepatic lipidosis are common. Microscopically there is hepatocellular degeneration and lipidosis with variable necrosis depending on severity and chronicity of disease; peracute disease can result in fulminant hepatic necrosis. Confirmation of pregnancy toxemia is based on clinical history and signs in combination with supportive lesions.

A condition termed **hyracoidea stress syndrome** was identified in hyrax in Europe.[108] Animals that were exposed to environmental stressors, such as excessively cold temperatures, developed pneumonia, gastric ulcerations, and degeneration and necrosis of the pituitary, thyroid, and adrenal glands.

Toxic

In a search of a potential form of pest management to protect fruit orchards from rock hyraxes, **brodifacoum**-laced apple bait has been fed in an experimental setting to hyraxes.[73] Brodifacoum caused death as early as 6 days and all treated animals were dead by 15 days following consumption of 7–8 mg/kg body weight. Hemorrhage from the gums and nares were the noted gross lesions. Early embryonic deaths occurred in pregnant females. Additional studies using second-generation anticoagulants have also been performed and produced lower mortality rates.[74]

Fluorosis has been reported in a single hyrax-fed guinea pig chow, suggesting that this species may be sensitive to dietary fluoride (D. Agnew, unpublished data). The hyrax had grossly apparent thickening of appendicular bones (Fig. 21.5A) and red-brown discoloration of the bones and teeth. Periosteal new bone formation, enlargement of the medullary cavity, and replacement with coarse woven cancellous bone were seen histologically (Fig. 21.5B). Bone fluoride levels were 2,982 µg/g, which is in the toxic range for domestic ruminants (3,000–5,000 µg/g).[90]

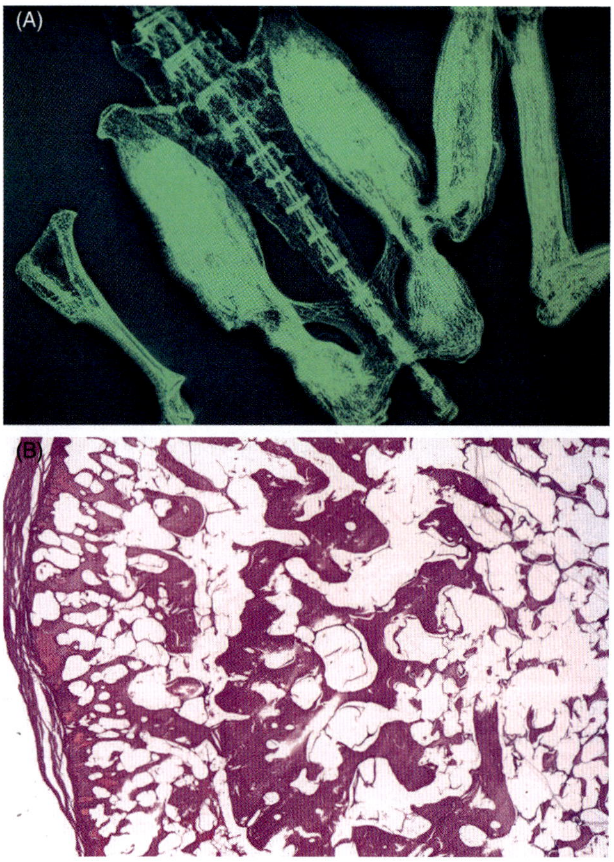

FIGURE 21.5 Fluorosis in a rock hyrax. (A) Intoxication is associated with hyperostosis of the appendicular bones and pelvis (note a normal control ilium is in the lower left). Xeroradiograph (B) A transverse section of the femur shows marked periosteal new bone formation, enlargement of the medullary cavity, and replacement with coarse woven cancellous bone.

Age-Related/Degenerative

A condition called "**Wobbly hedgehog syndrome**" (WHS) occurs in African and European hedgehogs. It is associated with progressive neurologic signs including ataxia and paraparesis to tetraparesis.[43,84] The onset of WHS commonly develops in animals under 2 years of age, and though they may survive for some time with supportive care, it is invariably fatal. Besides clinical neurologic signs, gross evidence of pronounced neurogenic muscle atrophy is also present. Histologically, there is extensive vacuolization of the white matter in the cerebrum, cerebellum, brain stem (Fig. 21.6A; eSlide 21.e3), and throughout the length of the spinal cord. Myelin likely degenerates first and is followed by secondary degeneration and loss of the axons with subsequent neuronal degeneration and necrosis. The corona radiata of the cerebrum is often the most pronounced area of demyelination (Fig. 21.6B). Degeneration of lower motor neurons in the ventral horns of the spinal cord and ventral rootlets is also apparent. Degeneration is not seen in the dorsal horn rootlets, spinal ganglia, or peripheral ganglia, and peripheral nerves are unaffected. Inflammation beyond mild astrocytosis is not a feature in African hedgehogs; mild meningitis and perivascular lymphocytic cuffing are reported as possible features of the early stages of the disease in European hedgehogs. Neurogenic atrophy of myocytes (angular, eosinophilic, and shrunken myocytes) without inflammation is seen in affected skeletal muscles. The cause is still under investigation, but proposed etiologies include nutritional, genetic, infectious, and neoplastic factors.[65,76] In African hedgehogs, there tends to be a strong familial correlation suggesting a hereditary component to the disease.[43,65]

Intervertebral disc disease in African hedgehogs must be differentiated from wobbly hedgehog syndrome.[91] Typical signs include lameness progressing to loss of proprioception and ataxia in the hindlimbs, as well as urinary tract dysfunction. Narrowed intervertebral cervical and lumbar disc spaces and spondylosis are noted radiographically. Histologically, there is neovascularization, fissures, mineralization, and dorsal extrusion of the annulus fibrosis and nucleus pulposus into the spinal canal. Spinal cord compression or traumatic damage due to disc herniation is associated with swollen axons (spheroids), edema within and around axons, and digestion chambers, particularly in lateral and ventral funiculi. Fibrocartilagenous emboli within venous sinuses may also be present.

Chronic renal disease is a common postmortem finding in sloths, armadillos, and to a lesser degree, anteaters.[8,71,104] Chronic tubulointerstitial nephritis with

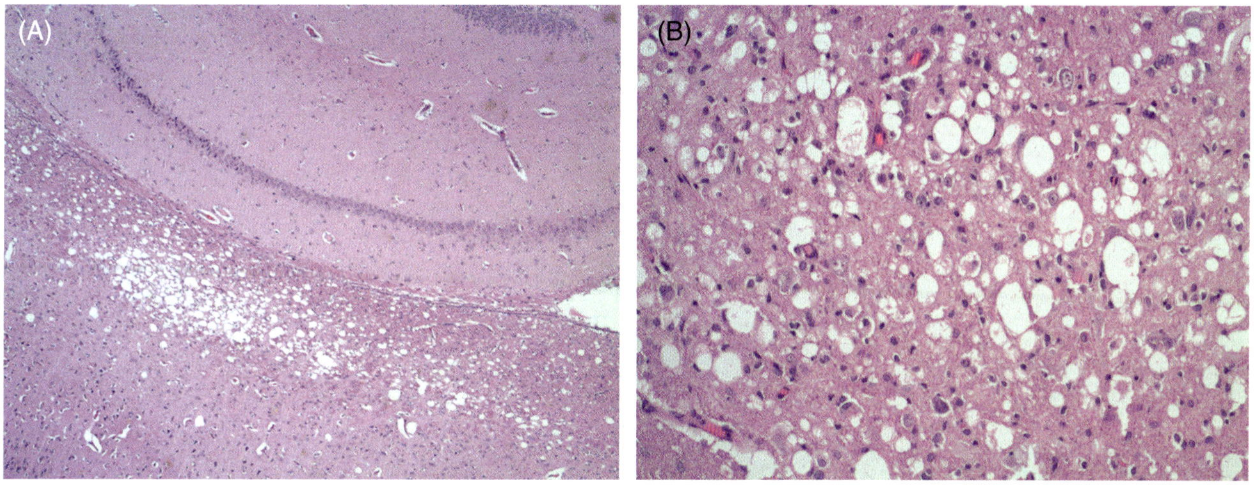

FIGURE 21.6 Wobbly hedgehog syndrome (WHS) in the brain of a hedgehog. (A) Histologically, extensive vacuolation of the white matter is seen in the cerebrum. The corona radiata of the cerebrum is often the most pronounced area of demyelination. Similar lesions can be found in the cerebellum, brain stem, and throughout the length of the spinal cord. (B) Myelin degenerates first, then secondary degeneration and loss of the axons occur with subsequent neuronal degeneration and necrosis characterized by prominent vacuolation and associated glial proliferation and collapse of parenchyma (see eSlide 21.e3).

lymphoplasmacytic inflammation, fibrosis, and variable degrees of membranoproliferative glomerulosclerosis are frequently cited as causes of death or as a complicating factor to some other cause. Nephrotoxic drugs, excessive dietary protein levels, and/or dehydration are likely causative factors, though infectious diseases have not been adequately investigated in these species. In a study of wild armadillos, leptospirosis was specifically ruled out.[104] Renal and urinary bladder **calculi** are frequently reported in sloths, though the specific composition varies. How these calculi might be related to chronic renal disease is not clear.

Metastatic mineralization is reported in captive rock hyrax with renal disease, renal failure, and subsequent hypercalcemia.[52] Gross lesions include chalky white gritty material in multiple tissues including skin (calcinosis cutis), soft tissues, and around joints and footpads (calcinosis circumscripta). Renal lesions will vary depending on severity of disease. Typically, there is decreased prominence of the renal corticomedullary junction and the cortices are pale tan. Mineralization is detected histologically within various tissues including the skin, footpads, soft tissues, vessels, lung, ovary, kidney, and intestinal tract (mucosa and muscular wall). Imaging along with supportive clinicopathologic data may be useful for antemortem diagnosis. Analysis of minerals within the urinary tract often demonstrates calcium carbonate.

Soft tissue mineralization has been reported in captive two-toed sloths of varying ages. Gross and radiographic lesions were most prominent in the aorta and arteries (Fig. 21.7).[46] Histologic changes include arterial mineralization with mural osseous and chondroid metaplasia and smooth muscle hyperplasia. Visceral mineralization can be seen in the gastric mucosa, kidneys, and lungs. Most affected animals have renal disease, which is likely an underlying cause of soft tissue mineralization; however, other causes of calcium and phosphorus imbalance (e.g., dietary) are likely to also be involved. Degenerative cardiac disease was commonly associated with vascular mineralization and myocardial necrosis. Similar lesions have been reported in giant and lesser anteaters.[6]

Adrenal cortical hyperplasia is reported in adult rock hyraxes and appears similar to other species with nodular accumulations of large (hypertrophied) cortical epithelial cells. Common concurrent diseases include pancreatic islet fibrosis and/or diabetes mellitus (see "Nutritional" section).[59]

Inflammatory, Non-Infectious

Gastrointestinal (GI) conditions including diarrhea, constipation, and anorexia are common in Xenarthra. Husbandry is likely responsible for many of these conditions due to an overreliance on moistened to liquid diets, insufficient or inappropriate dietary chitin or fiber, or a slow adaptation to captive diets in wild-caught animals. Few systematic studies examining the causes of GI disease have been performed, but moderate levels of inflammation have been reported at all levels of the GI tract.[50] Inflammation is nonspecific, lymphoplasmacytic, and often eosinophilic, and a variety of bacteria are often noted in the lumen (Fig. 21.8; eSlide 21.e4). Gram stains may show an abnormal balance of the enteric flora with a high percentage of gram-negative or gram-positive organisms suggesting dysbiosis. In addition, protozoa, such as trichomonads and ciliates, may become abundant, though whether these organisms are a cause or result of dysbiosis requires further investigation.[3]

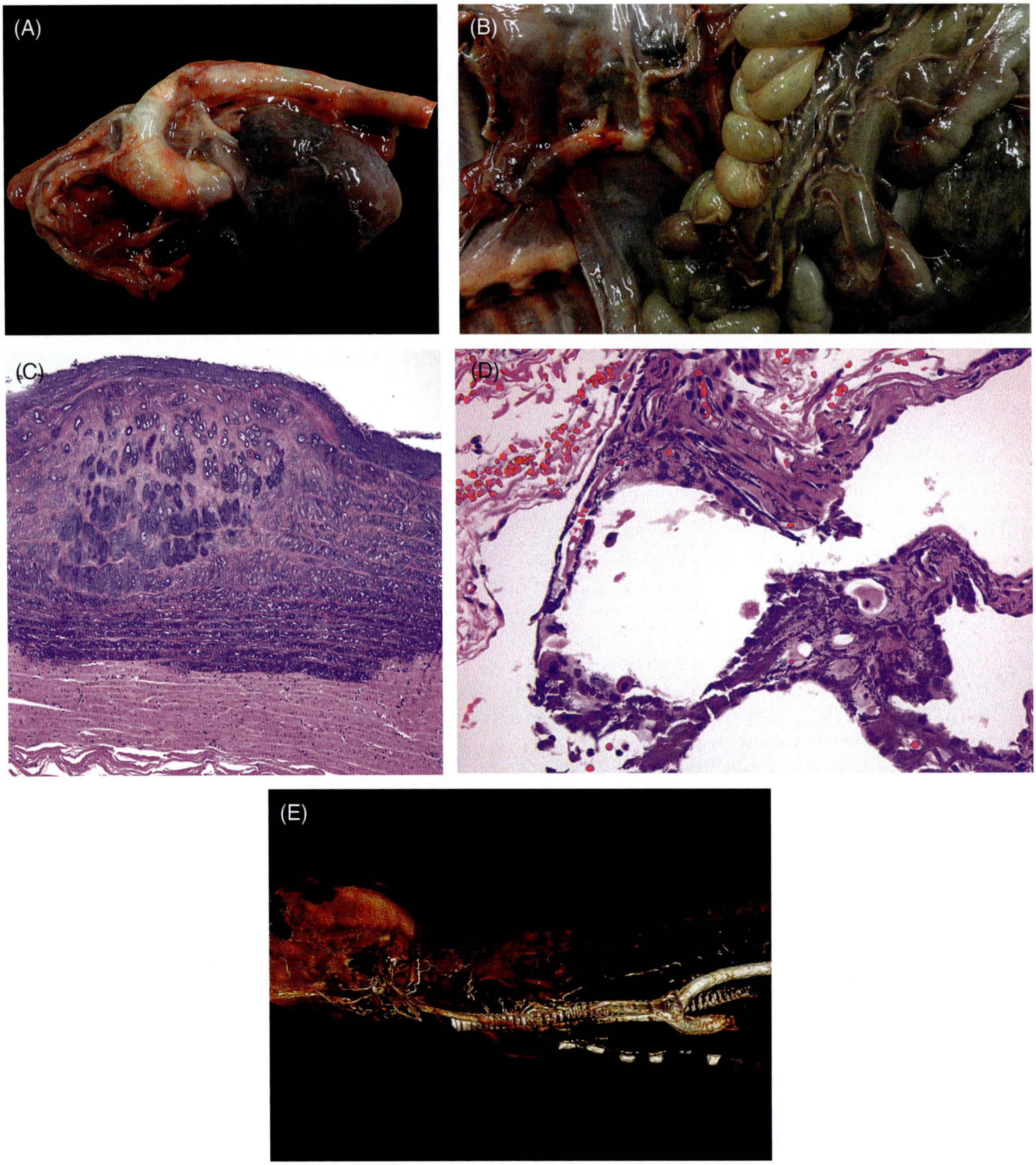

FIGURE 21.7 Arterial, arteriolar, and visceral metastatic mineralization in a Linne's two-toed sloth. (A) Mineralization of the aorta and cardiac arteries appears as bright white discoloration of the vessel walls. (B) Similar changes may be seen in mesenteric vessels. (C) Histologically, arterial mineralization, mural osseous and chondroid metaplasia, and smooth muscle hyperplasia are common findings. (D) Areas of mineralization in the pulmonary interstitium may appear more basophilic than unaffected tissue. (E) Three-dimesional volume rendering from computed tomography (CT) of a sloth with prominent mineralized aorta and vasculature. The skeletal structures have been digitally removed. *(Parts A–D: Courtesy of S. Han, Denver Zoological Foundation; Part E: Photo Courtesy of M. Ivančić and Brookfield Zoo Chicago Radiology Consulting Service.)*

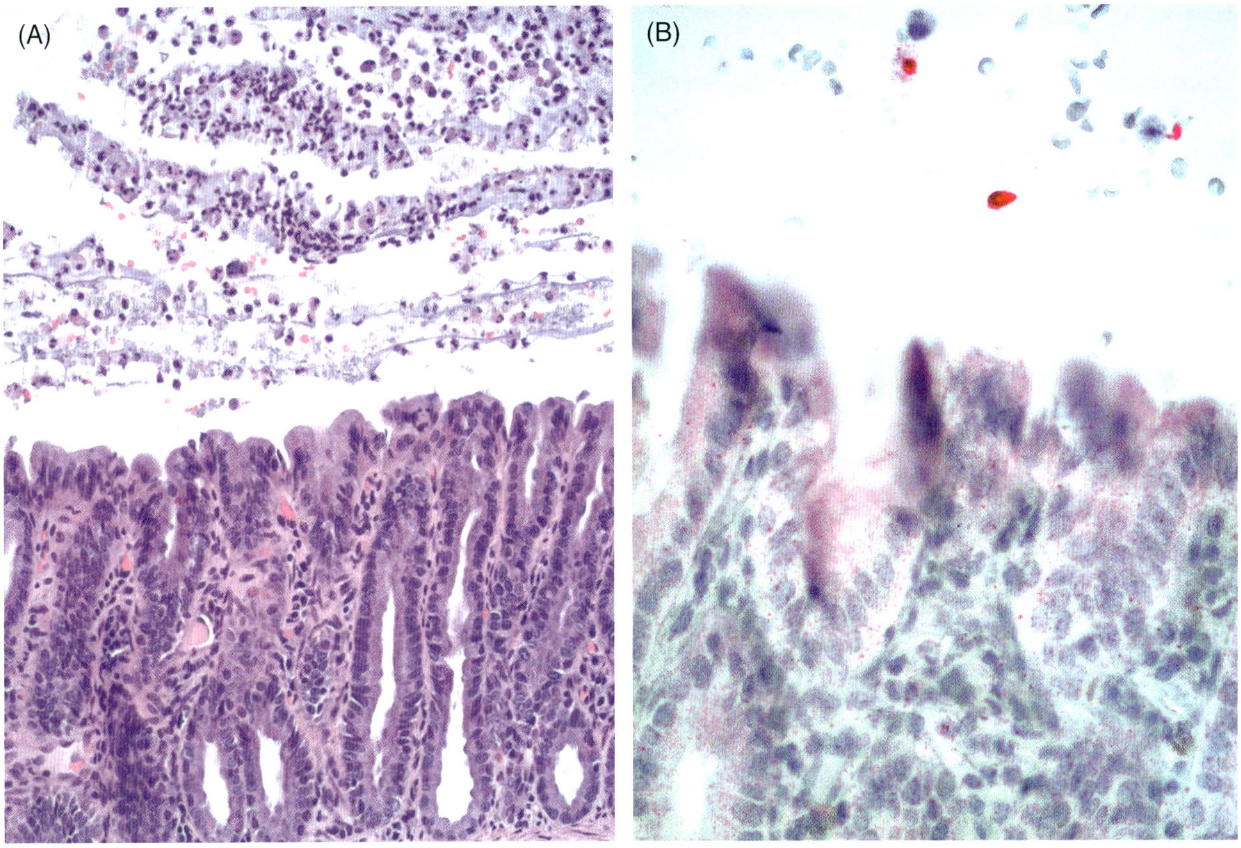

FIGURE 21.8 Colitis and colonic trichomoniasis in a giant anteater. (A) Chronic inflammation (lymphocytes and plasma cells along with neutrophils or eosinophils) is associated with variable mucus cell hyperplasia, ulceration, and luminal exudates. Inflammation is often associated with a predominance of either gram-negative or gram-positive bacteria (dysbiosis) (see eSlide 21.e4). (B) Immunohistochemical labeling (red) highlights *Tichomonad* spp. in the luminal contents of B. Trichomonad immunohistochemistry, red chromagen.

Miscellaneous/Multifactorial

Among the common findings that are seen in these species are fight wounds, as intraspecific and interspecific aggression are common. Wound infections are common due to the abundant bacterial flora in the oral cavity. Some species may have toxic elements in saliva that contribute to tissue damage. Giant anteaters can aggressively "slash" other anteaters or other species within the same or adjacent exhibits, leaving long deep lacerations, deep puncture wounds with extensive internal wounds, or even complete evisceration.

Due to the characteristic long tongue of anteaters, constrictions and strangulations of the tongue can occur. Entrapped hair, fascia from meat-based diets, or foreign fabric material have been identified at the base of gangrenous tongues. The tongue may also be entrapped in poorly designed caging material or traumatized by other animals (Fig. 21.9).

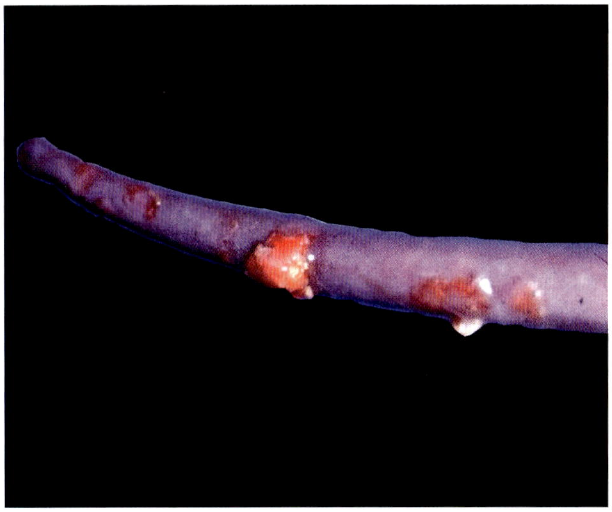

FIGURE 21.9 Tongue trauma in a giant anteater. Multifocal abrasions and lacerations on the tongue with proliferative scar tissue are due to trauma from an otter housed in an adjacent exhibit. Other tongue lesions may be due to constrictions and strangulations by entrapped hair, fascia from meat-based diets, or fabric. It may also be damaged in poorly designed caging. *(Photo Courtesy of A. Duncan, Detroit Zoological Institute.)*

Malnutrition, **wasting**, gastric ulcers, and gastritis are related syndromes associated with stress, inappropriate diet, and sand ingestion. Gross lesions include gastric ulceration and reddened mucosa, intraluminal sandy digesta, and atrophy of fat stores and skeletal muscle. Gastric lesions will have histologic evidence of chronic ulceration with mucosal atrophy and fibrosis. The intestinal lumen often has disproportionate populations of microflora (dysbiosis). Diagnosis is based on ruling out other systemic or gastrointestinal diseases and supportive gross and histologic lesions in the context of wasting. Wasting can also be seen in subordinate individuals in cases of overcrowding and insufficient food supplies.

Dental disease is common in hyrax and aardvarks at all ages, and other toothed insectivores as they age. Continuous growth of teeth may be a major contributor if feeds are not sufficiently coarse to wear the teeth down. Aggression and stereotypic behavior may lead to tooth fracture and root exposure. Mandibular and maxillary abscesses with bony proliferation are suggestive of a tooth fracture. Histologically, variable gingivitis, periodontitis, pulpitis, or osteomyelitis may be present depending on severity and chronicity of the disease. Oral exam, radiography, histologic lesions, and culture aid in the diagnosis. In aardvarks, tooth root abscesses occur most commonly in the mandible but may affect the maxilla and erode into the sinuses and nasal cavity.[17] Gingivitis, periodontal accumulations of calculus, and loosening of teeth are a common geriatric condition and can occur in younger animals with inappropriate husbandry. The role of inappropriate substrates and diets is unknown but likely significant in the development of dental diseases in many of these cases.

Pulmonary silicosis has been described in two captive rock hyraxes housed on play sand composed of quartz (silica polymorph). Lesions were identified incidentally with histology and included multifocal pulmonary granulomas intrahistiocytic, intracytoplasmic, and birefringent crystals.[88]

Reproductive diseases in these species have been poorly studied. Breeding and neonatal survival in captive settings are often poor, though it is likely that management practices are more to blame than specific disease entities.[105] Multiple reproductive conditions including **pyometra, metritis, retained fetal membranes, fetal retention, dystocia, and stillbirth** have been reported in a large breeding colony of Lesser Madagascar hedgehog tenrecs.[111] **Endometrial polyps** are frequently seen in African hedgehogs and must be differentiated from stromal neoplasms.[70] **Hydrometra**, characterized histologically by uterine lumen dilation and attenuation of endometrial glands, may occur in nulliparous captive hyraxes, and likely results in infertility.[75] **Endometritis** and **endometrial cystic hyperplasia** have also been reported in giant anteaters.[6]

Adenomyosis has been reported in a three-banded armadillo with a history of weakness and mild anemia.[66] Grossly, the uterus was enlarged and histologically, irregular but well-differentiated endometrial glands were observed in the myometrium. Many glands contained hemorrhage and the clinical signs resolved after ovariohysterectomy, suggesting a link between the adenomyosis and clinical signs.

Neoplastic

Neoplasia is not a significant factor in the Xenarthra or aardvarks; however, multiple neoplasms have been described in hedgehog tenrecs.[47,54] **Lymphosarcoma** (Fig. 21.10) and multiple carcinomas, particularly oral **squamous cell carcinoma**, are most common (Fig. 21.11). Table 21.e1 contains a list of tumors reported in these species.

Neoplasia is not common in rock hyraxes. Indeed, like their closest living relatives, elephants, hyraxes have multiple copies of the p53 gene (*TP53*), a cancer suppressor which is thought to regulate cell division and signal for apoptosis when a cell is irreparably damaged.[1] Single reports of various neoplasms in hyrax include ovarian granulosa cell tumor, gastric adenocarcinoma, metastatic seminoma, mammary adenoma, uterine leiomyosarcoma, thyroid myxoma, hepatic hemangioma, dermal hemangioma and keratoacanthoma, trichoepithelioma, trichofolliculoma, and oral squamous cell carcinoma.[49,59,80,109]

Multicentric skeletal sarcomas in African hedgehogs have been associated with a retrovirus.[86] In a report of two cases, one animal had a multinodular parosteal sarcoma located adjacent to the lumbar vertebra with osteomas/osteochondromas attached to the ribs and mandible. These tumors had abundant mesenchymal spindle cells mixed with varying amounts of chondroid and osteoid matrix. In some areas, the neoplasm invaded the lumbar vertebra and compressed the spinal cord. The other hedgehog had

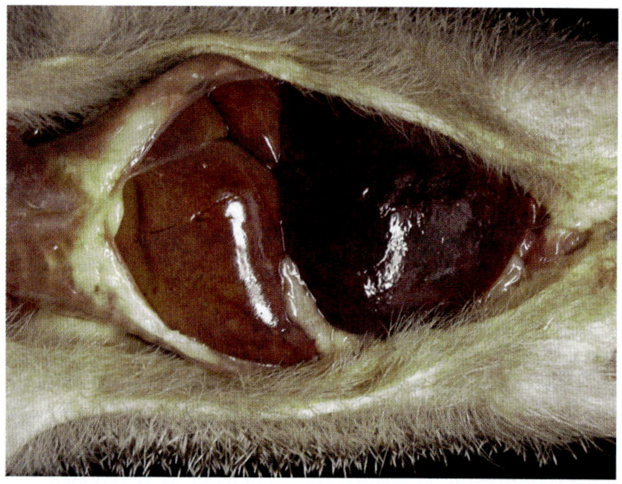

FIGURE 21.10 Hepatic and splenic lymphosarcoma in a tenrec. Tenrecs have a high incidence of lymphosarcoma, which is often characterized by parenchymal pallor, rounded organ margins, and massive hepatomegaly and splenomegaly. *(Photo Courtesy of San Diego Zoo and Wildlife Alliance.)*

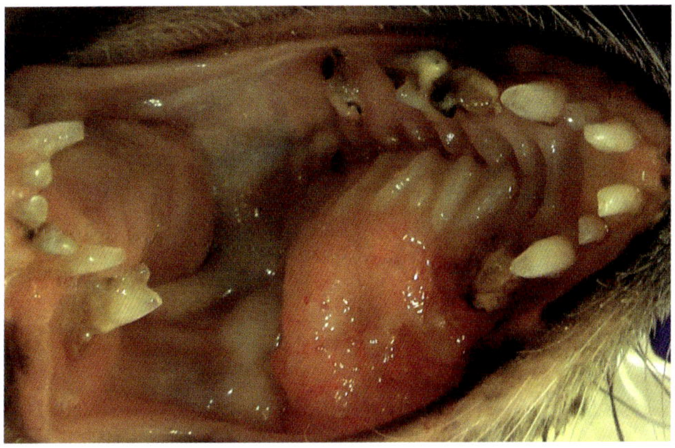

FIGURE 21.11 Squamous cell carcinoma in the oral cavity of a tenrec. These frequently present as a proliferative, often ulcerated, tumors that can invade the palate, nasal passages, and other adjacent structures. *(Photo Courtesy of D. McAloose, Wildlife Conservation Society.)*

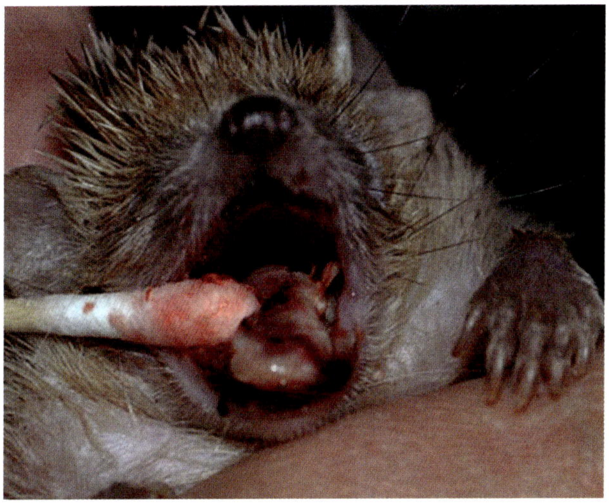

FIGURE 21.12 Ulcerative stomatitis and glossitis in a tenrec. Multiple deep ulcers, hemorrhage, and sloughed epithelium cover the tongue, gingiva, and inner lining of the cheeks. Herpesvirus was identified in multiple animals, which appeared to recrudesce during pregnancy. *(Photo Courtesy of T. Harrison, North Carolina State University, College of Veterinary Medicine.)*

multiple tumors of similar morphology arising from ribs in the thoracic cavity and one within the cranium. Very large numbers of 110 nm diameter enveloped viral particles were evident by transmission electron microscopy within the extracellular space. Particles had a centrally located electron-dense nucleoid averaging 55 nm in diameter and were often surrounded by an indistinct radio-dense line, compatible with core protein. Rare viral budding from the plasma membrane was also observed. Viral particles were not observed in other organs or lesions. These particles were most compatible with retroviruses, suggesting similarities between this condition in hedgehogs and retrovirus-induced osteochondromatosis, associated with feline leukemia virus (FeLV) infection in cats.

INFECTIOUS DISEASES

DNA Viruses

Herpesvirus (family *Herpesviridae*) was suspected as a cause of intermittent disease in a large colony of lesser Madagascar hedgehog tenrecs, in which occasional cases of necrotizing stomatitis and glossitis were observed; lesions often resolved with supportive care (Fig. 21.12). Viral inclusions were not observed in the lesions. The syndrome occurred in several pregnant animals that died and it is theorized that infection may recrudesce under the immune and metabolic stress of pregnancy. A novel "tenrec" gamma herpesvirus was identified in an oral lesion from one of these tenrecs via herpesviral PCR and subsequent sequencing (Harrison, T., unpublished data). Oral swabs from all tenrecs in the collection subsequently tested positive, suggesting this as a potential commensal herpesvirus.

An **alpha-herpesvirus** has been associated with disease in captive rock hyraxes, which included glossitis and oral ulceration, with high mortality in a group of zoo-housed rock hyrax.[39] A specific cause of death was not reported but many animals died after 2 weeks of debilitation. More moderate forms of this infection have also been reported. Histopathology of the ulcers showed amphophilic to basophilic intranuclear inclusion bodies in epithelial cells as well as cytomegaly, karyomegaly, and syncytia. Molecular analysis grouped the herpesvirus in the genus *Simplexvirus*. An alpha-herpesvirus was also reported as the cause of death in an African hedgehog, with multifocal hepatic necrosis, intranuclear inclusions in hepatocytes, and syncytia.[5] PCR identified the virus as **herpes simplex 1** and the owners reportedly had frequent "cold sores." Herpesviral infections with similar gross lesions and ultrastructural findings were also reported in European hedgehogs, but further molecular characterization of the viruses has not been performed.[117]

A primary gastric **T-cell lymphoma** was reported in one free-ranging giant armadillo from Brazil and a novel **gammaherpesvirus** (proposed as **Cingulatid gammaherpesvirus 1 (CiHV-1)**) was associated with the lesion. This virus has been detected in 14.3% of a small group of wild giant armadillos, but similar lesions were not noted.[79]

RNA Viruses

Foot and mouth disease, caused by *Aphthovirus* sp. in the family *Picornaviridae*, is reportable to the World Organization for Animal Health (WOAH) due to major impacts it has on livestock trade worldwide. Infection can also occur in hedgehogs. Though not common, the risk of infection is great enough that an awareness of the lesions is warranted. Natural infections during livestock outbreaks in the United Kingdom have been reported and experimental transmission studies have confirmed that the virus can successfully be passed between European hedgehogs and from hedgehogs to cattle.[69] Vesicular lesions are readily identified on the hairless parts of the feet, snout, lips, tongue, and perineum. These are typical intraepithelial vesicles that are easily disrupted or debrided. Systemic illness can also be a component and mortality can occur, likely from secondary bacterial infections. Definitive diagnosis is typically by PCR. Of the other species covered in this chapter, only armadillos and the eastern tree hyrax have been shown to carry foot and mouth disease, and then only under experimental conditions.[89,116] Testing of other species may reveal other carriers.

Influenza A, in the family *Orthomyxoviridae*, has been reported in a colony of giant anteaters. Infection caused severe nasal discharge, inappetence, and lethargy in all individuals except a neonate.[81] Simultaneously, staff keepers were also ill with the respiratory disease. Swabs of anteater nasal secretions were used to identify influenza A virus (H1N1). No mortality was reported in this incident, so lesion morphology is currently unknown.

Canine distemper virus (CDV) in the family *Paramyxoviridae*, genus *Morbillivirus*, has been reported in giant anteaters. Clinical signs included pulmonary congestion, lethargy, diarrhea, anorexia, and ataxia.[16,30,103] Histologic lesions typical of CDV in dogs and other carnivores included interstitial pneumonia with epithelial syncytia and eosinophilic nuclear and cytoplasmic inclusions. Ulcerative glossitis and compact hyperkeratosis on foot pads ("hard pad" disease) with opportunistic *Candida albicans* skin and pad infection occurred in a juvenile giant anteater. Encephalitis was not identified in any of the affected animals. Nuclear and/or cytoplasmic inclusions were present in epithelia of the lung, bile ducts, urinary bladder, renal pelvis, medullary tubules, intestine, skin, uterus, tongue, and exocrine pancreas. In another case described in a southern tamandua, brain, lung, urinary bladder, and skin were positive for CDV by fluorescent antibody, immunohistochemistry, and PCR. Immunohistochemistry demonstrated antigen in the necrotic neuronal cytoplasm in the cortex, spinal cord, and also in the urinary bladder epithelium.[63] Similarly, fatal CDV infections have been reported in a group of Linnaeus's two-toed sloths.[115] Clinical signs included hyporexia, lethargy, mucopurulent nasal discharge, and ulcers on the face and mucous membranes (Fig. 21.13A). All affected sloths had hepatic necrosis, lymphoid depletion, and bronchointerstitial pneumonia. Notably, no central nervous system lesions were present, but there was immunoreactivity in vessels of the brain. Epithelial cells and histiocytes in numerous organs contained eosinophilic intranuclear or intracytoplasmic inclusions; occasional syncytial cells were also seen (Fig. 21.13B).

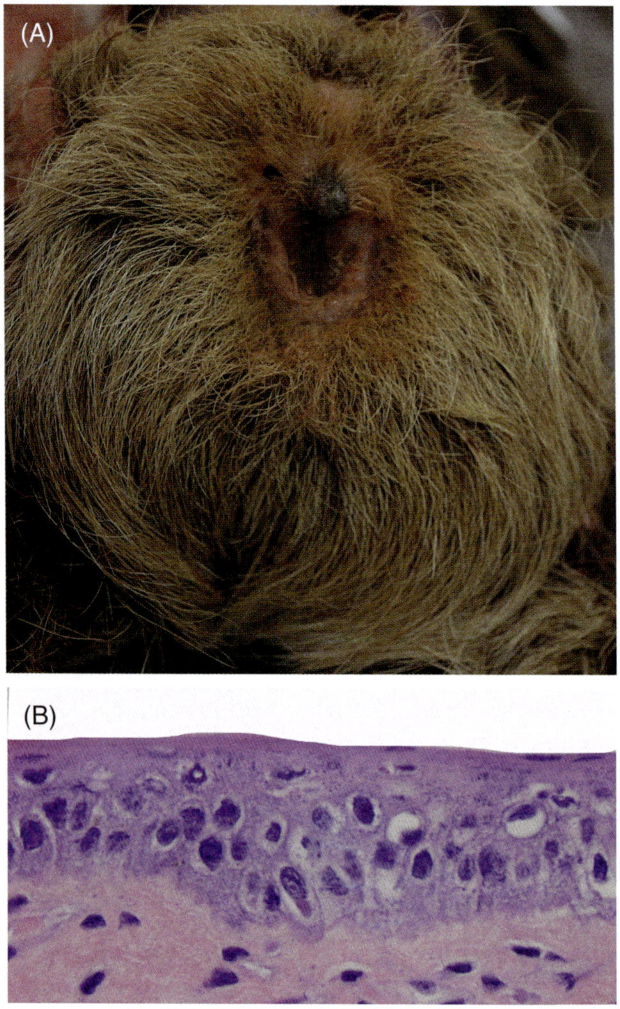

FIGURE 21.13 Canine distemper virus infection in the mucous membranes of a sloth. (A) Ulceration and crusting of the mucous membranes around the anus. (B) Intracytoplasmic and intranuclear eosinophilic inclusions in epithelial cells of the lip. *(Parts A, B: Photos Courtesy of L. Craig, University of Tennessee.)*

The role that Malayan pangolins may play in the epidemiology of **SARS-CoV-2-related coronaviruses** is a topic of considerable interest. Metagenomic sequencing has identified two sublineages of SARS-CoV-2-related coronaviruses in pangolins. The receptor-binding domain of one of these related viruses is very similar to that of SARS-CoV-2, suggesting that pangolins may be possible hosts of newly emergent coronaviruses. No lesions were reported in the pangolins in this study.[60] A case of SARS-CoV-2 infection was reported in a young giant anteater.[87] The animal was found with paralysis on a highway and sent to a rehabilitation center where multiple fractures were detected and it was euthanized. It had mild gross and histologic lesions in the lungs; however, there was no clinical suggestion of respiratory disease. Immunohistochemistry and molecular testing revealed the presence of the virus in lung, liver, spleen, lymph node, and large intestine.

Rabies was reported in a captive southern tamandua that was moved from a drive-through zoo in Virginia to a zoo in Tennessee, where the animal was housed indoors with one other tamandua and no access to native wildlife.[25,45] Two months after translocation, the animal displayed lethargy, anorexia, and diarrhea, and was euthanized. Neuronal cytoplasmic inclusions (Negri bodies) were detected, and rabies was confirmed by RT-PCR and immunohistochemistry that showed rabies antigen in the neuronal cytoplasm. Phylogenetic analysis and epidemiologic information suggested that the tamandua was infected through exposure to wild raccoons in Virginia. As it had not previously been reported in tamanduas and because the low basal body temperature of tamanduas was believed to reduce the susceptibility of the species to infection, rabies was not suspected during the clinical work-up or necropsy examination. The case resulted in 22 human exposures and is a reminder about the risk of rabies infection even in what are generally considered low-infection-risk species.

Bacteria

Leprosy (*Mycobacterium leprae*) is well documented in the armadillo (primarily the nine-banded armadillo).[96,102,112,113] It is endemic in wild armadillos in the southeastern United States and there are multiple reports of leprosy in armadillos in Central and South America.[102] Evidence supporting widespread dissemination of the disease outside the United States was initially less definitive but recent work suggests an over 40% prevalence in armadillos in the Andean region near where human cases also exist.[18]

The armadillo is one of a few species besides humans and some nonhuman primates that are susceptible to the *M. leprae* bacterium, making it a valuable experimental model for the disease and reservoir for zoonotic transmission in the wild.[10] In the Brazilian Amazon region, there is evidence of zoonotic transmission from the contact or consumption of armadillos.[26] As in humans, there appears to be a level of genetically based resistance to infection; approximately 20% of armadillos are resistant to *M. leprae* with sibling and dams responding similarly. Single nucleotide polymorphisms in toll-like receptors, similar to resistance in humans, have been identified in resistant armadillos.[2] *M. leprae* thrives at cooler temperatures (e.g., the extremities), which suggests a reason for the unique susceptibility in armadillos (their core body temperature (86–95°F (30–35°C)) is significantly lower than most mammals).

The lesions of *M. leprae* infection in armadillos are very similar to those in humans. After experimental (and likely natural infection), *M. leprae* localizes in peripheral nerves and the reticuloendothelial system. Subsequent intermittent and low-level bacteremia progresses to widespread systemic dissemination. Armadillos are rarely clinically ill until late in the disease and external lesions are subtle and difficult to identify. However, in susceptible armadillos, the bacillus proliferates rapidly and by 2 years postinoculation, most animals have developed severe and widespread infection.

Early in disease, the most common lesions are plantar ulcers and skin erosions and abrasions around the eyes, nose, and footpads. As disease progresses, foot ulcers are increasingly frequent, likely because of diminished neural sensitivity in the extremities. Other skin lesions are difficult to find due to cutaneous armoring, making it a challenging disease to monitor. A granulomatous response can be seen in multiple organs and the skin, but most damaging is the infection, inflammation, and ultimate destruction of nerves. Like humans, armadillo Schwann cells are susceptible to infection with the *M. leprae* bacillus, leading to neuritis and neurological dysfunction characteristic of the disease. In armadillos, the full spectrum of Ridley-Jopling responses from "tuberculoid" (TT) to three stages of "borderline" reaction [borderline tuberculoid (BT), intermediate borderline (BB), and borderline lepromatous (BL)] and finally to "lepromatous" inflammation (LL) are seen. In humans, tuberculoid leprosy is seen in individuals with intense cell-mediated immunity, and is characterized by small lesions with raised borders, papules, and plaques.[114] Lepromatous leprosy is seen when there is no effective reaction against the bacillus, which can then disseminate and produce the development of widespread granulomas. Borderline presentations are a mixture of the lepromatous and tuberculoid types, with larger plaques (a "Swiss cheese" appearance), increased reddening of the lesions, and more frequent small satellite lesions around larger skin lesions than the tuberculoid type lesions. The number of bacilli tends to be much higher in lepromatous inflammation than tuberculoid reactions and intermediate in the borderline lesions. Most armadillos have reactions toward the lepromatous end of the spectrum, with remarkably large numbers of leprosy bacilli in reticuloendothelial tissues.[10] The peripheral nerves are infected early and widely in the progression of the disease. Granulomatous interstitial neuritis involving the perineurium, epineurium, and endoneurium is characteristic (Fig. 21.14). Neural degeneration, myelin loss, and fibrosis are also present. Interestingly, due to the importance of the armadillo to leprosy research, there is a growing arsenal of armadillo-specific molecular and immunologic reagents available to study the immune response of these animals.

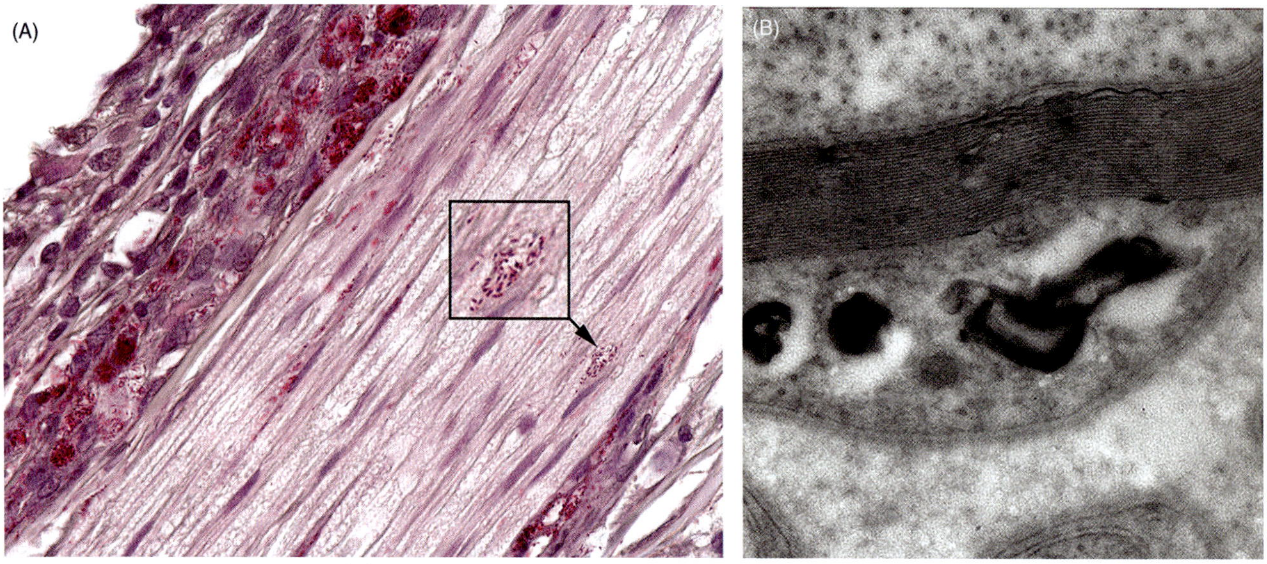

FIGURE 21.14 Leprosy due to *Mycobacterium leprae* in peripheral nerves of an armadillo. (A) Granulomatous interstitial neuritis involving the perineurium, epineurium, and endoneurium is characteristic. It is associated neural degeneration, myelin loss, and fibrosis. In this case, the armadillo's immune response is skewed toward the lepromatous end of the spectrum, with remarkably large numbers of leprosy bacilli in neural tissue and within macrophages *(see box inset)*. Ziehl-Neelsen acid fast. (B) Transmission electron microscopy (TEM) shows a typical twisted bacillus within the endoneurium. *(Parts A, B: Photos Courtesy of R. Truman, National Hansen's Disease Program.)*

Mycobacteriosis has been widely reported in hyrax. A specific variant of ***Mycobacterium tuberculosis* complex** called the "**dassie bacillus**"; ***M. africanum***, ***M. microti***, and ***M. microti*-like bacteria** have been reported.[64,77,85] Weight loss, lameness, pneumonia, and reproductive failure have been associated with these infections, and grossly, granulomas are found in the lungs, liver, spleen, and kidneys. Intralesional acid-fast bacilli are usually present in very small numbers. Spoligotyping (PCR and genotyping) can be performed on freshly collected material. Antemortem diagnostics are considered insensitive and nonspecific. The disease typically leads to death or euthanasia. Rock hyraxes may harbor ***M. avium* subspecies *paratuberculosis* (MAP)** in their intestines and feces, although it has not been associated with lesions of progressive enteropathy in this species.[36]

Bacterial septicemia can be a significant cause of death in captive rock hyraxes.[59] **Gram-negative sepsis** (e.g., **colibacillosis**) is the cause in most cases. Initial sites of infection include the lung, gastrointestinal tract, liver, uterus, kidneys, or external traumatic wounds. Comorbidities in affected individuals include ulcerative glossitis and esophagitis with intralesional yeasts consistent with *Candida* sp. Endometritis can be a major cause of septicemia in nulliparous, gravid, and postpartum females.

Bacterial pneumonia is another common cause of death in zoo-housed hyraxes.[59] **Pasteurellosis *(Pasteurella multocida)*** manifests as pneumonia with or without progression to sepsis in hyraxes. Grossly, lesions include pulmonary congestion, edema, and hemorrhage with cranioventral distribution. Bronchopneumonia to diffuse bronchointerstitial pneumonia is seen histologically. In some cases, there is also necrotizing interstitial pneumonia with variable vasculitis and splenic necrosis. Diagnosis is based on histologic lesions and bacterial culture.

Rock hyraxes may be an important host species for ***Borrelia persica***, a tick-transmitted bacterium which is associated with **tick-borne relapsing fever** in Israel; however, no lesions or clinical disease has been described.[56]

A survey in the Brazilian wetlands detected antibodies to ***Leptospira*** spp. in six-banded and giant armadillos.[27] Bacteria were not seen in tested samples and no disease was reported in these animals.

Enteritis associated with *Salmonella* spp., *Shigella* spp., *Campylobacter* spp., *Escherichia coli*, *Yersinia pseudotuberculosis*, and other enteric pathogens is relatively common in anteaters, and has also been reported in tenrec, hedgehogs, and armadillo. It is almost always associated with poor husbandry, contaminated feed (many diets may include uncooked meat), or immunocompromise due to stress, transport, or concurrent disease.

While lesions have not been attributed to **hemotropic mycoplasmas**, they have been identified in multiple species within Xenarthra. One individual isolate from a pale-throated sloth had a 99.7% identity with ***Mycoplasma wenyonii***, suggesting these organisms have disease potential.[29]

Fungi

Adiaspiromycosis caused by dimorphic Onygenales fungi (*Emmonsia crescens* and *Blastomyces parvus*) was identified in a large population of road-kill armadillos (six-banded and southern naked-tailed armadillos) in Brazil. Granulomatous pneumonia associated with adiaspores was noted, but the molecular analysis suggested possible novel species of adiaspores may be involved (Fig. 21.15).[78]

Sporotrichosis has been reported in a severely burned anteater. It was suggested that immunocompromise and damage to the skin barrier might be required for infection.[6]

Metazoa

In general, the members of the multiple orders covered in this chapter do not have significant endoparasite or ectoparasite diseases unless they are poorly managed, recently brought in from the wild, or if they are severely compromised.

Gastrointestinal parasites are likely common in wild and managed collections but have been rarely examined in detail. ***Grassinema procavia*** is a small white nematode found in the stomach and to a lesser extent in the intestinal tract of hyraxes. Gastric ulceration may be present in more severe infestations.[68] In the family *Tupidae* (shrews), **gongylonemiasis** is a very common finding leading to morbidity (e.g., weight loss, inappetence) and sometimes leading to death.[21] This parasite has been seen as an increasingly common finding in shrews in captivity, possibly due to increased use of mixed exhibits with natural substrates allowing more cockroaches and intermediate hosts to persist in the environment. These nematodes embed within the superficial epithelium of the tongue. ***Anoplocephala*** sp. is a cestode typically limited to the intestine while ***Hoplodontophorus*** sp. can be found in the small intestine, bile duct, and liver.[100] ***Physaloptera magnipapilla*** has caused severe hypochromic, microcytic anemia, hypoproteinemia, and grossly evident ascites, hydrothorax, pulmonary congestion, and severe catarrhal hyperplasic gastritis in anteaters (Fig. 21.16).[62] Mucous hyperplasia was also noted microscopically.

Ovarian filariasis was reported in a southern tamandua in which both ovaries were irregularly and multifocally nodular. The nodules contained adult nematodes (including gravid females with microfilaria) surrounded by mixed chronic inflammation.[37] Infection was first reported in 1914, suggesting it may be widespread in wild populations and may impact fertility.[53]

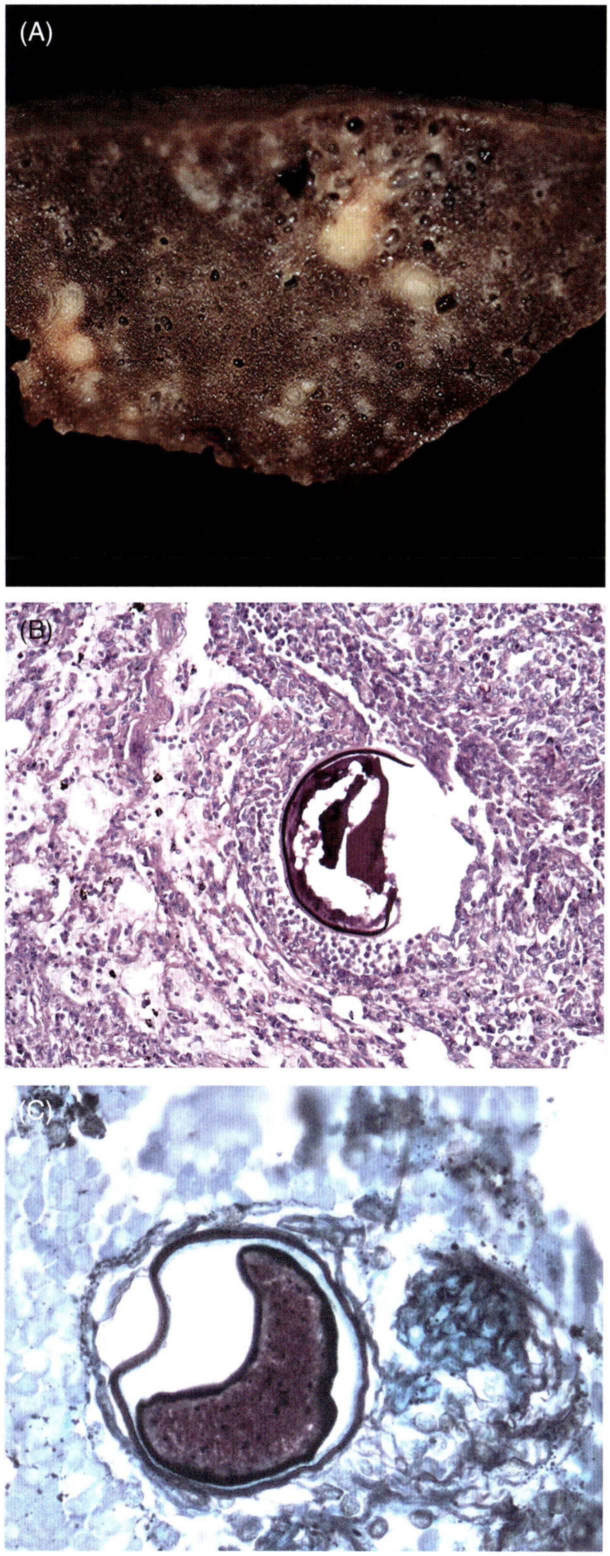

FIGURE 21.15 Adiasporomycosis in the lung of a nine-banded armadillo. (A) The parenchyma contains multifocal to coalescing, 0.5 to 1 mm diameter, moderately demarcated, pale tan to yellow firm nodules. (B) A pulmonary granuloma with a fragmented adiaspore in the center. PAS. (C) A fragmented adiaspore with a prominent trilaminar wall within a pulmonary granuloma. GMS. *(Parts A–C: Photos Courtesy of P. Navas-Suarez, Laboratory of Wildlife Comparative Pathology, University of São Paulo.)*

Protozoa

Protozoan diseases have been discussed earlier as it relate to dysbiosis in many of the insectivores and Xenarthra. However, other systemic protozoa can be a concern

FIGURE 21.16 *Physaloptera* spp. infection in the stomach of a giant anteater. Severe catarrhal hyperplasic gastritis with numerous intralesional nematodes. Gastritis and mucus hyperplasia can lead to hypochromic microcytic anemia, hypoproteinemia, and grossly evident ascites, hydrothorax, and pulmonary congestion. *(Photo Courtesy of P. Navas-Suarez, Laboratory of Wildlife Comparative Pathology, University of São Paulo.)*

(see "Inflammatory, Non-Infectious" section).[7] In particular, **Chagas disease** caused by ***Trypanosoma cruzi*** in hedgehogs and armadillos causes granulomatous and necrotizing myocarditis.[28,58,72] Antemortem diagnosis may be possible if the trypanosomes are identified in blood smears during the parasitemic stage, but this is only transient and may be missed.

Leishmaniasis is WOAH listed, emerging disease of concern in hyrax populations primarily as it relates to zoonotic transmission of ***Leishmania tropica*** and ***L. aeithopica***.[107] While *Leishmania* spp. do not commonly cause disease in rock hyraxes, studies suggest they may act as reservoir hosts and the origin of human cases of cutaneous leishmaniasis throughout their natural range in Ethiopia and the Middle East. Hyraxes capitalize on land-use changes, making use of shelters in urban areas, which increases contact with people and domestic animals.[11,97] *Leishmania* spp. cause cutaneous (***L. major*** and ***L. tropica***) and visceral (***L. infantum***) diseases in humans and animals, particularly dogs. The vector for *Leishmania* sp. is the female phlebotomine sandfly. Molecular identification and speciation have been useful to trace infections during epidemiological studies.

Fatal **toxoplasmosis** has been reported in pregnant and young rock (cape) hyraxes and a tree hyrax in captivity; titers to *Toxoplasma gondii* have been detected in free-ranging rock hyraxes in Africa.[31,33,83,95] Toxoplasmosis manifests similarly to diseases in other susceptible species with disseminated disease targeting the brain, liver, spleen, and adrenal glands. Encephalomalacia, hepatic, and/or splenic necrosis, as well as pneumonia may be seen on gross examination. Among affected organs, necrotizing lesions with or without substantial surrounding neutrophilic and lymphohistiocytic inflammation are typical. Within regions of necrosis, particularly along the margins of nonviable and viable tissue, intracellular and extracellular tachyzoites can be detected microscopically. Tissue cysts may be present in both severely affected and lesser-affected tissues. It can also cause abortion and stillbirth in pregnant hyrax. Diagnosis is made based on supportive gross and histologic lesions and confirmed with immunohistochemistry, PCR, or electron microscopy. Seropositivity has also been reported in armadillos.[58]

Babesia thomasi has been documented in hyraxes from Kenya.[51] Low numbers of erythrocytes contain protozoa (approximately 1 parasitized erythrocyte per 1,000). A novel blood piroplasm, ***Echinozoon hoogstraali***, is described in the bush hyrax. An interesting feature of this parasite is the presence of filament formation in infected erythrocytes.[42]

Ectoparasites

Ectoparasites, specifically mites and lice, appear to be a significant problem in quilled animals, such as the tenrec and hedgehog. Though commonly recognized, little work has been done to identify common species or assess the skin damage these parasites may cause. Lesions present as reddened areas and nodules with crusts, scabbing, and exudates among the quills of the head, dorsum, and hind limbs (Fig. 21.17). **Demodicosis (*Demodex* sp.)** has been reported in three sibling juvenile hyraxes.[106] Causative mites were identified histologically and characterized ultrastructurally. Lice (***Trichodectidae*** **sp.**), nasal mites (***Schoutedenichia schoutedeni***), and lung mites (***Pneumonyssus procaviens***) have all been reported in wild and less often in captive hyraxes.[77]

FIGURE 21.17 Acariasis in the skin of a tenrec. Myriad mites are present at the base of the quills on this tenrec. Mites and lice are commonly reported in tenrecs and hedgehogs.

FIGURE 21.18 Tungiasis in the skin of a giant anteater. *Tunga penetrans*, a zoonotic flea that burrows under the skin, causes white firm papules with dark centers where the flea neosomes (hypertrophied adult fleas) are present. *(Photo Courtesy of P. Navas-Suarez, Laboratory of Wildlife Comparative Pathology, University of São Paulo.)*

Tungiasis is caused by ***Tunga penetrans***, a zoonotic flea that burrows under the skin causing white firm papules with dark centers where the flea neosomes (hypertrophied adult fleas) are present (Fig. 21.18). Lesions in an anteater were primarily noted on both the plantar surface of the hindfoot and dorsal aspect of the forefoot. It has been described in a giant anteater, southern tamandua, and armadillos (P. Navas-Suárez, unpublished data).[32,35]

ADDITIONAL RESOURCES

Arenales A, Gardiner CH, Miranda FR, et al. Pathology of Free-Ranging and Captive Brazilian Anteaters. *J Comp Pathol*. 2020;180:55–68.

Arenales A, Silva FL, Miranda F, et al. Pathologic findings in 36 slots from Brazil. *J Zoo Wildl Med*. 2020; 51 (3):672–677.

Balamayooran G, Pena M, Sharma R, Truman RW. The armadillo as an animal model and reservoir host for *Mycobacterium leprae*. *Clin Dermatol*. 2015;33:108–115.

Harrison TM, Harrison SH. Evaluation of husbandry and mortality in lesser hedgehog tenrecs (*Echinops telfairi*). *J Zoo Wildl Med*. 2017;48 (2):440–445.

Kruse TN, Garner MM, Bonar CJ. A retrospective study of pathologic findings in captive rock hyrax (*Procavia capensis*) in the United States. *J Zoo Wildl Med*. 2015;46:798–805.

Superina M, Miranda F, Plese T. Maintenance of Xenartha in captivity. In: Vizcaino SF, Loughry WJ, ed. *The Biology of the Xenartha*. University Press of Florida; 2008.

Weaver GV, Domenech J, Thiermann AR, Karesh WB. Foot and mouth disease: a look from the wild side. *J Wildl Dis*. 2013;49:759–785.

References

Full references for this chapter can be found online at https://doi.org/10.1016/B978-0-443-18437-6.00026-2.

Chapter 22

Cetacea

Stephen Raverty[a], Judy St. Leger[b], Alexandria Mena[c], Ursula Siebert[d], Kathleen M. Colegrove[e], Maria Morell[d], and Antonio Fernandez[f]

[a]*Ministry of Agriculture and Food, Animal Health Center, Abbotsford, BC, Canada;* [b]*Department of Population Medicine and Diagnostic Sciences, College of Veterinary Medicine, Cornell University, Ithaca, NY, United States;* [c]*SeaWorld Parks and Entertainment, San Diego, CA, United States;* [d]*Institute of Terrestrial and Aquatic Wildlife Research, University of Veterinary Medicine Hannover, Büsum, Hannover, Germany;* [e]*Zoological Pathology Program, College of Veterinary Medicine, University of Illinois, Brookfield, IL, United States;* [f]*Institute of Animal Health and Food Safety, University of Las Palmas de Gran Canaria, Canary Islands, Spain*

INTRODUCTION

This chapter presents the pathology of cetaceans, a diverse group of mammals that are exclusive to aquatic habitats. The taxa include the largest mammals on earth, baleen whales (the mysticetes), as well as marine and freshwater-toothed whales, dolphins, and porpoises (the odontocetes). The taxonomy of the cetacean infraorder is presented in the Digital Content in Tables 22.e1 and 22.e2. There are approximately 89 living species of cetaceans. Of these, there are 15 species of baleen whales and more than 70 species of toothed whales. Odontocetes include species with diverse ecological (including foraging strategies, prey selection, and habitat use), anatomical, and physiological differences ranging from the freshwater river dolphins to deep-sea diving beaked whales. This subgroup includes sperm whales, porpoises, dolphins, beluga, narwhal, and beaked whales. These adaptations can have important implications when evaluating gross pathology and histopathologic findings. Free-ranging whales and dolphins may have global distributions, but many ecotypes have localized habitats with or without a specific migratory range. A limited number of odontocetes, primarily delphinids, are maintained in captivity. For this reason, the published material of cetacean pathology is heavily biased toward odontocetes using material from captive and stranded animals.

NON-INFECTIOUS DISEASES

Nutritional

Limitations in prey availability can result in **starvation** and **inanition** that causes the cessation of normal reproductive activity. Because changes in nutritional status are critical for management and from a population-level health perspective, blubber depths as well as multiple measurements of an animal's girth are critical information to collect during a health assessment or necropsy examination. However, it should be noted that alternations in body condition may reflect normal physiologic processes related to prolonged migration or reproductive status, or pathologic and attributed to a variety of conditions including impaired mobility, for example, due to musculoskeletal disease or verminous pneumonia. Vision or hearing deficits can also result in reduced foraging capacity. Inanition does not always mean that insufficient food was available, so both natural history considerations and disease processes should be thoroughly investigated to determine its underlying cause.

Iron overload disease has been described in captive bottlenose dolphins. These animals generally demonstrate high circulating levels of serum iron (values greater than 600–700 μg/dL; reference interval 92–300 μg/dL) and elevated transferrin saturation. Histologically, iron is seen in hypertrophied Kupffer cells and there is diffuse cytoplasmic iron accumulation in hepatocytes. The cytoplasmic granules are granular and brown with hematoxylin and eosin and blue with Prussian blue staining. In the absence of **hemochromatosis** (hepatic necrosis, hepatitis, and fibrosis), which is less common, a diagnosis of **hemosiderosis** is appropriate. The condition has been associated with suspected insulin resistance and a metabolic syndrome (a condition in humans that is associated with hepatic iron accumulation and resting hyperglycemia and has been suggested in dolphins). The cause and pathogenesis of this condition are poorly understood; however, chronic inflammation, iron sequestration secondary to acute sepsis, excessive dietary iron consumption or supplementation, emaciation, inborn errors of metabolism, and a maladaptive type syndrome have been suggested.[224]

Metabolic

Hepatic lipidosis, the accumulation of lipid in hepatocytes with nuclear displacement (Fig. 22.1), is a relatively

Pathology of Wildlife and Zoo Animals. https://doi.org/10.1016/B978-0-443-18437-6.00010-9

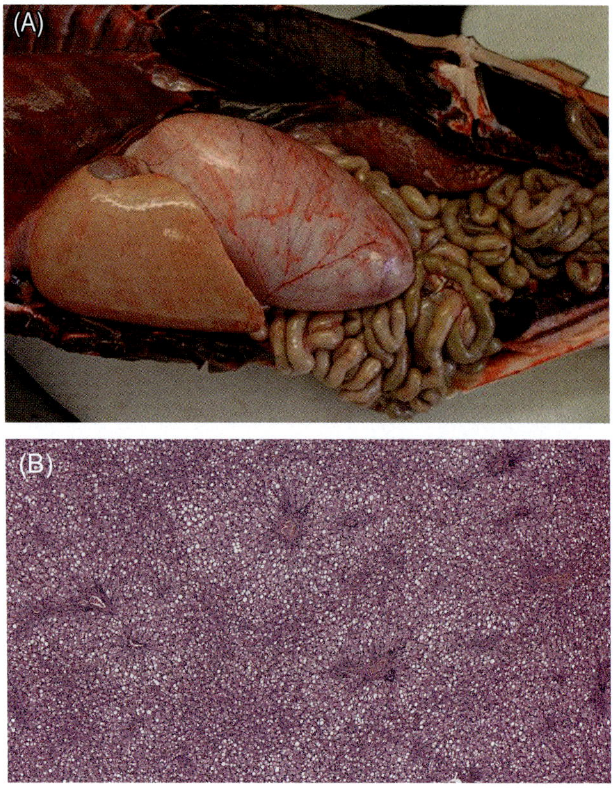

FIGURE 22.1 Hepatic lipidosis in a common dolphin. (A) Typical gross presentation is an enlarged, pale liver, often with rounded edges. (B) There is diffuse expansion of hepatocytes by clear, cytoplasmic vacuoles and nuclear displacement. This change can be seen as a pathologic change associated with inanition. Microvesicular or macrovesicular lipid accumulation in the absence of nuclear displacement may reflect a normal physiologic process (see eSlide 22.e1).

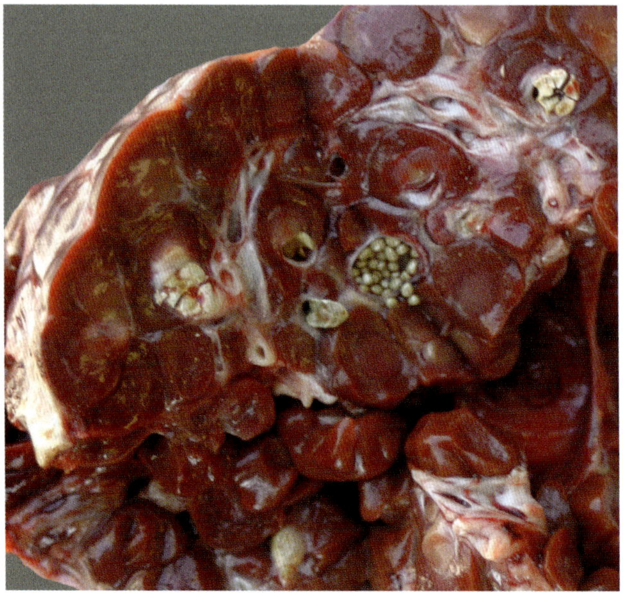

FIGURE 22.2 Ammonium urate nephroliths in the kidney from a captive pilot whale. The liths are present as miliary concretions scattered throughout the renal parenchyma. They can be large, up to 1–2cm in diameter, and are associated with hematuria, nephritis, ureteritis, and ureteral obstruction.

common finding in stranded and by-caught cetaceans, especially those with metabolic disorders produced by toxic injuries, hypoxia, or nutritional deficiencies. Stranded animals suffering from starvation or that are nursing calves that have a dietary intake high in carbohydrates accumulate excessive triglycerides in hepatocytes. A mild degree of hepatic lipid accumulation may be normal in young nursing calves due to the high fact content of milk (eSlide 22.e1), though hepatic lipidosis in young animals is usually an indication of inanition/starvation.[102] In cases of hepatic lipidosis, the liver is often enlarged, diffusely pale tan and has a waxy appearance. Microscopically, hepatocytes are swollen with clear cytoplasmic micro and macrovesicles that displace the nucleus and in more severe cases, there may also be compression of the sinusoids and bile canaliculi with cholestasis. Venn-Watson and colleagues[224] have associated mild cases of hepatic lipidosis in adult bottlenose dolphins in managed care with insulin resistance and metabolic syndrome.

Renal calculi associated with **ammonium urate nephroliths** (Fig. 22.2) have been described in bottlenose dolphins under managed care and have occasionally been seen in stranded, wild animals. Urolithiasis can result in chronic renal disease and end-stage kidneys in more severely affected animals. A study by Smith et al.,[201] suggests that dolphins are susceptible, in part, due to a high dietary load of purines that results in a transient but marked increase in the urinary supersaturation of the sparingly soluble ammonium urate salt. In advanced cases, nephrolithiasis is associated with clinical hematuria and nephritis that may progress to secondary renal atrophy. When dislodged from the kidney, stones can obstruct the ureters or urethra and cause subsequent hydroureter, hydronephrosis, and possible unilateral or bilateral (depending on the location of the stones) renal failure. Nephrolithiasis may predispose to urinary tract infection. In Europe, **vaginal stones** have been identified in young animals and the composition is similar to uroliths.

Massive urolithiasis of the penile urethra was reported in an adult pygmy sperm whale. The urolith was composed of 100% **struvite** (magnesium ammonium phosphate) and routine culture yielded *Klebsiella oxytoca*, a urease-positive bacterium occasionally associated with struvite urolith formation in domestic animals. Histologically, there was moderate multifocal to coalescing plasmacytic balanitis and penile urethritis. No associated bladder or kidney damage was identified.[91] While these cases are uncommon, they demonstrate the need for urolith composition analysis to better understand the pathogenesis of the condition. Struvite

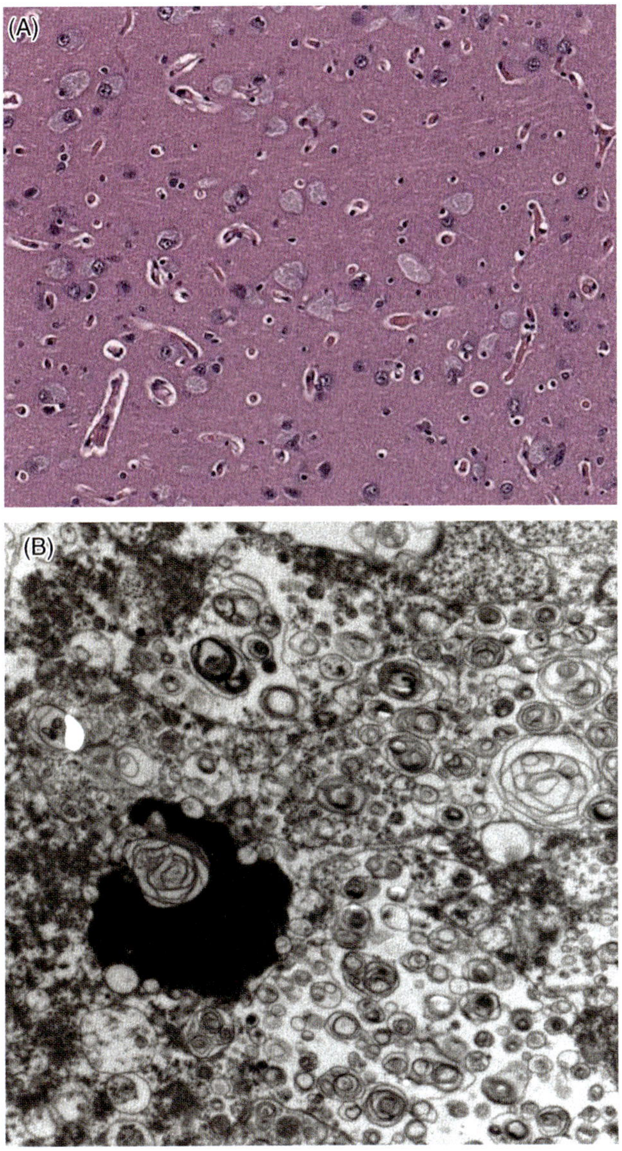

FIGURE 22.3 Lysosomal storage disease in a beluga whale. (A) The cytoplasm of the neurons is filled and distended by granular to vacuolated cytoplasmic content and there is nuclear margination. (B) Accumulation of incompletely degraded substrate within lysosomes appears as membrane-bound, concentric, lamellar inclusions. EM (see eSlide 22.e2).

urolithiasis is likely related to urinary tract infection as opposed to a metabolic etiology for the ammonium urate stones. **Vaginal struvite calculi** have also been described.

A **lysosomal storage disease** has been seen in multiple beluga whales (Figs. 22.3A, B; eSlide 22.e2). The condition is associated with reduced cognitive function. The lesions include neuron expansion and filling of the neurons with lysosomal lamellar bodies. To date, affected animals have all been adults. Investigations have focused on determination of the enzyme deficit and whether the condition is genetic or acquired.

Toxic

Harmful algal blooms (HABs) include a number of toxins that are elaborated by predominantly free-swimming unicellular dinoflagellates (Table 22.1). These organisms occur in deep offshore waters and near-shore bays, lagoons, and estuaries. As part of their life cycle, dinoflagellates can sexually reproduce, and then encyst for varying periods of time. Under appropriate environmental conditions, they reemerge and produce toxins. The geographic and ecologic distribution of these organisms parallels those of cetaceans. Blooms can result in significant mortality events involving marine mammals, birds, fish, and invertebrates.[120] Algal blooms occur globally and terrestrial species, including man, can be affected. It is important to note that toxin levels can vary between isolates within a bloom and not all algae recovered from a bloom may produce HABs. From a pathologic perspective, mass strandings of cetaceans alone or in combination with other aquatic or avian species should prompt an inquiry into an HAB event, and appropriate diagnostic and environmental samples must be collected for confirmation. There are few pathognomonic lesions related to HAB exposure in cetaceans and sublethal, long terms effects are largely unknown. Some dinoflagellates bioaccumulate or concentrate in invertebrates or fish up the food chain, resulting in more protracted strandings. Other toxins can be inhaled or ingested and are directly lethal, resulting in a pulse of mortalities. An increased frequency and intensity of HABs are suspected to be related to climate-change and associated ocean warming events.[61]

Saxitoxins are neurotoxic and affect voltage-sensitive sodium channels. This results in impaired nerve impulse propagation and produces paralysis. Saxitoxin has been implicated in the deaths of large whales. In addition to direct, lethal effects, sublethal exposure has been postulated to result in impaired reproductive success and impede an animal's ability to successfully forage, maintain nutritional condition, and possibly avoid marine vessel traffic. A disproportionate number of animals that present with evidence of ship strike or entanglement have detectable levels of saxitoxin and domoic acid. However, the contribution of these toxins to predisposing animals to vessel or other anthropogenic impact is unknown.[73]

Karenia brevis is geographically limited to the Gulf of Mexico, Atlantic coast of Florida, the Caribbean, and more recently has been identified in the waters off the Canary Islands. **Brevetoxicosis** or HABs involving this organism, called "red-tide events," are now annually reported off the Florida Gulf Coast. As the organism is unarmored, it tends to be relatively pleomorphic and labile. Toxins are released with rupture of the cell membranes. The pathogenesis of **brevetoxicosis** has been attributed to prolonged opening of voltage-gated sodium channels in

TABLE 22.1 Notable Algal and Toxic Pollutants

Disease	Toxin and Algae	Vectors or Exposure in	Clinical Signs and Gross Pathology	Histopathology	Confirmatory Testing	Mechanism
Amnestic shellfish poisoning	Domoic acid (DA) primary DA producing algal diatom *Pseudo-nitzschia australis*	Bivalves, scallops, mussels, clams, oysters and northern anchovy, benthic and pelagic fish, birds (pelicans and cormorants), and marine mammals	Mortality with peracute exposure may lack clinical signs and gross or histologic lesions. Piriform lobe malacia, myocardial necrosis, bronchopneumonia, reproductive loss (abortion).	In cetaceans, contribution to morbidity and mortality not fully resolved. Algae and toxins transit the bowel, whales tend to be asymptomatic, but there may be an association with ship strikes	LC–MS/MS on urine and feces. ELISA on serum, urine, and feces SEM of gastric content and feces to detect diatom frustules. Cytology, gastric fluid, and feces	Glutamate receptor binding and activation
Diarrheic shellfish poisoning	Okadaic acid, (DTX-1 and DTX-2) and dinophysis toxin	Bivalve shellfish, scallops, mussels, clams, oysters	Unknown	Unknown	Gastric fluid and feces—assays still to be validated	Inhibitor of protein phosphatase activity. Serine/threonine phosphatases 1 and 2A
Neurotoxic shellfish poisoning	Brevitoxin	Bivalve shellfish, scallops, mussels, clams, oysters, fish	Pulmonary and nasopharyngeal edema and hemorrhage, rhinitis, multiorgan hemosiderosis, nonsuppurative leptomeningitis	Unknown	Clinical history gross, histopathology. IHC on lungs and lymph nodes. ELISA, RBA, RIA, and LC–MS/MS on blood, liver, serum, urine, feces, and gastric content	Binds voltage-dependent sodium channels, results in prolonged activation
Paralytic shellfish poisoning	Saxitoxin	Scallops, mussels, butter clams, oysters, cockles, herbivorous fish	Humpback whales, birds, nonspecific effects	Unknown	ELISA, RBA, LC–MS/MS on urine, gastric content, and feces	Binds voltage-dependant sodium channels, inhibits channel conduction and blocks neuronal activity, respiratory paralysis

Ciguatera fish poisoning	Ciguatoxin, maitotoxin, scaritoxin *Gambierdiscus toxicus*	Large reef fish, grouper, red snapper, amberjack, barracuda	Unknown	Unknown	LC–MS/MS RBA, RIA on liver and feces	Similar to brevitoxin, bind to voltage-dependent sodium channels
Blue-green algae (cyanobacteria) Hepatotoxic shellfish poisoning (microcystin)	Anatoxin, microcystin, anabaena, nodularins (New Zealand and Baltic), cylindrospermopsin	Direct contact with water, aerosol inhalation, drinking water, shellfish bioaccumulation, and possibly fish	Jaundice with acute hepatic hemorrhage and necrosis	Hepatocellular vacuolation, necrosis and apoptosis, cholestasis, hemorrhage hepatocellular necrosis, hemorrhage, karyocytomegaly with pseudoinclusions, biliary ductular hyperplasia and renal tubular necrosis	Histopathology and IHC HPLC, LC–MS/MS, protein phosphatase inhibition assay, ELISA on fungal mats and water. Methodology in development for liver, and GI contents	Some toxins with "fast death factor" and peracute death, phospholipase inhibitor, phosphatase 1 and 2A inhibitor
POPs	PCBs, DDT, aldrin, chlordane, dieldrin, endrin, heptachlor, hexachlorobenzene, mirex, toxaphene, dioxin, polychlorinated dibenzofurans, and others	Ingestion, absorption, percutaneous, respiratory, in utero	Immune deficiency, potential reproductive failure	Thymic and splenic lymphoid depletion associated with starvation and chronic inflammation lipofuscinosis in the liver and kidney with hepatocellular hydropic change, central necrosis, and lymphocytic infiltrates	MS on blubber, liver, or serum	
Heavy metals	Lead, mercury, zinc, iron, manganese, copper, tin, vanadium	Ingestion, respiratory, in utero	Potential CNS and vascular disease. Concerns differ with differing metals		LC–MS on liver quantifies mercury with selenium levels	

CNS, central nervous system; *DA*, domoic acid; *DDT*, dichlorodiphenyltrichloroethane; *ELISA*, enzyme-linked immunosorbent assay; *GI*, gastrointestinal tract; *HPLC*, high-performance liquid chromatography; *IHC*, immunohistochemistry; *LC*, liquid chromatography; *MS*, mass spectroscopy; *PCB*, polychlorinated biphenyl; *POP*, persistent organopollutants; *RBA*, receptor binding assay; *RIA*, radioimmunoassay; *SEM*, scanning electron microscopy.

neurons. Sporadic large-scale stranding events of bottlenose dolphins have been associated with ingestion of heavily contaminated menhaden, aerosol exposure, and ingestion of contaminated fish during fish kills.[74] Aerosolization and inhalation of the toxin are associated with pulmonary and nasopharyngeal edema, hemorrhage, and multisystem congestion. Upper and lower respiratory tract lesions may be attributed to direct chemical irritation. Immunohistochemistry demonstrates brevetoxin in pulmonary alveolar lymphocytes and histiocytes as well as in pulmonary lymph nodes. Localized immunosuppression has been postulated as a contributing factor in secondary infections.[213] An unusual stranding event involved 12 rough-toothed dolphins found adrift or beach cast along the southwestern coast of Gran Canaria. Toxic levels of PbTx2 and PbTx3 brevetoxins were detected in gastric contents. This was the first report of brevetoxicosis in European waters.[70]

While **domoic acid** has been detected in the gastrointestinal contents and urine of several species of cetaceans, a direct causal link to clinical disease and characteristic pathologic lesions is less often reported when compared to lesions noted in pinniped species (see Pinnipediae). The hippocampal formation in cetaceans is extremely difficult to section consistently for histologic evaluation, especially when specimens are autolyzed, which is frequently encountered in stranded cetaceans. Thus, it should be expected that brain lesions associated with domoic acid toxicosis can be missed on histologic evaluation if appropriate sections of the hippocampus are not obtained. Based on extrapolation from California sea lions and a few cetaceans evaluated with lesions consistent with domoic acid toxicosis, acute to subacute gross lesions may include neuronal necrosis in the pyriform lobe, myocardial degeneration and necrosis, pulmonary edema with occasional pneumonia, gastrointestinal hemorrhage, and reproductive failure. Chronic lesions may include hippocampal atrophy and emaciation. Domoic acid degenerative cardiomyopathy has been noted in a few cetaceans. Microscopically, the acute phase is characterized by multifocal myocardial cytoplasmic vacuolation with variable interstitial accumulation of edema fluid. With more advanced lesions, there was more generalized myocardial vacuolation with multifocal necrosis, interstitial fibrosis, and scant mononuclear infiltrates with replacement of myocardial fibers by adipocytes.

Among the more naturally occurring toxic compounds are **cyanotoxins**, which are an example of land-to-sea toxin flow. Cyanobacterial blooms may occur in surface waters with eutrophication and have been associated with exposure to the neurotoxin, *B*-methylamino-L-alanine. *B*-amyloid plaques and neuronolysis, particularly in the auditory cortex, have been reported in dolphins recovered during stranding events in Florida and Massachusetts.[41] A history of cyanobacterial blooms in waterways, coupled with the microscopic findings, should prompt submission of brain tissue for analysis at a reference laboratory. The strandings are important sentinels for human and ecosystem health.

Cetaceans are particularly vulnerable to the effects of **oil spills**, which may occur as large-scale catastrophic events like the Deep Water Horizon or Exxon Valdez spills, or smaller and often recurrent localized spills such as those that are associated with industrial sites. Regardless of the size of the spill, even small regional spills can pose a risk to wildlife in general or if the spill affects critical habitats or threatened or endangered populations. The lack of baseline data on ecosystem components, animal health, demographics, and population trends within an oiled area can significantly impede interpretation of pathology and ancillary diagnostic findings and confound delineation of long-term lethal and sublethal effects of spills. Exposure can be through inhalation of oil products or volatile compounds, direct contact with the skin and mucous membranes, ingestion of contaminated prey, or through aspiration of compounds from the gastrointestinal tract into the lungs. A few laboratory studies focused on oil exposure were conducted in the 1980s; however, few studies have focused on the pathogenesis of exposure. Because petroleum products can include several inorganic and organic compounds, it is difficult to extrapolate the effects of one spill to another. As a result, the pathogenesis of oil exposure is often conjectured from a review of the findings in an oil spill event. In the Deep Water Horizon spill, **adrenocortical atrophy, pneumonia**, and **reproductive loss** were significant findings in exposed animals and long-term follow-up studies may reveal insights into the consequences of exposure and population recovery.[186] Detection of visible oil on the skin surface or within the upper respiratory tract or digestive system may indicate oil spill exposure and prompt notification of appropriate regional and national authorities. Large spill in the United States may illicit a national emergency response (e.g., Incident Command Activation).[225]

Although banned in many regions of the world, **persistent organic pollutants (POPs)** (Table 22.1) are still detected in many marine mammal populations.[76,106,173,203] Exposure and effects are due to aerosolization and widespread atmospheric dispersal, environmental persistence, and the longevity of many cetacean species.[14,221] Individual and synergistic effects of POPs may potentiate toxicity of some compounds. Although numerous hematology, clinical chemistry and in vitro and in vivo immune function studies have been conducted in cetaceans and surrogate pinnipeds, there are few descriptive gross or histologic descriptions. In addition to impacting immune function, endocrine disruption also occurs. Long-term necropsy results linked to contaminant levels in stranded cetaceans have identified a reduction in splenic and thymic lymphoid populations with increasing POP levels, and immunosuppression is inferred.

However, it is important to place contaminant loads in the context of postmortem state and nutritional status of the animal as artifactual increases can occur with peripheral mobilization of adipose tissue and postmortem dissolution of fat stores.

Cetaceans are exposed to **heavy metals** in the environment. Many have a significantly higher level of methylmercury relative to terrestrial species (Table 22.1). Nutrient upwellings and introduction of sequestered metals from depositions in substrates have been associated with El Niño events and have been implicated in exposure in cetaceans. Cetaceans have protein complexes (metallothioneins and possibly other chelators) that bind copper, zinc, mercury, and selenium, as well as sequester metals in the liver and skin with little to no negative effects. Elevated methylmercury has been implicated in peripheral and central neuropathies in terrestrial species and lab animals, but there is no indication of similar changes in cetaceans to date. In a case series of South Australia adult bottlenose dolphins, increased liver cadmium, copper, and zinc were associated with renal pathology.[121] Distention of Bowman's space and proteinuria, and bone lesions consistent with osteoporosis of the ribs were associated with elevated liver cadmium levels. Lower mineral densities of vertebrae were noted in some animals, but this was not statistically associated with liver metal concentrations and could reflect impaired renal function.

Congenital/Genetic

Multiple single cases of **leucism** (aberrant white coloration) in cetaceans have been observed but genetic analysis has not been reported. Overall, this has been documented in over 20 species of whales, dolphins, and porpoises. The etiology of this coloration is typically presumed to be genetically based. A single case of a white killer whale was attributed to congenital **Chediak-Higashi syndrome.**[210]

A number of **cardiac anomalies** have been reported in cetaceans. These include transposition of the pulmonary artery and aorta, ventricular septal defects (VSD), persistent ductus arteriosus (PDA), atrial septal defect (ASD), right ventricular hypertrophy (RVH) and subvalvular pulmonic stenosis with pulmonary artery and mitral valve hypoplasia.[161] Sequelae may include pulmonary hypertension and other systemic lesions which culminated in cardiopulmonary decompensation.

Congenital hyperplastic goiter has been associated with perinatal captive bottlenose dolphin deaths.[77] Histologic changes include reduced follicular luminal diameter, markedly reduced or absent luminal colloid, hypertrophy of follicular epithelium, micropapillary proliferation, and follicular dysplasia.

Skeletal malformations including **craniofacial abnormalities** such as **prognathism**, **brachygnathism**, **facial duplication**, and **scoliosis** (Figs. 22.4A, B) have been reported.[59] Care must be taken when diagnosing skeletal changes because early infections and traumatic events can cause skeletal changes in very young calves. Many alterations in bones such as incomplete fusion of the dorsal processes of cervical vertebrae occur with great variety in severity. **Spina bifida** in humpback whales has been associated with large gaps that span multiple cervical vertebrae.[85] In killer whales, the same lesion has been seen but has only been reported to impact single cervical vertebrae with dorsal gaps varying from less than 1 to 4 cm (Fig. 22.5). In both species, determination of clinical impact involved assessing the entire skeleton and determining the number of animals within populations that have these changes. Some severe lesions have been reported in humpback whales that lived to adult ages.

Age-Related/Degenerative

Multiple age-related and degenerative changes can be seen in cetaceans. Cardiac, dental, ocular, musculoskeletal, renal, endocrine, and geriatric conditions have been described. **Cardiac disease** is very common in both captive and free-ranging aged cetaceans (St. Leger, pers. obs.). Changes include multifocal myocardial degeneration and fibrosis, endocardiosis, valvular fibrosis and endocardiosis and left ventricular hypertrophy. **Arteriosclerosis** is uncommon but has been seen in both captive and free-ranging bottlenose dolphins. Degenerative **arthritis, spondylitis**, and **spondylosis** occur and are sometimes linked to inanition perhaps secondary to reduced mobility. Degenerative changes can develop independent of comorbidities and be unrelated to the proximate cause of death. A holistic evaluation of cases helps to contextualize pathologic findings.

Tooth wear, fracture, laxity, and **attrition** are associated with advanced age in many odontocetes (Figs. 22.6A, B). The limited or lack of enamel covering odontocete teeth increases their vulnerability to wear and damage. Wear intensity varies by species and diet, with killer whales and pseudorcas having the highest prevalence.[125] As patterns of dental wear in the killer whale can reflect dietary specializations (elasmobranchs versus salmonids or other marine mammal prey species) and learned behaviors; tooth wear is not always the best indicator of age in some odontocete species.

Both **nodular adenomatous thyroid gland hyperplasia** and **thyroid follicular cysts** are common in older bottlenose dolphins and are generally considered incidental, age-related changes. Mild to severe adenomatous thyroid hyperplasia is also commonly seen in beluga and harbor porpoises. In beluga, the total volume of the lesions is positively correlated with age.[115] These changes have not been sufficiently investigated to determine an etiology or the physiologic impact, but may have important implications

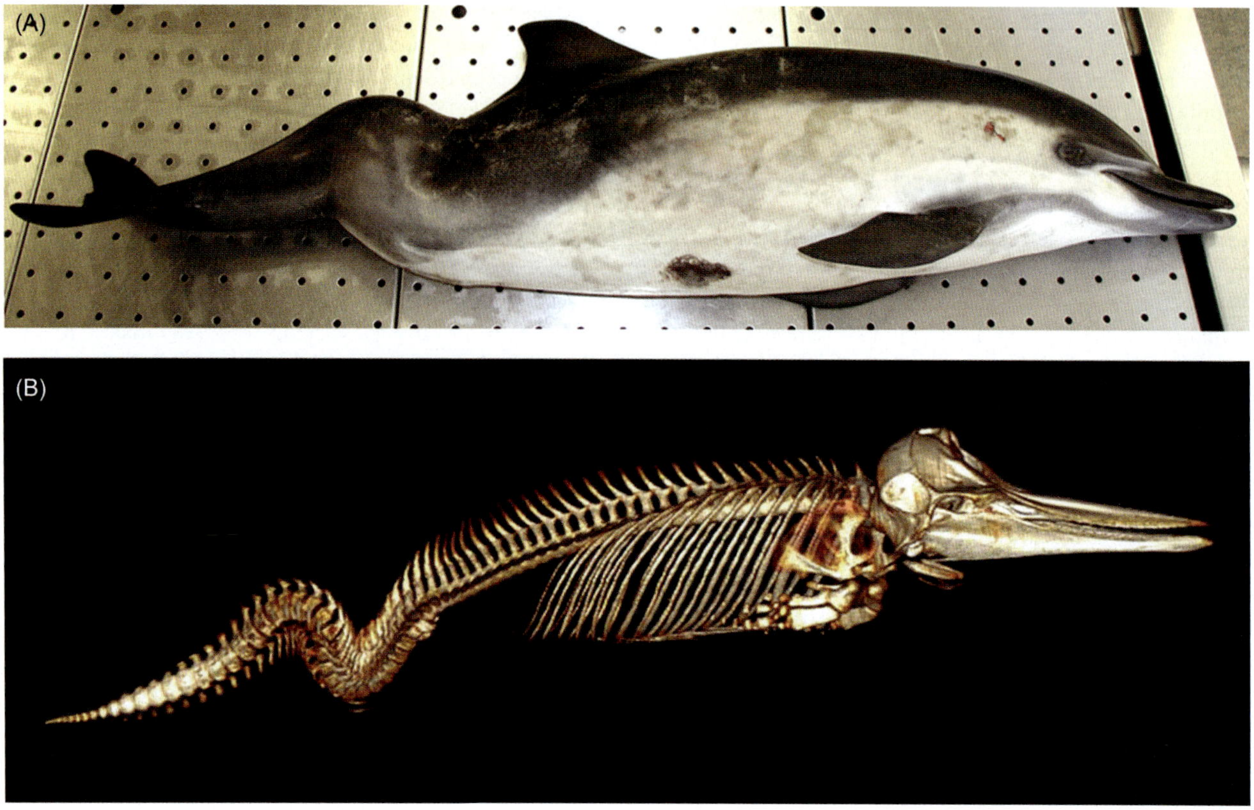

FIGURE 22.4 Scoliokyphosis in a young stranded common dolphin. (A) The animal has pronounced downward and lateral deviations in the vertebral column. The vertebral changes may be congenital but early trauma to the spine can cause similar vertebral lesions. (B) Computed tomography reconstruction of the vertebral column clearly demonstrates the dramatic malformation.

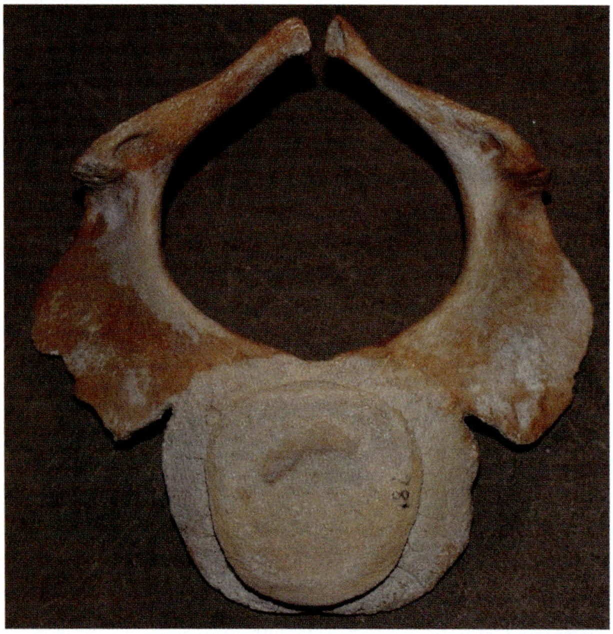

FIGURE 22.5 Incomplete fusion of the dorsal process in a cervical vertebra in a killer whale. This congenital defect is similar to the changes seen in occult spina bifida. In this case, it was not considered severe enough to have been associated with clinical effects.

when assessing contaminant loads on thyroid gland hormone levels or gene expression.

Adrenocortical hyperplasia is a common change in older delphinids (Fig. 22.7). It is seen in both free ranging and those in human care and consists of variation in the thickness of the adrenal cortex and an increase in ratio between the thickness of the cortex and medulla (corticomedullary ratios). Increased thickening can be diffuse or multifocal and nodular (cortical, medullary, or extracapsular) and is occasionally associated with cortical cyst formation (eSlide 22.e3).[115] The zona fasciculata and zona glomerulosa are most often expanded but all three cortical zones can be involved. Chlorinated hydrocarbons have been investigated as a cause for this lesion, but to date, no specific association has been identified.[113] In general, adrenal hyperplasia and cyst formation are not associated with clinical symptoms. They may represent compensatory physiology adaptations and typically appear incidental with few sequelae. This differs from adrenocortical atrophy, which has been reported in animals exposed to petroleum products (see "Toxic" section; Table 22.1), where profound effects associated with hypoadrenocorticism have been suggested. Adrenocortical atrophy may also be noted in animals on long-term steroid treatments.

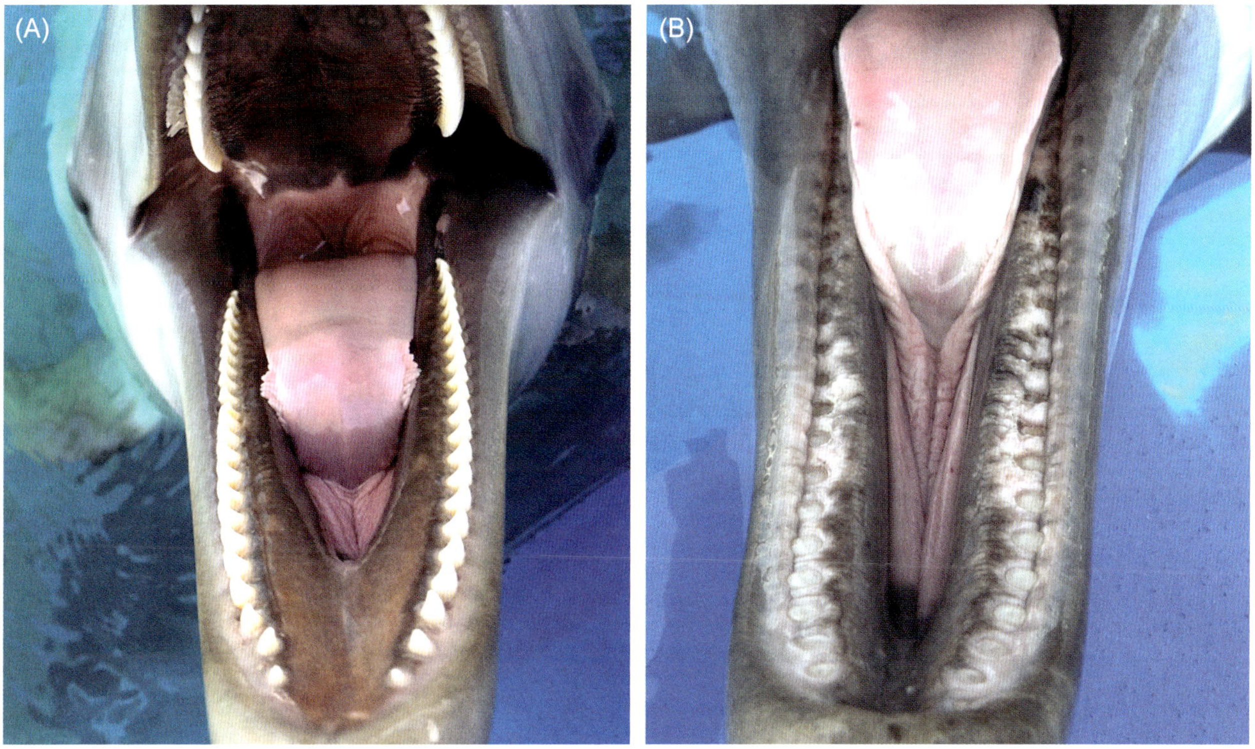

FIGURE 22.6 Tooth wear in a bottlenose dolphin. (A) Teeth are present in both the mandible and maxilla. They are conical. (B) Teeth wear as animals age. The teeth in this geriatric dolphin are worn to the level of the gingiva. Wear and tooth breakage are common in aged odontocetes.

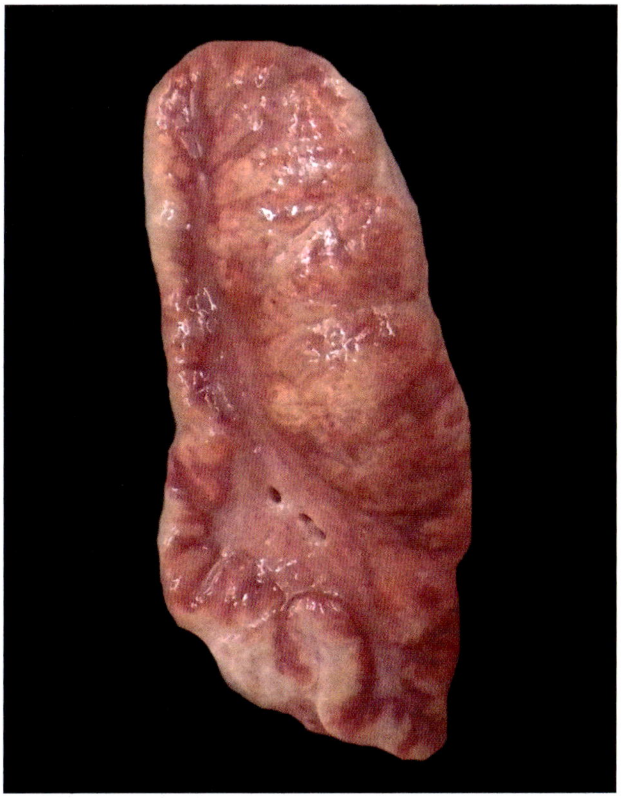

FIGURE 22.7 Adrenocortical hyperplasia in a killer whale. Irregular expansion of all layers of the adrenal cortex is present. It is most prominent in the upper right side of the gland in this image. The hyperplastic cortical tissue bulges from cut surfaces (see eSlide 22.e3).

Cardiomyopathy was first described in pygmy and dwarf sperm whales from the southeastern United States.[22] This condition is seen in animals from both the Atlantic and Pacific Oceans. Cardiomyopathy and **myocardial degeneration** are identified predominantly among adult whales. Males are overrepresented (up to 75%) in both species. The etiopathogenesis remains unknown. Case evaluations suggest this is a chronic and progressive disease that initially presents as myocardial degeneration that progresses to **dilated** or **hypertrophic cardiomyopathy**.[22] Gross findings include cardiac enlargement with right or biventricular dilation and thinning of the interventricular septum (Fig. 22.8). Right ventricular dilation can be impressive in the final stages of cardiac degeneration. Microscopically, lesions are characterized by moderate to extensive myocardial cellular degeneration, myofiber disarray, eosinophilic sometimes globular homogenization of myofiber sarcoplasm, anisokaryosis with karyomegaly, and myocardial atrophy and loss. There is concurrent interstitial edema and fibrosis. Infrequently, cases have limited associated mononuclear myocarditis. As expected, secondary concerns like pulmonary and hepatic congestion are common.

Stress cardiomyopathy has also been reported in cetaceans. It has most often been associated with bycatch, vessel or propeller strike, or live strandings.[29] Histopathology of a number of stranded cetaceans along the coast of the Canary Island and Andalusia revealed acute to subacute myocardial degeneration and necrosis with interstitial

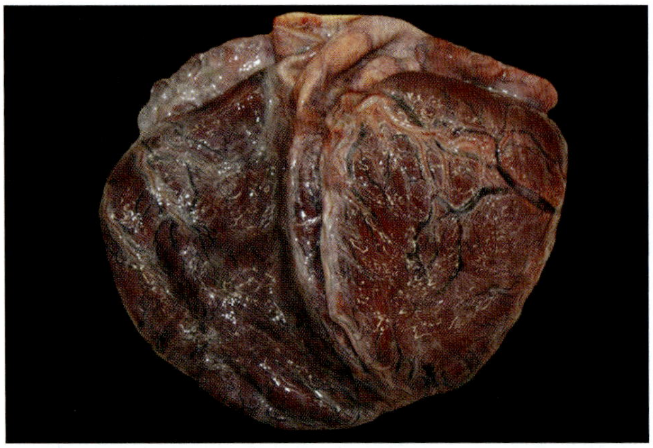

FIGURE 22.8 Dilated cardiomyopathy in a pygmy sperm whale. The heart is diffusely enlarged. Both ventricles are dilated with thin, flaccid walls. *(Photo Courtesy of Hubbs-SeaWorld Research Institute.)*

myoglobin deposition and nonsuppurative inflammation. Immunohistochemistry revealed depleted cardiac troponin 1, cardiac troponin C, and myoglobin, and clinical chemistry analysis showed an increase in circulating cardiac troponin 1.[28] The pathogenesis of the myocardial lesions was attributed to a cholinergic surge, transient coronary vasoconstriction, and subsequent ischemia-reperfusion type injury. Understanding the pathogenesis of these lesions is critical for medical management and recovery of live stranded cetaceans.

Inflammatory Non-Infectious

Gastritis and **gastric ulcerations** impact both captive and free-ranging cetaceans. Geraci and Gerstmann[79] associated nonparasitic gastric ulcerations with consumption of high levels of histamine from fish. Elevated histamine levels result from the freeze/thaw process, so it is unlikely that this mechanism occurs in free-ranging dolphins. The possibility of physiologic or psychogenic stressors related to social dynamics or environmental factors in the development of gastric ulcers in wild or display animals cannot be discounted.

Miscellaneous/Multifactorial

Anthropogenic factors including **entanglement** in fishing gear, **peracute underwater entrapment**, and sharp and blunt force trauma from **vessel strikes** are a significant contributor to mortality in marine mammals. Human interactions can result in long-term impacts on marine mammal population status, recruitment, and declines. Unfortunately, without direct field observations or photodocumentation for confirmation, many cases of human interactions are defined as only probable or suspect. This has important implications from a management, enforcement, and legal perspective. Moreover, postmortem change (autolysis), scavenging, and mutilation can impede recognition and interpretation of gross and microscopic findings in examined animals. However, advanced imaging studies, chemical analysis of gas emboli, and increasing knowledge about sharp and blunt force injuries, have all contributed significantly to improve detection and confirmation of anthropogenic insults. It is important to note that these categories are not mutually exclusive and detection and interaction of more than one entity may occur. For instance, large whales may have evidence of chronic or historic line or rope entrapment/entanglement and supervening acute injuries consistent with vessel strike. Conversely, a single insult may result in multiple pathologic manifestations. As these incidents may trigger regulatory enforcement or legal investigation, review and use of appropriate forensic diagnostic protocols is imperative (see Wildlife Forensic Pathology). Moore et al.[139] provide an excellent primer and includes downloadable forms as well as case definitions and gross and microscopic images of different serious injury entities.

Bycatch refers to the entanglement of cetaceans in fishing gear. It affects both small and large cetaceans worldwide.[164] It is considered one of the most significant threats to marine mammal populations. Critically endangered species including the vaquita and baiji have largely been extirpated by fisheries interactions that may ultimately result in their extinction. Derelict and active fisheries netting, commercial monofilament lines, crab-trap lines, and rope impact cetaceans directly through entanglement or entrapment. External evidence of net entanglement can include line or hatched gear impressions along the surface of the skin or oral mucosa (hatched or linear furrows), abrasions, incisions, lacerations, and subcutaneous hemorrhage (Fig. 22.9A). However, surface evidence of entanglement may not always be apparent, particularly when animals are entrapped in a trawl with large numbers of fish that prevent direct contact with the netting. With nets or longline and crab trap lines, and lines that are tethered to substrates, entangled small cetaceans may be trapped under water and die from asphyxiation. In these instances, animals may present in good nutritional condition with stable white tracheobronchial froth, hyperinflated lungs with generalized pulmonary edema, congestion, and subpleural emphysema. Pleural and pulmonary petechiae and ecchymoses may also be seen edema, often associated with hemorrhage may be observed in subcutaneous and muscular tissue in the mandibular, neck and axillary regions secondary to intense struggle.[101]

With large and smaller cetaceans that are released or that break the line or netting, subsequent wrapping or entanglement of the gear may occur. It is common for rope and net to entangle the peduncle or tail stock, fins, flippers, and through the oral cavity. In severe cases, ropes can become deeply embedded within and cause significant tissue damage with necrosis, fibrosis, and granulation tissue

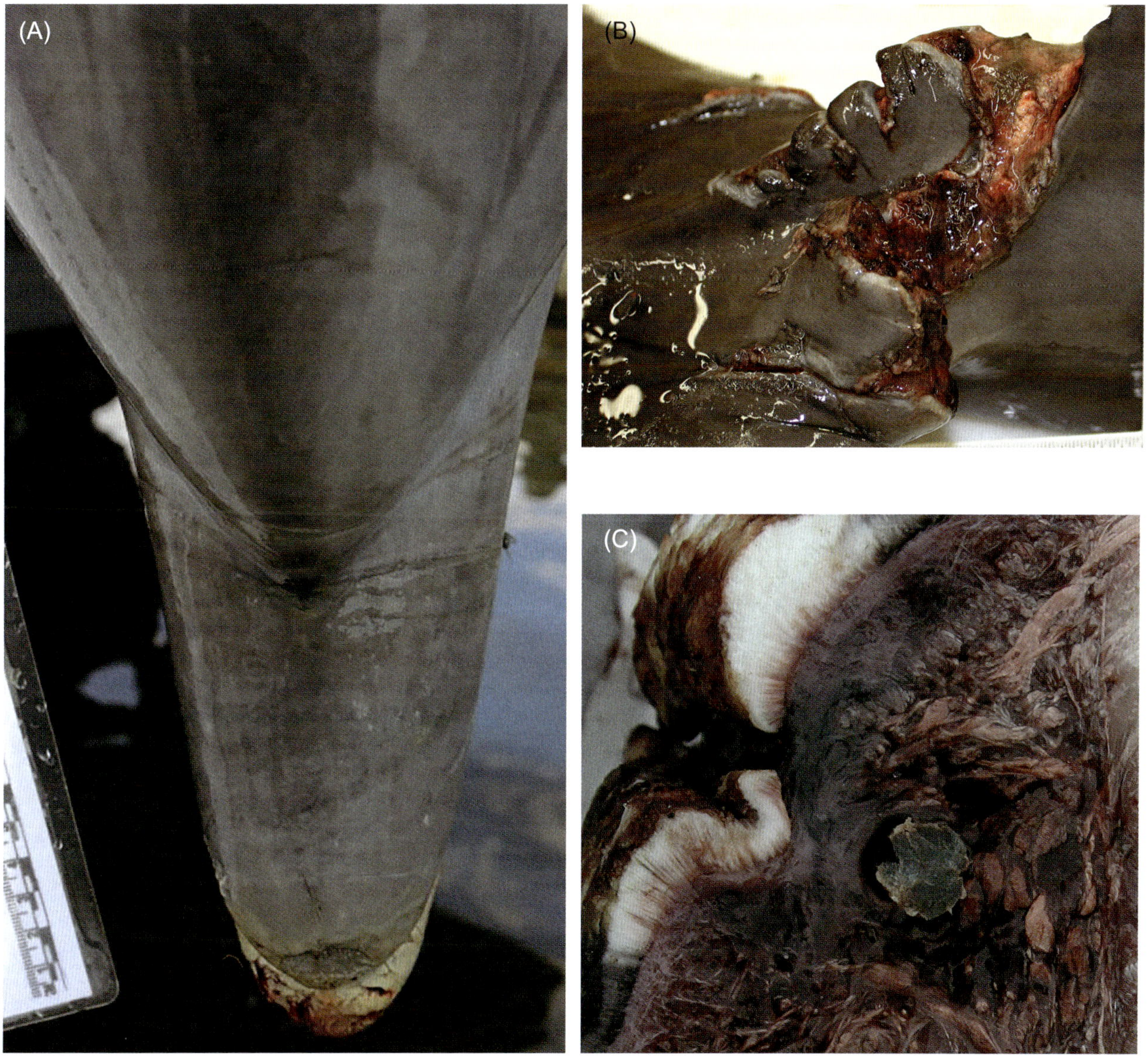

FIGURE 22.9 Fishing gear entanglement in dolphins (A, B) and a North Atlantic right whale (C). (A) Along the rostrum, melon and lateral aspect of the melon are multiple parallel linear wounds. The superficial furrows are attributed to line entanglement. (B) Fishing line laceration and net wounds can lacerate tissue and create deep wounds. The line may envelop the torso and with constriction, can extend deep into the underlying soft tissues. (C) In this chronic, fatal case of entanglement, rope suggestive of a lobster trap given an attached wire mesh fragment was wrapped around the head, through the mouth, and around both flippers. The image shows a transverse section through a deep groove cut into the skin and underlying tissues by the rope and the rope, which remained sequestered in the lower right lip and resulted in major scarring through the entry tract. *(Parts A, B: Virginia Aquarium & Marine Science Center; Part C: Photo Courtesy of the Florida Fish and Wildlife Conservation Commission [NOAA Research Permit 932-1905].)*

formation (Figs. 22.9B, C), and osseous metaplasia and periosteal bone proliferation can develop as a result of the embedded fishing gear.[128] Depending on the degree of constriction or compression and anatomic location, vascular perfusion may be impaired and hypoxia and ischemic necrosis of the distal extremities may occur. Amputation of an appendage may occur either acutely with incision on impact or due to the chronic ischemic effects. With more protracted entanglements, animals typically lose condition. If hemorrhage is present, a timeline may be inferred from observation of erythrocytes, hematoidin, or hemosiderosis, and margins of bone fractures should be closely evaluated to assess the extent of necrosis, inflammation, callous formation, and remodeling. Exertional myopathy and rhabdomyolysis, myoglobinuria, myoglobinuric nephrosis, and crystalline nephropathy can be seen, and adrenocortical hyperplasia with lipid degeneration can develop in chronic cases.[96,154] Once gear is released, wound resolution may occur with primary and secondary intention healing.

It is important to photodocument and diagram the course of the fishing line or rope throughout the carcass (wrapping pattern) and note any knots that may be present. In animals with chronic entanglement that are in an advanced state of decomposition with skin sloughing, the raised linear ridges of fibrosis associated with entanglement may be seen and associated with superficial plaques or plugs of adhered epidermis due to adhesions. Less common forms of entrapment may include interactions with aquaculture net-pen facilities, piers, and boat harbors.[168] Based on long-term data analysis, persistent organic pollutants and harmful algal blooms (toxins) have been associated with an increased risk of entanglement. If feasible, samples to screen for these compounds should be collected. Due to the significance of by-catch to cetaceans, pathologists should harbor a level of suspicion of entanglement and assess all carcasses for evidence of net or line entanglement. An indirect consequence of fisheries interactions is consumption of fish from long line hooks, a learned behavior recognized predominantly in killer, false killer, pilot, and sperm whales. In rare instances, hooks have been ingested and penetrated the oropharynx, resulting in secondary septicemia. This behavior may also result in whale harassment and possibly gunshot, further contributing to the loss of these animals.

Lesions consistent with **decompression sickness (DCS)**[68,104] are most often identified in deep-diving cetaceans, particularly beaked whales, but cases have also been reported in other species of odontocetes.[71] To date, DCS has not been identified in any stranded mysticetes (baleen whales). As with "the bends" in humans, changes in the dive profile under certain conditions is a prime contributor to interstitial or intravascular gas bubble formation, which can result in subsequent gas embolization, impaired vascular perfusion, ischemia, and infarction. The most common anthropogenic activities associated with this process include deployment of mid-frequency active sonar (MFAS) affecting primarily beaked whales (particularly *Ziphius cavirostris*). With acute presentations, multisystemic gas bubbles may be observed, particularly within in the mesenteric, renal, pulmonary, hepatic, coronary, meningeal, otic, and spinal epidural rete vasculature (Figs. 22.10A, B); subcutaneous, subserosal, and mesenteric vasculature should also be closely examined for intravascular gas. Associated hemorrhage is often identified in the visceral pleura, epicardium, meningeal areas, and acoustic fat (Fig. 22.10C). Histopathology typically features occlusion and variable vascular distention by clear microcavitations, acute perivascular hemorrhage, and occasional infarction (Fig. 22.10C). In the acute phase of severe decompression, fat emboli may also be identified histochemically in the pulmonary microvasculature as well as the epidural rete, subcapsular sinuses of lymph nodes, and less frequently, in the renal vasculature.[68] If animals survive the initial insult, lesions may resolve through vascular resorption and exhalation or offloading of gases.

A chronic form of gas bubble lesions that is not linked to MFAS but is associated with cystic lesions has been reported in several species of cetaceans (dolphins, whales).[69,104,105] The pathogenesis is still being characterized, but preexisting hepatic vascular disease (e.g., phlebothrombosis) and entrapment of decompression-related intravascular gas bubbles have been suggested as factors in its development. The histology resembles Budd-Chiari-like pathology.[69] Histologic features include intravascular fibrin thrombi and variably sized intravascular clear, cystic spaces that are associated with vascular dilation and sinusoidal congestion (Fig. 22.11). Hepatocellular atrophy and pressure or coagulation necrosis of adjacent hepatocytes can also be seen. The cystic spaces are clear (gas filled) or may contain or be lined by thin rims of peripheralized serum or fibrin. Based on several studies, this is a relatively uncommon but chronic, progressive process with the potential to contribute to individual morbidity and mortality in dolphins.[69] In the past, gas production attributed to putrefaction, and hemorrhage associated with impact with the substrate in live stranded animals, may have confounded differentiation between decompressive pathology and putrefaction. However, the advent of computed tomography imaging studies and chromatographic analysis of gas samples, facilitate differentiation of decompressive-related nitrogen from those gases attributed to decomposition artifact.[16–19]

Blast injury due to underwater detonation of explosives can have significant impacts on cetaceans.[193] Due to the unique physical properties of the aquatic environment and substrate, lesions are typically more profound in cetaceans than in terrestrial mammals. With detonation of explosives, there is rapid conversion of liquid or solid substrates to gas under high pressure and temperature, which generates a blast front and an immediate increase in underwater static pressure. This energy is transmitted and propagated radially as alternating positive and negative pressure waves.[98] On impact with an animal, there are implosive and pressure differentials, which can generate up to a ninefold increase in pressure. In contrast to detonations on the ground or in the air, underwater explosions are more lethal due to the incompressibility of water. This results in shock wave propagation up to three times farther in water than on land, and reflection of shock waves along the substrate and air surface interface resulting in more complex blast waves. In a recent study, the cause of death of 30% (8 out of 24 individuals) of harbor porpoises that stranded in the Baltic Sea after the underwater detonation of World War II munitions was assessed as blast injury.[193] In addition, one bycaught animal and another porpoise with signs of blunt force trauma also showed evidence of blast injury.

Pathology related to direct exposure to blast waves depends on the magnitude of the explosion, distance, and direction of travel of the animal from the blast, as well as

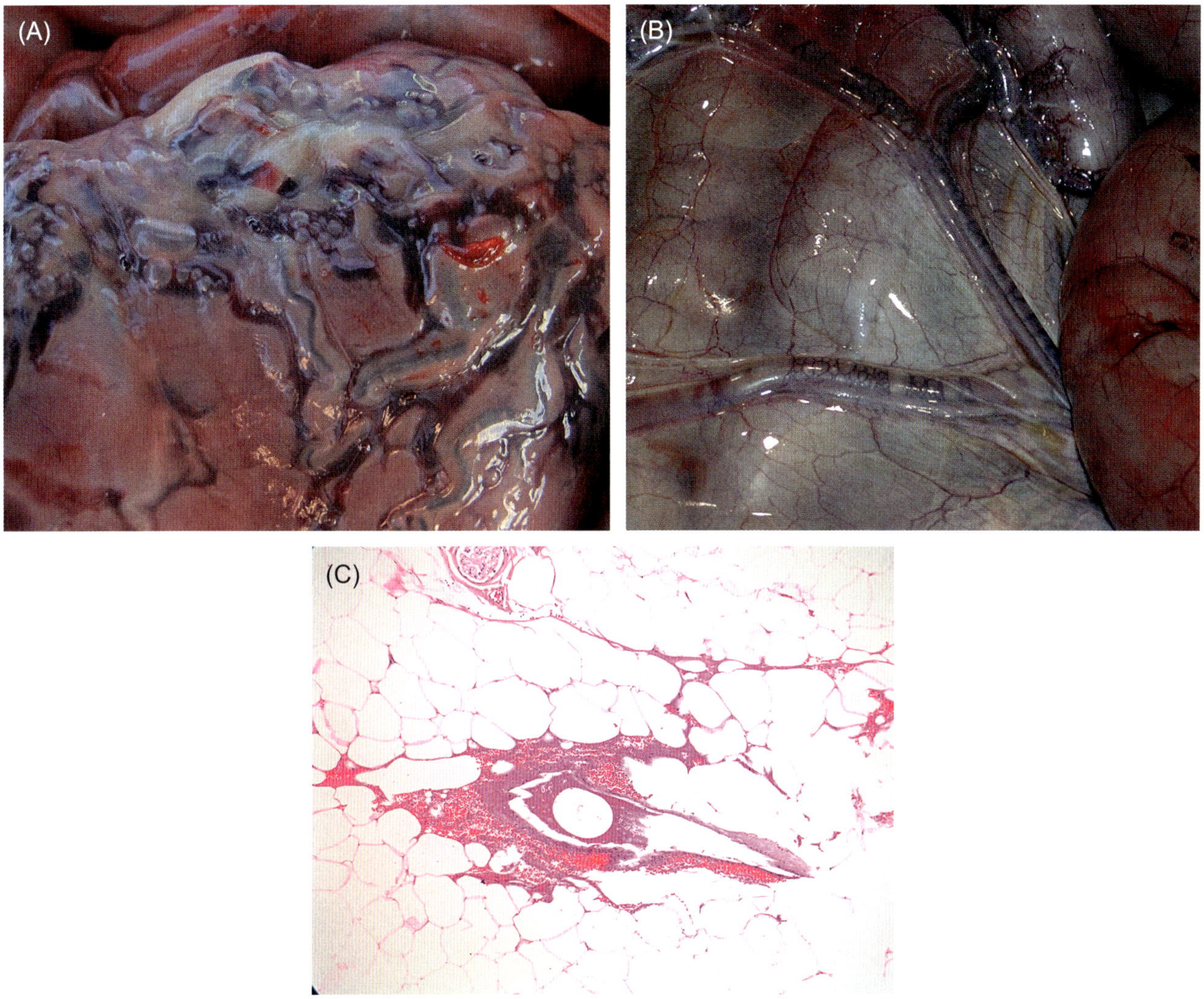

FIGURE 22.10 Gas bubble formation due to decompression phenomena in a beaked whale. Numerous, variably sized gas bubbles (gas emboli) are present in the coronary (A) and mesenteric (B) vasculature. (C) As in this image, intravascular gas embolus in the acoustic fat can be associated with acute perivascular hemorrhage. This is a common site for gas embolization. Acoustic fat hemorrhage may also be associated with impact of live stranding and entanglement, but gas emboli are not a related finding in these cases (see eSlide 22.e4, which features intravascular gas bubble accumulation secondary to blast injury). *(Part C: Photo Courtesy of the Institute of Animal Health and Food Safety, University of as Palmas de Gran Canaria.)*

environmental surroundings (sloping substrate versus underwater canyons). Typically, the respiratory, alimentary, circulatory, and auditory systems are most often and obviously impacted. On gross examination, hemorrhage may be apparent in the trachea and lungs. Pulmonary hematomas, edema, and emphysema tend to be more pronounced on the side of impact. Pulmonary rents with pneumothorax or hemothorax and acute intrabronchiolar hemorrhage and exfoliation of type I pneumocytes with detached and aspirated ciliated respiratory epithelia can be seen. Shearing of blood vessels in the lung and blubber can occur resulting in the formation of intravascular gas or fat emboli, which can compromise homeostasis. The gastrointestinal tract is particularly prone to injury due to gas content within its lumen. Gross lesions may be absent or include segmental hemorrhage and tearing or perforation of different levels of the digestive system. Lesions in the sound production and reception pathways in cetaceans include hemorrhages in the melon and acoustic fat around the lower jaw, hemorrhage in the peribullar sinus, middle and inner ear, dislocations and fractures of the middle ear ossicles (Fig. 22.12; eSlide 22.e4), and fractured tympano-periotic complex.[110,193]

Recent advancements in acoustic mapping of the cochlea have provided valuable insights into **acoustic injury**. If ears are harvested and perfused within 18–24 h after death, ultrastructural evaluation with electron microscopy and immunofluorescence studies of sensory cells and afferent innervation may provide evidence of hearing loss (Fig. 22.13).[81,142–145] In several cases of stranded cetaceans, lesions in the inner ear compatible with noise-induced hearing loss were identified.[140–142,177] These

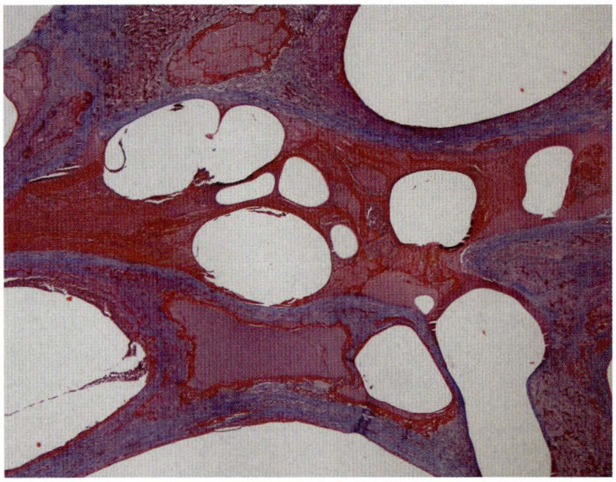

FIGURE 22.11 Budd-Chiari type lesions in a stranded striped dolphin. This condition, which contributes to hepatic venous outflow obstruction, is characterized by the formation of intrahepatic, intravascular encapsulated gas-filled cystic structures (gas bubble lesions). In this case, both fibrin thrombi (lightly eosinophilic aggregates) and cystic structures are present in the vessels. Associated changes can include fibrosis and hepatic cord atrophy. Trichrome. *(Reprinted with permission from Fernandez A, Jepson PD, Diaz-Delgado J, de Quiros YB, Sierra E, Mompeo B, Vela AI, Di Guardo G, Suarez-Santana C, de los Monteros AE, Herraez P, Andrada M, Caballero MJ, Rivero M, Consoli F, Castro-Alonso A, Quesada-Canales O, Arbelo M. Budd-Chiari-like pathology in dolphins. Sci Rep. 2022;12(1):12635.)*

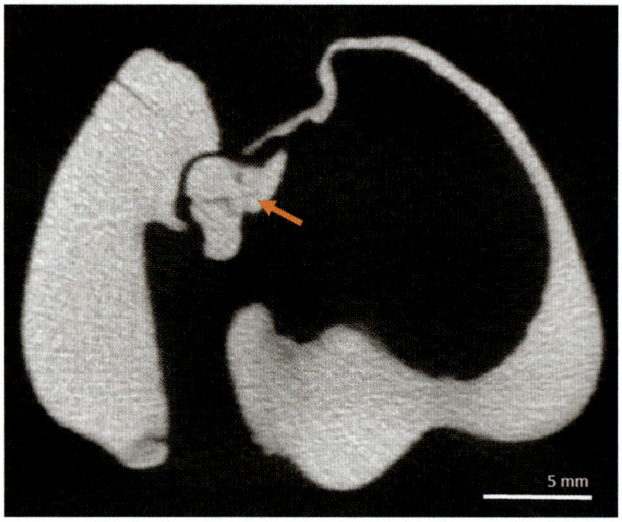

FIGURE 22.12 Blast injury in the middle ear ossicles of a harbor porpoise. A fracture line *(arrow)* is seen in the ossicle of the middle ear. *(Photo Courtesy of U. Siebert, Institute of Terrestrial and Aquatic Wildlife Research.)*

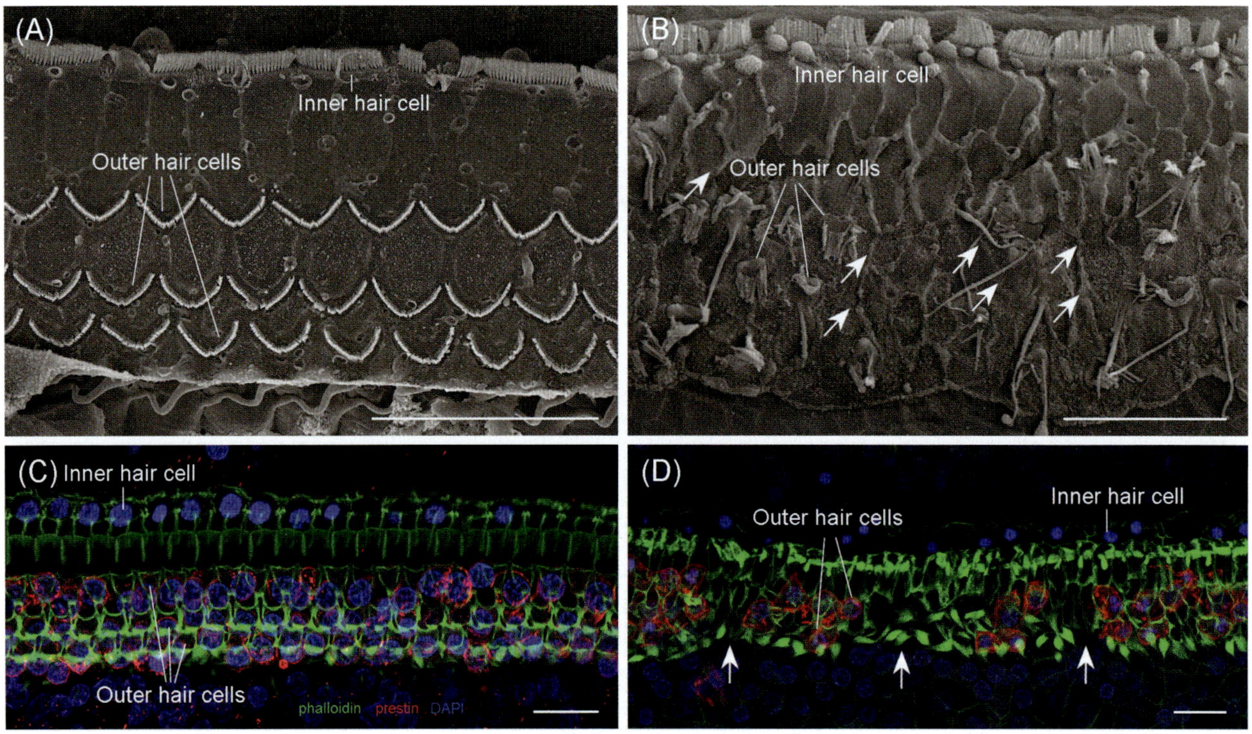

FIGURE 22.13 Damage in the organ of Corti of a beluga whale and harbor porpoise compatible with noise-induced hearing loss (SEM and immunofluorescence). (A) Normal arrangement of the sensory cells of the organ of Corti in one row of inner hair cells and three rows of outer hair cells. (B) The apex of the cochlea with missing outer hair cells, highlighted with *arrows*. (C) and (D) Maximum projections from confocal microscope images of the apex of the cochlea of the harbor porpoise showing the normal dispositions of the cells of the organ of Corti and evidence of lesions (*arrows* in D). The cochlear segments were labeled with anti-prestin antibody (red), phalloidin (green), and the nuclei were counterstained with DAPI (blue). Scale bars: 20 µm. *(Photo Courtesy of M. Morell, Institute of Terrestrial and Aquatic Wildlife Research: Part A: modified from Girdlestone C, Ng J, Piscitelli-Doshkov M, Ostertag S, Raverty SA, Morell M, Shadwick RE. Description of cochlear morphology and hair cell variation in the beluga whale. Arctic Sci. 2018;4(3):279–291; Part B: modified from Morell M, Ijsseldijk LL, Piscitelli-Doshkov M, Ostertag S, Estrade V, Haulena M, Doshkov P, Bourien J, Raverty SA, Siebert U, Puel J-L, Shadwick RE. Cochlear apical morphology in toothed whales: using the pairing hair cell-Deiters' cell as a marker to detect lesions. Anat Rec. 2022;305(3):622–642; Part C: modified from Morell M, Vogl W, Ijsseldijk L, Piscitelli-Doshkov M, Tong L, Ostertag S, Ferreira M, Fraija-Fernandez N, Colegrove K, Puel JL, Raverty SA, Shadwick RE. Echolocating whales and bats express the motor protein prestin in the inner ear: a potential marker for hearing loss. Front Vet Sci. 2020;7:429; Part D: modified from Rohner S, Morell M, Wohlsein P, Strznickel J, Reiter EB, Jahnke A, Prenger-Berninghoff E, Ewers C, Walther G, Striewe LC, Failla AV, Siebert U. Fatal aspergillosis and evidence of unrelated hearing loss in a harbor porpoise (Phocoena phocoena) from the German Baltic Sea. Front Mar Sci. 2022;15(159),107014, Prestin antibody for immunofluorescence Courtesy of Dr. J. Zheng, Northwestern University.)*

studies highlight the importance of collecting and analyzing the ears as part of the routine postmortem examination protocol.

With any suspected barotrauma, decompression sickness or acoustic injury, examination, dissection, and sampling of acoustic fat, auditory canals, peribullar sinuses, middle and inner ears, eyes, and nasal sinuses should be performed with care. Gas accumulation associated with putrefaction and autolysis can be differentiated from antemortem lesions by assessing the degree of autolysis and gas analysis of intravascular bubbles.[17,19]

Over the last 60 years, advances in vessel design and operating speeds of 14–15 knots or more have been associated with a significant increase in the numbers of dead floating and live moribund whales that have catastrophic lesions consistent with **ship (vessel) strike**. Although small cetaceans and larger whales can be impacted, larger slower moving whales tend to be more frequently involved, particularly when congregated in shipping lanes. Depending on the point of contact with the vessel, animals may present with blunt force or sharp injury that can be directly lethal or contribute to sustained injuries, which may resolve or can persist for extended periods of time. Case definitions and examples of traumatic lesions are available.[139] While some animals with blunt force injury may not have apparent external lesions, on reflection of the skin and blubber, subcutaneous hemorrhage, edema, and skeletal fractures at the impact site are usually identified (Figs. 22.14A, B). If the fibroelastic sheath is intact, focally extensive bulging may be observed and on incision, massive hematoma

FIGURE 22.14 Vessel injuries in gray whales (A, B) and a North Atlantic right whale (C). (A) Catastrophic closed and comminuted rostral fractures are present. Externally this appears as a prominent depression of the caudal half of the rostrum. (B) On reflection of the skin and blubber of the rostrum, there is a severe comminuted fracture with displacement of the bone fragments. (C) The regularly spaced and parallel incised wounds extend to varying depths into the underlying soft tissue. The lacerations extended into the thoracic cavity and caused fatal pneumothorax. (D) Myocellular degeneration and necrosis can be seen at sites of tissue damage. In this case, there is monophasic myofiber swelling and degeneration, contraction band necrosis, discoid degeneration and fragmentation, and loss of cross-striations. The inset highlights the myofiber changes with special staining. Phosphotungstic acid hematoxylin (PTAH). *(Parts A, B: Photos Courtesy of P. Duignan, The Marine Mammal Center; Part C: Photo Courtesy of the Florida Fish and Wildlife Conservation Commission [NOAA Research Permit 932-1489]; Part D: Photo Courtesy of the Institute of Animal Health and Food Safety, University of Las Palmas de Gran Canaria.)*

formation with shearing and necrosis of the subjacent skeletal musculature may be evident. Rupture of the sheath at the time of impact is more typically associated with extensive subcutaneous hemorrhage, which may track dependently in the hypodermis. Inspection of the abdominal and thoracic cavities may reveal organ herniation, transposition, rupture, and associated hemorrhage. Histologically, acute myocellular degeneration and necrosis with edema and hemorrhage at the impact site and in muscle samples remote to the site of impact are seen, and flocculent, granular, discoid, and hyalinized segmental myodegeneration are attributed to agonal catecholamine surge. Established reference intervals and creatinine and potassium levels from serum, vitreous humor, or urine collected from live-stranded and fresh carcasses may provide insights into the extent, duration, and severity of myopathy in vessel-struck animals.[235] Fat embolization with a tropism to the pulmonary microvasculature is frequently detected in lethal vessel/ship strike.[8,9] Depending on the circumstances, impacts with stationary structures, such as piers, moorings, wind farm structures, or tidal hydroelectric turbines, may present with similar gross lesions to those of vessel strike.[139]

In contrast to blunt force injury, animals with **propeller injuries** have incised percutaneous wounds that may extend to varying levels into the underlying soft tissue and be curvilinear to sigmoidal, linear, or fenestrated. Catastrophic injuries are associated with incision into the thoracic or abdominal cavities as well as in those cases with amputation of flippers, fins, or even the peduncle. Involvement of larger caliber blood vessels will result in hypovolemic shock and acute death (Fig. 22.14C). Histopathology of the wound margins typically features subcutaneous edema, hemorrhage, shearing at the site of impact, necrosis (Fig. 22.14D), and retraction. As these wounds are exposed to the marine environment, imbibed water results in dissolution of erythrocytes, tissue maceration, and leaching as well as microbial and ciliate colonization and proliferation. In nonlethal cases, wound healing may resolve with primary and secondary intention repair.

Southern right whales at Peninsula Valdés, Argentina, show wounds produced by **kelp gull predation** (Fig. 22.15). Kelp gulls normally scavenge sloughed whale skin. However, an unusual form of kelp-gull predation on the skin in living, free-swimming whales that was first noted in the 1980s has become a locally important problem. Adult whales have developed several strategies to avoid gulls, for example, by keeping their backs under water; calves that have not developed this behavior remain highly vulnerable. The attacks can last for hours at a time and interfere with normal calf resting and nursing. The kelp gull population near Peninsula Valdés has grown dramatically in recent decades, and the percentage of mothers and calves with gull lesions has correspondingly increased from an average of 2% in the 1970s to 99% in the 2000s. Gull harassment is considered a possible contributing factor in calf mortality.[133] Skin wounds can vary in size from small, focal lesions covering approximately 0.1% of the dorsal back surface to large, coalescing wounds over 8% of the dorsal body wall. The wound patterns vary from round to oval and often coalesce. Histologically, lesions in affected skin include dermal congestion, suppurative dermatitis and panniculitis, necrotizing vasculitis, and vascular thrombosis. Two calves with wounds forming discrete rhomboid shapes have been described. Bacterial culture and PCR confirmed *Erysipelothrix rhusiopathiae* from lesions in these two calves.[72] However, these were isolated cases and bacterial, viral, or other infectious diseases that predispose the skin to predation have, to date, not been identified. Physiologic and metabolic factors associated with extensive skin loss and energetic costs associated with prolonged flight/avoidance activities are topics of active investigation.

FIGURE 22.15 Kelp gull trauma in the skin of southern right whales. The adult female and calf have single or multiple, often circular, dorsal skin wounds typical of kelp gull predation. As seen in this image, the wounds are often much more extensive in calves than in adults. *(Photo Courtesy of M. Sironi, Instituto de Conservación de Ballenas/Ocean Alliance.)*

Also of note is the effect of **interspecies cetacean on cetacean trauma**. Attacks on porpoises by bottlenose dolphins have long been recognized in the United Kingdom[178] and are increasingly seen on the Pacific Coast of the United States. In these cases, harbor porpoises are pursued and killed but not eaten by bottlenose dolphins. Lesions include massive blunt force traumatic lesions, such as hemorrhage, bruising, and bone fractures with or without abundant skin lacerations from teeth (rakes). Infanticide has been reported in bottlenose dolphins and killer whales. Conspecific aggression has also been noted in some species, such as bottlenose dolphins and other cetacean species.[166] Gray seal predation events involving harbor porpoises have also been documented along the North Sea coast of Northern France/Belgium.[89] Porpoise carcasses presented with severe,

FIGURE 22.16 Gray seal predation in a harbor porpoise. The predatory wounds vary and in this case were associated with severe, extensive blubber and skin loss with sharp lacerations, abrupt margins, and extensive loss of large parts of the skin and underlying soft tissues. These are gross images of gray seal predation of harbor porpoises, a recent development in Europe and should be differentiated from propeller strike, flensing, or post-mortem scavenging. *(Photo Courtesy of U. Siebert, Institute of Terrestrial and Aquatic Wildlife Research.)*

parallel, linear, and/or fenestrated lacerations. The margins were abrupt and blubber segments were detached, reflected, or avulsed from the torso (Fig. 22.16). These events often involve solitary seals with aberrant behavior. Gray seal predation is usually performed by individual males that have recently attained sexual maturity. Their predatory behaviors may be accompanied by other aberrant behaviors including cannibalism and copulation with other seal species. Because of this, cases tend to cluster temporally and spatially with individual animals that exhibit this specialized behavior. Differentiating predatory attack wounds from other forms of traumatic injury (propeller strike, fishing gear, etc.) can be difficult. Use of a standardized necropsy protocol[223] and the advent of specific predator DNA detection protocols for suspect wound sites have advanced forensic investigations in these cases.[94]

As with other forms of anthropogenic impacts, ingestion of lost or discarded **marine debris and plastics** is increasingly recognized as a significant cause of morbidity in cetaceans. Foreign material can be found anywhere along the gastrointestinal tracts of marine mammals and be associated with generalized emaciation. Pica and ingestion of foreign debris have been reported in live, moribund-stranded dolphins and may be behavioral, while in ram or lunge-feeding baleen whales, ingestion of floating or suspended debris may be inadvertent. Cetaceans with more discriminate feeding habits occasionally selectively ingest material that can accumulate in quantities sufficient to impair normal digestive processes. Reported findings have included

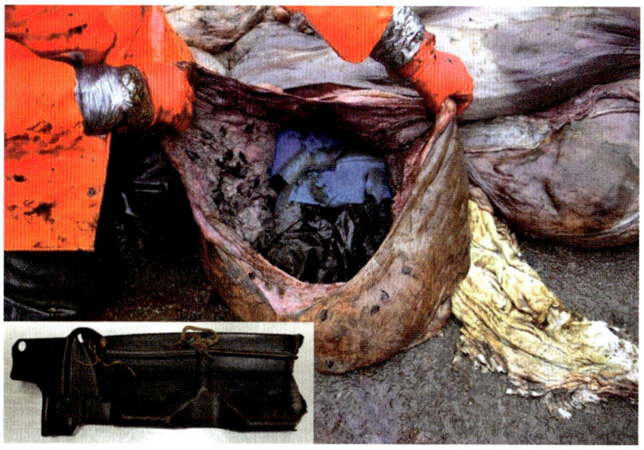

FIGURE 22.17 Foreign debris ingestion in a sperm whale. The contents of the stomach included a mixture of foreign materials including automobile parts (inset). *(Photo Courtesy of U. Siebert, Institute of Terrestrial and Aquatic Wildlife Research, University of Veterinary Medicine Hannover.)*

distention of the stomach by netting (up to 16 m^2 in sperm whales), rope, plastic bags, fishing line, metal filaments, glass fragments, and other materials, which in more severely affected animals has resulted in obstruction, localized ulceration, perforation, and secondary peritonitis (Fig. 22.17).[139,165] In addition to the physical consequences of ingested debris, manufactured plastics also contain polychlorinated biphenyls (PCBs), persistent organic pollutants, pesticides, heavy metals, and other additives from the environment, which over time, can be assimilated and have a direct immunotoxicologic consequence in animals.[229] This process may be exacerbated by the inadvertent ingestion of microplastics or may result from mechanical breakdown of larger plastic items into secondary microplastics in the marine environment.[199] The current understanding of the distribution and impacts of marine debris on cetaceans is poorly understood and global initiatives to better document affected species, stranding location, debris composition, and dispersal may help better define the source of marine debris.

As with terrestrial species, **asphyxiation** can occur in cetaceans, but establishing confirmatory findings can be challenging in animals that are found dead. Determining a cause of death can be further complicated as there are different types of asphyxiation. These include cellular, vascular, and ventilatory or mechanical obstruction. At present, there are no pathognomonic lesions or biochemical markers for asphyxiation and diagnosis typically relies on circumstances associated with recovery of the carcass. A presumptive diagnosis of asphyxiation in marine mammals is most commonly seen in bycaught individuals where animals present or are recovered entrapped in gill nets or trawls. An additional form of asphyxiation is more sporadic and related to ingestion of foreign debris, such as fishing line which can obstruct or displace the larynx. In some

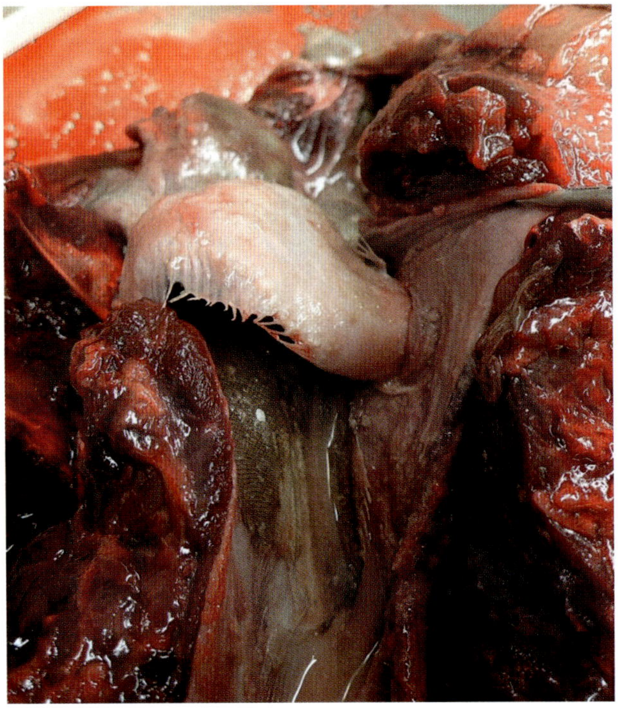

FIGURE 22.18 Laryngeal displacement, obstruction, and asphyxiation in a harbor porpoise. A prey item (common sole) has become lodged within the lumen and resulted in asphyxiation and the death of the animal. *(Photo Courtesy of U. Siebert, Institute of Terrestrial and Aquatic Wildlife Research, University of Veterinary Medicine Hannover.)*

cases, laryngospasm may be a related, life-threatening process. The same effect is achieved when cetaceans consume prey that is sufficiently large to cause displacement of the larynx and secondarily obstruct the exposed airway (Fig. 22.18).[86] Cases in animals in managed care have been seen when regurgitated objects became lodged in the displaced larynx and resulted in fatal asphyxiation. A recent case of phytobezoar asphyxiation in a bottlenose dolphin demonstrates that material need not be swallowed to be fatal.[233] Fatal blowhole obstructions have sporadically been reported in long-finned pilot whales, harbor porpoises, and more recently in common bottlenose dolphins. Laryngeal displacement and blowhole obstruction resulting in asphyxiation have also been described in a bottlenose dolphin that ingested a whip eel and shrimp eel.[83] These findings may have important implications with respect to prey shifts and feeding ecology.

Angiomatosis is a common finding among the stranded bottlenose dolphins in the Gulf of Mexico. The lesion has also been seen in stranded Atlantic bottlenose dolphins in eastern Florida and occasionally in common dolphins in Southern California. An evaluation of common dolphin strandings in the Canary Islands found that a majority of animals had pulmonary angiomatosis.[43] A correlation with lungworm infections was seen in these cases and is often noted in bottlenose dolphins with the condition. The condition is characterized by a proliferation of small, thick-

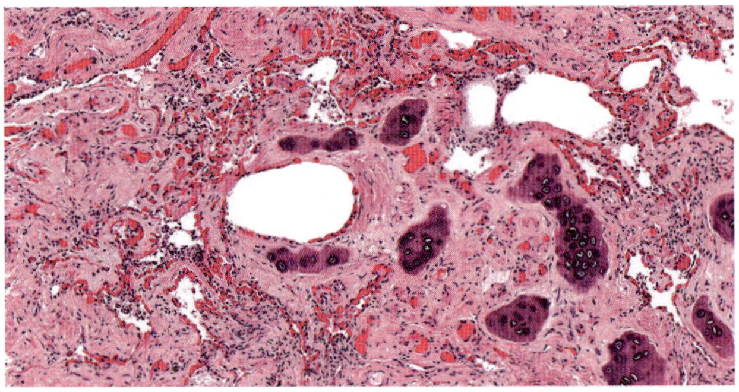

FIGURE 22.19 Pulmonary angiomatosis in the lung of a bottlenose dolphin. A moderate to marked proliferation of small vessels with thick walls arranged in a haphazard pattern disrupts the bronchiole and replaces the pulmonary parenchyma (see eSlide 22.e5).

walled blood vessels focally, multifocally, or diffusely throughout the lungs without inflammation, exudation, or alveolar hemorrhage (Fig. 22.19; eSlide 22.e5). The vascular proliferation also occurs in lung-associated lymphoid tissue and other visceral lymph nodes. Changes in the lungs, if severe, can impinge upon and compromise alveolar airspaces and may occlude small airways. Although the exact cause of this condition remains undetermined, the increasing frequency of this diagnosis suggests a complex multifactorial etiology and pathogenesis.

Lesions of **ocular disease** in cetaceans are similar to those of domestic animals but reports as scarce. Corneal ulceration commonly affects marine mammals, and corneal defects can develop secondary bacterial or fungal infections. Scarring is a common consequence and visual deficits can result. Corneal opacities that progress and are in which **keratomycosis** caused by ***Candida albicans*** develops have been described in bottlenose dolphins.[198]

Sunburn is a common sequel to stranding in many cetaceans. As with other species, darker pigmentation contributes to preventing sun damage. Grossly, the condition presents as dorsal body skin edema and sloughing with patchy to diffuse ulcerative dermatitis and pronounced erythema. In stranded animal in which the skin on their lateral body walls is exposed, the damage may be present on the upward side as opposed to the dorsal distribution. Skin sloughing due to postmortem decomposition is a differential for sun burn. Histologic changes in the epidermis in humans and many animals associated with sunburn include dyskeratotic, vacuolated, and apoptotic keratinocytes, so-called sunburn cells. Similar findings are seen in cetaceans with sunburns. This includes degeneration, cell contraction, and dissociation of the cells.[131] In these instances, the burn damage is coupled with damage from drying/dehydration of the skin. This results in the fragmentation and lifting of the stratum externum and superficial layers of the stratum spinosum. In more severe cases, clefts extend below the stratum germinativum, which leads to lifting of the entire epidermis. Dermal changes include hemorrhage, vascular dilation, and perivascular edema.

Damage commonly extends deep in the dermis. Healing occurs through epidermal migration and scarring. Areas of prior damage can often be identified grossly by white cicatricial change in the skin that remains for life.

Skin trauma can also result from changes in the osmolality or salinity in the aquatic environment such as can occur in animals that become entrapped or established in a hyposaline environment (typically less than 10 practical salinity units) or that are exposed to increases in freshwater runoff. Marine cetaceans with **fresh-water skin disease** can exhibit osmotic skin damage.[92] Gross lesions include patchy to diffuse pallor and discrete circular to coalescing areas of targetoid or elliptical skin thickening, sloughing, erosion, and ulceration (Fig. 22.20A). Overgrowth of brown to green algal mats may occur. Histologically, lesions can have varying degrees of epidermal thickening with acanthosis, marked intracellular edema, vesicle formation, epidermal and dermal inflammation, and pustule formation with epidermal necrosis and ulceration (Fig. 22.20B). Fungal hyphae can sometimes be seen superficially in affected skin but do not typically invade deeper into the tissue. Secondary bacterial infections are common in severe lesions, and some have superficial *Dermatophilus* spp.-like colonization. Dolphins exposed to freshwater may develop electrolyte abnormalities such as hyponatremia, hypochloremia, and hyposmolality and if these derangements are sufficiency severe, could lead to death.[50]

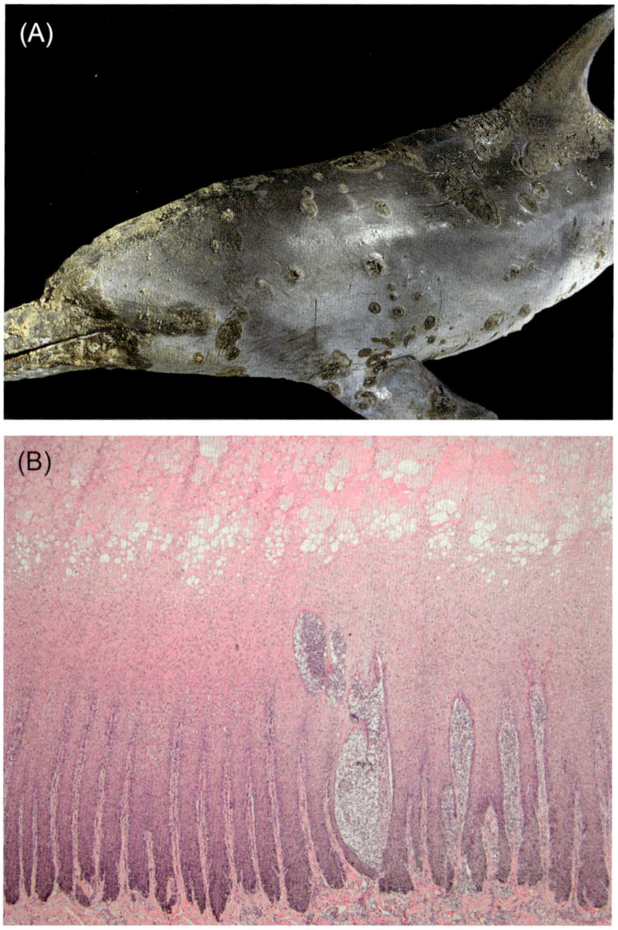

FIGURE 22.20 Fresh-water skin disease in a dolphin. (A) Lesions are characterized by extensive pallor of the skin which may progress to multifocal and coalescing targetoid or elliptical plaques that can be centrally eroded or ulcerated. (B) Histologic changes include hydropic degeneration with vesicle and pustule formation. *(Part A: Photo Courtesy of P. Coughran, Western Australia Department of Parks and Wildlife; Part B: Photo Courtesy of P. Duignan, The Marine Mammal Center.)*

Species-specific diagnostic keys to describe skin lesions have been developed for free-ranging cetaceans and are helpful in lesion interpretation, directing ancillary diagnostics, and in contextualizing lesions and potential population-level impacts.[62,78,212]

As with many animals, **amyloidosis** occurs in cetaceans. The product is most likely serum amyloid A (AA amyloid), an acute-phase reactant protein produced secondary to an inflammatory process. Amyloid can be seen in vascular walls and appears as acellular, amorphous, eosinophilic material that can segmentally or diffusely expand the vascular wall. In tissues, it can be seen anywhere in the body but has been specifically described in the corticomedullary regions of the kidneys, around acini of the palatal salivary gland and in the thyroid glands.[33] Stejneger's beaked whales appear to have a predilection for developing **systemic amyloidosis.**[147] In them, amyloid has been detected in the space of Disse in the liver and in the kidney, heart, spleen, adrenal glands and pancreas.

Neoplastic

A comprehensive review of marine mammal tumors was compiled by Newman and Smith[148]; additional reports of neoplasia in cetaceans published since his 2006 review are listed in Table 22.e3. Three neoplastic conditions that warrant special attention are **gastrointestinal neoplasia** in belugas from the St. Lawrence Estuary, papillomas in both free-ranging and captive dolphins, and oral squamous cell carcinoma in bottlenose dolphins in human care.

Long-term investigations into the health of the beluga whales in the St. Lawrence Estuary (SLE) have documented tumor rates that are among the highest in wild cetacean populations anywhere in the world.[116] Over a 30-year time period, 222 dead whales were recovered and available for necropsy. Of these, 39 neoplasms were diagnosed in 35 adults. Malignant tumors were identified as the cause of death in 14% of the examined belugas and **gastrointestinal adenocarcinoma** the most common, affecting 7% of mature adults. Eight cases of **mammary carcinoma** were identified in 10% of examined adult females. These tumors were often metastatic and in 6 of the 8 cases, mammary gland carcinoma was identified as the cause of death in affected whales. There were two fatal cases of **thyroid carcinoma**, one **lymphoma**, and two benign ovarian tumors.[116] No comparable prevalence of neoplasia has been found in other free-ranging cetacean populations. The hypothesis to explain this unusually high frequency of neoplasia is environmental exposure, persistence, and possible foodweb bioaccumulation of carcinogens, such as polyaromatic hydrocarbons.[130]

Oral papillomas (Figs. 22.21A, B; eSlide 22.e6) and **squamous cell carcinoma (SCC)** in bottlenose dolphins under managed care may reflect a progression along a spectrum. **Genital papillomas** have been identified in both sexes of captive and free-ranging bottlenose dolphins. These proliferative lesions are often called orogenital papillomas though not all dolphins with oral lesions have genital lesions and vice versa.[24] Oral papillomas present as raised, well-circumscribed, pale, irregular masses or plaques in the mucosa of the oral cavity, mucosa, mucocutaneous junction, or on the tongue. Histologically, they are composed of hyperplastic epithelium and they tend to recur after surgical excision. With time, and likely due to additional factors that are not yet defined, these masses may progress to carcinoma in situ or squamous cell carcinoma. Oral lesions that have progressed to carcinoma are often irregular, nodular to coalescing, poorly demarcated, and have ulcerated areas.

Oral SCCs are composed of nests of neoplastic epithelial cells that breach the mucosal basement membrane and extend widely and deeply from a central tumor into underlying or adjacent tissues as irregular islands and fingers of proliferating neoplastic cells (Figs. 22.22A, B; eSlide 22.e7). Lesions along the ventral frenulum and back of the mouth appear to progress to SCC more commonly. Oral SCCs tend to be slowly progressive but can be highly aggressive. The poor demarcation and wide tumor extension result in common recurrence after surgical excision. Both local aggression and metastasis characterize the biological behavior of these neoplasms, however, some animals live for many years following initial diagnosis of oral SCC without complete resolution of the lesion. Metastasis to tonsils as well as thoracic and prescapular lymph

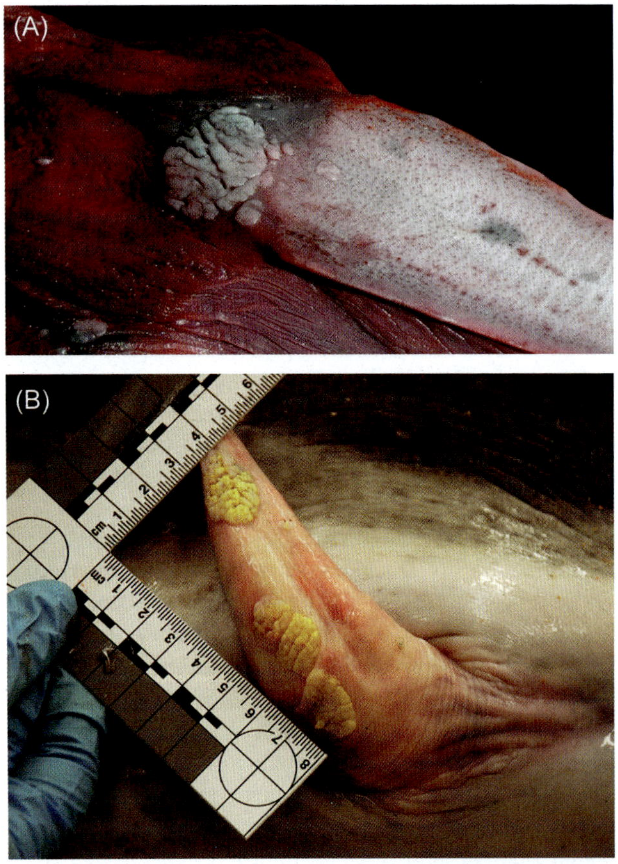

FIGURE 22.21 Oral papilloma in a common dolphin (A) and genital papillomas in a bottlenose dolphin (B). (A) A discrete, raised, white plaque is present in the mucosa at the base of the tongue. (B) Multiple, light yellow, well-circumscribed plaques are present along the mucosa of the penis (see eSlide 22.e6). *(Part B: Photo Courtesy of Hubbs-SeaWorld Research Institute.)*

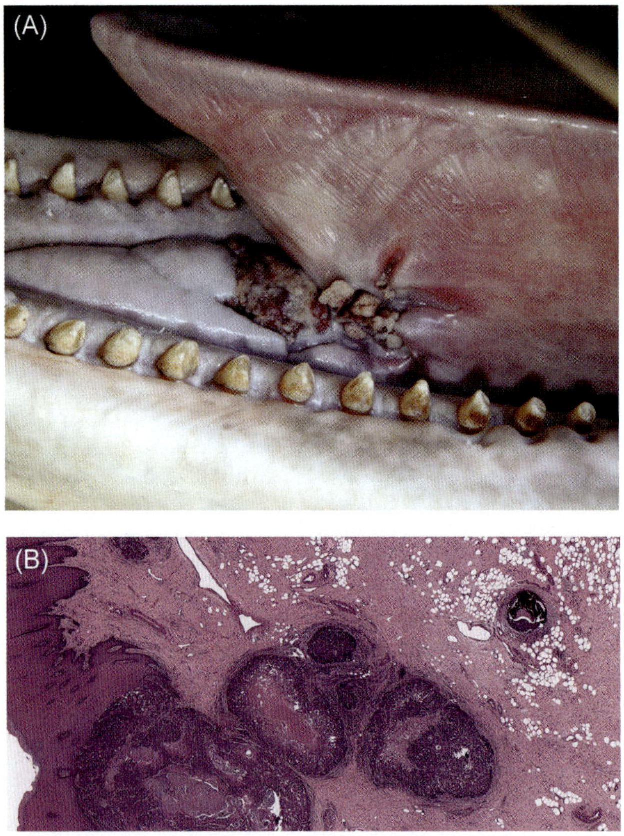

FIGURE 22.22 Oral squamous cell carcinoma in a bottlenose dolphin. (A) Irregular tissue proliferation with foci of ulceration is present on the floor of the oral cavity at the base of the frenulum. (B) Irregular proliferations of neoplastic epithelial cells form islands that invade beneath the basement membrane of the oral mucosa. Multifocal central necrosis is present, as is intravascular metastasis in a blood vessel (upper right) (see eSlide 22.e7).

nodes and lung, have been described. Prescapular lymph node involvement is important clinically because this tissue is accessible for both ultrasound and biopsy to screen for metastatic spread. The aggressive nature of these neoplasms is reported to be enhanced during pregnancy, suggesting a role of both hormones and immune modification in the pathogenesis of these tumors. Exposure of the oral cavity to ultraviolet light associated with "heads up" feeding of captive dolphins above the water surface is suggested as having a potential role in malignant transformation of the mucosal epithelium (N. Stedman, personal communication).

Both papillomaviruses and herpesviruses have been associated with genital and oral papillomas and oral squamous cell carcinomas in multiple cetacean species, both in captive and stranded animals (see "DNA Viruses" section).[170,202,219] Intranuclear viral inclusions have occasionally been noted in epithelial cells in oral or genital papillomas.

INFECTIOUS DISEASES

In addition to the descriptions below, see Tables 22.2, 22.e4, and 22.e5 for additional viral and parasitic infections in cetaceans.

DNA Viruses

A unique pattern of disease colloquially called "**tattoo**" typifies **poxvirus infection** in cetaceans. Phylogenetic studies have identified six separate species or clusters of cetacean poxviruses that are similar to each other and different from other orthopoxviruses, suggesting that cetacean poxviruses may be a separate genus. This unique phylogeny fits the unusual biologic behavior of cetacean pox in creating nonproliferative skin lesions that are present for extended periods of time. No pattern regarding species-specificity of viruses has been delineated. Both toothed and baleen whales can be infected, and typical lesions have been identified in bottlenose, short and long-beaked common, Commerson's, striped, white-beaked, Pacific white-sided, and dusky dolphins, harbor and Burmeister's porpoises, and killer, short- and long-finned pilot, Southern right and bowhead whales.[13]

The lesions are demarcated areas of hyperpigmented skin that give the condition its name "tattoo." They variably present as pinpoint, ring, or serpiginous patterns (Figs. 22.23A, B). Lesions are generally contiguous with the healthy skin surface. Pinpoint lesions have a dark center, which may be depressed relative to adjacent skin and may be surrounded by a pale margin forming a target lesion. Although lesions can occur anywhere on the body, the flippers and flukes are generally less affected. The lesions often wax and wane over many months and years. Studies in captive bottlenose dolphins show a marked reduction in lesion intensity to the point of clearing the skin with warming water temperatures (S. Di Rocco, personal communication). Work by Van Bressem et al.[220] with free-ranging animals strongly suggests anthropogenic factors as promoting expression of infection. Tattoo lesions

TABLE 22.2 Herpes Virus Infections in Cetaceans

Species	Herpes Subfamily	Location	Lesions
Beluga whale	Alpha	Genital/oral mucosa, skin	Necrotizing dermatitis[15,129]
Atlantic bottlenose dolphin	Alpha	Skin	No lesions described[202]
Atlantic bottlenose dolphin	Alpha	Multiorgan	Necrotizing inflammation[21]
Striped dolphin	Alpha	Lymphoid system	Lymphoid depletion[204]
Blainville's beaked whale	Alpha	Kidneys	Nephritis[4]
Cuvier's beaked whale	Alpha	Lymphoid system	Lymphoid necrosis[7]
Harbor porpoise	Alpha	Brain	Encephalitis[109,217]
Striped dolphin	Gamma	Genital mucosa/skin	Epithelial hyperplasia[195]
Atlantic bottlenose dolphin	Gamma	Genital and oral mucosa/skin	Epithelial hyperplasia[202,222]
Risso's dolphin, dwarf sperm whale, Blainville's beaked whale	Gamma	Genital mucosa/skin	Epithelial hyperplasia[109,202,222]

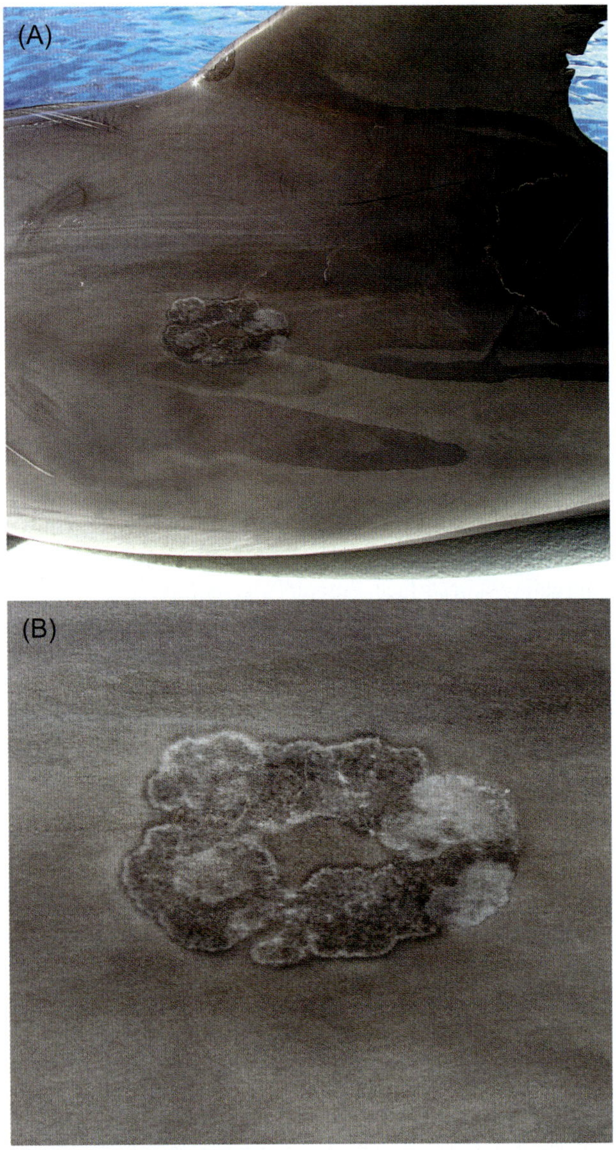

FIGURE 22.23 Poxvirus infection ("tattoo") in the skin of a bottlenose dolphin. (A) Infection takes the form of anastomosing cords and rings of pigmented skin that give the lesion its name (tattoo). (B) The patterns can have various forms and be irregular or create prominent ring lesions, as is present on the lateral trunk of this animal. The lesions typically come and go over time and are generally not raised (eSlide 22.e8).

commonly occur adjacent to or are directly associated with areas of skin damage from conspecific trauma (rakes) and may be seen more frequently in animals undergoing social stress. Histologic changes in the skin include an increase in melanocytes in the basal epithelium, cytoplasmic vacuolation in the stratum intermedium, and mild to moderate hyperplasia in the stratum externum (eSlide 22.e8). Small round, eosinophilic intracytoplasmic viral inclusions may be seen but are rare and the absence of inclusions should not rule out the possibility of poxvirus infection. Unless there is associated skin damage (as with rakes), inflammation is not a feature of these lesions, and despite hyperplasia of the stratum externum, these lesions are not raised. Noninvasive skin sampling may facilitate surveillance of free ranging and possibly, more susceptible cetacean populations.[189] Scanning and transmission electron microscopy (SEM and TEM, respectively) may identify particles with ultrastructural characteristics of both parapox and orthopoxviruses. Cetacean poxvirus particles are arranged haphazardly in the epithelial cytoplasm. They are oval with a central core and an outer membrane and measure approximately 150–200 µm by 250–300 µm, which is similar to parapoxviruses. However, the surface morphology resembles a "ball of yarn," which is characteristic of the disorderly structure of an orthopoxvirus rather than the orderly crisscross patterns of parapoxviruses.[13,80]

In both mysticetes and odontocetes, **gammaherpesvirus** infection has been associated with skin lesions, genital lesions, nephritis, encephalitis, and disseminated infections (Table 22.2).[197] Viral infections can also be detected in the absence of lesions. Cetacean herpesvirus infections can result in host immunosuppression, and latent viral infections may be reactivated and possibly exacerbated by generalized debilitation or localized immunosuppression. Both herpes and papilloma viruses have been identified in proliferative orogenital lesions (see "Neoplastic" section).

Alphaherpesvirus infections have been reported less frequently than gammaherpesviruses in cetaceans (Table 22.2). Associated lesions vary from incidental and asymptomatic to acute, systemic necrotizing inflammation in multiple organ systems.[4]

Based on viral sequencing, there are currently at least five papillomavirus variants in cetaceans. As with many viruses of cetaceans, **cetacean papillomaviruses (PVs)** are phylogenetically related to PVs of artiodactylids.[82] Serologic positivity to PV is high in free-ranging and is common in captive dolphins (up to 90% and 51%, respectively, in some serosurveys), and ELISA reactivity is higher in males than females. Cetacean papillomaviruses are associated with both mucosal and cutaneous papillomas in free-ranging and captive cetaceans (see "Neoplastic" section; Figs. 22.21A, B; eSlide 22.e6). The mean age of free-ranging dolphins with papillomas is ~11 years; it is ~30 years in captive dolphins. PV infection in bottlenose dolphins is common and the main route of transmission is likely horizontal with orogenital papilloma development early in life in certain free-ranging bottlenose dolphins.[170] PV-associated mucosal lesions are typical and consist of demarcated plaques or broad-based areas of thickened/proliferative light pink to tan mucosa. The orogenital tumors have a predilection for vulvar and vaginal mucosa, penile mucosa, and oral or esophageal mucosa and frequent involvement of the frenulum of the tongue is seen. While far less common, cutaneous papillomatous proliferations have been seen in a variety of cetaceans including harbor porpoises, killer whales, and pilot whales. Lesions can be warty and/or plaque-like, singular or multiple, and coalescing. Histologic findings consist of epithelial hyperplasia producing finger-like or warty projections along a fibrovascular core. Within the hyperplastic epithelium, there is variable acanthosis and thickening of the stratum spinosum. The basement membrane remains intact. With cutaneous or mucocutaneous tropisms, many viruses may be difficult to differentiate clinically or at necropsy[188] and comprehensive sampling and archiving of fresh tissue and formalin-fixed tissues is imperative.

RNA Viruses

Cetacean morbilliviruses (CeMV) are RNA viruses that were first recognized in 1988 following a series of mass cetacean and pinniped mortalities in Northwestern Europe. There are at least six distinct viral strains, each of which is associated with a particular cetacean species. They include **porpoise morbillivirus (PMV)**, first isolated from harbor porpoises, **dolphin morbillivirus (DMV)**, first isolated from Mediterranean striped dolphins, and **pilot whale morbillivirus (PWMV)**. Closely related to these three strains is **beaked whale morbillivirus (BWMV)**. Additional, less well-characterized strains include isolation of CeMVs from a long-finned pilot, Longman's beaked whale, Guiana dolphin, and Indo-Pacific bottlenose dolphin. Cetacean morbilliviruses have also caused several mass mortality events in odontocetes and mysticetes.[20,53,219]

Morbillivirus transmission is thought to occur mostly by horizontal dissemination after inhalation of aerosolized virus shed by infected individuals. The close association of animals in cetacean pods makes aerosol transmission within a group highly effective. The migratory and social behavior of cetaceans promotes wide geographical dissemination.[182] CeMV-infected females may also transmit the virus transplacentally to fetuses and to neonates during lactation. Chronically infected animals likely serve as reservoirs for the virus.[25] Morbilliviruses are lymphotropic, epitheliotropic, and neurotropic and they replicate in the lymphoid tissue prior to dissemination and infection of other cell types and organ systems. This group of viruses is highly contagious and generally causes pneumonia and encephalitis with marked immunosuppression. Infection rates vary among species and populations.

Episodic large, unusual mortality events related to cetacean morbillivirus infections have occurred in many areas including along the east coast of the United States and in the Mediterranean Sea. Mortality events may occur when significant portions of the population lack or have declining natural immunity to the virus or new viral strains are introduced to the population.[58,108,124] Outbreaks of DMV infection in the northern Gulf of Mexico (GoM) are sporadic and infection is present at low levels. Confirmation of DMV infections and existing DMV titers demonstrate continued exposure to morbillivirus among the northern GoM cetaceans.[64]

Gross lesions include pneumonia with pulmonary consolidation and congestion and lymphadenomegaly. Oral proliferative lesions and oral mucosal and skin ulceration have also been described. Opportunistic infections associated with profound immunosuppression are common and thus the gross presentations vary based on these secondary infections. These include opportunistic protozoal (e.g., *Toxoplasma gondii*), viral (e.g., herpesviruses), bacterial (e.g., *Photobacterium damselae*, *Streptococcus phocae*, *Brucella* spp.), ciliate, algal, and fungal (e.g., *Aspergillus* spp.) infections. Some animals have heavy parasite loads. Histologically, CeMV infection can be pancytopathic, but infection most often affects the respiratory, lymphoid, and central nervous (CNS) systems. Generalized lymphoid depletion and edema with germinal center necrosis are common and syncytial cells (Warthin-Finkeldey type) are often prominent (Fig. 22.24A). The cortex of lymph nodes is generally the best site to identify inclusions within syncytia; detection of virus-laden cells may be impeded by expansive inflammation and necrosis associated with supervening bacterial or fungal infections. Bronchointerstitial pneumonia is characterized by necrosis of type I pneumocytes and bronchiolar epithelial cells, interstitial edema, type II pneumocyte hyperplasia, and the formation of large syncytia in the alveolar and bronchiolar air spaces. Intracytoplasmic and intranuclear, eosinophilic viral inclusion bodies are found, but generally only rarely, in respiratory epithelia, bronchiolar gland epithelia, syncytia, and other epithelia within the body (Fig. 22.24B). CNS lesions include nonsuppurative meningoencephalitis with prominent perivascular cuffs, expansion of the meninges with edema, fibrin, hemorrhage, and a mix of lymphocytes and plasma cells. Eosinophilic viral inclusions may be seen (Fig. 22.24C). With chronic infection, multifocal areas of demyelination may be apparent. Immunohistochemical (IHC) labeling using monoclonal antibodies (MoAb) to the hemagglutinin (H) glycoprotein of phocine distemper or neuraminidase (N) protein of canine distemper virus, or a rabbit polyclonal antiserum to rinderpest virus have been successfully used to diagnose CeMVs and are valuable when tissue decomposition impedes histologic examination (Fig. 22.24D).[219]

In cases in which the epidemiology, clinical history, and gross necropsy and histologic examination are highly suggestive of CeMV infection, virus isolation using VERO or SLAM cells is confirmatory, and remains the gold standard diagnostic test. Reverse transcription (RT)-PCR using various universal primers sets against different conserved regions of morbillivirus genes followed by sequencing

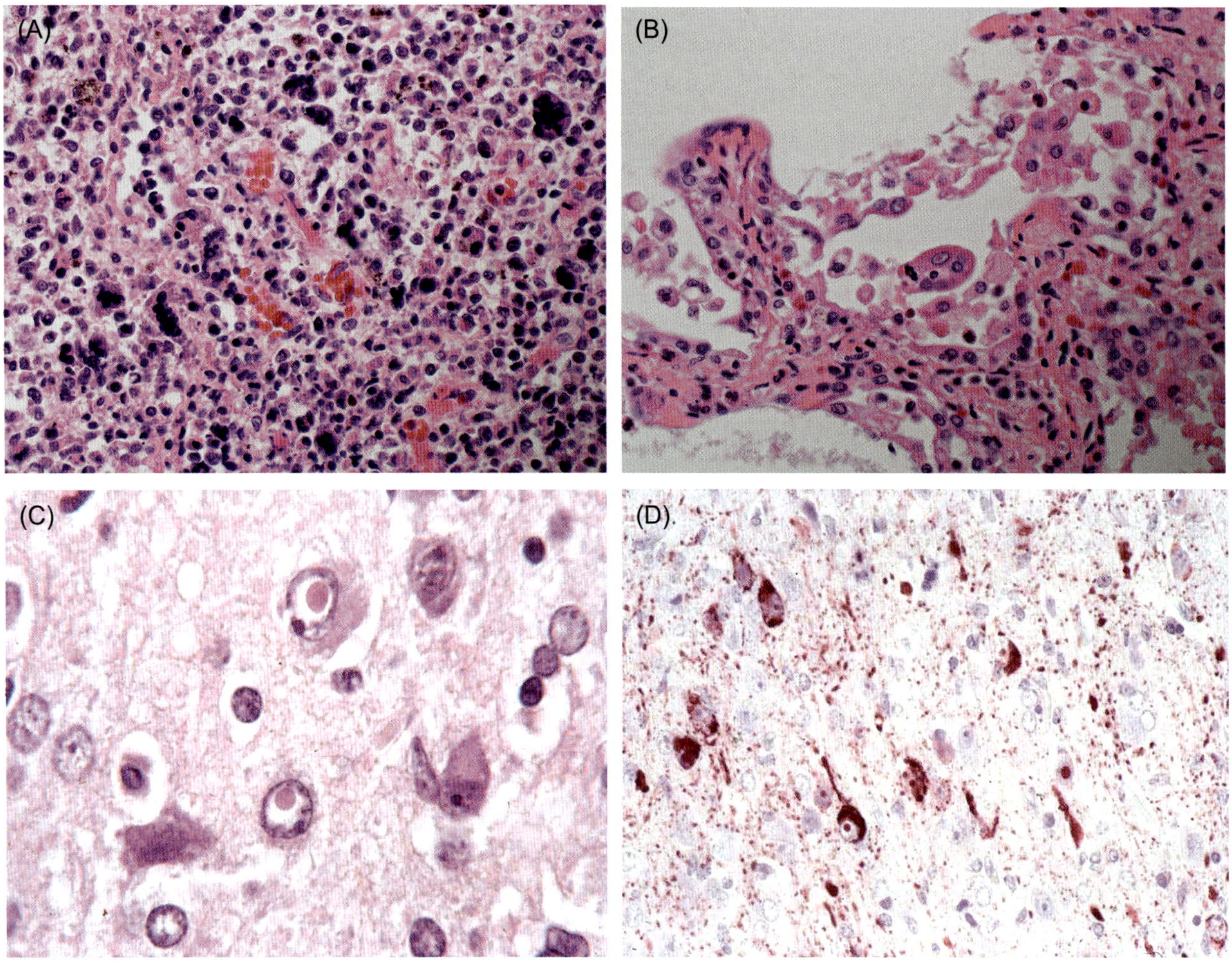

FIGURE 22.24 Morbillivirus infections in bottlenose (A, B) and striped (C, D) dolphins. Cetacean morbilliviruses can be pancytotrophic with lesions commonly seen in the lymphoid (A), respiratory (B), and nervous systems (C, D). (A) There is diffuse lymphocytolysis and disruption of the normal architecture. Multiple multinucleated syncytia are scattered throughout the tissue. (B) Pulmonary lesions include interstitial pneumonia with epithelial cell necrosis, cell sloughing, and the formation of syncytial cells. Intranuclear eosinophilic inclusions are present in two syncytial cells in the middle of this image. (C) Prominent and single, eosinophilic viral inclusion bodies are present in the nucleus of two neighboring neurons. (D) Widespread immunolabeling for morbilliviral antigen is present in cortical neuronal *somata*, axons, dendrites, intranuclear viral inclusions and in the surrounding neuropil (IHC for CDV antigen). *(Parts A, B: Photos Courtesy of D. Rotstein, Marine Mammal Pathology Service: Parts C, D: Photos Courtesy of G. Di Guardo, University of Teramo.)*

can provide rapid CeMV confirmation in a number of species with additional clade determination. Real-time RT-PCR (rtRT-PCR) targeting the hypervariable C terminal domain of the N gene can also provide rapid and reliable detection of species-specific dolphin and porpoise morbilliviruses.[219] Serological studies including virus neutralization (VN) tests, plaque reduction (PR) assays, and indirect enzyme-linked immunosorbent assays (iELISAs) are the main platforms for detecting antibodies against CeMV. Serology is useful for studying CeMV epidemiology, assessing population immune status before and after events and to identify new epidemics early in the disease course.

A form of **chronic CeMV** has been described and characterized by profound lymphoid depletion and fatal secondary infections without the typical morbillivirus lesions in the lungs and brain, but morbillivirus RNA can be amplified in tissues (e.g., lymph nodes) with PCR. Additionally, cetaceans that have cleared systemic DMV infection may develop lesions that localize only in the brain. Neuronal processes have patchy positive immunolabeling; some zones have almost no labeling. In these cases, CeMV spread may be the result of cell-to-cell transmission rather than blood-borne infection. This CNS form shares histological characteristics with **subacute sclerosing panencephalitis (SSPE)** and **old dog encephalitis (ODE)**, chronic latent localized infections in humans and dogs caused by defective forms of MV and CDV infections, respectively. Lesions are localized predominantly in the cerebral cortex, subcortical white matter, and the thalamus.

Perivascular cuffs, diffuse gliosis, and glial nodules with neuronophagia are the most prominent changes. Eosinophilic inclusions and syncytia are typically not seen. Demyelination may be apparent, but in some cases, it is less prominent than in dolphins with subacute infection. As in human and canine presentations, antigen and viral RNA can be detected in dolphin brains but the virus is difficult to isolate. The mechanism underlying the demyelination has not been well characterized, but delayed antigen and RNA clearance from the CNS may be related to the reduced immune surveillance.[219]

The close phylogenetic genetic relationship between cetacean and ruminant morbilliviruses has led to the hypothesis that these pathogens may be derived from a common source. Cetaceans belong to the phylogenetic clade Cetartiodactyla. Several species in this clade are susceptible to rinderpest and peste des petits ruminants (PPR). Therefore, it is possible that interspecies transmission between a cetacean and another member of the Cetartiodactyla led to the development of distinct virus species that cycle exclusively in the marine environment. However, the presence of similar host proteins and cell receptors in cetaceans and artiodactyls may favor cross-species transmission.

In addition to terrestrial mammals, **influenza A** can infect marine mammals as well. Some of the earliest reports in marine mammals are from the early 1930s, with more recent outbreaks documented in phocids (e.g., harbor seals, Caspian seals, and elephant seals) (see Chapter 23: Pinnipediae) and sporadic spillovers into cetaceans.[23] These viruses are members of the orthomyxovirus family which consists of types A, B, C, and D viruses with only types A and B reported in marine mammals. Several viral subtypes have been recovered from marine mammals, including H1, H3, H4, H7, H10, and H13, which appear to have originated from infected birds. During the ongoing (as of chapter preparation in 2023) **highly pathogenic avian influenza virus (HPAI)** pandemic, which has not had an impact on humans, large numbers of wild birds and production poultry flocks throughout Europe and North America have been infected. The circulating virus is H5N1 clade 2.3.4.4b. Between June and July 2022, increased numbers of harbor and gray seals presented with respiratory or neurologic signs or were found dead. These events often coincided with eider, gull, or tern die-offs. Individual strandings of harbor porpoises along the west coast of Sweden and the United Kingdom, and bottlenose dolphins in Florida, United States were also reported. Histopathology of porpoises and bottlenose dolphins revealed nonsuppurative meningoencephalitis with neuronolysis, gliosis, perivascular cuffing, and transmural vasculitis or acute necrosis throughout the neuropil (Figs. 22.25A, B). In contrast to pinnipeds, lesions in cetacean lungs were more subtle, consisting of alveolar histiocytosis and occasional

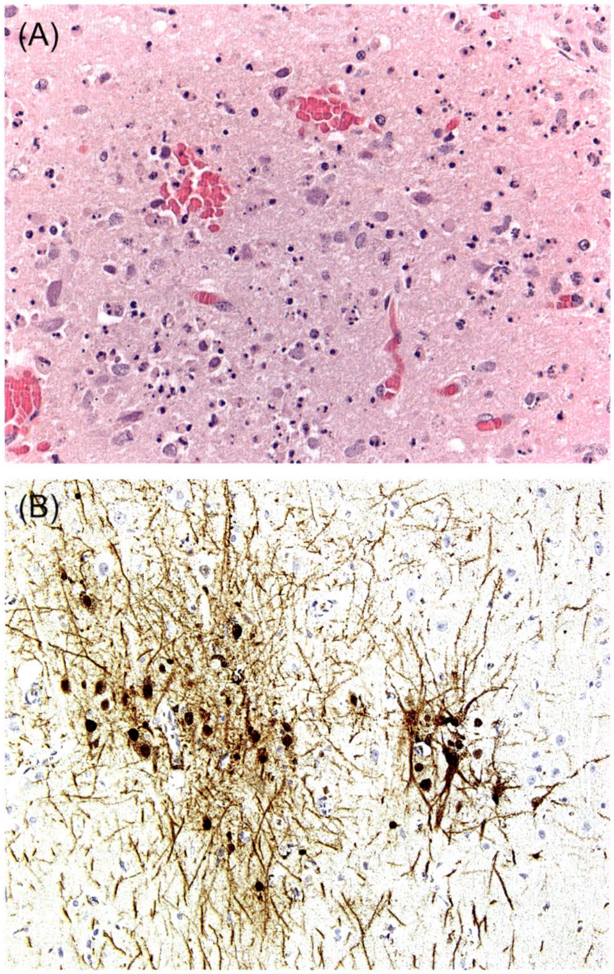

FIGURE 22.25 Highly pathogenic avian influenza (HPAI) infection in a bottlenose dolphin. (A) Infection is associated with widespread neuronolysis and necrosis with granular cellular and karyorrhectic debris throughout the neuropil. (B) Immunolabeling for influenza antigen is strongly positive (granular brown) in infected neurons (IHC). *(Part A: Photo Courtesy of R. Ossiboff, University of Florida; Part B: Photo Courtesy of P. Vogel, St Jude Children's Hospital.)*

mononuclear infiltrates within alveolar septae. Confirmation in the cetacean was accomplished by immunohistochemistry, molecular studies (RT-PCR for matrix IA and H5/H7, MAPAQ, and ACIA for confirmation), and phylogenetic analysis.[211] Live-stranded animals can be screened by collecting and screening blowhole, rectal, or conjunctival swab samples using virus isolation or molecular testing. The role of marine mammals as a wildlife reservoir and mammalian adaptation of an avian virus, as well as the mode of transmission have not yet been fully resolved.

Severe acute respiratory syndrome coronavirus 2 (SARS-CoV-2) is the cause of the ongoing coronavirus disease-2019 (COVID-19) global pandemic. SARS-CoV-2 is an RNA virus of putative animal origin. Although infections have not been reported in free-ranging or display cetaceans, due to homology between marine mammal and human angiotensin-converting enzyme 2 (ACE2) receptors and human fecal contamination of marine areas frequented by cetaceans, a risk of exposure and infection has been proposed.[10]

Retroviruses

Endogenous gammaretroviruses have been studied in both killer whales and bottlenose dolphins. While no direct disease associations have been made, anecdotal increases in the incidence of cancer and immunodeficiency suggest a potential for viral activation and associated pathology. A survey of multiple cetacean species by PCR for gag, pol, and env gene sequences showed homologs of this virus in the DNA of eight species of delphinids, pygmy and dwarf sperm whales, and harbor porpoises, but not in screened beluga or fin whales. Analysis of the bottlenose dolphin genome revealed two full-length proviral sequences with 97.4% and 96.9% nucleotide identity to the killer whale gammaretrovirus.[119]

Bacteria

Erysipelothrix rhusiopathiae, the agent of **erysipelas**, is a common clinical concern for captive cetaceans and has been identified both sporadically and in several case clusters in stranded free-ranging cetaceans.[40,181,185] In 2021, infection with *E. rhusiopathiae* was implicated in an unusual stranding event involving 190 dead porpoises in the Dutch Wadden Islands.[100] The animals presented in moderate to advanced autolysis and most were in a good plane of nutrition. Postmortem examinations were conducted in 22 porpoises and a few animals had lesions suggestive of septicemia. Bacterial culture was positive for *E. rhusiopathiae* in 16 individuals. This bacterium is readily isolated from the slime layer of food fish and the presumptive route of exposure is oral or perhaps though rake marks or ulcers in the skin.[216] Two forms of the disease exist, an acute fatal septicemic form and a cutaneous form.[216] The acute presentation is associated with sudden death and few, if any, premonitory signs. On gross examination, these animals have ascites along with mottled livers, enlarged spleens, multiorgan congestion, and occasional skin ulceration. Histologically, there is moderate multifocal necrotizing and neutrophilic lymphadenitis, vasculitis, splenitis, and hepatitis (Fig. 22.26A) with numerous intralesional and often intracellular gram-positive bacilli (eSlide 22.e9). Vasculitis may lead to ulceration in tissues such as the gastrointestinal tract, urinary bladder, and skin. The milder, cutaneous form of the disease has been seen in display beluga whales and bottlenose dolphin calves. In these cases, cutaneous lesions appeared as a rhomboid shaped patch of discolored skin with associated necroulcerative dermatitis (Fig. 22.26B). These lesions are associated with vasculitis and infarction, and clinically, there may be a loss of appetite and an inflammatory leukogram. These cases are responsive to antibiotics and the lesions resolve by regression with or without sloughing of the affected skin. Additionally, development of disease in calves from vaccinated dams in enzootic populations suggests a role of species susceptibility or antibody transfer that is protective against the acute, septicemic form of the disease.

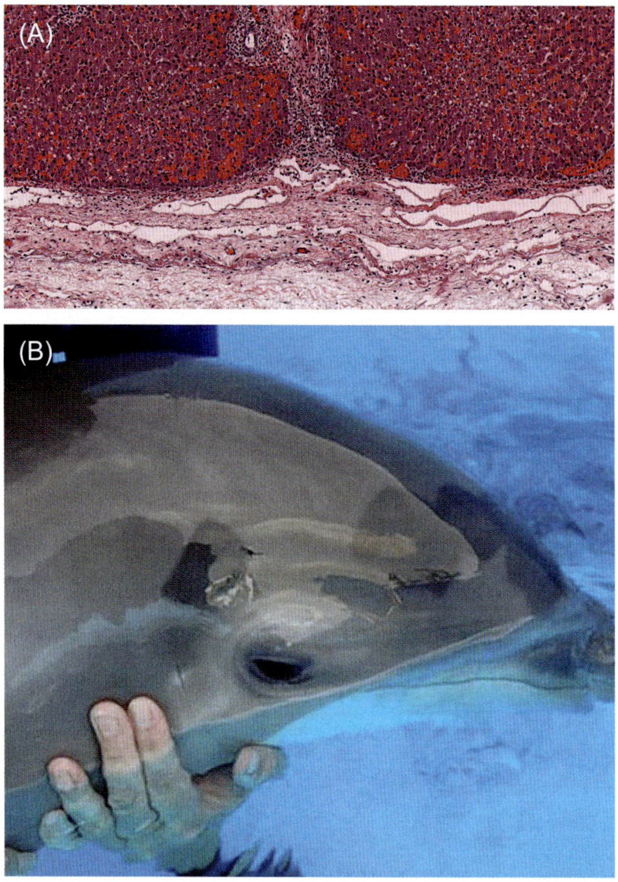

FIGURE 22.26 Acute septicemic (A) and cutaneous (B) erysipelas due to *Erysipelothrix rhusiopathiae* infection in the liver and skin of bottlenose dolphins. (A) Necrotizing periportal and capsular hepatitis due to infection is associated with marked capsular edema and expansion of the capsule and periportal regions by neutrophilic inflammation. This change is often associated with acute ascites. (B) A characteristic skin lesion is the formation of rhomboid discoloration and/or ulceration in the skin comparable to "diamond skin disease" due to *Erysipelothrix rhusiopathiae* infection in pigs (see eSlide 22.e9). *(Part B: Photo Courtesy of SeaWorld of Orlando.)*

Brucella ceti is a gram-negative bacterium that causes brucellosis. The bacterium has been isolated from a number of cetacean species, is thought to be widespread in marine environments, is zoonotic, and has caused bovine abortion by experimental challenge. *B. ceti* has been identified by culture or PCR in four cetacean families, and antibodies against *Brucella* spp. antigens have been detected in seven whales, dolphin, and porpoise families. The bacterium replicates in host macrophages and trophoblasts, and result in acute abortion and reproductive loss as well as more chronic disease in both baleen and odontocete cetaceans. *Brucella* spp. exposure is common in free-ranging populations, and a significant proportion of infected animals appear to overcome infection and may act as carriers and potential shedders.[87]

As with many pathogens, the population prevalence of brucellosis in cetaceans varies by species and geographic location, and understanding population impacts that may be due to this pathogen is important, given the ability of the bacterium to cause reproductive loss. In Costa Rica, six cetacean species have shown antibody positivity to *B. ceti* and the organism was cultured from 70% of stranded striped dolphins.[95] *Brucella* spp. have also been identified in bottlenose dolphins in the Gulf of Mexico and along the east coast of the United States,[32,36] and in two distinct beluga stocks in Alaska.

Transmission of cetacean *Brucella* spp. likely involves three pathways. Brucella are nonmotile, labile in the environment, and are unlikely to persist for extended time periods outside the host. Based on extrapolation from terrestrial species, *Brucella* spp. may be transmitted by close physical contact between susceptible hosts during coitus, through exhaled breath, by exposure to expelled fluids and placental tissues during delivery, or during nursing. *B. ceti* have been isolated from the reproductive tracts and milk of infected cetaceans, and vertical transmission has been documented in a fetus with in utero pneumonia.[32] *Brucella* spp. have also been recovered from fish mucus and the bacterium is capable of replication in these vertebrates. *B. ceti* also cycles in both nematodes and fish. This raises the possibility of transmission through vectors, such as *Halocercus* spp. and *Pseudalius* spp. lungworms in dolphins and porpoises, respectively, which can contain relatively large quantities of *Brucella* spp. Some of these parasites seem capable of crossing the placenta, which may provide another route of transmission between infected dams and fetuses.[87]

Gross and histologic lesions of cetacean brucellosis are quite variable and include chronic septicemia and abscessation. However, the most important lesions are identified in the reproductive, central nervous, and musculoskeletal systems.[196] *B. ceti* has been isolated or amplified via PCR from the reproductive organs of both male and female cetaceans. In females, endometritis, placentitis, and abortions due to brucellosis have been documented in bottlenose dolphins, harbor porpoise, and a striped dolphin. Placental lesions include multifocal necrosis, neutrophilic or mixed inflammation, submucosal edema, and multifocal, intralesional bacteria. Immunohistochemistry for brucella antigens highlights and localizes the bacteria; however, gram stains rarely demonstrate the bacteria. In bottlenose dolphins aborted due to *B. ceti* infection, a common finding is interstitial pneumonia that is often associated with prominent interstitial nodular lymphoid aggregates and increased numbers of amniotic squamous cells in the airways.[32] Metritis, placentitis, and fetal mortality due to a *Brucella* species most genetically similar to ***B. pinnipedialis*** infection have been reported in a pregnant Hector's dolphin,[27] and *Brucella* spp. have been isolated from the mammary glands of sperm whales and dolphins, suggesting infection of resident macrophages in these organs.

Brucella spp.-associated epididymitis and orchitis have been described in a number of cetacean species, including minke and Bryde's whales, harbor porpoise, beluga, and common dolphins. Infection is associated with testicular enlargement and a mix of abscesses and granulomas in the parenchyma. Histologically, granulomas are composed of macrophages admixed with epithelioid or giant cells and contain central necrosis and mineralization. Scattered between the granulomas are areas of necrosis, neutrophilic and mixed inflammation, and variable peripheral fibrosis (capsular formation).

In the CNS, there may be an increase in the volume and cellularity of the cerebrospinal fluid (CSF) with ependymal cells and mononuclear leukocytes noted on CSF cytology. There can be expansion and inflammation of the choroid plexus, leptomeninges, and the subependymal neuropil (periventriculitis) which can obstruct CSF flow and lead to secondary hydrocephalus (Figs. 22.27A, B). Occasional, multifocal Purkinje cell degeneration, gliosis, satellitosis, spongiosis, and perivascular edema is observed. There may also be mild to moderate perivascular cuffing in the neuropil and rarely, transmural fibrinoid degeneration and necrosis with perivascular hemorrhage and edema. Mononuclear choroiditis forming lymphoid follicles, periependymitis, and white matter and laminar necrosis and inflammation in the cerebrocortical gray matter have also been reported. Predilection sites are the cerebellum, brainstem, spinal cord, and medulla oblongata with less involvement of the cerebral cortex. Immunohistochemistry can highlight *Brucella* species antigens in phagocytic cells. Other confirmatory tests include PCR or bacterial culture on affected tissues, CSF, or joint fluid.

In addition to soft tissues, *B. ceti* has also been isolated from lesions in the bones and joints of cetaceans. Reported infections include **discospondylitis** in a harbor porpoise, **fibrinopurulent osteoarthritis** in a striped dolphin, and **degenerative joint changes** in a white-sided dolphin. Similar lesions are common in free-ranging bottlenose dolphins and in white-beaked dolphins. Cases of **blubber abscesses** and **pneumonia** have also been associated with cetacean brucellosis, and **vegetative valvular endocarditis** with intralesional *Brucella* spp. positive macrophages have been described in a striped dolphin with **neurobrucellosis** and concurrent severe endocarditis.[87]

Infections with **atypical *Mycobacterium*** spp. have been associated with panniculitis, pneumonia, and pleuritis in captive odontocetes; presumptive atypical mycobacterial lymphadenitis has also been reported in a stranded pygmy

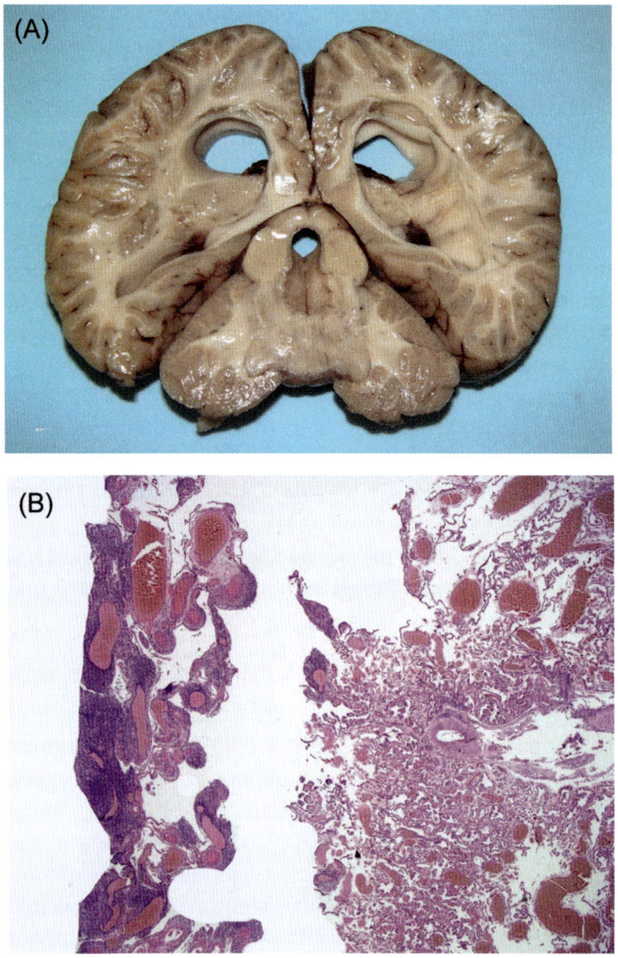

FIGURE 22.27 Brucellosis in the brain of a common dolphin. (A) Obstruction with increased intracranial pressures secondary to infection and inflammation can cause marked dilation of the ventricular system (hydrocephalus). (B) The meninges (left) and choroid (right) are markedly expanded by a mononuclear inflammatory cell infiltrate and edema. Blood vessels are markedly dilated.

sperm whale. This group of bacteria is ubiquitous in the environment and is associated with soil, detritus, and water. Common presumed routes of infection include percutaneous inoculation or secondary contamination of cutaneous lesions or via oral or respiratory routes; animal-to-animal or animal-to-human infections are uncommon. Infections may occur due to reduced host immunity or generalized debilitation, and the unique anatomy of the cetacean respiratory tract may predispose to and possibly exacerbate these infections. Multiple case reports of atypical mycobacteriosis in cetaceans currently exist but to date have not been compiled and reviewed. Notable cases include a beluga with extensive ulcerative dermatitis, panniculitis, and chronic proliferative pleuritis with marked thoracic histiocytic exudate due to ***Mycobacterium marinum*** infection[26]; panniculitis with blubber pyogranulomas, necrosuppurative pneumonia and lymphadenitis due to ***M. chelonae*** in a bottlenose dolphin[234]; ***M. abscessus*** pyogranulomatous pneumonia and associated coughing and bloody respiratory discharge in a bottlenose dolphin[30]; and an untyped mycobacterial pyogranulomatous lymphadenitis in a pygmy sperm whale and infection in bottlenose dolphin secondary to net entanglement.[114] The ecologic diversity and spectrum of lesions identified in cetaceans with atypical mycobacteriosis suggest a complex natural history of the microbe.

Nocardia spp. infections in wildlife and companion and production animals typically cause pyogranulomatous inflammation in a variety of organs, with an apparent predilection for the lung. In cetaceans, **nocardiosis** has been reported in Atlantic bottlenose dolphins, beluga whales, and killer whales. The most common presentation is the systemic form, which involves multiple organs. Organs most frequently affected are the lung and thoracic lymph nodes, but infections can occur anywhere; infection in the brain is the site most often associated with death. Molecular identification and bacterial isolation have identified a variety of pathogenic species in cetaceans including *N. asteroides*, *N. farcinica*, *N. brasiliensis*, *N. cyriacigeorgica*, and *N. levis*. Diagnosis is based on bacterial culture, gross and histologic findings, and molecular diagnostic tests. Cytology is particularly effective at identifying nocardial bacteria, which are branching, rod-shaped bacteria. The bacteria are gram-positive, and positive with modified acid-fast and Grocott's methenamine silver (GMS) special stains, the latter of which most consistently demonstrate the characteristic organisms (Fig. 22.28; eSlide 22.10). In both pinnipeds and cetaceans, juvenile animals are affected more often than adults.[46,206]

As with terrestrial animals, cetaceans harbor and occasionally develop fatal **salmonellosis**. Zoonotic transmission

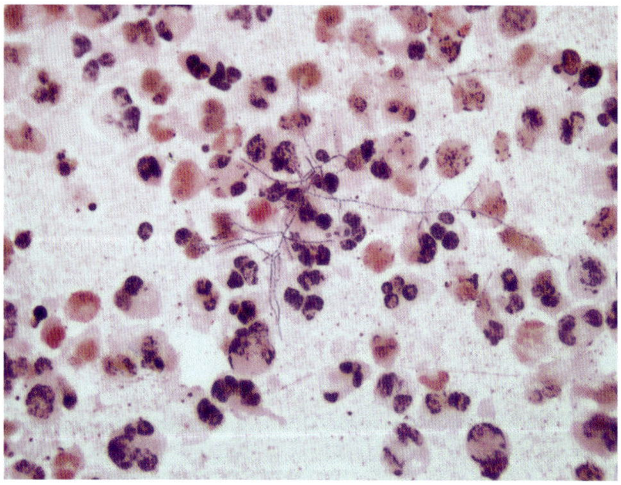

FIGURE 22.28 Nocardiosis in a beluga whale. Cytology from a tissue imprint (touch preparation) of intact and degenerate neutrophils and numerous beaded, branching filamentous bacteria consistent with *Nocardia* spp. Diff-Quik (see eSlide 22.e10).

can occur and is most common in people whose diets include consumption of porpoise, dolphin, or whale meat. Salmonella have been identified in the respiratory tracts of killer whales and harbor porpoises, and have been associated with mortality in a septic killer whale perinate.[31] In stranded animals off the Canary Islands, a dwarf sperm whale and common dolphin were emaciated and diagnosed with salmonellosis and enterotoxemia.[5] *Salmonella enterica* subspecies *enterica* is frequently found in the respiratory tract and other tissues of healthy and diseased harbor porpoises from northern European coasts. Ironically, this strain is most closely related to a strain identified in chickens in the United States in the 1950s. Reported infection rates are as high as 27% in harbor porpoises in the United Kingdom. Isolates are most commonly identified in lung tissue and transmission of via lungworm vectors has been proposed.[88]

Burkholderia pseudomallei, the agent of **melioidosis**, is indigenous to Southeast Asia and Northern Australia, and is sporadically isolated in Europe, Central and South America, Africa, the Middle East, and India. It is an important pathogen of managed cetaceans and pinnipeds in Asia. Infections in Pacific bottlenose dolphins, Indo-Pacific bottlenose dolphins, killer whales, pseudorcas, and Pacific white-sided dolphins have been reported. It is likely that any marine mammal can contract this infection. The disease presents in two forms, an acute septicemic form and a chronic, debilitating granulomatous form. Clinical presentation and pathologic lesions depend on exposure load and the immune status of the animal at the time of infection. The organism has been isolated from air samples and inhalation, ingestion, or contamination of open wounds are likely sources of infection. Reduced exposure to aerosolized bacteria in typhoon winds has been effective in mitigating clinical disease.[112]

The acute, disseminated form of melioidosis is by far the more common presentation in cetaceans and animals may die rapidly following initial clinical signs of disease such as lethargy and inappetence. Grossly, tan to yellow areas of necrosis that vary in size from pinpoint and miliary to several centimeters in diameter are present in multiple organs; concurrent splenomegaly and lymphadenomegaly with edema are also seen. These correspond histologically to small abscesses that contain a mixture of hemorrhage, necrosis, and intact or degenerate neutrophils, or poorly organized pyogranulomas with central necrosis, hemorrhage, fibrin, and edema admixed with intact and degenerate neutrophils surrounded by macrophages, lymphocytes and plasma cells. Acute infections in cetaceans most commonly affect the liver, spleen, and lung. In addition to acute disease, chronic melioidosis with lumbar vertebral osteomyelitis has been reported in a bottlenose dolphin. In all cases, slender, gram-negative bacteria are identified within the areas of necrosis.[112]

As with many animals, ***Staphylococcus*** spp. and ***Streptococcus*** spp. are common primary or opportunistic (secondary) infections in cetaceans. Of these, there are three pathogens that require a brief overview. ***Streptococcus phocae*** and ***S. iniae*** are pathogenic non-β-hemolytic species. In cetaceans, *S. phocae* can be cultured from many live-stranded and dead animals. It can be cultured from blowhole debris and feces of asymptomatic animals, and many isolates may reflect commensals, innocuous, or opportunistic infection. However, *S. phocae* has been associated with suppurative cervical lymphadenitis in bottlenose dolphins.[179] Other cases of septicemia due to *S. phocae* have been noted in species such as beluga, harbor porpoise, and a short-beaked common dolphin with lesions including suppurative and necrotizing inflammation in multiple tissues.[155]

Streptococcus iniae is a commensal and fish pathogen and exposure in free-living and display cetaceans is likely from contaminated prey/feed. *S. iniae* is associated with generalized sepsis as well as localized infections such as pneumonia, pleuritis, pyothorax, dermatitis, myositis, and panniculitis. Lesions consist of a mix of acute necrotizing and more chronic, pyogranulomatous inflammation.

Notable ***Staphylococcus aureus*** infections include sepsis in a killer whale and granulomatous inflammation in harbor porpoises. Lesions in stranded porpoises included pyogranulomatous myocarditis, necrotizing bronchopneumonia, pyelonephritis, osteomyelitis, leptomeningitis, and abscesses in lymph nodes and skeletal muscles. Other reports include fibrinous and suppurative epicarditis and pyogranulomatous myocarditis with abscessation in a captive porpoise and tooth root abscessation and suppurative pneumonia and nephritis in a dead-stranded killer whale.[162] Infection in the porpoises was suspected to have entered through skin lesions or the respiratory tract.[192] Because *S. aureus* has been isolated from blow samples from free-ranging killer whales, associated disease and pathologic findings to date likely represent opportunistic infection.

Mycoplasmosis has been associated with upper respiratory, oral, and otic infections in harbor seals and polyarthritis in sea lions. Reports of cetacean infections are rare and poorly described. ***Mycoplasma phocicerebrale*** was isolated from the lungs of three harbor porpoises and the liver of one of these animals. Novel *Mycoplasma* spp. have been isolated from the lungs of five harbor porpoises and kidney of another, and an isolate closely related to *Mycoplasma* species 13CL was obtained from the kidney of a Sowerby's beaked whale. No associated lesions were identified in any of these cases.[75]

Bartonellosis has been reported in multiple cetacean species.[90] These include free-ranging striped dolphins, harbor porpoises, Risso's dolphins, a pygmy sperm whale, captive bottlenose dolphins, and beluga whales. Sequencing identified a *Bartonella* spp. that is most similar to strains of ***B. henselae***. No specific pathology has been associated with these infections. **Angiomatosis**, an important pathologic manifestation of *Bartonella* spp. infection in immunocompromised humans, has been described in stranded bottlenose dolphins and hunter-harvested beluga. To date, a similar association in cetaceans has not been described.

Severe myositis due to infections with ***Clostridium*** spp. has been diagnosed in captive killer whales, pilot whales, and bottlenose dolphins. All marine mammals are likely susceptible to infections with bacteria in this genus; however, exposure is likely sporadic with a low incidence of clinical disease or pathology. The disease is characterized by acute swelling, muscle necrosis, and accumulation of gas in affected tissues. This is accompanied by severe clinical leukocytosis, serofibrinous exudate, and acute hemorrhage. Typical gross lesions include acute necrotizing myositis with dark discoloration and a general dryness of the muscle. Histologically, areas of acute muscle degeneration and necrosis often contain clear spaces due to gas production by the bacteria. Diagnosis of clostridial infection is based on detection of gram-positive bacilli in aspirates and confirmation with anaerobic culture, florescent antibody screening for the bacterium, and/or multiplex real-time PCR. As a reminder, microcavitations or clear bubbles in the tissues are not pathognomonic for clostridial myositis, and intramuscular and occasionally intravascular bubbles are often present in decomposed carcasses due to postmortem bacterial overgrowth and gas production by *Clostridium* spp. Detection of bacteria histologically or positive culture results must be interpreted along with carcass postmortem condition and evidence of associated hemorrhage, necrosis, and inflammation to differentiate postmortem bacterial overgrowth from an antemortem disease process.

Helicobacter cetorum has been identified in numerous delphinid species with gastritis.[42] Histopathology reveals mild to moderate lymphoplasmacytic gastritis in the superficial mucosa of the main and pyloric compartments with superficial to deep ulceration and occasional hemorrhage. Ulceration in the pylorus can perforate and result in fatal chemical or septic peritonitis. Spiral to fusiform bacteria have been detected in the gastric mucosa with Warthin Starry staining. Despite reports of infection, the role of *Helicobacter* spp. in gastric ulcerations has not been fully resolved.

Fungi

Lobomycosis in humans is caused by the fungus ***Lacazia loboi***. A morphologically similar infection in dolphins was historically attributed to this same pathogen. However, recent molecular studies have identified ***Paracoccidioides brasiliensis*** **(*Paracoccidioidomycosis ceti*)** as the causative organism.[228] This distinction may account for subtle ultrastructural distinctions in yeast morphology previously reported between dolphin and human cases. Dolphin infections are most commonly reported from Florida (United States) to the Gulf of Mexico and extend south into South America. Recent reports from Asia are suggestive of infections with a morphologically similar organism. Gross lesions tend to occur along the torso but may occur anywhere on the body, and consist of focally extensive to disseminated, smooth, cutaneous swellings that progress to papules, nodules, plaques, verrucae, and occasional ulcerations (Fig. 22.29A). Microscopically, the dermis is expanded and effaced by myriad, 6–10 μm diameter, round, thick-walled, PAS, and GMS positive (mucicarmine negative) yeast that form single to multiple budding cells or chains interconnected by a tubular isthmus. As with cryptococcosis, melanin can accumulate in or be associated with the cell wall and is considered a virulence factor that is protective against oxidative stress. Yeast are embedded within lymphohistiocytic inflammation that includes numerous foreign bodies and Langhan's type multinucleated giant cells, fewer plasma cells, and occasional neutrophils (Fig. 22.29B; eSlide 22.e11). The yeast are slow growing and depending on the stage of infection and degree of

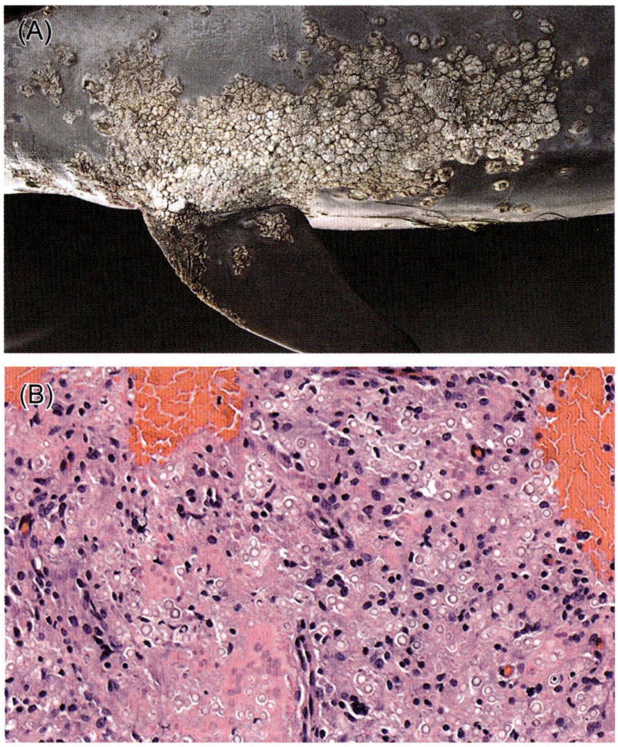

FIGURE 22.29 Cutaneous lobomycosis in a bottlenose dolphin. (A) Multifocal to coalescing patches of proliferative dermatitis are present along that body wall and flipper. (B) Numerous chains of budding yeast are present within mixed pyogranulomatous inflammation (see eSlide 22.e11). *(Photos Courtesy of Hubbs-SeaWorld Research Institute.)*

dermal invasion, pseudoepitheliomatous hyperplasia, acanthosis, and parakeratotic hyperkeratosis may also be seen. Expansile dermal lesions, atrophy of the overlying epidermis with attenuation or loss of rete ridges and ulceration may also occur. Opportunistic bacterial infection is associated with loss of epidermal integrity. Regional lymph node involvement has been reported, but infection is usually limited to the skin and can persist in a nonprogressive state for many years. The mode of infection is believed to be direct skin penetration and subsequent proliferation in the dermis.

Zygomycosis refers to infection with members of a diverse group of fungi (class Zygomycetes) in one of its two orders, the Mucorales and Entomophthorales. The most common fungi associated with **mucormycosis** are *Mucor* spp., *Rhizopus* spp., *Rhizomucor* spp., *Absidia* spp., *Apophysomyces* spp., and *Saksenaea* spp.[1,97,153] All are distributed worldwide and typically associated with soil and detritus. Infections are most commonly acquired through inhalation, ingestion, or percutaneous inoculation of spores. The fungi are angioinvasive and fungemia can result in disseminated infections. The hyphae are broad to ribbon like, have thin hyaline walls, are pauciseptate, nondichotomously branching, and prone to twisting and folding. Respiratory tract infections result in necrotizing and pyogranulomatous bronchopneumonia often with extensive necrosis of airways. Fungal hyphae can sometimes be found within bronchial cartilage. Vascular infections are associated with necrosis, vasculitis, perivascular hemorrhage, thromboembolism, ischemia, and infarction. An inflammatory response may be absent or consist of suppurative, pyogranulomatous, or granulomatous inflammation. Individual case reports of localized dermatitis, cellulitis or myositis, and disseminated mucormycosis (zygomycosis) have been reported occasionally in stranded wild and managed cetaceans, and mucormycosis with pulmonary dissemination has been reported secondary to satellite tag attachment in a southern killer whale.

Entomophthoramycosis is generally limited to tropical regions and involves one of two genera, ***Basidiobolus*** spp. and ***Conidiobolus*** spp. Infections are generally chronic, indolent, and can arise through inhalation or direct inoculation of spores from decaying plant material, soil, and leaves. Host animals are generally immunocompetent. In contrast to mucormycosis, fungi tend not to invade blood vessels and infections rarely disseminate from the primary site of infection. The hyphae are broad and pauciseptate with 90-degree branching. Skin and subcutaneous infections result in localized swelling, which may be fluctuant to firm. Microscopically, there is granulomatous inflammation with foreign body multinucleated giant cells and Splendore-Hoeppli material. Eosinophils often accompany the inflammation and eosinophilic microabscessation may occur.

The genus ***Aspergillus*** includes hundreds of species of molds, some of which are associated with sporadic respiratory infections in wild-stranded and display-managed cetaceans. In most cetacean cases, *Aspergillus fumigatus* has been isolated or identified by molecular studies, although *A. niger* and *A. terreus* have also been reported.[171] Infections are generally opportunistic and acquired by inhalation with deposition of spores deep in the respiratory tract. **Aspergillosis** in free-ranging cetaceans is often seen concurrently with morbillivirus infection. Initial stages of infection with airway-oriented fungal hyphae are associated with invasion and necrosis in subjacent parenchyma. Gross examination reveals well-delineated, gray nodules in the parenchyma with hemorrhagic margins or large cavitations composed predominantly of fungal hyphae. Dissemination to other tissues, such as the brain, kidney, and eye, occurs secondary to vascular invasion. Tracheal involvement with circumferential to segmental disruption and effacement of the respiratory mucosa by fibrosis and granulomatous infiltrates and variable luminal occlusion has been reported in four captive bottlenose dolphins (three with aspergillosis and one with possible zygomycosis). Severe mycotic otitis media due to *A. terreus* has also been described in a juvenile live-stranded harbor porpoise from the United Kingdom.[163] On gross exam, green-yellow purulent exudate and caseous debris were noted in both tympanic cavities and the periotic sinuses; histologically, a large fungal aggregate was present in the tympanic cavity. Osteolytic changes were evident in the adjoining periotic bone and stapes. This included lymphoplasmacytic and neutrophilic inflammation of the scala vestibuli with granulation tissue, serous exudate, and pseudocyst formation associated with the round and oval windows. The lesions, in conjunction with generalized emaciation, were considered sufficiently severe to account for the live stranding of the animal. A similar presentation was identified in a by-catch harbor porpoise. Recent developments in protein electrophoresis and metabolomics/proteomics may bolster serologic detection of aspergillosis in live stranded and display cetaceans.[51]

Obstructive tracheitis has been reported in four captive bottlenose dolphins.[49] The lesions consisted of severe, segmental, and circumferential fibrosing tracheitis with partial to total luminal obstruction. The tracheal cartilage, submucosa, and mucosa were distorted and replaced by extensive fibrosis and pyogranulomatous inflammation centered on fungal hyphae. The fungi associated with this lesion included *Aspergillus* spp. and zygomycetes (Fig. 22.30).

Cryptococcus is a genus of fungi that includes both yeasts and filamentous forms. The yeast have a prominent polysaccharide capsule. ***Cryptococcus neoformans*** and ***C. gattii*** are common pathogens. Infections are acquired predominantly via the respiratory tract with subsequent hematogenous dissemination. *C. neoformans* occurs

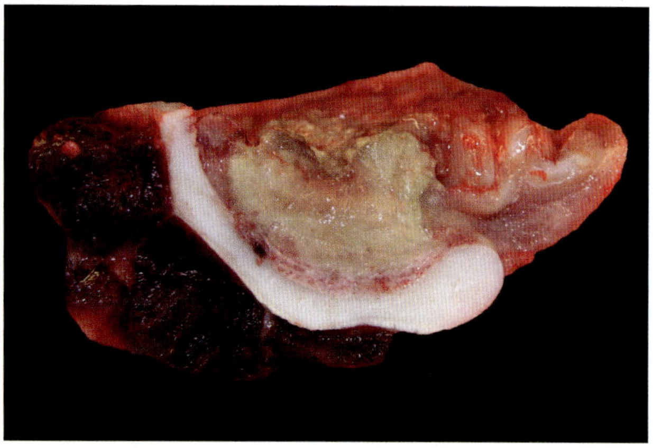

FIGURE 22.30 Obstructive tracheitis due to mixed fungal infection in a bottlenose dolphin. Tan/yellow, pyogranulomatous tracheal inflammation and fibrosis cause partial occlusion of the tracheal lumen (tracheal cartilage is white and crescent shaped).

worldwide and clinical disease tends to develop in immunosuppressed or debilitated individuals. *C. gattii* is limited to tropical and subtropical regions that largely parallel the distribution of eucalyptus trees. In contrast to *C. neoformans*, reported human infections in the northeastern Pacific tend to involve immunocompetent individuals with a history of outdoor recreational activities.[66,111] Dead-stranded cetaceans with **cryptococcosis** typically present with pneumonia and lymphadenitis.

The first multispecies outbreak of *C. gattii* was recognized in the Pacific Northwest and western Canada in 1998. Infections involved humans, companion animals, exotic birds, and terrestrial wildlife. Among the first animals diagnosed with *C. gattii* infection was a harbor porpoise that stranded along the southeast coast of Vancouver Island, BC. In the Pacific Northwest, *C. gattii* is one of the most common fungal pathogens isolated from stranded cetaceans. It is most often seen in harbor porpoises and Dall's porpoises and with sporadic infections of Pacific white sided dolphins. Gross lesions include generalized emaciation, pulmonary consolidation, and generalized lymphadenopathy. Pulmonary lesions vary from single large, tan-yellow to gray-red nodules that protrude above the pleural surface and extend deep into the parenchyma. In more severe infections, the nodules are miliary or become confluent and can efface entire areas. Secondary bacterial infections and concurrent verminous pneumonia have also been seen. Prominent rib impressions may be apparent. Regional lymph nodes are markedly enlarged (up to 5–10 times normal size) and on cut surface may be homogenous, glistening yellow-white, and ooze clear gelatinous material, or they may be multinodular, pale gray/white to brown/red and be associated with inflammation, necrosis, hemorrhage, and cavitation. Microscopically, in the more acute stages of infection, pulmonary bronchoalveolar spaces are distended and occluded by dense aggregates of encapsulated yeast with sparse to mild infiltrates of lymphocytes and histiocytes. With progression of inflammation to more chronic stages, there are nodular to diffuse accumulations of granulomatous inflammation with scattered Langhan's type multinucleated giant cells, intra- and extracellular yeast, and occasional fibrous encapsulation. Similar transition from small aggregates to variably extensive sheets of yeast is noted in the subcapsular and medullary sinuses of regional lymph nodes, where yeasts are often admixed with mild lymphohistiocytic infiltrates. In more fulminant disease, evidence of multisystemic fungemia may be seen in the brain, kidney, prostate, digestive tract, or mammary glands and in developing fetuses. The polysaccharide capsule of the yeasts are highlighted with methenamine silver and Fontana-Masson stains, and the capsule stains red with mucicarmine and blue with Alcian blue stains. Although fungal immunohistochemistry is available, fungal culture and isolation with molecular genotyping yield more valuable information.

In marine mammals, **coccidioidomycosis** is most often reported in California sea lions and southern sea otters. Sporadic cases in bottlenose and common dolphins have been identified in California. Notable is that *Coccidioides immitis* and *C. posadasii* are zoonotic. The most common findings are miliary nodules with central caseous necrosis in the lungs and perihilar lymph nodes. Pyogranulomatous pneumonia and lymphadenitis with intralesional fungal arthroconidia and immature and mature spherules, the latter of which are 60–80 μm diameter with a characteristic, double contoured, refractile wall and 5–7 μm diameter intralumenal endospores, area common histologic findings. Infection in some cases has extended to or been present in the brain where the fungus was associated with necrotizing encephalitis (Fig. 22.31).

Metazoa

As with all free-ranging wildlife, parasitic infections in cetaceans are common. Many infections are subclinical, and disease is often related to the extent of host–parasite adaptation and the health status of the host. The advent of gene sequencing (conventional nested or quantitative polymerase chain reaction) and restrictive length polymorphism (RFLP) has advanced the speciation and understanding of the natural history of many of these parasites.[107,117] In addition to the descriptions below, an overview of cetacean parasites is presented in Table 22.e5.

Anisakis simplex is a gastric nematode parasite of cetaceans. It is associated with volcano-like ulcers. The nonglandular forestomach is the most commonly reported site of infection in bottlenose dolphins and harbor porpoises. In addition to the forestomach, the parasites can be distributed within any of the three gastric compartments, overlying the mucosa or free within the gastric and

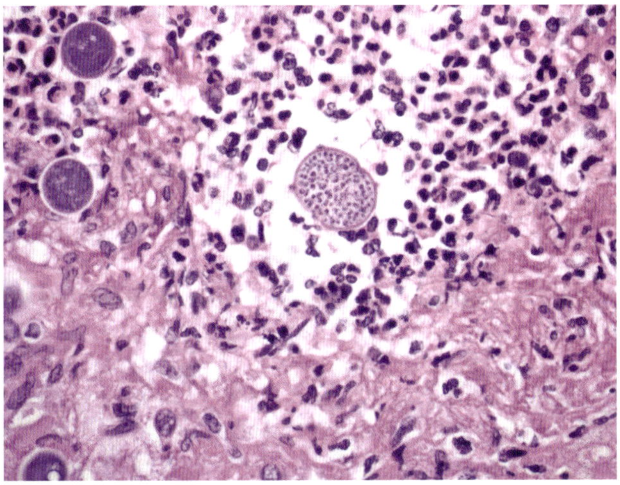

FIGURE 22.31 *Coccidioides immitis* infection in the brain of a common dolphin. The neuropil is disrupted by a focal area of suppurative inflammation and necrosis and intralesional fungal elements. These include darkly basophilic immature spherules (left) and a larger, more lightly basophilic mature spherule that contains numerous endospores (right).

intestinal lumina. Lesions vary in number and size. Larger reported lesions have been up to 3–5 cm in diameter and been associated with ulceration, fibrosis, and numerous intralesional nematodes, which can extend into the submucosa. Microscopically, marked gastritis with lymphocytic, plasmacytic, eosinophilic, and granulomatous inflammation, giant cells, hemosiderosis, fibrosis, and areas of necrosis are associated with the parasites. Degenerate and necrotic parasite remnants embedded within proliferative fibroplasia may be seen deep in the submucosa. Ulcers can be associated with gastric hemorrhage; perforation is uncommon. The larvae of this parasite are present within the muscles of numerous marine fish.

Another group of nematodes associated with increased morbidity and mortality of cetaceans are the sprurida, ***Crassicauda*** spp. These parasites have a tropism for the renal vasculature, mammary tissue, and subcutaneous tissue around the urogenital slit. In more severely affected animals, impaired renal function may contribute to significant loss within the population.[118] They can also migrate along the torso and have been associated with deep invasion into the tissues of the neck and underlying the spinal cord. Grossly, the parasites may be present in the renal calices, mammary ducts, and other tissues and, depending on the host immune status, they may be associated with mild to severe inflammation. With more chronic infections, granulomatous to fibrogranulomatous inflammation may develop and be associated with dense eosinophilic necrotic debris, dystrophic mineral deposition, and variable numbers of nematodes.

Braunina cordiformis is a digenic trematode of small delphinids. *Braunina* spp. trematodes are commonly found within a mat of mucus attached to the surface of the glandular compartment (second chamber) of the stomach (Figs. 22.32A, B; eSlide 22.e12). Occasionally, organisms will attach to the mucosa of the gastric pylorus and duodenal ampulla. The parasite produces a small focus of chronic gastritis at the site of mucosal attachment. While the attachment does not appear to be associated with significant disease, heavy parasitism can be so intense that extensive areas of the gastric mucosa (more than 50%) may be involved and covered by associated mucus. When parasites are primarily within the second gastric chamber,

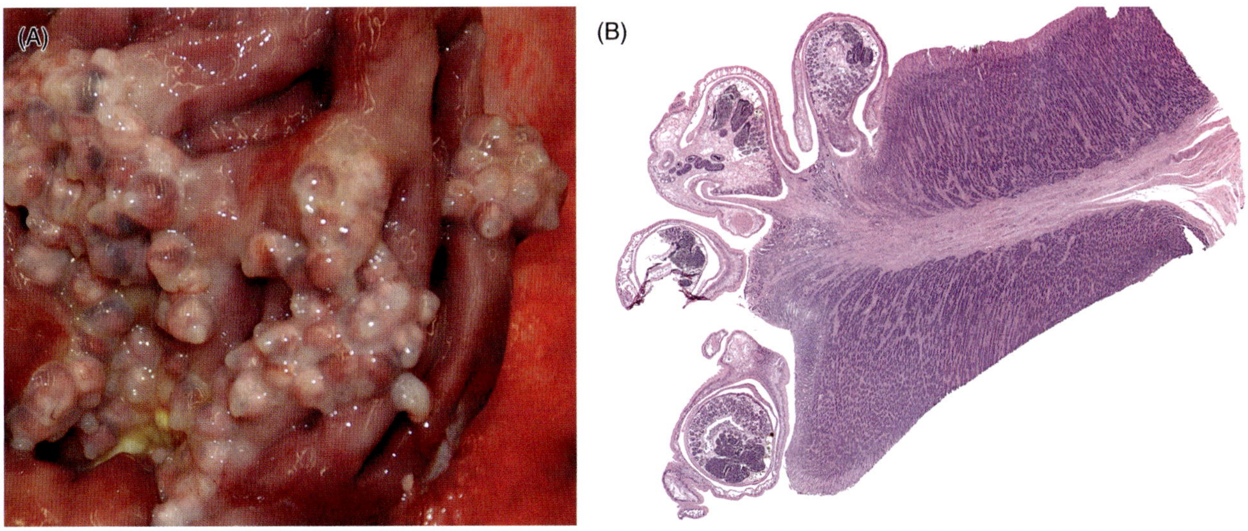

FIGURE 22.32 *Braunina* sp. infection in the glandular stomach of a bottlenose dolphin. (A) The mucosa is covered by numerous trematodes, which appear as translucent to opaque white nodules covered by an overlying mucous layer. (B) Numerous trematodes are attached to the gastric mucosa. Their presence is associated with very little, if any, associated inflammation (see eSlide 22.e12). *(Part A: Photo Courtesy of Hubbs-SeaWorld Research Institute.)*

animals should be evaluated for signs of maldigestion associated with a reduction in the effective gastric glands.

Hepatobiliary **trematodiasis** attributed to *Campulla* spp., *Oschmarinella* spp., and *Zalophatrema* spp. have been reported in a number of cetacean species and lesions typically include ductular hyperplasia, ectasia, periductular fibrosis, eosinophilic and granulomatous inflammation, fluke pigment, and variable numbers of intralesional adult trematodes with occasional dispersed triangular ova. Immunohistochemistry for inflammatory markers has identified distinct immune responses in harbor porpoises and Dall's porpoises infected with *Campula oblonga* and Hubbs' beaked whales with *Oschmarinella macrorchis*.[146] With ectopic migration, similar lesions with these species may occur within the pancreas. Eosinophilic and fibrosing pancreatitis due to Brachycladiidae trematodes has also been reported in bottlenose dolphins stranded along the east coast of the United States (Georgia).[187]

Nasitrema globicephalae and up to 9 other species of this genus are trematodes that inhabit the pterygoid sinuses and tympanic cavities of odontocetes.[65] The life cycle of this parasite is complex, and infection is most likely acquired through consumption of infected fish (prey items) that contain the larval stages. Aberrant migration of this parasite can result in otitis media, neuritis of the eighth cranial nerves, and marked meningoencephalitis (Fig. 22.33A). Migration tracts, which can be seen in both the gray and white matter of the brain, are associated with necrosis and cavitations, extensive areas of hemorrhage, edema, and inflammation with gitter cells, neutrophils, eosinophils, lymphocytes, plasma cells, and intralesional golden yellow to black birefringent material (fluke pigment). Spongiosis and gliosis may occur within the adjacent neural tissue, and the meninges and perivascular spaces are expanded by edema, fibrin, lymphocytes, plasma cells, eosinophils, and fewer neutrophils and hemosiderin-laden macrophages. Sections of adult trematodes and thick-walled, golden-brown, triangular eggs (approximately 50 μm in diameter) may be found and may be surrounded by multinucleated giant cells (Fig. 22.33B). Adult trematodes are 500 μm wide with a 15 μm thick, have a spiny tegument, and contain numerous vitellaria, paired ceca, and testes.

A study published in the mid-1980s found a high percentage of the cetaceans stranded along the Southern Californian coastline with massive infestation of *Nasitrema* spp., with mature, gravid (egg-bearing) helminths in the brain.[34] Because of the extent and severity of encephalitis and necrosis of the neuropil, researchers have inferred that the brain lesions were a contributing factor to the stranding of small cetaceans. A survey of cranial lesions of two coastal dolphin species in South Africa revealed that crassicaudiasis and cranial osteolysis were most profound in younger animals and suggested a possible relationship with the stage of bone ossification, exuberant inflammatory response or possible impaired immune response.[218] Interestingly, examinations of stranded cetaceans in this same region since 2000 have found a very low incidence of infection despite thorough neuropathologic examinations. This shifting prevalence may reflect a change in the dietary preferences, prey availability, or parasite distribution in this region.

Protozoa

Toxoplasmosis, due to ***Toxoplasma gondii***, is a sporadic, but significant cause of nonsuppurative meningoencephalitis, disseminated parasitemia, necrotizing placentitis,

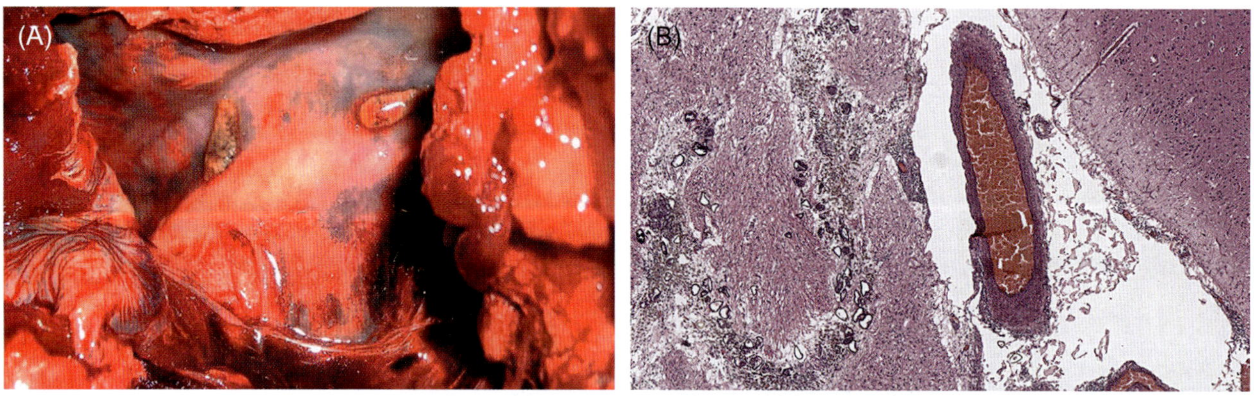

FIGURE 22.33 *Nasitrema* sp. in the nasal sinus of a striped (A) and *N. globicephalidae* meningoencephalitis in a common (B) dolphin. (A) *Nasitrema* sp. overlying the sinusoidal mucosa. The parasite likely features a complex life cycle and aberrant infections have resulted in neuritis of the eighth cranial nerve, otitis media, and meningoencephalitis. (B) The meninges are markedly expanded by mixed inflammation. Within the inflammation and extending into the subjacent neuropil are scattered brown, thick-shelled trematode eggs. The wide, curved pattern of inflammation and intralesional eggs suggests a migration track of an adult parasite through the brain. *(Part A: Photo Courtesy of the Institute of Animal Health and Food Safety, University of Las Palmas de Gran Canaria.)*

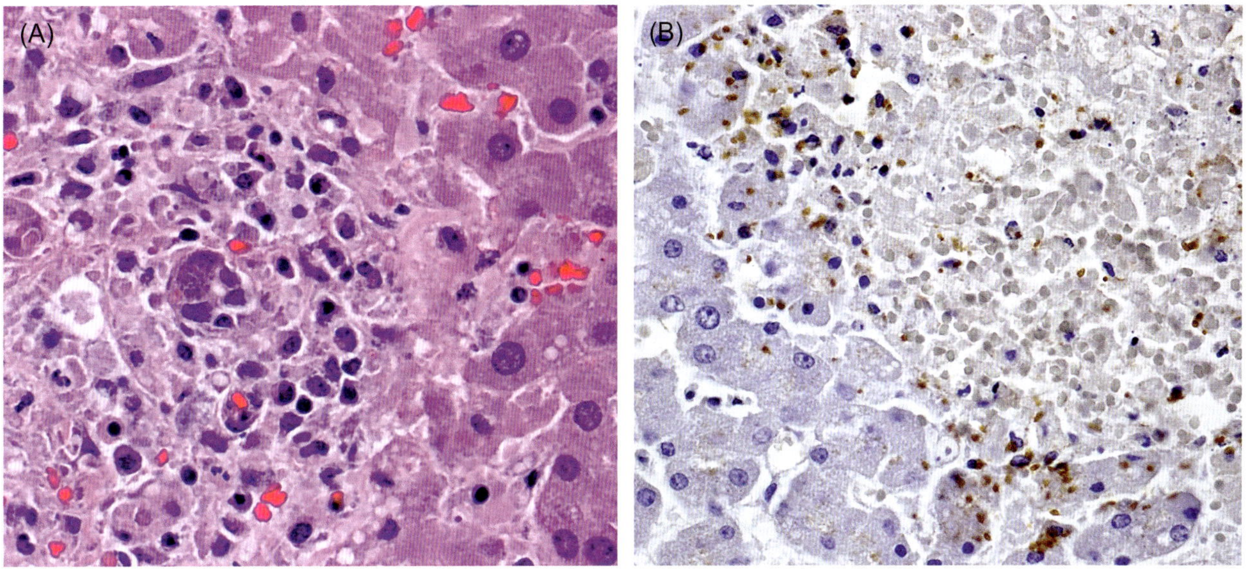

FIGURE 22.34 Hepatic toxoplasmosis in a Hector's dolphin. (A) Focal necrotizing hepatitis with mixed inflammation surrounds a single enlarged hepatocyte containing a protozoal cyst with multiple tachyzoites. (B) Immunohistochemical labeling at the edge of a necrotic focus contains moderate numbers of brown, reactive zoites. *(Parts A, B: Photos Courtesy of W. Roe, Massey University.)*

and abortion in cetaceans. Infections have been reported in a number of New Zealand Hector's and critically endangered Maui dolphins. One study identified death in 7 of 28 (25%) dolphins, including two of three Maui dolphins, due to disseminated infection. A further ten dolphins had one or more tissues that were positive for *T. gondii* DNA using PCR. Fatal cases had intralesional protozoal zoites and cysts in areas of necrotizing and hemorrhagic pneumonia, lymphadenitis, hepatitis, and adrenalitis (Figs. 22.34A, B). Tachyzoites and tissue cysts were also present in other organs including the brain, heart, stomach, and uterus with minimal associated inflammatory response. One dolphin had marked suppurative metritis with numerous intraepithelial tachyzoites.[176] Two of eight striped stranded dolphins in late 2007 on the Ligurian Sea coast of Italy had nonsuppurative meningoencephalitis characterized by prominent perivascular mononuclear cell cuffs and macrophage accumulations in the neuropil. These lesions were associated with mild lymphocytic and plasmacytic inflammation in the choroid plexus in one dolphin. *T. gondii* cysts and zoites, confirmed by immunohistochemical labeling, were scattered throughout the brain parenchyma.[54] Similarly, two Atlantic spotted dolphins were diagnosed with nonsuppurative meningoencephalitis due to toxoplasmosis in the Canary Islands.[5]

Several reports describe **sarcocystosis** in cetaceans. This includes ***Sarcocystis neurona*** infection with lesions and mortality in cetaceans from the Pacific Northwest, nonsuppurative and necrotizing meningoencephalitis with perivascular lymphoplasmacytic cuffs, scattered gliosis, and intralesional protozoal schizonts and dispersed merozoites in a Pacific white-sided dolphin and striped dolphin

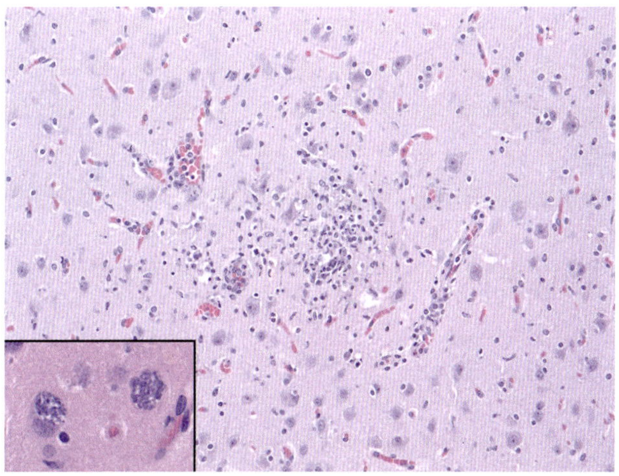

FIGURE 22.35 *Sarcocystis speeri* encephalitis in a stripped dolphin. Infection is associated with nonsuppurative encephalitis with perivascular lymphoplasmacytic cuffs and intralesional protozoa (inset). *(Photo Courtesy of R Ossiboff, University of Florida.)*

(Fig. 22.35). Interstitial pneumonia, peribronchial lymphofollicular hyperplasia and florid, predominantly intravascular accumulations of protozoa, mild nonsuppurative encephalitis, cholangiohepatitis, myocarditis, and tonsillitis were recorded in a harbor porpoise. *Sarcocystis* spp. cysts have also been seen as incidental findings in skeletal and cardiac muscle in Atlantic white-sided dolphins[61] and seen in a stranded beluga in the Alaska and Canadian Arctic. Immunohistochemistry and PCR are useful to confirm infection and establish the specific *Sarcocystis* species.

Ectoparasites

Cyamids (Fig. 22.36) are external amphipod parasites of cetaceans that infest both baleen whales and much less frequently, odontocetes. They are a common commensal and do not suggest specific disease. They are referred to as "whale lice" but they are not insects. They can be found anywhere on the skin. Typically, they reside in genital folds, ventral pleats, and nostrils, are associated with callosities and barnacles, and occur at the periphery of the eyes and within and around skin wounds. Each whale species generally has its own species of cyamid parasite, and in some, the species of cyamids change with the different life stages of the host. Cyamids do not have a free-swimming phase and infections are transferred by contact. The parasites spread from dams to their calves during birth, nursing, and general contact. In the sperm whale, the parasitic relationship is sex specific. *Cyamus catodontis* lives exclusively on the skin of the male, while *Neocyamus physeteris* is found only on females and calves. Cyamids consume the algae and sloughing superficial skin of their hosts. The rapid turnover of cetacean skin provides ample support for hundreds, even thousands of these organisms on individual whales.

FIGURE 22.36 Cyamids on the skin of a North Atlantic right whale. Often called "whale lice," cyamids are not insects. They are amphipod parasites that colonize the skin. *(Photo Courtesy of Hubbs-SeaWorld Research Institute.)*

ADDITIONAL RESOURCES

Dailey, M.D., 1985. Diseases of mammalia: Cetacea. Dis. Mar. Anim. 4 (2), 805–847.

Dierauf, L., Gulland, F.M. Whitman, L (Eds.), 2018. CRC Handbook of Marine Mammal Medicine: Health, Disease, and Rehabilitation. 3ed. CRC Press.

Dold, C., 2001. Cetacea (Whales, Dolphins, Porpoises). In: Fowler's Zoo and Wild Animal Medicine Current Therapy, Vol. 7. Elsevier Health Sciences, pp. 422–436.

Moore, M., van der Hoop, J., Barco, S., Costidis, A., Gulland, F., Jepson, P., Moore, K., Raverty, S., McLelland, W. Criteria and case definitions for serous injury and death of pinnipeds and cetaceans caused by anthropogenic trauma. Dis. Aquat. Org. 102, 229–243.

Morell, M Andre, M. Cetacean ear extraction and fixation protocol. https://www.zoology.ubc.ca/files/Ear_extraction_and_fixation_protocol_UBC.pdf.

References

Full references for this chapter can be found online at https://doi.org/10.1016/B978-0-443-18437-6.00010-9.

Chapter 23

Pinnipediae

Kathleen M. Colegrove[a], Kathy A. Burek-Huntington[b], Wendi Roe[c], and Ursula Siebert[d]

[a]Zoological Pathology Program, College of Veterinary Medicine, University of Illinois, Brookfield, IL, United States; [b]Alaska Veterinary Pathology Services, Eagle River, AK, United States; [c]Massey University, Palmerston North, New Zealand; [d]Institute of Terrestrial and Aquatic Wildlife Research, University of Veterinary Medicine Hannover, Büsum, Hannover, Germany

INTRODUCTION

The pinnipeds consist of 33 species of mammals inhabiting marine/terrestrial interfaces around the globe. They are the Otariidae (fur seals and sea lions), Phocidae (true seals), and Odobenidae (walrus) families. Common and scientific names of species described in this chapter are presented in Table 23.e1. Much of the knowledge on pathologic conditions of pinnipeds comes from necropsies of stranded animals and those housed in captivity. As such, disease knowledge is biased toward species frequently housed in zoos and aquaria, those that strand more commonly, or those in which free-ranging populations are more easily accessible. According to the IUCN (International Union for the Conservation of Nature) Red List of Threatened Species, 10% of pinniped species or subspecies are listed as vulnerable and 20% are considered endangered. The Caribbean monk seal and Japanese sea lion are extinct due to hunting. Current threats to pinniped species include anthropogenic factors, climate change, and habitat degradation. These threats are of particular concern in species or subspecies that inhabit small geographic ranges, have a low population abundance, and/or whose populations are currently or have recently undergone population declines.

In addition to descriptions of notable disease processes in these groups of animals, details about their unique anatomy can be found in the Digital Content, Figs. 23.e1–23.e5, and Table 23.e2.

NON-INFECTIOUS DISEASES

Nutritional

Thiamine deficiency can occur in pinnipeds in managed care, despite routine vitamin B supplementation due to one or more conditions such as high thiaminase activity in dietary fish, improper thawing procedures, inadequate vitamin administration, interactions with other medications that decrease vitamin absorption, and an increased thiamine demand caused by pregnancy and lactation. Affected seals can demonstrate neurologic signs such as ataxia, muscle fasciculations, vertical nystagmus, depression, and seizures. Lesions include **polioencephalomalacia**, neuronal necrosis in the cerebral cortex, and Purkinje cell necrosis in the cerebellum. Thiamine deficiency can be difficult to confirm using standard liver and serum concentrations, however, a diagnosis can be made by measuring thiamine-dependent enzyme activity in tissues such as brain and liver.[36]

Metabolic

Urolithiasis is occasionally diagnosed in pinnipeds, though an underlying cause for urolith development is unknown.[25,49]

Exertional rhabdomyolysis (capture myopathy) has been noted rarely in pinniped species and is most commonly associated with capture and restraint for health assessments. In several South American fur seal pups with suspected underlying stress due to inflammatory conditions, lesions associated with capture myopathy included myocardial degeneration and contraction band necrosis, especially in the subepicardial myocardium. Myoglobin immunostaining can be utilized to identify myoglobin depletion in affected myocytes, which can aid in differentiating artifactual changes from early necrosis. Acute necrosis of proximal tubular epithelium in the kidney may also be noted.[166]

Toxic

In addition to several notable toxins described below, see Table 23.1 for a list of vitamin deficiency diseases and additional toxins in pinnipeds.

Domoic acid (DA) is an amino acid excitatory neurotoxin produced by diatom species, especially in the genus ***Pseudo-nitzschia***. Domoic acid has a high affinity for the α-amino-5-hydroxy-3-methyl-4-isoxazole propionic acid (AMPA) and kainite subclasses of glutamate ionotropic receptors. Receptor binding of DA causes neuronal depolarization, endogenous glutamate release, and activation of

Pathology of Wildlife and Zoo Animals. https://doi.org/10.1016/B978-0-443-18437-6.00023-7

TABLE 23.1 Miscellaneous Toxic and Nutritional Conditions of Pinnipeds

Cause	Lesions	Species Affected
Vitamin E deficiency	Myopathy, steatitis	California sea lion
Scombroid poisoning[190]	Nonspecific respiratory tract congestion and edema, mucus membrane irritation	Multiple phocids and otariids
Lead toxicosis[192]	Erythrophagocytosis and hemosiderosis in lymph nodes and liver, myocardial necrosis, cerebral, and intramyelinic edema	Harbor seal
Organochlorines[10,143]	Claw deformities, uterine stenosis, still birth, thyroid abnormalities, skull bone lesions, adrenal hyperplasia	Gray seals, harbor seals, ringed seals
Oil/polycyclic aromatic hydrocarbons (PAHs)[143]	Fouling of pelage, impaired thermoregulation, epidermal acanthosis and hyperkeratosis, intramyelinic edema, neuronal necrosis, axonal swelling, conjunctivitis, keratitis	Harbor seals, fur seals, ringed seals

voltage-gated calcium channels resulting in cell dysfunction and death.[93] Toxicity due to DA exposure was first identified in California sea lions (CSLs) in 1998, following a bloom of ***Pseudo-nitzschia australis*** in central California.[162] Since then, there has been an increasing incidence of DA-producing diatoms along the California coast and increased frequency of strandings of CSLs with neurologic signs and lesions associated with exposure to DA.[71,79] Though CSLs are by far the most common species affected by DA, toxicosis has also been diagnosed in harbor seals and northern and Guadalupe fur seals.[109,125] Neurologic abnormalities with acute DA toxicosis may be continuous and include ataxia, head weaving, abnormal scratching, seizures, and coma. Acutely affected animals often strand in clusters temporally associated with blooms of toxin-producing diatoms.[71,162] Animals with chronic toxicosis have more intermittent seizures, unusual behavior, vomiting, and apparent blindness, and they may be clinically normal between neurologic events. Chronic cases often strand individually and may not be temporally associated with specific blooms. Eosinophilia and low serum cortisol can be noted in CSLs with both acute and chronic toxicosis.[79] Sea lions with chronic DA toxicosis may have abnormalities detected on electroencephalogram and hippocampal atrophy can be detected by magnetic resonance imaging.

Brain lesions associated with DA exposure are centered on the hippocampus and parahippocampal gyrus and can be unilateral or bilateral. Histologic sections of both sides of the ventral mid-hippocampus are therefore essential for diagnosis. In acute DA toxicosis, there is acute neuronal necrosis affecting the granular cells of the dentate gyrus, pyramidal neurons of the hippocampus, and neurons in the amygdala and piriform lobe (Fig. 23.1A; eSlide 23.e1). Within the hippocampus, pyramidal cells in cornu ammonis (CA) sectors CA1, CA3, and CA4 are affected often with apparent sparing of CA2 cells. Occasionally in animals dying very acutely following exposure, lesions are minimal with only rare neuronal necrosis or central hippocampal neuropil vacuolation. Neuropil vacuolation should be interpreted with caution as autolysis and specimen handling can cause artifactual vacuolation along the edges of the hippocampus. In chronic cases, grossly evident hippocampal atrophy may be seen unilaterally or bilaterally (Fig. 23.1B.). This corresponds histologically to parenchymal atrophy in the hippocampal formation, mild to severe loss of granular cells in the dentate gyrus and hippocampal neurons, astrocytosis, oligodendrogliosis, and mild lymphocytic perivascular cuffing (Figs. 23.1C, D; eSlide 23.e2). Scattered acute neuronal necrosis may be noted. In severe acute and chronic cases, there may be malacia in the adjacent piriform lobe associated with neuronal necrosis, loss, and gliosis. Occasional perivascular cuffing in the piriform lobe and scattered meningeal aggregates of lymphocytes and plasma cells can be seen in chronic disease.[71,171] In more recent years, occasional cases have been noted with gliosis and neuronal necrosis in the amygdala and only relatively mild or no lesions in the hippocampus, though the cause of these lesions is unknown. Also, some CSLs with DA-associated hippocampal lesions may have multifocal lymphohistiocytic encephalitis, though whether this is due to a concurrent infectious process or DA exposure has not been elucidated.

Cardiac lesions associated with DA toxicosis can manifest grossly as regions of myocardial pallor or streaks along the epicardial surface, pericardial effusion, or a globally flaccid heart. Histologically, lesions affect the interventricular septum and left ventricle with the earliest lesions appearing at the base of the septum. Acute lesions include interstitial edema, rowing of cardiomyocyte nuclei, cardiomyocyte vacuolation, and, less frequently, necrosis. In subacute to chronic lesions, there is myocyte loss and replacement by adipocytes or loose fibrous connective tissue and areas can contain occasional macrophages with

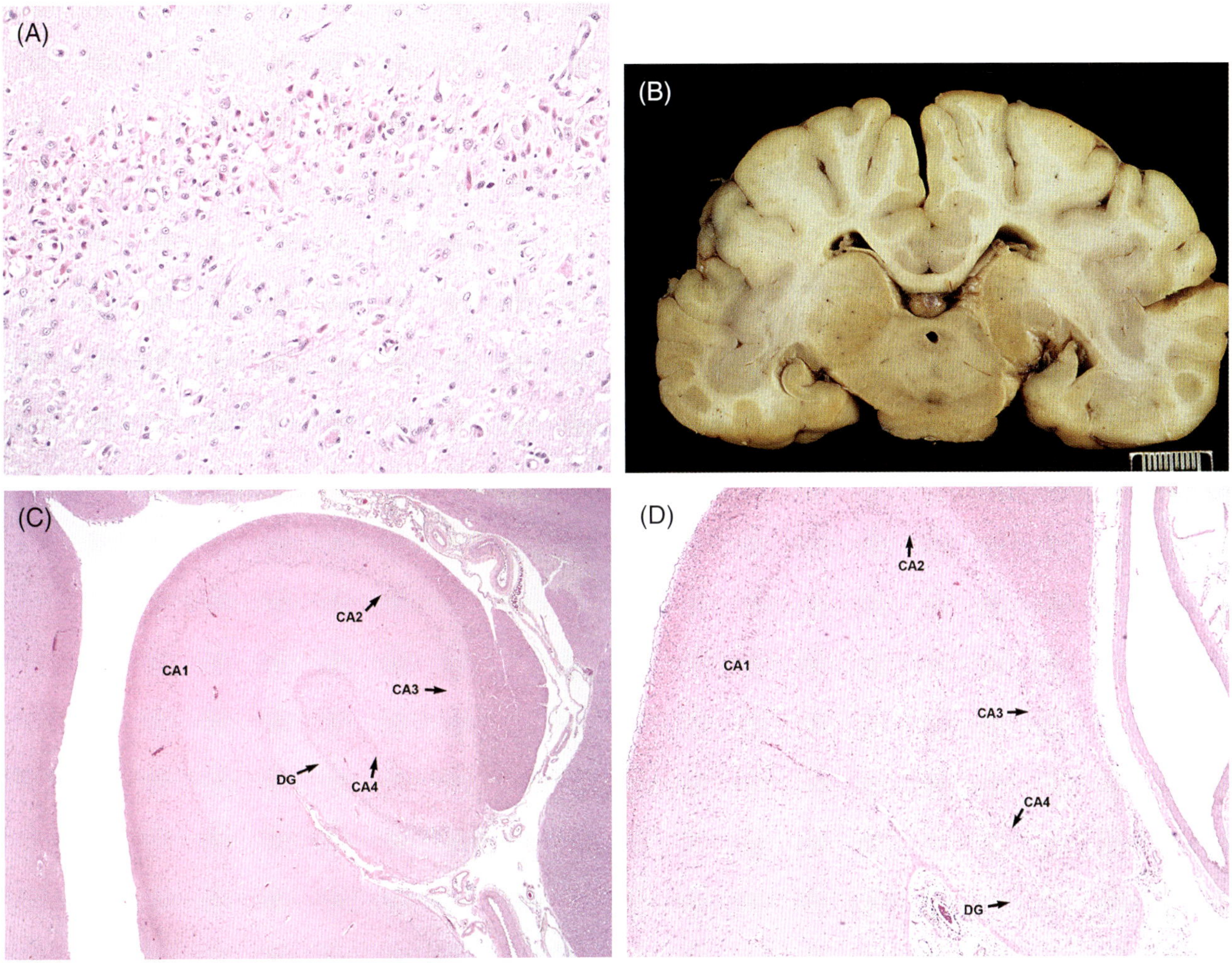

FIGURE 23.1 Domoic acid (DA) toxicosis in the brain of California sea lions. (A) Acute DA toxicosis is characterized by neuronal necrosis (angular, brightly eosinophilic neurons) and neuropil vacuolation in the pyramidal cell layer of the hippocampus. (see eSlide 23.e1) (B) Chronic DA toxicosis can present grossly as hippocampal atrophy (lower right; formalin fixed). (C) Normal hippocampus for comparison: dentate gyrus (DG); cornu ammonis sectors 1–4 (CA1–4). (D) Chronic lesions in the hippocampus due to DA toxicosis. There is parenchymal collapse, neuronal loss throughout the DG and CA sectors 1–4, and gliosis (see eSlide 23.e2).

cytoplasmic lipofuscin (Fig. 23.2; eSlide 23.e3). Lesions are suspected to be primarily due to direct toxin reaction with cardiac glutamate receptors causing myocyte apoptosis.[171,191]

Domoic acid may also cause abortion and premature parturition in affected pinnipeds and the toxin can be detected within amniotic fluid, fetal stomach contents, and urine. Cerebral edema is rarely identified in fetuses and placental abruption can occur in dams due to seizure activity.[72] Abnormal behavior can be noted in pup and yearling CSLs thought to have been exposed to DA in utero, though brain lesions in these animals are often not identified. Domoic acid can be detected in feces, serum, urine, and gastric contents using tandem mass spectrometry coupled with liquid chromatographic separation. Due to its short half-life, especially in serum, it will only be present in animals very recently exposed to the toxin.[56,171] Domoic acid has also been identified in samples from a number of other pinniped species including walrus in Alaska. It is often difficult to determine the significance of detection of low levels of the harmful algal toxins in marine mammal species.[108]

Congenital/Genetic

Congenital abnormalities have most often been identified in harbor and northern elephant seals and northern fur seals, with sporadic case reports in other species.[25,174,176,181] In harbor seals, the most common abnormalities identified are **hernias, skeletal malformations**, and proliferative conditions. In pups of multiple species including harbor seals there may be functional patency of the foramen ovale and ductus arteriosus up to 6–7 weeks of age, therefore, these should not be considered congenital defects in very

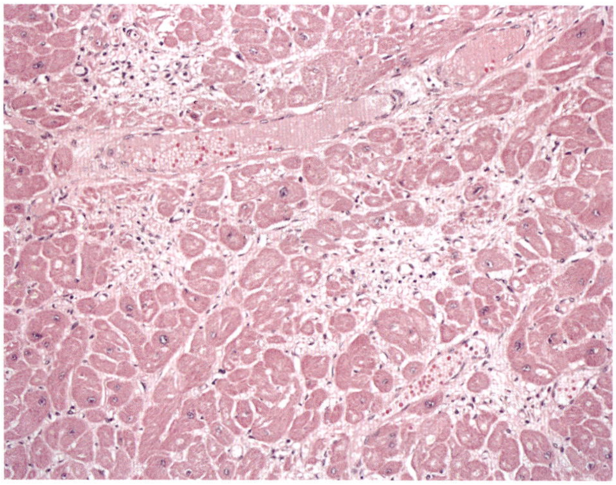

FIGURE 23.2 Domoic acid cardiomyopathy in a California sea lion. Histologically, there is loss of myocytes and replacement by loose fibrous connective tissue and adipocytes. Adjacent cardiomyocytes have enlarged hyperchromatic nuclei. Reactive cardiomyocyte nuclei surround the lesions (see eSlide 23.e3).

young pups without additional evidence of cardiac disease.[48] In elephant seals, **hydrocephalus** and **cardiac abnormalities** are most common. Low genetic diversity likely plays a role in the high prevalence of congenital defects noted in elephant seals.[181]

Age Related/Degenerative

Myocardial interstitial fibrosis is occasionally diagnosed in aged CSLs, most often in males. This condition has been associated with acute death during anesthetic procedures in several animals. Age-related **arteriosclerosis** can also be noted in pinnipeds. Acute death due to **cerebral infarction** has been reported occasionally in aged pinnipeds and may be related to age-associated arteriosclerosis or amyloid deposition in cerebral blood vessels. **Degenerative joint disease** can be noted in aged pinnipeds in captivity, especially in the hind limbs of otariid species.

Inflammatory Non-Infectious

Ocular disease affecting the lens and cornea is common in pinnipeds, especially those in managed care. In otariids, **progressive keratitis** is characterized by corneal edema, keratitis, corneal ulceration, pigmentation, vascularization, and in severe cases, stromal thinning and secondary bacterial and fungal infections (Fig. 23.3A). Exposure to low-salinity water, oxidants (ozone), concurrent lens disease (lens-induced keratitis), and exposure to ultraviolet (UV) radiation are also potential factors.[33] Cataracts and lens luxation also occur frequently in pinnipeds in captivity (Fig. 23.3B). Exposure to UV light, concurrent or previous keratitis, and fighting are risk factors for lens disease.[34]

Miscellaneous/Multifactorial

Trauma in pinnipeds can include **anthropogenic factors**, such as **gunshot, vessel strike, acoustic and blast trauma,** and **entanglement** in marine debris and fishing gear.[68,132] Entanglement in fishing gear, plastic, or metal rings can cause chronic infections and respiratory problems. **Peracute underwater entrapment (PUE)** or **by-catch** is a major source of mortality. Lesions in affected animals can include evidence of net or line marks or indentations in fur. Line marks may be especially notable around the mouth or appendages; however, the hair coat of many pinniped species can greatly reduce visibility. In some cases, associated acute hemorrhage, edema, or inflammation can be noted histologically in skin or mucosa, which can aid in diagnosing antemortem entrapment. Signs of struggle can also be noted and include hemorrhage or edema in

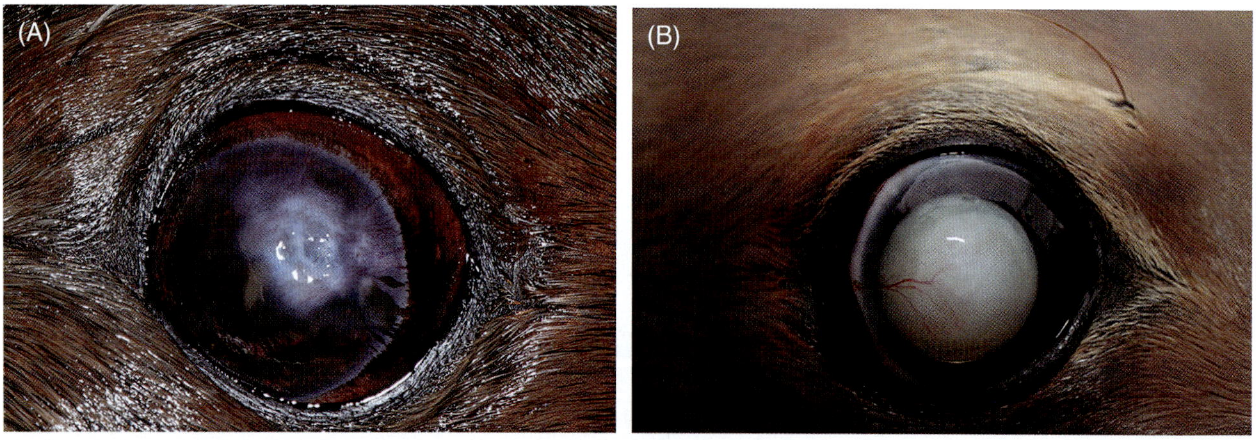

FIGURE 23.3 Ocular lesions in a California sea lion. (A) Active progressive keratopathy is characterized by corneal edema, fibrosis, stromal loss and ulceration, and vascular and pigment migration. (B) Cataract, anterior lens luxation, and keratopathy. The lens is diffusely opaque and cloudy and the cornea is slightly cloudy and has an ingrowth of congested blood vessels. *(Parts A and B: Photos Courtesy of C. Colitz.)*

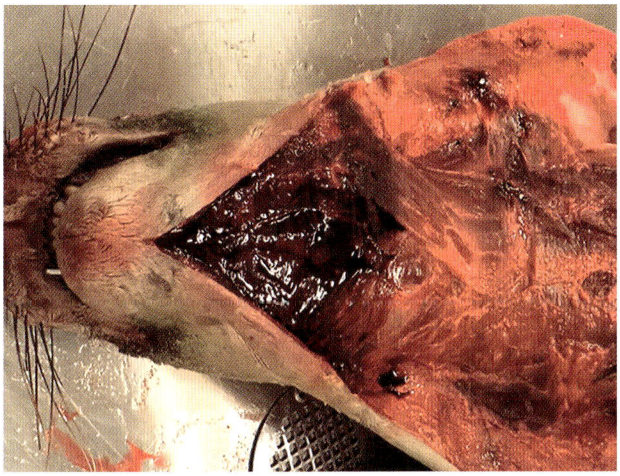

FIGURE 23.4 Lesions due to peracute underwater entrapment (PUE) in a Hawaiian monk seal. There is subcutaneous and muscular hemorrhage around the ventral neck region due to entrapment in fishing gear and subsequent struggle. The green algae on the fur around the mouth is normal for this species. *(Photo Courtesy of National Oceanic and Atmospheric Administration Hawaiian Monk Seal Research Program.)*

the subcutaneous or muscular tissue of the mandibular region, ventral neck, or axillary region (Fig. 23.4). Hyphemia can occasionally be identified, and some animals have stomachs full of nondigested prey suggestive of a recent meal and acute death. Nonspecific changes in the lungs suggestive of hypoxia include pulmonary congestion, edema, or emphysema. Rarely, gas-bubbles may be detected in multiple tissues.[131] Systematic, detailed, postmortem evaluation is essential in diagnosing anthropogenic trauma and criteria have been reviewed by Moore et al.[132]

Predation by large sharks and killer whales is a recognized major cause of trauma and in the case of some shark bites, the pattern of the bite wounds (Figs. 23.5A, B) or embedded tooth fragments can provide specific information about the predator. Spiral or cork-screw lacerations and wounds have been associated with gray seal predation in harbor seals, other gray seals, and harbor porpoises (Fig. 23.6).[16,185] Conspecific trauma can be associated with breeding and crowding on rookeries. In addition, fatal sexual interactions between gray seal males and pregnant female harbor seals have been documented.[159]

Malnutrition is commonly observed in young pinnipeds entering rehabilitation settings. Increased numbers of malnourished animals can strand when oceanic conditions, such as **El Niño events,** affect prey availability for both pups and dams.[26] In addition to a thinning and loss of blubber, lesions can include hepatocellular, muscle and gastric mucosal atrophy, and pancreatic zymogen depletion. Bile stasis and hepatic lipidosis can also be noted. Secondary bacterial infections are common. Poor juvenile survival related to emaciation and low prey availability threatens recovery of the endangered Hawaiian monk seal.

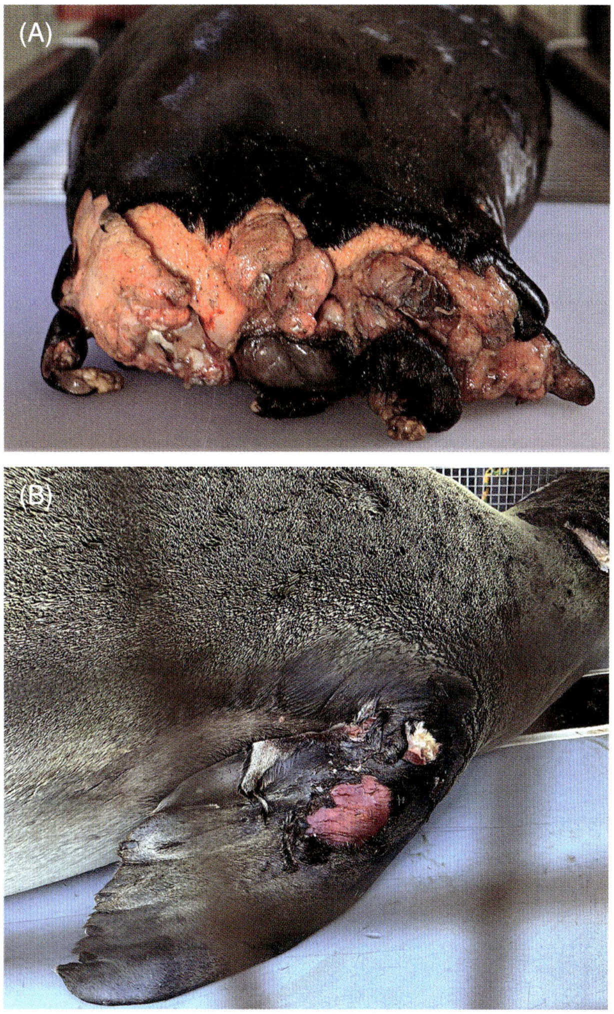

FIGURE 23.5 Traumatic lesions due to shark bite injuries in a California sea lion (A) and Hawaiian monk seal (B). (A) The caudal portion of the abdomen, pelvis, and hind flippers is completely amputated due to a suspected great white shark bite. Note the distinct triangular-shaped wound margin consistent with the dental arcade of a great white shark. (B) The torn skin and wounds with irregular, jagged edges are most consistent with bites from a tiger shark. *(Part A: Photo Courtesy of The Pacific Marine Mammal Center; Part B: Photo Courtesy of The Marine Mammal Center Ke Kai Ola/MMHSRP Permit 24356.)*

Systemic AA amyloidosis has been reported in CSLs and likely develops secondary to chronic inflammatory conditions. The kidney, thyroid gland, and blood vessels are most commonly affected. In the kidney, amyloid deposition occurs within the interstitium and glomeruli and often in a distinct band-like region along the corticomedullary junction (Fig. 23.7).[28]

Cutaneous microabscesses and **ulcerations** can occur due to sepsis and vasculitis. Folliculitis secondary to bacterial infection has been reported in captive walrus.[134] A skin disease characterized by **alopecia**, ulceration, and hyperkeratosis was described in northern elephant seals in the late 1990s, but diagnosis has been less frequent in more

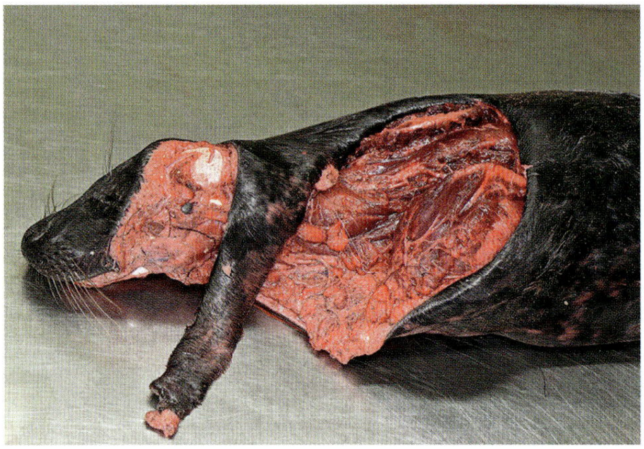

FIGURE 23.6 Conspecific predation wounds in a gray seal. Traumatic lesions are characterized by large, spiral-shaped, blubber defects and straight wound margins that are consistent with tearing of the tissue. These tears are adjacent to long strips of tissue that remain on the body. *(Photo Courtesy of the Institute for Terrestrial and Aquatic Wildlife Research.)*

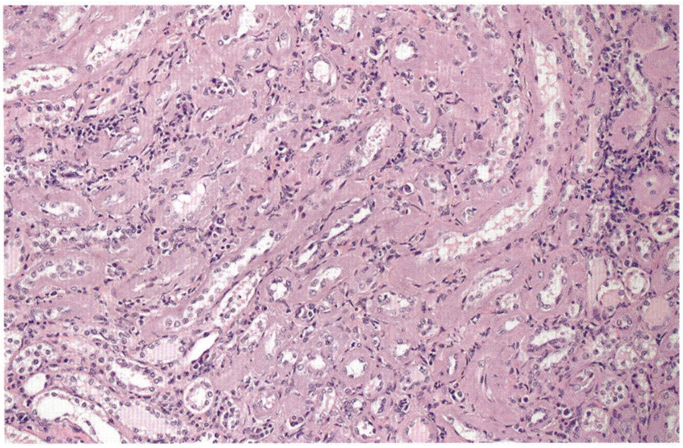

FIGURE 23.7 Amyloidosis in the kidney of a California sea lion. Pale, eosinophilic amyloid deposition is present within the intertubular interstitium. Small numbers of scattered lymphocytes and plasma cells are also present within the interstitium.

recent years.[9] Alopecia or **abnormal molts** can be noted on occasion, most commonly in phocids. This may be related to environmental, endocrine, dietary, or stress-mediated processes, however, pathogenesis in most cases is not elucidated. Alopecia, delayed molt, and ulcerative skin disease were primary findings of an unusual mortality event in ice seals and walruses in Alaska beginning in 2011. An exact etiology was not determined.[20]

Otitis media has been reported in stranded and captively managed harbor seals. Clinical signs include purulent ear discharge, and a number of different bacterial species can be isolated. Advanced imaging can be used to confirm the diagnosis and to guide medical and surgical treatments.[153]

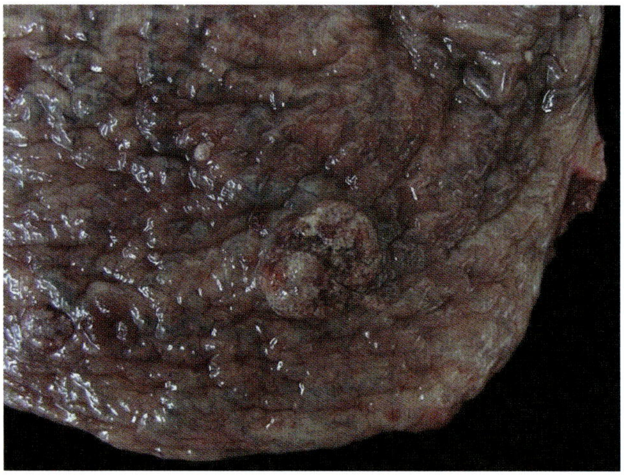

FIGURE 23.8 Urothelial hyperplasia in the urinary bladder of a California sea lion. Lesions may be single, multifocal, and/or papillary.

Myocardial necrosis can be secondary to hypoxia in pinnipeds that develop complications or die during or following anesthetic events. **Valvular endocarditis** due to bacterial infection has been reported in CSLs.[100]

Few scattered meningeal lymphocytes may be seen incidentally in free-ranging CSLs. Cerebral edema and meningitis are common lesions identified in septic phocid pups.

Multifocal papillary foci of **urothelial hyperplasia** (Fig. 23.8) are noted occasionally within the urinary bladder of CSLs and are not related to sea lion urogenital carcinoma (see below). The cause is unknown and lesions are not typically associated with significant cystitis.

Dental disease, abnormal tooth wear, gingivitis, secondary bacterial infections, and osteomyelitis have been reported in multiple pinniped species.[25] Tusk pulp bacterial infections have been an occasional problem in walruses in managed care secondary to wear. These infections can extend dorsally into the overlying maxillary bone causing osteomyelitis and fistulous tracts into the nasal cavity. Preventative measures aimed at preventing worn tusks have helped to manage this condition.[134]

Spiral bacteria morphologically consistent with ***Helicobacter*** sp. have been identified in the stomach of CSLs; however, often they are not associated with gastritis or significant glandular atrophy. Spiral to filamentous bacteria can be seen within colonic crypts of debilitated CSLs with dysbiosis. Protozoa is morphologically compatible with **trichomonads** and rarely can be found within gastric glands of free-ranging CSLs and also may be a form of dysbiosis.

Acute perforating proximal duodenal/pyloric ulceration is a condition identified in free-ranging, often malnourished pup and yearling CSLs. Lesions often lead to acute death. Histologically, there is inflammation and necrosis at the site of ulceration. Ulceration is not thought to be related to parasitism, bacterial infection, or any specific extra-gastrointestinal disease. Instead, ulcers may develop due to alterations in gastric emptying, abnormal gastrointestinal motility, or acid hypersecretion.[193]

Fatal **intestinal volvulus** and **intussusception** have been reported in multiple species and may be related to bacterial enteritis or dysbiosis.[170]

Hepatic hemosiderosis is seen frequently in several pinniped species including young northern elephants, harbor seals, and Hawaiian monk seals. In elephant seals, additional iron uptake by macrophages can occur secondary to vascular damage and leakage associated with ***Otostrongylus circumlitus*** infection (see Parasites). Hepatic hemosiderosis can also be seen in malnourished pinnipeds. In northern fur seals and CSLs, **iron overload syndrome** has been identified. Dietary factors such as rations with elevated proportions of high-iron-content fish (herring) and iron supplementation have been postulated to play a role in development of the syndrome.[123]

Mild chronic **cholecystitis** and **portal hepatitis** are common findings in wild pinnipeds secondary to trematode infection and trematode-associated pigment accumulation can occur. **Cystic hyperplasia of the gall bladder** is a common finding in older sea lions. This change may or may not have associated inflammation or infection. Small, randomly distributed granulomas in hepatic parenchyma can be noted secondary to parasite migration.

Neoplastic

A number of case reports describing neoplasia in pinnipeds and reviews detailing reported neoplasms have been previously published. Among the pinnipeds, neoplasms are most frequently diagnosed in free-ranging and captive CSLs.[25,136] In captive CSLs, **mammary carcinoma, lymphoma,** and **laryngeal squamous cell carcinoma** are among the most common. Neoplasia is less common in captive harbor and gray seals, though Flower et al.[57] documented esophageal squamous cell carcinoma in six aged harbor seals possibly associated with regurgitation. Uterine leiomyomas can be seen in aged pinnipeds.

Urogenital carcinoma (UGC) in CSLs is the most extensively studied neoplastic disease affecting male and female pinnipeds over the past four decades and is considered an important wildlife model of carcinogenesis.[15,80] A high prevalence of UGC has been diagnosed in adult CSLs that have stranded along the west coast of North America, with 26% of adult deceased CSLs affected at one rehabilitation facility over a 15-year period.[44] Occasional cases of UGC have been diagnosed in captive CSLs, most of which were wild-born. Carcinogenesis is considered multifactorial and the role of infectious agents, genetic factors, alterations in p53, endogenous hormones, and high levels of persistent contaminants have been studied in relationship to tumor development.[15,29,80] **Otarine herpes virus-1 (OtHV-1)**, a gammaherpesvirus that is sexually transmitted, was initially associated with UGC through PCR amplification of the virus from tumor and genital tissues.[17] Additional investigations utilizing a large case–control study of stranded CSLs over a 20-year period found that the odds of urogenital carcinoma were 43.6 times higher in sea lions infected with OtHV-1 and 1.5 times higher for every unit increase in blubber organochlorine concentration (i.e., PCBs and DDT).[80] In CSLs with urogenital carcinoma, high viral loads of OtHV 1 have been found in tumor tissues and viral oncogene expression has been documented in neoplastic cells using in situ hybridization.[45] Thus, OtHV-1 has been shown to play a critical role in the pathogenesis of urogenital carcinoma in CSLs with other cofactors, such as high burdens of persistent organochlorine contaminants, increasing the likelihood of tumor development.

Sea lions with UGC can have perineal edema, hind limb paresis, abdominal effusion, and penile prolapse with ulceration and necrosis.[82] Tumors originate in the cervix, vagina, penis, or prepuce and often there is widespread metastasis at the time of death. Subtly raised plaques, thickening, roughing, or dulling of the mucosa, and/or tan firm masses can be noted within affected urogenital epithelium (Figs. 23.9A, B). Metastatic tumors are often found in the inguinal and sublumbar lymph nodes as well as many other organs, such as liver, kidney, and lung. Hydronephrosis is a common sequela due to ureter compression by sublumbar lymph node masses and can be used to diagnose UGC antemortem via ultrasound.

Carcinoma in situ in the reproductive tract is often very difficult to discern grossly. In situ lesions may be seen incidentally in CSLs that die from other causes without metastatic disease. As such, it is critically important to routinely perform gross and histologic examination of the entire reproductive tract of male and female CSLs to identify these tumors.[29,82] Histologically, intraepithelial lesions may be observed multifocally along cervical, vaginal, penile, or preputial epithelium and often large regions of epithelium may be affected. Penile or prostatic urethral epithelium is rarely affected. Intraepithelial lesions may extend into the uterus along the endometrial surface; however, this is thought to be a form of intraepithelial metastasis and not primary transformation of uterine epithelium. Affected genital epithelium is often markedly thickened with dysplastic cells and disordered maturation involving up to the entire thickness of the epithelium (Fig. 23.9C). Increased numbers of mitotic figures can be seen. Direct infiltration of the underlying submucosa is difficult to identify, even in animals with widely metastatic disease. Multiple distinct morphologic patterns, including squamous, cribriform, comedone, or basaloid, can be observed in metastases. Neoplastic cells are polygonal with large nuclei and there is often a high mitotic index with

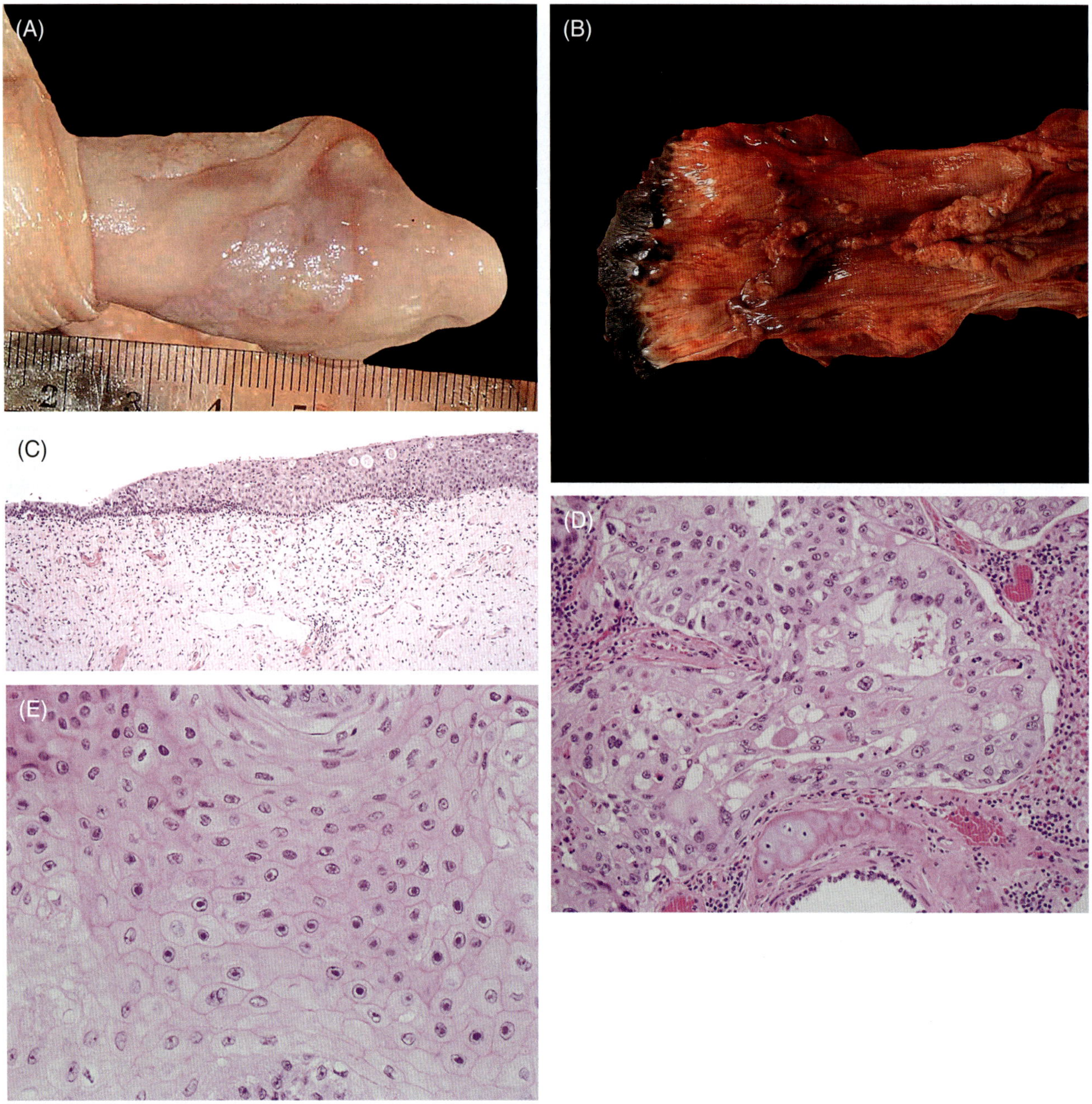

FIGURE 23.9 Urogenital carcinomas in California sea lions. (A) A well-demarcated, raised, roughened pale gray area of neoplastic mucosa is present on the penile mucosa. (B) Irregular thickening of the cervix and vagina with tan masses in a female sea lion. (C) The transition zone between normal cervical epithelium (left) and carcinoma in situ (right). There is dysplasia with disordered piling of neoplastic cells throughout the entire thickness of mucosa in the area of carcinoma in situ. (D) Metastatic urogenital carcinoma in the lung. A nodular mass of pleomorphic, polygonal neoplastic epithelial cells with distinct cell boarders, pale eosinophilic cytoplasm, and large nuclei replaces normal tissue (normal bronchiole in lower middle). Note the marked anisocytosis and anisokaryosis in tumor cells. (E) Numerous bright eosinophilic to magenta intranuclear viral inclusions can occasionally be noted in neoplastic genital epithelium. *(Part B: Photo Courtesy of The Pacific Marine Mammal Center.)*

bizarre mitotic figures or multinucleated neoplastic cells (Fig. 23.9D). Central necrosis within masses is common and lymphoplasmacytic inflammation is common adjacent to both intraepithelial lesions and surrounding metastases. Despite the association between UGC and OtHV-1, eosinophilic intranuclear inclusions are infrequently noted. When present, they are most common in genital intraepithelial lesions in which there is superficial ballooning degeneration (Fig. 23.9E). Inclusions are extremely rare in metastatic masses.

INFECTIOUS DISEASES

In addition to descriptions of notable infectious diseases that follow below, also see Tables 23.e3–23.e5 in the Digital Content for additional viral, bacterial, and parasitic diseases, respectively, of pinnipeds.

DNA Viruses

Phocine herpesvirus-1 (PhHV-1) is classified in the subfamily *Alphaherpesvirinae* and has been isolated from harbor and gray seals in Europe and from both coasts of North America. Serosurveys have documented titers in free-ranging phocids in both the southern and northern hemispheres.[67] PhHV-1 can be associated with high morbidity and mortality and outbreaks have been documented in rehabilitation facilities.[77] Clinical disease and fatal generalized infection are most common in neonates and young pups. Horizontal transmission via direct contact and aerosols is most common though vertical transmission has been identified.[70]

In Pacific harbor seals, multifocal adrenocortical and hepatic necrosis are the most common lesions associated with infection. Within and adjacent to areas of necrosis, cells may contain smudgy, eosinophilic, and intranuclear inclusions (Fig. 23.10A). Adrenal necrosis may be associated with mineralization (Fig. 23.10B). Thymic atrophy is often noted in affected pups and lesions secondary to bacterial infection and septicemia, including omphalophlebitis, pneumonia, and meningoencephalitis, may be concurrently identified.[77] Oral ulceration and small intestinal crypt dilation and necrosis are rare lesions. In infected European seals, interstitial pneumonia and massive liver necrosis have been reported.[13] Virus isolation, PCR, IHC, and serology (ELISA) are used to diagnose infections.

Otarine herpesvirus-1 (OtHV-1) is in the subfamily *Gammaherpesvirinae* and genus *Rhabinovirus*. The virus has associated CSLs with UGC and is considered a prime factor in tumor development (see Neoplasia). Viral DNA can be amplified from the urogenital tract and, less frequently, the oral cavity of sea lions with no clinical evidence of UGC and prevalence of virus in the genital tract is higher in adults than in juveniles. Sexual transmission is thought to occur and females may transmit the virus to pups during birth.[15,17] The virus has also been associated with a single case of genital carcinoma in a South American fur seal, and detected in a Steller sea lion without evidence of neoplasia.[37]

Pox viruses have been identified in skin and mucosal lesions from numerous pinniped species including CSLs, South American sea lions, Steller sea lions, northern fur seals, and harbor and gray seals. Most pinniped poxviruses are in the genus *Parapoxvirus* and viruses from Atlantic and Pacific pinnipeds are phylogenetically distinct.[138] There is widespread exposure in free-ranging pinnipeds and transmission occurs through direct contact, such as from conspecific rubbing or biting behaviors. Outbreaks have occurred in rehabilitation facilities, especially in young animals, and lesions may develop during times of stress or concurrent disease.[137] **Sealpox** is a zoonotic disease that causes nodular cutaneous lesions in people and transmission likely occurs when virus enters broken skin.[187]

Cutaneous poxvirus lesions appear as raised and firm, sometimes ulcerated nodules along the head, neck, and thorax (Fig. 23.11A). Nodular or plaque-like lesions may also occur along the tongue and oral commissure (Fig. 23.11B). Histologically, there is marked epidermal and follicular epithelial hyperplasia with ballooning degeneration in the stratum spinosum. Moderately sized, round to chunky eosinophilic cytoplasmic inclusions are noted

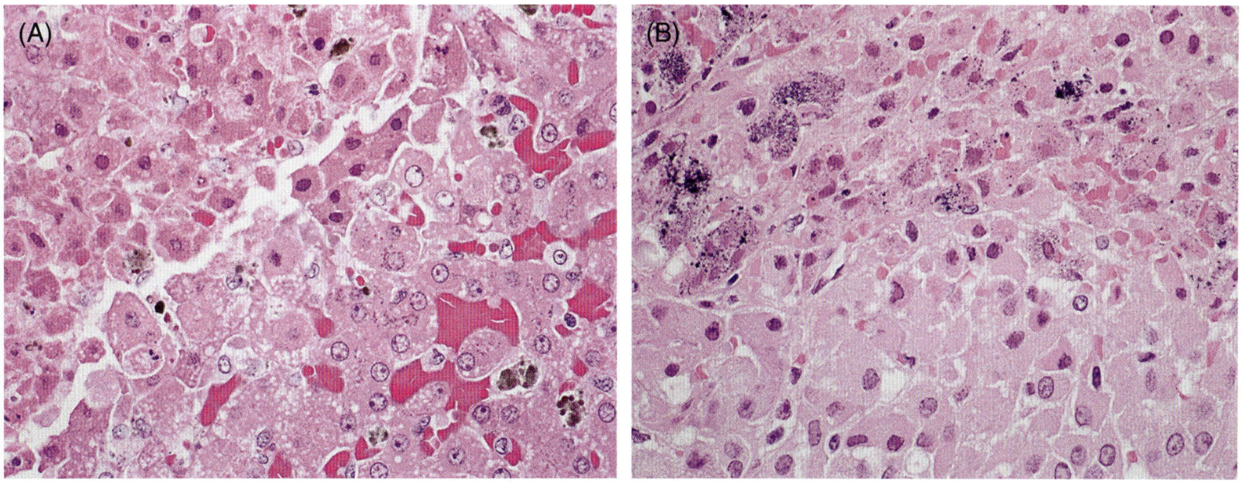

FIGURE 23.10 Phocine herpesvirus-1 infection in a harbor seal. (A) Hepatic necrosis (upper left) with smudgy, eosinophilic, intranuclear inclusions that often fill the nucleus. (B) Multifocal adrenocortical necrosis with numerous, smudgy, eosinophilic, intranuclear inclusions that fill the nucleus. Necrosis was associated with prominent mineralization.

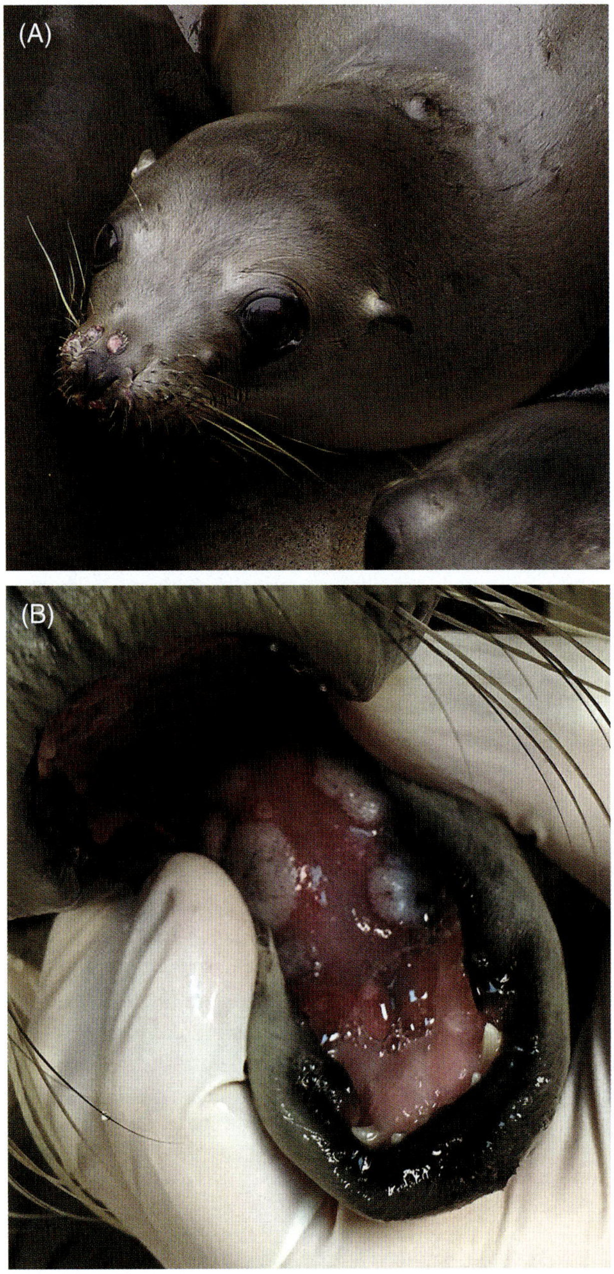

FIGURE 23.11 Poxviral lesions in pinnipeds. (A) Multifocal nodular and ulcerated lesions along the face and neck of a California sea lion. (B) Proliferative, gray/white lesions along the tongue of a young harbor seal. *(Part A: Photo Courtesy of J. St. Leger, SeaWorld; Part B: Photo Courtesy of The Marine Mammal Center.)*

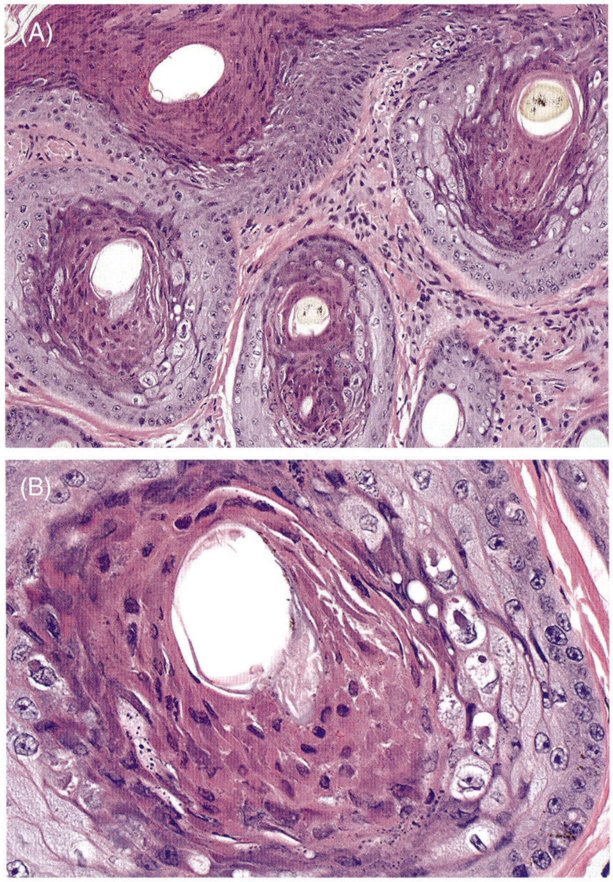

FIGURE 23.12 Poxviral infection in skin of a California sea lion. (A) Note the epidermal and follicular epithelial cell hyperplasia, hyperkeratosis, and eosinophilic intracytoplasmic inclusions. (B) Viral infection is associated with ballooning cellular degeneration and chunky, eosinophilic intracytoplasmic inclusions.

within epidermal and follicular keratinocytes (Figs. 23.12A, B). There is variable epidermal necrosis and ulceration, and often moderate associated mixed inflammation. Dermal fibrosis is occasionally noted.[139] Lesions will often spontaneously regress.

Adenovirus infections have been reported sporadically in CSLs and have recently been associated with an outbreak involving multiple otariids in managed care.[14,91] Viral sequence analysis has identified **California sea lion adenovirus-1 (CSLAdV-1)** and determined that it is unique from canine adenovirus-1 (CAV-1) and represents a new member of the genus *Mastadenovirus*.[65] Steller sea lions have antibodies to CAV-1 and canine adenovirus-2 (CAV-2).[19] Multifocal necrotizing hepatitis with hepatocellular intranuclear inclusions is the most common lesion reported in CSL infections (Fig. 23.13A). In one case report, endothelial infection was identified in the lung, lymph nodes, and eye in the absence of hepatic lesions. Ocular lesions included corneal edema, keratitis, and iridocyclitis. Several cases of necrotizing enteritis in CSLs have also been attributed to adenoviral infection based on histologic lesions (Fig. 23.13B); one CSL exhibited diarrhea prior to death. CSLAdV-1 has been amplified via PCR from liver and feces in affected individuals, and adenoviral particles have been identified via transmission electron microscopy (TEM) in several cases. The virus can also be amplified from feces of individuals without evidence of clinical disease, suggesting that some animals may be subclinical shedders of the virus.

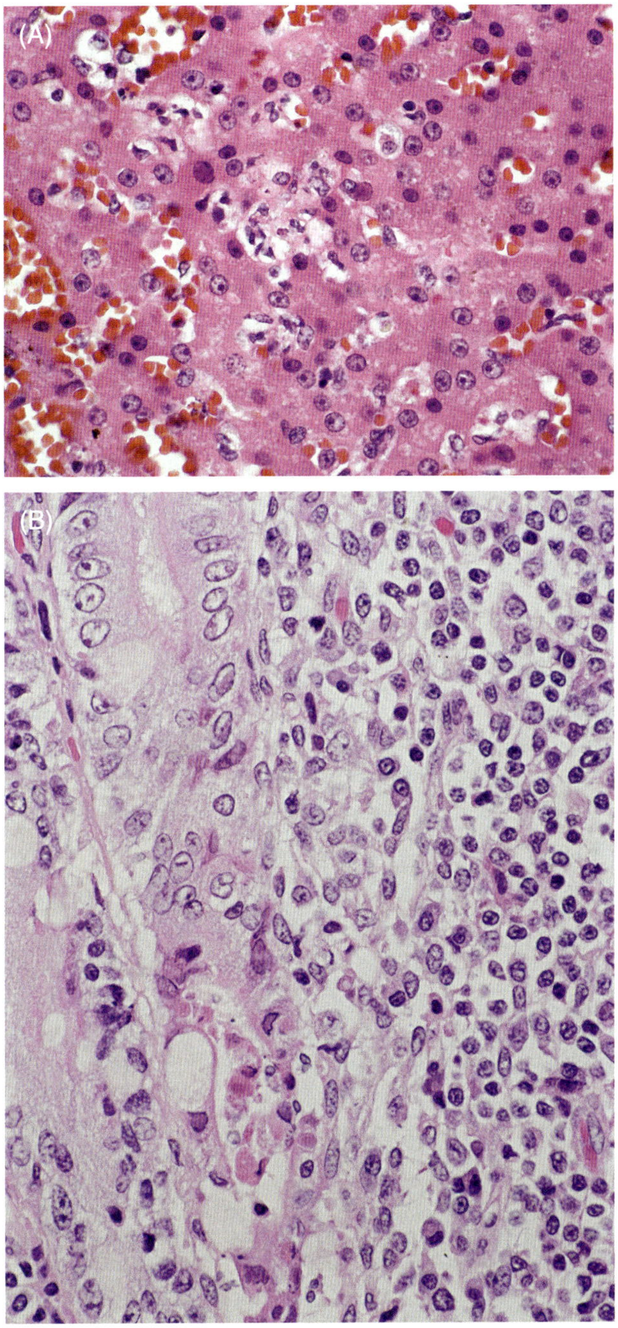

FIGURE 23.13 California sea lion adenovirus 1 (CSLAdV-1) infection in California sea lions. (A) Hepatic necrosis and large, smudgy, amphophilic, intranuclear inclusions that fill and sometimes enlarge the nucleus of the hepatocyte are characteristic findings. (B) Necrosis of intestinal crypt and mid-mucosal epithelium, mild to moderate inflammation, and smudgy eosinophilic to amphophilic intranuclear inclusions can also be seen.

RNA Viruses

Morbillivirus infection in pinnipeds was first observed in Baikal Seals in 1987. The infections were likely due to spread of a **canine distemper virus (CDV)** outbreak in terrestrial mammals.[74] In 1988, the first large-scale epizootic of **phocine distemper virus (PDV)** in harbor seals followed in European waters of the North and Baltic Seas causing 23,000 deaths, and in 2002, another outbreak occurred in the same region affecting 30,000 seals. Several smaller-scale mortality events in pinnipeds on the Atlantic coast of the United States have also been linked to PDV infections in harbor, harp, hooded, and gray seals. Positive morbillivirus antibody titers are regularly found in Arctic seal species and provide evidence of widespread distribution of PDV in seal populations.[50,84]

Morbilliviruses are most commonly transmitted through respiratory, nasal, or ocular secretions. Clinical findings with PDV include oculonasal discharge, conjunctivitis, keratitis, coughing, dyspnea, diarrhea, abortion, head tremors, convulsions, and increased buoyancy. Pathological findings include bronchointerstitial pneumonia, pulmonary atelectasis, congestion, and edema as well as pulmonary, mediastinal, and subcutaneous emphysema, nonsuppurative encephalitis, and lymphocyte depletion in lymphoid tissues (Fig. 23.14A). Intracytoplasmic and intranuclear eosinophilic inclusion bodies can be found in the respiratory epithelium, biliary and pancreatic ducts, renal pelvis, urinary bladder, gastrointestinal epithelium, neurons, and astrocytes (Fig. 23.14B).[50,96,168] Due to viral-induced immunosuppression, PDV is often associated with secondary parasitic (pulmonary nematodiasis, *Toxoplasma gondii* infection), bacterial (e.g., *Bordetella bronchiseptica*), and concurrent viral infections (herpesvirus, influenza virus). The diagnosis of CDV or PDV is accomplished by immunohistochemistry using cross-reactive monoclonal antibodies for the morbillivirus nucleoprotein and by the detection of morbillivirus-specific nucleic acid in tissues by RT-PCR. Antibody titers allow assessment of exposure and susceptibility to PDV.[50,117]

Mortality events associated with **Influenza A** infection have been documented since 1979, primarily in harbor seals in the Northeastern United States and Europe. A range of subtypes have been implicated (see Table 23.2).[3,21,63,88,188] In recent years **highly pathogenic avian influenza (HPAI; H5N8, H5N1)** mortalities in harbor and gray seals in Europe and Maine (United States) and South American sea lions in Peru have been temporally associated with HPAI outbreaks in wild birds and resulted in high numbers of dead pinnipeds in some areas.[110,149,152,167] A PCR survey for influenza virus in pinnipeds along the Pacific coast and Northeastern US has detected H1N1 infection in asymptomatic Northern elephant seals and circulating virus in gray seals. Antibodies to influenza have also been detected in various serosurveys. These data suggest that transmission of influenza viruses between aquatic avian species and pinnipeds that share habitat occurs with some frequency across many regions of the globe.[161] **Influenza B**, believed to be an exclusively

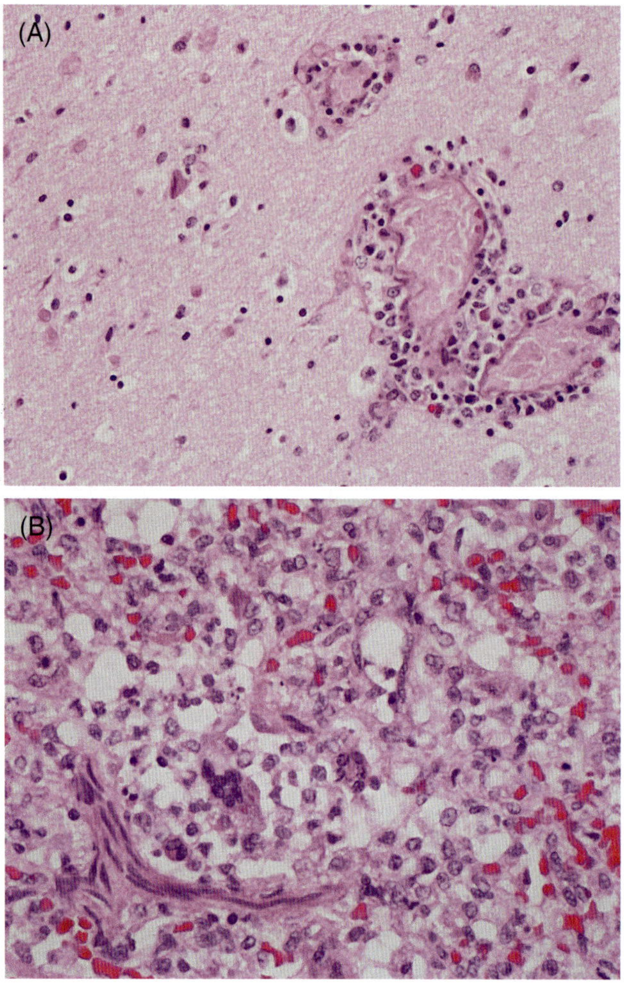

FIGURE 23.14 Phocine distemper virus infection in harbor seals. (A) Encephalitis characterized by perivascular lymphohistiocytic inflammation and neuronal necrosis can be noted. (B) Necrotizing and histiocytic bronchointerstitial pneumonia with syncytial cells. There are a few small intracytoplasmic pale eosinophilic inclusions within syncytial cells.

human pathogen, has been isolated in a single young seal from the Dutch Wadden Sea, though association with disease is unknown.[144]

Clinical findings of influenza are similar to PDV and include dyspnea, nasal discharge, lethargy, and seizures. Pathological findings of influenza include partially collapsed lungs, pulmonary congestion, emphysema, necrotizing bronchitis and bronchiolitis, hemorrhagic alveolitis, bronchial gland adenitis, and interstitial pneumonia (Fig. 23.15A). Lesions noted in harbor seals during some mortality events have been subtle or nonspecific, however, lesions noted with HPAI were striking and included severe necrosis of bronchiolar epithelium, necrotizing alveolitis, and epithelial dysplasia in more chronic lesions (Fig. 23.15B). Meningoencephalitis, neuronal necrosis, and occasional vasculitis have also been noted in some seals infected with HPAI. Additional concurrent respiratory tract lesions caused by parasitic or secondary bacterial infections are often present with associated inflammation.[11,63,88,168] Influenza was also detected in harbor seals from the North Sea with evidence of meningoencephalitis.[149] Diagnosis is accomplished by ELISA, immunohistochemistry using influenza A virus nucleoprotein-specific monoclonal antibody, and RT-PCR in lung and oral swabs.[11,42,183]

Marine **caliciviruses** are within the **vesivirus** genus and include over 40 serotypes including **San Miguel Sea Lion viruses (SMSV-1-17)** and **Walrus calicivirus (WCV).**[124] Isolates and serologic evidence of exposure are documented in multiple species and in both clinically affected and healthy individuals. An unusually wide variety of marine and terrestrial species, including fish, birds, and other mammals, may be infected, indicating that these viruses have the ability to infect many species of marine animals, some of which may act as reservoirs for others. Opal eye perch are a likely

TABLE 23.2 Influenza A Morality Events in Pinnipeds

Year	Species	Location	Influenza Subtype
1979–1980	Harbor seals	Northeast, USA	H7N7[188]
1982–1983	Harbor seals	Northeast, USA	H4N5[63]
1991–1992	Harbor seals	Northeast, USA	H4N6 and H3N3[21,161]
2014–2015	Harbor seals	Sweden, Denmark, Germany, Netherlands	H10N7[11]
2011	Harbor seals	Northeast, USA	H3N8[3]
2019–2021	Gray and harbor seals	Poland, Germany	H5N8[149]
2022	Harbor and Gray seals	Maine, USA	H5N1[152]
2022	South American sea lions	Peru	H5N1[110]

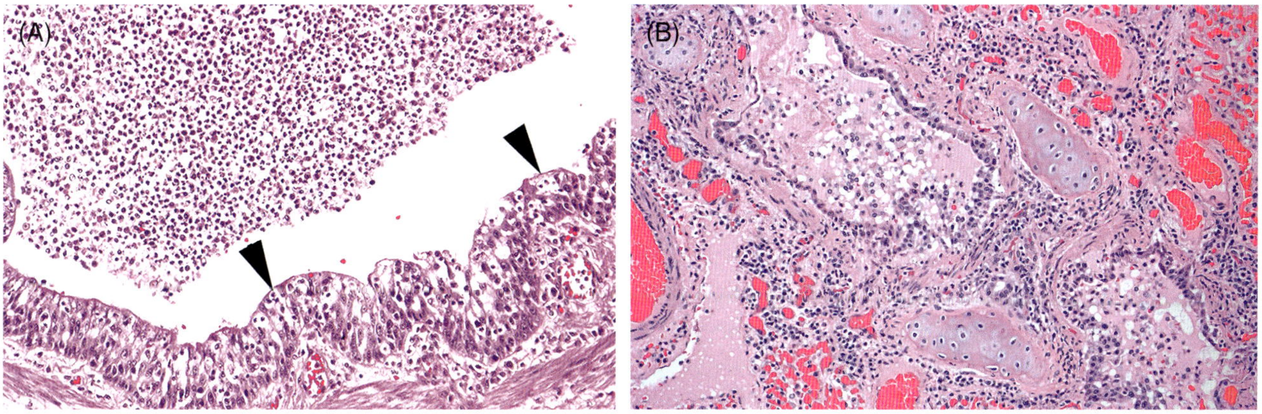

FIGURE 23.15 Influenza A infection in the lungs of harbor seals. (A) Infection with H10N7 characterized by acute single mucosal epithelial cell necrosis *(arrowheads)*. Neutrophilic bronchiolitis is present and may be due to secondary bacterial infection. (B) Infection with H5N1 was associated with necrotizing bronchiolitis and lymphoplasmacytic and neutrophilic inflammation in the surrounding interstitium. *(Part A: Photo Courtesy of P. Wohlsein, Department of Pathology, University of Veterinary Medicine Hannover, Germany.)*

natural host. These viruses are indistinguishable from the caliciviruses that caused **swine vesicular exanthema**, a disease occurring in the United States from 1932 to 1956. Feeding swine uncooked garbage containing marine mammals and fish was a likely route of introduction of the virus into the swine population. Caliciviruses are zoonotic, with rare cases of human vesicular dermatitis and flu-like illness after laboratory and field exposures.[172,187]

Gross findings of calicivirus infection in sea lions, include vesicular dermatitis that progresses to ulceration primarily of the nonhaired surfaces of the flippers (Fig. 23.16). Occasionally, necrosis of underlying digits can occur, usually associated with secondary bacterial infections. Ulcerative stomatitis and ulcerative and nodular dermatitis of the lips, nasal planum, and chin have also been reported in CSLs. Abortion and premature birth in CSLs have been attributed to calicivirus based on positive viral isolation and serologic evidence of exposure, but causation has not been proven. Lesions begin as spongiosis of the stratum spinosum followed by subcorneal vesicle formation without the formation of visible inclusion bodies.[130] The vesicles are very transient and ulcers with secondary bacterial infections are common. Diagnosis is by culture or RT-PCR.[130,172] Because of the importance in differentiating these viruses from foreign animal diseases, a real-time reverse transcription PCR has been developed that can identify most serotypes and differentiate them from reportable vesicular diseases.[124]

FIGURE 23.16 Calicivirus infection in the skin of a Stellar sea lion. There is a fluid filled, raised, smooth vesicle along the flipper skin. Vesicles may rupture and ulcerate.

Bacteria

Several notable bacterial infections are described below. Also see Table 23.e4 for additional important bacterial infections in pinnipeds.

Infection with ***Leptospira*** sp. is most commonly reported in free-ranging pinnipeds, including CSLs, harbor, and northern elephant seals along the Pacific coast of North America.[31,81] Sporadic infections and outbreaks have been reported in captive animals and exposure has been documented serologically in other geographic regions.[99,121] ***Leptospira interrogens*** **serovar** ***pomona*** has been implicated in disease in CSLs. Based on the long-term serosurveys utilizing the microscopic agglutination test, leptospirosis is thought to be enzootic in the CSL population. Periodic epizootics occur every 3–5 years and result in large numbers of CSLs stranding with clinical leptospirosis.[81,115] An asymptomatic, chronic carrier state may provide a mechanism for persistence of the bacterium in the population.[150] Though cases in free-ranging CSLs can occur throughout the year, they are most commonly noted between July and December.[81] Little is known about the epidemiology of leptospirosis in other pinniped species, where infections occur more sporadically or as small outbreaks in rehabilitation situations.[31] Leptospirosis may also cause abortion.[81]

Pinnipeds with leptospirosis have clinical signs and serum biochemistry values consistent with renal failure including polydipsia, and elevated blood urea nitrogen (BUN), creatinine, and phosphorus. Grossly, kidneys are swollen, pale tan, and exhibit lack of renule and corticomedullary differentiation (Fig. 23.17A). Histologically, there is subacute to chronic, lymphoplasmacytic, tubulointerstitial nephritis, typically with plasma cell heavy inflammation (Fig. 23.17B). *Leptospira*-related liver lesions are not commonly reported. In some phocids, tubular necrosis is more severe than inflammation suggesting that the animals die in the more acute stage of infection. Infection can be confirmed histologically by demonstrating spirochetes with silver stains or IHC. IHC is particularly useful in animals with chronic infections or those treated with antibiotics since the spirochetes may be difficult to visualize with routine hematoxylin and eosin staining; labeled antigen will be present in renal tubular epithelial cells and interstitial macrophages.[31,81] PCR can also be used for diagnosis. Given the zoonotic potential of the bacterium, appropriate personal protective measures should be undertaken when necropsying animals with known or suspect infection.[187]

Mycobacteriosis was first described in a group of captive pinnipeds at a marine park in Australia.[59] It was later classified as being caused by a previously undescribed member of the ***Mycobacterium tuberculosis* complex, *M. pinnipedii*.**[35] Since then, the disease has been reported in a number of captive otariid species worldwide, most frequently affecting Southern sea lions.[101] Infection in free-ranging pinnipeds is confined to the southern hemisphere and transmission between species is thought to be by sub-Antarctic fur seals whose range overlaps with eight other otariid species.[8] To date, mycobacteriosis due to *M. pinnipedii* has not been reported in any phocid species; however, the potential host range is broad and transmission from infected fur seals and sea lions has been described for zoo species, domestic cattle, and humans.[98,113,116,133,178,179] The most likely route of infection is aspiration of aerosolized bacteria shed in respiratory secretions of infected pinnipeds.

Pinnipeds with mycobacteriosis typically present in poor body condition. The majority of clinical infections involve the thoracic organs, supporting inhalation as the dominant route of infection. There are two main gross presentations. The first involves granulomatous to pyogranulomatous pleuropneumonia, thoracic lymphadenitis, and mediastinitis with pronounced fibrinous or serosanguinous effusion (Fig. 23.18). In the second type of presentation, multifocal caseating granulomas are seen in the lungs and thoracic lymph nodes. Several cases of respiratory mycobacteriosis have included concurrent granulomatous lesions in the mesenteric lymph nodes that might reflect spread into the gastrointestinal tract and abdominal cavity through swallowing infected respiratory secretions.[41,101,157]

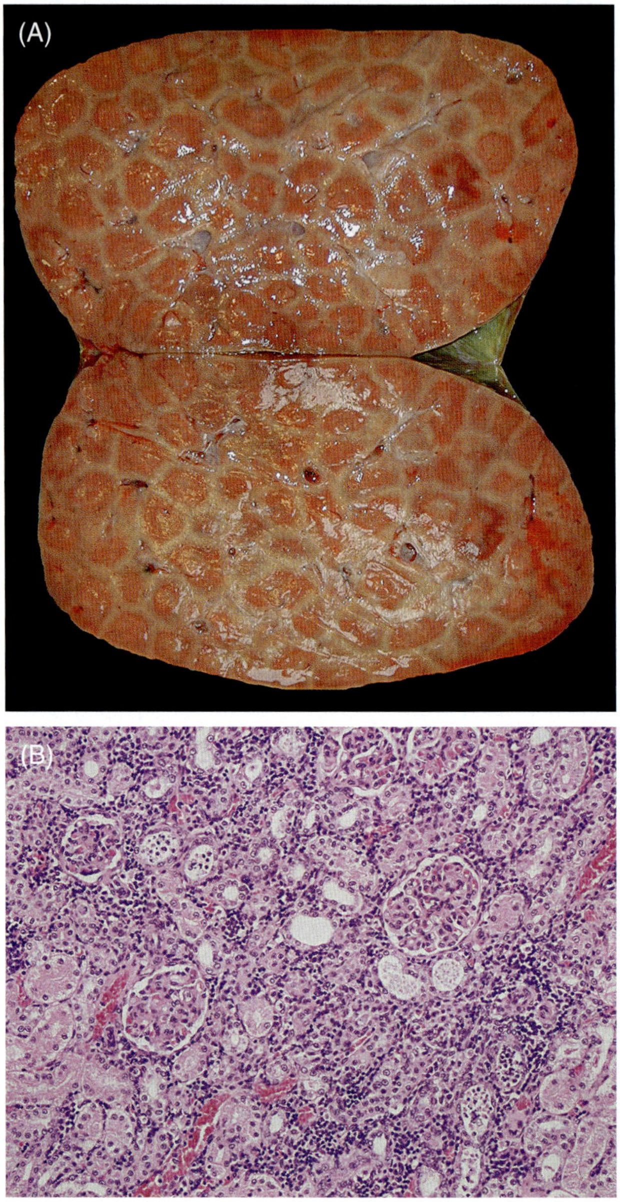

FIGURE 23.17 *Leptospira interrogens* serovar *pomona* infection in the kidney of a California sea lion. (A) The kidneys are enlarged, bulge from the capsule on cut section, and have swollen renules and pale cortices. (B) Tubulointerstitial nephritis with multifocal lymphoplasmacytic inflammation and multifocal tubular necrosis is present. Sloughed epithelial cells and few neutrophils are within necrotic tubules. *(Part A: Photo Courtesy of The Marine Mammal Center.)*

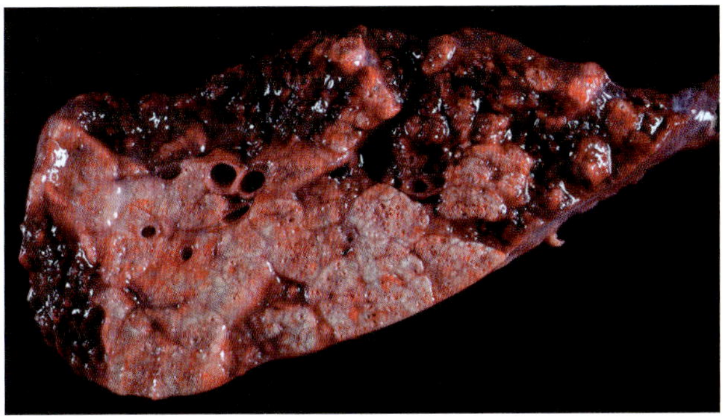

FIGURE 23.18 *Mycobacterium pinnipedii* infection in the lung of a New Zealand sea lion. Coalescing granulomas with central caseous necrosis replace the majority of the pulmonary parenchyma. *(Photo Courtesy of S. Hunter, Massey University.)*

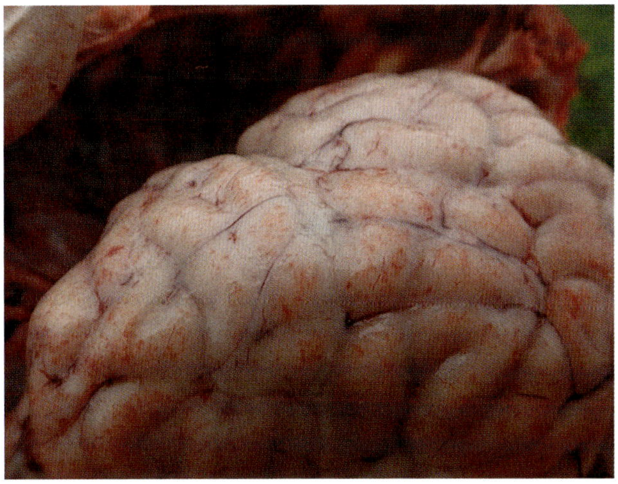

FIGURE 23.19 Meningitis due to hypermucoviscous *Klebsiella pneumoniae* infection in a New Zealand sea lion. Note the petechial hemorrhage over the gyri and meningeal opacity with fibrinosuppurative exudate in the subarachnoid space. *(Photo Courtesy of S. Michael, Massey University.)*

Meningeal, hepatic, and disseminated granulomas have also been reported.[59] Histologically, lesions are dominated by aggregates of epithelioid macrophages containing few to many intracellular, intracytoplasmic acid-fast bacilli. These cells are mixed with variable numbers of lymphocytes, plasma cells, and neutrophils. Caseous necrosis of the centers of these granulomas is common, mineralization is generally minimal, and multinucleated giant cells are rare. A presumptive diagnosis of mycobacteriosis can be made during postmortem examination based on the presence of intrahistiocytic, acid-fast positive bacteria within granulomas noted either on cytology or histopathology, though in some cases, bacteria are rare within lesions. Clinical diagnosis can be aided by PCR, culture of respiratory and gastric lavage samples, and radiographic findings. Population monitoring is best accomplished by necropsy of all dead pinnipeds and monitoring of live animals including lymph node cytology, thoracic radiographs, and bronchoalveolar lavage.[157]

Klebsiella pneumoniae are gram-negative bacteria in the Enterobacteriaceae family. They are ubiquitous in the environment and are common inhabitants of the gastrointestinal or respiratory tract of healthy mammals. Historically, most reports of *K. pneumoniae* in marine mammals were of sporadic respiratory infections or isolation from tissues of healthy individuals. However, a 2002 mass mortality event caused by *K. pneumoniae* occurred in New Zealand sea lion pups in the sub-Antarctic.[24] The causal isolate was later identified as a hypermucoviscous phenotype,[158] but appears not to be of human origin.[24] A similar isolate has also been reported in CSLs.[92,164] The clinical presentation of hypermucoviscous *K. pneumoniae* varies somewhat between host species. Affected New Zealand sea lion pups have bacteremia and fibrinosuppurative to histiocytic inflammation of the meninges, joints, lymph nodes, respiratory tract, peritoneum, and subcutaneous tissues (Fig. 23.19).[158] Disease has not been reported in juveniles or adults. Pleuritis, pyothorax, pneumonia, and abscesses are the most frequent lesions in CSLs and infection has been reported in all age groups[92,164]; septicemia and meningoencephalitis have also been identified in CSLs. Additionally, Spraker et al.[173] describe a syndrome in CSLs on San Miguel Island, California, where *K. pneumoniae* bacteria are transferred into the abdominal cavity on the cuticle of hookworms that penetrate the intestinal tract, resulting in septic peritonitis. This has not been reported at other locations.

Several different ***Mycoplasma*** spp. have been isolated from phocids and ottarids. These include ***M. phocicerebrale***, ***M. phocirhinis***, ***M. phocidea***, and ***M. zalophi***. *Mycoplasma* spp. are thought to be part of normal pinniped oral flora and infection is associated with skin wounds, abscesses, polyarthritis, lymphadenitis, and necrotizing pneumonia in CSLs and skin wounds in gray and harbor seals.[5,86,118] In utero infection with *M. phocicerebrale* is associated with myocarditis and pneumonia and has caused abortion in Australian fur seal fetuses.[118] *Mycoplasma* sp. have been isolated from the respiratory tracts of harbor seals during influenza outbreaks, though the role of the bacteria in respiratory tract disease was undetermined.[3,63] Infection is diagnosed via culture or PCR. *Mycoplasma* sp. is the cause of "seal finger" in humans, an important zoonotic disease most often associated with the bite of a pinniped to the extremities or contact with broken skin.[187]

Several new species in the genus ***Brucella*** have been recently described in marine mammals. ***Brucella pinnipedialis*** is the organism isolated from pinnipeds.[60,151,169] Serological studies show worldwide distribution in pinnipeds with a wide variation in prevalence between geographical locations and species, although a lack of consistency in the serological assays that are used makes valid comparisons difficult.[90,142] The route of infection for marine mammals has not been elucidated but a fish or invertebrate vector is possible. *Brucella*-infected lungworms can be found in the lungs of pinnipeds though whether the parasites serve as a primary route of transmission or are incidentally infected is unknown.[62] The majority of *B. pinnipedialis* infections in pinnipeds are thought to be subclinical. However, suppurative and necrotizing placentitis in a northern fur seal was confirmed to be due to *Brucella* spp. infection by immunohistochemistry and PCR.[52] While zoonotic transmission is a concern for humans who consume seal or whale meat as part of a traditional diet, none of the reported human cases have documented direct exposure to marine mammals.[187]

Nocardiosis is most commonly reported in juvenile stranded hooded seals and harbor seals; multiple different ***Nocardia*** species are associated with infection. Affected hooded seals stranded along the east coast of the United States and Canada typically outside of their normal geographic range. Underlying immunosuppression, environmental stressors, and decreased prey availability may increase susceptibility to infection. Lesions in affected seals include pyothorax, and abscesses or pyogranulomas in the lungs and thoracic lymph nodes. Infection is often systemic at death and pyogranulomatous inflammation in the skin, brain, or other tissues can occasionally be observed. Nocardiosis has also been identified in captively managed harbor seals. The primary lesion in these seals was chronic pyogranulomatous cellulitis (Fig. 23.20). Infection is diagnosed via culture or PCR, and organisms can be demonstrated using Gram, GMS, and acid-fast stains.[175]

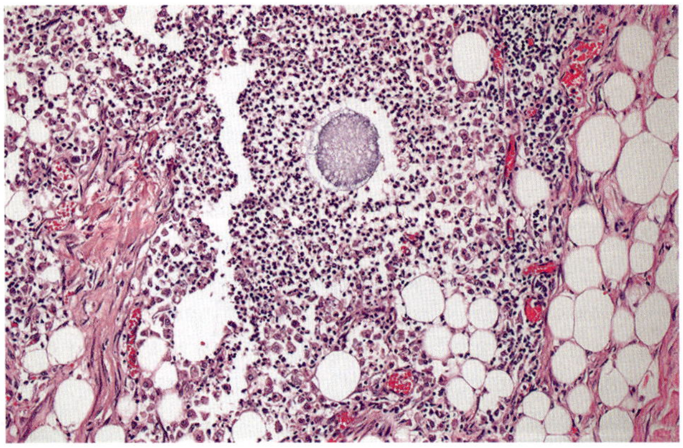

FIGURE 23.20 Chronic pyogranulomatous cellulitis in the subcutis and adipose tissue of a harbor seal due to *Nocardia* sp. infection. Inflammatory cells surround a central round cluster of pale basophilic filamentous bacteria.

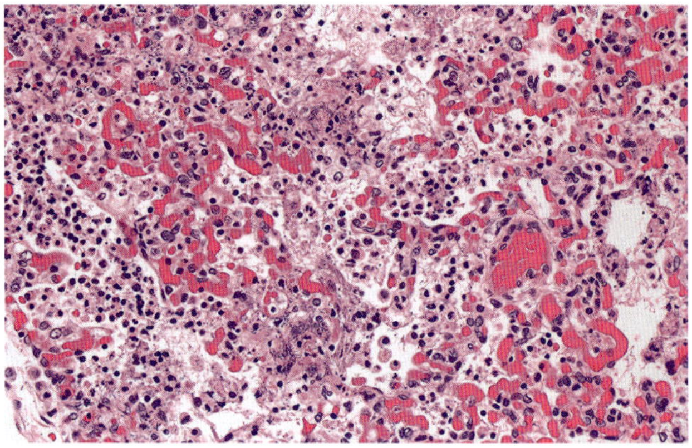

FIGURE 23.21 *Pseudomonas aeruginosa* infection in the lung of a harbor seal. Infection causes fibrinosuppurative and necrotizing pneumonia and vasculitis. A characteristic feature of infection is multifocal "swarms" of intralesional bacilli that are often centered on blood vessels.

Pseudomonas* sp.,** most commonly ***P. aeruginosa, can cause severe systemic infections in pinnipeds, most frequently in young animals in rehabilitation centers. Pneumonia is most common; however, meningoencephalitis, hepatitis, and splenitis can be noted with septicemia. Lesions are characterized by fibrinosuppurative inflammation with necrosis, interlobular edema, and fibrin accumulation in affected lungs. Histologically, the bacterium causes vasculitis with characteristic intralesional "swarms" of gram-negative bacilli (Fig. 23.21).[61] *P. aeruginosa* infection has also been associated with hemorrhagic pneumonia in a cluster of deaths in adult free-ranging harbor seals.[140]

Fungi

Fungal disease represents a relatively small proportion of infectious diseases that have been reported in pinnipeds. Systemic infections with opportunistic fungi have been reported sporadically and have been reviewed by Reidarson et al.[154] Exposure most often occurs through the respiratory system but can be due to implantation, such as spreading from the gastrointestinal tract or a skin wound. Distribution of lesions is widely variable but often involves pulmonary tissues with subsequent systemic spread. Cytology and histopathology can be used to identify characteristic fungal morphology; however, culture and/or PCR are often needed to differentiate/confirm the type and species of fungus. Burek[18] reviewed the epidemiology, gross and histopathologic features, and diagnosis of each fungal group.

Coccidioides immitis, an important human pathogen, is endemic in California and has been reported most commonly in CSLs and rarely in harbor and northern elephant seals. It is typically associated with disturbance of dry soil where it occurs in the mycelial phase and produces highly infectious arthroconidia. Transmission is generally through inhalation.

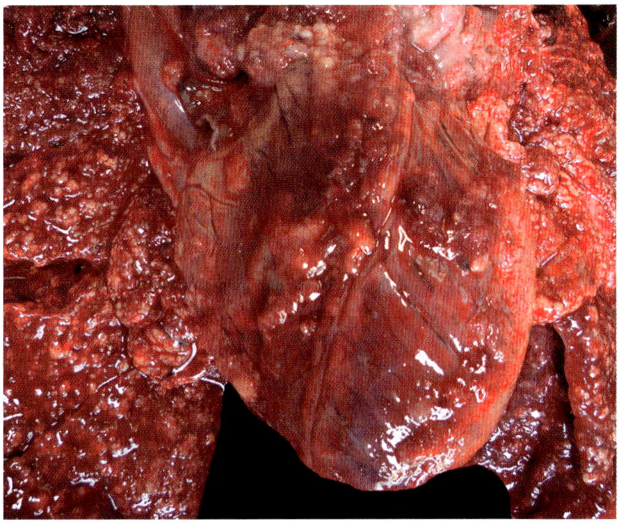

FIGURE 23.22 *Coccidioides immitis* infection in the lung of a California sea lion. Numerous fungal granulomas are scattered throughout the lungs and across the epicardial surface. *(Photo Courtesy of The Marine Mammal Center.)*

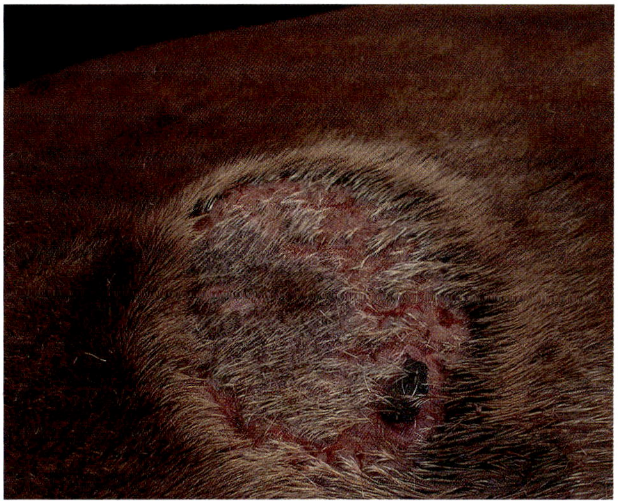

FIGURE 23.23 *Trichophyton* sp. dermatitis in the skin of a Stellar sea lion. Target-shaped areas of alopecia with ulceration are associated with infection.

In most animals, *C. immitis* causes respiratory disease but it can progress to disseminated fatal systemic infection. Animals with **coccidioidomycosis** are often emaciated and have pyogranulomatous pneumonia, pleuritis, peritonitis, and thoracic and abdominal lymphadenitis (Fig. 23.22). In animals with systemic spread, multiple miliary to large pyogranulomas, sometimes with indented necrotic centers, can be noted throughout the body. Histologically, there is intense pyogranulomatous inflammation centered on few to many, double contoured, 5–30 μm diameter immature spherules that mature into 30–200 μm diameter spherules containing a myriad of endospores. Mycelia with production of arthroconidia are rare in tissue. Because of the danger of exposure of necropsy personnel and laboratory workers to fungal spores, affected carcasses should be incinerated. Cytology, histology, and PCR are the preferred diagnostic methods.[89] Fungal pyogranulomas CSLs may be grossly difficult to distinguish from urogenital carcinoma metastatic masses, though heavy involvement of the lungs and granulomatous inflammation on impression smear cytology can aid in differentiation at necropsy.

Fungal dermatitis has been reported sporadically both in captive and free-ranging pinnipeds due to a wide variety of yeasts, nondermatophylic hyaline molds, and dermatophytes. In addition to previously described predisposing factors, mechanical disruption of the skin and water quality issues (low salinity, high temperatures) can predispose to infection. Alopecia, acanthosis, and hyperkeratosis can be associated with ***Candida albicans***, ***Fusarium*** spp. ***Microsporum*** sp. and ***Trichophyton*** **sp**. infections. Commonly affected areas include mucocutaneous junctions, nail beds, and the axillae.[146,154] In Steller sea lions in Alaska, "fungal patches" due to *Trichophyton* sp. distributed across the body are very commonly seen and can be present for extended periods of time. Affected skin is thickened and corrugated and often alopecic. Lesions sometimes have a target pattern with either healed or central patches of raw skin (Fig. 23.23). Histologically, there is chronic proliferative dermatitis with acanthosis, hyperkeratosis, and folliculitis, and in cases of dermatophyte infection, there may be intrafollicular or superficial epithelial fungal hyphae. Special stains, such as PAS or silver stains can aid in identification and characterization of the organisms. Ulceration and suppurative dermatitis are generally related to self-trauma and secondary bacterial infection. Diagnosis can be made by skin scraping, fungal cultures, and/or PCR; PCR is often needed to differentiate and confirm the fungal species.[18]

Metazoa

In addition to parasite descriptions below, a more extensive review of pinniped parasitic infections is presented in Table 23.e5.

There are two general categories of lungworms that infect pinnipeds, small lungworms, the *Parafilaroides* species that includes *P. decorus, P. gymnus*, *P. hispida, P. normani, P. measuresae, P. hydrurgae,* and *P. gullandae* and the large lungworms *Otostrongylus circumlitus*. They are metastrongyloid nematodes of the Filaroididae and Crenosomatidae families, respectively. Life cycles of marine mammal lungworms are not well understood, though fish species have been shown to be intermediate hosts.[126,127]

Parafilaroides spp. infect otariids and phocid hosts in both the northern and southern hemisphere and subclinical infections are very common. Adult parasites reside in the pulmonary parenchyma, have a thin, ridged cuticle, coelomyarian polymyarian musculature, and shed embryonated eggs into the small airways, where they develop into L_1 larvae. L_1 larvae are coughed up, swallowed, and shed in the feces. *Parafilaroides* spp. infect all age-classes at different intensities in different species, though infections are often more intense in younger animals. Lesions in uncomplicated infections appear grossly as small-raised nodules along the pleural surface and in the parenchyma adjacent to bronchioles (Fig. 23.24A; eSlides 23.e4, 23.e6). Parasites are fine and difficult to detect grossly; however, scraping the cut surface of lesions with a knife may enable visualization. Histologically, there are cross-sections of small nematodes within alveolar spaces and bronchioles. Infection in most animals results in minimal lymphocytic interstitial and peribronchiolar inflammation and occasional granulomas may develop around degenerating parasites or larvae (Fig. 23.24B; eSlides 23.e4 and 23.e6). Heavy infections and secondary bacterial pneumonia can result in more significant inflammation, especially in young or debilitated animals. Mild bronchiole epithelial hyperplasia and lymphocytic tracheitis are also common.[126]

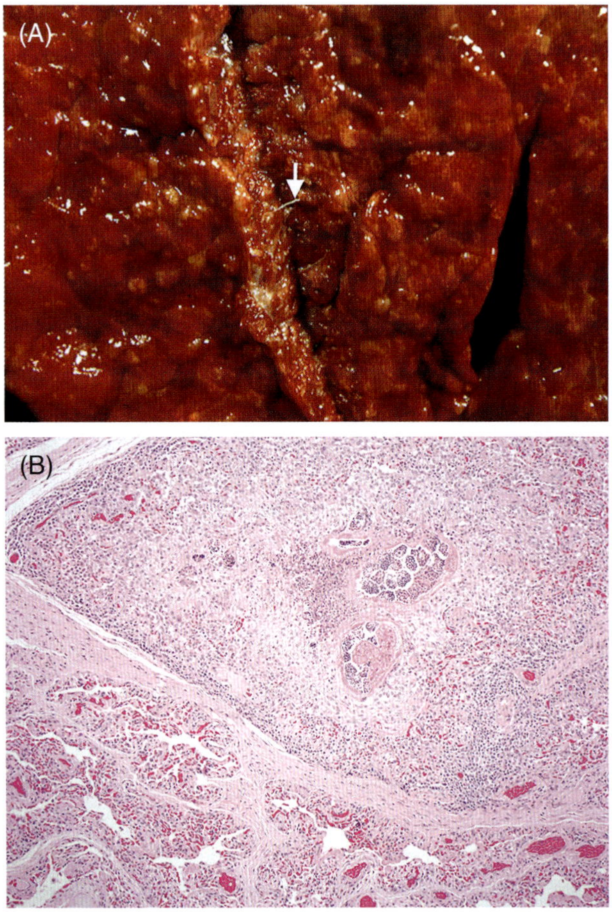

FIGURE 23.24 *Parafilaroides decorus* infection in a California sea lion. (A) Nodular areas of pneumonia are associated with thread-like metastrongyles *(arrow)*. (B) Granulomatous pneumonia is associated with intralesional larvae and necrotic adult parasites (see eSlide 23.e4, eSlide 23.e6). *(Part A: Photo Courtesy of The Marine Mammal Center.)*

Otostrongylus circumlitus is distributed in the circumpolar region and affects a variety of species. The primary hosts are ringed and harbor seals, but infections have also been reported in other phocids including gray, elephant, ribbon, and hooded seals.[111,126,127,170,182] After ingestion, larvae migrate to the right ventricle, through the wall of the pulmonary artery and into the lungs. Significant infection generally occurs only in young animals. Heavy infections may impact diving performance, health, and recruitment. Adult parasites reside in the bronchi and bronchioles (Fig. 23.25). The head of a worm is embedded in the parenchyma and the body and tail extend into the lumen. L1 larvae are released into airways. Infection induces variable amounts of mucus plugging of airways, mucosal cell hyperplasia, moderate lymphoplasmacytic and eosinophilic peribronchitis, and mild chronic arteritis with tunica intima hyperplasia.[126] Unlike other phocid species in which larval migration is not associated with lesions, in northern elephant seals reaction to larval parasite migration in the pulmonary arteries, especially the pulmonary arterioles, can cause severe disease and death in the prepatent period. Clinical signs in affected elephant seals include harsh lung sounds, depression, neutrophilia, and disseminated intravascular coagulation (DIC). Grossly, nematodes are found in the right ventricle and pulmonary arteries; they are often associated with extensive hemorrhage in lung parenchyma. Histologically, nematodes can be present within the pulmonary artery and major pulmonary artery branches.

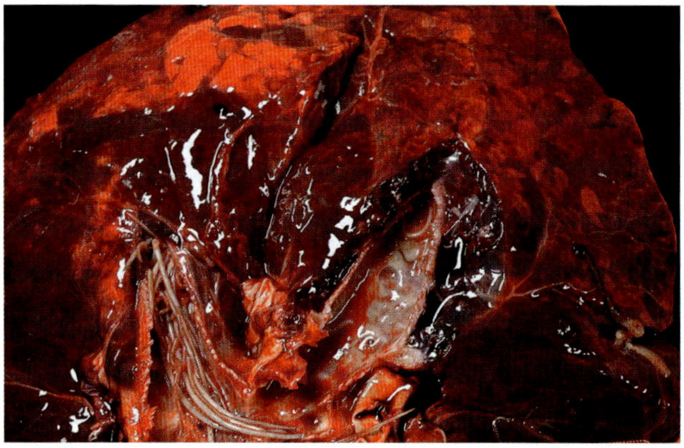

FIGURE 23.25 *Otostrongylus circumlitus* infection in the lungs of a gray seal. Numerous long, slender, white nematode parasites fill the lumen of a bronchiole. Note the associated mucous in the airways. *(Photo Courtesy of the International Fund for Animal Welfare.)*

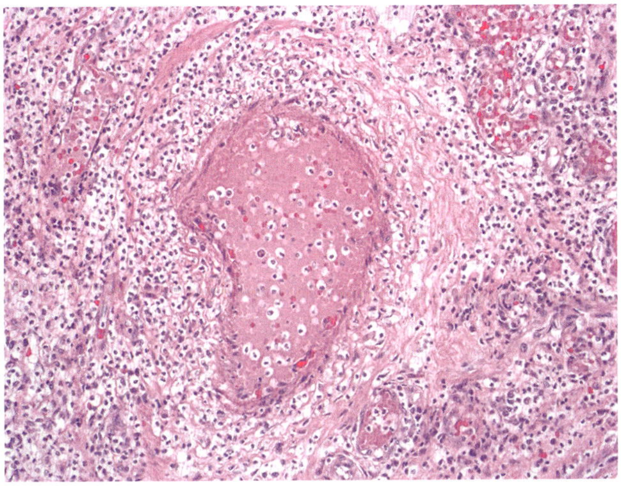

FIGURE 23.26 Vasculitis and interstitial pneumonia associated with *Otostrongylus circumlitus* infection in a northern elephant seal. Vasculitis and thrombosis can be noted in pulmonary arterioles distant from larger arteries containing the nematodes (see eSlide 23.e5).

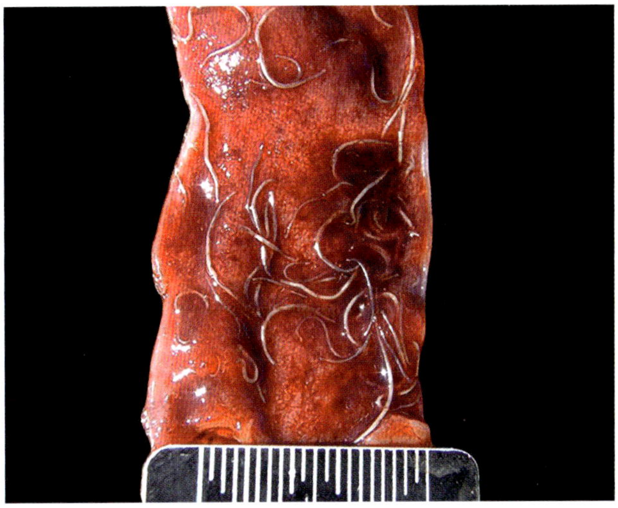

FIGURE 23.27 Hookworm infection in the small intestine of a New Zealand sea lion. Moderate numbers of hookworms are present within and attached to the small intestinal mucosa. Mucosal trauma caused by the parasites is associated with hemorrhage.

Severe neutrophilic and lymphocytic arteritis, thrombosis, and interstitial pneumonia with vasculitis and thrombosis are often present in smaller intrapulmonic vessel branches (e.g., arterioles) that do not contain the parasites (Fig. 23.26; eSlide 23.e5). DIC is a possible sequela. Lesions can be complicated by secondary bacterial infection.[78,170] The severe reaction in this species suggests a poorly adapted host.[126]

Hookworm infections caused by members of the genus ***Uncinaria*** have been described in a number of otariid and phocid species. Both Northern and Southern fur seals experience significant pup mortality associated with infections. In a recent study, 94% of pups examined had infections and infections were identified as the cause of death in 35% of pups.[164] While only two distinct hookworm species, ***U. lucasi*** and ***U. hamiltoni***, have been confirmed to infect pinnipeds, recent studies suggest that there is considerably more species diversity than is currently recognized.[135] The pinniped hookworm life cycle was largely characterized in northern fur seals in the 1960s and appears to be similar for other pinniped host species. Briefly, pups are infected via the transmammary route, with ingestion of L3 larvae in the milk. Larvae mature into adults in the pup's intestine and eggs are shed in feces. Larvae develop in the environment and L3 larvae penetrate the skin of a susceptible pinniped host or are transmitted via ingestion. Larva then migrate from the intestines or subcutaneous tissues to the host ventral blubber, where they lie dormant until, in pregnant females, they migrate into the milk during late pregnancy. Adult hookworms survive in the intestinal tract of infected pups for 6 weeks[120] to 8 months[87,94,119] depending on host species. During this time, they attach to the small intestinal mucosa and feed on blood. Clinical signs are generally only seen in pups and are associated with blood loss into the intestine from parasite-induced damage. Severely affected pups have reduced growth rate and can cause severe fibrinohemorrhagic enteritis, anemia, and death. Typical gross lesions include pallor of tissues due to anemia, nematodes admixed with hemorrhage in the small intestine, and numerous 2–4 mm diameter red foci on the serosal surface of the intestine that correspond to hemorrhagic and inflamed feeding sites (Fig. 23.27). Penetration of the wall of the intestine with subsequent peritonitis and bacteremia also occurs.[164]

Gastric and intestinal roundworm infection is extremely common in free-ranging pinnipeds, with ***Anisakis*** sp. and ***Contracaecum*** sp. being most commonly reported. Within the stomach, infection results in large, "volcanic" ulcers due to embedded parasites in the mucosa and adjacent mucosal hyperplasia and submucosal fibrosis (eSlide 23.e6). Infection is typically considered subclinical even in cases with very large parasite loads and multiple gastric ulcers.[127,165]

Heartworm, due to infection with ***Dirofilaria immitis*** and ***Acanthocheilonema spirocauda***, has been reported in multiple seal and sea lion species. Infection mirrors the condition in domestic dogs with congested, edematous, and hemorrhagic lungs, pulmonary emphysema, interstitial and exudative pneumonia, right ventricular hypertrophy, pulmonary catarrhal bronchitis, and hepatic congestion. Endarteritis and intimal proliferation of arteries and arterioles are observed microscopically. Worms can be found in the right ventricle, pericardial sac, pulmonary artery, and/or lung. Identification of microfilaria helps to differentiate between the two parasites; microfilaria of *D. immitis* are generally larger than *Acanthocheilonema* sp. Not all blood microfilaria reflect pathogenic heartworm infection. *A. odendhali* microfilaria is similar in size to *A. spirocauda*. *A. odendhali* infection is generally incidental with adults located in subcutaneous, intramuscular, and cavity spaces without associated cardiovascular disease. Captive seals and some sea lions in endemic regions are routinely screened for and placed on heartworm preventative to protect from this infection.[127]

Acanthocephalans can occur in the small and large intestines in a number of pinniped species. Lesions range from mild mucosal inflammation at the site of parasite attachment to severe ulceration and tunica muscularis hypertrophy. Rarely, transmural inflammation is noted when parasites invade into the submucosa and muscularis. Mild to moderate chronic lymphoplasmacytic peritonitis can be noted in some seals with more severe intestinal lesions.[104]

Protozoa

Toxoplasma gondii is an intracellular apicomplexan parasite that infects many warm-blooded species, including people, worldwide. Serologic evidence of exposure to *T. gondii* has been reported in a number of free-ranging and captive pinnipeds including artic and subarctic seals and walruses with low exposure to felid definitive hosts. Parasite-associated disease is most commonly identified in free-ranging Hawaiian monk seals, CSLs, and harbor seals.[128] It is a significant cause of mortality in the main Hawaiian islands, where there is a high feral cat population.[85] Hawaiian monk seals with fatal toxoplasmosis commonly have severe systemic disseminated infections that cause necrotizing inflammation in the brain, lung, adrenal gland, lymph nodes, and spleen. Severe steatitis is a characteristic feature of the infection and has been identified in the mesentery and epicardium but not within subcutaneous adipose tissue (Fig. 23.28A). Histologically acute necrotizing lesions are associated with myriad tachyzoites and less frequent cysts (Fig. 23.28B). Myocardial inflammation has been identified in seals dying with chronic infection.[6]

In CSLs with *T. gondii* infection, the most common lesions are necrotizing and lymphohistiocytic meningoencephalitis, myocarditis, and myositis with associated intralesional protozoa (Fig. 23.29). Neurologic signs, including ataxia, seizures, and abnormal mentation, have been observed. Infection has also been identified in aborted CSL pups, confirming vertical transmission of the parasite. Incidental tissue cysts can be occasionally seen within skeletal or cardiac myocytes.[22]

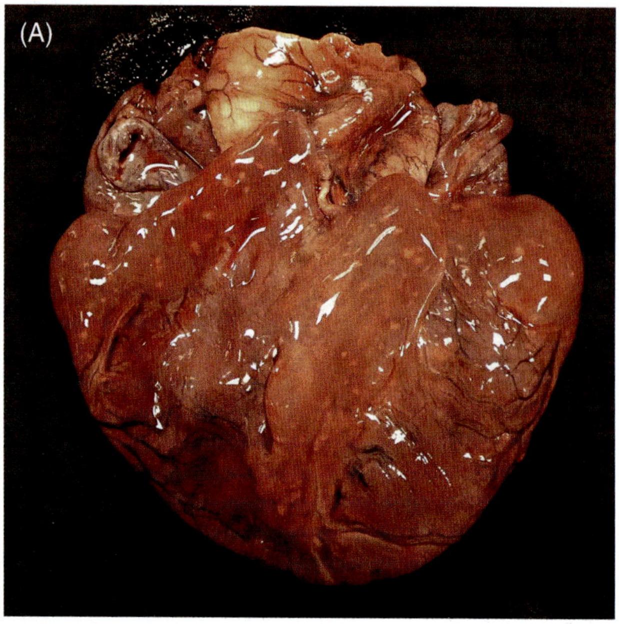

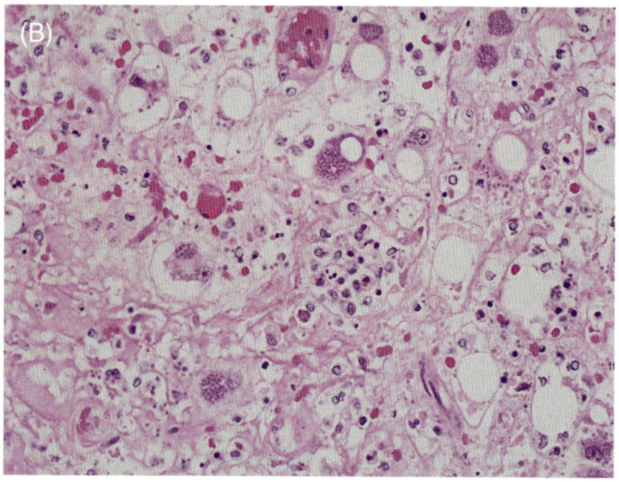

FIGURE 23.28 *Toxoplasma gondii* in a Hawaiian monk seal. (A) Multifocal small yellow nodular foci of steatitis are present throughout the epicardial adipose tissue. (B) Necrosis and inflammation of mesenteric adipose tissue with myriad 2–3 μm eosinophilic intra and extracellular tachyzoites. *(Part A: Photo Courtesy of National Oceanic and Atmospheric Administration Hawaiian Monk Seal Research Program.)*

In harbor seals, widespread meningoencephalitis is the most common lesion of toxoplasmosis and coinfection with ***Sarcocystis neurona*** has been identified in a number of seals. In one investigation in the Pacific Northwest of the United States, *T. gondii* infections were most commonly caused by Type I genotypes.[64]

PCR and IHC can confirm *T. gondii* infection. Occasionally, organisms are not observed with IHC in areas of protozoal-compatible inflammation, though *T. gondii* DNA can be amplified by PCR. Most commercially available IHC assays employ polyclonal antibodies to

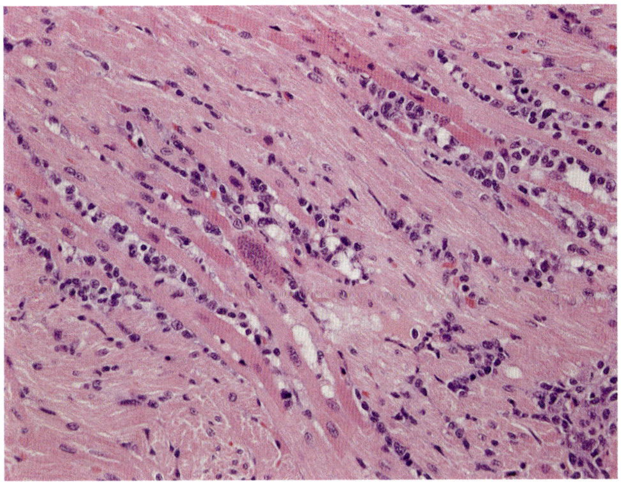

FIGURE 23.29 *Toxoplasma gondii* infection in the heart of a California sea lion. Lymphohistiocytic myocarditis is associated with intralesional protozoal tachyzoites.

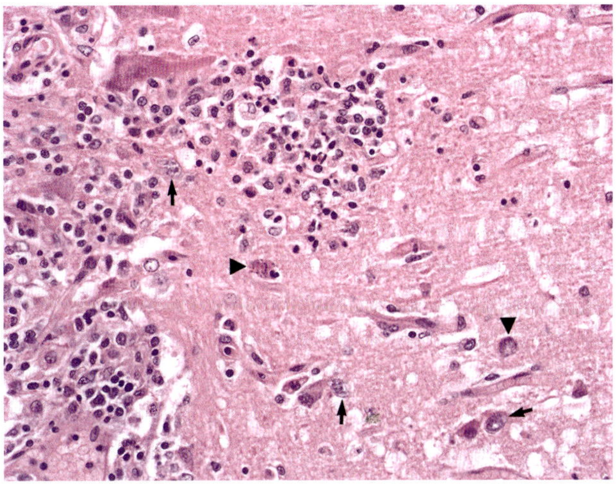

FIGURE 23.30 *Sarcocystis neurona* infection in the cerebellum of a harbor seal. Infection is associated with multifocal lymphocytic and histiocytic inflammation and necrosis. Within the areas of necrosis are a few mature and immature schizonts (*arrows*). Mature schizonts have radially arranged merozoites (*arrowheads*).

T. gondii, therefore, cross-reaction with other protozoa is possible.[27] Though serologic assays may be used to document exposure and rising titers have been associated with active infection, lack of pinniped-specific antibodies often makes defining positive titer ranges difficult.

Mortality due to ***Sarcocystis neurona***, an intracellular apicomplexan parasite, is most commonly identified in free-ranging pinnipeds from the Pacific coast of North America, and antibody titers reflecting exposure have been documented in harbor seals and CSLs.[23,76,128] Several cases have also been identified in captive phocids, otariids, and walrus.[102,105] Harbor seals are the most commonly reported species to develop fatal central nervous system infection and in California, subadults and adults are primarily affected.[7,128] Lesions include nonsuppurative meningoencephalitis which is often most severe in the cerebellum and can be associated with numerous organisms[105] (Fig. 23.30). Schizonts with radially arranged merozoites in a rosette pattern with occasionally visible central eosinophilic residual bodies can be used to differentiate organisms from *T. gondii*. Barbosa et al.[7] reported a novel genotype, Type XIII, that was associated with an increased severity of encephalitis in multiple marine mammal species.

While incidental sarcocysts can be identified in skeletal or cardiac myocytes in multiple species, infected CSLs can develop severe polyphasic rhabdomyositis. Multiple muscle regions are affected, especially the esophagus and diaphragm. Grossly, affected muscle may be atrophied and/or have pale gray to tan streaks (Fig. 23.31A). This corresponds histologically to multifocal lymphohistiocytic myositis with myonecrosis and regeneration (Fig. 23.31B). Rare sarcocysts can be noted in myocytes, but often are not directly associated with lesions, suggesting a possible immune-mediated pathogenesis.[23,163] Some may also develop secondary aspiration pneumonia and/or myoglobinuric nephrosis.

A novel ***Sarcocystis*** spp. closely related to ***S. canis*** and tentatively called ***S. pinnipedi*** has been associated with sporadic, fatal necrotizing hepatitis in several species including free-ranging and captive CSLs, Steller sea lions, monk, and gray seals. Affected sea lions were icteric, and had schizonts and merozoites in hepatocytes and Kupffer cells associated with foci of necrotizing hepatitis (Fig. 23.32). Immunohistochemistry and serology utilizing antibodies against *Sarcocystis neurona* failed to cross react with *S. pinnipedi* and diagnosis was made via PCR. Ringed seals have been identified with *S. canis*-like muscle cysts but no apparent hepatic lesions.[83,128,189] The intermediate host of *S. pinnipedi* has not been identified.

Eimeria phoca infection in young harbor seals results in enteritis ranging from mild self-limiting inflammation to severe fatal necrotizing enteritis.[184] Several novel apicomplexan parasites with closest homology to ***Neospora*** sp. and evidence of sexual replication in enterocytes have been identified within the intestinal tracts of CSLs. Though infection in CSLs causes only mild enteritis, one of these novel coccidian species was linked to fatal infection in a neonatal harbor seal, suggesting CSLs may be a definitive host for a protozoan pathogenic to other species.[27] Enteric coccidia have also been noted in elephant seals and South American fur seals. Both serologic evidence of exposure to *Neospora* sp. and infection diagnosed via PCR have been reported in several species.[64,128]

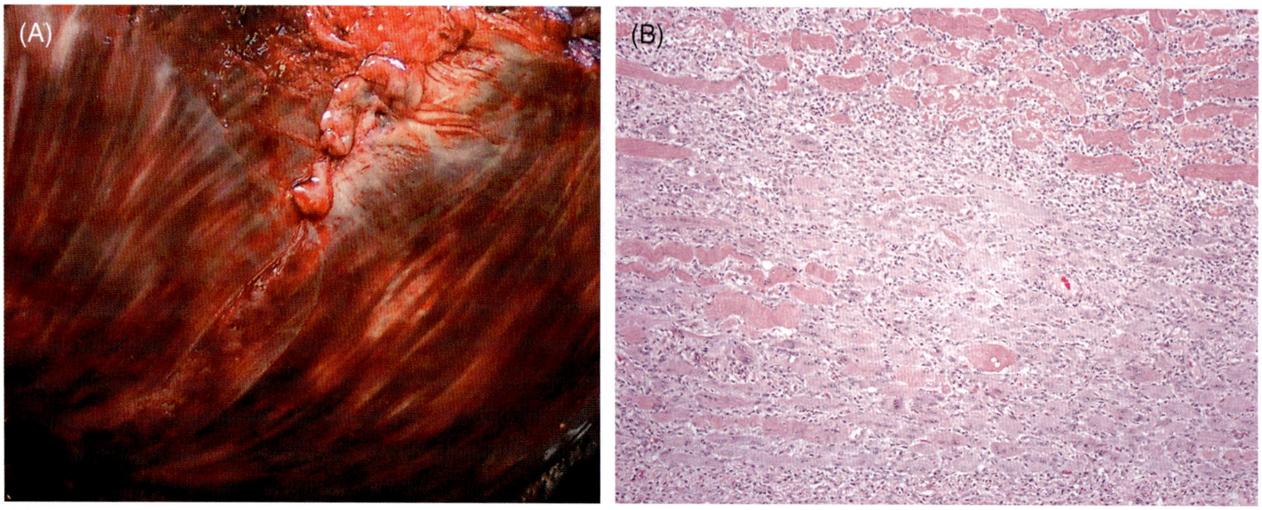

FIGURE 23.31 *Sarcocystis neurona*-associated myositis in a California sea lion. (A) Myositis causes pale tan streaks throughout the diaphragm. (B) Histologically, there is polyphasic rhabdomyositis with myonecrosis, mononuclear inflammation, and regenerating myocytes with pale basophilic cytoplasm and rowed nuclei. *(Part A: Photo Courtesy of The Marine Mammal Center.)*

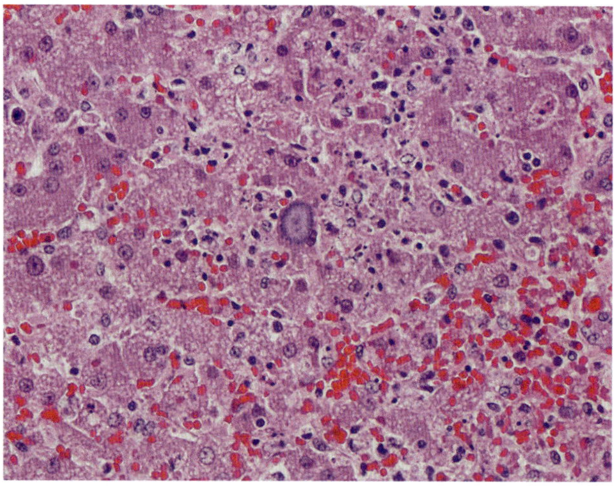

FIGURE 23.32 *Sarcocystis pinnipedi* infection in the liver of a California sea lion. Infection causes necrotizing hepatitis associated with schizonts and merozoites. Note the large mature schizont with radially arranged merozoites in the center of the image. *(Photo: Zoological Pathology Program, University of Illinois.)*

ADDITIONAL RESOURCES

Colegrove, K.M. 2018. Noninfectious diseases. *In*: Gulland, F.M.D., Dierauf, L.A. Whitman, K.L. (Eds). CRC Handbook of Marine Mammal Medicine, 3rd Edition. CRC Press, Boca Raton, pp. 255–284.

Duignan, P., Van Bressem, M.F., Baker, J.D., Barbieri, M., Kathleen, M., Colegrovem, K.M., De Guise, S., de Swart, R.L., Di Guardo, G., Andrew Dobson, A., Duprex, W.P., Early, G., Fauquier, D., Goldstein, T., Goodman, S.J., Grenfell, B., Groch, K.R., Gulland, F., Hall, A., Jensen, B.A., Lamy, K., Matassa, K., Mazzariol, S., Sinead, E., Morris, S.E., Nielsen, O., Rotstein, D., Rowles, T.K., Saliki, J.T., Siebert, U., Waltzek, T., Wellehan, J.F.X., 2014. Phocine distemper virus: current knowledge and future directions. Viruses 6, 5093–5134.

Gulland, F.M.D., Koski, M., Lowenstine, L.J., Colagross, A., Morgan, L., Spraker, T., 1996b. Leptospirosis in California sea lions (*Z. californianus*) stranded along the central California coast, 1981–1994. J. Wildl. Dis. 32, 572–580.

Gulland, F.M.D., Trupkiewicz, J.G., Spraker, T.R., Lowenstine, L.J. 1996a. Metastatic carcinoma of probable transitional cell origin in 66 free-living California sea lions (*Z. californianus*), 1979–1994. J. Wildl. Dis. 32(2), 250–258.

Measures, L.N. 2018. Helminths and parasitic arthropods *In*: Gulland, F.M. D., Dierauf, L.A., Whitman, K.L. (Eds.). CRC Handbook of Marine Mammal Medicine, 3rd Edition, CRC Press, Boca Raton, pp. 457–486.

Miller, M., Shapiro, K., Murray, M.J., Haulena, M., Raverty, S. 2018. Protozoan Parasites of Marine Mammals.

Moore, M.J., van der Hoop, J., Barco, S.G., Costidis, A.M., Gulland, F.M., Jepson, P.D., Moore, K.T., Raverty, S., McLellan, W.A., 2013. Criteria and case definitions for serious injury and death in pinnipeds and cetaceans caused by anthropogenic trauma. Dis. Aquat. Organ. 103, 229–264.

Silvangi, S., Lowenstine, L.J., Spraker, T., Lipscomb, T., Gulland, F.M.D., 2005. Pathology of domoic acid toxicity in California sea lions (*Zalophus californianus*). Vet. Pathol. 42, 184–191.

References

Full references for this chapter can be found online at https://doi.org/10.1016/B978-0-443-18437-6.00023-7.

Chapter 24

Sirenia

David Rotstein[a] and Martine de Wit[b]

[a]Marine Mammal Pathology Services, Olney, MD, United States; [b]Florida Fish and Wildlife Conservation Commission, Saint Petersburg, FL, United States

INTRODUCTION

Sirenians live in tropical and subtropical regions and include a single species of dugong (*Dugong dugon*) and three species of manatees, the African (*Trichechus senegalensis*), Amazonian (*Trichechus inunguis*), and West Indian (*Trichechus manatus*). The latter species includes two subspecies, the Antillean and the Florida manatee. Sirenia are generally not kept in zoological collections unless they are being rehabilitated or are permanently injured preventing release. With few examples from Sirenia in managed care, opportunities to study disease are limited. Additionally, disease is uncommonly reported in free-ranging Sirenia other than the Florida manatee due to limitations in animal examination. For these reasons, unraveling disease concerns is a complex process but is important as all Sirenia are listed as vulnerable in the IUCN Red List.[58] Manuals describing postmortem examination techniques are available.[10,40]

NON-INFECTIOUS DISEASES

Nutritional

Extensive **seagrass die-offs** have resulted in **starvation, emaciation**, and strandings of both manatees[34] and dugongs.[85] Depending on prior condition, animals may take 6–8 months to starve and may migrate large distances, so lack of proximity to an area of seagrass loss does not rule this out as a cause of emaciation.[85] The stomachs of affected animals often contain unusual material, such as algae, mangrove seeds, dead seagrass rhizomes, and anoxic sediment. Locally extensive areas of darkened skin have been noted in emaciated dugongs.[40] After a decade of seagrass losses in the Indian River Lagoon of Florida, mass morbidity and mortality from malnutrition were documented in manatees along Florida's Atlantic coast.[34] Manatees were emaciated with depleted fat stores and muscle. Digestive tracts often were relatively empty and gall bladders distended. Various pulmonary lesions (bullae, hematomas, infarction, and pneumonia) were identified in select animals. Microscopic findings (Fig. 24.1) were typical of starvation and included multiorgan atrophy of the liver, heart, skeletal muscle, pancreas, and hemidiaphragm. Atrophied hepatocytes had decreased cytoplasm and contained increased Schmorl's-positive lipofuscin. Hemosiderin was present in small amounts within hepatocytes and more moderate amounts within Kupffer cells. Periodic Acid-Schiff stains indicated diminished glycogen within hepatocytes.

Nephrolithiasis and suppurative pyelonephritis with abscessation have been reported in a Florida manatee and an Antillean manatee (Fig. 24.2). The stones were composed of calcium carbonate and it is not known whether the pyelonephritis predisposed to nephrolith formation.[59] In one case, there was renal contracture and in the other, renomegaly. Stones were observed with and/or without suppurative exudate and multiple different bacteria were cultured. Histopathologic findings included pyelonephritis, interstitial nephritis, glomerulonephritis, glomerulosclerosis, nephrocalcinosis, and cortical atrophy. One or several of the following bacteria were isolated including *Edwardsiella tarda*, nonhemolytic *Streptococcus* spp., *Aeromonas sobria*, α-hemolytic *Streptococcus*, and *Actinomyces renale*.

Metabolic

Cold stress syndrome (CSS) is usually a chronic disease process that can occur in manatees after approximately 1–2 weeks of continuous exposure to water temperatures below 20°C.[17] It is a relatively common cause of manatee mortality and catastrophic mortality events have been documented following periods of severe cold, with temperatures dropping to or below freezing.[47] Exposure to such extreme temperatures can also lead to acute hypothermia and presumed cold water drowning.[47]

Hematologic and clinical chemistry changes are observed with CSS and may have prognostic significance.[3,71] Hematologic indicators include low platelet count, hypothermic coagulopathies, hemoconcentration,

Pathology of Wildlife and Zoo Animals. https://doi.org/10.1016/B978-0-443-18437-6.00022-5

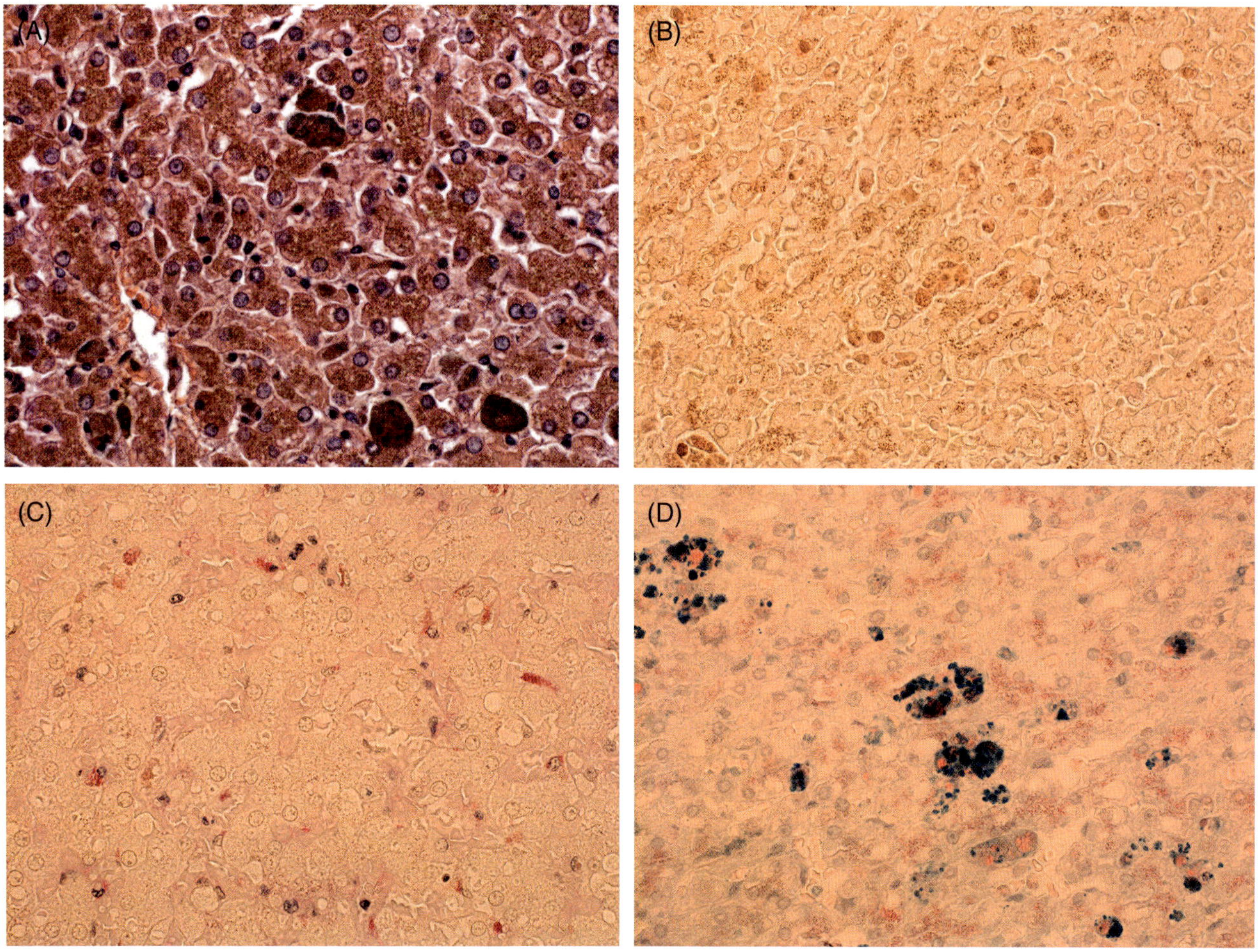

FIGURE 24.1 Hepatic atrophy from starvation in a Florida manatee. (A) There is collapse of the hepatic architecture. Hepatocytes contain diminished cytoplasm. Within the cytoplasm, there are abundant brown granules. Kupffer cells contain black-brown pigment (hemosiderin). (B) There is an accumulation of lipofuscin, a "wear and tear" pigment. Schmorl's stain. (C) There is scant glycogen within hepatocytes. Periodic Acid-Schiff. (D) Kupffer cells contain abundant hemosiderin. Perl's Prussian Blue.

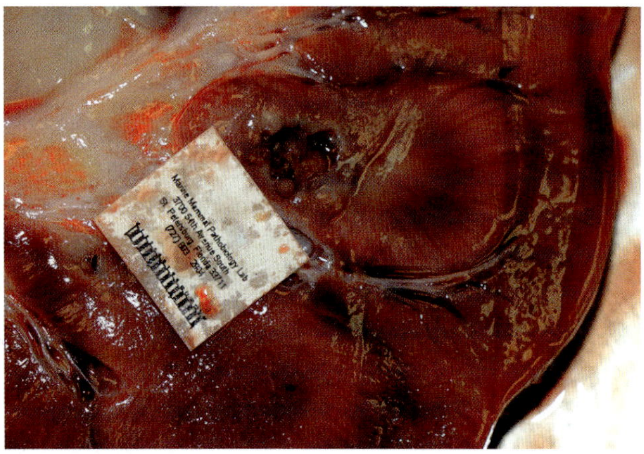

FIGURE 24.2 Nephrolithiasis in a Florida manatee. Within the renal pelvis, there is accumulation of variably sized and irregular stones.

and protein dyscrasias. Thromboelastographic (efficiency of blood coagulation) and coagulation panel changes may occur with CSS providing a possible indicator of thromboembolic complications including disseminated intravascular coagulation.[3] In comparison to healthy manatees for which hemostatic and coagulation factor profiles had been established,[4] reaction time was decreased, clot formation time increased, and increased percentage of clot lysed after 30 min in CSS manatees. Mean D-dimer concentration and coagulation factor XI activity were higher. Activated partial thromboplastin time was prolonged and mean antithrombin activity was decreased.

Emaciation, marked adipose depletion, and serous atrophy of internal fat is noted in most chronic cases of CSS. The findings are similar to starvation cases; however, atrophy of the liver and muscle have not been observed with

FIGURE 24.3 Early cold stress syndrome in a manatee, axillary region. Multifocal to locally extensive epidermal and dermal necrosis and sloughing.

CSS. Since there may be overlap in environmental conditions (e.g., cold temperatures in an environment with depleted seagrasses), diagnosis of CSS and/or starvation should be done with consideration of all factors. The upper gastrointestinal tract usually contains minimal or no ingesta, and the lower gastrointestinal tract contains dry hard feces indicative of dehydration and/or constipation. Pulmonary edema consistent with drowning is observed in acute CSS. Skin lesions are common. In more acute cases these can consist of spongy, edematous epidermis (Fig. 24.3). With chronicity, there is marked epidermal hyperplasia characterized by irregular, verruciform, pale gray plaques that are extensively fissured (Fig. 24.4). Multifocal to locally extensive cutaneous ulceration is also common with chronic cold stress. Cutaneous pustules or abscesses are reported in some animals. These skin lesions tend to be located on the flukes, flippers, and head with patchy distribution on the body.

Histologically, there is moderate-to-severe lymphoid depletion particularly in lymph nodes, spleen, gut, and respiratory mucosa-associated lymphoid tissue (Fig. 24.5; eSlide 24.e1). In addition, myocardial degeneration characterized by wavy, attenuated myofibers with loss of cross striations, karyomegaly, and occasional fibrosis is noted in some cases. Heterophilic and histiocytic colitis or enterocolitis is also relatively common, and smaller proportions of animals have chronic-active, suppurative bronchopneumonia.[17] Pancreatic zymogen granule depletion, a common finding with anorexia, is consistently observed. The most striking findings are multifocal superficial to full-thickness epidermal necrosis and spongiosis as a result of vasculitis, vascular fibrinoid necrosis, and intravascular fibrin thrombi (Fig. 24.6). Involved epidermis is covered by a thick layer of necrotic epithelial cells with abundant intraepidermal

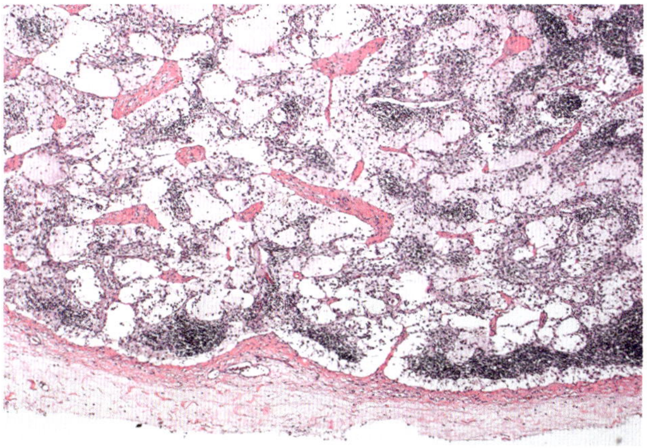

FIGURE 24.5 Cold stress syndrome in a manatee. Lymph node with severe lymphoid depletion in both follicles and paracortical areas (see eSlide 24.e1).

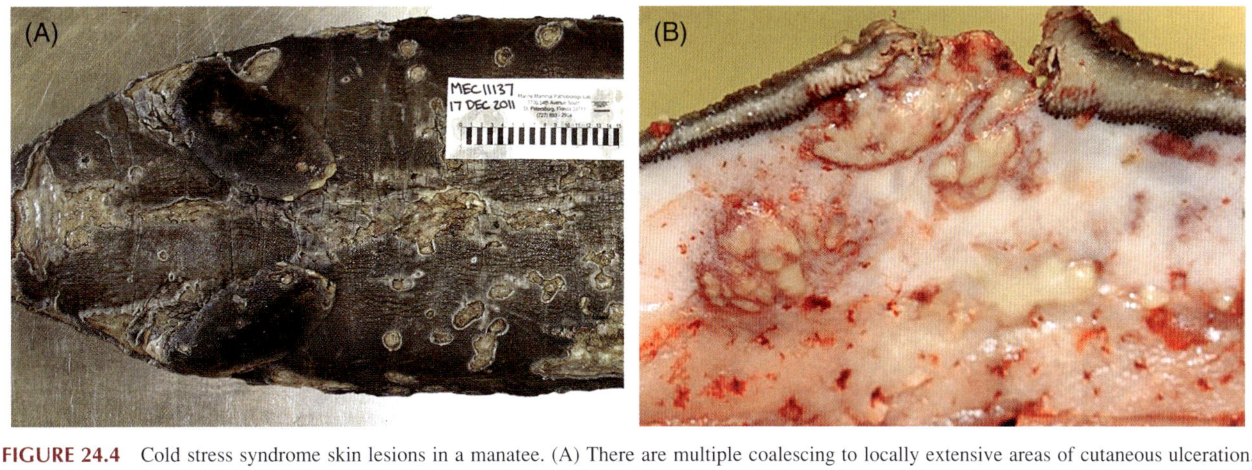

FIGURE 24.4 Cold stress syndrome skin lesions in a manatee. (A) There are multiple coalescing to locally extensive areas of cutaneous ulceration. (B) Cross-section through skin and subcutaneous fat with epidermal hyperplasia, ulceration, and multifocal abscessation superficially and within the dermis and adipose.

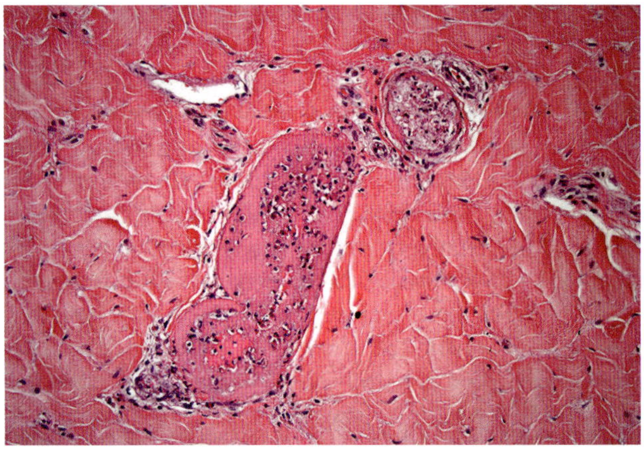

FIGURE 24.6 Acute cold stress syndrome in a manatee. Acute vasculitis in the dermis characterized by fibrinoid necrosis of vessel wall, perivascular and transmural heterophils, and fibrin thrombus formation. *(Photo Courtesy of L. Farina, University of Florida.)*

heterophilic pustules (Fig. 24.7; eSlide 24.e2). Dermal abscesses are also observed. There is epidermal hyperplasia at the ulcerated epithelial margins and pigment loss within basal epithelial cells (leukoderma). CSS remains a potential differential diagnosis for dugongs with a similar range of lesions and mortalities after periods of cold weather.[84]

Isolated cases of **fatal exertional myopathy** attributed to extensive mating pursuit by males have been observed in female Florida manatees[109] and in dugongs after pursuit for captures. The dugongs exhibit high serum lactate levels reflecting possible high muscle exertion and metabolic acidosis.[65] In manatees, the epidermis in the urogenital and peduncle region demonstrates abrasions, presumably from traumatic injury from the sandpaper-like texture of males' flippers. Histopathologic findings are observed in the skeletal muscle and kidney. There is myofiber degeneration with involved myofibers exhibiting sarcoplasmic clumping, vacuolation, and loss of cross-striations. Fragmented myofibers are also present. In the kidney, there is a myoglobinuric nephrosis with renal tubules containing brown intraluminal pigment (myoglobin) with sloughed tubular epithelial cells.

Toxic

Brevetoxicosis is caused by the marine dinoflagellate *Karenia brevis*, which produces several brevetoxins (PbTxs) that have potent neurotoxic and hemolytic properties. Brevetoxins can be harmful to aquatic animals, humans, and pets. With certain naturally occurring or anthropogenic factors influencing environmental conditions, these dinoflagellates can proliferate to the extent that they cause a "red tide," a form of harmful algal bloom (HAB). Red tides have been associated with mass mortalities in manatees, dolphins, and fish.[15,62] In many cases, multiple species are impacted in the same area at the same time.

Florida manatees are exposed to **brevetoxins** either through ingestion or inhalation. Ingestion occurs through consumption of filter-feeding organisms or toxins that are directly associated with seagrass.[41,42] This can occur even after a HAB has dissipated. In cases where the respiratory tract is affected, inhalation of the toxin is presumed; however, there is no indication that inhalation alone can result in death. Respiratory lesions associated with brevetoxicosis are nonspecific and variable. Animals may present with catarrhal inflammation of the nasal sinuses, which often contain copious exudate with submucosal congestion and hemorrhage. Microscopically, there are extensive lymphoplasmacytic infiltrates that thicken the nasal sinus submucosa.[15] Terminal pulmonary edema, congestion, and hemorrhage can be observed. Multiorgan hemosiderosis

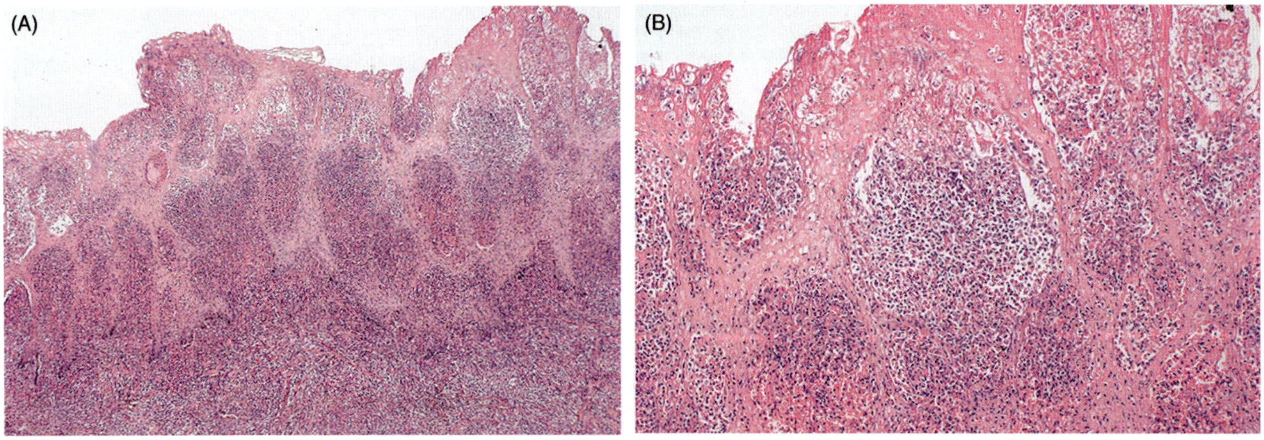

FIGURE 24.7 Chronic cold stress syndrome in a manatee. (A) Epidermal hyperplasia and erosion with intracorneal pustules and diffuse, marked, heterophilic dermatitis. (B) Higher magnification of the lesion in A (see eSlide 24.e2).

is observed in the liver, spleen, and central nervous system as a result of hemolytic toxins. The neurotoxin also produces nonsuppurative meningitis. Brevetoxins have an affinity for lymphocytes, macrophages, and microglial cells, and their pathogenesis may involve elevated circulating cytokines and toxic shock.[14]

A diagnosis of **brevetoxicosis** is typically based on history, necropsy findings, brevetoxin (PbTx) ELISA,[78] and exclusion of other etiologies. Toxins can be detected and quantified in body fluids, tissues, and gastrointestinal contents although the toxicity threshold is not known. Immunohistochemistry can be used to characterize the distribution of the toxin and associations between the toxin and lesions.[14,15,78] The presence of toxin alone is insufficient to warrant a diagnosis of brevetoxicosis. In live manatees, circulating brevetoxin can be found more than 1 month after exposure and neurological signs have resolved. Sublethal exposure to brevetoxins may also occur as the toxin can persist in seagrass after a bloom has dissipated. This sublethal exposure may compromise immune function; therefore brevetoxin could contribute to morbidity and mortality, even at lower concentrations.[105,106]

Congenital/Genetic

A number of congenital malformations have been described in Sirenia. These include uni- and bilateral **ectrodactyly of the pectoral flippers** in West Indian manatees.[110] Other reported malformations include **hydrocephalus** and **cardiac malformations** including right ventricle hypoplasia with tricuspid valve stenosis, ventricular septal defect, left ventricular hypertrophy, and pulmonary artery aneurysm.[26] In Florida manatees, malformations of the cardiac, pulmonary, renal, musculoskeletal, and gastrointestinal systems have been observed including enterocolonic strictures, scoliosis, midline deficits (umbilical hernia, omphalocele), and skull malformations such as open fontanelles.[107,111]

The etiology of these conditions has not been determined. The incidence of malformations in both manatees and dugongs requires a systematic approach to better understand the significance in populations.

Age-Related/Degenerative

Endocardiosis (Fig. 24.8) and **cardiomyopathies** have been described in manatees and dugongs.[43,44,83] The cause of cardiomyopathy is not always known; however, the lack of inflammation or causative agents makes an infectious cause less likely. One case of right-sided heart enlargement in a Florida manatee resulted from chronic traumatic diaphragmatic hernia.[43] Histological changes associated with cardiomyopathy vary from subtle to severe cardiomyocyte degeneration characterized by wavy, attenuated, variably sized myofibers, loss of cross striations, and anisokaryosis with karyomegaly. This can be associated with fibrosis and interstitial myocardial edema.[18,83] As similar lesions may be present with CSS, evaluation of environmental factors may be necessary to differentiate.

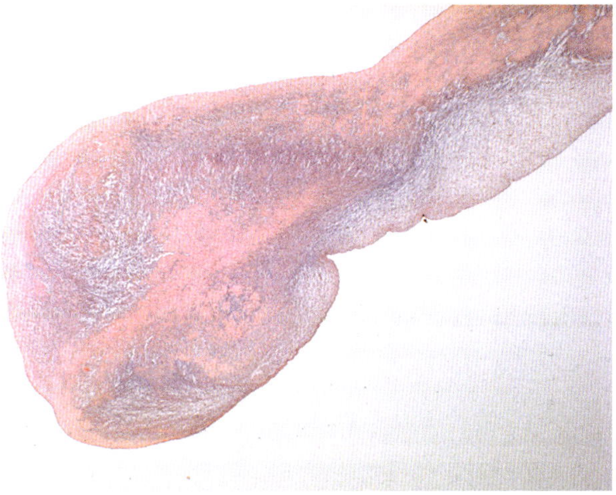

FIGURE 24.8 Valvular endocardiosis in the left ventricle of a manatee. The atrioventricular valve contains myxomatous material and collagen within the valve which is curled.

Miscellaneous/Multifactorial

Watercraft-related mortality (boat strike) is one of the most common causes of morbidity and mortality and is one of the major threats to the long-term conservation of the Florida manatee population.[93] This threat is likely to persist due to the overlap of nearshore manatee habitat and increasing coastal human populations that use waterways for commerce and recreation. A primary conservation action to reduce boat strike risk is the establishment of manatee protection zones limiting boat speeds in specific areas.[23] Small numbers of dugongs in Queensland, Australia have also died from boat strike injuries.[22,83] The lower number of incidents relative to the Florida manatee may be a function of the low human population in much of the dugong's range. Death following boat strike can result from sharp, penetrating trauma (e.g., from boat propellers), or from blunt trauma (e.g., boat hull or skeg impacts). Propellers often produce a parallel series of equidistant lacerations of the skin and deeper tissues on the dorsum with varying degrees of penetration (Fig. 24.9). However, in Florida manatees, the majority of lethal watercraft-related mortality is due to blunt force trauma from collisions with speeding and/or heavy vessels. These injuries can consist of minimal or underwhelming epidermal lesions with extensive trauma in deeper tissue.[68,92]

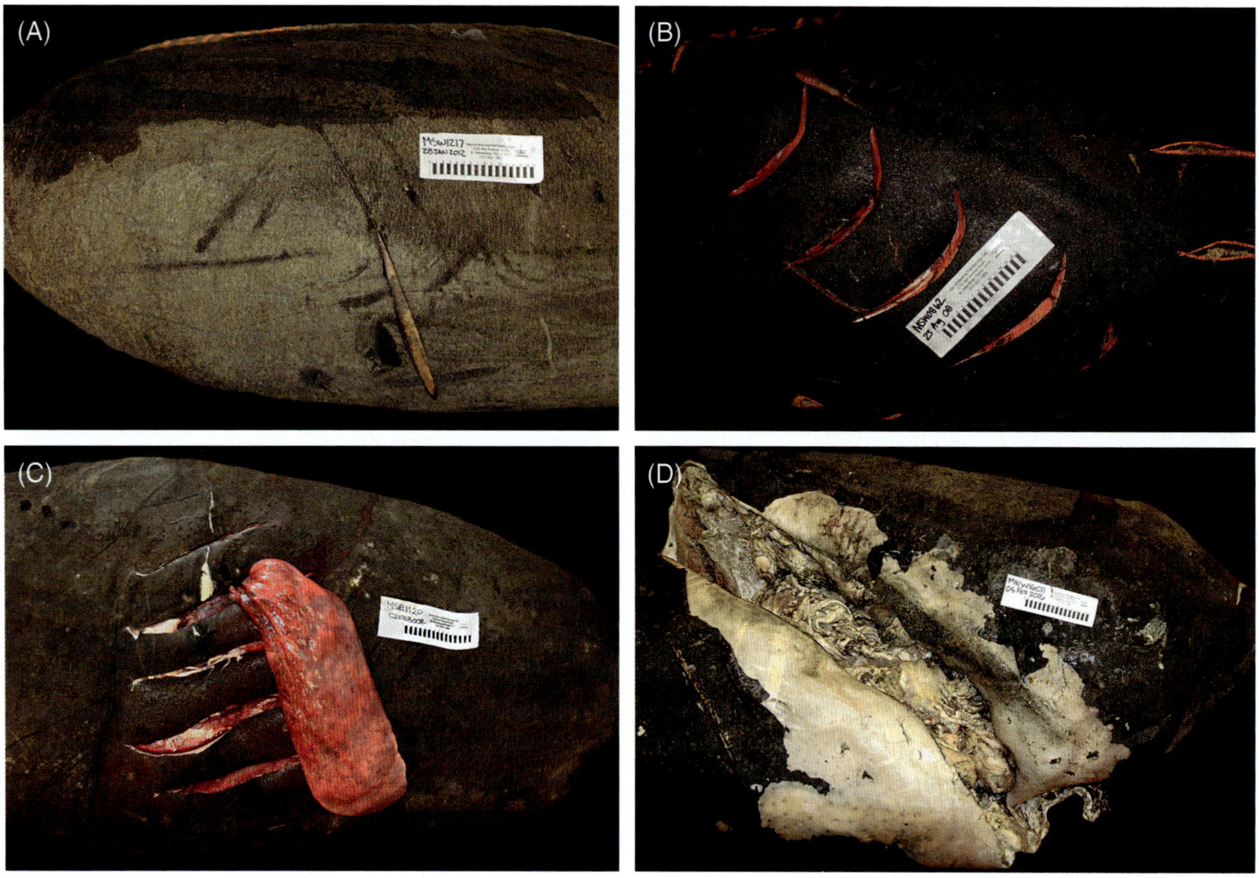

FIGURE 24.9 Boat strike injuries of increasing degrees of penetration in the skin of manatees. The linear, nonpenetrating injuries in A and B are due to blunt impact with parts of the boat or engine. Deep, penetrating wounds (C and D) can cause damage or extravasation of underlying viscera.

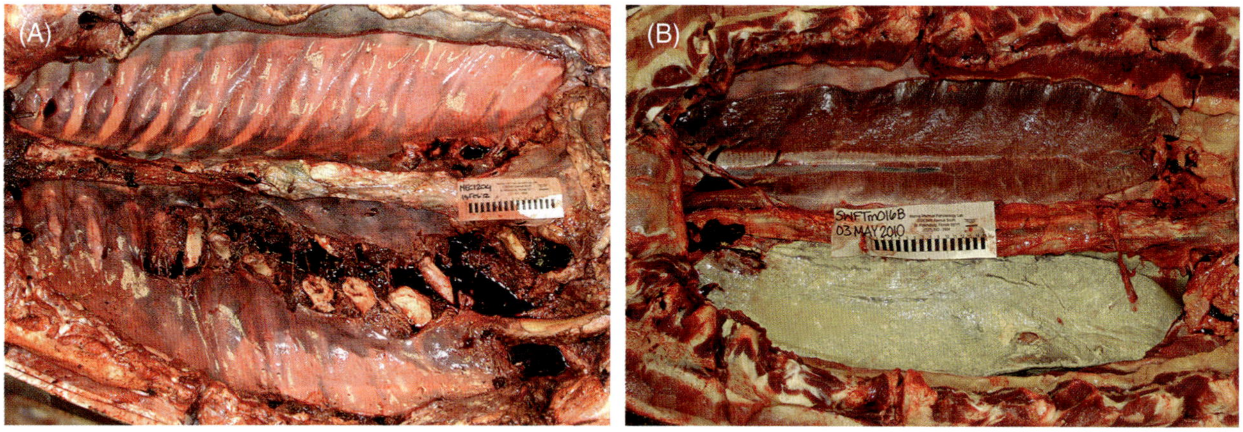

FIGURE 24.10 Boat strike injuries to the thoracic cavity of a manatee. (A) Acute boat strike injury to the thoracic cavity of a manatee with multiple rib fractures, and pleural and pulmonary lacerations. (B) Chronic boat strike injury with severe unilateral pyothorax resulting from primary dorsal pulmonary laceration and opportunistic infection.

Boat strike mortality can be acute or chronic resulting from sublethal injuries and secondary infections. The unusual anatomy of sirenians determines, to a degree, the lesions typically seen with boat strike. Their horizontal position in the water column with dorsal location of the lungs means that these are often injured with boat strike. Acute lesions to the thoracic cavity can result in soft tissue laceration from costovertebral fractures or luxations (Fig. 24.10A). Pulmonary sequelae can include hydrothorax, subcutaneous emphysema, and pneumothorax and secondary infections resulting in sepsis, pyothorax (Fig. 24.10B), pleuritis and peritonitis, osteomyelitis, and cellulitis.[13,22,50,68] The kidneys are also susceptible to injury because of their dorsal anatomic location. Renal lacerations can result in extensive hemorrhage and death due to hypovolemia. Diaphragmatic hernias have been observed.[43]

The most common consequences of boat strike are fractures, luxations, and subluxations of both the axial and appendicular skeleton. Sublethal skeletal injuries can also result in secondary osteomyelitis, periostitis, and osteitis with dramatic bone remodeling. Head injuries can lead to brain hemorrhage, with or without skull fractures. Manatees will often swallow blood following head or lung trauma; therefore the presence of blood in stomach contents provides evidence of antemortem injury.[68] Combined thoracic and abdominal injury usually results in extensive damage to multiple tissues, substantial hemorrhage, and acute mortality.[18,22] These findings aid in distinguishing between ante- and postmortem injuries as the cause of death, as it is also possible for watercraft to strike already dead and floating animals.

Other anthropogenic causes of mortality, such as **entanglement** in fishing and shark nets or **entrapment** in large-diameter drainage pipes and flood control gates or canal locks usually cause death by **drowning**.[22,55,81,83] When found in a drainage pipe, ventral recumbency of the carcass provides circumstantial evidence that the manatee may have entered when alive, as opposed to floating in after death because manatee carcasses usually float in dorsal recumbency. Flood control gates and canal locks can also cause death through crush injuries, which are often associated with extensively shredded muscle, vertebral separations, and/or bilateral trauma. Characteristic lesions of net entanglement are often absent but can occur (Fig. 24.11). Care must be taken not to attribute the presence of scars on dugong skin to net entanglement in all cases, as these findings are a common result of scratching against various materials and social interactions and could conceivably be scars from other types of superficial trauma.[40]

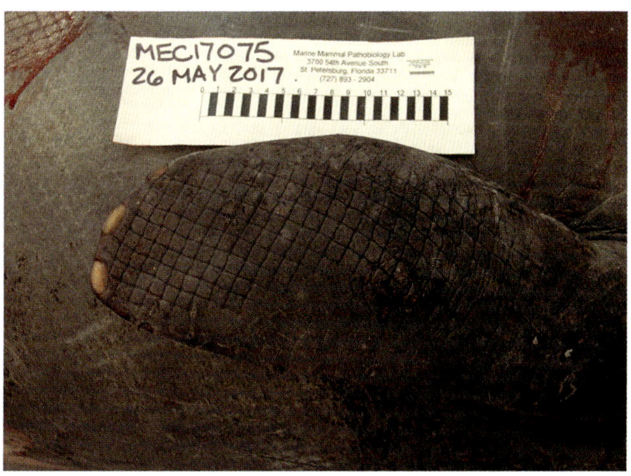

FIGURE 24.11 Net impressions in the skin of the flipper and body in a Florida manatee. This manatee drowned in a shrimp net. The skin contains cross-hatched lesions and impressions ($1.3\,cm^2$).

In a postmortem study performed on a small number of dugongs known to have died by drowning, investigators found no water, foreign material, or foam in the larynx, trachea, or primary bronchi. This finding supports the theory that dugongs drown due to asphyxiation following laryngospasm or the build-up of carbon dioxide in the blood causing paralysis of the closed larynx ("dry drowning"). Pulmonary abnormalities noted histologically included congestion, mild intraalveolar hemorrhage, edema in alveoli and the interstitium, and peripheral emphysema with peracute to acute alveolar septal myofiber degeneration.[114]

Fishery gear entanglement, typically around the flippers, and ingestion of discarded garbage can also cause lesions. Current and past flipper injuries due to entanglement include lacerations or scarring in the axillary region and partial or complete amputation of the limbs.[1] Mortality has been directly attributed to this entanglement in some cases. Although environmentally persistent debris is commonly found incidentally in the gastrointestinal tract, it can be significant if it causes impaction, intussusception, perforation, and secondary peritonitis.[7]

Calf mortality after maternal death or abandonment is common in Sirenia.[18,33] Neonates present with classic signs of negative energy balance including emaciation, serous atrophy of fat, pancreatic acinar atrophy (zymogen granule depletion), and hepatocellular atrophy. Perinatal death is suggested when the colons of calves contain meconium but there is no milk or vegetation in their digestive tracts. These animals also typically have an open or unhealed umbilicus and there is clotted blood in the umbilical artery.[40] In Florida manatees, death due to stillbirth or dystocia was the most common observation among natural death causes in small calves.[111]

Neoplastic

Neoplastic disease is rarely reported in manatees and no cases have been described in dugongs.[13,113] Reports of neoplasia are limited to **lymphoma** and **reproductive tumors**, which include uterine carcinoma, ovarian adnexal tumor, granulosa cell tumor, leiomyoma, and testicular interstitial cell tumor.[46,97] Other neoplasms observed in Florida manatees based on an unpublished review of medical records include disseminated histiocytic sarcomas, which were immunoreactive to vimentin and ionized calcium-binding adapter molecule (Iba-1), malignant testicular teratoma (Fig. 24.12), and hepatocellular carcinoma.

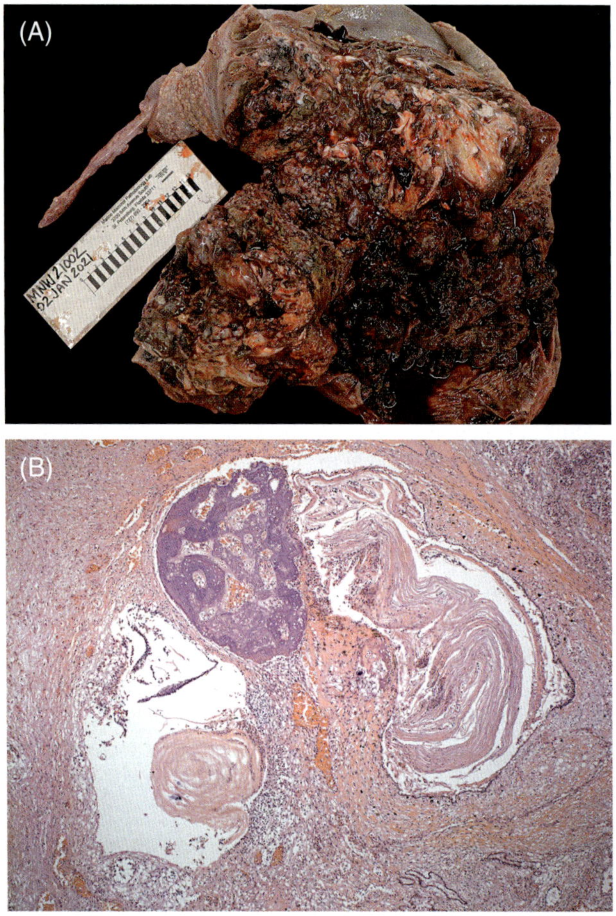

FIGURE 24.12 Teratoma in the testis of a manatee. (A) There is an infiltrative neoplastic mass with an irregular red to brown surface and interspersed cystic spaces. (B) The teratoma has epithelial and mesenchymal components. There is an island of cartilage within a dilated epithelial-lined cyst which contains lamellated keratin. The supporting stroma consists of variably dense spindloid cells with interspersed acini.

INFECTIOUS DISEASES

With the exception of a mass mortality event in Florida manatees associated with Clostridial infection, primary infectious disease is not a common cause of mortality in free-ranging sirenians.[14,46,63,70,83] Most reports are individual cases.

Viruses

***Trichechus manatus latirostris* papillomaviruses (TmPV)** cause a range of hyperplastic skin lesions in captive Florida manatees. Though extensively covered in literature, the disease is largely historical and associated cutaneous lesions have not been described in free-ranging animals. The lesions are variably multifocal, pedunculated, whitish-gray, and verruciform or filiform papillomas distributed over the head and leading contact regions of the pectoral flippers to more diffuse and numerous sessile, solid papillomas with a different distribution. Lesions sometimes appear to coincide with preexisting areas of superficial trauma, such as scratch marks.[16,37,86] Histologically, lesions are composed of hyperplastic and occasionally dysplastic keratinocytes. In some instances, stratum spinosum and granulosum keratinocytes have vacuolated cytoplasm and pleomorphic, vesicular nuclei typical of papillomavirus infection. Inclusion bodies have not been identified.

Certain features of disease presentation suggest that manifestation of infection may involve viral latency and immunologic suppression.[8,14] Diagnosis is confirmed by sequencing papillomavirus DNA from a clinically suggestive lesion. Immunohistochemistry for bovine papillomavirus cross-reacts and can be utilized when only histology tissues are available for diagnostics.

Antibodies to morbillivirus; pseudorabies, San Miguel sea lion virus, Eastern, Western, and Venezuelan equine encephalitis, and sequence of Trichechid herpesvirus 1 have been detected in free-ranging Florida manatee samples; associated clinical signs and lesions have not been described.[13,99,112]

Bacteria

Widespread bacterial **cellulitis** as well as sepsis and peritonitis can occur following sublethal boat strike and secondary infection.[68,83] In some cases, polyserositis and multifocal hemorrhages are noted similar to other mammals.[83] Sepsis can also occur with severe gastrointestinal and reproductive tract infections or perforations.[13,27,38,40,68] When bacterial infection is present in sirenians, the respiratory tract is commonly involved. Depending on the nature of the infection and route of exposure, infections of the respiratory tract can manifest as bronchopneumonia, granulomatous pneumonia, lung abscessation, pleuritis, or pleuropneumonia.[13,18,22,24,79,83]

Cases of **metritis** or **pyometra** resulting in sepsis and death have been reported in manatees and a dugong. Metritis in the manatees occurred following dystocia and subsequent fetal maceration, a similar pathogenesis was also suspected in the dugong.[13,27] *Aeromonas* spp. and *Pseudomonas* spp. were isolated from the dugong uterus. Rupture of a uterine abscess resulting in peritonitis and death has also been reported in a dugong.[83]

Mycobacteriosis due to *Mycobacterium marinum*, *M. fortuitum*, *M. kansaii*, and *M. chelonei* has been reported in captive Sirenia. Grossly, multiple coalescing caseous nodules were present throughout the lungs with pleural, lymph node, testicular, and cutaneous involvement in some cases. Bronchi contained purulent exudate. Histologically, lesions are typical of tuberculoid granulomas consisting of epithelioid macrophages rimmed by lymphocytes with intralesional acid-fast bacteria.[9,76,94]

Salmonella can cause gastroenteritis in young Sirenia. Severe necrotizing and granulomatous enteritis due to *Salmonella lohbruegge* has been described in a juvenile captive dugong. Infection was associated with focal granulomas (paratyphoid nodules) in the liver and reticuloendothelial cell hyperplasia in associated lymph nodes.[38] *Salmonella panama* isolated from an Antillean manatee calf caused gastroenteritis and septicemia.[103] An emaciated subadult male Florida manatee with *Salmonella enterica* serotype IV 50: z4, z23 had caseous exudate within the glenohumeral joint, intervertebral disk spaces, and costovertebral joints. Abscesses were present in the lung, kidney, and inguinal region. Microscopic findings include a pyogranulomatous nephritis, enteritis, osteomyelitis, pneumonia, myositis, and lymphadenitis.[104]

Clostridia have been associated with acute enterocolitis in Florida manatees.[63] Infection is precipitated by dysbiosis from a sudden shift to a predominantly macroalgal diet. *Paeniclostridium sordellii* and *Clostridium perfringens* with toxin genes and virulence factors plus detection of *Clostridioides difficile* toxin were identified. Affected manatees were in good nutritional condition. There was multiorgan congestion and thickening of the small and large intestine (Fig. 24.13A). Microscopic lesions included intestinal necrosis, transmural edema, hemorrhage, mixed inflammatory cell infiltrates, and mucosal-associated lymphoid depletion (Fig. 24.13B). Spore-forming rods were observed with Gram-stained sections within histologic and cytologic samples (Fig. 24.13C).

Orphaned manatees in rehabilitation settings can develop enterocolitis secondary to intestinal tract compromise from an artificial diet and concurrent immunosuppression. *Pseudomonas aeruginosa*, *Salmonella* spp., and *Clostridium difficile* are among the bacterial species that have been isolated from these cases. *Pneumatosis intestinalis* (PI), an accumulation of gas within the wall of the small or large intestine, is potential sequelae.[13] Grossly and microscopically, variably sized gas bubbles expand the submucosa with few surrounding mixed inflammatory cells. A case of multisystemic infection, possibly originating from a severe heterophilic and ulcerative enteritis and resulting in heterophilic meningoencephalitis in a captive neonatal dugong, has also been reported. *Enterococcus faecalis* was isolated from multiple body fluids.[79,83]

Necrotizing and eosinophilic **omphalitis** and hepatitis with fibrinous peritonitis in a free-ranging, juvenile, West Indian manatee have been reported. The inciting agent was not definitively identified.[108]

Cases of **bacterial encephalitis**, caused by both gram-positive and gram-negative organisms, have been described in Florida manatees.[21,22] In the case with gram-positive etiology, the source of the bacteria was thought to be an area of dermal erosion.

Antibodies to various serovars of ***Leptospira interrogans*** have been detected in Florida, Antillean, and Amazonian manatees and antibodies to ***Brucella*** spp. have been reported in Florida manatees. There are no reports of clinical disease or pathology associated with these infectious agents.[2,13,99]

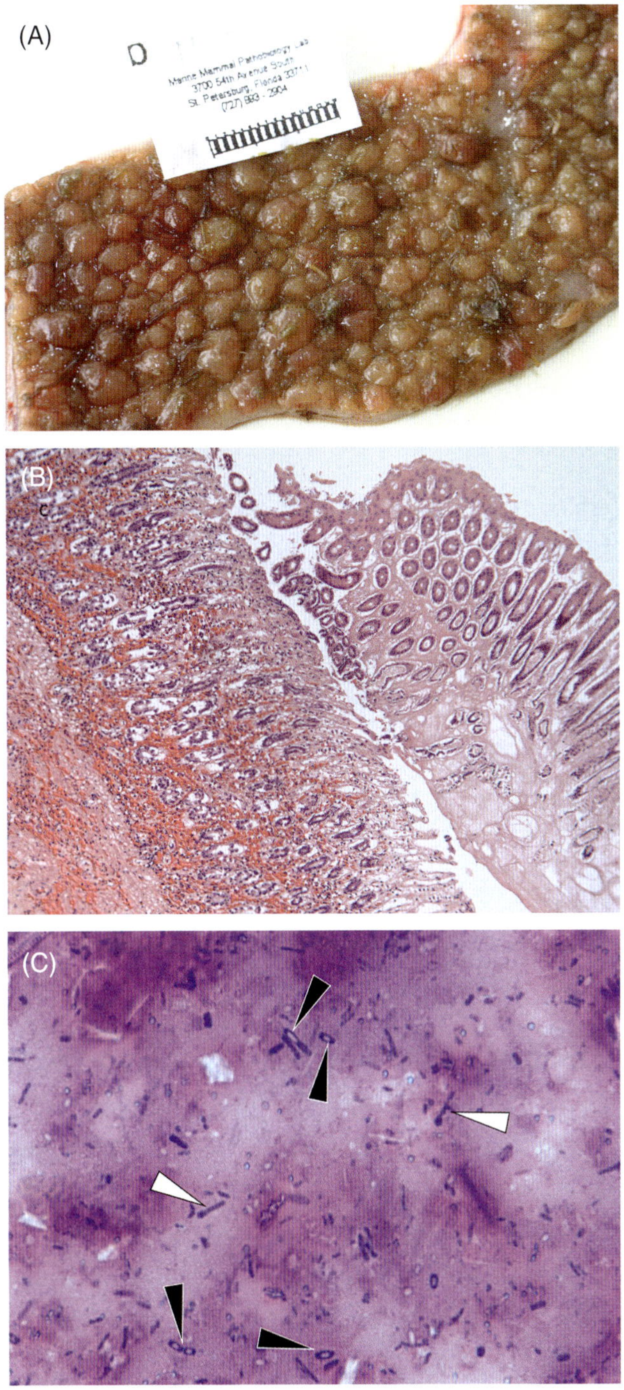

FIGURE 24.13 Clostridial infection in the proximal colon of a Florida manatee. (A) The mucosa is thickened with multiple mucosal proliferative lesions. (B) Edema and hemorrhage are present within the mucosa. (C) In a Gram-stained imprint, bacterial spore-forming rods are in two forms: matchstick like *(white arrowheads)* and safety pin shaped *(black arrowheads)*. *(Photo C: Courtesy of N. Stacy, University of Florida.)*

Fungi

Dermal mycosis has been reported in small numbers of Florida, Amazonian, and captive Antillean manatees.[13,35,70,95] Associated lesions range from multifocal areas of erosion to progressive, multifocal, white, gray, reddish-brown, or black plaques. A case with a draining sinus in the axillary region was also thought to have a fungal cause.[100] These lesions have variously yielded *Epidermophyton floccosum*, *Bipolaris hawaiiensis*, and likely *Cephalosporium*, *Cercospora*, and *Mucor* spp. on fungal culture. Pyogranulomatous pneumonia with *Aspergillus fumigatus* was found in a Florida manatee at a site of pulmonary bulla. Fungi were confined to the lung with no systemic dissemination.

A dermal condition described as "**cobblestone-like**" skin is seen in manatees from fresh water habitats or suffering from cold stress (Fig. 24.14). This is believed to be associated with fungi that are yet to be more fully characterized. Morphologically the fungi are aseptate, 20×5 to $10\,\mu m$, have nonparallel walls, and demonstrate irregular branching. Lesions are histologically characterized by epidermal hyperplasia with cytoplasmic swelling of keratinocytes overlain and occasionally invaded by a dense layer of the fungus/water mold. Rhabditiform nematodes are sometimes noted in the stratum corneum in these cases; their significance (if any) in development of the lesions is unclear.

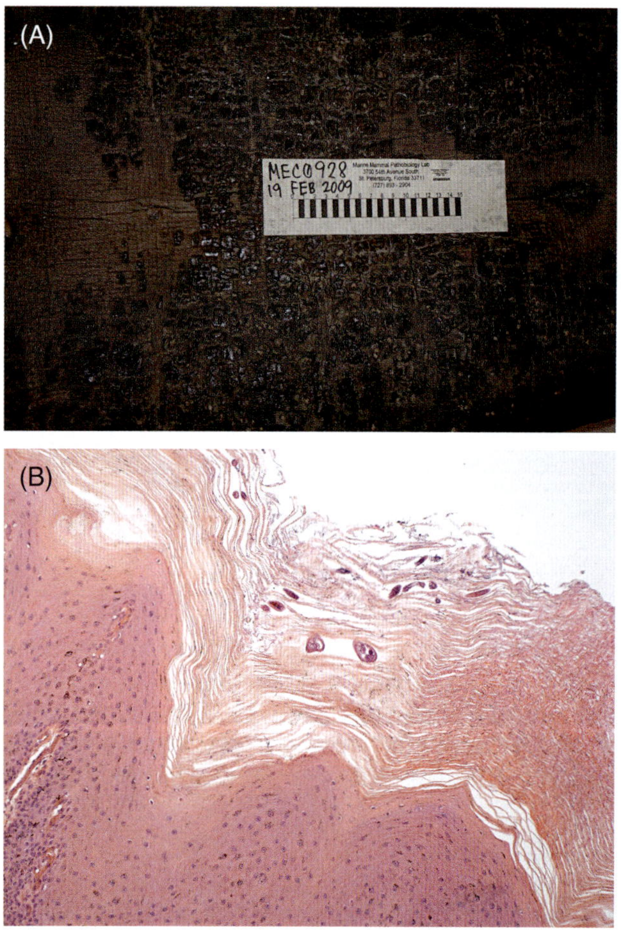

FIGURE 24.14 Cobblestone skin in a Florida manatee. (A) The skin is rugose with black discoloration. (B) Linear, C- or U-shaped fungal hyphae are present within increased layers of the stratum corneum. Nematodes are also present in the stratum corneum and do not elicit a host response.

Metazoa

Endoparasites are commonly seen in manatees and dugongs. The lesions they produce, if any, are usually of uncertain clinical significance. For these reasons, caution must be taken when attributing disease or death to the presence of parasites.

Enteric trematodiasis has been identified as a contributor to deaths in Florida manatee calves.[111] This disease is characterized by the gross presence of many small, white, pinpoint lesions within the intestinal mucosa and adult ***Moniligerum blairi*** or ***Nudacotyle undicola*** trematodes on mucosal scrapings when examined microscopically in direct association with significant gross and, where available, histological intestinal lesions. Overt gross lesions include intestinal wall thickening and increased mucosal rugosity, hemorrhage, and raised nodules with necrotic plaques or foci (Fig. 24.15A). Histological findings are characterized by adult trematodes embedded within intestinal mucosa with associated mixed inflammatory infiltrates, mucosal hyperplasia, erosion or ulceration, necrohemorrhagic and fibrinous debris, villus blunting, and crypt atrophy (Fig. 24.15B).

In dugongs, intestinal parasites have been associated with impaction (*Paradujardina halicoris*) and transmural abscessation with subsequent rupture (*Faredifex clavaa*). *Lankatrema mannarese* is associated with pyogranulomatous gastric nodules in dugong. Chronic rhinitis, tracheobronchitis, epiglottitis, and verminous pneumonia were observed in a Florida manatee and dugong associated with *Pulminicola cochleotrema*.

A brief summary of disease-associated sirenian parasites and their distribution is outlined in Table 24.1.

Protozoa

Disseminated **toxoplasmosis** has been rarely reported in manatees including two Florida manatees[98] and four Antillean manatees.[19] In the brain, there was a mild heterophilic and histiocytic inflammatory cell infiltrate with astrocytic hyperplasia, and approximately $60 \times 560\,\mu m$ protozoal cysts containing 2 to $3\,\mu m$ long bradyzoites. Affected parenchymal organs, including the liver, have

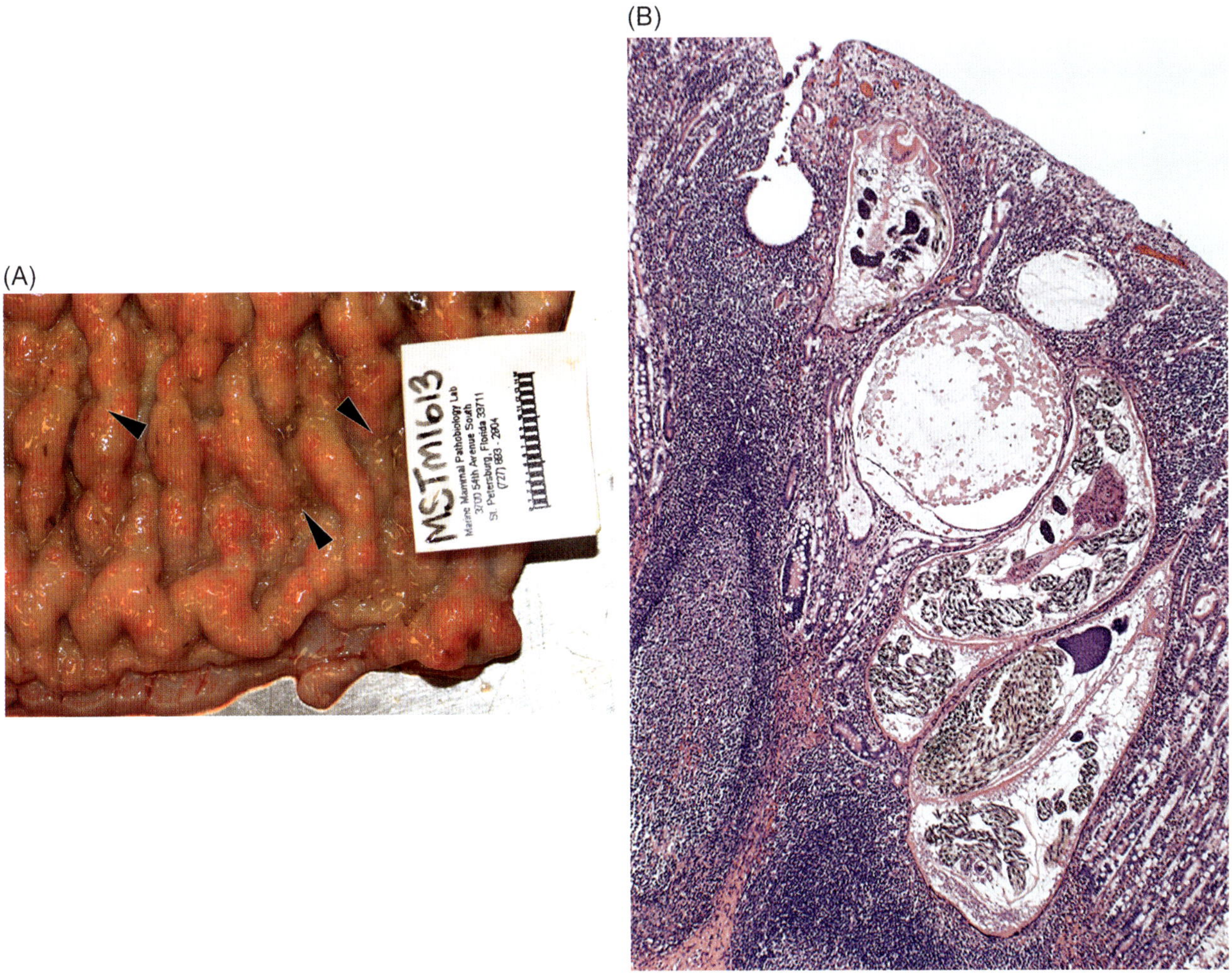

FIGURE 24.15 *Moniligerum blairi* enteritis in a Florida manatee. (A) The intestinal wall is rugose and hemorrhagic, with small 1 mm trematodes *(arrows)* in the luminal fluid and embedded in the wall. (B) Within the lamina propria of the small intestine are trematodes with aspinose tegument and oral sucker and uterus containing eggs.

TABLE 24.1 Endoparasites in Sirenia

Etiology	Location in Host	Associated Disease	Species Affected
Metazoan			
Labicola elongata[24]	Upper lip	Abscessation	Dugong
Lankatrema spp.[31]	Gastric and duodenal glands, pancreatic ducts	Gland dilation, pancreatic fibrosis and follicular lymphocytic inflammation	Dugong
Lankatrema mannarense[31]	Stomach	Caseous nodules	Dugong
Nudacotyle undicola[6,13]	Gastrointestinal tract	Hemorrhagic enteritis	Florida manatee
Moniligerum blairi[6,111] (Fig. 24.15)	Gastrointestinal tract	Hyperplastic, ulcerative, necrotic enteritis	Florida manatee
Faredifex clavata[40,114]	Gastrointestinal tract	Transmural abscesses, rupture and bacterial Peritonitis	Dugong
Paradujardinia halicoris[83]	Gastrointestinal tract	Impaction	Dugong

Continued

TABLE 24.1 Endoparasites in Sirenia—cont'd

Etiology	Location in Host	Associated Disease	Species Affected
Unidentified parasite spp.[114]	Bile ducts	Granulomatous inflammation and bile duct hyperplasia and fibrosis	Dugong
Pulmonicola cochleotrema[6,40,83]	Respiratory tract	Chronic rhinitis, ulcerative epiglottitis, tracheobronchitis, and verminous pneumonia	Florida manatee, dugong
Protozoan			
Cryptosporidium spp.[11,56,77]	Gastrointestinal tract	Heterophilic and lymphocytic enteritis	Antillean manatee, dugong
Toxoplasma gondii[19,21,83,98] (Fig. 24.16)	Disseminated	Necrotizing, multifocal to diffuse, chronic-active enteritis and myocarditis, pneumonia, lymphadenitis, hepatitis, meningoencephalitis	Antillian and Florida manatee, dugong

random to coalescing necrosis with fibrin exudation and heterophilic and histiocytic infiltrates (Fig. 24.16). Few to occasionally moderate numbers of 2 to 3 μm oval to crescent-shaped tachyzoites were present typically at the margin of the necrotic zones. Immunoreactivity is observed with commercially available reagents.[74] Seroprevalence studies of free-living or captive manatees in Florida, Puerto Rico, Belize, Peru, and Brazil ranged from 8% to 63%.[19,32,72,98,99]

Cryptosporidium species was found in the small intestine of an Antillean manatee[11] and dugong[77] associated with heterophilic and lymphocytic enteritis. *Cryptosporidium* oocysts were detected in the feces using the Kinyoun technique. Cryptosporidium immunofluorescence confirmed the diagnosis.

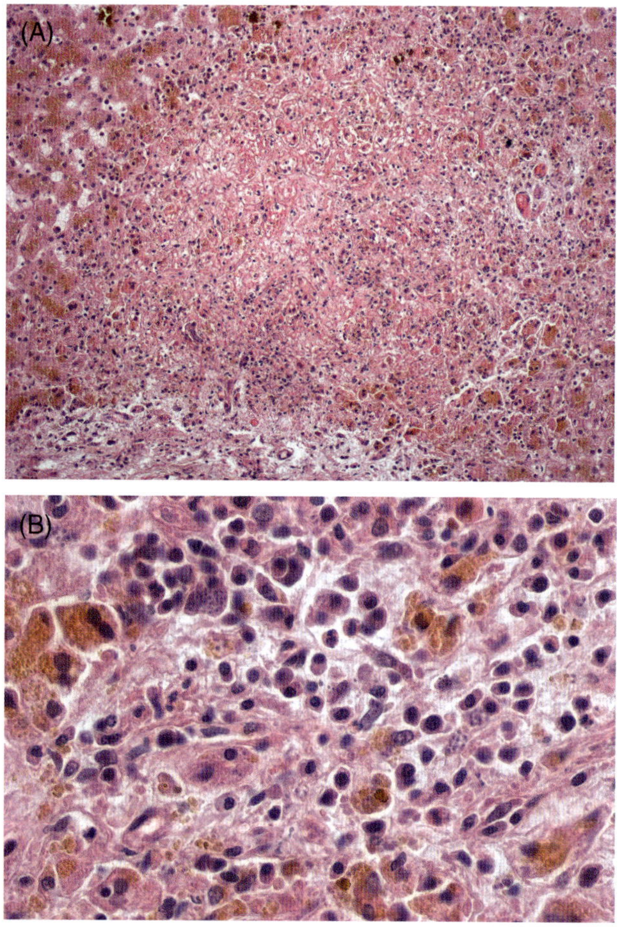

FIGURE 24.16 Toxoplasmosis in the liver of a dugong. (A) The liver contains multifocal, necrotizing hepatitis characterized by loss of hepatocellular detail and mild inflammation. (B) Necrotizing hepatitis with intra- and extracellular basophilic protozoal zoites. *(Photo Courtesy of H. Owen.)*

Ectoparasites

A large variety of **ectoparasites** can be found on sirenian skin; these are generally considered to be commensals.[70] Three species of nematodes have been identified on Florida manatees from southwest Florida and were not associated with skin lesions, suggesting that these are commensal organisms.[45] Species identified include *Cutidplogaster manati* and two unspecified species. The nematodes are sandwiched between layers of keratin with no inflammation or hyperplasia; nematodes do not extend into the epidermal layers. Mosquito species including Aedes, Anopheles, and Culex have been observed on the skin with penetration of the epidermis.[89] The role in transmission of vector-borne diseases is not known.

ADDITIONAL RESOURCES

Bossart GD, Meisner RA, Rommel SA, Ghim S, Bennett Jenson A. Pathological features of the Florida manatee cold stress syndrome. *Aquat Mamm*. 2002b;29:9–17.

Landsberg JH, Tabuchi M, Rostein DS, Subramaniam K, Rodrigues TCS, Waltzek TB, Stacy NI, Wilson PW, Kiryu Y, Uzal FA, de Wit M. 2022. Novel lethal clostridial infection in Florida Manatees (*Trichechus manatus latirostris*): Cause of the 2013 Unusual Mortality Event in the Indian River Lagoon. *Front Mar Sci*: 2022; 9:841857. https://doi.org/10.3389/fmars.2022.841857.

Lightsey JD, Rommel SA, Costidis AM, Pitchford TD. Methods used during gross necropsy to determine watercraft-related mortality in the Florida manatee (*Trichechus manatus latirostris*). *J Zoo Wildl Med*. 2006; 37: 262–275.

Owen H, Gillespie A, Wilkie I. Postmortem findings from dugong (*Dugong dugon*) submissions to the University of Queensland: 1997–2010. *J Wildl Dis*. 2012;48:962–970.

References

Full references for this chapter can be found online at https://doi.org/10.1016/B978-0-443-18437-6.00022-5.

Chapter 25

Chiroptera

Lisa L. Farina[a] and Julia S. Lankton[b]

[a]*College of Veterinary Medicine, University of Florida, Gainesville, FL, United States;* [b]*United States Geological Survey-National Wildlife Health Center, Madison, WI, United States*

INTRODUCTION

With more than 1,300 species identified, bats represent almost one-quarter of the world's mammals.[105] Bats provide important environmental services, such as insect control, seed dispersal, and pollination. They inhabit a wide variety of ecological niches on all continents except Antarctica. More than 200 species are listed as endangered or vulnerable to extinction,[160] primarily due to habitat degradation.

Despite their ubiquity and ecological importance, relatively little has been published on diseases of bats, whereas much has been written on bats' role as reservoirs in disease transmission. The decimation of certain bat populations in North America following the introduction of the fungus *Pseudogymnoascus destructans*, the cause of white-nose syndrome, has served to highlight both the importance and vulnerability of bats and has resulted in increased focus on disease threats to free-ranging bat populations. In addition to the threat of infectious disease, the proximity of bats to humans, including their use of human structures for hibernation and their role as a food source in many parts of the world, results in their susceptibility to anthropomorphic threats, such as trauma, toxicosis, and habitat loss. Ultimately, the interdependence of bats and humans means that an understanding of their health is critical for understanding and preserving human, domestic animal, and ecosystem health.

As many bats have multiple common names, the scientific names for species are used in this chapter (see Table 25.e1 in Digital Content for common and scientific names of species in this chapter).

NON-INFECTIOUS DISEASES

Nutritional

Iron overload (IO) has been reported in captive *Rousettus aegyptiacus*.[78,104,184] Overload is likely dietary, although the reason for the susceptibility of this species relative to other species on the same diet is unclear. Excess iron participates in the generation of free radicals, resulting in tissue damage.[304] The liver is the primary target organ, and excessive iron is observed in portal macrophages, Kupffer cells, and hepatocytes, as well as fibroblasts and bile duct epithelium. With chronicity, bats can develop portal to bridging fibrosis or cirrhosis (Figs. 25.1A, B; eSlides 25.e1 and 25.e2). Icterus and cavitary effusions can occur with liver failure.[78] Excessive hemosiderin is also identified in the spleen, and hemosiderin has also been identified in the heart, skeletal muscle, pancreas, kidneys, and various other tissues. Hepatocellular carcinoma and cholangiocarcinoma can be seen in association with IO in *R. aegyptiacus* with liver damage. Myocardial degeneration and/or fibrosis is also common in *R. aegyptiacus* with IO (occurs in about 45% of affected bats) but it is not correlated with the amount of iron present in the liver or heart.[184] Myocardial fibrosis is also observed in older fruit bats of other species, albeit at a lower rate than *R. aegyptiacus*. It is unknown whether myocardial fibrosis is related to IO or is simply common in this species, as a substantial number of *R. aegyptiacus* without IO have not been examined. Iron overload with hepatic fibrosis has also been observed in *R. lanosus* (unpublished data). While other fruit bats, including *Pteropus* spp.,[78] *Carollia perspicillata*, and *Cynopterus brachyotis*, have been observed with varying amounts of hepatic iron, associated damage has not been identified.

Hypovitaminosis C was suspected in an 8-week-old hand-reared frugivorous *Pteropus pumilus* and a captive colony of sanguinivorous *Desmodus rotundus*.[2,137] The *P. pumilus* presented with lethargy and joint enlargement that radiographically was associated with widening of the proximal physis and decreased bone opacity. In this case, the formula fed to the bat contained no vitamin C, and supplementation resolved clinical signs and radiographic abnormalities.[2] In *D. rotundus*, multiple deaths were seen in juvenile and adult bats. Lesions in adult bats included generalized subcutaneous hemorrhage and chronic nonresolving wing hematoma with associated necrohemorrhagic and suppurative dermatitis with ulceration. One adult bat

Pathology of Wildlife and Zoo Animals. https://doi.org/10.1016/B978-0-443-18437-6.00034-1

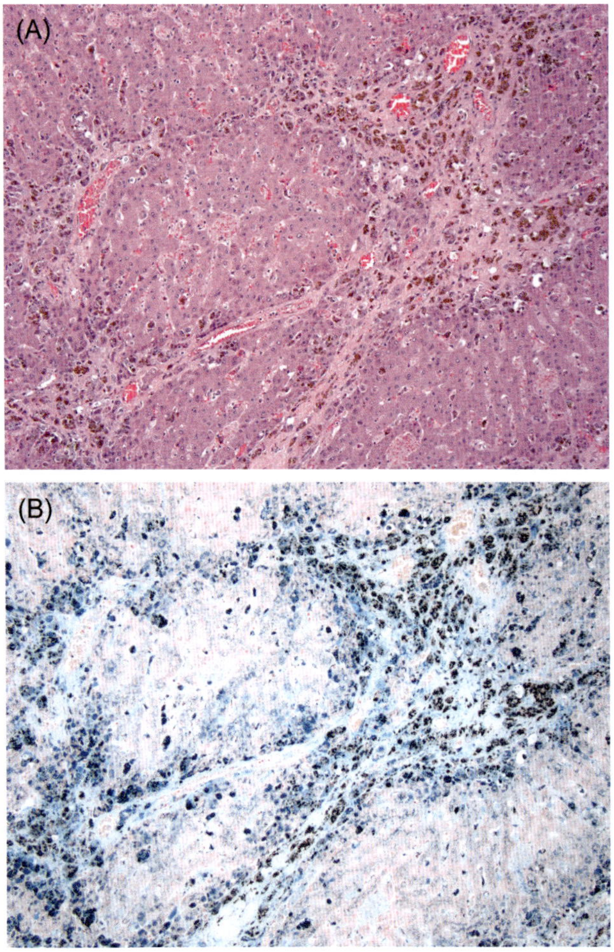

FIGURE 25.1 Iron overload in the liver of a *Rousettus aegyptiacus*. (A) Bridging fibrosis with iron overload. Abundant brown pigment is present within and at the margins of the connective tissue (see eSlide 25.e1). (B) Abundant hemosiderin is present in macrophages at the margins of and within the fibrous connective tissue, with lesser amounts in hepatocytes, bile ducts, and Kupffer cells. Prussian blue (see eSlide 25.e2).

euthanized for severe subcutaneous hemorrhage had a low serum vitamin C concentration. Gross lesions identified in juvenile bats at necropsy included single or multiple wing and/or leg bone fractures, fractures of the rib costochondral junctions, nonhealing fractures, and joint swelling and laxity. Histopathologic bone lesions in these cases included fractures, failure of endochondral ossification, osteopenia, microfractures, periosteal fibrosis, sequestra, and myelofibrosis. Multiple adult bats in the colony developed subcutaneous wing hematomas and gingival bleeding, which improved with supplementation with vitamin C. Hypovitaminosis was suspected but not confirmed in three young adult bats that had necrosis of the cerebellar granule layer, with two having mild Purkinje cell necrosis, and two with mild to moderate lymphocytic or lymphohistiocytic meningoencephalitis.[137] Although vitamin C is described to cause dysplastic changes in the cerebellum of developing fetuses,[52] and it was postulated that dams of these animals could have been deficient in vitamin C, these lesions were described as acute but affected bats were 2–5 years-old.[137] Therefore, the cerebellar lesions described in this study may have been unrelated to hypovitaminosis C in the colony.

Vitamin B_{12} deficiency has been induced in captive *Rousettus aegyptiacus* by feeding a diet consisting of washed, peeled bananas, oranges, pears, and papaya. After 200 days on this diet, bats became ataxic with hind limb proprioceptive deficits. Histologically, the spinal cord of the caudal cervical and cranial thoracic regions had patchy spongiosis in the lateral and ventrolateral white matter, consistent with early demyelination.[122] Lipid and fatty acid concentrations in the myelin of B_{12}-deficient bats were subsequently determined to be different than in control bats; bats in this study were fed washed, peeled banana and papaya with a supplement containing vitamins A, C, and D and four B complex vitamins (B_6, thiamin, riboflavin, niacin).[315] However, it should be noted that the experimental diets were likely deficient in multiple nutrients, including vitamin E, so reported nervous system lesions could have been the result of deficiencies other than vitamin B_{12} (Ellen Dierenfeld, Ellen S. Dierenfeld, LLC, verbal communication, 2016).

Additional nutritional conditions are included in the Digital Content.

Metabolic

Varying degrees of **hepatocellular lipid deposition**, suspected to result from negative energy balance, have been observed in sick or injured free-ranging bats in rehabilitation settings (Karrie Rose, Taronga Conservation Society Australia, written communication, 2016) and in captive bats.

Hepatic lipidosis, along with lipid deposition in the renal tubules and cardiac myocytes, has been documented in obese wild-caught, captive-held *Rhinolophus ferrumequinum* fed mealworms in captivity; the high dietary fat content may have been contributory.[121] Hepatic, renal tubular, and cardiomyocyte lipidosis with patagial ecchymoses, perirenal hemorrhage, and icterus has also been reported in a captive colony of *Eptesicus fuscus* secondary to decreased food intake associated with environmental disturbance, illness, and conspecific trauma.[293] In this group, it was unknown whether affected bats were overconditioned prior to the period of anorexia.

Toxic

Liver, kidney, and gastric contents are the most frequently utilized samples for toxicologic analysis. Contacting the laboratory that will perform the testing is recommended to get sample submission requirements and recommended tissues for testing for suspected toxins.

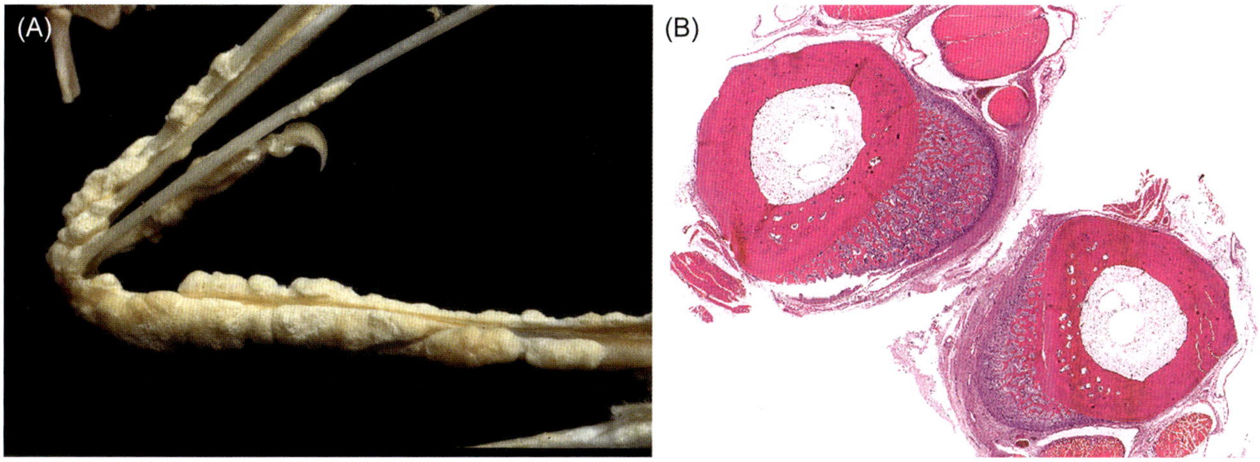

FIGURE 25.2 Fluorosis associated in the wing bones of a *Rousettus aegyptiacus* (A) and humerus of a *Pteropus giganteus* (B). (A) Gross lesions consist of nodular, bony proliferation along the surfaces of the bones. (B) This corresponds histologically to subperiosteal trabeculae of woven bone that is demarcated from the underlying cortex. *(Part A: Photo Courtesy of D. Smith, Toronto Zoo, Part B: Photo Courtesy of M. Duncan, St. Louis Zoo.)*

Fluorosis with hyperostosis in captive *Rousettus aegyptiacus*, *Pteropus giganteus*, and *P. poliocephalus* is associated with high dietary and elevated bone fluoride concentrations. The source for excess dietary fluoride was fluoride-rich shrimp meal and dicalcium phosphate. Multifocal to coalescing asymmetric bony proliferations, which are periarticular but do not involve the joints (Fig. 25.2A), are most frequently identified on the wings. However, lesions of the hind limbs, sternum, ribs, and mandible are sometimes seen. Histologically, subperiosteal nodules of new woven bone are clearly demarcated from the underlying cortex (Fig. 25.2B), with some resorption in the underlying cortical bone.[100] Bony proliferation is common in captive bats as a result of trauma, particularly on the leading edge of the wing, so lesions must be differentiated via histopathology and clinical correlations.

Pesticide residues, particularly **organochlorines**, such as **dichlorodiphenyldichloroethylene (DDE)**, **dichlorodiphenyltrichloroethane (DDT)**, **lindane**, and **polychlorinated biphenyls**, have been detected in many species of free-ranging bats in North America, Britain, Europe, Australia, and Africa, with occasional documentation of population-level lethal and sublethal impacts.[44,66,67,239,241,338] Insectivorous bats are thought to be particularly susceptible to organochlorine exposure due to their higher trophic level in food webs, high metabolic rate, and long life span, as well as the periodic mobilization of adipose tissue during migration and hibernation, at which time lipophilic compounds stored in fat are released.[118] Due to their propensity to roost in manmade structures, bats are often exposed to antifungal and insecticide compounds in treated lumber. Bats of all life stages are at risk for intoxication, as some organochlorine pesticides have been shown to cross through the placenta and pass to young in milk.[67]

Several species have been experimentally exposed to organochlorines at lethal or sublethal levels.[67,68] Pesticide-associated mortality has been widely documented; however, the effects of sublethal exposure are largely unknown. Some evidence indicates that sublethal exposure to organochlorines may increase metabolic rate, potentially interfering with adequate fat deposition for hibernation.[68,301] In an experimental study, fruit-eating *Artibeus lituratus* exposed to the organochlorine insecticide endosulfan developed hepatocellular degeneration, necrosis, steatosis, and fibrosis attributed to oxidative damage.[244]

Organophosphate toxicity resulting in brain cholinesterase inhibition has been reported in insectivorous bats in Europe and North America.[240] In one report, *Artibeus lituratus* exposed to fenthion developed hepatocellular cytomegaly and vacuolation, as well as increased muscle glycogen content.[7] Incoordination has been reported in *Eptesicus fuscus* and *Myotis lucifugus* exposed to acephate and methyl parathion.[67]

Additional toxicoses reported in bats are listed in Table 25.1 and in the text of the Digital Content.

Congenital/Genetic

Rare **congenital abnormalities** have been identified in captive and free-ranging bats, including albinism,[183] atresia ani, portosystemic shunts,[245] unilateral renal aplasia, and congenital cataracts.

In multiple years, several colonies of *Pteropus conspicillatus* in Australia had multiple births with congenital defects, including cleft palate and/or other craniofacial abnormalities and deformities of the toes and claws; a toxin or infection during gestation was suspected.[245]

TABLE 25.1 Additional Toxicoses Reported in Chiroptera

Toxin	Species Affected	Clinical Signs	Lesions
Ivermectin[92]	Captive *Cynopterus brachyotis*	Fell from perches, minimal movement in all parts of body	Renal tubular necrosis with regeneration (uncertain if directly related to ivermectin)
Rodenticides			
Diphacinone[93]	Free-ranging *Mystacina tuberculata*	Lethargy, death	Multisystemic hemorrhage
Zinc phosphide[156]	Free-ranging *Myotis lucifugus*	Tremors, death	Pathology not described
Heavy metals			
Lead[134,290,299,340]	Captive and free-ranging *Pteropus* spp.	Incoordination, inability to fly, aggression, muscle tremors, ptyalism, ataxia, separation from group, target of conspecific aggression	Renal tubular necrosis, proximal tubular karyomegaly, and nuclear pleomorphism; variably acid-fast eosinophilic intranuclear inclusion bodies in the proximal renal tubular epithelium and occasionally in hepatocytes
Copper[145]	Captive *Artibeus jamaicensis*	Death without prior clinical signs	Random acute hepatocellular necrosis with abundant copper pigment in hepatocytes and macrophages, hemoglobin casts in renal proximal tubules with tubular degeneration

Age-Related/Degenerative

Myocardial fibrosis is not uncommon in captive frugivorous bats. Affected bats are generally older but lesions can be present in young adult bats. A particular pattern or distribution for lesions has not been observed. Overall cardiomegaly and chamber dilation (most often the right or both ventricles) may be present; bats rarely have left atrial mural thrombi. Affected bats sometimes have pathologic evidence of heart failure, including pulmonary edema and intraalveolar hemosiderin-laden macrophages (heart failure cells). Bats with heart failure may have dependent, subcutaneous edema of the head and neck; similar cranial edema may also be observed in bats with protein-losing nephropathy and sepsis.

Several causes for myocardial fibrosis have been investigated including hypovitaminosis E in *Pteropus* spp.[138] and iron overload in *Rousettus aegyptiacus*[184]; however, not all bats with myocardial fibrosis have these conditions. Further studies into etiologies of myocardial disease in bats would be beneficial.

Unilateral or bilateral **testicular degeneration**, atrophy, and/or fibrosis have been observed in older captive pteropodid bats.

Degenerative joint disease is relatively common in older captive pteropodid bats; it also occurs in other captive bat species. Hip, stifle, interphalangeal, and metacarpophalangeal joints are most frequently affected. Increased incidence of falling from perches can be observed in affected animals, and injuries from falls may be the cause of death or may prompt euthanasia.

Vertebral spondylosis is sometimes seen in older captive pteropodid bats, and can affect the cervical, thoracic, and/or lumbar regions. These lesions are often associated with intervertebral disc degeneration, which can also be seen in the absence of spondylosis. Associated lesions include collapse of the vertebral disc space, endplate remodeling or sclerosis, and degenerative joint disease of the articular facets. Hansen type II disc herniations can be observed, associated with bony proliferations in some cases, but these are typically mild with evidence of cord compression only in rare cases. Animals with spondylosis and/or intervertebral disc disease may have histories of falling, being found down on the ground, and/or having mobility issues. As affected animals are typically older, concurrent limb degenerative joint disease is common, and may primarily cause or contribute to clinical signs.

Cataracts are relatively common in older frugivorous bats. While these are often incidental lesions at necropsy (unpublished data), lens luxation, lens-induced uveitis, and secondary glaucoma may be associated with cataracts.[314]

Inflammatory Non-Infectious

Allergic dermatitis has been observed in captive *Pteropus* spp., predominantly *P. vampyrus*, and cases have been reported that improved with allergen-specific immunotherapy. Cutaneous ulceration is common, and most often seen on the patagia, over the carpi or digits, or on the pinnae. Alopecia of the head, neck, shoulders, and/or ventral abdomen is also present in some cases. Histopathology typically shows mixed (lymphoplasmacytic and histiocytic, with variable eosinophils and neutrophils) perivascular to interstitial dermatitis with acanthosis, spongiosis, and hyperkeratosis.[61,120] Dermal mineral aggregates with associated fibrosis, granulation tissue, and granulomatous to mixed inflammation, are seen in some cases, and may appear grossly as patagial nodules.[61] Some bats also have blepharoconjunctivitis with chemosis in addition to skin lesions.[61,120] Conjunctival histopathology in one case identified submucosal, often perivascular, eosinophils with fewer lymphocytes and plasma cells, with mild mucosal hyperplasia with intraepithelial leukocytes.[120]

Ulcerative or alopecic lesions with eosinophilic and mixed dermatitis with necrosis, collagen degeneration, and/or panniculitis are occasionally seen, and are believed to represent insect bite reactions. Eosinophilic dermatitis in the lip of a captive *Pteropus poliocephalus* was associated with chronic ulceration and extensive collagen degeneration. The lesions were similar to collagenolytic dermatitis in cats (Karrie Rose, Taronga Conservation Society Australia, written communication, 2016).

Miscellaneous/Multifactorial

Trauma is a common cause of death in **free-ranging bats**.[147,221,222,265] **Wind turbine interactions** account for the highest number of traumatic mortality events in bats worldwide.[241] Migratory species that roost in trees are disproportionately affected, although high mortality can occur in nonmigratory species.[80,125] **Barotrauma** was initially proposed as the cause of death in many bats due to reports of intrathoracic and intraabdominal hemorrhage with a relative lack of external injuries. Other findings supporting barotrauma included pulmonary lesions such as alveolar bullae, hemorrhage, congestion, edema, and atelectasis.[14] Subsequent studies have concluded both that the magnitude of pressure changes associated with wind turbines are likely insufficient to result in mortality[182] and that injuries are more compatible with blunt trauma (Fig. 25.3). Commonly cited injuries include wing fractures, external lacerations, inguinal and diaphragmatic hernias, subcutaneous hemorrhage, and bone marrow emboli. Rupture of the tympanic membrane in the absence of associated mechanical injuries has been reported in a small subset (6%) of affected bats.[125,272] It is possible that both blunt force trauma and decompression play a role in wind turbine-related bat mortality, although the majority of injuries are likely attributable to blunt trauma (see Chapter 3: Wildlife Forensic Pathology).

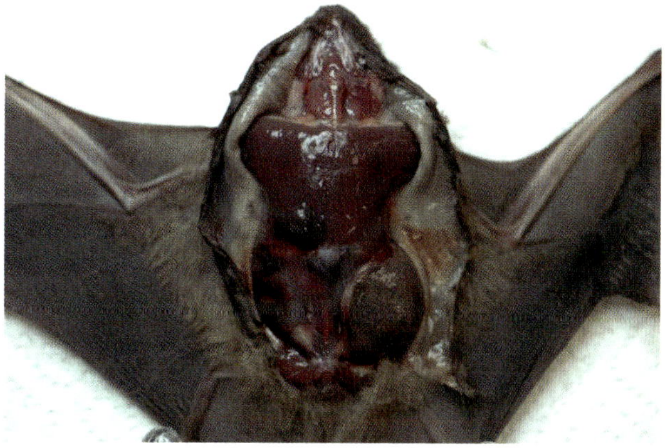

FIGURE 25.3 Wind turbine trauma in a *Nyctalus noctula*. Hemorrhage and abdominal hernia with prolapse of the stomach through the left abdominal wall were presumed to result from blunt-force wind turbine-related trauma. *(Photo Courtesy of G. Wibbelt, Leibniz Institute for Zoo and Wildlife Research (IZW), Berlin, Germany.)*

Cat predation is frequently documented, particularly in species roosting in buildings. Sequelae common to cat predation include wing lacerations and fractures, soft tissue injury including subcutaneous and intramuscular hematomas, and *Pasteurella multocida* infection.[221,222]

Bite wounds from conspecifics are also common in captive bats and may result in cellulitis or abscesses.

Free-ranging pteropodid bats are highly susceptible to **entanglement injuries** sustained from netting improperly installed around fruit trees (Karrie Rose, Taronga Conservation Society Australia, written communication, 2016).

Other reported causes of trauma in free-ranging bats include vehicular strike, collision with aircraft, building collision, and trauma secondary to entrapment in confined spaces.[221,241] Evidence of trauma has also been documented in live, apparently healthy, bats to varying degrees. In one survey of 3,525 live free-ranging bats in Poland comprising six species, 2.9% of bats exhibited evidence of damage to the wing or tail, although variation was wide in the prevalence of damage by species, which ranged from 0.3% to 21.74%; differences were attributed to behavioral and ecological factors. Loss of the edge of a wing membrane was the most common type of damage noted (29.3%).[64]

Electrocution from power lines or electric fences has been observed in free-ranging pteropodid bats in Australia (Karrie Rose, Taronga Conservation Society Australia, written communication, 2016) and India.[60] Acutely, there may be no lesions, or only charring of the fur or skin or a burnt smell may be identified. In some cases, a fine linear pattern of necrosis appears to represent electric current entry and exit sites (Fig. 25.4). Approximately 2–4 days after electrocution, the skin distal to the electrocution site becomes moist and friable and cutaneous defects appear,

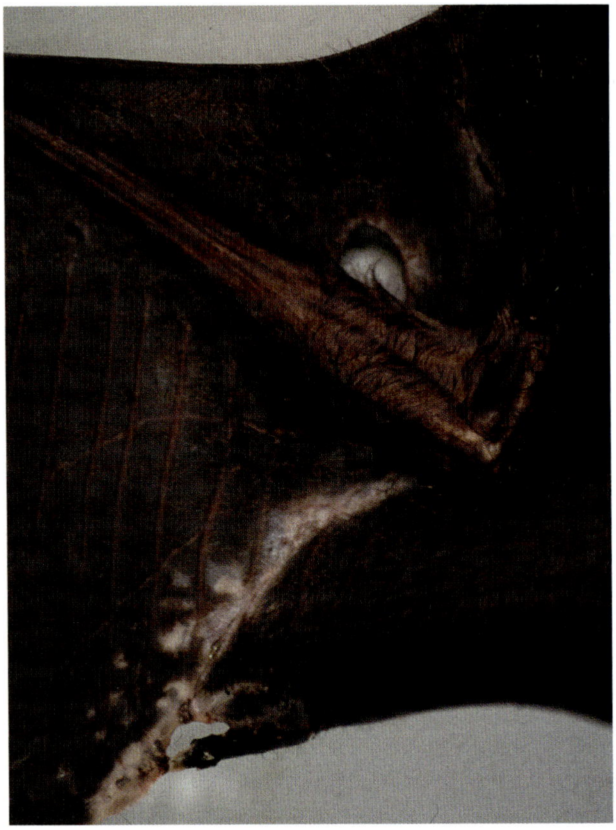

FIGURE 25.4 Electrocution injury of the patagium in a *Pteropus poliocephalus*. The linear pattern of patagial necrosis and full-thickness defect is consistent with an electrocution injury. *(Photo Courtesy of K. Rose, Australian Registry of Wildlife Health, Taronga Conservation Society, Australia.)*

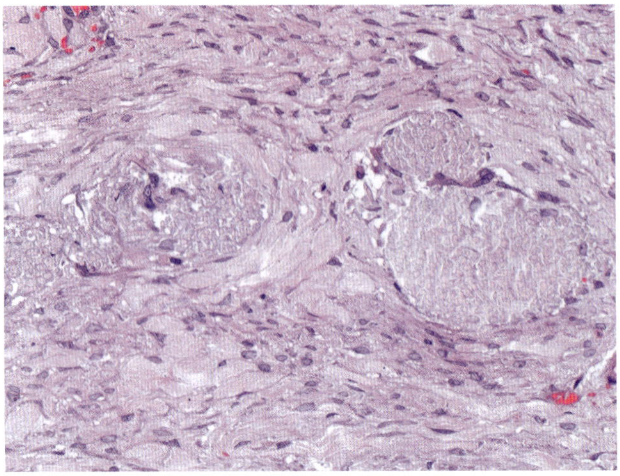

FIGURE 25.5 Extensive granulation tissue and fibroplasia associated with a chronic wound in the patagium of a *Pteropus vampyrus*. An irregular mass of fibroblasts expands the patagium and infiltrates into and around elastin bundles. Anisocytosis, anisokaryosis, and multinucleation are observed in some of the fibroblasts. While this lesion was initially diagnosed as a fibrosarcoma, the lesion subsequently regressed, indicating that this was reactive fibroplasia (see eSlide 25.e3).

most commonly on the wings. Histologically, these lesions consist of fibrinoid degeneration of blood vessels with epidermal, dermal, and adnexal coagulative necrosis. Secondary bacterial and fungal infection of these wounds is common. Edema of the meninges and brain is also observed in some acutely affected bats (Karrie Rose, Taronga Conservation Society Australia, written communication, 2016). Bats that recover from burns may have depigmentation, scarring, and alopecia of affected areas. Pups on dams that are electrocuted may survive due to the insulation provided by fur, but they may have oral burns sustained via their attachment to the teat.[245]

Trauma is also common **in captive bats**. Trauma to the wings, typically a result of wire enclosures, is described, and wounds often have secondary, opportunistic colonization with mixed bacteria, and sometimes fungi. Ulceration and hemorrhage or fatal hemorrhage can be associated with traumatic removal of metal identification bands. As soft tissue covering the wing bones is minimal, fractures, most often of digits, can occur, and bony proliferation and/or osteomyelitis may be sequelae to wing trauma.

Pteropodid bats can have **extensive granulation tissue (reactive fibroplasia) associated with chronic wounds**. Fibroblasts can be large with some cellular atypia, and the classic vascular arrangement of granulation tissue may not always be apparent, especially in the patagium. Lesions may sometimes be relatively discrete and nodular, and infiltration of normal structures by fibroblasts is occasionally observed (Fig. 25.5; eSlide 25.e3). Some lesions have been diagnosed as sarcomas and have subsequently regressed, so circumspection is needed in diagnosing cutaneous spindle cell proliferations in bats.

Other forms of trauma, for example, spinal fractures and luxations, long bone fractures, subdural hemorrhage, liver lacerations, hemoabdomen, and hemorrhage in the subcutis and skeletal muscle, have been observed sequelae from presumed falls in captive pteropodid bats.

In older captive insectivorous and carnivorous bats, **dental disease** is frequently a significant problem. Tooth loss, plaque, tartar, gingivitis, periodontitis, and osteomyelitis can be seen.[245] Similar lesions are observed sporadically in captive frugivorous bats.

Severe chronic periodontal disease of unknown cause characterized by inflammation of the gums, loss of the majority of teeth, and loss of bone around remaining teeth was reported in a free-ranging *Tadarida brasiliensis* in Texas, United States. Despite only eight teeth remaining, the bat was in adequate body condition.[310]

Traumatic damage has been associated with secondary gingivitis, pulpitis, and tooth root abscesses in captive pteropodid bats. Substantial dental attrition is common in older frugivorous bats but does not appear to cause difficulty with feeding.

Enamel defects and staining of the teeth are common in pteropodid bats.[245]

Alopecia of the chest, abdomen, and occasionally the back has been reported in 5%–5.6% of free-ranging *Artibeus jamaicensis* and *A. lituratus* in Mexico. Lesions are more prevalent in the dry season and in urban areas, and more females than males may be affected.[26] Alopecia is also seen in captive *A. jamaicensis* at multiple institutions, with males most affected in some groups, and females primarily affected in others, as well as in male and female captive *Carollia perspicillata*. In the latter, symmetrical hair loss typically starts on the dorsum or ventrum, extends to the limbs, and sometimes affects the entire body (Fig. 25.6A). Histologic lesions seen in *Artibeus* include loss of some hair follicles with rare catagen follicles (Fig. 25.6B); those in *Carollia* include dilation of follicular infundibula with attenuation of the infundibular epithelium, tricholemmal keratinization, and follicular atrophy/loss with retention of sebaceous glands. Nutritional or endocrine etiologies are considered possible differentials in these species but a specific cause remains to be identified[26] (Karen Terio, University of Illinois Zoological Pathology Program, written communication, 2016; Michael Garner, Northwest ZooPath, written communication, 2016; Denise McAloose, Wildlife Conservation Society, written communication, 2022). Alopecia has also been reported in captive bats due to inadequate dietary protein or fat-soluble vitamins, or feeding infant formula to bats past their normal weaning age.[245] Increased prevalence of alopecia, as well as abnormal tooth wear and increased prevalence of respiratory lesions, were reported in a free-ranging population of *Artibeus jamaicensis* on an island in the West Indies following volcanic activity. Alopecia was attributed to stress, high ectoparasite loads, or mineral deficiencies associated with ash ingestion.[248]

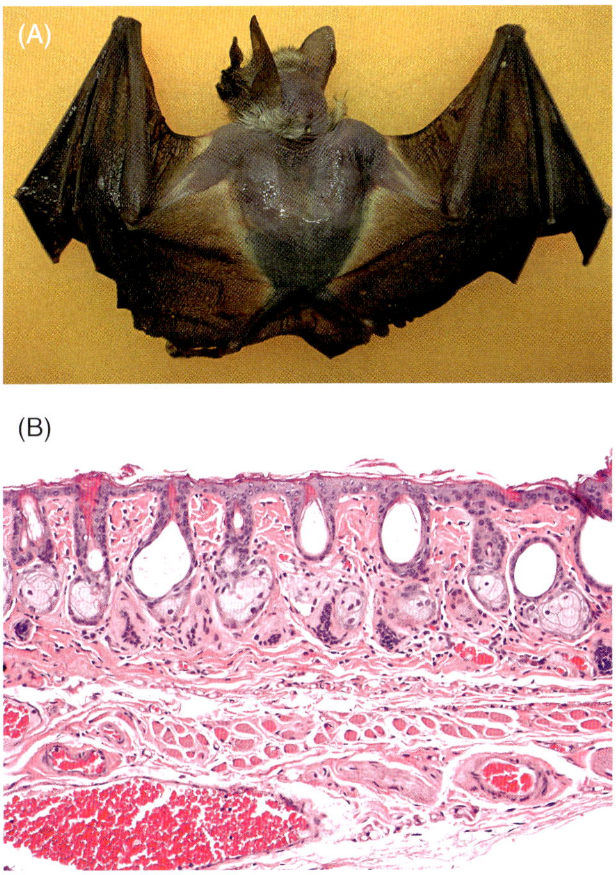

FIGURE 25.6 Alopecia in *Carollia perspicillata* (A) and *Artibeus jamaicensis* (B). (A) Alopecia is diffuse across the entire dorsum. (B) Histologic lesions include dilation of follicular infundibula, thinning of the infundibular epithelium, follicular atrophy and loss, and tricholemmal keratinization. Sebaceous glands are retained. *(Part A: Photo Courtesy of K. Conley, Wildlife Conservation Society; Part B: Photo Courtesy of K. Terio, University of Illinois Zoological Pathology Program.)*

Either as a primary process or as a secondary lesion in healing wounds, **depigmentation** can be observed in bat skin. Gross lesions of primary depigmentation (vitiligo) have been described in captive bats in a group of *Pteropus vampyrus* and *P. hypomelanus*,[297] a group of *Rousettus lanosus* (Michael Garner, Northwest ZooPath, written communication, 2016), and a single captive *Pteropus hypomelanus* (unpublished data). Lesions are typically patchy on the wings or face and limbs. Histologically, the majority of epidermal melanin is lost from affected areas with no inflammation (Michael Garner, Northwest ZooPath, written communication, 2016; unpublished data).[297] Hypovitaminosis D due to lack of sunlight exposure was proposed in the group of captive *Pteropus*,[297] but the etiology of these lesions has not been definitively determined.

Proliferative endometrial lesions have been observed in *Carollia perspicillata* in multiple zoo collections. Gross lesions include abdominal distention, marked uterine enlargement, and serosanguinous or pseudomembranous material in the uterine lumen. Histologically, there is multifocal to diffuse proliferation of glandular epithelium and stromal cells, with occasional mucosal epithelial erosion.[230]

Pteropus spp. occasionally have **choleliths** in the gall bladder or bile duct. In some cases, they are incidental findings. However, they may be associated with cholangitis/cholangiohepatitis and intrahepatic cholestasis.

Frostbite has been observed in captive pteropodid bats. Vasoconstriction with decreased blood flow typically causes necrosis of the distal extremities, in particular, the nail beds and wing tips.[107] Frostbite has been proposed as the cause of the so-called "square-eared anomaly" reported in free-ranging North American species of *Myotis*. Gross lesions are limited to the pinna, the height of which may be reduced by up to 50%. Microscopically, the edges of affected pinnae have a thinned epidermis, dermal granulation tissue and fibrosis, and disrupted, hypertrophic, and dystrophic cartilage consistent with frostbite.[170]

Multiple large-scale mortality events, with increasing numbers in recent years, have been associated with **high ambient temperatures** (above 42°C/107.6°F) in free-ranging pteropodid bat species in Australia and India,[241,326] and insectivorous bats (*Chaerephon plicatus* and *Taphozous theobaldi*) in Cambodia. Mild, multicentric hemorrhage in one of two necropsied Cambodian bats was the only lesion identified,[260] which is consistent with typical lesions of heat stress in other species.

Iatrogenic thermal burns, most commonly in the patagium, can occur in captive bats due to contact with heat lamps, hot packs, or heating pads/blankets.[139]

Rectal prolapses can occur secondary to constipation (usually due to feeding fruit bats a blended diet), diarrhea, colitis, and/or intestinal parasites.[245]

Bronchiolar metaplasia of the alveolar epithelium (Fig. 25.7) is occasionally observed in *R. aegyptiacus*. These are typically multiple lesions that are >1 mm in diameter and are adjacent to bronchioles. They can have an alveolar or papillary pattern, and may contain ciliated epithelial cells and increased amounts of fibrous stroma. Inflammation and/or fibrosis is present in some cases and indicates prior tissue damage. Bronchioloalveolar adenomas can also be seen in this species, sometimes concurrently with bronchiolar metaplasia, but these masses are typically discrete, expansile nodules (see "Neoplastic" section).

In older *Pteropus hypomelanus*, intranuclear inclusions are occasionally observed in meningothelial cells in the brain, sometimes in foci of meningothelial hyperplasia. Affected cells have peripheralization of the chromatin, with the central portion of the nucleus filled with a slightly irregular, eosinophilic, homogenous to slightly heterogeneous inclusion (Fig. 25.8) that does not stain with a Periodic acid-Schiff stain. Mild karyomegaly is sometimes observed associated with inclusions (unpublished data). These inclusions have a similar appearance to those reported in meningiomas in humans and a dog, which are pseudoinclusions as a result of cytoplasmic invagination into the nucleus.[135,158]

Additional miscellaneous conditions are included in the Digital Content.

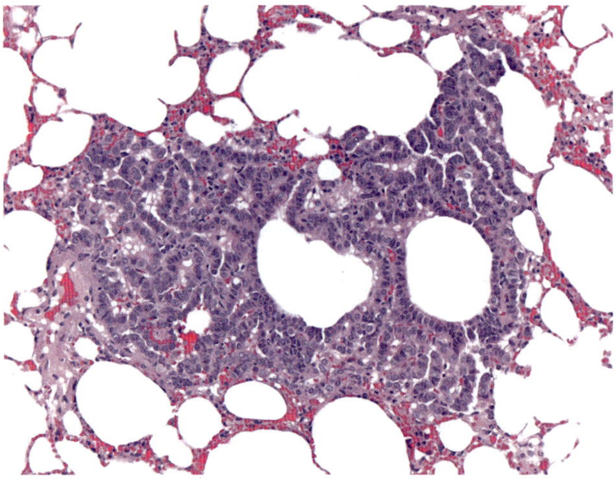

FIGURE 25.7 Bronchiolar metaplasia of the alveolar epithelium in the lung of a *Rousettus aegyptiacus*. Surrounding a small bronchiole, columnar epithelial cells line the alveoli and form papillary projections.

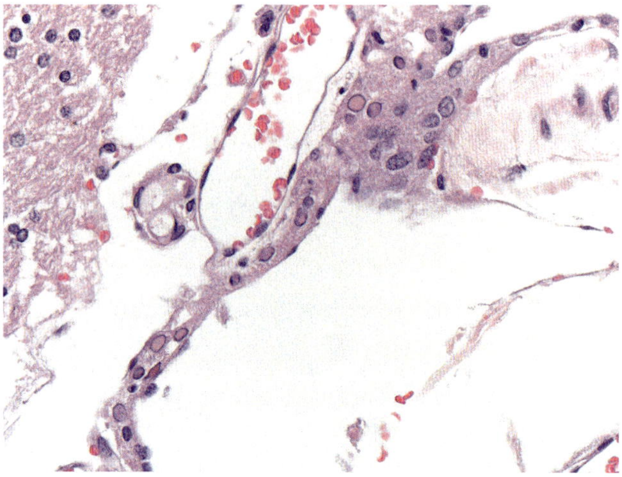

FIGURE 25.8 Meningothelial hyperplasia in the leptomeninges of the brain of a *Pteropus hypomelanus*. Eosinophilic intranuclear inclusions with peripheralization of the chromatin are occasionally seen in foci of meningothelial hyperplasia in geriatric *Pteropus hypomelanus*. These are presumed to be pseudoinclusions as a result of cytoplasmic invagination into the nucleus.

Neoplastic

A wide variety of neoplasms, including various soft tissue sarcomas, lymphoma, oral and nasal squamous cell carcinoma, cutaneous melanocytoma, osteosarcoma, and mammary adenomas and carcinomas, have been described in captive bats. Most are single case reports, with *Rousettus aegyptiacus* the most common species in these reports. The most commonly reported neoplasms in captive *R. aegyptiacus* are **hepatocellular carcinoma** and **cholangiocarcinoma**, thought to be associated with liver damage from iron overload (see Iron Overload above).[184]

Novel papillomavirus in a 5-year-old captive *Rousettus aegyptiacus* was associated with the development of multiple **cutaneous papillomas** and **basosquamous/squamous cell carcinomas**. Positive intranuclear immunostaining using antibodies against bovine papillomavirus identified viral antigen within the neoplasms.[206]

A subcutaneous interscapular **leiomyosarcoma** associated with a microchip was identified in a captive *R. aegyptiacus*. Neoplastic cells were elongate and arranged in bundles and streams, with moderate anisocytosis and marked anisokaryosis. Metastases were identified in the liver and peritoneum.[288]

Adrenal cortical lipomas and/or osseous metaplasia have been observed occasionally in captive *Pteropus* spp. Nodules of fat are relatively discrete but may be slightly infiltrative or contain entrapped clusters of cortical cells. Some masses have peripheral foci of osseous metaplasia, and osseous metaplasia with few or no adipocytes can also be seen.

Cutaneous **spindle-cell sarcomas** have been seen in captive bats; however, it is important to note that pteropodid bats can have nodular fibroplasia with some cellular atypia as a sequela of trauma. Some lesions that have been diagnosed as sarcomas in the skin have spontaneously regressed (see "Miscellaneous/Multifactorial" section).

Single or multiple bronchioloalveolar adenomas have been observed in *Rousettus aegyptiacus* but warrant differentiation from bronchiolar metaplasia (see "Miscellaneous/Multifactorial" section). Bronchioloalveolar adenomas are expansile nodules, while bronchiolar metaplasia does not significantly distort the architecture.

Of two known published reports of neoplasia in free-ranging bats, both were cutaneous tumors of smooth muscle origin. One case was a cutaneous **leiomyosarcoma** in a *Corynorhinus townsendii virginianus* in the United States.[74] The other was a **spindle cell tumor** in a hibernating wild female *Plecotus auritus* in England. The tumor infiltrated the dermis, epidermis, and underlying skeletal muscle and was also suspected to be of smooth muscle origin.[20]

INFECTIOUS DISEASES

Bats are host to many infectious organisms. Viral diseases of humans or other animal taxa for which bats are a proven or likely reservoir are summarized in Table 25.2. Many of these viruses are listed diseases (international diseases that are zoonotic and cause severe human disease and/or have significant impact on the health of wildlife and/or domestic animals) by the World Organization for Animal Health (WOAH; formerly OIE) and/or on the Blueprint list of priority diseases of the World Health Organization (WHO). Many molecular, serologic, and other surveys of infectious organisms in various bat species have been performed due to recognition of the role bats play as reservoirs for zoonotic pathogens.[41] These surveys have identified numerous viruses, including adenoviruses, astroviruses, bunyaviruses, circoviruses, coronaviruses, hepadnaviruses, herpesviruses, papillomaviruses, paramyxoviruses, parvoviruses, picornaviruses, polyomaviruses, retroviruses, and rotaviruses in bats. However, most of these investigations have not demonstrated any associated disease in host bats.[55,99,335]

The ability of bats to harbor a variety of agents, particularly viruses that are highly pathogenic to other species yet result in no apparent disease in the bat host, has resulted in many studies focusing on bat virus-host interactions and bat immunology.[163,236,246,305] The discussion that follows focuses on notable infectious organisms reported to cause disease in bats.

DNA Viruses

Poxviral infection in free-ranging *Eptesicus fuscus* is associated with joint swelling, necrosuppurative osteomyelitis, necrotizing tenosynovitis, and occasional vasculitis involving long bones and flat facial bones. Virus can be detected in the synovium by polymerase chain reaction (PCR) and electron microscopy. Genetic analysis has identified the virus (Eptesipox virus) as a member of a new genus of poxvirus most closely related to Cotia virus, an unclassified chordopoxvirus isolated from mice in Cotia County, São Paulo State, Brazil, in 1961.[102] A more typical cutaneous poxviral nodule was reported in a single free-ranging *Miniopterus schreibersii* in Australia. Microscopically, the lesion was typical of poxviral infection and characterized by focal epidermal thickening with ballooning degeneration and intraepithelial, eosinophilic intracytoplasmic inclusion bodies. Pox virions were confirmed by electron microscopy.[207]

Cytomegalovirus has been detected in the acinar cells of the principal submandibular salivary gland of two captive *Myotis lucifugus*. Although cytomegaly was evident microscopically, there was no evidence of clinically significant disease.[306]

A novel **gammaherpesvirus** has been identified in a captive *Pteropus vampyrus* with chronic lymphoplasmacytic blepharitis, Meibomian gland adenitis, and neutrophilic and lymphocytic conjunctivitis. A relationship between the virus and lesions has not been confirmed.[39]

RNA Viruses

In addition to their notoriety as reservoirs of **rabies virus** and other zoonotic **lyssaviruses**, bat species are susceptible to fatal lyssaviral infection. Lyssaviruses have high host specificity, so different species of bats are infected by different variants of the same Lyssavirus species.[75] Lyssaviruses documented to cause disease in humans or other animals, including bats, are included in Table 25.2. Aravan and Khujand lyssaviruses have each been isolated from a single bat in surveys in Central Asia, but natural disease does not appear to be documented.[10,172]

Although rabies virus is endemic in bat populations, estimates of subclinical infection vary from 0.1% to 2.5%

TABLE 25.2 Viruses Associated With Natural Disease in Humans and Animals for Which Bats Are a Proven or Likely Reservoir

Disease	Species Clinically Affected	Proven or Likely Main Bat Reservoir Species	Geographic Locations of Disease (Imported Diseases Reported as Country Where Infection Acquired)
Bunyaviruses			
Issyk-Kul virus[38,193]	Human—Fever, headache, myalgia, nausea	*Nyctalus noctula; Eptesicus nilssonii*	Kazakhstan, Germany
Kasokero virus[165]	Human (laboratory-acquired infection)—flu-like symptoms, abdominal pain	*Rousettus aegyptiacus*	Uganda
Coronaviruses			
Severe acute respiratory syndrome-related coronavirus (SARS-CoV)[98,155]	Human—flu-like symptoms, diarrhea, death	*Rhinolophus* spp.	China (no reports since 2003)
Severe acute respiratory syndrome coronavirus 2 (SARS-CoV-2)[12,194,285]	Humans and other mammals, particularly Mustelidae and Felidae—symptoms and clinical signs vary from absent to severe respiratory, nervous system, and gastrointestinal disease, death	*Rhinolophus* spp. are suspected primary reservoir; immediate animal progenitor to SARS-CoV-2 pandemic unknown	Worldwide
Middle East respiratory syndrome coronavirus (MERS-CoV)[195]	Human—flu-like symptoms, acute renal failure, death	*Taphozous perforatus*	Saudi Arabia, Jordan, United Arab Emirates, Qatar
Filoviruses			
Ebola virus[9,46,108,157,186,187,243,274]	Human—flu-like symptoms, abdominal pain, skin rash, encephalopathy, hemorrhagic symptoms, death Nonhuman primates—lethargy, abdominal pain, anorexia, hemorrhage, death Duiker—death (no pathology performed)	*Hypsignathus monstrosus, Epomops franqueti, Myonycteris torquata* suspected	Sudan, Democratic Republic of Congo, Uganda, Republic of Congo (not including disease known to be transmitted from primates)
Marburg virus[8,37,243]	Human—flu-like symptoms, nausea, vomiting, diarrhea, pharyngitis, dysphagia, skin rash, encephalitis, hemorrhagic manifestations, death	*Rousettus aegyptiacus*	Zimbabwe, Kenya, Democratic Republic of Congo, Angola, Uganda (not including disease known to be transmitted from nonhuman primates)
Lyssaviruses			
Rabies virus[17,76]	Most mammals at risk, rarely reported in poultry—progressive neurologic deterioration, coma and death	Multiple, including Vespertilionidae, Molossidae, Phyllostomidae, Mormoopidae, Noctilionidae, and Emballonuridae—especially important are *Eptesicus fuscus, Lasionycteris noctivagans, Perimyotis subflavus, Tadarida brasiliensis,* and *Desmodus rotundus*	Global; only found in bats in North, Central, and South America

TABLE 25.2 Viruses Associated With Natural Disease in Humans and Animals for Which Bats Are a Proven or Likely Reservoir—cont'd

Disease	Species Clinically Affected	Proven or Likely Main Bat Reservoir Species	Geographic Locations of Disease (Imported Diseases Reported as Country Where Infection Acquired)
Australian bat lyssavirus[22,106,109,286]	*Pteropus* spp., *Saccolaimus flaviventris*, human, horse—neurologic disease similar to rabies virus infection	*Pteropus* spp., *Saccolaimus flaviventris*	Australia
Bokeloh bat lyssavirus[101,292]	*Myotis nattereri*—grounded, aggression, weakness or paralysis, moribund or found dead *Pipistrellus pipistrellus*—found injured and died without other clinical signs	*Myotis nattereri*	Germany, France, Poland
Duvenhage virus[18,199,201,210,247,300]	Human—neurologic disease similar to rabies virus infection *Miniopterus schreibersii* (presumptive identification)—caught during the day by a cat	Unknown, virus isolated from a clinically normal *Nycteris thebaica*, and seropositivity is documented in this species and *Miniopterus natalensis*	Kenya, South Africa, Zimbabwe
European bat lyssavirus-1[11,49,82,224,279,283,287]	Multiple bat species—neurologic disease similar to rabies virus infection or found dead Human, domestic cat, stone marten (*Martes foina*), domestic sheep—neurologic disease similar to rabies virus infection	*Eptesicus serotinus*	Europe, especially the Netherlands, Germany, Denmark, Poland, France, Spain
European bat lyssa-virus-2[11,49,161,232,279,287]	*Myotis* spp., human—neurologic disease similar to rabies virus infection	*Myotis daubentonii*	Europe, especially United Kingdom and the Netherlands, also detected in Germany, Finland, Luxembourg, Switzerland
Gannoruwa bat lyssavirus[127]	*Pteropus giganteus*—found dead or clinically ill with death soon after capture	Unknown	Sri Lanka
Irkut lyssavirus[17,56,188]	Human—neurologic disease similar to rabies virus infection Dog—death with unknown history	*Murina leucogaster*	Russian Federation, China
Lagos bat virus[18,70,112,164,199,200,202,211,300]	*Epomophorus wahlbergi*—found sick or dead, sudden death, or neurologic disease similar to rabies virus infection Cat—neurologic disease similar to rabies virus infection or lethargy and paresis Dog, pet *R. aegyptiacus* in France (ex-Egypt or Togo), water mongoose (*Atilax paludinosus*)—neurologic disease similar to rabies virus infection Virus has been isolated from clinically normal/found dead/unknown history *Eidolon helvum*, *Micropteropus pusillus*, *Nycteris gambiensis*, *R. aegyptiacus*, *Epomophorus gambianus*, and *Epomops buettikoferi*	Unknown, neutralizing antibodies found in multiple bat families, most commonly in pteropodid bats including healthy *Eidolon helvum* and *Myonycteris torquata*. 30%–67% seropositivity in *Epomophorus wahlbergi* and *Rousettus aegyptiacus*	Central African Republic, Ethiopia, Ghana, Guinea, Kenya, Nigeria, Senegal, South Africa, Zimbabwe

Continued

TABLE 25.2 Viruses Associated With Natural Disease in Humans and Animals for Which Bats Are a Proven or Likely Reservoir—cont'd

Disease	Species Clinically Affected	Proven or Likely Main Bat Reservoir Species	Geographic Locations of Disease (Imported Diseases Reported as Country Where Infection Acquired)
Lleida bat lyssavirus[11,70,252]	*Miniopterus schreibersii*—lethargy and death or found dead	Unknown	Spain, France
Shimoni bat lyssavirus[71,164,173]	*Macronycteris vittatus*—unknown history; *Macronycteris commersoni*—found dead	Unknown, neutralizing antibodies found in healthy *Eidolon helvum* and *Macronycteris commersoni*	Kenya
Taiwan bat lyssavirus[152]	*Pipistrellus abramus*—loss of appetite and death or found dead	Unknown	Taiwan
West Caucasian bat lyssavirus[70,185,287]	Cat—neurologic disease similar to rabies virus infection	*Miniopterus schreibersii*	Russian Federation, Italy
Paramyxoviruses			
Hendra virus[69,131,169,181,329,333]	Horse—fever, respiratory or neurologic signs, death Human—encephalitis, flu-like pulmonary syndrome, death	*Pteropus alecto, P. poliocephalus, P. scapulatus,* and *P. conspicillatus*	Australia
Menangle virus[21,54,192,250,251]	Pig—sow infertility, embryonic death, stillborn pigs, abortion, skeletal and craniofacial deformities, pulmonary hypoplasia, brain/spinal cord degeneration Human—flu-like illness, skin rash	*Pteropus* spp. (neutralizing antibodies in multiple species, virus isolated from *P. alecto*)	Australia
Nipah virus[69,131,148,181,219,329]	Pig and human—fever, respiratory or neurologic disease, death Dog—fever, respiratory distress, mucopurulent oculonasal discharge, death Cat—generalized vasculitis, clinical signs not reported Horse—meningoencephalitis	*Pteropus vampyrus, P. giganteus,* and *P. hypomelanus*	Malaysia, Bangladesh, Singapore, India
Sosuga virus[4,179]	Human—fever, joint and muscle pain, headache, stiff neck, oropharyngeal ulcers, malaise, maculopapular rash	*Rousettus aegyptiacus*	South Sudan or Uganda
Reoviruses			
Nelson Bay virus[62,63,336]	Human—fever, acute respiratory disease	*Pteropus* spp.	Malaysia, Indonesia

in healthy bats. Prevalence is much higher in sick bats, with estimates ranging from 10% to 92%.[75,83] A high seroprevalence rate has been detected in apparently healthy bats, indicating that they may be able to survive infection.[5] Bats typically undergo a 2- to 25-week incubation period, although prolonged incubation of more than 1 year has been reported.[36] Cold temperatures inhibit disease progression in cold-adapted (i.e., hibernating) bats.[76] Virus may be present in saliva for up to 12 to 15 days before onset of clinical signs, and most bats die within 1 to 2 weeks of developing clinical signs.[76,140] Although direct contact with saliva, as through a bite wound, is typically required for transmission, both bats and humans can be infected by aerosolized virus.[75,85]

Bats more commonly display paralytic rather than "furious" clinical signs and are rarely reported to attack humans unless provoked. Clinical signs typically include neurologic deficits, such as paralysis, incoordination, disorientation, irritability, and increased vocalization.[75]

Microscopic lesions in bats with lyssaviral infection are similar to those reported in other mammalian species and include nonsuppurative meningoencephalomyelitis and ganglioneuritis. In a report of rabies infection in five *Molossus molossus* in Brazil that were unable to fly and seen out during the day, the presence of rabies virus was confirmed by immunohistochemistry, direct immunofluorescence, and mouse inoculation testing. However, no microscopic lesions were present, indicating that a lack of histologic lesions may not rule out lyssaviral infection in bats.[88] Neuronal intracytoplasmic eosinophilic inclusion bodies known as Negri bodies are typical, but their presence and frequency are highly variable. In one study of bats infected with Australian bat lyssavirus, the highest frequency and severity of microscopic lesions was detected in the hippocampus, thalamus and midbrain, and medulla oblongata and pons.[149] Lyssaviral antigen has been detected in bats in the central and peripheral nervous system as well as in other tissues. Reported affected sites include brainstem, cerebellar Purkinje cells, hippocampus, cerebrum, thalamus and midbrain, peripheral autonomic ganglia, nerve plexuses in the gastrointestinal tract, intramuscular nervous tissue, adrenal medulla, salivary gland, tongue, brown fat, lung, heart, kidney, bladder, stomach, intestine, and feces.[5,84,88,149,280,294]

Tacaribe virus, an arenavirus, has been reported in several *Artibeus* species in Trinidad. Clinical signs mimic rabies and include unkempt haircoat, isolation from other bats, and tremors.[97] Although a histologic review of naturally infected bats has not been performed, experimental inoculation of *A. jamaicensis* and *Carollia perspicillata* with Tacaribe virus resulted in multisystemic lesions including interstitial pneumonia, hepatocellular degeneration and necrosis, neutrophilic splenitis, and lymphocytic leptomeningitis or encephalitis. Development of significant clinical disease in these species may indicate they are unlikely to be reservoirs for Tacaribe virus.[72]

As with lyssavirus, bats can be subclinically or clinically affected by other viruses for which they serve as reservoirs. In addition to serving as a reservoir for **Marburg virus**, wild *Rousettus aegyptiacus* can develop mild hepatic lesions, including small areas of mononuclear cell inflammation and localized hepatocellular necrosis in association with viral antigen identified by immunohistochemistry.[312]

Chaerephon plicatus, in addition to serving as a reservoir for **Kaeng Khoi virus** (Bunyavirales), can also exhibit disease due to natural infection with this virus. Lesions include hepatomegaly, massive hepatocellular necrosis lacking inflammation, and interstitial pneumonia.[233]

Additional viral diseases are included in the Digital Content.

Bacteria

Yersinia pseudotuberculosis infections have been described in captive *Pteropus rodricensis*,[330] *Rousettus aegyptiacus*,[59,154,229] and *Carollia perspicillata*,[129] and in a free-ranging *Myotis myotis* in Germany.[223] In *R. aegyptiacus*, gross lesions include splenomegaly, hepatomegaly, mesenteric lymphadenopathy, and white to yellow-white nodules in the spleen, liver, kidneys, and lung. Histologic lesions include abscesses or neutrophilic to necrosuppurative inflammation in the liver, spleen, kidney, and lung; bacteria within renal and pulmonary blood vessels; thrombosis; necrosis of the bone marrow and large aggregates of gram-negative bacilli/coccobacilli within lesions[59,154,229] (Figs. 25.9A, B). Lesions in *C. perspicillata* include necrosuppurative inflammation in the liver, lung, spleen, lymph node, uterus, intestine, tongue, and oral cavity, with numerous gram-negative coccobacilli in lung and liver lesions.[129]

Infections with ***Pasteurella*-like bacteria** have been described in captive *Epomophorus wahlbergi* from one zoo collection with unilateral or bilateral pneumonia, pleuritis,[142] pleuropneumonia, suppurative or fibrinous epicarditis/pericarditis, and/or myocarditis with intralesional gram-negative pleomorphic coccobacilli. In some of these bats, type II pneumocyte and bronchiolar epithelial cell hyperplasia were prominent, possibly indicating that there could have been additional pathogens (unpublished data). Two *Pteropus pumilus* from the same institution had similar bacteria cultured from hind limb cellulitis and fracture with abscess formation in a wing digit; the respiratory and cardiac lesions seen in the *E. wahlbergi* were not identified in these bats.[142]

A mortality event involving approximately 100 free-ranging *Eptesicus fuscus* in Wisconsin, United States, was attributed to ***Pasteurella multocida*** serotype 1 as a cause of acute fatal septicemia with interstitial pneumonia. Gross lesions included pulmonary consolidation and splenomegaly. Microscopically, gram-negative coccobacilli were seen in blood vessels and endothelial cells of the lung, liver, and spleen. Additional lesions included alveolar thrombosis, endothelial necrosis, and neutrophilic interstitial pneumonia[33] (Fig. 25.10). Systemic pasteurellosis

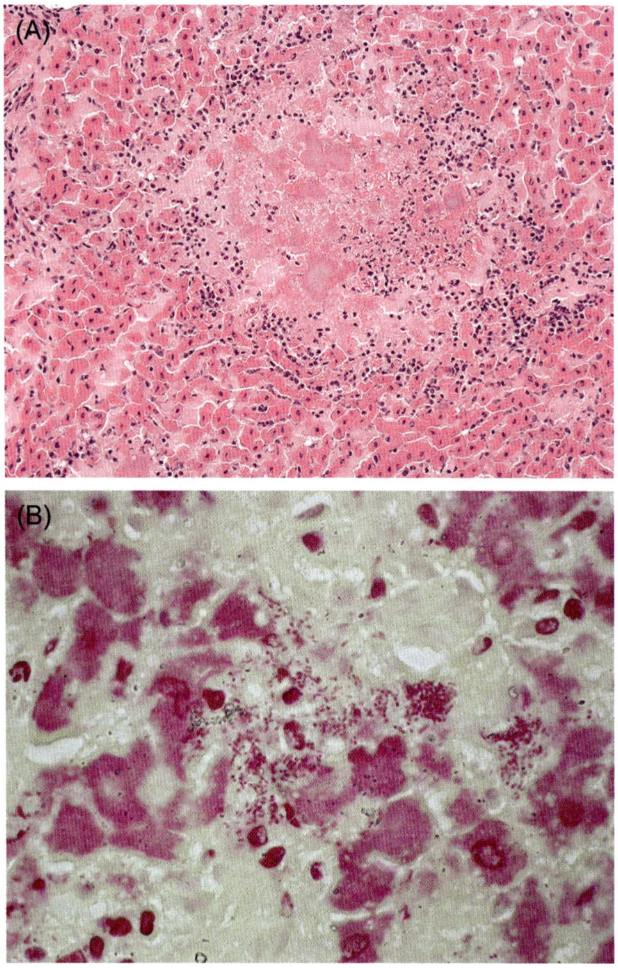

FIGURE 25.9 *Yersinia pseudotuberculosis* in the liver of a *Myotis myotis*. (A) Infection is associated with coagulative necrosis with a central accumulation of neutrophils and a rim of lymphocytes and histiocytes. (B) Abundant intralesional coccobacilli are highlighted with special staining. Gram. *(Parts A and B: Photos Courtesy of K. Meuhldorfer, Leibniz Institute for Zoo and Wildlife Research (IZW), Berlin, Germany.)*

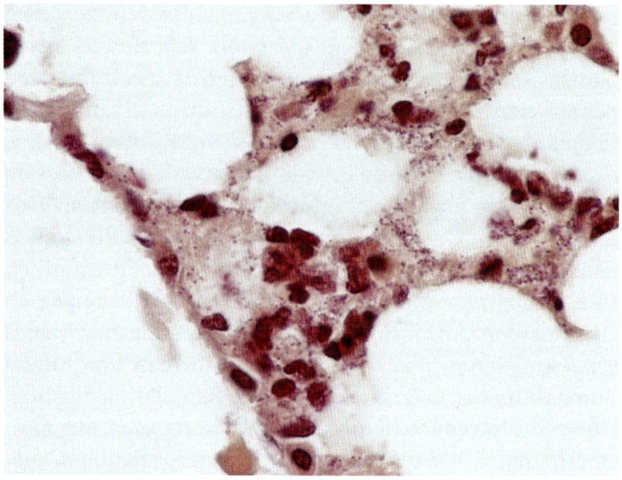

FIGURE 25.10 *Pasteurella multocida* pneumonia in the lung of an *Eptesicus fuscus*. Mild acute interstitial pneumonia is associated with numerous intracellular and extracellular gram-variable bacteria. Brown-Hopps.

is also strongly associated with cat predation. In one survey of European bats, most isolates of *Pasteurella* cultured from organs of 29 vespertilionid bats were identified as *P. multocida* subsp. *septica* capsular type A, a predominant strain found in the oral cavity of domestic cats.[221] *Pasteurella* spp. have also been reported as pathogens in individual free-ranging European bats.[220]

Salmonella sp. serovars capable of infecting a broad range of hosts have been identified in both healthy and septicemic bats. *Salmonella* serotype enteritidis and *Salmonella* serotype typhimurium have been isolated from viscera of free-ranging vespertilionid bats with interstitial pneumonia and suppurative meningitis.[220]

Clostridium perfringens and ***C. sordellii*** are reported as a cause of hemorrhagic diarrhea in European vespertilionid bat species.[130,222] *Clostridium perfringens* is thought to cause gastritis in captive pteropodid bats that consume inappropriate foods.[245]

In captive pteropodid bats, **α-hemolytic *Streptococcus*** spp. occasionally cause bronchopneumonia or subcutaneous abscesses.

Staphylococcus aureus has been identified in captive pteropodid bats with sepsis and osteomyelitis.

Multiple species of potentially pathogenic bacteria, including *Salmonella* spp., *Shigella flexneri*, *Campylobacter* spp., *Neorickettsia* spp. including *N. risticii*, *Escherichia coli*, *Yersinia enterocolitica*, *Borrelia* spp., *Bartonella* spp., hemotropic *Mycoplasma* spp., and *Enterococcus* spp. have also been detected in free-ranging bats in the absence of clinical disease.[40,43,198,203,220,221,223]

Additional bacterial diseases in bats are described in the Digital Content.

Fungi

The fungal infection known as **white-nose syndrome (WNS)**, caused by the fungus ***Pseudogymnoascus* (*Geomyces*) *destructans*** (***Pd***), was first detected in bats in a cave in upstate New York, United States, in 2006. Within a decade, an estimated six million bats had died of the infection, which may represent the largest mammalian wildlife mortality event in recorded history. By 2023, WNS had spread to 40 US states and 8 Canadian provinces throughout Eastern and Central North America with a cluster of detections in Washington State,[327] demonstrating the potential of the introduced fungal pathogen to impact bat populations across North America.[191] Certain hibernating North American bat species have been disproportionately affected, whereas others appear refractory to the disease. Particularly susceptible species include *Myotis lucifugus*, *M. septentrionalis*, and *Perimyotis subflavus*.[57,111] *Myotis lucifugus*, once among the most common species of bat in the Eastern United States, has undergone significant population declines and localized extinction with complete loss of populations in some affected caves.[113] Recent evidence indicates the possibility of adaptation and resistance in some *M. lucifugus* populations.[110,177] *Myotis lucifugus* is currently listed as endangered, *Perimyotis subflavus* is listed as vulnerable, and *M. septentrionalis* is listed as near threatened on the International Union for Conservation of Nature's Red List of Threatened Species.[160]

The fungus *P. destructans* is thought to be native to Europe and Asia, where it had not previously been recognized as a cause of mortality in native bats.[47,261] Since the North American outbreak began, both colonization by *Pd* and lesions diagnostic for WNS have been documented in several species of bats from Europe[16,50,117,254,328,341] and Asia[150,167] although significant mortality has not been reported in those populations. These differences are most likely due to environmental and host factors resulting from evolutionary host-pathogen evolution, rather than differences in fungal pathogenicity between isolates of *Pd* from Eurasia and North America.[16,342]

Pseudogymnoascus destructans is a psychrophilic and keratinophilic fungus. Like other *Pseudogymnoascus* species, it is saprophytic and does not require a host for survival. The fungus's ability to persist in cave sediment in the absence of living hosts is likely to be an important factor for environmental persistence and perpetuation of infection.[271] Bats are thought to become infected beginning in autumn, with transmission increasing throughout winter. Infection peaks in late winter but is cleared during summer in surviving bats,[176] although the fungus may be present on the skin of bats in the summer months.[15,51,153] The optimal temperature range for growth of *Pd* corresponds to the body temperature of most hibernating bats,[318] and hibernation and the associated decrease in body temperature during periods of torpor are significant factors in the pathogenesis of WNS.

As *Pd* infection progresses, hibernating bats arouse from torpor more frequently and exhibit abnormal energy-intensive behaviors, such as premature emergence from hibernacula during the winter period when insect prey is generally not available.[269] One experimental study indicated that fat energy utilization was twofold higher in infected bats even before the onset of increased arousal, indicating that other physiologic mechanisms of increased metabolism may be present.[319] Other systemic abnormalities, including increased evaporative water loss through damaged skin, acidosis, hyperkalemia, hypovolemia, hypotonic dehydration, and fever, have been detected in infected bats.[81,204,319,323,331]

The hallmark of WNS is white fungal growth on the muzzle, ears, and wing membranes of hibernating bats[32] (Fig. 25.11A). By the time a carcass is received for necropsy, however, the fungus may no longer be grossly evident. Wing damage may present as ~1–3 mm diameter, multifocal to coalescing white foci with a pinpoint black center or variably sized areas of depigmentation, tearing, or dryness of the patagia. Wings are best examined backlit as with a radiograph light box (Fig. 25.11B). Additionally, infected skin has a characteristic orange fluorescence under long-wave ultraviolet (UV) light[313] (Fig. 25.11C), although the specificity of this screening method has not been validated across a wide geographic area or temporal window. Examination under UV light can be a more sensitive means of detecting infection than gross lesions alone, as grossly unremarkable areas of the wing may demonstrate fluorescence and wing damage in otherwise seemingly uninfected bats.[256] Dead bats are frequently emaciated but there are no other systemic lesions, and visceral dissemination has not been reported.

Diagnosis of WNS requires both demonstration of *Pd* on the skin via PCR or fungal culture and histopathology to demonstrate the presence of characteristic lesions. The area of skin examined microscopically can be maximized by rolling excised portions of the patagia onto wax dowels, which are then cut transversely, resulting in multiple cross-sections of concentric rolls of skin[213] (Fig. 25.12A). Pathognomonic lesions include characteristic "cupping erosions" and ulcers filled with densely packed, PAS- and GMS-positive fungal hyphae with variable extension through the basement membrane (Figs. 25.12B, C). Hyphae commonly extend into hair follicles and sebaceous or apocrine glands. In tissue section, fungal hyphae exhibit mild morphological variability. Hyphae are septate and branching, with parallel to slightly irregular walls 2–5 μm in diameter. Conidia, when present, are curved and measure ~2.5 μm wide × 7.5 μm long[213] (Fig. 25.12C; eSlide 25.e4). The lack of inflammatory response to the invading hyphae is typical of the lesion in hibernating bats. However, neutrophilic inflammation can occur in hibernating bats when the fungal lesion is superinfected with bacteria.[77,213] Bats that have emerged from hibernation may exhibit a neutrophilic response to *Pd* (Fig. 25.12D), as well as edema and necrosis that are not present in hibernating bats.[212,215] A recent study provides evidence that, rather than creating true erosions, fungal hyphae proliferate within the intraepidermal space, after which the infection progresses by extending through the basement membrane, an invasion technique found in plant fungal pathogens but not previously described in animals.[214] To assist with diagnostic standardization, grading systems have been proposed for both the gross[270] and microscopic[253,269] lesions of WNS.

Although the mortality rate is often high among certain North American bat species, the disease is not invariably

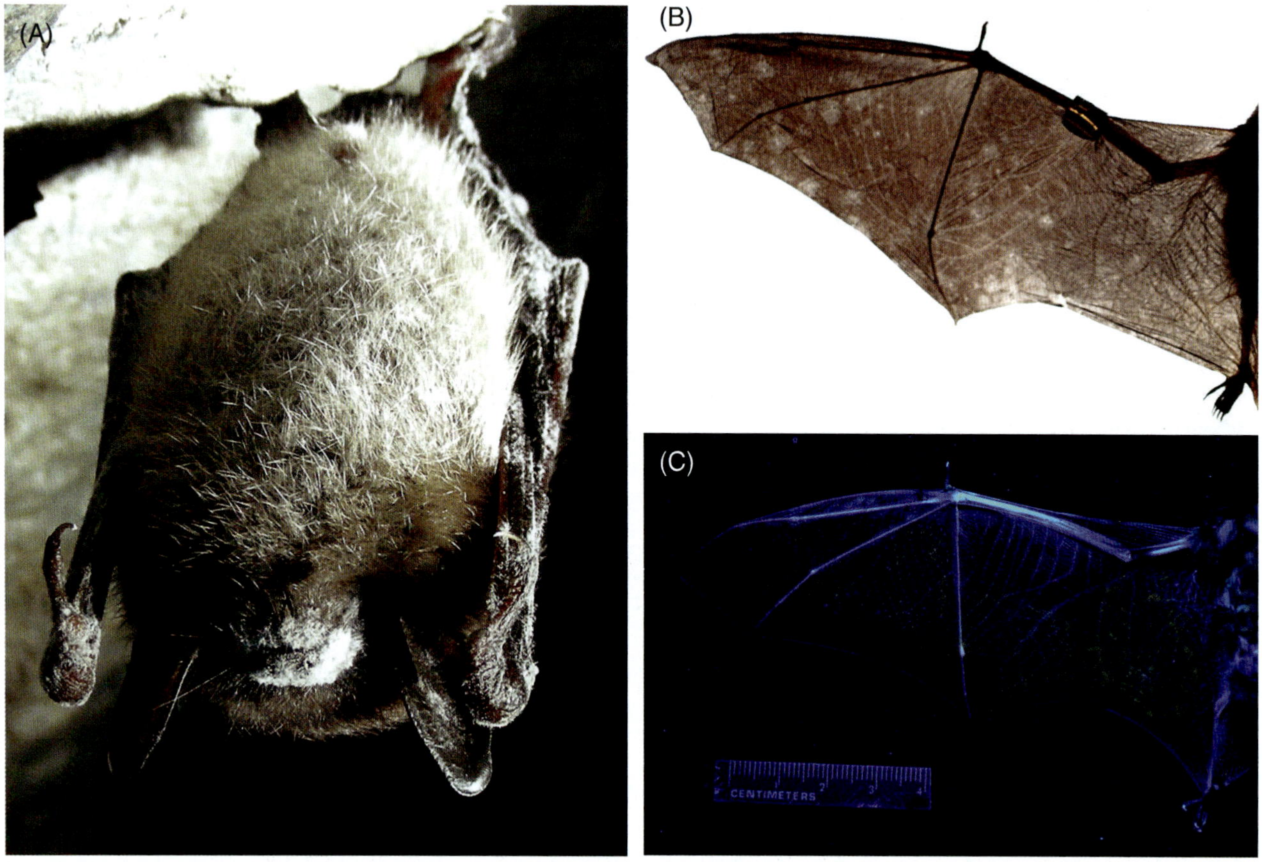

FIGURE 25.11 Gross lesions due to *Pseudogymnoascus destructans* infection, the cause of "white-nose syndrome," in a *Myotis lucifugus*. (A) White fungal growth is present on the muzzle, pinna, and wing. (B) Multifocal to coalescing circular areas of pallor are common on wing membranes of infected bats. (C) Ultraviolet illumination reveals characteristic orange fluorescence characteristic of infection. *(Part A: Photo Courtesy of U.S. Fish and Wildlife Service; Part B: Photos Courtesy of U.S. Geological Survey National Wildlife Health Center.)*

fatal. Infected bats can recover and the wing damage can heal.[114,215] In some cases, bats that survive into the summer may experience poor hunting or reproductive success due to postinfection scarring or elevated metabolic rate.[209,270]

The interaction between *Pd* and other etiologic agents in bats continues to be investigated. In one study, *M. lucifugus* coinfected with *Pd* and the *M. lucifugus* coronavirus (Myl-CoV) exhibited increased severity of WNS-associated lesions, increased presence of viral RNA in the intestinal tract, higher levels of coronavirus antibodies, and lower gene expression related to innate immune responses. These results indicate that WNS may down-regulate bats' antiviral responses and increase virus shedding.[86]

In addition to *P. destructans*, several other fungal species have been reported to cause mycotic dermatitis in bats. Most are limited to single case reports. One exception is a report describing gross and microscopic lesions due to infection with the dermatophyte ***Arthroderma* (formerly *Trichophyton*) *redellii*** in hibernating bats from Wisconsin, Indiana, and Texas, United States. *A. redellii* can cause gross lesions on the ears, wings, legs, and tail that may resemble WNS, but the muzzle is not noted to be affected. "Ring"-type lesions characteristic of human ringworm may also occasionally be seen (Fig. 25.13A). Microscopically, *A. redellii* invades the stratum corneum and deeper epidermal layers with a neutrophilic pustular response (Fig. 25.13B), but the aggregation, erosion, and deep dermal invasion common in WNS are not present. Conidia of *A. redellii* are radially symmetric and obovate to pyriform, in contrast to the asymmetrically curved conidia of *P. destructans*. Hyphae are septate, 2–6 μm diameter, and morphologically similar to *Pd*. Differentiation between WNS and *A. redellii* infection may require close examination of the pattern of invasion and inflammation and fungal culture or molecular identification.[190]

Hyphopichia burtonii has been reported as a cause of fungal dermatitis in a free-ranging *Barbastella barbastellus*. In this case report, gross lesions included skin thickening, crusting, ulceration, and nodule formation. Thin, branching, septate fungal hyphae, and masses of ~4 μm diameter ovoid spores were present in the stratum corneum and hair follicles. Infection was also associated with epidermal hyperplasia and follicular hyperkeratosis.[289]

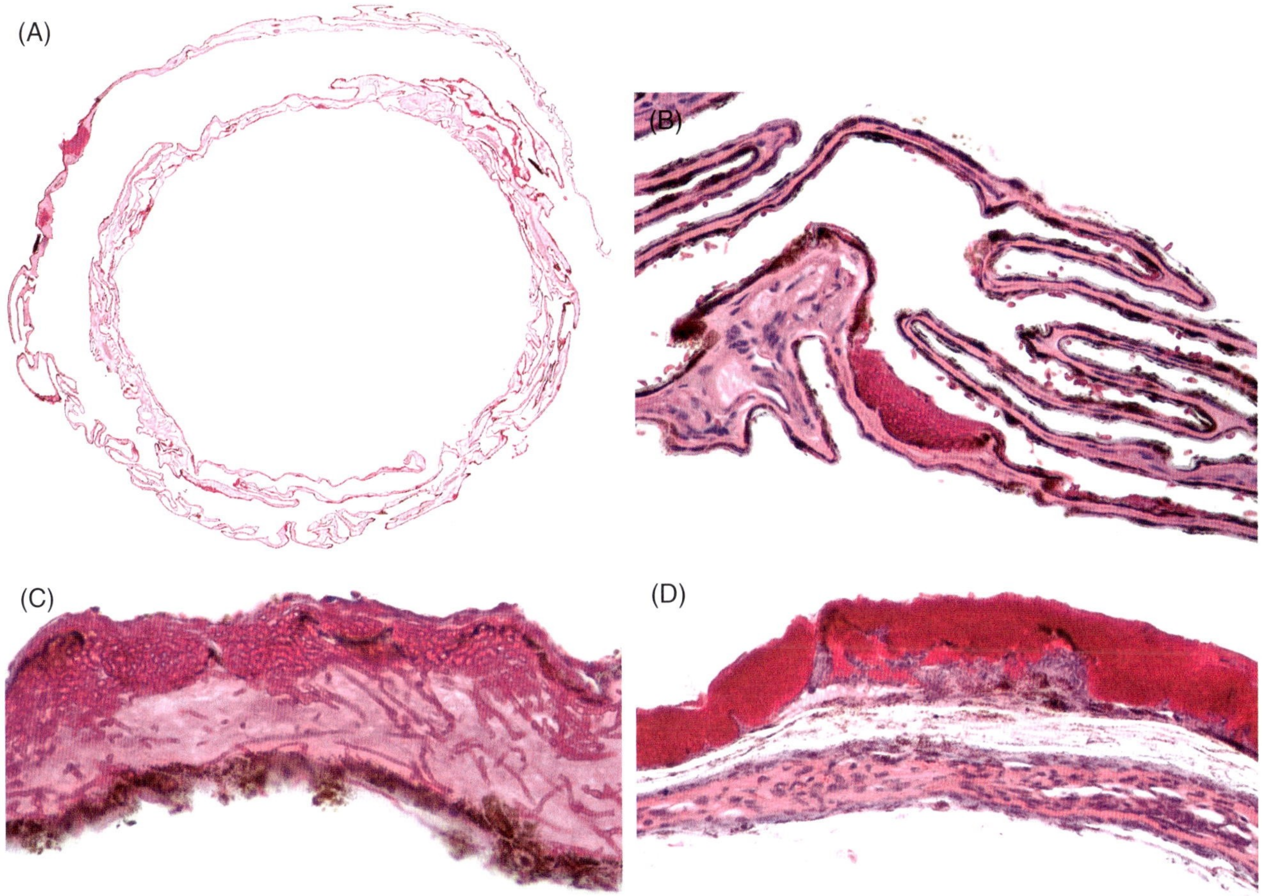

FIGURE 25.12 Histologic lesions due to *Pseudogymnoascus destructans* infection/white-nose syndrome (WNS) in the wing of a hibernating (A–C) and (D) euthermic *Myotis lucifugus* (see eSlide 25.e4). (A) Rolled wing membrane maximizes surface area of skin available for microscopic examination. Dense packets of PAS-positive fungi are easily identified at low magnification. (B) Characteristic cupping erosion in the epidermis. Scattered over the surface of the stratum corneum are many curved arthroconidia characteristic of *P. destructans.* (C) Severe *P. destructans* infection with ulceration and transdermal extension of fungal hyphae. Note the lack of inflammatory response. (D) This wing membrane was collected in May after emergence from hibernation. In contrast to the typical noninflammatory cupping erosion of WNS seen in hibernating bats (A–C), the patagium in this bat exhibits dermatitis, hyperkeratosis, and a serocellular crust containing myriad fungal hyphae. (A–D) Periodic acid-Schiff.

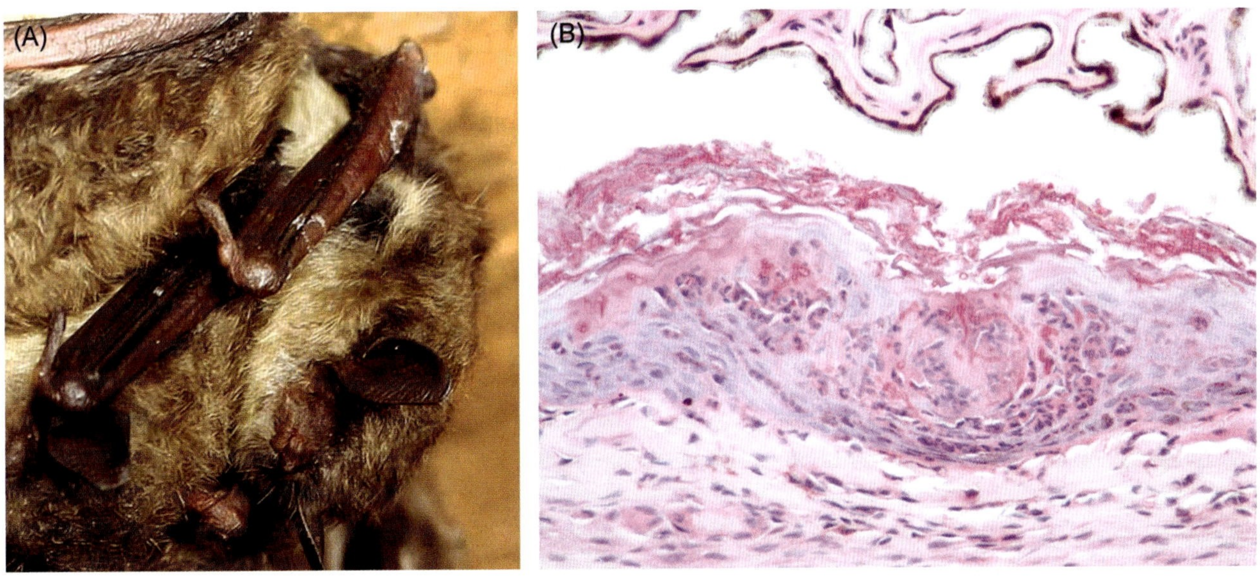

FIGURE 25.13 *Arthroderma* (formerly *Trichophyton*) *redellii* in the skin of *Myotis lucifugus* (A) and *Myotis velifer* (B). (A) White fungal growth is present on the wrist and wing but not on the muzzle, the latter a characteristic feature of *P. destructans* infection. A few lesions exhibit an outer zone of fungal growth with circular clearing typical of dermatomycosis in humans. (B) Fungal hyphae form a dense mat on the surface of the stratum corneum. Invasion of the epidermis is associated with neutrophilic inflammation, Periodic acid-Schiff. *(Part A: Photo Courtesy of H. Kaarakka, Wisconsin Department of Natural Resources.)*

Fungal dermatitis has also been associated with *Cladosporium* spp. in an euthermic *Vespertilio murinus*. Infection included white foci on wing membranes and severe necrotizing dermatitis with intralesional fungal hyphae.[19]

Bats can serve as r. servoirs for ***Histoplasma capsulatum***, although clinical disease in free-ranging bats is uncommon. Organisms are most often present in pulmonary alveolar macrophages but can also be found in circulating macrophages and in the mesentery, spleen, liver, adrenal gland, and intestine.[307,309] There is typically minimal to no cellular response to infection, which is thought to occur by the respiratory route. Infective fungi are excreted in feces.[208,307] There are no reports of mortality due to natural infections in free-ranging bats.[124] The ability of bats to survive disseminated infection, including intestinal infection with shedding of spores, likely allows them to serve as reservoirs. During an outbreak at 1 zoo, 12 *Rousettus aegyptiacus* and 1 *Artibeus jamaicensis* had systemic histoplasmosis. There was lung involvement in all cases except for four of the *R. aegyptiacus*.[311]

Juvenile pteropodid bats maintained under poor husbandry conditions or that have chronic illnesses can develop opportunistic cutaneous ***Candida albicans*** infections, most commonly on the wing membranes. Lesions initially present as erythema progressing to a gray, slimy pseudomembrane from which hair epilates in clumps. Concurrent cutaneous and oral candidiasis are sometimes identified. Oral candidiasis is common in hand-raised pteropodid bats, and gastric candidiasis may occur in pteropodid bats consuming inappropriately nutrient-dense diets, as well as in immunosuppressed or debilitated pups, especially after prolonged antibiotic administration.[245]

Oral and esophageal candidiasis is occasionally observed in captive bats. Infection is associated with neutrophilic infiltrates, erosions, ulcers, or necrosis of the mucosa and intralesional 2–6 μm diameter ovoid budding yeasts and 3–5 μm diameter hyphae and pseudohyphae. Often these animals have received antibiotics or have other concurrent disease processes. *Candida* infection has also been identified as the cause of keratitis and/or conjunctivitis in captive pteropodid bats.

Additional fungal diseases are described in the Digital Content.

Metazoa

Infections with various nematodes, trematodes, cestodes,[325] and acanthocephalans[291] have been described in bat species, but specific clinical diseases have only been described for a few metazoans.

Neurologic disease due to infection by ***Angiostrongylus cantonensis***, the rat lungworm, has been reported in free-ranging *Pteropus alecto* and *P. scapulatus* and captive and free-ranging *P. poliocephalus* in Australia[23,245,268]

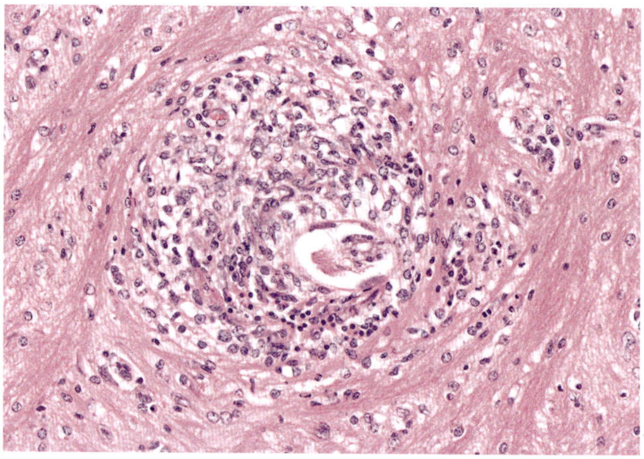

FIGURE 25.14 Encephalitis due to *Angiostrongylus cantonensis* infection in a *Pteropus poliocephalus*. A nodular infiltrate of predominantly mononuclear cells, with rare eosinophils, surrounds a transverse section of the nematode. *(Photo Courtesy of K. Rose, Australian Registry of Wildlife Health, Taronga Conservation Society Australia.)*

and a captive *P. giganteus* in the United States.[324] Related inflammation is eosinophilic and granulomatous or neutrophilic and lymphoplasmacytic and associated with meningitis or meningoencephalitis, foci of malacia, and/or hemorrhage and gliosis (Fig. 25.14). Larval or subadult nematodes with or without associated inflammation may be identified in the meninges, subarachnoid space, ventricles, or brain parenchyma.[23,268,324] Granulomatous endarteritis in *Pteropus alecto* associated with another rat lungworm, ***Angiostrongylus mackerrasae***, results from parasites in the pulmonary arteries. Severe, mixed inflammatory cell interstitial pneumonia with intralesional eggs and rare larvae in the parenchyma can also be seen.[196]

Infection with ***Toxocara pteropodis*** has been described in free-ranging flying foxes from Australia (*Pteropus alecto*, *P. scapulatus*, *P. poliocephalus*, and *P. conspicillatus*),[258,259] Sri Lanka (*P. giganteus*)[166] and the South Pacific (*P. tonganus*).[257] Infection has also been described in a captive flying fox in Florida (*P. hypomelanus*).[141] In adult bats, third-stage larvae are present in the liver, and larvae can be passed from lactating females to suckling bats via transmammary transmission. Adult nematodes are normally only identified in the intestine of juvenile bat,[257] but have rarely been identified in the stomach, esophagus, gall bladder, common bile duct, and pharynx/larynx.[166,259] Large numbers of intestinal nematodes can be negatively correlated with body condition, and volvulus, intussusception, and possible partial intestinal obstruction have been observed in some infected bats.[141,166] Failure to thrive may occasionally be observed in hand-reared *Pteropus* spp. with heavy ascarid loads.[245]

The nematode ***Riouxgolvania beveridgei*** has been identified as the cause of cutaneous granulomas in free-ranging *Miniopterus schreibersii bassanii*. Gross lesions include 1–2 mm diameter white nodules or ulcers on the dorsal aspect of the thoracic and pelvic limbs. Microscopically, dermatitis is granulomatous to pyogranulomatous with central degenerate nematodes. Males are more commonly affected than females.[207] Dermal nodules are also reported with ***R. kapapkamui*** in *Myotis* spp. in Japan.[136]

The filarioid nematode ***Molossinema wimsatti*** has been reported in the central nervous system of *Molossus rufus* (*M. ater*). These nematodes resemble *Litomosoides* and *Litosoma* spp., which have been described in the thoracic cavity of bats. Clinical signs are minimal, although ataxia was noted in one animal before death. From few to many 2–5 cm long, 100–200 μm diameter nematodes have been reported in the lateral, third, and fourth ventricles of the brain and in the subdural space of the cervical and thoracic vertebral canal. With the exception of ventricular dilation, tissue response was rare and limited to mild focal encephalitis and necrosis of adjacent parenchyma in one animal. Infection was considered an incidental finding.[237] Two free-ranging *Molossus molossus* in Brazil were also diagnosed with *M. wimsatti* infection based on the microscopic detection of adults, subadults, and eggs of filarioid nematodes in the cerebellum and ventricles of the brain; one of the two bats also had clinical signs of ataxia and mild lymphoplasmacytic meningoencephalitis.[89]

Physaloptera brevivaginata can cause gastritis in *Myotis* spp. Grossly, the site of attachment may be visible from the gastric serosa as a red focus, and there is mucosal erosion with occasional ulceration. Microscopically, parasites attach at the mucosal-submucosal junction with a mixed inflammatory response including mononuclear cells, polymorphonuclear cells, edema, and fibrosis.[35]

Additional metazoan diseases are included in the Digital Content.

Protozoa

Infection with ***Toxoplasma gondii*** in two captive juvenile Australian flying foxes (*Pteropus scapulatus* and *P. conspicillatus*) caused necrosis and granulomatous to pyogranulomatous inflammation in multiple tissues, including the lungs, brain, peritoneum, and gastrointestinal wall, as well as the heart and skeletal muscle in the *P. scapulatus*.[276]

Renal coccidiosis is reported in a few bat species, including *Eptesicus fuscus*, *Pipistrellus pipistrellus*, *Myotis mystacinus*, *Myotis nattereri*, *Myotis sodalis*, *Nyctalus noctula*, *Hipposideros caffer*, and *Rhinolophus* sp.[126,171,334] Early reports in *M. sodalis* presumptively identified the coccidia morphologically as *Klossiella* sp.[171] *Nephroisospora eptesici* is the causative agent in *E. fuscus*.[334] White foci up to 2 mm in diameter may be visible on the renal capsular surface (Fig. 25.15A). Tubules, particularly those in the deep cortex and medulla, are markedly dilated with hypertrophic and hyperplastic epithelium. Epithelial cells contain asexual and sexual stages of the organism, and tubular lumina contain schizonts, merozoites, macrogametocytes, microgametocytes, and sporulated and unsporulated oocysts[126,334] (Fig. 25.15B). Eosinophilic, histiocytic, and lymphoplasmacytic inflammation may occur but is rare. There has been no evidence of clinically significant impairment of renal function in reported cases. Marked cystic tubular dilation distinguishes renal coccidiosis of bats from that of waterfowl, amphibians, and other species.[126] Moreover, the entire life cycle is thought to be completed in a single host without requiring sporogony in the environment.[334]

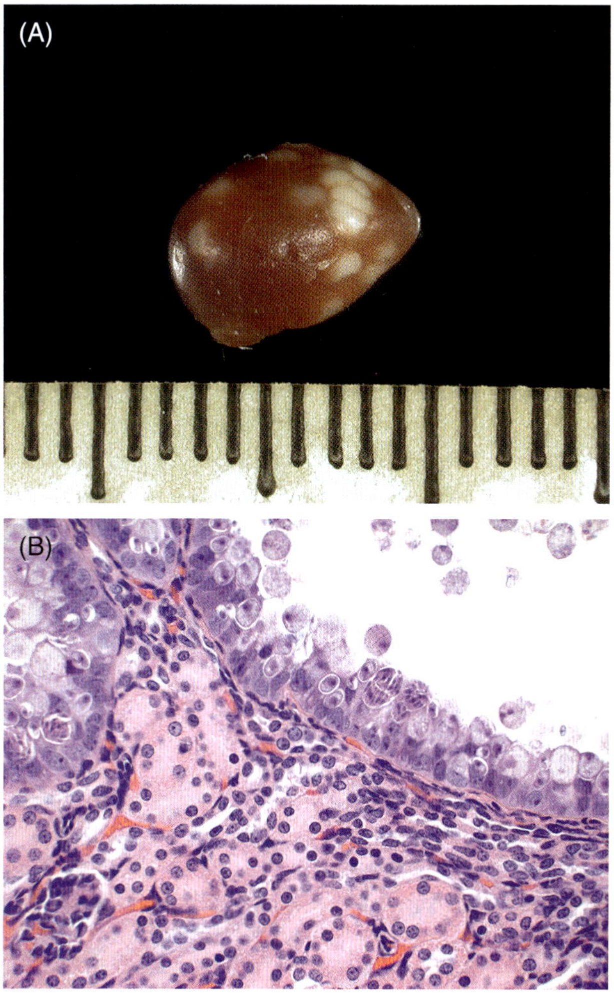

FIGURE 25.15 Renal coccidiosis in *Eptesicus fuscus*. (A) White foci up to 2 mm in diameter are present on the surface of the kidney. (B) Renal tubules are dilated and the epithelium is proliferative. Various life stages of coccidia, including macrogametocytes and meronts with merozoites, are present within the epithelium and lumen. *(Photos Courtesy of A. Wünschmann, University of Minnesota College of Veterinary Medicine.)*

Bats are thought to be potential reservoirs for ***Leishmania* spp.**, although there are few reports of associated disease.[273] In one report, *Leishmania* (*Leishmania*) *mexicana* was associated with a raised ulcerated mass on the leading edge of the patagium of a free-ranging *Artibeus* *lituratus* in Brazil that was found dead.[28] Cutaneous lesions not further described were also associated with *L.* (*Viannia*) *braziliensis* in two *Glossophaga soricina* in Brazil.[284] Other species of *Leishmania*, including *L.* (*Leishmania*) *amazonensis* and *L.* (*Leishmania*) *infantum chagasi*, have also been detected in Brazilian bats.[277]

Numerous **hemoparasites** are described in bats, although the clinical effect of these parasites is largely unknown, and most infections are likely subclinical. Free-ranging pteropodid bats may be infected with *Hepatocystis* sp., with gametocytes in the blood and schizonts in the liver.[175] Schizonts may be seen as small white foci in the liver,[116] but histologically there is no associated inflammation.[174] No clinical disease has been described. Several additional genera of hemoparasites, including *Trypanosoma*,[162] *Polychromophilus*, *Plasmodium*, *Nycteria*,[278] *Johnsprentia*, *Sprattiella*,[175] *Babesia*,[73] and *Hepatozoon*[255] have been reported in free-ranging bats. Clinical disease has not been reported.

Other apicomplexan parasites that have been identified in bats without evidence of significant lesions or clinical disease include *Eimeria* sp.[205,225] and *Cryptosporidium* sp.[225,322]

Additional protozoan diseases are included in the Digital Content.

Ectoparasites

Numerous **ectoparasites**, including flies (Streblidae and Nycteribiidae), chiggers (Trombiculidae), fleas, ticks, bed bugs, and mites have been described in bats.[94,218,303] This discussion will be limited to ectoparasites that have been associated with pathology.

Tick paralysis, caused by toxins of *Ixodes holocyclus*, is common in northern Australia in *Pteropus conspicillatus*. It typically affects females in maternity colonies. Animals present with ascending motor paralysis. Diaphragmatic involvement results in labored breathing and eventually respiratory failure. Diagnosis is based on clinical signs and the presence of the tick or an ulcer consistent with prior tick attachment; other gross or histologic lesions are not present.[45,245]

Free-ranging *Cynopterus brachyotis* infested with **teinocoptid mites** have acanthosis with dermal fibroplasia subtending the site of attachment of mites.[180] ***Demodex*** sp. infection has been reported in preputial hair follicles of *Eptesicus fuscus* and is associated with mild eosinophilic folliculitis and lymphoplasmacytic preputial gland adenitis.[178] Multiple 3- to 12-mm diameter raised cutaneous nodules on the face and ventrum of a captive adult female *Eidolon helvum* were histologically composed of cystic follicles containing numerous *Demodex* sp. mites.[31] Mites consistent with *Demodex* spp. are occasionally identified as an incidental finding at necropsy in the Meibomian gland ducts of *Rousettus aegyptiacus*, and can be associated with infiltrates of lymphocytes, macrophages, eosinophils, and neutrophils in ducts and surrounding affected glands.

Additional ectoparasitic diseases are included in the Digital Content.

OTHER INFECTIOUS ORGANISMS

Granulomatous/histiocytic inflammation and organisms consistent with the algae ***Prototheca*** were identified in the lymph nodes, spleen, meninges/brain, mesentery, pancreas, heart, skeletal muscle, and kidney of a captive *Pteropus lylei*.[216] Organisms consistent with *Prototheca* were also observed in the lung (associated with miliary white foci on gross examination), skin (associated with alopecia), spleen, heart, pancreas, peripancreatic lymph nodes, jejunum, and testicle of a captive *Rousettus lanosus*. This bat also had iron storage disease.[296]

ADDITIONAL RESOURCES

Calisher CH, Childs JE, Field HE, Holmes KV, Schountz T. Bats: important reservoir hosts of emerging viruses. *Clin Microbiol Rev*. 2006;19: 531–545.

Meteyer CU, Buckles EL, Blehert DS, et al. Histopathologic criteria to confirm white-nose syndrome in bats. *J Vet Diagn Invest*. 2009;21: 411–414.

Mühldorfer K. Bats and bacterial pathogens: a review. *Zoonoses Public Health*. 2013;60: 93–103.

Mühldorfer K, Speck S, Kurth A, et al. Diseases and causes of death in European bats: dynamics in disease susceptibility and infection rates. *PLoS One*. 2011;6: e29773.

O'Shea TJ, Cryan PM, Hayman DTS, Plowright RK, Streiker DG. Multiple mortality events in bats: a global review. *Mammal Review*. 2016;46: 175–190.

DISCLAIMER

Any use of trade, firm, or product names is for descriptive purposes only and does not imply endorsement by the US Government.

References

Full references for this chapter can be found online at https://doi.org/10.1016/B978-0-443-18437-6.00034-1.

Chapter 26

Apterygiformes, Casuariiformes, Rheiformes, Struthioniformes, Tinamiformes

Dale A. Smith[a] and Ilse H. Stalis[b]

[a]University of Guelph, Guelph, ON, Canada; [b]Independent Scholar, Long Island, NY, United States

INTRODUCTION

The infraclass Palaeognathae includes ratites (ostrich, emu, rhea, cassowary, kiwi), and tinamous, a group of related but flighted birds [see Table 26.e1 for a list of the members of the Infraclass (Clade) Palaeognathae]. Commercial ratite production for meat and leather (ostrich, emu, rhea), feathers (ostrich), and oil (emu) occurs worldwide. The South African ostrich industry is historic and legendary, and the source of considerable veterinary literature. Information on the kiwi and tinamou derives primarily from in situ conservation programs and zoological collections. Many ratite-related texts and detailed conference proceedings were produced in North America in the 1990s, a time when there was an explosive increase in the breeding of ostriches, emus and, to a lesser extent, common rheas on this continent. The product markets did not, however, expand at the same rate as the breeding programs and the heyday of the North American ratite industry ended within a decade or so. The management and diseases of these large ratites are similar. In addition to descriptions of non-infectious and infectious diseases of ratites described below, the Digital Content includes a list of reference texts describing the biology, management and diseases of ratites (Table 26.e2), comments on the application of clinical pathology, and text and a table describing major/unique anatomic features (Supplemental Text and Table 26.e3).

NON-INFECTIOUS DISEASES

Nutritional

Developmental **angular limb deformities** are a significant cause of mortality in ratite chicks.[34,156,180] These abnormalities are undoubtedly multifactorial, with nutritional and management factors intricately intertwined in the pathogenesis. Imbalances of Vitamin D_3, calcium, phosphorus, manganese, zinc, copper, and selenium; riboflavin deficiency; and excess protein have all been implicated in disease pathogenesis. The most described syndromes include unilateral or bilateral bowing or twisting of the leg bones, particularly of the tibiotarsus and/or tarsometatarsus (Fig. 26.1). Associated sequelae can include dislocation of the gastrocnemius tendon over the hock joint and lateral rotation of the major toe of the ostrich.

Rickets in ratite chicks has most often been associated with hypophosphatemia. It has been described in both rheas (called "rubber rhea" syndrome) and ostriches.[138,217] Characteristic lesions are pliability of long bones, angular limb deformities (Fig. 26.2A), and folding fractures. Also seen are enlargement and thickening of the metaphyseal plates due to elongation of the zone of hypertrophic cartilage, retained cartilage cores (to be differentiated from normal cartilage cones present in birds up to 6–8 weeks of age), subchondral microfractures, and reduced metaphyseal osteoid (Figs. 26.2B, C; eSlide 26.e1).

Atherosclerosis can develop in emu as a sequela to feeding a ration high in eggs (high cholesterol content) and fat for an extended period.[305] Histologic lesions include partial arterial occlusion by large confluent atherosclerotic plaques containing cholesterol clefts, lipid, and lipid-laden macrophages, fatty degeneration and calcium deposition in the media, fragmentation of the internal elastic lamina, and musculoelastic thickening. In a described case, the abdominal aorta and coronary arteries were the most severely affected.

Marked **pansteatitis** associated with low liver Vitamin E levels has been seen in adult kiwi.[41]

Sudden death due to a **dissecting aortic aneurysm** in ostriches and in a single emu has been suggested to be

Pathology of Wildlife and Zoo Animals. https://doi.org/10.1016/B978-0-443-18437-6.00045-6

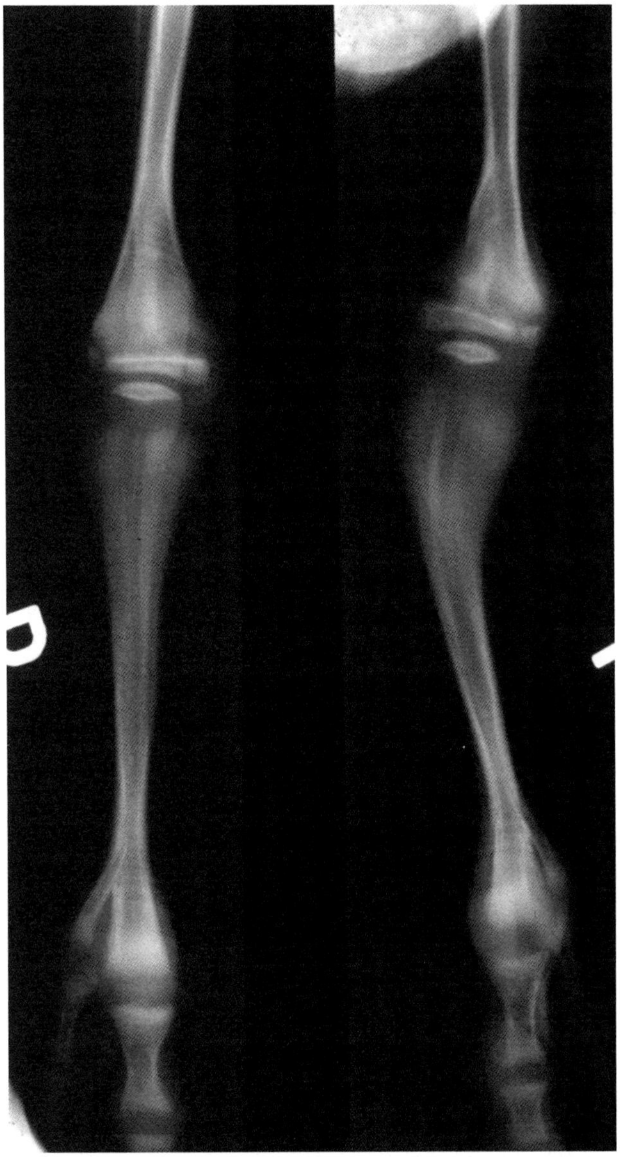

FIGURE 26.1 Angular limb deformity in an ostrich. Radiographs of the legs of a 1-month-old ostrich chick. Torsional and angular deformity of the distal tibiotarsus and the tarsometatarsus can be seen in the bones in the right image. The normal leg is shown for comparison in the left image.

related to copper deficiency.[103,314] Histologic aortic lesions include fragmentation and disruption of the elastic laminae, cystic extracellular spaces, and an accumulation of what have been interpreted to be strongly sulfated acid mucopolysaccharides based on staining with Alcian blue and periodic acid-Schiff (PAS). Copper deficiency was suspected based both on the character of the aortic lesions and the findings of low liver copper levels in a number of affected birds (liver copper: affected birds <3.5 ppm; unaffected >5 ppm). A copper-dependent enzyme, lysyl oxidase, is required for the cross-linking of collagen and elastin fibers that provides strength to the vascular wall.

Toxic

Monensin toxicity in ostriches, greater rhea, and an emu has occurred due to errors in feed formulation. It presents clinically as anorexia, ataxia, loss of control of the neck, recumbency and dyspnea prior to death, or death with no premonitory signs. Expected elevations in creatine kinase, aspartate aminotransferase, and lactate dehydrogenase and classic gross and histopathologic lesions occur. Characteristic findings of the latter include multifocal to coalescing areas of myocardial and skeletal muscle pallor that correspond to hyaline degeneration, vacuolization, and monophasic myodegeneration and myonecrosis with loss of cross striations.[23,241] Satellite cell proliferation and responsive lymphohistiocytic infiltrates may also be seen in animals with increased survival times.

The larger ratites are indiscriminate eaters, and ingestion of metallic foreign bodies has resulted in absorption of toxic levels of iron, lead, copper, and zinc. Pancreatic exocrine atrophy and fibrosis in association with regenerative ducts have been described in ostriches due to **zinc toxicity**.[56]

Poisoning due to the ingestion of certain **plants** and **plant toxins** has been reported in ratites. These include ingestion of parsley (*Petroselinum sativa*; ostrich, photosensitization); avocado (*Persea americana*; ostrich, myocardial degeneration); oak leaves (*Querca agrifolia*; double wattled cassowary, enteritis and nephrosis); wilted grass sprouts (ostrich, prussic acid poisoning); yew (*Taxus baccata*; emu, gastroenteritis and pulmonary edema); and a number of toxic southern African plant species [including ragwort (*Senecio scleratus*) and lantana (*Lantana camara*); ostrich].[70,156]

Mortality occurred in a flock of young emu chicks that ingested large numbers of **blister beetles** (*Pyrota insulata*) containing **cantharidin**, a vesicant. Resulting lesions included esophageal congestion and ulceration, submucosal vascular thrombosis and necrosis, and sloughing of the lining of the ventriculus.[26]

Levamisole toxicity has been reported in kiwi.[123] Described lesions and outcomes included acute bronchopneumonia with severe pulmonary edema and death in five of six affected birds. Hepatocellular degeneration and vascular thrombosis were also seen in one, and multifocal hepatic necrosis was seen in another of the birds.

Additional toxins that have been described as a cause of clinical disease and death in large ratite species include several compounds used therapeutically, in pest control, or that are found in feed and forage. These include colistin, dynamulin, furazolidone, lincomycin, lindane, mineral oil, morantel, furan, thiabendazole, streptomycin, warfarin, and gizzerosine (causes ventricular erosion), mycotoxins (aflatoxin, ochratoxin, T2 toxin, sporodesmin, zearalenone), salt, and selenium.[83,105,156,161,173,276]

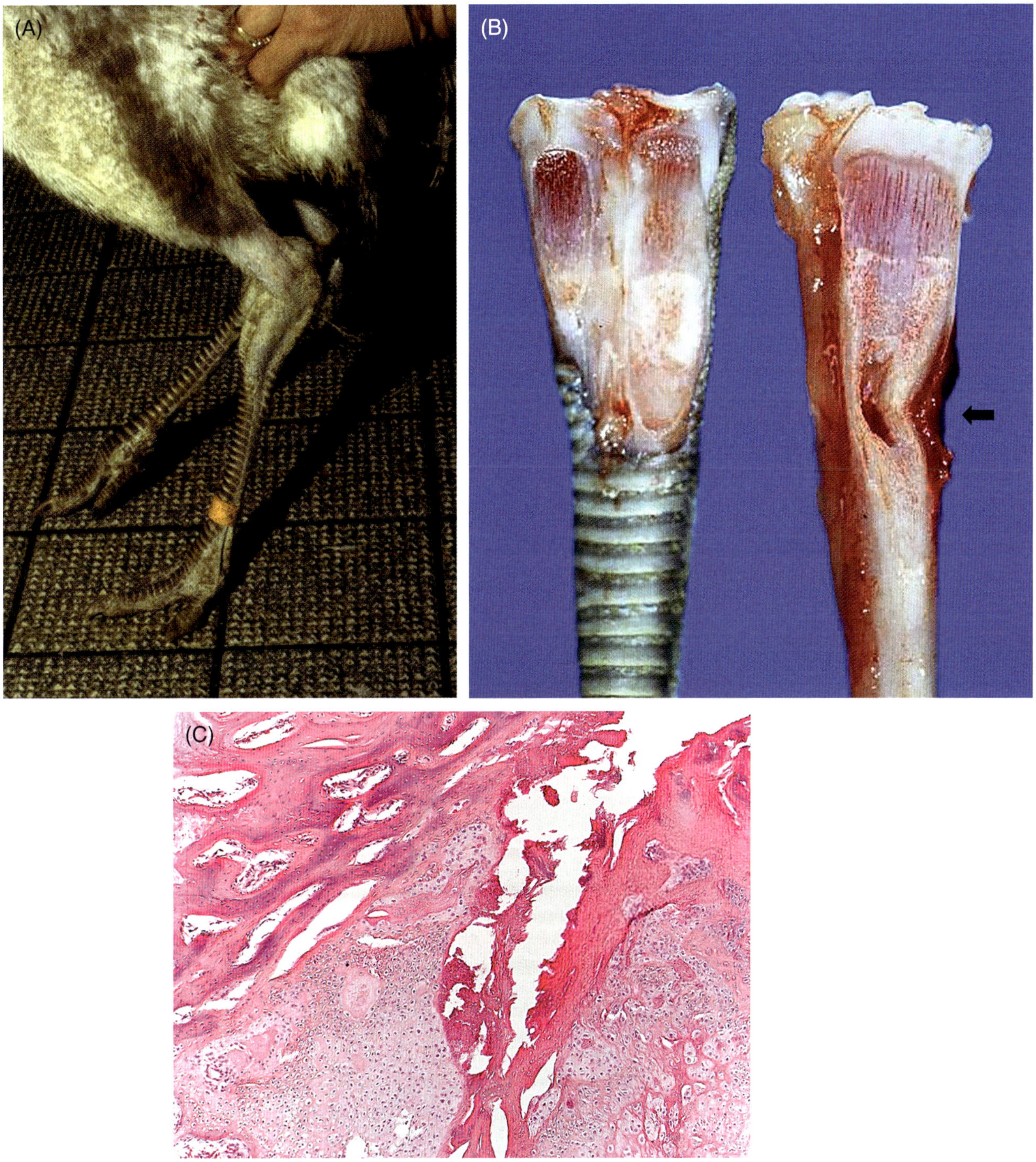

FIGURE 26.2 Phosphorus-deficiency rickets in young common rhea. Characteristic features include (A) tibiotarsal curvature and thickening at the proximal growth plate, (B) enlargement of the metaphyseal plates, elongation of the zone of hypertrophic cartilage, and pathologic fractures in the underlying metaphyseal bone *(arrow)*, and (C) pathologic fractures characterized by irregular, jagged-edged clefts in the bone surrounded by irregular rests of hypertrophic cartilage (see eSlide 26.e1). *(Part B: Photo Courtesy of H.L. Shivaprasad, University of California Davis, California Animal Health & Food Safety Laboratory.)*

Congenital/Genetic

A form of **mucopolysaccharidosis type IIIB**, an autosomal recessive, inherited **lysosomal storage disease** caused by a deficiency of lysosomal alpha-*N*-acetylglucosaminidase (NAGLU), has been described in emus.[19,125,229] Deficiency results in a failure of heparin sulfate degradation and the subsequent intracellular accumulation of this glycosaminoglycan. Juvenile birds develop progressive neurological signs; sudden death due to abdominal hemorrhage is also described. Characteristic histologic lesions include widespread neuronal swelling and fine, intracellular vacuolation in the brain, spinal cord, retina, and autonomic ganglia (Fig. 26.3A). Material in vacuoles stains variably positive with PAS and strongly positive with Luxol fast blue. Macrophages containing the material may be present in a variety of tissues including liver, spleen, intestine, and the tunica media of the aorta. On transmission electron microscopy (TEM), membrane-bound, electron-dense, lamellated cytoplasmic material (consistent with "zebra bodies") is seen in neurons and ganglial cells (Fig. 26.3B). Storage material in retinal cells may differ and consist of a central granular, highly electron-dense core surrounded by membranous whorls, while in macrophages, it may appear as well-circumscribed, electron-dense, homogenous circular bodies surrounded by stacks of rough endoplasmic reticulum.

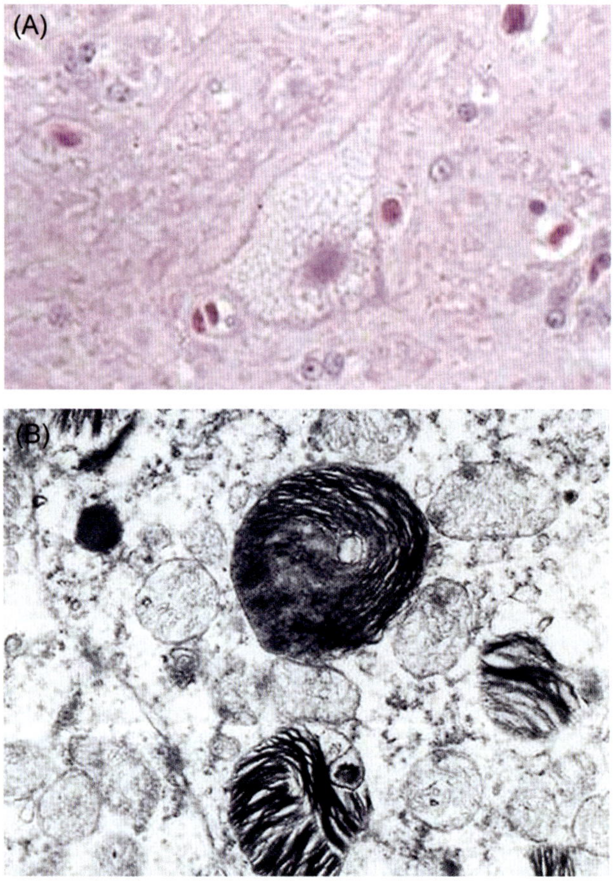

FIGURE 26.3 Mucopolysaccharidosis IIIB (lysosomal storage disease) in the brain of an emu. (A) Characteristic features are diffuse cytoplasmic vacuolization and cell swelling, both of which are seen in this cerebral neuron. (B) Storage material consists of accumulations of electron-dense, lamellar, membrane-bound cytoplasmic material due to defective enzymatic activity. TEM. *(Photos Courtesy of H.L. Shivaprasad, University of California Davis, California Animal Health & Food Safety Laboratory.)*

A **mechanobullous skin condition** with low prevalence occurred in ostrich chicks on a large farm in Israel over a 2-year period.[235] Lesions, present only in skin, consisted of dermolytic blistering, skin peeling, and feather loss. Disease resulted in death or euthanasia of affected chicks by 14 weeks of age. Microscopic and TEM examination of skin showed epidermolysis bullosa-like lesions, including subepidermal bullae, splitting of the sublamina densa, dermal edema, and with time, inflammation and crusting. Lesions were considered likely to be inherited, but the parentage of the affected birds could not be traced.

Miscellaneous/Multifactorial

Trauma, **misadventure**, and **exertional rhabdomyolysis** are leading causes of death in adult age group ratites. In the wild, the most commonly recognized causes of mortality for cassowaries and kiwis are trauma, and trauma and predation, respectively.[140] Tinamous are prone to injury and trauma when startled or frightened and must be handled carefully to avoid fractures.

Management-related factors are a major cause of mortality in captive ratite systems. In the hands of inexperienced ratite farmers, hatching and neonatal chick survival rates can be abysmal. Improper egg collection and storage, incubation, and hatching parameters result in embryonic mortality and failure to hatch, weak and edematous hatched chicks, and poor chick viability.[45,47,130] Acute generalized **skeletal myopathy and myonecrosis** involving the complexus (pipping) (eSlide 26.e2), hind limb, and costopulmonaris muscles has been described in newly hatched chicks and may be a result of exertion or physiologic difficulty at hatching.[240]

"**Fading syndrome**" and **gastric stasis** are major causes of stunted growth, wasting, and mortality in ostrich chicks up to 6 months of age; up to 70% of chicks can be affected to varying degrees. The cause is not fully understood but environmental parameters, maladaptation, and increased susceptibility to infectious agents are likely important factors.[51,156,223] Necropsy lesions are nonspecific but include emaciation and proventricular and ventricular changes, including either lack of contents or distension without impaction, and mucosal hyperplasia and degeneration.

Ratites are indiscriminate eaters and **gastrointestinal impaction** and **gastrointestinal foreign bodies** are, therefore, important causes of anorexia, wasting, and death

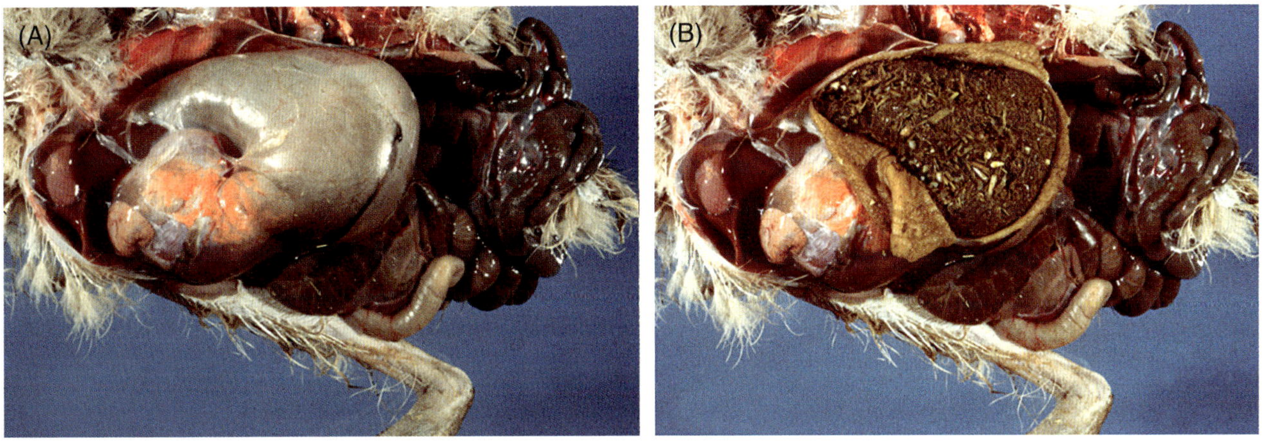

FIGURE 26.4 Impaction in an ostrich. The proventriculus and ventriculus are markedly distended and filled with coarse roughage-unopened (A) and opened (B). *(Photos Courtesy of M. Brash, Animal Health Laboratory, University of Guelph.)*

in birds of all ages.[180,257] Impaction with straw, shavings, coarse grass, and virtually any other ingestible substrate have been reported (Figs. 26.4A, B).

Amyloidosis has been reported in rhea and in the liver and/or spleen in young ostriches with mycotic airsacculitis and pneumonia.[7,72] Histologically, eosinophilic homogenous material positive with Congo red staining expands sinusoids and is associated with hepatocyte degeneration and atrophy. In the spleen, the material is within the tunica media and surrounds splenic blood vessels. The presence of amyloid A-type fibrils has been confirmed using immunohistochemistry.[7]

Pneumoconiosis is common in captive kiwi and lesions can be marked. The condition is thought to be associated with dry, dusty substrates and enhanced by the birds' sniffing behavior during feeding and substrate investigation.[286]

Miscellaneous conditions identified in tinamous held in zoos include a **myocardial infarct, aortic and carotid atherosclerosis, vegetative valvular endocarditis** associated with chronic sinus infection, **oviductal egg impaction**, and **nephritis.**[137] Conditions in birds raised under more intensive conditions are somewhat different with **uric acid nephrosis**, **amyloidosis**, **gastrointestinal foreign bodies**, renal trematodiasis (***Paratanaisia confusa***), and the presence of ***Capillaria penidoi*** in the upper digestive tract as the most significant findings in one study of red-winged tinamous.[211]

A **spongiform encephalopathy** similar to that seen with prion disease has been described in zoo-housed ostriches and emus in Germany. Neurological signs were present in most, but not all birds and other significant lesions were found in some of the birds. Lesions included marked vacuolation of the neuropil in the brain stem and medulla, as well as neuronal cytoplasmic vacuolation, particularly in the red and vestibular nuclei and the reticular formation. Mild gliosis, isolated necrotic neurons, and neuronophagia were present in the anterior brainstem. The prion protein PrP was not identified in these cases using IHC labeling.[30]

Neoplastic

Neoplastic diseases have rarely been reported in Struthioniformes. Multicentric lymphoid neoplasia in ostriches is most commonly described. Single cases of lymphoid leukemia and a monoclonal gammopathy have also been reported. There are no reports of immunohistochemical or other typing of neoplastic lymphocytes. Included in Table 26.e4 are brief descriptions of the gross and histologic details of these and several other reported cases of lymphoid neoplasia in ratites. Reports of nonlymphoid neoplasia include intrathoracic hemangiosarcoma[146] and capillary-type pulmonary hemangioma[274] in young ostriches, mucinous adenocarcinoma in the conjunctiva of a 2-year-old ostrich,[238] thyroid adenoma in adult ostrich,[30] telangiectatic osteosarcoma in a North Island brown kiwi,[50] salivary gland adenoma in a 7-month-old North Island brown kiwi,[65] fibrosarcoma in a 1-year-old ostrich,[127] and ovarian carcinoma with carcinomatosis and cutaneous squamous cell carcinoma in the foot of an ostrich and southern cassowary, respectively (personal communication, D. McAloose, Wildlife Conservation Society).

INFECTIOUS DISEASES

DNA Viruses

Viruses reported as frequent or significant causes of infection and/or disease are described below. Brief descriptions of additional viral agents reported in ratites are listed in Table 26.e5.

In 1992, an **adenovirus** outbreak caused 90%–100% mortality in ostrich chicks less than 2 months of age in Oklahoma, United States.[251] Egg transmission was highly

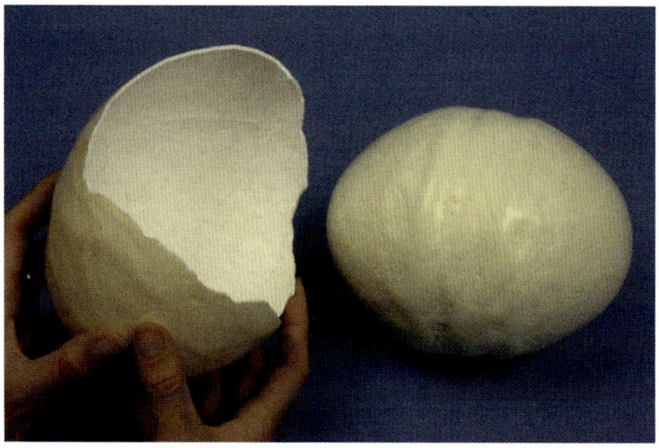

FIGURE 26.5 Shell abnormalities in ostrich eggs. Malformed, wrinkled eggshells. The causes of this nonspecific change are poorly understood but are postulated to include viral infection, particularly with adenovirus and Newcastle disease virus, bacterial salpingitis, nutritional deficiencies or imbalances, and abnormalities of the oviduct.

suspected. The most unique clinical symptom was foul-smelling, gray chalky feces. Gross necropsy findings were nonspecific and included pulmonary congestion, pale foci in the liver, a pale spleen, and airsacculitis. On microscopic examination, the most consistent lesions in chicks were splenic and bursal lymphoid depletion, hepatic necrosis, necrotic enteritis, and pulmonary congestion. Adult hens laid chalky or wrinkled eggs (Fig. 26.5) and one had evidence of hepatitis. An adenovirus was isolated from dead chicks, the yolk sac of a dead in-shell chick, and the trachea of healthy adult birds. Inclusion bodies consistent with adenovirus were not seen and the virus was not further characterized.

In the United Kingdom, six adenoviruses similar to **fowl adenovirus serotype 8** have been isolated from ostriches. The sources were the tracheas of an adult hen that laid eggs with chalky ridged shells and her mate, and the intestines of chicks with diarrhea.[133] In Italy, an adenovirus isolated from a 4-month-old ostrich chick was associated with an enlarged nodular pancreas and hemorrhagic enteritis. Inoculation studies in young guinea fowl chicks produced severe pancreatic lesions, including tissue swelling and hardening, loss of acini and inflammation, and death. Intranuclear viral inclusions were present in surviving and regenerating cells. Restriction endonuclease analysis showed the virus to be similar to **fowl adenovirus serotype 1**.[54]

An **avian pox virus** was identified histologically and by sequencing as the cause of transient nodular cutaneous lesions in the skin of the legs of two immature North Island brown kiwi and near the beak in one.[142] The histopathologic lesions were classic, with epidermal hyperplasia, epithelial swelling, and eosinophilic intracytoplasmic viral inclusions (Bollinger bodies).

A group of elegant-crested tinamous was found to be infected with a novel **hepadnavirus**.[163] Virus was identified in tissues of deceased birds and in serum of living birds without clinical signs. Liver lesions varied and were characterized as necrotizing, lymphohistiocytic, and pyogranulomatous. Intranuclear inclusions were seen in some birds.

Outbreaks of paralysis and diarrhea in 1- to 3-month-old ostrich chicks have recently been reported from China.[332] Affected birds had hemorrhage in multiple organs, especially muscle, intestine, and thymus, and swelling and effusion of the knee joint. Disease was attributed to a novel ostrich **parvovirus**.

RNA Viruses

Several of the RNA viruses that affect ratites cause diseases that are reportable to the World Organization for Animal Health (WOAH). These include avian influenza virus (AIV), avian orthoavulavirus 1 (Newcastle disease virus, avian paramyxovirus 1), and eastern and western equine encephalitis viruses (EEE and WEE, respectively).

Avian influenza virus (AIV) infection has been reported in ostriches, emus, rheas, and a cassowary. Infection with a variety of viral strains including those of both high (H5 and H7; **HPAI**) and low pathogenicity occur frequently in farmed ostriches in South Africa.[1,2] Outbreaks are thought to result from exposure to wild birds, but evidence of origin from chickens has been reported.[1] Clinical signs and lesions associated with infection in ostriches in natural outbreaks and with experimental infection are variable, and it is worth noting that there are no pathognomonic lesions. Mortality can be up to 100% and is dependent on age, virus strain, and secondary infections.[9,55,71] Chicks and juvenile birds are more susceptible to infection and develop more significant clinical disease than adults, who frequently appear unaffected. Clinical signs include anorexia, depression, and poor growth rates as well as green discoloration of the urine (biliverdinuria), swelling of the throat and neck, ocular discharge and sneezing, and evidence of respiratory difficulty. Gross necropsy findings are nonspecific and can include serous atrophy of fat and muscle atrophy; luminal bile and bile staining of the upper gastrointestinal tract; hepatic bile stasis; congestion and mottling of the liver, spleen, kidneys, trachea, and lungs; epicardial hemorrhages; and intestinal congestion to severe hemorrhagic enteritis. Histologic lesions also vary and include multifocal to diffuse necrosis of liver, spleen, and kidneys with a heterophilic response; pancreatic hemorrhage and necrosis with mononuclear cell inflammation; fibrinoid arteriolar necrosis and inflammation, particularly in spleen and brain; interstitial pneumonia; mild to marked airsacculitis; multifocal malacia and neuronophagia in the brain; and necrosis and hemorrhage in the intestine, particularly in the duodenum. Both gross and histologic lesions are affected by the presence of secondary infections, which have been prevalent in several reports. Infection with low pathogenic avian influenza viruses results in primarily a mild respiratory disease, with a reduction in egg production.[71]

In emus and rheas, respiratory disease is the predominant clinical presentation with HPAI H5N2 and H7N1 subtypes in the United States. Lesions are present in both the upper and lower respiratory tracts.[230,327] In China, an outbreak of low pathogenicity H9N2 influenza in emus was associated with 60% morbidity and 12.5% mortality.[164] Gross lesions included hyperemia and hemorrhage in the mucous membranes, edema and congestion of the lungs and cloaca, epicardial hemorrhage, hepatic and renal swelling, and widespread gastrointestinal hemorrhage. Histologic lesions included desquamation of the bronchial mucosa; myocardial necrosis; and liquefactive necrosis, satellitosis, neuronophagia, perivascular cuffing, and hemorrhage in the brain.

Standard methods of diagnosing avian influenza infections in ratites include PCR, virus isolation and typing, testing for pathogenicity in chickens, and standard serologic surveys. Tracheal swabs may be positive for virus earlier in disease than cloacal swabs.[9]

Avian orthoavulavirus 1 (**Newcastle disease virus**, **Avian paramyxovirus 1**) has been identified in ostriches in zoos and commercial operations in several countries. Like avian influenza viruses, this is a WOAH-listed reportable disease. Virus has been isolated from tissues of birds with and without a history of clinical disease. There are no pathognomonic gross or microscopic lesions. Chicks and juveniles appear most susceptible to disease, with sudden death as the most common presentation. In adult birds, notable clinical signs and lesions include edema of the head; mucoid sinusitis, tracheitis, and airsacculitis; splenomegaly; epicardial petechial hemorrhages; and neurologic abnormalities. Histologic lesions include neuronal chromatolysis, endothelial cell hyperplasia, perivascular lymphoid cuffs, and multifocal gliosis that is generally limited to the brainstem, and less frequently, splenic vascular necrosis with fibrin exudation or lymphoplasmacytic pancreatitis.[157] Cerebrospinal fluid has been suggested as a source of material for direct EM, hemagglutination inhibition, and virus isolation, although this last can be difficult.[9] An avirulent class I Avian orthoavulavirus was recently identified in ostriches in South Africa.[3]

Western equine encephalitis (**WEE**) is also a WOAH-listed reportable disease. Infection in emus is predominantly a neurologic disease with a reported morbidity of 10%–50% and mortality of approximately 10%.[20,254] Birds as young as 3 months of age can be affected. Clinical signs include depression, anorexia, and evidence of weakness and/or ataxia including muscle tremors, recumbency, and unnatural positioning of the head on the back (opisthotonus). Watery diarrhea, green urates, and pericardial effusion may also be present.[20] Microscopic lesions in the brain include mild to moderate infiltrates of lymphocytes, plasma cells, heterophils, and/or histiocytic cells perivascularly and in the meninges. Neuronal satellitosis, increased numbers of oligodendroglia, axonal degeneration, and empty axon sheaths can also be seen.[254] Extensive vasculitis with necrosis and inflammation in a wide range of tissues is common. Diagnosis can be made by virus isolation and PCR.

A single outbreak of WEE in rheas in the United States has been described. Diagnosis was based on seroconversion and response to vaccination.[253] Clinical signs were generally nonspecific and included anorexia, drowsiness, stupor, ataxia, hyperpnea, and loose stool for a period of 48–72 h prior to death. Twenty-two of the 90 birds died; at least 30 showed mild illness and recovered. Gross findings were restricted to mild hepato- and splenomegaly. On microscopic examination, there was mild multifocal subacute pericholangitis, diffuse hepatic sinusoidal leukocytosis, and moderate diffuse splenic histiocytosis and plasmacytosis.

Eastern equine encephalitis (**EEE**) is primarily a disease of emus. It can affect all ages of birds and has a mortality rate up to 87% in juveniles and adults. Death is frequently acute, but reported clinical signs include marked depression, bloody vomitus, and diarrhea. The predominant gross necropsy findings include large amounts of blood in the small and large intestines and systemic serosal petechiation to ecchymoses (Fig. 26.6). Histologically, consistent lesions include fibrinoid degeneration of splenic

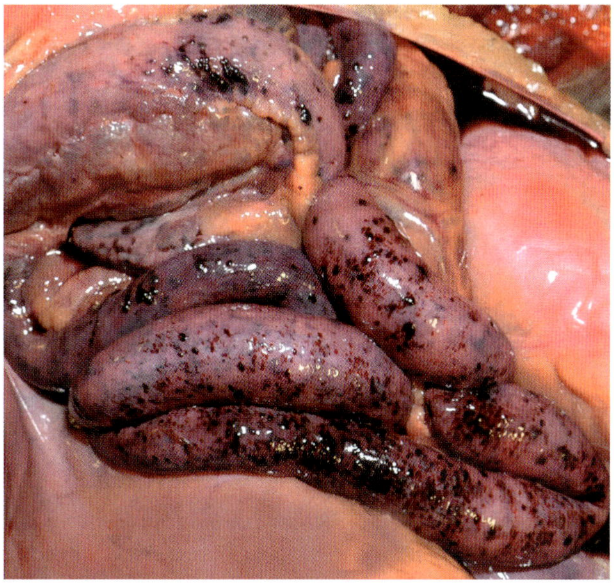

FIGURE 26.6 Eastern equine encephalitis virus in an emu. Infection causes vasculitis which can be seen as multifocal petechial and ecchymotic hemorrhages along the serosal surfaces of the intestinal tract. *(Photo Courtesy of L. Farina, College of Veterinary Medicine, University of Florida.)*

sheathed arterioles with necrosis of endothelial macrophages and diffuse lymphoid necrosis; hepatic congestion and hemorrhage with widespread hepatocellular necrosis; and changes in the intestinal lamina propria including the presence of large amounts of necrotic cellular debris, hemorrhage, a mild heterophilic infiltrate, and necrosis of lymphoid follicles. It is notable that lesions are not present in the central nervous system.[48,317] Emus experimentally infected as part of a vaccine trial developed severe clinical signs and shed virus in feces, oral secretions, and regurgitated material; infection spread horizontally in the absence of mosquito vectors.[302]

Naturally occurring EEE infection has also been described in six captive cassowaries in the United States.[141] Both juvenile and adult birds were affected. Clinical signs included sudden death, lethargy, weakness, ataxia, and convulsions. Gross necropsy lesions included coelomitis and diarrhea. The most consistent histologic lesion was vasculitis in the liver and spleen. Mild lymphocytic encephalitis and hepatitis, splenitis, pneumonia, and nephritis were also seen. The presence of intercurrent diseases had an impact on the range of pathologic changes. Notably, the affected cassowaries did not develop hemorrhagic enteritis.

The diagnosis of EEE can be confirmed by virus isolation from tissues including liver, spleen, blood, and brain; TEM; and PCR. Surviving birds develop both hemagglutinating and neutralizing antibodies. Antibodies can also result from maternal transfer, inapparent infection, and vaccination.[77]

A prolonged outbreak of **bornavirus**-associated neurologic disease occurred in ostrich chicks on several farms in Israel between 1988 and approximately 1993. Clinical signs included a brief period of incoordination followed by spastic paresis and eventually death due to secondary complications. Mortality varied but was up to 20%. There were no consistent gross lesions. Histologic lesions were restricted to the lumbosacral spinal cord and included neuronal degeneration and neuronophagia, glial satellitosis, and the formation of glial nodules. The disease was experimentally transmitted to ostrich chicks using the brains of infected chicks and on microscopic examination, notable perivascular cuffing developed. Extensive laboratory investigation into the cause of the syndrome provided convincing evidence of infection by **Borna disease virus**.[192]

Aquatic bird bornavirus-1 was identified as the cause of prolonged, subtle gastrointestinal, and neurologic clinical signs in an emu in a zoological park. Histopathologic lesions included marked lymphoplasmacytic encephalomyelitis with prominent perivascular cuffing (eSlides 26.e3 and 26.e4). Multifocal neuritis was also seen in the gastrointestinal tract. The disease was assumed to have been transmitted from wild Canada geese.[221]

Bacteria

Bacterial infection is a common cause of embryonic death and of disease and mortality in young hatched ratite chicks. Yolk sac infection, generally due to ascending infection through an open umbilicus, is common in incubated chicks and can lead to bacteremia (eSlide 26.e5). In one summary of findings from 59 tinamous of eight species in a North American zoo, bacterial omphalitis due to *Escherichia coli* and *Pseudomonas* spp. was among the noted, common conditions.[137]

Bacteria that cause localized (particularly respiratory and conjunctival), enteric, and systemic infection in ratites are no different from those that affect other species of birds and include a variety of gram-positive and gram-negative environmental and opportunistic organisms. Select bacterial diseases that are frequently reported in ratites or are of particular note are described in Table 26.e6. These include diseases due to *Bacillus anthracis*, *Clostridium perfringens* (eSlides 26.e6 and 26.e7), *C. difficile*, *C. chauvoei*, *Erysipelothrix rhusiopathiae*, *Brachyspira hyodysenteriae* (Fig. 26.7), *Bordetella avium*, *B. bronchiseptica*, *Campylobacter jejuni*, *Chlamydia psittaci*, *Lawsonia intracellularis*, *Hemophilus* spp., *Salmonella* spp., and mycobacterial (Fig. 26.8) and mycoplasmal infections.

Botulism due to ***Clostridium botulinum*** and associated with toxin types C and D has been described in ostriches and, presumptively, in a cassowary.[261] Diagnosis in these cases was related to clinical history and identifying toxins via mouse bioassay. Affected ostriches may be acutely ataxic, paretic, or completely paralyzed.[156]

Fungi

Infection of the upper and lower respiratory systems with ***Aspergillus* spp.**, particularly ***A. fumigatus*** and ***A. niger***,

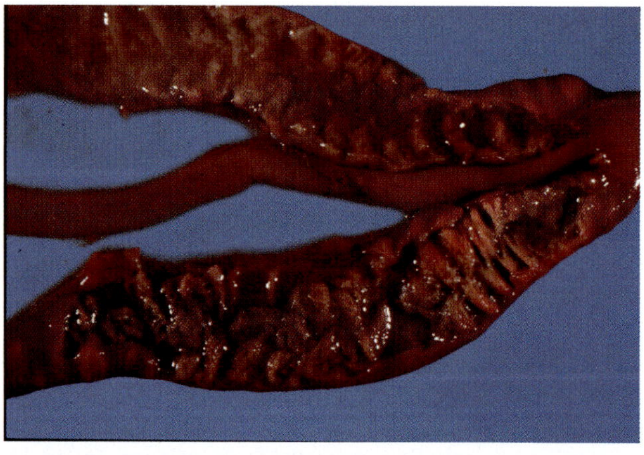

FIGURE 26.7 *Brachyspira hyodysenteriae* typhlitis in a common rhea. Fibrinous and necrotizing typhlitis is a common cause of mortality in young rheas. *(Photo Courtesy of M. Brash, AHL, University of Guelph.)*

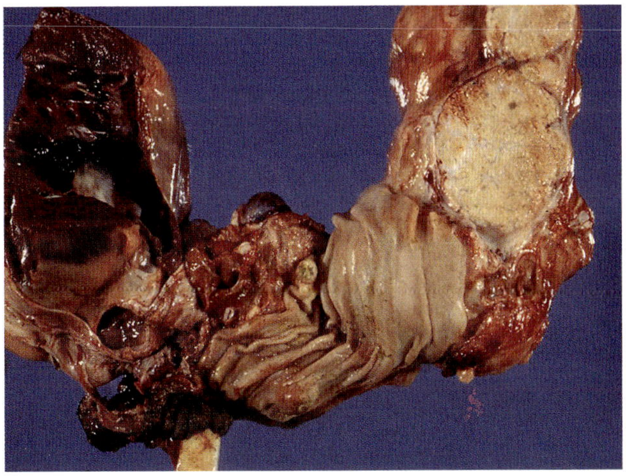

FIGURE 26.8 Mycobacteriosis in an ostrich. Esophageal and periesophageal granulomatous lesions are present in the thoracic region (heart on the left side of the image, esophagus centrally and on the right side). Large, caseous masses are seen more commonly with mycobacteriosis in ratites than in other frequently affected avian taxa. *(Photo Courtesy of M. Brash, AHL, University of Guelph.)*

is a common cause of disease in captive ostriches, emus, and rhea.[156] Outbreaks are frequent in young chicks (e.g., brooder pneumonia). Clinical signs and gross and pathologic lesions are similar to those in any avian species and include pale gray to yellow nodules and plaques in the lungs and air sacs (Fig. 26.9). Histopathologic lesions are typical for avian aspergillosis, with central areas of necrosis that contain parallel-walled, dichotomously branching, fungal hyphae and conidia (at air interfaces) that are surrounded by heterophilic and granulomatous inflammation with multinucleate giant cells and (in chronic lesions) an outer layer of fibroblasts or granulation tissue. Fungal elements are often easily identified with routine hematoxylin and eosin staining; silver or PAS stains highlight the organisms.

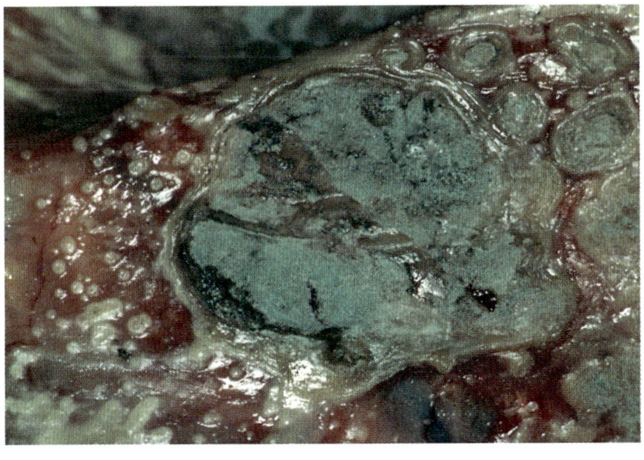

FIGURE 26.9 Mycotic airsacculitis in an emu. Opaque nodular and plaque-like thickenings are present on the serosal and air sac surfaces. Surrounding and underlying regions are covered by blue-green mats, which are characteristic of *Aspergillus* spp.

Aspergillus spp. have also been reported as the cause of dermatitis in ostriches. Lesions can be localized or extensive.[177] Regardless of site, fungal culture is confirmatory; PCR can be used for additional characterization and differentiation of species. Predisposing factors to infection include contamination of eggs and hatcheries, overwhelming exposure to fungal spores through contaminated feed and bedding, and immunosuppression; high humidity and/or poor hygiene are husbandry conditions that have been associated with cutaneous aspergillosis.

Infection of the upper digestive tract; the oral cavity, pharynx, and esophagus, by opportunistic yeasts is reported frequently in ostrich chicks and is primarily associated with ***Candida*** spp. Proliferative and necrotizing lesions containing intralesional 3–5 μm diameter yeast and interspersed nonseptate pseudohyphae or hyphae are characteristic. Organisms can be highlighted in histologic section with either silver or PAS staining. Overuse of antibiotics is a suspected predisposing factor[156] but infections can occur with no associated antibiotic administration.

Lesions of proventricular and ventricular **zygomycosis** in ostrich chicks include mucosal erosion and ulceration, subserosal edema, hemorrhage, and intralesional fungus. A thick layer of mucus covering the lining of the ventriculus, a result of mucosal mucous cell hyperplasia, appears to be a characteristic finding.[160,177] Fungal hyphae are broad (6–20 μm diameter), irregular in appearance, pauciseptate, and branch at 45–90 degrees. As with other fungal organisms, the fungi can be highlighted with silver or PAS staining. Infection is considered opportunistic or secondary to lesions from proventricular and/or ventricular impaction.

Macrorhabdus ornithogaster (avian gastric yeast) is an ascomycete yeast that is found in ostriches and common rhea. It is associated with a "failure to thrive" clinical syndrome with progressive emaciation despite an initially good appetite.[156,195] Mortality can be up to 100%. Gross necropsy findings include emaciation, a dilated but not impacted proventriculus, reduced digesta in the remainder of the gastrointestinal tract and, in some cases, erosion and ulceration of proventricular and ventricular mucosa. Organisms can be seen on fecal smears or proventricular or ventricular scrapings where they resemble oversized, elongate, rod-shaped bacteria. Histologically, they are present on the surface of the koilin where they appear as rigid, parallel, stacked rectangular structures. The organisms are Gram stain variable. Giemsa staining highlights the nuclei on cytology, and PAS and silver staining highlight the organisms in histologic sections. Associated inflammation is generally minimal.

Fungal dermatitis has been infrequently reported in ratites. A dermatophyte with an endothrix pattern in feather shafts, possibly a ***Trichophyton* sp.**, resulted in feather loss and proliferative crusting lesions on the head and upper neck in several ostrich chicks.[225] ***Microsporum gypseum*** was described as a cause of rows of small cutaneous lesions that resulted in downgrading of skins from commercially raised ostriches in South Africa.[156]

Four cases of infection with ***Cryptococcus bacillisporus*** (previously *C. neoformans* var. *gattii*) have been reported in North Island brown kiwi.[151,191] Disease was limited to the respiratory tract in two cases and was disseminated in two. Multifocal to diffuse necrosis, hemorrhage, and histiocytic inflammation with intralesional fungus was present in affected organs, which included the heart, liver, kidney, oviduct, pancreas, proventriculus, and intestinal serosa. Organisms were spherical, 8 µm in diameter, and surrounded by a 1.6 µm capsule that could be highlighted by staining with PAS or Best's mucicarmine. Narrow-necked budding, a characteristic feature of *Cryptococcus* spp., was present. The low body temperature of the kiwi, the use of eucalyptus mulch, and the species' habit of sniffing while probing, were all considered as predisposing factors.

Occasional cases of infection by a variety of opportunist agents have been reported in ratites. For example, two cases of cerebral and ocular infection by ***Ochroconis gallopava*** in elegant crested tinamou chicks.[227] Esophageal pythiosis, caused by the oomycete ***Pythium insidiosum***, was diagnosed in an ostrich in Brazil housed in an endemic area.[290]

Metazoa

A small number of parasitic diseases are or have been associated with substantial morbidity and/or mortality in ratites. As in any other species, intensive rearing of birds and environmental contamination increase the prevalence and severity of disease. Ratites will ingest the feces of other species, thus the presence of parasite eggs in their feces can reflect true or pseudoparasitism. Table 26.e7 contains a list of the major metazoan endoparasites that affect larger ratites along with their host/s, geographic distribution, and associated disease. Parasites of particular significance include *Libyostrongylus douglassii*, *Deletrocephalus dimidiatus*, *Codiostomum struthionis*, and *Houttuynia struthionis*. The endoparasites of the kiwi were tabulated by McKenna.[199] The ingestion of earthworms, which serve as intermediate and paratenic hosts for many nematode parasites, may predispose to the range of helminth infections described in kiwi.[63] A variety of helminth parasites have also been recorded in both captive and free-ranging tinamous, thickened oral and esophageal mucosa and fetid diarrhea being the most commonly associated pathologic changes/signs.[194,282]

Larval migrans is a concern in all ratite species. The presence of definitive hosts results in fecal contamination of food or the environment and subsequent parasite transmission. In many cases the causative nematode cannot be definitively identified. ***Baylisascaris procyonis*** (natural host: raccoon) is a significant cause of neural larval migrans in ratites in North America.[166] Characteristic features of *B. procyonis* larvae include paired lateral alae and a prominent intestine with paired lateral excretory columns (Fig. 26.10A; eSlide 26.e8). The parasite is most often seen in the brainstem or cerebellar white matter. Inflammation or tissue changes associated with parasite migration may be absent or minimal and can include neuropil vacuolation, hemorrhage, or tissue necrosis. Focal

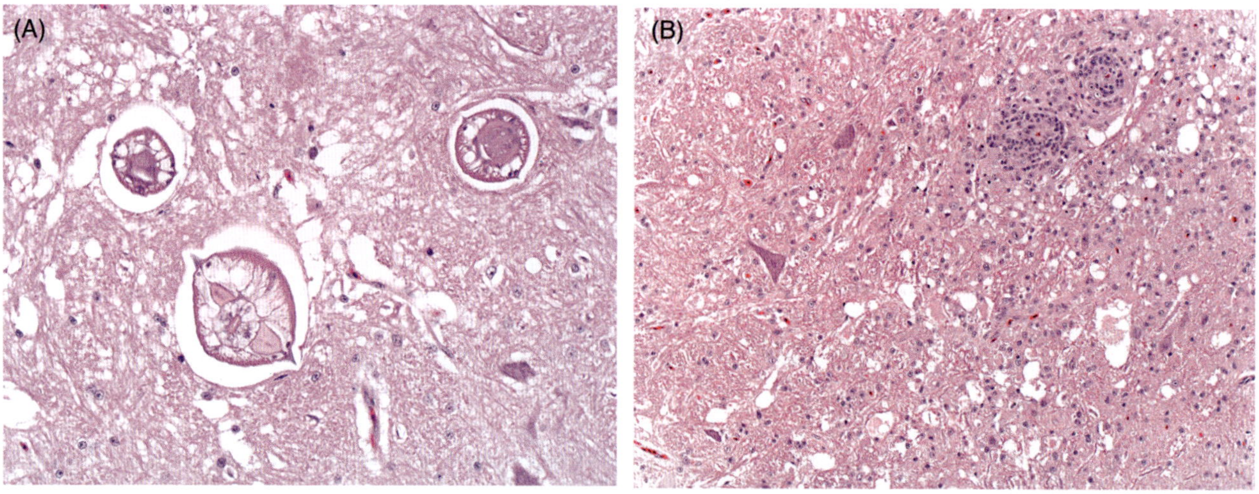

FIGURE 26.10 *Baylisascaris procyonis* larval migrans and associated encephalitis in the brain of an elegant crested tinamou. (A) Multiple cross-sections of nematode larvae consistent with *Baylisascaris procyonis* (the raccoon roundworm) are present in the brain. Notable features of the parasite in cross-section include a large body size (60–80 µm diameter), a prominent intestine with paired triangular lateral excretory columns, and paired lateral cuticular alae. (B) Perivascular lymphoplasmacytic infiltrates, gliosis, axonal degeneration, and neuropil vacuolation are typical sequela to neural larval migrans. These can be seen in the absence of in-section parasite larvae. Lesions are most commonly present in the brainstem and central cerebellar white matter (see eSlide 26.e8).

granulomas reflect previous parasite-related damage; unlike lesions in mammals, this is not eosinophil rich (Fig. 26.10B; eSlide 26.e8).

Visceral larval migrans is frequently described in kiwi, most often in the North Island brown kiwi and as an incidental finding with eosinophilic granulomas containing fragments of or intact nematode larvae particularly in lung and liver.[312] ***Toxacara cati*** (natural host: domestic cat) was identified by PCR in tissues from cases of visceral and neural larva migrans in North Island brown kiwi.[115] Adult and larval forms of *T. cati* were identified by morphologic features in the small intestine of a North Island brown kiwi.[61] *T. cati* has also been found in meat from farmed ostrich.[204]

Cutaneous larval migrans has been described in brown kiwi chicks held on a crèche island as part of a captive propagation program.[124] Infection was associated with feather loss, and dermatitis was present in the abdominal area and along the vent margin. Histologically, there was mild to moderate epidermal hyperplasia and hyperkeratosis with perivascular and nodular lymphoid infiltrates; lower numbers of heterophils and eosinophils; and in some cases, the presence of intraepithelial pustules, granulomas, and necrotic debris. Intralesional nematode larvae were identified and were closely related to ***Trichostrongylus axei*** using DNA sequencing. It was speculated that the parasites might be of sea bird or marine mammal origin.

Intraventricular ***Cyrnea apterycis*** (a spirurid nematode) and colonic ***Heterakis apterycis*** appear to be common incidental findings in kiwi.[61]

Protozoa

Most information pertaining to enteric protozoa in ratites is from fecal examinations. In the absence of correlation to clinical signs or histologic lesions, it is difficult to associate these findings with disease. Notable protozoal infections of ratites are described below. Several other intestinal coccidia that have been identified in the feces of ratites are listed in Table 26.e8.

Intestinal and extraintestinal **coccidiosis** are significant causes of disease in both free-ranging and captive kiwi. Birds are often concurrently infected with more than one species of parasite. Four ***Eimeria*** species have been described and additional species are suspected to occur in the North Island brown kiwi.[215] One of these species, ***Eimeria kiwi***, is present in the intestinal epithelium (Fig. 26.11). An unusual feature of infection with this organism is the presence of large macromeronts.[213] Other identified species include ***E. paraurii*** (suspected to play a role in the development of colorectal polyps), ***E. mantellii*** (has been identified in feces but has not yet been associated with clinical disease), and ***E. apteryxii*** (causes renal coccidiosis in the North Island brown kiwi, rowi, and tokoeka).[213,215,303] Additionally, asexual stages of unknown coccidial species have been found in the liver, spleen, lung, and pancreas. Periodic acid-Schiff staining of kidney to enhance parasite recognition is recommended as part of routine histopathologic examination.[214] **Eimeriosis** has also been described in red-winged tinamous. Infection with the causative agent, ***E. rhynchoti***, caused apathy and maldigested, fetid, pasty feces as well as duodenal hemorrhage and thickening.[75,194] Experimentally infected birds died between 1 and 6 days of patent infection.

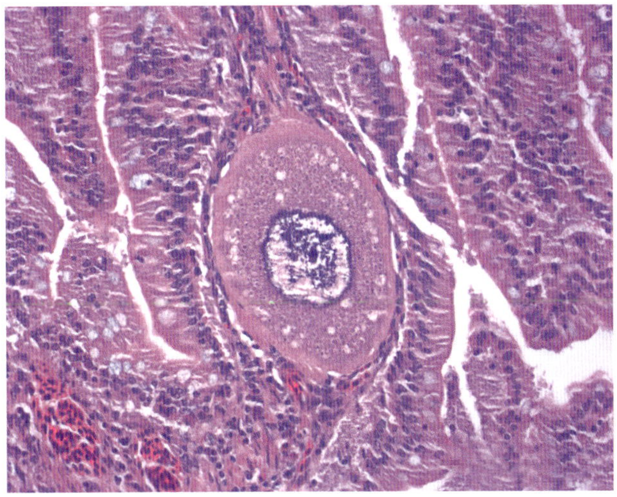

FIGURE 26.11 Enteric coccidiosis in a North Island brown kiwi. Coccidiosis is an important cause of morbidity and mortality in young kiwi. A number of incompletely described *Eimeria* species, both enteric and extraintestinal, have been described in the species. This image shows an *Eimeria* sp. megalomerozoite (macromeront) in the villar lamina propria of the small intestine. Macromeronts are an uncommon feature of avian coccidiosis. *(Photo Courtesy of K. Morgan, Massey University.)*

Cryptosporidia are frequent, incidental findings in the feces of apparently healthy ostriches; there is a greater prevalence in chicks and juvenile birds.[321] However, an associated syndrome of cloacal prolapse in birds with **cryptosporidiosis** is well recognized in ostrich chicks, particularly males. Infection of epithelial cells in the distal rectum and cloaca results in mucosal swelling, epithelial hyperplasia, and mixed heterophilic and mononuclear cell inflammation in the lamina propria.[234,267] Infection with necrosis and inflammation may also occur in the pancreatic ductal epithelium.[158] Molecular investigations have identified ***Cryptosporidium baileyi*** and ***Cryptosporidium* avian genotype II** in ostriches.[202,321] Disease has, thus far, only been associated with the latter organism. However, much of the literature predates the development of molecular methods for parasite identification.

Histomonas meleagridis is a cause of severe necrotizing typhlitis, localized peritonitis, and focal hepatic necrosis in ostrich and rhea chicks. Lesions mimic those seen in domestic chickens.[43,201]

A wide variety of **amoebae**, **ciliates**, and **flagellates** have been identified in the feces and, less often, in intestinal smears, from ratites without and occasionally with clinical enteric disease. Recent work has shown that what was originally described as ***Balantidium struthionis*** in a captive ostrich is morphologically identical to ***Balantidium (Balantioides) coli*** and has only minor differences in 18 s-rRNA sequences; it should therefore more correctly be described as "*B. coli*-like" until further characterized.[244] These organisms have been associated with typhlitis and colitis in ostrich chicks in South Africa,[156] but also appear to be part of the normal lower gastrointestinal tract fauna in healthy ostriches[297] as well as common and lesser rhea.[122] ***Entamoeba struthionis*** has been identified from the feces of healthy ostrich in Europe.[246,297]

Leukocytozoon struthionis infection is common in young ostriches in South Africa. Myocarditis has been described in association with the presence of megaloschizonts, and initial infection can lead to anemia prior to the development of stable low-level parasitemia. ***Plasmodium*** spp. have been identified in ratites without associated disease. Several protozoal **hemoparasites** have also been identified in kiwi. The death of a single wild-born juvenile North Island brown kiwi due to disseminated protozoal infection led to the identification of ***Plasmodium elongatum*** in North Island brown kiwi and rowi that were part of captive breeding and captive-rearing crèche programs.[24] Early trophozoites were present in blood smears collected at the time, with a prevalence of 68% by microscopy and 78% by PCR. Parasitemias were low, and subsequent testing did not show the continuing presence of the parasite. The most severe histologic lesion in the sentinel bird was severe interstitial pneumonia, with congestion, edema, and small blood vessel thrombosis. Hypertrophied pulmonary endothelial cells, hepatic Kupffer cells, and macrophages in the lung, liver, and spleen contained intracytoplasmic merozoites or hemozoin (hemoglobin breakdown product). ***Haemoproteus*** and ***Plasmodium*** have been found in wild great tinamou.[131]

Babesia kiwiensis, an intraerythrocytic parasite, has been identified in both captive and free-living North Island brown kiwi chicks.[159,233] Parasites were present as ring forms, schizonts, and merozoites within erythrocyte cytoplasm, without displacing the host nucleus. Single birds were coinfected with ***Hepatozoon kiwii***. The oval or elongated gametocytes of *H. kiwii* were present in the cytoplasm of monocytes and, less frequently, lymphocytes, where they distorted and compressed the nucleus of the host cell. ***Ixodes anatis***, also known as the kiwi tick, is commonly found on kiwi and is postulated as a vector for these organisms. The significance of these infections is currently unknown.

Toxoplasma gondii has been reported to have caused the death of a North Island brown kiwi.[154] Lesions included liver necrosis and protozoal organisms in hepatocytes and Kupffer cells. Exposure to *T. gondii* appears to be widespread in little spotted kiwi with more than 80% testing positive in a survey of a sanctuary population.[298] There are several reports of ratites (rhea, emu, ostrich) with antibodies to *T. gondii*[28,118,153] and mouse bioassays and PCR have been positive in ostrich.[76]

Blastocystis sp. has been identified in the feces or intestinal contents of asymptomatic ostriches in Europe, Japan, and most recently, Malaysia where the organism was characterized as Subtype 6.[58,245] This subtype has also been identified in humans.

Ectoparasites

Lice and **mites** are common ectoparasites of ostriches and rheas worldwide.[73,156,209] In addition to the descriptions that follow, see Table 26.e9 for a list of several additional notable ectoparasites.

Struthiolipeurus struthionis, the **ostrich louse**, feeds on the body scales and feather debris of ostriches and causes irritation and pruritis. Adult lice are 3–4 mm long and readily visible; eggs (nits) can be found on the underside of feathers attached to the barbs near the shaft. They live in the groove of the shaft and feed on the pulp of growing feathers. Parasite feeding and excessive grooming can lead to feather damage and loss.

Parasitic mites can also cause cutaneous lesions. For example, acute parasitic, cystic, lymphoplasmacytic dermatitis, **nodular trombiculinosis**, have been described in a juvenile ostrich in Brazil as a result of infestation with the trombiculid mite ***Apolonia tigipioensis***.[226]

In addition to lice and mites, international importations of ostriches and in-contact birds should be examined for the presence of **ticks** as a range of species of ectoparasitic arthropods have been identified on birds imported from Africa and Europe.[203] Feeding by **hippoboscid flies** and **black flies** (***Simulium* spp.**) can cause irritation and anemia, and severe dermatitis and conjunctivitis have been described in ostriches swarmed by thrips (***Limothrips denticornis***).[69,156]

Tinamous are commonly infected by one or more species of ectoparasite, with up to 15 species reported from a single great tinamou.[194,311] Kiwi can host several species of ticks; eight species of chewing lice and five species of feather mites have also been identified.[147] The significance of ectoparasitism in these two groups of birds is unclear.

ADDITIONAL RESOURCES

Brand, Z., Cloete, S.W., Malecki, I.A., Brown, C.R., 2017. Ostrich (*Struthio camelus*) embryonic development from 7 to 42 days of incubation. Br. Poult. Sci. 58(2), 139–143.

Deeming, D.C. (Ed), 1999. The Ostrich. Biology, Production and Health. CABI Publishing, Wallingford, Oxon, UK, 358 pp.

del Hoyo, J., Elliot, A., Sargatal, J, (Eds.), Handbook of the Birds of the World, Vol. 1. Lynx Edicions, Barcelona, Spain.

Huchzermeyer, F.W., 1998. Diseases of Ostriches and Other Ratites. Agricultural Research Council, Onderstepoort Veterinary Institute, Onderstepoort, Republic of South Africa, 296 pp.

Kummrow, M.S., 2014. Ratites or Struthioniformes: Struthiones, Rheae, Cassuarii, Apteryges (Ostriches, Rheas, Emus, Cassowaries, and Kiwis), and Tinamiformes (Tinamous). In: Miller, R.E., Fowler, M. (Eds.), Fowler's Zoo and Wild Animal Medicine, Vol. 8., Elsevier Health Sciences, St. Louis, Missouri, USA, pp. 75–82.

Silveira, L.F., Höfling, E., Gaglianone Moro, M.E., do Nascimento, A.A., Gouveia Arantes, I.M., 2001. Order Tinamiformes (Tinamous). In: Fowler, M.E., Cubas, Z.S. (Eds.), Biology, Medicine and Surgery of South American Wild Animals. Iowa State University Press, Ames, Iowa, USA, pp. 72–80.

References

Full references for this chapter can be found online at https://doi.org/10.1016/B978-0-443-18437-6.00045-6.

Chapter 27

Sphenisciformes, Gaviiformes, Podicipediformes, Procellariiformes, and Pelecaniformes

Daniela Denk[a] and Mark F. Stidworthy[b]

[a]SeaWorld Abu Dhabi, Yas Island, Abu Dhabi, United Arab Emirates; [b]International Zoo Veterinary Group, Keighley, West Yorkshire, United Kingdom

INTRODUCTION

The avian clade Aequorlitornithes consists of shorebirds, flamingos, grebes, gulls, tropicbirds, penguins, and other waterbirds. This chapter represents five orders of this clade: Sphenisciformes, Pelecaniformes, Gaviiformes, Podicipediformes, and Procellariiformes.

The Sphenisciformes (penguins) are pelagic birds with a lifespan of 25–40 years in captivity. The 6 genera and 18 currently recognized species range from cold-tolerant species populating Antarctica and sub-Antarctic areas to temperate species living near the equator.[164] Sexual maturity in these generally monogamous birds usually occurs around 3–5 years of age. Penguin neonates are altricial.

Pelecaniformes include pelicans (Pelecanidae), cormorants, and shags (Phalacrocoracidae), anhingas (darters; Anhingidae), boobies and gannets (Sulidae), frigate birds (Fregatidae), and tropicbirds (Phaethontidae). Shoebills and hamerkops are included under current classifications. These birds are long-lived (40 years for captive pelicans, 30 years for free-ranging frigate birds). All feed on fish, aquatic vertebrates, and invertebrates. Monogamous mating is typical. Apart from tropicbirds, eggs are incubated using foot webs, limiting clutch size to one or two eggs. Neonates are altricial. Gannets, boobies, frigate birds, and tropicbirds are pelagic, spending much of the nonbreeding season in extended flight. Most pelicans are surface feeders, while cormorants, shags, and anhingas are subsurface diving specialists. Brown pelicans, tropicbirds, gannets, and boobies are plunge divers. Frigatebirds capture their prey, including flying fish in aerial pursuit, from the water's surface. Shoebills and hamerkops hunt as waders in shallow marginal waters.

The Gaviiformes order comprises five species, distributed worldwide, known as loons in North America and as divers in the Old World. They are heavy, long-bodied birds adapted for underwater hunting. Neonates are precocial. They breed near freshwater but migrate and overwinter in marine environments where they may congregate.

Approximately 22 species of Podicipediformes exist worldwide. Like loons they are underwater hunters, but more varied in size. Grebes inhabit freshwater wetlands but may overwinter in marine environments, congregating in large numbers as they migrate. Neonates are precocial. Molting occurs away from breeding grounds. Mostly, they are solitary or live in small groups.

The order **Procellariiformes** includes four families: albatrosses and mollymawks (Diomedeidae), petrels and shearwaters (Procellariidae), storm petrels (Hydrobatidae), and diving petrels (Pelecanoididae). Although found worldwide, Procellariiformes predominantly inhabit the southern hemisphere. There is considerable size variation between the smallest storm petrels (adults 20 g) and the albatrosses (adults exceeding 10 kg). They are long lived: up to 20 years for the smaller petrels and 50–70 years for albatrosses. All Procellariiformes are pelagic, accomplished long-distance soaring fliers that spend months at sea feeding on fish, larger invertebrates, phytoplankton, and krill. Landfall is infrequent and occurs mainly for breeding and nesting. This is typically colonial on remote islands, on the ground or in burrows. Young are altricial to semialtricial with well-developed thermoregulatory capacity. Introduced predators, island habitat destruction, and intensive fishing pose threats. Many species, particularly albatross, are threatened or at risk of extinction. Wildlife disease surveillance highlights anthropogenic treats in Sphenisciformes and Pelecaniformes, while disease is more frequently reported in captive settings. Disease records of free-living members of the remaining orders (Gaviiformes, Podicipediformes, and Procellariiformes) stem predominantly from rehabilitation centers and wildlife surveillance efforts, as migratory patterns

Pathology of Wildlife and Zoo Animals. https://doi.org/10.1016/B978-0-443-18437-6.00017-1

and the pelagic lifestyle render members of these orders ill-suited for long-term captivity.

A table of all species included in this chapter, their conservation (IUCN) status and geographical distribution (Table 27.e1), as well as information on unique anatomical features and relevant clinical pathology, is provided in the Digital Content. A more detailed overview of the above groups is available in the literature.[189]

NON-INFECTIOUS DISEASES

Nutritional

Nutritional deficiencies are uncommon in captive penguins and are associated with feeding poor quality, improperly handled, or inadequately supplemented fish, shrimp, or krill. Poor parental care, miscalculation of the mineral formulation, in particular calcium, or lacking oral vitamin D_3 provision in combination with insufficient exposure to full-spectrum lighting in indoor housed animals may result in **nutritional metabolic bone disease.**[114] Affected chicks fail to thrive and develop folding fractures of the tibiotarsus, rib fractures, and luxation of the gastrocnemius tendon (Fig. 27.1).

Radiographic reduction in bone density is apparent. Histological findings are typical of marked diffuse osteopenia predominantly affecting long bones. Trabeculae are thin, irregular, and frequently contain closely spaced reversal lines, thin surface seams of osteoid, or both.[4] There are occasional reports of **rickets** resembling those of vitamin D-deficient domestic fowl in zoo-captive double-crested cormorant chicks receiving nonsupplemented smelt as sole food, and **fibrous osteodystrophy** caused by Ca:P imbalance in free-ranging royal northern albatross chicks receiving supplemental hand-feeding on their nests. In the latter, deterioration of fat-soluble vitamin D during frozen storage, feeding of boned fillets rather than whole fish, and peroxidation of stomach oil used as a supplement are contributing factors. Histologically, osteopenia and fibrous osteodystrophy predominate in these species, with rachitic elongation and widening of the metaphyseal cartilage in cormorants. Parathyroid glands may be grossly normal or enlarged and hypertrophied with chief cell vacuolation.

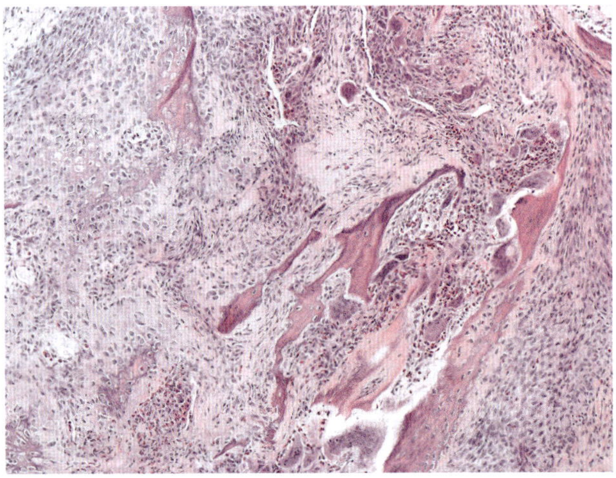

FIGURE 27.1 Metabolic bone disease with fibrous osteodystrophy in the tarsometatarsus and rib of a bank cormorant. Incomplete and attenuated trabeculae of cortical bone are associated with osteoclastic resorption, mesenchymal proliferation, collagen deposition, chondroid metaplasia, and marked periosteal thickening (lower right) in a 15-days-old chick.

Thiamine deficiency occurs in captive piscivores fed nonsupplemented diets or diets where supplementation allows for vitamin breakdown prior to administration. Torticollis, incoordination, and "stargazing" are observed in affected penguins. Histological features have not been described but polioencephalomalacia is reported from other susceptible bird species.[141] **Vitamin A deficiency** leading to periocular gland squamous metaplasia occurs infrequently in Sphenisciformes.[176] Vitamin A and essential amino acid deficiency have also been suggested as possible underlying factors for reproductive tract pathology in females.

Hypovitaminosis E following feeding of frozen-thawed fish with inadequate dietary supplementation is reported in pelicans, cormorants, and rarely penguins in managed collections.[228] Dietary antagonism by excess vitamin A may play a role. Birds present dead or with lethargy, weakness, cutaneous edema, heterophilic leukocytosis, and elevated plasma creatinine levels. Necropsy findings include pale and streaked striated musculature, pericardial effusion, necrotic adipose tissue, and subcutaneous hemorrhage in the absence of trauma. Hypovitaminosis E should be considered in all cases of nonspecific soft tissue hemorrhages in aquatic birds (Fig. 27.2A; eSlide 27.e1). Nutritional myopathy characterized by polyphasic multifocal rhabdomyolysis and myocardial necrosis, accompanied by necrotizing pansteatitis, is evident histologically (Fig. 27.2B).

Hypervitaminosis E due to excess dietary supplementation that caused severe coagulopathy with extensive subcutaneous, muscular, and intracoelomic hemorrhage in the absence of concurrent myopathy has been reported in a group of pink-backed pelicans.[180] Evaluation of vitamin and mineral levels from antemortem serum samples is a valuable diagnostic tool to determine nutritional deficiencies. Postmortem liver tissue analysis can be performed, but interpretation may be challenging in the absence of species-specific reference values.

Metabolic

Similar to other avian species, **gout** is characterized by the deposition of needle-like monosodium urate crystals on visceral or air sac surfaces, within the tissue parenchyma of viscera, or joint spaces.[164] In free-living seabirds, renal gout

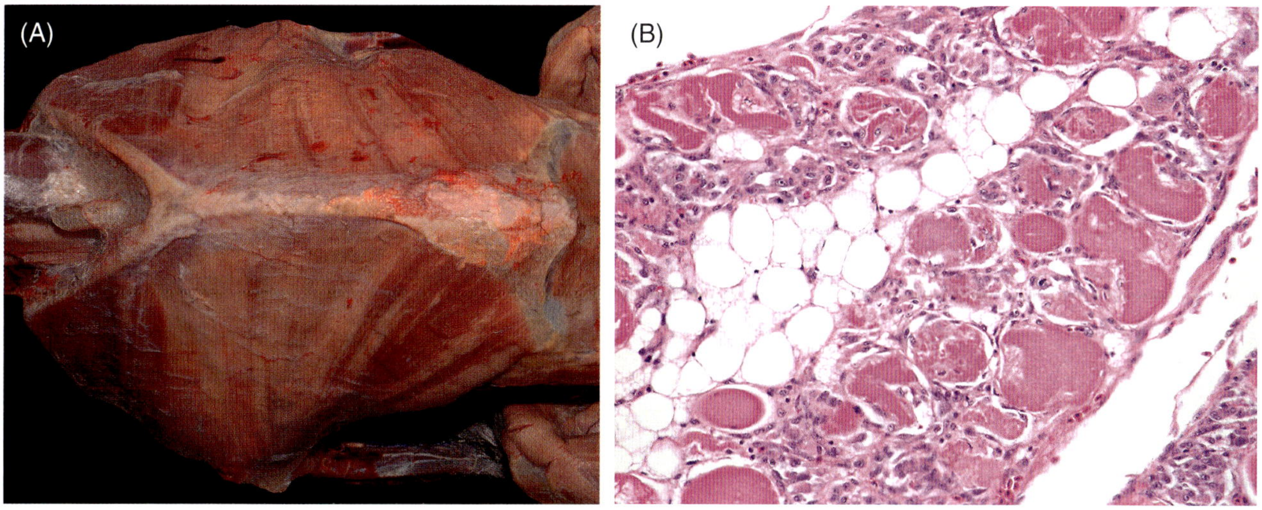

FIGURE 27.2 Hypovitaminosis E in the pectoral muscle of pelicans. (A) Gross lesions in a brown pelican include severe, multifocal to coalescing, intramuscular hemorrhage, and segmental marked pallor of the pectoral muscles. (B) Severe, multifocal, polyphasic myodegeneration with myofiber swelling, hypereosinophilia, loss of cross-striation, and early regenerative activity in a Dalmatian pelican (see eSlide 27.e1). *(Photo A: Courtesy of M. Busch, Wildlife Conservation Society.)*

is associated with cachectic condition and autumn stranding and correlated to metabolic imbalances, chronic dehydration, and overdoses of protein supplements following rehabilitation.[60]

Toxic

In addition to descriptions of several notable toxins below, algal and other toxins affecting these orders are listed in Table 27.e2.

Dose-dependent **voriconazole toxicity** occurs in multiple penguin species following antifungal treatment.[116] No specific gross lesions are identified. Central nervous system myelinic edema and acute pulmonary congestion are suggestive histopathologic findings, but death is primarily due to unsuccessful treatment of the fungal infection. Hepatotoxicity, as described in other avian species, is not a feature in penguins with voriconazole toxicity; however, hepatotoxicity resulting in anorexia, weight loss, and hepatic swelling, discoloration, and necrosis can occur in captive king penguins treated with **itraconazole** (R. Pizzi, personal communication).

Penguins and American white pelicans are susceptible to **fenbendazole toxicity**. Intoxication of African penguins has been induced accidentally by feeding fish dosed with 66–77 mg/kg of fenbendazole (S. Redrobe, personal communication). Death occurred 5–9 days posttreatment. Necropsy findings included empty gastrointestinal tracts, stomatitis (pelicans), and inconsistent hepatic, renal, or splenic enlargement. Intestinal crypt cell necrosis is seen histologically, with villous architectural collapse, superficial mucosal necrosis, and secondary bacterial colonization (Fig. 27.3; eSlide 27.e2). Necrotic or regenerating epithelial cells line remaining crypts. Intestinal lesions mirror fenbendazole toxicity in avian species including pigeons, storks, and vultures. Nonspecific radiomimetic bone marrow suppression and splenic lymphoid depletion are additional features.

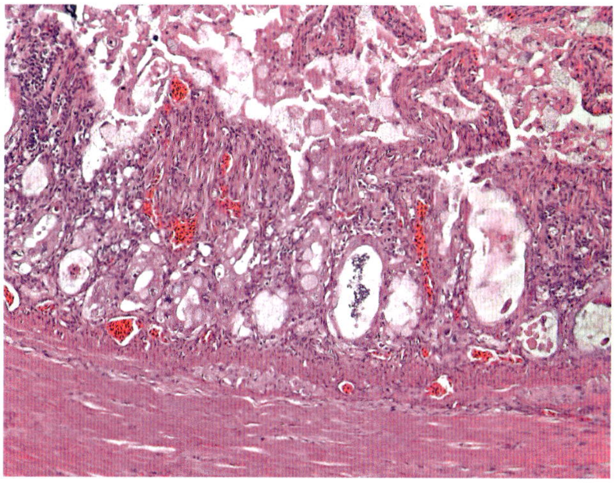

FIGURE 27.3 Fenbendazole toxicity in the small intestine of an African penguin. Throughout the small intestinal mucosa, there is diffuse villus distortion and blunting with widespread epithelial degeneration and regeneration, edema of the lamina propria, and mixed cellular inflammation. Clusters of indistinct basophilic bacteria (secondary opportunistic infection) are present within the crypt lumen (see eSlide 27.e2).

Toad poisoning may occur in zoo-captive penguins living in temperate climates in Europe, particularly when toads are migrating. Affected penguins die suddenly, in good condition and without clinical symptoms. Ingested toads are found in the upper gastrointestinal tract together with large quantities of mucus (sialorrhea). Biogenic

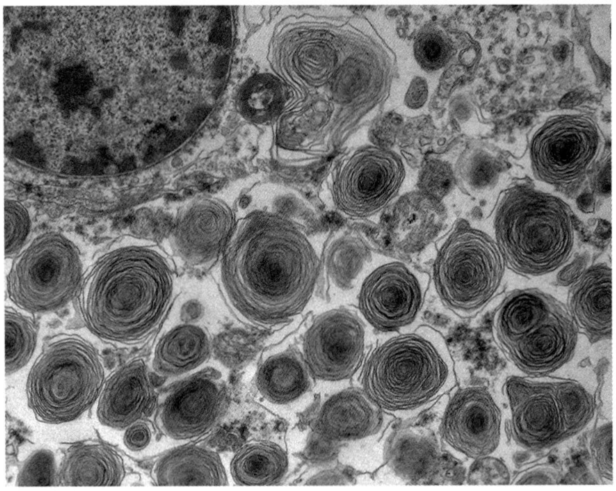

FIGURE 27.4 Sphingolipidosis in the spinal cord of a Humboldt penguin. Numerous multilayered concentric lamellar structures widely distend the perikaryon of a spinal cord motor neuron. Purkinje cells and neurons of brainstem nuclei are also affected. TEM. *(Photo Courtesy of A. Armien, University of Minnesota, Minnesota Veterinary Diagnostic Laboratory.)*

amines and steroid derivatives present in bufonid toad venom act similarly to digitalis and inhibit the sodium-potassium pump in cardiac myocytes. Histological abnormalities are not observed. Diagnosis can be aided by serum biochemical analysis to detect hyperkalemia and measure serum digoxin concentration.

A (possibly acquired) storage disease consistent with **sphingolipidosis** occurred in five related adult female captive Humboldt penguins. Clinical signs included slowly progressive lethargy, weakness, dysphagia, and ataxia.[285] Fine clear vacuolation within the perikarya and margination of the Nissl substance of swollen neurons, Purkinje cells, and motor neurons of the spinal cord was the histological hallmark. Axons contained fine clear vacuoles and infrequent spheroids. Multilayered concentric lamellar bodies resembling stored gangliosides were identified by transmission electron microscopy (Fig. 27.4). High cumulative doses of chloroquine, a lysosomotropic amine used for antimalarial treatment may have induced disease, raising concerns over long-term high dose administration.

Many reports documenting toxins involve surveillance detection of tissue contaminants in free-ranging species, particularly **persistent organic pollutants**, and **heavy metals**. Disease and/or pathology are frequently absent, but it has been suggested that such pollutants may contribute to hepatopathy in penguins. High levels of mercury and cadmium have been associated with proximal renal tubular epithelial necrosis in Manx shearwaters and northern fulmars. **Organochlorines** caused reproductive failure, including extinctions, in brown pelican colonies during the 1960s. Residues of **1,1-dichloro-2,2-bis (*p*-chlorophenyl) ethylene (DDE)** caused thinning of brown pelican eggshells leading to fragility and clutch failure. Regulations restricting the use of DDE in the 1970s have resulted in a decrease in environmental residues and restoration of normal shell thickness and reproductive success.[25]

Growing bodies of literature suggest an association between plastic/microplastic ingestion and persistent organic pollutant accumulation.[7,50,51] In the absence of long-term field data, there is early experimental evidence to indicate measurable sublethal toxic effects of ingested plastics, such as oxidative stress and alteration in reproductive hormone levels.[167,175] As with other species, no specific gross or histologic changes have been identified to date, but prospective studies may change this. Refer to the "Miscellaneous/Multifactorial" section for additional information related to plastic ingestion.

Free-ranging pelicans, cormorants, and particularly loons are susceptible to **lead intoxication**, mainly by ingestion of fishing tackle and ammunition. Similar items may be ingested by penguins. Occasional birds have exposure via tissue pellets from prior shooting incidents. Lead paint is a cause of intoxication in Laysan albatross chicks. Clinical signs include an inability to fly, drooped wing posture, and bright green diarrhea. Lead in the gastrointestinal tract can be identified radiographically. Necropsy lesions can be absent or include pectoral muscle atrophy, fat depletion, esophageal/proventricular impaction, ventricular erosions, gallbladder distension, and lead objects in gastric content. Toxic effects of lead (and zinc) on uterine smooth muscles may predispose to egg retention. Histologic lesions may include rare acid-fast positive intranuclear inclusions in renal tubular epithelial cells and hepatocytes, and Wallerian degeneration of peripheral nerves. Bench top analysis of tissue fluids shows good correlation with liver lead concentrations in loons.[137] Lead tissue concentrations and their clinical relevance in birds are listed in Table 27.1.

Botulism is caused by several toxins produced by the bacterium ***Clostridium botulinum***. **Type C botulism** causes large die-offs of free-ranging American white pelicans (and smaller numbers of brown pelicans) in southern California (United States).[224] Elsewhere, pelicans, cormorants, grebes, and occasionally northern fulmar and Laysan albatross have also contracted type C botulism. Piscivores (common loons, red-throated loons, and horned grebes) are more typically affected by **type E botulism**.[30] Outbreaks in loons typically occur in the fall (September) while pelican outbreaks may occur earlier in the summer. Risk factors include seasonally increased water temperatures, increased organic material in the sediment, low or fluctuating water levels, high water pH (7.5–9.0), and increased salinity (>2 ppt).[223] Dabbling and filter-feeding waterfowl, shorebirds that feed near waterbody surfaces, and sediment- or fish-eating birds are considered at higher risk of developing botulism. Increased bird mortalities may also trigger

TABLE 27.1 Diagnostic Significance of Blood and Tissue Concentrations of Lead in Birds[86,191]

Blood	Liver, Kidney (Wet Weight)	Bone (Dry Weight)	Clinical Significance
	<2 mg/kg ww		Background level
20 < 50 μg/dL	2–6 mg/kg ww	10–20 mg/kg dw	Subclinical poisoning
50–100 μg/dL	6–15 mg/kg ww	> 20 mg/kg dw	Clinical poisoning
>100 μg/dL	>15 mg/kg ww		Severe clinical poisoning and death
	27 and 107 mg/kg ww		Chronic lead poisoning

outbreaks as carcasses act as an environment for *C. botulinum* growth and toxin production. Botulinum toxins inhibit neurotransmission by blocking secretion of acetylcholine from peripheral cholinergic nerve terminals in the motor and autonomic nervous systems. Progressive (generally flaccid) paresis and paralysis lead to inability to fly, paddling wing movements across water, and paralysis of the nictitating membranes and the neck (limber neck). Birds lose the ability to raise wings and flap throat tissue. Secondary hyperthermia can be severe. Diagnosis is based on history, clinical signs, absence of other causes of paresis/paralysis on gross and histologic examination (e.g., trauma, inflammation), and laboratory confirmation of botulinum toxin in blood or tissue. Mouse bioassay remains the gold standard for diagnosis but antigen-capture ELISA testing of blood and serum for type C toxin has been used successfully in birds.

Outbreaks of ***Clostridium* enterotoxemia** have caused peracute and acute deaths in a variety of penguin species.[99] Birds are found dead in excellent body condition or may present with acute depression and brown, foul-smelling diarrhea.

Aquatic birds feeding predominantly on fish and invertebrates are susceptible to **algal biotoxins** generated during harmful algal blooms. Pathobiology and epizootiology have been reviewed in detail elsewhere.[143] Marine dinoflagellates and diatoms produce neurotoxic **saxitoxins**, **brevetoxins**, and **domoic acid**; freshwater cyanobacteria produce neurotoxic **saxitoxins** and **anatoxins**, hepatotoxic **microcystins**, and dermatotoxic ***Lyngbya* toxins**. Exposure is through bioaccumulation in prey species or drinking contaminated freshwater. Coastal piscivores hunting pelagic prey, such as cormorants and shags, are particularly at risk. Neurotoxins cause signs including lethargy, disorientation, paresis, paralysis, tremors and muscle fasciculations, pupillary constriction, torticollis, lost righting reflexes, seizures, and death by respiratory failure or drowning. Vomiting, diarrhea, excess oronasal and glandular discharges, dyspnea, and tachycardia may be seen. There are no specific gross or histologic lesions, but birds may be either emaciated or in good condition with stomachs full of prey. Concurrent, nonspecific findings include intestinal hemorrhage, catarrhal enteritis, and hemorrhage and necrosis of striated muscle. Toxins can be detected and quantified in supply water, gastrointestinal content, body fluids and tissues. **Microcystins** are the most common form of cyanobacterial toxicosis; affected birds may be jaundiced with evidence of massive acute hepatic necrosis and hemorrhage. Fatal pectoral and cardiac myonecrosis have been associated with microcystin intoxication in American white pelicans, in which a role for microcystin-induced vitamin E depletion was suggested.[160]

Congenital/Genetic

Congenital defects are uncommonly recorded in free-ranging or captive penguins. Congenital and genetic abnormalities in free-ranging populations of all bird species are likely overlooked, since long-term survival to allow identification of affected birds in the wild is unlikely. Occasional beak abnormalities and carpal rotation have been described in free-ranging double-crested cormorant chicks. A few case reports describe congenital cardiac anomalies in penguins, among which septal defects appear to be most common. **Ventricular septal defects (VSD)** have been reported in a 45-day-old female Humboldt penguin, a juvenile rockhopper penguin, a captive macaroni penguin and a 27-day-old Adélie penguin[144] (Fig. 27.5). Two cases of **atrial septal defects** occurred in gentoo penguin chicks with cardiomegaly (J. St. Leger and R. Pizzi, personal communications).

Although uncommon, abnormal plumage coloration occurs in penguins. Full or partial **leucism**, an incomplete partial albinism in which eyes, bill, and legs retain normal color while pigmentation of the feathers is absent or decreased, has been reported in several penguin species and free-ranging double-crested cormorant chicks.[78,260,262] **Melanism**, caused by increased amounts of eumelanin in feathers, has been reported in Adélie, royal, gentoo, macaroni, and chinstrap penguins, and partial melanism has been seen in king penguins.[23,78,260]

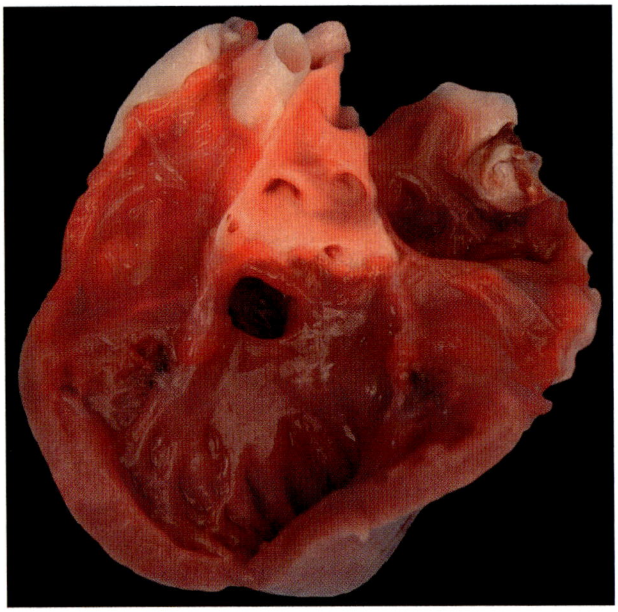

FIGURE 27.5 Congenital ventricular septal defect in the heart of an Adélie penguin. There is a full-thickness, round, subvalvular defect in the interventricular septum of this 27-day-old chick. Death was associated with marked pulmonary edema. *(Photo Courtesy of J. St. Leger, SeaWorld Parks & Entertainment.)*

Two cases of unilateral **micromelia** are described in free-ranging juvenile little penguins,[210] and **craniofacial deformities** occurred in free-ranging yellow-eyed penguin chicks in New Zealand.[37] **Malformed beaks** and the absence of flippers have been observed in little penguins.[215] These examples raise suspicion of teratogenic exposures in breeding colonies. Congenital **hemivertebrae** with spinal cord compression was diagnosed in a 10-week-old kyphotic African penguin.[29] **Polydactyly** is reported in a common loon.[204] A **conjoined twin** embryo was found in a Wedge-tailed shearwater during research egg collection. Duplication of the embryo was complete except for a common head missing one eye, and the brain was partially exposed.[199] An isolated case of oviduct duplication is recorded in a Humboldt penguin.[131]

Age-Related/Degenerative

Degenerative joint disease is uncommon in penguins. Among adults, 16% of rockhopper, 4% of gentoo, and 2% of king penguins examined at a zoo in the United Kingdom had osteoarthritic changes, mostly affecting the coxofemoral and stifle joints (R. Pizzi, personal communication). Bilateral degenerative joint disease was diagnosed in a free-ranging, juvenile, yellow-eyed penguin with abnormal stance and decreased mobility.[36] Necropsy identified bilaterally distended, thickened coxofemoral joints with increased laxity, and small, roughened, and angular femoral heads (Fig. 27.6). Histologically, the left femoral articular cartilage and subchondral bone were absent. The remaining femoral head consisted of trabecular bone overlain by fibrin and granulation tissue. Evidence of infection was absent. Osteonecrosis and osteochondrosis with subsequent bilateral, aseptic degenerative changes of the femoral head were the suspected pathogenesis.

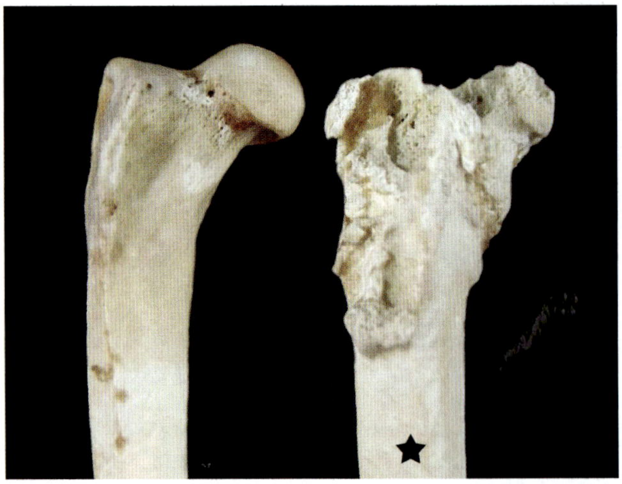

FIGURE 27.6 Degenerative joint disease in the hip joint of a yellow-eyed penguin. Affected right femur *(star)* and unaffected age-matched control. The abnormal femoral head is shrunken, angular, and roughened, with osteophyte formation on the femoral neck, and greater and lesser trochanters. *(Photo Courtesy of K. Buckle, Massey University Institute of Veterinary, Animal and Biomedical Sciences.)*

The importance of **atherosclerosis** (Figs. 27.7A, B; eSlide 27.e3) varies within captive penguin collections. In one review, a 2% incidence with no associated mortality was noted while in another, 37% of penguins had atherosclerosis, predominantly as an incidental finding. Adélie and emperor penguins were most commonly affected.[164] The reported prevalence may be influenced by the age of captive populations, species, dietary or climatic differences.

Despite their longevity, age-related/degenerative problems in zoo-captive Pelecaniformes are poorly documented. They include arthritis, arteriosclerosis, and hyperkeratosis of the tibiotarsus and tarsometatarsus in shoebills. **Cataracts** (Fig. 27.8) are the most commonly reported age-related ocular pathology in penguins, and recent evidence suggests that cataracts and vitreal degeneration may be underrecognized in zoo-captive adult pelicans.[24,185] Reports are scant in free-ranging or captive populations of Procellariiformes, Gaviiformes, or Podicipediformes. Most free-ranging aged birds die at sea rendering them unavailable for examination.

Inflammatory Non-Infectious

Due to their curious nature, penguins, particularly juveniles and nesting females, are prone to investigate and ingest novel items. These may include nesting and plant material,

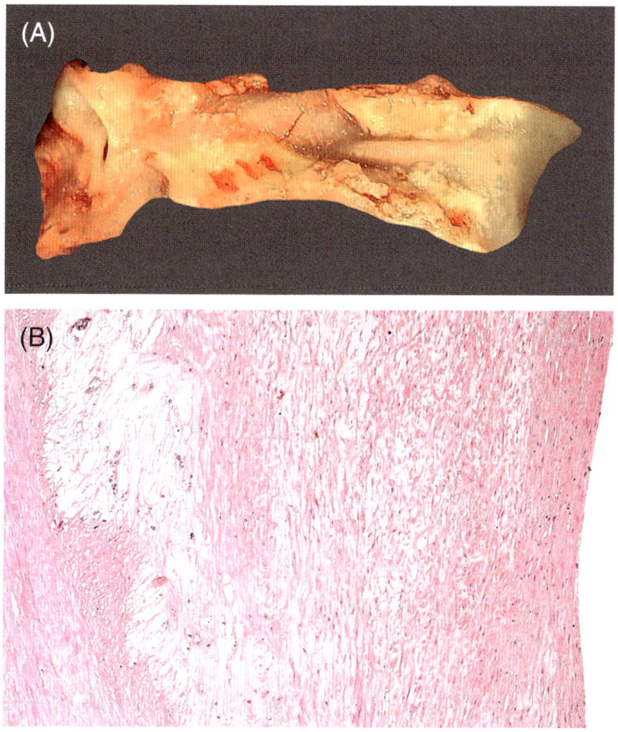

FIGURE 27.7 Atherosclerosis in the aorta of an African penguin (A) and an Eastern white pelican (B). (A) The intimal surfaces of the vessel are irregular, yellow, raised, and roughened due to expansion by atherosclerotic plaques. (B) Expansion of the vascular wall is due to marked transmural thickening by subintimal cholesterol clefts within a matrix of spindle cells, collagen, and occasional vacuolated macrophages. The vascular lumen lies to the right (see eSlide 27.e3). *(Photo A: Courtesy of A. Rich, International Zoo Veterinary Group.)*

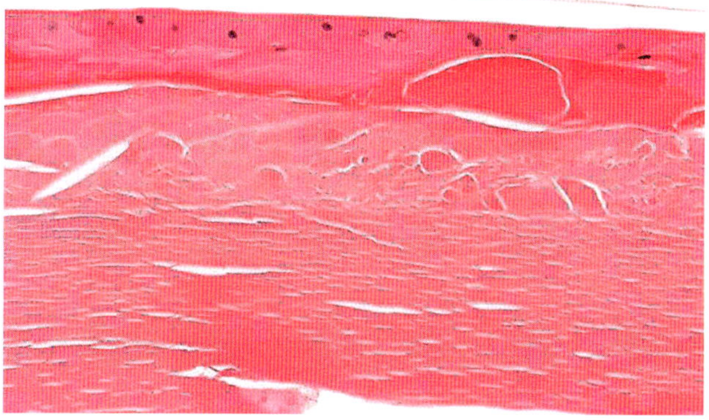

FIGURE 27.8 Early posterior cortical cataract in the eye of a Macaroni penguin. Multiple, roughly round aggregates of degenerate protein form Morgagni globules that disrupt the lens stroma.

such as sticks and stones or bristles from cleaning utensils, coins, fence clips, nails, lead pellets, molted tail feather shafts, enrichment items, and items introduced into enclosures by members of the public (Fig. 27.9A). **Gastrointestinal foreign bodies** are frequent in both captive and free-ranging penguins and cause significant morbidity and even death. Ingested objects are often retained in the stomach (Fig. 27.9B). Radiographically, it can falsely appear that objects are located in the distal intestine near the cloaca owing to the distensible stomach. Dehydration, lethargy, weakness, anorexia, coelomic distension, weight loss, regurgitation, vomiting, ill thrift, diarrhea, scant feces, and extended molt are presentations of acute and chronic gastric impaction or obstruction. Foreign bodies are readily identified at necropsy. Associated tissue changes range from mild mucosal irritation and ulceration to transmural foreign body penetration/perforation with localized to diffuse coelomitis and bacterial sepsis.

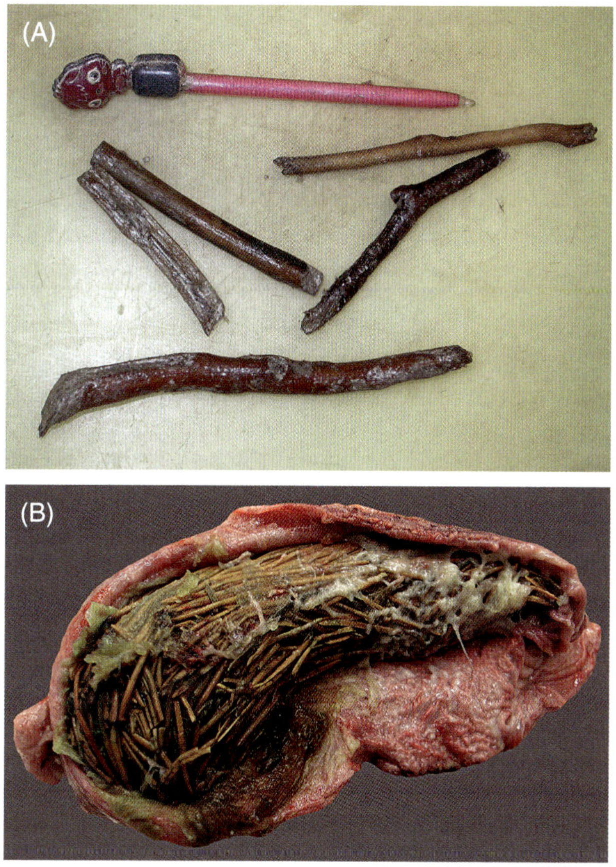

FIGURE 27.9 Gastric foreign bodies from Magellanic (A) and Humboldt (B) penguins. Birds can ingest many types of nondigestible man-made and natural products. (A) This assortment of ingested, gastric foreign objects includes twigs and a pen. (B) Ventricular distension by materials such as these compacted spruce needles can cause luminal obstruction. *(Photo B: Courtesy of C. Schulze, Berlin-Brandenburg State Laboratory.)*

Gastric impaction with feathers occurs in penguin chicks and ingestion of small rocks/pebbles from the environment is frequent. This behavior is benign if binding agents such as sand are not consumed at the same time. However, if binding agents turn the ingested material into a thick sludge that inhibits gastric grinding, impaction can develop from a very young age. **Phytobezoar** formation may lead to chronic obstruction in adult penguins.

Sporadic cases of gastrointestinal impaction and perforation due to foreign body ingestion are seen in zoo birds, particularly pelicans, and in free-ranging birds that tackle oversized or toxic prey items (e.g., puffer fish).

The specific underlying cause for the development of **uropygial gland infections** is unknown (for normal anatomy, see Fig. 27.e2). However, infection has not been recorded in free-ranging penguins, suggesting a multifactorial process related to captive environments. Predisposing factors include sedentary habits and reduced swimming activity, poor plumage, failure to preen, irregular molt, and nutritional deficiencies. Glands are enlarged and swollen and contain purulent or caseous material. Cultures often yield mixed bacteria. ***Candida*** spp. are commonly isolated, even following antifungal treatment. Rupture and secondary septicemia may develop. Histology may suggest underlying vitamin A deficiency, but supplementation does not resolve the condition. Early diagnosis and treatment may prevent impaction. Once preen gland infection has been established in a bird, future episodes are likely.[231]

Miscellaneous/Multifactorial

A disease complex termed **penguin diphtheria (PD)**, also referred to as **necrotizing gastroenteritis** or **diphtheritic stomatitis**, is recognized in wild and captive penguins of the genera Megadyptes und Spheniscus in New Zealand and Europe. It constitutes one of the most important causes of death in captive juveniles in outdoor facilities in these regions. Despite extensive research, a definitive etiology has not yet been established to date.[8,63] Proposed causes include underlying viruses, polymicrobial, protozoal (trichomoniasis) or fungal infections, and environmental factors including mucosal injury due to biotoxins or food item/foreign body trauma. Environmental hygiene is vital and prophylactic antibiotic and antiviral treatment appears to limit the incidence in some collections. Affected chicks frequently have outdoor access and present most commonly between 2 and 4 weeks (range from 1 to 10 weeks) of age with sudden death or nonspecific debilitation and anorexia. Once initiated, cases occur in clusters, with seasonality related to breeding.

Necropsy findings in PD include thick, yellow diphtheritic membranes on the hard palate, tongue, buccal mucosa, and beak commissures that overlie mucosal ulceration (Fig. 27.10A). Lesions may extend into the esophagus and gastrointestinal tract. Cases with unremarkable oral mucosa but multifocal diphtheritic and necrotizing enterocolitis occur. Reduced intestinal content indicates anorexia. Histologically, severe diphtheritic and necrotizing stomatitis, esophagitis, and/or enterocolitis predominate. Single-cell epithelial necrosis with sloughing of the esophageal mucosa lacking significant inflammation is seen in early PD stages (Fig. 27.10B; eSlide 27.e4). Mild lesions progress to mucosal hyperkeratosis, hyperplasia and

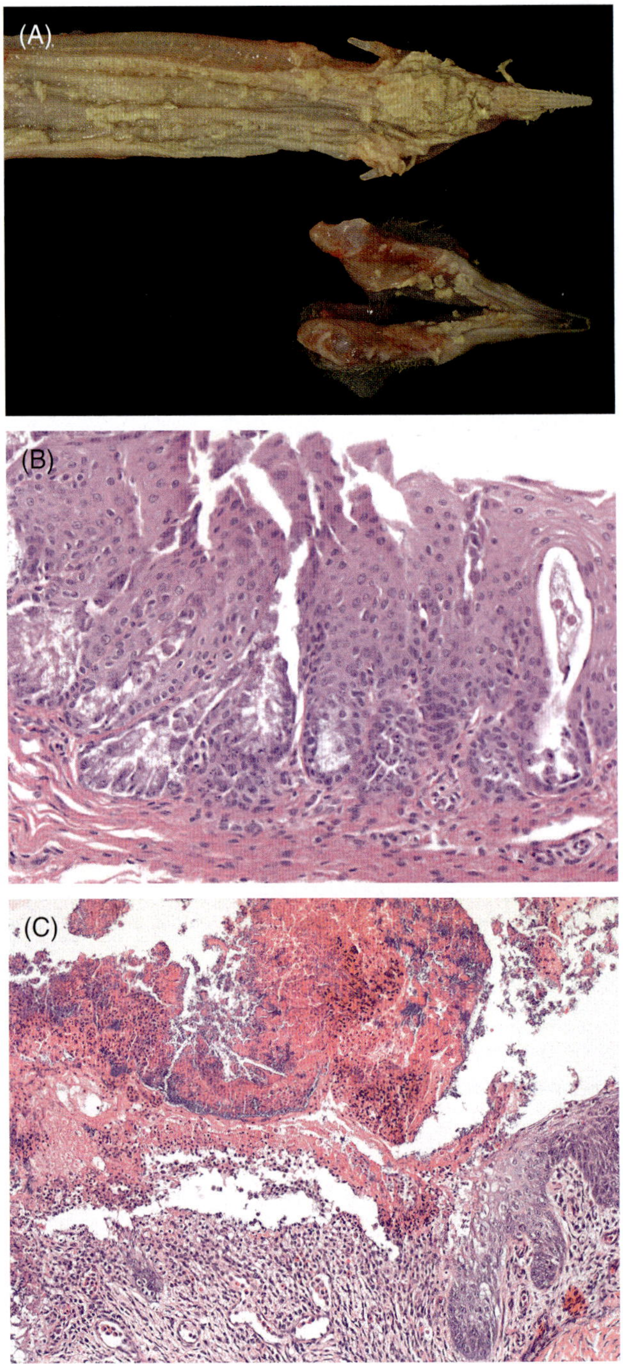

FIGURE 27.10 Penguin diphtheria (PD) in the oral cavity (A) and esophagus (B) of a Humboldt penguin and the pharynx of an African penguin (C). This syndrome has been reported as "diphtheritic stomatitis" and "necrotizing gastroenteritis" and primarily affects chicks between 2 and 4 weeks of age. (A) Severe multifocal to confluent plaques of soft light tan material cover the oral mucosa and extend into the esophagus. (B) Early lesions are characterized by multifocal epithelial single-cell drop out, necrosis, and sloughing of the esophageal mucosa in the absence of significant inflammation. (C) Changes in more advanced lesions include the formation of diphtheritic membranes composed of viable and degenerate granulocytes, necrotic debris, red blood cells, and colonies of bacteria. Adjacent remaining epithelium is spongiotic and mildly hyperplastic. Occasional indistinct viral eosinophilic intranuclear inclusion bodies may be present among epithelial cells (see eSlide 27.e4). *(Part C: Photo Courtesy of C. Schulze, Berlin-Brandenburg State Laboratory.)*

mucosal edema, intracytoplasmic vacuolation, and separation of the basement membrane. Intranuclear eosinophilic inclusion bodies are occasionally seen in oral mucosal epithelial cells at early disease stages. In advanced fatal lesions, thick serocellular crusts cover ulcerated mucosal and proprial surfaces (Fig. 27.10C). Extensive pleocellular inflammation extends into the submucosa where it is accompanied by granulation tissue. Survivors may develop chronic tracheal infections or diskospondylitis but this is rare.[22] Heavy colonization of diphtheritic lesions by mixed pleomorphic bacteria is common and secondary bacterial pneumonia may complicate the clinical picture. Bacterial cultures yield mixed nonspecific growth, but ***Plesiomonas shigelloides*** and ***Corynebacterium amycolatum*** are disproportionately frequent. *Clostridium* species are occasionally detected. The significance of isolated aquatic environmental organisms in disease development is uncertain, but secondary bacterial proliferation following initial mucosal damage (viral/environmental) is suspected. A novel avian *alpha herpesvirus*, designated ***Spheniscid herpesvirus*** **1 (SpHV-1)**, has been isolated and characterized in Humboldt and African penguins with typical oropharyngeal lesions, but further studies failed to consistently identify the virus in historic frozen case material ($n=42$).[200] Virus isolation from oral swabs of Humboldt penguin chicks proved more successful (unpublished results author). Recently, SpHV-1 was isolated from oral swabs of adult African penguins in managed care presenting with nonfatal oropharyngeal lesions and sneezing following intrazoo transfers in the United States (K.C. Gamble, personal communication). Continued monitoring of animals by oropharyngeal/choanal swab analysis in affected collections and attention to collection and transfer history is recommended to further establish the role of SpAVH-1 in penguin diphtheria and respiratory presentations.

There are multiple underlying causes of pododermatitis (bumblefoot) in birds. Predisposing factors in captive penguins include sedentary behavior, decreased swimming, and prolonged standing on abrasive, hard, moist, or fecal-contaminated flooring. Malnutrition, vitamin A deficiency, and exposure to allergens, strong cleaning products, or disinfectants in captivity have also been proposed. Similar considerations apply to pododermatitis in pelicans and shoebills. The primary lesion is ischemic pressure necrosis of soft tissues of the foot, or an initial fissure or puncture wound that compromises perfusion of the dermis and leads to erosion and ulcer formation (Fig. 27.11A). Affected animals present with abnormal gait, increased resting behavior, and discomfort on foot palpation. In nontraumatic lesions, unilateral or bilateral swelling and increased redness of one or multiple areas of the footpads followed by epithelial thinning, ulceration, hemorrhage, and scab or granulation tissue formation are seen on the plantar surfaces of the feet (Fig. 27.11B). Histological findings include skin ulceration, laminated parakeratosis, serocellular crusting with active inflammation, and variable bacterial colonization. Adjacent epidermis is hyperkeratotic and hyperplastic with irregular acanthosis or atrophic thinning. Early cases have mild pleocellular perivascular dermatitis. Advanced cases exhibit severe perivascular inflammation, edema, and collagen degeneration resulting in necrotizing vasculitis. Introduced foreign material incites granulomatous foreign body reactions with giant cell formation. Secondary opportunistic bacterial infections with *Escherichia coli, Proteus mirabilis* and *Staphylococcus*, *Enterococcus*, *Pseudomonas*, and *Clostridium* species, or invading yeasts and fungi, occur frequently. Soft tissue mineralization and ascending necrotizing osteomyelitis and/or tenosynovitis can be seen in untreated lesions (Figs. 27.11C, D) and may trigger the development of systemic **amyloidosis**. Chronic progression rarely results in vegetative valvular endocarditis affecting the aortic and mitral valves leading to vascular insufficiency, lethargy, and dyspnea. Prevention is paramount, focusing on environmental enrichment, encouragement of swimming, and avoidance of hard, rough, wet, or contaminated surfaces.

Plastic foreign bodies are commonly found in the gastrointestinal tracts of beached, by-caught, and free-ranging pelagic species trapped worldwide, particularly Procellariiformes. Plastic ingestion leads to lower fledging weight in Laysan albatross chicks but does not appear to affect chick survival. In one study, plastic was present in the stomach of up to 71% of monitored Manx shearwaters, and lighter birds had larger pieces of plastic than heavier birds, implying some physiological impact. Another recent study in flesh-footed shearwaters provides evidence for sublethal physical tissue damage due to the ingestion of macro- and microplastic resulting in "plasticosis," plastic-induced proventricular scar tissue formation. Masson's trichrome demonstrates widespread fibrosis and replacement of the proventricular mucosa and submucosa, not identified in animals ingesting natural items. There is increasing evidence to suggest an association between plastic ingestion and persistent organic pollutant accumulation (see "Toxic" section for additional details).[7,42,50,51]

Free-ranging aquatic bird species can drown due to entrapment in **marine debris and fishing gear**, including nylon drift nets, gill nets, and baited hooks. Closer to shore, nets catch pursuit-diving species including cormorants, loons, grebes, and penguins (Fig. 27.12).

Pelagic long-line fishing contributes to declines in albatrosses, petrels, boobies, and frigate birds. Most drowning cases occur in the absence of an underlying disease. In contrast to birds stranding or affected by pollution, drowned

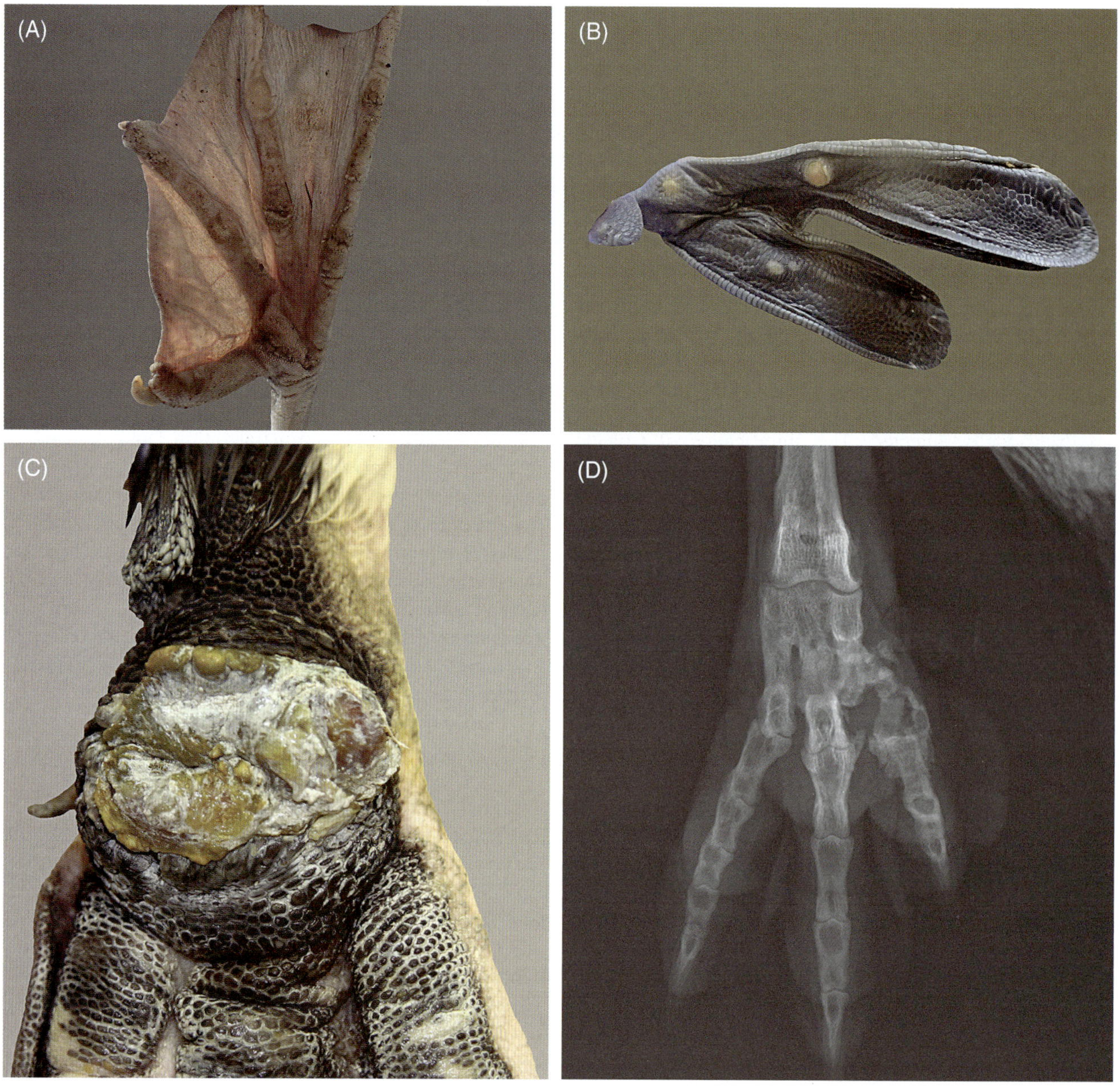

FIGURE 27.11 Pododermatitis (bumble foot) progression affecting the feet of a Dalmatian pelican (A), black-necked grebe (B), and Macaroni penguin (C, D). (A) Early lesions often present as superficial fissures in the skin overlying the digital joints of the feet. (B) Raised, proliferative, white lesions are the result of beginning granulation tissue development on the plantar aspect of the foot. (C) Chronic changes include focally extensive ulceration with exuberant granulation tissue formation and marginal dermal hyperplasia. (D) Ascending osteomyelitis with bone lysis can develop subsequent to chronic severe pododermatitis (same animal as in C). *(Parts A, B: Photos Courtesy of M O'Brien, Wildfowl & Wetlands Trust; Parts C, D: Photos Courtesy of J. St. Leger, SeaWorld Parks & Entertainment.)*

birds are in good body condition with subcutaneous fat and well-developed muscles, and their stomachs frequently contain recently ingested fish. Lesions in entangled animals can present as abrasions, bruising, wing erythema and edema, and multiorgan congestion. The air sacs or distal trachea may contain clear watery fluid (aspirated water). Lungs are congested, edematous, or hemorrhagic, and exude white frothy supernatant-rich fluid from cut surfaces. The heart and major veins, including the dural venous sinuses, are grossly distended. Histologic lesions are not pathognomonic but multiorgan congestion and pulmonary edema in conjunction with aquatic environmental debris in severely edematous lungs is supportive.[76,239]

Starvation is a significant problem in free-ranging penguins, particularly chicks and juveniles, and can result in mass mortality events. Free-ranging Procellariiformes,

FIGURE 27.12 Drowned Magellanic penguins entangled in an artisanal triple-layered drift gillnet. Entanglement in fishing gear can impact single, multiple, or many animals. Animals are generally in good body condition and lesions consistent with entanglement can include abrasions, bruising, erythema, wing edema, multiorgan congestion, jugular vein engorgement, pulmonary edema and hemorrhage, and serous bloody fluid in the trachea and lungs. *(Photo Courtesy of J. Catão-Dias, University of São Paulo.)*

including shearwaters and prions (genera *Pachyptila* and *Halobaena*) of all ages, frequently suffer **mass mortality events ("wrecks")** of hundreds to thousands of birds. These events occur predominantly during migration and are related to adverse weather and/or inadequate food. Most affected birds are emaciated juveniles with pectoral muscle atrophy, an empty stomach, complete fat exhaustion with serous atrophy of fat, and multiorgan atrophy. Investigations exclude infectious disease, heavy metal, or persistent organic pollutants but sepsis and dehydration can play a role. **Mass downing** of migrating, free-ranging eared grebes is seen periodically during the fall in North America; affected birds are typically adults in good body condition.[220] Events are associated with adverse weather or poor visibility as this species requires water for landing and take-off; downed birds may die on impact or remain trapped on the ground and subsequently die.

Winter mortality of free-ranging common loons is seen periodically in Florida (United States) between December and April.[6] Common findings include emaciation with muscle (pectoral muscle atrophy), fat exhaustion, and hemorrhagic enteritis. Synergistic etiologies include inadequate food supply, the high energetic cost of migration and molt, poor weather, and parasitism. Mass mortalities, sometimes involving thousands of birds, also occur from **idiopathic emaciation** in free-ranging nestling American white pelican breeding colonies.[221] Starvation and hypothermia due to adverse weather are implicated.

An iodine-containing disinfectant induced incidental **goiter** in captive penguins with renal failure and an association has been made between dioxin pollutants and the development of goitrous changes in great cormorants.[43,226] Histologically, cormorants showed irregularly shaped follicles with hypertrophic follicular epithelium and loss or deficiency of luminal colloid. Increased dioxin levels were detected in liver, muscle, and fat.

Penguin feather loss disorder is a syndrome characterized by aberrant premature loss of feathers that is associated with retarded growth rates and increased mortality in young birds. First identified in 2006, it has subsequently affected free-ranging and captive African, Magellanic, rockhopper, and Adélie penguins.[103,129] Aberrant feather loss exposes bare patches among normal feathered skin, ranging from a few centimeters in diameter to complete feather loss prior to usual molt. Retarded growth rates and increased mortality have been associated with the syndrome. Survivors remain featherless for several weeks but those that reach fledging regrow normal, complete juvenile plumage. A viral infection has been suggested but the etiology remains unproven and more complete investigation, including histology, will aid in characterizing this condition. The presentation is distinct from **molt abnormalities** seen in zoological collections, a spectrum of feather issues associated with increased patchy feather loss or poor feather growth following a molt. These cases may have a light-cycle, light-intensity, and/or nutritional basis.

Puffinosis is an enigmatic syndrome that causes annual seasonal mortality (late August/September) in free-ranging juvenile Manx shearwaters on Skomer and Skokholm Islands, Wales. Mortalities follow episodes of foot-web blistering (Fig. 27.13A), leg paralysis, locking of the tibiotarsotarsal joint, and occasionally conjunctivitis without systemic lesions. Histologic lesions include fluid-filled vacuoles in the epidermis (Fig. 27.13B). An etiology remains unproven but recent metagenomic studies have suggested that opportunistic bacterial infection may be the cause of blistering, facilitated by skin contact with caustic fecal ammonia while sitting in damp nesting burrows.[71] Similar blistering foot lesions have been described in European storm petrels, Leach's storm petrels, and northern fulmars from Scotland, California brown pelicans in the United States (J. St. Leger, personal communication) and in gentoo penguins with multiple ulcers on the dorsal surfaces of both feet. Poxviral infection should be considered as a differential diagnosis.

Predatory trauma is a significant cause of morbidity and mortality in many of the colonial nesting seabirds, particularly ground-nesting or burrowing island-dwelling Procellariform species. Predators may target penguins housed outdoors (Fig. 27.14). Introduction of feral rodents, cats, foxes, or primates can threaten extinction in the wild.

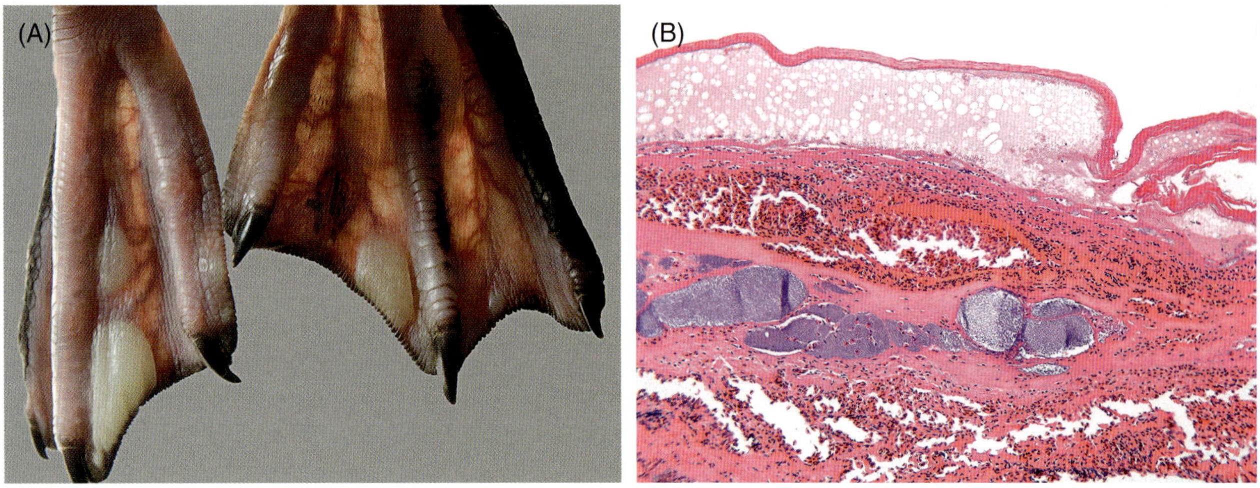

FIGURE 27.13 Puffinosis in the feet of a Manx shearwater. (A) Foot webs have characteristic fluid-filled vesicular lesions that are associated with seasonal mortality events. (B) The keratinized layer of the epidermis is markedly elevated from the basal lamina by fluid-filled vacuoles that form large blisters. The superficial dermis is expanded by edema fluid, hemorrhage, and heterophils. Dermal capillaries are dilated and markedly congested. Numerous bacterial colonies are present within the blister and dermal vessels. *(Part A: Photo Courtesy of C. Perrins; Part B: Photo Courtesy of R. Hanna, Agri-Food & Biosciences Institute.)*

FIGURE 27.14 Bite wound in the neck of an African penguin. The patterns of wounds can indicate predatory trauma. The two puncture wounds, hemorrhages and blood loss in this penguin were due to fox predation. *(Photo Courtesy of C. Schulze, Berlin-Brandenburg State Laboratory.)*

Neoplastic

The reported overall prevalence of neoplasms in captive penguins is low. **Malignant melanomas** are reported with the highest frequency, occurring in macaroni, Humboldt, and rockhopper penguins.[69,211] Predilection sites are the skin of the foot or hock, and subcutis and subjacent muscle near the beak and oral cavity. Humboldt penguins may have a unique presentation, with variably pigmented, cornified lesions in the inguinal area. Histologically, neoplastic cells are pleomorphic and heavily pigmented; greater than 50% contain melanin granules. Junctional activity is prominent. Surface ulceration correlates with systemic spread and metastases are reported in the liver, lung, adrenal gland, brain, and bone. Other prognostic factors include gross surface dimensions, mitotic index, and depth of neoplastic cell invasion. Amelanotic melanomas have not been reported. PNL-2 can be used for immunohistochemical confirmation; S-100, Melan-A, and Ki67 cannot. The average survival is 7 months. The pathogenesis underlying the high prevalence of melanocytic neoplasms in captive penguins is unknown. Advanced age is most probable. There is no evidence to suggest involvement of ultraviolet light or chlorine exposure, viral induction, or genetic predisposition.[69]

Other reported neoplasms in penguins include carcinomas (adenocarcinoma, squamous cell carcinoma, cholangiocarcinoma, renal tubular carcinoma, poorly differentiated hepatic carcinoma), sarcomas (anaplastic, mucinous) (Figs. 27.15A, B), malignant lymphoma, and an orbital round cell tumor.[44,64,80,128,214,216,217,230,235,279,287] **Adenocarcinomas** primarily affect the stomachs with a few cases of reported intestinal and urogenital origin. **Gastric adenocarcinomas** are firm, pale masses that expand the gastric wall. Transmural growth results in serosal adhesions. Gastrointestinal adenocarcinomas are histologically typical (Fig. 27.16; eSlides 27.e5 and 27.e6). Apparent predilection for the proventricular/ventricular junction can render it difficult to differentiate origin based on histomorphological features. Cytoplasmic positivity with alcian blue and Periodic acid–Schiff is helpful in confirming ventricular origin and positive galactose oxidase-Schiff staining can confirm proventricular origin. Ventricular adenocarcinomas may metastasize to the pancreas and intestines. **Uropygial gland squamous cell carcinoma** (SCC) should be considered an underlying problem in cases of chronic, recurrent disease of the uropygial gland.

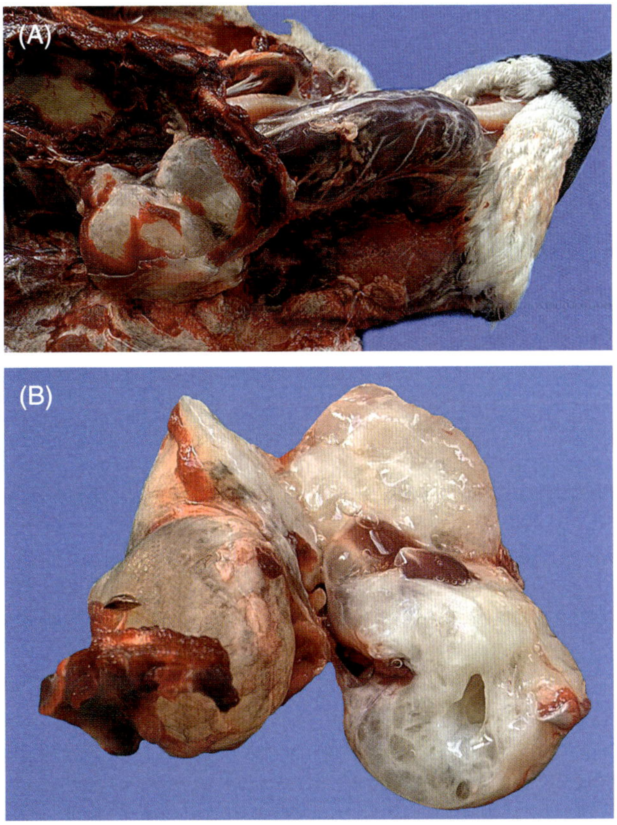

FIGURE 27.15 Myxoid sarcoma in the shoulder of an African penguin. (A) Surrounding the left shoulder joint is an expansile, multinodular, polycystic, soft mass. (B) Cut surfaces reveal numerous variably sized cystic cavities, filled with clear, gelatinous, lubricative fluid. *(Parts A, B: Photos Courtesy of S. Spiro, Zoological Society of London, London Zoo.)*

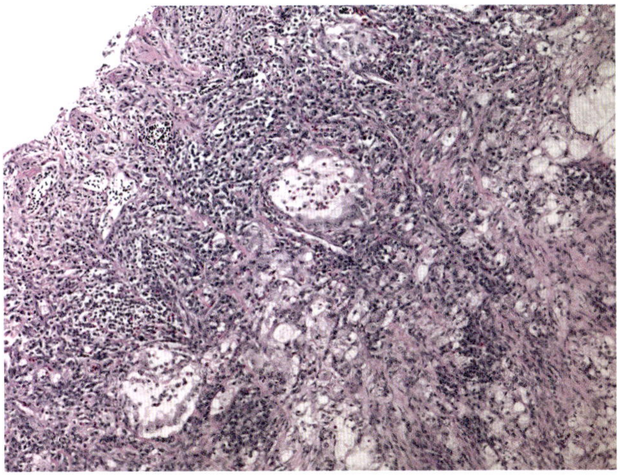

FIGURE 27.16 Mucinous adenocarcinoma in the cloaca of a Humboldt penguin. The mucosa is widely invaded and effaced by disorganized nests and tubuli of neoplastic epithelial cells containing pale basophilic mucin and cellular debris. Significant secondary inflammation dissects the fibrous stroma and further obscures the architecture (see eSlides 27.e5 and 27.e6).

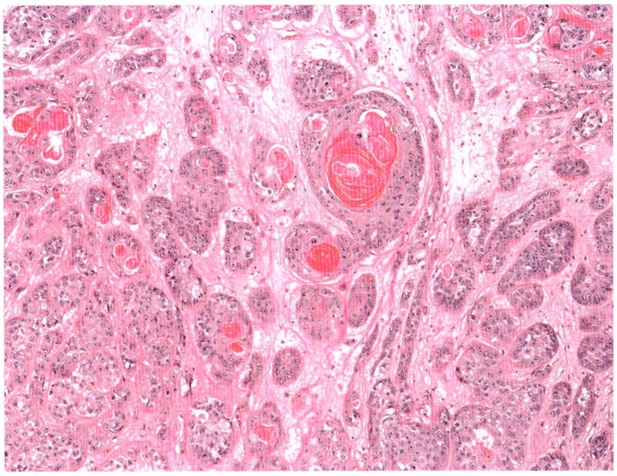

FIGURE 27.17 Squamous cell carcinoma in axillary skin of the left wing of Dalmatian pelican. Numerous cords of variably differentiated neoplastic squamous epithelial cells extend from the skin surface (not shown) into the underlying dermis. The tumor is associated with an edematous, scirrhous stroma. Central keratin pearls are present in the centers of some epithelial cords (see eSlide 27.e7).

Histomorphology is similar to SCC in other species. Chronic inflammation may predispose to neoplastic transformation, warranting aggressive early treatment. Underlying hypovitaminosis A should be excluded. Behavior is locally invasive with infrequent metastasis.[21,217] **Malignant lymphoma** may be multicentric or intestinal. Immunohistology can confirm T-cell origin (CD3). Commercial CD45 and CD79 antibodies are not cross-reactive.[128,230]

Despite their longevity, neoplasia is rarely documented in free-ranging or zoo-captive species in other orders. In captive pelicans, **squamous cell carcinomas** (Fig. 27.17; eSlide 27.e7) and **chondrosarcomas** of the beak are most common. Additional neoplasms of these orders are available in Table 27.e3. For further histology of neoplastic entities, also see eSlide 27.e8 (renal carcinoma), eSlide 27.e9 (cholangiocarcinoma), and eSlide 27.e10 (hepatic carcinoma).

INFECTIOUS DISEASES

In many of the free-ranging pelagic species in the clade Aequorlitornithes, in which sampling may be infrequent, relatively little is currently known about the prevalence and significance of infectious disease and associated pathology. A review of diseases in Procellariiformes is indicative of the wide range of sources and limits of methodology that form the body of current knowledge.[255] In addition to descriptions below, the Digital Content includes tables of notable viral (Table 27.e4), bacterial, fungal (Table 27.e5), and parasitic diseases (Table 27.e6).

DNA Viruses

Avipoxvirus infections have been documented in African, Humboldt, gentoo, rockhopper, and Magellanic penguins, but all species are probably susceptible. The disease affects both captive and free-ranging penguin populations and is more common in chicks than adults, suggesting adults have acquired immunity or more robust immune function. Increasing prevalence often coincides with weather-induced increases in vector populations but transmission also occurs by direct contact with abraded skin, conjunctiva, mucous membrane, contaminated air-borne particles, or infected tissues. At least two divergent avipoxvirus strains (*Chordopoxvirinae* subfamily) affect penguins.[186] Outbreaks manifest as cutaneous and diphtheritic ("wet") forms. Affected birds are often emaciated.

Cutaneous poxvirus infection induces typical raised and ulcerated wart-like papules and nodules in feathered and unfeathered skin around the eyelids, beak, neck, flippers, cloaca, and feet (Fig. 27.18A). Glossal lesions may develop. Ocular pathology resulting in blindness has been reported in Magellanic penguin chicks.[130] Histological findings are characteristic, with epidermal hyperplasia and ulceration, eosinophilic intracytoplasmic inclusions consistent with Bollinger bodies, and dermatitis (Fig. 27.18B; eSlide 27.e11). Although cutaneous poxvirus infection is rarely fatal, debilitation and secondary bacterial or fungal infections are common and can cause mortality.

The rarer diphtheritic form is more frequently fatal, and manifests as raised, yellow plaques in the oral cavity, throat, esophagus, trachea, and lungs. Infection in these locations can cause dyspnea or dysphagia. Lungs are often congested or hemorrhagic, and there may be splenomegaly and cardiomegaly. Histological findings include bronchopneumonia, nonsuppurative hepatitis, splenic necrosis, lymphopenia, and lymphocytolysis, as well as necrotizing enteritis, esophagitis, and airsacculitis. Typical cytoplasmic inclusion bodies are seen in affected epithelial cells. Concurrent infections with common pathogens (e.g., *Aspergillus* spp.) worsen the clinical course.

Avian poxvirus infections occur sporadically in Pelecaniformes (red-tailed and white-tailed tropicbirds, pink-backed pelican) and in Procellariiformes (southern giant petrel, Manx, flesh-footed, wedge-tailed and Audobon's shearwaters, shy albatross, Laysan albatross), but reports are rare or absent in Gaviiformes and Podicipediformes. Most infections are cutaneous, affect fledgling birds on the nest, and result in proliferative and ulcerative lesions on facial skin, beak, and legs. Diphtheritic lesions are documented in white-tailed tropicbirds and Laysan albatross. Prevalence in Laysan albatross fledglings can reach 88% in epizootic years of high rainfall.[288] Despite a range of severity (including proliferative lesions covering the eyes and that deform the bill and skull), most chicks recover, and overall fledging rate is not affected (Figs. 27.18C–E). Foot-web lesions due to avian poxvirus infection have also been reported in juvenile and adult Manx shearwaters. These are distinct from puffinosis (see "Miscellaneous/Multifactorial" section).

Herpesvirus-like infection is described in African penguins. Associated lesions resemble those of infectious laryngotracheitis (*Gallid herpesvirus* 1 syn: *Avian herpesvirus* 1). Debilitation and respiratory distress were clinical features in two adults and a chick from one zoo.[132] Histologically, inflammation and syncytial cells with eosinophilic Type A intranuclear inclusions were seen in respiratory epithelium. Findings bear some similarities to a reported outbreak of herpesvirus-like infection in eight African penguins in a rehabilitation center.[205] Affected chicks presented with poor weight gain, airsacculitis, and congested, edematous, and firm lungs. Tracheitis with epithelial necrosis, herpetic intranuclear inclusion bodies, and syncytial cell formation were characteristic histological features. Herpes-like virus particles were evident on transmission electron microscopy in both studies, but virus isolation and molecular testing could not confirm or characterize the virus further. Other reported herpesvirus infections include ***Magellanic penguin herpesvirus* 1 (MagHV-1)** which caused an outbreak of severe respiratory disease characterized by fibrinonecrohemorrhagic tracheitis with inclusion-bearing syncytial cells, fibrinous airsacculitis and heterophilic pneumonia in Magellanic penguins in Brazil. Subsequent surveillance revealed additional novel herpesviruses in asymptomatic free-living Magellanic penguins (MagHV-2) and Atlantic yellow-nosed albatross (Thalassarchid herpesvirus). Phylogenic analysis suggested an association between the MagHV-1 outbreak and previous viral seabird die-offs.[183] **Herpesvirus infections** has been demonstrated in cases of penguin diphtheria (see "Miscellaneous/Multifactorial" section).[200]

Gaviid herpesvirus 1 has been isolated from ulcerative tracheitis lesions in common loons. An **alphaherpesvirus** was isolated from magnificent frigatebirds with crusting proliferative hyperkeratotic lesions with ballooning epidermal generation of the facial skin and beak.

RNA Viruses

Several **arboviruses** affect birds, but **West Nile virus (WNV)** and **Eastern equine encephalitis (EEE)** infections are most relevant to penguins. Sphenisciformes experience high rates of morbidity and mortality during outbreaks in both free-ranging and zoo settings. Disease spreads via biting insects, usually mosquitoes, and wild birds act as virus reservoir.

Introduction of **WNV**, a *flavivirus*, into North America in 1999 caused fatal infections in many species of birds

FIGURE 27.18 Avian poxvirus infection of the skin of a Humbolt penguin (A, B) and infection progression and resolutions in a Laysan albatross (D–E). Typical raised and ulcerated wart-like papules and nodules can be present in feathered and nonfeathered skin. Clinical outcomes depend on site and severity of infection, and birds can survive infection despite transiently severe lesions. (A) Multifocal, raised, superficially ulcerated papules and nodules are present in nonfeathered skin around the eyelids and nares and to a lesser degree affect the feathered skin. (B) Marked epidermal hyperplasia with basal spongiosis and abundant brightly eosinophilic intracytoplasmic inclusions (Bollinger bodies) in superficial keratinocytes are characteristic of infection. (C) April 3, 2013. Proliferative epithelial lesions adjacent to the nostril, extending bilaterally to involve both eyelids. The chick exhibited unusual head posture and circling behavior. (D) April 7, 2013. Proliferative lesions have ulcerated, with crusting exudate. The blind chick wandered from its nest and was placed back by an observer, and its parents continued to feed it. (E) May 25, 2013. Skin lesions have largely resolved with return of normal vision and behavior. The chick later fledged successfully (see eSlide 27.e11). *(Part A: Photo Courtesy of C. Schulze, Berlin-Brandenburg State Laboratory; Parts C–E: Photos Courtesy of C. Granholm, Princeville, Kaua'I.)*

including Sphenisciformes, Gaviiformes, Podicipediformes, and Pelecaniformes. Occasional cormorant fatalities have been reported in Europe, where infection is considered enzootic. WNV is shed in respiratory secretions, and since avian-to-human transmission has been documented, it should be treated as a zoonotic pathogen. Infection and deaths have been reported in Humboldt and African penguins, common loons, pie-billed grebes, Clark's grebes, eared grebes, little grebes, American white and brown pelicans, and double-crested and Guanay cormorants. Major die-offs have occurred in eared grebes and pelicans, and WNV was implicated in at least 11 mortality events resulting in the deaths of over 9,000 American white pelicans. In penguins housed in outdoor exhibits, clinical signs can include anorexia, weakness, abnormal behavior, vomiting, and dyspnea due to excessive tracheal or pulmonary mucus secretion. There are no typical gross lesions. Histological findings are limited to multifocal lymphoplasmacytic meningoencephalitis with mild gliosis and satellitosis, primarily affecting the cerebellum. Immunohistochemistry can aid in the identification of infected Purkinje cells, neurons, and glial cells (Fig. 27.19).

In pelicans and cormorants, neurological signs include head tilt, ataxia, and disorientation. Brain hemorrhage may be apparent. Histological lesions in cormorants include meningeal or neuropil hemorrhage; neuronal vacuolation; lymphohistiocytic myocarditis, mineralization, and hemorrhage; splenomegaly with lymphoid depletion, fibrinous splenitis, and hemorrhage; renal and pancreatic lymphohistiocytic infiltrates; hepatocyte necrosis and intestinal crypt cell necrosis.[245] Encephalitis or meningoencephalitis and myocardial necrosis or myocarditis are the dominant lesions in American white pelicans; prominent findings in other species include encephalitis and myocarditis in eared grebes and myofiber drop out, nephritis and meningoencephalitis in little grebe.[91]

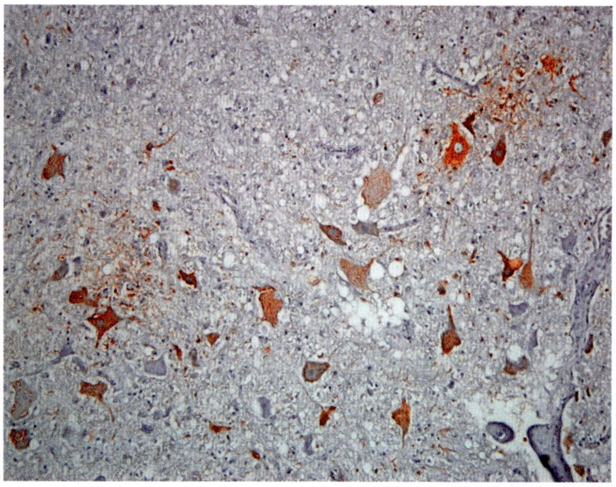

FIGURE 27.19 West Nile virus encephalitis in an African penguin. Numerous neuronal perikarya exhibit positive immunolabeling. There is a noticeable lack of lymphoplasmacytic inflammation in this case. IHC. *(Photo Courtesy of D. McAloose, Wildlife Conservation Society.)*

Eastern equine encephalitis (EEE) is an *alphavirus*. Clinical signs in African penguins include anorexia, antisocial behavior, mild lethargy, and intermittent vomiting.[254] Severely affected animals become recumbent and secondary infections may develop. Gross findings are minimal. Histologic examination reveals severe, chronic, multifocal, lymphoplasmacytic encephalitis with prominent perivascular cuffing. Lesions are notable in the cerebellum where mineralization of dendritic processes, focal infiltrates of mononuclear and glial cells, and loss of Purkinje cells are detected in folia; vacuolation, axonal degeneration, demyelination, and gitter cells may be seen in cerebellar white matter. Mild necrotizing and histiocytic myocarditis can be present. Antibodies to EEE have been detected in pelicans, but clinical disease has not been identified.

Recurrent high mortality events in double-crested cormorants, attributable to virulent **Newcastle disease virus (NDV); avian paramyxovirus-1** has occurred in breeding colonies in Canada and the United States since 1992.[269] The mortality of juveniles can exceed 90% in an affected colony. Spillover events in white pelicans are occasionally seen and large-scale mortality events in American white pelicans (>5,000 birds) have occurred concurrently with confirmed outbreaks in double-crested cormorants. Paramyxovirus seropositivity is widespread in free-ranging penguins but disease is rare. The major clinical sign in cormorants is an inability to fly, often with unilateral wing or leg paralysis. Pelicans are weak and unable to fly. Specific gross lesions are absent in pelicans or cormorants. Multifocal lymphoplasmacytic encephalitis and/or myelitis are found histologically. Lesions can be mild, focal, and include perivascular lymphocytic cuffing, focal gray matter gliosis with loss of Purkinje cells in the cerebellum, and white matter vacuolation in the cerebrum.[275] Foci of lymphocytic infiltration are seen inconsistently in other organs including the pancreas and liver.

Avian influenza viruses are common in free-living waterbirds including penguins. Serology or PCR-testing has confirmed exposure or infection in Pelecaniformes (cormorants) and Procellariiformes. Two spot-billed pelicans died during an outbreak of highly pathogenic, H5N1, avian influenza in Cambodia in late 2003. Mortalities associated with highly pathogenic H5N1 strains have been recorded in eared and great crested grebes in Europe. During the severe 2020–23 highly pathogenic avian influenza (HPAI) epidemic numerous seabird species, including Sulidae, Pelecanidae, and Phalacrocoracidae, have been severely affected by mass mortality events. In breeding colonies in Europe, HPAI virus detection in wild seabirds was seen in unprecedented numbers between June and August 2022 and similar patterns were recorded in many countries in Africa, Asia, and in the United States.[73,74,258,272] Grossly, lungs are severely congested and marked splenomegaly may be seen (Figs. 27.20A, B). Histological lesions include hepatic, myocardial, and pancreatic necrosis, interstitial pneumonia with necrotizing vasculitis, splenic lymphocytolysis, and nonsuppurative meningoencephalitis (Figs. 27.20C–E; eSlides 27.e12 and 27.e13). Low-pathogenic avian influenza infects North American red-necked grebes (six different strains) and grebes in Siberia (mainly H5 and H7).

Bacteria

As in all vertebrate species, the intestinal flora in healthy free-ranging penguins contains a variety of commensal bacteria, including diverse gram-negative and gram-positive aerobes and anaerobes. These include *Escherichia coli*, *Alcaligenes faecalis*, *Citrobacter freundii*, *Enterobacter*, *Paracolon*, and *Bacillus* species. *Firmicutes*, *Bacteroidetes*, *Proteobacteria*, *Fusobacteria*, *Actinobacteria*, and *Tenericutes* are the main components of the gastrointestinal flora in Chinstrap penguins, which demonstrate differences in the bacterial composition between juvenile and adult animals.[17] When studied, the fecal flora of Pelecaniformes and Procellariiformes is also found to be mixed (gram-negative bacilli, *Enterococcus*, *Staphylococcus*). Antibiotic resistance, including extended-spectrum β-lactamase (ESBL) strains, is present in isolates obtained from shearwaters and boobies living even in remote locations, such as Easter Island.[10]

Enteritis due to ***Escherichia coli*** and ***Klebsiella* species** occurs in captive penguins, but ***Salmonella*** and ***Clostridium* species** are the most significant enteric pathogens (see "Toxic" section for a discussion of clostridial enterotoxemia). *Salmonellae* (***S. typhimurium*** and ***S. anatis***) may cause clinical disease in penguins, and some infected birds are asymptomatic carriers. Mixed *Salmonella* serotypes, including antibiotic-resistant strains, are carried by double-crested cormorants and common loons, and occasionally contribute to overwintering mortality in loons. Occasional isolation is reported in grebes and albatrosses. Penguins may carry ***Campylobacter jejuni*** but clinical disease is not reported. Gross lesions of bacterial enteritis may include a combination of redness, exudation, and ulceration of the mucosa and diphtheritic membranes can be seen in severe cases. Intestinal lumina may be distended by gas or fluid. Histologically necrosis, fibrin deposition, and primarily heterophilic infiltrates characterize acute lesions, which may extend into the submucosa and muscularis. Bacteria may not be present in all lesions.

Confirmed disease due to ***Chlamydophila*** is rare in free-ranging penguins despite widespread seropositivity. Outbreaks have been reported in a captive colony of Magellanic penguins and African penguins in a mixed species exhibit. Affected birds were inappetant, lethargic, and produced light green urates. Necropsy revealed hepatomegaly and splenomegaly. Histologic lesions included necrotizing to granulomatous hepatitis, splenitis, and vasculitis with prominent ventricular vascular thrombosis[19,123] (Figs. 27.21A–C; eSlides 27.e14 and 27.e15). Nearly 90% of Magellanic penguins developed severe keratoconjunctivitis with serous to purulent conjunctival exudate, chemosis, and variable corneal edema; this was not a feature reported in the African penguins. Gimenez stain-positive, intracellular inclusions can be difficult to identify histologically, and definitive diagnosis requires immunohistochemistry or supplementary molecular testing.

Historically, human psittacosis (***Chlamydophila psittaci***) on the Faroe Islands was associated with processing of northern fulmar chicks for human consumption. A recent study suggests that the prevalence of infection in juvenile fulmars in the Faroes remains at approximately 10%.[111] Both northern gannets and short-tailed shearwaters are asymptomatic carriers of *Chlamydophila*.

Septicemic outbreaks of **avian cholera (*Pasteurella multocida*)**, leading to significant mortality events in free-ranging birds, are reported in rockhopper and Adélie penguins, brown and great white pelicans, double-crested and Cape cormorants, common loons, and pie-billed, horned, eared, and Western grebes. A single case was reported in a free-ranging southern giant petrel. Gross and histological findings are typical of *P. multocida* septicemia with hemorrhage, intestinal hyperemia, hepatomegaly, splenomegaly, and necrotizing hepatitis and splenitis with intralesional gram-negative coccobacilli. Other reported bacterial causes of mass mortality events in eared grebes include ***Erysipelas rhusiopathiae*** and ***Streptococcus zooepidemicus***.

In penguins, ***Plesiomonas shigelloides*** and ***Corynebacterium amycolatum*** have been commonly isolated from cases of penguin diphtheria. Polymicrobial and viral etiologies have been proposed as factors in the development of PD, though the pathogenesis remains to be fully characterized (see "Miscellaneous/Multifactorial" section).

Mycobacteriosis is rare in free-ranging birds and is more common in zoo collections. *Mycobacterium avium* complex species (*M. avium* sub. *Avium* and *M. avium* sub. *intracellulare*) and *M. genavense* are most frequently implicated. Sporadic infections have been recorded in African and little penguins and Pelecaniformes, and probably occur in other species.[138] Birds present emaciated, with typical necrogranulomas containing abundant intrahistiocytic acid-fast positive bacilli in multiple organs (particularly liver, spleen, lungs) (Figs. 27.22A, B; eSlide 27.e16). Rapid diagnosis is possible through review of impression

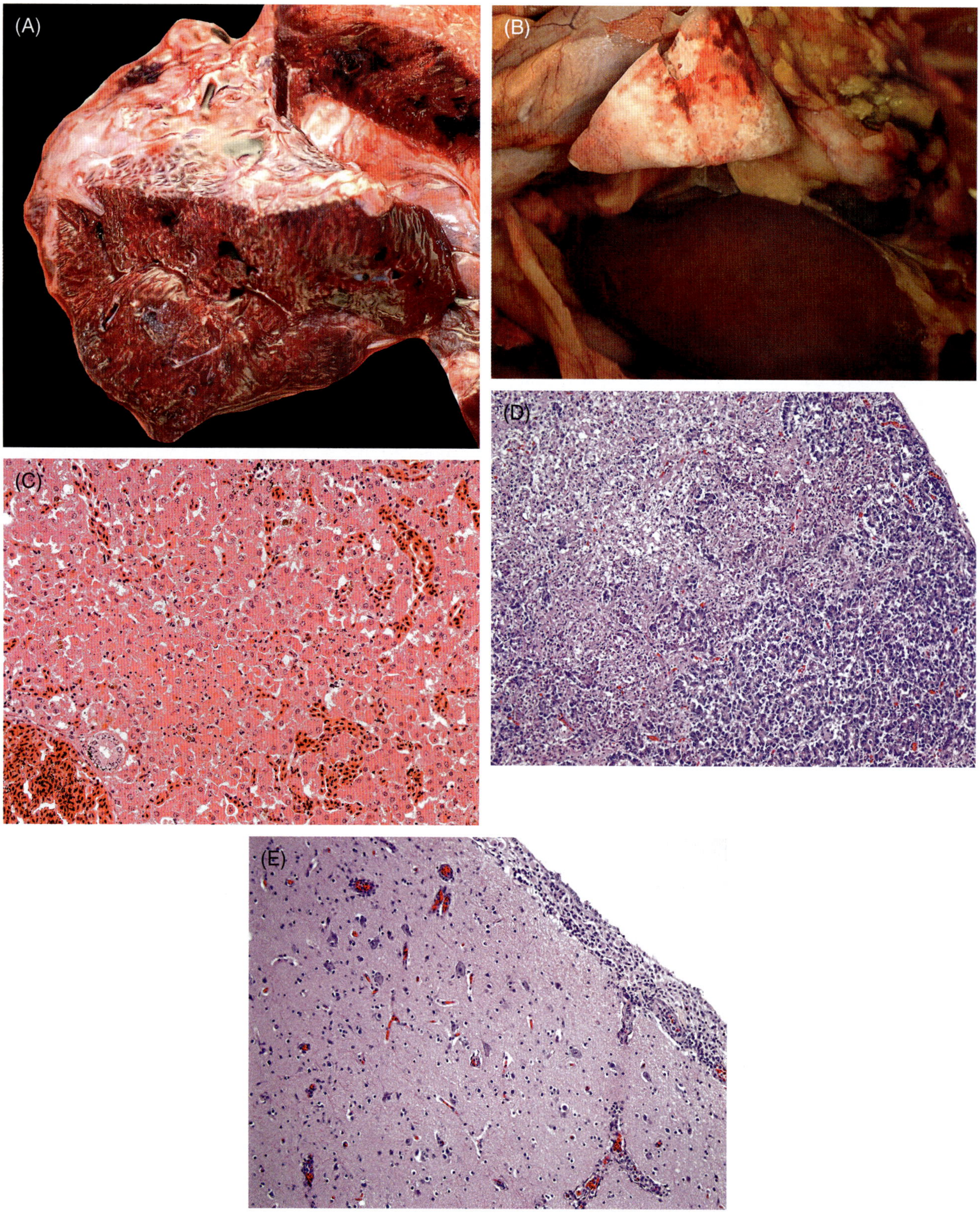

FIGURE 27.20 Highly pathogenic avian influenza (HPAI) in a Humboldt penguin (A, C) and a Dalmatian pelican (B, D, E). No pathognomonic lesions have been associated with highly pathogenic avian influenza virus infection. Gross lesions can include pulmonary congestion and edema, pancreatic hemorrhage, and splenomegaly. Infection can be associated with focal, multifocal, or multiorgan necrosis. Necrotizing vasculitis may also be seen. Lesions can be variable and may be subtle. (A) The lung tissue in this bird is markedly congested, diffusely edematous, and was rubbery with wet cut surfaces and rounded rib impressions. (B) The pancreas has confluent areas of hemorrhages and pale, friable areas of necrosis. Marked splenomegaly was also seen in this case (spleen is present in bottom of image). (C) The liver shows multifocal, random, coalescing, or diffuse (as seen in this image) acute hepatocellular necrosis. Necrotic foci contain shrunken, rounded, hypereosinophilic cells, karyorrhectic debris, occasional heterophils, mononuclear cells, and hemosiderin-laden macrophages. (D) The pancreas shows severe, regionally extensive, multifocal to coalescing, acute pancreatic necrosis. (E) There is nonheterophilic meningoencephalitis with severe inflammatory cell infiltrates in the meninges and multifocal perivascular lymphoid cuffs (see eSlides 27.e12 and 27.e13). *(Parts B, C, E: Photos Courtesy of D. McAloose, Wildlife Conservation Society.)*

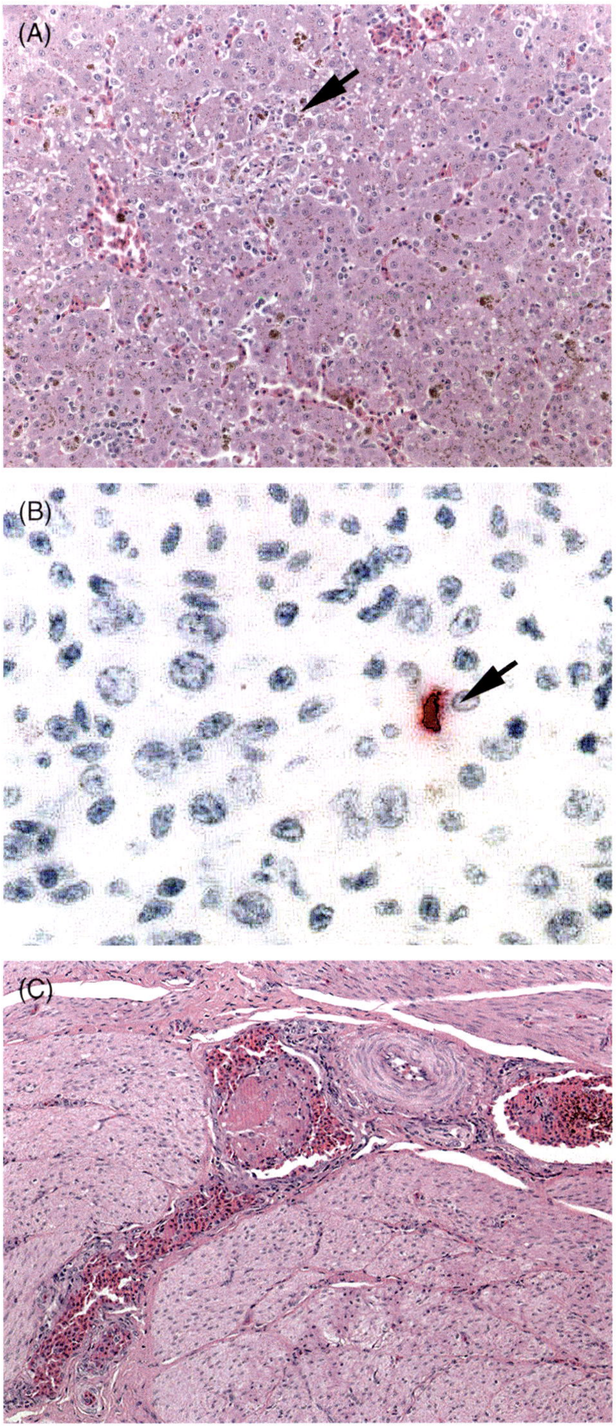

FIGURE 27.21 *Chlamydia* infection in an African penguin. (A) Infection is associated with multifocal individual hepatocellular necrosis *(arrow)*. (B) Occasional *Chlamydia* inclusions are seen and confirmed by immunohistochemistry *(arrow)*. (C) Vascular thrombosis in the ventriculus has been a characteristic in some species (e.g., African penguins in a mixed species outbreak) (see eSlides 27.e14 and 27.e15).

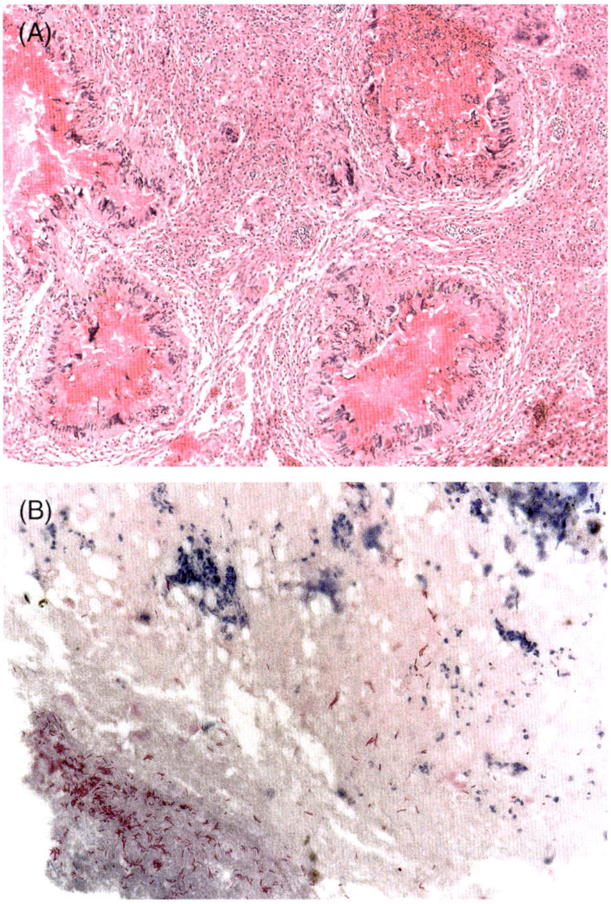

FIGURE 27.22 Necrogranulomatous rhinitis due to *Mycobacterium avium* infection in the nasal mucosa of a Dalmatian pelican. (A) Granulomatous inflammation and granulomas composed of necrotic foci rimmed by macrophages and multinucleated cells are typical of infection. The intervening mucosa contains dense infiltrates of macrophages, multinucleated cells, lymphocytes, and plasma cells. (B) Necrotic foci contain abundant, acid-fast positive, rod-shaped bacilli. Lesser numbers of bacteria are present in the cytoplasm of macrophages and multinucleated cells in the surrounding granulomatous inflammation. Ziehl-Neelsen (see eSlide 27.e16).

smear cytology with follow-up histology, and the identification of intra- or extracellular acid-fast positive bacteria. Diagnosis can be confirmed by mycobacterial culture or PCR.

Fungi

Aspergillosis is by far the most important fungal disease and most common cause of death in penguins under human management.[64] It is rare in free-ranging populations.[75] The causative agent is typically ***Aspergillus fumigatus. Aspergillus flavus*** also occurs but is more sporadic. Acute and chronic aspergillosis has been described. Acute cases associated with high spore inoculum and/or severe immunosuppression can progress to death within 24 to 48 h. More commonly, aspergillosis has a prolonged course over weeks or months. Predisposing factors in rehabilitation settings and captivity include high environmental spore loads, poor ventilation, high ammonia levels, physiologic, and social stressors, inadequate nutrition, temperature stress, age, and underlying disease. Acute and chronic aspergillosis differ grossly but may occur in the same bird. Acute aspergillosis targets the lungs, which are dark red, edematous, and contain multiple, small, white to tan, disseminated foci on serosal and parenchymal cut surfaces. Acute airsacculitis may occur. The more common chronic presentations are disseminated and incite variably severe air sac thickening with fungal plaques in interclavicular, thoracic, and abdominal air sacs (Fig. 27.23A). Plaques may appear blue- or olive-green, brown or black, and velvety if conidia develop. Compact mycelial masses (**aspergillomas**) may form in respiratory cavities. Pulmonary parenchyma is replaced by disseminated pearl-like pale white to yellow granulomas. Granulomas and granulomatous inflammation with fungal growth may extend along the coelomic serosa (Fig. 27.23B) and invade the liver, kidney, and adrenal glands. Histologically, visceral lesions have multifocal necrosis that encases radially arranged, lightly basophilic to transparent, 3–6 µm, regularly septate, thin- and parallel-walled fungal hyphae with dichotomous, acute-angled branching (Fig. 27.23C). Infection is accompanied by marked heterophilic inflammation. Necrotic cores are surrounded by dense infiltration with macrophages, multinucleated giant cells, heterophils, and fewer lymphocytes and plasma cells (Fig. 27.23D). In chronic lesions, degenerating fungal hyphae can be difficult to detect (eSlide 27.e17). Air sacs are markedly thickened and expanded by similar inflammatory infiltrates, and fibrosis may be evident. Mats of fungal hyphae and conidia frequently replace the lining of air sacs (Fig. 27.23E; eSlide 27.e18). Vasoinvasion in bird groups in this chapter is an inconsistent feature in comparison with other avian clades and is most commonly seen in acute *Aspergillus* spp. infection (eSlide 27.e19). Dual infection with avian malaria should be considered when splenomegaly and hepatomegaly are seen.

Adult birds are considered more resistant than chicks to aspergillosis yet are more commonly affected in captivity. Maternal antibodies may provide transient protection, but the role of humoral immunity is unclear. Long-term treatment is required and may be unsuccessful if begun late in the disease course. Antifungal drug resistance is increasingly recognized, and newer antifungal drugs may be associated with toxicity[116] (see "Toxic" section).

Aspergillosis has also been reported as a cause of death in free-ranging common loons and American white pelicans. Nutritional and migratory stress and high spore loads in wetland habitats are probably contributory. Sporadic aspergillosis-associated mortalities occur in free-ranging nestling pelicans, zoo-captive and free-ranging adult pelicans, and are occasionally seen in free-ranging gannets and albatross. Captive shoebills are highly susceptible. However, primarily a respiratory disease, a range of nonspecific clinical presentations can also be seen.

Candida spp. isolation is reported infrequently from rehabilitation settings and is of secondary relevance in seabirds. Recently, ***Candida palmioleophila*** was isolated in a case of respiratory candidiasis in a Magellanic penguin.[75]

Protozoa

A wide range of metazoan and protozoan parasites are reported in free-ranging populations of these species groups. These are summarized in Table 27.e6. This list should not be considered comprehensive. High parasite burdens, which may be facilitated by immune-suppressive diseases allowing unchecked parasite proliferation, may contribute to morbidity and mortality and parasites can act as vectors for other disease agents. Low parasite burdens are frequently not associated with clinical disease and are considered an incidental finding. Occasional case reports of parasite-related mortality are found in captive populations.

Avian malaria is the most significant parasitic disease in penguins. Disease is common in captive penguin colonies and rehabilitation centers and occurs in the wild. In some settings, mortality rates as high as 50%–80% have been reported.[101] Fatal infection is most commonly associated with the subgenera *Haemamoeba* (particularly *Plasmodium relictum*) and *Huffia* (particularly *P. elongatum*) which appear more pathogenic to penguins than other subgenera, but cases involving *P. cathemerium*, *P. juxtanucleare*, *P. tejerai*, *P. nucleophilum*, and *P. unalis* have been recorded.[49,117] In Europe, infection is predominantly associated with *P. relictum*, while both *P. relictum* and *P. elongatum* are common in captive penguins in North America, likely reflecting plasmodium prevalence in native wild bird populations.[264] Avian malaria affects a range of penguin species and reports are most common in Humboldt and African penguins. A bias introduced by their popularity in captivity is likely.[110]

Avian malaria is markedly seasonal ("**spring relapse**") and dependent upon the availability of mosquito vector populations. *Culex* species are significant vectors for avian malaria. Rising global temperatures facilitate the geographical expansion of these vectors, increasing the threat avian malaria poses to penguin populations.[117] Local wild avifauna serve as the reservoir host. Sporozoites in infected saliva of female mosquitoes are injected into a host and enter reticuloendothelial cells where they develop into

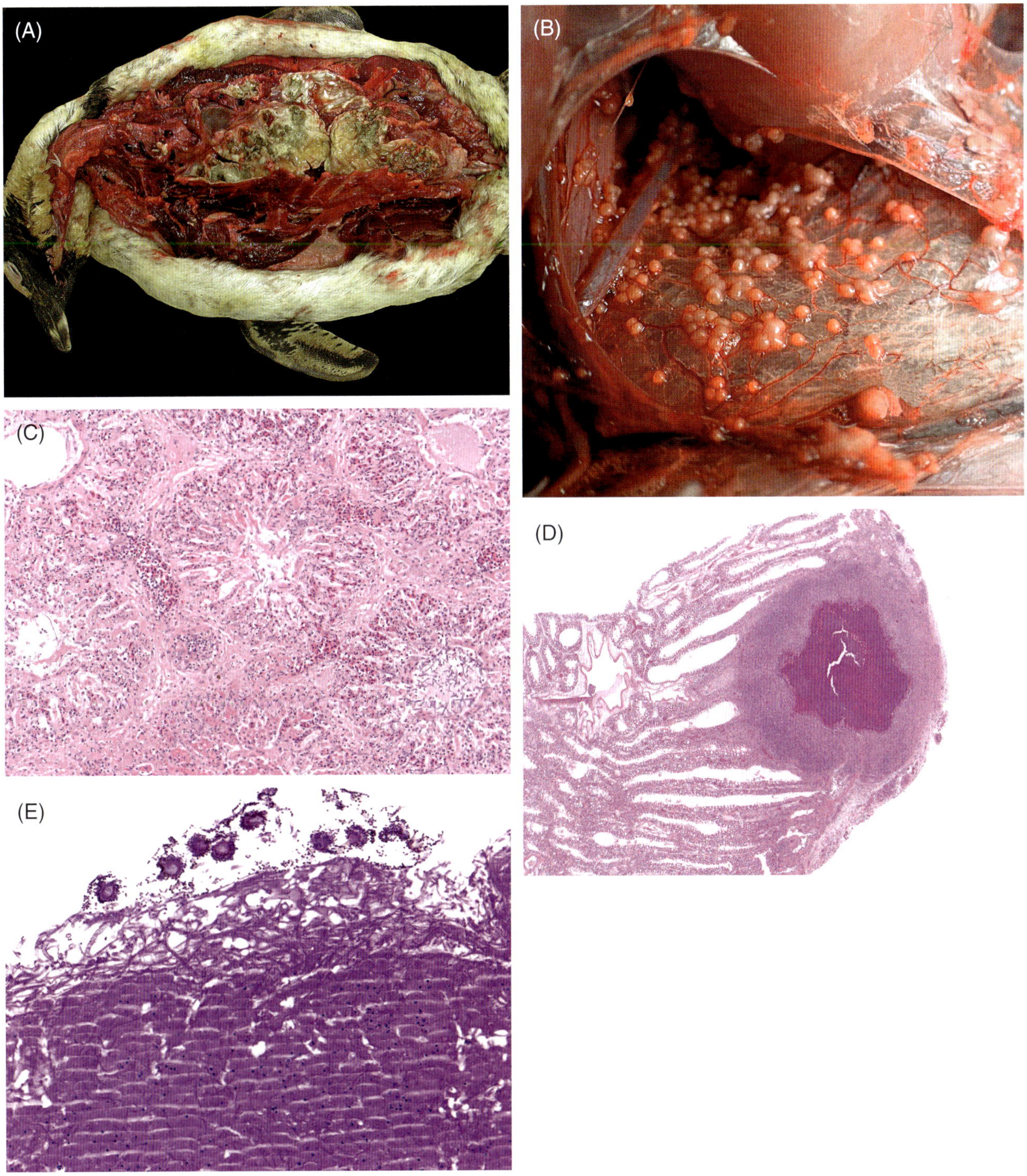

FIGURE 27.23 Aspergillosis in Humboldt penguins (A, C–E) and a Dalmatian pelican (B). (A) Chronic air sac aspergillosis can present with markedly thickened air sac walls and confluent green to gray velvety fungal plaques within air sac lumina. (B) Numerous pearl-like granulomas overly the coelomic serosal surfaces. Similar lesions can be present in the lungs of affected birds. The species-specific subcutaneous air sac diverticula are evident in the background. (C) Acute infection associated with marked confluent fibrinonecrotizing bronchopneumonia with intralesional fungal hyphae. Vascular thrombosis and intralesional and angioinvasive fungal hyphae may also be seen (not shown in this image). (D) Granuloma formation is seen in more chronic cases. Characteristic features include central necrosis with intralesional fungal hyphae surrounded by variably thick peripheral rims of fibrosis and histiocyte-rich mixed inflammation. Degenerating fungal hyphae can be difficult to identify in advanced cases. (E) The air sac is diffusely expanded by necrotic debris and myriads of thin (3–6 μm), regularly septate, parallel-walled fungal hyphae with acute-angle, dichotomous branching. Conidiophores and myriad conidia may develop at air interfaces and are present along this interface toward the top of this image. Periodic acid–Schiff. *(Part B: Photo Courtesy of M. O'Brien, Wildfowl & Wetlands Trust.)*

first-generation exoerythrocytic meronts that undergo asexual multiplication and form merozoites. Host cell rupture releases merozoites into the blood and results in additional infection of reticuloendothelial cells (undergoing new cycles of exoerythrocytic merogony) or infection of erythrocytes (proceeding to erythrocytic merogony or gametogony). Within erythrocytes, merozoites develop into trophozoites, which further differentiate into either erythrocytic meronts or gametocytes (macrogametocytes or microgametocytes). Gametocytes remain within erythrocytes until ingestion by an insect host in which sporogony takes place.

The cause of the extreme susceptibility of penguins to avian malaria is unclear. Penguins have not coevolved with hemosporidian parasites and immunologic naivety is considered significant.[152] Disease risk is highest in chicks, juveniles, and naive adults, and animals with outdoor access. Stressors (molt, chick rearing, concurrent disease, poor husbandry) associated with catecholamine/mediated immunosuppression increase mortality.[79] Clinical signs range from asymptomatic acute death to lethargy, anorexia, depression, vomiting, dyspnea, pale mucous membranes (anemia), and behavioral separation. Severe forms induce neurological signs including motor incoordination, seizures, and paralysis. Antemortem diagnosis is difficult as animals frequently succumb rapidly without detectable blood parasitemia. Intraerythrocytic meronts (syn. schizont) may be seen on Giemsa-stained blood smears but are often absent even in fulminant, acute cases. A serological test has been validated for African penguins and PCR-detection of malarial organisms in blood is available but it may fail to identify low-level parasitemia or mixed infections.

Rapid antemortem diagnosis can be achieved by detection of extraerythrocytic meronts in impression smears from affected animals (Fig. 27.24A). Typical gross findings include splenomegaly, hepatomegaly, pulmonary edema, and hydropericardium with granulocytic interstitial pneumonia, mononuclear hepatitis, granulocytic splenitis, and myocarditis. Vascular occlusion and rupture due to swollen, meront-filled endothelial cells can develop and result in tissue hypoxia and necrosis. Extramedullary hematopoiesis and hemosiderosis are seen in the spleen and liver, and deposition of birefringent hemozoin granules (malaria pigment) may occur. Exoerythrocytic meronts are most frequent in macrophages and endothelial cells in the lungs, heart, liver, and spleen (Fig. 27.24B; eSlide 27.e20). Endothelial changes (necrosis, regeneration) are present in many cases and are consistent with rupture of endothelial schizonts. In prolonged cases, erythrocytic stages (trophozoites, intraerythrocytic meronts, and gametocytes) are rarely detected and enable differentiation of *Plasmodium* species from morphologically similar ***Haemoproteus*** species, which have been shown to be competent in and able to complete their development and produce gametocytes in penguins.[118] Death is the culmination of circulatory shock from respiratory insufficiency, cardiac tamponade due to pericardial effusion, and massive release of inflammatory mediators (cytokine storms). Aspergillosis, enteritis, septicemia, chlamydiosis, *Avipoxvirus* and West Nile virus infections, enteric and visceral helminths, amyloidosis, and cholestasis have all been recorded as concurrent conditions in cases of avian malaria, which accounts for variability in clinical presentation. Prophylaxis is imperative and preventative measures include stringent vector control, such as introduction of infertile or malaria-resistant

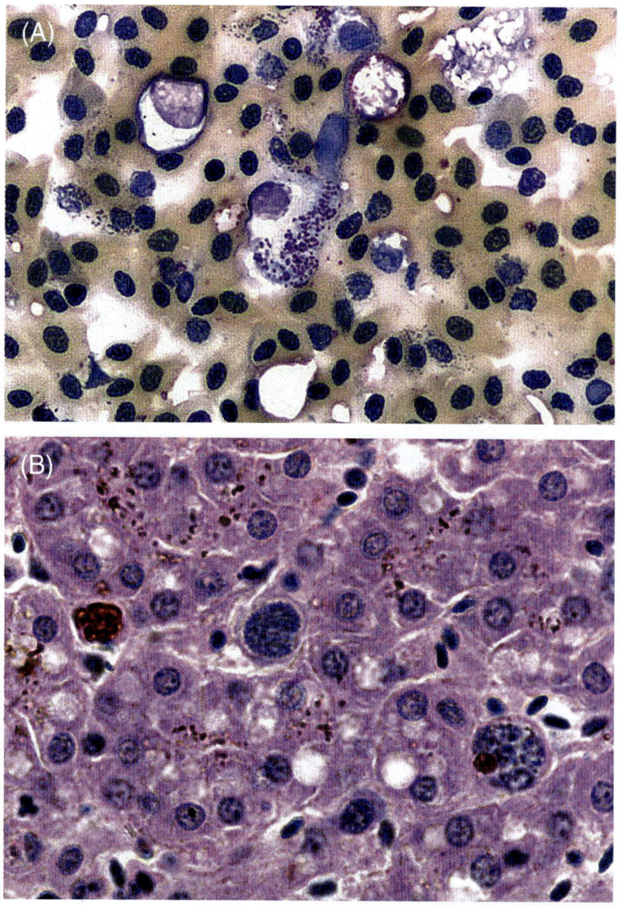

FIGURE 27.24 Avian malaria in the lung (A) and liver (B) of Humboldt penguins. (A) Cytological examination of impression smears of lung, liver, spleen, and heart blood is a useful diagnostic tool. Exo-erythrocytic intracellular zoites are seen in epithelial cells in a lung tissue imprint. The meront has an elongated form consistent with an intraendothelial location. (B) Exoerythrocytic meronts are present within hepatocytes. These appear as clusters of oblong basophilic protozoal organisms within the cytoplasm. The parasites are associated with the deposition of granular black/brown hemozoin granules (malaria pigment) and mononuclear infiltrates. Hemozoin crystals are birefringent under polarized light. Hemosiderin, which can look like hemozoin, is not birefringent but can be stained with Pearls Prussian blue (see eSlide 27.e20).

mosquitoes or mosquito-proofing of enclosures, and chemoprophylaxis. DNA vaccines trialed in canaries and captive penguins proved protective for the first year, but vaccinated birds were highly susceptible in the subsequent year, which was attributed to a failure to develop protective immunity.[102] Global warming and the anticipated increase in malaria exposure render appropriate malaria management critical to conservation efforts. Information on further blood parasite infections of penguins and other orders is available in Table 27.e6.

Renal coccidiosis due to ***Eimeria*** **species** may be encountered as an incidental finding in loons (***Eimeria gaviae***), European shags, and Australasian gannets. In double-crested cormorants, renal coccidia (***Eimeria auritusi***) are common in healthy birds, but coccidia-associated mortality events have occurred on breeding grounds or wintering areas. Renal coccidiosis is also seen in adult short-tailed shearwaters, flesh-footed shearwaters, and fluttering shearwaters (J. Hall, personal communication). Severe renal coccidiosis with ***Eimeria serventyi*** (**Limey disease**) affects short-tailed shearwater (Tasmanian mutton bird) nestlings.[171] Affected birds are malodorous, emaciated with yellow discoloration of fat and muscle and have cloacal impaction/vent caking with white urates and distal intestinal distension and hemorrhage. The kidneys may be enlarged and contain multifocal to coalescing white foci and chalky urate deposits, likely triggered by heterophilic to lymphoplasmacytic inflammation that is associated with gametocytes and oocysts in renal tubules, collecting ducts, and ureteric lumina.

See Table 27.e6 for an overview of other protozoan and metazoan parasites.

ADDITIONAL RESOURCES

Atkinson CT, Thomas NJ, Hunter BD. Parasitic Diseases of Wild Birds. Iowa, USA: Blackwell Publishing; 2008.

Brouwer K, Hiddinga B, King CE. Management and breeding of pelicans *Pelecunus* spp. in captivity. International Zoo Yearbook. 1994;33: 24–39.

Clarke JR, Knowles KR. Diseases and parasites of penguins. Korean J Polar Res. 1993;4: 79–96.

Friend M, Franson JC: Field manual of wildlife diseases. General field procedures and diseases of birds. Geological Survey Madison WI, Biological Resources Division, 1999.

Ings K, Denk D. Avian Malaria in Penguins: Diagnostics and Future Direction in the Context of Climate Change. Animals (Basel). 2022;12.

Ladds P. Pathology of Australian Native Wildlife. Collingwood, Australia: CSIRO Publishing; 2009.

Landsberg JH, Vargo GA, Flewelling LJ, Wiley FE. Algal biotoxins. In: Infectious and Parasitic Diseases of Wild Birds. Oxford: Blackwell Press; 2007:431–455.

Rocke T, Converse K, Meteyer C, McLean B. The impact of disease in the American white pelican in North America. Waterbirds. 2005;28: 87–94.

Thomas NJ, Hunter BD, Atkinson CT. Infectious Diseases of Wild Birds. Blackwell Publishing, Iowa, USA; 2007.

Wolfaardt A, Williams A, Underhill L, Crawford R, Whittington P. Review of the rescue, rehabilitation and restoration of oiled seabirds in South Africa, especially African penguins *Spheniscus demersus* and Cape gannets *Morus capensis*, 1983–2005. African Journal of Marine Science. 2009;31: 31–54.

References

Full references for this chapter can be found online at https://doi.org/10.1016/B978-0-443-18437-6.00017-1.

Chapter 28

Phoenicopteriformes

Judy St. Leger
Department of Population Medicine and Diagnostic Sciences, College of Veterinary Medicine, Cornell University, Ithaca, NY, United States

INTRODUCTION

Flamingos are hardy, long-legged, colonial birds that are able to thrive in harsh high salinity and/or alkalinity aquatic habitats and occupy a unique feeding niche. The six extant species are divided into three genera, *Phoenicoparrus*, *Phoeniconaias*, and *Phoenicopterus*, with members of the former two groups feeding on diatoms and algae and the latter group feeding mainly on arthropods and mollusks.[36] From a conservation standpoint, only Greater and American flamingos are listed by International Union for the Conservation of Nature (IUCN) as species of Least Concern. The Andean flamingo is considered Vulnerable and the Jame's/Puna, Chilean, and Lesser flamingo are all listed as Near Threatened.[29] Most of our knowledge of diseases in flamingos is based on the data accumulated through the study of captive birds. Data on free-ranging birds are sparse. Many of the diseases documented in captive birds are similar to those found in other avian species but a few conditions are particularly prevalent in flamingos. Common conditions can be roughly divided into infectious and non-infectious categories, recognizing that there is much overlap in these and one can potentiate the development of the other.

NON-INFECTIOUS DISEASES

Nutritional

Metabolic bone disease results in improper ossification and is usually due to improper dietary ratios of calcium, phosphorus, and vitamin D. These ratios can also be altered due to insufficient exposure to natural light. Multiple folding fractures, generalized bone fragility, or softening, particularly in the long bones and bones of the skull, indicate metabolic bone disease.

Diagnosis of metabolic bone disease can be made based on the gross and histologic examination of longitudinal sections through the physis of long bones (Fig. 28.1). Affected bones display widening of the epiphyseal plate, abnormal endosteal and periosteal osteogenesis in the metaphysis and diaphysis with thinning of trabecular and cortical bone, and sometimes, replacement by fibrous tissue. Epiphyseal changes most often correspond to expansion of the proliferating zone, indicating vitamin D deficiency, but determination of the exact cause of the lesion is best confirmed by dietary analysis or a history of prolonged exposure to artificial light.[20,21]

In addition to widespread metabolic bone disease, young flamingos are prone to development of a lesion resembling rachitic changes but confined to the tibiotarsus. This lesion is characterized by medial deviation of the upper one-third of the bone expansion of the epiphyseal plate. The lesion appears to be the result of abnormal ossification, causing bending of the bone as the birds grow.[22]

Long-legged wading birds are particularly prone to **exertional rhabdomyolysis** and animals with an underlying vitamin E/selenium deficiency are particularly prone to it. Gross and histologic changes are variable and examination of numerous muscles may be necessary to detect the changes. In peracute cases, birds die of shock or hyperthermia prior to the development of gross or histologically apparent lesions. In more advanced lesions, birds die of massive efflux of myoglobin into the kidneys resulting in renal failure. When gross lesions are present, affected muscles are soft and mottled pale pink to dark red. Kidneys may have a brown discoloration and appear swollen. Histologically, detectable lesions include rhabdomyolysis or overt necrosis of myofibers characterized by loss of cross striation, disruption of the nuclei, and disruption of the sarcoplasm. Brown pigment, consistent with myoglobin, is found in the renal tubules and renal tubular epithelial cells.[36] Because lesions are often absent, diagnosis of capture myopathy is often one of exclusion. In these cases, a history of capture, transport, or spurts of unusual intense physical activity should be used to support the diagnosis.

Nutritional factors, including high fat/energy, and advanced age are predisposing factors in the development of **atherosclerosis** in a number of species including birds. It is uncommonly reported in flamingos.[40] A recent report described atherosclerosis as the cause of sudden, unexpected death in an aged American flamingo.[24] Gross

Pathology of Wildlife and Zoo Animals. https://doi.org/10.1016/B978-0-443-18437-6.00044-4

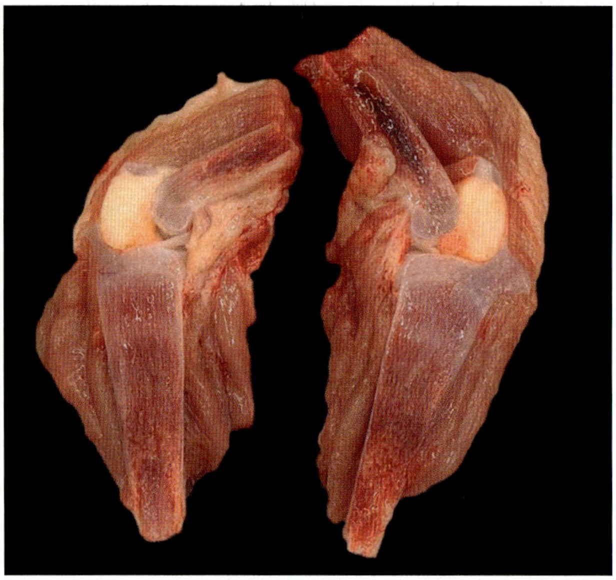

FIGURE 28.1 Metabolic bone disease in the long bones of the leg from a 1-month-old American flamingo. There is expansion of the epiphyseal cartilage and retention of cartilage cores in the trabecular bone. *(Photo Courtesy of N. Stedman, Busch Gardens.)*

findings were similar to those in other species and included mineralization and plaque formation in numerous vessels, including the aorta. Characteristic histologic features included expansion of vascular walls by acicular (cholesterol) clefts, histiocytic infiltrates, and mural fibrosis. Multifocal intralesional mineral and chondroid and osseous metaplasia were also seen. Death was due to aneurysmal dilation and rupture of an affected area of the aorta.

Metabolic

Few metabolic disorders have been described in flamingos. **Hepatic lipidosis** generally develops in overconditioned animals or secondary to underlying diseases, particularly those that lead to anorexia and mobilization of peripheral fat stores to maintain normal energy balance. Flamingos in managed settings tend to accumulate excessive fat stores, and **steatosis** of multiple organs, including the heart, liver, ventricular wall, and kidneys, may develop. The significance of this finding should be interpreted in light of other findings. Severe hepatic lipidosis can interfere with liver function, and fat infiltration around the Purkinje fibers of the heart can predispose the birds to fatal arrhythmias.

Because captive diets are intrinsically higher in iron content than the natural diet of the flamingos, birds can develop **iron storage (hemosiderosis)**, the accumulation of intracellular iron. This is most commonly seen in hepatocytes, Kupffer cells, renal tubular epithelia, and occasionally in cardiomyocytes. Relative levels of hepatic hemosiderin also increase secondary to inanition. With routine hematoxylin and eosin staining, hemosiderin appears as brown to gold intracytoplasmic granular material (eSlides 28.e1 and 28.e2). The granules will be dark blue when stained with a Prussian blue histochemical stain, and this can be used to differentiate the granules from lipofuscin.[63]

A rare form of **GM2-gangliosidosis**, a **lysosomal storage disease** resembling Tay-Sachs disease, has been described in juvenile flamingos. Clinically, the condition is characterized by progressive neurologic disease. The diagnosis is made based on the histologic lesions. Purkinje cells, cerebral neurons, astrocytes, bipolar retinal ganglion cells and cells of ganglia of the adrenal glands, and peripheral nerves are markedly swollen and have foamy, vacuolated cytoplasm containing material that stains with Luxol fast blue but not with Toluidine blue or periodic acid Schiff (PAS) histochemical stains. The ultrastructure of the material consists of membrane-bound, clear cytoplasmic vacuoles and lamellated inclusions. Additionally, the number of Purkinje cells may be reduced and cells that remain have decreased numbers of cytoplasmic processes when compared with normal cells.[70]

Toxic

Mass mortalities due to toxins are most often reported in free-ranging birds, but captive birds in contaminated environments are also at risk. A semiaquatic lifestyle and a feeding technique reliant on filtration of large amounts of silt and water, make flamingos particularly vulnerable to toxicosis from agents like **cyanobacteria**.[31,57] Suspicion of this toxin should be high in cases of multiple mortalities, especially those affecting multiple species in an aquatic area or exhibit. Growth is enhanced by increases in water temperature and there may or may not be visible blooms in affected areas. Sampling of the water and submission of the dead birds for necropsy is required for diagnosis. When present, gross findings include muscular, pulmonary, or intestinal hemorrhage. The most consistent histologic change is massive hepatocellular disassociation with hemorrhage. Detection of the organisms and toxins in the environment along with confirmation of the toxin in gastric contents are diagnostic. Toxins can sometimes be detected in the liver and feathers, but these tests are not fully validated.[1,13,33,35]

Lead toxicity can be a significant cause of mortality in birds in contaminated environments. Free-ranging birds are most at risk and toxicity can manifest in acute death due to short-term exposure to high levels of lead, or as a chronic disease due to long-term exposure to low levels of lead. Birds dying of acute exposure are generally in good body condition while those with long-term exposure may be emaciated. Ideally, lead material is detected in the ventriculus either by radiographs or at postmortem examination. In the absence of this specific finding, watery intestinal contents, bile staining of the ventricular mucosa, enlarged kidneys, and cardiac degeneration and congestion are suggestive lesions. In some cases, there are histologically detectable, small, acid-fast nuclear inclusions in the renal tubular epithelial cells.

In many instances of fatal lead intoxication, gross and histologic lesions are absent and diagnosis relies on detection of toxic levels of lead in the tissues. Liver and kidney are the preferred samples for evaluation. Bone should be evaluated if long-term exposure is a differential as lead is sequestered in bone. In some cases, lead levels in feathers can be evaluated, but care must be taken to ensure that the measured levels are present in the feather, and not the result of superficial environmental contamination.[2,10]

Environmental toxins, including perfluoroalkyl substances (PFAs), pesticides, polychlorinated biphenyls (PCBs), polycyclic aromatic hydrocarbons (PAHs), have also been directly or indirectly implicated as concerns for flamingos given their unique feeding behaviors and possible bioaccumulation in their food items or the environment.[15,17]

FIGURE 28.2 Pododermatitis in an American flamingo. Ulcerative and proliferative nodular lesions are present bilaterally on the plantar surfaces of the feet. Areas of yellow discoloration are areas of hyperkeratosis. Some of the ulcerated skin is covered by thick, dark crusts while others are crateriform with rounded, contracted edges (see eSlide 28.e3). *(Photo Courtesy of SeaWorld Parks and Entertainment.)*

Congenital/Genetic

Angular limb deformities, such as tibiotarsal rotation, must be differentiated from changes that occur secondary to metabolic bone disease. The etiology of these two lesions differs and requires different intervention strategies. Angular limb deformities lack the ossification defects in metabolic bone disease. Valgus and varus deformities may be minor or can severely alter mobility. They may also predispose birds to trauma of the affected limbs. The cause of the deformities is rarely identified. They are likely due to combinations of factors, including high fat or protein diets, lack of exercise, genetics, artificial incubation conditions, and inadequate light exposure.[71] If multiple unrelated, artificially incubated birds are affected, evaluation of the incubation protocols should be assessed.

Unilateral **renal agenesis**, with no evidence of contralateral renal hypertrophy or clinical evidence of renal dysfunction, was identified in a captive Chilean flamingo that died due to tracheal aspergillosis.[44]

Miscellaneous/Multifactorial

Pododermatitis (bumblefoot) is one of the most common clinical or postmortem findings in captive flamingos with 100% prevalence being reported in some populations. These lesions range from mild, clinically insignificant changes to those that severely impair limb function. Chronic inflammation associated with these lesions can potentiate fatal systemic diseases (see Amyloidosis). The term pododermatitis encompasses a wide array of gross and histologic changes and does not imply a specific etiology. As this condition is only documented in captive birds, conditions unique to the captive environment likely form the basis for lesion development. Housing on concrete floors, prolonged periods indoors, environmental temperatures of 15°C or below, and excessive body weight have all been implicated in pathogenesis.[38,41] An association with dietary vitamin A and E levels has been speculated, but evidence from a controlled study is lacking.[65–68] Lesions are broadly classified as inflammatory or proliferative (Fig. 28.2; eSlide 28.e3). Lesion severity is sometimes expressed by using scoring systems based on size, depth, and distribution of changes.[6,64] In general, inflammatory lesions are more common in younger birds and proliferative lesions are more common in older birds. Ulceration and heterophilic dermatitis, hydropic epithelial degeneration, areas of hyperplasia, formation of serocellular crusts, and heterophilic pustules characterize more acute lesions. As lesions become chronic, neovascularization, fibrosis, and infiltration by lymphocytes and macrophages occur. Older lesions are more likely to extend deep into the underlying bone. Regardless of the lesion timeline, opportunistic bacterial colonization and infection are usually present. Proliferative lesions include areas of hyperkeratosis, fissures, nodular lesions, and papillary growths. Histologically, these lesions are characterized by fibrosis, hyperplasia, and papilliferous projections of connective tissue covered by epithelium. Deep areas of mixed inflammation, surrounded by fibrosis, otherwise known as bumblefoot, can also develop.[65]

Various studies have analyzed pododermatitis lesions in flamingos for infectious agents. A hyphal-forming, micrococcus-like organism, *Arsenicicoccus dermatophilus*, has been identified in lesional and nonlesional feet in some young birds.[23] The microscopic appearance of some of the papillary lesions resembles viral-induced lesions from other species and there are anecdotal reports of intracytoplasmic epithelial inclusion bodies in affected tissues. However, fungal and bacterial agents are not consistently present and no virus has been isolated.[65] Hence, though infectious agents, particularly fungi and bacteria, may contribute to the pathogenesis of lesions, most are thought to be opportunists rather than primary drivers of the process. The primary differential for pododermatitis is poxvirus infection (see "DNA Viruses" section). Histology and ancillary diagnostics (e.g., PCR, DNA sequencing) can differentiate the two processes.

Amyloidosis describes a range of diseases related to the deposition of amyloid in tissues. In animals, AA amyloid is most common and often develops secondary to a variety of inflammatory conditions. The pathogenesis and severity vary widely and in severe cases, sequela can include organ failure and death.[47] Common predisposing factors include pododermatitis, pneumonia, and arthritis; an inciting cause is not always identified. Grossly, affected organs are often swollen, discolored (e.g., pallor, orange), and in the most severe cases, have a waxy consistency. The presence of amyloid can sometimes be demonstrated at the time of necropsy by treating the cut surface of the affected organ with Lugol's iodine and rinsing with sulfuric acid, which causes the amyloid deposits to turn dark brown or black. This technique is most effective on fresh tissue. Histologically, amyloid is homogenous, acellular, eosinophilic to amphophilic material that appears light pink with routine hematoxylin and eosin staining (Fig. 28.3; eSlide 28.e4). Accumulations are most common in hepatic sinusoids, renal glomeruli, and the spleen.[40] They can also occasionally be seen in vascular walls, including in the central or peripheral nervous system, and in the intestinal lamina propria.[45] Though the appearance with hematoxylin and eosin is distinctive, fibrin or collagen can be mistaken for amyloid. Congo red histochemical staining will highlight the amyloid as a red-orange material that has apple-green birefringence under polarized light; collagen is white under polarized light. Examination of the tissues in an environment of low ambient light may be necessary to detect the latter characteristic feature. Amyloid accumulation eventually results in physical and functional damage to organs or in rupture of organs due to disruption of the tissue by the abnormally folded protein.[72] Affected tissues are often friable and hepatic fractures with subsequent exsanguination can occur and cause acute death.

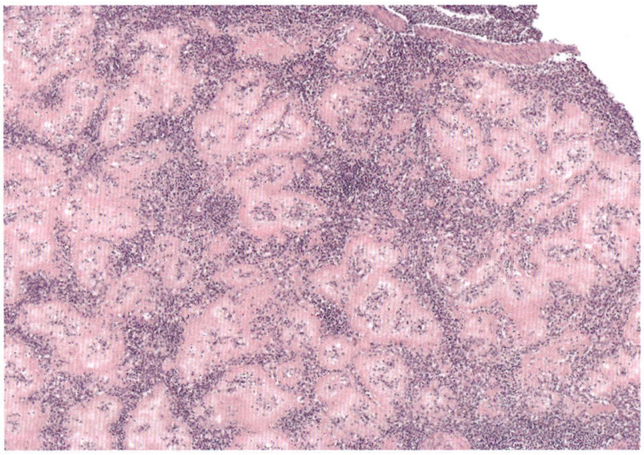

FIGURE 28.3 Amyloidosis in the spleen of an American flamingo. Extensive perivascular amyloid is present throughout the parenchyma. Deposition of the homogenous, acellular, eosinophilic material expands and replaces normal tissue (see eSlide 28.e4). *(Photo Courtesy of N. Stedman, Busch Gardens.)*

Trauma-induced fractures are the major cause of morbidity and mortality in captive flamingos. While severe acute fractures may result in death or euthanasia, fractures may also lead to the development of systemic complications during treatment. Amyloidosis and aspergillosis are common fatal sequelae. Long bone fractures are most common. Even when the fracture is easily palpated, or visible in postmortem survey radiographs, the area should be dissected to confirm the presence of associated tissues reaction, such as hemorrhage, ensuring that the fracture did not occur after death and to rule out the presence of preexisting conditions, such as osteomyelitis. Fractures, luxations, and subluxations of the cervical vertebrae can also occur. Severe damage is easily identified by hemorrhage with tissue and bone disruption, but acute compression fractures of the cervical vertebrae may be difficult to detect on postmortem examination. Knowledge of the conditions under which the bird was found, and radiographs can aid in diagnosis and careful dissection of the soft tissues of the neck may reveal subtle hemorrhage around the affected bone. In some cases, careful opening of the vertebral column and removal of the spinal cord may be needed to document related damage. Noncervical vertebral fractures are rare in birds, due to the fusion of most vertebrae. When vertebral fractures, luxations or subluxations occur, they are generally the result of massive crushing trauma, such as that occurs due to predation or vehicular strikes. Less dramatic fractures generally occur in the free vertebrae located between the notarium and synsacrum. If present, fractures in this area can indicate metabolic bone disease. Evaluation of the free vertebra or vertebrae is most easily accomplished with the bird in dorsal recumbency, after removal of the viscera.

In free-ranging birds, **predatory trauma**, including by dogs, has also been reported.[51] **Tongue** or **beak trauma** or other processes that prevent normal glossal movement result in failure to feed. Such damage may include ulcers, lacerations, foreign bodies, or bill fractures. In one report of **trauma-related obstruction**, captive birds developed oral impactions and lost condition and became hypoglycemic after ingestion of high-fat food that was meant for other species in the exhibit.[25] No mortalities resulted, but these cases illustrate the need for a thorough oral examination in animals with suboptimal body condition or with a history of anorexia. **Intestinal obstruction** and loss of condition can also occur secondary to the ingestion of inappropriate material. For example, one report describes erosive enteritis and intestinal obstruction in a group of flamingos that ingested decomposed granite from a recently renovated enclosure.[52] Toe and tarsal trauma was reported to be associated with nylon line (monofilament) and hook **entanglement** in a wild Austral flamingo in Argentina.[8] The case acts as a reminder that in addition to marine mammals, wild birds are also susceptible to injury and potential death due to abandoned, lost, or discarded fishing gear. And in an unusual environmental form of trauma, a severe hailstorm (hail up to 5 cm in diameter) in India in 2014 caused mass mortality across 35 species of birds, including nonlethal trauma in a greater flamingo.[39]

Neoplastic

Neoplasia is uncommon in flamingos but a variety of malignancies have been reported. These include **lymphoma, neuroendocrine** carcinoma, and **hepatocellular** and **ovarian neoplasms**. In captive birds, hepatocellular neoplasms are the most often reported. The gross and histologic appearances of these neoplasms are typical for the appearance of the tumors in other species. Immunohistochemistry can be helpful in determining histogenesis but species-specific positive and negative controls should be run in parallel with the test tissue to ensure the validity of the test.[60]

INFECTIOUS DISEASES

DNA Viruses

Poxvirus is a relatively common viral disease in captive flamingos. Reported infections include self-limiting lesions that regressed over time to more widespread disease. Complications are related to particularly intense lesions that interfere with feeding or sight, or the development of secondary bacterial or yeast infections.[4] Multiple poxviruses have been identified in flamingos. Molecular sequence data reveals that the viruses of flamingos have some sequence homology to pigeon, falcon, albatross, bustard, and condor poxviruses.[11,28,37]

Grossly, poxviruses appear as nodular lesions that generally develop on featherless areas of the skin, such as the beak, face, and legs (Fig. 28.4A). On the face, the lesions often develop near mucocutaneous junctions. Early lesions are raised and smooth. With time, the lesions develop crusts, become ulcerated or traumatized, and are colonized by opportunistic bacteria and fungus. Histologic lesions are pathognomonic. Characteristic features include marked epidermal hyperplasia with hypertrophy, prominent, cytoplasmic eosinophilic inclusion bodies (Bollinger bodies) and variable amounts of hyperkeratosis (Fig. 28.4B; eSlide 28.e5). The amount of associated inflammation varies depending on the amount of necrosis, ulceration, and secondary infection. Mixed dermal inflammation may be present and mixed morphology of bacteria and yeast may be seen in the areas of intense inflammation. The number of viral inclusion bodies varies depending on the stage of infection, and in cases where the lesion is heavily inflamed, inclusion bodies may only be detectable near the edge of the lesions. When present on the feet, gross pox lesions may be confused with the proliferative forms of pododermatitis, but the presence of inclusion bodies distinguishes the two entities.

Herpesvirus infection has been reported in flamingos but reports are scarce.[14,42] An alphaherpesvirus was detected in lesions from young Chilean and American flamingos.[14] The nonspecific findings include generalized

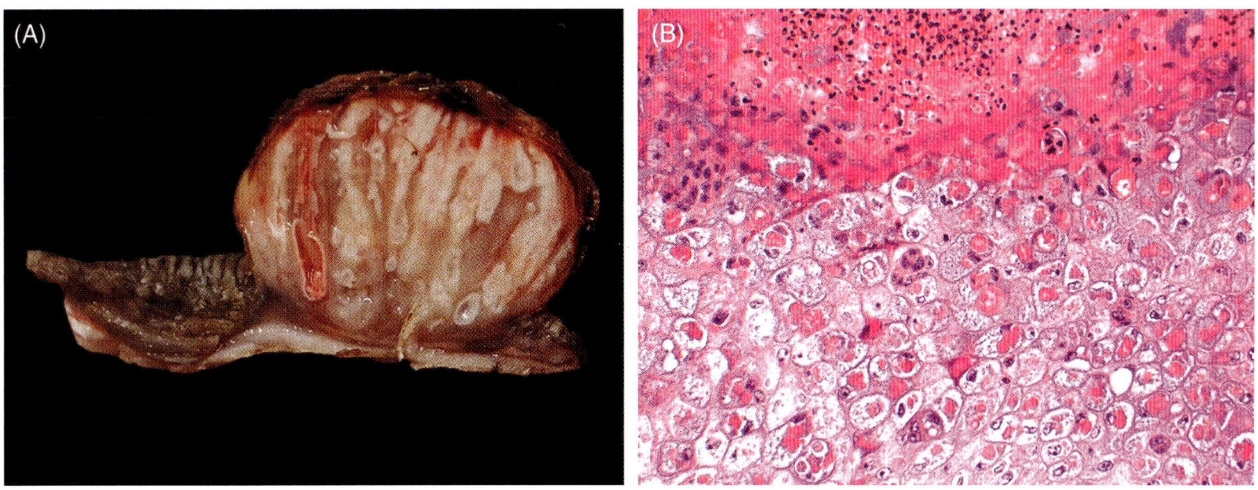

FIGURE 28.4 Cutaneous poxvirus infection in the toe of an American flamingo. (A) The skin is markedly hyperplastic due to marked expansion of the epidermis, which is overlain by a thick serocellular crust. (B) Markedly swollen keratinocytes contain distinctive, eosinophilic cytoplasmic viral inclusions. The proliferative tissue is covered by a serocellular crust (see eSlide 28.e5). *(Parts A and B: Photos Courtesy of N. Stedman, Busch Gardens.)*

pallor, mild hepatomegaly, splenic congestion, and pulmonary edema. Histologically, the liver lesions are most striking, consisting of zones of lytic necrosis, syncytial hepatocytes, and variable numbers of granulocytes along the edges of the areas of necrosis. Eosinophilic to amphiphilic intranuclear inclusion bodies (Cowdry type A) are found in a few hepatocytes. Hemorrhages and fibrinoid necrosis of splenic ellipsoids, lymphocytic, and granulocytic interstitial pneumonia may also be present as inclusion bodies within endothelial cells.

RNA Viruses

West Nile Virus is a single-stranded RNA virus in the *Flaviviridae* family. Found in many regions of the world, it was introduced into the western hemisphere in the late 1990s and since then has become endemic across the United States and its distribution extends from Canada to Central and South America. It is a significant viral infection in flamingos, and flamingos were some of the first birds to be diagnosed with infection following introduction of the virus into North America.[56] Since then, the virus has been associated with the disease in both captive and free-ranging birds, and mortality events remain a major cause of death. West Nile Virus should be considered a differential for any bird that dies in the summer and fall when mosquito vectors are most active. As this virus is zoonotic, suspect cases should be handled with appropriate levels of biosecurity.

Gross lesions of West Nile Virus are variable. Many birds die prior to the development of detectable morphologic changes. In cases where gross lesions are present, they may include congestion of the leptomeninges and brain, and/or irregular areas of pallor in the heart, liver, and spleen.[56] Affected areas of myocardium are often soft and lesions in the visceral organs may be accompanied by small areas of hemorrhage. West Nile Virus infection should be a top differential in birds with myocardial lesions. Histopathologic changes are equally variable. In some cases, no histologic changes are detectable with routine hematoxylin and eosin staining. In others, lesions can include necrotizing myocarditis with variable amounts of nonsuppurative inflammation (eSlide 28.e6). Changes in the liver and spleen are similar to those in the heart, and neural lesions range from perivascular lymphoid cuffs to areas of malacia and parenchymal collapse accompanied by nonsuppurative inflammation. Immunohistochemical labeling specific for West Nile Virus reveals abundant viral antigen in infected tissues.[3,5,56] The diagnosis can be confirmed via immunohistochemistry or by PCR on pooled tissues including brain, heart, and kidney. PCR is particularly helpful when autolysis precludes histologic evaluation or interferes with immunohistochemistry.

Though not well described, **avian paramyxovirus 3** was isolated from the carcasses of captive and free-ranging flamingos that died acutely with respiratory or neurologic signs.[36,53] Nonspecific findings included subcutaneous, hepatic, pericardial, splenic, renal, and intestinal hemorrhages, airsacculitis, hepatomegaly, and splenic necrosis. No histologic correlates are described, rather infection was inferred from the results of virus isolation and electron microscopy.

Like many other bird species, flamingos are susceptible to **avian influenza virus** infection. Infection and outbreaks due to **highly pathogenic avian influenza (H5N1)** have been reported in lesser, greater, and Puna flamingos.[18,30] Low pathogenic avian influenza has been identified on fecal RTPCR of Iberian Peninsula flamingos of the eastern Atlantic flyway.[48] Descriptions of lesions in the lesser and greater flamingos with HPAI infections included varying combinations of pulmonary congestion and edema, respiratory epithelial cell degeneration and loss, multifocal gliosis or lymphohistiocytic meningoencephalitis with perivascular cuffs, and congestion and multifocal necrosis in the liver, spleen, intestines and/or pancreas. Diagnosis relies heavily on virus isolation and molecular diagnostics (rapid immunoassays, avian influenza and HPAI-specific reverse-transcription PCR, DNA sequencing, etc.) since there are no pathognomonic gross or histologic lesions.

Bacteria

Flamingos can be infected by many different bacterial species. Some of the most commonly reported are opportunistic infections that develop secondary to trauma, pododermatitis, or poxvirus infection. These include environmental contaminants, such as members of the ***Enterobacteriaceae*** family and bacteria that are part of the normal skin microflora such as ***Streptococcus*** spp. and ***Staphylococcus*** spp. Spread of bacteria from traumatized or ulcerated skin, particularly from the feet, can lead to serious sequela including osteomyelitis, bacterial sepsis, and/or amyloidosis.

Mycobacterium avium infection is a concern in both captive and free-ranging birds. Drought, which causes reduction in the size of ponds and lakes and results in abnormally large congregations of birds, is thought to have played a role in two epizootics in free-ranging lesser flamingos. In captivity, mixed species exhibits and high population density can predispose birds to infection. Gross lesions include hepatomegaly and splenomegaly and tissue discoloration in areas of necrosis or petechia. Granulomatous inflammation, characterized by aggregates of epithelioid macrophages mixed with variable numbers of granulocytes and areas of necrosis similar to those described in other avian species, can be found in affected tissues including the liver and spleen, and occasionally also the bone marrow.[32,54] The intracellular bacteria do not stain with Wright-Giemsa or hematoxylin and eosin stains and may appear as negatively staining, rod-shaped organisms. They are positive with acid-fast stains (e.g., Ziehl-Neelsen or Fite-Faraco stains). More rapid diagnosis can be made by collecting impression preparations at the time of necropsy. PCR or bacterial culture is confirmatory.

Two new species in a new genus within the ***Chlamydiaceae*** family, *Chlamydiifrater phoenicpteri* sp. nov. and *Chlamydiifrater volucris* sp. nov. have recently been identified in wild, free-ranging, and zoo-managed greater flamingos in France.[62] Infection was not associated with disease in any of the birds.

The genus *Mycoplasma* contains many species of commensal and pathogenic bacteria. Among them is ***Mycoplasma synoviae***, an important poultry pathogen that causes respiratory and articular disease and production losses. Infection and associated catarrhal tracheitis and fibrinous airsacculitis have been reported in a lesser flamingo.[12] It was one of a group of 10 flamingos that died with similar clinical signs, which included lethargy, weakness, ruffled feathers, hypothermia, and evidence of severe enteritis with multifocal hemorrhage and ulcerative pododermatitis.

Fungi

Fungal infections in flamingos mirror the presentations typical in other birds. Infection with ***Aspergillus*** spp. is one of the most common isolates. Typical lesions include plaques and granulomatous and heterophilic inflammation in the trachea, air sacs, and lungs.[16] Infection and associated inflammation can occasionally also extend into other tissues, and the fungus can be angioinvasive and has been associated with aortic aneurysm and rupture.[61] Infection is often thought to develop opportunistically in birds that are compromised due to other disease processes. In flamingos, their propensity for traumatic injuries can predispose to the fungal infection as these can stress the bird during recovery or, in the case of open fracture of bone, provide direct passage for the fungus to enter the respiratory system. The fungus is parallel-walled, septate, has dichotomous branching and has distinctive conidia and arthroconidia (the latter develop at air interfaces). Though they are generally easily seen with hematoxylin and eosin staining, silver stains highlight the organisms, and fungal culture, PCR, and DNA sequencing can be used to confirm and characterize infections. Differentials include **candidiasis**, which was associated with granulomatous inflammation due to ***Candida albicans*** in the gnathotheca in a lesser flamingo.[50] Features of this fungus that help differentiate it from *Aspergillus* are that the fungi are pseudoseptate and are not parallel walled.

Another notable fungal infection is colonization of the foot webs by ***Geotrichum candidum***, which has been reported in captive Caribbean flamingos. This common saprophyte is known to opportunistically infect a variety of animals and plants. In flamingos, previous foot damage, prolonged submersion of the tissue in water, and older age of the birds are thought to predispose animals to infection. Histologic lesions include necrosis and mixed inflammation with large numbers of intralesional mycelia and arthrospores. In contrast to *Aspergillus*, the arthrospores of *Geotrichum* are elongate and rectangular.[55]

Metazoa

Parasites are common in flamingos, particularly in the wild. However, descriptions of related disease and pathology are limited. ***Schistosomes*** are reported in flamingos, but there is only a single case in which the presence of ova in multiple visceral organs and the brain was the cause of mortality.[46] **Cestodiasis** due to ***Flamingolepis liguloides*** has been reported and associated with intestinal necrosis in wild greater flamingos,[58] and ***Phoenicolepsis naukrensis*** was associated with necroulcerative enteritis in an American flamingo.[69] **Tetrameres** have been reported in association with proventricular atrophy.[43,58]

Protozoa

Though often incidental, significant disease and death have been reported in flamingos with **hemoprotozoal** infections. Gross lesions include multifocal subcutaneous and hemorrhage, hemocoelom, and multifocal areas of black to red discoloration in the liver (Fig. 28.5A). Histologically, numerous protozoal cysts (megaloschizonts) are present in the tissues. These are delineated by a thin membrane and hemorrhages and intense inflammation are associated with rupture of the cysts[19] (Fig. 28.5B). The organism has not been fully characterized, and the reason for some infections to be incidental while others can be fatal is unclear. It is possible that severity of lesions may be due to the flamingos being aberrant hosts.

Subclinical ***Plasmodium rouxi*** and ***P. vaughani*** infections have been detected by PCR in captive Chilean flamingos. Affected birds are sometimes anemic or lethargic.[22,49,59] **Trypanosome** and ***Sarcocystis*** species infections have been identified in wild greater flamingos.[58] Enteric **coccidiosis** with mucosal and glandular necrosis and hemorrhage was identified in the same study.

Ectoparasites

Various mites and lice have been found in and on birds, but only in rare cases are they considered significant.[49] In most cases, ectoparasites are asymptomatic or result in a nonspecific inflammation.

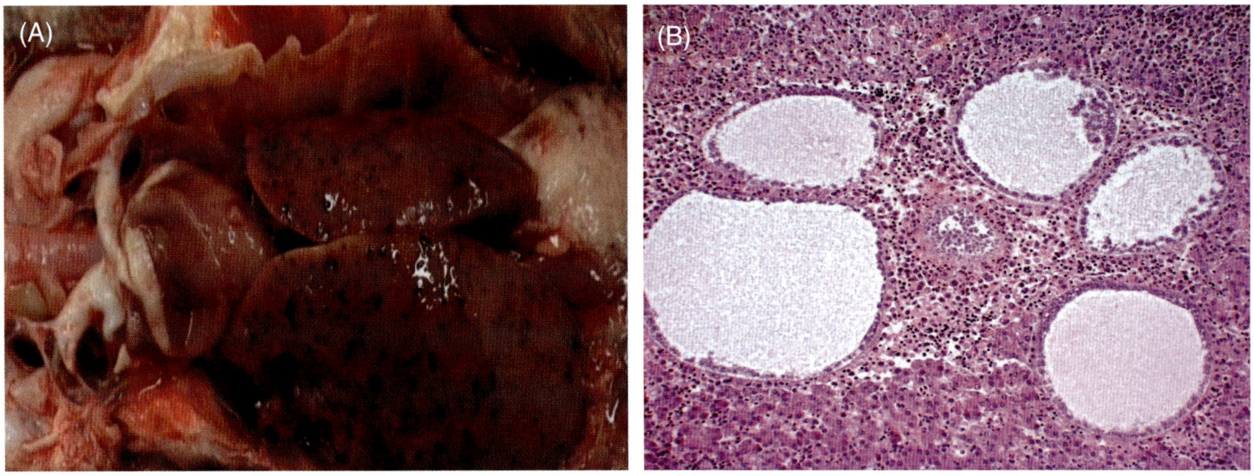

FIGURE 28.5 Hematoparasite infection of the liver of an American flamingo. (A) Dozens of small, randomly distributed dark red, round, discrete to coalescing areas of discoloration are present throughout the parenchyma. The remaining liver is discolored light brown. (B) Multiple megaloschizonts, the site of asexual reproduction (merogony), are present in the hepatic parenchyma. These large protozoal cysts contain central proteinaceous fluid and are lined by a peripheral rim of protoplasm containing developing merozoites. The surrounding tissue contains areas of necrosis, inflammation, and hemorrhage. *(Parts A and B: Photos Courtesy of N. Stedman, Busch Gardens.)*

ADDITIONAL RESOURCES

Abdul-Aziz, T., Barnes, H.J., Fletcher, O.J., 2016. Avian Histopathology, 4th edition. American Association of Avian Pathologists.

Cooper, J.E., Deacon, A.E., Nyariki, T., 2014. Post-mortem examination and sampling of African flamingos (Phoenicopteridae) under field conditions, Ostrich J Afric Ornith, 85(1): 75–83.

Fereidouni, S., 2024. Ecology of Wild Bird Diseases. CRC Press.

Krienitz, Lothar. The Lesser Flamingo. Lesser Flamingos. Springer, Berlin, Heidelberg, 2018. 3–18.

Owen, J.C., Hawley, D.M., Huyvaert, K.P. 2021. Infectious Disease Ecology of Wild Birds. Oxford University Press.

Randal, C.J., Reece, R.L., 1996. Color Atlas of Avian Histopathology. Mosby Ltd.

Schmidt, R.E., Struthers, J.D., Phalen, D.N., 2024. Pathology of Pet and Aviary Birds, 3rd edition. Wiley-Blackwell.

References

Full references for this chapter can be found online at https://doi.org/10.1016/B978-0-443-18437-6.00044-4.

Chapter 29

Anseriformes, Ciconiiformes, Charadriiformes, and Gruiformes

Heather Fenton[a], Rita McManamon[b], Elizabeth W. Howerth[b], and Aleksija Nemanis[c]
[a]Australian Registry of Wildlife Health, Taronga Conservation Society, Mosman, NSW, Australia; [b]University of Georgia, Athens, GA, United States; [c]Swedish Veterinary Agency, Uppsala, Sweden

INTRODUCTION

This chapter includes Anseriformes (three families: Anatidae [ducks, geese, and swans], Anhimidae [screamers], and Anseranatidae [magpie goose]); Ciconiiformes (five families: Ardeidae [herons, bitterns, egrets], Balaenicipitidae [shoebills], Ciconiidae [storks], Scopidae [hammerkops], and Threskiornithidae [ibises and spoonbills]); Charadriiformes (three suborders: Alcae [including auks], Charadrii [oystercatchers, plovers, stilts, others], and Lari [gulls, terns, skimmers]); and Gruiformes (families: Gruidae [cranes], Psophiideae [trumpeters], Rallidae [rails, crakes, coots, others], Sarothruridae [flufftails], Heliornithidae [sungrebes, finfoots], and Aramidae [limpkins]). The orders Otididiformes (bustards), Eurypygiformes (sunbitterns, kagus), Cariamiformes (seriemas), and Mesitonithiformes (mesites and monias) are also included in Gruiformes. At least 28 species pertaining to this chapter are listed as Critically Endangered and 43 species are listed as Endangered by the International Union for the Conservation of Nature (IUCN) (see Table 29.e1). Threats to members of this chapter are multifactorial and range from overharvesting, habitat loss, climate change, decreased availability of prey, toxicosis, and increasingly more frequently, infectious disease. Infectious and non-infectious diseases in these bird groups are the focus of this chapter.

NON-INFECTIOUS DISEASES

Nutritional

Angel wing is a permanent valgus malformation of the carpometacarpal joint of the wing that develops in young, captive, or free-ranging members of Anseriformes (Fig. 29.1). The condition results from rapid growth of the distal carpometacarpal bone and developing primary feathers when the developing proximal bones and tendons are not yet strong enough to support the increased weight. This mismatch results in lateral rotation (up to 360°) and deformity of the distal wing. Although the pathogenesis is not understood, implicated causes are high-protein or carbohydrate-rich diets (e.g., white bread).[124,232,297] Manganese deficiency has also been proposed in some cases. Wing deformities have also been reported in free-ranging cranes thought to be associated with previous traumatic injury, although nutritional or genetic factors have not been completely ruled out.[263]

Hypovitaminosis A is most often seen in hand-raised waterfowl and cranes. Free-ranging waterfowl may be affected during the winter if they feed on grain-based diets with limited access to green forage. Starvation can produce deficiencies in multiple vitamins, including vitamin A. Deficiency of vitamin A results in epidermal hyperkeratosis and squamous metaplasia of secretory epithelia.[156] Ataxia, poor growth, paresis, paralysis, blindness, xeropthalmia, and death are noted in ducklings and goslings.[67] In adult birds, gross lesions include blunting of choanal papilla, white nodules in the oral cavity, upper gastrointestinal tract, and cutaneous hyperkeratosis. Primary differentials for the gross oral cavity lesions are candidiasis (histology would demonstrate organisms) and pox virus infection (histology would reveal viral inclusions). Ureteral squamous metaplasia can result in intralumenal keratin plugs plus reduced mucin production, causing postrenal obstruction, and renal gout.[151] Histologic lesions show layers of keratinized epithelium replacing normal glandular secretory or lining epithelium, often plugging glandular or renal collecting duct and ureteral lumens.[105] Low retinoid (vitamin A) values in serum and/or liver confirm the diagnosis.

Nutritional myopathy is associated with inadequate dietary levels of vitamin E (α-tocopherol) and selenium due to free radicals causing cellular injury, as both compounds are required for the proper function of glutathione enzymes. Low dietary or decreased bioavailability of selenium resulting in **selenium deficiency (hyposelenosis)** can increase physiological demand for vitamin E. In

Pathology of Wildlife and Zoo Animals. https://doi.org/10.1016/B978-0-443-18437-6.00052-3

FIGURE 29.1 Angel wing in a Canada goose. Full body image of a subadult Canada goose with bilateral valgus of the carpometacarpal joints resulting in the deformity referred to as "angel wing." *(Photo Courtesy of the Swedish Veterinary Agency, Sweden.)*

domestic mammals, hyposelenosis is most often implicated, since selenium content can be low in forage in certain regions of the United States, but vitamin E deficiency is often implicated in zoological species and fish-eating birds.[42] No controlled descriptive studies have been performed in these avian species, to characterize lesions for either etiology, and they may overlap. Documented **vitamin E deficiency (hypovitaminosis E)** most often occurs in fish-eating birds (especially storks and herons); cases in captivity occur when birds are fed frozen whole fish without adequate vitamin E supplementation. Free-ranging birds are affected through the consumption of rancid fish, as might occur during fish mortality events. Causes of inadequate intake or absorption of vitamin E include inanition, reduced intake due to conspecific competition, high dietary polyunsaturated fats (oxidation of fatty acids releases lipid peroxides that increase demand for the antioxidant α-tocopherol), rancid fish oil, pelleted diets, or rapid postnatal growth. Some dietary mineral elements (e.g., sulfur, silver, arsenic, cadmium, copper, mercury, fluorine, zinc) or vitamins (e.g., vitamin C) can antagonize or synergize (vitamin E, cobalt, vitamin B6) selenium.

Gross findings of hypovitaminosis E include pale streaks or glistening smooth muscle bundles, often combined with fat necrosis and steatitis (see Chapter 27: Sphenisciformes, Gaviiformes, Podicipediformes, Procellariiformes, and Pelecaniformes). Histological findings in muscle include multifocal, polyphasic degeneration (hypereosinophilia, loss of cross striations, floccular degeneration, fragmentation, and myofibrillar loss) in skeletal, cardiac, and smooth myofibers. Acute necrosis may be associated with calcium salt release from damaged sarcolemma, or influx into damaged myocytes from the bloodstream.[42] Inflammation is typically absent, except for infiltration of macrophages during early stages, clearing debris. In skeletal muscle, attempts at regeneration and repair (e.g., satellite cells and internalized nuclei of myocytes) may be observed. When repair or regeneration is impossible in skeletal or cardiomyofibers, interstitial or replacement fibrous connective tissue is deposited. Cardiomyofibers are fragmented, lost, or replaced by interstitial fibrosis. Exertional myopathy and ionophore intoxication can cause similar myodegeneration and necrosis, but the pattern is monophasic; reparative processes and/or fibrotic sequelae are similar. **Fat necrosis (steatitis)**, which may implicate hypovitaminosis E, is recognized as hard yellow clumps of intracoelomic or subcutaneous adipose (Fig. 29.2A). Microscopic lesions include fat saponification, mineralization, cholesterol deposits, and infiltrates of low to moderate numbers of heterophils and histiocytes

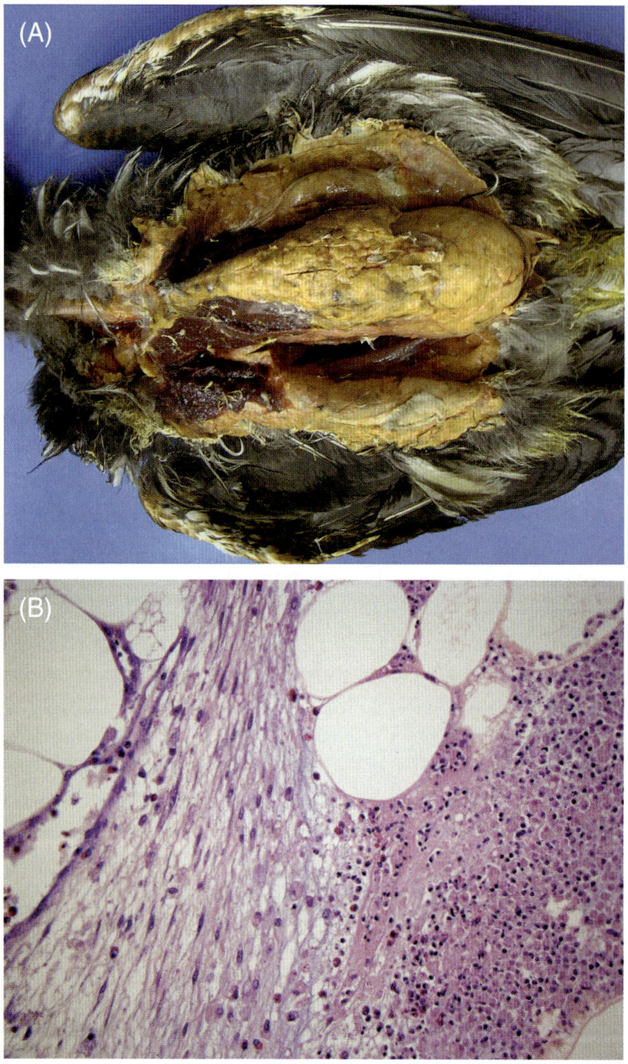

FIGURE 29.2 Fat necrosis and steatitis in a free-ranging great-blue heron. (A) The subcutaneous and visceral fat has a yellow-firm appearance. (B) Inflammation consists of periadipocyte infiltrates of heterophils and histiocytes. *(Part A: Photo Courtesy of M. Forzán, Canadian Wildlife Health Cooperative, Atlantic Region.)*

(Fig. 29.2B). The presence of concurrent steatitis also supports a nutritional etiology. In all cases of suspected nutritional myopathy, assays for vitamin E in serum or liver are useful ancillary tests but are not always sophisticated enough to differentiate the active form (α-tocopherol) from inactive stereoisomers. Serological assays for muscle enzymes, particularly creatine kinase (to confirm myofiber necrosis) and aspartate aminotransferase (not specific for muscle damage), may be informative. As with exertional myopathy, myoglobinuric nephrosis due to muscle necrosis is possible. Commercially available urine test strips do not differentiate hemoglobin from myoglobin; myoglobinuria in avian species can only be detected spectrophotometrically.[151]

Nutritional secondary hyperparathyroidism (metabolic bone disease) occurs under a variety of managed care and natural rearing conditions, in young waterfowl species including cattle egrets, cranes, spoonbills, marabou storks, and seriemas.[17,135,196,223,285] Long-legged and large species are particularly well represented in the literature. Risk factors involve complex interactions of diet components (calcium, phosphorus, vitamin D3) and ultraviolet B level exposure (influenced by weather variations or artificial lighting), particularly during rapid developmental growth phases, unusual handling, or trauma. Diagnosis most often relies on gross findings, which may include spontaneous or handling-induced rib or leg fractures, limb deviations, and/or suboptimal bone calcification. Evaluation of nutritional history (especially potential milling or supplement changes in managed diets; food sources in natural habitats), radiographic evaluation with comparison to established normal growth patterns, and parathormone (PTH) evaluation are helpful for diagnosis.[44,185,196,268,296]

Metabolic

For the species in this chapter, **gout** occurs in captive and free-ranging birds, particularly those with coastal to pelagic distributions.[125] The disease is characterized by the deposition of acicular (needle-like) monosodium urate (MSU) crystals in joints (chronic urate deposition), within tissue parenchyma (especially kidneys) or on muscle fascia, air sacs, or visceral serosal surfaces (acute urate deposition)[58] (see Chapter 31: Galliformes and Columbiformes). Lesions may appear grossly as white spots, streaks, or sheets on serosal surfaces and viscera (Fig. 29.3A). Articular forms of gout involve chalky white deposits in and around joints (Fig. 29.3B). Urate crystals are typically leached out during tissue processing, and their detection may require recognition of remnant, wispy, basophilic "ghost" crystals. Tissues fixed in alcohol rather than formalin and examined with the De Galantha method may improve detection. Histologically, fine radiating urate crystals may be present in the absence of inflammation or be associated with histiocytic, granulomatous, or mixed inflammation (Fig. 29.3C; eSlide 29.e1); chronic lesions may also include fibrosis.

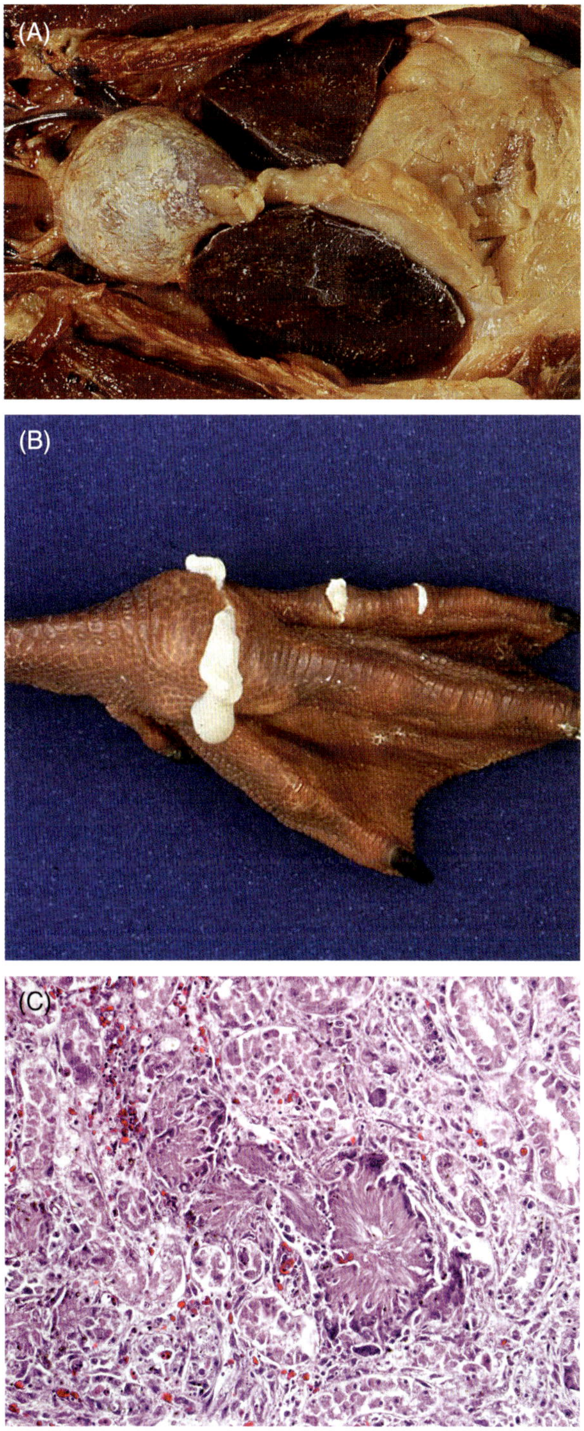

FIGURE 29.3 Acute visceral (A), chronic articular (B), and renal (C) gout in a Canada goose (A), snow goose (B), and common pochard (C). (A) Chalky white crystalline urate deposits cover the pericardial and hepatic surfaces. (B) Chalky white gritty urate crystalline deposits expand the articular space in the metatarsal and phalangeal joints. (C) Stellate, crystalline urate tophus formations that markedly fill and expand the renal tubules are surrounded by mild granulomatous inflammation containing scattered multinucleated giant cell macrophages (see eSlide 29.e1). *(Part A: Photo Courtesy of the Canadian Wildlife Health Cooperative, University of Saskatchewan; Part B: Photo Courtesy of Canadian Wildlife Health Cooperative, Atlantic Region; Part C: Photo Courtesy of D. McAloose, Wildlife Conservation Society.)*

Clinicopathologic confirmation of hyperuricemia provides supportive evidence for gout. Gout should be distinguished from pseudogout (e.g., nonurate, calcium pyrophosphate [CPP] crystals) or other forms of periarticular crystalline deposition (e.g., osteophytes). For joint fluid analysis, using a polarizing filter with a red compensator filter, MSU crystals are negatively birefringent (yellow when parallel to the slow axis of the red compensator; blue when aligned perpendicular to the direction of polarization); CPP crystals are positively birefringent (blue when parallel with the slow axis of the compensator; yellow when perpendicular). Calcium stains (e.g., Von Kossa) can distinguish mineral deposits from MSU.

Exertional rhabdomyolysis (capture myopathy) can occur in any avian species but is reported most frequently in long-legged birds (e.g., cranes). The condition is most commonly seen in captivity but can occur in any pursuit-capture situation with exertion, prolonged restraint, or struggling. Transportation in high environmental ambient temperatures, hyperthermia, hypoxemia, and dehydration, as well as preexisting or concurrent electrolyte abnormalities, have been proposed as contributory factors. Lameness and muscular atrophy can occur days to weeks after capture. Clinical pathology findings in acute to subacute cases may include: metabolic acidosis, myoglobinuria, oliguria, increased serum creatinine phosphokinase (CPK), aspartate aminotransferase (AST), and lactic dehydrogenase (LDH). Often no gross lesions are found in peracute or acute conditions, but pale streaks and swelling of large muscle groups in the neck, gluteals, leg muscles, and (to a lesser extent) pectoral muscles and/or cardiomyofibers may be present. The kidneys may be dark and swollen. In subacute to chronic cases, gross lesions may include shrinkage, atrophy, pallor, and streaking of myofibers in heart and skeletal muscle groups. Histologic lesions in acute to subacute cases of rhabdomyolysis include monophasic changes with rhabdomyofibers and/or cardiomyofibers appearing pale, edematous, hypercontracted, fragmented, degenerated, or necrotic with pyknotic nuclei. In some cases, there may be basophilic mineral deposits associated with the lesions. In more subacute to chronic cases, myofiber atrophy or necrosis with mineralization, variable macrophage infiltration, and replacement fibrosis are seen. Tubular necrosis associated with the deposition of myoglobin crystals may occur in the kidneys.[42]

Toxic

Toxicoses are common causes of large-scale mortality events in free-ranging water birds. Gross and microscopic lesions are often nonspecific or absent. Some detection methods are very sensitive and allow detection of small amounts of certain compounds, which may make interpretation of the significance of these findings challenging. Reference intervals are not always available for free-ranging wildlife or nondomestic species. Consultation with a toxicologist at referral laboratories can be an essential component of diagnostic case work up for proper test selection, appropriate sample collection, storage, and interpretation.

Avian botulism is caused by a neurotoxin produced by the gram-positive, saprophytic, anaerobic bacterium, ***Clostridium botulinum***. There are seven types of *C. botulinum* toxins: A, B, C1, D, E, F, and G. Most waterfowl and gull mortalities are associated with type C1, although type E is associated with piscivorous deaths in the Great Lakes, US and in France; outbreaks can also occur in zoo or other managed waterbird collections and at landfills.[181,207] Botulism outbreaks are associated with ingestion of preformed botulinum toxin produced under specific environmental conditions that allow spore germination. The bacterium grows most readily in the presence of decaying tissue that serves as a source of amino acids. This contributes to the "carcass-maggot cycle" that involves concentration of the toxin in fly larvae that feed on toxin-laden carcasses and outbreaks that are perpetuated by ingestion of maggots containing high levels of toxin. The botulinum toxin is thought to cleave synaptic receptor proteins essential for release of acetylcholine at neuromuscular junctions.[214]

Clinical signs of avian botulism include progressive weakness, paresis, and flaccid paralysis. Some birds drown due to an inability to hold their head above water. There are no characteristic gross or microscopic lesions of avian botulism, but the "limber neck" appearance can be seen prior to death. In some cases, birds may become infested with nasal leeches in their weakened state.[277] Feces can be bright green from high biliverdin content. Gastrointestinal tracts often are empty, and the colon may be dilated. Occasionally undigested/partially digested maggots (Fig. 29.4), or fish

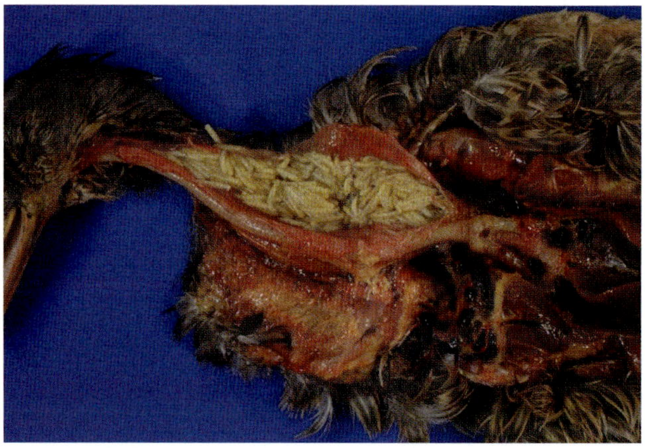

FIGURE 29.4 Avian botulism in a duck (unspecified species). The photo illustrates the classic "carcass-maggot" pathogenesis cycle: ingestion of blow fly maggots from dead waterfowl infected with *Clostridium botulinum* bacteria, distension of crop with ingested blow fly maggots containing concentrated botulinum toxin, and death due to complications of muscle paralysis due to botulinum toxicosis. *(Photo Courtesy of Canadian Wildlife Health Cooperative, University of Saskatchewan.)*

are observed within the upper gastrointestinal tract.[213] Animals that die from drowning may have aquatic microorganisms within air capillaries. Samples used for detection of botulinum toxin include blood or serum from moribund or freshly dead birds, liver, and gastrointestinal content. The gold standard test in the United States is mouse inoculation, although false negatives due to low levels of toxin and false positives due to the presence of bacteria or other contaminants are possible. The technique is expensive and faces criticism regarding the terminal use of live animals necessary for the test. Enzyme-linked immunosorbent assays (ELISA) have been developed and work best with fresh serum samples. An immunostick antigen-capture ELISA for type C botulinum toxin[215] has been effective in a research setting but is not yet commercially available. Additional diagnostic techniques include PCR detection of the C1 toxin gene.[63,72] However, concerns regarding false positive results due to the presence of dormant spores are associated with the use of those techniques. Additional ELISA, PCR, and endopep-mass spectrometry protocols have been established and are available in Europe[143,257] and Australia.[155] Liver is considered the sample of choice for PCR protocols and sample preservation at −18°C is recommended prior to testing. Liquid chromatography–mass spectrometry may be available at some laboratories. Botulinum toxin is regulated under the United States Federal Select Agent Program due to concerns related to potential use in bioterrorism, but neither the toxin nor botulism are a listed disease with the World Organization of Animal Health.

Necrotizing enteritis, most likely caused by ***Clostridium perfringens***, has been described in wild and captive waterfowl. Gross lesions of necrotizing enteritis are most severe in the ileum. Microscopically, the presence of large gram-positive bacilli in intestinal lesions can lead to a presumptive diagnosis; however, demonstration of the production of one or more disease-causing toxins (e.g., alpha, beta, epsilon, or iota) within intestinal contents is typically recommended as parts of a clostridial enteritis work up to differentiate from postmortem overgrowth.[289] Results often have to be interpreted in light of gross and histologic lesions.

Mycotoxins most likely to impact free-ranging and captive waterfowl are **aflatoxins** and **trichothecenes.** Mycotoxin exposure in waterfowl and cranes is suspected to occur from ingestion of fungal toxin on moldy feed left on pastures.[205] Mycotoxins have immunosuppressive properties and exposure may be a risk factor for avian cholera.[238] For example, neurologic disease and death in more than 9,000 sandhill cranes over a 5-year period in the United States was thought to result from exposure to mycotoxins produced by *Fusarium* spp.[272] Aflatoxin B1 is the most toxic of the **Aflatoxins** produced by *Aspergillus* spp. This group of toxins is commonly associated with moldy peanuts, corn, and other oil seeds, and mortality events have been associated with aflatoxin exposure in a variety of wild bird species.[76] Optimal environmental conditions for toxin production include high relative humidity, temperatures >25°C, and moisture content.[31] Aflatoxin is metabolized in the liver by cytochrome P450 enzymes into reactive metabolites that bind RNA and DNA and potentially alter cell division resulting in carcinogenic properties associated with DNA adduct formation.[293] Binding to DNA can also interfere with protein synthesis by blocking translation, which disrupts metabolic pathways, such as the Kreb's cycle, the clotting cascade, lipid metabolism, and the host's immune system due to altered globulin synthesis.[205] Toxicoses range from acute to chronic. Younger animals appear to be more susceptible to acute toxicosis. Reported gross lesions from experimental trials in ducklings up to three weeks of age include pallor and swelling of the liver and kidneys, as well as multiorgan petechiation. In slightly older ducklings, the liver is firm, shrunken, and has a reticular pattern; hydropericardium and ascites are also observed.[6] Clinical signs in free-ranging birds include blindness, unresponsiveness, depression, tremors, and an inability to fly. Free-ranging geese diagnosed with aflatoxicosis tend to have hepatic pallor, whereas ducks appear to have more severe hepatic lesions with hepatomegaly, hemorrhage, and generalized edema. Histologic lesions of acute aflatoxicosis include hepatocellular degeneration, intracellular lipid vacuolation, necrosis (often of periportal hepatocytes, but can be massive), and biliary proliferation associated with mononuclear cell infiltrates. Degenerative pancreatic and renal lesions have also been reported.[101] Chronic lesions include bridging hepatic fibrosis, biliary hyperplasia, karyomegaly, and nodular regeneration. Hepatocellular carcinomas due to chronic aflatoxin exposure are reported in domestic and suspected in wild ducks.[34] Diagnosis of aflatoxicosis can be challenging. The samples of choice are feed and gastrointestinal tract (GI) contents using ELISAs and radioimmunoassays. Detection of aflatoxin in tissue can be difficult due to its rapid degradation and requires high-performance liquid chromatography. Freezing feed samples is often recommended if they be immediately delivered to a lab in order to prevent further fungal growth and toxin production.[199]

Morbidity and mortality events associated with **trichocenes** have been reported in captive sandhill and whooping cranes.[180] Trichocenes (T2, HT-2, deoxynivalenol, and scirpenols) are produced primarily by *Fusarium* spp. and consist of sesquiterpene toxins that inhibit protein synthesis.[182] The resultant decrease in enzymatic activity can result in lipid peroxidation and associated cellular damage[239] that cause ulcerative and proliferative lesions that manifest as radiomimetic lesions in the intestine. Clinical signs from experimental infections of T2 intoxication in geese include vomition, thirst, depression, and head and leg tremors.[205] Affected sandhill cranes are reported to have a drooped posture of the head and neck in addition to wing and leg weakness that affect the ability of birds to fly. Trichocenes are immunosuppressive and can cause lymphoid necrosis and depletion, decreased immunoglobulin synthesis, and impaired macrophage function.[43] Low temperatures favor production of these toxins. Gross lesions of trichocene exposure include ulcerative lesions in the oropharynx and upper GI tract, esophageal impaction as well as the development of diphtheritic membranes in some cases. Subcutaneous edema and hemorrhage in affected muscle (e.g., cervical region in cranes), dermal irritation, anemia, and leukopenia are reported. Decreased size of the bursa of Fabricius and the thymus are also reported to be associated with HT-2 toxicosis.[24,97] Histologic lesions mirror the ulcerative gross lesions and include lymphoid depletion. Vascular thrombosis and ischemia have been reported in cranes and are associated with gross lesions in subcutaneous tissue and muscle.[217] Diagnosis of trichocene intoxication can be difficult as it involves confirmation of the toxin in feed in combination with appropriate lesions. Detection can be via thin-layer chromatography and ELISA methods. Testing multiple feed samples is recommended, as the distribution of the toxin is generally not uniform. Certain trichocenes are regulated under the United States Federal Select Agent Program and are often monitored as part of mycotoxin surveillance programs in cereal products globally (e.g., New Zealand).

Numerous large-scale mortality events in water birds have been associated with **harmful algal blooms (HABs).** These events typically occur in the summer months when temperatures and nutrient loads in water bodies are high. In general, piscivorous birds, such as cormorants and shags are commonly involved, but many species of ducks and gulls can be affected.[118,142] Several algal species, the toxins they produce, their mechanisms of action, and significant pathological findings in waterbirds are summarized in Table 29.1. Gastrointestinal content can provide clues as to a potential route of exposure (e.g., bright-green freshwater material, fish, or mollusks). Pulmonary edema, tracheal congestion, gastrointestinal irritation, and mucoid intestinal content are gross lesions in some HAB cases.[175] Animals that die from acute exposure are often in good body condition. Diagnosis of HABs can be challenging and requires exclusion of other potential causes of mortality (especially botulism). Mortality events frequently involve a variety of taxonomic groups, appropriate environmental conditions, detection of cyanobacteria, diatoms, or dinoflagellates, and toxin detection from environmental samples and/or gastrointestinal contents of affected animals. ELISA or HPLC are commonly used to detect toxins. An immunohistochemical test developed for tissue detection of

TABLE 29.1 Toxins Associated with Algal Blooms

Disease Name	Toxin Produced	Mechanism of Action	Algal Species	Pathology	Species Affected
Blue-green algal poisoning	Microcystin	Inhibits protein phosphatases 1 and 2B and causes hepatocyte apoptosis	*Microcystis aeruginosa, M. flosaquae, M. botrys, M vurudis, Anabaena flosaquae, Planktothrix agardhii*	Massive hepatic necrosis and hemorrhage[a]	Water birds including eared grebes, mallards, spot-billed ducks, Eurasian coot, purple gallinule, common moorhen, mew gulls, great-crested grebe, common golden eye[33]
	Unknown	Unknown	*Akashiwo sanguinea*	None reported	Surf scoters, white-wing scoters, common murres[118]
	Anatoxin-a	Postsynaptic depolarization	*Anabaena* spp.	None reported	Water birds including Franklin's gulls and coots[79]
	Anatoxin-a (s)	Inhibits acetylcholinesterase	*Anabaena* spp.	None reported	Diving birds, especially mergansers[98]
Paralytic shellfish poisoning	Saxitoxin	Blocks sodium channels	*Aphanizomenon* spp., *Alexandrium catenella, A. tamarense*	Enteritis	Herring gull, great black-backed gull, lesser black-backed gull, black-headed gull, black-legged kittiwake, western gull, kelp gull, Hartlab's gull, white-winged scoter, black scoter, common murre, tufted puffin, sooty shearwater, black oystercatcher, common eider, common tern, Arctic tern[175]

TABLE 29.1 Toxins Associated with Algal Blooms—cont'd

Disease Name	Toxin Produced	Mechanism of Action	Algal Species	Pathology	Species Affected
Neurotoxic shellfish poisoning	Brevetoxin	Binds sodium-gated channels and inhibits cathepsins	*Karenia brevis*	Mononuclear meningeal infiltrates	Numerous sea birds[65]
Amnesic shellfish poisoning	Domoic acid	Analog of glutamate	*Pseudonitzschia* spp.	Skeletal muscle hemorrhage and necrosis	Numerous sea birds[286]
Nodularin poisoning	Nodularin	Inhibits protein phosphatases 1 and 2B and carcinogen	*Nodularia spumigena*	Hepatic necrosis and hemorrhage	Multiple ducks[119]

[a]*Liver lesions due to microcystin intoxication are not well documented in avian species, although massive hepatic necrosis and hemorrhage is described in mammals. The toxin has been detected in a wide variety of avian species, but the pathology and overall significance remain unknown.*

Source: Adapted from Landsberg JH, Vargo GA, Flewelling LJ, Wiley FE. Algal biotoxins. In: Thomas NJ, Hunter DB, Atkinson CT, eds. *Infectious Diseases of Wild Birds*. Oxford: Blackwell Publishing; 2007:431–455.

brevetoxin is not commercially available.[25] Lethal doses of toxins are not well established for many wild bird species, and thus toxin detection does not always imply causality. Physiological assays, such as acetylcholinesterase activity, can be helpful if anatoxin-a(s) is suspected, and protein phosphatase inhibition can suggest microcystin exposure. Diagnosis can be complicated by multiple concurrent organisms and toxins. Toxins can persist in the environment and pose an exposure risk for weeks to months after an algal bloom has dissipated.

Avian vacuolar myelinopathy is a condition that affects primarily American coots and bald eagles in the United States (see also Chapter 30: Birds of Prey). It is thought to be due to exposure to a toxin produced by the cyanobacterium *Aetokthonos hydrillicola*, which is associated with the invasive aquatic plant *Hydrilla verticillata*.[270] The cyanobacterium produces a pentabrominated biindole alkaloid (aetokthonotoxin), which is lipophilic and capable of bioaccumulation (see Chapter 30: Birds of Prey).[21,27] Affected birds are unable to swim, walk, or fly properly. The disease appears to be associated primarily with specific sites with artificial reservoirs and large amounts of introduced aquatic vegetation in the southeastern United States.[70] There are no typical gross lesions, and affected birds are often in good body condition. Histologically, bilaterally symmetrical vacuolation (Fig. 29.5), intramyelinic edema, spheroid formation, and mild gliosis within the white matter of the brain and spinal cord. Myelin splitting detected by electron microscopy is characteristic. White matter tracts of the optic tectum are a consistent site for lesions, but they can also occur in other parts of the brain and spinal cord.[258]

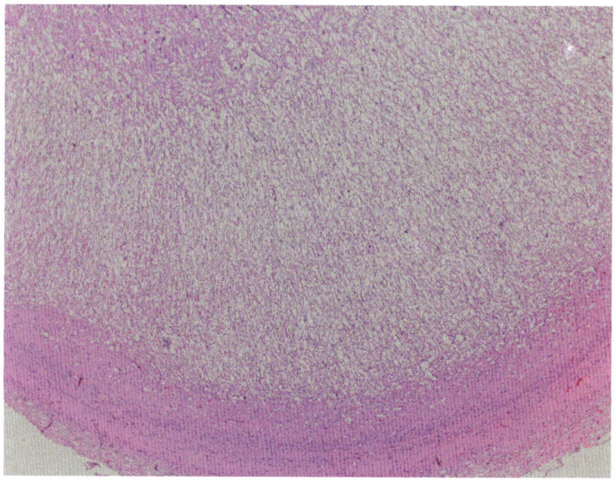

FIGURE 29.5 Avian vacuolar myelinopathy in a Canada goose. Marked vacuolation, which appears as elongated clear spaces, is present in the white matter tracts in the optic lobe of the brain.

Lead intoxication (plumbism) is a significant and common cause of morbidity and mortality in wild waterfowl worldwide. For this reason, lead ammunition has been partially or completely banned in waterfowl hunting in some countries. However, wild birds can still become exposed to lead through the ingestion of spent ammunition and fishing tackle.[76] Captive birds become exposed via ingestion of lead-based paint or other materials. Lead is quickly absorbed through the gastrointestinal tract and redistributed to the kidney, liver, and eventually bone. Clinical signs associated with lead intoxication include progressive weakness, green liquid feces (biliverdin), ataxia, and weight loss. Ingested lead-based materials can be differentiated from actual ballistic shot based on location (i.e., ingested shot is in the gastrointestinal tract) and external properties. Ingested shot tends to be brightly colored, eroded, pitted, and can be reduced in size; ballistic shot pellets are dull blue to gray, regular in size, and tend to be flattened.[277] Birds can survive shooting encounters where the pellets migrate through the tissues and are

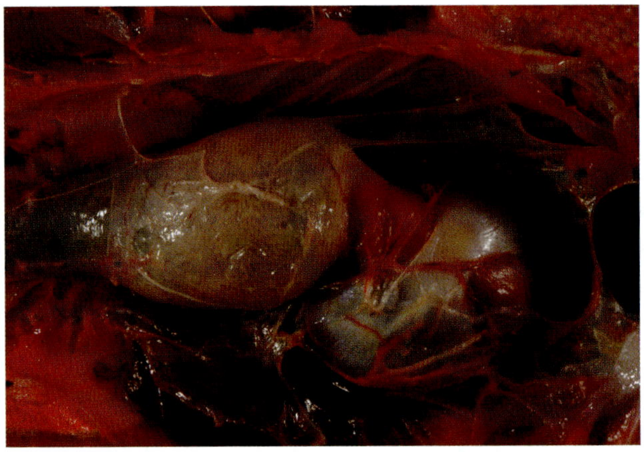

FIGURE 29.6 Lead intoxication in a Canada goose. Esophageal distension and impaction are common findings due to mechanical dysfunction (esophageal muscular degeneration, atrophy, and necrosis) or neurogenic paralysis induced by lead toxicosis. *(Photo Courtesy of B. Munk, Southeastern Cooperative Wildlife Disease Study.)*

generally unassociated with hemorrhage but can have surrounding fibrosis. Although some lead may be absorbed from embedded ammunition, exposure is primarily considered to be absorbed into the bloodstream from oral ingestion. Gross lesions of lead intoxication in waterfowl include green staining around the vent, proventricular or esophageal impaction (Fig. 29.6), gall bladder distension, myocardial pallor, and lymphoid atrophy. Lead-based materials may be present in the gastrointestinal tract (GIT). Subcutaneous edema around the head and mandible and hydropericardium, presumably associated with cardiotoxicity, are reported in geese and swans. Microscopically, hepatocellular degeneration, hepatic cord atrophy, hemosiderosis, and rarely, bile duct necrosis can be observed. Gastrointestinal effects can include arrest of mucosal regeneration and reduction in the height of proventricular plica often due to pressure necrosis in cases of proventricular impaction.[38] Renal tubular degeneration and necrosis can be seen; tubular epithelial cell, acid-fast positive, intranuclear lead inclusions are characteristic, but rarely observed finding. Focal degeneration, atrophy, and necrosis of skeletal muscle, smooth muscle of the gizzard, and cardiac muscle are also reported.[122] Foci of myocardial fibrosis may be present and are occasionally associated with vascular mural hyaline degeneration. Lymphoid depletion, myelosclerosis, and primary demyelination in central and peripheral nervous systems can be observed.

Liver is often the tissue of choice for diagnostic testing in suspected cases of lead intoxication; kidney can also be used. Chelation therapy and lead fragments from undigested shot (e.g., from the euthanasia method) may impact detection of lead in tissues.[73] Lead-based materials can be analyzed at specialized laboratories if necessary; however, a quick, inexpensive, but nonspecific screening method involves the use of a magnet. Lead-based material will not be magnetic, while alternative materials may be. Clinical pathology findings include hypochromic and hemolytic anemia in acute exposure, and leukocytopenia in chronic exposure. Basophilic stippling of erythrocytes seen in mammals is not reported in birds. Clinical conditions of lead poisoning can vary from acute to chronic depending on the dose ingested and other host factors, such as species, signalment, diet, concurrent infections, or exposure to other heavy metals (e.g., mercury).

Waterfowl that ingest zinc-containing objects are at risk of developing **zinc toxicosis.** It is generally considered a disease of captive birds but has also been documented (rarely) in free-ranging waterfowl.[236,295] The most commonly recognized scenario is the ingestion of pennies (coin currency), typically by diving ducks who mistake pennies thrown into water as food items. In the United States, pennies minted after 1983 contain a zinc core that, despite being covered by a thin layer of copper, is exposed as the copper erodes over time. Ingested pennies can persist in the ventriculus as they are too large to be passed. Caging, fencing or fence clips; nails, nuts, and bolts; and certain zinc-containing fertilizers and paints can all be potential sources of ingested zinc. Clinical signs can be absent or include weight loss, depression, weakness, and incoordination or paresis. Lesions are classically identified in the pancreas and are more commonly identified histologically. Variable pancreatic zymogen granule depletion, acinar cell atrophy, and/or necrosis with disruption of the normal acinar architecture are common acute lesions. With chronicity, damaged tissue is replaced by progressive interstitial and parenchymal fibrosis, and exocrine pancreatic regeneration and/or ductular hyperplasia are seen. Lesions in the ventriculus can include degeneration and necrosis of the koilin, glandular epithelium, and (occasionally) smooth muscle and heterophilic infiltration of the superficial mucosa. Hepatic hemosiderosis, perhaps secondary to low-level hemolysis, may also be present. Radiographic evidence of gastrointestinal metal object/s (antemortem), identification of an eroded, post-1983 penny (US) or other metal in the ventriculus (postmortem) in association with consistent pancreatic or ventricular lesions should raise the suspicion of zinc toxicity. Elevated serum and tissue zinc levels support the diagnosis.

Mercury exposure is common in birds that consume fish and shellfish, in which it bioaccumulates, and can be associated with coal combustion as well as other industrial activities. Historically, waterfowl were exposed through ingestion of mercury-treated seed grains, although this practice was abandoned in most countries after the 1970s. Animals with heavy diet of fish can accumulate large amounts of mercury in the absence of obvious clinical disease or pathology.[74,80] The form of mercury most relevant to biological systems is methyl mercury, which can be acutely toxic at species-specific doses. As the metal is known to bioaccumulate in aquatic systems without clinical effects in waterfowl, diagnosis of **mercury intoxication**, as opposed to exposure, can be a challenge. The methylated form of mercury is lipid-soluble, allowing it to cross the blood-brain barrier and impact the central nervous and reproductive systems. Clinical signs include weakness and inappetence, as well as difficulty flying and walking. Embryonic mortality and infertility are reported.[68] Subclinical impacts of mercury exposure may include behavioral changes (e.g., vocalization changes),[94] immunosuppression, reproductive alterations, and microscopic lesions of spinal cord demyelination.[224] Diagnostic testing for mercury can be altered by the presence of other metals, particularly selenium that can form complexes with mercury, facilitates systemic redistribution of mercury, and impacts enzyme activities in addition to consideration of nonmethyl forms of mercury that are considered biologically inactive. Although there are individual and species differences, kidney methyl mercury values >40 ppm have been associated with avian mortality.[69] However, tissue concentrations need to be interpreted in the context of clinical signs, pathological findings, and the amount of biologically active methyl mercury and other metals, particularly selenium that are detected.[199]

Selenium (Se) toxicosis occurs in free-ranging birds in regions with high soil selenium, potentially in association with industrial development. Toxicosis paradoxically causes similar muscle lesions to those described for selenium deficiency. Negative reproductive impacts, such as embryonic death, have been reported in wild birds. Gross lesions in young birds can include limb deformities, brachygnathia, fluid accumulation in the skull, and feather and claw abnormalities. Affected adult birds can be emaciated and have concurrent disease/s. Experimental intoxication of mallards and American coots causes coelomic fluid accumulation, subcutaneous edema, and cardiac flaccidity.[1] Histologic lesions are indistinguishable from selenium deficiency and include myofiber degeneration and necrosis, as well as hepatic necrosis and splenic, pancreatic, and muscular atrophy. Diagnosis can be confirmed by quantification of excessive selenium levels in tissue (e.g., liver). Elevated environmental selenium is also supportive.

Organophosphates (OP) and **carbamates** inhibit acetylcholinesterase, an enzyme necessary for acetylcholine recycling at nerve and neuromuscular junctions. These compounds are often sprayed on crops to control agricultural pests and, in some cases, used to intentionally poison nuisance wildlife. Their presence is often suspected when there is an acute, multispecies mortality event. Clinical signs of exposure are associated with overstimulation of muscarinic and nicotinic receptors in the central and peripheral nervous system including muscle tremors, hyperexcitability, diarrhea, vomition, excessive salivation and lacrimation, piloerection, respiratory difficulty, miosis or mydriasis, paralysis, epistaxis, depression, and opisthotonus.[76] Delayed neurotoxicosis that occurs days or weeks after the initial exposure is possible with some OPs.[199] Death is most often due to respiratory paralysis. Exposure may predispose birds to trauma or predation, interfere with thermoregulation, and reduce feeding activities.[76]

There are no characteristic gross lesions associated with OP or carbamate toxicosis. Nonspecific petechial hemorrhages and organ congestion may be observed. Animals are typically in good nutritional condition and have evidence of recent feeding with abundant feed material within their upper GI tract. There are also no characteristic microscopic lesions, although mild hemorrhage, especially in the brain, and gliosis have been observed (unpublished Southeastern Cooperative Wildlife Disease Study [SCWDS] data). Depression of brain cholinesterase activity can be used as a screening test for exposure to these compounds, and the cerebral cortex is the preferred test sample.[199] Prolonged frozen storage can artificially depress brain cholinesterase levels and should be considered when interpreting results.[203] Inhibition of cholinesterase activity is irreversible with OP exposure but can be reversible with carbamate compounds. Reference intervals for brain cholinesterase levels are species-specific.[99] Exposure is confirmed with detection of an OP or carbamate compound in shared feed or GI contents and in some cases detection of metabolites in the liver.

Organochlorines are a group of lipid-soluble, synthetic hydrocarbons that historically were used extensively for pest control. Examples include dichlorodiphenyltrichloroethane (DDT) and dichlorodiphenyldichloroethylene (DDE). Due to the lipid-solubility of the compounds, they bioaccumulate through trophic levels and have nontarget reproductive impacts on higher trophic-level species.[199] Most are now commercially unavailable in North America; however, illegal use can still occur and some countries have not banned the use of these compounds. Acute toxicosis is thought to occur due to negative impacts on the central nervous system. These compounds are not thought to play a significant role in morbidity and mortality of free-ranging waterfowl. However, due to recent interest in vector-borne zoonotic diseases and insecticide resistance, there is interest in bringing some of these compounds back to market, which could have nontarget effects on species at higher trophic levels.

Unfortunately, **oiling with petroleum products** continues to be a cause of morbidity and mortality for birds. Oiling can occur in large mortality events, such as the Ixtoc I (1979) in Mexico, Exxon Valdez (1989) and Deepwater Horizon (2010) oil spills in the United States, Erika (1999) in France, Prestige (2002) in Spain, Odyssey

(1988) in Canada, the Persian Gulf War (1991), and more frequent smaller spills associated with oil and gas development.[116] Direct mortality from oiling can occur from inhibition of normal thermoregulation, drowning from loss of buoyancy, direct intoxication from ingestion or inhalation, and embryonic mortality in eggs.[277] Subclinical effects of chronic exposure to oil are not well understood, but hematological abnormalities, such as anemia (potentially with the formation of Heinz bodies), reproductive effects, and immunosuppression have been recognized. The compounds in oil associated with toxic effects are **polycyclic aromatic hydrocarbons (PAHs)** and other hydrocarbons.[145] The toxicity of oil can vary widely based on chemical composition, environmental conditions, and composition changes associated with dispersal (weathering) within the environment. Gross lesions, aside from oil-covered plumage, are nonspecific and include emaciation, pulmonary congestion, pulmonary hemorrhage, mucosal congestion, or hemorrhage along the gastrointestinal tract, involution of lymphoid tissues, hepatomegaly, and pallor of the liver. Microscopic lesions include hepatic lipidosis, hemosiderosis, adrenal necrosis, urate deposition (gout), and lipid droplets in the lungs. Determination of exposure in chronic cases can be challenging and may be associated with concurrent illness, such as hemosiderosis, aspergillosis, ventricular ulceration,[10] or emaciation. Chemical analyses of feathers, skin, or other tissues can be used if necessary to confirm exposure to oil, PAHs, or other hydrocarbons.

Nontarget pesticide intoxication has been reported in many waterfowl species and associated with access to treated grain fields. Examples of such compounds include zinc phosphide,[20] fluoroacetamide (1081),[235] strychnine,[278] and alpha-chloralose.[92] The latter is also used as an avicide in some countries. Many of these compounds were used historically but their use is now illegal. Other rare intoxications reported in waterfowl include castor bean (*Ricinus communis*),[115] cyanide,[269] copper,[133] fluoride,[4] and white phosphorus.[266] In many cases, intoxication has been associated with site-specific industrial activity.[112] Newly developed toxins, such as sodium nitrite, proposed to assist with control of feral swine populations in the United States could be considered as differentials if historical information suggests exposure to such a compound.[240] Morbidity and mortality associated with exposure to the anticoagulant rodenticides brodifacoum, bromadiolone, chlorophacinone, coumachlor, difethialone, diphacinone, and warfarin have not been readily documented in members of the orders covered in this chapter, likely due to differences in life history strategies where rodents do not make up a large portion of the diet as it would in other avian orders. However, birds in this chapter that scavenge (e.g., Charadriiformes) could be exposed to these compounds with the potential for subclinical health impacts.[126]

Congenital/Genetic Diseases

Several reports of presumed "congenital" abnormalities present at hatching in wild and captive cranes and waterfowl are summarized in an online wildlife database.[271] The pathogeneses are unknown, but selenium intoxication, genetics, and conditions during incubation are suspected in some cases. Examples include limb deformities in mallards, premaxillary or maxillary deformities, and cranial, vertebral, or cardiac malformations. Developmental defects in captive ducklings and goslings include perosis (commonly called "hock disease," "slipped tendons," or "slipped hock") and bowed limbs. Perosis is characterized by medial luxation of the Achilles tendon and is thought to be associated with vitamin (biotin, choline) and/or mineral (manganese) dietary deficiencies. Bowed limbs include deformities of the tibiotarsus possibly associated with niacin deficiency.[8] Orthopedic conditions in hand-raised cranes include deformities in the proximal tibiotarsus and tarsometatarsus, distal tibiotarsus, and the intertarsal and tibiofemoral joints.[153] The pathogenesis is not known.

Age-Related/Degenerative

Degenerative joint disease (DJD) (osteoarthritis) of the pelvic joints is common in captive waterfowl. Nutrition, congenital conformational abnormalities, and rough surfaces have been proposed as potential risk factors for its development.[50] Gross lesions include swelling of the joints (commonly the stifle), erosion or loss of articular cartilage of affected distal and proximal condyles, as well as thickened joint capsules. Histologically, focal to severe erosions and cysts within the epiphysis, joint capsule, and synovial proliferation and fibrosis, as well as osteomyelitis may be observed (Fig. 29.7). No infectious agents have been associated with DJD in waterfowl.

Miscellaneous/Multifactorial

Amyloidosis is the descriptive term for a group of protein misfolding diseases that result in deposition of at least one of 20–25 distinct, homogenous proteinaceous substances. These insoluble substances aggregate and are deposited in tissues as β-pleated sheets of amyloid fibril protein.[169] Waterfowl are frequent models for spontaneous, induced, and potentially transmissible amyloidosis. The most common form of amyloid in birds is AA-amyloid, which is derived from serum A-amyloid (SAA), an acute phase lipoprotein that utilizes apoSAA as its precursor protein.[36] AA-amyloidosis includes secondary (reactive) amyloidosis associated with chronic inflammatory or neoplastic disease and familial AA-amyloidosis. A portion of the duck genome contains a single copy of the SAA gene, which can be expressed even in the absence of an acute phase

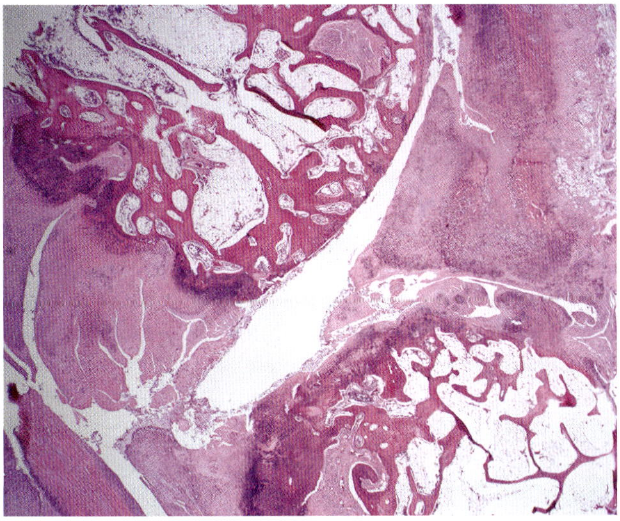

FIGURE 29.7 Degenerative joint disease in an Indian pygmy goose. There is degeneration of the articular cartilage and eburnation in the tibiotarsal-tarsometatarsal joint. Osteophytes with endochondral ossification are present in the attachment of the joint capsule.

inflammatory response.[93] Amyloidosis in birds is associated most often with chronic inflammatory conditions, such as avian mycobacteriosis and pododermatitis, although associations with other infections are also described.[233] Stress from captivity can also induce amyloidosis in free-ranging gulls.[102] Presumed mechanisms include increased expression of cytokines IL1 and IL6 that upregulate amyloid production by hepatocytes, potential defects in macrophage degradation of amyloid fibrils, and possible increased production of degradation-resistant fibrils. Transmissibility of amyloid has been demonstrated through experimental intravenous or oral administration within and between species, which may explain why outbreaks of amyloidosis can occur in chicken flocks, free-ranging waterfowl populations,[256] and gull populations.[113]

Gross lesions of amyloidosis most often include an enlarged, tan to tan/orange liver (Fig. 29.8A) and in some cases splenomegaly. Hepatic or splenic fracture with acute coelomic hemorrhage can occur. Identification of chronic inflammatory conditions warrants a search for evidence

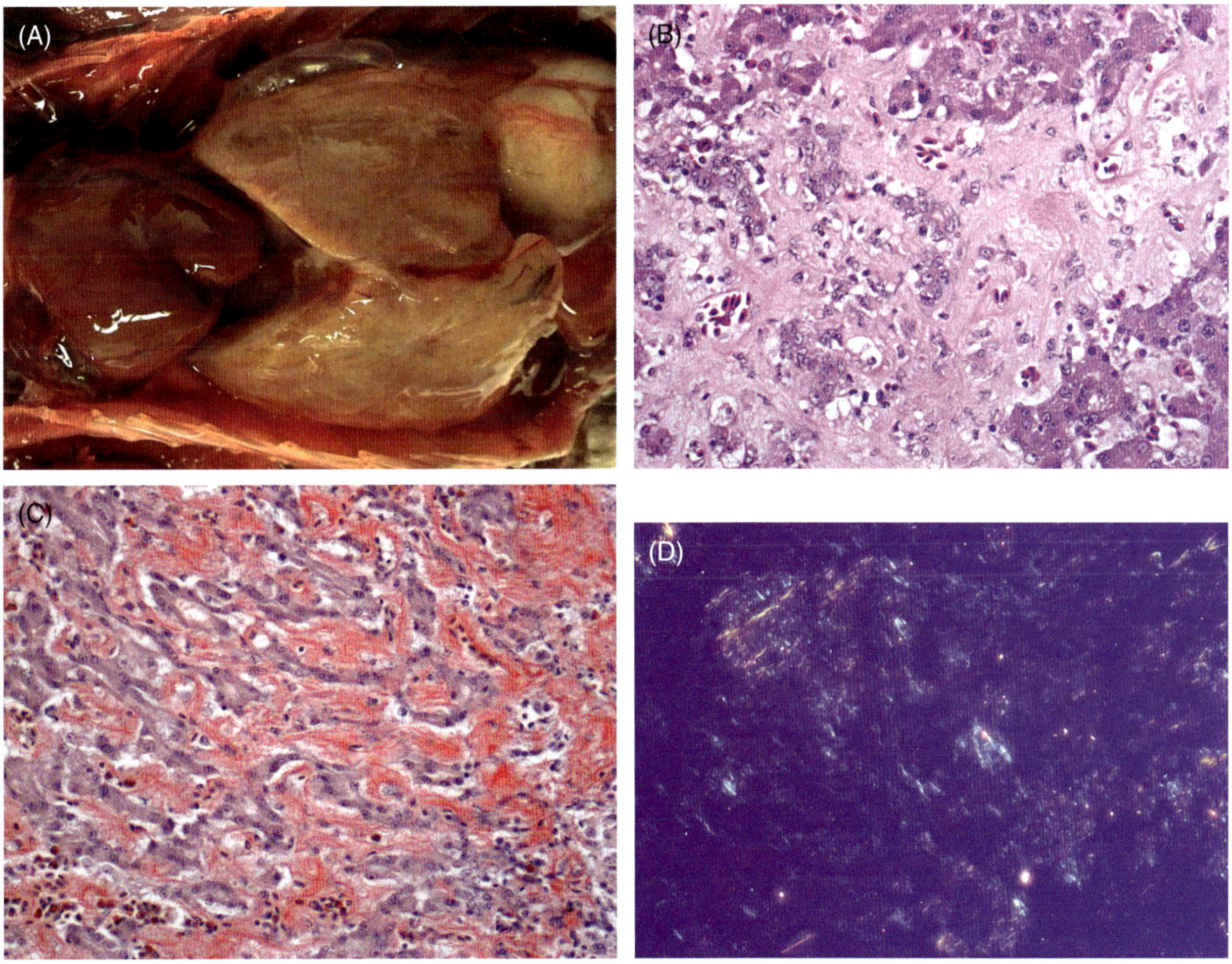

FIGURE 29.8 Amyloidosis in the liver of a redhead duck (A) and Brazilian teal (B–D). (A) Characteristic features of hepatic amyloidosis are hepatomegaly with rounded liver margins, tan to orange/tan discoloration, and a waxy texture. (B) Amorphous, acellular, eosinophilic, hyaline deposits markedly expand the space of Disse and sinusoids and compress and replace hepatic cords. (C) Amyloid deposits stain pale orange/pink with brightfield microscopy, and (D) have bright green birefringence with polarized light. Congo red (C, D) (see eSlides 29.e2 and 29.e3). *(Part A: Photo Courtesy of N. Stedman, Busch Gardens; Parts B–D: Photos Courtesy of K. Terio, University of Illinois Zoological Pathology Program.)*

of amyloidosis and vice versa. Characteristic light microscopic findings include deposits of pale, homogenous, and hyaline eosinophilic material within tissues, commonly the liver and spleen (Fig. 29.8B; eSlide 29.e2). Congo red staining highlights amyloid, which appears pink/orange with brightfield microscopy (Fig. 29.8C; eSlide 29.e3) and is birefringent and has apple-green birefringence under polarized light (Fig. 29.8D). Results of Congo red staining can be variable or negative in nonmammalian species. An additional option is thioflavin T stain which attaches to amyloid fibrils and is detectable with fluorescent microscopy. The diagnostic gold standard is ultrastructural identification of classic amyloid fibrils.

Pododermatitis (bumblefoot) has been reported in many avian species, including waterfowl and shorebirds. The pathogenesis is not completely understood and may be multifactorial. Precipitating conditions include suboptimal husbandry and sanitary conditions involving rough, uneven, or steep walking surfaces. Gross lesions include individual or coalescing nodules of thickened and hyperkeratotic epidermis with fissures, erosions, and/or ulcers overlying chronic granulation tissue and inflammation with described "abscesses" on the plantar foot surface (Fig. 29.9). Early lesions include irregular papillary epidermal proliferations with dyskeratosis. Chronic lesions feature hydropic degeneration of keratinocytes, epidermal hyperplasia, dyskeratosis, dermal neovascularization, increased intercellular matrix, diffuse lymphohistiocytic to granulomatous inflammation, discrete granulomas, and superficial to deep bacterial colonies.[288] Unresolved infection with granulation tissue can progress to chronic tenosynovitis and osteomyelitis. Secondary bacterial infections with organisms, such as *Fusobacterium* sp. and *Staphylococcus aureus*,[53] are common and exacerbate tissue damage and inflammation. Chronic inflammation associated with pododermatitis has been linked to systemic amyloidosis.[169]

FIGURE 29.9 Pododermatitis (bumblefoot) in a Chinese merganser. Chronic proliferative nodular lesions on the plantar surface of the foot. Lesions are often associated with superficial erosion or ulceration that microscopically overlie acute to chronic inflammation admixed with fibrosis or granulation tissue (depending on chronicity) and superficial bacterial infection. *(Photo Courtesy of C. Hollinger, Wildlife Conservation Society.)*

Esophageal impaction is frequently reported in free-living Canada geese in North America.[171] Underlying etiologic factors have not been established in all cases, but lead intoxication may decrease gastrointestinal peristalsis and precipitate impaction. Other potential contributing factors include consumption of dry plant material (soybeans, cowpeas, green plants), discarded fishing line, excess feed, ventricular nematode infection (*Amidostomum* sp.), and mycotoxin exposure. Grossly, a doughy tubular mass, with or without grit, is found in the mid-cervical esophagus. Histological findings in cases of proventricular impaction associated with avian bornavirus infection can include mild histiocytic and lymphoplasmacytic infiltrates in splanchnic autonomic ganglia.[52]

Ingestion of **marine debris** (micro- and macroplastics) is known to frequently occur in a variety of sea bird species with anticipated negative impacts on overall health that are poorly understood. Birds with plastic ingestion can be in good nutritional condition. Gastrointestinal obstruction, ulceration of the ventriculus, and bioaccumulation have been reported.[7,82] Although not reported in birds, lesions consistent with hepatotoxicosis (glycogen depletion, fatty vacuolation, and single-cell necrosis) have been reported in experimentally exposed fish, which raise health concerns for birds exposed to high levels of plastic.[212] Ingestion of plastic debris is associated with persistent pollutants such as polychlorinated biphenyls (PCBs) that are known to impact reproduction and have immunosuppressive impacts on hosts.[291]

Traumatic injuries are common in free-ranging birds. In waterfowl, common findings include fractures of the limbs or hemocoelom. Causes of traumatic lesions vary and include predation, intraspecific and interspecific aggression, acute collision with the ground associated with migratory exhaustion or other underlying factors, and anthropogenic causes, such as being hit by vehicles or hunting. Collisions with electrical wires occur in birds with large wingspans (e.g., swans)[192] and may also result in electrocution. Grossly, singed feathers or skin associated with lacerations are seen at the points of contact. Birds are subject to nontarget bycatch and become entangled or entrapped in fishing lines and nets or in traps for furbearing mammals.[9] In these cases, gross lesions seen in drowning may include distension of the heart and major veins, watery fluid in the air sacs, and heavy, edematous, congested lungs.[237] Ingestion of anthropogenic materials, such as fishing hooks, tackle, and plastics, is reported widely in wild birds. In some cases, these have been associated with gastrointestinal tract perforation and serositis or coelomitis, but in others, the material may be an incidental finding. As with lead-associated material, chemicals from plastics can be absorbed into the bloodstream and are thought to be immunosuppressive.

Waterfowl adapted to saline environments have supraorbital salt glands and other physiological adaptations to remove excess sodium and chloride.[189] Despite these adaptations, **salt intoxication** can occur with water deprivation and high sodium intake. Mortality events associated with drought conditions have been reported in Australian shelducks[210] and in waterfowl on highly saline lakes and wetlands in California, North Dakota, Saskatchewan, and New Mexico when freshwater sources were frozen or unavailable.[163,273,279] Precipitation of salt on the birds, thought to result from rapid environmental temperature drops, weighs down the birds, impairs their mobility, and prevents them from accessing freshwater. Younger animals are more susceptible to salt intoxication and water deprivation. Hyperosmolarity (hypernatremia) results in compensatory responses in the brain, which causes an influx of sodium, potassium, and chloride ions that results in brain shrinkage followed by an influx of organic osmolytes (amino acids, polyols, and methylamines). Upon rehydration, the brain swells as it takes on water to compensate for the increased osmolality. This causes neuronal damage.[294] Affected birds have limb paralysis and subcutaneous edema. In birds that succumb to the effects of severe dehydration without a rehydration phase, the cause of death is likely multiorgan failure due to decreased perfusion and hyperosmolality. Compounds with cholinergic activity, such as OPs and botulinum toxin, can impact salt metabolism and may predispose birds to salt toxicosis.

Gross lesions in salt intoxication include emaciation, external encrusting with salt, conjunctivitis, cataracts, dehydration with acute urate deposition (gout), mucoid GI content, sloughing of koilin from the ventriculus, multiorgan congestion (especially of the meninges), and small salt glands. When birds adapt to saline conditions, the glands hypertrophy. Microscopic lesions can be absent or include hemorrhage, congestion, encephalomalacia, glomerular capillary hyalinization, hepatic vacuolation, bile stasis, skeletal and cardiac muscle degeneration, and mineralization associated with mononuclear infiltrates, mononuclear conjunctivitis, and lens fiber necrosis with cataract formation. Renal tubular degeneration with cast formation may be present, indicating a state of dehydration. Diagnosis is suspected based on an appropriate history and gross lesions. The presumptive diagnosis can be supported by measurement of the sodium content in the brain. Reference intervals for each avian species are not available, but levels >1900–2000 ppm of sodium (wet weight) have been associated with salt toxicosis in free-ranging wild birds.[86] Comparison to presumably healthy control birds may be necessary.

Starvation can occur on an individual basis or can be associated with larger-scale mortality events, for example with certain weather events that prevent access to food.[280] Any factors that inhibit gastrointestinal motility or cause obstruction (e.g., esophageal impactions or ingested foreign bodies) can also result in mortality from starvation. In general, waterfowl that have lost approximately 45% of their body weight can die from starvation. As a differential, it is worth noting that some species (e.g., lesser snow geese) can undergo normal, physiological decreases of up to 42% of their body weight in association with reproduction.[5] Gross lesions of starvation include the complete absence of subcutaneous and visceral fat stores and marked pectoral muscle atrophy. Black, tarry material is often present along the mucosa of the gastrointestinal tract. The liver may be shrunken and the gall bladder distended. Serous atrophy of fat may be present at the heart base as well as in the bone marrow and is characterized by adipocytes with shrunken, pale, basophilic cytoplasm that are occasionally associated with lymphoplasmacytic and histiocytic inflammation. Atrophic changes may be observed in skeletal muscle, pancreas, and liver. A diagnosis of starvation is often based on a combination of factors including the history, clinical signs, poor nutritional condition, and consistent environmental factors; however, quantitative medullary bone marrow fat content <1% can be confirmatory. The ulna is the bone of choice for sampling.[211] This technique can be used on autolyzed and scavenged carcasses.

Adverse weather conditions, such as hailstorms[131] and heatwaves,[197] have been associated with mortality in large numbers of birds. Birds may die from hypo- or hyperthermia and often are emaciated. Changing climatic conditions alter availability of food and are associated with increased incidence of HABs. Although it is difficult to quantify, **habitat loss** worldwide likely contributes to overall morbidity and mortality in wild birds. In some cases, birds have had to adapt to shrinking available habitat by aggregating in large numbers, which reduces food availability and increases the potential transmission of density-dependent diseases.

Neoplastic

Reports of neoplasia in both free-ranging and captive water birds are relatively rare, typically representing <1% of diagnostic cases. These include **mesenchymal neoplasms (neurofibrosarcoma, liposarcoma** (Fig. 29.10), **fibromatosis, spindle cell sarcoma)** in captive ducks[11] and free-ranging ducks,[29] geese[47] and gulls (unpublished data Canadian Wildlife Health Cooperative, Atlantic region); **lymphoma** in ducks, swan, and geese; **leukemia** in a

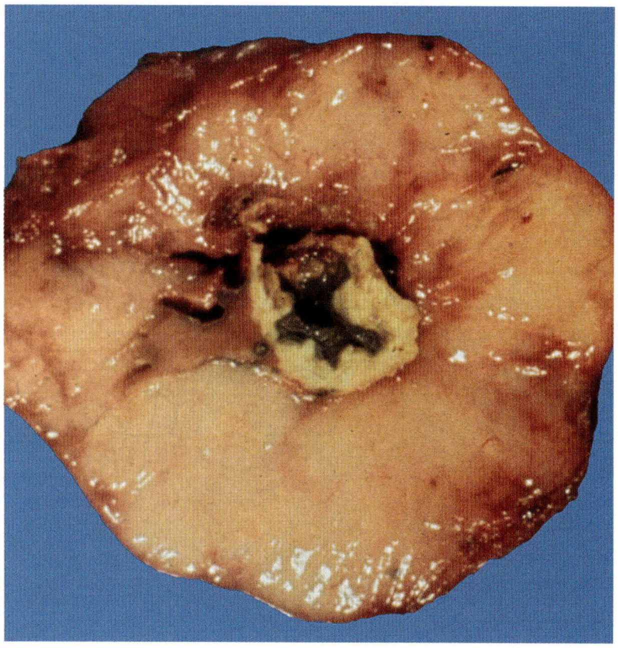

FIGURE 29.10 Liposarcoma in the intestine of a white-fronted goose. The infiltrative tumor is light tan and markedly expands and effaces the normal architecture of the intestinal wall. The lumen is filled with necrotic tissue debris. *(Photo Courtesy of P-Y. Daoust, Canadian Wildlife Health Cooperative, Atlantic region and F. Leighton, University of Saskatchewan, Western College of Veterinary Medicine.)*

moorhen.[56] **Testicular tumors,** the most common of which is the **seminoma,** have been reported in both captive and free-ranging ducks.[172] Seminomas are typically unilateral, encapsulated, homogenous, firm masses at the cranial pole of the kidney that bulge upon cross-section. Other reported germ cell tumors include **sex cord-stromal tumors** and **teratomas.**[144] A novel papillomavirus has been associated with spindle cell carcinoma in at least one case.[219] **Hepatocellular carcinomas** in domestic ducks in China are associated with both aflatoxin exposure and infection with the hepadnavirus hepatitis B virus (DHV).[57] The neoplasms present as multifocal to coalescing, raised hepatic nodules with occasional regions of umbilication. Metastasis can be observed. Microscopic lesions in ducks can vary from well-differentiated trabecular epithelial cells to markedly anaplastic cells with marked polymorphism.

INFECTIOUS DISEASES

DNA Viruses

Duck viral enteritis (DVE), also known as **duck plague (DP),** is caused by the alpha herpes virus **anatid herpesvirus 1** and is an acute, highly contagious disease of ducks, geese, and swans of all ages. It is a World Organisation for Animal Health (WOAH)-listed notifiable disease. Disease is characterized by sudden death and high mortality. Although disease is widespread in distribution and frequent in captive waterfowl, it is infrequently seen in wild waterfowl. Members of the family Anatidae are natural hosts, and there are species-specific differences in susceptibility with Muscovy ducks among the most susceptible.[26] Gross lesions include petechial and ecchymotic hemorrhages in the heart and mesentery, multifocal necrosis in the gastrointestinal tract and lymphoid tissue, and hepatomegaly with multifocal necrosis.[95] Gastrointestinal lesions, which occur anywhere from the oral cavity to cloaca, start as hemorrhage, progress to elevated crusted plaques, and eventually coalesce into large, ulcerated areas covered by diphtheritic membranes. Hemorrhage and necrosis of the gut-associated lymphoid tissue are characteristic and classically seen as hemorrhagic annular band necrosis in ducks and button-like ulcers in geese (corresponds to arrangement of intestinal lymphoid tissue) (Fig. 29.11A). These changes, along with pseudomembranous plaques at the esophageal-proventricular junction and cloaca, are almost pathognomonic for the disease.[277] Changes in lymphoid organs include a small spleen, petechiation, and atrophy of the thymus, as well as hemorrhage in the bursa of Fabricius. Lumina of the intestines and ventriculus are often filled with blood. The liver and pancreas may contain pinpoint hemorrhages mixed with white necrotic foci. Mature hens may have hemorrhages in deformed and discolored ovarian follicles as well as ruptured yolk and free blood in the abdominal cavity. Microscopic lesions include necrosis and hemorrhage in affected organs (Fig. 29.11B; eSlides 29.e4 and 29.e5). Eosinophilic, intranuclear, and sometimes cytoplasmic, inclusion bodies are seen in epithelial cells of the gastrointestinal tract (in some cases, they are most notable in epithelial cells at the esophageal/proventricular junction) (Fig. 29.11B; eSlide 29.e4), hepatocytes, lymphoid tissue, conjunctiva, and Harderian gland, and can occasionally be seen in bile duct epithelium.[95] Final diagnosis is made by virus isolation or PCR with liver being the sample of choice.

Gruid Herpesvirus 1 (GHv1) causes **inclusion body disease of cranes,** an acutely fatal disease of captive cranes characterized by hepatitis.[54] Many *Grus* spp., are naturally susceptible to the virus, and Pekin ducks and coots have been experimentally infected. Affected birds tend to separate from the flock, are lethargic and anorexic, and may have diarrhea prior to death. Grossly, birds usually have abundant fat stores; hepatosplenomegaly with pinpoint yellowish-white necrotic foci on the surface and throughout the parenchyma; intestinal and thymic hemorrhages; gastrointestinal erosions and ulceration; renomegaly with or without pinpoint necrotic foci; and miliary necrotic foci in other tissues and organs, such as bone marrow and thyroid glands.[229] Microscopically, the liver, spleen, gastrointestinal tract, and thymus contain foci of necrosis with

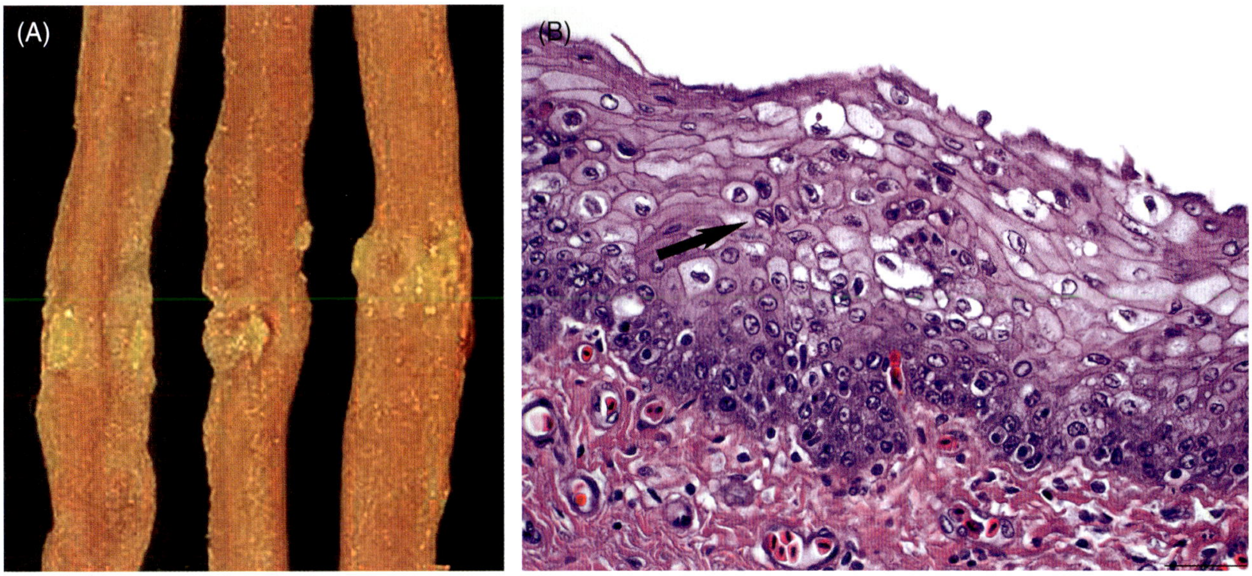

FIGURE 29.11 Duck plague (duck viral enteritis) caused by anatid herpesvirus-1 in a Canada goose (A) and (B) Muscovy duck. (A) Note the classic annular band necrosis in the gastrointestinal tract. (B) Numerous eosinophilic to amphophilic intranuclear viral inclusions (*arrow*) are present in infected epithelial cells; similar inclusions were present in subjacent esophageal glands (see eSlide 29.e4). *(Part A: Photo Courtesy of the Canadian Wildlife Health Cooperative, University of Saskatchewan; Part B: Photo Courtesy of D. McAloose, Wildlife Conservation Society.)*

heterophilic to mononuclear inflammatory cell infiltrates and intranuclear, eosinophilic inclusion bodies. Confirmation is made by virus isolation or PCR.

Avian pox, caused by viruses of the genus *Avipoxvirus*, has been reported in free-ranging species from the orders in this chapter. Characteristic gross lesions include cutaneous nodules on featherless skin (feet, legs, and face) (Fig. 29.12); diphtheritic oral lesions are occasionally seen. Microscopic lesions consist of epithelial cell hyperplasia and ballooning degeneration with large, eosinophilic, intracytoplasmic inclusion bodies (Bollinger bodies). Lesions may be traumatized, resulting in secondary bacterial and mycotic infections that typically do not spread systemically.[262] Microscopic lesions are diagnostic; electron microscopy, PCR, and virus isolation are confirmatory.

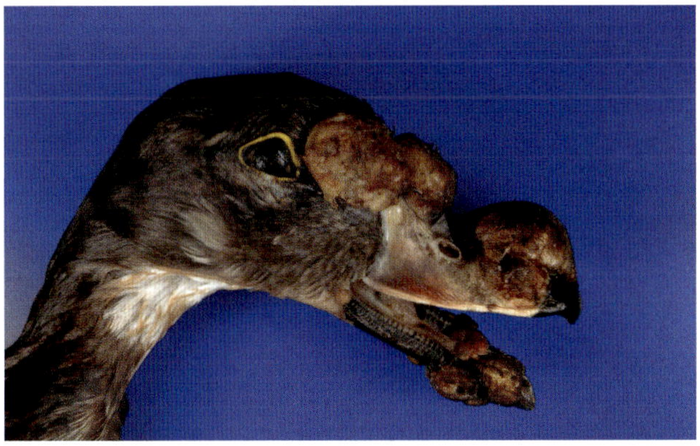

FIGURE 29.12 Avian pox virus infection in a wood duck. Gross lesions consist of multiple, crusty, and proliferative nodular lesions on the non-feathered skin of the head. In this case, they have caused considerable deformation of the bill. *(Photo Courtesy of M. K. Keel, Southeastern Cooperative Wildlife Disease Study.)*

Circovirus-like infections have been identified in several species of gulls including kelp gulls and ring-billed gulls. Bursal lesions (e.g., swelling and red mottling), emaciation, and aspergillosis are consistent opportunistic infections in circovirus-infected birds. Large intracytoplasmic, basophilic, botryoid inclusion bodies in macrophages and lymphocytes of the bursal follicles are a characteristic histologic finding. Care must be taken to not misinterpret the inclusions for phagocytized material. Bursal and splenic lymphoid depletion are also common. Ultrastructurally, inclusion bodies contain viral particles typical of circovirus. An underlying circovirus infection should be suspected in die-offs of gulls, particularly when juvenile birds are involved.[184] Molecular testing is confirmatory but not yet widely available.

A novel **adenovirus** isolated from common eiders caused a mortality event in the Baltic Sea near the coast of Finland in 1998.[104] Another adenovirus was associated with a mortality event affecting terns and gulls.[121] The most consistent lesions in these birds were impaction and mucosal necrosis of the small intestine.

Two viruses, **anseriform dependovirus 1 (syn. goose parvovirus; Derzsy's disease virus)** of domestic geese and Muscovy ducks[83] and **goose hemorrhagic polyomavirus,**[141] are important viral diseases of young birds. Goose parvovirus events can have hepatitis, enteritis and/or myocardial degeneration with fatty infiltration and intranuclear inclusion bodies. Cases of polyomavirus-associated mortality have hemorrhagic and necrotic enteritis, necrotizing nephritis and gout. The species range of susceptible hosts to these two viruses is unknown.

RNA Viruses

Wild waterfowl and shorebirds are frequently infected with many different types of **avian influenza A viruses;** they are considered reservoir hosts.[241] In general, infections with low pathogenic avian influenza viruses (LPAI) in wild waterfowl and shorebirds are subclinical and infected birds have no clinical signs. However, certain strains of highly pathogenic avian influenza (HPAI), notably recent H5 strains of the virus clade 2.3.4.4b, have caused mortality and outbreaks in over 100 species of wild swans, ducks, geese, terns, gulls, skuas, guillemots, shorebirds, coots, cranes, and storks. **Highly pathogenic avian influenza viruses** are often notifiable at state, provincial, and national levels, in addition to being a WOAH-listed notifiable disease. Clinical signs of HPAI in birds can be inapparent or include neurological signs, such as tremors, torticollis, opisthotonus, and depression. There are no pathognomonic gross lesions associated with LPAI or HPAI. Often birds are in good nutritional condition. Grossly, multifocal to coalescing hemorrhage with necrosis in the pancreas (Fig. 29.13A), severely congested lungs with edema, subepicardial and intramyocardial hemorrhages, and hepatosplenomegaly may be seen.[248] Microscopically, the predominant lesions are in brain, pancreas, heart, and liver and include severe multifocal lymphoplasmacytic encephalitis, neuronal necrosis, neuronophagia, and status spongiosis in the brain, multifocal necrosis in the pancreas and heart, and foci of necrosis with inflammation in the liver. Adrenal or lymphoid necrosis, pulmonary edema and hemorrhage, vasculitis, and myocardial hemorrhage may also be seen. HPAI viruses generally are neurotropic in wild birds. However, increasing frequency of lesions and presence of viral antigen in the intestinal tract of wild birds suggests that HPAI H5 viruses are becoming more enterotropic, thus becoming more adapted to wild birds.[32] The diagnosis must be confirmed by virus isolation or demonstrating viral nucleic acid or antigen within tissues. Cloacal and/or choanal swabs are recommended for qPCR testing. Brain, pancreas, heart, and liver are good samples for immunohistochemical confirmation ([28]; Fig. 29.13B).

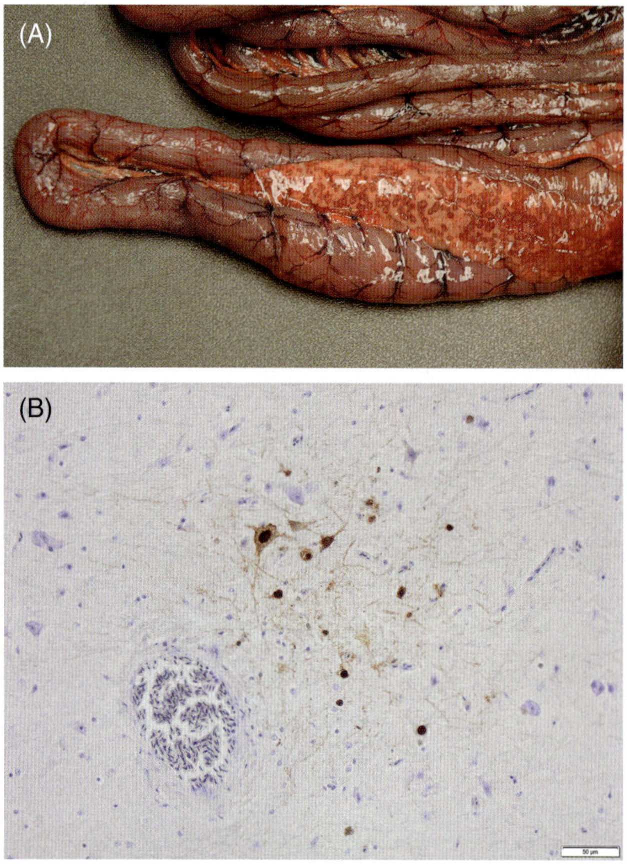

FIGURE 29.13 Lesions due to highly pathogenic avian influenza virus infection. (A) Gross image of the duodenum and pancreas of a whooper swan demonstrating multifocal to coalescing foci of centralized necrosis and hemorrhage located randomly throughout the parenchyma. (B) Immunohistochemical image of brain from a barnacle goose with intracytoplasmic and membranous positive immunohistochemical staining for Influenza A antigen within neurons, glial cells, and astrocytes that highlights dendritic and axonal processes. *(Photos Courtesy of the Swedish Veterinary Agency.)*

Newcastle disease (ND), also a WOAH-listed, notifiable disease, is caused by **avian paramyxovirus 1 (APMv1)**. The virus belongs to the genus *Avulavirus*, family *Paramyxoviridae*.[111] ND is caused by infections with ND virus (NDV) classified as virulent (vNDV). NDV of low virulence (loNDV) does not cause ND, although it can cause respiratory disease. NDV infects many species of birds, including ducks and geese; most infections are asymptomatic. However, frequent outbreaks of ND in domestic geese and several outbreaks in Muscovy ducks have been reported in China. Clinical disease is characterized by diarrhea, nervous signs, and mortality.[150,234] Gross and microscopic changes include hemorrhage and necrosis in the gastrointestinal mucosa, lymphoid organs, and pancreas with variable neuronal degeneration. Large-scale illness in wild birds has occurred only in double-crested cormorants and epornitics of ND in cormorants have occurred in Canada and in the Great Lakes region and north-central United States. Viral spread to poultry and mortality in pelicans, gulls, and terns has been reported in association with cormorant mortality.[107,137,281] Juvenile cormorants are typically affected and have nervous system signs, such as leg and/or wing paralysis, nystagmus, and circling. Gross lesions can include conjunctivitis, periocular edema, and erythema. Microscopically, characteristic

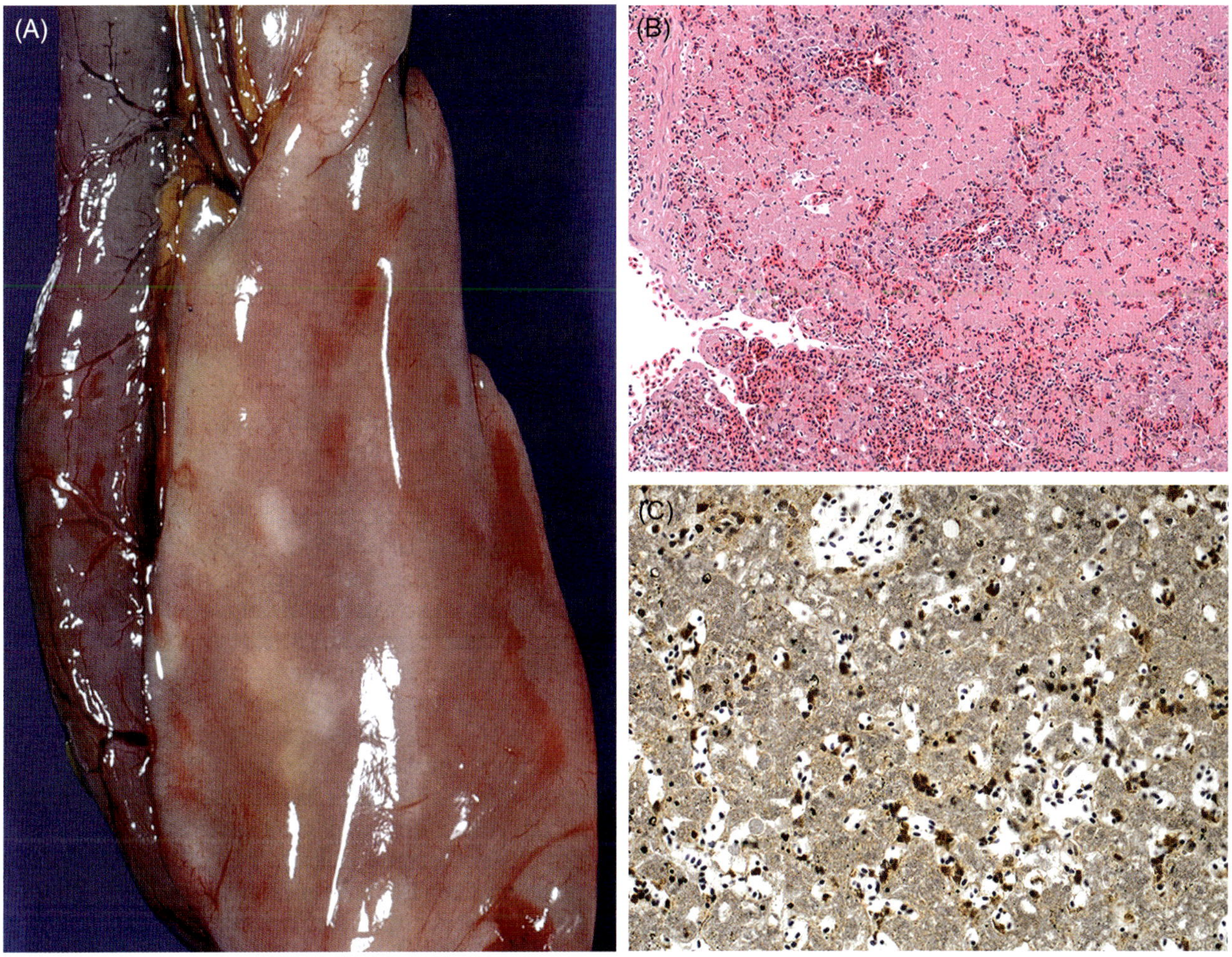

FIGURE 29.14 Wellfleet Bay virus infection in the pancreas (A) and liver (B, C) of a common eider. (A, B) Infection is associated with multifocal areas of coagulation necrosis and hemorrhage. (C) The virus can be highlighted with immunohistochemical labeling, here shown as intracytoplasmic and nuclear positivity in hepatocytes, Kupffer cells and leukocytes in regions of necrosis. *(Photos Courtesy of E. Howerth, J. Ballard and C. Bahnson, Southeastern Cooperative Wildlife Disease Study and the University of Georgia.)*

findings in the brain and spinal cord of cormorants include neuronal necrosis, perivascular infiltrations of mononuclear cells, and/or gliosis. Microscopically, gulls have mild, non-suppurative encephalitis consisting of small accumulations of lymphocytes around one or a few small vessels.[136,138] Immunohistochemistry is useful for detecting virus in Newcastle disease, especially in brain, spinal cord, and kidney, and infection with Newcastle disease virus or other paramyxoviruses can be confirmed by virus isolation and/or molecular testing. These viruses also cause hemagglutination when isolated in eggs.

Wellfleet Bay virus (WFBv), a member of the *Orthomyxoviridae* family, is associated with recurrent mortality in common eiders in Massachusetts (United States). This novel RNA virus has been classified within the *Quaranjavirus* genus.[2] The route of transmission is unknown, but a tick vector is suspected. Gross lesions include foci of pallor in the liver and pancreas, as well as splenomegaly (Fig. 29.14A). Characteristic microscopic lesions include hepatic, pancreatic, and splenic necrosis (Fig. 29.14B). A combination of histologic examination, virus isolation, nested, reverse-transcription polymerase chain reaction (RT-PCR), and serology is helpful for diagnosing WFBV. Immunohistochemical labeling may be useful for characterizing lesions (Fig 29.14C).

Aquatic bird bornavirus 1 (ABBv1) causes lymphoplasmacytic inflammation in the central, peripheral, and autonomic nervous systems of free-ranging and captive Canada geese and trumpeter swans.[52] Clinically, birds are somnolent, weak, lethargic, and may exhibit lameness, ataxia, hypermetria, torticollis, head tremors, stargazing, opisthotonos, and/or suspected blindness. Esophageal impaction is occasionally observed. Histologically, gliosis and lymphoplasmacytic perivascular cuffs in brain, spinal cord, peripheral nerves, and myenteric ganglia or nerves are reported. Lymphoplasmacytic infiltrates can also be found in the adrenal glands as serosal perivascular, perineural, and periganglional infiltrates or infiltrates throughout the muscularis of the gastrointestinal tract. Histologic lesions are suggestive of avian bornavirus infection but detection of viral antigen or nucleic acid by immunohistochemistry or RT-PCR, respectively, is needed for confirmation; brain is the diagnostic tissue of choice.

Eastern equine encephalitis virus (EEEV) is a zoonotic alpha virus in the family *Togaviridae*. It is an arbovirus transmitted by mosquitoes. EEEV can cause mortality in free-ranging Ciconiiformes (e.g., great egret) and certain species of Gruiformes (e.g., whooping cranes).[51,87] Gross findings may include multifocal myocardial pallor, pericardial effusion, ascites, or hepatosplenomegaly. Histologic lesions include necrotizing inflammation in liver, spleen, and lung. Lesions are suggestive of the disease, but detection of viral antigen or nucleic acid by immunohistochemistry or RT-PCR, respectively, is needed for confirmation. In the United States, the virus is considered a select agent and infections are reportable to the United States Department of Agriculture.

Although anseriform, gruiform, charadriiform, and ciconiiform birds are not thought to be particularly susceptible to **West Nile virus (WNV),** mortality due to the virus has been seen in a few species in each order in both captive and free-ranging birds. Hemorrhage, myocardial necrosis, and splenomegaly may be seen grossly. Pathology varies among different bird species but histologic lesions in most include degeneration and necrosis in the heart and lymphoplasmacytic inflammation in a variety of organs including the brain and spinal cord, heart, kidney, spleen, and liver.[77,100,231,249] Diagnosis is confirmed by immunohistochemistry, virus isolation, or RT-PCR.

The **Usutu virus (USUV)**, a type of Flavivirus, has been associated with mortality events in common scoters.[15] Hepato- and splenomegaly are seen grossly and necrotizing, lymphoplasmacytic hepatitis is seen microscopically.

The two duck ***avihepatoviruses*** (Family *Picornaviridae*) that have been described are **duck hepatitis virus 2 (DHV2)** and **duck hepatitis virus 3 (DHV3)**. DHV2 causes fatal hepatitis in domestic ducks.[134,149] Affected ducks have swollen kidneys and splenomegaly, as well as multifocal hemorrhage in the liver and sometimes other organs. Histologically, there is widespread necrosis of hepatocytes with minimal to marked perivascular and periportal infiltrations of mononuclear cells and heterophils. There also may be hemorrhage and biliary hyperplasia, but no inclusions are observed.[88] DHV3 has been associated with enteritis in ducks.[149] Astroviruses also have been associated with stunting and prehatch mortality in domestic ducks and geese.[19]

Woodcock reovirus has been isolated from and associated with die-offs of free-ranging American woodcock.[55] Birds are often found dead and emaciated but have no other gross lesions. Histopathology has not been described and diagnosis is by virus isolation. A mortality event in common eider ducklings in Finland in 1996 was also due to a **reovirus**. Poor body condition, hemorrhage in the liver, and necrosis of the liver and bursa of Fabricius were common findings.[103]

Bacteria

Fowl cholera (avian cholera; avian pasteurellosis), due to the gram-negative bacterium *Pasteurella multocida,* is a common and important disease of both captive and free-ranging birds. It has been described in various species in this chapter, occasionally in association with large disease outbreaks.[222] Acute and chronic disease can occur, but in wild waterfowl the disease usually presents as peracute septicemia with massive numbers of bacteria in blood collected shortly before death. A more chronic form of disease, characterized by wasting, may be seen in captive waterfowl. In the peracute phase, birds are in good condition and scattered petechial hemorrhage may be seen. With time, small foci of necrosis and petechiae may be seen in the hepatic (Fig. 29.15A) and splenic parenchyma, and there may be pulmonary consolidation. The intestinal tract may be empty or contain clear gelatinous material with large numbers of bacteria. The lesions can be similar to those of duck plague/DVE. Microscopic findings are not specific, but intralesional gram-negative coccobacilli are present (Fig. 29.15B). The organism can be isolated for definitive diagnosis; heart blood, liver, and bone marrow are the most suitable tissues to culture.

***Mergibacter septicus* (Bisgaard taxon 40),** a recently described bacteria in the Pasteurellaceae family, causes septicemic disease.[49] Mortality events have been reported in common and sandwich terns[174] and rhinoceros auklets.[132] Infection was suggested to have been precipitated by immunosuppression caused by brevetoxin exposure in the terns, as well as emaciation and poor body condition in the auklets. Gross lesions reported in the auklets included dark red, firm pneumonic lungs, and fibrinonecrotic pleuritis. Histological findings included multiorgan lymphoplasmacytic and heterophilic inflammation and necrosis associated with small gram-negative bacilli. The organism can be isolated for definitive diagnosis; lung, liver, and heart are suitable tissues to culture.

Avian mycobacteriosis is most commonly caused by the acid-fast bacteria *Mycobacterium avium* or *Mycobacterium genavense*, but other environment mycobacteria have been documented.[195] The condition is primarily a disease of captive birds. Gross lesions in infected birds include emaciation and solid to caseous nodules or inflammatory infiltrates in organs and tissues (Fig. 29.16A). The liver is most often affected in wild waterfowl, but spleen, lung, and intestine may contain

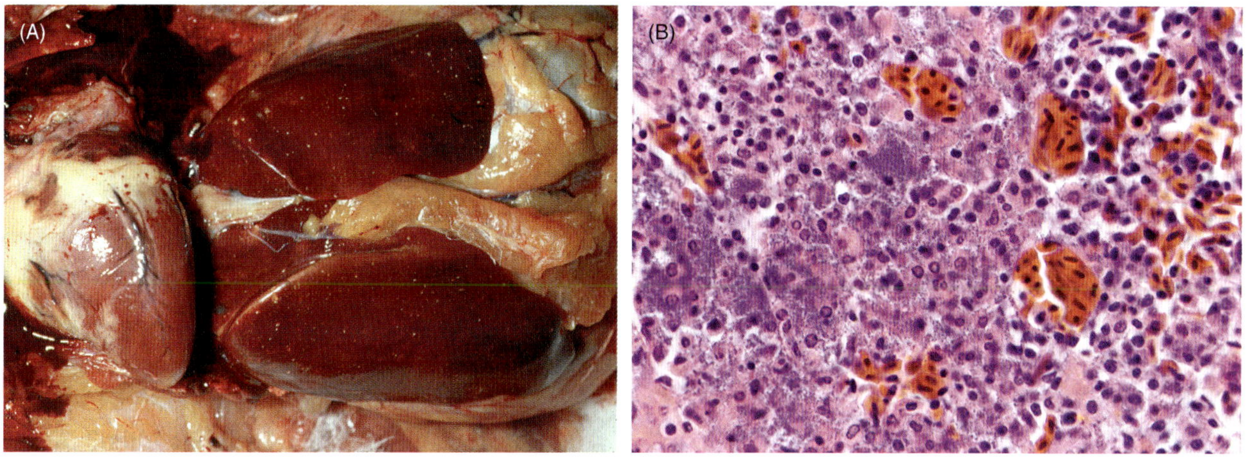

FIGURE 29.15 Avian cholera (avian pasteurellosis) in a snow goose. (A) Numerous, pinpoint white foci are present throughout the liver. (B) Myriad, intralesional coccobacilli are associated with fibrin deposits (pale pink amorphous material) and tissue necrosis. *(Photo Courtesy of the United States Geological Service, National Wildlife Health Center.)*

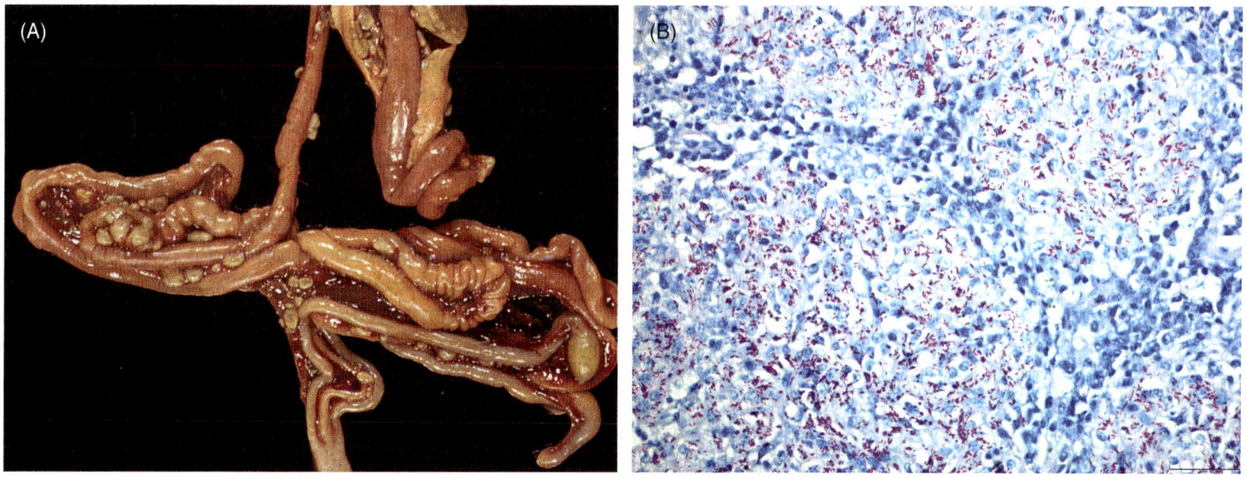

FIGURE 29.16 Avian mycobacteriosis in a duck (unspecified species) (A) and an Indian whistling duck (B). (A) Multiple, variably sized nodular granulomas and granulomatous inflammation expand the serosa along the intestinal tract. (B) Splenic parenchyma is replaced by multiple granulomas containing myriad intrahistiocytic, acid-fast positive bacilli. Ziehl-Neelsen. *(Part A: Photo Courtesy of the Canadian Wildlife Health Cooperative, University of Saskatchewan; Part B: Photo Courtesy of D. McAloose, Wildlife Conservation Society.)*

similar nodules or inflammation. Alternatively, there may be diffuse hepatosplenomegaly without nodule formation. Affected organs may be friable, due to secondary amyloid deposition.[216] Scattered foci of granulomatous inflammation with epithelioid macrophages and lymphocytes or heterophils, sheets of large macrophages with foamy cytoplasm, or well-defined caseous granulomas in advanced stages, are typical microscopic findings. Acid-fast bacilli, which are often numerous, are present within macrophages (Fig. 29.16B) or free among the necrotic debris and may be visible as negatively stained bacilli histologically or in cytologic preparations. Differentials include aspergillosis or granulomas due to infection with other bacteria. A tentative diagnosis can be made based on the presence of compatible gross and microscopic lesions coupled with the presence of intracellular acid-fast bacilli on cytology or histology. However, diagnosis should be confirmed by bacterial isolation (this is still the gold standard) or by PCR on fresh, fresh frozen, or formalin-fixed, paraffin-embedded tissue.

Die-offs due to **salmonellosis** in captive and free-ranging Anseriformes, Ciconiiformes, Charadriiformes, and Gruiformes (e.g., mute swans, ducks, American coots, gulls, and cattle egrets) are most often due to ***Salmonella typhimurium***.[46] In acute infections, gross lesions may be absent or there may be hepatosplenomegaly with tiny hemorrhagic or white necrotic foci in the parenchyma. As the disease progresses, small tan-white "paratyphoid nodules" (foci of lytic necrosis surrounded by mononuclear cells and macrophages) may develop in the liver, spleen, pectoral muscles, subcutis, and brain. Fibrinous to fibrinopurulent exudate may cover the pericardium, peritoneum, and air sacs. Hypopyon may be seen. Delayed yolk sac resorption may occur in infected young birds and contents of the yolk sac may have a creamy or caseous texture. Enteritis is more common in adult birds and in early infection, it is characterized by reddening and ulceration of the mucosa of the posterior small intestine, cecum, and colon. This may progress to the development of pseudomembranes and cecal cores. Enlargement and impaction of the rectum and arthritis may also occur. Microscopically, necrosis and inflammation with abundant gram-negative bacilli are characteristic. In the early stages of disease, inflammation is composed of fibrin and heterophils with infiltration of histiocytes, multinucleated giant cells, and lymphocytes as the disease progresses. Gross and microscopic lesions are similar to other bacterial diseases, so diagnosis must be confirmed by isolation and identification of the bacteria. **Pullorum disease** due to ***S. pullorum*** is less common.

Riemerella anatipestifer is a gram-negative bacillus that causes septicemic disease. Disease has been reported in various species of captive and free-ranging ducks, geese, and swans, and it has been responsible for die-offs in free-ranging black and whistling swans. The most prominent gross lesions are exudative and fibrinous polyserositis. Fibrinous sinusitis, tracheitis, bronchopneumonia, and salpingitis; splenomegaly, serosal petechiae, and swollen joints may also be seen. Microscopically, lesions consist of serosal fibrinous exudates with variable numbers and types of inflammatory cells. Myocarditis, bronchopneumonia, hepatitis, glomerulitis, fibrinous meningoencephalitis, and vascular thrombosis have been reported. Bacteria may be seen in large numbers, both free and within macrophages. Gross and microscopic changes, as well as the presence of gram-negative bacilli, are suggestive of *R. anatipestifer* infection, but confirmation via bacterial isolation is required. The organism may be difficult to isolate, so several organs, such as heart blood, brain, liver, spleen, kidney, lung, and air sacs, should be submitted; heart blood and brain are considered optimal samples.

Erysipelas, caused by the gram-positive bacillus ***Erysipelothrix rhusiopathiae,*** occurs sporadically in captive and free-ranging waterfowl. It has been described in a wide variety of birds including storks, cranes, coots, spoonbills, oystercatchers, and gulls.[283] Although rare, large-scale mortality events due to erysipelas have occurred in free-ranging birds, predominantly of other orders. Grossly, disease is characterized by subtle lesions suggestive of septicemia. These include generalized congestion, petechial and ecchymotic hemorrhage, and a swollen liver and spleen. Microscopical lesions may be subtle and contain intravascular clumps of gram-positive bacilli. Fibrin thrombi in small pulmonary vessels, glomeruli, and hepatic sinusoids may be accompanied by acute parenchymal degeneration and necrosis. Diagnosis depends on isolation or detection of the bacterium. Liver, spleen, heart, and kidney are the most appropriate organs for bacterial culture. Fluorescent antibody testing and immunohistochemistry can be used for antigen detection.

Mycoplasma anatis is the most commonly isolated species of *Mycoplasma* in Anseriformes and Gruiformes; associated disease has not been reported.[152] **Chlamydiosis** due to ***Chlamydia psittaci*** has been reported in free-ranging gulls, geese, ducks, herons, and terns. It has occasionally been associated with die-offs.[3] Grossly, infected birds are often thin and may have fibrinopurulent polyserositis, airsacculitis, and hepatosplenomegaly. Histopathologic findings have not been described for species in this chapter, but lesions are likely similar to those seen in other birds. These include basophilic, intracytoplasmic inclusions (reticulate bodies) in macrophages in the serosa, air sacs, liver, and spleen. Diagnosis of chlamydiosis can be made by fluorescent antibody testing, immunohistochemistry, PCR, or electron microscopy.

Fungi

The most common fungus isolated from cases of **aspergillosis** is ***Aspergillus fumigatus***, which has a ubiquitous environmental distribution worldwide. Factors associated with the development of disease include immunosuppression, malnutrition, concurrent disease, or other physiologic stressors, such as migration, oiling, and captivity. Additional host factors associated with disease include age, nutritional condition, or species susceptibility. Sea ducks, eiders, and trumpeter swans appear to be highly susceptible, although the reason is not known.[208] Outbreaks involving multiple birds at one time have been reported and likely result from environmental exposure to a common source (e.g., moldy crops) as the disease is not contagious and cannot be transferred bird to bird. Other environmental conditions that may affect susceptibility to disease include temperature, humidity, amount of dust particles, and any factors that could negatively impact the food supply.

Clinical signs of aspergillosis include weakness, inability to fly, or open-mouth breathing occasionally accompanied by rales, tachypnea, cyanosis, and dyspnea. Diarrhea, vomiting, and neurological signs are reported in later stages of disease and are associated with lesions in the affected system/s.[41] After inhalation of spores, the fungus grows in the lungs and air sacs, and because hyphae are angioinvasive they can enter the bloodstream and disseminate widely. Grossly, multiple granulomas are present in subacute stages of the disease (Fig. 29.17A), while thickening of the air sacs with raised, firm, flattened, white-green caseous plaques that resemble bread mold with a central

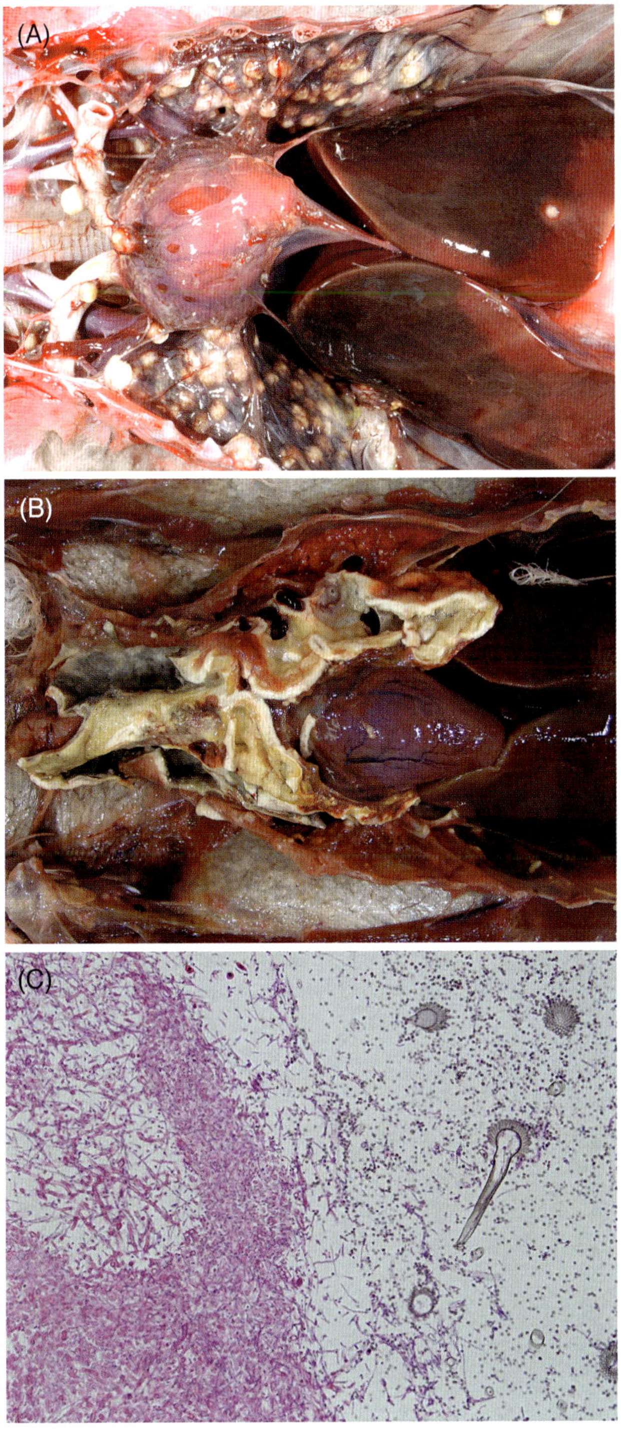

FIGURE 29.17 Aspergillosis in a Canada goose (A, C) and herring gull (B). (A) Numerous coalescing nodules are present throughout the lungs and air sacs. A single small, demarcated nodule is also present in the liver. (B) Marked thickening of the air sacs and a green velvety appearance along the inner surface are characteristic of fungal growth. (C) A thick mat of *Aspergillus fumigatus* fungal hyphae with conidiophore formation is present at the gas exchange surface. PAS. *(Part A: Photo Courtesy of M. Ruder, Southeastern Cooperative Wildlife Disease Study; Part B: Photo Courtesy of Canadian Wildlife Health Cooperative, Atlantic Region; Part C: Photo Courtesy of W. Yau, University of Georgia.)*

green, velvet appearance occurs in chronic stages (Fig. 29.17B). Affected birds can die from asphyxiation associated with blockage of airways with fibrinonecrotic material.[252] The typical microscopic appearance of the fungus is that of ascomycete with parallel, septate hyphae with dichotomous branching at 45° angles that will stain with both silver and Periodic acid-Schiff (PAS) stains. In some instances, large conidiophores can be observed, often at an air interface (Fig. 29.17C). Fungal hyphae may be associated with hemorrhage, fibrin, thrombosis, and necrosis if angioinvasion has occurred. Necrosis of pulmonary epithelium results in large accumulations of fibrinonecrotic debris within air spaces. In chronic cases, the hyphae are associated with caseonecrotic, fibrous, and granulomatous lesions (including numerous multinucleated giant cells), most prevalent at regions of gas exchange (e.g., luminal surface of air sacs). The gross and microscopic lesions with detection of characteristic fungal hyphae are almost pathognomonic. However, cytology, fungal culture, molecular testing, and immunohistochemistry can be performed for confirmation as infection with other bacterial or fungal pathogens (e.g., *Mycobacterium avium, Riemerella anatipestifer, Zygomycosis*) can produce similar lesions. Zygomycetes, such as ***Mucor* spp.,** have larger hyphae (5–10 μm), are aseptate, lack parallel walls, have an irregular appearance with bulging dilations, and branch at right rather than 45° angles.

Candidiasis, caused by the ubiquitous and opportunistic pathogen ***Candida albicans*,** has been reported rarely in waterfowl. It is usually associated with concurrent illness or traumatic injury. Typical gross lesions include ulcerative and proliferative plaques in the upper respiratory and gastrointestinal tracts. Microscopic lesions include necroulcerative lesions with intralesional yeast or pseudohyphae, fibrin, and necrotic debris potentially associated with opportunistic or concurrent bacterial infection. Candidiasis is diagnosed based on the microscopic appearance of ovoid yeast, hyphae, and pseudohyphae with routine hematoxylin and eosin (HE) and Grocott-methenamine silver (GMS) staining and can be confirmed with culture and molecular methods (Fig. 29.18).

Mortalities in a variety of ducks and great-blue herons from Western Canada have been reported in association with an intracellular, **yeast-like organisms** of an unknown species.[166] Disease appears to peak in autumn and be associated with respiratory distress. Although the species has not yet been identified, it may be in the *Saccharomycetales* family of yeast. Gross lesions include edematous, dark, congested and heavy lungs, hydropericardium, splenomegaly, and hepatomegaly.[206] Microscopic lesions include pneumonia, which is associated with large numbers of round to ovoid (0.5–1 × 1–4 μm), basophilic, intracellular, yeast-like organisms often found in aggregates in endothelial

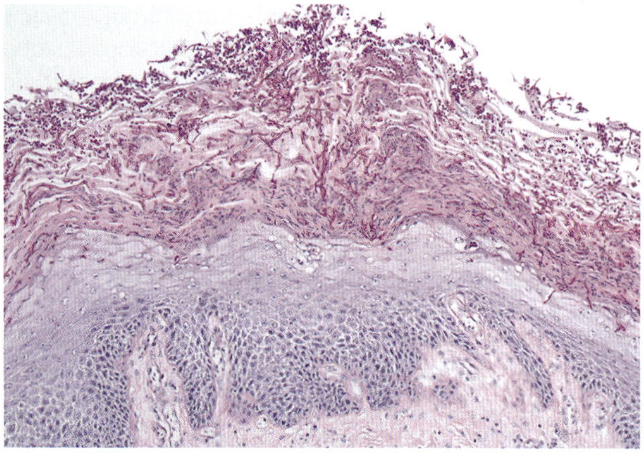

FIGURE 29.18 Candidiasis in the tongue of a red shoveler. The epithelium is hyperplastic with thick layers of parakeratotic and orthokeratotic hyperkeratosis infiltrated by yeast cells, elliptically contoured pseudohyphae, and septate hyphae with parallel walls. PAS. *(Photo Courtesy of E. Howerth, University of Georgia.)*

cells. Less commonly, hepatitis, splenitis, and myocarditis are reported. Organisms stain with Geimsa, PAS, and silver stains but are acid-fast negative. The characteristic ultrastructural appearance of the organism includes evidence of budding, a nucleus, a nuclear membrane, and the presence of mitochondria.

Although its phylogenetic status is currently controversial, there are reports of rhinitis and sinusitis associated with ***Rhinosporidium seeberi*** infection in geese, ducks, and swans.[127] Infection is most often associated with polypoid masses in the nasal cavity, conjunctiva, and eyelids.

Metazoa

In addition to the parasites described below, information related to several other notable metazoan parasites can be found in Table 29.e2. Among those in this table are: ***Amidostomum* spp, *Epomidiostomum spp.*, spirurid nematodes (*Streptocara, Echinuria, Synhimantus [Dyspharynx]*), *Capillaria* sp., tetrameres, *Trichostrongylys* sp., gapeworm (*Cyathostoma bronchialis*), *Avioserpensis spp.*,** and ***Diplotriaena* spp.**

Eustrongylidiosis, caused by infections with **nematodes** in the genus ***Eustrongylides*** (***E. tubifex*, *E. ignotus***, and ***E. excisus***), is common in fish-eating birds as the larvae of these nematodes utilize many fish species and oligochaete worms as intermediate hosts. Mortality can occur in individual cases and epizootics in wading birds; young birds are most susceptible.[244] Death is associated with coelomitis and bacterial sepsis from direct penetration of the nematodes through the gastrointestinal tract (Fig. 29.19A).[243] Fibrinous and fibrous channels on the serosal surfaces of the intestinal tract created by parasite migration (Fig. 29.19B; eSlide 29.e6) are associated with visceral and coelomic adhesions. Severe lesions affect the ventriculus in *E. ignotus* infections of Ciconiiformes, while *E. tubifex* infection tends to target the esophagus and proventriculus of mergansers. Dense fibrosis, multinucleated giant cells, eggs, cuticular remnants, and secondary bacterial colonies characterize chronic infections. Characteristic histologic features of the parasite include prominent lateral cords, a glandular esophagus, pseudomembranes, operculated eggs, and coelomyarian musculature

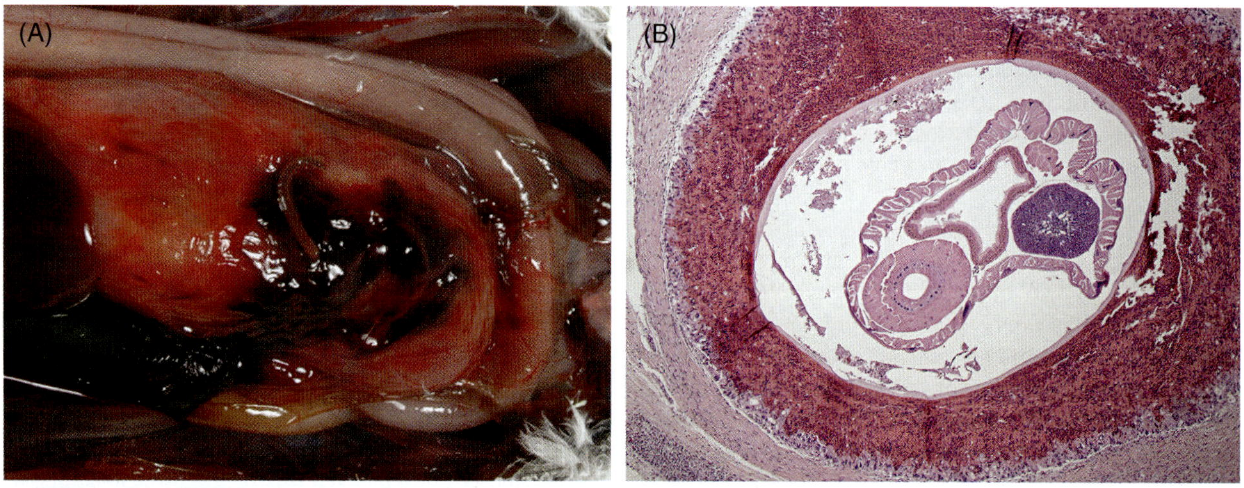

FIGURE 29.19 Eustrongylidiasis in a great egret. *Eustrongylides* spp. are thin red nematodes that create tunnels and channels on serosal surface of the gastrointestinal tract. (A) A slender red nematode is present on the serosal surface of the proventriculus where it is associated with regionally extensive hemorrhage. (B) A transverse section of the nematode shows it within a serosal channel surrounded by a thick rim of fibrinonecrotic and heterophilic debris and peripheral multinucleated giant cells and fibrous connective tissue. The parasite has a thick, smooth cuticle, coelomyarian/polymyarian musculature, a pseudocoelom, and gastrointestinal tract lined by tall columnar cells (see eSlide 29.e6). *(Part A: Photo Courtesy of M. K. Keel, Southeastern Cooperative Wildlife Disease Study.)*

(Fig. 29.19B; eSlide 29.e6).[78] Differentials include a penetrating foreign body or severe ascarid infection.

Ascarid nematodes have direct life cycles and are found within the host GI tract. ***Contracaecum* spp.** utilize crustacean intermediate and fish paratenic hosts. Heavy infections may be associated with poor nutritional condition, coelomitis, and anemia in juvenile birds.[64] Aberrant larval migrans of ascarid nematodes (e.g., ***Baylisascaris* spp.**) have rarely been reported in members of the *Anas* genus. Nematodes may be visible free in the gastrointestinal tract, and gross findings can vary from reddening of the mucosa to erosions, ulcers, or nodular inflammatory lesions. Infections with nematode species in aberrant hosts can be associated with proventricular rupture and coelomitis.[173] Microscopic lesions are minimal or associated with eosinophils, lymphocytes, plasma cells, and macrophages. Characteristic microscopic features of ascarids include lateral alae, lateral cords, coelomyarian/polymyarian musculature, and a digestive tract lined by uninucleate epithelium. In chronic lesions, fibrous tissue and inflammation surround degenerated nematode material referred to as hyaline caps.[136,138] Nematode identification is recommended in cases of rupture as it may indicate an aberrant host infection and potential expansion of geographical parasitic range.[245]

The life cycle of **filarial nematodes** involves a hematophagous invertebrate vector that ingests L1 (microfilariae) larva from the bloodstream. Microfilariae embed within peripheral vessels and are associated with subcutaneous inflammatory reactions. Adult nematodes are found in the heart, blood vessels, or venous sinuses where males and females mate (see Table 29.e2 for additional locations). Microfilariae often are considered incidental findings. Gross lesions typically are not observed, but cardiomegaly can result from significant infections in some species (e.g., *Splendidiofilaria* spp. in geese and swans).[76] In species that cause subcutaneous infections (e.g., *Pelecitus* spp.), edematous nodules can form in sites of infection involving joints and surrounding tendons.[140] Evidence of chronic inflammation may be present within arterial walls. Myocardial degeneration, myocarditis, and fibrosis can be seen in the heart.[62] In tissue sections, filarial parasites have coelomyarian musculature, small lateral cords, a small intestine, occasionally cuticular ridges, and thick-shelled eggs with microfilariae. In some cases, the commercial test kit for canine heartworm (*Dirofilaria immitis*) can be used to detect avian filarial infections.[60]

Trematodes are often considered host-adapted and in many bird groups in this chapter detection is considered an incidental finding. Those known to cause disease are summarized in Table 29.e2.

Adult **schistosome trematodes** live in the circulatory system or nasal sinuses and lay eggs that disseminate throughout the body via the bloodstream. These parasites are responsible for "swimmer's itch," a dermatitis associated with free-living cercaria embedding within the skin in humans. Eight genera infect Anseriformes (*Allobilharzia, Austrobilharzia, Bilharziella, Dendritobilharzia, Gigantobilharzia, Jilinobilharzia, Macrobilharzia*, and *Trichobilharzia*), three infect Ciconiiformes (*Austrobilharzia, Gigantobilharzia*, and *Trichobilharzia*), and three infect Charadriiformes (*Austrobilharzia, Gigantobilharzia*, and *Ornithobilharzia*).[109] Emaciation, dehydration, and pectoral muscle atrophy are reported in aberrant hosts. In some cases, thrombosis of the caudal mesenteric vein, fibrinohemorrhagic colitis, hepatomegaly, pulmonary congestion, gall bladder distension, and empty gastrointestinal tracts are reported.[282] Lymphoplasmacytic to granulomatous inflammation, as well as Splendore-Hoeppli-like material and mineral associated with eggs or adults, can be observed in affected organs. Adults and eggs can be found within the mesenteric veins or branches (Fig. 29.20). Lamina proprial edema, elongation of villi, and distention of crypts can be observed in the GI tract.[284] In some instances, proliferative vascular lesions and endophlebitis are observed. Unlike other trematodes, schistosomes are not hermaphrodites. Schistosome eggs do not have an operculum, but have terminal or lateral spines. Nasal swab cytology or Western blots can assist with antemortem detection of infections.[22,106]

Adult **cestodes** (**tapeworms**) are commonly found in the intestine and are often considered incidental findings. Notable cestodes of waterbirds are summarized in Table 29.e2. In a few cases, heavy cestode infections have been associated with malnutrition, mortality events,[114] diarrhea, hemorrhagic gastrointestinal content, and distention and occlusion of the intestine.[90] Histologic features

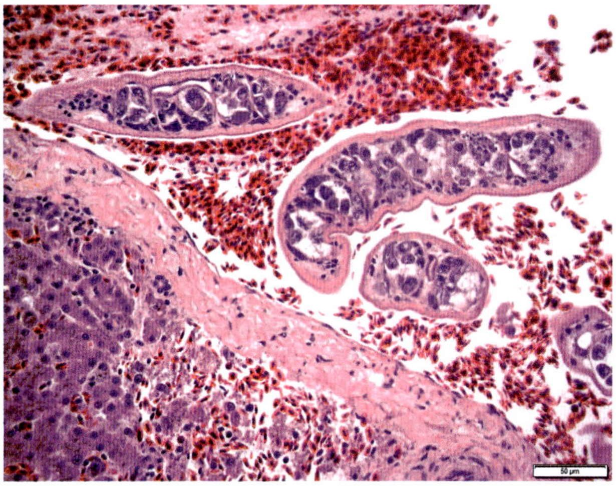

FIGURE 29.20 Adult schistosomes in an intrahepatic blood vessel of a blue-winged teal. The tegument of the parasites overlies a layer of muscle and subjacent parenchymal cells and sections of intestinal tract. Tegumentary papilla, a characteristic feature of adult schistosomes, is not apparent in this section.

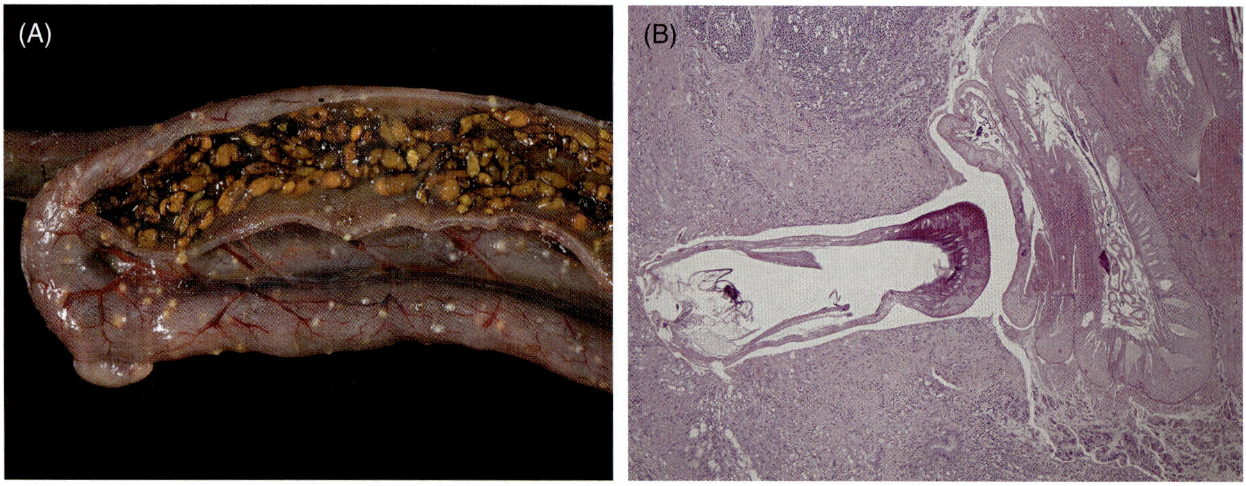

FIGURE 29.21 Acanthocephalans in a common eider. (A) The lumen of the small intestine is filled with plump, oblong acanthocephalan parasites. The small, demarcated white serosal nodules are related to mural damage caused by the parasites. (B) The proboscis is embedded in the submucosa (left in image) and associated with necrotizing transmural lymphoplasmacytic and granulomatous inflammation. Lacunar channels and the thick hypodermal layer of the acanthocephalan body are visible in the image. *(Part A: Photo Courtesy of M. K. Keel, Southeastern Cooperative Wildlife Disease Study.)*

include segmentation, lack of a gastrointestinal tract, a thin tegument, calcareous corpuscles, and a scolex.[78]

Acanthocephalans, known as the spiny-headed or thorny-headed worm due to the appearance of their proboscis, have no mouth or digestive tract and absorb nutrients from the host through their integument. Adult acanthocephalans are macroscopic and located almost exclusively within the small intestine of the definitive host[209] (Fig. 29.21A). These worms have an indirect life cycle that involves an arthropod intermediate host where larvae (cystacanths) develop.[225] Heavy infections have been associated with mortality[147,187] often with emaciation, fibrous gastrointestinal nodules, and serositis. The most common genera in Anseriformes are ***Corynosoma*** and ***Polymorphus***; in Charadriiformes are ***Arhythmorhynchus*** and ***Plagiorhynchus* spp.**; and in Ciconiiformes are ***Southwellina hispida***. For a complete list of acanthocephalan species, see Golvan.[85] Microscopic lesions at sites of attachment of the proboscis (Fig. 29.21B) include multifocal, transmural, granulomatous, ulcerative, and fibrosing enteritis and serositis.[23] Characteristic microscopic features of acanthocephalid parasites include prominent lacunar channels, a hypodermal layer that is thicker than the muscular layer, and a spined lemniscus that assists with eversion of the proboscis.

Protozoa

Disseminated visceral coccidiosis (DVC) is primarily a disease of captive, mostly juvenile, birds of *Grus* spp., including sandhill cranes, whooping cranes, and Eurasian, and Japanese white-naped cranes.[128,179] The disease is caused by either ***Eimeria reichenowi*** or ***E. gruis***. These are apicomplexan protozoal parasites. Birds may be found dead or lethargic with diarrhea and oocysts in the feces. Unlike many eimerian parasites which are found only in the gastrointestinal tract, DVC parasites can be found in numerous tissues and completion of their life cycle occurs in both the gastrointestinal and respiratory tracts. Distribution of the parasite includes the oral cavity, liver, lung, spleen, kidneys, heart, adventitia of blood vessels, gastrointestinal tract serosa and mucosa, muscles, and the subcutis. Infection often presents grossly as multifocal granulomas within and on the surfaces of affected tissues. Microscopically, lesions contain various developmental stages with asexual stages present in mononuclear cells and sexual stages in the gut and lungs. The parasites are associated with lymphohistiocytic and granulomatous inflammation as well as granuloma formation (Fig. 29.22).

Intestinal coccidiosis, caused by ***Eimeria***, ***Isospora***, ***Wenyonella***, or ***Tyzzeria* spp.**, is seen in all four orders, but is most frequent in Anseriformes; it is primarily a problem in domestic ducks and geese. *T. perniciosa* is considered the most pathogenic intestinal coccidia in ducks while *E. anseris* is most important in geese. Clinical signs may include sudden death, depression, blood-stained vent, diarrhea, emaciation, and pallor. In domestic ducks, infection with *T. perniciosa* may result in distension of the small intestine with necrohemorrhagic or caseous material. Disease is not common in free-ranging birds but epizootics due to *E. aythyae* have been reported in lesser scaup.[274] Acutely infected scaup have hemorrhagic enteritis; pseudomembranous plaque formation along the mucosal surface occurs in more chronic stages. Microscopic lesions involve multiple life stages of coccidia within mucosal epithelial cells and in the intestinal lumen.

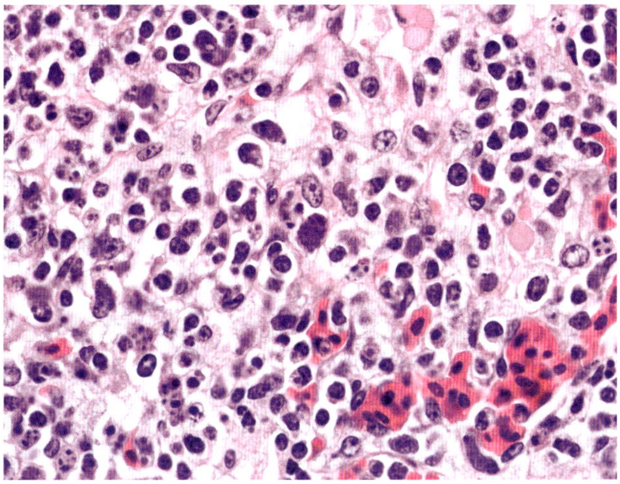

FIGURE 29.22 Disseminated visceral coccidiosis (*Eimeria reichenowi*) in a juvenile white-naped crane. Lymphohistiocytic inflammation is associated with intrahistiocytic asexual stages (zoites) of this apicomplexan coccidian parasite.

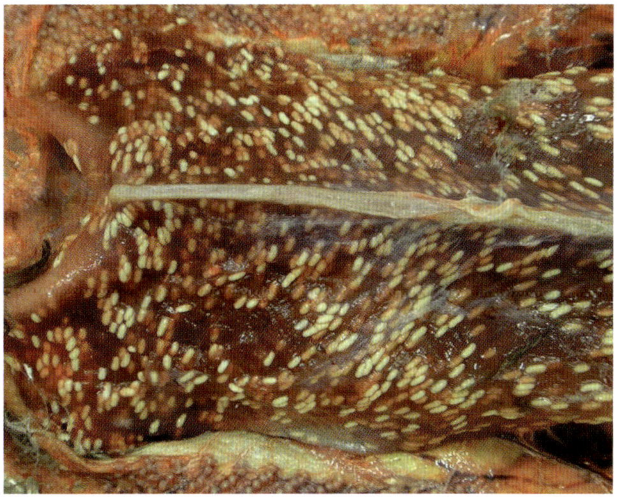

FIGURE 29.23 Sarcocystosis (*Sarcocystis* sp.) in an Indian runner duck. Severe sarcocystosis in which myriad short, white, intramyofibrillary, cystic structures that resemble grains of rice are present throughout the pectoral muscles. *(Photo Courtesy of A. Newton, Wildlife Conservation Society.)*

Renal coccidiosis occurs in many species of domestic and wild waterfowl; *Eimeria truncata* infections in geese are the best documented. Organisms typically mature and reproduce in the kidney or occasionally in the cloaca adjacent to the ureteral junction. Infection is typically asymptomatic in older birds, whereas infection in young birds can be pathogenic. Grossly, young birds may be emaciated with a prominent keel and have enlarged, pale kidneys with multifocal to coalescing white foci. Microscopically, there is extensive destruction of tubular epithelium, and tubules may be dilated and contain oocysts and inflammatory cells. Gametocytes are prominent in the epithelium and the interstitium is infiltrated with lymphocytes, histiocytes, and fewer heterophils.

Sarcocystosis is a nonfatal, typically asymptomatic infection in waterfowl seen most frequently in dabbling ducks. The most commonly described species in wild North American and European waterfowl is ***Sarcocystis rileyi,***[201] although *S. anasi*, *S. albifronsi*, *S. fulicae*, and *S. wobeseri* are also reported. Sarcocysts are well described in waterfowl worldwide, although molecular species differentiation is not always performed.[168,183,200] Sarcocysts are visible grossly within skeletal muscles and resemble rice grains (thus the name "rice breast disease") (Fig. 29.23). Microscopically, myocytes are expanded by large protozoal cysts containing multiple bradyzoites. The cysts are typically not associated with inflammation or tissue necrosis. Infection is typically considered an incidental finding unless severe. Confirmation of the species involved requires molecular testing.

Other protozoal infections in waterbirds include ***Cryptosporidia* spp**. in Anseriformes,[230] Ciconiiformes, Charadriiformes,[188] and Gruiformes. Infections have been reported in the bursa of Fabricius, cloaca, and respiratory tract. The most commonly identified organism is ***C. baileyi***. **Systemic *Isospora*/*Atoxoplasma*** has been reported as an incidental finding in an East African crowned crane.[190] Infection and exposure to ***Toxoplasma gondii*** have been reported in migratory waterfowl with mortality associated with a low percentage of infections, although infection may be associated with an increased susceptibility to trauma.[264,287] **Trichomonads** have been reported but are rare in waterfowl.

See the Digital Content and Table 29.e2 for more information on notable nematode, cestode, trematode, myxozoan, and hematozoan protozoa parasites (e.g., *Haemoproteus, Leukocytozoon,* and *Plasmodium* sp.).

Ectoparasites

Bloodsucking **nasal leeches** of the genus ***Theromyzon*** feed directly in the nasal passages, trachea, and beneath the nictitating membrane of many duck species that inhabit coldwater lakes. These may be seen attached to the mucous membranes of the eye where they are associated with exudate under the nictitans, a cloudy cornea, or even a collapsed globe. They may also be seen protruding from the nares or blocking the nasal passage, throat, or trachea[48] (Fig. 29.24). Mortality can occur with heavy infestations and heavy nasal leech infestations may be seen in birds that die of lead poisoning.

Almost all birds have host-adapted **chewing lice** (predominantly in the genus ***Mallophaga* spp.**).[37] Although many lice are easy to detect, others are small or can be hidden in the shafts of developing pin feathers. They may therefore not be easily seen and additional methods of

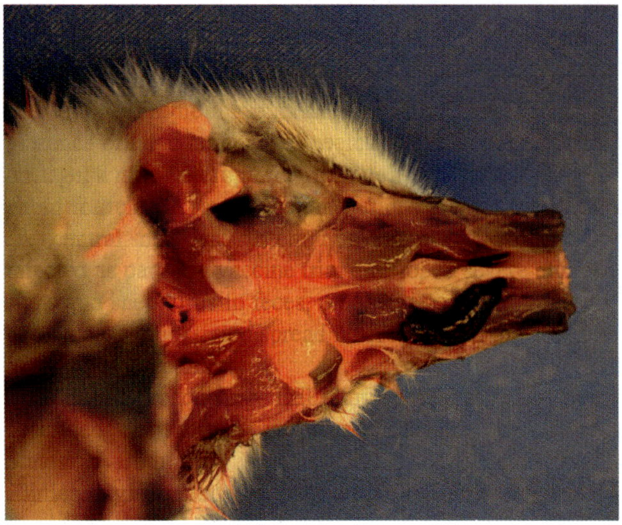

FIGURE 29.24 Nasal leeches. Horizonal section through the head, including the maxilla and the nasopharynx of an mute swan species demonstrating the presence of a large, *cylindrical*, dark red to black leech approximately 1.5 × 4 cm × 1 cm within the nasal cavity. *(Photo Courtesy of the Swedish Veterinary Agency.)*

detection must be employed (e.g., body washing and feather ruffling in dead birds). Louse eggs are usually white and tend to be found.

All birds may be infested with any number of mite species. **Mites** can infect feathers (e.g., ***Stibarokris annae*** in Maguari stork[84]), subcutaneous tissues (e.g., ***Hypodectes propus*** in herons[61]), and the nasal cavity (e.g., ***Rhinonyssus rhinolethrum*** in black-necked swan[154]). **Trombiculosis**, dermatitis due to the chigger mite ***Blankaartia sinnamaryi***, has been described in young Florida sandhill cranes.[246] Infestation is characterized by firm, raised papules with intralesional red-orange mites in a small depression on top of the papule. Microscopically, a stylostome extends into the dermis and is surrounded by pyogranulomatous inflammation. Dermatitis due to chiggers has been reported in mallards, black ducks, and a ruddy duck duckling.

Hematophagous myiasis in nestlings is common in nidicolous birds, but natural infestations are not reported in water or shore birds.[148] **Traumatic myiasis** due to nonhematophagous dipteran larvae has been reported in a wide variety of species. Fatal myiasis due to the flesh fly ***Wohlfahrtia opaca*** has been reported in newly hatched wild ducklings. In these ducklings, larvae entered via the umbilicus and burrowed extensively in the abdominal subcutis.

ADDITIONAL RESOURCES

Atkinson, C.T., Thomas, N.J., Hunter, D.B. (Eds.), Parasitic Diseases of Wild Birds. Wiley-Blackwell, Ames, pp. 289–315.

Fowler, M.E. (Ed.), Zoo and Wild Animal Medicine. W. B. Saunders Company, Philadelphia, pp. 178–181.

Friend, M. (Ed.), Field Guide to Wildlife Diseases. Vol. 1. General Field Procedures and Diseases of Migratory Birds. U. S. Department of the Interior Fish and Wildlife Service. Resource Publication 167, Washington, pp. 95–99.

http://wildpro.twycrosszoo.org/S/00dis/Miscellaneous/Congenital_Abnormalities.htmhttp://www.wildlife information.org.

Hubálek Z. 2021. Pathogenic microorganisms associated with gulls and terns (Laridae). J. Vert. Biol. 70:21009-1.

Kuiken, T., et al., 1999.b Pathology of Newcastle disease in double-crested cormorants from Saskatchewan, with comparison of diagnostic methods. J. Wildl. Dis. 35,8–23.

Thomas, N.J., Hunter, D.B., Atkinson, C.T. (Eds.), Infectious Disease of Wild Birds. Blackwell Publishing, Ames, pp. 108–130.

Wobeser, G., 1997. Diseases of Wild Waterfowl, second ed. Plenum Press, New York.

References

Full references for this chapter can be found online at https://doi.org/10.1016/B978-0-443-18437-6.00052-3.

Chapter 30

Birds of Prey

Arno Wünschmann[a], Anibal G. Armién[b], Ursula Höfle[c], Jörg Kinne[d], and H.L. Shivaprasad[e]

[a]University of Minnesota, St. Paul, MN, United States; [b]California Animal Health and Food Safety Laboratory System, Davis Branch, CA, University of California, Davis, CA, United States; [c]SaBio (Health and Biotechnology) Working Group, National Game and Wildlife Research Institute, Cuidad Real, Spain; [d]Central Veterinary Research Laboratory, Dubai, United Arab Emirates; [e]California Animal Health and Food Safety Laboratory System, Tulare Branch, University of California, Tulare, CA, United States

INTRODUCTION

All birds in the orders Falconiformes (hawk-like birds) and Strigiformes (owls) are considered birds of prey (raptors). The order Falconiformes contains five families: Accipitridae (hawks, eagles, kites, harriers, and Old World vultures), Falconidae (falcons and caracaras), Cathartidae (New World vultures), Pandionidae (ospreys), and Sagittriidae (secretary birds). The families Strigidae (typical owls) and Tytonidae (barn owls) make up the order Strigiformes. Falconiformes are diurnal and Strigiformes are nocturnal with some exceptions.

NON-INFECTIOUS DISEASES

Nutritional

Obesity primarily occurs in raptors kept as exhibit and interactive birds. In contrast, falconer-owned raptors tend to be lean and athletic. Obese birds have abundant adipose in the subcutis, at the heart base and within the coelom, and hepatic lipidosis may be an associated finding.

Pansteatitis has been described in red-tailed hawks and to a lesser degree in bald eagles during the winter months in northern latitudes (Ontario, Canada, and upper Midwest of the United States). The cause of the condition is unclear although oxidative damage due to vitamin E deficiency (hypovitaminosis E) is suspected.[144] Affected birds may have elevated creatine kinase values and their serum may be lipemic. At necropsy, internal and subcutaneous adipose tissue is discolored yellowish beige and has a chalky consistency (Fig. 30.1). Histologically, adipocyte necrosis is associated with heterophilic and granulomatous inflammation, including multinucleated giant cells (eSlide 30.e1). Cholesterol-like crystals and yellowish granular pigment may be present in the necrotic fat and in macrophages. The pigment stains with hematoxylin phloxine saffron and Ziehl-Neelsen stains and are Periodic acid-Schiff (PAS) positive, resembling ceroid. Hyaline degeneration and necrosis of myofibers and cardiomyocytes have been described but may not be present in all cases.

In recently hatched nestlings (up to approximately 1 week of age), a mild to moderate degree of hepatic lipid accumulation, characterized by diffuse, beige to yellowish discoloration of a slightly enlarged liver, and hepatocellular cytoplasmic vacuoles is associated with resorption of yolk and considered a normal, physiologic finding. A similar transient physiologic hepatic lipidosis occurs in egg-laying raptors as a result of altered lipid metabolism during vitellogenesis. On the other hand, marked diffuse **hepatic lipidosis (fatty liver)**, characterized by liver enlargement with rounded liver margins and diffuse yellowish discoloration, is a pathologic condition occasionally encountered in captive raptors. Samples of affected liver may float in formalin if severely affected. Pathologic hepatic lipidosis usually affects small-sized owls, such as screech owls and saw-whet owls but also has been described in larger owls, such as tawny owls and barred owls, merlins, kestrels, and other birds of prey.[51] Affected owls are usually obese and may have a history of recent weight loss. The sedentary lifestyle of captivity and high caloric intake associated with high-fat diets (such as a diet of day-old chicks) are considered the primary risk factors. Affected animals may have greasy plumage. Histologically, hepatocytes throughout the lobules are enlarged and contain one or multiple optically empty vacuoles. In merlins, similar vacuoles develop in renal tubular epithelial cells and in the spleen in addition to the marked hepatic lipidosis.

Although **atherosclerosis** has been reported as a cause of death in free-ranging birds of prey, it is more common in aged captive raptors including various falcon species, black kites, red-tailed hawks, eagles, and owls.[33] It primarily affects large arteries, which includes the brachiocephalic trunk, pulmonary artery, and aorta but can also affect myocardial arteries. Brahminy kites are suspected to have a genetic predisposition for atherosclerosis since individuals

Pathology of Wildlife and Zoo Animals. https://doi.org/10.1016/B978-0-443-18437-6.00015-8

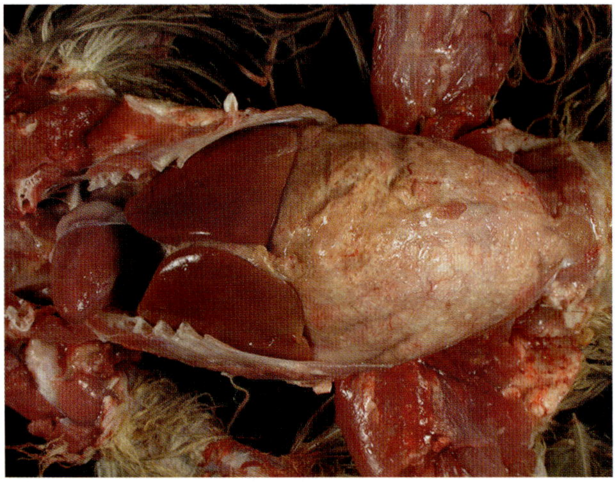

FIGURE 30.1 Pansteatitis in the coelom of a red-tailed hawk. The hawk is markedly obese. Intracavitary adipose has a chalky texture and patchy, beige discoloration (see eSlide 30.e1).

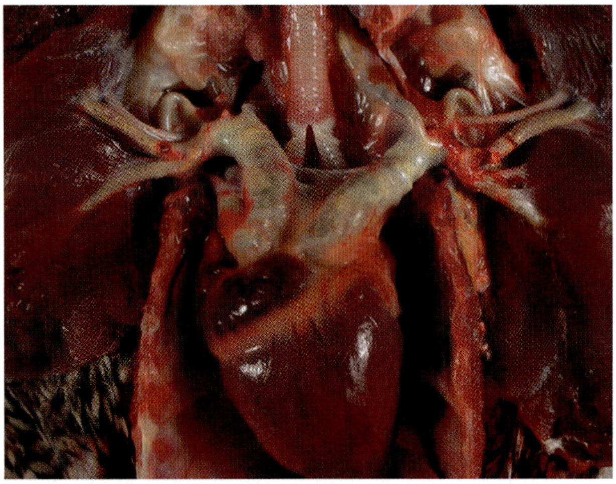

FIGURE 30.2 Atherosclerosis in the great vessels of a peregrine falcon. The walls of the subclavian arteries, brachiocephalic trunks, and aortic arch are multifocally opaque white due to expansion and hardening of the vascular walls by yellowish intimal plaques.

as young as 55 days have been diagnosed with moderate to marked atherosclerosis and lipidosis of the major organs.[102] Grossly, the brachiocephalic trunk, pulmonary artery, subclavian arteries, and aorta are thickened due to the presence of raised yellow atheromatous plaques that cause narrowing or occlusion of the arterial lumina (Fig. 30.2). Histological sections reveal pools of lipid within the intima and media that are associated with foamy macrophages and the presence of cholesterol clefts. Deposition of collagen fibers and osseous metaplasia are present in advanced cases. In rare cases, ischemic cardiomyopathy, characterized by beige discoloration of the myocardium, may be associated with the presence of atherosclerosis. Dissecting aneurysms with or without rupture represent complications of advanced atherosclerosis in falcons and kites.

Metabolic

The pathogenesis and pathology of **uric acid deposition disease** (or **gout**) do not differ in raptors from uric acid deposition in other birds (see Chapter 31: Galliformes and Columbiformes).

Toxic

Direct or indirect intoxications of birds of prey have been well reported, often as individual case reports or small case series. However, population declines associated with widespread intoxication, such as **dichlorodiphenyltrichloroethane (DDT)** or **diclofenac**, are also recognized in a variety of raptors and vultures.[96] Since vultures and condors feed communally on carcasses, even a single contaminated carcass has the potential to kill many individual birds. When many contaminated carcasses are present in the landscape, the effects can be catastrophic.

Carcass disposal by avian scavengers is important in environmental "recycling". However, pharmaceuticals and parasiticides in the carcasses that are ingested by scavengers can cause morbidity and mortality. One of the most striking examples is the marked population decline of *Gyps* spp. vultures in India and Pakistan. This decline, which began in the 1990s and continued into the 2000, was determined to be the result of residues of the nonsteroidal anti-inflammatory drug **diclofenac** in livestock carcasses scavenged by the vultures. Bird deaths were due to acute renal tubular necrosis with renal and visceral urate deposition (gout) (Fig. 30.3).[82] Experimental dosing of Asian *Gyps* vultures recreated the lesions, and experimental work in cell culture systems has defined the injury primarily as cell death by necrosis rather than caspase-mediated apoptosis. New World turkey vultures appear to be fairly

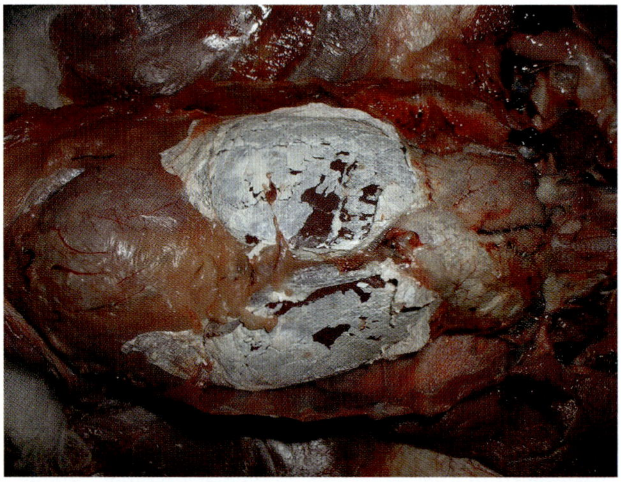

FIGURE 30.3 Visceral gout in a *Gyps* spp. vulture with diclofenac poisoning. Marked diffuse chalky, white urate deposits are present on the liver capsule. *(Photo Courtesy of Martin Gilbert, Cornell Wildlife Health Center, Cornell University.)*

resistant.[104] Though banned in some parts of the world, diclofenac continues to be used in others including some parts of Europe. Its continued use in livestock poses an ongoing threat to vultures in an expanding geographic distribution. In captive settings, accidental intoxication of birds under treatment with antiinflammatory drugs, which include carprofen, flunixin, diclofenac, ketoprofen, and phenylbutazone, has also been reported.[26]

Lead toxicity due to ingestion of carcasses or offal from animals hunted with lead-based ammunition is very common in eagles and vultures.[20,25,45] In fact, many vulture species and condors throughout the world are threatened by extinction in large part due to ingestion of lead.[101] Cases of lead toxicity often occur shortly after the hunting season.[106] In cold climates, a second peak of cases occurs after the snow melts in the spring, which allows access to lead in thawing gut piles. As a general rule, any detectable concentration of lead in blood or tissues is abnormal; there are species-dependent variations in toxic threshold concentrations. In bald eagles, blood lead concentrations of >1 ppm are considered lethal, and liver and kidney concentrations >20–30 ppm (dry weight) in eagles and >6 ppm (wet weight) in California condors, respectively, are interpreted as evidence of fatal intoxication.[44] Neurologic signs, such as severe apathy, hock sitting, ocular twitching, anemia, and gastrointestinal stasis, are common in eagles with lead poisoning; severe dyspnea can occur terminally. Eagles also frequently have lime-green mutes. In experimental lead poisoning of turkey vultures, clinical signs include weight loss, incoordination, and lethargy; clinical chemistry abnormalities in experimentally intoxicated Andean condors included anemia, hyperbilirubinemia, and elevated liver enzymes.[17,100]

Gross lesions of lead toxicity may be absent, particularly in cases of acute poisoning. The majority of bald eagles are emaciated and may have evidence of either one or a combination of the following: anemia, myocardial degeneration and necrosis (Fig. 30.4), hydropericardium, multifocal petechiae in the brainstem, mesencephalon, and cerebellar peduncles, erosive or ulcerative proventriculitis, distention of the gallbladder with dark-green concentrated bile, and bile staining of the gastrointestinal tract mucosa and perianal feathers.[38] Acute renal tubular degeneration and necrosis, edema, multifocal hemorrhage, and pan-necrosis in the brainstem and mesencephalon, or chronic myocardial fibrosis can also be seen.[94] In one study, slightly over 50% of the bald eagles with lead toxicity had grossly detectable evidence of myocardial degeneration and fibrosis while approximately 14% had gross evidence of brainstem hemorrhages and necrosis.[76] Histologically, fibrinoid necrosis usually affecting small- or mid-sized arteries in the heart, brain, and choriocapillaris of the eyes is relatively common in affected eagles (eSlides 30.e2–e4).[76] Ulcerative ventriculitis may be associated with the presence of irritating lead

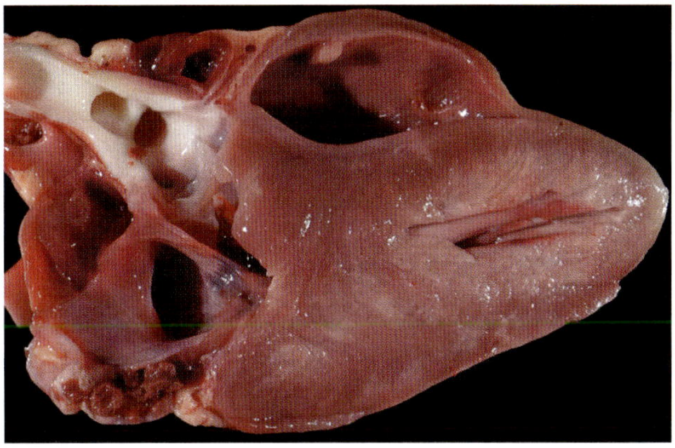

FIGURE 30.4 Heart of a bald eagle with lead poisoning. The myocardium contains multifocal, poorly defined areas of pallor particularly in the septum and left ventricular free wall. Discoloration is due to lead-induced myocardial degeneration and necrosis (see eSlides 30.e2–e4).

particles. In California condors, Kupffer cell erythrophagocytosis and hemosiderosis are the most common findings. Cerebellar perivascular and perineuronal edema, cerebellar status spongiosus, and Purkinje cell pyknosis, as well as intra- and extra-hepatic hemosiderosis are seen in experimentally induced lead intoxication in Andean condors. Acid-fast positive and intranuclear lead inclusions have not been documented in eagles or California condors but have been seen in experimentally intoxicated turkey vultures and Andean condors in renal tubular epithelial cells.

Liver lead concentrations are important in confirming intoxication.[73] Bone or feather analyses are useful in chronic cases or when soft tissues are no longer available due to decomposition. Analysis of different segments of the feathers provides information about past nonlethal as well as lethal lead exposure. Detection of metallic, lead-containing fragments in the proventriculus and ventriculus is uncommon since the animals frequently cast them out.[76] Nevertheless, careful examination of proventricular and ventricular contents for nonmagnetic, lead-based metallic fragments is recommended. Radiographic examination of the proventriculus and ventriculus at necropsy may facilitate detection of small fragmented projectiles.

Exposure of birds of prey to **anticoagulant rodenticides (AR)** is very common and is usually due to the ingestion of rodents poisoned with second-generation compounds, particularly brodifacoum.[56] Despite extremely high exposure rates of up to 92% in great horned owl populations, actual cases of toxicity with widespread hemorrhage are infrequently reported. Intoxication with ARs was considered the cause of death in only 8% of scavengers in a study in California despite 84% of birds having detectable hepatic concentrations of anticoagulant rodenticides. Similarly, the death of 4% of eagles with analytically confirmed AR exposure was attributed to the exposure itself.[95] Clinical diagnosis of impaired coagulation is challenging due to the relatively ineffective intrinsic clotting cascade in birds.[90] Raptors with anticoagulant rodenticide intoxication may present with anemia and bleeding from relatively minor wounds, such as injection sites. Widespread hemorrhage is considered uncommon, but if extensive subcutaneous hemorrhage, hemorrhage into the lungs, air sacs, or body cavities, and/or hemorrhage into the gastrointestinal tract are present in the absence of trauma-related lesions, anticoagulant rodenticide toxicity should be considered. Turquoise or blue-green staining of crop or gastric mucosa by tainted baits or finding stained ingesta heightens suspicion. Toxic concentrations of the rodenticide brodifacoum are considered to be >0.2 mg/kg in the liver but the concentration of anticoagulant rodenticides correlates poorly with the presence or severity of lesions.

Vultures and other scavengers are at significant risk for intoxication with a number of **pesticides.** Target species that have eaten poisoned bait but are still alive present risks for other birds of prey. In Europe, the most important pesticide intoxications in birds of prey have been cholinesterase inhibitors (e.g., aldicarb, fenthion), strychnine, and nitrofurans. Birds are usually found dead with no specific gross or histologic lesions. The temporally related, sudden death of groups or large numbers of birds raises the suspicion of intoxication. Diagnosis relies on the detection of bait in the proventriculus and ventriculus and chemical analysis.

There is great concern for bioaccumulation of **persistent organic pollutants (POP)** in several vulture species, especially California condors feeding on marine mammal carcasses.[63] Tissues suggested for analysis include feathers, dried blood spots, livers, and eggs. Consequences of POP exposure are largely reproductive with eggshell thinning in coast-dwelling condors and other apex-predator species.[57] The pathogenesis of eggshell thinning due to DDT and its metabolites is thought to involve interference with prostaglandin synthetase and/or carbonic anhydrase activity in the distal oviduct (shell gland). Decreased numbers of eggshell capillaries containing carbonic anhydrase have been correlated with a decreased number of mammillary cones in thin-shelled eggs. This effect is found in eggs from females that were exposed during the egg-laying cycle and in eggs laid by females of the F1 generation, indicating intergenerational effects.[46] Contaminants with estrogen mimicry may act synergistically with POPs compounding their negative effects; however, if this is the case, it is likely species specific.

Avian vacuolar myelinopathy (AVM) was first described in bald eagles in the early to mid-1990s in the southeastern United States.[131] The lesions of AVM have been experimentally reproduced in red-tailed hawks. The condition is considered to be site-specific, occurring and recurring at particular lakes or bodies of standing water with introduced aquatic plants. Outbreaks have been limited to the winter months. The disease is caused by aetokthonotoxin. This toxin is produced by the cyanobacterium *Aetokthonos hydrillicola* that colonizes the invasive aquatic macrophyte *Hydrilla verticillata*.[10,141] Waterfowl and American coots ingest the toxin while feeding on aquatic plants and develop AVM. Consumption of intoxicated birds likely causes the disease in raptors. Affected eagles show neurologic signs. Gross lesions are absent. Histologically, vacuolation of the white matter throughout the central nervous system is the hallmark of this condition (Figs. 30.5A, B; eSlide 30.e5). Variably sized, optically empty spaces are particularly apparent in the medulla oblongata, cerebellar folia, and (most notably) the optic tecta. The optic nerves are also affected; the retina is unaffected. Gliosis, neuronal degeneration, and occasional swollen axons may serve as corroborative evidence in

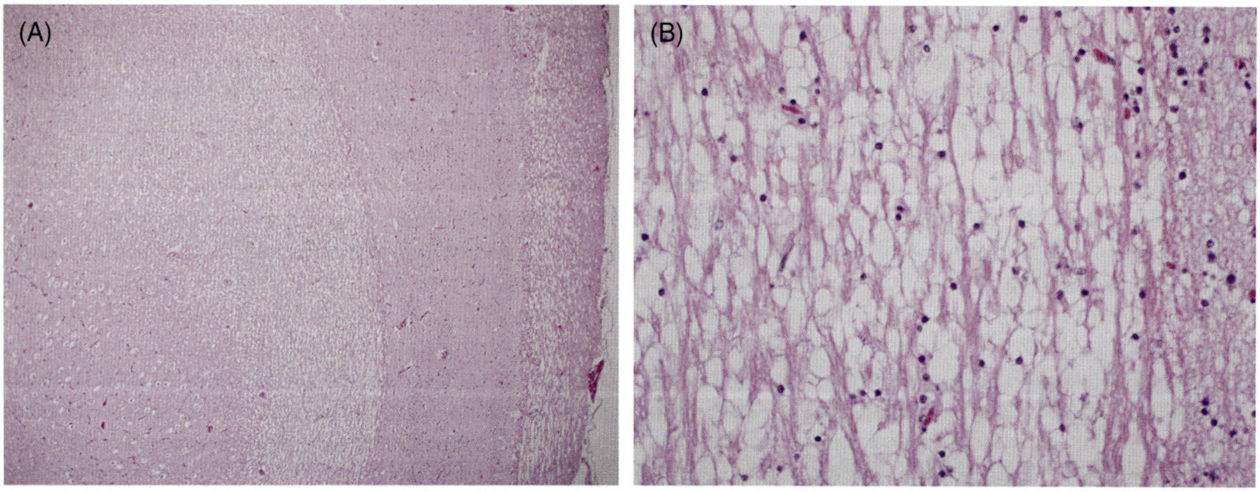

FIGURE 30.5 Avian vacuolar myelinopathy in a bald eagle. (A) The white matter of the mesencephalon is pale with numerous clear spaces. (B) Vacuolar change is characterized by variably sized, optically empty spaces in the white matter tract of the optic tectum (see eSlide 30.e5).

distinguishing avian vacuolar myelinopathy from artifactual vacuolation. Electron microscopy reveals intramyelinic vacuolation within axon sheaths and splitting of one or more of the myelin lamellae at the intraperiodic line. This lesion is considered pathognomonic in eagles submitted under epidemiologically appropriate circumstances; however, other toxins, such as bromethalin and hexachlorophene, are known to cause exactly the same lesion. Bromethalin and hexachlorophene can be detected by gas chromatography and mass spectroscopy in brain tissue although detection is challenging.

Environmental contaminants, such as **DDT** and chlorinated hydrocarbons, such as **polychlorinated biphenyls (PCB)** have been associated with reproductive disorders and malformed offspring. **DDE (dichlorodiphenyldichloroethylene),** a metabolite of DDT, was shown experimentally to cause eggshell thinning, which is the suspected mechanism of reproductive failure in bald eagles and other birds of prey in the mid-20th century; its use in the United States was banned in 1972 and since then populations of birds-of-prey have been slowly recovering. Developmental deformities induced by PCBs in free-ranging birds include crossed bills, clubbed feet, hip dysplasia, dwarf appendages, hydrocoelom and subcutaneous edema, gastroschisis, and craniofacial deformities.[9] Shortened mandibles and limbs, exophthalmia, underdevelopment, and hematomas have been induced experimentally in American kestrel offspring after long-term consumption of a PCB-spiked diet by the parents.[35]

Congenital/Genetic

Developmental abnormalities have been infrequently reported in raptors. The causes are unknown for many malformations (see Digital Content Table 30.e1). Affected birds, whether in natural domains or habitats heavily modified by humans, are rarely discovered thus the prevalence is probably underestimated.

Miscellaneous/Multifactorial

Physical injuries/trauma are the most common cause of raptor admission to rehabilitation centers worldwide. Mortality varies with geographic location, ecology, time of year, climatic conditions, and species. Causes of physical injuries include gunshot, collision with vehicles, stationary objects, blades of wind turbines, and power lines, as well as entrapment and interspecies and intraspecies aggression. Recognizing and documenting prevalence and cause of physical injuries are important to direct mortality and criminal investigations and in the enforcement of conservation policies.[42] Focal traumatic injuries include wing amputation, laceration, fractures, luxation, and hematomas. In owls, hepatic laceration with fractures and extensive necrosis are common after blunt force trauma. It has been speculated that the configuration of the rib cage of owls may predispose these animals to hepatic injuries during collisions.

Neurologic signs due to trauma may occur concurrently and mask underlying nontraumatic infectious or noninfectious disease processes. Paralysis of the legs is common due to acute or chronic spinal cord compression or transection due to vertebral fractures and remodeling. Acute or chronic trauma that affects the synsacrum/caudal vertebral segments can lead to urinary, intestinal, and/or cloacal autonomic dysfunction with coprostasis and marked intestinal/cloacal distension. Acute axonal injury with Wallerian degeneration is a common consequence of nerve and spinal cord trauma. It develops distal and proximal to the site of nerve transection and can extend proximally, affecting sensory as well as motor nerve and spinal ascending tracts. Brachial plexus avulsion (Fig. 30.6A) is another important

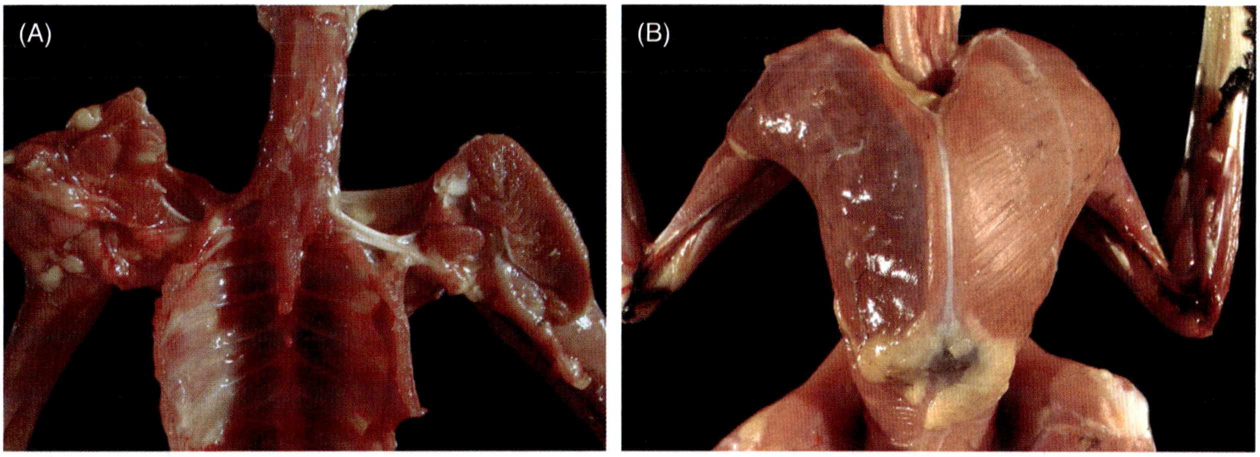

FIGURE 30.6 Brachial plexus avulsion in a northern saw-whet owl. (A) The nerves of the right brachial plexus (left side of image) are transected. (B) The right pectoral muscle (left side of image) is markedly atrophied due to denervation.

consequence of nervous system trauma.[124] Avulsions are most commonly unilateral. Clinical signs include lack of pain, proprioceptive deficits, paresis, and paralysis. Neuromuscular degeneration followed by severe muscular atrophy affects wing and pectoral muscles (Fig. 30.6B).

Head trauma, and in particular blunt force trauma to the eyes and their adnexal structures, is common in raptors.[120] Lesions in the posterior segment of the eye are more common than lesions in the anterior segment. This may be due to horizontal intraocular volume displacement and stretching and tearing of tissue that occurs after frontal trauma in the tubular-shaped eyes of owls. Commonly observed lesions include corneal erosion or perforation, hemorrhage and/or hyphema, lens luxation, secondary cataracts, tear and rupture of the pecten, chorioretinal rupture, retinal detachment (e.g., due to fluid exudation or hemorrhage into the subretinal space or countercoup forces), and posterior scleral rupture. Chronic lesions include superficial ulcers, anterior and posterior uveitis, irido- and cyclodialysis, posterior synechia, preretinal fibrous membranes, chorioretinitis, and glaucoma. Müller cell hypertrophy and hyperplasia are commonly associated with retinal detachment or atrophy. A unique chronic lesion in raptors with ocular injury is the formation of lentoid bodies (Fig. 30.7).[150] Lentoids are intraretinal, lens-like structures produced by Müller cells or other components of the neuroretina subsequent to retinal trauma. Histologically, lentoids are clusters of lens fibers and cells that are PAS positive and immunoreactive with crystalline alpha-A marker. Horner syndrome, mainly characterized by ptosis of the upper eyelid, has been documented in individuals of various raptor species, most of which had evidence of trauma.[65]

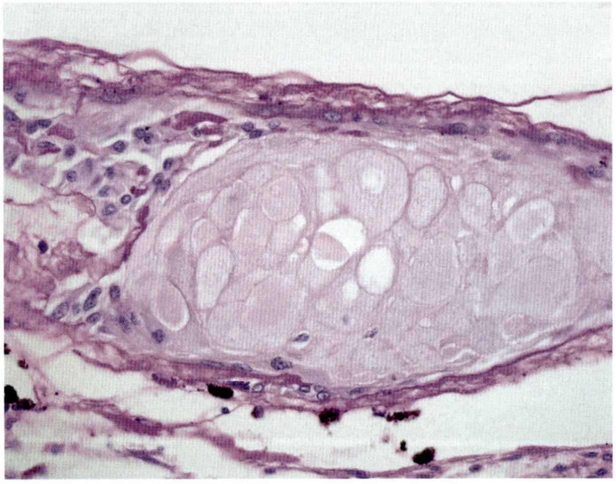

FIGURE 30.7 Lentoid body in the eye of a red-tailed hawk. The lentoid body is a cluster of large globoid cells resembling "bladder cells" of the lens. In this case, they are embedded in retinal scar tissue. The hawk had phthisis bulbi, presumably due to ocular trauma. PAS.

Macrotrash and microtrash ingestion is the single most important cause of death in nestling California condors.[36,106] It is postulated that plastics and other trash are mistaken by parents for bone fragments that would normally provide calcium for growing nestlings. Besides the obvious hazards of entanglement in macrotrash, such as fishing gear, obstruction of the gastrointestinal tract by macrotrash is common. Furthermore, components of microtrash, such as metal objects, can lead to zinc and lead toxicosis and there is concern that ingested microplastics may be toxic to the animals.

Emaciation is common, particularly in free-ranging, young raptors during their first winter. Gross and histologic lesions are similar to those of other species.

Systemic amyloidosis, which occurs secondary to serum amyloid A protein (SAA; an acute phase reactant protein) deposition in tissues, has been recognized as a major cause of death in captive falcons in the Middle East since the mid-1990s; the condition has not been reported in North America or Europe.[41] Gyrfalcons are particularly susceptible while saker falcons are relatively resistant. Affected animals are frequently emaciated. Severely affected livers (and other organs) are enlarged, have a waxy appearance (Fig. 30.8), and fracture readily upon handling. Greenish-brown discoloration of the liver, when seen, is attributed to bile stasis. The liver but also kidney, spleen, adrenal gland, or intestine are commonly affected. The earliest deposition of amyloid is present beneath the endothelial cells lining the space of Dissé in the liver and in blood vessels in histologic sections. Amyloid is eosinophilic, acellular, and positive with Congo red staining. Bumblefoot and chronic visceral uric acid deposition disease (gout) are the most common comorbidities in falcons with fatal SAA amyloidosis, but aspergillosis, mycobacteriosis, and poxvirus infection are also frequently documented.

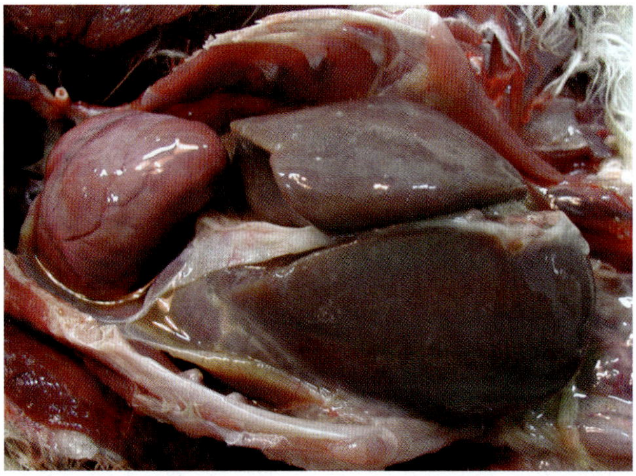

FIGURE 30.8 Amyloidosis in the liver of a falcon. The liver is markedly enlarged, has a waxy appearance, and is discolored light brown.

Neoplastic

Despite the long life of raptors (reported to be up to 55 years in eagles) compared with mammals and other bird groups, references to neoplasia in birds of prey are infrequent. Limited data are available to calculate the incidence and prevalence of neoplasia and species, sex, or age predisposition. Tumor nomenclature is inconsistent and etiology, oncogenesis, and biological behavior are basically unknown. Worldwide, neoplasms have been reported in more than 40 species among the families Accipitridae, Falconidae, Strigidae, and Cathartidae, and a wide variety of neoplasms have been described. The most commonly reported neoplasms are **soft tissue sarcoma, lymphosarcoma, adenocarcinoma**, and **squamous cell carcinoma**. **Synovial osteochondromatosis**, a tumor-like lesion, occurs in great horned owls and has been reported in a red-tailed hawk.[126] It is characterized by chondral or osteochondral nodules in synovial tissue of the joints and tendon sheaths (Figs. 30.9A–C; eSlide 30.e6). Nodules may become pedunculated and break off forming loose bodies in the joint. Lesions are typically bilateral and most frequent in the scapulohumeral joints, less common in the sternocoracoid and femoroacetabular joints, and occasional in the stifle and hock joints. The etiopathogenesis of osteochondromatosis is unknown although chronic joint disease and cartilage metaplasia with fragment release and chondrocyte reattachment and proliferation in the synovial membrane are suspected.

INFECTIOUS DISEASES

DNA Viruses

Falcons, owls, and hawks are susceptible to infection with **alphaherpesviruses**. In the past, it was suspected that closely related raptor-specific herpesviruses infected the various raptor species. However, molecular analysis determined that **columbid herpesvirus-1 (CoHV-1)** is by far the most common cause of clinical herpesvirus infections in raptors worldwide.[108] Raptors become infected by ingesting CoHV-1-infected pigeons. When tested against various raptor herpesvirus isolates, the seroprevalence is low to moderate among owls and falcons and very low among Accipitridae in the English raptor population, suggesting that latent infections occur in all three of these groups.[153] Clinical signs are frequently absent, but lethargy

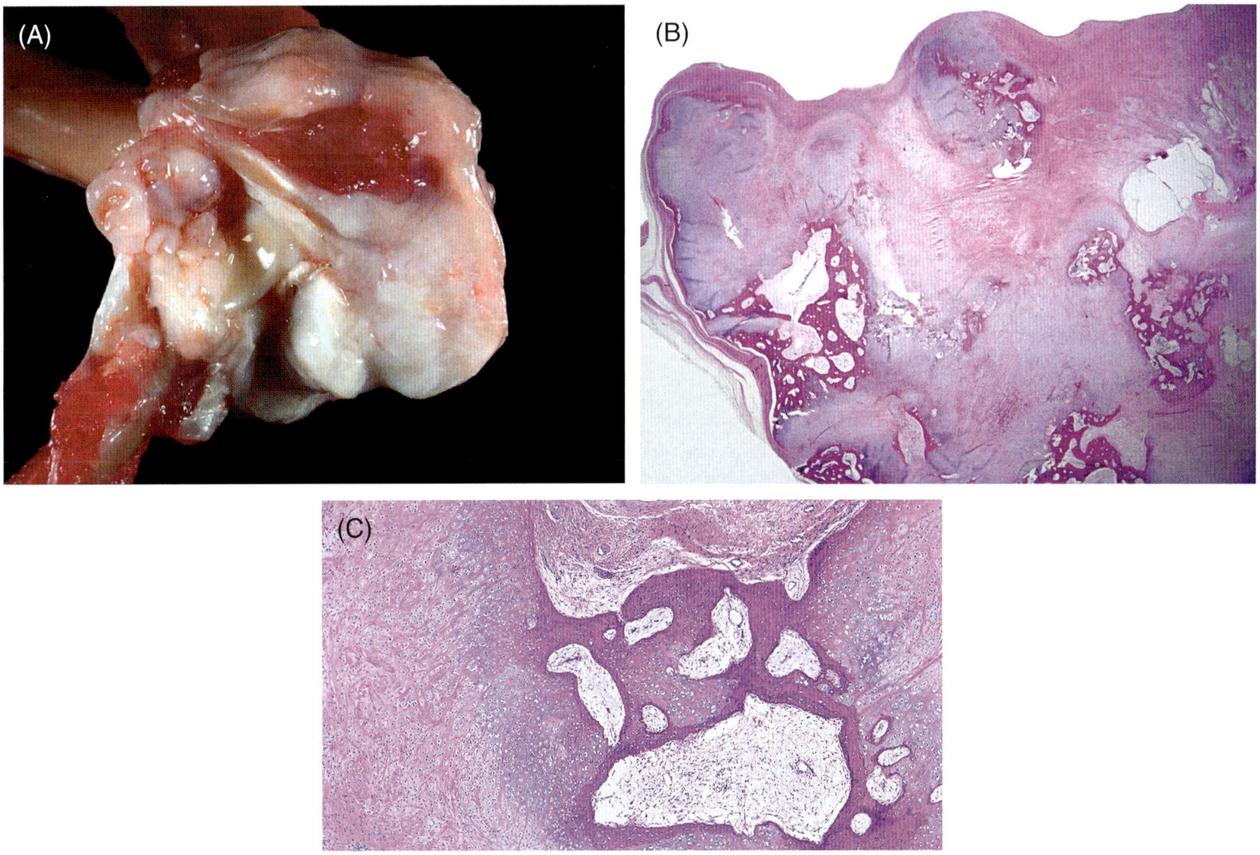

FIGURE 30.9 Synovial osteochondromatosis in a great horned owl. (A) Marked synovial proliferation with osseous and cartilaginous metaplasia replaces and distorts the normal configuration of the shoulder joint. (B, C) The synovium is expanded by multiple, variably sized, coalescing, unencapsulated proliferative nodules composed of hyaline cartilage and trabecular bone with marrow spaces (see eSlide 30.e6).

and anorexia of short duration (hours or days) may occur. Common gross lesions include multifocal to coalescing, pinpoint and variably sized, tan foci in the liver, spleen, and bone marrow (Figs. 30.10A, B; eSlide 30.e7) that correspond histologically to miliary areas of lytic necrosis. Necrotic foci may be present in other tissues including the pharynx, intestine, thymus, cecal tonsils, bursa of Fabricius, thyroid, and parathyroid glands. Hepatocytes, splenic reticuloendothelial cells, lymphocytes, bone marrow macrophages, pharyngeal epithelial cells, and enterocytes at the junction of the necrotic foci may contain eosinophilic, intranuclear viral inclusions with a distinct halo and hyperchromasia of the nuclear membrane (Fig. 30.10C; eSlide 30.e7). Herpesviral particles can be detected in the feces. Intranuclear, 100–120 nm diameter nucleocapsids are detectable by electron microscopy. Virus isolation is frequently unsuccessful using egg inoculation; PCR with subsequent sequencing can confirm and characterize the causative herpesvirus.

Fatal **adenovirus** infections are most commonly reported in falcons but have also been described in hawks, kites, and owls.[28,97,119] Various adenoviruses infect falcons. **Falconid adenovirus-1**, a member of the **aviadenovirus** genus, has been associated with mortality in Aplomado falcons, Taita falcons, orange-breasted falcons, peregrine falcons, a gyrfalcon, and a merlin. The virus is widespread in healthy peregrine falcons based on serological investigations. Juvenile falcons appear to be particularly susceptible, and interspecies infections can occur. Falconid adenovirus-1 is epitheliotropic and lymphotropic causing potentially systemic infections. Infection in natural hosts is usually subclinical or mild, while infection in closely related species can be fatal. Alternatively, falcons may develop inclusion body hepatitis after ingestion of prey items infected with **fowl adenovirus 2** and **fowl adenovirus 6**, both belonging to the aviadenovirus genus.[84]

In adenovirus-infected falcons, clinical signs include lethargy and anorexia or death without premonitory signs. Gross lesions include hepatomegaly and pallor (Fig. 30.11A), and splenomegaly with discrete white foci. Acute hemorrhage into the gastrointestinal tract, lungs, or the reproductive tract of reproductively active females and (rarely) renal swelling and pallor may also be seen. Histologically, multifocal to widespread or diffuse hepatocellular necrosis with rare to frequent, hepatocellular, and occasionally biliary epithelial cell, intranuclear inclusion bodies are hallmarks of infection (Figs. 30.11B, C; eSlide 30.e8); inclusions may also be seen in macrophages and lymphocytes in the spleen or (rarely) in renal tubular epithelial cells. The inclusion bodies are quite characteristic and may be large and fill the entire nucleus (pan-nuclear) or be surrounded by a clear halo and marginated chromatin with hyperchromasia of the nuclear membrane. While classically basophilic, inclusion bodies can, on occasion, also be amphophilic or even eosinophilic. Other findings may include mild to moderate lymphoplasmacytic hepatitis, splenic reticuloendothelial cell hyperplasia, lymphocytolysis and necrosis, and lymphoplasmacytic interstitial

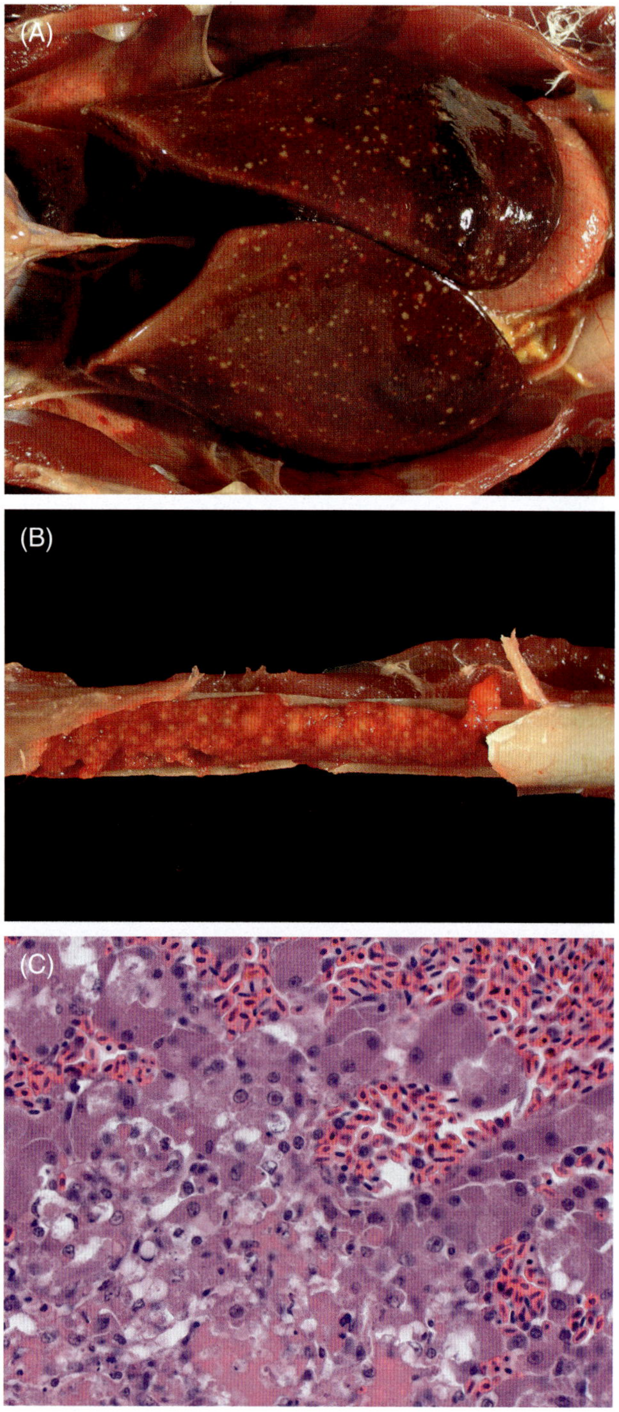

FIGURE 30.10 Columbid herpesvirus-1 in a gyrfalcon (A, B) and a gyrfalcon/peregrine falcon hybrid (C). (A) The liver is diffusely enlarged (note the rounded liver lobe edges) and contains numerous, distinct tan foci of necrosis that are typical of herpesvirus infection. (B) The tibiotarsal bone marrow contains numerous, coalescing tan foci of necrosis. (C) Eosinophilic to amphophilic, hepatocellular intranuclear inclusion bodies in hepatocytes adjacent to an area of hepatic necrosis (see eSlide 30.e7).

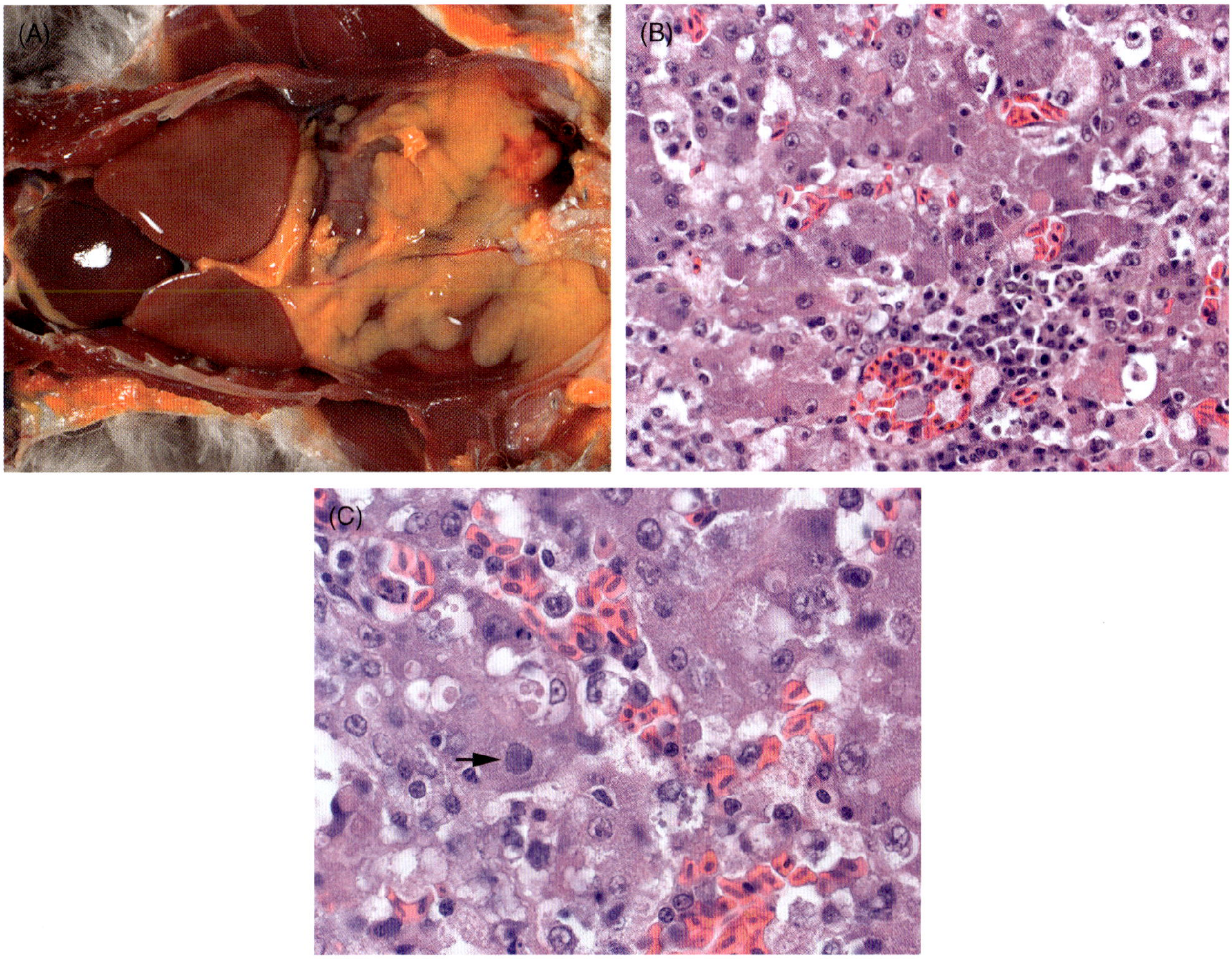

FIGURE 30.11 Falconid adenovirus-1 infection in a taita falcon (A) and a peregrine falcon (B, C). (A) The liver is diffusely pale brown and mildly swollen with rounded edges. (B, C) Widespread hepatocellular necrosis is present. Few hepatocytes contain large, basophilic, intranuclear inclusion bodies (*arrow*). The liver is multifocally infiltrated by plasma cells and lymphocytes (see eSlide 30.e8).

nephritis. Ventricular epithelial cell inclusions may be present in the absence of other lesions in adenovirus-infected falcons.

Relative to falcons, adenovirus infections appear to be rare in hawks and owls though several cases have been described. Red-tailed hawks infected with a **siadenovirus**, most closely related to **raptor adenovirus-1**, had hepatomegaly and splenomegaly grossly.[132] Histologically, widespread bone marrow necrosis, hepatocellular necrosis and splenic necrosis with occasional intranuclear inclusion bodies in hematopoietic cells, macrophages, and hepatocytes were the most consistent findings in affected birds (Fig. 30.12; eSlide 30.e9). Ulcerative ventriculitis and proventricular, epithelial cell inclusions have also been described in an eagle owl infected with a novel raptor adenovirus belonging to the siadenovirus genus.

By electron microscopy, adenoviral particles are intranuclear, isometric, icosahedral, approximately 75–100 nm in diameter, and arranged in a paracrystalline array. **Dependoviruses**, adenovirus-associated parvoviruses, may be detected by electron microscopy or PCR, and viral particles are identified in cloacal contents in some cases. Virus isolation with standard egg inoculation methods is usually unsuccessful. The use of specific raptor cell lines increases the likelihood of a positive result. In situ hybridization and PCR with subsequent sequencing are used to confirm the diagnosis and characterize the causative adenovirus.

As in many other bird groups, **avian pox virus** infection occurs in many different captive and free-ranging raptor species.[113] Disease is more frequently reported in captive than free-living birds. Both individual cases and epiornitics in rehabilitation centers have been described. Outbreaks in groups of captive raptors, including peregrine falcons in a reintroduction program, suggest direct and fly or mosquito-mediated transmission. Only a few, generally captive owls, have been diagnosed with pox.[39] The cutaneous (dry) form of pox most commonly involves the

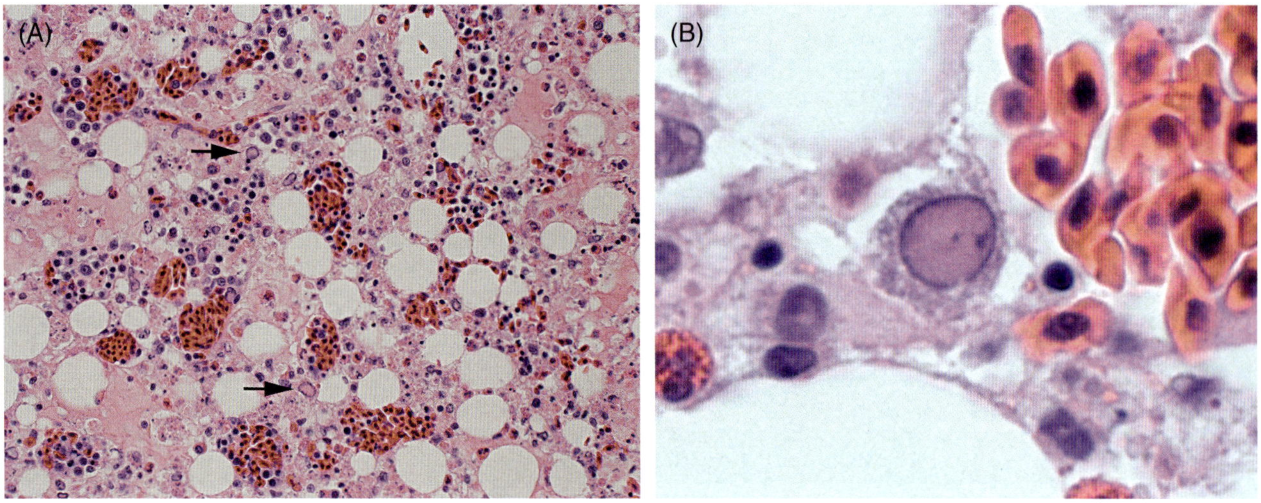

FIGURE 30.12 Siadenovirus infection in the bone marrow of a red-tailed hawk (A, B). (A) The bone marrow is multifocally necrotic. The nuclei of multiple hematopoietic cells are expanded by amphophilic inclusion bodies (*arrows*). (B) The nucleus of a hematopoietic cell is expanded by an amphophilic inclusion body (see eSlide 30.e9).

unfeathered skin of feet and legs (especially toes and palmar surface of the feet and the shanks) and face (cere, eyelids, commissures of beak, and skin). Lesions on the feet are often bilateral and can adversely affect the bird's ability to grasp perches or prey. The cutaneous form is often considered self-limiting but may become complicated by bacterial (*Staphylococcus* sp., *Escherichia coli*, *Clostridium perfringens*, and others) or fungal (*Candida albicans* and saprophytic fungi) infections. Lesions can be solitary, multifocal, or coalescing proliferative nodules or plaques that are often ulcerated and infected. Exudation and crusting are frequent. Similar to the microscopic appearance in other avian species, cutaneous pox lesions are characterized by epidermal hyperplasia with intra- and intercellular edema and pathognomonic large eosinophilic granular cytoplasmic inclusion bodies (Bollinger bodies) in which poxviral particles can be identified by electron microscopy. Inclusions are more common in superficial layers of the epidermis. Extension of cutaneous lesions to the mucosa of the oropharyngeal cavity and esophagus (diphtheritic or 'wet' form) may occur. Fatal diphtheritic forms of avian pox are observed less frequently in raptors than some other avian families but have been recorded in peregrine falcons, bald eagles, and white-tailed sea eagles. Microscopically, lesions of the mucous membranes principally resemble those of the cutaneous form. Systemic pox is infrequent but described in a juvenile turkey vulture and a juvenile Andean condor. The vulture had cutaneous (dry) lesions on face and feet and mucocutaneous (wet) pox lesions in the oropharynx, esophagus, and trachea, as well as pulmonary nodules and splenomegaly due to the proliferation of large mononuclear cells containing cytoplasmic inclusions.

RNA Viruses

A wide variety of raptors are susceptible to infections with **West Nile virus (WNV)** of lineages 1 and 2.[30,151] This RNA virus is a member of the Flaviviridae family and Japanese encephalitis complex. In some species, experimental studies demonstrate sufficient virus amplification within 10 days of infection to suggest that they may be competent vectors. Some infected individuals die within 1–2 weeks while others rapidly clear the virus. In free-ranging raptors in the United States, the seroprevalence of WNV is moderate to high in various populations of adult hawks, owls, caracaras, and kestrels suggesting that infections are commonly survived or are frequently subclinical.[92] The predominant route of infection is through mosquito bites, but raptors can also become infected through consumption of infected prey or scavenged carcasses.[50] The disease manifests itself differently in the various groups of raptors.

Accipitridae: Hatch year birds are more likely to develop fatal West Nile (WN) disease than older animals, although fatal disease has been documented in adults. Clinical signs in hawks are dominated by vision impairment and a wide variety of neurologic signs; trauma secondary to ocular and neurologic dysfunction is common. Birds are frequently undernourished but may present in good nutritional condition. Feather abnormalities, such as "pinched off feathers" and stunted feathers, occur in Swainson's hawks, red-tailed hawks, bald eagles, and barn owls.[93] These birds presumably survive acute infection but impaired feather growth persists for several molts. Other gross lesions include trauma (common in WNV-positive raptors), a combination of patchy tan discoloration of the myocardium (Fig. 30.13A), white discoloration (fibrin

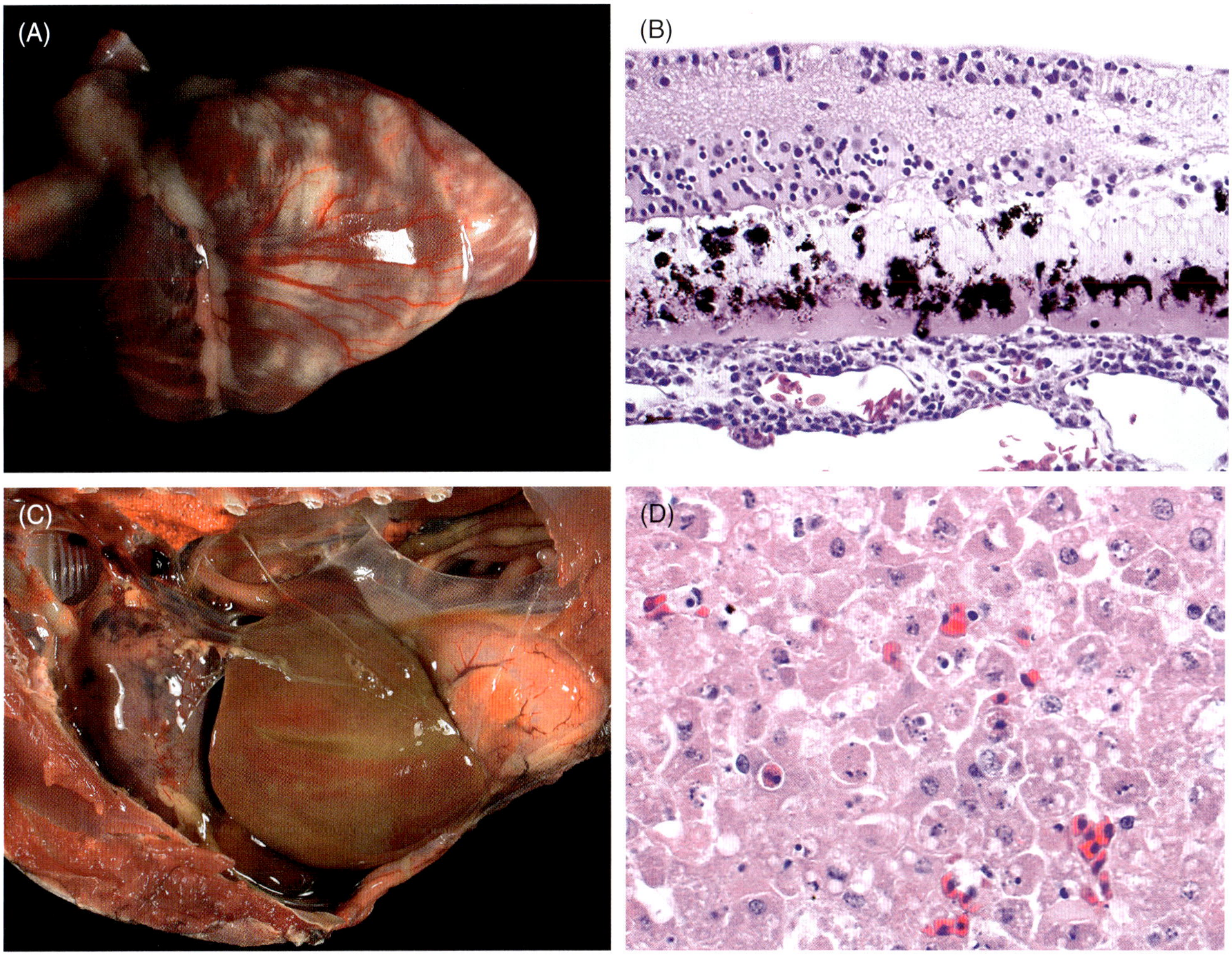

FIGURE 30.13 West Nile virus infection in a red-shouldered hawk (A), a red-tailed hawk (B), and a great gray owl (C, D). (A) Approximately 50% of the myocardium contains multifocal to coalescing areas of tan discoloration due to coagulative necrosis with dystrophic calcification. (B) In the eye, the choroid is infiltrated by lymphocytes and plasma cells. The retina is partially necrotic. The retinal pigmented epithelial cells are rounded and frequently detached. Photoreceptors are no longer apparent and the retinal layers are jumbled. The ganglion cell layer is infiltrated by a few plasma cells. (C) The liver is diffusely abnormally ochre-colored and is slightly swollen with rounded edges. (D) The liver is largely necrotic as evidenced by the individualization and rounding of the hepatocytes. The hepatocellular nuclei are karyorrhectic (see eSlides 30.10e and 30.11e).

deposition) in the fundus of the eye, and unilateral or bilateral collapse of the cerebral hemispheres.[145,148] Bald eagles may present with splenomegaly and/or cardiac dilation with a rounded, beige-discolored apex.

A triad of histologic lesions, which include myocarditis (eSlide 30.e10), endophthalmitis (Fig. 30.13B), and meningoencephalomyelitis (possibly including optic neuritis), is the hallmark of WNV disease (eSlide 30.e11). Multifocal to coalescing mononuclear, lymphoplasmacytic inflammation is typical in all sites. In the heart, up to 50% of the myocardium may be inflamed. In some hawk species, for example, red-shouldered and red-tailed hawks and goshawks, inflammation may be associated with myocardial necrosis and dystrophic calcification. The lesions of WNV infection in the eye are largely limited to the retina and choroid.[148] They range from subtle, focal to multifocal infiltration of the choroid with lymphocytes and necrosis of superjacent pigmented retinal epithelial cells to more striking inflammatory infiltrates with full-thickness retinal necrosis. In later stages, segments of retinal scarring over parts of the choroid that are heavily inflamed are present. Lymphoplasmacytic meningoencephalomyelitis affects all parts of the brain and spinal cord. Extensive cerebellar necrosis with hydrocephalus ex vacuo and parenchymal collapse primarily occurs in bald eagles and red-tailed hawks.[148] The lesion is usually bilateral and symmetrical and is characterized by rarefaction of the brain parenchyma with infiltration by gitter cells and perivascular lymphoplasmacytic infiltrates (eSlide 30.e11). Glial nodules are present in most cases, particularly in the molecular layer of the cerebellar gray matter. Multifocal necrosis of exocrine pancreatic cells may be seen. Splenitis and multifocal interstitial lymphoplasmacytic nephritis are also described in raptors with WNV infection; however, these lesions are nonspecific and potentially related to concurrent disease.

Infected neurons in the brain (e.g., Purkinje cells), spinal cord and retina, and glial cells in the brain and spinal cord are commonly WNV antigen positive. Cardiomyocytes may be widely or sparsely positive or may be negative, and occasional epithelial cells in distal tubules and collecting ducts are positive. Other cells that may be antigen-positive include exocrine cells of the pancreas, dendritic cells of splenic follicles, arterial smooth muscle cells, adrenal cells, and keratinocytes. Detecting WNV-specific antibodies in the cerebrospinal fluid is supportive of a diagnosis of WNV infection, even in the absence of detectable antigen in tissue sections or negative PCR results.

Strigiformes: A wide range of clinical signs, lesions, and antigen distribution is seen in owls infected with WNV. Adult and juveniles appear to be equally susceptible. Species with a more northern distribution, such as snowy owls and great gray owls, frequently die from infection without showing clinical signs or after a short (e.g., day-long) period of lethargy.[74] Gross lesions are characterized by subepicardial and subendocardial petechiae and ecchymoses, enlarged ochre-colored livers (Fig. 30.13C), and splenomegaly with pinpoint white foci. Widespread hepatocellular necrosis (Fig. 30.13D), multifocal splenic ellipsoid necrosis, and multifocal renal tubular epithelial cell necrosis are seen histologically. Intestinal mucosa may be multifocally necrotic. Rare glial nodules are present in the cerebrum and cerebellar gray matter. Viral antigen is abundant and may be present in endothelial cells of numerous organs (including brain), macrophages, and hepatocytes.

Owl species with a more southern range, including great horned owls and barred owls, have a prolonged course of disease. Great horned owls with WNV disease may develop acute head tremors that resemble rapid head bobbing and become persistent without improvement or worsening over time. Gross lesions are frequently absent except for emaciation. Histologic lesions include mild, but in few cases moderate to marked, lymphoplasmacytic myocarditis. Brain lesions are dominated by rare glial nodules and occasionally by subtle lymphoplasmacytic perivascular infiltrates (occasionally with heterophils) in the brain (particularly in the molecular layer of the cerebellum) and spinal cord. Necrosis of individual neurons may be seen. Mild lymphoplasmacytic pectenitis and mild iridocyclitis are seen but retinal lesions are absent. Some owls have subtle lymphoplasmacytic infiltrates in the pancreatic and renal interstitium. WNV-infected great horned owls occasionally present clinically with unilateral or bilateral wing or leg paralysis. In these owls, lesions may be concentrated in the spinal cord or peripheral nerves. WNV antigen is very sparse in great horned owls in general. Occasionally as few as one individual WNV antigen-positive neuron (e.g., Purkinje cells or large neurons of the brainstem) may be present in a brain section. The highest likelihood of detecting WNV antigen-positive cells in great horned owls is in cardiomyocytes and epithelial cells lining collecting ducts in the medullary cones of the kidney.

Falconidae: Published reports of naturally occurring WN disease in falcons are rare.[143] Falcons may have similar lesions and antigen distribution as those described in hawks. A lesion that appears to be more common and prominent in falcons (e.g., American kestrels and merlins) than in other raptors is fibrinoid necrosis of medium caliber arteries and arteritis in the spleen or heart. Myocardial necrosis and necrosis of brain parenchyma may be a sequela of arteritis or arterial necrosis. WNV antigen can be detected in similar cells as are seen in hawks and in arterial smooth muscle cells.

Cathartidae: Although many California condors (both vaccinated and unvaccinated) are seropositive for WNV, the virus has been implicated as the cause of death in only a single juvenile with non–suppurative encephalitis.[106] This bird had concurrent aspergillosis.

Usutu virus (USUV), another virus in the Japanese encephalitis complex of Flaviviridae, is implicated in the death of captive, nonnative owls in Europe.[152] In great gray owls, primary gross lesions include splenomegaly, hepatomegaly, and congestion of the lungs and brain. Encephalitis with neuronal necrosis and the formation of glial nodules, coagulative necrosis in the liver and hyperplasia of the red pulp in the spleen are seen histologically. Fatal infections have also occurred in captive Northern hawk owls, Tengmalm's owl, pygmy owls, and a free-living long eared owl. Antibodies to USUV have been detected in a Eurasian eagle owl, snowy owl, and Ural owl and in an Egyptian vulture.[12]

Raptors have been infected naturally and experimentally with **avian influenza viruses**, including **highly pathogenic avian influenza virus (HPAI)** strains. Virus-associated mortality in raptors has been a prominent feature of recent HPAI outbreaks throughout the world, presumably due to predation of or scavenging on waterfowl or other infected birds. There are no pathognomonic lesions in HPAI-infected raptors. Differences in the distribution of macroscopic and microscopic lesions in various species, outbreaks, and natural versus experimental infections might be due to different tissue tropism of the various viral strains.[64] Gross lesions are only inconsistently present. Hemorrhagic pancreatic necrosis has been the most prominent gross finding in HPAI-infected falcons during outbreaks caused by H5N1 strains in the United Arab Emirates (UAE) and in experimental infections of kestrels with H5N1 strains (Fig. 30.14A).[91] Experimental

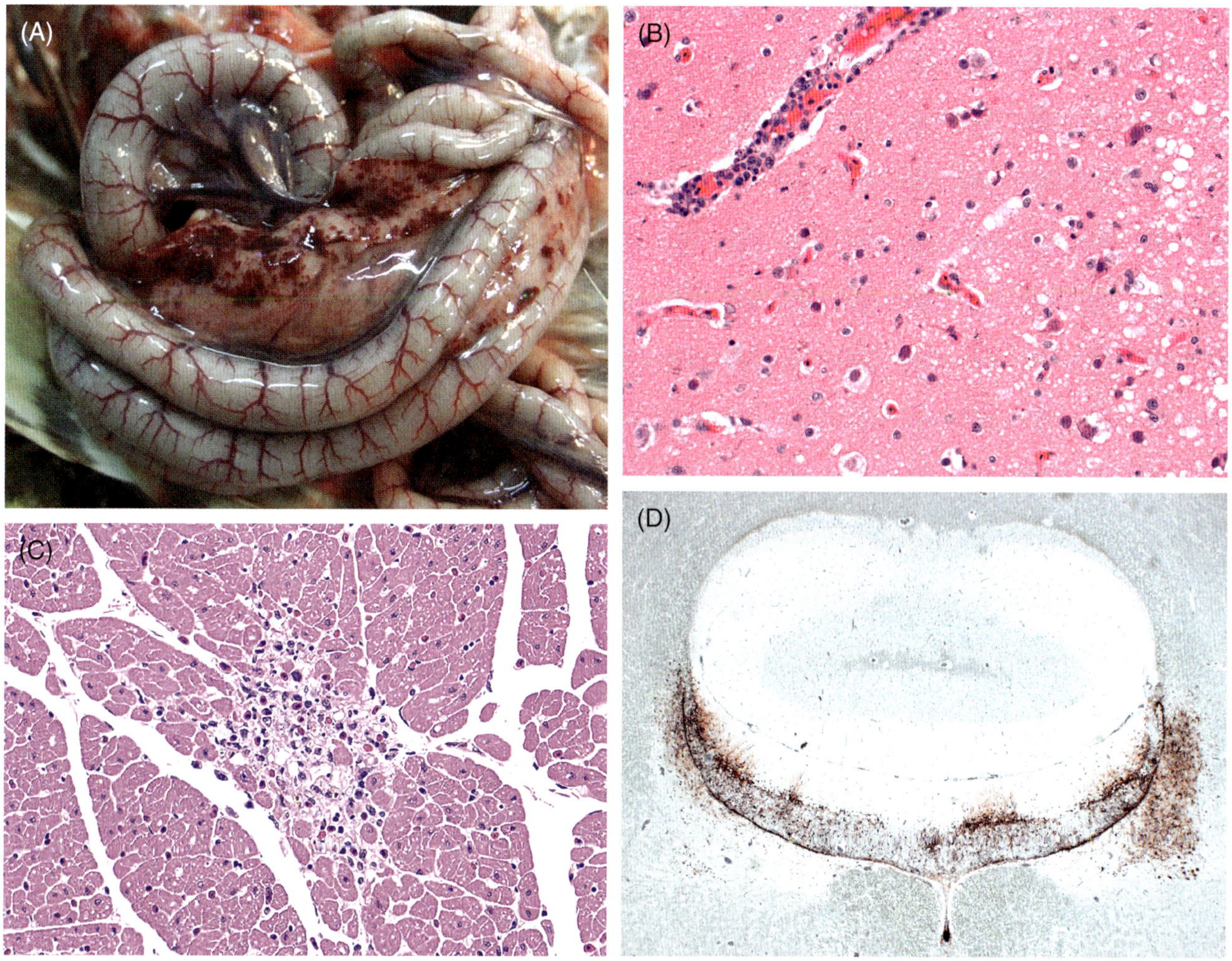

FIGURE 30.14 Highly pathogenic avian influenza virus in the pancreas of a falcon (A), the cerebrum of a red-tailed hawk (B), the heart of a bald eagle (C), and the brain of a red-tailed hawk. (A) The pancreas is swollen and contains numerous foci of hemorrhage and necrosis. There are no pathognomonic HPAI lesions; however, this lesion is a relatively consistent finding in some species. (B) A capillary is surrounded by multiple layers of mononuclear inflammatory cells including mononuclear cells and heterophils. Numerous brightly eosinophilic (necrotic) neurons are scattered in the adjacent neuroparenchyma. (C) A small cluster of cardiomyocytes is necrotic. The focus of necrosis is infiltrated by few lymphocytes and macrophages. (D) Ependymal cells of the 4th ventricle and neurons of the adjacent brainstem and cerebellum (including Purkinje cells) are immunopositive for nucleoprotein antigen of influenza A virus (see eSlides 30.e12 and 30.e13). *(Part A: Reprinted with permission from Naguib MM, Kinne J, Chen H, Chan KH, Joseph S, Wong PW, Woo PCY, Wernery R, Beer M, Wernery U, Harder TC. Outbreaks of highly pathogenic avian influenza H5N1 clade 2.3.2.1c in hunting falcons and wild birds in Dubai suggest progressive intercontinental virus spread. J Gen Virol. 2015;96:3212–3222.)*

nasochoanal infection of gyrfalcon-saker falcon hybrids with H5N1 produces similar pancreatic necrosis and causes pancreatic exocrine atrophy, multifocal mucosal petechiae at the proventriculus-ventriculus junction, pulmonary congestion with focal hemorrhage, and mild catarrhal tracheitis.[7] While gross lesions were uncommon in white-tailed eagles and common buzzards that died of H5N8 virus infection in 2016–17 and in 2020–21 in Europe, a combination of reddened firm lungs, pancreas necrosis, splenomegaly, epicardial hemorrhage, and myocardial necrosis were fairly common gross findings in various raptor species that died during the 2014–15 H5N2 and H5N8 outbreaks in the United States.[16,61,121]

Microscopic nervous system lesions in natural infections in raptors can be subtle consisting of mild multifocal gliosis with only minor mononuclear inflammation. However, mild to marked necrotizing meningoencephalitis with neuronal necrosis, most severely affecting the forebrain and also affecting mesencephalon and brainstem with only minor involvement of the cerebellum, has been described (Fig. 30.14B; eSlide 30.e12). The combination of fibrinoid necrosis of the walls of brain capillaries (with or without thrombosis), mononuclear meningoencephalitis, selective acidophilic neuronal necrosis, focal or multifocal pannecrosis with granulocytic infiltration, and mixed-cellular ependymitis was described in great horned owls, red-tailed hawks, and bald eagles during the HPAI outbreak that began in North America in the winter of 2021. When present, this combination of brain lesions can help to differentiate HPAI virus-induced encephalitis from WNV-induced encephalitis on a microscopic level.[146] Experimental infection causes multifocal necrosis, neuropil spongiosis, chromatolysis, gliosis, and karyolysis in the cerebrum, and Purkinje cell chromatolysis, necrosis, and nonsuppurative inflammation. Similar neurologic lesions were seen in free-ranging common buzzards and peregrine falcons during an H5N1 outbreak in Germany in 2006.[133] Besides the microscopic brain lesions, myocardial necrosis, and myocarditis (Fig. 30.14C) and necrosis of the exocrine pancreas (eSlide 30.e13) were prominent microscopic features in various raptors that died during a 2014–15 outbreak in the United States and 2020–21 outbreak in Europe.[121]

Avian influenza viruses can be detected using oropharyngeal swab samples and are fairly consistently present in the brain (Fig. 30.14D) and heart by immunohistochemistry. Viral antigen is less consistently expressed in the pancreas, adrenal gland, kidney, spleen, liver, and eye. Vaccination of gyrfalcon x saker hybrid falcons using an inactivated H5N2-specific vaccine has been shown to prevent clinical disease following experimental infection with highly pathogenic H5N1 virus.[67] Natural transmission of HPAI to raptors may be via several mechanisms. Wild birds used as prey for the hunting falcons may be a source of infection. For example, gyrfalcons in the United States were infected with HPAI H5N8 after consumption of an infected American wigeon.[49] The Arabian Peninsula lies along the central Asian flyway, a major migration route for birds from breeding grounds in central Asia to African wintering sites. The timing of an outbreak of HPAI in falcons of the Arabian Peninsula in the fall of 2014 supports a scenario of introduction of the virus by migrating birds. HPAI virus was detected in hooded vultures found dead with respiratory or neurologic signs during a poultry outbreak in Burkina Faso in 2006.

Newcastle Disease viruses (NDV) (World Organization for Animal Health [WOAH] listed reportable disease) cause one of the most important diseases of domestic poultry worldwide. They belong to the *Avulavirus* genus and *Paramyxoviridae* family, and are recognized as **avian paramyxovirus serotype-1 (PMV-1)**. NDVs are divided into lentogenic, mesogenic, and velogenic strains based on their pathogenicity (least, moderate, and most, respectively) in chickens. Many different paramyxoviruses can infect birds. Over a period of 20 years, 255 paramyxoviruses were isolated and identified as the cause of death in 1167 birds of prey at the Central Veterinary and Research Laboratory in Dubai. This included infections in several different raptor species: 247 from falcons, 5 from hawks and 3 from secretary birds. Because low, moderate, and high virulence NDV strains exist, outcomes of infection differ, and attempts should be made to characterize and differentiate them. In the Middle East, hundreds of captive falcons died from infection with velogenic NDV before a killed vaccine containing strains isolated from four local avian species was produced and became available in the year 2000.[72] Raptors usually become infected after ingestion of infected prey.

Infections, particularly with velogenic NDV, produce fatal neurotropic and viscerotropic disease in falcons.[114] Anorexia, vomiting, and paralytic ileus are common acute signs; bloody diarrhea may also be seen. Central nervous system signs include ataxia, head tilt, tremors, and wing and leg paralysis. Torticollis develops late in the disease. Mucosal petechiae in the proventriculus or hemorrhage and congestion in the respiratory tract, typical of infection in poultry, are rare in birds of prey.[138] Besides a swollen pancreas with petechiae (Fig. 30.15A), no gross lesions are detected in raptors showing neurological signs.

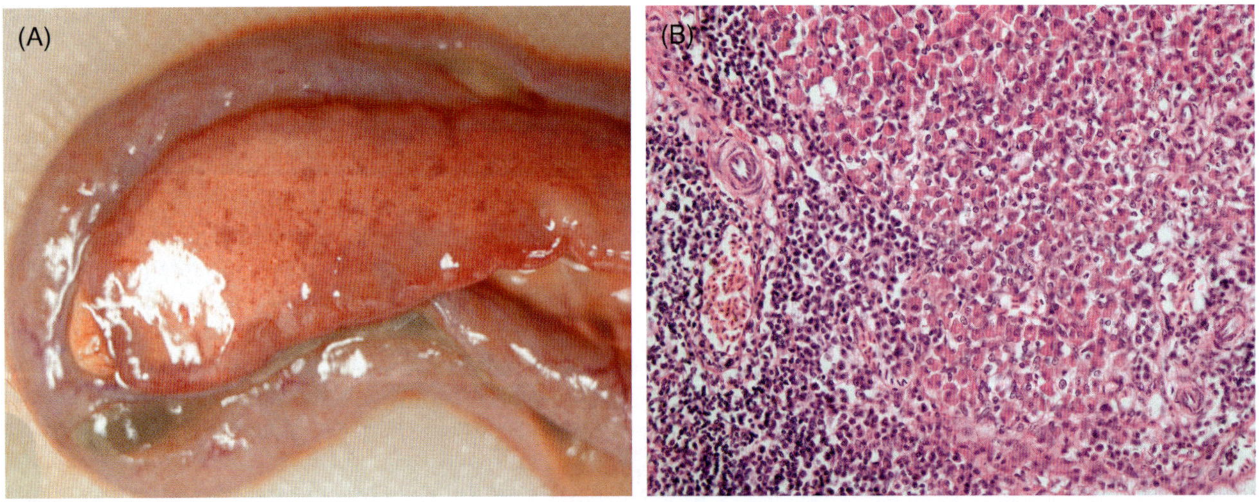

FIGURE 30.15 Newcastle disease virus in the pancreas of a Saker falcon (A) and a peregrine falcon (B). (A) The pancreas is swollen and contains numerous pinpoint red foci (petechiae). (B) The interstitium contains focally extensive inflammation composed of mononuclear cells.

Lymphoplasmacytic and histiocytic pancreatitis is the microscopic lesion in falcons with the viscerotropic form (Fig. 30.15B); falcons with the neurotropic form often only have subtle nonsuppurative encephalomyelitis with mild to severe demyelination in the cerebrum. Velogenic NDV has also been reported in an adult female bearded vulture found dead in a zoo. Gross lesions included intestinal hyperemia with hemorrhagic content, hepatomegaly, and pulmonary and renal congestion. Histiocytic hepatitis and duodenal and jejunal mucosal erosions were seen histologically.[75] Infection was confirmed with virus isolation. The detection and characterization of lentogenic strains of NDV from raptors has been reported in the United States.[53] Virus can be isolated from the spleen, brain, or lungs among other organs.

Bacteria

Anecdotally, vultures appear to be resistant to pathogens encountered in carcasses, such as *Bacillus anthracis*. Factors that might contribute to resistance include anatomic features, such as the relatively bald head and neck in carcass feeders, low gastric pH, gut microflora that is inhospitable to certain pathogens, and hypothesized functional differences in vulture receptors for specific molecule patterns in pathogens such as toll-like receptors.[27] Toll-like receptors are important components of the innate immune system that are involved in antimicrobial responses. African vultures have also been shown to carry antibodies to *B. anthracis*, and are implicated in both containment (ingestion and killing of the vegetative forms) and spread (regurgitation or fecal release of undigested spores) of *B. anthracis* during scavenging. Despite the ability to resist diseases found in carrion, vultures are not immune to infectious bacterial diseases, though intoxications and trauma are far more significant causes of mortality.

With the exceptions of *Pasteurella multocida* and *Mycobacterium avium* subspecies *avium* (described later), bacteria that cause disease in raptors are not well documented.[142] Infrequently described bacteria in raptors include *Chlamydia psittaci*, *Mycoplasma* spp., *Staphylococcus* spp., and *Salmonella* spp.; those that are rarely reported include *E. coli*, *Borrelia* spp., *B. anthracis*, *Streptococcus faecalis* and *S.* spp., *Clostridium* spp., *Erysipelothrix rhusiopathiae*, *Listeria monocytogenes*, *Corynebacterium* spp., and *Ornithobacterium rhinotracheale*. See Table 30.e2 in the Digital Content for brief summary information on *E. coli*, *Borrelia* spp., and *O. rhinotracheale* infections in raptors.

Avian cholera is caused by ***P. multocida***. Serotype 1 followed by 3 and 4 are the most common *P. multocida* serotypes isolated from eagles, hawks, falcons, and owls.[87] The bacterium is not only transmitted orally through ingestion of contaminated food, water, and infected waterfowl but can also be transmitted through inhalation or by cat scratches and bites. Pathogenesis is due to endotoxemia, which results in hyperemia and fibrinous polyserositis (Fig. 30.16). Acute lesions include hepatomegaly and splenomegaly with pale foci of necrosis and fibrinoheterophilic inflammation with numerous, intralesional bacteria. Similar inflammation can be seen in the pericardium, air sacs, lungs, coelomic cavity, intestine, and the irises and ciliary bodies of the eyes, as well as the skin and skeletal muscle in areas of infected cat scratches and bites. An interesting chronic form of avian cholera characterized by oropharyngeal and esophageal granulomas is described in *Buteo* hawks. Histologically, granulomas contain multinucleated giant cells and intralesional bacteria. A tentative diagnosis of *P. multocida* can be made by reviewing blood films and/or organ impression smears stained with methylene blue or Wright's Giemsa stains which highlight the bipolar nature of the bacteria. Confirmation of the diagnosis is by bacterial isolation. Serotyping is performed only in a few laboratories.

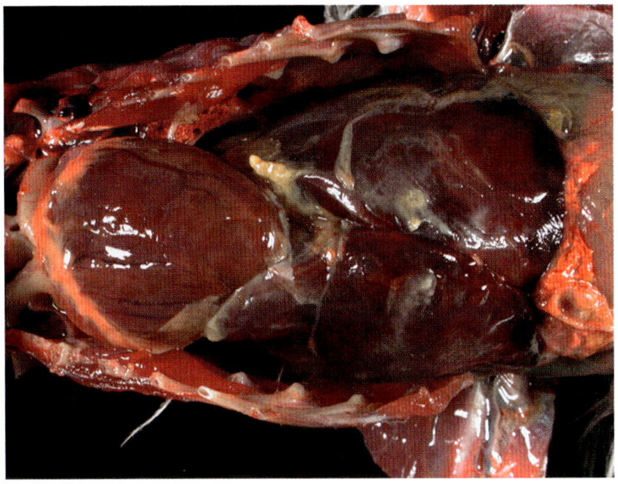

FIGURE 30.16 *Pasteurella multocida*, serotype 1 in the liver of a gyrfalcon. The liver is partially covered by thin, opaque white, fibrinous membranes caused by the bacterial infection and associated inflammation.

Mycobacterium avium ssp. ***avium*** causes **avian mycobacteriosis**, a chronic debilitating disease in various species of birds including raptors. Disease is described in various species of hawks, falcons, eagles, kestrels, and owls.[111,129] Affected raptors are generally weak and emaciated. The most common lesions involve the liver, spleen, and intestine. The liver and spleen can be enlarged and contain yellow/tan nodular foci that often coalesce into larger granulomas (Fig. 30.17). The intestine can contain similar mural nodules. Occasionally, the lungs, air sacs, kidneys, vertebrae, skin, joints, brain, spinal cord, and/or eyes are similarly involved. Histologically, granulomas are typical and composed of central necrosis and necrotic eosinophilic and heterophilic debris surrounded by multinucleated giant cells and epithelioid macrophages. This, in turn, can be

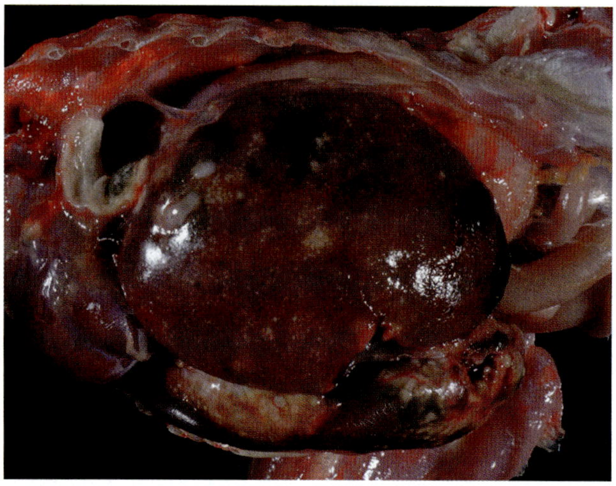

FIGURE 30.17 *Mycobacterium avium* infection in the liver of a red-tailed hawk. The liver is markedly enlarged and contains multifocal to coalescing, pinpoint to variably sized areas of yellow/tan discoloration that correspond histologically to granulomas. Concurrent fungal infection, seen as a plaque covered by a fungal mycelium (aspergilloma), is present in left cranial thoracic air sac in this bird.

surrounded by lymphocytes, plasma cells, heterophils, and fibrosis in chronic infections. Acid-fast staining (e.g., Ziehl-Neelsen or Fite's acid-fast) is usually positive for large numbers of intrahistiocytic acid-fast-positive bacteria. This is the easiest and fastest method of diagnosing mycobacteriosis in birds. Isolation or identification of *M. avium* ssp. *avium* by PCR is confirmatory.

Chlamydia psittaci are Gram-negative, obligate intracellular bacteria that cause **chlamydiosis** in various species of raptors. Despite having a high prevalence of antibodies to *C. psittaci*, disease has been infrequently described in hawks, falcons, eagles, kestrels, and owls.[54] Transmission is through respiratory, oral, and conjunctival routes. Gross pathology includes hepatosplenomegaly as the most common lesion, and fibrinous exudate in the pericardium, air sacs, or on the capsular surface of the liver. Microscopic lesions vary in severity and distribution but most commonly involve the liver and spleen, which contain few to large numbers of macrophages or histiocytes with or without necrosis and fibrin exudation. Some macrophages contain small, basophilic coccoid organisms. Other histologic lesions include fibrinous pericarditis and airsacculitis with infiltration of macrophages containing intracytoplasmic organisms. Pleuritis, pneumonia, and meningoencephalitis may also be seen in red-tailed hawks. Recently, *C. buteonis*, an intermediate between *C. psittaci* and *C. abortus*, was isolated from hawks and falcons that had clinical signs of upper respiratory disease and/or diarrhea.[125]

Various serotypes of ***Salmonella*** have been isolated from normal and diseased raptors, but only sporadic reports associate the bacteria with disease.[85,139] *S. typhimurium* was isolated from the liver, spleen, and intestine of six peregrine falcons and two gyrfalcons that had hepatomegaly and splenomegaly with miliary necrosis. *S. havana* and *S. virchow* were isolated from embryos and neonates of the European eagle owl, peregrine falcon, and lanner falcon. The neonates infected with *S. havana* had congestion in various organs and histologically there was severe vasculitis in the liver, lungs, and brain. *S. virchow* infection caused polyserositis, duodenal abscessation, and gelatinous edema in the cervical region. *S. enteritidis* phage type 21b isolated from the liver, spleen, and intestine of a Eurasian eagle-owl was associated with small whitish nodules in the liver and hemorrhagic exudate in the proventriculus and proximal ventriculus. Intestinal salmonellosis is characterized grossly by serosal petechiae and fibrinonecrotic mucosal exudate, and enterohepatic spread leads to periportal hepatic coagulative necrosis and inflammation and portal vein thrombosis.

Many species of **mycoplasma** have been isolated from healthy and ill raptors.[69] In one study of debilitated or injured birds of prey in Germany, *Mycoplasma* spp. were isolated from 82% of nestlings, 26% of juveniles, and 55% of adults. Some of these birds had mild airsacculitis from which *M. buteonis*, *M. gypis*, or *M. falconis* were isolated. In a study summarizing the literature on Mycoplasma isolations, *M. anatis*, *M. gallisepticum, M. gallinaceum, M. gallinarum*, *M. iners*, and *M. corogypsi* were isolated (typically) from the respiratory tract of various species of raptors, some of which had respiratory signs and sinusitis. Transmission of mycoplasma in birds is primarily through the respiratory tract; vertical transmission can also occur.

In vultures, *M. corogypsi* has been associated with polyarticular swelling and distension of tendon sheaths in American black vultures.[109,135] Gross lesions include thickening, opacification, and proliferation of synovial membranes with pannus formation and varying degrees of cartilage damage ranging from mild fibrillation to cartilaginous ulceration with extension into the subchondral bone. Histologically, synovial inflammation is lymphoplasmacytic with fibrinonecrotic debris in joint spaces. Bacterial isolation or PCR is confirmatory; a PCR technique has been published for detection and differentiation of mycoplasmas, including *M. corogypsi* in birds of prey.[68]

Staphylococcus spp. are gram-positive, coccoid bacteria. *S. aureus* is the most common and pathogenic species isolated from pododermatitis lesions (**bumblefoot**) in raptors, but septicemic forms with valvular endocarditis also occur (Figs. 30.18A–C).[142] Pododermatitis occurs primarily in captive raptors as a sequela to suboptimal perching substrate/or underlying trauma. Skin abrasion, ulceration, inflammation, and swelling of the metatarsal or a digital pads are characteristic lesions. Depending on the severity of the lesions 5 grades of pododermatitis are distinguished by clinicians. Microscopically, acute inflammation is typically fibrinoheterophilic; macrophages and multinucleated giant cells and severe fibrosis with granulation tissue formation are present in subacute and chronic

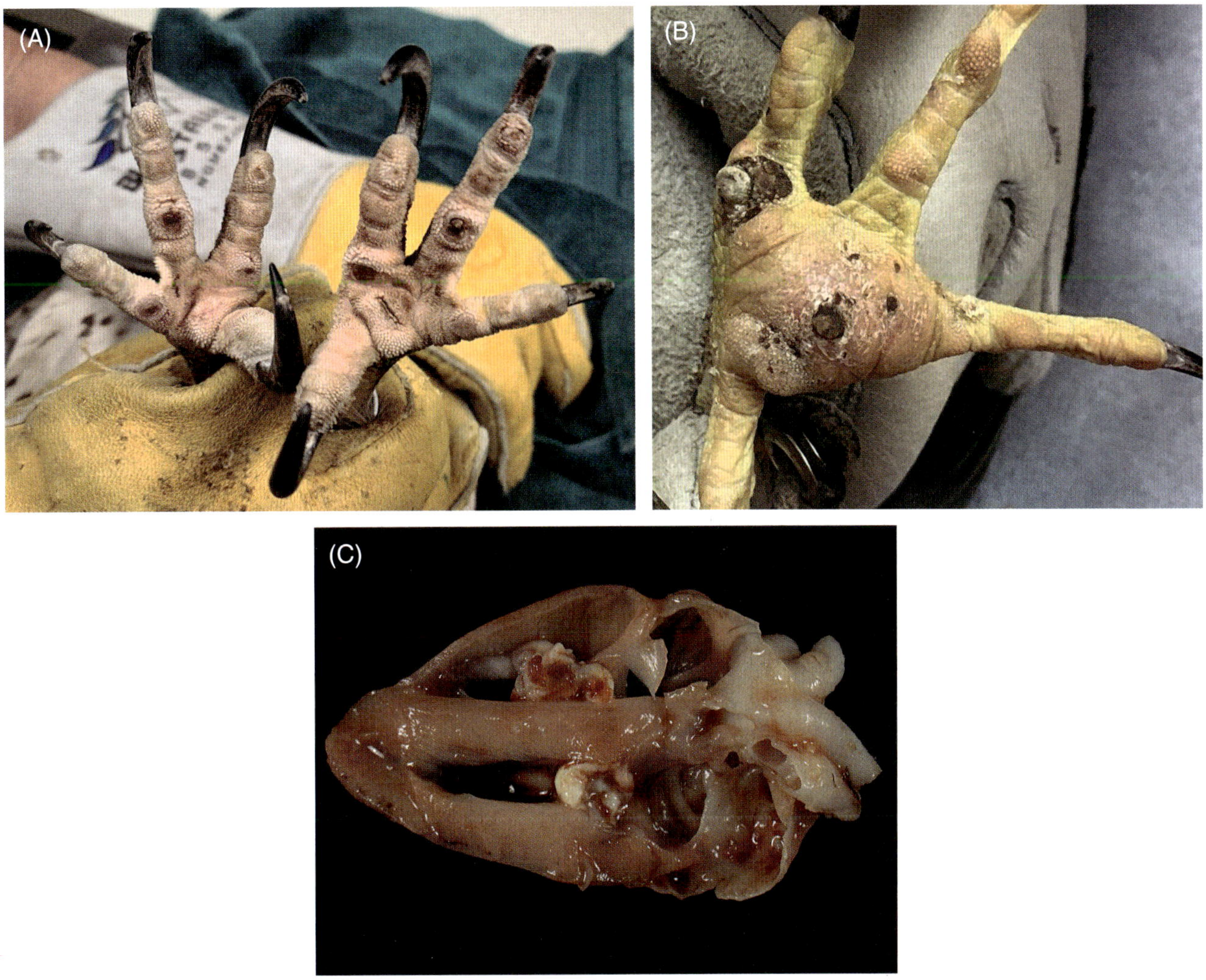

FIGURE 30.18 *Staphylococcus aureus* infection in areas of ulcerative pododermatitis in the feet of a red-tailed hawk (A) and red-naped Shaheen falcon (B), and the heart of a Cooper's hawk (C). (A) Ulcerated lesions with proliferative inflammation, commonly called "bumblefoot," are present in the skin of multiple digits of both feet over pressure points. (B) The foot is markedly swollen and has multiple areas of ulceration. (C) Ulcerated skin provides a portal for entry of the bacteria into the systemic circulation, which can result in valvular endocarditis. Characteristic lesions are white/tan material/thrombi on the valve leaflets, in this case, on both the right and left atrioventricular valves. *(Parts A, B: Photos Courtesy of Dana Franzen-Klein, The Raptor Center, University of Minnesota.)*

lesions. Superficial to deep colonies of gram-positive cocci are common. Colonies of similar bacteria and associated inflammation and necrosis affect and occasionally expand the affected cardiac valves in septicemia.

Fungi

Aspergillosis is one of the most important causes of death in captive raptors worldwide. The disease is similar to that in other birds with respect to pathogenesis and pathology. Predisposing factors include environmental and husbandry conditions, especially high humidity and temperature or dry air containing a high amount of dust, or stress associated with excessive fasting in falconry birds. A single exposure to ***A. fumigatus*** caused acute aspergillosis in juvenile falcons after experimental infection. Aspergillosis is far less frequent in free-ranging raptors. In them, it has been related to chronic physiologic (fasting) and psychologic (agonistic) behavior during migration.[55,105] Trauma or other chronic, debilitating disease conditions, such as lead toxicosis, predispose to aspergillosis in free-ranging birds of prey. Among them, gyrfalcons, *Accipiter* species, eagles, merlins, and some owls (e.g., barn owls) appear to be the most susceptible species. Aspergillosis is commonly caused by inhalation of spores of ***A. fumigatus***, but ***A. niger*** or ***A. flavus*** inhalation also can result in aspergillosis.

Lesions can be found throughout the respiratory tract (Fig. 30.19) and can extend to any other organ. Lesions of primary aspergillosis are often found in the caudal thoracic air sac, as this is where spores first settle due to the anatomy of the avian respiratory tract. Acute aspergillosis is characterized by miliary lesions in lungs and air sacs and is thought to be related to aspiration of high number of spores. Lesions of chronic aspergillosis may be restricted

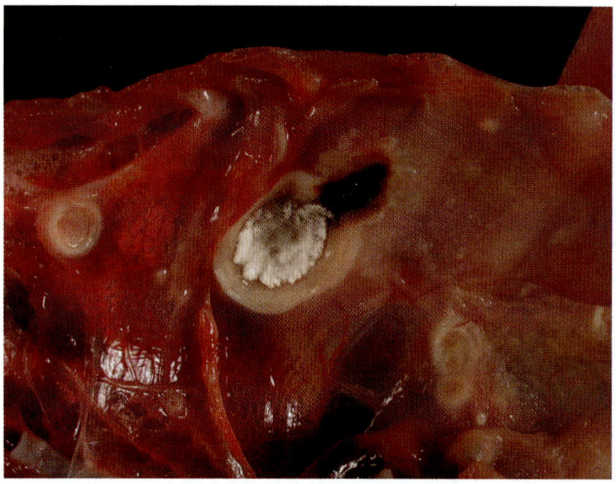

FIGURE 30.19 Aspergillosis in the respiratory tract of a red-tailed hawk. Multifocal fungal plaques, one of which is covered by a white, fungal mycelium, are present in the air sacs and the lung.

to the respiratory tract or affect a variety of organs. The extension, type, and severity of lesions are largely related to the susceptibility of the species and the immune reaction of the individual host. Aspergillosis lesions may consist of well-delineated granulomas that range in size from pinpoint to many centimeters in diameter.[6] Granulomas may be located in the trachea, nasopharynx, air sacs, or lung parenchyma but can also affect the serosa or parenchyma of other organs. Granuloma formation in the upper respiratory tract, especially the trachea or nasal turbinates, or debris and fibrinous plaques can, in some cases, lead to obstruction of the trachea or syrinx and death from asphyxia. Nongranulomatous lesions may include air sac thickening or severe fibrinous airsacculitis with abundant fungal growth on the serosal surface of the air sacs and in pneumatized bones.

Microscopically, chronic aspergillosis lesions in birds of prey typically consist of granulomas with a central core of degenerate heterophils surrounded by multinucleated giant cells and epithelioid macrophages within which are embedded radially organized, septate, parallel-walled, dichotomously branching fungal hyphae and characteristic conidiophores (the latter are most often found at air interfaces). A fibrous capsule may be seen in chronic cases and fungus may invade blood vessels of the lung and occasionally other organs. Severe chronic infection and inflammation may lead to the replacement of lung parenchyma by fibrotic scar tissue and produce secondary right ventricular dilatation and cor pulmonale. Acute aspergillosis lesions are usually found in nestlings that are exposed to large numbers of spores. In these cases, the lesions are dominated by a fibrinous and heterophilic exudate plugging parabronchi with abundant fungal hyphae. The fungus is positive with PAS and Grocott methenamine silver (GMS) stain. For confirmation of presumptive aspergillosis, fungal culture, PCR, or immunohistochemistry are recommended.

The primary cause of **candidiasis** in birds of prey is ***Candida albicans***, which is present as part of the normal flora of the upper digestive tract in healthy individuals. Lesions of candidiasis are often associated with predisposing conditions that have an adverse effect on the mucosal immunity and microflora composition of the host, such as vitamin A deficiency, prolonged antibiotic treatment, or concomitant infections; systemic infection can occur secondary to predisposing infections that impair the mucosal barrier, such as avian poxvirus infection.[122] It is more commonly observed in captive animals than in free-living individuals. Candida lesions in the oral mucosa, soft palate, and upper esophagus may be present as whitish-gray foci or plaques that, upon direct microscopy with methylene blue/lactophenol blue stain, contain numerous pear-shaped budding organisms. Crop candidiasis can lead to severe thickening and focal necrosis of the crop epithelium resulting in a rugal appearance or pseudomembranes. Histologically, lesions are characterized by hyperplasia of the epithelium, abundant lymphoplasmacytic inflammation of the submucosa and muscularis, and intralesional yeast and pseudohyphae (highly elongated filamentous structures with parallel walls and septal constrictions).

Enterocytozoon bieneusi genotype D, an obligate, intracellular, **microsporidian** organism recently classified as a fungus, causes fatal infections in various species of falcons in the Middle East. Disease is characterized by the presence of yellowish, well-demarcated, raised nodules in liver, kidneys, spleen, and intestine (Fig. 30.20); some

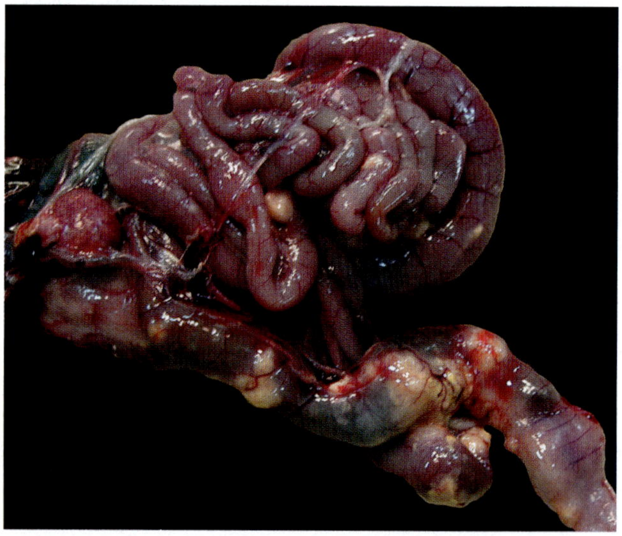

FIGURE 30.20 *Enterocytozoon bieneusi* infection in the intestine and spleen of a falcon. The intestinal wall is multifocally expanded by numerous, yellowish nodules, and plaques that correspond histologically to transmural, necrotizing enterocolitis with diphtheritic membrane formation. The spleen is mildly enlarged and has poorly defined white foci. *(Reprinted with permission from Müller MG, Kinne J, Schuster RK, Walochnik J. Outbreak of microsporidiosis caused by Enterocytozoon bieneusi in falcons. Vet Parasitol. 2008;152:67–78.)*

birds may be heavily coinfected with *Caryospora* spp.[89] Histologically, hepatic, renal, and splenic lesions consist of multifocal necrosis and pyogranulomatous inflammation. In the intestine, there is transmural, necrotizing enterocolitis with diphtheritic membrane formation. Organisms are oblong, up to 2 μm long, intracytoplasmic, and are generally rare and only detectable histologically in a small number of cases. Immunohistochemistry with cross-reacting commercially available polyvalent rabbit antiserum against *Encephalitozoon cuniculi* aids in detection of the intracytoplasmic microsporidia. PCR targeting microsporidian DNA can be used to confirm the diagnosis. Detection of microsporidia in fecal samples is challenging due to the small size of the organism. Infected feces from ducks, cranes, and pigeons may be the source of *E. bieneusi* infection in falcons.

Metazoa

A wide variety of helminths, including nematodes, trematodes, cestodes, and acanthocephalans, infect the alimentary tract and other organ systems of birds of prey. It is neither the intent nor purpose of this textbook to provide a complete list of the parasites of raptors. However, several notable helminths that are associated with morbidity and mortality are described below.

Spirurid nematodes in the genus ***Serratospiculum*** and ***Serratospiculoides*** infect the air sacs of falcons and less commonly of hawks. Infection of the thoracic and abdominal air sacs and body cavity with *Serratospiculum amaculata* is common in prairie falcons and may cause clinical disease and death on rare occasions.[137] *Serratospiculum seurati* causes serratospiculiasis in the Middle East.[128] Adult worms colonize the air sacs where they lodge in the mesenchymal layer (Fig. 30.21A). Adult female nematodes are large, up to 20 cm in length, and lay embryonated eggs in the air sacs. Eggs travel to the lungs and are coughed up, swallowed, and passed in the feces. Beetles ingest the embryonated eggs and infect raptors via ingestion. Birds are suspected to serve as paratenic hosts for avivorous raptors. L3 larvae are thought to penetrate the proventricular and ventricular wall and lodge in the air sacs, completing the life cycle. Adult nematodes in the air sacs can cause clinical signs; however, large numbers can be found in apparently healthy, well performing falcons. Clinically affected falcons may have poor stamina, labored breathing, and may be "poor doers". The air sac membranes are grossly thickened (Fig. 30.21A). Complications include secondary bacterial infections and/or aspergillosis. Other findings include edema, hemorrhage, mild necrosis, and heterophilic and histiocytic inflammation of the air sacs (Fig. 30.21B). Aberrant migration of this spirurid nematode has resulted in neurologic signs.[134]

Infection with opisthorchiid **liver flukes**, such as ***Opisthorchis* sp.**, ***Metorchis bilis***, or ***Erschoviorchis*** sp., is common in bald eagles and white-tailed sea eagles.[62,79] These infections also occur in harriers. Infection with *Erschoviorchis* sp. can be severe in hatch-year bald eagles but the clinical significance is uncertain. Disease and mortality are attributed to massive infection with *M. bilis*. Opisthorchiid flukes usually have two intermediate hosts that may include an aquatic snail primary intermediate host and an amphibian or fish secondary intermediate host. Adult *M. bilis* flukes reside in the biliary system where they lay eggs. Massive infection with *M. bilis* can be associated with granulomatous cholangitis, pericholangitis, and necrosis of adjacent hepatocytes ultimately leading to significant fibrosis. Thin-walled, adult flukes in North American bald

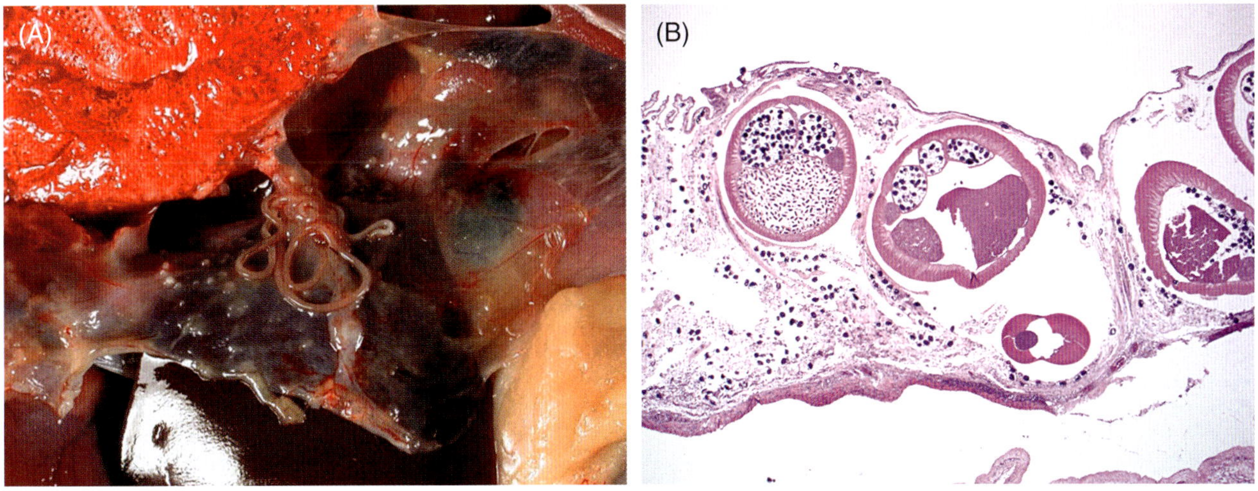

FIGURE 30.21 Spirurid nematodes (*Serratospiculum* sp.) in a gyrfalcon (A) and prairie falcon (B). An adult spirurid nematode is present in the membranes of a thoracic air sac. The air sac is slightly opaque and contains multiple, pinpoint white foci of inflammation. (B) Cross-sections of female spirurid nematodes are present in the air sac membrane. The nematodes have a thick cuticle, lateral cords, polymyarian-coelomyarian musculature, and gravid uteri with embryonated eggs. Heterophilic granulocytes infiltrate the air sac membrane adjacent to the nematodes.

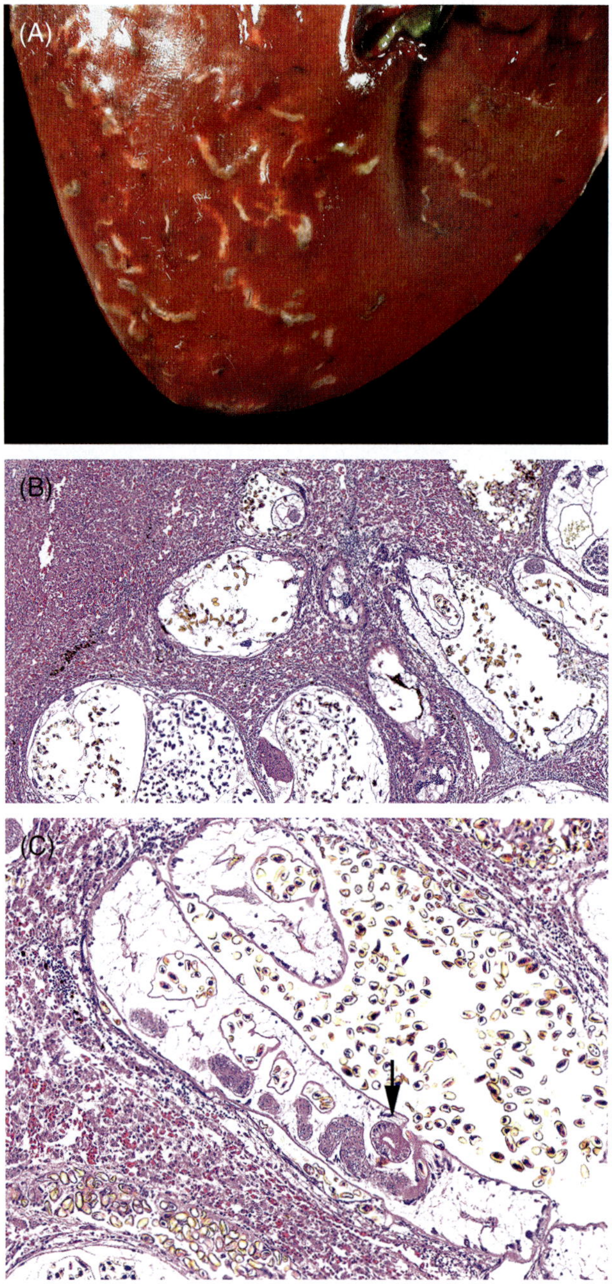

FIGURE 30.22 Liver flukes (resembling *Amphimerus* sp.) in a bald eagle. (A) Numerous, slightly raised, translucent to opaque white, serpentine parasite migration tracts that contain slender flukes run beneath the liver capsule and within the parenchyma. (B) A large percentage of the liver parenchyma is replaced by flukes that contain numerous pigmented ova. A degenerate fluke in the center of the image is associated with granulomatous inflammation. (C) A single fluke compressing hepatic parenchyma and surrounded by pigmented ova is present. A ventral sucker can be seen (*arrow*) (see eSlide 30.e14).

eagles (*Erschoviorchis* sp.), resembling *Amphimerus* sp., extensively migrate within the liver parenchyma causing slightly raised serpentine migration tracts and granulomatous hepatitis (Figs. 30.22A–C, eSlide 30.e14).

For information on additional notable helminths in raptors, see Table 30.e3 in the Digital Content.

Protozoa

Species of ***Sarcocystis*** (including the former ***Frenkelia* sp.**) have a heteroxenous life cycle with a predatory final host and a prey species intermediate host. Sporulated oocysts of *Sarcocystis* species (or *Frenkelia* species) are commonly found in the lamina propria mucosae of the small intestine in raptor species. Intestinal infection is considered insignificant except in massive infection in nestlings that may lead to malabsorption. Raptors may be the definitive or intermediate hosts for a variety of *Sarcocystis* species (e.g., *S. calchasi*, *S. glareoli*, and *S. jaypeedubeyi*, formerly *Frenkelia microti*).[78,98] The intermediate host status of a raptor is characterized by the presence of tissue cysts in skeletal muscle and the myocardium of otherwise healthy raptors belonging to various families. In the case of ***Sarcocystis falcatula***, muscle tissue cysts are elongate and conform to the shape of host myofibers. The cysts are 30–40 μm × 250 μm with a cyst wall up to 3–4 μm thick that has fine villous projections. Numerous bradyzoites are visible within cysts, which may lack or contain internal septation. Meronts and schizonts of *S. falcatula* may be present in endothelial cells of pulmonary capillaries, neurons (Fig. 30.23), and glial cells and have the potential to cause both morbidity and mortality.[147] Protozoal schizonts in pulmonary capillary endothelial cells are compressed resulting in characteristic, elongated serpentine profiles. Some schizonts are mature and contain numerous, mature merozoites whereas others are immature and contain large basophilic serpentine nuclei. Intraendothelial protozoa in the pulmonary

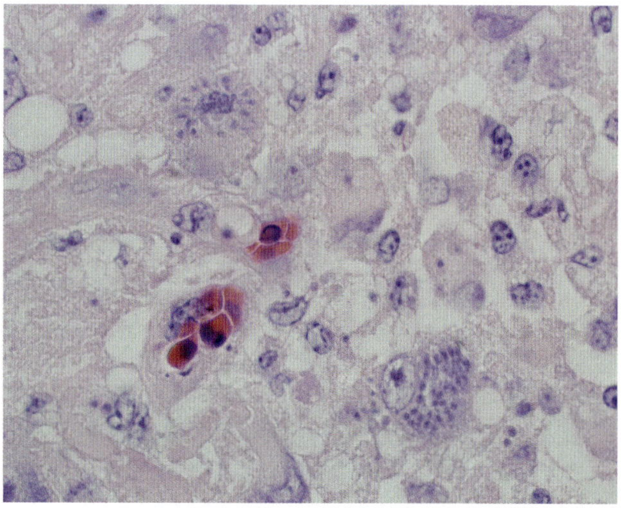

FIGURE 30.23 *Sarcocystis falcatula* in the cerebrum of a great horned owl. Schizonts are present in two neurons. The infected neuron in the upper left contains an intermediate stage schizont and the merozoites are clustered around a residual body in a rosette pattern. The second neuron (lower right) contains a single mature schizont with randomly scattered merozoites. *(Reprinted with permission from Wünschmann A, Rejmanek D, Cruz-Martinez L, Barr BC. Sarcocystis falcatula-associated encephalitis in a free-ranging great horned owl (Bubo virginianus). J Vet Diagn Invest. 2009;21:283–287.)*

vasculature can be associated with pulmonary edema and widespread capillary thrombosis. Affected lungs may contain mild, multifocal lymphoplasmacytic, histiocytic, and polymorphonuclear interstitial inflammation. Neuronal and glial infection can lead to necrotizing and granulomatous, polioencephalitis, and myelitis with meningitis. Lesions may be present in the brainstem, cerebellum, and cerebrum. Glial nodules are frequently present in the gray matter. Meningeal and cerebral blood vessels may be surrounded by macrophages, lymphocytes, and plasma cells. Protozoal organisms in the brain are rare in some cases. Schizonts in various stages of development may be present. Schizonts range from immature with large central round, oval, or multilobulated nuclei and peripheral pale amphophilic cytoplasm, to mature with clusters of small elliptical merozoites and central nuclei. Merozoites within mature schizonts may cluster randomly or occasionally arranged in a rosette pattern around central pale residual bodies.

Caryospora sp. have a monoxenous life cycle, although some species may be facultative heteroxenous. All sexual and asexual stages of the life cycle of most *Caryospora* sp. occur in the epithelial cells of the small intestine in raptors (or snakes). A variety of *Caryospora* sp. have been described in falcons and owls.[37,99] Oocysts are readily identifiable by fecal floatation or smears of intestinal content.[43] Oocysts are large, sporulated, and have one sporocyst with eight sporozoites. While infection is typically clinically insignificant, disease can occur. For example, *C. neofalconis* can cause catarrhal enteritis with petechial hemorrhage in young falcons, particularly in captivity. Concurrent infection with *Clostridum perfringens* type A/B can result in fatal hemorrhagic enteritis. Histologically, protozoa are located in the cytoplasm of enterocytes, beneath the nucleus.

Trichomoniasis, known in falconer circles as "**frounce**" is due to infection with a virulent strain of the flagellate ***Trichomonas gallinae***. Rock pigeons are considered the primary host for *T. gallinae* but mourning doves and songbirds can also be infected (see also Chapter 31: Galliformes and Columbiformes). Infection rates among pigeons and raptors vary regionally. Disease is common in captive raptors fed pigeons and in free-ranging raptors, particularly in nestlings of avivorous raptor species in urban settings or areas where other prey has become rare. The most common lesions of trichomoniasis in raptors are beige to yellowish mucosal plaques in the oropharynx and upper esophagus (Fig. 30.24A). Lesions frequently extend deep into subjacent soft tissues, underlying bone, and the nasal cavity.[115] Crop fistulas have been described in short-toed eagles. Trichomoniasis lesions are uncommon in vultures. However, a unique trichomonad, ***T. gypaetinii***, has been isolated from the oropharynx of normal Egyptian and cinereous vultures in Spain and causes trichomoniasis lesion in juvenile cinereous vultures.[77] DNA sequences from the trichomonads of these vultures are similar to those from a bearded vulture.

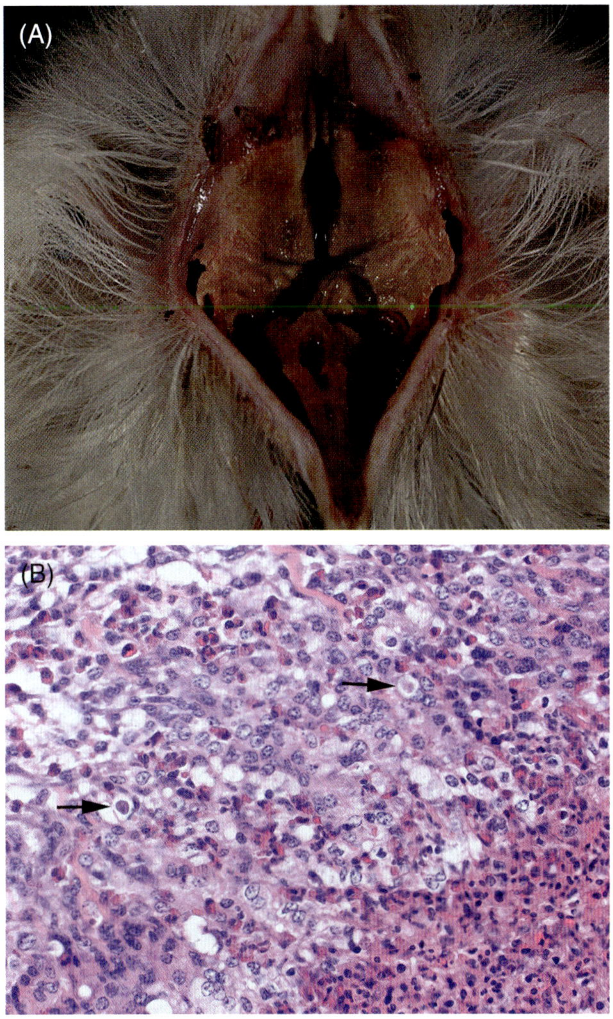

FIGURE 30.24 Trichomoniasis in a barn owl (A) and a barred owl (B). (A) The mouth is open to reveal the nasopharynx which contains a massive amount of necrotic debris and diphtheritic inflammation that extends into the underlying soft tissue and bone due to infection with *Trichomonas gallinae*. (B) Two slightly basophilic, approximately 6 μm in diameter protozoa (*arrows*) are present among the histiocytic and heterophilic inflammation at the leading edge of a focus of lytic necrosis in the oral mucosa (lower right). The characteristic flagella are not recognizable in this image (see eSlide 30.e15). *(Part A: Photo Courtesy of William R. Pritchard, Veterinary Medical Teaching Hospital.)*

In clinical and diagnostic settings, the diagnosis of the upper alimentary and upper respiratory infection is made by identification of characteristic motile, flagellated protozoa on microscopic examination of wet mounts. Histologically, fibrinoheterophilic and necrotizing (diphtheritic) inflammation is characteristic. The protozoa may be found at the leading edge of a lesion in well-preserved specimens. However, the absence of microorganisms does not rule out trichomoniasis, as they are commonly not seen amidst the exudate in histological preparations. Organisms are round to ovoid and approximately 5–7 μm in diameter (Fig. 30.24B; eSlide 30.e15) with a small, condensed nucleus. The flagella are almost impossible to detect in histological sections. Concurrent fungal (e.g., *Candida* sp.) and bacterial infections (e.g., *Staphylococcus aureus*) are common. Testing of swabs or tissue for *T. gallinae* using culture and PCR is more sensitive than smears or histology. Differential diagnoses for trichomoniasis in the upper alimentary tract of raptors include avian poxvirus infection, herpesvirus infection, candidiasis, salmonellosis, and capillariasis.

Cryptosporidium baileyi infections occur in owls and falcons. Infection causes upper respiratory tract disease that may extend to the lungs via the trachea or middle ears via the Eustachian tube, and may involve conjunctival tissue (possibly ascending along the nasolacrimal duct).[8] Grossly, clear mucoid exudate may be seen in the trachea (Fig. 30.25A), epicardium, or in the middle ears. Fibrinous airsacculitis may also occur. Histologically, lymphoplasmacytic and heterophilic mucosal and submucosal inflammation and epithelial hyperplasia of affected sites are present. Round, intracellular, extracytoplasmic cryptosporidial organisms are 2–6 μm in diameter and are present in the brush border along the apical surface of respiratory epithelial cells. Organisms can be identified in smears from the trachea, air sac, or conjunctiva. PAS reaction and acid-fast stains help visualize the organisms (Fig. 30.25B). PCR allows not only for the diagnosis of cryptosporidial infection but also for differentiating various cryptosporidial species.

In general, **intracellular hematozoa** are considered to be nonpathogenic in raptors. The exception is avian *Plasmodium* infections to which raptors are highly susceptible; high nestling infection rates can be seen. Examination of blood films or impressions smears of lungs, liver, and spleen is useful when hemoparasitosis is suspected based on history or necropsy findings, such as hepatomegaly and/or splenomegaly. The classification of the **hemosporidian parasites** is rapidly changing and has been in flux since the advent of molecular methods. For the purposes of this chapter, we follow established classification schemes and recommend recent and future specialty literature for additional consideration.

Plasmodium* sp., *Hemoproteus* sp.,** and ***Leucocytozoon* sp.** are closely related hemosporidian blood parasites. *Plasmodium* sp. are the cause of **avian malaria**. Parasites are transmitted by culicine and anopheline mosquitoes.[4,58] The disease is sporadic but may occur on a larger scale in naïve, captive raptors introduced to a new region and exposed to suitable mosquito vectors. Among the raptors, gyrfalcons, gyrfalcon hybrids, and snowy owls are particularly susceptible to fulminant disease. A variety of *Plasmodium* species infect birds, but avian malaria in raptors is usually caused by ***Plasmodium relictum. In cases of fulminant avian malaria, affected birds are anemic and the liver and spleen are swollen and discolored dark red to black

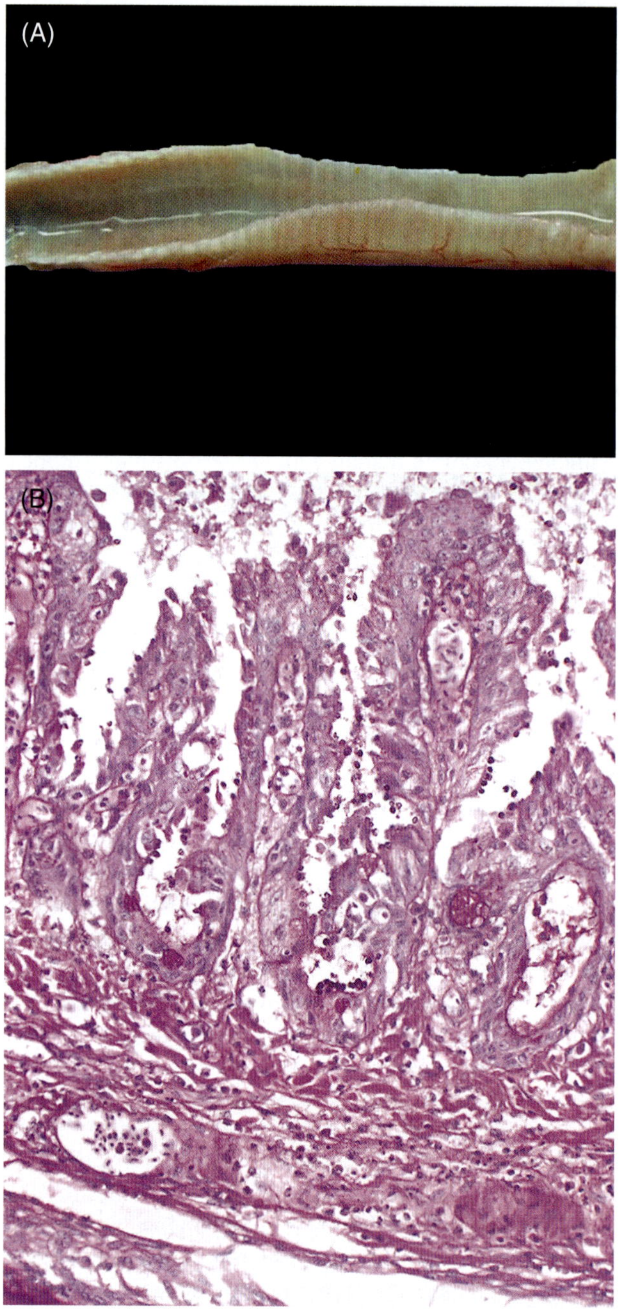

FIGURE 30.25 *Cryptosporidium baileyi* in the trachea of a falcon (A, B). (A) Clear, mucoid exudate in the trachea is a common finding. (B) Infection is associated with marked mucosal hyperplasia and villous proliferation of the tracheal mucosal epithelium. Numerous small, round, PAS-positive, protozoa are present along the apical epithelial border. PAS.

(Fig. 30.26A). The bone marrow is diffusely dark red. The lungs are markedly edematous. In some cases, the cerebellum has petechiae. Touch preparations (impression smears) of lung, liver, and spleen are highly useful in identifying parasitized erythrocytes and extraerythrocytic schizonts and meronts. In the case of *Plasmodium*

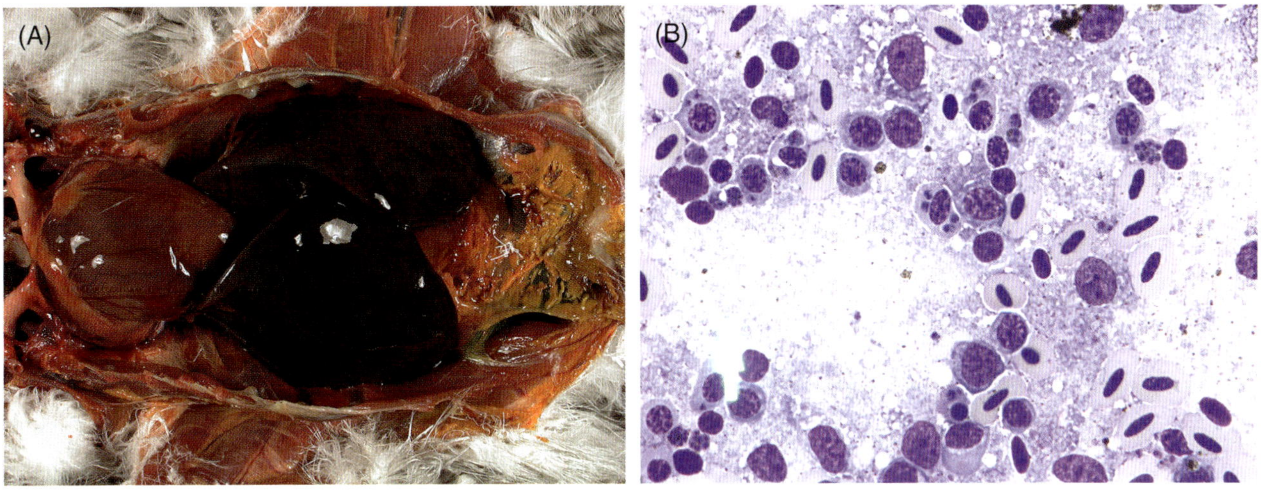

FIGURE 30.26 Avian malaria in the liver of a gyrfalcon. (A) The liver is markedly enlarged and dark red/brown due to infection with *Plasmodium* sp. (B) The organisms may be more easily seen in touch preparations (impression smear) than in histologic sections. *Plasmodium* sp. gametocytes are basophilic, spherical to irregular, and up to 2 µm in diameter. They are intracytoplasmic and can be adjacent to or surround the nucleus of erythrocytes. Brownish black malarial pigment (hemazoin) is present within parasitized erythrocytes and is present extracellularly. Modified Romanowsky stain. *(Part B: Photo Courtesy of J. Overmann, College of Veterinary Medicine, University of Minnesota.)*

infections, one to multiple, intracytoplasmic basophilic spherical to irregular gamonts, up to 2 µm in diameter, and hemozoin (brownish black malarial pigment) are present in the cytoplasm of erythrocytes. Histopathology may reveal intraerythrocytic parasites, particularly in pulmonary and brain capillaries. The bone marrow is hypercellular due to marked erythroid hyperplasia. The sinusoids of the liver and the spleen are expanded by erythroid hematopoietic cells. Necrosis of hepatocytes and splenic cells may occur. Spleen, liver, and bone marrow contain multifocal, individual, or clustered macrophages with intracytoplasmic hemozoin, which is refractile with polarized light - this feature allows hemozoin to be differentiated from hemosiderin (hemosiderin does not polarize). Extraerythrocytic meronts may be present in endothelial cells and tissue macrophages, particularly in lungs, kidney, liver, spleen, and brain, although birds dying from certain *Plasmodium* sp. infections may lack extraerythrocytic meronts. Finding extraerythrocytic meronts depends on a thorough scanning of tissue sections.

Identification of the *Plasmodium* species is based on molecular analysis and morphologic features of the intraerythrocytic gametocytes and meronts, such as shape and size, and the association of these forms with the erythrocyte nucleus. Morphologic identification requires careful antemortem examination of blood in a live bird or cytologic examination of touch preparations from the lung, liver, and/or spleen at necropsy (Fig. 30.26B). *Plasmodium* species undergo asexual reproduction with the formation of schizonts in erythrocytes. When seen, this feature distinguishes *Plasmodium* from other hemosporidians, including *Hemoproteus* and *Leucocytozoon*. Also, only *Plasmodium* and *Hemoproteus* species digest and convert hemoglobin to hemozoin (malarial pigment) or a closely related heme polymer pigment that is recognizable as refractile, golden brown, or black granules. The pigment is present in the cytoplasm of the intraerythrocytic parasite and can be seen in the blood or in macrophages in the bone marrow, spleen, and liver. Asexual reproduction of *Hemoproteus* sp. occurs exclusively in endothelial cells of various tissues (including liver and kidney) and not in circulating erythrocytes. The histological identification and differentiation of tissue stages and intraerythrocytic stages of *Plasmodium* sp., *Hemoproteus* sp., and *Leucocytozoon* sp. is frequently challenging. PCR followed by DNA sequencing are common tools for confirming infection and speciating the organisms.

Seroprevalence for ***Toxoplasma gondii*** is high in birds of prey; however, clinical disease and death appear to be very rare.[15,83] ***Eimeria*** sp. infect the intestine of raptors but have not been associated with morbidity or mortality. Various life stages of a novel *Eimeria* sp., most closely related to ***E. reichenowi***, can be found in ductular epithelial cells of the kidney of great horned owls.[52] This renal protozoal infection is considered to be an incidental finding since the intracytoplasmic microgametocytes, macrogametocytes, and oocysts do not appear to cause inflammation (Fig. 30.27; eSlide 30.e16). For information on additional notable protozoa in raptors, please see Table 30.e4 in the Digital Content.

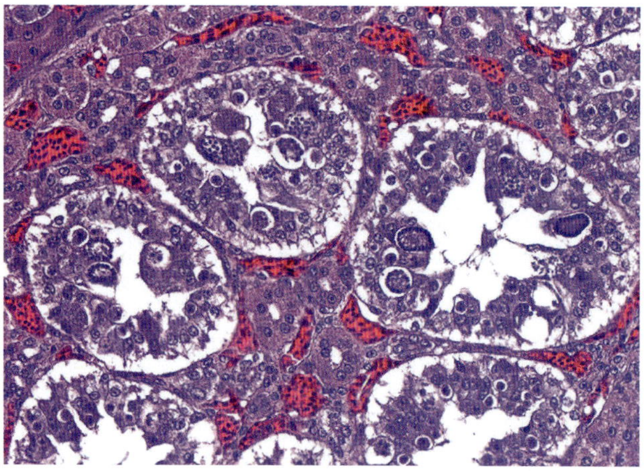

FIGURE 30.27 Coccidian parasites in the kidney of a great horned owl. The ductular epithelium of the medullary cones is hyperplastic. The cytoplasm of numerous ductular epithelial cells is expanded by the parasites. Various life stages in this image include trophozoites, microgametocytes, macrogametocytes, and schizonts (see eSlide 30.e16).

Ectoparasites in Raptors

For information on notable ectoparasites, please refer to Table 30.e5 in the Digital Content.

References

Full references for this chapter can be found online at https://doi.org/10.1016/B978-0-443-18437-6.00015-8.

Chapter 31

Galliformes and Columbiformes

Rocio Crespo[a], Monique S. França[b], Heather Fenton[c], Simone Stoute[d], and H.L. Shivaprasad[e]
[a]North Carolina State University, Raleigh, NC, United States; [b]Pathology at Johnson and Johnson Innovative Medicine, San Diego, CA, United States; [c]Australian Registry of Wildlife Health, Taronga Conservation Society, Mosman, NSW, Australia; [d]California Animal Health and Food Safety Laboratory System, Trulock Branch, University of California, Turlock, CA, United States; [e]California Animal Health and Food Safety Laboratory System, Tulare Branch, University of California, Tulare, CA, United States

INTRODUCTION

Galliformes are a diverse order of heavy-bodied, ground-feeding birds with short, rounded wings for short-distance flight. The order includes approximately 290 species divided into five families distributed worldwide.[43,124] Phasianidae (e.g., chicken, quail, partridges, pheasants, turkeys, peafowl, and grouse), Odontophoridae (New World quails), Numididae (guineafowl), Cracidae (e.g., chachalacas and curassows), and Megapodiidae (e.g., malleefowl, maleo, and brush-turkeys). Galliformes vary in size from the small king quail that weighs 28–40 g to the large North American wild turkey weighing up to 14 kg. Some species have been domesticated and are raised for human consumption of meat and eggs. Grouse, quail, partridges, pheasants, and turkeys are hunted regularly in all parts of the world. The majority of the birds in the group are granivorous to slightly omnivorous.

Columbiformes consists of two families, the Columbidae (pigeons and doves) and Raphidae (Dodos). There are about 310 extant species in the Columbidae family. All species of the Raphidae family are extinct. The terms pigeon and dove are used interchangeably. These birds are common and occur worldwide. Pigeons vary in size and color, from small, sparrow-sized ground doves to large turkey-sized crowned pigeons. Pigeons are kept for various purposes, such as for racing (homing pigeons), show, sports, acrobatics, and pets and as meat (e.g., squab). Some breeds are used in biomedical research.

Nondomestic galliformes and columbiformes are susceptible to many of the same diseases that occur in domestic chickens and turkeys. A majority of the literature on these groups of birds relates to the latter, which, in this chapter, will be used as models for many of the shared, common diseases in both domestic and nondomestic species. Specific references and descriptions of diseases unique to nondomestic species will also be discussed.

NON-INFECTIOUS DISEASES

Nutritional

Galliformes and columbiformes are susceptible to a number of nutritional deficiencies; diagnosis for all of these conditions requires evaluation of lesions in combination with feed analysis.

Inadequate intake or lack of absorption of calcium, vitamin D_3, or phosphorus causes **rickets** in young growing birds and **osteomalacia** in adults.[156] Rickets is characterized by enlargement or beading at the costochondral junctions of the ribs with the sternum and vertebrae (rachitic rosary), bending of the ribs with lateral flattening of the thorax, pathologic bone fractures with callus formation, lateral deviation of the sternum, and spinal deformities (Fig. 31.1A). Rickets caused by hypocalcemia and/or vitamin D_3 deficiency is characterized by a thickened zone of proliferation in the growth plate of bones (Fig. 31.1B) and enlarged parathyroid glands. Similar growth plate changes in the absence of parathyroid gland enlargement are associated with phosphorus deficiency or calcium excess. With osteomalacia, cortical and trabecular bone are thinner than normal and cortical bone may be porous (osteoporosis).

Vitamin E and selenium are antioxidants that protect cell membranes against oxidative damage caused by reactive oxygen species. **Vitamin E deficiency** (**hypovitaminosis E**) develops secondary to consumption of diets low in vitamin E and may be accompanied by deficiency in selenium or sulfur-containing amino acids. Accumulation of peroxides and deficiency of antioxidants results in a spectrum of diseases, which includes encephalomalacia, nutritional myopathy, and exudative diathesis.[156] **Encephalomalacia,** also known as "crazy chick disease," is a disease of the central nervous system in young gallinaceous birds fed diets low in vitamin E; it is more commonly seen in 2- to 4-week-old birds.[19] Peroxidative damage to

Pathology of Wildlife and Zoo Animals. https://doi.org/10.1016/B978-0-443-18437-6.00042-0

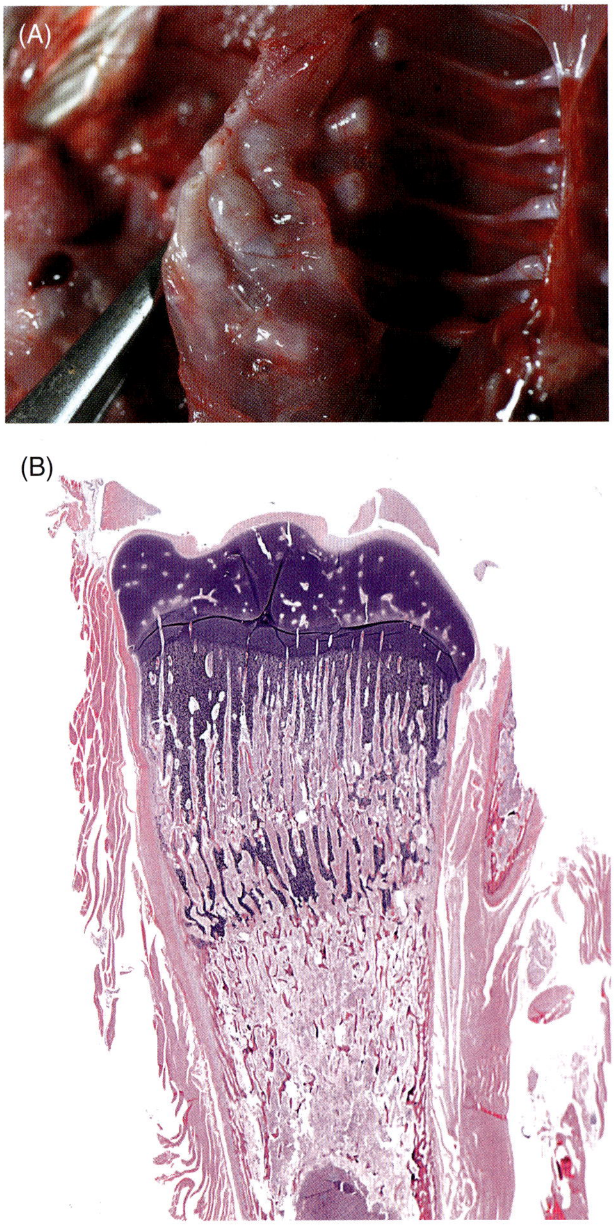

FIGURE 31.1 Rickets in turkey poults. (A) Enlargement at the costochondral junctions of the ribs is caused by calcium and/or vitamin D_3 deficiency. (B) Rickets caused by phosphorus deficiency causes increased width of the zone of hypertrophy with vascular invasion. The zone of proliferation beneath the epiphysis is relatively normal. There is a compression fracture at the margin of the zone of hypertrophy in the growth plate. *(Part A: Photo Courtesy of R. Chin, California Animal Health and Food Safety Laboratory – UC Davis; Part B: Photo Courtesy of P. Cortes, USDA.)*

capillary walls results in edema, hemorrhage, and malacia of the affected areas. Macroscopically, the cerebellum is swollen and has petechial to diffuse hemorrhage producing a "cherry red" appearance in more severe lesions (Fig. 31.2). Microscopically, there is necrosis of the cerebellar folia that affects all layers, capillary thrombosis, edema, and axonal

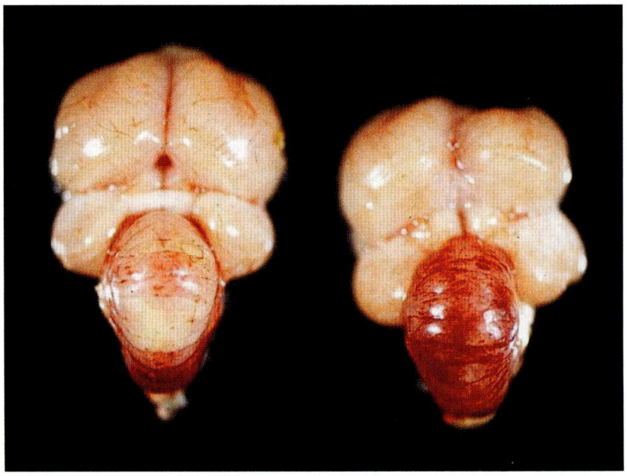

FIGURE 31.2 Vitamin E deficiency encephalomalacia in turkey poults. Lesions can include swollen and edematous cerebellum with petechiae *(left)* and/or diffuse cerebellar hemorrhage *(right)*.

degeneration and hemorrhage in the cerebellar folia and meninges. Birds with **nutritional myopathy** ("**white muscle disease**") have white streaks in muscle fibers of the breast, leg, and gizzard. Microscopically, there is degeneration and necrosis of myofibers, myofiber mineralization, satellite cell proliferation, and heterophilic cell infiltrates in acute cases with subsequent fibroplasia. Hypovitaminosis E can also cause **exudative diathesis** and is characterized by subcutaneous edema and hemorrhage; a green-blue gelatinous appearance of the subcutis is commonly seen in the ventral aspect of the thorax and abdomen and less frequently in the mandible and periorbital region.[20]

Curled-toe paralysis is a disease of young gallinaceous poultry caused by **vitamin B2 (riboflavin) deficiency**, which is believed to be required for myelin synthesis in peripheral nerves. Deficiency causes generalized demyelinating polyneuropathy. The sciatic, brachial, cervical, and lumbar nerves and large and medium intramuscular nerves are commonly affected and may be swollen and soft.[23] The sciatic nerves may be four to five times larger than normal.[156] Microscopically, there is myelin and axonal degeneration, edema, mild lymphocytic infiltrates, gliosis, and hyperplasia of Schwann cells in later stages.

Vitamin A is critical for differentiation of epithelial cells into cuboidal, columnar, or mucous-producing cells. **Hypovitaminosis A** produces epithelial squamous metaplasia and hyperkeratosis.[31,156] Macroscopically, the mucosa of the tongue, choana, and salivary glands in the oropharynx and esophageal glands are thickened and form "pustule-like" nodules (Fig. 31.3) due to hyperkeratosis and distension of glands and ducts with keratin, secretions, and cellular debris. The conjunctiva, bursa of Fabricius, nasal passages, and sinuses may contain caseous exudate, mucous membranes may be dry, corneas may be opaque and dry, and plantar surfaces may be thickened and hyperkeratotic.

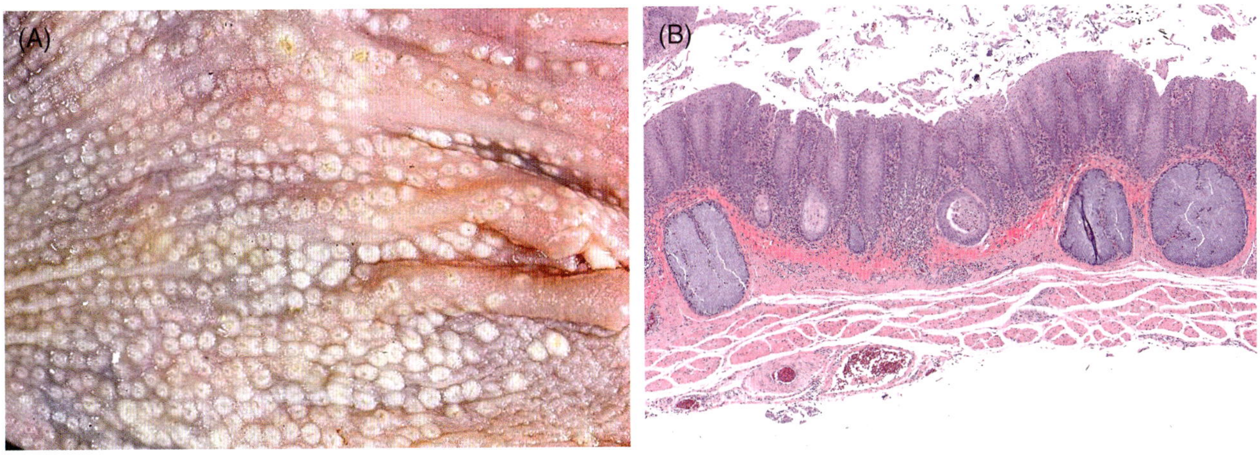

FIGURE 31.3 Hypovitaminosis A in a chicken (A) and pheasant (B). (A) Deficiency is most often associated with distended mucosal glands in the esophagus. (B) Esophageal glands are enlarged and lined by squamous epithelium, a metaplastic response.

Choline or **manganese deficiency** causes **perosis** or **chondrodystrophy**, which are diseases of immature gallinaceous birds that result in impaired endochondral bone growth.[156] Affected birds are small for their age and have thickened, short, twisted, and bent tarsometatarsi, and widened, deformed articular cartilage. In more severe cases, gastrocnemius tendon dislocation occurs. Microscopically, there is a narrow physeal zone of proliferation with disorganized chondrocytes.[53]

Thyroid hyperplasia (goiter) is considered a common problem in birds and is common in pigeons.[143,167] The most common causes include dietary deficiency or excess of iodine, consumption of goitrogenic substances (e.g., spinach, cassava, peanuts, soybeans, kale, broccoli, Brussels sprouts, cabbage, canola, cauliflower, mustard greens, radishes, and rapeseed), goitrogenic drugs (e.g., sulfonamides) and a defective negative feedback control by the pituitary.

Metabolic

Hyperuricemia (gout) is a metabolic condition in which white chalky or semifluid-like urates accumulate in soft tissues or joints of various organs in the body. In birds, uric acid is the end-product of protein and purine metabolism (uricotelic), whereas in mammals, urea is the end-product (ureotelic). Gout in birds occurs in acute (visceral) and chronic (articular) forms (Fig. 31.4). These two forms differ in age of onset, frequency, sex predilection, gross

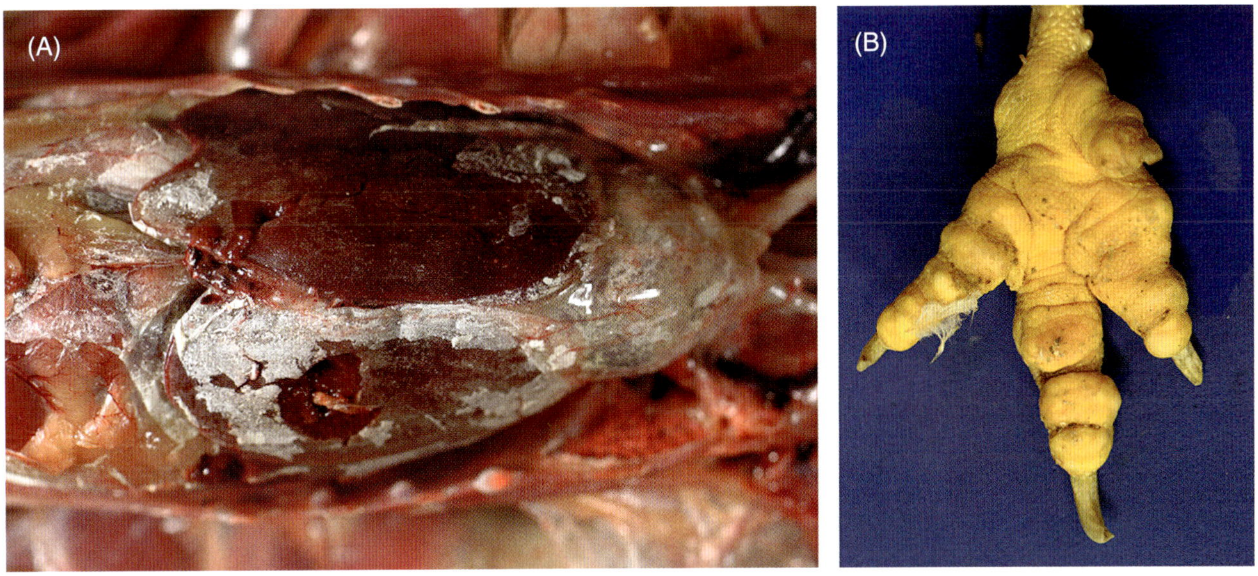

FIGURE 31.4 Acute/visceral (A) and chronic/articular (B) uric acid (urate) deposition in a chicken. (A) Fine, chalky-white crystalline material is present on the pericardium, across the capsular surface of the liver, and in the soft tissues of the coelom. (B) The joints of the toes are enlarged and deformed due to uric acid deposition and accumulation.

and microscopic lesions, pathogenesis, and cause/s (Table 31.1). A great deal of confusion exists between the two syndromes because urate deposition in joints can occur in both disease forms. For clarity, commonly used terms "visceral gout" and "articular gout" should be avoided and replaced with "acute urate deposition" and "chronic urate deposition", respectively. Histologically, feathery crystals may be seen within tissues; however, much of the urate deposits are lost when tissues are processed. In chronic cases, granulomatous inflammation is observed.

Exertional myopathy is caused by excessive muscle activity during restraint. It commonly affects the wings and legs. Histologic lesions are characterized by swelling of the myofibers with striation loss, fragmentation of myofibrils, and subsarcolemmal nuclear pyknosis. **Deep pectoral myopathy** is an **exertional compartment syndrome** of the supracoracoid muscle characterized by ischemic necrosis in the absence of other muscular lesions. It occurs in domesticated heavy breeds of chickens and turkeys.[148] Deep pectoral myopathy has been reported in zoo white-headed pigeons but not in free-ranging birds.[71] Deep pectoral myopathy can be induced by holding a heavily muscled bird by its legs and allowing it to flap its wings[149] or by occlusion of the subclavian artery with muscle stimulation.[175] Compartmental myopathy produces high levels of lactic acid and myonecrosis. Further complications include lactic acidosis, renal failure, arrhythmias, and death. Initially, the supracoracoid muscle is swollen, pale, and edematous. Later, muscle becomes green, dry, and friable. Histologically, there is necrosis with no evidence of inflammation. Blood vessels are filled with lysed red blood cells.

Toxic

Galliformes and columbiformes are curious animals that may unintentionally ingest toxic compounds from their environments. Most compounds considered toxic for mammals and other avian groups are likely to be toxic to galliformes and columbiformes (see Table 31.e1). As occurs in other avian species, galliformes and columbiformes are more susceptible to inhaled toxins than mammals because of their rapid metabolic rate and highly efficient respiratory system. Good ventilation should be maintained in birds kept in captivity.

Pesticides and **anticoagulant rodenticides** are known to cause mortality in nontarget species, including wild pigeons, turkeys, and other galliformes.[52,126,140] Though more commonly used historically, nonanticoagulant

TABLE 31.1 Differences Between Acute and Chronic Urate Deposition in Gallinaceous Birds

	Acute Urate Deposition	Chronic Urate Deposition
Frequency	Common	Rare or sporadic
Age	Any age	Generally mature birds
Renal lesions	Almost always involved, with deposition of white, chalky precipitates	Normal
Soft tissue lesions	Visceral organs like liver, myocardium, spleen, or serosal surfaces like pleura, pericardium, air sacs, mesentery, etc., are commonly involved	Soft tissues other than synovium are rarely involved. Occasionally comb, wattles, and trachea are involved
Joint lesions	Soft tissues around the joints may or may not be involved. Surfaces of muscles, synovial sheaths of tendons, and joints are involved in severe cases	Soft tissues around the joints are always involved, especially feet. Other joints of the legs, wing, spine, and mandible are also commonly involved
Microscopic lesions	Generally no inflammatory reaction in synovium or visceral surfaces. Kidney and viscera have an inflammatory reaction around tophus	Granulomatous inflammation in synovium and other tissues
Pathogenesis	Generally due to renal failure	Probably due to a metabolic defect in the secretion of urates by the kidney tubules
Causes	Dehydration Nephrotoxicity Nephrotropic infectious agents, Vitamin A deficiency Urolithiasis Neoplasia (lymphoma, primary renal tumors) Immune-mediated glomerulonephritis	Genetics High protein diet

rodenticides (e.g., zinc phosphide or strychnine) can also cause toxicity. Some species, for example, feral pigeons, may be purposely poisoned if they are considered a nuisance. The only avicide registered in the United States to poison birds is 4-aminopyridine, the active ingredient in **Avitrol** baits (4-aminopyridine). Other compounds, such as alpha-chloralose, thallium, or ethylene glycol (antifreeze), may be used as poison in other parts of the world; some require special permits for their use. Avitrol baits and alpha-chloralose affect the nervous system. Gross findings in the former may include extensive subcutaneous congestion in the region of the crop and a full crop containing baited seeds. Thallium and antifreeze (ethylene glycol) trigger acute nephrosis with accumulation of oxalate crystals in the renal tubules and secondary uric acid deposition in many organs.

Heavy metal intoxications, mainly associated with lead and zinc, are some of the most commonly reported and clinically recognized poisons in domestic and free-ranging birds. These metals are included in many man-made products, such as metal tools, coins, batteries, and gun pellets. Clinical signs include weight loss, lethargy, depression, anemia, paralysis, head tremors, convulsions, and death. Microscopically, the most diagnostic lesions are demyelination of peripheral nerves and focal areas of vascular damage in the cerebellum. Heavy metal poisoning should be differentiated from botulism, by analyzing the plasma or tissues (e.g., kidney and liver) for heavy metal concentrations. The primary treatment for lead or zinc poisoning is removal of the toxic metal. Chelation therapy (e.g., calcium ethylenediaminetetraacetic acid [CaEDTA]) may also be beneficial in some cases.

Mycotoxicosis is caused by toxic metabolites of fungi. Hundreds of mycotoxins are known. The most common are **aflatoxins** and **fusariotoxin** (e.g., T-2 toxin and deoxynivalenol [DON]).[9,59] Because galliformes and columbiformes consume primarily seed-based diets, they have a high risk of exposure. Susceptibility to mycotoxins varies among species. All mycotoxins cause immunosuppression. Clinical signs of aflatoxins may include anemia, hemorrhage, or paralysis; histologic findings include hepatocellular vacuolation, cytomegaly karyomegaly, degeneration, and necrosis with biliary hyperplasia and, in more chronic cases, fibrosis. Fusariotoxins are associated with feed refusal, extensive necrosis of oral mucosa and skin in contact with toxins, and hyperplastic bone marrow. Embryonic exposure to mycotoxins is also known to be associated with developmental defects.[98]

Inflammatory Non-Infectious

Amyloidosis is a well-recognized pathologic disorder in birds (see Chapter 29: Anseriformes, Ciconiiformes, Charadriiformes, and Gruiformes). Deposition of amyloid protein occurs commonly in the subsinusoidal space in the liver and spleen but can accumulate in other organs. In chickens, turkeys, and pigeons, it is commonly associated with chronic infections including *Mycobacterium avium*, *Mycoplasma* spp., and *Enterococcus faecalis*. Amyloid in the tissues is highlighted with Congo red histochemical staining.

Xanthomas are rare nonneoplastic, white to yellow, nodular masses that occur in the skin, subcutaneous tissues, and internal organs.[89] Histologically, xanthomas are composed of sheets of large, foamy, and lipoprotein-laden macrophages. Occasionally, cutaneous xanthomas develop over lipomas.[49,156] The pathogenesis is poorly understood, but it has been associated with contamination of feed fat with hydrocarbons or abnormal lipid metabolism. The majority of xanthomas that are surgically removed do not recur.[132]

Several metabolic diseases that are commonly seen in commercial poultry are only occasionally diagnosed in wild gallinaceous species and pigeons. These include hemorrhagic **fatty liver syndrome, aortic rupture, ascites syndrome, dilated cardiomyopathy (round heart disease), tibial dyschondroplasia,** and **femoral head degeneration**. The reader is referred to other publications for information related to these conditions.[19]

Neoplastic

Virus-associated neoplasms tend to be the most common tumors observed in gallinaceous and columbiform species and are summarized in the infectious disease section. The etiopathogenesis of infectious neoplasms is well described for many tumors in domestic chickens (i.e., retrovirus-associated lymphoma) and may be similar, even though uninvestigated, in nondomestic poultry.[72,156,168]

Tumors described in gallinaceous birds and pigeons include oral squamous cell carcinomas in peafowl; fibromas, lipomas, and fibrosarcomas in the skin, skeletal muscle, and visceral tissues of free-ranging pigeon, pheasant, and quail; leiomyomas (often in the mesosalpinx), leiomyosarcomas, rhabdomyomas, and rhabdosarcomas in domestic chickens, quail, and a number of pigeon species; and ovarian or oviductal adenocarcinomas with or without widespread carcinomatosis in domestic and nondomestic chickens.[72,134,156]

INFECTIOUS DISEASES

There are a number of important viral infections of domestic galliformes that nondomestic galliformes are presumed susceptible to but that, to date, have not been reported to cause disease (see Table 31.e2).

DNA Viruses

The family *Adenoviridae* is divided into five genera: *Mastadenovirus*, *Aviadenovirus*, *Atadenovirus*, *Siadenovirus*, and *Ichtadenovirus*. The majority of adenoviruses in birds are classified as *Aviadenoviruses*.[67]

Quail bronchitis, caused by an *Aviadenovirus* of the species fowl aviadenovirus A (FAdV-A), is a respiratory disease of bobwhite and Japanese quail.[156] Quail under 4 weeks of age are more susceptible, and mortality rate may reach 100% in very young birds. In older birds, infections are usually milder or subclinical.[19] Gross lesions include catarrhal tracheitis (Fig. 31.5A), corneal clouding, conjunctivitis, congestion in the mucosa of the nasal passages and infraorbital sinuses along with pulmonary congestion, consolidation, airsacculitis, dilated intestines with frothy luminal contents, erosion in the gizzard mucosa,[93] pinpoint necrotic foci in the liver, and a swollen and mottled spleen. Microscopically, there is fibrinoheterophilic or lymphoplasmacytic rhinitis, tracheitis and bronchitis with loss of cilia, epithelial cell necrosis and hyperplasia, and large, basophilic, intranuclear inclusions in intact or sloughed epithelium (Figs. 31.5B, C). Other microscopic lesions include necrosis in liver, spleen, and

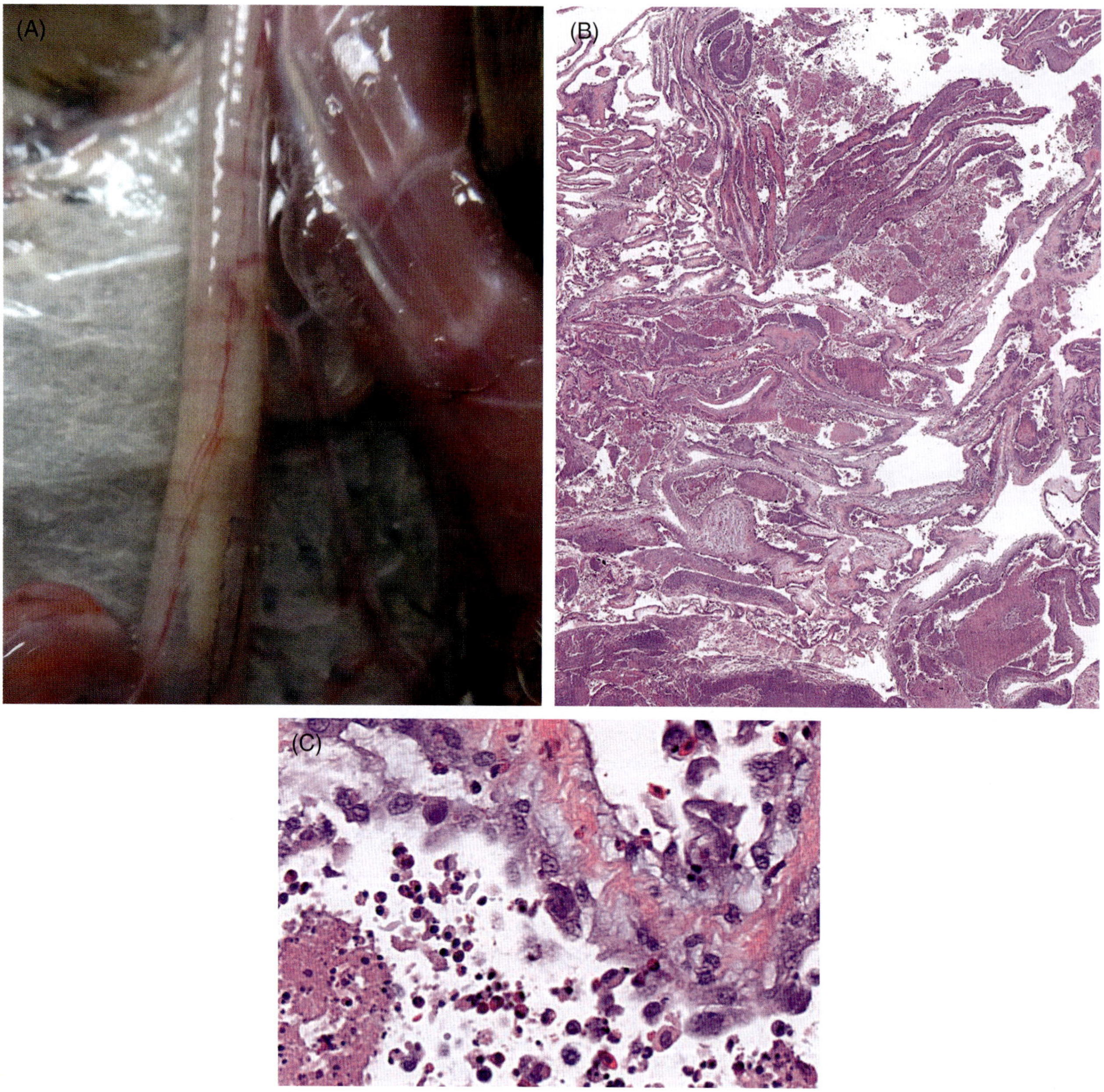

FIGURE 31.5 Quail bronchitis. (A) Catarrhal tracheitis in a quail. The lumen of the trachea is filled with necrotic caseous exudate. (B) Numerous inflammatory cells infiltrate the walls of the air sac and are present with necrotic material within the lumen. (C) Higher magnification of the air sac with large basophilic intranuclear inclusion bodies, consistent with Adenovirus, are present in the epithelium of the air sac. *(Part A: Photo Courtesy of E. Pitman, Georgia Poultry Laboratory Network.)*

bursa of Fabricius, as well as epithelial degeneration in the gizzard with fragmentation of the koilin layer. Diagnosis is made by evaluating lesions and the presence of amphophilic to basophilic intranuclear inclusions in respiratory epithelium, hepatocytes, and epithelial cells of the bursa of Fabricius. Virus isolation can be performed in samples of trachea, lungs, and air sacs for confirmation. Additionally, PCR for aviadenovirus can be performed in liver, respiratory tract, or other affected tissues.

Inclusion body hepatitis (IBH) is caused by an *Aviadenovirus* of the species fowl aviadenovirus D (FAdV-D) that occurs worldwide. It primarily affects chickens, though guinea fowl and turkeys can also develop disease.[29,106] Immunosuppressive viruses, such as chicken anemia virus or infectious bursal disease virus, predispose chickens to IBH. The disease is typically seen in 2- to 3-week-old birds but it has been diagnosed in chickens as young as 4 days and as old as 7 weeks.[19] Gross lesions include an enlarged, friable, and pale liver that in some cases contains multifocal foci of necrosis and hemorrhage. Other lesions include hydropericardium and swollen, pale, mottled, and hemorrhagic kidneys. Microscopically, there is multifocal hepatic necrosis, basophilic to amphophilic intranuclear inclusions in hepatocytes, and inflammation composed of lymphocytes, heterophils, macrophages, and plasma cells. Pancreatic necrosis with intranuclear inclusions has been reported in guinea fowl.[20] Diagnosis is made by detecting characteristic hepatocellular intranuclear inclusions within lesions (Figs. 31.6A, B). Virus isolation, immunofluorescence, and PCR can confirm the diagnosis.

Pigeon aviadenovirus occurs in many parts of the world. They are also relatively common in wild and feral pigeons and there is evidence of host switching between pigeons and other wild bird species.[164] This adenovirus causes two disease syndromes: classical adenovirus and necrotizing hepatitis.[165] Classical adenovirus occurs in young pigeons (<1 year) and causes diarrhea, vomiting, and weight loss with morbidity approaching 100%. This disease is associated with fowl adenoviruses belonging to various serotypes.[100] Gross lesions include watery intestinal contents and microscopic villous atrophy with basophilic, intranuclear inclusions in enterocytes. Necrotizing hepatitis occurs in pigeons of all ages and results in vomiting, watery yellow diarrhea, and death within 24–48 h. Mortality ranges from 30% to 100%. Grossly, the liver is enlarged and pale yellow. Microscopically, there are multifocal to coalescing, random foci of hepatocyte necrosis with heterophilic inflammation and hemorrhage. Karyomegaly may be seen and is due to large, round, and basophilic intranuclear inclusion bodies with occasional margination of chromatin. Diagnosis is made by histopathology and identification of 75–85 nm diameter, icosahedral, and nonenveloped viral particles by transmission electron microscopy. Pigeon aviadenovirus isolation is difficult. PCR can be performed in liver samples.[172]

Hemorrhagic enteritis, caused by a *Siadenovirus*, affects turkeys, Guinea fowl, and chukar partridges.[102,147] Lesions are characterized by small intestinal hemorrhage as well as an enlarged, friable, and mottled spleen with reticuloendothelial cell hyperplasia and basophilic/amphophilic intranuclear inclusions in mononuclear phagocytic system cells (Fig. 31.7). **Marble spleen disease**, also caused by a *Siadenovirus* is a disease of pheasants. It affects 3- to 8-month-old birds.[22,104] Grossly, the spleen is swollen and mottled, and the lungs are congested and edematous. Microscopically, basophilic/amphophilic intranuclear inclusions are present in macrophages and lymphocytes in the spleen and Kupffer cells in the liver.[50,51] In addition

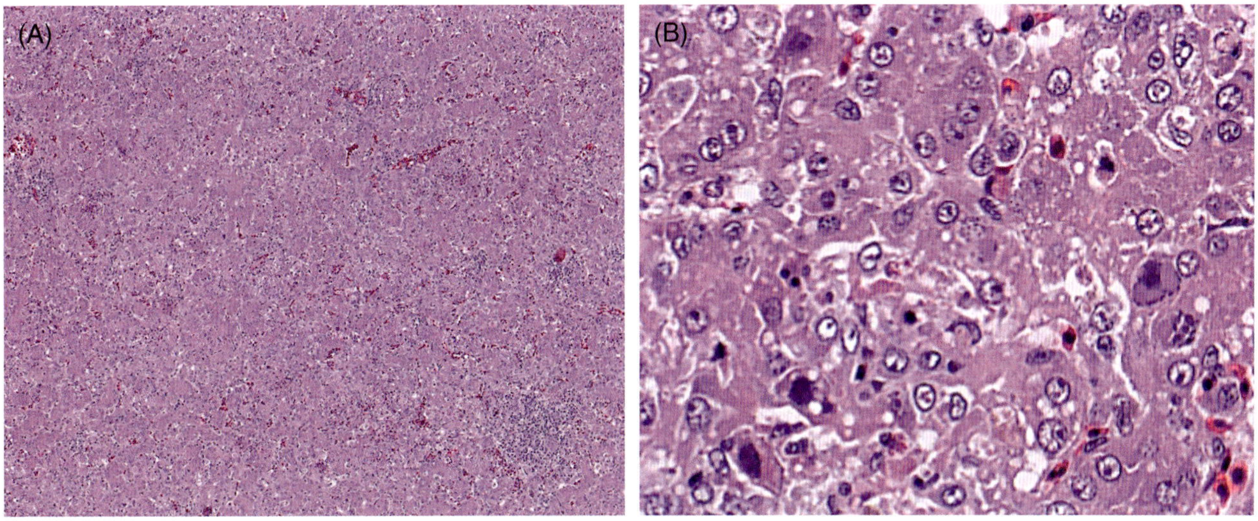

FIGURE 31.6 Inclusion body hepatitis in a chicken. (A) Multifocal hepatic necrosis and inflammation. (B) Higher magnification of the previous, showing basophilic intranuclear inclusions in multiple hepatocytes.

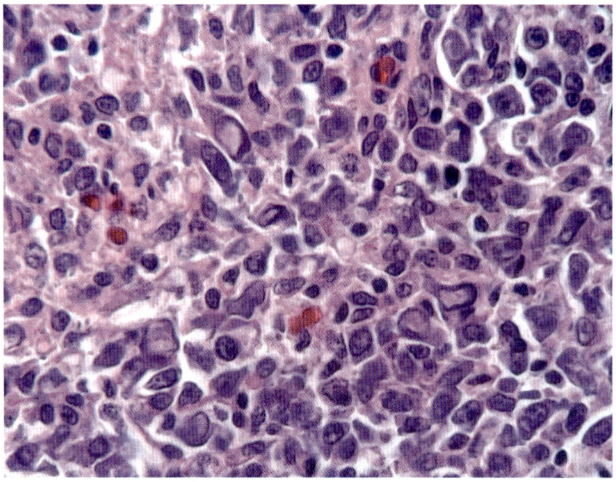

FIGURE 31.7 Siadenovirus infection in the spleen of a turkey. Large glassy intranuclear inclusion bodies are present in the reticuloendothelial cells of the spleen.

to gross and microscopic lesions, agar gel immunodiffusion or PCR assays can be performed for confirmation.

Pigeon circovirus (PiCV) is a small, nonenveloped, circular, single-stranded DNA virus. It is common in pigeons in the United States of America, Europe, Australia, South Africa, Japan, and China and is probably distributed worldwide. PiCV causes a complex, multifactorial disease primarily in young pigeons (<4 months of age) called young pigeon disease syndrome; adults can be carriers. PiCV can be found in healthy, asymptomatic pigeons (by PCR or ELISA) and has an incidence of 70% or higher.[153,156] Infection causes immunosuppression due to apoptosis of bursal lymphocytes. Sequela include severe acquired immunosuppression[1] and various secondary infections. Transmission is primarily horizontal but vertical transmission can also occur. Clinical signs vary and often depend on secondary or concurrent infections. Mortality is variable but can approach 100%. Occasionally, abnormal feathers can also be observed. Gross lesions include an enlarged or small bursa of Fabricius with or without luminal, caseous exudate, and splenomegaly. Microscopically, variable degrees of lymphoid necrosis are seen in the bursa of Fabricius: lymphoid depletion is seen in acute disease and cystic follicles are seen in subacute and chronic disease. Lymphoid depletion can also be seen in the spleen, cecal tonsils, and bone marrow. Intranuclear and intracytoplasmic basophilic globular or characteristic botryoid inclusion bodies are present in macrophages and, less commonly, epithelial cells of the bursa of Fabricius (Figs. 31.8A, B; eSlide 31.e1). Similar inclusions can also be observed in the macrophages of the thymus, spleen, cecal tonsils, lungs, and duodenum. Ultrastructurally, inclusion bodies are composed of 15–20 nm in diameter, nonenveloped, icosahedral viral particles that are either loosely arranged or in a paracrystalline array. Diagnosis is based on a combination of PCR (feces or tissues), ELISA, and identification of characteristic inclusion bodies in the bursa of Fabricius and/or other organs, immunohistochemistry, and electron microscopy.

The family **Herpesviridae** is divided into three subfamilies. All avian herpesviruses are in the *alphaherpesvirinae* subfamily.

Infectious laryngotracheitis (ILT) caused by ***Gallid herpesvirus 1*** **(GaHV-1**; also known as ***Avian herpesvirus*** **1**, genus *Iltovirus*) is a WOAH (World Organization for Animal Health) notifiable disease. It occurs primarily in chickens, but pheasants, pheasant-chicken crosses, peafowl, and turkeys are susceptible.[32,86,127] Virus infects and replicates in respiratory tract epithelium and conjunctiva. Latency is established in the trigeminal ganglia[12] and viral infection is reactivated in immunosuppressed birds. In mild cases, the larynx, trachea,

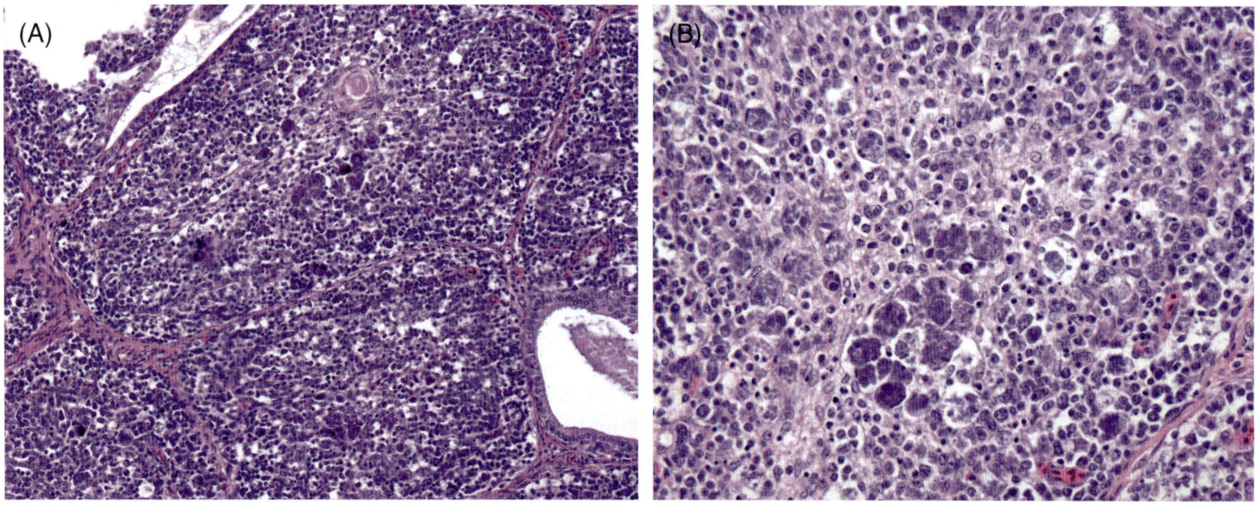

FIGURE 31.8 Circoviral infections in the bursa of a pigeon. (A) Characteristic, basophilic botryoid inclusion bodies are present in epithelial cells of the bursa of Fabricius. (B) Higher magnification of the previous image (see eSlide 31.e1).

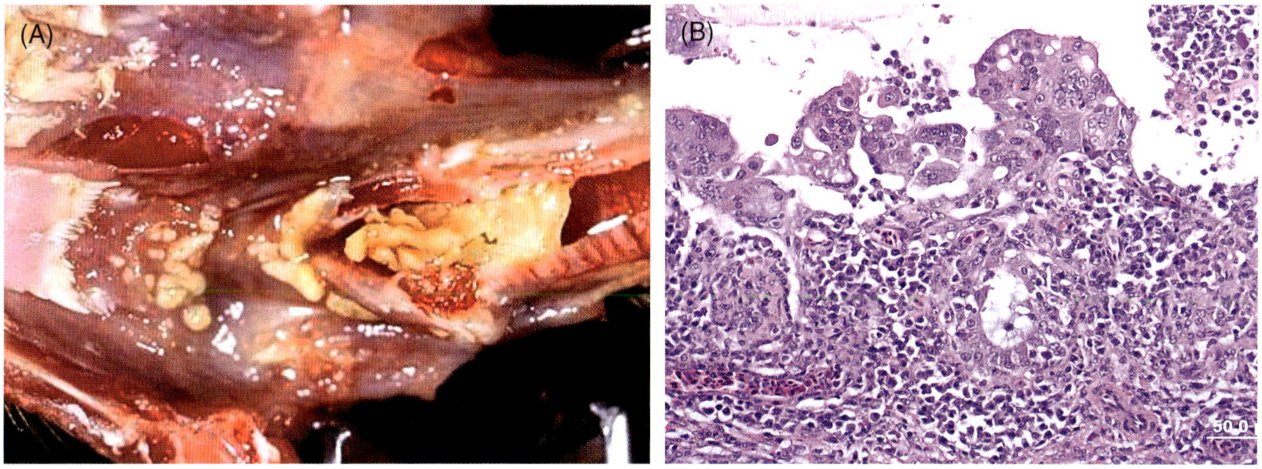

FIGURE 31.9 Infectious laryngotracheitis (Gallid herpesvirus-1) in a chicken. (A) Accumulation of fibrinonecrotic debris in the larynx of a chicken infected with Gallid herpesvirus-1. (B) Severe loss of cilia, necrosis of epithelium, and numerous syncytial cells with eosinophilic intranuclear inclusions on the luminal surface (see eSlides 31.e2 and 31.e3).

and primary bronchi contain excess mucus and have hyperemic mucosa with mild petechiae. In more severe lesions, intraluminal accumulation of fibrinonecrotic material and/or abundant hemorrhage, which occludes the laryngeal opening and can cause suffocation, are seen (Fig. 31.9A; eSlide 31.e2). Affected birds may also have oculonasal discharge, conjunctivitis, and swelling of the infraorbital sinuses. Microscopic lesions in the respiratory tract and conjunctiva consist of epithelial erosion, necrosis, and ulceration with formation of syncytial cells containing eosinophilic, intranuclear inclusion bodies (Fig. 31.9B; eSlide 31.e3). Syncytial cells may be sloughed into the lumen and are only evident in the early stages of infection (1–5 days).[63] The mucosa and submucosa are edematous, congested, and infiltrated by mononuclear cells and heterophils. In addition to the pathologic changes, diagnosis can also be made by virus isolation, immunofluorescence, immunohistochemistry, and PCR for GaHV-1 on samples of larynx, trachea, and lungs.[25,162]

Gallid herpesvirus 2 (**Marek's disease virus** [**MDV**], genus *Mardivirus*) causes **Marek's disease (MD)**, one of the most common diseases of chickens. Quails, pheasants, and turkeys are also susceptible, and disease in Japanese quail is relatively common.[156] The disease is more often seen in unvaccinated and sexually immature chickens that are 2–7 months old, but can occur at any age.[19] Infection causes mononuclear infiltration and/or neoplasia in multiple sites, most often the skin, peripheral nerves, gonads, eyes, muscle, and viscera. The virus replicates in feather follicular epithelium and is shed into the environment. Marek's disease pathogenesis has four phases.[18,156] In the earliest phase, cell-free virus in feather dander, chicken-house dust, feces, or saliva is inhaled and infects respiratory epithelial cells and local macrophages. Viremia with cytolytic viral replication in lymphocytes and immunosuppression subsequently develops. This is followed by a latency phase in CD4+ T cells and results in systemic viral dissemination. During this phase, cutaneous viral infection, replication, and shedding occur. The third phase of MDV infection is characterized by reactivation in CD4+ T cells. During this phase, a late cytolytic stage and immunosuppression occur. The final, proliferative phase is characterized by reactivation of the virus and neoplastic transformation of CD4+ T cells.

Gross lesions include bursal and thymic atrophy; tumor infiltration with enlargement of peripheral nerves (Fig. 31.10A), white discoloration of the iris with an irregularly shaped pupil, nodular lesions in feather follicles, splenomegaly, hepatomegaly, and the presence of pale white tumors in various organs. Peripheral nerve lesions may be unilateral or bilateral and involve the sciatic, brachial, and vagus nerves and spinal root ganglia. There are two main types of lesions in peripheral nerves, type A (neoplastic) and B (inflammatory) lesions and characterized by an infiltration of neoplastic CD4+ T cells (Fig. 31.10B; eSlide 31.e4) and mixed inflammation with edema, respectively.[21,156] Peripheral nerve lesions are rare in quail and turkeys.[68,123,156,176] Microscopic changes in the central and peripheral nervous system include perivascular lymphocytic cuffing, vasculitis, edema, gliosis, and lymphocytic meningitis. Visceral organs contain pleomorphic and neoplastic T lymphocytes (Fig. 31.10C; eSlide 31.e4). Skin lesions may be inflammatory or neoplastic but usually surround feather follicles. Eosinophilic intranuclear inclusions may be present in cells of the transitional zone and stratum corneum of the feather follicle epidermis.[53] Malignant lymphoma can be present in the anterior and posterior uvea. Detection of lymphoma in peripheral nerves,

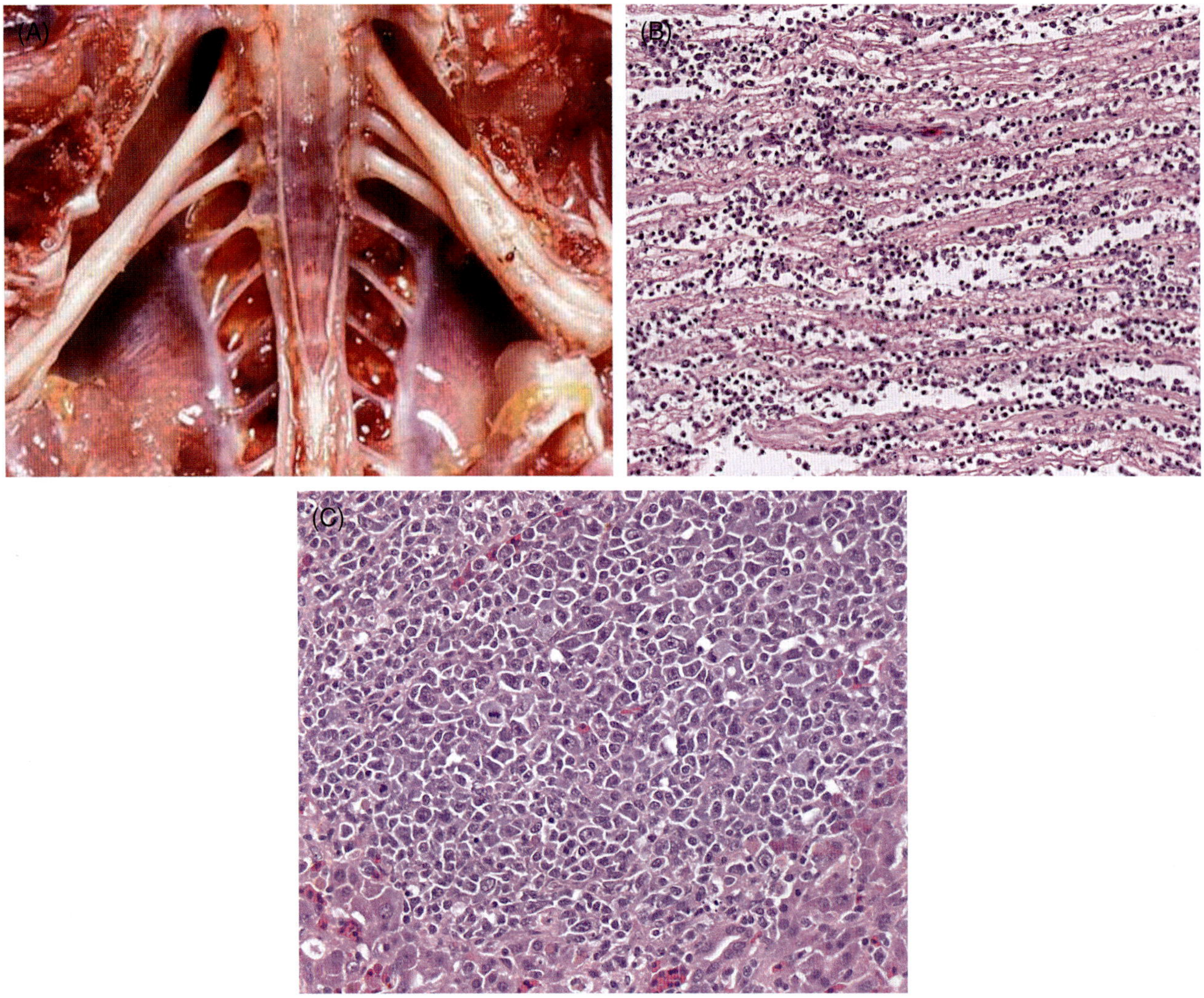

FIGURE 31.10 Gallid herpesvirus 2 (Marek's disease) in the sciatic nerve (A, B) and liver (C) in chickens. (A) A characteristic finding is enlargement and edema of the sciatic nerves due to neoplastic CD4+ T cell infiltrates (Type A) or mixed inflammation and edema (Type B). (B) Expansion of the tissue is due to diffuse infiltration of pleomorphic, anisokaryotic, neoplastic lymphocytes between the axons. (C) The normal parenchyma is expanded and replaced by sheets of pleomorphic, neoplastic CD4+ T lymphocytes (see eSlide 31.e4).

brain, and eye supports a diagnosis of Marek's disease. As MDV is ubiquitous, viral detection alone is not confirmatory. Diagnosis therefore relies on characteristic lesions combined with detection of Marek's disease virus antigens (pp38 and meq) in tissues by immunohistochemistry, virus isolation, or PCR of tumor from feather tips, blood, or tissues.

A novel **phasianid herpesvirus**, member of the genus *Mardivirus* and closely related to Gallid herpesvirus-3, was detected in three species of pheasants vulnerable to extinction.[145] Birds were found dead or presented nonspecific clinical signs including lethargy and anorexia prior to death. Microscopically, affected birds had necrotizing hepatitis, splenitis, and enteritis associated with intranuclear eosinophilic inclusions in hepatocytes and intestinal epithelial cells. The virus was detected by PCR using degenerate herpesvirus primers followed by sequencing. In situ hybridization labeled viral DNA in the nuclei of hepatocytes.

Pigeon herpesvirus 1 (syn. **Columbid herpesvirus 1**) is worldwide distributed and relatively common in wild and feral pigeons.[125] It causes generalized infection in young pigeons; disease can occur in adults, who can be asymptomatic carriers. Stress or immunosuppression caused, for example, by pigeon circovirus infection predisposes birds to infection and disease.[156] Columbid herpesvirus 1 also causes disease in raptors including owls (see Chapter 30: Birds of Prey).[55,177] Virus sources include contaminated environments, food, and water, as well as crop milk, which the adults regurgitate and feed their young. The virus initially infects the mucosa of the conjunctiva, respiratory, and digestive tracts, after which the birds develop viremia.

Clinical signs include lethargy, depression, excess lacrimation, nasal discharge, diarrhea, neurological signs, and sudden death. Morbidity and mortality are variable in a flock but are usually low. Squabs that have maternal immunity are protected against severe disease; they survive but become asymptomatic carriers.[101] Gross lesions include conjunctivitis, fibrinonecrotic exudate in the pharynx, esophagus, and crop and most commonly, hepatomegaly with pale foci of necrosis. The spleen can be enlarged, and the pancreas may contain pale, patchy areas of necrosis. Microscopically, there is necrosis of the pharyngeal, esophageal, and ingluvial epithelium with ulceration and inflammation. Epithelial cells subjacent to ulcers often contain eosinophilic intranuclear inclusions. The liver contains multifocal to coalescing areas of coagulative necrosis with mixed inflammatory cell infiltrates. Eosinophilic intranuclear inclusion bodies are present in hepatocytes (Fig. 31.11), pancreatic acinar cells, and splenic mononuclear cells. Syncytia, some with inclusions, may also be present. Other lesions described less frequently include stomatitis, sinusitis, tracheitis, nonsuppurative encephalomyelitis, myocarditis, sialoadenitis, and proventriculitis. Virus isolation, transmission electron microscopy, or PCR for the viral DNA polymerase gene in liver, chicken egg embryos, chorioallantoic membrane, oropharyngeal swab samples, or cell culture are confirmatory when combined with characteristic histologic lesions.[41]

Avian poxvirus infections are caused by *Avipoxviruses* in the family of *Poxviridae*. Within the *Avipoxvirus* genus are numerous species, including fowlpox, pigeonpox, quailpox, and turkeypox. Infections are reported in a wide host range with a degree of host-specificity; morbidity can approach 100% but mortality is low. Avian poxviruses encode a protein similar to epidermal growth factor (EGF), which is believed to be associated with epithelial cell hyperplasia in infected tissues.[156] Gross lesions of cutaneous avian pox consist of proliferative tan nodules primarily on the unfeathered portions of the head (Fig. 31.12A) and feet. Nodules may become ulcerated, dark brown, and hemorrhagic as a result of localized trauma. The "wet" or diphtheritic form of avian pox is characterized by proliferative and ulcerative lesions with diphtheritic membranes in the mouth, tongue, esophagus, conjunctiva, and upper respiratory tract (Fig. 31.12B). Sometimes, both forms can occur simultaneously. Characteristic microscopic lesions include irregular acanthosis, spongiosis, and ballooning degeneration with brightly eosinophilic, intracytoplasmic inclusions (Bollinger bodies) in epidermal (Fig. 31.12C), and mucosal epithelial cells. The surface epithelium may be ulcerated and contain secondary bacterial and/or fungal infections. Mononuclear cells are commonly present in the surrounding submucosa or dermis. The microscopic appearance of Bollinger bodies is typically considered diagnostic, but molecular testing is available. Pox virus infection occurs via ocular, digestive, and respiratory routes by exposure to contaminated environments, food, and water. Biting insects aid in the transmission of the virus, which results in breaks in the epidermis that facilitate transmission. The incubation period is between 4 and 10 days and birds of all ages are susceptible. Virus isolation in embryonated chicken egg chorioallantoic membranes, PCR, and electron microscopy (especially in the absence of obvious viral inclusions) (Fig. 31.12D) support the histologic diagnosis of poxviral infection.

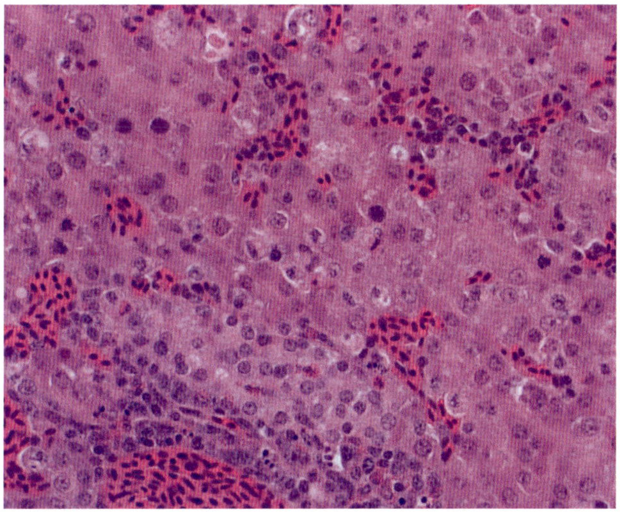

FIGURE 31.11 Pigeon herpesvirus 1 (syn. Columbid herpesvirus 1) in a pigeon. Characteristic findings of infection are periportal lymphoplasmacytic and heterophilic inflammation and hepatocellular intranuclear eosinophilic inclusions.

Parvoviruses in the *Chaphamaparvovirus* genus, within the newly created *Hamaparvovirinae* subfamily, have been associated with high mortality and hepatitis in gray partridges and in pheasants.[103] The livers of deceased birds are swollen, mottled, yellowish-brown, or they may be congested with ecchymotic lesions. These gross lesions look like inclusion body hepatitis in chickens. Histologically, the disease is characterized by multifocal to coalescing necrotizing hepatitis, with large amphophilic to acidophilic intranuclear inclusion bodies in degenerative hepatocytes. Although this condition was described over 20 years ago,[137] the etiology has not been elucidated until recently.

RNA Viruses

There are four types of influenza viruses: A, B, C, and D. **Avian influenza viruses** (**AIV**) are influenza A viruses. They have been detected in more than 105 bird species from different taxa. Wild waterfowl in the orders Anseriformes and Charadriiformes are the natural reservoirs.[152] Traditionally, AIVs have been classified by their hemagglutinin (HA) and neuraminidase (NA) surface glycoproteins. Based

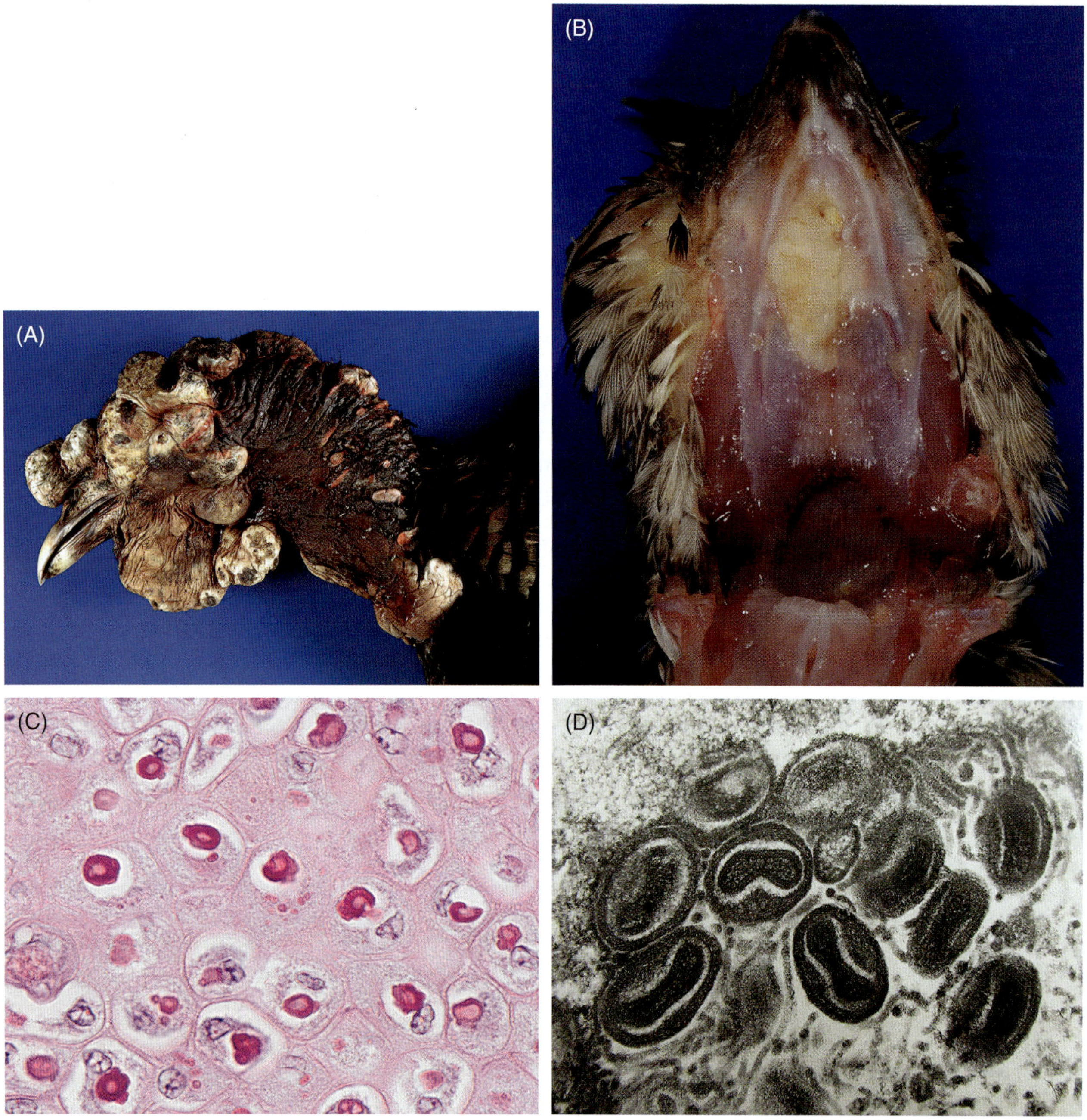

FIGURE 31.12 Poxviral infection in a turkey (A), quail (B), and chicken (C). (A) Multiple proliferative nodules are present on the unfeathered skin of the head. (B) Proliferative lesions with diphtheritic membranes within the oral cavity are typical of the "wet" or diphtheritic form of avian pox. (C) The epidermis is thickened due to ballooning degeneration of infected epithelial cells, many of which contain characteristic, large, eosinophilic intracytoplasmic inclusions. (D) Virus particles are large (250–300 nm) and ovoid with a biconcave core. TEM.

on the clinical signs and mortality caused experimentally in chickens, AIVs are classified into two pathotypes, low and highly pathogenic AIV (LPAIV and HPAIV, respectively). For domestic species, all AIVs are notifiable to the WOAH; for wild birds only HPAIVs are notifiable.[120] The same AIV may induce a wide range of clinical signs depending on host species, age, or immune status. LPAIV infections may be subclinical or associated with diseases affecting the respiratory, gastrointestinal, urinary, and reproductive tracts. These viruses require trypsin-like enzymes for the cleavage of the HA protein, thus viral replication is usually limited to mucosal epithelial cells of the respiratory and intestinal tracts. Circulation of H5 and H7 subtypes of LPAIV in gallinaceous poultry may result in the acquisition of multiple basic amino acids in the HA protein cleavage site and the emergence of highly pathogenic AIV strains.[155]

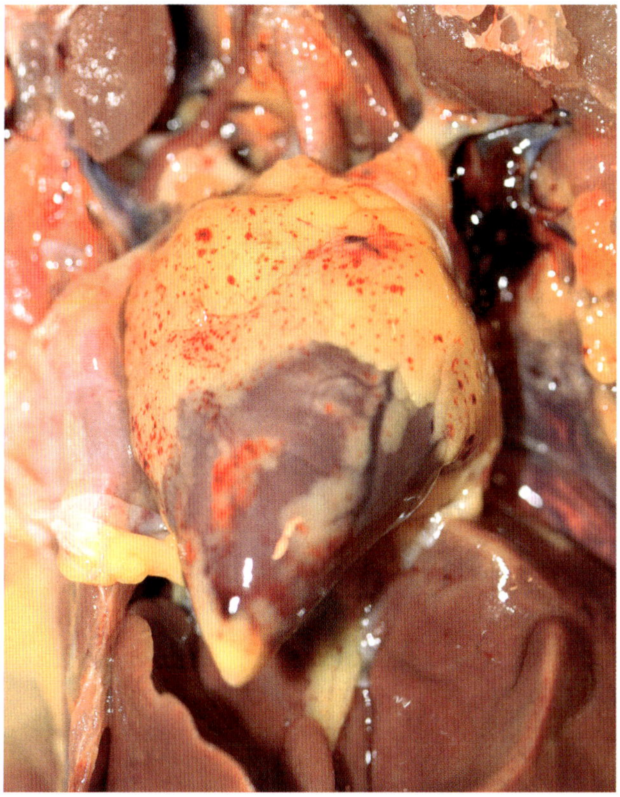

FIGURE 31.13 Highly pathogenic avian influenza virus (HPAI) infection in a chicken. Petechial hemorrhages on the heart and abdominal adipose tissue due to infection with the H5N1 stain of HPAI. Though there are no pathognomonic lesions of HPAI infection, petechial hemorrhage has been reported in some species.

HPAIV infection causes systemic disease with high mortality in chickens and turkeys. Other gallinaceous poultry species, such as bobwhite quail, ring-necked pheasants, and chukar partridges may survive longer and develop more severe neurological disease than chickens. It has long been thought that pigeons are resistant to HPAI; however, certain strains of HPAIV can cause disease.[77,91]

Clinical signs of HPAIV infection include depression, paresis, paralysis, and mortality. Affected birds may have edema and cyanosis of the head and comb, hemorrhage in the skin of the legs, edema and hemorrhage in the lungs, hemorrhage in the gastrointestinal mucosa and serosa, and necrotic foci and hemorrhage in visceral organs (Fig. 31.13), such as the pancreas.[154,156,157] Microscopic lesions of HPAIV infection vary and involve multiorgan necrosis, hemorrhage, and inflammation. Confirmation of disease in suspected cases is through virus isolation, antigen capture immunoassays, and RT-PCR or qRT-PCR in oropharyngeal and cloacal swabs, feces and in samples of affected organs. Rapid diagnosis can be achieved by screening oropharyngeal swabs using a matrix gene for influenza A by qRT-PCR and subsequent testing of positive samples with H5 and H7 subtype-specific qRT-PCR.[151] WOAH influenza reference laboratories perform confirmatory testing and final identification of these viruses.

Newcastle disease virus (**NDV**), a variant of **avian paramyxovirus 1** (**APMV-1**), causes **Newcastle disease** (**ND**). Infection with NDV is WOAH notifiable.[120] NDV has been detected in at least 241 bird species.[82] NDV strains are categorized as velogenic (highly virulent), mesogenic (moderately virulent), and lentogenic (nonpathogenic). Velogenic strains cause severe nervous and respiratory disease and high mortality (up to 100%); mesogenic strains primarily affect egg production, can produce mild respiratory disease, and cause low mortality; lentogenic strains usually are asymptomatic in adult birds but may cause respiratory disease in the young. Chickens are highly susceptible to velogenic NDV. Pathogenesis of ND involves viral replication in mucosal epithelium of the respiratory and intestinal tracts followed by viremia and viral dissemination to multiple visceral organs and the brain. Gross lesions include facial edema, ventricular and/or proventricular hemorrhage, necrosis and hemorrhage in the intestinal mucosa, especially at cecal tonsils (Figs. 31.14A, B) and Peyer's patches, and splenomegaly with congestion and mottling.[156] Microscopically, affected birds have nonsuppurative encephalitis with gliosis and neuronal degeneration, disseminated vasculitis and thrombosis, and necrosis in lymphoid tissues as well as necrosis and hemorrhage in the mucosa of the conjunctiva, trachea, and gastrointestinal tract. Infections with low virulence NDV can cause mild conjunctivitis, rhinitis, tracheitis, pneumonia, and airsacculitis. Secondary bacterial infections enhance the severity of the respiratory disease caused by low virulence NDV. Virus isolation, RT-PCR, and DNA sequencing of the fusion protein cleavage site on tracheal and cloacal swabs and samples of affected organs support the gross and histological diagnosis; the intracerebral pathogenicity index (ICPI) test is performed to determine the virulence of the virus.[119,156]

Pigeon paramyxovirus 1 (**PPMV-1**) is a variant of APMV-1 that is common in pigeon populations worldwide. Pigeons of all ages are susceptible. The APMV-1 variant affecting pigeons is commonly known as pigeon PMV-1 or PPMV-1. It has a distinct binding pattern with monoclonal antibodies and, by phylogenetic analysis, the majority of PPMV-1 isolates are placed in lineage 4b (VIb).[3] Disease is transmitted most commonly by ingestion of contaminated food and water but also through ocular and respiratory routes. The incubation period is 10–12 days. Lesions are due to viremia and noted in various organs including the brain, inner ear, and viscera. Most commonly, affected pigeons have neurological signs, such as ataxia, paralysis of legs and wings, torticollis, and opisthotonos (Fig. 31.14C), but clinical signs can also include anorexia, greenish watery diarrhea, lethargy, and depression. Morbidity can be high but mortality in a flock is generally

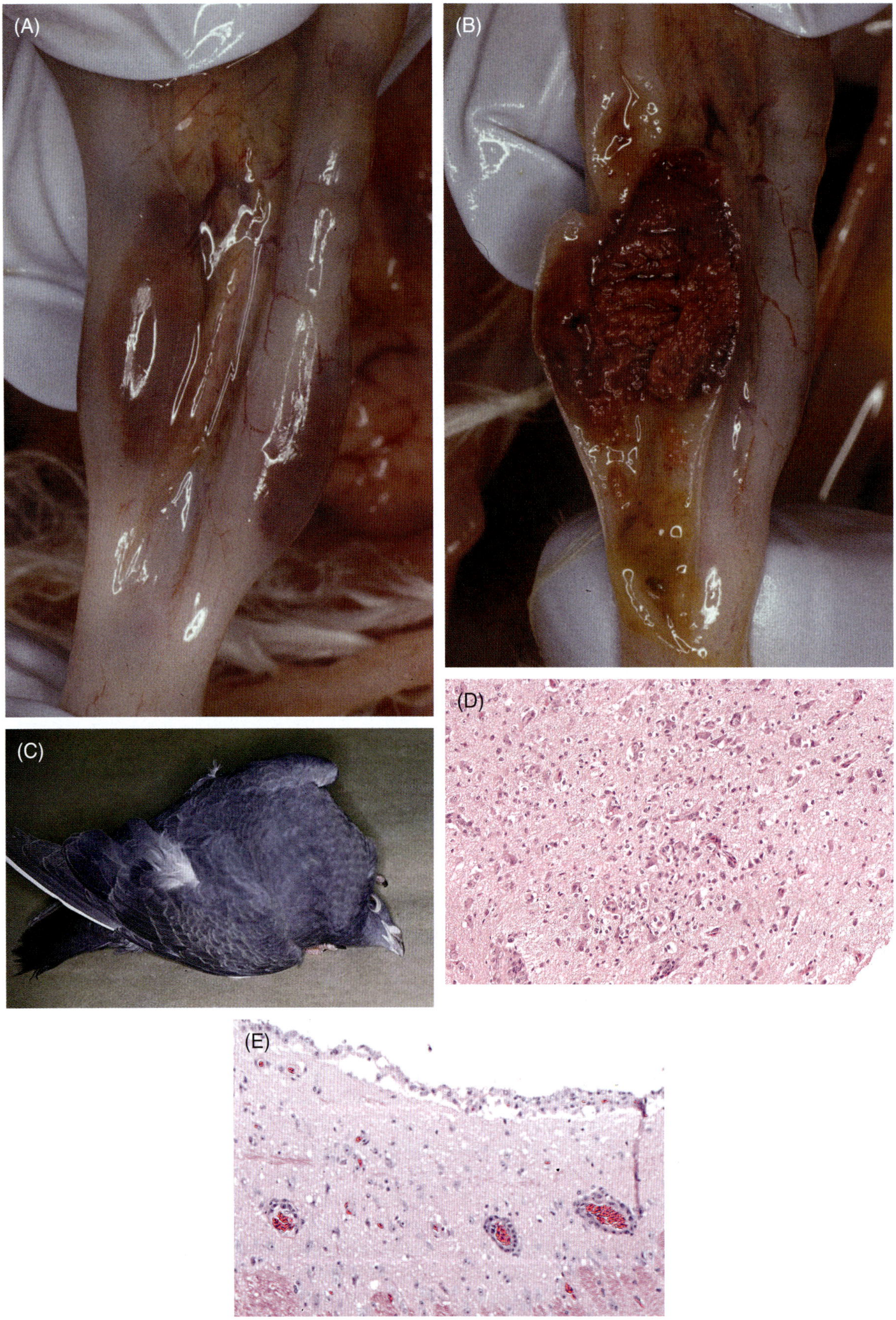

FIGURE 31.14 Paramyxovirus-1 infection. Velogenic Newcastle disease (vND) in a chicken with (A) characteristic lesions include focal discoloration of the cecal wall due to hemorrhage of the cecal tonsils that can be observed through the serosal surface. (B) With vND, accumulation of red/brown necrohemorrhagic luminal content is observed when the cecal wall is open. (C) Clinical signs of paramyxovirus-1 infection in pigeons can include opisthotonus and torticollis. (D) Histologically, paramyxovirus infection in a pigeon causes nonheterophilic, necrotizing meningoencephalitis and (E) lymphocytic perivascular cuffing.

low. However, infection has been associated with mortality events in wild Eurasian collared and rock doves in the United States. Gross lesions include greenish watery intestinal contents, pale tan and enlarged kidneys with increased urates, enlarged and mottled pale tan or red pancreas, and splenomegaly.[144] Microscopically, the most common lesions are lymphoplasmacytic tubulointerstitial nephritis, necrotizing pancreatitis, enteritis and splenitis, nonsuppurative meningoencephalitis (Figs. 31.14D, E), and occasionally otitis interna.[76] PPMV-1 can be diagnosed using the same methods as for any other APMV-1, including virus isolation and RT-PCR.

***Avian Paramyxoviruses*-2 to -11** are occasionally isolated from wild birds; however, it is unclear if they are responsible for clinical signs and death because experimental infection usually does not induce disease. Serotypes 2, 3, 6, and 7 have been isolated from turkeys. Serotype 7 has been isolated from dead collared doves.

A **coronavirus** similar to infectious bronchitis virus of chickens is associated with clinical signs of respiratory disease and pancreatitis in pigeons.[128] Clinical signs include depression, polydipsia, ocular and nasal discharge, and tracheal rales. Necropsy findings include an enlarged and mottled pancreas, and edema and congestion in the lungs.

Eastern equine encephalitis (**EEE**) (family *Togaviridae*; genus *Alphavirus*) is a zoonotic alphavirus that causes neurological signs and encephalitis in ring-necked pheasants, chukar partridges, ruffed grouse, and turkeys.[8,17,156] Mortality can be high (up to 80%) in naturally occurring outbreaks; young birds are most susceptible. Affected ruffed grouses have been reported to have poor body condition with atrophied breast musculature.[8] Gross lesions have not been observed in pheasants. Chukar partridges may have pale tan focal areas in the heart and an enlarged, mottled spleen.[130] Microscopically, affected birds have lymphocytic meningoencephalitis with perivascular cuffing, vasculitis, gliosis, neuronal degeneration, and satellitosis. Myocardial necrosis and lymphocytic myocarditis may also be seen. Lymphoid depletion and necrosis in the bursa of Fabricius and thymus are reported in experimentally infected turkeys.[64]

Highlands J virus (**HJ**) is an alphavirus that causes disease in turkeys and chukar partridges. Known distribution is in North and South America.[44,48,169] Affected birds have a mottled, enlarged spleen, myocardial pallor due to necrotizing myocarditis, intestinal serosal hemorrhage, enteritis with necrosis of gut-associated lymphoid tissue, lymphocytic meningoencephalitis, and lymphoid depletion with lymphoid necrosis in the bursa of Fabricius, spleen, and thymus.

West Nile virus (**WNV**), a type of *flavivirus*, is not considered a significant pathogen of commercial poultry or columbiformes despite recognized low-level viremia in infected birds.[6] On the other hand, natural disease has been reported and associated with population-level declines in some species of Galliformes, including in threatened species such as the greater sage-grouse.[159,171,178] Lymphocytic encephalitis, myocarditis, pancreatitis, ventriculitis, and myositis with bone marrow hyperplasia have been reported in a single wild turkey that was clinically disoriented and failed to flee when approached.[180] Disease in Impeyan pheasants was associated with hemorrhage at the proventricular-ventricular junction, cecal ulcers, necrotizing splenitis, and mild myocardial necrosis, while chukar partridges had more significant myocardial necrosis.[178] Experimental infection in red-legged partridges resulted in mortality with inflammation and necrosis that was associated with WNV antigen in the heart, brain, liver, lung, spleen, intestine, and bursa of Fabricius.[57] In ruffed grouse, experimental WNV infection resulted in mortality with severe necrotizing myocarditis and mild encephalitis.[115]

Israel turkey meningoencephalitis virus (**ITV**) and **Bagaza virus** (**BAGV**), also flaviviruses, have been proposed as single virus species.[47] ITV is reported in turkeys,[156] and BAGV has been detected in free-ranging gallinaceous birds, including partridges, pheasants, and common wood pigeons.[56] Disease causes nervous signs and high mortality. Gross lesions vary from none to pallor of the pectoral muscle, pancreas, and the myocardium. Congestion of encephalic and epicardial vessels may be present. Microscopically, affected birds have nonsuppurative meningoencephalitis and splenitis, as well as myocardial necrosis.[56,75,92]

Diagnosis of arboviral diseases is through evaluation of lesions and by performing virus isolation (in level 3 biocontainment facilities), RT-PCR, and immunohistochemistry.

Turkey rhinotracheitis (**TRT**) and **swollen head syndrome** (**SHS**) are highly contagious respiratory diseases caused by **avian metapneumoviruses (aMPV).** TRT is primarily a disease of turkeys, while SHS is primarily a chicken disease. aMPV infections similar to SHS have also been detected in pheasants[60,118,173] and guinea fowl.[97] The viruses infect and replicate in epithelial cells of the respiratory tract causing ciliostasis, which predisposes to secondary bacterial infections.[26,78,80] Chickens, pheasants, and guinea fowl with SHS may present with swelling of the infraorbital sinuses, periorbital region, eyelids and subcutaneous tissues of the mandible, neck, and wattles due to accumulations of yellow gelatinous to caseous material (Fig. 31.15). Turkeys with TRT have watery to mucoid exudate in the nasal passages and trachea[80] and periorbital and infraorbital edema. Fibrinous polyserositis with airsacculitis, pericarditis, perihepatitis, salpingitis, and oophoritis occur in birds with secondary bacterial infections. Microscopically, there is a loss of epithelial cilia in the respiratory tract and the lamina propria is expanded with edema and inflammatory cells. Small eosinophilic intracytoplasmic inclusions may be seen in epithelium of the nasal

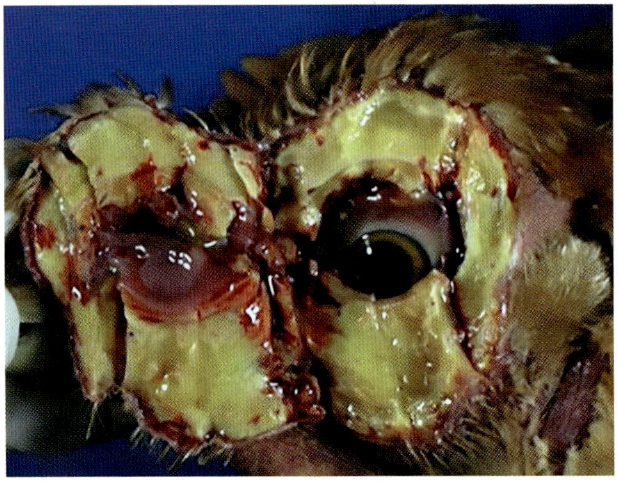

FIGURE 31.15 Swollen head syndrome due to avian metapneumovirus infection in a chicken. Infection can result in an accumulation of yellow gelatinous to caseous material in the subcutaneous tissue of the periorbital region.

turbinates and trachea in the acute stage of the disease.[20] Conjunctivitis, osteomyelitis of cranial bones, and otitis may also occur. Diagnosis is made by evaluating lesions and by virus isolation and/or RT-PCR on upper respiratory tract tissues, swabs, or mucus.

Most **enteric viral infections** occur in young birds. The clinical signs and lesions induced by different viruses may be similar. Diarrhea and distention of the gastrointestinal tract with gas and liquid are common manifestations. Histologically, there is degeneration of enterocytes, villous blunting, and leukocytic infiltration of the lamina propria (Fig. 31.16). The etiology of many poultry enteric disease syndromes is difficult to elucidate, as many of these viruses cannot be cultured. **Astroviruses, reoviruses, coronaviruses, and even parvoviruses (*Aveparvovirus* genus, ssDNA)**, to mention a few, are implicated as causative agents for enteric disease because they have been isolated from or identified in the intestines and intestinal contents of affected poultry.[122,156] However, many of these enteric viruses can also be detected in otherwise healthy birds.

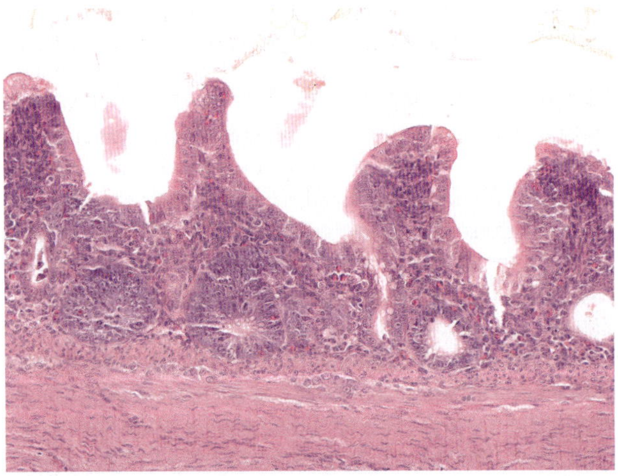

FIGURE 31.16 Coronavirus enteritis in a turkey. Infection is associated with degeneration of enterocytes, villous blunting, and leukocytic infiltration of the lamina propria.

Avian encephalomyelitis (AE) is a neurological disease of chickens, turkeys, quail, and pheasants caused by **avian encephalomyelitis virus** (family Picornaviridae; genus *Tremorvirus*). Clinical disease with neurological signs most commonly occurs in 1- to 3-week-old birds. Pathogenesis involves infection of the intestinal epithelium followed by viremia and viral dissemination to other visceral organs, skeletal muscle, and the brain.[156] Gross lesions are often absent. Pale discoloration in the muscular layers of the ventriculus may be the only gross lesion seen in birds with clinical disease; grayish foci may be seen in the pancreas of pigeons.[161] Cataracts may be observed in birds that survive infection. Microscopically, encephalomyelitis with lymphocytic perivascular cuffs, gliosis, glial nodules, central chromatolysis of neuronal cell bodies, and neuronal satellitosis are often seen in the brain and spinal cord. Multifocal lymphocytic infiltration may be present in the myocardium, muscular layer of the gastrointestinal tract (Fig. 31.17A), pancreas, skeletal muscle, and peripheral nerves.[20,156] Diagnosis is based on a combination of clinical signs, disseminated nonpurulent inflammation of the central nervous system, and identification of characteristic central chromatolysis in neuronal cell bodies in the brain stem (Fig. 31.17B). Virus isolation performed on brain samples confirms the diagnosis. Fluorescent antibody tests can be performed to demonstrate viral antigen in tissues.

Avian reoviruses are ubiquitous viruses among birds and infection is usually present without disease (an estimated 85%–90% of isolated avian reoviruses are nonpathogenic); one exception being viral arthritis/tenosynovitis in chickens and turkeys.[79,156] Reoviruses may be involved in several other disease syndromes, including respiratory, enteric, and systemic infections, and the stunting/malabsorption syndrome of poultry. However, reoviruses are not considered the main cause in any of these conditions. Reoviruses have been isolated from tissues or feces of sick pigeons, pheasants, and bobwhite quail; in all these cases, other pathogens (e.g., cryptosporidium and *Staphylococcus aureus*) have also been identified in lesions.

A novel **pigeon rotavirus** has been reported in racing and feral pigeons from Australia.[105] Infection causes high rates of morbidity and mortality. Clinical signs are characterized by regurgitation, diarrhea, severe lethargy, and depression. Necropsy findings are frequently unremarkable. Histologic lesions are characterized by extensive hepatic necrosis with minimal inflammatory response. The pathogenesis is poorly understood. Although avian and mammalian rotavirus sequences are noticeably different, this novel virus shares genes with other pigeon rotaviruses and a rotavirus isolated from a fox. It is capable of infecting mice.[108]

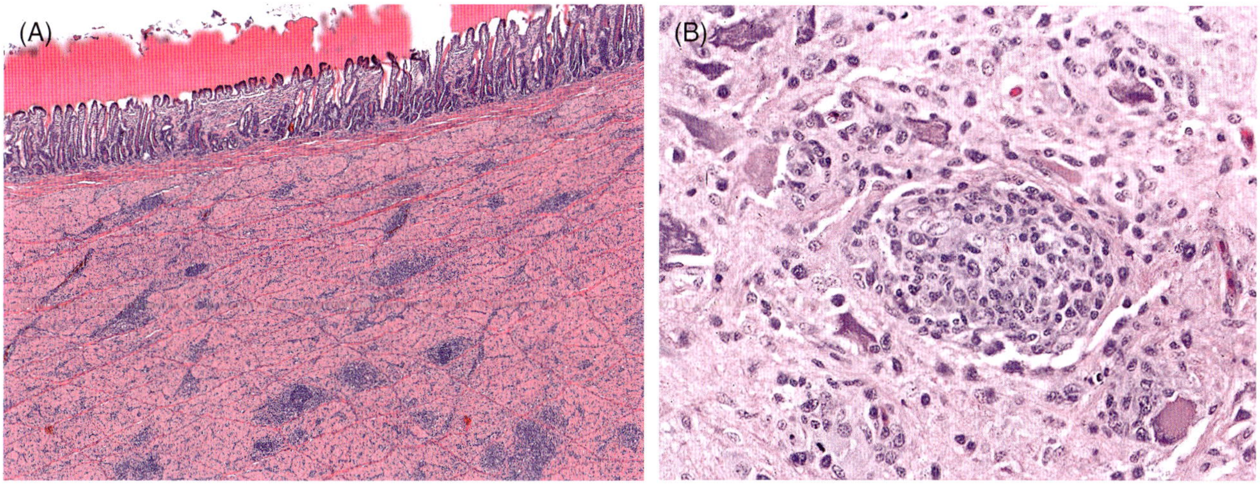

FIGURE 31.17 Avian encephalomyelitis virus infection in a chicken. (A) Infection is associated with multifocal lymphocytic infiltration in the muscular layer of the gizzard and (B) nonheterophilic inflammation with glial nodules and central chromatolysis of neuronal cell bodies in the brain.

Retroviruses

Lymphoid leukosis (LL) is a disease of semimature to mature chickens that is caused by **avian leukosis sarcoma viruses (ALVs)**. These viruses are classified into 10 subgroups: A, B, C, D, E, F, G, H, I, and J. ALVs may cause a variety of tumors, including lymphoid leukosis, soft tissue sarcomas, epithelial cell tumors (primarily renal tumors), and tumors of endothelial cell origin (hemangioma and hemangiosarcoma among others).[156] Lymphoid leukosis is the most common of these neoplastic conditions. Disease is more frequently seen in chickens older than 16 weeks of age. ALVs infect and replicate in various tissues, but the bursa of Fabricius is the target organ for neoplastic transformation in chickens with LL. ALVs cause insertional activation of *c-myc* in transformed follicles by placing this gene under the control of viral LTR, which causes maturation arrest and proliferation of bursal stem cells.[113,114] Gross lesions of LL consist of splenomegaly, hepatomegaly, and grayish-to-white soft tumors in various organs that are most common in the liver, spleen, and bursa of Fabricius. Microscopically, affected organs have multifocal to coalescing nodules containing a uniform population of large lymphocytes (lymphoblasts) that efface normal tissue architecture and compress adjacent parenchyma. Lymphoma in the bursa of Fabricius has an intrafollicular growth pattern. Diagnosis can be difficult because LL tumors are similar to those caused by Marek's disease virus and reticuloendotheliosis virus (REV; below)[19] Lymphoma with an intrafollicular pattern in the bursa of Fabricius is typical for LL, but bursal lymphoma caused by REV needs to be ruled out. Immunohistochemistry for B-cell and IgM surface markers, virus isolation from peripheral white blood cells and tumor homogenates, and standard or quantitative RT-PCR can be used for confirmation. Because ALVs are ubiquitous, viral detection alone is not confirmatory.

Reticuloendotheliosis, primarily a disease of chickens and turkeys, is caused by **reticuloendotheliosis virus (REV)**, a retrovirus in the *Gammaretrovirus* genus. REV is known to cause lymphoma, acute reticular cell sarcoma (reticuloendotheliosis), and various nonneoplastic lesions in various species of gallinaceous birds. In addition to chickens and turkeys, lymphomas caused by REV occur in quail, pheasants, peafowl, and Attwater's prairie chickens,[156] and have recently been associated with disseminated lymphoma in young and adult pigeons.[179] Clinical signs include reduced reproductive performance, poor weight gain, and increased flock mortality. Lesions coupled with virus isolation and PCR-positive results from blood and tumor homogenates are diagnostic.

Lymphoproliferative disease (LD), caused by an exogenous retrovirus called **lymphoproliferative disease virus (LPDV)**, occurs in domestic turkeys in Israel and several countries in Europe, and in wild turkeys in North America.[5,69,160] LD is characterized by multifocal to coalescing tan, raised, firm skin nodules composed of pleomorphic lymphoid cells in the dermis of unfeathered skin on the eyelids, head, neck, or legs (Fig. 31.18)[45,99] and in visceral organs. Birds with systemic LD may have hepatomegaly, splenomegaly, and multifocal to coalescing tan, raised nodules or infiltrates in many organs, including the liver, kidney, lung, spleen, gastrointestinal tract, and brain; neoplastic lymphocytic infiltrates often also surround blood vessels.[116] While infection with LPDV is common in wild turkeys in the United States, LD lesions are rare. Diagnosis of LD requires histological confirmation of neoplastic and nodular infiltrates of pleomorphic lymphocytes in the skin or visceral organs. The most sensitive tissue for molecular detection of LPDV is bone marrow, but whole blood, visceral organs, and spleen can also be used.[4,160] The method of virus transmission in nature is not known.

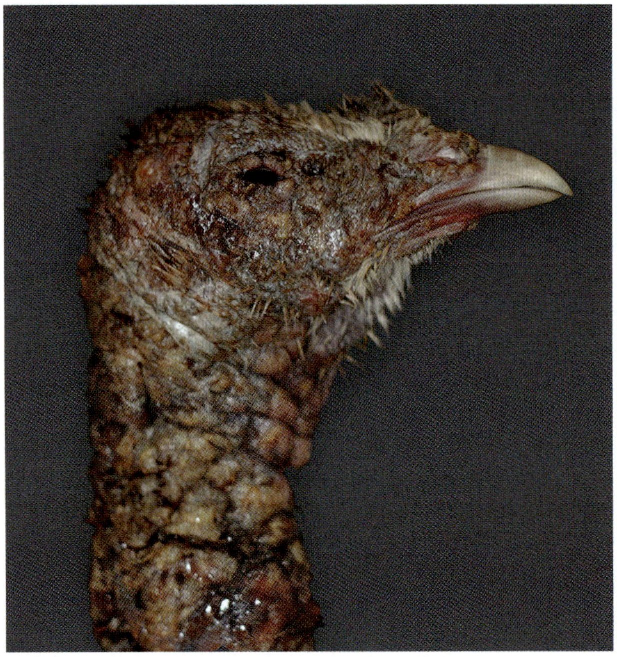

FIGURE 31.18 Retroviral lymphoproliferative cutaneous nodules in a turkey. Nodules in the unfeathered skin on the head are composed of pleomorphic lymphoid cells. *(Photo Courtesy of the Southeastern Cooperative Wildlife Disease Study.)*

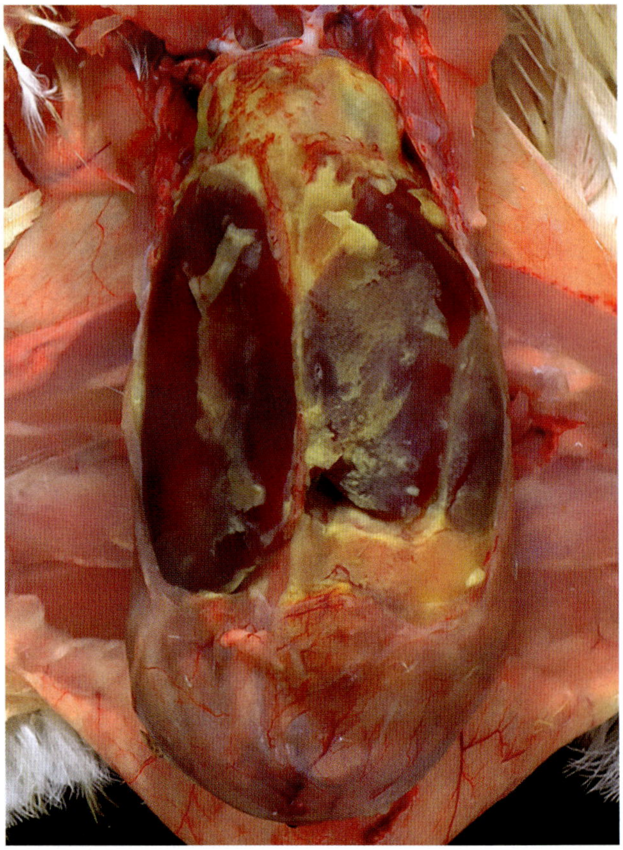

FIGURE 31.19 Enterobacterial pericarditis, perihepatitis, and peritonitis in a chicken. Infection is associated with fibronecrotic exudate over the Glisson's membrane, the pericardium, and acute fibrinous coelomitis. *(Courtesy of H. John Barnes, NC State University.)*

Bacteria

Members of the **Enterobacteriaceae** family are ubiquitous and normal intestinal flora. Under certain circumstances, they may become pathogenic. They are not part of the normal respiratory or reproductive tract flora. Efforts should be made to avoid contamination during prosection, so that accurate culture results are obtained from these or other sites.

Colibacillosis refers to localized or systemic infection caused by the gram-negative bacillus ***Escherichia coli***. It occurs in all types and age groups of poultry and pigeons. *E. coli* is present in the intestine of birds and excreted in feces. It may be a primary or secondary pathogen. Clinical signs vary from absent, when lesions are mild or localized, to lethargy, anorexia, or peracute death in severely affected birds. Chronically infected birds are often stunted and unthrifty. Colisepticemia or systemic *E. coli* infection is the most common pathogenic presentation (Fig. 31.19). Localized infections include omphalitis in hatchlings, salpingitis in laying hens, cellulitis caused by the secondary infection of skin wounds, synovitis as a sequela to systemic infections, and airsacculitis, pericarditis, and perihepatitis. Chronic granulomatous infections with *E. coli* are called coligranulomas or **Hjarre's disease**. Enteritis caused by pathogenic **attaching and effacing *E. coli* (AEEC)** has been reported in gallinaceous birds and pigeons.[20,166]

Infections with ***Salmonella* spp**. (gram-negative bacillus) can cause acute and chronic disease. Gallinaceous species are natural reservoirs of the bacteria. ***S. enterica*** subsp. ***enterica* serotypes Pullorum** and **Gallinarum** are responsible for the septicemic diseases known as **pullorum disease** and **fowl typhoid**, respectively. These bacteria are nonmotile, generally host specific and reportable to WOAH.[120] Clinical signs include anorexia, dehydration, huddling, depression, diarrhea, and death. Mature birds may be asymptomatic or show reduced feed consumption, depression, and decreased egg production, fertility, and hatchability. Other lesions may include swelling of the joints, dyspnea, and blindness. Histology is characterized by fibrinoheterophilic inflammation, usually in multiple organs.

The majority of the other serotypes in *S. enterica* subsp. *enterica* are motile and nonhost specific, and diseases caused by these bacteria are referred as **paratyphoid salmonellosis**. Infections are more important as a public health issue than as a disease concern for galliformes. Although all ages are susceptible to infection, clinical signs are usually seen in young and immunosuppressed birds. Clinical signs and lesions are nonspecific and identification of the pathogen is needed to differentiate it from other bacterial infections. Birds look depressed and dehydrated, have diarrhea, and increased mortality, mainly in the first 2 weeks of age or at the onset of laying. Adult birds typically do not develop clinical signs. Bacteria are localized to the intestine or gallbladder of carriers and intermittently shed in feces.

In pigeons, one of the most common diseases is **wing paralysis** caused by ***S. enterica*** **subsp.** ***enterica*** **serovar Typhimurium** var. **Copenhagen**, a host-adapted pathogen.[129] Clinical signs can be variable ranging from acute death to depression, listlessness, anorexia, greenish diarrhea, corneal opacity, swollen joints, ataxia, opisthotonus, and torticollis and occasionally respiratory signs. Gross lesions include hepatomegaly with green discoloration, splenomegaly and renomegaly in acute cases, and pale yellow, dry exudates in the peritoneum, gonads, oviduct, heart, lungs, joints, intestine, brain, and the eye in chronic cases. Microscopically, lesions can vary from hepatocellular necrosis with fibrin exudation and heterophilic inflammation in acute cases to caseous granulomas with lymphocytic and histiocytic inflammation with multinucleated giant cells in chronic cases.

S. enterica subsp. ***arizonae*** causes **Arizonosis** in primarily young turkey poults. Infection is egg transmitted. Disease can be similar to other *Salmonella* infections, mostly paratyphoid infections.[20] Clinical signs are nonspecific and include depression, weakness, anorexia, diarrhea, paralysis, torticollis, opisthotonus, and blindness. Gross necropsy findings are nonspecific and include yolk sac retention, peritonitis, cecal cores, enlarged and mottled liver, splenomegaly, cloudy meninges, and cornea.

Diagnosis of salmonellosis is achieved by isolation and identification of the bacteria from internal organs. Isolation of *Salmonella* spp. from the intestines or feces of adult gallinaceous species should be interpreted carefully and in the context of history, clinical signs, and lesions. Gallinaceous birds are the natural host of these bacteria and may carry *Salmonella* without clinical disease.

As are other members of the Enterobacteriaceae family, ***Yersinia*** spp. are ubiquitous in the environment and water and can multiply at low temperatures. ***Yersinia pseudotuberculosis*** is the most important avian pathogen in this genus; it causes septicemic disease. Clinical signs depend on disease chronicity with the chronic form being more common.[19] Clinical signs include lethargy, anorexia, loss of weight, diarrhea, dyspnea, dehydration, and sudden death. In chronic infections, signs include weight loss, swollen joints, and paresis. Gross changes in acute cases are associated with swelling of the liver and spleen and bloody to fibrinous exudates in the coelom. Gross lesions in chronic cases consist of granulomas in the skeletal muscles and organs. The primary differential is avian mycobacteriosis. Microscopically, lesions range from acute hepatic coagulative necrosis, fibrin exudation, and heterophilic and histiocytic inflammation with multinucleated giant cells. Large numbers of intrahistiocytic and intralesional gram-negative coccobacilli are highlighted with special stains. Similar lesions can be present in other organs, including the spleen, kidneys, and lungs (Figs. 31.20A, B; eSlides 31.e5 and 31.e6).

Pasteurella multocida, the causative agent of **fowl cholera**, occurs frequently as acute septicemic disease (Fig. 31.21A) with high mortality and morbidity. All birds may be susceptible under appropriate conditions. Chronic,

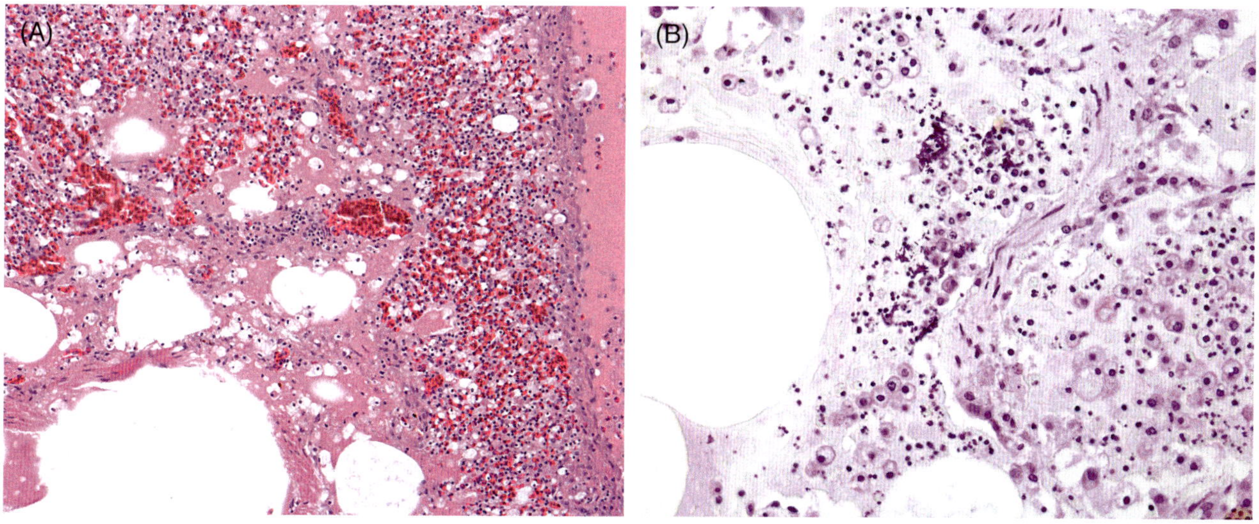

FIGURE 31.20 *Yersinia pseudotuberculosis* pneumonia in a chicken. (A) Infection is associated with acute coagulative interstitial pneumonia, with fibrin exudation and heterophilic inflammation. (B) Numerous intralesional gram-negative bacilli with fibrin accumulation are present in the parabronchial lumen (see eSlides 31.e5 and 31.e6).

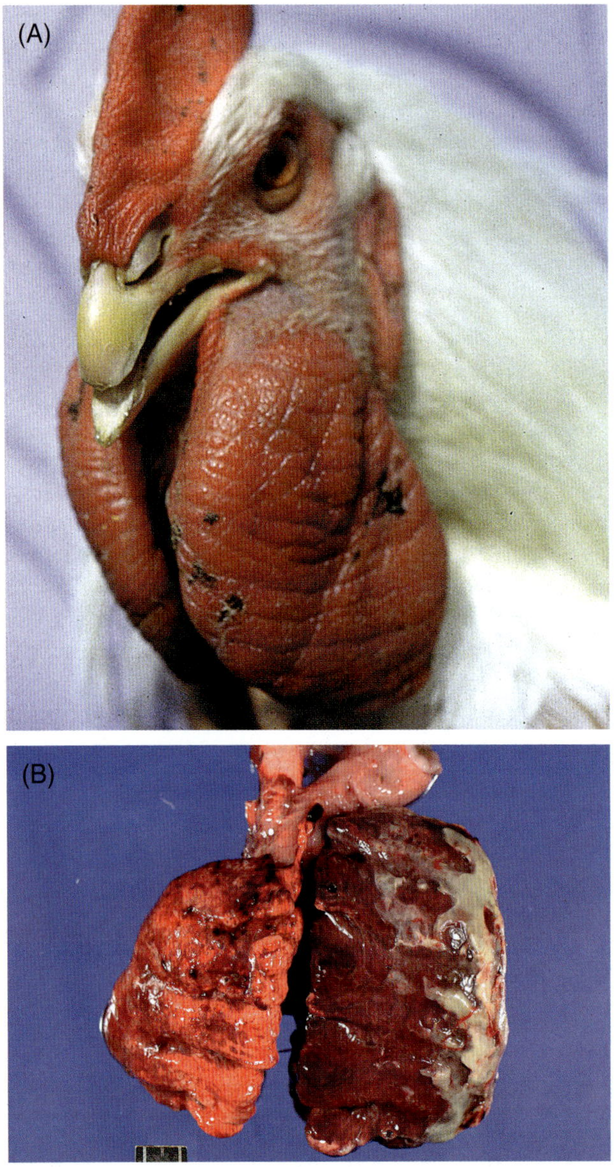

FIGURE 31.21 *Pasteurella multocida* infection in the wattles of a chicken (A) and lungs of a turkey (B). (A) The wattles are markedly swollen due to accumulation of caseous exudate and inflammatory cell infiltration. (B) Infection can also be associated with acute pneumonia. In this case, it was unilateral and the affected lobe (*right*) was enlarged, discolored dark red, and congested.

localized disease may occur following acute disease. Localized infections are found mainly in joints, head (including subcutaneous tissue, sinus, and middle ear) (Fig. 31.21B), oviduct, and the respiratory tract. Transmission and infection between flock mates includes cannibalism of sick and dead birds. Furthermore, many recovered flocks may include subclinical carriers that spread *P. multocida* to susceptible birds. Carriers may harbor the organism in the choanal cleft leading to contamination of feed, water, and the environment. Cats, dogs, rodents, and pigs are reservoirs of *P. multocida* strains that are virulent for gallinaceous birds.

Infectious coryza is an acute respiratory disease caused by ***Avibacterium paragallinarum*** (gram-negative bacillus). Domestic chickens of all ages are susceptible and susceptibility increases with age. Pheasant, guinea fowl, and turkeys have also been reported to develop disease.[109,170] Transmission is by direct contact, aerosol, and contamination of drinking water. Recovered birds may become carriers. Disease is characterized by upper respiratory clinical signs, including nasal discharge, sneezing, and swelling of the face (Fig. 31.22A). There is also a drop in egg production. Lesions include sinusitis, rhinitis, conjunctivitis, tracheitis, bronchitis, and even airsacculitis and peritonitis (Fig. 31.22B). Histologically, there is epithelial hyperplasia, edema, and infiltration of heterophils, macrophages, and mast cells. An experimental study demonstrated that quail are less susceptible to developing clinical signs caused

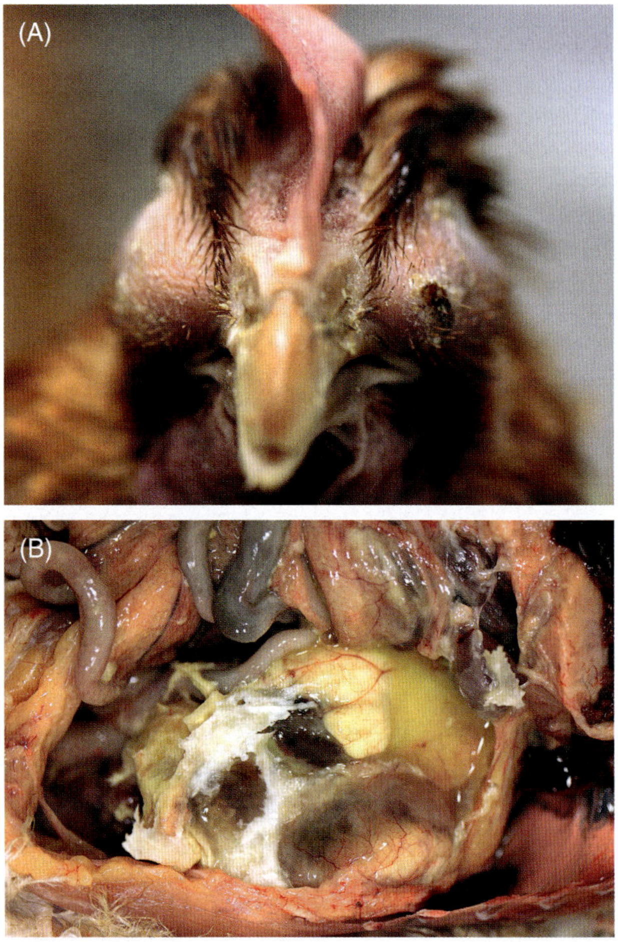

FIGURE 31.22 Infectious coryza due to *Avibacterium paragallinarum* infection in chickens. (A) Severe swelling of the infraorbital sinus is a characteristic lesion. (B) Infection can also cause pneumonia and fibrinous accumulation of caseous exudate in the peritoneum and air sacs.

by *A. paragallinarum* infection, possibly due to their faster and stronger immunity response to this pathogen.[40] Diagnosis is achieved by isolation of this gram-negative organism or by PCR. Specialized media are required with culturing *A. paragallinarum* because of its nutritional requirement for NAD (V-factor).

Ornithobacterium rhinotracheale is a gram-negative bacterium that is associated with acute, highly contagious respiratory disease. The bacterium has been isolated from domestic and wild gallinaceous birds, including chicken, chukar partridge, guinea fowl, partridge, pheasant, quail, and turkey. In young birds, the bacterium causes relatively mild respiratory signs. In older birds, especially turkeys, *O. rhinotracheale* can cause acute pneumonia with high mortality rates.[38,41] In pheasants the most common lesions are redness of the mucosa of the upper respiratory tract and excess mucus in the nasal cavities, infraorbital sinuses, larynx, and trachea.[58] The severity of disease may increase with the presence of other infectious agents, especially *Mycoplasma gallisepticum*.[13] It may also cause paralysis through arthritis, osteitis, and osteomyelitis, as well as a drop in egg production and quality.[38,163]

Bordetellosis is a highly contagious upper respiratory tract disease, primarily of young turkeys, quail, and partridges, caused by ***Bordetella avium***, a gram-negative coccobacillus.[156] Disease is characterized by conjunctivitis, sneezing, and coughing, with moist tracheal rales in young birds and dry coughs in older ones. Birds may be dyspneic with open-mouth breathing due to accumulation of mucus in the trachea, and softening and deformation of the tracheal rings. Because *B. avium* damages the epithelium of the upper respiratory tract, affected birds are susceptible to secondary infections. The severity of disease varies with the strain virulence and the presence of other infectious agents, such as Newcastle disease virus, *Mycoplasma*, *Pasteurella*, and *E. coli*.

The genus ***Clostridium*** are gram-positive, spore-forming, catalase-negative, anaerobic bacilli. Many have been recovered in environmental and clinical specimens from wild and domestic avian species. Pathogenicity is mediated through potent toxins and disease may occur as soft tissue infections, intoxications, and toxico-infections. There are four clostridial diseases of importance in gallinaceous birds. They are described below.

Ulcerative enteritis is caused by ***Clostridium colinum*** and occurs worldwide in many bird species, including quail, grouse, turkeys, chickens, and partridges.[156] It is an acute bacterial infection, primarily of young birds. Lesions are characterized by multiple ulcers with or without hemorrhage throughout the intestinal tract (Fig. 31.23) and foci of hepatic necrosis and inflammation. Ulcers can perforate the wall of the intestine causing peritonitis. Birds that recover naturally from the disease develop immunity but remain carriers and shed the bacterium in their feces.

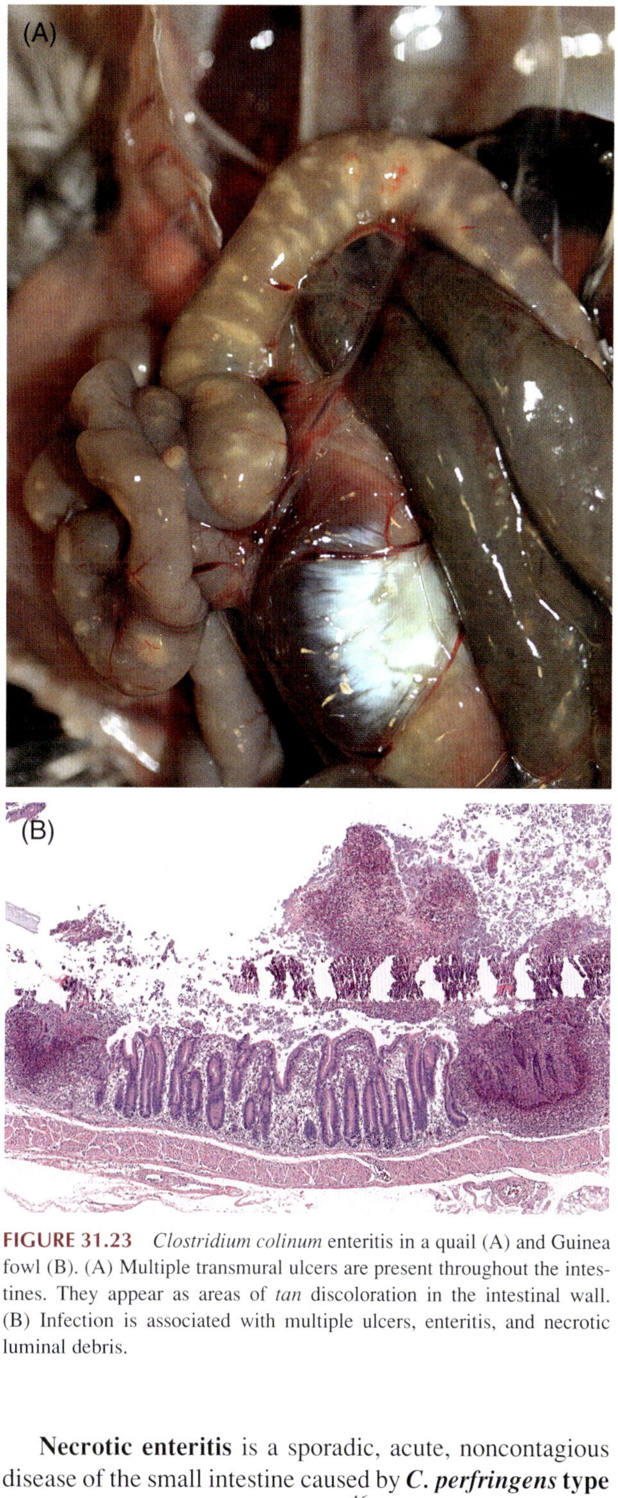

FIGURE 31.23 *Clostridium colinum* enteritis in a quail (A) and Guinea fowl (B). (A) Multiple transmural ulcers are present throughout the intestines. They appear as areas of *tan* discoloration in the intestinal wall. (B) Infection is associated with multiple ulcers, enteritis, and necrotic luminal debris.

Necrotic enteritis is a sporadic, acute, noncontagious disease of the small intestine caused by ***C. perfringens* type A**, and to a lesser extent, **type C**.[46] The novel NetB toxin, not alpha-toxin, as previously believed, is critical in the pathogenesis of necrotic enteritis.[87] Lesions are characterized by severe fibrinonecrotic enteritis with the formation of diphtheritic pseudomembranes. Mortality rates are high. Among game fowl, quail are the most susceptible.[54]

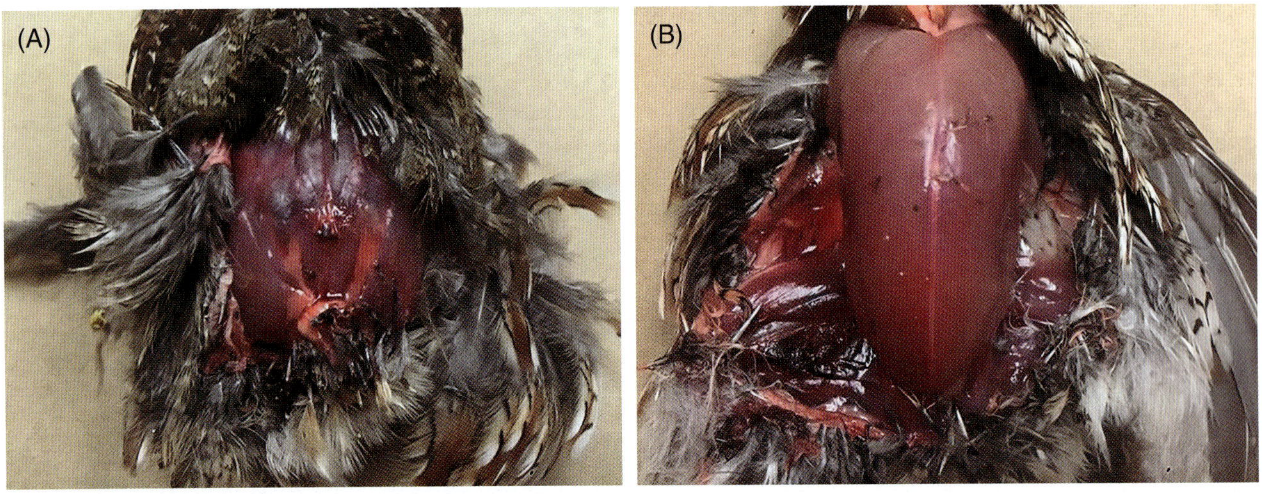

FIGURE 31.24 Gangrenous clostridial dermatitis in quail. Toxins produced by the bacterium result in the production of (A) serosanguineous fluid that in this case is present in the subcutaneous space above the tail head and (B) along the inside of the thigh muscles and the pectoral muscles. *(Photo Courtesy of M. Ficken, Texas A&M Veterinary Medical Diagnostic Laboratory.)*

Gangrenous dermatitis is a peracute, fatal disease, caused by ***C. perfringens*** or ***C. septicum***. It occurs more frequently in warm and humid environments. It is believed that the bacteria are introduced through cutaneous wounds. Immunosuppression enhances susceptibility to disease. Affected birds are depressed or found dead with darkened and gangrenous skin, with emphysematous or serosanguinous cellulitis (Figs. 31.24A, B).

Botulism, caused by an exotoxin of ***C. botulinum*** that produces progressive flaccid paralysis, occurs more often in captive pheasants and occasionally in chickens.[19] It occurs more frequently in warmer months. Of the eight known types of botulism toxins, type C is the most common in gallinaceous birds.[156] Birds with type C botulism lack characteristic gross or microscopic lesions. Definitive diagnosis requires detection of toxin in serum, crop, or gastrointestinal contents of moribund birds by mouse bioassay (MBA), ELISA, or PCR. MBA is the most sensitive assay and is capable of detecting a wide variety of botulinum toxins. *C. botulinum* can be found in the gut of normal chickens and toxin can be produced after a bird dies. Therefore, finding toxin in tissues of dead birds does not confirm botulism.

Mycoplasma species are small (approximately 200 nm), prokaryotic organisms that lack a cell wall and are phylogenetically related to gram-positive bacteria.[131,141] Mycoplasmas tend to be host-specific. Of the many *Mycoplasma* species that infect birds, only the species capable of causing disease in commercial poultry have received attention by researchers. *Mycoplasma* spp. can be transmitted vertically or horizontally by contact or by airborne dust or droplets.

Mycoplasma gallisepticum has been reported in chickens, turkeys, pheasants, partridges, peafowl, guinea fowl, and quail.[109] In turkeys, infection is more prevalent in domestic flocks, but the bacteria are found in wild populations.[142] Infection is often associated with respiratory signs and swelling of the infraorbital sinus. The condition is frequently found as a coinfection with other respiratory pathogens such as *E. coli*, *Avibacterium* sp., *Aspergillus* sp., or in Newcastle disease. Signs of ***M. synoviae*** infection include a pale comb, stunting, lameness, and swollen joints. Swellings of the tissues adjacent to the ventral edge of the keel bone or sternal crest (referred as breast blisters) are frequently observed. The infection in the respiratory tract is mostly subclinical. ***M. meleagridis*** and ***M. iowae*** infections affect mainly turkeys, and are responsible for lower egg production and hatchability with minimal or no respiratory signs.

Three species of *Mycoplasma*, ***M. columbinum***, ***M. columbarale***, and ***M. columbinasale***, have been isolated from clinically healthy pigeons and pigeons with respiratory disease.[88] Clinical signs can include ocular and respiratory disease; however, most sick birds have concomitant infections and it is difficult to determine whether or not the signs are due solely to mycoplasmas. *M. columbinum* has also been associated with arthritis of the shoulder joint in a juvenile racing pigeon that had a droopy wing and was unable to fly.[70]

Histologically, lymphoplasmacytic inflammation of the conjunctiva, sinuses, trachea, and occasionally air sacs and bronchi are the principal lesions of mycoplasmosis. Diagnosis is through isolation on a suitable culture media or by PCR.

Erysipelas is an acute septicemic disease caused by ***Erysipelothrix rhusiopathiae***. In Galliformes, it occurs primarily in turkeys, occasionally in pheasants, and rarely in chickens. The onset of disease can be sudden with birds

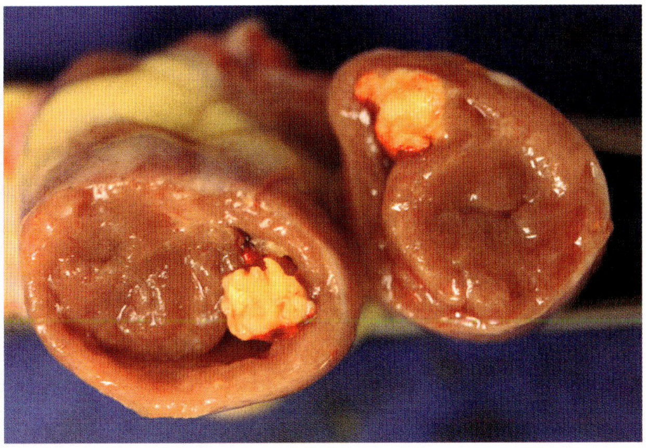

FIGURE 31.25 *Staphylococcus aureus* valvular endocarditis in a chicken. A large yellow granuloma is attached to the right atrioventricular valve. The granuloma almost occupies the entire space of the right ventricle.

found dead or dying after a short period of illness. Serosal, cutaneous, and muscular hemorrhage and splenomegaly are characteristic. A chronic form, characterized by polyarthritis or endocarditis, occasionally occurs. Infection is zoonotic.

Staphylococcus aureus, the cause of systemic disease **staphylococcosis,** is widespread in the environment and causes a variety of conditions, including arthritis, tenosynovitis, osteomyelitis, endocarditis (Fig. 31.25), and omphalitis. Skin injuries are a key point of entry, and *S. aureus* is a common opportunistic infection in ulcerative pododermatitis in many bird species, including galliformes and columbiformes. The respiratory tract is another important portal of entry.

Mycobacterium avium is the causative agent of avian mycobacteriosis. Susceptibility studies suggest that among gallinaceous birds, chickens and partridges are highly susceptible; pheasants, guinea fowl, and domestic turkeys are less susceptible. Epidemiological surveys classify pigeons as moderately susceptible to highly resistant to *M. avium* infection.[39,42,59] Avian mycobacteriosis caused by ***M. genavense*** has been reported in captive doves.[66,138] Contaminated fecal material is the most important means of dissemination of *M. avium*. Respiratory secretions are also a potential source of environmental contamination and infection. The bacteria can survive for more than 4 years in the environment, when not under direct sunlight.[42] Infected birds lose weight, their feathers look dull, and birds have a marked leukocytosis due to monocytosis. Abnormal gait or paralysis can occur with joint infection. The most frequent gross lesions are individual or coalescing granulomas that vary from pinpoint to several centimeters and can be found in many tissues including the liver, spleen, intestine, lung, and bone marrow. Histologically, lesions vary from mild to severe and consist of multifocal granulomatous inflammation or organized granulomas that, in some cases, replace significant portions of parenchymal organs; multifocal caseous necrosis is also seen. Granulomas are characterized by large accumulations of macrophages with abundant cytoplasm surrounded by multinucleated giant cells and fibrosis. Acid-fast stains (e.g., Ziehl-Neelsen) highlight large numbers of bacteria within the cytoplasm of macrophages and multinucleated giant cells. These findings are usually sufficient for presumptive diagnosis of mycobacteriosis. Other bacterial conditions, such as coligranuloma, *Salmonella* infections, and fowl cholera, must be ruled out as differentials. *Mycobacterium* can be cultured in suitable media and the species can be identified by PCR and sequencing.

Species of ***Pseudomonas*** are ubiquitous bacteria that are found in soil, water, and humid environments. They can cause respiratory infections, keratitis, omphalitis, or septicemia in young birds.

Riemerella anatipestifer occasionally affects turkeys. See Chapter 29: Anseriformes, Ciconiiformes, Charadriiformes, and Gruiformes for a more detailed description.

Members of the genus ***Chlamydia*** infect and cause disease in a wide range of species including humans, for whom wild and companion birds can act as reservoirs and sources of infection.[84] **Chlamydiosis**, also called **ornithosis or psittacosis**, is most commonly caused by ***Chlamydia psittaci*** genotype B. Other species of chlamydia, such as ***C. pecorum***, ***C. trachomatis***, and a novel species ***C. avium,*** have also been detected in a variety of bird species, but their role in causing the disease is not known.[181] Chlamydiosis is a common disease of pigeons of all ages but young birds are more susceptible. One study showed that 19%–96% of tested pigeons were positive by the complement fixation test, 0%–57% by isolation in chicken egg embryos and cell culture systems, and 3%–50% by PCR. Young birds are most susceptible, and adults can be carriers. Chlamydiosis, on the other hand, is rarely detected in galliforms.[83] Turkeys are the most common species affected.[7] One report from China demonstrated that *C. gallinacea*, not *C. psittaci* was common in clinically healthy chickens.[62] The disease characteristics, transmission, and diagnosis are essentially the same as in Psittaciformes (see Chapter 32: Psittacines, Coliiformes, Musophagiformes, Cuculiformes).

Both ***Streptococcus*** spp. and ***Enterococcus*** spp. are ubiquitous bacteria found in the intestines of many animals, including galliformes. The bacteria can enter the blood stream through intestinal lesions, resulting in sepsis. *Streptococcus* sp. are mainly associated with septicemia, endocarditis, and necrotic hepatitis and splenitis. ***Streptococcus gallolyticus* (previously *S. bovis*)** is an important pathogen in some parts of Europe that causes septicemia and increased mortality as high as 78% in some commercial poultry flocks. The disease is seen more commonly in young pigeons and clinical signs include acute death, anorexia, inability to fly, lameness, swollen joints, loss of weight, and greenish diarrhea. Grossly, the liver and spleen are enlarged and mottled, the kidneys are enlarged, and there are occasional pale foci in the heart and consolidation of lungs. Microscopically, fibrinoheterophilic inflammation in the liver, spleen, kidneys, heart, lungs, and synovium, and occasional encephalitis or meningoencephalitis are associated with intralesional coccoid bacteria.

Enterococcus cecorum is associated with infections of the skeleton, including spondylitis, femoral head necrosis, and osteomyelitis. ***E. faecalis*** has been associated with endocarditis, granulomatous hepatitis, and amyloid arthritis. Clinical signs and lesions are not specific to these bacteria and other bacterial infections need to be ruled out.

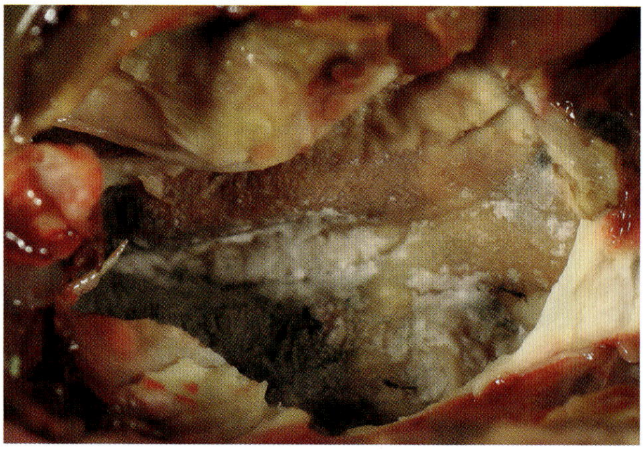

FIGURE 31.26 *Aspergillus fumigatus* airsacculitis in a pigeon. The right air sac is opaque and thickened. A fungal mat completely overlies the air sac adjacent to the lung.

Fungi

Aspergillosis is a common infection caused by ***Aspergillus fumigatus*, *A. flavus*,** or ***A. niger*,** saprophytic fungi associated with decaying organic matter[28] (also see Chapter 27: Sphenisciformes, Gaviiformes, Podicipediformes, Procellariiformes, and Pelecaniformes; Chapter 30: Birds of Prey). Birds typically become infected through the inhalation of fungal spores from the environment. The spores germinate in the respiratory tract and incite a strong inflammatory response.[74] In experimentally infected chickens, proinflammatory cytokines such as IL-1, TNF-α, Cxcl-8, and IFN-γ are up-regulated through a Toll-like receptors (TLR-1 and TLR-2) mediated innate immune response that recruit heterophils and macrophages. The fungi can continue to spread within the respiratory system via production of new spores under aerobic conditions, by direct seeding into the coelomic cavity and extension to pneumatized bones can also occur.[30,135] Systemic spread is via macrophages or, because *Aspergillus* is vasotropic, via hematogenous routes. Disease, occasionally termed "brooder pneumonia," tends to be more severe in younger birds (7–10 days after hatching in domestic chickens) and associated with concurrent immunosuppression, infection, or stress.[121] In captive settings, disease is commonly associated with poor husbandry, moldy feed or litter, damp bedding, high stocking densities, poor ventilation, or breed susceptibility.[11] Aspergillosis causing morbidity and mortality has also been reported in a free-ranging wild turkey[35] and Gambel's quail.[61]

Gross lesions include mild to severely thickened air sacs and pleura coated with white exudate (mycelia) or green powder-like (conidia) material (Fig. 31.26). The lungs contain small, pale yellow nodules to larger areas of consolidation. Birds in chronic stages of the disease tend to be in extremely poor body condition and may have underlying or concurrent disease. Plugs of caseonecrotic material may be observed within the lumen of the trachea or syrinx. Ascites is frequently reported in chickens and is presumably associated with vascular failure.[81] Deformities of the vertebrae due to osteomyelitis have been reported in domestic chickens with chronic infections.[16] Ocular aspergillosis can be subdivided into two conditions based upon the underlying pathogenesis. Topical exposure results in keratoconjunctivitis with caseous plaques beneath the nictitating membranes (more common in chickens), while hematogenous spread results in posterior ophthalmitis or panophthalmitis (more common in turkey poults).[156]

Microscopic lesions consist of giant cell granulomas with intralesional septate, branching hyphae that have 3–6 μm in diameter parallel walls with dichotomous branching. Fungi stain with both Gomori's methenamine silver (GMS) and periodic acid-Schiff (PAS) techniques.[27] In some instances, large conidiophores (radiating fruiting bodies) are observed at an air interfaces[30] and calcium oxalate crystals can be observed in some infections with *A. niger*. In acute–subacute cases, inflammatory cells are less organized but organization into granulomas occurs with chronicity as does surrounding fibrosis. Necrotic lesions associated with hyphae can be found in the brain in some cases, often with limited inflammation. In posterior ocular infections, granulomas are present within the pectin and hyphae can be found in the vitreous humor and the retina.

While the gross and microscopic lesions are almost pathognomonic, infection can be confirmed with culture, molecular techniques,[85,174] and immunohistochemistry.[24] A combination of PCR with galactomannan assays can increase the reliability of molecular results.[112] In live birds, tracheal washes can be cultured, the granulomas can be visualized and sampled via cytologic preparations or endoscopic methods,[133] and an indirect enzyme-linked immunosorbent assay (ELISA) can be used on serum for detection of antibodies and to monitor treatment.[33] Radiographs can also suggest aspergillosis. As *Aspergillus* spp. can be ubiquitous in the environment, caution should be used to maintain sterility in sample collection. For cytology, nodules can be digested in 20% potassium hydroxide (KOH) and characteristic conidiophores and hyphae are visualized with the assistance of ink dye; fungi can also be seen with Romanowsky or Wright Giemsa stains.

Infection with **zygomycete** fungi, such as ***Mucor, Rhizopus, Asidia, Rhizomucor***, and ***Mortierella,*** can present clinically and cause similar lesions to aspergillosis. Infections are often secondary to concurrent illness or stress. Hyphae of zygomycete fungi are larger (7–20 µm), aseptate, lack parallel walls, have an irregular appearance with bulging dilations, and branch at right angles.[27] Like *Aspergillus*, zygomycete hyphae will stain positively with GMS and PAS. Diagnosis can be confirmed with fungal culture or molecular methods.

Candidiasis, **crop mycosis**, or **thrush** is caused by the ubiquitous and opportunistic yeast ***Candida albicans***. Infection is often associated with concurrent illness or traumatic injury. Found worldwide, this yeast can be isolated from the gastrointestinal tracts of clinically healthy birds. Other species of *Candida* have also been isolated but are rare.[41] Candidiasis is one of the most common mycotic diseases in pigeons, especially neonates that are not immunocompetent. Pigeons with concurrent circovirus infections are especially susceptible. Disease also occurs in gallinaceous birds including chickens, turkeys, guinea fowl, pheasants, and quail.[10,107,156] Large mortality events have been described in young domestic turkeys and chickens. Younger birds are more susceptible to disease than older birds. In pigeons, transmission of *Candida* occurs through ingestion of crop milk, contaminated feed and water, and from the environment. Clinical signs include anorexia, diarrhea, depression, regurgitation, crop distension, and increased morbidity and mortality.

Typical lesions of candidiasis include ulcerative and proliferative white plaques. Pseudomembranes that can be easily peeled off may be present. Lesions are common in the oral cavity, crop mucosa, and the esophagus. Less commonly, candidiasis may be observed in the proventriculus and ventriculus and can be seen in association with hemorrhage and ulceration. Rare systemic, dermatologic, and comb infections are reported. Microscopically, lesions include necrosis and epithelial ulceration, intralesional yeast, hyphae and pseudohyphae, fibrin, sloughed epithelial cells, and necrotic debris that are often limited to the stratum corneum (Fig. 31.27A) or mucosa, though deeper infections can also develop. Intraepithelial edema, hyperkeratosis, and secondary bacterial colonization may also be observed. Oval *Candida* spp. yeast are 3–6 µm. Hyphae are parallel-walled, septate, and 3–5 µm in diameter while pseudohyphae include chains of elongated yeast that resemble hyphae with intermittent constrictions. *Candida* spp. stain with GMS (Fig. 31.27B) and PAS and are gram-positive. Focal periportal necrosis and vascular changes are suspected to be associated with an endotoxin. Candidiasis is diagnosed based on the microscopic appearance of the yeast, hyphae, and pseudohyphae. Infection can be confirmed with culture and molecular methods.

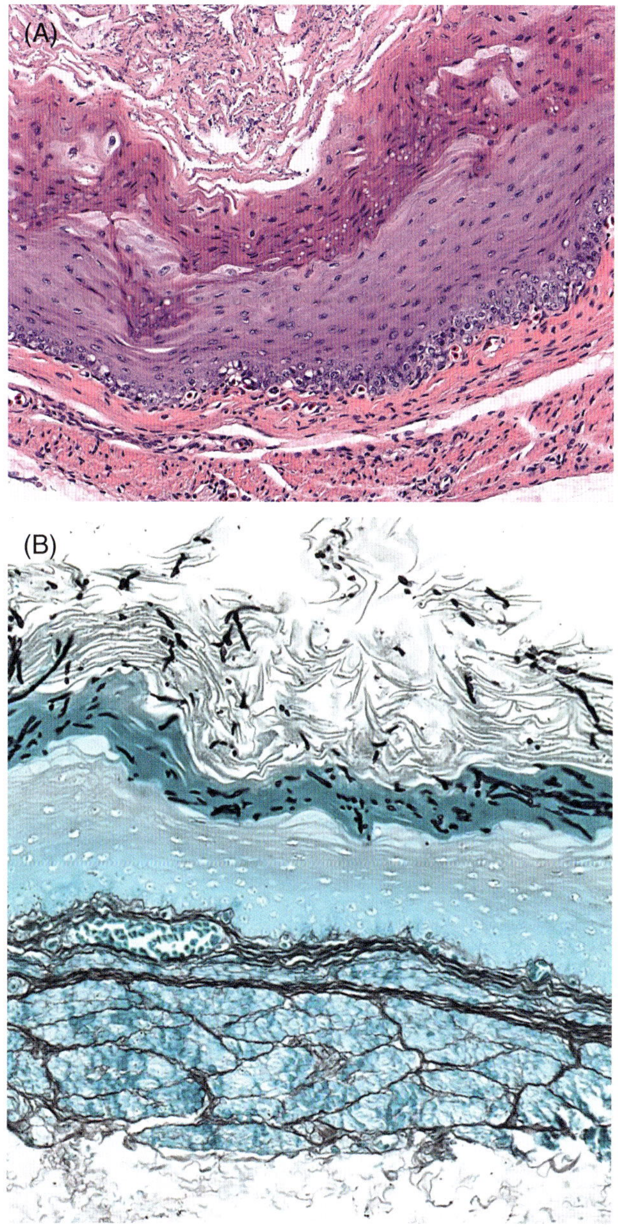

FIGURE 31.27 *Candida albicans* ingluvitis in quail. (A) Infection is associated with hypertrophy of the mucosal epithelium and koilin of the crop. Numerous yeast and pseudo-hyphae are present in the epithelium. (B) Numerous yeast and pseudohyphae that are present in the mucosa are highlighted with special staining, GMS.

Ochroconosis (syn. **dactylariosis**), due to ***Ochroconus gallopava***, causes encephalitis in young domestic turkeys, chickens, and quail. Mortality events involving a large number of birds are often associated with contaminated litter.[146,156] Gross lesions described include focally extensive, firm, gray or red foci in the cerebellum or cerebrum, as well as rare pulmonary granulomas. Histologic lesions include foci of necrosis associated with numerous heterophils, macrophages, and multinucleated giant cells, including vasocentric lesions. The hyphae of *Ochroconis* spp. are pigmented yellow to light brown, septate, irregularly branched, and 1.2–2.4 μm wide.[156] Hyphae in tissue section stain positively with GMS. Diagnosis is based on fungal isolation or PCR.

Macrorhabdosis (previously megabacteriosis) is caused by infection with the opportunistic yeast ***Macrorhabdus ornithogaster*** (see Chapter 33: Passeriformes, Caprimulgiformes, Coraciiformes, Piciformes, Bucerotiformes, and Apodiformes) that localizes to the proventricular-ventricular junction. Infection has been documented in domestic turkeys, chickens, guinea fowl, quail, and partridges. Gross lesions may not be present or can be associated, including those seen with a chronic, debilitating disease of the gastrointestinal tract such as emaciation and thickening of the proventriculus. Histologic lesions vary from none to lymphoplasmacytic and histiocytic inflammation in the lamina propria of the proventriculus and ventriculus. Organisms are often arranged in parallel bundles within superficial mucus, proventricular crypts, koilin, and rarely, within the mucosa.[156] Diagnosis is based on detecting the organism in fecal smears, cytology of gastric mucus, or histology of the proventricular-ventricular junction. Organisms are Gram variable but will stain positively with PAS.

Feather abnormalities in wild turkeys are associated with colonization by numerous fungal species including ***Aspergillus, Curvularia, Cladosporium, Dactyella, Exophiala, Helminthosporium, Trichophyton,*** and ***Streptomyces***. The pathogenesis is unknown. Gross lesions include cracks in the ventral median groove of the rachis with visible mold and broken tail feathers. Histologic lesions include fungal hyphae on the surface of the rachis and within the feather pulp and associated separation of the middle Malpighian layer. Diagnosis is based on the microscopic appearance of the fungal hyphae and fungal culture.[36]

Metazoa

Common internal parasites, associated lesions and life cycles are summarized in Tables 31.2 and 31.3. Most nematodes that affect galliformes and columbiformes infect the digestive tract and shed eggs in droppings.

TABLE 31.2 Common Intestinal Parasites

Parasite	Location in the Host	Clinical Signs/Lesions	Life Cycle
Nematodes			
Capillaria	Crop, small intestine, and ceca	Thickened mucosa, emaciation, hemorrhage/anemia	Direct. May be ingested by earthworms
Dispharynx, Tetrameres, Cyrnea	Proventriculus	Diarrhea, emaciation. Mucosa ulceration, necrosis	Indirect: grasshoppers, cockroaches, sowbugs, and pillbugs are intermediate host
Cheilospirura	Gizzard	Damaged gizzard wall	Indirect: grasshoppers, beetle
Ascarids sp. (Figs. 31.28A, B)	Lumen of small intestine	Weight loss, potential intestinal blockage, and blood loss	Direct: *Eggs may be ingested by insects*
Heterakis	Ceca	Cecal mucosa inflammation and thickening	Direct: *Eggs may be ingested by earthworms*
Cestodes	Small intestine	Mild or unknown, except *Raillientina* sp. that causes severe lesions and weight loss	Indirect: insect, earthworm, snail
Trematode	Small intestine	Mild or unknown	Indirect
Protozoa			
Eimeria (Coccidia)	Intestine, region depends on species	Varies with species, from diarrhea to bloody feces and death; from whitish-white foci on serosa to hemorrhage and necrosis	Direct
Trichomonas	Esophagus, crop	White-yellow plaques (ulcers)	Direct
Histomonas	ceca, liver	Necrosis	Indirect: *Heterakis gallinarum* is the initial intermediate host Direct (?)
Cryptosporidium	Small intestine, bursa of Fabricius	Diarrhea, villus atrophy	Direct

TABLE 31.3 Common Nonintestinal Parasites

Parasite	Location in the Host	Clinical Signs/Lesions	Life Cycle
Nematodes			
Syngamus (Fig. 31.29)	Respiratory system: trachea	Gasping, dyspnea, nodules in trachea	Direct or indirect: earthworm, slugs, snails
Oxyspirura	Special senses: eye	Conjunctivitis, found under nictating membrane	Direct or indirect: cockroaches
Cestodes			
	Reproductive tract: oviduct	Weight loss and lower egg production	Indirect: insect, earthworm, snail
Protozoa			
Leucocytozoon	Circulatory system	Sudden death, hepatomegaly, splenomegaly. Schizogony in tissues. Gametogony in erythrocytes and leukocytes	Indirect: blackflies and culicoides midges
Plasmodium	Circulatory system	Schizonts in lung and blood endothelium; merozoites: erythrocytes (displace nucleus to the side)	Indirect: mosquitoes and biting midges
Hemoproteus	Circulatory system	Schizonts in lung and blood endothelium; merozoites: erythrocytes (around nucleus)	Indirect: biting flies and midges
Cryptosporidium	Respiratory system	Bronchopneumonia	Direct

Capillaria nematodes infect the crop and esophagus but can also infect the intestine. They cause extensive damage to their hosts because adults feed directly on blood and tissue. Eggs are bioperculate and easily recognized with cytology or histology. Some *Capillaria* species have direct life cycles, while others require intermediate hosts, such as earthworms, beetles, and cockroaches.[95,111]

Worms of the proventriculus and gizzard belong to the genera ***Dispharynx***, ***Tetrameres***, and ***Cheilospirura***; all have indirect life cycles.[111,156] The intermediate host of *D. nasuta* is the sowbug. *Tetrameres* utilize amphipods, grasshoppers, cockroaches, and earthworms as intermediate hosts. Beetles and grasshoppers serve as intermediate hosts to *Cheilyspirura*. These worms infect a variety of galliformes. All three genera embed their head/mouthparts in the mucosa causing tissue proliferation and deterioration. Affected birds may appear emaciated.

Ascarid parasites, or large roundworms, are found commonly in the intestines of many galliformes. Heavy infestations cause growth retardation, constipation or diarrhea, loss of condition, and sometimes intestinal blockage (Figs. 31.28A, B). The ascarid life cycle is simple and direct with no intermediate host.[95] It is also possible that the eggs are ingested by grasshoppers or earthworms and in turn infect an avian host when the invertebrate is eaten. After ingestion, eggs hatch in the proventriculus or upper intestine and larvae cause damage during migration.

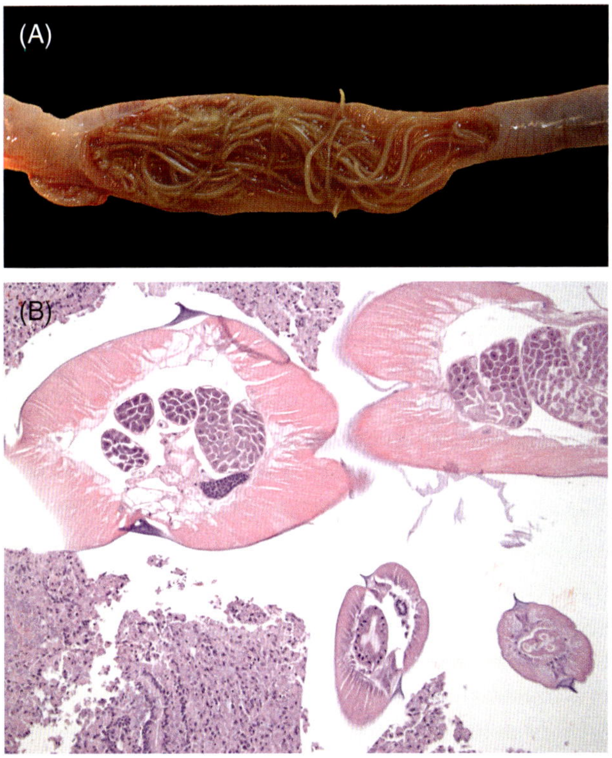

FIGURE 31.28 Intestinal ascarid infections in a chicken. (A) The small intestine is obstructed due to impaction of the lumen with *Ascaridia galli*. (B) Ascarid nematodes have prominent lateral alae, thick coelomyarian musculature, a pseudocoelom, a uterus filled with eggs, and a central gastrointestinal tract lined by uninucleate cells.

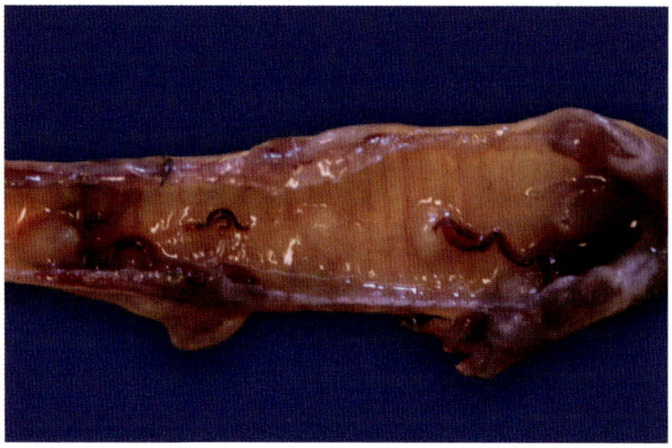

FIGURE 31.29 *Syngamus trachea* tracheitis in a pheasant. Also known as "gapeworm," these are bright red nematodes that attach themselves to the tracheal wall and form granulomas. Several granulomas are protrude from the mucosal surface in this case.

Members of the genus ***Heterakis*** are parasites mainly of the large intestine (cecum). The parasites live on intestinal contents and do not migrate during their development, so damage to the host is minimal. But one species, ***Heterakis gallinae***, plays an important role in the transmission of the protozoan parasite ***Histomonas meleagridis***, which causes **histomoniasis** (**blackhead disease**; see below) in Galliformes.

Syngamus trachea or **gapeworm** has been reported to affect a broad range of captive and wild galliformes and columbiformes.[111,156] The parasites are usually encountered in the trachea, and occasionally in the bronchi. The worms attach to the tissue and feed on blood (Fig. 31.29). Clinical signs include gaping, coughing, and pneumonia-like symptoms. The lumen of the trachea may become obstructed, resulting in dyspnea or suffocation. Eggs are shed with the feces. *Syngamus* has a direct cycle, although earthworms may serve as intermediate host.

Oxyspirura mansoni, the common eye worm, infects numerous Galliformes, including chicken, turkey, grouse, guinea fowl, peafowl, quail, and prairie chicken.[111,156] This parasite is rare in pigeons. Adult worms are found beneath the nictitating membrane of the eye, causing lesions which vary from mild conjunctivitis to severe ophthalmia. The cockroach is the intermediate host.[111,156]

Well over 1,000 species of cestode belonging to 10 different genera (including *Davainea*, *Raillentina*, and *Metroliasthes*) have been identified in domestic and exotic Galliformes.[111,156] Infection occurs through ingestion of an intermediate host. Most of them are not very pathogenic; however, weight loss has been noted in severely affected birds.

Trematodes (flukes) are small (usually less than 1 cm), flat, leaflike parasites of the gastrointestinal tract. They have an indirect life cycle. Numerous genera of flukes have been reported in domestic and exotic galliformes species and occasionally in pigeons. They are often missed if only fecal flotations, and not fecal sedimentation, are performed. Necropsy may detect specimens as incidental findings. Mortality has been reported in association with heavy infestation of ***Morishitium*** in guineafowl.[110]

Protozoa

Protozoans are a diverse group of unicellular organisms. Most species are free-living but some parasitize animals, including birds. Infections range from asymptomatic to life-threatening, depending on the species and strain of the parasite and the resistance of the host.

Eimeria is a genus of intestinal coccidian parasites with a direct life cycle and fecal-oral transmission. Coccidiosis has been reported in numerous galliformes, including chickens, turkeys, quail, grouse, pheasants, chukar, guinea fowl, and peafowl. In general, the different species are host-specific but some species that affect domestic birds may also cause disease in exotic Galliformes.[109] Coccidiosis is also one of the most prevalent parasitic diseases in pigeons.[150] The parasites invade intestinal epithelial cells. This is followed by leakage of blood, severe mucosal edema submucosal necrosis, and loss of villi (Fig. 31.30). These changes trigger diarrhea, and often dehydration. Disease occurs in young and newly acquired animals. Immunity increases with exposure and older individuals often serve as reservoirs.

Trichomonads are flagellated protozoa. Most species are commensals of the avian digestive tract. Yet the species ***Trichomonas gallinae*** affects the upper digestive tract and causes the condition known as **trichomoniasis**, **canker (pigeons)**, or **frounce (see Chapter 30: Birds of Prey)**. Although pigeons are the primary carriers,[139] *T. gallinae* also parasitizes chickens, turkeys, and other domestic and wild galliformes. Outbreaks occur more frequently in warmer months. Outbreaks of trichomoniasis in wild birds typically occur in passerines in the family Fringillidae,[136] but some mortality events have been reported in free-ranging columbiformes, such as the woodpigeon[65] and the endangered pink pigeon.[158] Birds become infected by ingesting water or feed contaminated with oral secretions from infected pigeons (e.g., supplemental feeding sites). Young pigeons can become infected from crop milk that is regurgitated and fed to young birds by their parents. The protozoa invade the mucosal surface and cause stomatitis, esophagitis, and ingluvitis (Fig. 31.31).[156] The affected mucosa becomes necrotic with profuse caseous exudate. Diagnosis of *T. gallinae* infection is more reliably made by demonstrating the organisms in wet mounts or tissue imprints obtained from scraped material from a bird's mouth and crop than by histology. The parasite can also be cultured in Diamond's medium or in commercially available kit pouches.

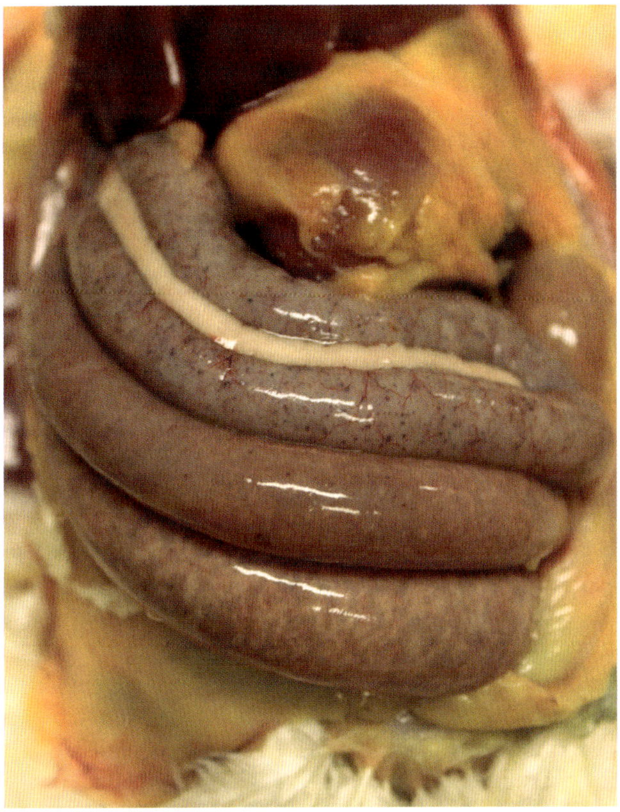

FIGURE 31.30 *Eimeria* enteritis in the small intestine of a chicken. *E. necatrix* causes intestinal inflammation, dilation, hemorrhage, and necrosis. Note that the walls of the small intestine are severely swollen and multifocal white foci (necrosis) and hemorrhage on the intestinal serosal surface.

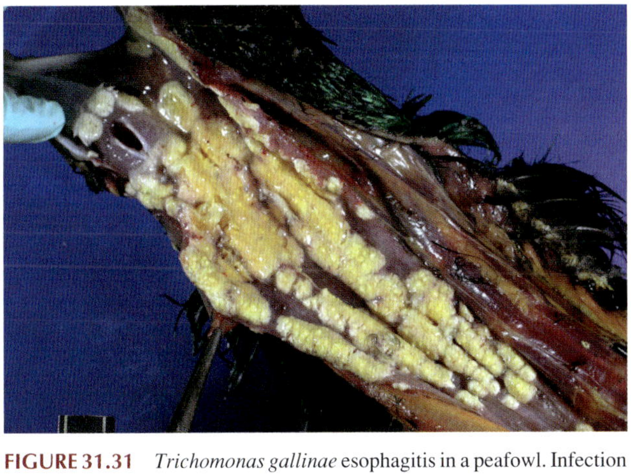

FIGURE 31.31 *Trichomonas gallinae* esophagitis in a peafowl. Infection is associated with tissue necrosis that is overlain by caseous exudate.

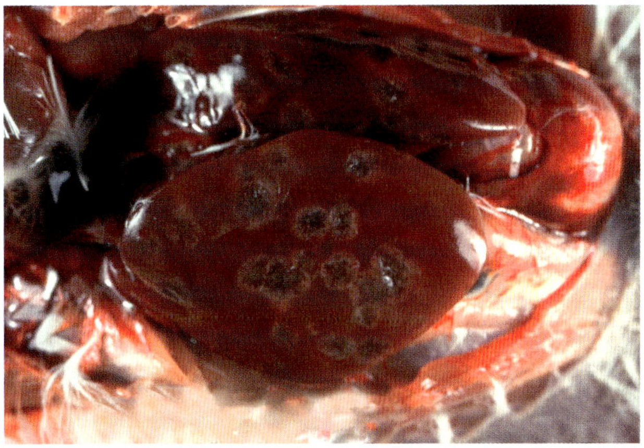

FIGURE 31.32 *Histomonas meleagridis* necrotizing hepatitis in a turkey. Multiple demarcated to coalescing areas of necrosis in the liver have characteristic bull's-eye appearance (see eSlide 31.e7).

Histomonas meleagridis is a well-known and economically important protozoal parasite of gallinaceous birds that causes the disease **histomoniasis (syn. Blackhead disease)**. Clinical signs include yellow-colored droppings, depression, and death in affected birds. Lesions are characterized by multifocal necrosis of the liver that have a bull's-eye appearance (Fig. 31.32; eSlide 31.e7) and necrotizing typhlitis with luminal necrotic cecal cores. Domestic chickens may carry this parasite and not demonstrate any clinical signs; however, turkeys, pheasants, and peafowl are highly susceptible. Guinea fowl, chukar, grouse, quail, and partridges can also be infected.[111,156] Cases of histomoniasis typically occur in birds that have access to the outdoors and can encounter arthropod intermediate hosts as well as ova within the environment. In captive birds, the organism is usually passed from bird to bird within the ova of the nematode ***Heterakis gallinae***, and one study mentioned speculation about an increased likelihood of transmission to turkeys based on a farm's proximity to litter-raised chicken farms. Mechanical transmission may occur via earthworms, flies, grasshoppers, sowbugs, and crickets. In a recent survey, blackhead was not evident in any of the wild turkeys surveyed despite identification of *H. gallinae* in the intestine of several birds.[117] *H. meleagridis* infection is reported periodically in free-ranging wild turkeys,[73] but histomoniasis has a low prevalence in wild gallinaceous birds.[34,37]

Hemosporidia are blood-borne, apicomplexan protozoa that include parasites of three main genera: ***Plasmodium*** (the cause of **avian malaria**), ***Hemoproteus***, *and* ***Leucocytozoon***. These parasites invade red blood cells and endothelial cells of many domestic and wild species, including galliformes and columbiformes. The life cycle requires two hosts: sporogony (asexual phase) occurs in insects, and schizogony (merogony; asexual phase) and gametogony (sexual phase) occur in tissues or blood cells, respectively, of the vertebrate host. Plasmodium insect hosts are sucking mosquitoes (*Culicidae*), while *Hemoproteus* uses biting midges (*Ceratopogonidae*) and louse flies (*Hippoboscidae*) and *Leucocytozoon* uses blood-sucking blackflies (*Simuliidae*) as the intermediate host. Avian hosts and their hemosporidian parasites share an evolutionary history in their native geographic distributions and most hemosporidian parasites are not lethal to birds in the wild. Many of the examples of mortality in zoos caused by hemosporidia involve hosts challenged by parasites not found in their native habitats. On the other hand, hemosporidian parasites may act as population modulators because they reduce fitness.

Plasmodium spp. have been reported in pheasants and guinea fowl;[109] *Hemoproteus* spp. have been reported in guans, curassows, chachalacas,[15,20] and free-living pigeons and doves.[90,95,96]

Differences in the prevalence, geographic distribution, and hosts of hemosporidia are associated with the habitat preference of bird hosts. The abundance and feeding habits within habitats of suitable insect vector and innate physiological differences make some avian hosts more susceptible than others. Most of these parasites are not very pathogenic unless parasite loads are high. Following initial infection and an acute stage, most infections become chronic with occasional relapses when the host is immunosuppressed. Giemsa staining of blood films allows observation of parasites in the erythrocytes. Splenomegaly and hepatomegaly are commonly observed in birds that die from these infections; hemozoin (**malarial pigment**) may be seen histologically in some infections. PCR may be necessary for confirmation of infection, especially in low-level infections, or to speciate the organisms.

Ectoparasites

Lice (*Menacanthus* sp.) on the feathers are highly host-specific (Fig. 31.33) and the most commonly reported ectoparasites in wild turkeys and pigeons.[2,94] The entire life cycle occurs on live birds. Lice do not live for more than a few days away from the host. Lice have chewing mouthparts and feed primarily on feathers or scavenged debris from feather surfaces; however, they may also take blood, serum, and skin tissues. The mouthpart of each louse species is adapted to different body parts, thus common names applied to louse species originate mainly from the preferred site of infestation of the species.[156] Transmission is by direct contact. Clinical signs are associated with weight loss and restlessness due to constant irritation. Eggs are laid along the feather, and usually at the base of the shaft. Harm to the host is minimal, except in very heavy infections. Bathing, preening, and other grooming behaviors help to eliminate lice.

Mites are not host specific, although they are more successful on their preferred host.[156] They are divided into two groups: (1) mange mites, which feed on skin cells and cause skin irritation and scabbing, and (2) blood-sucking mites. ***Knemidocoptes mutans***, also known as scaly leg mite, is the most commonly diagnosed mange mite in birds; it has a wide host range. Mites burrow into the skin of the birds' shanks and feet causing swelling and deformation of scales (Fig. 31.34). In severe infections, loss of digits has been reported.[109] ***K. laevis* (syn. *Neocnemidocoptes gallinae*)**, also known as the **depluming mite,** burrows into the skin at the base of feathers, resulting in severe pruritus and feather pulling. In severe infestations, the mite may cause scaly, thickened, and wrinkled skin, mainly along the neck.

Among the blood-sucking mites, ***Dermanyssus gallinae***, the poultry red mite, is one of the best known and a common parasite of birds.[20,109] It feeds at night and hides during the day. *D. gallinae* can cause weight loss, anemia, and even death. Another mite that sucks blood and can cause similar serious problems is ***Ornithonyssus sylviarum*** or Northern fowl mite. It visits the host only for feeding and therefore might be missed. In heavy infestations, feathers of the birds appear blackish, especially in the vent region, due to mite droppings (Fig. 31.35).

FIGURE 31.33 Lice on the feathers of a chicken. The body louse (*Menacanthus stramineus*) is one of the most common parasites found on the body of many gallinaceous birds. This yellowish, chewing lice have a flat body and are about 3 mm long.

FIGURE 31.34 *Knemidocoptes mutans* (the scaly leg mite, infestation of the foot and shank in a chicken. The mites burrow into the skin of the shanks and feet causing swelling and deformation of scales.

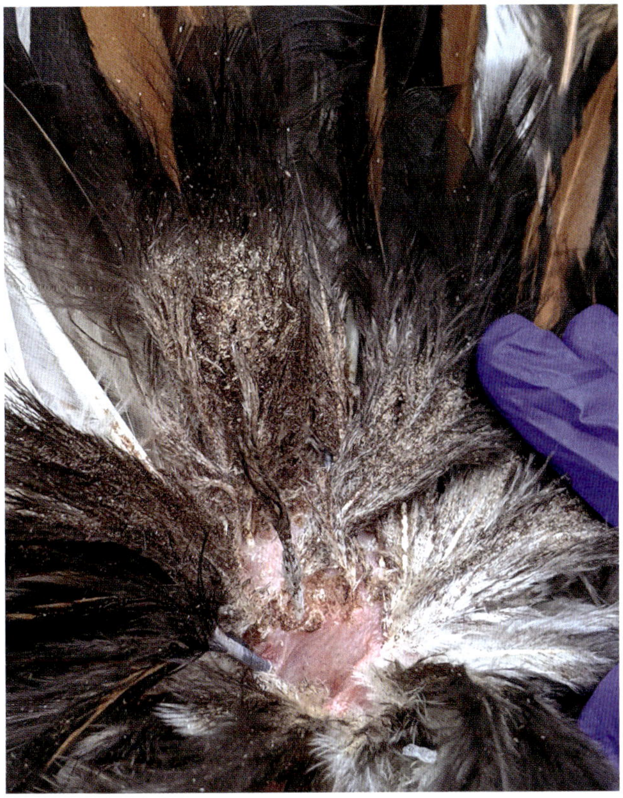

FIGURE 31.35 *Ornithonyssus sylviarum* (Northern fowl mite) infestation in a chicken. Heavy infestation of mites is present on the feathers around the vent region. The fine mite droppings make the feathers appear dirty or covered by dust.

The main flea in Galliformes is the **stick-tight flea**, ***Echidnophaga gallinacean***. It infects chickens, pheasants, and quail.[109,156] Females embed themselves in the skin of the host, usually on or around the head; males remain mobile. Adults feed on blood and can cause irritation, anemia, and even death in heavy infections.

Ticks, **biting flies**, and **bedbugs** feed periodically on blood of birds. Heavy infestation can lead to severe host anemia, nest desertion, and even death. In addition, they serve as important vectors for other parasites including black flies with *Leucocytozoon*; biting midges with *Hemoproteus*; and several genera of mosquitoes with *Plasmodium*.[156] The major concern with these bugs is that they may be transferred from birds to infest human houses, when carried in clothing or infected materials.

ADDITIONAL RESOURCES

Boulianne M, Brash ML, Fitz-Coy SH, et al.: Avian Disease Manual. American Association of Avian Pathologists, Jacksonville, FL, 2013.

Doufour-Zavala L, Swayne DE, Pearson JE, Reed WM, Jackwood MW, Woolcock PR: Isolation, identification, and characterization of avian pathogens, 5th ed. The American Association of Avian Pathologists, Athens, GA, 2008.

Fletcher O: Avian Histopathology, 3rd ed. American Association of Avian Pathologists, Jacksonville, FL, 2008.

Greenacre CB, Morishita TY, eds. *Backyard Poultry Medicine and Surgery: A Guide for Veterinary Practitioners*. Ames, IA: Wiley; 2015.

Morishita TY. Galliformes. In: Miller RE, Fowler ME, eds. *Fowler's Zoo and Wild Animal Medicine*. St. Louis, Missouri: Elsevier; 2015:143–155.

Swayne DE, Boulianne M, Logue CM, McDougald LR, Nair V, Suarez DL: Diseases of Poultry, 14th ed. Wiley-Blackwell, Ames, IA, 2020.

Thomas NJ, Hunter DB, Atkinson CT, eds. *Infectious Diseases of Wild Birds*. Oxford: Blackwell Publishing; 2007.

References

Full references for this chapter can be found online at https://doi.org/10.1016/B978-0-443-18437-6.00042-0.

Chapter 32

Psittacines, Coliiformes, Musophagiformes, Cuculiformes

Neel Aziz[a], Jeann Leal[b], Gerry Dorrestein[c], and Drury R. Reavill[d,†]

[a]*Smithsonian National Zoological and Conservation Biological Institute, Washington, DC, United States;* [b]*Universidade Federal da Paraíba UFPB, Areia, Paraíba, Brazil;* [c]*Pathology Laboratory NOIVBD, Vessem, The Netherlands;* [d]*Zoo/Exotic Pathology Service, Citrus Heights, CA, United States*

INTRODUCTION

Birds from the order Psittaciformes, also known as psittacines or parrots and parakeets, are popular display animals and pets due to their colorful behaviors and widely recognized intelligence. Birds in this order are common among those wildlife that are illegally traded and trafficked. Mousebirds (order Coliiformes) are a small group of birds that are related to the clade Eucavitaves [includes the cuckoo roller (Leptosomatiformes) and trogons (Trogoniformes)]. This is the only bird order whose natural distribution is limited entirely to sub-Saharan Africa. The turacos make up the bird family Musophagidae ("banana-eaters"; order Musophagiformes) which includes plantain-eaters and go-away-birds from southern Africa. This chapter discusses normal, unique anatomy (see Digital Content) and diseases in these groups of birds. In addition to the descriptions that follow, also see Tables 32.e1–32.e3; Figs. 32.e1 and 32.e2 for several additional infectious diseases that have been reported in Coliiformes, Cuculiformes, and other Psittacines.

NON-INFECTIOUS DISEASES

Nutritional

The term "vitamin A" includes a family of metabolically related retinoids (retinal, retinol, retinoic acid, and a storage form-retinyl ester) and several provitamin A carotenoids (e.g., beta-carotene). Vitamin A is an essential vitamin that is required throughout life and participates in numerous cellular activities involved in reproduction, embryonic development, vision, growth, cellular differentiation and proliferation, the integrity of mucus membranes, enhancing immunity to infections, and lipid metabolism.[9,22] Retinoid must be obtained through the diet, where it undergoes various pathways for metabolism that involve specific carrier proteins and enzymes. Most of the body's retinoid is stored in the liver. Vitamin A, stored as retinyl esters, is delivered from liver stores into the bloodstream as retinol bound to retinol-binding protein. This process results in a constant plasma retinol level.

Hypovitaminosis A (vitamin A deficiency) in birds results in epithelial squamous metaplasia of glandular epithelium that manifests grossly as hyperkeratosis of the oral cavity, conjunctiva, nasolacrimal duct, upper alimentary tract (esophagus), and respiratory tract (nasal passages, sinuses, trachea, and syrinx). The metaplastic changes are clinically most prominently seen in the glossal and mandibular salivary glands (Fig. 32.1). In large parrots, epithelial changes appear when vitamin A concentrations in the liver decrease below 50 IU/g (normal 500 IU/g liver).[26] Keratin plugs from hyperkeratosis can block the ducts of the submucosal mucus glands. The secretions build up and the glands will enlarge and become secondarily infected resulting in large keratin granulomas within the nasal or oral cavity. Some birds, especially African gray parrots, may develop massive keratin rhinoliths that distort the nares and nasal sinus. Squamous metaplasia of the mucosal epithelium also alters the mucosal defenses, predisposing birds to fungal and bacterial infections, commonly in the respiratory tract. In advanced cases, intraluminal accumulation of keratin due to squamous metaplasia of the epithelium of the ureters and collecting ducts can result in partial or complete obstruction.[117] Vitamin A deficiency in parrots is commonly due to deficiencies in the diet. Seeds and cereal grains of commercial psittacine diets are typically low in vitamin A or carotenoids. With the increasing use of formulated diets, the incidence of vitamin A deficiency is decreasing. In addition to diet, enteritis can also lead to and be a differential for hypovitaminosis A since the intestinal mucosa functions to convert carotenoids to vitamin A.

[†]Deceased.

Pathology of Wildlife and Zoo Animals. https://doi.org/10.1016/B978-0-443-18437-6.00012-2

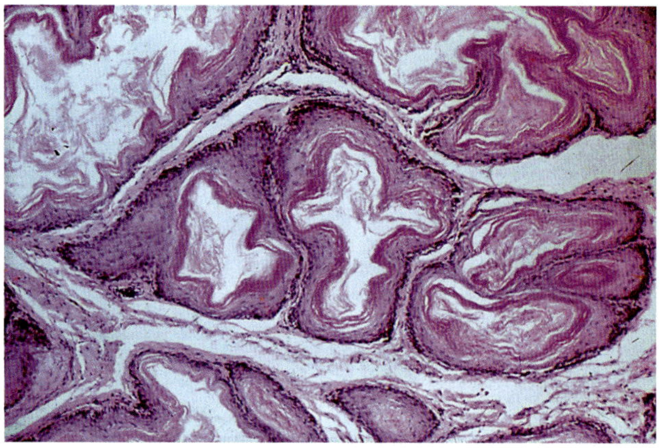

FIGURE 32.1 Squamous metaplasia due to hypovitaminosis (vitamin A deficiency) in the mandibular salivary gland in a sulfur-crested cockatoo. Hypovitaminosis A is associated with squamous metaplasia in several locations including the oral cavity and glandular tissues. In this case, glandular epithelial squamous metaplasia is associated with luminal accumulation of keratin.

Hypocalcemia syndrome in African gray parrots has been historically characterized by decreased levels of blood calcium associated with weakness, ataxia, tetany, or seizures.[96] These clinical signs are also seen with atherosclerosis, and avian circovirus and bornavirus infections in African gray parrots. The birds are generally adults (2–15 years) and most are on nutritionally poor diets. Seed diets (peanuts, corn) are typically low in calcium and vitamin D_3, and birds fed a high seed diet without proper nutritional supplementation can develop nutritional secondary hyperparathyroidism. Most affected African gray parrots will have low plasma ionized calcium levels (0.96–1.22 mmol/L), or plasma calcium levels less than 0.15 mmol/L (6.0 mg/dL) and low 25-hydroxycholecalciferol levels (14.50–20.0 nmol/L).[130] The primary function of vitamin D is to maintain serum calcium at optimum levels. This is achieved by increasing intestinal absorption of calcium, parathyroid-dependent renal tubular resorption of calcium, and parathyroid hormone-dependent mobilization of calcium from the bone. Birds acquire vitamin D_3 from the diet and can also endogenously synthesize D_3 from provitamin D_3. Total calcium is representative of the active ionized calcium, protein-bound calcium, and calcium chelated with anions. In African gray parrots, a significant correlation has been found between total calcium and albumin concentration in the plasma, which has not been demonstrated in most other psittacine groups. Clinically, this is important when calculating total calcium in birds with low albumin.[69]

Although all psittacines may develop lesions of **osteopenia** (reduced bone mass in adults) and **osteomalacia** (in young birds with a growing skeleton), adult African gray parrots are unique as they present with neurological signs when abnormalities of calcium, vitamin D_3, and/or phosphorus are present.[129] Lesions are poorly described, with only a suggestion of enlargement and/or **secondary hyperparathyroidism** by gross evaluation (Figs. 32.2A–C).

Goiter is a noninflammatory, nonneoplastic enlargement of the thyroid gland. It can be functional as with **hyperthyroidism** or have subnormal function as with **hypothyroidism**. In general, there are three mechanisms that lead to goiter and hypothyroidism: iodine deficiency (selenium deficiency will make it worse), ingestion of goitrogens (Brassica and Cruciferae plants), and hereditary biosynthetic defects. All result in inadequate thyroxine (T_4) synthesis and decreased blood levels. Decreased thyroxine triggers the hypothalamus and pituitary gland to increase the production of thyrotropin-releasing hormone (TRH) and thyroid-stimulating hormone (TSH), respectively, which results in hypertrophy and hyperplasia of follicular cells of the thyroid gland.

Thyroid control in birds is different from mammals. For one, thyroxine (T_4) levels in birds are approximately one-tenth the concentrations found in mammals, which complicates clinical attempts to measure baseline levels in birds. Briefly, the hypothalamus of birds produces TRH that acts on the anterior pituitary to release growth hormone (GH), instead of TSH as in mammals. Avian GH then stimulates the monodeiodination of thyroxine (T_4) creating an increase of triiodothyronine (T_3). This increase of T_3 then stimulates the anterior pituitary to release TSH; the TSH in turn stimulates the thyroid gland to release T_4.[72,73]

Hyperplastic goiter and **colloid goiter** have been described in a number of domestic and nondomestic species including birds. Hyperplastic goiter was historically most common in budgerigars, but it occurs in a wide variety of pet and nonpet birds. Grain grown on iodine-deficient soils is the primary cause for its development. Budgerigars clinically present with respiratory clicks. Bilateral hyperplasia results in enlarged, red-brown or purple glands that put pressure on the trachea, esophagus, and other soft tissues of the neck and can cause collapse of the interclavicular air sac. The enlarged gland may displace the trachea dorsally or ventrally. Associated clinical signs include inspiratory squeak, dyspnea, and regurgitation. Clinically, goiter does not cause hypothyroidism which makes thyroid function tests or diagnostics ineffective.[70,81] A hyperplastic goiter syndrome is seen in young blue and gold and occasionally other large macaws. The etiology has not been determined. These birds may unexpectedly die with no other lesions to explain a cause of death.[116] Histologically, affected glands in animals with **hyperplastic goiter** consist of numerous small- to medium-sized follicles lined by enlarged (hypertrophied) epithelial cells. There is no colloid and often no apparent follicular lumen (Fig. 32.3). **Colloid goiter** is considered the involution phase of thyroid gland hyperplasia. Histologically, the gland is composed of large follicles containing abundant colloid and lined by epithelial cells that may be cuboidal or columnar initially, but eventually become flattened and atrophic (Fig. 32.4).

Emaciation and dehydration are common issues in psittacines. In birds and mammals, three distinct levels of

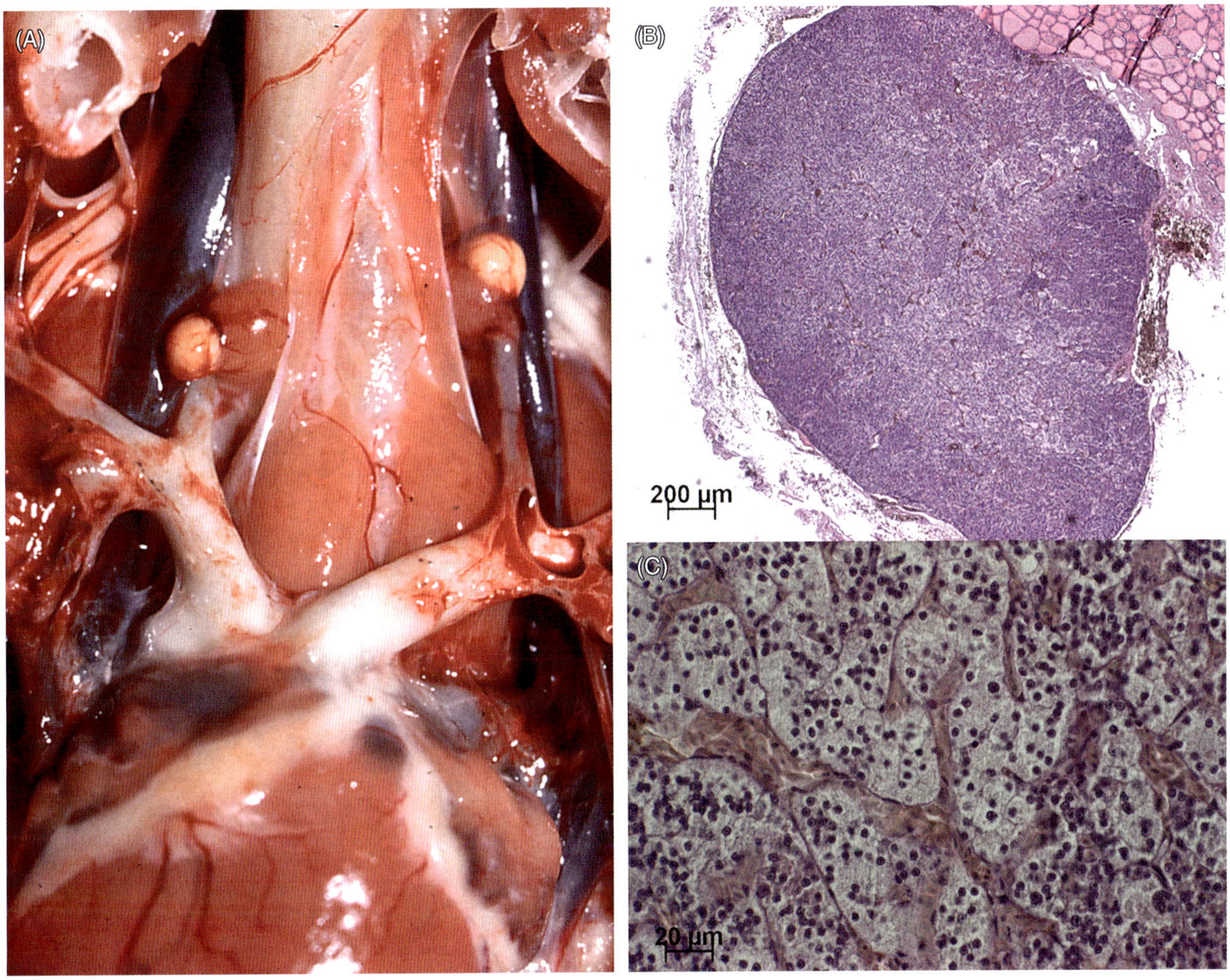

FIGURE 32.2 Hyperparathyroidism in an African gray parrot. (A) The thyroid and parathyroid glands are located cranial to the heart. The parathyroid glands in this bird are dramatically enlarged and are light tan. The much smaller adjacent normally sized red/brown thyroid glands are immediately caudal to the parathyroid glands. (B) Diffuse enlargement of the gland (upper right). (C) Hypertrophic chief cells (lower right).

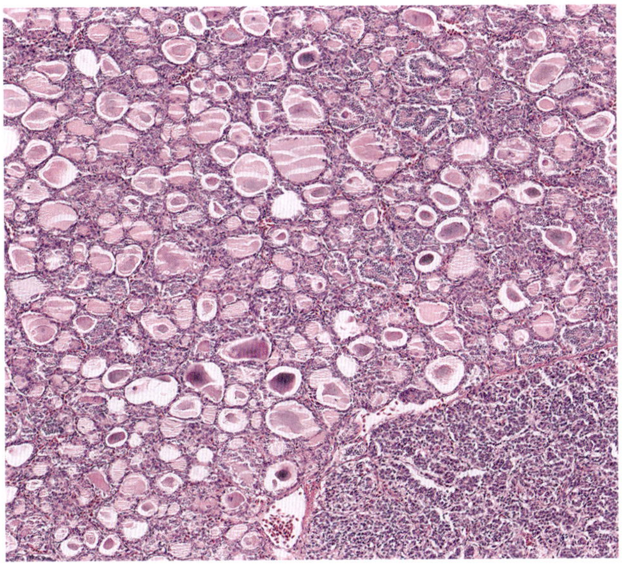

FIGURE 32.3 Hyperplastic goiter in a blue and gold macaw. The thyroid gland contains numerous small follicles, many of which contain little to no colloid. The parathyroid gland is in the lower right of this image.

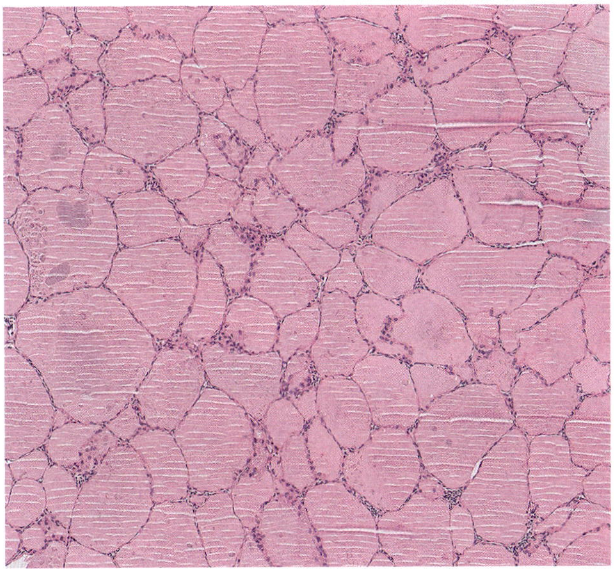

FIGURE 32.4 Colloid goiter in a Goffin's cockatoo. The thyroid gland is greatly enlarged and the follicles are large, lined by attenuated epithelium, and filled with colloid.

energy depletion during **anorexia** or **prolonged fasting** have been described.[42,43] The first phase lasts only a few hours and is a rapid period of adaptation marked by the flushing of the gut contents. Phase two is characterized by intestinal mucosal atrophy following a period of starvation during which the fat stores are mobilized for energy expenditure. During phase two, apoptosis is present at the tips of the intestinal villi, the number of mitoses in crypts shows a 30% decrease, and a drop in cell migration is observed resulting in defects of the mucosal cell lining. In phase three, after a long fast, mucosal atrophy worsens and is marked by an increase in protein catabolism; fat stores are progressively exhausted. Changes of phase three fasting are common in necropsies of ill psittacines. Due to small body size, there is a high energetic need to maintain body temperature. Anorexia, inanition, or starvation leads quickly to hypoglycemia resulting in, among other things, defects in the mucosal lining and **hemorrhagic diathesis** (leaking of blood into the intestines). Hemorrhagic diathesis, seen frequently in small birds (budgerigars, lovebirds, or nestlings) with low fat stores that are anorexic for 24h or longer, is often mistaken for hemorrhagic enteritis. Clinical history and necropsy findings consistent with hemorrhagic diathesis include recent or prolonged anorexia and an empty crop with reflux of denaturated (black) blood from the intestines into the ventriculus.

Hepatic lipidosis (Fig.32.5) is also a relatively common finding in psittacine birds, especially those kept under erroneous high-fat diets. In these cases, the liver usually has rounded margins and it is diffusely pale tan to yellow. Samples collected from a severely affected liver may float in formalin. Microscopically, hepatocytes have large cytoplasmic vacuoles and peripheralized nuclei. **Cirrhosis** (Fig. 32.6), in which the liver is small and the normal parenchyma is replaced by multiple pale yellow-tan firm hyperplastic nodules of variable sizes separated by thick fibrous tissue, can occur as a result of severe chronic disease processes including chronic hepatitis, biliary obstruction, and severe hepatic lipidosis.

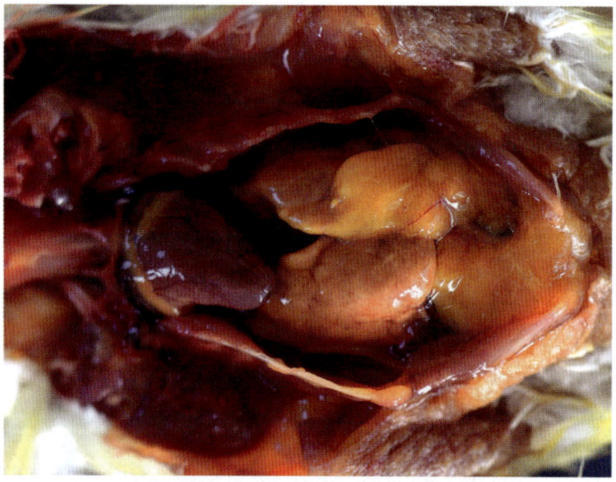

FIGURE 32.5 Hepatic lipidosis in a cockatiel. The liver is enlarged and diffusely discolored yellow.

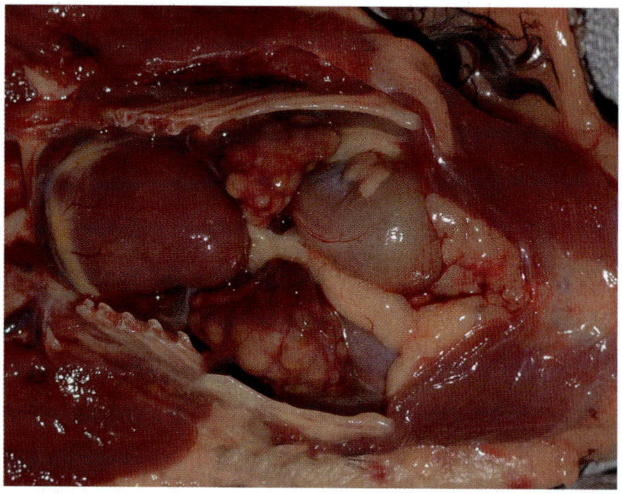

FIGURE 32.6 Cirrhosis in a cockatiel. The liver is small and much of the normal parenchyma is replaced by multiple pale yellow-tan firm hyperplastic nodules of variable sizes.

Metabolic

Similar to other species, iron pigment can accumulate in the liver and in more severe cases be associated with hepatocellular damage and **iron overload** (see Chapter 33: Passeriformes, Caprimulgiformes, Coraciiformes, Piciformes, Bucerotiformes, and Apodiformes). It is seen infrequently in parrots, but when seen, it is most common in lories and lorikeets.[156] Hemosiderosis has been reported in captive Cuculiformes and go-away birds.[154] Cases of iron overload have also been described in captive turacos.

The predominant nitrogenous waste product of birds is uric acid. Elevations in plasma uric acid can result in **gout**. The majority (90%) of uric acid is actively secreted by the proximal renal tubules. The remaining 10% is filtered by the glomerulus. Therefore, severe dehydration, extensive damage to the proximal tubules, lesions that obstruct urine outflow, or congenital malformations of the kidney predispose to hyperuricemia and the precipitation of uric acid crystals on mesothelial surfaces and within the kidney. These lesions may be nodular (urate tophi), such as those that are found on synovial surfaces or in periarticular tissues, or they may be diffuse, such as those found on the pericardium, liver capsule, air sacs, and peritoneum. The urate deposits can either have a finely granular (Fig. 32.7A) or a homogenous shiny (Fig. 32.7B) aspect. In the kidney, gross lesions are characterized by multifocal white to yellow-white foci or streaks that represent urate

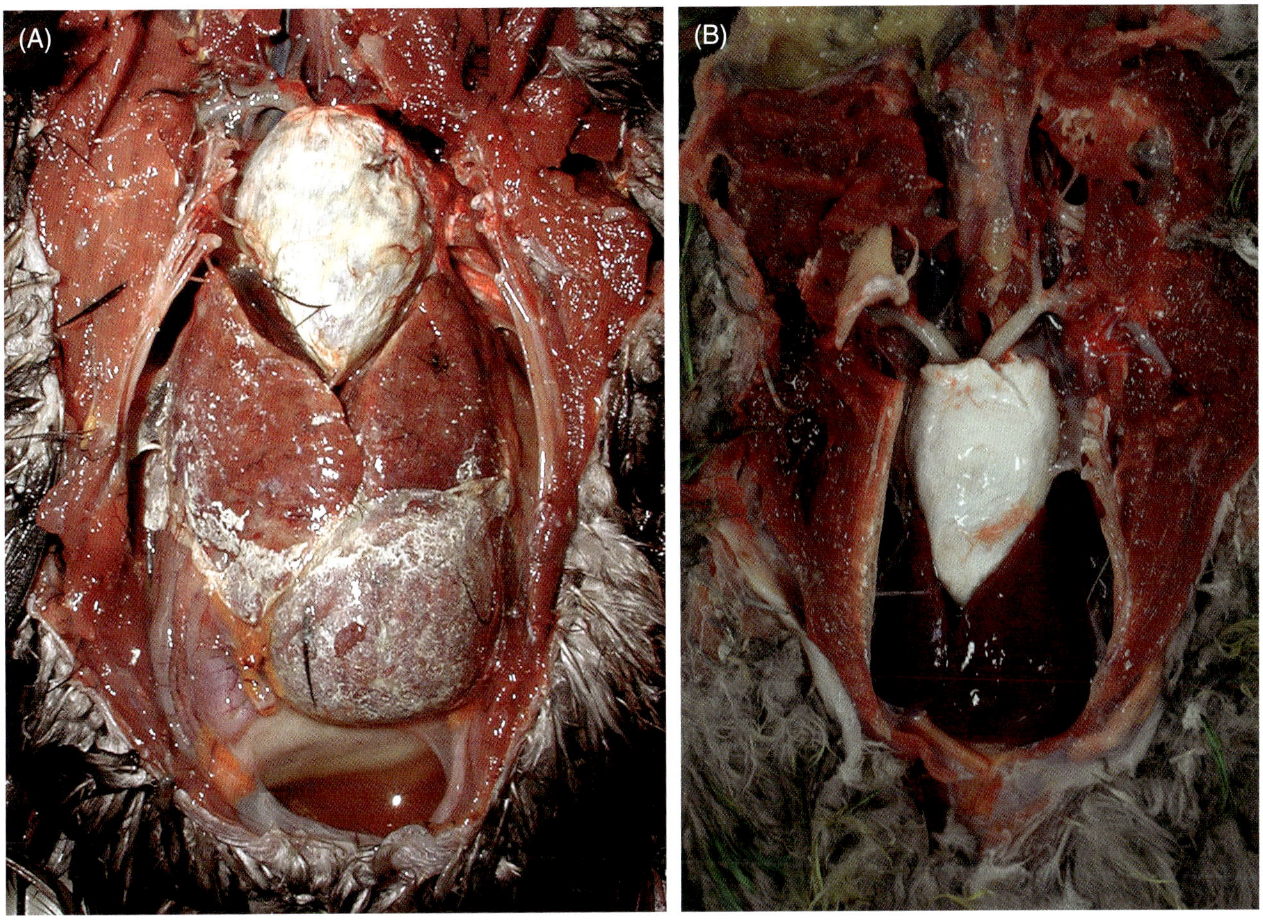

FIGURE 32.7 Visceral gout in a black lorry (A) and blue-fronted Amazon parrot (B). (A) Uric acid deposition on visceral surfaces can appear dry and chalky. In this case, the pericardium and visceral surfaces throughout the coelom appear white due to precipitate deposition. (B) In some cases, visceral gout has a smooth, shiny, almost wet rather than chalky appearance as in (A).

deposits in tubules and the interstitium.[117] Microscopically, urates dissolve during fixation but leave behind characteristic needle-shaped (acicular) and amorphous clefts and spaces surrounded by an eosinophilic protein matrix where the crystals had been. In the kidney, secondary epithelial necrosis occurs due to mechanical damage from the urate crystals, which elicits a heterophilic to histiocytic inflammatory response (urate tophi) (Fig. 32.8). Along serosal and visceral surfaces, acellular, finely fibrillar, lightly basophilic to amphophilic material is more common than the acicular clefts that are present in the kidneys. Necrosis with a primarily heterophilic inflammatory response is dependent on the duration of urate deposition in these locations.

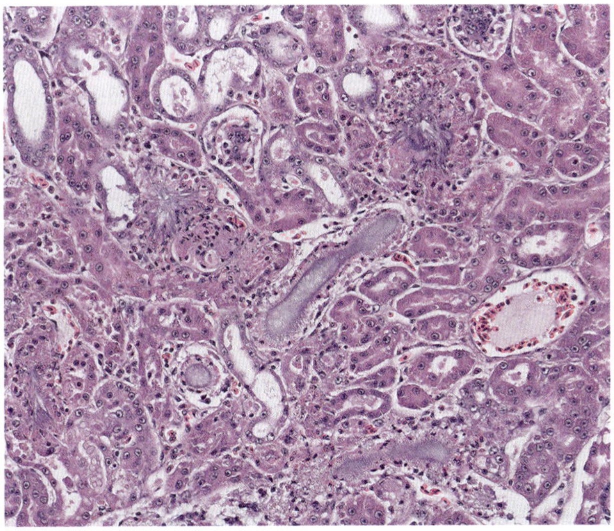

FIGURE 32.8 Renal gout in an eclectus parrot. Urate tophi in the kidney (upper right and in tubules) surrounded by acute inflammatory cells and necrosis.

Toxic

Sources of lead are numerous and include lead shot, colored coatings on electrical/electronic wires, lead-based paints, lead putty, solder, plaster impregnated with lead, linoleum, hardware cloth, some types of foils, some types of welds, lead weights (curtains, fishing lures), bells with lead clappers, frames of stained glass in any form, improperly glazed ceramics, batteries, weighted bird toys, costume jewelry, mirror backs, leaded gasoline fumes, and some vinyls and plastics that are stabilized with lead. Lead may also be present in plastics, steel pipes lined with zinc that contains lead, or ceramics used as dishes or cage furnishings.[66] Clinical signs of **lead toxicity** involve the hematopoietic, gastrointestinal, renal, and nervous systems. Effects are dosage, exposure, and time dependent. Chronic low-level exposure can cause immunosuppression as well as tissue build-up and eventual manifestations similar to that of an acute exposure. Psittacines may show depression, weakness, vomiting, polyuria/polydipsia, seizures or tremors, hemoglobinuria, and diarrhea.[66] Lead-induced neuropathy is initiated primarily by Schwann cell damage.[89] Myelin sheath damage and lead-induced vascular endothelial damage result in endoneurial edema. Lead interferes with enzymes of the cellular oxidative processes by inhibiting delta amino levulinic acid dehydrase (ALAD) and heme synthesis.[68] Histologically, findings are nonspecific but hepatocyte necrosis and hemosiderosis may be present in the liver. Erythrophagocytosis, as well as accumulation of hemosiderin pigments, are common findings due to the removal of the damaged red blood cells by hepatic Kupffer cells. There may be impairment of liver function due to direct lead effects on the hepatocytes as well as functional impairment of bile flow due to lead-caused vagal neuropathy. Renal nephrosis and accumulation of hemosiderin pigments within the cytoplasm of the renal tubular epithelial cells can also occur.[76] Some birds with lead toxicity have acid-fast, intranuclear inclusion bodies within renal tubular epithelial cells.[117]

Lead is likely the most commonly ingested toxin diagnosed in pet psittacine species. Confirmation of lead toxicosis requires all or a combination of the following: a clinical history of presumptive lead consumption, radiopaque foreign bodies in the crop and gastrointestinal tract, increased aspartate aminotransferase (AST), lactate dehydrogenase (LDH), and creatine phosphokinase/creatine kinase (CPK/CK), low red blood cell count, hypochromasia, cytoplasmic vacuolization, ballooning and basophilic stippling of erythrocytes. Blood lead levels of 20 mg/dL are indicative of lead toxicity and levels of 50 mg/dL or more confirm the diagnosis. Liver and kidney levels of 3–6 ppm or more indicate lead toxicity, and levels above 6 ppm confirm the diagnosis.[113,147]

Acute signs of **zinc toxicity** are similar to those seen with lead toxicosis; some birds may have concurrent toxicity. Zinc plasma concentrations above 4 mg/L are suggestive but not diagnostic for intoxication.[93,111] Sources of zinc exposure include coins, galvanized containers or caging, staples, nails, fertilizers, some paints, zinc pyrithione shampoos, zinc oxide, zinc undecylenate (Desenex cream), costume jewelry, and other metallic items. Zinc toxicity can cause hepatic hemosiderosis and erythrophagocytosis. The associated pancreatic histological and ultrastructural lesions include disruption of the normal acinar architecture of the exocrine pancreas, loss of zymogen granules and basophilic staining, the formation of hyaline bodies and other electron-dense aggregates, cellular atrophy and necrosis of individual acinar cells and interstitial fibrosis. The pancreatic islets are generally spared.[93] Zinc toxicity may also cause erosive ventriculitis. Histologically, the koilin layer is disrupted, there is ulceration of the underlying mucosa, and the ventricular glands are dysplastic. In the kidneys, zinc toxicity produces acute tubular necrosis.

The best-documented reports of airborne pulmonary toxins are of those caused by inhalation of pyrolysis products produced from overheated **polytetrafluoroethylene (PTFE** or **Teflon)**-coated cooking pans, stove tops, coated heat lamps, and lightbulbs. Toxicity results from the fumes emitted as the polymers undergo pyrolysis at 536°F (280°C) and release PTFE fumes. The toxic products are made up of both gaseous and particulate materials. In combination, the fumes and particulates cause direct damage to type I pneumocytes and capillary endothelial cells.[155] Other airborne toxins include aerosol sprays, cooking gas, carbon monoxide, and tobacco smoke. Although the specific toxin is unknown, the operation of self-cleaning ovens has also resulted in acute deaths[132] and compounds emitted from burned foods, cooking oils, and other materials can also be toxic. Clinical signs are usually reported as respiratory problems leading to acute death. Often there is sudden death of more than one bird in a group along with a history of cooking with or warming a nonstick pan. Mild exposures may present with moist rales, dyspnea, ataxia, depression, or restless, anxious behavior. These signs are similar to other respiratory diseases or stress.

The lesions of airborne toxins are nonspecific and can be easily overlooked. Affected birds consistently have severely congested or hemorrhagic red and wet lungs. Death is from acute pulmonary edema, hemorrhage, and shock. Watery red fluid may exude from the trachea and nares. In the lungs, the lumen of the tertiary bronchi, atria, and air capillaries are filled with abundant edema fluid. Multifocal to coalescing hemorrhage in the pulmonary interstitium often extends into air spaces. Pale-staining edema fluid accumulates around most arteries and veins. Transmission electron microscopy after PTFE exposure shows tracheal mucosal and tertiary bronchiolar epithelium degeneration and ulceration with necrosis of air capillary membranes. Cardiac lesions include multifocal degeneration of myocytes with deeply eosinophilic cytoplasm, occasional vacuolization, and deeply basophilic nuclei.

Definitive diagnosis of airborne toxicity from any cause is limited in clinical cases. The lesions as described by TEM may provide more definitive diagnosis for PTFE exposure; however, an appropriate history and consistent gross and histologic lesions are usually all that will be available.

Vitamin D_3 toxicity usually results from excess dietary supplementation. Cockatiels, blue and gold macaws, and hyacinth macaws seem more prone to disease. Renal and soft tissue mineralization is a characteristic feature of hypervitaminosis D_3. Affected kidneys are enlarged, firm, and contain numerous yellow-white foci. Histologically, damage is variable and includes tubular necrosis and basement membrane mineralization. Necrotic epithelial cells and secondary urate deposition (urate tophi) may be seen. Liver lesions are characterized by mineralization of sinusoidal basement membranes and the tunica media of blood vessels.

Although a variety of drugs and chemicals are potentially cardiotoxic in birds, there are very few documented cases. Natural and experimental **avocado toxicity** has been reported in canaries, cockatiels, and budgerigars.[12] The common avocado varieties on the market in the United States are the Fuerte and Hass cultivars. The leaves, fruits, and seeds are toxic.[13] The toxic agent is persin,[45] a long-chain, unsaturated compound that has been isolated from the leaves.[77] Gross lesions include subcutaneous edema and hydropericardium (Fig. 32.9). The primary histologic lesion is myofiber degeneration; inflammation is uncommon in psittacines. There may be generalized congestion of tissues, especially the lungs. Diagnosis is based on history, lesions, and identifying the plant material in the digestive tract.

Congenital/Genetic

Choanal atresia is described in African gray parrots and an umbrella cockatoo.[38,46] A persistent membrane or bony plate at the palate in the nasal cavity results in a closed choanal slit. This blocks the normal drainage of nasal secretions into the oral cavity. Young birds present with a chronic nasal or ocular discharge and in some cases, the infraorbital sinuses will be distended with clear secretions. The cause is unknown.

Age Related/Degenerative

Atherosclerosis, vascular intimal thickening with mural lipid accumulation, is reported most often in three psittacine genera: *Psittacus*, *Nymphicus*, and *Amazona*.[6] In one report, the incidence of atherosclerosis in histologically examined parrots was 91.9% in African gray parrots and 91.4% in Amazon parrots.[29] Many risk factors are associated with atherosclerosis including increased age, a diet rich in fat, and reproductive disease. Affected birds can be of any age, but most are 8 or more years old and many are over 15 years of age.[87,117] The frequency increases dramatically in birds 20–30 years of age, but disease in birds as young as 6 months of age has also been reported. In addition, there is a bias toward females developing more severe lesions.[6]

Atheromatous plaques, raised lesions with a fibrous cap, can be found in any vessel. In psittacines, the aorta, brachiocephalic trunks, and pectoral and carotid arteries are common sites; coronary artery involvement is relatively rare. The arterial wall is variably thickened with roughened yellow intimal plaques (Fig. 32.10; eSlides 32.e1–32.e3). Histologically, this corresponds to vacuolated smooth muscle cells and macrophages (foam cells that contain cytoplasmic cholesterol and cholesterol esters) within the intimal layer of the aorta and large arteries. Microhemorrhage, chondroid metaplasia, and fibrosis can also be seen. Mural mineralization is less common in birds than in mammals. In severe cases, atherosclerosis causes increased

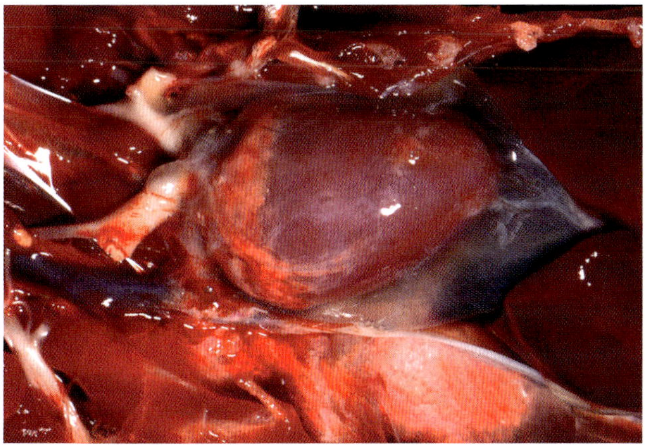

FIGURE 32.9 Avocado toxicity in an African gray parrot. Hydropericardium has been reported in association with avocado toxicity but is a nonspecific finding.

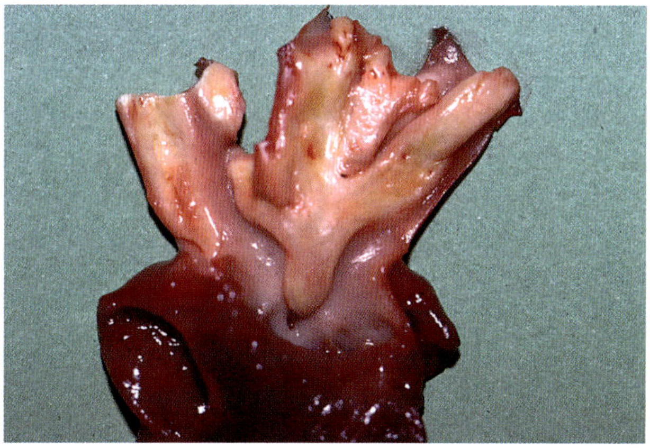

FIGURE 32.10 Atherosclerosis in the great vessels of an African gray parrot. Thick, yellow plaques are present along the intimal surfaces of the great vessels (see eSlides 32.e1–32.e3).

arterial resistance that affects the heart. Early changes in the heart include hypertrophy of the left ventricle followed by left ventricular dilation, dilation of the left atria, right heart dilation, and right heart failure. Right heart failure can lead to congestion, atrophy, and subsequent hepatic fibrosis. Ischemic changes in the heart are rare. In chronic cases, birds may present with loss of body condition, although many die unexpectedly in excellent or obese body condition. Death can also develop secondary to neurologic complications due to decreased cerebral blood supply and anoxia due to the severely narrowed carotid arteries. There may be a history of the bird going through periods of disorientation in the days or weeks prior to death. Atherosclerosis can also produce aneurysmal dilation and aortic rupture in psittacines.

Chronic pulmonary interstitial fibrosis (CPIF) is a syndrome best described in older captive Amazon parrots (average age 20.5 years). It typically presents as exercise intolerance. Most birds will have a long history of respiratory disease. The condition has been more frequently described in psittacines in European populations, although it has also been seen in the United States. In affected birds, there is loss of functional lung tissue due to pulmonary interstitial fibrosis (Fig. 32.11), and secondary cardiomyopathy consisting of right ventricular hypertrophy or right ventricular dilation develops as sequela to increased pulmonary resistance. Severe cases have an associated elevated packed cell volume as a result of an increase in erythrocyte size and an increased hemoglobin mass per erythrocyte related to the progressive reduced oxygen exchange surface in the lungs with normal blood flow. Venous blood gas analysis may reveal hypoxia and hypercapnia.[157] In live birds, computed tomography (CT) is a valuable diagnostic tool (dilation of bronchi and variable thickened walls of the bronchi).[3] The cause of this syndrome has not been identified. Characteristic histologic changes include mild to severe interstitial fibrosis with filling or replacement of air and blood capillaries (Fig. 32.11). In the functional lung tissue, there will be variable amounts of smooth muscle hypertrophy of the tertiary bronchi, edema, and congestion.

Chronic obstructive pulmonary disease (macaw pulmonary hypersensitivity) with polycythemia appears most common in the blue and gold macaw, although other macaw species can be affected.[138] Early diagnosis is difficult due to the reserve capacity of avian lungs and relative inactivity of macaws in aviaries. Generally, there is severe exertional expiratory and inspiratory dyspnea, cyanosis of facial skin, and a dry, nonproductive cough. The lung lesions are generally advanced when polycythemia occurs. Grossly, the lungs have a firm, rubbery texture and are moderately congested. Histologically, there is extensive consolidation and a thickening of the interstitium with eosinophilic material and fibrous tissue with a mixed cellular infiltrate. There is partial or complete obliteration of the parabronchi lumen. The prominent lesion is the pulmonary atrial smooth muscle hypertrophy (Fig. 32.12) and some atrial loss due to fusion and epithelial bridging. Uncommonly, there is a proliferation of parabronchial lymphoid tissue and in severe cases lymphoid nodule formation.

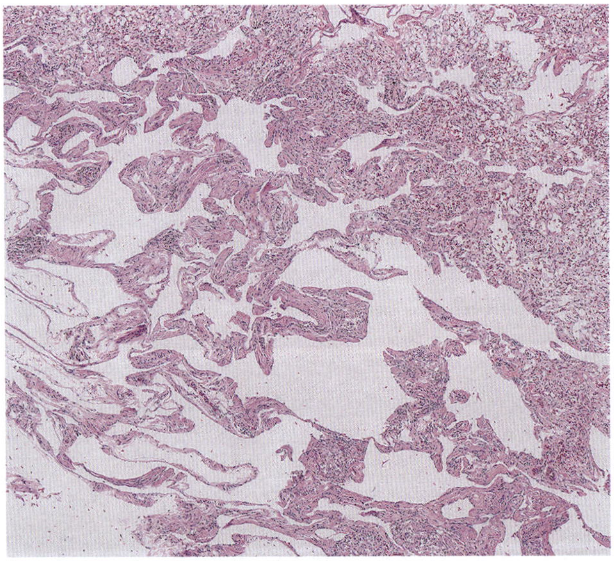

FIGURE 32.11 Chronic pulmonary fibrosis in the lungs of an aged yellow-headed Amazon parrot. The interstitium is diffusely mildly to moderately thickened due to collagen deposition/fibrosis. There is congestion of the more normal lung (upper right). Foci of pneumoconiosis are also present.

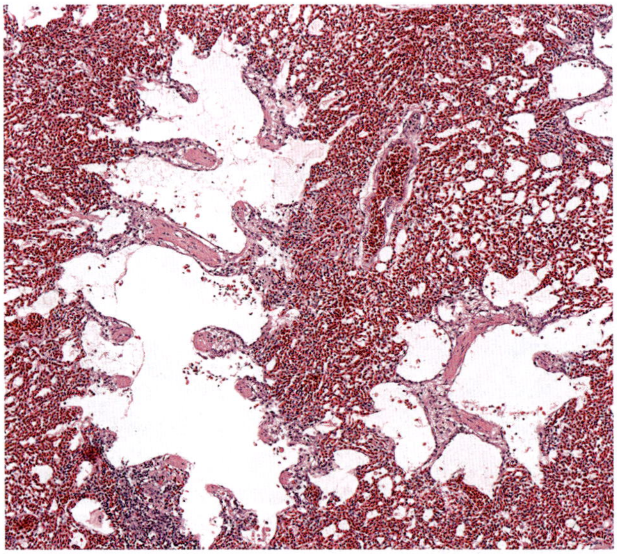

FIGURE 32.12 Macaw pulmonary hypersensitivity (chronic obstructive pulmonary disease). The prominent lesion is the pulmonary atrial smooth muscle hypertrophy and some atrial loss due to fusion and epithelial bridging.

Inflammatory Non-Infectious

Egg yolk coelomitis (EYC) is a common sequela of chronic reproductive tract disease. It can occur after salpingohysterectomy secondary to ovulation directly into the coelomic cavity. Other causes are ectopic ovulation, salpingitis, neoplasia, cystic hyperplasia, or oviduct rupture. Egg yolk with or without bacterial infection (commonly *E. coli*, *Staphylococcus* sp.) causes a severe inflammatory reaction and can lead to egg-related pancreatitis or yolk emboli. This occurs most commonly in cockatiels. The yolk material (proteins and fats), which in small amounts can be gradually resorbed, generally elicits a mild histiocytic response along the serosa of the intestines and oviduct. Large amounts of yolk material and the presence of bacteria can result in severe adhesions between the loops of intestines and variable organ dysfunction.

Acute EYC is characterized by large amounts of refractile, orange/pink to amphophilic or basophilic, variable-sized protein globules deposited on the serosal surfaces. These yolk globules can also be found in circulation. Most birds with EYC present in respiratory distress with a distended, fluid-filled coelom.[117] Respiratory distress is related to retrograde yolk inhalation (RYI) leading to an embolic pneumonia. Histologically, emboli consist of eosinophilic, globular aggregates in pulmonary capillaries and arteries. They may be associated with widespread, multifocal hemorrhage in tertiary bronchi and air capillaries. In these cases, chronic RYI pneumonia can develop and lead to interstitial dystrophic mineralization.[110]

Xanthomas are nonneoplastic, local depositions of yellowish cholesterol-rich material. They are usually identified in older (range 3–30 years; average age 10 years) psittacine birds and are more commonly reported in budgerigars and cockatiels.[102] They are most commonly found in the skin over the ventral abdomen, thighs, wings, eyelids, and around the face, and are rarely described in internal organs or bone marrow. Lesions may be pruritic and are often associated with other findings, such as lipomas, hernias, and sites of chronic irritation. The underlying cause is uncertain but high-fat or high-cholesterol diets, genetic predispositions, or trauma may contribute to their formation. Dietary improvement, including assuring sufficient vitamin A consumption, has been curative in less advanced cases.

Xanthomas are composed of numerous large macrophages and multinucleated giant cells that contain abundant foamy cytoplasm skin.[100] The cytoplasmic material is lipid. Extracellular lipid and cholesterol cleft formation may also be seen. Xanthomas can contain multifocal necrosis and variable heterophilic and lymphocytic inflammation (Fig. 32.13; eSlides 32.e4 and 32.e5). Superficial ulceration may occur at sites of skin lesions; internal masses may efface normal parenchyma.

Wing web dermatitis of the prepatagium and axilla of African gray parrots and lovebirds is a frustrating syndrome that is difficult to manage and resolve. The potential etiologies are complex and poorly understood, although hypersensitivity, self-induced or hormonal prepatagial feather loss, or excessive moisture of the wing web may initiate the early events. Chronic cases can result in scarring leading to prepatagial contracture, habitual self-mutilation, and recurrent secondary infections. Grossly, there is a moist, exudative dermatitis of the ventral prepagium and axillary areas with loss of downy and contour feathers. Multifocal areas of skin are ulcerated and inflammatory exudates are found on the surface or at the margins of the affected area (Fig. 32.14). One or both wings can be affected. The histopathological findings include superficial necrosis, epithelial hyperplasia, perivascular inflammation, and dermal fibrosis. There does not appear to be a sex predilection.[105] African gray parrots appear to be overrepresented, although the condition has been observed in other species, including lovebirds, hyacinth macaws, blue-throated macaws, cockatiels, cockatoos, and rarely other psittacine species.[91]

Hepatic fibrosis and bile duct hyperplasia (biliary reaction) in Psittacines is a confusing syndrome that is accompanied by chronic hepatitis of various severity and/or hepatic lipidosis. It can mimic neoplasia and there is not a known etiology or pathogenesis. Proposed causes include chronic toxicities (e.g., microcystin), infections (*Helicobacter pullorum*, psittacosis, viruses), immune-mediated disorders, or chronic malnutrition.[17,61,62] Grossly, there is diffuse hepatic pallor, firmness, hepatomegaly, and the surface of the liver may have a micronodular appearance.[17]

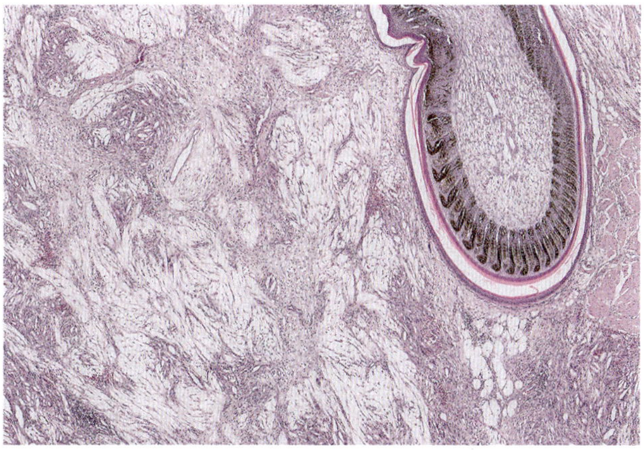

FIGURE 32.13 Xanthoma in the dermis of a cockatiel. Xanthomas are characterized by stacked, acicular clefts surrounded by multinucleated giant cells and macrophages. The intervening tissue is expanded by numerous, finely vacuolated histiocytes. (see eSlides 32.e4 and 32.e5).

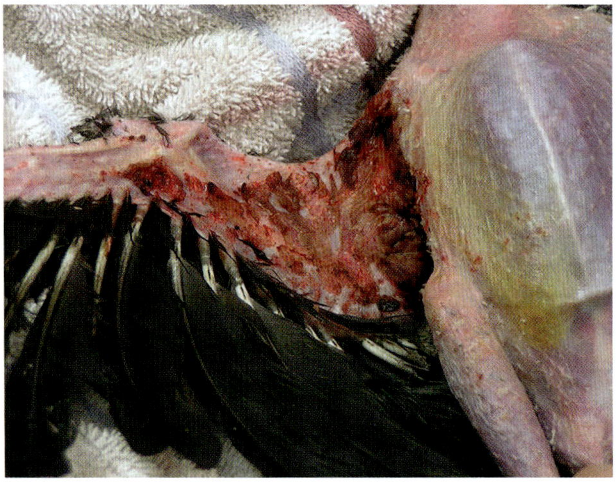

FIGURE 32.14 Wing web dermatitis in an African gray parrot. Chronic lesions include feather loss, moist dermatitis, and ulceration. Disease is often bilateral.

Miscellaneous/Multifactorial

Self-mutilation is a common issue for psittacine under managed care. The condition can vary from mild wing trimming and feather plucking to extensive **automutilation of the sternum (split sternum)**. This condition occurs over the cranial portion of the sternum along the point of the keel and may appear as an open wound of the skin and muscle. Continued self-trauma can lead to tissue necrosis and opportunistic infection, which extends into the deeper muscles and bone of the keel (osteomyelitis). Histologically, there will be variable epithelial hyperplasia, compact orthokeratotic hyperkeratosis, dermal fibrosis, and perivascular inflammation.[34] Mixed opportunistic bacterial infection (*Staphylococcus aureus, Escherichia coli* [*E. coli*], and/or *Streptococcus*, *Pseudomonas*, and/or *Klebsiella* species) and occasionally fungi can be isolated from open wounds. The pruritis typical of these lesions may be due to inflammation and histamine release that are stimulated by the secondary infections. Improper wing trimming in heavy-bodied birds (e.g., African gray parrots, Amazon parrots) resulting in traumatic landing on the sternum is suspected to be a predisposing factor. In birds with full flight ability and no history of a traumatic event, the reason for self-mutilation is unknown.[100]

Amyloidosis is better described in other avian species (see Chapter 29: Anseriformes, Ciconiiformes, Charadriiformes, and Gruiformes). It has been described in captive parrots or Coliiformes and is more common though poorly described in some Musophagiformes (e.g., red crested, Hartlaub's, and Livingstone turacos).[121]

Perivascular dermatitis associated with **feather picking/destructive behavior** is a common, nonspecific lesion in psittacines. The behavior has been associated with numerous underlying disease conditions and inappropriate social interactions between owners and their birds. The gross appearance includes missing and frayed feathers, reddened skin, ulcers, and variable pruritus. Histologically, mild to marked perivascular inflammation in the superficial or deep dermis is present in skin samples collected from clinically affected as well as unaffected areas. The inflammation is generally lymphocytic with fewer plasma cells, rare histiocytes, and few heterophils (Fig. 32.15). Varying degrees of edema, epidermal and follicular hyperkeratosis, and perivascular pulpitis can also be present. The pattern and the cellular constituents of the inflammation may suggest delayed-type hypersensitivity.[34,117]

Amazon foot skin necrosis is a syndrome of foot skin erythema and necrosis with automutilation in Amazon parrots, primarily the yellow-naped Amazon parrot. It occurs in birds of all ages with no sex predilection. Affected birds typically act very pruritic and will attack affected areas of skin. No etiologic agent has been identified; a delayed hypersensitivity reaction has been speculated. Proposed initiating causes include staphylococcal dermatitis or contact with nicotine or some element in tobacco smoke.[21] The clinical lesions appear as an irregular pattern of erythema or brown to black discoloration on the unfeathered skin of legs and feet. Self-trauma will result in ulcerated skin lesions with varying amounts of edema. Some birds have recurring episodes.[55,99]

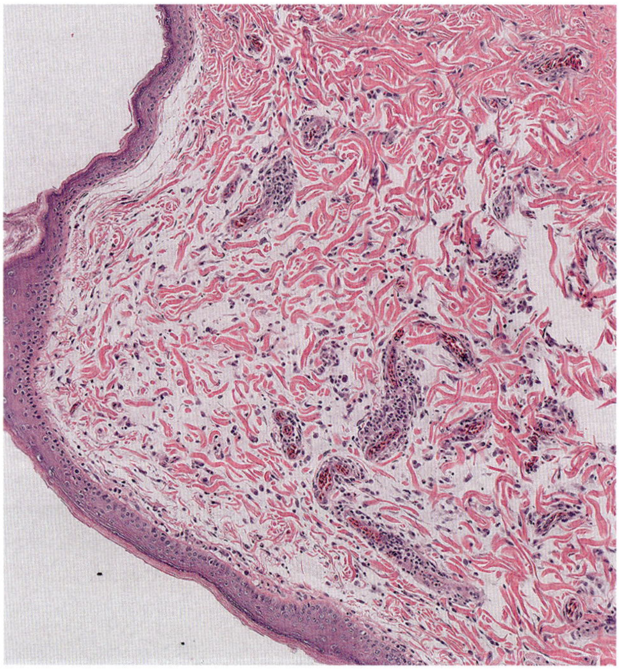

FIGURE 32.15 Lymphocytic plasmacytic perivascular dermatitis in a caique. This is a nonspecific, but common inflammatory lesion in the skin of blue-and-gold macaws and blue-fronted Amazon parrots. It can be associated with a number of different conditions including chronic feather picking.

FIGURE 32.16 Lateral beak deviation in a macaw. Deformities and malocclusion vary in severity but result in a 'scissors' effect of the beak. *(Graphic Courtesy of P. Latas.)*

Lateral beak deviation (scissors beak) is most often diagnosed in macaws[127] (Fig. 32.16). The cause is unknown and may include low or unbalanced calcium in the diet, viral diseases, trauma, improper handfeeding techniques, and alterations in the mandibular occlusal surface. Scissors beak deformities are characterized by variably severe deformation of the upper beak rhinothecal keratin and/or bones with resultant overgrowth of the opposing lower gnathotheca. This deformity is progressive, ultimately generating a "scissors" effect. The problem may be corrected by trimming the lower beak and reshaping the occlusal surface of the rhamphotheca. If that is not successful, various hardware can be surgically applied.

Constricted toe syndrome can be caused by fibers, scabs, or necrotic tissue and may result in avascular necrosis of the distal toe. In juvenile and nestling macaws (*Ara* species), African gray parrots, eclectus parrots, cockatoos, and conures, some cases develop in the absence of constricting foreign material.[150] Proposed causes include low humidity, egg-related strictures, ergot-like intoxication, or digit fractures. This deformity consists of an annular ring of constriction that eventually causes swelling and necrosis of the distal segment of the toe. The histology of amputated toes is nonspecific and can include dermal edema of the distal toe, and dermal fibroplasia with neovascularization and mixed inflammation and perivascular dermatitis sites of constriction. Toe constrictions can often be prevented by keeping susceptible species on nondesiccating surfaces and increasing brooder humidity. If lesions are noticed before necrosis is evident, surgically "cutting" the constriction may prevent necrosis of the distal toe.

Neoplastic

Age has an important influence on the likelihood of tumor development in many animals, including birds.[102] This is due to an increase in genetic damage from the action of environmental agents and errors in repair. Given the increasingly improved care and the long lives of many psittacines, there is a growing need for in-depth studies of their tumors using routine and advanced diagnostic modalities and therapies.[158] Tumor prevalence is relatively high in some species, such as budgerigars and cockatiels. This may be related to in-breeding and potential viral tumor induction, as occurs in psittacid herpesvirus papillomas. The following tumors are a few that are seen more commonly in psittacines than other avian species. See the eSlide collection for several other less common tumors (thyroid adenoma, rhabdomyosarcoma, fibrosarcoma, osteosarcoma, perivascular wall tumor, granular cell tumor, cholangiocellular carcinoma in eSlides 32.e6–32.e12, respectively).

Gastric carcinomas and **gastric adenocarcinomas** are reported in several species of birds, but in pet birds they are most common in budgerigars, gray-cheeked parakeets, lovebirds, cockatiels, various conures, and Amazon parrots.[101] The tumors are uncommon in macaws and African gray parrots and rare in cockatoos. Reported clinical signs include weight loss, weakness, inability to perch, head tilt, melena, undigested seeds in the feces, beak overgrowth, and polyuria. Common hematologic changes are anemia and hypoproteinemia. The tumors are often found at the proventricular-ventricular junction. Tumors present grossly as a thickening in the gastric wall. Tumors may contain multifocal inflammation, hemorrhage, and necrosis. Extension of the neoplasm to the serosal surface may be associated with peritonitis, fibrin deposition, and adhesion to the liver or other organs. Tumor cells may be poorly differentiated and tend to form infiltrative nests, tubules, and cords, but may be individualized. Goblet cells may be present. Moderate numbers of mitotic figures and associated desmoplasia may be seen. Metastases are uncommon but have been reported in the lung and liver.[14]

Papillomatosis is the most common neoplasm of the oral cavity and cloaca in New World psittacines.[63,134] This disease is recognized in macaws, conures, Amazon parrots, and hawk-headed parrots. Oral lesions are particularly common macaws but are relatively rare in Amazon parrots. They are most commonly located along the margins of the choanal fissure, at the base of the tongue, and on the glottis. They can occasionally grow large enough to cause obstruction of the airways. Papillomatous lesions are white to pink, raised, and focal to locally extensive. Most have a typical cauliflower-like appearance. These lesions wax and wane and smaller, slightly raised discolored lesions may be overlooked. Thickening of the choanal margin and discoloration and blunting of the choanal mucosa are other subtle changes that may be seen. In the cloaca, tumors occur near the junction with the skin but may extend into the cloaca (eSlides 32.e13 and 32.e14). Lesions may be focal or diffuse. Early lesions may have an irregular "cobblestone" pattern. Acetic acid (vinegar) is handy when clinically evaluating avian species for cloacal papillomas as topical application will blanch the papillomatous growth white. Large papillomas may prolapse and ulcerate. Chronic soiling of the surrounding skin leads to inflammation, fibrosis, and stricture of the vent.

Papillomas have a highly vascular connective tissue core overlain by proliferative mucosal epithelium. Lymphoplasmacytic inflammation may be present in adjacent connective tissue. Research links these tumors to infection with **psittacid herpesvirus (PsHV) genotypes 1, 2, and 3**.[35,63] In some species, such as macaws and Amazons, these viruses have also been associated with hyperplastic to neoplastic changes of the bile or pancreatic ducts.[35,63]

Testicular tumors are best described in budgerigars and pigeons. Reported tumors include **seminoma**, **Sertoli cell tumor**, **interstitial cell tumor (Leydig cell tumor)**, **lymphosarcoma (lymphoma)**, **undifferentiated sarcoma**, and **teratoma**. Seminomas and Sertoli cell tumors are the most common.[58] Tumors of the testicle frequently result in testicular enlargement and coelomic cavity distension. Dyspnea, lethargy, anorexia, and ascites are the common clinical signs.[103] Seminomas are tumors of immature germ cells, grossly soft, yellow-red, and cause enlargement of the testis. Histologically, neoplastic cells are somewhat pleomorphic with indistinct cytoplasmic margins and large vesicular nuclei. There is minimal stroma, and there may be remnants of tubular architecture within the tumor. There are rare reports of metastasis to the liver, spleen, and local spread of the tumor to the kidney. Sertoli cell tumors are tumors of the gonadal stroma that arise from the Sertoli (sustentacular) cells. These are generally firm, gray-white, nodular neoplasms. Cystic spaces may form within the tumor. Tumors contain numerous tubules lined by pleomorphic polyhedral cells. Neoplastic tubules are separated by abundant stroma. Some Sertoli cell tumors are reported to be functional, resulting in feminization with an associated change in cere color in budgerigars and feather changes in a Nyasa lovebird.[104] Hepatic metastasis has been reported.

Primary renal tumors of pet birds include **embryonal nephromas (nephroblastomas)**, **adenocarcinomas**, **carcinomas**, and **adenomas**. All are reported more frequently in young- to middle-aged (approximately 4 years) male budgerigars. Common clinical signs include weight loss, vomiting or regurgitation, and unilateral lameness or paresis[75]; paresis is believed to be due to compression of the root of lumbar and sacral nerve plexi as they pass through or dorsal to the kidney or from tumor growth into and adjacent to the synsacrum and ilium.[115] Skeletal muscle atrophy and osteopenia may also be seen.[115] Although it has been speculated there may be a viral etiology associated with renal tumors, this has not been proven.[125]

Most primary renal tumors are described in the cranial division of the kidneys and grow by expansion and local infiltration. Embryonal nephromas, adenocarcinomas, and carcinomas do not appear to compromise renal function. Metastasis to the liver, oviduct, lung, and subcutis have been reported but are rare. Radiographically, there may be coelomic cavity enlargement and the ventriculus may be displaced ventrally against the body wall. Grossly, **renal adenocarcinomas** and **carcinomas** appear as irregular, nodular, red-brown or gray, occasionally cystic masses that fill most of the coelomic cavity. Microscopically, they are composed of anaplastic epithelial cells forming tubules, nests, and sheets, usually with minimal stroma. **Renal adenomas** may be identified only by histologic examination as they seldom result in gross lesions. Histologically, they are comprised of tubular structures lined by fairly well-differentiated epithelial cells (eSlide 32.e15). There is usually slight to moderate stroma. **Embryonal nephromas (nephroblastomas)** are unusual tumors believed to develop from the metanephric blastema. They are typically large, unilateral, multilobular masses. Nephroblastoma in budgerigars presents clinically similar to renal adenocarcinoma, although leg paresis is not common. Histologically, they are composed of tubules, glomerulus-like structures, and sheets of undifferentiated epithelial cells (Fig. 32.17).[47,94] Some tumors have abundant stroma.

Air sac carcinomas are rare tumors that are often difficult to definitively diagnose as arising from air sacs. The few described cases have been in large psittacines, although there are scattered reports in other avian species including a captive roadrunner.[33] The initial presentation can vary and may include gross or radiographic evidence of cystic masses in soft tissues or bone. Affected birds may have a drooped wing or may be reluctant to move

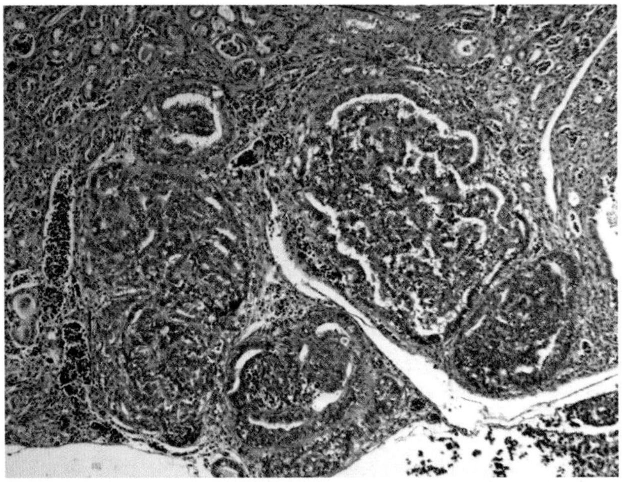

FIGURE 32.17 Nephroblastoma in a budgerigar. The tumors are composed of tubules, glomeruli, and immature mesenchyme. This image contains tubular and glomerular components.

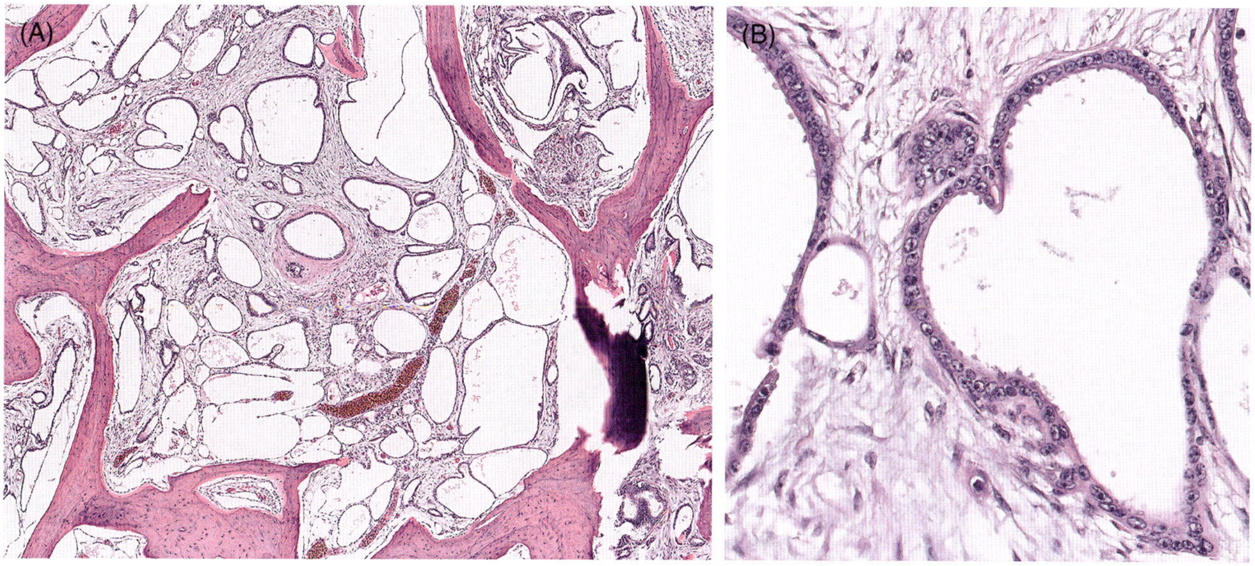

FIGURE 32.18 Air sac carcinoma in the humerus of a yellow-naped Amazon parrot. (A) Neoplastic cystic structures are present within the bone. (B) Note the apical cilia on the lining of the neoplastic epithelium.

the wing. The tumor may develop in other air sac diverticula, such as the sternum and humerus (Fig. 32.18A).[67] Tumors may be fluid filled and have large, intraluminal, friable, gray-brown, mottled polypoid growths/masses. Radiographs may demonstrate deformed and proliferative new bone with areas of osteolysis. Tumors are poorly demarcated, unencapsulated, and composed of variably sized acini, tubules, and papillary structures supported and separated by fine fibrous connective tissue. Neoplastic cells are cuboidal to low columnar and rudimentary cilia may be present on apical cell surfaces (Fig. 32.18B). Finding areas where the tumor is continuous with the air sac reflection in the bone or identifying it within the pulmonary parenchyma helps determine tumor origin. Neoplastic cells in one report of a Quaker parrot were cytokeratin positive and vimentin and calretinin negative by immunohistochemistry.[67] Both endostosis and exostosis of the involved bone may be present.

INFECTIOUS DISEASES

In addition to those described below, descriptions of several additional infectious diseases can be found in the Digital Content (Figs. 32.e1 and 32.e2; Tables 32.e1–32.e3).

DNA Viruses

Psittacine beak and feather disease (PBFD) is caused by **psittacine beak and feather disease virus (PBFDV)**, a type of circovirus. The virus is common in many species of captive and wild African and Indopacific birds. The infection poses a threat to many endangered psittacines.[25,95] In Australia, repeat spillover of beak and feather disease virus into endangered wild parrots from captive-bred released birds highlights the risk associated with endemic pathogen loss in endangered species.[23] Viral proliferation occurs in germinal basal epithelial cells in feather follicles and the rhamphotheca and results in feather and beak deformities. On histologic examination, circovirus inclusions are intracytoplasmic (can be intranuclear, but rare), 3–6 μm, basophilic, and have a botryoid appearance. They are present in feather follicular epithelium and feather pulp; they may also be seen in dermal and epidermal macrophages. PBFDV also causes necrosis and lymphoid depletion of the bursa of Fabricius and the thymus as well as circulating cells of the immune system, causing immunosuppression and predisposing to opportunistic infections; hepatic necrosis has also been reported in a few species.

PBFD in nestlings is a generalized, peracute to acute disease that affects all of the growing feathers as well as lymphoid organs. Flight feathers take longer to develop than contour feathers. If nestlings are somewhat older when infected, only the flight feathers may be affected. Nestlings typically die within 2 weeks of disease onset. The thymus and bursa of Fabricius should be examined in all young birds for the typical lesions: profound lymphoid depletion with variable necrosis and basophilic botryoid intracytoplasmic viral inclusions (Fig. 32.19A; eSlide 32.e16). The inclusions can also be found in follicular epithelium of abnormal feathers.

Chronic PBFD is generally seen in birds 8 months to 3 years of age or is first recognized when birds go through their first molt. Dystrophic feathers replace normal ones during the molt. The feathers have thickened sheaths, hemorrhage within their shafts, and pinch off at their base. Powder-down feathers may be the first affected in cockatoos, resulting in a clean, shiny beak. Cockatoos classically

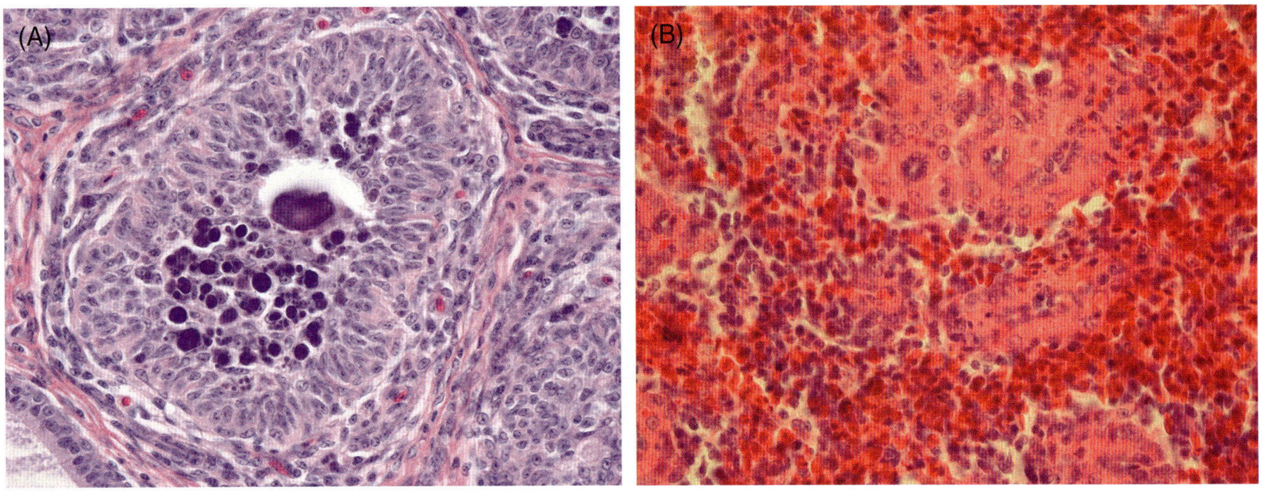

FIGURE 32.19 Circovirus infection in the bursa of Fabricius (A) and spleen (B) of an African gray parrot. (A) Typical viral inclusions are basophilic, intracytoplasmic, and are globular/botryoid. (B) Infection is often associated with lymphoid depletion (see eSlide 32.e16).

present with progressive, generalized feather dystrophy, delayed molt, and beak elongation and fracture.[109] Because of the feather damage, chronic feather picking is a differential for these changes.

Circovirus infection in lovebirds and eclectus parrots presents in one of two ways. The first is a peracute to acute infection associated with pneumonia, enteritis, feather deformities, and/or death. The second, a chronic form of infection, is characterized by progressive feather degeneration and occasionally beak deformity. Many affected lovebirds show only thin, dull plumage or have an increased number of broken or worn feathers. Advanced cases may show some feather dysplasia, or new feathers may simply not develop, leaving portions of the bird unfeathered. A significant percentage, perhaps the majority of lovebirds infected with circovirus, will have no signs of disease.

Circovirus infection in African gray parrots is characterized by significant immunosuppression; the progressive feather lesions that develop in other species are rarely seen. In addition to an absence of feather abnormalities, disease in African gray parrots is typically peracute and characterized by an acute onset of nonspecific clinical signs, severe leukopenia, and anemia. Varying degrees of hepatic coagulation-necrosis can be seen histologically.[119] Infection is associated with secondary bacterial or fungal infections resulting in sepsis, pneumonia, enteritis, and death. Severe leukopenia and nonregenerative anemia are present and most likely due to bone marrow infection that produces pancytopenia. Lymphocytolysis and lymphoid depletion are seen in the spleen and bursa of Fabricius (Fig. 32.19B; eSlide 32.e16). In adult African gray parrots, red zonal discoloration of the contour feathers may occur in those that survive earlier circoviral infection (Fig. 32.20).

FIGURE 32.20 Abnormal plumage due to circovirus infection in an African gray parrot. Zonal, red discoloration of the contour feathers is a typical finding in African gray parrots that survive infection.

PBFDV is shed in the feces, crop secretions, and feather dust. The minimum incubation period is 21–25 days with a maximum period of up to several years. Diagnosis is made by histopathology and PCR. Avian polyomavirus infection in nestling budgerigars can cause identical signs to those of PBFDV infection (see below) and can only be differentiated with PCR-based diagnostics or histopathology.

Avian polyomavirus (APV) is a nonenveloped, DNA virus in the family *Polyomaviridae*. APV primarily infects and causes disease in captive parrots; it has also been identified serologically in wild Australian cockatoos. First described in budgerigars, it was originally referred to as **budgerigar fledgling disease (BFD)**. Before circovirus was fully characterized, budgerigars with PBFDV infections were often incorrectly diagnosed as BFD given the similar clinical signs.[108] Transmission is most likely by respiratory exposure. Once birds are infected, they become viremic, and virus can be found in nearly all organs in both symptomatic and asymptomatic birds. Infected birds shed virus in droppings, oral secretions, and feather and skin dander.[88] It is speculated that vertical transmission occurs in budgerigars; however, this has not been confirmed. Many cases of APV infections are asymptomatic.

Disease and death due to APV predominantly occur in nestling parrots that are less than 14 weeks old. The most susceptible birds are budgerigars, lovebirds, macaws, conures, caiques, eclectus, and ring-necked parrots. APV disease in adult birds has been reported in eclectus parrots, caiques, and lovebirds. Coinfection with circovirus (see above) may predispose some adult birds to APV disease due to the immunosuppressive effects of circovirus infection. Disease in nestling cockatoos and Amazon and African gray parrots is relatively uncommon.

Gross lesions vary with the species affected. Nestling budgerigars 10–20 days old will be stunted and have delayed and/or abnormal feather development (Fig. 32.21A; eSlides 32.e17 and 32.e18). Additional findings can include hydropericardium, myocardial hemorrhage, perihepatic effusion and ascites, hepatomegaly, hepatic necrosis, renal swelling, petechia, and general tissue pallor. Budgerigars that survive acute infection will often fail to develop their primary wing and tail feathers, or these feathers will be dystrophic. Histologically, the nuclei of virally infected cells in budgerigars are enlarged (karyomegaly), have marginated chromatin, and contain a finely granular, basophilic to amphophilic inclusions. Inclusions occur in multiple organs and tissue types including, but not limited to, liver, spleen, kidney (mesangial and tubular cells), feather follicles, skin, esophagus, brain, and heart. Moderate, multifocal, random, hepatic necrosis is common. Inclusion bodies are often abundant in the spleen and are accompanied by necrosis of perivascular histiocytes and other phagocytic cells. Massive infection of the germinal epithelium of feather follicles is also common in budgerigar APV infections. Inflammation is generally minimal.

Nonbudgerigar nestling psittacines with APV infection typically develop normally and have substantial body fat. Gross necropsy findings in fatal APV infections are considered to be fairly classic and involve subcutaneous and subserosal hemorrhage and splenic and hepatic enlargement, both of which are typically enlarged and friable. Less commonly, ascites and pericardial effusion are present. Similar lesions are also seen in adult birds with APV disease. Histologically, some degree of hepatic necrosis and often massive necrosis is present. Liver necrosis is multifocal to coalescing and, when severe,

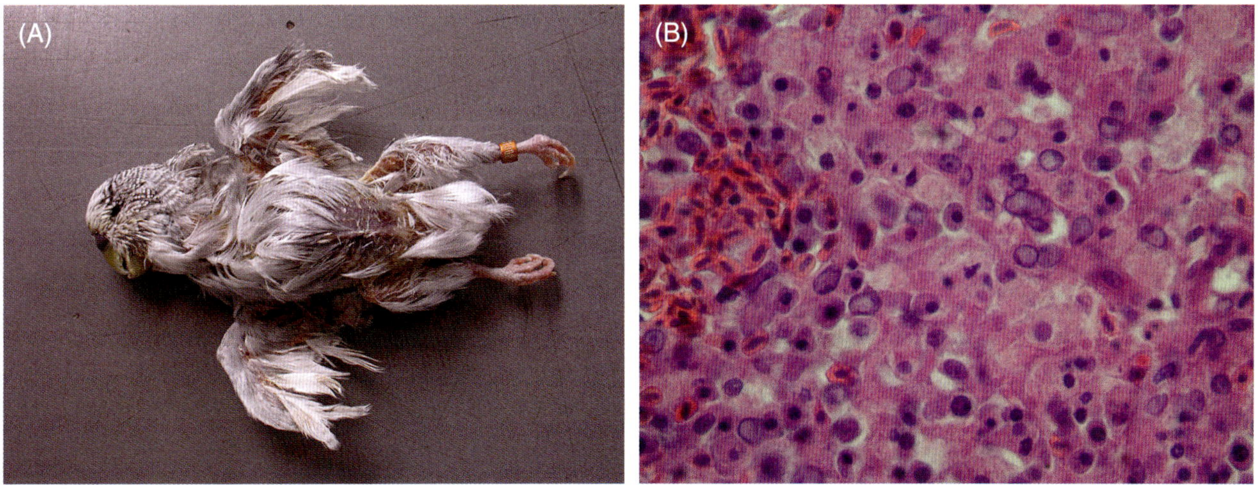

FIGURE 32.21 Polyomavirus infection in the feathers/skin of a budgerigar (A) and spleen of a lovebird (B). (A) This young bird is stunted and has diffuse abnormal feather development. (B) Infected cells are often karyomegalic and contain characteristic intranuclear inclusion bodies that are large and amphophilic to basophilic (see eSlides 32.e17 and 32.e18).

spares only the periportal hepatocytes. Characteristic inclusion bodies as described in the budgerigar may be seen in many organs, most often the kidneys, the Kupffer cells in the liver, and the spleen (Fig. 32.21B; eSlides 32.e17 and 32.e18). The degree and severity of necrosis and hemorrhage and the presence of inclusion bodies are variable and species dependent.

In infected birds, APV can be found in large concentrations in all tissues and the blood by PCR. Infection is confirmed with in situ hybridization of paraffin-embedded sections of the spleen or liver or immunofluorescent staining of impression smears.

Pacheco's disease (PD) is a peracute to acute cause of fatal hepatitis in psittacines. It may occur in an individual bird or as an explosive outbreak within a collection. The cause is **psittacid herpesvirus type 1 (PsHV-1)**,[140] an alphaherpesvirus that is closely related to the infectious laryngotracheitis virus in poultry. Based on UL16 gene sequencing, PsHV-1 has been subdivided into 4 genotypes. Amazon parrots are most commonly diagnosed with PD and have been infected with all 4 PsHV-1 genotypes. The 4 PsHV-1 genotypes have also been found in birds from the Pacific region, such as cockatiels and cockatoos; however, these species are somewhat more refractory to the disease. PsHV-1 genotypes found in African gray parrots include genotypes 2, 3, and 4, but not genotype 1. Genotype 4 is most commonly associated with Pacheco's disease in macaws and conures while genotype 3 is rarely found and genotype 1 has not been found in these species.[144,146]

Clinical signs of Pacheco's disease are often minimal to absent. If noted, they include depression, anorexia, diarrhea, tremors, and weakness. The most common history is peracute death. Necropsy specimens are typically well muscled with adequate body fat. The liver may be enlarged and friable with yellow-gray mottling and areas of hemorrhage. Histologically, acute multifocal to submassive hepatic necrosis is a consistent finding. Inflammation is generally minimal. Syncytial cell formation may occur but is uncommon. Intranuclear inclusion bodies (INIB) in hepatocytes, biliary epithelium, and other epithelia are typically numerous (Fig. 32.22; eSlide 32.e19). Inclusion bodies (Cowdry type A) are typically deeply eosinophilic but may be lightly basophilic.[117] INIB can be seen cytologically in touch preparations of the liver. Extrahepatic necrosis and INIB in the spleen and pancreas may also be seen. The crop is another commonly affected organ. Multifocal ballooning degeneration of the basal cell layer in the crop may be associated with erosions, ulcers, and rarely vesicles, possibly with inclusions. Intestinal lesions are relatively uncommon and are generally mild and often limited to crypt epithelial cells that contain nuclear inclusions. Severe, necrotizing tracheitis may also, but uncommonly, be observed.[117] The differential diagnoses based on the gross lesions and clinical signs include bacterial hepatitis, chlamydiosis, lead poisoning, and avian polyomavirus, psittacine adenovirus, and reovirus infections.[51]

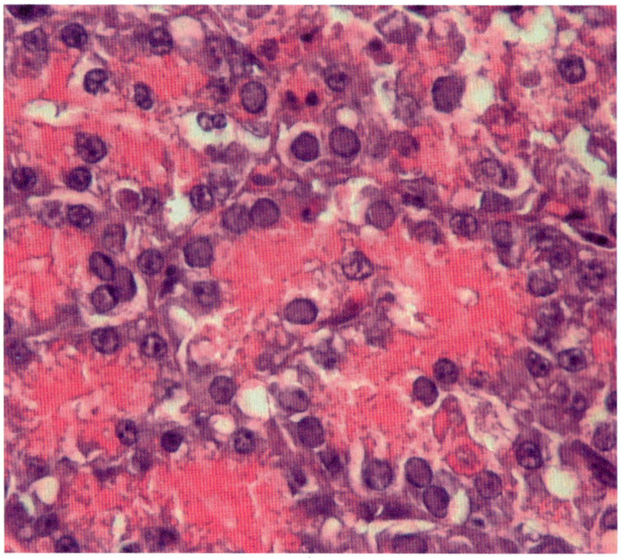

FIGURE 32.22 Pacheco's disease due to herpesvirus infection in the pancreas of a blue-fronted Amazon parrot. Viral inclusion bodies are intranuclear and eosinophilic to amphophilic (see eSlide 32.e19).

PsHV-1-associated mucosal papillomas (papillomatosis; see Neoplastic above) have been detected in New World parrots.[63,134] Not all PsHV-1 infections cause disease and some parrots are asymptomatic carriers. Latently infected parrots, generally Amazon parrots and certain species of conures and macaws, are potential sources of virus shedding (PsHV-1 has been identified in oral and cloacal mucosa at a prevalence as high as 92% in some groups).[146] Ingestion, inhalation, and conjunctival exposure have experimentally been shown to result in infection and disease. Psittacid herpesvirus, **PsHV-2**, has been identified in cutaneous papillomas in three African gray parrots,[135] and PsHV-1 genotypes 1, 2, and 3 (but not 4) have been identified in mucosal papillomas.[134]

Avian adenoviruses are classified into three genera: *Aviadenovirus*, *Atadenovirus*, and *Siadenovirus*.[7] Most of the psittacine adenoviruses that have been characterized are classified into the genus *Aviadenovirus*. The adenovirus genome is a linear, double-stranded DNA and is estimated to be 25–45 kbp in length. Adenoviruses are nonenveloped and have an icosahedral capsid with a diameter of 70 nm. The capsid is made up of two types of capsomeres: 12 vertex capsomeres composed of a fiber attachment protein and its penton base, and 240 hexons. The hexon protein is the major viral capsid protein and interacts with the immune response of the host. The hexon protein is known, therefore, to possess family-, genus-, species-, and type-specific determinants. Infections with adenovirus or adenovirus-like particles have been described in a variety of psittacine birds, including budgerigars, macaws, Amazon parrots, and cockatoos, but little is known about the genomic organization of these viruses. Those that have been recently described include **psittacine adenovirus (PsAdV)** in Senegal parrots, **psittacine adenovirus HKU1** in mealy parrots (related to duck adenovirus A [*Atadenovirus* genus])[142] and **budgerigar adenovirus type 1 (BuAdV-1)**, a siadenovirus that is associated with ruffled feathers.

Adenovirus infections in psittacine birds have been associated with depression, anorexia, diarrhea, and cloacal hemorrhage. Adenovirus outbreaks have been described as having a clinical progression similar to psittacid herpesvirus 1 (PsHV-1) infection, in which normal-appearing birds are found dead in their enclosures. Gross lesions associated with adenovirus infections in birds include hepatomegaly, splenomegaly, dilatation of the duodenum and proventriculus, swollen kidneys, and edema, congestion, and pulmonary hemorrhage. Enlarged friable livers may be hemorrhagic, pale, or mottled. Liver lesions are the most consistent gross change in infected birds. Basophilic intranuclear inclusion bodies are routinely seen in association with necrosis in the liver and spleen[51] (Figs. 32.23A, B). The diagnosis is based on the histopathology and confirmed by electron microscopy (EM) and/or PCR.

RNA Viruses

Proventricular dilatation disease (PDD; also called neuropathic gastric dilatation, myenteric ganglioneuritis, and splanchnic neuropathy) is a progressive and often fatal disease of the central (brain and spinal cord) and/or peripheral nervous systems caused by **parrot bornaviruses (PaBVs)**. To date, nine pathogenic psittacine genotypes have been identified.[82] The virus has also been isolated from clinically healthy psittacines.[65] It has been reported in many species of captive psittacine birds and has been described in other species of birds as well.[32,40,53] African gray parrots, macaws, conures, *Poicephalus* spp., and cockatoos are the most commonly affected species. It has not yet been reported in free-ranging psittacines.

Many birds with PDD develop ileus and as a result are severely emaciated with marked pectoral muscle and adipose atrophy. The primary gross lesions are flaccidity and dilatation of any portion of the gastrointestinal tract with the proventriculus, ventriculus, and crop being most commonly affected.[141] The proventriculus may be so dilated as to fill much of the left side of the celomic cavity with concomitant displacement of the ventriculus to the right and cranially (Fig. 32.24A). If the bird has been on a seed diet, the proventriculus and ventriculus will be packed with seeds. Atrophy of the muscles of the ventriculus and thinning of the proventricular wall are common. The lack of muscular grinding of seeds leads to

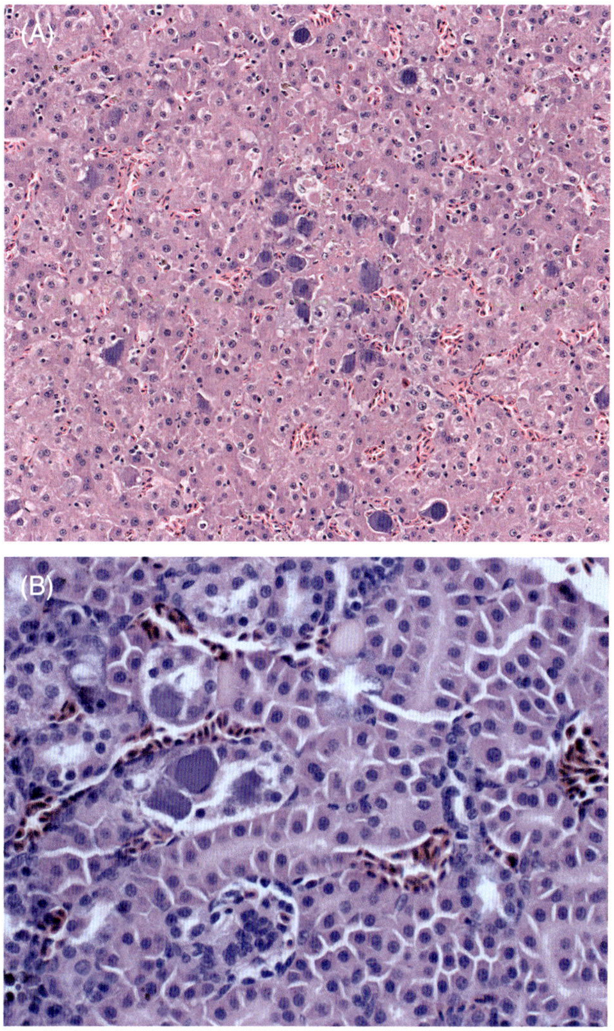

FIGURE 32.23 Adenovirus hepatitis in a Senegal parrot (A) and nephritis in a budgerigar (B). (A) Infection is associated with areas of hepatic necrosis. Large angular basophilic inclusions greatly expand the nucleus of infected cells in both the (A) liver and (B) kidney.

poor absorption of nutrients, hypoglycemia, and hemorrhagic diathesis. Some birds with neurologic lesions or blindness may present for necropsy with no gross lesions in the digestive system. It has been recently demonstrated that epicardial ganglioneuritis and myocarditis associated with PaBV infection may be linked to cardiac disease and sudden death in psittacine birds.[59]

The characteristic histologic PDD lesion is variably severe, lymphoplasmacytic inflammation in the myenteric plexus in any part of the esophagus, crop, proventriculus, or gastrointestinal tract (Fig. 32.24B; eSlides 32.e20–32.e25). The mononuclear inflammatory infiltrate may also affect the conduction fibers of the heart, medullary tissue of the adrenal glands, smooth muscle tunics, the brachial, vagus, and sciatic nerves, and/or the central nervous system. Crop biopsy is a useful antemortem diagnostic sample in

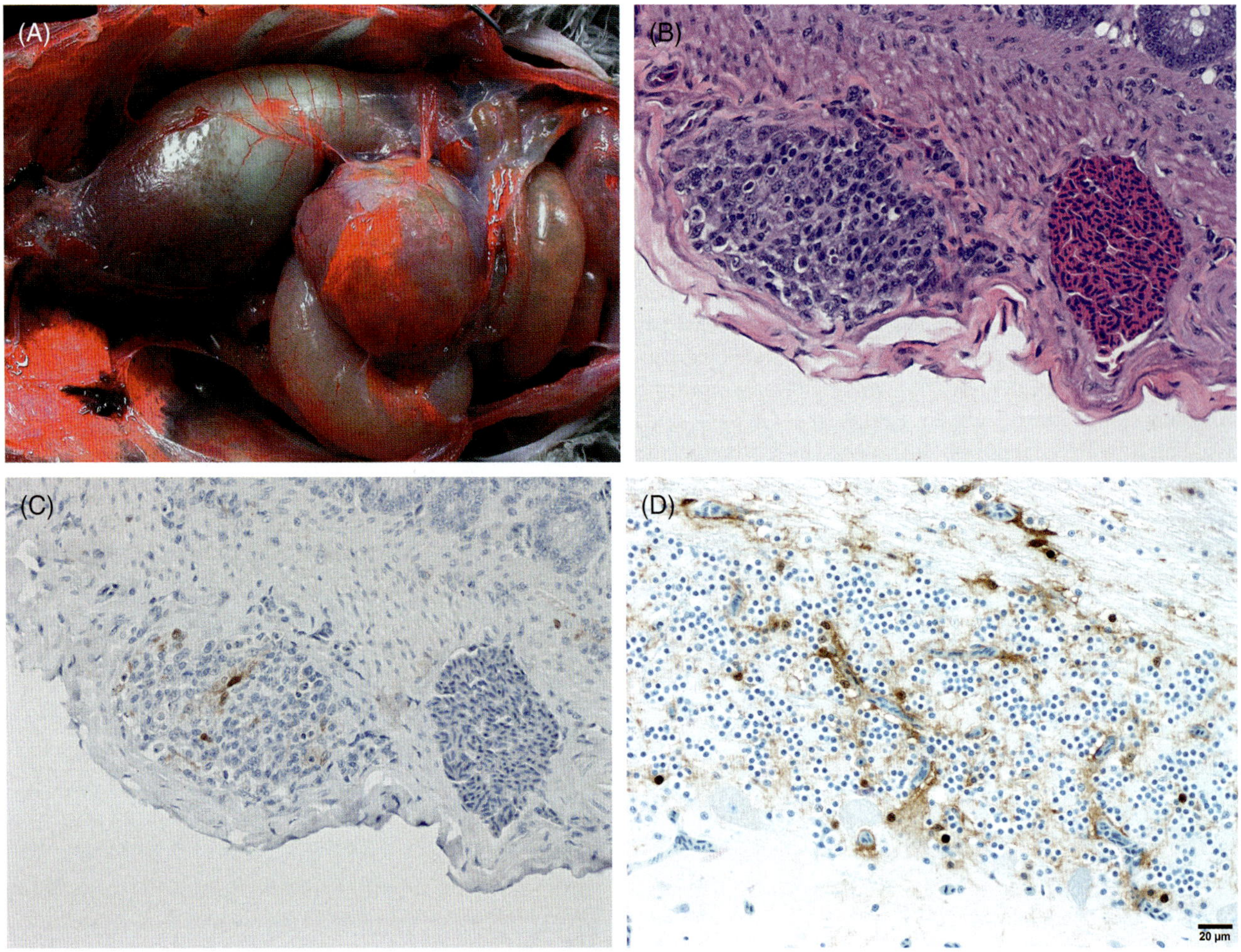

FIGURE 32.24 Proventricular dilatation disease due to parrot bornavirus infection in the proventriculus of an African gray parrot (A) and the proventriculus (B, C) and brain (D) of a cockatiel. (A) The proventriculus can become dilated and impacted and can displace the ventriculus to the right and cranially. (B) Inflammation in the gastrointestinal tract is typically lymphoplasmacytic and associated with the myenteric plexus. (C, D) Intranuclear and intracytoplasmic immunolabeling (brown in these images) for parrot bornavirus infection can confirm infection. IHC-N protein chromogen (see eSlides 32.e20–32.e25).

birds with clinical signs involving the crop; if the crop is not dilated, biopsy may not be diagnostic. Real-time PCR from various samples can be diagnostic for the presence of parrot bornaviruses. However, not all PCR PaBV-positive birds will develop clinical disease. Serological tests (ELISA) can be used to detect birds that have seroconverted after being infected with PaBV. Immunohistochemistry against PaBV-2 N protein can be used to diagnosis PaBV infections in both the peripheral (Fig. 32.24C) and central (Fig. 32.24D) nervous system.

While proventricular dilation is characteristic, it is not pathognomonic for PDD. Any disease that causes partial or complete obstruction of the intestines will result in proventricular dilation. The differentials for gastric dilation can include any number of factors, including clostridial or eastern equine encephalomyelitis (EEE) infections, gastric mycobacteriosis, fungal gastritis, lead poisoning, serositis and ulcerative or erosive gastritis, and/or peritonitis.[97] The most common cause for proventriculus/ventriculus dilation on necropsy is postmortem autolysis.

Bacteria

Salmonella are gram-negative bacilli in the Enterobacteriaceae family. Infection by members of the different species and subspecies causes a variety of diseases, notably gastroenteric and septicemic disease. ***Salmonella typhimurium*** infection can cause high mortality in young lories and African gray parrots.[114] The most frequent isolate is ***Salmonella enterica* serotype typhimurium**.[8] Some serotypes are invasive and can penetrate the mucosal barrier while noninvasive serotypes may result in a carrier state. Transmission is by ingestion of contaminated food or water or by direct contact with aerosolized bacteria from feces or feather dust. Feces from chronically infected carrier birds are one of the most common sources of infection. Diagnosis is made by fecal or affected organ culture, although antemortem diagnosis by fecal culture is complicated by intermittent shedding. Control of infection depends on strict isolation, vector control, and appropriate disinfection. Infection in birds can pose a public health risk.

The disease course of salmonellosis in birds is highly variable and based on host and organism interaction. Clinical signs include depression, lethargy, anorexia, weight loss, diarrhea, pasty vents, lameness, abscess formation, convulsions, poor hatching or excessive fledgling mortality, dehydration and crop stasis, meningitis, osteoarthritis, and sudden death. Characteristic gross lesions are hepatomegaly, splenomegaly, pneumonia, and catarrhal to hemorrhagic enteritis. Histologic lesions include multifocal random necrosis and inflammation in the gastrointestinal tract, liver, spleen, or other sites that range from heterophilic to lymphohistiocytic. In birds that develop bacterial septicemia, the number of bacteria visible in organ impressions and tissue sections is remarkably low.

Chlamydia psittaci, the cause of **chlamydophilosis**, **chlamydiosis**, **psittacosis**, or **parrot fever**, is an obligate intracellular, zoonotic bacterium.[4,137] The organism is gram-negative, contains DNA, RNA, and prokaryotic ribosomes, synthesizes its own proteins, nucleic acids, and lipids, and is susceptible to antibiotics. It has a biphasic life cycle.[153] Aerosol inhalation and ingestion are the primary routes of transmission/infection. Egg transmission has been described but is believed to be rare.[1] In some species, parent-to-young transmission via feeding is thought to be important, although contamination of nesting sites may play a larger role. The development of clinical signs postinfection ranges from 5 days to several weeks, and in cockatiels, chlamydia can be shed for 16 days after inoculation. Carriers can excrete the organism intermittently over several months. Immunity is usually short-lived in recovered cases, permitting rapid reinfection.

Clinical signs of **chlamydiosis** vary tremendously and are affected by management practices, immune competence of the bird, pathogenicity of the particular bacterial strain, and the presence of co-pathogens. Young birds are more susceptible than mature birds. Common signs include sinusitis, dyspnea, conjunctivitis, polyuria, diarrhea, lethargy, anorexia, yellow to dark green droppings, poor feathering, flaccid paralysis of the legs, chronic weight loss and emaciation, opisthotonos, tremors, or convulsive movements. Increased embryonic death, decreased fertility, and early chick death may be seen in breeding birds. Serious disease has not been noted among wild populations, possibly due to host-adapted strains.

Hepatic and splenic enlargement combined with pericarditis and airsacculitis are common findings in birds with **chlamydiosis**. Affected livers may be enlarged, discolored, and may contain numerous gray-yellow foci of necrosis. Histologically, lymphoplasmacytic and histiocytic inflammation, and multifocal to confluent necrosis can also be seen (eSlides 32.e26 and 32.e27). Giminez's staining will highlight the small, 0.5 μm, intracellular organisms, which may be found in macrophages and/or hepatocytes. They may be easier to find in cytologic rather than histologic preparations of fresh tissue (liver, spleen, or air sacs). In chronic disease, portal fibrosis and bile duct hyperplasia occur. Electron microscopy can confirm the diagnosis and will highlight the life stages (infective elementary body and replicative reticulate body) of the bacterium; PCR is an additional diagnostic option. It is important to remember that no antemortem test will certify a bird is chlamydia-free.

Mycobacteria are acid-fast positive bacilli. Several species that infect animals, including birds, have zoonotic potential. All species of birds are likely susceptible to mycobacterial infections, although it is more commonly reported in pet and zoo gray-cheeked and canary-winged parakeets and in *Pionus* species, Amazon parrots, and turacos.[128] In these birds, **avian mycobacteriosis** is most commonly caused by ***Mycobacterium avium***; ***Mycobacterium genavense*** is also reported.[83] Infection is a slowly spreading, usually chronic disease of subadult to adult birds. The main port of entry is typically the gastrointestinal tract, and the liver, spleen, and bone marrow are common sites of lesion development.[139] Other sites of infection, including the skin, can occur.

Typical gross lesions of avian mycobacteriosis include hepatosplenomegaly, regional thickening of the wall/villi of the intestines, and multiple tan to white nodules in affected sites. Granulomas and granulomatous inflammation are commonly found in the liver, intestine, spleen, lung, and air sacs (Fig. 32.25A); they are uncommonly found in the bone marrow and kidney. Microscopically, infections take on two primary forms but may contain elements of both. In the first, granulomas contain a central necrotic (caseous) region surrounded by macrophages, multinucleated giant cells, histiocytes, and plasma cells. Bacteria may be scarce in these lesions. More commonly in pet species, early hepatic lesions consist of portal and multifocal random aggregates of heterophils and small macrophages. Older lesions consist of small foci and eventually extensive sheets of bacteria-laden, epithelioid macrophages. Confirmation of infection is via a combination of histology, positive acid-fast bacterial staining (Fig. 32.25B), bacterial culture, and PCR.

A rare form of **tuberculosis** is found in pet parrots in intimate contact with people (transmission from humans to pet birds). Chronic lesions are seen in the skin, oral cavity around the tongue, and in the bones/joints with systemic involvement as the disease progresses. ***Mycobacterium tuberculosis*** is the common isolate.[118] Natural infection with ***Mycobacterium bovis*** seems to adapt to "colder" parts of the bird; experimental infection has resulted in systemic involvement.[2,60]

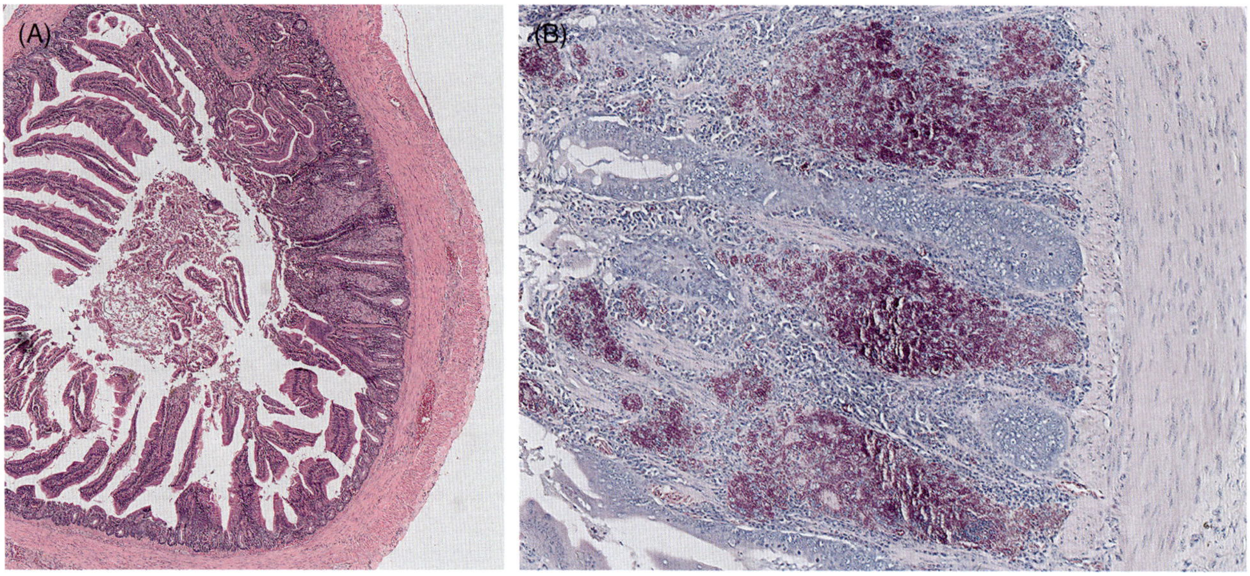

FIGURE 32.25 Mycobacterial infection in the intestine of an Amazon parrot. (A) The mucosa is segmentally severely expanded by granulomatous inflammation including macrophages and fewer multinucleated giant cells that are admixed with lymphocytes and plasma cells. (B). Myriad intralesional bacteria are highlighted with acid-fast staining. Fite's.

Yersinia pseudotuberculosis infections are rare in psittacines but do occur.[19,44,133] Infection is characterized by multifocal, miliary, caseous lesions in the liver, and other organs (Fig. 32.26); mycobacteriosis is a differential diagnosis. Large colonies of gram-negative bacilli that histologically are often associated with remarkably little host inflammation are characteristic findings. Because these bacteria can replicate at low environmental temperatures, outbreaks can occur in the winter months.

Fungi

Macrorhabdus ornithogaster, a yeast that in the past was referred to as **"megabacteria"** or **"avian gastric yeast"** is commonly found on the mucosal surface of the isthmus of the ventriculus of small psittacines, especially budgerigars, parrotlets, lovebirds, cockatiels, and rosellas; infection has also been reported in a lesser sulfur-crested cockatoo. Phylogenetic research identifies this organism as anamorphic ascomycetous yeast representing a novel genus.[145] The organisms are relatively large (2×20–$40\,\mu m$), gram- and periodic acid-Schiff (PAS)-positive bacilli that stain strongly with Calcofluor White M2R (a chitin stain). Organisms can be identified by examining cytologic or histologic preparations of samples from the ventricular isthmus. The organism is pathogenic and can be associated with chronic wasting disease; however, many infected birds do not have clinical signs or histologic lesions. Underlying stress may be necessary for the infection to become clinically significant. Concurrent bacterial infections, poor husbandry, and gastric lesions, such as gastric carcinomas that lead to elevated pH changes, can result in yeast proliferation and fecal shedding.[57] The primary gross lesion associated with macrorhabdiosis is excessive mucus production. Histologically, goblet cell hyperplasia and variable mononuclear cell inflammation are present in the mucosa of the proventriculus. Yeasts are found densely packed and stacked on the mucosal surface (Fig. 32.27A; eSlides 32.e28 and 32.e29) and often extend down into the ducts and glands. They may also extend into the ventriculus, where they can be found in the koilin layer. In live birds, multiple wet mount preparations of the feces may detect infection, but not all infected birds shed the organism. Budgerigars may have nonspecific hematologic changes including anemia, leukocytosis, heterophilia, monocytosis, lymphocytosis, basophilia, or

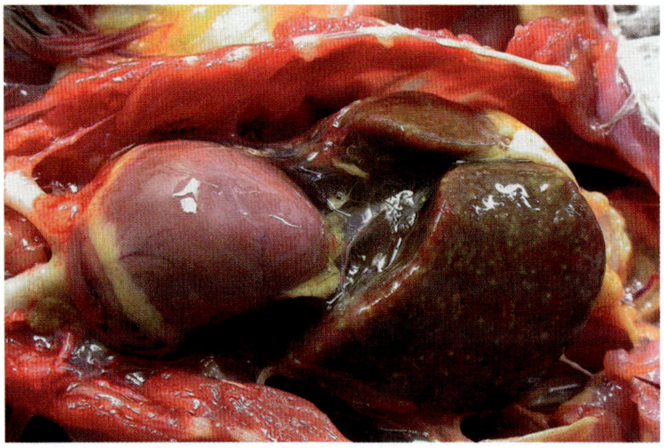

FIGURE 32.26 *Yersinia pseudotuberculosis* in the liver of a Galah. Miliary light tan foci of inflammation are typical of infection.

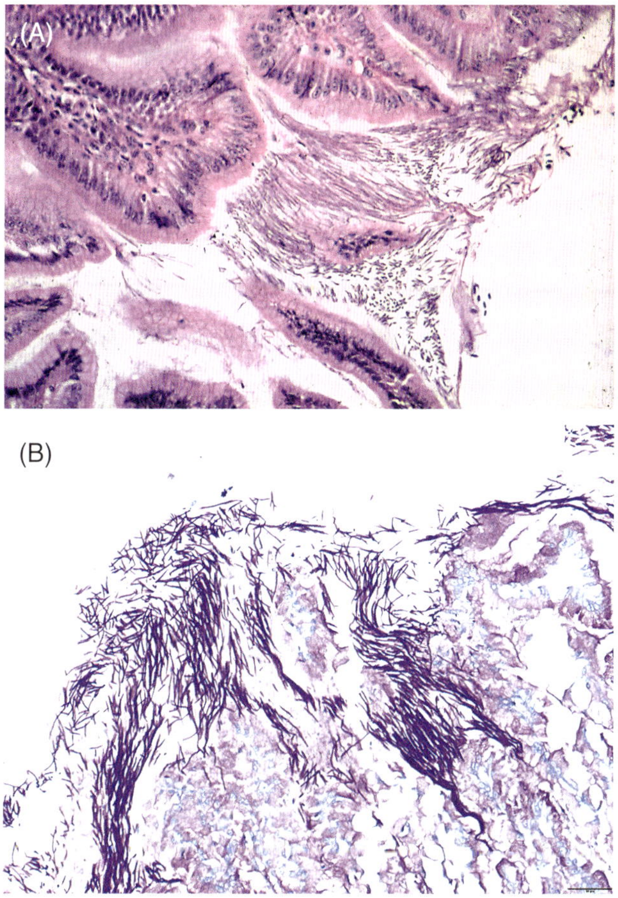

FIGURE 32.27 *Macrorhabdus ornithogaster* infection in a budgerigar. (A) Long, rod-shaped, stacked yeast are present on the mucosal surface at the proventricular-ventricular junction. This is a common location for infection. (B) The organism can be highlighted with special stains. PAS (see eSlides 32.e28 and 32.e29).

thrombocytosis. PAS can be a useful stain for histological preparations (Fig. 32.27B).

Members of the genus ***Aspergillus*** are ubiquitous, opportunistic fungi that grow readily in moist environments and on substrates, such as wood shavings, corncob bedding, and seed hulls. They are capable of causing localized and/or disseminated granulomatous infections. Captive birds are thought to be more prone to the infection than wild birds due to a number of factors including poor husbandry, malnutrition (hypovitaminosis A from a deficient diet), use of immunosuppressive drugs, and/or concurrent infections.[151] The most common route of exposure is by inhalation of the spores, and pneumonia and/or airsacculitis are the most common lesions. ***Aspergillus fumigatus***, ***A. flavus***, ***A. glaucus***, ***A. oryzae***, ***A. niger***, and ***A. nidulans*** have all been isolated from pneumonia and other lesions in birds; *A. fumigatus* is the most common isolate. Infection can be diagnosed by classic imaging or endoscopy, but antemortem diagnosis is often difficult and infection frequently does not respond even to aggressive therapy.[20]

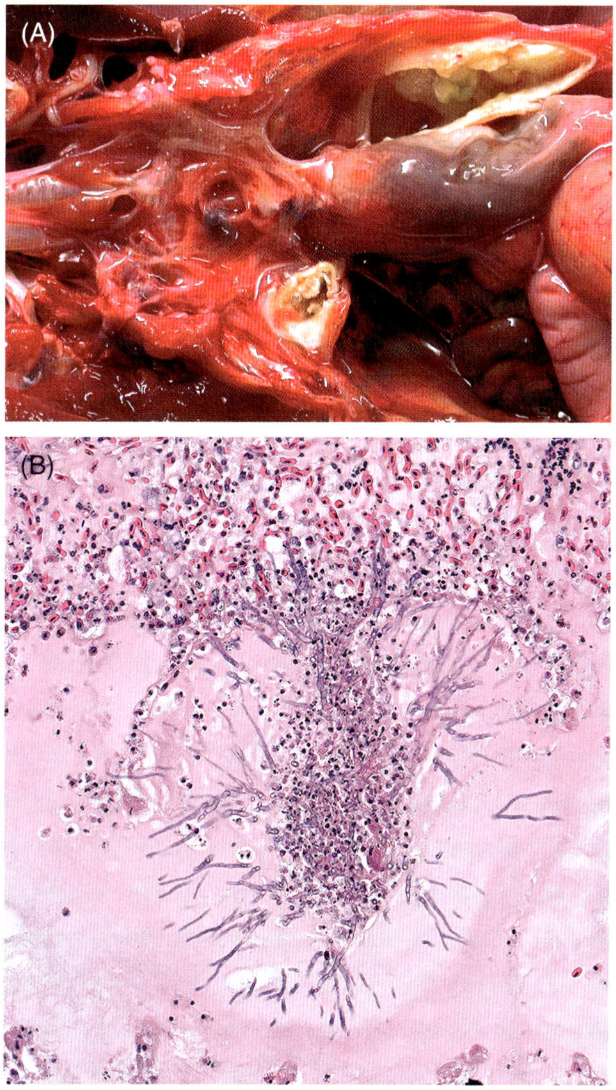

FIGURE 32.28 Aspergillosis in blue-fronted Amazon. (A) Thick yellow pseudomembranes can fill the air sacs. (B) Fungal infection is often associated with necrosis, hemorrhage, fibrin deposition, edema, variable numbers of degenerate heterophils and intralesional, parallel-walled, septate fungal hyphae.

Aspergillosis presents in several forms in pet birds. The lesions depend on the number of spores inhaled and the chronicity of infection. Colonization may be limited to the site of primary infection. The acute form results in multiple, miliary granulomas in the lungs and air sacs. Histologically, these are multifocal to coalescing foci of cellular necrosis, hemorrhage, fibrin deposition, and edema. Areas of infection contain large numbers of degenerate heterophils and variable numbers of characteristic intralesional, parallel-walled, septate, dichotomously branching fungal hyphae. The chronic pulmonary and air sac form is a slowly progressive disease. Grossly, the lesions may include white-to-yellow plaques (scutula) in the trachea, syrinx, bronchi, thickening of the air sacs (Fig. 32.28A); nodules in the lung parenchyma with caseous, consolidated, or necrotic centers; mycelial formation on air sacs or other serosal surfaces; and caseous nodules within organ parenchyma. Angioinvasion and dissemination can produce lesions in the coelomic cavity, central nervous system, liver, intestines, kidney, bone, adrenal glands, or spinal column. Central necrotic cores surrounded by degenerate inflammatory cells and a variably thick capsule of multinucleated giant cells and epithelioid macrophages characterize the granulomas. Intralesional fungal hyphae are seen histologically (Fig. 32.28B) and can also be seen cytologically in touch impressions. Conidiophores, conidia, and conidial heads (fruiting bodies) can be seen in lesions at air interfaces (e.g., nasal passages, air sacs, cavitary lesions.[90] Confirmation of the diagnosis can be made through a combination of cytology, histology, special stains, fungal culture, and/or PCR with DNA sequencing (to confirm species or strain).

Birds that develop primary aspergillosis in the syrinx will present clinically abnormal vocalizations, depression, vomiting, polyuria/polydipsia, weight loss, and recent dyspnea, especially when handled. The lumen of the trachea, syrinx, or bronchi will be partially to completely occlude with a white to cream-colored fibrinocaseous plug. Histologically, fungal mycelia in the trachea and bronchi penetrate the walls and combine with inflammatory cells to form caseogranulomatous nodules. Tracheal washes and brush cytology may be diagnostic. Mycotic rhinitis may be associated with invasion of the sinuses, blood vessels, turbinate cartilage, and nasal bones. Abnormal beak growth can be the result of this destruction.

Penicillium species and at least 11 species of the fungal class **Zygomycetes** are regarded as medically important. The class includes the order **Mucorales**, which contains the genera ***Mucor***, ***Rhizopus***, and ***Absidia***. All have been reported in birds. Infections are usually opportunistic and associated with other underlying disease. This order is unique in that hyphae are nonseptate or rarely septate and do not dichotomously branch (eSlide 32.e30). They are widespread in nature, thriving in organic material. They are also common laboratory contaminants. In psittacines, tissue hyphae are associated with pneumonia, airsacculitis, and granulomatous inflammation. Granulomas or granulomatous inflammation can efface the normal pulmonary architecture and appear similar to lesions caused by *Aspergillus* species.

Candida albicans is a normal commensal organism in the alimentary tracts of many species, including birds. Alteration of the normal bacterial flora, the integrity of the alimentary mucosa, or immunosuppression can permit overgrowth (and possible dissemination) of resident yeast. Local tissue invasion or systemic dissemination, and the formation of mycelial forms (pseudohyphae), especially on mucosal surfaces, may be seen. Factors predisposing birds to infection include prolonged antibiotic therapy, hypovitaminoses (especially hypovitaminosis A), feeding spoiled, stale, or sour foods, a stressful environment with moist floors, dirty nests, and fecal contamination, malnutrition, and coexisting bacterial or viral infections. Human error (overfilling the crop or feeding formula at an improper temperature) can predispose to crop stasis and *Candida* overgrowth in handfed psittacines. Cockatiels, parakeets, and lovebirds appear to be highly susceptible; yeast infection appears to be a primary disease in these species.[122] Candidiasis is more prevalent in young birds and may progress to a fatal systemic disease. The high sugar content of nectarivorous species (lorikeets and lories) diets may predispose these species to yeast/fungal infection if there are suboptimal hygiene practices. On exfoliative cytology, increased numbers of organisms (>1/hpf) are diagnostic. The oral, esophageal, and/or ingluvial mucosa may be thickened and covered by gray-white plaques. The crop wall is often significantly thickened and thrown into broad folds. Histologically, there may be excessive keratinization and few to numerous, 3–6 μm, oval to round yeasts with narrow-based budding with or without short chains of blastoconidia (pseudohyphae), and slender, 3–4 μm wide, septate, parallel walled hyphae are present in the mucosa and keratin (Fig. 32.29). Mucosal and submucosal inflammation is variable and primarily consists of lymphocytes, plasma cells, and macrophages.

Encephalitozoon hellem is an obligate, intracellular, single-celled microsporidial parasite. Infections have been documented in a variety of captive psittacines, primarily young birds.[15,92,112] ***E. bieneusi***, ***E. cuniculi***, and ***E. intestinalis*** have also been reported, but less commonly. *Encephalitozoon* species are zoonotic.[5,61,62] Encephalitozoonosis in birds generally involves the eye and/or more frequently the kidneys, liver, and intestines. Immunosuppression or poor husbandry, especially for lovebirds and

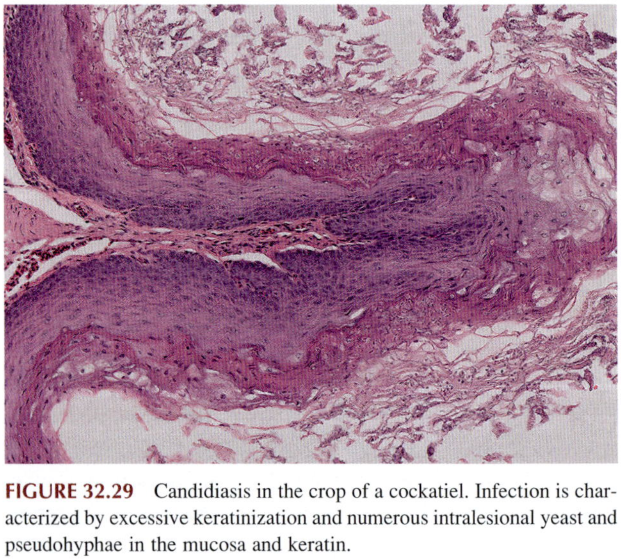

FIGURE 32.29 Candidiasis in the crop of a cockatiel. Infection is characterized by excessive keratinization and numerous intralesional yeast and pseudohyphae in the mucosa and keratin.

budgerigars, can result in clinical disease. Coinfection with circovirus in lovebirds or polyomavirus, cryptosporidia, giardia, *Macrorhabdus ornithogaster*, and *Chlamydia* spp. in budgerigars is reported. Grossly, the intestines may be distended with fluid and/or gas. Irregular pale foci may be present in the liver and kidney. Birds with ocular infections present with a keratitis and conjunctival swelling. Histologically, there is subacute hepatitis and lymphoplasmacytic interstitial nephritis. Organisms cluster in the cytoplasm of infected cells. The most commonly infected cells are renal tubular and collecting duct epithelium, hepatocytes, and intestinal epithelial cells. Infection results in cytoplasmic expansion and cell rupture. Organisms can also be found in the spleen, the lamina propria of the intestines, and the ocular conjunctiva. Ocular infections cause subacute keratitis and conjunctivitis. *E. hellum* stains poorly with hematoxylin–eosin, but they are gram-positive and can be highlighted with modified trichrome and Calcofluor White M2R stains. The diagnostic single nucleus and helical polar filaments with 5–7 turns within the spherical spores can be confirmed by electron microscopy.[126]

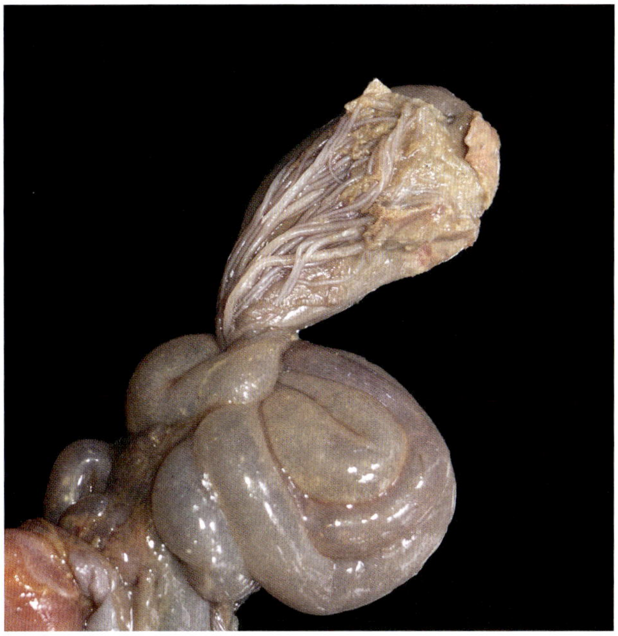

FIGURE 32.30 Intestinal ascarid infection in a blue-fronted Amazon parrot. The intestinal lumen is obstructed by numerous roundworms consistent with ascarids.

Metazoa

It seems reasonable to assume that all of the members of the species discussed in this chapter, in the wild or under managed care, can be infected by nematodes. This reportedly occurs more often when they are housed with access to the ground. In parrots, seven **ascarid** species have been identified in wild birds and in birds in captivity. Species specific to parrots include ***Ascaridia hermaphrodita***, ***A. sergiomeirai***, ***A. ornata***, ***A. nicobarensis***, ***A. sprenti***, and ***A. platyceri***.[74] Two species, ***A. galli*** and ***A. columbae***, which are parasites of gallinaceous and columbiform birds, respectively, have also been described in parrots.[50] All species of ascarids have a direct life cycle involve fecal shedding of eggs into the environment and ingestion of infected materials. Signs of infection include anorexia, diarrhea, stunted growth, listlessness, behavioral changes, enteritis, and/or death. Heavy infections can cause partial or total obstruction of the gastrointestinal tract. Antemortem infection is diagnosed by fecal examination. At postmortem examination, single to numerous juvenile and adult worms can be found in the lumen of the intestinal tract (Fig. 32.30). Once in the intestine, *Ascaridia* may migrate to the liver through the portal vein and trigger inflammatory lesions in the hepatic parenchyma (Fig. 32.31).

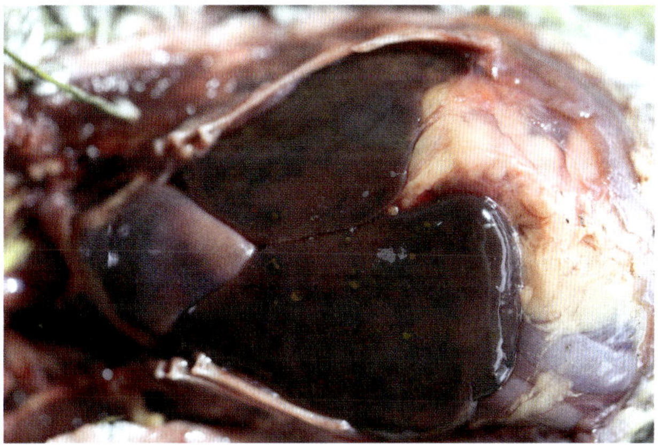

FIGURE 32.31 Hepatic ascarid infection in a blue-fronted Amazon Parrot. Multifocal dark red and tan foci represent parasite migration.

Species of ***Capillaria*** are tiny (10–15 mm long), threadlike nematodes infect the gastrointestinal tract. They have a direct life cycle; embryonation requires approximately 2 weeks, and eggs can remain infectious in the environment for several months. Infections appear to be more common in canaries, pigeons, and gallinaceous birds than in psittacines, and have been reported in macaws, cockatoos, budgerigars, Amazon and regent parrots, sun conures, and *Neophema* species. Clinical signs include depression, dysphagia, regurgitation, diarrhea, melena, anemia, and weight loss. *Capillaria* that infect the crop, esophagus, and oral cavity burrow into the mucosa, creating tracts that may fill with blood, producing hyperemic streaks. Luminal hemorrhage can be seen in the upper intestinal tract in heavily parasitized animals. Diphtheritic lesions can develop in the mouth, pharynx, esophagus, and crop. In psittacines, the duodenum is the most common site for infection and mild to severe enteritis with hemorrhage can be seen.

Histologically, sections of larvae and adult worms with eggs can be seen between the villi and in the mucosal lamina propria (Fig. 32.32). Inflammation is mixed and primarily heterophilic and mononuclear. Scrapings of suspect lesions or fecal flotation can be used to detect the characteristic bipolar eggs. Microscopic wet mounts may be necessary to see the adults.

Cestodes are very rarely seen in psittacines because they are granivorous, frugivorous, or nectarivorous, which reduces the risk of contacting the intermediate hosts (mostly insects). Infections are uncommon in birds that do not have access to the ground. It is for this reason that you can expect cestode infections in Cuculiformes that feed mainly on insects.

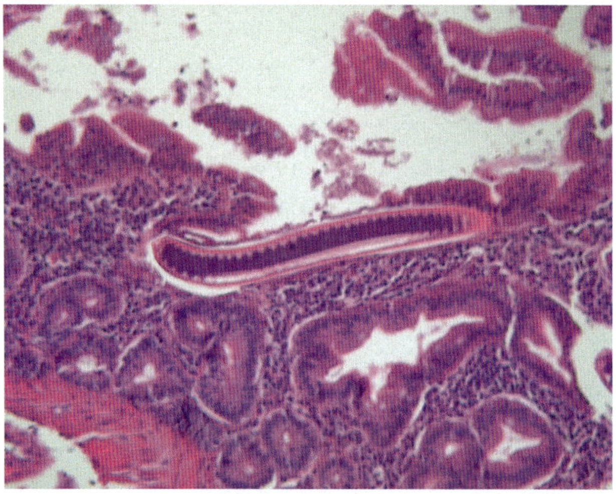

FIGURE 32.32 Intestinal capillariasis in a Fischer's lovebird. A longitudinal section of the parasite is present in the superficial mucosa of this bird.

Protozoa

Sarcocystis* spp**. are apicomplexan, protozoa that have an obligate two-host life cycle. At least six species of *Sarcocystis* infect birds. ***Sarcocystis falcatula appears to be the most significant pathogen in susceptible psittacines. The North American opossum is the definitive host and cowbirds and grackles are the intermediate hosts. Asexual reproduction of the protozoa occurs within the intermediate host's endothelial cells, which results in endothelial damage and serious or fatal disease in aberrant hosts. In natural hosts, schizogony is followed by the formation of sarcocysts in muscles. American or neotropical (Mexico, South, and Central America) psittacines are usually resistant to disease as adults, but clinical disease and significant clinical disease and death can occur. In contrast, Old World psittacines from Australia, Asia, and Africa are highly susceptible to fatal respiratory disease, as are mousebirds kept in endemic areas within the Americas.[117] The primary gross lesion and cause of death in susceptible species are severe pulmonary congestion, edema, or hemorrhage. Histologically, the lungs are congested and contain fibrin deposition, edema, and hemorrhage. Lymphocytes and plasma cells accumulate around blood vessels and bronchi. Multiple aggregates or clusters of small elliptical or crescent-shaped protozoal merozoites are present throughout the pulmonary vessels (Fig. 32.33A; eSlide 32.e31). Clusters of

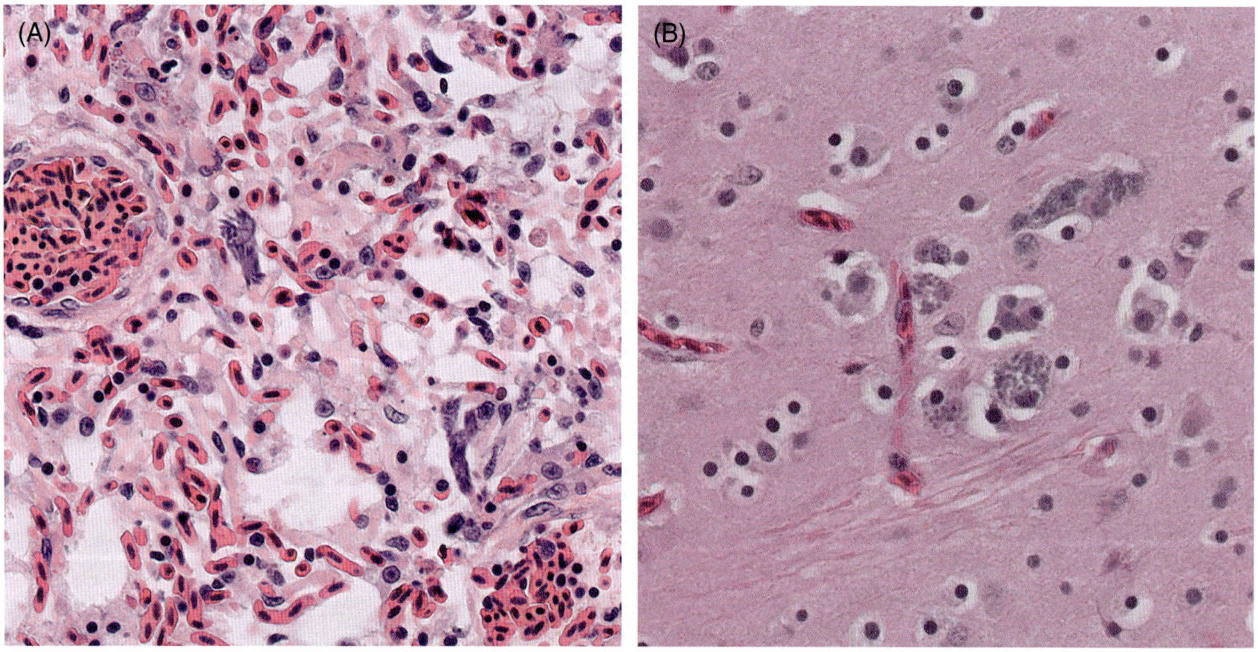

FIGURE 32.33 Sarcocystosis in the lung of an African gray parrot (A) and brain of a lory (B). (A) Multiple aggregates and clusters of small elliptical or crescent-shaped organisms compatible with protozoal merozoites can be seen in pulmonary vessels. (B) Intracellular zoites in the infection in this bird were not associated with inflammation (see eSlide 32.e31).

intraendothelial cell zoites form long and sinuous structures that may resemble microfilaria. The merozoites do not stain well with Brown & Brenn, PAS, or Giemsa stains.[117]

Sarcocystosis can also be seen in muscle and nervous tissue. In the muscular form, birds initially present with severe lethargy and weakness. They also have elevations of muscle enzymes (creatine phosphokinase, aspartate aminotransferase) and beta and gamma globulins. Protozoal cysts may be found in the skeletal muscle and be associated with lymphoplasmacytic myositis.[152] These birds may be positive with immunofluorescent serologic testing.[21] In the neurologic form, birds present with severe ataxia and uncoordinated head movements. Intralesional, intracellular zoites and schizonts may be seen in the absence of inflammation (Fig. 32.33B) or be associated with lymphoplasmacytic meningoencephalitis.[123,152] ***Sarcocystis*** causes encephalitis in pigeons and has also been reported as a cause of encephalitis in a cockatiel, rose-breasted cockatoo, long-billed corella, and a princess parrot.[79,106] Histologic lesions include lymphoplasmacytic and histiocytic encephalitis with mixed histiocytic, lymphocytic, and plasmacytic inflammation. Parenchymal gliosis, edema, and hypertrophy of small vessels associated with immature to mature protozoa can also be seen.

Spironucleus is a genus of flagellated, enteric, diplomonad protozoa that inhabit the gastrointestinal tract. They have been identified in a variety of small, free-ranging, and captive psittacines. ***Spironucleus meleagridis*** (formerly *Hexamita meleagridis*) has been identified in cockatiels by PCR. Clinical signs are nonspecific and include lethargy, weight loss, diarrhea, melena, poor feathering, and death in affected birds. Concurrent infections (e.g., *Macrorhabdus ornithogaster*) and age (cockatiels less than 6 months) are associated with more significant disease.[98] Diagnosis can be made by identifying motile trophozoites and cysts in a saline wet mount of fresh feces, intestinal luminal contents, or intestinal mucosal scrapings. It can be misidentified as *Giardia*. The two flagellates can be differentiated by the "falling leaf" motion of *Giardia* spp. trophozoites as compared to the forward motion of *Spironucleus* spp. trophozoites in wet mount preparations. Grossly, birds may have general loss of body condition and feces pasted around the vent. The intestines are variably reddened and distended. Thin, pyriform protozoa that are approximately 5–12 μm in length and have internal, eccentrically located, dense basophilic structures, and multiple flagella are present along the mucosal epithelium and between the villi histologically (Fig. 32.34).[64] The organisms are associated with variably severe, mixed lymphocytic, plasmacytic, histiocytic, and heterophilic inflammation in affected mucosa, though some cases have little to no associated inflammation.

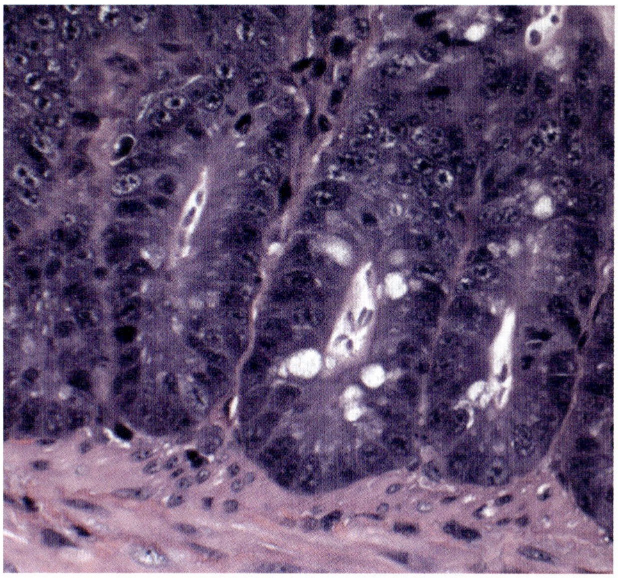

FIGURE 32.34 *Spironucleus meleagridis* infection in a cockatiel. The thin, pyriform flagellated protozoa (flagella are often not apparent in histologic sections) aggregate along the surface and within glands of the intestinal mucosa.

Trichomonas gallinae is common in many avian species and sporadic infections occur in budgerigars, especially in juvenile birds.[159] Infection has also been described in a juvenile military macaw and nestling African gray parrots. Common clinical signs include respiratory symptoms, regurgitation, nasal discharge, and emaciation. Antemortem diagnosis can be made in wet mounts from a crop swab. The most common gross lesion is thickening of the crop wall. In some cases, aspiration of crop contents causes pneumonia. Histologically, there is necrosis and ulceration of the crop wall. The flagellates can be identified, even when the bird is not very fresh, in crop smears stained with Hemacolor or other rapid stains (e.g., Romanowsky stain variants).

Ectoparasites

Scaly face and **Scaly leg** disease are relatively common diseases of caged psittacine birds, especially budgerigars, lovebirds, and cockatiels. In psittacines, these conditions are caused by ***Knemidocoptes pilae***. These mites burrow into the skin and feed on epidermal keratin in the stratum corneum. The mites induce marked hyperkeratosis and swelling of the featherless regions of the skin (Fig. 32.35), especially on the face, beak, and legs.

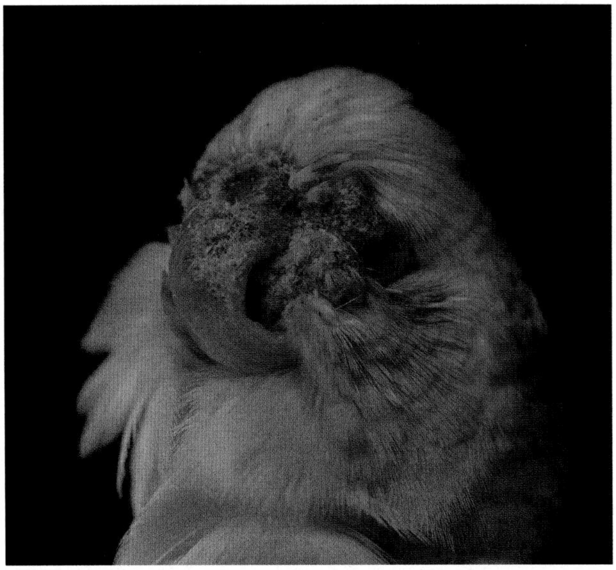

FIGURE 32.35 *Knemidokoptes* infection in the beak of a budgerigar. *Knemidocoptes pilae* infection causes severe proliferative hyperkeratosis.

Additional Resources

Abdul-Aziz T, Fletcher OJ, Barnes JH editors: Avian Histopathology, 4th edition. 2016. Madison, WI, AAAP.

McVey SD, Kennedy M, Chengappa, Wilkes R, editors: Veterinary Microbiology, 4th edition. 2002. Hoboken, NJ, Wiley-Blackwell.

Schmidt RE, Reavill DR, Phalen DN: Pathology of Pet and Aviary Birds 2nd edition. 2015. Ames, Iowa, John Wiley & Sons.

References

Full references for this chapter can be found online at https://doi.org/10.1016/B978-0-443-18437-6.00012-2.

Chapter 33

Passeriformes, Caprimulgiformes, Coraciiformes, Piciformes, Bucerotiformes, and Apodiformes

Michael M. Garner[a], Carles Juan-Sallés[b], Rachel E. Burns[c], John Trupkiewicz[a,d], and Denise McAloose[e]

[a]*Northwest ZooPath, Monroe, WA, United States;* [b]*Noah's Path, Elche, Alicante, Spain;* [c]*San Diego Zoo Wildlife Alliance, San Diego, CA, United States;* [d]*School of Medicine, Johns Hopkins University, Baltimore, MD, United States;* [e]*Pathology Department, Zoological Health Program, Wildlife Conservation Society, Bronx, NY, United States*

INTRODUCTION

The Aves Group includes more than half of all tetrapods on the planet. They are diverse in size and morphology, and inhabit a wide range of habitats and ecological niches. The Orders in this chapter include some of the most numerous and most endangered bird species. In the wild, conservation threats include habitat loss, anthropogenic factors, and introduced diseases. Under managed care in zoos or as pets, husbandry-related issues, breeding situations, and infectious diseases are substantial challenges. The focus of this chapter is on several of the most common or important diseases in these species and settings.

The largest and most diverse Order of birds among those in this chapter is Passeriformes, the perching birds. More than half of all bird species are classified in this Order. Its many Suborders and Infraorders include (but are not limited to) babblers and pseudobabblers, bellbirds, birds-of-paradise, bowerbirds, buntings, chickadees, cotingas, crows, finches, flycatchers, grass finches, honeycreepers, jays, kinglets, larks, lyrebirds, bulbuls, manakins, orioles, ovenbirds, oxpeckers, pittas, robins, shrikes, sparrows, tanagers, tits, vireos, warblers, wattlebirds, waxwings, weavers, white eyes, whydahs, woodcreepers, and wrens.

The Order Apodiformes consists of three families: Trochilidae (hummingbirds), Apodidae (swifts), and Hemiprocnidae (treeswifts). Caprimulgiformes (oilbirds, owlet-nightjars, frogmouths, potoos, nighthawks, and night jars) are stocky birds with exceptionally large and wide mouths. The mouth may be fringed with fine setae, which assist in capturing insects for food. Coraciiformes (bee-eaters, kingfishers, motmots, rollers, and todies) are a somewhat diverse group with syndactyly and partial fusion of the three anterior toes. Piciformes (aracaris, barbets, honeyguides, jacamars, nunbirds, nunlets, piculets, puffbirds, toucans, woodpeckers, and wrynecks) are notable for their exceptionally large bills, which are used for feeding on fruits and seeds. Bucerotiformes are the hornbills, birds of variable size with a characteristic casque modification of the rhamphotheca. These birds are native to subtropical Africa, Asia, and Melanesia.

NON-INFECTIOUS DISEASES

Nutritional

Iron is a biologically essential element in many physiologic processes, notably as a component of hemoglobin. It is absorbed in the intestinal tract and stored as hemosiderin or ferritin in the cytoplasm of hepatocytes and Kupffer cells in the liver, reticuloendothelial cells in the spleen, siderophages, and in the bone marrow. It is transported via transferrins and within the body is recycled and conserved. Iron bioavailability and absorption from the gut are facilitated by dietary vitamin C and reduced by dietary tannins, phytates, and other plant chemicals. **Iron overload** occurs when the rate of iron uptake exceeds use, excretion, or both. It can be due to heritable conditions, diet, or disease (e.g., infectious disease, toxins, diseases that compromise mucosal integrity and allow increased iron uptake). In zoo settings, iron overload is most commonly related to diet. In the absence of observable biochemical, cellular, or tissue damage, this condition is called **hemosiderosis**, while **hemochromatosis (iron storage disease)** indicates functional or morphologic evidence of tissue damage.[33,34,96,116] The diagnosis can be made using a combination of tests including clinical chemistry and liver biopsy in live birds and necropsy and tissue iron levels in dead birds. For the latter, comparison of results to normal levels may not always be possible as reference intervals are not available for all species, and caution is advised as seasonal variation in hepatic iron levels is seen

Pathology of Wildlife and Zoo Animals. https://doi.org/10.1016/B978-0-443-18437-6.00037-7

in some species such as common starlings.[62,121] Histological review of the liver, spleen, kidneys, and other target organs is used to make the diagnosis.

Hemosiderosis is common in toucans, mynahs, hornbills, aracaris, Micronesian kingfishers, birds-of-paradise, tanagers, jays, and hummingbirds.[96,116] These and other species that predominantly eat fruits or insects in the wild are at increased risk when they receive or have access to iron-rich diets or supplements. This can be a particular challenge in mixed species exhibits with multiple food options. In addition, fruits that are high in vitamin C promote intestinal absorption of iron and can cause overload in the absence of elevated dietary iron. The primary site of hemosiderin accumulation is typically the cytoplasm of hepatocytes and Kupffer cells in the liver. In mild cases, birds may be clinically normal and the liver may be grossly normal. In more severe cases, the liver may be enlarged and discolored dark brown, orange, or bronze (Fig. 33.1A; eSlide 33.e1). Histologically, hemosiderin appears as granular to globular golden to brown aggregates with hematoxylin and eosin (HE) staining (Fig. 33.1B) and is blue to blue-green with Prussian blue (also known as Perl's) reaction (Fig. 33.1C; eSlide 33.e2), which can be used to differentiate iron pigment from bile, lipofuscin, copper, or other deposited materials.[23,25,33,34] Ferritin is not apparent in HE or Prussian blue stained sections.

Characteristic lesions of **hemochromatosis** include cytoplasmic deposition of iron in hepatocytes and Kupffer cells in association with acute to chronic hepatocellular degeneration or necrosis, disruption of hepatic cords, periportal fibrosis, biliary or nodular hyperplasia, and cirrhosis (Fig. 33.1D). Clinical evidence of hepatic dysfunction or failure may occur in severe or prolonged cases. Nodular hepatocellular regeneration and transformation to hepatocellular or biliary adenocarcinoma has been reported to occur in mynahs[63] but is also seen in other species. Iron deposition in other sites, such as the intestinal lamina propria, myocardium, spleen, pancreas, or kidneys, is typically not associated with cellular or architectural changes.

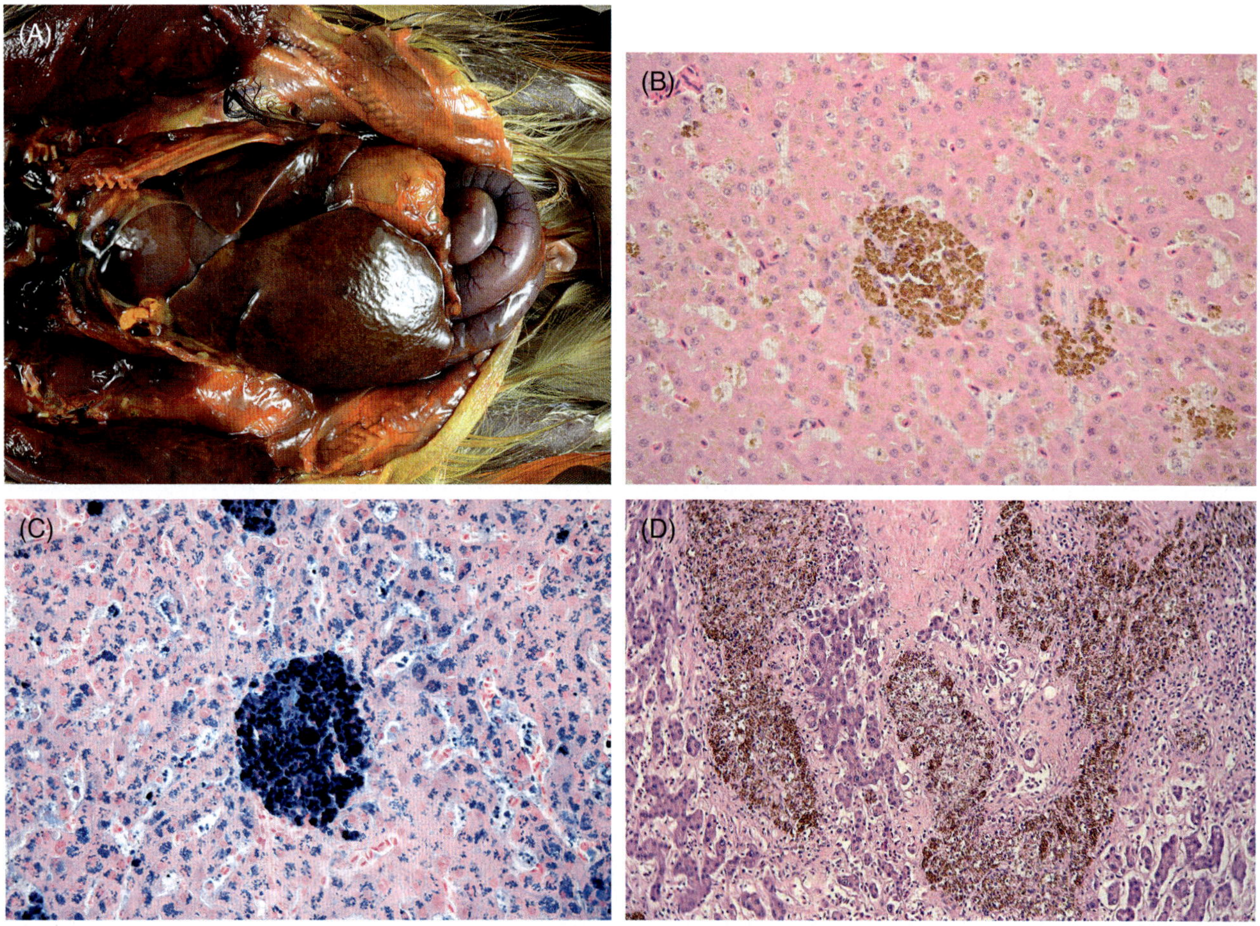

FIGURE 33.1 Hemosiderosis in a bird-of-paradise (A) and Micronesian kingfisher (B, C), and hemochromatosis in a bird-of-paradise (D) due to iron overload. (A) The liver can appear light brown/orange (as in this image), dark brown, or bronze due to iron accumulation. (B) Hepatocytes and clusters of macrophages and sinusoidal Kupffer cells contain moderate, intracytoplasmic deposits of granular brown pigment in cases of hemosiderosis. (C) Intracellular iron stains blue. Prussian blue. (D) Characteristic features of hemochromatosis, in addition to intracellular iron, include hepatocellular degeneration, drop-out, and replacement by fibrosis (see eSlides 33.e1 and 33.e2).

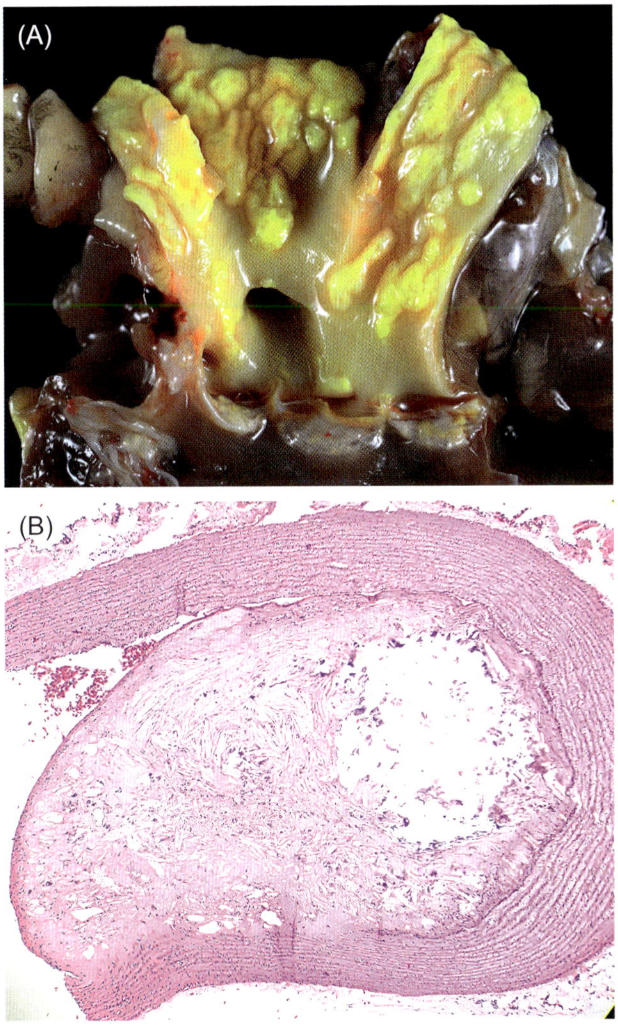

FIGURE 33.2 Atherosclerosis in the great vessels of a great hornbill (A) and motmot (B). (A) The vascular walls are markedly expanded by thick, yellowish tissue that protrudes from the intimal surface into the lumen. (B) Characteristic changes include expansion of the vascular wall by lipid-laden macrophages with intralesional cholesterol clefts.

Atherosclerosis is the accumulation of lipid and lipid-laden macrophages in the intima and media of blood vessels. In advanced cases, mineralization and ossification of the vascular wall may ensue and produce partial to complete occlusion of vessel lumina (Fig. 33.2A). Dissection with aneurysmal dilation and rupture can occur in advanced cases. Characteristic histologic findings include expansion of the intima and media of vessels by foamy macrophages and acicular cholesterol clefts, and associated inflammation and/or fibrosis may also be seen (Fig. 33.2B). Avian atherosclerosis has been linked to high-fat diets, aging, and confinement or lack of exercise. Affected birds may be hypercholesterolemic. Atherosclerosis is especially common in finches, hornbills, tawny frogmouths, and rhamphastids. In one retrospective study of necropsy specimens, Coraciiformes had the highest prevalence of atherosclerosis.[65] The most commonly affected sites are the great vessels, aortic valves, myocardial arteries, spleen, and glomeruli. Lesions may also be seen less commonly in vessels of the brain or skin. Secondary cardiac changes may include myocardial fibrosis, dilated or hypertrophic cardiomyopathy, and endocardiosis. Birds that die with atherosclerosis are often found dead unexpectedly or develop signs of cardiopulmonary or central nervous system disease. The cause for the central nervous system signs in these birds is not always apparent histologically, but it may be related to hypoxia or cephalic atheromatous emboli. Those with cardiac insufficiency may have ascites. In birds that die unexpectedly, histologic evidence of pulmonary hemorrhage is a common finding, although the source may be difficult to find.

Metabolic

Gout, or the deposition of uric acid precipitates, is common in birds. The pathogenesis is complex but generally related to dehydration, inflammation, or any renal insult that causes renal tubular obstruction or necrosis (see Chapter 31: Galliformes and Columbiformes). Deposition sites include visceral serosal surfaces and the pericardium (visceral gout; Fig. 33.3A), renal tubules, glomeruli and the renal interstitium (renal gout; Fig. 33.3B), and articular surfaces and periarticular tissues (articular gout; Fig. 33.3C). The different forms of gout may be present concurrently and the cause for their development may be related or independent. Deposits appear grossly as soft to gritty white aggregates or plaques that are (generally) soluble in formalin and may appear histologically as acicular or stellate clefts of dissolved urate crystals (urate tophi) surrounded by foci of heterophilic to granulomatous inflammation, or fibrillar, acellular, slightly basophilic material along serosal surfaces of the viscera. Acute renal gout is most commonly associated with dehydration or exposure to toxins or nephrotoxic drugs, and appears histologically as tubular dilation with urate stasis and urate tophi within tubules. It is occasionally associated with tubular rupture and mild mixed-cell interstitial inflammation. Renal gout is not necessarily a primary renal lesion and can develop secondary to renal insufficiency or be associated with other disease processes such as parasitism, age-related degenerative disease (e.g., nephrosclerosis), or other deposition disorders such as mineralization or amyloidosis.

Multiple cases of **systemic mineralization** have been seen in adult female ‘alala under managed care (unpublished data, San Diego Zoo Wildlife Alliance). Metastatic mineralization is presumed based on affected tissues, most commonly the lung, heart, kidney, and proventriculus, but serum calcium and phosphorus levels have generally not been available for confirmatory testing. Sensitivity to increased dietary calcium and/or vitamin D provided during active egg laying is suspected, but underlying renal disease and other causes are also possible. Affected birds

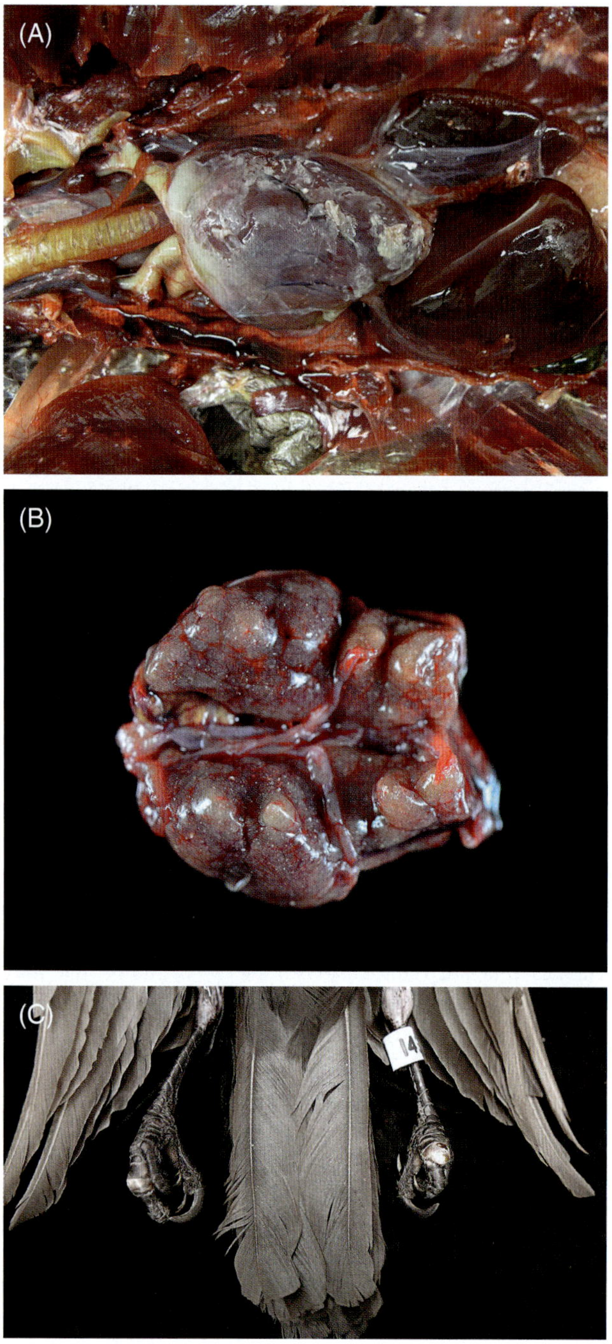

FIGURE 33.3 Visceral gout in a red-knobbed hornbill (A) and renal (B) and articular (C) gout in an ‘alala. (A) Uric acid precipitates are deposited as whitish plaques on the surfaces of organs such as the heart, (B) can cause pinpoint white speckling throughout the renal parenchyma, and (C) can form nodular periarticular thickenings that may be associated, as in this bird, with light discoloration of the overlying skin.

commonly have concurrent renal gout of varying severity and duration. Grossly, the mineralization can appear as pale tan streaking in the myocardium, areas of pallor in the lungs, and pallor and white stippling of the kidneys. Prominent mineralization in the pulmonary vasculature and renal tubular basement membranes, cardiomyocytes, and proventricular glands is seen histologically. Metastatic mineralization is a common histologic finding in birds with underlying severe acute or chronic renal disease.

Hepatic lipidosis is the accumulation of excessive lipid (fatty change) in the cytoplasm of hepatocytes. Lipidosis can be associated with obesity, inanition, metabolic derangements, exposure to hepatotoxins, and active reproduction (females). It has also been described in association with hepatic fracture and hemocoelom, and over time can progress to cirrhosis. In passerines, hepatic lipidosis is generally associated with obesity and appears to be common in blue-crowned motmots, honeyeaters, and tawny frogmouths but sporadic in many other species. Lipidosis may also be seen in the spleen and lamina propria of the intestinal tract in some birds with severe hepatic lipidosis.

Diabetes mellitus has been described in toco and keel-billed toucans.[116] Clinical signs include weight loss, polyuria, polydipsia, polyphagia, hyperglycemia, and glucosuria. Reported histologic lesions include pancreatic islet cell hyperplasia, islet cell eosinophilia, and iron accumulation in the acinar and ductal cells of the pancreas. However, there are often no histologic changes associated with the clinical condition. Although toucans can develop reactive systemic amyloidosis, primary islet amyloid deposition (islet amyloid polypeptide) that is seen in several other species, is not a feature of diabetes mellitus development in these birds. Glucagon appears to function as the primary glucose regulatory hormone in birds.

Toxic

Rodenticides typically do not target specific species and the availability of these products can place wild and captive nontarget species at risk for primary (direct ingestion) or secondary (ingestion of poisoned prey) intoxication. Second-generation anticoagulants, such as **brodifacoum** and **difethialone**, pose a great risk to birds, while first-generation anticoagulants (e.g., **warfarin** and **diphacinone**), while still toxic, pose less risk. Anticoagulant rodenticides inhibit vitamin K_1 2,3-epoxide reductase, which interferes with the production of vitamin K and the clotting factors II, VII, IX, and X. The increased potency and significant effects of long-acting anticoagulant rodenticides are attributed to their: (1) greater affinity for vitamin K_1 2,3-epoxide reductase, (2) ability to disrupt the vitamin K_1 2,3-epoxide cycle at more than one point, (3) hepatic accumulation, and (4) unusually long biological half-lives due to high lipid solubility and enterohepatic circulation.[169] The most common findings in birds with rodenticide toxicosis are subcutaneous or cavitary hemorrhage and tissue pallor due to blood loss. Diagnosis is based on gross evidence of ingested bait, gross, and histologic evidence of hemorrhagic diathesis, and analysis of liver for suspect toxins.

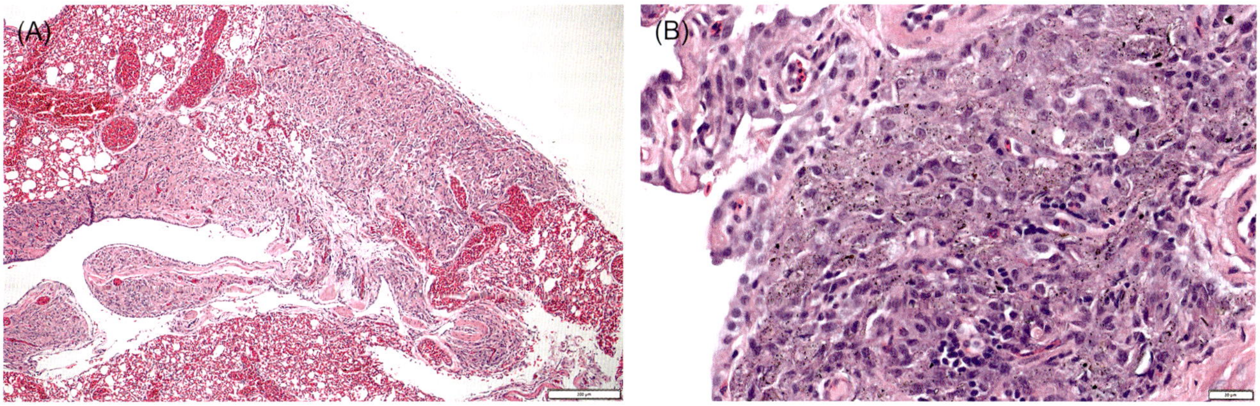

FIGURE 33.4 Pneumoconiosis in the lungs of a starling (A) and babbler (B). (A) Tissue adjacent to parabronchioles is expanded by aggregates of histiocytes that can, as in this bird, also extend into the adjacent interstitium. (B) Histiocytes contain material that may be finely granular, black (carbon), or refractile and depends on the type of inhaled particulate material.

The nonanticoagulant **zinc phosphide** is also used as a rodenticide. The chemical reacts with gastric acids to produce highly toxic phosphine gas that is rapidly distributed throughout the body, causing hypoxia and death. Formulations include its incorporation into wheat and oat baits. Ingestion can kill not only to rodent targets but nontarget birds, mammals, and fish that eat the bait or that ingest poisoned prey or carcasses.[42]

Primary **organophosphate** or **carbamate** toxicoses in wild birds are usually associated with consumption of insect baits or agricultural products, such as corn, wheat, or nuts, that are sprayed with these compounds. Organophosphates (e.g., dichlorvos, chlorpyrifos, diazinon, fenthion, and famphur) and carbamates (e.g., carbaryl, Sevin®) inhibit cholinesterase activity resulting in the accumulation of acetylcholine and persistent stimulation of nerve synapses. Death is due to respiratory paralysis. Passerines, including songbirds, woodpeckers, starlings, sparrows, and corvids, may be poisoned accidentally or intentionally (and sometimes illegally) in agricultural areas; secondary poisoning in raptors or other birds that feed on target primary species is common. Affected birds develop seizures and die within hours of exposure. Histologic lesions are generally mild and limited to the effects of seizure activity rather than direct toxicosis, and include trauma-induced coelomic or cephalic hemorrhage and seizure-induced rhabdomyolysis. Diagnosis of toxicosis is based on history, absence of other histologic causes of death, identification of suspect compounds in gastrointestinal contents, and identification of elevated blood or brain anticholinesterase activity.[52] Sublethal doses of cholinesterases can cause alterations in thermoregulation, behavior, and reproduction that could potentially impact bird populations.[69]

Due to their densely packed air capillary system, avian lungs have twice the surface area for gas exchange as mammals. This anatomic feature has been proposed as a cause for the increased susceptibility of some bird species to **inhalant toxins**. For example, historically, canaries were used as sentinels for toxic fumes in mine shafts due to their low tolerance for **carbon monoxide**. Inhalation of toxic chemicals, dust, smoke, oil, or other aerosolized pollutants can induce **pneumoconiosis**, which appears histologically as variable but often histiocytic inflammation and fibrosis (Figs. 33.4A, B). Susceptibility to inhaled toxins varies between species. For example, woodpeckers, granivorous, or burrowing birds may tolerate pneumoconiosis better than arboreal birds exposed to a smoky environment.

Polytetrafluorethylene (Teflon™), a common non stick coating on cookware, can cause acute pulmonary hemorrhage and necrosis in birds exposed to toxic fumes released from Teflon™-coated pans heated to temperatures over 260°C.[17] Teflon™ is also used as a coating on some light bulbs to decrease the chance of breakage and may be an alternative source of toxic fumes when used in aviaries. Grossly, affected lungs are diffusely red, wet and heavy, and ooze red fluid on cut surface. Histologically, severe congestion and hemorrhage are present in parabronchi, air capillaries, and around larger blood vessels. Air capillaries may be atelectatic or contain edema. Endothelial cells lining the capillaries and parabronchial epithelium may be necrotic. Birds generally die acutely before other tissue reactions are observed. Lesions are often severe but nonspecific and are similar to hemorrhagic diatheses due to other causes such as blunt trauma and rodenticide toxicosis. History, which might include the death of one, or a number of birds, of a single or multiple species and identification of Teflon™-coated light bulbs or a history of cooking with Teflon™ coated pans are critical factors in making this diagnosis as there are no tests available for the compound. Full histologic tissue examination to rule out other disease processes is also helpful in establishing a diagnosis of Teflon™ toxicosis.

Heavy metal toxicosis, particularly due to ingestion of **lead** and **zinc**, while more common in other species, such as waterfowl, raptors, or pet and zoo psittacines (see Chapter 29: Anseriformes, Ciconiiformes, Charadriiformes, and Gruiformes; Chapter 30: Birds of Prey; Chapter 31: Galliformes and Columbiformes; Chapter 32: Psittacines, Coliiformes, Musophagiformes, Cuculiformes), has been described in song birds foraging at a shooting range.[93]

Polysulfated glycosaminoglycans are used to treat degenerative arthritis in various avian species. Oral formulations are generally well tolerated. However, cases of possible high dose-related hemorrhagic diathesis have been reported following intramuscular administration of this drug in a hornbill, two raptors, and a cockatiel. Affected birds developed large, fatal hematomas at injection sites. Visceral tissues reflect hypoperfusion and hypoxic damage due to blood loss.[3]

Congenital/Genetic

Genetic disorders are infrequently described in free-ranging birds. Some studies suggest an association between **inbreeding** in insular populations and decreased reproductive success in song sparrows.[83,130,158] Breeding between related individuals in these closed populations results in decreased annual survival rates, reduction in egg-hatching success, and a marked reduction in chick survival to maturity. Inbreeding in wild sparrows and captive populations of zebra finches can reduce humoral antibody and cell-mediated immune responses.[75,130]

A **lipid storage disease** with a presumed genetic basis has been seen in a highly inbred captive population of superb birds-of-paradise.[111] Affected birds were subadult to adult, female, full siblings from different clutches from a breeding pair that shared multiple common ancestors. Gross lesions included an enlarged pale tan liver and pale heart, and severely affected tissues that floated in formalin due to the accumulation of intracellular lipids. Pronounced lipid-type vacuolar change was present histologically in cardiomyocytes (Fig. 33.5A) and hepatocytes; variable accumulation was seen in the renal tubular epithelium, parathyroid glands, gastric glandular epithelium, and pancreatic islets of Langerhans. The lipid composition of the deposits was confirmed with oil red O staining (Fig. 33.5B), electron microscopy, and lipidomic profiling, the last demonstrating primarily triglyceride accumulation. A hereditary enzyme deficiency is presumed to be the cause and adipose triglyceride lipase is a possible candidate due to similarities between the condition in the birds and neutral lipid storage disease with myopathy that is seen in humans.

A **lysosomal storage disease** of undetermined etiology has been described in captive Costa's hummingbirds.[127] Affected birds were found dead with no premonitory signs or presented with neurologic signs including rhythmic motion of the head and erratic or backward flight. Gross lesions were not seen but neuronal cytoplasmic microvesicles, neuroaxonal degeneration, and spheroids were seen histologically in the brain, spinal cord, root ganglia, optic tract, and inner nuclear layer of the retina. Vesicles stained with Luxol fast blue, indicating the presence of cell membrane material or myelin. Cytoplasmic electron-dense bodies consisted primarily of lamellar material arranged in concentric myelin figure-type whorls or lamellar stacks. Other bodies contained primarily electron-dense, granular material. Many of the bodies contained heterogenous mixtures of these two components. The features were consistent with previously described lysosomal storage diseases. The abnormal storage material and aberrant metabolic pathway in hummingbirds have not been identified.

A fatal genetic anomaly resulting in severe **ocular defects** has been described in an isolated population of red-billed choughs in Scotland.[158] Lesions consist of anterior segment abnormalities and corneal fibrosis. Some features are similar to those of Peter's anomaly in humans (anterior segment dysgenesis with iridocorneal synechiae and corneal opacification). Pedigree analysis suggests a single-locus, autosomal lethal recessive defect. Interestingly, carrier parents show higher fecundity than noncarrier pairs, suggesting that the genetic mutation may spread rapidly in the population over time.

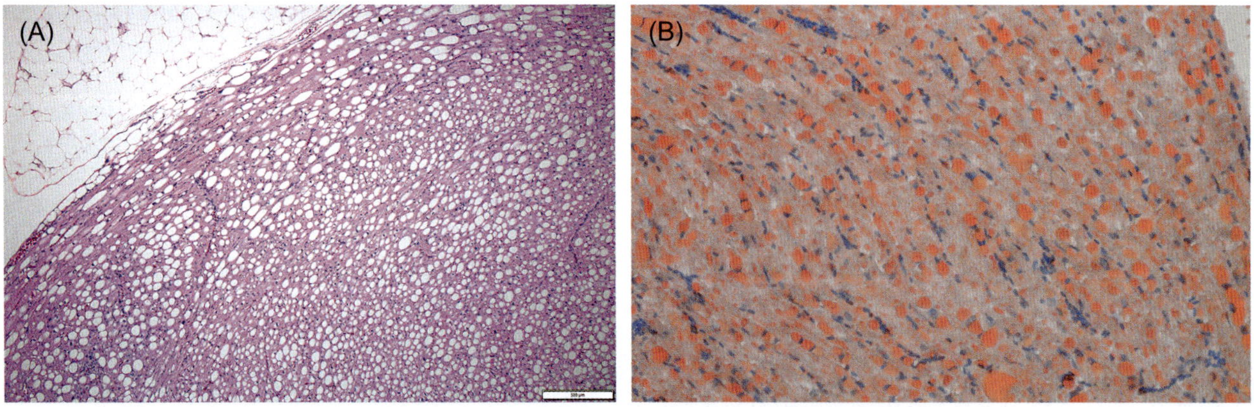

FIGURE 33.5 Lipid storage disease in the heart of a superb bird-of-paradise. (A) The sarcoplasm of cardiomyocytes throughout the heart is expanded by single, large, lipid-type vacuoles. (B) This is due to lipid accumulation, which is highlighted with special staining. Oil red O.

Congenital or genetic factors may also contribute to the development of **feather dysplasia**, which is seen sporadically in various species. Affected birds are often young and present prior to the first molt. Alopecia may be present over the body and wings, and can affect down, contour, and primary feathers. Affected follicles include blood feathers with twisted shafts, reflection of the barbules, and retained follicular sheaths. In the authors' experience, histologic manifestations of feather dysplasia in birds include growth of the bulb parallel to the epidermis, folding of the shaft within the dermis, retroflexion of the barbules, and multiple bulbs within a single follicular sheath. Interestingly, this condition may resolve in some affected birds following subsequent molts. Rarely, unusual leucystic or **amelanistic (albino) color morphs** occur. This change appears to be particularly common in wild northern cardinals.[30]

Inflammatory Non-Infectious

Aspiration and **aspiration pneumonia** are common causes of morbidity and mortality in passerines and piciformes. Numerous predisposing factors include oral trichomoniasis or candidiasis, assist feeding, alimentary tract disease, trauma, or beak anomalies. Because piciformes lack a crop, modified programs for hand feeding are required to prevent aspiration of food materials. The gross pulmonary pattern of aspiration in birds is generally cranial and unilateral or bilateral. Histologically, aspirated material is present in parabronchi and is associated with varying degrees of hemorrhage, inflammation, and sometimes secondary, opportunistic bacterial or fungal infection. Aspiration of amniotic fluid during pipping is a common cause of mortality in perinatal birds, especially those with malformations or those that are malpositioned. Yolk may also be aspirated into the lungs or ingested by embryos that are malpositioned, or it can occur subsequent to yolk sac rupture during internalization.

Xanthomas are nodular or raised, plaque-like, tan to yellow, irregular, and often corrugated masses in the skin and subcutis. **Xanthomatosis** refers to multiple cutaneous and visceral masses. Xanthomas can occur anywhere on the body but are most common in the wings, sternum, or pubic area of birds.[78,151] The etiopathogenesis of xanthoma development in birds is unclear but may involve dietary factors, abnormal lipid metabolism or trauma. Hypercholesterolemia is not necessarily present. Histologically, xanthomas are composed of sheets of vacuolated macrophages, multinucleated giant cells, and acicular (cholesterol) clefts (Figs. 33.6A, B). Evidence of chronic hemorrhage and scarring is often present. Ulcerated xanthomas can be secondarily infected with bacteria or fungus. The lesions tend to be invasive and are notoriously difficult to completely excise; wing lesions sometimes require amputation. Xanthomas and xanthomatosis are somewhat common findings in many avian Orders (especially Psittaciformes), but are not common lesions in Passeriformes.

Miscellaneous/Multifactorial

Chronic inflammation can predispose individuals of many bird species to reactive systemic (secondary) **amyloidosis**, a disease caused by the abnormal tissue deposition of amyloid derived from the acute phase reactant serum amyloid A protein (SAA: see Chapter 29: Anseriformes, Ciconiiformes, Charadriiformes, and Gruiformes). Common associated inflammatory conditions include mycobacteriosis, aspergillosis, and pododermatitis. Within the Passeriformes, amyloid deposition is especially common in finches. Common sites of amyloid deposition are vascular walls and adventitia, the space of Disse in the liver, adrenal glands, spleen, and glomeruli. Grossly, the liver may be enlarged, pale

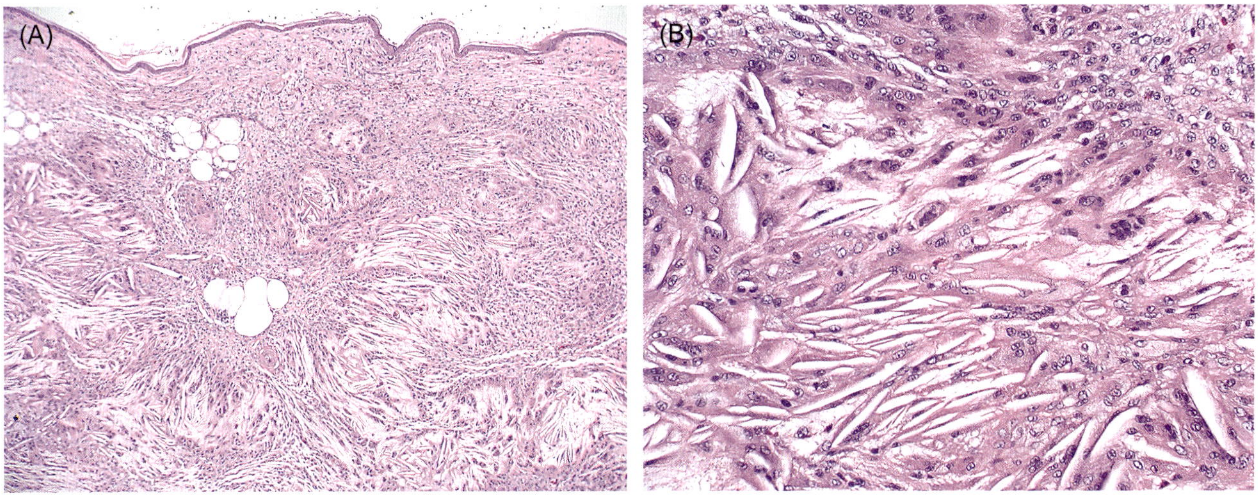

FIGURE 33.6 Xanthoma in the skin of a frogmouth. (A, B) The skin is markedly expanded and normal tissue is replaced by (B) numerous acicular (cholesterol) clefs surrounded by epithelioid macrophages and multinucleated giant cells.

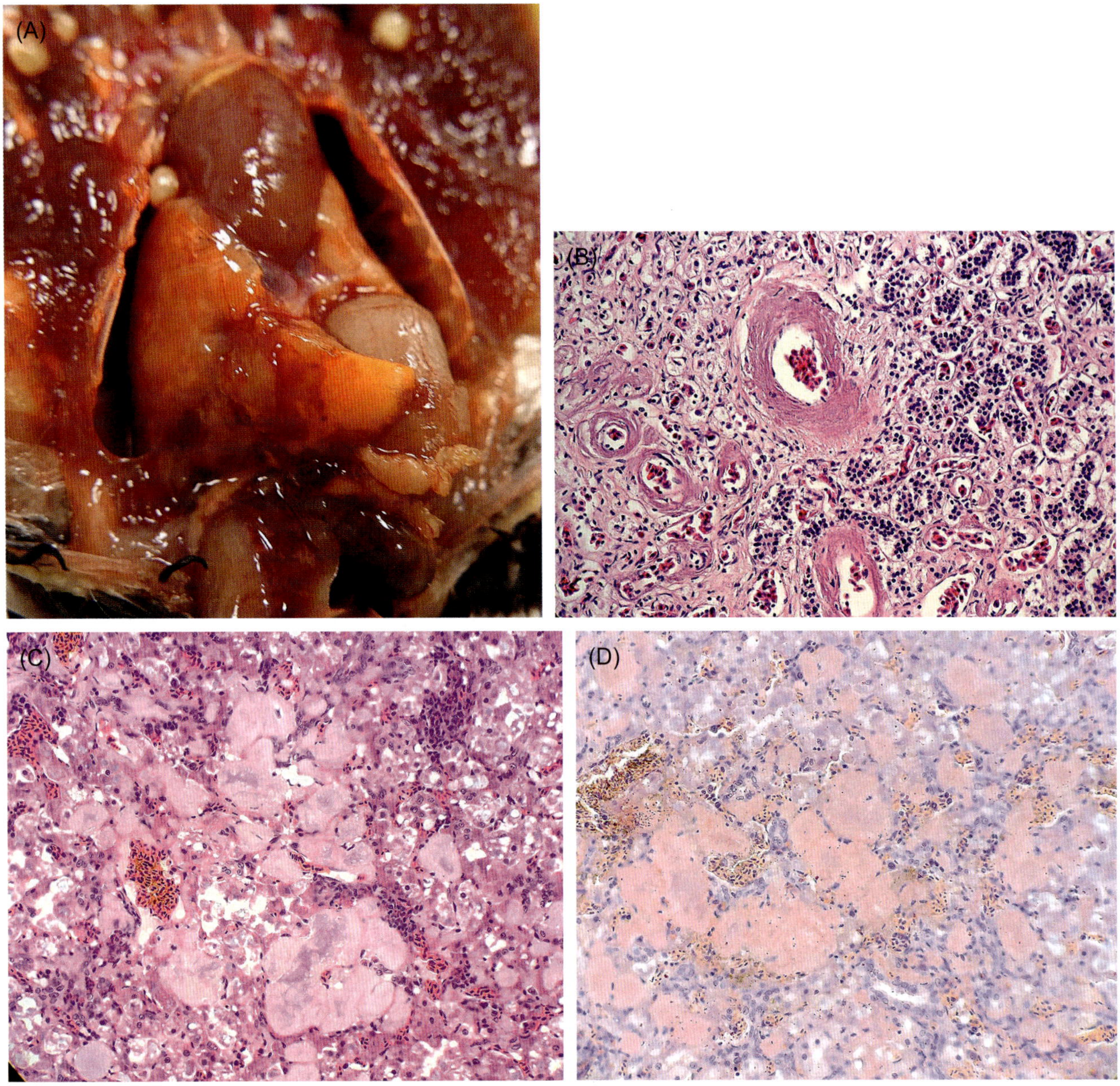

FIGURE 33.7 Amyloidosis in a whydah (A), frogmouth (B), and society finch (C, D). (A) The liver is enlarged, discolored orange/tan, and has a slightly waxy appearance. Amyloid is acellular and lightly eosinophilic with routine hematoxylin and eosin staining. Deposition can cause expansion of (B) vascular walls and perivascular connective tissues and in the (C) liver, it can expand the space of Disse causing occlusion of sinusoids and compression of hepatic cords. (D) Amyloid appears light pink/orange with Congo red staining and brightfield illumination. It is birefringent with polarized light (not shown). Congo red.

tan/orange, and have a characteristic waxy appearance (Fig. 33.7A). Amyloid appears histologically as extracellular deposits of faintly fibrillar, eosinophilic material with routine hematoxylin and eosin staining (Figs. 33.7B, C). With Congo red staining, it is light pink/orange when viewed with brightfield illumination (Fig. 33.7D) and green birefringence may be seen when viewed with polarized light. In the liver, deposition can coalesce and efface the parenchyma, causing hepatic atrophy and in severe cases, hepatic failure. Effaced adrenal glands may be associated with some degree of metabolic derangement or adrenal insufficiency; however, this is generally not documented prior to death.

Avian keratin beak disorder, which produces dramatic, debilitating overgrowth and malformation of the beak, was first noted in 1998 in black-capped chickadees and Northwestern crows in Alaska (United States).[161,178] Since the initial reports, similar lesions have been noted in over 80 species of birds and the geographic distribution

FIGURE 33.8 Black-capped chickadee with a beak deformity characteristic of avian keratin disorder. The upper bill is overly long, crossed, and laterally deviated. *(Reprinted with permission from van Hemert C, Handel CM. Blood serum chemistry of wild Alaskan black-capped chickadees (Poecile atricapillus) with avian keratin disorder. J Wildl Dis. 2016;52:927–930.)*

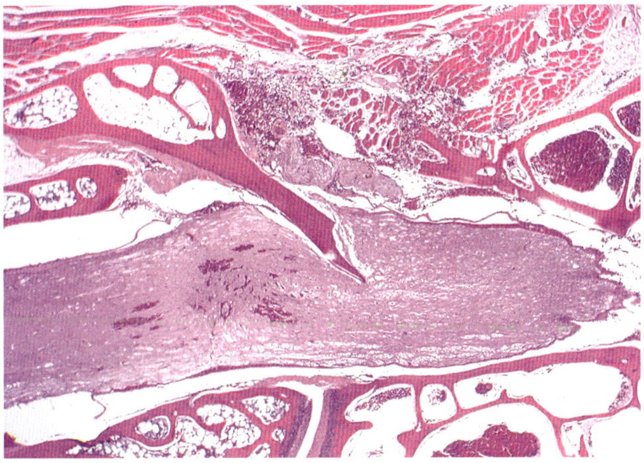

FIGURE 33.9 Blunt-force trauma in the cervical spine of a Taveta golden weaver. Lesions associated with trauma in this bird include a displaced fracture of a cervical vertebral body and multifocal paravertebral and spinal cord hemorrhage.

has expanded regionally to include parts of the states of Washington and Oregon. Beak deformities are described as elongation, crossing, and lateral deviation of the bill (Fig. 33.8).[162] In some cases, roughening or flaking of the skin of the head and nails of the digits has also been described. This corresponds histologically to dysplastic changes in the stratum corneum and marked epithelial hyperplasia and hyperkeratosis. Additionally, accumulations of sloughed cells and eosinophilic debris with areas of hemorrhage are present along the stratum corneum. Low numbers of eosinophilic, intranuclear, and intracytoplasmic inclusion bodies are present in keratinocytes. Bone lysis with proliferation of woven bone is present in the underlying bone of the beak. To date, the cause and pathogenesis have not been definitively identified, although a virus is suspected. One metagenomic study has identified a possible association with a novel picornavirus (***Poecivirus; Picornaviridae***).[178] Other studies show no evidence of an underlying infectious etiology. Nutritional and toxicologic analyses have failed to show significant differences between affected and unaffected birds.

Feather cysts are common in passerine birds, especially canaries. Solitary or multicentric cysts may involve the wings, head, neck, back, or uropygial gland region. Grossly, the cysts are raised and smooth, dark brown to black, superficially featherless, and are occasionally ulcerated due to cyst rupture or excoriation. Histologically, cyst walls are lined by stratified squamous epithelium and luminal contents include stratified keratin and feather barbules. Secondary infection and hemorrhage are common in traumatized cysts, and the release of keratin or other feather components from ruptured cysts incites severe granulomatous or xanthogranulomatous inflammation in the surrounding dermis.

Trauma is an important cause of mortality in many passerine species. **Collision with window glass (window strike)** is particularly common, especially during migration or during the breeding season. It may account for the deaths of as many as 600 million birds annually in the United States.[43] Blunt force trauma due to window strike can also occur in captive settings in glass-walled enclosures. Gross lesions vary and can include head trauma and intracranial hemorrhage[84] or fracture of the coracoid bone, cervical vertebrae (Fig. 33.9) or liver, laceration of the aorta, and acute hemocoelom. In some instances, there are no gross lesions despite circumstantial evidence of a glass strike. Histologically, birds may have hemorrhage at sites of impact in the pectoral muscles, parabronchi, coelom, air cells of the skull, meninges, brain, and/or at sites of trauma. **Wind turbines** used for power generation have been estimated to cause 134,000 to 230,000 passerine mortalities annually in the United States and Canada.[43] Window and **wall strikes** are common in passerine collections kept in human care, especially in mixed species exhibits. Other forms of trauma that can be seen in exhibit species include **entrapment** in netting, pipes or enrichment devices, or **impalement** on branches. Injuries are dependent on the manner and site of the trauma.

Predation is common in passerines, and predators include feral or outdoor cats, rats (wild and captive settings), raptors (e.g., at bird feeders or through enclosure wire), and other mesocarnivores and carnivores. Estimates of annual mortality due to cat predation in wild birds range from 110 to 350 million birds in Canada[14] to more than 1 billion birds per year in the United States.[10,14,29] Gross lesions can include bite wounds and perforations, including penetrating wounds into the coelomic cavity. An additional form of predation is **ant bite** predation, which is common in ground-nesting birds. Among ant species, fire ant predation has been reported as a substantial cause of mortality in hatchlings in the southern United States. Lesions may include partial consumption of carcasses, evisceration,

FIGURE 33.10 Snake predation in a Taveta golden weaver. The anterior portion of the bird in the lower half of the image is wet and covered by a thin film of mucoid material and blood. This is due to regurgitation of the incompletely ingested, regurgitated bird. The bird in the upper part of the image is a normal control.

and intracoelomic ants. In less obvious presentations, histologic evaluation may identify bite wounds. Lesions may be associated with necrosis or ulceration of the epidermis and subjacent zones of hemorrhage, necrosis, and collagen degeneration in the dermis.[64] **Snake predation** can also occur in small passerines. Birds are generally ingested head first. Regurgitation of incompletely ingested birds can result in a bird being found dead with a mucoid matt over the anterior body (Fig. 33.10). The snake may not be found but the gross appearance of the regurgitated bird is characteristic.

An inflammatory **vasculopathy** of undetermined cause has been seen sporadically in a large captive flock of New Guinea metallic starlings. Gross lesions included hydropericardium, epicardial hemorrhage, and areas of myocardial pallor. The condition was characterized histologically by fibrinonecrotic to proliferative arteritis (primarily the coronary arteries) with associated myocardial degeneration, necrosis, hemorrhage, mineralization, atrophy, and fibrosis. Lesions were also occasionally seen in arteries in the proventriculus, ventriculus, and mesentery. Affected birds were usually 3–4 weeks old. Due to colony nesting behavior, the parentage and relatedness of affected individuals were unknown. In the absence of detectable infectious agents, such as aspergillosis or West Nile virus infections, an immune-mediated or toxic cause was considered. All birds had varying degrees of hepatic hemosiderosis, but iron levels did not correlate with the occurrence of this syndrome.

Neoplastic

Neoplasia is common in many species of birds; however, with a few notable exceptions (see below), passerines are infrequently affected.[63]

Lymphoma is a relatively common tumor in many species of birds. Among passerines, finches and canaries may be overrepresented when compared to other species, and lymphoid infiltrates suggestive of neoplastic transformation and intestinal lymphoma have been reported in house sparrows and American goldfinches infected with systemic *Isospora* (see Systemic Isosporosis below).[26] Lymphoid malignancies in finches typically present as high-grade, multicentric tumors. As with other species, gross changes include nodular to diffuse enlargement of affected organs, which have patchy to diffuse tan discoloration (Fig. 33.11A). Cells are blast-like with large vesicular nuclei (Fig. 33.11B). Mitotic figures are often numerous and the tumor may be associated with necrosis or hemorrhage. Tumors can potentially be classified into T or B cell

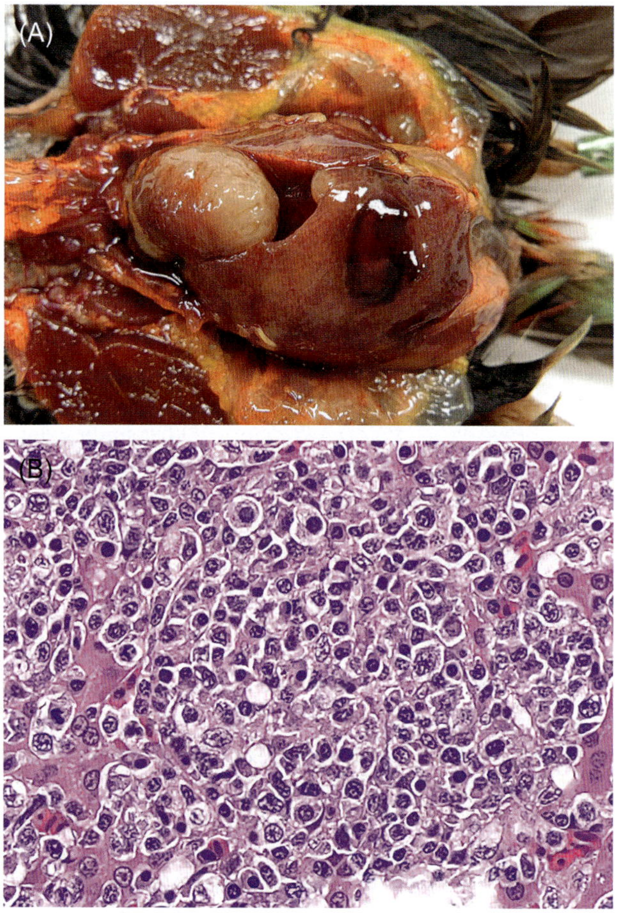

FIGURE 33.11 Lymphoma in the heart of a tricolored parrot finch (A) and liver of a barbet (B). (A) Diffuse enlargement and light tan discoloration are seen in the liver and heart. (B) This corresponds to infiltrates of neoplastic lymphocytes with moderate anisocytosis and anisokaryosis that enlarge the tissue and efface the normal hepatic parenchyma.

lineage using CD3 or PAX5 immunohistochemical labeling, respectively, based on the use of these markers in other species.[67]

A relatively common tumor in great Indian hornbills is **squamous cell carcinoma** of the casque. Affected birds are generally middle aged to older (25 to 40 years), and clusters of cases have been identified. Tumors typically appear at the interface of the rostroventral casque and dorsocaudal maxillary rhamphotheca; metastasis is reported. Affected birds develop softening, flaking, and/or discoloration of the outer casque (Fig. 33.12A), and tumor masses may be inapparent or appear as fleshy white/pink nodules that grow through the choanal slit. Tumors are nodular to plaque-like and may extend locally into the maxillary sinus and bone (Fig. 33.12B; eSlide 33.e3). As in other species, tumors may be well to poorly differentiated and consist of variably thick cords of invasive, neoplastic stratified squamous epithelial cells that extend from the skin or mucosal surface into the underlying tissues. Keratin "pearl" formation (brightly eosinophilic aggregates of concentrically laminated keratin) may be seen in well-differentiated tumors. Neoplastic cells are polygonal, spindle, or stellate and contain moderate to abundant amphophilic to eosinophilic cytoplasm, may have prominent desmosomes, and show variable pleomorphism, anisocytosis, and anisokaryosis. Superficial ulceration, multifocal intratumoral necrosis, and mixed inflammation may be seen. Superficial bacterial or mycotic infection and hyperplasia or dysplasia of the mucosa may be seen in superficial biopsies; deep biopsies and imaging are often necessary to make an antemortem diagnosis. A cause for the apparent predilection for tumor development in the casque in Indian hornbills has not been identified. However, one study has suggested that affected birds may be deficient in uropygial gland carotenoids, which may be protective against the oxidative damage of ultraviolet light.[59]

Folliculomas, benign feather follicle tumors, are common in canaries, especially densely feathered or curled-feather breeds.[54,174] Predilection sites include the wing, dorsum, and breast. Grossly, these are discrete, nodular to multinodular, freely movable dermal tumors that are yellow to yellow-orange and have a cystic center. Histologically, they are unencapsulated, well-demarcated, and composed of multiple, variably sized cystic spaces lined by variably thick, well differentiated to somewhat dysplastic and disorganized proliferative follicular basal epithelium that shows progressive maturation to stratified, keratin-producing, squamous epithelium. A defining feature that distinguishes these lesions from feather cysts is intraluminal feather differentiation (e.g., pulp, barbs), which may not be apparent in all sections. Aggregates of compact keratin are common in the central lumina. Mural inflammation is common and typically consists of mixed inflammatory cells or in the case of rupture or trauma, granulomatous inflammation.

Hepatocellular and **biliary neoplasms** in passerines are often associated with iron overload, although they can occur sporadically, especially in old birds. The role of iron overload in the development of hepatic neoplasia is not clear, but iron-related hepatocellular damage, for example, in cases of hemochromatosis (see above), is thought by some to be a predisposing factor. Hepatic neoplasms are often lighter than adjacent normal parenchyma and are variably well-demarcated and irregularly nodular. Hepatocellular carcinomas typically have a trabecular pattern composed of enlarged hepatocytes with moderate

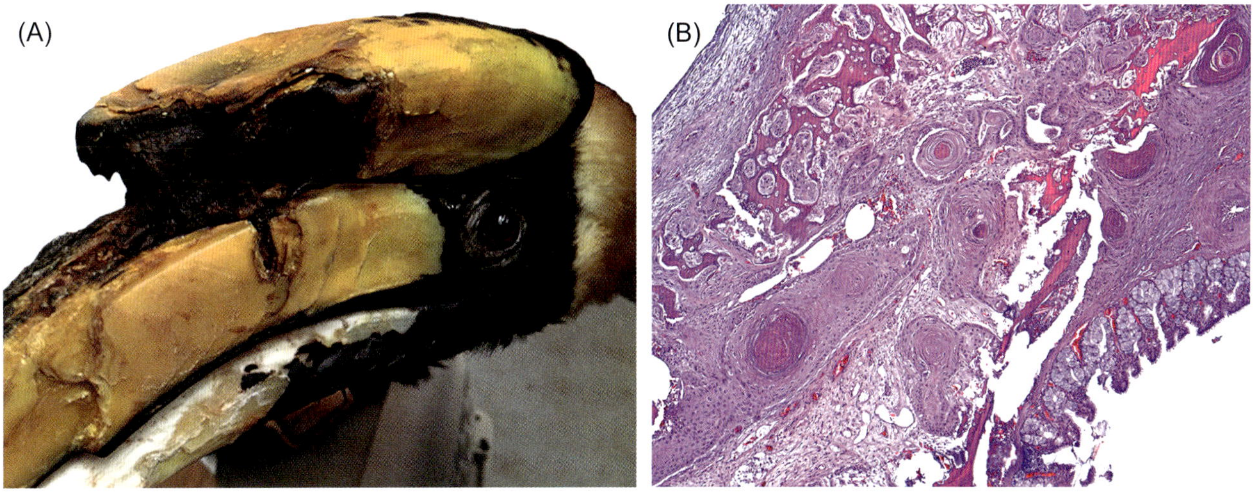

FIGURE 33.12 Squamous cell carcinoma in the casque of an Indian hornbill. (A) As in this case, the neoplastic process is often locally aggressive and causes extensive destruction of the normal architecture of the casque. Damaged tissue is dry to fleshy, discolored brown/black, and extends into the underlying tissues. (B) Characteristic neoplastic features typically include the formation of "keratin pearls" and often involve invasion into underlying soft tissue and destruction, lysis, or fracture of bone of the rhinotheca. Note the normal respiratory epithelium in the lower left of the image (see eSlide 33.e3). *(Part A: Photo Courtesy of S. Bartlett, Zoo New England).*

anisocytosis and anisokaryosis, large singular nucleoli, and occasional mitotic figures. Hepatocellular carcinomas typically are not encapsulated and invade the surrounding hepatic parenchyma. Well-differentiated hepatocellular carcinomas can be difficult to distinguish from hepatocellular adenomas, especially in small biopsy specimens where the interface of the tumor and surrounding hepatic parenchyma is not present. Histologically, biliary adenocarcinoma is easily differentiated from hepatocellular carcinoma due to the formation of ductular or tubular structures one to several cells thick in the former. Neoplastic cells may show crowding, loss of nuclear polarity, anisokaryosis, and frequent mitotic figures. In contrast to hepatocellular carcinomas, biliary adenocarcinomas are often associated with a scirrhous response.

Renal tubular adenomatosis is characterized by the formation of multicentric benign renal tubular adenomas. It has been described in rhamphastids and has also occasionally been reported in hornbills and various species of waterfowl and raptors.[63,98] The tumors are well circumscribed, occasionally encapsulated, and are often present in multiple renal lobes. The tumors may be associated with slight displacement and compression of surrounding renal parenchyma. Tubular epithelium is well differentiated, and tubules may be mildly dilated and contain proteinaceous fluid or urates. The tumors are generally incidental and not likely to substantially impair renal function in most cases.

INFECTIOUS DISEASES

DNA Viruses

Poxviruses are large, enveloped, double-stranded DNA viruses. Infections with **avipoxviruses** are relatively common among free-ranging passerines. Numerous strains of poxvirus have been identified and there appears to be some degree of species specificity. Transmission is most commonly by direct contact with infected birds or from biting insect vectors (primarily mosquitos) and fomites. Transmission of some strains can also occur by inhalation or infection through open skin wounds. Disease may be more severe when introduced into naïve populations, such as isolated island species.[149,168]

Avipoxvirus infection classically presents as one of two forms, cutaneous ("dry") or diphtheritic ("wet"), though both forms can be present in the same bird and systemic pox has also occasionally been described. The **cutaneous form** is more common in passerines and produces nodular growths on nonfeathered skin of the feet, digits, and cere (Figs. 33.13A, B; eSlide 33.e4).[156] In many cases, lesions are self-limiting and spontaneously regress over several weeks. However, impaired vision and interference with the ability to eat or perch may occur. Opportunistic bacterial and fungal infections in affected skin may contribute to increased morbidity and mortality. Histologically, skin lesions consist of marked epithelial cell hyperplasia and hypertrophy throughout the stratum spinosum or the full thickness of epidermis and follicular epithelium. Infected epithelial cells may contain one large or few to numerous variably sized, eosinophilic, intracytoplasmic inclusion bodies often with central pallor or clearing (Bollinger bodies) (Fig. 33.13C; eSlide 33.e4). Epithelial cells may appear swollen with large cytoplasmic vacuoles due to loss of the inclusion during processing or diffuse cytoplasmic edema (ballooning degeneration). Cytologic examination of skin lesions will sometimes also reveal intraepithelial inclusions that do not stain or stain slightly blue with Romanowski stains.[18] Hyperkeratosis, ulceration, necrosis, and subacute inflammation may occur, possibly secondary to lesion excoriation. In canaries, typical epidermal lesions can be accompanied by marked lymphoplasmacytic and histiocytic inflammation that extends into soft periocular and leg tissues (Fig. 33.13D). Periocular cutaneous and conjunctival infection is often severe and fatal in this species due to inanition/anorexia and secondary localized or embolic bacterial and fungal infections.

The diphtheritic form of avian poxvirus infection, which is common in canaries, produces proliferative and fibrinonecrotic, plaque-like or pseudomembranous lesions.[132,145,171] Severe lymphoplasmacytic inflammation and intraepithelial, eosinophilic, and intracytoplasmic inclusions are seen in affected tissues. The oral cavity (Fig. 33.13E), trachea (Figs. 33.13F, G), lungs, and air sac are most commonly affected. Infection can also involve the nasal cavity, lymphoid tissues (e.g., bursa of Fabricius, thymus), uropygial gland, esophagus, and/or proventriculus. Outbreaks can cause mortality as high as 100% in canaries, and substantial mortality has been reported in sparrows and finches.

A systemic form of avian poxvirus has also been described. It is likely an extension from the cutaneous or diphtheritic types.[148] Reported primarily in zoo-housed passerines but also in kingfishers and several other orders, this form has two histologic manifestations that often occur concurrently with diphtheritic and cutaneous lesions. One pattern consists of extension of histiocytic inflammation from severe epithelial lesions deep into underlying connective tissues, including bone. Macrophages may contain poxviral-like cytoplasmic inclusions. The second pattern involves histiocytic inflammation and intracytoplasmic inclusions can be seen in various organs including the liver, spleen, kidney, thyroid gland, bone marrow, adrenal glands, and the gastrointestinal tract. Poxviral DNA can be detected in histiocytes in both types of lesions with in situ hybridization and PCR. Comparison of gene sequences from poxviruses causing both the cutaneous/diphtheritic and systemic forms of disease showed that those causing systemic lesions grouped in cluster 2 of the B1 subclade of canarypox viruses. A combination of viral and host factors is therefore suspected to be involved in the manifestation of the systemic forms of disease.

The gross oral lesions of avian poxvirus infection closely resemble those of candidiasis, trichomoniasis, hypovitaminosis A, and capillariasis. Histologic identification of intracytoplasmic **Bollinger body inclusions** is characteristic of avian poxvirus infection but these may be inapparent or ambiguous in severely inflamed, necrotic, or autolyzed lesions, especially in the diphtheritic form of the disease. A number of ancillary diagnostic tests, including PCR, electron microscopy, or viral culture, are confirmatory.

Polyomaviruses are a diverse group of double-stranded DNA viruses that includes 8 genera and more than 115 species.[167] Avian polyomavirus infections have been described in Passeriformes including finches, canaries, crows, butcherbirds, and green aracaris.[13,21,71,74,87,107] The authors have also identified an enteric form of polyomavirus infection in a tawny frogmouth. Lesions in passerines are more typical of the psittacine variant of systemic polyomavirus infection. Disease is relatively common in finches (especially zebra finches) and canaries, and most often occurs in neonates and juveniles. These viruses can potentially be spread through secretions and feather dust, and evidence suggests vertical transmission may occur. Hepatomegaly and occasionally splenomegaly with foci of acute hemorrhage in the subcutis and serosal surfaces of the heart and liver are typical gross lesions. Areas of necrosis and lymphocytolysis are typical histologic lesions in the liver (Fig. 33.14A) and spleen and may be seen in other sites such as the kidneys, lung, thymus, or intestine.

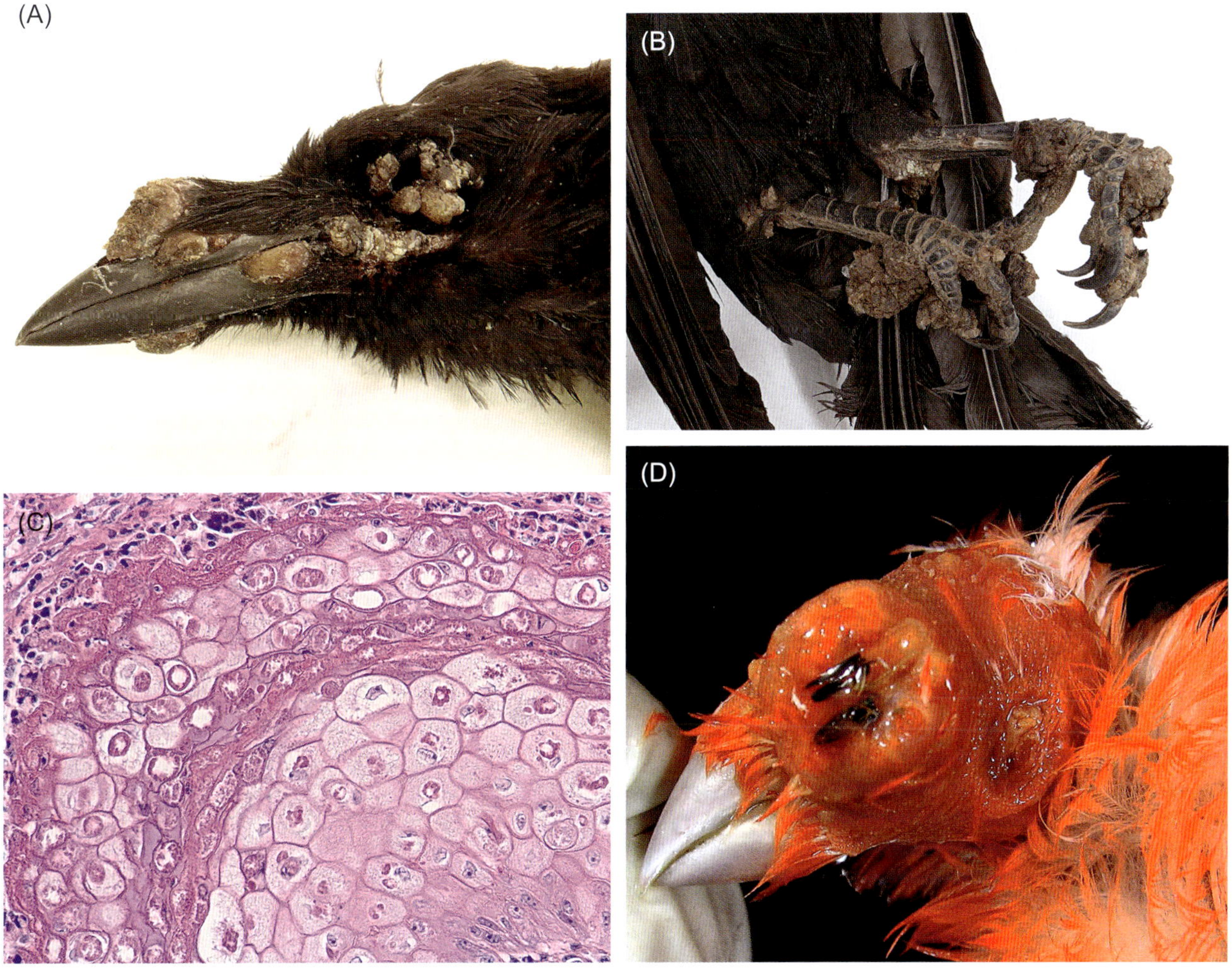

FIGURE 33.13 Cutaneous avian poxvirus infection in a crow (A), raven (B), blue jay (C), and canary (D); poxviral stomatitis in a blue jay (E); and poxviral tracheitis in a canary (F, G). (A) Characteristic cutaneous lesions include proliferative lesions with superficial crusts on nonfeathered skin, especially on the face (A) and legs (B). (C) The follicular epithelium is intensely hyperplastic and ballooning degeneration of most keratinocytes is striking. Eosinophilic intracytoplasmic inclusion bodies, often with central pallor, are present in numerous keratinocytes. (D) While respiratory poxviral infection is most common in canaries, localized infections (e.g., lingual, nasal, palpebral, or cutaneous) and mucocutaneous presentations involving the skin and one or more mucosal surfaces, particularly in the head, can be observed. This canary has severe palpebral and periocular cutaneous involvement.

(Continued)

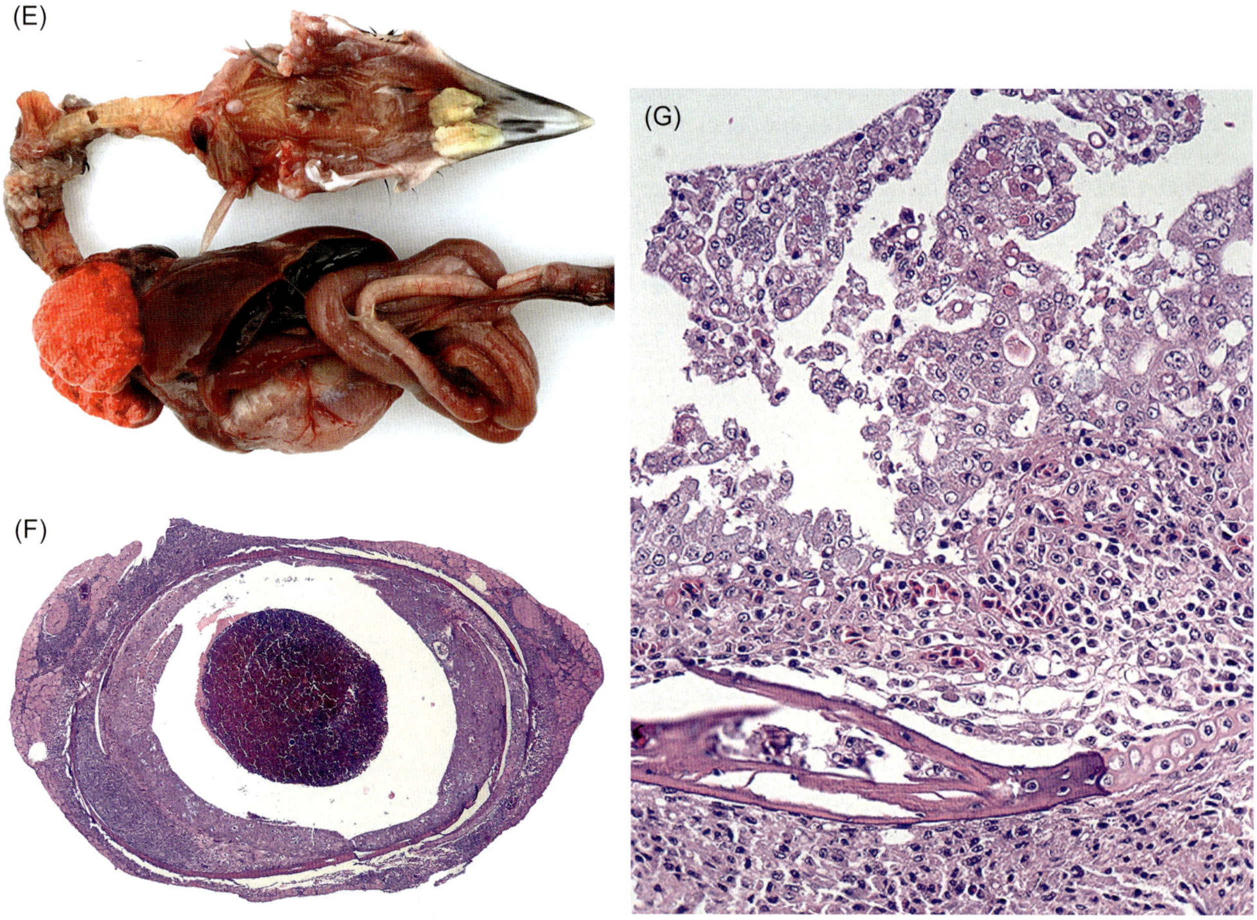

FIGURE 33.13, CONT'D (E) Thick, yellow diphtheritic plaques composed of necrotic tissue and cell debris are present on the dorsal oral mucosa. This bird had concurrent mucocutaneous and cutaneous poxviral infection. (F) Infection is associated with marked expansion of the mucosa and peritracheal tissues due to epithelial hyperplasia and inflammation. Lymphoplasmacytic inflammation extends through the tracheal rings into the peritracheal skeletal muscle. (G) The tracheal mucosa is markedly hyperplastic and epithelial cells contain eosinophilic, intracytoplasmic inclusion bodies often with a central pallor (see eSlide 33.e4).

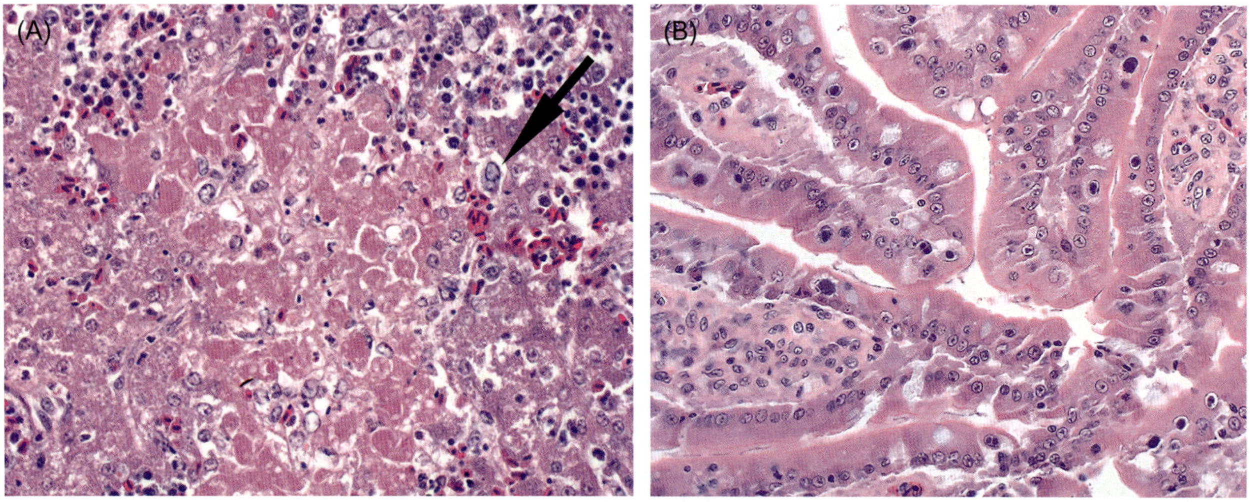

FIGURE 33.14 Polyomavirus infection in the liver of a Cordon-bleu finch (A) and intestine of a frogmouth (B). (A) Infection can be associated with necrosis, which in this case is seen as brightly eosinophilic discoloration and rounding and dissociation of hepatocytes (center of image). (A, B) Viral inclusions are large, intranuclear, basophilic, and can be found in hepatocyte (A; *arrow*) and enterocyte (B) nuclei. Karyomegaly is a characteristic feature of infection.

Karyomegaly with large, pale, Cowdry B type, basophilic, or clear intranuclear inclusion bodies (Figs. 33.14A, B) can be seen in epithelial and endothelial cells. Inclusions in the absence of necrosis or other tissue reaction can also be seen in various tissues in canaries and finches. A Sertoli cell tumor associated with polyomavirus infection and intralesional inclusions in neoplastic cells has been described in a Gouldian finch.[137] The role of the virus in tumor development was not identified, but the polyomavirus group is known to contain several oncogenic viruses. Confirmation of polyomavirus infection is by PCR, in situ hybridization (ISH), or transmission electron microscopy (TEM).

Herpesviruses cause disease in many species of passerines. Fatal outbreaks of alphaherpesvirus respiratory disease have been identified in Gouldian finches,[173] canaries, cardinals, weavers, and starlings. Affected birds present with severe dyspnea, lethargy, and occasionally diarrhea or conjunctivitis. Gross lesions are variable but some birds have luminal tracheal debris or exudates. Characteristic histologic lesions in the respiratory system include epithelial cell hyperplasia and exfoliation (cell sloughing), syncytial cell formation, erosion, ulceration, and necrosis with intraluminal cellular debris. Lesions and inclusions are typically most prominent along the tracheobronchial and parabronchial mucosa. Eosinophilic, intranuclear viral inclusion bodies are present in hyperplastic and hypertrophic (cytomegalic) respiratory epithelial cells. The tracheal lumen may contain necrotic cell debris, degenerate heterophils, and macrophages.

Free-ranging tawny frogmouths in eastern Australia can develop **inclusion body hepatitis**. Clinical signs in affected birds include weakness, lethargy, and (frequently) acute death. Gross examination reveals an enlarged, friable liver with miliary white foci. Histologic lesions consist of acute hepatic necrosis with eosinophilic intranuclear inclusion bodies. This is presumed to represent an adenoviral (or possibly herpesviral) infection based on electron microscopy.[128] A novel alphaherpesvirus was detected in a live tawny frogmouth by sequencing and phylogenetic analysis. It is unclear if this virus causes disease or is related to the previously described inclusion body disorder.[2]

RNA Viruses

West Nile Virus (WNV) is an arthropod-borne flavivirus in the Japanese encephalitis virus complex. Flaviviruses are enveloped, single-stranded RNA viruses that are ~40–65 nm in diameter with an icosahedral nucleocapsid. Other members of this complex include Saint Louis encephalitis virus, Kunjin virus, Usutu virus (see below), and Murray Valley encephalitis virus. WNV is endemic in parts of the Middle East, Africa, India, Europe, Asia, and Australia. Since its introduction into New York State in 1999, WNV infection has become endemic across the continental United States and from Canada to Central and South America.[106] Wild and domestic birds serve as reservoirs for the virus, and mosquitoes (primarily *Culex* and *Aedes* genera) are virus vectors and the common source of transmission. Transmission by consumption of infected carrion has been demonstrated experimentally in crows, and fecal-oral transmission is also a suspected mode of transmission.[105] Birds in the orders *Passeriformes*, *Charadriiformes*, *Strigiformes*, and *Falconiformes* develop sufficiently high titers to allow transmission of virus to mosquitoes during a blood meal; birds in the *Piciformes*, *Columbiformes*, and *Anseriformes* orders develop lower titers that are thought to be insufficient for transmission.[105]

Affected birds are found dead or present with depression, anorexia, or neurologic signs including ataxia, torticollis, incoordination, and an inability to fly. Gross lesions may include calvarial and meningeal hemorrhage (the former should not be mistaken for congestion or blood pooling in the sinus), myocardial pallor, splenomegaly, hepatomegaly, and/or intestinal hemorrhage and congestion. The kidneys are generally swollen and may contain pinpoint white foci. Histologically, there is mild to marked lymphoplasmacytic meningoencephalitis and myelitis with perivascular cuffing, gliosis, and microglial nodule formation. Lesions in the heart can include multifocal myocardial necrosis, histiocytic myocarditis, and epicarditis (Fig. 33.15A; eSlide 33.e5). Similar inflammatory infiltrates with or without necrosis may be present in the kidney, spleen, or other lymphoid tissues, liver, intestine, adrenal gland, and pancreas. Acute pancreatic acinar cell degeneration or necrosis is a consistent finding in affected passerines. Differential diagnoses include other viral encephalitides (e.g., eastern equine encephalitis, avian influenza virus).

Confirmation of the diagnosis is by immunohistochemical labeling (Fig. 33.15B; eSlide 33.e6), virus isolation, or PCR testing of heart, kidney, or brain samples (or other apparent sites of infection); feather pulp or skin are also useful samples in live or dead birds.[35,60,119] In most bird species, Purkinje cells, the molecular layer of the cerebellum, and neurons in the brain stem show strong immunohistochemical labeling, however, this may be less prominent in corvids. In crows, tissue macrophages in the spleen, Kupffer cells in the liver, and blood monocytes typically stain strongly for West Nile Virus antigen. Recent studies in humans indicate that the virus may persist following clinical recovery, and may contribute to chronic illnesses including chronic nephritis.

In the summer of 2001 and continuing through 2002, an epornitic of **Usutu virus** caused a mortality event in Eurasian blackbirds in Austria.[7] Like WNV, Usutu virus is a

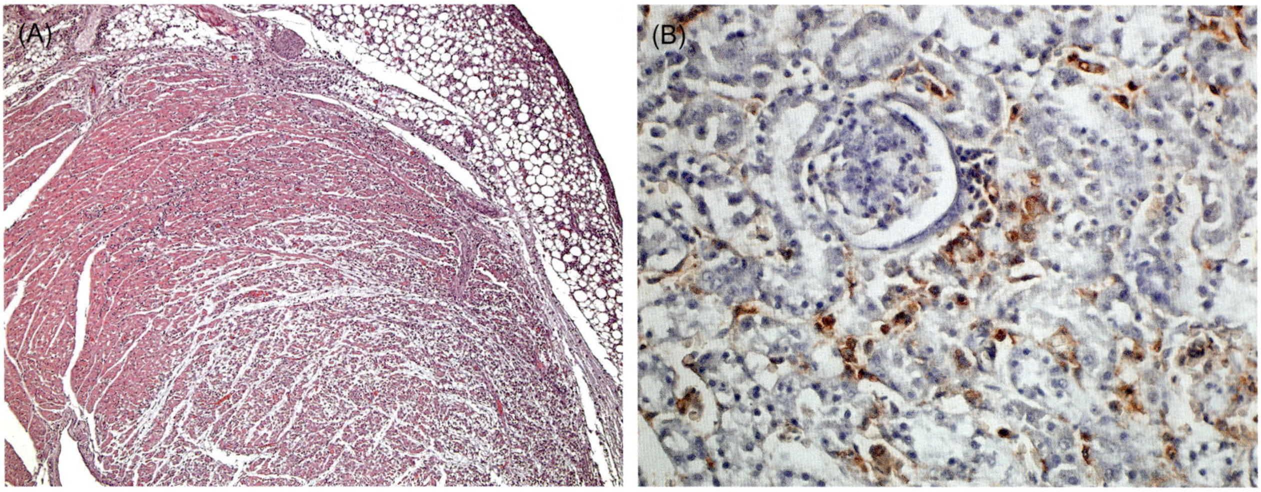

FIGURE 33.15 West Nile virus infection in the heart (A) and kidney (B) of an American crow. (A) Common changes in the heart include myocardial degeneration, edema, and/or inflammation. Necrosis in this case is associated with drop-out of myocardiocytes, increased clear space due to edema, and mild subacute, pancardial inflammation. (B) Kidney is a commonly affected site of infection in crows. Renal lesions may be subtle and associated with mild to moderate lymphoplasmacytic, interstitial nephritis, and occasionally tubular epithelial cell necrosis. In suspect cases with minimal or mild nonspecific lesions, viral antigen can be highlighted with immunohistochemical labeling. IHC (see eSlides 33.e5 and 33.e6).

flavivirus in the Japanese encephalitis group. Grossly, the primary lesion was hepatosplenomegaly. Histologically, neuronal necrosis, myocarditis, and hepatic and splenic necrosis were observed. Immunohistochemical labeling showed cross-reactivity with WNV antibodies in affected tissues. However, in situ hybridization using Usutu virus-specific oligonucleotide probes indicated the causative agent was Usutu virus. In subsequent years, additional cases have been seen in Germany, Italy, Spain, Greece, Poland, and Belgium.

Eastern, Western, and **Venezuelan equine encephalitis viruses (EEE**, **WEE**, and **VEE**, **respectively**) are members of the *Togaviridae* family (*Alphavirus* genus) and are mosquito-borne arboviruses. Ratites and passerines are susceptible to infection and passerines act as reservoir hosts. EEE is perhaps the most virulent of the group and is an important cause of encephalitis in humans and horses. In the United States, EEE is occasionally seen in passerines, ciconiiformes, penguins, and ratites, and is typically reported in the southern states.[6,70] Eastern equine encephalitis virus has caused deaths in assurance colonies of Florida grasshopper sparrows, in the United States.[64] Experimentally infected passerines develop leukopenia, suggesting the agent may be immunosuppressive.[122] Affected birds may have a short period of nonspecific clinical illness prior to death. Gross lesions are nonspecific and may include hepatosplenomegaly. Necrosis is a prominent histologic feature. Fibrinoid necrosis of the splenic ellipsoids with extension of fibrin into the surrounding red pulp is a relatively common finding and should raise the level of suspicion (Fig. 33.16). Random foci of acute lytic necrosis and parenchymal hemorrhage may be seen in the liver, and acute multifocal myocardial and systemic vascular necrosis may also be seen. Encephalitis is not a consistent finding in birds. Confirmation is by PCR or culture for EEE in suspect cases.

Infection with **avian influenza virus** has historically been infrequently documented in Passeriformes. However, natural infection, including infection with highly pathogenic H5N1 avian influenza virus **(HPAI)**, has been reported (HPAI is a World Organization for Animal Health [WOAH] reportable disease). Prior to the early 2020s, the latter included reports in bee-eaters, mynahs, and magpie robins in southeast Asia.[147] More recently, numerous passerine species have been infected during an ongoing (at the time or chapter preparation in 2023) highly pathogenic H5N1 pandemic. Affected passerine species have included American and fish crows, American robins, black-billed magpies, boat-tailed, common and great-tailed grackles, common ravens, dark-eyed juncos, red-winged blackbirds, tree swallows, and house sparrows in the United States; reports in passerines have been lower than what has been seen in several other bird groups such as gulls and pelicans. HPAI infections have also been experimentally induced in several species (e.g., common reed buntings, brown-eared bulbuls, thrushes, blackcaps, red-billed queleas, zebra finches, house finches, house sparrows).[15,57,123] Outcomes of HPAI infections can include high mortality and appear to be dependent on several factors including avian species, virus strain, and dose. Gross lesions may be absent or

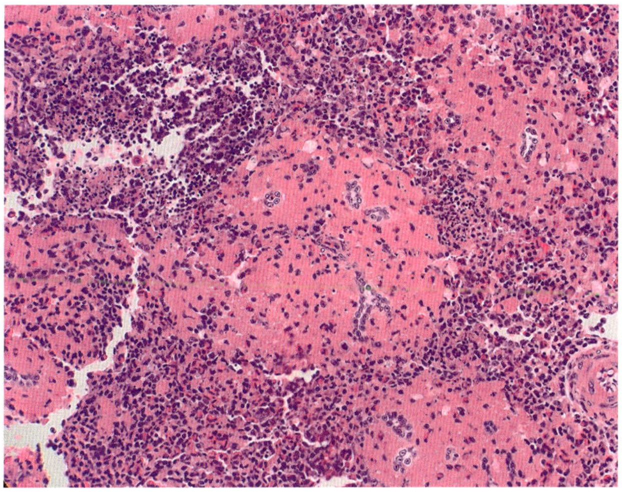

FIGURE 33.16 Eastern equine encephalitis virus infection in the spleen of a grasshopper sparrow. Infection is associated with multiorgan necrosis. In the spleen, fibrinoid necrosis of splenic ellipsoids, characterized by amorphous cellular and tissue degeneration, is a characteristic finding.

nonspecific (e.g., pulmonary congestion and edema), and demarcated red or white areas of discoloration may be seen in tissues (e.g., pancreas). Histologic lesions vary in severity and distribution and can include neuronal necrosis, neuronophagia, and neuropil vacuolization, as well as moderate to severe pancreatic acinar necrosis with heterophilic and histiocytic inflammation. Necrotizing rhinitis and sinusitis, lymphocytolysis (lymphoid patches, spleen), mild heterophilic monocytic hepatitis, myocarditis, and interstitial pneumonia can also be seen. Ancillary diagnostic tests such as IHC, ISH, virus isolation, and/or PCR are necessary to confirm the diagnosis and differentiate it from infection with other viruses such as Newcastle disease and West Nile viruses.

An orthoreovirus has been implicated in multiple large mortality events in American crows across North American.[50] Reoviruses are nonenveloped, double-stranded RNA viruses, and the newly described corvid orthoreovirus is distinct from another avian orthoreovirus that causes arthritis and tenosynovitis in domestic poultry. Corvid orthoreovirus-associated mortality events most often occur in the winter months ("winter mortality of crows") but can also occur at other times of the year. In one large outbreak in Southern California, juvenile and subadult birds in poor body condition were primarily affected, so immunosuppression, debilitation, and/or concurrent diseases were thought to predispose to infection and clinical disease. Gross and histologic findings commonly include multifocal to coalescing, hemorrhagic and necrotizing enterotyphlocolitis, and splenitis (Fig. 33.17A). Necrosis and hemorrhage in the intestines involve the full-thickness mucosa, including the crypts, resulting in ulceration and fibrinonecrotic

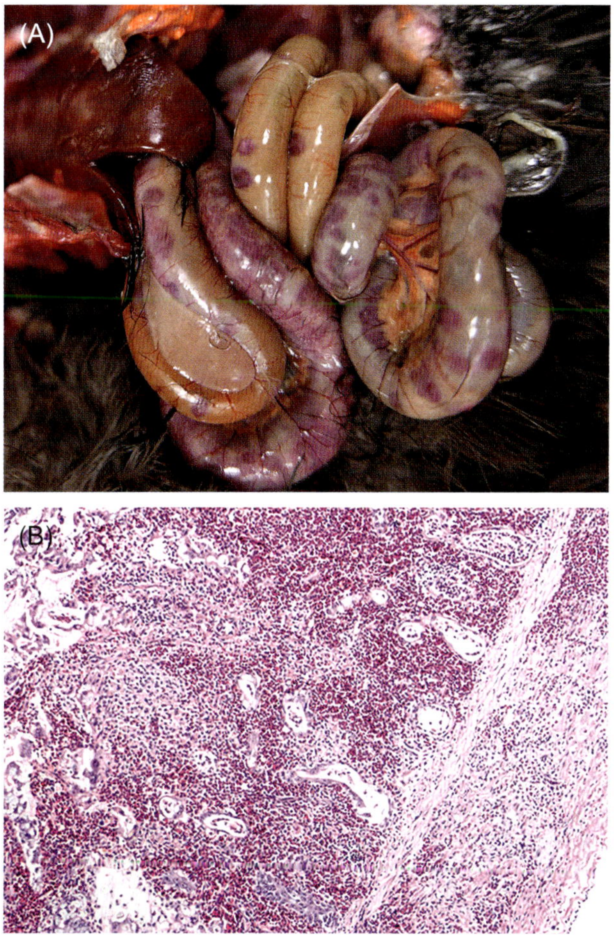

FIGURE 33.17 Orthoreovirus in the small intestine of a crow. (A) Characteristic gross lesions include multifocal to coalescing areas of red/purple discoloration in the intestinal wall. (B) This corresponds histologically to transmural hemorrhage and superficial to deep crypt necrosis and associated inflammation.

pseudomembrane formation (Fig. 33.17B). Heterophilic and lymphohistiocytic inflammation can extend transmurally. Rarely, intestinal epithelial cell syncytia may be seen. Lytic necrosis, fibrin, and cellular debris may be seen in the spleen. Confirmatory diagnostic tests include virus isolation, in situ hybridization, and PCR.

Avian bornavirus (ABV) infection in passerine birds was first described in 2009 in a canary in Austria that died with ganglioneuritis and encephalitis.[170] Viral antigen was detected by immunohistochemistry in neural and extraneural tissues, and sequencing revealed a novel avianbornavirus genotype. Subsequent work showed avian bornaviruses to be widespread in canaries in Germany, both in patients with neurological and gastrointestinal signs and in clinically normal individuals.[139] Sequencing revealed several distinct genotypes that were different from those involved in psittacine infections. In the same study, canaries that were infected experimentally developed cerebral and systemic infections that were subclinical during the 5-month observation period, and the virus was transmitted horizontally to sentinel birds. Embryonated eggs of infected canary hens contained specific ABV viral RNA, but the virus was not reisolated from embryos. So far, productive vertical transmission of ABVs in birds has not been demonstrated.[138] Canary bornaviruses in central Europe include CnBV 1, CnBV 2, and CnBV 3.[138] Other distinct passerine bornaviruses have been identified in estrildid finches in Germany and in munias in Japan.[138] More recently, canary bornavirus infection was detected by PCR and sequencing in captive and wild passerine birds in Brazil; phylogenetic analysis confirmed that these infections were due to CnBV-1.[125] In addition, a recent study has suggested but not confirmed a relationship between concurrent CnBV (1–3) and circovirus infections in the development of neurological and gastrointestinal signs (proventricular dilation disease-like signs) in canaries in Germany.[134] In Spain, cases of systemic ganglioneuritis, myocarditis, gastrointestinal leiomyositis, endocarditis and in some birds arteritis, pansteatitis, proventriculitis, and/or ventriculitis in canaries and other passerine species including tanagers have been seen since the early 2000s (Figs. 33.18A, B). Some that were tested for ABV infection were positive with immunohistochemistry (Juan-Sallés, unpublished data). Regardless of the strain or tissue distribution, inflammation is typically nonheterophilic, and inflammation of mural plexi in the gastrointestinal tract is a characteristic, though not pathognomonic, finding.

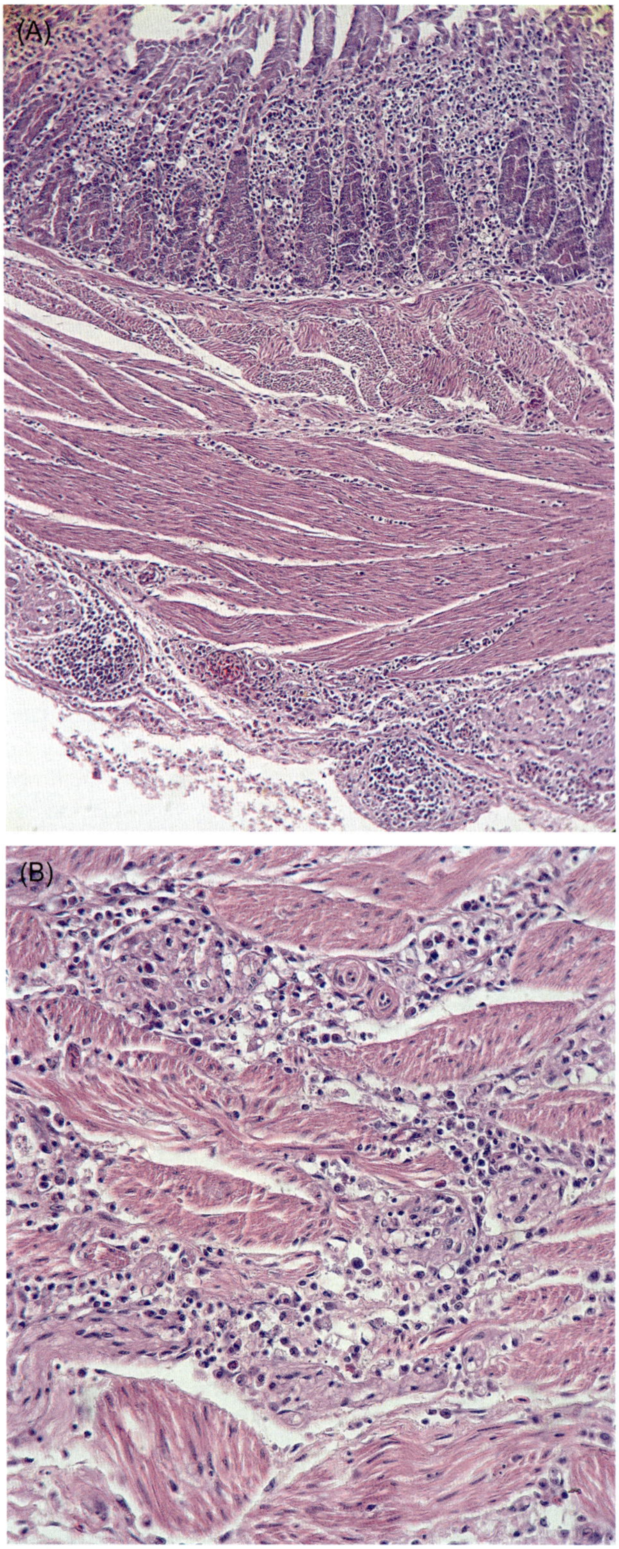

FIGURE 33.18 Avian bornavirus ganglioneuritis in the ventriculus of a seven-colored tanager. (A, B) Inflammation surrounds and infiltrates nerves in the muscular wall. It is often most apparent in the outer layers (myenteric plexus). (B) Inflammation is typically nonheterophilic.

Bacteria

Avian **mycobacteriosis** is a frequent finding in Passeriformes and zoo-housed Guam kingfishers. Mycobacteriosis in birds is generally caused by infection with ***Mycobacterium avium***, ***M. intracellulare***, or ***M. genavense***.[103,124] *M. avium* and *M. intracellulare* are two of the three nontuberculous mycobacteria grouped into the ***M. avium* complex (MAC; also called *M. avium-intracellulare* complex, MAI)**. Numerous serotypes exist within the MAC; virulence varies widely depending on the serotype and affected species.[143] Other mycobacterial species, including ***M. tuberculosis*** (canary), ***M. peregrinum*** (Gouldian finches), and ***M. simiae*** (kingfishers), have also been reported in passerines.[79,80,159,166]

Mycobacteria are ubiquitous in the environment, existing as saprophytic organisms in soil and water. Fecal shedding by infected birds can contribute to environmental contamination and a higher incidence of infection in some populations. In general, it is not a highly contagious disease. Mycobacteria are nonmotile, nonspore forming, Gram variable (due to variable staining of cell surface), acid-fast positive bacilli. Routine histologic stains, including HE and Wright-Giemsa stains, fail to stain the organisms, and non-staining, ghostlike bacilli (negatively stained) may be seen in the cytoplasm of macrophages, multinucleate giant cells, or be extracellular but intralesional (strain dependent) in cytologic (Fig. 33.19A) or histologic preparations. Special acid-fast stains such as the Ziehl-Neelsen or Fites are necessary to clearly identify the organisms. PCR testing from fresh or FFPE tissue may identify mycobacteria to the species and sometimes subspecies level. Whole genome sequencing has recently been used to more completely characterize strains.[124] Despite this, mycobacterial culture is still currently considered the "gold standard" diagnostic test for mycobacterial identification. A radiometric culture technique (BACTEC) has also been developed that is more rapid and may be more sensitive in identifying atypical species than traditional culture techniques.[155]

Infection can be seen in any tissue and can be local or widely disseminated. Hematological changes and clinical signs vary widely and depend on the affected species and infected organ/organ system. Chronic weight loss, ill thrift, hepatomegaly, intestinal thickening and dilation, and lethargy may be seen. Many birds are found dead with no premonitory signs. Diarrhea, dyspnea, and other respiratory signs or skin lesions, including thickenings, plaques, or distinct nodules, can also be seen but are less common. Heterophilic and monocytic leukocytosis may occur and can be dramatic. Common necropsy findings include hepatomegaly and splenomegaly (Fig. 33.19B). Others include pneumonia or airsacculitis, renomegaly, nodular lesions, or a thickened pale intestinal wall. In some cases, discrete granulomas with necrotic centers surrounded by macrophages, epithelioid macrophages, and multinucleated giant cells may develop, however, these are more commonly seen in waterfowl than in passerines. In passerines, inflammation is more often diffuse and histiocytic or is mixed histiocytic and heterophilic without granuloma formation. In HE-stained histologic sections, infected macrophages and multinucleate giant cells may have bluish-to-gray cytoplasm and pale-staining bacilli. Myriad intrahistiocytic bacilli are best highlighted with acid-fast stains (Fig. 33.19C). In the gastrointestinal tract, fusion of the villi and marked expansion of the mucosa and submucosa by histiocytes that contain numerous, acid-fast positive bacilli is common. Transmural infection and inflammation may occur in severe cases. An interesting and somewhat unusual manifestation in some species, especially small passerines (e.g., finches), is thickening of the elastic arteries at the heart base due to infection and inflammation (Figs. 33.19D, E; eSlide 33.e7). Mycobacteria in low numbers can sometimes be found in foci of pneumoconiosis, especially if the latter is more pronounced than expected for the species of affected bird. Acid-fast staining of these large foci of pneumoconiosis is recommended.

Salmonellosis is a common cause of mortality in passerines (so-called songbird salmonellosis). Most infections are due to ***Salmonella enterica* serovar Typhimurium** (commonly abbreviated *Salmonella typhimurium*). Several definitive phage types (DT) have been identified in outbreaks of songbird salmonellosis, including DT 40, 56v (56 variant).[56,109] Multiple mortality events have been documented across North America, Europe, Great Britain, Japan, and New Zealand. Documented outbreaks include those in greenfinches, house sparrows, tree sparrows, and pine siskins but a large variety of species have been affected in epornitics. Outbreaks appear to be more common in colder months or during migration, and some are associated with crowding of birds at feeding stations. Affected birds may be found dead or may show nonspecific signs of illness, including lethargy, fluffed feathers, and weight loss. Pasting of feces at the vent may be noted. Commonly described gross lesions include pinpoint white foci in the liver, spleen, and/or the gastrointestinal tract. Caseous exudates that form raised nodules in the crop and esophagus may be seen in more chronic cases. Hepatomegaly or splenomegaly is seen in some birds and in others, caseonecrotic sinusitis and wet, red lungs may be seen. Gross lesions correspond histologically to foci of necrosis with infiltrates of heterophils and macrophages, and few to numerous and occasionally large colonies of intralesional, Gram-negative bacilli.[58,129] Evidence of septicemia with multiorgan necrosis and subacute inflammation may also be seen.

Mycoplasma spp. play a role in many acute and chronic diseases in birds and mammals including humans. Mycoplasmas that frequently cause infection and disease in birds include ***Mycoplasma meleagridis***, ***M. synoviae***, and ***M. gallisepticum***. A diverse range of passerine species can carry or show evidence of exposure to *M. gallisepticum*, and mycoplasma-associated outbreaks have occurred in a number of wild passerine species.[31,32] ***Mycoplasma sturni*** has also been identified as a cause of conjunctivitis in blue jays, mockingbirds, and a European starling.[53,94,172] Affected birds present with periocular (eyelid and infraorbital sinus) swelling (Fig. 33.20A), clear to slightly cloudy ocular discharge, and crusting on the periocular skin. Some affected birds present with serous nasal discharge. Swelling and crusting may impair vision, and affected animals may be found on the ground near feeders or collide with objects. Histologically, lesions consist of lymphoplasmacytic conjunctivitis, rhinitis, and sinusitis with variable heterophilic inflammation, mucosal epithelial hyperplasia, erosions, and loss of cilia; lesions can also, but generally only rarely, extend into the lower respiratory tract (tracheitis, pneumonia, airsacculitis) (Fig. 33.20B). Corneal lesions are infrequently seen and consist of edema and epithelial hyperplasia. Gram stains are unrewarding as the organism does not stain with routine histochemical stains because it lacks

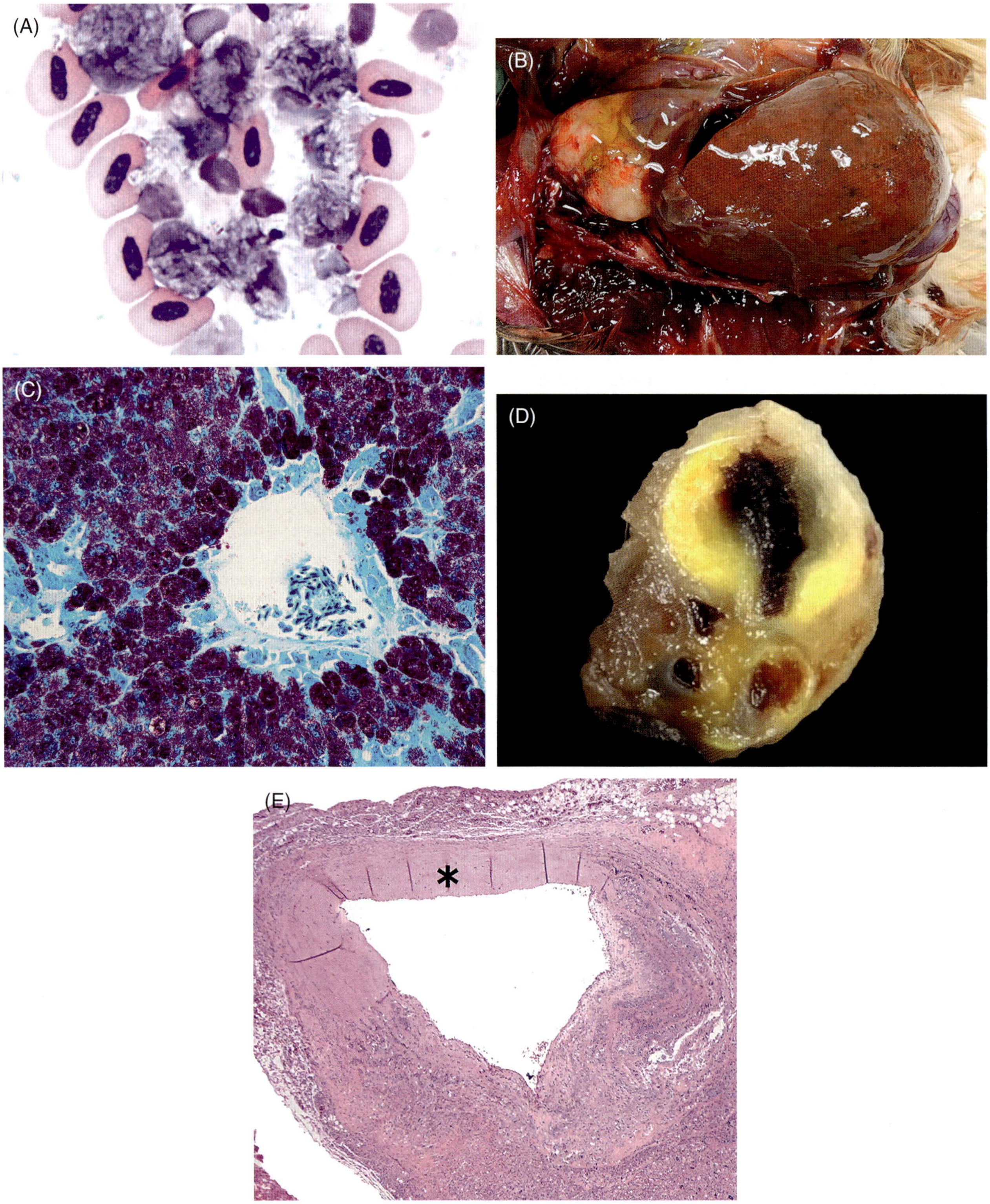

FIGURE 33.19 Avian mycobacteriosis in the liver of a bowerbird (A, B), liver of a Micronesian kingfisher (C), and the great vessels of a finch (D, E). (A) The bacteria can appear as unstained, elongate, ghost-like (negatively stained) structures in cytologic preparations. (B) Severe diffuse hepatomegaly (note the rounded liver lobe margins) and light brown discoloration reflect diffuse hepatic infection. Expansion and white discoloration over the heart base were due to regionally extensive infection. (C) The hepatic sinusoids are expanded by numerous macrophages containing myriad acid-fast bacilli, typical of avian mycobacteriosis. Ziehl-Neelsen. (D) Marked thickening with luminal narrowing and yellow discoloration of the arteries at the heart base are due to mycobacterial arteritis, medial necrosis, and arteriosclerosis with lipid deposition and mineralization. (E) Marked thickening of the arterial wall *(star)* is due to intense, transmural granulomatous infiltration with mycobacteria-laden macrophages (see eSlide 33.e7).

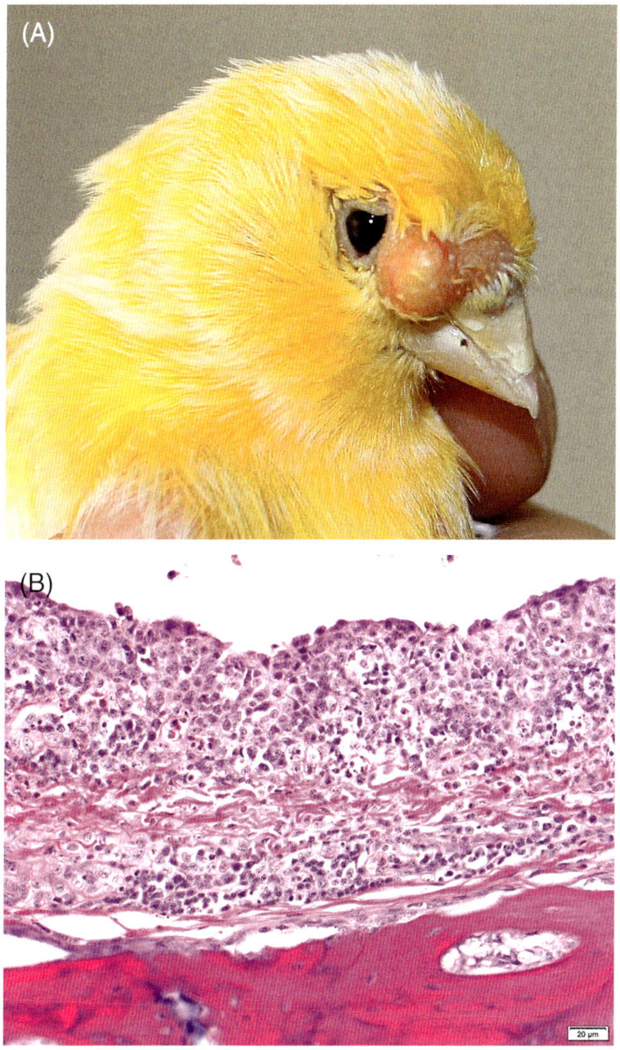

FIGURE 33.20 Infraorbital sinusitis in a canary (A) and tracheitis in a crow (B) due to mycoplasmosis. (A) The infraorbital sinus is markedly distended. Differential diagnoses include infection with *Mycoplasma* spp. or other bacteria, and more rarely poxvirus, flagellated protozoa, fungi, or mixed infections. (B) Inflammation can be mild to severe (the latter is transmural in this case) and is typically lymphoplasmacytic. *(Part A: Photo Courtesy of M. Máinez, Centro Veterinario La Marina Exóticos.)*

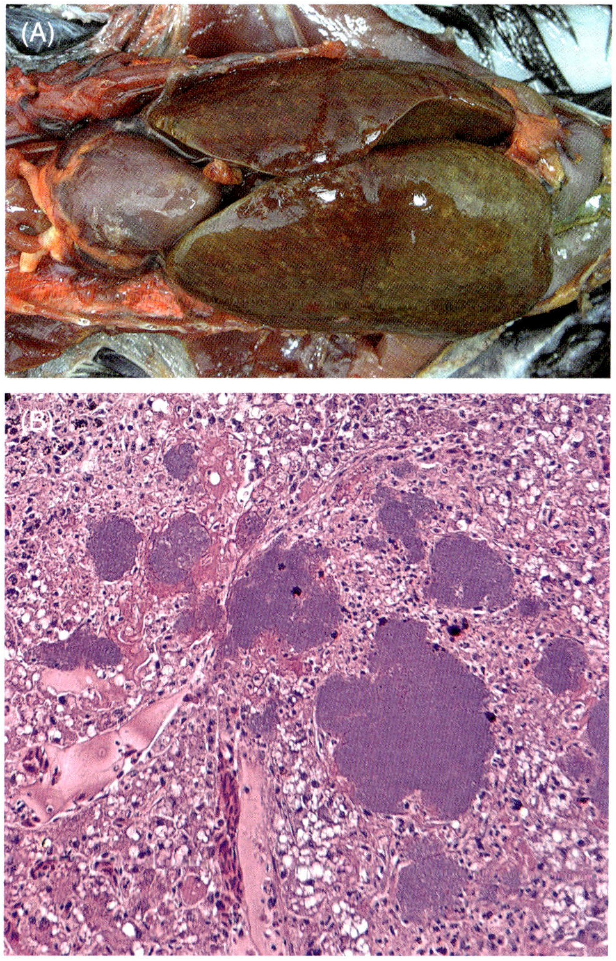

FIGURE 33.21 Pseudotuberculosis due to *Yersinia pseudotuberculosis* infection in the liver of a toco toucan. (A) The liver contains miliary yellow/tan foci and is enlarged due to infection and concurrent iron overload. (B) The miliary foci correspond histologically to multifocal hepatocellular necrosis and fibrin deposition with large, intralesional, multilobed, basophilic colonies of bacteria.

a cell wall. Immunohistochemistry is available at some laboratories and there may be cross-reactivity between mycoplasma species. Bacterial culture (using techniques specific to mycoplasma), PCR, or other molecular techniques coupled with characteristic lesions are used to confirm the diagnosis.

Pseudotuberculosis, caused by the gram-negative bacillus ***Yersinia pseudotuberculosis***, occurs in many species and has been reported in several passerine and toucan species in zoo settings.[22,118] Typical gross lesions across species include hepatosplenomegaly with small yellow foci throughout the parenchyma (Fig. 33.21A), and/or enteritis with enlargement of cecal tonsils. Pneumonia is less frequent and may produce intraluminal exudates. Areas of necrosis in the spleen and liver, and more rarely in the intestine, lung, and kidney, contain numerous large, multilobed colonies of gram-negative bacilli that may be surrounded by a narrow zone of degenerative heterophils (Fig. 33.21B). Bacterial emboli may be seen in capillaries of other tissues, especially the lung and heart. Bacterial culture or PCR is used to confirm the diagnosis; immunohistochemical labeling is also available in some laboratories or research settings.

Necrotic enteritis associated with ***Clostridium perfringens* subtype A** has been seen in toucans and hornbills. Infection can result in outbreaks or peracute death with no premonitory signs. The intestines may be distended by gas, the intestinal wall is thickened, and the mucosa is covered by red/brown to tan pseudomembrane (Fig. 33.22). This corresponds to fibrinonecrotic enteritis with abundant,

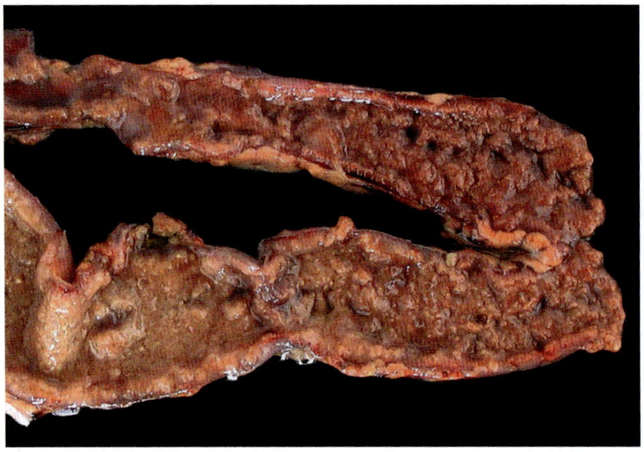

FIGURE 33.22 Necrotizing enteritis associated with *Clostridium perfringens* subtype A in a toco toucan. The intestinal wall is diffusely thickened and red/brown necrotic debris coats the mucosal surface.

intralesional pleomorphic gram-positive bacilli that often palisade in thick colonies along the denuded tips of necrotic villi. Foci of acute necrosis and hemorrhage with intralesional bacilli may also be seen in the liver. Affected birds are usually in good body condition. Anaerobic bacterial culture of intestine and PCR for *C. perfringens* toxins is needed to confirm the diagnosis. Possible predisposing factors have not been identified.

Clostridium piliforme **(Tyzzer's disease)** has occasionally been reported as a cause of encephalitis in young wild and captive birds from several orders, including passerines (starling, jay, weaver, finchbill), piciformes (woodpecker), and psittacines.[114] Nestling to fledgling birds are affected and present clinically with neurologic signs such as torticollis. Lesions are often limited to the brain and are characterized histologically by multifocal to regionally extensive areas of necrosis and mixed inflammatory cell infiltrates. Within the lesions and visible with silver stains, such as Steiner's or Warthin-Starry, are long argyrophilic rod-shaped bacteria in characteristic "haystack" arrangements (Figs. 33.23A, B). The bacteria can be confirmed as *C. piliforme* using PCR assays on fresh or paraffin-embedded brain tissue.

Though more commonly a pathogen of psittacines and columbiformes, ***Chlamydophila psittaci*** infection has been associated with high mortality in colonies of violaceous euphonia and members of the Trochilidae family (e.g., emerald hummingbirds).[115] Birds typically die after a brief period of illness or are found dead. Gross lesions consist primarily of hepatomegaly and splenomegaly; muscle atrophy and weight loss are also reported. Histologically, areas of necrosis and lymphohistiocytic inflammation are present in the liver, and sinusoidal macrophages may be prominent. Macrophages contain basophilic elementary bodies. Foci of necrosis and histiocytic and/or plasmacytic infiltrates may be seen in the spleen. Organisms are best identified with

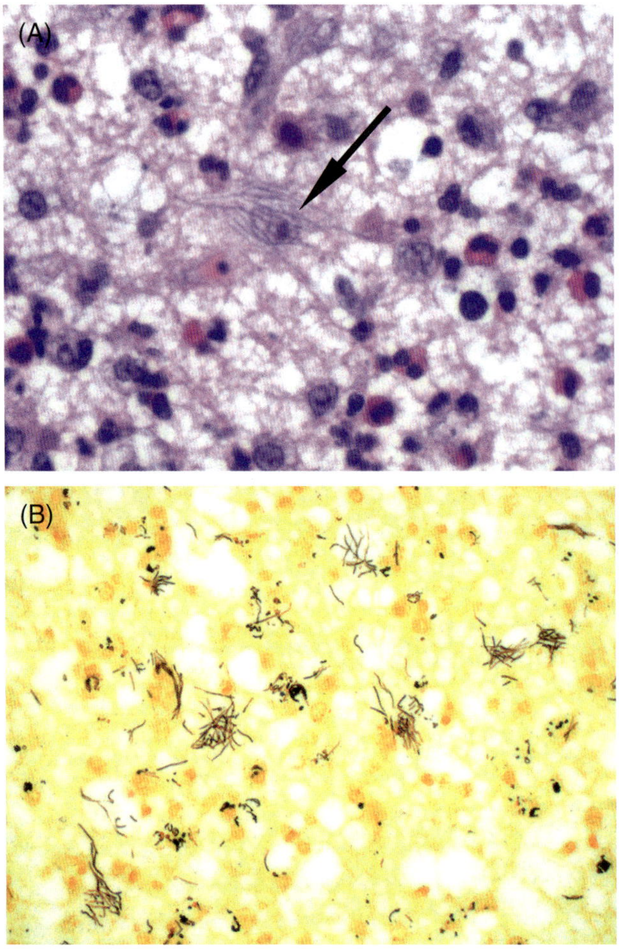

FIGURE 33.23 *Clostridium piliforme* (Tyzzer's disease) in the brain of a starling. (A) The bacteria are slender, intracellular, and often have a "stacked" appearance *(arrow)*. (B) They can be highlighted with silver stains. Steiner's.

Pierce-Vanderkamp (PVK; also called Gimenez) or Macchiavello stains and confirmed by PCR or immunohistochemical labeling.

Fungi

As occurs in other bird Orders, **aspergillosis** is a relatively common fungal infection in passerines. Its pathogenesis and the features of the disease are, generally, as described in other avian species and chapters in this book but with several somewhat unique manifestations. For example, vasoinvasive infections in the arterial branches at the heart base have been associated with thrombosis and complete, unilateral, pectoral muscle infarction in passerines. Birds may present clinically with a wing droop, and affected pectoral muscle is markedly pale relative to the contralateral muscle, and is often somewhat swollen, soft, and dry (Fig. 33.24A). Examination of the arteries in the cranial coelomic cavity on the affected side is likely to reveal a

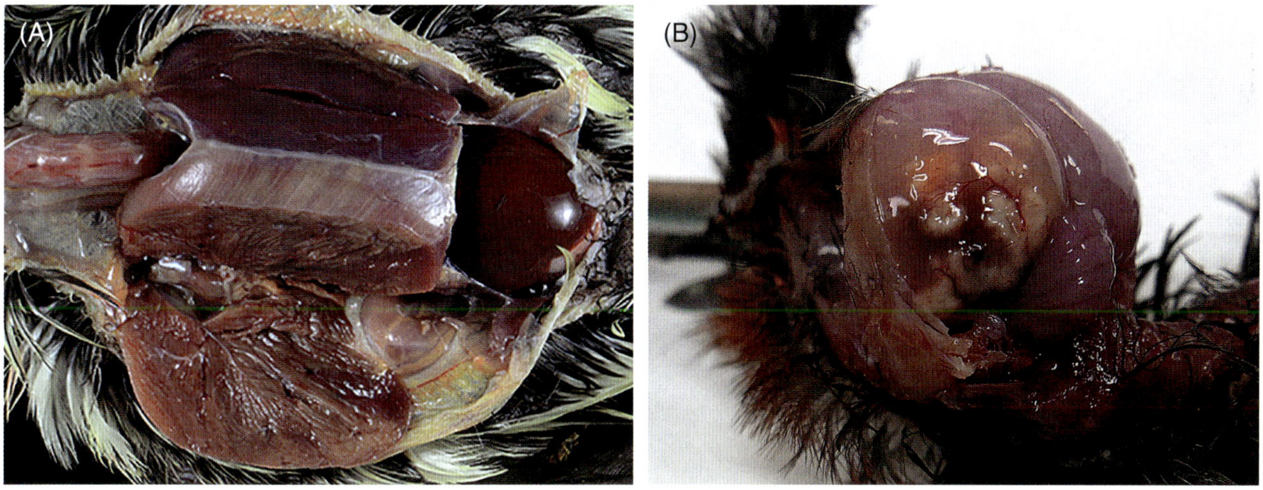

FIGURE 33.24 Aspergillosis-associated disease in the pectoral muscle of a metallic starling (A) and brain of a liocichla (B). (A) Thrombotic myonecrosis can present as unilateral, diffuse tan discoloration of the pectoral muscle. Muscle infarction is the result of vascular occlusion due to vasoinvasive fungal infection. (B). Infection and associated inflammation can be found in any tissue and result in multifocal to regionally extensive areas of necrosis (light tan in this image) that may be associated with congestion and/or hemorrhage.

thickened, discolored, or occluded vascular segment. Confirmation can be made by histologic examination. Mycotic angioinvasion can also result in arterial aneurism or rupture with hemocoelom or hemopericardium, valvular endocarditis, infarction of other tissues, such as liver, and hematogenous dissemination of infection. Spread to the brain resulting in large areas of necrosis and inflammation is sometimes seen, especially in small passerines (Fig. 33.24B). Due to their small size, occlusion of the trachea, syrinx, or bronchi may also be more common in passerines and hummingbirds than in larger birds. These lesions can be solitary and localized, and may be easy to miss if the entire trachea and its branches are not examined.

Candidiasis, infection with ***Candida* spp.**, is common in passerine birds; in piciformes infection may be associated with the high sugar content in diet formulations. Infection is often opportunistic. Factors that predispose to disease are concurrent clinical illness (e.g., bacterial or viral disease), vitamin deficiency (e.g., hypovitaminosis A), stress, poor husbandry conditions, malnutrition, or prolonged antibiotic treatment for other diseases.[141] In the upper alimentary tract, common lesions include mucosal hyperplasia and hyperkeratosis with crust formation, erosion, and ulceration. Large numbers of *Candida* yeasts and pseudohyphae may be present on the surface or through the mucosal epithelium. Yeasts are 2–6 μ, round to oval, and show narrow-based budding; pseudohyphae are 3–4 μ wide, septate, and parallel-walled. Special stains are useful for highlighting the yeast and pseudohyphae (Figs. 33.25A, B). Associated inflammation may be heterophilic in acute lesions or when the mucosa is intact, but as disease progresses and fungi invade, there is often an influx of histiocytes, lymphocytes, and plasma cells. Granulomas or microgranulomas oriented around fungi and eosinophilic necrotic debris may be seen in the deeper layers of the affected gut wall. Similar lesions occur in the proventriculus and ventriculus. At low magnification, the most notable feature is often degeneration of the koilin layer in the ventriculus. Infection in the lower alimentary tract may cause glandular and crypt hyperplasia, mucosal necrosis, erosions, or ulcers. Direct extension from the gastrointestinal tract may be associated with infection and inflammation in the liver or on visceral coelomic surfaces. When present, granulomatous, necrotizing bronchopneumonia is likely due to aspiration of organisms from lesions in the upper alimentary tract. Systemic dissemination can be associated with inflammation and granuloma formation in multiple organs, even in the absence of typical mucosal lesions. Less commonly, candidiasis may involve keratin layers of the skin, feather follicles, and beak.

Infection with the **avian gastric yeast**, ***Macrorhabdus ornithogaster*** (previously megabacteria), is not common in passerine species except in finches and canaries.[141] It is generally not associated with clinical signs or gross lesions, and is often considered an incidental histologic finding. However, it can be associated with wasting or excessive mucus production, particularly in some finch species such as Gouldian finches. Affected birds may also have goblet cell or glandular hyperplasia, or mild lymphoplasmacytic inflammation in the lamina propria. The bacillus-like organisms are most often identified in the mucus and glands in the proventricular isthmus (proventricular/ventricular junction) in cytologic preps (smears of superficial mucosal sample) or with histology (Figs. 33.26A, B; eSlides 33.e8 and 33.e9). The organisms are relatively large (2×20–$40\,\mu m$) and gram- and PAS-positive. They are also strongly positive with Calcofluor White MR2 staining.

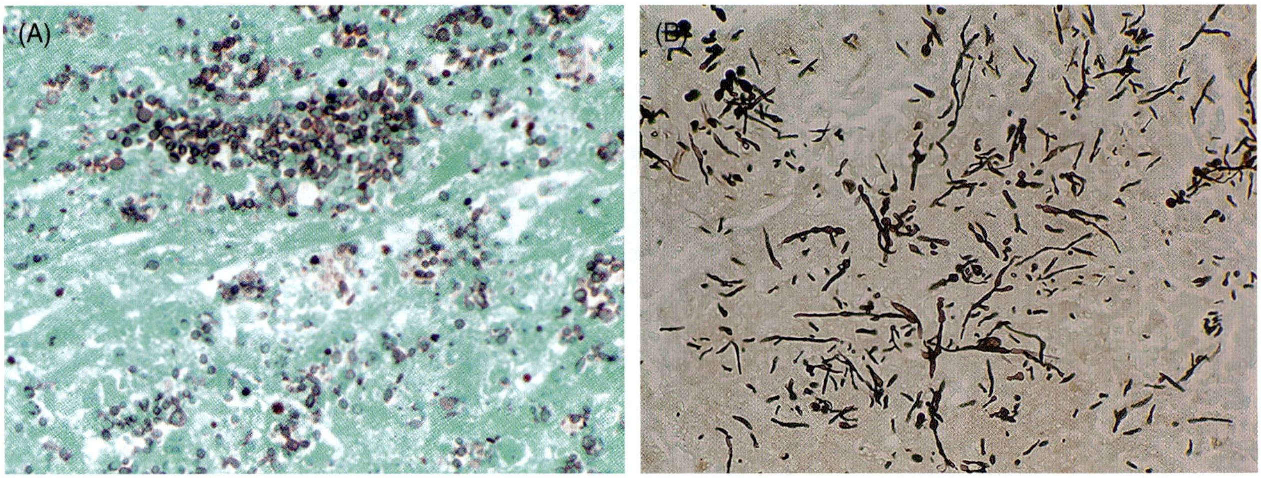

FIGURE 33.25 Candidiasis in a saffron finch (A) and Taveta golden weaver (B). (A) Myriad, round, intralesional yeast, and (B) septate, parallel-walled pseudohyphae are highlighted with special stains. A: GMS; B: silver stain.

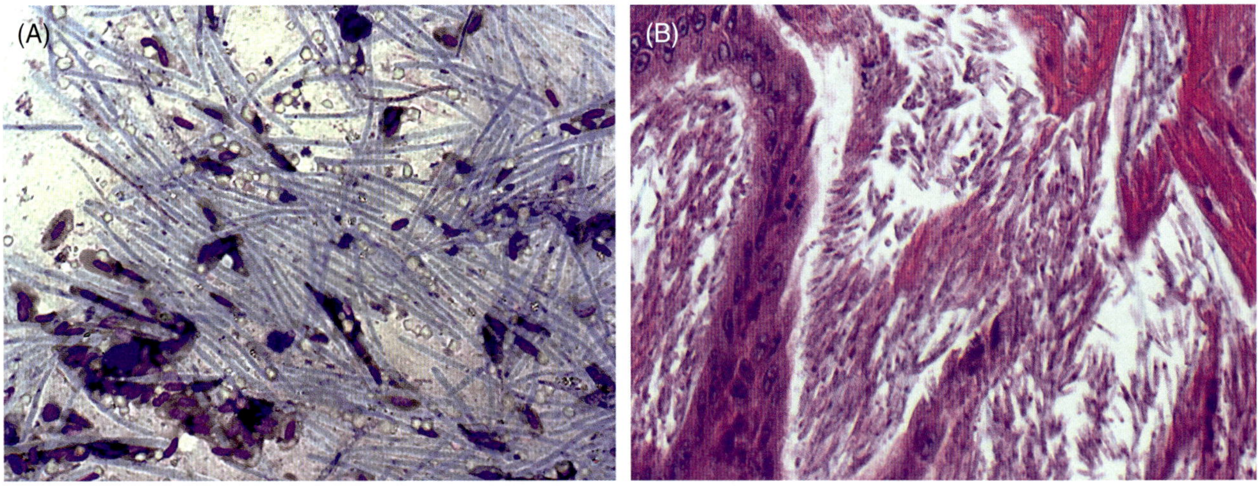

FIGURE 33.26 *Macrorhabdus ornithogaster* infection in the proventriculus of a European goldfinch. (A) These uninucleate yeasts are relatively large, bacillus-like organisms. Note their size relative to red blood cells in this cytologic preparation. The presence of a nucleus helps distinguish these from large bacteria. Modified Romanowsky stain. (B) The fungi are elongate, typically oriented perpendicular to the mucosal surface, and can form thick mats in heavy infections (see eSlides 33.e8 and 33.e9).

Malassezia yeasts are normal commensals on the skin of mammals and birds. The yeast has a characteristic "footprint" or "peanut-shaped" morphology. They are rarely seen in birds with clinically and histologically normal skin. Malasseziosis in passerines is most commonly seen in canaries and a variety of finches and other fringillids.[81] Disease can occur in pet, aviary, and zoo birds as isolated cases or outbreaks. Juvenile and adult birds are most commonly affected but *Malassezia*-associated skin disease has also been described in neonates. The most common lesions are feather loss and flaky/crusty skin, particularly on the head and neck (Fig. 33.27A). This corresponds histologically to epidermal and follicular hyperkeratosis, and perivascular dermatitis with numerous, round, oval, or morphologically characteristic peanut-shaped, intralesional yeasts (Fig. 33.27B). Affected birds often have concurrent diseases such as mycobacteriosis or other bacterial infections, systemic isosporosis, macrorhabdiosis, invasive candidiasis, or avian pox. Factors influencing *Malassezia* proliferation in affected birds and the role of *Malassezia* as a primary or secondary pathogen are not known but hypersensitivity to yeast antigens, hypovitaminosis A (and any conditions causing hyperkeratosis), and underlying immunosuppression are potentially involved.

Microsporidia are obligate, intracellular, spore-forming, 1–4 μm diameter, unicellular organisms that have recently been reclassified from protozoa to fungi. Spores are generally oblong with a thick, double-layered cell wall (refractile with polarized light and positive with trichrome or PAS staining), a polar cap, and a coiled polar filament

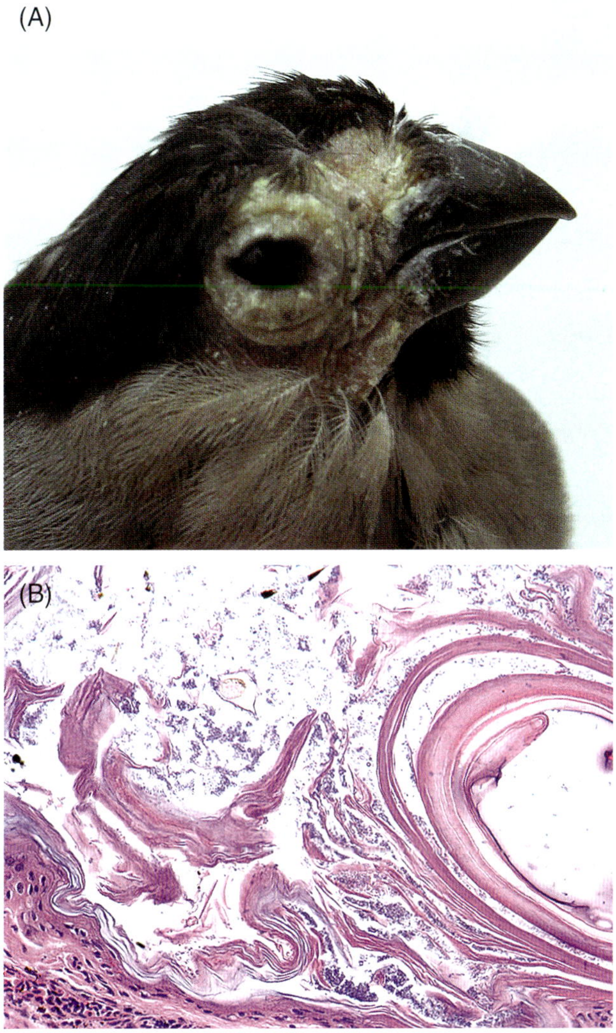

FIGURE 33.27 Malasseziosis in the skin of a bullfinch (A) and canary (B). (A) Infection can be associated with feather loss, cutaneous thickening, and hyperkeratosis. (B) Prominent epidermal and follicular orthokeratotic hyperkeratosis is associated with marked proliferation of *Malassezia* yeasts, which are small, basophilic, round to peanut-shaped organisms in the stratum corneum and on the surface of skin. *(Part A: Photo Courtesy of M. Máinez, Centro Veterinario La Marina Exóticos.)*

(the latter two are ultrastructural features). Cell-to-cell infection occurs when a spore extrudes its polar filament, inserts it through the cell membrane of an uninfected host cell, and injects its sporoplasm. Spores develop within an infected cell. Cell rupture releases spores to complete the cycle. In birds, pathogenic microsporidia have been documented in tissues, or more commonly feces, of a variety of species. In passerines, microsporidiosis has been reported in finches, canaries, and hummingbirds.[19,66,95,136,150] ***Encephalitozoon hellem*** has been identified by molecular methods in most cases. Infection occurs primarily in enterocytes in the small intestine. Tissue response varies from none to chronic, proliferative inflammation. Additional sites of infection in passerine birds include the conjunctival mucosa and bursal (Fig. 33.28A) and renal tubular epithelium. Expansion of intestinal villi due to epithelial cell hypertrophy and identification of morphologically consistent, intraepithelial organisms are the primary histologic findings. Tricolored parrot finches infected with an unidentified, intrahistiocytic microsporidian have been reported to develop granulomatous inflammation of the serosal surfaces of the gastrointestinal tract, air sacs, bone marrow, and conjunctiva.[66] Infected cells are usually recognizable in HE-stained sections of intestines and Gram, Stamp, or Luna staining may be helpful in identifying the intracellular organisms. Mature spores are gram-positive (Fig. 33.28B) but may be gram-variable depending on life stage. The spores stain magenta with the Luna technique.

Metazoa

Internal parasites are relatively common in free-ranging passerines. Given the great diversity of species and worldwide distribution, an exhaustive review of all parasites, internal and external, in this group of birds is beyond the scope of this text. The reader is referred to other published reference works for detailed lists.[9,51,144] Several notable parasites are discussed below.

Synhimantus (Dispharynx) nasuta is a proventricular spirurid nematode that contributes to debilitation and mortality in many species, most notable among groups in this chapter is infection in passerines and frogmouths. The primary lesion is ulcerative and proliferative proventriculitis, which can lead to anemia, anorexia, and death. Small (5 to 10mm) coiled white nematodes are present in the thickened mucosa. They are best visualized with a dissecting microscope (Fig. 33.29). Hyperplasia of the proventricular mucosa with dilated glands and mild to moderate lymphocytic and histiocytic inflammation with or without hemorrhage is seen histologically. Numerous thick-walled, embryonated ova are found in the proventricular glands. Transmission is indirect via terrestrial isopods (e.g., pillbugs and sowbugs).

Necrotizing ventriculitis with koilin degeneration and thickening due to infection with the spirurid ***Acuaria skrjabini*** has been reported in finches in Australia.[77,112] Birds may present with lethargy, inappetence, diarrhea, and weight loss. Nematodes burrow into the interface between mucosal glands and koilin. Ventricular spiruridiasis in passerines can also be caused by nematodes in the genus ***Procyrnea*** (Superfamily Habronamatidae), which has also been reported in woodpeckers and toucans.[40,120,146] Clinical signs may include emaciation, lethargy, and perching low or on the ground. Koilin fragmentation and thickening with ulcerative ventriculitis and intralesional nematodes are common histologic findings.

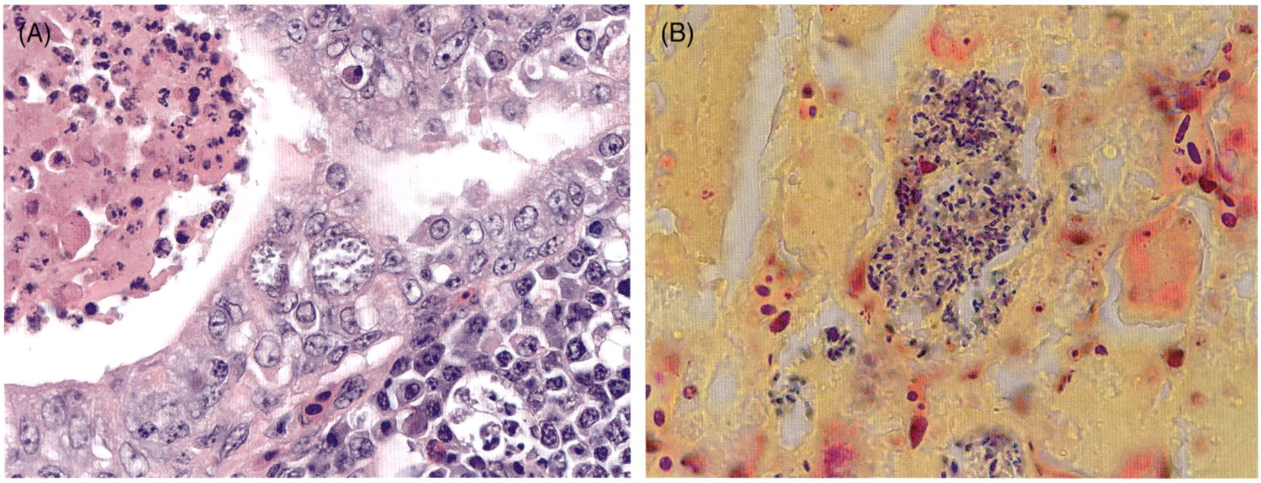

FIGURE 33.28 Microsporidiosis in the bursa of Fabricius of a green jay (A) and kidney of a waxbill (B). (A) Numerous small, oblong, basophilic, intracytoplasmic organisms cause marked expansion of infected bursal epithelial cells. (B) The organisms are positive with gram-positive. Brown and Brenn.

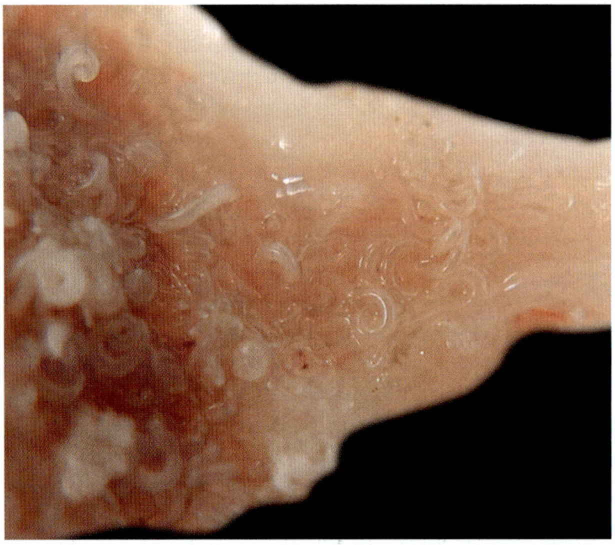

FIGURE 33.29 *Synhimantus* (*Dispharynx*) *nasuta* in the proventriculus of a saffron finch. The proventriculus is moderately thickened by numerous, slender, coiled, white nematode parasites that are partially embedded in the wall. Associated tissue damage includes multifocal erosions, ulcers, and red discoloration of the mucosal surface. *(Photo Courtesy of C. Hollinger, Wildlife Conservation Society.)*

Chandlerella quiscali, also called the grackle nematode, is a filarid worm that resides in the ventricles of the brain of its reservoir host, the common grackle.[4,11] The parasite is transmitted by mosquitoes, causes verminous encephalitis in numerous species of wild passerines in the United States, and has also been described in emus and a caracara.[39,89] It is noteworthy that the microfilaremic stages of this parasite and of ***Eufilaria*** sp. potentially have a role in the facilitation and amplification of arbovirus transmission, specifically West Nile virus, by mosquito vectors.[164]

Geopetitia aspiculata is a tetramerid spirurid that burrows into the serosal surfaces (Fig. 33.30) of the esophagus, proventriculus, and ventriculus of Passeriformes, Charadriiformes and Coraciiformes.[12,55] Intermediate hosts include crickets and cockroaches. Weight loss and distention of the coelomic cavity may be noted clinically. Lesions consist of granulomatous inflammation around encysted nematodes. Nematodes may penetrate the capsule of the adjacent liver in some birds.

Other relevant spirurids in these avian taxonomic groups include ***Monopetalonema*** (Diplotriaenoidea superfamily), which occurs in the air sacs of kingfishers and can cause airsacculitis and pneumonia,[16] and ***Thelazia anolabiata***, which targets the periocular tissues and may cause keratoconjunctivitis and corneal opacity in some affected birds.[41]

Calodium* spp**. (previously ***Capillaria) are commonly seen in free-ranging and zoo-housed passerines and are reported in captive frogmouths and toucans. The parasite is typically found in the mucosa of the crop, esophagus, and intestine in association with areas of epithelial hyperplasia, hyperkeratosis, erosion, and ulceration. Inflammation is often absent or minimal in the absence of ulceration, but it can include mucosal infiltrates of heterophils, lymphocytes, and plasma cells in some cases. Characteristic features are bacillary/hypodermal bands in the adult aphasmid nematodes and thick-walled, bioperculate intralesional ova.

In tawny frogmouths, severe neurologic signs with high mortality have been reported in free-ranging birds due to infection with the rat lungworm ***Angiostrongylus cantonensis***. The nematode is transmitted by slugs and snails, and ingestion by an aberrant host (including passerines and humans) may result in neural larval migrans. Histologic lesions in the brain include malacia, gliosis, and eosinophilic meningoencephalitis.

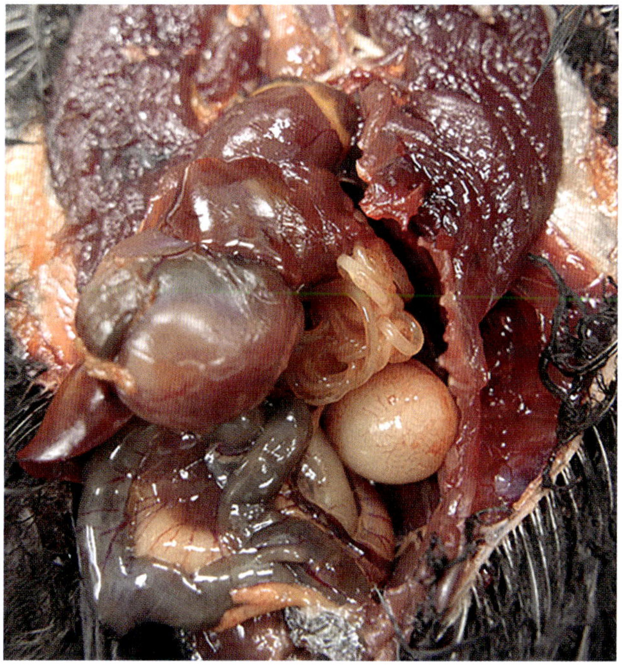

FIGURE 33.30 *Geopetitia aspiculata* infection in the coelom of a blackbird. These slender, white nematodes burrow into serosal surfaces. Infection with single individuals or large aggregates of the parasite, as in this case, can be seen.

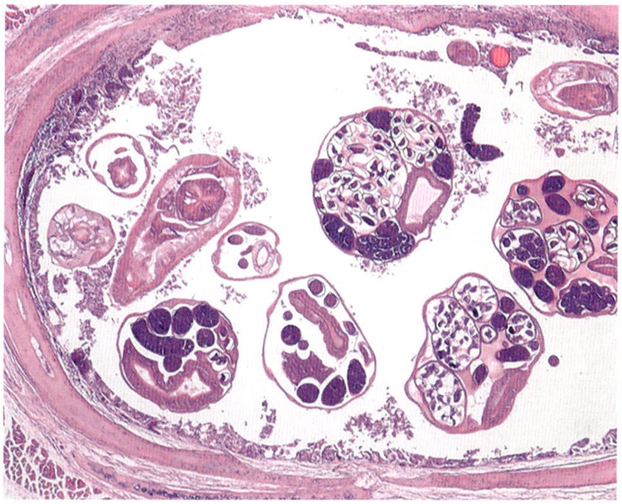

FIGURE 33.31 *Syngamus* sp. tracheitis in a common hill mynah. Luminal nematodiasis is associated with extensive ulceration, necrosis, and inflammation of the tracheal mucosa. Characteristic features of this nematode include platymyarian subcuticular musculature, an intestine with a prominent brush border, vacuolated lateral cords, and uteri distended with numerous thick-shelled, embryonated eggs.

Several species of ***Syngamus*** are found in the respiratory tract of passerines; infection is more commonly associated with disease in gallinaceous birds (see Chapter 31: Galliformes and Columbiformes).[82] When present in large numbers, they may cause obstruction and asphyxia or produce inflammation, necrosis, ulceration, and intraluminal hemorrhage (Fig. 33.31). Intraluminal hemorrhagic exudate admixed with nematodes may cause tracheal obstruction and be noted grossly as segmental red discoloration. Nematodes may be attached to mucosal surfaces by their buccal apparatus at sites of ulceration. Eggs that are deposited in the lungs may cause granulomatous inflammation in the parabronchi.

Proventricular nematodiasis with visceral larval migrans and associated granulomas in the coelom and coelomic viscera caused morbidity and subsequent death (or euthanasia) in three wrinkled hornbills in a zoo collection. The identity of the larval nematode was not determined by morphologic or molecular methods, although gravid tetramerids were present in the proventricular mucosa.[44]

Intestinal cestodiasis is rarely a clinical problem in passerine birds but parasitism with ***Choanotaenia*** sp. was found to be a cause of disease and death in captive populations of Australian finches.[113] Young adult birds were primarily affected and grossly had small intestinal distension with numerous luminal cestodes and watery content. Cestodiasis has also been described in a colony of tricolored parrot finches. Infection was associated with intestinal impaction and may have contributed to debilitation when large numbers of parasites were present. Definitive identification was not made, but cestodes were morphologically consistent with those in the family Davaineidae. Intestinal cestodes are a common incidental finding in free-ranging hummingbirds in California (United States) (Fig. 33.32).[100]

Protozoa

Coccidia are apicomplexan protozoal parasites with worldwide distribution. Those in the genus ***Isospora*** are common in many birds. *Isospora* spp. with an extraintestinal life

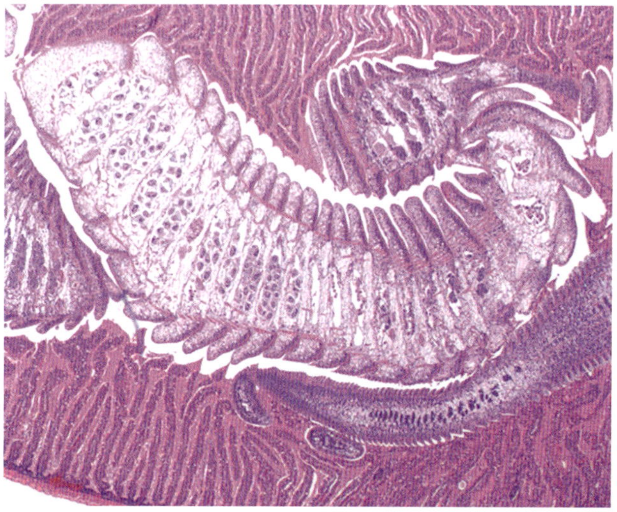

FIGURE 33.32 Cestodiasis in the intestine of a hummingbird. Characteristic features of the parasite include a segmented body (proglottids), thick tegument, and reproductive organs within each proglottid.

stage can cause **systemic isosporosis** (formerly referred to as **Atoxoplasmosis**). Systemic isosporosis is a major protozoal disease of passerine birds. Controversy and uncertainty still exist as to its exact taxonomic classification and molecular studies have questioned the traditional classification of these coccidian parasites.[142] Additionally, there have been differing findings regarding pathogenesis, cellular targets, and the nature of the associated common lesions.[26] The terminology in some texts and references adds confusion. Systemic isosporosis has been traditionally attributed to infection with apicomplexan protozoans characterized by a life cycle with sexual reproduction (gametogony) in the enteric epithelium and production of oocysts with *Isospora* morphology after sporulation, as well as an asexual replication (merogony) phase within the cytoplasm of circulating mononuclear leukocytes. Monocytes are often mentioned along with lymphocytes as targets of the merogonic phase of the life cycle of these parasites, and immunohistochemical studies have confirmed infection in lymphocytes.[26] Infection is assumed to occur after ingestion of sporulated oocysts, and a recent study suggests the possibility of vertical transmission but no conclusive evidence of this has been shown.[117]

Infection has been reported in a wide range of passerine birds from house sparrows and canaries to a variety of finches, starlings, siskins, tanagers, weavers, jays, and endangered species such as the Bali mynah, blue-crowned laughing thrush, and Florida grasshopper sparrow.[1,26,45,47,64,68,86,108,117,140,142] Phylogenetic analysis of genetic diversity in the cytochrome C oxidase gene (subunit I) of *Isospora* from a large variety of passerines showed that sequences generally grouped by host species or family, indicating a relatively strong degree of host-specificity.[86] Disease is reported in wild and captive settings, and in the latter, infection and disease may be influenced by certain husbandry conditions including crowding, exposure to infected free-range birds, suboptimal sanitation with accumulation of infective oocysts, and stress.[47] Mortality is most common in fledglings but disease can also occur in adults. Clinical signs may include anorexia, weight loss, ruffled feathers, diarrhea, hepatomegaly with distention of the coelomic cavity, and muscular atrophy. Cytological evaluation of percutaneous fine-needle aspirates of the liver in live birds with hepatomegaly can be used for a rapid diagnosis,[99] and postmortem lung, liver or splenic touch preparations for cytologic examination (eSlide 33.e10) are recommended as organisms can be difficult to see in histologic sections.

Severe diffuse splenomegaly is a common gross finding in birds with systemic isosporosis (Fig. 33.33A). Hepatomegaly and atrophy of adipose tissue and skeletal muscles may also be seen. In fringillids, including canaries and a variety of finches, marked thickening of the duodenum is the most consistent gross finding (Fig. 33.33B). Body condition may be normal in acute cases. In cases with associated myositis, streaking in the cardiac and peripheral skeletal muscle may be seen. In some cases, gross lesions may resemble those of mycobacteriosis and cytological evaluation of the duodenal mucosa, liver, and spleen is useful in differentiating the two diseases.

Lymphohistiocytic inflammation in the liver, intestine, and splenic lymphoid hyperplasia are common (eSlides 33.e10 and 33.e11). However, histologic lesions may be more apparent and severe in other organs; this may differ by host species and stage of infection. In the intestine and other tissues, lymphocytic infiltration can be intense (Fig. 33.33C) and lymphocytes can be atypical or blast-like. Interestingly, the blast-like lymphocytes may contain meronts. Neoplastic transformation has been suspected in some cases, but conclusive evidence of neoplastic transformation to lymphoma, such as the clonality of the infiltrative cells, has not yet been demonstrated.[26] Marked cellular infiltration in the mucosal lamina propria is commonly accompanied by infiltration of similar cells in the muscular tunics, visceral nervous ganglia, and serosal surfaces; severe transmural inflammation may be accompanied by marked thickening and loss of normal architecture of villi (Fig. 33.33C). As in cytologic preparations from tissue imprints and aspirates (Fig. 33.33D), infected monocytes in HE stained sections contain one to multiple, round, approximately 2 μm in diameter, intracytoplasmic protozoal organisms (merozoites; asexual stage) that cause characteristic indentation of the cell nucleus (Fig. 33.33E). In many cases, asexual (mononuclear cell) and sexual (enterocyte) stages are seen. Margination of nuclei in infected mononuclear leukocytes, most notably in the microcirculation, is often prominent and may be widespread in tissues particularly the intestine, liver, spleen, lung, and bone marrow. In the liver, cholangiohepatitis may be accompanied by foci of necrosis that is scattered randomly in the parenchyma. A unique manifestation of isosporosis in white-rumped shamas is characterized by infiltration of various tissues with nodular aggregates of atypical epithelioid macrophages containing intracytoplasmic protozoa (Fig. 33.33F).[175] The most consistently involved tissue is skin (dermis), but these lesions also occur in liver, spleen, and gastrointestinal tract.

Enteroepithelial coccidiosis, primarily with various *Isospora* spp., can also occur in passerine birds particularly canaries and finches. Clinical signs in affected birds include diarrhea, weight loss, and dilation of the intestinal loops. Coinfection with systemic isosporosis may be noted in some affected birds. Sexual stages of coccidian parasites are observed within the cytoplasm of enterocytes and oocysts may be seen in the intestinal lumen. In severe cases, diffuse infection with high numbers of coccidia may be seen (Fig. 33.34).

Cryptosporidiosis, a disease caused by different species of the apicomplexan parasite ***Cryptosporidium***, is common in passerine birds (over 60% in one study).[135] The proventriculus is the most common site of infection, and infection in the intestine and cloaca may also be seen (Fig. 33.35). ***Cryptosporidium galli*** has been identified as

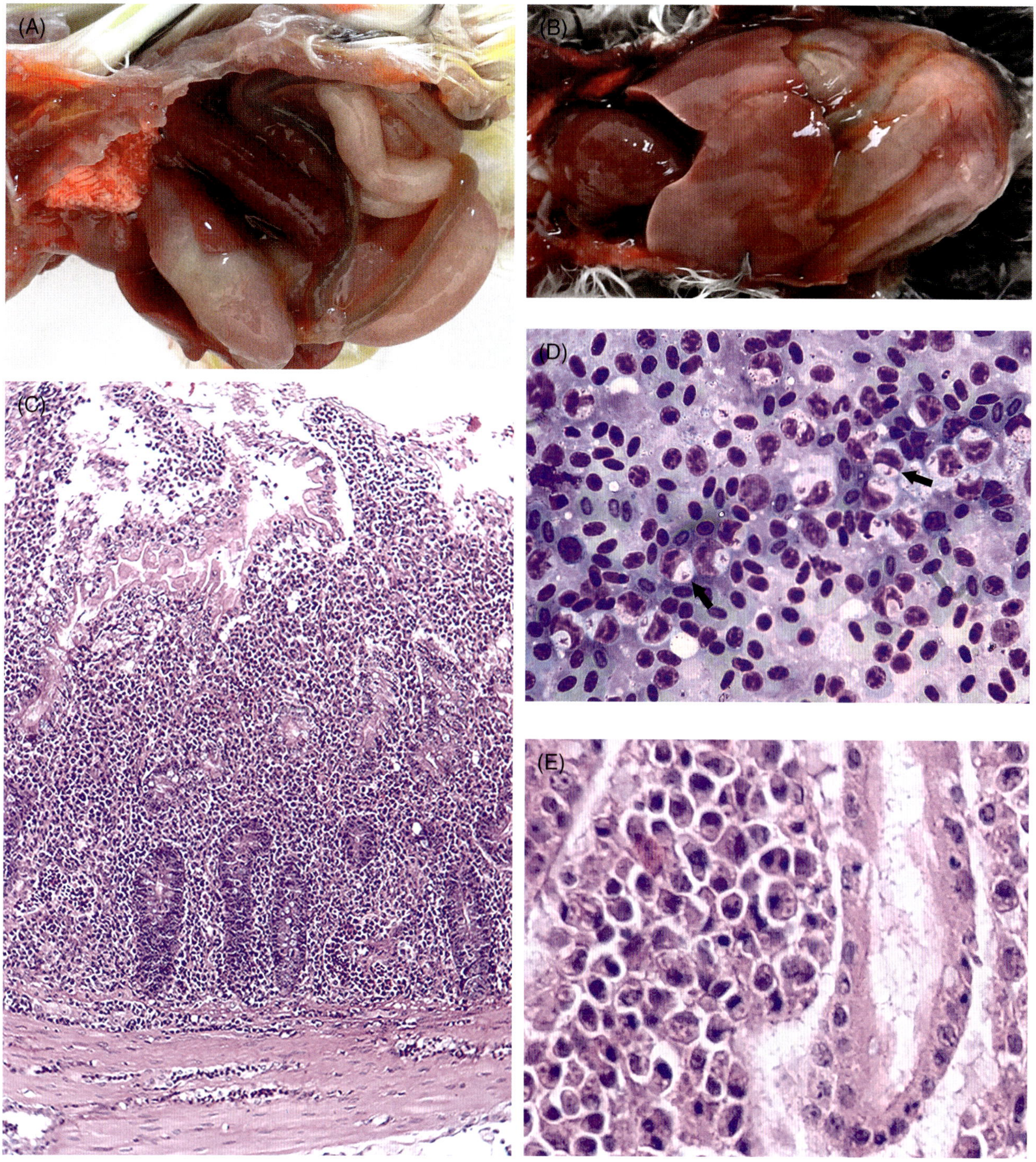

FIGURE 33.33 Systemic isosporosis in finches and canaries (A–E) and cutaneous isosporosis in a shama thrush (F). (A) Severe diffuse splenomegaly is a common finding in birds with systemic infection. (B) In some species, severe thickening of the intestinal wall is a common finding. (C) Severe enteritis is associated with infiltration of the lamina propria and expansion and replacement of normal tissue architecture by numerous lymphocytes and histiocytes. Similar inflammation is also present multifocally in the muscular layers and ganglia in the muscular tunics. (D) Large, round to oval, intracytoplasmic, protozoal zoites with a central basophilic nucleus cause characteristic indentation of the nucleus (arrows) of infected cells and are easily seen in touch preparations from necropsy samples. Modified Romanowsky stain. (E) Most inflammatory cells in the lamina propria of the duodenum contain one to multiple round, amphophilic to basophilic cytoplasmic merozoites.

(Continued)

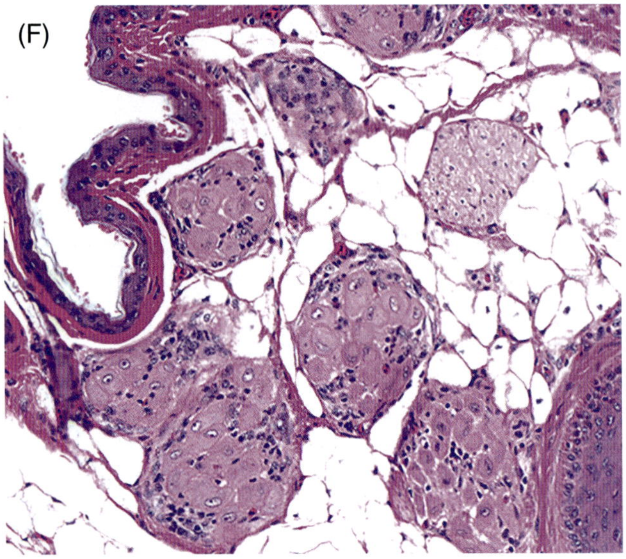

FIGURE 33.33, CONT'D (F) Aggregates of atypical epithelioid macrophages containing one or multiple cytoplasmic zoites are an unusual manifestation of isosporosis in white-rumped shama thrushes (see eSlides 33.e10 and 33.e11).

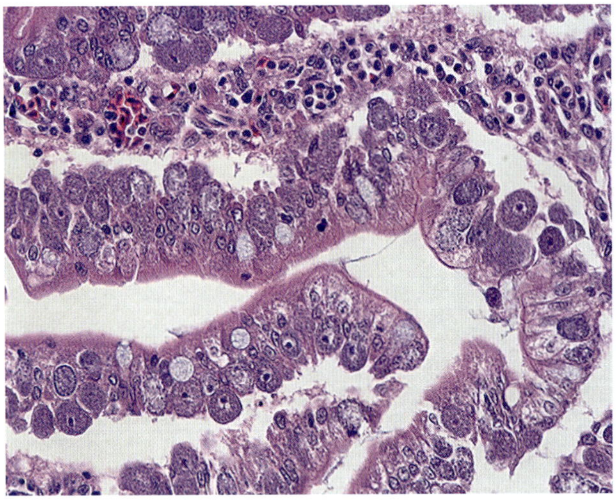

FIGURE 33.34 Enteroepithelial coccidiosis in a finch. Severe thickening of the intestinal mucosa is due to expansion of epithelial cells by large, intracytoplasmic coccidian parasites. Numerous macrogamonts (large central nucleus) and fewer microgramonts (numerous peripheral small nuclei or microgametes) are present in this image.

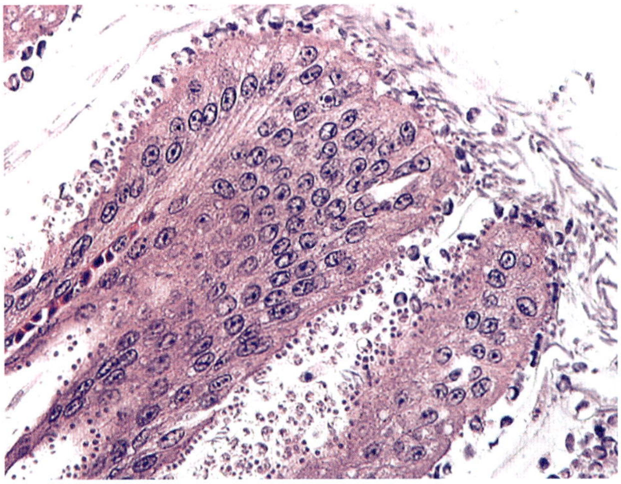

FIGURE 33.35 Cryptosporidiosis in the proventriculus of a canary. The mucosal epithelium is hyperplastic and heavily colonized by various sexual stages of this protozoal parasite, which appear as small, dense to larger, basophilic structures along the mucosal surface. Concurrent luminal infection with *Macrorhabdus* (see Fig. 33.26) is also present within glandular lumina and on the mucosal surface.

the cause of proventricular cryptosporidiosis in some studies.[5,28] In the proventriculus, cryptosporidia may be associated with mucosal hyperplasia, lymphoplasmacytic inflammation, mucous metaplasia of proventricular glands, and luminal hemorrhage. Atrophy of adipose tissue may also be seen; however, most of these lesions appear to be more commonly seen in psittacine birds and passerine infections may have no associated architectural changes in the proventricular mucosa. Intestinal lesions more commonly consist of atrophy and fusion of villi. Cryptosporidiosis can frequently present with one or more concurrent infectious diseases including proventricular macrorhabdiosis, systemic isosporosis, candidiasis, mycobacteriosis, and poxvirus.[135]

Infections with flagellated protozoa, most commonly ***Trichomonas gallinae***, occur in captive and free-ranging passerine birds. **Trichomoniasis** caused by a clonal strain of *T. gallinae* emerged in passerine birds, primarily finches, in 2005 in the United Kingdom.[61,90] It rapidly spread to continental Europe, likely by migratory birds, and common chaffinches are suspected to be the main vector. The common practice of supplemental feeding of wild birds using backyard bird feeders is suspected to be an additional factor in pathogen transmission.[20,91,177] In North America, trichomoniasis was first noted in wild finches in Canada in 2007.[49,110] Lesions most commonly involve the oropharynx, esophagus, and crop and consist of necrosis, ulceration, and epithelial hyperplasia with mucosal infiltration of lymphocytes, plasma cells, heterophils, and macrophages. Colonies of flagellated protozoa and bacteria may be seen in foci of necrosis but can also be difficult to distinguish from necrotic histiocytic cells. In Spain, flagellated protozoal infections occur occasionally in aviary canaries, particularly as outbreaks of respiratory tract disease with high mortality due to tracheitis and pneumonia. The gross lesions clinically mimic poxviral infections. The flagellates in the canaries have not been molecularly identified (Juan-Sallés, personal observation). Other presentations include infraorbital sinusitis and more sporadically, sialoadenitis (Figs. 33.36A, B), conjunctivitis, or cervical infections without tracheitis or pneumonia. High numbers of intralesional flagellated protozoa are typically seen, usually palisading over the surface of infected epithelia or mucosa. Protozoal emboli are occasionally also seen.

Hemoparasites, including species of ***Plasmodium*** (the cause of **avian malaria**) and the related hemosporidians ***Leukocytozoon*** and ***Haemoproteus***, are a diverse group of protozoal pathogens that infect a wide range of species including many avian species. In Passerines, infection is common and disease outcomes are dependent on affected host and protozoan species. For example, infection with ***P. relictum*** is a cause of notable population declines in several populations of Hawaiian forest birds.[9,88,163,168] Mortality is highly variable between species. For example, iiwi and kiwikiu show high mortality, apapane appear to be resistant, and resistance may be developing in the amakihi.[8,9,88] Some species of migrating songbirds do not appear to be adversely affected by blood stages of some *Plasmodium* or *Haemoproteus* infections,[24] while in others, asexual tissue (megaloschizont) stages of *Haemoproteus* can cause severe hepatic and renal lesions.[36] There is also speculation that some species of passerines may be reservoir hosts for virulent hemoparasitic infections in other avian species.[92]

The parasites are transmitted by biting insects, including mosquitoes and blackflies (Simulidae). Replication occurs in the insect gut and gametogenesis occurs in the stomach. Fertilization results in the production of motile ookinetes that develop into oocysts in the epithelium of the stomach. Following rupture of the oocyst, released sporozoites travel to the salivary glands and are subsequently injected during feeding into the avian host. Hematogenous stages of *Leucocytozoon* and *Haemoproteus* are usually subclinical, although heavy infections can cause anemia and clinical morbidity, especially in young birds.

Clinical signs in birds with avian malaria include weakness, lethargy, and a fluffed appearance; some birds are found dead with no premonitory signs. Anemia due to damage to erythrocytes may be severe. Grossly, hepatosplenomegaly and pulmonary pallor are common. Impression smears of the lung or smears of peripheral blood may identify plasmodial life stages in erythrocytes, including schizonts; intraerythrocytic schizonts are not seen in infections with *Leucocytozoon* or *Haemoproteus* (Fig. 33.37A). The organism and various life stages are often more easily seen in blood films or cytology of tissue imprints than with histologic examination of tissue sections. Infection may be associated with necrosis and inflammation in the spleen and liver, and malarial schizonts, which are approximately 16–25 μm by 11–16 μm, may be seen in any tissue but are most commonly found in the spleen, liver, or lung (Fig. 33.37B). Pulmonary hypoperfusion, edema, moderate histiocytic interstitial pneumonia with cellular phagocytosis

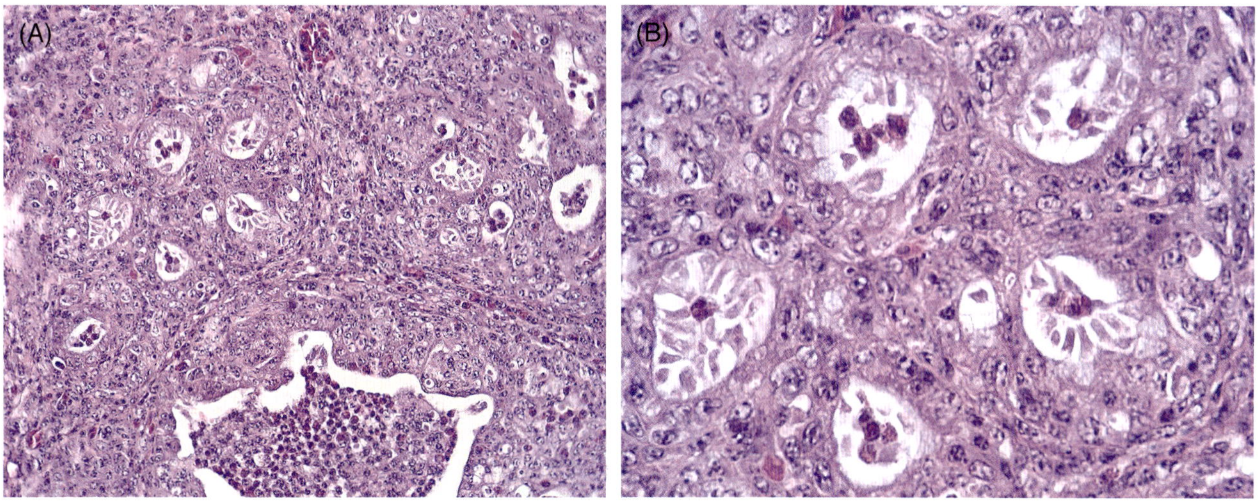

FIGURE 33.36 Flagellated protozoa in the salivary glands of a canary. (A) Numerous elongate protozoa are present within many of the glands. (B) Infection can be associated with epithelial hyperplasia and mixed histiocytic and heterophilic inflammation.

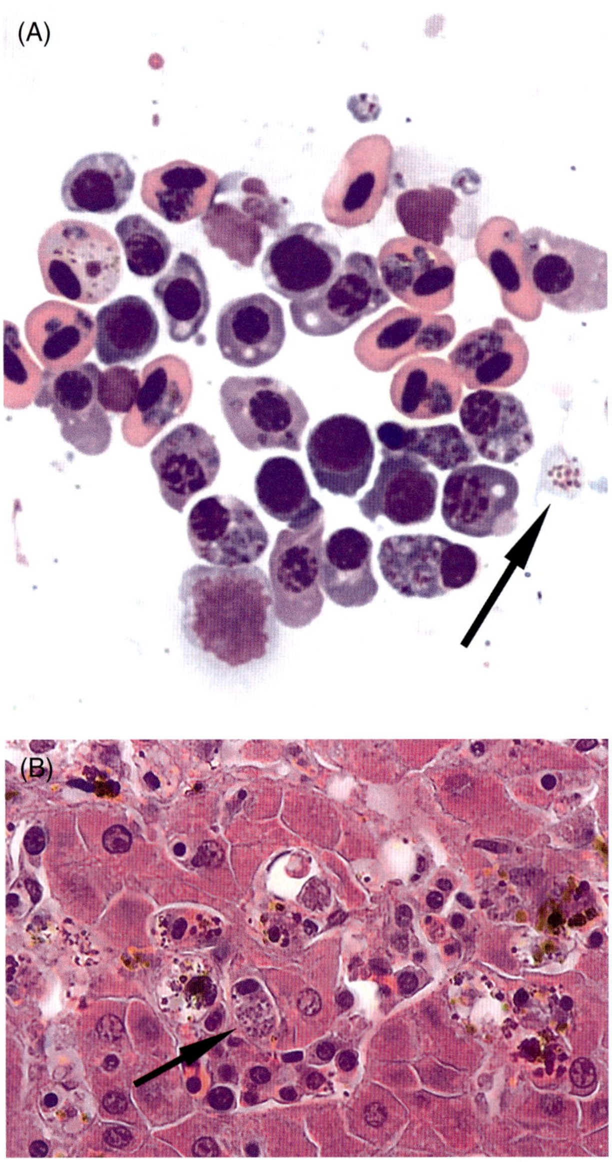

FIGURE 33.37 Avian hemoparasite infections in the blood of a Maui parrotbill (kiwikiu) (A) and liver of a bulbul (B). (A) One to multiple small intracellular magenta organisms at different life stages, including a schizont *(arrow)*, are present in erythrocytes. Intraerythrocytic schizonts are seen in *Plasmodium* spp. but not *Leucocytozoon* spp. or *Hemoproteus* spp. infections. Erythrocyte basophilia (reticulocytosis) reflects a regenerative response to the infection. (A, B). Hemozoin, a granular brown crystalline material produced through red blood cell destruction by *Plasmodium* and *Hemoproteus* hemoparasites, is present in erythrocytes (A) and macrophages (B). It is a characteristic finding in birds infected with these parasites. (B) Intracellular hemoparasites, in this case a schizont *(arrow)*, can be seen histologically but they are more easily seen in cytologic preparations. Part A: Touch preparation, lung. Wright's stain.

in the lungs, and microvascular injury and thrombosis in the lungs or other sites may also be seen. Erythrocytes and tissue macrophages may contain large amounts of brown crystalline material, hemozoin (Figs. 33.37A, B), which is formed as a secretory breakdown product of hemoglobin (seen with *Plasmodium* and *Hemoproteus* but not *Leucocytozoon* infections). This pigment is birefringent with polarized light, which is useful in diagnosis of infection and differentiation from hemosiderin, which is not birefringent. PCR can be used to confirm infection and differentiate between the various parasites; coinfections are common.

Infection with a promiscuous species of ***Haemoproteus*** (or ***Parahaemoproteus* sp.**) has been associated with sudden death in a number of passerine birds, especially oropendolas and other South American species, in zoological collections in North America.[36,45] This organism has not been identified to species. Infection is associated with multifocal hepatic hemorrhage (Fig. 33.38A) and the formation of large cyst-like structures (megaloschizonts or macroschizonts) containing merozoites (Fig. 33.38B) in infected tissue. Formation of these unusual structures is suspected to be the result of infection in aberrant hosts. The schizonts eventually rupture, which results in foci of intra-hepatic hemorrhage and fatal intracoelomic exsanguination. The gross lesions of multifocal hepatic hemorrhage and hemocoelom are highly characteristic. Circulating intracellular forms are not seen in these cases, suggesting the parasite is unable to complete its life cycle in these hosts.

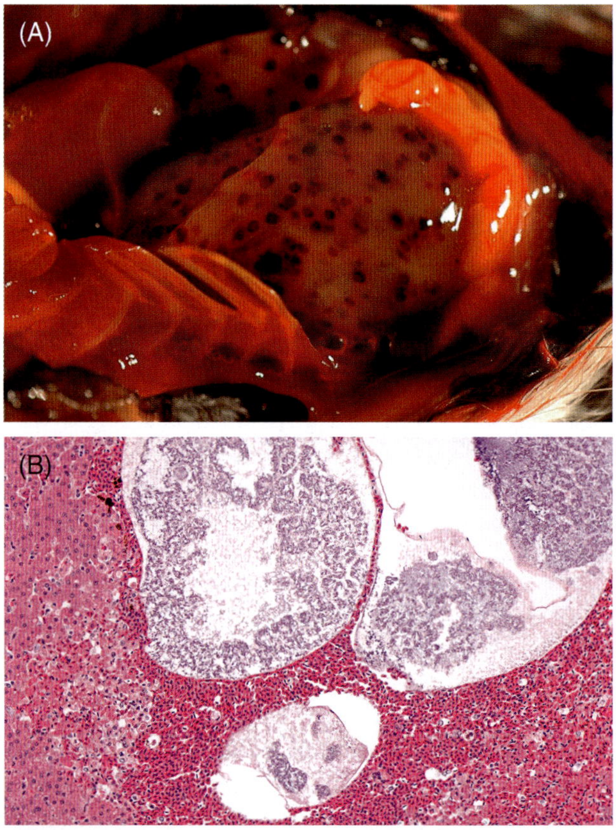

FIGURE 33.38 Atypical *Hemoproteus* infection in the liver of a red-capped cardinal. (A) Multifocal hemorrhage in the liver is a characteristic gross finding. (B) A characteristic histologic finding is the development of large protozoal megaloschizonts containing numerous merozoites.

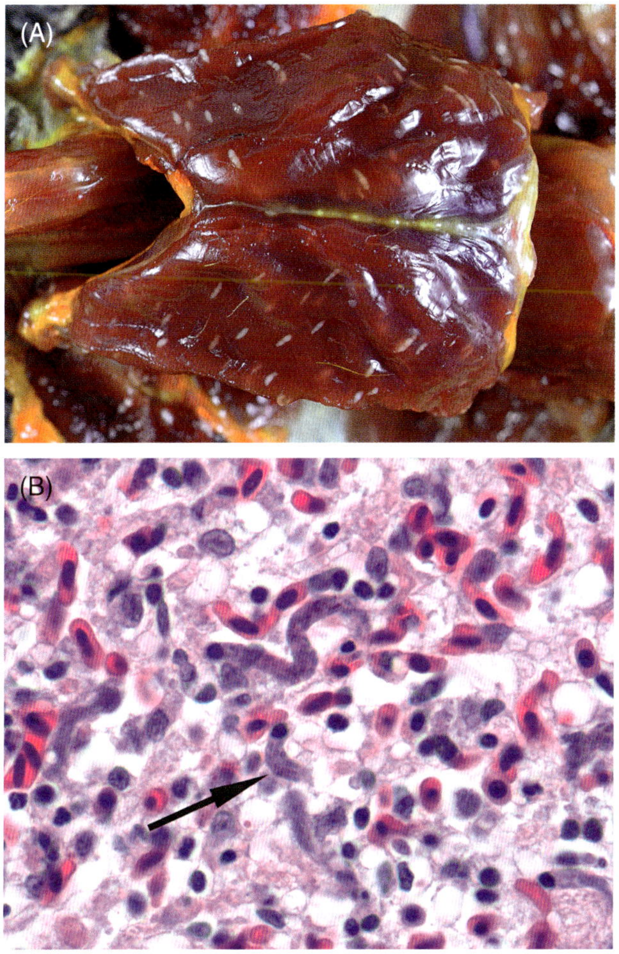

FIGURE 33.39 *Sarcocystis* sp. infection in the skeletal muscle of a weaver (A) and lung of a bee-eater (B). (A) Intramuscular tissue cysts are elongate and white. (B) A characteristic feature of intraendothelial infection in the lungs is expansion and elongation of the cells, which conform to the serpiginous contour of the capillaries *(arrow)* and contain numerous zoites.

Although more commonly noted in psittacines, ***Sarcocystis*** spp. can infect and cause disease in Passeriformes and Coraciiformes.[37,38] ***Sarcocystis falcatula***, spread by the Virginia opossum in North America, is most commonly implicated.[38] After ingestion of infective sporocysts from feces of the definitive host, sporozoites excyst, invade the intestinal mucosa, and infect endothelial cells where they form schizonts. Characteristic elongate, white tissue cysts can be seen grossly in the skeletal muscle (Fig. 33.39A) but are not as common in passerines as other groups of birds. In the lungs, schizonts often conform to the contour of the capillary and can be elongate and tortuous (Fig. 33.39B). Release of merozoites causes endothelial damage resulting in edema, inflammation, and less commonly hemorrhage. In the lung, a common lesion is pulmonary edema, which can be fatal. Organisms are most commonly seen in tissues with prominent capillaries (e.g., lung, spleen, liver) and the presence of red, wet lungs should raise suspicion of infection in the pulmonary form of disease. ***Sarcocystis calchesi*** can infect hepatocytes and causes severe hepatic necrosis in mousebirds and Taveta golden weavers.

Infection by the apicomplexan protozoal parasite ***Toxoplasma gondii*** is infrequently identified in passerines but has been documented as an important cause of mortality in Hawaiian forest birds and canaries. Canaries present with conjunctivitis that, when severe, can progress to blindness.[165] Crusting of the eyelids, cataracts, and collapse of the globe have been seen grossly. Histologic lesions include choroiditis, retinitis, and retinal detachment. In these cases, tachyzoites of *T. gondii* may be present in the retina and vitreous humor. In the ʻalala, a Hawaiian corvid, gross findings include splenomegaly, hepatomegaly, and pericardial effusion. Areas of necrosis and mononuclear cell inflammation are present in the liver, spleen, and brain, and protozoal tissue cysts can be seen in any of these sites especially in the brain.[176]

Ectoparasites

Hippoboscid flies (*Pseudolynchia* sp., *Ornithoica podargi*, *Ornithomya fuscipenni*) are frequently found in Caprimulgiformes (frogmouths and nightjars) and swifts. These biting flies can transmit *Haemoproteus* sp. and be a vector for the mite ***Microlichus americanus***, which is associated with dermatitis and folliculitis in Saffron finches.[131]

Larvae of dipteran flies of the genus **Philornis** are a cause of mortality in nestling passerines, including several finch species in South America and the Galapagos archipelago.[46,104] Early instars are found embedded in the skin of the face and nares (Fig. 33.40). Later instars are free-living in the nesting material and feed on blood from nestling birds. Mortality is due to anemia and tissue damage. Infection with this parasite can have population effects

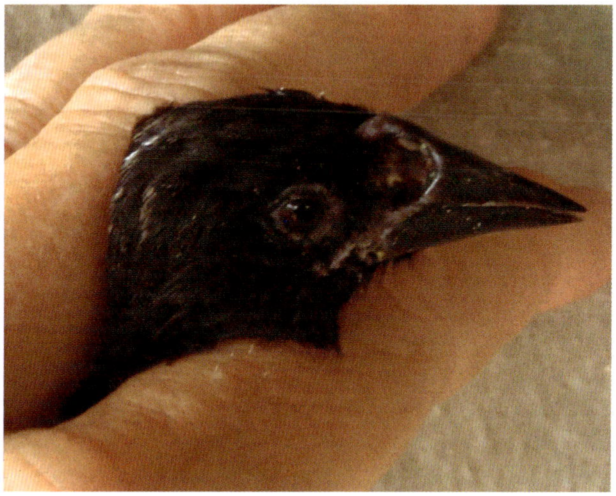

FIGURE 33.40 *Philornis downsi* in the beak and naris of a common cactus finch. The parasite burrows into the tissues of the beak and naris creating a large, open wound. *(Photo Courtesy of S. Deem, St. Louis Zoo.)*

that may increase the risk of local extinction (e.g., medium ground finches on the island of Santa Cruz in the Galapagos).[85]

Like other species, passerines can be infected by numerous species of **mites**. **Blood-sucking mites**, including ***Ornithonyssus*** sp. and ***Dermanyssus* sp.**, can cause mortality in nestling passerines. Some live in nest material and only feed at night, complicating detection if they are no longer present on the body during postmortem examination. Severe infestations can result in respiratory disease and anemia, and cutaneous feeding sites can lead to secondary infections. **Nectar mites** are transferred between flowers by hummingbirds and may be found in the nares and sinuses. These mites are considered nonpathogenic. **Feather mites** also occur in hummingbirds, nightjars, potoos, and numerous other species of passerines. Treatment in emerald hummingbirds may be complicated by an apparent sensitivity of this species to the anthelminthic drug ivermectin diluted with propylene glycol, which has been reported to cause seizures and central nervous system depression.[144]

Harpirhynchid mites cause cysts in the skin and feather follicles of birds. A number of species, including *Harpirhynchus nidaluns H. brevis*, and *H. quasimodo*, have been described in birds in North America. The latter, ***H. quasimodo***, has been reported in free-ranging brown-headed cowbirds.[101] Lesions consist of large, occasionally dramatic, nodular, firm, yellow masses, most often in the axillary and inguinal regions, but they can be found in other parts of the skin as well. On section, the masses are multiloculated epidermal cysts filled with granular yellow to tan material. Cytologically and histologically, the contents contain adult and juvenile mites, mite eggs, and keratin and cellular debris. A characteristic feature of the mites is a tripartite, humplike projection. The life cycle and pathogenesis of these mites are poorly understood.

The tracheal mite ***Sternostoma tracheacolum*** is common in canaries and finches, especially Gouldian finches.[133,157] Infection has been documented in wild populations and captive collections. Adult mites reside in the major airways, parabronchi, and air sacs. With close inspection and magnification, adults can be seen in the trachea or air sacs at necropsy (Fig. 33.41A). Lesions are primarily induced by eggs and mite excretion, resulting in moderate to severe heterophilic to granulomatous tracheitis, bronchopneumonia, and airsacculitis. Characteristic features of the mites include a chitinized exoskeleton, jointed appendages, and striated skeletal muscle (Fig. 33.41B).

The subfamily Knemidocoptinae includes the scaly leg mites ***Knemidokoptes mutans*** and ***K. jamaicensis***. Infections with *Knemidokoptes* species have been reported in many species of free-ranging and captive passerines including common chaffinches, Eurasian tree sparrows, black-faced and little buntings, and eastern great reed warblers.[27,102] Typical lesions include marked hyperkeratosis and scale on the featherless parts of the skin (Fig. 33.42A). Characteristic features of the mites are an eosinophilic chitinous exoskeleton, a hemocoelom, and striated muscle (Fig. 33.42B).

Ticks found in free-ranging passerines include ***Haemaphysalis longicornis*** (the longhorn tick), ***Argas* species** including ***A. giganteus*** and ***A. americanum***, ***Ixodes scapularis*** and ***I. brunneus***, and ***Haemaphysalis cordeilis***. The longhorn tick and *Ixodes* species ticks are vectors for several important pathogens including *Anaplasma*, *Coxiella burnetii*, *Rickettsia japonica*, *Theileria mutans*, and *Theileria orientalis*.[160] ***Hyalomma rufipes*** is also common on passerines in some regions and is a known vector for *Babesia*, *Rickettsia*, and *Anaplasma* and Crimean-Congo hemorrhagic fever virus.[48,72,73] The most common tick-associated disease in passerines is tick paralysis. Neurotoxins

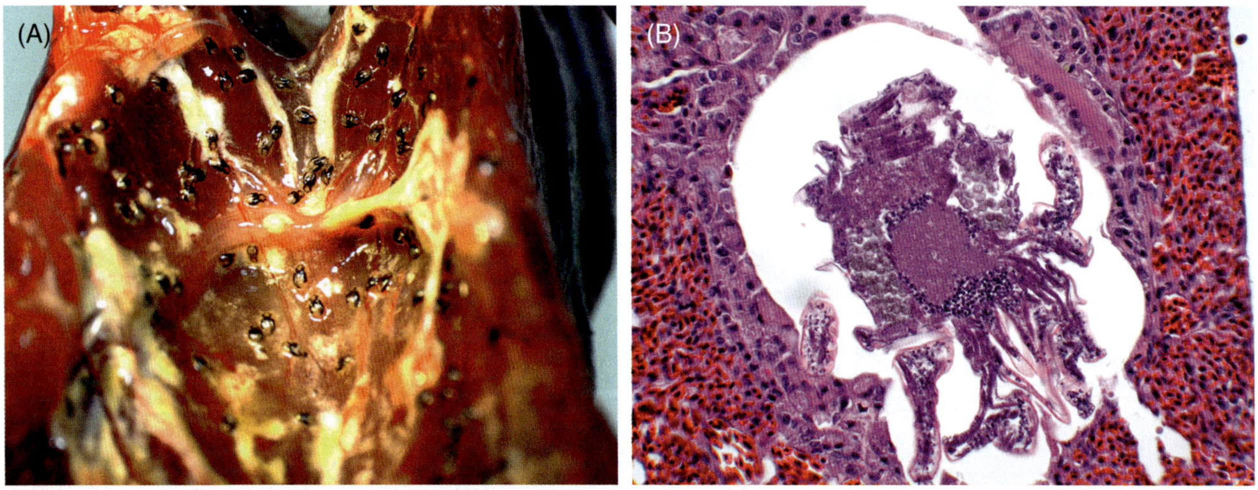

FIGURE 33.41 *Sternostoma tracheacolum* in the air sacs of a Gouldian finch. (A) Numerous small, black/tan mites are present in the respiratory system of this finch. (B) Characteristic features of these large arthropod parasites include a chitinized exoskeleton, jointed appendages, and striated muscle. *(Part B: Photo Courtesy of R. Berlinski.)*

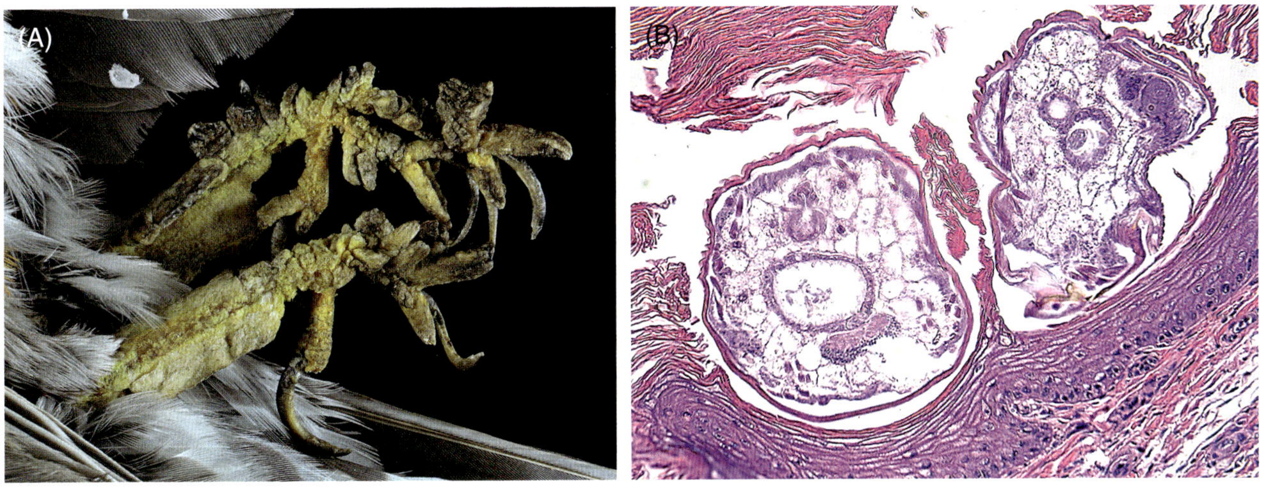

FIGURE 33.42 *Knemidocoptes* sp. infection in the nonfeathered skin of an American robin. (A) Marked, bilateral, thick, crusty skin is present on the digits and tarsometatarsus. (B) Infection is associated with marked compact, orthokeratotic hyperkeratosis, and acanthosis with intracorneal mites. (B) Characteristic features of the mites are an eosinophilic chitinous exoskeleton, a hemocoel, and striated muscle.

produced by hard and soft tick species have been associated with disease, which occurs globally and has been reported in humans, domestic animals, wild mammals, and birds.[97] Toxin produced by *Ixodes brunneus* and *I. holocyclus* is known to cause disease in passerines. Single or multiple ticks are found primarily on the feet, legs, keel, and inguinal regions. Clinical signs including sternal recumbency, polyneuropathy, distal to proximal ascending paralysis, anemia, dehydration, respiratory distress, lethargy, ataxia, dysphagia, and death.[76,97] Clinical signs of progressive flaccid paralysis can be seen with infection by just a single tick and can lead to death.[76] Edema, ecchymoses, and petechiae can be found at the attachment sites but there are no other direct gross or histologic lesions. Successful outcomes are best achieved with identification and removal of the tick/s, provision of supportive care, and where available, administration of antivenom.[76] Provisional diagnosis is based on morphologic identification of a tick species known to cause tick paralysis, absence of other disease processes, and response to treatment.

ADDITIONAL RESOURCES

Atkinson, C. T., Thomas, N. J., Hunter, D. B., eds. 2008. Parasitic Diseases of Wild Birds. Ames, Iowa. Wiley-Blackwell: A John Wiley & Sons, Ltd. Pp. 595.

Miller, E. R., Lamberski, N., Calle, P. P., eds. 2022. Fowler's *Zoo and Wild Animal Medicine* Current Therapy, volume 10. St. Louis, MO. Saunders. Pp. 832.

Owen, J., Hawley, K., Kuyvaert, K., eds. 2021. *Infectious Disease Ecology of Wild Birds*. Oxford, UK. Oxford University Press. Pp. 280.

References

Full references for this chapter can be found online at https://doi.org/10.1016/B978-0-443-18437-6.00037-7.

Chapter 34

Chelonia

Carlos E. Rodríguez[a], Ana María Henao Duque[b], and Jennifer Steinberg[c]

[a]Disney's® Animals, Science and Environment, Lake Buena Vista, FL, United States; [b]Zoopath Veterinary Pathology Laboratory-Cali Zoo, Cali, Colombia; [c]Mid-Atlantic Veterinary Pathology, LLC, Baltimore, MD, United States

INTRODUCTION

Chelonians are poikilothermic tetrapod vertebrates distinguished by a unique, specialized bony shell that covers the body. They are members of the class Reptilia, order Testudines that includes approximately 350 extant species of terrestrial, fresh water and marine turtles, tortoises, and terrapins[3,142] (see Table 34.e1).

Chelonians vary greatly in size, from the diminutive 100 g speckled Cape tortoise to the leatherback sea turtle, which can weigh over 800 kg.[43,89,100,117] Most chelonians are ectotherms that rely on ambient temperature gradients to maintain optimal body temperature. The notable exception to this is the leatherback sea turtle, which can reg ulate its body temperature through a series of vascular plexuses deep within the musculature of the limbs. This helps prevent excessive heat loss while swimming in cooler water or from overheating while nesting on land.[29,43]

Increasing pressures from habitat loss, emerging infectious diseases, worldwide anthropogenic activities such as hunting for food, the collection of turtle parts and products for use in traditional medicines, fashion, or ornaments, and the increasing demands of the pet trade have placed many chelonian species in peril. The stresses inherent with animal harvesting, including those related to removal from the wild, transport, less than optimal housing and substandard nutritional conditions for prolonged periods of time, place these animals at a notably increased risk for intractable debilitation, immunosuppression, and disease emergence.[31,121] Nowhere has this been more significant than in Asia, where 75% of the 90 Asiatic chelonian species are facing extinction.[89] Of the 270 species of chelonians on the 2022–23 IUCN Red List of Threatened Species, 170 (62.9%) are listed as critically endangered, endangered, or vulnerable, including six of the seven marine turtle species. Eight turtle species are considered currently extinct. This places chelonians among the most threatened vertebrate groups in the world.[43,44,70,142] The possibility of infectious disease transmission within and between comingled species at any point along the trade chain can have severe implications for the reintroduction and/or translocation of animals in conservation projects. For example, the introduction of, or an outbreak of mycoplasmosis associated with the translocation of desert tortoises under the governmental population management and sustainability plans could be catastrophic to vulnerable wild populations.[141] A number of governmental, nongovernmental, and nonprofit organizations support invaluable in situ and ex situ research, capacity building, rehabilitation, and reintroduction/release projects and lobby for the enactment of legislation to monitor and protect these species throughout the world. In addition, many private and public institutions provide critical logistical and financial support for the rescue, rehabilitation, and release of many species of sea turtles.

NON-INFECTIOUS DISEASES

Nutritional

In chelonians, as in other reptiles, **metabolic bone disease (MBD)** can be of nutritional or renal in origin. **Nutritional MBD (NMBD)** is more common and seen almost exclusively in chelonians under managed care as the result of suboptimal diets and husbandry practices. Diets deficient in calcium or vitamin D, imbalances in the dietary calcium–phosphorus ratio, suboptimal (e.g., low) ambient temperatures, and inadequate exposure to ultraviolet B (UV-B) light are thought to play a role in disease development.[58,89,91] NMBD is also commonly seen in aquatic turtles that are housed indoors and do not receive appropriate exposure to UV-B spectrum as they cannot synthesize calcitriol, the active form of vitamin D3 that is necessary for intestinal calcium absorption.[89] Hypocalcemia triggers a physiologic increase in the production of parathyroid hormone (PTH) that results in enhanced resorption of calcium from the skeletal system and subsequent osteopenia. In young animals, this is known as **rickets** and is characterized by abnormally thickened and expanded growth plates in which the physeal cartilage is

Pathology of Wildlife and Zoo Animals. https://doi.org/10.1016/B978-0-443-18437-6.00041-9

distorted by retained, roughly parallel, elongated columns of poorly organized, hypertrophic chondrocytes that span the expanded metaphysis. These retained cartilage columns are surrounded by a smudged, poorly mineralized osteoid matrix, and osteoblasts can often be found along the scalloped margins of the matrix. Ineffective mineralization and bone remodeling result in abnormally widened physeal plates. In many cases, the periosteum is notably expanded by fibrosis and when present, cortical bone is notably thinned. By contrast, **osteomalacia** in adult animals is characterized by failure of endochondral mineralization during normal remodeling of mature bones in response to normal daily stressors on the skeletal system. This is characterized by abnormal retention of disorganized, poorly mineralized osteoid and cartilage. In general, this occurs away from the physis and is often associated with thinning of preexistent cortical and trabecular bone and replacement with poorly mineralized hypertrophied cartilage and reactive woven bone. This results in brittle bones that are prone to pathological fractures. Periosteal fibrosis may or may not be present but can extend into the medullary space, replacing adipose tissue and hematopoietic elements. Serum calcium remains at the low end of the normal range in most cases, and phosphorus can be normal or slightly low due to increased renal excretion. **Renal MBD (RMBD)** is a sequela to primary chronic renal disease and as such, the pathophysiology between NMBD and RMBD differ, even though the end result is similar. Decreased glomerular filtration leads to **hyperphosphatemia**, which is the hallmark of RMBD. Hyperphosphatemia can induce profound hypocalcemia, resulting in osteoclastic bone resorption and thinning of cortical and medullary bone. Failed mineralization leads to retained irregular islands of chondrocytes with unmineralized osteoid matrix, particularly noticeable as osteoid seams along medullary bone and periosteum. Varying degrees of medullary and periosteal fibrosis may also develop.[26] **Fibrous osteodystrophy** can be a sequela to osteopenia in all its forms. As in other vertebrates, there is overlap in the gross appearance of the lesions of osteomalacia, rickets, and fibrous osteodystrophy; however, fibrous osteodystrophy is not as commonly reported in chelonians as it is in other reptile species.[26,82,89]

Regardless of the cause, clinical signs of MBD in adults include generalized muscle weakness, cloacal or rectal prolapse, tremors, ataxia, reproductive stasis, and growth retardation. In juveniles, softening of the carapace and plastron, pyramiding (abnormal and abrupt vertical growth of individual scutes into pyramid-like structures) and other carapacial deformitics (Fig. 34.1), separation of the scutes, thickening of the rhamphotheca, growth retardation, and bowing of the limbs are common gross lesions.[89,91] Aquatic or semiaquatic animals can present with erosive shell lesions that are often complicated by secondary infections.[26]

FIGURE 34.1 Metabolic bone disease in a red-footed tortoise. The carapace is deformed by pronounced pyramiding of the scutes, which is characterized by abnormal and abrupt vertical development of individual scutes into pyramid-like structures.

Secondary nutritional hypothyroidism and subsequent **goiter** have been neither sufficiently studied nor adequately described in chelonians. The condition is seen in animals under managed care and is associated with diets low in iodine or high in goitrogenic elements such as kale, broccoli, and cabbage. Iodine toxicity can also result in goiter.[89] Anecdotally, Galapagos and Aldabra tortoises seem to be more susceptible.[89,91] Although the exact pathogenesis of this remains unclear, goiter is also common in sea turtles kept in artificial salt water aquaria over prolonged periods of time (B. Stacy, personal communication). Grossly, the thyroid gland is visibly enlarged, and follicles are engorged with either clear to yellow-tinged fluid or coagulated, yellow-orange gelatinous colloid (Fig. 34.2). Histologically, the follicular epithelium is typically attenuated and compressed. Selenium deficiency, a known cause of hypothyroidism in other vertebrate species, has not been described in chelonians.[89]

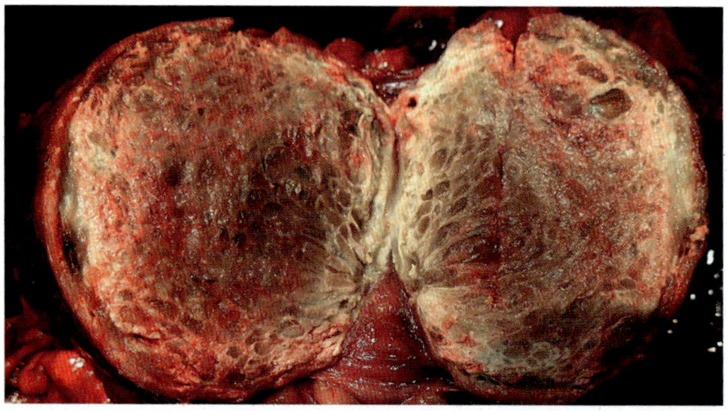

FIGURE 34.2 Goiter in the thyroid gland of a loggerhead sea turtle. The enlarged thyroid gland is bisected and has been distorted by numerous, variably sized ectatic thyroid follicles filled with inspissated, gelatinous colloid.

FIGURE 34.3 Obesity in a loggerhead sea turtle. Excess subcutaneous adipose tissue is typically deposited along the soft tissue areas around the limbs and neck, causing bulging and stretching of the skin. Excessive fat deposition can hinder mobility by limiting the effective range of motion of the limbs.

FIGURE 34.4 Bilateral aural abscesses related to hypovitaminosis A in a painted turtle. The pronounced areas of swelling on either side of the head just caudal to the eyes are caused by an accumulation of keratinaceous and inflammatory debris in the tympanic cavity. This is often associated with opportunistic colonization by gram-negative anaerobic bacteria. *(Photo Courtesy of N. Mylniczenko, Disney's Animals, Science and Environment.)*

Obesity is common in chelonians under managed care. It is often due to a combination of excessive caloric intake, an inappropriately balanced diet for the species, and a sedentary lifestyle.[89] Excess adipose stores can predispose an animal to **hepatic lipidosis/fatty change** in the face of starvation; however, this must be interpreted with caution as the fat content of the liver can vary dramatically depending on the metabolic state of an animal, as occurs with hibernation or during the reproductive season. In obese animals, excessive subcutaneous fat deposits can be seen around the neck and limbs and can sometimes hinder mobility (Fig. 34.3).[91]

Prolonged **anorexia**, regardless of the cause, can result in severe protein malnutrition and cachexia. This can manifest as loss of muscle mass, weakness, and lethargy, and grossly as reduction in subcutaneous adipose tissue deposits and loss of coelomic fat bodies.[89] Gallbladder ectasia with normal bile consistency may be indicative of recent anorexia, whereas a distended gallbladder filled with thick, mucoid, inspissated bile can be associated with prolonged anorexia.[45] Mobilization of fat and subsequent hepatic lipidosis (see "Miscellaneous/Multifactorial" section) may develop as a sequela to anorexia and can be clinically significant.

Scattered foci of granulomatous inflammation associated with oxalate crystals, also referred to as **renal oxalosis**, are not uncommon in the kidneys of some herbivorous marine and terrestrial chelonians and are considered incidental.[44,132]

Chronic **hypovitaminosis A** has been associated with the development of **squamous metaplasia** of the epithelial lining of the Eustachian tubes and middle ear in chelonians. This presents as the accumulation of keratinized debris and progressive swelling of the middle ears. Vitamin A is essential for cellular differentiation and maintenance of the specialized ciliated epithelium of the nasopharynx. Without proper dietary supplementation of vitamin A, sensitive epithelial tissues revert to a more resilient stratified squamous keratinized epithelium. Over time bacterial colonization from the nasopharynx occurs and the resultant heterophilic and histiocytic inflammation manifests as the accumulation of caseous debris and uni- or bilateral aural abscesses with swelling and lateral displacement of the tympanic membrane (Fig. 34.4). In chronic cases, multinucleated giant cells and a fibrous capsule around the abscess are common findings.[80] Hypovitaminosis A is also associated with squamous metaplasia of the lacrimal and Harderian gland ducts, and can lead to the development of cysts in the third eyelid that interfere with closing of the eyelids.[71,91]

Metabolic

Hypothermic (cold) stunning occurs in free-ranging marine turtles when there are abrupt drops in the water temperature outside the normal, typical seasonal variations or when animals are carried by currents into areas of colder water. Juveniles tend to cool down faster due to their smaller bodies and higher surface-to-volume ratio. Hypothermia leads to reduced activity, leaving animals prone to predation. In a retrospective study of cold-stunned Kemp's ridley sea turtles in a rehabilitation setting, manifestations of disease were most common in the gastrointestinal (GI) tract and respiratory system and associated with fungal and/or bacterial pneumonias and mixed gram-positive and -negative bacteria ulcerative enteritis and gastritis.[67] GI and visceral granulomas associated with a variety of nematode, cestode, and trematode parasite larvae and adults were also common but these may have been concurrent, independent infections. Bacterial and fungal infections are typically considered opportunistic and related to immune system dysfunction secondary to debilitation and hypothermia.[45,67,72,89]

Hepatocellular fatty change can be the end result of physiologic as well as pathologic processes that result in the cytoplasmic accumulation of triglycerides within hepatocytes. Hepatic lipid metabolism is a normal physiologic process that involves an energy-dependent, complex pathway that culminates in the hepatic release of very low-density lipoproteins (VLDL) into circulation. Physiologic hepatocellular fatty change reflects normal metabolic processes that can be associated with events, such as preparation for hibernation or reproductive seasonality, particularly in vitellogenic females (see Digital Content). Grossly, the liver is slightly enlarged with rounded lobe edges and light brown to pale tan.[38,45,89,91] As hibernation progresses, hepatocytes undergo atrophy and lack cytoplasmic lipid droplets when compared with hepatocytes of nonhibernating chelonians.[45] Early stages of pathologic hepatocellular fatty change **(lipidosis)**, associated with anorexia, protein malnutrition, hepatic injury, and systemic diseases can be grossly indistinguishable from physiologic fatty change. Microscopically, early stages of both physiologic and pathologic fatty change manifest by the presence of multiple clear, crisply defined cytoplasmic microvesicles. At this stage, this process is reversible if the stimulus for increased lipid mobilization is resolved. As pathologic fatty change/hepatic lipidosis progresses, lipid microvesicles coalesce into a single or several large vesicles that peripherally displace the nucleus (Fig. 34.5). In severe cases, lipid vacuoles from adjacent hepatocytes coalesce either by rupture or fusion of adjacent hepatocytes and form large foamy lipid-laden masses, referred to as fatty cysts. In severe hepatic lipidosis, the edges of the liver lobes are rounded and the liver has a friable texture. Cut sections of liver may float in fixative or water due to abundant hepatocellular lipid accumulation. The increased levels of fatty acids within hepatocytes increase the oxidative stress on cells and can result in single-cell or regional hepatocellular necrosis, fibrosis, melanomacrophage and pigment accumulation, and nodular hyperplasia.[28]

Gout occurs frequently in managed populations of terrestrial chelonians.[89,91] **Hyperuricemia** is the most common cause in chelonians and is associated with renal disease, dehydration, inappropriate diet, reduced renal perfusion, and the use of diuretics. Uric acid circulates in blood primarily as insoluble monosodium urate that is excreted directly by the renal tubules.[37,89] Hyperuricemia results in precipitation of urates and its salts as insoluble crystals.

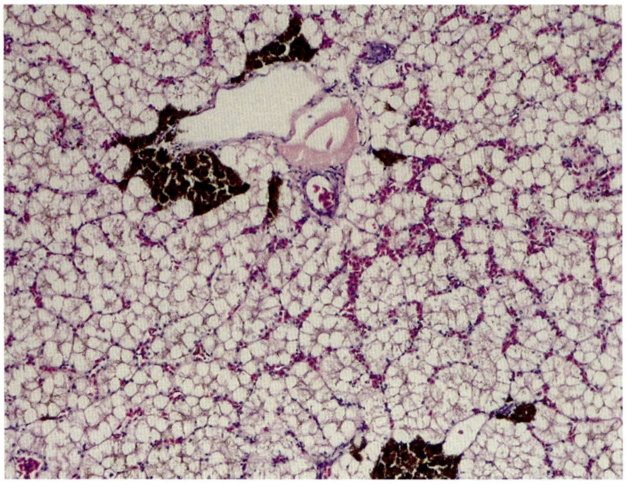

FIGURE 34.5 Pathologic hepatic lipidosis/fatty change in a red-footed tortoise. The hepatocyte cytoplasm is expanded by single large vacuoles that appear to be "empty" and displace the nucleus toward the periphery of the cell. Displacement of the nucleus can be an indicator to differentiate pathologic hepatic lipidosis from nonpathologic physiologic fatty change. There are scattered melanomacrophage aggregates throughout the sinusoids.

This is common in the kidneys as **renal gout,** and in joints as **articular gout** (Fig. 34.6A). It can also occur on visceral surfaces as **visceral gout** (Fig. 34.6B). Grossly, the kidneys can be swollen and mottled with myriad disseminated pinpoint to linear, chalky, white deposits. In articular gout, the soft tissues surrounding the joints and joint spaces can be swollen and hinder mobility. Microscopically, renal gout is characterized by renal tubular ectasia with intralumenal, acicular, cleft-like clear spaces that aggregate into single or coalescent stellate inflammatory masses. Renal tubular damage occurs via mechanical abrasion and occlusion of the tubules and is associated with intense heterophilic and histiocytic to granulomatous inflammation. Where crystals bind with cellular, inflammatory, and organic debris, they may form stellate aggregates called tophi. Urate crystals dissolve readily in alcohol during histologic processing and what remains are cellular infiltrates, debris, and the organic matrix that surrounds the outlines of the dissolved crystals. Unfixed, unstained cytologic imprints can aid in the diagnosis as the needle-like crystals readily polarize. Along visceral surfaces, stellate tophi are generally not formed as there is often minimal inflammation associated with these deposits and only the organic, protein-rich matrix that accompanies the urate deposits is typically visible.[37] Gross lesions in acute, terminal cases include bright white, chalky deposits on the surfaces of the pericardial sac, liver, spleen, lungs, and other viscera. In the kidneys and joints, tophi formation tends to occur in chronic disease.

Pseudogout has been reported sporadically in several species, including red-bellied turtles and a painted tortoise. It is caused by the deposition of crystals other than urate

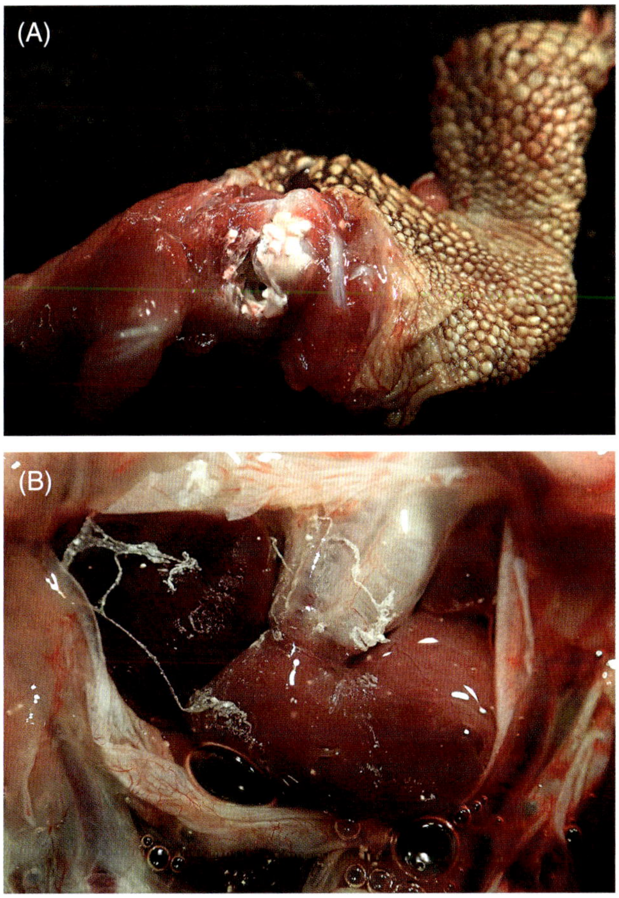

FIGURE 34.6 Articular gout (A) and visceral gout (B) in a pancake tortoise. (A) Urate deposition (gout) in the shoulder joint of a pancake tortoise. The articular surfaces are partially covered by thick deposits of chalky, bright white material. This is the typical appearance of uric acid precipitates (tophi) when complexed with cellular and inflammatory debris. (B) Urate deposition (gout) on the epicardial surface of the heart of a pancake tortoise. Wispy to amorphous deposits of chalky, bright white material consistent with urate deposits are present multifocally on the epicardial surface.

salts, such as calcium hydroxyapatite or calcium pyrophosphate dehydrate, typically in and around joints. Histologically, variably sized periarticular and intraarticular nodular deposits of basophilic granular matrix with pleomorphic, faintly birefringent crystals, histiocytic to granulomatous inflammation, mild heterophilic infiltrates, and frequent multinucleated, foreign body-type giant cells can be seen. The hallmark of pseudogout is the destruction of articular cartilage secondary to the deposition of calcium crystals around individual chondrocytes. With progressive disruption of the cartilage matrix, the crystalline material reaches the joint space, inciting the inflammatory reaction. At this point, the gross lesions of gout and pseudogout can be identical.[17,89,91]

Uroliths are relatively common in terrestrial chelonians. A particularly high incidence is seen in desert tortoises. The majority of the cases are documented in

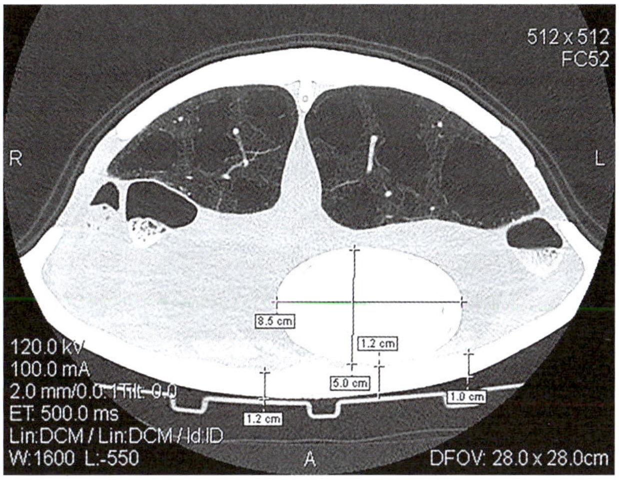

FIGURE 34.7 Urolith in the bladder of a Malaysian river terrapin. The lith appears as a single, large, oval, radiodense object in the lower right quadrant. Computed Tomography. *(Photo Courtesy of G. Pye, Disney's Animals, Science and Environment.)*

animals under managed care, but bladder stones are also reported sporadically in free-ranging individuals.[71,89] Uric acid is the most common component. It is usually complexed with calcium, potassium, and/or ammonium salts.[89] These stones are radiodense and easily identified radiographically or, where available, with advanced imaging techniques such as computed tomography (Fig. 34.7).[53,64,89] Dehydration is the most common predisposing factor. Other factors include deficiencies in vitamins A and D and calcium, excess dietary protein or oxalates, bacterial infections, and complications related to foreign bodies (e.g., suture remnants from prior cystotomies). Uroliths are most common in the urinary bladder as single smooth, lamellated concretions but multiple stones can be present. Uroliths can also be present in the ureters and in the cloaca. Concretions within the kidneys have not been reported in reptiles. Serum levels of uric acid are typically not elevated unless there is renal disease and do not correlate with the presence of uroliths. Small stones usually cause mild mucosal irritation and mild hematuria. Larger stones can cause mucosal erosions and ulceration, thickening and hypertrophy of the bladder wall and, in severe cases, transmural inflammation, pressure necrosis, and coelomic/visceral adhesions. Histologic lesions include urothelial erosion, ulceration, and hyperplasia accompanied by varying degrees of heterophilic inflammation. Retropulsed eggs from the urodeum can occasionally become trapped in the urinary bladder and develop into pseudocalculi.[89]

Miscellaneous/Multifactorial

Traumatic shell fractures are one of the most common causes for submission of free-ranging terrestrial and aquatic chelonians for necropsy examination.[119] Impacts from vehicular trauma in terrestrial species and boat strike in marine species are the most common causes for traumatic fractures of the shell. Predation can also present as shell fractures but in this type of trauma there are often other injuries, such as limb amputation and soft tissue lacerations. Typical presentation in terrestrial species includes multiple, variably fragmented, and displaced carapacial fractures accompanied by abrasions and ulcerations on the plastron and skin of the limbs and head (Fig. 34.8A). In marine species, either linear fractures from impact with the keel of a boat or a series of parallel cuts caused by the propeller are most common (Fig. 34.8B).[45] If the fractures involve or bisect the vertebral column, the prognosis for recovery decreases substantially due to spinal cord damage. Deep fractures of the shell can also expose the lungs and/or the coelomic cavity, resulting in the development of deep-seated bacterial or fungal infections. Chelonians do not require negative pressure within the coelomic cavity to maintain respiration, so exposure of the coelom will not necessarily result in respiratory failure.[89,91]

Despite the use of turtle excluder devices in some parts of the world, **entrapment or entanglement in fishing nets or gear** during fishing operations remains a significant threat to sea turtles, particularly in identified hotspots, such as the southwest Atlantic and Mediterranean sea.[84] Even with postcapture releases, animals may be significantly compromised due to water aspiration or injuries from hooks and fishing gear, or from systemic lactic acidosis and stress related to overexertion.[146] Entrapment, entanglement, and other interactions with fishing gear or floating debris can hinder the animal's swimming and surfacing ability, especially if the affected animals are cold-stunned and weak. Entanglement with fishing gear or plastic debris can result in strangulating lesions (Fig. 34.9), which may become opportunistically infected or result in limb amputation. Drowning is a common sequela to entanglement but it is a difficult postmortem diagnosis to establish and often based on circumstantial evidence and lack of other identifiable causes of death. Airway-associated fluid, foreign material, and environmental debris may not be present in the majority of cases. **Exertional myopathy/exertional**

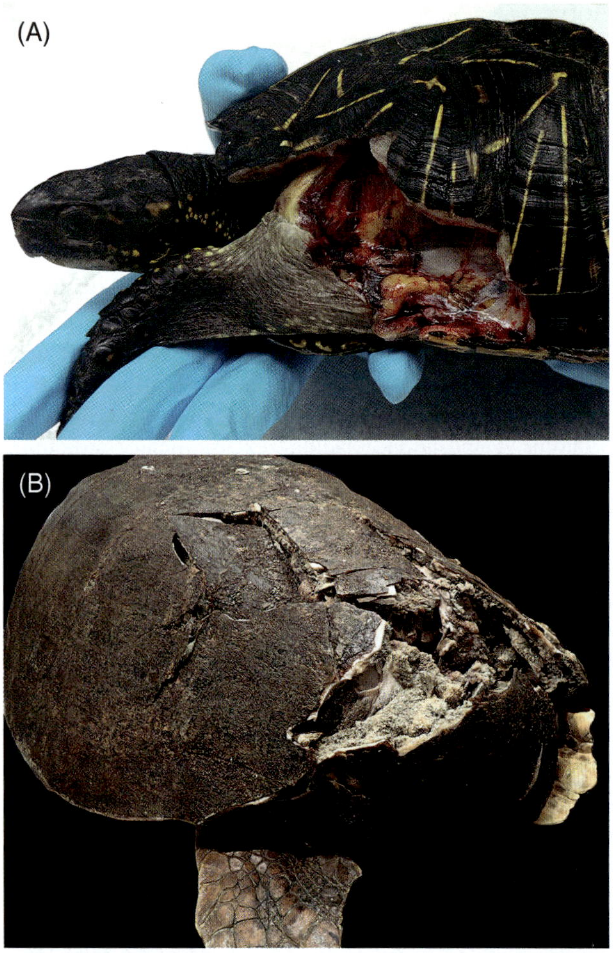

FIGURE 34.8 Traumatic shell fractures in a Florida box turtle (A) and loggerhead sea turtle (B). (A) Automobile trauma in a Florida box turtle with a severe comminuted carapacial fracture with loss of an extensive section of the carapace and exposure of the underlying soft tissues and musculature. (B) Motorboat trauma tends to produce, as in this Loggerhead sea turtle, a unique fracture pattern characterized by a single, long linear fracture due to the impact with the keel of the boat and/or multiple, shorter, regularly spaced parallel fractures at an angle caused by propeller strike. *(Part A: Photo Courtesy of J. Holmes, Disney's Animals, Science and Environment.)*

FIGURE 34.9 Trauma due to fishing gear entanglement in a green sea turtle. Monofilament fishing line wrapped tightly around the flipper caused a progressive stricture, cutting the skin and soft tissues almost to the bone. This occurred over time, as scarring of the exposed tissue is extensive. An algae-covered metal identification tag is present on the flipper near the top of the image. *(Photo Courtesy of N. Mylniczenko, Disney's Animals, Science and Environment.)*

rhabdomyolysis with severe metabolic acidosis and lactic acidemia has been reported in an entangled juvenile green sea turtle.[114] Hypermagnesemia is a consistent finding in sea turtles and humans that drown or near-drown in salt water, presumably from aspiration of sea water.[67]

Predation can be seen in both free-ranging and privately held aquatic and terrestrial chelonians. Typically, the limbs and tail are affected. It is not uncommon for limb injuries to be complicated by joint infections and osteomyelitis.

In chelonians, **burns** can be caused by thermal (most common), electrical, chemical, or radiant energy. In animals under managed care, malfunction or inappropriate placement of exogenous sources of heat, such as hot rocks, heat lamps, or thermal mats, are the most common cause. In free-ranging chelonians, burns are common during fires, either natural or anthropogenic.[89,91] Depending on the duration and intensity of exposure, burns are classified as first-, second-, third-, and fourth-degree burns, although some authors favor a more recent classification scheme of partial-thickness and full-thickness burns. In first-degree or partial-thickness burns, the skin is erythematous. Blistering is rare in chelonians with first-degree burns. Histologic changes in first-degree burns involve only the epidermis and changes are mild, including desquamation of keratinocytes with dermal congestion and edema. In second-degree or deeper partial-thickness burns, there is coagulative necrosis of the epidermis and involvement of the superficial dermis. Blistering, swelling, discoloration, and formation of serocellular crusts are common. Lesions are characterized by loss of differential staining, hypereosinophilia and karyorrhexis throughout the epidermis accompanied by vesicle and bulla formation both within the epidermis as well as at the dermoepidermal junction. In third-degree burns, there is destruction and loss of the epidermis and dermis, and the tissue takes on a white-gray to charcoal-black appearance. The remaining necrotic tissue is hypereosinophilic and partially covered by a coagulum of serous exudate. Fourth-degree burns are similar in severity to third-degree burns but extend deep into muscle and bone. Severe osmotic and electrolyte imbalances as well as sepsis are common sequela in burn patients. Inhalation of smoke and super-heated gasses can also damage the respiratory mucosa, resulting in the following three concurrent types of injury. Thermal burns, which typically affect the upper respiratory tree, chemical injuries that extend throughout the entire respiratory track, and the systemic effects of toxic gasses such as carbon monoxide. Bronchoconstriction develops secondary to the release of inflammatory mediators following tissue injury, eventually leading to the development of pulmonary edema. Grossly, the lungs are wet, heavy, and congested and sections may sink or sink then float or float beneath the surface in fixative solutions. Histologically, black or pigmented particulate debris may be present throughout the mucosa of the respiratory tree or alveolar walls but this is not always the case. Heterophilic inflammation may not be present in acute cases.[56,89]

Unilateral or bilateral **aural abscesses** are a common finding in chelonians kept under suboptimal husbandry conditions; they have been well documented and described in Eastern box turtles.[23,89] The etiology is multifactorial and involves factors such as improper housing, unsanitary conditions, and poor nutrition. Mixed gram-negative, aerobic bacteria that are part of the normal flora of the oral cavity are most commonly implicated and gain access to the tympanic cavity via the Eustachian tube.[80,89] Squamous metaplasia associated with chronic hypovitaminosis A, although not essential for the development of all aural abscesses, predisposes to the accumulation of keratinized debris and facilitates bacterial colonization and development of aural abscesses.[71,80,91]

Disseminated **hepatic fibrosis** and **hepatocellular nodular hyperplasia (cirrhosis)** can occur as a consequence to several chronic disease processes, including hepatic inflammation, heart failure, hepatic or posthepatic biliary obstruction, and severe hepatic lipidosis. At necropsy, subcutaneous edema and ascites can be seen. In cirrhosis, the liver is small and composed of variably sized, pale yellow-tan, firm, hyperplastic nodules (Fig. 34.10A). Histologically, the remaining hyperplastic cords of the hepatocytes are surrounded and separated by variably thick fibrous connective tissue septae often populated with hyperplastic biliary ducts and pigment-laden macrophages (Fig. 34.10B). Hepatocytes in the center of the regenerative nodules often exhibit macrovesicular lipidosis (fatty change) with eccentric nuclei.[55]

Melanomacrophages are pigmented phagocytic cells thought to participate in the innate as well as adaptive immune system. These are commonly seen in the kidneys, spleen, and livers in many species of fish, amphibians, and reptiles (Fig. 34.11A). The accumulated pigment in these specialized cells consists of melanin with various amounts of lipofuscin, hemosiderin, and other phagocytized materials.[136] The presence of single to few small melanomacrophage aggregates in tissues is considered normal and the size of normal melanomacrophage aggregates can vary greatly between species and due to a number of factors (Fig. 34.11B). A diagnosis of **melanomacrophage hyperplasia** can therefore be challenging and should take into consideration species, age, body condition (hepatic atrophy can make normal, resident melanomacrophage aggregates appear closer together), and immunocompetency. Formation of discrete, expansile melanomacrophage nodules that prominently distort the preexistent tissue architecture and compress surrounding hepatic cords may offer supportive evidence for a diagnosis of melanomacrophage nodular hyperplasia (Fig. 34.11C).

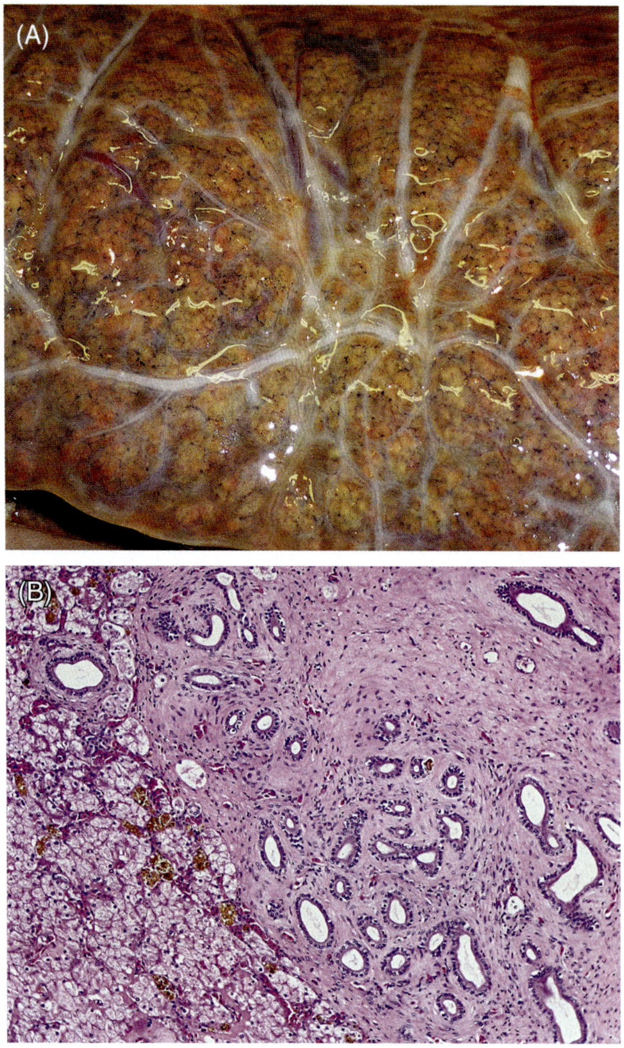

FIGURE 34.10 Hepatic cirrhosis in a Galapagos tortoise (A) and South American snapping turtle (B). (A) The liver of this Galapagos tortoise is subdivided into irregular nodules by bands of portal and bridging fibrosis. This gives the liver surface a cobblestone appearance. Hepatocytes within the nodules are distended with lipid and there is an increase in melanomacrophages (pinpoint pigmented dark brown/black foci). The white vascular structures along the capsular surface are dilated lymphatic vessels secondary to fibrosis-related compromised vascular flow. Cut sections of this liver floated when placed in a liquid fixative. (B) Histologically, the liver of this South American snapping turtle is characterized by displacement of the hepatic cords by thick bands of mature, dense fibrous connective tissue with entrapped hepatocytes as well as few to many hyperplastic bile ducts.

Neoplastic

Other than herpesvirus-induced cutaneous fibropapillomas and internal fibromas (see "DNA Viruses" section below), myxosarcomas, and fibrosarcomas in marine turtles, neoplasia is infrequently reported in chelonians. Individual reports of neoplasia in chelonians include dermal fibroma, cutaneous mast cell tumor, squamous cell carcinoma, thyroid adenoma and carcinoma, parathyroid adenoma, pulmonary

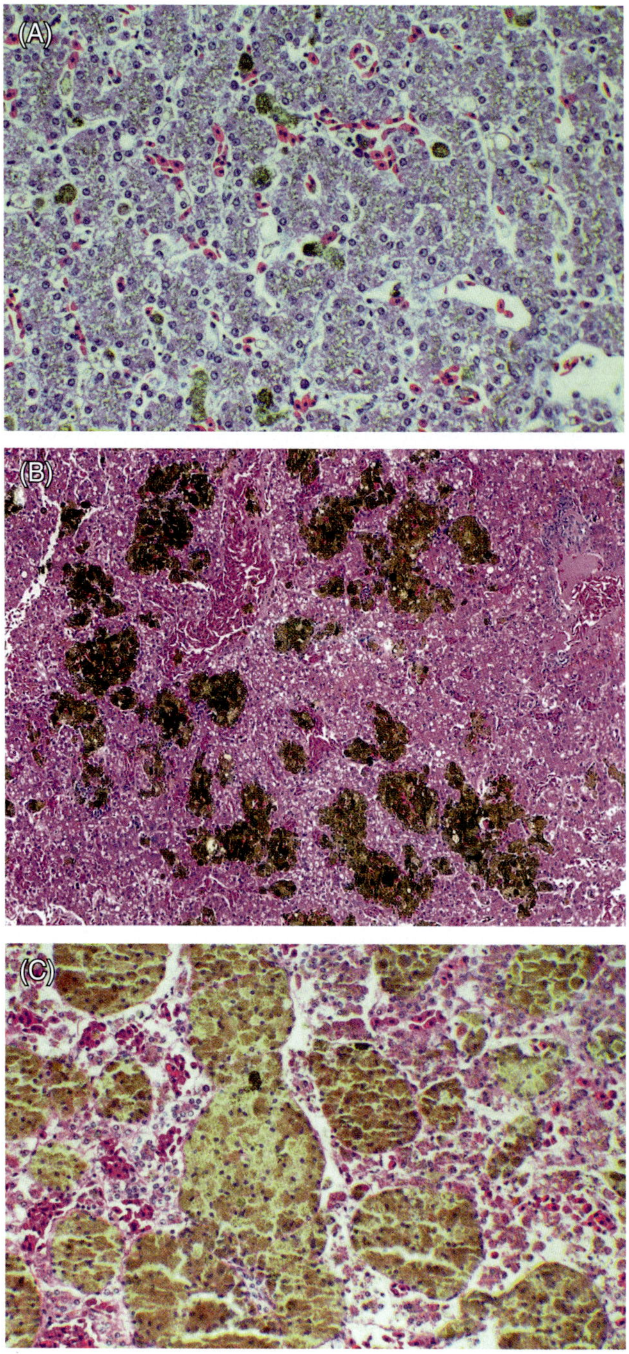

FIGURE 34.11 Melanomacrophages in the liver of a green sea turtle (A), a white-lipped mud turtle (B), and an Eastern box turtle (C). (A) The presence of individual and small melanomacrophage aggregates in the liver of this green sea turtle is considered normal. (B) Melanomacrophage nodular hyperplasia can be challenging to diagnose, as in the liver of this white lipped mud turtle. There are coalescent melanomacrophage nodules that distend the sinusoids but this may represent variations of normal based on species, age, gender, physiologic state, and immune status. (C) Liver of an Eastern box turtle with prominent melanomacrophage nodular hyperplasia characterized by pronounced distortion of the hepatic cords by large, coalescent melanomacrophage nodules that compress and distort adjacent hepatic cords.

fibroma and fibroadenoma, gastric carcinoma, intestinal lymphosarcoma, gallbladder papilloma, testicular interstitial cell tumor, and renal adenocarcinoma. Lymphoid leukemias occur rarely but several cases have been reported in confiscated, siadenovirus infected Sulawesi tortoises.[11,54,59,86,89,121,124]

INFECTIOUS DISEASES

DNA Viruses

All reptile **herpesviruses** belong to the subfamily *alphaherpesviridae*.[9,92,143] Since the early 1980s, the presence of herpesviruses has been documented in an ever-increasing number of species.[9] The majority of reports describe herpesviruses in privately held collections but reports of herpesviruses in free-ranging chelonians with and without clinical disease are increasing. Several of these virus isolates have been sequenced but due to differences in diagnostic methodologies and inconsistency in molecular follow-up, the nomenclature of chelonian herpesviruses is at best confusing and likely to change as more phylogenetic information becomes available. The most commonly reported and best-described herpesviruses are discussed below. For a more complete list of chelonian herpesviruses see Table 34.e2. Several additional notable chelonian viruses are listed in Table 34.e3.

The first herpesvirus isolated from terrestrial chelonians was identified in several species of privately held tortoises presenting with necrotizing stomatitis and glossitis with diphtheritic plaques in the oral cavity, cranial esophagus, and trachea (several of these individuals were coinfected with *Mycoplasma* sp.). The virus was tentatively named **tortoise herpesvirus 1 (THV-1)** and since then, it has been identified in a number of other species of terrestrial chelonians.[57,101] Rhinitis, hepatomegaly, enteritis, and conjunctivitis have also been reported.[9,72] Histologically, lesions are characterized by mucosal ulceration and necrosis with diphtheritic membrane formation, heterophilic, and histiocytic inflammation, and opportunistic, usually gram-negative bacterial infections.[72,77] Characteristic eosinophilic to amphophilic glassy intranuclear inclusions are present in the mucosal epithelium of the oral cavity, tongue, upper digestive tract, nasal and respiratory tract, and urinary and genital systems. Inclusions are only present during the acute phase of the disease.[102]

Tortoise herpesvirus 2 (THV-2) was initially isolated from a California desert tortoise held in a private collection that presented with severe necrotizing stomatitis and glossitis with numerous eosinophilic to amphophilic glassy intranuclear inclusions in the mucosal epithelium (eSlide 34.e1).[77] The endotheliotropism of the virus was evidenced by vasculitis, edema, and thrombosis in several of the cases.[139] Electron microscopic features included enveloped intranuclear hexagonal viral particles with a 110–120 nm electron-dense core, consistent with herpesvirus.[72,76]

Fibropapillomatosis (FP) is an emerging, panzootic infectious disease that affects primarily green sea turtles around the world but has also been documented in all marine turtle species.[79,83] FP was initially recognized in 1938 and though it has been the subject of intense research, the pathogenesis and epidemiology have not been completely elucidated and the causative agent has not been definitively identified.[36,79] Currently, the putative agent is believed to be **chelonid herpesvirus 5 (ChHV-5)** [previously known as chelonid fibropapilloma-associated herpesvirus (CFPHV)], an *alphaherpesvirus* in the genus *Scutavirus*. There are at least four identified genetic variants, although to date, the virus has not been successfully cultured.[79] Interestingly, the presence of virus in nonaffected animals has also been documented and suggests that other factors may be important in the development of lesions.[83] FP is most common in postbenthic juveniles migrating to coastal waters and shallow feeding areas, regardless of gender. This suggests a horizontal mode of transmission, possibly through intraspecies aggression or during mating. FP has not been documented in pelagic, deep ocean-dwelling turtles. A number of factors may predispose newly arrived juveniles recruited into the neritic, near-shore populations to the development of FP and include stressors, such as recent migration, adapting to new environments, and social dynamics with the resident turtle populations. External factors, including mechanical vectors, such as marine leeches, elevated water temperatures, increased local anthropogenic influences, and toxic algal blooms have also been suggested but causation remains to be established. Spontaneous regression of the tumors has been documented in some geographic locations. External tumors in natural and in experimentally induced infections are identical and pathognomonic.[59,79,83]

Tumors vary from flat, plaque-like lesions to raised, exophytic, gray to black, verrucous masses that can grow to over 30 cm in diameter (Fig. 34.12A; eSlide 34.e2). Tumors arise from the skin around the eyes, mouth, and limbs but can also develop on the cornea and the surface of the carapace and plastron. External tumors can affect vision or hinder feeding if they are associated with the eyes or mouth or interfere with swimming. Early stages of the disease are characterized by localized fibroblast proliferation within the superficial dermis. As tumors grow, the epidermis becomes hyperplastic, up to 30 cells thick, and expands into exophytic nodules composed of papilliferous, hyperkeratotic fronds (Fig. 34.12B). Acanthosis, basal cell necrosis, and cleft formation along the dermo-epidermal junction are common histologic features. Subepidermal cleft formation, spongiosis, and occasional epidermal inclusion cysts are also seen. Prominent interdigitating rete pegs may extend deep into the dermis. Eosinophilic

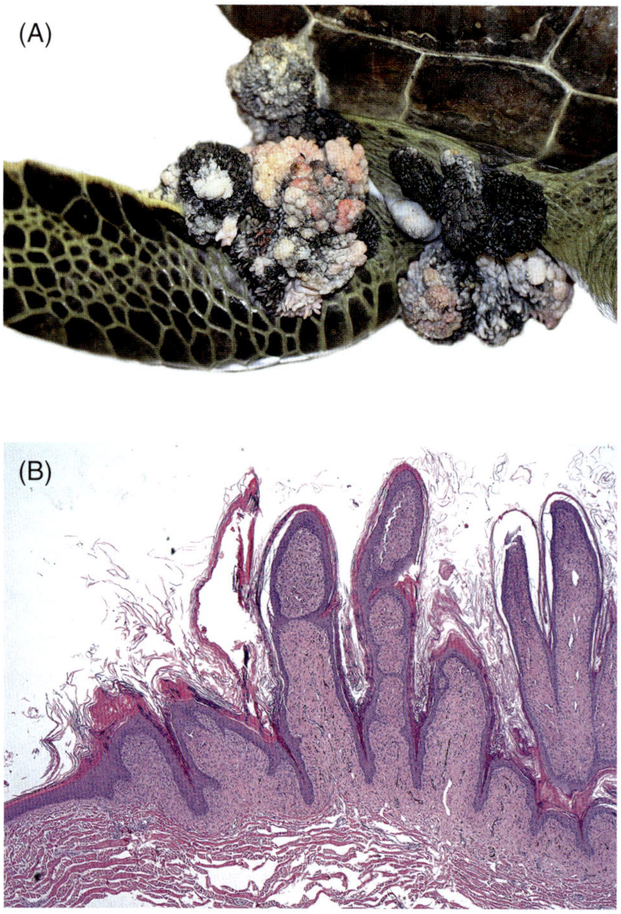

FIGURE 34.12 Fibropapillomatosis in the skin of a green sea turtle. (A) Multiple coalescing, verrucous, exophytic, tan to black nodular masses are present on the flipper. (B) Proliferative fibrovascular connective tissue stroma within the superficial dermis supports multiple papilliferous fronds of thickened, hyperkeratotic epidermis. Multiple foci of perivascular melanomacrophage aggregates are evident throughout the proliferating dermal connective tissue (see eSlide34.e2). *(Parts A and B: Photos Courtesy of B. Stacy, National Oceanic and Atmospheric Administration.)*

intranuclear inclusions are rare in most cases and are typically transient. Inflammation is generally sparse except in areas of cleft formation, in which intense granulocytic aggregates can evolve into intraepidermal pustules. Mild mononuclear superficial perivascular dermal inflammatory infiltrates can be present.[59,79] Internal tumors, which have been described in the oral cavity, glottis, lungs, kidney, heart, liver, spleen, and GI tract, tend to develop late in the disease process.[59,83] When present, they are typically well-demarcated, smooth, firm, nodular white masses. Histologic features of the internal tumors in early lesions include extensive fibroblast proliferation associated with the production of a myxomatous matrix followed by a more mature, fibrillary collagenous stroma in later stages.[59,79,150] These have been diagnosed as fibromas, fibrosarcomas, and myxosarcomas. Mitoses are rare. Peritumoral fibrosis is common and entrapment of normal structures has been reported. Inflammation is sparse and can consist of granulocytic to lymphocytic infiltrates.[79] Foreign body granulomas with multinucleated giant cells associated with spirorchiid trematode ova or epidermal inclusion cysts are not uncommon in FP lesions in free-ranging animals. Areas of ulceration are often associated with bacterial and/or fungal colonization.[45,59,79,150]

Two genetically distinct **papillomaviruses**, ***Caretta caretta papillomavirus (CcPV)*** and ***Chelonia mydas papillomavirus (CmPV)***, have been identified in epidermal lesions in a free-ranging loggerhead sea turtle and a free-ranging green sea turtle, respectively.[90] Both turtles presented for unrelated reasons and were housed at different times for treatment and rehabilitation. Both turtles developed few to many raised, smooth, soft, white nodules over the head and flippers. The lesions consisted of well-demarcated areas of epidermal hyperplasia associated with ballooning degeneration and keratinocyte anisokaryosis, nuclear swelling, irregular chromatin clumping, and eosinophilic intranuclear inclusions surrounded by a clear halo. Electron microscopy confirmed viral particles consistent with papillomavirus.[60,90] Papillomavirus-like particles have also been identified in a lung wash of a Horsfield's Russian tortoise and intranuclear, papillomavirus-like paracrystalline arrays have been described in a Bolivian side-necked turtle with papular skin lesions.[72]

Adenoviruses in chelonians were, until recently, unrecognized.[42,50] The first report involved a single, moribund leopard tortoise coinfected with an uncharacterized herpesvirus and adenovirus-like viral particles. In 2006, adenoviruses in chelonia garnered increased attention due to a mortality event involving 105 confiscated Sulawesi tortoises that died due to systemic disease that was associated with a novel adenovirus.[121,126] This virus, identified as **Sulawesi tortoise adenovirus-1 (STAdv-1),** is the only reptilian adenovirus within the *Siadenovirus* genus.[121] Fatal siadenoviral infections were subsequently identified in impressed tortoises and in one Burmese star tortoise housed in the same facility as the confiscated Sulawesi tortoises, suggesting that interspecies transmission and infection are possible.[50,126] A novel adenovirus has also been described in a single ornate box turtle from Europe that was coinfected with a distinct, uncharacterized *Mycoplasma* sp. Although the gross and histologic lesions in this animal were consistent with those caused by STAdv-1, phylogenetic analysis failed to cluster this virus with any of the known adenovirus families.[42]

In the confiscated Sulawesi tortoises, affected tortoises were in poor body condition and many presented with anorexia, lethargy, ulcerative stomatitis, nasal and ocular discharge, and diarrhea. The most common microscopic lesions were necrotizing and pseudomembranous enterocolitis, individual to multifocal and massive hepatocellular necrosis, myeloid necrosis, and splenic necrosis.

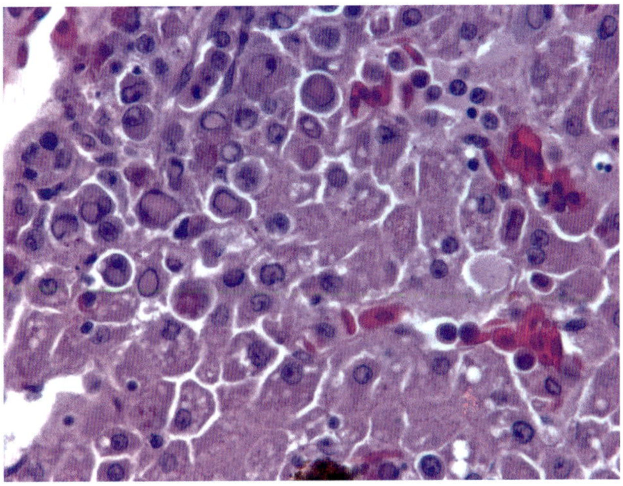

FIGURE 34.13 Siadenovirus infection in the liver of a Sulawesi tortoise. Viral infection is characterized by hepatocellular degeneration and hepatic cord dissociation, multifocal individual cell necrosis, and large, glassy, amphophilic to eosinophilic intranuclear viral inclusions that peripherally displace nuclear chromatin (see eSlide 34.e3). *(Photo Courtesy of D. McAloose, Wildlife Conservation Society.)*

Intranuclear inclusions were present in a number of sites including intestinal epithelial cells, hepatocytes, Kupffer cells, cardiac and vascular endothelial cells, bone marrow and splenic myeloid and stromal cells, pancreatic acinar cells, stromal and germinal cells of the testes and ovaries, renal tubular epithelial cells, respiratory epithelium, cerebral glia, and the choroid plexus. The intranuclear inclusions are glassy, amphophilic to eosinophilic, variably sized and discretely marginate the nuclear chromatin (Fig. 34.13; eSlide 34.e3). Electron microscopy confirmed 70–90 nm, nonenveloped viral particles with occasional formation of paracrystalline arrays in the nuclei of infected cells. Adenovirus-related disease most often occurs in young and immunocompromised animals, and naïve animals in the pet trade may be at higher risk of infection with novel pathogens due to overcrowding and mixing of geographically distant species, inappropriate husbandry, suboptimal nutrition, and other physiologic stressors. Virus-associated myelonecrosis in the Sulawesi tortoises and immunosuppression may have predisposed the affected tortoises to a number of opportunistic coinfections that included amoebiasis, nematodiasis, and *Escherichia coli*, *Aeromonas hydrophila*, and *Chlamydophila* sp. infections.[50,121,126,143]

Iridoviridae are a diverse group of viruses that infect vertebrate and invertebrate species. Members include viruses in the ***Iridovirus*** and ***Ranavirus*** genera, which are listed as reportable diseases by the World Organization for Animal Health (WOAH) and have been identified and implicated as the cause of disease and mortality in turtle and tortoise populations worldwide.[32,50] Infections and mortality have been diagnosed in privately owned Mediterranean tortoises, gopher tortoises, and a mortality event involving Eastern box turtles and red-eared sliders held at a zoo.[5,32,50] A virus recovered from one of the Eastern box turtles was initially named **Turtle virus-3 (TV-3)** but sequence was nearly identical to **Frog virus-3 (FV-3)**, a type of ranavirus.[32] To date, all ranaviral infections described in chelonians have been caused by FV-3 or FV-3-like viruses and infections have been documented worldwide, affecting over 20 species in seven countries.[3] Environmental temperatures may have a significant effect on virus-associated morbidity and mortality.[5] The mechanism of transmission is unknown and outbreaks in free-ranging populations tend to be characterized by high levels of morbidity and mortality.[3] Because iridoviral particles can be observed in circulating leukocytes (Fig. 34.14A), it is postulated that transmission by blood-feeding parasites or biting insects may be possible. Other possible transmission routes include consumption of infected animals, environmental contamination, and exposure to infected fomites, amphibians, or fish.

Lesions vary somewhat by species and are nonspecific. In fact, the clinical presentation is virtually identical to that of herpesvirus and *Mycoplasma* spp. infections and can manifest as oculonasal discharge, lethargy, anorexia, dyspnea, oral plaques and ulcers, necrotizing glossitis, and tracheitis. Gross lesions can include oral, esophageal and GI ulceration, multifocal gray foci and/or hemorrhages throughout the liver (Fig. 34.14B) and spleen, and coelomic effusion.[3,32,50] Histologically, fibrinoid vasculitis is a consistent finding in infected chelonians and lesions can include cutaneous and mucosal ulceration and necrosis, stomatitis, glossitis and pharyngitis, necrotizing pneumonia, conjunctivitis, esophagitis, tracheitis, enteritis, pancreatitis, and cystitis.[32,50,78] Necrotizing splenitis with hemorrhage has consistently been seen in red-eared sliders, whereas in Eastern box turtles raised and often ulcerated skin abscesses appear to be common.[3,5,32] Intracytoplasmic basophilic inclusions can be seen in epithelial cells, including renal tubular epithelium (Fig. 34.14C) and hepatocytes (eSlide 34.e4), although they may be uncommon and difficult to identify.[78] On electron microscopy, hexagonal to icosahedral viral particles that measure approximately 150 nm in diameter cluster in the cytoplasm.[72,76] Paired blood and oral/cloacal swabs are recommended for PCR testing.[3]

RNA Viruses

Tortoise Picornavirus (ToPV), also known as **Virus X**, has been associated with skeletal lesions as well as kidney disease in juvenile spur-thighed tortoises, Hermann's tortoises, and Indian star tortoises. Clinically, this syndrome was characterized experimentally and infected individuals developed softening of the shell and plastron and delayed

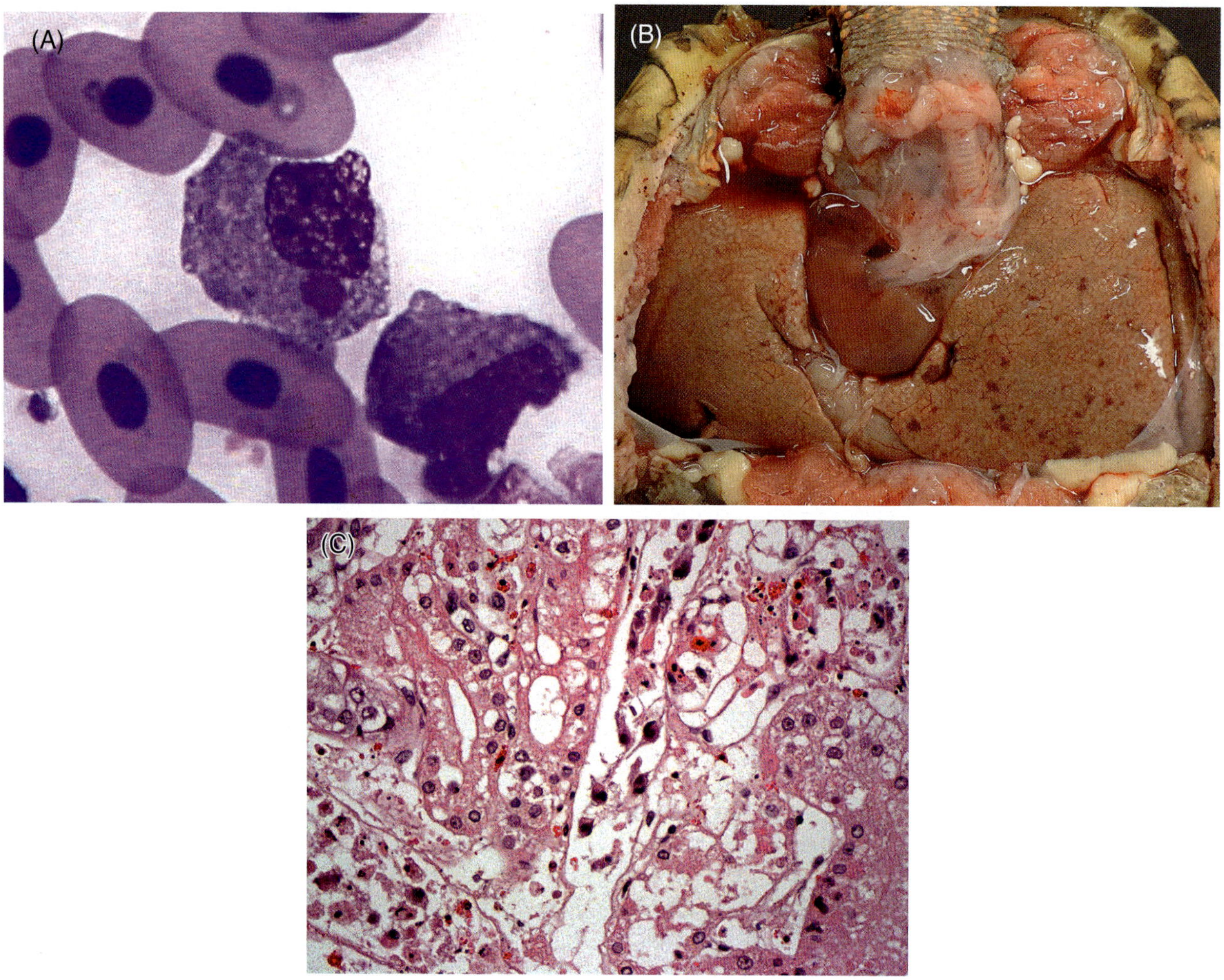

FIGURE 34.14 Ranavirus infection in an Eastern box turtle. (A) A circulating monocyte contains an amphophilic ranaviral cytoplasmic inclusion. Wright-Giemsa. (B) Multiple, randomly scattered foci of hemorrhage admixed with punctate focal to coalescent areas of hepatic pallor (necrosis) may be present throughout the liver (the liver is also discolored light tan due to hepatocellular lipid deposition). (C) Infection causes acute renal tubular epithelial necrosis and heterophilic inflammation with dissociation and sloughing of cells into the tubule lumena. Intracytoplasmic basophilic inclusions are present in several necrotic renal tubular epithelial cells (see eSlide 34.e4). *(Parts A–C: Photos Courtesy of the University of Illinois Zoological Pathology Program.)*

growth compared to noninfected animals. Histologically, the bony lesions featured a lack of primary bone trabecular remodeling and formation of dense fibrous connective tissue, the hallmarks of osteodystrophy. The pathophysiology that leads to the development of these lesions is not completely understood but in many of these cases there was concurrent evidence of injury to the kidneys characterized by renal tubular vacuolization and interstitial edema. Few studies have been carried out but virus shedding was documented in pharyngeal and cloacal swabs from infected individuals throughout the duration of the study (up to 165 days postinfection) and virus was successfully cultured from a number of infected tortoises.[63,110] A novel member of the ***Picornavirus*** family has also been recently discovered and sequenced in samples of liver and kidney from Sulawesi tortoises involved in the mortality event associated with STAdv-1. The designation of **Tortoise Rafivirus-A (Tor-Raf-A)** has been proposed. At this time, the host range of this virus is restricted to Sulawesi tortoises from the mortality event and the virus has not been successfully cultured. Its role in the mortality event or as a cause of disease in reptiles remains unknown and warrants investigation.[98]

Bacteria

Ulcerative lesions of the skin and shell that are secondarily infected by single or mixed bacterial infections are primarily reported in managed populations of fresh water chelonians. Lesions are often associated with suboptimal housing, nutrition, and management practices. Lesions typically start as abrasions or as the result of conspecific aggression. If disease becomes systemic, a syndrome known as **septicemic cutaneous ulcerative disease (SCUD)** develops.[23,72] Animals may present with lethargy, muscle weakness, limb paralysis, cutaneous abscesses, and/or necrosis of the digits or shell.[72] The syndrome has been reported in North America and Europe, and an identical presentation has been documented in terrapins in Africa.[89] The most commonly isolated organism reported in the literature is *Citrobacter freundii,* but *Serratia* sp. and *Beneckea* sp. have also been implicated and may be synergistic.[72] Gross lesions include cutaneous erosions and ulcers on the plastron, carapace, and skin with loosening of the scutes (Fig. 34.15). Disseminated infection can produce necrotic foci in many sites, in particular in the liver, spleen, kidney, and heart. Multiple caseous yellow masses may also be seen throughout the coelomic cavity.[72] Histologically, multifocal, variably extensive foci of necrosis are present on the plastron, carapace, and skin. Acute lesions are characterized by intense heterophilic and histiocytic inflammation with necrosis. Bacteria may or may not be evident with routine hematoxylin and eosin (HE) stains. More chronic cases have an increased mononuclear inflammatory cell component and can be surrounded by granulation tissue and fibrosis. When present on the shell, lesions often penetrate the keratin layer and can result in periosteal inflammation, localized osteonecrosis, and bone remodeling.[72] The recent identification and characterization of *Emydomyces testavorans* (see "Fungi" section below) in some shell lesions can be a complicating factor and a reminder that not all shell lesions are due to bacterial infection. Histologic examination of the shell with appropriate special staining and ancillary diagnostic testing is critical for accurate diagnosis and treatment.

FIGURE 34.15 Septicemic cutaneous ulcerative disease (SCUD) in a Colombian wood turtle. Extensive necrotizing dermatitis and abscess formation is characterized by necrosis within and beneath the scutes. The affected scutes were on the plastron and were loose and easily removed during necropsy, exposing the underlying tissues covered by necrotic tan material.

Upper respiratory tract disease (URTD) is a clinical syndrome characterized by rhinitis and oculonasal discharge. It was first described in free-ranging populations of desert tortoises in California and has been associated with significant declines in free-ranging populations of this endangered tortoise.[46,71,73,89,122] URTD has subsequently been identified in several other species of tortoises, some of which are highly endangered, in both free-ranging and managed populations. At least five species of *Mycoplasma* have been identified in tortoises. Of these, *M. agassizii* and *M. testudineum* have been associated with clinical URTD in desert tortoises and gopher tortoises, respectively. Mycoplasmas are thought to be largely commensal organisms; however, they can be opportunistic pathogens and development of clinical disease is likely due to a complex interplay of factors, such as infective dose and strain virulence, external stressors, immunocompetency, and coinfections with other pathogens.[122] Acute high mortality, though rare, has been reported in an experimental study with gopher tortoise hatchlings. Spontaneous clearance of infection is also rare. More commonly, animals develop chronic infections that follow cyclical periods of recrudescence. Transmission is thought to be horizontal between individuals in close proximity and latent infections have been reported.[74,89] Animals usually present with palpebral edema, conjunctivitis, rhinitis, and nasal and ocular discharge (Fig. 34.16A). Chronic cases may have partial or complete obstruction of the nasal passages with caseous plugs. Erosion and depigmentation of the nares with formation of vertical fissures ventral to the nares can occur in chronic cases. Animals that die usually are cachectic and lethargic with multisystemic disease due to secondary infections. Lesions in the nasal cavity can vary dramatically between individuals and include necrosis of ciliated epithelium with erosion, basal cell and ciliated epithelial hyperplasia, mucus gland metaplasia, mixed heterophilic, histiocytic and lymphoplasmacytic inflammation, and subepithelial lymphoid aggregates (Fig. 34.16B; eSlides 34.e5 and 34.e6). Cytoadherence of *Mycoplasmas,* at least *M. pneumoniae,* to the respiratory epithelium is facilitated by a specialized attachment structure coated with adhesins. Once adhered to the cell membrane, accessory proteins promote the intracellular translocation of the organism. Cytotoxic effects leading to epithelial cell necrosis are the result of oxidative damage and cytokine-mediated inflammation.[21]

Salmonellosis is recognized as the most important zoonotic disease of reptile origin.[72,89,95] Turtle-associated human salmonellosis (TAHS) was first reported in 1943. Over the subsequent two decades, an increasing number of TAHS cases were identified but in the late 1960s and early 1970s an epidemic of salmonellosis in children linked

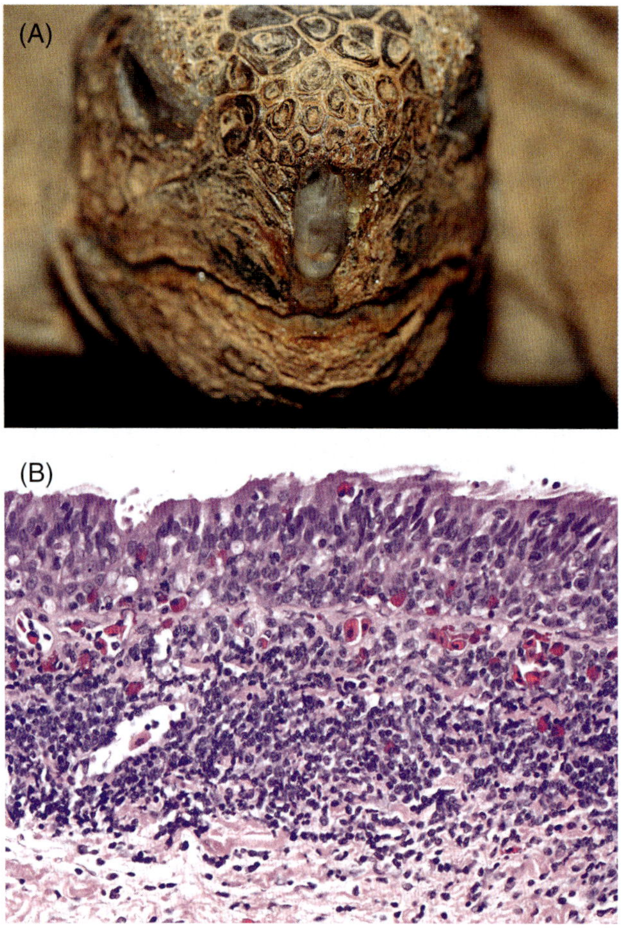

FIGURE 34.16 Mycoplasmosis in (A) a desert tortoise and (B) an Eastern box turtle. (A) Rhinitis with mucopurulent nasal discharge along with conjunctivitis and ocular discharge are the most common clinical signs and are classic symptoms of infection. (B) Conjunctivitis is associated with epithelial hyperplasia and intense leukocytic exocytosis that, as in this case, can obscure the basal lamina. The inflammatory reaction is characterized by a mixed heterophilic, histiocytic, and lymphoplasmacytic infiltrate with extensive subepithelial lymphoid aggregates (see eSlides 34.e5 and 34.e6). *(Part A: Photo Courtesy of B. Rideout, San Diego Zoo Wildlife Alliance; Part B: Photo Courtesy of K. Terio, University of Illinois Zoological Pathology Program.)*

to pet ownership of red-eared sliders and other aquatic turtles resulted in legislation in the United States affecting the sale of aquatic turtles in the pet trade. ***Salmonella*** spp. are routinely isolated from clinically asymptomatic chelonians in free-ranging and managed populations around the world and are generally considered part of the normal intestinal microbiota.[72,95] Up to 85% of chelonians may be infected with as many as five different serotypes. In a survey of the chelonian collection at one institution, *S. durham* was the most common isolate identified among over 30 serotypes.[95] Intermittent shedding of *Salmonella* has been reported in healthy hosts. Under the right conditions and in a susceptible, debilitated animal, pathologic infections can develop, although clinical disease is far less commonly reported in chelonians than in other reptile groups. Transmission can be from animal to animal and via feces and urine, and vertical transmission to eggs has been documented. Virulence factors vary greatly between *Salmonella* species and serovars. In one report, the death of two Galapagos tortoises was putatively attributed to salmonellosis but lesions were not described.[72] In another study, clinical signs in a group of ill, free-ranging, flattened musk turtles included prolonged basking (when compared to apparently healthy animals) and multiple foci of shell necrosis with loosening of the scutes, consistent with bacterial septicemia. Sick animals were thin to emaciated. Histologic lesions consisted of focal areas of epidermal necrosis throughout the shell as well as foci of hepatic necrosis, airway-associated heterophilic and histiocytic pulmonary inflammation, and submucosal edema in the intestinal tract. *S. arizonae*, among a number of other microorganisms, was cultured from these animals.[72]

A novel species of ***Helicobacter*** has been implicated in several cases of rhinitis in gopher tortoises in the Southeastern United States. Affected individuals presented lethargic with opaque, watery to mucoid nasal discharge and dyspneic. Cytologic preparations of nasal discharge from affected animals contained heterophilic and histiocytic inflammation and Warthin-Starry (WS) stain-positive spiral bacteria. Histologic lesions were centered on the nasal mucosa and conjunctiva. These included edematous mucosa with ulcero-erosive heterophilic rhinitis, submucosal lymphocytic infiltrates, and mucoid glandular ectasia with intralumenal accumulations of cellular and secretory debris. Chronic lymphoplasmacytic and heterophilic inflammatory infiltrates were present in the conjunctival submucosa. The organism has not been successfully cultured. Ongoing and further studies are needed to assess the role of this novel organism in morbidity and mortality in tortoises as well as the potential negative impact on wild populations.[31]

Mycobacteria are slender, aerobic, generally acid-fast, and variably gram-positive bacilli. Among reptiles, infections are most common in aquatic chelonians. This is not surprising as the mycobacteria that infect chelonians are ubiquitous in aquatic environments.[94,99,120] *Mycobacterium marinum* and *M. chelonae* are common isolates.[72] Other species, such as *M. kansasii* have also been isolated and there are few reports of *M. avium* in free-ranging and managed groups of green sea turtles.[72,89,96] While more commonly reported in previously free-ranging animals brought under managed care or being housed for rehabilitation, infections have occurred in both free-ranging and managed populations of chelonians. Most mycobacterial infections in chelonia tend to be chronic; however, acute disease has been reported. Acute mycobacteriosis can present as fulminant disease with lethargy, petechial hemorrhages on the limbs, and bloody nasal discharge. Internal lesions can include hemorrhagic coelomic effusion, petechiae throughout the intestinal serosa, and pulmonary hemorrhage. Bacterial emboli and disseminated intravascular coagulopathy have been reported.[96,99] Chronic mycobacterial infections, which are by far more commonly reported than acute infections, are associated with granulomatous disease. Lesions are often present in the lungs, coelomic viscera (Fig. 34.17A), and skin but can also involve bones and joints. Histologically, granulomas are composed of dense aggregates of heterophils with or without a central core of liquefactive necrosis surrounded by histiocytes, epithelioid macrophages, and multinucleated giant cells. Chronic granulomas may be surrounded by fibrous connective tissue (Fig. 34.17B). Mycobacteria typically stain positive with acid-fast stains, such as Ziehl-Neelsen and the Fite's method; atypical mycobacteria may not stain.[94] Diagnosis of mycobacteriosis is based on detection of acid-fast positive bacilli in exudates, fine-needle aspirates, or tissue biopsy or necropsy specimens, bacterial culture, and via molecular diagnostics such as PCR.

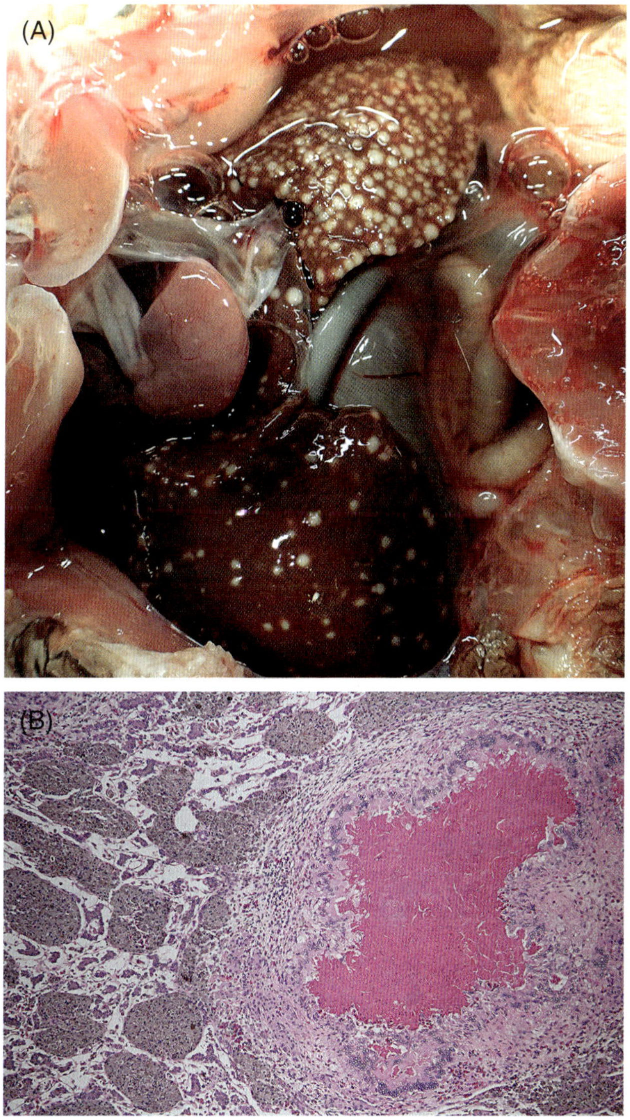

FIGURE 34.17 Mycobacteriosis in an Eastern box turtle. (A) Disseminated, well-demarcated, raised, light tan nodules are scattered randomly throughout the liver. There is moderate ascites throughout the coelom. (B) Granulomatous hepatitis consists of a central core of intensely eosinophilic necrotic cellular and inflammatory cell debris rimmed by multinucleated giant cells and surrounded by heterophils and histiocytes. Prominent perilesional fibrosis with lymphocytic infiltrates partially displaces the adjacent parenchyma. Adjacent hepatic cords are compressed by broad sinusoidal nodular aggregates of melanomacrophages.

Austwickia (previously ***Dermatophilus***) ***chelonae*** is a gram-positive Actinobacterium that forms branching filamentous structures and produces motile, coccoid zoospores. Lesions in reptiles associated with *A. chelonae* include cutaneous, subcutaneous, and muscle abscesses, hepatic granulomas, granulomatous coelomitis, pneumonia, and septic arthritis.[72,85,143] The distribution and mode of transmission are poorly understood but prolonged exposure to an abnormally wet environment may be a predisposing factor.[10,23] External lesions are typically yellow, raised, hyperkeratotic skin nodules around the neck and head.[22] Microscopic lesions consist of superficial cutaneous crusts with necrosis of the underlying epidermis and transmigrating heterophils, or poorly encapsulated intradermal nodules composed of a core of intensely eosinophilic necrotic material surrounded by heterophilic to granulomatous inflammation, and intralesional gram-positive filamentous bacteria.[10,22,143] Recent studies have identified production of a diphtheria-like toxin gene associated with *A. chelonae*. This is concerning as diphtheria toxin is one of the most potent toxins known and capable of interfering with protein synthesis. Furthermore, the isolation of *A. chelonae* in a free-ranging tortoise in North America could have implications for conservation and release of rehabilitated animals.[85]

Fungi

Chelonians have diverse cutaneous and intestinal mycobiota. Mycoses are largely considered opportunistic in chelonians and most reports are from animals under managed care.[34,72,109] Mycoses are usually associated with suboptimal housing and management conditions, such as overcrowding, inappropriate temperature gradients, and poor hygiene. These conditions allow for organic and environmental debris to build up and promote fungal overgrowth.[72,125] In free-ranging chelonians, mycoses have been associated with exposure to adverse environmental conditions, such as unusual changes in temperature.[72,89] Both **hyalohyphomycoses** and **phaeohyphomycoses** have been reported in chelonians. Based on the distribution of lesions, infections can be generally classified as superficial shell and skin lesions or **superficial dermatomycoses**

(SD), and visceral or **systemic mycoses (SM)**.[72,89,125] SD manifests as lesions on the shell and less commonly on the skin. SD is more common in marine and fresh water turtles than in terrestrial chelonians. Lesions develop as white to lightly discolored blemishes on the carapace that can progress to scute necrosis and pitting. These can become ulcerative lesions as the fungi invade and disrupt the keratinized layer. Minimal to no inflammation is seen in these cases but if the lesions become ulcerated, variably intense heterophilic and histiocytic to granulomatous inflammation develops.[89] Fatal SD infections are rare but outbreaks of dermatomycosis associated with *Mucor* and *Paecilomyces* species have been reported in young Florida softshell turtles and in Fly River turtles, respectively.[72,89] In privately held terrestrial chelonians, secondary fungal pododermatitis can develop, putatively the result of inappropriate substrate.

Fusarium incarnatum has been isolated from free-ranging and managed populations of Texas tortoises with **necrotizing scute disease (NSD)**. NSD manifests as blemishes on the shell that are confined to the superficial epidermal keratin layers; NSD differs from SCUD in that lesions, although disfiguring, do not result in systemic disease or death.[72,89]

Recently reported in privately held Hermann's tortoises, **microsporidiosis** has been associated with granulomatous and necrotizing hepatitis and pneumonia with intralesional basophilic spores. Necrotizing to pseudomembranous enteritis is less common but is described. Histologically, spores are surrounded by a thin, fibrous capsule, are $1.5\,\mu m \times 2.5\,\mu m$ in diameter, and are positive with Grocott-Gomori's methenamine silver (GMS) and Periodic acid-Shiff (PAS) stains. On electron microscopy, the spores are characterized by an electron-dense outer sheath and an electron-lucent inner sheath that surrounds an electron-lucent vacuole with eight polar tubes arranged in two rows.[41]

Phaeohyphomycosis has been reported in a number of chelonians. In general, most of the pigmented fungi are considered opportunistic organisms and infections in chelonians have been associated with adverse housing or environmental conditions and immunosuppression.[109] Infections with dematiaceous fungi, such as ***Ochroconis*** and ***Exophiala*** species have been associated with granulomatous to ulcerative skin lesions on the feet and tail of box turtles. *Exophiala* sp. has also been identified as the cause of an oral granuloma in a radiated tortoise, deep ulcerative shell lesions in an Aldabra tortoise, and unilateral ocular infection and systemic disease in a Galapagos tortoise.[72,137] *Veronea* sp. has been identified in fatal cases of obstructive tracheitis in stranded green sea turtles.[34]

A unique presentation of mycotic shell and skin disease associated with a recently described keratinophilic fungi in the order Onygenalean has been reported in a variety of institution-housed and free-ranging aquatic and semi-aquatic chelonian species. This organism, originally grouped with the *Chrysosporium* anamorphs of *Nannizziopsis vriesii*, has been recently definitively identified and named ***Emydomyces testavorans***.[3,147,149] Gross lesions consist of multifocal to coalescing areas of scute ulceration, keratin discoloration and deformation, and necrotizing osteomyelitis with softening of the scutes. This is accompanied by formation of single to coalescent nodular and cystic bone lesions filled with tan to gray caseous material that protrude into the coelomic cavity (Fig. 34.18A). These nodules can often compress or displace internal organs and can be seen radiographically as lytic lesions within the bony structure of the carapace and/or plastron. Histologically, lesions are characterized by ulcerative dermatitis, necrotizing osteomyelitis, and effacement of dermal bone by multilocular cystic structures lined by stratified squamous epithelium and filled with necrotic keratinaceous debris (epithelial inclusion cysts) and may contain fragments of necrotic bone (Fig. 34.18B). Hyphae are readily visible within the keratin debris and superficial epithelial layers of the affected shell as well as within the cystic lesions with a GMS stain (Figs. 34.18C, D) but may be inapparent with routine HE staining. Morphologically, hyphae are 3–5 μm diameter, are thin-walled and regularly septate, and occasionally produce acute to right-angle branching. Arthroconidia formation is not a feature of this fungus (eSlide 34.e7).[3,134,148,149] Another member of the Onygenalean family, ***Aphanoascella galapagosensis***, has been described in shell lesions from a subspecies of Galapagos tortoise.[138]

In **systemic mycoses**, the respiratory tract is the primary route of infection and usually involves the lungs.[72,89] Dissemination of the infection to other organs is possible. Fungal pneumonias associated with ***Fusarium*** sp. are prevalent in free-ranging sea turtles and have been associated with cold-stunning events.[45,72,103,104] Fungal pneumonia in giant tortoises is also relatively common and is often associated with sporulating fungi, such as ***Aspergillus***, ***Paecilomyces***, and ***Penicillium*** species.[62,72,89] Fungal pneumonias associated with *Aspergillus* spp. usually involve the lower respiratory tract and present as caseous and firm nodular masses within the lung and as fungal plaques on mucosal (e.g., trachea) or coelomic surfaces. Acute lesions are characterized by heterophilic and histiocytic inflammation and exudates. In chronic cases, single to multiple coalescent granulomas are typical, often with central cores of intensely eosinophilic necrotic debris, heterophilic, and histiocytic inflammation with multinucleated giant cells, and variable degrees of fibrosis.[67,72,125] Dissemination of infection to the liver, spleen, kidneys, or other viscera is possible.[72] More recently, pulmonary and systemic mycoses caused by entomopathogenic fungi ***Metarhizium*** and ***Beauveria*** sp. have been documented in several chelonian species.[65] These fungal organisms are primary

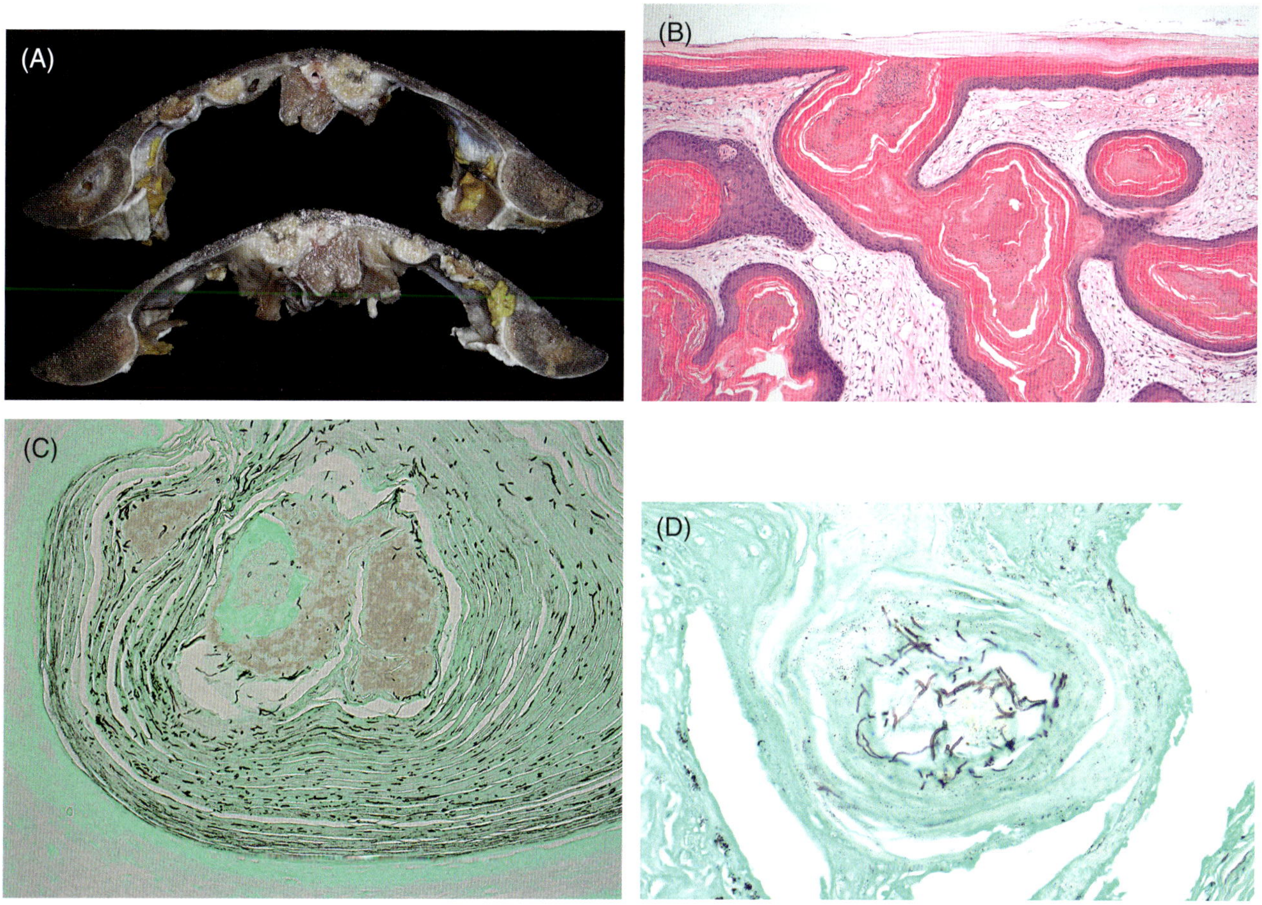

FIGURE 34.18 *Emydomyces testavorans* infection in the carapace of a (A, C) mata mata, (B) Savanna side-necked turtle, and (D) a yellow-spotted river turtle. (A) Lesions consist of necrotizing osteomyelitis and multifocal ulceration along the dorsal surface of the carapace. Broad, lytic, tan to white "inverted" nodular masses with central caseous debris are present subjacent to the ulcer. These lesions markedly distort the contour of the shell and bulge ventrally into the coelomic cavity. (B) The normal architecture of the dermal bone is effaced and replaced by broad cavernous cysts that are lined by stratified squamous keratinized epithelium and filled with necrotic lamellated, keratinaceous, and cellular debris. (C and D) Within the cystic structures and admixed within the keratin and cellular debris are numerous intralesional 3–5 μm, septate, parallel-walled, dichotomously branching hyphae. GMS (see eSlide 34.e7). *(Parts A–D: Photos Courtesy of the University of Illinois Zoological Pathology Program.)*

pathogens of insects, have a worldwide distribution in terrestrial as well as freshwater and marine environments, and are believed to be opportunistic vertebrate pathogens in compromised hosts. Lesions have been predominantly centered on the respiratory system; however, at least one case of granulomatous coelomitis has been described.[65] Grossly, these lesions feature the formation of necrotic granulomas filled with caseous debris scattered throughout the lungs. Histologic lesions are characterized by heterophilic and histiocytic to granulomatous inflammation that surround abundant lamellated necrotic debris. The fungal organisms, which can be identified on routine as well as silver stains, display nonparallel, branching walls and in some cases are accompanied by the formation of birefringent oxalate crystals. These fungal organisms can exhibit angioinvasive behavior with vascular necrosis and thrombosis.[65]

Yeasts, in particular ***Candida albicans***, are widespread in the GI tract of chelonians but pathologic infections are rare. Oronasal ulcers in a radiated tortoise, pneumonia in a Greek tortoise, and enteritis and pneumonia in a stranded loggerhead sea turtle have been associated with *Candida* spp.[72] Fungal organisms can be difficult to see with routine HE; however, GMS or PAS stains highlight the organisms (Fig. 34.19).[72,89,125]

Metazoa

At least 10 different genera of **spirorchiid trematodes** have been documented in free-ranging marine sea turtles. They are considered to be one of the most widespread and common chelonian cardiovascular parasites around the world and are thought to play an important contributory role in sea turtle stranding and mortality.[19,25,45,130] The life cycle is unknown but is generally thought to involve a gastropod and at least one motile cercarial stage; vermetid gastropods have been found to be infected with sporocysts and

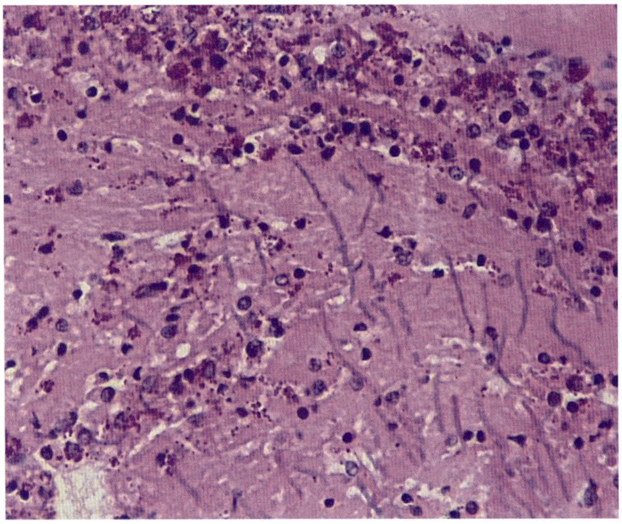

FIGURE 34.19 Pulmonary candidiasis in a South American snapping turtle. Extensive necrosis, fibrin, and proteinaceous exudate accompanied by intense heterophilic and histiocytic inflammation are associated with numerous slender pseudohyphae that have occasional bulbous dilations.

cercaria of *Amphiorchis* sp. and *Learedius learedi* and identified as the key intermediate host for spirorchiid infections in loggerhead and green sea turtles at aquarium facilities. Cercaria of *Neospirorchis* sp. have been documented in several species of marine polychaetes.[27,123] Recent advances in the application of molecular tools have helped in the understanding of parasitic biodiversity and epidemiology, host burden, and intermediate hosts. Furthermore, recent studies suggest not only host specificity but also strong organ system specificity, as exhibited by loggerhead turtle infections by *Neospirorchis* species exquisitely targeting the submucosal vasculature of the intestinal tract.[19,123] Infection is extremely common in free-ranging marine turtles but trematode burden is not always correlated with clinical disease or death. However, increased tissue ova burden may correlate with animals in declining body condition, especially in cases where ova and associated lesions are present in the brain. In addition, concurrent infections with multiple species of spirorchiid nematodes are common.[19,45,130] Spirorchiid infections often colocalize anatomically with other disease processes, such as fibropapilloma-associated cutaneous and internal tumors, coccidia-associated enteritis, and systemic gram-negative bacterial infections.[45,59,79,150]

Adult trematodes are typically found in the cardiac chambers and lumina of the great vessels (Fig. 34.20A) but can also be seen in blood vessels in many organs[25,45,130] including the lungs and intestinal tract (Figs. 34.20B, C).[45] Gross cardiac and vascular lesions include endocarditis and arteritis, thickening of the tunica muscularis, mural mineralization, intimal nodular proliferation, thrombosis, and aneurysms. Heterophilic and histiocytic to granulomatous endocarditis and arteritis, muscular hypertrophy, proliferative endarteritis, perivascular edema, and thrombosis with recanalization are common histologic vascular lesions.[45] The round to elongated ova are approximately $50 \times 100\,\mu m$, can be deposited within the vascular lumen, wall, or at perivascular sites, and are often associated with granulomatous inflammation (Fig. 34.20D). Perivascular deposition of ova in the spleen is very common and often accompanied by hypertrophy of the arterial and arteriolar tunica media. However, parasitic granulomas can be seen in virtually any organ.[19,25,45,130]

Protozoa

Infection of chelonians with **intranuclear coccidia** has been reported sporadically since the 1990s. All cases to date involve chelonians in private collections or recently captured free-ranging individuals that were confiscated after transitioning through the international pet trade. The causative agent is a yet-to-be-named coccidian organism that is associated with the syndrome known as **Testudiniid intranuclear coccidiosis (TINC)**. Fecal–oral transmission has been identified as the route of infection.[3] Species in which the coccidia has been identified include radiated tortoises, leopard tortoises, Forstern's tortoises, Sulawesi tortoises, giant tortoises, Eastern box turtles, flat-tailed tortoises, spider tortoises, Arakan forest turtles, bowsprit tortoises, and impressed tortoises.[48,50,68] Clinical presentation is nonspecific and variable, including clear mucoid nasal and/or ocular discharge, lethargy, weakness, anorexia, anemia, leukocytosis, hyponatremia, hyperuricemia, and hyperglycemia. The life cycle of the parasite and mode of infection and transmission have not been elucidated. Concurrent mycoplasmosis, amoebiasis, and bacterial septicemia can be seen. Intranuclear coccidial infections are, in general, systemic and widespread.[48]

Severity and distribution of gross lesions can be variable and include mucous membrane pallor, rhinitis with oronasal fistulas, generalized subcutaneous edema, coelomic effusion, pseudomembranous enteritis, epicardial and intestinal petechiae, tan discoloration of the liver, splenic congestion, gray pallor to red mottling throughout the kidneys, urinary bladder distention, pulmonary congestion, and skin ulceration.[48,50] Intranuclear stages are more commonly seen in renal tubular epithelial cells, hepatocytes, the biliary tree, enterocytes, pancreatic ductular epithelium and acinar cells, urinary bladder and ureteral epithelium, nasal mucosa, tracheal epithelium, and pneumocytes. Ovary, testes, spleen (Fig. 34.21A; eSlide 34.e8), lymphoid tissue, thyroid and adrenal glands, and epidermis can also be infected. The most common histologic lesion is single-cell necrosis associated with intranuclear, or less frequently, intracytoplasmic stages of the organism. This can be accompanied by mild lymphocytic or lymphoplasmacytic

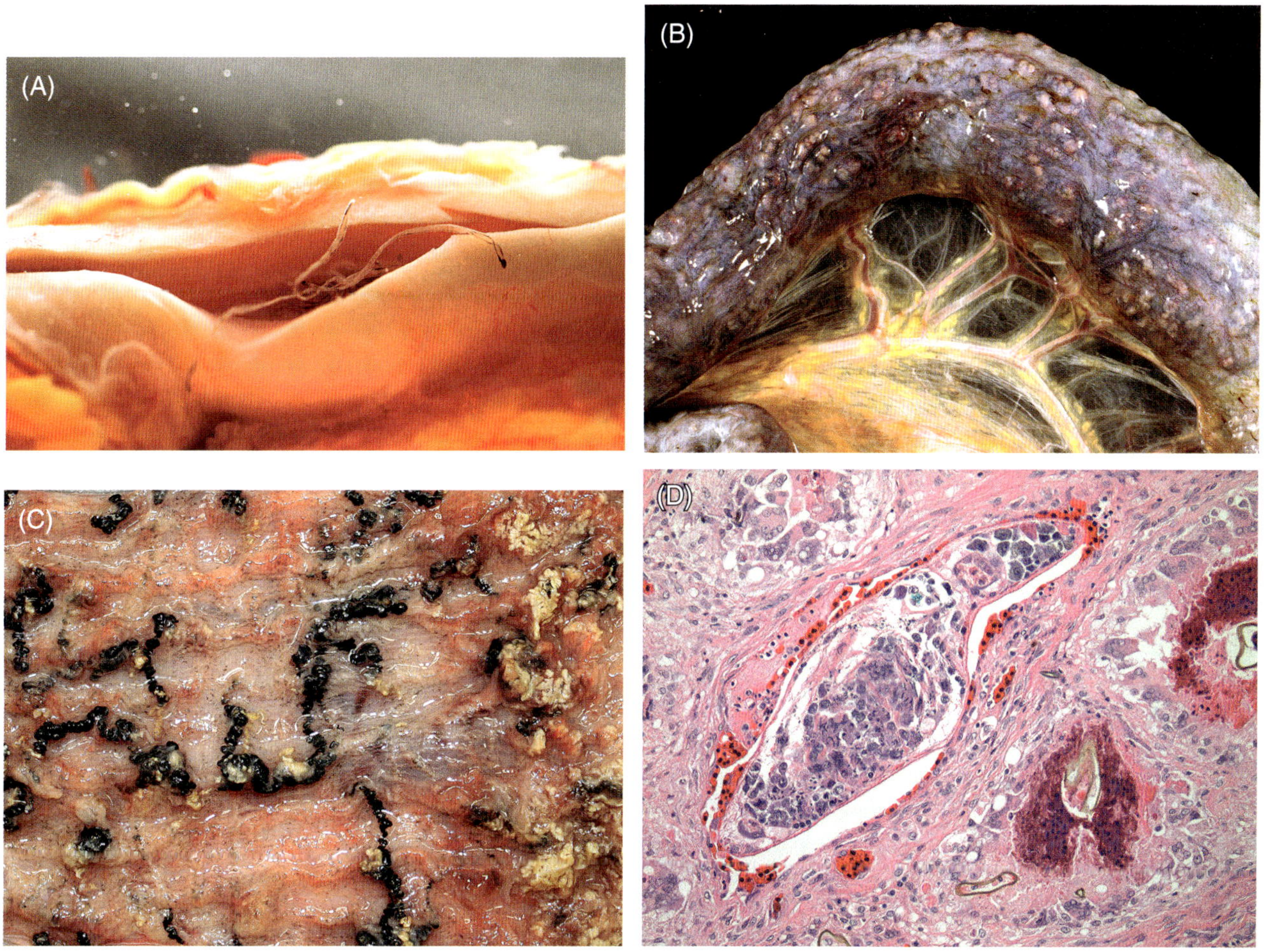

FIGURE 34.20 Spirorchiidiasis in the aorta of a green sea turtle (A) and intestine of loggerhead sea turtles (B–D). (A) Vascular spirorchiidiasis in a green sea turtle. Multiple flattened, filariform, unsegmented, adult *Hapalotrema* sp. spirorchiid trematodes are present in the lumen of the aorta. (B) Intestinal spirorchiidiasis in a loggerhead sea turtle. Throughout the wall of the small intestine are myriad, demarcated, raised, tan nodules that consist of *Amphiorchis* sp. ova-associated granulomatous inflammation. (C) Numerous coiled to serpiginous black trematodes are present in the colonic mucosa and submucosa. Spatially associated with the adult trematodes are irregular aggregates of fragmented to coalescent, yellow-tan material consistent with ova-associated granulomatous inflammation. (D) Intravascular adult trematodes are characterized by a dorso-ventrally flattened body wall with a thin-walled, unadorned tegument, a thin flat musculature, and a parenchymatous body. The intestinal tract is difficult to discern in this image, but bilateral male and female gonads are present. The extravascular ova are roughly elongated to triangular and contain the miracidium encased in a 2–3 μm thick, yellow-brown pigmented wall. Ova are typically associated with an intense, localized heterophilic to granulomatous response with a central core of brightly eosinophilic, necrotic cellular and inflammatory debris surrounded by multinucleated giant cells, *Amphiorchis* sp. *(Parts A–D: Photos Courtesy of B. Stacy, National Oceanic and Atmospheric Administration.)*

inflammation.[76] Infected mucosal surfaces in the GI, respiratory, or genitourinary tracts may be opportunistically colonized by bacteria and develop intense heterophilic and histiocytic inflammation.[50] Although organisms have not been identified in the central nervous system, mild perivascular lymphoplasmacytic infiltrates have been reported throughout the brain and spinal cord in some affected animals.[48] Gamete stages are spherical, 6 μm diameter intranuclear structures with slightly granular basophilic cytoplasm and an eccentric nucleus. Meronts are spherical with a central residual body surrounded by several to more than a dozen merozoites. The distinct apical conoid characteristic of apicomplexan protozoa, numerous rhoptries, and the nucleus of the banana-shaped, up to 4 μm merozoites are evident on electron microscopy (Fig. 34.21B). More than one gamete can be present in an infected cell. The organisms can be difficult to identify on routine HE stains and do not selectively stain with Fite's acid-fast or PAS methods. Cytology of nasal exudates stained with Wright's-Giemsa may highlight the organisms.[68,76] Quantitative PCR can be performed on swabs from the conjunctiva, oral, choanal mucosa, and cloaca for confirmation in suspect cases.[3,48,50]

Several GI protozoal parasites infect the intestinal tract of chelonians. Most cases reported come from animals in managed populations of terrestrial tortoises or in stranded,

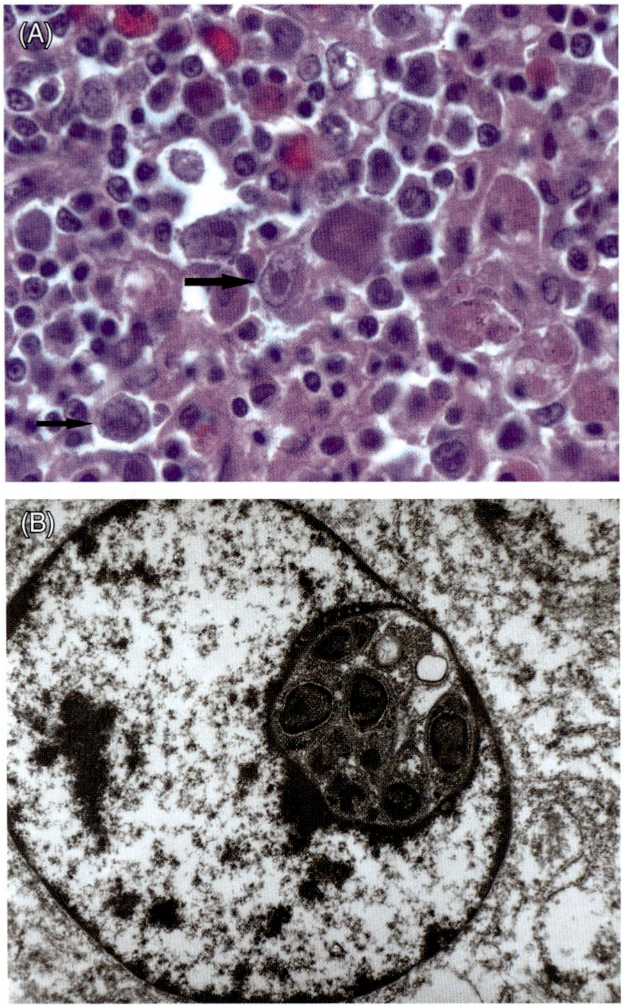

FIGURE 34.21 Intranuclear coccidia in the spleen of a star tortoise. (A) One or occasionally multiple, round to polygonal, amphophilic to eosinophilic organisms (*arrows*) are present in the nuclei of infected cells. (B) Multiple life stages can be present. This nucleus contains an intranuclear meront with residual body and multiple electron-dense oval to elongated merozoites, TEM (see eSlide 34.e8). *(Parts A, B: Photos Courtesy of D. McAloose, Wildlife Conservation Society.)*

free-ranging marine sea turtles. Although ***Eimeria*** spp. are the most common coccidia in chelonians, ***Cryptosporidium* spp.**, ***Eimeria caretta***, and ***Caryospora cheloniae*** are also reported. The latter has been associated with mortality events in mariculture-raised green sea turtles in Grand Cayman and with a mass mortality event in free-ranging green sea turtles in Australia.[20,45,48] The main lesion of *Caryospora cheloniae* infection is heterophilic to granulomatous meningoencephalitis and fibrinous or fibrinonecrotizing enteritis. Meronts measuring 30–80 μm packed with numerous ≤1 μm basophilic merozoites can be present within the cytoplasm of intestinal epithelial cells and within leukocytes in the intestinal mucosa and lamina propria, thyroid gland, and brain.[20] Organisms can also be found in the renal tubular epithelium.[20,45]

At least three species of ***Cryptosporidium*** have been identified in the GI tract of chelonians. *Cryptosporidium testudinis*, or *Cryptosporidium* sp. tortoise genotype I, shows affinity for infecting gastric epithelium, while *Cryptosporidium ducismarci* or *Cryptosporidium* sp. tortoise genotype II, is more commonly found in the intestinal tract. A third species, *Cryptosporidium* sp. tortoise genotype III has also been identified but little is known about this organism at this time. Clinical signs are not always apparent but can include diarrhea, pica, weight loss, undigested food in feces, and anorexia. Organisms, characterized by round, 1–5 μm amphophilic bodies with an eccentric, densely basophilic nucleus can be seen along the apical surface of the gastric or intestinal mucosal epithelial cells. Organisms are intracellular but extracytoplasmic. In cases of gastric cryptosporidiosis, the stomach can be dilated and filled with fluid. Gastric epithelial hyperplasia is variable and often accompanied by mild lamina proprial lymphoplasmacytic and heterophilic inflammation. Intestinal cryptosporidiosis may manifest as fibrinonecrotic lesions, ulceration, and hemorrhages. Inflammation can be variably intense and mixed lymphohistiocytic to heterophilic, especially in areas of mucosal ulceration.[3,50]

Amoebiasis attributable to ***Entamoeba invadens*** has been reported in a variety of chelonian species. Pathogenesis is similar to that in other reptiles (see Lacertia and Serpentis).[75,88] Lesions consist of fibrinonecrotic enteritis or colitis and, in some cases, it is also associated with necrotizing hepatitis (Fig. 34.22A) via dissemination of organisms through the portal vasculature or common bile duct. Amoebae are readily apparent on HE-stained sections of affected tissue and are characterized by an eccentric basophilic macronucleus and multiple cytoplasmic vacuoles (Fig. 34.22B; eSlides 34.e9 and 34.e10). Special stains, such as PAS, highlight the organisms.

Ectoparasites

Epibionts are organisms that live on the surface of other living organisms. They include barnacles, leeches, algae, and other marine life that can attach to the shell of sea turtles. Increased burden can be an indicator of weakness or disease, especially if found adhered to the skin in addition to the carapace and plastron.[89] Excessive algae loads may be an indicator that an animal is consistently floating near the surface and an excessive number of barnacles may indicate weakness or an inability to scrape them off on rocks. This must be interpreted with caution as the normal epibiont load varies between species.[45]

Free-ranging terrestrial chelonians are host to a number of ectoparasites, including **ixodid** (hard bodied) and **argasid** (soft bodied) **ticks**. Heavy parasite burdens can cause skin wounds that may become secondarily infected

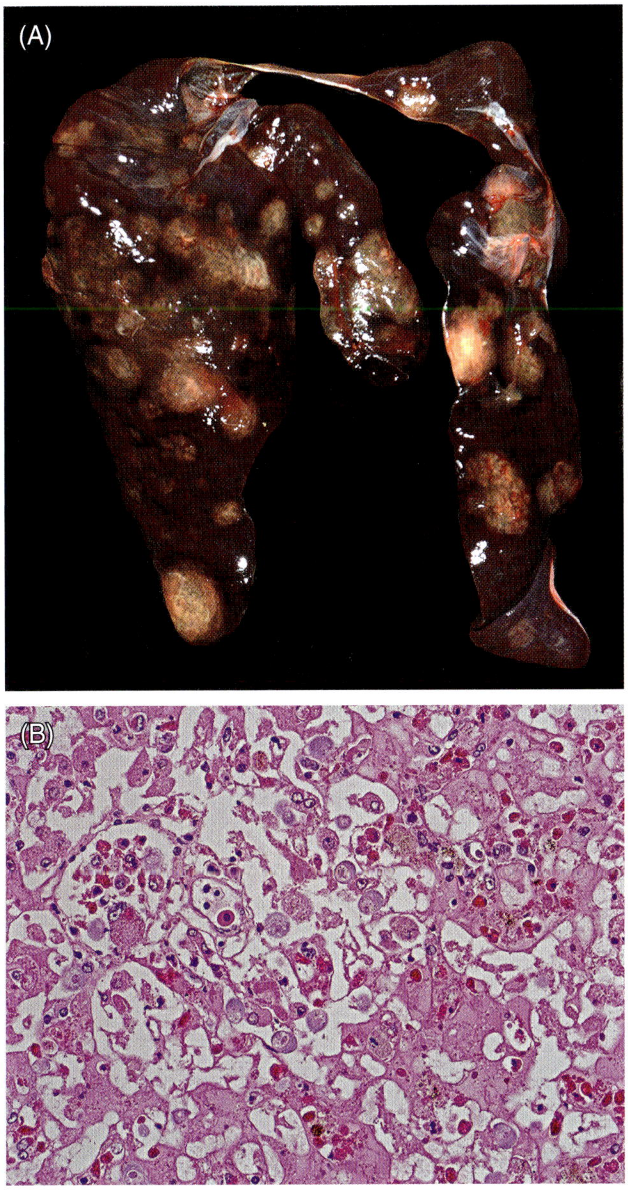

FIGURE 34.22 Hepatic amoebiasis in a gopher tortoise. (A) Multifocal, variably sized, coalescing, nodular, tan foci expand and distort the contour of the liver. (B) Infection is characterized by necrotizing hepatitis with intense heterophilic inflammation and numerous intralesional amoeba. Amoeba can be distinguished by their pale-staining basophilic cytoplasm, an eccentric basophilic macronucleus, and multiple cytoplasmic vacuoles. Cytoplasmic edges may exhibit irregular pseudopod formation (see eSlides 34.e9 and 34.e10). *(Parts A, B: Photos Courtesy of the University of Illinois Zoological Pathology Program.)*

and may cause anemia if the parasite burden is severe enough. Ticks are often attached to the soft skin of the limbs and neck but some species will localize to the seams between scutes.[72,89] As a reptile disease vector, there are no reports of tick-borne disease in chelonians and there are very few reports of tick-related disease in reptiles overall.[89] Ticks are, on the other hand, an important vector for many mammalian pathogens and are the most commonly reported parasite on internationally traded exotic animals. *Amblyomma* ticks recovered from turtles and tortoises are known vectors for pathogens, such as *Ehrlichia ruminantium* (previously *Cowdria ruminatium*), the cause of heartwater disease in domestic ruminants in sub-Saharan Africa. Many such ticks have been recovered from imported leopard tortoises from Africa. Although there is no evidence to suggest that leopard tortoises are susceptible to infection with *E. ruminantium* or that they would serve as an amplification or reservoir host, they could serve as the vehicle to introduce an exotic disease to a new area.[23,72,89,93] Careful examination and removal of ticks from animals being locally or internationally traded is therefore important in preventing inadvertent tick and pathogen introduction or disease transmission.

ADDITIONAL RESOURCES

Divers SJ, Stahl, SJ. Reptile and Amphibian Medicine and Surgery, 3rd ed. Elsevier, 2019.

Flanagan, J.P., 2015. Chelonians (Turtles, Tortoises). In: Miller, R.E., Fowler, M.E. (Eds.), Fowler's Zoo and Wild Animal Medicine. Vol 8. Saunders Elsevier, St. Louis, pp. 27–38.

Mader, D.R., 2006. Reptile Medicine and Surgery, second ed. Saunders Elsevier, St. Louis.

Raphael, B.L., 2003. Chelonians (Turtles, Tortoises). In: Fowler, M.E., Miller, R.E. (Eds.), Zoo and Wild Animal Medicine. Fifth ed. Saunders, St. Louis, pp. 48–58.

Reavill, D.R., Griffin, C., 2014. Common pathology and diseases seen in pet store reptiles. In: Mader, D.R., Divers, S.J. (Eds.), Current Therapy in Reptile Medicine & Surgery. Elsevier Saunders, St. Louis, pp. 13–19.

Steinberg JS, Divers SJ. Hematology of Testudine. In: Brooks MB, Harr KE, Seelig DM, Wardrop KJ, Weiss DJ., eds. Schalm's Veterinary Hematology, 7th ed. Wiley, 2022;1219–1227 U.S. Fish and Wildlife Service. 2019. Health Assessment Procedures for the Mojave Desert Tortoise (*Gopherus agassizii*): A Handbook Pertinent to Translocation. Desert Tortoise Recovery Office, U.S. Fish and Wildlife Service, Reno, Nevada.

Wyneken, J., 2009. Normal reptile heart and function. In: Hatley, J.J. (Ed.), Vet Clinics of North America: Exotic Animal Practice (Cardiology). W. B. Saunders, Philadelphia, pp. 51–64, vol. 12 (Abou-Madi and Jacobson, 2003).

References

Full references for this chapter can be found online at https://doi.org/10.1016/B978-0-443-18437-6.00041-9.

Chapter 35

Crocodilia

Kenneth J. Conley[a] and Catherine M. Shilton[b]

[a]*Disney's® Animals, Science and Environment, Lake Buena Vista, FL, United States;* [b]*Northern Territory Department of Industry, Tourism and Trade, NT, Australia*

INTRODUCTION

The order Crocodilia includes approximately eight genera and 24 species in three families—Crocodylidae (crocodiles), Alligatoridae (alligators and caiman), and Gavialidae (gharial and false gharial). As a group, crocodilians are immediately recognizable, largely aquatic reptiles. They range in size from the relatively small, 2 m long dwarf caiman,[80] to the estuarine (saltwater) crocodile, which can reach 6–7 m in length.[124] Crocodilians are tropical to subtropical animals found on all continents except Antarctica and Europe. In addition to free-ranging and zoo-held animals, there is extensive commercial crocodilian production, a somewhat unique occurrence within the Reptilian class. Production animals are either from captive-bred, farmed stock or are reared from eggs or juveniles collected from the wild (ranching); they are harvested predominantly for meat and skin. Due to the breadth of farming across the world, much of the crocodilian literature is focused on those species most widely held in captivity, including the American alligator, Nile crocodile, freshwater crocodile, and estuarine crocodile (Table 35.e1).

Unique anatomic features of crocodilians, including the integumentary sensory organs, adaptations for their amphibious nature, and unique cardiovascular and respiratory anatomy, are discussed in the Digital Content.

NON-INFECTIOUS DISEASES

Nutritional

Nutritional diseases in crocodilians are largely reported in captive settings. **Hypovitaminosis E**, associated with **(pan) steatitis** and **fat necrosis**,[47,58,75] can be caused by diets containing fish high in polyunsaturated fatty acids, feed rancidity, or a combination of both.[24] Grossly, this condition results in discrete nodular or diffuse, firm, yellow/brown adipose that will fluoresce yellow/orange under ultraviolet light.[57] Lesions are often most severe in the tail (Fig. 35.1) and can result in immobility. Additionally, impingement by masses of hardened fat can negatively affect gastrointestinal tract motility. Histologic appearance includes necrosis, hemorrhage, saponification, inflammation (composed of predominantly macrophages, lymphocytes, and multinucleated giant cells), and bands of fibrous connective tissue. Pigment consistent with ceroid can be found in phagocytic cells in primary lesions, adjacent stromal tissue and skeletal muscle, and splenic reticuloendothelial cells (eSlide 35.e1). This pigment is presumably responsible for autofluorescence.

Pansteatitis has also been diagnosed in free-ranging Nile crocodiles.[74] Multiple complex ecological changes were reported during the time of the event; a specific cause was not confirmed. However, one of the species of prey fish was similarly affected, and ingestion of these fish by crocodiles was considered as a possible contributory cause for the disease. Selenium deficiencies have been implicated in disease processes similar to those caused by hypovitaminosis E in other animals; this factor has not been extensively studied in reptiles and the role that selenium plays in the development of crocodilian pansteatitis remains uncertain.

Vitamin A (retinol) is an essential fat-soluble vitamin that is stored principally in the liver, and meat-only diets can be deficient if not adequately supplemented.[3] **Hypovitaminosis A** is predominantly associated with development of **squamous metaplasia**. In crocodiles (Nile and freshwater), marked squamous metaplasia with retention of abundant sloughed keratin within glandular ostia can occur in the lingual glands (Fig. 35.2A). This presents grossly as numerous nodules on the dorsal surface of the tongue (Fig. 35.2B). Squamous metaplasia can also affect large collecting ducts in the kidneys, resulting in renal and visceral gout and secondary tubulonephritis.[3] Squamous metaplasia of conjunctival and corneal epithelia is another potential sequela (Fig. 35.2C).

Thiamine (vitamin B1) is a water-soluble vitamin present in meat, fish, and plant sources.[35] As in mammalian carnivores, **thiamine deficiency** in crocodilians results in neurologic disease. This can present as abnormal buoyancy,

Pathology of Wildlife and Zoo Animals. https://doi.org/10.1016/B978-0-443-18437-6.00008-0

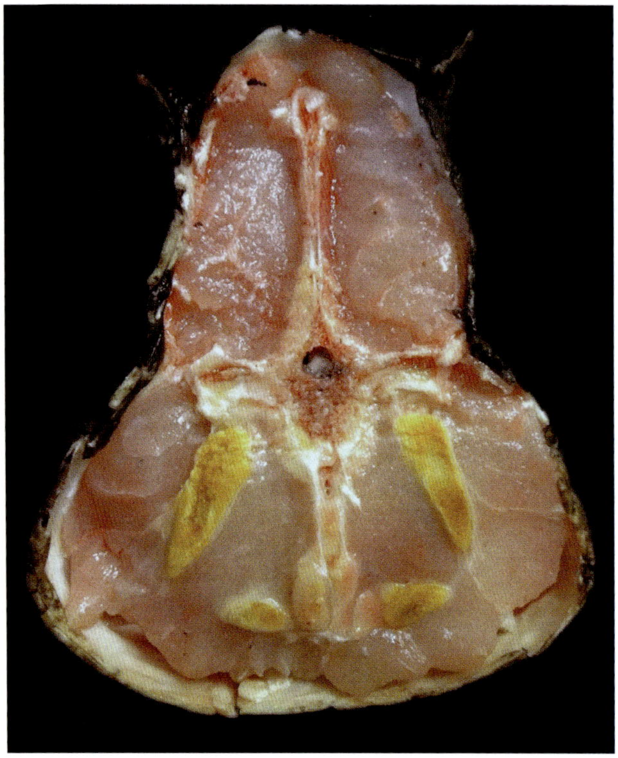

FIGURE 35.1 Pansteatitis in the tail of a dwarf caiman. Multifocal, dark yellow discoloration, and hardening of the intermuscular adipose stores is secondary to presumed hypovitaminosis E (see eSlide 35.e1). *(Photo Courtesy of the Wildlife Conservation Society.)*

head tilt, limb paresis, anisocoria, lethargy, lack of vocalization, and dorsal extension of the head and neck (stargazing).[45,110] In most cases, thiamine deficiency in crocodilians is the result of one of two mechanisms of thiamine destruction, or both in combination: degradation of the compound due to repeated freezing and thawing (e.g., in food items), and enzymatic destruction by thiaminases. Thiaminases are normally present in fish at varying levels depending on fish species. Crocodilians at risk include those with diets consisting predominantly of fish with high thiaminase levels, such as gizzard shad and smelt, especially if the fish are not maintained frozen.[43,57] There are no gross lesions in thiamine-deficient crocodilians. Histologically, American alligators exhibit multifocal gray matter necrosis. The dorsal ventricular ridge and torus semicircularis are the most consistently affected areas but lesions can extend into the subpallium, thalamus, and tegmentum and affect the cerebellum and brainstem. Lesions are characterized by neuropil rarefaction, neuronal necrosis, and endothelial cell hypertrophy. Denervation atrophy within skeletal muscles of the hindlimb and endoneurial fibrosis within the sciatic nerve are possible sequelae.

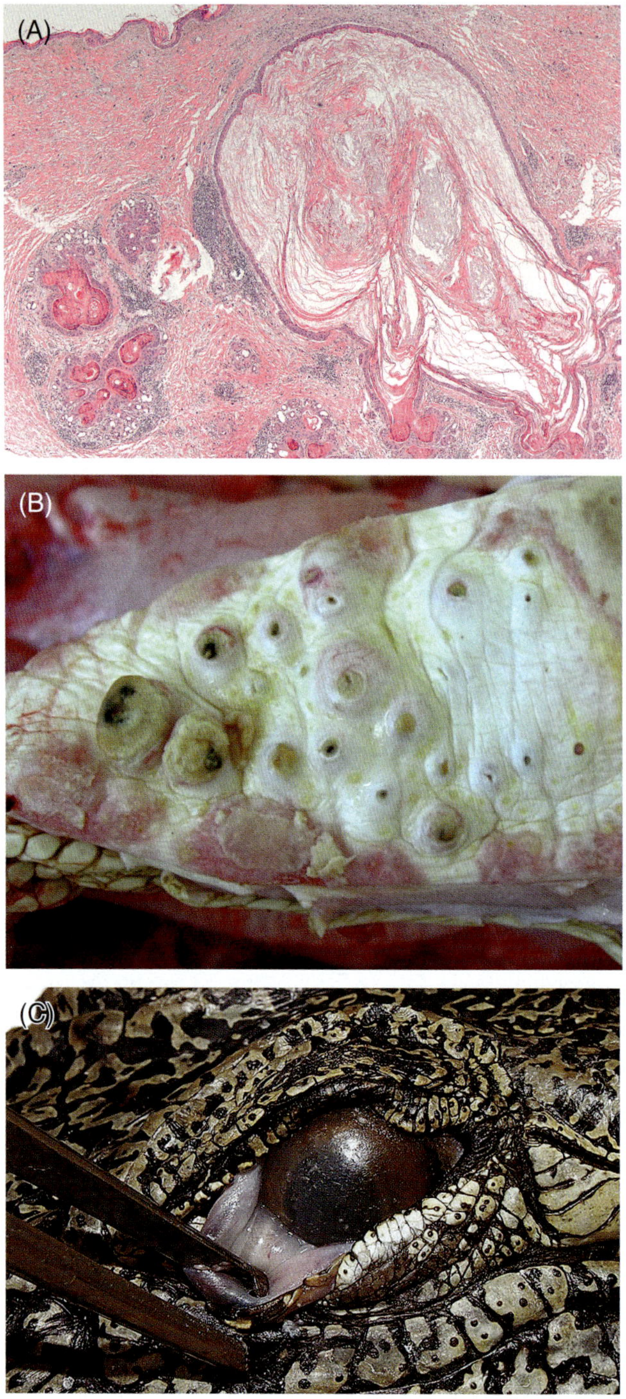

FIGURE 35.2 Hypovitaminosis A resulting in squamous metaplasia of the lingual glands (A, B) and corneal epithelium (C) of estuarine crocodiles. (A) Deposits of laminated keratin cause marked expansion of a central lingual gland duct and mild expansion of smaller surrounding ducts by accumulated keratin. (B) Grossly, this appears as enlargement of the glands with accumulation of keratin and caseous exudate, which appears as tan to black material in this image. (C) In the eyes, squamous metaplasia and keratinization of the corneal epithelium result in increased opacity and a dry appearance, along with mild neovascularization and fibrosis of the superficial corneal stroma.

Metabolic

Crocodilians are ammono-uricotelic and produce predominantly liquid nitrogenous waste as ammonia. Increased uric acid excretion occurs during times of fasting. Uric acid levels surpassing the renal threshold for clearance result in **gout**, the precipitation of uric acid in tissues. In crocodilians, these precipitates form within the kidneys, on serosal surfaces, and within and around joints. There is no indication that the form of gout (**visceral, renal, or articular**) correlates with the underlying cause in crocodilians. Any disease process that affects the kidneys and leads to reduced uric acid excretion can result in gout. Predisposing factors include low environmental temperatures, dehydration, bacterial nephritis,[13] sepsis, squamous metaplasia of collecting ducts, and nephrotoxicosis.[3,58] Uric acid accumulation can also have underlying nutritional and environmental causes. Farmed crocodilians can develop gout when overfed without intervening fasts, suggesting that protein overload may be contributory.[58]

Uric acid accumulation (gout) presents as deposits of white, chalky to gritty material, most commonly on serosal surfaces, predominantly the pericardium, and in and around joints (Figs. 35.3A, B). Within the renal parenchyma, gout is most commonly identified grossly as numerous pinpoint

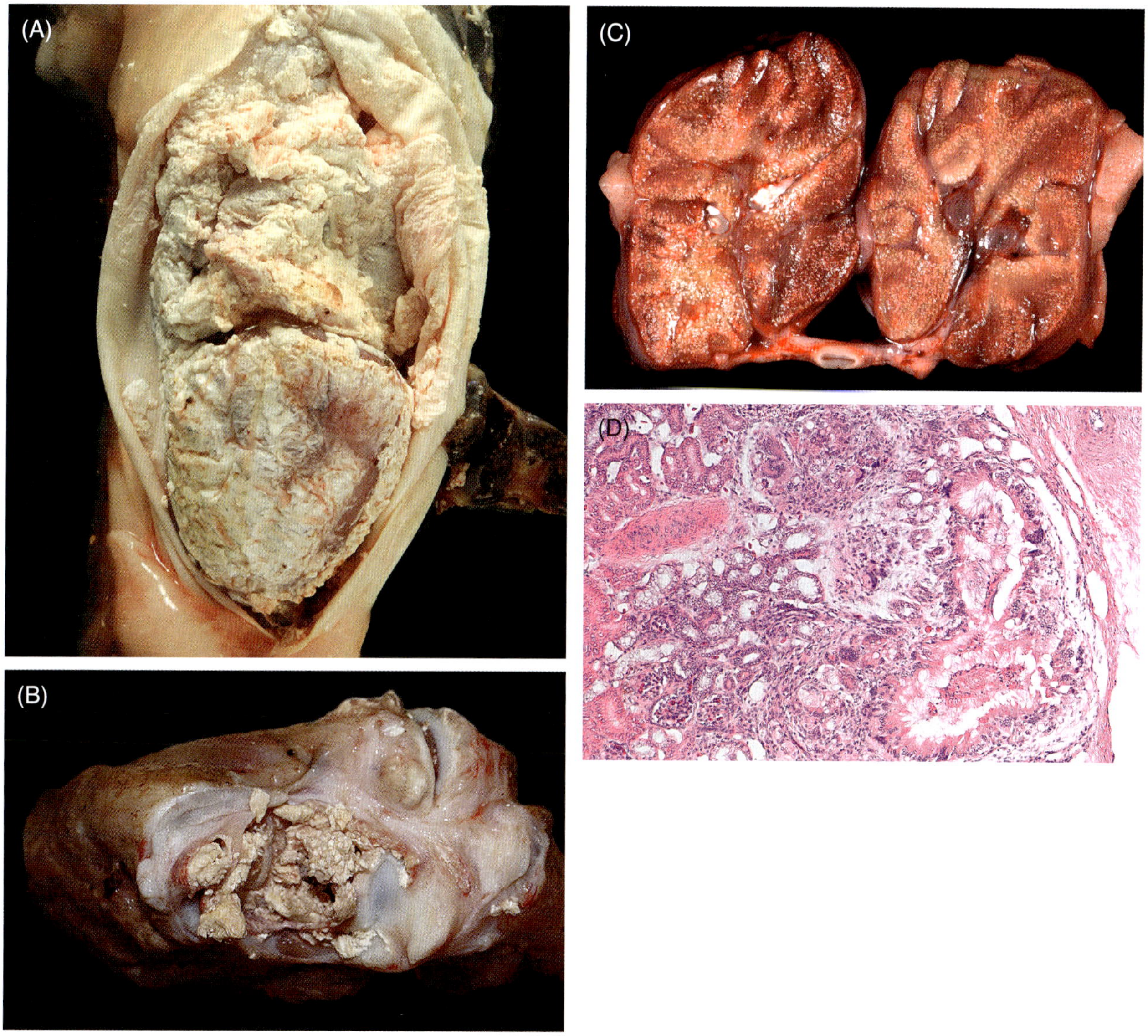

FIGURE 35.3 Pericardial gout in an African dwarf crocodile (A), articular gout in the distal limb of an estuarine crocodile (B), and renal gout in an American alligator (C) and African dwarf crocodile (D). (A) Pericardial gout develops with the accumulation of uric acid admixed with inflammatory exudates within the pericardial sac. (B) Similar material can accumulate in and around joints in articular gout. (C) Chalky, white foci of uric acid accumulation are characteristic of renal gout. (D) Histologically, renal gout is characterized by distention and replacement of renal tubules by urate tophi surrounded by multinucleated giant cells and granulocytes. Clear, acicular clefts within the tophi comprise the typical histologic appearance. *(Part A: Photo Courtesy of R. Ossiboff, University of Florida College of Veterinary Medicine; Part C: Photo Courtesy of the Wildlife Conservation Society.)*

to slightly larger white foci throughout the tissue (Fig. 35.3C). Cytologic confirmation of uric acid accumulation typically requires wet mounts of material scraped from serosal surfaces or touch or crush preparations of affected parenchymal tissue. Disease-associated uric acid crystals are acicular (needle-like) and often form radiating arrays. Histologically, **gout tophi**, the accumulation of uric acid crystals with an associated inflammatory cell response, consist of central eosinophilic material that often has a "star-burst" appearance, with radiating clear acicular clefts (the crystals are typically lost during processing leaving behind only the inflammatory response and remnant clear spaces and clefts) (Fig. 35.3D). Inflammation is often mixed and includes epithelioid macrophages and multinucleated giant cells immediately surrounding the crystals, as well as more peripheral heterophils. Lymphocytes and plasma cells are present in lower numbers and are seen more commonly with increasing chronicity. In the kidney, tophi are present within tubules or within effaced parenchymal tissue. In many cases, associated renal tubular epithelial cells are destroyed and inflammation extends from tubules into the adjacent interstitium. In viscera other than the kidney, tophi are typically scattered randomly with little destruction of tissue outside of the directly affected area. Inflammation associated with articular gout can be severe with destruction of cartilage, bone, and surrounding soft tissues.

As in other carnivorous reptiles, **metabolic bone disease** is a condition of captive crocodilians fed inappropriately formulated diets. The pathophysiology in this taxon has not been specifically elucidated but it is presumably due to calcium and phosphorus imbalance similar to what occurs in other reptiles. The importance of ultraviolet light relative to metabolic bone disease in crocodilians is not clear but lack of vitamin D may exacerbate the condition.[46] Gross findings include softening and increased elasticity of the bones, including the long bones, maxilla, and mandibles (Fig. 35.4A). Dental changes include a diaphanous appearance, loss and malpositioning of the teeth, and changes in the skeleton, including shortening of the axial skeleton and long bone malformations. Axial skeletal lesions also include kyphoscoliosis in young animals and spinal fracture (Fig. 35.4B), the latter resulting in tetany and hind end paresis.[28,99] Osteoporosis, osteomalacia, and fibrous osteodystrophy are reported, though detailed histologic and gross descriptions are generally sparse in the literature. **Fibrous osteodystrophy**, characterized by the proliferation of fibrous connective tissue replacing bone, has been described in hypocalcemic juvenile broad-snouted caimans, accompanied by bone marrow depletion.[33] The mandibles and long bones were evaluated and subperiosteal bone was most significantly affected. Histopathologic changes in the physes of affected individuals included disorganized and hypertrophic chondrocytes, but detailed descriptions suggesting true rickets-like changes were not included. In juvenile farmed estuarine crocodiles, although bones are pliable, they are otherwise morphologically normal, with only mild histologic osteoporosis in decalcified sections. Additional lesions associated with metabolic bone disease include cystic degeneration and hyperplasia of the parathyroid gland.

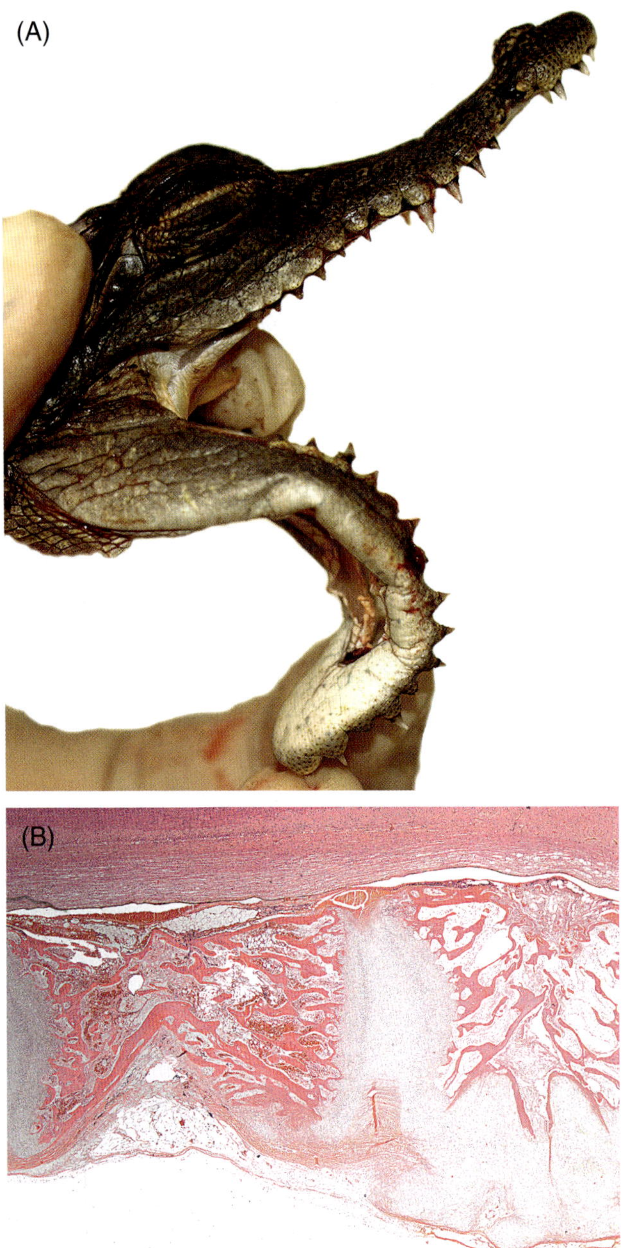

FIGURE 35.4 Metabolic bone disease in the jaw (A) and spine (B) of estuarine crocodiles. (A) Metabolic bone disease results in softening and elasticity of the bones, as in this crocodile's mandible, which is otherwise grossly normal. (B) Spinal fracture is commonly identified in young crocodilians with metabolic bone disease. The vertebral body on the left side of this image is folded and the vertebral body on the right exhibits a large, primarily cartilaginous callus enveloping a previous fracture site. Compression and Wallerian degeneration of the ventral aspect of the spinal cord are present dorsal to the fracture site.

Toxic

Acute toxic disease processes are uncommon in crocodilians, and chronic effects of potential toxicants are not well-investigated. **Lead toxicity** as a clinical condition is not commonly reported in reptiles, and there are varying reports regarding lead toxicity both as a clinical entity and as an insignificant occurrence in crocodilians.[22,40,123] There may be a variable lead tolerance between species and age groups. Relatively high tissue and blood lead levels have been reported in both wild and captive crocodilians but associated clinical conditions or pathologic lesions are rare. Recent research into a population of Nile crocodiles exposed to lead for several decades suggested that increased blood and tissue lead levels were associated with anemia and increased tooth loss, presumably related to impaired heme synthesis and osteoblast and osteoclast function, respectively.[53] Elevated lead in captive American alligators is typically the result of consuming prey items killed with lead shot. Histologic findings include individual renal tubular epithelial cell necrosis and enlarged tubular cell nuclei with marginated chromatin.[17] Reduced reproduction due to embryonic death may be the only indication of lead toxicity in farmed American alligators.[73]

An investigation into potential toxicity of **polyfluoroalkyl substances (PFAS)** in wild American alligators revealed possible immune dysfunction in alligators with evidence of elevated blood PFAS in a contaminated river system. Compared to a relatively uncontaminated reference population, Alligators from the contaminated river system had increased skin lesions, infected and unhealed wounds, slightly decreased blood clotting, increased red cell schistocytes, elevated lysozyme activity (a marker of innate immunity), markedly elevated interferon gamma responsive gene expression, and elevated autoantibodies.[36]

Neoplastic

Neoplasia is rarely described in crocodilians. Tumors that have been reported in the literature and case material from captive crocodilians are listed in Table 35.e2 and include squamous cell carcinoma (Figs. 35.5A, B). Similar to other reptiles, crocodilians are particularly adept at hyperplastic responses that may be interpreted as neoplastic in a mammal. One example is the ability to produce exuberant fibroblastic tissue at sites of previous or chronic soft tissue injury that may be mistaken for fibromas (Fig. 35.6) and disorganized, hyperplastic chondroid responses at the site of fractures. Another example is discrete masses of lymphocytes within nonhematopoietic tissues that are likely manifestations of generalized lymphoid hyperplasia rather than malignant transformation (see the discussion on lymphonodular skin lesions in the Miscellaneous/multifactorial section in this chapter).

INFECTIOUS DISEASES

DNA Viruses

The first viral disease to be identified in crocodilians was caused by **poxvirus** in the skin of captive spectacled caimans.[64] Following this initial description, multiple poxviral diseases have been recognized. **Caiman poxvirus** infection, occurring largely in young animals (<1 year), results in gray-white circular papules that are flat or slightly depressed. Commonly affected sites include the dorsal surfaces of the head and body, the oral mucosa, palpebrae, and limbs. These lesions can regress uneventfully or progress to large coalescent areas. Histologically, papules consist of areas of markedly hypertrophied epidermis (acanthosis) and intracellular edema, accompanied by hyperkeratosis

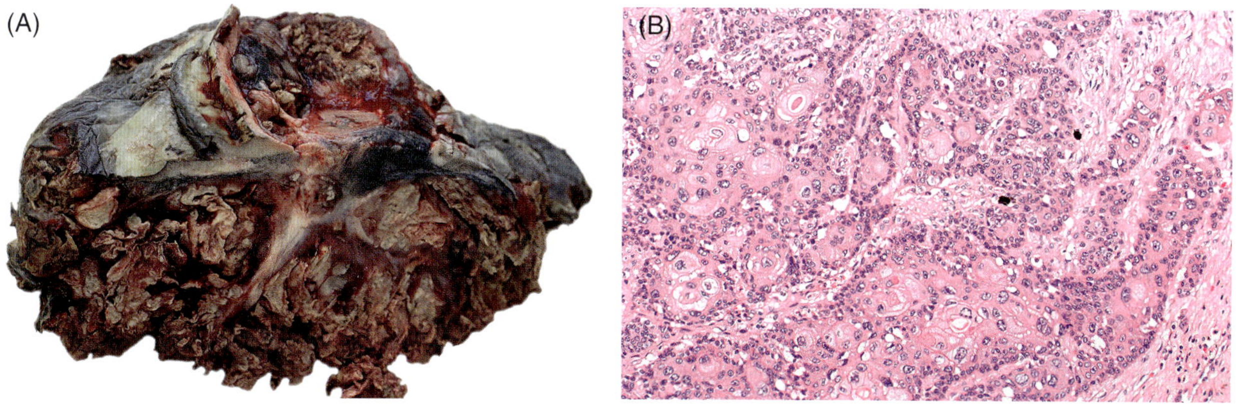

FIGURE 35.5 Squamous cell carcinoma in a hindlimb toe of an estuarine crocodile. (A) The digit is viewed from the lateral aspect, with the surgical margin at the top of the image and a cut surface at the bottom. The digit is markedly expanded by a bulbous, ulcerated mass of friable tissue with an accumulation of laminated keratin. (B) The histologic presentation is similar to that in other taxa, with coalescing nests and cords of neoplastic squamous epithelial cells with disorderly maturation, dyskeratosis, and frequent bizarre and karyomegalic nuclei.

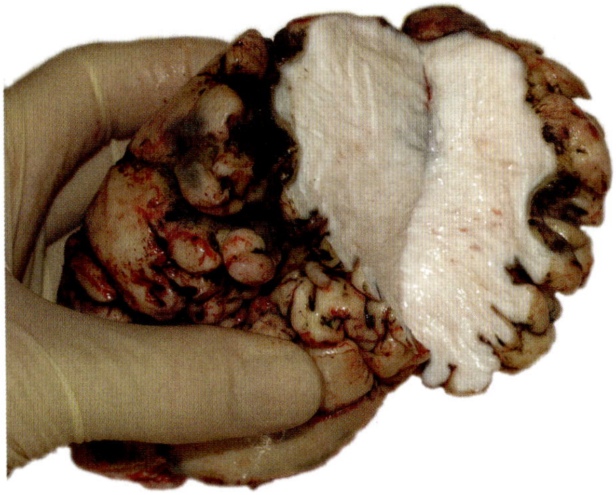

FIGURE 35.6 Proliferative callous from the foot of an estuarine crocodile. Uniform, firm, white dermal tissue mimicking a neoplasm is present on the plantar surface of the skin. The proliferative tissue was confirmed histologically to consist of massive collagen proliferation, mature granulation tissue, and epidermal hyperplasia.

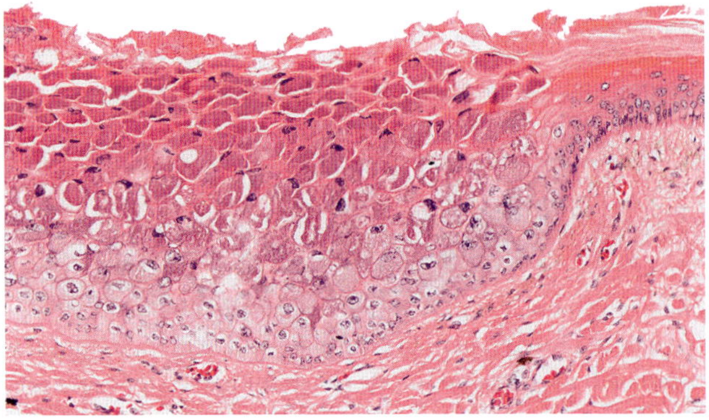

FIGURE 35.7 Poxvirus infection in the ventral abdominal skin of an estuarine crocodile. There is depression of the lesion below the skin surface and thickening of the affected epidermis due to ballooning degeneration, acanthosis, and large, intracytoplasmic, eosinophilic viral inclusion body formation.

and multifocal necrosis. Intracytoplasmic inclusion bodies are present in keratinocytes with ballooning degeneration. Inclusions are variably sized with small, intensely eosinophilic forms and large hyaline forms.[98]

Crocodile poxvirus of young (<1 year) farmed Nile crocodiles presents differently. The most characteristic findings are circular, brown, raised, ulcerated lesions rather than the gray-white lesions that are seen in poxvirus infections in caiman. Lesions in Nile crocodiles can occur anywhere on the body but are consistently reported on the head and ventrum.[44,50] Brown, plaque-like lesions can also occur in the oral mucosa.[28] Lesions around the eyes and around the nares can affect vision and respiration, respectively. Histologically, there is acanthosis, marked parakeratotic hyperkeratosis with erosions, and typical intracytoplasmic, eosinophilic inclusion bodies in hypertrophied keratinocytes with ballooning degeneration; lesions in the oral cavity may lack inclusion bodies.[44] While morbidity in affected groups can be very high, mortality is typically low with few exceptions.[2] A secondary presentation consists of lesions referred to as "pit holes" consisting of small, deep holes in the scales of the ventral coelom.[52] Histologically, the lesions progress from debris-filled cysts in hatchlings to collapsed cysts in 2- and 3-year-old animals. Typical intracytoplasmic eosinophilic inclusions are restricted to the hyperplastic epithelium lining the early, debris-filled cysts. Estuarine and freshwater crocodile poxvirus infections present with gross lesions similar to the relatively mild, focal, flat, or depressed lesions in caiman, and a histologic appearance consistent with the other poxvirus infections already described (Fig. 35.7).[11] To date, only the Nile and estuarine crocodile poxviruses have been fully classified molecularly.[2,108]

Until recently, **herpesvirus** infection in crocodilians was limited to a single report of dermatitis in laboratory-held estuarine crocodiles. In addition to extensive poxviral lesions in these animals, one individual also had areas of hyperkeratosis with degeneration of superficial keratinocytes and focal serocellular crust formation. Degenerate epithelial cells contained large amphophilic intranuclear inclusion bodies, and intranuclear virions consistent with herpesvirus were identified ultrastructurally.[84] More recently, three new herpesviruses, **crocodyline herpesvirus 1, 2, and 3** (CrHV-1, 2, 3, respectively), have been identified in farmed estuarine and freshwater crocodiles.[54,112] CrHV-1 and CrHV-2 are associated with three disease syndromes in estuarine crocodiles. These are **systemic lymphoproliferative syndrome with encephalitis** (SLPE), **conjunctivitis/pharyngitis** (CP), and **lymphonodular skin lesions** (LNS). CrHV-3 is associated with a lymphoproliferative syndrome in freshwater crocodiles. Of these, only SLPE is strongly linked with the virus (see Bacteria for additional details related to CP and next for additional details of LNS). SLPE affects juveniles and is recognized in pens experiencing increased mortalities, poor growth, and ill thrift. Gross lesions beyond poor body condition are minimal; when present, they may include conjunctival hyperemia, splenomegaly, and/or pulmonary edema. Histologically, the syndrome is characterized by lymphohistiocytic infiltrates in numerous tissues including the pulmonary and pancreatic interstitium, portal tracts in the liver, submucosa throughout the gastrointestinal tract, the cardiovascular system (epi/pericardium and

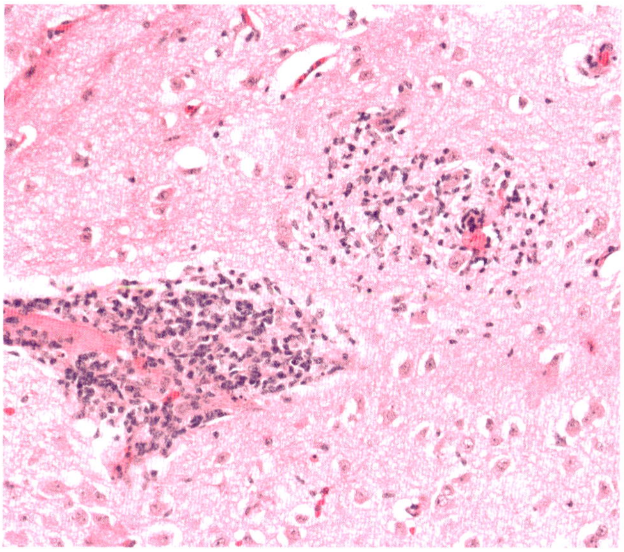

FIGURE 35.8 Systemic lymphoid proliferation and encephalitis syndrome (SLPE) associated with crocodyline herpesvirus infection in an estuarine crocodile. Nonsuppurative encephalitis characterized by thick perivascular cuffs of lymphocytes and fewer macrophages, along with multifocal gliosis and glial nodules.

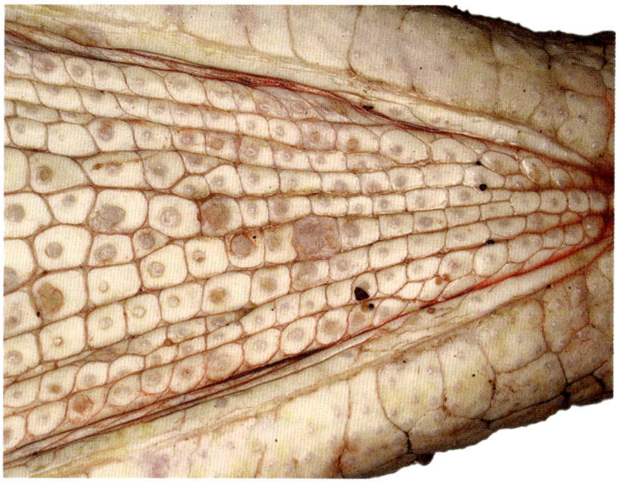

FIGURE 35.9 Lymphonodular skin syndrome (LNS) in a farmed estuarine crocodile. Characteristic lesions are raised, discolored lesions on the scales of the ventral jaw, primarily affecting the integumentary sensory organs.

vasculature), brain (meninges, choroid plexus, and Virchow-Robin spaces), and eye (conjunctiva and iris). Additional findings include pulmonary edema and severe lymphoid proliferation in the spleen and tonsil with a prominent histiocytic population (Fig. 35.8). Viral and molecular investigation of these cases has not been exhaustive and although other potential agents have not been identified, definitive association between SLPE and CrHV requires additional research.

Lymphonodular skin syndrome (LNS) is reported in 2- to 4-year-old estuarine crocodiles.[112] A herpesvirus (CrHV-2) is the suspected etiology, but this has not been confirmed. Subtle, soft, pale, sessile nodules up to 2 cm in diameter most frequently affect the skin of the lateral body wall. Lesions in other areas of the skin and the lingual mucosa can also be seen but are less common. Larger lesions can ulcerate. When only a portion of a scale is affected, it often involves the integumentary sensory organ (Fig. 35.9). Additional gross lesions include a fine multinodular appearance or multifocal white foci in the tonsils, conjunctiva, cloacal mucosa, heart, liver, and/or kidney. Histologically, the skin and oral nodules and white foci in internal organs consist of large dense aggregates of variably sized lymphocytes and macrophages. In the skin, these efface the dermal collagen. Marked lymphoid hyperplasia is present in the tonsils.

Adenovirus infection in young (<1 year, often <5 months) farmed Nile crocodiles results in hepatic or intestinal disease.[28,29,62,95] In both forms, animals are typically found moribund or dead without specific premonitory signs. Those with hepatic disease may exhibit mottling and enlargement of the liver. The enteric form can present with a thickened intestinal wall with liquid to bloody contents, or can have no gross lesions. Histologically, there is acute degeneration and necrosis of hepatocytes and enterocytes with lymphocytic to histiocytic (liver) or mixed (intestine) inflammatory cell infiltrates. Basophilic intranuclear inclusion bodies with margination of chromatin and variable karyomegaly are present in hepatocytes in affected areas or in viable crypt epithelial cells underlying areas of villous loss and necrosis. Additional findings reported with adenoviral disease in Nile crocodiles include pancreatitis with or without intranuclear inclusion bodies, peripancreatic steatitis and chronic inflammation with fibrosis, biliary hyperplasia, and parenchymal atrophy. Further classification of the causative agent(s) has not been reported; adenovirus is implicated in these diseases based on the light microscopic and ultrastructural findings.

RNA Viruses

West Nile virus is a potential cause of mortality in farmed American alligators and Nile crocodiles. Mortality rates in alligators in the Southeastern United States range from around 3% farm-wide to up to 60% in individual pens.[52,85,87,115] Hatchlings and juveniles are most severely affected. Mortality occurs acutely or may be preceded by neurologic disease and/or hematochezia in some instances. Gross lesions, while not consistent, can include coelomic or pericardial effusion, oropharyngeal fibrinonecrotic inflammation, fibrinonecrotic colitis, intestinal hemorrhage, and tan foci or mottling of the liver, spleen, kidneys, and/or heart. Histologically, American alligators have heterophilic to lymphoplasmacytic meningoencephalitis with variable spinal cord involvement, which includes perivascular cuffs in both the gray and white matter in affected areas. Hepatitis, pancreatitis, interstitial nephritis, and gastrointestinal inflammation from the oral cavity to the intestines may consist of heterophilic, lymphocytic, and/or histiocytic infiltrates, or be predominantly necrotizing. Other variably present lesions include myocarditis and myocardial degeneration and necrosis, interstitial pneumonia, necrosis of palatine tonsils, and heterophilic and necrotizing adrenalitis.

A direct association between West Nile virus infection and fatal disease is less clear in Nile crocodiles. In one study, the virus was identified via PCR of whole blood in 2 of 11 Nile crocodiles exhibiting clinical neurological signs, bloody diarrhea and scoliosis, and pulmonary congestion, tracheal and intestinal hemorrhage, and hydropericardium at necropsy. Detection of the virus by PCR in whole blood suggested viremia at the time of death, although further diagnostics to substantiate clinical signs and lesions as being associated with West Nile virus infection were not pursued.[115]

West Nile virus infection (WNV) has also been associated with dermal lymphohistiocytic aggregates in alligators infected with the New York strain of WNV, and estuarine crocodiles infected with the endemic Australian Kunjin strain of WNV.[54,87,88] Grossly, this syndrome presents as spherical opaque lesions, approximately 1 mm in diameter in fresh skin, and as pit-like scars in tanned skins ("Pix").[121] In younger animals (typically less than 2 years of age), multifocal, discrete nodular aggregates of lymphocytes and macrophages replace and compress the collagen of the superficial dermis. Unlike American alligators and Nile crocodiles, overt clinical signs (aside from cutaneous lesions) and internal lesions are not found in West Nile virus-infected estuarine crocodiles,[38,55] which may be related to the endemicity of the infecting Kunjin strain found in Australia.

Transmission of West Nile virus amongst crocodilians likely occurs via multiple routes. High viral shedding from the cloaca of infected alligators has been identified[63] and experimental infections of estuarine crocodiles confirmed virus in water and mucosal samples.[38] These findings suggest horizontal transmission may play a role in local propagation of disease. Recent research has confirmed that alligators can both be infected by, and serve as a source of transmissible infection to, mosquito vectors.[15] The level of viremia in infected crocodilians is high and appropriate personal protective equipment should be considered for prosection of these animals.

Bacteria

There are few specific bacterial disease conditions that affect crocodilians in either captive or free-range settings. Those that do occur include **mycoplasmosis**, **chlamydiosis**, dermatitis associated with ***Dermatophilus*-like bacteria**, and **mycobacterial infections**. Gram-negative septicemia, caused by a variety of organisms, occurs with some frequency, especially in farming situations. ***Providencia rettgeri***, one organism implicated in septicemia, affects the central nervous system more commonly than others. **Necrotizing fasciitis** associated with ***Streptococcus agalactiae*** is a rare example of a significant gram-positive bacterial infection in this group.

Mycoplasma (syn. **Mycoplasmopsis**[37]) infections in captive Nile crocodiles and American alligators cause arthritis and pneumonia in both species, and pericarditis and polyserositis in alligators.[21,86] **Mycoplasmosis** in farmed Nile crocodiles is caused by ***M. crocodyli***.[68] The disease affects only young animals (1–3 years); morbidity and mortality rates are less than 10%. Infection is associated with marked enlargement of the joints resulting in lameness and paresis. Grossly, swelling is the result of fluid accumulation within the joints and edema of the joint capsule and surrounding tissues. Intraarticular effusions begin as turbid, mucoid exudates that progress to inspissated, yellow-tinged fluid in more chronically affected animals. Bacterial culture may only be successful in the earlier stages of infection. Histologically, lesions consist of fibrinonecrotic synovitis with lymphocytic infiltrates. Pneumonia, characterized by consolidation with mononuclear and heterophilic infiltrates, hemorrhage, and edema, may also be present. Mycoplasmosis in American alligators causes a more fatal and widespread disease. It has only been reported in experimentally induced cases and in zoological collections. The causative agent is ***M. alligatoris***.[9] Gross lesions consist predominantly of cloudy joint effusions containing fibrinous strands, edema of periarticular soft tissues, fibrinous epicarditis with effusion and adhesions between the epicardium and pericardial sac and serositis/capsulitis affecting coelomic viscera (Fig. 35.10A). Histologically there is fibrinonecrotic synovitis and fibrinous epicarditis and serositis with lymphocytic and heterophilic to granulomatous inflammation (Fig. 35.10B, eSlide 35.e2). Other potential lesions include fibrinous hepatic capsulitis, portal hepatitis, splenic necrosis, and lymphocytic and heterophilic to granulomatous meningitis.[10] Pneumonia is variable and may be complicated by bacterial infection.

In crocodilians, **chlamydiosis** is a suspected cause of disease in farmed Nile, estuarine, and Siamese crocodiles,[46,48,51,107] and American alligators.[19] Disease may occur as large-scale outbreaks or sporadic isolated cases. The precise causative agent has not been elucidated in most cases, but is suspected to be a novel species, which has been supported by recent molecular and microbiological investigations.[19,20] Upper respiratory and systemic infection, which can occur concurrently, have been attributed to ***Chlamydiaceae*** in crocodilians, as has conjunctivitis, although the relationship of the latter manifestation to chlamydial infection is tenuous (see the Miscellaneous/multifactorial

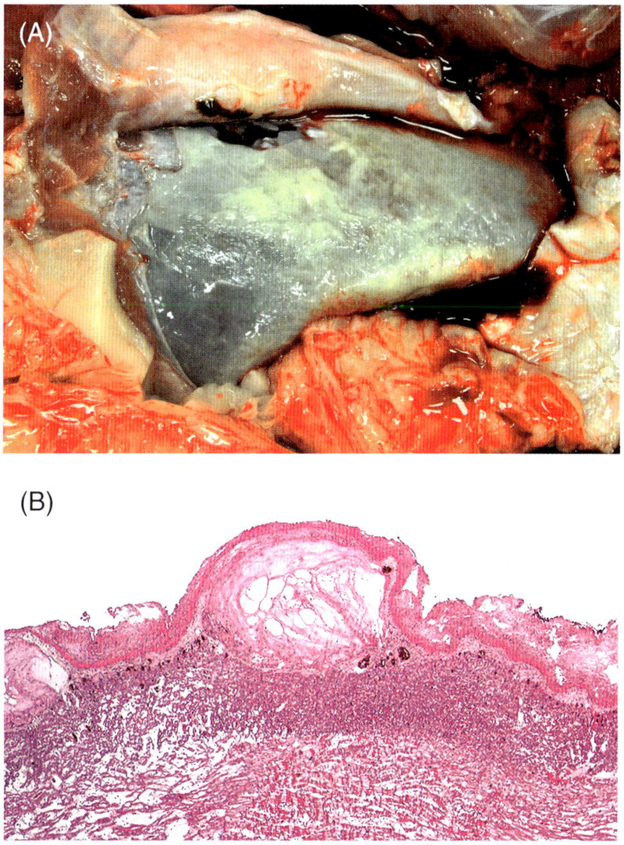

FIGURE 35.10 Liver (A) and heart (B) of American alligators with mycoplasmosis. (A) The hepatic capsule is markedly thickened and opaque due to fibrinous inflammation and there is adhesion of the capsule to the adjacent coelomic body wall. Infection in this case was due to Infection with *M. alligatoris*. (B) Infection is characterized histologically by the formation of a thick, fibrinous exudate containing predominantly lymphocytic and heterophilic infiltrates (see eSlide 35.e2). *(Part A: Photo Courtesy of R. Ossiboff, University of Florida College of Veterinary Medicine; Part B: Photo Courtesy of the University of Florida College of Veterinary Medicine.)*

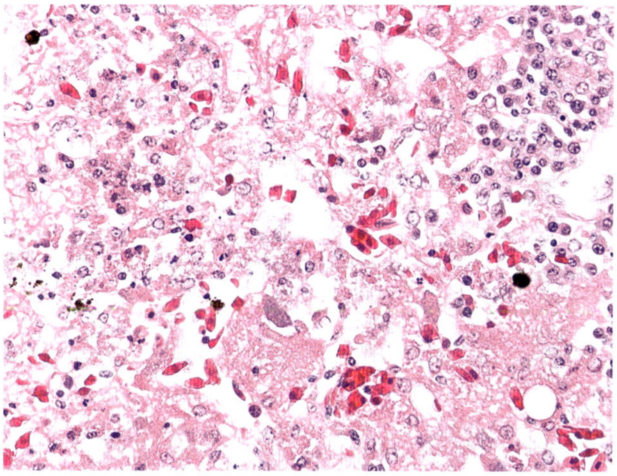

FIGURE 35.11 Hepatic necrosis and inflammation due to chlamydiosis in an estuarine crocodile. Hepatic necrosis and mixed lymphoplasmacytic, histiocytic, and heterophilic inflammation are associated with intralesional clusters of smudgy, basophilic material. Identification of *Chlamydiaceae* was confirmed by polymerase chain reaction and sequencing of the omp2 gene.

section in this chapter). In Nile and estuarine crocodiles, hepatitis is the predominant lesion of systemic chlamydiosis, while more widespread disease is present in Siamese crocodiles and American alligators. Grossly, the liver of affected animals is enlarged, pale, yellow to orange to gray, and mottled (similar to that seen with gram-negative sepsis). The spleen is less severely enlarged, and fluid accumulation may be present in the pericardial sac and the coelom. Mucoid pulmonary exudates, necrotizing tracheitis and stomatitis, fibrinous inflammation of the body cavities, and pinpoint white foci on the epicardial surfaces can be present with more widely disseminated disease. Alligators may additionally exhibit fibrinous enteritis.[19,120] Cytologically, organisms consistent with *Chlamydiaceae* can be detected in touch preparations of the liver, providing a potential differentiating diagnostic feature from gram-negative sepsis. The hepatic lesion ranges from coalescing necrotizing, heterophilic, and granulomatous inflammation to lymphoplasmacytic and histiocytic inflammation in a portal to diffuse pattern with multifocal hepatocellular necrosis and intermingled heterophils (Fig. 35.11). Groups of round organisms consistent with *Chlamydiaceae* can be identified within hepatocyte cytoplasm histologically [hematoxylin and eosin (HE), Macchiavello and Gimenez stains] and confirmed immunohistochemically and ultrastructurally.[19,48] Additional lesions may include histiocytic interstitial to heterophilic bronchopneumonia, lymphohistiocytic to necrotizing and heterophilic myocarditis, individual cell necrosis and lymphoid depletion in the spleen, enterocolitis, vasculitis, pulmonary edema, and mild multifocal necrosis in the coelomic fat body. Of note, in many of the described mortality events attributed to chlamydiosis, other pathogens, including *Aeromonas* spp., adeno- and herpesviruses, have been identified concurrently and may contribute to observed disease. This is especially true for Siamese crocodiles with vasculitis and more widespread disease in which ***Aeromonas sobria*** was cultured concurrently, and for American alligators with enterocolitis in which ***Salmonella*** spp. was cultured concurrently.

Conjunctivitis with or without pharyngitis (CP) in Nile, Siamese, and estuarine crocodiles is well described but lacks a clear cause. Classically described as a lesion associated with *Chlamydiaceae*, recent evidence suggests this may be a multifactorial condition in which herpesviruses may also play a role.[93,112] In some cases, conjunctivitis presents concurrently in animals with confirmed chlamydial hepatitis but confirmation of *Chlamydiaceae* in the ocular lesions has been inconsistently reported.[47,48,51,107] Unlike in systemic (hepatic) chlamydiosis in which typical organisms are readily histologically seen, chlamydial organisms are not histologically evident in conjunctival lesions that are positive by PCR, and immunohistochemical studies in conjunctival lesions have not been performed.[51,93,112] Additionally, a previous investigation into conjunctivitis/pharyngitis in farmed estuarine crocodiles showed poor correlation between the disease and identification of chlamydial organisms, as PCR detection of chlamydial organisms was common in some clinically and grossly normal farmed populations.[65] Shilton et al.[112] describe a syndrome of conjunctivitis and pharyngitis unassociated with systemic lesions in estuarine crocodiles that is associated with both *Chlamydiaceae* and herpesviruses (CrHV-1, 2) based on PCR testing. Grossly, lesions are typically bilateral and consist of thickening and edema of the nictitating membrane with mucopurulent to fibrinocaseous exudate (Fig. 35.12), and gray discoloration and opacification of the cornea; corneal rupture occurs in severe cases. Histologic lesions consist of edema and necrosis of the conjunctival epithelium with erosion or ulceration, overlying fibrinonecrotic debris, and infiltration by heterophils and macrophages; inflammation in chronic cases is predominantly lymphoplasmacytic and there is prominent epithelial hypertrophy and hyperplasia. The corneal changes include edema, erosion, ulceration, and keratitis (lymphocytic and/or heterophilic). Most cases have lymphohistiocytic inflammation of the iris, which can be marked, and associated with intermingled heterophils. Pharyngitis consists of multifocal erosion or ulceration with adherent fibrinocaseous debris in the caudal oral cavity around the base of the tongue and gular flap; lesions may extend into the larynx. The histologic presentation is similar to that described for the conjunctiva. Neither chlamydial organisms nor viral inclusion bodies have been described in any of these lesions.

FIGURE 35.12 Conjunctivitis/pharyngitis syndrome in a farmed estuarine crocodile. The typical presentation of the eyes is closure of the eyelids and marked mucopurulent exudate.

Skin lesions caused by bacteria morphologically consistent with ***Dermatophilus*** spp. have been identified in (predominantly farmed) American alligators, estuarine crocodiles, and freshwater crocodiles.[14,59] Grossly, **dermatophilosis** can present as red or brown flat maculae up to 4 mm diameter; irregularly shaped brown ulcerations up to 2 cm diameter; extensive erosions up to 5 cm long, most commonly identified on the jaw; and subcutaneous granulomas that occasionally underlie ulcerations.[12] Lesions are most frequent on the ventrum and neck and occasionally occur on the feet and tail. Lesions are often located between scales although the center of the scale is commonly affected in crocodiles. Lesions can also involve the integumentary sensory organs (Fig. 35.13A). Histologically, dermatophilosis lesions range from parakeratotic hyperkeratosis with accumulation of karyorrhectic debris to ulceration.

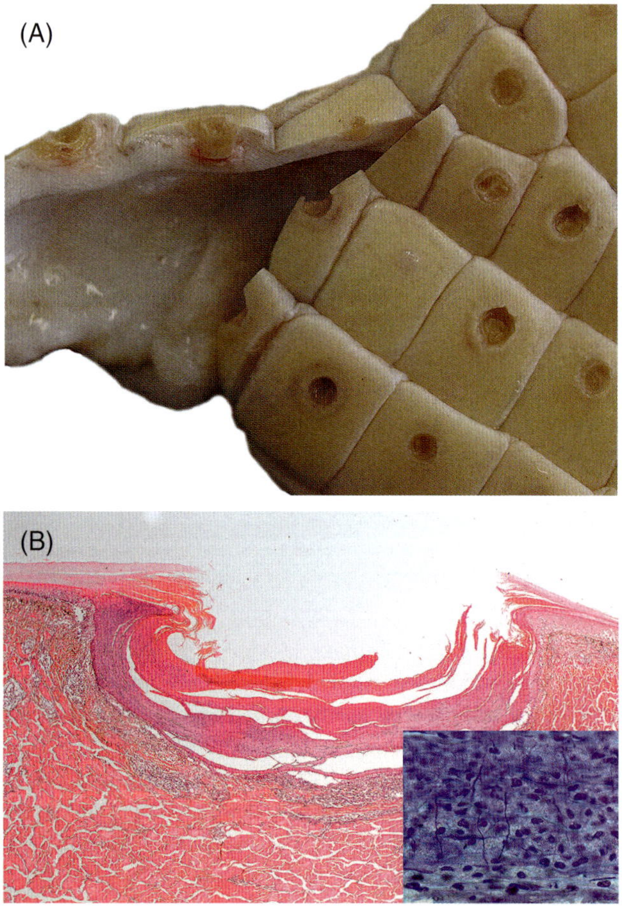

FIGURE 35.13 Cutaneous dermatophilosis in the skin of an estuarine crocodile. (A) Light brown pitting lesions result in widening and depression of the integumentary sense organs. (B) Histologically, the integumentary sensory organ is effaced by ulceration and an overlying thick, laminated serocellular crust in which there are numerous long, filamentous, beaded gram-positive bacteria (inset). Gram (inset).

Epidermal hyperplasia with formation of rete pegs and keratinocyte dysplasia is typically present. Dermal inflammation progresses with epidermal lesions and is primarily lymphohistiocytic and perivascular; granulomatous inflammation is rare. Gram-positive filaments and coccoid structures (presumed zoospores) are most often present in the stratum corneum but can extend into the deeper aspects of the intact epidermis (Fig. 35.13B). With ulceration, organisms can extend into the dermis, subcutis, and less frequently, underlying musculature. *Dermatophilus* spp. may also occur in ulcerative skin lesions along with various other agents, raising the likelihood that under some circumstances it is an opportunistic secondary invader.

As with other reptile taxa, **mycobacteriosis** in crocodilians is caused by nontuberculous mycobacteria and typically presents as a granulomatous disease. It can be restricted to the skin or be multisystemic with or without skin involvement. Nontuberculous mycobacteria are considered ubiquitous and most cases in crocodilians likely represent opportunistic infection. Table 35.e3 provides specific mycobacterial species that have been identified in crocodilians.

Gram-negative septicemia in crocodilians can be caused by a wide range of bacteria, most of which are considered opportunists. Temperature-associated stress (both natural and enclosure-related), other forms of stress in the captive environment and trauma are frequently reported as contributors or underlying causes for the overwhelming infections. ***Aeromonas hydrophila*** is one of the most frequently isolated organisms; other genera involved include ***Salmonella*, *Edwardsiella*, *Providencia* (*Proteus*), *Klebsiella*, *Citrobacter*, *Morganella*, *Enterobacter*,** and ***Serratia***.[13,28,46,71,89,100,104] ***A. hydrophila*** has been identified as part of the normal flora of wild American alligators, and is considered ubiquitous in the aqueous environment.[58] Crocodile mortalities due to *A. hydrophila* septicemia following improper thermoregulation support the opportunistic nature of this infection. In farmed crocodiles, *A. hydrophila* septicemia commonly affects hatchlings, and is associated with fibrinous polyserositis, hepatitis, and pneumonia, as well as subcutaneous edema and ventral skin erosions. Microscopically, liver lesions consist of necrosis with intralesional gram-negative bacteria as well as lymphocytic and granulocytic inflammation; similar infiltrates are present in the lungs. Less common lesions include necrotizing enterocolitis, pancreatitis, and nephritis. Similar lesions, with a predominance of hepatic infection and inflammation, are typical with other listed gram-negative bacteria. Of note regarding ***Salmonella* spp.**: similar to the situation in many other reptiles, while infection resulting in lesions typical of gram-negative sepsis does occur,[105] it is uncommon despite high intestinal carriage rates of the organism.[79,82]

Providencia rettgeri (previously ***Proteus rettgeri***) is a motile, gram-negative bacillus that, like *A. hydrophila,* is ubiquitous in the environment and can be a significant cause of gram-negative septicemia in farmed crocodile hatchlings and juveniles (2–8 months of age).[7] Lesions are similar to those noted earlier for other gram-negative septicemias with fibrinous to heterophilic and granulomatous polyserositis, hepatitis, and splenitis predominating (Fig. 35.14A);

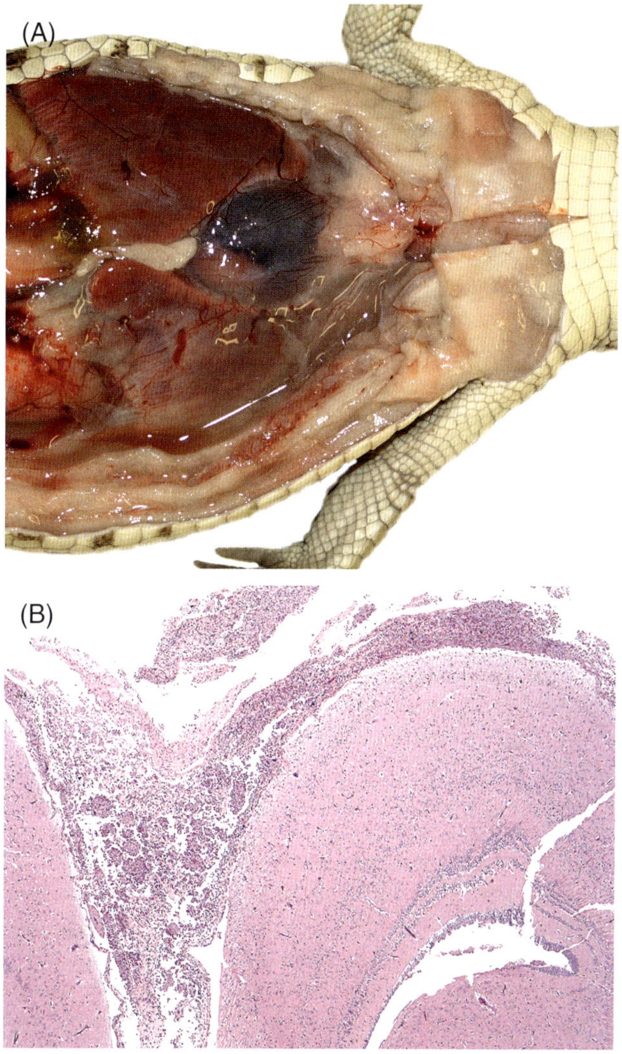

FIGURE 35.14 Bacterial septicemia in an estuarine crocodile (A) and meningitis (B) in an American alligator caused by *Providencia rettgeri*. (A) Fibrinous polyserositis, seen in this image as thickened, wet, colorless to lightly opaque thickening on or of serosal surfaces, is a common finding secondary to bacterial septicemia caused by a variety of gram-negative bacteria. (B) Infection with this organism is more commonly associated with meningoencephalitis than other causes of gram-negative septicemia in crocodilians. Histologically, this presents as heterophilic infiltrates expanding the meninges and occasionally invading the underlying cerebral tissue (see eSlide 35.e3). *(Photo Courtesy of R. Ossiboff, University of Florida College of Veterinary Medicine.)*

subcutaneous erythema and edema and/or swelling involving an entire limb is also common. Central nervous system disease as a component of septicemia is seen more consistently with *P. rettgeri* infection than with other bacteria and acute, heterophilic meningitis with occasional intralesional or intravascular bacilli is the most common histologic lesion (Fig. 35.14B, eSlide 35.e3).[16,69] Heterophilic and fibrinous ventriculitis, heterophilic perivascular cuffing, necrosis, and inflammation of the choroid plexus and heterophils scattered throughout the neuropil are less consistent findings.

Yokenella regensburgei is a gram-negative bacillus that causes predominantly respiratory lesions and acute mortality in young farmed American alligators; up to 10% daily mortality rates have been reported.[6] Disease is primarily respiratory, with pleuritis being a consistent finding amongst other pulmonary lesions. Vasculitis with thromboembolism, coelomitis, and myocarditis are also common. But as with other gram-negative infections, inflammation may be widespread, affecting numerous organ systems.[6]

While gram-negative bacteria certainly predominate as causes of infection in crocodilians, one notable exception is necrotizing fasciitis associated with an aggressive clone of ***S. agalactiae*** in farmed juvenile estuarine crocodiles.[8] Grossly, areas of necrosis and full-thickness ulceration involve the body wall and/or limbs and cause exposure of underlying muscles and soft tissue (Fig. 35.15A). Histologically, there is necrosis, edema, and fibrin deposition in the subcutis that dissects through fascial planes and contains heterophilic and granulomatous inflammation admixed with intralesional gram-positive cocci (Fig. 35.15B). Fasciitis can be accompanied by polyarthritis, which presents as multiple swollen joints filled with caseous exudate. This corresponds to heterophilic and granulomatous arthritis with abundant intralesional gram-positive cocci.

Infection with ***Burkholderia pseudomallei*** **(melioidosis)** is generally rare in crocodilians living in areas where the bacterium is endemic, however, there are a few case reports of infection. These include **yolk sacculitis** and systemic infection in two hatchling-farmed crocodiles in which an identical strain of *B. pseudomallei* was detected in the incubator in which the crocodiles hatched.[97] The yolk sac and systemic lesions in the hatchlings were typical of those caused by gram-negative bacteria generally.

Wounds are often infected with mixed bacteria. Exudate may be caseous, as is typically the case in most reptile infections, to viscous depending on the organisms involved. Yolk sac infections, sometimes associated with poor yolk scar closure, are common in farmed crocodiles in the first week after hatching, especially when held in unhygienic conditions.

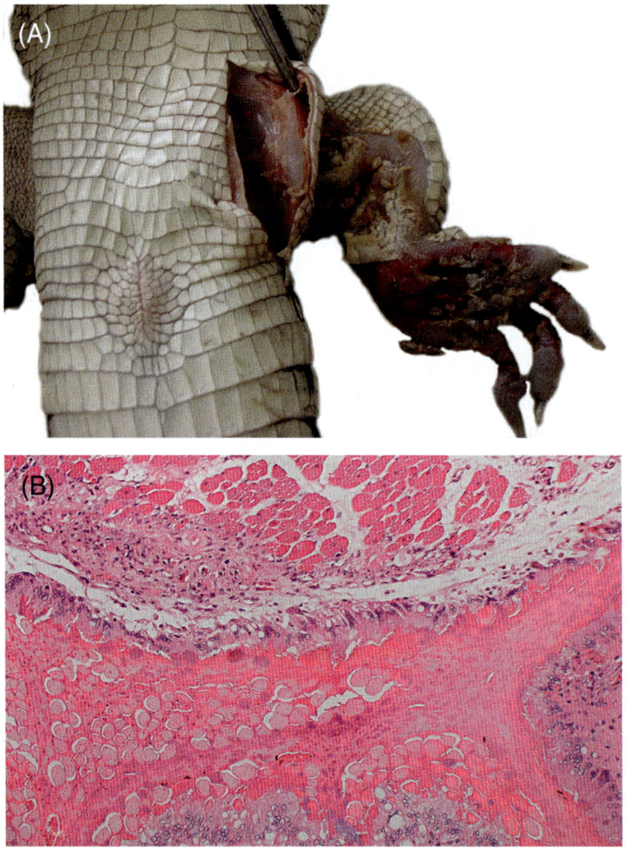

FIGURE 35.15 Necrotizing fasciitis caused by *Streptococcus agalactiae* in an estuarine crocodile. (A) Caseous exudates can be found in the subcutis and along fascial planes. In this case, lesions are present in the muscle of a hindleg and adjacent coelomic body wall. This is accompanied by regionally extensive sloughing of overlying skin. (B) This corresponds histologically to skeletal muscle and connective tissue necrosis with numerous intralesional colonies of coccoid bacteria surrounded by multinucleated giant cells. Note viable skeletal muscle and connective tissue in the upper portion of the image.

Fungi

Fungal infections can occur in any tissue in crocodilians. They most commonly affect the skin or lungs but can be widely disseminated and affect numerous organ systems.[46,92] In captive animals, suboptimal environmental conditions predispose to fungal infection, presumably due to suppression of immune system function. Isolated fungi are often ubiquitous environmental organisms and represent opportunistic infection.[13,30]

Cutaneous infection typically presents grossly as an alteration in the color and texture of the skin surface. Superficial cutaneous mycotic infections can appear dull,

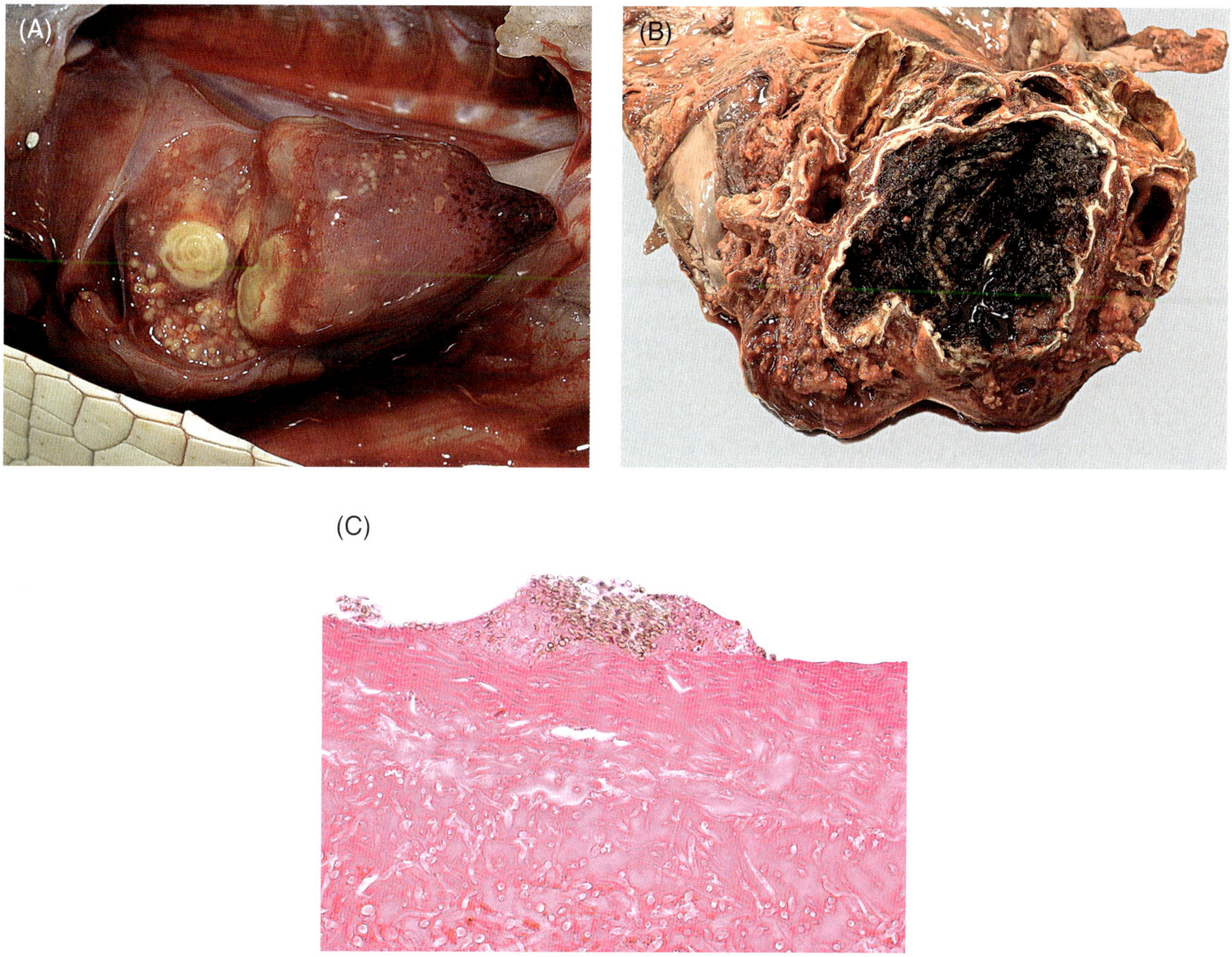

FIGURE 35.16 Mycotic pneumonia in an estuarine crocodile (A) and American alligators (B, C). (A) As is often identified in crocodilians with mycotic pneumonia, the pulmonary parenchyma is expanded by caseous exudates in the airways, which appear as both large, lamellated foci, and abundant miliary yellow foci. An *Acremonium* spp. was cultured from these lesions. (B) An alternative appearance of mycotic pneumonia in crocodilians is the lining of airways by thick plaques composed of dense fungal mats and associated inflammation, as is present in this *Metarhizium robertsii* infection. (C) In some cases, adventitious sporulation may be seen and used for morphological identification of fungal species. In this case, the conical, pigmented conidia forming chains were typical of *Metarhizium anisopliae* var. *anisopliae*. Hyalohyphomycosis within the inflammatory debris at the bottom of the image lines the pulmonary airway. *(Part B: Photo Courtesy of R. Ossiboff, University of Florida College of Veterinary Medicine.)*

thickened, moist, and gray, while deeper infections appear as eroded or ulcerated, red to brown, crusty lesions. An alternate presentation is proliferative granulomatous inflammation on the feet of crocodilians housed on abrasive surfaces.[58] The histologic appearance of cutaneous mycotic disease involves epidermal changes, such as hyperplasia, acanthosis, inter- and/or intracellular edema along with erosion or ulceration. The gross and histologic appearance of lesions also varies with lesion depth and concurrent bacterial infection(s). Heterophilic inflammation predominates, especially in ulcerated tissues but mononuclear to granulomatous inflammation, including discrete granulomas, can be present.

In most parenchymal organs, mycoses typically present grossly as granulomatous disease with firm, pale-colored nodules (Fig. 35.16A). A central core of lamellated, brown, or tan material may be apparent, while others may be more homogenous and solid on section. In the lungs, fungal infection can also present as thick plaques lining the airways (Fig. 35.16B). Histologically, inflammation is predominantly granulomatous, including multinucleated giant cells and peripheral lymphocytic aggregates; heterophilic inflammation is variable and can be severe.

Many fungal species affecting crocodilians have similar morphologic features and are narrow, septate, and nonpigmented with parallel walls (**hyalohyphomycosis**), so morphology alone is often of limited use for identification. Fungal culture and/or molecular diagnostics, in combination with histologic appearance, are used for positive identification. That said, some morphologic features, such as pigmentation, calcium oxide crystal formation (***Metarhizium anisopliae, Beauveria bassiana***, or ***Aspergillus* spp.**)[30,39] or the presence of reproductive propagules (adventitious sporulation) (Fig. 35.16C) can be useful in categorizing some infections. Grossly, tissue colonies of ***Purpureocillium lilacinum*** (previously ***Paecilomyces lilacinus***) may impart a violet hue to infected tissues, ***Metarhizium anisopliae*** dark green and ***Beauveria bassiana*** white,[92] although the latter two are less specific changes.

Nannizziopsis crocodili, initially described as the ***Chrysosporium* anamorph of *Nannizziopsis vriesii***,[113,119] is one of the few specific and possibly nonopportunistic fungal diseases in crocodilians, though it has, to date, only been identified in captive hatchling and juvenile animals.[42,119] It causes a dull to leathery and plaque-like or focal caseous appearance on the skin of affected estuarine crocodile hatchlings. Various presentations are possible in juvenile freshwater crocodiles, ranging from ulceration to gray to black scale discoloration and loss of scale definition (Fig. 35.17). Histologic features are as generally described earlier for other fungal infections and include intralesional hyphae that exhibit typical features of hyalohyphomycosis in areas of crusting or epidermal necrosis and occasionally invade into the underlying dermis. Adventitious sporulation may be present with characteristic arthroconidia (1.5–2 μm × 2–3 μm cylindrical), which are the reproductive propagules, near the skin surface. *Nannizziopsis crocodili*-related mortalities can be due to secondary bacterial septicemia or can be the direct result of cutaneous infection, although the pathogenesis of the latter is unknown.

There are numerous single case reports and small case series of crocodilian mycotic disease in the literature including one report of **microsporidiosis**. These are summarized in the Table 35.e4.

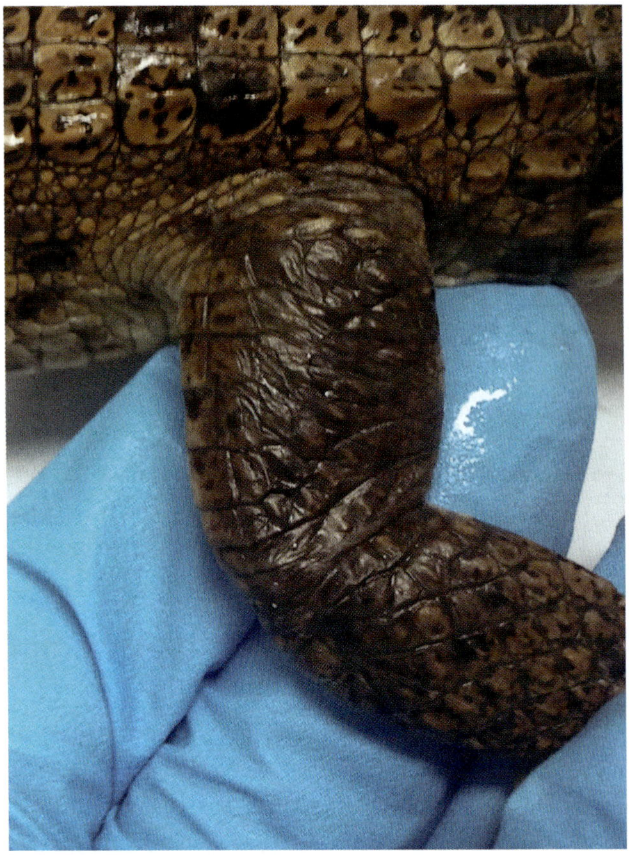

FIGURE 35.17 Mycotic dermatitis in a juvenile freshwater crocodile caused by *Nannizziopsis crocodili* infection. Characteristic changes of infection are locally extensive loss of scale definition and discoloration, which are present on the proximal aspect of the hindlimb in this animal. *(Photo Courtesy of A. Hill, Currumbin Wildlife Sanctuary.)*

Metazoa

A wide array of internal parasites has been identified in captive and wild crocodilians, and a comprehensive checklist of host–parasite interactions is available.[118] Many of these are clinically insignificant. Several that result in clinical disease and/or mortality are described below.

Gastric ascaridiasis, caused by ***Dujardinascaris, Ortleppascaris***, or ***Brevimulticaecum*** species, does not always cause clinical disease. However, severe infestations can cause gastric ulceration and granuloma formation in the stomach of wild and captive crocodilians.[18,46,71,126,127]

Lingual trematodiasis in the American alligator is caused by ***Odhneriotrema incommodum***. Gross lesions in wild alligators can be numerous and include circular, umbilicated foci accompanied by mucosal hemorrhage and edema. Histologically, mucosal ulceration is subtended by extensive mixed inflammation that includes granulocytes, macrophages, and lymphocytes and is associated with neovascularization, fibrosis, hemorrhage, and edema.[125] While thought to be of little clinical concern, Woodyard et al.[125] suggest that the potential impact of lingual trematodiasis on wild American alligators may be underestimated.

Cutaneous **paratrichosomiasis** occurs in multiple farmed and free-ranging crocodilian species.[46] ***Paratrichosoma crocodylus*** affects New Guinea, estuarine, and (likely) freshwater crocodiles and ***P. recurvum*** affects American and Morelet's crocodiles; similar undescribed

FIGURE 35.18 Paratrichosomiasis in the skin of an estuarine crocodile. Dark, serpiginous tracts traverse the scales of the lateral abdominal skin due to tunneling of the nematode within the epidermis.

parasites also occur in Nile, Orinoco, and mugger crocodiles. Infection is not clinically significant but lesions affect the appearance of the hide and its commercial value. Infection is characterized grossly by serpentine, tightly curved to hairpin, darkly colored, "zigzag trails" on the lateral or ventral skin of affected animals (Fig. 35.18). Acute lesions are often subtle, appearing as less distinct slightly erythematous tracts. Histologically, clear cores created by migrating nematodes are present within the stratum corneum and underlying layers of the epidermis. These cores can contain operculated and larvated eggs but rarely contain the adult parasite.[12] Presumably, the eggs are elevated toward the surface with maturation of the epidermis, eventually being released into the environment upon shedding of superficial keratin layers. Inflammation is typically dermal, perivascular, and lymphocytic but can be more diffuse and granulocytic when migration occurs deeper in the epidermis.

Pentastomiasis is a common parasitic infection in wild and captive crocodilians around the world and according to Riley,[101] most if not all species of crocodilians harbor a pentastomid. Pentastomes are hematophagous, vermiform arthropods with a thin, chitinous cuticle, annulations, and two pairs of hooks surrounding the mouth. The pentastome life cycle is indirect and includes intermediate and definitive hosts. Eggs remain viable in water for long periods of time and hatch upon ingestion by an intermediate host (commonly a fish for those species infecting crocodilians). After multiple molts, most larvae encyst in the intermediate host's tissues and excyst upon the death of that host. They then migrate from the gastrointestinal tract of the definitive host to the site of final maturation. Multiple different pentastome species have been identified in crocodilians.[66] Adults of most species that occur in crocodilians live in the lungs. Exceptions are ***Leiperia*** spp. and ***Subtriquetra*** spp. that inhabit the trachea and bronchi and the nasopharynx, respectively.[46,102] *Leiperia* spp. also are unique in that maturation stages occur in the cardiovascular system (aorta and pulmonary artery) prior to migration to the respiratory tree.[91]

Significant disease is not typically associated with pentastome infection in the definitive host. However, they can induce or predispose crocodilians to bacterial sepsis and cause respiratory tract disease or death.[1] For example, ***Sebekia oxycephala*** has caused fatal pulmonary hemorrhage in American alligators.[23] In general, respiratory tract disease is most often associated with those species that are embedded in tissue (e.g., *Leiperia* spp.) rather than with those that move freely in airways. When present, histologic changes in the lungs include hemorrhage and granulomatous to lymphoplasmacytic inflammation.[13,60] Lesions in estuarine and New Guinea crocodiles can also include extensive interstitial pneumonia with bronchiectasis and bronchiolar epithelial hyperplasia.[71]

Protozoa

A potentially fatal form of **coccidiosis** occurs in farmed crocodilians. Disease is caused by an organism morphologically consistent with a ***Goussia* spp.**[29,31,46,71] However, in many descriptions, identification is based solely on the presence of oocysts containing four sporocysts, each with two sporozoites, which is insufficient for genus-level identification.[78,83] The life cycle of the parasite is currently unknown. A systemic form of infection occurs in Nile, estuarine, and New Guinea crocodiles. Sporulating coccidia occur in the intestine, spleen, liver, lung, kidney, and stomach, presumably within histiocytic cells and possibly pulmonary endothelial cells. Associated inflammation is rare, though focal heterophilic pneumonitis can occur.[71] Intestinal infection may be incidental or can cause villous loss and fusion, mononuclear cell inflammation, and/or necrotizing or fibrinous enteritis that in estuarine crocodiles can be transmural (Figs. 35.19A, B).[29,46] Histologically, mural fibrosis, marked lymphohistiocytic and plasmacytic enteritis, and multifocal granulomas are seen in severe cases. Oocysts are present throughout the intestinal wall, within the dense mononuclear infiltrates, and within granulomas.

A rarely described condition in farmed estuarine crocodiles in Australia called **giant cell enteritis** is characterized grossly by thickening, reddening, and ulceration of the proximal intestinal tract. This corresponds histologically to massive transmural accumulation of multinucleated giant

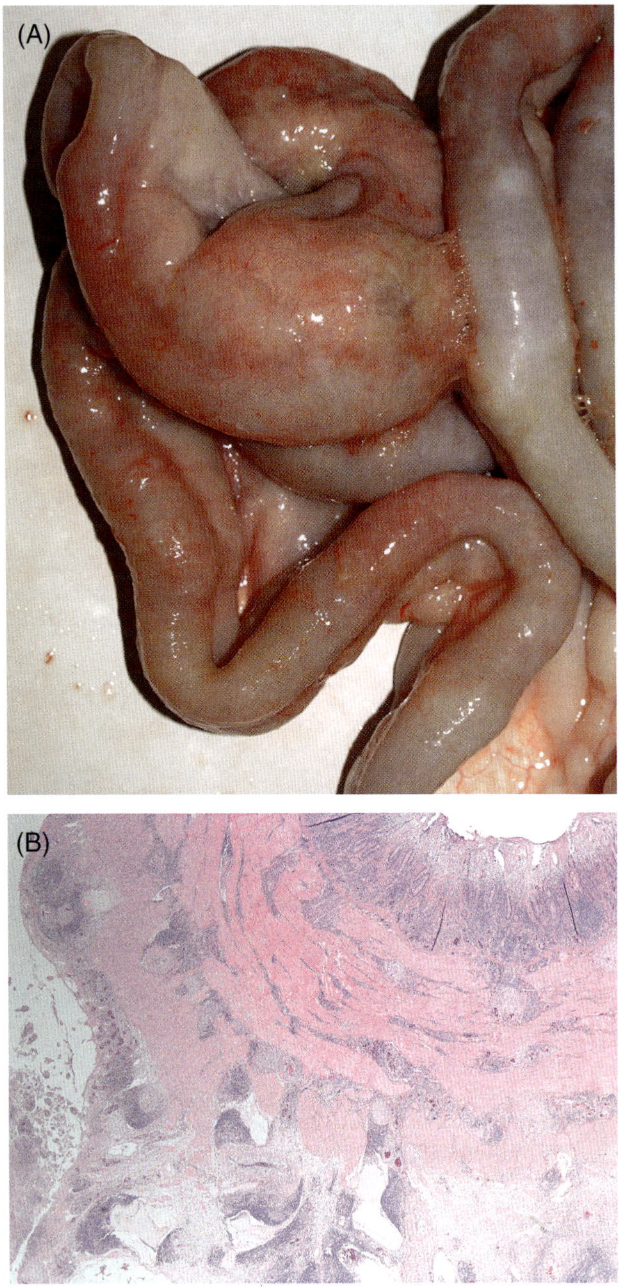

FIGURE 35.19 Intestinal coccidiosis in an estuarine crocodile. (A) Infection is associated with segmental thickening and hyperemia in the intestinal wall and an irregular, finely multinodular serosal surface. In severe infections, as in this image, fine, fibrinous adhesions may develop between affected and adjacent, normal tissue. (B) Chronic infection in this case includes transmural granulomatous inflammation with abundant lymphoid aggregates. Clusters of coccidial sporozoites are present within granulomatous foci. Note that this is a particularly severe case of coccidiosis and that infections are more commonly limited to the mucosa.

cells that markedly expand the lamina propria and submucosa.[70] Numerous 1 μm in diameter, round to ovoid intracellular organisms are present in vacuoles within the giant cells, resulting in a foamy appearance. Further classification of these organisms has not been performed although they are presumed to be protozoan based on histologic appearance.

Other Infectious Organisms

Achyla species are oomycetes, protistans often referred to as water molds that include the more commonly identified *Saprolegnia* spp. in fish. Infection of hatchling American alligators by an *Achyla* spp. can result in gross lesions similar to those described in fish, including gelatinous, off-white growths overlying ulcerative lesions on multiple areas of the skin.[76] The lesions consist of epidermal necrosis and ulceration with accumulation of keratin (orthokeratotic hyperkeratosis) admixed with necrotic debris, hyphal-like structures, and mixed bacterial populations. As with other oomycetes, hyphal-like structures are relatively wide, negatively staining with HE (GMS-positive) and have nonparallel walls.[76] Confirmation of oomycetes and differentiation from some true fungi (e.g., zygomycotic agents) can be challenging and, as with other fungal infections, culture and/or molecular diagnostics may be required to facilitate identification.

ADDITIONAL RESOURCES

Buenviaje, G., Hirst, R.G., Summers, P.M., 2000. Skin Diseases of Farmed Crocodiles. Rural Industries Research and Development Corporation, Canberra, Australia.

Huchzermeyer, F.W., 2003. Crocodiles: Biology, Husbandry and Diseases. CABI Publishing, Oxon, UK.

Jacobson, E.R., Garner, M.M., 2021. Infectious Diseases and Pathology of Reptiles. 2nd Edition. CRC Press, Boca Raton.

Tellez, M., 2013. A Checklist of Host–Parasite Interactions of the Order Crocodylia. University of California Publications in Zoology, Berkeley and Los Angeles, vol. 136.

Van der Merwe, N.J., Kotze, S.H., 1993. The topography of the thoracic and abdominal organs of the Nile crocodile (*Crocodylus niloticus*). Onderst. J. Vet. Res. 60, 219–222.

References

Full references for this chapter can be found online at https://doi.org/10.1016/B978-0-443-18437-6.00008-0.

Chapter 36

Lacertilia

Francesco C. Origgi
Institute of Microbiology, Department of Environment Construction and Design, University of Applied Sciences and Arts of Southern Switzerland, Mendrisio, Switzerland; Department of Veterinary Sciences, University of Messina, Messina, Italy

INTRODUCTION

The suborder *Lacertilia* (lizards) in the order *Squamata* includes more than 11,000 species; it accounts for the largest group of living reptiles (see Table 36.e1 for more detailed taxonomic information).[143] At least 110 species of this suborder are considered critically endangered. Fossil records confirm the presence of lizards on the earth for hundreds of millions of years and the crocodile-lizard lineage split is estimated to have occurred approximately 250 million years ago.[82] Lizards are the most diverse living group of reptiles, and as such they have colonized most of the surface of the globe with the exclusion of the polar regions.[143] Lizards are generally tetrapods, however, several species are reduced to complete loss of limbs (e.g., legless lizards). The members of the suborder *Lacertilia* can greatly vary in weight and length from a few centimeters for the genus *Sphaerodactylus* (ball finger geckos) to more than 3 m and about 100 kg for Komodo dragons. Most lizards are oviparous, although viviparous and parthenogenetic lizards exist. Very limited information is available on free-ranging lizard pathology, whereas a significant body of literature has documented the occurrence of a variety of diseases in captive lizards. Much of the early literature describes diseases associated with suboptimal care and husbandry practices. Highlighted in this chapter are the most relevant and/or unique morphological features of lizards and a comprehensive review of the most important diseases affecting this group of reptiles.

NON-INFECTIOUS DISEASES

Nutritional

Metabolic bone diseases include a group of conditions (rickets, fibrous osteodystrophy, osteoporosis, and osteomalacia) characterized by an overwhelming exacerbation, disturbance, or derangement of physiological bone processes that result in severe, systemic bone changes and can be fatal if left untreated.[176] Of the different forms of metabolic bone diseases, **fibrous osteodystrophy** occurs most frequently in reptiles. It is an overrepresented disease in captive reptiles and is almost invariably secondary to nutritional imbalances and/or renal hyperparathyroidism. Associated imbalances include one or more of the following: inappropriate calcium to phosphorus ratio or low calcium (intake, but also absorption or retention) and/or vitamin D; or inappropriate exposure to ultraviolet (UV) B light. Renal disease resulting in abnormal and progressive loss of calcium is also a significant predisposing condition. The main driver of the process is chronic, limited availability of calcium in the blood, which stimulates the production of parathyroid hormone and subsequent progressive resorption of calcium from the bones. Three different stages of the disease can be defined: a preclinical disease, when bone resorption is occurring but the affected animal is not showing obvious clinical signs; an early clinical stage when the animal is showing anorexia, difficult movements and lethargy with or without muscle twitching, tremors, and neurological signs, and calcium levels drop under physiological levels; and a late clinical stage with characteristic severely thickened and swollen limbs (Fig. 36.1A). Bowing of the limbs is also common especially in chameleons. Long bones (femur, humerus, tibia, fibula, and radius) are most commonly affected[13] and may undergo pathologic fractures, generally in proximity to the metaphysis. Other affected bones include the mandible which due to demineralization and fibrous replacement, softens and becomes pliable. This produces a condition commonly referred as "rubber jaw." Vertebra can also fracture leading to neurological complications. Kyphosis, lordosis, and scoliosis are common findings in lizards that develop nutritional or renal secondary hyperparathyroidism at a young age and represent a permanent scar (Fig. 36.1B). Microscopically, there is progressive reduction to complete loss of the cortical bone (Fig. 36.1C; eSlide 36.e1) and loss of bony trabeculae along with proliferation of fibrous vascular tissue in the marrow cavities.[12,13,127] In animals that are still capable of a certain degree of movement, the mechanical forces promote the exuberant deposition of cartilage and fibrous tissue along the

Pathology of Wildlife and Zoo Animals. https://doi.org/10.1016/B978-0-443-18437-6.00020-1

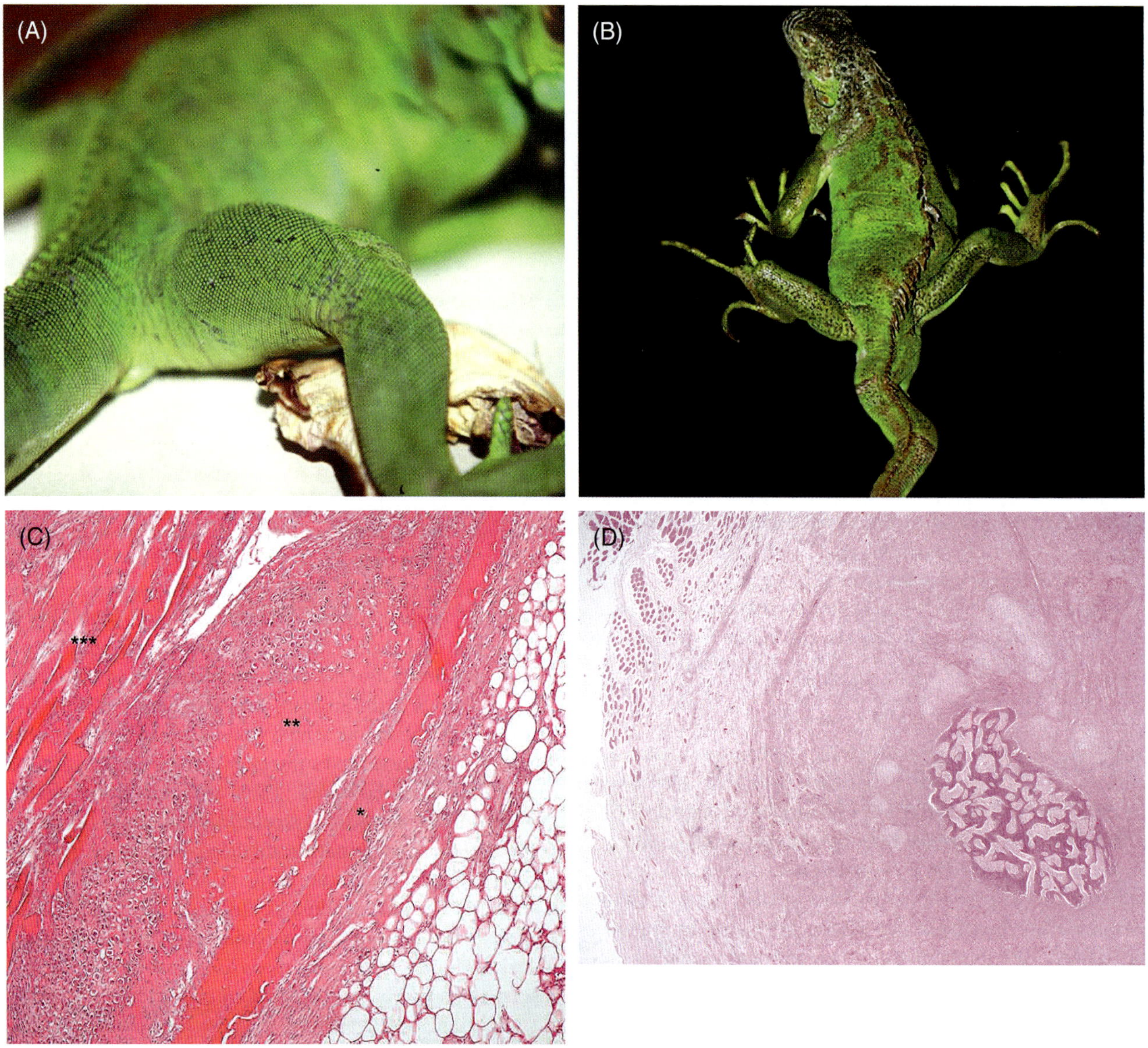

FIGURE 36.1 Fibrous osteodystrophy and metabolic bone disease in green iguanas. (A) The thigh of this green iguana is markedly swollen giving the false impression of prominent muscular hypertrophy. The enlargement is due to severe cartilage and fibrous proliferation secondary to severe bone demineralization and consequent regional compression and atrophy of the muscles. (B) Rhoecosis describes the combination of scoliosis, kyphosis, and lordosis, all of which can occur secondary to fibrous osteodystrophy. (C) Characteristic changes associated with fibrous osteodystrophy (metabolic bone disease) include thinning of the cortical bone *(single asterisk)* with a scalloped edge secondary to active osteoclastic resorption. Severe fibrocartilaginous proliferation along the periosteal surface *(double asterisks)* and compression and degeneration of adjacent skeletal myofibers *(triple asterisks)* are also seen. (D) In this cross-section of the femur, there is a complete loss of the cortical bone, which is replaced by exuberant fibrous proliferation along with the seeding of a few cartilage cores. The trabecular bone of the femur is still visible surrounded by the massive fibrosis. More than 90% of the skeletal muscles have been replaced by the fibrosis and the remaining muscle fibers have been compressed and pushed to the periphery of the section (see eSlide 36.e1). *(Part B: Photo Courtesy of M. Bielli, Novara, Italy; Parts C, D: Photos Courtesy of the University of Bern, Centre for Fish and Wildlife Health-FIWI.)*

periosteal surfaces (Fig. 36.1D; eSlide 36.e1), which may dramatically expand the bone profile leading to progressive pressure atrophy of the surrounding muscles and the pseudohypertrophy of the limbs observed grossly.[12,258]

In growing animals, the hypertrophic cartilage at the physis forms variably extensive irregular clumps of chondrocytes.[12,13,127] Shortening and widening of the metaphyseal trabeculae and replacement of the primary spongiosa (lacking cartilage cores) by woven bone[127] and/or its progressive loss together with the secondary spongiosa is also described.[12,13]

Hypo- and **hypervitaminoses** are common conditions in captive lizards. Vitamin D is a secosterol and together with its metabolites 25-hydroxyvitamin D_3 (25(OH)D_3) and 1,25 dihydroxyvitamin D_3 (1,25(OH)$_2D_3$) has a major role in calcium and bone metabolism. **Hypovitaminosis D_3**

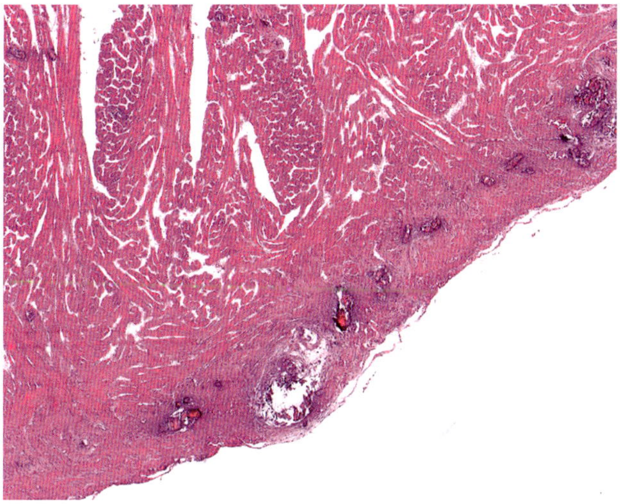

FIGURE 36.2 Metastatic myocardial mineralization in a chameleon. The subepicardium is multifocally thickened by the deposition of mineral and associated fibrosis. *(Photo Courtesy of the University of Bern, Centre for Fish and Wildlife Health-FIWI.)*

in the diet of captive lizards has been associated with the development of fibrous osteodystrophy[258] (see also metabolic bone disease-fibrous osteodystrophy above), whereas massive deposition of calcium salts in the arterial media and terminal nephritis is reported in green iguanas undergoing **excessive supplementation of vitamin D** together with calcium and phosphorus.[278] However, metastatic calcification in green iguanas (Fig. 36.2) and dystrophic mineralization in *Uromastyx* sp. have been associated with low plasma or serum vitamin D_3 concentration.[127] Tubular mineralization has also been observed in the kidney of veiled chameleons fed unsupplemented experimental diets lacking vitamin D, A, and calcium along with lack of UVB exposure. The breakdown of inhibitors of soft tissue mineralization secondary to contextual hyperparathyroidism and hypovitaminosis D have been suggested as a possible pathogenetic process, consistent with the existence of a complex etiology behind these tissue changes.[127]

Vitamin A is an important cofactor for the integrity of the epithelia and for the preservation of their normal differentiated state. A number of conditions consistent with **hypovitaminosis A** resulting in squamous metaplasia have been documented in lizards. Tissues particularly sensitive to hypovitaminosis A are the pharyngeal mucosa, conjunctiva, urogenital tract, and upper and lower respiratory tracts. Toxic compounds, including organochlorine pesticides, may inhibit vitamin A metabolism causing similar changes.[38] Suspected hypovitaminosis A in a group of green anoles was associated with irregular thickening and red/brown discoloration of the lower lip, ulcerative cheilitis, blepharedema, anorexia, lethargy, depression, and weight loss. Histological lesions were consistent with squamous metaplasia of the oral mucous glands, palpebral conjunctiva, and upper respiratory mucosa.[192] Hypovitaminosis A in old world chameleons may also predispose to stomatitis,[84] and geckos and chameleons are also reported to be susceptible and develop multisystemic lesions.[178] In panther chameleons, gross changes include ocular discharge, swelling of the lips, gular and cervical edema, tail tip necrosis, and skeletal abnormalities.[84] However, Hoby et al.[127] suggested that vertebral abnormalities would more likely be related to hypovitaminosis D and hypocalcemia. In chameleons, soft tissue mineralization has been described with **vitamin A oversupplementation.**[84] Similarly, hypervitaminosis A has been associated with segmental calcification in the colon of chameleons experimentally fed a diet supplemented with calcium and vitamin A.[127]

Vitamin E is an important cofactor whose reduced availability may lead to significant diseases including muscle degeneration and necrosis.[165] Captive-bred leaf-tailed geckos with presumptive **vitamin E/selenium deficiency** had multifocal areas of cutaneous red-to-white discoloration, myocardial mineralization, and severe multifocal skeletal muscle degeneration. Granulomatous lesions were observed in the dermis and corresponded to the gross skin lesions. **Nutritional myopathy** has also been seen in veiled chameleons with progressive inability to use their tongue along with circling and weakness,[178] and in monitor lizards (M. Garner, personal communication).

Vitamin B1 (thiamine) is an important cofactor whose absence or low level in the diet has been associated with severe brain lesions and subsequent central nervous system (CNS) signs in several species. Presumptive **thiamine deficiency** has been documented in a colony of anoles[83] with reluctance to move, tremors of the distal appendages, and abnormal posture of the axial skeleton including hyperextension or twisting and serpentine distortion. Gross findings at necropsy were nonspecific. No histologic examination was carried out in the affected animals.

Impaction is a relatively common occurrence in reptiles; it is essentially a management-related issue. Impaction can occur following the accidental or voluntary ingestion of excessive amounts of food or indigestible foreign material, for example, ingestion of bedding material. Impactions are usually fatal if left untreated.

Metabolic

Gout is a relatively common metabolic disorder in captive reptiles including lizards. Interestingly, presumptive occurrence of gout has been reported in dinosaurs.[234] The disease consists of the deposition in tissues, on tissue surfaces, or in joints of characteristic basophilic, needle-like (acicular), monosodium urate crystals (tophi) that are often arranged in a radial pattern; the deposits can also be amorphous and eosinophilic to amphophilic (Fig. 36.3; eSlide 36.e2). Under polarized light, gout tophi are birefringent.[150]

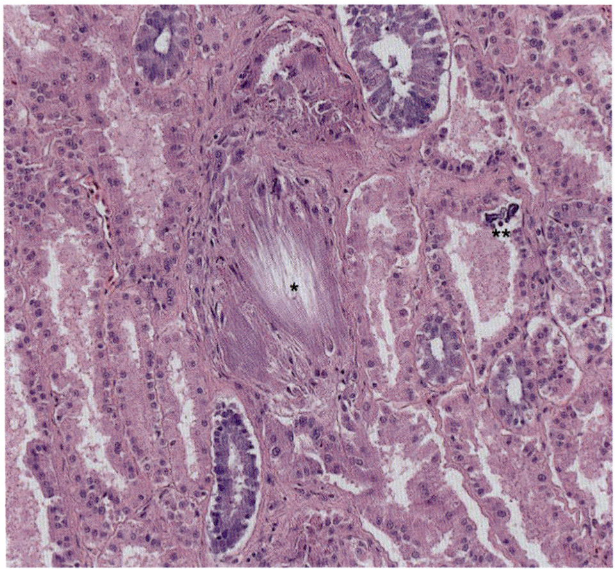

FIGURE 36.3 Renal gout in a chameleon. An intratubular tophus is surrounded by few mononuclear cells and occasional heterophils (see eSlide 36.e2). *(Photo Courtesy of the University of Bern, Centre for Fish and Wildlife Health-FIWI.)*

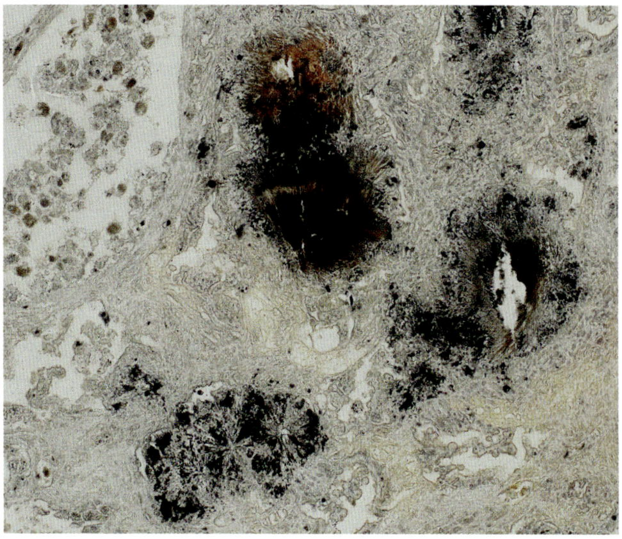

FIGURE 36.4 Renal pseudogout in a chameleon. Several tophi with prominent calcium component expand and efface multiple renal tubules. Associated interstitial fibrosis is also observed. Von Kossa (see eSlides 36.e3 and 36.e4). *(Photo Courtesy of the University of Bern, Centre for Fish and Wildlife Health-FIWI.)*

Predisposing conditions include any direct or indirect factor that contributes to the increased production of urates or its decreased excretion including (commonly) dehydration and renal disease.[191] The end result of the process is the deposition of urate crystals that can be seen grossly as chalk-like precipitates. Based on the location, of deposition of urate precipitates, gout is referred to as articular, visceral, or renal. In visceral gout, deposition can occur on the surfaces of any tissue within the coelomic cavity. With articular gout, swollen and deformed joints are a relatively common finding in chronic cases. With renal deposition, white to gold foci may be evident grossly. These correspond microscopically to tophi within renal tubules. Crystals may cause renal tubular necrosis and secondary foreign body reaction with variably extensive fibrosis and nephromegaly in chronic cases. Visceral and articular lesions may occur in the same individual. Fixation of tissues in 10% neutral buffered formalin often results in a significant solubilization and loss of the urate crystal, whereas fixation in 100% alcohol preserves the tophi in affected tissues.[16] The outcome of chronic and untreated gout is generally fatal. Gout is one of the most common deposition diseases reported in both Gila monsters and beaded lizards.[177]

A similar disease with partially overlapping macroscopic and microscopic changes is **pseudogout**. In pseudogout, tissue deposits consist of crystals that are not composed of urates but instead are generally composed of calcium. Histologically, pseudogout crystals present as pale brown to light basophilic deposits. Pseudogout crystals are often variably birefringent under polarized light. Using Von Kossa or Alizarin red S stains on nondecalcified sections allows differentiation of calcium in pseudogout crystals from true gout crystals (Fig. 36.4; eSlides 36.e3 and 36.e4). Gout and pseudogout may also occur at the same time.[150]

The deposition of fat within the liver and more frequently in the hepatocytes is a common finding in captive reptiles. The collection of lipid droplets in hepatocyte cytoplasm might not reflect a disease and its presence can be influenced by a number of factors including species, season, sex, and nutritional status. However, when the collection of lipids within the hepatocytes affects the entire cytoplasm, displaces the nucleus, and is observed in more than 20% of the parenchyma, this condition is called **hepatic lipidosis** (Figs. 36.5A–C). This metabolic derangement is referred as moderate when it affects 80% of the liver parenchyma and marked when the entire organ is involved. Excess physiologic fat reserves are more common in females just before hibernation, whereas minimal reserves are reported in males at the end of hibernation. Accordingly, females might show changes consistent with hepatic lipidosis under specific physiologic conditions, however, this would not necessarily reflect a pathologic change. Overfeeding and lack of exercise are the most common causes of hepatic lipidosis, which can be life-threatening in severely and chronically affected individuals. Gross changes include pale yellow to light tan enlarged livers with rounded profiles. Histologically, the classic enlarged hepatocytes with clear cytoplasm and a peripherally located nucleus are common findings.[74] Hepatic lipidosis (fatty change) was one of the most common tissue changes observed in bearded dragons

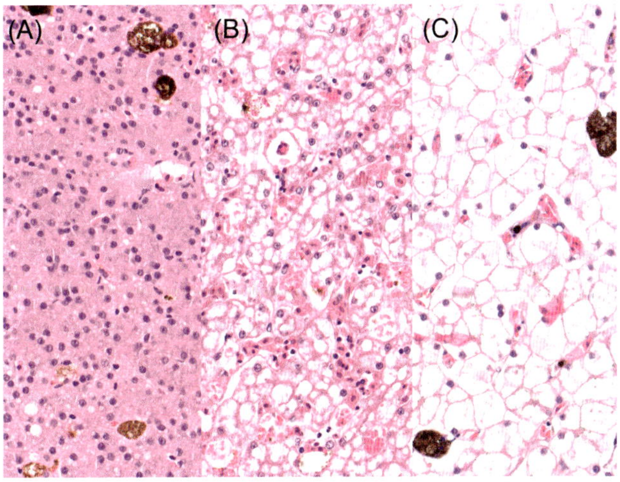

FIGURE 36.5 Normal liver (A) and hepatic lipidosis (B, C) in lizards. (A) Normal liver. Occasional clear vacuoles (lipid deposits) are observed in hepatocytes. (B) Moderate lipidosis. Hepatocellular cytoplasmic lipid vacuoles are present in the majority of the hepatocytes. (C) Severe lipidosis. The cytoplasm of all of the hepatocytes is completely replaced by lipid deposits (clear vacuoles) with peripheralization of the nuclei. *(Parts A–C: Photos Courtesy of the University of Bern, Institute of Animal Pathology, ITPA.)*

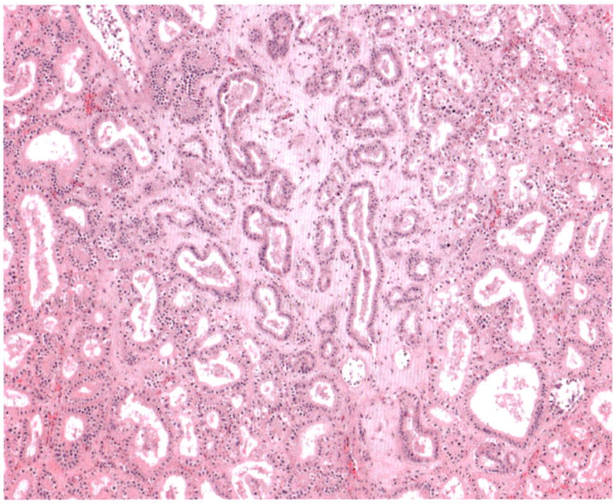

FIGURE 36.6 Chronic nephrosis in a green iguana. The renal interstitium is replaced by severe fibrosis. Tubules are variably ectatic with irregular serpiginous profiles and have segmental hyperplasia of the tubular epithelium. *(Photo Courtesy of the University of Bern, Centre for Fish and Wildlife Health-FIWI.)*

according to a recent retrospective study.[63] Similarly, hepatic lipidosis was among the most common deposition diseases in a recent report in Gila monsters and beaded lizards.[177]

Toxic

A number of tissue changes secondary to the effect of toxic compounds, mostly pesticides, have been described in lizards. The most frequently affected tissues are the gonads and liver. See Table 36.e2 in the Digital Content for specific toxin-associated lesions in lizards.

Congenital/Genetic

A variety of congenital defects have been documented in several lizard species. They include among others, head duplication, scoliosis, hydrocephalus,[175] ankyloblepharon,[230] and atrial septal defect.[221] Recently, ocular neural heterotopia was reported in a Moroccan uromastyx.[53] Overall, anecdotal reports of congenital defects are not uncommon, however, to the best of our knowledge, there are no comprehensive studies that provide solid numbers or a clear epidemiological picture. The great majority have been described in captive animals and accordingly, the conservation significance of these defects in free-ranging individuals remains unclear.

Age-Related/Degenerative

Renal diseases are relatively common in lizards and primarily include degenerative changes associated with fibrosis, gout, and mineralization along with nonspecific inflammatory changes (Fig. 36.6). Anecdotal reports suggest that the degenerative changes may be linked to husbandry. The most common outcome of the chronic renal changes is nephromegaly, which has often been reported in captive green iguana. Renal enlargement can cause compression of the distal colon and intestinal obstruction. A common sequela of chronic renal insufficiency is renal secondary hyperparathyroidism and subsequent fibrous osteodystrophy (see above). A recent investigation attempted to clarify the nature of the nephromegalic changes in green iguanas. Histologic examination revealed that the enlarged kidneys consisted mainly of tubular hyperplasia, resembling poorly developed proximal segments. Additionally, a reduction in density of the glomeruli was observed along with a reduction in the number of distal segments.[294] A classification of reptile glomerulonephritis has been proposed by Zwart.[292,293] In it, acute glomerular disease is characterized by moderate thickening of glomerular capillary basal membranes by deposition of PAS-positive, eosinophilic material, and hypertrophy of the capillary endothelial cells. Proliferation of the visceral epithelium of Bowman's capsule, thickening of Bowman's capsule, and synechiae formation are also seen. Acute glomerular disease is also subdivided into membranous, exudative, and mesangioproliferative according to specific hallmarks characterizing each of the distinct subtype inflammatory processes. The scheme also differentiates intra- and extracapillary glomerulonephritis according to the presence or absence of changes affecting the capillaries. Chronic glomerular disease is characterized by fibrosis, glomerular tuft sclerosis, and tubular nephrosis. Additional changes observed in the tubular epithelial cells include the deposition of brown pigment. The intracellular material has been described as melanin degradation product by Jacobson.[138] The deposition of pigment within the tubular epithelial cells of the renal tubuli was found to be a common finding in bearded dragons.[63] Deposition of amyloid-like material that is Congo red positive but not birefringent and does not stain with crystal violet, PAS, or Von Kossa stains has also been reported in the kidneys of Komodo dragons.[97]

The deposition of birefringent, eosinophilic to amphophilic, hyaline-like material forming globular aggregates filling faveolar spaces, was found to be overrepresented in the lung of bearded dragons in the context of a retrospective study. No inflammatory reaction was found associated with this material, whose origin remains unclear.[63]

A number of **cardiovascular diseases** have been described in lizards; inland bearded dragons appear to be overrepresented. The etiology of most of these diseases remains obscure. Vascular diseases include atherosclerosis[241,243] aortic aneurysm,[261] and pseudoaneurysm. The latter appear as large, fluctuant to firm swellings up to 5 cm in diameter that cause marked expansion and bulging of the dorsolateral neck region just caudal to the skull.[24] Lesions associated with atherosclerosis, include first-degree atrioventricular block, arteriolosclerosis of the cerebral arterioles, and multifocal spongiosis of the brain, were documented in a bearded dragon,[243] and arteriosclerosis has recently been described in bearded dragons.[63] Hemopericardium has also been reported in lizards.[197]

Inflammatory Non-Infectious

Folliculostasis and **egg binding** (Fig. 36.7) are a relatively common occurrence in captive reptiles. The pathogenesis is not clear, however, failure to provide appropriate environmental conditions (e.g., nesting material) for female lizards may negatively impact ovulation and/or egg laying. Similar reproductive disturbances (not obstructive dystocia) have been reported in varanids[96,191] and iguanids.[186,256] **Obstructive dystocia** includes conditions that physically prevent the passage of eggs through and from the oviduct with subsequent stasis of eggs or deposition of follicles in the coelomic cavity, which can result in chronic compression of the viscera. A possible sequela of folliculostasis and egg binding is **yolk coelomitis**. This process is caused by the release of yolk into the coelomic cavity with subsequent foreign body-type inflammation, mesothelial cell proliferation, and possible secondary bacterial complications.[256] This condition is often fatal. In addition to a traumatic etiology, Stacy et al.[256] have proposed an additional pathogenesis. In these cases, mature vitellogenic follicles that fail to ovulate would undergo atresia. Atretic follicles subsequently rupture ("bursting atresia"—a normal physiologic process) and release yolk into the ovary and/or the coelomic cavity. The yolk released by these large follicles overwhelms the resorption capacity of the host resulting in secondary mesothelial proliferation, inflammation, and potentially, bacterial colonization. Yolk coelomitis has been reported as a major cause of death in captive, sexually mature female Fiji Island banded iguanas.[256]

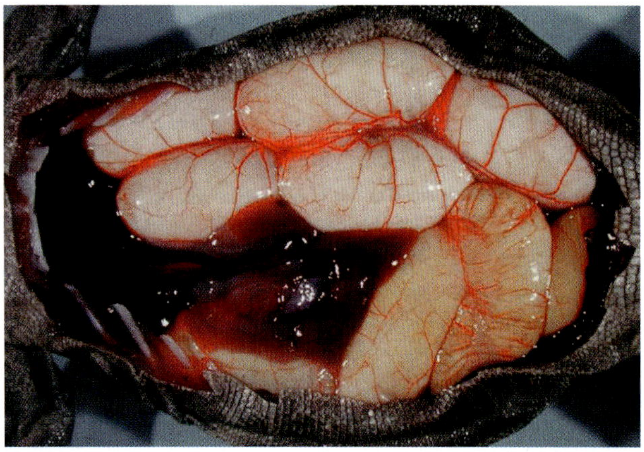

FIGURE 36.7 Egg binding in a green iguana. The oviduct is markedly distended and filled with a large number of eggs. The enlarged oviduct fills the coelomic cavity. *(Photo Courtesy of the University of Bern, Centre for Fish and Wildlife Health-FIWI.)*

Miscellaneous/Multifactorial

Relatively common lesions observed in captive lizards are those secondary to **thermal burns**. Lesions can range from superficial to severe full-thickness necrosis of the skin and underlying tissues with variably extensive and deep ulceration (Fig. 36.8). Mycotic spectaculitis secondary to thermal burns has been described in gold-striped geckos in a zoological institution in New Zealand. The lizards showed extensive areas of ulceration of the spectacle with serocellular crusts and intralesional, embedded fungal structures.[100] Ultraviolet-associated photodermatitis has also been recently described in a blue-tongued skink.[95]

Cloacal and **penile prolapse** are relative common occurrences in lizards (Fig. 36.9). The etiologies that have been proposed for both lesions include nutritional and/or renal secondary hyperparathyroidism and infections. Dystocia, urolithiasis, and neoplasia have also been described as inciting factors for cloacal prolapse, while trauma, bite wounds, traction during copulation, myopathy, neurological disease, and gastrointestinal foreign bodies have been reported as potential causative events for hemipene prolapse.[25] Independent of the inciting cause, cloacal and penile prolapse are considered relevant clinical entities and their severity and prognosis largely depend on the chronicity of the lesion. Common sequelae are regional edema, passive congestion, and necrosis, which can predispose to ascending infections and septicemic events.

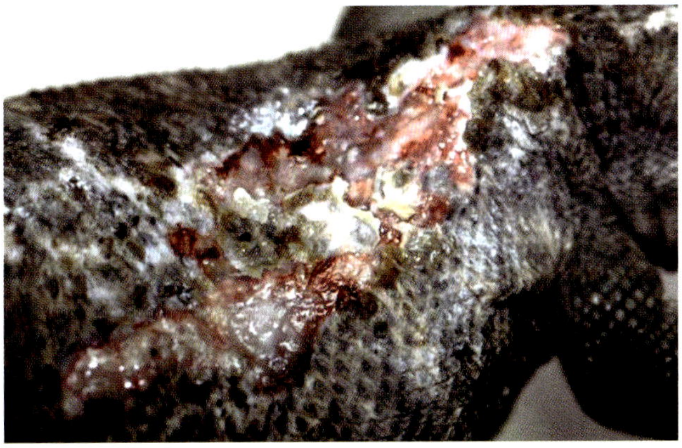

FIGURE 36.8 Thermal burns in the skin of a monitor lizard. Damage to the skin can result in extensive skin ulceration and subsequent granulation tissue and crust formation. Multifocal hemorrhage surrounds the wound. *(Photo Courtesy of the University of Bern, Centre for Fish and Wildlife Health-FIWI.)*

FIGURE 36.9 Cloacal prolapse in a leopard gecko. The typical appearance is that of a red, round to ovoid nodular structure located lateral (as in this image) or ventral to the tail at the level of the cloacal opening. *(Photo Courtesy of the University of Bern, Centre for Fish and Wildlife Health-FIWI.)*

Xanthomatosis refers to the deposition of cholesterol crystals associated with foreign body reaction in different body sites. Xanthomatosis has been documented in green (Chinese) water dragons and geckos.[99,162] Cholesterol granulomas in the brain of affected animals have been associated with displacement and pressure of the adjacent brain and involvement of the spinal cord. Nutritional factors or trauma have been suggested as being involved in the pathogenesis of this condition.[162] The geckos also had lesions on the coelomic surfaces.[99]

Proliferative spinal osteopathy is a bone disease with unknown etiology that has been described only occasionally in lizards[252]; it is more frequently described in snakes.[86] Affected animals have variably lytic and proliferative lesions along the vertebral column that can result in morphologic defects in the column, including kyphosis, lordosis, scoliosis, or a combination of all three features, a condition known as rhoecosis (Fig. 36.1B). Proposed etiologies include trauma, infectious diseases,[65] nutritional deficiencies, neoplasia, immune-mediated diseases, and reduced physical exercise in captivity. Additional differentials include vertebral instability, degenerative joint disease, periosteal reactivity or inflammation, proliferative responses, and neoplasia. Variably extensive bone proliferation and remodeling with reduced or lost mobility may follow.[86] Under the same umbrella of proliferative spinal lesions is **spondyloarthropathy**, which, in one report, was described in 41 of 11,138 reptile skeletons, including those of several lizard families and crocodilians. The author described this as the major nontraumatic osseous disease in lizards, with an overrepresentation in varanids. *Salmonella* spp. and *Shigella* sp. have been linked to the disease in primates, however, no correlation between these bacterial agents and spondyloarthropathy has been established in lizards.[233] A novel mycoplasma (***Mycoplasma iguana* sp. nov**) has been described in the vertebrae of green iguanas with granulomatous osteomyelitis and hindlimb paralysis, meningitis, and demyelination of the spinal cord.[37,39,187]

Skin lesions and overall poor body condition are often associated with focal or diffuse **skin darkening/hyperpigmentation**. The basis of the process is unknown, but it is likely to involve changes affecting either the distribution or the arrangement or the number of chromatophores in the skin of the affected individual.[60]

A cystic (urinary bladder) **calculus** was recently reported in a laboratory housed green anole.[30]

Avascular necrosis of the tail is commonly observed finding in a wide range of reptiles. The etiology may vary, but trauma is often a primary differential. Iatrogenic damage following venipuncture is also considered as a possible causative event.[24]

Rostral trauma is relatively common in lizards housed in glass-sided enclosures (e.g., terrariums). Sequela can range from minute erosions to deep ulcerations with bone exposure. Secondary bacterial infection and associated complications, including stomatitis, periodontitis, osteomyelitis, calculus formation, abscesses, and gingival recession, especially in acrodont lizards, are reported.[24]

Compressive myelopathy secondary to subluxation of the cervical spine has been described in Komodo dragons. Trauma was considered as one of the possible etiologies. Affected cervical vertebrae included C1–C4. Histological changes included articular cartilage degeneration, necrosis, and axonal degeneration with mild multifocal hemorrhage, demyelination, diffuse Wallerian degeneration, and oligodendrocyte necrosis.[291] Crouch et al.[63] described bilateral symmetric degenerative myelopathy characterized by dorsolateral and ventrolateral myelin swelling, vacuolation, and degeneration of the spinal cord in two bearded dragons. The lesions were considered suggestive of either a nutritional or toxic etiology.

Neoplastic

Several tumors have been described in lizards (Figs. 36.10 and 36.11; eSlide 36.e5). In one study, of 1,297 Saurian submissions to a private diagnostic laboratory, 81 (6%) were diagnosed with neoplastic disease. The most common lizard neoplasms were in the hematopoietic system, liver, and skin, or were of musculoskeletal origin.[98,126] A second study identified neoplasia in 22.9% of the submissions. Bearded dragons and panther chameleons were the species most commonly affected; geckos were underrepresented. Testicular Sertoli cell tumor, testicular granulosa cell tumor, splenic hemangiosarcoma, gastric adenocarcinoma and basal cell carcinoma have also been reported.[160] The most common neoplasm described in *Heloderma* was recently reported to be renal adenocarcinoma.[177] See Table 36.e3 for a list of neoplasms that have been reported in lizards.

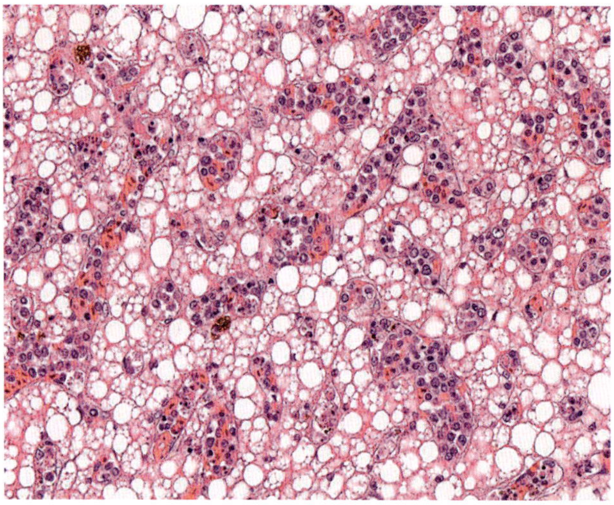

FIGURE 36.10 Leukemia in a bearded dragon. The sinusoids of the liver are filled and expanded with a monomorphic population of pleomorphic mononuclear cells with large nuclei and scant cytoplasm. *(Photo Courtesy of the University of Bern, Centre for Fish and Wildlife Health-FIWI.)*

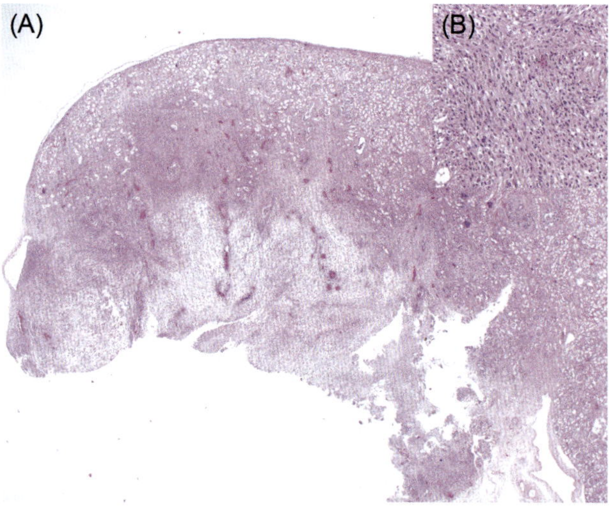

FIGURE 36.11 Renal sarcoma in a lizard. (A) The normal renal parenchyma is expanded and replaced by an infiltrative, densely cellular mass composed of streams and bundles of spindloid cells embedded in scant eosinophilic matrix. (B) The most affected portion of the kidney (medulla) shows feature of a myxoid-like component (see eSlide 36.e5). *(Photo Courtesy of the University of Bern, Institute of Animal Pathology, ITPA.)*

INFECTIOUS DISEASES

DNA Viruses

Adenoviruses are large, nonenveloped, double-stranded DNA viruses belonging to the family *Adenoviridae*. Five genera are currently recognized in the family, including *Aviadenovirus*, *Mastadenovirus*, *Atadenovirus*, *Ichtadenovirus*, and *Siadenovirus*. A new genus tentatively named *Testadenovirus* has also been recently proposed in testudinoid hosts.[76] The large majority of reptilian adenoviruses are included in the genus *Atadenovirus*. Adenoviruses have been documented in a number of lizards including eastern bearded dragon,[151] inland bearded dragon,[75,142,156,196,242,281] Rankin's dragon lizard,[90] savannah monitor,[141] central netted dragon, western bearded dragon,[131] green striped tree dragon, green anole, Jackson's chameleon,[21] brown anolis,[222] *Anolis* spp.,[17] common agama,[22] Gila monster, beaded lizard, emerald monitor,[205,281] tropical girdled lizard,[20] *Lacertoidea,*[263] leopard gecko, fat-tail gecko, blue-tongued skink, Tokay gecko, and mountain chameleon.[156,157,281] The infection is overrepresented in bearded dragons and a single adenovirus (**bearded dragon adenovirus 1—*Lizard atadenovirus B***) is probably responsible for most of the infections in these lizards both in Europe and in the United States.[21,206,281] Clinical signs vary from none to wasting, anorexia, and central nervous system (CNS) signs including circling and head-tilt. Gross findings are frequently unremarkable.[131] Histologically, prominent basophilic intranuclear inclusions, most often within the hepatocytes but also in bile duct epithelial cells, enterocytes, endothelial cells, renal tubular epithelial cells, glomeruli, pancreatic acini, and oral mucous membranes, are a hallmark of disease (Fig. 36.12). The nature of the lesions is essentially necrotizing but it can also be proliferative. Necrotic lesions are commonly reported as being more severe in the liver, but other affected organs can include the heart,[141] kidney, pancreas,[196] and intestine.[156]

Recently, **Helodermatid adenovirus 2 (HeAdV2)**, the only reptile adenovirus currently officially classified by the ICTV, as ***Lizard atadenovirus A***,[181] was described in a captive central bearded dragon with heterophilic and

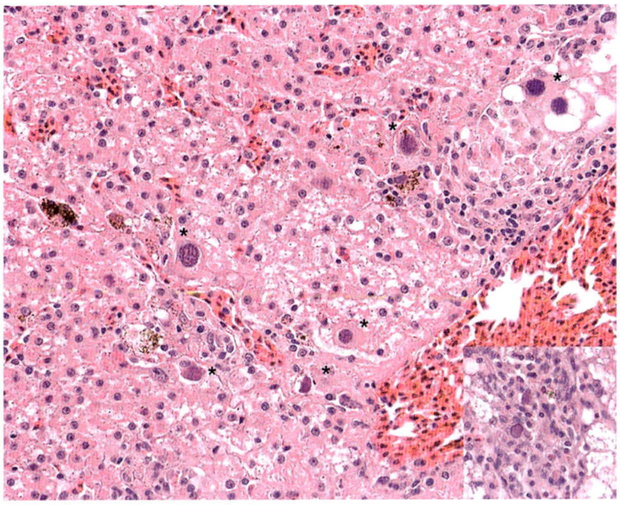

FIGURE 36.12 Adenoviral hepatitis in a bearded dragon. Several hepatocytes contain prominent, basophilic, intranuclear viral inclusions. *(Photo Courtesy of the University of Bern, Centre for Fish and Wildlife Health-FIWI.)*

lymphocytic interstitial pneumonia with occasional amphophilic intranuclear inclusions in pneumocytes and type II pneumocyte hyperplasia.[62] The dragon was coinfected with a novel mycoplasma, tentatively named ***Mycoplasma pogonae***, however, its clinical relevance was unclear. Helodermatid adenovirus 2 has also been reported in Gila monsters.[28] Occasionally, proliferative lesions have been reported affecting the tracheal and esophageal mucosa in Jackson's chameleon.[181] Bile duct proliferation is also often observed in the liver of the affected lizards. Coinfection with dependoparvoviruses (formerly dependovirus),[142,156] ranavirus and invertebrate iridovirus,[27] coccidians (*Isospora* sp.; *I. amphiboluri*),[131,156] and nematodes[157] has been reported. Adenovirus infection is not always associated with disease.[20,159]

Iridoviruses are large, enveloped or nonenveloped, double-stranded DNA viruses in the family *Iridoviridae*. The family *Iridoviridae* comprises two subfamilies, the Alphairidovirinae and Betairidovirinae. The former infect vertebrate poikilotherms and the latter infect insects and other invertebrates.[181] The *Alphairidovirinae* include the genera: *Ranavirus*, *Megalocytivirus*, and *Lymphocystisvirus*. *Hemocytivirus* is a newly proposed genus. *Betairidovirinae* includes the genera *Iridovirus*, *Chloridovirus*, and *Decapodairidovirus*. The genus *Ranavirus* includes viruses that infect amphibians, fish, and reptiles (vertebrate poikilotherms). **Ranavirus** infections are currently WOAH (World Organization for Animal Health, the former OIE) reportable in amphibians but not reptiles. Susceptible lizards include brown anoles, Asian grass lizards, green anoles, green iguanas and central bearded dragons, green striped tree dragons, and leaf-tailed geckos. The hallmark of infection is the presence of intracytoplasmic, basophilic inclusions that are associated with necrosis and commonly affect the skin (necrotizing, ulcerative, and hyperkeratotic dermatitis). However, inclusions may be elusive in affected tissues. Endoparasitism, myositis, osteomyelitis, hemorrhage and edema of the gastrointestinal tract, granulomatous dermatitis, sinusitis, muscle necrosis, vacuolar tubulonephrosis, and hyperemia with hemorrhage and necrosis in the liver have been reported in affected lizards. Adrenal tissue and bone marrow necrosis was reported in a fatal ranavirus outbreak in a group of Meller's chameleons.[217] A transmission study carried out in juvenile eastern water dragons inoculated with Bohle iridovirus (genus Ranavirus) revealed that the progression of the viral infection is first detected in the spleen, followed by the liver. Later affected organs include kidney, pancreas, oral mucosa, and skin.[174] Ranavirus infection in lizards has often been reported in association with infection of other microorganisms including other viruses (invertebrate iridovirus, reovirus, and adenovirus), bacteria, and/or fungus. Consequently, it is not clear if all the lesions described in the literature are caused primarily or solely by ranavirus. Differential diagnosis of the observed skin lesions should include suboptimal environmental factors, infection with *Nannizziopsis* sp. (formerly *Chrysosporium*) secondary gram-negative bacteria, *Dermatophilus* sp., invertebrate iridovirus, or parasites, and trauma.[27,179,260]

Invertebrate iridoviruses have also been isolated in the lung, liver, and intestine of two bearded dragons and a four-horned chameleon, and from the skin of a frill-necked lizard. One of the bearded dragons had pneumonia and the frill-necked lizard had pox-like skin lesions.[152] Invertebrate iridoviruses have also been described in a green iguana and a spiny-tailed lizard with hyperkeratosis.[27]

In addition to the five genera described earlier, there is a group of **unclassified iridoviruses** (proposed genus *Hemocytivirus*). This group includes viruses that were originally misidentified as intraerythrocytic parasites (*Pirhemocytozoon*, *Toddia*) and later named "lizard erythrocytic viruses" and "snake erythrocytic viruses."[181,268] These are now referred to as "reptile erythrocytic viruses."[113] Infections with **intraerythrocytic iridoviruses** have been reported in several species of lizards including *Gekkonidae*, *Phyllodactylidae*, *Scincidae*, *Cordylidae*, *Lacertidae*, *Varanidae*, *Iguanidae*, *Phrynosomatidae*, *Agamidae*, *Chamaleonidae* and more recently in captive central bearded dragons[113] and in *Lacerta*.[68,69] Associated clinical signs are inconsistent, nonspecific and include anorexia, dehydration, and blepharospasm. Grossly, multifocal white plaques on the dorsal aspect of the tongue, pale mucous membranes, and weight loss are observed. Histologically, intracytoplasmic inclusions can be observed in up to 40%–50% of the mature and immature red blood cells.[113] Infection has been shown to be restricted to the erythrocytes

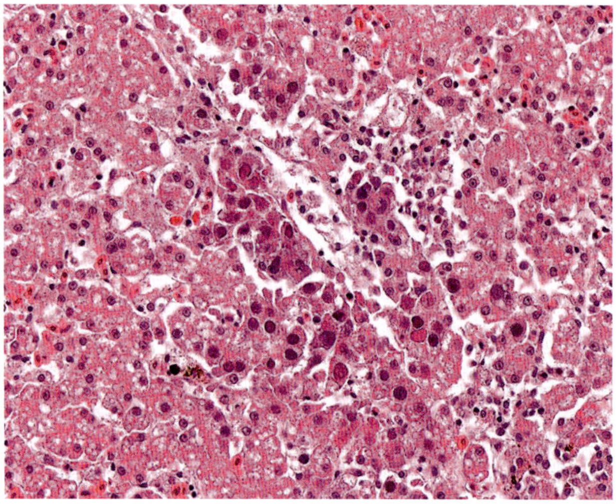

FIGURE 36.13 Herpesviral hepatitis in a common agama. A cluster of hepatocytes contains prominent intranuclear, darkly eosinophilic to basophilic inclusions. Associated hepatocellular necrosis is also seen. *(Photo Courtesy of the University of Bern, Centre for Fish and Wildlife Health-FIWI.)*

and occurrence of disease appears to be temperature dependent.[69]

Herpesviruses are large, enveloped, double-stranded DNA viruses. All of the partially characterized reptilian herpesviruses cluster within the subfamily of the *Alphaherpesvirinae* in the family *Orthoherpesviridae*. Lizard herpesviruses are associated with lesions in the skin, oral cavity, or visceral organs of the affected host. The hallmark of herpesvirus-associated infection is the presence of intranuclear eosinophilic inclusions (Fig. 36.13) and associated necrotizing and/or proliferative lesions. Herpesvirus-associated papillomas have been repeatedly observed in the genus *Lacerta* and occasionally in other lizards.[111,126,167,225] Other viruses, including reovirus and members of the former family *Papovaviridae,*[61] have been detected in similar lesions with or without detectable coinfecting herpesvirus particles.[225,273] Papillomas in male European green lizards cluster around the base of the head. In females, they are most common in the caudal aspect of the dorsum near the base of the tail. Histologically, in green lizards, there is epithelial hyperplasia with intralesional, enlarged nuclei with marginalization of the chromatin, and intranuclear inclusions were observed. The presence of viral particles consistent with herpesvirus, reovirus, and papovavirus was confirmed by transmission electron microscopy (TEM).[225] Papillomas that were TEM negative but PCR and DNA sequence positive for herpesvirus have also been described.[167]

Herpesvirus with or without notable lesions has been repeatedly reported in iguanas. Iguanid herpesvirus 1, the first herpesvirus isolated from a green iguana, was obtained from an individual with no detectable lesions and experimental inoculation in other juvenile lizards was not associated with a consistent lesional pattern.[55,181] In another report, liver necrosis was documented in a male green iguana that died 5 days after being transferred to a room with a temperature of 3–5°C cooler than the one where it was previously housed. Intralesional intranuclear eosinophilic inclusions were observed with light microscopy and viral particles consistent with herpesvirus were seen through electron microscopy.[286] In other reports, an uncharacterized herpesvirus was associated with multisystemic granulomatous inflammation in iguanas,[91] hepatic and splenic necrosis with intranuclear eosinophilic inclusions and mononuclear interstitial myocarditis and myositis was reported in a San Esteban chuckwalla (Iguanid herpesvirus 2-associated),[284] and liver necrosis with none or pinpoint white foci in the liver, hepatic and splenic necrosis, and exudative pneumonia with intranuclear amphophilic inclusions was seen in red-headed (common) agamas.[279] A herpesvirus closely related to Varanid herpesvirus 2 and tentatively named Varanid herpesvirus 3 was detected in monitor lizards with hepatitis, enteritis with coagulative necrosis, and intralesional intranuclear eosinophilic inclusions.[130] Also described are herpesvirus-associated oral lesions, including recurrent stomatitis and cheilitis in plated lizards (Gerrhosaurus herpesvirus 1–3)[282]; proliferative stomatitis, gingivitis, and rare squamous cell carcinoma in green tree monitors (Varanid herpesvirus 1-associated)[280]; and a herpesvirus associated odontogenic tumor in a banded Gila monster (Helodermatid herpesvirus 1).[102] The most recently described lizard herpesvirus was named Chamaeleonid herpesvirus 1. It was associated with cheilitis in panther chameleons.[121] The affected chameleons showed severe cheilitis and histologically there was epidermal hyperplasia of the epidermis with hyperkeratosis. Basophilic intranuclear inclusions were seen in the affected epidermis along with spongiosis and mild heterophilic exocytosis. Phylogenetic analysis revealed the closest relatedness with Varanid herpesvirus 2.

Recently, two DNA viruses, specifically a **circovirus** and a **parvovirus**, were detected in a colony of central bearded dragons tentatively named **bearded dragon circovirus (BDCV)** and **bearded dragon chaphamaparvovirus (BDchPV)**, respectively.[48] Affected lizards showed proliferative lung lesions. More specifically, the epithelium of the interfaveolar septa underwent proliferation and metaplasia ranging from columnar to pseudostratified columnar to ciliate. Basophilic intranuclear and eosinophilic intracytoplasmic inclusions were observed in some of the affected lizards. Rare binucleated cells and syncytia were also observed. A novel **poxvirus** was recently documented in a crocodile tegu.[250] The captive lizard showed obvious nodular skin lesions and weight loss. Following euthanasia, gross nodular lesions extended over the entire dorsal aspect of the animal but did not reach the underlying muscle. Histologically, there was extensive epidermal hyperplasia with ballooning degeneration and eosinophilic intracytoplasmic inclusions. Phylogenetic analysis revealed a close relatedness with avipoxviruses.[250]

RNA Viruses

Paramyxoviruses are enveloped, single-stranded, negative-sense RNA viruses. All the partially characterized lizard paramyxoviruses are included in the genus *Ferlavirus* in the family *Paramyxoviridae*. The genus Ferlavirus includes the genogroups A, B, and C and a distinct "Tortoise" lineage.[132] Lizards, together with snakes, are the reptilian groups more frequently affected by paramyxoviruses. Infection in lizards is rarely associated with disease.[132] The only reported paramyxovirus-associated diseases in lizards occurred in Caiman lizards[144] and a water dragon.[36] Infected caiman lizards were either found dead or were observed anorexic before death. Grossly, yellow/gray mucoid fluid and yellow flocculent material were present in the central airway of the lungs. Microscopically, proliferative heterophilic and histiocytic interstitial pneumonia was present.[181] Diffuse, marked hyperplasia and hypertrophy of the respiratory epithelial cells lining the faveoli (type II pneumocyte hyperplasia) with syncytial cells was observed. Occasional intracytoplasmic, 1–2 µm eosinophilic inclusions were noted in epithelial cells.[144] Similarly, a proliferative interstitial pneumonia was also observed in a water dragon.[36] More recently, pneumonia in a green iguana infected with paramyxovirus was reported but no gross or histological description of the lesions was provided.[182] No lesions or diseases have been reported in other lizards that have either been infected or exposed to paramyxoviruses including wild-caught Mexican lizards,[180] false tegu,[5] wild iguanas, a monitor lizard,[110] rhinoceros iguanas, prehensile-tailed skink, blotched blue-tongued skink, and beaded lizard.[169]

Rhabdoviruses are single-stranded, negative-sense, enveloped RNA viruses within the order *Mononegavirales*. Rhabdoviruses of reptiles have primarily been reported in teiid lizards from South America, more frequently in the giant ameiva. However, Australian skinks and geckos can also be infected.[283] Five rhabdoviruses from South American lizards (Marco, Chaco, Timbo, and Sena Madureira) and two from Australia (Almpiwar and Charleville) have been reported. A novel rhabdovirus was recently described in a caiman lizard (another South American lizard) with pale, basophilic intracytoplasmic inclusions in Wright-Giemsa stained blood smears. Sequencing identified the presence of a previously undescribed rhabdovirus tentatively named **Caiman lizard virus.**[283] Further studies are needed to determine the ecology and clinical significance of these viruses.

Reoviruses have been isolated and/or detected from/in several lizards including green,[31,32] Baker's spiny-tailed, and black spiny-tailed iguanas,[110] Knob-scaled lizards (serologically),[180] spiny-tailed lizards,[77] and green lizards.[225] Descriptions of lesions are not available, with the exception of associated papillomas in green lizards. Reovirus coinfections have been reported in brown anoles with concurrent ranavirus[260] (see also iridovirus discussed earlier) and juvenile veiled chameleons with concurrent **Serpentovirus (Nidovirus)** infection. The latter exhibited respiratory signs, including wheezing and vertical head tilting with gasping, increased mucus in the oral cavity, anorexia, and reduced water intake. All affected chameleons died within 0.5–1.5 months after their arrival at a research facility.[129] Similarly, respiratory disease was observed in a shingleback lizard infected with a nidovirus.[198]

Bacteria

Bacteria-associated diseases have been extensively documented in reptiles.[214] Characteristic lesions include conjunctivitis, abscesses, granulomas (Fig. 36.14), stomatitis, pneumonia, and other visceral diseases. Bacteremia is a frequent outcome and can result in death. An association between stomatitis and pneumonia is relatively common and it has been suggested that pneumonia might occur following inhalation/aspiration of exudate from the oral cavity.[214] A variety of gram-negative bacteria have been consistently isolated from these lesions. ***Pseudomonas*** spp. and ***Aeromonas*** spp. are common isolates.[214] However, the primary role of these bacteria in disease

FIGURE 36.14 Bacterial granulomatous inflammation in a green iguana. Infection and associated inflammation cause marked expansion of the periocular soft tissues of the lower lid of the left eye. *(Photo Courtesy of M. Bielli, Novara, Italy.)*

and death may historically have been overestimated, especially in the past, when bacterial sampling was rarely supported by histological examination. In more recent work, the presence of powerful virulence factors, including the type III secretion system (T3SS) of *Aeromonas* species, suggests that some strains of these bacteria might not necessarily always be secondary invaders. The information in the following paragraphs describes several of the most common and significant bacterial infections in lizards.[214]

Salmonella are a component of the normal bacterial flora in both captive and free-ranging wild reptiles.[214] *Salmonella* bacteria are commonly isolated from cloacal and fecal samples in healthy captive lizards,[213] and can be a source of infection for humans. In addition, several reports describe *Salmonella*-associated disease in reptiles.[212] These include systemic salmonellosis with associated lesions often occurring in debilitated individuals with concurrent infections.[259] *Salmonella* spp.-associated lesions in lizards, similar to what is described in other reptiles, may affect the skeletal system, although spinal lesions do not appear as common in Lacertilians. Infections may affect skin, visceral organs, joints, periarticular tissues, and bones. Lesions include, among others, ovarian granulomas, salpingitis, dermatitis, diphtheritic enteritis with diphtheronecrotic and heterophilic exudate, purulent osteomyelitis, and arthritis or periarthritis with periarticular abscesses and secondary involvement of the adjacent subcutaneous tissue.[153,166,202,212] In a recent study, the risk of developing disease was higher in carnivorous species and overall was lower in lizards compared to other taxa.[54]

Devriesea agamarum is a gram-positive bacillus that was described for the first time in 2008.[72,122,183] The bacterium appears to be part of the oral microbiota of bearded dragons where it may also be associated with disease. Disease has also been described in desert-dwelling lizards, although not all desert species show the same sensitivity to the bacterium.[122,123] Lizards of the genus *Uromastyx* appear to be particularly susceptible. Agama and collared lizards have been reported to be susceptible with bearded dragons serving as a reservoir.[122] Infection has also been described in the leopard gecko, jeweled curly-tailed lizard, frilled lizard, and Philippine sailfin lizard.[26,232] Associated disease includes proliferative and hyperkeratotic dermatitis (Fig. 36.15A) and cheilitis characterized by a crusty appearance and brown coloration of the affected skin[122,123]; septicemia can also develop. Limb swelling is described in an affected Saharan spiny-tailed lizard,[171] plumed basilisk, and Chinese (green) water dragons.[232] The water dragons and basilisk also developed multifocal granulomas, panniculitis, hepatitis, pneumonia, osteitis, myositis, skin wounds, and subcutaneous to coelomic fistulae (Fig. 36.15B).[232] A combined infection with *D. agamarum* and ***Nannizziopsis* (formerly *Chrysosporium*) *guarroi*** (see below) was recently reported in an inland bearded dragon.[247] A recent article describes the likely spillover of this bacterium in a free-ranging colony of the endangered Lesser Antillean iguana in the French Caribbean associated with severe skin disease.[124] Interestingly, the same bacterium was observed in clinically healthy anoles present on the same island, further supporting the hypothesis of a different sensitivity to this bacterial agent among lacertilian. More recently, cheilitis associated with *D. agamarum* infection was documented in a North African spiny-tailed lizard.[93]

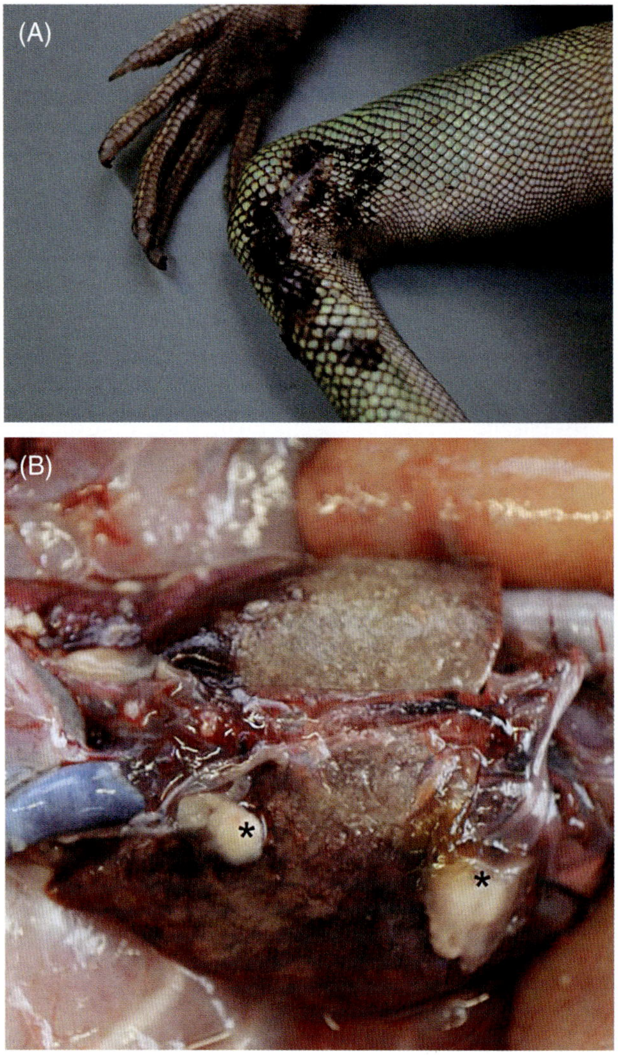

FIGURE 36.15 Bacterial dermatitis (A) and hepatitis (B) due to *Devriesea agamarum* infection in a water dragon. (A) Multifocal to coalescing nodules elevate the skin. The nodules are covered by dark brown to black crusts. (B) Necrotizing hepatitis is associated with multiple granulomas *(asterisk)* and a thick layer of fibrin that partially covers the liver surface. *(Parts A, B: Photos Courtesy of the University of Bern, Institute of Animal Pathology, ITPA.)*

Dermatophilus sp. is a gram-positive bacterium with zoonotic potential. It is associated with orthokeratotic and parakeratotic hyperkeratosis in several animal species

(A)

(B)

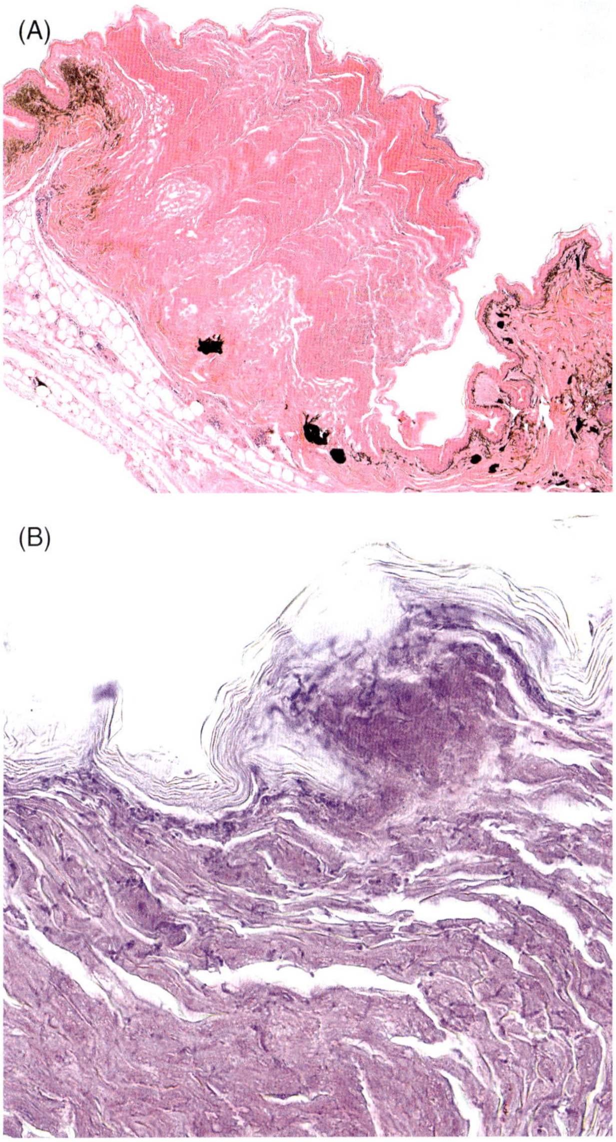

FIGURE 36.16 Dermatophilosis in the skin of a chameleon. (A) Exuberant hyperkeratosis is associated with large numbers of intracorneal microorganisms. (B) The bacteria have a characteristic arrangement in segmented rows. PAS. *(Parts A, B: Photos Courtesy of the University of Bern, Centre for Fish and Wildlife Health-FIWI.)*

including reptiles (Fig. 36.16A). The bacterium has a unique, characteristic growth, and morphologic pattern that resembles "railroad-tracks" in which the coccoid bacteria are arranged in parallel, serpiginous lines (Fig. 36.16B). Dermatophilosis has been observed in a variety of lizards including bearded dragons,[195,253] marbled lizards,[15] bush anoles, collared lizards, Senegal chameleons, green iguanas, savannah monitors, and panther chameleons.[136,137,200,235] The characteristic lesions consist essentially of (often) markedly proliferative/hyperkeratotic dermatitis. Development of subcutaneous nodules in affected lizards has also been reported.[51] An outbreak of dermatophilosis has been reported in captive bearded dragons that had concurrent ranavirus infection.[265] The affected lizards developed multifocal to coalescing, 1–10mm in diameter, crusty nodular lesions scattered mainly over the ventral aspect of the body including the head, tail, and the limbs. Histologically, lesions were ulcerative and necrotizing with epidermal hyperplasia, parakeratosis, and granulation tissue. Large numbers of intralesional, gram-positive bacteria consistent with *Dermatophilus* sp. and positive with anti-*D. congolensis* antibody by immunohistochemistry were observed. Molecular diagnostics confirmed the bacteria were consistent with ***Austwickia chelonae*** (formerly ***Dermatophilus chelonae***).

Leptospira species are significant infectious agents in mammals and they are zoonotic. Several reports describe the presence of antibodies against this bacteria in a number of reptiles. Recently, an investigation carried out on wildlife in Nigeria revealed the presence of kidney lesions in two Nile monitors infected with *Leptospira* spp. The monitors had interstitial nephritis and tubular nephrosis together with interstitial fibrosis and tubular casts. Warthin Starry stain revealed the presence of bacteria in the renal tubuli of the affected animals.[6]

Mycobacteria are acid-fast positive bacilli that are divided into two main groups, tuberculous bacteria in the ***Mycobacterium tuberculosis*** complex and **nontuberculous *Mycobacterium* species**. Most mycobacteria isolated from reptiles belong to the nontuberculous *Mycobacterium* group and cluster in Runyon groups I and IV.[214] Mycobacteria reported in reptiles include ***M. chelonae***, ***M. fortuitum***, ***M. intracellulare***, ***M. marinum***, ***M. phlei***, ***M. smegmatis***, ***M. kansasii***, ***M. thamnopheos***, ***M. haemophilum***, ***M. ulcerans***, ***M. confluentis***, ***M. haemophilum***, ***M. hiberniae***, ***M. neoaurum***, and ***M. nonchromogenicum.***[78,214,255] The most commonly reported mycobacterium in reptiles is *M. marinum*[214]; however, there are exceptions to this rule.[78] In general, reptilian mycobacteriosis is a chronic disease.[214] Lesions may develop anywhere in the body and range from extracellular proliferations of bacteria with associated necrosis and heterophilic to histiocytic inflammation in early infections and classic granuloma formation in chronic mycobacterial infections (Fig. 36.17A; eSlides 36.e6 and 36.e7) with intralesional acid-fast organisms (Fig. 36.17B). Systemic mycobacteriosis with mycobacterial myocarditis has been reported in a frilled lizard[197] and granulomatous osteomyelitis due to atypical mycobacteriosis has been documented in a bearded dragon.[158] Soldati et al.[255] reported that diagnosis of mycobacterial infection based on the association between organisms and histiocytic granulomas might not always be reliable, and that PCR is more sensitive than histopathology and immunohistochemistry.

Chlamydia are intracellular bacteria that are being more and more commonly detected both in amphibians

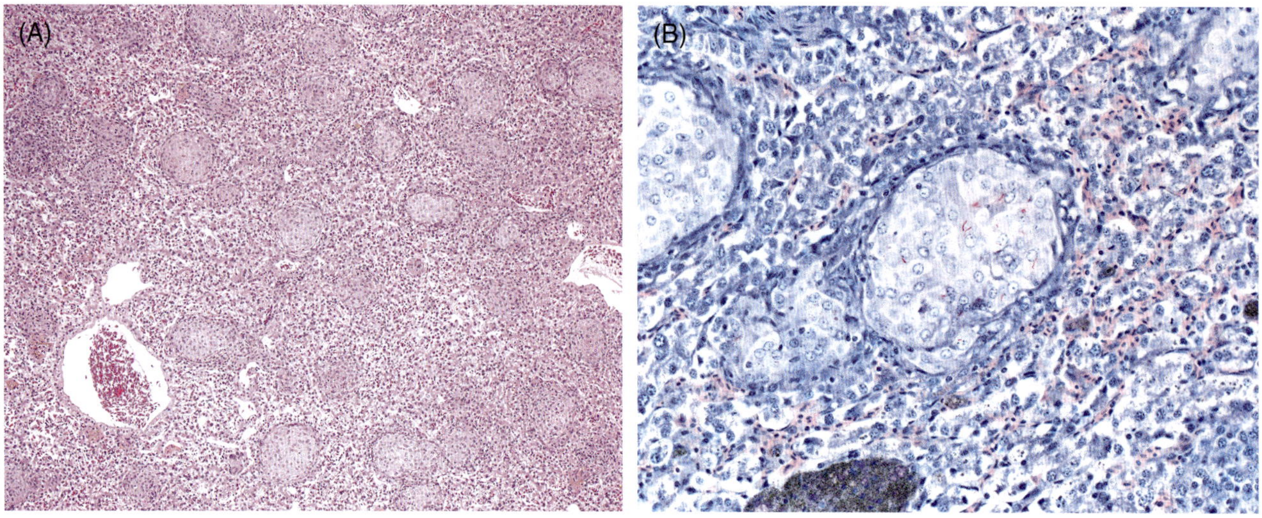

FIGURE 36.17 Mycobacteriosis, liver of a lizard and chameleon (B). (A) Scattered in the liver parenchyma are large numbers of histiocytic granulomas. (B) Histiocytes contain rare acid-fast positive bacilli. Ziehl-Neelsen (see eSlides 36.e6 and 36.e7). *(Parts A, B: Photos Courtesy of the University of Bern, Centre for Fish and Wildlife Health-FIWI.)*

and reptiles. Classic-associated lesions include histiocytic granulomas with intralesional organisms, which in some reptile species may cluster in such large numbers that they might be visible histologically. Less organized lesions can be seen in chelonians. Coinfection of the bacteria with other pathogens has been described. Cases include concurrent ***Chlamydia pneumoniae*** and poxvirus infection in a flap-necked chameleon[33,145] and *C. pneumonia* and ***C. felis*** in farmed green iguanas with diffuse necrosis of the intestinal mucosa.[33] A ***Chlamydia*-like organism** in lizards with granulomatous lesions has also been described.[255] Disease and death may occur in a collection at relatively distant points in time, making an association between individual cases more difficult. Detection of *Chlamydia* spp. in lizards is relatively uncommon or less frequent than in other reptilian taxa, such as snakes. However, a more thorough testing, especially when classic histiocytic granulomas are found in the tissue of the examined lizards, might provide a different epidemiological picture in the future.

Other notable bacterial infections in lizards include ***Burkholderia pseudomallei***, the agent of the zoonotic disease **melioidosis**, detected in abscesses in captive green iguanas in Europe and the United States[79,125,290]; ***Neisseria iguanae*** isolated from oral abscesses of healthy and affected, cage-mate, green and rhinoceros iguanas (Splendore-Hoeppli-like formations were observed in the lesions)[214]; and ***Serratia*** spp. in pyogenic arthritis in a tegu lizard, tumor-like lesions in knight anoles, and subcutaneous abscesses in a green and a northeastern spiny-tailed iguana.[214]

Helicobacter sp. was reported in Grand Cayman blue iguanas and introduced green iguanas. The clinical significance of infection in either species is not yet clear.[59]

Anaerobes have been reported relatively infrequently in reptiles and are even more rarely reported as the cause of disease. ***Clostridium difficile***, ***C. innocuum***, ***C. perfringens***, and ***C. sordelli*** were frequently observed in bacterial cultures from clinically healthy captive monitor lizards.[214] ***Clostridium perfringens***-associated bacteremia was reported in a rough-necked monitor with spondylosis.[65]

Fungi

Mycoses in reptiles include superficial and deep/systemic infections. Infections can also transition from superficial to deep/systemic. A number of fungal organisms belonging to the order ***Onygenales*** (genera ***Nannizziopsis***, ***Paranannizziopsis***, and ***Ophydiomyces***) have been shown to be clinically relevant for several reptilian groups including lizards. Those more frequently detected in lizards include ***N. dermatitidis***, ***N. guarroi***, ***N. chlamydospora***, ***N. barbatae***, ***N. draconii***, ***N. arthrosporioides***, and ***N. pluriseptata***. Other members of the genus include ***N. vriesii***, ***N. hominis***, ***N. obscura***, ***N. infrequens*** and ***N. crocodile***.[210,211,251] These fungi are present on the skin of clinically healthy reptiles less than 1% of the time.[209] *N. dermatitidis* has been shown to be a primary pathogen, with the fulfillment of Koch's postulates in veiled chameleons.[207] This same fungus has been reported in Day geckos and Parson's, carpet, and Jackson's chameleons.[210] Infections generally cause superficial skin lesions. Bullae or small cutaneous vesicles may occur in the early stages of the disease and their rupture is followed by formation of serocellular crusts and hyperpigmentation. Epidermal hyperplasia and hyperkeratosis with necrosis and sloughing of affected skin followed by exposure of the underlying dermis are additional common findings (Fig. 36.18A). Granulomatous and nodular dermatitis, cellulitis, and myositis, along with deep granulomas in visceral organs and intralesional fungal hyphae, are also reported[207,208,248,272]; granulomatous inflammation appears to be particularly prominent in bearded dragons.[41]

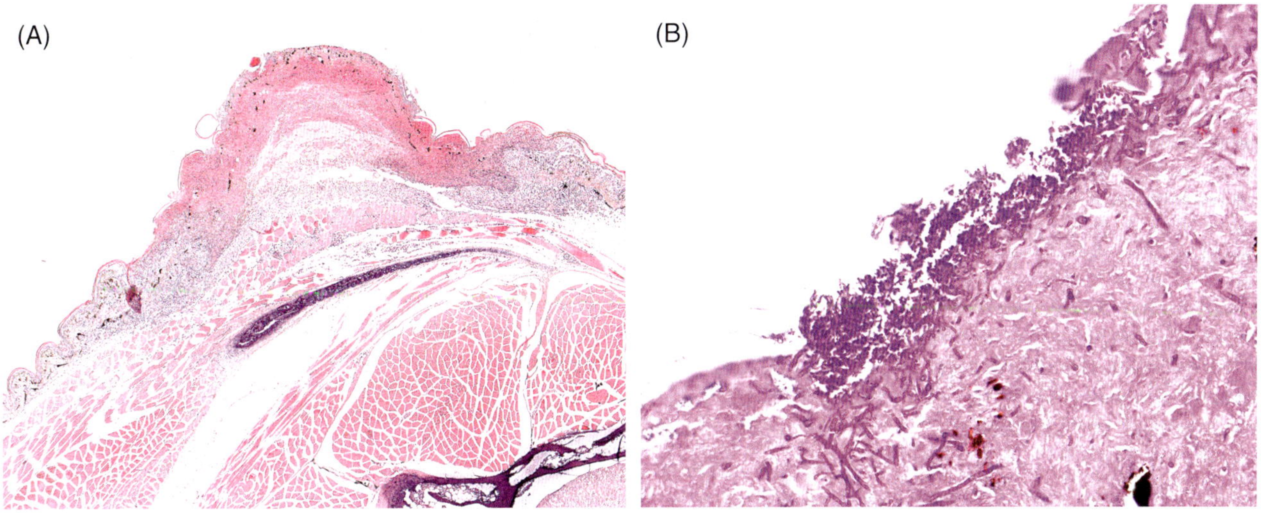

FIGURE 36.18 *Nannizziopsis* spp. dermatitis and myositis in a bearded dragon. (A) The epidermis is focally thickened, necrotic, and elevated by underlying inflammation and edema. Inflammation separates the deep dermis from the underlying skeletal muscle, which is also necrotic. (B) A dense mat of fungal hyphae covers the surface of the skin and extends into subjacent deeper tissues. Typical arthroconidia consistent with *Nannizziopsis* spp. are visible. PAS. *(Parts A, B: Photos Courtesy of the University of Bern, Centre for Fish and Wildlife Health-FIWI.)*

Conjunctivitis and rhinitis are less commonly described.[148] Histologically, the fungi are 2–4 μm in diameter, hyaline, generally parallel-walled, septate hyphae with irregular branching and arthroconidia (Fig. 36.18B). Infection in bearded dragons, which was originally named **yellow fungus disease (YFD)**,[35,209,210] is most commonly associated with ***N. guarroi.***[211,251] *N. guarroi* has also been identified in infections in iguanas, agamas, giant girdled lizards, and central bearded dragons.[1–3,35,116,119,120,251] *N. dermatitidis* has been linked to infections in day geckos and chameleons,[207–210,272] and *N. barbatae* (previously *N. barbata*) has been identified in infected coastal bearded dragons,[148,251] Eastern water dragons, Tommy roundhead dragons, and Eastern blue tongue, shingleback and Centralian blue-tongue skinks.[219] *N. chlamydospora*[248] and *N. draconii* were recovered from inland bearded dragons with lesions typical of YFD; *N. pluriseptata* was detected in a southeastern five-lined skink and *N. arthrosporioides* was obtained from a captive water dragon.[211] A fungus taxonomically related to *N. vriesii* and *N. dermatitidis* was isolated from leopard geckos that were part of commercial breeding colony in Florida.[272] Among the fungi in the genus *Paranannizziopsis*, *P. australiensis*, was reported in captive tuataras and a costal bearded dragon in the same facility.[185] The infection was fatal in the bearded dragon, whereas the tuataras survived following antimycotic therapy.[211] PCR and DNA sequencing of the internal transcribed spacer region of the rRNA gene has been reported as confirmatory.[163]

Several reports describe **dermatophytoses** in lizards. These include systemic infection with ***Microsporum canis*** in a green iguana with a nodular, dark-pigmented skin lesion and multiple granulomas in the liver, lung, and cardiac apex[52]; ***Trichophyton mentagrophytes*** var. ***interdigitale*** in green iguanas with patchy areas of darkened skin that histologically were acanthotic, hyperkeratotic, and contained intralesional fungal hyphae[154]; ***Trichophyton*** sp. infection in a Tenerife lizard with ulcerative and heterophilic pustular dermatitis[201]; nuchal hematomas in green anoles infected with ***Trichosporon cutaneum***[208] and trichosporonosis in a *Uromastyx*.[209] Additionally, the ascomycete ***Chamaleomyces* sp.** (now ***Metarhizium* sp.**) is associated with fungal dermatitis, glossitis, systemic, and fatal infection in chameleons, and ***C. granulomatis*** and ***C. viridis*** have been associated with dermatitis and granulomatous lesions in the tongue and throat in veiled, panther, and carpet chameleons.[220,245,246] Fatal cutaneous and systemic mycoses associated with ***Lecanicillium*** spp. have been reported in captive Guthega skinks.[240] Affected skinks showed erythema and ulcerations of the skin with crusts and raised scales. The skin lesions were reported to occur predominantly on the ventral surface of the skin near the cloaca and the proximal ventral tail base. Histologically, full-thickness epidermal necrosis with hemorrhage, edema, fibrin, and intralesional 4–7 μm thick septate and periodically branching hyphae were observed. Fungal hyphae and necrotic foci extended into the skeletal muscle, and one individual showed evidence of systemic spread with hyphae infiltrating into the liver sinusoids.

Systemic **microsporidiosis** in bearded dragons can cause death in the absence of premonitory signs. Both intracytoplasmic and extracellular organisms are observed in hepatocytes in association with liver necrosis, in renal epithelial cells with cytoplasmic vacuolation, and in pulmonary and gastric mucosal epithelial cells, enterocytes, capillary endothelial cells, and ventricular ependymal cells in the brain. Granulomatous inflammation with intrahistiocytic organisms may cause massive expansion of the colonic wall and be present in the adrenal glands and ovaries. Organisms are 2 to 3 μm in diameter and Gram and acid-fast positive with a PAS-positive small polar granule. Affected lizards may have concurrent diseases including coccidiosis, adenovirus infection, pneumonia, or hepatobiliary disease.[140] Microsporidiosis has also been reported in a central knob-tailed gecko, an eastern water dragon. Multisystemic granulomatous inflammation associated with a previously undescribed member of the genus ***Encephalitozoon*** has also been described in inland bearded dragons; the organism shared strong sequence similarity but was distinct from ***E. cuniculi.***[226] Reports of microsporidiosis include *Encephalitozoon* sp. in an African skink, *Encephalitozoon lacertae* in the common wall lizard, and *Pleistophora* sp. in tuatara.[277] Bearded dragon microsporidia has recently been proposed as a novel species, named ***Encephalitozoon pogonae*** and this agent was documented in two juvenile central bearded dragons with granulomatous arteritis and fatal hemopericardium; one had restrictive epicarditis.[288] Both individuals had intralesional microsporidians. Paratesticular granulomas with intralesional organisms were also seen.[288] Microsporidiosis associated with *E. pogonae* has also been reported in an inland bearded dragon in Australia that developed large ovarian granulomas with intralesional microsporidia.[168]

Metazoa

A large number and variety of parasites have been either detected or isolated from reptiles. However, few publications describe associated pathology. Most clinically significant lizard pathogens are protozoa, whereas metazoan parasites, including ***Foleyella* sp.**, a filaroid metazoan commonly observed in the subcutis and visceral surfaces of chameleons (Fig. 36.19), appear to play a more marginal role.

FIGURE 36.19 Coelomic and visceral filariasis in a panther chameleon. Multiple slender, yellow, filarial nematodes (*Foleyella* sp.) are scattered over the surface of the muscle wall of the coelomic cavity.

Summarized in Table 36.e4 in the Digital Content are the main parasitic diseases in lizards.

Protozoa

Protozoa are among the most clinically relevant parasites reported in lizards. Some are nonhemoparasitic (amoeba, cryptosporidia, isospora, eimeria, microsporidia, and caryospora), while others are hemoparasitic (hemogregarines and plasmodium).

Amoebic parasitic disease has been reported occasionally in lizards. In most cases, infection is presumed due to ***Entamoeba invadens***; however, definitive species identification is rarely performed. Infection is generally associated with intestinal and liver disease that varies from insignificant to fatal (Figs. 36.20A, B; eSlide 36.e8).[277]

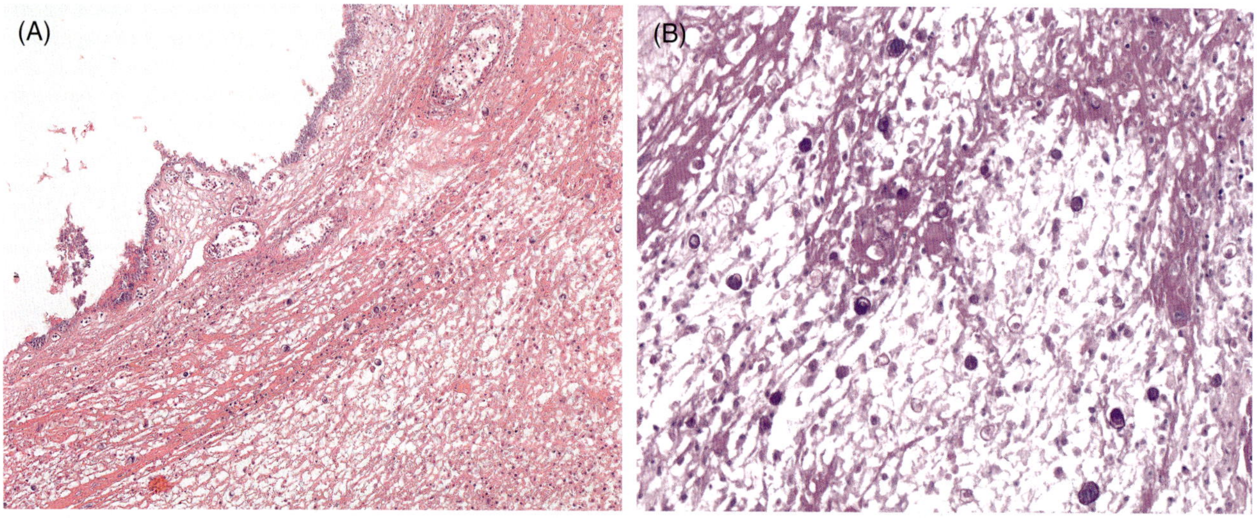

FIGURE 36.20 Amoebiasis in the intestines of a skink. (A) The intestine is ulcerated and necrotic and large numbers of basophilic bacterial colonies coat the ulcerated mucosa. A large amount of eosinophilic fibrillar material (fibrin) and numerous round-to-ovoid basophilic organisms are seen in the necrotic debris in the wall of the intestines. (B) Amoebic trophozoites are round with a spherical nucleus and granular endoplasm. PAS (see eSlide 36.e8). *(Parts A, B: Photos Courtesy of the University of Bern, Centre for Fish and Wildlife Health-FIWI.)*

Reports of **amoebiasis** have most often been in monitor lizards with hemorrhagic colitis, enteritis, and hepatitis[277] or multisystemic infection with ulcerated skin wounds and granulomatous myositis.[50] Amoebiasis has also been observed in blue-tongued skinks and (more rarely) green iguanas with systemic disease.

At least 57 species of reptiles have been reported to be susceptible to the infection with ***Cryptosporidium* sp.**[163,199] geckos appear to be overrepresented. Infection is typically associated with proliferative gastritis in snakes and the intestinal tract in lizards.[57,204,269] Two distinct cryptosporidial species, ***C. serpentis*** and ***C. varanii* (syn. *C. saurophilum*)**, have been associated with snakes and lizards, respectively.[163] However, this separation is not strict.[216,227] Cryptosporidia other than *C. serpentis* and *C. varanii* exist. For example, a *Cryptosporidium* species "lizard genotype"[26] and a species with 18S RNA sequence similarity to a *Cryptosporidium* species that infects cockatiels have been reported in lizards.[155]

Leopard geckos with cryptosporidiosis may present clinically with anorexia, lethargy, and diarrhea. Gross lesions include a marked reduction of the fat deposits and thickening and reddening of the intestinal wall that corresponds histologically to mucosal hyperplasia and mononuclear cell infiltration in the small intestine (Figs. 36.21A, B).[269] In green iguanas, lesions include cystitis and colitis with recurrent cloacal prolapse.[155] Other species in which intestinal infection is described include panther chameleons, savannah monitors, white tegus, freckled monitors, giant day geckos, and Schneider's skinks.[277] Gastric cryptosporidiosis with mucosal hyperplasia has been described in a free-ranging male frilled lizard,[203] a chameleon,[73] two ocellated lizards, and a savannah monitor.[277] *Cryptosporidium* spp. was recently documented to be the most common cause of enterocolitis in Gila monsters but not in beaded lizards[177] and was associated with enteritis, villous atrophy, and edema of smooth muscle of the intestinal wall.[161]

Cryptosporidium sp. infection has been associated with ear lesions in green iguanas. At least two distinct reports describe the development of **aural polyps** in affected lizards.[87,274] In both cases, affected iguanas developed aural polyps with protrusion of the tympanic membrane extension from the ear canal. Histologically, nests of cystic glands are surrounded by abundant fibrous connective tissue that is lined by hyperplastic cuboidal to pseudostratified columnar epithelium that has moderate to severe cryptosporidial colonization. Nonalimentary cryptosporidiosis has been reported in two green iguanas and a Parson's chameleon.[89] One of the affected iguanas and the chameleon had renal cryptosporidiosis, whereas the other iguana had cryptosporidial sialoadenitis.[89]

Isospora amphiboluri has been described in captive-bred colonies of inland[188] and coastal bearded dragons.[44] Infection is associated with mortality in neonatal and juvenile lizards. The infection begins in the duodenum and progresses to the colon.[163] A combined infection together with adenovirus and dependoparvovirus has also been reported in inland bearded dragons.[156] More recently, a new coccidian, ***Eimeria pogonae*** (genus *Choleoeimeria*) that infects the biliary and gallbladder epithelium, has been described in bearded dragons.[204,262] The gallbladder of affected lizards was markedly dilated with focal thickening and tan discoloration of the wall. Oocysts were present both in the intestinal contents and the biliary epithelium, where they were aggregated in dense clusters. Coccidian development occurs in the epithelium of the gallbladder and bile ducts. Epithelial metaplasia, hypertrophy, and cell degeneration were observed together with numerous heterophils in the subepithelial tissue.[262]

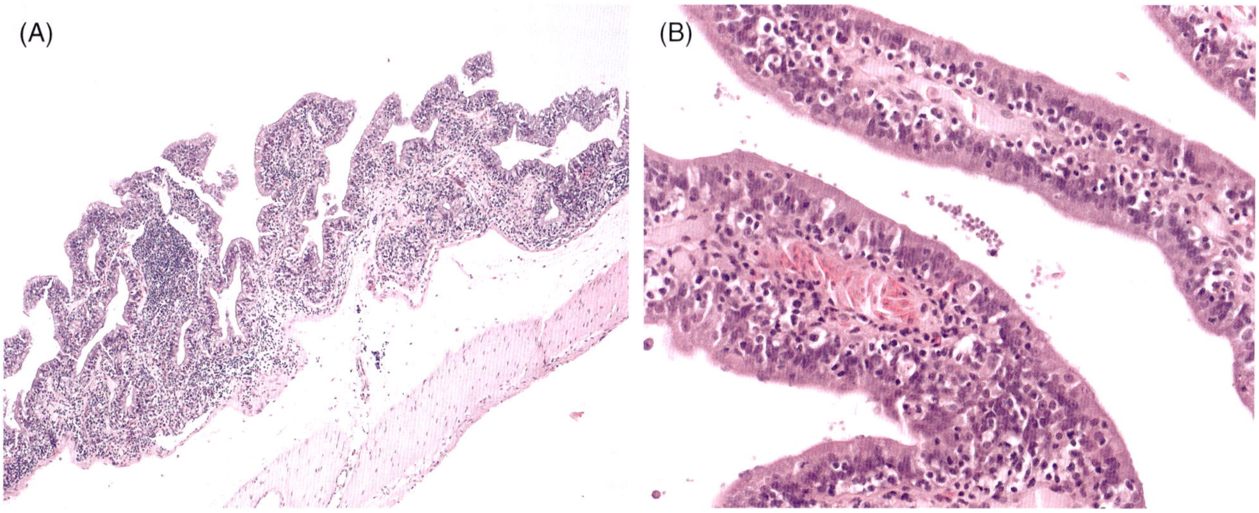

FIGURE 36.21 Intestinal cryptosporidiosis in a leopard gecko. (A) The intestinal villi are multifocally severely hyperplastic with branching, anastomosis, fusion, and blunting. (B) Large numbers of 2–3 μm in diameter, round, intracellular, extracytoplasmic protozoal organisms are present along the mucosal surface. *(Parts A, B: Photos Courtesy of the University of Bern, Centre for Fish and Wildlife Health-FIWI.)*

Sarcocystis-like organisms in Fischer's chameleon were associated with swollen skeletal muscles and epaxial muscle necrosis in a black-headed monitor. ***Caryospora*** sp. has been reported in a few lizards but no lesions were mentioned in the reports.[277]

Hemogregarines are rarely associated with lesions in reptiles.[277] However, a high parasite load has been suggested as a possible cause of reduced individual animal fitness in infected European green lizards.[194] Additionally, free-ranging lizards infected with hemogregarines have been shown to have a smaller home range.[34]

Experimental infection of Easter fenced lizards, Great Plains skinks, and crevice spiny lizards with **hepatozoon** caused severe lethargy and anorexia with leukocytosis and increased aspartate aminotransferase (AST) plasma activity. Random hepatocellular necrosis with heterophilic infiltration centered on ***Hepatozoon mocassini*** meronts was observed in all the lizards. Interstitial heterophilic pneumonia centered on meronts was also seen.[277,287] Great genetic diversity has been observed in *Hepatozoon* spp. infections in lizards in the Mediterranean.[277]

Lizards can be parasitized by apicomplexan parasites of the genera ***Plasmodium*** (erythrocytes), ***Fallisia*** (leukocytes and thrombocytes), and ***Saurocytozoon*** (lymphocytes).[277] Infections associated with variable levels of anemia have been reported in Japanese grass lizards and in green grass lizards infected with ***Plasmodium sasai,***[266,267] in Amazon lava lizards experimentally infected with ***P. tropiduri***, and in grass anoles naturally infected with ***P. colombiense.***[18] Infection of western fenced lizards with ***P. mexicanum*** was associated with anemia, decreased oxygen consumption, and running stamina.[238] Similar findings were also observed in the common agama.[239] Striped forest whiptail lizards infected with *Plasmodium* spp. had systemic intraerythrocytic life stages in the liver, lung, and heart. The liver of the infected lizards showed mild to moderate Kupffer cell and melanomacrophage hypertrophy/hyperplasia with sinusoid leukocytosis.[85]

Ectoparasites

Mites and **ticks** are the most common ectoparasites observed in lizards. Mites are ranked as the sixth most common etiology in lizard skin disease.[285] ***Ophyonissus* sp.** (*Macronissidae*) is the most common genus of mites infesting lizards. ***Ophyonissus natricis*** is probably the most common lizard and snake mite and it can be associated with severe infestations possibly leading to anemia.[277]

Parasites in free-ranging, side-botched lizards included the trombiculid ***Neotrombicula californica***, the pterygosomatid mite ***Gekobiella texana***, the macronyssid mite ***Ophyonissus natricis***, and the ixodid tick ***Ixodes pacificus***.

Necrosis of the epidermis with occasional serocellular crust formation may be seen at sites of mite attachment. Inflammation associated with mite and tick bites consists of chronic mixed inflammation composed of histiocytes, heterophils, and lymphocytes along with granuloma formation.[107]

Ticks, in particular those of the genus ***Amblyomma***, can be found parasitizing lizards, especially varanids, as well as iguanids and teiids. Some are species specific, whereas others have a broader host range.[277] Ticks of the genus ***Aponomma*** are also commonly detected in reptiles.[189]

The site of attachment of tick nymphs of ***Ixodes pacificus*** and ***I. ricinus*** was examined in the skin of a sand lizard. The rostrum of the ticks was surrounded by a fibrin cone and reactive collagen capsule that was associated with granulocytic and eosinophilic inflammation.[112] *I. scapularis* larvae and nymphs were found preferentially attached to the axillae/front leg of the five-lined skink.[101] Argasid ticks have also reported parasitizing lizards.[277]

ADDITIONAL RESOURCES

Barten, S.L., 2006a. Lizards. In: Mader, D.R. (Ed.), Reptile Medicine and Surgery. second ed. Elsevier, St Louis, MO, pp. 59–76, Chapter 6.

Jacobson, E.R., 2007. Overview of Reptile anatomy, physiology and histology. In: Jacobson, E.R. (Ed.), Infectious Diseases and Pathology of Reptiles: Color Atlas and Text. CRC Press, Boca Raton, pp. 1–130, Chapter 1.

Jacobson, E., R., 2007. Parasites and Parasitic Diseases of Reptiles. In: Jacobson, E.R. (Ed.), Infectious Diseases and Pathology of Reptiles: Color Atlas and Text. CRC Press, Boca Raton, pp. 571–665, Chapter 12. Parasites.

Jacobson, E.R., Lillywhite, H. B., Blackburn, D. G., 2021. Overview of Biology, anatomy and histology of Reptiles. In: Jacobson, E.R., Garner, M. M. (Eds.), Infectious Diseases and Pathology of Reptiles: Color Atlas and Text. 2nd edition CRC Press, Boca Raton, pp. 1–214, Chapter 1.

Marschang, R. E., Origgi F. C., Stenglein, M. D., Hyndman, T. M., Wellehan, J. F. X., Jacobson, E. R., 2021. Viruses and viral diseases of reptiles. In: Jacobson, E.R. Garner, M. M. (Eds.), Infectious Diseases and Pathology of Reptiles: Color Atlas and Text. 2nd edition CRC Press, Boca Raton, pp. 575–704, Chapter 10.

References

Full references for this chapter can be found online at https://doi.org/10.1016/B978-0-443-18437-6.00020-1.

Chapter 37

Serpentes

Robert J. Ossiboff

Aquatic, Amphibian, and Reptile Pathology Service, College of Veterinary Medicine, University of Florida, Gainesville, FL, United States

INTRODUCTION

All snakes are members of the suborder Serpentes, a monophyletic clade deeply embedded within the phylogeny of the Squamata (lizards). They are distinguished from other legless lizards by morphologic features of the skull and dentition that first appear in the fossil record during the Middle Jurassic–Lower Cretaceous periods. It is hypothesized that changes to the cranium preceded the loss of limbs, and that the first snakes were likely short-bodied lizards with limbs.[18]

Ophidian taxonomy, like that for much of herpetofauna, is fraught with disagreement and is in a near-constant state of flux. It is generally accepted that the basal division of extant species occurs between the Scolecophidia (blind and thread snakes) and the Alethinophidia (the "advanced" snakes). The Alethinophidia contains at least five superfamilies, including the Acrochordoidea (file snakes), the Uropeltoidea (pipe and shield-tailed snakes), the Pythonoidea (pythons and sunbeam snakes), the Booidea (boas), and the incredibly diverse Colubroidea (colubrids, elapids, viperids, and numerous others).[116,156]

There are now more than 4,000 identified species of snakes, an increase of over 13% since this chapter was completed for the inaugural edition of this text. Despite continued identification of novel ophidian species, substantial threats to snake biodiversity persist. Approximately 15% of all snake species are estimated to be threatened with extinction, and an additional 19.5% of species are data deficient.[26] The most highly threatened snakes include uropeltid (shield-tailed snakes) and tropidophiid (dwarf boas) species, though a variety of species inhabiting freshwater environments, tropical regions, and oceanic islands are also considered to be of great extinction risk.[12,26] While one study identified habitat destruction due to agricultural expansion as the most frequent threat to squamate reptiles, climate change, over collection, invasive species, and infectious diseases are all contributors to species declines.[26]

This chapter aims to introduce snake anatomy and histology (see Digital Content), highlight the most important diseases of wild and captive snakes, and provide important diagnostic features of these processes to aid pathologists and clinicians in diagnosing ophidian diseases. In addition, particular emphasis will be placed on ophidian infectious diseases as they are of paramount importance to the health of both captive and free-ranging snakes. As the popularity of keeping snakes and other reptiles as pets increases so does the widespread potential of disease spillover from snakes maintained in captivity to free-ranging snakes, and vice versa. Ophidian virology and mycology continue to be rapidly evolving areas of reptile medicine and pathology. Ophidian parasitic diseases are now also an area of increasing significance for both captive and free-ranging snakes, which in some instances is due to the introduction of invasive parasites. Pathogens from each of these arenas can have direct implications for conservation of free-ranging species, ex situ conservation, and individuals within zoological and captive collections, and as such it is important for veterinary pathologists and diagnosticians to be familiar with the major disease entities discussed herein.

NON-INFECTIOUS DISEASES

Nutritional

The most significant nutritional issue observed in captive snakes is **obesity**. In captivity, snakes are most often kept in small enclosures that offer limited opportunities for movement and they are fed calorically dense rodents at a frequency that is much higher than needed to meet their metabolic/physiologic needs. The end result is over conditioning. This is particularly true for either very large snake species (such as reticulated and Burmese pythons) or snakes in intense breeding facilities where animals may be very frequently "power-fed" larger meals to foster rapid growth. While the risk obesity poses to snakes as it relates to the development of other disease processes is not clear, it is likely that obesity has long-term deleterious effects on many aspects of animal health, including longevity.

Necrotizing steatitis, more commonly associated with piscivorous species such as crocodilians or birds, can also

Pathology of Wildlife and Zoo Animals. https://doi.org/10.1016/B978-0-443-18437-6.00027-4

occur in snakes maintained on a fish diet. In tentacled snakes, characteristic histologic lesions of fat necrosis and granulomatous inflammation were observed secondary to the long-term feeding of frozen, thawed lesser sand eels low in vitamin E and high in peroxides.[111]

The accumulation of cytoplasmic iron is common in snake hepatocytes (**hemosiderosis**; Fig. 37.e7), renal tubular epithelial cells, and occasionally enterocytes of both captive and free-ranging snakes. This change is most often secondary to advanced age, chronic disease states, or in the case of the liver, chronic negative energy balance and hepatocellular atrophy. Issues with iron overload are not recognized in snakes.

Other nutritional issues have not been identified as a major source of disease in either captive or free-ranging snakes. This is likely a testament to the relatively simple nutritional requirements of most snakes and the evolution of an incredibly efficient gastrointestinal system. While exposure to ultraviolet light (UV) is essential for many other reptiles, including the lacertilian squamates (see Lacertilia: Chapter 36), the majority of captive snakes are able to absorb appropriate calcium from their diet in the absence of natural or artificial UV exposure. Moreover, studies of ultraviolet exposure in captive snakes and resulting alterations in blood plasma vitamin D_3 often show either conflicting or clinically ambiguous results.[1,15,55] However, other potential benefits to ultraviolet light exposure, particularly as it relates to behavior, are poorly characterized and there may still be benefit to UV supplementation in captive snakes.

Metabolic

Gout, the deposition of uric acid in tissues, is commonly encountered in the kidney and viscera of snakes. The kidneys can be swollen and contain multifocal to miliary, white-tan, pinpoint nodules (Fig. 37.1A; eSlide 37.e1). On gross examination, care must be taken to differentiate tubular urate stasis and the sexual segment of the nephron in males, seen as radiating white or tan streaking, from true renal gout. Impression smears or squash preps of the kidney can be used to differentiate urate stasis and renal gout. In true renal gout, characteristic radiating, monosodium urate tophi will be present (Fig. 37.1B; eSlide 37.e1); in instances of urate stasis, round to amorphous urates will be seen. Histologically, radiating, acicular clefts (gout tophi) that often appear as empty spaces (the crystals dissolve during routine histologic processing) are surrounded by epithelioid macrophages, multinucleated giant cells, and heterophils (Fig. 37.1C; eSlide 37.e1). In snakes, secondary gout associated with dehydration is most common and uric acid deposition in the kidneys typically precedes that of other

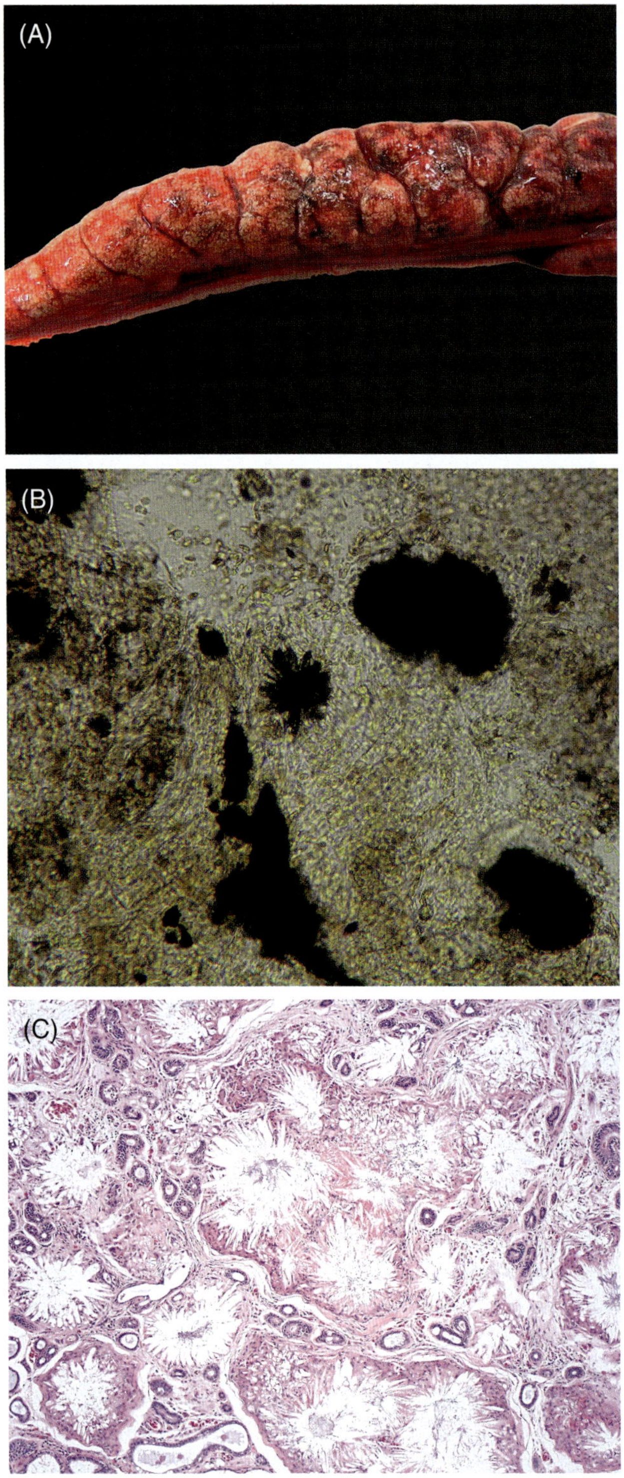

FIGURE 37.1 Renal gout in a corn snake. (A) Tophi composed of crystallized uric acid and associated inflammation are grossly visible as white, pinpoint foci that disrupt the normal parenchyma. (B) In renal cytologic squash preps, gout can be confirmed by the presence of radiating crystallized aggregates of uric acid. (C) Histologically, granulomatous inflammation surrounds empty, radiating, acicular clefts. The precipitated uric acid dissolves during routine processing (see eSlide 37.e1).

tissues (commonly the pericardial sac, pleura, and the serosa of the liver and other coelomic organs). Articular (e.g., intervertebral) accumulations of uric acid are not as commonly seen in snakes, but do occur and may be underreported due to the additional time and steps required for histologic processing of the spinal column.

In captive snakes, the presence of macrovesicular, hepatocellular lipid accumulation is a common, and most often incidental finding. Often times these are likely a result of over conditioning as discussed above (see Obesity). However, degenerative hepatic changes associated with lipid accumulation, including intra- and intercellular lipid vacuolar fusion, nuclear peripheralization, and ultimately intrahepatic cholestasis and/or hepatocellular necrosis can be seen in snakes. This is particularly true for over conditioned snakes following an extended period of anorexia. Despite this, changes associated with hepatocellular lipid accumulation very infrequently manifest as clinical hepatic lipidosis. Histologic interpretation of **hepatocellular lipid vacuolation** can be challenging and diagnosis and conclusions should align with other historical, clinical, and medical information.

Congenital/Genetic

Certain commonly kept snake species, such as ball pythons and corn snakes, are intensively and selectively bred for the production of unique color morphs to supply the private herpetofaunal market. But as can occur with selective breeding of any captive animal, selection of deleterious alleles associated with clinical disease has also occurred. For example, the **"spider" morph** of ball python has a highly desirable unique pattern and coloration. However, animals produced of this lineage often display neurological deficits, such as abnormal head movement/trembling that leads to feeding difficulties and is termed **wobble syndrome**. Studies using detailed, micro-computed tomography and magnetic resonance imaging identified malformations of the sacculus, ampullae, and semicircular canals of spider morph ball pythons not present in wild-type ball pythons. These changes are likely responsible for the abnormal neurologic behavior exhibited in "spider" morph ball pythons.[128,137] Complete characterization of the genetic underpinnings of selected traits in snakes is only beginning, and much will likely be learned in future studies.[16]

Though most often an incidental finding, hepatic changes indicative of embryological **ductal plate malformations** occur in captive individuals of a number of species including colubrids, vipers, and boids. The malformations consist of well-demarcated nodules of irregular bile ducts surrounded by fibrous connective tissue. In some snakes, the malformations can progress to cystic liver disease.[110]

Age-Related/Degenerative

Chronic vertebral osteopathy, chronic proliferative, degenerative, and/or inflammatory lesions in the vertebral bodies, intervertebral joints, and ribs is a common finding in snakes (Fig. 37.2). Lesions can include a spectrum of changes, which include bony remodeling, proliferative new bone and fibrocartilage formation, sclerosis, vertebral and costal spondylosis and ankylosis, degeneration of the articular cartilage, osteomyelitis, osteoarthritis, osteonecrosis, and pathologic fractures.[40,109] Chronic, noninflammatory lesions can include sclerotic changes of both cortical and trabecular bone that are characterized by widening and confluence of trabeculae by disorganized, irregular patches of woven and lamellar bone with mosaic reversal lines; the latter of which can be quite dramatic. Fibrocartilaginous proliferation can also be a prominent feature. An increase in osteoclastic activity is variably reported. In other cases, chronic noninflammatory, degenerative changes may be admixed with small foci of chronic inflammation. This suggests that many cases may fall on a spectrum of disease progression from initial inflammatory lesions to chronic, aseptic, degenerative, and proliferative changes. Some comparisons have been made between chronic vertebral osteopathy in snakes and osteitis deformans (Paget disease) in humans.[115] However, features of vertebral osteopathy in snakes do not follow the same hypothesized pathologic progression as the human disease and the designation of reptile lesions as osteitis deformans should be avoided.[109] Several differential etiologies have been proposed for chronic vertebral osteopathy in snakes

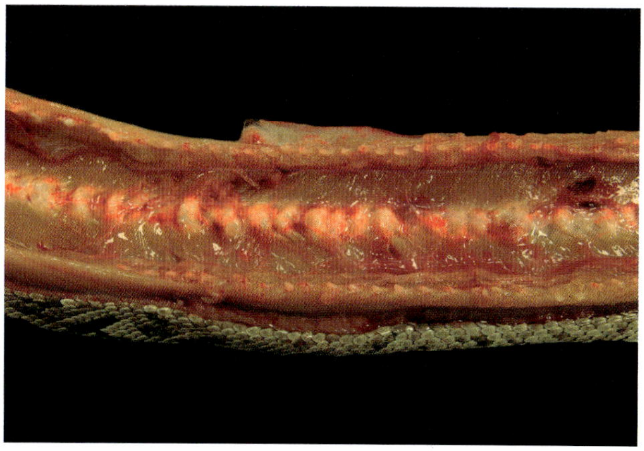

FIGURE 37.2 Chronic vertebral osteopathy in a Cuban boa. Chronic vertebral osteopathy is a common finding in snakes, and can include bony remodeling, proliferative new bone formation, spondylosis, and ankylosis. While a common finding, representative samples from such lesions should be collected and examined microscopically to rule out a chronic bacterial osteomyelitis, which can be grossly indistinguishable from sterile lesions.

including trauma, viral infection, nutritional deficiencies (hypovitaminosis A and hypervitaminosis D), chronic inactivity secondary to cage confinement, and chronic or resolved bacterial infections (see "Bacteria" section).[40]

Miscellaneous/Multifactorial

In captivity, **dysecdysis** (incomplete ecdysis and shed retention) is common (Fig. 37.3). Dysecdysis can be a manifestation of chronic dehydration, malnutrition, underlying disease, and/or improper husbandry conditions such as low humidity and inappropriate temperatures. Chronic dysecdysis can be associated with dermatitis either as a sequela or an inciting factor; the latter is particularly true for scale retention around the margins of the oral cavity. Retention of the spectacle, the modified scale overlying the cornea, can result in secondary bacterial infections and inflammation. Edema, hyperkeratosis, and granulocyte infiltration of the spectacle (spectaculitis) may be associated with bacterial and/or fungal infections.[27]

Ovoretention and **dystocia** are common in captive oviparous snakes and, in severe cases, can be associated with oviductal prolapse.[135] **Salpingitis** can either be a cause or sequela of ovoretention/dystocia. Ovoretention of nonviable ova (slugs) can also occur (Fig. 37.4). Chronic ovoretention is most often seen as a sterile condition, though bacterial salpingitis may occur and be associated with sepsis or rupture and release of infected yolk material into the coelom. While **yolk coelomitis** can occur in snakes, it is far less common than in other squamates or other reptiles. As an aside, given the close anatomic proximity of the ovary and the adrenal gland, close examination of any antemortem biopsy submissions of the squamate female reproductive tract for the presence of adrenal tissue is recommended as unintended removal of this tissue could result in negative endocrine effects.

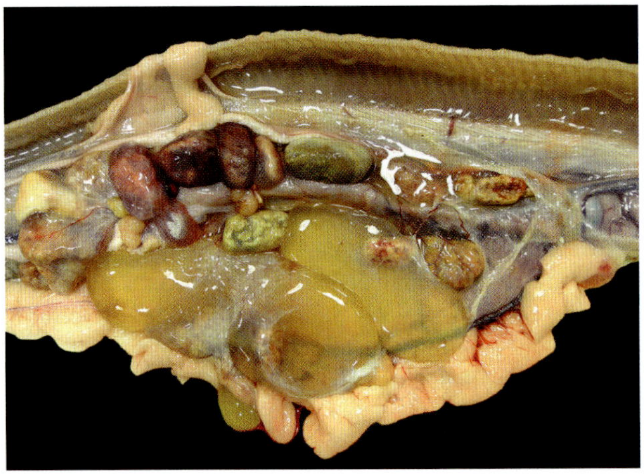

FIGURE 37.4 Chronic sterile ovoretention (oviductal slugs) in an Aruba Island rattlesnake. Yellow to green aggregates of inspissated yolk material (slugs) within the oviduct are variably associated with congestion and hemorrhage and/or surrounded by abundant translucent, yellow fluid. *(Photo Courtesy of R. Ossiboff, Wildlife Conservation Society.)*

FIGURE 37.3 Dysecdysis and retained spectacle in a corn snake. Dull, slightly opaque, retained keratin (shed) is present on the surface of the skin and multifocally evident as flaking over the scales. There is also a retained, slightly opaque spectacle over the eye with evident wrinkling.

The male hemipenes are sac-like extensions inverted within the base of the tail. **Paraphimosis** with prolapse and inflammation of the hemipenes is not uncommon. Prolonged hemipenal prolapse may result in extensive tissue necrosis. Inflammation of the hemipenes and the hemipenal sac are often associated with bacterial infections in the presence or absence of concurrent cloacitis. **Squamous metaplasia of the hemipenis** and hemipenal sac mucosa is typical with chronic hemipenal inflammation.

Husbandry-related trauma can be an important cause of morbidity and mortality in captive snakes. **Thermal burns** can occur on the dorsum (due to heat lamps and/or radiant heat panels) or ventrum (due to heating pads, cable, and/or tape) and can be associated with extensive epidermal, dermal, and muscular necrosis. Localized **facial trauma** secondary to repetitive injury (nose-rubbing) on either habitat furniture or the enclosure itself is also common. Such facial trauma can also be the hallmark of a variety of changes seen in wild-caught, imported reptiles, including snakes, as a result of facial injury during the prolonged shipping process. Lastly, **bite wounds** are a common finding if the snake is being fed live prey (typically larger rodents). Secondary bacterial infections are a common complication of husbandry-related traumatic injuries.

Neoplastic

Neoplasia is a common finding in captive snakes and in retrospective studies of reptile neoplasia, snakes are more often diagnosed with neoplasia than other reptile species.[138]

While certain neoplasms are more frequently encountered than others, a great diversity of ophidian tumors have been reported.[20,44,94,142]

Hematopoietic neoplasms, most notably **lymphoma** (Figs. 37.5A–C and 37.6), are common in snakes. Neoplastic aggregate distribution is most often multicentric, though solitary tumors, for example, in esophageal mucosa-associated lymphoid tissue (MALT) (Fig. 37.6) and leukemic extension and vascular tropisms can all be observed. **Leukemia** can also occur in the absence of solitary or multicentric lymphoma (Fig. 37.7).[94] Neoplastic lymphocytes can have variable morphology but are most commonly large and blast-like.[44] Immunohistochemical characterization of neoplastic lymphocytes can be difficult due to a lack of antibodies that have been validated for use in snakes. CD3+ T-cell lymphomas seem to be most common in snakes (Figs. 37.5C and 37.6). However, this observation may be biased as commercially available CD3 antibodies are broadly reactive and useful in reptile tissues, while B-cell markers (BLA36, CD79a, and CD20) are of variable diagnostic utility (with substantial laboratory variation), this observation may be biased. **Mast cell tumors** and **histiocytic origin round cell tumors** can also occur.[44,94,129] Histochemical stains, such as toluidine blue, can enhance the metachromatic granules of snake mast cells.[103] IBA1 is a highly evolutionarily conserved protein expressed in the cytoplasm of activated macrophages and microglial cells, and is an effective marker of cells of histiocytic lineage in snakes as well as other nonmammalian vertebrates.

Neoplasms are common throughout the ophidian gastrointestinal tract. Oral cavity tumors are observed frequently, and solitary or multifocal **squamous cell carcinoma** is perhaps the most common.[94] Neoplastic cells can be quite invasive and locally destructive, and while metastasis is reported, it is uncommon.[138] Squamous cell carcinoma can arise de novo or in the face of chronic stomatitis. An association between chronic inflammation and neoplasia is possible but has not been fully examined. In smooth green snakes, Opheodrys herpesvirus 1 is associated with the development of oropharyngeal squamous cell carcinoma (see "DNA Viruses" section).[85] A variety of other epithelial and mesenchymal tumors, particularly **fibrosarcoma**, can occur in the oral cavity.[94] Commercially available pan-cytokeratin (AE1/AE3) antibodies are often useful in immunostaining snake epithelial cells, while vimentin immunoreactivity is inconsistent and therefore of limited diagnostic utility. In a report of an oral **amelanotic melanophoroma** in a *boa constrictor*, neoplastic cells were positive for melan-A immunohistochemical labeling.[147] **Oral tumors of odontogenic origin** are sporadically reported in captive snakes.[94] Though there are few reports of neoplasia in free-ranging snakes, an **ameloblastoma** was documented in wild black rat snake.[25]

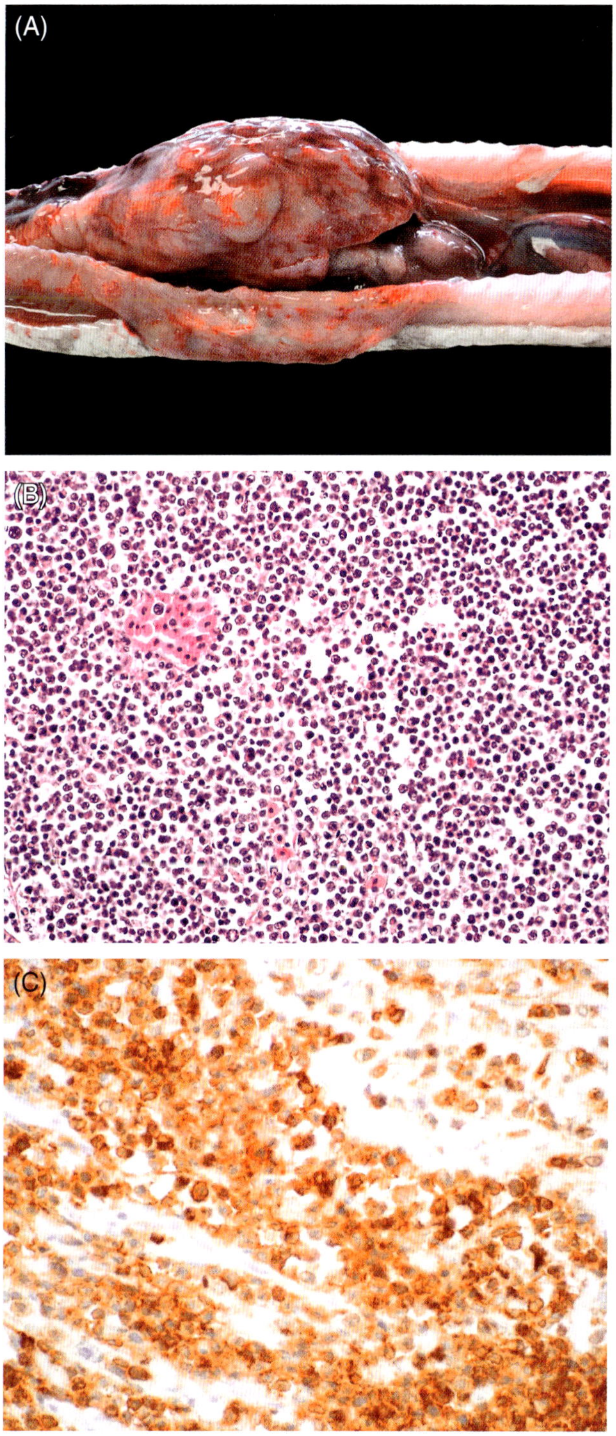

FIGURE 37.5 Thymic lymphoma in a ball python. (A) A large, mottled, light tan/red, multinodular mass effaces the thymus and expands the soft tissues cranial to and surrounding the heart. (B) The tumor is composed of sheets of neoplastic round cells with moderate anisocytosis and anisokaryosis. (C) Neoplastic cells are characterized by strong, membrane-positive immunolabeling for CD3, consistent with T-lymphocyte origin.

Gastric adenocarcinomas are locally invasive and expansile mural masses composed of tubulopapillary arrangements of columnar to cuboidal neoplastic epithelial

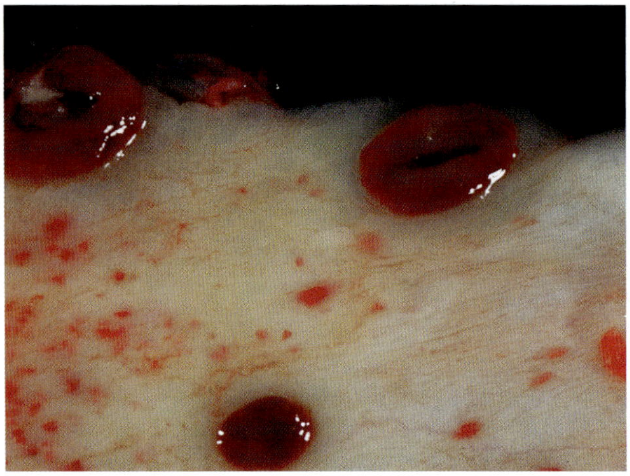

FIGURE 37.6 T-cell lymphoma in esophageal mucosal-associated lymphoid tissue (MALT) aggregates in a *boa constrictor*. MALT aggregates expanded by neoplastic lymphocytes are markedly congested and project from the mucosal surface. There are also multifocal areas of mucosal petechiation and ecchymosis. *(Photo Courtesy of William R. Pritchard Veterinary Medical Teaching Hospital.)*

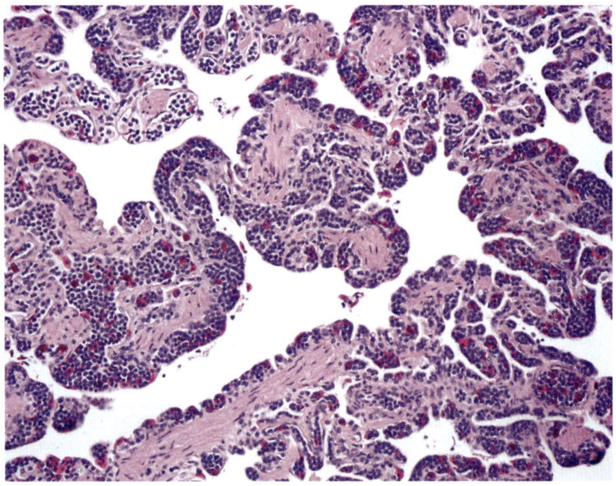

FIGURE 37.7 Leukemia in a western hognose. Sheets of neoplastic round cells expand and fill the lumina of faveolar capillaries and interstitial vessels.

cells with a prominent scirrhous response.[7,92] **Intestinal adenocarcinomas** are common, particularly in colubrids. Both the distal small intestine and colon are common sites, though cloacal adenocarcinomas can also occur. In some tumors, neoplastic epithelial cells may contain a single, large, cytoplasmic vacuole (signet ring morphology); in others, cells may contain abundant, PAS-positive cytoplasm with distension of glandular lumina by wispy, basophilic material (mucinous morphology).[83] Intestinal adenocarcinomas can be exophytic or mural, and, like their gastric counterparts, locally invasive with a prominent scirrhous reaction (Figs. 37.8A, B; eSlide 37.e2).[71,83,94] Visceral metastasis of intestinal adenocarcinomas is uncommon but reported.[105] **Benign and malignant smooth muscle, endothelial**, and **unclassified mesenchymal origin neoplasms** are also reported in the gastrointestinal tract.[94]

Benign and malignant hepatic and biliary tumors are also a common diagnosis (Fig. 37.9); colubrids appear to be at higher risk for these neoplasms than other snakes.[44,142] Both can be solitary or multifocal, and biliary neoplasms are often cystic. Carcinomas of the biliary duct and/or gall bladder have been reported in several viperids.[94] Hepatocellular adenomas and/or foci of nodular hepatocellular hyperplasia are a common finding in aged snakes. **Adenocarcinomas of both the exocrine pancreas** and **of pancreatic ducts** have been reported in multiple species of colubrids and viperids; the tumors, while well differentiated, can be extremely aggressive and metastasize.[44] **Exocrine pancreas adenomas and nodular exocrine hyperplasia** are a frequent, generally incidental, finding.

Renal adenocarcinomas are frequently diagnosed in snakes; colubrids are overrepresented (Figs. 37.10 and 37.11; eSlide 37.e3). The tumors are most often large and solitary, and composed of variably sized tubules lined by well-differentiated epithelial cells surrounded by scirrhous stroma that expands and disrupts the normal parenchyma. Urate tophi are often present within the neoplastic tubules and the adjacent desmoplastic stroma (Fig. 37.11; eSlide 37.e3).[44] Large, cystic fluid-filled cavities containing urine and urates can also be seen. **Renal tubular adenomas** can be an incidental finding during postmortem evaluation and consist of solid, nodular foci of tubular proliferations that may compress the adjacent tissue but lack an associated desmoplastic response.

Tumors of the female reproductive tract are more common than testicular tumors. **Ovarian carcinoma**, **sarcoma**, and **benign granulosa cell tumors** have all been reported.[94] **Oviductal adenocarcinomas** are exophytic, papillary, and composed of acini or cords of neoplastic cells that may have areas of hemorrhage, necrosis, and inflammation. The biologic behavior of these tumors appears to be variable. In one retrospective study, no metastases were noted while in another case series visceral metastases were noted in all cases.[44,113]

Tumors of the integument are commonly encountered in snakes. **Squamous cell carcinomas** are most commonly associated with the cloacal region and can originate from the hemipenes, cloacal glands, or cloacal skin.[44] A papillomavirus has been associated with a squamous cell carcinoma in a Dumeril's boa (see "DNA Viruses" section).[36] **Chromatophoromas** are neoplasms arising from the pigmented cells in either the epidermis (melanophores) or dermis (xanthophores, iridophores, and melanophores). **Melanophoromas** are most common, and are generally grossly gray or black and composed of neoplastic spindled cells with variably prominent black cytoplasmic granules (melanosomes).[36]

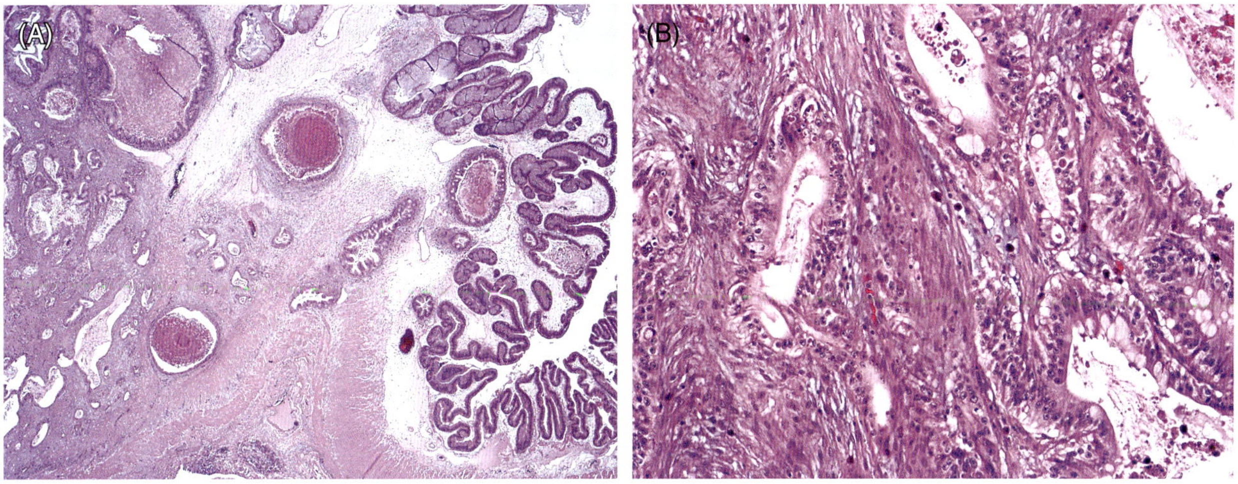

FIGURE 37.8 Colonic adenocarcinoma in a gopher snake. (A) There is abrupt neoplastic transformation of the colonic mucosa with the formation of papilliferous projections. Neoplastic infiltrate of the colonic wall. The submucosa is segmentally expanded by colorless space (edema). (B) Islands and tubuloglandular arrangements of neoplastic epithelial cells invade the colon transmurally and are associated with a prominent scirrhous response and heterophilic inflammation (see eSlide 37.e2).

FIGURE 37.9 Hepatocellular carcinoma in a Florida kingsnake. A moderately demarcated, nodular, hepatocellular neoplasm expands the hepatic parenchyma regionally. In this case, the neoplastic parenchyma is darker than the surrounding extant liver; however, the color of hepatocellular carcinomas in snakes can vary from tan, to red-brown, to brown-black, depending on the amount of lipid, pigment, and vascularity of the neoplasm.

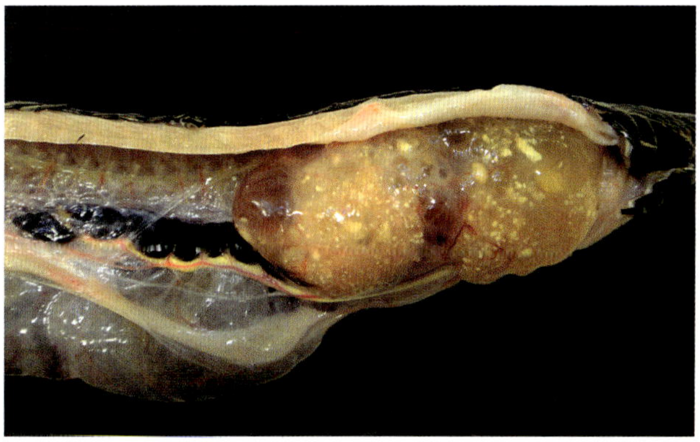

FIGURE 37.10 Renal adenocarcinoma in a Mexican black kingsnake. The caudal pole of the kidney is expanded by a cystic, multilobulated neoplasm. The neoplasm is composed of nodules of light brown neoplastic renal tissue admixed with large, fluid and urate-filled cavities (see eSlide 37.e3). (The colon has been reflected ventrally to expose the kidney.) *(Photo Courtesy of J. Rodriguez-Ramos Fernandez, Wildlife Conservation Society.)*

Both melan-A and S100 positive immunohistochemical labeling has been reported in melanophoromas. These tumors can exhibit variable biological activity, with some being invasive with intravascular and visceral metastases. **Iridophoromas** are grossly white to light gray, dermal masses composed of neoplastic polygonal to spindle cells with aggregates of golden brown to silver tan cytoplasmic pigment (Fig. 37.12A), that is, refractile under polarized light (Fig. 37.12B). Iridophoromas in reptiles are most often benign, though malignant forms occur.[36] Definitive diagnosis of **xanthophoromas** is more difficult due to the absence of clear, distinguishing light microscopic features between neoplastic xanthophores and dedifferentiated melanophores and iridophores. Additional diagnostics including electron microscopy may be necessary to confirm a diagnosis. The cytoplasmic pigment-containing pterinosomes of xanthophores may exhibit a positive histochemical reaction with Fontana-Masson stains.[47] While further characterization of chromatophoromas to the cell of origin is often ideal, it can prove quite challenging given available diagnostic modalities. Furthermore, ultrastructural examination of chromatophoromas can further complicate diagnosis, as tumors with clear gross or histologic features of one pigment tumor type may consist of more than one pigment cell based on ultrastructural features, consistent with mixed chromatophoromas.

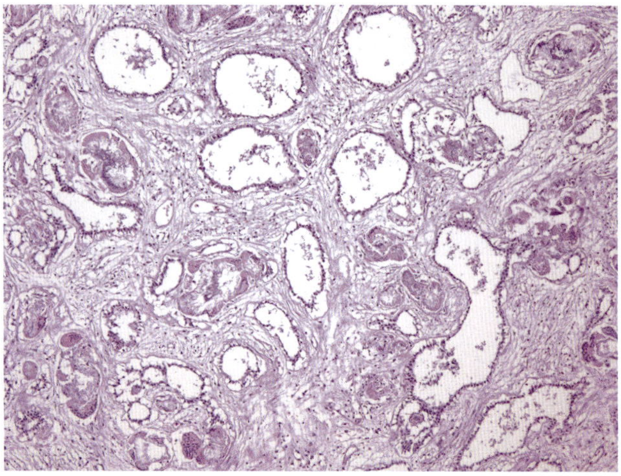

FIGURE 37.11 Renal adenocarcinoma in a gray-banded kingsnake. Variably sized, infiltrative neoplastic tubules lined by well-differentiated epithelial cells are surrounded by a desmoplastic stroma. Urate tophi are present predominantly within neoplastic tubules but also occasionally within the stroma (see eSlide 37.e3).

Cutaneous soft tissue sarcomas are also commonly found in snakes.[31] Locally invasive sarcomas composed of interlacing bundles and streams of neoplastic spindle cells, interpreted as **fibrosarcomas** are most common; colubrids and viperids are overrepresented.[44] Due to a lack of appropriate reagents and diagnostic criteria, further characterization of most soft tissue sarcomas in reptiles is limited.

Endocrine tumors in snakes are rare. **Thyroid and parathyroid adenomas** are reported. Both **pheochromocytomas** and **interrenal (adrenocortical) adenocarcinomas** are also documented. Neoplastic interrenal cells from a woma python were positive for melan A, an immunohistochemical marker also used for neoplastic adrenocortical cells in mammals.[74]

Neoplasms of the respiratory, cardiovascular, musculoskeletal, and central nervous systems are rare, though sporadically reported. **Intratracheal chondromas** resulting in airway obstruction occur in ball pythons (Fig. 37.13).[30,33] **Hemangiosarcomas** of the heart and spleen, respectively, have been described in colubrids.[130,152] Neoplastic cells in a case of cardiac hemangiosarcoma in a Madagascar giant hognose snake were positive for factor VIII-related antigen.[130]

INFECTIOUS DISEASES

DNA Viruses

Reports of **herpesvirus** infections in snakes are uncommon. In smooth green snakes, a novel herpesvirus, Opheodrys herpesvirus 1, the first to be genetically characterized from a snake, was identified in multiple individuals with oropharyngeal squamous cell carcinoma (Fig. 37.14).[85] In situ hybridization localized viral nucleic acid to neoplastic keratinocytes. Herpesviruses of the venom gland have been reported in cobras and kraits. In cobras with herpesvirus infection, inflammation, degeneration, and necrosis of the venom gland epithelium is associated with decreased venom production.[133] Hepatocellular necrosis with amphophilic intranuclear hepatocellular inclusions was observed in a clutch of captive *boa constrictors*, with concurrent pancreatic, renal, and adrenal intranuclear inclusions. While a herpesviral infection was not proven, the findings were strongly suggestive of a herpesvirus infection.[50] Herpesviruses have also been found in mixed infections of snakes with gastrointestinal disease.[89]

A novel **ranavirus** has been isolated and characterized from a group of green tree pythons with systemic

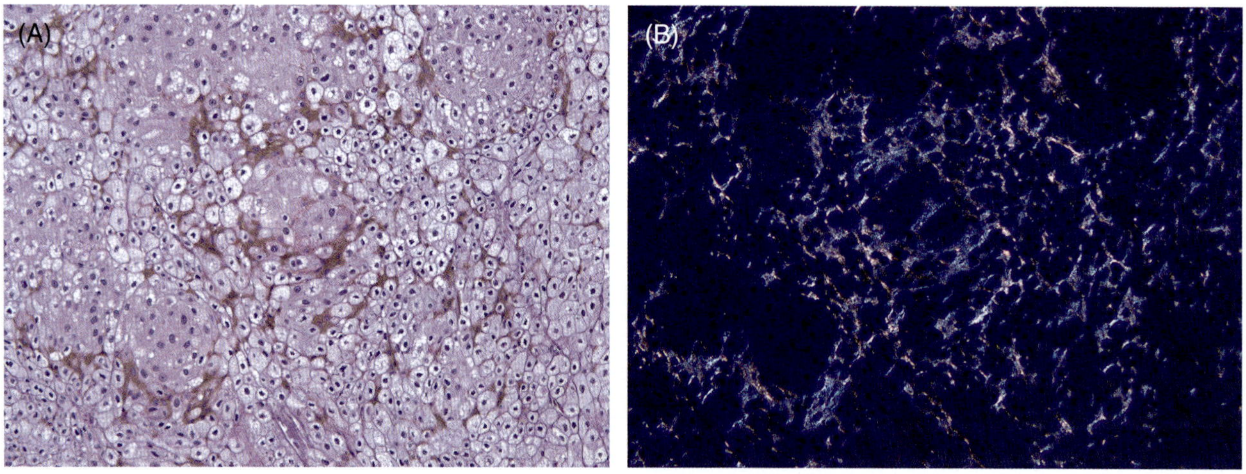

FIGURE 37.12 Metastatic iridophoroma in a carpet python. (A) Neoplastic polygonal chromatophores contain variable amounts of golden brown to silver tan, crystalline cytoplasmic pigment. (B) Under polarized light, the crystalline pigment is refractile and bright white to silver. In iridophoromas, particularly in aggressive tumors, not all neoplastic cells will contain refractile pigment.

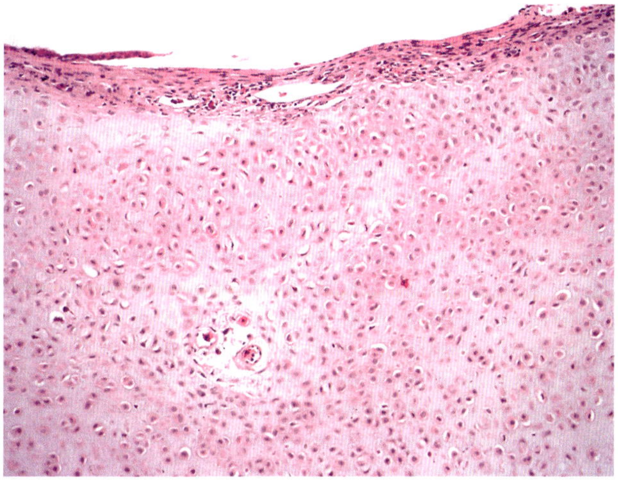

FIGURE 37.13 Intratracheal chondroma in a ball python. A nodular mass composed of well-differentiated chondrocytes amid chondroid matrix projects into the lumen of the trachea. There is squamous metaplasia and lymphocytic and heterophilic inflammation of the overlying mucosa.

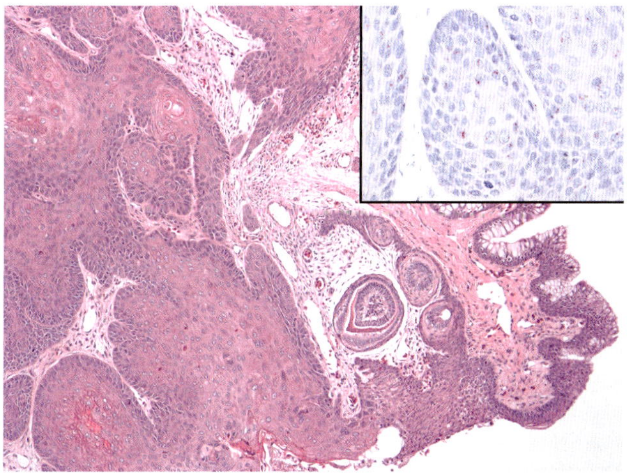

FIGURE 37.14 Squamous cell carcinoma due to Opheodrys herpesvirus 1 in a smooth green snake. Islands and interconnected trabeculae of neoplastic keratinocytes expand the oropharyngeal mucosa and invade into the underlying submucosa. Inset: Neoplastic cell nuclei are positive by in situ hybridization for Opheodrys herpesvirus 1. *(Photo Courtesy of R. Ossiboff, Zoological Pathology Program, University of Illinois Urbana-Champaign.)*

disease. Lesions in affected snakes included ulcerative and necrotizing rhinitis, stomatitis, pharyngitis, and hepatic and splenic necrosis; viral inclusion bodies were not seen.[62] **Erythrocytic iridoviruses** are seen in many species of squamates and chelonians. Erythrocyte inclusions can be seen in free-ranging North American northern water, plains garter, and eastern ribbon snakes and are attributed to assembly sites of snake erythrocytic iridoviruses (Fig. 37.15).[90,134] Historical reports of snake erythrocytic inclusions were often attributed to protozoal organisms (*Toddia* or *Pirhemocyton*) in the absence of ultrastructural confirmation. However, in the absence of this confirmation, some cases of erythrocytic iridoviruses may have been misdiagnosed and prevalence may be higher than currently thought. Iridoviral erythrocytic inclusions can be seen in both healthy and anemic snakes.[72,90]

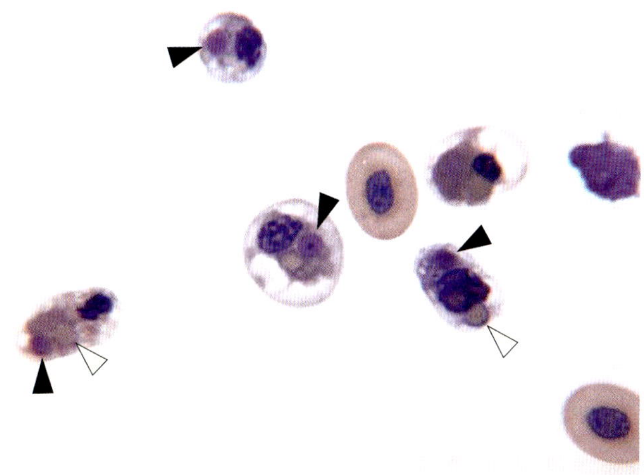

FIGURE 37.15 Snake erythrocytic iridovirus in a peninsula ribbon snake. Erythrocytes in a blood smear contain basophilic, round, well-demarcated viral inclusion bodies *(black arrowheads)* consistent with a snake erythrocytic iridovirus infection. A subset of erythrocytes also contain cytoplasmic aggregates of tan-yellow material *(white arrowheads)* that likely represents either hemoglobin crystals or viral-associated protein. Wright-Giemsa. *(Photo Courtesy of N. Stacy, Aquatic, Amphibian, and Reptile Pathology Service, University of Florida.)*

Scattered reports of other viral infections of snakes are present in the literature. These include several **adenoviruses** that are associated with sporadic gastrointestinal disease. In *boa constrictors*, adenoviruses can cause hepatocellular necrosis and large, characteristic basophilic intranuclear hepatocellular viral inclusions may be seen.[119] Adenoviral-like inclusion bodies can also be seen in enterocytes of a variety of colubrids with enteritis.[90,166] While reptile adenoviruses are generally considered to be host-adapted, spillover of a helodermatid adenovirus 2-like virus, first reported in beaded lizards, has been reported in a death adder with proliferative interstitial pneumonia.[9] However, the clinical significance of the adenovirus in this adder was undetermined.

Papillomaviruses have been associated with pigmented, papilloma-like lesions in a diamond python, and a cutaneous squamous cell carcinoma in a Dumeril's boa.[36,54] A **circovirus** was identified by metagenomic analysis in black-headed pythons, Boelen's pythons, and an annulated tree boa in gastrointestinal and liver tissue. Many of the study animals had spinal arthropathy lesions, though the clinical significance of virus identification in either the spinal lesion or the health of the animals is unclear.[6]

RNA Viruses

Inclusion body disease (IBD) is a global, transmissible, and progressively fatal disease of almost exclusively captive boids (boas and pythons). IBD was first identified in the 1970s in North America but has since been reported in captive snakes in Europe, Australia, and Africa.[22] Recently, both free-ranging and captive boids native to Costa Rica were confirmed to be infected with reptarenavirus. Archival analysis showed the virus was present in wild Costa Rican *boa constrictors* as early as the 1980s. This was the first time the virus was identified in wild snakes.[3] While the extent of natural infections of free-ranging snakes has yet to be determined, the identification of the virus in wild boas in Costa Rica likely represents a potential wild reservoir of the viruses, and movement of animals via the pet trade may have facilitated viral spread.

Clinical signs of snakes with IBD include central nervous system abnormalities, such as flaccid paralysis, stargazing, and torticollis, as well as chronic regurgitation and loss of body condition. Some individuals may succumb to the disease in weeks while others may survive for months making it a challenging disease to manage in captive collections. IBD was initially suspected to be caused by a retrovirus. However, subsequent research has identified divergent **arenaviruses** (family *Arenaviridae*; genus *Reptarenavirus*) in snakes with IBD. **Arenaviruses** are enveloped, segmented, single-stranded, negative sense RNA viruses that were previously believed to infect only rodents and humans. However, a number of snake arenaviruses have now been identified. Those identified from healthy and diseased IBD snakes are categorized in the genus *Reptarenavirus*. Recent studies of reptarenaviruses in snakes have identified variability in the disease susceptibility of pythons and boas to the viruses.

The most striking diagnostic feature of inclusion body disease is the presence of variably sized, eosinophilic to amphophilic, intracytoplasmic inclusions (Fig. 37.16; eSlide 37.e4). Inclusions are composed primarily of the ~68 kDa reptarenaviral nucleoprotein (also referred to as inclusion body disease protein) and cells with inclusions can frequently be observed in the absence of associated inflammation.[22,57,165] Within the central nervous system, inclusions may be observed in neurons and glial cells (Fig. 37.16A; eSlide 37.e4). In snakes (particularly pythons) with a more rapid progression of neurologic disease, inclusions are often limited to the central nervous system. In snakes with a more chronic manifestation of disease, such as has been reported in *boa constrictors*, inclusions can be seen throughout the body including in enteric, respiratory, and renal epithelial cells (Figs. 37.16B, C; eSlide 37.e4), hepatocytes, pancreatic acinar cells, and mononuclear cells.[90] Inclusions in lymphocytes are prominent within lymphoid aggregates in the esophagus (esophageal tonsils), and they can be observed systemically and in circulating leukocytes in blood films (Fig. 37.17; eSlide 37.e4).[21] Secondary infections (bacterial, fungal, and protozoal) are common and a significant factor in IBD mortalities; concurrent lymphoid neoplasia has also been reported.[90]

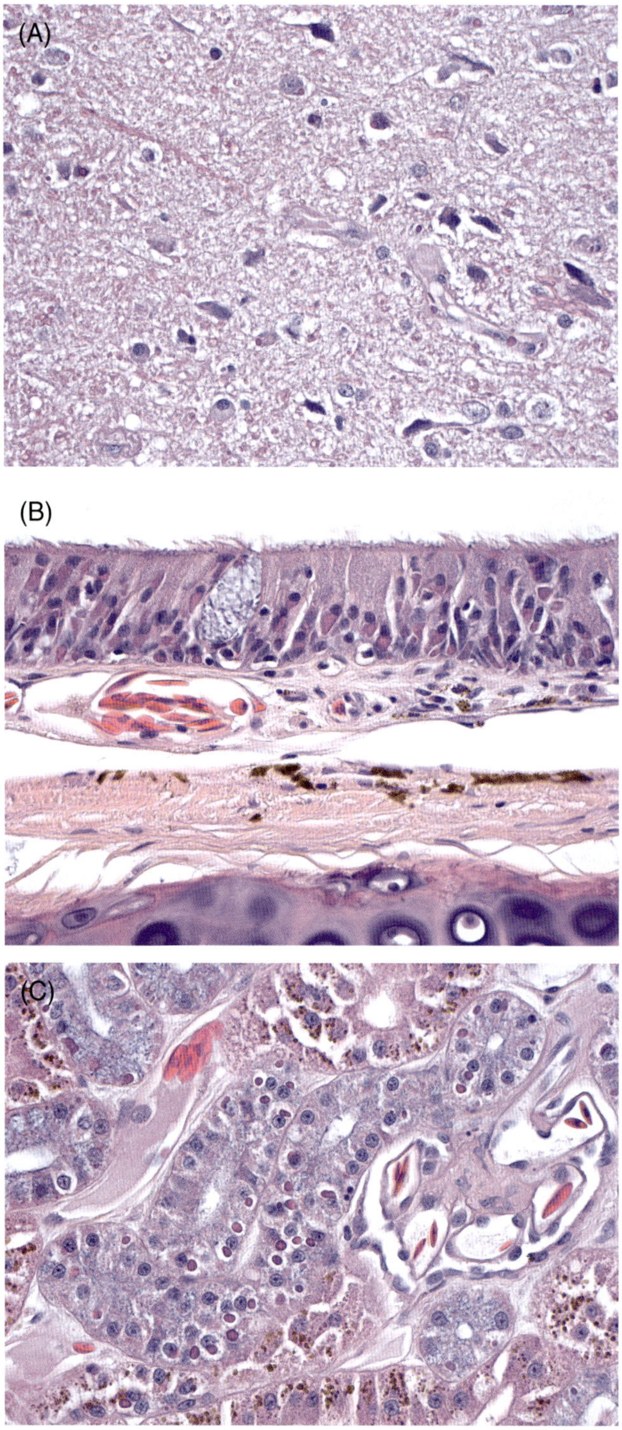

FIGURE 37.16 Inclusion body disease in a *boa constrictor*. Prominent, eosinophilic, intracytoplasmic, viral inclusion bodies are present within neurons and glial cells (A), respiratory epithelial cells of the trachea (B), and renal tubular epithelial cells (C). A subset of tubular epithelial cells also contain yellow-brown, cytoplasmic pigment, a common incidental finding in snakes (see eSlide 37.e4).

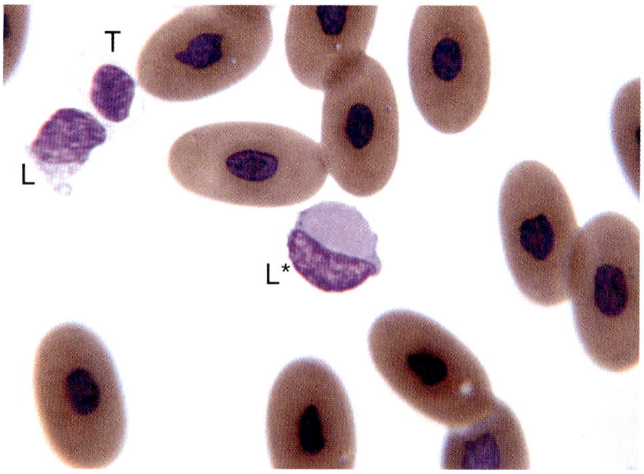

FIGURE 37.17 Inclusion body disease in a rainbow boa. Within a single lymphocyte (L*) of a blood smear, the nucleus is peripheralized by a pale basophilic, cytoplasmic reptarenaviral inclusion body. In contrast to the striking, hypereosinophilic inclusions noted on histology, reptarenaviral inclusions on blood smears are often basophilic. Also labeled are an unaffected lymphocyte (L) and a thrombocyte (T). Wright-Giemsa (see eSlide 37.e4). *(Photo Courtesy of N. Stacy, Aquatic, Amphibian, and Reptile Pathology Service, University of Florida.)*

Experimental inoculation of ball pythons with multiple reptarenaviruses resulted in severe and rapid (within 2 months) neurologic disease. Viral replication was only detected in the brain. In contrast, *boa constrictors* infected with the same viruses displayed no clinical signs for 24 months despite the presence of high-level systemic virus replication in the blood, brain, and other viscera and characteristic inclusion bodies.[140] This is consistent with other studies that have identified reptarenaviruses in overtly healthy snakes, often at a high prevalence.[21,63] While additional studies over a longer time course are needed to understand the pathogenesis and implications of reptarenavirus infections in boas, it is essential for pathologists to realize that a diagnosis of IBD or reptarenavirus infection in a boid snake may not cause immediate, significant morbidity or mortality. As such, do not overlook concurrent disease processes that may also have significance to the health of the individual or colony.

In addition to horizontal transmission, reptarenaviruses can also be transmitted vertically, and young boas born to infected dams can show inclusion bodies as early as 2 months after birth.[75] Reptarenaviruses are best characterized in boas and pythons. They have also been detected in other snakes, including captive Wagler's vipers, by molecular screening.[32] The clinical significance of such "aberrant" infections remains unclear.

The **serpentoviruses** (family *Tobaniviridae*; subfamily *Serpentovirinae*), previously referred to as **nidoviruses**, are an important, emerging group of reptile viruses, particularly in pythons. Serpentoviruses are large, enveloped, negative sense, single-stranded RNA viruses that were first reported in 2014 in ball pythons and an Indian python with fatal respiratory disease.[14,139,153] A number of serpentoviruses have since been identified in Australian and Asian python species in captivity in which they are often associated with significant respiratory disease[61]; green-tree pythons, carpet pythons, and other Australian pythons in captivity are all overrepresented.

Pythons with serpentoviral infections most often present with signs of respiratory disease including open-mouth, labored, and audible breathing and copious, mucoid to exudative, oral, and glottal/tracheal discharge. On gross examination, catarrhal and necroexudative stomatitis, tracheitis, and pneumonia are common (Fig. 37.18A; eSlide 37.e5). Aggregates of mucus and exudate frequently expand faveoli and occlude the vorbronchus. Similar material swallowed from the oral cavity can also be present along the length of the gastrointestinal tract. Histologically, severe epithelial hyperplasia and necrosis accompanied by lymphoplasmacytic and heterophilic inflammation can be seen in the oral and nasal cavities, glottis, trachea, proximal esophagus, and the lungs; pneumocyte hyperplasia can be profound (Fig. 37.18B; eSlide 37.e5). Abundant viral RNA can be observed within oral, tracheal, esophageal, and pulmonary epithelial cells (Fig. 37.18C; eSlide 37.e5). Secondary gram-negative bacterial stomatitis and bronchopneumonia are common.

Successful isolation of serpentoviruses on snake cell lines has permitted further investigation of these viruses, including fulfilling Koch's postulates for ball python nidovirus 1 in ball pythons. Lesions in experimentally infected ball pythons are similar to those of natural infections and include rhinitis, stomatitis, tracheitis, esophagitis, and proliferative pneumonia.[60] The diversity of identified serpentoviruses is ever increasing.[61] Serpentoviruses have been detected pythons, boas, colubrids, and viperids. Moreover, they have been identified in a variety of clinically healthy captive snakes, including pythons, and have recently been identified in free-ranging pythons and colubrids.[10,117,148] As such, it is clear that serpentoviral infection in snakes more often than not is subclinical and does not result in fatal respiratory disease. Other factors, including captive husbandry and stress, may play a role in clinical outcomes. There also appears to be a significant correlation between the detection of serpentoviruses and *Mycoplasma* spp. infection by molecular diagnostic screening, and it is possible that coinfections play an important role in disease.[117] While much work on serpentoviral infection and pathogenesis remains to be done, it is without question that these viruses represent a significant threat to the health of captive pythons and potentially other captive and even free-ranging snakes.

Ferlaviruses (paramyxoviruses) are another important group of viruses responsible for respiratory disease in snakes. Ferlaviruses are enveloped, negative sense, and single-stranded RNA viruses. There is substantial ferlaviral diversity, and four genogroups are recognized in the genus

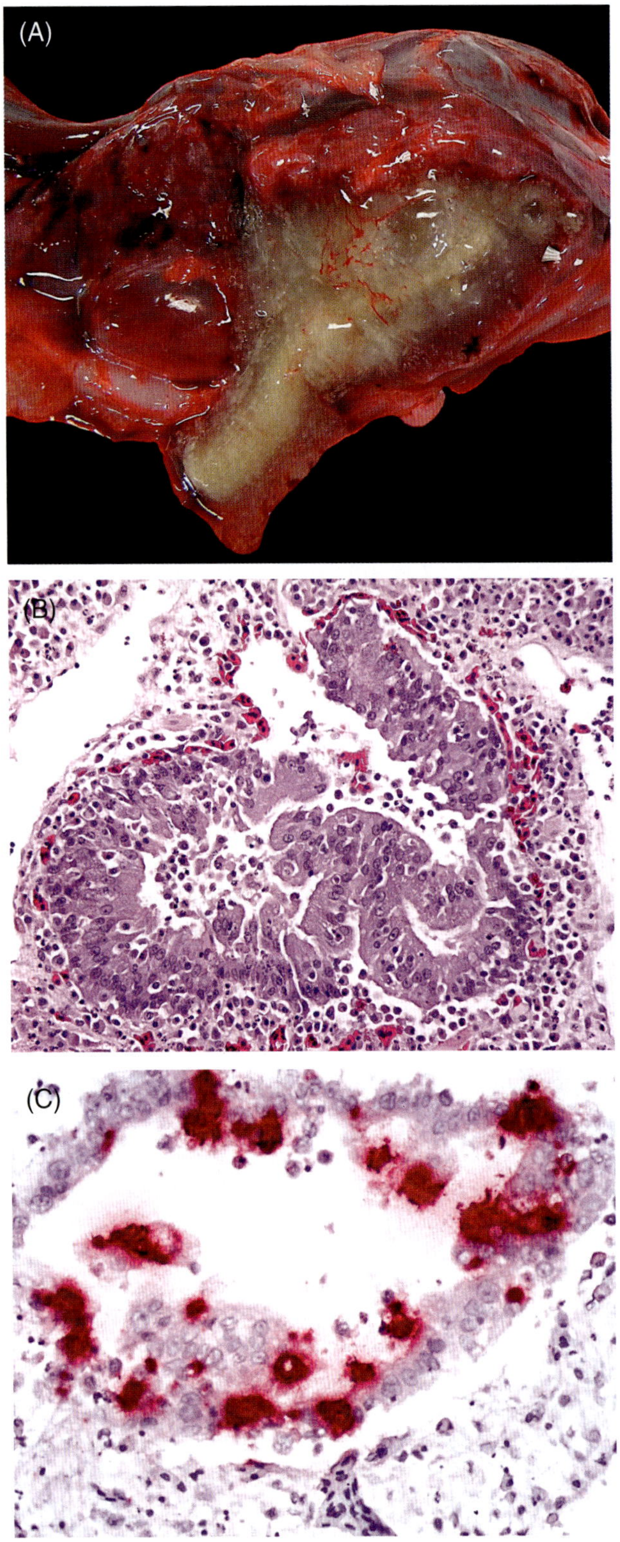

FIGURE 37.18 Serpentoviral pneumonia in a ball python. (A) The vorbronchus and faveolar lumina are expanded and obscured by accumulations of mucus, inflammatory exudate, and necrotic debris; the pulmonary parenchyma is edematous and congested. (B) Severely hyperplastic faveolar pneumocytes surround sloughed and exudative cellular debris; numerous lymphocytes, plasma cells, and fewer granulocytes are present within the interstitium. (C) Positive in situ hybridization to ball python nidovirus 1 RNA colocalizes viral RNA with pulmonary epithelial hyperplasia and inflammation (see eSlide 37.e5).

Ferlavirus (A, B, C, and "tortoise").[91,163] In snakes, ferlaviruses are primarily reported in captive viperids and crotalids, though colubrids, elapids, and boids are also susceptible.[65] Ferlavirus infections also occur in free-ranging North American snakes. Gross lesions are primarily noted in the lungs, which are often thickened and edematous. Hemorrhage and necroexudative material can accumulate in airways along the length of the lung, including the caudal, saccular nonrespiratory portion. Hemorrhage and exudate in the oral and nasal cavities, glottis, and esophagus, as well as coelomic cavity hemorrhage, may also be present. Histologic pulmonary changes are similar to those in other viral pneumonias of snakes, with prominent pneumocyte hyperplasia and variable lymphocytic, interstitial inflammation. Concurrent heterophilic bronchopneumonia associated with secondary gram-negative bacterial infections is common. The presence of both pulmonary epithelial syncytia and eosinophilic, intracytoplasmic inclusion bodies within pneumocytes can be diagnostic features of ferlavirus infection (Fig. 37.19), though inclusion bodies may be inconspicuous. The presence of the syncytia and/or inclusion bodies and the concurrent absence of a segmental esophagitis that is typical in serpentovirus infections can help differentiate between the two etiologies. Pancreatic inflammation and necrosis may be present in ferlavirus infection, and in some snakes, may be striking.[65,90] Snakes with ferlaviruses may also have concurrent neurologic disease with meningoencephalitis, lymphocytic perivascular cuffing, demyelination, axonal and neuronal degeneration, and/or intracytoplasmic glial and intranuclear neuronal viral inclusion bodies.[65] Ferlavirus has also been localized to the epididymal and deferent duct tubular epithelium, suggesting the potential for vertical transmission in snakes.[114]

Experimental inoculation of corn snakes with ferlavirus isolates representing genogroups A, B, and C resulted in variable lesion severity. Snakes inoculated with a genogroup B isolate showed the fastest, most widespread, and ultimately most severe disease, while the genogroup A isolate was the least virulent to infected snakes. Following ferlaviral inoculation, increased numbers of interstitial heterophils and lymphocytes, loss of type I pneumocytes, and type-II pneumocyte hyperplasia all occurred in the early stages of the disease.[136] With disease progression, type II pneumocyte proliferation continued and became multilayered, effectively reducing pulmonary function. Though the genogroup B isolate was the most pathogenic in these experimental studies, it was accompanied by the weakest ophidian immune response.[102] This indicates that tissue damage may be primarily due to viral cytopathic effect and not the host immune response, and that viral determinants of pathogenicity are not yet entirely understood.

A number of **reoviruses** (family *Reoviridae*; genus *Orthoreovirus*) have been reported in snakes. Reovirus infections are a diagnostic challenge due to the overlap of respiratory and neurorespiratory lesions with those of

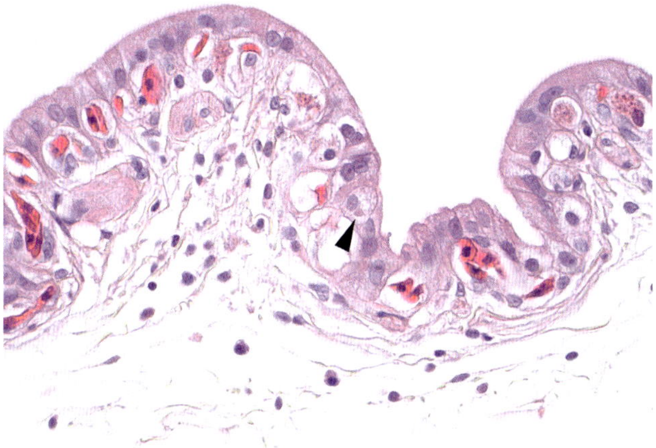

FIGURE 37.19 Ferlaviral pneumonia in a pygmy rattlesnake. Faveolar pneumocytes are hypertrophic and small numbers of binucleate (syncytial) cells are present. Small numbers of pneumocytes also have pale eosinophilic, intracytoplasmic viral inclusion bodies *(arrowhead)*. Low numbers of granulocytes and mononuclear cells are present within the interstitium that is also expanded by low amounts of colorless space (edema).

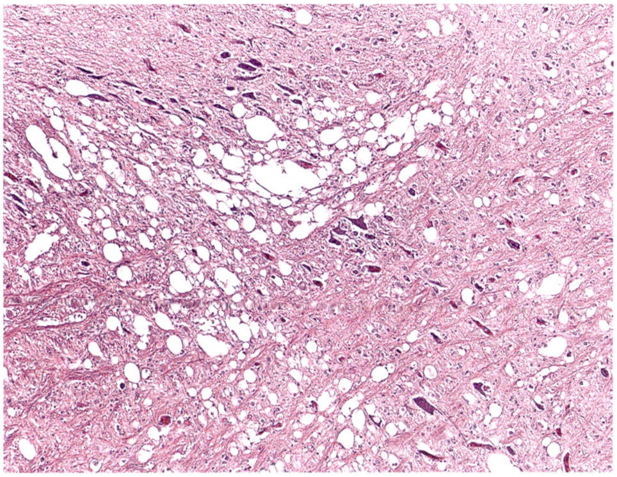

FIGURE 37.20 Sunshine virus infection in the brain of a carpet python. Severe white matter spongiosis and mild to moderate gliosis in the cranial medulla of a carpet python with neurologic disease secondary to a sunshine virus infection. *(Photo Courtesy of C. Shilton and K. Dyrting, Berrimah Veterinary Laboratories.)*

serpentovirus and paramyxovirus infections, respectively. Reports of reoviral disease in snakes are characterized by mild proliferative pneumonia and tracheitis.[80,90] Inclusion bodies, epithelial syncytia, and hemorrhage, which can be diagnostic features of ferlavirus infection, and segmental esophagitis, a common diagnostic finding in serpentovirus infections are not present in reoviral infections. In cases of suspected viral respiratory disease, a combination of diagnostic testing, including virus isolation on ophidian and select mammalian (Vero and CV-1) cell lines and targeted PCR testing for reovirus, ferlavirus, serpentovirus, or other respiratory and/or neurologic pathogens is recommended. Next-generation sequencing may be needed for confirmation.

Sunshine Coast virus (family *Sunviridae*; genus *Sunshinevirus*) has been associated with neurorespiratory disease in a number of Australian python species. The virus was originally believed to be a paramyxovirus but was reclassified to its own family in the order *Mononegavirales*.[2] Gross lesions are generally unremarkable. Histologically, central nervous system changes are characterized by spongiosis and gliosis of the hindbrain white matter (Fig. 37.20); extension into the gray matter and neuronal necrosis can be seen in severe cases. Mild bronchointerstitial pneumonia may also be seen.[64]

Divergent **bornaviruses** have also been identified genetically from the brains of carpet pythons with neurologic disease.[66]

Reports of other RNA viruses in snakes are sporadic. **Picornaviruses** have been reported in cases of gastrointestinal disease in snakes and **retrovirus** particles have been identified in neoplasms from snakes.[90] While reptile medicine and research has made substantial progress characterizing respiratory RNA viruses in snakes, much work needs to be done to characterize enteric viral infections.

Bacteria

Many gram-negative organisms, including *Pseudomonas*, *Aeromonas*, *Serratia*, *Morganella*, and *Providencia* spp., while considered to be part of the normal oral flora of healthy snakes can also be opportunistic pathogens.[123] **Bacterial stomatitis** is a common disease in captive snakes that is often associated with other problems (e.g., poor husbandry, concurrent disease, dysecdysis). Ulcerative and heterophilic inflammation of the oral mucosa, particularly at the junction between the labial scales, is a common finding early in disease. Inflammation, exudation, and accumulation of necrotic debris can progress to involve large areas of the oral and pharyngeal mucosa, surround the teeth, and extend into the choana. Heterophilic exudate can accumulate within the lingual sheath and with chronicity can manifest as severe necrotizing glossitis. Chronic oral inflammation may extend to the bones of the dentary and skull causing extensive osteomyelitis and bone loss. Septicemia and bronchopneumonia are common sequelae.

Salmonella is a component of the normal gastrointestinal flora of reptiles and a common pathogen in snakes. In a retrospective study of reptile submissions from a zoological collection, *Salmonella* was cultured from one-third of all submitted snake fecal samples and approximately one-third of all *Salmonella* positive samples were from clinically ill snakes.[23] *Salmonella enterica* **subspecies enterica (I), arizonae (IIIa), and diarizonae (IIIb)** are all commonly reported in snakes.[23,109] Septicemia, heterophilic dermatitis, necrotizing inflammation of visceral organs (liver, spleen, pancreas) and the gastrointestinal tract (stomach, small intestine, large intestine), and osteomyelitis (Figs. 37.21A, B; eSlide 37.e6) can all be observed in cases of ophidian salmonellosis. A specific serotype of *Salmonella enterica arizonae* (**56:Z4, Z23**) has been associated with osteotropism in captive ridgenose rattlesnakes.

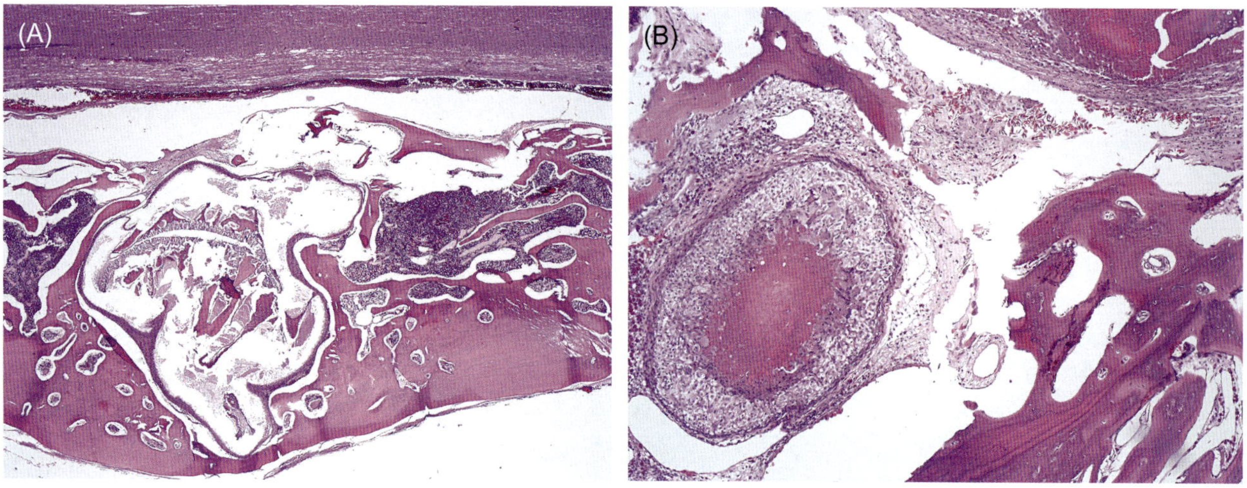

FIGURE 37.21 Necrotizing vertebral osteomyelitis in a corn snake due to *Salmonella enterica arizonae*. (A) Well-defined, foci of heterophilic and granulomatous inflammation surround necrotic bone, cartilage, and cellular debris and disrupt the normal architecture of the vertebral body. (B) Foci of heterophilic and granulomatous inflammation are associated with bone loss and active bone lysis with evident bony remodeling (see eSlide 37.e6).

Infection produced heterophilic and granulomatous osteomyelitis, osteonecrosis, sequestra, and pathologic fractures.[120] *Salmonella enterica arizonae* or other subspecies can often be isolated from cases of ophidian vertebral osteomyelitis when culture is productive (Figs. 37.21A, B; eSlide 37.e6). However, such lesions are not pathognomonic for *Salmonella* spp., as a number of other organisms, including *Pseudomonas* spp. and *Staphylococcus* spp., can cause vertebral infections and active granulomatous and heterophilic inflammation in snakes.[109] Moreover, as gross lesions of *Salmonella* osteomyelitis are often indistinguishable from chronic spinal vertebral osteopathy (Fig. 37.2), histologic characterization is often necessary for accurate lesion interpretation. Finally, because spinal bacterial osteomyelitis can be a primary source of septicemia, close examination of the spinal column for any irregularities or foci of bony proliferation is a necessary component of a snake necropsy.

Mycobacterial infections are encountered in captive snakes, albeit not as commonly as in some other reptiles and amphibians. Lesions may be localized (e.g., oral, pulmonary, oviductal, or dermal) or disseminated. Pulmonary mycobacteriosis is particularly common (Figs. 37.22A–C), and while boids seem to be overrepresented, it can be seen in most snake families. Typical reptilian granulomatous inflammation, with central cores of necrotic debris and often mineral surrounded by epithelioid macrophages, multinucleated giant cells, and variable numbers of heterophils with an outer rim of lymphocytes and plasma cells, is characteristic for the disease (Fig. 37.22B). Bacteria are usually readily identified with acid-fast histochemical stains (Fig. 37.22C). In reptiles and amphibians, Fite-Faraco staining may be more sensitive at detecting mycobacterial organisms than Ziehl-Neelsen staining.

There has been recent and extensive revision of the genus *Mycobacterium* based on phylogenetic analysis and growth characteristics in culture.[48] The historical genus *Mycobacterium* has been emended to include only slow-growing members of the "Tuberculosis-Simiae" clade, and four novel genera have been created: *Mycobacteroides* (rapid-growing, "Abscessus-Chelonae" clade), *Mycolicobacterium* (rapid-growing, "Fortuitum-Vaccae" clade), *Mycolicibacter* (slow-growing, "Terrae" clade), and *Mycolicibacillus* (slow-growing, "Triviale" clade). Both rapid- and slow-growing mycobacterial species can be identified in reptiles. *Mycobacterium marinum*, *M. haemophilum*, *M. kansasii*, *M. leprae*, and *Mycobacteroides* (*Mycobacterium*) *chelonae* are all reported in snakes.[56,101,121]

Chlamydia are obligate intracellular bacteria that can infect reptiles and typically cause granulomatous and granulocytic inflammation of the visceral organs including the spleen, heart, lung, and liver. In snakes, *Chlamydia pneumonia* is most commonly identified and proliferative pneumonia is a typical finding.[146] However, other incompletely characterized *Chlamydia* spp. can also cause disease. On microscopic examination, the identification of intrahistiocytic or intraepithelial, small, basophilic, punctate inclusions (Fig. 37.23) can be extremely helpful for making a diagnosis of chlamydiosis. However, ultrastructural, immunohistochemical, and/or molecular testing are often required for confirmation.[13,67–69] While reports of snake chlamydiosis are limited, it should be included as a differential etiology in cases of granulomatous and granulocytic inflammation with no clear evidence of a mycobacterial pathogen. The pathogenic potential of *Chlamydia pneumonia* and other *Chlamydia* spp. in snakes is not clear-cut, however, as they can be detected in choanal and cloacal swabs from healthy, captive snakes.[145,146]

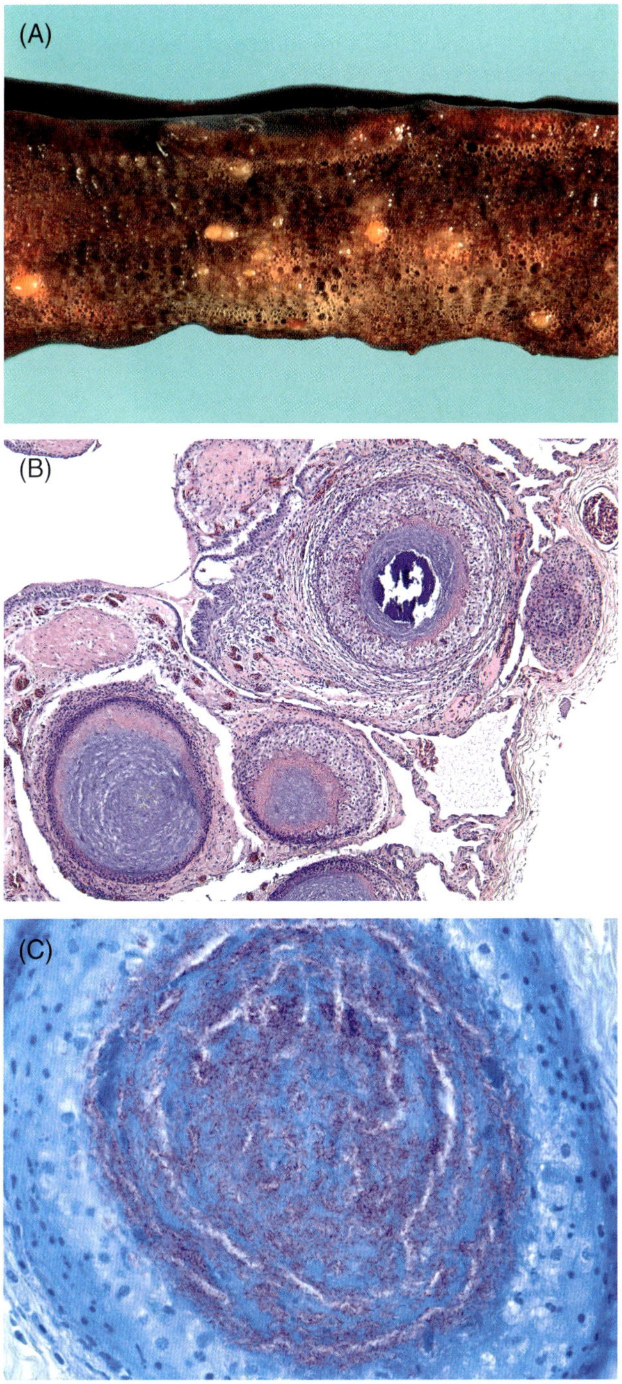

FIGURE 37.22 Pulmonary mycobacteriosis in a green tree python. (A) The pulmonary parenchyma is diffusely congested and edematous. Multifocally, faveoli are filled or obscured by nodular, tan masses (granulomas). (B) Microscopically, classic reptile granulomas composed of sheets of macrophages that often surround cores of hypereosinophilic to basophilic debris along and variable mineralization distort the pulmonary architecture. Granulomas are rimmed by variable numbers of lymphocytes and plasma cells. (C) Abundant acid-fast-positive bacilli are present within granulomatous foci. Ziehl-Neelsen.

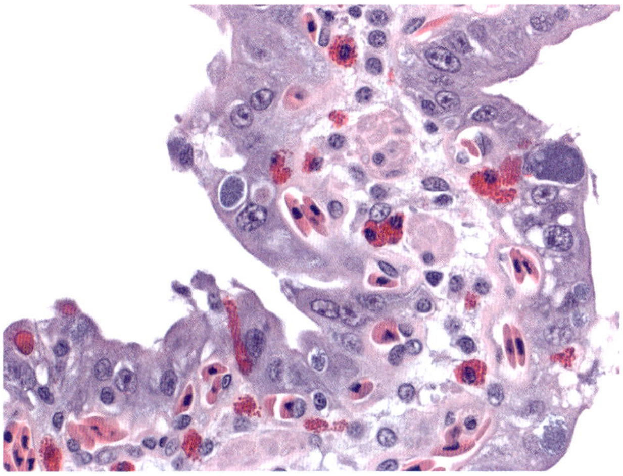

FIGURE 37.23 Pulmonary chlamydiosis in a corn snake. Diffuse pneumocyte hypertrophy and hyperplasia is associated with mixed granulocytic and mononuclear interstitial inflammation. Occasionally, the cytoplasm of pneumocytes is expanded by well-demarcated, granular, basophilic inclusions, consistent with developmental stages of *Chlamydia*.

There is limited literature on the clinical significance of ophidian mycoplasma infections. Mycoplasma infections have been detected multiple times in snakes, from both diseased and healthy pythons, boas, and vipers. A ***Mycoplasma*** sp. isolated from a Burmese python with proliferative tracheitis and pneumonia was associated with the surface of the tracheal and pulmonary airways by ultrastructural examination.[112] As mentioned earlier (see "RNA Viruses" section), there is also a strong correlation between detection of *Mycoplasma* spp. and serpentoviruses in python samples.[117] Similar to the revisions described previously for the genus *Mycobacterium*, extensive revisions to the phylum Tenericutes have been proposed. The creation of multiple novel genera to characterize organisms previously all considered to be members of the genus *Mycoplasma* has been proposed.[49] These genera include *Mycoplasmopsis* (containing reptile pathogens *M. agassizi*, *M. alligatoris*, and *M. crocodyli*), *Mesomycoplasma*, *Metamycoplasma*, and *Mycoplasmoides* (containing the reptile pathogen *M. testudinis*). Though a more thorough characterization of ophidian mycoplasmas is still needed, it is likely that *Mycoplasma* spp. previously identified in snakes represent novel specie(s) in a subset of the newly created genera mentioned above.

Granulomatous mucosal disease, affecting the oral cavity, cloaca, and hemipenes of captive ball pythons has been associated with a novel ***Actinomyces*** bacterium (Figs. 37.24A, B).[149] The disease predominantly affects male snakes, and granulomas of the hemipenes are one of the earliest noted lesions. Disease progression involves extensive granulomatous inflammation of the cloaca with occasional

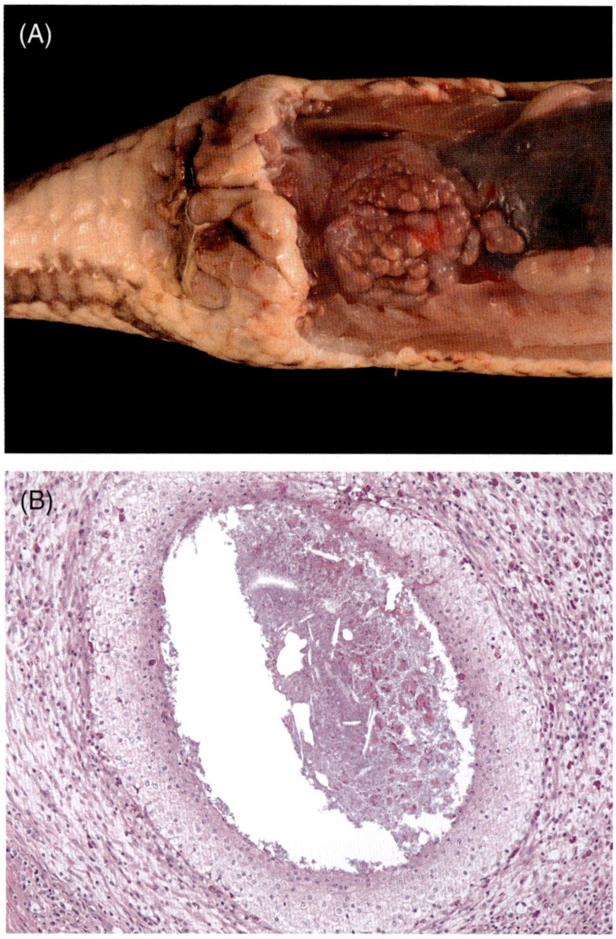

FIGURE 37.24 Actinomycosis in a ball python. (A) Numerous, well-demarcated, nodular granulomas are present within the dermis surrounding the cloaca, extending down along the hemipenes (not seen in photo), and extending along the serosal viscera of the caudal coelomic cavity. (B) A typical reptile granuloma surrounds a core of necrotic debris, with mixed morphology bacteria within granuloma cores, and characteristic cholesterol clefts. On high magnification (often requiring oil and a 100× objective), filamentous, gram-positive bacteria may be seen at the edges and within the wall of the granulomas (not shown).

local extension into the coelomic cavity (Fig. 37.24A); oral granulomas can also be seen in some snakes. The granulomas most often have a characteristic core of cellular debris, cholesterol clefts, and variable mineralization (Fig. 37.24B). Mixed morphology bacteria are often present within the granulomas. Among them are often variable numbers of filamentous, gram-positive bacilli. The filamentous bacilli are most reliably identified using histochemical stains and are seen along the margins of granulomas, at the edge of necrotic cores, and extending into the granuloma wall. Culture of these lesions is often unrewarding given the mixed bacterial populations and location of lesions at mucocutaneous junctions. PCR may be required for confirmation. As for other instances of actinomycosis in animals, trauma of the oral cavity (by food or substrate) or genitals (through intensive breeding) is suspected in the pathogenesis. Lesions are rarely observed in female ball pythons.

***Dermatophilus*-like bacteria** have been associated with subcutaneous masses in a *boa constrictor* and a king cobra.[109,164] Numerous mixed aerobic and anaerobic bacteria have been isolated from foci of dermal inflammation in snakes, including, but not limited to *Morganella*, *Fusobacterium*, and *Serratia*.[109]

Care must be taken when interpreting the identification of bacteria within the blood of snakes by way of blood culture or postmortem histologic examination. When bacteria are identified with other changes consistent with septicemia, such as embolic inflammation, disseminated fibrin thrombi, or haphazard multiorgan parenchymal necrosis, the interpretation of the bacteria as a significant component of the disease process is straightforward. However, it has been shown that a number of bacteria, including *Morganella*, *Citrobacter*, and *Proteus* spp., can be cultured from the blood of healthy snakes antemortem.[161] Moreover, gram-negative bacteria can often be seen in the vessels of snakes with even short postmortem intervals in the absence of significant localized or systemic inflammation. While their presence likely indicates a true antemortem bacteremia, it may not always indicate bacterial disease and should be interpreted in the context of the findings of the entire examination.

Fungi

Ophidiomycosis (snake fungal disease) (Figs. 37.25–37.27A, B; eSlide 37.e7) is an important disease of wild and captive snakes caused by the ascomycete *Ophidiomyces ophidiicola*. First reported in free-ranging snakes of the eastern and midwestern United States in the late 2000s, ophidiomycosis has now been documented in free-ranging and captive snakes throughout North America, Great Britain, Europe, and Australia.[28,52] Initially, ophidiomycosis was reported primarily in wild crotalids (timber, eastern Massasauga, and pygmy rattlesnakes) and sympatric colubrids including but not limited to garter, water, and milk snakes. The disease has now been documented in at least 60 snake species including boas, colubrids, vipers, and pythons.[5,52,107]

The most consistent and striking feature of ophidiomycosis is fungal dermatitis. In mild infections or early in the disease course, typical lesions consist of superficial foci of dermatitis and scale loss (Fig. 37.25; eSlide 37.e7). Severe and chronic infections can progress to cause widespread deep dermatitis, occasionally with substantial facial disfiguration (Fig. 37.26; eSlide 37.e7). Ultraviolet light (365 nm) may be a useful tool to quickly identify ophidiomycosis cases in the field or on the necropsy floor. Ophidiomycosis cutaneous lesions exhibit yellow-green fluorescence not seen in the surrounding, unaffected skin, though additional

FIGURE 37.25 Ophidiomycosis (*Ophidiomyces ophidiicola infection*) in a timber rattlesnake. Early or mild *O. ophidiicola* infection often presents as one or multiple area/s of mild, superficial dermatitis with localized scale loss (see eSlide 37.e7). *(Photo Courtesy of R. Ossiboff, Cornell University Wildlife Health Program.)*

FIGURE 37.26 Ophidiomycosis (*Ophidiomyces ophidiicola infection*) in a Florida kingsnake. Chronic *O. ophidiicola* infections can cause extensive, severe, and crusting dermatitis with facial and oral disfiguration (see eSlide 37.e7).

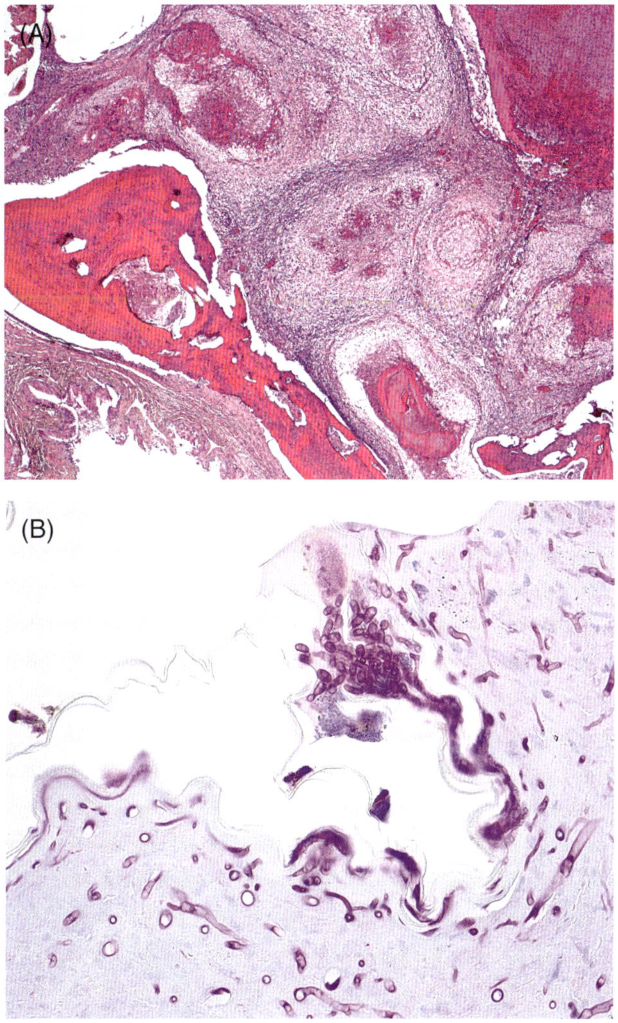

FIGURE 37.27 Ophidiomycosis (*Ophidiomyces ophidiicola infection*) in a timber rattlesnake. (A) Microscopically, severe and chronic cases of ophidiomycosis are characterized by foci of heterophilic and granulomatous inflammation that can extend deep to the level of underlying bone, where they are associated with bone resorption and loss. (B) A unique feature of *Ophidiomyces* and other, related Onygenalean fungi with pathogenic potential in reptiles is the formation of cylindrical air-lesion interface overlying invasive, septate, branching hyphae that extend throughout the inflammatory foci. Small numbers of cocci are also present along the surface of the lesion. Periodic acid-Schiff (see eSlide 37.e7). *(Photo Courtesy of R. Ossiboff, Cornell University Wildlife Health Program.)*

testing is needed to determine the specificity of lesion fluorescence and infection with *O. ophidiicola*.[157]

Histologically, the lesions are characterized by necroulcerative, heterophilic, and granulomatous dermatitis (Fig. 37.27A; eSlide 37.e7). Overlying serocellular crusts are composed of abundant necrotic debris and proteinaceous material that subtend the stratum corneum and frequently contain myriad profiles of up to 5 μm diameter, parallel walled, branching, septate hyphae. Along the surface of the crusts, cylindrical fission arthroconidia may be present (Fig. 37.27B; eSlide 37.e7). Histochemical stains [Grocott-Gomori methenamine silver (GMS); Periodic acid-Schiff (PAS)] may assist with arthroconidial identification. However, the absence of arthroconidia does not rule out ophidiomycosis. Moreover, arthroconidia are a diagnostic feature common to many members of the fungal order Onygenales, and their presence cannot differentiate between other closely related fungi, such as *Nannizziopsis* and *Paranannizziopsis* spp. In some cases, the superficial dermatitis will resolve with successive rounds of ecdysis. In severe cases, however, deep, chronic granulomatous and heterophilic inflammation is associated with severe osteolysis and osteomyelitis (Fig. 37.27A; eSlide 37.e7). Disseminated infections characterized by widespread visceral granulomas with intralesional fungal elements also occur.

Given that a number of important reptile fungi are morphologically indistinguishable from *O. ophidiicola* in tissue section, additional ancillary testing is required for a definitive diagnosis of ophidiomycosis. Both fungal culture and quantitative PCR specific for *O. ophidiicola* are recommended as fungal culture alone may result in the isolation of saprophytic organisms or concurrent, opportunistic copathogens. For field surveillance, the examination of shed skins can be quite useful in identifying the presence of *Ophidiomyces*. Snake sheds can be examined for the presence of focal or localized thickenings of the corneum or crusts, particularly along the ventral scales. Histology, fungal culture, and nucleic acid extraction for molecular testing can all successfully and reliably be performed on shed skins.

The exact origins of *O. ophidiicola* remain undetermined. Retrospective analysis of museum specimens has identified infections in free-ranging snakes in North America as early as 1945 and in free-ranging snakes in Europe as early as 1959.[84,104] Recent genomic analysis of *O. ophidiicola* strains suggests that the fungus was introduced into North America multiple times from a presumed single source population during the past few hundred years.[79] Strain analysis also indicates that North American and European strains form distinct clades and may have different source populations. The actual impacts of ophidiomycosis on free-ranging snake populations are undetermined. Ophidiomycosis does not appear to pose a significant threat to many wild snake populations in North America, though it is very possible that certain populations are experiencing declines in the face of ophidiomycosis and other concurrent infections or stressors.[52] In wild European snakes, ophidiomycosis cases are much milder in nature and are often limited to multifocal mycotic dermatitis.[42] Despite this, ophidiomycosis can have significant impacts on the health of individual, free-ranging snakes and captive snakes and collections.

Fungal dermatitides histologically similar to ophidiomycosis but caused by other fungi previously classified using the now obsolete terminology *Chrysosporium anamorph of Nannizziopsis vriesii* (CANV) can also be seen, particularly in aquatic snakes. In captive tentacled snakes, fungal dermatitis is a common finding (often due to suboptimal water quality and parameters) and two species of ***Paranannizziopsis*** (*P. californiensis* and *P. crustacea*) have been identified.[132] Likewise, *P. australasiensis* has been reported as a cause of necrotizing dermatitis in aquatic file snakes.[132] A novel *Paranannizziopsis* species, *P. tardicrescens*, was recently described from skin lesions in captive Wagler's vipers, housed in a facility alongside tentacled snakes and a rhinoceros rat snake with cutaneous fungal infections due to the same agent.[118] In general, **dermatomycoses** are common in snakes and opportunistic hyalohyphomycotic fungi, such as *Paecilomyces*, *Fusarium*, *Aspergillus*, and *Penicillium*, can also be causative. For this reason, accurate description of fungal morphology and additional ancillary diagnostics are very important for proper diagnosis.

Cases of visceral mycotic infections due to hyalohyphomycotic fungi occur sporadically. *Purpureocillium* (both *P. lilacinum* and *P. lavendulum*) have been reported as potential primary respiratory agents in green tree pythons, causing fatal granulomatous pneumonia.[97,127] Reports of infections by yeast and dimorphic fungi in snakes are rare. Granulomatous **cryptococcal pneumonia** and **meningoencephalitis** have been documented in a green anaconda.[96] *Coccidioides immitis-associated pneumonia* with mixed inflammation surrounding round, thick-walled spherules was reported in a Sonoran gopher snake.[150]

Metazoa

Metazoan parasitism in snakes is primarily an issue in free-ranging or wild-caught animals in captive collections, though there are some exceptions.

Rhabditid nematodes are of particular clinical significance in captive snakes. *Rhabdias* and *Strongyloides* are direct developing and free-living parasites that can also enter a parasitic, parthenogenetic phase. The larvae of both genera initially target pulmonary tissues. *Rhabdias* spp. remain in the lung for maturation while *Strongyloides* spp. complete their development in the intestinal tract. Snakes with severe infections or extensive larval migration can develop proliferative pneumonia and secondary bacterial infections. *Strongyloides* infections can result in proliferative disease of the oral cavity, esophagus, stomach, and intestinal tract. Given the life cycles of these nematodes, hyperinfections can occur following repeated exposure to infective larvae in the environment. Rhabditid nematodiasis is also a cause of ocular disease in captive ball pythons (Figs. 37.28A–C). Snakes infected with *Serpentirhabdias dubielzigi* can present with opaque spectacles and swelling of the oral, facial, and periocular tissues (Fig. 37.28A). Histologically, the subspectacular space, the space between the cornea and the overlying spectacle, is expanded by inflammatory and proteinaceous debris (Fig. 37.28B) with abundant nematode parasites at multiple life stages including ova, larvae, and adults (Fig. 37.28C). Adult *Serpentirhabdias* have characteristic double lateral alae, and ova are commonly larvated or embryonated. Parasites and mixed granulocytic and granulomatous inflammation and edema can extend into the connective tissues of the head.[51,86]

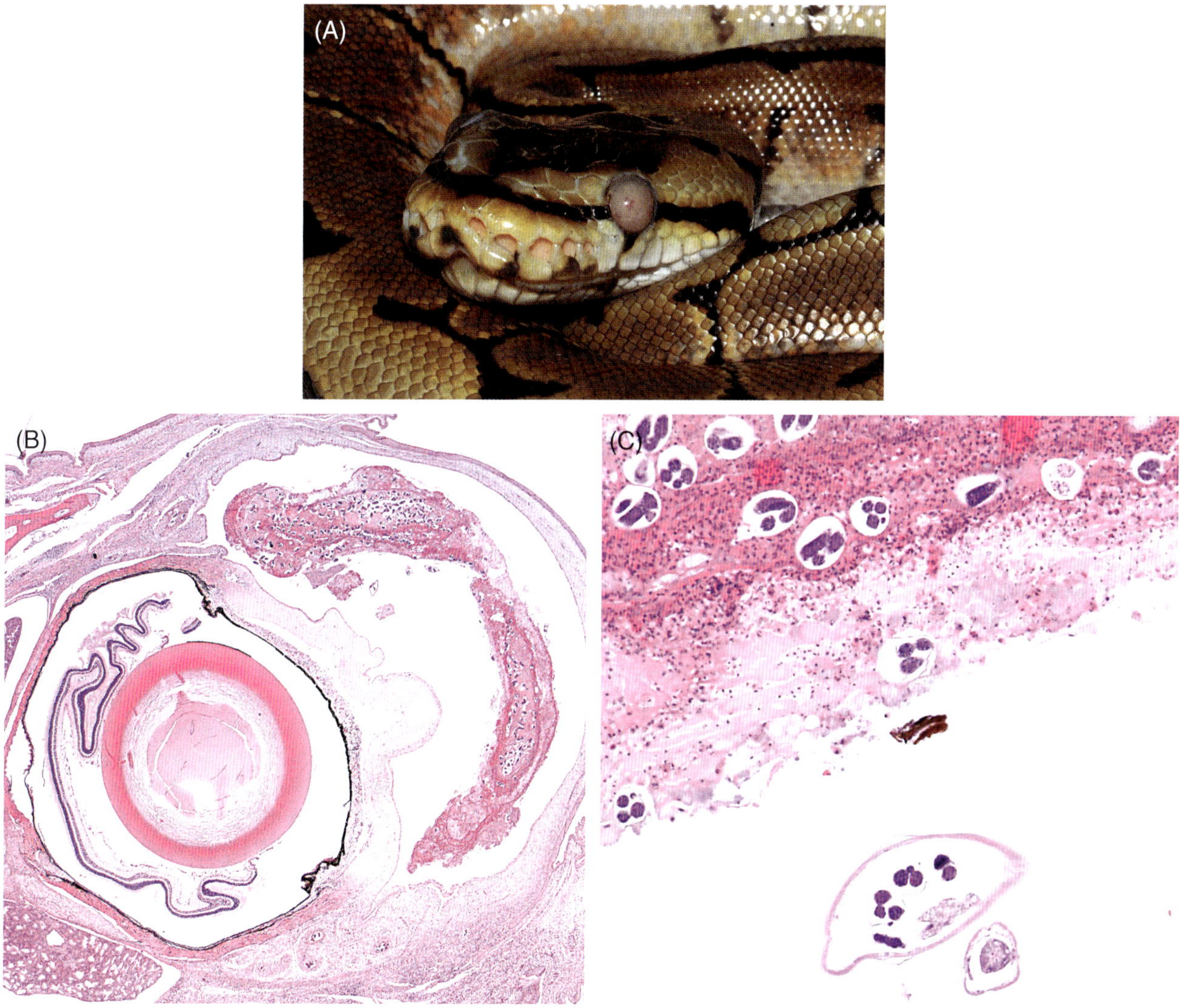

FIGURE 37.28 Ocular rhabdiasis in a ball python. (A) The spectacle bulges out from the globe due to the accumulation of fluid, exudate, and hemorrhage within the subspectacular space. There is also periorbital and subcutaneous swelling of the right lateral side of the head. (B) The subspectacular space, the potential space between the cornea and the overlying spectacle, is expanded by inflammatory and proteinaceous debris. (C) Numerous transverse sections of larval and mature nematode parasites (the type host of *Serpentirhabdias dubielzigi*) expand the subspectacular space. Double lateral alae are a characteristic feature of adults, and the presence of abundant larvated and/or embryonated ova is a common finding. *(Part A: Photo Courtesy of C. Mans, University of Wisconsin School of Veterinary Medicine and A. Lucio-Forster and D. Bowman, Cornell University; Parts B, C: Photos Courtesy of University of Wisconsin School of Veterinary Medicine Pathology Service and A. Miller, A. Lucio-Forster and D. Bowman, Cornell University.)*

Ascarids are common in free-ranging snakes and often associated with gastric lesions. Adults of ***Ophidascaris*** spp. may embed in the caudal esophageal and gastric mucosa and wall, resulting in ulceration, inflammation, and fibrosis. Visceral larval migration can cause multifocal granulocytic and granulomatous inflammation. ***Kalicephalus***, the snake hookworm, has a direct life cycle, and can be maintained in wild-caught snakes housed in captivity. In severe infections, gastritis, ulcerative enteritis, and hemorrhage or impaction may be observed.[76,158] *Kalicephalus* can also be seen in captive-born corn snakes, where the parasites exhibit a tropism for the esophagus and are associated with necrotizing stomatitis and esophagitis. Adult parasites can be found in the oral cavity and esophageal lumen, and embryonated ova can be seen in the esophageal mucosa.[93]

Larvae of **Dracunculoidea** nematodes can cause raised, pustular-like lesions in a wide variety of snakes. ***Macdonaldius***, a filarid of ophidians, is most often an incidental finding. Intravascular adults and microfilariae are seen histologically and can be associated with vascular obstruction, localized necrosis, and inflammation; microfilariae can also be observed in blood smears.[158]

The majority of cestode infections in snakes are subclinical. ***Bothridium*** and ***Bothriocephalus*** are parasites of boid snakes that are rarely associated with disease, even in heavy infections. **Tetrahydria**, larval cyclophyllidean cestodes of the genus *Mesocestoides*, are commonly found in snakes and develop following the ingestion of a cysticercoid-containing arthropod. Tetrahydria can be found within white nodules in the intestine, liver, or serosal surfaces of the coelomic viscera.[158]

Renifers (digenetic trematodes of the families Ochetosomatidae and Plagiorchiidae) are common inhabitants of the oral cavity, proximal esophagus, and lungs of free-ranging snakes. Adults migrate through the oral cavity into the glottis, trachea, and lungs, where they attach to the mucosal epithelium and can produce focal lesions and predispose to secondary bacterial infections. Plagiorchiid trematodes of the genus ***Styphlodora*** reside in renal collecting ducts and ureters and can cause tubular dilatation and chronic interstitial nephritis.[158]

Pentastomids are superficially segmented parasitic crustaceans that are commonly found in reptiles (Figs. 37.29 and 37.30). In snakes, *Armillifer*, *Kiricephalus*, *Porocephalus*, and *Raillietiella* are reported. The parasites are ingested as larvae and migrate extensively before attaining maturity in the lung (*Armillifer*, *Porocephalus*, and *Raillietiella*) or subcutis (*Kiricephalus*).[158] While adults in the lungs (Fig. 37.29) are not often associated with significant lesions other than obstruction, there can be significant inflammation associated with parasite migration and molting. A pentastome of increasing concern for free-ranging snakes in the southeastern United States is the invasive parasite, ***Raillietiella orientalis***. First documented in invasive Burmese pythons and sympatric native squamates in the Everglades in southern Florida, the range of *R. orientalis* has expanded north of the documented range of the Burmese python in Florida and has infected at least 16 native snake species.[38,99,159] The parasite is able to use both invasive brown anoles and native Florida toads as intermediate hosts to successfully complete its life cycle.[106] The parasite has also been detected in the pet trade due to the sale of wild-caught Florida reptiles.[39] Though population studies to determine the exact impact the parasite may have on native snake populations are ongoing, parasite loads in affected snakes can be very high (Fig. 37.30), and coinfections with other ophidian pathogens, such as *Ophidiomyces ophidiicola*, are common.

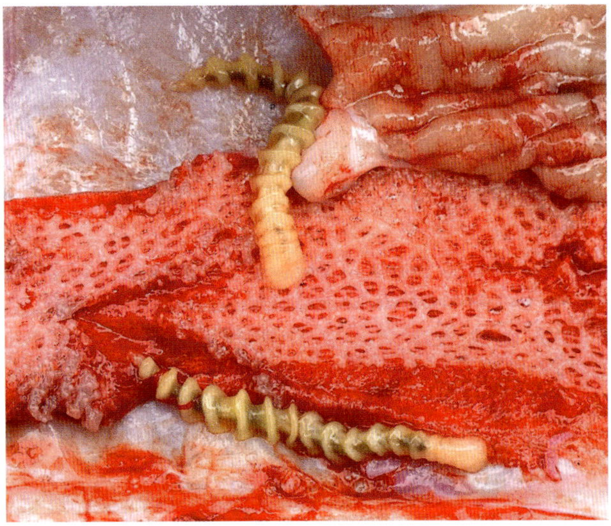

FIGURE 37.29 *Armillifer armillatus* in the lung of an African rock python. *Armillifer*, a pentastome parasite, attaches to the luminal surface of the lung in this free-ranging African rock python by way of compound hooks that surround its mouth. *(Photo Courtesy of B. Cossic, Cornell University.)*

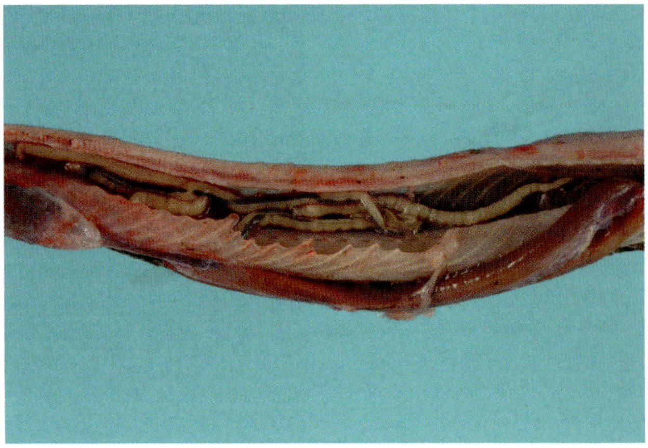

FIGURE 37.30 *Raillietiella orientalis* in the lung of a pygmy rattlesnake. Numerous adult *R. orientalis* are present within the lung of this free-ranging pygmy rattlesnake.

Protozoa

Cryptosporidial infections are an important cause of gastric and enteric disease in snakes. While generally only a clinical issue in captive snakes, ***Cryptosporidium*** spp. have been identified in free-ranging snakes as well.[155] ***Cryptosporidium serpentis*** infections can present as enlarged and palpably firm stomachs that may cause bulging of the body wall. Grossly, the gastric mucosa is thickened with prominent rugae and mucus accumulation (Fig. 37.31; eSlide 37.e8). Histologically, the gastric mucosa is hyperplastic with increased amounts of connective tissue in the lamina propria and submucosa (Fig. 37.32A; eSlide 37.e8). Gastric glands are dilated and there can be significant atrophy of granular, serous cells with hyperplasia of mucus cells. Variable numbers of intracellular, extracytoplasmic cryptosporidial oocysts (3–6 μm) are seen along the microvillar surface (Fig. 37.32B; eSlide 37.e8) and can be highlighted with Giemsa histochemical staining.[17] Though more common in lizards, chronic enteritis associated with ***Cryptosporidium varanii*** (and other less characterized intestinal *Cryptosporidium* spp. such as that reported in the Japanese grass snake) can also be seen in snakes.[78]

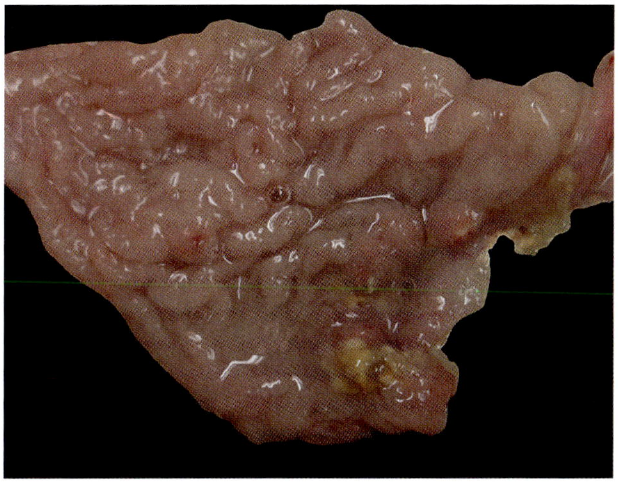

FIGURE 37.31 Gastric cryptosporidiosis in a corn snake. The gastric mucosa is rugose (severely hyperplastic) with abundant luminal mucus (see eSlide 37.e8). *(Photo Courtesy of W. Yau, University of Georgia Zoo and Exotic Animal Pathology Service.)*

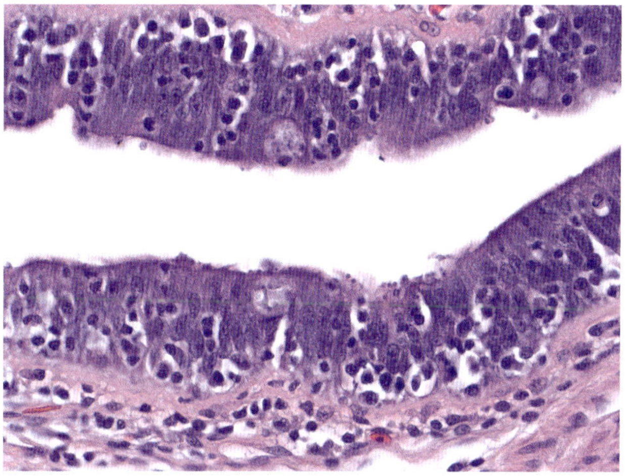

FIGURE 37.33 Intestinal cryptosporidiosis in a corn snake. Intestinal cryptosporidiosis is characterized by enterocyte hyperplasia, mucosal and lamina proprial lymphoplasmacytic and granulocytic inflammation with cryptosporidial organisms studding the apical margin of enterocytes. This case was confirmed to be *Cryptosporidium varanii* by PCR.

Intestinal lesions include mucosal epithelial hyperplasia and disorganization accompanied by mixed lymphoplasmacytic and granulocytic inflammation (Fig. 37.33). *C. varanii* can also cause ascending infections of the epithelium of the gall bladder and biliary ducts, resulting in proliferative cholecystitis and choledochitis. Definitive identification and speciation of *Cryptosporidium* requires molecular characterization. Frozen feces, stomach, and/or small intestine are the best samples for molecular characterization; formalin-fixed paraffin-embedded tissues are more limited in their diagnostic utility, though can be useful if formalin-fixation time is short (ideally <48 h).

Entamoebiasis is an important disease of captive snakes. Infections are often seen in mixed collections of herpetofauna that include subclinical reservoir animals/species, such as herbivorous lizards and turtles or amphibians. While many snake species are susceptible, some snake species, including garter snakes and northern black racers, seem to be more resistant to infection though individuals can still succumb to amoebiasis. ***Entamoeba invadens*** is encountered most commonly in a clinical setting, and causes necrotizing hepatitis and enterocolitis. Ingested quadrinucleate cysts develop into invasive trophozoites within the alimentary tract and cause segmental to diffuse mucosal necrosis of the large intestine though the small intestine and stomach can also be affected. Widespread mucosal necrosis permits amoebae access to the

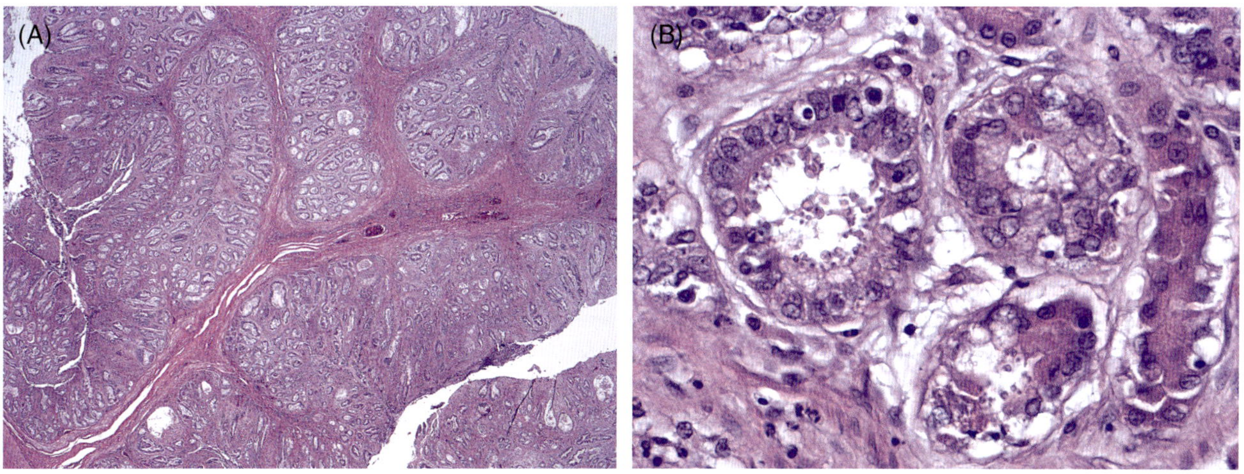

FIGURE 37.32 Gastric cryptosporidiosis in an indigo snake. (A) Gastric glands are branching, tortuous, and distended with moderate amounts of luminal debris. The glands are surrounded by increased amounts of fibrous connective tissue and admixed inflammation. The combination of mucosal proliferation, fibrosis, and inflammation is the cause of the rugose gross appearance of the mucosa in snakes with gastric cryptosporidiosis. (B) Numerous cryptosporidial organisms stud the surface of the gastric epithelial cells within gastric glands and the surface mucosa (not shown) (see Fig. 37.31; eSlide 37.e8).

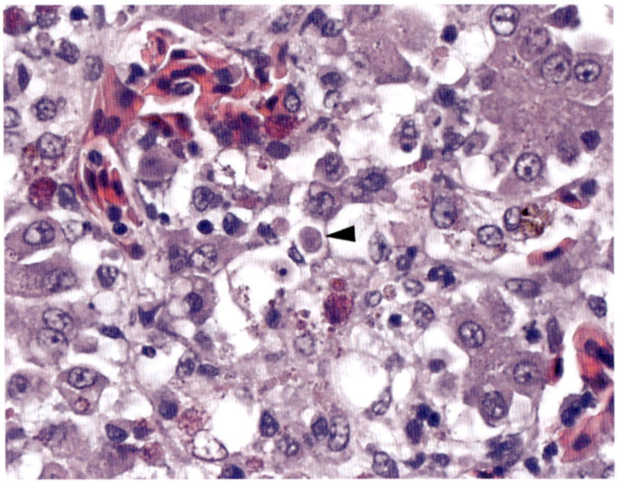

FIGURE 37.34 Hepatic amoebiasis in a garter snake. Within foci of hepatocellular degeneration and necrosis are small numbers of extracellular, eosinophilic, amoebic trophozoites *(arrowhead)* with a round to ovoid amphophilic nucleus (*Entamoeba invadens*, presumptive). Amoeba are accompanied by moderate numbers of intact and degenerate heterophils and fewer lymphocytes.

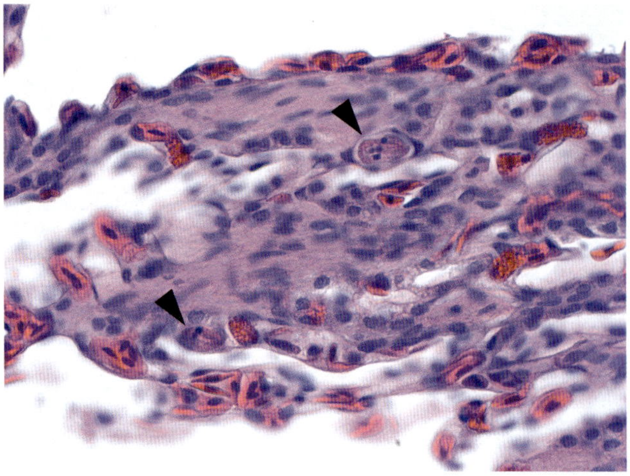

FIGURE 37.35 Pulmonary endothelial hemogregarines in an eastern coral snake. Multifocal endothelial cells of the lung are expanded by hemogregarine meronts (*Hepatozoon* sp., presumptive) and are contained within a parasitophorous vacuole *(arrowheads)*.

portal circulation and the liver. Necrotizing hepatitis (Fig. 37.34) is common due to ascending infection with both amoebic trophozoites and alimentary bacteria.[158] ***E. ranarum*** infections can also be seen in snakes. In contrast to *E. invadens*, these parasites are restricted to the mucosa and superficial submucosa of the large intestine and are associated with necrotizing and ulcerative colitis.[95,98,131] *E. ranarum* does not spread to the liver, instead it exhibits a pattern more similar to *E. histiolytica* infections in mammals. Periodic acid-Schiff staining is useful to highlight *Entamoeba* trophozoites in tissue section, while for tissue imprints or mucosal scrapes, Masson's trichrome staining of polyvinyl alcohol fixed samples is best.

Monocercomonas-like trichomonads are commonly seen in the gastrointestinal tract of snakes and are most often an incidental finding, though sporadic individual reports associate the organisms with pathology of a variety of tissues (gall bladder, oviducts, stomach, and intestine). *Eimeria*, *Isospora*, and *Choleoeimeria* also can be found in snakes, though their presence is most often incidental.

Hemogregarine apicomplexans are the most common intraerythrocytic parasite of snakes and even severe infections are typically clinically incidental.[158] These parasites are most often members of the genus *Hepatozoon*. Gamonts can be seen in erythrocytes on blood films; in tissue sections, meronts surrounded by a parasitophorous vacuole can most commonly be found in endothelial cells of the lung (Fig. 37.35) and adjacent to sinusoids in the liver. Vertical transmission has been documented in the viviparous garter snake. Infections may therefore develop in captive-born offspring produced in captivity from wild-caught parents.[73]

Ectoparasites

Mites are a common finding in free-ranging snakes and can inadvertently be introduced into naïve environments and populations of herpetofauna through the international movement of animals. They are also common in captive snakes with suboptimal husbandry and are frequently associated with dysecdysis. Severe ***Ophionyssus natricis*** acariasis can result in anemia and debilitation. *O. natricis* is also suspected of having a role in the transmission of disease in captive snakes, including divergent arenaviruses associated with IBD and certain bacterial diseases due to *Aeromonas hydrophila*. *Entonyssus* and *Hamertonia* spp. are snake lung mites that are reported as incidental findings in free-ranging and recently wild-caught snakes.[158]

ADDITIONAL RESOURCES

Boon A, Iredale ME, Tillis SB, Ossiboff RJ. Ophidian Serpentoviruses: A Review and Perspective. J Herp Med and Surg. 2023; 33 (4):1–12.

Campbell TW. Clinical pathology. In: Mader DR, Divers SJ, eds. Current Therapy in Reptile Medicine & Surgery. Elsevier Saunders 2014: 70–92.

Cox N, Young BE, Bowles P, Fernandez M, Marin J, Rapacciuolo G, Böhm M, Brooks TM, Hedges SB, Hilton-Taylor C, Hoffmann M, Jenkins RKB, Tognelli MF, Alexander GJ, Allison A, Ananjeva NB, Auliya M, Avila LJ, Chapple DG, Cisneros-Heredia DF, Cogger HG, Colli GR, de Silva A, Eisemberg CC, Els J, Fong GA, Grant TD, Hitchmough RA, Iskandar DT, Kidera N, Martins M, Meiri S, Mitchell NJ, Molur S, Nogueira C de C, Ortiz JC, Penner J, Rhodin AGJ, Rivas GA, Rödel M-O, Roll U, Sanders KL, Santos-Barrera G, Shea GM, Spawls S, Stuart BL, Tolley KA, Trape J-F, Vidal MA, Wagner P, Wallace BP, Xie Y. A global reptile assessment highlights shared conservation needs of tetrapods. Nature. 2022;605(7909): 285–290.

Garner MM, Hernandez-Divers SM, Raymond JT. Reptile neoplasia: A retrospective study of case submissions to a specialty diagnostic service. Vet Clin North Am - Exot Anim Pract. 2004;7 (3):653–671.

Haynes E, Allender M. History, Epidemiology, and Pathogenesis of Ophidiomycosis: A Review. Herpetol Rev. 2021;52:521–536.

Hyndman TH, Shilton CM, Marschang RE. Paramyxoviruses in reptiles: A review. Vet Microbiol. 2013;165 (3, 4):200–213.

Jacobson ER, Lillywhite HB, Blackburn DG. Overview of reptile biology, anatomy and histology. In: Jacobson ER, Garner MM, eds. Infectious diseases and pathology of reptiles, Color atlas and text. Taylor & Francis 2021:1–214.

Marschang RE, Origgi FC, Stenglein MD, Hyndman TH, Wellehan JFX, Jacobson ER. Viruses and viral diseases of reptiles. In: Jacobson ER, Garner MM, eds. Infectious diseases and pathology of reptiles, Color atlas and text. 2021:575–704.

Pasmans F, Martel A, Jacobson ER. Bacterial diseases of reptiles. In: Jacobson ER, Garner MM, eds. Infectious diseases and pathology of reptiles, Color atlas and text. Taylor & Francis 2021:705–794.

Walden HDS, Greiner EC, Jacobson ER. Parasites and parasitic diseases of reptiles. In: Jacobson ER, Garner MM, eds. Infectious diseases and pathology of reptiles, Color atlas and text. Taylor & Francis 2021:859–968.

References

Full references for this chapter can be found online at https://doi.org/10.1016/B978-0-443-18437-6.00027-4.

Chapter 38

Amphibia

Allan P. Pessier
Washington State University, Pullman, WA, United States

INTRODUCTION

The vertebrate Class Amphibia contains 3 Orders and approximately 8,600 species that are distributed globally with the exception of Antarctica. The largest Order is the Anura (frogs and toads) with over 7,300 species. This is followed by the Order Caudata (salamanders and newts) with over 700 species, and the legless caecilians in the Order Gymnophiona with over 200 species. Estimates of amphibian species numbers change frequently because hundreds of new species are still being described annually. Evolutionarily, Amphibia is a monophyletic group and should not be considered as transitional between fish and reptiles. The greatest amphibian biodiversity is found in the neotropics. There is no generic "amphibian" and species are found in habitats ranging from desert to rainforest and they have life histories that range from entirely aquatic to completely terrestrial. Most caecilians are fossorial and are only rarely observed in nature. From a conservation perspective, the Amphibia are the most imperiled class of vertebrates with 40.7% of known species listed as globally threatened in the 2022 IUCN Global Amphibian Assessment and an estimated extinction rate over 200 times historical numbers.[31,103,171] Factors contributing to amphibian population declines include habitat loss, climate change, and introduced infectious diseases, especially the amphibian chytrid fungus *Batrachochytrium dendrobatidis*.

The role of infectious diseases in declines of free-ranging amphibians was recognized in the late 1990s, and since then heightened interest in the study of amphibian disease has resulted in an explosion of new scientific literature. In addition, the development of amphibian conservation programs that bring endangered species into captive survival assurance populations has encouraged the investigation of infectious as well as non-infectious diseases, the latter often related to nutrition and husbandry.[51] The vast diversity of amphibians in terms of lifestyle and habitat means that knowledge about the natural history of individual species often greatly assists morbidity and mortality investigations. And because the majority of amphibians spend at least their larval stages in an aquatic environment, a background in fish medicine and pathology is helpful because interpretation of water quality parameters (e.g., ammonia) is essential and many pathogens are similar (e.g., microsporidiosis, mycobacteriosis). Finally, because amphibian skin has unique physiologic attributes, evaluation of the integument is essential in amphibian disease investigations.[135]

NON-INFECTIOUS DISEASES

Nutritional

Nutritional diseases are common in captive amphibians because captive-reared insects (e.g., domestic crickets) without adequate supplementation are deficient in key nutrients, such as calcium or vitamin A.[51] Metabolic bone disease (MBD) is one of the most common nutritional diseases and usually results from dietary calcium deficiency, vitamin D_3 deficiency or combinations of these factors. Based on their natural history, amphibians vary in their requirement for exposure to ultraviolet B (UVB) light to meet their requirements for vitamin D_3. For instance, fossorial species (e.g., many caecilians) require very little UVB exposure, whereas species with natural basking behaviors, such as the mountain chicken frog, need supplemental UVB light in captivity.[3,162] Amphibians also acquire nutrients, such as calcium or phosphorus, through the skin and therefore environmental water composition (e.g., high levels of phosphorus in a municipal water supply) is a consideration when evaluating altered intake ratios of calcium and phosphorus that could result in MBD. Recently, the possibility of **fluorosis** was considered as a contributor to MBD-like lesions in frogs exposed to a fluoridated water supply; however, these animals also had other risk factors for MBD, such as lack of adequate UVB light exposure.[150] MBD is also occasionally observed in some anurans with advanced kidney disease and the pathogenesis is presumably similar to renal secondary hyperparathyroidism in mammals.[48,133] Clinical signs in animals with MBD include tetany (often induced in response to handling or stress) and skeletal deformities of the long bones, spinal column, and cranium.

Pathology of Wildlife and Zoo Animals. https://doi.org/10.1016/B978-0-443-18437-6.00024-9

Radiographically, there is diffusely reduced bone mineral density and folding fractures. On histologic examination, MBD lesions are best described as **osteomalacia** with wide seams of unmineralized osteoid that line the outer cortex of affected bones. In addition, the cortex is frequently expanded by exuberant nodular proliferation of hyaline cartilage, especially at sites of folding fractures (Fig. 38.1A). Trichrome stains are useful for highlighting widened osteoid seams (Fig. 38.1B).

Hypovitaminosis A is a more recently recognized nutritional disease of captive anurans and caudates fed insect-based diets.[142] By themselves, captive-reared insects lack significant amounts of preformed vitamin A and the plant-based carotenoids found in wild insects that are converted to active vitamin A. Initially, hypovitaminosis A was described in the Wyoming toad as "short tongue syndrome" because affected animals had progressive difficulty in prehending prey items. There are usually no gross lesions; however, brown to black pinpoint foci were described on the mucosal surface of the tongue in Puerto Rican crested toads. Other toads presented with facial nodules and oronasal fistulae arising from keratinizing cysts in cutaneous mucus glands. Histologically, lesions are composed of squamous metaplasia of mucus-producing glands or ciliated epithelium of the tongue (Figs. 38.2A, B), oral mucosa, esophagus, cloaca, and urinary bladder. Less frequently, squamous metaplasia is observed in renal tubules, cutaneous mucus glands, and the oviduct. Lesions of squamous metaplasia are often multifocal and are not present in every affected animal within a population or in every potential anatomic site in affected animals. A presumptive diagnosis is made based on histologic findings in the absence of chronic inflammatory processes that might promote squamous metaplasia in the absence of deficiency. Confirmatory diagnosis requires determination of hepatic retinol levels with levels less than 5–10 ppm being suspicious for deficiency.

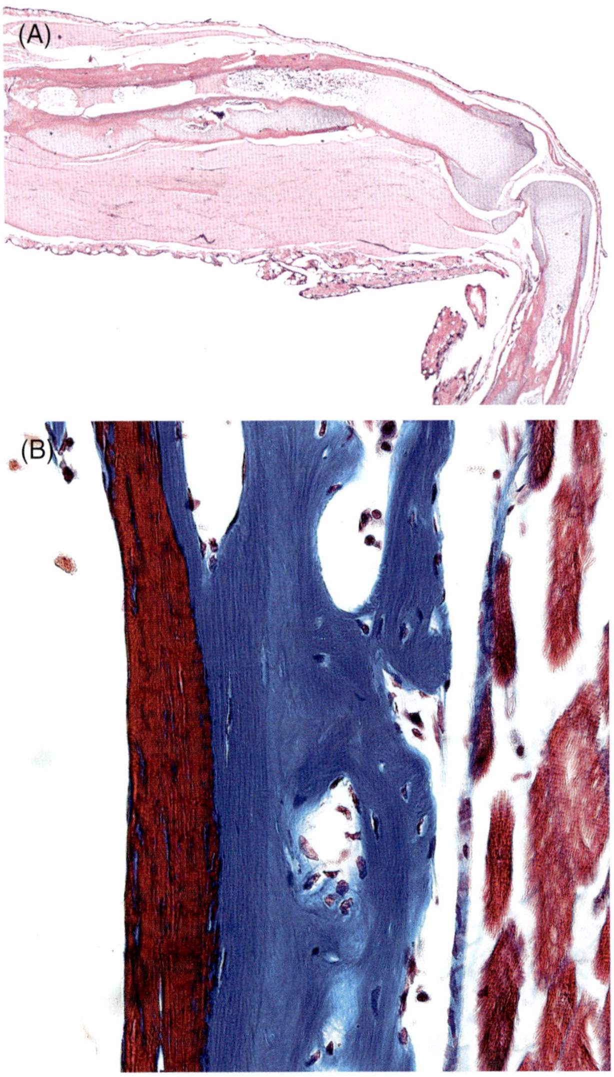

FIGURE 38.1 Metabolic bone disease in a coqui frog. (A) The femur is misshapen (*curved*) and the cortices are expanded by poorly mineralized osteoid and nodules of proliferative cartilage. (B) Special stains can highlight abnormal tissue. Poorly mineralized osteoid stains blue and preexisting cortical bone (*left*) stains red with trichrome staining (normal skeletal muscle is on the right). Masson's trichrome.

Corneal lipid deposition, also referred to as **lipid keratopathy** or **corneal lipidosis,** is a common condition in captive anuran amphibians. It is suspected to be associated with captive diets that differ in cholesterol content or other lipid composition compared to the diets of free-ranging animals.[154] Affected animals frequently have elevated serum cholesterol concentrations. Although several published reports describe this condition in the Cuban tree frog, it is also seen in older captive anurans of a variety of different species. Lesions are usually bilateral and begin as a hazy white discoloration at the limbus with extension over time to completely cover the cornea with thick white nodules and plaques (Fig. 38.3A). Histologically, early lesions consist of small stromal deposits of cholesterol with mild infiltrates of foamy macrophages (Fig. 38.3B). In advanced lesions, cholesterol clefts are surrounded by moderate to severe granulomatous inflammation with multinucleated giant cells, neovascularization, and fibrosis. In some affected animals, **xanthomas** have been observed in the brain, pituitary gland, stomach, lung, ovary, nerves, arteries, and periarticular soft tissues.[30]

Renal oxalosis occurs in tadpoles and recently metamorphosed captive ranid frogs[52,165] fed oxalate-rich plants, such as spinach. It has also been observed in postmetamorphic waxy monkey frogs that consumed crickets fed an oxalate-containing plant and in tadpoles of wild anurans feeding on oxalate-containing plants. Clinically, severely affected animals have evidence of renal dysfunction, including hydrocoelom or subcutaneous edema, as well as white foci (renal calculi) within the kidneys. Histologically, renal tubules are filled with radially arranged crystals

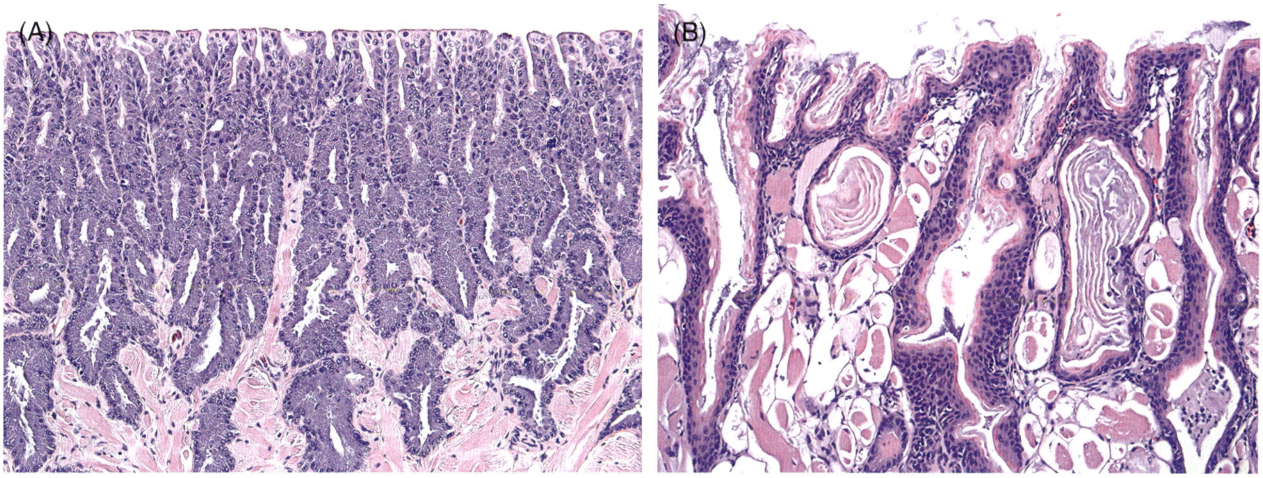

FIGURE 38.2 Hypovitaminosis A in a Wyoming toad. (A) Normal tongue. The mucosa is comprised of large mucus-producing glands. (B) Hypovitaminosis in amphibians is associated with glossal squamous metaplasia. In this case, the normal mucosa is replaced by squamous epithelium and multiple glands are distended with laminated keratin.

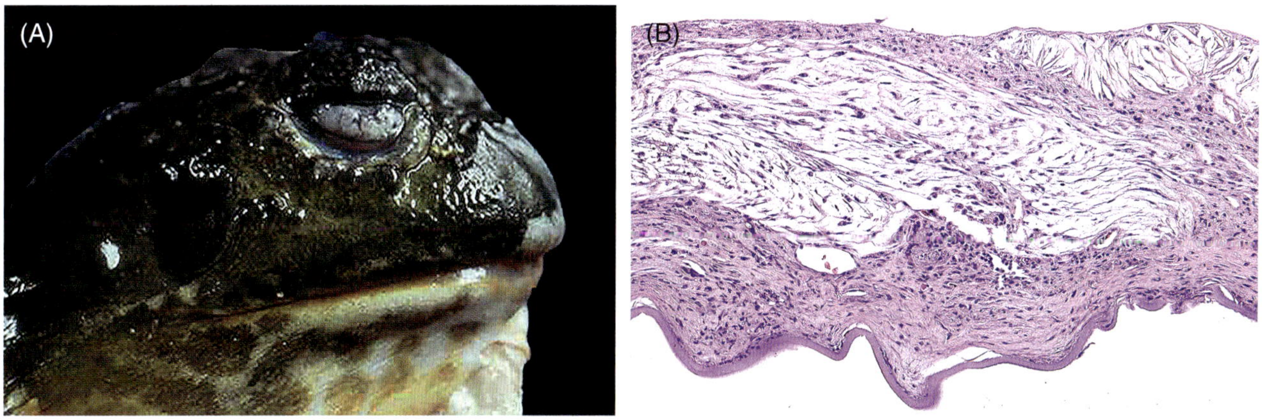

FIGURE 38.3 Corneal lipidosis in an African bullfrog. (A) The cornea is diffusely opaque white and multinodular due to advanced corneal lipid accumulation. (B) Histologically, the cornea is ulcerated and the stroma is expanded by cholesterol clefts and infiltrates of macrophages.

that are birefringent with polarized light. Diagnosis is presumptive based on histologic findings. Oxalate crystals in published reports are positive with Alizarin Red S stain at pH 7.0 but negative at pH 4.2.

Ammonium urate urolithiasis with the formation of large urinary bladder stones (Fig. 38.4) is seen in captive waxy monkey frogs.[4] Factors contributing to the development of these stones include the unique uricotelic metabolism of this species, high-protein captive diets, and dehydration.

Cholelithiasis occurs frequently in captive mountain chicken frogs and sporadically in other anurans.[107,133] Choleliths are discovered by clinical ultrasonography or radiography or at necropsy.

Visceral **metastatic mineralization**, such as in the heart and kidneys, with varying degrees of granulomatous inflammation, has been observed in captive frogs with dietary hypervitaminosis D.

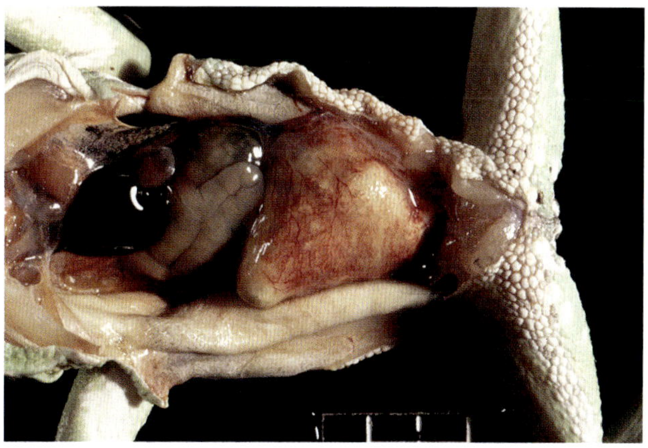

FIGURE 38.4 Urolithiasis in a waxy monkey tree frog. A large ammonium urate urolith distends the urinary bladder and displaces other viscera cranially. This species is one of the very few uricotelic amphibians.

Toxic

Calcinosis circumscripta, suspected to be due to exposure to endocrine-disrupting chemicals or cholecalciferol-containing plants, was described in tadpoles living near a wastewater treatment facility.[93] Affected tadpoles had grossly visible raised tan to yellow skin nodules containing gritty to pasty material in the tail and gular regions. Histologically, nodules were composed of calcium phosphate mineral deposits surrounded by macrophages, giant cells, and fibrosis (consistent with calcinosis circumscripta). Tadpoles had whole-body elevations of cholecalciferol.

The occurrence of free-ranging, **intersex (hermaphrodite)** anuran amphibians has been associated with natural and experimental exposure to **endocrine disrupting chemicals (EDCs)**, such as the herbicide atrazine, polychlorinated biphenyls (PCBs), bisphenol A, and estrogenic compounds found in domestic wastewater.[1] Lesions can be complicated and there have been attempts to standardize the nomenclature used in the histologic examination of amphibian gonads for defining intersex conditions.[74] Intratesticular oocytes, small numbers of oocytes dispersed within the testes, are a common gonadal lesion (Fig. 38.5). An ovotestis is a testicle that contains more than 30% oocytes. If a discrete ovary and testis are found in the same animal, that individual is said to have mixed gonadal tissue. Not all intersex amphibians are the result of exposure to EDCs and normal background levels of intersex animals can result from plasticity in amphibian gonadal differentiation. For instance, in some anurans, there is developmental hermaphroditism wherein genetic males first develop either intersex gonads or ovaries that after metamorphosis differentiate into testes.

Miscellaneous/Multifactorial

Because of the physiologic importance of the skin, histologic changes considered minor or insignificant in other vertebrates assume greater importance in amphibian histopathology. Furthermore, the relative lack of protective structures (e.g., hair) increases the susceptibility of the skin to injury, especially in captivity where trauma from flaws in enclosure design, aggressive enclosure-mates, and suboptimal water quality (e.g., elevated environmental ammonia) is more common. In most cases, the lesions associated with **environmental skin disease** are nonspecific and consist primarily of epidermal hyperplasia and hyperkeratosis. Epidermal hyperplasia is usually characterized by varying degrees of increased thickness of the epidermis, disorganization of keratinocytes, anisokaryosis, and increased numbers of mitotic figures. Epidermal thickness varies between species and is usually greater in those with a terrestrial lifestyle (e.g., toads). For most amphibians, a stratum corneum of greater than 1–2 cell layers will justify a designation of hyperkeratosis. Secondary colonization of the stratum corneum by bacteria and fungi is common in cases of environmental skin disease and clinically significant dermatitis can result (Fig. 38.6).

Brown skin disease is a dermatosis of unknown etiology in captive Puerto Rican crested toads.[35] The clinical course can be protracted and severely affected animals fail to thrive. On gross examination, affected animals have a darkened leathery appearance with abnormal shedding of brown flaky fragments of skin (dysecdysis). Histologically, findings are nonspecific and overlap with those described earlier for environmental skin disease.

Rostral abrasions (Fig. 38.7) are common ulcerative skin lesions on the rostral maxilla in captive amphibians.

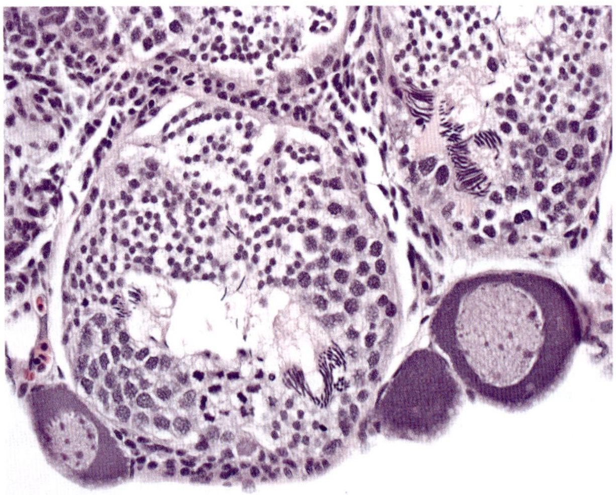

FIGURE 38.5 Gonad of a White's tree frog. Intratesticular oocytes are present along the lower edge of the testicular tissue.

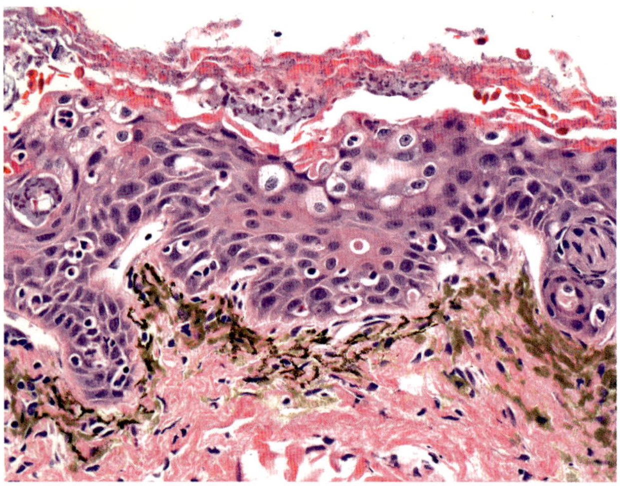

FIGURE 38.6 Severe environmental skin disease in a Wyoming toad. Shown are marked hyperkeratosis and epidermal hyperplasia. These lesions are frequently accompanied by secondary bacterial or fungal dermatitis.

FIGURE 38.7 Rostral abrasion in a green tree frog. Two pink-gray cutaneous ulcers are present on the nasal rostrum and maxillary skin. These lesions occur secondary to trauma and can be complicated by secondary bacterial infections.

These lesions occur following repeated trauma from striking enclosure glass, screens, or shipment containers. Secondary bacterial infections can result in bacteremia (see red leg syndrome discussed in the subsequent section). Chronic rostral abrasions have significant granulation tissue or fibromyxomatous tissue proliferation that must be distinguished from a neoplastic process (e.g., myxoma).

Gas bubble disease causes high rates of morbidity and mortality in aquatic amphibians housed in water supersaturated with gases (e.g., nitrogen). Most cases occur in captive settings secondary to malfunction of municipal or facility water handling and delivery systems. However, occasional mass mortality events have been observed in free-ranging animals.[19] In African clawed frogs, affected animals present with buoyancy problems characterized by floating in a head-down position.[168] Other clinical or gross necropsy findings can include cutaneous hyperemia, macroscopic air bubbles in the skin especially in the toe webbing, bloating (pneumocoelom), retinal detachment, and prolapse of the tongue or stomach from the mouth. Histologically, gas bubbles in the skin appear as discrete clear spaces in the dermis or epidermis that sometimes extend to form fissures that separate layers of the epidermis. Gas bubbles in blood or lymphatic vessels in the skin or viscera are associated with ectasia, congestion, and hemorrhage.

Cutaneous **chromatophore hyperplasia** presents as flat to raised, multifocal to coalescing areas of hyperpigmentation at sites of skin injury, such as those associated with penetrating nematode parasites.[82,160] The color of these lesions ranges from white-yellow to black depending on the predominant chromatophore present (e.g., iridophores and melanophores, respectively).

Considerable attention has been paid to a high prevalence of **limb malformations** occurring over expected background levels in populations of free-ranging anuran amphibians.[98,101] Etiologies that have been proposed for these phenomena include exposure to excess UV (ultraviolet) radiation, exposure to chemical contaminants (e.g., pesticides, insecticides, and retinoids), trauma, and parasitism. Other factors that could influence amphibian limb development include environmental pH and temperature. In nature, the spectrum of malformations is broad and ranges from missing limbs (amelia) and shortened limbs (ectromelia) to supernumerary limbs (polymelia) and toes (polydactyly).[112] The trematode parasite *Ribeiroia ondatrae* is a frequent cause of polymelia in free-ranging anurans from the United States and the prevalence of infection is enhanced by environmental eutrophication.[143] Malformations are attributed to disruption of the embryonic limb bud by the encysting metacercaria. Amelia and ectromelia are among the most common malformations observed in free-ranging amphibians, and may be attributable to sublethal trauma from predators, such as dragonfly nymphs.[8,21] The morphology of posttraumatic limb malformations shows considerable variation and complexity because of the ability of many amphibian species, including anurans, to partially regenerate amputated appendages. Anuran limb regeneration is incomplete with formation of tapering cartilaginous spikes (Fig. 38.8) that can resemble congenital malformations.[8]

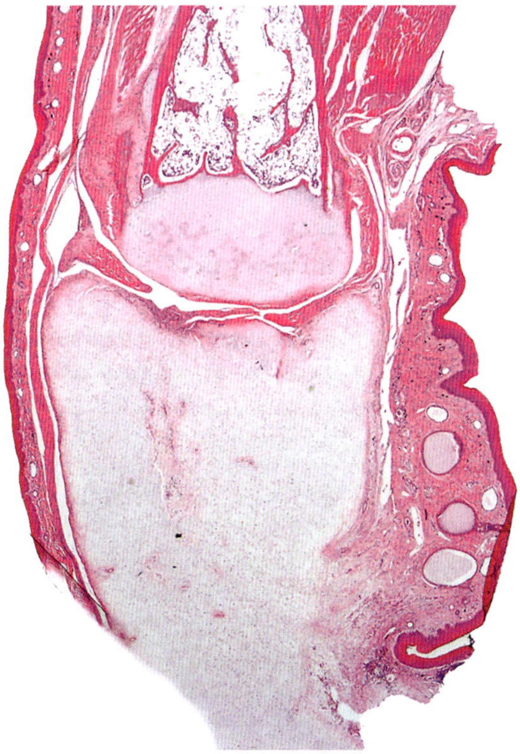

FIGURE 38.8 Limb regeneration in a Cururu toad. Incomplete regeneration of a digit. The distal phalanx is composed of cartilage that forms a tapering distal spike.

FIGURE 38.9 Spindly leg syndrome in a recently metamorphosed harlequin frog. The forelimbs are thin and underdeveloped. *(Photo Courtesy of B. Gratwicke, Smithsonian National Zoological Park.)*

Limb malformations are also observed in captive amphibians and the most common is **spindly leg syndrome (SLS)** characterized on gross examination by thin underdeveloped forelimbs that are present from the time of metamorphosis (Fig. 38.9). There is little published information on histologic findings in SLS, but lesions have included reduced numbers of forelimb myofibers and incomplete joint development.[70] The cause of SLS is likely multifactorial with nutritional and environmental factors suspected. Experimental studies in harlequin frogs demonstrated a relationship to overfeeding and environmental calcium and phosphorus availability.[99] In the dendrobatid frog, *Andinobates geminisae,* water supplementation with vitamin B complex or folic acid and increased calcium availability reduced the prevalence of SLS.[76]

Effusion within the coelomic cavity **(hydrocoelom** or **hydrops)** and the large subcutaneous lymphatic sacs of anurans (**lymphedema**) referred to as "**edema syndrome**" is a common nonspecific clinical presentation in captive and free-ranging amphibians[34,132] (Fig. 38.10). There are non-infectious and infectious causes of edema syndrome and the tendency of amphibians to develop generalized edema reflects their specialized anatomy and physiology in regards to osmoregulation and water balance. Edema can be physiologic if animals are housed in low solute water (e.g., distilled or unreconstituted deionized water) that creates an osmotic gradient favorable to cutaneous water absorption. In captive anurans, one of the most common causes of edema syndrome is chronic renal disease. The underlying etiology is only rarely determined, in part, because most cases evaluated by pathologists are end-stage with extensive fibrosis and tubular loss. Glomerular disease (e.g., membranous glomerulopathy or glomerulonephritis) is also common and comprehensive investigations using

FIGURE 38.10 Severe subcutaneous edema (lymphedema) in an American bullfrog with inflammation of the caudal lymph heart.

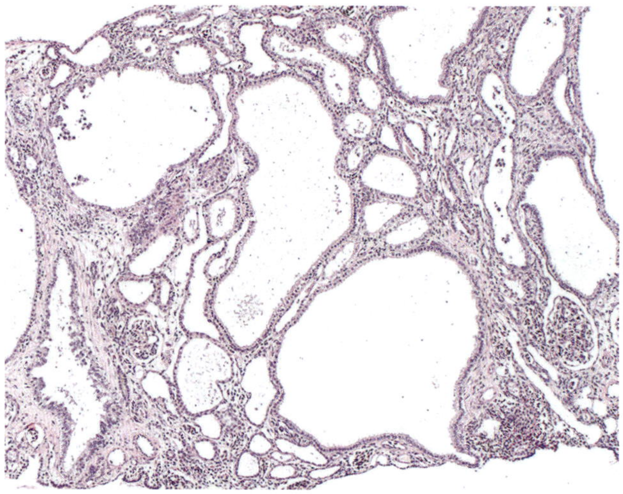

FIGURE 38.11 Severe polycystic nephropathy in a harlequin frog. The proximal renal tubules are markedly ectatic and tubules are lined by a plump cuboidal epithelium.

techniques, such as electron microscopy of well-preserved material, can be helpful to determine etiology.

Acquired **polycystic nephropathy** occurs in four species of captive harlequin frogs (e.g., Panamanian golden frog) and rusty robber frogs from Panama (Fig. 38.11). The cause is unknown but speculated to be secondary to environmental water composition.[133] Similar lesions occurred in a single clutch of captive Panamanian golden frogs in the United States.[48] In Rio Cauca caecilians, polycystic nephropathy was associated with significant interstitial fibrosis not observed in affected anurans.[91] Finally, polycystic kidneys have also been observed as a precursor to the development of renal cell carcinoma in hybrids of Japanese and Chinese toads.[108] An important differential for polycystic changes in the kidney is tubular infection with myxozoan parasites (see "Other Infectious Organisms" section).

Other common non-infectious causes of edema syndrome include hypocalcemia (see Nutritional-metabolic bone disease) with secondary lymph heart failure, hypoproteinemia, gastrointestinal disease, heart failure, renal toxicity (see Nutritional-oxalate nephrosis), and congenital anomalies of lymph hearts. Edema syndrome has also been observed in African clawed frogs with **ovarian hyperstimulation syndrome** following treatment with human chorionic gonadotropin. Common infectious causes of edema syndromes include *Ranavirus* infection, systemic bacterial infections (see red leg syndrome in "Bacteria" section), mycobacteriosis, and infections with metacercariae of echinostome trematodes (see "Metazoa" section).

Neoplastic

A wide variety of neoplasms have been described in amphibians and there are several comprehensive published reviews.[65,78,160] Most reports of neoplasia originate from captive anurans and caudates and less frequently from free-ranging animals. The best-known neoplastic disease in amphibians is **Lucke's renal adenocarcinoma** of free-ranging and captive northern leopard frogs caused by **ranid herpesvirus 1** (see "Viruses" section). Renal adenocarcinomas that are morphologically similar but not known to be associated with a viral infection are sporadically observed in other species including caecilians (Fig. 38.12).[91] Recently, there have been several reports of intestinal adenocarcinomas in captive anurans including Amazon milk frogs, mountain chicken frogs, and marine toads.[7,102]

Neoplasms of the skin are among the most commonly observed and arise from the epidermis and cutaneous glands, pigment cells, and elements of the cutaneous stroma. **Squamous papillomas** occur most often in caudates and appear grossly as single or multiple sessile plaques or pedunculated masses that sometimes spontaneously regress over several months. Histologically, papillomas are composed of discrete foci of marked epidermal hyperplasia that can progress to lesions with multiple papillary projections. Papillomas in free-ranging Japanese fire-bellied newts are influenced by environmental temperature, UV light exposure, and possibly an uncharacterized virus.[5] Sessile papillomas are difficult to distinguish from discrete areas of epidermal hyperplasia (pseudo-epitheliomatous hyperplasia) as described in tiger salamanders from a chemically contaminated environment.[65]

Squamous cell carcinoma (SCC) is commonly observed in anurans[9,22,83,170] and is histologically distinguished from squamous papilloma by cellular atypia and evidence of invasive biological behavior. Cutaneous masses with histologic evidence of glandular differentiation (e.g., formation of tubules or papillary projections) including adenomatous hyperplasia, adenoma, cystadenoma, and adenocarcinoma arise from both serous and mucous skin glands.[15,149,170]

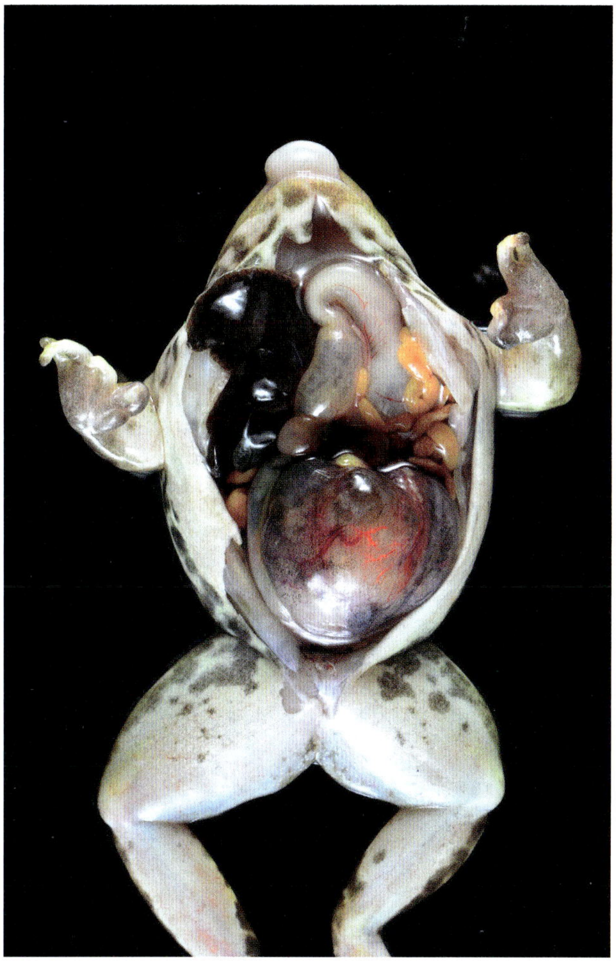

FIGURE 38.12 Renal adenocarcinoma in a mountain yellow-legged frog. A large, mottled tan, well-vascularized renal mass displaces the viscera cranially.

Chromatophoromas are neoplasms arising from cutaneous pigment cells (e.g., melanophores, erythrophores, xanthophores, and iridophores) and the majority are melanin-producing **melanophoromas**. Grossly, melanophoromas range from discrete flat to slightly raised areas of black discoloration to larger nodular masses that are invasive to underlying soft tissues. Like melanocytomas and melanomas in other vertebrates, there is variation in the morphology of individual neoplastic cells including epithelioid, spindloid, and dendritic cell forms. The degree of pigmentation also varies widely, sometimes even within the same neoplasm, and ranges from amelanotic to heavily pigmented neoplasms that require bleaching of sections to adequately evaluate cellular morphology. The biological behavior of melanophoromas ranges from self-limiting and benign to malignant neoplasms with local invasion of lymphatic sacs (anurans) and skeletal muscle, as well as occasional visceral metastases. **Iridophoromas** are rare neoplasms arising from crystalline purine-producing chromatophores. On gross examination, iridophoromas

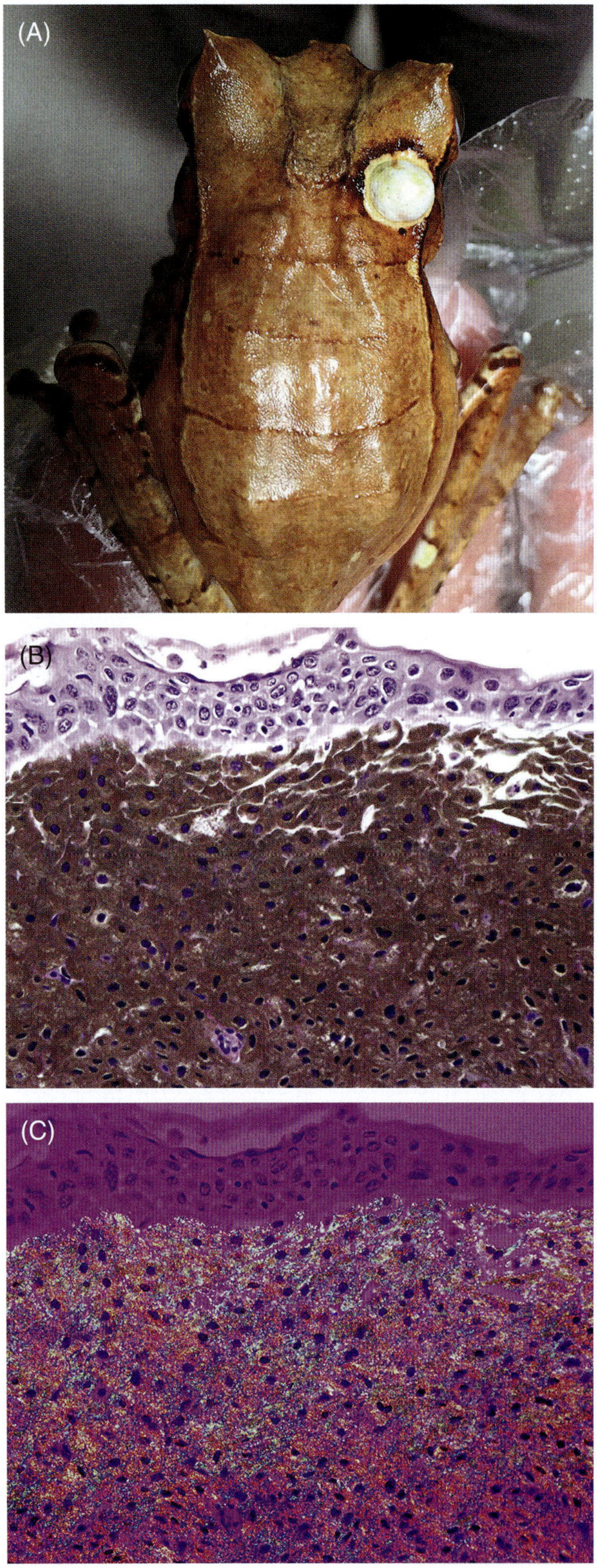

FIGURE 38.13 Cutaneous iridophoroma in a marsupial frog. (A) A well-demarcated, exophytic, white mass is present in the skin caudal to the eye. (B) The cytoplasm of neoplastic cells is filled with myriad refractile brown-yellow pigment granules. (C) Cytoplasmic pigment granules are birefringent with polarized light.

are yellow to white cutaneous masses (Fig. 38.13A). Histologically, they are similar to melanophoromas, but neoplastic cells have refractile brown intracytoplasmic pigment that is anisotropic when examined with polarized light (Figs. 38.13B, C). Both melanophoromas and iridophoromas must be distinguished from chromatophore hyperplasia (see "Miscellaneous/Multifactorial" section).

Lymphoma and other round cell tumors, including histiocytic sarcoma and cutaneous mast cell tumor, can resemble macrophagic inflammatory responses to infectious disease, especially mycobacteriosis.[6,83] Acid-fast staining of round cell tumors, especially those with histiocytic morphology, is helpful to avoid misdiagnosis. **Lymphoid leukemia** is increasingly recognized in Houston toads, Wyoming toads, and Puerto Rican Crested toads bred in captivity for species conservation programs (Fig. 38.14).[43,83] **Epitheliotropic lymphoma** in captive American bullfrogs presents clinically with patchy cutaneous erythema that can be confused with environmental skin disease (see above) or red leg syndrome (see below). Immunohistochemistry (IHC) with the mammalian lymphocyte markers CD3, CD20, and CD79a show inconsistent results but none have been systematically validated. Cutaneous **mast cell tumors** (MCT) are described in axolotl, tiger salamanders, and Puerto Rican crested toads.[72,83] Neoplastic mast cells in amphibians have a varied appearance ranging from round to polygonal or spindle-shaped cells (Fig. 38.15). Eosinophils are observed in MCT of salamanders, but are not a prominent feature in crested toads. Metachromatic cytoplasmic granules can be demonstrated with Giemsa or toluidine blue stains; some tumors are paucigranular. In Puerto Rican crested toads, there was variable immunoreactivity with tryptase IHC, and neoplastic cells

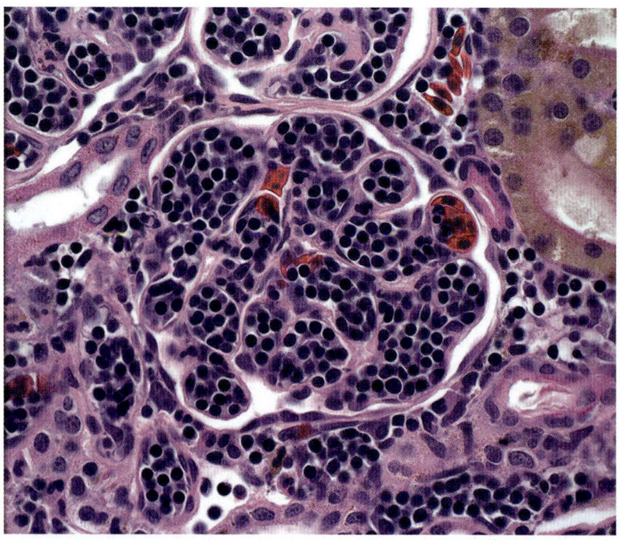

FIGURE 38.14 Lymphoid leukemia in a Wyoming toad. Glomerular (center) and interstitial capillaries are distended by a uniform population of neoplastic lymphocytes.

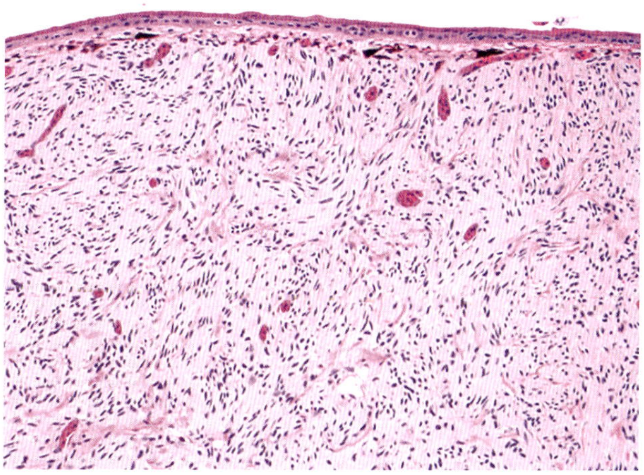

FIGURE 38.15 Cutaneous mast cell tumor in a Puerto Rican crested toad. Neoplastic cells are round to spindloid. Differentials can include histiocytic sarcoma and granulomatous inflammation due to mycobacteriosis. *(Photo Courtesy of M.A. Delaney, University of Illinois Zoological Pathology Program.)*

were nonimmunoreactive for ionized calcium-binding adapter molecule (IBA-1) which aids in differentiation from histiocytic sarcoma. MCT in crested toads was disseminated and present in organs including the liver and kidney.

INFECTIOUS DISEASES

DNA Viruses

Iridoviruses in the genus *Ranavirus* are a significant cause of mortality events in both free-ranging and captive anurans and caudates. Concerns about the potential for ranaviruses to contribute to declines of free-ranging populations resulted in listing of these viruses as notifiable to the World Organization for Animal Health (WOAH). Several species of ranaviruses are associated with amphibian disease. These include **Frog Virus 3** (FV3; predominantly affecting anurans), ***Ambystoma tigrinum*** **virus** (ATV; predominantly affecting salamanders), **Bohle iridovirus** (BIV; anurans), and **Common Midwife Toad virus** (CMTV; anurans and caudates). Some of these viruses can also cause disease in fish and reptiles.[42,172] In anurans, disease is most frequently observed in tadpoles and recently metamorphosed frogs, but experimental and natural disease is increasingly recognized in adults. Animals subclinically infected with ranaviruses may be important as disease reservoirs.[146] Infections are multisystemic with a tropism for hematopoietic tissues, endothelial cells, and a variety of epithelial cells. On gross examination, common lesions are erythema and petechial and ecchymotic hemorrhages of the skin and viscera (Fig. 38.16), subcutaneous and coelomic cavitary edema, and cutaneous ulcers. Additional gross findings in tiger salamanders and Chinese giant salamanders include raised cutaneous plaques or polyps.[18] Histologic lesions of systemic ranaviral infections can

FIGURE 38.16 Erythema in the ventral skin of a smoky jungle frog. Differential diagnoses for cutaneous erythema in amphibians include *Ranavirus* infection, chytridiomycosis, and bacterial infections (red leg syndrome).

include: (1) necrosis of hematopoietic tissues in the bone marrow and visceral organs that have significant extramedullary hematopoiesis, such as liver, spleen, and kidney (Fig. 38.17A); (2) multicentric hemorrhage with varying degrees of fibrinoid vasculitis and endothelial cell necrosis (Fig. 38.17B); and (3) necrosis of epithelial cells in the skin, liver, and kidney, among other sites.[33,53,87,114] Characteristic basophilic intracytoplasmic inclusion bodies are frequently observed in ranaviral lesions (Fig. 38.17C). Experimental infections of wood frogs with FV3 are associated with neutrophilia, circulating undifferentiated blast cells, and intracytoplasmic inclusion bodies in lymphocytes, monocytes, neutrophils, and eosinophils.[54]

A presumptive diagnosis of ranaviral infection can be made from histologic lesions and observation of characteristic cytoplasmic inclusions. However, secondary bacterial infections with organisms, such as *Aeromonas hydrophila*, are common in affected animals and these lesions overlap and obscure primary lesions of ranaviral infection.[38,114] In addition, inclusion bodies are sometimes absent or

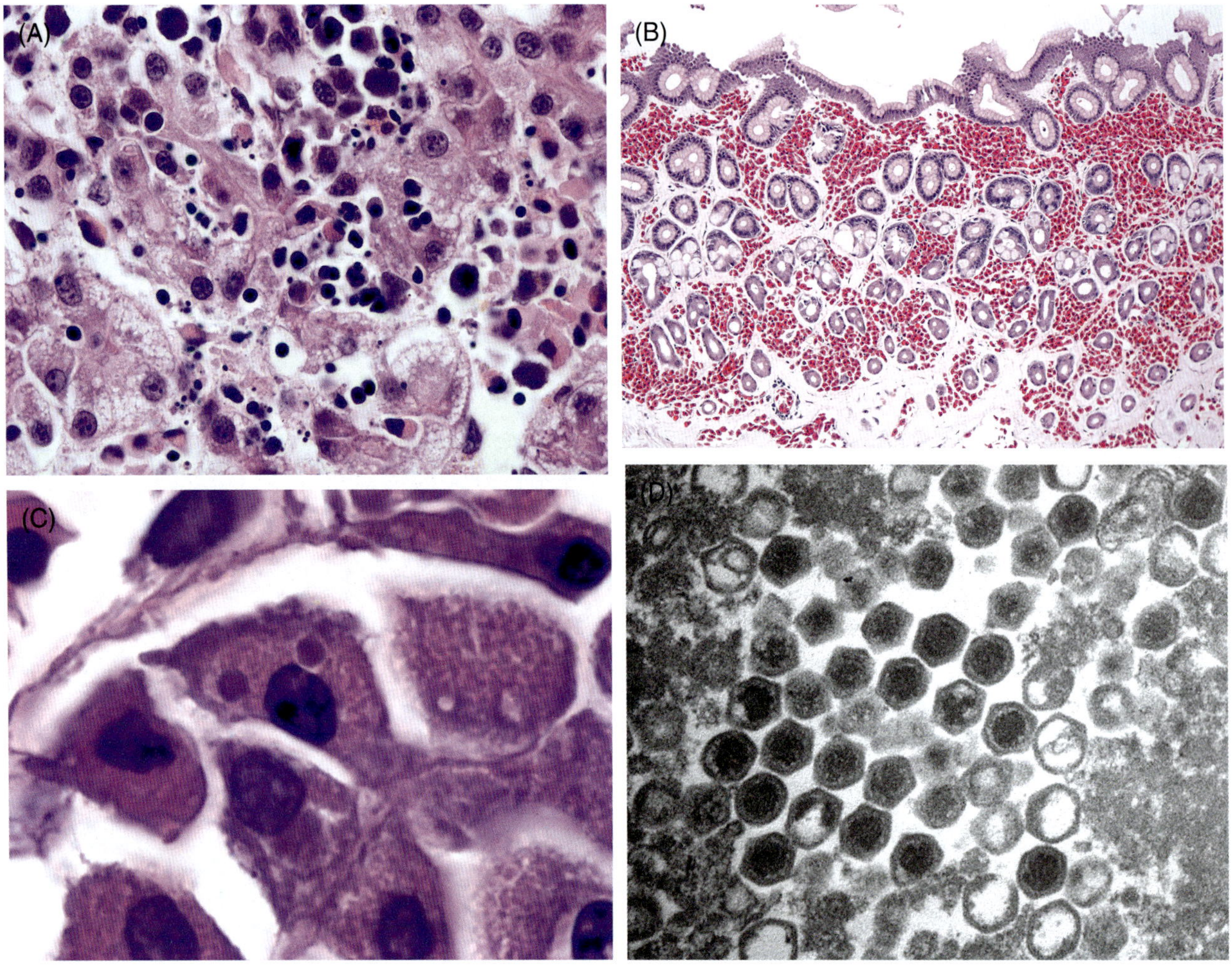

FIGURE 38.17 Ranavirus infection in the liver of a green frog tadpole (A), stomach of a spadefoot toad (B), liver of a leopard frog (C), and kidney of a western toad (D). Infection can be associated with cellular necrosis (A, C) and hemorrhage (B). (A) In the liver, sinusoids are filled with karyorrhectic debris representing necrosis of preexisting hematopoietic tissue. This change is sometimes mistaken for autolysis. (B) Infection can be associated with diffuse mucosal hemorrhage. (C) In addition to individual hepatocyte necrosis, one hepatocyte (center) contains two characteristic intracytoplasmic amphophilic inclusion bodies. (D) Cytoplasmic arrays of icosahedral virions are typical of iridoviruses. Transmission electron microscopy (TEM).

present in low numbers and in cases with substantial necrosis, they may be confused with karyorrhectic debris. Therefore, confirmation of infection using ancillary diagnostic tests, such as PCR or virus isolation, should be run in cases of systemic bacterial infection (i.e., red leg syndrome) or those with compatible histologic lesions. Immunohistochemistry using polyclonal antibodies as well as in-situ hybridization can be used to demonstrate ranaviruses in histologic section.[39,127,146] Transmission electron microscopy is a useful diagnostic tool for demonstrating iridovirus virions in affected tissues, even in some cases that lack obvious inclusion bodies. *Ranavirus* virions have icosahedral symmetry and range from 120 to 175 nm in diameter (Fig. 38.17D).

Erythrocytic iridoviruses produce intracytoplasmic acidophilic to basophilic intracytoplasmic inclusion bodies in red blood cells and splenic reticuloendothelial cells of free-ranging anurans.[69,159] Previously, these inclusions were interpreted as protozoa in the genera *Pirhemocyton* or *Toddia*. Infections are subclinical in most cases, but anemia was observed in heavily infected American bullfrogs. **The *Rhinella marina* erythrocytic-like virus (RMELV)** in the *Betairidovirinae* subfamily of the Iridoviridae was recently characterized.[145] Most other members of the *Betairidovirinae* are pathogens of invertebrates. Similar erythrocytic viruses occur in reptiles and fish and they are unrelated to the ranaviruses (*Alphairidovirinae*) that cause systemic disease in amphibians.[68] Ultrastructurally, virions of the amphibian erythrocytic viruses range from 300 to 450 nm in diameter and therefore are considerably larger than ranaviruses.

First described in 1934, infection with **Ranid herpesvirus 1 (RaHV-1)** is the cause of **Lucké's renal adenocarcinoma** in free-ranging and captive Northern leopard frogs (eSlide 38.e1).[29] During the algid phase of infection, cool winter temperatures (<7°C) are permissive for viral replication but proliferation of neoplastic cells is minimal. In contrast, when temperatures increase, the calid phase begins and is associated with increased proliferation of neoplastic cells and greatly reduced viral replication. On gross examination, there are single to multicentric, smooth to multilobular, white to yellow masses in the kidney that in advanced stages fill the coelomic cavity and displace viscera.[110] Metastasis occurs during the calid phase and is most often to the liver. Histologically, the neoplasms are unencapsulated but well-demarcated from unaffected kidney, and histologically there is formation of variably sized tubules and papillary projections. During the algid phase, Cowdry type A intranuclear inclusion bodies typical of herpesvirus infections are observed within neoplastic cells. To date, RaHV-1 associated neoplasms have only been described in the Northern leopard frog; however, grossly and histologically similar renal adenocarcinomas are sporadically observed in other amphibians including a caecilian. Whole genome sequences of **RaHV-1** and **RaHV-2**, a herpesvirus of uncertain clinical significance, show close relationships to Ictalurid herpesvirus-1 of fish.

Ranid herpesvirus skin disease occurs in free-ranging agile frogs, common frogs, and common toads in Europe. Originally described from observation of herpesvirus-like particles by transmission electron microscopy, molecular characterization identified the viruses as **Ranid herpesvirus 3 (RaHv-3)** in frogs and **Bufonid herpesvirus 1** in toads.[47,119,121,120] Disease is observed seasonally near the end of hibernation and extending into the breeding season. Mortality has not been observed. Infection with RaHv-3 occurs in both adult frogs and tadpoles, but gross and histologic lesions have only been observed in postmetamorphic animals.[122]

Grossly, affected frogs have multifocal to coalescing, gray to white areas of thickened skin over the dorsum and inguinal region; in agile frogs, the lesions have been described as vesicular. Affected toads have multifocal to coalescing areas of brown to black hyperpigmentation that progress to raised, round cutaneous papules that are sometimes described as "cauliflower-like". Histologically, there is epidermal hyperplasia that in mature lesions has a line of demarcation between the superficial and deep epidermal layers. In frogs, there is intracellular edema of the superficial keratinocyte layer with subsequent necrosis and sloughing of the superficial layer. Lesions extend into cutaneous glands with dermal edema. In toads, lesions are limited to the epidermis, lack keratinocyte degeneration and necrosis, and have overlying orthokeratotic hyperkeratosis. In both frogs and toads, there are eosinophilic to amphophilic intranuclear inclusion bodies in affected superficial keratinocytes. Associated infiltrates of granulocytes and mononuclear cells are minimal to mild.

In captive Mexican giant tree frogs, a herpesvirus (not yet further characterized) identified by electron microscopy has been associated with intranuclear inclusion bodies and necrosis in multiple organs including the liver, heart, pancreas, heart, kidney, and skin.[57]

Other DNA and RNA Viruses

Surprisingly, few other significant viral infections have been reported in amphibians. Metatranscriptomic and metagenomic surveys of the Cane Toad microvirome revealed infections from the Circoviridae, Hepesviridae, Nackednaviridae, Papillomaviridae, and Picornaviridae families.[145] Necrosis and lymphohistiocytic myositis in striated muscle of the tongue with eosinophilic intranuclear inclusion bodies was described in captive spring peepers. Transmission electron microscopy of these inclusions revealed 20–24 nm in diameter, nonenveloped, icosahedral virus-like particles suggestive of a **parvovirus.**[139] A suspected poxvirus-associated dermatitis was described in European common frogs, but the putative viral particles were later determined to be melanosomes.[37,38] Experimental infections of American Bullfrogs with **West Nile Virus** resulted in low-level viremia but no mortality was described.[95] The **rhabdovirus** responsible for **spring viremia of carp** was isolated from Chinese fire-bellied newts, but no histologic lesions were described.[84] Similarly, **adenoviruses** and **caliciviruses** have been isolated from frogs but are of unknown clinical significance.

Bacteria

Red leg syndrome or **bacterial dermatosepticemia** is one of the first described and most widely recognized diseases of amphibians. Although commonly attributed to the gram-negative bacillus, *Aeromonas hydrophila*, a wide variety of other gram-negative bacteria, such as *Pseudomonas aeruginosa*, *Elizabethkingia meningoseptica* (*Flavobacterium meningosepticum*), *Klebsiella pneumoniae*, and *Chryseobacterium indologenes* as well as some gram-positive organisms, such as *Streptococcus iniae* have been isolated from affected animals.[67,100,109,147] These are systemic bacterial infections that are clinically manifested by cutaneous hyperemia and hemorrhage distributed on the limbs, digits, and ventral abdomen (Fig. 38.16). Red leg syndrome is most often observed in individual animals or as outbreaks in captive amphibians that have been stressed by overcrowding, recent shipment, poor water quality, low environmental temperatures, or cutaneous injuries. Occurrence of disease solely attributable to bacterial infection is rare in free-ranging amphibians and most published accounts fail to exclude concurrent or predisposing primary infectious diseases that have overlapping clinical signs (e.g., *Ranavirus* infection).[64] In addition to cutaneous hyperemia, gross findings in red leg syndrome can include cutaneous and visceral petechiae, cutaneous ulcers, ascites, and subcutaneous edema. Other manifestations of bacteremia in amphibians are otitis media and panophthalmitis. Infection with *Elizabethkingia miricola* is an emerging cause of meningitis in farmed anurans that present clinically with torticollis.[80,167] Histologic findings are typical of those seen in bacterial infections in other animals and include multicentric congestion, hemorrhage or fibrin thrombi, and multifocal random necrosis in the liver, spleen, or kidney. Inflammation is granulocytic (neutrophilic) in early lesions with later infiltration by macrophages to form granulomas. Intralesional bacteria, which sometimes form visible colonies, can be observed within thrombi or within areas of necrosis and inflammation. Diagnosis of red leg syndrome is by demonstration of compatible histologic lesions with bacterial culture. However, consideration should be given to exclusion of other potential or concurrent causes of cutaneous hyperemia and hemorrhage, such as ranavirus infection (see "Viruses" section).

Mycobacteriosis, caused by numerous species of nontuberculous mycobacteria not limited to *Mycobacterium marinum*, *M. chelonae*, *M. fortuitum*, *M. xenopi*, and *M. ulcerans* ecovar Liflandii, is frequently observed in captive anurans and less often in salamanders. Reports in free-ranging amphibians are rare. These bacteria are common in aquatic environments and in most cases are probably opportunistic pathogens. In experimental studies of *M. marinum* infection in leopard frogs, chronic subclinical infections could be converted to lethal infections by immunosuppression with corticosteroids.[138] There is a single report of tuberculosis caused by *M. bovis* in a captive American bullfrog.[81]

The clinical presentation of mycobacteriosis ranges from sporadic in individual animals to population outbreaks that occur in association with husbandry factors, such as stress, overcrowding, or suboptimal environmental temperature. Mycobacteriosis is usually seen in adult postmetamorphic animals but cases are occasionally observed in juveniles and tadpoles. Infections can be limited to a single organ system (e.g., the integument or bone) or widely disseminated. On gross examination, nodules in the skin and subcutaneous lymphatic sacs are common and may become ulcerated or form abscesses (Fig. 38.18A). In disseminated infections, white to yellow nodules ranging from numerous and miliary to large, solitary, and discrete are commonly observed in the liver, spleen, and kidney but can be found in virtually any tissue. Coelomic effusion is observed in animals with mycobacterial coelomitis or widely disseminated infections. Histologically, mycobacterial lesions are composed predominantly of infiltrates of macrophages arranged either as discrete, multifocal to coalescing granulomas or as poorly defined sheets (Fig. 38.18B). In some cases, macrophages are admixed with small numbers of neutrophils, eosinophils, and lymphocytes. Chronic granulomas are surrounded by fibroblasts and fibrosis. Central necrosis in granulomas is uncommon but is observed in some chronic lesions. However, infections caused by mycolactone-producing organisms, such as *M. ulcerans* ecovar Liflandii, can have lesions with extensive necrosis (Fig. 38.18C). Early or subclinical lesions of visceral mycobacteriosis appear as aggregates of minimally pigmented macrophages that expand melanomacrophage centers in the liver and spleen. Macrophages in mycobacterial lesions frequently have a uniform appearance with small to moderate amounts of cytoplasm, and multinucleated giant cells are present in low numbers or absent. This morphology sometimes leads to misdiagnosis of amphibian mycobacteriosis as a lymphoma or other round cell neoplasm.

The presumptive diagnosis of mycobacteriosis is made by demonstration of acid-fast bacteria within inflammatory lesions. Although Ziehl-Neelsen acid-fast staining is sufficient for many cases of amphibian mycobacteriosis, use of alternative methods, such as Fite's, are most reliable for demonstrating aquatic mycobacteria in histologic section (Figs. 38.19A, B). Bacteria are most often intracellular within macrophages, but are sometimes present extracellularly in colonies large enough to be observed on routine hematoxylin and eosin (HE)-stained sections. The appearance of individual organisms ranges from short bacilli to long filamentous beaded forms. Confirmation of the diagnosis requires either isolation of bacteria in culture from fresh tissues or use of molecular methods.

Gram-negative, facultatively intracellular bacteria in the family ***Brucellaceae*** are emerging causes of soft tissue and bone infections in a wide variety of captive and free-ranging anuran amphibians worldwide.[60,75,116,152] Although first reported in cane toads as *Ochrobactrum anthropi*, these infections are now attributed to several genetically diverse ***Brucella inopinata*-like** organisms. Common gross lesions are subcutaneous abscesses on the dorsum or limbs that respond to surgical treatment and antibiotic therapy. Other reported lesions include chronic spinal arthropathy with cartilage proliferation and joint fusion, limb arthritis with bone loss, abscesses or nodules in the coelomic cavity and viscera and oophoritis. Histologically, there is pyogranulomatous inflammation and necrosis with infiltrates of macrophages, fewer neutrophils, and occasionally eosinophils, lymphocytes, and plasma cells. Organisms are difficult to detect in histologic section but have been observed as coccobacilli in smears of abscess exudate. Cases of **brucellosis** must be differentiated from other common causes of subcutaneous abscesses, disseminated granulomatous inflammatory lesions, or osteomyelitis in amphibians, especially mycobacterial and fungal infections. *B. inopinata*-like organisms are readily isolated using routine bacteriological methods, however, they can be misidentified (e.g., *Ochrobactrum anthropi*) using some commercially available biochemical test systems.

Infections with gram-negative intracellular pathogens in the family *Chlamydiaceae* are increasingly recognized in both captive and free-ranging amphibians. The original descriptions of amphibian **chlamydiosis** in captive African clawed frogs identified the causative agent as *Chlamydia*

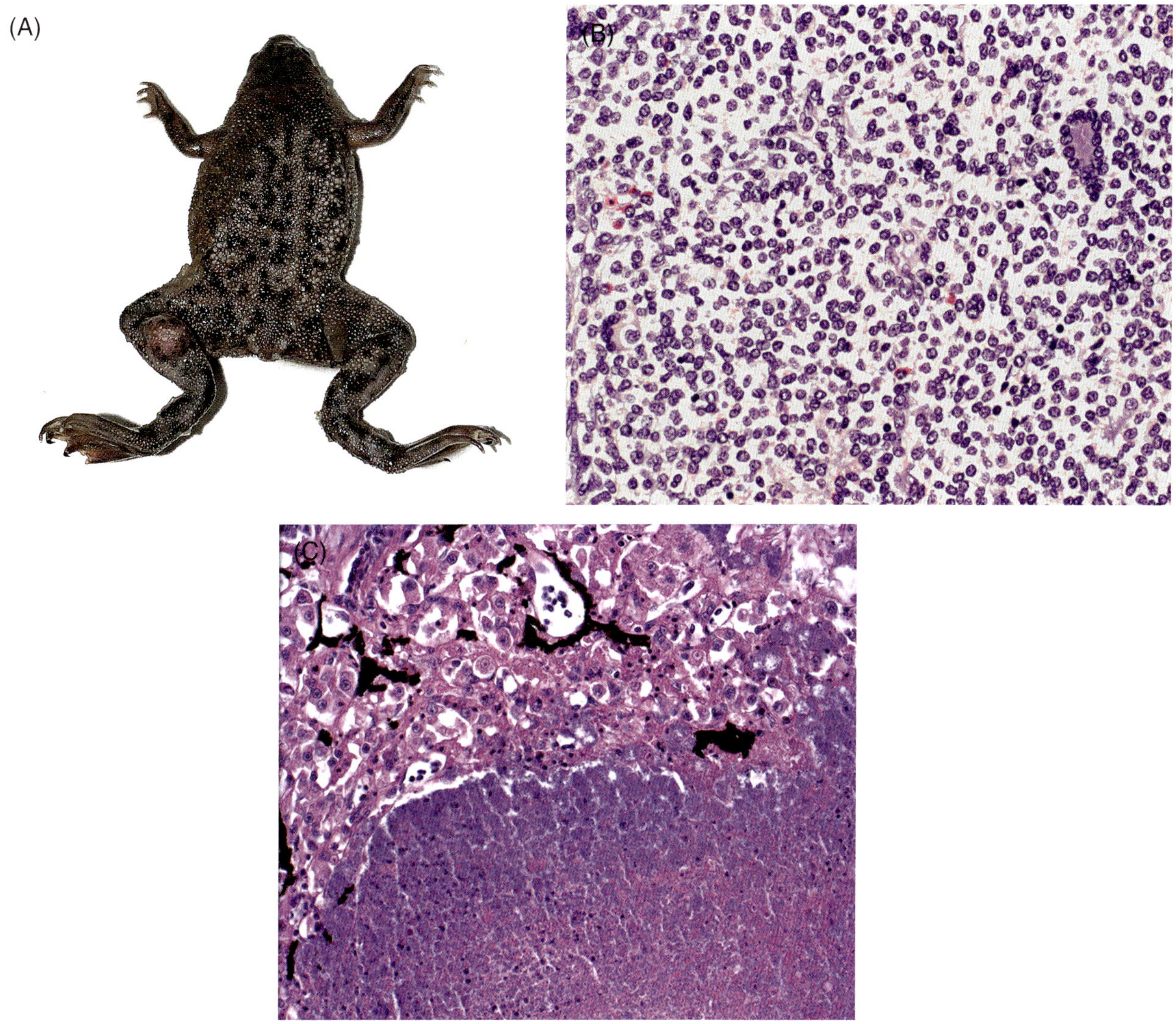

FIGURE 38.18 Mycobacteriosis in an African dwarf frog (A, C) and a Western toad (B). (A) Cutaneous mycobacteriosis is associated with the formation of a discrete granuloma on the skin of the leg of this frog. (B) Granulomatous inflammation consists of a near-uniform population of macrophages that can be mistaken for neoplastic round cells. There are occasional multinucleate giant cells. (C) *Mycobacterium ulcerans* ecovar *liflandii* in the liver. Infection is associated with regionally extensive lytic necrosis with bacteria (bottom of image).

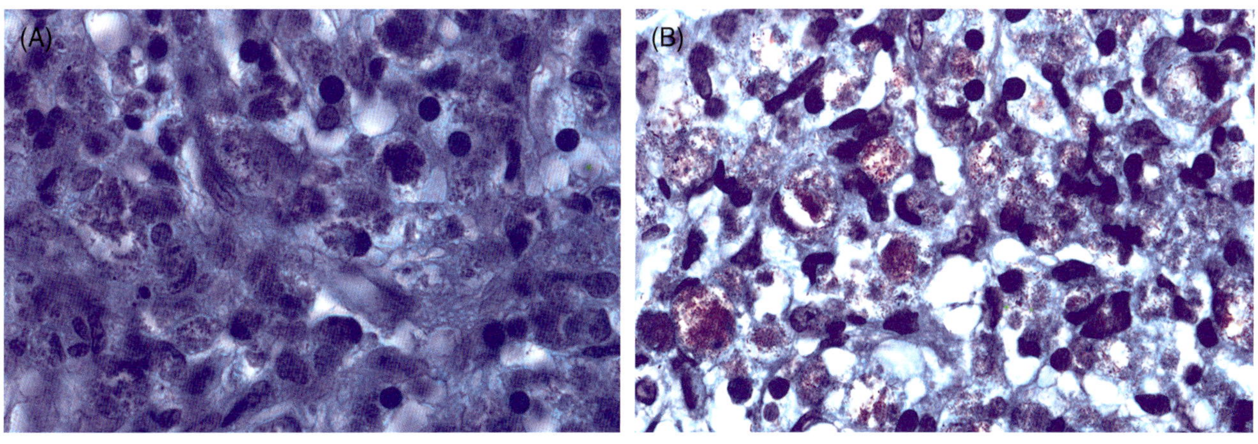

FIGURE 38.19 Comparison of acid-fast stains for mycobacterial infections. *Mycobacterium chelonae* infection granuloma in a Houston toad tadpole. Compare the Ziehl-Neelsen acid-fast stained section with the same section stained with Fite's acid-fast stain. There is greatly improved staining of bacilli with Fite's (B) compared with Ziehl-Neelsen (A).

psittaci,[79,174] but subsequent reports of chlamydiosis in anurans have usually implicated ***C. pneumoniae***. Disease in European newts was attributed to a novel taxon ***Candidatus Amphibiichlamydia salamandrae*** and a related organism ***Amphibiichlamydia ranarum*** was identified at a high prevalence in asymptomatic American Bullfrog tadpoles.[104,105] Immunohistochemical and PCR surveys of free-ranging and captive frogs from Switzerland revealed an unexpectedly high prevalence of subclinical infections suggesting that amphibians could act as reservoir hosts for these organisms.[17] Clinical and gross necropsy findings in sick animals are nonspecific and can include cutaneous hyperemia, excessive shedding of skin (sloughing), subcutaneous edema, and hepatosplenomegaly. Histologically, in systemic infections, there is hepatocellular and splenic necrosis with infiltrates of macrophages and fewer lymphocytes and plasma cells. Interstitial nephritis, epicarditis, and myocarditis with similar lymphohistiocytic infiltrates have also been described. In a free-ranging giant barred frog from Australia, the predominant lesion was a mononuclear interstitial pneumonia.[13] In captive Houston toads, necrotizing and histiocytic polioencephalomyelitis and ganglionitis were the predominant histologic findings.[55] In most cases of chlamydiosis, large intracytoplasmic granular basophilic inclusions are found in infiltrating macrophages or in hepatocytes, respiratory epithelial cells, or neurons of affected tissues (Fig. 38.20A). These inclusions are reliably positive using a Gimenez stain. Transmission electron microscopy of inclusions demonstrates characteristic elementary and reticulate bodies of the *Chlamydiaceae* (Fig. 38.20B). Diagnosis of chlamydiosis is usually made by observation of suggestive histologic lesions and cytoplasmic inclusion bodies combined with immunohistochemistry or PCR and DNA sequencing of chlamydial 16 s RNA and ompA genes.

Infection with an unnamed genus of intracellular gram-negative bacteria in the Order **Rickettsiales** and family ***Anaplasmatacea*** causes clear spherical cytoplasmic inclusions in the erythrocytes of anurans and caudates.[40] These organisms had previously been placed in the genus *Aegyptianella*. The clinical significance is unknown.

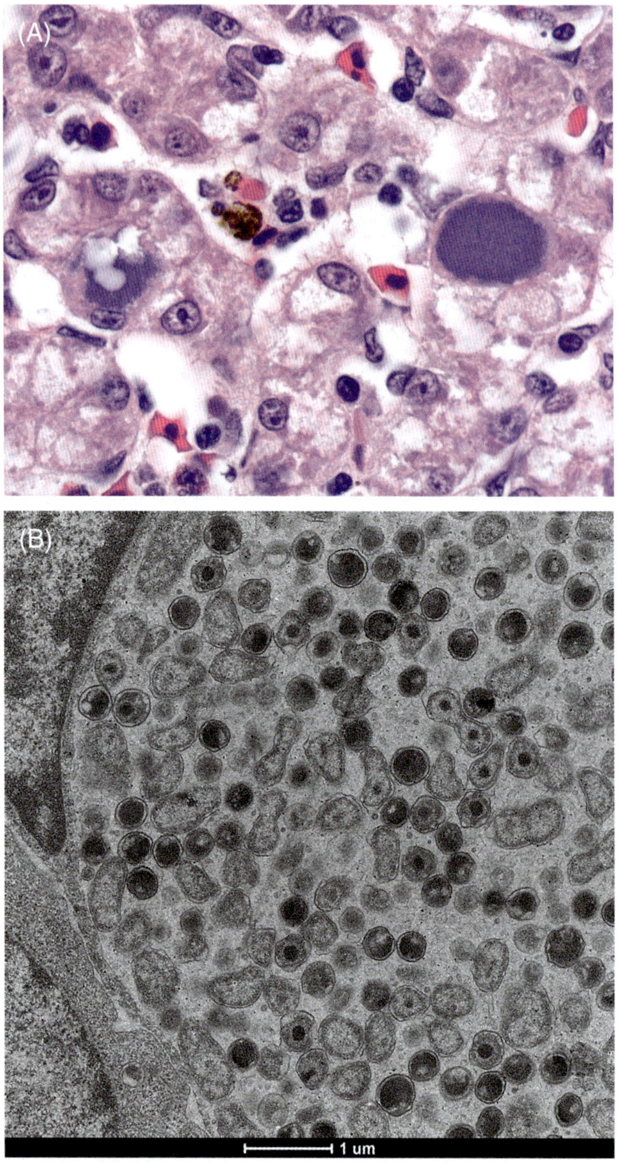

FIGURE 38.20 Chlamydiosis in the liver of an evergreen toad (A) and kidney of a harlequin frog (B). Scattered hepatocytes contain large, intracytoplasmic basophilic granular inclusions. This animal was infected with *Chlamydia pneumonia*. (B) The cytoplasm of an interstitial macrophage is filled with numerous reticulate and elementary bodies (electron-dense structures) characteristic of *Chlamydia* sp. Transmission electron microscopy (TEM).

Fungi

The chytrid fungi ***Batrachochytrium dendrobatidis* (*Bd*)** and ***Batrachochytrium salamandrivorans* (*Bsal*)**, the cause of ***Bd* chytridiomycosis** and ***Bsal* chytridiomycosis** respectively, have emerged as important causes of morbidity and mortality events and population declines in free-ranging and captive amphibians worldwide.[12] *Bd* is a generalist pathogen that infects anurans, caudates and caecilians and has been found on every continent with extant amphibians. *Bsal* was described as a new *Batrachochytrium* species causing mortality events of free-ranging and captive salamanders in Europe.[105,106] *Bsal* predominantly infects salamanders but experimental infections and disease have been described in a small number of anuran species.[62,166] *Bd* and *Bsal* are notifiable to the World Organization for Animal Health (WOAH) because of concerns that international movement of amphibians for trade contributes to the dissemination of the chytrid fungi. *Bd*-associated population declines are caused by prevalent hypervirulent genotypes (i.e., the *Bd* Global Panzootic Lineage [*Bd*-GPL]), but ancient geographically restricted and less pathogenic *Bd* lineages are also recognized.[25]

Both *Bd* and *Bsal* are intracellular pathogens of keratinocytes in the skin of postmetamorphic animals. *Bd* is also found in the keratinized mouthparts of anuran tadpoles and infection moves to the keratinized skin that develops during metamorphosis. The life cycle is simple with intracellular fungal bodies (thalli) that mature to form zoosporangia containing numerous infective, posteriorly flagellated zoospores. Death due to *Bd* infection is attributed to interference with cutaneous Na^+ channel function resulting in loss of electrolyte homeostasis. Severely affected animals become hyponatremic, hypokalemic, and hypochloremic and die due to terminal cardiac arrest.[26] In *Bsal* chytridiomycosis, clinically affected newts had decreased plasma Na^+ and Cl^- and increased K^+ compared to subclinically infected and uninfected newts.[151] Other clinicopathologic findings included increased plasma protein levels and a left shift (i.e., increased numbers of circulating immature neutrophils) with toxic changes and Dohle bodies. These changes were attributed to possible dehydration and development of secondary bacterial infections. The outcome of infection with *Bd* or *Bsal* ranges from subclinical carrier states to severe disease associated with substantial population morbidity and mortality. Resistant species, such as the American bullfrog and the African clawed frog, develop minimal skin lesions and low infection intensity after exposure. Others, such as the Pacific chorus frog, are tolerant of high fungal infection intensity but rarely develop disease that results in mortality.[140] Highly susceptible species like the endangered Panamanian golden frog have a high infection intensity, severe skin disease, and significant population mortality after exposure to *Bd*.[44]

Pathogen, host, and environmental factors that influence the outcome of *Bd* infection have been identified. These include the composition of antimicrobial peptides secreted by the cutaneous granular glands, innate immunity conferred by cutaneous bacterial flora that secretes antifungal metabolites (e.g., *Janthinobacterium lividum*) and environmental temperature and moisture. The role of adaptive immunity in the outcome of *Bd* infection is uncertain and while *Bd*-specific antibodies can be found in serum and cutaneous mucus, vaccination trials have had mixed results. Investigations of naturally and experimentally infected frogs have shown enhanced survival with specific MHC class II genotypes.

Gross findings in *Bd* chytridiomycosis include brown to red (erythema) skin discoloration, rough skin texture, excessive skin shedding (dysecdysis) (Fig. 38.21A), and occasional cutaneous erosions or ulcers. Dysecdysis caused by chytridiomycosis is characterized by shedding of numerous small discolored skin fragments rather than the large translucent to white sheets of skin observed with normal ecdysis. In salamanders infected with either *Bd* or *Bsal* there may be multifocal areas of black hyperpigmentation in areas of infection. Gross lesions of *Bsal* chytridiomycosis in susceptible salamanders include dysecdysis, but epidermal erosions and ulcers are the predominant feature in later stages of the disease (Fig. 38.21B). In anuran tadpoles, depigmentation of the keratinized toothrows and jaw sheaths is common with *Bd* infection. Chytridiomycosis caused by either *Bd* or *Bsal* can present as sudden death without gross lesions.

Histologic lesions of both *Bd* and *Bsal* chytridiomycosis are limited to areas of keratinizing stratified squamous epithelium in the skin of postmetamorphic animals or the oral disc of tadpoles. In aquatic neotenic salamanders, infections are limited to small areas of keratinized epithelium on the feet. Lesions of *Bd* chytridiomycosis are most reliably present on the ventral body, feet, and within skin folds, but can be found anywhere, especially in totally aquatic species. In *Bd* chytridiomycosis, characteristic histologic lesions include epidermal hyperplasia and orthokeratotic hyperkeratosis with numerous characteristic intracellular chytrid thalli (fungal bodies) within the stratum corneum (Fig. 38.21C). In some species, there is marked retention (up to 20 cell layers) of the expanded stratum corneum, but other species more frequently shed the infected keratinized layers. In the latter instances, only epidermal hyperplasia may be observed and a detailed search through several histologic sections is required to demonstrate chytrid thalli. Inflammatory cell infiltrates of neutrophils or lymphocytes are uncommon and more consistently observed in *Bd*-tolerant species or with secondary bacterial or fungal infections. Occasionally, intracellular edema of superficial keratinocytes or even epidermal erosion and ulceration are present.

In contrast to *Bd* chytridiomycosis, *Bsal* chytridiomycosis causes multifocal keratinocyte necrosis that is associated with numerous chytrid-type fungal thalli (Fig. 38.21D). In the early stages of infection or at the margins of an established lesion, *Bsal* chytridiomycosis resembles the hyperplastic lesions of *Bd* chytridiomycosis. In these instances, *Bsal* thalli are intracellular within keratinocytes and marginate host cell nuclei. Necrotic foci of *Bsal* chytridiomycosis extend partially or fully through the epidermis and when lesions are mature the necrotic epidermis is sloughed leaving areas of erosion and ulceration.

In histologic section, multiple developmental stages of *Bd* and *Bsal* thalli can be identified and are useful for diagnosis.[11,134] Zoosporangia are intracellular within keratinocytes and range from ~5 to 15 μm in diameter. The earliest form is round and uninucleate with progression to multinucleate forms that have stippled to microvacuolated cytoplasm and ultimately formation of a mature zoosporangium with multiple discrete 2–3 μm in diameter basophilic spherical zoospores. Occasionally, zoosporangia are observed with discharge papillae that lead to the skin surface that give the thallus a flask-like appearance (Figs. 38.22A–C). In diagnostic practice, empty zoosporangia that have already discharged their zoospores are the predominant form available for examination. The empty zoosporangia easily allow for identification of internally septate colonial thalli (Fig. 38.23), which are highly characteristic of the genus *Batrachochytrium*. Extending from some mature thalli are

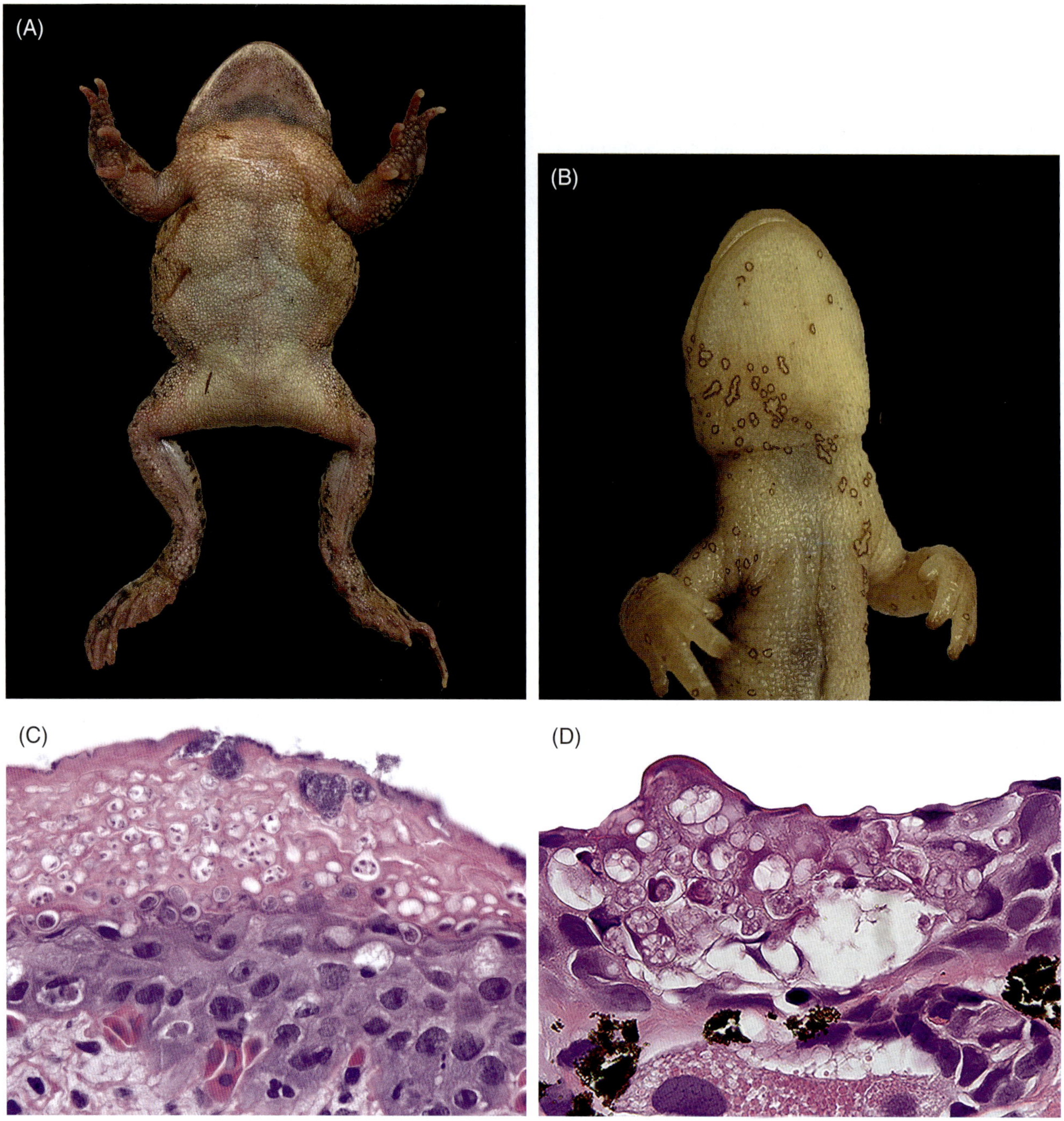

FIGURE 38.21 Chytridiomycosis in amphibians. *Batrachochytrium dendrobatidis* (*Bd*) in a Southern toad (A) and Wyoming toad (C) and *Batrachochytrium salamandrivorans* (*Bsal*) in a rough-skinned newt (B) and an eastern newt (D). (A) In *Bd* infection, dysecdysis is characterized by multiple sheets of brown shedding skin. This reflects hyperkeratosis typical of *Bd* chytridiomycosis. (C) This corresponds to severe epidermal hyperplasia and hyperkeratosis. The thickened stratum corneum contains myriad spherical chytrid thalli and superficial bacterial colonies. (B) In *Bsal* infection, multifocal to coalescing, circular to irregular foci bounded by brown pigmentation are seen. The centers of these lesions contain necrotic keratinocytes and *Bsal* fungal thalli. Cutaneous ulceration occurs as the necrotic centers are sloughed. (D) This corresponds to near full-thickness epidermal necrosis with myriad spherical chytrid thalli. Most of the thalli are empty with numerous fine internal septa (colonial thalli).

fine root-like structures termed rhizoids that are best visualized with silver stains, such as GMS (Fig. 38.22D).

Diagnosis of chytridiomycosis is by recognition of characteristic chytrid fungal thalli in wet mounts, stained cytologic preparations of skin scrapings or by histopathology. Recognition of flask-shaped thalli, thalli containing discrete zoospores, and empty internally septate colonial thalli are the most useful forms for increasing confidence in the morphologic diagnosis. Zoosporangia that contain zoospores are sometimes confused with protozoal cysts and some early publications attributed lesions and chytrid thalli to the zygomycete fungus *Basidiobolus ranarum*.[117]

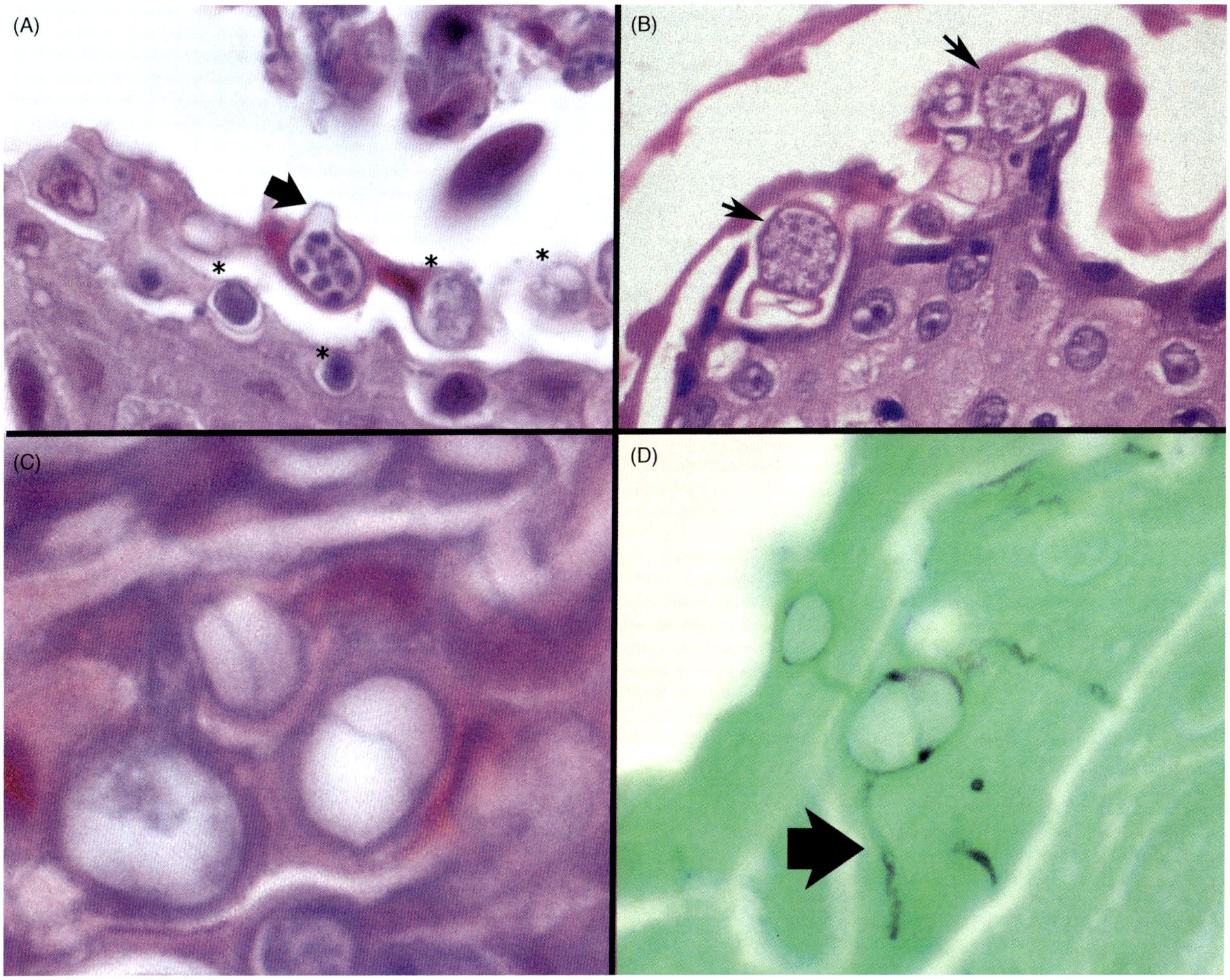

FIGURE 38.22 *Batrachochytrium dendrobatidis*. Detailed morphologic features of thalli useful for histologic diagnosis. (A) Mature zoosporangium with discrete basophilic zoospores and a discharge tube (*arrow*). (B) Developing multinucleated thalli (*arrows*). (C) Three "empty" thalli that have previously discharged their zoospores. The thalli in the center are colonial thalli as indicated by fine internal septation that appears to divide the thallus in half. (D) The rhizoids (*arrow*) are thin root-like projections from the thallus that are intermittently observed in silver-stained histologic sections. GMS. *(Reprinted with permission Mader and Divers. Current Therapy in Reptile Medicine and Surgery.)*

The lesions of *Bd* and *Bsal* chytridiomycosis overlap especially in early stages of infection and with low-level infections. Histologic findings of prominent necrosis with development of erosions and ulcers as well as the frequent presence of larger (up to 20 μm) colonial thalli with multiple internal septa should raise suspicion for *Bsal* chytridiomycosis.[173] Thalli are highlighted by GMS and PAS stains (Fig. 38.22D). Immunohistochemistry using a polyclonal antibody was described to aid in identification of *Bd* chytrid thalli, but cannot distinguish between thalli of *Bd* and *Bsal*.[10,106] In situ hybridization can be used to distinguish between thalli of *Bd* and *Bsal* in histologic section and for diagnosis of mixed *Bd* and *Bsal* infections.[123] PCR is useful to diagnose infection in clinical settings, for confirmation of histologic diagnoses and for differentiation between *Bd* and *Bsal* infections. Routine clinical fungal culture methods are not helpful for the diagnosis of chytridiomycosis.

Infections with several pigmented (dematiaceous) fungi, including *Cladosporium*, *Philalophora, Exophiala,* and *Fonsecaea* species, among others, cause **chromoblastomycosis** and **phaeohyphomycosis** in amphibians.[90] Most published reports describe infections in captive anurans, but sporadic cases have been observed in free-ranging animals and caudates. These fungi are common saprophytes in soil and plant materials worldwide and infection follows traumatic inoculation of the skin with environmental organisms. On gross examination, there are cutaneous papules and nodules that may be ulcerated or pigmented brown to black (Figs. 38.24A, B). In some cases, visceral dissemination occurs with nodules most frequently found in the liver, spleen, and kidneys. Histologically, cutaneous and visceral nodules are granulomas with prominent multinucleated giant cells and fewer eosinophils and neutrophils. Granulomas of chromoblastomycosis contain characteristic spherical, 10–15 μm, internally septate, brown-pigmented sclerotic

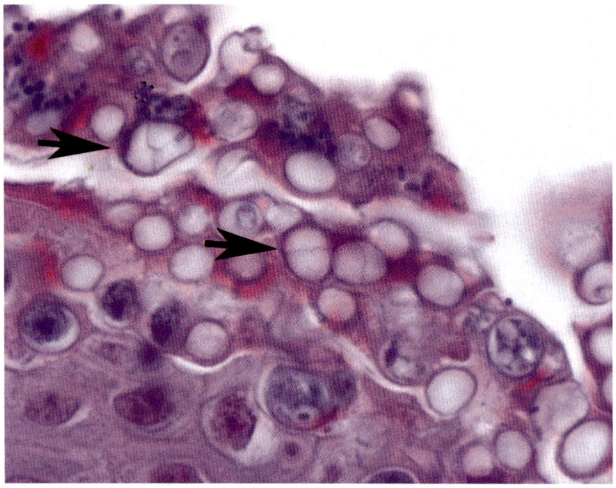

FIGURE 38.23 Chytridiomycosis caused by *Batrachochytrium dendrobatidis* in a smoky jungle frog. There are multiple spherical chytrid thalli in the stratum corneum. Most of the thalli are empty with occasional fine internal septa (colonial thalli; *arrows*). *(Reprinted with permission Mader and Divers, Current Therapy in Reptile Medicine and Surgery, Elsevier, 2014.)*

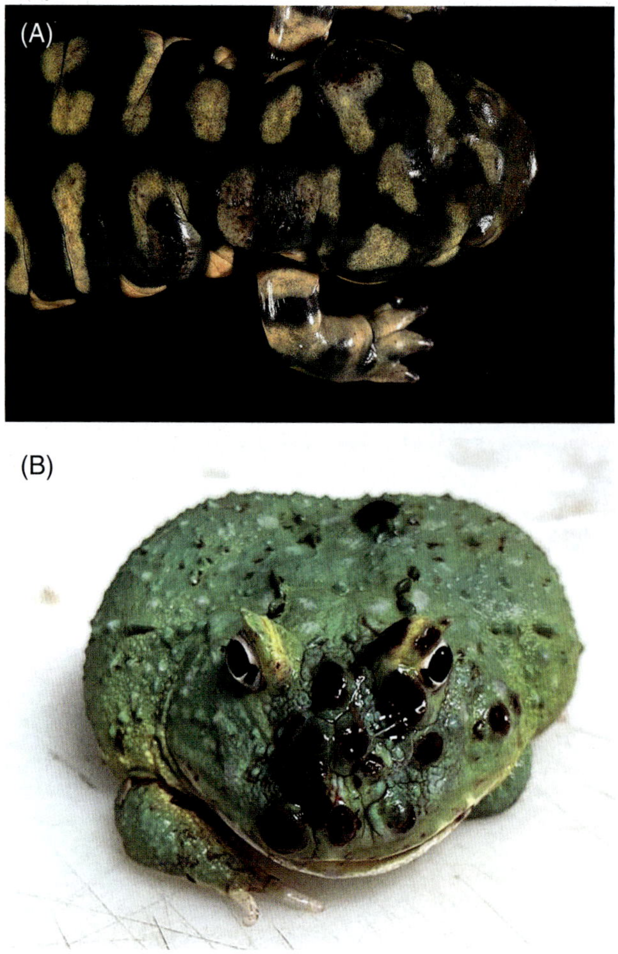

FIGURE 38.24 Chromomycosis in the skin of a tiger salamander (A) and phaeohyphomycosis in the skin of an ornate horned frog (B). (A) Infection is associated with multifocal ulceration. Differentials for cutaneous ulcers and nodules include mycobacteriosis, neoplasia, and chytridiomycosis caused by *Batrachochytrium salamandrivorans*. (B) Infection can be associated with discoloration of the skin. In this case, it was associated with multifocal to coalescing black-pigmented cutaneous nodules.

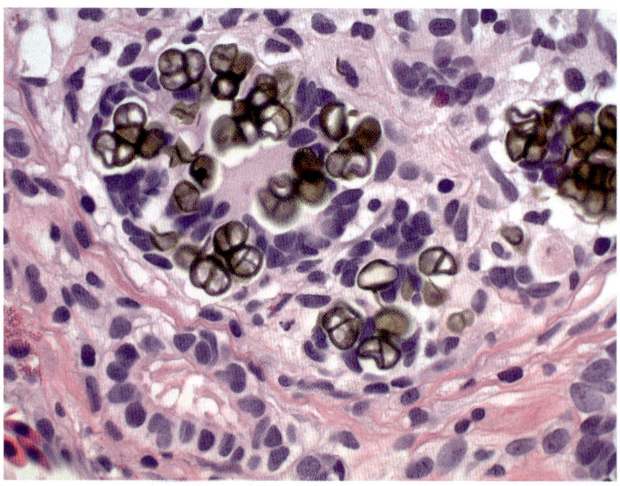

FIGURE 38.25 Chromoblastomycosis in the kidney of a harlequin frog. Granulomatous inflammation surrounds numerous internally septate brown-pigmented sclerotic bodies.

bodies (Fig. 38.25) that are sometimes accompanied by septate fungal hyphae. In phaeohyphomycosis, only pigmented hyphae are observed.

The dimorphic zygomycete fungus ***Mucor amphibiorum*** causes a distinctive disseminated mycosis in free-ranging and captive Australian anurans.[36,158] Disease has also been described in captive anurans and caudates from Europe exposed to an imported Australian green tree frog. *M. amphibiorum* has a mycelial stage in soil and infection of amphibians is thought to occur through ingestion with subsequent hematogenous dissemination. In amphibian hosts, the fungus forms a unique nonhyphal spherule. Gross lesions include multifocal to coalescing pale cream to yellow nodules on serosal surfaces and in the parenchyma of multiple organs, including the liver, spleen, and kidney, among others. Skin lesions may be present as papules with variably ulcerated centers. Histologically, lesions are granulomas predominantly composed of epithelioid macrophages, including prominent multinucleated giant cells, with some lesions containing smaller numbers of neutrophils, eosinophils, and lymphocytes, as well as areas of central necrosis. Fungal spherules are found both extracellularly and intracellularly within macrophages, and are between 2 and 40 μm in diameter with a refractile outer wall. Small "daughter" spherules develop within larger "mother" spherules before rupturing through the outer wall. Other lesions described in some affected animals include rhinitis with nasal granulomas and transmural granulomatous and necrotizing enteritis. Diagnosis is by demonstration of characteristic spherules in granulomas by squash preparations or histology and by fungal culture. Morphologically, the spherules of *M. amphibiorum* must be distinguished from endosporulating fungi, such as *Coccidioides immitis* or algae (e.g., *Prototheca* spp.).

Saprolegniasis, caused by infections with Oomycete water molds in the genera *Saprolegnia*, *Aphanomyces*, and *Achyla*, is observed in both captive and free-ranging larvae and aquatic postmetamorphic amphibians.[91,131] Mortality of egg masses (embryos) due to infections with water molds is a concern for amphibian conservation in association with environmental factors, such as UVB radiation. Although the mycelial forms of water molds morphologically resemble true fungi, their cell walls are composed of cellulose rather than chitin and genetically they are most closely related to brown algae and diatoms. In captive amphibians, saprolegniasis is usually an opportunistic infection and frequently lesions develop following skin trauma or exposure to low environmental temperatures, high environmental organic loads, or poor water quality (e.g., high environmental ammonia). On gross examination, saprolegniasis has a characteristic appearance with tufts of tan to white or gray cottony material on the skin that is best appreciated underwater (Fig. 38.26A). Outside of water, tufts collapse to form soft granular plaques that overlie erosions or ulcers. In some cases, only cutaneous ulcers will be observed. In larvae, lesions are often located on the keratinized mouthparts, nostrils, or gills.[14] Histologically, lesions are composed of mats of thin-walled, nonseptate, usually nonbranching hyphae (Fig. 38.26B) that stain poorly with PAS and well with GMS. The underlying epithelium is necrotic or ulcerated and hyphae can penetrate into the deeper soft tissues or body cavity. Infiltrates of associated inflammatory cells are usually minimal, even in lesions known to have been present for several days. Diagnosis of saprolegniasis is usually presumptive by identification of the characteristic gross and histologic morphology of the lesions. Water molds are ubiquitous in aquatic environments and will grow as saprophytes on dead amphibians. Therefore, postmortem overgrowth should be considered in autolyzed animals. Species identification of water molds is by specialized cultured techniques or molecular methods.

Fungal dermatitis caused by infections with true fungi that form hyphae visible on skin scrapings or by histology is common in captive amphibians. Similar to saprolegniasis, these infections are frequently opportunistic following skin irritation from suboptimal water quality or environmental conditions. Many different fungi can be isolated in culture from clinical samples, reflecting the opportunistic nature of most infections. Commonly isolated species are *Fusarium* spp. and zygomycetes like *Basidiobolus ranarum*, *Mucor* spp., and *Rhizopus* spp.[129] Gross findings overlap with chytridiomycosis and chromoblastomycosis, as well as red leg syndrome, ranaviriosis, and non-infectious conditions, such as brown skin syndrome. Observations include brown to red skin discoloration, excessive skin shedding, cutaneous nodules, and ulcers. Infections with a *Mucor*

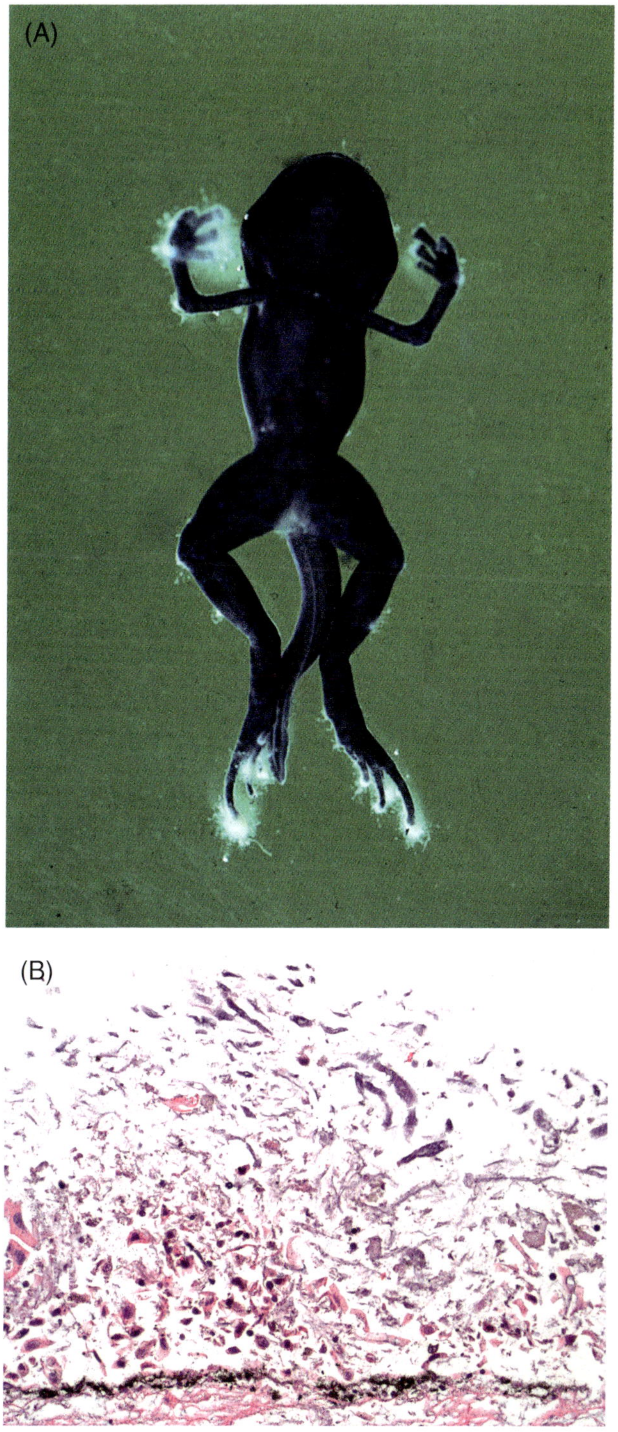

FIGURE 38.26 Saprolegniasis in the skin of a metamorphosing blue poison dart frog tadpole (A) and Kaiser newt (B). (A) This photograph was taken with the frog underwater to show the characteristic white tufts of hyphal growth on the feet. (B) There is full-thickness necrosis of the epidermis with large numbers of thin-walled, pleomorphic hyphae characteristic of the Oomycete water molds.

spp. in Wyoming toads[163] and a *Fusarium* spp. in harlequin frogs caused cutaneous nodules with grossly visible fuzzy mycelial growth similar to saprolegniasis. Histologically, lesions range from proliferative with epidermal hyperplasia, hyperkeratosis, and intracorneal fungal hyphae to necrotizing with infiltration and destruction of the epidermis by fungal hyphae. In proliferative lesions, epidermal microabscesses with neutrophils and macrophages surround the penetrating fungal hyphae and there is mixed dermal inflammation that includes neutrophils, macrophages, and lymphocytes. The specific identification of fungi within lesions of fungal dermatitis is by fungal culture or by PCR and DNA sequencing. Because skin samples from amphibians are frequently contaminated with environmental fungi, attempts should be made to match the morphology of the hyphae observed in tissue section with any fungal isolates. For instance, the hyphae of *Fusarium* spp. are typically slender with parallel walls and frequent septation, whereas those of *Mucor* spp. are wide and infrequently septate. Infections initially attributed to *Basidiobolus ranarum* in both dwarf African clawed frogs and Wyoming toads have subsequently been determined to be chytridiomycosis caused by *B. dendrobatidis*.[117]

Although traditionally classified as protozoa, the microsporidia are now known to be obligate intracellular parasites within the Kingdom Fungi. **Microsporidiosis** occurs in both captive and free-ranging anurans and caudates. ***Pleistophora myotrophica*** caused myositis and high mortality in captive European common toads.[27] Affected animals became emaciated and on gross examination had white discoloration and streaking of the pectoral, abdominal, and femoral musculature. Histologically, myriad refractile 3.2–4.5 μm spores were observed within the sarcoplasm of skeletal myocytes and in the cytoplasm of macrophages. Necrosis and regeneration were seen in affected myocytes. A ***Pleistophora*-like microsporidian** was the cause of an epaxial myositis with secondary kyphosis in a San Marcos salamander.[56] Infections with ***Alloglugea bufonis*** cause intestinal xenomas (enlarged host cells containing microsporidial spores) in free-ranging cane toad tadpoles and recently metamorphosed juveniles from Brazil.[125] On gross examination, affected tadpoles had white nodules or cysts on the intestinal serosa, spleen, and kidneys. Histologically, xenomas expanded the lamina propria of the intestine and spores were also found within macrophages in the spleen and kidney. Infections with unidentified microsporidians were the cause of necrotizing hepatitis in free-ranging mountain yellow-legged frog tadpoles and histiocytic enteritis and coelomitis in White's tree frog tadpoles. Other unidentified microsporidians have been observed in the ovary and less frequently the testis or Bidder's organ of several anuran and caudate species. In these cases, there are aggregates of spores (sporophorous vacuoles) that when ruptured are associated with granulomatous inflammation (oophoritis) (Fig. 38.27A). The clinical and reproductive significance of these infections has not been established. The initial diagnosis of microsporidiosis is by identification of xenomas and spores by histopathology. Spores can be highlighted by several histologic stains and are usually positive with Gram, Luna, Ziehl-Neelsen, and Fite acid-fast methods (Fig. 38.27B). Confirmation of the diagnosis is usually by transmission electron microscopy to identify ultrastructural features of spores, such as the characteristic coiled polar filaments or by PCR and DNA sequencing of affected tissues.

Metazoa

A wide variety of metazoan and protozoan parasites occur in free-ranging and captive amphibians, and amphibians are intermediate and paratenic hosts for innumerable nematodes, cestodes, trematodes, acanthocephalans, and pentastomes. However, relatively few of these parasites are known to cause disease.[96] The following discussion does not attempt to be comprehensive but rather focuses on

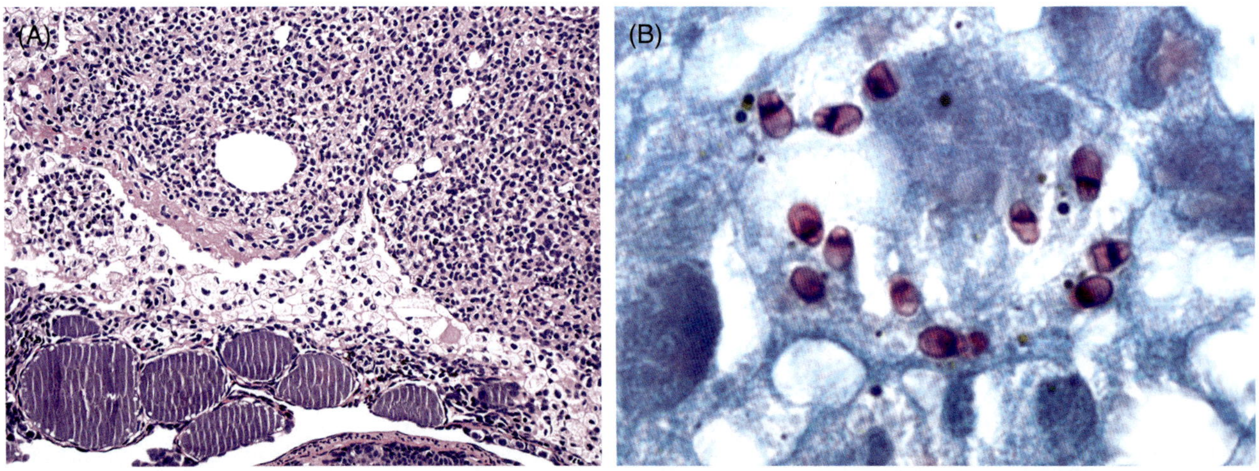

FIGURE 38.27 Microsporidiosis in the ovary of a San Marcos salamander. (A) There is granulomatous oophoritis with dense infiltrates of macrophages. (B) Special staining can highlight the oval microsporidial spores, which are pink with this stain. Fite's.

parasites problematic for captive amphibians or those that are commonly encountered as incidental findings in diagnostic examinations.

Over 40 different species of the amphibian lungworm ***Rhabdias*** (family *Rhabdiasidae*) are common parasites of both free-ranging and captive anuran and caudate amphibians worldwide. Many *Rhabdias* species are host-specific; some are generalists. These nematodes have a direct life cycle with reproductive stages that occur both within the host and in the environment. Hermaphroditic adults in the lung produce eggs or larvae that are swallowed and pass through the gastrointestinal tract to be shed in feces. In the environment, larvae develop into male and female adults that mate to produce infective larvae that penetrate the skin of the amphibian host and migrate internally to the lung. In captive animals, this life cycle can result in clinically significant hyperinfection, especially in situations with suboptimal hygiene; infections in free-ranging amphibians are usually well tolerated. On gross examination, *Rhabdias* spp. are mottled white-black worms within the airway lumen of the lung (Fig. 38.28A). Low parasite loads have no associated histologic lesions. In hyperinfections, there is pneumocyte hypertrophy and hyperplasia with mild interstitial infiltrates of lymphocytes, macrophages, or occasional eosinophils. Characteristic histologic features of *Rhabdias* spp. are scant platymyarian musculature, vacuolated lateral chords, a uterus with developing larvae, and an intestinal tract containing abundant dark-pigmented material (Fig. 38.28B). Histologic lesions are also observed as the result of larval penetration of the skin with subsequent soft tissue and coelomic cavitary migration and associated mixed cellular infiltrates of varying proportions of macrophages, neutrophils, eosinophils, and lymphocytes surrounding migrating larvae. Diagnosis is by the identification of morphologically compatible nematodes on gross examination or histopathology. Embryonated eggs or larvae can be recovered from fecal samples or tracheal wash specimens. Diagnostic differentials for metazoan parasites in amphibian lungs include digenetic trematodes in the genus *Haematoloechus* and monogenean trematodes in the genus *Pseudodiplorchis* (see later).

Intestinal nematodes are common in free-ranging and captive amphibians and while nonspecific clinical signs of weight loss or declining condition are often attributed to nematodiasis, there are very few published reports of confirmed disease. ***Strongyloides*** spp. nematodes cause proliferative enteritis in both free-ranging and captive anurans.[126,133] These nematodes have a direct life cycle similar to that described for *Rhabdias* sp. (see above) and hyperinfections are a significant problem for captive animals with in settings with suboptimal hygiene. Histologically, there is small intestinal mucosal hyperplasia with thick blunted villi and intramucosal tunnels containing

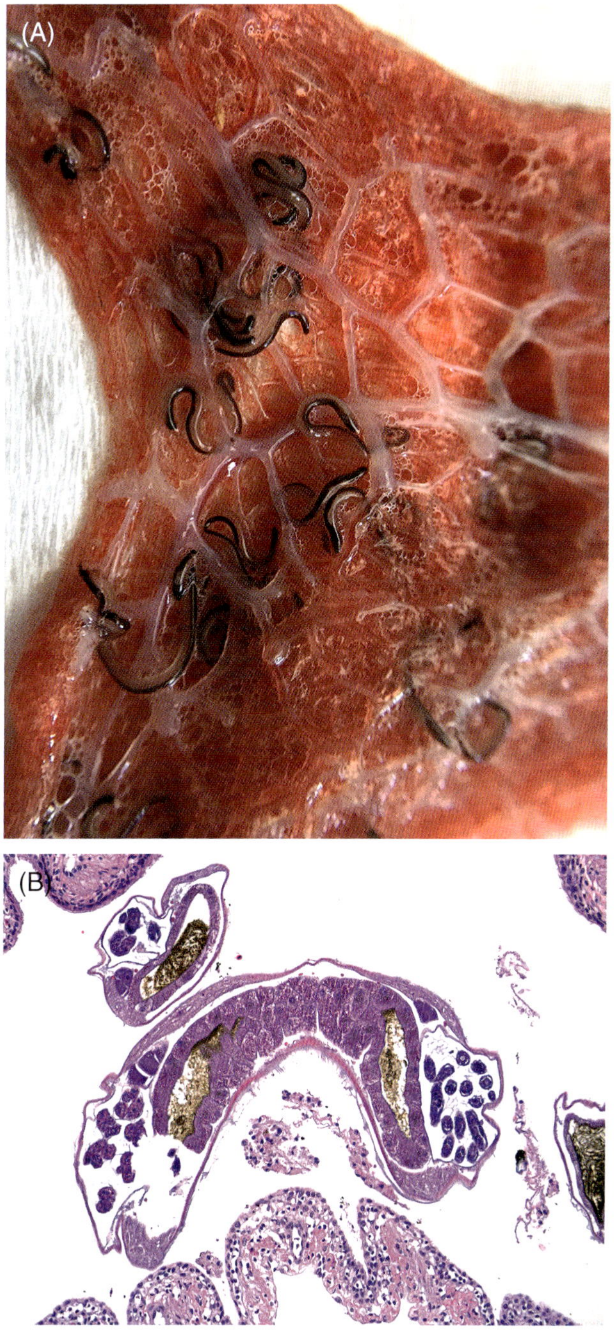

FIGURE 38.28 Pulmonary nematodiasis in a cururu toad (A) and reed frog (B). Numerous, dark, coiled *Rhabdias* spp. nematodes are present on the parenchymal surfaces of the opened lung. (B) Histopathology of *Rhabdias* sp. in the lung of a tree frog. Histologic features of the worms include larvae within the uterus and brown-pigmented material within the intestinal tract. Associated lesions include mild pneumocyte hyperplasia. *(Reprinted with permission Mader and Divers, Current Therapy in Reptile Medicine and Surgery, Elsevier, 2014.)*

cross-sections of nematodes (Fig. 38.29). There are increased numbers of lymphocytes and macrophages within the lamina propria. Presumptive diagnosis is by observation of suggestive lesions by histopathology and definitive diagnosis

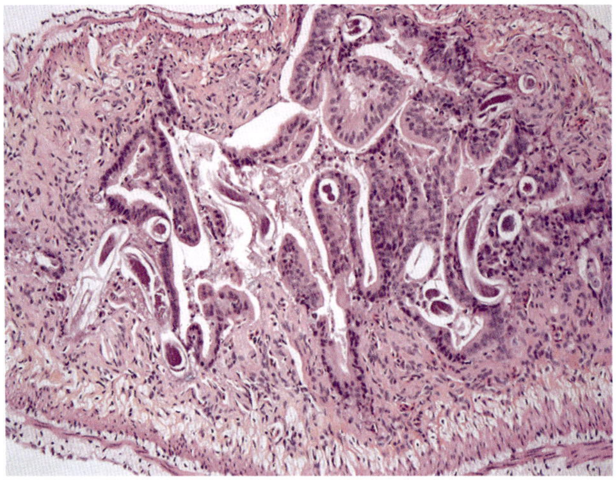

FIGURE 38.29 Strongyloidiasis in a reed frog. The mucosa of the small intestine contains numerous profiles of a rhabditiform nematode parasite.

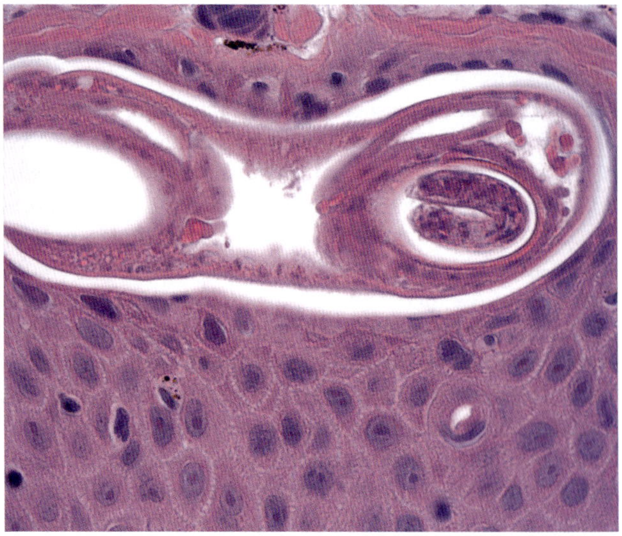

FIGURE 38.30 *Pseudocapillaroides xenopi* in the epidermis of an African clawed frog. An intraepidermal parasite migration tract contains a longitudinal section of a nematode containing a larvated egg.

requires examination of whole adult worms from affected intestinal tracts. Larvated eggs or adult worms can be observed in fresh fecal samples examined by wet mount or floatation methods. There are few clear guidelines for differentiation of the eggs of *Strongyloides* sp. from other intestinal nematodes (usually of low or no pathogenicity) that have larvated eggs, such as *Aplectana, Cosmocerca*, and *Cosmocercoides* spp. or the lungworm *Rhabdias.*[48] Some intestinal nematodes, such as pinworms in the genus *Gyrinicola,* are considered to be beneficial commensal organisms.

The capillarid nematode ***Pseudocapillaroides xenopi*** infects the epidermis of free-ranging and captive African clawed frogs.[49] On gross examination, there is excessive skin shedding and roughening, discoloration of the skin surface, and multifocal cutaneous erosions or ulcers. Heavy infections cause a wasting syndrome. Histologically, there is epidermal hyperplasia and hyperkeratosis with intraepithelial tunnels or subepidermal clefts that contain adult worms, larvae, or eggs. Like other capillarids, the eggs are bioperculate and a feature unique to this species is the presence of larvated eggs within the uterus of the female (Fig. 38.30). Diagnosis is by demonstration of characteristic eggs or adult nematodes by histopathology or wet mount examination of shed skin fragments or skin scrapings. A quantitative PCR assay is described for detection of *P. xenopi* in aquarium sediment.[49]

Intermediate stages of many nematode parasites can be found encysted within soft tissues, viscera, and the coelom in free-ranging amphibians. On gross examination, there may be up to several hundred small (1–2 mm) white to yellow nodules on serosal surfaces. Although visually arresting, these lesions are usually clinical insignificant. Histologically, there is minimal active inflammation with encysted nematodes being surrounded by thin layers of attenuated macrophages and fibrous connective tissue. Examples of some well-known nematodes that encyst in amphibians include ***Eustrongylides* sp.** (common in the proventriculus of birds) and ***Angiostrongylus*** *cantonensis.*[32] Definitive identification of encysted nematodes is rarely attempted in diagnostic practice. Histologically, encysted nematodes must be differentiated from encysted cestodes and trematodes that are also commonly observed in free-ranging amphibians.

Amphibians are second intermediate hosts for **cestodes** in the genus ***Spirometra*** that cause **sparganosis** in mammals. The plerocercoids (sparagana) encyst in the subcutis, skeletal muscle, and coelomic cavity of free-ranging amphibians.[13] On gross examination, affected animals have multicentric cutaneous nodules, especially on the ventral abdomen and legs with a thin membrane encasing the encysted worms. Histologically, worms have features typical of cestodes (e.g., calcareous corpuscles and lack of a coelomic cavity or digestive tract) and are surrounded by varying degrees of fibrosis and granulomatous inflammation.

Free-ranging amphibians are intermediate hosts for many species of digenetic **trematodes** and the encysted metacercariae are sometimes clinically significant. Environmental changes (e.g., eutrophication) related to agricultural practices and urbanization contribute to increased abundance and impact of trematode infections.[143] Species in genera such as *Clinostomum* (yellow grub), *Diplostomum,* and *Alaria* encyst in sites including the subcutis, skeletal muscle, eye, and brain. Large numbers of encysted ***Clinostomum marginatum*** in the head of red-spotted newts can interfere with feeding and the associated skin nodules must be differentiated clinically from those caused by the mesomycetozoan parasites *Ichthyophonus* and *Amphibiocystidium* that occur in this species. Although these parasites are space-occupying the inflammatory response associated with the encysted metacercariae is usually minimal (Fig. 38.31), but is severe in some cases.[130] Heavy infections with echinostome trematodes (e.g., ***Echinostoma trivolis***) result in slow growth, generalized edema, and increased mortality in leopard frog and green frog tadpoles.[77] Metacercariae localize to the kidney in tubules or encysted within granulomas composed of a few layers of attenuated macrophages. Encysted metacercariae of ***Ribeiroia ondatrae*** within limb buds of anurans is associated with limb malformations (see above).

Amphibians are also definitive hosts for both monogenetic and digenetic trematodes. Digeneans in the genus ***Haematoloechus*** are common in the lung of free-ranging anurans worldwide (Fig. 38.32). Adult flukes inhabit the airways and there is usually little host response (e.g., inflammation or pneumocyte hyperplasia). Infections with *Haematoloechus* sometimes occur concurrently with *Rhabdias* sp. nematode infections. The monogenetic trematode ***Pseudodiplorchis americanus*** inhabits the lungs and urinary bladder of Couch's spadefoot toad and chronic infections have caused mild histiocytic interstitial pneumonia and fibrosis.[164] **Gorgoderid digenetic trematodes** of the genera ***Gorgoderina*** and ***Phyllodistomum*** inhabit the Wolffian ducts and urinary bladder of anurans and caudates. Infections are usually subclinical but mild mucosal hyperplasia and lymphohistiocytic cystitis can be seen. Rarely, heavy infections cause perforation of the Wolffian duct. Several species of ***Gyrodactylus*** colonize the skin of frogs and salamanders in North America.[124] Heavy infections can result in cutaneous depigmentation and erosion.

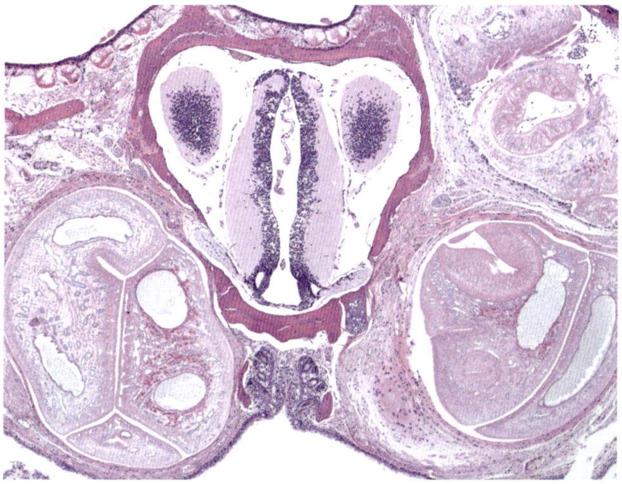

FIGURE 38.31 Trematodiasis in the head of an eastern newt. The soft tissues of the head contain sections of three encysted metacercariae of a *Clinostomum* spp. trematode.

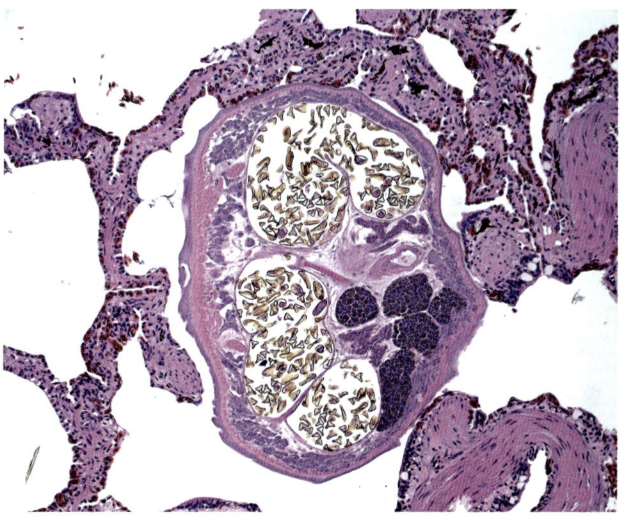

FIGURE 38.32 *Haematoloechus* sp. trematodiasis in the lung of a leopard frog. The ovaries of the trematode contain myriad brown-pigmented eggs.

Free-living **oligochaete worms** (Phylum Annelida; Family Naididae) in the genus ***Dero*** that inhabit bromeliad water cups are opportunistic parasites of the cloaca, urinary bladder, Wolffian ducts (ureter), and Harderian glands of free-ranging and captive anurans.[2] Infections are mostly subclinical, but occasionally heavy worm burdens result in Wolffian duct rupture with secondary peritonitis. If the parasites extend into renal tubules, mild lymphohistiocytic interstitial nephritis develops. Histologically, oligochaete worms are distinguished from nematode parasites by the presence of a cellular epidermis rather than an acellular cuticle, foamy cells within the coelomic cavity (coelomic corpuscles) and refractile hair-like chaetae that exit through the body wall (Fig. 38.33).

Myxozoa are microscopic metazoan parasites in the Phylum Cnidaria that are increasingly recognized as pathogens of free-ranging and captive amphibians. The life cycle of most myxozoans uses a vertebrate intermediate host (e.g., amphibian or fish) and an aquatic invertebrate definitive host (e.g., annelid worm). Myxozoa in the genus ***Cystodiscus*** (formerly *Myxidium* spp.) develop spore-forming (sporogonic) plasmodia in the gallbladder of anurans, caudates, and caecilians worldwide. In most instances, these are incidental findings without histologic evidence of inflammation or other host response (Fig. 38.34A). However, infections with ***C. australis*** and ***C. axonis*** have been described as emerging diseases in native and introduced anurans from Australia.[73] In addition to typical sporogonic plasmodia in the gallbladder, nonsporogonic plasmodia of *C. axonis* are found in the brain, spinal cord, nerve roots, and occasionally the peripheral nerves of tadpoles and postmetamorphic frogs. Plasmodia are associated with increased numbers of glial cells (gliosis) and hemorrhage; necrosis, and vasculitis with lymphocytic perivascular cuffs are variably present in lesions. Plasmodia localize within axons and are 5–25 µm with numerous small basophilic nuclei. Similar lesions have been observed in Cururu toads from Brazil

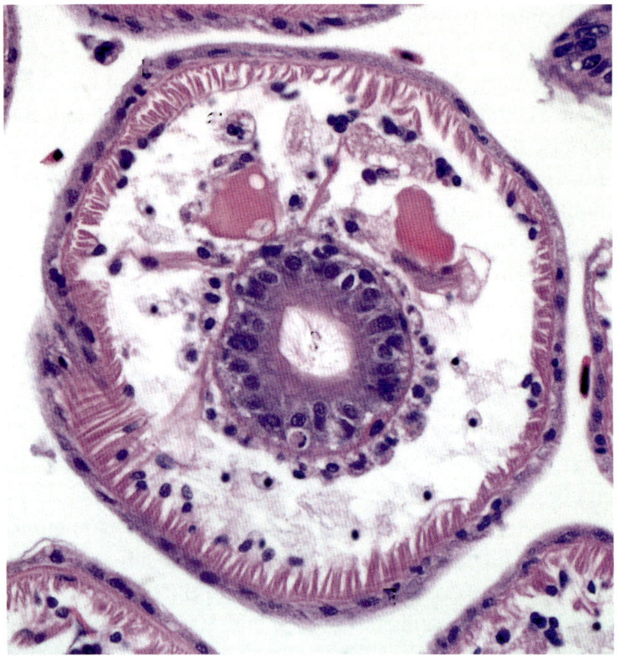

FIGURE 38.33 An oligochaete worm in the cloaca of a poison dart frog. Characteristic histologic features in this transverse section include a cellular epidermis and foamy coelomic corpuscles. *(Reprinted with permission Mader and Divers, Current Therapy in Reptile Medicine and Surgery, Elsevier, 2014.)*

infected with ***C. immersus*** and in toads from Uruguay with an unidentified intraaxonal organism in the spinal cord.[161] Both *C. axonis* and *C. australis* also form nonsporogonic plasmodia in the bile ducts of tadpoles and some postmetamorphic frogs where infection is associated with biliary hyperplasia, portal fibrosis, and lymphoplasmacytic portal hepatitis. Diagnosis of sporogonic *Cystodiscus* spp. infections is by observation of characteristic plasmodia and spores within the lumen of the gallbladder of infected animals. The nonsporogonic stages in nervous tissues and the liver strongly resemble apicomplexan protozoal cysts (e.g., *Toxoplasma gondii*). PCR assays are available for amphibian *Cystodiscus* spp. and in combination with DNA sequencing are helpful for species determination. Morphologic examination of the size and the arrangement and number of transverse ridges in spores by light microscopy or scanning electron microscopy for features, such as may be helpful in species determination.

Other myxozoa that infect amphibians occur in the kidney, testes, and skin. Organisms in the genera ***Chloromyxum***, ***Sphaerospora***, and ***Hoferellus*** inhabit renal tubules and glomeruli of anurans and caudates.[88] Most renal myxozoa are incidental findings when present in low numbers and without appreciable inflammation or evidence of tubular injury. In wild-caught Asian horned frogs, *C. careni* was associated with mild to severe multifocal lymphocytic and histiocytic interstitial nephritis that may have contributed to death.[45] Pseudoplasmodia and plasmodia expanded Bowman's space and renal tubules and contained large numbers of 6–8 μm spherical to pyriform spores with four 2–3 μm polar capsules (Fig. 38.34B). **Frog kidney enlargement disease** caused by ***Hoferellus anurae*** occurred in wild-caught African hyperoliid frogs.[118] On gross examination, the kidneys were enlarged and bright yellow to orange with a granular surface. In some frogs, renal cysts containing yellow to green-blue fluid were present. Myxozoan spores and plasmodia were present within ectatic renal tubules.

Protozoa

Protozoa including opalinids, flagellates, and ciliates are normal commensal organisms in the lumen of the intestinal

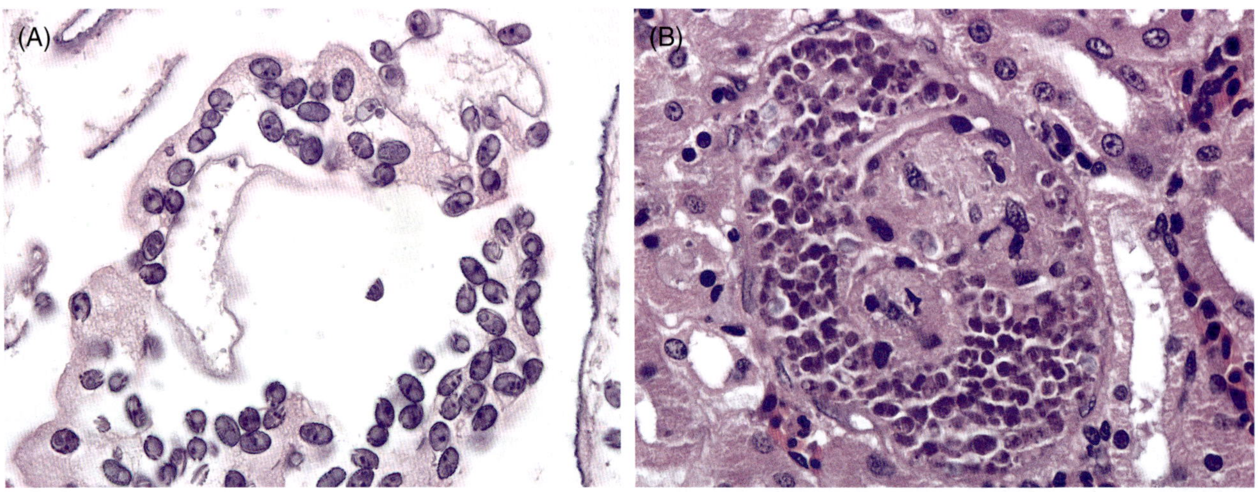

FIGURE 38.34 Myxozoa in the gallbladder of a Panamanian golden frog (A) and kidney of an Asian horned frog (B). (A) Plasmodia within the gallbladder lumen contain numerous basophilic spores of a *Cystodiscus* spp. myxozoan parasite. (B) Plasmodia and numerous basophilic spores of *Chloromyxum careni* are present in a glomerulus.

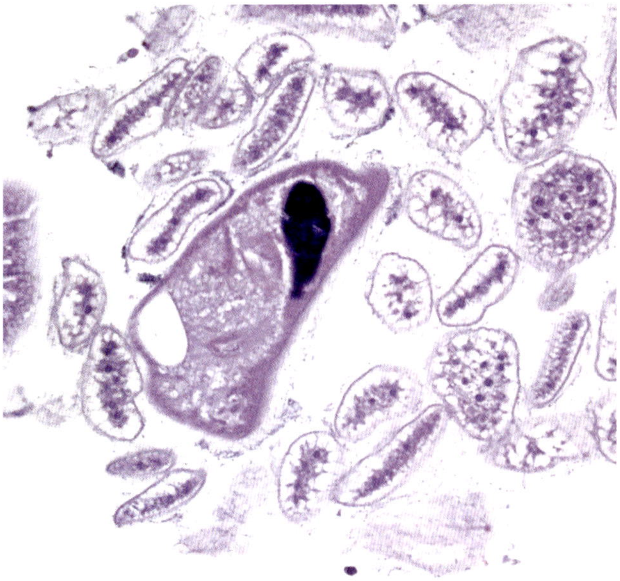

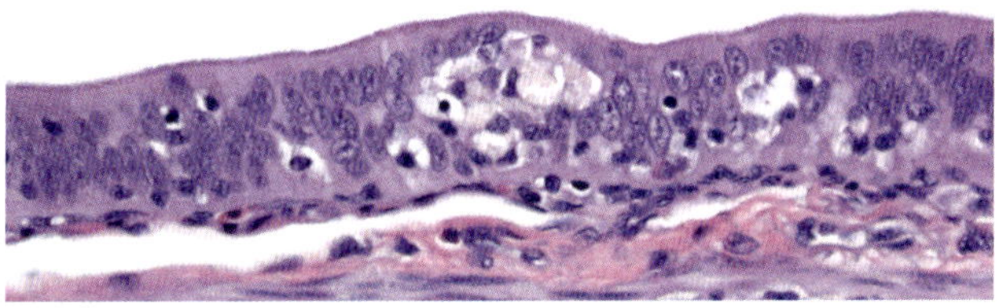

FIGURE 38.35 Commensal protozoa in the colon of a dusky salamander. A ciliate protozoan with a characteristic dark macronucleus is present in the center. The ciliate is surrounded by numerous opalinid-type protozoa.

tracts of amphibians.[136] These organisms can be seen in wet mounts of feces or histologically (Fig. 38.35). Colonization is not clinically significant unless the organisms are invasive or are present in very high numbers. Several species of the ciliate ***Trichodina*** are incidental findings on mucosal surfaces of the skin, gills, and urinary bladder. **Intravascular ciliated protozoa** (not further characterized) have been associated with mortality in the endangered Kihansi spray toad in an assurance colony.[58] Disseminated infection with high numbers of protozoa can cause obstruction of lymphatic vessels, hepatic sinusoids, and glomerular capillaries. Sequela can include regional edema, cellulitis, and hepatitis. Aquatic ciliate protozoa in the genus ***Tetrahymena*** can invade the skin and muscle and disseminate to visceral organs in tadpoles. Trypanosome flagellates (e.g., ***Trypanosoma ranarum***) are occasionally seen as incidental findings in blood or tissue impression smears from free-ranging amphibians.

Cryptosporidiosis caused by apicomplexan protozoa in the genus ***Cryptosporidium*** has occasionally been described in captive anurans. ***Cryptosporidium fragile*** has been identified in the superficial gastric epithelium of black-spined toads.[89] Infected toads had a history of weight loss and histologically there was mild gastric epithelial hyperplasia without significant associated inflammation. An African clawed frog with a history of weight loss and emaciation was infected with an unidentified *Cryptosporidium* spp. On histologic examination, there was marked intracellular, extracytoplasmic gastric epithelial hyperplasia (proliferative gastritis) with myriad superficial 0.5–3.0 µm developmental stages.[66] Gastric mucosal colonization with *Cryptosporidium*-like protozoa has also been observed as an incidental finding in a free-ranging dusky salamander. Another unidentified *Cryptosporidium* spp. infect the ventral ciliated epithelium of the tongue in the Wyoming toad. Infections are subclinical with mild mucosal hyperplasia and lymphocytic glossitis (Fig. 38.36).

A variety of other apicomplexan protozoa are observed in diagnostic samples from amphibians, but are rarely associated with disease. Gamonts of ***Hepatozoon*** spp. occur in erythrocytes of free-ranging anurans, caudates, and caecilians. Sporozoites of ***Lankesterella*** spp. are found in the erythrocytes of many anurans and intracytoplasmic tissue stages have been observed in many tissues including leukocytes, melanomacrophages, endothelial cells, and hepatocytes. Stages of *Lankesterella* have cross-reacted on immunohistochemistry with polyclonal antibodies to *Toxoplasma gondii*.[59] Intestinal **coccidiosis** associated with ***Eimeria, Isospora***, and ***Goussia*** spp. is observed in anuran and caudate larvae, but is usually not of clinical significance. Oocysts of *Goussia* spp. have been reported in melanomacrophage centers of tadpoles.

Amoebiasis, caused by a novel species of *Entamoeba* related to ***E. ranarum*** and ***E. histolytica*** caused an outbreak

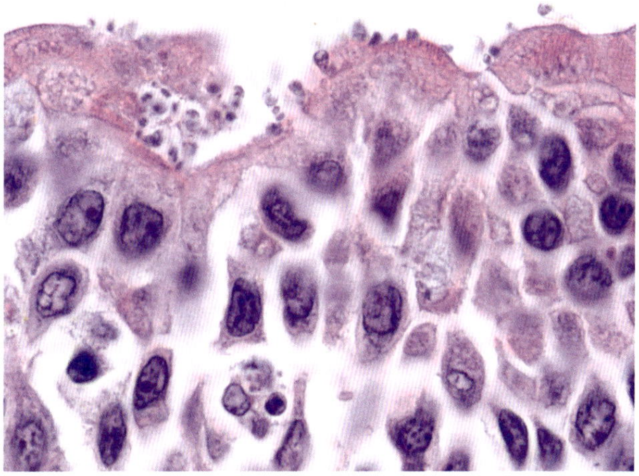

FIGURE 38.36 Cryptosporidiosis in the ventral tongue of a Wyoming toad. Numerous small basophilic protozoal organisms are present along the mucosal surface. *(Reprinted with permission Mader and Divers. Current Therapy in Reptile Medicine and Surgery.)*

of hyperplastic and multifocally ulcerative colitis in free-ranging cane (marine) toads from Australia.[153] A subset of affected animals had enteritis and gastritis. On gross examination, toads had variable serosal congestion, thickening of the colonic wall, and hemorrhagic colonic contents. Histologically, trophozoites within the epithelium and lamina propria were highlighted with PAS. Another suspected *Entamoeba* sp. was associated with lymphohistiocytic nephritis in marine toads.[169] Many amphibians harbor *Entamoeba ranarum* within the gastrointestinal tract, but associated disease is rare.

Mass mortality events in free-ranging tadpoles of anuran species in the family Ranidae (e.g., American bullfrogs, leopard frogs, and green frogs) from the United States are caused by a pathogenic lineage of **alveolate protozoa** in the Phylum Perkinsea.[41,85,86] A suspected mortality event was also described in captive European tree frogs (family Hylidae) from the United Kingdom.[157] Molecular evidence suggests that infection with similar organisms occurs in the absence of gross or histologic lesions in anurans besides just ranids from North America, Central America, South America, Africa, and Europe.[92,157] On gross examination, the liver, spleen, and kidney of affected tadpoles are markedly enlarged and discolored white to yellow. Less frequently, pancreatomegaly and thickened white intestinal loops are observed.[85] Histologically, affected tissues are infiltrated by large numbers of 6–9 μm spherical basophilic spores (Fig. 38.37A). In a single report of infection in an adult Southern leopard frog, infection presented as a soft tissue mass on the hindlimb. Histologically, this was composed of similar basophilic spores with mild granulomatous inflammation. Spores are positive with Periodic acid-Schiff (PAS), Grocott's methenamine silver (GMS), and Congo Red stains. On transmission electron microscopy, the spores have a trilaminar cell wall that is centrally electron dense and mitochondria with tubular cristae. Using scanning electron microscopy, the spores have a characteristic polyhedral shape with square to hexagonal interconnecting plates (Fig. 38.37B). The diagnosis of alveolate protozoal infection is made by recognition of spores in tissue wet mounts or histologic sections. If necessary, confirmation can be obtained by electron microscopy or molecular methods.[156]

Ectoparasites

Metazoan ectoparasites are common findings in free-ranging amphibians. Cutaneous **trombiculosis** caused by **trombiculid mites** (chiggers) in genera such as *Hannemania* (North and South America), *Endotrombicula* (Africa), and *Vercammenia* (Australia) occurs in many species of anurans and caudates.[111,155] Larval mites penetrate the skin and encyst within the dermis for up to several months. Grossly, there are distinctive multifocal to coalescing, orange to red, 1–4 mm in diameter cutaneous papules and vesicles that are most commonly distributed on the feet, legs, ventral body surfaces, and head including the nasolabial groove of plethodontid salamanders. Histologically, mites are located in the superficial dermis and are surrounded by mild fibrosis and minimal to mild infiltrates of macrophages and lymphocytes. Heavy parasite burdens may be associated with secondary bacterial infections. Diagnosis is by microscopic examination of mites recovered from skin scrapings or by histopathology. Other mites that might be encountered in clinical samples from amphibians include *Xenopacarus africanus* described from the nasal cavity of African clawed frogs and environmental mites, such as

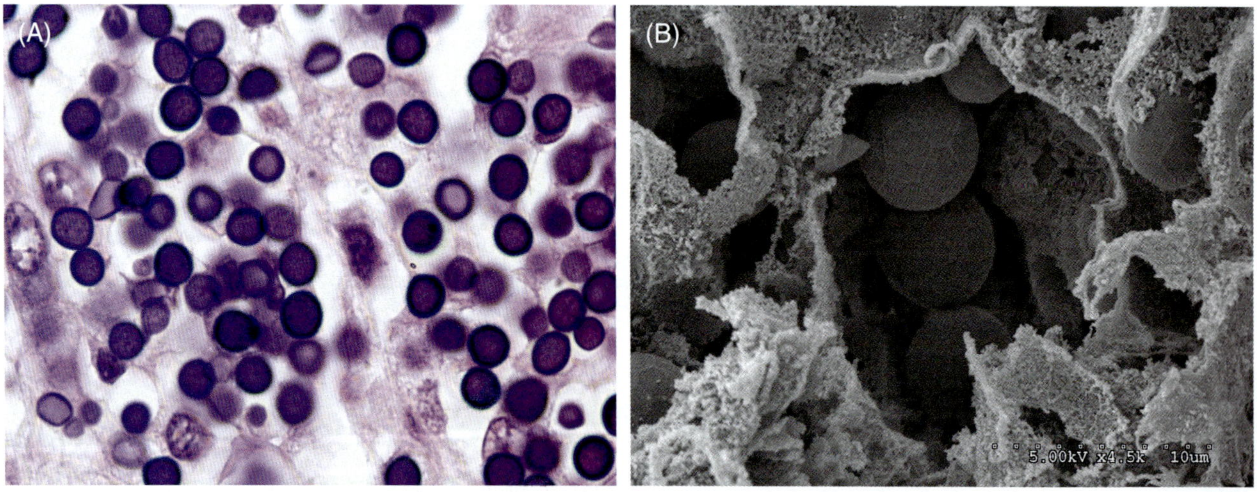

FIGURE 38.37 Alveolate protozoa in a Southern leopard frog. (A) Basophilic spores of an unidentified alveolate protozoan are present in the skeletal muscle. (B) The spores are square to hexagonal plates on the surface. Scanning electron microscopy (SEM).

Rhizoglyphus spp. that can incidentally colonize animals. **Ticks**, usually *Amblyomma* spp., are sometimes associated with granulomatous and eosinophilic inflammation and granulation tissue at chronic attachment sites. **Anchor worms,** such as *Lernaea cyprinacea*, are copepod parasites of fish that occasionally infect free-ranging anuran tadpoles and juveniles.[97] The head of the parasite penetrates the skin and anchors within soft tissues, and the elongate abdomen with paired egg sac sacs projects from the skin surface. Fish **lice** (*Argulus* spp.) are blood-feeding brachyuran crustaceans that can be found on aquatic anurans and caudates. **Myiasis** caused by larvae of several species of **dipteran flies** from genera, such as *Notochaeta*, *Batrachomyia*, and *Bufolucillia*, occurs in the subcutaneous lymph sacs or nasal cavities of anurans worldwide.

Other Infectious Organisms

Organisms in the **protist** class **Mesomycetozoea** arose at the animal-fungal divergence and have been previously classified as either protozoa or fungi. These organisms can be found in both marine and freshwater environments as commensals or parasites of invertebrates, fish, amphibians, as well as birds and mammals (e.g., *Rhinosporidium seeberi*). In contrast to mesomycetozoal disease in fish, where systemic infections with organisms like *Sphaerotheceum destruens* are associated with significant mortality events, infections in amphibians are often incidental. Very little is known about the lifecycle or mode of transmission of amphibian mesomycetozoans, which have been reported in free-ranging anurans and caudates from North and South America, Europe, and Africa. Amphibian mesomycetozoans from the Order Dermocystida have recently undergone taxonomic revisions and new species have been described. Historical reports in amphibians usually refer to the genera *Dermosporidium*, *Dermocystidium*, or *Dermomycoides*, which have been replaced by *Amphibiocystidium* spp., *Amphibiothecum* spp., and a new species of *Rhinosporidium* spp. based on ultrastructural examination and phylogenetic analyses of 18S rRNA sequences.[128,137,148] On gross examination, infections appear as multicentric skin nodules that range from oval to "C" or "U"-shaped and often occur on ventral body surfaces. In some cases, nodules have a depressed or ulcerated central region. Infections with *Rhinosporidium rwandae* in African reed frogs were multifocal to coalescing with many spherical nodules with a central black spot present on dorsal and ventral skin surfaces. Histologically, skin nodules are composed of large intradermal cysts containing myriad 2–12 μm in diameter endospores (Fig. 38.38). Cyst walls range from thin to thick and refractile, and some have evidence of internal septation. Cysts are surrounded by a minimal to marked inflammatory cell infiltrate that can include macrophages, eosinophils, and lymphocytes. Diagnosis is by histopathology or by recognition of endospores in wet mounts or cytologic preparations of skin nodule aspirates. On gross examination, skin nodules must be differentiated from those caused by mycobacteriosis, fungal granulomas, and encysted parasites. Histologically, the morphology of the spores can overlap with unidentified alveolate protozoa observed in ranid frogs (see above). Determination of individual species is by observations of cyst and spore morphology by light and transmission electron microscopy and by PCR and DNA sequencing.

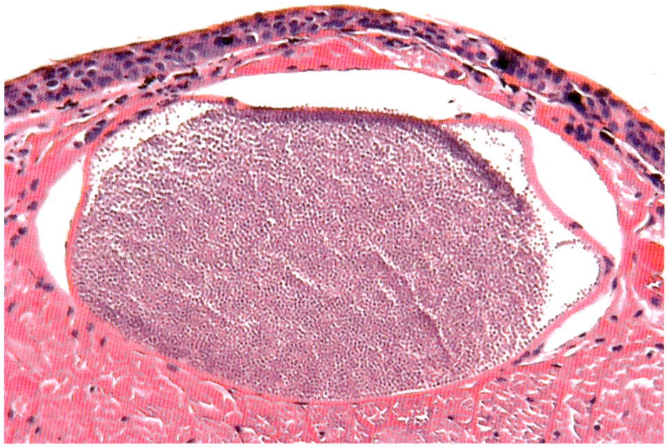

FIGURE 38.38 An *Amphibiocystidium* spp. cyst in the skin of an eastern newt. The cyst expands the dermis and contains myriad eosinophilic endospores. *(Photo Courtesy of T. Raffel, Oakland University.)*

Infections with ***Ichthyophonus*-like mesomycetozoan** organisms in the Order Ichthyophonida occur in free-ranging anurans and caudates in North and South America.[20,113] Although the organisms morphologically resemble *Ichthyophonus hoferi* infections of fish, the phylogenetic relationships of the organisms infecting amphibians have not been determined. The lifecycle is unknown, but leeches (*Plactobella picta*) have been suggested as a possible vector in the red-spotted newt.[137] Mortality of individual animals or sometimes groups of animals is associated with heavy infections, but infection is often incidental. On gross examination, heavily infected animals have swellings or nodules on the body, rump (region of urostyle in anurans), and the tail in caudates. Histologically, there is a granulomatous myositis associated with parasitic cysts within skeletal myocytes and surrounding connective tissues. Active cysts have a thin outer wall, are 100–220 μm in diameter and are present in the sarcoplasm of infected myocytes. In contrast, passive cysts are 40–100 μm in diameter, occur outside of myocytes, and have a thick refractile wall that is often wrinkled (Fig. 38.39). A neurotropic *Ichthyophonus*-like organism has been observed in

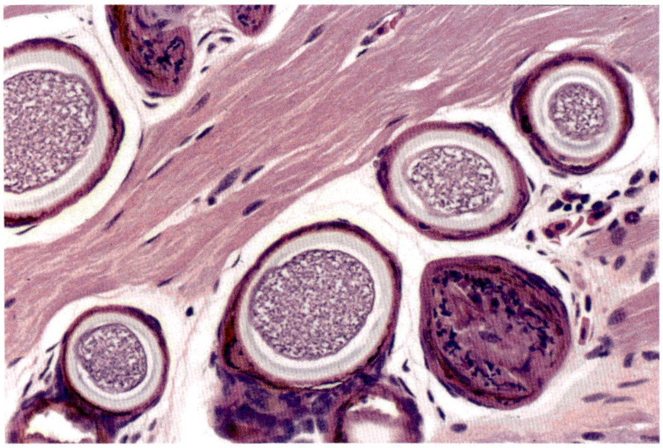

FIGURE 38.39 *Ichthyophonus*-like organisms in the skeletal muscle of an eastern newt. Granulomas contain thick-walled cysts. *(Reprinted with permission Mader and Divers, Current Therapy in Reptile Medicine and Surgery, Elsevier, 2014.)*

tadpoles of southern toads[94] and southern leopard frogs from the southeastern United States. Parasitic cysts ranging from 10 to 35 μm in diameter were observed histologically in the brain, spinal cord, retina, and gills of infected animals.

ADDITIONAL RESOURCES

Densmore, C.L., Green, D.E., 2007. Diseases of amphibians. ILAR J. 48, 235–54.

Forzan, M., Heatley, J., Russell, K.E., Horney, B., 2017. Clinical pathology of amphibians: a review. Vet. Clin. Pathol. 46, 11–33.

Green, D.E., 2001. Pathology of amphibia. In: Wright, K.M., Whitaker, B.R. (eds): Amphibian medicine and captive husbandry. Malabar, FL, Krieger. Pp 401–485.

Green, S.L., 2010. The laboratory *Xenopus* sp. Boca Raton, FL, CRC Press.

McDiarmid, R.W., Altig, R., 1999. Tadpoles. The biology of anuran larvae. Chicago, IL, University of Chicago Press.

Mutschmann, F., 2010. Erkrankungen der amphibian. Stuttgart, Enke Verlag.

Pessier, A.P., 2021. Amphibians. In: Sharkey, L.C., Radin, M.J., Seelig, D. (eds): Veterinary cytology. Hoboken, NJ, John Wiley and Sons. Pp 869–875.

Wells, K.D., 2007. The ecology and behavior of amphibians. Chicago, IL, University of Chicago Press.

References

Full references for this chapter can be found online at https://doi.org/10.1016/B978-0-443-18437-6.00024-9.

Chapter 39

Osteichthyes

Salvatore Frasca, Jr.[a], Abigail Armwood[b], Emily M. Reinhardt[a], Eric D. Lombardini[c], and Alvin C. Camus[d]
[a]University of Connecticut, Storrs, CT, United States; [b]North Carolina State University, Raleigh, NC, United States; [c]Pre-Clinical Global, Boston Scientific Corporation, Marlborough, MA, United States; [d]University of Georgia, Athens, GA, United States

INTRODUCTION

A fish, from an anatomic and morphological perspective, refers to a water-dwelling poikilothermic vertebrate possessing gills throughout its life and fins, if present, as limbs. Osteichthyes, i.e., bony fish, represent the largest taxonomic class of vertebrates in the modern world. According to Nelson et al.[327], over one-half of the extant vertebrate species are species of fish. Two subclasses are recognized within the class Osteichthyes. These are the Sarcopterygii and the Actinopterygii. The former are the lobe-finned fishes and are represented by eight extant fish species that include lungfishes and coelacanths. The latter are the ray-finned or spiny-finned fish that are represented by some 30,500 species. Subclass Actinopterygii includes the infraclass Chondrostei, which includes paddlefishes and sturgeons, and the clade Neopterygii, which includes the infraclass Holostei, i.e., gars and bowfins, and the subdivision Teleostei, which includes a reported 96% of all extant fishes spread throughout freshwater and marine habitats globally.[327] It is from among the Actinopterygii, particularly the Teleostei, also known informally as "teleosts," that the veterinary pathologist will encounter the vast majority of fish. A comprehensive description of the taxonomy and species diversity of fish is provided by Nelson et al.[327]

The biological diversity of the Osteichthyes is truly awesome. The Osteichthyes includes species that have been recognized for their importance to the commercial aquarium industry (e.g., seahorses, sea dragons, goldfish, clownfish, and tangs), the production aquaculture industry (e.g., salmon, cod, catfish, and tilapia), the ecological and environmental scientific communities (e.g., winter flounder[322]), as well as the biomedical community (e.g., zebrafish and medaka). The range of species in the Osteichthyes is as vast as the environments they occupy such that no one species or environmental context, whether natural or artificial, can adequately represent the entirety of the bony fish when considering a description of their diseases. Hence, this chapter has not been prepared to serve as a summary of the extant knowledge of fish diseases. Rather, it has been prepared so that the text might be representative of the breadth of diseases a veterinary pathologist or clinician might encounter from the most common sources of fish cases, for example, commercial and laboratory aquariums, coldwater aquaculture, warm water aquaculture, and environmental population studies. A number of excellent and comprehensive fish disease reference texts for both pathologists and clinicians currently exist. These are the foundation for the accumulated knowledge base of this chapter. It is hoped that this chapter will serve as a primer on fish pathology. Those whose interest is piqued are encouraged to seek out any number of these other texts, so they might serve the reader as they have the authors. Some notable reference texts embraced by the authors of the first and second editions are provided (Box 39.1); others may be noted in the references of this chapter.

While the diversity of bony fish and their environments is vast, there are themes that contribute to an understanding of disease in fish across contexts. Principal among these is the concept that the health of fish is intimately linked to the health of the aquatic environments they inhabit.[403] The aquatic environment is immensely complex and provides the physical and chemical features that influence and maintain the physiological and immunological homeostasis of fish. Perturbations in environmental parameters can disrupt physiological and immunological homeostasis, which directly or indirectly results in disease and death. In aquaculture and aquarium settings, fish often do not have the opportunity to migrate to avoid deleterious changes in their environment, and the effects of environmental stresses can manifest more intensely as a consequence, at times affecting high numbers of fish.[403] The pathogenesis of disease in bony fish, and other aquatic species for that matter, is therefore a direct reflection of the pathobiology of destabilized yet inextricably linked biological systems.

Pathology of Wildlife and Zoo Animals. https://doi.org/10.1016/B978-0-443-18437-6.00002-X

Box 39.1 Reference Textbooks for Fish Pathology

- Bruno, D.W., Poppe, T.T., 1996. *A Color Atlas of Salmonid Diseases.* Academic Press, San Diego, US.
- Ferguson, H.W. (Ed.), 2006. *Systemic Pathology of Fish, A Text and Atlas of Comparative Tissue Responses in Diseases of Teleosts,* 2nd ed. Scotian Press, London, UK.
- Noga, E.J., 2010. *Fish Disease: Diagnosis and Treatment,* 2nd ed. Wiley-Blackwell, Ames, US.
- Roberts, R. J. (Ed.), 2012. *Fish Pathology,* 4th ed. W.B. Saunders, Philadelphia, US.
- Stoskopf, M.K. (Ed.), 1993. *Fish Medicine.* W.B. Saunders, Philadelphia, US.
- Woo, P.T.K. (Ed.), 2006, *Fish Diseases and Disorders,* Vol. 1, *Protozoan and Metazoan Infections,* 2nd ed. CAB International, Wallingford, UK.
- Woo, P.T.K. (Ed.), 2010, *Fish Diseases and Disorders,* Vol. 2, *Non-Infectious Disorders,* 2nd ed. CAB International, Wallingford, UK.
- Woo, P.T.K. (Ed.), 2011, *Fish Diseases and Disorders,* Vol. 3, *Viral, Bacterial and Fungal Infections,* 2nd ed. CAB International, Wallingford, UK.

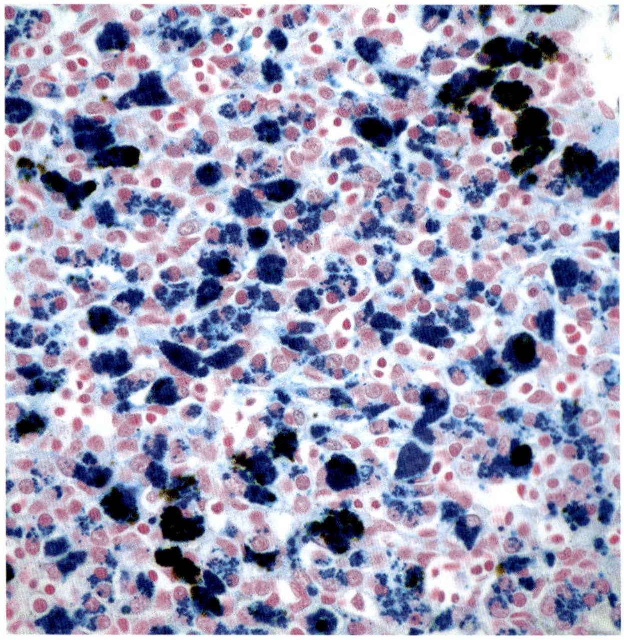

FIGURE 39.1 Iron overload in the liver of a lake sturgeon. Iron, which is stained blue in this preparation, has accumulated in the cytoplasm of hepatocytes, and in melanomacrophages. Prussian blue reaction.

NON-INFECTIOUS DISEASES

Nutritional

Hypovitaminosis C due to dietary deficiency can affect any cultured or display teleost, as the ability to synthesize vitamin C, while a trait of cartilaginous fish and nonteleostean bony fish, appears not to be a trait of teleosts.[111] Hypovitaminosis C is most frequently reported in catfish and salmonid production. Pathogenesis in fish is similar to that in other species as vitamin C is a cofactor for prolyl and lysyl hydroxylases, and deficiency leads to both impaired collagen crosslinking and decreased overall collagen production. In growing fish, improper vertebral development with spinal deformation and growth retardation are typical. Increased risk of fracture with impaired callus formation as well as impaired wound healing is also common.[259] Scoliosis and/or lordosis and deformations of opercula, branchial arches, and gill filaments may be grossly evident. Fractures may be grossly observable, but many are only evident radiographically. Histologically osteopenia, cartilaginous and osseous dysplasia, fractures, and reduced callus formation may be observed. Abnormal wound healing is characterized by epithelial restitution with reduced or absent subtending fibroplasia. Observation of typical gross and histologic lesions, particularly if occurring in multiple individuals or an entire age class, is diagnostic. Assessment of vitamin C levels in tissue and/or the provided ration may be of utility in supporting gross and histologic diagnoses. Production systems losses are readily prevented with adequate feed formulation. In aquaria, skeletal deformities and fractures can lead to loss of display value and deaths.

Iron overload occurs in a variety of wild and aquarium-held freshwater and marine species. Whether hepatic iron accumulation is due to environmental and or dietary factors similar to iron storage disease in other zoo species, or if it is due to specific metabolic defects, is not understood. Hepatocellular iron accumulation in fish exposed to high environmental iron levels has been documented experimentally and has also been associated with exposure to heavy metal- and PCB-contaminated sites in wild fish.[36,71] Grossly, the liver may be orange-brown. Histologically, variable quantities of iron will be evident in the cytoplasm of hepatocytes, and sinusoidal macrophages, without distortion of hepatic architecture (Fig. 39.1). Confirmation is made by visualizing stainable iron within hepatocytes. The clinical significance of iron overload is likely low, except as a potential indicator of another process, such as contaminant exposure.

Metabolic

Nephrocalcinosis is best described in rainbow trout production systems, where it has been known for many years, and it has been noted in a wide variety of display species. The disease in trout has been primarily associated with high CO_2 levels, but it is also seen or experimentally induced with dietary manganese deficiency and selenium and arsenic toxicity.[248,453] The mechanism by which elevated CO_2 leads to nephrocalcinosis is obscure. While best described in trout, identical lesions have been observed in fish maintained in

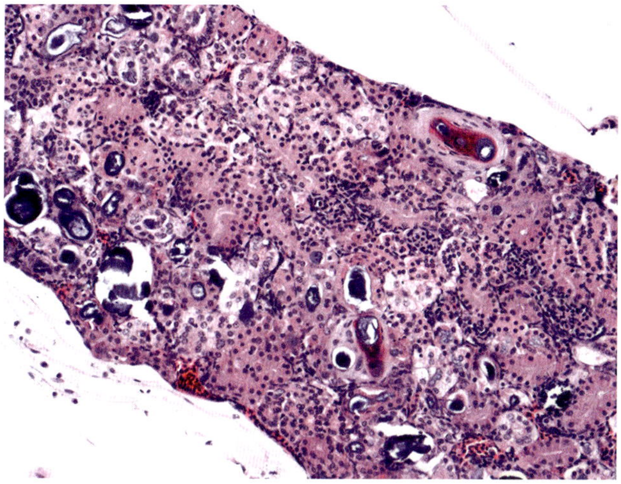

FIGURE 39.2 Nephrocalcinosis in the kidney of a teleost. Basophilic deposits of mineral are present in tubules and the interstitium. Mineral deposits in the interstitium elicit a granulomatous inflammatory reaction.

aquaria. These include variably sized, hard, off-white renal nodules that may be visible grossly. Microscopically, cases may range from small renal tubular accumulations of mineral without apparent epithelial cell damage to large aggregates of mineral surrounded by granulomatous inflammation and fibrosis (Fig. 39.2). In severe cases, extensive distortion and loss of normal renal parenchyma may occur with resultant compromise of renal function.

Hepatic lipid accumulation is common in bony fish and is influenced by species, age, sex, reproductive season, nutritional status, and concurrent infectious disease or toxicities. Bony fish have a remarkable capacity for hepatocytic lipid storage without apparent ill effects and at levels far exceeding those considered life-threatening in higher vertebrates. Histologically, hepatocytic lipid accumulation is characterized by discrete, clear, variably sized, cytoplasmic vacuoles (Fig. 39.3). Failing to recognize the capacity of bony fish for hepatic lipid storage as a physiologic reality too often leads to an erroneous diagnosis of **hepatic lipidosis**.[524] Hepatic lipidosis is a pathologic condition associated with abnormalities in lipid metabolism, incumbent morbidity, or even mortality. The diagnosis of hepatic lipidosis should be assiduously avoided unless clear indicators to the contrary are noted. The transition of lipid accumulation from physiologic to pathologic is poorly understood and probably varies by species, further underscoring the value of a conservative approach to diagnosing lipidosis. Lipid accumulation in bony fish is frequently dietary in origin and reflects feeding practices in confinement systems (Figs. 39.4A, B) such as aquaculture operations where rapid maximal growth is desirable. In mixed species aquarium exhibits, more aggressive species often overconsume, while least aggressive specimens may not receive adequate nutrition. Additionally, some teleosts lack swim bladders and

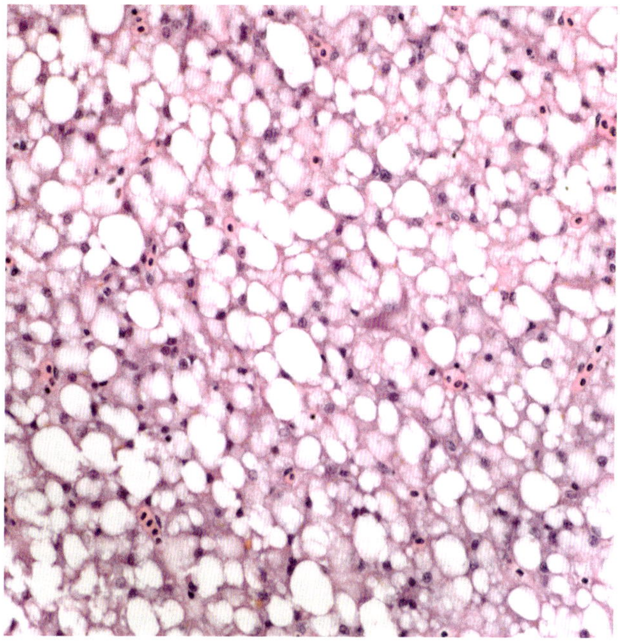

FIGURE 39.3 Lipid accumulation in the liver of a teleost. The pathologic significance of hepatic lipid content is subject to consideration of a number of factors such as diet, feeding frequency, reproductive status, and behavior. There is marked lipid accumulation in hepatocytes that is not pathologic. Hepatocytes contain one to several, large, clear, sharply marginated vacuoles in their cytoplasm without loss of cellular margins, evidence of pyknosis, or inflammation.

utilize hepatic lipid storage for buoyancy similar to elasmobranchs (see Chapter 40: Chondrichthyes).

Histologic criteria warranting a diagnosis of hepatic, i.e., pathologic, lipidosis include: (1) coalescing hepatocellular lipid vacuoles indicating loss of membrane integrity, (2) marked disparity in hepatocellular size, (3) lipid granulomas and macrophages containing ceroid suggestive of increased lipid processing, and/or (4) concurrent lesions indicative of hepatic insult, such as biliary hyperplasia and fibrosis in the absence of other inciting causes (Fig. 39.5). Excessive lipid accumulation can predispose to oxidative injury related to the abundance of polyunsaturated fats in fish, which are prone to peroxidation caused by diets containing oxidized (rancid) fats, particularly frozen fish, or low in antioxidants such as vitamin E and selenium. Attempts to mimic natural diets for aquarium-housed fish may overlook appropriate antioxidant supplementation necessary to deter oxidative injury. This is often manifested by the accumulation of pigmented lipid residues represented by ceroid within the cytoplasm of hepatocytes and macrophages.[100] The accumulation of pigmented lipid residues in macrophages and fat saponification may be seen in coelomic adipose. Historically, a problem in commercial aquaculture, pathologic lipidosis, so-called **lipoid liver disease**, was largely eliminated by formulated diets fortified with appropriate vitamin and mineral supplements.[187,454] Regarding reporting, a diagnosis of

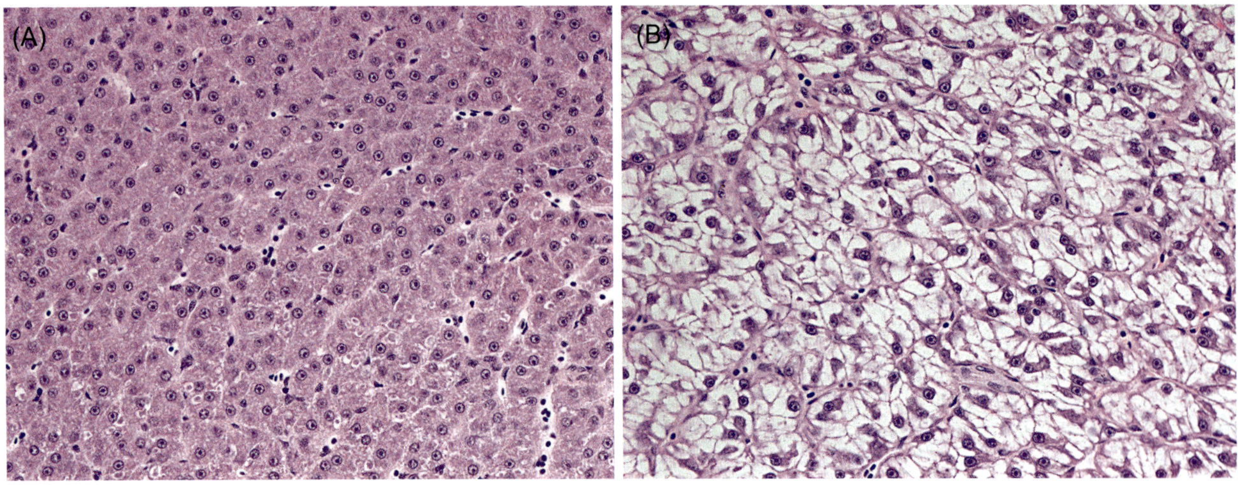

FIGURE 39.4 Normal livers from channel catfish fingerlings with physiologic variation in lipid accumulation. (A) Hepatocytes are diffusely devoid of discrete lipid vacuoles. (B) Hepatocytes diffusely contain multiple, clear, discrete lipid vacuoles. Note the absence of lipid vacuole coalescence, variation in hepatocellular size, lipid granulomas, ceroid accumulation, and concurrent lesions of hepatic insult, which would support a diagnosis of pathologic lipidosis.

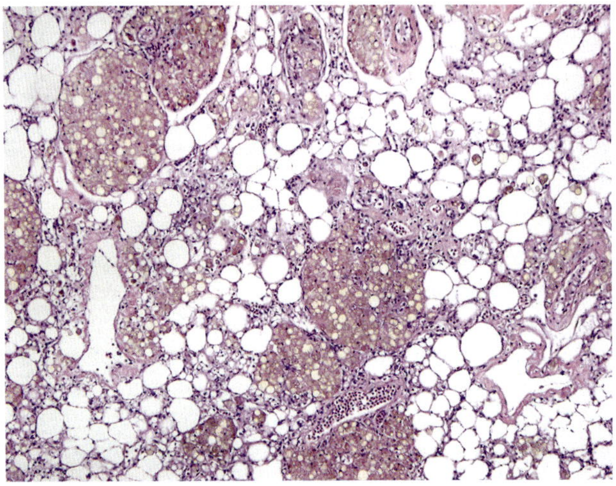

FIGURE 39.5 Pathologic hepatic lipid accumulation, Lionfish, liver. Hepatocytes are distended by large, clear, well-defined, cytoplasmic lipid vacuoles leading to marked disparity in hepatocellular size and coalescing lipid vacuoles. Hepatocytes occasionally contain golden-brown intracytoplasmic pigment. Numerous macrophages form aggregates and frequently contain lipid droplets and golden-brown pigment granules.

hepatic lipid accumulation with a semiquantitative (minimal, mild, etc.) or qualitative (e.g., excessive) assessment is preferred, as low hepatic lipid stores may also be informative, suggesting a fish is inappetent, stressed, or malnourished. In the case of elasmobranchs, hepatocytic lipid accumulation is expected in well-conditioned sharks, skates, and rays, while depletion of hepatic lipid is indicative of morbid nutritional imbalance (see Chapter 40: Chondrichthyes). Perhaps, then, the misdiagnosis of hepatic lipidosis in bony and cartilaginous fish is of greatest significance, as it may obviate the search for the actual cause of morbidity or mortality.

Toxic

Ammonia toxicosis, "**new tank syndrome**," is due to ammonia inputs in excess of the ammonia-oxidizing capacity of filter media bacteria and archaea within a system. Common scenarios include adding fish to a new tank prior to the development of appropriate numbers of filter microorganisms; filter changes where media containing microorganisms are removed in excess; high stocking densities, especially when coupled with marginal filter microorganism capacity; overfeeding, where protein breakdown results in generation of ammonia; and any combination of the earlier. Toxicosis during shipping also occurs with some frequency. Small water volumes, high animal densities, and lack of appropriate microorganisms are all risk factors. Assessment of water ammonia levels ($NH_3{-}N$) is diagnostic, and results are immediate. Ammonia concentration is affected by pH and temperature and should be assessed in context with these parameters using readily available tables that indicate the percent of total ammonia present in the unionized (toxic) form. Levels of unionized ammonia greater than 0.05 mg/L are considered elevated and should be treated; levels greater than 1 mg/L can be rapidly fatal.[333] High and rapid mortality can occur when levels are substantially elevated. Sublethal effects can be difficult to definitively ascertain. In production systems, decreased growth and increased susceptibility to infection are reported, and similar morbidity would be expected in display species.[81]

Morphologic changes in gills associated with ammonia toxicosis can range from no visible changes to excess mucus production in acute cases and a wide range of alterations in subacute to chronic exposures including interstitial edema, epithelial degeneration, epithelial hyperplasia, and fusion of lamellae.[63,472] As in any process that affects the gills, critical assessment of cellular and architectural preservation along with associated artifacts, which can mimic pathologic changes, particularly edema and degeneration, is necessary (Fig. 39.6). Although high ammonia levels can lead to convulsions and coma, presumably due to alterations in neurotransmitter activity, morphologic changes similar to those of mammals have not been documented in the brains of affected teleosts.[471]

Most situations that result in ammonia toxicosis can also lead to **nitrite toxicity,** since oxidation of nitrite is similarly dependent on the existence of appropriate microorganisms in the filter bed. Nitrite toxicosis is largely a disease of freshwater fish due to differences in marine versus freshwater electrolyte physiology and mechanisms of NO_2^- uptake.[209] As for other species, a primary effect of elevated nitrite is the oxidation of hemoglobin, limiting the oxygen-carrying capacity of the blood. Hemolytic anemia and derangements in chloride and potassium homeostasis also occur and contribute to morbidity. Sudden death of single or multiple fish is common, and significant morbidity and mortality can occur. Gills and blood may be grossly brown. Serum biochemical and complete blood count abnormalities reflecting nitrite-associated derangements, including anemia, methemoglobinemia, low chloride, and acidosis, may be noted.[209] Diagnosis is dependent on the documentation of methemoglobinemia in affected individuals coupled with elevated nitrite levels in the water (Figs. 39.7A, B). As susceptibility is also variable by species, Noga[333] suggests that freshwater nitrite levels greater than 0.10 mg/L should be considered elevated for warm water aquarium species. Even in systems where treatment is employed, lingering effects, such as anemia and electrolyte imbalances, especially of potassium, may result in extended periods (weeks) of elevated sensitivity to stressors and increased rates of infectious disease.[290]

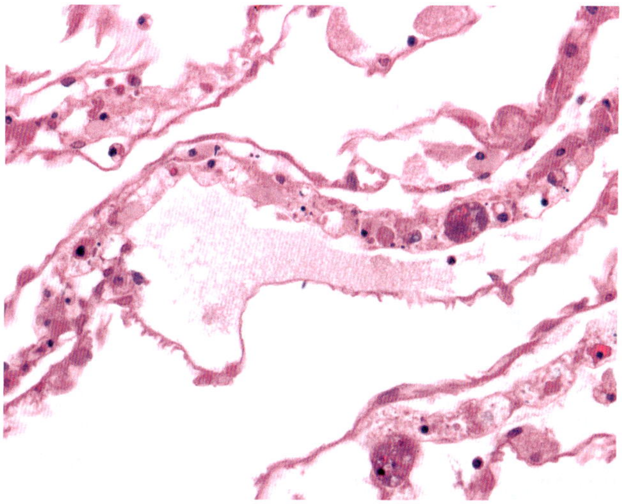

FIGURE 39.6 Lamellar edema in the gill of a weedy seadragon. The lamellar epithelium is elevated away and separated from the lamellar vascular core by lightly eosinophilic edema fluid. Note the presence of several scuticociliates in the lamellar vasculature.

Toxicosis and death of fish are associated with several other soluble chemicals that can achieve noxious levels under certain environmental conditions, most notably chlorine and hydrogen sulfide. **Chlorine toxicosis** commonly occurs after contamination of water systems from the release of treated water by sewage treatment plants or from the use of chlorinated municipal water, i.e., "city water," in home and commercial aquariums. Fish exposed to chlorine may swim down concentration gradients, if possible, or exhibit erratic swimming behaviors or die acutely, depending on the concentration and duration of exposure. Experimental intermittent exposure of bluegill and rainbow trout to chlorine led to branchial lamellar epithelial hyperplasia and swelling of lamellar epithelial cells, which was severe at lethal concentrations. Additionally, hepatic necrosis was reported in those fish exposed to high

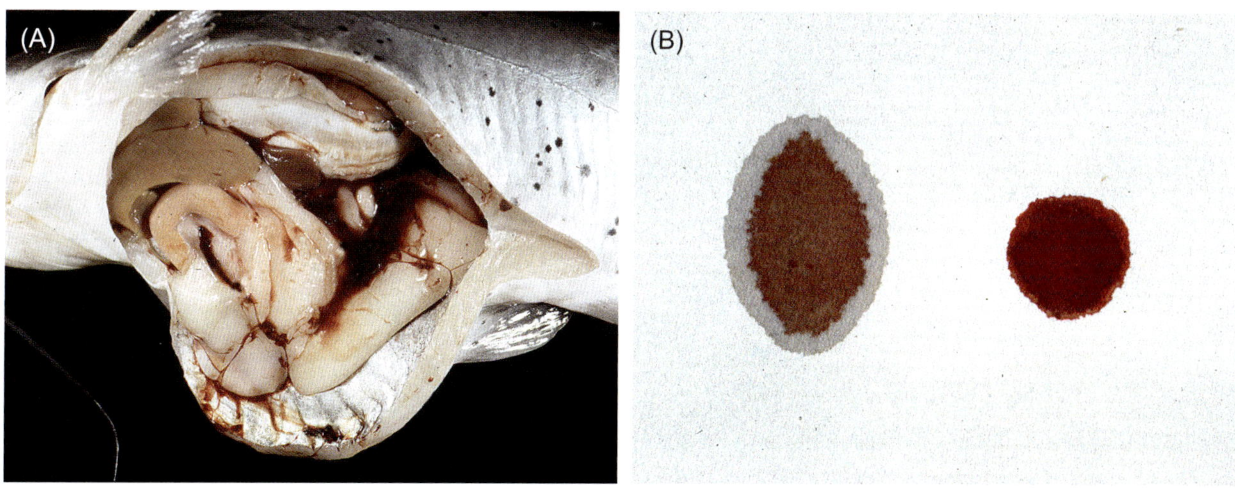

FIGURE 39.7 Methemoglobinemia in a channel catfish. (A) Blood that is brown is present in the coelomic cavity. There is a generalized pallor to the viscera. (B) Drops of blood from a catfish with methemoglobinemia (brown, left) and a normal channel catfish (red, right).

levels of chlorine.[22] Water from municipal supplies or treatment plants should be treated with sodium thiosulfate or allowed to stand for several days in order for the chlorine to outgas.[379] **Hydrogen sulfide toxicosis** is typically associated with sewage pollution or organic enrichment of anaerobic bottom sediments. Hydrogen sulfide is highly soluble in water and highly toxic to fish in its undissociated state. Turnover of bottom sediments can release hydrogen sulfide, resulting in mass mortalities of fish. This environmental scenario is exacerbated in acidic waters.[379,403]

Aflatoxicosis results from the ingestion of **Aflatoxin B1** produced by fungi such as ***Aspergillus flavus*** and ***A. parasiticus*** growing on plant-based dietary protein sources in fish feed. The parent compound is biotransformed via the cytochrome P-450 monooxygenase system into the biologically relevant 8,9 epoxide, which is subsequently detoxified via glucuronidation.[423] Aflatoxin has both cytotoxic and mutagenic properties leading to both direct hepatocellular damage and neoplastic transformation. Exposures can be acute to chronic and responses dose dependent. Fish are considered highly sensitive to the toxin, but even in the teleosts, variable species susceptibility is observed. Differential sensitivity among salmonids is related to the efficiency of monooxygenase conversion of the parent compound more so than the detoxification and excretion of the active compound.[16]

The importance of aflatoxicosis in fish has been recognized since the 1960s when large numbers of commercially raised trout presented with **hepatocellular carcinoma**. Neoplastic transformation is species dependent, probably again reflecting monooxygenase efficiency. In highly susceptible species, the incidence of neoplasia is high and tumors are characteristically malignant, while in less susceptible species incidence is lower and hepatomas predominate (Fig. 39.8).[16] Aflatoxin-associated hepatocellular neoplasia and hepatotoxic effects have subsequently been noted in a variety of species, the best studied being those fish produced commercially for human consumption.[423] Elevations in hepatic transaminases, hepatocellular lipid accumulation, megalocytosis, megalokaryosis, hepatocellular necrosis, biliary hyperplasia, hepatoma, and hepatocellular and biliary carcinoma have all been reported.[16,491,542] Since hepatocellular lipid accumulation is common in display fish, it is probably of limited diagnostic utility in most instances. Compatible histologic lesions raise suspicion of toxicosis, and testing of feed for Aflatoxin B1 is of great utility, particularly in suspected acute exposures. Liver from suspect exposed fish can also be tested as bioaccumulation can occur.

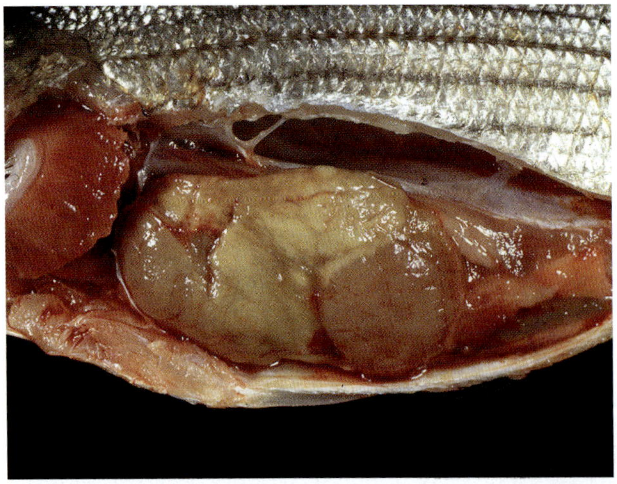

FIGURE 39.8 Hepatocellular carcinoma in the liver of a hybrid striped bass with exposure to aflatoxin. The liver is enlarged and has irregular margins, areas of nodularity, and tan-yellow mottling.

Miscellaneous/Multifactorial

Water quality plays a pivotal role in fish health and is integral to an understanding of the pathogenesis of fish disease and to any diagnostic investigation. Both physical and chemical parameters directly and indirectly affect the occurrence of disease in fish. Water temperature can have a dramatic impact on fish health and water quality. Many fish have upper and lower limits of thermal tolerance as well as optimum temperature ranges for essential physiological processes, such as growth, immune cell responsiveness, metabolism of nutrients, and reproduction.[131,403] Changes in water temperature can have varied effects on cellular elements of the adaptive immune response (e.g., lymphocytes) as well as on acellular and cellular elements of the innate immune response. Such changes in the activity of immune cells can alter immunosurveillance, resulting in the development and regression of some neoplasms, e.g., walleye dermal sarcoma.[37,39] Many fish pathogens, e.g., koi herpesvirus, only cause disease within specific temperature ranges. Temperature can also affect the chemical properties of water causing environmental stress. As water temperature increases, the solubility of dissolved oxygen decreases, which has resulted in large-scale fish die-offs due to hypoxia (Fig. 39.9). Conversely, the solubility of toxins, such as crude oil, some pesticides, and certain heavy metals, increases with increasing water temperature.[403] Similarly, water pH can have broad effects on fish health. In general, extremes in pH have deleterious effects on osmoregulation, blood pH, and respiration. Water pH can also have a significant effect on the toxicity of certain compounds, such as ammonia; the amount of unionized ammonia (NH_3), which is highly toxic to fish (see "Toxic" section), increases with increasing temperature, and increasing pH above pH 7.5.[403] Low water pH increases the risk of toxicity from elevated levels of CO_2, chlorine and heavy metals.

The presence of gas supersaturation can lead to a condition known as "**gas bubble disease**." Gas supersaturation occurs when the total pressure of dissolved gases in an

FIGURE 39.9 The role of water quality in fish health. Acute mortality of Atlantic menhaden *Brevoortia tyrannus* related to hypoxia in an estuary. *(Photo Courtesy of Dr. Jan Lovy, USGS Western Fisheries Research Center.)*

aquatic system exceeds barometric pressure. Common scenarios are those in which gases are drawn or injected into the water. These may include impeller cavitation in pumps, holes on the low pressure (intake) sides of pumps, and natural or artificial waterfalls. Other scenarios include utilization of water from deep wells when paired with heating to higher temperatures and/or without providing for off-gassing (passive or active) at lower surface air pressure.[467] Supersaturation can result in gas embolism in any vessel or emphysema in any organ. The most commonly reported lesions are branchial gas embolism, which can be seen in gill wet mounts (Fig. 39.10A), exophthalmos (due to retrobulbar emphysema), and intraocular gas accumulation (Fig. 39.10B). Histologically, the most striking lesions may be colorless spaces indicative of gas accumulation in the choroidal rete and/or retrobulbar tissues and, in chronic cases, granulomatous inflammation in affected tissues. High morbidity and mortality can occur acutely when total pressures are high. In surviving animals, neurologic signs, blindness, generalized debilitation, and persistent exophthalmos are concerns. In lower-pressure exposures, elevated morbidity and mortality without significant lesions and increased infectious disease, presumably subsequent to gas-related tissue damage, have been reported.[377] Accurate assessments of physical conditions and insightful historical information together with characteristic gross lesions can make the diagnosis straightforward. Evaluation of gas saturation of the water from the enclosure can provide confirmation. In chronic mild supersaturation, gas emboli may not be seen, and diagnosis is difficult in the absence of evaluation for gas saturation.

Current evidence points to an association of **head and lateral line erosion** (**HLLE**) syndrome with use of activated carbon in the filtration cycle. Stamper et al.[476] showed rapid development of characteristic lesions in ocean surgeonfish upon experimental exposure to activated charcoal; lesions resolved when affected individuals were subsequently transferred to enclosures without carbon. In another experimental setting, sharpnose sea bream raised in saline well water developed lesions while those raised in seawater did not. Analysis of water suggested that the only significant difference was a fourfold greater CO_2 concentration in well water.[220] Lesions can vary from depigmentation to pitting to large erosions or ulcerations centered on lateral line structures of the head or other portions of the body, such as the opercula and body wall lateral line (Fig. 39.11A). Histologically, loss of dermis, hypodermis, and bone may all be present; however, lesions of epidermal erosion with minimal or, at most, mild chronic dermatitis are more common in the experiences of some

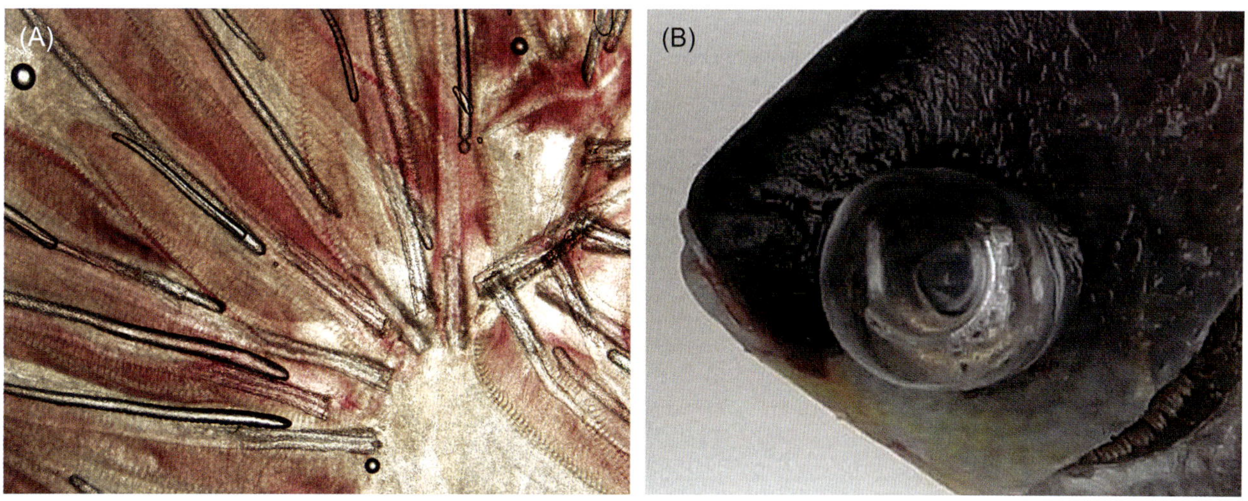

FIGURE 39.10 (A) Gas bubble disease in a gill clip wet mount from a teleost housed in an enclosure that suffered from a gas supersaturation event. Note the elongate gas bubbles within branchial vessels. (B) Gas bubble disease in a creole wrasse from an aquarium. There is an outward bulging of the cornea from the accumulation of gas in the anterior chamber. *(Part A: Photo Courtesy of University of Illinois Zoological Pathology Program. Part B: Photo Courtesy of University of Illinois Zoological Pathology Program.)*

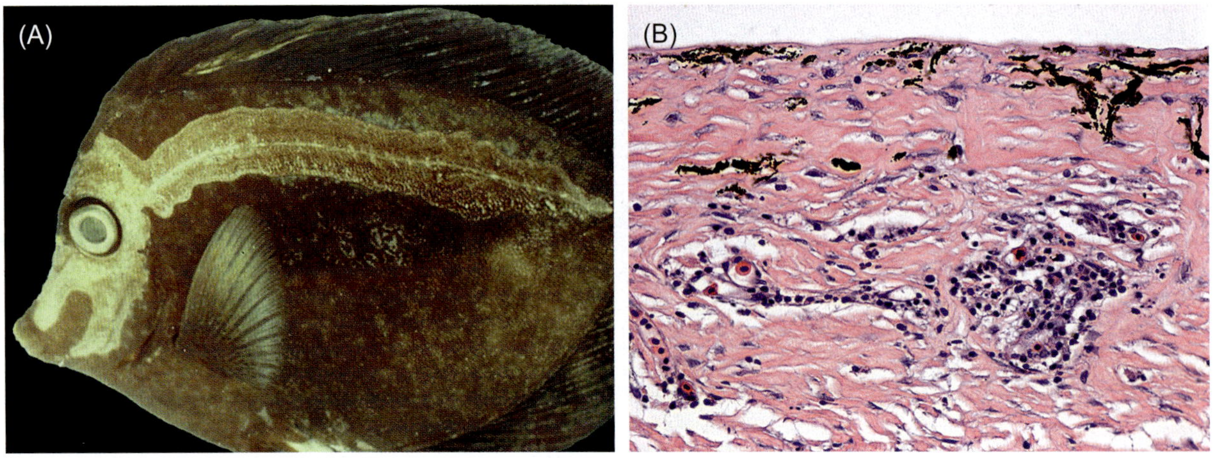

FIGURE 39.11 (A) Head and lateral line erosion syndrome in a tang. There is erosion of the skin bordering the length of the lateral line system and its cephalic extensions. (B) Head and lateral line erosion syndrome in the skin of a teleost. The epidermis is eroded, leaving a denuded basement membrane. In the dermis, a mild perivascular infiltrate of lymphocytes and macrophages is present.

pathologists. (Fig. 39.11B). Characteristic lesions can be readily detected grossly and histologically. Fish can and do survive long term with HLLE. Significance is centered on animal welfare concerns (minimizing opportunistic infections and concurrent disease) and managing the concerns of the public regarding readily visible lesions (maintaining exhibit-worthy specimens).

Goiter associated with **iodine deficiency** is a concern in managed fish populations. Iodine is an essential element required for thyroid hormone production. Iodine in the form of iodide (I^-) is more efficiently absorbed from the water across the gill than across the gut in fish.[386] Elevated nitrates and ozonation, frequent challenges in closed aquarium systems, indirectly contribute to goiter development. Nitrates inhibit I^- uptake by the thyroid.[315] Ozonation converts I^- to iodate (IO_3) which is not biologically available.[440] Even though dietary supplementation and iodine additions to water are common in display aquaria, goiter still occurs, inferring that such inputs may not be sufficient to fully counter the above effects. Hyperplastic thyroid tissue is often relatively localized in branchial arches and/or glossal regions and appears as one or more discrete, expansile masses, which may appear invasive (Fig. 39.12A). As thyroid tissue does not occur as an encapsulated organ in teleosts (see Unique Features, Digital Content), hyperplasia can result in grossly or microscopically evident hyperplastic thyroid tissue in various parenchymatous organs, i.e., ectopic thyroid follicular hyperplasia. Hence, interpretation of proliferative thyroid lesions requires consideration of these anatomic features in conjunction with criteria that distinguish hyperplasia from neoplasia (see Neoplastic later). Diagnosis of goiter is based primarily on the documentation of characteristic gross or histologic lesions. Histologically, thyroid follicular cell hyperplasia is characterized by the formation of well-differentiated follicles of various sizes and shapes that contain colloid and are lined by a simple cuboidal to columnar epithelium (Fig. 39.12B). Follicular cell hyperplasia can be simple, i.e., an increase in the number of follicles within the hypobranchial stroma of the pharynx, or nodular, i.e., an increase in the number of follicles in a localized region of the hypobranchial stroma. Nodular thyroid follicular cell hyperplasia typically results in goiter and often is multinodular.[142] Identification of low I^- levels in water and feed can support the diagnosis. Total water iodine determination is not a sufficient test modality as it will not differentiate biologically available (I^-) and unavailable (e.g., IO_3) forms. Serum thyroxine levels in affected individuals could also be assessed (baseline and response to treatment). The importance of thyroid hormone for normal metabolic function and the significance of substantial derangements of normal metabolic function in iodine deficiency are not well understood in fish.

Neoplastic

Neoplastic disease in fish significantly impacts a wide variety of economic ventures, including commercial and sport fisheries, aquaculture, mariculture, and public aquaria. Equally, neoplasia in fish has contributed significantly not only to our understanding of fish health but also to our understanding of environmental pollution and human cancer. Fish serve as both sentinels of carcinogenic environmental contamination and subjects for monitoring the effects of remediative efforts. In biomedical research, fish serve as models for human carcinogenesis and as screening tools for antineoplastic agents. Consequently, the detection and accurate morphologic diagnosis of fish tumors has been and always will be a worthwhile endeavor.

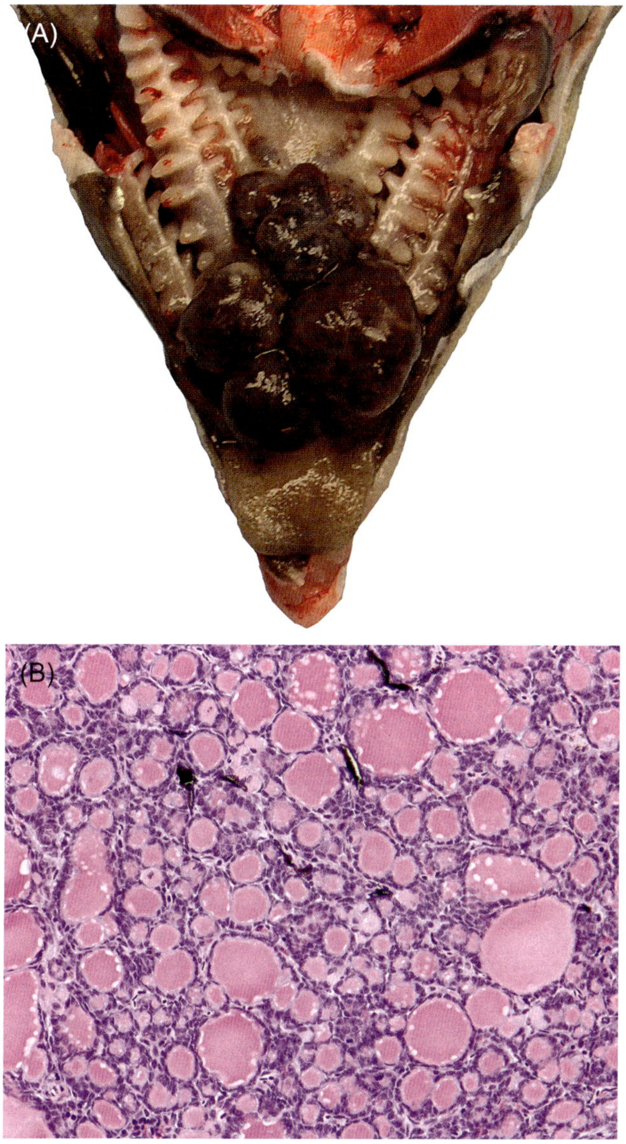

FIGURE 39.12 (A) Goiter in the branchial arches and glossal tissues of a sablefish. This multinodular mass is composed of hyperplastic thyroid tissue. (B) Follicular cell hyperplasia in mummichog. This goiter was composed of densely packed and varyingly sized follicles containing abundant colloid. Follicles are lined by simple, cuboidal epithelium. *(Part A: Photo Courtesy of University of Illinois Zoological Pathology Program.)*

Fish serve as models for carcinogenesis and research into the molecular basis of cancer through investigations of naturally occurring neoplasms as well as transgenic applications in experimental studies. Whereas no naturally occurring models of melanoma formation exist in rodents, the occurrence of melanomas in F1 backcrosses of swordtail and platyfish (*Xiphophorus* spp.) hybrids is well established and has served as a model for research into melanoma carcinogenesis.[153] The variety of tumor models that currently can be produced by genetic alteration of laboratory fish species, such as zebrafish, is almost limitless, making a comprehensive list of fish neoplasms something of a Sisyphean task. Some tumor types in wild and captive fishes have been reported to occur naturally, that is, spontaneously, although certain of these tumors actually may have environmental causes. Alternatively, certain other neoplasms have been induced experimentally (Figs. 39.13A, B and 39.14). For more about experimental fish tumor models, the article by Bunton[52] remains a valuable resource for understanding the chemical induction of tumors in fish, and more recently, Barriuso et al.[21] reviewed various transgenic zebrafish cancer models.

The causes of neoplastic disease in fish parallel those of mammals. Evidence for chemical tumor induction is based on prospective controlled experimentation (e.g., intentional exposure of laboratory fish species to established carcinogens, such as diethylnitrosamine or *N*-methyl-AT-nitro-*N'*-nitrosoguanidine). In other cases, carcinogenesis has been linked convincingly to inadvertent environmental chemical exposure, for example, hepatic and exocrine pancreatic neoplasms in mummichog from creosote-contaminated waters.[141] Foci of hepatocellular alteration are recognized in fish; such foci are considered putatively preneoplastic lesions and can be associated with contaminant exposure. Foci of hepatocellular alteration are named based on their tinctorial properties and include basophilic, eosinophilic, vacuolated cell, and clear cell types.[27] The margins of these lesions are discrete, and these foci merge seamlessly with the adjacent parenchyma. Foci of cellular alteration can be distinguished from hepatocellular neoplasms based on their retention of hepatic architecture, lack of cellular atypia, and absence of compression of the neighboring parenchyma (Fig. 39.15). The association between tumorigenesis and infectious agents such as retroviruses, herpesviruses, and papillomaviruses ranges from weak, e.g., detection of reverse transcriptase activity or ultrastructurally evident viral particles in tumor samples, to strong, e.g., virus isolation followed by sequencing and successful generation of the characteristic neoplastic disease by exposure of fish to cell-free filtrates derived from homogenized tumor tissue.[77] In this context, retroviruses have been associated with the epizootic and seasonal occurrence of several tumors in fish, e.g., walleye dermal sarcoma, northern pike lymphosarcoma, and Atlantic salmon swim bladder fibrosarcoma; the retroviruses implicated appear highly adapted to both their hosts and their host's environment, as evidenced by the presence of a temperature-sensitive reverse transcriptase in *Walleye dermal sarcoma virus* and the *Pike lymphosarcoma retrovirus*.[286] More can be found regarding walleye dermal sarcoma later in this chapter (see "RNA Viruses" section). Associations have also been noted between intestinal parasitism by *Pseudocapillaria tomentosa* and the formation of hepatic or ductular tumors and malignant neoplasms of the intestine in zebrafish (eSlides 39.e1 and 39.e2), and a malignant ovarian tumor in a longjaw mudsucker was

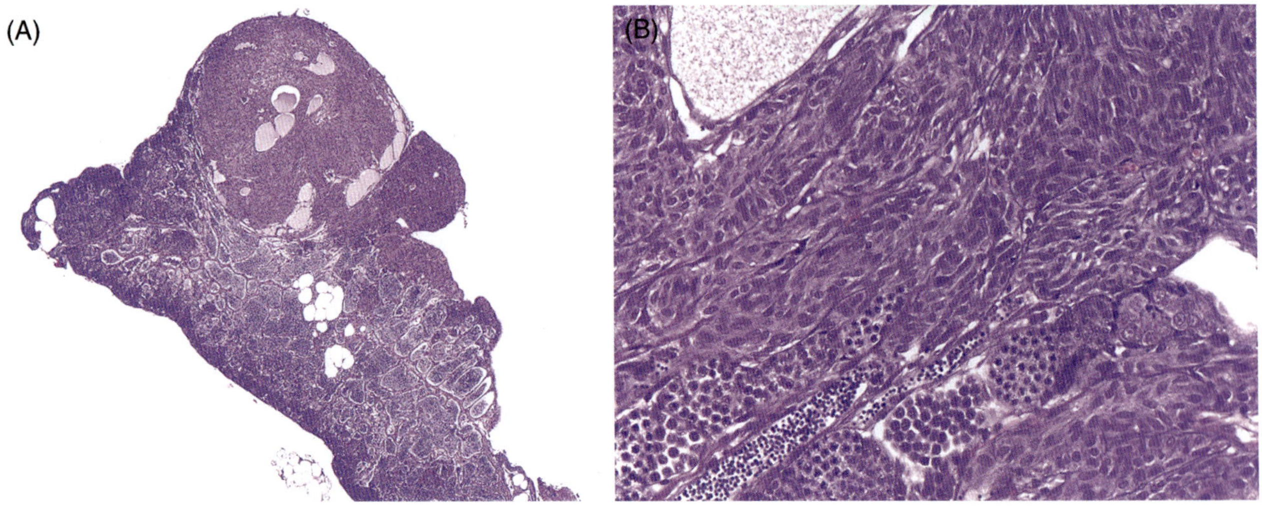

FIGURE 39.13 Sertoli cell tumor in the testis of a medaka. (A) A large spherical nodule with irregular extensions and internal cystic spaces expands the margin of the testis. (B) In this high magnification view, the tumor tissue is characterized by cells with carrot-shaped nuclei and moderate amounts of cytoplasm that form bundles divided by fine fibrous septa. Near the bottom of the image, portions of the Sertoli cell tumor surround several spermatocysts, each of which contains a clonal group of cells in a stage of spermatogenic development. *(Part A: Photo Courtesy of Dr. Jeffrey C. Wolf, EPL, Inc. Part B: Photo Courtesy of Dr. Jeffrey C. Wolf, EPL, Inc.)*

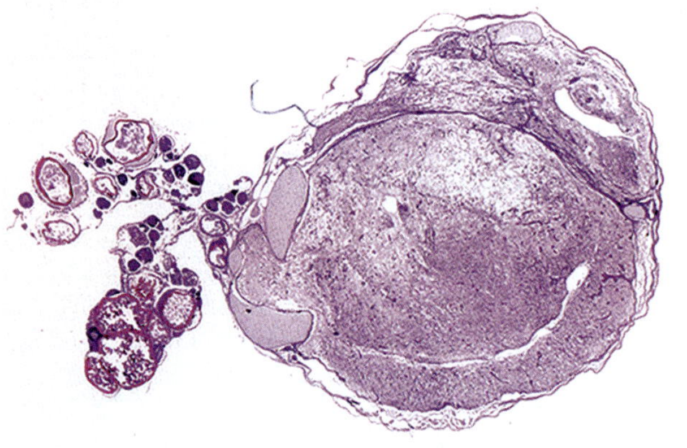

FIGURE 39.14 Teratoma in the ovary of a striped bass. Attached to the nonneoplastic ovary on the left is a large ovoid mass that consists primarily of poorly organized tissue that resembles brain neuropil and cartilage. *(Photo Courtesy of Dr. Jeffrey C. Wolf, EPL, Inc.)*

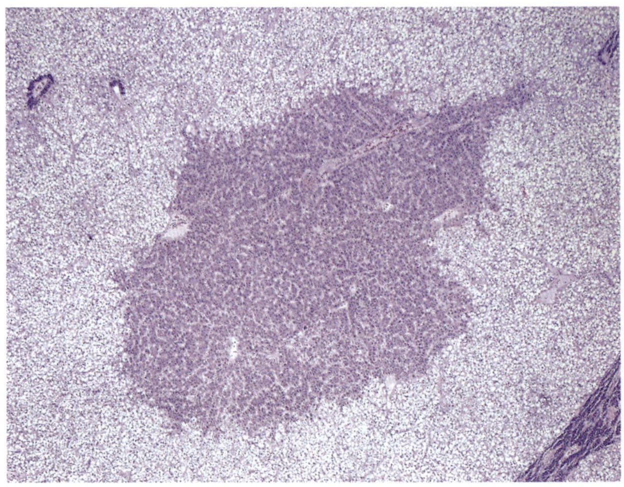

FIGURE 39.15 Focus of cellular alteration in the liver of a striped bass. A discrete focus of eosinophilic, nonvacuolated hepatocytes blends inconspicuously into the adjacent hepatic parenchyma. Note the retention of the hepatic architecture and lack of neighboring tissue compression.

considered a sequela of granulosa cell hyperplasia induced by a microsporidian infection.[33,227]

Fish are subject to a diverse array of tumor types, the variety of which seems to increase as captive fish achieve longer lifespans in public and private aquaria, the number of piscine models of carcinogenesis multiply, and awareness grows of the potential for environmental contamination to induce tumor formation. Diagnostic criteria for tumors from fish largely have followed fundamental principles applied to mammalian tumor diagnosis. Many of the characteristic features of well-recognizable tumors such as squamous cell carcinomas, fibrosarcomas, peripheral nerve sheath tumors, hepatocellular carcinomas, and **primitive neuroectodermal tumors (PNETs)** (Fig. 39.16; eSlide 39.e3) are seen in fish with minor variations in a few tumor types. For example, a hallmark of mammalian squamous cell carcinomas (SCC) is the formation of structures known as keratin pearls; however, these are typically not observed in SCC of fish because keratinized epithelium is not anatomically present in most fish species, one exception being the chewing pad located in the pharynx of some species. Comparable to mammals, fish develop tumors of epithelial cell origin, mesenchymal cell origin, and tumors that are embryonal, that is, develop from pluripotential cells that give rise to histologically complex neoplasms with epithelial and mesenchymal cell types and

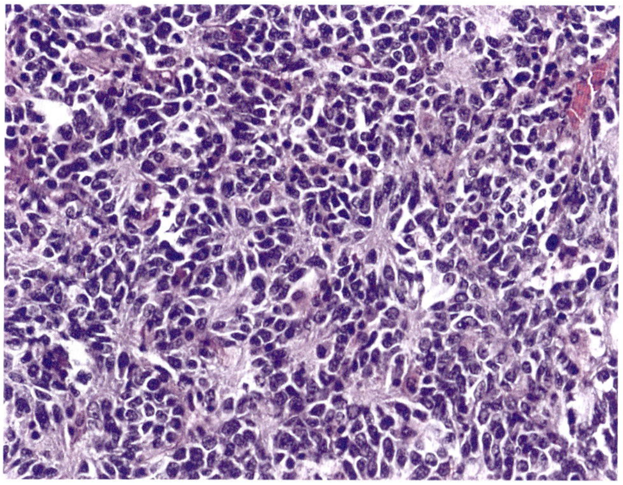

FIGURE 39.16 Retinoblastoma in an African cichlid. The neoplasm is composed of triangular (carrot-shaped) to round cells often arranged in rosettes supported by a fine fibrovascular stroma. Rosettes consists of neoplastic cells surrounding central areas of fibrillar eosinophilic cytoplasm (see eSlide 39.e3).

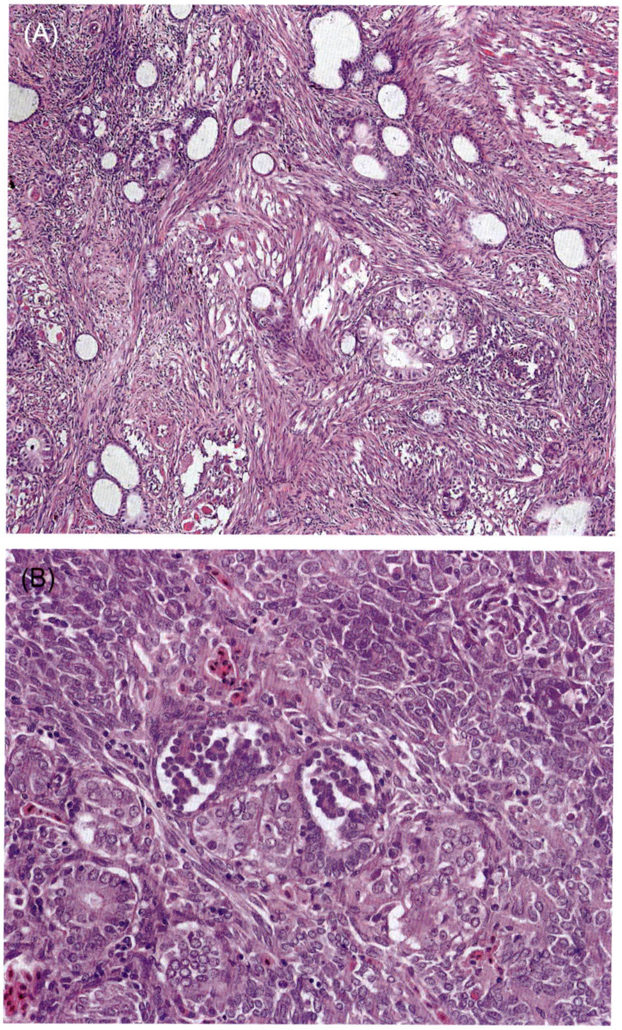

FIGURE 39.17 Nephroblastoma in a Japanese eel. Nephroblastomas have morphologic features stemming from their constituent cells attempting to recapitulate nephrogenesis; as such, three constituent cell types are classically seen, i.e., blastemal, stromal, and epithelial. (A) The emphasis of this photomicrograph is on the stromal component, which consists of intersecting streams of fibroblastic cells and collagen interrupted by tubular arrangements of epithelial cells occurring individually or in clusters. (B) The emphasis of this photomicrograph is on the blastemal and epithelial components, represented by solid aggregates of oval basophilic blastemal cells and epithelial cells arranged in tubules and glomeruloid structures, which consist of luminal spaces within which are tufts of cells. *(Photos Courtesy of Dr. Jeffrey C. Wolf, EPL, Inc.)*

tissues, e.g., **nephroblastoma** (Figs. 39.17A, B) and **branchioblastoma** (Figs. 39.18A, B). Also akin to mammals, tumors have been reported from nearly all organ systems and tissues of fish. An excellent description of common neoplasms in fish is provided by Martineau and Ferguson.[286] Epithelial tumors of the skin of fish include **papillomas** (Figs. 39.19A, B), which can be exophytic or plaque-like, and squamous cell carcinoma. Skin tumors of mesenchymal origin include **chromatophoromas**, such as **melanophoromas (melanomas)**, **erythrophoromas**, **iridophoromas** (Figs. 39.20A–C), and spindle cell tumors, which may have histologic features typical of fibrosarcoma or **peripheral nerve sheath tumor** (Fig. 39.21), and tumors of vascular origin, for example, hemangioma, hemangiosarcoma, and vascular wall tumors. Tumors of the liver in fish include **hepatoma** and **hepatocellular carcinoma**, which have been associated with aflatoxin exposure, particularly in rainbow trout (see Toxic Diseases section), and **cholangioma** and **cholangiocarcinoma**. Tumors of the reproductive tract include germ cell tumors, e.g., **seminomas** (Fig. 39.22) and **dysgerminomas** (Figs. 39.23A, B) and sex cord-stromal tumors, e.g., **Sertoli cell tumors** (Figs. 39.13A, B). Hematopoietic neoplasms of fish are represented primarily by **lymphosarcoma** (Fig. 39.24) and lymphocytic or lymphoblastic **leukemia**, although epizootics of plasmacytoid neoplasia in Chinook salmon and histiocytic tumors in northern pike have been reported.[77,486] As in mammals, it may not always be possible to identify the precise cell of origin in poorly differentiated tumors, and in such cases, the goal may be to at least narrow the diagnosis to a broad morphologic category, e.g., malignant round cell tumor.

Given that a comprehensive listing of all known fish tumors would be impractical, Table 39.e1 provides a compilation of tumor types in fish that is intended to be representative and, as such, affirming of the aforementioned themes. Tumor types are accompanied by an example and literature reference that describes that neoplasm in a particular fish species, although in reality, most tumor types are not limited to a single species. For the diagnostician, it can be helpful to have an idea of the relative frequency of tumor occurrence. For example, some neoplasms, such as peripheral nerve sheath tumors, are comparatively common in fish, whereas tumors such as pituitary adenomas appear

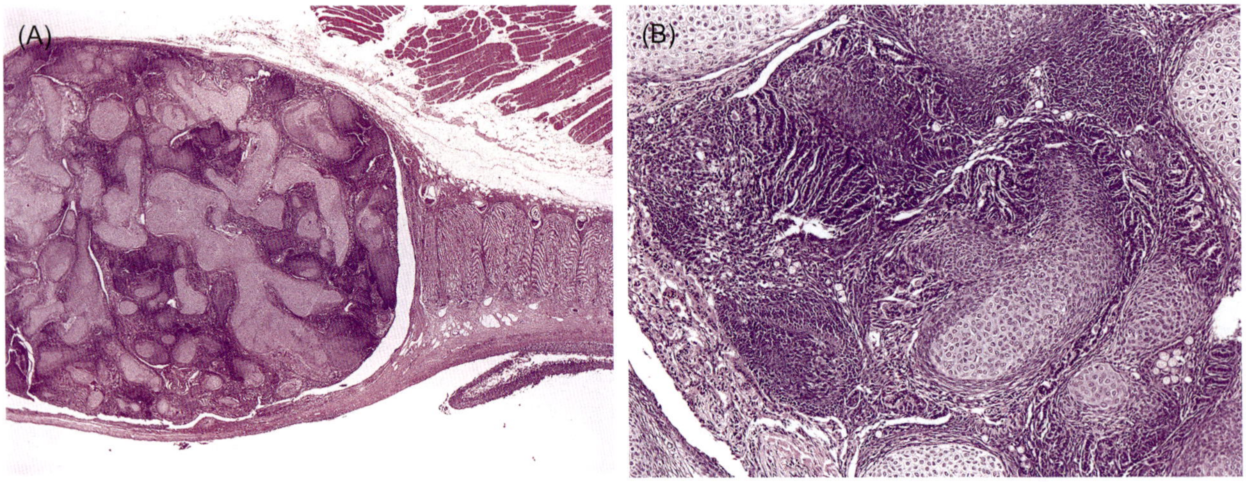

FIGURE 39.18 Branchioblastoma in a salmon. (A) A well-demarcated, expansile, nodular neoplasm effaces and replaces the normal architecture of the gill. (B) There are attempts by the poorly organized tumor tissue to replicate the filamental and lamellar gill architecture. *(Part A: Photo Courtesy of Dr. Jeffrey C. Wolf, EPL, Inc. Part B: Photo Courtesy of Dr. Jeffrey C. Wolf, EPL, Inc.)*

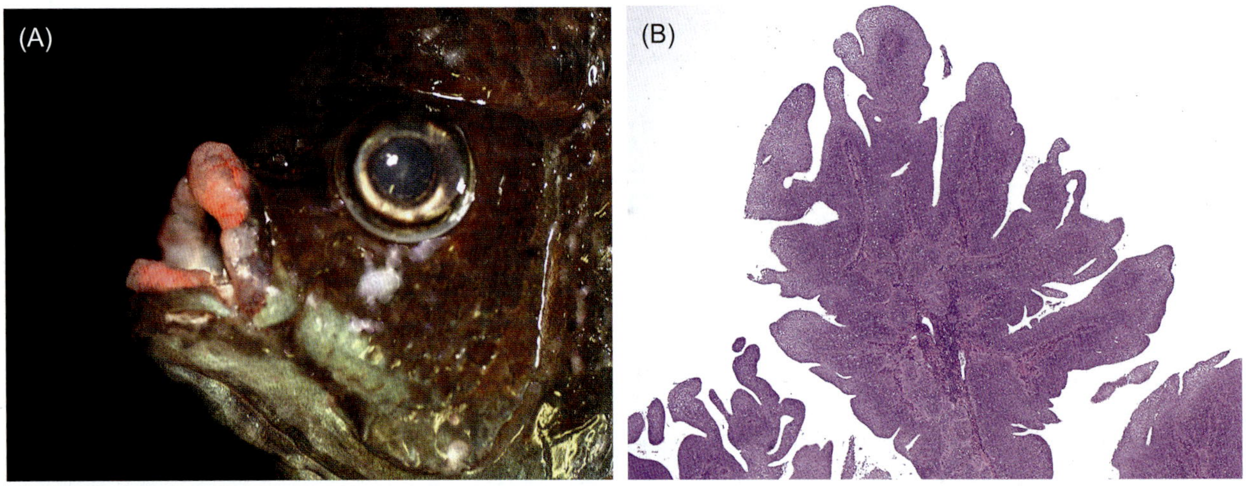

FIGURE 39.19 Papilloma in a bluegill. (A) Linear, raised masses with reddened, beaded surfaces extend from the mucosa at the mucocutaneous junctions of the mouth. (B) The mass consists of arborizing fronds lined by variably wide epithelium supported by thin stalks of fibrovascular connective tissue.

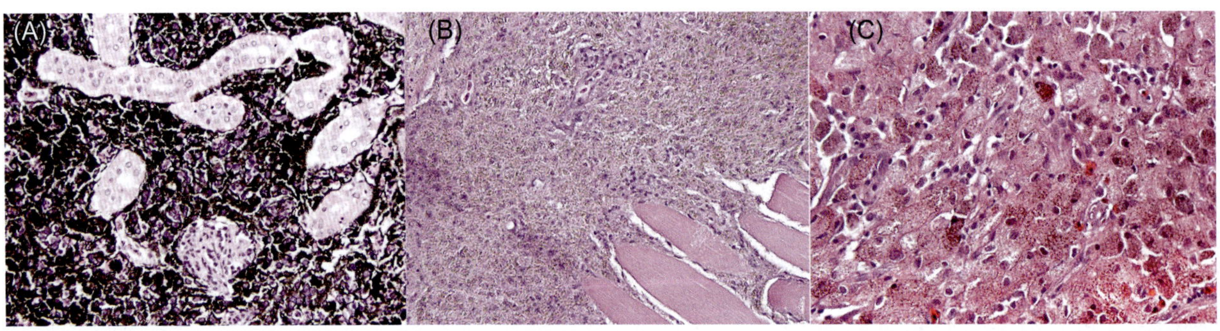

FIGURE 39.20 Chromatophoromas in teleosts. (A) Melanophoroma in the posterior kidney of a striped eel catfish. Pleomorphic neoplastic cells replace the hematopoietic tissue and contain abundant, cytoplasmic, black to brown pigment granules. (B) Iridophoroma in the skin and underlying skeletal muscle in a betta fish. Note the neoplastic cells contain abundant, refractile, elongate, golden-brown crystals, which are birefringent with polarized light. (C) Erythrophoroma in the skin of a kelpfish. Round to polygonal cells contain red-brown, cytoplasmic pigment granules.

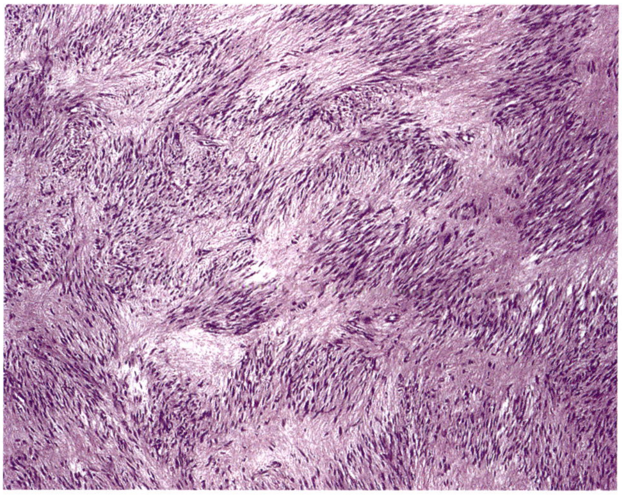

FIGURE 39.21 Peripheral nerve sheath tumor in the dermis of a goldfish. A feature that is highly characteristic of nerve sheath tumors (e.g., schwannoma and neurofibrosarcoma) is the organization of spindle cells into palisading bundles or fascicles. *(Photo Courtesy of Dr. Jeffrey C. Wolf, EPL, Inc.)*

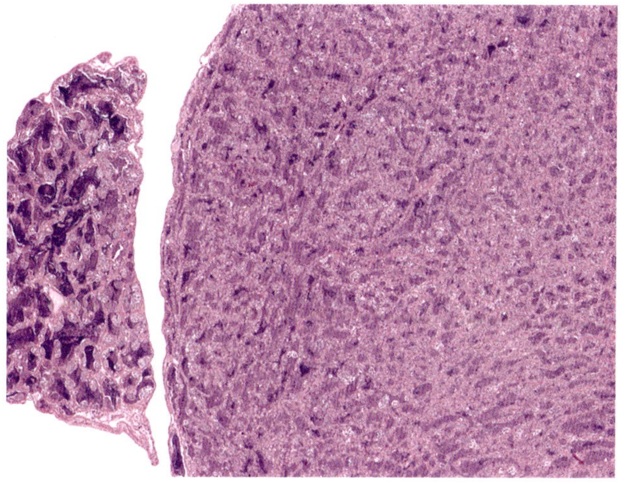

FIGURE 39.22 Seminoma in the testis of a *Synodontis* catfish. The tumor is the large mass in the right half of the image, and a portion of residual nonneoplastic testis is present on the left. As is the case for many teleost seminomas, this tumor is spermatocytic, that is, there is a progression of spermatogenesis from spermagonia to more mature stages. *(Photo Courtesy of Dr. Jeffrey C. Wolf, EPL, Inc.)*

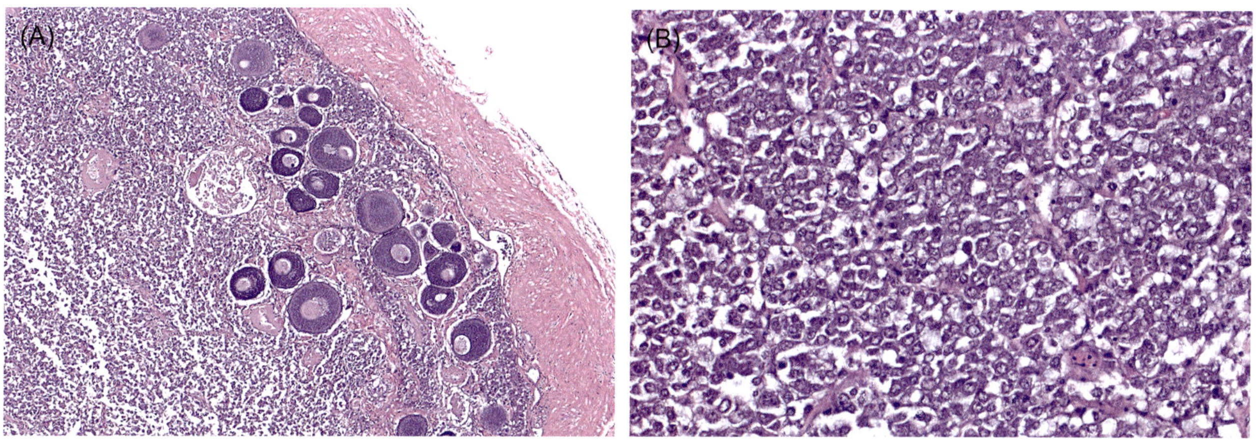

FIGURE 39.23 Dysgerminoma in the ovary of a pipefish. (A) Ova and follicles are replaced by sheets of round cells. (B) Neoplastic cells are arranged in variably wide cords separated by fine fibrovascular stroma, and individual cells have moderate amounts of amphophilic cytoplasm and a round to oval nucleus with coarse chromatin and a single nucleolus, which is typical of tumors of germ cell origin whether they arise from ovary or testis.

to be exceedingly rare; however, true tumor prevalence in a given species is often difficult to determine, especially for wild fish in which the development of cancer may impair survival and therefore impact collection. In Table 39.e1, the relative frequency of tumor occurrence is represented by the number of cases of each neoplastic diagnosis that can be found among the 7,768 case submissions contributed over 42 years in the currently inactive Registry of Tumors in Lower Animals archival collection.[415]

Although definitive evidence of a genetic etiology is difficult to establish in naturally occurring (versus experimentally induced) neoplasms, a number of species predispositions for certain tumors appear to exist. Prominent examples include: renal tubular neoplasms of oscars and nephroblastomas in Japanese eels (Figs. 39.17A, B); testicular seminomas and ultimobranchial gland tumors in zebrafish; nerve sheath tumors in gray snappers and gonadal stromal tumors in goldfish and carp; chromatophoromas in butterflyfish and Pacific rockfish; **ameloblastomas** in Chinook salmon (Figs. 39.25A, B) and primitive neuroectodermal tumors in coho salmon.[93,171,230,269,287,352,354,427,512] As noted by Dawe[93], the tendency for certain fish species to develop particular types of tumors may have more to do with habitat than genetics; consider the liver and exocrine pancreatic tumors in Elizabeth River mummichog and the hepatocellular neoplasms in English sole of Puget Sound, WA, USA as examples of epizootic tumorigenesis associated with environmental contamination.[93,141,283] On the other

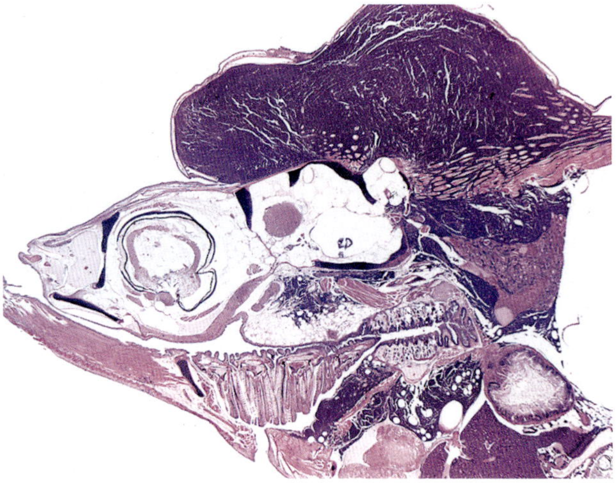

FIGURE 39.24 Lymphosarcoma in a medaka. A large irregular mass comprised of dark lymphoid cells massively expands the hypodermis of the dorsal neck region and infiltrates the adjacent epaxial musculature. In normal medaka, a population of resident lymphoid cells can be found in the dorsal hypodermis, which is the likely site of tumor origination in this case, although extension from the nearby thymus is also possible. *(Photo Courtesy of Dr. Jeffrey C. Wolf, EPL, Inc.)*

hand, the early presumption that oral and cutaneous papillomas in bullhead catfish were contaminant-related has been difficult to confirm.[38] One myth that has been entirely debunked, however, is the idea that elasmobranchs (sharks, skates, and rays) do not develop cancer (see Chapter 40: Chondrichthyes).[360,415]

Several types of tumors are essentially exclusive to fish by virtue of their unique anatomy. These include the branchioblastoma, an embryonic tumor of gill (Figs. 39.18A, B), **lepidocytoma**, a tumor-forming scale (Fig. 39.26), and tumors of the swim bladder and gas gland, an organ responsible for swim bladder inflation in physoclistous teleosts (Fig. 39.27). Tumors of the same type and degree of differentiation generally behave similarly in fish and mammals, although distant spread of malignant tumors by metastasis tends to be far less common in fish. The reason for this disparity is not entirely clear. Additionally, there are proliferative lesions in fish that are suggestive of metastasis but actually are not; examples include ectopic gonad tissue, ectopic thyroid tissue, and hyperplastic or neoplastic proliferations of such ectopic tissues. Notable in this context are the proliferative lesions of the thyroid in fish. It is important to remember that the thyroid of teleosts is a diffuse, unencapsulated organ located in the hypobranchial stroma. In multinodular goiter (see Miscellaneous/Multifactorial diseases section, Goiter), hyperplastic thyroid tissue can be recognized in the surrounding connective tissue, muscle, or gill, suggestive of local invasion, or in ectopic sites, such as the kidney or spleen, suggestive of metastasis. True neoplastic transformation leads to thyroid follicular adenoma or carcinoma. Follicular cell adenoma can be distinguished from hyperplasia by the formation of solid or papillary growth patterns that compress the adjacent stroma or follicles. Follicular cell carcinoma exhibits cellular pleomorphism, nuclear atypia, high numbers of mitotic figures, loss of nuclear polarity, piling of cells along segments of follicles, and scant colloid (Fig. 39.28).[142]

Fish may also present with a variety of nodular lesions that are not neoplasms. Examples of pseudo-tumors in fish include parasite-induced granulomas or large parasitic cysts, e.g., the protozoan-induced "X-cell pseudotumors" of certain marine fishes (Fig. 39.29), viral-induced cellular hypertrophy, e.g., the extreme cytomegaly of dermal fibroblasts caused by the iridovirus responsible for lymphocystis

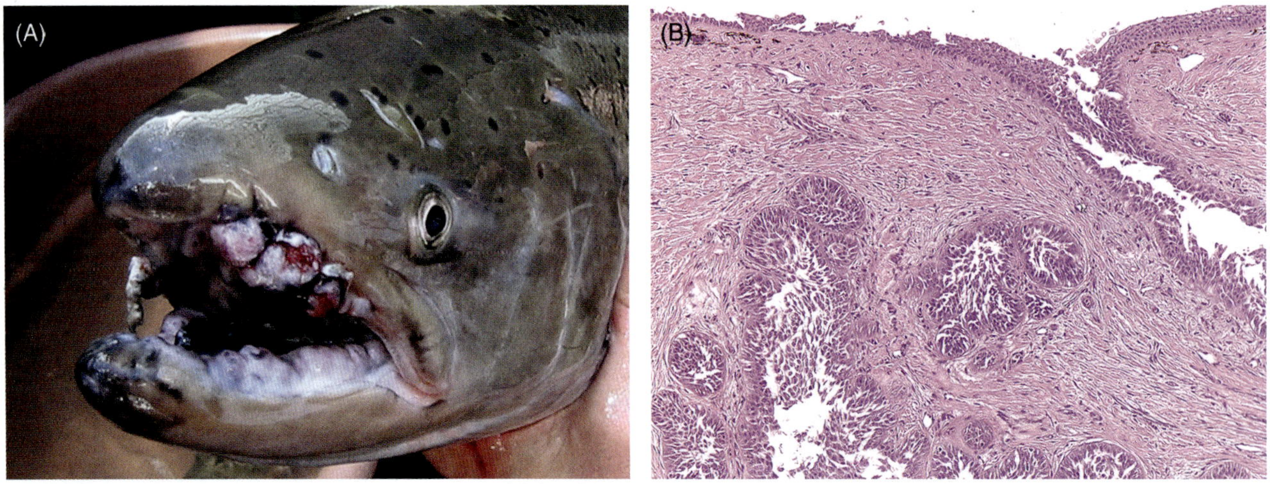

FIGURE 39.25 Ameloblastoma in a Chinook salmon. (A) Multiple, raised, and coalescent, nodular masses are present along the orocutaneous junction and along the roof of the mouth. (B) Islands of tumor tissue are present within the propria mucosa, and there is continuity between the neoplastic tissue and the mucosal surface epithelium. In areas of this specimen that contained teeth, the tumor tissue was observed to arise from the enamel-forming epithelium. This neoplasm resembles squamous cell carcinoma but has more of a tendency to dissociate centrally to form cleft-like structures. *(Part A: Photo Courtesy of Dr. Jeffrey C. Wolf, EPL, Inc. Part B: Photo Courtesy of Dr. Jeffrey C. Wolf, EPL, Inc.)*

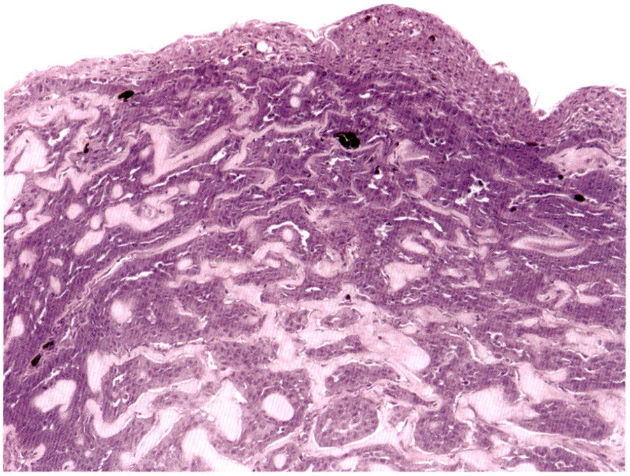

FIGURE 39.26 Lepidocytoma in the skin of a teleost. This neoplasm is characterized by anastomosing linear islands of osteoid and intervening osteoblast-like cells that appear to recapitulate scale formation within the dermis and hypodermis. *(Photo Courtesy of Dr. Jeffrey C. Wolf, EPL, Inc.)*

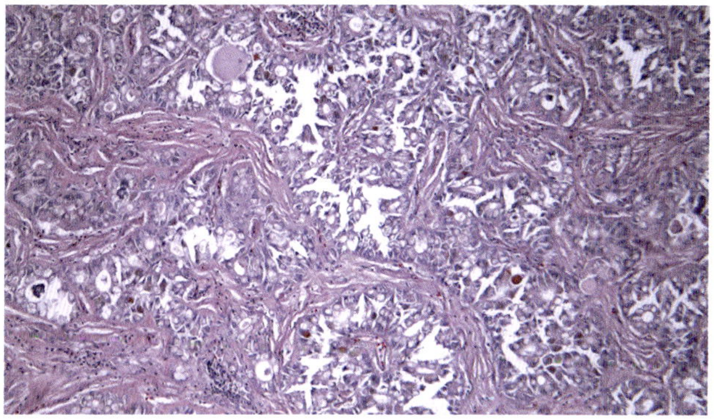

FIGURE 39.28 Thyroid follicular cell carcinoma in a French angelfish. There is marked cellular pleomorphism of follicular epithelium, a loss of nuclear polarity, increased numbers of mitotic figures, piling of follicular cells, and the formation of papillary-like projections into the follicular lumen. Minimal colloid or no colloid is seen within follicular epithelial cells. In some cases, neoplastic transformation can occur within a mass that otherwise has histologic features of follicular cell hyperplasia.

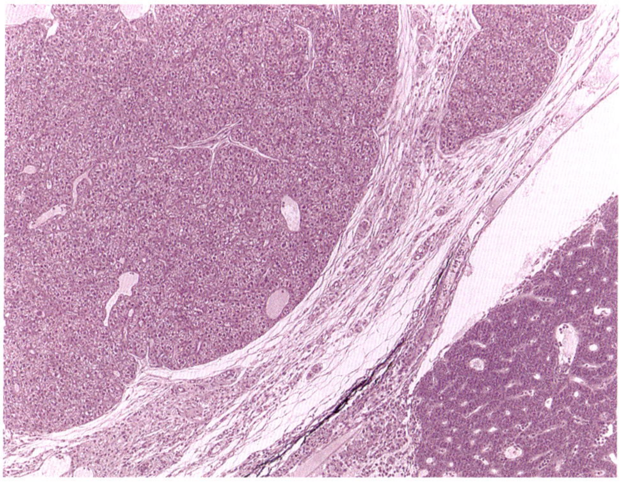

FIGURE 39.27 Gas gland adenoma in the swimbladder of a medaka. The neoplasm, which is comprised of sheets and folded cords of polygonal cells, occupies the upper left half of the image, whereas a portion of unaffected liver is evident at the bottom right. *(Photo Courtesy of Dr. Jeffrey C. Wolf, EPL, Inc.)*

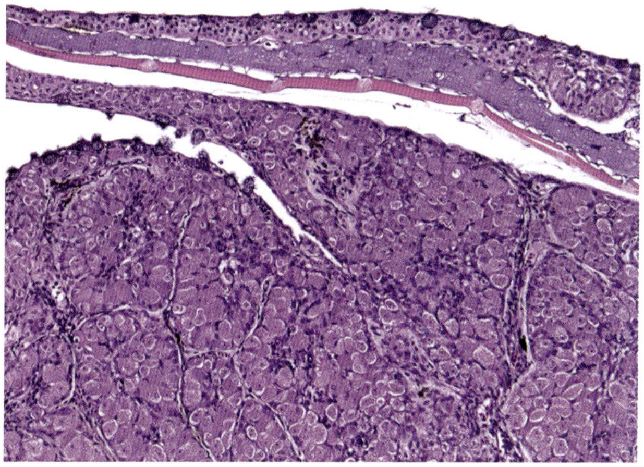

FIGURE 39.29 X-cell pseudotumor in the skin of a wrasse. Sheets of ovoid cells occupy the hypodermis. Although the pattern resembles that of a neoplasm, genetic analysis has indicated that the mass-like tissue consists of parasitic protozoa. *(Photo Courtesy of Dr. Jeffrey C. Wolf, EPL, Inc.)*

disease (see "DNA Viruses" section), and the regenerative hyperplasia of hepatocytes that may occur as a response to certain hepatotoxicants, such as polycyclic aromatic hydrocarbons.[79,309]

INFECTIOUS DISEASES

DNA Viruses

Lymphocystis is an infectious disease of cultured and wild, freshwater, and saltwater teleosts that affects a remarkably wide variety of fish from more than 30 different families.[253,333] The disease is one of the evolutionarily advanced teleosts to date, not having been recognized in salmonids, catfish, or cyprinids. The etiologic agent of lymphocystis is *Lymphocystis disease virus* (LDV), an iridovirus. First isolated in 1966, the LDV virion is icosahedral with a diameter of approximately 200 nm.[25,528] There are currently two known lymphocystis viruses, LDV-1 of flounder and plaice, and LDV-2 of dab; iridoviruses isolated from other species have not been formally characterized.[333] The incubation period is thought to be long, i.e., weeks or months, and fish in which the disease occurs typically have a clinical history of related and/or recent stress. Transmission of the virus is horizontal, presumably from abrasion of lesions and subsequent release of iridovirus into the environment. Disease is self-limiting and often regresses spontaneously.

Gross lesions of lymphocystis consist of variably sized, raised, white to tan masses that are distributed across the skin, oral mucous membranes, and gills. Lesions may be pinpoint, discrete and nodular, and/or coalescent and multinodular.[333] Microscopically, there is extreme cytomegaly of dermal fibroblasts accompanied by dermal infiltration by lymphocytes and histiocytes, the influx of which has been associated with healing.[132] Affected dermal fibroblasts are severely enlarged, have a distinct, thick, eosinophilic, hyaline capsule, and contain basophilic cytoplasmic viral inclusions (Fig. 39.30; eSlide 39.e4).[132,390] Diagnosis is by microscopic examination of histologic sections or wet mount preparations of skin or gill and identification of cytomegalic dermal fibroblasts typical of the disease.[333] Lymphocystis is quite possibly the most common disease of aquarium fish.[333]

Systemic iridoviruses have caused significant mortality and economic loss in cultured food and ornamental fish.[152,475,511,515] The responsible iridoviruses, members of the genera *Megalocytivirus* and *Ranavirus* of the family *Iridoviridae*, infect wild and cultured fish from freshwater and marine environments and are distributed globally.[73,515] Of the megalocytiviruses, the ***Red sea bream iridovirus*** (RSIV) and ***Infectious spleen and kidney necrosis virus*** (ISKNV) are the best known. Infection with RSIV is listed by the World Organization for Animal Health (WOAH) as notifiable (Aquatic Code Online Access—WOAH—World Organisation for Animal Health). Noteworthy as well among the megalocytiviruses is ***Banggai cardinalfish iridovirus*** (BCIV), which has been implicated in mortalities of Banggai cardinalfish and orbiculate batfish and is the only megalocytivirus reported in marine ornamental fish. Significant among the ranaviruses for their ability to cause disease in finfish are ***Epizootic hematopoietic necrosis virus*** (EHNV), ***European catfish virus*** (ECV), ***Singapore grouper iridovirus*** (SGIV), and ***Frog virus-3***. Disease occurrence is influenced by environmental factors including temperature and stressors, such as transport, crowding, mating behavior, and malnutrition. Iridoviruses are very stable in aquatic environments and transmission is suspected to occur by ingestion of infected tissues and contaminated water. Unlike the megalocytiviruses, which are pathogens of fish, the ranaviruses, such as *Frog virus-3*, infect animals of diverse taxonomic classes (e.g., finfish and amphibians), which may contribute to their ability to persist in particular habitats and to spread geographically.

Fish with systemic megalocytivirus disease are lethargic and anemic. Gross examination may demonstrate branchial hemorrhages and splenomegaly. Histologically, cytomegalic cells are present in multiple organs, which include the kidney, spleen, liver, intestine, eye, pancreas, heart, gill, brain, and connective tissue and are accompanied by necrosis of individual cells and areas of tissue in more severe disease. Cytomegalic cells possess granular basophilic viral inclusions that distend the cytoplasm and are often located beneath the endothelium of blood vessels (Fig. 39.31; eSlide 39.e5).[152,511]

Ranaviruses in fish cause systemic, necrotizing disease. An example is multifocal necrosis of the renal hematopoietic interstitium, liver, and spleen caused by EHNV in redfin perch and rainbow trout. Foci of necrosis are often centered on blood vessels and include necrosis of endothelial cells.[515] Round, eosinophilic intracytoplasmic viral

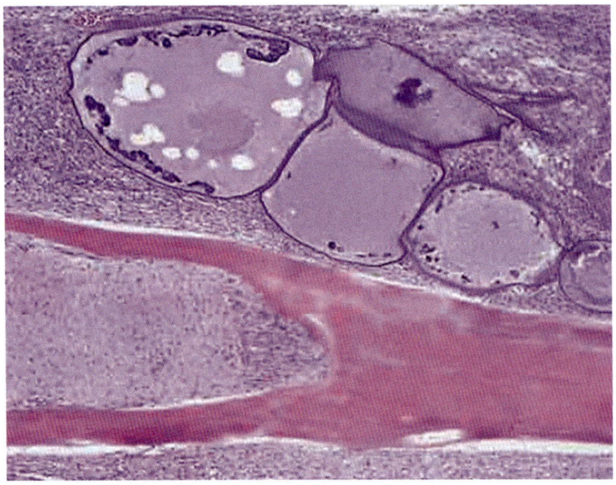

FIGURE 39.30 Lymphocystis in the skin of a teleost. Cytomegalic fibroblasts in the dermis adjacent to a fin ray (bone) have basophilic, granular viral inclusion material throughout their cytoplasm and are surrounded by a hyaline capsule. One cell retains central remnants of a degenerate nucleus (see eSlide 39.e4).

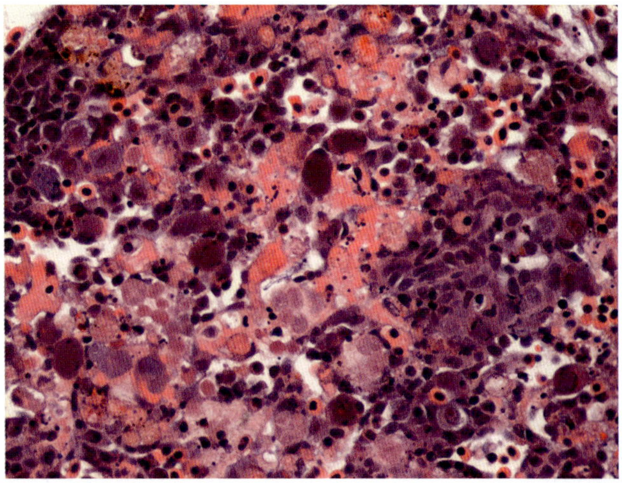

FIGURE 39.31 Systemic megalocytivirus disease in the cranial kidney of a Banggai cardinalfish infected with Banggai Cardinalfish Iridovirus. There is necrosis of hematopoietic cells in the renal interstitium. Basophilic granular viral inclusions distend the cytoplasm of cells protruding into sinusoids (see eSlide 39.e5).

inclusions that tend not to distend the cytoplasm can be seen in hepatocytes and, less frequently, in interstitial cells of the kidney and spleen. Other ranaviruses that cause fatal systemic disease incite lesions similar to those of EHNV and cause multiorgan necrosis with apparent endotheliotropism.

The diagnosis of systemic iridoviral diseases is achieved by identification through virus isolation, immunofluorescence staining of virus in tissue imprints and cytologic preparations, or PCR. PCR-based tests can be performed on DNA extracted from cell cultures with evidence of cytopathic effect or from affected tissues.[511,515] Tissues recommended for diagnostic testing include spleen, kidney, and liver.

Megalocytiviruses and ranaviruses are emerging pathogens of fish and other aquatic vertebrates. Severe morbidity and mortality, the broad range of fish species that these iridoviruses infect, and their global distribution are keys to the ecologic and economic impacts of these pathogens.[73,511,515] The global movement of fish and fish tissues through commercial activities of the ornamental and food fish industries facilitates and potentiates the transboundary spread of these diseases, which poses a significant threat to aquaculture operations worldwide.

Over 21 **alloherpesviruses** have been identified in bony fish (Table 39.e2). Typically, these viruses do not infect or have limited pathogenicity for nonnatural hosts. Associated diseases vary widely between viruses, and latent or recrudescent viruses may cause different signs than during the original infection. Lesions typically include epithelial proliferation of either skin or gill leading to hyperplasia or neoplasia, or necrosis of internal viscera, often with prominent intranuclear inclusions.

Common carp and koi are affected by *Cyprinid herpesvirus-1* (**CyHV1**; **carp pox**) and *Cyprinid herpesvirus-3* (**CyHV3**; **koi herpesvirus, carp interstitial nephritis and gill necrosis virus**), while *Cyprinid herpesvirus-2* (**CyHV2**; **goldfish herpesvirus, herpesviral hematopoietic necrosis virus**) causes disease primarily in goldfish.[183] Distribution of all three is virtually worldwide, and CyHV1 and CyHV3 have become endemic in many captive and wild fish populations. Disease outbreaks are strongly influenced by temperature, fish age, and environmental stressors. Elevating water temperatures and avoiding stress can help prevent and control outbreaks.[184]

"Carp pox" is usually a benign skin disease of adult common carp and koi. In contrast, CyHV1 can cause high mortalities in young fry and systemic disease characterized by hepatocellular, renal tubular, and intestinal necrosis. Clinical signs are nonspecific and include anorexia, darkening of skin, exophthalmia, petechiation, coelomic distension, and abnormal swimming. Months later, surviving fish may develop friable, plaque-like skin lesions with a translucent, mucoid appearance variably described histologically as epidermal hyperplasia or flat papillomas. Cowdry type A inclusions are described in hepatocytes of fry and epidermal cells of papillomas.[422] Highest mortalities occur in fry. Skin lesions in adult fish correspond with periods of highest viral replication at temperatures of 15–20°C and decline or regress at 25–30°C.[421] While lesions can recur, because they are usually self-limiting, carp pox is primarily a cosmetic problem for koi producers and fanciers, but it can also affect the marketability of common carp.

CyHV2 causes outbreaks with severe losses primarily in juvenile goldfish but can affect all ages. External signs can be few, and lesions are limited to patchy foci of pallor in the gills resulting from epithelial hypertrophy and hyperplasia with lamellar fusion, necrosis, and inflammatory cell infiltration.[175,213] Internally, renomegaly, splenomegaly, and pallor correspond to the principal lesions involving necrosis of renal and splenic hematopoietic tissues. The intestinal lamina propria and submucosa, exocrine pancreas, and epithelium of the oropharynx, opercular cavity, and integument are less consistently affected.[162,175,213] Fine, pale, homogenous, amphophilic to basophilic inclusions that enlarge nuclei and marginate chromatin can be numerous in affected hematopoietic and epithelial cells (Fig. 39.32).[162,175] Temperature optima for outbreaks of CyHV2 are similar to those of CyHV1.

CyHV3 is a WOAH notifiable disease agent (Aquatic Code Online Access—WOAH—World Organisation for Animal Health) of common carp, koi, and certain hybrid crosses. The virus has been detected in additional fish species, including noncyprinids. Gross and microscopic lesions are similar to those produced by the closely related CyHV2 in goldfish, primarily affecting the gills and hematopoietic organs, and mortalities in naïve fish of all ages can be high in the ornamental koi and common carp food fish

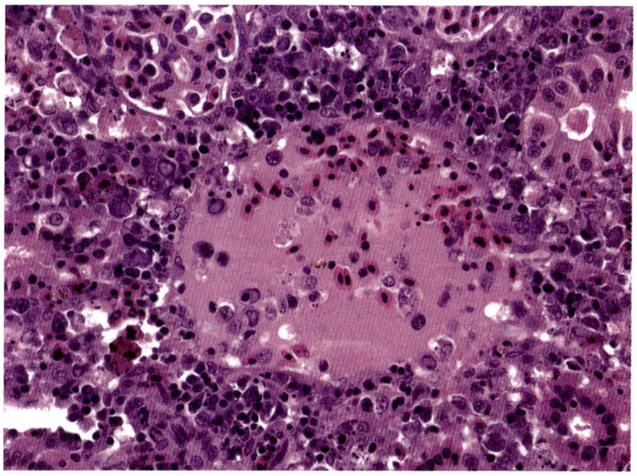

FIGURE 39.32 Herpesviral nephritis in a goldfish infected with Cyprinid herpesvirus-2 (CyHV2), also known as goldfish herpesvirus and herpesviral hematopoietic necrosis virus. There is focal necrosis of hematopoietic tissue with eosinophilic proteinaceous material and necrotic debris. Large amphophilic intranuclear viral inclusions are present within precursor cells. *(Photo Courtesy of Dr. Jeffrey C. Wolf, EPL, Inc.)*

industries.[196] In situ hybridization studies demonstrate wide distribution of the virus, although corresponding lesions may not always be present. Additionally, ulcerative dermatitis and notably encephalitis with the formation of glial nodules and lymphocytic perivascular cuffing occurs in some fish, and inclusion bodies can be seen in these lesions as well as in the heart, liver, renal tubular epithelium, and olfactory epithelium.[308,311] Outbreaks of CyHV3-related disease occur optimally at 22–26°C, are typically caused by introduction of the virus into naïve populations by persistently infected carrier fish, and often involve stressful events. Tentative diagnosis is based on history, gross lesions, and histopathology. Confirmatory diagnosis and surveillance of apparently healthy carriers is accomplished primarily by quantitative and other PCR methods (https://www.woah.org/fileadmin/Home/eng/Health_standards/aahm/current/2.3.06_KHV.pdf).

Poxviruses are a group of viruses well known for the size and complexity of their linear, double-stranded DNA genomes and enveloped virions, which can be large enough to be visible by light microscopy.[455] The large genome encodes for viral polymerases, which allow poxviruses to replicate within the host cell cytoplasm rather than entering the cell nucleus.[317] Members of *Poxiviridae* family infect a diverse array of insect (*Entomopoxivirnae*) and vertebrate (*Chordopoxivirnae*) hosts including reptiles, birds, and mammals. Over the past decade, two members of *Chordopoxivirinae*, ***Carp edema virus*** (CEV) and ***Salmon gill poxvirus*** (SGPV), have been found responsible for disease in farmed and collection fish species.[156]

CEV infects a relatively narrow number of fish species, i.e., carp, koi, and carp hybrids. The virus causes disease that severely limits production. Together with CyHV3 (koi herpesvirus), ranavirus, and carp sprivivirus (*spring viremia of carp virus*), CEV challenges carp production with high mortality events and impacts on potential growth rates of fish.[276] CEV was first reported in ornamental koi in Japan in 1974 and has since been identified as a cause of outbreaks in common carp for food production in the USA and in a rapidly growing number of countries in Europe and Asia.[195,321,510] The global distribution of CEV is tied to international trade of ornamental fish.[4]

CEV has a large, ovoid, 250–400 nm mature virion with a tubular unit on its membrane surface.[272,358] Currently, three distinct genogroups of CEV have been identified using a region of gene 4a which encodes a major core protein; genogroup I viruses infect common carp exclusively and have been identified in the US and Europe, genogroup II viruses are split into two subgroups, IIa and IIb, both of which have been isolated from koi carp and common carp in Asia and Europe.[3,272,288]

There is an incomplete understanding of the transmission of both CEV and SGPV, but recent identification of subclinical viral carriers of CEV in common carp and koi and reports of recurring SGPV infection could provide evidence of viral persistence as seen in pox viruses in terrestrial species.[156,288] Horizontal transmission of CEV and SGPV and the potential for vector transmission are under current study. Cross-species viral transmission of CEV has been shown experimentally between cyprinid species, and CEV has been detected on the gill or skin of noncyprinid species cohabited with CEV-infected carp.[3,289] Water temperature seems to be the most important determinant of mortality from CEV, with disease progressing at water temperatures of 15–25°C and most outbreaks occurring in spring and summer.[258,310]

CEV has a predilection for gill where it causes a progressive proliferative response leading to lamellar fusion that limits the gill surface area for respiratory and metabolic exchange, including ammonia excretion and osmoregulation. Age and water temperature impact the manifestation of disease, resulting in two distinct forms: **carp edema** in small fry in winter and **koi sleepy disease** in juveniles in summer. Fish with CEV may be found floating on their sides and are poorly responsive to stimuli, giving this disease the name "koi sleepy disease".[310] Other commonly reported signs include enophthalmos, hemorrhages in this skin around the mouth and at the base of fins, and the characteristic finding of edema of the gill and skin.[258,362,510] Histopathologic changes of CEV include branchial epithelial hyperplasia with lamellar fusion, interstitial edema, and epithelial cell apoptosis in interlamellar regions of the gill (Figs. 39.33A–C).[258,272,310,362] Lethargy may result from hypoxia along with metabolic derangements, i.e., elevated blood ammonia levels and alterations in blood sodium concentrations, which have proposed effects on nervous tissue.[4]

History, clinical signs, gross lesions, histologic changes, and transmission electron microscopy (TEM) may be supportive of CEV infection, but gross and histologic lesions are not specific to CEV, and the success rate of TEM identification of viral particles can be highly variable.[272,510] Definitive diagnosis of CEV requires identification of viral DNA in tissues achieved by real-time PCR or nested PCR.[362,510] CEV has been shown not to be detectable in gills following abatement of clinical signs, necessitating testing during active infections.[272] There is no known treatment for CEV. Mortality of infected koi can be significantly decreased using long-term 0.5% saltwater baths, although this could be considered symptomatic treatment as viral infection is maintained.[432]

Poxvirus infection in Atlantic salmon has been suspected of contributing to high mortality events since the 1990s, but *Salmon gill poxvirus* (SGPV) was not fully characterized until recent advances in molecular diagnostics allowed identification and sequencing of the virus.[155,342]

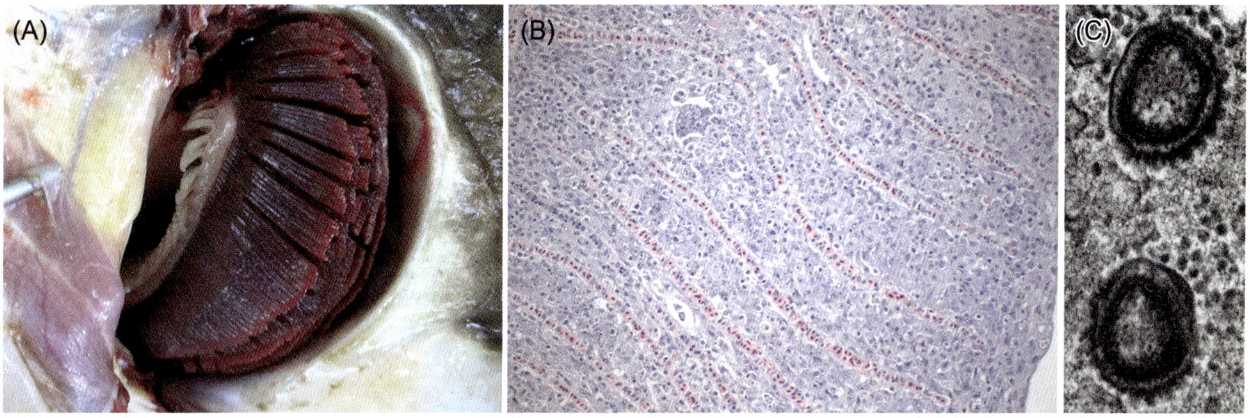

FIGURE 39.33 Branchial lesions of carp edema virus (CEV) infection in common carp *Cyprinus carpio*. (A) Gills are relatively normal in appearance despite there being significant lamellar pathology evident microscopically. (B) There is severe lamellar epithelial hyperplasia with lamellar fusion; foci of necrosis are present in the hyperplastic epithelium, together with scattered individual apoptotic cells that are rounded, have eosinophilic cytoplasm, and possess a shrunken, round, heterochromatic nucleus. (C) CEV virions occur within branchial epithelial cells. *(Photos Courtesy of Dr. Jan Lovy, USGS Western Fisheries Research Center.)*

SCPV has been identified throughout the production cycle of farmed salmon, causing disease in freshwater and saltwater life stages.[156] Infection by SPGV is often associated with acute mortality and can result in major economic losses due to failure of smoltification.[156] Similar to CEV, SGPV also has a predilection for gill and has recently been identified as a common denominator in recurrent and multifactorial gill disease in salmon.[156] In experimental infections in salmon, the major clinical feature is lethargy. Gross lesions include swollen and pale gills, splenomegaly, and pallor of internal organs. The characteristic histologic lesion of SGPV is apoptosis of lamellar epithelial cells in acute stages, followed by epithelial hyperplasia in later stages of disease. Hemophagocytosis in hematopoietic tissue may be associated with severe disease.[155] Diagnosis of SGPV is possible by immunohistochemistry, TEM, and PCR testing. There is no known treatment for SGPV.

RNA Viruses

Infectious pancreatic necrosis (IPN) due to *Infectious pancreatic necrosis virus* (IPNV) is a highly contagious disease of young salmonids worldwide that has had a highly significant impact on salmonid aquaculture.[346,407] IPNV is a bisegmented, double-stranded RNA virus in the *Birnaviridae* family that has significant antigenic diversity and a spectrum of virulence. IPNV most commonly affects rainbow trout, brook trout, brown trout, Atlantic salmon, and a range of Pacific salmon species.[18,407,416,533] The virus is also suspected to be the cause of clinical disease in some farmed marine species, such as halibut, southern flounder, cod, yellowtail, and turbot; fish in a wide range of other families are known asymptomatic carriers.[292,316,416] Species susceptibility is reported to decrease with age and to correlate with rearing intensity, temperature, and transfer from fresh to saltwater, with epizootics typically occurring in young fish, especially in species that have a more rapid initial growth phase. Mortality may range from less than 10% to 90% of a single population based on a range of potential factors, such as viral strain, host species, temperature, water quality, and stocking density.

In freshwater, affected fry and parr frequently display a corkscrew or spiral swimming behavior, dyspnea, severe abdominal swelling, and pigmentary darkening of the body.[416] Posttransfer to salt water, smolts may become anorexic, lethargic, and display abnormal swimming patterns with an inability to maintain buoyancy. Gross findings in fry include darkening of the skin along the tail and abdomen, pale yellow liver, and pale yellow catarrhal exudate within the gastrointestinal tract. Histologic lesions typically consist of severe pancreatic acinar cell necrosis, epithelial necrosis of the intestinal mucosa, and hepatocellular necrosis (Figs. 39.34A, B). A characteristic feature of the intestinal lesion is production of McKnight cells, apoptotic mucosal epithelial cells that are extruded from the intestinal mucosa.[296] With survival and chronicity, pancreatic tissue will be replaced by fibrous connective tissue.[46] Examination of infected tissue using TEM has revealed cytoplasmic viral-specific tubules and paracrystalline arrays of virions.[84]

When sampling for virus isolation from healthy or subclinical animals, liver, spleen, kidney, and brain or ovarian fluid from broodfish during spawning are preferred specimens. Molecular diagnostics can confirm an outbreak that is suspected based on history and gross and histologic findings.[346,416] IPNV can be isolated from asymptomatic fish that may be reservoirs of infection.

The economic impact of IPNV on the farmed salmon industry has been sizeable. A single outbreak of IPNV that occurred in Ireland in 2006 resulted in the deaths of

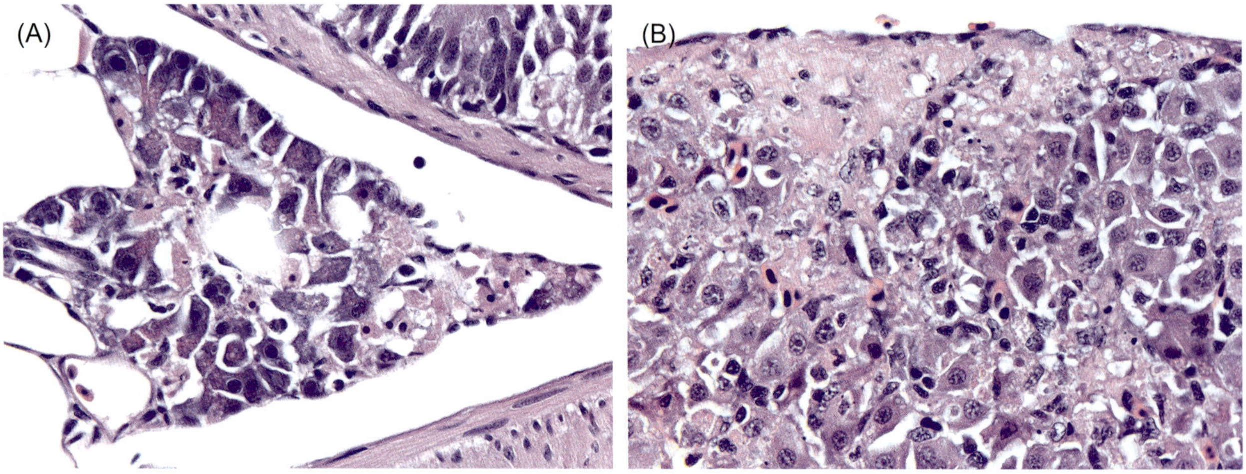

FIGURE 39.34 Infectious pancreatic necrosis virus (IPNV) infection in a rainbow trout. (A) Pancreatic acini in the mesentery are disrupted, and necrotic acinar epithelial cells have eosinophilic cytoplasm lacking zymogen with or without a pyknotic nucleus. (B) There is focal hepatic necrosis with disruption of the normal arrangement of hepatocytes and replacement by eosinophilic protein accompanied by degeneration and necrosis of individual hepatocytes. *(Part A: Photo Courtesy of Dr. David Groman, Aquatic Diagnostic Services, Atlantic Veterinary College. Part B: Photo Courtesy of Dr. David Groman, Aquatic Diagnostic Services, Atlantic Veterinary College.)*

1,650,000 salmon and was estimated to cause losses of up to USD 43 million.[416] In addition to the commercial cost, there is an obvious concern about the impact of the virus on wild populations, especially in regions in which native salmonids are being reintroduced.

Infectious salmon anemia virus (**ISAV**), which causes **infectious salmon anemia** (**ISA**), is primarily considered a disease of marine, farm-raised Atlantic salmon. The disease causes significant pathology and high mortality. Following the initial description of the disease in Norway, the virus has since been identified in salmon farmed in both freshwater and saltwater.[451]

ISAV is an RNA virus, the only species in the genus *Isavirus*, and it is the only fish Orthomyxovirus described to date.[451] It is an enveloped virus with a genome comprised of eight single-stranded RNA segments with negative polarity. The virus has been demonstrated to agglutinate erythrocytes from a range of different fish species.[126] Viral virulence is associated with the highly polymorphic region (HPR) of the hemagglutinin-esterase (HE) gene, and there is a marked difference in the virulence of geographically different isolates. For example, outbreaks in the Bay of Fundy in Canada associated with HPR4 isolates resulted in mortality ranging up to 92%, while an event linked to the HPR2 isolate was noted to result in mortality between 27% and 40%.[451] Infection with pathogenic HPR-deleted variants or nonpathogenic HPR0 strains of infectious salmon anemia virus is listed as notifiable by WOAH (Aquatic Code Online Access—WOAH—World Organisation for Animal Health). Sea-run brown trout, rainbow trout, and Coho salmon typically develop subclinical disease and are theorized to act as disease reservoirs; high mortality can occur in experimentally infected rainbow trout, while experimentally infected herring, Arctic char, and cod do not develop disease.[233,339,457] Transmission is primarily horizontal, and the virus is carried in seawater contaminated with infected urine or feces.[208] Additionally, sea lice have been implicated as a vector of the disease.[341] Disease outbreaks with high mortality appear to correlate with stressors, such as overcrowding and environmental conditions, including significant changes to water temperature and poor water quality.[84] Similarly, seasonality appears to play a role, insomuch as the temperature changes in spring and winter are posited to parallel epizootics. Mortality can range from background levels to greater than 90% of a given population based on the isolated strain, environmental stressors, fish age, and population density. Primary mortality events typically occur between 6 and 10 months following the introduction of naïve salmon to saltwater pens.[451] Acutely infected animals typically display evidence of abnormal swimming behavior and marked lethargy.

At necropsy, affected animals have ascites, exophthalmia, and severe pallor of the gills and viscera (Fig. 39.35A). Visceral adipose may contain petechiae and ecchymoses (Fig. 39.35B), while the liver and spleen are often markedly congested.[158,451] With chronicity, petechiation extends into the subcutaneous adipose. Necrohemorrhagic hepatitis, renal interstitial hemorrhage, and noninflammatory tubular epithelial degeneration and necrosis (Fig. 39.35C) are present histologically. There is also congestion of the branchial filaments, gastric and foregut lamina propria, as well as the spleen. Marked erythrophagocytosis can be observed.[158] For rapid diagnosis, especially in diseased animals, viral antigen can be identified in tissue smears using immunofluorescent antibody tests (IFAT). For carriers, there are

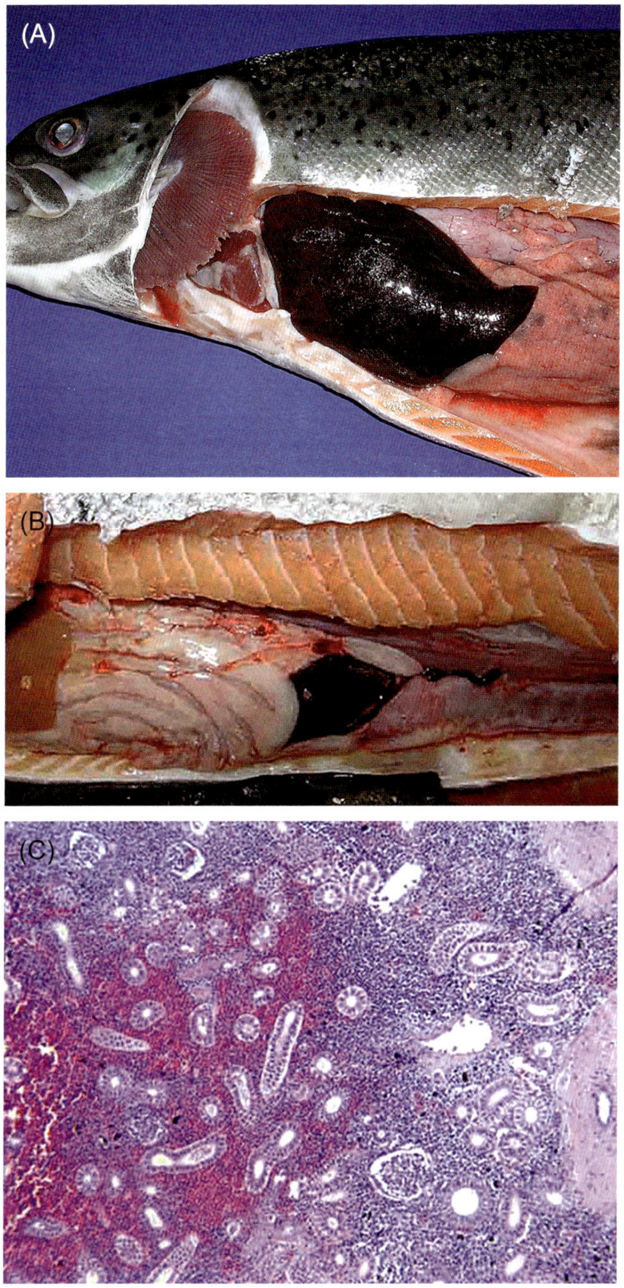

FIGURE 39.35 Infectious salmon anemia (ISA) in a salmon. (A) Gill pallor is a characteristic feature and is indicative of the severity of the anemia that occurs in ISA. (B) Petechiae are present in the serosae, mesentery, and muscle. (C) Hemorrhage is present in the hematopoietic interstitium between tubules. *(Part A: Photo Courtesy of Dr. Marcos G. Godoy. Part B: Photo Courtesy of Dr. Marcos G. Godoy. Part C: Photo Courtesy of Dr. David Groman, Aquatic Diagnostic Services, Atlantic Veterinary College.)*

effective RT-PCR assays. RT-PCR is the most sensitive diagnostic test, followed by virus isolation and IFAT.[328]

Outbreaks of ISA have an extraordinary economic impact that can be measured in the tens of millions of dollars, and in some parts of the world, the impact of significant epizootics with high mortality can be devastating.

Viral hemorrhagic septicemia virus (**VHSV**) is considered one of the most significant pathogens affecting specific species of farmed fish, such as rainbow trout and turbot.[84,370] As a consequence, **viral hemorrhagic septicemia** is listed as a WOAH notifiable disease (Aquatic Code Online Access—WOAH—World Organisation for Animal Health). It was first recognized in the early 1900s and has since been described in over 50 species of marine and freshwater fish in the northern Atlantic and Pacific Oceans, the North Sea, the Baltic Sea, Great Lakes region in North America, and in Japanese and Korean waters.[124] Affected species include salmonids, such as brown and rainbow trout, muskellunge, smallmouth bass, walleye, lake whitefish, freshwater drum, and Atlantic herring. VHSV is a rhabdovirus in the genus *Novirhabdovirus*. There are four genotypes and multiple subgenotypes. Some VHSV strains are capable of infecting multiple fish hosts and infrequently can cross species barriers, which has been described experimentally between turbot and rainbow trout.[428] Transmission is thought to occur through predation or scavenging of diseased or dead fish, through the contamination of water by infected urine or feces, mechanically by human activity or piscivorous birds, or vectored by parasites, such as leeches, or prey species, such as amphipods.[123,125]

VHSV disease may present in one of three forms: acute, chronic, and nervous. Acute infection may result in sudden death in up to 100% of fry. Additional clinical symptomology includes darkening of the skin, lethargy, exophthalmia, ascites, and abdominal distention. Petechiae and coalescing ecchymoses at the base of the fins, gills, mouth, eyes, and multifocally along the skin, as well as branchial pallor and abnormal swimming patterns, are also observed.[370,451] With chronicity, swimming behavior is described as more lethargic, while in the nervous form, fish will be hyperactive, with abnormal spiraling swimming patterns, flashing and jumping from the water.

Gross lesions in infected fish with the acute form typically include hemorrhage within the mesentery, abdominal adipose, skeletal muscle, and in some species, the heart. There is often marked renal congestion and renomegaly as well as hepatic pallor. Animals that survive to the chronic disease stage have similar renal and hepatic changes; however, mortality is typically decreased.[8,451] The endothelium in the liver, kidney, spleen, and skeletal muscle together with circulating leukocytes appear to be the primary target cells, and lytic parenchymal necrosis can be seen in the liver, kidney, and heart.[8,10,451] Other findings include congestion of the hepatic sinusoids, degranulation of melanomacrophages in the head kidney, and intramuscular extravasated erythrocytes. Vasculitis has been a consistent finding in specimens from mortality events, while necrotizing enteritis, nephritis, and splenitis may also be seen. In muskellunge, crappie, and smallmouth bass, there is evidence of a diffuse aerocystitis.[8] Diagnosis is based on histopathology and electron microscopy, with identification of bullet-shaped virions. Confirmation can be done by immunohistochemistry with the most significant immunoreactivity occurring in the endothelial cytoplasm, especially in the meninges, reproductive tract, epicardium, spleen, gastrointestinal tract, hepatopancreas, and the choroidal rete.[8] Additionally, virus isolation and RT-PCR, IFAT on fixed tissue smears, and direct ELISA can be useful in making the diagnosis.

Betanodavirus is a genus comprised of four confirmed species of nonenveloped positive-strand RNA viruses in the family Nodaviridae.[417] *Nervous necrosis virus* (NNV) causes **viral nervous necrosis** (VNN), also known as **viral encephalopathy and retinopathy** (VER), in a wide range of teleost fish.[319] Over 50 species of fish, primarily marine but also freshwater, are reported to be affected.[349] The four recognized species of NNV are *Barfin flounder nervous necrosis virus* (BFNNV), *Red-spotted grouper nervous necrosis virus* (RGNNV), *Striped jack nervous necrosis virus* (SJNNV), and *Tiger puffer nervous necrosis virus* (TPNNV), with RGNNV having the widest geographic and species distribution. An additional genotype, turbot betanodavirus strain (TNV), has been proposed.[210] The original identification occurred in barramundi in Australia; however, infections have been reported in Australian, African, European, Asian, and North American fish, as well as fish in the South Pacific. Different viral species within the genus each infect a wide range of species.[507] Betanodaviruses are considered highly resistant, surviving for long periods in marine environments at low temperatures; however, at 25°C or higher, viral survivability is significantly decreased. Water temperature is considered a critical factor in the development of clinical signs. The impact of temperature is well described in farmed sea bass, which was originally known as **summer disease**.[349] Transmission is thought to be through direct contact or passive diffusion and is suspected to occur through vertical transmission, as virus has been detected in the gonads of infected fish, especially in times of stress.[154] The four species of virus primarily infect larval and juvenile stages but also adult fish, and in particular Atlantic halibut, sevenband grouper, and European sea bass.[325] In certain species, such as striped jack, European sea bass, barfin flounder, sevenband grouper, and Atlantic halibut, brood fish are considered to be the primary viral reservoir and the key route of infection for larval and juvenile fish.[349] Betanodavirus is recognized as a significant concern to fish culture operations with catastrophic losses reported worldwide ranging from 15% to 100% mortality. One report observed that in striped jack, mortality most frequently occurred within 10 days after hatching while the earliest observation of clinical disease is at day two posthatching, and in that case resulted in almost complete loss of the larvae.[349] Susceptibility varies, although one study evaluated a range of species experimentally, and noted that differing species can harbor the disease subclinically, with Celebes rainbowfish, dwarf rainbowfish, threadfin rainbowfish, and medaka being most susceptible to RGNNV. Celebes rainbowfish, threadfin rainbowfish, and medaka, demonstrated the highest morbidity rates and viral titers in brain. The authors determined that these three species appear to be appropriate animal models for experimental infections with betanodavirus.[148]

VNN presents clinically as anorexia, lethargy, circular swimming, and alterations in pigmentation. There are few signs on the body surface and gills apart from a progressive change in pigmentation. Anecdotally, European sea bass have been occasionally reported with erosive dermatitis on the head and mandible, which has been theorized to be due to trauma, potentially as a result of retinal lesions. While pathology is variable between species and age groups, the virus generally infects the central nervous system, targeting the brain and spinal cord, as well as the retina. In the CNS and retina, microscopic lesions consist of neuronal vacuolar degeneration and necrosis accompanied by microglial and histiocytic infiltration of the neuroparenchyma (Figs. 39.36A, B).[330] Intracytoplasmic inclusions were observed in the brain of European sea bass, Asian sea bass, Japanese parrotfish, and brown-spotted grouper.[319] The retina often had degenerative lesions in the bipolar cells of the inner nuclear layer and in the ganglion cell layer. In persistently infected Atlantic halibut, multifocal aggregates of melanin-containing cells were observed in the sensory retina (SR) and retinal pigment epithelium (RPE).[154] Ultrastructurally, 25 nm subspherical viral particles have been observed in brain, spinal cord, and retina of heavily infected fish. Virions may be found in the cytoplasm or in association with membranes of the endoplasmic reticulum and are observed in neurons, astrocytes, oligodendrocytes, and microglia.[349]

Conventional RT-PCR is the most efficient method for diagnosing clinically infected fish, while nested PCR or real-time RT-PCR protocols are appropriate for diagnosing subclinically infected and reservoir fish. NNV shows a clear predilection for brain and retina, suggesting that these are the primary locations for viral replication, while cranial kidney is recommended as the best tissue for possible NNV detection by nonlethal biopsy.[242]

The economic impact of outbreaks of betanodavirus in farmed fish is significant, and the wide geographic distribution as well as the susceptibility of such a diverse range of species means that the disease represents an increasing risk to global fish populations. The presence of reservoirs in wild populations is a persistent threat for disease in expanding farmed populations, while trade and transport of subclinically infected juveniles likely represent the primary mechanism for global distribution.

Between 2013 and 2014, four different picornaviruses, single-stranded positive-sense RNA viruses, were isolated

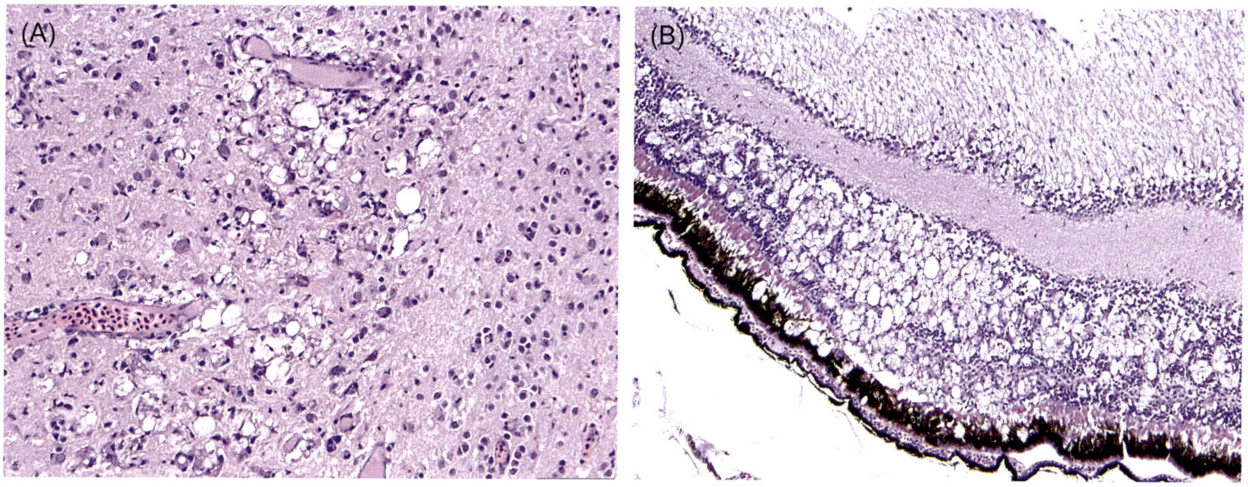

FIGURE 39.36 Betanodavirus infection in an anthias. (A) There is a focus of neuronal vacuolation in the brain with infiltration by low numbers of glial cells. (B) Marked vacuolation of neuronal cell bodies is present in this segment of the retina.

from European eels, common carp, bluegill, and fathead minnow. The eel picornavirus is a member of the genus *Potamipivirus*, while the latter three fish had viruses belonging to the genus *Limnipivirus*.[20,426] Subsequent studies using a metagenomic approach described 29 novel fish picornaviruses assigned to six newly recognized fish-associated genera of picornaviruses (*Limnipivirus*, *Potamipivirus*, *Fipivirus*, *Symapivirus*, *Rajidapivirus*, and *Danipivirus*) with several more picornaviruses identified in saltwater and freshwater fish remaining to be taxonomically determined.[151,441] These included exemplars derived from both wild marine and freshwater fish, and comprised specimens of lobe-finned, ray-finned, and cartilaginous fishes. Around the same time, a picornavirus was discovered and described in an aquacultured juvenile clownfish as part of two outbreaks of disease. That virus has been named *Clownfish picornavirus* (CFPV). A study was conducted on a population of cultured gilthead seabream in a hatchery subsequent to a disease outbreak resulting in 70% mortality of 50-day posthatching juveniles, and a separate picornavirus was identified, *Gilthead seabream picornavirus* (GSPV).[271] In addition to CFPV and GSPV, multiple other fish picornaviruses have been associated with mortality; however, the majority of those reports are from individual animals and, as such, do not address population-wide lesions, susceptibility, or geographic distribution.

Clinical signs observed in GSPV infection included anorexia and abnormal swimming behavior, while clinical signs of CFPV included changes in pigmentation, lethargy, loss of body condition, tachypnea, and inability to regulate position in the water column. Gross lesions in both ranged from no signs to erythema and hemorrhage at the base of fins and multifocal to coalescing superficial hemorrhages. Histopathological findings from multiple animals from both CFPV and GSPV outbreaks were similar. These included minimal to moderate individual cell necrosis characterized by cytoplasmic fragmentation and karyorrhexis admixed with mononuclear cell inflammation localized to the mucosal epithelia of the branchial cavity, pharynx, esophagus, and/or stomach. In one instance, there was pronounced necrosis of gastric glands and abundant sloughed debris in the proximal intestine. In several clownfish, 10–15 μm diameter, round to oval basophilic inclusions are present within mildly hyperplastic pharyngeal mucosal epithelium. Under light microscopy, it may be difficult to localize viral inclusions to the cytoplasm or nucleus.[426] Conventional RT-PCR assays are the most rapid screening tool to test tissues for CFPV. Histopathologic findings combined with ultrastructural identification of picornavirus-like particles in clownfish support the diagnosis of CFPV. For the diagnosis of subclinically infected fish and reservoir species, metagenomic analysis or RT-PCR testing of fecal samples can be conducted without requiring euthanasia of target animals.[188]

While certain picornaviruses are important viral pathogens of fish and our collective knowledge is expanding rapidly, little is known about the diversity, epidemiology, and pathogenesis of fish picornaviruses.[313] Clownfish are important marine fishes for the international commercial pet trade, and it is estimated that greater than 90% of all traded clownfish species are raised in captivity. A common practice in equatorial scuba diving is the repopulation of coral reefs with farmed clownfish. The release of farmed clownfish, while with good intentions, poses a risk of transmission of CFPV into wild populations.

Tilapia lake virus **(TiLV)** is an enveloped, negative-sense, single-stranded RNA virus that has been classified as a relative of the *Orthomyxoviridae* family of viruses. The virus has been associated with outbreaks of epizootic mortality in wild, farmed, and commercial hybrid tilapia.[121] Blue tilapia, cichlids, redbelly tilapia, tinfoil barb, river barb, giant gourami, and Tvarnun simon are also reported to be susceptible either naturally or experimentally.[6,14,102,329] It is listed as a notifiable disease by the WOAH (Aquatic Code Online Access—WOAH—World Organisation for Animal Health).

Tilapia lake virus disease, or **syncytial hepatitis of tilapia**, was first described in Ecuador and Israel in 2013 and 2014, respectively. However, it is thought to have been responsible for significant mortality events in Israeli-farmed tilapia since 2009. The virus has subsequently been described in multiple sites around the globe with spread presumed due to translocation of live tilapia for aquaculture.[6,14,102,329] A study from Thailand assessed a partial genome which showed relatively high nucleotide identity (i.e., approximately 97%) to strains from Israel.[102] Several reports noted very high mortality, e.g., 90%, in fingerlings, whereas mortality rates are significantly lower, e.g., 9%, in medium- to large-sized, older, Nile tilapia. Disease has been associated with transfer between ponds and may be associated with stress, while temperature and salinity have not been correlated as potential risk factors. Direct horizontal transmission is thought to be the primary mechanism of infection. While infective virus has been identified in gonads and fry in natural outbreaks and experimentally exposed fish, there is no evidence to date of vertical transmission.

In fish infected with TiLV, clinical signs include abnormal swimming behavior, pallor, anorexia, loss of condition, severe anemia, bilateral exophthalmia, skin erythema, scale protrusion, and ascites. Lens opacification and rupture of the lens with chronicity as well as dermal erosion, endomeningeal hemorrhage, and splenic congestion may be evident on gross examination.[121] Histologic lesions primarily have been observed in brain, eye, and liver.[121] Encephalic lesions consist of edema, focal endomeningeal hemorrhage, and diffuse capillary congestion, as well as neuronal degeneration. Occasionally, there are glial nodules and sporadic perivascular lymphocytic cuffing. Lesions in the eye include ruptured lenticular capsule and cataractous changes. Within the liver, syncytial cell formation and extensive hepatocellular necrosis with pyknosis and karyolysis as well as hepatocellular swelling, are often noted (Fig. 39.37). Splenic lymphoid hyperplasia is typically observed. There is hyperplasia of melanomacrophage centers in both spleen and liver. Examination by TEM revealed multinuclear hepatocytes with markedly swollen mitochondria and infected cells with numerous, round-shaped, viral particles measuring approximately 100nm in diameter.[102]

High mortality in tilapine species with the aforementioned clinical signs, grossly apparent ocular changes such as lens opacity, and postmortem findings of dermal erosion, leptomeningeal hemorrhage, and moderate congestion of kidney and spleen should elevate suspicion of TiLV.

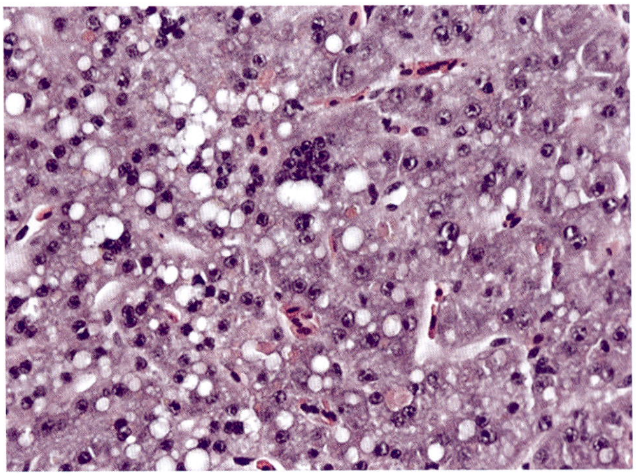

FIGURE 39.37 Tilapia Lake Virus infection of liver of a Nile tilapia. Syncytial cell formation is evident in the liver of this Nile tilapia after experimental challenge with the virus.

While liver and brain have been the principal targets of molecular assays due to the presence of histologic lesions, an ISH assay with a TiLV-specific probe identified viral RNA in kidney, gill, and spleen, suggesting these tissues should also be considered targets for molecular diagnosis.[102]

Retroviruses

Walleye dermal sarcoma (WDS) was first reported in walleyes in Oneida Lake in New York State (USA) and subsequently has been described in fish in Canada.[508] The causative agent, *Walleye dermal sarcoma virus* (WDSV), is a type C retrovirus that is one of the newest members of the genus *Epsilonretrovirus* and induces dermal sarcomas.[412] In Oneida Lake, greater than 27% of the walleye population are infected and develop WDS; 10% are diagnosed with **walleye epidermal hyperplasia (WEH)**, which is similarly induced by an epsilonretrovirus.[412] A fascinating aspect of these two proliferative diseases is found in their seasonality. Peak incidence occurs during the late autumn, and lesions last up to spring spawning when they are observed to regress naturally. During summer when ambient and water temperatures are higher, these proliferative diseases typically do not occur.[37,39] Transmission of WDSV is theorized to occur horizontally from contaminated water or through direct contact between infected fish during spawning.[40]

WDS is a mesenchymal neoplasm that arises within the dermis and occurs along the external surface of affected fish. Tumors are white to pink, often have a smooth or cobblestoned surface, range in size from 0.2 to 1.0cm in diameter, and frequently coalesce. Microscopic examination of the neoplastic tissue reveals unencapsulated, multinodular masses arising from the stratum spongiosum.

These are composed of interlaced streams, whorls, and bundles of spindle cells supported by a dense fibrous matrix that elevate the scales and compress the underlying dermis. Neoplastic cells have a small to moderate amount of vacuolated cytoplasm and oval nuclei with a central nucleolus. Mitoses are rare. There is frequent ulceration and secondary inflammation. Rare cases of WDS are locally invasive, infiltrating beyond the dermis into the musculature and bone of the operculum; there are no reports of visceral metastasis. Definitive diagnosis requires histopathology and identification of the retroviral C particles via TEM or RT-PCR analysis.[412,451]

In 1978, farmed salmon in Scotland were diagnosed with **swim bladder leiomyosarcomas and fibrosarcomas.**[112,297] An incidence rate of approximately 4.6% with a mortality rate of less than 3% was determined. C-type retroviral particles were identified in the tumors.[77] Subsequent epizootics were reported in Atlantic salmon in New England, and an exogenous retrovirus was isolated and characterized from neoplastic tissue.[369] Affected salmon are lethargic, moribund, and have distended abdomens. At necropsy, abdominal cavities contain variably sized, multinodular masses associated with both the external and internal surfaces of the swim bladder (Fig. 39.38A). Microscopically, masses arise from and efface the inner smooth muscle layer as well as the swim bladder areolar tissue zone. Tumors are well-circumscribed, unencapsulated, multinodular masses composed of interlacing streams, and bundles of spindle cells supported by scant collagenous matrix (Fig. 39.38B). There is mild to moderate anisocytosis and anisokaryosis and a high mitotic rate. Positive immunohistochemical labeling with alpha-smooth muscle actin and desmin confirms the diagnosis of leiomyosarcoma.[77,369] TEM may identify the C-type retroviral particles. PCR amplification is also possible as a confirmatory assay. The incidence of disease and mortality in farmed salmon is low, and disease does not present a significant concern to the population dynamics or to the economics of salmon farming.

Bacteria

A suite of **gram-negative bacteria** cause septicemia in a wide variety of wild, cultured, and display freshwater and marine species. *Vibrio alginolyticus*, *Vibrio* (*Listonella*) *anguillarum* (**red pest of eels**), *Vibrio ordalii*, *Allivibrio salmonicida* (**Hitra disease**), *Vibrio vulnificus*, and *Photobacterium damselae* subsp. *damselae* are pathogens of warm and coldwater marine species, and *Aeromonas hydrophila* (**red pest**), *Aeromonas veroni*, *Aeromonas caviae*, *Aeromonas sobria*, *Pseudomonas fluorescens*, and other motile gram-negative bacteria are common in freshwater and some brackish water species, particularly warm water species. Most of these bacteria exist in free-living forms in the environment, populate surfaces or are constituents of biofilms, and/or are part of the flora of apparently healthy fish, but are capable of acting as opportunistic pathogens.

For many vibrios and aeromonads, rising environmental temperature is an important factor in pathogenesis. Aeromonads are generally considered reliant upon stressors and preexisting conditions, such as high stocking density, altered water quality, poor host condition, and skin lesions to establish infections, but bacterial virulence as a significant factor in disease is increasingly recognized.[24,75] A highly virulent strain of *A. hydrophila* has caused significant losses in commercial catfish aquaculture in the southeastern US since 2009.[23] Spring epizootics are not uncommon due to rising temperatures conducive to infection coupled with seasonal marginal host condition and

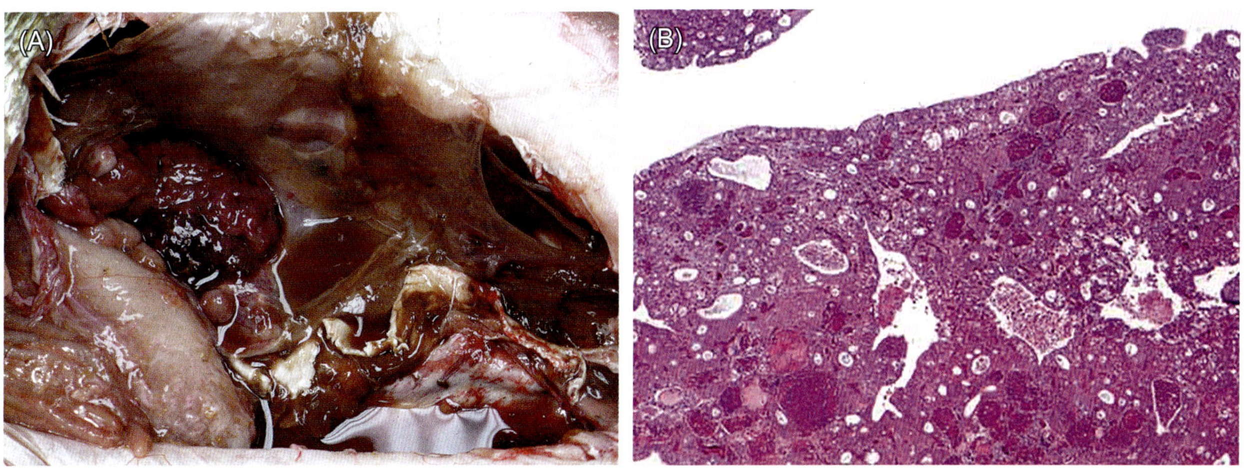

FIGURE 39.38 Swim bladder sarcoma in a bass with swim bladder sarcoma virus infection. (A) Effacing the swim bladder and extending into the coelomic cavity is a multinodular, highly vascular mass. (B) An invasive, multinodular, highly vascular mass replaces and effaces the normal architecture of the swim bladder mucosa. *(Part A: Photo Courtesy of Dr. John Hawke, Louisiana State University. Part B: Photo Courtesy of Dr. Wes Baumgartner.)*

physiologic stressors, such as reproductive activity.[75] Vibrio infections are dependent on host and environmental conditions, but infection can occur in the absence of recognized predisposing factors.[2,143] In display settings, gram-negative septicemias commonly occur in recently shipped individuals, reflecting the enhanced capacity for infection associated with stressful events.

Gross and histologic lesions are those typical of septicemia. Acute mortalities may have no gross lesions. Multifocal hemorrhage of oral cavity, skin, fin bases, and visceral organs, and nephro- and splenomegaly may be observed. Subacute to chronic cases may have cutaneous ulcers. Histologically, acute cases may have renal and splenic reticuloendothelial cell accumulations of bacteria and widespread intravascular bacteria (Fig. 39.39). Necrotizing dermatitis and vasculitis are common and characteristic; progression to deep ulceration or cavitary necrotizing myositis as well as necrotizing lesions in visceral organs may also occur. Multicentric granulomas are the typical presentation for chronic *P. damselae* subsp. *piscicida* infection (pseudotuberculosis). Diagnosis is dependent on visualization of typical lesions of septicemia, along with identification of the organism by bacterial culture and/or molecular methodologies. Several of the listed organisms are important pathogens in production systems and can cause high morbidity and mortality. In display settings, cases tend to be sporadic except for circumstances such as recent shipment and quarantine, which may result in epizootics.

Mycobacteriosis and **nocardiosis** are common, generally chronic, progressive, debilitating, systemic diseases that may affect wild or captive individuals of virtually all freshwater, euryhaline, or marine fish, although species susceptibility may vary.[525] Both causative agents are gram-positive, acid-fast (or in the case of *Nocardia* spp., partially acid-fast), short- to medium-length bacilli, that are thought to cause disease by initiating a foreign body-type host response while simultaneously evading intraphagocytic lysosomal destruction. Causative agents most commonly isolated from fish include *Mycobacterium marinum*, *Mycobacterium fortuitum*, *Mycobacterium chelonae*, *Mycobacterium peregrinum*, *Mycobacterium abscessus*, *Nocardia asteroides*, *Nocardia seriolae* (previously *Nocardia kampachi*), *Nocardia crassostreae*, and *Nocardia salmonicida*. Many of these bacteria are ubiquitous in aquatic environments, and transmission can occur by exposure to infected water, biofilms, sediment, and contaminated fish carcasses. Infections may spread systemically within phagocytes or extracellularly via the lymphatic and/or blood circulatory systems.

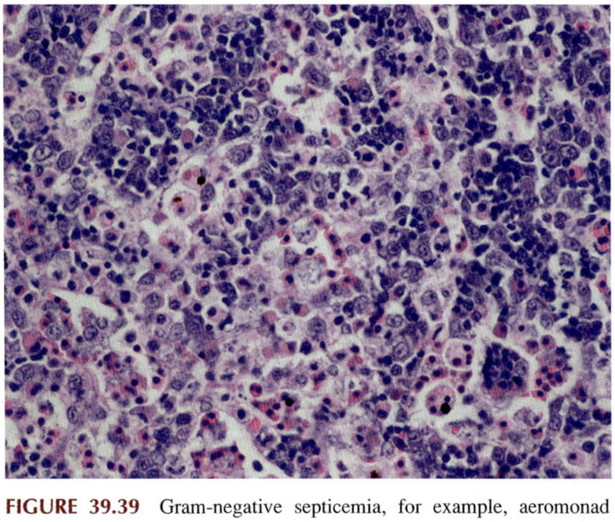

FIGURE 39.39 Gram-negative septicemia, for example, aeromonad septicemia, in an aquarium housed teleost. The cranial kidney has both extracellular and intrahistiocytic bacteria; a few macrophages are necrotic.

The presentation of acid-fast bacterial infections may vary according to factors such as host species, virulence of the particular bacterial strain, exposure magnitude, robustness of the host immune system, length of infection, and ambient water temperature. Infected fish are often thin despite having visceral organomegaly, and cutaneous ulcers may be present. Frequently, the parenchyma of the spleen and kidney are expanded and studded with multiple gray-white nodules (Fig. 39.40A). Essentially any tissue can be affected, but other commonly affected tissues include the liver, mesentery, gill, gonads, and choroid rete (eSlide 39.e6). Depending on some of the aforementioned factors, the inflammatory response may present as discrete granulomas consisting of layers of flattened macrophages (with or without central necrosis) (Fig. 39.40B), or it may manifest as nonencapsulated, nodular, and coalescing foci of granulomatous inflammation (Fig. 39.40C). The occurrence of multinucleated giant cells is variable and may be rare or not at all in some instances. Acid-fast bacilli within and external to granulomas may be rare to numerous (Fig. 39.40D). Mycobacterial disease is common in syngnathids; however, unlike other bony fish, syngnathids rarely produce discrete granulomas. Lesions are typically characterized by necrotic foci with massive numbers of free bacteria, sheets of macrophages with abundant intracytoplasmic bacteria, and/or aggregates of bacteria in vascular lumens (Fig. 39.40E).[138] Common diagnostic tools include the visualization of acid-fast bacilli in impression smears or histologic sections, bacterial culture, and PCR. Culture utilizes special agar and often requires a long incubation period, during which overgrowth by nonpathogenic bacteria may occur, although some species, including *M. marinum* and *M. chelonae*, will grow on blood agars in less than 7 days.

Because of their persistence in biofilms, *Mycobacterium* and *Nocardia* spp. can be especially problematic in artificial systems in which captive fish are raised or maintained. Recirculating systems seem to be particularly prone to recurring patterns of acid-fast bacterial disease, which characteristically presents as chronic, low-level morbidity and mortality that primarily affects older specimens. Sport and commercial fisheries (e.g., mycobacteriosis in

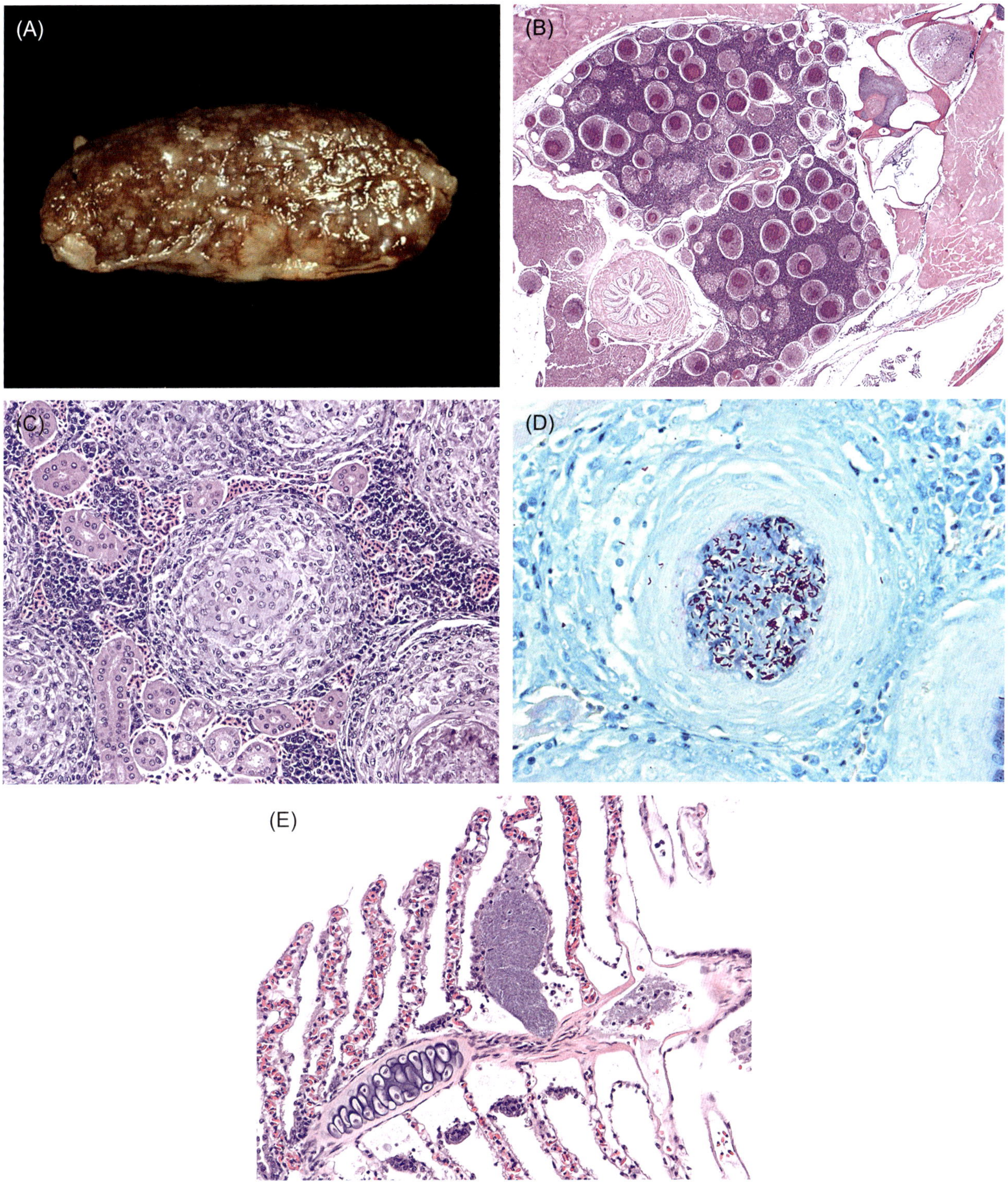

FIGURE 39.40 Mycobacteriosis. (A) Mycobacteriosis in a hybrid striped bass. Severe splenomegaly is present with extensive formation of granulomas, which are represented by multiple, sometimes coalescent, gray-white nodules. (B) Mycobacteriosis in a tetra. Numerous, discrete, occasionally coalescent granulomas with cores of hypereosinophilic necrotic macrophages are present throughout the cranial kidney. A few granulomas also are present in the liver. (C) Mycobacteriosis in a rainbow fish due to *Mycobacterium marinum*. There is granulomatous nephritis characterized by the formation of nonencapsulated whorls of epithelioid macrophages. (D) Mycobacteriosis in a rainbow fish. The granuloma in this section is composed of a collar consisting of multiple layers of flattened macrophages that surround a core of necrotic debris in which are located numerous, acid-fast bacilli. Ziehl-Neelsen. (E) Mycobacteriosis in an alligator pipefish. In syngnathids, blood vessels may contain bacteria free in circulation, particularly in lamellar capillaries. Note the absence of inflammation (see eSlide 39.e6).

striped bass of the Chesapeake Bay; e.g., nocardiosis in Japanese yellowtail) and environmentally significant species (e.g., menhaden of the Chesapeake Bay) may also be impacted.[215,247] Atypical acid-fast bacterial infections can be zoonotic, especially for immune-compromised individuals.

Multiple species of **gram-positive lactic acid bacteria** from multiple genera are pathogenic to fish. Most significant are *Lactococcus garvieae*, *Streptococcus iniae*, *Streptococcus agalactiae*, *Streptococcus parauberis*, and *Streptococcus dysgalactiae*. Several, particularly *S. iniae*, have zoonotic potential. The incidence of disease and number of pathogens belonging to the lactic acid group of bacteria has increased steadily over the past few decades. Collectively, they cause significant losses in aquaculture, and both *S. iniae* and *S. agalactiae* are responsible for major wild fish kills. In contrast to many fish pathogens, large fish may be impacted as severely as juveniles, compounding economic losses.[5,500] A detailed description of the convoluted phylogeny and taxonomy of the group or of specific aspects of individual agents is beyond the scope of this chapter, so the included material will focus on common features of the assemblage.

L. garvieae, first described in Japanese yellowtail, causes serious losses in Asian mariculture and rainbow trout in many parts of the world.[246,500] Disease caused by *S. iniae* dates to the early 1980s in Japan. Since then, it has emerged as a major pathogen responsible for losses in a diverse group of cultured and wild fish species in fresh, brackish, and salt water.[5] *S. agalactiae* is best known as a cause of multispecies wild fish kills in brackish water habitats but has also impacted coastal aquaculture facilities.[380,463] *S. parauberis* and *S. dysgalactiae* are the most recently recognized pathogens in the group. First reported from cultured turbot in Spain, *S. parauberis* now affects cultured marine species in Asia and wild striped bass in the Chesapeake Bay, USA.[101,180,531] *S. dysgalactiae*, originally described from Japanese yellowtail and amberjack, was thought to be a marine fish pathogen but has since been isolated from Nile tilapia in Brazil.[329,335]

Primarily warm water pathogens, disease occurs in many species under a range of culture conditions at temperatures above 15°C. Epizootics are often precipitated by and overall mortalities influenced by environmental stress. With *L. garvieae* and *S. iniae*, asymptomatic carriers are believed to play important roles in transmission. Infected wild fish may also be present in proximity to culture operations. Epizootics begin when environmental conditions, particularly temperature, are optimal, and transmission occurs horizontally by fecal–oral routes in water and possibly by cannibalism. In general, there is predilection for the central nervous system, including the eyes, and coelomic mesothelial surfaces, often with comparatively mild parenchymal involvement. Some species are known to produce potent hemolytic and cytolytic toxins that damage tissues, and some have surface molecules and capsules that aid in avoidance of phagocytosis.[5,500]

Disease is primarily acute to subacute, with typical signs of septicemia, but some, e.g., *S. iniae*, may cause chronic disease and produce granulomas.[463] Common signs and gross changes include anorexia, lethargy, loss of orientation, erratic swimming, exophthalmia, corneal opacity, skin darkening, and hemorrhages. Internally, turbid to hemorrhagic ascites, fibrin deposition, organomegaly with congestion and hemorrhages, and meningeal congestion with cloudy cerebral spinal fluid may be seen. Microscopic changes frequently include severe pyogranulomatous meningoencephalitis, hemorrhagic panophthalmitis, epicarditis, coelomic polyserositis, and cellulitis (Fig. 39.41A). Serosal surfaces may be expanded by mixed inflammatory exudates dominated by macrophages, many containing phagocytized cocci possibly in chains, and smaller numbers of neutrophils.

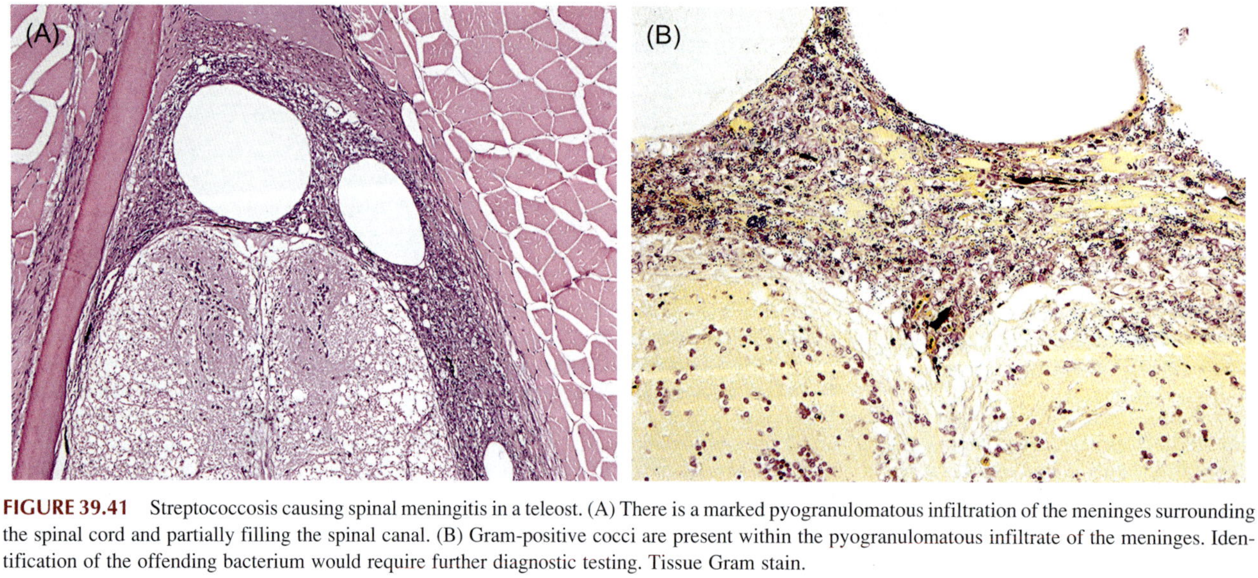

FIGURE 39.41 Streptococcosis causing spinal meningitis in a teleost. (A) There is a marked pyogranulomatous infiltration of the meninges surrounding the spinal cord and partially filling the spinal canal. (B) Gram-positive cocci are present within the pyogranulomatous infiltrate of the meninges. Identification of the offending bacterium would require further diagnostic testing. Tissue Gram stain.

Lesions in parenchymal organs and other tissues are primarily necrotizing with free and phagocytized cocci and similar inflammatory infiltrates. In some instances, bacteria may be present in vascular spaces and attached to erythrocytes. Enteritis is reported with *S. agalactiae* and *S. parauberis* infections, and hemorrhagic enteritis is a prominent feature of L. *garvieae* infection.[101,119,463] Muscle abscesses may be the primary lesion seen with *S. dysgalactiae*, but these also occur with *S. iniae*.[335]

The generic term streptococcosis is applied to cases where gram-positive cocci are present in tissues, but the offending agent is not identified (Fig. 39.41B). Although some authors link specific lesions to individual organisms, some reports are contradictory, and the opinion here is that the respective diseases cannot be reliably discriminated based solely on histopathologic changes.[119,335,531] Complicating matters, diseases caused by more than one of these bacteria overlap and affect multiple fish species in certain areas.[531] Members of the group are readily cultured from most organs, which should include the brain, on common and selective bacterial media. Blood agars are recommended to be observed for hemolysis. Serologic and molecular methods provide for accurate species identification.

Bacterial gill disease (BGD) is a commercially important disease of cultured fish maintained in freshwater throughout the world. It is a major infectious disease problem for freshwater aquaculture, most notably in farmed salmonid species where the disease is most common and most severe in intensively reared fry and fingerlings.[160,468] It is primarily a husbandry and management disease in salmonid hatcheries. Affected fish are lethargic, anorexic, and display signs and behaviors associated with dyspnea, such as flared opercula and crowding at water inlets. Morbidity and mortality have been reported as explosive. The causative agent of BGD is a gram-negative, nonmotile, slender, filamentous bacterium, *Flavobacterium branchiophilum*.[505] Disease has been reproduced in rainbow trout under laboratory conditions by horizontal transmission and by exposure to *F. branchiophilum* inoculated into tank water.[134] Additionally, when BGD was produced in rainbow trout by manipulation of the tank environment (i.e., without bacterial challenge), the dominant bacterial species on gills of affected rainbow trout was shown to be *F. branchiophilum*.[160] Environmental factors, such as overcrowding, high levels of nitrogenous wastes, and low levels of dissolved oxygen, are implicated as factors that may promote BGD in intensive culture operations; however, retrospective epidemiological studies indicate that specific management practices in hatcheries and specific species, such as brook trout, are more consistently affected by BGD.[161] A seasonal pattern of BGD occurrence has been suggested, with a rise in cases in spring and summer months.[468] Rising temperature and higher levels of precipitation during the spring have been posited as risk factors; however, the significance of these environmental factors needs to be tempered by the knowledge that populations of higher-risk age groups (e.g., fry and fingerlings) peak in the spring.[161]

According to a study of the sequential development of branchial lesions of BGD by Speare et al.,[470] bacterial colonization of gill filaments is immediately preceded by ultrastructural changes to the superficial epithelium of the filament that include cytoplasmic blistering and degeneration of microridges together with irregularities at the tips of filaments. Bacterial colonization begins at irregular filament tips and proceeds proximally along the filament, extending over the lamellae. Very high morbidity and mortality coincide with extensive bacterial colonization of lamellar surfaces, hydropic degeneration and necrosis of lamellar epithelial cells, and development of lamellar edema, which occur during the acute phase (0–2 days) of the disease. Epithelial cell hyperplasia and mucous cell metaplasia, particularly along the distal one-third of filaments with mucous cells on lamellar surfaces, characterize the subacute phase (days 3–5); severe and extensive epithelial hyperplasia and mucous metaplasia, leading to lamellar and filamental fusion with intralesional filamentous bacteria in clumps of mucus along the branchial surfaces, characterizes the chronic phase (days 7–14) (Fig. 39.42).

The diagnosis of BGD can be obtained by microscopic examination of whole mount preparations or histologic sections of gill leading to the recognition of numerous gram-negative filamentous bacteria located in interlamellar spaces and along branchial lamellar epithelium.[468] Methods for the specific detection of *F. branchiophilum* include ELISA, IFAT, and PCR.[160] Failure to obtain a rapid, accurate diagnosis and provide appropriate treatment can

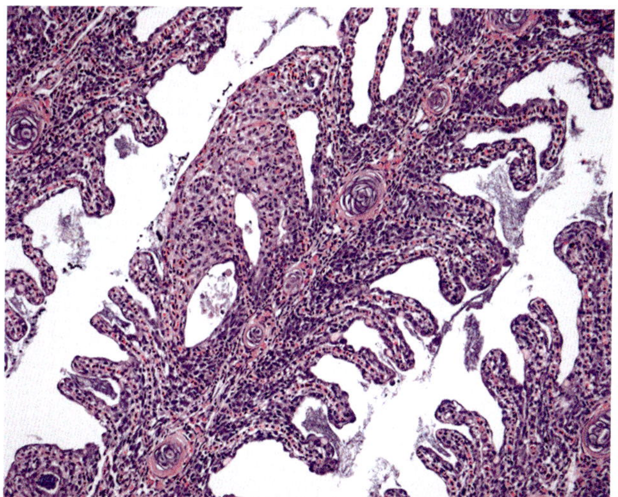

FIGURE 39.42 Bacterial gill disease in a rainbow trout. There is diffuse hyperplasia of branchial lamellar epithelium with foci of lamellar fusion and synechiae. Mats of filamentous bacteria occupy interlamellar spaces and extend across the tips of lamellae in the interfilamental space.

lead to high levels of mortality and pronounced economic impacts.

Similar to salmonids, gill disease associated with poor husbandry and environmental conditions can predispose bacterial branchitis in ornamental fish. Injuries to gill epithelium resulting from chemical or physical irritants, such as elevated ammonia or suspended solids, induce epithelial and goblet cell hyperplasia, often with subsequent partial to complete lamellar fusion. Poor water circulation within interlamellar pseudocystic spaces created by lamellar synechia, along with trapped mucus and cellular debris, creates conditions conducive to bacterial proliferation and branchial necrosis (Figs. 39.43A, B).

Epitheliocystis is an infectious disease that affects fish across a remarkably broad range of environments and species. The condition has been observed in over 50 species of wild and cultured, freshwater and saltwater fish from warm and coldwater environments.[336] It is also seen in one chondrostean (white sturgeon) and in several chondrichthyans (see Chapter 40: Chondrichthyes).[56,174,381] Epitheliocystis affects the epithelium of the gill and skin; however, gill infections overwhelmingly make the most clinical and pathological impacts. The agents of epitheliocystis are gram-negative bacteria that occur intracellularly, typically in epithelial cells but also in chloride, mucous, and pillar cells and in macrophages. In the vast majority of cases, bacteria have been identified as members of the order Chlamydiales, i.e., chlamydia-like bacteria, e.g., *Candidatus* Piscichlamydia salmonis, *Candidatus* Clavochlamydia cysticola, *Candidatus* Syngnamydia salmonis, and *Candidatus* Parilichlamydia carangidicola.[110,217,343,483] It is important to note that in several instances β-proteobacteria, e.g., *Candidatus* Branchiomonas cysticola and the BJC epitheliocystis agent, as well as a γ-proteobacterium, *Endozoicomonas elysicola*, have also been identified in cases of epitheliocystis.[82,245,300,306,488] Epitheliocystis may be incidental or may be associated with mortality of up to 100% in cultured fish.[336] A direct, causal link between infection and disease has not been established. *E. elysicola* is the only epitheliocystis-associated bacterium to be isolated in vitro; nevertheless, at the time of this publication, Koch's postulates have not been fulfilled for any recognized epitheliocystis-associated bacterium.[82,336] Further complicating the issue of disease causation is the fact that mixed infections of epitheliocystis-associated bacteria have been detected in the same population of fish.[306,429,488] Environmental, management, and host factors (temperature, season, stocking density, age-class of fish, etc.) are risk factors for epitheliocystis.

Epitheliocystis is characterized histologically by the presence of hypertrophied cells that contain intracytoplasmic, membrane-bound, vacuolar inclusions with basophilic, granular contents that displace the host cell cytoplasm, often including the nucleus, to the cell periphery.[82,110,336,364,366] Host proliferative and inflammatory responses can differ considerably, and in many cases, cytoplasmic inclusions are associated with minimal or no such responses. In the gill, proliferative responses vary in severity and extent, and include hyperplasia of lamellar epithelium, lamellar fusion, and bridging of hyperplastic epithelium across the tips of consecutive lamellae (Figs. 39.44A, B). Hyperplastic responses are accompanied by varying degrees of epithelial cell necrosis, goblet cell metaplasia, and inflammation. The bacteria associated with epitheliocystis demonstrate differences in histochemical staining patterns and differences in the ultrastructure of intracytoplasmic vacuolar inclusions in different host fish.[109,110,336] Determination of the specific agent(s) of epitheliocystis requires PCR amplification and analysis of

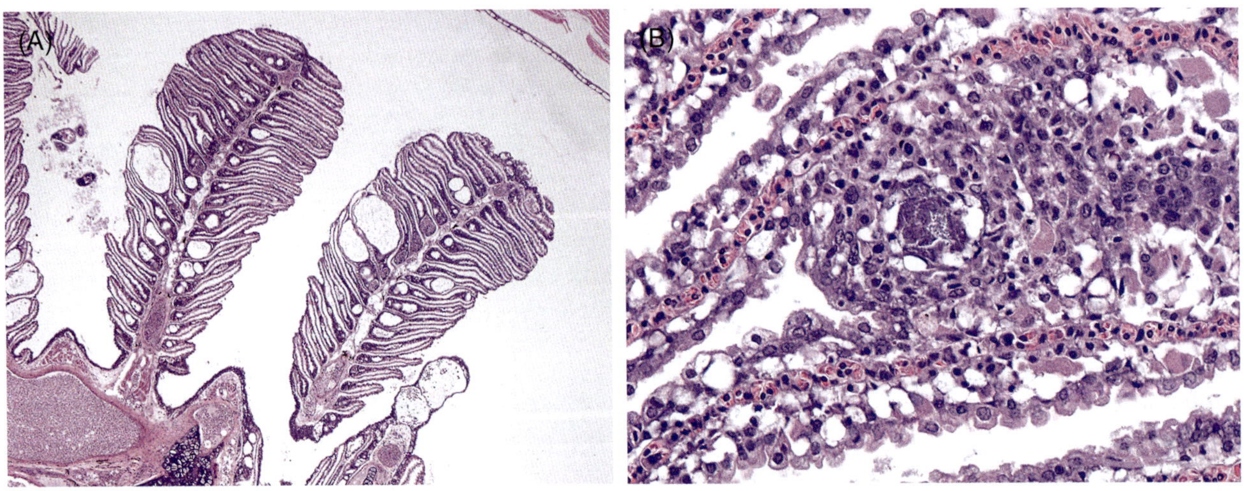

FIGURE 39.43 Bacterial branchitis in the lined seahorse. (A) Synechiae at the tips of lamellae or along their lengths create interlamellar cavities in which water circulation is limited, accompanied by hyperplasia of lamellar epithelium. (B) Focal necrosis surrounding a bacterial colony within an area of hyperplastic epithelium and lamellar fusion.

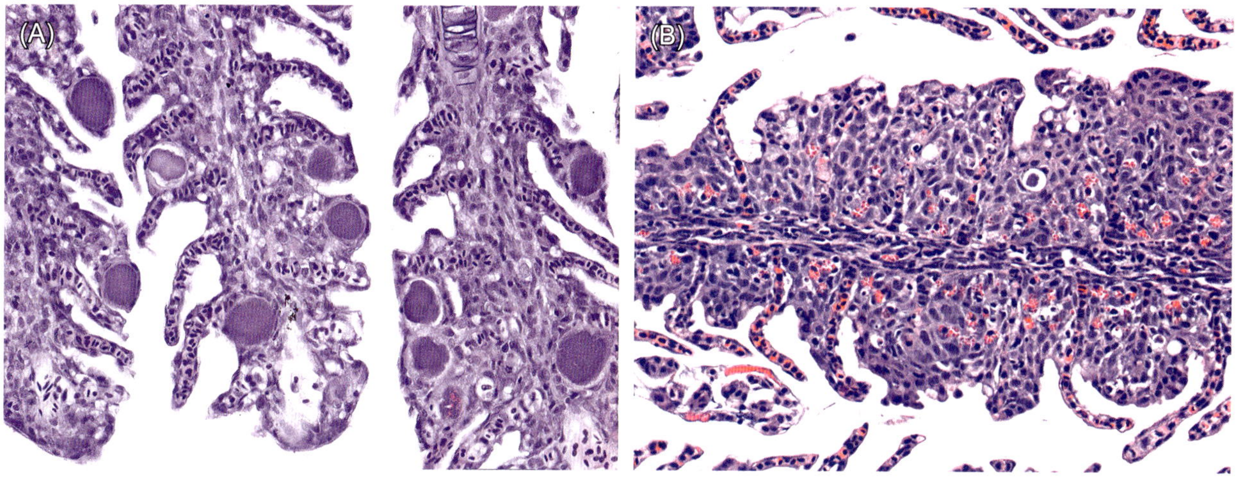

FIGURE 39.44 Epitheliocystis in the gill of a barramundi (A) and Arctic char (B). (A) The cytoplasm of many epithelial cells lining lamellae is expanded by inclusions that contain granular basophilic contents representing developmental stages of the bacterium. While there is mild hyperplasia of the epithelium of a few of the lamellae in this field, the epithelium lining many lamellae is flat and the interlamellar spaces are open, i.e., no hyperplasia. (B) An inclusion with basophilic granular contents surrounded by a clear space halo and displacing the nucleus to the periphery is present in the cytoplasm of an epithelial cell, accompanied by severe hyperplasia of lamellar epithelium resulting in fusion of consecutive lamellae. Note that eosinophilic granular cells, which are normally observed in the gill of salmonids, infiltrate the hyperplastic epithelium. Contrast the proliferative response in this specimen with the markedly less proliferative, i.e., nonproliferative, response of the specimen in (A), exemplifying the range of host responses in this disease.

phylogenetically relevant genetic sequences (e.g., 16S rRNA), which is best coupled with in situ hybridization to localize amplified sequences to bacteria in epitheliocystis inclusions in tissue section.[110,336,488]

Piscirickettsiosis is the systemic disease caused by *Piscirickettsia salmonis*, a gram-negative, nonmotile, aerobic, nonencapsulated, fastidious, typically coccoid bacterium that is an intracellular pathogen of fish, particularly farmed salmonids in seawater.[145,146,414] It is a cause of significant mortality in salmonids reared under culture in marine net pens. Outbreaks of disease occur 6–12 weeks after transfer to seawater net pens, and cumulative mortality ranges from 30% to 90%.[45] There is variability in the susceptibility of salmonids. For example, coho salmon are considered more susceptible while Atlantic salmon are more resistant.[145] The causative agent and, hence, the disease have a global distribution in aquaculture-reared salmonids, occurring in the Pacific Ocean in Chile, the Pacific northwest, and the northeastern and northwestern Atlantic Ocean.[145] Disease has also been identified in white seabass.[69] Transmission of *P. salmonis* is horizontal between infected and noninfected fish; a reservoir of *P. salmonis* has not yet been identified, although a marine reservoir is suspected since outbreaks of piscirickettsiosis occur after transfer to seawater.[88,145,414] On the basis of nucleotide sequence and secondary 16S rRNA structure, *P. salmonis* has been assigned to the class Gammaproteobacteria, which includes the genera *Francisella*, *Coxiella*, and *Legionella*.[144]

The characteristic lesion of piscirickettsiosis is systemic, septicemic disease with lesions in multiple organs, the most severe occurring in the liver, kidney, spleen, and intestine.[44,88,145] The most consistent gross finding is pallor of the gills, a consequence of severe anemia. The hematocrit of *P. salmonis*-infected fish may be as low as 2%, with an average reduction of 18.5% (normal range is 35%–50%). Peripheral blood smears have high numbers of macrophages, most of which contain cytoplasmic bacteria or cellular debris.[88] Gross internal lesions commonly include renomegaly, splenomegaly, ascites, and hemorrhages in the visceral fat, stomach, swim bladder, and skeletal muscle. Severely affected fish may have circular, cream-colored lesions throughout the liver. Necrosis is the principal pathologic change during the acute phase of disease followed by granulomatous inflammation as disease progresses into the chronic phase. Foci of necrosis occur in the renal hematopoietic interstitium and spleen, which is the cause of the characteristic anemia.[88,145,414] Granulomatous inflammation replaces renal hematopoietic tissue and splenic lymphoid tissue as the disease process becomes chronic. Necrotizing hepatitis with necrotizing vasculitis and thrombosis is commonly seen in the liver, as is necrosis, hemorrhage, and granulomatous inflammation in the lamina propria of the intestine. Inflammation and thrombosis are observed in the brain, heart, pancreas, adipose tissue, and ovary, along with lamellar hyperplasia with necrosis and thrombosis in the gill.[88]

The diagnosis of piscirickettsiosis is based on recognition of typical gross and histologic lesions together with observation of gram-negative intracellular bacteria and confirmatory identification of *P. salmonis* by one of several testing methods. These include immunohistochemistry, culture of bacteria followed by labeling with specific fluorescent antibody, or PCR of culture isolates or tissues.[145,216,251,305] Isolation of *P. salmonis* has been accomplished using fish-derived cell lines and in vitro utilizing cysteine-enriched agar.[305] Tissues of choice for isolation of *P. salmonis* include kidney, liver, and blood.[414]

Francisellosis in farmed, aquarium displayed, and wild fish caused by bacteria in the genus *Francisella* are increasing in incidence worldwide and affect a number of marine and freshwater fish, as well as the giant abalone *Haliotis gigantea*.[26,57,80] Once referred to as Piscirickettsia-like, the small, pleomorphic, intracellular bacteria are distinct but genetically related and morphologically similar to *P. salmonis*.[80,145] While new species are being identified, most *Francisella* spp. isolates obtained from warm water fish species are *F. orientalis*, whereas those from coldwater fish species are *F. noatunensis*.[80,392,460] Transmission of the bacteria is horizontal with the skin, gill, and gastrointestinal tract serving as portals of entry, and evidence indicates transmission can occur through contaminated water without direct contact between fish.[80,462]

Francisellosis in fish is a systemic granulomatous disease. Findings are the same irrespective of bacterial or host fish species. Formation of grossly visible granulomas is extensive, occurring principally throughout the spleen, kidney, and liver, but can affect virtually all tissues including the heart, gill, testis, muscle, brain, and eye (Fig. 39.45A). Best visualized with Giemsa stains, the gram-negative bacteria are observed microscopically within large clear vacuoles in the cytoplasm of macrophages (Fig. 39.45B).[26,80,462] Bacteria are numerous early in lesion development, but numbers decline over time as centers of granulomas become progressively necrotic. In addition to recognition of gross and microscopic lesions, diagnosis of francisellosis is achieved by isolation of the bacteria in culture on media such as modified Thayer Martin agar. Species-specific PCR tests are capable of discriminating *F. orientalis* from *F. noatunensis*.[113] Vaccines exploiting components of the pathogenicity island necessary for intracellular survival and others hold promise for the development of preventative vaccines against piscine francisellosis.[436]

Principal disease-causing representatives of the *Flavobacteriaceae* group of gram-negative bacteria include *Flavobacterium columnare* (**columnaris disease, saddleback**), *Flavobacterium psychrophilum* (**bacterial coldwater disease, peduncle disease, rainbow trout fry syndrome**), *F. branchiophilum* (**bacterial gill disease**), and *Tenacibaculum maritimum* (**tenacibaculosis, saltwater columnaris,** and **black patch necrosis**). Columnaris disease affects many freshwater species, including numerous tropical ornamentals. Coldwater disease affects salmonid and several nonsalmonid species in freshwater. Bacterial gill disease occurs primarily in young hatchery-reared salmonids. Tenacibaculosis impacts many marine fish and is a major impediment to commercial mariculture. *F. columnare* and *T. maritimum* are notable for their apparent lack of host specificity. As a group, these pathogens share common morphology as long slender bacterial rods that adhere to epithelial surfaces, exhibit gliding motility, and produce yellow flexirubin or carotenoid pigments.[262,294]

Despite their prevalence, virulence mechanisms used by the flavobacteria and their interactions with host immune systems remain unknown. Tissue damage and invasion is likely promoted by extracellular proteases such as chondroitin AC lyase, which degrades the ground substance of connective tissues.[94,477] These bacteria can be persistent in the environment, and epizootics are often associated with stressors, such as handling, transport, overcrowding, and poor water quality. Diseased and dead fish are important sources of infection, and vertical transmission is suspected

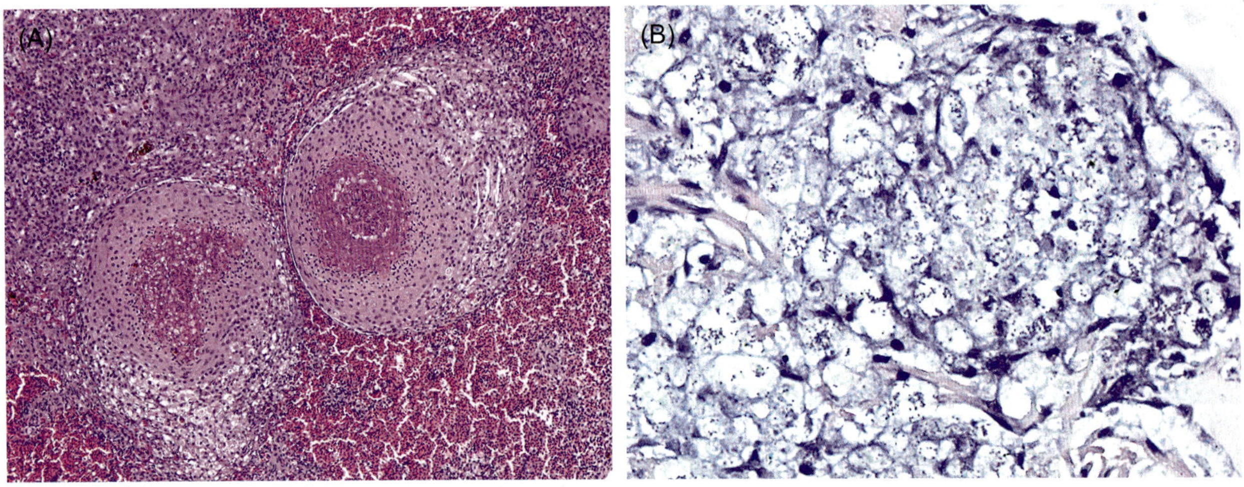

FIGURE 39.45 Francisellosis in a Nile tilapia due to *Francisella orientalis*. (A) Granulomas observed in francisellosis consist of cores of necrotic debris surrounded by wide collars of epithelioid macrophages and must be distinguished from those observed in mycobacteriosis. (B) Early in lesion development, small, gram-negative, coccobacilli are numerous. Best visualized using Giemsa stains, bacteria are frequently found in macrophages within a single large cytoplasmic vacuole. Giemsa.

with *F. psychrophilum*.[294,477] All ages are susceptible, but signs and mortality vary with age; young fish are most susceptible. Disease outcome and lesion progression are also influenced by temperature and lesion distribution. Outbreaks of columnaris and tenacibaculosis typically occur above 15°C, while coldwater disease occurs at 15°C and below.[13,94,477] Bacterial gill disease can occur at temperatures ranging from 5°C to 20°C.[492] Virulent strains of *F. columnare* can cause death in less than 24h with few external signs, while less virulent strains result in a more protracted course, often with extensive external lesions. Similarly, internal infection by *F. psychrophilum* and gill infection by *F. columnare* are often rapidly fatal.[94,477]

Columnaris, coldwater disease, and tenacibaculosis cause ulcerative dermatitis and necrotizing myositis. The characteristic lesion is pale yellow discoloration of the skin that microscopically corresponds to large numbers of long slender bacteria in dermal connective tissues and along epithelial margins (Figs. 39.46A–C; eSlide 39.e7). Focal depigmentation is followed by necrosis and often rapid lesion expansion. Inflammatory responses are mononuclear and relatively mild, particularly with columnaris, and may be accompanied by dermal edema and hemorrhage. All three affect multiple skin sites. Columnaris frequently affects the caudal peduncle, mouthparts, fins, and dorsum of the back, creating the so-called saddleback lesion. Coldwater disease commonly involves the peduncle, as well as the base of the skull, jaw, dorsum, and perianal area. *F. columnare* and less frequently *T. maritimum*, also cause necrotizing branchitis. Columnaris gill lesions are well-demarcated yellow-tan areas of necrosis that grossly resemble those of CyHV3 in koi and consist of a mixture of epithelial proliferation and necrosis with mats of bacteria within debris and between gill components. In freshwater, lesions are often secondarily colonized by *Saprolegnia* spp.[13,94,477]

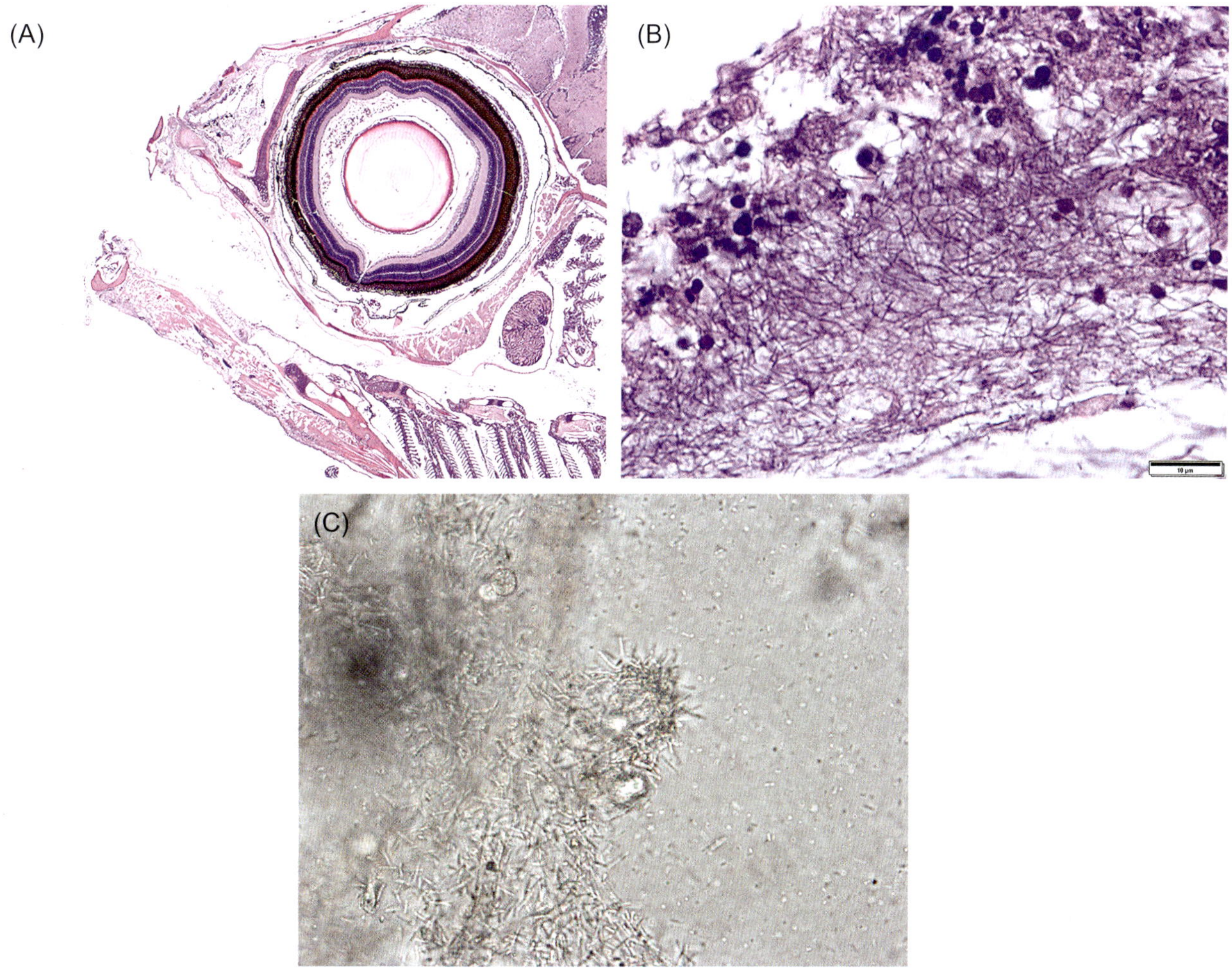

FIGURE 39.46 Columnaris disease in a Cardinal tetra. (A) There is regionally extensive ulceration of the skin of the snout and along the margins of the lips, with edema and necrosis of subcutaneous tissues. (B) Mats of innumerable long slender bacteria are present within the necrotic dermis in the ulcerated skin. (C) Wet mount preparation from the affected skin of a teleost. Intertwined filamentous bacteria in raised sheave-like arrangements ("haystacks") are present (see eSlide 39.e7).

Unlike other members of the group, *F. psychrophilum* causes severe systemic disease, resulting in visceral hemorrhagic necrosis. Muscle necrosis can also be severe, resulting in complete loss of the caudal fin and exposure of vertebral bone. The bacterium has a predilection for connective tissues, which includes hyaline cartilage and immature bone. Chronically infected fish may develop spiral and erratic swimming, blackening of the tail and spinal deformities, reminiscent of whirling disease, in association with meningitis, osteitis, and periosteal proliferation. Similar, severe, internal changes with mortality exceeding 50% are associated with **rainbow trout fry syndrome** in sac fry and early feeding stages.[477]

The gross and microscopic lesions of columnaris, coldwater disease, and tenacibaculosis are virtually pathognomonic. However, columnaris can be mistaken for coldwater disease in salmonid species. The flavobacteria require special media and do not grow on typical diagnostic agars. Confirmation can be achieved by various serologic- or molecular-based methodologies.

Aeromonas salmonicida is a gram-negative, nonmotile, short rod-shaped obligate fish pathogen. Subspecies *salmonicida* is best known and is the cause of **furunculosis** in salmonids. A diverse group of "atypical" *A. salmonicida* including three additional subspecies (subspecies *smithia*, subspecies *achromogenes*, and subspecies *masoucida*) and a number of additional strains cause disease in salmonids, cyprinids, and an expanding list of freshwater and marine species. Recognized disease entities associated with atypical *A. salmonicida* subspecies include **goldfish ulcer disease**, **carp erythrodermatitis**, and **ulcer disease of flounder**.[517] Carrier fish are crucial to maintenance of the agents. Transmission can occur via direct contact, contact with contaminated water, and also apparently ingestion of infected material.[75,410] Disease severity and lesion development are influenced by age, host resistance, and stressors, such as crowding and transport. Bacterial virulence factors also play a significant role in the establishment of infection and lesion progression, and multiple mechanisms have been delineated or proposed.[498]

Peracute to acute septicemia is the typical outcome of *A. salmonicida* subsp. *salmonicida* infections in fingerlings through adults, while the classic, subacute to chronic form of the disease, furunculosis, is most frequently observed in older individuals with partial immunity or in species with greater innate resistance.[75] Atypical infections generally are associated with ulcerative rather than septicemic disease in a variety of species. This may in part be due to differing virulence factors. Cases of acute septicemia may have fin base and visceral organ hemorrhages as well as splenomegaly. Grossly, furunculosis is characterized by extensive cavitary or hemorrhagic lesions in skin and skeletal muscle that visibly distort the external contours of the body (Fig. 39.47). In cases with cutaneous ulceration, ulcers will

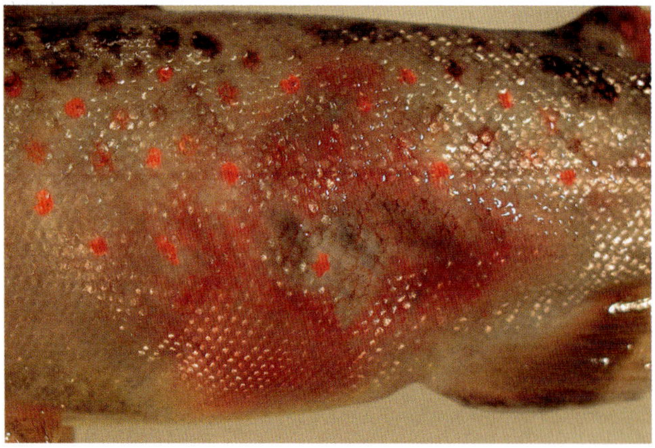

FIGURE 39.47 Furunculosis in a brown trout due to *Aeromonas salmonicida*. There is an extensive area of the lateral body wall over which the surface contour is distorted, accompanied by hemorrhage in the skin with erosion and ulceration centrally. Beneath the skin at the site of this lesion, there is necrosis of the muscle of the trunk. *(Photo Courtesy of Dr. Spencer Russell, Vancouver Island University.)*

be contiguous with deep cavitary lesions. Histologically, in peracute and acute cases, septicemia presents as large intravascular aggregates of bacteria with little to no inflammation in a variety of organs but particularly kidney, gill, heart, and spleen. Subacute cases will have multicentric necrosis, hemorrhage, and inflammation. Classic furunculosis lesions are deep regions of liquefactive necrosis and hemorrhage of body wall muscle, often with cutaneous ulceration as an extension of the necrotizing inflammatory process. Atypical infections, common in koi, are characterized by cutaneous ulcers of various severity, sometimes accompanied by extensive dermal edema. Typical gross and histologic lesions may be an indicator of disease, but identification of the causative organism using selective bacterial culture or PCR is necessary for definitive diagnosis.

Furunculosis is one of the most important bacterial diseases in captive salmonids. Morbidity and mortality are typically high, economic losses can be extreme, and prevention is an important endeavor in production systems.

The gram-negative bacterial genus, *Edwardsiella*, family Hafniaceae, contains the facultative intracellular fish pathogens, *Edwardsiella ictaluri* (**enteric septicemia of catfish**; **ESC**), *Edwardsiella tarda* (**emphysematous putrefactive disease**, **fish gangrene**, **red disease**), *Edwardsiella piscicida*, and *Edwardsiella anguillarum*. *Edwardsiella* spp. cause disease in wild and cultured fish from cool, temperate, and warm waters.[169] Isolates previously identified as *E. tarda* represent three separate taxa, including two new species, *E. piscicida* and *E. anguillarum*.[1,437] *E. tarda* thus must be diagnosed cautiously, as current MALDI-TOF databases, 16s rRNA sequencing, and commercial biochemical test panels do not differentiate the new taxa, necessitating use of species-specific PCRs or sequence analysis of appropriate genes.[167,170,394] *Edwardsiella* spp. readily grow on common media incubated at 28–30°C (*E. ictaluri*) or 28–37°C (*E. tarda*, *E. piscicida*, *E. anguillarum*).

E. ictaluri, an obligate pathogen, initially emerged with channel catfish farming and now affects over 15 species, including zebrafish, tilapia, and, recently, ornamental fish.[169,191,192,459] Infection occurs via ingestion or olfactory epithelium invasion. *E. ictaluri* causes hemorrhagic sepsis in the acute stage, with internal and external petechiation, organomegaly, and multiorgan necrosis with abundant bacteria histologically (Fig. 39.48). Ascites is often prominent in catfish. Ulcerative dermatitis likely results from bacterial embolization to the skin. Inflammation becomes progressively granulomatous, predominated by macrophages containing phagocytized bacteria. Olfactory epithelium infection results in olfactory neuritis and subsequent granulomatous meningoencephalitis, which in channel catfish, leads to the classic "hole-in-the-head" lesion (Fig. 39.49; eSlide 39.e8). Descriptions of the inflammatory response differ somewhat but may reflect chronicity rather than variation among fish species. In catfish aquaculture, naïve fish up to market size can be affected, but disease primarily involves fingerlings in late summer or fall, when water temperatures are 18–28°C. Immune survivors become carriers and perpetuate disease, particularly in US channel catfish operations where all-in-all-out culture is rare.[239] Management techniques include live attenuated vaccination, feed restriction, and antimicrobials.[522,523]

While pathogenic to over 20 fish species, *E. piscicida* is increasingly prevalent in US catfish farming, emerging concurrently with the increase in channel x blue hybrid catfish.[168,169] Lesions parallel those caused by *E. ictaluri*, including dermal ulcerations, hemorrhagic branchitis, mononuclear meningoencephalitis, splenitis, granulomatous interstitial nephritis, hepatitis, hemorrhagic enteritis, and the "hole-in-the-head" lesion. In contrast to *E. ictaluri*, *E. piscicida* predominantly affects large market-sized hybrid catfish.[168] Outbreaks have involved largemouth bass, farmed barramundi, and farmed whitefish, among others, generally causing lesions characteristic of gram-negative sepsis.[58,139,263,434]

Edwardsiella anguillarum was originally described as a hemorrhagic sepsis in eels but has an expanding host range, including cultured tilapia, grouper, and sharpsnout sea bream.[11,220,437,494] In tilapia, *E. anguillarum* causes widespread acute necrosis in highly vascular tissues with abundant bacteria, advancing to granulomatous inflammation and discrete granulomas with progressively fewer intrahistiocytic bacteria.[11] An important differential in tilapia is *Francisella orientalis*.[461]

Though *E. tarda* is historically a major fish pathogen, its recent separation into three taxa has minimized its significance, especially in US catfish.[393] Published descriptions and present-day diagnoses should be interpreted considering current *Edwardsiella* classification. *E. tarda* causes opportunistic infections via horizontal transmission and infrequent outbreaks coinciding with stress from elevated temperatures, high stocking densities, and high organic waste concentrations.[167,303] Original reports of emphysematous putrefactive disease in channel catfish describe large raised cutaneous bullae that rupture to release malodorous gas with cavernous areas of liquefied muscle.[303] In most species, *E. tarda* commonly presents as generalized septicemia.[91,169]

Photobacteriosis, or **fish pasteurellosis** and **pseudotuberculosis**, refers to disease caused by the bacterium *P. damselae* subsp. *piscicida*, of the family *Vibrionaceae*.

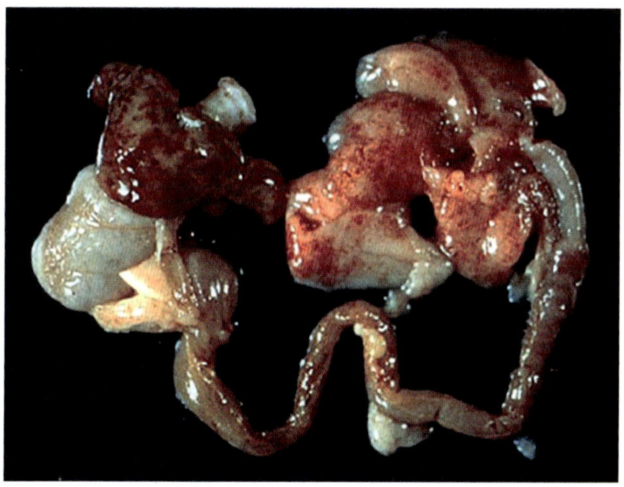

FIGURE 39.48 Enteric septicemia of catfish in a channel catfish. The acute presentation of disease is that of hemorrhagic septicemia characterized by widespread petechial and ecchymotic hemorrhages involving the coelomic viscera.

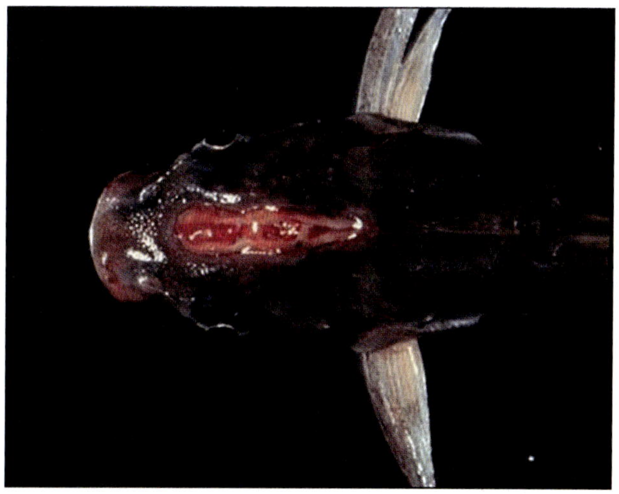

FIGURE 39.49 Hole-in-the-head lesion in a channel catfish due to *Edwardsiella ictaluri*, the agent of enteric septicemia of catfish. Chronic infection is associated with bacteria translocating the olfactory epithelium and ascending the olfactory nerve, resulting in granulomatous meningoencephalitis, lysis of the dorsal cranium, and ultimately ulceration of the overlying skin, giving rise to the characteristic lesion (see eSlide 39.e8).

This bacterium is a gram-negative, nonmotile, halophilic, and facultative intracellular pathogen that may show pleomorphism and filament formation.[404] This pathogen is considered a significant threat to aquaculture of marine fish due to its wide host and geographic range.[409] Previously identified as *Pasteurella piscicida*, *P. damselae* subsp. *piscicida* is capable of infecting as many as 20 warm water marine fish species worldwide, resulting in high mortality events of cultured and wild marine fish in the USA, Mediterranean, Europe, and Japan.[333] High mortality events typically occur at water temperatures above 18–20°C and salinities of 5–15 ppt with epizootic events occurring in summer and more serious outbreaks occurring at water temperatures above 20–25°C and in brackish waters.[193,404] At lower water temperatures, infected fish may serve as subclinical carriers for extended periods of time.[278] Age of fish and maturity of the host immune system have been identified as major determinants in disease susceptibility in some fish species.[337] Size may be related to the functionality of macrophages and neutrophils with larger fish showing better ability to phagocytize and kill this bacterium.[337,450] *P. damselae* subsp. *piscicida* shares phenotypic characteristics with several important fish pathogens including *Vibrio anguillarum* and *A. salmonicida*, and its historical classification has been contentious.[333] Analysis of 16S rRNA gene sequences and DNA–DNA hybridization studies of *P. damselae* subsp. *piscicida* informed the current classification and revealed high similarity to *P. damselae* subsp. *damselae,* an emerging fish pathogen in its own right and similar in its lack of host specificity and global distribution.[402]

As a facultative intracellular pathogen, *P. damselae* subsp. *piscicida* is capable of residing within macrophages and epithelial cells.[280,337] A number of virulence factors have been identified that contribute to infection and survival within phagocytes, including a polysaccharide capsule, plasmids, serum resistance, mechanisms of iron uptake involving iron-binding siderophores, plasmid-encoded apoptosis-inducing proteins, and a number of extracellular products with enzymatic activities.[98,106,107,277,279,324] Multidrug transferable genetic elements (R plasmids) have been identified in *P. damselae* subsp. *piscicida* conferring resistance to many commonly used antibiotics.[97,234]

Infection by *P. damselae* subsp. *piscicida* in susceptible fish results in bacterial septicemia, which in turn leads to one of two forms of disease. In the acute form, affected fish may be dark and inappetant or may show no clinical signs. Gross lesions include small hemorrhages around the operculum and base of fins together with enlargement of kidney and spleen. Histologically, there is necrosis of spleen, liver, and pancreas without attendant inflammation.[193] In the chronic form of the disease, granulomas centered on bacteria form in the kidney and spleen, and these may be grossly apparent as 1–2 mm, white nodules.[238] The formation of multicentric granulomas in hematopoietic tissues has earned the disease the name, "pseudotuberculosis," as its gross appearance mimics that of mycobacteriosis. Diagnosis of *P. damselae* subsp. *piscicida* is based on the presence of gross and histologic lesions together with microscopic evidence of intralesional gram-negative bacteria and confirmatory bacterial culture or molecular testing.[333]

Erysipelothrix rhusiopathiae, a known pathogen of marine mammals (see Chapter 22: Cetacea), was first associated with disease in fish in 2015 as a cause of hemorrhagic septicemia in Australian eels *Anguilla* spp.[74] However, the phenotypic and limited molecular methods used made the species-level identification uncertain. In the United States, bacteria identified phenotypically as *E. rhusiopathiae* have caused low to moderate mortalities and unique pathologic changes in ornamental barbs and tetras of multiple genera since 2012.[383] Infections have also occurred in mosquitofish *Gambusia affinis* present naturally in commercial catfish ponds.[480] Discriminating molecular analyses of the inciting agent characterized the novel species, *Erysipelothrix piscisicarius*.[382] Nile tilapia and channel catfish are refractory to experimental challenge, indicating limited species susceptibility to infection.

Clinical signs and gross changes of **piscine erysipelosis** due to *E. piscisicarius* are nonspecific, while histopathologic lesions are pathognomonic with bacteria exhibiting distinct tropism for collagenous tissue.[68] Infections begin as an ulcerative dermatitis and stomatitis followed by wide distribution of microcolonies and macrophages replete with small, gram-positive, rod-shaped bacteria extending from the dermis throughout the interstitium of skeletal muscle (Fig. 39.50A). Bacteria invade vascular adventitia, connective tissue components of organs such as capsule, lamina propria, submucosa, and serosa, and connective tissue surrounding bone in sites that may include the peritoneum, pharynx, gill arches, heart, swim bladder, spinal canal, mesentery, and intestine (Fig. 39.50B). Bacteria are notably absent from vascular lumina.[68] Necrosis becomes increasingly severe after several days but generally spares parenchymal organs. In fish that survive acute infection, bacterial numbers decline precipitously as cellular responses become progressively granulomatous and fibrosing, replacing extensive areas of skeletal muscle and enveloping bone.[68,383]

Fungi

Pigmented fungi are well-known pathogens in food and aquarium-housed bony fish, with an increasing diversity of implicated fungal genera/species and affected fish hosts. Factors contributing to their pathogenicity include presence of melanin and carotene, formation of thick cell walls, thermo- and osmotolerance, assimilation of aromatic hydrocarbons, and production of extracellular polysaccharides,

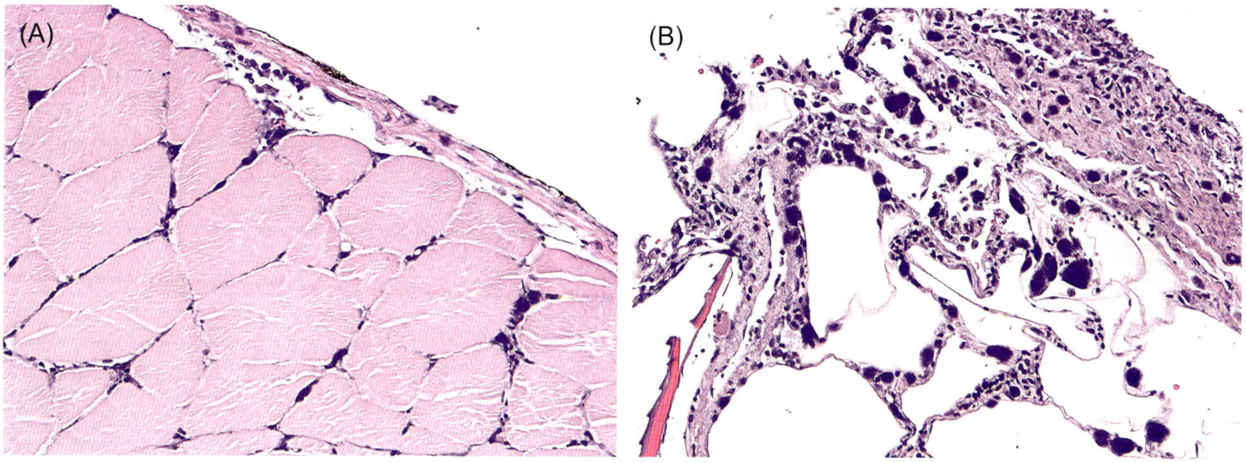

FIGURE 39.50 Acute erysipelosis in a tetra. (A) Bacterial colonies are present within interstitium of skeletal muscle, characteristic of the acute presentation. Necrosis and inflammatory changes are minimal during acute infection. (B) Necrotizing dermatitis with bacterial colonies distributed in the vascular walls and interstitium supporting adipocytes in the subtending hypodermis.

siderophores, and acidic or alkaline secondary metabolites.[95,433] Temperature is a critical microenvironmental element, influencing the susceptible host range. Melanized fungal infections occur in three patterns: chromoblastomycosis, eumycetoma, and phaeohyphomycosis. Differences are based on the fungal morphology in tissue section and the accompanying histopathologic findings.[295,400,466] **Chromoblastomycosis** is a chronic subcutaneous mycosis characterized by verrucous plaques possessing muriform fungal cells known as sclerotic bodies. **Eumycotic mycetoma** is a chronic subcutaneous nodular granulomatous inflammatory condition that may develop draining tracts. Lesions are characterized by "grains" or sclerotia consisting of mycelial aggregates surrounded by an extracellular matrix containing a melanin-like compound. **Phaeohyphomycosis** refers to localized or systemic disease caused by melanized fungal elements, for example, yeasts, pseudohyphae, hyphae, or any combination of these.

Phaeohyphomycoses may be cutaneous, subcutaneous, or systemic, and can incite a range of host inflammatory responses.[295,400,466] Gross lesions vary, ranging from none to hemorrhagic to dark green or black nodules. Skin including fin, skeletal muscle, and kidney are the most frequently affected tissues in ornamental fish, where both localized and systemic presentations are common.[12] Histologic lesions vary in the extent and intensity of host response but are typically necrotizing and granulomatous, with or without multinucleated giant cells.[12] In some species, inflammation may be granulocytic.[361,478] Angioinvasion with associated vascular necrosis and tissue infarction is common (Fig. 39.51).

Phaeohyphomycosis is diagnosed generically using wet mount preparations, impression smears, or histopathology, while fungal identification relies on culture and appropriate discriminatory molecular analysis.[60] Molecular findings should be correlated with fungal morphology to determine clinical and pathologic relevance. Genera of pigmented fungi implicated in fish disease include *Exophiala*, *Neodevriesia*, *Ochroconis/Verruconis*, *Phoma*, and *Veronaea*.[12,122,478] Pigmented fungal genera cannot be reliably differentiated using histopathology alone, as all the fungi demonstrate similar morphologic characteristics in tissue sections with regard to their septation, alignment of hyphal walls, branching pattern, and golden-brown pigmentation.[12]

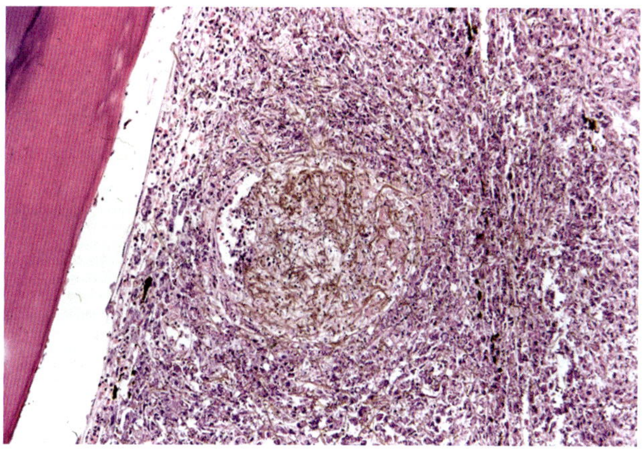

FIGURE 39.51 Phaeohyphomycosis in a seadragon due to *Exophiala* sp. Innumerable, intertwined, slender, melanized fungal hyphae occupy the lumen of a vein and course through the vessel wall and into surrounding necrotic parenchyma of the kidney.

Exophiala spp. are recognized with increasing frequency in aquarium and farmed fish. Their associated disease in bony fish is summarized by Armwood et al.[12], Nyaoke et al.[338], and Seyedmousavi et al.[433] The nonthermotolerant, waterborne *Exophiala* species, for example, *E. salmonis, E. aquamarina,* and *E. pisciphila*, cluster in a genetic clade known as the salmonis-clade. Together they account for most of the recognized *Exophiala* spp. infections of aquatic poikilotherms (e.g., fish, toads, and crabs).[96,433] Phaeohyphomycosis caused by *Exophiala* spp. is relatively common in syngnathids, inducing localized skin lesions or systemic necrotizing lesions with angioinvasion and often skin involvement.[12,338] Phaeohyphomycosis due to *E. pisciphila* in Atlantic salmon is characterized by widespread granulomas and/or hyphal invasion of cranial structures, including the semicircular canals and lateral line, accompanied by granulomatous inflammation.[249,401]

Veronaea botryosa is a significant, emerging cause of systemic phaeohyphomycosis in white and Siberian sturgeon aquaculture.[478] The disease, referred to as "fluid belly" by producers based on the severe coelomic effusion, affects valuable mature animals, causing significant economic losses.[478] Affected animals orient abnormally or are abnormally buoyant, have coelomic distension, exophthalmos, cutaneous erythema, and skin and eye ulcerations.[478] Gross findings include raised to ulcerative dark black-green cutaneous lesions, exophthalmia, yellow coelomic effusions, and organomegaly with nodules.[176] Histologically, multiple organs have extensive granulomatous to pyogranulomatous inflammation, with occasional multinucleated giant cells, phagocytized melanized hyphae, and angioinvasion. Tissue tropisms include the liver and secondary lymphoid organs, such as the spleen and kidneys.[78] Elevated temperatures (≥18°C) used in sturgeon culture to promote rapid growth are conducive to disease development.[78]

Phomamycosis is a phaeohyphomycosis of hatchery-reared fish. The disease is most notable in intensive salmonid aquaculture, especially Chinook salmon, coho salmon, rainbow trout, and ayu.[67,122,190] Mortality is variable in fry and fingerlings (e.g., 2%–67%).[54] The etiologic agent, *Phoma herbarum*, is ubiquitous in the environment and can be isolated from soil, water, and food.[30] A recognized saprophyte and plant pathogen, it can also be pathogenic to fish.[411] Transmission has been postulated to occur by inhalation or ingestion of conidia; however, experimental transmission of *Phoma* sp. by oral or airborne routes has met with varied success.[54,67,411] Phomamycosis is perhaps best recognized for its ruinous effect on salmonid restoration efforts. Minimizing stress in culture conditions and proper storage of food to limit contamination are keys to its control. The primary target organ is the swim bladder, with potential spread into adjacent tissues, angioinvasion and hematogenous dissemination.[122,411] Gross lesions include caseous material and fungal plaques in the swim bladder lumen with possible hemorrhage in the adjacent kidney and hypaxial muscle.[67,122] Microscopically, lesions consist of necrotizing and granulomatous aerocystitis, nephritis, myositis, gastroenteritis, and coelomitis, with necrotizing vasculitis and intravascular, intralesional fungal hyphae. Definitive identification of the fungus can be obtained via culture, ideally followed by multilocus PCR and sequencing.

Microsporidia are highly specialized, obligate intracellular, eukaryotic parasites of a wide range of hosts, including invertebrate and vertebrate, mammalian and nonmammalian, ectothermic and endothermic organisms.[222,223] Taxonomic classification of microsporidia has varied over time, but the most recent molecular phylogenetic data place them within or as a sister group to the fungi, although exact subclassification is still contentious. The most common of the varied taxonomic groups that host microsporidia are fish and insects. Fish are hosts to over 150 species of microsporidia assigned to 14 genera.[268] Several of these microsporidial infections cause ecologically and economically significant infections in fish, e.g., salmon.[222] Given the widespread adoption of the zebrafish in research, perhaps the most significant microsporidial infection of cultured fish, however, is caused by ***Pseudoloma neurophilia***, which has become the focus of control measures and preventive practices in zebrafish laboratory facilities worldwide. *Pseudoloma neurophilia* infects the spinal cord, hindbrain, and ventral nerve roots of zebrafish, with spores also found in other tissues, most notably muscle, kidney, and ovary (Fig. 39.52A).[228,323]

The vast majority of fish microsporidia have a simple, single-host life cycle characterized by infection, merogony, sporogony, spore release, and presumed direct transmission.[268,408] The spore serves as both the dispersal and infective stage of the microsporidial life cycle; it is also the stage most relevant to diagnostics as it is the only stage present outside of the host cell and the only stage that allows species differentiation.[114,223,265] Microsporidial spores have thick cell walls, a cytoplasm (i.e., sporoplasm) containing a nucleus and a coiled, extrudable polar tube, but lack mitochondria, Golgi complex, and peroxisomes. Extrusion of the polar tube occurs in response to environmental stimuli. The polar tube provides a conduit for the passage of the sporoplasm into the host cell cytoplasm, thereby initiating infection while also avoiding interaction with host immune cells or immune molecules.[222,408] Once inside the host cell, the sporoplasm matures into a meront. Development within the host cell occurs by two recognized phases, merogony or the proliferative phase that produces more meronts, and sporogony or the spore development phase during which meronts undergo conversion to spores.[268] Meronts lie in the host cytoplasm, with the exceptions being those of *Nucleospora* that lie in the host nucleoplasm, and grow into multinucleate plasmodia (uninucleate meronts in some species); sporogonial plasmodia arise from meronts. Infected cells may develop at sites of primary infection or be transported to other tissues for development, although the process of transport within the host is not well characterized.[408]

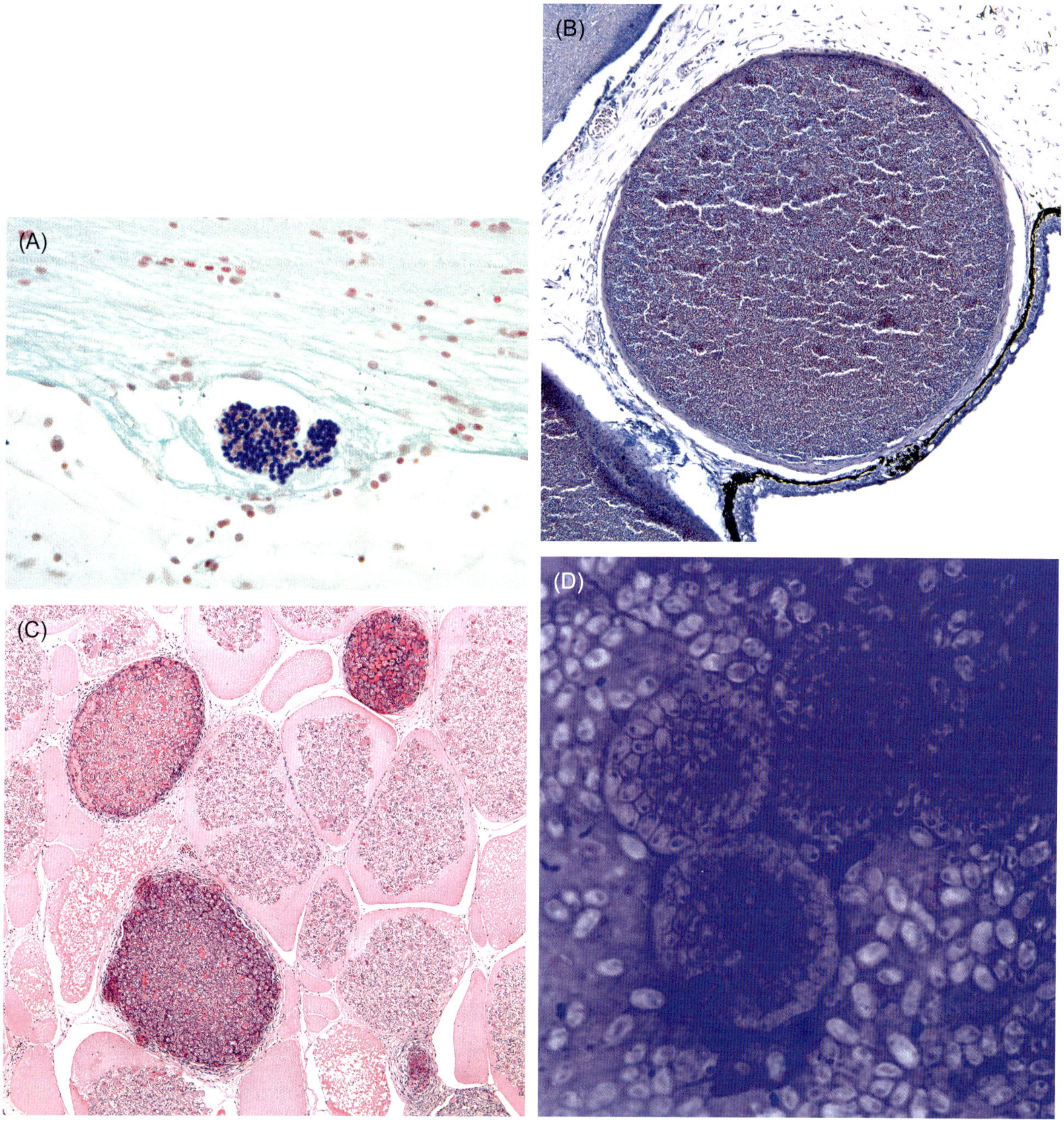

FIGURE 39.52 Microsporidial lesions in multiple species. (A) *Pseudoloma neurophilia* infection of the spinal cord of a zebrafish. There is an aggregate of gram-positive spores located between nerve fibers of the spinal cord. Twort's Gram. (B) Cutaneous microsporidiosis in a seahorse resulting in xenoma formation. The xenoma is surrounded by a thin layer of fibrous connective tissue, slightly compresses the surrounding dermis, and elevates the overlying epidermis. The mature spores of this microsporidian are fuchsinophilic when stained using an acid-fast technique. Ziehl-Neelsen. (C, D) Microsporidial myositis in a shorthorn sculpin. (C) Granulomas, necrotic myofibers, and meronts and spores developing within the sarcoplasm of myofibers are present without hypertrophy of infected myofibers or overt xenoma formation. (D) Microsporidial myositis in a wet mount from a shorthorn sculpin. Sporophorous vesicles, round meronts, and spores are present. *(Photo Courtesy of Dr. Kathryn Tuxbury, New England Aquarium.)*

Lesions associated with microsporidian infections vary by species of microsporidian, infected host cell, anatomic tissue distribution within the host, the temporal phase of the infection, and the host response incited by the microsporidian (eSlide 39.e6). Inoculation of microsporidia into the host cell cytoplasm, or rarely the nucleus, either induces hypertrophy of the infected target cell and its transformation into a xenoma, or it causes the destruction of the cytoplasm resulting in degeneration and necrosis of the cell.[115,268] Microsporidia are therefore categorized by their ability to induce xenoma formation or not.[115,268,408] The xenoma is comprised of an enormously hypertrophied host cell that is replete with microsporidian stages, for example, meronts and spores, and takes on the appearance of a cyst-like structure distinct from the host tissue.[115,408] In xenoma-inducing microsporidians, e.g., ***Glugea* spp.**, in the weakly reactive phase to xenoma development, there is compression of the surrounding tissue, pressure atrophy, and mild fibrosis so that xenomas become surrounded by a thin layer of fibrous connective tissue (Fig. 39.52B).[115] Mature xenomas, i.e., those that continue to enlarge and become filled with spores, elicit granuloma formation. Healing can occur as spores at the centers of granulomas are destroyed by macrophages and granulomas become surrounded by granulation tissue and are eventually replaced by fibrous connective tissue. Infections by microsporidians that do not induce xenomas are characterized by the destruction of host cell cytoplasm and its replacement by developmental stages without cell hypertrophy. This type of reaction is characteristic of ***Pleistophora*** spp. infections, e.g., those that occur in muscle due to *P. hyphessobryconis*.[115] In these infections, development of meronts and spores destroys the sarcoplasm and results in degeneration and necrosis of myofibers. The subsequent release of spores elicits a granulomatous inflammatory response (Fig. 39.52C).[115]

Lesions of mycobacteriosis, lymphocystis, ichthyophonus, metacercarial granulomas, cyst-like myxozoan plasmodia, and neoplastic masses may appear similar to xenomas and xenoma-induced granulomas.[333] Confirmation of microsporidiosis is achieved by the identification of spores in tissue through examination of wet mount preparations (Fig. 39.52D) or histologic sections, which can be stained using acid-fast, Gram, and Luna techniques (Figs. 39.52A, B). Genus- or species-level identifications may in some instances be possible through morphological evaluations of properly prepared preparations of spores by knowledgeable, experienced individuals. PCR assays are available and can provide rapid identification of some fish microsporidians.[268]

Metazoa

Monogenea are in the order Platyhelminthes. They are not trematodes but may incorrectly be referred to as "monogenean trematodes". Monogenea are characterized by the opisthaptor, the posterior holdfast organ. Monogenea consists of two groups, monopisthocotyleans and polyopisthocotyleans. Most, but not all, of the important monogenean parasites of captive-held fish are monopisthocotyleans, with dactylogyrids, gyrodactylids, and capsalids most frequently implicated in severe disease. Pathogenesis is considered related to irritation and tissue damage associated with opisthaptor hooks, as well as loss of epithelium due to consumption by the parasite. Gyrodactylids infect both fresh and marine species, dactylogyrids occur in freshwater species, and capsalids are seen in marine species.[50,333] Cutaneous and branchial infestations are common with few species infecting mucosal surfaces of internal organs, such as esophagus and urinary tract. Capsalids and gyrodactylids commonly colonize epithelium of the body and fins and sometimes the cornea; dactylogyrids and some gyrodactylids are found on gill. Some, especially capsalids, readily infect a variety of host species while others exhibit high host fidelity.[50] Examples of important parasite–host pairs include *Dawestrema cycloancistrium* in arapaima, *Gyrodactylus salaris* in Atlantic salmon, and *Benedenia* and *Neobenedenia* spp. in a large variety of marine species. While monogenean infections of importance can occur in free-living fish (e.g., *G. salaris* in salmon), significant infections most commonly occur in captive-held fish. Infection with *G. salaris* is listed as notifiable by the WOAH (Aquatic Code Online Access—WOAH—World Organisation for Animal Health).

The monogenean life cycle is direct. Most gyrodactylids are viviparous; capsalids and dactylogyrids are oviparous. In the viviparous gyrodactylids, juveniles are fully functional, and autoinfection and direct transfer to new hosts maintain and expand the infection. Oviparous species can deposit large numbers of eggs that can accumulate in closed systems with resultant high numbers of infective larvae.[514] High captive stocking densities coupled with rapid monogenean reproduction rates can lead to heavy parasite burdens across most or all susceptible individuals in an enclosure.

Mild cutaneous lesions associated with monogenean infections may include discoloration, small erosions or ulcerations, and mild inflammation. In heavier infestations, large erosions or ulcers with significant inflammation may be noted. Secondary bacterial, fungal, and/or oomycete infections are possible. In gill, mild to severe lamellar epithelial hyperplasia (Fig. 39.53A) with or without necrosis, inflammation, and secondary infections can be observed. Mild infections may be subclinical, but detection of even low numbers on an individual fish may represent a significant group health concern. Loss of display value, generalized debilitation, osmoregulatory imbalance, respiratory distress, secondary infections, and death are all possible with severe infestations.

While few species are large enough to be seen grossly and organisms can be detected histologically, evaluation of skin scrapes and gill wet mounts on fresh samples are of primary diagnostic utility (Fig. 39.53B). Identification to group can often be accomplished using morphologic criteria (e.g., holdfast characteristics, presence of eyespots). Submission of specimens for morphologic and/or molecular characterization by a parasitologist may also be desirable.

The **Myxozoa** are metazoan parasites of invertebrates and vertebrates. With more than 2,000 recognized species,

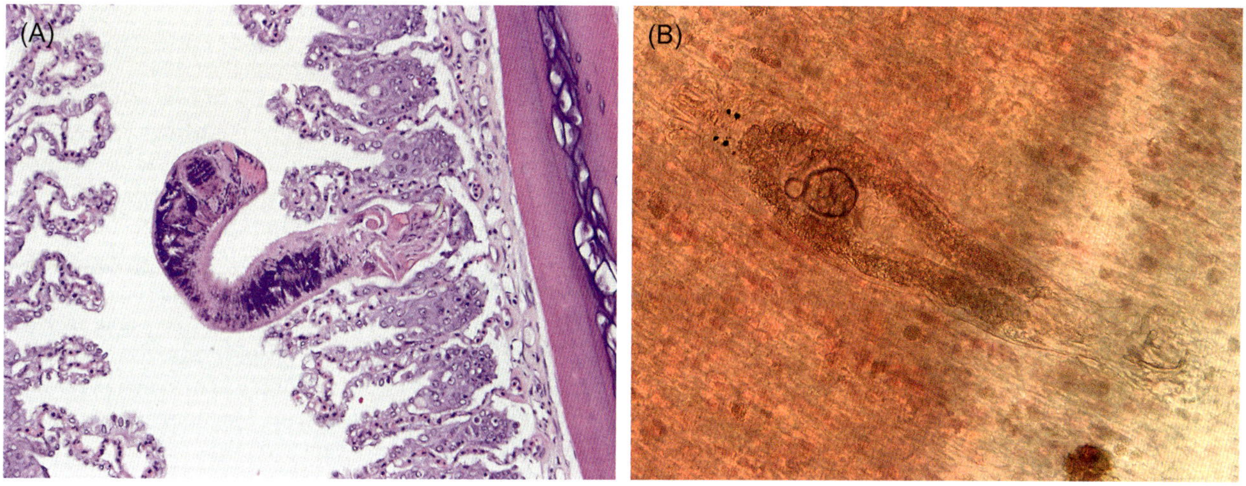

FIGURE 39.53 (A) Monogenean infestation in the gill of a white bass. A monogenean attaches at the interlamellar sulcus by its haptor, which has hamuli, i.e., hooks or anchors. (B) Monogenean infestation in a wet mount of a gill clip from a koi. A monogenean is present between the lamellae.

the Myxozoa not only represent perhaps the most extensive assemblage of metazoan fish parasites that exists but also include ecologically and economically impactful fish parasites, for example, *Myxobolus cerebralis*, *Sphaerospora* spp., *Kudoa thyrsites*, *Ceratomyxa* spp., *Tetracapsuloides bryosalmonae*, and *Henneguya ictaluri* (Table 39.e3; Figure for Table 39.e3).[226,267] Myxozoa are characterized by multicellular spores that form when 1–7 constituent cells come together to form outer shell valves that enclose one to many amoeboid germ cells (sporoplasms) and one to multiple nematocyst-like polar capsules; the polar capsules house an extrudable filament that allows attachment to host target cell(s) for subsequent infection by the myxozoan sporoplasm.[265–267] The multicellular nature of myxozoan spores with specialization of constituent cells for particular functions corroborates molecular evidence of the metazoan nature of the Myxozoa.[61,443]

Since elucidation of the life cycle of *M. cerebralis*, the causative agent of **whirling disease of salmonids**, it is now generally accepted that the life cycle of the Myxozoa involves alternation between a vertebrate and an invertebrate host with sporogony occurring in each.[226,267,529] The phylum Myxozoa contains two classes, the Myxosporea and the Malacosporea. In the Myxosporea, the myxospore phase takes place typically in fish, with the rare exceptions that amphibians and reptiles (or perhaps birds or mammals) may serve as the vertebrate host, and produce a myxospore that is infective to the invertebrate host; the invertebrate host is an annelid, rarely a sipunculid, and supports the actinospore phase, which involves a sexual process and generation of the actinospore that is infective to the vertebrate, e.g., fish, host.[267] This complex life cycle with bisporogony and alternation between a vertebrate and an invertebrate host becomes vitally important when considering the infective or amplification potential of myxozoan infections in aquarium settings. Although the spores of Myxozoa are morphologically complex, the vegetative, trophic stages within hosts are much simplified, consisting of syncytial plasmodia and single cells.[61,265–267] Myxozoan infections of fish can be categorized as coelozoic, wherein vegetative stages occur in body or organ cavities, e.g., the lumen of the gall bladder or lumina of bile ducts or renal tubules, or they can be histozoic, wherein vegetative stages occur in tissues either intercellularly or intracellularly.

Most myxozoans cause little or no inflammatory reaction in fish; this is true of the many coelozoic myxozoans, in which vegetative stages and spores are restricted to luminal spaces (Fig. 39.54A).[449] Some, however, do incite severe chronic inflammation, the most common morphologic expression of which is granuloma formation. Others are associated with lytic changes of host tissues, such as cartilage (Figs. 39.54B, C).[66,449,493] The presence of vegetative stages of histozoic myxozoans can also be associated with pressure atrophy of host tissues as a consequence of their expansion.

Microscopic examination of histologic sections with observation of vegetative stages can provide a simple and straightforward diagnosis of myxozoan infection, which can be augmented by the use of special stains (eSlides 39.e9 and 39.e10).[481] However, identification of the specific myxozoan requires detailed morphologic examination of myxospores and/or ancillary diagnostic testing.[226,267] Depending on the myxozoan, antibody-based and lectin-based assays have been developed; however, the inherent limitation of these techniques in the detection and diagnosis of myxozoan infections is that these may exhibit binding affinities to particular developmental stages.[226] PCR and in situ hybridization avoid this limitation and have been developed to detect several of the economically important myxozoa.[226]

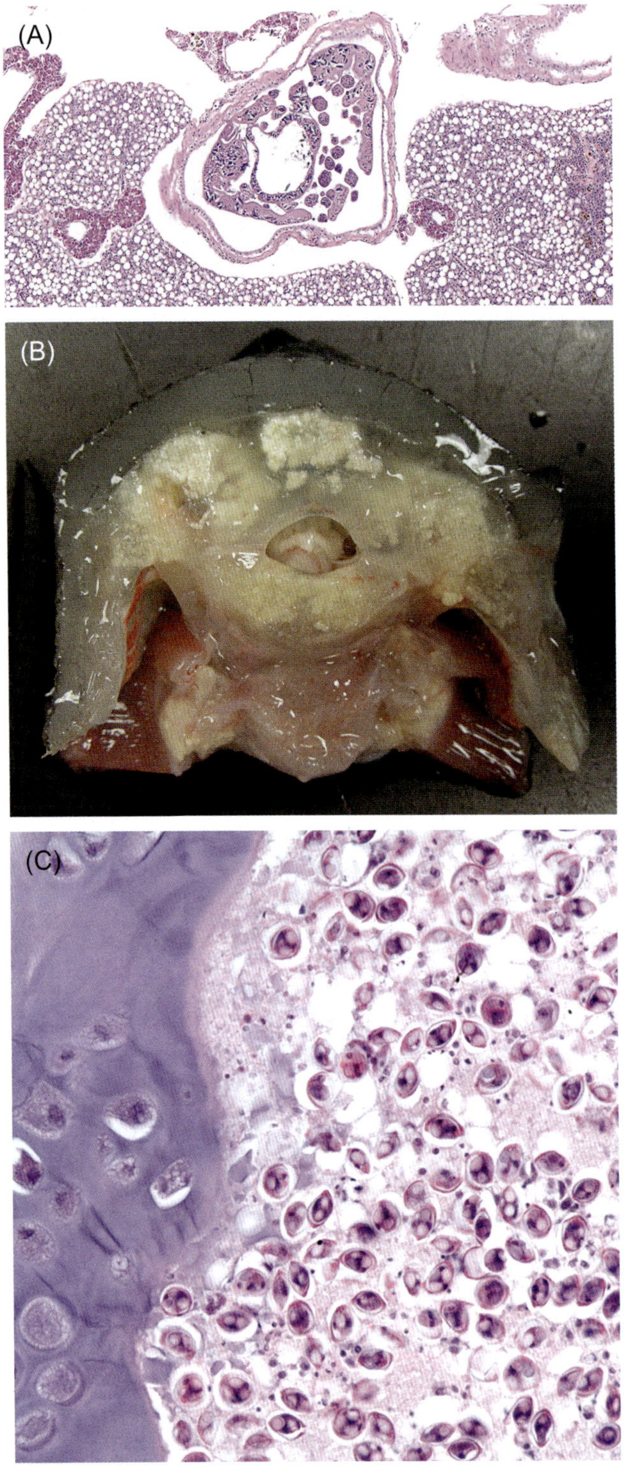

FIGURE 39.54 Coelozoic myxozoan in the bile duct of a lined seahorse. (A) Developmental stages of a coelozoic myxozoan are present in the lumen of a bile duct and include smaller, round, 5–10 μm in diameter stages containing multiple cells and larger, ribbon-like plasmodia containing multiple round cells and basophilic bar-shaped spores. Myxozoanosis in a lumpfish due to *Myxobolus albi*. (B) Multiple, expansile, myxozoan pseudocysts (plasmodia) with *white* granular contents are located in the cranial cartilage. (C) At the margin of a myxozoan plasmodium in the scleral cartilage, there is chondrolysis without an apparent granulomatous inflammatory infiltrate. Numerous spores and other developmental stages are present within the plasmodium. *(Part B: Photo Courtesy of University of Illinois Zoological Pathology Program.)*

Protozoa

Ichthyophthirius multifiliis (**Ich**) and *Cryptocaryon irritans* (**marine Ich**) are large ciliated protozoa that parasitize the cutaneous, oral, and branchial epithelium in freshwater and marine fish, respectively. *Ichthyophthirius multifiliis* can be up to 1,000 μm in diameter; *Cryptocaryon irritans* is somewhat smaller, up to 450 μm in diameter.[265] Both are widespread, warm water protists that have a similar life cycle with a parasitic stage (trophont) and an environmental amplifying stage (tomont). Reproduction can be rapid, and high stocking densities typical in display aquaria can support disease at epizootic levels. Recovery from infection can result in limited (cryptocaryonosis) to long-term (ichthyophthiriosis) immunity.[53,99] Cryptocaryonosis can be endemic in systems and cause epizootics when numbers of susceptible individuals rise due to loss of immunity, additions of naïve stock, and immunomodulation due to stressors. Grossly, discrete, raised, white lesions in gill, oral cavity, and/or skin may be evident. These correspond histologically to embedded, intraepithelial organisms that may be associated with epithelial erosion or ulceration as well as epithelial hyperplasia. Gross lesions, if evident, are indicative of infection, but the presence of trophonts with characteristic nuclear morphology should be confirmed in wet mounts or histologic section for definitive diagnosis (Figs. 39.55A, B; eSlide 39.e11). Individuals with partial immunity may develop self-limiting disease. Otherwise, any untreated infection should be considered as potentially life-threatening, and heavy infection at initial diagnosis carries a guarded to grave prognosis.

Trichodiniosis refers to disease caused by peritrichal ciliated protozoans of the family Trichodinidae Raabe, 1959, which includes the genera *Trichodina*, *Hemitrichodina*, *Paratrichodina*, *Tripartiella*, *Trichodinella*, *Dipartiella*, and *Vauchomia*.[205,264] Disease occurs worldwide in freshwater and saltwater environments, is essentially unrestricted in host range, and affects all cultured and feral fish.[264] Ciliates of the family Trichodinidae have a characteristic basal attachment disc possessing a ring of chitinoid denticles that mediates firm attachment to epidermal or branchial epithelial surfaces.[205,264] The oral end of trichodinids is directed away from the host epithelium, and the ciliate feeds on organic matter and bacteria from the water column and sloughed cells and cellular debris from the surface of the skin. When environments are healthy, trichodinids exist in low numbers on fish and can be considered commensal; however, when environments are stressed, trichodinids occur in high numbers on fish and can be pathogenic.[264] Transmission occurs directly between hosts.[264] Lesions typically occur on the gill and skin, where there can be excessive mucus production leading to a bluish film in severe infestations.[264,293] Hyperplasia of the epidermis and of branchial lamellar epithelium may compromise respiration and osmotic balance; severe and extensive erosion of branchial lamellar and filamental epithelium and destruction of lamellae with hemorrhage may be present.[264,293,307] Diagnosis involves cytological or histological examination of

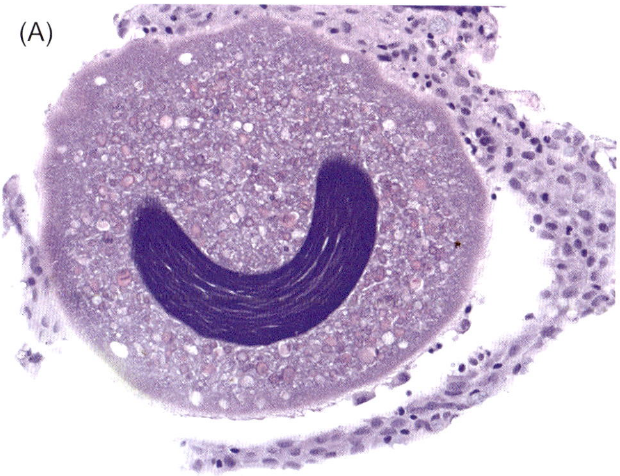

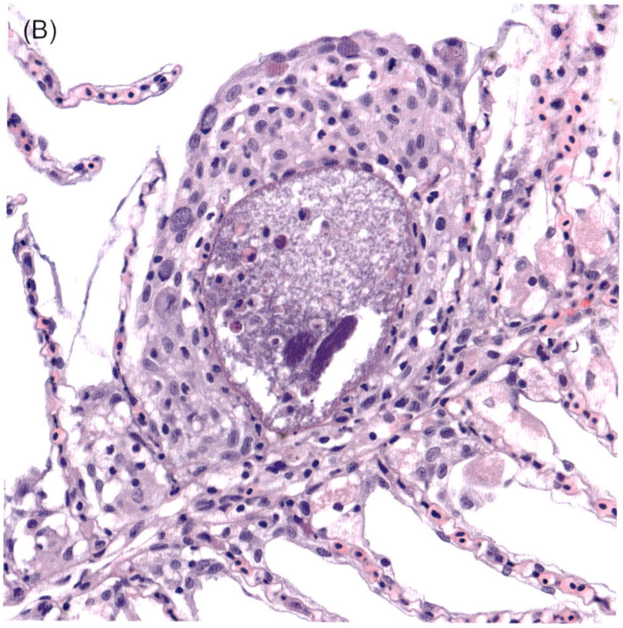

FIGURE 39.55 Freshwater and marine Ich. (A) Ichthyophthiriosis in a freshwater fish. The trophont of *Ichthyophthirius multifiliis* infection is very large, intraepithelial, and characterized by a large C-shaped macronucleus and cilia coating the entire organism. Trophonts of *I. multifiliis* can measure up to 1 mm in diameter. (B) Cryptocaryonosis in the gill of a marine teleost. A trophont of *Cryptocaryon irritans* is located within hyperplastic epithelium in a focus of lamellar fusion. The trophont of *C. irritans* has a segmented macronucleus with four lobes; two lobes of the segmented macronucleus are visible within the cytoplasm of the trophont in this image (see eSlide 39.e11).

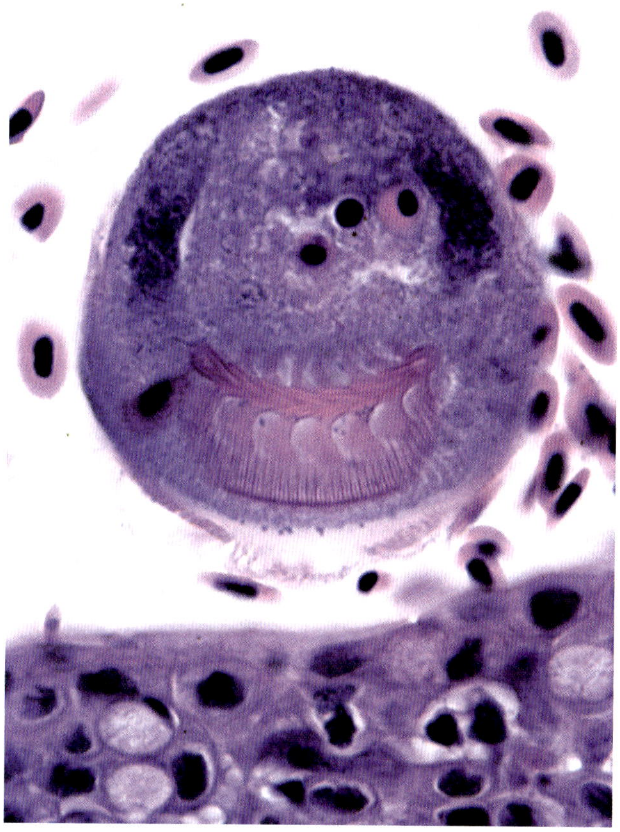

FIGURE 39.56 Trichodiniosis in a teleost. Trichodinid ciliates possess a ring of denticles that are brightly eosinophilic in histologic sections stained with hematoxylin and eosin. Typically, differentiation of trichodinids to genus is not diagnostically relevant because the same, or very similar, treatment approaches are used for all the trichodinids (see eSlide 39.e12).

affected areas of skin and gill leading to identification of ciliates possessing the characteristic basal attachment disc with a ring of denticles (Fig. 39.56; eSlide 39.e12).

Scuticociliatosis is the generic term applied to disease of marine fish, crustaceans, and mollusks caused by histophagous ciliated parasites (subclass Scuticociliatia), the taxonomy of which has historically been confused and complicated.[275] The best known are ***Uronema marinum, U. nigricans,*** and ***Philasterides dicentrarchi*** (syn. ***Miamiensis avidis***).[458] Common in aquarium species, particularly syngnathids, disease is problematic in cultured fish, such as black sea bass, sea bream, and several pleuronectids.[212,485] *U. nigricans* causes severe central nervous system disease in cultured bluefin tuna.[320] Disease occurs over a broad range of salinities (15–35 ppt) and water temperatures (8–28°C), and it is strongly associated with environmental stress.[70,87] Once established, there is potential for extensive systemic invasion. *Uronema marinum* and *P. dicentrarchi* are ovoid, holotrichous ciliates, 32–50 μm and 40–50 μm in length, respectively, with longitudinal rows of kineties and a single long caudal cilium. The anterior pole of *U. marinum* is flattened, while that of *P. dicentrarchi* is tapering.[265] Scuticociliate-induced skin lesions begin as areas of epidermal pallor, followed by sloughing and ulceration. Invasion of skeletal muscle and widespread systemic tissue dissemination may then occur. Despite the presence of large numbers of ciliates, inflammatory responses are often minimal. When tissue responses do occur, lesions are primarily necrotizing, hemorrhagic, and may have pronounced edema. In tissue sections, the organisms are round, oval, or pyriform, with a large, round, central to slightly eccentric macronucleus and abundant, foamy, basophilic cytoplasm frequently contain ingested host erythrocytes (Figs. 39.57A, B). Organisms are readily recognized in wet mount preparations of lesions and histologic sections (Fig. 39.57C). However, speciation based on morphologic features is difficult.

Two groups of gill amoeba are responsible for branchial ectoparasitism in salmonids. The most economically important are infections caused by *Neoparamoeba* spp. These affect Atlantic salmon, other salmon species, and rainbow trout reared in saltwater, and infections have been termed "**amoebic gill disease**" (**AGD**). Of the three amoeba species implicated in this disease, a recent study suggests that *N. perurans* may be more relevant than *N. branchiphila* or *N. pemaquidensis*.[539] The other syndrome, "**nodular gill disease**" (**NGD**), occurs primarily in rainbow trout from freshwater habitats and is associated with various amoeba of the family Cochliopodidae.[118] AGD has a worldwide distribution and causes recurrent outbreaks in seawater-farmed salmonids. NGD appears to be a much more sporadic disease of wild and captive trout.

The macroscopic and microscopic appearance of amoebic infections is remarkably similar. Fish present with rapid respiration and mortality, and gills have poorly defined, multifocal to coalescing, gray-white, flattened nodular areas. Microscopically, gill lesions are characterized by severe lamellar epithelial hyperplasia that leads to segmental lamellar fusion (Fig. 39.58). Some areas of gill may be markedly thickened, while neighboring portions of the same filament are minimally affected. In addition to a modest inflammatory response, other frequent findings include lamellar capillary thrombosis and lamellar adhesions (synechiae), the latter of which may form interlamellar pseudocystic spaces in which amoebic trophozoites can be found. The other location at which trophozoites are most frequently observed is along the distal surface of fused lamellar tips where they can be mistaken for exfoliated pavement cells;

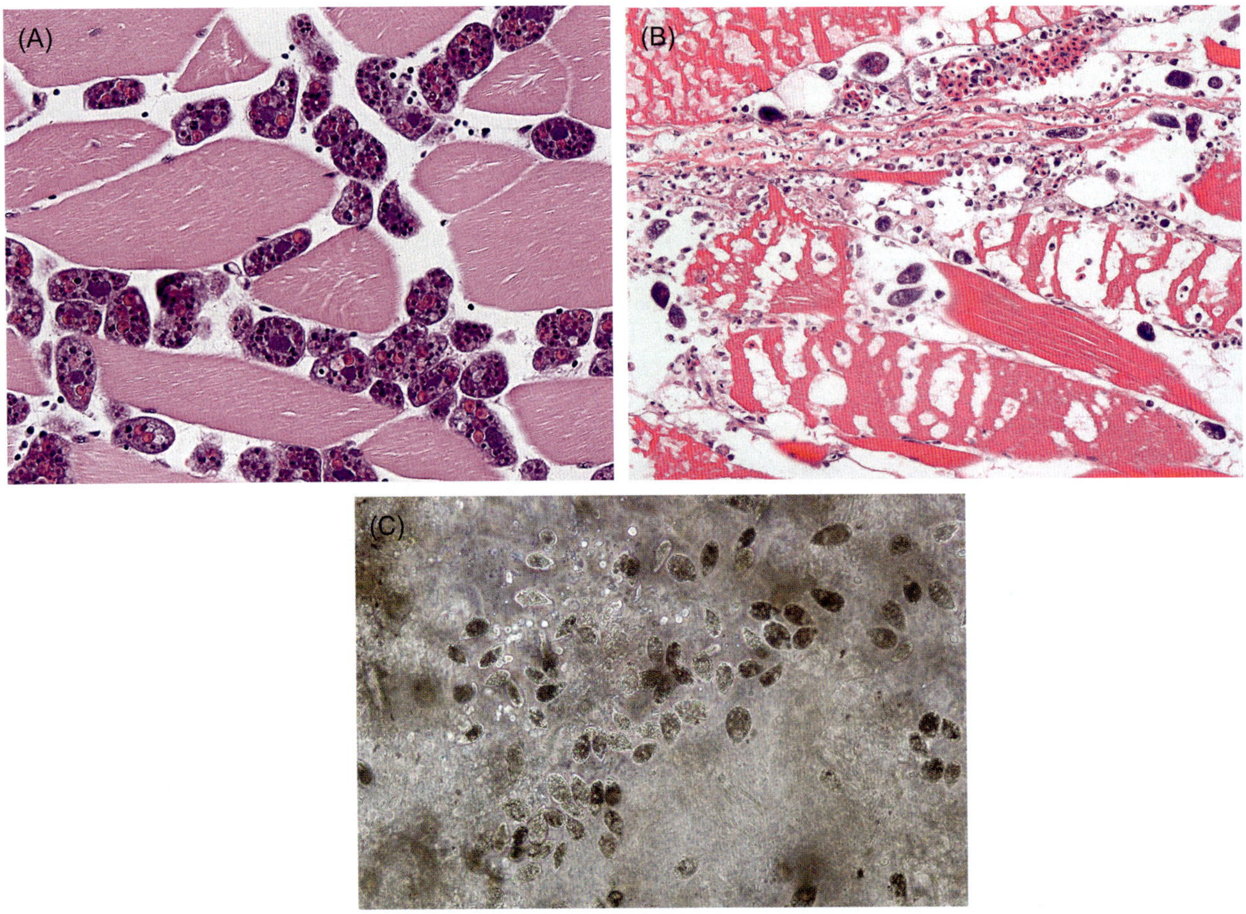

FIGURE 39.57 Scuticociliatosis in teleost fish. (A) Scuticociliatosis in a dispar anthias. Large numbers of ciliates with abundant cytoplasmic granules, ingested erythrocytes, and ovoid macronuclei are present within endomysium of skeletal muscle. Note the lack of necrosis and inflammatory response and compare to (B). (B) Scuticociliatosis in skeletal muscle from a teleost. Muscle fibers in this section are necrotic and an inflammatory cell infiltrate composed predominantly of macrophages is present in the interstitium, where there is congestion of blood vessels and edema. Scuticociliates are present in the interstitium and in a necrotic muscle fiber. (C) Scuticociliatosis in a wet mount from a teleost. There are numerous, pyriform ciliates in the field. *(Part B: Photo Courtesy of University of Illinois Zoological Pathology Program. Part C: Photo Courtesy of University of Illinois Zoological Pathology Program.)*

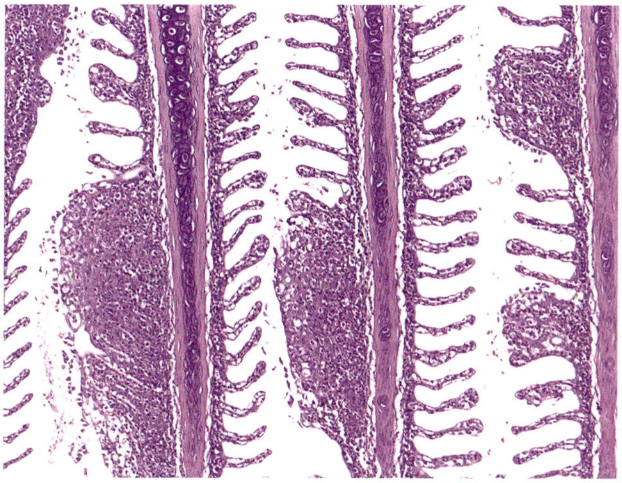

FIGURE 39.58 Amoebiasis in the gill of a rainbow trout. The reason for the common name of "nodular gill disease" is obvious here. This host response is one of multifocal epithelial hyperplasia with lamellar fusion. Cochliopodid amoebae line themselves along the apical surfaces of hyperplastic lamellar epithelium, where the organisms can easily be mistaken for exfoliated cells. *(Photo Courtesy of Dr. Jeffrey C. Wolf, EPL, Inc.)*

fewer organisms are evident in less affected gill areas. Diagnosis is usually made by observation of amoebae in wet mounts of gill biopsies or in histologic sections. Although they can be detected in hematoxylin and eosin-stained sections, visualization of trophozoites may be enhanced by Giemsa staining. In situ hybridization or PCR can be used to determine the precise etiologic agent, and PCR has also been advocated as a monitoring tool.[108]

Ichthyobodo spp. (currently referred to as *Ichthyobodo necator*, and previously as *Costia necatrix*) are small flagellates that parasitize the cutaneous and branchial epithelium of freshwater and marine teleosts, and species of octopus (see Chapter 41: Invertebrates). Rather than a solitary species, these have been shown to be a group of closely related but distinct species that colonize different hosts.[487] *Ichthyobodo* spp. are approximately 10 × 5 μm, round to pyriform, with two flagella.[265] Organisms can be free swimming or attached to the host epithelium. Organisms are more characteristically pyriform and flagella inapparent when attached to the host.

Mild *Ichthyobodo* cutaneous lesions may be limited to discoloration and/or excess mucus production. Epithelial hyperplasia and degeneration can also be evident. In the gills, lamellar epithelial hyperplasia and synechiae are common. Synechiae at lamellar tips with creation of pseudocystic spaces is also common, and the presence of this lesion heightens suspicion for ichthyobodosis (Figs. 39.59A, B).[469] In overwhelming infections, mortalities may occur with little lesion development. Heavy *Ichthyobodo* infections can lead to high morbidity and rapid mortality and may occur with little lesion development. Nonfatal chronic infection can result in respiratory compromise, associated generalized debilitation, and secondary infection.

Cryptobia spp. are flagellate protozoa closely related to trypanosomes. Like trypanosomes *Cryptobia* have two flagella, one of which is recurrent with an incomplete, undulant membrane (those of trypanosomes are complete).[532] Cryptobia can be free-living commensals in the aquatic environment but are also external, gastrointestinal, and hematozoic parasites in teleosts. Hematozoic parasites utilize leeches as vectors and are uncommon in display aquaria. Several species of *Cryptobia,* especially *C. branchialis*, are known to colonize the skin and gill of freshwater and marine teleosts. Of the known species that parasitize the gastrointestinal tract, the most important in display and aquaculture situations is *C. iubilans*, which affects a variety of cichlids and can cause high morbidity and mortality in cichlids both in display and aquaculture situations. In discus, severity of disease is reportedly influenced by factors such as

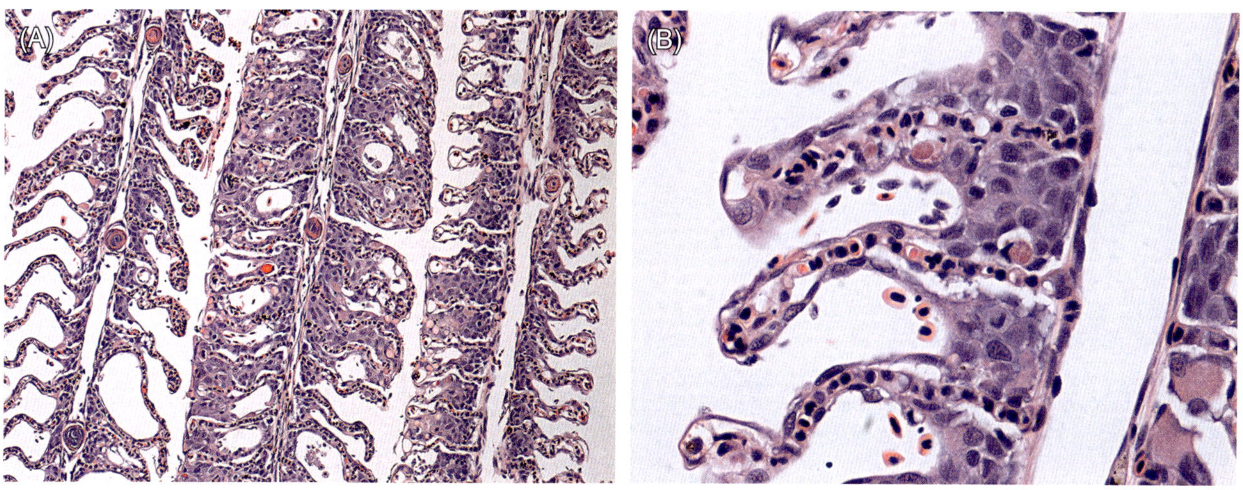

FIGURE 39.59 Branchial ichthyobodosis in a teleost. (A) This section of gill has areas of lamellar epithelial hyperplasia with fusion and frequent lamellar synechiae leading to the formation of pseudocystic spaces commonly observed in *Ichthyobodo* infections. (B) Pyriform ~10 μm long flagellates are attached to the lamellar epithelium.

age, concurrent infection, and water quality.[538] There also appears to be some variation in host resistance.[165] *Cryptobia branchialis* infections cause similar, though milder, branchial changes as do *Ichthyobodo* spp.

Cryptobia iubilans infection causes variably severe granulomatous and necrotizing gastritis (Fig. 39.60A). Inflammation can be extensive, and grossly large transmural defects together with intracoelomic gastric contents may be observed. Gastritis can be accompanied by significant hemorrhage and anemia, which may be evidenced by pale gills. Dissemination of infection to kidney, liver, and spleen may occur in severe cases, resulting in similar granulomatous inflammation. In chronic and resolving cases, granulomas and fibrosis may be all that remains, though the distribution of lesions may still implicate *C. iubilans*.

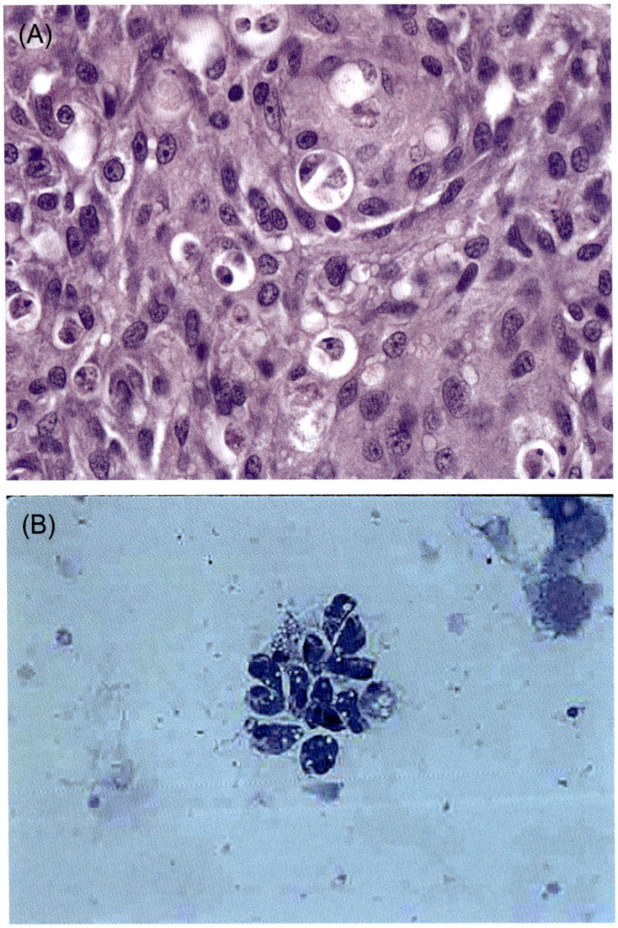

FIGURE 39.60 (A) Cryptobiosis in the stomach of a cichlid with *Cryptobia iubilans* infection. Submucosal granulomatous inflammation is present with scattered individual or clustered, 10–20 μm, pyriform to crescentic flagellates, with a small basophilic nucleus and anterior kinetoplast. (B) Cryptobiosis in a squash preparation of kidney from a cichlid with disseminated *Cryptobia iubilans* infection. Clusters of typical flagellated organisms are present. Such clusters can be readily observed in cytologic preparations of tissues from fish with active infections. Giemsa.

Differentiation of *Ichthyobodo* spp. and *C. branchialis* is reliant upon visualizing morphologically compatible organisms in wet mounts or histologic sections. Noga[333] describes the motility of *Ichthyobodo* as "flickering" and that of *Cryptobia* species as "flowing." Additionally, when attached to the gill, *Cryptobia* are not closely associated with the epithelium and are round to oval, while *Ichthyobodo* are tightly attached and have a pyriform shape. As for most external parasitic diseases, fresh specimens are essential for an accurate diagnosis, as organisms may leave the host upon death.

Definitive diagnosis of *C. iubilans* is reliant on visualizing intralesional organisms cytologically or histologically. Giemsa-stained squash preparations of granulomatous lesions are useful for rapid diagnosis, and organisms often are easier to recognize cytologically than histologically (Fig. 39.60B).

Other Infectious Organisms

Oomycetes

Oomycetes are eukaryotic microorganisms that act as parasites or saprophytes in aquatic environments.[375] They are collectively referred to as **water molds**. Oomycetes are not true fungi and are classified as Stramenopiles (Heterokonts), which includes the golden-brown algae and diatoms.[17] Oomycetes are separated into three subclasses: Saprolegniomycetidae, Rhipidiomycetidae, and Peronosporomycetidae. The Saprolegniomycetidae contains two orders, the Saprolegniales and the Leptomitales. Within the Saprolegniales are the genera *Saprolegnia*, *Branchiomyces*, *Achlya*, and *Aphanomyces*, to which several of the well-known fish and shellfish pathogens are assigned.[92,499]

Saprolegnia parasitica is one of the most destructive of the oomycete fish pathogens. It infects wild and cultured freshwater fish, is endemic to freshwater habitats over the world, is responsible for substantial economic losses in salmon and catfish aquaculture, and has been implicated in population declines of wild salmonids.[499] *S. parasitica* is a serious pathogen of eggs maintained in culture, e.g., cultured salmon eggs, where dead eggs promote colonization, mycelial growth results in more egg deaths, and the cycle becomes self-perpetuating.[499]

Gross lesions of saprolegniasis in fish are characterized by focal or focally extensive, slightly raised, white to gray, cotton-like plaques occurring on the external surfaces of the body including the fins and/or gills and opercula (Fig. 39.61A). Microscopically, lesions consist of focal to focally extensive epidermal necrosis, erosion, and ulceration with hyphal invasion of the dermis and occasionally the underlying necrotic muscle (Fig. 39.61B). In the gills, hyphae cover filaments and lamellae, which are necrotic and hyperplastic.[49] Inflammation is minimal and often

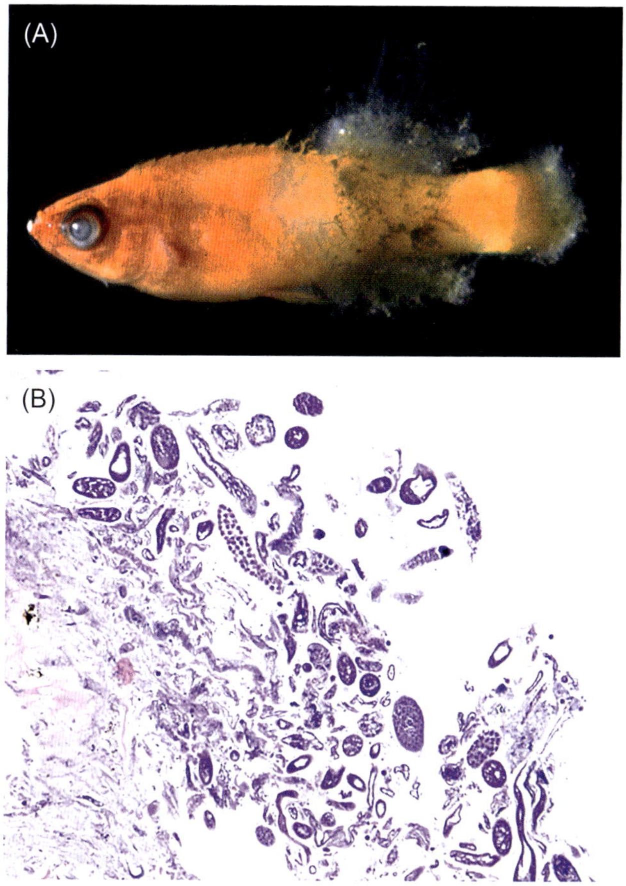

FIGURE 39.61 (A) Saprolegniasis in a medaka. A locally extensive, white, cotton-like plaque involves the skin of the caudal half of the body. (B) Saprolegniasis in a teleost. A mat of filamentous hyphae and terminal, elongate, club-shaped zoosporangia is present along a margin of necrotic tissue. Zoosporangia contain round, biflagellated zoospores. The thick-walled, nonseptate hyphae typically do not invade viable tissue and elicit minimal inflammation.

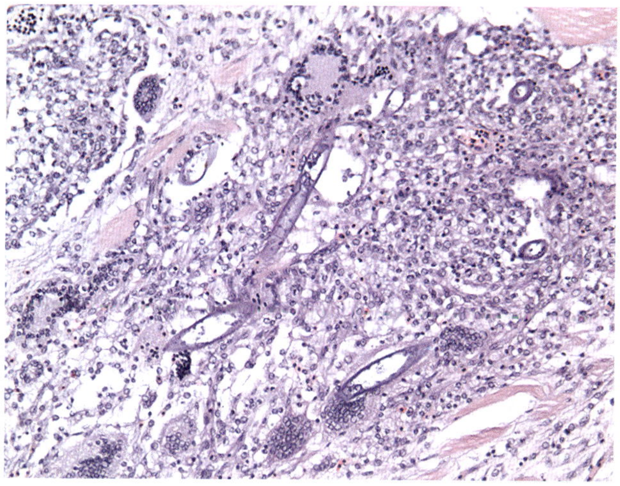

FIGURE 39.62 Aphanomycosis (epizootic ulcerative syndrome) in a channel catfish due to *Aphanomyces invadans*. Unlike saprolegniasis, hyphae of this oomycete are highly invasive and induce intense granulomatous inflammatory reactions, including multinucleated giant cells.

absent unless lesions are complicated by bacteria.[376,499] On wet mount preparations, hyphae appear nonseptate, multinucleate, variably branching, and transparent.[49] Injury to the skin or gill that compromises its innate barrier function predisposes those tissues to infection by *Saprolegnia* spp., which often cause secondary infections but can act as primary pathogens. Death is a consequence of hemodilution and osmoregulatory failure.[376,499]

Aphanomyces invadans is recognized as the agent of **epizootic ulcerative syndrome (EUS)**, a WOAH-listed notifiable disease of fish (Aquatic Code Online Access—WOAH—World Organisation for Animal Health). The disease is also known as **red spot disease (RSD)**, **mycotic granulomatosis (MG)**, and **ulcerative mycosis (UM)**. *Aphanomyces invadans* causes disease in wild and farmed, freshwater and estuarine fish worldwide and has been reported in 94 fish species.[132,347,405] Disease occurs seasonally at water temperatures of 18–22°C and often after heavy rainfall. In juvenile and young adult fish, morbidity and mortality are high (>50%). Infections of eggs, fry, and larvae have not been reported. Genetic analyses have demonstrated that all strains of *A. invadans* are identical regardless of fish host or geographic source.[405] Lesions attributed to the toxin of *Pfiesteria piscicida* in populations of estuarine fish were shown to be lesions of aphanomycosis.[28]

In the early stages, gross EUS lesions consist of red spots on the skin of the head, operculum, body, and caudal peduncle. Lesions progress to focally more extensive, shallow red to gray skin ulcers that may have brown, necrotic centers and are associated with waves of acute mortality. Lesions in some fish, e.g., menhaden, progress to circumscribed, deep ulcers, most often in the ventral perianal region, that penetrate the coelomic cavity. In fish that typically take longer to die, such as snakeheads, skin ulcers can be regionally extensive, necrotizing, and ulcerative and involve the caudal portion of the body or the head, exposing the brain. The predominant histopathologic pattern in EUS is one of extensive ulceration overlying deeply penetrating myonecrosis. Microscopically, early lesions consist of erythematous dermatitis, which progresses to produce severe, locally extensive, necrotizing, and granulomatous dermatitis and myositis.[28,347,405] Hyphae of *A. invadans* penetrate the dermis and extend deeply into skeletal muscle (Fig. 39.62). PCR assays for the identification of *A. invadans* are available.[347]

Branchiomyces spp. are responsible for **branchiomycosis**, a.k.a. **gill rot**, a disease of cultured and wild freshwater fish from Europe, Taiwan, Egypt, India, Israel, Brazil, and the United States.[304,333] The disease has been reported in fish of 17 families, which include the Anguillidae, Salmonidae, Ictaluridae, Cyprinidae Centrachidae, Cichlidae, and Percidae.[333] It is considered a major problem in European aquaculture.[304] Two species are recognized, that is, *B. sanguinis* and *B. demigrans*; each affects different fish and has different geographic distributions.[333] Disease occurs at water temperatures typically exceeding 20°C, develops rapidly, is associated with high mortality (e.g., 50% or 100%), and can be epizootic. The course of disease can be exacerbated by crowding, algal blooms, or increased levels of unionized ammonia.[232,304]

Lesions observed in **branchiomycosis** occur in the gill and consist of regions of infarctive necrosis resulting from invasion of branchial blood vessels. Microscopically, partially to completely occlusive intravascular hyphae in branchial blood vessels, thrombosis, vascular necrosis, and regional necrosis of the gills are accompanied by hyperplasia of lamellar epithelium, lamellar fusion, and telangiectasis (Fig. 39.63). Granulomatous reaction to hyphae has been reported in channel catfish.[232] Hyphae of the *Branchiomyces* spp. are branched and nonseptate, with *B. sanguinis* reported to be 8–30 µm in diameter and those of *B. demigrans* larger.[405]

Diagnosis of oomycete infections is by microscopic identification of hyphae in wet mount preparations and examination of histologic tissue sections. The oomycetes can be isolated in culture, but culture can be difficult when lesions are heavily coinfected with bacteria.

Algae

Oodiniosis is caused by five genera of dinoflagellates highly modified to a parasitic lifestyle. The most significant are *Amyloodinium ocellatum* (**amyloodiniosis and marine velvet disease**) in brackish and marine environments and *Piscinoodinium* spp. (**velvet or rust disease**) in freshwater. *Piscinoodinium limneticum* occurs in North America, while *P. pillulare* is reported from Europe and Asia.[435] The parasites feed as sessile trophozoites (trophonts) on gill, skin, and pharyngeal epithelia. With structural variations, both genera have pedunculated attachment discs equipped with filiform feeding structures that embed in and damage host cells. Trophozoites detach to become cyst-like tomonts that undergo asexual reproduction and release hundreds of biflagellated dinospores that attach to a host and begin the cycle again.[265] *Amyloodinium ocellatum* does not appear to negatively impact wild fish, but it affects hundreds of marine teleosts, as well as elasmobranchs, and it is one of the most important pathogens to affect global mariculture. It is also problematic in public and home aquariums, as it is capable of producing mortalities of 100% within days.[244,252,365] *Piscinoodinium* spp. are less pathogenic, affecting tropical and temperate freshwater fish, primarily in aquaria and less commonly in aquaculture facilities.[334]

Gill is the most important predilection site, with less frequent skin and occasional corneal involvement. Heavy parasitism with *Piscinoodinium* spp. gives skin the gross appearance of dusting by fine grains of sand and a rusty hue.[334] Mucus production is excessive, with darkening and extensive erosion.[135] Histologic changes include variable degrees of epithelial hyperplasia, inflammation, hemorrhage, and necrosis in association with trophozoites, which are attached by a short stalk or peduncle with associated rhizoids (Fig. 39.64). In the gill, lamellar synechia and fusion may progress to hyperplastic clubbing of entire filaments. The ovoid, flask- or sac-like parasites are present in the interlamellar and interfilamental spaces. Their abundant basophilic cytoplasm contains numerous vacuoles

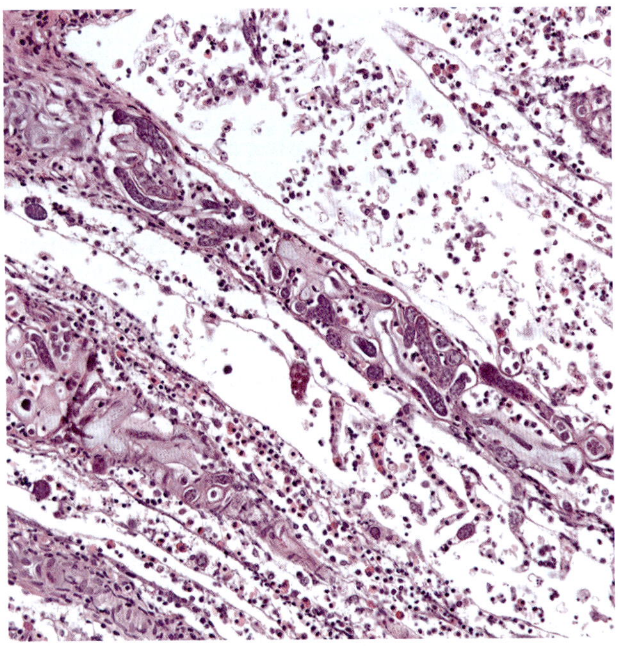

FIGURE 39.63 Branchiomycosis in the gill of an alewife. Hyphae invade blood vessels and the interstitium along the lengths of filaments. *(Case material provided by Dr. Rodman Getchell, Cornell University.)*

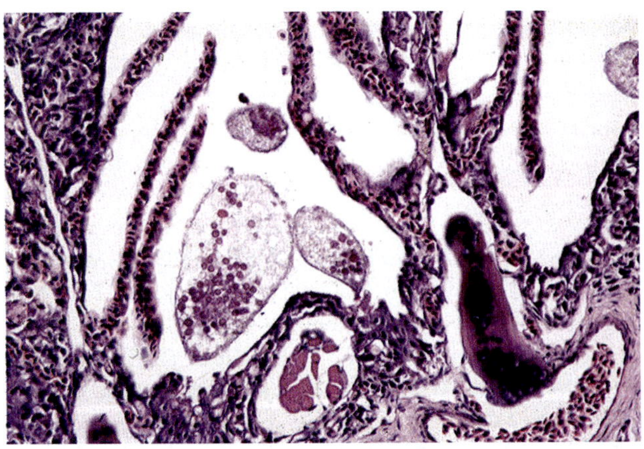

FIGURE 39.64 Oodiniosis in the gill of a marine teleost. Trophonts are located at the interfilamental sulcus. Extending toward the surface of the epithelium is the rhizoid of one trophont. There is little or no inflammatory cell infiltrate in the branchial epithelium, which often can be the case.

and refractile granules. Macronuclei are large, up to 40 μm, and approximately centrally located, with conspicuously condensed chromosomes.[135,365,435] Trophozoites of *A. ocellatum* are 150–350 μm and pyriform to ovoid, while those of *Piscinoodinium* spp. are pyriform to sac-like, yellowish-brown, and up to 160 μm.[265] In both genera, the cytoplasm contains starch granules and digestive vacuoles. *Piscinoodinium* spp. also have peripheral cytoplasmic chloroplasts. Diagnosis is achieved by identification of the parasites in wet mounts of gill biopsies and skin scrapes or in histological sections. Trophozoites detach shortly after the death of the host and stain dark brown in fresh preparations stained with Lugol's iodine.[334]

Mesomycetozoea

Dermocystidiosis is a globally distributed, parasitic disease of fish caused by microorganisms taxonomically assigned to the genus *Dermocystidium* of the order Dermocystida within the class Mesomycetozoea, a clade comprised of microorganisms that lie phylogenetically at the animal-fungal boundary.[299] Members of the order Dermocystida infect fish, apart from species of the genus *Rhinosporidium* which infect mammals and birds. Currently 18 *Dermocystidium* species have been described in fish.[261] Members of the genus *Dermocystidium* have been reported to infect a growing number of cultured freshwater and marine fish worldwide including eels, salmonids, juvenile catfish, cyprinids, tilapia, and ornamental fish.[117,282,298,378,447,464,535,541] Disease can occur in adult and juvenile fish, resulting in chronic or acute forms.[189] Proposed factors contributing to disease include the presence of environmental pollutants in water and lower water temperature.[218,237,357,496]

Species of the genus *Dermocystidium* form cysts in host tissue within which endospores are produced.[299] In vitro, endospores give rise to motile uniflagellated zoospores which infect a host, encyst, and begin the creation of new endospores by the formation of sporonts within plasmodia or tubular-like fungoid hyphae.[117,270] Water-borne transmission by zoospores resulting in proliferative gill lesions has been demonstrated, and experimental studies have shown that infection is possible by ingestion of spores.[282,357]

Infection of gill and skin is common in dermocystidiosis, although systemic infection has been reported.[47,298,326] Gross lesions include small, white, spherical to oval cysts on skin, fin, and gill; in systemic infections, cysts may be present in kidney, spleen, and liver. In some infections, cysts become elongate or vermiform.[147] Microscopically, cysts measure 200–400 μm in diameter and can contain hundreds of 3–10 μm endospores. The spherical endospores of *Dermocystidium* spp. characteristically have a central cytoplasmic vacuole or inclusion and a peripheral nucleus.[47,270,357] Reportedly, the walls of endospores are strongly PAS-reactive, while the large central vacuoplast is refractile and not PAS-reactive.[47] Infection of gill has been associated with lamellar epithelial hyperplasia that can be severe and fatal.[357] Intact cysts in visceral organs may not elicit a significant host inflammatory response; however, free endospores elicit a progressive granulomatous response with phagocytosis of spores by macrophages.[47] Diagnosis relies on visualization of cysts and endospores in fresh mount preparations and histopathologic examination of infected tissue. In vitro culture has been reported, although its utility in routine diagnostics is debatable.[282]

Ichthyophonosis is a systemic granulomatous disease of freshwater and marine fish that is widely distributed geographically. The etiologic agent is *Ichthyophonus hoferi*, which is taxonomically classified as a member of the class Mesomycetozoea, order Ichthyophonida.[299] *Ichthyophonus hoferi* is an obligate pathogen, and disease is endemic in wild coldwater marine fish populations, e.g., Atlantic herring, in which epizootics have occurred; it is rarely seen in cultured fish.[333,465] Transmission appears to be direct and via the oral route, and infection is presumed to occur by the ingestion of contaminated feed, infected fish, and/or infected copepods.[405,465] *Ichthyophonus hoferi* was the only recognized fish pathogen of the genus until the relatively recent discovery of *Ichthyophonus irregularis*, a genetically distinct isolate in yellowtail flounder.[299]

The lesions of ichthyophonosis occur principally in highly vascularized organs, e.g., the heart, spleen, kidney, and liver. The stage recognized most in tissues of affected fish is the spore. Round, double-walled, multinucleate spores, or flask-shaped germinating spores are characteristic and considered pathognomonic; hyphae may also be seen. Spores elicit the formation of granulomas composed of variably wide collars of epithelioid macrophages and multinucleate giant cells surrounded by thin rims of fibrous connective tissue; spores are occasionally surrounded only by a thin rim of fibrous connective tissue.[333,405,471] Disease is often systemic, and granulomas form around spores in all anatomic locations (Fig. 39.65). In the skin, fish may develop a roughened surface texture akin to that of sandpaper, the "sandpaper effect," which is the result of elevation of the epidermis and superficial dermis overlying granulomas disseminated throughout the deeper layers of the dermis.[333,405] Involvement of the brain and spinal cord may lead to neurologic signs, which has been referred to as "**swinging disease**" in salmon.[534] Fish with ichthyophonosis may also have spinal curvature and darkening of the skin.[333]

Diagnosis of ichthyophonosis is by culture, or by examination of wet mount preparations or histologic sections of the skin or viscera.[333,405] Unlike some of the Mesomycetozoea, *I. hoferi* can be cultured in vitro using fungal media, such as Sabouraud dextrose agar with 1% serum or tissue culture media supplemented with 10% fetal bovine serum.[333,350,465]

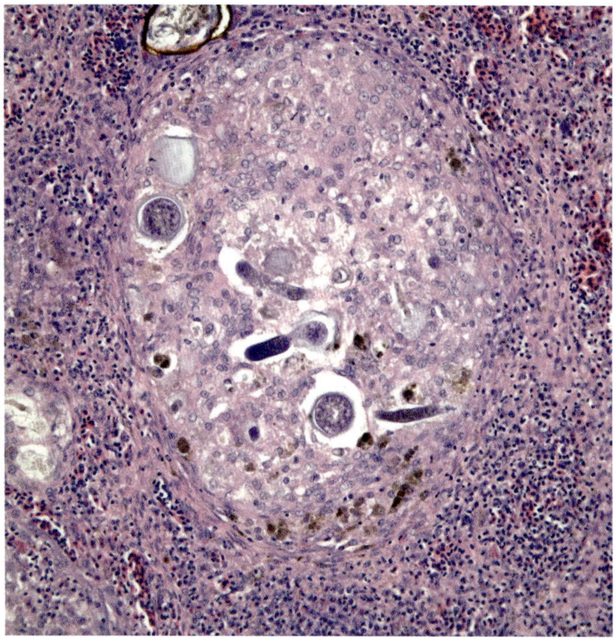

FIGURE 39.65 Ichythophonosis in a teleost. A granuloma in the spleen contains "resting" and germinating spores, the latter distinguished by their flask-like shape. A few hyphae are also present.

Ichthyophonus hoferi has a global distribution, a wide host range involving more than 80 fish species, and it has been associated with epizootic mortalities of fish.[465] There is no treatment, and lesions in muscle render fillets unmarketable, which further adds to the economic impact of this disease.[333]

ADDITIONAL RESOURCES

Refer to Box 39.1 Reference Textbooks for Fish Pathology.

References

Full references for this chapter can be found online at https://doi.org/10.1016/B978-0-443-18437-6.00002-X.

Chapter 40

Chondrichthyes

Nancy L. Stedman[a], Michael M. Garner[b], and Mark S. Okihiro[c]

[a]*Busch Gardens Tampa, Tampa, FL, United States;* [b]*Northwest ZooPath, Monroe, WA, United States;* [c]*California Department of Fish and Wildlife, Wildlife and Fisheries Division, Fisheries Branch, Vista, CA, United States*

INTRODUCTION

The class Chondrichthyes includes approximately 1268 known extant species divided into subclasses Elasmobranchii (sharks, skates, and rays) and Holocephali (chimaeras; 52 described species).[145] Elasmobranchii contains Selachii (sharks; 548 described species) and Batoidea (skates and rays; 668 described species).[145] Many chondrichthyans, particularly sharks, are rapidly declining in their natural habitats with an estimated 100 million sharks killed annually, mostly for the fin trade.[197] At least 37.5% of chondrichthyans are considered threatened or endangered because of overfishing followed by habitat loss, climate change, and pollution which is compounded by low reproductive rates and prolonged generation time.[45] As chondrichthyans decline in their natural habitats, managed care and propagation become increasingly important, mandating a need for pathologists to work ably with these species. A summary of the unique anatomical features of chondrichthyans is provided in the digital content.

NON-INFECTIOUS DISEASES

Nutritional

Abnormal thyroid gland morphology, presumably from **iodine deficiency**, occurs in free-ranging chondrichthyans and chondrichthyans in managed care. Causes for iodine deficiency in free-ranging chondrichthyans are unknown. In managed care, iodine deficiency develops because of difficulty in providing adequate iodide in treated natural and artificial seawater. Ozone, used to disinfect water in closed systems, converts iodide to less absorbable iodate. Nitrates, from bacterial breakdown of ammonia, are absorbed and inhibit iodide uptake by thyroid epithelium. Iodine supplementation is difficult in small chondrichthyans that may spit out capsules hidden in prey items. Grossly, thyroid-deficient chondrichthyans may have thyroid enlargement that appears as a mass on the ventral midline caudal to the mouth. The thyroid gland may be thickened or rounded and pale orange, and may increase up to 40 times its normal size (Fig. 40.1A). Histologic findings vary and have been termed diffuse hyperplastic goiter, diffuse colloid goiter, and multinodular colloid goiter based on similarities with iodine deficiency in mammals.[33] In diffuse hyperplastic goiter, follicles contain little colloid and are lined by densely arranged cuboidal to columnar cells with pseudostratification and occasional papillary projections (Fig. 40.1B). In diffuse colloid goiter, follicles are also lined by hyperplastic epithelium but contain abundant colloid. In multinodular colloid goiter, features of both hyperplastic and colloid goiter are often present, and the hyperplastic thyroid tissue is arranged in nodules separated by fibrous stroma. Iodine-deficient chondrichthyans do not consistently have thyroid gland enlargement or histologic lesions, and similar lesions may be seen with iodine excess. Thyroid gland morphology may vary with age, further complicating the histologic interpretation. Antemortem diagnosis of iodine deficiency is difficult because of a lack of species-specific reagents to measure thyroid hormones. Measurement of thyroid iodine content is a reliable indicator of iodine status in deceased chondrichthyans and can be used to assess the efficacy of iodine supplementation. Thyroid iodine ranges from approximately 3000–7000 ppm dry weight in healthy adult chondrichthyans[19] but may be lower in juveniles (Stedman, unpublished data). The clinical effects of iodine deficiency in chondrichthyans are largely unknown, but severe deficiency may result in poor reproduction and pup survival. **Oversupplementation with iodine** may cause similar gross and microscopic findings in the thyroid gland. Its clinical importance is uncertain; one author has seen several elasmobranchs with thyroid iodine concentrations up to 39,000 ppm dry weight with no overt clinical impact (Stedman, unpublished data).

Metabolic

Hypoxia may be caused by environmental factors such as low dissolved oxygen secondary to equipment failure or high bioload in a system, or may be caused by reduced

Pathology of Wildlife and Zoo Animals. https://doi.org/10.1016/B978-0-443-18437-6.00043-2

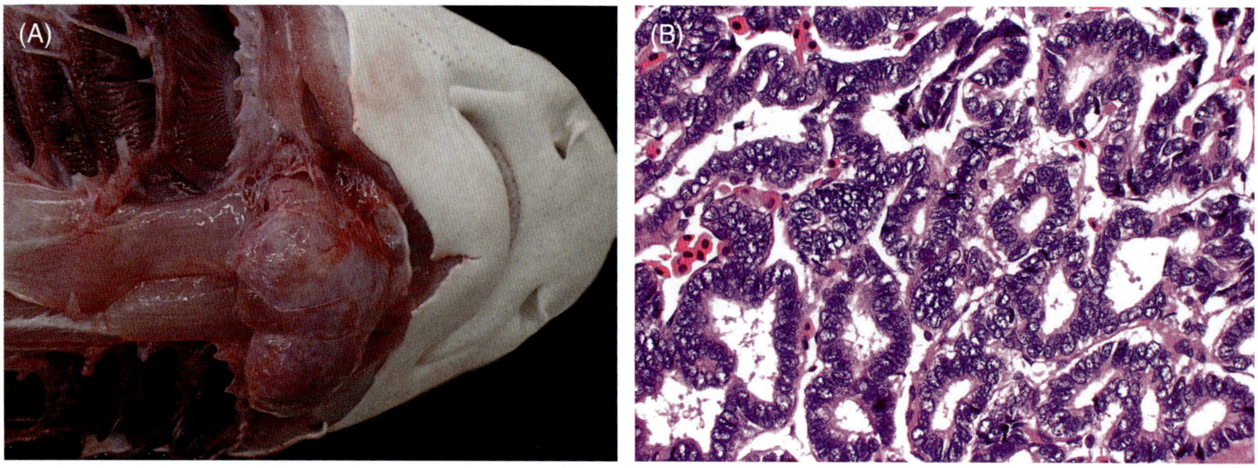

FIGURE 40.1 Iodine deficiency in a leopard shark. (A) Severe enlargement of the thyroid gland is present. Thyroid iodine content was 9.9 ppm dry weight in contrast to the reported range of approximately 3000–7000 ppm dry weight in healthy chondrichthyans. (B) Diffuse severe hyperplastic goiter is present. Follicles are lined by densely packed epithelial cells with pseudostratification, and contain little colloid.

oxygen absorption secondary to acute or chronic gill damage. There is tremendous variation in susceptibility to hypoxia among chondrichthyans. Some species, such as the epaulette shark, have evolved to survive in a hypoxic environment. Method of ventilation affects hypoxia susceptibility. While most chondrichthyans ventilate by buccal pumping and compensate for hypoxia by decreasing activity and increasing respiratory rate, obligate ram ventilators, such as white, mako, hammerhead, and sandbar sharks, must increase swim speed and mouth gape to compensate for hypoxia.[23] Presumably, death in hypoxic chondrichthyans is associated with anaerobic glycolysis including lactate accumulation and metabolic acidosis. Chondrichthyans that die from acute hypoxia are typically in good nutritional condition with no gross lesions or nonspecific visceral and meningeal congestion. Histologic lesions may be absent, or animals may have acute necrosis in viscera, particularly gill lamellae and renal tubules. Because microscopic lesions may be minimal, documentation of low dissolved oxygen in the water is helpful to support the diagnosis.

Chondrichthyans, particularly sharks, are susceptible to **stress and mortality associated with capture and handling**, including when they are caught as bycatch.[168] Adverse effects of capture and handling may be mediated by catecholamine release and the inhibition of respiration during restraint, particularly in obligate ram ventilators. Consistent findings are decreased blood pH and increased blood lactate related to anaerobic glycolysis (metabolic acidosis). Inhibition of respiration in some sharks also causes hypercapnia, resulting in concurrent respiratory acidosis. Some sharks, such as dogfish, may have hyperglycemia, and the magnitude may be related to the duration of stress.[55] Increased plasma potassium in bull, tiger, sandbar, blacknose, blacktip, and spinner sharks and decreased blood potassium in catsharks may also be indicators of stress.[160,195] There are no gross or histologic lesions and the diagnosis is confirmed by documentation of decreased blood pH and elevation of pCO_2 and lactic acid in an antemortem blood sample. Generally, capture-related mortality tends to be higher in species that are obligate ram ventilators, particularly hammerhead and thresher sharks.[50] Smaller sharks and sharks dwelling in shallower waters may be more susceptible than deep-dwelling sharks.[151] Capture-related mortality in sharks with similar degrees of acidosis and increased plasma potassium differs substantially with species suggesting factors other than those indicated by measurable physiologic parameters affect mortality.[195] Decreases in blood pH and increases in blood lactate and glucose can also occur with prolonged or repeated tonic immobility in sharks, presumably related to impaired ventilation.[17] Capture-and-release fishing also increases body temperature in the dorsal musculature of blue sharks and tiger sharks, suggesting body temperature is an indicator of stress in some chondrichthyans.[71] Chondrichthyans may develop **exertional rhabdomyolysis** of skeletal and cardiac muscle following stress and handling.[66,130] Exertional rhabdomyolysis can be severe enough to result in grossly visible myoglobinuria.[137]

Toxic

Free-ranging chondrichthyans, particularly elasmobranchs, are indicator species for environmental pollutants, and although numerous studies have documented effects on specific biomarkers, documentation of effects on health is limited.[2] Chondrichthyans bioaccumulate heavy metals such as mercury, selenium, and cadmium in various tissues but are generally resistant to **heavy metal toxicosis**. However, sublethal copper exposure is associated with decreased erythrocyte numbers and hematocrit, decreased plasma urea, increased lactate, and acidosis in dogfish.[36,186] High concentration of mercury in the liver is associated with less lipid deposition in neonatal blacktip sharks and increased numbers of pigmented macrophages in juvenile blacktip sharks.[134] Hepatic cobalt accumulation is associated with increased serum ALT in multiple shark species.[198] Exposure to tin as tributyltin oxide, an antifouling chemical used in marine paint, produced branchial epithelial necrosis secondary to membrane peroxidation in yellow stingrays.[47]

Female round rays naturally exposed to **polychlorinated biphenyls (PBCs)** lost hepatic lipid stores, hepatic and muscle glycogen stores, and muscle protein during gestation resulting in smaller embryos with lower embryonic hepatic weights compared with unexposed gravid females.[101,103] PCB-exposed round rays also failed to show increased plasma glucose following stress.[102] Findings suggest PCBs may have important adverse effects on the metabolism of chondrichthyans. Atlantic stingrays experimentally exposed to **crude oil** at concentrations similar to those during the Deepwater Horizon oil spill in 2010 had decreased olfactory organ responses to amino acids relative to control stingrays, although the specific mechanism of the decrease was not determined.[25]

At therapeutic doses, the anthelminthic **fenbendazole** may have adverse effects in some chondrichthyans, particularly sharks.[66] Clinical findings can consist of lethargy, inappetence, increased respiratory rate, leukopenia, and/or anemia. Gross lesions are nonspecific and may include ventral erythema and necrotic margins on the tail and pectoral fins. Histologically, necrosis is seen in hematopoietic tissues, particularly the epigonal organ, and gastrointestinal mucosal basal cells (Fig. 40.2). Animals may die from sepsis that originates from the necrotic gastrointestinal mucosa. Documentation of lesions and history of recent administration of fenbendazole support the diagnosis of fenbendazole toxicosis.

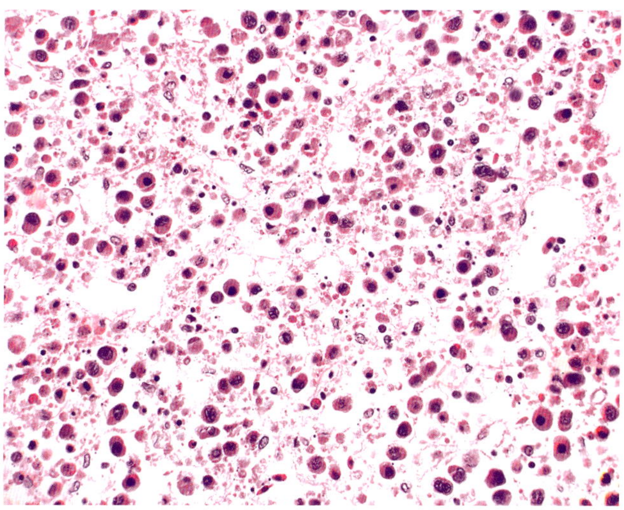

FIGURE 40.2 Fenbendazole toxicity in the epigonal organ of a sand tiger shark. Massive necrosis of hematopoietic precursors is present.

Multiple **volatile organic chemicals** associated with paint fumes have been associated with acute mortality in sand tiger sharks under managed care.[152] Gross lesions consisted of branchial congestion. Histologic lesions were acute branchial epithelial cell necrosis with thrombosis, lamellar collapse, and congestion. Confirmation of the presence of volatile organic chemicals in the exhibit water and blood of affected sharks supported the diagnosis. Interestingly, southern stingrays in the same exhibit survived.

Ozone toxicosis occurs with equipment failure resulting in excessive ozone output into the water. Ozone is an oxidizing agent that is toxic to branchial epithelium and erythrocytes, presumably by membrane peroxidation, and causes formation of methemoglobin. The proximate cause of death may be hypoxia secondary to these effects. Species and individual susceptibility to ozone toxicosis is variable. Grossly, chondrichthyans that die from ozone toxicity are typically in good nutritional condition with no gross lesions, nonspecific visceral and meningeal congestion, or small foci of necrosis in viscera. Histologically, gill epithelium may be acutely necrotic with or without congestion and lamellar atrophy. Acute necrosis may be present in other viscera. Because lesions are often subtle and nonspecific, documentation of a sudden increase in ozone is necessary to support the diagnosis.

Mortality in blacktip sharks and Atlantic sharpnose sharks has been reported during ***Karenia brevis*** **blooms** ("**red tide**") off the coast of Florida.[61] Specific lesions have not been reported, but demonstration of high brevetoxin concentrations in liver, gill, and skeletal muscle support the diagnosis. Because free-ranging chondrichthyans will naturally bioaccumulate brevetoxins in tissues, particularly the liver, the diagnosis of brevetoxicosis can be challenging. Sharks appear to be resistant to the toxic effects of domoic acid.[162]

Congenital/Genetic

Multiple morphological abnormalities suspected to be developmental have been described in elasmobranchs, although no genetic bases have been proven.[29,53,59,111,123,128,148,158,166] Abnormalities reported in elasmobranchs that survived to maturity are partial or complete albinism; separation of the cranial aspects of the pectoral fins from the torso in batoids; asymmetric, incomplete, or missing fins; shortened claspers or tail; anophthalmia; and head deformities. Abnormalities reported in embryos that would presumably be fatal in utero or within the early postnatal period are synophthalmia with and without albinism, dicephaly, and diprosopia. A bonnethead shark born to parents in managed

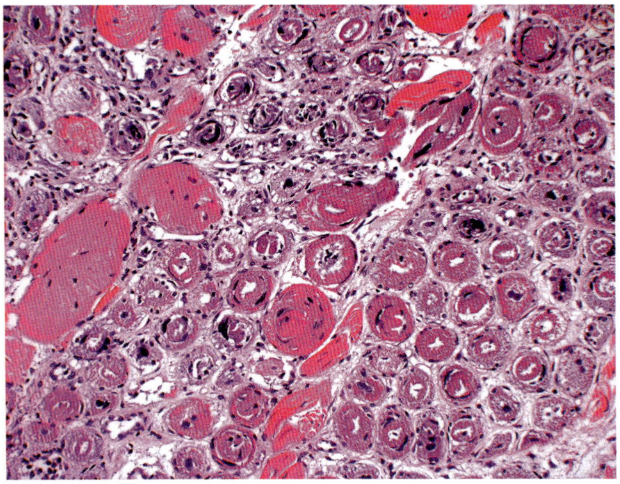

FIGURE 40.3 Skeletal muscle from a bonnethead shark pup. Myofibers have marked hypoplasia or atrophy with central clear spaces and/or a concentric fiber arrangement. Collagenous fascia between myofiber bundles is discontinuous. Findings are suggestive of dystrophy.

care had worsening abnormal swim patterns necessitating euthanasia at 6 months of age and was found to have small myofibers in all epaxial muscles with frequent splitting and a whorled pattern suggestive of muscular dystrophy (Stedman, unpublished data; Fig. 40.3).

Age Related/Degenerative

Coronary arteriosclerosis occurs in free-ranging sharks and sharks in managed care.[15,63] Lesion severity correlates with larger body size, and, as sharks grow continuously, coronary arteriosclerosis may be an age-related change. Lesions are not apparent grossly. Histologic findings are loss of the internal elastic lamina and proliferation of connective tissue and smooth muscle in the intima with reduction of the vascular lumen. Lesions are most common at branch points of coronary arterioles. **Arteriosclerosis** may also be seen in the meningeal and ocular choroidal vessels of geriatric chondrichthyans, and also is presumably an age-related, degenerative change. Meningeal arteriosclerosis may be visible grossly as thickened, rope-like vessels along the lateral aspects of the caudal cerebral hemispheres, cerebellum, and brainstem. Histologic findings are proliferation of dense collagenous connective tissue within and around meningeal and ocular choroidal vessels, sometimes with cartilaginous metaplasia. In the heart, **myocardial fibrosis** may be seen in geriatric chondrichthyans and may contribute to stress-related mortality. Myocardial fibrosis in chondrichthyans is typically not visible grossly. Histologically, fibrosis occurs specifically at the junction of the inner spongy myocardium and outer dense myocardium of the ventricle. The role of hemodynamic factors has not been investigated in the development of these lesions.

Glomerulosclerosis with increased mesangial stroma, increased glomerular cellularity, or both occurs in chondrichthyans, especially with age. Findings do not have a consistent clinical correlation. Renal tubular atrophy with interstitial fibrosis may be visible grossly as pale, firm, and/or nodular kidneys. The pathogenesis of these findings in chondrichthyans is unknown.

Degeneration of the notochord may be seen in geriatric chondrichthyans. Notochord cells are replaced by acellular matrix and intact and degenerate hyaline cartilage (Fig. 40.4; eSlide 40.e1). Ventral tesserae of the centra around the notochord are replaced by poorly organized sheets of proliferating cartilage that may be visible grossly. The process appears somewhat analogous to intervertebral disc degeneration and spondylosis in geriatric mammals. Cataracts may also be seen in geriatric chondrichthyans.

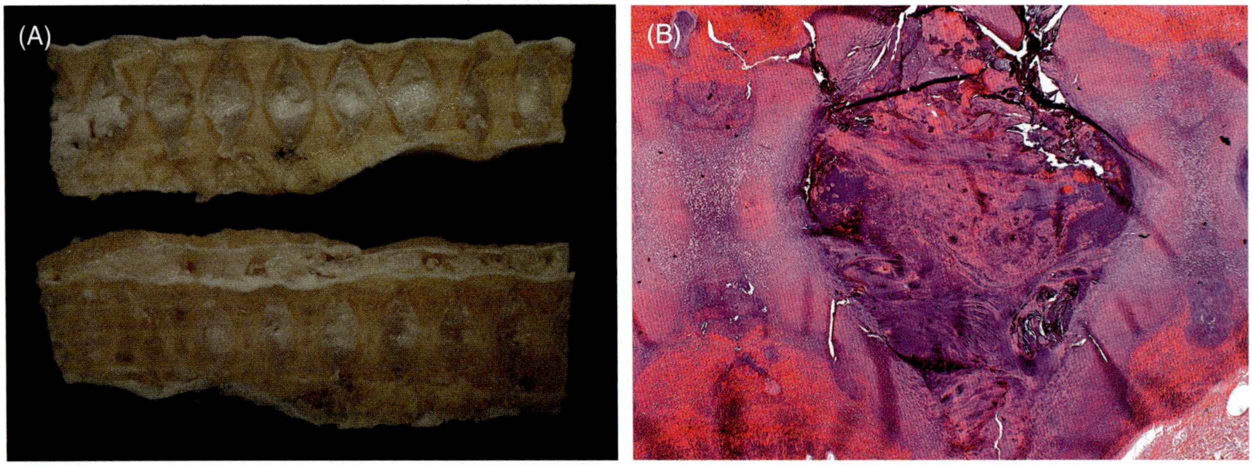

FIGURE 40.4 Degeneration of the notochord in a geriatric southern stingray (A) and cownose ray (B). (A) Two abnormal segments of vertebral bodies are shown. The notochord segments are off white with central fragmentation. Some are misshapen. Cartilage proliferation is present on the ventral aspect of the vertebral centra. (B) The vacuolated cells of the notochord remnant are replaced by poorly organized cartilage. The surrounding centra are intact (see eSlide 40.e1).

Inflammatory Non-Infectious

Bottom-dwelling chondrichthyans may develop **ventral torso cutaneous inflammation** in response to irritants in environmental substrates. Affected chondrichthyans have erythema and sometimes an irregular, pitted skin texture limited to the ventral torso. Histologic lesions are epidermal hyperplasia with dermal perivascular and interface infiltrates of lymphocytes and fewer granulocytes that may extend into the epidermis. Epidermal necrosis with erosion and ulceration and secondary infection may be present in more severe cases. The irritant in the substrate may be difficult to identify and may be physical, chemical, or biological. The ventral cutaneous reddening must be distinguished from the agonal pooling of blood commonly encountered in deceased chondrichthyans.

Miscellaneous/Multifactorial

Soft tissue mineralization occurs in chondrichthyans in managed care and is most common in batoids. It may be an incidental finding at necropsy or may be a progressive disorder resulting in weight loss and death. The chondrichthyan skeleton is not a substantial source or reservoir of calcium. Calcium is absorbed through the gastrointestinal tract and gills. Soft tissue mineralization may be related to exposure to high concentrations of calcium in water and/or diet, resulting in excessive absorption. Fatal soft tissue mineralization in black tip reef sharks at one institution was associated with oversupplementation with a commercially available vitamin and mineral supplement for elasmobranchs.[142] Grossly, affected chondrichthyans with soft tissue mineralization have firm, expansile lesions on the skin, particularly the distal pectoral fins, ventral torso, dorsal midline, and tail (Fig. 40.5A). Lesions may be domed or flat and ulcerated, and may exude chalky, white deposits that mimic inflammatory exudates caused by infection. Internally, similar chalky, white deposits may be seen in the small intestine, proximal spiral valve, uterus, heart, conus arteriosus, ventral aorta, kidneys, ovarian interstitium, mesentery, cornea, and oral cavity, and smooth muscle of the branchial corpus cavernosum (Fig. 40.5B). Basement membranes of the proximal branchial lamellae may also be affected. Histologically, basophilic mineral deposits of variable size and shape are present in affected connective tissues, smooth muscle, myocardium, and basement membranes, and they may be surrounded by granulomatous inflammation and fibroplasia, resembling the lesions of calcinosis circumscripta (Fig. 40.5C). A Von Kossa stain can confirm the presence of mineral. Death is presumably related to the effects of mineralization on these tissues, particularly in the cardiovascular system.

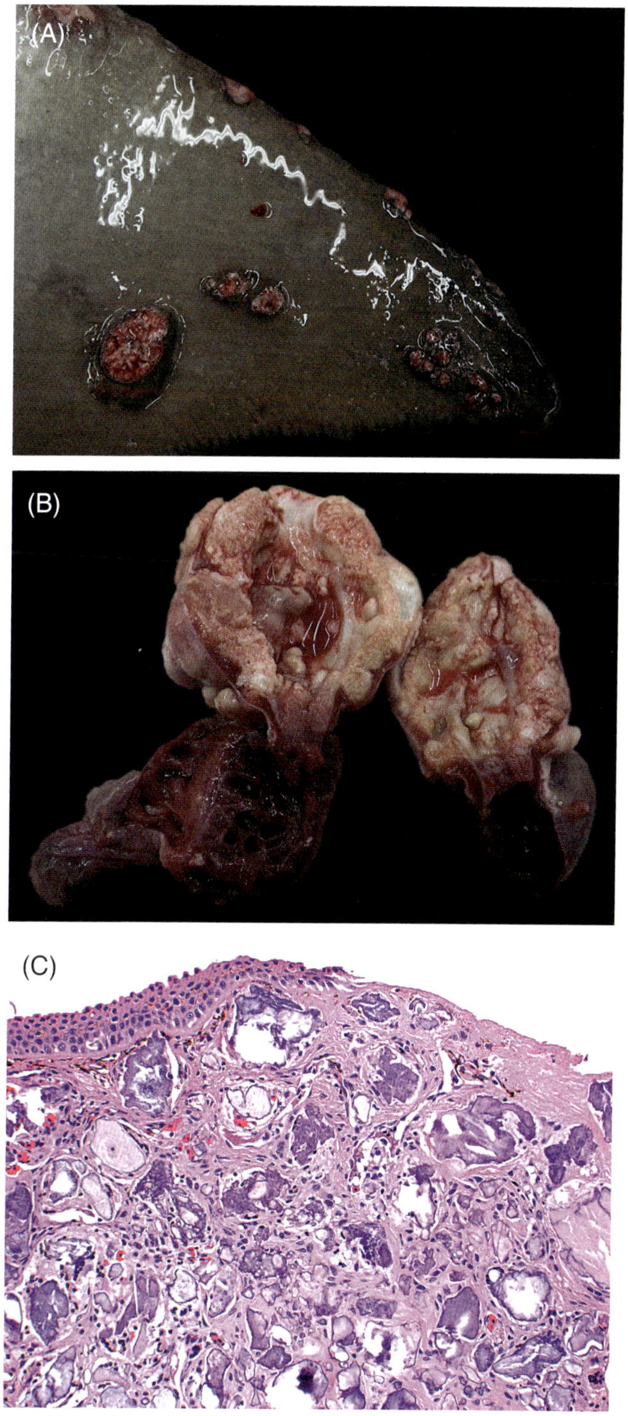

FIGURE 40.5 Mineralization of the pectoral fin (A, C) and conus arteriosus in a cownose ray. (A) The dermis has multifocal raised, nodular, well-demarcated, ulcerated lesions composed of off-white to yellow mineral admixed with inflammation. Lesions are on the leading edge and dorsal aspect of the fin. (B) Severe transmural mineralization of the conus is present. (C) The dermis is filled with and expanded by irregularly shaped deposits of mineral surrounded by epithelioid macrophages. The deposits are forming a raised dermal lesion with associated attenuation, erosion, and ulceration of the epidermis.

Chondrichthyes are susceptible to **gas embolism** (gas bubble disease) when exposed to supersaturated water (**supersaturation**). Gross lesions in chondrichthyans may be absent or fish may have white foci in the skin and gas accumulation in the heart or gills.[120] Histologically, gas emboli are present in vascular spaces, including branchial capillaries and corpora cavernosa (eSlide 40.e2). They may be associated with acute branchial epithelial necrosis.[66]

While the effects of **ocean acidification** associated with rising atmospheric carbon dioxide on chondrichthyans are largely unknown, experimental exposure of puffadder shy-sharks to hypercapnic conditions simulating ocean acidification resulted in corrosion of the surfaces and blunting of the edges of placoid scales.[48]

Trauma is common in both free-ranging chondrichthyans and chondrichthyans in managed care. Anthropogenic trauma is the most common trauma in free-ranging elasmobranchs. Mortality can result from retained fish hooks that penetrate the oropharynx and enter the cardiac coelom.[1] Additionally, large hooks can also severely damage teeth, jaws, and jaw musculature. Elasmobranchs may also ingest retained fish hooks in prey items. In one case, an ingested fish hook that perforated the cardiac stomach of a sand tiger shark and migrated into the spiral valve resulted in enterolith formation around the hook, with subsequent cachexia and death.[138] Microplastics have been found in the stomachs of free-ranging sharks but to date no health effects have been documented in these fish.[74,121] Elasmobranchs may become entangled in fishing line and netting or sustain gunshot injuries. Collision with boats resulting in blunt force trauma and propeller wounds are a leading cause for mortality in juvenile and subadult white sharks off the coast of California (Okihiro, unpublished data; Fig. 40.6). Collision with ships is also a cause for mortality in free-ranging whale sharks.[196] Sharks such as sevengill sharks that prey on venomous fish such as scorpionfish or rays may sustain puncture wounds in the oropharynx and esophagus with associated mural esophageal abscess formation (Okihiro, unpublished data). Although sharks are anecdotally considered resistant to the venom of scorpionfish and rays, systemic effects have been seen including perirenal hemorrhage, renal tubular necrosis, and necrosis of gastrointestinal smooth muscle and cardiac muscle.

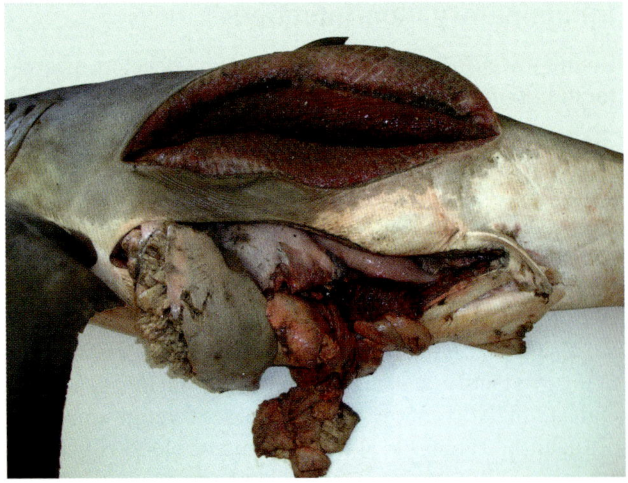

FIGURE 40.6 Watercraft trauma in a white shark. Two parallel propeller wounds are present. One extends deeply into skeletal muscle and the other has opened the coelomic cavity with incise wounds present in multiple viscera.

Conspecific or interspecific aggression, predation, and self-trauma are the most common causes for trauma in elasmobranchs in managed care. Male aggression in managed care is seen most commonly with smaller sharks such as leopard sharks, reef sharks, and bamboo sharks. Bite wounds present most commonly on the fins, especially the pelvic fins, and sometimes the claspers, but wounds can be seen anywhere. Clasper bite wounds may be limited to the dorsal aspect of the clasper and may not be apparent by observing a free-swimming animal. These wounds may result in infection of fin and clasper cartilage, often with *Enterococcus faecalis*. Male sharks bite the female before and during copulation, which results in bite wounds on the head, cranial torso, pectoral fins, and less frequently other fins of the female. Bite wounds associated with breeding are generally shallow and heal without complications. Female rays may sustain bite wounds to the trailing edges of the pelvic fins from pursuing males. Blunt force trauma to the torso during pursuit and capture by a male may cause fatal hepatic rupture and hemocoelom in female chondrichthyans with abundant hepatic lipid storage. Hemorrhage exuding from the coelomic pores can be seen with intracoelomic hemorrhage secondary to hepatic rupture. **Self-trauma** may be related to exhibit design. While it commonly presents as ulcers on the face, ulcers or other forms of trauma can occur on the trunk, fins, and on specialized appendages such as the cephalofoil of sphyrnids, cephalic lobes of some rays, and elongated rostrum of the Pristiformes. Batoids that hit their pectoral fins against structures in an exhibit develop erosive and ulcerative lesions on the underside of the distal fins. Secondary bacterial and less commonly fungal or ciliate infection may be present at these sites. Elasmobranchs in managed care may be injured during capture and handling and care should be taken when netting, restraining, and positioning elasmobranchs to avoid trauma to the fins, viscera, and spine.

Ulceration of the ventral pectoral girdle may result from excessive bottom resting and ischemic necrosis of the skin (Fig. 40.7). These ulcers may indicate chronic underlying disease. Histologically, lesions extend deeply into the dermis and sometimes to pectoral cartilage. Inflammatory cell infiltrates are often minimal, and secondary bacterial infection may be present.

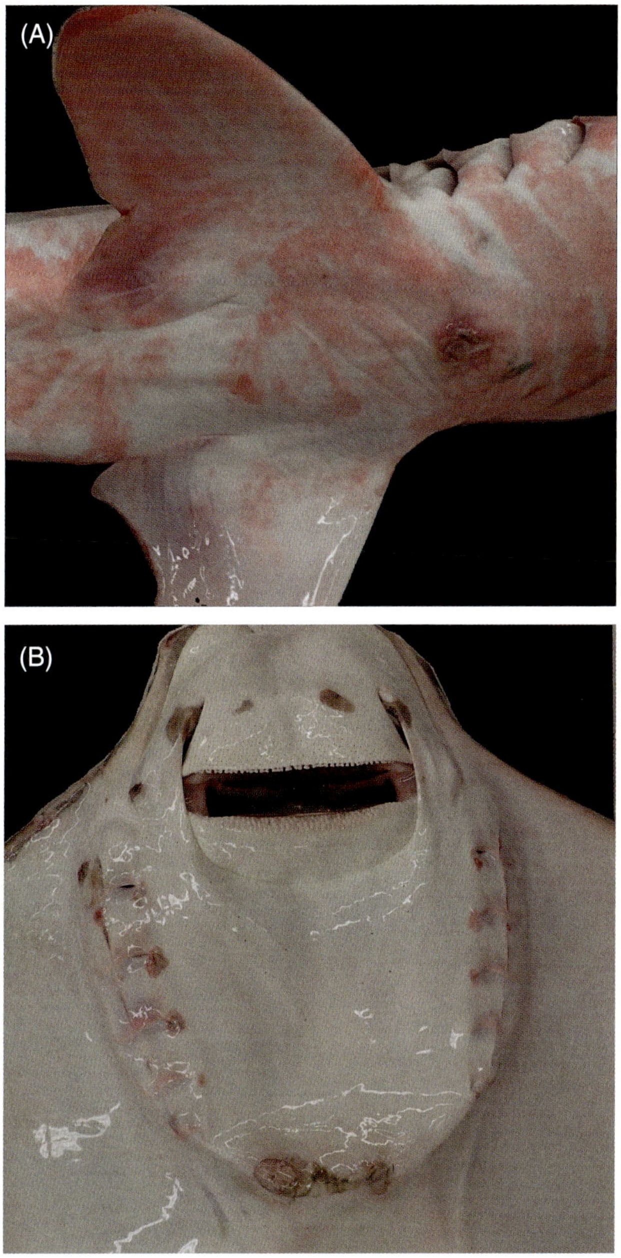

FIGURE 40.7 Ventral pectoral ulceration in a bonnethead shark (A) and cownose ray (B). (A) A single circular ulcer is present on the ventral midline over the pectoral girdle. (B) The ray is in poor nutritional condition. Ulceration is more linear and follows the cartilage around the gill slits and nares.

Some chondrichthyans are able to quickly evert the glandular stomach through the mouth and evert the caudal spiral valve, rectum, and cloaca through the vent, and then replace these organs. This behavior is for voiding and cleaning. Thin, loose mesenteric attachments of the gastrointestinal tract facilitate the behavior. **Gastric eversion** occurs in free-ranging sharks and elasmobranchs in managed care. It can be experimentally induced in skates and is rarely documented in batoids naturally.[167] Gastric

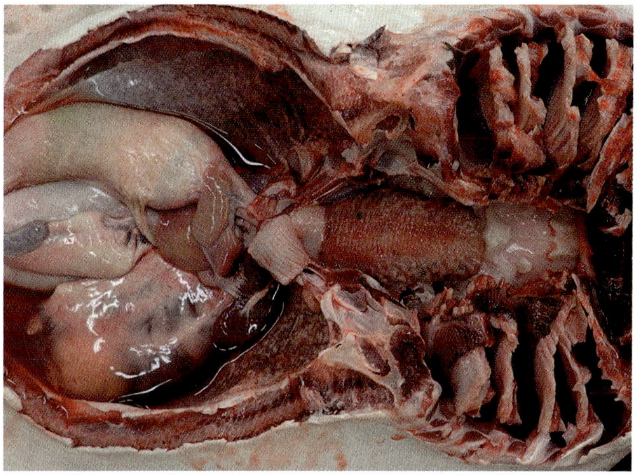

FIGURE 40.8 Gastric eversion in a cownose ray. The ventral body wall, liver, cardiac coelom, and ventral pharyngeal wall have been removed to show the everted cardiac stomach in the pharynx. The proximal spiral valve and pancreas are deviated medially.

eversion can be fatal if the stomach is not swallowed and replaced, becomes prolapsed through the gill slits, or is entrapped in the oropharyngeal cavity (Fig. 40.8).[188] The specific cause for failure to swallow and replace the stomach following eversion is unknown but possibilities include trauma during eversion causing tissue swelling and prolapse/entrapment through the gill slits from buccal pumping. In severe cases, portions of the pyloric stomach and spleen may also be displaced into the oropharyngeal cavity. Death is often related to sepsis originating in the prolapsed tissue wall. Bite wounds to the spiral valve from other elasmobranchs may occur during **spiral valve eversion**, and are most common in chondrichthyans with a scroll-type spiral valve.[32] These wounds may cause sepsis and death but may not be apparent until necropsy if the traumatized tissue was retracted prior to death. **Gastric bloating** occurs in young epaulette sharks in managed care (Fig. 40.9). Sharks are suspected of swallowing air for unknown reasons. Affected sharks have variable coelomic distension related to gastric distension with air or fluid. Concurrent hepatic atrophy is common with little to no hepatic lipid stores. Affected sharks are also often stunted and have soft tissue pallor. Other conditions of the gastrointestinal tract are rarely described in chondrichthyans. Fatal **gastrointestinal volvulus** may occur in free-ranging sharks and sharks in managed care, possibly associated with rolling movements while struggling or during handling (Fig. 40.10). Because of limited mesenteric attachments of the gastrointestinal tract, the entire tract may undergo volvulus with twists present at the esophagus and rectum. An enterolith formed around an octopus beak in the spiral valve of a sand tiger shark in managed care, resulting in severe weight loss.[122]

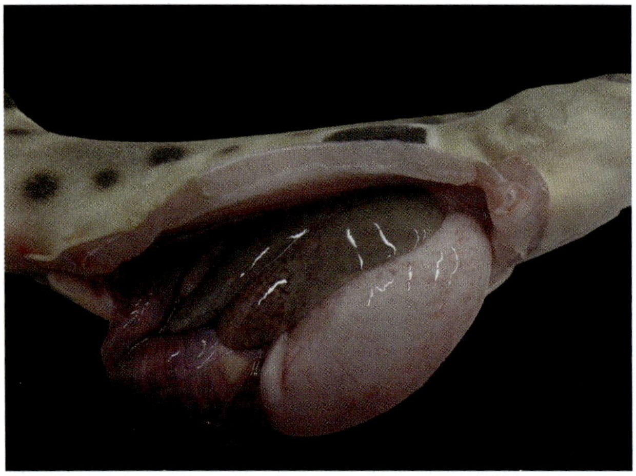

FIGURE 40.9 Gastric bloat in an epaulette shark pup. The coelomic cavity has been opened to show the enlarged stomach. The pup is also in poor nutritional condition with a small dark liver.

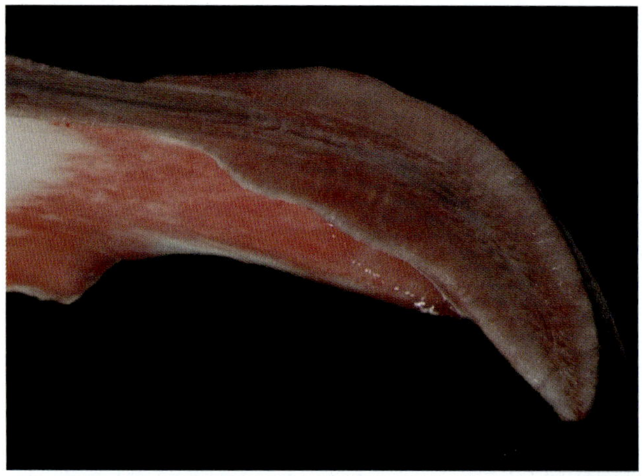

FIGURE 40.11 Pectoral fin curvature in an Atlantic blacktip shark. Fin curvature is present at the distal fin, and is most severe along ceratotrichia. There is marked thickening of the fin at the curvature related to increased connective tissue.

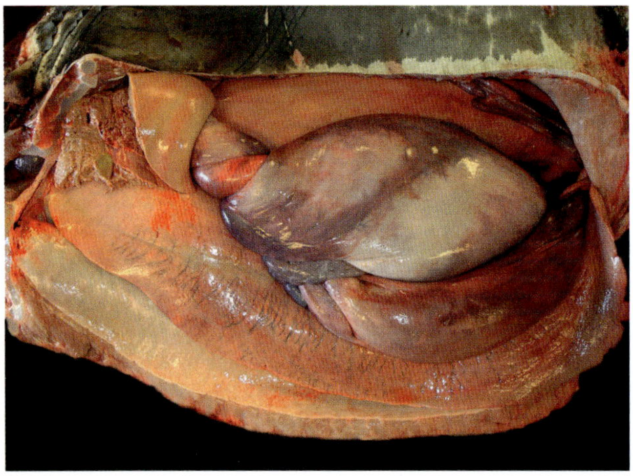

FIGURE 40.10 Gastrointestinal volvulus in a white shark. The left lobe of the liver has been removed to show the gastrointestinal tract. Twists are present at the esophagus and distal aspect of the spiral valve.

Gingival masses are occasionally reported in captive sand tiger sharks. Their cause is uncertain, although trauma may be involved because their protruding teeth are easily snagged during capture and handling. Masses are polypoid projections composed of hyperplastic epithelial cells, often with abundant cytoplasmic mucus, overlying dense or variably mature fibrovascular stroma.

Bilateral downward **curvature of the pectoral fins** occurs in captive large sharks of the genera *Carcharias* and *Carcharhinus* that rely partially or completely on ram ventilation. The specific cause is uncertain, but may be related to forces on the pectoral fins associated with redundant swimming patterns in an exhibit. Sharks with curved pectoral fins are unable to swim and ventilate efficiently, and can die of hypoxia. Grossly, the distal pectoral fins are curved ventrally and medially. Traumatic erosions and ulcers may be present on the dorsal aspects of the distal fins. The curved portions of the fins are thickened by connective tissue (Fig. 40.11). Lesions appear to be predominantly within distal portions of the fins containing collagenous ceratotrichia. Histologically, increased mucinous and collagenous connective tissue between the dermis and curved ceratotrichia is present. In severe cases, fracture and fragmentation of the ceratotrichia are present.

Spinal deformities occur in sand tiger sharks in managed care and occasionally other chondrichthyans. The cause is unknown but spinal trauma; deficiencies in zinc, potassium, or vitamins C or E; short pool length; or redundant swimming behaviors are possibilities.[4,179] Affected sharks have progressive difficulty swimming that may necessitate euthanasia. Grossly affected sharks have scoliosis, kyphosis, subluxation, compressed and deformed vertebrae, and spondylosis between the pectoral girdle and dorsal fin.[150] Histologically, poorly organized proliferation of hyaline cartilage extends from the centra of the affected vertebrae. Surrounding epaxial muscle may have necrosis, hemorrhage, inflammation, atrophy, fibrosis, and mineralization. Affected sharks have progressive difficulty swimming that may necessitate euthanasia. Spinal deformities were also noted in two free-ranging common angel sharks (*Squatina squatina*) and were suspected to be the result of trauma.[140]

Congestive heart failure is described in a mature sand tiger shark with increased respiratory effort, anemia, and edema of the claspers and proximal pelvic fins.[76] The shark had dilation of the atrium, ventricle, sinus venosus, and hepatic veins, and the condition responded to treatment. A mature female cownose ray with atrial thrombosis similarly had edema of the caudal trunk body wall, proximal pelvic fins, ovarian interstitium, and connective tissue around the thyroid gland, presumably related to chronic passive congestion (Stedman, unpublished data). Notably, neither elasmobranch was reported to have coelomic fluid accumulation.

A mature female roughtail stingray in managed care but recently obtained from the wild had increased eosinophilic glomerular mesangial stroma and thickened capillary basement membranes that stained light red with Congo red and ultrastructurally contained fibrils consistent with amyloid.[21] **Amyloid** accumulation may also be seen in the acini and interstitium of the pituitary gland of mature elasmobranchs possibly related to aging and of unknown clinical importance.

Reproductive disorders, best described in Southern stingrays, occur in female elasmobranchs housed without males.[131] The cause for these conditions is uncertain, although they may have a physiologic basis related to chronic nulliparity. Grossly, affected stingrays have an unusually large left ovary and a large left uterus with prominent trophonemata and histotroph accumulation (Fig. 40.12). Enlargement may cause clinically apparent coelomic distension. The left ovary contains numerous mature follicles and may contain cystic and inflamed follicles. The right oviduct, and sometimes the vestigial left oviduct, may contain mucosal cysts and fluid accumulation. Chronically affected rays are frequently in poor nutritional condition. Histologically, ovarian tissue is often unremarkable. Stromal hyperplasia may occur in the left ovary and vestigial right ovary and may mimic a mass in the latter (Fig. 40.13). Granulocytic inflammation and mural necrosis may be present within cystic follicles. Uterine trophonemata are elongated and may be hyperplastic (eSlide 40.e3). One or both oviducts may have cystic mucosal hyperplasia. Affected rays may have coelomic mesothelial hyperplasia. Generally, increasing estrone, 17 beta-estradiol, and progesterone are associated with increasing severity of ovarian disease and increasing estrone and 17 beta-estradiol are associated with increasing severity of uterine disease in southern stingrays, although findings are not consistent for all stingrays.[131]

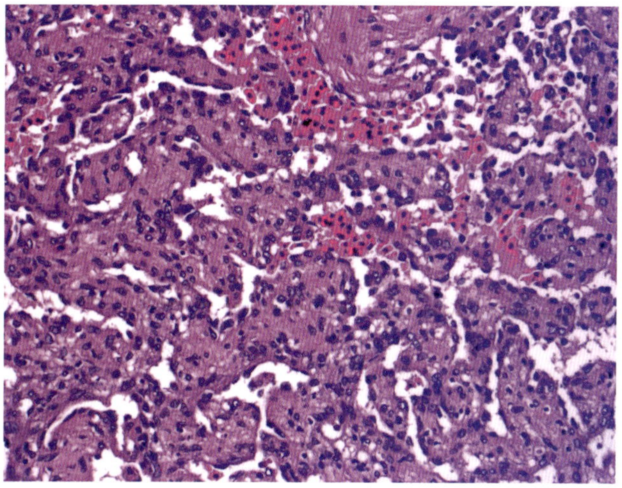

FIGURE 40.13 Ovarian stromal proliferation in a southern stingray with reproductive disease. The proliferating stroma can be in the left or vestigial right ovary and can mimic neoplasia.

Dystocia may occur in viviparous chondrichthyans and result in death of the pup(s), the dam, or both. The cause is often undetermined. In batoids it may be related to large pup size. It is important to be familiar with the reproductive biology of any female chondrichthyan prior to diagnosing dystocia and other disorders related to pregnancy and parturition. For example, cownose rays typically have a single pup in the left uterus that can present head or tail first at parturition. Right uterine pregnancies and twin pregnancies are less common but known to occur, and they may not cause dystocia. Grossly, in female chondrichthyans that die from dystocia, the head or tail of the intrauterine pup may be visible through the cloaca and vent. Pup size may appear large relative to dam size, and the uterine wall is attenuated

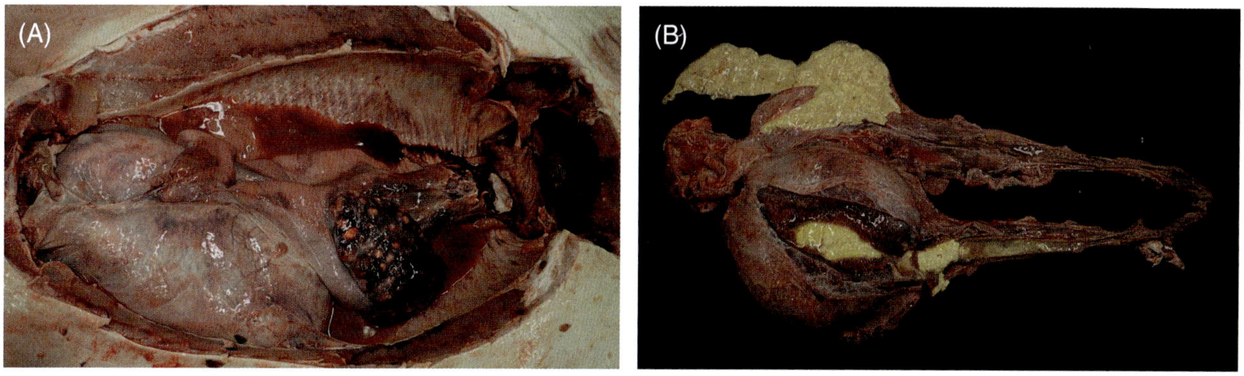

FIGURE 40.12 Reproductive disease in a female southern stingray. (A) The reproductive tract is shown in situ. The liver, gastrointestinal tract, spleen, pancreas, and rectal gland have been removed. The ovarian stroma is congested. The left uterus has marked enlargement. The vestigial right uterus is also enlarged. (B) Reproductive disease in a female southern stingray. Both uteri contain ovarian follicle material. The left uterus has elongated trophonemata. Several small cysts are present in the mucosa of both oviducts (see eSlide 40.e3).

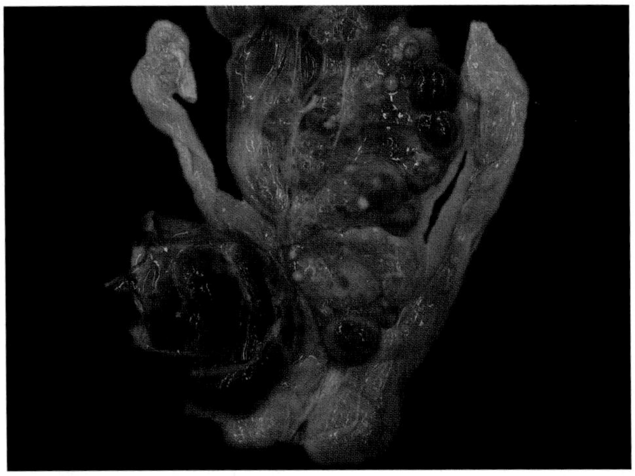

FIGURE 40.14 Egg impaction in a white spotted bamboo shark. The reproductive tract is shown ex situ. The ovary has numerous maturing follicles. The right uterus is opened to show a mass of egg cases.

and hyperemic. Parthenogenesis occurs in chondrichthyans, and disorders relating to pregnancy and parturition are possible in females in single-sex groups.

Uterine impactions with egg cases may occur in oviparous chondrichthyans. Uterine infection is commonly present, but it is uncertain if infection precedes and causes impaction or is a consequence of it. During reproduction, oviparous chondrichthyans should have one or two eggs in each uterus. Grossly, an impacted uterus is distended with multiple egg cases (Fig. 40.14). If the uterus is sufficiently enlarged, it may cause coelomic distension and rupture of the adjacent hepatic lobe, presumably by pressure necrosis. If infection is present, the uterus may also contain caseous exudate, although most chondrichthyans with uterine impaction and infection do not survive long enough to accumulate sufficient exudate to be grossly visible. Mucosal ulceration and variable mural granulocyte infiltrates are common. Similar to other oviparous species, uterine infection often spreads to the ovary and systemically. Captive female sand tiger sharks that do not breed accumulate numerous ovarian follicles within the uteri, causing marked distension and mimicking impaction but of doubtful clinical importance. Mummification of two in utero embryos accompanied by 12 viable embryos was reported in a free-ranging scalloped hammerhead shark.[112]

Neoplastic

Despite the common misconception that sharks do not get cancer, numerous benign and malignant tumors occur in chondrichthyans (Table 40.1). Immunohistochemistry is often of little value for chondrichthyan neoplasms because of limited cross-reactivity with currently available reagents. Commercially available antibodies against smooth muscle actin, MART1, and S-100 may be of value

TABLE 40.1 Neoplasms Reported in Free-Ranging and Managed-Care Chondrichthyes by Origin[5,14,34,66,75,80,132,136,176,191]

Epithelial neoplasms	Mesenchymal neoplasms	Melanocytic, neural, and neuroendocrine neoplasms
Cholangiocarcinoma	Chondroma, cutaneous	Choroid plexus papilloma
Ductus deferens carcinoma	Chondroma, skeletal	Chromaffinoma
Fibropapilloma, gingival	Fibroma	Cutaneous melanoma*
Hepatoblastoma	Fibrosarcoma	Ependymoma
Hepatocellular adenoma	Fibrous histiocytoma, gingiva	Neurofibroma
Hepatocellular carcinoma	Hemangioma, cutaneous and subcutaneous	Olfactory neuroblastoma
Intestinal adenocarcinoma	Hemangioma, visceral	
Mesothelioma	Lipoma	**Germ cell neoplasms**
Pancreatic adenocarcinoma	Lymphosarcoma*	Seminoma
Papilloma, cutaneous*	Myelocytoma	Ovarian dysgerminoma
Papilloma, gingival	Myxofibroma	
Pituitary adenoma	Myxofibrosarcoma	
Renal carcinoma	Myxosarcoma	
Squamous cell carcinoma	Osteoma, cutaneous	
Thyroid carcinoma	Odontoma, cutaneous	
	Sarcoma of soft tissue, origin undetermined	

More common neoplasms are indicated by an asterisk.

in identifying neoplasms of smooth muscle, melanocytic, and neural origins. There is little information on the biological behavior of most malignant neoplasms in chondrichthyans.

Multiple **cutaneous papillomas** have been reported in both free-ranging and captive sharks. These include cutaneous papillomas of the torso in bamboo, sand tiger, leopard, and white-tipped reef sharks and on the gill slits and conjunctiva of nurse sharks. A viral cause is suspected because lesions are multifocal and multiple animals in the same tank can be affected; however, to date, viral particles have not been identified in lesions. Grossly, lesions are raised and polypoid or plaque like with a smooth surface. Histologically, lesions are composed of multiple projections of well-differentiated hyperplastic epidermis with orderly maturation, overlying fibrovascular stroma (Fig. 40.15). Stroma may contain lymphocytic infiltrates. Squamous metaplasia and atypia suggestive of carcinoma in situ may be seen. Malignant transformation has not been reported.

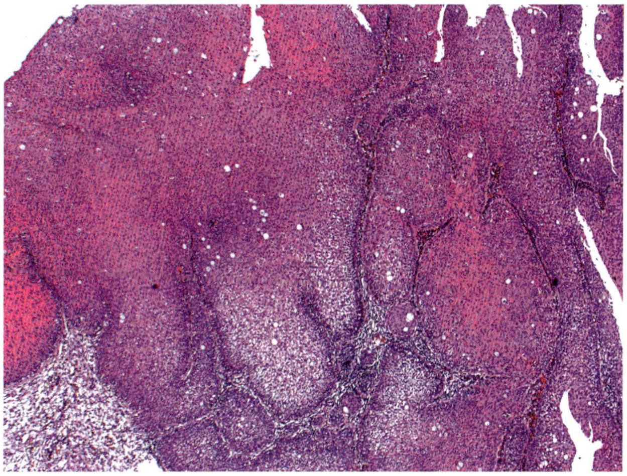

FIGURE 40.15 Cutaneous papilloma of gill slit epithelium in a nurse shark. Severely hyperplastic epithelium forms papillary projections and rete pegs. There is loss of epithelial mucus production.

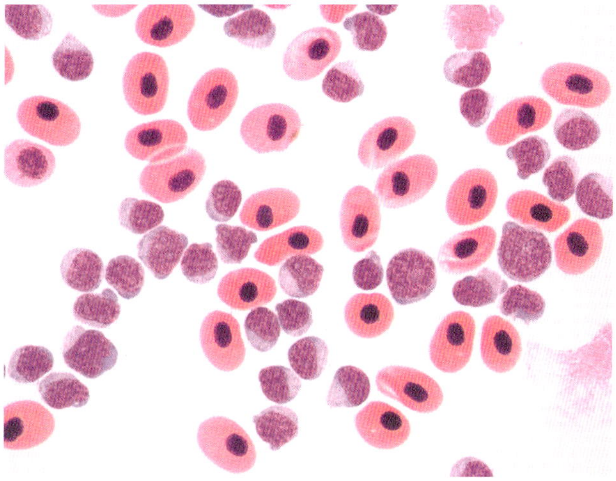

FIGURE 40.16 Lymphocytic leukemia in a sand tiger shark. Numerous circulating monomorphic neoplastic lymphocytes are present. The total white cell count in this case was 130,500 cells per microliter with 93% neoplastic lymphocytes. Wright stain.

Lymphosarcoma occurs in both free-ranging and captive chondrichthyans. The cause is uncertain. Lymphosarcoma may present as disseminated disease with widespread visceral involvement and in some cases, leukemia (Fig. 40.16). Gross and histologic lesions are similar to those in other species and are composed of homogeneous, off-white soft tissue with blastic lymphocytes (Fig. 40.17). Thymic lymphosarcoma may present as a raised, subcutaneous mass dorsal to the gill arches and may mimic a primary skin or subcutaneous tumor. Immunohistochemistry using antibodies against mammalian CD20 may assist in diagnosis of B cell lymphosarcoma.[106]

Cutaneous melanomas are described in free-ranging and captive chondrichthyans of multiple species, most commonly in Rajiformes.[136] Some chondrichthyans are known to increase cutaneous melanin content in response to ultraviolet (UV) light exposure, but a conclusive link between chondrichthyan melanomas and UV light exposure has not been made. Grossly, cutaneous melanomas are plaque-like or raised, sessile, or multilobular (Fig. 40.18A). Gross pigmentation is variable. Histologically, neoplastic cells are spindled or round with variable cytoplasmic melanin content (Fig. 40.18B). Junctional activity may be seen.

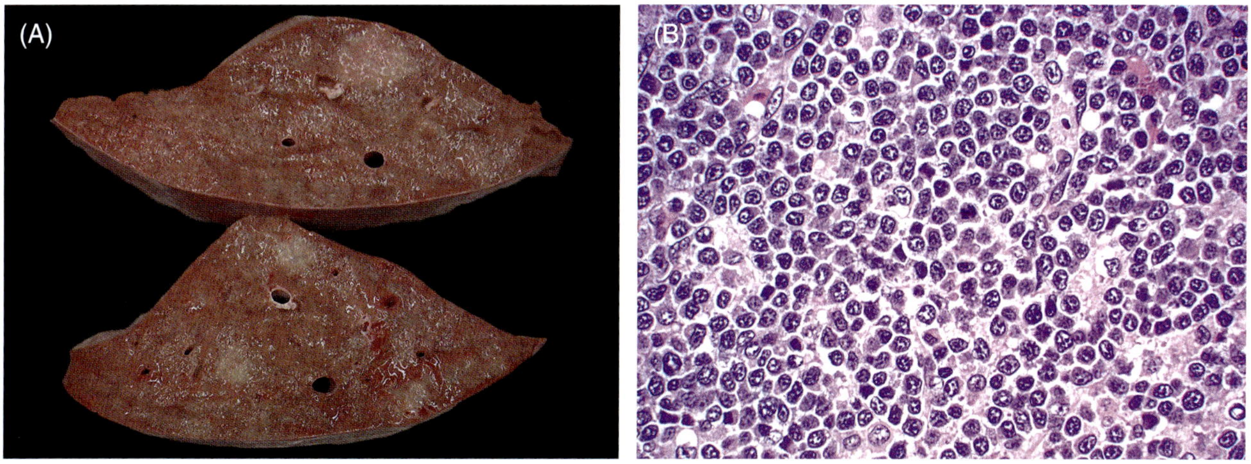

FIGURE 40.17 Lymphosarcoma in the liver of a sand tiger shark. (A) Multifocal off-white infiltrates of metastatic lymphosarcoma are present in the hepatic parenchyma. The shark still has adequate lipid storage in the liver despite having leukemia and lymphosarcoma in multiple viscera. (B) Sheets of large neoplastic lymphocytes are present within the thymus. Cells have sparse, lightly basophilic cytoplasm and large, round to oval nuclei with a vesicular chromatin pattern and multiple nucleoli. A mitotic figure is present.

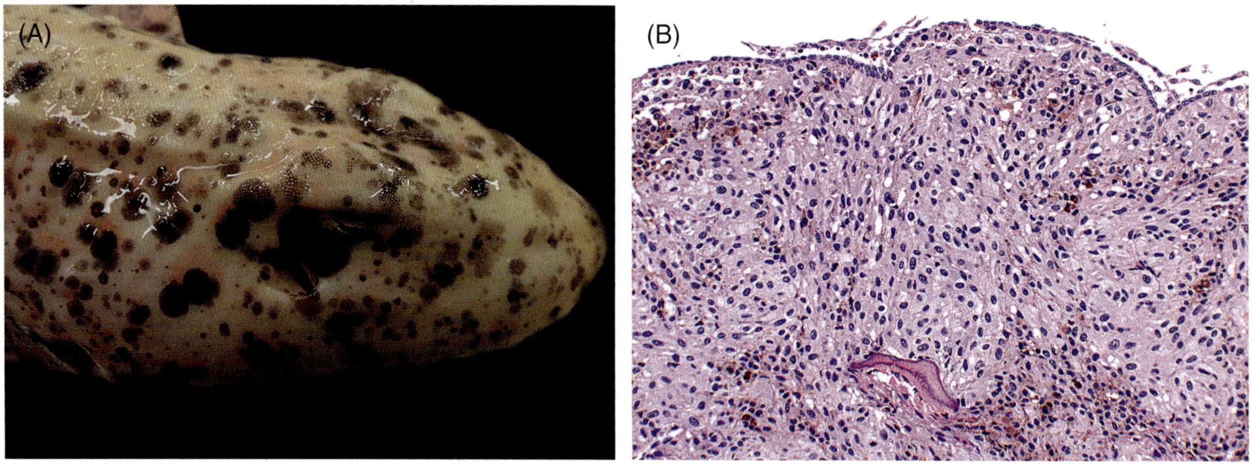

FIGURE 40.18 Cutaneous melanomas in a white spotted bamboo shark. (A) This individual is leucistic and has numerous proliferating plaques of hyperplastic and neoplastic melanocytes. (B) This neoplasm is from the skin of the spiracle. The epidermis is raised and attenuated. The dermis is effaced by poorly organized sheets of malignant melanocytes with variable melanin content. The neoplastic cells entrap the remnants of a denticle.

INFECTIOUS DISEASES

DNA Viruses

Little is known about viral infections in chondrichthyans and most data are from case reports. **Viral dermatitis** has been described in dusky smoothhounds.[90] Grossly, sharks had raised, round to oval, white to gray, depigmented skin lesions ranging in size from 1 mm to over 1 cm in diameter. Lesions had a slight central depression and a reddish margin and were most numerous on the fins, including the tail. Histologically, epidermal cells had intracellular edema, ballooning degeneration, necrosis, and eosinophilic intranuclear and cytoplasmic inclusions. Necrosis caused epidermal vesicles, erosion, and ulceration with dermal congestion and hemorrhage. Ultrastructurally, inclusions contained hexagonal virions with a bilayered envelope that were 150 to 185 nm in diameter suggestive of herpesvirus, but the virus was not characterized further. Lesions regressed spontaneously to foci of hyperpigmentation. A similar condition has been seen in a spiny dogfish with intranuclear inclusions in epidermal and endothelial cells.[66]

A juvenile tiger shark obtained from the wild and introduced into a tank with two captive-born sandbar sharks developed necrotizing branchitis with gill epithelial cell necrosis and capillary thrombosis of the gill lamellae. Gill epithelial cells contained large cytoplasmic and intranuclear amphophilic to basophilic inclusions. Similar inclusions were also present in esophageal epithelial cells. Ultrastructurally, inclusions contained capsulated virions 148 nm in diameter with an icosahedral nucleocapsid and an electron-dense core. DNA and protein sequence homology were indicative of a divergent **alloherpesvirus**.[8]

A giant guitarfish in managed care developed pink to red raised cutaneous lesions. Histologically, the epidermis formed papillary projections with numerous amphophilic intranuclear inclusions and marginated chromatin in epithelial cells (Fig. 40.19). Ultrastructurally, 75 nm icosahedral viral particles were present. Sequencing identified a 22 kb circular dsDNA genome representing a novel clade named guitarfish **adomavirus**.[42] Metagenomic analysis also identified a polyomavirus in the skin designated giant guitarfish polyomavirus 1.[43] However, polyomavirus virions were not seen ultrastructurally, and in situ hybridization confirmed adomavirus nucleic acid in the inclusions. Lesions regressed spontaneously. Adomavirus infection also caused red to pink, raised skin lesions in multiple sand

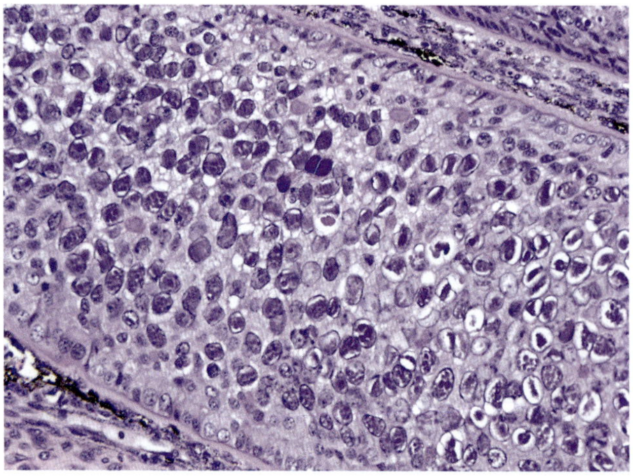

FIGURE 40.19 Adomavirus infection in the skin of a giant guitarfish. Numerous intranuclear inclusion bodies are present in epithelial cells of the epidermis. *(Photo Courtesy of Al Camus, Department of Pathology, University of Georgia.)*

tiger sharks housed together in managed care.[147] Histologically, lesions consisted of proliferating epithelium and misshapen denticles without intranuclear inclusions. Viral particles were not seen ultrastructurally. The adomavirus infecting the sand tiger sharks differed from the virus infecting the guitarfish based upon sequence data. Partial regression of the lesions occurred in one shark.

Iridovirus infection of dusky smoothhounds, leopard sharks, and possibly other chondrichthyan species causes erythrocytic necrosis.[83,85] Affected animals are anemic and may have soft tissue pallor. Cytoplasmic inclusions in erythrocytes can be identified in blood smears.

A group of young dusky smoothhounds obtained from the wild developed fatal ulcerative dermatitis and branchitis with large intranuclear inclusions that ultrastructurally contained virions resembling **adenovirus**.[66] Histologically, dermal ulcers were coinfected with flagellates and ciliates, and gill lesions consisted of atrophy of lamellae with large intranuclear inclusions in lamellar endothelial cells.

RNA Viruses

Clinically important RNA virus infections have not been described in chondrichthyans. A hepacivirus (Wenling shark virus; Flaviviridae) has been detected in hepatic tissue from a graceful catshark (*Proscyllium habereri*), but disease associated with the virus was not reported.[165] Diverse endogenous retroviral sequences have been identified in elasmobranchs.[70,193]

Bacteria

In one review, bacterial infections were the most common infectious disease in chondrichthyans, and **bacterial sepsis** was a common sequela.[66] *Vibrio* species, particularly those that produce urease, are common pathogens, although it is unknown if bacterial urease production plays a role in pathogenicity, especially in chondrichthyans that have high circulating urea.[67] *Vibrio harveyi*, a urease-producing species, was recovered from 62% of elasmobranch sepsis cases at one institution (Stedman, unpublished data). *V. harveyi* is also the most commonly recovered bacterium from the blood of healthy free-ranging electric rays.[178] Thus it can be difficult to determine if chondrichthyan sepsis is a novel infection or overgrowth of normal circulating bacteria. *V. alginolyticus*, a species that does not produce urease, was responsible for an outbreak of sepsis with 25% mortality in dark spotted stingrays and Tahitian stingrays recently added to an exhibit with multiple established shark species that were not affected, suggesting factors such as recent stress and stocking density may also promote sepsis in elasmobranchs.[51] Cutaneous and branchial lesions are commonly associated with sepsis and may be portals of bacterial entry. Septic chondrichthyans may have no gross lesions, cutaneous necrosis and ulceration, or foci of necrosis in the liver, spleen, pancreas, and kidneys. Meninges may be reddened and thickened. Histologically, bacteria are identified within blood vessels or foci of necrosis in affected viscera, sometimes accompanied by inflammatory cell infiltrates. Fibrin thrombi or aneurysms may be present in capillary beds of the branchial corpus cavernosum and lamellae. Diagnosis of acute sepsis is challenging because bacteria can be recovered from the blood of healthy chondrichthyans, and autolyzed specimens frequently have substantial postmortem bacterial overgrowth. Histologic confirmation of necrosis and inflammation in association with bacteria or frequent and numerous intravascular bacteria in tissue with minimal autolysis can support the diagnosis (Fig. 40.20; eSlide 40.e4).

Epitheliocystis, gram-negative organisms within the order Chlamydiales, can infect the gills of chondrichthyans.[22,144] In heavy infections, the gills may appear pale and mottled. Histologically, epithelial cells of brachial lamellae are enlarged by intracellular cysts that vary from 10 to 200 μm in diameter (Fig. 40.21). Cysts are filled with granular, lightly basophilic material surrounded by a thin, eosinophilic membrane. In heavy infections, mixed inflammatory cell infiltrates, lamellar epithelial cell hyperplasia, and fusion of lamellae may be present. The organisms stain pink to red with the Gimenez stain. Demonstration of lesions and identification of organisms by electron microscopy can confirm the diagnosis, but molecular methods such as PCR amplification and sequencing of16S rRNA are necessary to definitively place the causative agent within the order Chlamydiales. Epitheliocystis can be an incidental finding in chondrichthyans infected with low numbers of organisms.

Carnobacterium maltaromaticum causes meningoencephalitis and otitis in free-ranging juvenile salmon,

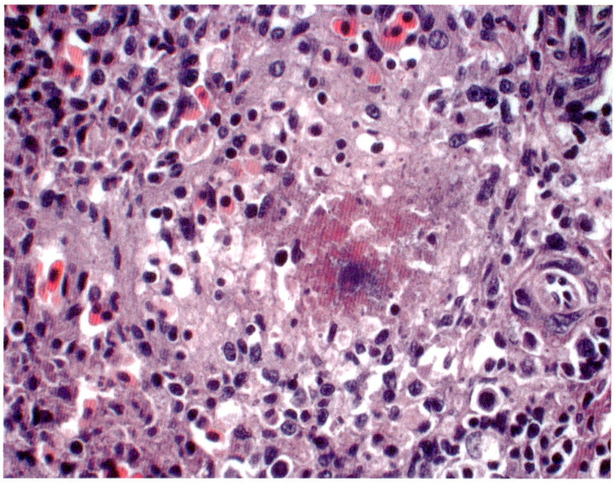

FIGURE 40.20 *Vibrio harveyi* sepsis in the spleen of a cownose ray. Acute necrosis is associated with colonization by numerous bacteria (see eSlide 40.e4).

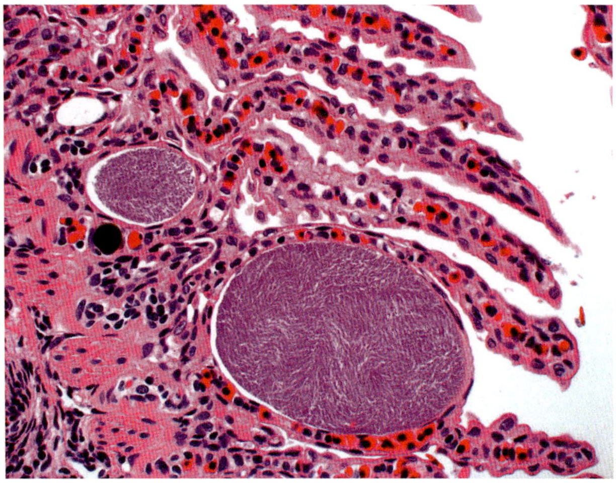

FIGURE 40.21 Epitheliocystis of the gills of a spotted eagle ray. Two branchial epithelial cells are filled with cyst-like aggregates of granular basophilic organisms. Slight lymphocytic inflammation is also present.

common thresher, and shortfin mako sharks along the California coast,[161,174] and in one Pacific electric ray (Okihiro, unpublished data). Because affected sharks are often less than one year of age and *C. maltaromaticum* has been recovered from the uterus of thresher sharks, the infection in some sharks may begin in utero.[174] The endolymphatic ducts of chondrichthyans open at the endolymphatic pores on the dorsal aspect of the head and are likely portals of entry for this agent. The bacteria travel through the endolymphatic ducts to the ears and subsequently into the cranial vault, presumably along the meninges of the auditory nerves (Fig. 40.22A). Grossly, there may be erythema, caseous exudate, or both between the skin and endolymphatic fossa of the chondrocranium. Exudate at this site may cause a focal domed lesion covered by intact skin on the dorsal midline of the head at the endolymphatic pores. Infected ears are filled with translucent or opaque perilymph. Inflamed meninges appear erythemic and extradural fluid may be translucent or opaque. Histologically, mixed mononuclear and granulocyte infiltrates surround endolymphatic ducts with epithelial pseudostratification. Similar inflammatory cell infiltrates are present in the perilymph of the ear and endolymphatic space of the sacculus. Meningeal inflammation is mostly mononuclear with fewer granulocytes and is most severe over the telencephalon and cerebellum (Fig. 40.22B). Meningeal inflammation may be accompanied by proteinaceous edema, congestion, hemorrhage, and necrotic cell debris. Small gram-positive coccobacilli may be seen among inflammatory cell infiltrates in all locations. If the infection becomes systemic, splenomegaly and foci of hepatic necrosis and mononuclear cell inflammation may be present. Chondrichthyans in managed care may also acquire other bacterial, mycobacterial, and fungal infections through the endolymphatic ducts resulting in similar lesions. These infections are most common in sharks but also can occur in batoids. Fungal infections acquired through the endolymphatic ducts can cause severe lesions with rupture of the ducts, lysis of the membranous labyrinth of the ears, and extension of infection into the chondrocranium. Free-ranging chondrichthyans with lesions similar to those caused by *C. maltaromaticum* may also have infections with other pathogens. *Streptococcus iniae* infection caused meningoencephalitis, otitis, and marked subcutaneous inflammation around the endolymphatic ducts in a large free-ranging common thresher shark that stranded in California (Okihiro unpublished data). The infection also caused multifocal hepatic necrosis.

Streptococcus agalactiae septicemia was responsible for clusters of mortality in multiple batoid species.[16] Grossly, splenomegaly, hepatomegaly, swelling of the dorsal cranium, and cerebral meningeal hyperemia were seen. Histologically, granulocytic and mononuclear leukocytic myocarditis was the most common finding, followed by meningoencephalitis, splenitis or splenic congestion,

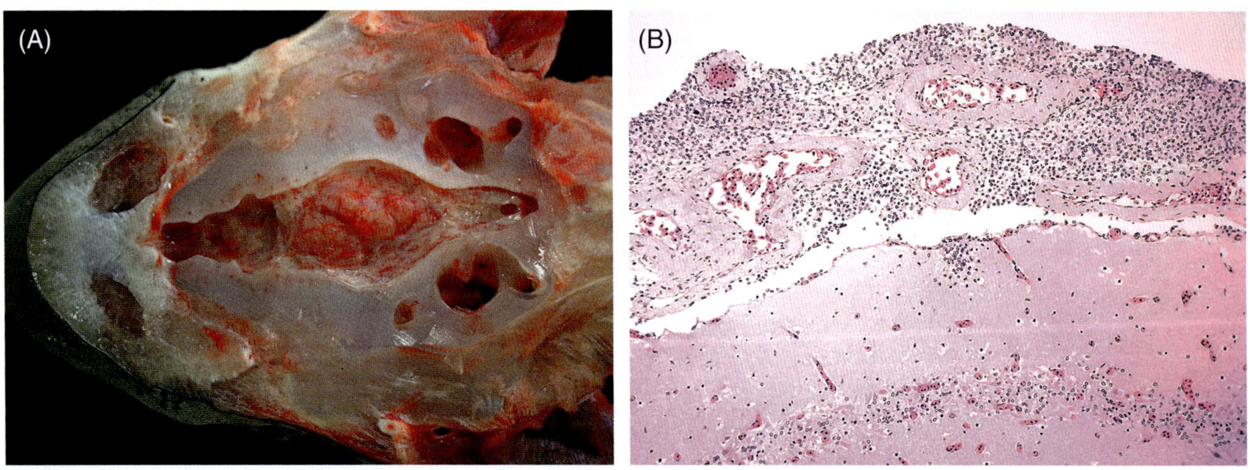

FIGURE 40.22 *Carnobacterium maltaromaticum* infection of the inner ears and meninges of a salmon shark. (A) The ears are hyperemic and severe meningitis is present. (B) Lymphocytic inflammation is present in the meninges.

branchitis, interstitial nephritis, hepatitis, and inflammation in other viscera. Gram-positive cocci were easily identified in blood vessels and inflamed tissues. One ray had been collected from an area where *S. agalactiae* was causing mortality in wild Queensland grouper and this ray was speculated to be the source for the outbreak. *S. agalactiae* infection was also reported in a deceased wild female Sandbar shark.[126] The shark had grossly visible granulomatous oophoritis, epicardial hemorrhage, and atrial emphysema. Histopathologic examination identified granulocytic, lymphocytic, and histiocytic epicarditis and meningoencephalitis accompanied by necrotizing vasculitis with intralesional bacteria.

Enterococcus faecalis infection of cartilage of the claspers and cartilage and ceratotrichia of the fins occurs in sharks and is most common in small sharks, particularly male bamboo sharks (Stedman and Garner, personal observations). Because *Enterococcus* is part of the normal flora of the oral cavity of some chondrichthyans,[177] the lesions are likely a result of bite wounds from conspecifics introducing bacteria. Grossly, nodular enlargements of the claspers and fins are present, and may ulcerate. The affected cartilage is yellow and may be friable. Caseous exudate may be visible grossly within and around the affected cartilage. Histologically, the infected cartilage has matrix eosinophilia and chondrocyte necrosis (Fig. 40.23). Microcolonies of gram-positive cocci are often seen in the necrotic cartilage. Necrotic cartilage may be infiltrated by histiocytic inflammation and fibroplasia and aggregates of macrophages are present around the cartilage and in the surrounding soft tissue. Gram-positive cocci may also be present in the cytoplasm of macrophages or in aggregates among the macrophages. Infected ceratotrichia of the fins

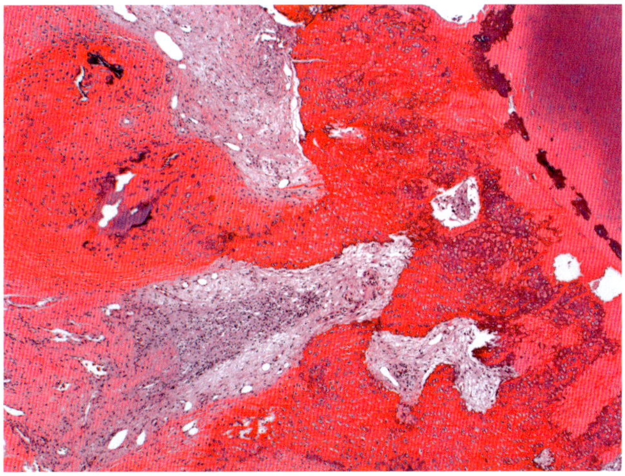

FIGURE 40.23 Chronic *Enterococcus. faecalis* infection of the pelvic fin cartilage of a white spotted bamboo shark. Proliferating immature cartilage and fibroplasia are extending from necrotic tessellated cartilage (right side of image). A microcolony of bacterial cocci is present in the proliferating cartilage (upper left of image).

are fragmented and colonized by gram-positive cocci with surrounding histiocytic inflammation. Fin lesions may have degenerate denticles embedded in the dermis from the initial bite wound. Affected sharks may be bacteremic and *E. faecalis* can spread to cartilage and soft tissues at other sites. Infection with other Gram positive cocci can cause similar lesions (eSlide 40.e5). *E. faecalis* infection was also reported to cause an abscess-like lesion in the rostral cranial vault of a spotted eagle ray that resolved with medical management.[38]

Fatal ***Citrobacter freundii*** meningitis occurred in a lemon shark in managed care.[155] Grossly, the shark had a subcutaneous abscess over the dorsal chondrocranium, a tract extending through the chondrocranium, and yellow exudate in the meninges that was most severe over the cerebellum. Histologically, lymphoplasmacytic granulocytic, and fibrinonecrotizing meningoencephalitis with bacteria and vascular thrombosis was present. Other lesions were lymphoplasmacytic heterophilic coelomitis, interstitial nephritis, and myocarditis. The same *Citrobacter freundii* was recovered from the intestinal tract and hematogenous spread of the organisms from this site was suspected. However, the presence of a cutaneous abscess over the dorsal chondrocranium is suggestive of the endolymphatic ducts as a possible portal of entry for the bacteria.

Tenacibaculum maritimum caused necrotizing skin lesions on the dorsum of an adult female sand tiger shark (*Carcharias taurus*) in managed care.[62] Lesions were suspected to have occurred at sites of skin trauma and responded to antimicrobial treatment. Lesions appeared white grossly with numerous filamentous gram-negative bacteria present cytologically. Histologically, skin lesions in sharks seen by one author (Stedman) have consisted of dermal infiltration by numerous granulocytes and macrophages that surround necrotic cellular and other debris admixed with mattes of filamentous gram-negative bacteria. The overlying epidermis was hyperplastic with denticle loss.

Mycobacterial infections are uncommon in chondrichthyans. Some have been confirmed to be caused by *Mycobacterium chelonae* or *M. marinum* with a single report of *M. avium* infection in an epaulette shark.[3,28,78,79,82] Cutaneous or visceral granulomas may be present; however, acute embolic disease or cutaneous lesions with necrosis and exudate may also be present. Benthic species and particularly guitarfish may be predisposed.[189] One study reported a 30% overall prevalence of mycobacteriosis in elasmobranchs in managed care with most cases manifesting as localized urogenital infections in benthic shark species, suggesting substrates as a possible source of infection.[79] Although most animals lacked gross lesions, granulomas with intralesional acid-fast bacteria were detected microscopically in the kidneys and reproductive tract including gonads, epigonal organ, and spiral valve. Additionally, acid-fast bacteria were present in the cytoplasm of epithelial cells of the kidney, epididymis, and seminal vesicles.

Fungi

Fungi of the genus ***Fusarium***, particularly ***Fusarium solani* species complex (FSSC)**, are the most common fungal pathogens of chondrichthyans. Trauma to the skin may be an important predisposing factor. Infection is most common along the cephalofoil of bonnethead and scalloped hammerhead sharks in managed care.[58,143] Infection also commonly involves the lateral line system of sphyrnids. Well-demarcated, ulcerative, and/or hemorrhagic lesions, sometimes containing off-white exudate, are observed in areas of the lateral line organs of the head and torso (Fig. 40.24A). Lesions may be confined to the dermis or may extend into subjacent skeletal muscle and cartilage. Histologically, cutaneous necrosis and ulceration with infiltration by granulocytes and macrophages are present. Angioinvasion with thrombosis and infarction may be seen.[143] Intact lateral line organs have vacuolation, necrosis, and hyperplasia of epithelial cells and intraepithelial and luminal inflammatory exudate (Fig. 40.24B). Nonpigmented branching septate fungal hyphae can be easily demonstrated with fungal stains but are difficult to see in routine hematoxylin and eosin (HE) stained sections (Fig. 40.24C). Dissemination of angioinvasive fungi may cause visceral necrosis and hemorrhage that can be visible grossly. Characteristic lesions and confirmation of the presence of the organisms by culture and molecular methods are diagnostic. Although *Fusarium* is the most common pathogen infecting the lateral line system of chondrichthyans, bacterial infection of the lateral line system can also occur (eSlide 40.e6). *Fusarium* also caused ulcerative dermatitis in a rough-tailed stingray without lateral line involvement; the infection responded to treatment.[73] *Fusarium* may also cause infection of the endolymphatic ducts and ears.

Infection with other fungi such as ***Exophiala*** and ***Paecilomyces*** can occur in captive chondrichthyans and present as skin and/or gill necrosis, erosion, and ulceration, infection of the endolymphatic ducts and ears, meningoencephalitis, and/or systemic infection. Fungal infection of deeper tissues and systemic infections may be present without apparent cutaneous lesions.[66,107] Infection of the endolymphatic ducts and ears or other structures of the head with pigmented fungi that spread to the surrounding cartilage may cause gray to black discoloration of the cartilage (Fig. 40.25).[52] Infection of a free-ranging white shark with *Mortierella alpina* also caused severe meningoencephalitis and hepatic necrosis; the infection appeared to have started in the endolymphatic ducts and ears with extension into the cranial vault and angioinvasion (Okihiro, unpublished data).

Microsporidiosis is rare in chondrichthyans but is reported in captive leopard sharks.[66] Histiocytic inflammation is associated with cytoplasmic microsporidians in the gills, spleen, pancreas, brain, and blood vessels. ***Dasyatispora levantinae*** infects free-ranging common stingrays. Infection presents as spindle-shaped, raised, white lesions filled with thick, caseous material arranged parallel to muscle fibers around the margin of the disc.[41] Histologically, sporophorous vesicles 30 to 40 μm in diameter and

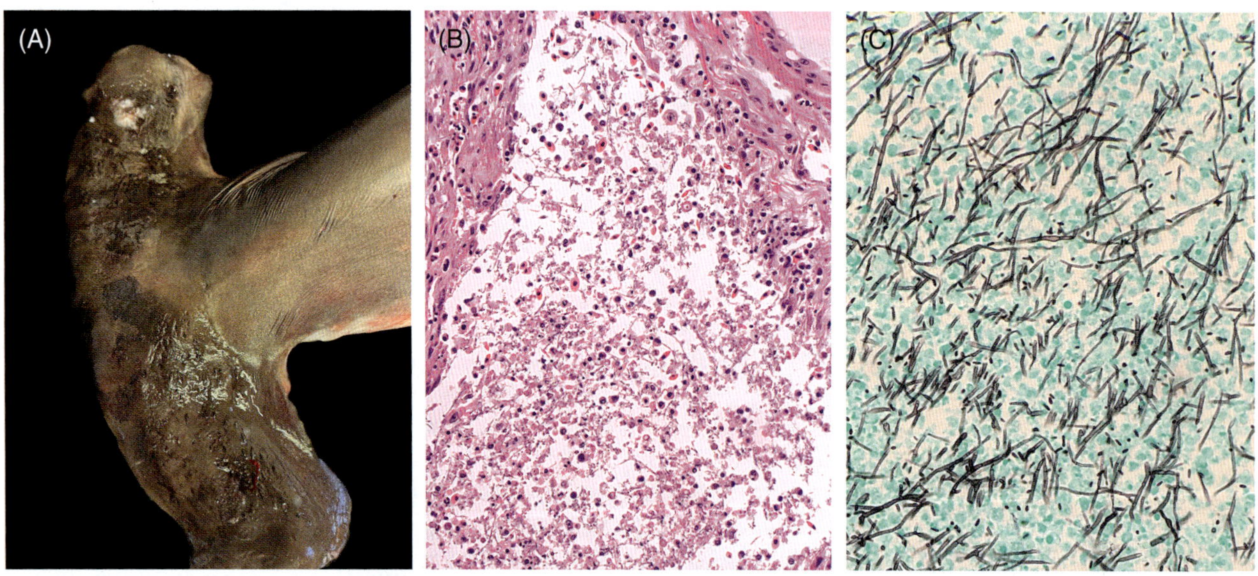

FIGURE 40.24 *Fusarium* infection of the lateral line system of a scalloped hammerhead shark. (A) Ulceration and off-white exudate are present on the dorsal cephalofoil. (B) A lateral line canal is filled with inflammatory cells and numerous fungal hyphae. (C) Hyphae are easily seen with a fungal stain. Grocott's methenamine silver. *(Part A: Photo Courtesy of Kristen Dubé, Phillip and Patricia Frost Museum of Science.)*

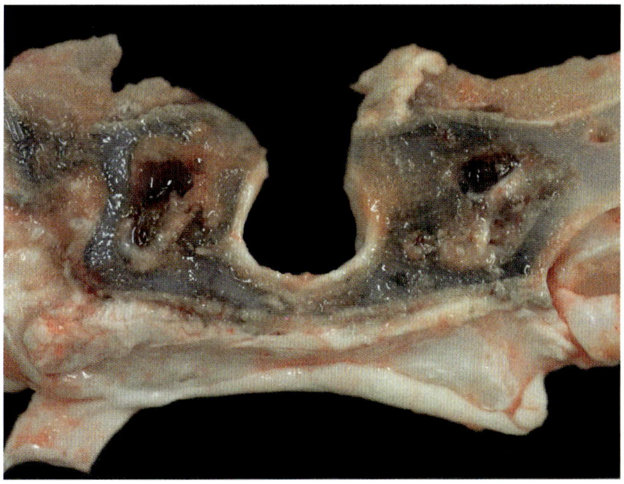

FIGURE 40.25 *Cladosporium* otitis and chondritis in a spotted wobbegong. A transverse section of the chondrocranium through the ears shows replacement of maculae with exudate and massive green discoloration of the surrounding cartilage.

containing different stages of the parasites are present in hypertrophied and necrotic skeletal myofibers and as extracellular spores released from ruptured myofibers. Phagocytosed spores are also present in macrophages. Severe infection may cause mortality.

Metazoa

Monogeneans parasitize the skin, gills, nasal tissues, urogenital tract, gallbladder, or coelom of free-ranging and captive chondrichthyans. Many of these parasites are species specific. Free-ranging chondrichthyans typically harbor low numbers of these parasites with few apparent adverse effects. In managed care, the direct life cycle of these parasites and relative ease with which newly hatched larvae find a host promote hyperinfection. Infections may be more severe in warmer water. Spotted eagle rays and bonnethead sharks in managed care are particularly predisposed to fatal branchial monogenean infections. Monogeneans have a posterior haptor organ for attachment with hooks or muscular suckers and a smaller anterior prohaptor for feeding and attachment. Both structures cause varying degrees of mixed inflammation, necrosis, cellular exfoliation, and reactive epithelial cell hyperplasia at the sites of attachment. Little is known about the biology and feeding behavior of chondrichthyan monogeneans. Some cutaneous monogeneans are large enough to be seen grossly and skin scrapes and gill clips can be useful for cytologic diagnoses. Parasitism of the skin causes grossly visible, multifocal, thick, gray to white lesions. Erosion and ulceration may develop due to rubbing and flashing, and lesions may become secondarily infected with bacteria, fungi, or other parasites. Histologically, epidermal hyperplasia, erosion, and ulceration are present. Chondrichthyans with chronic branchial monogenean parasitism have grossly thickened gill filaments with occasional yellow exudate, and they are often thin with ventral pectoral ulceration (from bottom resting) and soft tissue pallor (Fig. 40.26A). Many branchial monogeneans are too small to see grossly. Gill epithelial cell hyperplasia is characteristic and may obliterate the interlamellar spaces due to fusion of lamellae (Fig. 40.26B; eSlide 40.e7). Epithelial necrosis, erosion, and ulceration and infiltration by granulocytes and lymphocytes may also be present. Chronic branchial infections may have mucoid and alarm cell metaplasia, partial collapse or squamous metaplasia of distal filament, and cartilaginous nodular metaplasia in filaments. Adult monogeneans can

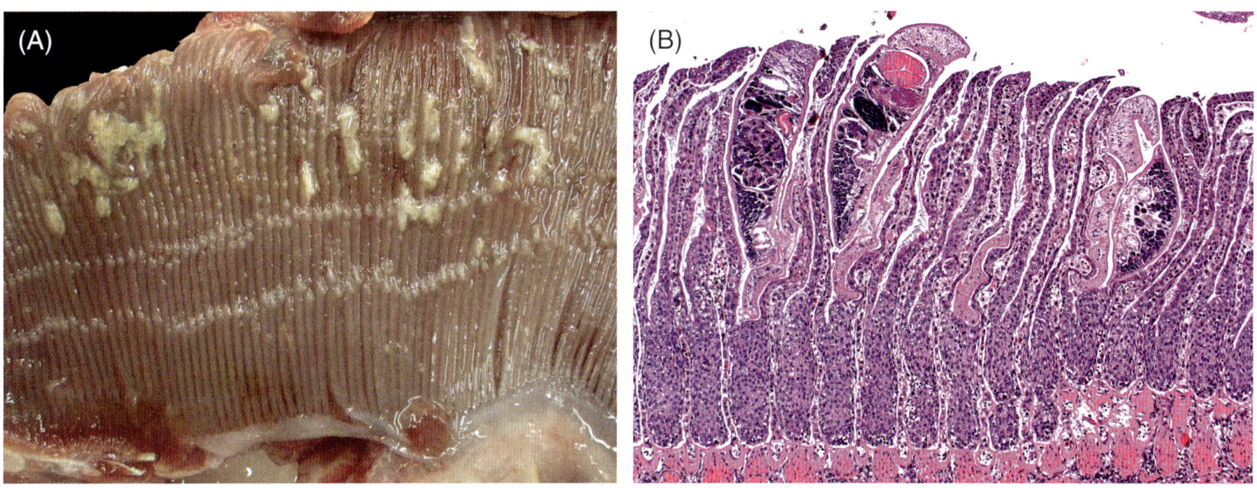

FIGURE 40.26 Gill monogenean parasitism in the gills of a spotted eagle ray. (A) Multifocal accumulations of mucus and parasites are present. Gill filaments are diffusely thickened. (B) Severe gill epithelial hypertrophy and hyperplasia with fusion of secondary lamellae are present related to heavy monogenean parasitism. Four organisms are present in the field, and have a large haptor organ attached to the gill epithelium. The haptor organ of a fifth monogenean is also present (see eSlide 40.e7).

be identified by the large haptor organ, thick eosinophilic cuticle, abundant vitellaria, and paired ceca.

Digenean trematodes are uncommon parasites of the gastrointestinal tract, coelom, cardiovascular system, rectal gland, oral cavity, cloaca, gills, and viscera of chondrichthyans. Rectal gland infections are especially common in black-tipped reef sharks (Garner, pers. observation). Severe infections in the rectal gland may disrupt osmoregulation. Digenean trematodes of the family Aporocotylidae have been found in the heart, conus arteriosus, dorsal aorta, postcardinal vein around the kidney, and mesenteric arteries of elasmobranchs.[194] Digenean trematodes and their ova have been associated with vascular necrosis, vasculitis, and hemorrhage in the brain of chondrichthyans.[66]

Cestodes are the most common helminths of chondrichthyans with approximately 1000 species described; almost all parasitize the spiral valve of free-ranging chondrichthyans and are species specific.[20] Cestodes may also be found in the stomach, small intestine, cloaca, gallbladder, and urinary tract of chondrichthyans.[117] Cestodes may cause epithelial necrosis, ulceration, and hemorrhage at the site of attachment. Mural plerocerci occur most commonly in the stomach and may be surrounded by granulomatous inflammation forming nodules visible grossly.[13] Encysted cestode larvae may also be seen in the mesentery and viscera.[66] Cephalopods are intermediate and/or paratenic hosts for some elasmobranch cestodes.[180]

Nematodes are uncommon parasites of the spiral valve of free-ranging chondrichthyans with no adverse effects. Ova of *Huffmanela* (Nematoda: Trichosomoididae) occur in the skin and oral cavity of free-ranging sharks and sharks in managed care. The life cycle of these parasites is unknown but may require an intermediate host. Affected sharks have black, serpiginous tracts or discrete black foci in the skin or oral cavity.[104] Histologically, larvated eggs with bipolar plugs and a thick, dark brown shell are embedded within the epithelium, which may be hyperplastic and accompanied by granulocytic inflammation and subjacent lymphocytic dermatitis. Adults are not present within lesions and their location within the chondrichthyan host is uncertain.

Meningoencephalitis with intralesional **spirurid** larvae occurs in free-ranging nurse sharks.[30] The location of adult nematodes in the host is unknown. Although not grossly apparent, histologically the meninges contain nodular aggregates of coiled nematode larvae with variable numbers of granulocytes, lymphocytes, and plasma cells. Meningeal vessels are cuffed by inflammatory cells and the subjacent cerebral parenchyma contains perivascular mononuclear inflammation and gliosis. Larvae may spread to the meninges by vascular routes. Similar verminous meningitis was seen in five free-ranging soupfin sharks and one soupfin shark in managed care (Okihiro, unpublished data). Grossly, meninges were thickened and congested, and extradural fluid was cloudy. Histologically, meninges had fibrosis, vascular proliferation, and infiltrates of lymphocytes and macrophages with giant cell formation. Larval nematodes and less frequent ova and adults were present in the inflamed meninges. Intracardiac nematodes of uncertain identity were seen in the atrium of a free-ranging Pacific electric ray (Okihiro, unpublished data). Grossly, the atrium was enlarged, thin walled, and filled with nematodes admixed with a thrombus. Histologically, cardiac myocytes had anisokaryosis, karyomegaly, and vacuolar degeneration accompanied by infiltrates of macrophages. Larval nematodes and thrombi were present in the cardiac lumen. A granuloma surrounding larval nematodes was found adjacent to one kidney and both kidneys had membranoproliferative glomerulonephritis. Nematodes in the soupfin sharks and electric ray were not characterized further. Intravascular nematode larvae may also occur in the brain, skin, gastrointestinal tract, liver, mesentery, and gills of chondrichthyans as an incidental finding or in association with capillary occlusion, necrosis, and inflammation.[66] Incidental granulomas surrounding nematode larvae may also be seen in the gastric wall and other tissues in healthy, free-ranging caught chondrichthyans.

Acanthocephalans are occasionally present in the stomach or spiral valve of free-ranging chondrichthyans and may be accidental infections.[86]

Protozoa

Eimeria spp. are major parasites of chondrichthyans. Most *Eimeria* are parasites of the small intestine and spiral valve, where they cause little disease except in heavy infections, which can result in spiral valve epithelial necrosis, sloughing, and hemorrhage. In contrast, ***Eimeria southwelli*** infects the coelomic epithelium of cownose rays. It is associated with high mortality in captivity.[173] This parasite is also common in free-ranging cownose rays but whether it contributes to morbidity or mortality is unknown. The life cycle is likely direct through environmental contamination from infected animals. Grossly, infected animals have coelomic distension due to turbid effusion, weight loss, and prominent cutaneous blood vessels. The visceral mesothelium is thickened due to both mesothelial hypertrophy and hyperplasia with necrosis and mononuclear infiltrates. Mesothelial cells contain sporulated and unsporulated oocysts and meronts (Fig. 40.27). Organisms can be identified in coelomic fluid obtained at necropsy or through the coelomic pores in live animals.

Apicomplexan hemoparasites of the genus *Haemogregarina* can be seen in the blood cells of chondrichthyans but have not been associated with clinical disease. Trypanosomes have been seen in the blood of free-ranging catsharks and shysharks; their clinical importance is unknown.[149]

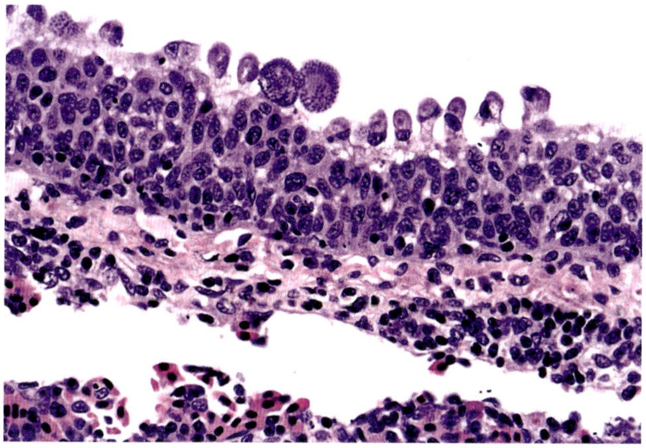

FIGURE 40.27 Coelomic coccidiosis in a cownose ray. Mesothelium has marked hyperplasia with lymphocyte exocytosis and submucosal lymphocyte infiltrates. Superficial epithelial cells contain intracytoplasmic coccidian organisms. *(Photo Courtesy of S. Frasca, Connecticut Veterinary Medical Diagnostic Laboratory, Department of Pathobiology and Veterinary Science, University of Connecticut.)*

Fatal **scuticociliate** infections have been reported in multiple species of sharks; infections are occasionally seen in other chondrichthyan species.[66,95,133,154,175] Most infections are caused by two morphologically similar unicellular ciliates, *Miamiensis avidus* and *Philasterides dicentrarchi*, that share over 95% small subunit ribosomal ribonucleic acid (SSU rRNA) gene identity and had been considered the same organism prior to 2017.[37] In chondrichthyans in managed care, scuticociliatosis can be systemic or limited to the central nervous system, and may occur as an isolated case or an outbreak. Zebra sharks may be predisposed. The portal of entry may determine if the infection is systemic or limited to the central nervous system. Grossly, sharks with systemic infection have multifocal necrohemorrhagic hepatitis, cutaneous ecchymoses, turbid intracranial and coelomic fluid, and/or necrotizing pancreatitis. Histologically, necrotizing vasculitis and intravascular fibrin thrombi are associated with necrosis in affected organs, most commonly the liver, pancreas, and brain (eSlide 40.e8). Intravascular organisms are approximately $20\,\mu m \times 50\,\mu m$ in diameter with a single hyperchromatic round to oval macronucleus and cytoplasmic eosinophilic globules.

In free-ranging chondrichthyans, scuticociliatosis manifests as necrotizing meningoencephalitis without systemic dissemination.[133,154] Epizootics of meningoencephalitis caused by *Miamiensis avidus* in leopard sharks in San Francisco Bay, California (USA) are highly correlated with wet winters. Other wild elasmobranchs affected during epizootics in this geographic region are bat rays, brown smoothhound sharks, shovelnose guitarfish, and one sevengill shark. Ciliates enter through the nares and invade the olfactory organs. Because the olfactory bulbs of chondrichthyans are located near or within the nasal cavities adjacent to the olfactory lamellae, ciliates are able to colonize the meninges and olfactory bulb parenchyma and migrate into other areas of the brain. Gross lesions may not be apparent, or extradural fluid may be cloudy and/or hemorrhagic and the meninges may be thickened (Fig. 40.28). Olfactory lamellae may appear hemorrhagic or severely congested. Histologically, necrosis is present in the olfactory lamellar mucosa accompanied by infiltrates of lymphocytes, macrophages, and granulocytes within the hyperplastic epithelium, submucosa, and olfactory nerve branches. Meningitis and necrosis are present in the olfactory bulbs and peduncles and can extend to other areas of the brain. The rostral cerebral hemispheres are commonly affected, likely because of their proximity to the olfactory bulbs. Affected areas have necrosis of inflammatory cells, neurons, and neuropil. Ciliates may be present

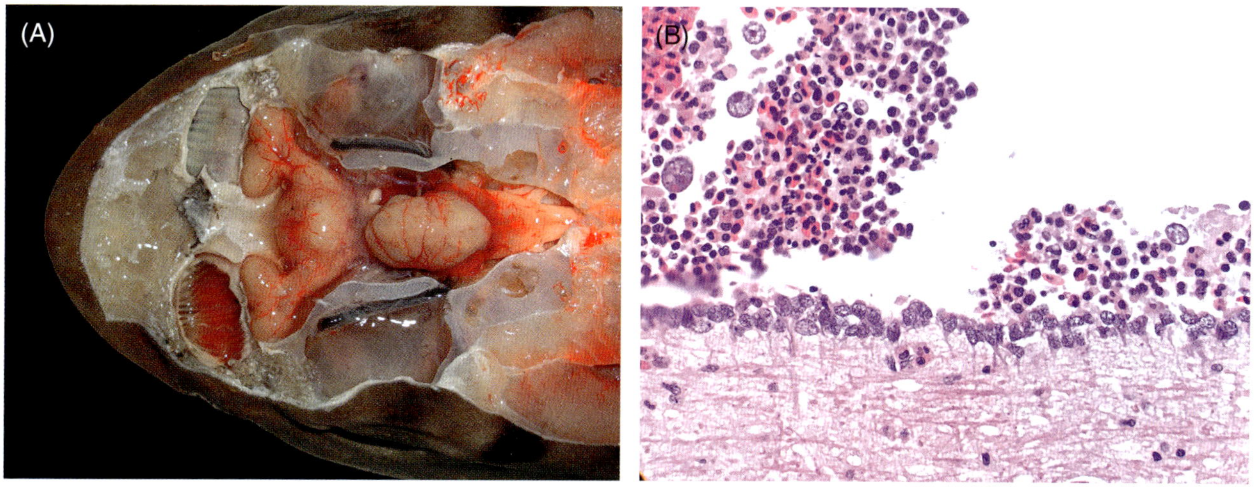

FIGURE 40.28 *Miamiensis avidus* infection of the olfactory organ and brain of a leopard shark. (A) The left olfactory organ is hyperemic. The meninges are congested. (B) The lateral ventricle contains inflammatory cells and scuticociliates.

in areas of necrosis in the olfactory lamellae, olfactory bulbs and peduncles, and other areas of the brain (Fig. 40.28B). Organisms are variable in number and may be difficult to detect microscopically. The lateral ventricles may contain inflammatory cells and ciliates. The meninges and parenchyma are also often congested. In chondrichthyans in managed care with scuticociliatosis of the central nervous system, gross lesions are often mild or inapparent, and only one olfactory organ may be involved.

Other ciliate infections are occasionally described in captive chondrichthyans as opportunistic infections of skin lesions with *Uronema*-like and *Cryptocaryon*-like organisms in the gill interstitium.[66]

Ectoparasites

Barnacles, amphipods, brachyurans, isopods, and copepods may parasitize the skin, corneas, nasal tissues, spiracles, gills, branchial chambers, buccal cavity, stomach, or cloaca of chondrichthyans. Most are of no clinical significance. Larvae of *Gnathia* spp. isopods feed on blood and tissue fluids, and heavy infestations may cause anemia. The isopod *Cirolana borealis* can penetrate the body wall and skeletal muscle of sharks, enter the body cavities, and feed on cardiac and other tissues.[12] Affected sharks have grossly visible isopods within and around the heart, and histologic lesions include myocardial necrosis and mild pericardial lymphocytic inflammation. The condition is severe enough to cause death and may cause periodic declines in free-ranging shark populations. Copepods may cause epithelial erosion, ulceration, hemorrhage, inflammation, and fibroplasia at attachment sites, although their clinical importance may be minimal. Some copepods that attach to the cornea, such as *Ommatokoita elongata*, can cause keratoconjunctivitis.[11] Grossly, affected corneas are multifocally opaque. Histologically, corneal hyperplasia, erosion, ulceration, inflammation, stromal fibrosis and mineralization, and neovascularization are associated with copepod bullae. Anterior uveitis and conjunctival erosion, ulceration, and inflammation may also occur.

Leeches are occasionally found attached to the skin, gills, claspers, oral mucosa, and cloaca of chondrichthyans. They may transmit hemoparasites but are otherwise incidental. Heavy infestations may be clinically important. Experimental infection with the marine leech *Branchellion torpedinis* caused anorexia, cutaneous ulceration, blood loss anemia and decreased total solids, and death in yellow stingrays.[108] Leeches may enter the uterus and attach to the skin of intrauterine embryos.[127]

Other Infectious Organisms

Myxozoan spores or spore-forming plasmodia are most commonly found in the gallbladder, bile ducts, glomeruli, meninges, and skeletal muscle of free-ranging and captive chondrichthyans and are generally incidental.[66,96] Plasmodia in the cytoplasm of the host cell contain numerous developing myxospores, often with no other histologic abnormalities in the host cell or associated tissue inflammatory response. Larger spore size is correlated with host size and habitat depth for myxozoa of the genus *Ceratomyxa*.[96]

ADDITIONAL RESOURCES

Garner MM. A retrospective study of diseases in elasmobranchs. *Vet Pathol*. 2013;50:377–389.

Mylniczenko ND, Sumigama S, Wyffels JT, Wheaton CJ, Guttridge TL, DiRocco S, Penfold LM. Ultrasonographic and hormonal characterization of reproductive health and disease in wild, semiwild, and aquarium-housed southern stingrays (*Hypanus americanus*). Am J Vet Res. 2019;80:931–942.

Pirarat N, Sahatrakul K, Lacharoje S, Lombardini E, Chansue N, Techangamsuwan S. Molecular and pathological characterization of *Fusarium solani* species complex infection in the head and lateral line system of *Sphyrna lewini*. *Dis Aquat Organ*. 2016;120: 195–204.

Retallack H, Okihiro MS, Britton E, Van Sommeran S, DeRisi JL. Metagenomic next-generation sequencing reveals *Miamiensis avidus* (Ciliophora: Scuticociliatida) in the 2017 epizootic of leopard sharks (*Triakis semifasciata*) in San Francisco Bay, California, USA. J Wildl Dis. 2019;55: 375–386.

Steele LM, Okihiro MS, Berlemont R, Dillon JG, Young KA, Hesami S, Sommeran SV, Lowe CG. *Carnobacterium maltaromaticum* associated with meningoencephalitis and otitis in stranded common thresher sharks (*Alopias vulpinus*). *Vet Pathol*. 2022;59:850–859.

References

Full references for this chapter can be found online at https://doi.org/10.1016/B978-0-443-18437-6.00043-2.

Chapter 41

Invertebrates

Alisa L. Newton[a,b], Roxanna Smolowitz[c], and Elise LaDouceur[d]

[a]ZooQuatic Laboratory, LLC, Baltimore, MD, United States; [b]OCEARCH, Park City, UT, United States; [c]Roger Williams University, Bristol, RI, United States; [d]Joint Pathology Center, Veterinary Pathology Service, Silver Spring, MD, United States

INTRODUCTION

More than 99% of all animal species and over 1 million named and an estimated 30 million unnamed species are invertebrates.[193] They suffer from a variety of diseases similar to their vertebrate counterparts, but disease investigations have primarily focused on infectious diseases of economically important marine species. Invertebrate conservation efforts lag woefully behind those in vertebrates. The International Union for the Conservation of Nature (IUCN) red list excludes small-bodied species, those with narrow distribution ranges and low dispersal abilities, leading to a taxonomic bias.[34] With the most recent update, the Red List has assessed 80% of described vertebrate species but only 2% of described invertebrate species. Within the 2% of invertebrates evaluated, 23% are considered threatened (Critically Endangered, Endangered, or Vulnerable) (Table 41.1).[110] By necessity, this chapter will focus on a small number of diseases in only a fraction of all invertebrates and aims to provide an approach to assessing disease processes in other invertebrates. The four major phyla (Cnidaria, Mollusca, Arthropoda, Echinodermata), subphyla, and diseases selected in this chapter were due to their ecological and commercial value, and pertinence to species in managed care. These factors have driven the need for diagnostic pathology and ancillary diagnostic testing in invertebrates to advance veterinary care.

Invertebrate diseases are frequently complex multifactorial processes, and the common terrestrial model of one-pathogen one-disease simply does not apply in many cases. The fulfillment of Koch's postulates, where a single species of disease-causing pathogen is successfully (1) isolated from an infected animal, (2) grown in pure culture, (3) reinoculated into a healthy individual, and (4) produces the same disease, is particularly challenging in invertebrates due to limited invertebrate cell lines and the difficulty of culturing many invertebrate pathogens. The marine environment is also particularly complex due to the higher taxonomic diversity of hosts and parasites, differences in life histories, the more open nature of recruitment, differences in anthropogenic impacts, and the potential for different modes of disease transmission.[141] It is important as we approach disease investigations in invertebrates to consider the significant contribution of environmental and other factors to susceptibility and pathogenesis, and to employ methods of investigation that include detailed anatomic descriptions, ancillary diagnostic testing, and controlled experimental design to arrive at the case description, etiology, and pathogenesis of a disease.[240]

NON-INFECTIOUS DISEASES

Nutritional

Molt death syndrome (MDS) of larval decapod crustaceans is related to lipid nutrition, specifically low dietary phospholipids, cholesterol, and n-3 highly unsaturated fatty acids. Gross findings are best described in American lobster and include animals dying in a partially molted condition or animals molting and having missing or deformed appendages. Multifocal calcium deposits are associated with the inner surface of the exuvial skeleton and are visible grossly with the aid of a dissecting microscope and scanning electron microscopy. These range in size up to 0.2 mm diameter, are pink to colorless, and are primarily found on the inner surface of the claws and carapace. In severe cases, there is mottling of the carapace and the cuticle of the thoracic limbs, tail segments, uropods, and telson. Deposits are similar to calcite ($CaCO_3$) and may adhere to the new inner exoskeleton and cause deformities that make shedding of the exuvium difficult and may lead to death.[26,230]

Impaction has been reported anecdotally in terrestrial invertebrates in managed care due to factors such as dehydration and ingestion of inappropriate/nonnative feed or substrate. Histologic lesions include distention, impaction, ulceration, and hemocytic inflammation of the midgut with associated melanization.[32]

Pathology of Wildlife and Zoo Animals. https://doi.org/10.1016/B978-0-443-18437-6.00030-4

TABLE 41.1 Number and Percentage of Described Invertebrate Species Evaluated by the IUCN Red List 2022, Including the Number of Threatened Species by Major Group

	Estimated Number of Described Species (2022)	Number Evaluated by IUCN (2022)	% of Described Species Evaluated (2022)	Number of Threatened Species (2022)
Insects	1,053,578	12,161	1.2%	2,291
Mollusks	84,528	9,017	11%	2,384
Crustaceans	80,122	3,197	4%	745
Corals	5,574	846	15%	232
Arachnids	110,615	441	0.40%	251
Velvet worms	210	11	5%	9
Horseshoe crabs	4	4	100%	2
Other	157,543	904	0.57%	152

Adapted from IUCN Red List version 2022-1: Table 1a.

Toxic

A wide array of organic and inorganic toxins are described in invertebrates; many are intentionally used against invertebrates in pest control. These agents are summarized in other texts (see Invertebrate Pathology: Noncommunicable Diseases[212] and Pathobiology of Marine and Estuarine Organisms[83]). One toxin of note in aquatic invertebrates is **copper**. Copper is an essential micronutrient but toxic concentrations of the bioavailable forms result in adverse effects on aquatic and semiaquatic invertebrates. Environmental factors, such as dissolved organic and inorganic ligands, and competition with dissolved cations (water hardness) all significantly impact bioavailability and therefore total copper levels in water samples are insufficient to predict toxicity. Sensitivity, morbidity, mortality, and tissue response vary significantly between species and are intimately related to species physiology and life stage.[7,91] For example, adult oysters and clams are largely resistant to copper toxicity and can sequester copper in hemocyte clusters in the gills and mantle resulting in green discoloration in these areas, while larval oysters, adult and larval mussels (genus *Mytilus*), and some echinoderms (sand dollar, sea urchin) are considered some of the most copper sensitive taxa.[7,212]

The proposed pathogenesis of **copper toxicity** includes oxidative stress, DNA damage, altered ion regulation, and metabolic derangements due to impaired oxygen uptake. In invertebrates, lesions described with copper toxicity are similar to those described with other heavy metals and pollutants and are not specific to copper. Gross lesions include **"distress syndrome"** in freshwater pulmonate snails, which is characterized by immobility, muscular spasms, extension of the cephalopedal mass, inability to attach the foot to substrate, swelling of the tentacles, and epidermal sloughing. Gross findings in sea stars are described as muscular weakness, loss of righting ability, lack of tube feet coordination, mouth relaxation, and gastric eversion. Histologic lesions described in invertebrates include zooxanthella expulsion in cnidarians; branchial epithelial necrosis, hemocytic inflammation in the connective tissue, and necrosis and sloughing of the mucosa in gastropods; nephridial necrosis in sea hares; mucosal necrosis in the stomach and digestive gland atrophy in oysters; inflammation, necrosis, and vacuolation in the digestive diverticula, kidney, and intestinal tract of clams; branchial necrosis in crabs; and epidermal ulceration and irregular epidermal contour in numerous aquatic invertebrate species. The latter is believed to be the result of copper-induced peroxidation of membrane lipids and impaired epidermal cellular membrane structure and function.[61,83,125]

Age Related/Degenerative

Senescence occurs at the end of a brief, single reproductive event in many invertebrate species.[148] It is hypothesized to be the end result of the energetic costs of gamete production and the act of mating.[71] However, some species are able to mate over an extended period of time and moderate the effects of reproduction, resulting in an uncoupling of reproduction and senescence. Among invertebrates in managed care, senescence is most often seen in cephalopods after spawning. Histologic lesions associated with this process include degeneration and atrophy of the gonads, digestive glands, salivary glands, and accessory sex glands. Senescent animals also often have lesions of negative energy balance (due to terminal anorexia) and concurrent infections with particularly high levels of parasitism (LaDouceur, unpublished data). In senescent animals, bacteria can often be cultured from the hemolymph or renal sac (Smolowitz, unpublished data).

Inflammatory Non-Infectious

Eversion syndrome is an idiopathic condition of captive scyphomedusa jellies that results in bell eversion into a concave position. This results in exposure of the manubrium and gastric cavity, which often leads to death. The primary histologic finding is striated muscle degeneration (loss of striation and swelling) and necrosis, and mesogleal degeneration characterized by thinning, fiber loss, and disarray. Affected species include moon jellies, fried egg jellies, Pacific sea nettle, Atlantic sea nettle, Northern sea nettle, and purple striped jellies. Affected age groups are either young growing animals (1–2 months old) or older animals (greater than 6 months). A common etiology has not been identified despite investigations into husbandry practices, environmental, and nutritional factors.[72]

Miscellaneous/Multifactorial

The most common non-infectious disease processes encountered in invertebrates are direct trauma or organism stress. The etiologies vary by situation but frequently overlap. Environmental stress (e.g., temperature, pH, and anthropogenic), handling/husbandry, and social interactions are the most frequent factors. If a limb is damaged due to trauma, invertebrate animals may induce **autotomy**, which is a voluntary action to remove tissue. Autotomy occurs in arachnids, echinoderms, cephalopods, crustaceans, and other species. Following autotomy, invertebrates may regrow the body part, often a limb. Gross lesions consistent with trauma vary from areas of epidermal or exoskeletal discoloration, tissue loss, and a loss of surface features due to exoskeletal fractures. Damage to the cuticle, epidermis, or exoskeleton can result in loss of osmoregulatory capacity, hemolymph loss and shock, opportunistic infection, and in many cases death. Even minor trauma in arachnids can cause death, particularly if located on the dorsal prosoma, due to exsanguination. Trauma with exsanguination is a common cause of captive spider mortality.[174]

The histologic response to injury has been described in detail in many invertebrate species including soft and hard corals,[144,165] black abalone,[6] oyster,[54] freshwater mussel,[167,168] Drosophila,[18] wax moths,[191] horseshoe crabs,[33] sea cucumbers,[48,143] penaeid shrimp,[69,134]

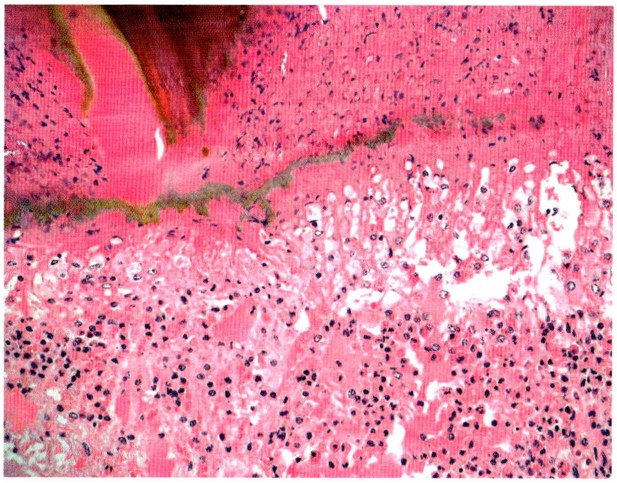

FIGURE 41.1 Scab formation at a site of carapace trauma on a walking leg of a goliath bird-eating spider. Eosinophilic lakes of coagulated hemolymph (clot) bridges a site of carapace fracture. Granular hemocytes infiltrate the clot and aggregate with signs of degeneration at the surface of the clot. A line of melanization seals the hemocoel.

leeches,[103,104,221] and octopus.[66,176] In most invertebrates, **wound healing** begins with the formation of a **hemolymph clot**, which rapidly prevents further hemolymph loss and entraps microbial pathogens before they can enter and spread through the hemocoel.[112] Clotting initiation is either the result of a primary proteolytic cascade or is initiated by degranulation of local hemocytes resulting in the formation of a protein aggregate or "**soft clot**." These proteins become crosslinked in some species (the "**hard clot**") and later infiltrated by other types of hemocytes (scab formation; Fig. 41.1) that seal the hemocoel from the outside environment and serve as the scaffolding over which the epidermis can regenerate, cover the wound site, and replace the scab. Additional in-depth description of the invertebrate immune response, inflammation, and special responses such as melanization are discussed in the Digital Content.

The epidermis and mantle of cephalopod mollusks are common sites of injury. The simple columnar epithelium is very fragile, easily damaged, and heals poorly. Opportunistic bacterial infections secondary to epidermal damage are common and often result in septicemia and death. Bacteria of the genus *Vibrio* sp. are the most common opportunistic agents. **Trauma to the mantle apex** (Fig. 41.2) due to collision with tank walls after startling or while trying to escape perceived threats is a frequent occurrence in managed care.[199,201] These lesions can be acute or chronic (following multiple episodes of collision) and can result in deep ulcerations that extend into the subjacent connective tissues and in squid and cuttlefish may also extend to involve the pen or cuttlebone sac. In cuttlefish, mantle abrasions and lacerations may occur secondary to crowding and territorial competition between males. Cuttlefish have also been reported to develop gas pockets beneath the mantle

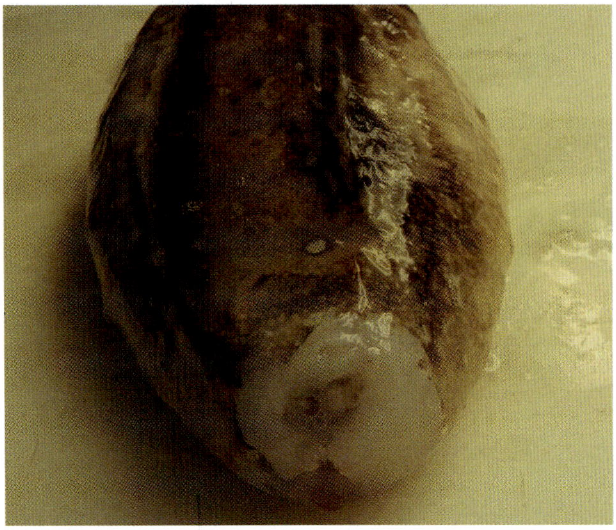

FIGURE 41.2 Mantle trauma in a common cuttlefish. The posterior tip of the mantle demonstrates locally extensive epidermal ulceration with exposure of the mantle dermis, skeletal muscle, and (centrally) the distal tip of the intact cuttlebone. *(Photo Courtesy of K. Conley, Disney's Animals, Science & Environment.)*

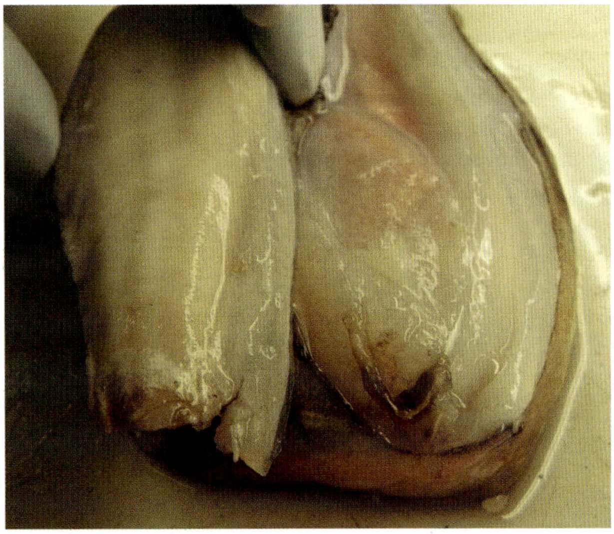

FIGURE 41.3 Cuttlebone fracture in a common cuttlefish. The posterior margin of the cuttlebone is deformed, irregular, and fractured with exposure of the interior lamella and white discoloration of the cuttlebone sac. *(Photo Courtesy of C. Rodriguez, Disney's Animals, Science & Environment.)*

epidermis secondary to hovering in the stream of bubbles from air stones. These gas pockets later rupture resulting in a focal ulcer leading to similar complications to primary traumatic injuries.[198]

Cuttlebone fracture (Fig. 41.3) is a frequent additional injury seen with trauma in cuttlefish. The cuttlebone consists of cuttlebone sac and epithelium that secretes lamellae of calcified chitin arranged in parallel layers separated by pillars that are sealed along the lateral and anterior margins by a thick calcified layer. These layers are added sequentially, with new chambers added ventrally. Ultimately, what is created is a series of chambers filled with either gas (ventrally) or fluid (dorsally) that allow for near-neutral buoyancy of cuttlefish.[130,242] Fracture is a common occurrence in both free-ranging and managed cuttlefish due to external trauma by predators or intraspecies aggression, compression fracture due to excess hydrostatic pressure, or self-trauma (e.g., collisions with tank walls).[22] Nutritional deficiencies and stressors, such as abnormal water temperatures or elevated nitrogen cycle metabolites, may impact normal cuttlebone mineral deposition in animals in managed care leaving it brittle and more easily fractured.[201] Fractures can be either longitudinal or transverse. Gross lesions are most often evident along the anterior rim near the nuchal mantle edge. They consist of mild thickening of the dorsal aspect of the bone and characteristic dark lines along the ventral surface due to melanization. Healing of lateral shell fractures may result in an asymmetrical cuttlebone shape, as recovery of normal shell growth is not always possible after healing.[22,158] Chronic low-impact trauma can cause prolapse and proliferation of the cuttlebone sac connective tissue and adjacent mantle dermis reminiscent of a neoplasm. Occasionally, the stomach and cecum will herniate, or the dorsal mantle will detach from the cuttlebone. These complications often lead to death.[158] Healing is facilitated by secretions of the cuttlebone sac epithelium.

Idiopathic self-trauma (automutilation syndrome) has been described in three species of octopus in managed care: giant Pacific octopus, California two-spot octopus, and Mexican four-eyed octopus.[57,182] Animals develop circular ulcers on the mantle, proximal arms, or both. Mantle lesions begin as punctate ulcers that gradually enlarge over several weeks. They typically occur on the anterior surface between the eyes or immediately above the origin of the arms. Limb lesions include ulcers on the dorsal surface, loss of suckers, or loss of the tips of multiple arms. These lesions are described as occurring acutely (often overnight) with no history of exposure to live prey or conspecifics. In many cases, the suckers and arm tips are later recovered from the stomach of the animal, confirming self-mutilation. Histopathology has identified arm nerve cord (axial nerve) abnormalities including edema, neuritis, and necrosis in many cases. Vascular lesions, including thrombosis within the arms and mantle, and myositis adjacent to areas of mantle damage, have also been noted. A mixture of common opportunistic bacteria has been isolated from these lesions including *Aeromonas* spp., *Vibrio* spp., and *Staphylococcus* spp. The etiology of this syndrome is unknown, but hyperesthesia secondary to the neural and vascular lesions is suspected.

In cnidaria, **sedimentation** and increased turbidity from dredging, drilling, and agricultural runoff interferes with coral autotrophy and heterotrophy by decreasing available light and photosynthetic activity, which causes polyp retraction and interferes with prey capture. This is believed to produce negative energy balance that may be further complicated by increased energy demand to clear excess sediment. Sediments can also cause local and systemic tissue injury.[187] Gross lesions include polyp swelling, tissue thinning, discoloration, pallor, and tight polyp retraction, along with distention and discoloration of the oral disc.[227] Early histologic changes include multifocal to diffuse changes in the number and size of mucous secretory cells (mucocytes) in the epidermis and gastrodermis. Mucocytes may show variable uptake of pentachrome dyes with special staining, suggesting a change in the composition of the mucus. At 3–4 weeks of exposure, colonies develop attenuation of the epidermis and gastrodermis, atrophy of epidermal mucocytes, decreased mucus secretion, and eventually tissue damage and necrosis.

Ulcerative lesions of the umbrella (bell rot) have been reported in a number of jellyfish species in managed care including moon jellies, Pacific sea nettles, Atlantic sea nettles, lagoon jellies, blubber jellies, Sandaria jellies, and white-spotted jellies (Fig. 41.4; eSlide 41.e1). Although a specific etiology has not been identified, trauma and environmental stress are considered significant contributing factors. Husbandry practices resulting in irregular tank flow may allow for collisions with tank walls, and increased tank density can allow for conspecific collisions. Gross lesions vary from erosion to ulceration of the peripheral and dorsal aspects of the bell and can range in size 1–2 mm and up to 6 mm diameter. Lesions can progress to full-thickness tears with secondary herniation of the internal viscera. Histologic lesions include epidermal necrosis (cell swelling, pyknosis, and lysis) of the exumbrella and on occasion the subumbrella with ulceration and local hemocyte infiltration. Cnidocytes are spared. Secondary algal and bacterial invasion may be present.[123,215]

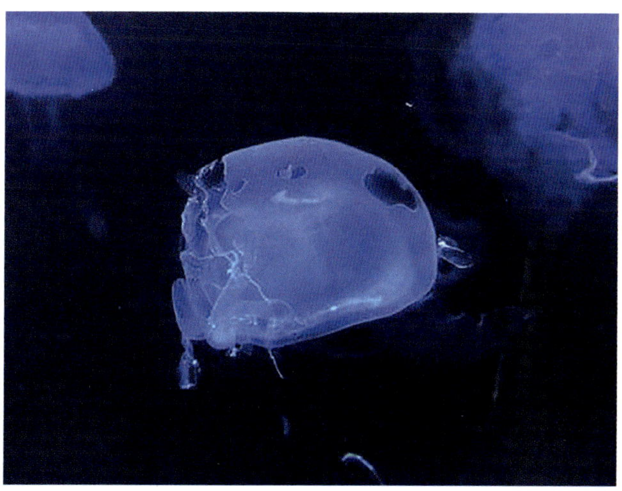

FIGURE 41.4 Ulceration of the umbrella of moon jelly. Along the aboral surface of the bell, there are numerous irregular dark areas of cavitation at sites of exumbrella ulceration along with sites of extensive umbrellar tissue loss and deformity where ulcers are coalescent. *(Photo Courtesy of N. Mylniczenko, Disney's Animals, Science & Environment.)* (see eSlide 41.e1).

Emaciation and **dehydration** in terrestrial invertebrates are common issues in managed care. Gross manifestations are most apparent within the digestive organs, specifically the digestive gland. These organs often serve as sites of energy/nutritional storage, similar to what is seen in the liver and adipose stores of vertebrates. Gross lesions are apparent externally in arthropods and hexapods, and consist of a pronounced decrease in the size of the abdominal segment or opisthosoma due to the reduced size of the digestive system (Figs. 41.5 A, B). Internally, the small size of the digestive gland and/or digestive system will be grossly apparent in most species. Histologically, the digestive gland is typically composed of pyramidal cells with abundant cytoplasm containing eosinophilic secretory products, fine granulation, or vacuolation. With emaciation, digestive gland cells are reduced in size and demonstrate reduced vacuolation, eosinophilic cytoplasmic products, and granulation. Emaciation can be due to a variety of causes and can be compounded by debilitation from other illnesses. Husbandry should always be investigated, particularly nutrition, prey appropriateness (species, size) and availability, water sources, water quality, and humidity levels. Lighting needs to be investigated in photosynthetic species and overventilation may contribute to dehydration in terrestrial species.

The majority of coral diseases are complex processes, and a single primary etiologic agent has been confirmed in only a handful of diseases to date. Even in those diseases, other contributory factors are likely to be significant. A thorough summary of the current understanding of coral diseases was published in Diseases of Coral by Woodley et al.[236] and exceeds what can be covered in this chapter. A few important disease examples are highlighted below.

Bleaching in cnidaria, particularly scleractinian corals, is the result of a breakdown of coral-algal symbiosis due to loss of symbiotic photosynthetic algae (*Symbiodinium* or zooxanthella) or a reduction in algal per cell pigment concentrations. Temperature extremes and solar radiation have received the most focus, but other environmental factors, which include salinity extremes, sediment, low temperature, desiccation, low light, ultraviolet light, and reduced pH are considered as risk factors. Contaminants, such as herbicides, copper, cyanide, oil, and sunscreens, may also serve as initiators. The primary gross lesion associated with bleaching is patchy to diffuse pallor of the affected colony.[11] Three different tissue changes result in reduced zooxanthellae densities in affected corals. These include degeneration of the zooxanthellae within gastrodermal and epidermal cells, release of zooxanthellae from the gastrodermis lining the mesenterial filaments, and sloughing of

FIGURE 41.5 Dehydration in a goliath bird-eating spider. (A) Normal opisthosoma. (B) The opisthosoma in a spider with dehydration. Note the overall decrease in size of the opisthosoma as well as the irregular contour (wrinkling) of the cuticle in the animal with dehydration.

the gastrodermis and secondary loss of the zooxanthellae.[28] By light microscopy, intact zooxanthella should be circular with a distinguishable nucleus, pyrenoids, and assimilation bodies. Degeneration of zooxanthella is characterized by a loss of circular profile, cytoplasmic vacuolation, and in some cases extreme enlargement. Zooxanthella loss from gastrodermal cells is believed to be related to the changes in coral metabolism associated with stress. Environmental bleaching is due to a mixture of these three tissue changes. Additional associated lesions seen with temperature stress include epidermal and gastrodermal atrophy and erosion, retractor muscle atrophy, and ultimately loss of tissue architecture and tissue necrosis secondary to zooxanthella necrosis.[89] If prolonged and severe, bleaching results in coral mortality; milder bleaching events can be reversible in weeks or months.

Colony collapse disorder (CCD) is a complex syndrome affecting honey bee colonies, the hallmark of which is rapid loss of adult worker bees. CCD is discussed here, prior to other infectious diseases of bees, as this disorder is likely multifactorial and some of the identified contributory factors will be discussed in more detail later in the text. Large-scale losses (30%–90%) of honey bee colonies were first reported in the fall of 2006 in the United States. Symptoms in affected colonies were not consistent with known bee pathogens.[62] Characteristic findings in affected colonies included (1) rapid loss of adult worker bees; (2) brood populations in excess relative to remaining adult bee populations; (3) lack of dead worker bees within or surrounding affected hives; (4) delayed invasion of common hive pests, such as small hive beetles and wax moths; and (5) kleptoparasitism from neighboring honey bee colonies.[226] In colonies that had not yet completely collapsed, other findings included (1) an insufficient number of bees to maintain the number of brood in the colony; (2) a workforce composed predominantly of younger adult bees; (3) persistence of the queen; and (4) remaining bees that were reluctant to consume food offered by bee keepers.[62] The three major hypotheses regarding the cause of CCD include (1) novel or resurgent pathogens; (2) poor nutrition and resultant developmental disorders; and (3) environmental contaminants, which include agricultural pesticides or chemicals used to control common bee colony pests.[47,62] Other factors suggested in the literature include management stress and poor genetic biodiversity in the queen bee source.[62] Potential synergistic interactions between two or more of the aforementioned hypotheses (and potentially others) are likely.

To date, no single disease agent has been associated with CCD, but infectious diseases are believed to play critical roles. Epidemiologic studies of CCD have shown a higher incidence and abundance of infectious agents in CCD colonies, which suggests they either are exposed to greater numbers of pathogens, that coinfections can act synergistically, or that colonies are immune compromised.[47,226]

In particular, RNA viruses of the family **Dicistroviridae [Israeli acute paralysis virus (IAPV), acute bee paralysis virus (ABPV), Kashmir bee virus (KBV), black queen cell virus (BQCV)],** and **Iflaviridae [deformed wing virus (DWV)]** and **microsporidia (*Nosema* spp.)** have been implicated. Factors that can affect colony immunocompetence include poor nutrition and chronic sublethal exposure to agricultural or bee-keeping pesticides. Evidence for low-level exposure to environmental toxins is equivocal and neither individual chemicals nor chemical load have been associated with increased risk of CCD. Nutritional resources can affect overwintering survival and bee longevity but analysis of bee colonies for body weight, protein components, and lipid stores has not shown a direct link between food resources and collapse.[226]

Epizootics of a process termed **wasting disease** have been reported across all subphyla of echinoderms, including sea stars, sea urchins, brittle stars, and sea cucumber species, with descriptions going back to 1898.[13,59] **Sea star wasting (SSW)** events, often resulting in regional mortality, have been reported globally including the North American Pacific Coast, Atlantic Coasts of North America and Europe, the Mediterranean Sea, East China Sea, Yellow Sea, and Port Philip Bay, Australia.[162] The most recent event, which began in 2013 resulted in the loss of billions of sea stars of at least 20 species and devastated populations along the Pacific coast of North America from southeastern Alaska to southern California. At least five Pacific coast public aquaria that utilized "open" water systems were also impacted. *Pisaster*, *Pycnopodia*, and *Evasterias* were the most impacted genera and mortality approached 100% in select species at many locations. Clinical and gross findings associated with wasting disease include (1) loss of body turgor (deflation) and weakness; (2) foci of body wall pallor and tissue loss; (3) sloughing of multiple rays and/or rupture of the body wall with evisceration; and (4) death (Fig. 41.6). Consistent histologic changes include epidermal degeneration, necrosis, and ulceration with dermal separation, necrosis, and inflammation. Community fingerprinting and bacterial metagenomics performed during the recent mortality event identified three candidate disease-associated bacterial families including Bacteroidetes, Gammaproteobacteria, and Spirochaetes. Viral metagenomes prepared from symptomatic and asymptomatic sea stars identified several candidate disease–associated metazoan viruses including a sea star-associated densovirus (Parvoviridae).[92] Links between the histologic lesions and candidate disease-associated pathogens remain under investigation but to date no infectious agents have been demonstrated by light or electron microscopy within affected sea star tissues or lesions. Small changes in temperature, both increased and decreased have been reported to influence the progression and intensity of disease in different species.[118,142] Analysis of sea surface temperature and precipitation patterns present during the Pacific Coast wasting event demonstrated an inconsistent temperature relationship and regional associations with reduced precipitation. A relationship between environmental conditions and RNA viruses was demonstrated, but the association between densoviral infection was only apparent in some locations and specific species, making it unlikely to be the sole cause of the wide-ranging Pacific coast wasting event.[98] Seasonal changes in reproductive potential, fluctuations in temperature (both increase and decrease), host condition/acclimation, and anthropogenic factors may impact the dynamics of the disease.[13] SSW likely represents a common suite of gross and histologic changes that echinoderms elicit in the face of heterogeneous pathogens and environmental conditions.[98]

FIGURE 41.6 Sea star wasting in an ochre sea star. Characteristic lesions begin on the ray near the junction with the disc. The white coloration may be due to pallor of the intact epidermis or may represent areas of tissue loss with dermal and skeletal exposure. *(Photo Courtesy of M. Miner, University of California, Santa Cruz.)*

Neoplastic

Neoplasia is a documented occurrence across all invertebrate taxa, but challenges remain with regard to tumor diagnosis, identification, and treatment. Spontaneous, hereditary, or acquired (e.g., radiation, chemical, or viral) genetic alterations that are well described in vertebrates, also occur in invertebrates.[161,188] Reports of neoplasia are frequent in the phylum Mollusca and subphylum Hexapoda (class Insecta), infrequent in phylum Cnidaria and subphylum Crustacea, and are not yet documented in phyla Porifera or Echinodermata. This distribution is likely biased by the intensive monitoring that occurs in invertebrates of economic importance.

There are many challenges to documenting neoplasms in invertebrates. Images and descriptions of gross and microscopic findings in the older literature, which would allow for comparison or peer evaluation of tumor identification, are inadequate.[196] The classifications that exist for mammalian tumors are difficult to apply to invertebrate neoplasms and may not be appropriate. Ancillary testing to facilitate cell identification, such as immunohistochemistry, is generally unavailable. Therefore, it is critical for pathologists to be familiar with the identification of invertebrate tissues, normal tissue reaction, inflammation, and wound repair in a given species when considering a diagnosis of neoplasia. The characteristics used to define neoplasms in vertebrates, such as self-sufficiency in growth signals, insensitivity to antigrowth signals, evasion of apoptosis, unlimited replicative potential, sustained angiogenesis, tissue invasion, and metastasis, may be challenging to fulfill across invertebrate taxa. Organs, organ systems, cell types, and metabolic reactions are different, and the anatomy of most invertebrate circulatory systems (partially open or lacking organization) makes documenting angiogenesis extremely difficult.[95,153]

The fundamental characteristics used to differentiate benign from malignant neoplasms in vertebrates include (1) degree of cellular differentiation and anaplasia; (2) rate of growth; (3) local invasion; and (4) evidence of local or distant metastasis.[120] These features should be applied with caution as invertebrate neoplasms have rarely been studied over time, mitotic figures can be transient, cell division may be under heavy hormonal control, and evidence of local invasion and metastasis can be difficult to assess in species with open circulatory systems.[153] The majority of neoplasms described in invertebrates have been reported to be benign, largely due to a lack of variation in cell size and shape (anaplasia), lack of mitotic figures, and a lack of organ invasion or metastatic spread.[188] The establishment of the Registry of Tumors in Lower Animals (RTLA), a cooperative project in the United States between the National Institutes of Health's (NIH), National Cancer Institute (NCI), and the Smithsonian Institution's National Museum of Natural History in 1965 greatly advanced our understanding of the occurrence, gross and microscopic pathology, and etiology of invertebrate neoplasms. It is notable that of the 1550 invertebrate submissions made to the RTLA database and evaluated by the RTLA pathologists between 1965 and its closure in 2007, only 30% were ultimately designated to be neoplasms.[153]

Disseminated neoplasia of presumed hematogenous (hemic) origin (also termed **hemic neoplasia, hemic sarcoma, hematopoietic neoplasia**, and **hemocytic leukemia**) is one of the most commonly described neoplastic conditions of invertebrates. This condition is described in 15 species of bivalve mollusks as well as white shrimp, and mutant strains of Drosophila.[64,65,133,203] The cytologic and histologic features are similar across species and include the presence of large, round, anaplastic cells in connective tissues, blood vessels, sinuses of the visceral mass, muscle, and mantle tissue (Fig. 41.7). These cells are two to four times the diameter of a normal hemocyte, with a nuclear to cytoplasmic volume ratio of 1:1. They contain one or more prominent nucleoli and one to several vesicles filled with triglycerides and lipids. Mitotic figures are common. Infiltrating cells cause displacement, compression, and necrosis of the gill, gonad, and connective tissue, and arrest gametogenesis by causing gonad atrophy and degeneration. Death is the ultimate outcome of this tumor due to the effects of organ disruption and dysfunction of the immune system. Gross findings in affected individuals are nonspecific and considered general signs of poor doing in bivalves including emaciation, pallor of the digestive gland, and mantle recession.[12,63]

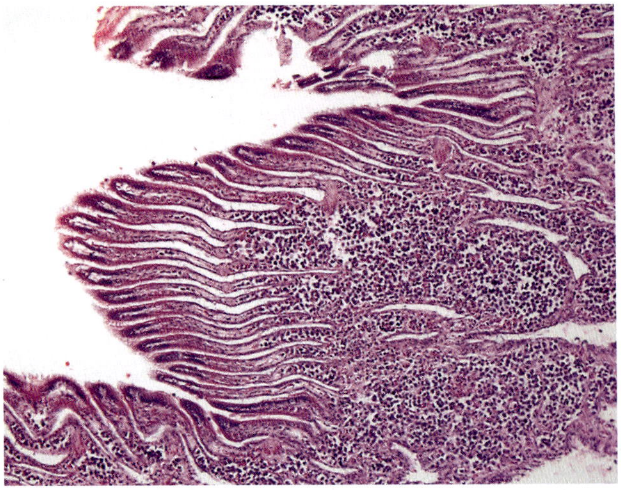

FIGURE 41.7 Disseminated hemic neoplasia in the gill of a softshell clam. Neoplastic hemocytes fill and expand the vascular spaces of the gill resulting in distortion of the tissue architecture.

An etiology for disseminated neoplasia has not been identified in bivalves or crustaceans. A clear consensus has not been reached regarding the exact cell of origin in most tumors. Most authors favor a hemic origin due to similarities between the neoplastic cells and granular or hyaline hemocytes and detection of neoplastic cells first in the circulatory system; a gonadal origin has also been proposed.[12,64] A relationship between prevalence and environmental carcinogens, biotoxins, stress, and genetics has been suggested but not proven.[12,25,229] There is variation in the occurrence of disseminated neoplasia between species, locations, and time of year suggesting that tumor development is likely a complex, multifactorial process.[229] In a single case report of disseminated neoplasia of shrimp, lesions consistent with **infectious hypodermal and hematopoietic necrosis virus (IHHN)** were present in the neoplastic hematopoietic tissue suggesting a possible viral infection.[133] Metzger et al.[145] have recently shown that these tumors in bivalves are most likely the result of a transmissible neoplastic cell that in most cases are species specific or at least only infect closely related

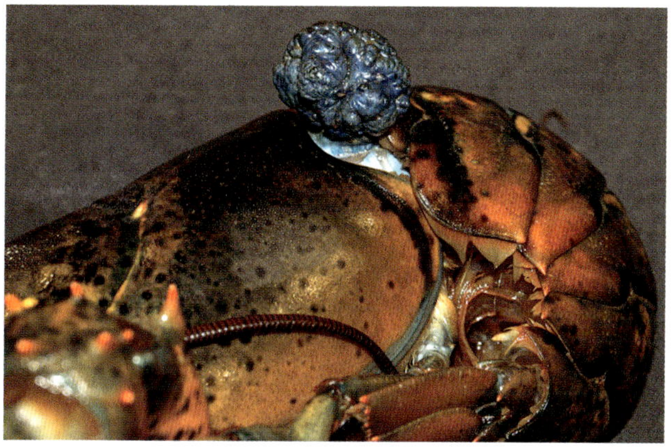

FIGURE 41.8 Hamartoma on the carapace of an American lobster. A bright blue, broad-based, exophytic papilliform mass protrudes from the arthrodial membrane between the thorax and the abdominal somites. *(Photo Courtesy of J. Shields, Virginia Institute of Marine Science.)*

species. A retrotransposon, Steamer, is expressed in high numbers only in neoplastic cells.

Tumor-like lesions of the carapace have been described in crustaceans and classified as **papillomas** or **hamartomas**, a proliferation of disorganized tissue indigenous to the examined site.[153,202] Carapace tumors arising from the abdominal somites occur in brown shrimp, California brown shrimp, American lobster, and other lobster species.[153,202,213] Affected crustaceans are often adults at an intermolt stage. Carapace tumors are broad-based, firm exophytic proliferations with a papilliform to rugose surface and increased apical pigmentation (Fig. 41.8). On cut section, tumors are fibrous with a superficial convoluted cuticle of variable thickness covering the surface. Histologically, the epidermis and subepidermis are continuous with the adjacent unaffected areas, but epithelial cells within the mass demonstrate altered histomorphology described as hypertrophy and hyperchromasia in shrimp, and columnar with globose apical vacuoles in lobster. Adnexa (tegmental glands, setal inserts, and pigment cells) are locally to multifocally absent or in shrimp adnexa can be increased in number and in an abnormally deep location relative to the epidermis. Mitotic figures are not observed but the strong hormonal control of cell division in crustaceans may limit the ability for mitoses to occur outside times of ecdysis even in neoplastic lesions. Tumor stroma is composed of fibrous connective tissue with scattered muscle fibers, few arterioles, and small numbers of infiltrating hemocytes. The etiology of these carapace lesions is currently unknown. Infectious agents have not been identified. In lobster, local trauma, bleeding, and wound repair have been a suggested cause, but the lack of adnexa suggests the epidermal disruption is more complex than simply failing to unfold during ecdysis and is more consistent with a neoplasm. It is not clear if animals affected with these lesions can survive molting due to the lack of tegmental glands, which provide lubrication during ecdysis, and the size of the lesions, which would likely adhere to the instar during molting.[202,213]

Growth anomalies in cnidaria are reported in more than 40 species of scleractinian corals, 5 gorgonians, and 1 hydrozoan.[56] Despite having a similar external appearance and gross features, there are significant interspecies differences in histologic morphology and not all growth anomalies have been classified as neoplasms.[31,139,173,237] Growth anomalies present as discrete, nodular protrusions from the colony and share the gross characteristics of fewer polyps per surface area, decreased pigmentation (fewer zooxanthella per polyp), and finer (more porous) skeletal structures than normal. These foci also grow more rapidly than the adjacent normal appearing tissues.[239]

Calicoblastic epitheliomas, tumors resulting from abnormal proliferation of the ectodermal cells that produce the aragonite skeleton, are described in both *Acropora* spp. and *Montipora* spp. corals.[153,239] Histologically, there is proliferation of gastrovascular canals and the associated calicoblastic epidermis (calicodermis) with a loss of normal polyp structures and decreased numbers of gastrodermal zooxanthellae, mesenteries, and gonads. Affected calicoblasts, instead of having a squamous morphology, are cuboidal to columnar and resemble those found near the rapidly growing regions in corals.[173] The designation of these lesions as neoplastic and specifically malignant is controversial but is based on the histologic features that suggest rapid, uncoordinated growth of the tissue and skeleton, and destruction of adjacent normal tissue.[56,173] Lack of anaplasia and mitotic figures are cited suggesting that these may be growth anomalies or hyperplasias rather than true neoplasms.[188,238] In areas of anomalous growth, the loss of polyps and mucus secretory cells makes affected corals susceptible to sediment accumulation, which causes tissue necrosis and filamentous algae overgrowth in the coral skeleton and can cause regional polyp or entire colony death.[153] Although the normal aragonite crystalline structure is retained in areas of growth anomalies, the skeletal density is reduced, further predisposing the affected area to erosion and predation. In *Montipora* spp., similar anomalies reduce fecundity of affected colonies and in some cases nearly 70% of the colony is replaced by growth anomalies.[31] The etiology of these lesions is not known. Viral infections and UV light exposure have been proposed but remain to be proven.[237]

A number of different benign external tumors, **papillomas**, **polypoid tumors**, or **polypoid growths**, affect the body wall in gastropod and bivalve mollusks. The locations and species affected include the foot of freshwater mussels, oysters, and a butter clam, and the base of the siphon of soft shell clams and a horse clam.[55,166,168,169] These tumors are focal, exophytic, and firm with a texture and color similar to the adjacent tissue of the foot and siphon; some cases have a firm but flexible stalk.[166] Histologically, the tumors are covered by normal tall columnar epithelium that is often thrown into convoluted folds. Multifocal hyperplasia of the basophilic glandular cell layer may occur. The neoplastic portion of these growths consists of a compact, disorganized proliferation of smooth muscle cells originating from the muscle beneath the epithelium of the foot or siphon. Mitotic figures are not a feature of any of the described tumors and all are classified as benign. The etiology of these neoplasms is unknown, although some authors have suggested environmental contaminants while others have described them as spontaneous.[157,168]

Similar exophytic lesions have been described in apple snail and giant African snail, Pacific oyster, Sydney rock oyster, and multiple planaria species. Histologically, these tumors are composed of large altered epithelial cells forming convoluted acini and covering a fibrous connective tissue stalk containing bands of muscle. The altered epithelial cells indicate a different cell of origin and tumor type and the term **adenopapilloma, epithelioma,** or **epidermal papilloma** is applied to these lesions.[146,153,220] All are similarly described as benign and of unknown etiology.

Gonadal tumors are described in 14 species of marine bivalve mollusks from 3 continents. Species include ocean quahog, bay scallop, cockle, Pacific oyster, Eastern oyster, chalky macoma, Northern quahog, hybrid quahog, soft shell clam, blue mussel, bluff oyster, and razor clam.[51,97,172] Three histologic patterns are documented in these neoplasms: **germinomas**, tumors of germ cell origin; **fibroma/myofibroma**, tumors of stromal/connective tissue origin; and **gonadoblastoma**, mixed tumors containing both components.[172] Germinomas are composed of undifferentiated germ cells that fill individual follicles, efface the gonadal area, or spread to other adjacent tissues. Stromal tumors are composed of neoplastic cells that are myxoid to vesicular to spindle-shaped and efface normal reproductive tissue as they grow. In mixed tumors, there is a central region that consists of dense fibrous tissue and the surrounding margins are neoplastic germ cells with either ovarian or spermatogenic differentiation. The latter group of tumors is reported infrequently but has an aggressive pattern of growth with evidence of extension through the follicular basement membrane, invasion of the adjacent connective tissue of the gill, and mass formation along the wall of the branchial vein.[172] Small foci of isolated neoplastic cells in the gill microvasculature in a soft shell clam were consistent with metastasis. Gonadal neoplasia has also been described in a male limpet.[35] The etiology of gonadal neoplasms is currently undetermined. Environmental pollutants and/or viruses are proposed but causal relationships remain to be confirmed.[153,172]

INFECTIOUS DISEASES

Viral-like particles (VLPs) have been reported in the plumose anemone, a cnidarian that lacks symbiotic zooxanthella, and the temperate sea anemone.[233,234] In corals, VLPs are reported in *Pavona danai*, *Acropora formosa*, *Acropora muricata*, *Zoanthus* spp., *Montastraea faviolata*, and their zooxanthella.[235] It remains undetermined if this association is mutualistic, opportunistic, or pathogenic.

DNA Viruses

Large DNA viruses of the families *Ascoviridae*, *Baculoviridae*, *Herpesviridae*, *Hytrosaviridae*, *Iridoviridae*, *Malacoherpesviridae*, *Nimaviridae* (*Whispovirus*), *Nudiviridae*, and *Poxviridae* (*Entomopoxvirinae*), and small DNA viruses including family *Bidnaviridae*, *Circoviridae, and Parvoviridae* (*Densoviriae*), are important pathogens of aquatic and terrestrial invertebrates. Due to their economic importance three DNA viruses that affect farmed shrimp and one DNA virus that affects abalone are currently listed as reportable to the World Organization for Animal Health (WOAH): **white spot syndrome virus (WSSD), infectious hypodermal and hematopoietic necrosis virus (IHHNV), decapod iridescent virus (DIV1)** in shrimp, and **abalone herpesvirus (AbHV).**[105,129,131] Notable DNA viruses of invertebrates organized alphabetically by family are discussed below.

Baculoviruses are a large group of host-specific DNA viruses, the majority of which infect lepidopteran caterpillars. These viruses occur naturally but are also used for biological control of pest species. Due to their high host specificity and high mortality, these viruses make excellent biological control agents. Most baculoviruses are spread horizontally by ingestion, after which they infect midgut epithelial cells and spread to the rest of the body. Vertical transmission is also possible. The virus produces large intranuclear inclusions (also called occlusion bodies) throughout the midgut. These are accompanied by nuclear hypertrophy and chromatin margination. Viral protease and chitinase enzymes cause liquefaction of the host and release of millions of virions into the environment for subsequent infections (eSlide 41.e2).[44] Although mortality can be high in certain insect hosts, covert infections, when there is low host density and limited horizontal transmission, do occur.[232]

Monodon baculovirus (MBV) disease (also ***Penaeus monodon*-type baculovirus**) (USDA reportable disease) occurs in various species of cultured penaeid shrimp and is found in global shrimp aquaculture.[24] It was also recently identified via PCR and histopathology in wild-collected juvenile and subadult mud crab.[180] The virus can be horizontally transmitted in water contaminated with broodstock

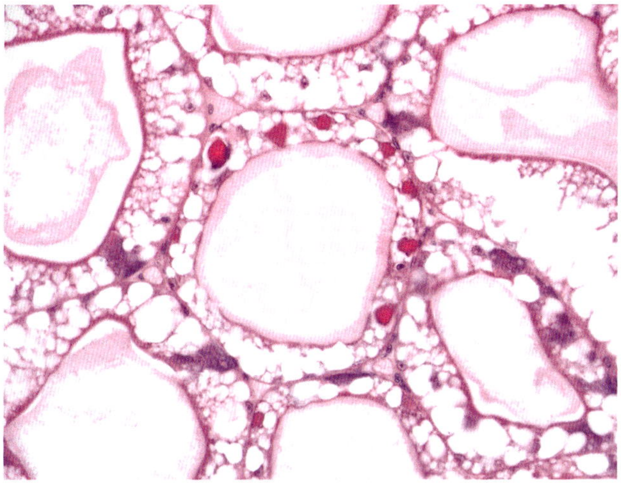

FIGURE 41.9 Monodon baculovirus disease in the hepatopancreas of a shrimp. Centrally, the glandular epithelial cells of hepatopancreatic tubules contain characteristic eosinophilic triangular intranuclear inclusions.

feces and also appears to be vertically transmitted.[115] Infection causes slow growth, decreased feeding, lethargy, and gill fouling. Mortality can be high in late larval to post-metamorphic juvenile shrimp.[24] Adults generally do not express disease. Heavily infected larvae and early juveniles exhibit a white midgut that is visible through the thin abdominal carapace and have an enlarged hepatopancreas. Histologically, the enlarged hepatopancreatic or midgut epithelial cell nuclei contain eosinophilic triangular inclusions (paracrystalline occlusion bodies) with marginated chromatin (Fig. 41.9). Necrosis and sloughing of the epithelium follows. These paracrystalline inclusions may also be detected in squash preparations of the dissected hepatopancreas or in the feces stained with malachite green.[115]

Herpes-like viral infections of larval and juvenile Pacific oyster caused by **oyster herpes virus 1 microvariant (OsHV-1μ-var)** have resulted in disease and high mortality in aquacultured animals, first in France, then in other locations on the Atlantic coast of Europe and in Italy.[183] Disease outbreaks are associated with warmer summer temperatures (>16°C). Infected animals do not show disease at lower temperatures (13°C).[170] A different strain of OsHV has also been implicated as the cause of mortality in aquacultured juvenile Pacific oysters in California and other locations along the west coast of the United States. Eosinophilic inclusion bodies are noted in the hypertrophic nuclei of hemocytes, connective tissue, and epithelial cells in oysters from Europe but not in oysters from the United States. Despite these differences, OsHV-associated mortality has been confirmed in Europe and the United States using molecular methods.[74]

Abalone viral ganglioneuritis due to **abalone herpesvirus** (**AbHV**) (WOAH and USDA reportable disease) is reported in farmed and wild greenlip, blacklip, and hybrid abalone in Australia and cultured diversicolor abalone cohoused with black abalone in Taiwan. Mortality rates across all age groups of up to 90% and 70%–80% have been reported in Australian outbreaks and in adult and juvenile abalone in epizootic and experimental infections of diversicolor abalone, respectively. Horizontal transmission has been confirmed experimentally. Affected animals demonstrate curling of the foot, excessive mucus production, swelling and protrusion of the mouth parts, eversion of the radula, and abnormal spawning or bloating. Foot abnormalities may include irregular peripheral concave elevation of the foot, lack of foot movement, poor righting reflex, and reduced pedal adhesion to substrates. Death occurs within 1–2 days. Histologically, there is individual cell necrosis of the nervous tissue including cerebral, pleuropedal, and buccal ganglia, branches of the pedal nerve and peripheral nerves. Margination of the chromatin and central pallor suggestive of intranuclear inclusion bodies may be seen in neurons.[101] Conventional and real-time PCR testing and in situ hybridization are available for confirmation.[45]

Invertebrate iridescent viruses (IIVs) infect numerous invertebrates with members of genera *Iridovirus* and *Chloriridovirus* infecting pest insects, and *Decapodiridovirus* infecting decapod crustaceans. **Decapod iridescent virus 1 (DIV1)** is a recently listed WOAH-reportable disease of penaeid shrimp, which first emerged as an issue in aquaculture in 2014.[131] These viruses cause slow, low-level infections resulting in either covert (sublethal) or patent infections; patent infections cause an iridescent color change and are usually fatal to larvae and pupae.[131,192] In shrimp, characteristic gross lesions include white to yellow discoloration of the head due to hepatopancreatic atrophy and an empty gastrointestinal tract. Some species show a focal white area under the carapace at the base of the rostrum. The viral target organs include lymphoid and hematopoietic tissues, hemocytes, hepatopancreas, muscle, and periopods with sporadic infection documented in the subcuticular epithelium. Affected cells contain dark eosinophilic cytoplasmic inclusions with an adjacent smaller basophilic body. Nuclear pyknosis is apparent in the hematopoietic tissue, branchial hemocytes, hepatopancreas, and periopods.[131]

Caribbean spiny lobsters, an important commercially fished species, are endemically infected with a DNA virus, ***Panulirus argus*** **virus 1 (PaV1)**. It was recently suggested that PaV1 be placed into a new viral family: Mininucleoviridae.[218] Infection with PaV1 occurs by water exposure and contact is potentiated by the behavior of these gregarious lobsters. High mortality (>40%) can develop in infected juveniles. Uninfected lobsters can detect infected juveniles and behaviorally segregate themselves from infected animals. Adults appear to be resistant to the disease but that may be partially behaviorally based. Grossly, juvenile lobsters are lethargic before death and have milky hemolymph that does not clot (Fig. 41.10A).

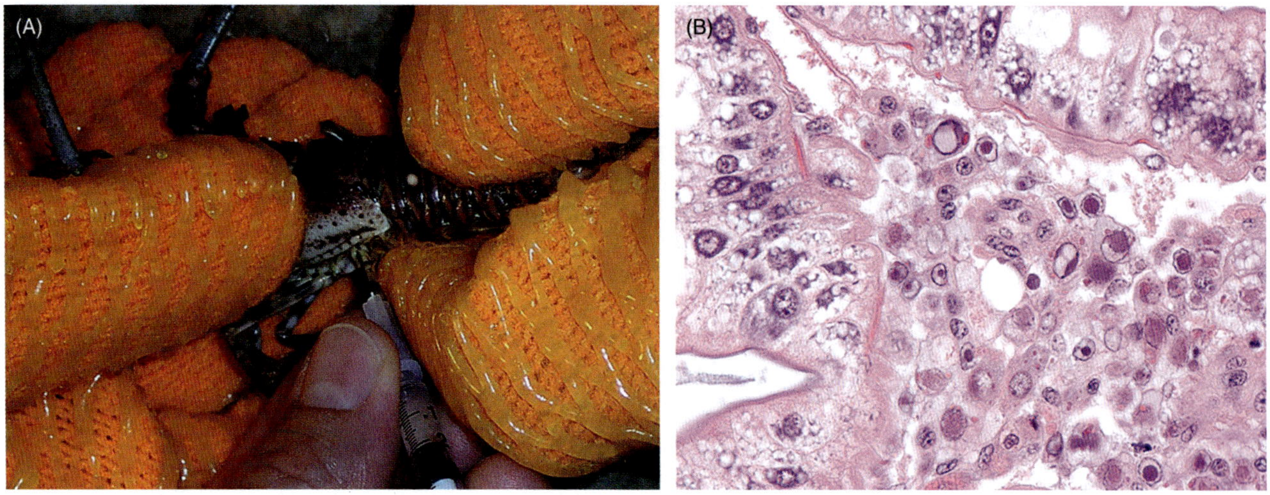

FIGURE 41.10 *Panulirus argus* virus 1 infection in a Caribbean spiny lobster. (A) The characteristic cloudy discoloration of the hemolymph. (B) Between two hepatopancreatic tubules, the fixed hemocytes demonstrate margination of the chromatin and large eosinophilic intranuclear inclusion bodies. *(Photos Courtesy of J. Shields, Virginia Institute of Marine Science.)*

Microscopically large eosinophilic intranuclear inclusions with marginated chromatin can be found in the fixed hemocytes of the hepatopancreas (Fig. 41.10B) and connective tissues throughout the body as well as in circulating hemocytes. Infection causes destruction and hemocytopenia as the disease progresses.[16]

The agent of **white spot disease (WSD),** also termed **white spot syndrome virus** (**WSSV**), is a species of *Whispovirus*. The virus, a DNA virus and the only member of the *Nimaviridae* family, can be found in all decapod crustaceans in many locations throughout the world, but is best known for causing disease with significant morbidity and mortality in several species of aquacultured penaeid shrimp. Characteristic gross lesions are white spots in the carapace. Depending on the severity, the spots can coalesce to form extensive areas of white discoloration across the carapace.[195] Microscopically, the virus infects ectodermal and mesodermal tissues and forms large, eosinophilic to basophilic inclusions in hypertrophic nuclei resulting in necrosis of infected cells (Fig. 41.11). White spots in the cuticle are due to infection and necrosis of the underlying cuticular epithelium. The virus is very stable in dead tissues and survives freezing.[184] Heavy infections have been reported in other decapods without marked disease. Gross and microscopic findings combined with disease outbreaks and molecular identification of the virus are used in diagnosis. In addition to cultured shrimp, the virus has been responsible for disease outbreaks in semicultured crayfish in Louisiana.[14] Numerous other crustaceans and aquatic insects are thought to be carriers/reservoirs of the virus.[195]

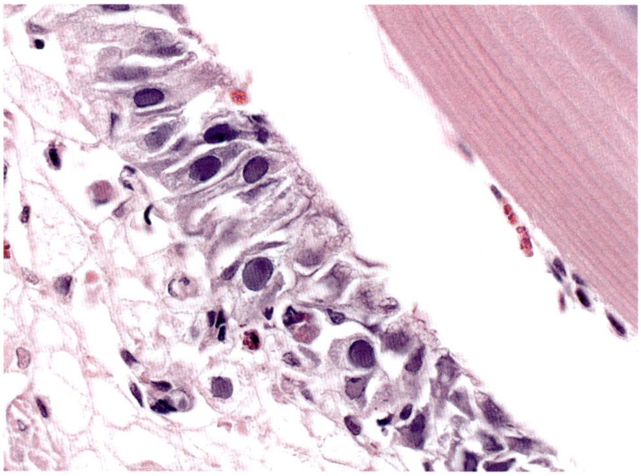

FIGURE 41.11 White spot disease in the cuticular epithelium of a shrimp. The cells contain hypertrophied nuclei that contain large eosinophilic to basophilic intranuclear viral inclusions. There is an artificial separation between the epithelium and the overlying carapace.

All parvoviruses that infect insects are from the subfamily ***Densovirinae***. These viruses most commonly infect lepidopterans but have also been reported in hemipterans, dipterans, hymenopterans, orthopterans, and aquatic arthropods. Densoviruses in insects can cause mass mortalities and devastating outbreaks in mass-reared insect facilities, particularly cricket and wax moth-rearing facilities.[192]

Infectious hypodermal and hematopoietic necrosis (**IHHN**) (WOAH-reportable disease) is a **densovirus** (**brevidensovirus**) in the *Parvoviridae* family. IHHN causes acute disease with high mortality in juvenile and subadult aquacultured blue shrimp. Infected animals exhibit abnormal behavior in the water column and irregular white to buff-colored spots are seen through the carapace. Chronic infections resulting in poor growth and deformities, such as bent bodies, deformed rostrums, and wrinkled antennae, cause **runt-deformity syndrome** (**RDS**) in white shrimp.[82,129] Histologically, eosinophilic intranuclear

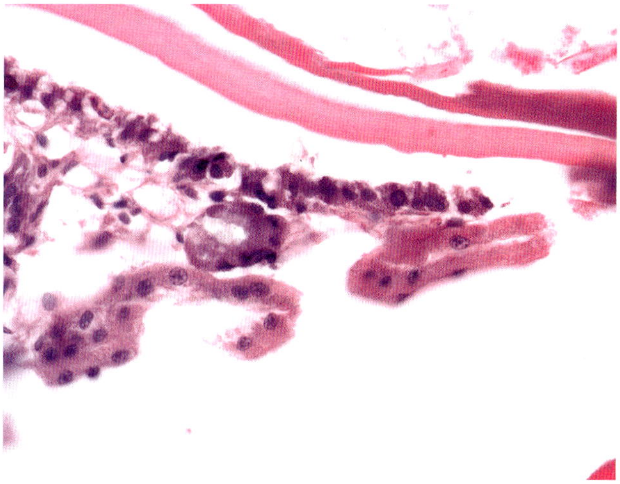

FIGURE 41.12 Infectious hypodermal and hepatopancreatic necrosis virus in the cuticular epithelium of a shrimp. Numerous eosinophilic intranuclear viral inclusions are present in the cuticular epithelial cells. The separation between the epithelium and the carapace is artifactual.

inclusions are noted in the cuticular epithelium and other ectodermal and mesodermal origin tissues in acute outbreaks (Fig. 41.12). Inclusions are not identified in chronic disease and diagnosis in these cases is usually accomplished using molecular methods.

Until recently, there were no described viruses of echinoderms. Metagenomic studies in three species of healthy urchins from Hawaii and five species of sea stars recovered viral sequences matching single-stranded DNA densoviruses (family *Parvoviridae*). Densoviruses were initially believed to be associated with sea star wasting disease (discussed previously).[92,99] However, subsequent studies revealed this densovirus was widely distributed in the water column and in both healthy and affected animals. No studies have confirmed a location of viral replication or the presence of densoviral particles in association with lesions of wasting disease.

Viruses from subfamily **Entomopoxvirinae** infect dozens of species of coleopteran, dipteran, lepidopteran, and orthopteran insects. These viruses infect midgut epithelial cells then migrate via hemocytes to the fat body, which is the primary site of replication. Time from entomopoxvirus infection to death is typically prolonged.[96]

RNA Viruses

Many RNA viral families are known to infect invertebrates, including double-stranded RNA families *Birnaviridae* and *Reoviridae*, and single-stranded RNA families *Bunyavirida*, *Discitroviridae*, *Iflaviride*, *Nodaviridae*, *Orthomyxoviridae*, *Rhabdoviridae*, *Roniviridae*, and *Totiviridae*, among others. RNA viral pathogens of invertebrates include four WOAH-reportable diseases of penaeid shrimp, **infectious myonecrosis virus (IMNV), yellow head virus genotype 1 (YHV genotype 1), Taura syndrome virus (TSV), white tail disease (WTD)**,[129] and significant viral diseases of honey bees including **acute bee paralysis virus (ABPV), Kashmir bee virus (KBV), Israeli acute paralysis virus (IAPV), deformed wing virus (DWV)** and **chronic bee paralysis (CBP)**.[225] RNA viruses of note are presented in alphabetical order by family.

Taura syndrome virus [TSV; family **Dicistroviridae,** genus ***Aparavirus*** (WOAH-reportable disease)] infects several shrimp species. It causes disease primarily in newly metamorphosed, postlarval (PL) aquacultured blue and white shrimp; however, larger shrimp may develop disease if not previously exposed. The virus infects the cuticular epithelium, hematopoietic tissues, and connective tissue but not tissues of endoderm-derived organs (hepatopancreas and gut epithelium, etc.). Histologically, it causes multifocal acute necrosis of the cuticular epithelium and dermis (Fig. 41.13).[132] The disease first occurred in 1992 in Ecuador and is now found in shrimp farms throughout the world; however, evidence indicates some strains of the virus were acquired from wild broodstock in waters not contiguous with the Americas.[219]

Viruses from the order *Picornavirales*, specifically those from *Iflaviridae* and *Dicistroviridae,* exclusively infect invertebrates. Several are of particular importance in pollinators such as **deformed wing virus (DWV), sacbrood virus**, and **acute bee paralysis virus**. DWV can be transmitted vertically, horizontally, or by the vector *Vorra* mites, but clinical disease only occurs when the vector infects pupae, resulting in gross wing and abdomen deformities in adult bees. Detailed histologic lesions in the wings, legs, brain, eye, and glands have been reported in bees with clinical DWV infection and include degenerative changes in the hypopharyngeal glands and flight muscles

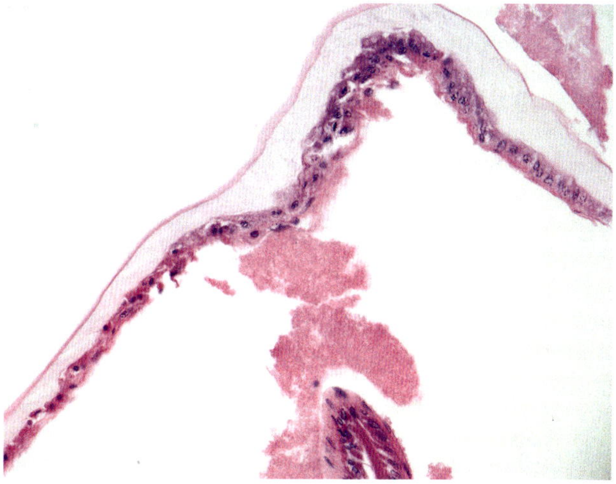

FIGURE 41.13 Taura syndrome virus in the cuticular epithelium of a shrimp. There is regional acute necrosis of the cuticular epithelium characterized by hypereosinophilic, nuclear pyknosis, and a loss of cellular detail.

and inflammation in the midgut and hemocele.[10,119,178] The hypopharyngeal glands produce the protein fraction of the worker and royal jelly fed to developing larvae and queens which plays a critical role in hive health. Degenerative changes include tall irregular acinar cells with hyperchromatic nuclei, depletion of cytoplasmic vacuoles, and eosinophilic cytoplasmic granules.[178]

Two agents, ***Macrobranchium rosenbergii* nodavirus (MrNV)** and **extra small virus (XSV)** cause **white tail disease of shrimp**, a WOAH-reportable disease that causes severe mortalities in larval and postlarval giant freshwater prawn. Early juveniles are also susceptible while adults are resistant and can act as carriers. Target organs include gill, muscles of the head, heart and abdomen, ovaries, pleopod, and tail muscle. Transmission via horizontal routes is reported in hatcheries and a vertical route of infection is suspected from carrier animals. Hosts include artemia and aquatic insects. Clinical signs include lethargy, opaque abdominal musculature, degeneration of the uropods, and telson with up to 100% mortality in four days. Histologically, there is acute necrosis of striated muscle of the cephalothorax, abdomen, and tail characterized by hyaline degeneration, lysis, and intramuscular edema. Large oval or irregular basophilic cytoplasmic inclusion bodies, 1–4 μm diameter, are present in the muscles of the abdomen, cephalothorax, and interstitium of the hepatopancreas. Diagnostic confirmation includes RT-PCR, dot-blot hybridization, in situ hybridization, and enzyme-linked immunosorbent assay (ELISA) testing.[194,214]

Cypovirus is a reovirus that infects a wide range of insects predominantly dipterans and hymenopterans. It has low host specificity and predominantly infects midgut epithelial cells of larvae. Infected animals may fail to pupate or have low weight gain.[96] Histologic lesions include cytoplasmic inclusions in midgut epithelial cells and fat body atrophy.[122]

Yellow head disease of shrimp (YHD) (WOAH-reportable disease) is caused by **yellow head virus (YHV)**, **gill-associated virus (GAD)**, and six other closely related viral strains. YHV, related to nidoviruses in the *Coronaviridae* and *Arteriviridae* families, and GAD are ssRNA viruses in the genus *Okavirus*, family *Roniviridae*.[129,135] YHV is primarily seen in Asia while GAD is reported in Australian shrimp farming facilities. There are no confirmed cases of YHV in the Americas. YHV is potentially lethal to most cultured penaeid shrimp strains. Mass mortality occurs in the early to late juvenile stages. Gross lesions are characterized by bleaching of the carapace and yellow discoloration of the cephalothorax due to the abnormally yellow hepatopancreas. The hepatopancreas is frequently soft in affected animals when compared to healthy shrimp. YHV targets the lymphoid organ, hemocytes, hematopoietic tissue, gill lamellae, subcutaneous connective tissue, gut, antennal gland, gonads, and nervous system (nerve tracts and ganglia). Histologic lesions include characteristic deeply basophilic cytoplasmic inclusions approximately 2 μm in size or smaller. The lymphoid organ, stomach, and gills are the best locations for detection of inclusions. Lymphoid organ necrosis may also be seen. Confirmation is through in situ hybridization along with RT-PCR and sequencing.

Infectious myonecrosis of shrimp (WOAH-reportable disease) is caused by a double-stranded RNA totivirus and causes significant mortalities in juvenile and adult-farmed Pacific white shrimp.[129] Pacific blue shrimp and black tiger shrimp have been infected experimentally. Disease outbreaks may follow stressful events, such as feeding, capture, or sudden environmental changes (salinity, temperature). Gross and histologic findings are similar to those seen with white tail disease of shrimp. Animals present acutely with focal to locally extensive white areas in the striated muscles that correspond to areas of necrosis. The distal abdominal segments and tail fan are the most severely affected tissues and can become necrotic and red in some shrimp. The primary target organs are striated muscle (skeletal and less often cardiac), connective tissue, hemocytes, and lymphoid organ parenchyma. Histologic lesions include coagulation to lytic necrosis frequently accompanied by edema and hemocyte infiltration in affected striated muscle. With chronicity, inflamed muscle fibers are replaced by fibrocytes and connective tissue interspersed with hemocytes and regenerating muscle fibrils. A consistent finding is hypertrophy of the lymphoid organ caused by accumulation of lymphoid organ spheroids and spheroids in ectopic locations (gills, heart, antennal gland, and ventral nerve cord). Transmission can occur horizontally through cannibalism. Vertical transmission is suspected in carriers. Confirmation and differentiation from white tail disease requires a combination of nested PCR or ISH and histopathology.[136,154,177]

Chronic bee paralysis (CBP) is caused by a unique, unclassified virus with a fragmented positive-stranded RNA genome composed of two major RNAs. It shares some characteristics with *Nodaviridae* and *Tombusviridae* family viruses, but is considered as the type species of a new group of positive-stranded RNA viruses.[185] Infection results in two distinct syndromes, both of which can be present in affected colonies to varying degrees. Bees with type 1 syndrome are unable to fly, often crawl on the ground or up plant stems, and may form masses of thousands of affected animals that huddle together on top of the cluster in the hive. Affected animals develop severe abdominal swelling secondary to distention of the honey sac with fluid. Symptomatic animals die within a few days. In severely affected colonies, there can be sudden colony collapse at the height of summer leaving only the queen and a few workers remaining. Type 2 syndrome has several names (black robbers, little blacks, and hairless black syndrome) that all reflect the characteristic loss of hair on the body, which causes the animals to appear dark to almost black and shiny or greasy. They are smaller than healthy animals, have a broad abdomen, and can fly initially but are attacked by healthy bees in the colony and become flightless, develop body tremors, and die. The virus is neurotropic and in situ hybridization has demonstrated infection and replication within neurons of somata and neuropile regions of the protocerebron and deuterocerebron. Targeted neurons in portions of the brain include the mushroom bodies and central complex, which are involved in sensory processing, learning, memory storage, control of motor patterns, locomotor control, and orientation behavior, and the antennal lobes which have primary sensory function.[159,185] Histopathologic lesions other than localization of viral replication in these regions are not described. A two-step RT-PCR test with a fluorescent probe to quantify viral load has been described.[19,20] Viral spread is horizontal and associated with host density within the hive. Confinement and crowding of bees due to poor weather conditions, or increased bee colony density within areas of limited foraging promote viral spread. Increased bee contact due to crowding results in cuticle abrasion and transmission.[85]

Bacteria

Bacterial infections in invertebrates are common particularly in situations where other environmental factors (abnormal temperature, abnormal pH, biofouling, and toxins) contribute to debilitation. Further, microbial gene expression can be rapidly altered by changes in environmental conditions, which along with plasticity in biochemical substrates may alter microbial behavior creating opportunities for infection by organisms that would otherwise be normal constituents of the host microbiome.[109] The respiratory apparatus and digestive system are primary targets in hexapods and arthropods. The epidermis and body wall are most frequently affected in echinoderms. Gram-negative bacteria, often mixed populations, are frequently identified. However, filter-feeding bivalves can accumulate large numbers of bacteria from surrounding seawater and can act as passive carriers, therefore, it is important to consider that a positive culture does not necessarily correlate with disease.[127] Conversely, pathogenic bacteria of aquatic animals may prove difficult to demonstrate in culture, and a negative culture result does not rule out a bacterial infection. Hence the importance of establishing histologic correlates.[240] Gross lesions of bacterial infections are often not apparent in some arthropods and mollusks. Insects may present with color changes, soft texture, or dark spots on the exoskeleton. Histopathology associated with invertebrate bacterial infections may include multifocal lamellar necrosis in the gills or book lungs, multifocal necrosis and hemocyte infiltration in the digestive gland, multifocal

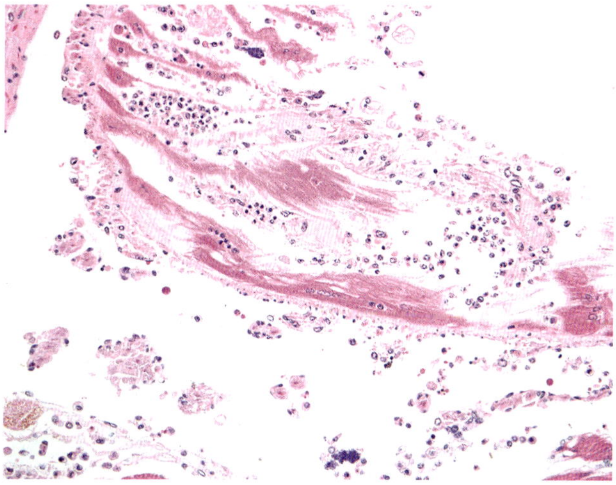

FIGURE 41.14 Bacterial sepsis in the heart of a greenbottle blue tarantula. Circulating hemocytes within the pericardium and ventricular lumen contain cytoplasmic aggregates of small bacilli consistent with circulating bacteremia.

necrosis and hemocyte infiltration in the connective tissue, and in cases of sepsis, bacteria are present within circulating hemocytes (Fig. 41.14).

In echinoderms, particularly sea urchin, multifocal loss of spines, pedicellaria, and tube feet, epidermal ulceration with swelling of the epidermis at the tissue margin, and black discoloration of the exposed test are evident in a condition known as **bald urchin disease**.[15,140] Histologic lesions associated with bald urchin disease include multifocal epidermal necrosis, disorganization of the skeletal muscle, coelomocyte infiltration, and the presence of mats of bacteria with frequent secondary invaders (i.e., ciliates, algae, and fungi).[87]

Shell disease is a general term used in reference to any crustacean with carapace erosions. In the American lobster, **epizootic shell disease (ESD)** has recently been identified (Fig. 41.15) and is common in wild-caught American lobsters along the northeast US coast (specifically Massachusetts and Rhode Island).[208] The disease begins as deep erosions on the dorsal cephalothoracic carapace in late summer (a few months after molting). Lesions progress both laterally across the carapace surface and into deeper layers of the carapace, and can result in ulcerations into the underlying carapace epithelium and dermis with significant dermal inflammation. Ulceration may result in adhesions of the old shell to the new, inability to completely molt, and mortality. If an animal can molt successfully, there are generally no indications it was affected before molting. No significant internal tissue changes are noted in animals without ulcerations. Affected berried females, who usually carry eggs (berries) on the ventral abdomen for 9–12 months, will molt early; early molt and egg loss are partly responsible for decreased lobster production in diseased areas. The cause of ESD has not been determined

FIGURE 41.15 Epizootic shell disease in an American lobster. Throughout the dorsal cephalothoracic carapace and extending along the somites are extensive, frequently coalescent areas of severe cuticular erosion.

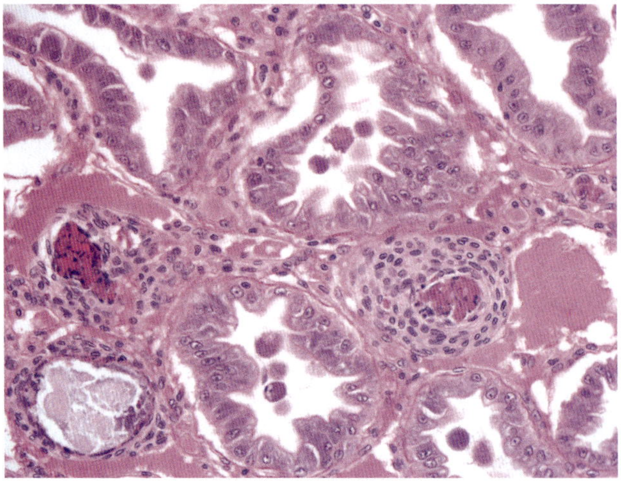

FIGURE 41.16 Necrotizing hepatopancreatitis in shrimp. Within the hepatopancreatic interstitium are numerous hemocytic infiltrates that surround aggregates of the Rickettsiales-like organism. Adjacent tubules are dilated and contain luminal hemocytes.

but evidence suggests that elevated water temperatures at the time of molting[88] and the occurrence of specific types of bacteria (***Aquimarina homaria*** and others) in the lesions are important in the development of the disease.[42,109]

Other forms of shell disease described in American lobsters are **enzootic shell disease** (**EnSD**), which has gross and histologic features similar to ESD, but is rarely found in fished populations and is of unknown cause, **black spot disease** which presents as focal to focally extensive deep black erosions that develop secondary to carapace trauma, and **impoundment shell disease** (**ISD**). ESD can be grossly distinguished in early stages from ISD when ISD has not yet spread across the surface of the shell.[211] Early ISD begins around the bilaterally symmetrically positioned tegmental gland ducts and sensory hairs and therefore lesions are bilaterally symmetrical; ESD lesions are not. In all three of these diseases, bacteria are cultured from lesions, but only bacterial culture of ESD lesions consistently shows *Aquimarina homaria* as at least a partial cause of the disease.

Necrotizing hepatopancreatitis (**NHP**) is a WOAH-reportable disease that causes high mortalities in penaeid shrimp. It is caused by a gram-negative, dimorphic intracellular **Rickettsiales-like** (**RLO**; unclassified alphaproteobacterium) that targets the tubular cells of the hepatopancreas. Based on genome sequencing information, the name *Hepatobacterium penaei* has been proposed. Early in infection, small, pleomorphic, intracytoplasmic bacteria are present in the apical cytoplasm or fill hepatopancreatic tubular epithelial cells resulting in cellular hypertrophy of individual to small patches of cells. Later in infection, there is more prominent tubular epithelial cell hypertrophy or epithelial attenuation with tubular dilation. Hemocytic infiltrates are present and tubular necrosis is mild and typically affects distal tubular segments. At the end stage of disease, there are linear areas of tubular epithelial necrosis that extend to the medial and proximal segments. Necrotic tubules are collapsed, contain luminal hemocytes, and have a hyalinized and fragmented basement membrane. Melanization and lamellar peritubular interstitial fibrosis are present and remaining tubules demonstrate moderate to severe dilation and epithelial attenuation (Fig. 41.16). The RLO are gram-negative but discrete organisms are best seen in Steiner's stains.[73] Elevated temperature (near 30°C) and elevated salinity [30 ppt (part per thousand)] contribute to disease occurrence. Diagnosis is confirmed by PCR.[138,156]

Withering syndrome (**WS**) is a WOAH-reportable, fatal disease of abalone caused by the **Rickettsiales-like prokaryote** (**WS-RLP**) *Candidatus Xenohaliotis californiensis*.[75,84] All abalone of the genus *Haliotus* in free-ranging, cultured, and laboratory settings can develop infection, although susceptibility varies by species. Black and white abalone populations suffer severe mortality, with only moderate mortality in red abalone and little to no mortality in free-ranging pink and green abalone.[77,149,150] The disease is most prominent in free-ranging and cultured populations along the Pacific coast from central California through Baja California, Mexico, but the geographical range is broad due to global culture of red abalone and secondary exposure of native species. Infected abalone are known to have been transported to Chile, China, Taiwan, Iceland, Ireland, Israel, Spain, Thailand, and Japan. External gross lesions include poor body condition and foot muscle atrophy following a clinical period of anorexia and lethargy. Internal gross lesions include retraction of the visceral tissues and digestive gland atrophy. Histologically,

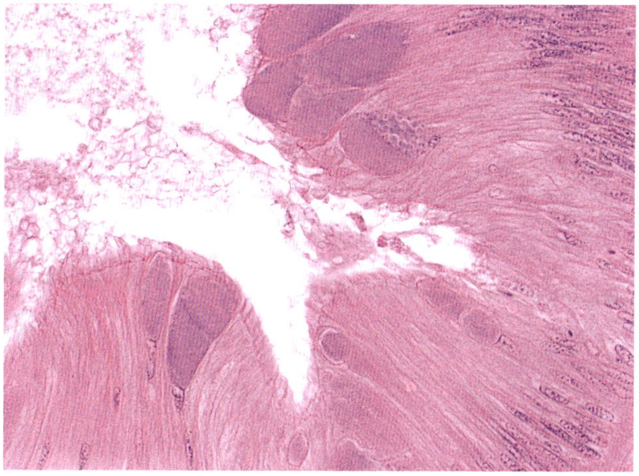

FIGURE 41.17 Intestine of an abalone with withering syndrome. Rickettsiales-like prokaryotes forming basophilic inclusions are present in the apical cytoplasm of the intestinal epithelium.

organisms are present within the epithelium of the posterior esophagus and digestive gland causing epithelial hypertrophy, necrosis, and low cuboidal to squamous epithelial metaplasia. Deeply basophilic, apical cytoplasmic bacterial colonies form inclusions that range from approximately 10–20 μm diameter and are evident in affected digestive tissue on routinely stained sections (Fig. 41.17). WS-RLP inclusions are also detectable on squash preparations of digestive tissues stained with the nucleic acid-specific fluorochrome Hoechst 33258. Pedal atrophy is characterized by a diffuse decrease in the size of skeletal muscle fibrils with secondary wide separation of the fibrils by extracellular matrix material. Transmission is likely fecal–oral and early infection is located in the posterior esophagus and to a lesser extent in the intestine. The pathophysiology of the disease is physiologic starvation secondary to digestive gland changes that result in depletion of glycogen reserves, catabolic changes in the pedal musculature, muscular and digestive gland atrophy, and ultimately death.[75]

Diagnosis of WS requires consistent gross lesions with confirmation of RLP infection either by histopathology or in situ hybridization (method of choice) coupled with PCR (conventional and quantitative) and sequence analysis.[4,50,75] Definitive diagnosis of WS must be conducted according to World Organization of Animal Health (WOAH) standards in the Manual of Diagnostic Tests for Aquatic Animals.[107]

A second RLP, the **stippled RLP (ST-RLP)** has been identified in California abalone but at this time it is considered nonpathogenic and is found in tissues only at low levels. A phage hyperparasite of the WS-RLP has been detected that may reduce the pathogenicity of the bacterium. WS-RLP and ST-RLP the phage variant can be differentiated by light microscopy based on their differential staining properties on routine hematoxylin and eosin (HE) staining and the cytoplasmic location of the inclusion.[50,78]

Roseovarius oyster disease (ROD, also known as **juvenile oyster disease, JOD)** is caused by an alphaproteobacteria, *Roseovarius crassostrea*. The bacterium infects juvenile eastern oysters and causes significant accumulation of abnormal (flaky/mucoid) conchiolin, the organic matrix of the shell within which calcium carbonate is deposited, between the membranous mantle and the nacre of the shell. Affected animals show cupping of the left valve and shortening of the right valve with death of affected animals (often >90%) within 6 weeks of initial signs of infection. The disease was first identified in 2002 and is noted sporadically in aquacultured juvenile oysters in the northeastern United States from Maine to New York. Histologically, mantle epithelium progresses from a hyperplastic/hypertrophic condition to ulceration of the shell epithelium of the mantle with buildup of conchiolin between the ulcerated mantle and the shell. Increases in hemocyte numbers are noted throughout all tissues in early to mid-disease stages. This agent proliferates in juvenile oysters less than 25 mm in shell height and only in seawater greater than 21°C and salinity between 25 and 32 ppt. Larger juveniles show signs of infection but survive disease. Diagnostics include characteristic gross lesions, histological evaluation of tissues, and PCR identification of the bacteria from infected oyster tissues. Resistance to the disease appears to develop in the F1 generation by using survivors as broodstock.[21]

A recently described infection of emperor scorpions with ***Rickettsiella scorpionisepticum*** results in hemocyte proliferation and widespread multiorgan infiltration of hemocytes that have abundant, granular, pale basophilic cytoplasm. Ultrastructural studies revealed that these hemocytes are distended with numerous bacteria of multiple forms (eSlide 41.e3).[94]

Rickettsiales-like organisms (RLO) cause large, intracytoplasmic, bacterial inclusions in branchial epithelial cells of cephalopods. These are likely an incidental infection, but there is speculation that in high numbers these organisms may cause respiratory compromise.[86]

Species of ***Vibrio*** are gram-negative, motile bacilli that are widespread in coastal and estuarine environments. Some species are serious pathogens of aquatic invertebrates. Vibrio-induced bleaching is one of the few coral diseases for which Koch's postulates have been fulfilled. Two examples include **bacterial bleaching** of *Oculina patagonica* due to ***Vibrio shiloi*** and **coral bleaching and lysis** in *Pocillopora damicornis* and *Paramuricea clavata* due to ***Vibrio coralliilyticus***. Environmental temperature and nutrients are risk factors for the development and progression of disease. *V. shiloi* causes the death of endosymbiotic zooxanthella within the gastrodermis.

V. coralliilyticus causes tissue lysis due to the synthesis of a potent metalloproteinase at temperatures above 26°C. Gross lesions in *V. shiloi* infection include multifocal areas of loss of pigmentation particularly in colonies in shallow water at depths of 1–6 m. Histologic lesions confirm large aggregates of bacteria at the border between affected and unaffected tissue and surrounding released zooxanthellae on the tissue surface. Transmission electron microscope (TEM) images demonstrate intracellular bacteria within the ectoderm, vacuolar change in the gastrodermis, and degeneration of zooxanthellae. Gross lesions in *V. coralliilyticus* infection include white patches, typically starting on the outgrowth/polyp extensions of the branch and later becoming confluent. Tissue lysis results in exposure of the underlying skeleton. Scanning electron microscopy (SEM) confirms high numbers of rod-shaped bacteria on the surface, below the necrotic tissue, and on the zooxanthellae.[9]

Acute hepatopancreatic necrosis disease (syn. **Early mortality syndrome**) of penaeid shrimp is caused by a unique strain of ***Vibrio parahaemolyticus*** that produces a plasmid-encoded toxin. Grossly, the hepatopancreas, which usually appears brown, appears white when examined through the carapace. Microscopically, infection results in acute necrosis of the hepatopancreatic tubular epithelium, sloughing of degenerate cells into the lumen of the tubules, marked intratubular hemocytic inflammation, and late development of a secondary massive bacterial infection.[222] In addition to gross and microscopic findings, molecular methods to detect the gene responsible for toxin production have been developed.

Brown ring disease caused by ***Vibrio tapetis*** **predominant 1 (Vibrio P1)** causes severe disease and mortality in aquacultured Manila clams in Europe. Cockles and other bivalves in affected areas are infected but do not develop disease. Grossly, brown rings composed of abnormal conchiolin are observed at the edges of the pallial line and the edge of the shell. Microscopically, the brown rings are composed of a mixture of necrotic periostracum and inner poorly formed layers of the prismatic and nacreous shell along with hemocytes and other cell debris from the mantle. It appears that the bacteria attach at the palladial line, an area where the shell epithelium of the outer mantle lobe detaches from the shell.[164] Disease is associated with colder temperatures (4°C) and is not noted at temperatures above 22°C. In animals that recover, nacre covers the inner surface of the shell lesion forming an irregularity in the inner shell surface, a process termed nacrezation, and circumferential growth can then continue. While soft tissue lesions are not a feature of this disease, a decrease in glycogen occurs and may be the cause of mass mortalities in the stressful winter period. Disease caused by *V. tapetis* has been noted in fish in areas where infected clams are found.[163]

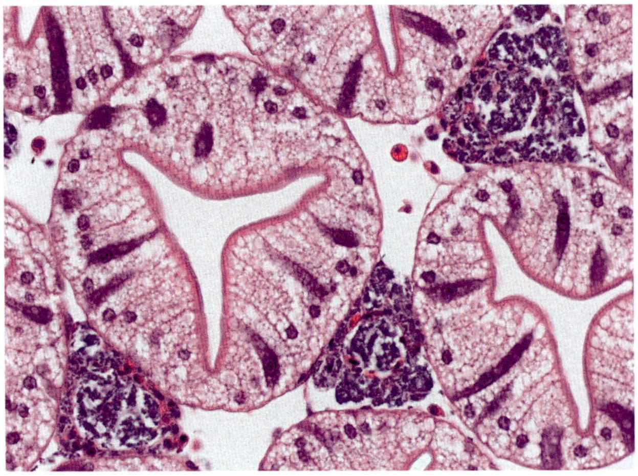

FIGURE 41.18 Hepatopancreas of an American lobster with Gaffkemia (red tail) due to *Aerococcus viridians* infection. The vascular spaces (arterioles especially) throughout the hepatopancreas are distended and filled with bacteria.

Gaffkemia (red tail) in lobsters is caused by the bacterium, ***Aerococcus viridans***. This gram-positive, polysaccharide capsule-producing coccus forms tetrads and infects American and European lobsters. Most infections occur through breaks in the carapace during molting. The bacteria proliferate in the open vascular system of the lobster (Fig. 41.18). Grossly, infected animals are lethargic and the arthrodial membranes on the ventrum of the tail as well as hemolymph samples are pink to red. Hemocytopenia eventually occurs and the vascular system is filled with tetrads of bacteria. The bacteria do not cause direct organ necrosis in chronic disease, rather thrombosis of vessels causes lack of nutrients to the tissues with tissue necrosis and atrophy. The increased density of captured lobsters maintained in impoundment/holding areas at >16°C is considered a contributory factor to disease development. The disease is not seen in recently caught wild animals. Gram stains of hemolymph and histologic examination are commonly used diagnostic methods as is culture of the bacteria. Oxytetracycline has been approved by the United States Food and Drug Administration (FDA) for use in treatment of this disease.

Two of the most economically important bacterial diseases in honeybees are American and European foulbrood, both of which are reportable to WOAH.[70] **American foulbrood** is a bacterial infection of the larval stages of the honey bee (*Apis mellifera* and other *Apis* species) by the gram-positive, spore-forming bacillus ***Paenibacillus larvae***. It occurs worldwide. Gross lesions consist of a mottled appearance of the colony due to a mixture of healthy capped, unhealthy uncapped, and empty brood cells.[108] Capped cells containing infected larvae are moist and dark colored and become concave and punctured as infection progresses. Affected larvae change color from creamy white to brown, and develop a semifluid, glue-like consistency (ropy stage) that can be drawn out as a thread with a probe or matchstick, and can serve as a method of field diagnosis. A month after the ropy phase, the larval material dries to a hard, dark scale (foulbrood scale) that is tightly adhered to the lower cell wall.[53] The scale contains millions of spores that are a source of disease transmission between colonies. Spores can remain infectious for more than 35 years and withstand moderate heat, cold, drought, and humidity. Bacterial spores are transmitted orally by nurse bees or via spore contamination at the base of a brood cell.[108] Transmission via hive contamination necessitates hive destruction by high heat (burning) to prevent subsequent transmission from colony to colony. Spores are infectious only to larvae, which are most susceptible 12–36 h after hatch. These spores pass through the foregut and germinate in the midgut, where they colonize the epithelium and massively proliferate, eventually filling the lumen, penetrating the peritrophic membrane, and passing through the epithelium to the hemocoel. Histologic lesions include initial vacuolation and eventual lytic necrosis of the midgut epithelium. Diagnosis is based on confirmation of the bacteria and clinical signs either through direct testing of colony material or following specialized culture with *P. larvae* agar (PLA), MYPGP agar, BHIT agar, and Columbia sheep agar. A conventional PCR test for rapid confirmation and identification after culture, as well as a nested PCR that allows for direct analysis of spore solutions are also useful in confirming infection. Colony samples should consist either of a section of comb (20 cm^2) containing affected brood cells or collection of affected larval remains with a sterile swab. Smears made at the apiary can be used to confirm diagnosis through the presence of characteristic small gram-positive bacilli both singly and in chains, or with carbol fuchin staining of larval smears which highlights the spores. Appropriate sampling techniques, storage, shipping, and testing are described in the *WOAH Manual of Diagnostic Tests and Vaccines for Terrestrial Mammals*.[106]

European foulbrood of honey bees caused by ***Melissococcus plutonius*** is the most widespread brood disease in England and has broad distribution that includes most locations where apiculture is practiced except New Zealand.[70] The name is derived from the foul odor associated with infected colonies. This disease results in the death of larvae at 4–5 days of age predominantly in unsealed broods. Gross diagnostic features during colony assessment include dead larvae that are not in their normal coiled position but are instead stretched lengthwise within the cell or twisted along the walls, and a distinct larval color change from normal white-tan to yellow, progressing to brown, and finally black-gray as larvae decompose. Some larvae die in sealed cells resulting in sinking of the cap, similar to American foul brood. *M. plutonius* is a microaerophilic to anaerobic, gram-positive, lanceolate coccus found singly, in pairs, and in chains. Secondary/concurrent bacterial infection is common and can lead to misdiagnosis of other bacterial etiologic agents. Pathogenesis includes ingestion of the bacterium with contaminated food followed by rapid proliferation within the larval midgut. Although not all infected larvae may be killed, those that survive infection are stunted compared to uninfected animals. Confirmation of infection can be made through smear preparations of larvae showing gross lesions or with a rapid detection kit.

Bacillus thuringiensis **(*Bt*)** is a ubiquitous, gram-positive bacterium originally identified in diseased silkworms (*Bombyx mori*) in 1901. It can infect a wide variety of insects, primarily lepidopterans. Insects ingest *Bt* spores from the soil. The bacteria produce toxins in the midgut, resulting in midgut cell necrosis, translocation of the bacteria to the hemolymph (sepsis), and death, usually within a few hours of ingestion depending on the host species. Many *Bt* toxins have been identified in insect hosts. *Bt* is also formulated as a biopesticide that has advantages due to its rapid mortality and relatively high host specificity.

Cyanobacterial gill disease due to filamentous blue-green bacteria (***Oscillatoria* spp.**) can also colonize the thin chitinous book gill lamellae. Gross lesions include swelling and rupture of gill leaflets and extensive tissue necrosis and loss[206] (Fig. 41.19A). Lesions in fungal and cyanobacterial gill disease are histologically similar. Gill leaflets are expanded by increased numbers of hemocytes that frequently form large aggregates obstructing the lumen of the leaflet (Fig. 41.19B). Distal to these aggregates, leaflets often show signs of regional necrosis similar to embolic disease. With fungal disease, hyphae are present both within the sinuses and penetrating through the gill leaflets. With cyanobacterial gill disease, thick mats of organisms frequently coat both surfaces of individual leaflets.

Fungi

Fungal infections, similar to bacterial infections, are common in situations where other environmental disruption or toxin exposure also contributes to debilitation. Even small erosions in the cuticle of terrestrial invertebrates can lead to primary or secondary fungal infection. Cuticular erosions are often associated with hyphal invasion in captive arachnids and insects (eSlide 41.e4), and are often considered secondary to cuticular or environmental disruption. Primary fungal infections in insects are caused by **entomopathogenic fungi**, most of which infect insects via the cuticle, proliferate inside the body, kill the host, and then sporulate and erupt from the cuticle. Phyla with entomopathogenic fungi include Blastocladiomycota, Zoopagomycota, Ascomycota, and Basidiomycota. Infective

FIGURE 41.19 Cyanobacterial gill disease in an Atlantic horseshoe crab. (A) Gross lesions include black discoloration of the gill lamellae, lamellar swelling, and multifocal lamellar necrosis. (B) Histologically, a thick mat of cyanobacteria covers the lamellar surfaces and the lamellar vascular spaces are infiltrated and obstructed by hemocyte aggregates.

stages are typically zoospores and conidia. Proliferative stages are often hyphae or hyphal bodies (yeast-like structures that proliferate with budding) (eSlide 41.e5 of *Massaspora cicadina*).

Fungal infections in horseshoe crabs are a significant issue in managed care particularly when animals are housed without sand substrate (Fig. 41.20). Infection targets the book gills, operculum, and carapace resulting in thickening and white to yellow discoloration of the book gill leaflets, yellow discoloration and thickening of the opercular cuticle, and cavitation, softening and discoloration globally of the carapace. Sites of infection are associated with necrosis, hemocytic infiltrates, and intralesional hyphae and may result in direct extension into the subjacent viscera (hepatopancreas).[124] Fungal cultures commonly grow *Fusarium* spp. particularly ***Fusarium solani***, but other species have been identified.[124,223]

Crayfish plague (also **Crayfish aphanomyciasis, La Peste**) is a fungal disease primarily of the carapace (cuticle) of many species of crayfish.[60,137] It is caused by the Oomycete (water mold), ***Aphanomyces astaci***. The fungus infects through small breaks in the arthrodial membranes primarily at molting and proliferates within the carapace tissues. Gross lesions include dark brown-black (melanized) or white to brown spots and patches on the arthrodial membranes of the ventral cephalothorax and abdomen. The fungus is highly branched, approximately 7 μm to 9 μm in width, and nonseptate. Spore formation is absent until late stages of infection. Examination of wet mounts with and without Giemsa staining helps with identification. Histological sections are also diagnostic. Molecular methods are available but due to the many strains of fungus, may not be accurate.

Aspergillosis of sea fan corals (*Gorgonia* spp.) due to infection with the ***Aspergillus sydowii*** causes mild morbidity to severe localized mortality events in the Caribbean.[3,116] Gross lesions affect primarily the fan portion of the colony and include multifocal tissue loss, distinct purple discoloration of the viable tissue at lesion margins, and large nodular growths (galls) on the fan of otherwise normally pigmented and intact colonies. Lesions characteristically occur away from the colony edge along major veins. The purple discoloration is the result of increased production of purple sclerites (skeletal elements), a nonspecific response to exposure to biotic agents. Histologic lesions include fungal infiltration, tissue necrosis, and extension of fungal elements into the axial skeleton. Granular amoebocyte infiltration and melanization often surround fungal hyphae. The margin between infected and healthy tissue is devoid of coral polyps and has a lower density of zooxanthella.[205] Stressors, such as elevated temperature and nutrient pollution likely contribute to disease development.[117]

Chalkbrood of honey bees due to the fungus ***Ascosphaera apis*** affects colonies in the spring during cool and humid weather conditions and affects larval stages exclusively. Larvae typically die before pupation, which affects colony populations. Infection is spread through the colony by food contaminated with fungal spores (ascospores). Ascospores germinate in the midgut of infected

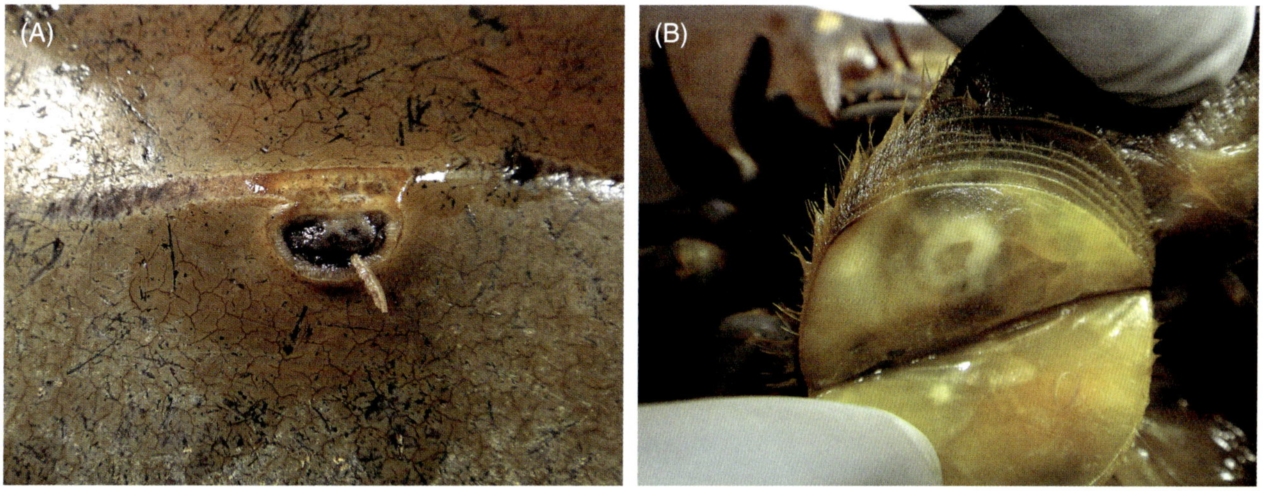

FIGURE 41.20 Fungal infection in an Atlantic horseshoe crab. (A) The ocellus (eye) and surrounding carapace are multifocally eroded and an irregular white-tan plaque covers the surface. (B) Within the book gill lamellae, there are multiple foci of white discoloration, regional lamellar thickening, and necrosis.

larvae and the fungus invades the viscera. White masses of mycelia are visible through the skin of infected larvae and eventually break through the posterior end of the larvae after death. Following death, hyphae cover the entire cadaver growing from the posterior end toward the head.[8]

Microsporidia are important pathogens of invertebrates, infecting host groups from all major taxa in all environments. Of the genera identified to date, nearly half infect aquatic organisms including 50 genera in aquatic arthropods and 21 genera in aquatic, nonarthropod invertebrates.[58,216] They cause significant morbidity and mortality in beneficial insects (pollinators) and farmed Crustacea and in some cases are used as biological control for insect pests. Horizontal and vertical transmission are both possible. Though most major organ and tissue types can be affected, infection is typically restricted to one organ, tissue, or cell type in an individual. Regardless of tissue type, histologically microsporidia are evident in large numbers in the host cell cytoplasm. Some form sporophorous vesicles, a barrier between the parasite and the host cell cytoplasm. Host cells are frequently hypertrophied. **Xenoma** formation occurs in numerous hosts and tissue types.

Metazoa

Invertebrates serve as intermediate and final hosts for a wide variety of metazoan parasites.

Pleurocercoids (intermediate forms of tapeworms) of **Calliobothrium verticillatum** are found in the anterior midgut ceca of *Pagurus* spp. The final host is the dogshark. Pleurocercoids cause significant damage to the blind-ended tubules of the glands and cause severe necrosis, rupture, and secondary bacterial infections that lead to death.[210] Numerous cestode larvae can also be found in the cecal lumen of squid and other cephalopods; tissue invasion into the intestines, cecum, and digestive gland with necrosis and inflammation is uncommon, but can occur. ***Licnophora* sp.** (phylum Ciliophora, class Spirotichea) infest several genera of animals including gastropods with and without shells (limpets and sea hares) as well as seahorses, corals, clams, and others. Their importance as a disease-causing agent varies by host species and environmental conditions, which may promote their sometimes opportunistic nature.[67]

Dicyemida (**Rhombozoa**) are a phylum of unusual organisms that may be related to roundworms, although their life cycle is not similar. All life stages, vermiform embryos, nematogens, rhombogens, sexually produced larvae, and infusoriform embryos are found in the renal sacs surrounding the appendages (kidney) of cephalopods. The number and maturity of the organisms parallel the age/growth of the cephalopod. Dicyemids are host specific but more than one species of dicyemid may be found in the same host.[79,80] The method of infection is not known, although it is suspected that the infusoriform larvae may both exit and enter via the renal pore that empties into the mantle cavity. Dicyemids attach to the renal tubule/crypts of the renal appendages that fill the lumen of the renal sac into which urine is deposited (Fig. 41.21). Some species have also evolved to only live in the sac surrounding the branchial heart appendage. Dicyemids tend to infect cephalopod species that spend time in the benthic layer (primarily octopi and some species of cuttlefish). These are considered a commensal organism that may promote urine excretion and flow.[81] An inflammatory response is not typically associated with infection, but one of the authors has seen renal necrosis and increased debris in the renal sac with heavy infection (Smolowitz, unpublished data).

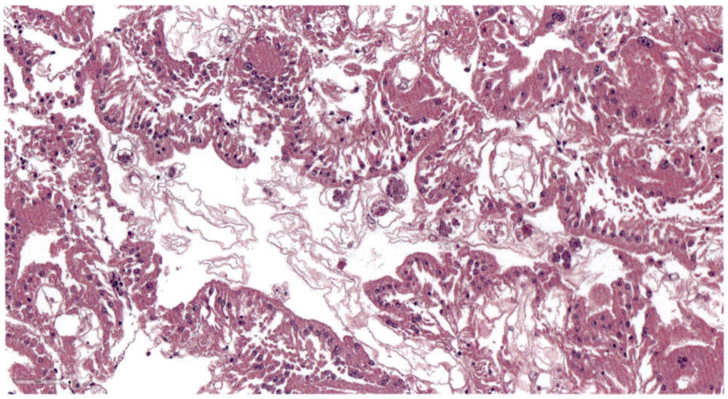

FIGURE 41.21 Renal dicyemids in a two-spot octopus. The tubular lumina contain multiple life stages of metazoan organisms with peripheral cilia, thin membranes, and internal cell clusters.

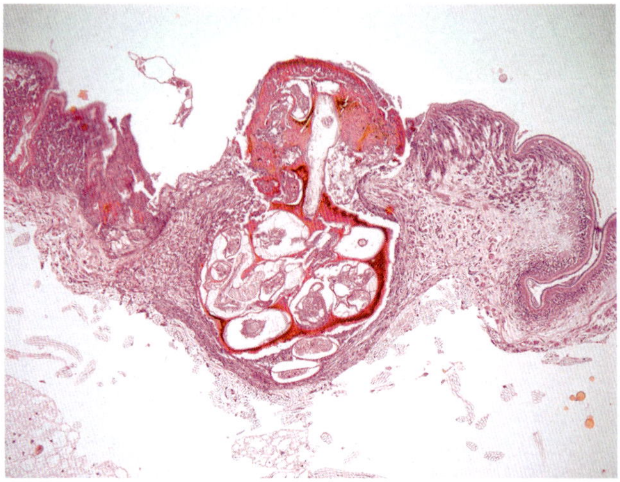

FIGURE 41.22 Rhabditiform nematode infection in the midgut of a millipede. Extending from the mucosa deep into the submucosa is a dense hemocytic infiltrate surrounding multiple cross-sections of adult rhabditiform nematodes.

Nematode infections of arachnids include those from families **Mermithidae** and **Panagrolaimidae**. At least 51 species of spiders and harvestmen have been recorded as mermithid nematode hosts across various habitats. Poinar provides a detailed summary of recorded instances of mermithid infection in arachnids.[175] Symptoms include abnormalities in body size and shape including swelling of the opisthosoma, altered epigynum, malformed palpi, legs that are shorter and thicker than normal, and poorly developed male secondary sexual characteristics. Coils of the parasite may be visible through the integument as the nematode eventually occupies the entire abdomen and occasionally the cephalothorax. This results in digestive gland reduction or absence, and loss of the prosomatic muscles, the entire digestive system, or the reproductive system.[174,175] **Panagrolaimid nematodes** infect the oral cavity of spiders, and have been recorded in both captive and wild spiders from multiple continents. Clinical signs include anorexia, decreased activity, and often death after a period of weeks to months. White, viscous discharge may be apparent between the mouth and chelicerae late in infection. Histologically, nematodes may be associated with secondary bacterial infections, inflammation, and necrosis.[174]

Nematodes are also common endoparasites of millipedes and are found across a similarly wide range of species.[102] **Rhabditiform** and **rhigonematid nematodes** have been recovered from fecal samples of myriapods.[43] Although previously not thought to be pathogenic, nematodes with morphology consistent with rhigonematids have been documented to cause midgut ulceration and transmural hemocytic inflammation in Puerto Rican white and giant African millipedes (Fig. 41.22). At necropsy, affected animals are in thin body condition with evidence of digestive gland atrophy.

Numerous nematodes infect insects, but a special group of nematodes called **entomopathogenic nematodes (EPN)** have bacterial symbionts that are rapidly lethal to the insect host. Once inside the host, EPN molt and release bacteria that are the primary killing agent for the host. Due to their ability to cause mass mortality in their hosts, EPNs have been commercialized and used for biological control of pest species.[192]

Turbellarids (flatworms) of many different genera infest and infect several types of invertebrates. A commonly encountered example is the **Triclad turbellarid (*Bdelloura* spp.)** found on the gills and body of Atlantic horseshoe crabs. Large white flatworms are visible on the gills, ventrum, and arthrodial membranes. Adults are not believed to cause disease, but the egg cases in the gills can cause pressure necrosis of the lamellae causing inflammation and opportunistic infections (Fig. 41.23). Turbellarid infections with members of the Order **Rhabdocoela** in the opalescent nudibranch used in research, involve a flatworm that invades the coelom through the digestive tract prior to collection. The turbellarid develops there, and eventually lyses the body wall of the host causing death. Once released, it forms cocoons from which the resulting embryo then invades a new slug host. Turbellarids also infect the gills and digestive tracts; they potentially cause disease in various bivalves.[29,224]

Flatworms of the order **Polycladida**, including ***Amakusaplana acroporae***, are of particular concern and affect acroporid corals. Gross lesions in affected *Acropora* consist of white bands of tissue loss which begin at the base and move dorsally along the branches. Diagnosis can be made through scrapings at the margins of tissue loss or through identification with a dissecting microscope (Fig. 41.24).

Trematode infections are uncommon in insects.[192]

Protozoa

Eukaryotic organisms including those that are pathogenic for invertebrates have recently been reclassified into

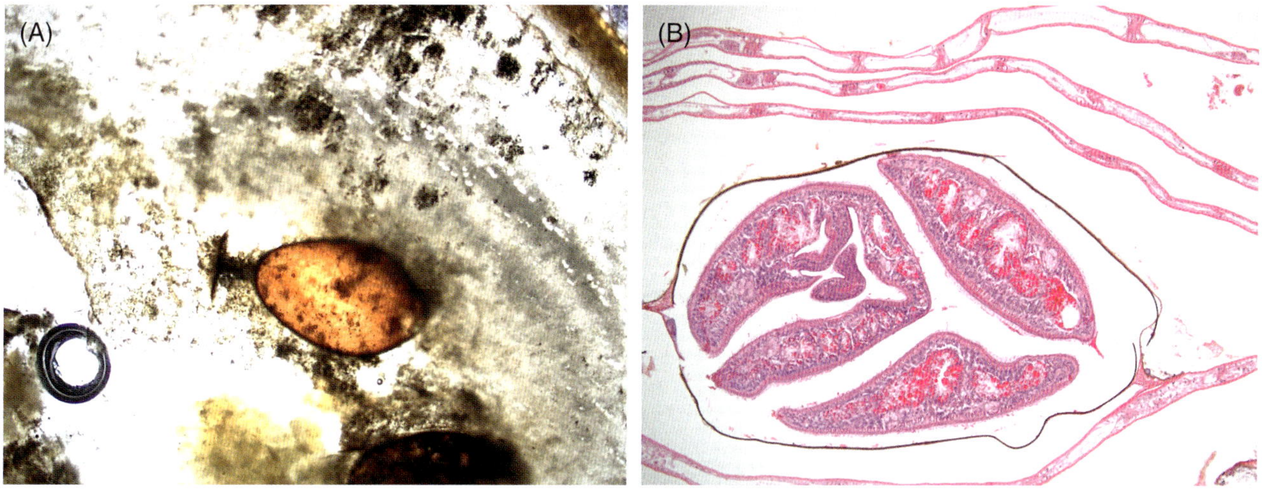

FIGURE 41.23 *Bdelloura* infection in the gill of an Atlantic horseshoe crab. (A) Wet mount cytology of a gill leaflet contains a focal egg casing from a Triclad turbellarid. (B) Histologically, adult *Bdelloura* are embedded within the gill lamella causing focal distortion.

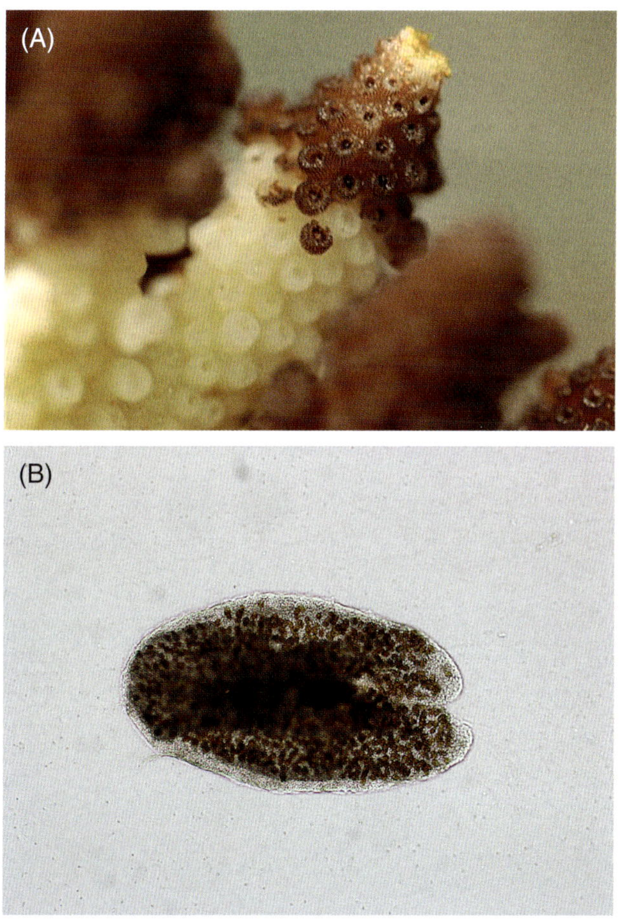

FIGURE 41.24 Acropora eating flatworm infection in *Acropora* spp. coral. (A) A single branch of an affected coral demonstrating tissue loss beginning at the base of the branch. (B) Wet mount cytology of the margin of the viable tissue recovered characteristic polyclad flatworm.

supergroups based on molecular similarities instead of morphological appearance.[1,30] While many organisms have been reclassified, they still retain their original phylum name. The phyla of pathogens described here will be identified.

Paramoebiasis in decapod crustaceans has been responsible for both endemic and epizootic disease and mortality in blue crabs and American lobsters. ***Paramoeba perniciosia*** (phylum Amoebozoa) infection occurs at molting. Lobsters may sustain a background infection without progression to mortality. Infections of blue crabs lead to death and most mortality is noted in high-density holding tanks in the fall and winter.[114,152] Grossly, crabs show gray and lobsters red discoloration along the ventrum when the disease is most severe, and hemolymph is white (crabs) or pink (lobsters). Early in disease development, *P. perniciosa* organisms are histologically found in the connective and neural tissues and are associated with hemocyte micronodulation and encapsulation (Fig. 41.25). In the late stages of the disease, hemocytopenia and lack of response to the proliferation of organisms in the vascular system are noted. In lobsters with chronic low-grade infections, fewer, melanized (chronic), micronodular foci containing paramoeba are found in neural and other tissues.[151] Since the range of these two crustaceans overlap, it is possible that lobsters may be a reservoir host.

Protozoal infections consistent with amoeba are also occasionally seen in other invertebrates, including cephalopods and American horseshoe crabs. Histologically, there is consistently substantial necrosis and inflammation centered on amoeba that are approximately 30-60 μm in diameter with amoeboid shape, round nuclei with dense chromatin, and granular cytoplasmic contents. Identification of species and genera has not been performed.[124,199]

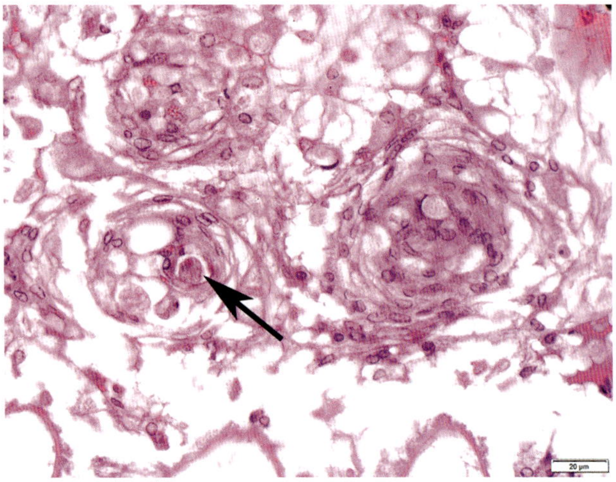

FIGURE 41.25 *Paramoeba perniciosi* (gray crab disease) in blue crabs. The organisms (*arrow*) are surrounded by an encapsulating inflammatory response.

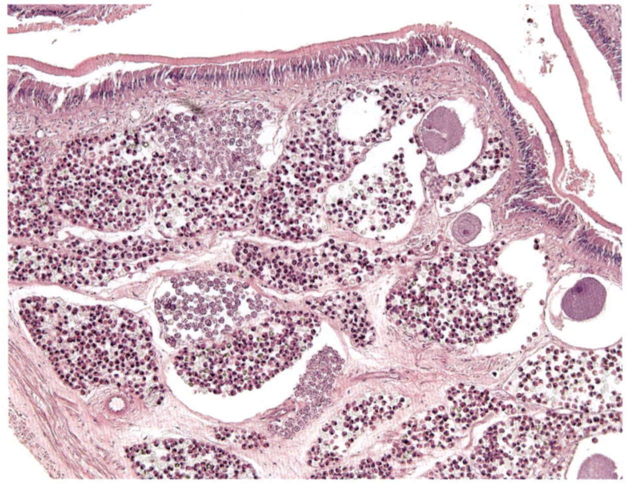

FIGURE 41.26 *Aggregata* spp. in the esophagus of a cephalopod. Expanding the submucosa are nodular aggregates of coccidia in which both gametogeny and sporogony can be seen.

In gastropod mollusks, the renal coccidian ***Merocystis kathae*** (phylum Apicomplexa) has been reported in dog whelks. ***Margolisiella haliotis*** is also often seen in the renal epithelium (nephridium) of black, red, pink, green, flat, and pinto abalone concurrently affected with wasting syndrome.[90] These coccidia do not typically result in gross lesions or morbidity. All developmental stages include oocysts, macrogametes, microgametocytes, and meronts, which are evident on histopathology with none to moderate hemocytic nephridial infiltrates present.[76,90] All life stages exist within a parasitophorous vacuole. Transmission is direct and horizontal; the life cycle is completed within a single host.

Many benthic-oriented cephalopods, including most octopi and several species of cuttlefish, are infected by ***Aggregata* spp.** (phylum Apicomplexa), highly host-specific coccidians that develop unusual forms in the gastrointestinal tract of the host.[86] Both gametogony and sporogony occur in the submucosa of all parts of the tract, but are usually most notable in the cecum. Grossly, infection is characterized by white, sometimes confluent, nodules (1 mm and less) throughout the gastrointestinal tract. Microscopically, in severe infections, numerous parasites in various life stages are seen (Fig. 41.26). Severely affected animals show a loss in body condition.[86] A second less well-known, ciliated parasite, ***Chromidina* spp.**, is sometimes found in the renal sacs of epi- and mesopelagic squids and octopi.[80]

Many species of decapod crustaceans can be infected with histophagous, holotrich ciliates (phylum Ciliophora) causing **bumper car disease**, so called due to the shape and activity of the organism. The most commonly diagnosed infection occurs in American lobster and is caused by ***Anophryoides haemophila*** but other genera (Mugardia sp., *Paranophrys* sp., *Anophrys* sp., etc.) occur in other decapods including several crabs.[100] The disease can be found in low prevalence in several types of decapods and only rarely causes severe disease in wild animals, typically in late winter/early spring. The method of entry appears to be through cuticular breaks and may occur most commonly in the gills.[37] Clinically, animals are lethargic. Histologically, when present in low abundance, the parasite is primarily located in the sinusoids around the hepatopancreas. As disease progresses, small granulomas/encapsulations occur in several areas (including the gills) and in late stages, hemocytopenia occurs and abundant ciliates may be found free in the vascular system throughout the animal, especially in areas underlying the carapace (Fig. 41.27). Parasites can be identified by direct examination of the hemolymph for the characteristic, active, ciliated, bumper car-shaped protozoa, by Giemsa-stained examination of

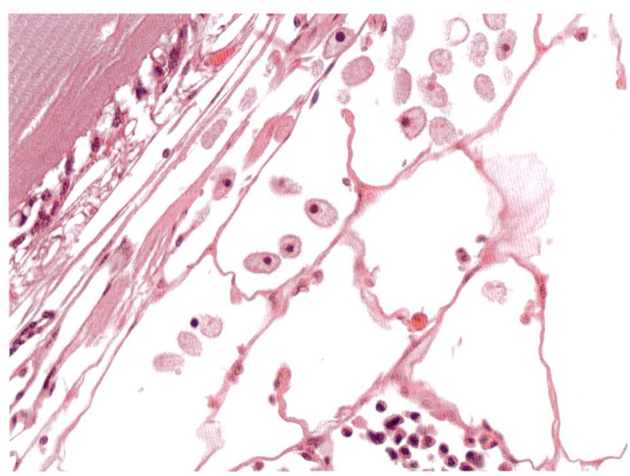

FIGURE 41.27 Bumper car disease due to *Anophryoides hemophilia* infection in the carapace of an American lobster. Ciliates are present in the sinusoids underlying the carapace.

hemolymph or by histologic examination. Direct or stained imprints of the hepatopancreas may detect early/light infections in lobsters.

Scuticociliatosis is seen in numerous marine invertebrates, particularly cephalopods, cnidarians, and tridacnid clams.[46,86,181] In most cases, the pathogenicity of the ciliates is unknown as these ciliates commonly ingest necrotic tissue, but some species are believed to represent primary pathogens and drive tissue necrosis. Generally speaking, invasion of viable tissue at the leading edge of a lesion is suggestive of a pathogenic infection; inflammation is typically mild. Tridacnid clams can develop overwhelming systemic infections, with scuticociliates filling the hemolymphatic spaces along with epidermal ulceration (Newton, unpublished data).

Hematodiniosis in several species of crustaceans, including copepods and amphipods, is caused by parasitic dinoflagellates (phylum Dinoflagellate) in the family Syndiniceae.[217] Infection by members of the genus *Hematodinium* produces the most economically important diseases in decapod crustaceans. These include **gray crab disease** of the blue crab, **bitter crab disease** of snow crabs, and **postmolt syndrome** in Norway lobsters. Ingestion and infection of the hemolymph through the gastrointestinal tract is the primary method of infection.[228] Severely infected animals take on a cooked appearance and the hemolymph becomes milky.[204] Histologically, as the disease progresses, abundant mononucleated or binucleated amoeboid trophonts can be seen in the hemolymph along with occasional vermiform plasmodia. Small encapsulations in several organs are seen in early infections. In late infections (1–2 weeks after infection), there is wasting of the tissues and hemocytopenia with abundant dinoflagellate forms filling the vascular spaces (Fig. 41.28). Single nucleated amoeboid trophonts can be easily mistaken for hemocytes histologically and on examination of hemolymph samples. Diagnosis is by gross and histologic evaluation and molecular methods. *H. perezi* can be diagnosed by staining a sample of hemolymph with Oil Red O, which results in lysosome staining in the parasite but not in the hemocytes.[217]

Perkinsosis (also called Dermo) is caused by the dinoflagellate ***Perkinsus marinus*** and is an important, WOAH-reportable disease in bivalves. *P. marinus* (phylum Perkinsozoa) is one of the two diseases that have caused great reductions in the population of eastern oysters (*Crassostrea virginica*) on the East Coast of the United States. The infectious form, a trophozoite, enters primarily through the gastric epithelium and digestive gland tubules but can also enter through the mantle and gill epithelium. The parasite is most active in oysters in seawater at a salinity of 12–28 ppt and at temperatures greater than 18°C. The trophozoites are engulfed by the hemocytes in which they proliferate and form signet ring profiles consisting of a hemocyte containing a 2–4 μm round parasitic vacuoles. Occasionally, a refringent vaculoplast can be identified in the vacuole. The mature trophozoite undergoes karyokinesis, then cytokinesis to form a rosette within the cells. Mature rosettes, called sporangia, are filled with trophozoites causing swelling of the hemocyte (Fig. 41.29). Evidence strongly suggests that the agent can inhibit apoptosis of the hemocyte until it has finished its replication cycle, at which time the host cell lyses and the newly formed trophozoites are released to infect naïve hemocytes. Infected hemocytes circulate throughout the open vascular system of the oyster. In early infections, moderate mixed hemocytic (granulocytic/agranulocytic) inflammation is noted, especially in and adjacent to the gastric epithelium or digestive gland tubules. In late infections, few normal

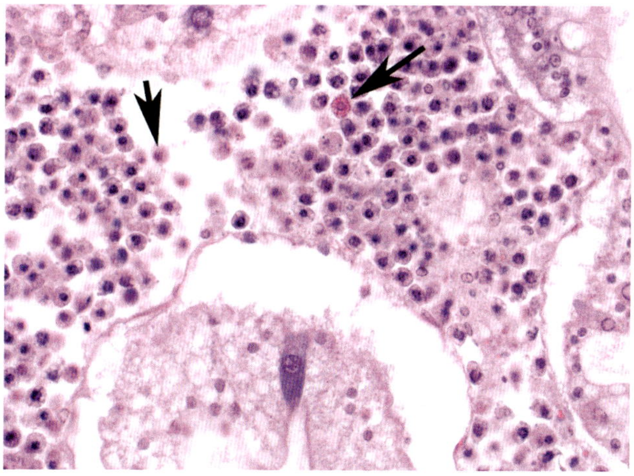

FIGURE 41.28 *Hematodinium* sp. in the digestive gland of a blue crab with gray crab disease. Dinoflagellates fill the sinusoids surrounding the hepatopancreatic tubules. *Arrows* indicate two normal hemocytes which are of similar size to the myriad dinoflagellates.

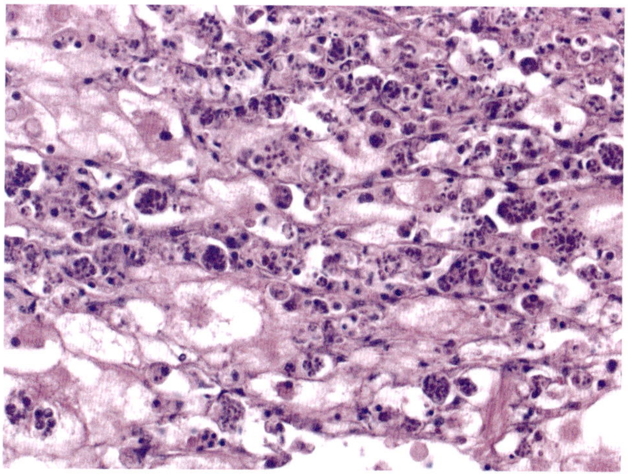

FIGURE 41.29 *Perkinsus marinus* (Dermo) infection in the hemocytes of an eastern oyster (*arrows*). Most hemocytes are distended and contain various maturation stages of the organism.

hemocytes remain in the tissues, and the tissues appear thinned and edematous. The agent is directly infective and trophozoites are released from dead, decaying animals as well as with the feces from live animals. This slowly progressive disease has spread and is now found in aquacultured and natural populations of eastern oysters in the United States from Maine to Louisiana, where it usually causes death in the fall. *P. marinus* has also recently been identified in other oyster species in the Gulf of Mexico, and the Pacific oyster (*Crassostrea gigas*) which is cultured on the west coast of the United States. These oyster species can become infected but do not exhibit disease. Histopathological examination is not as sensitive or specific a diagnosis as PCR/qPCR, and is not as sensitive as the Ray's fluid thioglycollate media (RFTM) method.[207]

Other *Perkinsus* species have been isolated from various bivalve species around the world. The most important of these is ***P. olseni***, which is also a WOAH-reportable disease. Grossly, *P. olseni* and other *Perkinsus* spp. infections result in small to large, white to brown nodules in the tissues of the body, gill and mantle. Microscopically, foci of infection are characterized by hemocytic inflammation that contain and surround or attempt to surround signet ring and rosette forms of the parasite (Fig. 41.30; eSlide 41.e6). Trophozoites of *P. olseni* can be up to 40 µm in diameter, much larger than those of *P. marinus*. *P. olseni* has been identified in Tridacnid clams (displayed in aquariums) and other cultured bivalves, such as Palourde clams, and infection has caused major mortality in these and other bivalves. It is also a directly infective agent. *Perkinsus* species can be cultured in RFTM, but culture is nonspecific if more than one *Perkinsus* species is present.[200]

Bodonid flagellates (phylum Euglenozoa) of the genus ***Ichthyobodo*** have been intermittently reported in cephalopods, particularly octopus, but the host–parasite interaction has not been studied in detail. The parasite is free-swimming in the water column and can attach to epithelial surfaces, particularly targeting the gills in octopus. There are no gross lesions and diagnosis is primarily made by histopathology. Histologically, there is swelling, degeneration, necrosis, and erosion of the branchial epithelium with hemocytic infiltration of the submucosa. *Ichthyobodo* species are pyriform, 6–10 µm long, 3–6 µm wide and are attached to the apical epithelium frequently in large numbers that are densely packed (eSlide 41.e7).[199]

Haplosporidium nelsoni, also named **multinucleated sphere unknown (MSX)**, is in the phylum Haplosporidia. MSX produces spores but they are considered more primitive than other Cercozoa. It was first identified in eastern oysters in late 1950s to early 1960s in cultured animals from New Jersey to Virginia (United States). It is now found in eastern oysters throughout the northeast United States and into Canada and along with *P. marinus*, and is responsible for decimating populations of aquacultured and wild eastern oysters along the eastern United States. Its most common form in tissues is a multinucleated plasmodia. The infectious form is a spore that is first found between epithelial cells of the gill, where it develops into a multinucleated plasmodia (Fig. 41.31; eSlide 41.e8). The plasmodia later breaks through the basement membrane of the branchial epithelium and enters the open vascular system, where it continues to proliferate through multiple serial divisions that produce more multinucleated plasmodia. Inflammation at sites of infection in the gills is intense and localized to areas of infection. The inflammatory response is ineffective in destroying or isolating the plasmodia, which proliferate in the open vascular system. As disease progresses, the inflammatory response

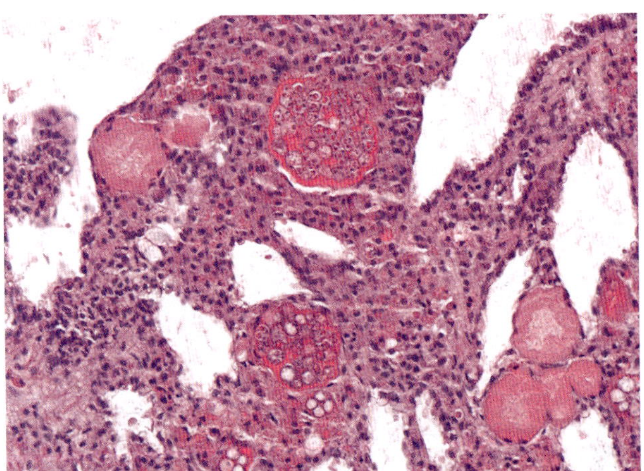

FIGURE 41.30 *Perkinsus* sp. infection in the gill of a soft-shelled clam. There is prominent granular hemocytic inflammation throughout the gill leaflets. Hemocytes are distended and contain sporangia filled with trophozoites (see eSlide 41.e6).

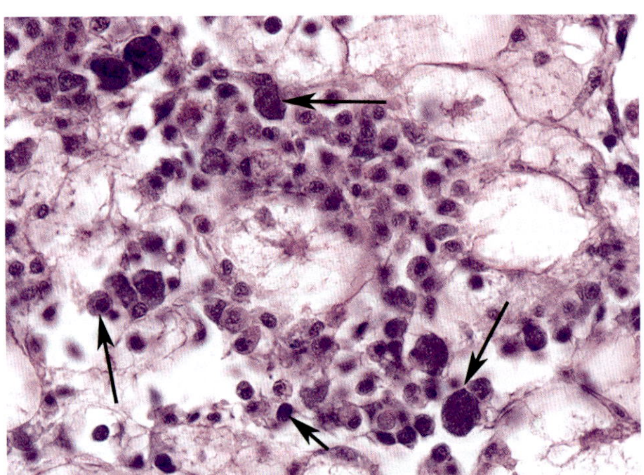

FIGURE 41.31 Multinucleated sphere unknown (MSX; *Haplosporidium nelsoni*) infection in the digestive gland of an eastern oyster. Multinucleated (and some immature single-nucleated) plasmodia in the sinusoids (*arrows*) and sporulated plasmodia in the digestive gland epithelium. (also see eSlide 41.e8).

is greatly muted. Terminally, connective tissue and cells are shrunken and animals develop generalized edema with increased "brown cells" (spent tissue hemocytes) in the tissues. Mortality is most severe in the late summer/early fall. The parasite grows best in oysters in salinities greater than 15 ppt and in warm water up to 20°C. *H. nelsoni* is reported to die in oyster tissues at temperatures greater than this. Spores are formed in the digestive tubules of juvenile oysters in some circumstances, but the reason/conditions for this are not understood. While sporulation kills the juvenile oysters, the spores are not infective to other adults or juveniles indicating that this is an indirect infection that requires another host in the parasite's life cycle. To date, other hosts have not been identified but movement of infected animals to areas without disease results in transmission indicating that the other hosts are commonly found in many areas. The agent and a mild form of the disease in low prevalence have been identified in Pacific oyster. Transport of Pacific oyster to the east coast of the United States is believed to be responsible for the introduction of MSX into eastern oyster populations. Histopathological examination of tissues (especially the gills in early infections) is the most common method of diagnosis. Molecular methods have been published but are not commonly used.

Haplosporidium costale, also called **seaside organism (SSO)**, is a close relative of MSX and also infects eastern oysters. The disease was first identified in 1962 in oysters grown in high-salinity waters on the Atlantic coast of Virginia (United States). Recently, the disease has been found in aquacultured oysters in more northern states. The portal of entry is primarily through the digestive tract epithelium, unlike MSX. Early SSO infections are identified when multinucleated plasmodia are present in the sinusoids in connective tissues around the stomach and digestive gland (Fig. 41.32). As in MSX, plasmodia proliferate over time and fill the vascular system of the oyster. Inflammation is moderate in early infections but as the disease progresses, the number of viable hemocytes decreases and inflammation is not seen late in the disease. Unlike MSX, plasmodia of SSO synchronously sporulate throughout the sinusoids in late May to early June in adult oysters. If abundant plasmodia are present in the sinusoids, the animals will die during sporulation. If few plasmodia are present, the animals will survive the event. Those that survive often appear thin and edematous. Microscopically, the only remaining sign that an animal was infected are numerous "brown cells" in circulation. Infection only rarely causes severe disease in wild animals, typically in late winter/early spring. Transmission using spores from infected oysters does not cause infection in naïve oysters so indirect infection methods are suspected but no intermediate host has been identified. The parasite grows best in oysters held in salinities 25 ppt or above. Diagnosis of SSO is hampered by the similar appearance of the

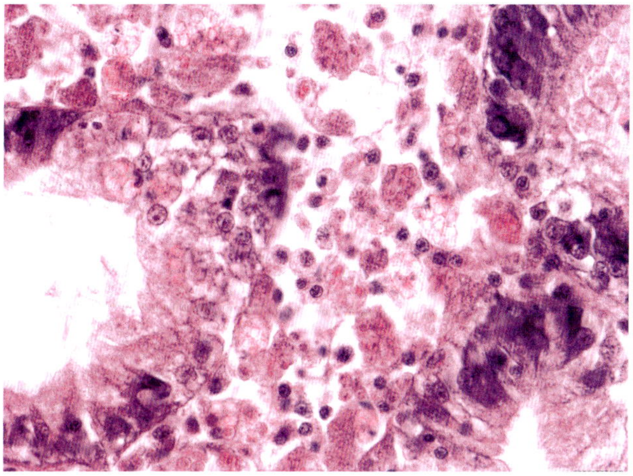

FIGURE 41.32 Sporulating seaside organism (SSO; *Haplosporidium costale*) in the connective tissue of an eastern oyster. Multinucleate plasmodia are undergoing sporulation (*arrow*) within the connective tissue surrounding the digestive tubules.

plasmodia to that of MSX. MSX plasmodia are usually slightly bigger and more eosinophilic than SSO, but that is not consistent in all cases. Molecular confirmation is very helpful in these cases.

Bonamia spp. parasites are considered members of phylum Haplosporidia, although no spore form has been identified. The disease, also termed **microcell disease** and **hemocytic disease**, was first reported in the United States during the mid-1960s, but disease caused by *Bonamia* spp. occurs worldwide. Several species of bivalve can be infected by the two most important species, which produce varying morbidity and mortality, especially during warmer months. ***B. ostreae*** is most problematic in European flat oyster. ***B. exitiosa***, and the genetically similar ***B. roughleyi***, are found primarily in European flat oyster in New Zealand, Spain, and Italy. Importantly, some species of oysters, including Pacific oyster, a commonly aquacultured oyster in Europe and the United States and Suminoe oyster, which was considered as a replacement species in aquaculture for Eastern oyster, become infected but do not develop disease and are considered carriers. *Bonamia* spp. infect the hemocytes of oysters (primarily adults) and form few to many 2–5 μm round cells within the cytoplasm of the hemocyte. It is directly infective. Grossly, yellow discoloration of the gills may be noted. Histologically, in early stages of the disease, an increase in circulating hemocytes is noted in affected animals and multifocal hemocyte aggregates/small encapsulations contain few to many *Bonamia* organisms within the cytoplasm of some of the hemocytes (Fig. 41.33). Interestingly, da Silva[52] found a correlation between high percentage of agranulocytes in the hemolymph and increased probability of infection of that strain of oyster. Histology and molecular methods can be used to diagnose the disease. Additionally,

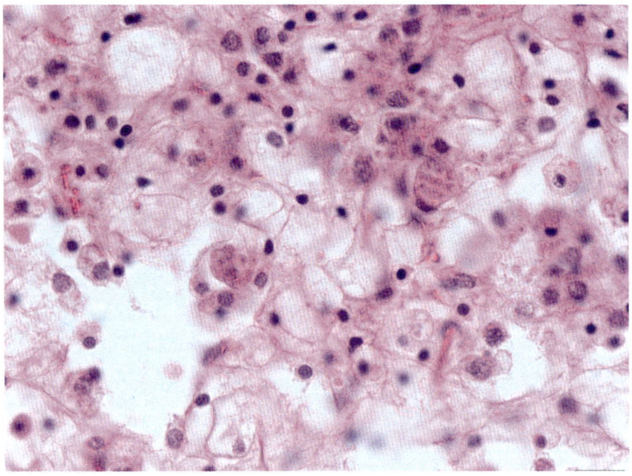

FIGURE 41.33 *Bonamia* spp. infection in the digestive gland of a Suminoe oyster. The cytoplasm of enlarged hemocytes within the connective tissues of the digestive gland contains numerous 2–5 μm round, intracytoplasmic parasites.

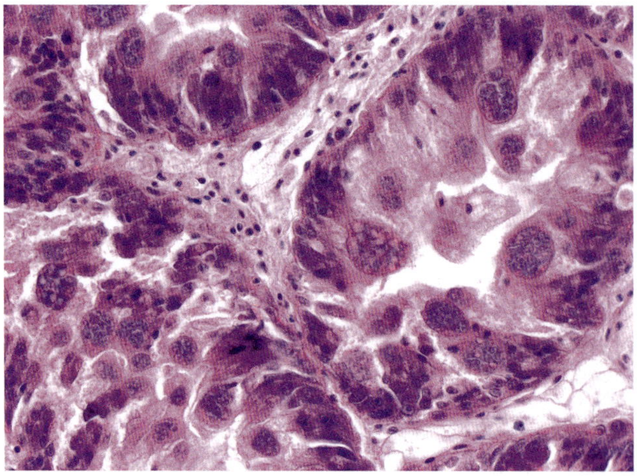

FIGURE 41.34 Digestive gland of a European oyster infected with *Marteilia refringens*. Digestive gland epithelial cells form a "nurse cell" containing 8–16 sporonts per cell. Each contains four spores and a few refringent bodies.

imprints of the heart on coated slides stained with Giemsa stain can be diagnostic. *B. ostreae* and *B. exitiosa* are both WOAH reportable.

Marteilia refringens belongs to the phylum Endomyxa and causes **Marteiliosis**, a WOAH-reportable disease. It causes significant disease in aquacultured and wild populations of European flat oysters on the west coast of Europe. However, other species of bivalve, including clams, scallops, and mussels, can be infected with *M. refringens*, or a closely related species, in various parts of the world. Associated morbidity and mortality vary greatly, and some bivalve species may be carriers. Tissues of clinically ill animals are thin and edematous. Infection occurs via ingestion when a primary/stem cell enters the digestive system and infects digestive gland epithelial cells forming a "nurse cell." Nurse cells undergo consecutive internal cleavage to produce 8–16 intracellular sporonts, each containing 4 spores and a few refringent bodies that can be seen histologically (thus its species name) (Fig. 41.34). Histology is used to diagnose the infection but smears/digestive gland imprints on Wright-Giemsa stained slides can be used in moderate to severe infections. Molecular diagnostic methods are also available.

Quahog parasite unknown (**QPX**), recently named ***Mucochytrium quahogii***, is a protozoan belonging to the phylum Labrinthomorpha and class Thraustochytrid. It is an opportunistic pathogen that causes disease in hard clams (*Mercenaria mercenaria*) on the East Coast of the United States. It was first identified in clams in Canada in the late 1940s then in aquacultured clams in the United States in 1996. It has caused disease in New York and Virginia, however, the prevalence and severity of the disease were highest in areas of Massachusetts. The protozoa infect the mantle of clams close to the base of the siphon; almost all infected animals show nodules and swellings in this area. This is the area where pseudofeces are deposited and it is hypothesized that the increased contact time between the parasite, pseudofeces, and the mantle allows for infection into the tissues of that area. More advanced infections show spread of the organisms throughout the clam vascular system. The inflammatory reaction to the invading parasite is seasonally dependent. In the early spring and late fall, the inflammatory response is poor to moderate and characterized by numerous proliferating organisms within an often thick coating of tenacious mucus (produced by the parasite) surrounded by only a few layers of responding hemocytes. In the summer (when metabolism of the clam peaks), the inflammatory response is intense with many layers of hemocytes forming hemocytic nodules with the occasional occurrence of hemocyte-derived multinucleated giant cells and histologic evidence of destruction and phagocytosis of the invading organism (Fig. 41.35). Diagnosis can be accomplished with stained or unstained squash preparations of the nodules in the mantle or with histological examination. Molecular diagnostic tools are also available.[209]

Ectoparasites

The most common arthropod parasites of terrestrial invertebrates include mites, wasps, and flies, while those of marine invertebrates include amphipods and copepods. Many of these are ectoparasites, though endoparasitism (parasitoid wasps and flies) also occurs.

Several honeybee mite infections are WOAH-reportable diseases, including **acarapisosis, tropilaelapsis**, and **varroosis**. **Acarapisosis** (also **acariosis** or **acarine disease**) of honey bees is caused by the Tarsonemid mite ***Acarapis***

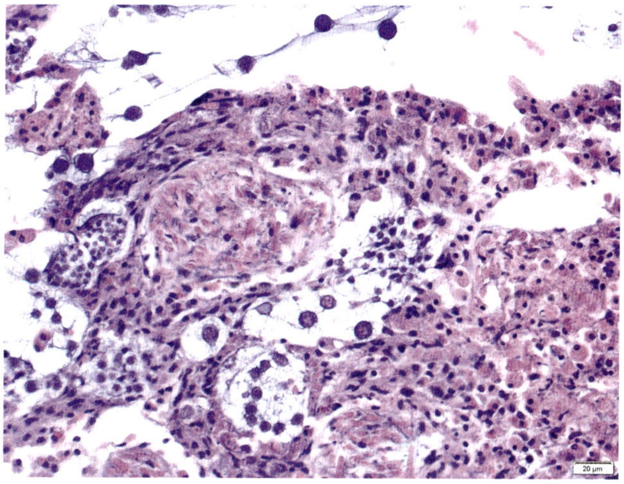

FIGURE 41.35 Portion of a QPX infection in the mantle of a hard clam (*Mercenaria mercenaria*) including a QPX thallus, recently lysed sporangia releasing endospores, hemocytic inflammation and remnants of mucus that surround the parasite in living tissue (mucus removed during tissue processing).

woodii, an internal pathogen of the respiratory system of the adult honey bee *A. mellifera* and other *Apis* species. Disease has been documented in North and South America, Europe, and the Middle East. These mites reproduce within the respiratory tract and feed upon the hemolymph of host bees. They are approximately 150 μm in diameter and are found predominantly in the prothoracic trachea but can be present in the head, thoracic, and abdominal air sacs. Infection causes mechanical damage to the airways, physical obstruction of the respiratory tract by mites or lesions, and hemolymph loss. With heavy parasitism, gross lesions, specifically opaque brown-black foci in the tracheal walls, develop due to inflammation and melanosis. Mortality from this infection is moderate to high and can be a cofactor in colony losses.[38] Diagnosis is by microscopy, ELISA, or the observation of guanine, a mite waste product. Microscopic confirmation requires identification of the mites first within the trachea and then after removal for positive identification. Spread is through direct contact with newly hatched (less than 10 days old) bees who are the most susceptible to novel infection.

Tropilaelaps of honey bees is due to infection with a few different species of *Tropilaelaps* mites, most frequently ***Tropilaelaps clareae*** and ***T. koenigerum***. The mites feed on the brood (larvae and pupae). Gross lesions include an irregular pattern of sealed and unsealed brood and deformities in adult bees. Spread is by direct contact, either bee to bee or by movement of brood frames. Diagnosis is by positive morphologic identification, although molecular testing is available.[106,108]

Varroosis of honey bees is due to infection with an ectoparasitic mite that affects both adult bees and the brood. Of the four species of mite, ***Varroa destructor*** is the most important. These mites have nearly a worldwide distribution with the exclusion of Australia and the southern island of New Zealand. Gross findings include shrunken abdomens in adult bees and visible external mites. Spread is by direct contact between adult bees and movement of infested bees or bee brood. In addition to direct effects, this mite is also an important vector for viral diseases of honey bees, including transmission of a virus that causes wing deformity.[106,108]

Endoparasitoids are typically wasps or flies that inject their eggs into an invertebrate host (or, less commonly, a host consumes eggs). The parasitoid larvae develop in the hemocoelom of the host, who is typically killed upon emergence or pupation of the endoparasitoid. **Endoparasitoid wasps** transmit **polydnaviruses** when the female injects the eggs into this host. These polydnaviruses inhibit the host immune response and retard growth, which prevents the host immune response from killing the parasitoid wasp eggs and larvae.[192]

Other arthropod parasites include amphipods, copepods, and beetles. **Hyperiid amphipods** are obligate parasites of gelatinous zooplankton that cause ectoparasitism in jellyfish and comb jellies in managed care. During larval development, gravid females select host medusa, bite holes into the tissue, and place larvae into the openings near the gonadal and gastric tissues. Developed females feed on host tissues and on prey captured by the medusa.[23,49] *Dugesia, Waminoa,* and *Convolutriloba* species flatworms can cause significant ectoparasitism in coral reef tanks. The copepod ***Tegastes acroporanus*** is an ectoparasite of acroporid corals and actively feeds on coral tissues causing reduced polyp extension, loss of coral tissue, and zooxanthellae pigments, reduced growth, and death.[36] The common name for this disease is "red bug" and the copepods are visible with a dissecting microscope (Fig. 41.36).

Small hive beetle infestation is caused by ***Aethinia tumida***, a beetle native to Africa that was introduced to the United States, Egypt, Canada, and Australia by commercial bee movement. Adult and larval beetles consume larvae, pollen, honey, and bee brood. The adult females lay eggs in the hive and the beetle larvae feed within the hive, then leave to pupate in the soil where the adults hatch and can fly to new hives. Spread can be rapid and over a range of several kilometers. If infestation is severe, bees will abandon the hive. Diagnosis is made by positive identification of beetles within the hive.[106,108]

Other Ectoparasites

Nudibranchs can be ectoparasites in coral reef aquaria, feeding on Montipora plate corals and earning the moniker **Montipora-eating nudibranchs** (Fig. 41.37). These organisms also cause tissue loss in affected coral colonies

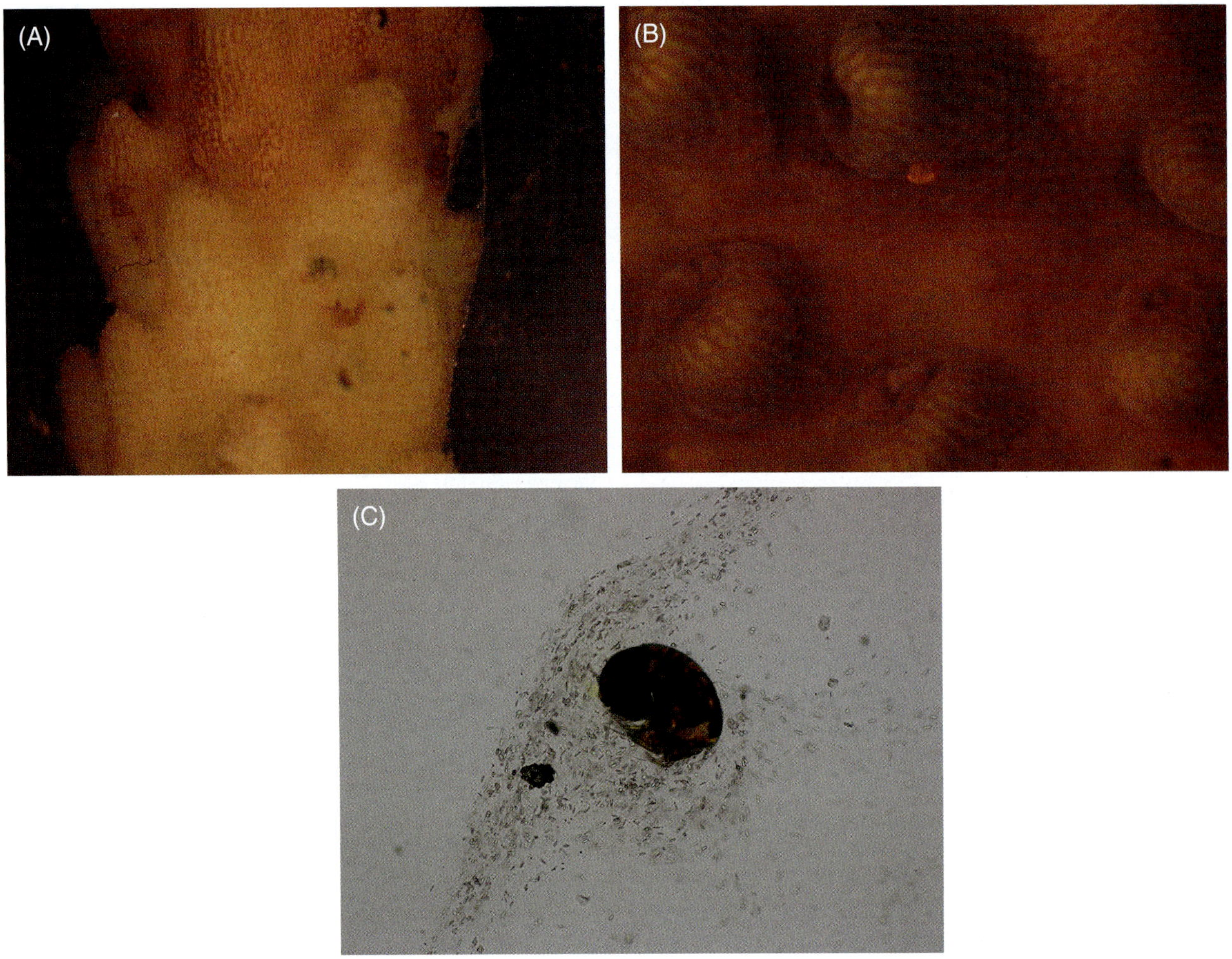

FIGURE 41.36 *Tegastes acroporanus* (red bug) infection in an *Acropora* spp. coral. (A) The affected branch demonstrates regionally extensive pallor and multifocal tissue loss. (B) Superficial copepod parasites with a yellow body and prominent red focus consistent with *Tegastes* are visible on the surface using a dissecting microscope. (C) Wet mount preparation of the crustacean parasite recovered in a mucus sample.

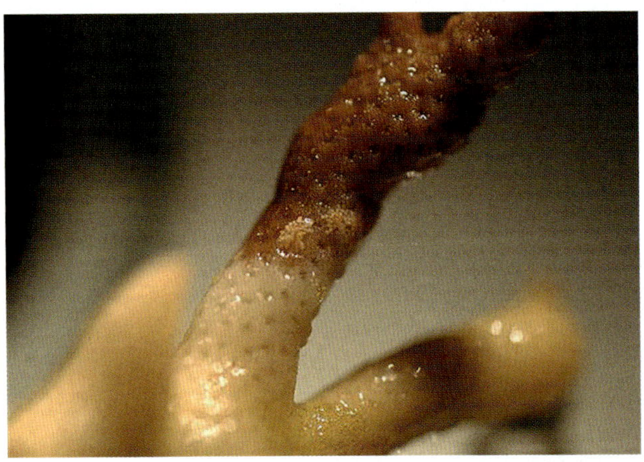

FIGURE 41.37 *Montipora* eating nudibranchs on a *Montipora* spp. coral. A representative branch from an affected coral shows tissue pallor and loss. Multiple light tan, leaf-shaped nudibranch parasites are present at the border between viable dark brown tissue and white to tan regions of tissue loss and skeleton exposure.

and are visible with the naked eye or by dissecting microscope, though they can be difficult to detect during daylight hours.[36]

Shells of various invertebrates are often infested by various **shell-boring organisms**. Some types of sponges (*Cliona* spp.) infest the shells of bivalves, dissolving the shell and establishing colonies on the shell surface. Colonization results in brittle, easily broken shells or pinpoint round holes in the shell when the sponge is no longer present. The shells of most bivalves and some other invertebrates, such as abalone, are infested by **Polychaeta** (*Polydora* spp. and *Boccardia* spp.) worms that borrow into the matrix of the shells and form tunnels. Sand and mud are often found in the tunnels along with the worms and forms "mud blisters" when the tunnel is viewed from the inner surface of the bivalve shell. These tunnels occasionally serve as homes for other invertebrates.

Other Infectious Organisms

Algae

Algae are common causes of exoskeletal and respiratory disease in Atlantic horseshoe crabs. **Green algal (chlorophycophytal)** infection of the carapace is one of the most common diseases in wild and captive animals. It causes severe pitting and deformity of the dorsal exoskeleton, eyes (ocelli), arthrodial membranes, and the base of the telson resulting in shell deformity, abnormal molting, end-stage eye disease, and full-thickness perforation of the carapace. Lesions have a characteristic dark green color. Histologically, there is multifocal erosion of the cuticle with rhizoidal processes of algal zygotes extending between the chitin laminae of the carapace. These lesions can be full thickness, penetrating the carapace, and infiltrating the internal viscera and creating portals for secondary bacterial infections. Histomorphology in one study identified the algae as being from the family Ulvaceae.[27]

Algal infections are reported in echinoderms. ***Coccomyxa ophiura*** infects two species of brittle stars (ophiuroids) causing multifocal patches of subepidermal green discoloration along the aboral surface of the disc and rays. Histologically, the algae infiltrate the dermal and mutable connective tissues of the body wall in the space between the dermal ossicles. Lesions progress and coalesce resulting in epidermal necrosis, dissolution of the dermal ossicles, infiltration of the connective tissue, and ossicles with algal cells and eventually ray autotomy. A similar disease process caused by ***C. astericola*** occurs in sea stars.[111]

Algae also comprise a large component of the **endolithic community** of reef-building corals. The endolithic community is part of the coral microbiome. It consists of algae, fungi, and prokaryotes that form a symbiotic consortium within the coral skeleton. The role of these microbial communities in coral health and disease is still being investigated.[171]

ADDITIONAL RESOURCES

See Table 41.2.

TABLE 41.2 Selected References for Normal Anatomy, Necropsy, and Histology of Invertebrates[a,b]

Phyla	Anatomy and Necropsy References	Histology References
Nematoda, Cestoda, Platyhelminthes	Low P, Molnar K, Kriska G. 2016. Atlas of Animal Anatomy and Histology. Springer, 413 pp.	Gardiner CH and Poynton SL. 1999. An Atlas of Metazoan Parasites in Animal Tissues. Registry of Veterinary Pathology. Armed Forces Institute of Pathology. Washington, DC, 63 pp.
Echinodermata	Newton AL, Dennis MM. 2021. Echinodermata. In: Invertebrate Histology, LaDouceur EEB, ed. Wiley and Sons. 1-18.	Newton AL, Dennis MM. 2021. Echinodermata. In: Invertebrate Histology, LaDouceur EEB, ed. Wiley and Sons. 1-18.
Porifera	Ereskovsky A, Lavrov A. 2021. Porifera. In: Invertebrate Histology, LaDouceur EEB, ed. Wiley and Sons. 19-54.	Ereskovsky A, Lavrov A. 2021. Porifera. In: Invertebrate Histology, LaDouceur EEB, ed. Wiley and Sons. 19-54.
Cnidaria	Peters, E. 2015. Anatomy. In: Diseases of Coral, Eds. Woodley CM, Downs CA, Bruckner AW, Porter JW, Galloway SB. Wiley-Blackwell, 85-107.	Peters, E. 2015. Anatomy. In: Diseases of Coral, Eds. Woodley CM, Downs CA, Bruckner AW, Porter JW, Galloway SB. Wiley-Blackwell, 85-107.
Mollusca: Gastropoda	*Terrestrial:* Low P, Molnar K, Kriska G. 2016. Atlas of Animal Anatomy and Histology. Springer, 413 pp. *Marine:* Tiley K, Freeman MA, Yen I, Dennis MM. 2019. Histology atlas and systematic approach to postmortem examination of the queen conch *Lobatus gigas*. *J of Shellfish Research*, 38(1):131-148.	*Terrestrial:* Low P, Molnar K, Kriska G. 2016. Atlas of Animal Anatomy and Histology. Springer, 413 pp. *Marine:* Tiley K, Freeman MA, Yen I, Dennis MM. 2019. Histology atlas and systematic approach to postmortem examination of the queen conch *Lobatus gigas*. *J of Shellfish Research*, 38(1):131-148.
Mollusca: Cephalopoda	Gestal C, Pascual S, Guerra A, Fiorito G, Vieites JM. 2019. Handbook of Pathogens and Diseases in Cephalopods. Springer, 234 pp.	Gestal C, Pascual S, Guerra A, Fiorito G, Vieites JM. 2019. Handbook of Pathogens and Diseases in Cephalopods. Springer, 234 pp.

Continued

TABLE 41.2 Selected References for Normal Anatomy, Necropsy, and Histology of Invertebrates—cont'd

Phyla	Anatomy and Necropsy References	Histology References
Mollusca: Bivalvia	Howard DW, Lewis EJ, Keller BJ, Smith CS. 2004. Histological techniques for marine bivalve mollusks and crustaceans. NOAA Technical Memorandum NOS NCCOS 5, 218pp. Low P, Molnar K, Kriska G. 2016. Atlas of Animal Anatomy and Histology. Springer, 413pp. Smolowitz R. 2021. Mollusca: Bivalvia. In: Invertebrate Histology, LaDouceur EEB, ed. Wiley and Sons. 163-184.	Low P, Molnar K, Kriska G. 2016. Atlas of Animal Anatomy and Histology. Springer, 413pp. Smolowitz R. 2021. Mollusca: Bivalvia. In: Invertebrate Histology, LaDouceur EEB, ed. Wiley and Sons. 163-184.
Annelida	Low P, Molnar K, Kriska G. 2016. Atlas of Animal Anatomy and Histology. Springer, 413pp.	Low P, Molnar K, Kriska G. 2016. Atlas of Animal Anatomy and Histology. Springer, 413pp.
Arthropoda: Arachnida	Foelix R. 2010. Biology of Spiders, 3rd ed. Oxford University Press. 432pp.	Kennedy B, Trim SA, Laudier D, Ladouceur EEB, Cooper JE. 2021. Arthropoda: Arachnida. In: Invertebrate Histology, LaDouceur EEB, ed. Wiley and Sons. 221-246.
Arthropoda: Merostomata	Bergdale KJ. 2017. Thesis: Illustrations for Health Assessment Techniques of the Atlantic Horseshoe Crab, *Limulus polyphemus*. Baltimore. John Hopkins University.	Bergdale KJ. 2017. Thesis: Illustrations for Health Assessment Techniques of the Atlantic Horseshoe Crab, *L. polyphemus*. Baltimore. John Hopkins University.
Arthropoda: Myriapoda	Newton AL, LaDouceur EEB. Myriapoda. In: Invertebrate Histology, LaDouceur EEB, ed. Wiley and Sons. 263-274.	Newton AL, LaDouceur EEB. Myriapoda. In: Invertebrate Histology, LaDouceur EEB, ed. Wiley and Sons. 263-274.
Arthropoda: Decapoda	Howard DW, Lewis EJ, Keller BJ, Smith CS. 2004. Histological techniques for marine bivalve mollusks and crustaceans. NOAA Technical Memorandum NOS NCCOS 5, 218pp. Smolowitz R. 2021. Arthropoda: Decapoda. In: Invertebrate Histology, LaDouceur EEB, ed. Wiley and Sons. 277-300.	Bell T, Lightner DV. 1988. A Handbook of Normal Penaeid Shrimp Histology. World Aquaculture Society. 114pp. Smolowitz R. 2021. Arthropoda: Decapoda. In: Invertebrate Histology, LaDouceur EEB, ed. Wiley and Sons. 277-300.
Arthropoda: Insecta	Stell I. 2012. Understanding Bee Anatomy: A Full Color Guide. The Catford Press. 199 pp. Low P, Molnar K, Kriska G. 2016. Atlas of Animal Anatomy and Histology. Springer, 413 pp.	Stell I. 2012. Understanding Bee Anatomy: A Full Color Guide. The Catford Press. 199 pp. Low P, Molnar K, Kriska G. 2016. Atlas of Animal Anatomy and Histology. Springer, 413 pp.

[a]*Microscopic Anatomy of Invertebrates, edited by Frederick W. Harrison and others, is a 14 volume set of invertebrate microanatomy with histology and electron microscopy images. Each volume is dedicated to a different taxon. All the taxa above are covered in this work. The work is out of print, but available at some large university libraries.*
[b]*Invertebrate Histology, edited by Elise E. B. LaDouceur, includes gross necropsy, normal anatomy, and normal histology of the above taxa with the exception of Nematoda/Cestoda/Platyhelminthes.*

References

Full references for this chapter can be found online at https://doi.org/10.1016/B978-0-443-18437-6.00030-4.

Index

Note: Page numbers followed by *f* indicate figures, *t* indicate tables, *b* indicate boxes, and *e* indicate online pages.

A

C

D

F

I

J

K

L

M

N

O

Q

R